TABLE A-4 COMMON ELEMENTS

Name	Symbol	Approx. mass	Common ox. nos.	Name	Symbol	Approx. mass	Common ox. nos.
aluminum	Al	27.0	+3	magnesium	Mg	24.3	+2
antimony	Sb	121.8	+3, +5	manganese	Mn	54.9	+2, +4, +7
arsenic	As	74.9	+3, +5	mercury	Hg	200.6	+1, +2
barium	Ba	137.3	+2	nickel	Ni	58.7	+2
bismuth	Bi	209.0	+3	nitrogen	N	14.0	−3, +3, +5
bromine	Br	79.9	−1, +5	oxygen	O	16.0	−2
calcium	Ca	40.1	+2	phosphorus	P	31.0	+3, +5
carbon	C	12.0	+2, +4	platinum	Pt	195.1	+2, +4
chlorine	Cl	35.5	−1, +5, +7	potassium	K	39.1	+1
chromium	Cr	52.0	+2, +3, +6	silicon	Si	28.1	+4
cobalt	Co	58.9	+2, +3	silver	Ag	107.9	+1
copper	Cu	63.5	+1, +2	sodium	Na	23.0	+1
fluorine	F	19.0	−1	strontium	Sr	87.6	+2
gold	Au	197.0	+1, +3	sulfur	S	32.1	−2, +4, +6
hydrogen	H	1.0	−1, +1	tin	Sn	118.7	+2, +4
iodine	I	126.9	−1, +5	titanium	Ti	47.9	+3, +4
iron	Fe	55.8	+2, +3	tungsten	W	183.8	+6
lead	Pb	207.2	+2, +4	zinc	Zn	65.4	+2

TABLE A-5 COMMON IONS

Name	Symbol	Name	Symbol
aluminum	Al^{+3}	lead(II)	Pb^{+2}
ammonium	NH_4^+	magnesium	Mg^{+2}
barium	Ba^{+2}	mercury(I)	Hg_2^{+2}
calcium	Ca^{+2}	mercury(II)	Hg^{+2}
chromium(III)	Cr^{+3}	nickel(II)	Ni^{+2}
cobalt(II)	Co^{+2}	potassium	K^+
copper(I)	Cu^+	silver	Ag^+
copper(II)	Cu^{+2}	sodium	Na^+
hydronium	H_3O^+	tin(II)	Sn^{+2}
iron(II)	Fe^{+2}	tin(IV)	Sn^{+4}
iron(III)	Fe^{+3}	zinc	Zn^{+2}
acetate	$C_2H_3O_2^-$	hydrogen sulfate	HSO_4^-
bromide	Br^-	hydroxide	OH^-
carbonate	CO_3^{-2}	hypochlorite	ClO^-
chlorate	ClO_3^-	iodide	I^-
chloride	Cl^-	nitrate	NO_3^-
chlorite	ClO_2^-	nitrite	NO_2^-
chromate	CrO_4^{-2}	oxide	O^{-2}
cyanide	CN^-	perchlorate	ClO_4^-
dichromate	$Cr_2O_7^{-2}$	permanganate	MnO_4^-
fluoride	F^-	peroxide	O_2^{-2}
hexacyanoferrate(I)	$Fe(CN)_6^{-4}$	phosphate	PO_4^{-2}
hecacyanoferrate(II)	$Fe(CN)_6^{-3}$	sulfate	SO_4^{-2}
hydride	H^-	sulfide	S^{-2}
hydrogen carbonate	HCO_3^-	sulfite	SO_3^{-2}

MODERN CHEMISTRY

ANNOTATED TEACHER'S EDITION

HOLT, RINEHART AND WINSTON

AUSTIN NEW YORK SAN DIEGO CHICAGO TORONTO MONTREAL

THE HRW MODERN CHEMISTRY PROGRAM

Modern Chemistry (Student Text)
Modern Chemistry (Annotated Teacher's Edition)
Laboratory Experiments (Consumable Version)
Laboratory Experiments (Nonconsumable Version)
Laboratory Experiments (Teacher's Edition)
Section Reviews
Section Reviews (Teacher's Edition)
Tests (Blackline Masters and Answer Key)
A Student Guide to Problem Solving

Cover: A view of light fuel oil using the technique of phase contrast photography. In phase contrast photography, transmitted light is refracted in the same way that a prism breaks white light into a full spectrum of colors. (See Section 4.1.) Light fuel oil is one of the many products of fractional distillation of crude oil. (See Chapter 21.) Other products in the distillation of crude oil include hydrocarbon polymers, the building blocks of plastics.

Photo: Manfred Kage/Peter Arnold, Inc.

For permission to reprint copyrighted material, grateful acknowledgment is made to the following sources:

CRC Press, Inc.: From *CRC Handbook of Chemistry and Physics*, 69th Edition, page B-88; Editor-in-Chief: Robert C. Weast, Ph.D. Copyright © 1988 by CRC Press, Inc., Boca Raton, Florida.

Doubleday, a division of Bantam, Doubleday, Dell Publishing Group, Inc.: From *Radiation Alert, A Consumer's Guide to Radiation* by David I. Poch. Copyright © 1985 by Energy Probe Research Foundation.

Printed in the United States of America

ISBN 0-03-014503-1

90123456 036 987654321

CONTENTS

PE TE

Introducing *Modern Chemistry* T6
Using *Modern Chemistry* T14
Curriculum Planner T18
Sources for Audiovisual Resources
 and Computer Software T28
Demonstration Supplies T29
Pupil's Edition Table of Contents iv
Discovering *Modern Chemistry* ix

**Unit 1 Introduction to Chemistry
 and Matter** 2

Chapter 1 Matter, Energy, and Change
 Chapter Planner and Teaching Strategies 4A
 Answers and Solutions 4F
 Chapter Opener **4**
1.1 What is Chemistry 5
1.2 Matter and Energy 10
1.3 Classification of Matter 19
1.4 The Chemical Elements 23

Chapter 2 Measurements and Solving Problems
 Chapter Planner and Teaching Strategies 30A
 Answers and Solutions 30D
 Chapter Opener **30**
2.1 Units of Measurement 31
2.2 Heat and Temperature 40
2.3 Using Scientific Measurements 46
2.4 Solving Quantititative Problems 55
 Careers Fire Chemist, Science Writer 64
 Cumulative Review 65
 Intra-Science The Process in Chemistry:
  The Portable Power of the Battery 66

Unit 2 Organization of Matter 68

Chapter 3 Atoms: The Building of Matter
 Chapter Planner and Teaching Strategies 70A
 Answers and Solutions 70D
 Chapter Opener **70**
3.1 The Atom: From Philosophical Idea to
  Scientific Theory 71
3.2 The Structure of the Atom 75
3.3 Weighing and Counting Atoms 82

Chapter 4 Arrangement of Electrons in Atoms
 Chapter Planner and Teaching Strategies 96A
 Answers and Solutions 96D
 Chapter Opener **96**
4.1 Refinements of the Atomic Model 97
4.2 Quantum Numbers and
  Atomic Orbitals 108
4.3 Electron Configurations 112

Chapter 5 The Periodic Law
 Chapter Planner and Teaching Strategies 126A
 Answers and Solutions 126D
 Chapter Opener **126**
5.1 History of the Periodic Table 127
5.2 Electron Configuration and
  the Periodic Table 132
5.3 Electron Configuration and
  Periodic Properties 143

PE TE

Chapter 6 Chemical Bonding
 Chapter Planner and Teaching Strategies 160A
 Answers and Solutions 160E
 Chapter Opener **160**
6.1 Introduction to Chemical Bonding 161
6.2 Covalent Bonding and
  Molecular Compounds 164
6.3 Ionic Bonding and Ionic Compounds 175
6.4 Metallic Bonding 179
6.5 The Properties of Molecular
  Compounds 182
 Careers Laser Chemist, Chemical
  Technician 194
 Cumulative Review 195
 Intra-Science The Question in Chemistry:
  How to Replace Asbestos 196

Unit 3 Language of Chemistry 198

Chapter 7 Chemical Formulas and Chemical Compounds
 Chapter Planner and Teaching Strategies 200A
 Answers and Solutions 200E
 Chapter Opener **200**
7.1 Chemical Names and Formulas 201
7.2 Oxidation Numbers 212
7.3 Using Chemical Formulas 216
7.4 Determining Chemical Formulas 224

Chapter 8 Chemical Equations and Chemical Reactions
 Chapter Planner and Teaching Strategies 232A
 Answers and Solutions 232E
 Chapter Opener **232**
8.1 Chemical Equations 233
8.2 Types of Chemical Reactions 244
8.3 Activity Series of the Elements 251

Chapter 9 Stoichiometry
 Chapter Planner and Teaching Strategies 260A
 Answers and Solutions 260D
 Chapter Opener **260**
9.1 Introduction to Stoichiometry 261
9.2 Ideal Stoichiometric Calculations 265
9.3 Limiting Reactants and Percent Yield 272
 Careers Analytical Pharmaceutical
  Chemist, Chemical Plant Manager 282
 Cumulative Review 283
 Intra-Science The Process of Chemistry:
  The Amazing Polyhexamethylenea-
  dipamide 284

Unit 4 Phases of Matter 287

Chapter 10 Representative Gases
 Chapter Planner and Teaching Strategies 288A
 Answers and Solutions 288D
 Chapter Opener **288**
10.1 Oxygen and Ozone 289
10.2 Hydrogen 294
10.3 Nitrogen and Ammonia 298
10.4 Carbon Dioxide and Carbon Monoxide 303

Chapter 11 **Physical Characteristics of Gases**
Chapter Planner and Teaching Strategies 314A
Answers and Solutions 314C
Chapter Opener **314**
11.1 The Kinetic Theory of Matter 315
11.2 Qualitative Description of Gases 320
11.3 Quantitative Description of Gases 322

Chapter 12 **Quantitative Behavior of Gases**
Chapter Planner and Teaching Strategies 342A
Answers and Solutions 342C
Chapter Opener **342**
12.1 Volume–Mass Relationships of Gases 343
12.2 The Ideal Gas Law 350
12.3 Stoichiometry of Gases 356
12.4 Effusion and Diffusion 360

Chapter 13 **Liquids and Solids**
Chapter Planner and Teaching Strategies 370A
Answers and Solutions 370C
Chapter Opener **370**
13.1 Liquids 371
13.2 Solids 377
13.3 Changes of State 383
13.4 Water 390
Careers   Occupations in the
Compressed-Gas Industries, Chemical
Patent Attorney 398
Cumulative Review 399
Intra-Science   The Problem in Chemistry:
Preserving the Ozone Layer 400

Unit 5 **Solutions and Their Behavior** 402

Chapter 14 **Solutions**
Chapter Planner and Teaching Strategies 404A
Answers and Solutions 404D
Chapter Opener **404**
14.1 Types of Mixtures 405
14.2 The Solution Process 410
14.3 Concentration of Solutions 420
14.4 Colligative Properties of Solutions 426

Chapter 15 **Ions in Aqueous Solutions**
Chapter Planner and Teaching Strategies 438A
Answers and Solutions 438D
Chapter Opener **438**
15.1 Ionic Compounds in Solution 439
15.2 Molecular Electrolytes 447
15.3 Properties of Electrolyte Solutions 450

Chapter 16 **Acids and Bases**
Chapter Planner and Teaching Strategies 460A
Answers and Solutions 460E
Chapter Opener **460**
16.1 Acids 461
16.2 Bases and Acid–Base Reactions 470
16.3 Relative Strengths of Acids and Bases 475
16.4 Oxides, Hydroxides, and Acids 479
16.5 Chemical Reactions of Acids,
Bases, and Oxides 483

Chapter 17 **Acid–Base Titrations and pH**
Chapter Planner and Teaching Strategies 490A
Answers and Solutions 490D
Chapter Opener **490**
17.1 Concentration Units for
Acids and Bases 491
17.2 Aqueous Solutions and
the Concept of pH 496
17.3 Acid–Base Titrations 505
Careers   Food Chemist, Aroma
Chemist 520

Cumulative Review 521
Intra-Science   The Problem in Chemistry:
Detergents and the Environment 522

Unit 6 **Chemical Reactions** 524

Chapter 18 **Reaction Energy and Reaction Kinetics**
Chapter Planner and Teaching Strategies 526A
Answers and Solutions 526E
Chapter Opener **526**
18.1 Thermochemistry 527
18.2 Driving Force of Reactions 536
18.3 The Reaction Process 541
18.4 Reaction Rate 549

Chapter 19 **Chemical Equilibrium**
Chapter Planner and Teaching Strategies 562A
Answers and Solutions 562D
Chapter Opener **562**
19.1 The Nature of Chemical Equilibrium 563
19.2 Shifting Equilibrium 569
19.3 Equilibrium of Acids, Bases, and Salts 576
19.4 Solubility Equilibrium 583

Chapter 20 **Oxidation–Reduction Reactions**
Chapter Planner and Teaching Strategies 595A
Answers and Solutions 595E
Chapter Opener **596**
20.1 The Nature of Oxidation
and Reduction 597
20.2 Balancing Redox Equations 602
20.3 Oxidizing and Reducing Agents 608
20.4 Electrochemistry 611
Careers   Industrial Technician,
Environmental Chemist 626
Cumulative Review 627
Intra-Science   The Problem in Chemistry:
Chemiluminescence 628

Unit 7 **Carbon and Its Compounds** 630

Chapter 21 **Carbons and Hydrocarbons**
Chapter Planner and Teaching Strategies 632A
Answers and Solutions 632F
Chapter Opener **632**
21.1 Abundance and Importance of Carbon 633
21.2 Organic Compounds 639
21.3 Hydrocarbons 643
21.4 Representative Hydrocarbons
and Polymers 655

Chapter 22 **Substituted Hydrocarbons**
Chapter Planner and Teaching Strategies 664A
Answers and Solutions 664E
Chapter Opener **664**
22.1 Alcohols 665
22.2 Halocarbons 670
22.3 Ethers 673
22.4 Aldehydes and Ketones 674
22.5 Carboxylic Acids 676
22.6 Esters 679

Chapter 23 **Biochemistry**
Chapter Planner and Teaching Strategies 684A
Answers and Solutions 684D
Chapter Opener **684**
23.1 The Chemistry of Life 685
23.2 Proteins 686
23.3 Carbohydrates 689
23.4 Lipids 691
23.5 Nucleic Acids 695
23.6 Biochemical Pathways 700

T4

Careers Agricultural Chemist,
  Biochemist 706
Cumulative Review 707
Intra-Science The Process in Chemistry:
  Custom-Designed Plants 708

Unit 8 Descriptive Chemistry 710

Chapter 24 The Metals of Groups 1 and 2
Chapter Planner and Teaching Strategies 712A
Answers and Solutions 712D
Chapter Opener **712**
24.1 The Alkali Metals 713
24.2 The Alkaline–Earth Metals 717
24.3 Sodium: One of The Active Metals 720

Chapter 25 The Transition Metals
Chapter Planner and Teaching Strategies 728A
Answers and Solutions 728D
Chapter Opener **728**
25.1 The Transition Elements 729
25.2 Iron, Cobalt, and Nickel 736
25.3 Copper, Silver, and Gold 744

Chapter 26 Aluminum and the Metalloids
Chapter Planner and Teaching Strategies 750A
Answers and Solutions 750C
Chapter Opener **750**
26.1 Introduction to Aluminum and
  the Metalloids 751
26.2 Aluminum 753
26.3 Representative Metalloids 758

Chapter 27 Sulfur and Its Compounds
Chapter Planner and Teaching Strategies 764A
Answers and Solutions 764C
Chapter Opener **764**
27.1 Elemental Sulfur 765
27.2 Important Compounds of Sulfur 770

Chapter 28 The Halogens
Chapter Planner and Teaching Strategies 778A
Answers and Solutions 778C
Chapter Opener **778**
28.1 The Halogen Family 779
28.2 Fluorine 781
28.3 Chlorine 782
28.4 Bromine 787
28.5 Iodine 790
Careers Metallurgist and Materials
  Specialist, Chemical Supplier 794
Cumulative Review 795
Intra-Science The Process in Chemistry:
  The Miracle Fluorocarbon Resin 796

Unit 9 Nuclear Reactions 798

Chapter 29 Nuclear Chemistry
Chapter Planner and Teaching Strategies 800A
Answers and Solutions 800C
Chapter Opener **800**
29.1 The Composition and Structure
  of the Nucleus 801
29.2 The Phenomen of Radioactivity 804
29.3 Applications of Radioactivity 816
29.4 Energy from the Nucleus 817
Appendix A Useful Tables 824
Appendix B Logarithms 834
Glossary 836
Index 845
Credits 854
Answers to Cumulative Review
  Questions and Problems 856A

DESKTOP INVESTIGATIONS
Physical and Chemical Changes 18
Density of Pennies 38
Constructing a Model 74
The Wave Nature of Light: Interference 102
Designing Your Own Periodic Table 131
Models of Chemical Compounds 215
Balancing Equations Using Models 249
Limiting Reactants in a Cookie Recipe 273
Properties of Carbon Dioxide 307
Diffusion 361
Surface Tension of Liquids
  and Water Solutions 374
Solutions, Colloids, Suspensions 408
Acids and Bases in Your Kitchen 476
Testing for Acid Precipitation 509
Factors Influencing Reaction Rate 555
Redox Reactions 600
Enzyme Action on Starch 690
Paramagnetism of Nickel 730
Silver Halides 789

CHEMISTRY NOTEBOOK
Interconversion of Matter and Energy 15
Millikan's Oil-Drop Experiment 79
Heisenberg's Uncertainty Principle 107
Exceptions to the Octet Rule 168
Resonance 174
Atomic Symbols of Molecular Formulas 221
Scientific Journals: Getting the Word Out 264
The Nitrogen Cycle 300
Carbon Monoxide Poisoning 309
Molecular Motion 316
Electrolytes in Your Body 447
Boiling Liquids 451
"Acid Stomach" and Antacids 481
Acid Rain and Lakes 504
Not-So-Blue Jeans 610
Chlorohydrocarbons 640
Octane Ratings 646
Polyunsaturated Fats and
  Pour-on Margarines 668
Group 1 Ions in Medication and Diet 715
Strontium: Fireworks and Fallout 719
Gallium Arsenide, Lasers,
  and Computer Chips 761
Medicines Made from Sulfur 769
The Smell of Sulfur Compounds 772
Discovering How to Discover Elements 791
Radium—Cause of and Cure for Cancer 802
Measuring Radiation Exposure 818

TECHNOLOGY
"Looking" at Atoms 81
Hydrazine, a Liquid Rocket Fuel 254
Measuring and Reaching
  Low Temperatures 322
Separating Isotopes 363
Liquid Crystals 376
Superconductors 382
Fresh Water From the Sea 428
Fuels and Energy Content 531
Swimming Pool Chemistry 570
Toward Human Gene Transplants 699
Can a Can Become a Plane? 754
PVC Pipes 784

Modern Chemistry
The Right Text For Today's World

*I*n a complex world that relies increasingly on sophisticated technology, students and teachers need a chemistry text that is accessible, current, and complete. Holt, Rinehart and Winston's new edition of *Modern Chemistry* continues a tradition of scientifically accurate and up-to-date content in a new organizational framework that makes it the most functional and informative chemistry text available.

The *Pupil's Edition* has been completely redesigned to capture students' attention with four-color photographs and illustrations. New and interesting features help maintain and deepen students' interest in chemistry by showing them the relationship between technology and their world.

The *Annotated Teacher's Edition* has also been restructured to provide teachers with a more useful teaching tool. **Chapter Planning Guides** and annotations that appear directly on the text page give teachers the support they need in a practical format.

*Discovering **Modern Chemistry*** is a feature of the *Pupil's Edition* designed to give students insight into the structure of the text and help them achieve greater success in their study of chemistry.

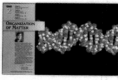

Discovering *Modern Chemistry*

The world we live in is complex and diverse. It contains the hardness of steel and the softness of cotton, the brilliance of diamonds and the blackness of coal, the heat of flames and the chill of ice, the tranquility of a slow moving river and the din of a busy city street. All the colors, textures, objects, animals, and plants that exist are too numerous to count. Yet scientists have learned that there are approximately 90 different ingredients, or elements, that make up all the material things we see and experience.

By studying chemistry we can understand how the diversity and complexity of our world can arise from so few ingredients. As you will discover, the science of chemistry ranges from a thought-provoking academic subject to a powerful tool for innovation in our rapidly changing society.

These pages will show you how *Modern Chemistry* is designed to guide you through your discovery of the science of chemistry.

Modern Chemistry is divided into nine units. On the unit opening pages, men and women who have devoted their careers to science retell what sparked their interest as young people. Maybe you share some of the thoughts and experiences of these scientists. These Unit Openers introduce you to the chapters to follow and help you to understand the ways in which chemistry is important in our lives.

Modern Chemistry contains 29 chapters. The first two pages of each chapter set the stage for learning and discovery. Introduction and Looking Ahead help you frame questions about what you are going to read. Each chapter is divided into numbered sections that begin with learning objectives. These objectives help you anticipate the major concepts of each section. Study Hints tell you where to refresh your memory when previously covered topics are needed to understand new material. Margin notes offer short comments that give greater insight or simply point out items of special interest.

ix

Engaging Introductions

Attractive **Unit** and **Chapter Openers** capture students' attention and help foster interest in upcoming material.

UNIT OPENER

Dramatic photographs accompany a letter of introduction to the students from a renowned chemist working in a field related to the material in the unit.

In these letters, the scientists explain what sparked their interest in chemistry as young people and discuss the value and usefulness of chemistry in the real world.

In each **Unit Opener**, a list of the chapters previews upcoming content.

CHAPTER OPENER

The **Introduction** ties the chapter content to material studied in previous chapters.

Looking Ahead helps students set goals for reading by providing them with learning objectives.

The **Section Preview** identifies the major instructional focus of each lesson.

Intriguing photographs and accompanying captions illustrate the impact that chemistry has on society.

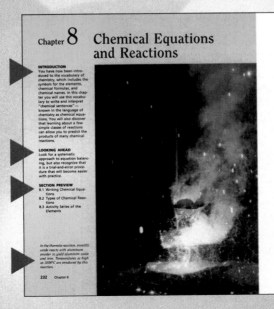

Modern Chemistry —clearly the right text for today's students and teachers.

Well-Organized Chapters

Study Hints provides students with review information, interesting facts, and mnemonic devices to assist them in the lesson.

Sample Problems familiarize students with an easily mastered four-part problem-solving process.

Practice Problems give students the opportunity to practice newly acquired problem-solving skills. Answers in the *Pupil's Edition* allow them to check their work.

*C*hapters are divided into smaller, sequential lessons that make the study of chemistry more accessible.

Section Objectives help students set goals for learning by providing them with a list of pre-reading objectives.

Tables, charts, and photographs visually organize data and reinforce abstract concepts.

Caution alerts students to possible hazards that may arise when working with chemicals.

Important vocabulary words are clearly identified through boldface type.

Thought-Provoking Features

Interesting features enliven the text by providing students with real-life applications for chemistry.

Chemistry Notebook

Chlorohydrocarbons

Chlorohydrocarbons are hydrocarbons in which chlorine atoms are substituted for one or more of the hydrogen atoms. A wide variety of chlorohydrocarbon compounds were used extensively in industry and agriculture until it was found that use of many of these compounds may cause cancer, damage the kidney and liver, and result in ailments of the circulatory and respiratory systems.

The members of one group of chlorohydrocarbons, the chlorinated forms of methane, were widely used until their health hazards were recognized. Carbon tetrachloride (tetrachloromethane) was used in dry cleaning because it dissolves fats, oils, and greases. Chloroform (trichloromethane) is still used as a solvent in many paint strippers. Methyl chloride (chloromethane) was stored under pressure as a liquid and then sprayed onto skin as a local anaesthetic.

Many other chlorohydrocarbons are known to be dangerous to living organisms. DDT (dichlorodipenyltrichloroethane) is a highly effective pesticide, but is now banned in many parts of the world. DDT becomes increasingly concentrated as it moves up the food chain and accumulates in animal tissues. It poisons fish and many other living organisms and interferes with bird reproduction. PCBs (polychlorinated bipenyls) were used in the United States as coolants and insulating fluids for electrical equipment. A herbicide called Agent Orange was used as a defoliant in Vietnam by the United States during its participation in the Vietnam War. A byproduct of the manufacture of PCBs and Agent Orange, the dioxins are a group of chlorinated aromatic hydrocarbons that are known to cause a wide variety of health problems. The town of Times Beach, Missouri, shown in the photograph below was abandoned in 1983 because of dioxin contamination of the soil.

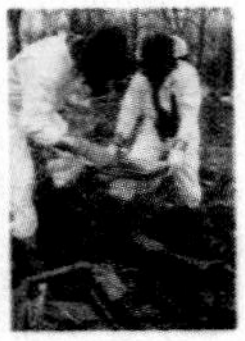

or groups behave similarly, we are able to classify them in terms of their functional groups.

Structural Formulas And Bonding

When you introduce structural formulas, use ball-and-stick models to help students understand the three-dimensional shapes of organic molecules and how these shapes are translated into the two-dimensional structural formulas used in the text.

The molecular formula of a compound shows the exact number of atoms of each type of element present in a molecule. While the formula for sulfuric acid (H_2SO_4) gives enough information for most purposes in inorganic chemistry, a molecular formula such as that for sucrose ($C_{12}H_{22}O_{11}$) is not satisfactory in organic chemistry because there are over 100 different compounds that have the same molecular formula as sucrose.

For this reason, organic chemists use structural formulas to represent organic compounds. *A* **structural formula** *indicates the exact number and types of atoms present in a molecule and also shows how the atoms are bonded*

640 Chapter 21

Technology explains how chemistry has been applied to the world's needs, bringing about worthwhile innovations.

Technology

PVC Pipes

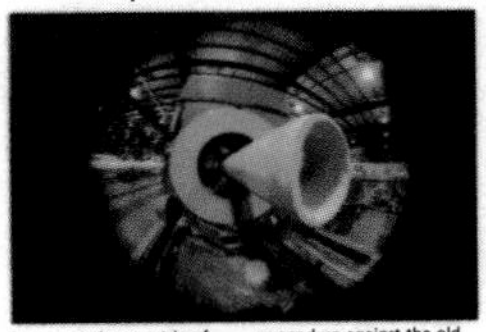

Lead was once used extensively for water pipes and tanks because it was inexpensive and slow to corrode. When the hazards of lead poisoning were recognized, clay was used instead. As our cities age, replacement of those aging clay pipes has become a major problem. How can the pipes be replaced without interrupting the water and sewage service? Engineers developed an interesting process using PVC.

PVC stands for "polyvinyl chloride," a polymeric form of $CH_2—CH—CH_2Cl$. This plastic is durable, cheap, and easy to manufacturer into pipes that do not corrode. The PVC pipes are created from a continuous strip of PVC with special clamping edges. The PVC is fed as a spiral directly into the old pipes, forming a sealed tube that is flexible enough to expand up against the old pipe. This lightweight PVC tube is supported by the old pipes, so it doesn't have to be as thick as other kinds of pipe.

Physical and Chemical Properties of Chlorine

The odor of chlorine is familiar to everyone as the odor of chlorine bleaches and of swimming-pool disinfectant.

Chlorine is a toxic, highly reactive element. Its chemical properties, however, make it an extremely useful element.

Physical Properties At room temperature, chlorine is a greenish-yellow gas with a disagreeable, suffocating odor. It is about 2.5 times as dense as air. It is moderately soluble in water, forming a pale greenish-yellow solution.

Intra-Science shows how chemists work with biologists, physicists, geologists, and other scientists to solve important questions that cross the boundaries of any one science.

Intra-Science
How the Sciences Work Together

The Process in Chemistry: Custom-Designed Plants

Custom-designed plants that are able to flourish in salty soil, resist drought, or survive deadly diseases are gradually becoming a reality. Tobacco plants that are resistant to herbicides are being developed. Strawberry plants sprayed with genetically altered bacteria are able to resist frost damage.

Ever since agriculture began, about 10 thousand years ago, farmers have tried to create improved breeds of plants. Parent plants with desirable characteristics were bred to each other to produce superior offspring. It was not until the mid 1800s, however, that a systematic study of inheritance

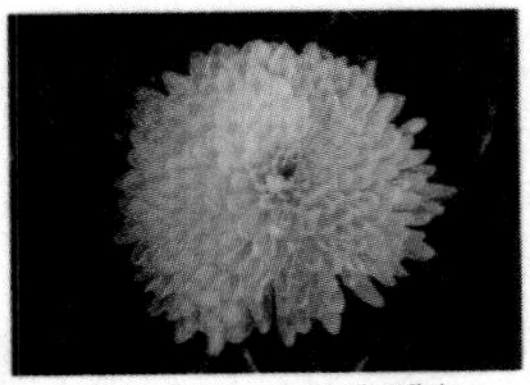

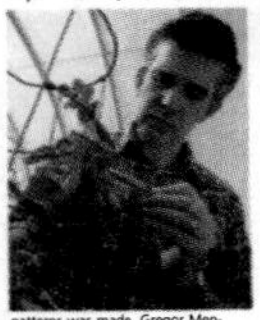

patterns was made. Gregor Mendel, an Austrian monk, clarified the

tion of the DNA (deoxyribonucleic acid) molecule. Their work in the 1950s began a revolution in genetics, leading to the creation of artificial genes and new forms of life.

The molecular basis of heredity is the DNA molecule. DNA is made of long chains of nucleotides. A nucleotide consists of a molecule of the sugar deoxyribose, a phosphate group, and one of four bases—adenine, thymine, cytosine, or quinine. The sequence of bases acts as a code that regulates the sequence of amino acids in a protein. DNA controls the characteristics of an organism by controlling the synthesis of specific proteins.

Every plant has thousands of genes (a gene is a section of DNA). By changing the genes or inserting new ones, plants with new characteristics—custom-designed plants—can be created.

The Connection to Physics
The chemical nature of genes already present in an organism can be altered with radiation. Ultraviolet light, alpha-particles, beta-particles, high-energy electrons,

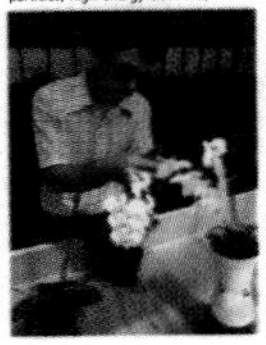

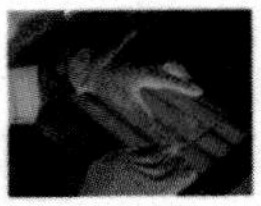

X rays, and other types of high-energy radiation can dislodge one or more electrons from the DNA of a gene. The ionized molecule is more reactive than normal and may combine with other substances. A nucleotide may be added or deleted, and this changes the sequence of bases. As a result, the sequence of amino acids in the

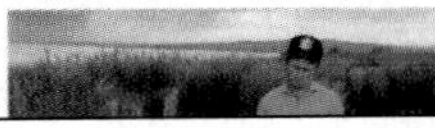

inserted. This process, electroporation, has been used to transfer new genes into corn cells. The biggest stumbling block has been the difficulty of regenerating whole plants from genetically altered cells. If this problem can be solved, electroporation will enable scientists to transfer genes between unrelated species. More nutritious corn, or wheat that can survive floods, may be produced.

The Connection to Biology
Viruses and bacteria are used to transport new genes into the cells of plants. Bacteria of the genus *Agrobacterium* are most commonly used. These bacteria normally cause diseases in plants by inserting a plasmid, a ring of bacterial DNA, into plant cells. The plasmid

the synthesis of an enzyme, EPSP synthetase, which is needed for making three essential amino acids. A mutant gene that produces a slightly different EPSP synthetase enzyme was inserted into cells of tobacco plants, using *Salmonella* bacteria. When fields of genetically altered tobacco plants are sprayed by glyphosate, the weeds are killed. The tobacco plants survive because they are still able to produce the essential amino acids.

Plant pathologists isolated the

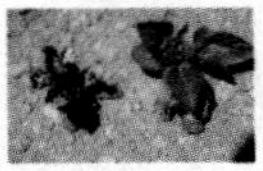

gene of bacteria that causes ice crystals to form and destroy plant cells. They altered the gene that codes for the synthesis of the pro-

Chemistry Notebook is an in-depth look at historical anecdotes, a current advance in chemistry, or an interesting topic related to material in the chapter

Desktop Investigations provides students with experiments that can be done at home or in class and do not require a laboratory setting or complex equipment.

Desktop Investigation

Testing for Acid Precipitation

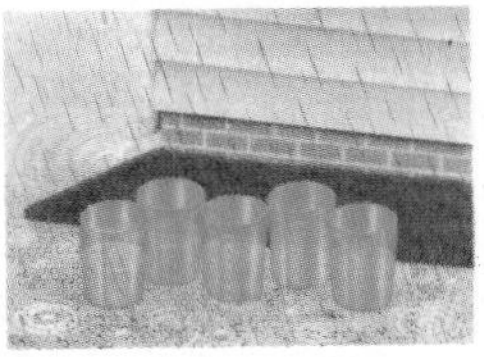

Question
Do you have acid precipitation in your area?

Materials
rainwater
distilled water
500-mL glass containers
thin, transparent plastic metric ruler (±0.1 cm)
hydrion-pH test paper: narrow-range, ±0.2–0.3, or pH meter.

Procedure
Record all your results in a data table.

1. Each time it rains, set out five clean glass containers to collect the rainwater. If the rain continues longer than 24 hours, put out new containers at the end of each 24-hour period until the rain stops. (The same procedure can be used with snow if the snow is allowed to melt before measurements are taken. You may need to use larger containers if a heavy snowfall is expected.)
2. After the rain stops or at the end of each 24-hour period, measure with a thin plastic ruler the depth of the water collected to the nearest 0.1 cm, and test the water with the hydrion paper to determine its pH to the nearest 0.2–0.3.
3. Record the following information: *(a)* the date and time the collection was started; *(b)* the date and time the collection was ended; *(c)* the location where the collection was made (town and state); *(d)* the amount of rainfall in cm; and *(e)* the pH of the rainwater.
4. Find the average pH of each collection that you have made for each rainfall, and record it in the data table.
5. Make collections on at least five different days. The more collections you make, the more reliable your data becomes.
6. For comparison, determine the pH of pure water by testing five samples of distilled water with hydrion paper. Record your results in a separate data table and then calculate an average pH for distilled water.

Discussion

1. What is the pH of distilled water?
2. What is the pH of normal rainwater? How do you explain this?
3. How is acid precipitation defined?
4. What are the drawbacks of using a ruler to measure the height of collected water?
5. Does the amount of rainfall or the time of day the sample is taken have an effect on its pH? Explain any variability among samples.
6. What conclusion can you draw from this investigation? Explain how your data supports your conclusion.

Careers highlights a variety of professions that utilize chemistry to make work easier and more productive.

Careers

Occupations in the Compressed-Gas Industries

The major method of producing pure gases is by compressing and cooling air until it liquefies. This liquid is then fractionally distilled to extract the desired gases. The process requires special equipment but generally does not require chemical expertise from the equipment operators.

The liquefied gases are extremely cold. People who work with compressed gases must understand the changes of pressure and volume that occur as the gases warm and the changes that occur in materials that come in contact with cold, liquified gases. Pipes can shatter, since many metals get brittle when they are cold. Joints will leak if they have not been sealed with materials that can withstand extreme temperature changes.

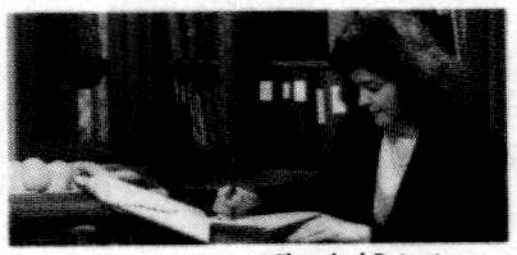

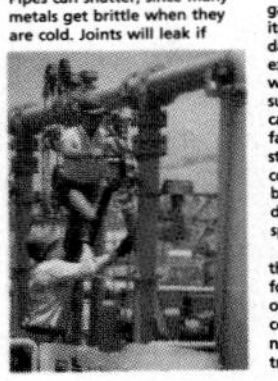

The manufacture and handling of nonatmospheric gases require chemical knowledge as well as common sense. The chemical reaction used to produce the gas must be carefully monitored to avoid potentially dangerous side reactions. For example, chlorine gas reacts with nearly any metallic surface it touches. Leaks of carbon monoxide can lead to fatalities because, in its early stages, CO poisoning is easily confused with the onset of a bad cold. Plans must be developed for coping with spills and leaking pipes.

Opportunities abound in the compressed-gas industries for people with a background or interest in chemistry. A college degree is usually not necessary, but some technical training is essential.

Chemical Patent Attorney

Students who have an interest in both the law and chemistry may find that chemical patent law is an attractive career. Chemical patent attorneys are becoming increasingly important to industry.

In addition to doing the painstaking work needed to establish a valid patent, chemical patent attorneys help decide whether a product is sufficiently different to justify filing a patent application. Deciding if an idea is new enough to patent or whether it infringes on existing patents is often difficult. Even if a chemical compound is already known, it may be possible to patent a new method of making it.

To become a chemical patent attorney, you must study chemistry as part of your undergraduate college program and then obtain a law degree.

398

Comprehensive Reviews

Chapter 8 *Review*

Chemical Equations and Reactions

Vocabulary

activity series
chemical equation
coefficient
combustion reaction
decomposition reaction
double-replacement reaction (ionic reaction)
electrolysis
formula equation
hydrocarbons
reversible reaction
single-replacement reaction (displacement reaction)
synthesis reaction (composition reaction)
word equation

Questions

1. What information is provided by a chemical equation?
2. List the three requirements for a correctly written chemical equation.
3. *(a)* What is meant by the term coefficient in relation to a chemical equation? *(b)* How does the presence of a coefficient affect the number of atoms of each type in the formula that it precedes?
4. Give one example each of a word equation, formula equation, and chemical equation.
5. What limitations are associated with the use of both word and formula equations?
6. Define each of the following: *(a)* aqueous solution *(b)* catalyst *(c)* reversible reaction.
7. Write formulas for each of the following compounds: *(a)* potassium hydroxide *(b)* calcium nitrate *(c)* sodium carbonate *(d)* carbon tetrachloride *(e)* magnesium bromide *(f)* sulfur dioxide *(g)* ammonium sulfate.
8. List the there steps used in balancing an equation.
9. What four guidelines are useful in balancing an equation?
10. How many atoms of each type are represented in each of the following? *(a)* $3N_2$ *(b)* $2H_2O$ *(c)* $4HNO_3$ *(d)* $2\ Ca(OH)_2$ *(e)* $3Ba(ClO_3)_2$ *(f)* $5Fe(NO_3)_2$ *(g)* $4Mg_3(PO_4)_2$ *(h)* $2(NH_4)_2$ *(i)* $6Al_2(CO_3)_3$.
11. Cite two common mistakes often encountered when attempting to balance chemical equations.
12. Define and give general equations for the five types of chemical reactions.
13. Complete each of the following general patterns for synthesis reactions by identifying the type of product usually formed: *(a)* element + oxygen *(b)* metal + nonmetal (other than oxygen) *(c)* metal oxide (of active metal) + water *(d)* nonmetal oxide (selected) + water *(e)* metal oxide (selected) + nonmetal oxide (selected).
14. What relationship exists between synthesis and decomposition reactions?
15. How are most decomposition reactions initiated?
16. What is electrolysis?
17. *(a)* In what environment do most single replacement reactions occur? *(b)* How do single-replacement reactions compare with synthesis and decomposition reactions in terms of the amount of energy generally involved?
18. What are hydrocarbons?
19. Complete the general pattern for the complete combustion of hydrocarbons by identifying the types of products usually formed:

    hydrocarbon + oxygen

20. *(a)* What is meant by the "activity" of an element? *(b)* What is the relationship between the activity of an element and the ease with which it reacts with other substances?
21. *(a)* What is an activity series of elements? *(b)* What is the basis for the ordering of the elements in the activity series?
22. *(a)* What is the relationship between the activity of the halogens and their arrangement on the Periodic Table? *(b)* List the halogens in order of decreasing activity.
23. *(a)* What is the chemical principle upon which the activity series of metals is based? *(b)* What is the significance of the distance between two metals in the activity series?

256 Chapter 8

The **Chapter Review** reexamines important vocabulary words, concepts, and problem-solving skills from the chapter. Enrichment activities enhance the content and reinforce science process skills.

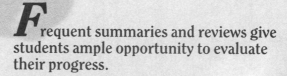

***F**requent summaries and reviews give students ample opportunity to evaluate their progress.*

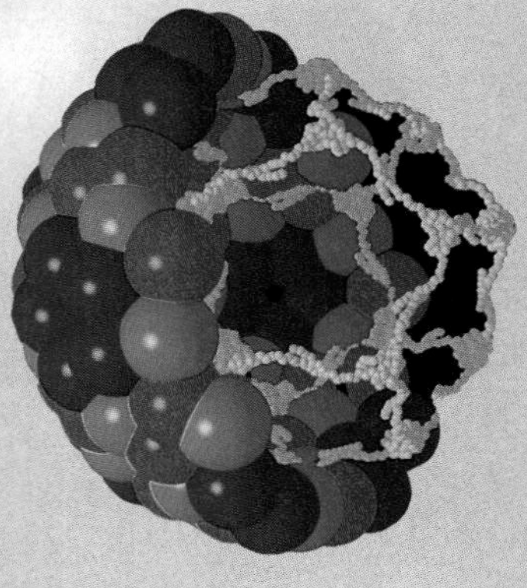

Cumulative *Review*

Unit 1 • Introduction to Chemistry and Matter

Questions

1. Define each of the following: *(a)* science *(b)* chemistry *(c)* scientific method *(d)* inertia *(e)* matter *(f)* energy *(g)* characteristic property *(h)* precipitate. (1.1)
2. List the five major areas of chemistry. (1.1)
3. Identify and describe the four main parts of the scientific method. (1.1)
4. Distinguish between each of the following pairs of terms: *(a)* qualitative and quantitative information *(b)* weight and mass *(c)* kinetic and potential energy *(d)* physical and chemical properties of matter *(e)* physical and chemical changes *(f)* reactants and products *(g)* exothermic and endothermic processes *(h)* heterogeneous and homogeneous matter *(i)* pure substances and mixtures *(k)* groups and periods on the periodic table *(j)* elements and compounds. (1.1, 2, 3, 4)
5. State each of the following: *(a)* the law of conservation of matter *(b)* the law of conservation of energy *(c)* the law of definite composition. (1.2)
6. List and define the three states of matter. (1.2)
7. Classify each of the following as a physical or chemical change: *(a)* freezing water *(b)* dissolving sugar in iced tea *(c)* making a tossed salad *(d)* burning wood *(e)* tearing paper *(f)* tarnishing silver. (1.2)
8. Identify three observable changes that generally indicate a chemical reaction has occurred. (1.2)
9. List and distinguish among the three general classes of elements. (1.4)
10. What are the fundamental SI units for length, mass, and time? (2.1)
11. Distinguish between *(a)* temperature and heat *(b)* accuracy and precision *(c)* direct and inverse proportion between two variables. (2.2)
12. List the four steps in the stepwise method of problems solving. (2.4)
13. Identify the SI unit(s) that would be most appropriate for measuring each of the following: *(a)* the thickness of a dime *(b)* the mass of a vitamin pill *(c)* the volume of a glass full of orange juice *(d)* your mass *(e)* the distance from New York City to Chicago *(f)* the volume of an olympic-size swimming pool. (2.1)
14. Round off 125.0965 cm to the indicated number of significant figures: *(a)* 6 *(b)* 5 *(c)* 3 *(d)* 2 *(e)* 4 *(f)* 1. (2.3)
15. Express each of the following in scientific notation: *(a)* 4500 *(b)* 370 000 *(c)* 0.053 *(d)* 0.000 081 *(e)* 650×10^4 *(f)* 206×10^{-7}. (2.3)
16. Write each of the following in the usual, long form: *(a)* 7.2×10^4 *(b)* 1.3×10^{-5} *(c)* 3×10^{10}. (2.3)

Problems

1. Express each measurement in the SI unit indicated: *(a)* 3.5 m in cm *(b)* 425 g in kg *(c)* 7.8 L in ml *(d)* 1.5 km in mm *(e)* 1.26 l in cm^3. (2.1)
2. Determine the density of an unknown substance if 15.0 grams of the material occupy a volume of 4.5 cm^3. (2.1)
3. Find the volume of an object having a density of 8.62 g/cm^3 and a mass of 19.40 g. (2.1)
4. Convert each of the following temperature readings: *(a)* 20°C to K *(b)* −65°C to K *(c)* 350K to °C *(d)* 100K to °C. (2.2)
5. Convert 350.0 calories to joules. (2.2)
6. How much heat would be absorbed by 20.0 g of silver when heated from 20°C to 35°C? (2.2)
7. Solve each of the following:
   (a) $4.2 \times 10^3 + 2.9 \times 10^4$
   (b) $(3.12 \times 10^5)(8.1 \times 10^{-9})$
   (c) $\frac{(5.20 \times 10^{-7})(7.0 \times 10^4)}{1.2 \times 10^{-5}}$ (2.3)
8. A rectangular tank has a length of 2.40 m, a width 14.0 dm, and a height of 80.0 cm. *(a)* Find its volume in cm^3. *(b)* How many liters of water could this tank hold? (2.4)

65

The **Cumulative Review** helps students reevaluate all the information in the unit through a variety of vocabulary, short-answer, and problem-solving questions.

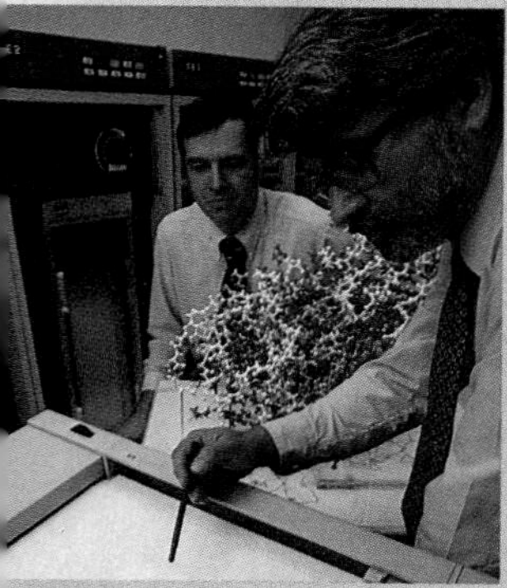

Sample Problem 4.4 *continued*

2. *(a)* Identify the element whose electron-dot symbol contains five dots in its fifth and highest-energy level. Write its shorthand electron-configuration notation. *(Ans.)* Sb, $[Kr]4d^{10}5s^25p^3$
   (b) Name the fourth-period element whose atoms have seven dots in their electron-dot notation. Which element would have that same number of outermost electrons in its sixth main-energy level? *(Ans.)* Br, At

Section Review

1. *(a)* What is electron configuration? *(b)* What three principles guide the development of the electron configuration of an atom?
2. What three methods are used to represent the arrangement of electron in atoms?
3. What is an octet of electrons?
4. Write both the complete and shorthand electron configurations, as well as the orbital and electron-dot notation for *(a)* C *(b)* Ne *(c)* S.
5. Identify each of the indicated elements: *(a)* $1s^22s^22p^63s^23p^3$ *(b)* $[Ar]4s^1$ *(c)* contains four electrons in its third and outer main energy level *(d)* contains one paired and three single electrons in its electron-dot symbol in the fourth and outer main-energy level *(e)* the first element to contain a *d* electron.

1. *(a)* The arrangement of electrons in atoms *(b)* The Aufbau principle, Hund's rule, and the Pauli exclusion principle
2. Orbital notation, electron-configuration notation, and electron-dot notation
3. The filling of the *s* and *p* sublevels of an atom's highest main energy level with eight electrons
4. The indicated responses are as follows:
   (a) $1s^22s^22p^2$, $[He]2s^22p^2$
   1s 2s 2p 2p 2p, ·Ċ·
   (b) $1s^22s^22p^6$, $[H]2s^22p^2$
   1s 2s 2p 2p 2p, :N̈e:
   (c) $1s^22s^22p^63s^23p^4$, $[Ne]3s^23p^4$
   1s 2s 2p 2p 2p
   3s 3p 3p 3p, :Ṡ:
5. The indicated elements are: *(a)* P *(b)* K *(c)* Si *(d)* As *(e)* Sc

Chapter Summary

- Quantum theory was developed early in this century to explain such observations as the photoelectric effect and the emission spectrum of hydrogen.
- Quantum theory states that electrons can exist in atoms only at specified energy levels called orbitals.
- When an electron moves from one orbital to an orbital with lower energy, a photon is emitted whose energy equals the energy difference between the two orbitals.
- An electron in an atom can move from one orbital to a higher energy orbital within the atom only by absorbing an amount of energy exactly equal to the difference between the two orbitals.
- The four quantum numbers that describe the properties of electrons in atomic orbitals are the principal quantum number, the orbital quantum number, the magnetic quantum number, and the spin quantum number.
- Electrons occupy atomic orbitals in the ground state of an atom according to the Aufbau principle, Hund's rule, and the Pauli exclusion principle.
- Electron configurations can be written using three different types of notations. They are orbital notation, electron-configuration notation, and electron-dot notation.
- Some electron configurations of larger atoms such as chromium do not strictly follow the Aufbau principle, but the entire grounds x state configuration that results is thought to be of a lower energy than that predicted by the Aufbau principle.

Arrangement of Electrons in Atoms 121

The **Chapter Summary** provides students with a succinct restatement of major concepts.

Figure 10-13 When a mixture of ammonium chloride and calcium hydroxide is heated in a test tube, ammonia is given off. The ammonia is collected by the downward displacement of air because it is less dense than air and very soluble in water. Litmus paper turns blue in the presence of ammonia, a base.

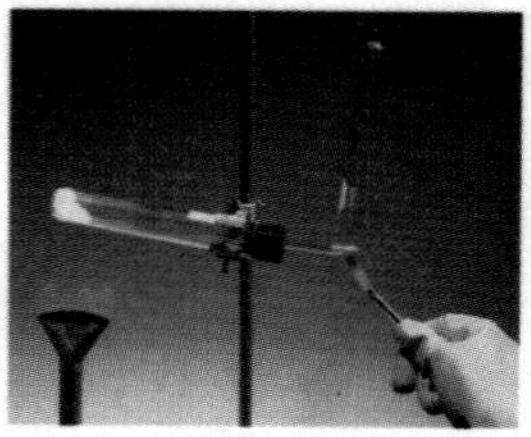

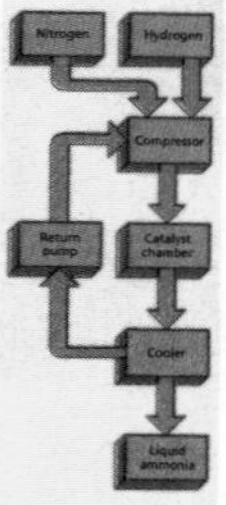

Figure 10-14 *(a)* A flow chart of the Haber process. Ammonia gas produced in the catalyst chamber is condensed into a liquid in the cooler. The uncombined nitrogen and hydrogen are recirculated through the compressor and catalyst chamber.

Industrial preparation of ammonia Ammonia is formed as an industrial by-product in the production of coal gas by heating bituminous coal in the absence of air. Most ammonia is produced industrially by the Haber process, however. *The* **Haber process** *is the catalytic synthesis of ammonia from nitrogen gas and hydrogen gas.*

$$N_2(g) + 3H_2(g) \rightleftharpoons 2NH_3(g)$$

This reaction is reversible and is carried out at temperatures of 400°–500°C and at pressures as high as 1000 atm (101 MPa) using a catalyst that is a mixture of porous iron and oxides of other metals. Under these conditions, about 40 to 60 percent of the reacting gases are changed to ammonia in each pass through the catalyst chamber, as shown in the chart in Figure 10-14. The ammonia is separated from the unreacted nitrogen and hydrogen by dissolving it in water or by cooling and liquefying it. The unreacted nitrogen and hydrogen are recompressed and again passed over the catalyst. Eventually, all the nitrogen and hydrogen are converted to ammonia in this way. More than 10 billion kilograms of ammonia are produced each year in the United States.

Section Review

1. Explain the importance of nitrogen to life on earth.
2. Compare the structures of nitrogen (N_2) and ammonia (NH_3).
3. Compare the physical properties of nitrogen and ammonia.
4. What is photochemical smog?
5. How is nitrogen prepared industrially?
6. Explain what is meant by the nitrogen cycle.

302 Chapter 10

The **Section Review** occurs at the end of every lesson and reinforces the **Section Objectives** through a series of questions.

A Functional *Teacher's Edition*

***C*hapter Planning Guides** and on-page annotations give teacher's the support they need in a practical format.

The **Chapter Planner** helps teachers pick topics, features, and review materials appropriate for the ability level of their students and the focus of the class.

The **Introduction** clarifies the teaching goal of the chapter and suggests ways to involve students.

The **References and Resources** section is a complete list of books, periodicals, audiovisual aids, and computer software that can be used to enrich the chapter content.

Explicit directions for the **Desktop Investigation** and follow-up **Discussion Questions** provide a useful resource for teachers who want to perform the experiment as a class activity.

Detailed teaching suggestions for each lesson recommend a teaching strategy and feature a hands-on **Demonstration**.

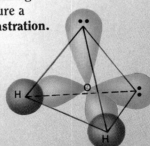
Water

Annotated Pupil's Page

Helpful annotations supply insightful background information, interesting science facts, safety notes, useful teaching suggestions, and answers for the Section Reviews.

Realistic Laboratory Experiments

*N*umerous laboratory experiments reinforce material presented in the text and provide hands-on opportunities to sharpen science process skills.

SKILL-BASED INVESTIGATIONS

Clearly labeled **Objectives** identify the concepts and science process skills being addressed.

Real applications for each experiment are emphasized in the **Introduction** and **General Conclusions** sections of the lab.

Seventeen of the labs offer a **Micro** experiment format that uses minute amounts of chemicals. The **Micro** experiments are safer, less expensive, and reduce chemical disposal problems.

Strategy notes provide helpful hints about doing the experiment and reinforce the process skills involved.

FLEXIBLE ORGANIZATION

▶ A total of forty-four experiments provides a variety of labs from which to choose. All labs are specifically correlated to the content of the pupil's text.

▶ The *Pupil's Edition* of the *Laboratory Experiments* is available in both a consumable and nonconsumable format.

Complete Ancillary Materials

*T*he *Teacher's Resource Organizer* provides for additional practice, process skill development, and testing.

TEACHER'S RESOURCE ORGANIZER

▶ The *Guide to the Teacher's Resource Organizer* provides a quick guide to *Modern Chemistry* and its components.

▶ The *Pupil's Edition* of the *Laboratory Experiments* provides forty-four consumable lab investigations with accompanying data sheets. The *Teacher's Edition* contains answers to all questions.

▶ *Section Reviews* contains questions correlated to the objectives in each lesson that can be used as quizzes, additional practice, or homework. The *Teacher's Edition* supplies answers to all questions.

▶ *A Student Guide to Problem Solving* helps familiarize students with the four-part problem-solving process used in *Modern Chemistry.* Solutions to all problems are included in the booklet.

▶ The *Tests* for *Modern Chemistry* include two versions of each Chapter and Unit test and the Final Exam. Answers for each test are included.

ALSO AVAILABLE

▶ The *TestBank* ™ is a computerized testing system with 5,000 test items that can be organized according to the objectives presented in the chapters. A *Teacher's Guide* and program disc are included.

▶ The non-consumable *Laboratory Experiments* contains the same experiments as the consumable version but does not include the data sheets for student response.

USING MODERN CHEMISTRY

Modern Chemistry is a complete introductory chemistry course for students in college-preparatory or general chemistry programs. *Modern Chemistry* is a program designed to help students see how chemical principles and concepts are developed from experimental observations and data, and how these principles can be used to explain phenomena in daily life as well as in the laboratory. In this text, special attention is given to problems we face today and the attitudes, understandings, and skills that will help students analyze carefully and act wisely on issues that will confront us all as citizens in our technological world.

Many concepts in an up-to-date chemistry program are abstract and may be difficult to grasp. In this text, these concepts are introduced when students need to know about them and only after they have the background needed to understand them. Concepts are first introduced qualitatively and in simple terms, and then developed quantitatively in a sequence of steps to a depth that gives them meaning and significance. There are many Sample Problems to illustrate concepts in the text. Each Sample Problem consists of a statement of the problem and a four-part solution that includes a discussion of the answer. Following each Sample Problem are Practice Problems with answers; encourage students to use them to build problem-solving skills. In *Modern Chemistry*, concepts are developed, reinforced, and extended in numerous ways to help students experience successful learning.

The *Modern Chemistry* text contains 29 chapters organized into nine units. The text contains many features designed to enhance learning by making the study of chemistry interesting, understandable, and relevant. For example, each unit begins with an exciting photograph, a thought-provoking essay addressed to the students, and a list of the chapters in the unit. Each chapter introduces its subject with a full-color photograph that highlights the relevance of the topic in that chapter. The Introduction section of each chapter gives an overview of the ideas and sequence of topics developed within the chapter, and sets the stage for the material that follows. The initial Looking Ahead section alerts students to what to look for as they work through the chapter. Each chapter is divided into sections. The Section Objectives list the tasks that students should be able to do after studying the section. After each section, Section Review Questions give students an opportunity to assess their understanding. Each chapter contains special features that enrich the study the chemistry. Chemsitry Notebook features describe advanced topics, instrumentation, related topics, and historical topics, and provide concrete reinforcement and extension of concepts developed in the text. Technology features describe new and break-through technologies that relate to the chapter material. Desktop Investigations suggest

brief, interesting, thought-provoking investigations, projects, experiments, or activities that can be done in class or at home. Finally, study hints and margin notes reinforce concepts, offer additional information, and assist students in developing effective study techniques.

With each chapter of *Modern Chemistry* students will experience the excitement of making discoveries and decisions, of thinking creatively, and solving problems in chemistry. The clearly written text will help students grasp ideas easily. New terms are highlighted in boldface type and defined at their first appearance. Charts, tables, and hundreds of detailed full-color illustrations and photographs organize scientific information in an easy-to-use fashion. The Chapter Summaries focus on the relative importance of the topics and connect them with the themes of the book. The Vocabulary section highlights key terms introduced in the chapter. Review Questions and Problems provide ongoing reinforcement and extension of critical-thinking skills. Application Questions and Problems require students to use problem-solving skills, understand information in the text, generate new ideas, and combine ideas in new ways. Enrichment activities invite students to explore concepts beyond those introduced in the text through library research and hands-on projects that extend learning.

Eight of the nine units of *Modern Chemistry* close with a Careers feature that exposes students to the diverse job opportunities in chemistry and provides descriptions of and educational requirements for employment in a variety of fields. Intra-Science Features show you how the sciences work together, and present interesting aspects and useful applications of chemistry to fields such as biology, physics, and engineering. The Cumulative Review tests varying levels of understanding and requires the application of concepts covered in the unit and to think critically by reinforcing skills such as inferring, hypothesizing, prediction, and problem solving.

Finally, the Appendices contain many important and useful tables and a discussion of logarithms. The Glossary lists important terms with their definitions and page references. The Index will help students quickly locate information in the text.

For more information on adapting *Modern Chemistry* to meet your specific curricular needs, please refer to the curriculum planning section that follows.

In sum, the authors feel that *Modern Chemistry* is a friendly, interesting, and readable book that will help students master the principles of chemistry.

Nicholas D. Tzimopoulos, Ph.D.

Planners

The *Modern Chemistry* system of lesson planning contains two types of planners: The Curriculum Planner, which begins on page T17, and the Chapter Planners, which appear before each chapter.

The Curriculum Planner displays the entire academic year at a glance, with all major topics shown, so that the chemistry teacher may get an overall view of the *Modern Chemistry* course. By consulting the Curriculum Planner, the teacher can design a complete year of lessons, choosing topics to be covered according to their importance in a first-year chemistry course and the abilities of students.

After a topic is chosen from the Curriculum Planner, the teacher may use the Chapter Planner to find out which activities increase understanding of that topic. The Chapter Planners provide unusually detailed information: teacher demonstrations, experiments from *Lab Experiments in Modern Chemistry*, Desktop Investigations, features, and end-of-chapter questions and problems are all broken down by topic *and* by degree of importance to understanding the topic.

A portion of the Curriculum Planner and a sample Chapter Planner are provided here to illustrate their use.

Curriculum Planner

Chapter 18 Reaction Energy and Reaction Kinetics		5–7 days
18.1 Thermochemistry	Heat of Reaction	
	Heat of Formation	
	Stability and Heat of Formation	
	Heat of Combustion	
	Calculating Heats of Reaction	
	Bond Energy and Reaction Heat	
18.2 Driving Force of Reactions	Enthalpy and Reaction Tendency	
	Entropy and Reaction Tendency	
	Free Energy	
18.3 The Reaction Process	Reaction Mechanisms	
	Collision Theory	
	Activation Energy	
	The Activated Complex	
18.4 Reaction Rate	Rate-Influencing Factors	
	Rate Laws for Reactions	

Coded Topics. Each topic is coded to indicate its importance in a first-year chemistry course:

Fundamental: topics that are fundamental to an understanding of chemistry.

Important: topics that are important to an understanding of chemistry and should be added if time allows.

Additional: topics that are suitable for accelerated classes or classes in which there is particular interest in a topic.

Guides for Time Allocation. The number of class days that should be devoted to each chapter is suggested, taking into account the time necessary for demonstrations, laboratory work, and tests. This pacing guide is presented as a range of days, as you see in the example above, in which the range is 5–7 days. In addition, a blank column is provided to allow the teacher to make his or her own time allocations.

180-Day Curriculum: The first number of the range is the number of days we suggest to be spent on the chapter if the teacher is required to prepare a curriculum in which all areas of chemistry are covered in a 180-day academic year. The first numbers of the ranges add up to 180 days.

Extended Curriculum: The second number of each range is the number of days we suggest a class of average scholastic ability spend for in-depth coverage of the chapter. The total number of days for the entire course adds up to greater than 180.

Chapter Planner

18.1 Thermochemistry
Heat of Reaction
- Demonstration: 1
Questions: 1-4
Problems: 1, 2
- Application Questions: 4, 5

Heat of Formation
- Question: 5
- Application Question: 3

Stability and Heat of Formation
- Application Question: 1
- Application Problem: 2

Heat of Combustion
- Technology
Experiment: 25
Problems: 5-9
- Question: 6
Application Question: 2
- Application Problem: 1

Calculating Heats of Reaction
- Experiment: 26
Questions: 7, 8a
Problems: 3, 4
- Question: 8b

Bond Energy and Reaction Heat
- Question: 9

18.2 Driving Force of Reactions
Enthalpy and Reaction Tendency
- Application Question: 7

Entropy and Reaction Tendency
- Demonstration: 2
Application Question: 6

Free Energy
- Questions: 10-15
Problem: 10
- Application Question: 8
Application Problems: 3, 4

18.3 The Reaction Process
Reaction Mechanisms
- Application Question: 9

Collision Theory
- Question: 16
Application Questions: 10, 11

Activation Energy
- Question: 17

The Activated Complex
- Questions: 18-20
Problems: 10, 11
- Application Problem: 5

18.4 Reaction Rate
Rate-Influencing Factors
- Demonstrations: 3, 4
Desktop Investigation
Experiment: 27
Questions: 21, 22
Application Question: 13
- Application Question: 12

Rate Laws for Reactions
- Questions: 23-24
Problem: 12
Application Problem: 6
- Application Questions: 14-16

Coded Topics. Each topic is coded to indicate its importance to an understanding of the topic:

Fundamental: Activities that are fundamental to an understanding of the topic.

Important: Activities that are important to an understanding of the topic and should be added if time allows.

Additional: Activities that are suitable for accelerated classes or classes in which there is particular interest in the topic.

CURRICULUM PLANNER

Chapter 1 Matter, Energy, and Change		9–11 days
1.1 What Is Chemistry?	Chemistry is a Physical Science	
	Chemistry in Modern Society	
	What is a Chemical?	
	Branches of Chemistry	
	The Scientific Method	
1.2 Matter and Energy	Definition of Matter	
	Definition of Energy	
	Law of Conservation of Energy	
	States of Matter	
	Properties and Changes of Matter	
1.3 Classification of Matter	Mixtures	
	Pure Substances	
1.4 The Chemical Elements	Introduction to the Periodic Table	
	Symbols for the Elements	
	Types of Elements	
	Metals	
	Two Typical Elements	

Chapter 2 Measurements and Solving Problems		10–12 days
2.1 Units of Measurement	The SI Measurement System	
	Fundamental SI Units	
	Units of Measurement in Calculations	
	Derived SI Units	
2.2 Heat and Temperature	Definitions of Heat and Temperature	
	Units of Heat	
	Heat Capacity and Specific Heat	
2.3 Using Scientific Measurements	Accuracy and Precision	
	Significant Figures	
	Scientific Notation	
2.4 Solving Quantitative Problems	Four Steps for Solving Quantitative Problems	
	Proportional Relationships	

Chapter 3 Atoms: The Building Blocks of Matter		5–6 days
3.1 The Atom: From Philosophical Idea to Scientific Theory	Dalton's Atomic Theory	
	Law of Multiple Proportions	

Chapter 3 Atoms: The Building Blocks of Matter (Continued)		
3.2 The Structure of the Atom	Discovery of the Electron	
	Discovery of the Atomic Nucleus	
	Composition of the Atomic Nucleus	
	The Isotopes of Hydrogen	
	The Sizes of Atoms	
3.3 Weighing and Counting Atoms	Atomic Numbers and Mass Numbers	
	Relative Atomic Masses	
	Average Atomic Masses of Elements	
	The Mole, Avogadro's Number, and Molar Mass	

Chapter 4 Arrangement of Electrons in Atoms		5–6 days
4.1 Refinements of the Atomic Model	Wave–Particle Nature of Light	
	Bohr Model of the Hydrogen Atom	
	Spectroscopy	
	Quantum Model of the Atom	
4.2 Quantum Numbers and Atomic Orbitals	Quantum Numbers	
	Electrons in Each Main Energy Level	
4.3 Electron Configurations	Rules Governing Electron Configurations	
	Representing Electron Configurations	
	Elements of the Second and Third Periods	
	Elements of the Fourth and Fifth Periods	
	Elements of the Sixth and Seventh Periods	

Chapter 5 The Periodic Law		4–6 days
5.1 History of the Periodic Table	Mendeleev and Chemical Periodicity	
	Moseley and the Periodic Law	
	The Modern Periodic Table	
5.2 Electron Configuration and the Periodic Table	Groups, Periods, and Blocks of the Periodic Table	
	The s-Block Elements: Groups 1 and 2	
	The d-Block Elements: Groups 3–12	
	The p-Block Elements: Groups 13–18	
	The f-Block Elements: Lanthanides and Actinides	
5.3 Electron Configuration and Periodic Properties	Atomic Radii	
	Ionization Energy	
	Electron Affinity	
	Ionic Radii	
	Valence Electrons	
	Electronegativity	
	Periodic Properties of the d- and f-Block Elements	

Chapter 6   Chemical Bonding | 8–10 days

6.1	Introduction to Chemical Bonding	Types of Chemical Bonds	
		Why Chemical Bonding Occurs	
6.2	Covalent Bonding and Molecular Compounds	Formation of Covalent Bonds	
		The Octet Rule	
		Lewis Structures	
		Multiple Covalent Bonds	
		Polyatomic Ions	
6.3	Ionic Bonding and Ionic Compounds	Formation of Ionic Bonds	
		A Comparison of Ionic and Molecular Compounds	
6.4	Metallic Bonding	Formation of Metalic Bonds	
		Nature of Metals	
6.5	The Properties of Molecular Compounds	VSEPR Theory	
		Hybridization	
		Intermolecular Forces	

Chapter 7   Chemical Formulas and Chemical Compounds | 22–24 days

7.1	Chemical Names and Formulas	Significance of a Chemical Formula	
		Monatomic Ions	
		Binary Ionic Compounds	
		Compounds Containing Polyatomic Ions	
		Binary Molecular Compounds	
		Acids and Salts	
7.2	Oxidation Numbers	Assigning Oxidation Numbers	
		Using Oxidation Numbers for Formulas and Names	
7.3	Using Chemical Formulas	Formula Masses and Molar Masses	
		Molar Mass as a Conversion Factor	
		Percent Composition	
7.4	Determining Chemical Formulas	Calculation of Simplest Formulas	
		Calculation of Molecular Formulas	

Chapter 8   Chemical Equations and Reactions | 12–13 days

8.1	Chemical Equations	Reading and Writing Chemical Equations	
		Significance of a Chemical Equation	
		Balancing Chemical Equations	
8.2	Types of Chemical Reactions	Synthesis Reactions	
		Decomposition Reactions	
		Single-Replacement Reactions	
		Double-Replacement Reactions	
		Combustion Reactions	
8.3	Activity Series of the Elements	Principles of the Activity Series	
		Useful Generalizations Based on the Activity Series	

Chapter 9 Stoichiometry		4–5 days
9.1 Introduction to Stoichiometry	Applications of Stoichiometry	
	Reaction-Stoichiometry Problems	
9.2 Ideal Stoichiometric Calculations	Mole–Mole Calculations	
	Mole–Mass Calculations	
	Mass–Mole Calculations	
	Mass–Mass Calculations	
9.3 Limiting Reactants and Percent Yield	Limiting Reactant	
	Percent Yield	

Chapter 10 Representative Gases		3–4 days
10.1 Oxygen and Ozone	Occurrence of Oxygen	
	Structure of the Oxygen Molecule	
	Physical Properties of Oxygen	
	Chemical Properties of Oxygen	
	Uses and Preparation of Oxygen	
	Formation and Properties of Ozone	
10.2 Hydrogen	Occurrence of Hydrogen	
	Physical Properties of Hydrogen	
	Chemical Properties of Hydrogen	
	Uses and Preparation of Hydrogen	
10.3 Nitrogen and Ammonia	Occurrence of Nitrogen	
	Physical Properties of Nitrogen	
	Structure and Chemical Properties of Nitrogen	
	Uses and Preparation of Nitrogen	
	Occurrence of Ammonia	
	Structure and Properties of Ammonia	
	Uses and Preparation of Ammonia	
10.4 Carbon Dioxide and Carbon Monoxide	Occurrence of Carbon Dioxide	
	Structure of Carbon Dioxide	
	Physical Properties of Carbon Dioxide	
	Chemical Properties of Carbon Dioxide	
	Uses and Preparation of Carbon Dioxide	
	Occurrence of Carbon Monoxide	
	Properties and Uses of Carbon Monoxide	

Chapter 11 Physical Characteristics of Gases		5–6 days
11.1 The Kinetic Theory of Matter	The Kinetic-Molecular Theory of Gases	
	The Kinetic Theory and the Nature of Gases	
	Deviations of Real Gases from Ideal Behavior	

Chapter 11 Physical Characteristics of Gases (Continued)		
11.2 Qualitative Description of Gases	Pressure and Volume at Constant Temperature	
	Temperature and Volume of Gases at Constant Pressure	
	Pressure and Temperature	
	Relationship Between Pressure and Moles	
	Relationship Between Volume and Moles	
11.3 Quantitative Description of Gases	Pressure	
	Boyle's Law: Pressure–Volume Relationship	
	Charles' Law: Temperature–Volume Relationship	
	Gay-Lussac's Law	
	The Combined Gas Law	
	Dalton's Law of Partial Pressures	

Chapter 12 Molecular Composition of Gases		9–10 days
12.1 Volume–Mass Relationships of Gases	Measuring and Comparing the Volumes of Reacting Gases	
	Avogadro's Principle	
	Molar Volume of Gases	
	Gas Density	
	Finding Molar Mass From Volume at STP	
12.2 The Ideal Gas Law	Derivation of the Ideal Gas Law	
	The Ideal Gas Constant	
	Using the Ideal Gas Law	
12.3 Stoichiometry of Gases	Volume–Volume Calculations	
	Volume–Mass and Mass–Volume Calculations	
12.4 Effusion and Diffusion	Graham's Law of Effusion or Diffusion	
	Applications of Graham's Law	

Chapter 13 Liquids and Solids		5–6 days
13.1 Liquids	Kinetic Theory Description of the Liquid State	
	Properties of Liquids and the Particle Model	
13.2 Solids	Kinetic Theory Description of the Solid State	
	Properties of Solids and the Particle Model	
	Crystalline Solids	
13.3 Changes of State	Equilibrium	
	Equilibrium Vapor Pressure of a Liquid	
	Boiling	
	Freezing and Melting	
	Phase Diagrams	
13.4 Water	Structure of Water	
	Physical Properties of Water	

Chapter 14 Solutions		9–10 days
14.1 Types of Mixtures	Solutions	
	Suspensions	
	Colloids	
14.2 The Solution Process	Factors Affecting the Rate of Dissolving	
	Solubility	
	Factors Affecting Solubility	
	Heats of Solution	
14.3 Concentrations of Solutions	Percent by Mass	
	Molarity	
	Molality	
14.4 Colligative Properties of Solutions	Vapor-Pressure Lowering	
	Freezing-Point Depression	
	Boiling-Point Elevation	
	Determination of Molar Mass of a Solute	

Chapter 15 Ions in Aqueous Solutions		4–6 days
15.1 Ionic Compounds in Aqueous Solution	Theory of Ionization	
	Dissolving Ionic Compounds	
	Solubility Equilibria	
15.2 Molecular Electrolytes		
15.3 Properties of Electrolyte Solutions	Conductivity of Solutions	
	Colligative Properties of Electrolyte Solutions	

Chapter 16 Acids and Bases		4–6 days
16.1 Acids	General Properties of Aqueous Acids	
	Definitions of Acids	
	Names and Structures of the Common Acids	
	Some Common Acids	
16.2 Bases and Acid–Base Reactions	General Properties of Aqueous Bases	
	Definitions of Bases and Acid–Base Reactions	
	Types of Bases	
16.3 Relative Strengths of Acids and Bases	Brønsted Acid–Base Pairs	
	Relative Strengths of Acids and Bases in Chemical Reactions	
16.4 Oxides, Hydroxides, and Acids	Basic and Acidic Oxides	
	Amphoteric Oxides and Hydroxides	
	Hydroxides, Acids, and Periodic Trends	
16.5 Chemical Reactions of Acids, Bases, and Oxides		

Chapter 17   Acid–Base Titration and pH | 7–9 days

17.1	Concentration Units for Acids and Bases	Chemical Equivalents	
		Normality	
		Relationship of Normality to Molarity	
17.2	Aqueous Solutions and the Concept of pH	Self-Ionization of Water	
		The pH Scale	
		Calculations Involving pH	
17.3	Acid–Base Titrations	Indicators	
		The Principle of Titration	
		Molarity and Titration	
		Normality and Titrations	

Chapter 18   Reaction Energy and Reaction Kinetics | 5–7 days

18.1	Thermochemistry	Heat of Reaction	
		Heat of Formation	
		Stability and Heat of Formation	
		Heat of Combustion	
		Calculating Heats of Reaction	
		Bond Energy and Reaction Heat	
18.2	Driving Force of Reactions	Enthalpy and Reaction Tendency	
		Entropy and Reaction Tendency	
		Free Energy	
18.3	The Reaction Process	Reaction Mechanisms	
		Collision Theory	
		Activation Energy	
		The Activated Complex	
18.4	Reaction Rate	Rate-Influencing Factors	
		Rate Laws for Reactions	

Chapter 19   Chemical Equilibrium | 5–7 days

19.1	The Nature of Chemical Equilibrium	Reversible Reactions	
		Equilibrium, a Dynamic State	
		The Equilibrium Constant	
19.2	Shifting Equilibrium	Reactions that Run to Completion	
		Common-Ion Effect	
19.3	Equilibria of Acids, Bases, and Salts	Ionization Constant of a Weak Acid	
		Ionization Constant of Water	
		Hydrolysis of Salts	
19.4	Solubility Equilibrium	Solubility Product	
		Calculating Solubilities	
		Precipitation Calculations	

Chapter 20 Oxidation–Reduction Reactions		9–10 days
20.1 The Nature of Oxidation and Reduction	Oxidation	
	Reduction	
	Oxidation and Reduction as a Process	
20.2 Balancing Redox Equations	Oxidation-Number Method	
	Ion-Electron Method	
20.3 Oxidizing and Reducing Agents		
20.4 Electrochemistry	Electrochemical Cells	
	Electrolytic Cells	
	Rechargeable Cells	
	Electrode Potentials	

Chapter 21 Carbon and Hydrocarbons		7–9 days
21.1 Abundance and Importance of Carbon	Structure and Bonding of Carbon	
	Allotropic Forms of Carbon	
	Amorphous Forms of Carbon	
21.2 Organic Compounds	Structural Formulas and Bonding	
	Differences Between Organic and Inorganic Compounds	
21.3 Hydrocarbons	Alkanes	
	Alkenes	
	Alkynes	
	Benzene and Aromatic Hydrocarbons	
21.4 Representative Hydrocarbons and Polymers	Natural Gas and Petroleum	
	Petroleum Substitutes	
	Rubber	

Chapter 22 Substituted Hydrocarbons		4–5 days
22.1 Alcohols	Preparation of Alcohols	
	Reactions of Alcohols	
	Properties, Preparation, and Uses of Alcohols	
22.2 Halocarbons	Preparation of Halocarbons	
	Reactions of Halocarbons	
	Properties, Preparation, and Uses of Several Halocarbons	
22.3 Ethers	Preparation of Ethers	
	Reactions, Properties, and Uses of Ethers	
22.4 Aldehydes and Ketones	Preparation and Uses of Aldehydes	
	Preparation and Uses of Ketones	
22.5 Carboxylic Acids	Preparation of Carboxylic Acids	
	Reactions of Carboxylic Acids	
	Preparation, Properties, and Uses of Carboxylic Acids	
22.6 Esters	Preparation of Esters	
	Reactions of Esters	

Chapter 23 Biochemistry		3–4 days
23.1 The Chemistry of Life	Inorganic Molecules and Water	
	Macromolecules	
23.2 Proteins	Amino Acids	
	Protein Structure	
	Enzymes	
23.3 Carbohydrates	Simple Sugars	
	Disaccharides and Polysaccharides	
23.4 Lipids	Fatty Acids	
	Waxes	
	Fats and Oils	
	Phospholipids	
	Steroids	
23.5 Nucleic Acids	Nucleotides	
	The Genetic Code	
	DNA Replication	
	RNA Synthesis	
	Protein Synthesis	
	Genetic Engineering	
23.6 Biochemical Pathways	Glycolysis	
	Organizing Metabolism	

Chapter 24 The Metals of Groups 1 and 2		4–5 days
24.1 The Alkali Metals	Structure and Properties	
	Chemical Behavior	
24.2 The Alkaline-Earth Metals	Structure and Properties	
	Properties of Alkaline-Earth Metal Compounds	
24.3 Sodium: One of the Active Metals	Preparation of Sodium	
	Properties and Uses of Sodium	
	Important Compounds of Sodium	

Chapter 25 The Transition Metals		4–5 days
25.1 The Transition Elements	General Properties of the Transition Elements	
	Transition Metal Groups	
	Variable Oxidation States	
	The Formation of Colored Compounds	
	The Formation of Complex Ions	
25.2 Iron, Cobalt, and Nickel	General Characteristics	
	Occurrence of Iron	
	Pure Iron	

Chapter 25　The Transition Metals (Continued)		
25.3　Copper, Silver, and Gold	General Characteristics	
	Copper	

Chapter 26　Aluminum and the Metalloids		3–4 days
26.1　Introduction to Aluminum and Metalloids	The Nature of Aluminum	
	The Nature of Metalloids	
26.2　Aluminum	Recovery of Aluminum	
	Properties of Aluminum	
	The Thermite Reaction	
	Some Compounds of Aluminum	
26.3　Representative Metalloids	Silicon	
	Arsenic	
	Antimony	

Chapter 27　Sulfur and Its Compounds		2–3 days
27.1　Elemental Sulfur	The Production of Sulfur	
	Physical Properties of Elemental Sulfur	
	Chemical Properties of Elemental Sulfur	
	Uses of Elemental Sulfur	
27.2　Important Compounds of Sulfur	Sulfur Dioxide	
	Sulfuric Acid	

Chapter 28　The Halogens		3–4 days
28.1　The Halogen Family	General Characteristics of the Halogens	
	The Physical Phases of the Halogens	
28.2　Fluorine	Preparation and Properties of Fluorine	
	Compounds of Fluorine and Their Uses	
28.3　Chlorine	Preparation of Chlorine	
	Physical and Chemical Properties of Chlorine	
	Uses of Chlorine	
	An Important Chlorine Compound: Hydrogen Chloride	
28.4　Bromine	Preparation of Bromine	
	Physical and Chemical Properties of Bromine	
	Compounds of Bromine and Their Uses	
28.5　Iodine	Preparation of Iodine	
	Physical and Chemical Properties of Iodine	
	Uses of Iodine and Its Compounds	

Chapter 29 Nuclear Chemistry		6–8 days
29.1 The Composition and Structure of the Nucleus	Mass Defect and Nuclear Binding Energy	
	Relationship Between Nuclear Stability and the Neutron/Proton Ratio	
	Types of Nuclear Reactions	
29.2 The Phenomenon of Radioactivity	Naturally Occurring Radioactive Nuclides	
	Artificially Induced Radioactive Nuclides	
29.3 Applications of Radioactivity	Radioactive Dating	
	Radioisotope Uses in Medicine and as Tracers	
29.4 Energy from the Nucleus	Nuclear Fission	
	Nuclear Fusion	

DEMONSTRATION SUPPLIES

1,1,2-trichloro-1,2,2-trifluoroethane [21,22]
1-butanol (*n*-butyl alcohol) [22]
2-butanol (*sec*-butyl alcohol) [22]
2-methyl-2-propanol (*tert*-butyl alcohol) [22]
acetic acid [1]
acetic acid. 0.1-*M* [15]
acetic acid, glacial [15,22]
acetone [9,22]
alcohol, ethyl or isopropyl, 70% [14]
alligator clips [15]
aluminum foil [3,24]
aluminum foil, heavy strip [26]
aluminum nitrate, 0.1-*M* [8]
aluminum soda can [11]
aluminum sulfate, 0.1-*M* [16]
aluminum, sample [3,8,9]
ammeter, 0-200 mA [15]
ammonia water (conc.) [11,12,15,16]
ammonia water (dil.) [1,16,21]
ammonia water, 0.1-*M* [15,17]
ammonia, household [16]
ammonium chloride [14,21]
ammonium thiocyanate [18]
apples (2) [20]
baby food jars (49) [8]
balance [1,9]
balloon, round [4]
balloons, 15"-diameter (3) [9]
balloons, oblong (2) [6]
balloons, oblong—different colors (3) [4]
balloons, oblong—same color (4) [4]
balloons, small (2) [1,18]
balloons, thin oblong (2) [4]
barium chloride [24]
barium hydroxide [18]
barium oxide [24]
batteries, 6-V (2) [15]
battery charger, 6 V/12 V d-c [20]
battery, 9-V [25,28]
beakers, 50-mL (49) [1,7,8,20,21,22]
beaker, 100-mL [27]
beakers, 125-mL (2) [21]
beakers, 150-mL (4) [16,17]
beakers, 250-mL (10) [3,13,14,15,16,20,25]
beaker, 400-mL [1,22]
beakers, 600-mL (2) [10,24]
beakers, 1000-mL (12) [1,2,11,12,14,17,18,23]
beral micropipets [8]
block of wood [18]
bottle caps (12) [9]
bottles (10) [9]

bottles of club soda (2) [14]
bottles of club soda, 12-oz (4) [14]
bromphenol blue [17]
bromthymol blue [8,17]
Bunsen burner [6,9,11,14,21,24–27]
burets (2) [9][17]
burette stand [9]
butyric acid [22]
cafeteria tray [6]
calcium (metal) [5,24]
calcium carbonate [16,25]
calcium chloride [14][24]
calcium hydroxide [16,24]
calcium oxide [11,24]
candle [11,18]
cardboard [4]
celsius thermometers (4) [2,14]
ceramic cup, porous, unglazed [12]
ceramic pad [1]
charcoal, sample [3]
cobalt, sample [8]
cobalt nitrate [26]
cobalt sulfate [25]
cobalt(II) chloride [25,26]
cobalt(II) chloride, 0.1-*M* [8]
cobalt(II) nitrate, 0.1-*M* [8]
coffee can, 3-lb. [6]
commercial molecular model kit (optional) [21]
conductivity detector or meter [17]
connecting wires (5) [20]
construction paper, white and colored [25]
cooking oil [22]
copper, sample [3,7,8]
copper nitrate (or copper sulfate), 0.1-*M* [20]
copper shot [1]
copper strips (5) [20]
copper wire [9]
copper wire, No. 20, insulated [15]
copper wires (2) [20]
copper(I) oxide [3]
copper(II) chloride crystals [26]
copper(II) nitrate, 0.1-*M* [8]
copper(II) oxide [3]
copper(II) sulfate pentahydrate [22,25]
copper(II) sulfate, 0.1-*M*, 20 mL [8]
copper(II) sulfate, 0.1-*M* [25]
cork [24]
cornstarch [21]
cotton [12]
cranberry juice [16]
crucible and crucible cover [24]
crucible tongs [25]

cyalume light sticks (4) [1,18]
cyclohexane [21]
cyclohexene [21,22]
decanoic acid [22]
decigram balance [3,14]
deflagrating spoon [6]
detergent solution [24]
diffraction grating [4]
distilled water [1,5,6,14–17,21–24]
distilled water bottle [18]
drain cleaner [16]
drinking straw [24]
dropping bottles [11]
dropping bottles, 125-mL (2) [16]
dry ice [17]
dry yeast [21]
egg, boiled [4]
electrodes (2) [20]
electrodes, carbon [25][28]
ethanol, 95% [6,14,15,22]
ethanol, absolute [15]
evaporating dish, porcelain [15, 25]
eyedroppers (2) [6,16]
felt, colored [3]
filter paper [16,22,24]
filters (colored cellophane) [25]
flashlight [25]
flask, Dewar [10]
flask, Erlenmeyer, 125-mL [14]
flask, Erlenmeyer, 250-mL (3) [9,10,18,17,22]
flask, Erlenmeyer, 500-mL (3) [1,9,12,17]
flask, Florence [9]
flux [20]
fresh pineapple [23]
fructose solution, 1% [23]
fruit-flavored gelatin [14,16]
funnel [10,16,24]
gas-collecting bottles [10]
gelatin (3 pkgs.) [23]
gelatin capsules (pharmaceutical) [25]
glass bend, 5 × 20 cm [16]
glass bends, 5 × 5 cm (3) [14]
glass container, 2-L [27]
glass plate [14]
glass tubing [12]
glucose [22]
glucose solution, 1% [23]
glycerin [15]
graduated cylinder, 10-mL [6,16]
graduated cylinder, 25-mL [22]
graduated cylinder, 100-mL (4) [2,14,16,17]
graduated cylinder, 500-mL [1]

graduated cylinder, tall [26]
grape juice [16]
grapefruit juice [16]
graphite, sample [3]
"grow beast" [1]
gumdrops, large—3 colors [6,15,21]
gumdrops, small—2 colors [15]
gumdrops, small—same color [6]
hard water [24]
heat lamp [9]
heat source [13]
heater, with temperature control [7]
helium [1]
Hoffman apparatus [8]
hot plate [1,9,22,25]
hydrochloric acid (conc.)
   [11,12,15,16,25]
hydrochloric acid (dil.) [1,16]
hydrochloric acid, 0.1-M [7,8,9,15,17]
hydrochloric acid, 1-M [16,20,24]
hydrochloric acid, 12-M [9]
hydrochloric acid, 6-M [7,10,17,25,27]
hydrogen [1,12,18]
hydrogen peroxide, 20% [3]
hydrogen peroxide, 3% [10]
ice [14]
ice cubes [7,9]
ice, 100-g blocks (2) [2]
iodine [21,22]
iodine crystals [7,9,14]
iron, sample [3,8]
iron nail [16]
iron ring [21,24]
iron(II) sulfate [23,25,26]
iron(III) chloride [26]
iron(III) chloride, 0.1-M [16,25]
iron(III) chloride, 0.5-M [21]
iron(III) nitrate, 0.1-M [8]
iron-enriched cereal [25]
iso-amyl alcohol [22]
kerosene [21]
kits composed of 15 large
   gumdrops—one color, 36 small
   gumdrops—different colors, 51
   toothpicks, the gumdrop-toothpick
   model from Demonstration 1 [21]
lamp holder, threaded [15]
lead, sample [8]
lead chloride, 0.1-M [8]
lead(II) nitrate, 0.1-M [8]
leads with alligator clips [20,25,28]
lemon [20]
lemon juice [16]
lemon lime soda [14,16]
Life Savers®, wintergreen [4]
lightbulb, 6.3-V/150-mA [15]
lime [16]
limewater [16]
liquid antacid medication [16]
liquid nitrogen [10]

lithium chloride [14]
lithium hydroxide (dil.) [1]
litmus paper, red and blue [16,24]
long-stem glass [10]
Lugol's solution [23]
lycopodium powder [6]
lye [16]
magnesium (metal) [24]
magnesium chloride [24]
magnesium oxide [24]
magnesium ribbon [5,6,24]
magnesium turnings [9]
magnet [10,25]
magnetic stirrer and stirring magnet
   [24,25]
magnifying lens [4]
manganese, sample [25]
manganese sulfate [25]
manganese(IV) oxide [3,25,26]
marker, felt-tip, permanent [14]
marshmallows, large white [6]
marshmallows, small colored [6]
masking tape [18]
matches [1,10,18]
medicine droppers [8]
mercury vapor lamp [4]
mercury(II) oxide [9]
meter stick [1,18]
methyl alcohol [22]
methyl orange [17]
methyl red [17]
methylene blue [23]
metric ruler, clear plastic [3]
microplate, 24-well [8]
milk [16]
milk of magnesia [14,16]
milk of magnesia tablets [24]
mineral oil [21]
modeling clay, 4 colors [15]
Modern Chemistry textbook [3,11]
mustard [16]
n-amyl alcohol [22]
nails (2) [3,4]
naphthalene [21]
nickel, sample [8]
nickel(II) nitrate [26]
nickel(II) nitrate, 0.1-M [8]
nitric acid, conc. [1]
nitric acid, dil. [1,16]
nitric acid, 0.1-M [7]
octyl alcohol [22]
oleic acid [6]
orange juice [16]
oven cleaner [16]
overhead projector
   [5,8,20,21,22,23,24,25]
oxalic acid, 0.300-M [20]
oxygen [1,10,18]
pH meter [17]
paint can and lid (1-pt.) [18]

pan, 2-L [14]
paper [1,4]
paper cup, 5-oz [22]
paper towels [2]
pennies, minted after 1982 (100) [3,25]
pennies, minted before 1982 (100)
   [3,25]
petri dish, 100-mm [8,20,23]
petri dish, glass, 100-mm [9]
phenol red [17]
phenolphthalein [1,5,8,9,17,20,24,28]
phosphoric acid, (dil.) [16]
phosphoric acid, 0.1-M [7]
pickle juice [16]
piezo sounder [20]
pipet, 1-mL [6]
plastic milk container (1-gal.) [11]
plastic ruler [6]
plastic sandwich bags [3,13,22]
plastic soda bottles, 2-L (2) [14]
plastic soft drink bottle, 2-L [24]
plastic storage bag, sealable (1-gal.) [16]
platinum or nichrome wire [26]
pliers [4]
pneumatic trough (or plastic dishpan)
   [14]
popsicle stick, wooden or plastic [15]
porous cup, 35 × 80 mm [20]
poster board, black [23]
potassium chloride [24]
potassium hydroxide (dil.) [1]
potassium iodide, 0.1-M [20]
potassium nitrate [14,15,21]
potassium permanganate [10,21,22,25]
potassium permanganate, 0.000 40-M
   [7]
potassium permanganate, 0.005-M [7]
potassium permanganate, 0.01-M [20]
potassium permanganate, 0.05-M [7]
potassium sodium tartrate [22]
potassium thiocyanate, 0.01-M [25]
power supply, 6 V/12 V d-c [8]
prism [4]
raisins [1]
reagent bottles (4) [16]
Red Hots® [6]
right-angle glass bend [10]
ring stand [12,14,21,24]
rings stand clamps (2) [12]
rock salt [14]
rubber delivery tubes (3) [14]
rubber hose [18]
rubber stopper, 1-hole (3) [12,14]
rubber stopper, 2-hole [10,12]
rubber stopper, solid [16,18]
rubber stoppers (8) [21]
rubber stoppers, #3 solid (4) [14]
rubber stoppers, #5 (3) [22]
rubber tubing [10]
salad dressing, oil and vinegar [14]

salicylic acid [22]
salt solution [1]
sand paper [5]
scissors [21]
sharp file [25]
silicon, sample [3]
silver nitrate, 0.1-M [21,25]
slide projector [4,14,27]
small battery clock [20]
soap solution [24]
soda bottles, 1-L (3) [14]
soda pop, clear [1]
soda straws, long [17]
soda-straw balance [25]
sodium bisulfite, 0.1-M [7]
sodium carbonate [24]
sodium chloride [7,25,28]
sodium chloride, 0.1-M [15,21]
sodium chromate, 0.1-M [8]
sodium hydroxide [8,9,15,17,21,22]
sodium hydroxide pellets [24]
sodium hydroxide, 0.1-M [15]
sodium hydroxide, 1-M [16]
sodium hydroxide, 2-M [16]
sodium hydroxide, 6-M [25]
sodium hydroxide, 8-M [7]
sodium nitrate [7,24]
sodium phosphate [7]
sodium silicate solution [22,26]
sodium sulfate [7,8]
sodium tetraborate [26]
sodium thiosulfate pentahydrate [14,17]
sodium thiosulfate, 0.1-M [9]
sodium-lead alloy (90% Pb, 10% Na)
  [24]

solder [20]
soldering gun [20]
sour milk or buttermilk [16]
spray bottle [24]
spray room deodorizer [13]
starch liquid (from pasta) [14]
starch solution [20]
starch solution, 1% [23]
steel wool [20,25]
stirring rods (7) [1,14,16,21,22,25]
stoppers, cork or rubber [12]
straight pins (2) [4]
string [18]
strontium chloride [24]
sucrose [14,15,21,27]
sucrose solution, 1% [23]
sulfur, sample [3,6]
sulfur, powdered [27]
sulfuric acid (conc.) [22,27]
sulfuric acid (dil.) [1,16]
sulfuric acid, 0.1-M [7,9]
sulfuric acid, 3-M [10,23]
superabsorbent diaper [1]
switch, push-button [15]
syringe, 50-mL or 100-mL [11]
talcum powder [6]
tape [4,6,18,21]
tea bag [14]
test tube holder [22]
test tube rack [22]
test tube, 10-mL [8]
test tube, large [27]
test tube, large pyrex [10]
test tubes (5) [9,17,23,24,25]
test tubes, large (8) [16,21]

test tubes, medium (7) [3,5,14,16,21,22]
thermometer [13,24]
thistle tube [10]
thread [25]
thumbtack [11]
thymolphthalein [17]
tongs [6,11,27]
toothpicks [15,21]
toothpicks, round or heavy [6]
transparencies [5,21]
triangle [24]
trough or large pan [10]
U-tube [28]
universal indicator [7,17,24]
urea [14]
vinegar [16]
voltmeter, 0–5-V [20]
watch glasses (7) [7,9,16,22,25]
water bath [7]
Waterlock J-500 [1]
Waterlock J-550 [1]
wide bore glass tubing [12]
wide-mouth container [26]
wire gauze [21]
wire gauze with ceramic center [14]
wooden splints [9,10,16,22]
yogurt [16]
zinc nitrate [26]
zinc nitrate, 0.1-M [8,20]
zinc strips (5) [20]
zinc, granular [25]
zinc, mossy [7,8,10]
zinc, sample [3,8]
zinc-coated nail [20]

SOURCES FOR AUDIOVISUAL RESOURCES AND COMPUTER SOFTWARE

The Aluminum Association
Educational Services
818 Connecticut Avenue, NW
Washington, D.C. 20006

Barclay School Supplies
166 Livingston Street
Brooklyn, NY 11201

Cambridge Development Laboratory
42 4th Avenue
Waltham, MA 02154

Central Scientific Company
11222 Melrose Avenue
Franklin Park, IL 60131–1364

Chemistry According to ROF
Richard O. Fee
5645 South Kensington Avenue
Lagrange, IL 60525

COMPress
(see Queue, Inc.)

Coronet Films and Video
108 Wilmot Road
Deerfield, IL 60015

Dow Corning Corporation
P.O. Box 1767, Dept. No. 01–58
Midland, MI 48640

Educational Images, Ltd.
P.O Box 3456, West Side
Elmira, NY 14905

Films for the Humanities and Sciences
Box 2053
Princeton, NJ 08543

Fisher Scientific Company
Educational Materials Division
4901 West LeMoyne Street
Chicago, IL 60651

Flinn Scientific, Inc.
P.O. Box 219
Batavia, IL 60510–0219

Frey Scientific Company
905 Hickory Lane
Mansfield, OH 44905

HRM
(see Queue, Inc.)

Human Relations Media
175 Tomkins Avenue
Pleasantville, NY 10570–9973

Intellectual Software
(see Queue, Inc.)

Journal of Chemical Education
Software
Project Seraphim/JCE
Department of Chemistry
Eastern Michigan University
Ypsilanti, MI 48197

Knowledge Factory
(see Queue, Inc.)

Learning Arts
P.O. Box 179
Wichita, KS 67201

Learning by Practice
(see Queue, Inc.)

Modern Talking Picture Service
5000 Park Street North
St. Petersburg, FL 33709

National Science Teacher's Association
NSTA Special Publications
1742 Connecticut Avenue, NW
Washington, D.C. 20009

Prentice Hall Media
150 White Plains Road
Tarrytown, NY 10591

Programs for Learning
(see Queue, Inc.)

Project Seraphim
NSF Science Education Project
Headquarters
Department of Chemistry
Eastern Michigan University
Ypsilanti, MI 48197

Queue, Inc.
562 Boston Avenue
Bridgeport, CT 06610
Also available through Queue:
COMPress, HRM, Intellectual Software, Knowledge Factory, Learning by Practice, Programs for Learning

Random House Media
Department 119
P.O. Box 126
Stamford, CT 06904

Richard O. Fee
(see Chemistry according to ROF)

Sargent Welch
7300 N. Linden Avenue
Skokie, IL 60077

TV Ontario/TVO Video
143 West Franklin St., Suite 206
Chapel Hill, NC 27516

University of Akron Audiovisual
Services
302 East Buchtel Avenue
Akron, OH 44325

Walch Science
J. Weston Walch, Publisher
321 Valley Street
P.O. Box 658
Portland, ME 04104–0568

Ward's Modern Learning Aids
Ward's Natural Science Establishment,
Inc.
5100 West Henrietta Road
P.O. Box 92912
Rochester, NY 14692

MODERN CHEMISTRY

ANNOTATED TEACHER'S EDITION

HOLT, RINEHART AND WINSTON

AUSTIN NEW YORK SAN DIEGO CHICAGO TORONTO MONTREAL

THE HRW MODERN CHEMISTRY PROGRAM

Modern Chemistry (Student Text)

Modern Chemistry (Annotated Teacher's Edition)

Laboratory Experiments (Consumable Version)

Laboratory Experiments (Nonconsumable Version)

Laboratory Experiments (Teacher's Edition)

Section Reviews

Section Reviews (Teacher's Edition)

Tests (Blackline Masters and Answer Key)

A Student Guide to Problem Solving

Cover: A view of light fuel oil using the technique of phase contrast photography. In phase contrast photography, transmitted light is refracted in the same way that a prism breaks white light into a full spectrum of colors. (See Section 4.1.) Light fuel oil is one of the many products of fractional distillation of crude oil. (See Chapter 21.) Other products in the distillation of crude oil include hydrocarbon polymers, the building blocks of plastics.

Photo: Manfred Kage/Peter Arnold, Inc.

Printed in the United States of America

ISBN 0-03-014502-3

90123456 036 987654321

AUTHORS

Nicholas D. Tzimopoulos, Ph.D.
Teacher of chemistry and Chairman of the Science Department of the Public Schools of the Tarrytowns, North Tarrytown, New York, and Adjunct Professor of Chemistry at PACE University, New York, New York. He is currently chairman of the Westchester (NY) Section of the American Chemical Society.

New material appearing in this edition of *Modern Chemistry* is the work of Dr. Tzimopoulos.

H. Clark Metcalfe
Formerly teacher of chemistry at Winchester-Thurston School, Pittsburgh, Pennsylvania, and Head of the Science Department, Wilkinsburg Senior High School, Wilkinsburg, Pennsylvania.

John E. Williams
Formerly teacher of chemistry and physics at Newport Harbor High School, Newport Beach, California, and Head of the Science Department, Broad Ripple High School, Indianapolis, Indiana.

Joseph F. Castka
Formerly Assistant Principal for the Supervision of Physical Science, Martin Van Buren High School, New York City, and Adjunct Associate Professor of General Science and Chemistry, C.S. Post College, Long Island University, New York.

EDITORIAL ADVISORY BOARD

CONTENTS

	Discovering Modern Chemistry	ix
UNIT 1	**INTRODUCTION TO CHEMISTRY AND MATTER**	**2**
Chapter 1	**Matter Energy and Change**	**4**
1.1	What Is Chemistry	5
1.2	Matter and Energy	10
1.3	Classification of Matter	19
1.4	The Chemical Elements	23
Chapter 2	**Measurements and Solving Problems**	**30**
2.1	Units of Measurements	31
2.2	Heat and Temperature	40
2.3	Using Scientific Measurements	46
2.4	Solving Quantitative Problems	55
Careers	Fire Chemist, Science Writer	64
Cumulative Review		65
Intra-Science	The Process in Chemistry: The Portable Power of the Battery	66
UNIT 2	**ORGANIZATION OF MATTER**	**68**
Chapter 3	**Atoms: The Building of Matter**	**70**
3.1	The Atom: From Philosophical Idea to Scientific Theory	71
3.2	The Structure of the Atom	75
3.3	Weighing and Counting Atoms	82
Chapter 4	**Arrangement of Electrons in Atoms**	**96**
4.1	Refinements of the Atomic Model	97
4.2	Quantum Numbers and Atomic Orbitals	108
4.3	Electron Configurations	112
Chapter 5	**The Periodic Law**	**126**
5.1	History of the Periodic Table	127
5.2	Electron Configuration and the Periodic Table	132
5.3	Electron Configuration and Periodic Properties	143
Chapter 6	**Chemical Bonding**	**160**
6.1	Introduction to Chemical Bonding	161
6.2	Covalent Bonding and Molecular Compounds	164
6.3	Ionic Bonding and Ionic Compounds	175
6.4	Metallic Bonding	179
6.5	The Properties of Molecular Compounds	182
Careers	Laser Chemist, Chemical Technician	194
Cumulative Review		195
Intra-Science	The Question in Chemistry: How to Replace Asbestos	196
UNIT 3	**LANGUAGE OF CHEMISTRY**	**198**
Chapter 7	**Chemical Formulas and Chemical Compounds**	**200**
7.1	Chemical Names and Formulas	201
7.2	Oxidation Numbers	212

7.3	Using Chemical Formulas	216
7.4	Determining Chemical Formulas	224
Chapter 8	**Chemical Equations and Chemical Reactions**	**232**
8.1	Chemical Equations	233
8.2	Types of Chemical Reactions	244
8.3	Activity Series of the Elements	251
Chapter 9	**Stoichiometry**	**260**
9.1	Introduction to Stoichiometry	261
9.2	Ideal Stoichiometric Calculations	265
9.3	Limiting Reactants and Percent Yield	272
Careers	Analytical Pharmaceutical Chemist, Chemical Plant Manager	282
Cumulative Review		283
Intra-Science	The Process in Chemistry: The Amazing Polyhexamethyleneadipamide	284
UNIT 4	**PHASES OF MATTER**	**287**
Chapter 10	**Representative Gases**	**288**
10.1	Oxygen and Ozone	289
10.2	Hydrogen	294
10.3	Nitrogen and Ammonia	298
10.4	Carbon Dioxide and Carbon Monoxide	303
Chapter 11	**Physical Characteristics of Gases**	**314**
11.1	The Kinetic Theory of Matter	315
11.2	Qualitative Description of Gases	320
11.3	Quantitative Description of Gases	322
Chapter 12	**Quantitative Behavior of Gases**	**342**
12.1	Volume–Mass Relationships of Gases	343
12.2	The Ideal Gas Law	350
12.3	Stoichiometry of Gases	356
12.4	Effusion and Diffusion	360
Chapter 13	**Liquids and Solids**	**370**
13.1	Liquids	371
13.2	Solids	377
13.3	Changes of State	383
13.4	Water	390
Careers	Occupations in the Compressed-Gas Industries, Chemical Patent Attorney	398
Cumulative Review		399
Intra-Science	The Problem in Chemistry: Preserving the Ozone Layer	400
UNIT 5	**SOLUTIONS AND THEIR BEHAVIOR**	**402**
Chapter 14	**Solutions**	**404**
14.1	Types of Mixtures	405
14.2	The Solution Process	410
14.3	Concentration of Solutions	420
14.4	Colligative Properties of Solutions	426
Chapter 15	**Ions in Aqueous Solutions**	**438**
15.1	Ionic Compounds in Solution	439

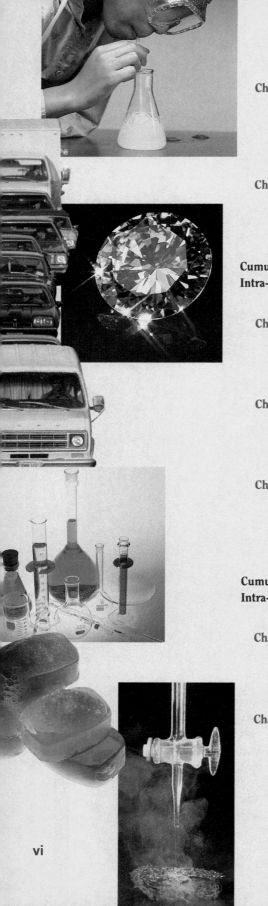

15.2	Molecular Electrolytes	447
15.3	Properties of Electrolyte Solutions	450
Chapter 16	**Acids and Bases**	**460**
16.1	Acids	461
16.2	Bases and Acid–Base Reactions	470
16.3	Relative Strengths of Acids and Bases	475
16.4	Oxides, Hydroxides, and Acids	479
16.5	Chemical Reactions of Acids, Bases, and Oxides	483
Chapter 17	**Acid–Base Titration and pH**	**490**
17.1	Concentration Units for Acids and Bases	491
17.2	Aqueous Solutions and the Concept of pH	496
17.3	Acid–Base Titrations	505
Careers	Food Chemist, Aroma Chemist	520
Cumulative Review		521
Intra-Science	The Problem in Chemistry: Detergents and the Environment	522
UNIT 6	**CHEMICAL REACTIONS**	**524**
Chapter 18	**Reaction Energy and Reaction Kinetics**	**526**
18.1	Thermochemistry	527
18.2	Driving Force of Reactions	536
18.3	The Reaction Process	541
18.4	Reaction Rate	549
Chapter 19	**Chemical Equilibrium**	**562**
19.1	The Nature of Chemical Equilibrium	563
19.2	Shifting Equilibrium	569
19.3	Equilibrium of Acids, Bases, and Salts	576
19.4	Solubility Equilibrium	583
Chapter 20	**Oxidation–Reduction Reactions**	**596**
20.1	The Nature of Oxidation and Reduction	597
20.2	Balancing Redox Equations	602
20.3	Oxidizing and Reducing Agents	608
20.4	Electrochemistry	611
Careers	Industrial Technician, Environmental Chemist	626
Cumulative Review		627
Intra-Science	The Problem in Chemistry: Chemiluminescence	628
UNIT 7	**CARBON AND ITS COMPOUNDS**	**630**
Chapter 21	**Carbons and Hydrocarbons**	**632**
21.1	Abundance and Importance of Carbon	633
21.2	Organic Compounds	639
21.3	Hydrocarbons	643
21.4	Representative Hydrocarbons and Polymers	655
Chapter 22	**Substituted Hydrocarbons**	**664**
22.1	Alcohols	665
22.2	Halocarbons	670
22.3	Ethers	673
22.4	Aldehydes and Ketones	674
22.5	Carboxylic Acids	676
22.6	Esters	679

Chapter 23	**Biochemistry**	**684**
23.1	The Chemistry of Life	685
23.2	Proteins	686
23.3	Carbohydrates	689
23.4	Lipids	691
23.5	Nucleic Acids	695
23.6	Biochemical Pathways	700
Careers	Agricultural Chemist, Biochemist	706
Cumulative Review		707
Intra-Science	The Process in Chemistry: Custom-Designed Plants	708

UNIT 8	**DESCRIPTIVE CHEMISTRY**	**710**
Chapter 24	**The Metals of Groups 1 and 2**	**712**
24.1	The Alkali Metals	713
24.2	The Alkaline-Earth Metals	717
24.3	Sodium: One of The Active Metals	720
Chapter 25	**The Transition Metals**	**728**
25.1	The Transition Elements	729
25.2	Iron, Cobalt, and Nickel	736
25.3	Copper, Silver, and Gold	744
Chapter 26	**Aluminum and the Metalloids**	**750**
26.1	Introduction to Aluminum and the Metalloids	751
26.2	Aluminum	753
26.3	Representative Metalloids	758
Chapter 27	**Sulfur and Its Compounds**	**764**
27.1	Elemental Sulfur	765
27.2	Important Compounds of Sulfur	770
Chapter 28	**The Halogens**	**778**
28.1	The Halogen Family	779
28.2	Fluorine	781
28.3	Chlorine	782
28.4	Bromine	787
28.5	Iodine	790
Careers	Metallurgist and Materials Scientist, Chemical Supplier	794
Cumulative Review		795
Intra-Science	The Process in Chemistry: The Miracle Fluorocarbon Resin	796

UNIT 9	**NUCLEAR REACTIONS**	**798**
Chapter 29	**Nuclear Chemistry**	**800**
29.1	The Composition and Structure of the Nucleus	801
29.2	The Phenomenon of Radioactivity	804
29.3	Applications of Radioactivity	816
29.4	Energy from the Nucleus	817

Appendix A	Useful Tables	824
Appendix B	Logarithms	834
Glossary		836
Index		845
Credits		854

TECHNOLOGY

"Looking" at Atoms 81
Hydrazine, a Liquid Rocket Fuel 254
Measuring and Reaching Low Temperatures 322
Separating Isotopes 363
Liquid Crystals 376
Superconductors 382
Fresh Water From the Sea 428
Fuels and Energy Content 531
Swimming Pool Chemistry 570
Toward Human Gene Transplants 699
Can a Can Become a Plane? 754
PVC Pipes 784

DESKTOP INVESTIGATIONS

Physical and Chemical Changes 18
Density of Pennies 38
Constructing a Model 74
The Wave Nature of Light: Interference 102
Designing Your Own Periodic Table 131
Models of Chemical Compounds 215
Balancing Equations Using Models 249
Limiting Reactants in a Cookie Recipe 273
Properties of Carbon Dioxide 307
Diffusion 361
Surface Tension of Liquids and Water Solutions 374
Solutions, Colloids, Suspensions 408
Acids and Bases in Your Kitchen 476
Testing for Acid Precipitation 509
Factors Influencing Reaction Rate 555
Redox Reactions 600

Enzyme Action on Starch 690
Paramagnetism of Nickel 730
Silver Halides 789

CHEMISTRY NOTEBOOK

Interconversion of Matter and Energy 15
Millikan's Oil-Drop Experiment 79
Heisenberg's Uncertainty Principle 107
Exceptions to the Octet Rule 168
Resonance 174
Atomic Symbols and Molecular Formulas 221
Scientific Journals: Getting the Word Out 264
The Nitrogen Cycle 300
Carbon Monoxide Poisoning 309
Molecular Motion 316
Electrolytes in Your Body 447
Boiling Liquids 451
"Acid Stomach" and Antacids 481
Acid Rain and Lakes 504
Not-So-Blue Jeans 610
Chlorohydrocarbons 640
Octane Ratings 646
Polyunsaturated Fats and Pour-on Margarines 668
Group 1 Ions in Medication and Diet 715
Strontium: Fireworks and Fallout 719
Gallium Arsenide, Lasers, and Computer Chips 761
Medicines Made from Sulfur 769
The Smell of Sulfur Compounds 772
Discovering How to Discover Elements 791
Radium–Cause of and Cure for Cancer 802
Measuring Radiation Exposure 818

Discovering *Modern Chemistry*

The world we live in is complex and diverse. It contains the hardness of steel and the softness of cotton, the brilliance of diamonds and the blackness of coal, the heat of flames and the chill of ice, the tranquility of a slow moving river and the din of a busy city street. All the colors, textures, objects, animals, and plants that exist are too numerous to count. Yet scientists have learned that there are approximately 90 different ingredients, or elements, that make up all the material things we see and experience.

By studying chemistry we can understand how the diversity and complexity of our world can arise from so few ingredients. As you will discover, the science of chemistry ranges from a thought-provoking academic subject to a powerful tool for innovation in our rapidly changing society.

These pages will show you how *Modern Chemistry* is designed to guide you through your discovery of the science of chemistry.

Modern Chemistry is divided into nine units. On the unit opening pages, men and women who have devoted their careers to science retell what sparked their interest as young people. Maybe you share some of the thoughts and experiences of these scientists. These Unit Openers introduce you to the chapters to follow and help you to understand the ways in which chemistry is important in our lives.

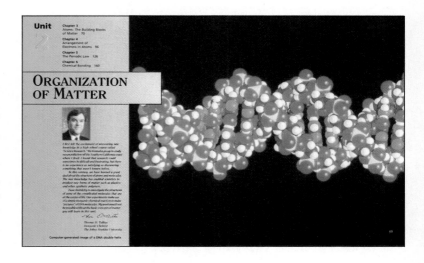

Modern Chemistry contains 29 chapters. The first two pages of each chapter set the stage for learning and discovery. Introduction and Looking Ahead help you frame questions about what you are going to read. Each chapter is divided into numbered sections that begin with learning objectives. These objectives help you anticipate the major concepts of each section. Study Hints tell you where to refresh your memory when previously covered topics are needed to understand new material. Margin notes offer short comments that give greater insight or simply point out items of special interest.

Throughout the text, *Modern Chemistry* introduces a systematic four-step method for approaching and solving "word problems." Study the four-step method in Sample Problems and use the method when solving problems on your own. Test your understanding of each new type of "word problem" by solving the Practice Problems that follow each Sample Problem. See if you get the same answers as in the book. At the end of each section, Section Review questions help you check your understanding of each section.

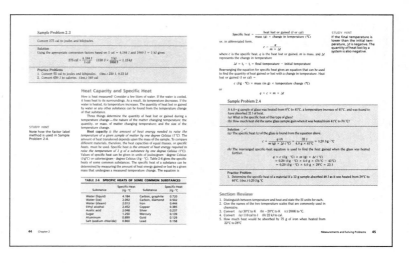

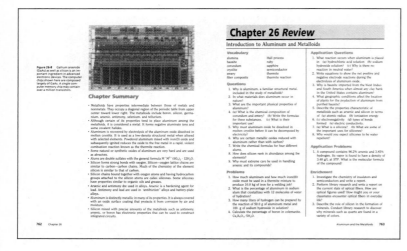

Chapter Summaries and the Chapter Reviews help you look back on each chapter. Many of the key concepts developed in each chapter are listed in the summary for quick review and reference. All of the new words defined in each chapter are listed. The questions and problems will enable you to practice and test your understanding of each chapter.

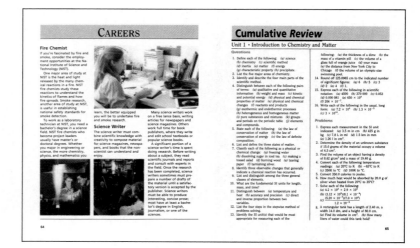

The Careers features highlight 18 different jobs and describe what the people working in these jobs do and how chemistry helps them to do it.

Cumulative Reviews bring together a large body of knowledge from each unit in the form of review questions and problems.

Desktop Investigations demonstrate the basics of experimental science using everyday materials. There are no right or wrong results. Your careful experimental technique and observations are the key to successful experimenting.

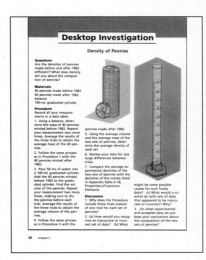

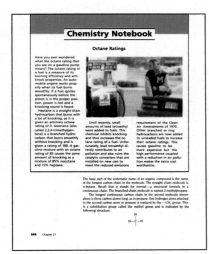

Chemistry is more than a collection of facts. Chemistry Notebook features take an in-depth look at special topics in chemistry or simply look at the story of a scientific discovery. These features reveal the richness and vitality of chemistry. Chemistry is a powerful tool for innovation. Technology features explain how chemistry applied to real world needs has brought about changes in the way we live.

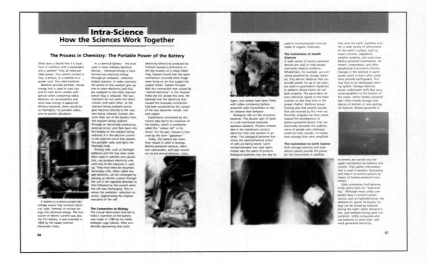

Science is a cooperative adventure. Chemists work with biologists, physicists, geologists and others to answer many important and intriguing questions. Intra-Science features pose a question in science and examine how scientists work together to find solutions.

Unit

1

Chapter 1
Matter, Energy, and Change   4

Chapter 2
Measurements and
Solving Problems   30

INTRODUCTION TO CHEMISTRY AND MATTER

At the NASA Lewis Research Center, I am involved in developing computer codes that use the laws of thermodynamics to model combustion processes. These codes have become useful tools for solving a wide range of problems in chemistry and engineering.

Space exploration depends on the science of chemistry. Chemists, for example, helped produce the rocket fuels that provide 4.4 million pounds of thrust to the space shuttle at liftoff.

Scientists also apply chemical principles to meet our society's need for food, clothing, building materials, medicines, and many other essentials. This unit introduces you to chemistry, a science that is an integral part of our lives.

Bonnie J. McBride

Bonnie J. McBride
Physical Chemist
NASA Lewis Research Center

The Space Shuttle *Discovery* at liftoff

Chapter 1 Matter, Energy, and Change

Chapter Planner

1.1 What is Chemistry?
Chemistry is a Physical Science
- Demonstrations: 1–3, 5
- Questions: 1a, 2a
- Experiment: SE-1

Chemistry in Modern Society
- Questions: 1b, 2,

What is a Chemical?
- Question: 3

Branches of Chemistry
The Scientific Method
- Questions: 5–7
- Demonstration: 6

1.2 Matter and Energy
Definition of Matter
- Experiment: 1
- Questions: 11, 12
- Application Question: 2
- Questions: 8–10
- Application Question: 1

Definition of Energy
- Questions: 13–15

Law of Conservation of Energy
- Chemistry Notebook
- Question: 16

States of Matter
- **Demonstration: 4**
- **Question: 17**
- **Question: 18**

Properties and Changes of Matter
- Demonstration: 7
- Desktop Investigation
- Experiment: 2
- Questions: 20, 22–27
- Application Question: 3
- Question: 19
- Application Questions: 4, 7
- Questions: 21
- Application Questions: 5, 10, 11

1.3 Classification of Matter
Mixtures
- Questions: 28, 29
- Application Question: 6

Pure Substances
- Questions: 30–33, 37
- Application Questions: 8, 9

1.4 The Chemical Elements
Introduction to the Periodic Table
- Question: 34

Symbols for the Elements
Types of Elements
Metals
Two Typical Elements
- Questions: 35, 36

Teaching Strategies

Introduction

This first chapter is perhaps the most important chapter because it will set the tone and lay the intellectual basis needed to teach chemical principles throughout the year. It introduces chemistry as a physical science, the scientific method, matter and energy, and the Periodic Table. Mathematical operations are left to the second chapter. You should emphasize that the chemistry learned in this chapter is used repeatedly in future chapters.

Critical to the success of the course is the need to stimulate and captivate your students' interest in chemistry, to make them believe that chemistry is exciting, and, by example, to establish the proper safety precautions and procedures for the laboratory. Interest can be aroused by a spectacular chemistry show the first day of class.

Let students experience the "wonder" of chemistry by giving them a glimpse of a few of the reactions they will study during the course of the year. Encourage them to think of explanations, but leave detailed explanations until later. You are simply whetting their appetites. The demonstrations will also give you a chance to teach safety in the laboratory by example. You will need to explain to the students why you are wearing goggles, apron, gloves (if needed) and why you are following the safety precautions that you outline.

Demonstration 1: The Ira Remsen Reaction

PURPOSE: To introduce a chemical change using a historical anecdote

MATERIALS: At least one pre-1982 penny, 500-mL Erlenmeyer flask, concentrated nitric acid.

PROCEDURE: While reading the text that follows, place the penny in the Erlenmeyer flask and add 20 mL of concentrated nitric acid. Place the flask on a table where all may view the reaction.

While reading a textbook of chemistry, I came upon the statement, "nitric acid acts upon copper." I was getting tired of reading such absurd stuff, and I was determined to see what this meant. Copper was more or less familiar to me, for copper cents were then in use. I had seen a bottle marked "nitric acid" on a table in the doctor's office where I was then "doing time." I did not know its peculiarities, but the spirit of adventure was upon me. Having nitric acid and copper, I had only to learn what the words "act upon" meant. The statement "nitric acid acts upon copper" would be something more than mere words. All was still. In the interest of knowledge, I was even willing to sacrifice one of the few copper cents then in my possession. I put one of them on the table, opened the bottle marked nitric acid, poured some of the liquid on the copper and prepared to make an observation. But what was this wonderful thing which I beheld? The cent was already changed, and it was no small change either! A green-blue liquid foamed and fumed over the cent and over the table. The air in the neighborhood of the performance became colored dark red. A great colored cloud arose. This was disagreeable and suffocating. How should I stop this? I tried to get rid of the objectionable mess by picking it up and throwing it out of the window. I learned another fact. Nitric acid not only acts upon copper, but it acts upon fingers. The pain led to another unpremeditated experiment. I drew my fingers across my trousers and another fact was discovered. Nitric acid acts upon trousers. Taking everything into consideration, that was the most impressive experiment and, relatively, probably the most costly experiment I have ever performed. It was a revelation to me. It resulted in a desire on my part to learn more about that remarkable kind of action. Plainly, the only way to learn about it was to see its results, to experiment, to work in a laboratory.

Have the students pay special attention to the discussion questions that follow the Desktop Investigation.

SAFETY CONSIDERATIONS: Point out to your students that ALL colored gases are poisonous. Use of the large flask will concentrate the gas and not allow it to diffuse through the room. This demonstration should be done under a hood or using an improvised funnel vent connected to an aspirator on a faucet. Handle the nitric acid carefully. Point out that it is a key laboratory acid, but it must be handled with extreme caution. Do not dispose of the solution. After adding the 200 mL of distilled water, point out that blue is a characteristic color of dissolved copper. Pour the solution into a clear glass bottle and put it on display. You will be able to refer to it many times in future lessons.

Demonstration 2: Cyalume Light Stick

PURPOSE: To demonstrate that some chemical reactions give off light and that the rate of the reaction depends on the temperature of the system
MATERIALS: cyalume light stick, 1000-mL beaker with ice water, a 1000-mL beaker with hot (boiling) water
PROCEDURE: Remove the light stick from its protective wrapper and ask the students if they have ever seen or used one. Explain the construction of the light stick—two chemicals one of which is kept in a sealed glass vial. The reaction begins by bending the light stick and breaking the vial allowing the two chemicals to mix. The reaction produces an intense light. The color depends on the type of light stick. Yellow-green, red, blue, and white light sticks, are available. Turn off the room lights and place the light stick in the beaker of cold water. Note the dimming of the light evolved. Remove the light stick and place it in the beaker with the boiling water. Note the increase in the intensity of the light. Point out that temperature is critical to the speed at which a reaction occurs. Low temperatures slow down reactions while high temperatures speed up reactions.

Demonstration 3: Superabsorbent Polymers

PURPOSE: To demonstrate a relatively new product with amazing properties and a variety of uses
MATERIALS: Distilled water, 400-mL beaker, Waterlock J-550 or Waterlock J-500, a "grow beast," and a superabsorbent diaper

PROCEDURE: Add 200 mL of distilled water to the 400-mL beaker. Sprinkle one half teaspoon of the polymer into the beaker. Swirl the contents; wait a few seconds, and turn the beaker over. The contents should have solidified and the gelled water remains in the beaker.

Students are amazed by this reaction. Ask the students if they have ever seen or used such a product. If no one volunteers, show them a diaper and a "grow beast." The product is the main ingredient in the new varieties of superabsorbent diapers. The toy industry uses the polymer to make the "grow beasts"—the plastic toys that expand to 200 times their original size when placed in water. Commercial airlines use the polymers to make fuel-filter elements. The polymer takes up any water that might have mixed with the fuel. The petroleum industry uses them as "superslurpers" to clean up oil spills.

After allowing the students to observe the beaker, sprinkle some salt (sodium chloride) on the gel. It immediately begins to liquefy. Swirling the beaker will show the change in properties. In place of the solid salt, 20 mL of a concentrated salt solution may be poured into the beaker. The salt interferes with the absorbing capacity of the polymer.

Ask students about the absorbing capacity of the polymer. Does it make a difference if tap water were used instead of distilled, or if hot water were used instead of cold? How well is urine absorbed in the diapers since urine contains dissolved salts? You might at this time place a "grow beast" in a wide-mouth gallon jar filled with distilled water and observe its growth during the coming week. The jar and "beast" may then be put on display.

SAFETY AND DISPOSAL: The gelled Waterlock polymer may be kept as a display. It may be safely washed down the drain after it has been poisoned with the salt.

Demonstration 4: Dancing Raisins

PURPOSE: To stimulate thought and to demonstrate density and buoyancy
MATERIALS: Cold, clear sodapop, raisins, a large beaker or 500-mL graduate cylinder
PROCEDURE: Pour the cold, clear soda into the container. Add a dozen raisins. Observe. The raisins begin to rise to the surface where they appear to "dance" and then sink, only to rise once again, repeating the process many times.

Have the students attempt to explain what they are observing. But, again, do not spend much time on the discussion. (Gas bubbles are trapped by the convoluted surface of a raisin, and thus lower the density of the whole. As the raisin reaches the surface, it begins to loose the gas bubbles, the density increases, and it sinks to the bottom again.)

Demonstration 5: Bursting Balloons

PURPOSE: To demonstrate the synthesis of water from hydrogen and oxygen gas while demonstrating that some reactions take place violently
MATERIALS: Balloons, a source of helium, hydrogen, and oxygen gas
PROCEDURE: Fill three balloons with (1) helium, (2) pure hydrogen, and (3) a mixture of hydrogen and oxygen. Tie a string to each balloon and allow them to float over the desk. Ask the students if they know what gas(es) might be in the balloons. Most are familiar with helium balloons. Few will have seen a

hydrogen balloon. Ask if any know the story of the German zeppelin the Von Hindenberg destroyed by lightning in the late 30's in Virginia. Dim the room lights and with a candle taped to the end of a meter stick, ignite the gas in each of the balloons. CAUTION: Be sure and have the students cover their ears. You should use ear protectors yourself. The helium balloon will pop with very little noise. The hydrogen balloon will burst with a muffled explosion and a ball of flame. The hydrogen/oxygen balloon will explode with a very loud noise.

SAFETY AND DISPOSAL: Have the students cover their ears. Small balloons should be used and should not be overfilled. They will make a loud noise.

1.1 What is Chemistry?

The first day demonstrations should have whetted the appetites of the students. After distributing the textbooks during the second day of class, discuss the introduction to chemistry. Point out how chemistry is everywhere. Ask the students for examples where they have come in contact with chemistry in action. The text lists a number, but the students should be able to come up with their own. Define chemistry and relate the definition to the previous day's demonstrations.

To help students see the wide scope of chemistry in their daily lives, display several household chemicals—a variety of plastics, fibers, fertilizers, motor oil, car polish, medicines, and vitamins. Chemistry not only has provided us with many useful products, but manufacturing processes have left us a legacy of problems of waste disposal, ozone depletion, water contamination, etc. Point out that solutions to the problems facing society will be found by chemists themselves. Note the importance of chemistry as a career option.

Encourage students to bring in newspaper and magazine articles that bear on chemical products or issues. You may wish to make this an assignment.

Discuss the traditional branches of chemistry, but emphasize that these boundaries are not as clear today as they were in the past. As biology has moved to the molecular level, the border between biology and chemistry is disappearing. Chemistry is becoming ever more important.

Point out that scientific discoveries are made in two ways: (1) by accident and (2) through the use of scientific methods. Since there are so many fascinating stories surrounding accidental discoveries, discuss a few important discoveries made by accident, such as the discoveries of penicillin, saccharin, dynamite, quinine, petroleum jelly, and teflon. During your discussion, use Pasteur's comment that "Chance favors the prepared mind" to emphasize that unless a scientist recognizes the importance of a discovery, it will be meaningless.

Define Scientific Method. Ask the students to give you examples of the scientific method. Most will give the example of experimentation. Prior to discussing generalizations, you should cover order in the universe, the principle that allows broad generalizations to be made from a rather limited number of observations because of the belief that everything in the universe follows a given pattern. Call to the students' attention the fact that they have been generalizing for many years. For example, they learned early in life that every time they dropped something heavier than air, it fell to the floor. Ask what would happen if a beaker were dropped in the lab. All would agree that

it would fall to the floor and break even though no one had ever seen that particular beaker before. This points out that after a limited number of trials, one can accurately predict future outcomes because of order in the universe. Remind students that scientists are always looking for patterns and keep pointing this out throughout the course.

In discussing the scientific method, stress the importance of observations in collecting data. Observations are simple statements of facts and do not imply anything that cannot be measured. These observations or information can be quantitative or qualitative. To distinguish between the two, show your students a candle and have them list observations. Describing the candle as a white, cylindrical solid makes use of qualitative information. A meter stick can provide quantitative information such as length and width of the candle. Qualitative information may become quantitative if it can be compared to other measurable information. For example, to say that the candle is made up of a white solid soft enough to be scratched by a fingernail adds a quantitative element to the observation. To say that the lit candle gives off light is a qualitative observation, but to add that it gives off enough light to be visible in a well-lit room makes the information quantitative.

In making observations, encourage the students to pretend that they are radio broadcasters. Their verbal descriptions must be so complete and graphic that a person hearing or reading the descriptions will be able to visualize the object.

Next it is important to help students distinguish between a theory and a law. Make sure that you make the distinction that a law tells what happens and a theory explains why it happens.

Demonstration 6: Effect of phenolphthalein on acids and bases

PURPOSE: To elicit a generalization from students
MATERIALS: Dilute solutions of several different acids and bases (concentration is not critical) LiOH, NaOH, KOH, NH_4OH, HCl, HNO_3, H_2SO_4, $HC_2H_3O_2$ in eight different beakers; phenolphthalein; stirring rod
PROCEDURE: Tell the students the names of each of the solutions. Add a few drops of phenolphthalein to all but the LiOH and HCl solutions. Explain that the first four are classed as bases and the final four as acids. Explain that acids and bases will be studied later. Ask the students to make a broad generalization concerning the observation they have just made. Have them predict what will happen if phenolphthalein is added to LiOH, a base, and to the HCl, an acid.

1.2 Matter and Energy

Define matter. Use various objects in the room as examples of matter. Solids and liquids should pose no problems. To demonstrate that gases are also a form of matter, determine the mass of an empty balloon. Blow up the balloon and determine the new mass. Although the difference in mass does not reflect the actual mass of the air in the balloon, the increase in mass should convince the students that gases are composed of matter.

Define inertia. Ask the students for examples of inertia from their own experiences. Action of cars on wet or icy roads is a good example. Others may have seen video of the astronauts pushing objects while in orbit. Use this last example to distinguish between mass and weight. Stress the fact that weight depends on the gravitational attraction. The mass, however, remains the same. Stress that chemists are concerned with determining the mass of substances. Point out that on a platform balance, mass is determined by comparing an unknown mass to a set of known masses. Next show the students a spring balance. Bathroom and/or kitchen scales are usually spring balances. Gravity pulling on the object being weighed determines the reading on the scale. If the same object could be weighed on the moon, the platform balance would give the same mass, but the spring balance would give a smaller reading because of the weaker gravitational field on the moon.

Ask the students to state the law of conservation of matter. It should have been covered in earlier courses. The law applies to "ordinary chemical and physical changes," not to nuclear reactions which are beyond the scope of the high school lab.

Define energy. Point out that scientists have some difficulty with this definition because it is not precise. Since all energy can be classified as kinetic or potential, use a transparency or the chalkboard and have the students classify the forms of energy listed in the chapter. The chart might look as follows:

Type of Energy	Kinetic	Potential
Chemical energy		X
Mechanical (including sound)	X	X
Electrical	X	X
Radiant (heat, light, etc.)	X	
Nuclear	X	X

As before, have the students state the law of conservation of energy. It too applies to ordinary laboratory situations and must be modified when discussing nuclear reactions.

Chemistry Notebook: Interconversion of Matter and Energy

This feature on the interconversion of matter and energy gives you the opportunity to tie the two conservation laws into one more general law. Break the law into two parts: (1) matter and energy are interchangeable and (2) the total amount of matter and energy in the universe is constant. Although Einstein formulated his theory in 1905, it was not confirmed in the lab until nuclear reactions were carried out. The power of a nuclear bomb arises from the conversion of a small amount of mass into a very large supply of energy.

List the three common states of matter. Some students may list a fourth, the plasma state. Point out that plasmas require temperatures too high (1×10^7 °C) to be reached in a high school lab. Plasmas are a gaseous system composed of positive ions and electrons. Have sample of solids and liquids as well as some empty containers on the desk. Show that the solids do not change shape when moved from container to container. The liquids do not change volume but they do conform to the shape of the container. To confirm that gases fill the container in which they are placed, spray some cologne or deodorizer into a beaker covered with a watch glass. Remove the cover and have students sitting near the desk indicate when the odor is noticeable. The cologne or deodorant will eventually fill the room.

Two distinctions must be made regarding the properties of matter. Students must be able to distinguish between physical and chemical properties; between intensive and extensive physical properties, and between physical and chemical changes.

Demonstration 7: Physical-chemical properties/changes

PURPOSE: To help students distinguish between physical and chemical properties and changes, and intensive physical and extensive physical properties

MATERIALS: Paper, matches, large watch glass or ceramic pad, balance, copper shot (any metal available in shot form), graduated cylinder, hot plate, 1-L beaker, salt solution

PROCEDURE: To distinguish between physical versus chemical properties and changes:

(1) Show students a piece of white paper. Have them describe it in terms of color, physical state, size, etc.

(2) Cut the paper in half. Ask if any of the physical properties have changed? Note that a physical change, cutting the paper, does not change the identity of the material.

(3) Holding one half of the paper with a pair of tongs over the ceramic pad or large watch glass, light a match and bring the match to the paper. Is burning a physical or chemical change? Is the ability of paper to burn a physical or chemical property? In this case, a new substance does form.

To distinguish between intensive physical and extensive physical properties:

(1) Prior to class measure approx. 25 ml of metallic shot. (2) Show the shot to the students and have the students describe it. (3) Place the shot on the balance, and determine its mass. (4) Pour the shot into a dry graduated cylinder, and determine its volume. (5) Divide the sample into two unequal portions. (6) Ask the students to list the properties that have changed. Intensive properties such as color and melting point have not changed. Mass and volume have changed. The extensive properties have changed; the intensive properties did not. (7) If time is available, you can determine the densities of a small and a large sample by (a) determining the mass of each and (b) determining the volume by pouring each into a graduated cylinder half-filled with water. Show the students how to determine the volume of the solid and show that the ratio of mass to volume, i.e. the density does not change. This may be used to introduce density before actually discussing it in the text, or, if you wish, you may delay this part of the demonstration until later and use it to reinforce the distinction between intensive and extensive properties.

Physical changes associated with a phase change.

(1) Place the 1-L beaker on a hot plate. Add about 200-mL of water and cover with a watch glass. (2) Place several ice cubes on the watch glass. (3) Turn on the hot plate and allow the water to boil and the steam to condense on the cold watch glass. (If this the first time your students are seeing a hot plate, explain its use and any safety precautions that must be taken.) (4) Point out that a phase change is a physical change, whereas burning is a chemical change. The physical change can be reversed by simply changing the temperature while the chemical change results in new compounds.

SAFETY CONSIDERATIONS: Wear your goggles. Be careful when burning the paper. Be sure that the large watch glass or ceramic plate are below the burning paper. The other materials may be reused.

Indications of chemical reaction are best presented in the context of the first Desktop Investigation. It also serves as the ideal time to introduce you students to the laboratory and to the required safety rules.

Desktop Investigation: Physical and Chemical Changes

Go over the procedure with your students. Stress the importance of recording all observations in the data table. After they have performed the exercise, elicit from them any observations that indicated a chemical change. Relate these to the ones listed in the text. Define endothermic and exothermic changes in the context of their observations.

1.3 Classification of Matter

Introduce this section using as many examples as you can. Have available on your desk a variety of samples of matter. For mixtures you might use a bottle of salad dressing, a piece of granite, a candy bar, and a copper-sulfate solution. For pure substances include water, sugar, salt, copper, and other element samples. Have the students describe each and begin to place them into the classification scheme for matter.

Although not discussed in this part of the text, some students will be aware of colloidal suspensions. Explain that if particles can be made small enough, the motion of the water molecules will keep them suspended. Milk is the best example.

Care must be taken in distinguishing the various types of mixtures and pure substances. You may wish to evaporate the liquid from two beakers. One beaker might contain distilled water (nothing remains), the second, a salt solution (the salt remains). If an electrolysis apparatus is available, electrolyze the water and show that it can be decomposed into simpler components. Stress that once one reaches the atomic or element level, one may no longer subdivide the matter in question with the chemical means available in the high school lab.

1.4 The Chemical Elements

As an introduction to the Periodic Table, if a recording of the song, "The Elements" from the album *An Evening Wasted* (Reprise Records) by Tom Lehrer is available, use it to introduce the section on the elements.

In this section you are introducing one of the most important tools a chemist has. Understanding how to use the Periodic Table is one of the skills your students must develop. Hand out Periodic Tables on which the students may write and take notes. First divide the table in two. Show them where the metals, nonmetals, and metalloids are found. Further subdivide the table into the main group or representative elements, the transition metals, and the rare earths. Distinguish between periods and families, and have them write down the names of the most common families, i.e. alkali metals, alkaline earths, halogens, and noble gases. Do not get bogged down with the group numbering system. Use the 1–18 adopted internationally.

List on the board those elements whose symbols are not related to their modern names. Encourage the students to begin learning them. The symbols of the elements are analogous to the letters of the alphabet. Just as we use letters to make up words, chemists use the symbols of the elements to write the formulas of compounds, and, finally, just as words are used to make up sentences, chemists use the formulas in writing chemical equations.

A Periodic Table with photographs of each elements would be a great aid in distinguishing types of elements. List properties of metals and then contrast the properties of nonmetals. Show how metalloids fall in between.

Have available samples of the two typical elements discussed in the text. **CAUTION:** White or yellow phosphorus is very dangerous and may not be allowed in your school. Use only red phosphorus and show pictures of the other form. If phosphorus is not available, use roll sulfur as your nonmetal sample.

Stress with the students that this has been a first look at the Periodic Table and the wealth of information that it contains. As the year goes on, each chapter will unlock more secrets and make the Periodic Table ever more useful.

As a special exercise, you may wish to have your students pick and report on a favorite element. If your students are seated alphabetically, point out that they are seated according to a classification system. At some future date, you may wish to seat your students according to some other property, e.g. home telephone numbers, and have them try to discover the pattern. Anything that you do to impress on the students the importance of the Periodic Table will bear fruit in the long run.

References and Resources

Books and Periodicals

Alyea, Hubert and Frederic B. Dutton. *Tested Demonstrations in Chemistry*, Easton, PA: Journal of Chemical Education, 1965. (Facsimile edition available from Flinn Scientific).

Borgford, Christie L. and Lee R. Summerlin. *Chemical Activities: A Sourcebook for Science Teachers*, Washington, D.C.: The American Chemical Society, 1988.

Gensler, Walter J. "Physical Versus Chemical Change," *Journal of Chemical Education*, Vol. 47, No. 2, Feb. 1970, pp. 154–156.

Redlich, Otto. "Intensive and Extensive Properties," *Journal of Chemical Education*, Vol. 47, No. 2, Feb. 1970, pp. 154–156.

Shakashiri, Bassam Z. *Chemical Demonstrations*, Vol. 1, Madison: University of Wisconsin Press, 1983.

Shakashiri, Bassam Z. *Chemical Demonstrations*, Vol. 2, Madison: University of Wisconsin Press, 1985.

Strong, Laurence E. "Differentiating Physical and Chemical Changes," *Journal of Chemical Education*, Vol. 47, No. 10, Oct. 1970, pp. 689–690.

Summerlin, Lee R., et al. "Introducing Chemistry," "Physical Changes," and "Reactions of Some Elements," *Chemical Demonstrations—A Sourcebook for Teachers*, Vol. 2, Washington, D.C.: American Chemical Society, 1987.

Talesnick, Irwin. *Idea Bank Collation—A Handbook for Science Teachers*, Kingston, Ont.: S17 Science Supplies and Services Company, Inc., 1984.

Webb, Michael J. "Physical and Chemical Change: What's the Difference?," *The Science Teacher*, Vol. 49, No. 3, March 1982, pp. 39–40.

Answers and Solutions

Questions

1. *(a)* Science is a systematic and organized collection of facts. *(b)* Scientific facts are used to provide explanations, to make predictions, and to manage events.

2. *(a)* Chemistry is the study of the composition of materials and changes in the composition of materials. *(b)* Chemistry is involved in the manufacture of synthetic fabrics, in the growing of fruits and vegetables, and in the purification of water.

3. The dictionary defines *chemical* as a substance produced by or used in a chemical process. According to this definition, everything is made of chemicals. As used in the advertising slogan cited, however, the word "chemical" refers to particular substances associated with undesirable circumstances. In fact, although some chemicals (in the dictionary sense) are harmful, many are not.

4. The scientific method is a logical approach to the solution of problems that lend themselves to investigation.

5. *(a)* In science, a law is a generalization that describes behavior in nature. *(b)* Laws in science are generally expressed by mathematical equations or concise statements.

6. Numerical information is quantitative, whereas nonnumerical information is qualitative.

7. *(a)* Measuring is the collecting of numerical information about the object, material, or process under study. *(b)* Experimenting is carrying out sequences of observations under controlled conditions. *(c)* A hypothesis is a testable statement. *(d)* A theory is a broad generalization that accounts for a class of known facts or phenomena.

8. *(a)* Inertia is resistance to change in motion. *(b)* A moving object resists a change in direction or speed, whereas an object at rest resists being moved. *(c)* Inertia varies directly with the quantity of matter to be moved.

9. Weight is a measure of the earth's attraction for matter, whereas mass is a measure of the quantity of matter.

10. *(a)* The weight of an object varies with the distance of the object from the planet or other very large object that attracts it. *(b)* The mass of an object is an unchanging property of that object.

11. *(a)* Matter is anything that has mass and occupies space. *(b)* Sand, paper, air, and chalk are examples of matter.

12. Matter cannot be either created or destroyed in ordinary chemical or physical changes.

13. *(a)* Energy is the ability to cause change. *(b)* Chemistry deals mainly with the chemical energy stored in matter and either the effects on matter or the production from matter of thermal, radiant, or electrical energy.

14. The two major types of energy are kinetic energy (the energy of an object in motion) and potential energy (the energy an object has by virtue of its position or composition).

15. Options *(a)* and *(d)* illustrate kinetic energy, whereas *(b)* and *(c)* illustrate potential energy.

16. Energy can be converted from one form to another, but it cannot be either created or destroyed in ordinary chemical or physical changes.

17. The three states of matter are the solid state (the state of matter characterized by definite shape and volume), the liquid state (the state of matter characterized by definite volume, but not definite shape), and the gaseous state (the state of matter characterized by neither definite shape nor definite volume).

18. The particles of matter in a solid are closely packed and rigidly held in fixed positions, those in a liquid are closely packed but are not bound in fixed positions, and those in a gas are very far apart and have no ordered positions at all.

19. Three examples of intensive properties of matter that are qualitative include color, hardness, and odor; three that are quantitative include density, melting point, and boiling point.

20. *(a)* A physical property is one that can be observed or measured without altering the identity of a material. *(b)* Color, melting point, and boiling point are examples of physical properties.

21. *(a)* Extensive properties depend on the amount of matter present, intensive properties do not. *(b)* Extensive: mass, length, volume; Intensive: density, ductility, color

22. A chemical property is the ability of a substance to undergo a change that alters the identity of a substance.

23. *(a)* Physical changes do not result in a change in identity. Chemical changes involve conversions into different substances. *(b)* Three examples of physical changes are cutting a wire, grinding a solid, and melting a solid. Examples of chemical changes include the burning of wood, the rusting of iron, the tarnishing of silver.

24. The substances that undergo a chemical reaction are the reactants, whereas the products are the new substances produced by a chemical reaction.

25. Examples *(a)*, *(b)*, *(d)*, and *(e)* are physical changes, whereas *(c)* is a chemical change.

26. Observable changes that generally indicate that a chemical reaction has occurred include the evolution of heat and light, the production of a gas, and the formation of a precipitate.

27. *(a)* In exothermic processes, heat is released, whereas in endothermic processes, heat is absorbed. *(b)* exothermic: hydrogen and oxygen → water and heat; endothermic: calcium carbonate + heat → calcium oxide + carbon dioxide

28. Any sample of matter can be classified as being either a pure substance or a mixture. Pure substances are either elements or compounds. An element is a substance that cannot be separated into other substances by any ordinary chemical change. A chemical compound is a pure substance that can be made to decompose chemically into two or more simpler substances. A mixture is a combination of two or more kinds of matter, each of which retains its own characteristic composition and properties.

29. *(a)* Heterogeneous matter is not uniform in composition and properties, whereas homogeneous matter is uniform in composition and properties. *(b)* Examples of heterogeneous matter include vegetable soup, whole-wheat bread, and granite. Examples of homogeneous matter include water, mercury, and glass.

30. Homogeneous matter is either a pure substance or a solution. A pure substance is a homogeneous material that has the same composition and characteristic properties, whatever its source. A solution is a homogeneous mixture.

31. Pure substances differ from mixtures in the following ways. *(1)* Every sample of a given pure substance has exactly the same characteristic properties; this is not the case in a given mixture. *(2)* Every sample of a given pure substance has exactly the same composition, whereas mixtures have variable compositions. *(3)* A pure substance cannot be separated into other substances without changing its identity and properties. In a mixture, the components do not lose their identity and can be separated by physical means.

32. *(a)* An element cannot be separated into other substances by ordinary chemical changes, but a compound can undergo decomposition in this way. *(b)* Aluminum, oxygen, and copper are elements, whereas sugar and water are compounds.

33. The law of definite composition states that every chemical compound has a definite composition by mass.

34. Groups (or families) are the vertical columns of the Periodic Table, and are numbered, left to right and consecutively, from 1 to 18. Periods are the horizontal rows of the table, and are numbered 1 to 7 from the top down.

35. *(a)* Copper has a metallic luster, high boiling and melting points, can be formed into wire, can be hammered into sheets, and exhibits high electrical conductivity. *(b)* Copper is used for electrical wiring, water pipes, and roof flashing.

36. *(a)* Red phosphorus is a dark-red powder that remains unchanged in air, is not toxic, and can be handled with reasonable safety. White phosphorus is a waxy, low-melting solid that ignites spontaneously in air, burns the skin, and is extremely toxic. *(b)* Phosphorus compounds are used in agricultural fertilizers, cleaning agents, and matches.

37. (a) A total of 109 elements are presently known.
    (b) Roughly 90 elements exist naturally on Earth.

Application Questions

1. *(a)* At the top of Pike's Peak, the mass of an object would be the same as its mass at sea level, but its weight would be slightly lower because it is farther from the earth's surface. *(b)* In the airplane, the object's mass would be the same as

at sea level, but its weight would be even lower than on Pike's Peak because the object is even farther from the earth's surface.

2. *(a)* It is not possible for an object to be massless because it will always consist of matter. *(b)* It is possible for an object to be weightless if it is far enough from any planet so as not to experience the effects of that object's attraction.

3. Examples *(b)*, *(e)*, and *(g)* indicate exothermic processes, whereas *(a)*, *(c)*, *(d)*, and *(f)* are endothermic.

4. The reactants in an exothermic reaction are analogous to a boulder at the top of a hill. Like the boulder, when they react ("fall down the hill"), their potential energy will be converted into an equal amount of energy released as heat. On the other hand, in an endothermic reaction, the reactants are like a boulder that must be moved from the bottom to the top of a hill. This sort of change necessitates the input of energy.

5. *(a)* The kinetic energy of object A is twice that of B.
   (b) Object A has a kinetic energy four times that of B.
   (c) Object A has a kinetic energy eight times that of B.

6. Examples *(c)*, *(e)*, and *(f)* illustrate heterogeneous matter, where *(a)*, *(b)*, and *(d)* represent homogeneous matter.

7. *(a)* Texture, color, and density could possibly be used.
   (b) Color, density, and melting point could possibly be used.
   (c) Color, density, and texture could possibly be used.
   (d) Density and viscosity could possibly be used.

8. The 12 earliest-known elements generally occur in nature in free form and are thus more easily identified. The vast majority of the remaining naturally occurring elements are found in nature in combined form, where they are more difficult to isolate, and thus are not readily identifiable. Gold, silver, and copper generally occur in uncombined form in nature, and therefore were more obvious than elements that occur in combined form.

9. The law of definite composition allows analytical chemists to identify the types and quantities of elements combined in any chemical compound.

10. Metals have a characteristic luster (they shine), most are ductile (they can be drawn out into fine wire), most are malleable (they can be hammered or rolled into thin sheets), and many have high tensile strength (they resist breaking when pulled upon).

11. Nonmetals are brittle and are neither malleable nor ductile.

Teacher's Notes

Chapter 1 Matter, Energy, and Change

INTRODUCTION

An understanding of chemistry and the methods of science is essential to many aspects of life in modern society. Chemistry is the study of matter, changes in matter, and the energy associated with those changes. Following an overview of chemistry and the processes of science, this chapter introduces the classification of matter and one of the most important tools in chemistry, the Periodic Table of the elements.

LOOKING AHEAD

As you study this chapter, think of ways in which chemistry touches everyday life and how the methods of science might be used to solve everyday problems. Become familiar with the Periodic Table. It is a valuable resource for organizing what you will be learning in later chapters.

SECTION PREVIEW

1.1 What is Chemistry?
1.2 Matter and Energy
1.3 Classification of Matter
1.4 The Chemical Elements

Energy is produced by the burning of wood—a chemical change. Heat produced by the fire melts the ice in the pan—a physical change.

1.1 What Is Chemistry?

SECTION OBJECTIVES

- Define chemistry.
- Identify some applications of chemistry in everyday life.
- List the major areas of chemistry.
- Describe the scientific method.

Human beings are naturally curious. We wonder and ask questions about what we see and experience. The phenomena of nature provide a never-ending source of questions such as: What is fire made of? What happens to the air we breathe? Why do objects fall toward the earth? While these three questions and many more like them have been answered, countless questions remain: How was the universe formed? Can heart disease be cured? Is there a material that superconducts at room temperature?

Posing questions about nature and carefully gathering the facts to provide answers is the work of natural scientists. Over the centuries, natural scientists have contributed to a greater understanding of our world and our experiences. This knowledge has enabled us to make tremendous changes in the way we live. Today, for example, laser-disk players are widely used to reproduce music with stunning clarity. Positron-emission tomography is used in hospitals to harmlessly probe the human brain for signs of disease. By using genetic engineering, bacteria can be designed to produce life-saving insulin for diabetics. None of these were possible when most of today's high school seniors were born.

These changes bring many benefits, but they also provide challenges to all of us. One of these challenges is cost. Many new and sophisticated medical techniques are expensive. Can society afford to make them available for everyone? Another challenge is the risk that genetically engineered life forms may pose to humans. Attempting to understand words and phrases like *laser*, *superconductivity*, *positron-emission tomography*, *pollution*, and *genetic engineering* is another challenge.

To successfully evaluate and meet these challenges, members of today's society must have a greater understanding of natural science than ever before.

Have students list other benefits and contributions of science, as well as other potential problems and hazards attributed to science.

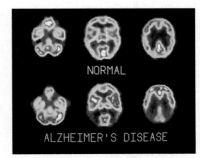

Figure 1-1 Color enhanced images produced by *Positron Emission Tomography* show differences between normal brains and brains affected by Alzheimers disease.

Chemistry is a Physical Science

The natural sciences fall into two general categories—the biological sciences and the physical sciences. The biological sciences are concerned primarily with living things, and the physical sciences are concerned primarily with nonliving things, including rocks, the stars, electricity, the weather, energy from the sun, and the composition of all materials. Chemistry is a physical science.

Chemistry *is the study of the composition and structure of materials and the changes they undergo.* This definition may suggest to you that chemistry has little to do with everyday life. This is not true. Your way of life would be radically different without the practical applications of chemistry. Imagine the contents of your closet if all clothing made of synthetic fabrics were removed. Chances are, over half your wardrobe would be gone. Imagine a supermarket offering only fruits and vegetables grown without manufactured fertilizers or pesticides. The quantities and varieties offered would be far fewer. Imagine drinking water from your tap that had not been purified. The unpurified water would probably make you sick. Try to imagine a world without gasoline or heating oil. It would be very different from the world we live in.

Chemistry in Modern Society

In generations past, the design and construction of houses, public buildings, and vehicles was limited by the available materials. Physicians had medicines that were discovered only by accident or by laborious trial and error. Today, new substances can be created to meet specific needs in architecture, aircraft and automobile design, medicine, and many other human endeavors.

The use of chemistry to manipulate the composition and behavior of substances is also important in meeting the challenges of our complex society. To feed an expanding world population and keep it healthy is an ever-present challenge. New ways must be found to generate and use energy efficiently. Dwindling supplies of natural resources must be replaced with other substances. Ways to recycle materials such as plastics (Figure 1-3) rather than allowing them to clutter and foul our environment are needed.

Figure 1-2 The scientists pictured above are studying the structure of renin; a chemical produced by the kidneys that causes blood pressure to rise. Their hope is to find a blood-pressure medicine that works by inhibiting the action of renin.

Figure 1-3 Plastic bottles (left) can be recycled and turned into raw material (right) for new plastic products.

Other examples are: new adhesives, pharmaceuticals, TV sets, insecticides, fertilizers, textiles, etc.

Encourage students to bring in newspaper and magazine articles that deal directly or indirectly with chemistry. Try to find those that place chemistry in a favorable light.

What is a Chemical?

Have you ever seen an advertisement that says, "All natural—no chemicals added." Perhaps it was in reference to a fruit drink or a snack of some kind. What, exactly, is a chemical and what does the advertisement mean? One dictionary defines *chemical* as a substance produced by or used in a chemical process. Sugar is produced by plants from carbon dioxide and water—definitely a chemical process. We separate the sugar from the plants, purify it, and add it to foods. Is sugar a chemical? It certainly is to a chemist, and he or she can describe its exact composition. But what about water? It can be found everywhere. It can also be made in a laboratory by combining hydrogen and oxygen. Water is indeed a chemical. The point is that chemistry is the study of any and all materials. With sufficient investigation, the composition of every material can be described in terms of chemistry—everything is made of "chemicals."

Unfortunately, the word *chemical* has become associated with undesirable circumstances—pollution, cancer, poisonings, and accidents that result in spills of harmful substances. Some chemicals do, indeed, have unpleasant and dangerous properties, and contact with them must be avoided. Other chemicals cure diseases, maintain the food supply, and contribute to our comfort. One outcome of your study of chemistry should be an ability to evaluate uses of the word *chemical*. A substance that is a "chemical" is not, by definition, harmful. Nor can anything else be "free of chemicals."

Branches of Chemistry

Chemistry is a very broad subject. Most chemists would describe themselves as working in one of the following five major areas of the science:
1. *Organic chemistry*—the study of substances containing carbon and hydrogen, and their derivatives
2. *Inorganic chemistry*—the study of all substances not classified as organic chemicals, which includes the chemistry of all substances containing elements other than carbon
3. *Physical chemistry*—the study of the properties and transformations of matter in terms of fundamental physical processes
4. *Biochemistry*—the study of all substances and processes that occur in living things
5. *Analytical chemistry*—the identification of materials and the qualitative and quantitative determination of the composition of materials

Chemists also work in numerous specialized branches of chemistry. Nuclear chemists study the properties and transformations of matter that involve the atomic nucleus. Polymer chemists concentrate on understanding and making polymers—the raw materials for, among many other things, plastic objects. Some chemists become specialists in interdisciplinary fields, such as bioinorganic chemistry, which is the study of the role of inorganic materials in living things, or pharmaceutical chemistry, which is the chemistry of drugs, medicines, and vitamins.

Chemistry is applied to the large-scale production of useful materials by chemical engineers who are involved in the design, construction, and operation of equipment required in the chemical manufacturing processes.

Originally, "organic" chemicals were thought to be made only within living things. We now know this to be incorrect, but the name remains in use.

Current boundaries among the major areas of science are not as clear as they were in the past.

The atomic nucleus and nuclear chemistry are described in later chapters.

The Scientific Method

The general approach to investigations in chemistry is no different from that in any other natural science. Some important scientific discoveries come about by accident. Others are the result of brilliant new ideas. Most scientific knowledge, however, is the result of carefully planned investigations carried out by trained scientists, like the analytical chemist pictured in Figure 1-4.

The aim of specific investigations is usually to solve a problem or answer a question about a natural phenomenon. For instance, an analytical chemist may carry out an investigation to determine the exact composition of certain substances that have shown the ability to kill cancer cells in rats while leaving normal cells unharmed.

The ultimate goal of scientific investigations is to explain and predict natural phenomena. Scientists investigating cancer, for example, would like to explain how cancers develop and predict what substances will kill human cancer cells and leave normal cells unharmed.

The way scientists carry out their investigations is often referred to as the "scientific method." *The* **scientific method** *is defined as a logical approach to the solution of problems that lend themselves to investigations by observing,*

Figure 1-4 A chemist at Lamont-Doherty Geological Observatory, NY carries out a carefully planned experiment.

generalizing, theorizing and testing. The processes involved in observing, generalizing, theorizing, and testing are explained below.

Observing Observing substances and events is essential to all natural sciences. Practical chemistry began with observations of materials found in nature, such as sand that could be converted to glass and rocks that yielded copper or iron when heated.

Collecting data goes hand in hand with observation. The data may be written descriptions of what is observed. The data may also be the result of *measuring* the object, material, or process under study. *Numerical information is referred to as* **quantitative information**. *Nonnumerical information is referred to as* **qualitative information**. Modern science relies on instruments of many kinds to extend the boundaries of what our senses can detect and measure. Figure 1-5 shows an image made by an electron microscope that allows details invisible to the naked eye to be "seen."

Figure 1-5 The scanning electron microscope (SEM) at the left can achieve magnifications of 100 million times the size of a specimen. A false color image (right) of *Corynebacterium* magnified 5400x. This bacterium causes diptheria.

Scientists know that observing is most productive when conditions that affect the observations are brought under their control. Chemists do most of their observing in the controlled environment of the laboratory. *Experimenting* is carrying out sequences of observations under controlled conditions. Experimentation is the foundation upon which modern science is built.

Before new experiments are planned and carried out, most scientists turn to the scientific literature, which is extensive. They study the reported observations of others working in the same area of science. In this way, new experiments can be designed to extend knowledge and understanding. When experiments are completed, each scientist has the responsibility of *communicating* results and new information to others.

Generalizing Once observations have been made and data collected, the process of generalizing gets under way. *Organizing data* by making graphs, charts, or tables may be necessary. *Analyzing data* by using statistics, mathematical relationships, or other techniques is often necessary. Frequently, computers are used to organize and analyze data. If possible, general principles will be developed that help in *classifying* the data.

The goal is to find relationships among the accumulated data. Scientists hope that determining such relationships will enable formulation of a **hypothesis** *which is a testable statement.* A hypothesis is the basis for making predictions and carrying out further experiments. A generalization that describes a wide variety of behaviors in nature is known as a **law**. Laws may be expressed by mathematical equations or concise statements. Scientific laws describe, but do not explain, natural phenomena. If a scientific law is observed to be violated, then the law must be changed because it is not a valid description of nature. Two of the laws you will study in chemistry are *Boyle's law* and the *periodic law*.

Theorizing   Once successful predictions can be made in an area of science, it becomes possible to move on to the task of determining *how* something occurs. To do so, scientists often engage in *constructing models*. Such models are usually not physical objects, like airplane models. Rather, they are visual, verbal, or mathematical models that show relationships among data or events. Frequently, a scientific model relates observed behavior to familiar and well-understood phenomena. Modern chemistry is based upon a model that states that all matter is composed of tiny particles.

In many cases, a model that is successfully applied may become part of a **theory**, *which is a broad generalization that explains a body of known facts or phenomena*. Theories can never be proven but they are considered successful if they can predict the results of many new experiments. Two major theories in chemistry that you will study in later chapters are the atomic theory and the kinetic–molecular theory.

Figure 1-6   A flow chart of the scientific method

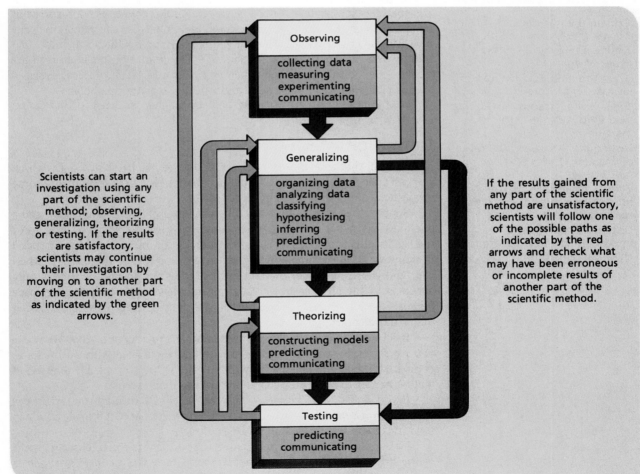

Scientists can start an investigation using any part of the scientific method; observing, generalizing, theorizing or testing. If the results are satisfactory, scientists may continue their investigation by moving on to another part of the scientific method as indicated by the green arrows.

If the results gained from any part of the scientific method are unsatisfactory, scientists will follow one of the possible paths as indicated by the red arrows and recheck what may have been erroneous or incomplete results of another part of the scientific method.

1. *(a)* Biological and physical *(b)* Physical
2. *(a)* A substance produced by or used in a chemical process. *(b)* Many are possible. Sugar (sucrose) and salt (sodium chloride)
3. Organic, inorganic, physical, biochemistry, and analytical
4. Observing, generalizing, theorizing, testing

Testing Through every stage of scientific investigation, testing takes place continuously. Scientists test data and the results of experiments by carrying out new experiments. Frequently, testing is based upon *predicting* the outcome of experiments based upon hypotheses, laws, or theories. No hypothesis, law, or theory is *ever* free from testing. All are constantly open to modification or abandonment based on new evidence. A summary of the scientific method and its processes is given in Figure 1-6.

Section Review

1. *(a)* What are the two categories of natural sciences? *(b)* Into which category does chemistry fall?
2. *(a)* What is a chemical? *(b)* Give two examples of chemicals.
3. List the five major areas of chemistry
4. What are four major activities of the scientific method?

1.2 Matter and Energy

SECTION OBJECTIVES

- Explain the difference between mass and weight.
- Define energy and list some types of energy.
- State the laws of conservation of mass and energy.
- Explain the gaseous, liquid, and solid states in terms of particles.
- Distinguish between physical and chemical properties of matter.
- Classify changes in matter as physical or chemical and give reasons for your choices.
- Distinguish between endothermic and exothermic chemical reactions.

Everything that you see or sense is either matter or the interaction of energy and matter. Pick up this book. You are holding a chunk of matter in your hands. You can read these words because light energy is reflected off the page in distinctive patterns. After reading this book for a while, you might get hungry and eat a peanut butter sandwich. The peanut butter and bread are matter, too. Within your body they will be converted into other types of matter plus energy that can help you continue reading this book.

We have used the terms *matter* and *energy,* but we must further define exactly what they mean.

Definition of Matter

Every object that you can see is matter, and gases like those in the air, which you can't see, are also matter. To define "matter" requires recognizing some properties common to all samples of matter.

Suppose we examine two very different kinds of matter such as air and a pile of bricks. What do they have in common? First, they occupy space. For the bricks, this is obvious. It is less obvious for air, but can be demonstrated in many ways. For example, water will not rise in a tube (Figure 1-7) when you hold your finger over the end because the space inside the tube is fully occupied by the air. Removing your fingers allows water to enter the space because some of the air can be pushed out and replaced by water, showing that water also occupies space.

Bricks can be used to demonstrate another property of matter. What can you expect to observe in pushing a wheelbarrow loaded with 20 bricks as compared to pushing one loaded with 60 bricks? As you probably guessed, it will be harder to start the second one rolling than the first.

If you throw a brick across the yard, it will fly through the air in the direction you aimed it. You would be very surprised indeed if it made a sudden turn in the air and headed off in a new direction.

The resistance of the wheelbarrows to moving and the unchanging path of the brick through the air demonstrate the property of **inertia**, *which is resistance to change in motion.* Inertia is a property of all matter. A moving object resists a change in direction or being stopped; an object at rest resists being moved.

Inertia varies directly with the quantity of matter to be moved. To explain the greater effort needed to move the larger load of bricks, however, it is doubtful one would say "Sixty bricks have more inertia than 20 bricks." Most likely, you would say "Sixty bricks weigh more than 20 bricks." Now we must find out what this means in the specific language of science.

Weight *is a measure of the earth's gravitational attraction for matter.* Hanging an object on a spring scale shows how the earth pulls that object down (Figure 1-8). The more matter in the object, the stronger the force exerted on it and the further it is pulled down. Comparing the weight of a vessel containing a vacuum and one containing a gas would demonstrate that gases, like bricks, can be weighed.

Scientists make a distinction between weight and mass that is usually ignored in everyday life. An astronaut weighs less on the moon than on earth. But on earth, on the moon, or in orbit, the amount of matter in the astronaut's body is the same. Here is where mass comes in. **Mass** *is a measure of the quantity of matter.*

The interchangeable use of the terms *mass* and *weight* seems unavoidable and ordinarily it causes no problems. If the masses of two objects are the same, they have equal weights while in the same location.

Now that you know something about the properties of matter, the usual definition should be more meaningful to you: **Matter** *is anything that has*

Figure 1-7 Air trapped in the above tube will not allow water to rise. When air is allowed to escape, below, water will rise in the tube.

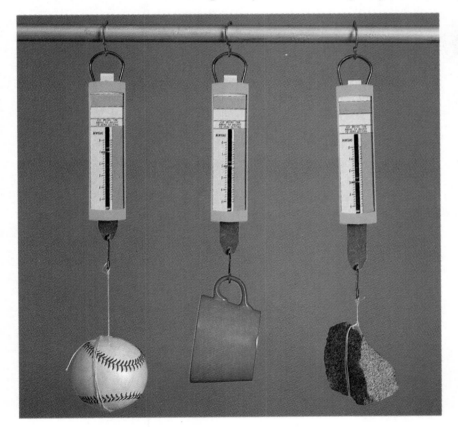

Figure 1-8 These spring scales show how strongly each object is attracted by the gravitational pull of the earth. Which object is attracted the most and which is attracted the least?

weight and occupies space. The **law of conservation of matter** is fundamental to chemistry. It states: *Matter cannot be either created or destroyed in ordinary chemical or physical changes.* Determining the mass of a sample of matter before and after a change allows this law to be tested.

Imagine using a spark to ignite a match in a glass jar that is so tightly sealed that no matter can enter or escape. As the match burns, flammable chemicals in the head of the match and the wood of the matchstick are transformed into water, carbon dioxide, and other substances. Some oxygen from the air in the jar is "used up" as it combines with materials from the match. If matter is being created or destroyed, the mass of the closed jar plus its contents must change. Measurements taken before and after the match has burned would show that the total mass remains the same, however.

The phrase "in ordinary chemical and physical changes" must be added to the law of conservation of mass because one circumstance is known in which matter is changed into energy, as described in the Chemistry Notebook on page 15. In your experiences in everyday life and in the chemistry lab, however, you can expect that matter will not be created or destroyed.

The word "ordinary" is very important. Only in nuclear reactions does one encounter a change in mass. The "lost" mass is known as "the mass defect."

Definition of Energy

Energy is available from a variety of sources. Consider an automobile. It uses chemical energy stored in the battery and in gasoline as its primary sources of energy. The stored energy is converted to other forms of energy as it is needed. Some of the energy conversions in starting a car and driving away, as diagrammed in Figure 1-9, are the following:

Chemical energy in battery → *electrical energy* in starter and spark plugs.
Electrical energy in starting motor → *mechanical energy* to crank engine.
Electrical energy in spark plug ignites gasoline/air mixture in piston.
Chemical energy from gasoline/air mixture → mechanical energy as gasoline explodes and drives pistons down.
Mechanical energy in engine → mechanical energy in wheels.
Mechanical energy from engine → electrical energy in alternator.
Electrical energy from alternator → *sound energy* for horn, *light (radiant) energy* for headlights.

Look over Figure 1-9. Note the change that is occurring in each transformation. **Energy** *is the ability to cause change or the ability to do work.* Chemistry deals mainly with the chemical energy stored in matter and with either the effects of energy on matter or the production of thermal energy, radiant energy, or electrical energy from matter.

All energy is classified as either kinetic energy or potential energy. *The energy of an object in motion is* **kinetic energy.** A baseball in flight, a car rolling down the street, and a book falling off the edge of the desk all have *kinetic energy.* The kinetic energy of a moving object has depends on the mass of the object and its velocity. *The energy that an object has because of its position, or composition is classified as* **potential energy.** Potential energy can be thought of as stored energy, waiting to be released. The energy in gasoline is chemical potential energy. Water held behind a dam has gravitational potential energy. When water is released from the dam, its potential energy is converted to kinetic energy as it falls to a lower level. At the lower level it possesses less potential energy.

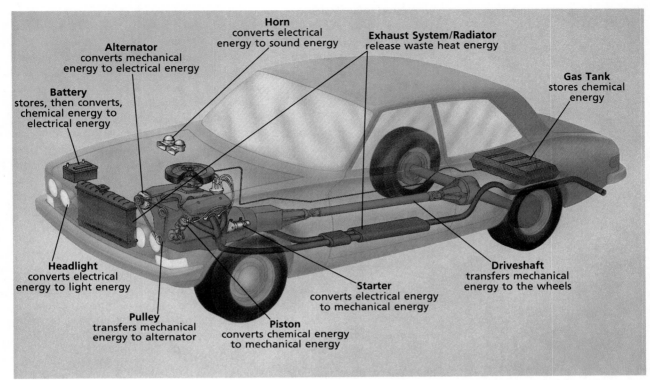

Figure 1-9 Numerous energy conversions take place when a car is started.

Labels in figure:

Alternator converts mechanical energy to electrical energy

Horn converts electrical energy to sound energy

Exhaust System/Radiator release waste heat energy

Gas Tank stores chemical energy

Battery stores, then converts, chemical energy to electrical energy

Headlight converts electrical energy to light energy

Pulley transfers mechanical energy to alternator

Piston converts chemical energy to mechanical energy

Starter converts electrical energy to mechanical energy

Driveshaft transfers mechanical energy to the wheels

Law of Conservation of Energy

Accounting for all of the energy present before and after an event is not as easy as keeping track of mass by weighing. But scientists have done the necessary "energy bookkeeping" are confident in the **law of conservation of energy**: *Energy can be converted from one form to another, but it cannot be created or destroyed in ordinary chemical or physical changes.*

States of Matter

A piece of wood placed on a table retains its shape and volume. The change either the shape or volume of the piece of wood would require applying considerable external force. *The **solid state** is the state of any matter that has a definite shape and volume.*

Suppose some water is poured onto the table. It flows out over the surface because it does not retain its shape. A liquid takes the shape of its container. The following example demonstrates that liquids retain their volume. Suppose you attempt to pour a quart of milk into a pint bottle you will discover that the volume of the milk does not change. The pint bottle holds only half the milke *The **liquid state** is the state of any matter that has a definite volume but an indefinite shape.*

When an automobile tire is inflated with air, the air assumes the shape and volume of the fully inflated tire. If a blowout occurs, the air escapes and immediately expands in volume. *The **gaseous state** is the state of any matter that has neither a definite shape nor a definite volume.*

The volume of an object is the amount of space it occupies.

Solid

decreasing ↑↓ increasing
temperature ↑↓ temperature

Liquid

decreasing ↑↓ increasing
temperature ↑↓ temperature

Gas

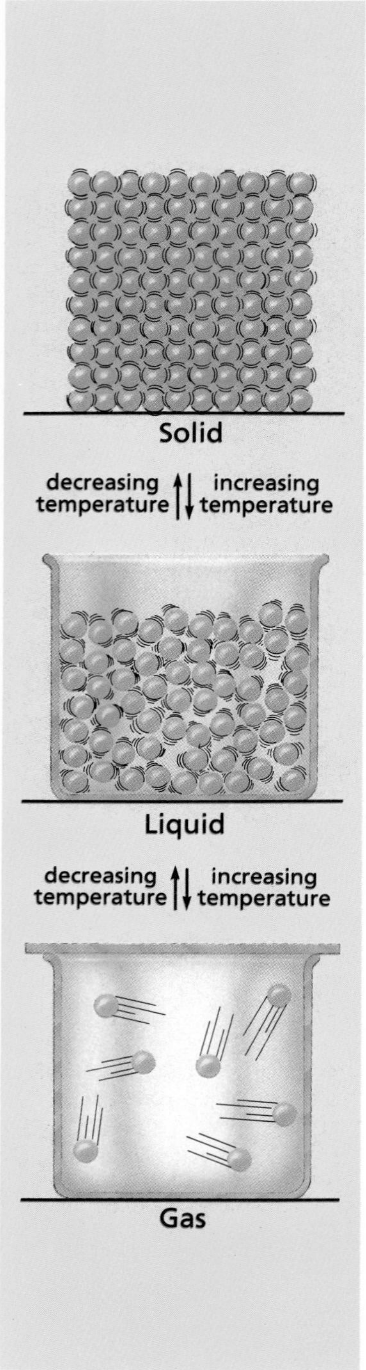

Figure 1-10 The distance between molecules in a substance and the rigidity with which they hold their positions determine whether the substance is a solid, a liquid, or a gas.

Water is one of many substances that can exist in all three common states of matter. A block of ice is a solid. When it melts, it becomes a liquid. When liquid water evaporates, it becomes a gas.

Figure 1-10 illustrates how chemists explain the differences between solids, liquids, and gases according to a model that states that matter is composed of tiny particles. The particles in a solid are closely packed and rigidly held in fixed positions, giving a solid its definite volume and shape.

The particles in a liquid are also closely packed. Because they remain close together, liquid volume changes very little with pressure or temperature. The particles in a liquid are not bound in fixed positions, however. They can "slide" past and around each other, which allows a liquid to take the shape of its container and flow when poured.

The particles in a gas under ordinary conditions of temperature and pressure are widely separated. Their positions have no order at all and, because they are constantly in motion, they quickly expand into any space available.

Properties and Changes of Matter

Most chemical investigations are related to the properties of matter. Before a material is used in an experiment or in the design of a product, data is gathered on the known properties of the material. The purpose of many experiments is to gain new data on properties. **Properties** *are characteristics that enable us to distinguish one kind of matter from another.*

Physical properties   Melting point and boiling point are examples of physical properties. A **physical property** *can be observed or measured without altering the identity of a material.* Many physical properties can be described both qualitatively and quantitatively. Qualitatively, for example, a box can be described as "large." With the aid of a ruler, the size can be given a quantitative description by specifying the height, width, and length of the box. Measuring the size of the box causes no change in its identity. It is the same box after it is measured.

Physical properties are either intensive or extensive. **Extensive physical properties** *depend on the amount of matter present and include mass, length, and volume.* **Intensive physical properties** *do not depend on the amount of matter present and include melting point, boiling point, density, ductility, malleability, color, crystalline shape, and refractive index.*

Any change in a property of matter that does not result in a change in identity is called a **physical change**. Cutting wire, grinding a solid, or allowing a gas to expand are physical changes. Observations would show that the substances changed in these ways retain their own characteristic properties, which is evidence that their identities are unchanged.

Changes of state *are changes between the gaseous, liquid, or solid states.* Changes of state are physical changes. Melting and boiling are changes of state, as are the reverse processes, freezing and liquefaction of a gas. These changes do not alter the identity of the substance.

Chemical properties   In contrast to a physical property, a **chemical property** *refers to the ability of a substance to undergo a change that alters it identity.* Chemical properties are observed when substances are converted to new substances with different characteristic properties. The tendency of wood

Chemistry Notebook

Interconversion of Matter and Energy

During the early 1900s, a revolution occurred in how scientists view matter and energy. Experiments on the speed of light that were then in progess provided results that could not be explained by the theories and laws of the time. Albert Einstein (1879–1955) is the best known of many scientists who rose to the challenge of finding new explanations for the observed behavior of matter and energy.

In 1905, Einstein suggested that matter and energy are related. His now-famous equation

$$E = mc^2$$

states this relationship. In the equation, E represents

energy, m represents the amount of matter (or mass), and c is the speed of light. Many experiments have established the validity of this equation.

The equation shows that matter can be converted to

energy and energy to matter. The conversion factor, c^2, has a very large value, 9×10^{16} meters² per second², which means that a very small amount of matter can be converted to a very large amount of energy. The world has seen evidence for this conversion in nuclear weapons and nuclear power plants as shown in the above photographs.

The evidence for the interconversion of matter and energy is taken into account in the *law of conservation of matter and energy,* which states: Matter and energy are interchangeable, and the total matter and energy in the universe is constant.

---

to burn in air is a chemical property. We find out that a material has this property by observing the material burning and being converted into different substances. When a match burns completely, the materials in the match no longer exist—the particles of which each was composed are transformed into particles of different substances. Rusting iron, leaves changing color in the fall, silver that is tarnishing, milk turning sour—all these commonly observed phenomena are evidence of chemical changes.

Any change in which one or more substances are converted into different substances with different characteristic properties is described as a **chemical change** *or a* **chemical reaction**. *The substance or substances that undergo a chemical reaction are the* reactants. *The new substance or substances produced by a chemical reaction are the* products. In the chemical reaction between sodium, an explosive metal, and chlorine, a poisonous gas, sodium chloride—known to everyone as "salt"—is formed. The vastly different characteristic properties of the reactants and product are apparent in Figure 1-11.

reactants → product

sodium + chlorine → sodium chloride (table salt)

Arrows like those above are used to indicate *chemical* or, sometimes, *physical* changes and are read as "yields" or "gives."

Take a wooden safety match and break it into two pieces. This is a physical change. Strike the match and allow the wood to burn. This is a chemical change.

Figure 1-11 Sodium and chlorine (left) can be combined chemically (middle) to form sodium chloride (right). Both sodium and chlorine are poisonous while sodium chloride is essential in your diet.

Figure 1-12 Sodium reacts in chlorine gas to form sodium chloride.

Figure 1-13 Bubbles of carbon dioxide form when vinegar is poured on sodium bicarbonate (baking soda).

A solution in which water is the solvent is referred to as an *aqueous solution.*

Indications of chemical reaction   To know for certain that a chemical reaction has taken place requires evidence that one or more substances have been changed in identity. Absolute proof of such a change can be provided only by chemical analysis of the products. Certain easily observed changes strongly suggest that a chemical reaction has occurred, however.

1. Evolution of heat and light. A change in matter that releases energy as both heat and light is good evidence that a chemical reaction has taken place. If you burn gas for cooking in your house, you see evidence of the reaction of natural gas with oxygen every day. Figure 1-12 shows the result of the reaction between the elements sodium and chlorine.

Heat evolution without accompanying light also indicates a possible chemical reaction. But heat alone is not as certain an indication of chemical change because some physical changes also release heat.

2. Production of a gas. The evolution of gas bubbles when two substances or solutions are mixed is a possible indication of a chemical reaction. When sodium bicarbonate (baking soda) is added to vinegar, for example, bubbles of carbon dioxide gas form immediately, as shown in Figure 1-13.

3. Formation of a precipitate. Many interesting and useful chemical reactions take place between substances that are dissolved in water. If a solid appears after two solutions have been mixed, a chemical reaction most likely has occured. *A solid that separates from a solution is known as a* **precipitate**. Figure 1-14 shows the result of mixing a solution containing sodium chloride (table salt) with a solution containing silver nitrate.

Energy and changes in matter   Chemical and physical changes are accompanied by energy changes. Energy may be either released or absorbed. The reaction of hydrogen with oxygen to form water produces a large amount of heat, which can be thought of as another reaction product:

$$\text{hydrogen} + \text{oxygen} \rightarrow \text{water} + \text{heat}$$

The change of state from liquid to solid water releases heat and can be represented as

$$\text{liquid water} \rightarrow \text{ice} + \text{heat}$$

A process that releases heat is **exothermic**. The reaction shown in Figure 1-15 is exothermic.

According to the law of conservation of energy, the energy released as heat in an exothermic process must come from somewhere. Prior to the process, the energy exists in the reactants. In the reaction of hydrogen with oxygen, chemical potential energy stored in the reactants was released as kinetic energy. In the freezing of water, the water particles give up some of their potential energy to the surroundings in the form of heat.

Figure 1-14 A precipitate of silver chloride forms when a solution of silver nitrate is poured into a solution of sodium chloride.

Figure 1-15 Sulfur (yellow) and zinc (grey) are mixed at left. When the mixture is heated, an exothermic reaction takes place which gives off heat and light (right).

One way to think of energy changes in chemical reactions is to picture a boulder at the top of a hill, as shown in Figure 1-16. The boulder has potential energy because of its position; this is converted to kinetic energy as the boulder rolls downhill. The reactants in an exothermic chemical reaction also "fall down a potential energy hill," converting potential energy into an equal amount of energy released as heat.

Figure 1-16 As a boulder rolls down a hill (left), potential energy is released as shown in the diagram (middle). When hydrogen and oxygen combine to form water, potential energy is released as shown in the diagram (right).

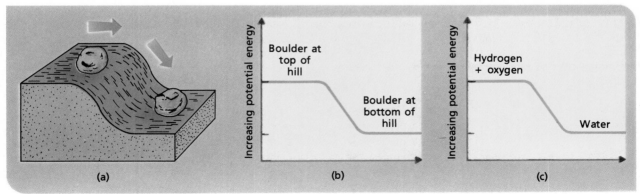

Desktop Investigation

Physical and Chemical Changes

Question

How can you distinguish a physical change from a chemical change?

Materials

matches
candle
chalk (calcium carbonate)
vinegar (dilute acetic acid)
soluble starch
water
tincture of iodine
Alka-Seltzer® tablet
phenolphthalein solution
milk of magnesia
    (suspension of magnesium
    hydroxide)
dilute ammonia solution
iron(III) chloride solution
four 250-mL beakers
mortar and pestle
two test tubes

Procedure

Record all of your results in a data table. **CAUTION:** Be careful when lighting the match; be sure to dispose of it properly. Romove all combustible materials from the work area. Remember to wear safety glasses and an apron.
1. Light a match.
2. Use the lighted match to

ignite a candle sitting in a 250-mL beaker.

3. Powder a 3-cm piece of chalk with a mortar and pestle.

4. React the powdered chalk with 10 mL of vinegar in a second beaker.

5. Add 0.5 g of soluble starch to a third beaker containing 100 mL of water.

6. Add two drops of tincture of iodine to the starch solution.

7. Drop one Alka-Seltzer® tablet into a fourth beaker containing 200 mL of water.

8. Add three drops of phenolphthalein to a test tube containing 5 mL of milk of magnesia.

9. Add 5 mL of ammonia solution to a second test tube containing 5 mL of iron(III) chloride solution.

Discussion

1. List four different types of evidence you observed in this lab that suggest a chemical change occurred.

2. List three types of physical changes you observed in this lab.

Many chemical reactions do not take place unless they are able to absorb heat from their surroundings. For example, limestone is composed primarily of a chemical know as calcium carbonate. It is obvious that limestone remains unchanged for centuries when it is left undisturbed. Heating limestone causes the chemical reaction in which limestone is converted to lime (calcium oxide) plus carbon dioxide. This chemical reaction is used in industry for the production of *lime*, a chemical that is used in many processes such as water and waste treatment and paper production.

$$\text{limestone (calcium carbonate)} + \text{heat} \rightarrow$$
$$\text{lime (calcium oxide)} + \text{carbon dioxide}$$

Heat is also required to melt ice.

$$\text{ice} + \text{heat} \rightarrow \text{liquid water}$$

A process that absorbs heat is **endothermic**. An endothermic reaction is equivalent to moving a boulder from the bottom to the top of a hill. It only happens with the input of energy.

Section Review

1. What is inertia?
2. Distinguish between *(a)* weight and mass *(b)* kinetic and potential energy *(c)* extensive and intensive physical properties of matter *(d)* physical and chemical properties of matter *(e)* physical and chemical changes *(f)* exothermic and endothermic processes.

1.3 Classification of Matter

Matter exists in countless forms, however each sample of matter can be classified as one of two types: a pure substance or a mixture. Pure substances are either elements or compounds, both of which have uniform composition and properties throughout a given sample and from one sample to another. Some mixtures also have compositions and properties that are uniform throughout a given sample, but the composition and properties of a mixture can vary from one sample to another. The composition and properties of many mixtures are not uniform, that is, they vary from one part of the mixture to another. All matter can thus be classified according to the uniformity of composition and properties of a given sample.

Mixtures

Most samples of matter are mixtures. The air is a mixture of oxygen, nitrogen, carbon dioxide, other gases, and water. Brass is a solid mixture of copper and zinc. Gasoline is a mixture that may contain only a few or as many as 20 different substances. Grapefruit is a mixture of about 2000 different substances. Some other familiar examples of mixtures are granite, vegetable soup, milk, soft drinks, and sugar syrup.

A **mixture** *is a combination of two or more kinds of matter each of which retains its own composition and properties.* The different kinds of matter in a mixture can be separated by physical means. The properties of a mixture are a combination of the properties of the different kinds of matter it contains. Unlike a pure substance, the composition of a mixture must be specified because it can contain verying amounts of different kinds of matter.

Mixtures may be either heterogeneous or homogeneous. *In a* **heterogeneous mixture**, *the composition and properties are not uniform—they differ*

Figure 1-17 Granite is a mixture of different minerals such as quartz, mica, and feldspar.

Figure 1-18 The components of a mixture can be separated by using differences in their physical properties. Here a mixture is separated by filtration after mixing it in a liquid in which one component (but not the other) is soluble.

from point to point in the mixture. Vegetable soup and granite are heterogeneous mixtures. Granite, shown in Figure 1-17, is a mixture of different minerals. Physical examination reveals that granite consists of nonuniformly distributed matter identified as the minerals quartz, mica, and feldspar. Crystals of each mineral have their own appearance, composition, and properties. Some materials must be examined quite closely to reveal that they are not uniform. Milk, for example, appears to be the same throughout, but under a microscope, suspended globules of fat can be seen.

Portions of matter that have both the same chemical properties and the same physical properties are called a **phase**. Phases might be different states of matter, such as ice and liquid water. Phases also might have the same state, such as the different crystalline phases in granite.

Other mixtures, such as the air we breathe or the beverages we drink, do not have distinct regions. These are homogeneous mixtures. *In a* **homogeneous mixture**, *the composition and properties are uniform throughout the mixture.* Such mixtures are the same from point to point. Consider the result of stirring some sugar into a glass of water. As the sugar dissolves, it spreads throughout the water and seems to disappear. Examination under a microscope would fail to reveal phases of different substances. Each teaspoonful of the mixture contains the same amount of dissolved sugar and will taste as sweet as the next teaspoonful.

Homogeneous mixtures are also called **solutions**. Solutions are found in the gaseous and solid states as well as in the liquid state. A sample of air is a gaseous solution; a sample of sea water is a liquid solution; a sample of brass is a solid solution.

The components of any mixture can be separated from each other by physical methods. That is, no chemical changes take place when mixtures are separated. The separated substances have the same characteristic properties and identities as before they were mixed together. Separating the components of a mixture may not always be as easy as it is for the mixture shown in Figure 1-18, but it is always possible. A sugar and water solution can be separated by carefully evaporating the water and collecting it. The crystalline sugar would be left behind, as shown in Figure 1-19. How many ways can you think of to separate a mixture of salt and pepper?

Pure Substances

Some kinds of matter display the same properties throughout. A teaspoonful of sugar taken from any spot in a large sugar bowl has the same properties, such as melting point and sweet taste, as a sample taken from any other spot in the bowl. The same properties would be found in any teaspoonful of pure sugar, no matter what its source. Chemical analysis shows that these kinds of matter have the same composition throughout a sample. Such kinds of matter are known as pure substances. A **pure substance** is a *homogeneous sample of matter that has the same composition and properties, whatever its source.* Examples include water, sugar, oxygen, silver, the graphite in a pencil, and the mercury in a thermometer. Each of these substances differs from one another in composition and can be identified by their physical and chemical properties. Pure substances are similar to homogeneous mixtures in that they both have uniform compositions throughout. But each pure substance always has the same composition in every sample, while the composition of a homogeneous mixture can vary from sample to sample. Pure substances differ from mixtures in three major ways.

Figure 1-19 Sugar is added to water (left) to form a homogenous mixture or solution (middle). When the water is evaporated the sugar is left behind (right).

1. Every sample of a given pure substance has exactly the same characteristic properties. Pure substances can be identified by determining their characteristic physical and chemical properties. The properties of a mixture of two substances vary with the amounts of each that are present in the mixture.

2. Every sample of a given pure substance has exactly the same composition. Unlike mixtures, which have variable compositions, pure substances have fixed compositions. Pure water is always composed of 11.2% hydrogen combined with 88.8% oxygen by mass.

3. A pure substance cannot be separated into other substances without changing its identity and properties. A pure substance cannot be broken down by any process that is based on a physical property. For example, when water is evaporated, filtered, melted, or frozen, it remains water. To separate water into the elements hydrogen and oxygen that compose it, a chemical change is necessary. As shown in Figure 1-20, hydrogen and oxygen— substances totally unlike water in their properties—can be produced when electrical current is passed through a sample of water.

Elements and chemical compounds Pure substances are either elements or compounds. Among the first questions a chemist asks in identifying an unknown material are, "Is this a mixture or a pure substance?" and "What is the composition of this material?" Experiments must be performed to determine whether the substance can be divided into other substances by either physical or chemical means.

Early in the history of chemistry, it was discovered that certain substances could not be separated into other substances by *any* means. These substances became known as "elements." A few elements, including gold, silver, copper, and sulfur have been known since ancient times because they are found as pure substances in nature. Eventually, about ninety elements were found to exist naturally on our planet. In recent times, other elements have been made artificially.

Figure 1-20 Water can be decomposed into hydrogen and oxygen by electrolysis. In the apparatus shown above hydrogen forms at the negative electrode (left) and oxygen forms at the positive electrode (right).

Figure 1-21 Sugar decomposes into carbon and water when heated. A small quantity of sugar heated in a flask melts and turns brown. Eventually only a black carbon residue and water droplets remain.

Figure 1-22 The classification of matter.

At the time this book is being written, a total of 109 elements are known. *An* **element** *is a substance that cannot be separated into other substances by any ordinary chemical change.* An element is composed only of one type of matter. Pure gold, for example, is 100% gold. All matter found on earth is composed of the elements.

Most pure substances are not elements, but chemical compounds. Water is a chemical compound. In Figure 1-20 the process by which water can be broken down by a chemical reaction to the elements hydrogen and oxygen is shown. We say that, in water, hydrogen and oxygen are "chemically combined," indicating that they can be separated only by a chemical reaction. *A* **chemical compound** *is a pure substance that can be decomposed into two or more simpler substances by an ordinary chemical change.*

Sugar is also a chemical compound. If a small quantity of sugar is heated in a flask as shown is Figure 1-21, the sugar first melts and changes color. Eventually, only a black residue remains and a few drops of a clear, colorless liquid can be seen near the opening of the flask. The black substance is carbon, an element, and the colorless liquid is water. The sugar has been broken down chemically into two simpler substances, an element and a compound.

A landmark in the history of chemistry was the formulation of the **law of definite composition**: *A chemical compound contains the same elements in exactly the same proportions by mass regardless of the size of the sample or source of the compound.* Experiments show that pure sugar is composed of 42.1% carbon, 51.4% oxygen, and 6.5% hydrogen by mass. *Every* pure chemical compound consists of two or more elements combined chemically in definite percentages. The law of definite composition was formulated in

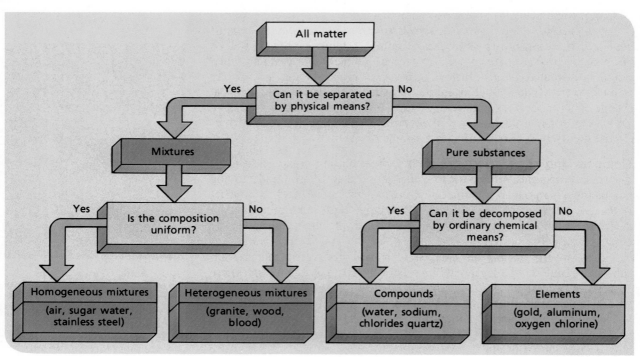

1799 by Joseph Louis Proust (1754–1856), who showed that copper carbonate ($CuCO_3$), whether prepared in the laboratory or obtained from natural sources, always contains the same elements—copper, carbon, and oxygen—in the same proportion by mass. Analytical chemists are able to identify the types and quantities of elements combined in any chemical compound. Figure 1-22 summarizes the classification of matter, and Table 1-1 summarizes the differences between mixtures and chemical compounds.

Section Review

1. Distinguish between: *(a)* heterogeneous and homogeneous matter *(b)* a pure substance and a mixture.
2. Identify two types of homogeneous matter.
3. State the law of definite composition.

1.4 The Chemical Elements

The chemical elements are the building blocks of matter. Anyone who studies or uses chemistry, therefore, must be familiar with their properties. Each element has its own characteristic properties. Early chemists made many attempts to organize the elements into groups with similar properties. Eventually, a table was devised in which all elements having similar properties could be grouped together in the same column.

Introduction to the Periodic Table

Figure 1-23 is a modern version of this table, known as the Periodic Table. Each block contains the name of an element and the symbol for its name. *The vertical columns of elements in the Periodic Table are referred to as* **groups** *or* **families** *and are numbered consecutively from 1 to 18.* The elements of Group 1, for example, are lithium (Li), sodium (Na), potassium (K), rubidium (Rb), cesium (Cs), and francium (Fr). Elements within a group have many similar properties.

The horizontal rows of elements in the Periodic Table are referred to as **periods** *and are numbered 1 to 7 from the top down.* elements close to each other in the same period are more similar than those further apart. For instance, the properties of the elements of Group 15 vary significantly from those of Group 1 but are somewhat similar to the properties of Group 16 elements. The two rows of elements placed apart from the others that appear below the table belong to periods 6 and 7. They are placed at the bottom simply to keep the table conveniently narrow.

Symbols for the Elements

The majority of elements have one- or two-letter symbols derived from their names, such as H for hydrogen, O for oxygen, or Si for silicon. Table 1-2 is a list of the names and symbols of some common elements.

In period 7 of the table there are some three-letter symbols. The elements represented by these symbols are all synthetic and have been created for the first time in recent years. These symbols are temporary until the International Union of Pure and Applied Chemistry (IUPAC) gives them official names.

The Swedish chemist Jöns Jakob Berzelius (1779–1848) introduced a system in which elements were symbolized by letters. In earlier times, pictorial symbols were used for elements, or what were thought to be elements.

In 1799, the French chemist Joseph Louis Proust (1754–1826) first stated the law of constant proportions, or the law of definite composition.

1. *(a)* Homogeneous matter is uniform throughout; heterogeneous is not. *(b)* A mixture can be subdivided into two or more components by physical means; a pure substance cannot.
2. Solutions and pure substances.
3. A chemical compound contains the same elements in exactly the same proportions by mass regardless of the size of the sample or the source of the compound.

SECTION OBJECTIVES

- Name the common elements, given their symbols.
- Write the symbols of the common elements, given their names.
- Describe the arrangement of the periodic table.
- Discuss the differences between metals, nonmetals, and metalloids.

TABLE 1-2 ELEMENTS WITH SYMBOLS BASED ON OLDER NAMES

Modern Name	Symbol	Older Name
antimony	Sb	stibium
copper	Cu	cuprum
gold	Au	aurum
iron	Fe	ferrum
lead	Pb	plumbum
mercury	Hg	hydrargyrum
potassium	K	kalium
silver	Ag	argentum
sodium	Na	natrium
tin	Sn	stannum
tungsten	W	wolfram

Periodic Table of the Elements

	Group 1																		Group 18

Metals ☐ **Metalloids** ☐ **Nonmetals** ☐

*Name not official

Lanthanide Series

58 Ce Cerium	59 Pr Praseo-dymium	60 Nd Neo-dymium	61 Pm Prometh-ium	62 Sm Samarium	63 Eu Europium	64 Gd Gadolinium	65 Tb Terbium	66 Dy Dyspro-sium	67 Ho Holmium	68 Er Erbium	69 Tm Thulium	70 Yb Ytterbium	71 Lu Lutetium
90 Th Thorium	91 Pa Protactin-ium	92 U Uranium	93 Np Neptunium	94 Pu Plutonium	95 Am Americium	96 Cm Curium	97 Bk Berkelium	98 Cf Californium	99 Es Einsteinium	100 Fm Fermium	101 Md Mendelev-ium	102 No Nobelium	103 Lr Lawrenc-ium

Actinide Series

Figure 1-23
The Periodic Table.

Types of Elements

General similarities among the properties of different elements provide one way of classifying them. Three general classes of elements are the metals, the nonmetals, and the metalloids. See Figure 1-23.

Metals

You would be able to recognize many metals by characteristic properties with which you are already familiar. Most metals have a characteristic *luster*—that is, they shine like silver. Copper and gold have a metallic luster and they also have distinctive colors. Metals are good reflectors of heat and light. Most metals are *ductile*; they can be drawn out into fine wire. As illustrated in Figure 1-24, most are also *malleable*; they can be hammered or rolled into thin sheets. Many have a high *tensile strength*; they resist breaking when pulled. Different metals have the metallic properties of ductility, malleability, and high tensile strength to varying degrees.

The property that most clearly distinguishes metals from other types of elements is the ease with which they conduct heat and electricity. *A* **metal** *is an element that is a good conductor of heat and electricity.* All metals except for mercury are solids under ordinary conditions. Mercury is a liquid at room temperature and pressure.

Figure 1-24 Because aluminum foil is ductile, it resists breaking when reshaped and can be used to wrap sandwiches.

Nonmetals In contrast to metals, nonmetals are very poor conductors of heat and electricity. Under ordinary conditions, some nonmetals—such as hydrogen, oxygen, chlorine, and nitrogen—are gases. Bromine, however, is a dark red liquid. Those nonmetals that are solids, carbon and sulfur, for example, are neither malleable nor ductile but instead are brittle.

A **nonmetal** *is an element that is a poor conductor of heat and electricity.* As shown in Figure 1-23, there are far fewer nonmetals that metals.

Metalloids In the Periodic Table, the metals appear at the left and the nonmetals are on the right. Between them lies a small class of elements known as the "metalloids," which are somewhat like metals, but not entirely like them. In appearance, metalloids such as arsenic and antimony, shown in Figure 1-25, are lustrous metals. But in many properties, they are more like nonmetals. Under ordinary conditions, all are solids.

The ordinary conditions referred to are room temperature (25°C) and the pressure exerted by the atmosphere at sea level (1 atm). You will learn more about pressure and "standard" conditions in Chapter 11.

Gallium and cesium are two low melting metals. Under normal lab conditions on a warm day, both would be liquids. Cesium melts at 28.5°C and gallium at 29.78°C. Since body temperature is 37°C, samples of the two metals (sealed in glass vials) would melt if held in one's hand.

Figure 1-25 Elemental antimony (left) and arsenopyrite (right) a mineral which contains the element arsenic

Metalloids conduct electricity, but do so in a significantly different manner than metals. For example, silicon and germanium are much better conductors of electricity than the nonmetal sulfur, but much poorer conductors than the metals silver and copper. Specifically, silicon, germanium, and the other metalloids are, in varying degrees, *semiconductors.*

A **metalloid** *is an element that has some properties characteristic of metals and others characteristic of nonmetals, and is a semiconductor.* The microelectronics revolution, which has resulted in desktop computers, hand-held calculators, digital watches, and hundreds of other items that are changing our lifestyles, is based upon devices made from semiconducting elements.

The metalloids are also referred to as *semimetals* or *semiconducting elements.* The property of semiconductivity is discussed in Section 26.3.

Two Typical Elements

The properties of the elements and their compounds are the basis for what is sometimes called "descriptive chemistry." Descriptive chemistry is involved more with observable facts than explanations. What does an element look like? How does it behave when combined with other substances? What are its physical properties? What are some of the ways in which its properties can be put to use? Answers to questions such as these are at the heart of the descriptive chemistry of the elements.

As an introduction to descriptive chemistry, we will look at two elements in the following paragraphs. They are chosen from two of the three general classes of elements shown in the Periodic Table in Figure 1-23. Copper is a well-known metal. Phosphorus is a nonmetal that has widespread applications.

Copper—a metal   Copper is found naturally noth as a free element and in minerals such as those shown in Figure 1-26, in which it is combined with sulfur (chalcocite and chalcopyrite) or with oxygen or a compound of oxygen and carbon (cuprite, malachite). Although it is needed in extremely small amounts, copper is an essential element in the human diet. In elemental form, it is not known to be toxic.

Figure 1-26   Copper is found naturally in a number of different forms. Two forms, free element (left) and combined with oxygen as *malachite* (right), are shown here.

Copper has a characteristic reddish color, a metallic luster, high boiling and melting points, and other properties considered typical for metals. It is readily formed into wire, sheets, and tubing. Copper allows electricity to pass through it with little loss of energy. This makes copper the metal of choice for electrical wiring, its principle commercial application. The resistance of copper to weathering and corrosion makes it desirable for water pipes, roofing materials, and other applications in plumbing and construction.

Copper remains unchanged in pure, dry air at room temperature. When heated, it reacts with the oxygen in the air. Copper also reacts easily with sulfur and the elements of Group 17 in the Periodic Table, which includes chlorine and bromine. Copper that is exposed to the weather for a long time reacts with oxygen, carbon dioxide, and sulfur compounds to produce a green coating on the copper surface. This coating is responsible for the green color of the Statue of Liberty and other copper monuments. Copper is mixed with other metals to form many useful alloys, including both brass and bronze.

Phosphorus—a nonmetal   Phosphorus is one of the five nonmetals that are solid rather than gaseous at room temperature. Pure phosphorus is known in two common forms that are quite different from each other in their properties. Red phosphorus is a dark red powder with a melting point (mp) of 597°C, remains relatively unchanged in air, is not toxic, and can be handled with reasonable safety. White phosphorus is a waxy, low-melting solid with an mp of 44°C, ignites spontaneously in air, causes serious burns when it comes into contact with the skin, and is extremely toxic. White phosphorus is usually stored under water, as shown in Figure 1-27.

Phosphorus is much too reactive to be found free in nature. It is present in huge quantities in deposits known as phosphate rock, that contain chemical compounds composed of phosphorus, oxygen, and calcium. All living things contain phosphorus, and require it as a nutrient. One of the major uses of phosphorus compounds is in agricultural fertilizers. For this purpose, phosphates can be manufactured directly from phosphate rock. In living things, biochemical compounds containing phosphorus are essential in the storage and production of energy.

Phosphates, phosphoric acid, and other chemical compounds containing phosphorus have many applications. Many phosphorus compounds are not toxic and are added to food products, including soft drinks and baking powders. Cleaning agents such as laundry detergents, dishwasher detergents, and soaps contain phosphorus compounds. Phosphorus compounds also play important roles in water softening and water purification. The flammability of phosphorus and certain phosphorus compounds allows them to be used in matches and fireworks. Interestingly, some phosphorus compounds are also excellent flame retardants and are used in fabric for clothing.

Section Review

1. In reference to the Periodic Table, what are *(a)* groups? *(b)* periods?
2. Write the chemical symbols for: *(a)* nitrogen *(b)* calcium *(c)* zinc *(d)* oxygen *(e)* sulfur *(f)* radon.
3. Name the element represented by the symbol: *(a)* He *(b)* Al *(c)* Br *(d)* Ni *(e)* Ag *(f)* Sn.
4. What are the three general classes of elements?

Chapter Summary

- Chemistry is a physical science with five major branches: organic chemistry, inorganic chemistry, physical chemistry, biochemistry, and analytical chemistry.
- The scientific method involves observing, generalizing, theorizing, and testing.
- Mass and weight measure different quantities. Weight is a measure of the earth's gravitational attraction for matter. Mass is a measure of the quantity of matter.
- The law of conservation of matter and energy says that the total matter and energy in the universe is constant.
- A physical property can be observed or measured without altering the identity of a material.
- A chemical property refers to the ability of a substance to undergo a change that alters the identity of the substance.
- A chemical reaction is likely to have taken place if one or more of the following changes occur: evolution of heat and light, production of a gas, formation of a precipitate.
- A quantity of matter may be classified as either a pure substance or a mixture.
- Mixtures are either homogenous or heterogenous.
- Pure substances are either compounds or elements.
- The Periodic Table arranges the elements according to their properties. Elements with similar properties are found in the same column.

Figure 1-27 Red and white phosphorous actually has a yellowish color and is stored under water since it ignites spontaneously in air.

1. *(a)* Vertical columns
   (b) Horizontal rows
2. *(a)* N   *(b)* Ca   *(c)* Zn
   (d) O   *(e)* S   *(f)* Rn
3. *(a)* Helium   *(b)* Aluminum
   (c) Bromine   *(d)* Nickel
   (e) Silver   *(f)* Tin
4. Metals, nonmetals, and metalloids

In some areas, the use of phosphates in household detergents has been limited to protect natural bodies of water. Waste phosphates that find their way into streams and lakes can cause excessive growth of algae and aquatic plants. This can harm other living things in these bodies of water.

Chapter 1 *Review*

Introduction to Chemistry and Matter

Vocabulary

changes of state	law
chemical change or chemical reaction	liquid state
	mass
chemical compound	matter
chemical property	metal
chemistry	metalloid
element	mixture
endothermic	nonmetal
energy	period
exothermic	phases
extensive physical property	physical change
	physical property
gaseous state	precipitate
groups or families	potential energy
heterogeneous mixture	products
	properties
homogeneous mixture	pure substance
hypothesis	qualitative information
inertia	
intensive physical property	quantitative information
	reactants
kinetic energy	scientific method
law of conservation of energy	solid state
	solution
law of conservation of matter	theory
	weight
law of definite composition	

Review Questions

1. *(a)* What is science? *(b)* How does science affect the way we live?
2. *(a)* What is chemistry? *(b)* Give three examples of the many applications of chemistry in everyday life.
3. Distinguish between the dictionary definition of the word *chemical* and the definition implied in such advertising slogans as, "All natural—no chemicals added."
4. What is the scienfific method and its major components?

5. *(a)* What is the meaning of the term *law* as used in science? *(b)* How are laws in science generally expressed?
6. Distinguish between quantitative and qualitative information.
7. Define: *(a)* measuring *(b)* experimenting *(c)* hypothesis *(d)* theory.
8. *(a)* What is inertia? *(b)* Explain how inertia is demonstrated in both moving objects and objects at rest. *(c)* What is the relationship between inertia and the quantity of matter in an object?
9. Distinguish between weight and mass.
10. What is the relationship between the location of an object and its *(a)* weight? *(b)* mass?
11. *(a)* What is matter? *(b)* Among the following, which are examples of matter: sand, paper, heat, air, sound, light, and chalk?
12. State the law of conservation of matter.
13. *(a)* What is energy? *(b)* Describe chemistry in terms of energy.
14. List and define the two major types of energy.
15. Identify the following as examples of either kinetic or potential energy: *(a)* a moving car *(b)* a coiled spring *(c)* food to be eaten *(d)* water flowing downstream.
16. State the law of conservation of energy.
17. List and define the three states of matter.
18. Describe the model that explains the differences between solids, liquids, and gases.
19. List three examples of intensive properties of matter that are qualitative in nature, and three that are quantitative.
20. *(a)* What is a physical property? *(b)* List three examples of physical properties.
21. *(a)* Distinguish between intensive and extensive physical properties. *(b)* Give three examples of each.
22. What is a chemical property?
23. *(a)* Distinguish between physical and chemical changes. *(b)* List three examples of each.
24. Distinguish between the reactant and products in a chemical reaction.

25. Classify each of the following as a physical or chemical change: *(a)* melting ice *(b)* chopping wood *(c)* burning paper *(d)* dissolving salt in water *(e)* boiling water.

26. List three observable changes that would generally indicate that a chemical reaction has occurred.

27. *(a)* Distinguish between exothermic and endothermic processes. *(b)* Give an example to illustrate each.

28. List and define the two basic categories into which all matter can be classified.

29. *(a)* Distinguish between heterogeneous and homogeneous matter. *(b)* Give three examples of each.

30. List and define the two categories of homogeneous matter.

31. List three ways in which pure substances differ from mixtures.

32. *(a)* What is the basic difference between an element and a compound? *(b)* Identify each of the following as elements or compounds: aluminum, sugar, oxygen, copper, and water.

33. State the law of definite composition.

34. Distinguish between groups and periods in the Periodic Table.

35. *(a)* List five properties of copper that indicate its metallic character. *(b)* Identify three commercial applications of copper.

36. *(a)* Identify and distinguish between the two common forms of phosphorus. *(b)* List three general uses of phosphorus compounds.

37. *(a)* How many elements are presently known? *(b)* Of that number, how many exist naturally on earth?

Application Questions

1. How would the mass and weight of an object at sea level compare with its mass and weight *(a)* at the top of Pike's Peak? *(b)* in an airplane 35,000 feet above the earth?

2. Is it possible for an object to be *(a)* massless? *(b)* weightless? Explain.

3. Identify each of the following as exothermic or endothermic processes: *(a)* the melting of ice *(b)* the burning of charcoal *(c)* photosynthesis *(d)* the vaporization of water *(e)* the explosion of a firecracker *(f)* the decomposition of water by electricity *(g)* the freezing of water.

4. Use the boulder-and-hill example discussed in the text to explain the basic difference between exothermic and endothermic reactions.

5. The formula for determining the kinetic energy of a moving object is: $K \cdot E \cdot = \frac{1}{2} mv^2$, where m = the mass of the object and v = its velocity. Based on this equation, compare the relative kinetic energies of objects A and B given the following information:
   (a) Object A has a mass twice that of B, but both are moving with the same velocity.
   (b) Object A has a velocity twice that of B, but both have the same mass.
   (c) Both the mass and velocity of object A are twice that of object B.

6. On the basis of general appearance, identify each of the following examples of matter as heterogeneous or homogeneous: *(a)* notebook paper *(b)* pencil lead *(c)* fresh-squeezed lemonade *(d)* silver *(e)* chocolate-chip cookies *(f)* oil-and-vinegar salad dressing.

7. Given samples of the following pairs of substances, name two physical properties that could be used to distinguish between each of the paired materials: *(a)* sand and salt *(b)* copper and silver *(c)* cork and lead *(d)* water and motor oil.

8. Despite the fact that 88 elements occur naturally, only 12 were known in ancient times. *(a)* How can this be explained? *(b)* Why do you think gold, silver, and copper were among those initial 12 elements?

9. Explain why the law of definite composition is important to analytical chemists.

10. List and define four intensive physical properties of metals.

11. List three intensive physical properties of nonmetals.

Enrichment

1. Design an experiment to show that the law of conservation of matter is not violated even though the ashes that remain after a given quantity of wood is burned have a mass considerably less than that of the wood.

2. Design and conduct an experiment to separate a mixture of sugar, sawdust, iron filings, rubbing alcohol, and water.

3. Prepare a report concerning the role of semi-conducting elements in the microelectronics revolution.

Chapter 2　Measurements and Solving Problems

Chapter Planner

2.1 Units of Measurement
The SI Measurement System
- ▫ Questions: 1–3
- ▪ Question: 4
- Application Question: 1

Fundamental SI Units
- ▫ Questions: 5, 6b, 7, 8b, 9, 10b
- ▪ Problem: 2
- ▫ Questions: 6a, 8a, 10a

Units of Measurement in Calculations
- ▫ Questions: 11, 12
- Problem: 3
- ▪ Problem: 1

Derived SI Units
- ▫ Desktop Investigation
- Questions: 13–16, 19, 20
- Problems: 4–9
- ▪ Questions: 17, 18
- Application Question: 2
- Application Problems: 1–5, 7
- ▫ Application Problem: 6

2.2 Heat and Temperature
Definitions of Heat and Temperature
- ▫ Demonstration: 1
- Questions: 21, 22
- Problem: 11
- ▪ Questions: 23, 24
- Application Question: 3
- ▫ Application Questions: 4, 5
- Application Problem: 8

Units of Heat
- ▫ Experiment: 5
- Question: 25
- Problems: 12, 13

Heat Capacity and Specific Heat
- ▫ Experiment: 6
- Questions: 26, 27
- Problems: 14, 15
- Application Question: 6
- Application Problems: 10, 11

2.3 Using Scientific Measurements
Accuracy and Precision
- ▫ Experiment: 3
- Questions: 28–30
- Problems: 16, 17

Significant Figures
- ▫ Questions: 31–37
- Problems: 16, 17
- Application Questions: 7

Scientific Notation
- ▫ Question: 38
- Problems: 18, 19, 21
- ▪ Problem: 20

2.4 Solving Quantitative Problems
Four Steps for Solving Quantitative Problems
- ▫ Question: 39
- Problem: 22
- ▪ Application Problem: 13
- ▫ Application Problems: 12, 14

Proportional Relationships
- ▫ Questions: 40, 41a
- Problem: 23
- ▪ Questions: 41b-c
- ▫ Application Problem: 15

Teaching Strategies

Introduction

Chapter 2 presents the mathematical foundations for the remainder of the course. SI units are introduced. These will be similar to, but not identical with, the metric system students have encountered in other science courses. The factor-label method of problem solving is introduced in the context of converting SI units. This method represents an algorithm that students must learn if they hope to be good problem solvers. Heat and temperature and attendant specific heat and heat-capacity problems are introduced. Significant figures, rounding off, scientific notation and the mathematical operations involving significant figures and scientific notation are discussed. The chapter concludes with an introduction to the four steps for solving quantitative problems.

2.1　Units of Measurement

Bring the metric system home to the students by displaying a variety of household products. Most household products now carry both English and metric equivalents. Have the students examine their pantries and medicine cabinets at home and bring in other examples.

Stress from the beginning that chemists do not work with pure numbers. To report 125.3 is meaningless unless one adds a unit of measurement, 125.3 grams, for example. Any number measured or derived in the chemistry course will consist of a number *and* a unit of measurement. Students must get used to always writing the units.

If time permits you may wish to give a historical background to the metric system and to the International System of Units. In any case, list and use each of the three fundamental SI units discussed. Have the students relate the size of each unit to something with which they are familiar. For example, a 2 meter-tall basketball player would probably play center on the school's basketball team.

Introduce the factor-label method by having students convert between units they are accustomed to using, e.g. yard to inches. Most can tell you that there are 120 in. in 10 ft., but they don't know how to set it up in the factor-label form. Do it for them. Move on to a conversion from days to seconds. Show that if the problem is set up correctly, the unit found in a numerator is cancelled by the unit in the following denominator. The unit remaining is the unit requested by the problem. For example, to convert 1 day to seconds, set up the following:

$$1 \text{ day} \times \frac{24 \text{ hours}}{1 \text{ day}} \times \frac{60 \text{ minutes}}{1 \text{ hour}} \times \frac{60 \text{ s}}{1 \text{ minute}} = 86\ 400 \text{ s}$$

Not only are you demonstrating the factor-label method, you are also introducing the concept of exact conversion factors and building a case for the four-step quantitative problem-solving method that will be introduced in Section 2.4. Explain also the international system used for writing numbers. Commas are no longer used to separate groups of three digits. A space is used instead. Many students will resist having to write the units. They consider it a waste of time when problems are relatively simple. Require them to write a unit after every number.

To illustrate the importance of units, write the following on the chalkboard: 36 + 16 = 1. When students protest say, "Oh, I left out the units, but that doesn't matter, does it?" Then add the units: 36 weeks + 16 weeks = 1 year.

Continue practicing conversions as you describe the derived SI units. Explain to the students that the text is using a set of modified SI units. Some of the SI units are too large to be of practical use in chemistry. For example, few chemists ever encounter volumes measured in cubic meters. The following calculation may bring this home. For water:

$$1 \text{ m}^3 = 1\ 000\ 000 \text{ cm}^3 = 1\ 000\ 000 \text{ g} = 1\ 000 \text{ kg}$$

$$1000 \text{ kg} = 1 \text{ metric ton} = 2200 \text{ lbs.}$$

Have the students set up the proper conversion factors. Most students will need help in converting cubic meters to cubic centimeters. Explain that everything in the conversion factor must be cubed, both the number and the unit.

$$1 \text{ m}^3 = 100 \text{ cm}^3$$
$$(1 \text{ m})$$

$$= 100 \text{ cm} \times 100 \text{ cm} \times 100 \text{ cm} = 1\ 000\ 000 \text{ cm}^3$$
$$(1 \text{ m}) \qquad (1 \text{ m}) \qquad (1 \text{ m})$$

The same problem arises when discussing density. The SI unit of kg/m³ is not commonly used. We shall use g/cm³ for solids and liquids and g/L for gases.

Desktop Investigation: Density of Pennies

Proceed with this as a way of bringing home the concept of density. Pennies minted prior to 1982 were made of an alloy containing 95% copper and 5% zinc. Penny hoarding as the price of copper rose in the 1970's forced Congress to adopt a change in which pennies would be made of a barrel composed of an alloy of 99.2% zinc and 0.8% copper onto which was electroplated a minimum thickness of 0.000 2 inches of copper. The overall composition of these pennies is 97.6% zinc and 2.4% copper. The pre-1982 pennies have a mass of 3.1 grams while the post-1982 pennies only a mass of 2.5 grams. When doing the experiment, stay away from 1982 pennies. These vary in composition.

For students who have difficulty with density problems, introduce the following mnemonic aid.

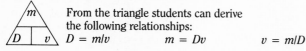 From the triangle students can derive the following relationships:

$$D = m/v \qquad m = Dv \qquad v = m/D$$

To work a density problem, two of the three variables must be known. The third, the unknown, can then be calculated.

A density column of ordinary substances, such as corn syrup, glycerin, water, salad oil, alcohol, and solids such as rubber erasers, plastic, cork, etc. would be good to use. Another simple, yet effective density "interest center" is a cylinder (1000 mL) 3/4 filled with a saturated NaCl solution. Float an egg in the salt water, and then carefully fill with tap water. The egg will appear to be suspended and will remain in position for weeks.

2.2 Heat and Temperature

To help students distinguish the difference between heat and temperature, perform the following demonstration as an introduction to the topic.

Demonstration 1: Heat vs Temperature

PURPOSE: To help students distinguish between heat and temperature

MATERIALS: Four 1–L beakers half filled with water at approximately the following temperatures: beaker 1 at 0 °C (ice water), beakers 2 and 3 at 20 °C (approx. room temperature), and beaker 4 at 40 °C; four Celsius thermometers, paper towels, two graduated cylinders, two 100–g blocks of ice.

PROCEDURE: Invite two or three volunteers to come to the front of the room. Do not tell them the temperature of the water in each beaker. Have each student place one hand in beaker 1 and their other hand in beaker 4. Next have them remove their hands from beakers 1 and 4 and place the hand that was in beaker 1 into beaker 2 and the hand that was in beaker 4 into beaker 3. On the basis of this data, ask them how they think the temperatures of beakers 2 and 3 compare with one another. Most will indicate that beaker 2 is warmer when in fact they are the same. Use the thermometers to show that the temperatures of beakers 2 and 3 were indeed the same.

Pour 10 mL of boiling water (100°C) onto one of the blocks of ice, pour 100 mL of boiling water on the other block. Though the water is the same temperature in both instances, the effect on the blocks of ice is much different.

In discussing the difference between heat and temperature, point out that temperature is an intensive physical property since it measures the kinetic energy of particles while heat is an extensive property because it is the sum total of the kinetic energy of the sample and depends on the size of the sample. Using the examples in the text, define temperature, heat, and heat flow, especially as the latter applies to the body. Heat flowing into the body is sensed as warm, while heat flowing away from the body is sensed as cool.

To explain the various temperature scales, draw three thermometers on the board or on a transparency. Label each as Fahrenheit, Celsius, and Kelvin. Draw two horizontal lines, one representing the freezing point of water and the other the boiling point of water. Show that there are 100 divisions between the freezing and boiling points on the Celsius and Kelvin scales and 180 divisions on the Fahrenheit scale. Then 100 Celsius degrees must equal 180 Fahrenheit degrees.

Although the students are most familiar with the calorie, the text will be using the SI unit of heat, the joule. Emphasize that the calorie is a very small unit and the joule an even smaller unit.

Remind the students that temperature is an intensive property while heat is extensive. Heat lost or gained by objects can be determined if one knows the change in temperature, the mass of the object, and a new intensive property known as the specific heat of the object. Metals which conduct heat very well have low specific heats while non-conductors, like the plastic handles on frying pans, have high specific heats.

2.3 Scientific Measurements

Use the dartboard analogy on page 46 to distinguish between precision and accuracy. Follow this discussion with a set of data representing masses of a beaker. Show how precision and accuracy apply to scientific measurements.

To help students comprehend significant figures, have them read measurements from meter sticks, graduated cylinders, and thermometers. In each case, they have to estimate between the two smallest marks on the instrument if they are to report the measurement to the limits of the measuring tool. Two students may read that final place differently.

The greatest problem students encounter in determining significant digits is deciding whether or not beginning or trailing zeros are significant. The following mnemonic may help:

If the number contains a decimal point, draw an arrow from the decimal point LEFT to RIGHT stopping at the first NON-ZERO digit. It and the remaining digits are significant. Add to the digits to the left of the decimal point. (To include a geographical twist: If a decimal point is *present*, draw an arrow from the *P*acific Ocean.)

If the number does not contain a decimal point, draw an arrow from RIGHT to LEFT stopping at the first NON-ZERO digit. It and the remaining digits are significant. (Continuing the geographical twist: If a decimal point is *a*bsent, draw an arrow from the *A*tlantic Ocean.)

Examples: 60.300 grams (5 significant figures)
86 000 miles/s (2 significant figures)

The rounding rule presented in the text will most likely be new to your students. They are accustomed to rounding number 5 or larger up to the next number. The rule that they will find strange applies only to rounding when the final digit is a 5. Rather than automatically rounding up, students must look to the digit preceding the 5, if it is an even digit drop the 5, if it is an odd digit round up. The following example may help clarify the issue.

If the numbers 10.5 and 11.5 were known to three significant digits, their average would be (10.5 + 11.5)/2 or 22.0/2 or 11.0. If, on the other hand, 10.5 and 11.5 were known only to two significant digits, should one round off the numbers and then take the average, the following would occur.

(a) Using the recommended system 10.5 would round to 10 and 11.5 would round to 12. The average of 10 and 12 is 11.

(b) Using the old system of always rounding up, 10.5 would round to 11 and 11.5 would round to 12. The average of 11 and 12 is 11.5 or, rounding, 12.

The method recommend by the text (rules 4 and 5) keeps large amounts of data from being skewed too high.

Go over the rules for adding and subtracting and multiplying and dividing numbers with significant figures. The students must be able to classify the problem they are working on as one (adding/subtracting) or the other (multiplying/dividing) before they can apply the correct rule. When adding and subtracting, the decimal points must line up. Perform the required operation and make sure that the answer is rounded off to the leftmost uncertain digit appearing in the original numbers. Multiplication and division is a bit easier. Simply determine the number of significant digits in each of the numbers being multiplied and/or divided. Make sure to round the answer to the number representing the fewest number of significant digits. When performing such calculations students will run across exact conversion factors, and they will try to round off numbers incorrectly. Point out the presence of exact conversion factors and that such factors do not limit the number of significant digits in the answer.

Students with a poor algebra background will usually encounter difficulties when working with exponential notation. You may wish to give a pre-test as a diagnostic tool. If students ask why they have to use scientific notation, write a number like Avogadro's number (6.02×10^{23}) on the board including all of the zeros. Write a very small number, one outside the range of most calculators, e.g. the mass of an electron (9.1×10^{-28} g). The only practical way to use such numbers is by expressing them in scientific notation.

2.4 Solving Quantitative Problems

The text proposes a four step method:
(1) analyze the problem
(2) plan the route to solve the problem
(3) compute the answer
(4) evaluate the answer

Stress that steps *(1)* and *(2)* are the key steps. They demand an understanding of the chemistry required to find the route to the solution. The final two steps are little more than arithmetic, yet they are important if one is to report the correct answer. Most students initially find the process too time consuming and too difficult. Work some examples for them and then allow them to work some of the practice problems while you observe. The more practice they get, the less difficult will later problem-solving chapters appear.

Conclude the unit by distinguishing between direct and inverse proportions. The definitions may appear strange to them. Two variables are directly proportional if dividing them gives a constant value, and two variable are inversely proportional if multiplying them gives a constant value.

References and Resources

Books and Periodicals

Asten, A.V. "Standards of Measurement," *Scientific American*, Vol. 218, No. 6, 1968.
Calder, Lord Ritchie. "Conversation to the Metric System," *Scientific American*, Vol. 233, No. 1, 1970.

Fields, Lawrence D. and Stephen J. Hawks. "Minimizing Significant Figure Fuzziness," *Journal of College Science Teaching*, Vol. XVI, No. 1, Sept./Oct. 1986, pp. 30–34.

Guymon, E. Park, et al. "Teaching Significant Figures Using a Learning Cycle," *Journal of Chemical Education*, Vol. 63, No. 9, Sept. 1986, pp. 786–787.

Norris, A.C. "SI Units in Physico-Chemical Calculations," *Journal of Chemical Education*, Vol. 48, No. 12, Dec. 1971, pp. 797–800.

Pinkerton, R.C. and C.E. Gleit. "The Significance of Significant Figures," *Journal of Chemical Education*, Vol. 44, No. 232, 1967.

Whitmer, Joan C. "Are Your Students Proportionality Literate?," *The Science Teacher*, Vol. 54, No. 8, Nov. 1987, pp. 37–39.

Computer Software

Groves, Paul, Science Skills—Disk 1 "Uncertainties and Measurements?," The Mole Company, Flinn.

Smith, S.G., R. Chabay, and E. Kean. Introduction to General Chemistry, Disk 9, The Metric System, COMPress.

Answers and Solutions

Questions

1. A unit of measurement is a physical quantity of a defined size.
2. *(a)* Scientists around the world use the International System of Units. *(b)* The abbreviation "SI" is generally used. *(c)* Many SI units were taken from the metric system.
3. *(a)* Standards of measurement are objects or natural phenomena of unchanging size equal to the size of the units. *(b)* Standards are necessary so that everyone will know the exact size of a reported measurement.
4. *(a)* 6500 *(b)* 78 420 *(c)* 207 891 *(d)* 0.3500 *(e)* 0.920 400 *(f)* 127.800 00 *(g)* 42 600.3500
5. *(a)* A fundamental unit is one that is defined by a physical standard of measurement.

   (b)

Quantity	Unit	Symbol
length	meter	m
mass	kilogram	kg
time	second	s
temperature	kelvin	K
amount of substance	mole	mol
electric current	ampere	A
luminous intensity	candela	cd

6. *(a)* One meter is the distance that light travels in a vacuum in 1/299 792 458 of a second. *(b)* Examples of approximately one-meter distances include the width of a doorway, the length of a baseball bat, and the distance that an adult can cover in one large step.
7. *(a)* cm or dm *(b)* m or dam *(c)* dm *(d)* mm *(e)* m or dm *(f)* km *(g)* Gm or Mm *(h)* μm
8. *(a)* The international standard for the kilogram is a platinum-iridium cylinder kept at the International Bureau of Weights and Measures in Sèvres, France. *(b)* One kilogram is approximately equal to the mass of this textbook.
9. *(a)* kg *(b)* g *(c)* mg *(d)* g
10. *(a)* The SI standard for the second is the time required for 9 192 631 770 vibrations of radiation of a cesium-133 atom to occur. *(b)* The second is commonly defined as $\frac{1}{60}$ of a minute.
11. The factor-label method is a problem-solving method based upon treating calculation units like algebraic factors.
12. A conversion factor is a ratio that is derived from the equality between two different units and that can be used to convert from one unit to the other.
13. *(a)* A derived unit is a unit that can be obtained from combinations of fundamental units. *(b)* Two examples of important derived units in chemistry are volume and density.
14. *(a)* The derived SI unit for area is the square meter (m^2). *(b)* The derived SI unit for volume is the cubic meter (m^3).
15. *(a)* Volume is defined as the amount of space occupied by an object. *(b)* The SI unit for volume, the cubic meter, corresponds to a cube whose sides are each one meter (the SI length unit) long.
16. *(a)* A liter is a unit of volume equal to 1000 cm^3. *(b)* One cubic centimeter is equivalent to a volume of one milliliter.
17. *(a)* mL or cm^3 *(b)* L *(c)* mL *(d)* m^3
18. The terms "light" and "heavy" are misleading because they do not take into account the volumes of the objects being compared.
19. *(a)* The density of a material is its mass divided by its volume, or its mass per unit volume.

    (b) Density = $\frac{mass}{volume}$, or $D = \frac{m}{V}$

    (c) The SI unit for density is kg/m^3; other metric units for density include g/cm^3 and g/mL.
20. Since many substances increase slightly in volume as their temperatures increase, densities determined at different temperatures will vary somewhat.
21. *(a)* Temperature is a measure of the average kinetic energy of the particles in a sample of matter. *(b)* The greater the energy of the particles, the higher the temperature of the matter.
22. *(a)* Heat is energy transferred between two systems at different temperatures, and is the total temperature related energy of the particles in a sample. *(b)* Heat always flows spontaneously from matter at a higher temperature to matter at a lower temperature.
23. Most materials expand when heated and contract when cooled. The liquids used in thermometers do so uniformly with changing temperature.
24. *(a)* The two fixed points on which the Celsius scale is based are the freezing point (0°C) and boiling point (100°C) of water. *(b)* One degree Celsius is the unit of temperature on the Celsius scale.
25. This means that the banana would provide 130 kilocalories of heat if it were completely burned (metabolized) by the body.
26. The quantity of heat lost or gained during a temperature change depends on the nature of the matter changing temperature, the quantity of matter changing temperature, and the size of the temperature change.
27. *(a)* Heat capacity is the quantity of heat needed to raise the temperature of a sample of matter by 1°C. *(b)* Specific heat is the quantity of heat required to raise the temperature of 1 g of a substance by 1°C.

28. The quality of a measurement depends on the measuring instrument and the skill of the person using it.

29. Uncertainty in measurements can result from limitations in accuracy or precision. Accuracy is the closeness of a measurement to a true or accepted value; precision is agreement among the numerical values of a set of measurements made in the same way.

30. The measurements in (a) and (c) are precise. Those in (b) are not.

31. (a) The significant figures in a measurement consist of all digits known with certainty plus one final digit, which is uncertain or estimated. (b) The location of the estimated digit depends upon the smallest subdivision of the measurement scale being used.

32. (a) 4 (rule 1, 1, 1, and 1) (b) 3 (rule 1, and 1) (c) 2 (rules 4, 1, and 1) (d) 4 (rules 1, 2, 1, and 1) (e) 6 (rules 1, 2, 2, 1, 1, and 1) (f) 1 (rules 1, 3, and 3) (g) 3 (rules 1, 1, and 3) (h) 2 (rules 4, 1, and 1) (i) 2 (rules 4, 4, 4, 1, and 1) (j) 2 (rules 1 and 5) (k) 4 (rules 1, 1, 5, and 5) (l) 2 (rules 4, 4, 4, 1, and 5) (m) 7 (rules 1, 2, 2, 1, 2, 1, and 5) (n) 5 (rules 1, 1, 3, 5, and 5)

33. (a) 500 g (b) 500. g (c) 500.0 g (d) 500.000 g

34. (a) 2.7 g (b) 47.37 mL (c) 8.4 kg (d) 4.16 L (e) 20 km (f) 0.049 mg (g) 0.064 mL

35. (a) 140 g (b) 140. g (c) 139.6 g (d) 139.58 g (e) 100 g

36. (a) The initial sum or difference should be rounded off so that the final digit is in the same place as the leftmost uncertain digit. (b) The initial product or quotient should be rounded off to the same number of significant figures as in the measurement that has the fewest significant figures.

37. (a) Exact conversion factors are those that have no uncertainty and do not limit the number of digits in a calculation. (b) Examples of exact conversion factors are
$$\frac{100 \text{ cm}}{1 \text{ m}}, \frac{1000 \text{ g}}{1 \text{ kg}}, \text{ and } \frac{4.184 \text{ J}}{1 \text{ cal}}.$$

38. Scientific notation is a shorthand technique in which numbers are written in the form $M \times 10^n$ where M is a number greater than or equal to 1, but less than 10, and n is an integer.

39. The stepwise method of problem solving consists of the following: Step 1: Analyze the problem; Step 2: Plan a route to solve the problem; Step 3: Compute the answer; Step 4: Evaluate the answer.

40. (a) A variable is a quantity that can change in value. (b) Examples are temperature, mass, and volume.

41. (a) Two variables are directly proportional to each other if dividing one by the other gives a constant value. Two variables are inversely proportional to each other if their product has a constant value.

(b) direct: $\frac{x}{y} = k$ or $x = ky$ inverse: $xy = k$ or $y = \frac{k}{x}$

(c) Values of directly proportional quantities produce a linear graph; inversely proportional quantities produce a hyperbola.

Problems

1. (a) $2 \text{ m} \times \dfrac{10 \text{ dm}}{\text{m}} = 20 \text{ dm}$

$2 \text{ m} \times \dfrac{100 \text{ cm}}{\text{m}} = 200 \text{ cm}$

$2 \text{ m} \times \dfrac{1000 \text{ mm}}{\text{m}} = 2000 \text{ mm}$

$2 \text{ m} \times \dfrac{\text{km}}{1000 \text{ m}} = 0.002 \text{ km}$

$2 \text{ m} \times \dfrac{1\,000\,000\,000 \text{ nm}}{\text{m}} = 2\,000\,000\,000 \text{ nm}$

(b) $1.5 \text{ km} \times \dfrac{10 \text{ hm}}{\text{km}} = 15 \text{ hm}$

$1.5 \text{ km} \times \dfrac{100 \text{ dam}}{\text{km}} = 150 \text{ dam}$

$1.5 \text{ km} \times \dfrac{1000 \text{ m}}{\text{km}} = 1500 \text{ m}$

$1.5 \text{ km} \times \dfrac{10\,000 \text{ dm}}{\text{km}} = 15\,000 \text{ cm}$

$1.5 \text{ km} \times \dfrac{100\,000 \text{ cm}}{\text{km}} = 150\,000 \text{ cm}$

$1.5 \text{ km} \times \dfrac{1\,000\,000 \text{ mm}}{\text{km}} = 1\,500\,000 \text{ mm}$

$1.5 \text{ km} \times \dfrac{1\,000\,000\,000 \text{ } \mu\text{m}}{\text{km}} = 1\,500\,000\,000 \text{ } \mu\text{m}$

$1.5 \text{ km} \times \dfrac{1\,000\,000\,000\,000 \text{ nm}}{\text{km}} = 1\,500\,000\,000\,000 \text{ nm}$

(c) 48 cg = 480 mg
= 4.8 dg
= 0.48 g
= 0.000 48 kg
= 480 000 μg

(d) 7.5 L = 75 dL
= 750 cL
= 7500 mL
= 7 500 000 μL

2. (a) First, express each in comparable units, such as meters:
1.2 m = 1.2 m

$750 \text{ cm} \times \dfrac{1 \text{ m}}{100 \text{ cm}} = 7.50 \text{ m}$

$0.005 \text{ km} \times \dfrac{1000 \text{ m}}{1 \text{ km}} = 5 \text{ m}$

$65 \text{ dm} \times \dfrac{1 \text{ m}}{10 \text{ dm}} = 6.5 \text{ m}$

$2000 \text{ mm} \times \dfrac{\text{m}}{1000 \text{ mm}} = 2 \text{ m}$

Thus, the values ranked from largest to smallest are: 750 cm, 65 dm, 0.005 km, 2000 mm, and 1.2 m.

(b) Convert each into comparable units, such as grams:

$450 \text{ mg} \times \dfrac{\text{g}}{1000 \text{ mg}} = 0.45 \text{ g}$

$3.8 \text{ cg} \times \dfrac{\text{g}}{100 \text{ cg}} = 0.038 \text{ g}$

$0.27 \text{ dg} \times \dfrac{\text{g}}{10 \text{ dg}} = 0.027 \text{ g}$

$0.50 \text{ g} = 0.50 \text{ g}$

$0.000 47 \text{ kg} \times \dfrac{1000 \text{ g}}{\text{kg}} = 0.47 \text{ g}$

Thus, the values ranked from largest to smallest are: 0.50 g, 0.000 47 kg, 450 mg, 38 cg, and 0.27 dg.

(c) Convert each into comparable units, such as liters:

$$2.5 \text{ L} = 2.5 \text{ L}$$

$$22 \text{ cL} \times \frac{\text{L}}{100 \text{ cL}} = 0.22 \text{ L}$$

$$13 \text{ dL} \times \frac{\text{L}}{10 \text{ dL}} = 1.3 \text{ L}$$

$$870 \text{ mL} \times \frac{\text{L}}{1000 \text{ mL}} = 0.87 \text{ L}$$

$$175 \text{ cm}^3 \times \frac{\text{mL}}{\text{cm}^3} \times \frac{\text{L}}{1000 \text{ mL}} = 0.175 \text{ L}$$

Thus, the values ranked from largest to smallest are: 2.5 L, 13 dL, 870 mL, 22 cL, and 175 cm³.

3. apples: $3 \times 125 \text{ g} \times \frac{\text{kg}}{1000 \text{ g}} = 0.375 \text{ kg}$

 bananas: $5 \times 115 \text{ g} \times \frac{\text{kg}}{1000 \text{ g}} = 0.575 \text{ kg}$

 pears: $2 \times 75 \text{ g} \times \frac{\text{kg}}{1000 \text{ g}} = 0.15 \text{ kg}$

 peaches: $4 \times 65 \text{ g} \times \frac{\text{kg}}{1000 \text{ g}} = 0.26 \text{ kg}$

 melon: $1750 \text{ g} \times \frac{\text{kg}}{1000 \text{ g}} = 1.75 \text{ kg}$

 grapes: $5.5 \text{ hg} \times \frac{\text{kg}}{10 \text{ hg}} = 0.55 \text{ kg}$

 sum $= 3.66 \text{ kg}$

4. $\dfrac{88 \text{ km}}{\text{h}} \times \dfrac{10 \text{ h}}{\text{day}} \times 5 \text{ days} = 4400 \text{ km}$

5. (a) $30 \text{ L} \times \dfrac{1000 \text{ mL}}{\text{L}} = 30\ 000 \text{ mL}$

 (b) $12 \text{ dL} \times \dfrac{100 \text{ mL}}{\text{dL}} = 1200 \text{ mL}$

 (c) $4.5 \text{ cL} \times \dfrac{10 \text{ mL}}{\text{cL}} = 45 \text{ mL}$

 (d) $6.4 \text{ cm}^3 \times \dfrac{\text{mL}}{\text{cm}^3} = 6.4 \text{ mL}$

6. (a) $22.4 \text{ L} \times \dfrac{1000 \text{ mL}}{\text{L}} = 22\ 400 \text{ mL}$

 (b) $22.4 \text{ L} \times \dfrac{1000 \text{ mL}}{\text{L}} \times \dfrac{\text{cm}^3}{\text{mL}} = 22\ 400 \text{ cm}^3$

7. (a) $D = \dfrac{m}{V} \qquad D = \dfrac{240.00 \text{ g}}{60.0 \text{ cm}^3} = 4.00 \text{ g/cm}^3$

 (b) $D = \dfrac{65.48 \text{ kg}}{12.5 \text{ m}^3} = 5.24 \text{ kg/m}^3$

 (c) $D = \dfrac{8.25 \text{ g}}{3.6 \text{ mL}} = 2.3 \text{ g/mL}$

8. (a) $D = \dfrac{m}{V}$

 $m = D \times V$

 $m = 11.3 \dfrac{\text{g}}{\text{cm}^3} \times 4.6 \text{ cm}^3 = 52 \text{ g}$

(b) $m = 7.31 \dfrac{\text{kg}}{\text{m}^3} \times 8.73 \text{ m}^3 = 63.8 \text{ kg}$

(c) First, express volume in cm³:

$$24 \text{ L} \times \frac{1000 \text{ cm}^3}{\text{L}} = 24\ 000 \text{ cm}^3$$

$$m = 0.090 \frac{\text{g}}{\text{cm}^3} \times 24\ 000 \text{ cm}^3 = 2200 \text{ g}$$

9. (a) $D = \dfrac{m}{V} \qquad V = \dfrac{m}{D}$

 $V = \dfrac{14.50 \text{ g}}{13.6 \text{ g/mL}} = 1.07 \text{ mL}$

 (b) $V = \dfrac{28.73 \text{ g}}{7.29 \text{ g/cm}^3} = 3.94 \text{ cm}^3$

 (c) $12 \text{ kg} \times \dfrac{1000 \text{ g}}{\text{kg}} = 12\ 000 \text{ g}$

 $V = \dfrac{12\ 000 \text{ g}}{1.43 \text{ g/cm}^3} = 8400 \text{ cm}^3$

10. (a) $T(\text{K}) = t(°\text{C}) + 273$
 $T(\text{K}) = 0°\text{C} + 273 = 273\text{K}$
 (b) $T(\text{K}) = 27°\text{C} + 273 = 300\text{K}$
 (c) $T(\text{K}) = 72°\text{C} + 273 = 345\text{K}$
 (d) $T(\text{K}) = -50°\text{C} + 273 = 223\text{K}$

11. (a) $t(°\text{C}) = T(\text{K}) - 273$
 $t(°\text{C}) = 300\text{K} - 273 = 27°\text{C}$
 (b) $t(°\text{C}) = 350\text{K} - 273 = 77°\text{C}$
 (c) $t(°\text{C}) = 273\text{K} - 273 = 0°\text{C}$
 (d) $t(°\text{C}) = 150\text{K} - 273 = -123°\text{C}$

12. (a) $25.0 \text{ cal} \times \dfrac{4.184 \text{ J}}{\text{cal}} = 105 \text{ J}$

 (b) $105 \text{ J} \times \dfrac{\text{kJ}}{1000 \text{ J}} = 0.105 \text{ kJ}$

13. (a) $1.42 \text{ kJ} \times \dfrac{1000 \text{ J}}{\text{kJ}} = 1420 \text{ J}$

 (b) $1420 \text{ J} \times \dfrac{\text{cal}}{4.184 \text{ J}} = 339 \text{ cal}$

 (c) $339 \text{ cal} \times \dfrac{\text{kcal}}{1000 \text{ cal}} = 0.339 \text{ kcal}$

14. Sp. ht. $= \dfrac{\text{ht.}}{m \times t \text{ change}}$

 Sp. ht. $= \dfrac{34 \text{ J}}{25 \text{ g} \times (33°\text{C} - 21°\text{C})}$

 Sp. ht. $= \dfrac{34 \text{ J}}{25 \text{ g} \times 12°\text{C}} = 0.11 \dfrac{\text{J}}{\text{g} \cdot °\text{C}}$

15. Ht. $=$ Sp. ht. $\times m \times t$ change

 Ht. $= \dfrac{0.39 \text{ J}}{\text{g} \cdot °\text{C}} \times 55 \text{ g} \times (40°\text{C} - 24°\text{C})$

 Ht. $= \dfrac{0.39 \text{ J}}{\text{g} \cdot °\text{C}} \times 55 \text{ g} \times 16°\text{C} = 340 \text{ J}$

16. (a) $4.07 \text{ g} + 1.863 \text{ g} = 5.933 \text{ g} = 5.93 \text{ g}$
 (b) $3127.55 \text{ cm} - 784.2 \text{ cm} = 2343.35 = 2343.4 \text{ cm}$
 (c) $0.067 \text{ mL} + 1.01 \text{ mL} + 2.5 \text{ mL} = 3.577 \text{ mL} = 3.6 \text{ mL}$
 (d) $36.427 \text{ m} + 12.5 \text{ m} + 6.33 \text{ m} = 55.257 \text{ m} = 55.3 \text{ m}$

(e) $20.2 \text{ L} + 7.6342 \text{ L} + 0.21 \text{ L} - 0.0758 \text{ L}$
$= 27.9684 \text{ L} = 28.0 \text{ L}$

(f) $600. \text{ km} + 29.45 \text{ km} + 0.0778 \text{ km} = 629.5278 \text{ km}$
$= 630. \text{ km}$

17. (a) $9.40 \text{ cm} \times 2.6 \text{ cm} = 24.44 \text{ cm}^2 = 24 \text{ cm}^2$
(b) $8.08 \text{ dm} \times 5.3200 \text{ dm} = 42.985\ 6 \text{ dm}^2 = 43.0 \text{ dm}^2$
(c) $1.50 \text{ g} / 2 \text{ cm}^3 = 0.75 \text{ g/cm}^3 = 0.8 \text{ g/cm}^3$
(d) $0.084\ 21 \text{ g} / 0.640 \text{ mL} = 0.131\ 578\ 1 \text{ g/mL} = 0.132 \text{ g/mL}$
(e) $4.00 \text{ m} \times 0.020 \text{ m} \times 1.57 \text{ m} = 0.1256 \text{ m}^3 = 0.13 \text{ m}^3$
(f) $\dfrac{21.50 \text{ g}}{4.06 \text{ cm} \times 1.8 \text{ cm} \times 0.905 \text{ cm}} = 3.250\ 808\ 2 \text{ g/cm}^3$
$= 3.3 \text{ g/cm}^3$

18. (a) $900 = 9 \times 10^2$
(b) $750\ 000 = 7.5 \times 10^5$
(c) $93\ 000\ 000 = 9.3 \times 10^7$
(d) $0.000\ 082 = 8.2 \times 10^{-5}$
(e) $0.000\ 403 = 4.03 \times 10^{-4}$
(f) $6500 \times 10^7 = 6.5 \times 10^{(7+3)} = 6.5 \times 10^{10}$
(g) $0.000\ 015 \times 10^8 = 1.5 \times 10^{(8-5)} = 1.5 \times 10^3$
(h) $320\ 000 \times 10^{-9} = 3.2 \times 10^{(-9+5)} = 3.2 \times 10^{-4}$
(i) $0.0097 \times 10^{-4} = 9.7 \times 10^{(-4-3)} = 9.7 \times 10^{-7}$

19. (a) $5 \times 10^3 = 5000$
(b) $7.8 \times 10^4 = 78\ 000$
(c) $2 \times 10^{-5} = 0.000\ 02$
(d) $3.81 \times 10^{-4} = 0.000\ 381$
(e) $4.07 \times 10^7 = 40\ 700\ 000$
(f) $6.09 \times 10^{-9} = 0.000\ 000\ 006\ 09$

20. (a) 6.7×10^2
(b) 9.5×10^{-5}
(c) 2.34×10^4
(d) 8.07×10^{-6}
(e) 3.7×10^4

21. (a) 6×10^{12}
(b) $15.96 \times 10^7 = 1.596 \times 10^8 = 1.6 \times 10^8$
(c) $12.504 \times 10^{-10} = 1.2504 \times 10^{-9} = 1.3 \times 10^{-9}$
(d) $18.06 \times 10^{12} = 1.806 \times 10^{13} = 1.8 \times 10^{13}$
(e) $4.2 \times 10^2 = 4 \times 10^2$ (f) $1.875 \times 10^8 = 2 \times 10^8$

22. *Step 1. Analyze*

Given: speed $= 40.0 \text{ km/1.0 h}$

time $= 5.5 \text{ h}$

Unknown: distance traveled

Step 2. Plan

speed, time $\rightarrow$ distance

speed $\left(\dfrac{\text{km}}{\text{m}}\right) \times$ time (h) $=$ distance (km)

Step 3. Compute

$\dfrac{40.0 \text{ km}}{1.0 \text{ h}} \times 5.5 \text{ h} = 220 \text{ km}$

Step 4. Evaluate

The answer is close to an estimated value of 200 km (40×5). Units have canceled to give kilometers, and the answer is correctly rounded to two significant figures.

23. *Step 1. Analyze*

Given: mass $= 8.50 \text{ g}$

density $= 0.24 \text{ g/cm}^3$

Step 2. Plan

mass, density $\rightarrow$ volume

$$D \times \frac{g}{\text{cm}^3} = \frac{m \text{ (g)}}{V \text{ (cm}^3)} \qquad V = \frac{m \text{ (g)}}{D \times \dfrac{g}{\text{cm}^3}}$$

Step 3. Compute

$$V = \frac{8.50 \text{ g}}{0.24 \text{ g/cm}^3} = 35 \text{ cm}^3$$

Step 4. Evaluate

The value of the answer is close to an estimated value of 40 cm³ $\left(\dfrac{8}{0.2}\right)$. Units have canceled as desired and the answer is correctly rounded to two significant figures.

Application Questions

1. (a) SI is a decimal (base-10) system, so larger and smaller units for each type of measurement can be created by multiplying the fundamental unit by the appropriate power of ten. Secondly, the prefixes added to the name of a unit to indicate the power of ten by which the unit has been multiplied can be used with any SI unit.
(b) Outside the scientific community, SI is not familiar to most Americans. Considerable reindustrialization would have to take place because many of our factories make parts and products that conform to the English system of units; the changeover would thus take a considerable period of time, during which our economy might suffer somewhat.

2. The statement could be considered true if reference is being made to a very-high-volume sample of feathers being compared to a low-volume sample of iron, since iron's density is much greater.

3. The hands are able to provide an indication of heat because they can detect the direction of energy being transferred between two systems. Temperature cannot be accurately detected by the hands because they are not able to measure objectively the average kinetic energy of the particles in a system.

4. Since the temperature of the copper is higher than that of the water, heat will flow away from the copper to the water, thereby decreasing the temperature of the copper and increasing the temperature of the water. When the two temperatures are equal, heat will no longer flow between the two objects.

5. (a) Mercury is generally used in thermometers because it expands and contracts in a nearly linear manner with changing temperatures.
(b) Water could possibly be used within a narrow range of temperatures, but at or below 0°C it would freeze, and at or above 100°C, it would boil.

6. Heat capacity depends on the mass of the sample, whereas specific heat depends only on the composition of the sample.

7. (a) Expressing the answer in four significant figures is misleading because one of the measurements has only two significant figures. (b) The correct density is 32 g/cm³.

Application Problems

1. (a) $\dfrac{65 \text{ km}}{\text{h}} \times \dfrac{1000 \text{ m}}{\text{km}} \times \dfrac{100 \text{ cm}}{\text{m}} = 6.5 \times 10^6 \text{ cm/h}$

   (b) $\dfrac{6.5 \times 10^6 \text{ cm}}{\text{h}} \times \dfrac{\text{h}}{60 \text{ min}} \times \dfrac{\text{min}}{60 \text{ s}} = 1.8 \times 10^3 \text{ cm/s}$

2. First, convert each dimension to cm:

   $125 \text{ cm} = 125 \text{ cm}$

   $0.750 \text{ m} \times \dfrac{100 \text{ cm}}{\text{m}} = 75.0 \text{ cm}$

   $500. \text{ mm} \times \dfrac{\text{cm}}{10 \text{ mm}} = 50.0 \text{ cm}$

   Next, find the volume in cm^3:
   $V = l \times w \times h$
   $V = 125 \text{ cm} \times 75.0 \text{ cm} \times 50.0 \text{ cm}$
   $V = 4.69 \times 10^5 \text{ cm}^3$

   Converting into liters,
   $4.69 \times 10^5 \text{ cm}^3 \times \dfrac{\text{L}}{1000 \text{ cm}^3} = 469 \text{ L}$

3. $1 \text{ m}^3 = 1 \text{ m} \times 1 \text{ m} \times 1 \text{ m} = 100 \text{ cm} \times 100 \text{ cm} \times 100 \text{ cm} = 1 \times 10^6 \text{ cm}^3$

4. $V = l \times w \times h$
   $V = 12.00 \text{ cm} \times 8.50 \text{ cm} \times 0.50 \text{ cm}$
   $V = 51 \text{ cm}^3$

   $D = \dfrac{m}{V}$

   $D = \dfrac{457.00 \text{ g}}{51 \text{ cm}^3} = 9.0 \text{ g/cm}^3$

5. $D = \dfrac{m}{V}$

   $V = \dfrac{m}{D}$

   $V = \dfrac{25.00 \text{ g}}{1.83 \text{ g/mL}} = 13.7 \text{ mL}$

6. $D = \dfrac{m}{V}$

   $V = \dfrac{m}{D}$

   $V = \dfrac{521.10 \text{ g}}{19.3 \text{ g/cm}^3}$

   $V = 27.0 \text{ cm}^3$

   Since $V_{\text{cube}} = l \times w \times h = l^3$,

   $L^3 = 27.0 \text{ cm}^3$

   $L = \sqrt[3]{27.0 \text{ cm}^3} = 3.00 \text{ cm}$

7. $D = \dfrac{m}{V}$

   $D = \dfrac{32.50 \text{ g}}{3.10 \text{ cm}^3}$

   $D = 10.5 \text{ g/cm}^3$

   Based on the density-table information, the unknown could be silver.

8. $t(°C) = \dfrac{5}{9} \times [t(°F) - 32F]$

Rearranging, $t(°F) = \dfrac{9}{5}°C + 32$

$t(°F) = \dfrac{9}{5}(37.0°C) + 32 \qquad t(°F) = 98.6°F$

9. First, determine the specific heat of the unknown:

   $\text{Sp. ht.} = \dfrac{\text{Ht.}}{m \times t \text{ change}}$

   $\text{Sp. ht.} = \dfrac{20.4 \text{ J}}{6.67 \text{ g} \times 7.0°C} \qquad \text{Sp. ht.} = 0.44 \text{ J/g} \cdot °C$

   Based on the table of specific heats, the substance is most likely iron.

10. $\text{Ht.} = \text{Sp. ht.} \times m \times t \text{ change}$

    $\text{Ht.} = \dfrac{0.13 \text{ J}}{\text{g} \cdot °C} \times 45 \text{ g} \times (20.0°C - 38.0°C)$

    $\text{Ht.} = \dfrac{0.13 \text{ J}}{\text{g} \cdot °C} \times 45 \text{ g} \times (-18.0°C)$

    $\text{Ht.} = -110 \text{ J}$ (the negative sign indicates heat loss)

11. $\text{Ht.} = \text{Sp. ht.} \times m \times \triangle t$

    $\triangle t = \dfrac{ht}{\text{Sp. ht.} \times m}$

    $\triangle t = \dfrac{46.5 \text{ J}}{\dfrac{0.24 \text{ J}}{\text{g} \cdot °C} \times 20.0 \text{ g}}$

    $\triangle t = \dfrac{46.5 \text{ J}}{4.8 \dfrac{\text{J}}{°C}} \qquad \triangle t = 9.7°C$

    Thus, the final temperature will be: $28.5°C + 9.7°C = 38.2°C$.

12. (a) $\dfrac{3.00 \times 10^{10} \text{ cm}}{\text{s}} \times \dfrac{\text{km}}{10^5 \text{ cm}} \times \dfrac{3600 \text{ s}}{\text{h}} \times \dfrac{24 \text{ hours}}{\text{day}} \times$

    $\dfrac{365 \text{ days}}{\text{y}} = 9.46 \times 10^{12} \text{ km/y}$

    (b) $1.50 \times 10^8 \text{ km} \times \dfrac{\text{light-year}}{9.46 \times 10^{12} \text{ km}}$

    $= 1.59 \times 10^{-5} \text{ light-year}$

13. (a) $60. \text{ Å} \times \dfrac{1 \times 10^{-8} \text{ cm}}{\text{Å}} = 60. \times 10^{-8} \text{ cm}$

    $= 6.0 \times 10^{-7} \text{ cm}$

    (b) $6.0 \times 10^{-7} \text{ cm} \times \dfrac{\text{m}}{10^2 \text{ cm}} = 6.0 \times 10^{-9} \text{ m}$

    (c) $6.0 \times 10^{-7} \text{ cm} \times \dfrac{10 \text{ mm}}{\text{cm}} = 6.0 \times 10^{-6} \text{ mm}$

    (d) $6.0 \times 10^{-7} \text{ cm} \times \dfrac{10^4 \text{ } \mu\text{m}}{\text{cm}} = 6.0 \times 10^{-3} \mu\text{m}$

14. $(2.00 \text{ mg B}_2) \dfrac{\text{g B}_2}{1000 \text{ mg B}_2} \dfrac{\text{g of cheese}}{5.2 \times 10^{-6} \text{ g B}_2} = 380 \text{ g, or}$

    $0.38 \text{ kg of cheese}$

15. (a) Direct relationship
    (b) As Celsius temperatures increase, Kelvin temperatures also increase.

Teacher's Notes

Chapter 2 Measurements and Solving Problems

INTRODUCTION

In Chapter 1, chemistry is described as a quantitative science. In this chapter you will find out about the quantitative language of chemistry—the system of measurement called *SI*, the units of measurement common in chemistry, and a useful method for writing numbers in science. You will also be introduced to a procedure that will help you solve quantitative problems.

LOOKING AHEAD

As you study this chapter, concentrate on developing problem-solving skills. These will be invaluable to you throughout the course. Practice using metric and SI units of measurement as well as the factor-label method as you solve quantitative problems.

SECTION PREVIEW

2.1 Units of Measurement
2.2 Heat and Temperature
2.3 Using Scientific Measurements
2.4 Solving Quantitative Problems

Measurement is a commonplace feature of our lives. The work of science, however, requires a special standardized system of measurement.

2.1 Units of Measurement

You have probably had the experience of ordering a large pizza in a restaurant only to discover that what is called "large" on the menu is smaller than you expected. A less ambiguous way of ordering would have been to ask for a 16-inch pizza.

A scientific description should be unambiguous, that is, it should mean the same thing to people all over the world. This section introduces the system of measurement scientists have agreed to use in order to avoid confusion and facilitate communication.

The SI Measurement System

Many systems of measurement are used throughout the world. In the United States, most people are familiar with the English system that uses units of measurement such as inches, ounces, and pounds. Units of measurement are used to compare whatever is to be measured with something of a previously defined size. For instance, to measure the length of this book, you could compare its length with a ruler marked off in centimeters. However, before the ruler could be made, the length of the centimeter had to be defined. A **unit of measurement** *is a physical quantity of a defined size.* Scientists from all countries have agreed to use the measurement system called *Le Systéme International des Unités (the International System of Units), usually abbreviated* SI. This system, which includes hundreds of units, was adopted in 1960 by the General Conference of Weights and Measures.

Science is now in a period of transition between the use of non-SI units and SI units. Therefore, some non-SI units are used in this book. The reasons for this are explained as the units are introduced in the text.

Various international and national organizations have the responsibility for defining the sizes of units and governing their use. Certain units are defined by choosing **standards of measurements**, *which are objects or natural phenomena of constant value, easy to preserve and reproduce, and practical in size.* In the United States, the National Bureau of Standards in Washington, D.C. plays a major role in maintaining standards of measurement (Figure 2-1). Standards are necessary so that everyone will know the exact size of a reported measurement.

If it were not for the use of standards as a basis for comparison, references to sizes of objects might mean completely different things to different people. Suppose the unit *foot* was not based on a standard, universally accepted length, but instead depended on the size of the feet of the person making the measurement. A person with very large feet might measure the length of a field as 100 feet. A person with very small feet might measure the same length as 200 feet.

It is also important that standards remain unchanged over time. Standards are chosen because they do not change and their dimensions are exactly reproducible whenever measurements are made.

The result of every measurement is a number plus a unit. As you will see throughout this chapter, units are usually represented in writing by symbols that are abbreviations of the unit names. For example, "ten centimeters" is written 10 cm.

Figure 2-1 A copy of the standard kilogram mass is kept at the National Bureau of Standards.

A number without a corresponding unit is of little use in chemistry.

Figure 2-2 A man and a boy use their feet as measurement standards. How can their measurements be made more useful?

This marks a departure from previous editions.

To calibrate a measuring device means to adjust its scale reading to agree with the size of a standard.

In this book, we follow an international system for writing numbers. Commas are not used in numbers. Counting from the decimal point, numbers with five or more digits on either side of the decimal point have a space rather than a comma after each group of three digits. For example, 45 000 (*not* 45,000) is used to represent "forty-five thousand." In this example, note the space after three digits to the left of the implied decimal point. Take another example. The number "four hundred and fifty thousand" is expressed as "450 000" (*not* 450,000). For numbers less than 1, spaces follow every three digits counting to the right from the decimal point—for example, 0.435 98 and 0.045 237.

Fundamental SI Units

Seven SI units are fundamental units, or base units. A **fundamental unit** *is one that is defined by a physical standard of measurement.* The seven SI units are shown in Table 2-1. All other SI units can be derived from these fundamental units. A discussion of the fundamental units for length, mass, and time follows. Later in this chapter, the SI unit for temperature is discussed. Another fundamental unit, the mole, is discussed in Chapter 3.

TABLE 2-1 SI FUNDAMENTAL UNITS		
Quantity	Unit	Symbol
Length	meter	m
Mass	kilogram	kg
Time	second	s
Temperature	kelvin	K
Amount of substance	mole	mol
Electric current	ampere	A
Luminous intensity	candela	cd

SI is a decimal (base 10) system. Therefore, larger and smaller units for each type of measurement can be created by multiplying the fundamental unit by the appropriate power of 10. A prefix is then added to the name of the unit, indicating the power of 10 by which the unit has been multiplied. These prefixes can be used with any SI unit. Refer to Table 2-2 for their meanings. For example, a length of 1 centimeter (cm) is equal to 10^{-2} meter, or 0.01 meter.

TABLE 2-2 SI PREFIXES

Prefix	Symbol	Exponential Factor	Meaning	Example
tera	T	10^{12}	1 000 000 000 000	1 terameter (Tm) = 1×10^{12} m
giga	G	10^{9}	1 000 000 000	1 gigameter (Gm) = 1×10^{9} m
mega	M	10^{6}	1 000 000	1 megameter (Mm) = 1×10^{6} m
kilo	k	10^{3}	1000	1 kilometer (km) = 1×10^{3} m
hecto	h	10^{2}	100	1 hectometer (hm) = 100 m
deca	da	10^{1}	10	1 decameter (dam) = 10 m
deci	d	10^{-1}	1/10	1 decimeter (dm) = 0.1 m
Centi	c	10^{-2}	1/100	1 centimeter (cm) = 0.01 m
Milli	m	10^{-3}	1/1000	1 millimeter (mm) = 0.001 m
micro	μ	10^{-6}	1/1 000 000	1 micrometer (mm) = 1×10^{-6} m
nano	n	10^{-9}	1/1 000 000 000	1 nanometer (nm) = 1×10^{-9} m
pico	p	10^{-12}	1/1 000 000 000 000	1 picometer (pm) = 1×10^{-12} m

Length *The SI standard unit for length is the* **meter.** The international standard for the meter (m) is the distance light travels in a vacuum during a time interval of 1/299 792 458 of a second. Most of us use meter sticks or other measuring devices with lengths that have been set on the basis of this standard. To estimate a length in meters, you can remember that one meter is about the width of an average doorway.

When measuring lengths shorter than a meter, it is often convenient to use centimeters. For longer distances, the kilometer (km) is more useful. One kilometer is equal to 1000 m. The dimensions of this book would most conveniently be measured in centimeters, whereas meters would be more useful for measuring the dimensions of your classroom. For measuring distances between cities, as shown in Figure 2-3, kilometers are usually the most practical unit.

Mass *The SI standard unit for mass is the* **kilogram.** The international standard for the kilogram (kg) is a platinum-iridium cylinder kept at the International Bureau of Weights and Measures in the French town of Sèvres. Copies of this standard are used to calibrate precision balances all over the world. You may find it helpful to think of a kilogram as approximately equal to the mass of this textbook.

The gram (g) is a more convenient unit than the kilogram for measuring small masses. One gram is equal to 10^{-3} kg (0.001 kg). Figure 2-4 shows the mass of a slice of bread and the mass of a paper clip are both about 1 g. For smaller objects, the milligram (mg) is useful. One milligram is equivalent to 0.000 001 kg or 0.001 g.

Figure 2-3 The distance between two cities is conveniently expressed in kilometers, although in the United States most people still use the more common unit of miles.

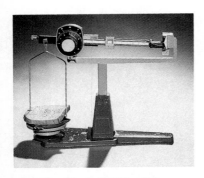

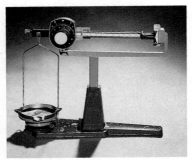

Figure 2-4 The gram is a convenient unit of mass for relatively small items, such as a slice of bread or a paper clip.

Students have a difficult time associating the units with algebraic factors. Use of letters other than *a, b, c; m, n;* or *x, y, z* (the letters usually encountered in math problems) seems to confuse them.

Time *The SI standard unit for time is the* **second.** The second (s) is commonly defined as $\frac{1}{60}$ of a minute. The definition of a minute depends on the definition of an hour, which in turn is based on the duration of a day. A day is the length of time it takes the earth to rotate exactly once on its axis. Because this time varies slightly, a more precise standard is needed for the SI unit. The SI standard for the second is the duration of 9 192 631 770 periods of a particular radiation emitted by cesium-133 atoms. This interval of time is always exactly the same. In everyday life, where such precise time measurements are unnecessary, the SI second can be considered equal to the common second.

Units of Measurement in Calculations

Factor-label method   The units given for a measured quantity are a very important part of that measurement and must be included in all calculations. Two times a 6-kg mass is 12 kg, not 12. This calculation should be written

$$2 \times 6 \text{ kg} = 12 \text{ kg}$$

Units in calculations are treated exactly like algebraic factors. If you are calculating the area of a rectangle 2 m on one side and 6 m on the other, the calculation should be written

$$2 \text{ m} \times 6 \text{ m} = 12 \text{ m}^2$$

The proper use of units in calculations is an extremely valuable aid to solving problems. Understanding how to use units and using them in every calculation in chemistry (and other subjects) will greatly decrease mistakes.

The use of units in calculations is referred to as the factor-label method. *The* **factor-label method** *is a problem-solving method based upon treating units in calculations as if they are algebraic factors.* The factor-label method is used in sample problems throughout this book. In the next section, the use of factor-label method in unit conversion is illustrated. You should use this valuable problem-solving technique throughout your study of chemistry.

Unit conversion   As an introduction to unit conversion, consider a very simple, everyday type of problem. Suppose you need to know how many quarters there are in 75 dollars (Figure 2-5). We are going to solve the problem by the factor-label method.

You know that four quarters equal one dollar. You can express this as a simple mathematical equation: 4 quarters = 1 dollar. Because these two quantities are equal, their ratios are both equal to 1:

$$\frac{4 \text{ quarters}}{1 \text{ dollar}} = 1 \qquad \frac{1 \text{ dollar}}{4 \text{ quarters}} = 1$$

Each of these expressions is a conversion factor. *A* **conversion factor** *is a ratio derived from the equality between two different units and can be used to convert from one unit to the other.* Because conversion factors like those above are equal to 1, they can be multiplied by other factors in equations without changing the validity of the equations. The answer will just be in different units. When using conversion factors, set the problem up as follows:

$$\text{quantity sought} = \text{quantity given} \times \text{conversion factors.}$$

Figure 2-5 Each of these two students has a different approach to a quantitative problem. Which way would you choose to find the number of quarters in $75?

For example, to find out how many quarters there are in 75 dollars, we must carry out the unit conversion from dollars to quarters.

$$\text{? quarters} = 75 \text{ dollars} \times \text{conversion factor}$$

The appropriate conversion factor is the one that allows cancellation of the unwanted unit and gives the answer in the desired unit. Set up the calculation with both conversion factors and cancel units as in algebra.

$$75 \text{ dollars} \times \frac{4 \text{ quarters}}{1 \text{ dollar}} = 300 \text{ quarters}$$

$$75 \text{ dollars} \times \frac{1 \text{ dollar}}{4 \text{ quarters}} = \frac{75 \text{ dollars}^2}{4 \text{ quarters}}$$

This shows that the first conversion factor is the correct one. The second gives the meaningless units of dollars²/quarters.

Sample Problem 2.1

Express a mass of 92 cg in milligrams and grams.

Solution
Solving this problem requires conversion factors that relate centigrams to milligrams and grams. Table 2.4 shows that

$$1 \text{ cg} = 10 \text{ mg} \quad \text{and} \quad 1 \text{ g} = 100 \text{ cg}$$

which give the conversion factors

$$\frac{1 \text{ cg}}{10 \text{ mg}} \quad \frac{10 \text{ mg}}{1 \text{ cg}} \quad \text{and} \quad \frac{1 \text{ g}}{100 \text{ cg}} \quad \frac{100 \text{ cg}}{1 \text{ g}}$$

Choosing in each case the conversion factor that leads to an answer in the desired units gives

$$92 \text{ cg} \times \frac{10 \text{ mg}}{1 \text{ cg}} = 920 \text{ mg} \qquad \text{and} \qquad 92 \text{ cg} \times \frac{1 \text{ g}}{100 \text{ cg}} = 0.92 \text{ g}$$

Practice Problems

1. Express a distance of 25 meters in centimeters and kilometers. *(Ans.)* 2500 cm, 0.025 km
2. Express a mass of 72.0 grams in milligrams, micrograms, and megagrams. *(Ans.)* 72 000 mg, 72 000 000 ug, 0.000 072 Mg
3. Express one week in terms of seconds. *(Ans.)* 604 800 s

Derived SI Units

The majority of SI units are derived units. *A **derived unit** is a unit that can be obtained from combinations of fundamental units.* Examples of some common SI derived units are given in Table 2-3. The units for pressure and energy are discussed in detail later in this book.

TABLE 2-3 DERIVED SI UNITS

Quantity	Unit	Symbol	Derivation	Units
Area	square meter	m^2	length × width	$m \times m$
Volume	cubic meter	m^3	length × width × height	$m \times m \times m$
Density	kilogram per cubic meter	kg/m^3	$\dfrac{mass}{volume}$	$\dfrac{kg}{m^3}$
Speed (velocity)	meter per second	m/s	$\dfrac{length}{time}$	$\dfrac{m}{s}$
Force	newton	N	$\dfrac{length \times mass}{time \times time}$	$\dfrac{m \times kg}{s^2}$
Pressure	pascal	Pa	$\dfrac{force}{area}$	$\dfrac{N}{m^2} = \dfrac{kg}{m \times s^2}$
Energy	joule	J	force × length	$N \times m = \dfrac{m^2 \times kg}{s^2}$

Examine the units given in the last column of Table 2-3. In each case, the quantity has units that are a combination of fundamental SI units. For example, speed, which is the distance in meters traveled in a given time in seconds, has the units of meters per second. The combinations of fundamental units for force, pressure, and energy include several different units. To simplify matters, therefore, each of these unit combinations is given a name of its own. For example, for the pascal (an SI pressure unit), 1 Pa = 1 kg/(m × s²).

Volume *The amount of space occupied by an object is called its **volume**.* The SI unit for volume is the cubic meter (m³). As shown in Table 2-3, this

1m³ of water has a mass of approx. 1000 kg or 2200 lbs. It represents a very large mass.

unit is derived from the fundamental SI unit of length, the meter, and corresponds to a cube whose sides are each 1 m long. Because this volume is often too large to be useful in normal laboratory measurements, the cubic centimeter (cm^3) is used more frequently in chemistry than the cubic meter.

For liquid and gas volume, an older metric unit, the liter, is often used, although it is not the official SI unit of volume. Imagine a cube with edges that are 10 cm long. The volume of this cube is 10 cm × 10 cm × 10 cm, or 1000 cm^3. *The* **liter (L)** *is a volume equal to 1000 cm^3.* One liter of water, or any other liquid, would fill the 1000-cm^3 cube. One milliliter (1 mL) is equivalent to 0.001 liter and is thus equivalent to a volume of 1 cm^3. These units—milliliter and cubic centimeter—are used interchangeably.

Most laboratory volumetric glassware, such as that shown in Figure 2-6, is calibrated and marked in liters or milliliters. These units are used throughout this book in references to liquid or gas volumes.

Density What do we mean by "light" and "heavy"? Figure 2-7 shows what happens when a piece of lead and a piece of cork are placed into a beaker of water. We commonly say that lead sinks in water because "lead is heavy." Likewise, we say that cork floats on water because "cork is light." Figure 2-7 shows the same pieces of lead and cork on opposite pans of a balance. As you can see, the lead and the cork have the same mass. It would appear that the piece of cork is just as "heavy" as the piece of lead. However, the volume of the cork is much greater than that of the lead.

When we speak of lead as "heavy" and cork as "light," we are actually referring not to their masses but to their densities. *The* **density** *of a material is its mass divided by its volume, or its mass per unit volume.* Density is expressed mathematically as

$$\text{density} = \frac{\text{mass}}{\text{volume}} \quad \text{or} \quad d = \frac{m}{V}$$

where m is the mass of a material, V is its volume, and d is its density.

In SI units, density is expressed in kilograms per cubic meter (kg/m^3). More commonly, however, the density of a solid is expressed in grams per cubic centimeter (g/cm^3), the density of a liquid in grams per milliliter (g/mL), and the density of a gas is expressed in grams per liter (g/L). Because the volumes of solids and liquids change slightly with temperature and the volumes of gases change greatly with temperature, the densities also change.

Figure 2-6 Most of the glassware used to measure volume in the chemistry laboratory is calibrated in liters or milliliters.

The liter is also equivalent to a cube with 1-decimeter sides. The volume of the cube, 1 dm^3, is equivalent to 1000 cm^3 or to 1 L.

Note that one liter is equal to one cubic decimeter (1 L = dm^3).

The volume of most kinds of matter changes with temperature. To be accurate, devices for measuring volume must be calibrated at a given temperature. Room temperature (25°C) is usually used for this purpose.

Figure 2-7 Cork floats in water because it is less dense than water; lead sinks because it is denser than water (left). Comparison on a double-pan balance shows that the cork and lead have equal masses, although their volumes are different (right).

Desktop Investigation

Density of Pennies

Questions

Are the densities of pennies made before and after 1982 different? What does density tell you about the composition of pennies?

Materials

40 pennies made before 1982
40 pennies made after 1982
balance
100-mL graduated cylinder

Procedure

Record all your measurements in a data table.

1. Using a balance, determine the mass of 40 pennies minted before 1982. Repeat your measurement two more times. Average the results of the three trials to obtain the average mass of the 40 pennies.

2. Follow the same process as in Procedure 1 with the 40 pennies minted after 1982.

3. Pour 50 mL of water into a 100-mL graduated cylinder. Add the 40 pennies minted before 1982 to the graduated cylinder. Find the volume of the pennies. Repeat your measurement two more times, making sure to dry the pennies before each trial. Average the results of the three trials to obtain the average volume of the pennies.

4. Follow the same process as in Procedure 3 with the

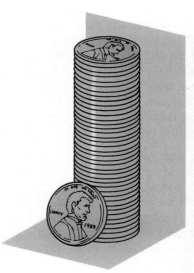

pennies made after 1982.

5. Using the average volume and the average mass of the two sets of pennies, determine the average density of each set.

6. Review your data for any large differences between trials.

7. Compare the average experimental densities of the two sets of pennies with the densities of the metals listed in Appendix Table A-18, Properties of Common Elements.

Discussion

1. Why does the Procedure include three trials instead of one trial for each set of pennies?

2. (a) How would you recognize an inaccurate or incorrect set of data? (b) What

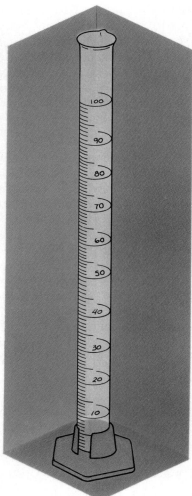

might be some possible causes for such faulty data? (c) What would a scientist do with sets of data that appeared to be inaccurate or incorrect? Why?

3. On what experimental and accepted data do you base your conclusions about the composition of the two sets of pennies?

TABLE 2-4 DENSITIES OF SOME COMMON MATERIALS

Solids		Liquids		
Substance	Density at 20°C (g/cm³)	Substance	Density (g/mL)	Temperature (°C)
Cork	0.24*	Gasoline	0.67*	20
Butter	0.86	Ether	0.736	0
Ice	0.91	Alcohol, ethyl	0.781	20
Sugar	1.59	Kerosene	0.82	20
Bone	1.85*	Turpentine	0.87	20
Clay	2.20*	Water	1.00	4
Glass, common	2.60*	Sea Water	1.025	15
Granite	2.70*	Milk	1.031*	20
Diamond	3.26*	Chloroform	1.489	20
Lead	11.35	Mercury	13.6	20

*Average density

Thus, when the density of a substance is given, the temperature at which it was measured must also be given. The densities of some common materials are listed in Table 2-4. Compare the densities of cork and lead. As shown by the experiments pictured in Figure 2-7, lead is denser than cork—in fact, about 47 times denser. The density of a material can be determined by measuring the mass and volume of a sample of any size. Although lead is dense element, there are elements that are even denser. Osmium (Os), for example, is a metal that has a density of 22.57 g/cm³, nearly twice that of lead. Osmium is the densest of all elements. Whatever type of material is involved, and whatever the size sample that is available in a density determination, the procedure is still essentially the same: simply measure the mass and volume of the sample, and divide the mass by the volume.

The change in gas volume with temperature is discussed in chapter 11.

Osmium, Os, is the densest element with a density of 22.57 g/cm₃.

Sample Problem 2.2

Find the density of a piece of aluminum with a volume of 4.0 cm³ and a mass of 10.8 g. Is aluminum more or less dense than lead? Than cork? You can calculate the density of the sample of aluminum by substituting the given values for mass and volume in the equation for density:

$$\text{density} = \frac{\text{mass}}{\text{volume}} \qquad \text{density} = \frac{10.8\ \text{g}}{4.0\ \text{cm}^3} = 2.7\ \text{g/cm}^3$$

The results of this calculation show that aluminum is denser than cork, but that it is less dense than lead, which has a density of 11.35 g/cm³.

Practice Problems

1. What is the density of a block of marble with a mass of 552 g and a volume of 212 cm³? (Ans.) 2.60 g/cm³
2. A 30 cm³ sample of quartz has a density of 2.65 g/cm³. What is the mass of the quartz sample? (Ans.) 80 g
3. The density of a sample of cork is 0.22 g/cm³. What is the volume of this sample if it has a mass of 35 g? (Ans.) 160 cm³

Section Review

1. What are the fundamental SI units for length, mass, and time?.
2. Express 24 meters in   *(a)* centimeters   *(b)* kilometers.
3. Express 5.6 kilograms in   *(a)* grams   *(b)* milligrams.
4. What volume of water, in liters, would be required to fill a tank 25 cm long, 18 cm wide, and 13 cm deep?
5. A piece of granite has a mass of 55 g and a volume of 20.0 cm³. What is the density of this piece of granite?

2.2 Heat and Temperature

What do you mean when you say something feels hot or cold? Nerves in your skin send signals to your brain that seem to tell you whether something you touch is hot, cold, or warm. Whether something feels hot or cold does not indicate its temperature, however. You can prove this to yourself by a simple experiment. Put your left hand briefly into a container of cold water. At the same time, put your right hand into a container of very warm (*not* boiling) water. Then place both hands into a container of water that is at room temperature. Your left hand, which is colder than the room temperature water, will signal "warm." Your right hand, which is warmer than the room temperature water, will at the same time signal "cold."

The experiment just described demonstrates the difference between heat and temperature. The warm right hand signaled "cold" because energy was flowing away from it into water colder than your hand. The cold left hand signaled "warm" because energy was flowing into it from water that was warmer than the hand. Your hands detect the transfer of energy as heat. Yet, as indicators of temperature, the senses are clearly unreliable.

Definitions of Heat and Temperature

Temperature *is a measure of the average kinetic energy of the particles in a sample of matter.* Temperature indicates how hot or cold something is. The greater the kinetic energy of the particles in a sample of matter, the higher the temperature of the matter. To assign a numerical value to temperature, it is necessary to define a temperature scale.

Heat *or* **heat energy** *can be thought of as the sum total of the kinetic energies of the particles in a sample of matter.* Heat always flows spontaneously from matter at a higher temperature to matter at a lower temperature, as is shown in Figure 2-8. The temperature of the cold water in the right-hand beaker will increase as heat flows into it. Likewise, the temperature of the hot water in the left-hand beaker will decrease as heat flows away from it. When the temperature of the water in each beaker and the water in the pan are equal, heat will no longer flow in the system.

To understand the difference between temperature and heat, consider a bathtub full of warm water and a cup full of boiling water. The temperature of the water in the tub is lower than that of the water in the cup. Although the kinetic energy of each particle in the cup is greater than that of each particle in the tub, the total heat energy in the tub will likely be greater than that in the cup, which has far fewer particles.

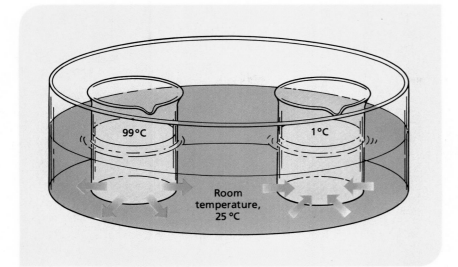

Figure 2-8 The direction of heat flow is determined by the temperature differences between the water in the two beakers and the basin. The transfer of heat out of the beaker on the left into the beaker on the right will continue until the water in all three containers has reached the same temperature.

Measurement of temperature and temperature units Some properties of matter vary with temperature and therefore can be used to measure temperature. For example, most materials expand when heated and contract when cooled. The most familiar instruments for measuring temperature are thermometers containing mercury or alcohol (colored red). The measurement is based on the uniform expansion or contraction of the liquid with changing temperature.

If the thermometer comes into contact with a system at a higher temperature, heat is transferred to the thermometer. This heat transfer raises the temperature of the liquid and increases its volume, causing the liquid to rise in the thermometer. The liquid rises until the temperatures of the system and the thermometer are equal. Similarly, if the thermometer comes into contact with something at a lower temperature, the column of liquid falls as the liquid contracts. The liquid rises or falls next to a set of numbered marks, or calibrations, from which you can read the temperature on a specific temperature scale. The ability to measure temperature is thus based on heat transfer.

Several temperature scales have been developed. One of these, the Celsius scale, was developed by the Swedish astronomer Anders Celsius (1701–1744). Celsius established his temperature scale by defining two fixed points, the freezing and boiling points of water. He defined the normal freezing point of water as 0°C. He defined the normal boiling point of water as 100°C. Both of these fixed points are under normal atmospheric pressure (1 atm). He then divided the interval between the freezing point and the boiling point into 100 equal parts, or degrees, each of which represents a temperature change of 1°C. *One* **degree Celsius** *is the unit of temperature on the Celsius scale.* By extending the same scale divisions beyond the two fixed points, temperatures below 0°C and above 100°C can be measured.

Another temperature scale was developed by William Thomson, Lord Kelvin (1824–1907), and bears his name. *The* **kelvin (K)** *is the unit of temperature on the Kelvin scale and is the fundamental SI unit for temperature.* The kelvin temperatures at which water freezes and boils are

"Normal" freezing and boiling points are those at 1 atm pressure.

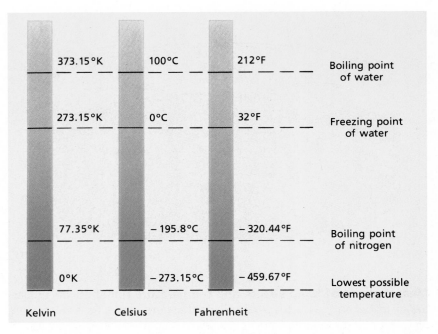

Kelvin	Celsius	Fahrenheit	
373.15°K	100°C	212°F	Boiling point of water
273.15°K	0°C	32°F	Freezing point of water
77.35°K	−195.8°C	−320.44°F	Boiling point of nitrogen
0°K	−273.15°C	−459.67°F	Lowest possible temperature

273.15 K and 373.15 K, respectively. Figure 2-9 includes a comparison of the Kelvin and Celsius temperature scales. Note that on both scales, there are 100 units between the freezing and boiling points of water and the sizes of the degree divisions on both scales are equal. The following equation can be used to convert temperatures from the Celsius (t) to Kelvin (T) scales.

$$T(\text{K}) = t(°\text{C}) + 273.15$$

The equation shows that 273.15 must be added to any Celsius temperature to obtain the Kelvin temperature. For example, converting a room temperature of 25.00°C to the Kelvin scale gives

$$T(\text{K}) = 25.00°\text{C} + 273.15 = 298.15 \text{ K}$$

Subtracting 273.15 allows conversion of a Kelvin temperature to a temperature on the Celsius scale. The equation is

$$t(°\text{C}) = T(\text{K}) - 273.15$$

For example, converting 315.00 K and 210.00 K to temperatures on the Celsius scale gives

$$t(°\text{C}) = 315.00\text{K} - 273.15 = 41.85°\text{C}$$
$$t(°\text{C}) = 210.00\text{K} - 273.15 = 63.15°\text{C}$$

In most calculations in this book, it is acceptable to use 273° rather than 273.15. This book uses both Celsius and Kelvin scales.

Another temperature scale with which you are probably more familiar is the Fahrenheit scale, which is also illustrated in Figure 2-9. In the United States, temperatures are frequently expressed in Fahrenheit degrees (°F). Water freezes at 32°F and boils at 212°F. As shown in Figure 2-9, 100 Celsius degrees are equal to 180 Fahrenheit degrees (the divisions between freezing and boiling). The Fahrenheit degree therefore, is equal to $\frac{100}{180}$ or $\frac{5}{9}$ the size of a Celsius degree. Figure 2-9 also shows that 0°C is equal to 32°F. Thus, the

Gabriel Daniel Fahrenheit (1686–1736), born in Danzig and settled in Holland, invented the mercury thermometer in 1714. For his zero point he chose the temperature produced by mixing equal amounts of snow and ammonium chloride. So that body temperature should be 100°, he chose 212° to be the boiling point of water. The absolute scale based on the Fahrenheit scale is known as the Rankine Scale.

relationship between Celsius and Fahrenheit temperatures is expressed by the equation

$$t\,(°C) = \frac{5}{9} \times [t\,(°F) - 32]$$

The conversion of, for example, the temperature on a hot summer's day of 95°F to a Celsius temperature gives

$$t\,(°C) = \frac{5}{9} \times (95°F - 32)$$

$$= \frac{5}{9} \times 63°F$$

$$= 35°C$$

Units of Heat

The SI unit of heat energy and of all forms of energy is the **joule** (pronounced "jool"). The joule (J) is not a fundamental unit. However it is derived from the units for force and length, as you can see by looking at Table 2-3.

An older unit of heat, the calorie (cal), which is not an SI unit, was originally defined as the quantity of heat required to raise the temperature of 1 g of water from 14.5°C to 15.5°C. The calorie is now defined in terms of the SI unit: *one* **calorie** *is equal to 4.184 J.* The calorie and the joule are both rather small units for the quantities of heat typically dealt with in chemistry. Thus, the kilocalorie (kcal) and the kilojoule (kJ) are also commonly used as heat units.

The energy value of foods is reported in kilocalories, often (and somewhat confusingly) called Calories (with a capital C). For example, if a glass of whole milk contains 150 Calories, this means that the glass of milk will provide 150 kcal of energy if it is completely burned (metabolized) by the body. Table 2-5 lists the number of calories in some common foods.

TABLE 2-5 AVERAGE ENERGY VALUES OF SOME COMMON FOODS

Food	Measure	Energy (kcal)	(kJ)
apple	1 large	100	419
banana	1 medium	130	545
bread	1 slice	60	250
cottage cheese	25 cm³	93	390
cupcake	1 average	200	838
custard	100 mL	125	524
hamburger	113 g (1/4 lb) patty	420	1760
hot dog	1 average	138	578
honey	10 mL	67	280
ice cream	100 g	185	775
orange juice	200 mL	83	350
peach	1 medium	50	210
peanut butter	32 g (1 tbsp)	190	796
potato	1 medium	100	419
skim milk	200 mL	74	310
sugar (sucrose)	4 g (1 tsp)	16	67
whole milk	200 mL	133	557

Sample Problem 2.3

Convert 275 cal to joules and kilojoules.

Solution

Using the appropriate conversion factors based on 1 cal = 4.184 J and 1000 J = 1 kJ gives

$$275 \text{ cal} \times \frac{4.184 \text{ J}}{1 \text{ cal}} \quad 1150 \text{ J} \times \frac{1 \text{ kJ}}{1000 \text{ J}} = 1.15 \text{ kJ}$$

Practice Problems
1. Convert 55 cal to joules and kilojoules. *(Ans.)* 230 J, 0.23 kJ
2. Convert 650 J to calories. *(Ans.)* 160 cal

Heat Capacity and Specific Heat

How is heat measured? Consider a few liters of water. If the water is cooled, it loses heat to its surroundings. As a result, its temperature decreases. If the water is heated, its temperature increases. The quantity of heat lost or gained by water or any other substance can be found from the temperature change of that substance.

Three things determine the quantity of heat lost or gained during a temperature change—the nature of the matter changing temperature; the quantity, or mass, of matter changing temperature; and the size of the temperature change.

Heat capacity *is the amount of heat energy needed to raise the temperature of a given sample of matter by one degree Celsius (1°C).* The amount of heat transferred depends upon the mass of the sample. To compare different materials, therefore, the heat capacities of equal masses, or *specific heats,* must be used. Specific heat *is the amount of heat energy required to raise the temperature of 1 g of a substance by one degree Celsius (1°C).* Values of specific heat can be given in units of joules/gram · degree Celsius (J/g°C) or calories/gram · degree Celsius (J/g · °C). Table 2-6 gives the specific heats of some common substances. The specific heat of a substance can be determined by measuring the amount of heat energy gained or lost by a given mass that undergoes a measured temperature change. The equation is

STUDY HINT

Note how the factor label method is used in Sample Problem 2-4.

TABLE 2-6   SPECIFIC HEATS OF SOME COMMON SUBSTANCES

Substance	Specific Heat $J/g \cdot °C$	Substance	Specific Heat $J/g \cdot °C$
Water (liquid)	4.184	Carbon, graphite	0.720
Water (ice)	2.092	Carbon, diamond	0.502
Water (steam)	2.013	Iron	0.444
Ethyl alcohol	2.452	Copper	0.385
Acetic acid	2.048	Silver	0.237
Sugar	1.250	Mercury	0.139
Aluminum	0.899	Gold	0.129
Salt (sodium chloride)	0.860	Lead	0.158

$$\text{Specific heat} = \frac{\text{heat lost or gained (J or cal)}}{\text{mass (g)} \times \text{change in temperature (°C)}}$$

or, in abbreviated form,

$$c = \frac{q}{m \times \Delta t}$$

where c is the specific heat, q is the heat lost or gained, m is mass, and Δt represents the change in temperature

$$\Delta t = t_f - t_i = \text{final temperature} - \text{initial temperature}$$

Rearranging the equation for specific heat gives an equation that can be used to find the quantity of heat gained or lost with a change in temperature. Heat lost or gained (J or cal) =

$$c \text{ (J/g} \cdot \text{°C)} \times \text{mass (in g)} \times \text{temperature change (°C)}$$

or

$$q = c \times m \times \Delta t$$

Sample Problem 2.4

A 4.0=g sample of glass was heated from 0°C to 41°C, a temperature increase of 41°C, and was found to have absorbed 32 J of heat.
(a) What is the specific heat of this type of glass?
(b) How much heat did the same glass sample gain when it was heated from 41°C to 70.°C?

Solution
(a) The specific heat (c) of the glass is found from the equation above.

$$c = \frac{q \text{ (J)}}{m \text{ (g)} \times \Delta t \text{ (°C)}} = \frac{32 \text{ J}}{4.0 \text{ g} \times 41\text{°C}} = 0.20 \text{ J/g} \cdot \text{°C}$$

(b) The rearranged specific-heat equation is used to find the heat gained when the glass was heated further.

$$q = c \text{ (J/g} \cdot \text{°C)} \times m \text{ (g)} \times \Delta t \text{ (°C)}$$
$$= 0.20 \text{ (J/g} \cdot \text{°C)} \times 4.0 \text{ g} \times (70.\text{°C} - 41\text{°C})$$
$$= 0.20 \text{ (J/g} \cdot \text{°C)} \times 4.0 \text{ g} \times 29\text{°C} = 23 \text{ J}$$

Practice Problem
1. Determine the specific heat of a material if a 12-g sample absorbed 48 J as it was heated from 20°C to 40°C. *(Ans.)* 0.20 J/g.°C

Section Review

1. Distinguish between temperature and heat and state the SI units for each.
2. Give the names of the two temperature scales that are commonly used in chemistry.
3. Convert   *(a)* 30°C to K   *(b)* −20°C to K   *(c)* 200K to °C.
4. Convert   *(a)* 110 cal to J   *(b)* 22 kJ to cal.
5. How much heat would be absorbed by 75 g of iron when heated from 22°C to 28°C

1. Temperature is a measure of the average kinetic energy of the particles in a sample of matter. Heat is energy transferred between two systems at different temperatures. The SI unit for temperature is the kelvin (K); that for heat is the joule (J).
2. Celsius and Kelvin
3. *(a)* 303K *(b)* 253K *(c)* −73°C
4. *(a)* 460 J *(b)* 5300 cal
5. 200 J

2.3 Using Scientific Measurements

If a reported measurement is to be useful, it must include some indication of its uncertainty. The complete expression of a measured quantity must include the number value and the unit and also show how reliable the number is. This section discusses how measurements are reported and used so that the reliability of the numbers is understood by everyone.

Accuracy and Precision

The quality of a measurement depends on the measuring instrument and the skill of the person making the measurement. Uncertainty in measurements can result from limitations in accuracy or limitations in precision. The limitations may be those of the instrument or the experimenter, or both.

Accuracy *refers to the closeness of a measurement to the true or accepted value of the quantity measured.* An electronic balance that is often calibrated with a standard mass is likely to be more accurate than an old, mechanical balance that has been dropped many times.

Precision *refers to the agreement among the numerical values of a set of measurements of the same quantity made in the same way.* A chemist who frequently carries out a complex experiment is likely to have more precise results than someone just learning the experiment.

One way to visualize the difference between accuracy and precision is with some darts and a dartboard, as shown in Figure 2-10. The center of the bull's-eye represents the true or accepted value. The closer a dart is to the bull's-eye, the more accurate the throwing of the dart. The closer the darts in a group are to each other, the more precise that series of throws. A good darts player might produce a result like that in Figure 2-10a—both accurate and precise. On the other hand, the group of darts shown in Figure 2-10b was thrown with good precision, but poor accuracy. The player who threw the darts shown in Figure 2-10c had neither good accuracy nor good precision.

Suppose it is necessary to determine the density of a sample of chloroform ($CHCl_3$) at 20°C. Upon repeated measurements with the same balance and volumetric flask, one chemist obtained values of 1.495 g/mL, 1.476 g/mL, and 1.485 g/mL for the density of the sample. These measurements vary widely; the precision is poor. Uncertainty due to poor

SECTION OBJECTIVES

- Distinguish between accuracy and precision.
- Determine the number of significant figures in measurements.
- Perform mathematical operations involving significant figures.
- Use scientific notation to write numbers and carry out arithmetical operations.

Figure 2-10 The center of the dartboard is the "accurate" location for a dart thrown at the board. The three groups of darts illustrate the distinction between accuracy and precision.

precision may result from limitations of either the measuring devices or the experimenter's ability to use those devices, or from both of these factors.

Another chemist repeated the measurements three times using the same equipment as the first chemist and found densities of 1.487 g/mL, 1.490 g/mL, and 1.488 g/mL. This group of measurements has good precision. Comparing these values with the accepted value of 1.489 g/mL for chloroform at 20°C shows that this group of measurements also has good accuracy.

Significant Figures

Suppose you were asked to measure the length of the nail shown in Figure 2-11 with the ruler drawn next to it. How long is the nail? It is longer than 3.6 cm, but how much longer? The tip of the nail appears to be less than halfway between 3.6 and 3.7, meaning that the nail is less than 3.65 cm long. Is it 3.61, 3.62, 3.63, or 3.64 cm long? The third digit in the length must be estimated. This estimated digit tells anyone reading the measurement that the nail is longer than 3.6 cm.

When scientists record and report the numerical values of their measurements, they want to indicate as much information as possible. They do not, however, want to give more information than is actually available and dependable. The compromise between these two goals is made by reporting only significant figures. *The **significant figures** in a measurement consist of all digits known with certainty plus one final digit, which is uncertain or estimated.*

In your study of chemistry, you will often make use of rulers, thermometers, graduated cylinders, or other instruments that have linear, graduated scales. For such measuring devices, the location of the estimated digit depends upon the smallest subdivision of the scale that appears on the device. An illustration of correctly reported measurements using thermometers with three different scales is shown in Figure 2-12. In each case, the final digit is an estimate of a digit between graduations on the scale. You can see that in correctly reported measurements, all digits retained are significant, but the last digit is uncertain.

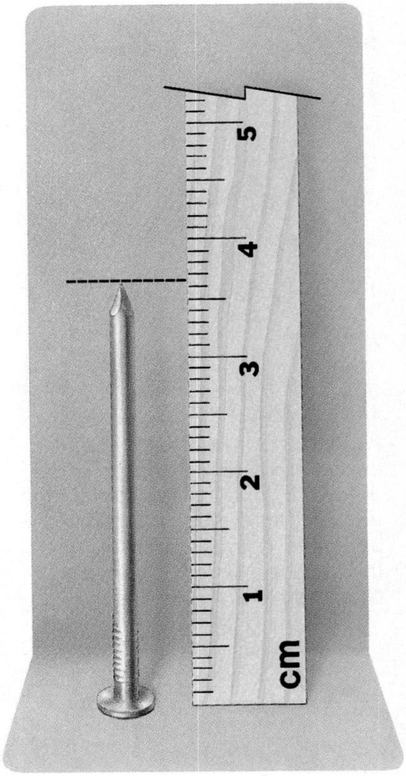

Figure 2-11 Measuring the length of a nail.

It is unfortunate that the term *significant* was chosen. It leads students to believe that *nonsignificant* digits are not important, when, in fact, they indicate the size of the number. For example, 600 000 000.

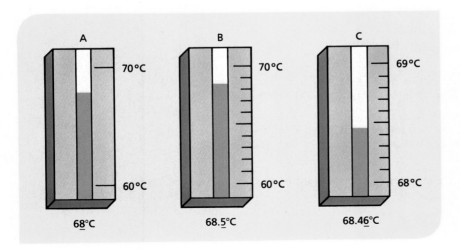

Figure 2-12 Thermometer A has 10-degree calibrations, thermometer B has 1-degree calibrations, and thermometer C has 0.1-degree calibrations. The temperature can therefore be read from thermometer A to two significant figures, from thermometer B to three significant figures, and from thermometer C, at temperatures higher than 10 degrees, to four significant figures.

Figure 2-13 An electronic balance with a digital scale readable to 1 mg. Just place an item on the pan in such a balance and read its mass to 1 mg in an instant. The readout provides the final, estimated digit.

As a student of chemistry, you will use significant figures in three different ways:

1. You will make measurements and report them to others. Here, the number of significant figures is easy to determine. It will usually be the number of digits you are able to read with certainty from the measuring device, plus one digit that you estimate. Figure 2-13 shows a digital readout, which requires no estimate.
2. You will interpret measurements reported by others.
3. You will need to keep track of significant figures when measurements are used in calculations.

The following sections explain how to deal with the kinds of situations described in items 2 and 3.

Determining the number of significant figures Determining the number of significant figures in the number 4.643 is easy—just count the digits. This number has four significant figures. Some rules are needed to count significant figures in numbers containing zeros because the zeros can be significant or not, depending on their location. Table 2-7 gives the complete set of rules for counting significant figures, together with examples of the use of each rule.

TABLE 2-7 RULES FOR DETERMINING SIGNIFICANT FIGURES INVOLVING ZEROS

Rule	Examples of Measurements	Number of Significant Figures
1. All nonzero digits are significant.	438 g	3
	26.42 m	4
	1.7 cm	2
	0.653 L	3
2. All zeros between two nonzero digits are significant.	506 dm	3
	10 050 mL	4
	900.43 kg	5
3. Zeros to the right of a nonzero digit, but to the left of an understood decimal point, are not significant. If such zeros are known to have been measured, however, they are significant and should be specified as such by inserting a decimal point to the right of the zero.	4830 km	3
	60 g	1
	4830. L	4
	60. K	2
4. In decimals less than one, zeros to the right of a decimal point that are to the left of the first non-zero digit are never significant. They are simply placeholders.	0.06 g	1
	0.0047 L	2
	0.005°C	1
5. In decimals less than 1, the zero to the left of the decimal is never significant. It is there to make sure the decimal point is not overlooked. overlooked.	0.8 g	1
6. All zeros to the right of a decimal point and to the right of a nonzero digit are significant.	8.0 dm	2
	16.40 g	4
	35.000 L	5
	1.60 sec	3

Sample Problem 2.5

How may significant figures are contained in each of the following measurements? Which significant-figure rule from Table 2-7 applies to each digit? *(a)* 37.4 *(b)* 6070. dm *(c)* 0.009 03 km *(d)* 0.0540 cm

Solution

(a) With no zeros in the number, rule 1 applies to each digit.

$$\begin{array}{ll} 111 & \text{Rule} \\ 37.4\,\text{g} & \text{three significant figures} \end{array}$$

(The numbers of the rules that apply to each digit are written above the digits.)

(b) The first zero is between two nonzero digits. The second zero is shown to be significant by the decimal point.

$$\begin{array}{ll} 1213 & \text{Rules} \\ 6070.\,\text{dm} & \text{four significant figures} \end{array}$$

(c) The first three zeros are not significant because they precede the first nonzero digit. The next zero is significant because it is between nonzero digits.

$$\begin{array}{ll} 5\,441\,21 & \text{Rules} \\ 0.009\,03\,\text{km} & \text{three significant figures} \end{array}$$

(d) The first two zeros are not significant because they precede the first nonzero digit; the last zero has been added to show uncertainty and is significant

$$\begin{array}{ll} 5411 & \text{Rule} \\ 0.0540 & \text{three significant figures} \end{array}$$

Practice Problems

1. Determine the number of significant figures in each of the following:
   (a) 605.03 g *(Ans.)* 5   *(c)* 0.006 03 mL *(Ans.)* 3   *(e)* 450 m *(Ans.)* 2   *(g)* 300. dm *(Ans.)* 3.
   (b) 0.8030 L *(Ans.)* 4   *(d)* 20.00 cm *(Ans.)* 4   *(f)* 0.000 070 kg *(Ans.)* 2   *(h)* 350.0 K *(Ans.)* 4.
2. Write "two hundred meters" in numerical form with the indicated number of significant figures: *(a)* 1 *(Ans.)* 200 m   *(b)* 3 *(Ans.)* 200. m   *(c)* 4 *(Ans.)* 200.0 m.

Rounding off   The measurements to be used in a calculation often contain different numbers of significant figures. The answer must not have more significant figures than are justified by the measurements. Especially if you use an electronic calculator, the number of digits in your answer often suggests a greater degree of certainty than really exists. For example, the density of an object with a mass of 1.5 g and a volume of 4 mL can be calculated using the formula for density:

$$\text{density} = \frac{\text{mass}}{\text{volume}} = \frac{1.5\,\text{g}}{4\,\text{mL}} = 0.375\;\text{g/mL}$$

This answer, although arithmetically correct, contains three digits that, according to the rules, would be interpreted as significant. The initial measurements of 1.5 g and 4 mL contained, respectively, two and one significant figures. In division and multiplication, the answer must not have

Figure 2-14   Rounding off is especially important when using an electronic calculator. It would be misleading to report all digits in the calculator answer to 96 divided by 7.

more significant figures than the number in the calculation with the fewest significant figures. The answer of 0.375 given by the calculator must therefore be rounded off to only one significant figure, or 0.4 g/mL.

In order to reflect the true uncertainty of calculations with measurements, the number of significant figures in the answers should be determined according to specific rules discussed in the next section. First, however, it will be helpful to review the rules for rounding off numbers. These rules are given in Table 2-8. Note that the procedure for rounding off is governed by the digit to the immediate right of the last digit to be retained. The use of these rules is illustrated in Sample Problem 2.7. Rounding off should be done only when all calculations are finished. Otherwise, a significant error may be introduced.

TABLE 2-8   RULES FOR ROUNDING OFF NUMBERS

If the digit immediately to the right of the last significant digit you want to retain is:	Then the last significant digit should:	Example (each rounded off to 3 significant figures):
1. Greater than 5	Be increased by 1	42.68 g → 42.7 g
2. Less than 5	Stay the same	17.32 m → 17.3 m
3. 5, followed by nonzero digit(s)	Be increased by 1	2.785 1 cm → 2.79 cm
4. 5, not followed by nonzero digits(s), and preceded by an odd digit	Be increased by 1	4.635 kg → 4.64 kg (since 3 is odd)
5. 5, not followed by nonzero digits(s), and the preceding significant digit is even	Stay the same	78.65 mL → 78.6 mL (since 6 is even)

Sample Problem 2.6

Round off each of the following measurements to the indicated number of significant figures:

(a) 35.27 g to 3 significant figures
(b) 0.414 kL to 2 significant figures
(c) 87.257 dm to 3 significant figures
(d) 1.35K to 2 significant figures
(e) 6250 cm to 2 significant figures

Solution
(a) 35.27 g is rounded to 35.3 g because 7 is a number larger than 5 (Rule 1)
(b) 0.414 kL is rounded to 0.41 kL because the final 4 is a number less than 5 (Rule 2).
(c) 87.257 dm is rounded to 87.3 dm because the 5 is followed by a nonzero digit (Rule 3).
(d) 1.35 K is rounded to 1.4 K because the 5 is preceded by an odd number (Rule 4).
(e) 6250 cm is rounded to 6200 cm because the 5 is preceded by an even number (Rule 5). Note that zeros, though not significant, must be used here to retain the size of the number.

Practice Problems
1. Round off each of the following measurements to the indicated number of significant figures:
   (a) 6.42 g to 2 significant figures (Ans.) 6.4 g   (b) 7.535 mL to 3 significant figures (Ans.) 7.54 mL   (c) 4.681 cm to 2 significant figures (Ans.) 4.7 cm   (d) 56.45 kg to 3 significant figures (Ans.) 56.4 kg
2. Round off 279.55 m to the indicated number of significant figures:
   (a) 2 (Ans.) 280 m   (b) 3 (Ans.) 280. m   (c) 4 (Ans.) 279.6 m   (d) 1 (Ans.) 300 m.

Addition and subtraction involving significant figures   If 3 000 000 (1 significant figure) and 4 192 378 (7 significant figures) are added, the result is 7 192 378. This answer indicates 7 significant figures. All the zeros in the first addend, 3 000 000, are not significant and are therefore uncertain. *In addition or subtraction, the arithmetic result should be rounded off so that the final digit is in the same place as the leftmost uncertain digit.*

For example, in the following subtraction (the uncertain digit in each measurement is underlined)

$$213.6\underline{7}$$
$$\underline{9\underline{8}} \quad \text{leftmost uncertain digit}$$
$$115.\underline{6}7, \quad \text{rounded to } 11\underline{6}$$

the final digit in the answer must be in the ones place because in the numbers subtracted the leftmost uncertain digit is the 8 in 98.

Multiplication and division involving significant figures   *The arithmetic product or quotient should be rounded off to the same number of significant figures as in the measurement with the fewest significant figures.* Suppose you want to find the following product: 12.0 cm × 4.3 cm. The answer should have two significant figures, the same number as in 4.3:

$$12.0 \text{ cm} \times 4.3 \text{ cm} = 51.6 \text{ cm}^2, \text{ rounded to } 52 \text{ cm}^2$$

Sample Problem 2.7

Perform each of the following calculations and express the answers with the correct number of significant figures.   *(a)* 6.43 mL + 2.015 mL   *(b)* 2.50 g / 0.04 cm³.

Solution

(a) $6.4\underline{3}$

       $+ \ 2.01\underline{5} \text{ mL}$

       $8.\underline{4}45 \text{ mL}, \quad$ rounded to $8.\underline{4}4$ mL

(b) $\dfrac{2.50 \text{ g}}{0.04 \text{ cm}^3} = 62.5 \text{ g/cm}^3$, rounded to $\underline{60} \text{ g/cm}^3$

Practice Problems

Perform the following calculations and express answers in the correct number of significant figures:

1. 6.821 m + 2.0 m + 0.45 m *(Ans.)* 9.3 m
2. 107.38 L − 65 L *(Ans.)* 42 L
3. 26.50 dm × 0.062 dm *(Ans.)* 1.6 dm²
4. 1.30 g / 0.02 cm³ *(Ans.)* 60 g/cm³

Exact conversion factors   Certain conversion factors do not limit the number of digits in the answer to a calculation. Consider the relationships

$$7 \text{ days} = 1 \text{ week} \qquad 100 \text{ cm} = 1 \text{ m}$$

There is no uncertainty in either of these statements because they are the result of definitions, not measurements. There are *exactly* 7 days in 1 week and *exactly* 100 cm in 1 m. *Exact conversion factors* have no uncertainty and do not limit the number of digits in a calculation. Exact numbers can be considered to have an unlimited number of significant figures. For example, in converting 7866 cm to meters

$$7866 \text{ cm} \times \frac{1 \text{ m}}{100 \text{ cm}} = 78.66 \text{ m}$$

STUDY HINT

Conversion factors based on units modified by SI prefixes are all exact conversion factors.

STUDY HINT
1 cal = 4.184 J is an exact conversion factor.

STUDY HINT
In algebra: $2a^5 + 6a^5 = 8a^5$.

the answer correctly has four significant digits because it is not limited by the 100-cm factor.

Suppose that, by counting, you find that there are 25 desks in a classroom. The conversion factor 25 desks/classroom would also be exact. Exact conversion factors result from definitions, or counting.

Scientific Notation

In science, numbers can range from very small to very large, depending on what is being measured. A very large number that is extremely important in chemistry, called *Avogadro's number*, is

$$602\ 252\ 000\ 000\ 000\ 000\ 000\ 000$$

The mass of an electron, on the other hand, is a very small number:

$$0.000\ 000\ 000\ 000\ 000\ 000\ 000\ 000\ 000\ 000\ 910\ 7\ \text{kg}$$

Because such numbers are tedious to read and write, scientists use a shorthand technique called scientific notation.

In **scientific notation**, *numbers are written in the form*

$$M \times 10^n$$

where M, the coefficient, is a number greater than or equal to 1 but less than 10 and n, the exponent, is an integer. For example, the speed of sound in water is about 1500 m/s. Written in scientific notation, this quantity is

$$1500\ \text{m/s} = (1.5 \times 1000\ \text{m/s}) = 1.5 \times 10^3\ \text{m/s}$$

An influenza virus is about 0.000 12 mm in diameter. In scientific notation, this quantity is

$$0.000\ 12\ \text{mm} = 1.2 \times \frac{1}{10\ 000}\ \text{mm} = 1.2 \times 10^{-4}\ \text{mm}$$

Avogadro's number and the mass of an electron are written in scientific notation as $6.022\ 52 \times 10^{23}$ and 9.107×10^{-31}, respectively.

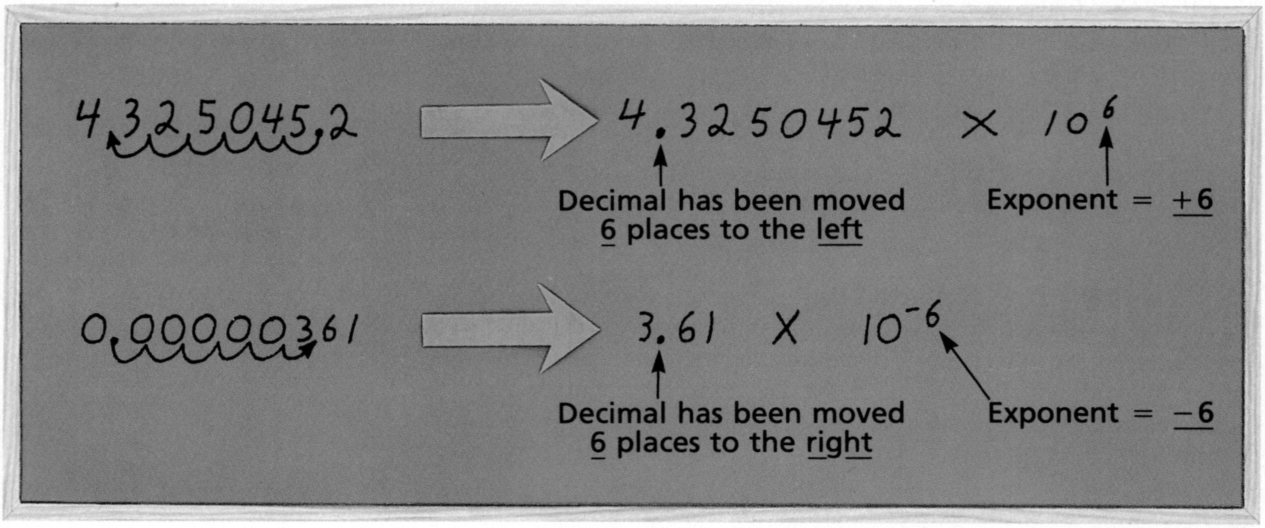

4,325,045.2
4.3250452 × 10^6
Decimal has been moved 6 places to the left
Exponent = +6
0.00000361
3.61 × 10^-6
Decimal has been moved 6 places to the right
Exponent = −6

Figure 2-15 An illustration of how to express numbers in scientific notation. The number of places the decimal point must be moved gives the exponent for 10.

Scientific notation makes it easy to indicate the correct number of significant figures. In $M \times 10^n$, all digits, both zero and nonzero, expressed in M, are significant. For example, notice that the volume, 10 050 mL, in Table 2-7 is assumed to have four significant figures as indicated by the scientific notation: 1.005×10^4 mL. If this measurement was known to have five significant figures, it would be written $1.005\ 0 \times 10^4$ mL. The distance from the earth to the sun is 149 740 000 km and has five significant figures. In scientific notation, it is $1.497\ 4 \times 10^8$ km. Rounded·off to three significant figures, the distance is 1.50×10^8 km.

Refer to Figure 2-15 as you read how to change a number from the long form to scientific notation:

1. Determine M by moving the decimal point in the original number to the left or to the right so that only one nonzero digit remains to the left of it.

2. Determine n by counting the number of places that the decimal point was moved. If moved to the left, n is positive; if moved to the right, n is negative.

To express 100 g as a two-significant-digit number, one would have to use scientific notation. As written, 100 g has only one significant figure. If one includes a decimal point, 100. g, then the number has 3 significant figures. To show two significant figures one would have to write it as 1.0×10^2 g.

Sample Problem 2.8

Write each of the following numbers in scientific notation. *(a)* 85 000 000 *(b)* 0.000 9

Solution
(a) The decimal point in 85 000 000 must be moved 7 places to the left.

$$85\ 000\ 000 = 8.5 \times 10^7$$

(b) In 0.0009, the decimal point must be moved 4 places to the right, so the sign of the exponent must be negative.

$$0.000\ 9 = 9 \times 10^{-4}$$

Practice Problems
Write each of the following numbers in scientific notation:

1. 74 000 *(Ans.)* 7.4×10^4
2. 0.000 005 *(Ans.)* 5×10^{-6}
3. 30 000 000 000 *(Ans.)* 3×10^{10}
4. 864 000 *(Ans.)* 8.64×10^5
5. 0.000 602 *(Ans.)* 6.02×10^{-4}
6. 4700 $\times 10^5$ *(Ans.)* 4.7×10^8

To write out a number in the usual long form that is given in scientific notation, the above procedure is reversed.

Sample Problem 2.9

Write each of the following numbers in the usual long form. *(a)* 7×10^4 *(b)* 2.31×10^{-7}.

Solution
(a) The positive exponent shows that the decimal point must be moved to the right. Changing this to long form gives

$$7 \times 10^4 = 70\ 000$$

(b) The negative exponent shows that the decimal point must be moved to the left to give a number smaller than 1. Moving the decimal place to the left seven places gives

$$2.31 \times 10^{-7} = 0.000\ 000\ 231$$
$$765\ 432\ 1$$

Note that the zero before the decimal point is not included in counting the zeros that must be added.

Practice Problems

Express each of the following numbers in the usual, long form:

(1.) 5.3×10^4 *(Ans.)* 53 000   *(2.)* 7×10^{-5} *(Ans.)* 0.000 07   *(3.)* 4.21×10^{10} *(Ans.)* 42 100 000 00

STUDY HINT

In algebra:
$2a^7 \times 3a^{10} = 6a^{17}$.

Stress the difference in handling exponents when adding and subtracting and when multiplying and dividing.

STUDY HINT

In algebra: $\dfrac{8a^7}{2a^5} = 4a^2$

STUDY HINT

To restate numbers in scientific notation: if M is made larger, the exponent n becomes smaller; if M is made smaller, the exponent n becomes larger. The size of the change in n is equal to the number of places the decimal point is moved.

Mathematics involving scientific notation   Because many measurements in science are expressed in scientific notation, arithmetic operations involving such numbers are frequently necessary. The rules governing such operations are the same as the rules for exponents that you learned when you studied algebra.

In each case, the M part of the answer must be rounded to the correct number of significant figures. Then, if necessary, the decimal point can be moved and the exponent changed so that M is greater than or equal to 1 and less than 10.

1. Addition and subtraction of expressions in scientific notation can be performed only if the expressions have the same exponent. The coefficients themselves can then be added or subtracted, but the exponent stays the same. If the initial expressions do not have the same exponent, they must first be rewritten so that they do have the same exponent. Consider the following example:

$$\begin{array}{lll} 6.3 \times 10^4 & \text{is written as} & 0.63 \times 10^5 \\ +2.1 \times 10^5 & & +2.1 \times 10^5 \\ \hline & & 2.73 \times 10^5 \end{array}$$
$$\text{rounded to } 2.7 \times 10^5$$

2. In the multiplication of expressions involving scientific notation, the coefficients themselves are multiplied and the exponents are added. Thus $(4.5 \times 10^8) \times (3.17 \times 10^5) = 14.265 \times 10^{13}$, which is rounded to 14×10^{13}, and restated as 1.4×10^{14}. Note that, to restate the number, the decimal point was moved to the left one place and the exponent was increased by 1.

3. In the division of expressions involving scientific notation, the coefficients themselves are divided and the exponent in the denominator is subtracted from that in the numerator. As shown in the second of the following two examples, the exponent can change sign in division.

$$\frac{6.85 \times 10^7}{2.0 \times 10^3} = 3.425 \times 10^4 \text{, rounded to } 3.4 \times 10^4$$

$$\frac{9.9 \times 10^5}{3.0 \times 10^{10}} = 3.3 \times 10^{-5}$$

Sample Problem 2.10

Perform the indicated operations. Express all answers in scientific notation and with the correct number of significant figures. *(a)* $(3 \times 10^7) + (5 \times 10^6)$ *(b)* $(4.3 \times 10^8) \times (2.51 \times 10^{-14})$ *(c)* $(6.81 \times 10^{10}) \div (1.2 \times 10^{-4})$.

Solution

(a) Before performing the indicated addition, both expressions must be written in terms of the same exponent. For example, 5×10^6 can be rewritten as 0.5×10^7. Then $(3 \times 10^7) + (0.5 \times 10^7)$ $= 3.5 \times 10^7 = 4 \times 10^7$.

(b) The coefficients must be multiplied and the exponents added:

$(4.3 \times 10^8) \times (2.51 \times 10^{-14}) = 10.793 \times 10^{-6}$, rounded to 11×10^{-6} and restated as 1.1×10^{-5}

(c) The coefficients must first be divided and the exponents then subtracted:
$(6.81 \times 10^{10}) \div (1.2 \times 10^{-4}) = 5.675 \times 10^{14}$, rounded to 5.7×10^{14}

Practice Problems

Perform the indicated operations and express all answers in scientific notation with the correct number of significant figures:
1. $(4.6 \times 10^5) + (3.2 \times 10^5)$ *(Ans.)* 7.8×10^5
2. $(5.7 \times 10^6) \times (2.8 \times 10^9)$ *(Ans.)* 1.6×10^{16}
3. $(7.7 \times 10^{-12}) \div (2.5 \times 10^5)$ *(Ans.)* 3.1×10^{-17}

Section Review

1. A student records volume readings of 8.42 mL, 8.41 mL, and 8.40 mL. What can you say about the precision of the student readings?
2. How many significant figures are indicated in each of the following measurements? *(a)* 49.35 g *(b)* 104.07 cm *(c)* 0.0082 kL *(d)* 20.00 m *(e)* 150 mL *(f)* 0.050 60 kg *(g)* 200 K.
3. Round off 785.6553 cm to the indicated number of significant figures: *(a)* 5 *(b)* 3 *(c)* 2 *(d)* 4 *(e)* 1.
4. Perform each of the following calculations and express all answers in the correct number of significant figures: *(a)* 430.62 m + 6.1 m + 10.5300 m *(b)* 750 cm × 24.32 cm *(c)* 804.00 g / 20 cm³.
5. Write each of the following numbers in scientific notation: *(a)* 45 000 *(b)* 0.000 002 7 *(c)* 301×10^{21}.
6. Perform the following calculations. Express all answers with the correct number of significant figures and in scientific notation: *(a)* $(6.8 \times 10^5 \text{ g}) - (3.1 \times 10^4 \text{ g})$ *(b)* $(7.32 \times 10^4 \text{ dm}) \times (4.1 \times 10^{-10} \text{ dm})$ *(c)* $(9.05 \times 10^{-8} \text{ g}) \div (5.3 \times 10^{-3} \text{ mL})$.

1. These measurements are precise.
2. *(a)* 4 *(b)* 5 *(c)* 2 *(d)* 4 *(e)* 2 *(f)* 4 *(g)* 1
3. *(a)* 785.66 cm *(b)* 786 cm *(c)* 790 cm *(d)* 785.7 cm *(e)* 800 cm
4. *(a)* 447.2 m *(b)* 18 000 cm² *(c)* 40 g/cm³
5. *(a)* 4.5×10^4 *(b)* 2.7×10^{-6} *(c)* 3.01×10^{23}
6. *(a)* 6.5×10^5 g *(b)* 3.0×10^{-5} dm *(c)* 1.7×10^{-5} g/mL

2.4 Solving Quantitative Problems

Much of your work in chemistry will focus on solving quantitative problems. Many such problems will be word problems. Learning to solve a variety of problems with ease requires practice. The steps outlined below will help you analyze and solve even complicated word problems correctly.

SECTION OBJECTIVES

- Use the suggested four-step method to solve quantitative problems.
- Define and state equations for directly and indirectly proportional relationships.
- Plot graphs showing direct and indirect proportionality between two variables.

Four Steps for Solving Quantitative Problems

Step 1. Analyze
Read the problem carefully at least twice. Then list all measurements and other data *given* in the problem. Next, identify the unknown, with units.

Step 2. Plan
The route for solving a problem shows how the given information can be used to determine the unknown. To choose the route, decide which conversion factors or algebraic equations relate the given to the unknown. Then write out the complete setup needed to solve the problem, including units.

The route might consist of a single calculation using one conversion factor. It might use one equation, such as that for density, or it might consist of both equations and conversion factors, used together or in separate steps. In most Sample Problems in this book, we will summarize the route with arrows showing the pathway from the given to the unknown. (You may choose to do the same when you solve problems.) We will then explain the route and show the complete setup. The factor-label method is the most useful tool that we have for deciding how to set up the calculation. If units cancel to give the answer in desired units, the setup is probably correct.

Step 3. Compute
Substitute the given information, conversion factors, and other necessary values into the setup from Step 2. Cancel units, calculate the answer, and round off to the correct number of significant figures.

Step 4. Evaluate
After performing the calculations and obtaining an answer, evaluate the answer. To do so, first make an estimate of the answer's expected value. Is the answer close to your estimate of what it should be? Is the answer realistic and reasonable? If you solve the problem a second time, do you obtain the same answer? Has the setup allowed units to be canceled correctly so that your answer has the correct units? Does your answer have the correct number of significant figures? Have you answered all parts of the question?

Sample Problems 2-11 to 2-13 illustrate the use of the four-step problem solving method. This method will be applied to quantitative problems throughout the rest of this book.

Sample Problem 2.11

How many seconds are there in one year? Assume that there are exactly 365 days per year.

Solution

Step 1. Analyze   Given: 1 year
                      Unknown: seconds

Step 2. Plan   year$\longrightarrow$days$\longrightarrow$hours$\longrightarrow$minutes$\longrightarrow$seconds

Solving this problem requires using four conversion factors. Such a series of conversion factors can be combined in a single setup.

$$1 \text{ year} \times \frac{\text{days}}{\text{year}} \times \frac{\text{hours}}{\text{day}} \times \frac{\text{minutes}}{\text{hour}} \times \frac{\text{seconds}}{\text{minute}}$$

Sample Problem 2.11 *continued*

Step 3. Compute	$1 \text{ year} \times \dfrac{365 \text{ days}}{1 \text{ year}} \times \dfrac{24 \text{ hr}}{1 \text{ day}} \times \dfrac{60 \text{ min}}{1 \text{ hr}} \times \dfrac{60 \text{ s}}{1 \text{ min}} = 31\ 536\ 000. \text{ s or } 3.153\ 600\ 0 \times 10^7 \text{ s}$
Step 4. Evaluate	An answer of about 10^7 seems reasonable ($300 \times 20 \times 60 \times 60$ is approximately 2×10^7). Units have canceled to give seconds, the desired unit. Note that each of the conversion factors is based upon a definition and therefore the answer is not limited in number of significant figures.

Practice Problems
1. A clock gains 0.25 seconds per minute. How many seconds will the clock gain in exactly 180. days? *(Ans.)* 6.4800×10^4 s
2. How many kiloliters are contained in 62.5 cm³? *(Ans.)* 6.25×10^{-5} kL

Sample Problem 2.12

What is the volume of a piece of aluminum that has a mass of 22.5 g? The density of aluminum is 2.7 g/cm³.

Solution

Step 1. Analyze	Given: mass = 22.5 g aluminum, density = 2.7 g/cm³
	Unknown: volume of aluminum
Step 2. Plan	mass, density $\longrightarrow$ volume

The mass and volume of a specific substance are related by the density of that substance, which is mass per unit volume. To find volume, the equation for density must be rearranged:

$$d \ (\text{g/cm}^3) = \frac{m \ (\text{g})}{V \, (\text{cm}^3)} \qquad V \ (\text{cm}^3) = \frac{m \ (\text{g})}{d \ (\text{g/cm}^3)}$$

Step 3. Compute	$V = \dfrac{2.25 \text{ g}}{2.7 \text{ g/cm}^3} = 8.3 \text{ cm}^3$
Step 4. Evaluate	The value of the answer is reasonable compared to an estimated answer of 7 cm³ (21/3). Units have canceled as desired. The answer is correctly given with two significant figures, as in the factor 2.7 g.

Practice Problem
Find the mass of an object with a density of 1.8 g/cm³ and a volume of 4.5 cm³. *(Ans.)* 8.1 g

Sample Problem 2.13

What quantity of heat in calories is needed to increase the temperature of 35 g of copper by 7.2 °C.

Solution

Step 1. Analyze	Given: mass = 35 g copper
	temperature change = 7.2°C
	specific heat = 0.385 J/g · °C
	Unknown: heat gained or lost in calories

Step 2. Plan

$$m, \text{T}, c \longrightarrow q \text{ (joules)} \longrightarrow q \text{ (calories)}$$

Solving this problem requires using specific heat. The equation for specific heat must be rearranged and solved for heat. A conversion factor is also needed for converting joules to calories. The equation for specific heat can be multiplied by a unit conversion factor to convert joules into calories.

$$q = c \times T(°C) \times m\,(\text{g}) \times \frac{\text{cal}}{\text{J}}$$

Step 3. Compute

$$\frac{0.384 \text{ J}}{\text{g} \times °C} \times 7.2\ °C \times 35 \text{ g} \times \frac{1 \text{ cal}}{4.184 \text{ J}} = 23 \text{ cal}$$

Step 4. Evaluate

The answer is reasonable in comparison with an estimated value of 30 cal $(0.5 \times 8 \times 30 / 4)$. Units have canceled to give the answers in calories, as desired. The answer is correctly limited to two significant figures by the factors 7.2 and 35.

Practice Problem

What quantity of heat, in kilocalories, is required to raise the temperature of 22.5 g of lead from 18°C to 30.°C? The specific heat of lead is 0.158 J/g · °C. *(Ans.)* 10×10^{-3} kcal

Proportional Relationships

Solving quantitative problems often involves using proportional relationships between variables. *A* **variable** *is a quantity that can change in value.* For example, temperature is a variable.

Direct proportion *Two variables are* **directly proportional** *to each other if dividing one by the other gives a constant value.* In other words, directly proportional values have a constant ratio. For example, for a given substance, mass and volume are directly proportional to each other. The mass of a given substance divided by its volume is the density of that substance. The general equation for a direct proportion between two variables is

$$\frac{y}{x} = k$$

where x is one variable, y is the other variable, and the constant value of the ratio, known as the proportionality constant, is represented by k. A direct proportion can also be represented as

$$y = kx$$

In the equation for density, mass is equivalent to y, volume is equivalent to x, and k is equivalent to density:

$$\frac{\text{mass}}{\text{volume}} = \text{density} \quad \text{or} \quad \text{mass} = \text{density} \times \text{volume}$$

At 3.98 °C, the density of water is equal to 1.00 g/mL. Figure 2-16 gives some values of the mass and volume of different samples of water at 3.98 °C and a graph of these values. If the variables in a directly proportional relation are plotted, a straight line, or linear graph/is formed.

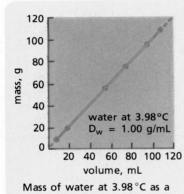

water at 3.98°C
$D_w = 1.00$ g/mL

Mass of water at 3.98 °C as a function of its volume

Figure 2-16 A graph of the mass of water at 3.98 °C as a function of its volume. At this temperature, the density of water has the constant value of 1.00 g/mL. Mass and volume (at a constant temperature) are directly proportional variables. All directly proportional variables give straight-line graphs.

Inverse propoertion *Two variable are* **inversely proportional** *to each other if their product has a constant value.* For example, the time it takes to travel a certain distance is inversely proportional to the speed at which you travel. If your speed increases, the time required to travel the fixed distance decreases.

The general equations for an inverse proportion between two variables are

$$xy = k \quad or \quad y = \frac{k}{x}$$

where x and y are the variables and k is the proportionality constant. The relationship between the speed and time it takes to travel a fixed distance is;

$$\text{speed} \times \text{time} = \text{distance} \quad or \quad \text{time} = \frac{\text{distance}}{\text{speed}}$$

Figure 2-17 gives some values of the time taken to travel a constant distance of 600 m at different speeds and a graph of these values. A graph of inversely proportional quantities is always of the shape shown. This shape is known as a hyperbola.

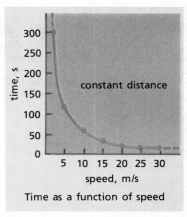

Figure 2-17 A graph of the time to travel a constant distance of 600 m as a function of the speed of travel. The time and speed of travel for a constant distance are inversely proportional variables.

Section Review

1. List and explain the four steps in problem solving.
2. Calculate the number of seconds in exactly one week.
3. What is the volume of a 12 ≅ g. (d = 19.3 g/cm³) piece of gold with a mass of 12 g?
4. An object travels at a speed of 7500 cm/s. How far will it travel in a day?
5. What type of relationship is illustrated between the variables A and B in each of the following: *(a)* As A increases, B decreases *(b)* As A increases, B increases.

Chapter Summary

- All chemists have agreed to use the SI measurement system.
- The SI system has seven fundamental units: meter, kilogram, second, kelvin, mole, ampere, and candela.
- The factor-label method is a valuable problem solving tool that treats units in calculations as if they are algebraic factors.
- Temperature is the average kinetic energy of the particles in a sample.
- All measurements are subject to some uncertainty.
- Accuracy refers to the closeness of a measurement to the true or accepted value. Precision refers to the agreement among the numerical values of a set of measurements made in the same way.
- Significant figures convey the reliability or certainty of measurements.
- Scientific notation is a shorthand way of writing numbers with many decimal places and also a convenient way of indicating significant figures.
- Solving quantitative problems is simpler when done in a methodical way. This book uses a four-step method. The steps are: *(1)* Analyze *(2)* Plan *(3)* Compute and *(4)* Evaluate.
- Many quantities, or variables, that you will encounter in chemistry are related to each other by either a direct proportion or an inverse proportion.

1. *Step 1. Analyze* In this step, the problem is carefully read, and the given and unknown variables are identified.
   Step 2. Plan Appropriate conversion factors or algebraic equations are selected to relate the given to the unknown.
   Step 3. Compute All necessary values are substituted into the setup and the answer is calculated.
   Step 4. Evaluate The appropriateness of the answer is determined.
2. $6.048\,00 \times 10^5$ s
3. $V = m/D = 0.62$ cm³
4. 6.5×10^8 cm
5. *(a)* Inverse proportion
   (b) Direct proportion

STUDY HINT
A variable with assigned values is referred to as an independent variable. A variable with determined values is referred to as a dependent variable. It is customary to plot the independent variable on the x axis and the dependent variable on the y axis.

Chapter 2 *Review*

Measurement and Solving Problems

Vocabulary

accuracy	liter
calorie	meter
conversion factor	precision
degree Celsius	second
density	SI
derived unit	scientific notation
directly proportional	significant figures
factor-label method	specific heat
fundamental unit	standards of
heat	measurement
heat capacity	temperature
inversely proportional	unit of
joule	measurement
kelvin	variable
kilogram	volume

Questions

1. What is a unit of measurement?
2. *(a)* What system of measurement is used by scientists around the world? *(b)* What abbreviation is used for the name of this system? *(c)* From what system were many of its units derived?
3. *(a)* What is meant by standards of measurement? *(b)* Why are standards necessary?
4. Write each of the following according to the international system for writing numbers:
   (a) 6500   *(b)* 78420   *(c)* 207891
   (d) 0.3500   *(e)* 0.920400   *(f)* 127.80000
   (g) 42600.3500
5. *(a)* What is meant by a fundamental unit?
   (b) List the seven SI fundamental units, identify the quantity measured by each, and write the corresponding symbol for each unit.
6. *(a)* What is the international standard for the meter?   *(b)* Give three examples of a distance that is approximately one meter in length.
7. Identify the SI unit(s) that would be most appropriate for measuring each of the following: *(a)* the length of a new pencil   *(b)* the width of your classroom   *(c)* the length of your arm   *(d)* the diameter of pencil

lead   *(e)* your height   *(f)* the distance from your home to the next town   *(g)* the distance from the earth to the sun   *(h)* the diameter of a bacterial cell.
8. *(a)* What is the international standard for the kilogram?   *(b)* What is an everyday approximation of the mass?
9. Identify the SI unit(s) that would be most appropriate for expressing the mass of each of the following: *(a)* a person   *(b)* a banana   *(c)* your daily vitamin C requirement   *(d)* a slice of cheese.
10. *(a)* What is the scientific basis for the SI unit of time?   *(b)* What is the common basis for this unit?
11. What is the factor-label method?
12. What is a conversion factor?
13. *(a)* What is a derived unit?   *(b)* Give two examples of derived units that are important in chemistry.
14. What is the derived SI unit and corresponding symbol for   *(a)* area   *(b)* volume?
15. *(a)* Define the term *volume*.   *(b)* What is the relationship between the SI unit for volume and that for length?
16. *(a)* What is a liter?   *(b)* What is the relationship between cubic centimeters and millimeters?
17. Identify the unit(s) that would be most appropriate for expressing the volume of each of the following: *(a)* a glass of milk   *(b)* the amount of water required to fill a bathtub   *(c)* a child's dosage of cough syrup   *(d)* a truckload of sand.
18. Why are the terms "light" and "heavy" misleading when describing an object's mass?
19. *(a)* What is meant by the density of a material?   *(b)* What formula can be used to find the density of an object?   *(c)* What are three SI or metric units that can be used to express density?
20. Why do temperatures usually accompany the density values of many substances in charts or tables?

21. *(a)* What is temperature? *(b)* How is temperature related to the energy of the particles in a sample of matter?
22. *(a)* What is heat? *(b)* What is the normal direction of heat flow?
23. What is the fundamental principle upon which the thermometer is based?
24. *(a)* What are the two fixed points that form the basis of the Celsius temperature scale? *(b)* What is meant by "one degree Celsius"?
25. Explain what is meant when we say that a banana contains 130 Calories.
26. What three factors determine the quantity of heat lost or gained by a sample of matter during a temperature change?
27. *(a)* What is meant by the heat capacity of a material? *(b)* What is specific heat?
28. What factors determine the quality of a measurement?
29. Identify and define the two limitations that contribute to uncertainty in measurements.
30. Describe each of the following measurements in terms of precision. *(a)* 458.43 m, 458.41 m, and 458.40 m *(b)* 462.78 m, 451.33 m, and 475.89 m *(c)* 468.49 m, 468.51 m, and 468.52 m
31. *(a)* What are significant figures in a measurement? *(b)* What determines the location of the estimated digit in making a measurement?
32. Determine the number of significant figures in each of the following measurements. Cite the rule that applies to each digit, in order, from left to right: *(a)* 65.42 g *(b)* 385 L *(c)* 0.14 mL *(d)* 709.2 m *(e)* 5006.12 kg *(f)* 400 dm *(g)* 260. mm *(h)* 0.47 cg *(i)* 0.0068 km *(j)* 7.0 cm^3 *(k)* 36.00 g *(l)* 0.0070 kg *(m)* 100.6040 L *(n)* 340.00 cm
33. Write "five hundred grams" in numerical form to indicate the following numbers of significant figures: *(a)* 1 *(b)* 3 *(c)* 4 *(d)* 6
34. Round off each of the following measurements to the indicated number of significant figures:
(a) 2.68 g to 2 significant figures
(b) 47.374 mL to 4 significant figures
(c) 8.35 mL to 2 significant figures
(d) 4.165 L to 3 significant figures
(e) 24 km to 1 significant figure
(f) 0.048 51 mg to 2 significant figures
(g) 0.063 50 mL to 2 significant figures

35. Round off 139.575 g to the indicated number of significant figures: *(a)* 2 *(b)* 3 *(c)* 4 *(d)* 5 *(e)* 1.
36. What is the rule governing the number of significant figures in calculations involving *(a)* addition and subtraction? *(b)* multiplication and division?
37. *(a)* What is meant by exact conversion factors? *(b)* Give three examples of such factors.
38. What is scientific notation?
39. List the four steps used in the suggested method of problem solving.
40. *(a)* What is a variable? *(b)* Give three examples of variables.
41. Distinguish between two variables that are directly proportional, as opposed to inversely proportional, in terms of *(a)* the definition of each *(b)* the general equation for each *(c)* the type of graph that illustrates each.

Problems

1. Express each of the following measurements in the SI units indicated: *(a)* 2 m into dm, cm, mm, km, and nm *(b)* 1.5 km into hm, dam, m , dm, cm, mm, μm, and nm *(c)* 48 cg into mg, dg, g, kg, and μg *(d)* 7.5 L into dL, cL, mL, and μL.
2. Rank the measurements within each set from largest to smallest: *(a)* 1.2 m, 750 cm, 0.005 km, 65 dm, and 2000 mm *(b)* 450 mg, 3.8 cg, 0.27 dg, 0.50 g, and 0.000 47 kg *(c)* 2.5 L, 22 cL, 13 dL, 870 mL, and 175 cm^3.
3. Find the total mass, in kilograms, of the following grocery purchases: 3 apples (125 g each), 5 bananas (115 g each), 2 pears (75 g each), 4 peaches (65 g each), 1750 g of melon, 5.5 hectograms (hg) of grapes.
4. If an automobile travels at a speed of 88 km/hr for 10 hours each day, what distance will it have covered at the end of five days?
5. Express each of the following in milliliters: *(a)* 30 L *(b)* 12 dL *(c)* 4.5 cL *(d)* 6.4 cm^3
6. Express the volume of a 22.4 L container in *(a)* milliliters *(b)* cubic centimeters.
7. Based on the given information, determine the density of each of the following samples: *(a)* mass = 240.00 g; volume = 60.0 cm^3 *(b)* mass = 65.48 kg; volume = 12.5 m^3 *(c)* mass = 8.25 g; volume = 3.6 mL.

8. Based on the given information, find the mass of each of the following samples:
   (a) density of lead = 11.3 g/cm^3; volume = 4.6 cm^3
   (b) density of tin = 7.31 kg/m^3; volume = 8.73 m^3
   (c) density of hydrogen = 0.090 g/L; volume = 24 L.
9. Based on the given information, find the volume of each of the following samples:
   (a) density of mercury = 13.6 g/mL; mass = 14.50 g   (b) density of zinc = 7.29 g/cm^3; mass = 28.73 g   (c) density of oxygen = 1.43 g/cm^3; mass = 12 kg.
10. Convert the following temperatures from the Celsius to Kelvin scales: (a) 0°C   (b) 27°C   (c) 72°C   (d) −50°C.
11. Convert the following temperatures from the Kelvin to Celsius scales: (a) 300K   (b) 350K   (c) 273K   (d) 150K.
12. Convert 25.0 calories to   (a) J   (b) kJ.
13. Convert 1.42 kJ to   (a) J   (b) cal   (c) kcal
14. Determine the specific heat of a substance if a 25 g sample absorbed 34 J as it was heated from 21°C to 33°C.
15. How much heat would be absorbed by 55 g of copper when heated from 24°C to 40°C.
16. Perform each of the following calculations and express all answers in the correct number of significant figures:
   (a) 4.07 g + 1.863 g
   (b) 3127.55 cm − 784.2 cm
   (c) 0.067 mL + 1.01 mL + 2.5 mL
   (d) 36.427 m + 12.5 m + 6.33 m
17. Perform each calculation and express all answers in the correct number of significant figures:
   (a) 9.40 cm × 2.6 cm
   (b) 8.08 dm × 5.3200 dm
   (c) 1.50 g / 2 cm^3
   (d) 0.084 21 g / 0.640 mL
   (e) 4.00 m × 0.020 m × 1.57 m
   (f) $\dfrac{21.50 \text{ g}}{4.06 \text{ cm} \times 1.8 \text{ cm} \times 0.905 \text{ cm}}$.
18. Write each of the following numbers in scientific notation: (a) 900   (b) 750 000
   (c) 93 000 000   (d) 0.000 082
   (e) 0.000 015 × 10^8 (f) 6500 × 10^7
   (g) 0.000 015 × 10^8 (h) 320 000 ×10^{-9}
   (i) 0.097 × 10^{-4}.

19. Write each of the following numbers in the usual, long form: (a) 5 × 10^3
   (b) 7.8 × 10^4 (c) (d) 3.81 × 10^{-4}
   (e) 4.04 × 10^7 (f) 6.09 × 10^{-9}.
20. Find each sum or difference. Express all answers in scientific notation and in the correct number of significant figures.
   (a) 2.6 × 10^2 + 4.1 × 10^2
   (b) 8.3 × 10^{-5} + 1.2 × 10^{-5}
   (c) 7.43 × 10^4 − 5.09 × 10^4
   (d) 9.30 × 10^{-6} − 1.23 × 10^{-6}
   (e) 1.2 × 10^3 + 3.6 × 10^4
21. Find each product or quotient. Express all answers in scientific notation and in the correct number of significant figures:
   (a) (3 × 10^5) (2 × 10^7)
   (b) (4.2 × 10^4) (3.8 × 10^3)
   (c) (5.21 × 10^{-6}) (2.4 × 10^{-4})
   (d) (6.02 × 10^{23}) (3.0 × 10^{-11})
   (e) (8.4 × 10^5) / (2 × 10^3)
   (f) (7.5 × 10^6) / (4 × 10^{-2})
22. A cyclist is able to maintain an average speed of 40.0 km per 1.0 hour. What distance could he travel in 5.5 hours? Use the suggested method of problem solving and label each step.
23. Use the suggested method of problem solving to determine the volume occupied by 8.50 g of cork. The density of cork is 0.24 g/cm^3.

Application Questions

1. (a) Give at least two advantages of SI over the English system used in the United States.
   (b) What are some of the considerations that would have to be taken into account if SI were to replace the English system?
2. Consider the following statement: Feathers are heavier than iron. Could such a statement be true?
3. Explain why your hands, though they may be good indicators of heat, are unreliable as indicators of temperature.
4. A cube of copper at 60°C is placed in water at 20°C. In terms of temperature and heat, explain what must occur in order for the copper and water to come to the same temperature.
5. (a) Why is mercury generally used in thermometers?   (b) Could water be used instead of mercury? Explain.
6. What is the major difference between heat capacity and specific heat?

7. *(a)* Why would it be misleading to express the final answer to the density problem below as 32. 43 g/cm^3? *(b)* What would be the correct density?

$$D = \frac{m}{V} = \frac{64.86 \text{ g}}{2.0 \text{ cm}^3} = 32.43 \text{ g/cm}^3$$

Application Problems

1. The speed limit on a highway is 65 km/h. *(a)* What is this speed in cm/h? *(b)* What is it in cm/s?
2. Find the volume, in cubic centimeters, of a rectangular aquarium that is 125 cm long, 0.750 m wide, and 500. mm deep. What is its volume in liters?
3. Determine the number of cubic centimeters in one cubic meter.
4. A piece of copper 12.00 cm long, 8.50 cm wide, and 0.50 cm thick has a mass of 457.00 g. What is its density?
5. Sulfuric acid has a density of 1.83 g/mL at room temperature. If 25.00 g of this acid is required for an experiment, what volume of the acid should be measured out?
6. A cube of tungsten has a mass of 521.10 g. If tungsten has a density of 19.3 g/cm^3, what are the dimensions of the cube?
7. A 3.10 cm^3 sample of an unknown element is found to have a mass of 32.50 g. Based on the information in the table of densities, which of the following elements could the unknown be: platinum, iron, gold, magnesium, silver, or copper?
8. Given the formula, $t(°C) = \frac{5°C}{9°F} [t(°F) - 32°F]$, convert 37.0°C (normal body temperature) to °F.
9. If 20.4 J of heat are required to raise the temperature of 6.67 g of an unknown substance from 35.°C to 42.0°C, determine which of the following would be the unknown by using the table of specific heats: water, ice, aluminum, copper, glass, iron, silver, or lead?
10. How much heat would be lost when a 45 g sample of lead is cooled from 38.0°C to 20.0°C?
11. A 20.0 g sample of silver initially at 28.5°C absorbs 46.5 J of heat. What will its final temperature be?
12. In astronomy, the non-SI unit frequently used to express intergalactic distance is the *light-year:* the distance that light travels in one year.

If light travels at a speed of 3.00×10^{10} cm/s, and a year is defined as exactly 365 days, determine each of the following: *(a)* the distance of a light-year in kilometers *(b)* the distance from earth to the sun in light-years if the sun is 1.50×10^8 km away.

13. A non-SI unit often used for expressing extremely small dimensions is the angstrom Å. It represents a distance of 1.00×10^{-8} cm. If a soap bubble has a thickness of 60. Å, express this dimension in *(a)* centimeters *(b)* meters *(c)* millimeters *(d)* micrometers.
14. The average adult daily requirement of riboflavin (vitamin B$_2$) is 2.00 mg. If the only source of riboflavin in a particular adult's diet is cheese that contains 5.2×10^{-6} g of riboflavin per gram of cheese, how many grams of cheese would have to be consumed daily to meet that requirement?
15. Construct a graph based on the following information with Celsius readings on the horizontal axis and the corresponding Kelvin readings on the vertical axis. *(a)* What type of relationship is illustrated? *(b)* Explain the meaning of this relationship.

Celsius scale (°C)	Kelvin scale (K)
0	273
27	300
50	323
100	373
−73	200
−100	173

Enrichment

1. Write a report that describes the history of the SI system.
2. For each room in your home, compile a list of the various items whose measurements are expressed in SI units. Include storage rooms, basements, laundry rooms, and garages. Compare your list with those of your classmates.
3. Prepare a presentation advocating the nationwide replacement of the English system of units with the SI system of units.
4. Trace the historical development of the mercury thermometer.
5. The meter is defined as the distance light travels in 1/299 792 458s. Why didn't scientists simply pick 1/3s?

CAREERS

Fire Chemist

If you're fascinated by fire and smoke, consider the employment opportunities at the National Institute of Science and Technology (NIST).

One major area of study at NIST is the heat and light released by the many chemical reactions in a fire. NIST fire chemists study these reactions to understand the kinetics of flames and how fire spreads. Smoke research, another area of study at NIST, is useful in establishing national safety standards for smoke detection.

To work as a laboratory technician at NIST, you need a bachelor's degree in a science field. NIST fire chemists who become project leaders usually have master's or doctoral degrees. Whether you major in engineering or science, the more chemistry, physics, and mathematics you

learn, the better equipped you will be to undertake fire and smoke research.

Science Writer

The science writer must combine scientific knowledge with creativity to compose material for science magazines, newspapers, and books that the nonscientist can understand and enjoy.

Many science writers work on a free lance basis, writing articles for newspapers and science magazines. Others work full time for book publishers, where they write and edit school textbooks or popular science books.

A significant portion of a science writer's time is spent doing research. Before writing about a subject, they read scientific journals and reports and consult with experts in the field. Once the research has been completed, science writers sometimes must prepare a number of drafts of the material until a satisfactory version is accepted by the publisher. Science writers must be able to produce interesting, concise prose; most have at least a bachelor's degree in English, journalism, or one of the sciences.

Cumulative *Review*

Unit 1 • Introduction to Chemistry and Matter

Questions

1. Define each of the following: *(a)* science *(b)* chemistry *(c)* scientific method *(d)* inertia *(e)* matter *(f)* energy *(g)* characteristic property *(h)* precipitate. (1.1)
2. List the five major areas of chemistry. (1.1)
3. Identify and describe the four main parts of the scientific method. (1.1)
4. Distinguish between each of the following pairs of terms: *(a)* qualitative and quantitative information *(b)* weight and mass *(c)* kinetic and potential energy *(d)* physical and chemical properties of matter *(e)* physical and chemical changes *(f)* reactants and products *(g)* exothermic and endothermic processes *(h)* heterogeneous and homogeneous matter *(i)* pure substances and mixtures *(k)* groups and periods on the periodic table *(j)* elements and compounds. (1.1, 2, 3, 4)
5. State each of the following: *(a)* the law of conservation of matter *(b)* the law of conservation of energy *(c)* the law of definite composition. (1.2)
6. List and define the three states of matter. (1.2)
7. Classify each of the following as a physical or chemical change: *(a)* freezing water *(b)* dissolving sugar in iced tea *(c)* making a tossed salad *(d)* burning wood *(e)* tearing paper *(f)* tarnishing silver. (1.2)
8. Identify three observable changes that generally indicate a chemical reaction has occurred. (1.2)
9. List and distinguish among the three general classes of elements. (1.4)
10. What are the fundamental SI units for length, mass, and time? (2.1)
11. Distinguish between *(a)* temperature and heat *(b)* accuracy and precision *(c)* direct and inverse proportion between two variables. (2.2)
12. List the four steps in the stepwise method of problems solving. (2.4)
13. Identify the SI unit(s) that would be most appropriate for measuring each of the following: *(a)* the thickness of a dime *(b)* the mass of a vitamin pill *(c)* the volume of a glass full of orange juice *(d)* your mass *(e)* the distance from New York City to Chicago *(f)* the volume of an olympic-size swimming pool. (2.1)
14. Round off 125.0965 cm to the indicated number of significant figures: *(a)* 6 *(b)* 5 *(c)* 3 *(d)* 2 *(e)* 4 *(f)* 1. (2.3)
15. Express each of the following in scientific notation: *(a)* 4500 *(b)* 370 000 *(c)* 0.053 *(d)* 0.000 081 *(e)* 650×10^4 *(f)* 206×10^{-7}. (2.3)
16. Write each of the following in the usual, long form: *(a)* 7.2×10^4 *(b)* 1.3×10^{-5} *(c)* 3×10^{10}. (2.3)

Problems

1. Express each measurement in the SI unit indicated: *(a)* 3.5 m in cm *(b)* 425 g in kg *(c)* 7.8 L in ml *(d)* 1.5 km in mm *(e)* 1.26 l in cm^3. (2.1)
2. Determine the density of an unknown substance if 15.0 grams of the material occupy a volume of 4.5 cm^3. (2.1)
3. Find the volume of an object having a density of 8.62 g/cm^3 and a mass of 19.40 g. (2.1)
4. Convert each of the following temperature readings: *(a)* 20°C to K *(b)* −65°C to K *(c)* 350K to °C *(d)* 100K to °C. (2.2)
5. Convert 350.0 calories to joules. (2.2)
6. How much heat would be absorbed by 20.0 g of silver when heated from 20°C to 35°C? (2.2)
7. Solve each of the following:
   (a) $4.2 \times 10^3 + 2.9 \times 10^4$
   (b) $(3.12 \times 10^5)(8.1 \times 10^{-9})$
   (c) $\dfrac{(5.20 \times 10^{-7})(7.0 \times 10^4)}{1.2 \times 10^{-5}}$ (2.3)
8. A rectangular tank has a length of 2.40 m, a width 14.0 dm, and a height of 80.0 cm. *(a)* Find its volume in cm^3. *(b)* How many liters of water could this tank hold? (2.4)

Intra-Science
How the Sciences Work Together

The Process in Chemistry: The Portable Power of the Battery

What does a World War II U-boat have in common with a pacemaker and a satellite? They all need portable power. You cannot connect a ship, a person, or a satellite to a power cord. You need batteries.

Batteries provide portable, stored energy that is used to start cars and to track Arctic wolves with special collars containing radios. Batteries run wristwatches and store solar energy in spacecraft. Without batteries, there would be no flashlights, no portable radios, and no pocket calculators.

A battery is a direct-current (dc) voltage source that converts chemical, solar, thermal, or nuclear energy into electrical energy. The first source of electric current was also the first battery. It was invented in 1800 by the Italian scientist Alessandro Volta.

In a chemical battery—the kind used in most ordinary electrical devices—chemical energy is transformed into electrical energy through an oxidation–reduction (redox) reaction. In redox reactions the atoms of one reactant give up one or more electrons (and thus are oxidized) to the other reactant (which thus is reduced). The two reactants do not come into direct contact with each other, so the reactant being oxidized cannot give electrons directly to the reactant being reduced. Instead, electrons flow out of the battery from the reactant being oxidized through an electrically conducting external circuit, and then back into the battery to the reactant being reduced. It is the electron current in the external circuit that powers the portable radio and lights the flashlight bulb.

Primary cells, such as flashlight batteries and the tiny silver oxide discs used in watches and calculators, can produce electricity only until one of the reactants is used up. They must then be discarded. Secondary cells, often called storage batteries, can be recharged by passing an electric current through the cell in the opposite direction to that followed by the current when the cell was discharging. This reverses the oxidation–reduction reaction, regenerating the original reactants of the cell.

The Connection to Biology
The crucial observation that led to Volta's invention of the battery was made in 1786 by the Italian biologist Luigi Galvani. After accidentally discovering that static electricity (electricity produced by friction) caused a contraction in the leg muscles of a newly killed frog, Galvani found that the same contraction occurred when frogs were hung on an iron support by copper hooks. Galvani thought that this contraction was caused by "animal electricity" in the muscles. Volta did not accept this view. He believed that the electricity that caused the muscular contraction had been produced by the contact of the two dissimilar metals: iron and copper.

Experiments stimulated by this correct idea led to his invention of the battery, which is sometimes called the "voltaic cell" in his honor. For his part, Galvani is honored by the term "galvanize."

Today, the battery has more than repaid its debt to biology. Battery-powered cameras, television camcorders, and tape recorders record animal behavior. Lions,

tigers, and wolves have been fitted with collars containing battery-powered radio transmitters to better observe their behavior.

Biological cells act like miniature batteries. The double layer of lipids in a cell membrane separates aqueous solutions. Proteins embedded in the membrane conduct electrons from one solution to another. Tiny biological batteries that mimic the electrochemical action of cells are being tested. Sandwiched between two lipid layers, metals take the place of proteins. Biological batteries may one day be used in microcomputers that are made of organic molecules.

The Connection to Health Sciences

A wide variety of battery-powered devices are used to help people overcome medical problems. Wheelchairs, for example, are commonly powered by storage batteries. Tiny electric batteries that can provide power for up to ten years are used in pacemakers implanted in patients whose hearts do not beat properly. The pacemaker delivers electrical signals to the heart muscles so that they beat in the proper rhythm. Batteries power hearing aids that amplify sounds that are received by the inner ear. Recently, progress has been made toward the development of battery-powered devices that can electrically stimulate the auditory nerve of people who otherwise could not hear sounds, no matter how strongly they were amplified.

The Connection to Earth Science

Both storage batteries and solar-electric panels provide the power for the instruments in satellites that orbit the earth. Satellites monitor a wide variety of phenomena on the earth's surface, such as ocean currents, vegetation, weather patterns, and snow lines. Battery-powered strainmeters, tilt-meters, creepmeters, and other geophysical instruments monitor changes in the bedrock in earthquake zones to learn what conditions precede earthquakes. This may lead to an earthquake warning system. Storage batteries power underwater craft that carry oceanographers to the bottom of the ocean, where battery-powered light often reveals strange new species of animals or new geological features. Battery-powered in-

struments are carried into the upper atmosphere by balloons and rockets. They gather information that is used in weather forecasting and relay it to control stations by means of battery-powered transmitters.

Utility companies find batteries to be useful tools for "load leveling." Although many utility companies have a constant power source, such as hydroelectricity, the demand for power fluctuates. Energy can be stored by batteries during the night, when demand is low, and released during peak consumption. Utility companies also use batteries to store solar- and wind-generated electricity.

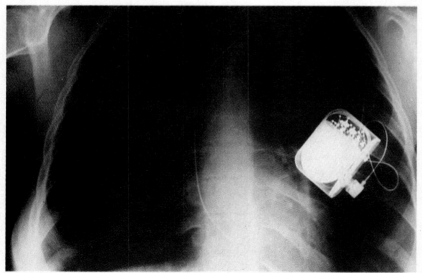

Unit

2

Chapter 3
Atoms: The Building Blocks
of Matter   70

Chapter 4
Arrangement of
Electrons in Atoms   96

Chapter 5
The Periodic Law   126

Chapter 6
Chemical Bonding   160

ORGANIZATION OF MATTER

I first felt the excitement of uncovering new knowledge in a high school course called "Science Research." We formed a group to study ocean pollution off the Southern California coast where I lived. I found that research could sometimes be difficult and frustrating, but there is no experience as satisfying as discovering something that wasn't known before.

In this century, we have learned a great deal about the structure of atoms and molecules. The new knowledge has enabled scientists to produce new forms of matter such as plastics and other synthetic polymers.

I use chemistry to investigate the structures of some of the complicated molecules that are at the center of life. Our experiments make use of a simple inorganic chemical reaction to make "pictures" of DNA molecules. My work would not be possible without the basic concepts of matter you will learn in this unit.

Thomas D. Tullius
Inorganic Chemist
The Johns Hopkins University

Computer-generated image of a DNA double helix

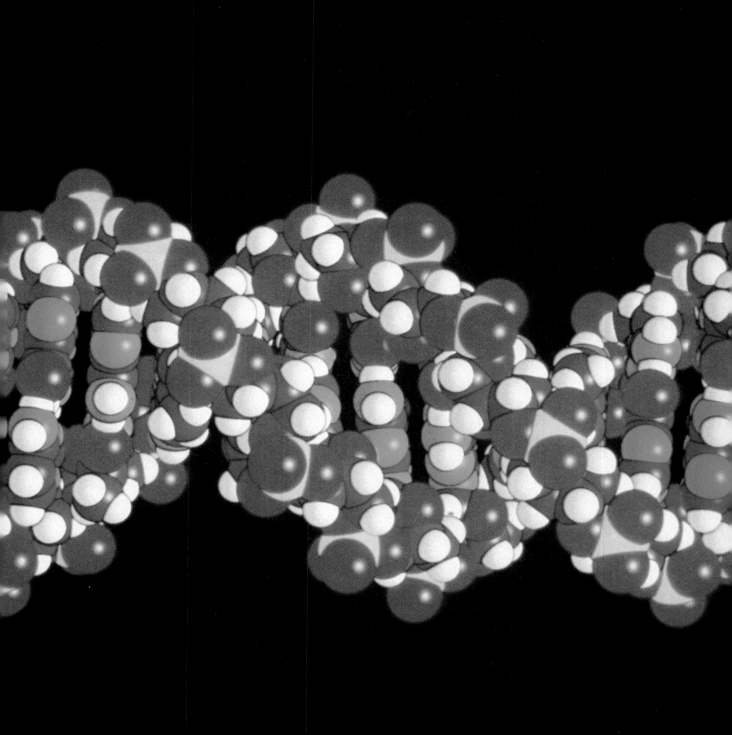

Chapter 3 Atoms: The Building Blocks of Matter

Chapter Planner

3.1 The Atom: From Philosophical Idea to Scientific Theory
- Desktop Investigation Experiment: 4
- Questions: 1, 5a
- Questions: 4, 5b
- Questions: 2, 3

Dalton's Atomic Theory
- Application Question: 1
- Problems: 2, 3

Law of Multiple Proportions
- Demonstration: 1
- Question: 5a
- Questions: 4, 5b

3.2 The Structure of the Atom
Discovery of the Electron
- Questions: 6, 7, 12, 13
- Questions: 8, 11
- Chemistry Notebook Questions: 9, 10

Discovery of the Atomic Nucleus
- Question: 14
- Application Questions: 2, 3
- Application Question: 4

Composition of the Atomic Nucleus
- Questions: 15, 16
- Questions: 17

The Isotopes of Hydrogen
- Questions: 18, 19
- Question: 20

The Sizes of Atoms
- Demonstration: 2
- Technology

3.3 Weighing and Counting Atoms
- Questions: 21, 23, 24
- Question: 22

Relative Atomic Masses
- Questions: 25, 26, 28
- Question: 27

Average Atomic Masses of Elements
- Demonstration: 3
- Questions: 29, 30
- Application Question: 6
- Application Problem: 1
- Application Problems: 2–4

The Mole, Avogadro's Number, and Molar Mass
- Demonstration: 4
- Questions: 31–33
- Problems: 1–7
- Application Question: 6
- Application Problems: 5–7
- Problems: 8, 9
- Application Question: 7
- Application Problem: 8
- Application Problems: 9–11

Teaching Strategies

Introduction

Start this chapter by pointing out that atoms are invisible particles that no one has ever seen except for the very large atoms that have been photographed under a transmission-electron microscope. Even in this case, they only appear as tiny blobs of light. Yet this entire course in chemistry is based on their existence and their structure. With something this small, how can we begin to determine not only that they exist but also what they are made up of and how they are put together? To help your students understand how they can accept the existence of invisible things, ask the following questions. Does wind exist? Have you ever seen wind? Then how do you know it exists? If you go to a lake or pond, how might you know that deer, raccoons, or dogs have been there if you did not see them? Information that results from effects on or tracks left on other substances is known as indirect evidence. Much of the evidence that accounts for atoms, subatomic particles, and their relationship to one another is based on similar types of data. This chapter uses many examples of indirect evidence in discussing the atom and its structure.

A basic goal in teaching this chapter is to familiarize the students with the structure of an individual atom and show them how atoms of different elements and even certain atoms (isotopes) of the same element differ from one another. The approach used is a historical one, which makes it important to place a special emphasis on what the presently-accepted theory for each area is so students do not get the information included in original theories confused with that in modern theories. A second goal deals with indirect measurement, both in the explanation as to how historical developments were made and in counting out atoms using molar mass.

3.1 The Atom: From Philosophical Idea to Scientific Theory

After discussing Dalton's atomic theory from a historical viewpoint, go over the Modern atomic theory, which is stated as follows:

1. All matter is made up of very small particles called atoms.
2. Atoms of the same element have the same chemical properties, while atoms of different elements have different chemical properties.
3. While individual atoms of a given element may not all have the same mass (due to differences in nuclear structure), any

natural sample of the element will have a definite average mass that is characteristic of the element and different from that of any other element.

4. Compounds are formed when atoms of two or more elements unite, with each atom losing its characteristic properties as a result of this combination.

5. Atoms are not subdivided in physical or chemical reactions.

Help students compare and contrast Dalton's theory with the Modern theory. It is important for students to know what the presently accepted theory is. This also helps emphasize the changing nature of a theory as new data is discovered.

Demonstration 1: Compounds Exemplifying the Law of Multiple Proportions

PURPOSE: To show that compounds composed of the same elements, but in different proportions, are different compounds with different properties

MATERIALS: Copper(I) oxide, copper(II) oxide, 20% hydrogen peroxide*, manganese dioxide, water, colored felt, 2 250–mL beakers, 2 medium test tubes

PROCEDURE: Have students examine copper (I) oxide, Cu_2O, and copper(II) oxide, CuO, and note how they differ from one another in physical properties. (Copper(I) oxide is red and copper(II) oxide is black.)

Have students examine water, H_2O, and hydrogen peroxide, H_2O_2, and note how they are similar to one another in physical properties. (Both are colorless liquids.)

Show students that water and hydrogen peroxide differ greatly from one another in chemical reactions by doing the following:

(1) Put a piece of colored felt** in a beaker half filled with water, and put a second piece in a beaker half filled with 20% hydrogen peroxide. Allow it to stand overnight and observe results. (The hydrogen peroxide will bleach the felt, but the water will not affect it.)

(2) To 5 mL of hydrogen peroxide in a test tube, add manganese dioxide equal to the quantity of a rice grain. Repeat with 5 mL of water. (The H_2O_2 will bubble vigorously; the water does not.)

 * 20% hydrogen peroxide is used to bleach hair and can be purchased at a drugstore.
** You can experiment with other pieces of material or yarn that might bleach equally well. It all depends upon the permanency of the dye. Organic dyes and organic substances such as hair work best.

Desktop Investigation: Constructing a Model

This Desktop could be done prior to the presentation of the materials in Section 3.2 so that students will have something to relate to as you discuss experiments regarding the discovery of electrons and the nucleus. Or it could be done after the completion of the discussion on the structure of the atom and then related to that discussion.

To make the sealed container, cut off the foot of a sock and attach the cut end of the sock to the top of an empty two-pound coffee can using duct tape (see Figure in text). Put one or more objects of various sizes, shapes, masses, and materials into the can and seal the cuff of the sock tightly with duct tape.

Try to use unfamiliar objects to reduce the student's tendency to attempt to guess the identity of the object rather than focusing on developing a model of the object. (A good source of such objects is old junk shops and antique stores. You can collect odd and unusual objects throughout the year, or ask members of your faculty to help you.) Number the cans, and make a key as to their contents. To save time, assign each student the task of making one "mystery can" according to the above suggestions. Have them turn it in to you with its contents identified prior to the lab.

You can provide the students with a data table or have them design their own to record their observations. Estimation of accuracy will improve considerably with practice, but use practice objects different from those in the sealed cans. Making mass/size estimations prior to the activity would probably work best. Remind the students of the importance of not looking into the can or removing the object(s) from the can by emphasizing that models are the visualizations of phenomenon that *cannot* be seen based on scientific data. When students are finished with their "mystery can," reseal the sock end with tape so it will be ready for the next class.

A fourth discussion question you might use is:

4. Scientists often gather data or figure out answers by comparing their observations with known quantities. How did you improve your estimations of size or mass using a comparison method?

ANSWERS TO DISCUSSION QUESTIONS

1. In Part 1, all observations were made without contact with the object itself and, thus, were examples of indirect evidence (gathered through the sense of hearing). In Part 2, the observations were made by using a seldom used sense: touch; but since contact with the object occurred, this would be an example of direct observations for qualitative data. However, because of the constraints in Part 2, all measurements still had to be estimations. Scientists gather and use both indirect evidence and estimations as data in their attempts to develop models.

2. Determination of mass, size, and number of objects resulted in quantitative data which was generally estimated. Determination of shape, material, and texture involved the gathering of qualitative data.

3. This will vary from student to student depending upon the nature of the object.

4. People can learn to use their fingers as crude devices for measuring the length of objects. As one works with objects of varying masses, one becomes more aware of how different masses "feel." The more one works with estimation, the more accurate the estimations become.

3.2   The Structure of the Atom

Demonstration 2: Indirect Measurement

PURPOSE: To help students understand how the size of small objects, like atoms and subatomic particles, can be determined by using indirect measurement

MATERIALS: *Modern Chemistry* textbook, clear plastic metric ruler

PROCEDURE: How could you determine the thickness of one page of your textbook using an ordinary ruler? After a student has suggested the solution of measuring the thickness of all the pages of the book with the ruler and dividing that amount by the total number of pages in the book, have all the students measure the book and make the necessary calculations. Then look at everyone's data, to determine if it all seems reasonable; and take an average of that which is determined to be acceptable. Those that look like serious errors can be re-checked. This is an excellent opportunity to learn the concept of indirect measurement and at the same time look at the degree that measurements can vary and how serious errors can be detected.

Answers should be in the area of 0.016–0.018 cm. You can use a micrometer to measure the thickness of a page directly, and compare the results with the indirect measurements.

When discussing Rutherford's experiment, caution students that what might at first appear to be a significant deviation from the norm in an experiment might not be error at all but very useful data; that is, in Rutherford's experiment, the few alpha particles that were deflected slightly and the few that were deflected back. A scientist has to be able to distinguish between error and meaningful deviations.

3.3   Weighing & Counting Atoms

Prior to your explanation of the individual atomic mass and average atomic mass, explain that *relative* means compared to something else. To aid in your discussion of individual atomic masses of isotopes and average atomic mass, you might refer to "Table of Isotopes" in Section B of the *CRC Handbook of Chemistry and Physics*. The individual atomic mass and percentage of natural abundance for every known nuclide is listed there. Students often ask, "How do you know what the atomic mass of a given nuclide is?" This a good time to let them look at this section in the handbook. They should find it interesting, and it is a chance to introduce them to the *CRC Handbook* as an excellent resource.

Students tend to get the following terms confused. A good way to help them is to put the following summary and table on the board or on a transparency.

Individual atomic mass—Exact mass in atomic mass units of one atom

Average atomic mass—Average mass in atomic mass units of all natural isotopes of an element

Mass number—Total number of protons and neutrons in a nucleus of an isotope OR the individual atomic mass rounded off to a whole number

Atomic number—number of protons in an atom of an element

Below is a table which helps explain each term:

Isotope	% in Nature	Individual Atomic No.	Atomic Mass	Mass No.	Average Atomic Mass
Ne-20	90.51%	10	19.99244 u	20	
Ne-21	0.27%	10	20.395 u	21	20.183 u
Ne-22	9.22%	10	21.99138 u	22	

A demonstration similar to the marble analogy discussed in the *Average Atomic Masses of Elements* section in the text follows.

Demonstration 3: Isotopes and Average **Atomic Mass**

PURPOSE: To show how different percentages of natural isotopes of a given element would affect the average atomic mass

MATERIALS: 100 pennies minted before 1982, 100 pennies minted after 1982, a decigram balance [Note: Pennies minted in 1982 are a mixture of both copper and zinc pennies; therefore, do not use them.]

PROCEDURE: Using the decigram balance, determine the mass of 100 pennies minted before 1982. To determine the mass of a single penny, divide the total mass by 100. (The mass of each penny is about 3.1 g. Answers of actual mass measurements should be in terms of four significant figures.) Is this direct or indirect measurement? (Indirect) Repeat the procedure with 100 pennies minted after 1982. (The individual mass is about 2.5 g.)

Have students predict (through mathematical calculations) the average mass of an "isotopic" mixture of 50% pre-1982 and 50% post-1982 pennies. Then do an actual weighing with 50 pennies minted before 1982 and 50 after 1982 and determine their average. (It should be about 2.8 g.) Repeat with a mixture of 25% before and 75% after (about 2.65 g.), 75% before and 25% after (about 2.95 g), or any other percentage you might wish to select.

Demonstration 4: Counting Out Atoms

PURPOSE: To determine the number of atoms in a few samples of different elements

MATERIALS: Aluminum, charcoal, copper, graphite, iron, silicon, sulfur, zinc, reclosable plastic bags, a penny minted before 1982, a nail, aluminum foil, decigram balance

PROCEDURE: Prior to class, weigh out 1 mol samples of aluminum (27.0 g.), carbon (12.0 g), copper (63.5 g), iron (55.8 g), silicon (28.1 g), sulfur (32.1 g), zinc (65.4 g), or any other element that you might have available. Place each sample in a reclosable plastic bag and label what element it contains. Have a few extra empty bags available for use in class.

Locate objects composed of the above elements such as a penny for Cu, a nail (Fe), a lump of charcoal (C), a piece of foil (Al), a chunk of S, a lump of Si, or a small piece of Zn.

When class begins, tell students that they are going to help you determine how many atoms there are in samples of various elements.

Start by holding up the bags of elements containing one mole of an element. (Do not tell them each bag contains one mole. That is for them to determine.) Ask the students: "How can I figure out how many atoms I have in these bags?" When a student suggests that you weigh them, you might ask some of them to determine the mass of the bag and its contents. Again ask "How can we determine the mass of the contents of this bag without removing it from the bag?" When someone suggests weighing several plastic bags to determine an average mass for the plastic bag, have other students do those mass determinations. Calculate the mass of each substance by subtracting the average mass of the plastic bag from the mass of the bag and its contents. The results will be very close to the molar mass of each element, and students should quickly conclude that each bag contains approximately 6.022×10^{23} atoms.

Ask your students: How many atoms are in this penny? Nail? Piece of aluminum foil? etc. Have them make weighings and calculate the number of atoms.

This demonstration can be easily changed into an inquiry lab in which you give each student a one mole bag of an element and a small piece of an element and ask them to devise a way to determine the number of atoms each contains. They can consult with one another as scientists often do; but do not tell them how to solve the problem—let them do that!

A poster on moles is available from the American Chemical Society, 1155 16th St. NW, Washington, D.C. 20036

References and Resources

Books and Periodicals

Bent, Henry A. "Should Atoms Be X-rated?" *Journal of Chemical Education*, Vol. 63, No. 10, Oct. 1986, pp. 878–9.

Herron, Dudley. "Rutherford and the Nuclear Atom," *Journal of Chemical Education*, Vol. 54, No. 8, Aug. 1977, p. 499.

Kolb, Doris. "But If Atoms Are So Tiny," *Journal of Chemical Education*, Vol. 54, No. 9, Sept. 1977, pp. 543–7.

Poskozim, Paul, et al. "Analogies for Avogadro's Number," *Journal of Chemical Education*, Vol. 63, No 2, Feb. 1986, pp. 125–6.

Rayner-Canham, Geoffrey, and Marelene Rayner-Canham. "The Shell Model of the Nucleus," *The Science Teacher*, Vol. 54, No. 1, Jan. 1987, pp. 18–21.

Records, Roger. "Developing Models: What is the Atom Really Like?" *Journal of Chemical Education*, Vol. 59, No. 4, April 1982, pp. 307–9.

Weinberg, Steven. *The First Three Minutes*, New York: Bantam Books, 1977.

Audiovisual Resources

The Atom and Atomic Mass—Nuclear Physics Series, filmstrip or slides, Educational Images, Ltd.

Experiments in Subatomic Particles, filmstrip, Educational Images, Ltd., 1988.

You in the Universe—The Atomic Universe, filmstrip or VHS and Beta cassettes, Educational Images, Ltd.

Computer Software

Atomic Structure, Apple 48K. Chemistry according to ROF. Project Seraphim, 1988. Apple II Disks, vol. 2, Disk nos. 1, 2, 4, 5 and vol 3, Disk nos. 4, 5, 6.

SEI Atomic Structure, Apple 64K and IBM PC 128K. Queue, 1988.

Smith, S. G., R. Chabray, and E. Kean, *The Elements*, Apple 48K and IBM PC 128K; *Atomic Weights*, Apple 48K and IBM PC 128K. Queue, 1988.

Answers and Solutions

Questions

1. *(a)* The idea of the atom originated as early as 400 B.C. in ancient Greece. Democritus is generally credited with the first use of the term *atom*. *(b)* The word *atom* is from the Greek word meaning "indivisible."

2. The five essential points of Dalton's atomic theory are: (1) All matter consists of extremely small particles called atoms. (2) Atoms of a given element are identical in size, mass, and other properties; atoms of different elements differ in size, mass, and other properties. (3) Atoms cannot be subdivided, created, or destroyed. (4) Atoms of different elements can combine in simple, whole-number ratios to form chemical compounds. (5) In chemical reactions, atoms are combined, separated, or rearranged.

3. *(a)* Since all chemical reactions simply involve the rearrangement of indivisible atoms, then matter, and its mass, can be neither created nor destroyed in such changes. *(b)* If atoms of each element have their own characteristic masses, then compounds consisting of these elements should always have the same composition by mass. *(c)* Since only whole atoms combine in chemical compounds, different compounds between the same two elements must result from the combination of different relative whole numbers of atoms.

4. Element AB would be expected to have mass of 5 mass units.

5. *(a)* The law of multiple proportions states that if two or more different compounds are composed of the same two elements, the masses of the second element combined with a fixed mass of the first element can be expressed as ratios of small whole numbers. *(b)* John Dalton is credited with the discovery of this law.

6. *(a)* An atom is the smallest particle of an element that can exist either alone or in combination with other atoms. *(b)* Atomic structure refers to the identity and arrangement of smaller particles within atoms.

7. All atoms consist of a very dense positively charged nucleus at the center, and a comparatively large region occupied by negatively charged particles surrounding the nucleus.

8. In the late 1800s, numerous experiments were performed in glass tubes containing various gases at very low pressure. When an electric current was passed through such tubes from the negative electrode to the positive electrode, the current flowed readily. These and other observations eventually led to the hypothesis that cathode rays consist of negatively charged particles.

9. The cathode-ray-tube experiments led to the establishment of the following facts. (1) Different gases glow with different colors as current passes through the tube. (2) The glass tube directly opposite the cathode glows. (3) An object placed between the cathode and the opposite end of the tube casts a shadow on the glass. (4) A paddle wheel between the electrodes rolls along on its rails from the cathode toward the anode. (5) Cathode rays are deflected by a magnetic field in the same manner as a wire carrying electric current, which is known to have a negative charge. (6) The rays are deflected away from a negative electrode.

10. Thomson found that the charge-to-mass ratio of the cathode-ray particles was always the same, regardless of the metal used to make the cathode or of the gas in the tube. He concluded, after study of other cathode-ray properties, that all cathode rays were composed of identical, negatively charged particles, subsequently called electrons.

11. According to Thomson and Millikan, an electron is negatively charged, has a mass approximately 1/2000 that of a hydrogen atom, has the smallest possible negative electric charge, has a fixed charge-to-mass rate, and is present in atoms of all elements.

12. An electron is a negatively charged subatomic particle.

13. Two inferences could be made about atomic structure: (1) Because atoms are electrically neutral, they must contain positive charge to balance the negative charge of the electrons. (2) Because electrons are so much lower in mass than atoms, atoms must contain additional particles that account for most of their mass.

14. (a) Rutherford is credited with the discovery of the atomic nucleus. (b) The nucleus of an atom is its positively charged, dense, central portion, which contains nearly all of its mass but takes up only an insignificant fraction of its volume. (c) According to Rutherford, an atom consists of a densely packed bundle of matter with a positive charge, called a nucleus, surrounded by mostly empty space. This nucleus contains nearly all of the mass of the atom but takes up only an insignificant fraction of its volume. He further proposed that the negatively charged electrons surround the positively charged nucleus like the rings around Saturn, but he could not explain what kept the electrons in motion around the nucleus.

15. Every atomic nucleus is made up of protons and neutrons. Protons are subatomic particles that have a positive charge of the same magnitude as the negative charge of an electron and are present in atomic nuclei. Neutrons are electrically neutral, subatomic particles found in atomic nuclei.

16.

Particle	Symbol	Mass	Actual Mass	Relative Charge
Electron	e^-	0	9.110×10^{-28} g	-1
Proton	p^+	1	1.673×10^{-24} g	$+1$
Neutron	n^0	1	1.675×10^{-24} g	0

17. The nuclear forces within an atom are the short-range proton-neutron, proton-proton, and neutron-neutron forces that hold the nuclear particles together.

18. The identity of an atom is determined by the number of protons in its nucleus.

19. (a) Isotopes are atoms of the same element that have different masses. (b) Isotopes of a particular element have the same number of protons. (c) Isotopes of a particular element have different numbers of neutrons.

20.

Isotope	Name	No. of Protons	No. of Electrons	No. of Neutrons
H-1	protium	1	1	0
H-2	deuterium	1	1	1
H-3	tritium	1	1	2

21. (a) The atomic number of an element is the number of protons in the nucleus of the atoms of that element. (b) The

mass number of an isotope is the total number of protons and neutrons in its nucleus. (c) The atomic number is at the lower left, whereas the mass number is at the upper left. (d) The hyphen notation for this isotope is hydrogen-2.

22. A nuclide is a general term for any isotope of any element.

23. The hyphen notation for each isotope is: (a) helium-4 (b) oxygen-16 (c) potassium-39.

24.

Nuclear Symbol	Atomic No.	Mass No.	No. of Protons,	Electrons,	Neutrons
a) $^{4}_{2}$He	2	4	2	2	2
b) $^{9}_{4}$Be	4	9	4	4	5
c) $^{32}_{16}$S	16	32	16	16	16
d) $^{28}_{14}$Si	14	28	14	14	14
e) $^{65}_{30}$Zn	30	65	30	30	35
f) $^{40}_{18}$Ar	18	40	18	18	22

25. (a) The standard is the carbon-12 atom, or $1.660\ 565\ 5 \times 10^{-24}$ g. (b) The relative atomic mass of an atom is the mass of that atom expressed in atomic mass units.

26. (a) One atomic mass unit is exactly 1/12 the mass of the carbon-12 atom, or $1.660\ 565\ 5 \times 10^{-24}$ g. (b) The relative atomic mass of an atom is the mass of that atom expressed in atomic mass units, as compared to the atomic mass of C-12.

27. The indicated relative atomic masses would be: (a) 4 u (b) 6 u (c) 24 u (d) 54 u.

28. The relative atomic masses are: (a) 1.007 276 u (b) 0.000 548 6 u (c) 1.008 665 u.

29. (a) The average atomic mass of a naturally occurring element is the weighted average of the masses of the naturally occurring isotopes of the element. (b) The average atomic mass depends on both the mass and the relative abundance of each isotope of the element. (c) The average atomic mass of an element is calculated by multiplying the relative atomic mass of each isotope by its relative abundance, and adding the results.

30. The indicated atomic masses are: (a) 15.999 u (b) 26.982 u (c) 63.546 u (d) 196.967 u.

31. (a) A mole is the amount of a substance that contains a number of particles equal to the number of atoms in exactly 12 grams of carbon-12. (b) The symbol for the mole is *mol*. (c) There are $6.022\ 045 \times 10^{23}$ particles in one mole. (d) The name given to the number of particles in one mole is Avogadro's number.

32. (a) The molar mass of an element is the mass in grams of that element that is numerically equal to its mass on the relative atomic mass scale. (b) The indicated molar masses are (1) 12.011 g (2) 20.179 g (3) 55.847 g (4) 238.029 g.

33. The responses are: (a) 27 (b) 59 (c) 27, 27, 32 (d) 59 u (e) 58.933 u (f) 58.933 g (g) 6.022×10^{23} atoms.

Problems

1. (a) 6.94 g Li   (b) 27.0 g Al   (c) 40.1 g Ca
   (d) 55.8 g Fe   (e) 12.01 g C   (f) 107.9 g Ag

2. (a) $3.00 \text{ mol Al} \times \dfrac{27.0 \text{ g Al}}{\text{mol Al}} = 81.0 \text{ g Al}$

(b) $4.25 \text{ mol Li} \times \dfrac{6.94 \text{ g Li}}{\text{mol Li}} = 29.5 \text{ g Li}$

(c) $1.38 \text{ mol N} \times \dfrac{14.0 \text{ g N}}{\text{mol N}} = 19.3 \text{ g N}$

(d) $8.075 \text{ mol Au} \times \dfrac{197.0 \text{ g Au}}{\text{mol Au}} = 1591 \text{ g Au}$

(e) $6.50 \text{ mol Cu} \times \dfrac{63.5 \text{ g Cu}}{\text{mol Cu}} = 413 \text{ g Cu}$

(f) $2.57 \times 10^8 \text{ mol S} \times \dfrac{32.1 \text{ g S}}{\text{mol S}} = 8.25 \times 10^9 \text{ g S}$

(g) $1.75 \times 10^{-6} \text{ mol Hg} \times \dfrac{201 \text{ g Hg}}{\text{mol Hg}} = 3.52 \times 10^{-4} \text{ g Hg}$

3. (a) $40.1 \text{ g Ca} \times \dfrac{\text{mol Ca}}{40.1 \text{ g Ca}} = 1.00 \text{ mol Ca}$

(b) $11.5 \text{ g Na} \times \dfrac{\text{mol Na}}{23.0 \text{ g Na}} = 0.500 \text{ mol Na}$

(c) $5.87 \text{ g Ni} \times \dfrac{\text{mol Ni}}{58.7 \text{ g Ni}} = 0.100 \text{ mol Ni}$

(d) $150 \text{ g S} \times \dfrac{\text{mol S}}{32.1 \text{ g S}} = 4.7 \text{ mol S}$

(e) $2.65 \text{ g Fe} \times \dfrac{\text{mol Fe}}{55.8 \text{ g Fe}} = 0.0475 \text{ mol Fe}$

(f) $0.007 \, 50 \text{ g Ag} \times \dfrac{\text{mol Ag}}{108 \text{ g Ag}} = 6.94 \times 10^{-5} \text{ mol Ag}$

(g) $3.25 \times 10^5 \text{ g Pb} \times \dfrac{\text{mol Pb}}{207 \text{ g Pb}} = 1.57 \times 10^3 \text{ mol Pb}$

(h) $4.50 \times 10^{-12} \text{ g O} \times \dfrac{\text{mol O}}{16.0 \text{ g O}} = 2.81 \times 10^{-13} \text{ mol O}$

4. (a) $6.022 \times 10^{23} \text{ atoms Ne} \times \dfrac{\text{mol Ne}}{6.022 \times 10^{23} \text{ atoms Ne}} = 1.000 \text{ mol Ne}$

(b) $3.011 \times 10^{23} \text{ atoms Mg} \times \dfrac{\text{mol Mg}}{6.022 \times 10^{23} \text{ atoms Mg}} = 0.5000 \text{ mol Mg}$

(c) $2.25 \times 10^{15} \text{ atoms Zn} \times \dfrac{\text{mol Zn}}{6.02 \times 10^{23} \text{ atoms Zn}} = 3.74 \times 10^{-9} \text{ mol Zn}$

(d) $50.0 \text{ atoms Ba} \times \dfrac{\text{mol Ba}}{6.02 \times 10^{23} \text{ atoms Ba}} = 8.31 \times 10^{-23} \text{ mol Ba}$

5. (a) $1.50 \text{ mol Na} \times \dfrac{6.02 \times 10^{23} \text{ atoms Na}}{\text{mol Na}} = 9.03 \times 10^{23} \text{ atoms Na}$

(b) $6.755 \text{ mol Pb} \times \dfrac{6.022 \times 10^{23} \text{ atoms Pb}}{\text{mol Pb}} = 4.068 \times 10^{24} \text{ atoms Pb}$

(c) $0.250 \text{ mol Si} \times \dfrac{6.02 \times 10^{23} \text{ atoms Si}}{\text{mol Si}} = 1.50 \times 10^{23} \text{ atoms Si}$

6. (a) $3.011 \times 10^{23} \text{ atoms F} \times \dfrac{\text{mol F}}{6.022 \times 10^{23} \text{ atoms F}} \times \dfrac{19.0 \text{ g F}}{\text{mol F}} = 9.50 \text{ g F}$

(b) $1.50 \times 10^{23} \text{ atoms Mg} \times \dfrac{\text{mol Mg}}{6.02 \times 10^{23} \text{ atoms Mg}} \times \dfrac{24.3 \text{ g Mg}}{\text{mol Mg}} = 6.05 \text{ g Mg}$

(c) $4.50 \times 10^{12} \text{ atoms Cl} \times \dfrac{\text{mol Cl}}{6.02 \times 10^{23} \text{ atoms Cl}} \times \dfrac{35.5 \text{ g Cl}}{\text{mol Cl}} = 2.65 \times 10^{-10} \text{ g Cl}$

(d) $8.42 \times 10^{18} \text{ atoms Br} \times \dfrac{1 \text{ mol Br}}{6.02 \times 10^{23} \text{ atoms Br}} \times \dfrac{79.9 \text{ g Br}}{\text{mol Br}} = 1.12 \times 10^{-3} \text{ g Br}$

(e) $25.0 \text{ atoms W} \times \dfrac{\text{mol W}}{6.02 \times 10^{23} \text{ atoms W}} \times \dfrac{184 \text{ g W}}{\text{mol W}} = 7.64 \times 10^{-21} \text{ g W}$

(f) $1.00 \text{ atoms Au} \times \dfrac{\text{mol Au}}{6.02 \times 10^{23} \text{ atoms Au}} \times \dfrac{197 \text{ g Au}}{\text{mol Au}} = 3.27 \times 10^{-22} \text{ g Au}$

7. (a) $5.40 \text{ g B} \times \dfrac{\text{mol B}}{10.8 \text{ g B}} \times \dfrac{6.02 \; 10^{23} \text{ atoms B}}{\text{mol B}} = 3.01 \times 10^{23} \text{ atoms B}$

(b) $8.02 \text{ g S} \times \dfrac{\text{mol S}}{32.1 \text{ g S}} \times \dfrac{6.02 \; 10^{23} \text{ atoms S}}{\text{mol S}} = 1.50 \times 10^{23} \text{ atoms S}$

(c) $1.50 \text{ g K} \times \dfrac{\text{mol K}}{39.1 \text{ g K}} \times \dfrac{6.02 \; 10^{23} \text{ atoms K}}{\text{mol K}} = 2.31 \times 10^{22} \text{ atoms K}$

(d) $0.025 \, 50 \text{ g Pt} \times \dfrac{\text{mol Pt}}{195.1 \text{ g Pt}} \times \dfrac{6.022 \times 10^{23} \text{ atoms Pt}}{\text{mol Pt}} = 7.872 \times 10^{19} \text{ atoms Pt}$

(e) 1.00×10^{-10} g Au $\times \dfrac{\text{mol Au}}{197 \text{ g Au}} \times \dfrac{6.02 \times 10^{23} \text{ atoms Au}}{\text{mol Au}} = 3.06 \times 10^{11}$ atoms Au

8. *(a)* 0.500 mol C $\times \dfrac{6.02 \times 10^{23} \text{ atoms C}}{\text{mol C}} = 3.01 \times 10^{23}$ atoms C

0.750 mol Li $\times \dfrac{6.02 \times 10^{23} \text{ atoms Li}}{\text{mol Li}} = 4.52 \times 10^{23}$ atoms Li

The Li sample thus contains the greater number of atoms. Note: Since the number of units in one mole of any substance is constant, this answer could have been determined without any calculations.

(b) 19.0 g F $\times \dfrac{\text{mol F}}{19.0 \text{ g Ca}} \times \dfrac{6.02 \times 10^{23} \text{ atoms F}}{\text{mol F}} = 6.02 \times 10^{23}$ atoms F

19.0 g Ne $\times \dfrac{\text{mol Ne}}{20.2 \text{ g Ne}} \times \dfrac{6.02 \times 10^{23} \text{ atoms Ne}}{\text{mol Ne}} = 5.66 \times 10^{23}$ atoms Ne

The F sample thus contains the greater number of atoms.

(c) 0.250 mol Si $\times \dfrac{6.02 \times 10^{23} \text{ atoms Si}}{\text{mol Si}} = 1.50 \times 10^{23}$ atoms Si

10.0 g Cl $\times \dfrac{\text{mol Cl}}{35.5 \text{ g Cl}} \times \dfrac{6.02 \times 10^{23} \text{ atoms Cl}}{\text{mol Cl}} = 1.70 \times 10^{23}$ atoms Cl

The Cl sample thus contains the greater number of atoms.

(d) 15.00 g Ca $\times \dfrac{\text{mol Ca}}{40.08 \text{ g Ca}} \times \dfrac{6.022 \times 10^{23} \text{ atoms Ca}}{\text{mol Ca}} = 2.254 \times 10^{23}$ atoms Ca

40.00 g Ag $\times \dfrac{\text{mol Ag}}{107.9 \text{ g Ag}} \times \dfrac{6.022 \times 10^{23} \text{ atoms Ag}}{\text{mol Ag}} = 2.232 \times 10^{23}$ atoms Ag

The Ca sample thus contains the greater number of atoms.

9. 12.0 g Al $\times \dfrac{\text{mol Al}}{27.0 \text{ g Al}} = 0.444$ mol Al

Since 0.444 mol Al contains the same number of atoms as 0.444 mol of any other element:

(a) 0.444 mol O $\times \dfrac{16.0 \text{ g O}}{\text{mol O}} = 7.10$ g O

(b) 0.444 mol Cl $\times \dfrac{35.5 \text{ g Cl}}{\text{mol Cl}} = 7.00$ g Cl

(c) 0.444 mol Ca $\times \dfrac{16.0 \text{ g Ca}}{\text{mol Ca}} = 17.8$ g Ca

(d) 0.444 mol Ni $\times \dfrac{58.7 \text{ g Ni}}{\text{mol Ni}} = 26.2$ g Ni

(e) 0.444 mol Pb $\times \dfrac{207 \text{ g Pb}}{\text{mol Pb}} = 91.9$ g Pb

Application Questions

1. In spite of the limitations of Dalton's theory that are now known, the concepts that all matter is composed of atoms and that atoms of different elements differ in properties from one another remain unchanged. The exceptions have simply led to an expansion of the theory to explain the new observations.

2. According to Rutherford, the possibility that any of the fast-moving positively charged particles he fired at the metal foil could have been repelled back toward the source was as remote as having a 15–inch shell fired at tissue paper bound back toward the source. Until this deflection was observed, it had been assumed that the mass and the positive charge of atoms were uniformly distributed, but the assumption of such a distribution could not be used to explain Rutherford's findings.

3. *(a)* Most of the fast-moving particles pass through the foil with only slight deflection. *(b)* Few of the fast-moving particles were deflected by a nucleus. *(c)* The positively charged particles were deflected by, rather than attracted to, the nucleus. *(d)* He had no experimental evidence to support this proposal.

4. In the nucleus of an atom, short-range attractive proton-neutron, proton–proton, and neutron–neutron forces serve to hold the nuclear particles together, rather than to allow the protons to repel one another, as might otherwise have been expected.

5. *(a)* These atoms represent isotopes of the same element. *(b)* These atoms represent different elements that simply have the same mass numbers.

6. Since the average atomic mass of an element is the weighted average of the relative atomic masses of its isotopes, only for those elements that occur in single isotopic form or for which the combination of mass and relative abundance would result in a whole number sum could average atomic mass be a whole number.

7. The atomic mass of a nuclide is the relative atomic mass of atoms of that nuclide, whereas the atomic mass of an element is the weighted average atomic mass of the naturally occurring mixture of isotopes of the element.

Application Problems

1. Average atomic mass $= (0.003\ 37 \times 35.968\ u) + (0.000\ 63 \times 37.963\ u) + (0.996\ 00 \times 39.962\ u)$
   Average atomic mass $= 0.1212\ u + 0.0239\ u + 39.8022\ u$
   Average atomic mass $= 39.947\ u$

2. Expressing this relationship algebraically, and solving for x, the atomic mass of the other isotope,
   $10.811\ u = (0.8020 \times 11.009\ u) + (0.1980 \times x)$
   $10.811\ u = 8.829\ u + (0.1980 \times x)$
   $1.982\ u = 0.1980 \times x$
   $10.010\ u = x$

3. Expressing this relationship algebraically and solving for x,
   $35.453\ u = (x \times 34.969\ u) + (1.000 - x)(36.966\ u)$
   $35.453\ u = (34.969\ u \times x) + 36.966\ u - (36.966\ u \times x)$
   $1.997\ u \times x = 1.513\ u$
   $x = 0.757\ 64$
   Thus, Cl-35 has a relative abundance of 75.764%, and Cl-37 has an abundance of 24.236%.

4. $\text{mass C-12}_{\text{old}} = 12.000\ 00\ u \times \dfrac{16.000\ 00\ u}{15.9994\ u}$

   $\text{mass C-12}_{\text{old}} = 12.000\ 45\ u$

5. (a) $0.0750\ \text{mol Zn} \times \dfrac{65.4\ \text{g Zn}}{\text{mol Zn}} = 4.90\ \text{g Zn}$

   (b) $3.40 \times 10^5\ \text{mol N} \times \dfrac{14.0\ \text{g N}}{\text{mol N}} = 4.76 \times 10^6\ \text{g N}$

   (c) $4.62 \times 10^{26}\ \text{atoms Pb} \times \dfrac{\text{mol Pb}}{6.02 \times 10^{23}\ \text{atoms Pb}} \times \dfrac{207\ \text{g Pb}}{\text{mol Pb}} = 1.59 \times 10^5\ \text{g Pb}$

   (d) $50.0\ \text{atoms Hg} \times \dfrac{201\ \text{g Hg}}{6.02 \times 10^{23}\ \text{atoms Hg}} = 1.67 \times 10^{-20}\ \text{g Hg}$

6. (a) $1.00\ \text{g C} \times \dfrac{\text{mol C}}{12.0\ \text{g C}} = 0.0833\ \text{mol C}$

   (b) $7.50\ \text{kg Fe} \times \dfrac{10^3\ \text{g Fe}}{\text{kg Fe}} \times \dfrac{\text{mol Fe}}{55.8\ \text{g Fe}} = 134\ \text{mol Fe}$

   (c) $1.25 \times 10^{20}\ \text{atoms Ni} \times \dfrac{\text{mol Ni}}{6.02 \times 10^{23}\ \text{atoms Ni}} = 2.08 \times 10^{-4}\ \text{mol Ni}$

   (d) $1000.\ \text{atoms Au} \times \dfrac{\text{mol Au}}{6.022 \times 10^{23}\ \text{atoms Au}} = 1.661 \times 10^{-21}\ \text{mol Au}$

7. (a) $3.25\ \text{mol S} \times \dfrac{6.02 \times 10^{23}\ \text{atoms S}}{\text{mol S}} = 1.96 \times 10^{24}\ \text{atoms S}$

   (b) $4.3 \times 10^4\ \text{mol H} \times \dfrac{6.02 \times 10^{23}\ \text{atoms H}}{\text{mol H}} = 2.6 \times 10^{28}\ \text{atoms H}$

   (c) $350.\ \text{g Al} \times \dfrac{6.02 \times 10^{23}\ \text{atoms Al}}{\text{mol Al}} \times \dfrac{\text{mol Al}}{27.0\ \text{g Al}} = 7.80 \times 10^{24}\ \text{atoms Al}$

   (d) $5.25\ \text{kg Zn} \times \dfrac{10^3\ \text{g Zn}}{\text{kg Zn}} \times \dfrac{6.02 \times 10^{23}\ \text{atoms Zn}}{\text{mol Zn}} \times \dfrac{\text{mol Zn}}{65.4\ \text{g Zn}} = 4.83 \times 10^{25}\ \text{atoms Zn}$

8. $4.75 \times 10^{23}\ \text{atoms Ag} \times \dfrac{\text{mol Ag}}{6.02 \times 10^{23}\ \text{atoms Ag}} \times \dfrac{108\ \text{g Ag}}{\text{mol Ag}} = 85.2\ \text{g Ag}$

   $85.2\ \text{g Au} \times \dfrac{6.02 \times 10^{23}\ \text{atoms Au}}{\text{mol Au}} \times \dfrac{\text{mol Au}}{197\ \text{g Au}} = 2.60 \times 10^{23}\ \text{atoms Au}$

9. (a) $\dfrac{7.21\ \text{g Mg}}{\text{cm}^3\ \text{Mg}} \times 12.0\ \text{cm}^3\ \text{Mg} = 86.5\ \text{g Mg}$

   $86.5\ \text{g Mg} \times \dfrac{\text{mol Mg}}{24.3\ \text{g Mg}} = 3.56\ \text{mol Mg}$

   (b) $3.56\ \text{mol Mg} \times \dfrac{6.02 \times 10^{23}\ \text{atoms Mg}}{\text{mol Mg}} = 2.14 \times 10^{24}\ \text{atoms Mg}$

10. 1.25×10^{18} atoms Si $\times \dfrac{\text{mol Si}}{6.02 \times 10^{23} \text{ atoms Si}} \times \dfrac{28.1 \text{ g Si}}{\text{mol Si}} \times \dfrac{\text{cm}^3 \text{ si}}{2.33 \text{ g Si}} = 2.50 \times 10^{-5} \text{ cm}^3 \text{ Si}$

11. $1.500 \text{ carat} \times \dfrac{200.0 \text{ mg}}{1.000 \text{ carat}} \times \dfrac{\text{g}}{10^3 \text{mg}} \times \dfrac{\text{mol}}{12.01 \text{ g}} \times \dfrac{6.022 \times 10^{23} \text{ atoms}}{\text{mol}} = 1.504 \times 10^{22} \text{ atoms}$

Teacher's Notes

Chapter 3 Atoms: The Building Blocks of Matter

INTRODUCTION
This chapter discusses the ancient Greek idea of the atom and the development of the atomic theory of matter. All matter is now viewed as composed of atoms, and therefore atomic theory is the basis for modern chemistry.

LOOKING AHEAD
As you study this chapter, pay particular attention to how theories and models are proposed, developed, and modified to account for experimental observations.

SECTION PREVIEW
3.1 The Atom: From Philosophical Idea to Scientific Theory
3.2 The Structure of the Atom
3.3 Weighing and Counting Atoms

Individual atoms may be seen in this image of the surface if gallium arsenide, produced by a scanning tunneling microscope. The blue spots are gallium atoms, and the red are arsenic atoms.

3.1 The Atom: From Philosophical Idea to Scientific Theory

The idea that matter is made up of simple, indivisible particles originated a long time ago. As early as 400 B.C., some Greek thinkers had the idea that matter could be divided into smaller and smaller particles until a basic particle of matter that could not be divided further was reached. Democritus (460–370 B.C.) called these particles "atoms." The word *atom* is from the Greek word meaning indivisible. However, Aristotle (384–322 B.C.), a Greek whose ideas had a large impact on Western civilization, did not believe in atoms and his view was accepted until the 18th century, when experimental evidence proved him wrong. Neither Democritus' nor Aristotle's theories were based on experimental evidence.

When you crush a lump of sugar, you can see that it is made up of many small particles of sugar. You may grind these particles into a very fine powder, but each tiny piece is still sugar. Now suppose you dissolve the sugar in water. The tiny particles disappear completely. Even if you look at the sugar-water solution through a microscope, you cannot see any sugar particles. But, if you taste the solution, you would know that the sugar is still there.

If you open a perfume bottle, you can smell the escaping perfume. Yet you cannot see gas particles in the air of the room, even if you use a powerful microscope. These observations and many others like them have led scientists to believe that the basic particles of matter must be very small.

Dalton's Atomic Theory

In the late 1700s, invention of the chemical balance gave chemists a tool for studying the composition of pure substances quantitatively. They discovered that a small number of elements are the components of many different substances. Investigations revealed how elements combine to form compounds and how compounds can be broken down into their constituent elements.

Observations made during the period resulted in the discovery of the law of conservation of matter and the law of definite composition, which were introduced in Chapter 1. An English schoolteacher, John Dalton (1766–1844), was the first to recognize that these laws and many experimental results could be explained by the existence of atoms.

In 1803, Dalton proposed an atomic theory that greatly expanded upon the Greek concept. Dalton was able to relate atoms to the measureable property of mass. Dalton's atomic theory can be summarized by the following statements:

1. All matter is composed of extremely small particles called atoms.
2. Atoms of a given element are identical in size, mass, and other properties; atoms of different elements differ in size, mass, and other properties.
3. Atoms cannot be subdivided, created, or destroyed.
4. Atoms of different elements can combine in simple, whole-number ratios to form chemical compounds.
5. In chemical reactions, atoms are combined, separated, or rearranged.

- Summarize the five essential points of Dalton's atomic theory.
- Explain the relationship between Dalton's atomic theory and the laws of conservation of matter and definite composition.
- Explain the law of multiple proportions.

Figure 3-1 John Dalton *(top)* and Democritus

John Dalton was interested
in the composition and
properties of gases in the
atmosphere. He kept a daily
record of the weather from
1787 until 1844.

In John Dalton's own words:
"No new creation or destruction of matter is within the
reach of chemical agency.
We might as well attempt to
introduce a new planet into
the solar system, or to anni-
hilate one already in existence, as to create or destroy
a particle of hydrogen. All
the changes we can produce
consist in separating particles already in a state of cohesion or combination, and
joining those that were previously at a distance."
—John Dalton: *A New Sys-
tem of Chemical Philosophy*
(London, 1842).

STUDY HINT

Isotopes are defined and discussed in Section 3-3

Dalton's atomic theory was accepted and became the basis for further experiments because it successfully explained the laws of conservation of mass and definite composition, and other observations. Although exceptions to Dalton's atomic theory are now known, atomic theory has not been discarded. We now know that atoms are divisible and that a given element can have atoms with different masses, called *isotopes*. Atomic theory has been expanded to explain the new observations. The concepts that all matter is composed of atoms and that atoms of different elements differ in properties from atoms of other elements remain unchanged.

Atoms and the conservation of mass   If atoms are indivisible and atoms of different elements can combine in chemical reactions then it *must* be that mass is conserved in chemical reactions. Let's apply these ideas to two imaginary elements, element A and element B. Suppose atoms of B weigh three times as much as atoms of A. If we assign a mass of 1 mass unit to atoms of A, then atoms of B will have 3 mass units.

In a chemical reaction, let's assume one atom of A and one atom of B combine to give a chemical compound, which we represent as AB:

$$A + B \rightarrow AB$$

According to atomic theory, neither atom can be subdivided in this reaction. Therefore, the total mass of the product can only be the sum of the masses of A and B

$$
\begin{array}{ccccc}
A & + & B & \rightarrow & AB \\
1 \text{ mass unit} & + & 3 \text{ mass units} & = & 4 \text{ mass units}
\end{array}
$$

The reverse also would hold true in a reaction in which AB is separated into its constituent elements

$$
\begin{array}{ccccc}
AB & \rightarrow & A & + & B \\
4 \text{ mass units} & = & 1 \text{ mass unit} & + & 3 \text{ mass units}
\end{array}
$$

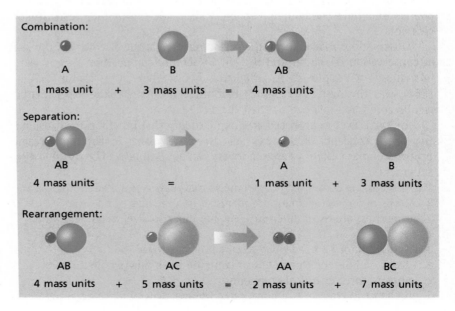

Figure 3-2   The combination, separation, and rearrangement of atoms are depicted, showing the conservation of mass in chemical reactions.

If all chemical reactions involve the combination, separation, or rearrangement of indivisible atoms, then matter and its mass can be neither created nor destroyed in such changes. (See Figure 3-2)

Atoms and the definite composition of chemical compounds   Figure 3-2 also illustrates how atomic theory accounts for the definite composition of chemical compounds by mass. If atoms of A and atoms of B always have their own characteristic masses, then compound AB must always have the same composition by mass. Whatever mass units are used and no matter how much of compound AB you have or where it came from, the mass ratio of A to B in the compound will be 1 to 3. Four grams of AB will contain 1 g of A combined with 3 g of B. One hundred grams of AB will contain 25 g of A and 75 g of B. In other words, compound AB will always have the composition of 25% by mass of substance A and 75% by mass of substance B.

Law of Multiple Proportions

Certain elements can combine to form two or more different chemical compounds. Carbon and oxygen provide a familiar example. They are known to form carbon monoxide (CO), which is the poisonous gas in automobile exhaust, and carbon dioxide (CO_2), which is given off by most living things and taken up by plants.

How does atomic theory explain the formation of different compounds by the same two elements? Quite simply, only whole atoms combine in chemical compounds between the same two elements. Therefore, different compounds formed by the same two elements must result from combinations of different relative numbers of atoms. The atoms of our elements A and B might combine not only one-to-one, as in AB, but in other ways. One atom of A and two atoms of B could form a compound written as AB_2. They might combine one-to-three in the compound AB_3.

The ratio of the masses of one element combined with a certain fixed mass of a second element in two compounds must be the same as the ratio of the numbers of atoms of the first element in the two compounds. Consider our two elements, A and B. Above, we noted that 1 g of A combines with 3 g of B to give 4 g of AB. If instead, these elements form AB_2, then 1 g of A would have to combine with twice as much, or 6 g of B, to give AB_2. In these two compounds the masses of B have the ratio of 1 to 2. This whole-number ratio is the same as the ratio of the numbers of atoms of B in the two compounds.

John Dalton made many similar observations and proposed the **law of multiple proportions** which states: *If two or more different compounds are composed of the same two elements, the masses of the second element combined with a certain mass of the first element can be expressed as ratios of small whole numbers.*

If carbon monoxide and carbon dioxide were each produced by the reaction of 12 g of carbon with oxygen, the resulting compounds would contain 16 g and 32 g of oxygen, respectively. The ratio of 16 g of oxygen to 32 g of oxygen $= \frac{1}{2}$. This indicates that, in carbon monoxide, there are half as many oxygen atoms for every carbon atom as there are in carbon. This result reflects the known compositions of carbon monoxide and carbon dioxide. Carbon monoxide contains one carbon atom and one oxygen atom. Carbon dioxide contains one carbon atom and two oxygen atoms.

Proposes this problem. When magnesium burns in oxygen, the ratio is 3 mass units of magnesium to 2 mass units of oxygen to form 4 mass units of magnesium oxide. How many grams of oxygen are needed to react with 6 g of magnesium? (4 g) What would happen if there were 6 g of oxygen? (When the reaction is over, 2 g of oxygen would be left.)

For a given mass of carbon combined with oxygen to form carbon dioxide (CO_2: mass ratio, 1 C: 2.66 O) the mass of oxygen (O), in carbon dioxide is always twice that in carbon monoxide (CO: mass ratio, 1 C : 1.33 O). Every gram of carbon combines, therefore, with either 2.66 g or 1.33 g of oxygen.

Desktop Investigation

Constructing a Model

Question
How can you construct a model of an unknown object by *(1)* making inferences about an object that is in a closed container and *(2)* touching the object without seeing it?

Materials
sealed container
one or more objects that fit in the container
cm ruler
balance

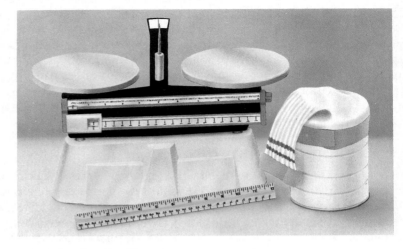

Procedure
1. Your teacher will provide you with a can covered by a sock and sealed with tape. Record all your observations in a data table. Without unsealing the container, try to determine the number of objects, their mass, shape, size, plus the kind of material they are made of and their texture. To do this, you may carefully tilt and shake the can.
2. Remove the tape from the top of the sock. Do *not* look inside the container. Put one hand through the opening and make the same observations as in Procedure 1 by touching and holding the objects. To make more accurate estimations, practice estimating the size and mass of some known objects outside the can and then compare your estimates of the objects outside the can with actual measurements using a cm ruler and a balance.

Discussion
1. Scientists often use more than one method to gather data. How was this illustrated in the investigation?
2. Which of the observations you made were qualitative and which were quantitative?
3. Using the data you gathered, draw a model of the unknown object(s) and write a brief summary of your conclusions.

---

1. *(a)* Democritus is generally credited with the use of the term *atom* to describe the smallest particles of matter that can exist. *(b)* John Dalton proposed an atomic theory that related atoms to the measurable property of mass.
2. See bottom of page 71.
3. Laws of conservation of matter, definite composition, and multiple proportions
4. If two or more different compounds are composed of the same two elements, the masses of the second element combined with a fixed mass of the first element can be expressed as ratios of small whole numbers.

The two compounds are represented by combining the atomic symbols as follows:

Compound Name	Compound Symbol
Carbon Monoxide	CO
Carbon Dioxide	CO_2

Section Review

1. Cite the major contributions of each of the following toward the present theory of the atom: *(a)* Democritus *(b)* John Dalton.
2. List the five essential points of Dalton's atomic theory.
3. What laws can be explained through the use of Dalton's theory?
4. State the law of multiple proportions.

3.2 The Structure of the Atom

You have thus far seen evidence that an **atom** *is the smallest particle of an element that can exist either alone or in combination with other atoms.* Experiments about the nature of matter continued into the 1800s as Dalton's atomic theory gained acceptance. The next developments on the pathway to modern atomic theory came with observations showing that atoms are divisible. Gradually, a model of atomic structure emerged. **Atomic structure** *refers to the identity and arrangement of smaller particles within atoms.*

Although atoms of various elements differ in the arrangements and numbers of particles that they contain, all atoms consist of two main regions. Within the atom, there is a very small region called the *nucleus* that, in all but one case, is occupied by positively charged particles called *protons* and neutral particles called *neutrons.* Surrounding the nucleus is a region occupied by negatively charged particles called *electrons.* The region occupied by electrons is very large compared to the size of the nucleus.

SECTION OBJECTIVES

- Summarize the observed properties of cathode rays that led to the discovery of the electron.
- Summarize the experiment conducted by Rutherford that led to the discovery of the nucleus.
- Describe the properties of protons, neutrons, and electrons.
- Define "atom" and "isotope."
- Describe the atomic structures of the isotopes of hydrogen.

It was not until 1939 that Lise Meitner figured out that an atom could indeed be subdivided. She and her associate Otto Hahn bombarded U-235 atoms with neutrons. Although Hahn suspected the uranium atoms were being split, it was Meitner who officially published the conclusion naming the process fission.

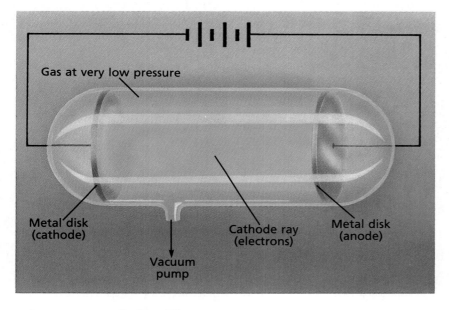

Figure 3-3 A simple cathode-ray tube.

Discovery of the Electron

The first evidence for the existence of smaller particles within atoms came in studies of the relationship between electricity and matter. Numerous experiments were performed in the late 1800s in which electrical current was passed through different gases at very low pressures. These experiments were carried out in glass tubes like those shown in Figure 3-3. An electrical current passes through the tube from the *cathode,* the electrode (plate) connected to the negative terminal of the battery, to the *anode*, the electrode (plate) connected to the positive electrode. Such tubes, called cathode-ray-tubes, were the forerunners of television tubes.

Cathode rays and electrons Gases at atmospheric pressure do not conduct electricity very well. At the very low pressures in cathode-ray-tubes,

In a television tube, a cathode ray constantly undergoes deflections that result from the varying magnetic field within the tube. The beam strikes a coated screen, where it creates a luminescent image.

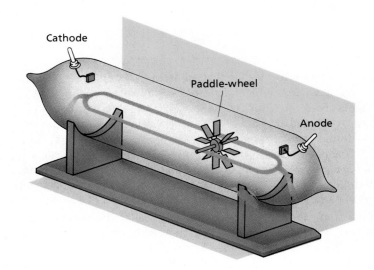

Figure 3-4 A paddle-wheel cathode-ray tube

Cathode

Paddle-wheel

Anode

The different colors result from the different energies that electrons have in different kinds of atoms. The electrons, however, are identical in all atoms.

current flows readily through a gas. A series of experiments established a number of facts about what happens in such tubes: *(1)* Different gases glow with different colors as current passes through the tube. *(2)* The part glass tube directly opposite the cathode glows. *(3)* An object placed between the cathode and the opposite end of the tube casts a shadow on the glass. *(4)* A paddle wheel between the electrodes rolls along on its rails from the cathode toward the anode (Figure 3-4).

The cathode ray is composed of particles (as shown by the paddle wheel experiment) that pass from the cathode to the anode. Further experiments gave additional information: *(5)* Cathode rays are deflected by a magnetic field in the same manner as a wire carrying electrical current, which was known to have a negative charge. *(6)* The rays are deflected away from a negative electrode. The observations of cathode-ray behavior led to the hypothesis that cathode rays consist of negatively charged particles. A brilliant series of experiments carried out in 1897 by the English physicist Sir Joseph John Thomson (1856–1940) strongly supported the hypothesis that cathode rays consist of negatively charged particles. Thomson used the apparatus shown in Figure 3-5 to measure the ratio of the charge of cathode-ray particles to their mass. He found that this ratio was always the same, regardless of the metal used to make the cathode or of the gas in the cathode-ray tube. Thomson concluded that all cathode rays were composed of identical negatively charged particles, which came to be called electrons. **Electrons** *are negatively charged subatomic particles.*

Charge and mass of the electron The ratio that Thomson obtained in his experiments was extremely high. This ratio, indicated that the electron had a very small mass compared to its charge. In 1909, experiments conducted by the American physicist Robert A. Millikan (1868–1953) showed that the mass of the electron is in fact approximately one two-thousandth the mass of a hydrogen atom, which is the smallest atom known. This experiment is discussed in detail in the Chemistry Notebook beginning on page 79. More accurate experiments conducted since then indicate that the electron has a mass of 9.109×10^{-28} g, which is $\frac{1}{1837}$ the mass of the simplest type of hydrogen atom.

Millikan's experiments also confirmed that the electron has the smallest possible negative electric charge, which is what Thomson and others had hypothesized. All other negative charges are whole-number multiples of the charge of the electron. Thus, an electric charge can be twice or three times that of an electron but never one-half or three-halves of that charge. Because cathode rays always have the same properties, regardless of the element used to produce them, it was concluded that electrons are present in atoms of all elements.

The electrons in cathode rays could come only from within the atoms of the gas or the metal cathode. Cathode-ray experiments had provided, therefore, clear-cut evidence that atoms are divisible. Two other inferences could be made about atomic structure based on what was learned about electrons. These are: *(1)* Because atoms are electrically neutral, they must contain a positive charge to balance the negative electrons; and *(2)* Because electrons are so much smaller in mass than atoms, atoms must contain additional particles that account for most of their mass.

Chapter 4 is devoted to the arrangement of electrons in atoms. This arrangement plays a very important role in determining the properties of matter.

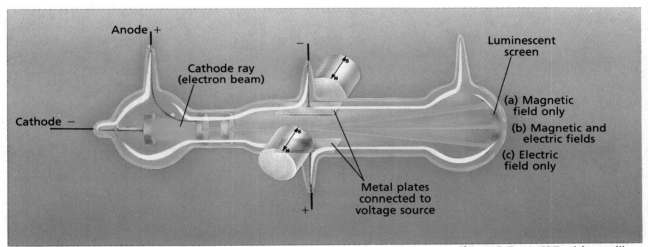

Discovery of the Atomic Nucleus

In 1908 and 1909, New Zealander Ernest Rutherford (1871–1937), working in England with his associates Hans Geiger (1882–1945) and Ernest Marsden (1889–1970), studied the bombardment of thin metal foils with fast-moving positively charged particles. The experiments were carried out using the setup shown in Figure 3-6. Almost all the fast-moving particles passed through the foil with only a slight deflection. This result is to be expected in bombarding the 10 000-atom-thick foil if the mass and positive charge of the atoms is uniformly distributed.

Geiger and Marsden also checked for the possibility of wide-angle deflections. Much to their surprise, they found that roughly one in eight thousand particles actually ricocheted back toward the source, as shown in Figure 3-6 As Rutherford later exclaimed, it was "as if you had fired a 15-inch [artillery] shell at a piece of tissue paper and it came back and hit you."

Rutherford thought about this experimental result until, after two years, he finally saw the solution. He reasoned that the fast-moving particles must be repelled by some powerful force within the atom. He reasoned further that

Figure 3-5 A CRT with a calibrated luminescent screen can measure the charge-to-mass ratio of a charged particle.

After Rutherford received the 1908 Nobel Prize in chemistry for his work with radioactivity, he discovered the proton. He was also the first person that was able to carry out the alchemist's dream of transmutation when he changed nitrogen atoms into oxygen atoms via nuclear reactions.

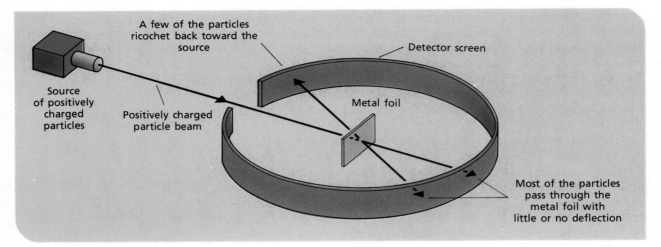

Figure 3-6 Rutherford used an appraratus like this in experiments that led him to consider the existence of the atomic nucleus.

whatever caused this repulsion must occupy a very small amount of space, since only a very few particles shot into the foil ran into it. Rutherford concluded that this force must be caused by a very densely packed bundle of matter with a positive electric charge. As shown in Fig 3-7, this bundle of matter is the nucleus. *The* **nucleus** *is the positively charged, dense central portion of the atom that contains nearly all of its mass but takes up only an insignificant fraction of its volume.* If the nucleus had the diameter of a dime, the distance to the outer edge of the atom would be about one-half the length of a football field! Thus, the atom is mostly empty space.

Where, then, were the electrons? Rutherford suggested, although he had no supporting evidence, that the electrons surrounded the positively charged nucleus like the planet around the sun. He could not explain, however, what kept the electrons in motion around the nucleus.

Rutherford discovered the proton in 1919 as a particle produced in the bombardment of atoms of several elements.

Composition of the Atomic Nucleus

Except for the nucleus of the simplest type of hydrogen atom (discussed in the next section) every atomic nucleus is made up of two kinds of particles,

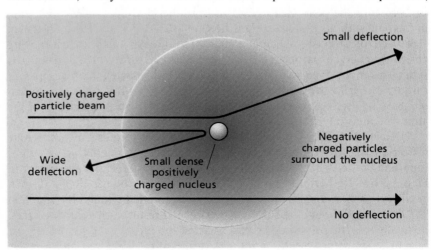

Figure 3-7 Rutherford bombarded metal foil with a beam of positively charged particles. Something deflected them; Rutherford suggested that it was a dense, positively charged nucleus.

Chemistry Notebook

Millikan's Oil-Drop Experiment

The problem of determining the actual mass and charge of the electron was solved in 1909 by Robert A. Millikan of the University of Chicago using the apparatus shown in the figure to the right.

A mist of fine droplets of oil was allowed to fall slowly through the air in the chamber. Some of these droplets were negatively charged, having acquired extra electrons dislodged from gases in the air by X rays. A microscope outside the chamber allowed Millikan to observe the relatively few droplets that fell through the small central hole in the upper metal plate.

When there was no voltage between the two metal plates, the falling droplets were acted on only by the force of gravity and by friction with the air. Millikan carefully adjusted the voltage between the metal plates so that the downward gravitational force on a droplet was countered by an upward force between the charged droplet and the charged metal plates. At this

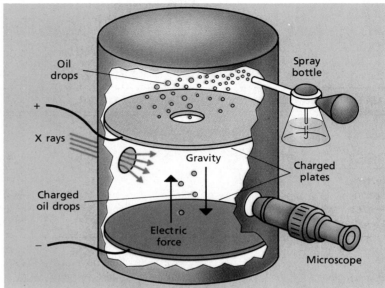

voltage, the drop is held motionless in the air. The upward force was proportional to the product of the charge of the droplet and the voltage between the two plates. The charge on each droplet could then be calculated from the voltage between the two plates. It is proportional to the charge required to balance the gravitational force on the droplet.

Millikan found that all the negative charges on the falling droplets are whole-

number multiples of the same negative charge. He assumed that this was the smallest possible negative charge, and, therefore, the charge of a single electron. Using this value for the charge of the electron, he then used Thomson's charge-to-mass ratio for the electron to calculate the mass of the electron by using the following relationship:

$$\text{mass of electron} = \frac{\text{charge of electron}}{\text{charge-to-mass ratio of electron}}$$

protons and *neutrons*. Protons have a mass of 1.673×10^{-24} g, which is 1836 times larger than the mass of an electron and $\frac{1836}{1837}$ of the mass of the lightest and simplest type of hydrogen atom, which consists of a single-proton nucleus with a single electron moving about it. Most of the mass of such a hydrogen atom is due to the mass of the proton. **Protons** *are subatomic particles that have a positive charge equal in magnitude to the negative charge of an electron and are present in atomic nuclei.* The electrical neutrality of atoms is due to the presence of equal numbers of protons and electrons.

Neutrons were discovered in 1932 by the English scientist James C. Chadwick (1891–1974), who had worked with Rutherford.

Protons alone do not account for the masses of atoms of elements other than the simplest hydrogen atom. These atoms also contain **neutrons,** *which are electrically neutral, subatomic particles found in atomic nuclei.* The mass of a neutron is 1.675×10^{-24} g—about the same mass as that of a proton.

Physicists have now identified many other subatomic particles. However, particles other than electrons, protons, and neutrons play little role in the properties of matter that are of interest in chemistry. Table 3-1 summarizes the properties of electrons, protons, and neutrons.

TABLE 3-1 PROPERTIES OF SUBATOMIC PARTICLES

Particle	Symbol	Relative Electric Charge	Mass Number	Relative Mass*	Actual Mass
electron	e^-	−1	0	$0.000\,548\,6u$	9.110×10^{-28} g
proton	p^+	+1	1	$1.007\,199\,u$	1.673×10^{-24} g
neutron	n^o	0	1	$1.008\,\,94\,u$	1.675×10^{-24} g

*$1\,u = 1.660\,565\,5 \times 10^{-24}$ g

STUDY HINT

Chemical changes do not involve the creation or change in identity of atoms.

Changes in which atoms are changed in identity are referred to as nuclear reactions, or nuclear changes and are discussed in Chapter 30.

An *element* is a substance composed of atoms all of which contain the same number of protons.

Tritium, which is found in the atmosphere, might be formed when nitrogen is bombarded by neutrons from cosmic rays.

$$^{14}_{7}\text{N} + ^{1}_{0}\text{n} \rightarrow ^{12}_{6}\text{C} + ^{3}_{1}\text{H}$$

It is used as a tracer, to make luminous paint, and in the production of hydrogen bombs. Both deuterium and tritium may be used as nuclear fuels for controlled nuclear fusion reactions in the future.

Particles that have the same electric charge generally repel one another. Nevertheless, over 100 protons can exist close together in a nucleus. When two protons are very close to each other, there is a strong attraction between them. Proton–proton attractive forces, neutron–neutron attractive forces, and proton–neutron attractive forces exist when these pairs of particles are very close together. *These short-range proton–neutron, proton–proton, and neutron–neutron forces hold the nuclear particles together and are referred to as* **nuclear forces.**

The nuclei of atoms of different elements differ in numbers of protons and therefore in positive charge. The nuclei of atoms of different elements also have different masses, although the difference in mass is sometimes very slight.

The Isotopes of Hydrogen

The number of protons in the nucleus determines the identity of an atom. All hydrogen atoms contain only one proton. An atom that contains more than one proton in its nucleus is an atom of an element other than hydrogen.

Like those of all naturally occurring elements, however, hydrogen atoms can contain varying numbers of neutrons. Three types of hydrogen atoms are known. The most common type of hydrogen is sometimes called *protium.* The nucleus of a protium atom consists of *one proton only,* and has one electron moving about it.

In addition to protium, which makes up 99.985% of naturally occurring hydrogen, there are two other known types of hydrogen. One is *deuterium,* which occurs in 0.015% of all hydrogen in nature. Each deuterium atom has a nucleus containing one proton and one neutron. The third form of hydrogen is known as *tritium,* which is radioactive. It exists in very small amounts in nature, but can be prepared artificially in a nuclear reaction. Each

Technology

"Looking" at Atoms

Scientists have long believed that atoms exist because the atomic theory successfully explains observable properties of matter. Now various modern techniques provide images of atoms on a computer screen or on film. When transmission electron microscopy (TEM) was first developed in the 1930s by Ernst Ruska, it could magnify objects 12 thousand ×. Since then, TEM has achieved magnifications of 2 million × and can be used to show the arrangement of relatively large atoms as illustrated below. A thin specimen (about 1000 atoms thick) is illuminated from the rear by an electron beam. Electrons emerging through the front of the sample are focused by a magnetic lens onto camera film or a fluorescent screen.

Images of atoms on the surface of a material are provided by a technique in microscopy called scanning-tunneling microscopy (STM).

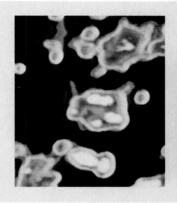
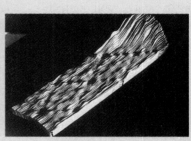

The STM can achieve magnifications of up to 100 million. It was developed in 1981 by Gerd Binnig and Heinrich Rohrer. A precise mechanism causes an extremely thin needle-like probe to move along the surface of the sample in parallel, closely spaced lines. A high-voltage electric current is maintained between the sample surface and the needle tip. This current is related to the distance between the tip and the atoms on the sample surface.

The same distance is always maintained between the tip and the surface atoms. In this way, the needle precisely follows the uneven contour of the surface. The up-and-down motion of the tip is processed by a computer, which generates an image of the sample surface like that shown in the STM image of silicon on the right. Ernst Ruska, Gerd Binnig, and Heinrich Rohrer shared the 1986 Nobel Prize in physics for their work on microscopy.

tritium atom contains one proton, two neutrons, and one electron, as shown schematically in Figure 3-8.

Protium, deuterium, and tritium are described as *isotopes* of hydrogen. Atoms of each have one proton, but they have one, two, and three neutrons, respectively, and therefore different total masses. **Isotopes** *are atoms of the same element that have different masses.* In all three types of hydrogen atoms, the charge of the single proton is balanced by that of one electron.

The Sizes of Atoms

The nuclei of atoms have radii of about 1×10^{-13} cm. They have a density of about 20 metric tons/cm^3. In a unit that is more convenient for the sizes of atoms, nuclei have radii of about 0.001 pm (picometers—1 pm $= 10^{-12}$ m). By contrast, atomic radii range from about 40–270 pm. To get some idea of the extreme smallness of the picometer, consider the fact that 1 cm is the same fractional part of 10^3 km (about 600 miles) as 100 pm is of 1 cm.

The radius of an atom is the distance from the center of the nucleus to the outer edge of the region occupied by electrons, which are rapidly moving around the nucleus. It is convenient to think of the region occupied by the electrons as an electron cloud—a cloud of negative charge. In the next chapter, you will learn more about the nature of the electron cloud.

1. (a) The smallest particle of an element that can exist either alone or in combination with other atoms (b) A negatively charged subatomic particle (c) The positively charged, dense, central portion of an

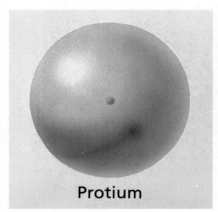
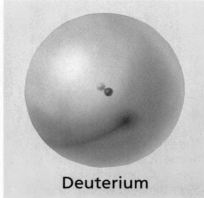
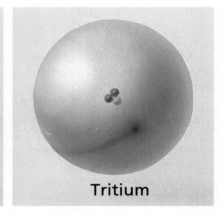

Protium   **Deuterium**   **Tritium**

Figure 3-8 Hydrogen and its isotopes have different nuclei.

atom that contains nearly all of its mass but takes up only an insignificant fraction of its volume (d) Subatomic particles of the nucleus that have a positive charge of the same size as the negative charge of an electron (e) Electrically neutral, subatomic particles found in atomic nuclei (f) Atoms of the same element that have different masses
2. (a) Thomson's work provided strong support for the hypothesis that cathode rays are negatively

Section Review

1. Define each of the following: (a) atom (b) electron (c) nucleus (d) protons (e) neutrons (f) isotopes.
2. Cite at least one contribution of each of the following toward the development of the present atomic theory: (a) Thomson (b) Millikan (c) Rutherford.
3. Compare and contrast the three types of subatomic particles in terms of location in the atom, mass, and relative charge.
4. How do the three isotopes of hydrogen compare in terms of the numbers of subatomic particles in each?

3.3 Weighing and Counting Atoms

Consider neon (Ne) the gas in many illuminated signs. Neon is one of the minor components of air, since dry air contains only about 0.002% neon. Yet about 5×10^{17} atoms of neon are present in each breath you inhale. Atoms are much too small to be weighed individually or counted in most experiments. How, then, can chemists quantitatively study elements and the chemical compounds made from them?

In this section, you will be introduced to some of the concepts used to link observed chemical behavior with tiny atomic and subatomic particles.

SECTION OBJECTIVES

- Define "atomic number" and "mass number" and describe how they apply to isotopes and nuclides.
- Determine the number of protons, neutrons, and electrons in a nuclide, given the identity of the nuclide.

charged particles (b) Millikan's work confirmed that the electron has the smallest possible negative electric charge

(c) Rutherford discovered the nucleus and described its contents.

Atomic Numbers and Mass Numbers

The **atomic number** *(Z) of an element is the number of protons in the nucleus of each atom of that element.* An element consists of atoms all of which have the same number of protons in their nuclei. Atoms of different elements contain different numbers of protons.

In the Periodic Table (inside the back cover), the elements are placed from left to right in order of increasing atomic numbers. At the top of the table is hydrogen (H), which has atomic number 1. Atoms of the element hydrogen, whether they be protium, deuterium, or tritium atoms, have one proton in each nucleus. Next in order are helium (He) with two protons in each nucleus; lithium (Li) with three protons, beryllium (Be) with four protons; and so on.

The atomic number alone identifies an element. If you want to know which element has, for example, atomic number 47, consulting the Periodic Table would show that it is silver (Ag). All silver atoms contain 47 protons in their nuclei. Because atoms are neutral, we know from the atomic number that all silver atoms must also contain 47 electrons.

Atoms of different isotopes have different masses. Identifying an *isotope* requires knowing both the name or atomic number of the element and the mass of the isotope. *The* **mass number** *is the total number of protons and neutrons in the nucleus of an isotope.* The three isotopes of hydrogen described earlier have mass numbers 1, 2, and 3, as shown below.

	Atomic Number (Number of Protons)	Number of Neutrons	Mass Number
protium	1	0	$0 + 1 = 1$
deuterium	1	1	$1 + 1 = 2$
tritium	1	2	$1 + 2 = 3$

Two methods are used to designate isotopes in writing. In the first method, the mass number is added with a hyphen to the name of the element. Tritium, for example, is described as hydrogen-3. We will refer to this method as "hyphen notation." The uranium isotope used in fuel for nuclear power plants has a mass number of 235 and is known as uranium-235.

To determine the composition of, for example, a uranium-235 nucleus, one first find from the Periodic Table the atomic number of uranium, which is 92. The number of neutrons is then found by subtracting the atomic number from the mass number:

$$\text{mass number} - \text{atomic number} = ?$$
$$235 \text{ (protons + neutrons)} - 92 \text{ protons} = 143 \text{ neutrons}$$

A uranium-235 nucleus contains 92 protons and 143 neutrons.

In the second method of designating isotopes, the atomic number is added to the symbol for the element as a subscript on the left and the mass number is added as a superscript on the left. For hydrogen-3 and uranium- 235, these nuclear symbols are

$$\overset{\text{mass number}}{\underset{\text{atomic number}}{{}_{1}^{3}\text{H}}} \qquad {}_{92}^{235}\text{U}$$

- Distinguish between *relative* atomic mass and *average* atomic mass.
- Calculate the average atomic mass of an element given the relative abundances of each isotope of the element.
- Define a mole in terms of Avogadro's number; define "molar mass."
- Solve problems involving mass, mole, and the number of atoms of an element.

Section Review Answers
continued

3. Electrons are located in energy levels surrounding the nucleus of an atom, they have a mass of 9.110×10^{-28} g and a relative charge of -1. Protons are located in the nucleus, have a mass of 1.673×10^{-24} g, and have a relative charge of $+1$. Neutrons are also in the nucleus, have a mass of 1.675×10^{-24} g, and have no charge.
4. All contain one proton and one electron, but protium contains no neutrons, deuterium contains one neutron, and tritium contains two.

Table 3-2 gives the names, symbols, and composition of the isotopes of hydrogen and helium. **Nuclide** *is a general term for any isotope of any element*. For example, we could say that Table 3-2 lists the composition of five different nuclides.

TABLE 3.2   ISOTOPES OF HYDROGEN AND HELIUM

Isotope	Symbol	Number of Protons	Number of Electrons	Number of Neutrons
hydrogen-1 (protium)	^1_1H	1	1	0
hydrogen-2 (deuterium)	^2_1H	1	1	1
hydrogen-3 (tritium)	^3_1H	1	1	2
helium-3	^3_2He	2	2	1
helium-4	^4_2He	2	2	2

Sample Problem 3.1

How many protons, electrons, and neutrons are there in an atom of chlorine-37?

Solution

Step 1. Analyze   Given: name and mass number of chlorine-37
Unknown: numbers of protons, electrons, and neutrons

Step 2. Plan   atomic number ⟶ no. of protons and electrons
mass number ⟶ no. of neutrons

The atomic number of chlorine can be found in the Periodic Table, and this is equal to the number of protons and also the number of electrons. The number of neutrons must be found by subtracting the atomic number from the mass number.

mass number of chlorine-37 − atomic number of chlorine
= number of neutrons in chlorine-37

Step 3. Compute   mass number − atomic number = 37 (protons plus neutrons) − 17 protons
= 20 neutrons

An atom of chlorine-37 contains 17 electrons, 17 protons, and 20 neutrons.

Step 4. Evaluate   The number of protons equals the number of electrons and the sum of the proton and neutrons equals the given mass number as they should.

Practice Problems
1. How many protons, electrons, and neutrons are in an atom of bromine-80?   *(Ans.)* 35 protons, 35 electrons, and 45 neutrons
2. Write the nuclear symbol for carbon-13.   *(Ans.)* $^{13}_6\text{C}$
3. Write the hyphen notation for the element that contains 15 electrons and 15 neutrons. *(Ans.)* phosphorus-30

Relative Atomic Masses

Masses of atoms measured in grams are very small. An atom of oxygen-16, for example, has a mass of 2.657×10^{-23} g. It is only in this century that scientists have been able to measure masses this small. Although we now know the actual masses of atoms, for most chemical calculations it is more convenient to use *relative* masses.

As you learned in Chapter 2, scientists use standards of measurement that are constant and the same everywhere. In order to set up a relative scale of *atomic mass,* one atom is arbitrarily chosen as the standard and assigned a relative mass value. The masses of all the other atoms are then expressed in relation to this defined standard. The carbon-12 atom has been chosen as the standard by the international organization of scientists that governs units of measurement. A single atom of the nuclide carbon-12 is arbitrarily assigned a mass of exactly 12 atomic mass units. *The **atomic mass unit** is symbolized by u.* **One atomic mass unit,** *1 u, is exactly $\frac{1}{12}$ the mass of a carbon-12 atom, or 1.660 565 5 $\times$ 10^{-24} g.* The atomic mass of a carbon-12 is exactly 12 *u. The mass of an atom expressed in atomic mass units is called the **atomic mass** of the atom.*

The hydrogen-1 atom actually does have a relative atomic mass of *about* $\frac{1}{12}$ that of the carbon-12 atom, or about 1 *u.* The precise value of the relative atomic mass of a hydrogen-1 atom is 1.007 825 *u.* An oxygen-16 atom has about $\frac{4}{3}$ the mass of a carbon-12 atom. Careful measurements show the relative atomic mass of oxygen to be 15.994 915 *u.* The mass of a magnesium-24 atom is found to be slightly less than twice that of a carbon-12 atom. Its atomic mass if 23.985 042 *u.* The atomic mass of any nuclide is determined by comparison with the mass of the carbon-12 atom.

Some additional examples of the relative atomic masses of the naturally occurring isotopes of several elements are given in Table 3-3. Isotopes of an element may occur naturally, or they may be made in the laboratory (*artificial isotopes*). Although isotopes have different masses, they do not differ significantly in their chemical behavior. (See Appendix Table 2 for a list of the masses of natural and artificial isotopes.

The masses of subatomic particles can also be expressed on the atomic mass scale (see Table 3-1). The atomic mass of the electron is 0.000 548 59 *u.* That of the proton is 1.007 276 *u* and that of the neutron is 1.008 665 *u.* Note that the proton and neutron masses are close to but not equal to 1. You have learned that the mass number is the total number of protons and neutrons in the nucleus of an atom. You can now see that the mass number and relative atomic mass of a given nuclide are quite close to each other. They are not identical because the proton and neutron masses deviate slightly from 1 *u* and the atomic masses include electrons. Also, as explained in Chapter 30, a small amount of mass is changed to energy in the creation of a nucleus from protons and neutrons.

Frédéric and Irène Joliot-Curie received the 1935 Nobel Prize in chemistry for discovering a way to make the first artificial isotopes by bombarding stable atoms with neutrons or deuterons. As a result of their work, radioactive isotopes can be made for all elements.

Average Atomic Masses of Elements

As illustrated for the elements listed in Table 3-3, most elements occur naturally as mixtures of isotopes. The percentage of each isotope in the naturally occurring element is nearly always the same, no matter where the element is found.

STUDY HINT
The isotopes of hydrogen are unusual in having distinctive names. Most isotopes are identified only by their mass numbers.

Isotope	Mass Number	Percent Natural Abundance	Relative Atomic Mass (u)	Average Atomic Mass* of Elements (u)
hydrogen-1	1	99.985	1.007 825	1.007 94
hydrogen-2	2	0.015	2.0140	
carbon-12	12	98.90	12 (by definition)	12.011
carbon-13	13	1.10	13.003 355	
oxygen-16	16	99.762	15.994 915	
oxygen-17	17	0.038	16.999 131	15.9994
oxygen-18	18	0.200	17.999 160	
copper-63	63	69.17	62.939 598	63.546
copper-65	65	30.83	64.927 793	
cesium-133	133	100	132.905 429	132.905
uranium-235	235	0.720	235.043 924	238.024
uranium-238	238	99.280	238.050 784	

* Values, established by international agreement, may differ slightly from calculated averages.

A *nonweighted* average of the marble masses is (2g + 3g) ÷ 2 = 2.5 g.

Atomic masses for the elements, which are included in the Periodic Table are averages for these naturally occurring mixtures of isotopes. *Weighted averages* account for the percentages of each isotope of a given element that is present.

As a simple example of a weighted average, suppose you had a box containing two sizes of marbles. If 25% of the marbles weigh 2.00g each and 75% of the marbles weigh 3.00 g each, how could the weighted average be found? You could count the marbles, calculate the total mass of the mixture, and divide by the total number of marbles. Suppose you had 100 marbles. The calculations would go as follows: 25 marbles × 2.00 g = 50 g and 75 marbles × 3.00 g = 225 g. Adding these masses gives 50 g + 225 g = 275 g. Dividing by 100 gives an average marble mass of 2.75 g. A simpler method is to multiply the mass of each marble by the decimal fraction representing its percentage in the mixture.

$$2.0 \text{ g} \times 0.25 + 3.0 \text{ g} \times 0.75 \text{ g} = 2.75 \text{ g}$$

Average atomic mass *is the weighted average of the atomic masses of the naturally occurring isotopes of an element.* As illustrated, the average atomic mass depends on both the mass and the relative abundance of each isotope. For example, naturally occurring copper consists of 69.17% copper-63, which has an atomic mass of 62.939 598 *u*, and copper-65, which occurs with a proportion of 30.83% and has an atomic mass of 64.927 793 *u*. The average atomic mass of copper can be calculated by multiplying the atomic mass of each isotope by its relative abundance (expressed in decimal form) and adding the results:

$$0.6917 \times 62.939 \ 598 \ u + 0.3083 \times 64.927 \ 793 \ u = 63.55 \ u$$

The calculated average atomic mass of naturally occurring copper is 63.55 *u*.

STUDY HINT
You may sometimes see the term *atomic weight* used synonymously with *atomic mass.*

The average atomic masses of the elements are important quantities to chemists because they indicate relative mass relationships in chemical reactions. You can find the average atomic masses of the elements in the Periodic Table inside the back cover of this book. What is the average atomic mass of silicon (Si)? What is the average atomic mass of lead (Pb)?

The average atomic masses are included for the elements listed in Table 3-3. In common use, we do not bother to distinguish between relative and average atomic masses, for it is clear from the context which is intended. The **atomic mass** of a nuclide is the relative atomic mass of atoms of that nuclide; the atomic mass of an element is the average atomic mass of the naturally occurring mixture of isotopes of the element.

As illustrated in Table 3-3, most atomic masses are known to four or more significant figures. In calculations in this book (other than addition), atomic masses of the elements are rounded before the calculation is carried out to the full number of digits that are known.

The Mole, Avogadro's Number, and Molar Mass

The relative atomic mass scale makes it possible to know how many atoms of an element are present in a given sample of the element with a measurable mass. To understand the composition of chemical compounds and what happens in chemical reactions, however, relative numbers of atoms and other chemical species must be noted.

Imagine that you could place a single carbon atom on one side of a balance. How many helium atoms would be needed on the other side to balance the carbon atom? The atomic mass of the element carbon is 12.01 u (to two decimal places) and that of helium is 4.00 u. Three helium atoms would just about balance the carbon atoms, as shown schematically in Figure 3-9. If a second carbon atom were added to the first, then six helium atoms would be needed to achieve balance. No matter what number of carbon atoms were placed on the balance, three times as many helium atoms would be needed. Also, whatever mass of carbon were to be balanced, an equal mass of helium would contain three times as many atoms.

Figure 3-9 The relative masses of carbon and helium atoms

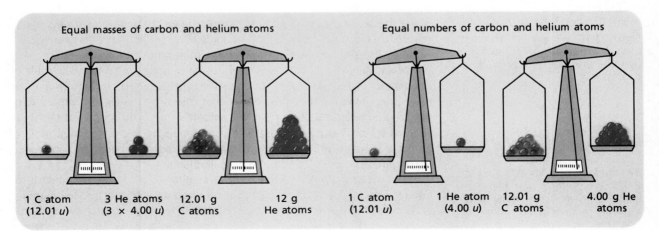

Equal masses of carbon and helium atoms				Equal numbers of carbon and helium atoms			
1 C atom (12.01 u)	3 He atoms (3 × 4.00 u)	12.01 g C atoms	12 g He atoms	1 C atom (12.01 u)	1 He atom (4.00 u)	12.01 g C atoms	4.00 g He atoms

Figure 3-10 One molar mass of each of three elements: sulfur *(top)*, aluminum *(center)* and carbon.

Next, imagine that you have placed 12.01 g of carbon on the scale. But instead of equal masses of helium atoms, now you would like to have the number of helium atoms present equal to the number of carbon atoms present. Note that 12.01 g is a mass in grams equal to the relative atomic mass of carbon, 4.00 g. Because the mass in grams of helium atoms and carbon atoms on the scale is equal to their relative atomic masses, you would have an identical number of helium and carbon atoms by 4.00 He on the scale.

The mole Three very important concepts—the mole, Avogadro's number, and molar mass—provide the basis for relating masses in grams to numbers of atoms. The first, the mole, is the SI unit for amount of substance. *The **mole** is the amount of a substance that contains the same number of particles as the number of atoms in exactly 12 g of carbon-12. The abbreviation for the mole is* **mol.** The mole is a counting unit, just like one dozen. We don't usually order 12 or 24 ears of corn; we order one dozen or two dozen. In writing, a chemist might refer to 1 mol of carbon, or 2 mol of iron, or 2.567 mol of calcium. In this chapter you will see how the mole applies to chemical compounds.

Avogadro's number The number of particles in a mole has been experimentally determined in a number of ways. The best modern value is $6.022\ 045 \times 10^{23}$ particles. This number is of such importance to chemistry that it is named in honor of the Italian scientist Amadeo Avogadro (1776–1856), whose ideas were crucial in the early understanding of chemistry. **Avogadro's number**—*$6.022\ 045 \times 10^{23}$—is the number of particles in exactly one mole of a pure substance.* Exactly 12 g of carbon-12 contains an Avogadro's number of carbon-12 atoms. For most purposes, Avogadro's number is rounded to 6.022×10^{23}.

To get a sense of how large Avogadro's number is, imagine this situation: Everyone living on earth (5 billion people) is to help count the atoms in one mole of an element. If each person counted continuously at the rate of one atom per second, it would require about 4 million years for all the atoms to be counted.

Molar mass Can you suggest how many grams of helium contain an Avogadro's number of helium atoms? If you decided it is 4.00 g, you are correct. Similarly, 6.941 g of lithium (Li) or 200.59 g of mercury (Hg) would contain the Avogadro's number of atoms. These convenient quantities are molar masses of these elements. *The **molar mass** is the mass in grams of one mole of an element.* The mass of one mole of atoms, in grams, is numerically equal to the atomic mass of the element in atomic mass nuits.

Molar masses of elements contain equal numbers of atoms. One mole of a substance is one molar mass of that substance. An alternate definition of the *mole* is as the amount of a substance that contains an Avogadro's number of particles, or chemical units. Here the chemical unit is the atom. For elements such as those shown in Figure 3-10, one molar mass contains one mole of atoms of the naturally occurring isotopes. For individual nuclides, one molar mass contains one mole of atoms of the nuclide.

The molar mass is used like a conversion factor in chemical calculations. For a specific substance, a known number of grams can be converted to moles or a known number of moles can be converted to grams. In such calculations, mclar mass has the units of grams per mole. For example, to find how many grams of helium is equal to 2 mol of helium, you would use the dimensional method introduced in Chapter 2. The molar mass of helium is taken as 4.0 g/mol He (to be required number of significant figures).

$$2.0 \text{ mol He} \times \frac{4.0\text{g}}{\text{mol He}} = 8.0\text{g He}$$

Figure 3-11 shows the conversions of molar mass, moles, and Avogadro's number.

Sample Problem 3.2

What is the mass in grams of 3.50 mol of the element copper (Cu)?

Solution

Step 1. Analyze Given: 3.50 mol Cu
Unknown: mass of Cu in grams

Step 2. Plan moles $\longrightarrow$ mass

The molar mass of copper is the necessary factor for converting from moles of copper to grams of copper. It is the mass in grams numerically equal to the atomic mass for copper given in the periodic table.

$$\text{mol Cu} \times \frac{\text{grams Cu}}{\text{mol Cu}} = \text{grams Cu}$$

Step 3. Compute $$3.50 \text{ mol Cu} \times \frac{63.5 \text{ g Cu}}{\text{mol Cu}} = 222 \text{ g Cu}$$

Step 4. Evaluate The number of moles of copper was given to three significant figures; the molar mass was therefore rounded to three significant figures. The size of the answer is about right, for it is somewhat more than 3.5 times 60

Practice Problems
1. What is the mass in grams of 2.25 mol of the element iron (Fe)?   *(Ans.)* 126 g Fe
2. What is the mass in grams of 0.375 mol of the element potassium (K)?   *(Ans.)* 14.7 g K

Sample Problem 3.3

A chemist produced 11.9 g of aluminum (Al). How many moles of aluminum has been produced?

Solution

Step 1. Analyze Given: 11.9 g Al
Unknown: moles of Al

Step 2. Plan mass $\longrightarrow$ moles

The molar mass of aluminum is the conversion factor for finding moles of aluminum from a known mass of aluminum in grams. As shown in Figure 3-11, going from mass in grams to number of moles is done by dividing by molar mass. In using the dimensional method, it is easiest to do this by inverting the factor to mole per grams and multiplying.

$$\text{grams Al} \times \frac{\text{mole}}{\text{grams}} = \text{moles Al}$$

Step 3. Compute

$$11.9 \text{ g Al} \times \frac{\text{mol Al}}{27.0 \text{ g Al}} = 0.441 \text{ mol Al}$$

Step 4. Evaluate   The molar mass and the answer are correctly rounded to three significant figures. The answer is reasonable because 11.9 g is somewhat less than half of 27.0 g.

Practice Problems
1. How many moles of calcium (Ca) are contained in 5.00 g of calcium? *(Ans.)* 0.125 mol Ca
2. How many moles of gold (Au) are contained in 3.60×10^{-10} g of gold? *(Ans.)* 1.83×10^{-12} mol Au

Figure 3-11 shows that Avogadro's number can be used to find the number of atoms of an element from number of moles or to find the number of moles from the number of atoms of an element. While these types of problems are less common in chemistry than converting between moles and grams, they are useful in demonstrating the meaning of Avogadro's number. Note that, in these calculations, Avogadro's number is expressed as units of atoms per mole.

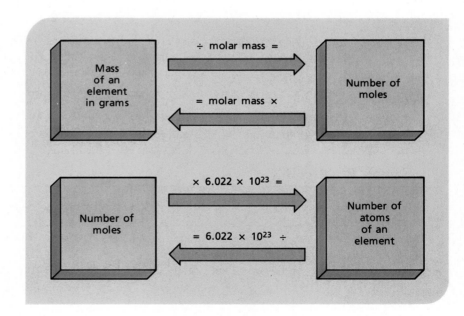

Figure 3-11 Relationships between mass, the number of moles, and the number of atoms of an element in a sample

Sample Problem 3.4

How many moles of silver (Ag) are in 3.01×10^{23} atoms?

Solution

Step 1. Analyze   Given: 3.01×10^{23} atoms of silver
Unknown: moles of silver

Step 2. Plan

atoms $\longrightarrow$ moles

Number of atoms is converted to number of moles by dividing by Avogadro's number (see Figure 3.13). In setting up the problem, we will use the inverted Avogadro's number factor so that we can multiply and cancel units.

$$\text{Ag atoms} \times \frac{\text{mol Ag}}{\text{Avogadro's number of Ag atoms}}$$

Step 3. Compute

$$3.01 \times 10^{23} \text{ Ag atoms} \times \frac{\text{mol Ag}}{6.02 \times 10^{23} \text{ Ag atoms}} = 0.500 \text{ mol Ag}$$

Step. 4. Evaluate   The answer is correct—units cancel correctly and the number of atoms is exactly one-half of Avogadro's number.

Practice Problems
1. How many moles of lead (Pb) are equivalent to 1.50×10^{12} atoms?   *(Ans.)* 2.49×10^{-12} mol Pb
2. How many moles of tin (Sn) are equivalent to 2500 atoms?   *(Ans.)* 4.2×10^{-21} mol Sn
3. How many atoms of aluminum (Al) are contained in 2.75 mol?   *(Ans.)* 1.66×10^{24} atoms Al

Sample Problem 3.5

What is the mass in grams of 1.20×10^8 atoms of copper?

Solution

Step 1. Analyze   Given: 1.20×10^8 atoms of copper
Unknown: mass in grams

Step 2. Plan

atoms $\longrightarrow$ moles $\longrightarrow$ mass

As indicated in Figure 3.13, the given number of atoms must first be converted to moles by dividing by Avogadro's number. The unknown mass in grams can then be found by multiplying by the molar mass of copper.

$$\text{Cu atoms} \times \frac{\text{mol Cu}}{\text{Avogadro's no. of Cu atoms}} \times \frac{\text{grams Cu}}{\text{mol Cu}} = \text{grams Cu}$$

Step 3. Compute

$$1.20 \times 10^8 \text{ Cu atoms} \times \frac{\text{mol Cu}}{6.02 \times 10^{23} \text{ Cu atoms}} \times \frac{63.6 \text{ g Cu}}{\text{mol Cu}} = 1.27 \times 10^{-14} \text{ g Cu}$$

Step 4. Evaluate   Units cancel correctly to give the answer in grams. The size of the answer is reasonable—10^8 has been divided by 10^{23}.

Practice Problems
1. What is the mass in grams of 7.5×10^{15} atoms of nickel (Ni)?   *(Ans.)* 7.3×10^{-7} g Ni
2. How many atoms of sulfur (S) are contained in 4.00 g?   *(Ans.)* 7.5×10^{22} atoms S
3. What mass of gold (Au) contains the same number of atoms as 9.0 g of aluminum (Al)?
   (Ans.) 66 g Au

1. (a) The number of protons in the nuclei of atoms of that element   (b) The total number of protons and neutrons in a nucleus.   (c) The mass of an atom expressed in atomic mass units.   (d) The weighted average of the masses of the naturally occurring isotopes of an element.   (e) The amount of a substance that contains a number of particles equal to the number of atoms in exactly 12 grams of carbon-12   (f) The number of particles in exactly one mole of a pure substance   (g) The mass in grams of an element that is numerically equal to its mass on the relative atomic mass scale

2. (a) 11, 11, 12   (b) 20, 20, 20   (c) 29, 29, 35   (d) 47, 47, 61

3. (a) $^{28}_{14}$ Si; silicon-28   (b) $^{56}_{26}$ Fe; iron-56   (c) $^{183}_{56}$ Ba; barium-138

4. (a) 39.098 u (b) 39.098 g

5. (a) 28.0 g N   (b) 17.8 g Cl

6. (a) 0.500 0 mol Mg   (b) 0.249 mol F

7. (a) 1.50×10^{24} atoms Zn   (b) 7.52×10^{22} atoms C

Section Review

1. Define each of the following: (a) atomic number   (b) mass number   (c) relative atomic mass   (d) average atomic mass   (e) mole   (f) Avogadro's number   (g) molar mass.

2. Determine the number of protons, electrons, and neutrons in each of the following isotopes: (a) sodium-23   (b) calcium-40   (c) $^{64}_{20}$ Cu   (d) $^{108}_{47}$ Ag.

3. Write the nuclear symbol and hyphen notation for each of the following neutral isotopes: (a) mass number of 28 and atomic number of 14   (b) 26 protons and 30 neutrons   (c) 56 electrons and 82 neutrons.

4. To three decimal places, what is the   (a) relative atomic mass and   (b) molar mass of the element potassium, K?

5. What is the mass in grams of: (a) 2.00 mol N   (b) 3.01×10^{23} atoms Cl?

6. How many moles are represented by: (a) 12.15 g Mg   (b) 1.50×10^{23} atoms F?

7. How many atoms are contained in: (a) 2.50 mol Zn   (b) 1.50g C?

Chapter Summary

- Scientists since the ancient Greeks have created theories of atoms. In the nineteenth century, John Dalton proposed a five-point theory of atoms that can still explain properties of many chemicals and compounds today.
- Mass can be neither created or destroyed in chemical reactions.
- Cathode ray tubes supplied the first evidence of the existence of *electrons*, which are negatively charged subatomic particles having relatively little mass.
- Rutherford found evidence for the existence of the atomic nucleus—a positively charged, very dense object within the atom—by using a positively charged particle beam aimed at metal foil.
- Atomic nuclei are composed of protons, that have an electric charge of +1, and, in all but one case, neutrons, that have an electric charge of −1.
- Isotopes of an element differ by the number of neutrons in their nuclei.
- Atomic nuclei have a radius of about 0.001 pm (pm = picometers; 1 pm = 10^{-12} m), while atoms have a radium of about 40–270 pm.
- The atomic number of an element is the number of protons in the nuclei of the atoms of that element.
- The atomic mass number is equal to the total number of protons and neutrons in the nucleus.
- The relative atomic mass unit (u) is based on the carbon-12 atom and is a convenient unit for measuring the mass of atoms.
- The average atomic mass of an element is found by calculating the weighted average of the atomic masses of the naturally occurring isotopes of the element.
- Avogadro's number is equal to $6.022 045 \times 10^{23}$ particles per mole of atoms or molecules. It is the number of atoms in 12 grams of carbon-12. A sample that contains a number of particles equal to Avogardo's number contains a mole of those particles.

Chapter 3 *Review*

Atoms: The Building Blocks of Matter

Vocabulary

atom	law of multiple
atomic mass	proportions
atomic mass unit	mass number
atomic number	molar mass
atomic structure	mole
average atomic mass	neutrons
Avogadro's number	nuclear forces
electrons	nucleus
element	nuclide
isotopes	protons

Questions

1. *(a)* When, where, and from whom did the idea of the atom originate? *(b)* What does the word *atom* mean?
2. Summarize the five essential points of Dalton's atomic theory.
3. Explain each of the following in terms of Dalton's atomic theory: *(a)* the law of conservation of mass *(b)* the law of definite composition *(c)* the law of multiple proportions.
4. According to the law of conservation of mass, if element A has a mass of 2 mass units and element B has a mass of 3 mass units, what mass would be expected for element AB?
5. *(a)* State the law of multiple proportions. *(b)* Who is credited with this law?
6. *(a)* What is an atom? *(b)* What is meant by the atomic structure of an atom?
7. What two major regions make up all atoms?
8. Briefly describe the types of experiments that eventually led to the discovery of the electron.
9. List six major facts that were established as a result of the cathode-ray tube experiments.
10. Briefly summarize Thomson's findings in support of cathode rays as negatively charged particles.
11. Cite at least four properties of electrons that evolved from the experiments of Thomson.
12. What is an electron?
13. Beyond that concerning electrons, what additional information about atomic structure was obtained from the cathode ray experiments?
14. *(a)* Who is credited with the discovery of the atomic nucleus? *(b)* What is the nucleus of an atom? *(c)* Summarize Rutherford's description of an atom.
15. Identify and define the two kinds of particles contained in the nucleus of an atom.
16. Copy and complete the following table concerning the properties of subatomic particles:

Particle	Symbol	Mass Number	Actual Mass	Relative Charge
Electron	_____	_____	_____	_____
Proton	_____	_____	_____	_____
Neutron	_____	_____	_____	_____

17. What are the nuclear forces within an atom?
18. What one factor determines the identity of a particular atom?
19. *(a)* What are isotopes? *(b)* How are the isotopes of a particular element alike? *(c)* How are they different?
20. Copy and complete the following table relative to the three isotopes of hydrogen (H-1, H-2, and H-3):

Isotope	Name	Number of Protons	Number of Electrons	Number of Neutrons
H-1	_____	_____	_____	_____
H-2	_____	_____	_____	_____
H-3	_____	_____	_____	_____

21. *(a)* What is the meaning of the term *the atomic number of* an element? *(b)* What is the mass number of an isotope? *(c)* In the nuclear symbol for deuterium ($_1^2H$), identify the atomic number and the mass number. *(d)* Write its hyphen notation.
22. What is a nuclide?
23. Use the Periodic Table and the information given below to write the hyphen notation for each isotope described: *(a)* atomic number = 2 mass number = 4 *(b)* atomic number = 8 mass number = 16 *(c)* atomic number = 19 mass number = 39

24. Copy and complete the following table:

	Nuclear Symbol	Atomic Number	Mass Number	Number of Protons, Electrons, Neutrons
(a)	^4_2He	___	___	___
(b)	^9_4Be	___	___	___
(c)	___	16	32	___
(d)	___	14	___	14 neutrons
(e)	___	___	65	30 protons
(f)	___	___	___	18 electrons 22 neutrons

25. (a) In the development of the relative scale for atomic masses, what atom is used as the standard? (b) What is its relative mass assignment?

26. (a) What is an atomic unit (symbol u)? (b) What is the relative atomic mass of an atom?

27. (a) What would the atomic mass of an atom be if its mass is approximately (a) $\frac{1}{3}$ that of C-12? (b) $\frac{1}{2}$ that of C-12? (c) 12 × that of C-12? (d) 4.5 times as much as C-12?

28. What are the atomic masses of each of the following subatomic particles? (a) proton (b) electron (c) neutron.

29. (a) What is the average atomic mass of a naturally occurring element? (b) Upon what two factors is the value of the average atomic mass dependent? (c) How is this value calculated?

30. What is the atomic mass of each of the following to 3 decimal places? (a) oxygen (b) aluminum (c) copper (d) gold.

31. (a) What is the SI definition of the mole? (b) What is the abbreviation for mole? (c) How many particles are contained in a mole? (d) What name is given to the number of particles in a mole?

32. (a) What is the molar mass of an element? (b) To three decimal places, write the molar masses for: (1) carbon (2) neon (3) iron (4) uranium.

33. Determine each of the following relative to cobalt: (a) atomic number (b) mass number of Co-59 (c) numbers of protons, electrons, and neutrons in Co-59 (d) relative atomic mass of Co-59 (e) the average atomic mass of all isotopes of cobalt to 3 decimal places

(f) molar mass of all isotopes of cobalt 1 to 3 decimal places (g) the number of atoms contained in 58.9 g of cobalt.

Problems

1. What is the mass, in grams, of each of the following? (a) 1.00 mol Li (b) 1.00 mol Al (c) 1.00 molar mass Ca (d) 1.00 molar mass Fe (e) 6.022×10^{23} atoms C (f) 6.022×10^{23} atoms Ag

2. Determine the mass, in grams, of each of the following: (a) 3.00 mol Al (b) 4.25 mol Li (c) 1.38 mol N (d) 8.075 mol Au (e) 6.50 mol Cu (f) 2.57×10^8 mol S (g) 1.75×10^{-6} mol Hg.

3. How many moles of atoms are represented by each of the following? (a) 40.1 g Ca (b) 11.5 g Na (c) 5.87 g Ni (d) 150 g S (e) 2.65 g Fe (f) 0.007 50 g Ag (g) 3.25×10^5 g Pb (h) 4.50×10^{-12} g O

4. How many moles of atoms are equivalent to each of the following? (a) 6.022×10^{23} atoms Ne (b) 3.011×10^{23} atoms Mg (c) 2.25×10^{25} atoms Zn (d) 50.0 atoms.

5. How many atoms are contained in each of the following? (a) 1.50 mol Na (b) 6.755 mole Pb (c) 0.250 mol Si.

6. What is the mass, in grams, of each of the following? (a) 3.011×10^{23} atoms F (b) 1.50×10^{23} atoms Mg (c) 4.50×10^{12} atoms Cl (d) 8.42×10^{18} atoms Br (e) 2.50 atoms W (f) 1.00 atom Au

7. Determine the number of atoms contained in each of the following: (a) 5.40g B (b) 8.02 g S (c) 1.50 g K (d) 0.025 50 g Pt (e) 1.00×10^{-10}g Au.

8. Within each of the following pairs, which contains the greater number of atoms? (a) 0.500 mol C or 0.750 mole Li (b) 19.0 g F or 19.0 g Ne (c) 0.250 mol Si or 10.0 g Cl (d) 15.00 g Ca or 40.00 g Ag.

9. Among the elements listed, determine the mass of each that would contain the same number of atoms as 12.0 g Al: (a) O (b) Cl (c) Ca (d) Ni (e) Pb.

Application Questions

1. Although exceptions to Dalton's atomic theory are now known, why is it still important to study it?

2. Explain the significance of Rutherford's statement that it was "as if you had fired a 15-inch [artillery] shell at a piece of tissue paper and it came back and hit you."

3. What specific evidence led Rutherford to come to each of the following conclusions? *(a)* an atom is mostly empty space *(b)* the nucleus occupies very little space in the atom *(c)* the nucleus is positively charged *(d)* the electrons surround the nucleus like the rings around Saturn.

4. Particles that have the same charge generally repel each other. As many as 100 protons can exist together in the small nucleus of an atom, however. Explain this seeming contradiction of observed phenomena.

5. *(a)* What is the relationship between an atom containing 8 protons, 8 electrons, and 8 neutrons, and one containing 8 protons, 8 electrons, and 10 neutrons? *(b)* What is the relationship between an atom containing 8 protons, 8 electrons, and 10 neutrons, and one containing 10 protons, 10 electrons, 8 neutrons?

6. Explain why the average atomic masses for most elements are rarely whole numbers.

7. Distinguish between the atomic mass of a nuclide and the atomic mass of an element.

Application Problems

1. Three isotopes of argon occur in nature,—$^{36}_{18}Ar$, $^{38}_{18}Ar$, and $^{40}_{18}Ar$. If the relative atomic masses and abundances of each of these isotopes are as follows, calculate the average atomic mass of argon to three decimal places argon-36 (35.978 u; 0.337%), argon-38 (37.963 u; 0.063%), and argon-40 (39.962 u; 99.600%).

2. If naturally occurring boron is 80.20% B-11 (atomic mass = 11.009 u) and 19.80% of some other isotopic form of boron, what must the atomic mass of this second isotope be in order to account for the 10.811 u average atomic mass of boron? (Express to three decimal places.)?

3. If the only two naturally occurring isotopes of chlorine, Cl-35 and Cl-37, have relative atomic masses of 34.969 u and 36.966 u respectively, what must be the approximate abundance of each in nature if the average atomic mass of chlorine is 35.453 u? (Express the final percentages to three decimal places.)?

4. Prior to 1961, the relative atomic mass scale was based on oxygen-16, the assigned atomic mass of which was 16.000 00 u. On the present scale, oxygen-16 has a relative atomic mass of 15.994 u. Using oxygen-16 as the standard, what would the relative atomic mass of carbon-12 have been?

5. Determine the mass in grams of:
   (a) 0.0750 mol Zn   *(b)* 3.40×10^5 mol N
   (c) 4.62×10^{26} atoms Pb   *(d)* 50.0 atoms Hg.

6. How many moles are represented by:
   (a) 1.00 g C   *(b)* 7.50 kg Fe   *(c)* 1.25×10^{20} atoms Ni   *(d)* 1000. atoms Au?

7. How many atoms are in: *(a)* 3.25 mol S
   (b) 4.3×10^4 mol H   *(c)* 350. g Al

8. If 4.75×10^{23} atoms of silver are placed on one side of a platform balance, how many atoms of gold should be placed on the other side to balance this quantity of silver?

9. Magnesium (Mg) has a density of 7.25 g/cm^3. Given 12.0 cm^3 of Mg   *(a)* how many moles of Mg atoms does this represent?   *(b)* how many atoms of Mg are contained in this volume?

10. The density of silicon (Si) is 2.33 g/cm^3. What volume of silicon would contain 1.25×10^{18} atoms of Si?

11. How many carbon atoms are contained in a 1.500 carat diamond (carbon) if one carat = 200.0 mg?

Enrichment

1. Write a report on the development and perfection of the laboratory balance and its significance in the study of chemistry.

2. In addition to chemistry, what other scientific areas were studied by John Dalton? List some of his major contributions in each area.

3. Observe a cathode ray tube in operation and write a description of your observations.

4. Prepare a report on the series of experiments conducted by Sir James Chadwick that led to the discovery of the neutron.

5. What are the functions of cyclotrons, synchrotrons, and linear accelerators? What subatomic particles have been discovered through the use of these devices?

7. Write a report on the contributions of Amedeo Avogadro that led to the determination of the value of Avogadro's number.

8. Trace the development of the electron microscope and cite some of its many uses.

Chapter 4   Arrangement of Electrons in Atoms

Chapter Planner

4.1 Refinements of the Atomic Model
Wave-Particle Nature of Light
- ▨ Demonstration: 1
  Questions: 1, 4, 7, 10b, 12, 13
- ▨ Desktop Investigation
  Questions: 2, 3, 5, 9, 10a, 15
- ▨ Questions: 6, 8, 11
  Problems: 1–5
  Application Questions: 1–5
  Application Problems: 1–9

Bohr Model of the Hydrogen Atom
- ▨ Questions: 16, 18
- ▨ Question: 17

Spectroscopy
- ▨ Demonstration: 2
  Experiment: 7
  Question: 14

Quantum Model of the Atom
- ▨ Chemistry Notebook
  Questions: 19–22

4.2 Quantum Numbers and Atomic Orbitals
Quantum Numbers
- ▨ Demonstration: 3
  Questions: 23–25, 26a-b, 27, 28a
  Application Question: 6
- ▨ Question: 28b-c
- ▨ Question: 26c

Electrons in Each Main Energy Level
- ▨ Demonstration: 4
  Question: 29
  Application Question: 7

4.3 Electron Configurations
Rules Governing Electron Configurations
- ▨ Questions: 30–32

Representing Electron Configurations
- ▨ Questions: 33–35

Elements of the Second and Third Periods
- ▨ Questions: 36–40
- ▨ Question: 44

Elements of the Fourth and Fifth Periods
- ▨ Questions: 41, 42
  Application Questions: 8–11
- ▨ Application Question: 13
- ▨ Application Question: 12

Elements of the Sixth and Seventh Periods
- ▨ Question: 45
- ▨ Question: 43

Teaching Strategies

Introduction

Remind students that scientists are always looking for patterns in nature (order in the universe). This chapter is an excellent example of such a pursuit—looking for the patterns followed in the electron arrangement of neutral atoms in the ground state. Mention that atoms can become charged ions which are not neutral and not always in the ground state, but that this chapter deals mainly with the electron configurations of unexcited neutral atoms. Tell the students that this is an essential chapter; the rest of the course is based on understanding it.

This is a good time to explain that although experimentation is the most well-known scientific method used in research, two equally important methods are observation and mathematical calculation. Distinguish between mathematics as a tool for expressing relationships when working with the experimental or observation method and mathematics calculations as a scientific method to figure out answers to problems, such as Einstein used in the development of the theory of relativity and Schrödinger and Heisenberg in the development of quantum mechanics. Since quantum mechanics is used to explain much of the electron structure of the atom, a good portion of the information in this chapter was gained through the mathematical calculation method and then supported by experimental evidence.

This is a difficult chapter with many concepts that students may not be capable of fully grasping. However, to determine the electron structure of the different atoms, it is necessary to cover concepts such as Heisenberg's uncertainty principle, Hund's rule, Pauli's exclusion principle, and the wave-particle duality theory. They are also important because they add excitement, beauty, and intrigue to chemistry. Hopefully, they help students understand that everything in science is not concrete and it sometimes needs to be looked at in different ways and with different rules than they are accustomed to!

Two basic goals in teaching this chapter are to help students understand *(1)* how to determine the electron configurations of atoms and *(2)* how scientists initially were able to figure out this information using experimentation and mathematical calculations. It is also important to tell students *why* this is important; that is, that ultimately we are interested in the number of electrons in the outermost energy level so we can figure out how atoms combine with one another when they form compounds.

4.1   Refinements of the Atomic Model

Desktop Investigation: The Wave Nature of Light: Interference
ANSWERS TO PROCEDURE QUESTIONS
    4. Yes, there are light and dark circular areas on the screen.

1. Actually, two phenomena are responsible for forming the pattern on the screen. They are *(1)* diffraction which is the spreading of a wave disturbance into a region behind an obstruction and *(2)* interference which is the mutual effect of two light beams that results in the loss of intensity in certain regions and the reinforcement of intensity in other regions resulting in the pattern of concentric circles of light and dark areas seen on the screen. Since diffraction and interference are properties shown by waves, one would conclude that light is acting like a wave in this experiment.

2. Light shows the wave property of interference when a beam of light is projected through a pinhole onto a screen because it produces a pattern of concentric circles (similar to that seen in a bull's eye) on the screen.

Demonstration 1: Exciting Atoms by Mechanical Means

PURPOSE: To show that atoms give off light when excited by mechanical means

MATERIALS: A pair of pliers, wintergreen Life Savers®

PROCEDURE: Crush a wintergreen mint suddenly between the jaws of a pair of pliers in a dark room. What did you see happen? (Flashes of blue-white light were produced.)

Why do you think this happened? (The electrons of atoms in the candy were excited as they were crushed. As the electrons of the excited atoms dropped back to their ground state, they gave off energy in the form of blue-white light. What kind of energy was used to excite these atoms? (Mechanical energy)

An additional interesting note is that similar results occur if you watch someone chew wintergreen Life Savers® in the dark. An interesting experiment that students could design and carry out at home would be to test to see *(1)* if other brands of wintergreen-flavored mints produce light when they are crushed and *(2)* if spearmint- or peppermint-flavored candies produce light when they are crushed.

Demonstration 2: Continuous and Line Spectra

PURPOSE: To show students how to produce a continuous spectrum and a bright-line spectrum

MATERIALS: Magnifying lens, mercury vapor lamp, piece of cardboard with a 1 mm slit, prism or diffraction grating, screen, slide projector

PROCEDURE: To produce a continuous spectrum, pass a beam of light from the slide projector through a narrow slit in a piece of cardboard, and project the image of this slit on the screen. Then put a prism or diffraction grating between the cardboard and the screen, and adjust the angle of projection so the spectrum appears on the screen.

What kind of spectrum is this? (Continuous) Why wasn't this spectrum produced the first time? (The prism or diffraction grating was not there to spread out the different colors in the light.) How does the prism spread these colors out? (Light of different frequencies is bent different amounts as it passes through a prism or diffraction grating.)

To produce a line emission spectrum, substitute a mercury vapor lamp and a magnifying lens for the slide projector. CAUTION: Since the mercury vapor lamp produces ultraviolet rays, it should be shielded so that no one can look directly at the light source. What kind of spectrum is this? (A line emission spectrum—sometimes called a bright-line spectrum.) What element does this line spectrum identify? (If possible, show the students pictures of line spectra of several elements including mercury so they can match the spectrum on the screen with the mercury spectrum in the picture.) Many supply houses have spectroscopy kits which contain several inexpensive spectroscopes and liquids to test. These can be used for flame tests and for seeing the characteristic line spectra produced by atoms of different elements. Once you purchase the kits, you can make your own solutions to replace the ones in the kit and continue to use the spectroscopes, making the purchase of the kit a good investment.

Chemistry Notebook: Heisenberg's Uncertainty Principle

Douglas Rickard's article—"A Model of Uncertainty" in **References and Resources** (at the end of this chapter), is an excellent way to explain and demonstrate Heisenberg's Uncertainty Principle.

4.2   Quantum Numbers and Atomic Orbitals

Students often have difficulty understanding the three-dimensional nature of orbitals. Many of them were taught Bohr's orbit theory in lower grades and have a hard time giving up that idea for the more complex orbital theory. The following demonstration will help.

Demonstration 3: Shapes of Orbitals

PURPOSE: To help students visualize the 3-D shapes of the *s*, *p*, *d*, and *f*-orbitals and of the *s*- and *p*-sublevels

MATERIALS: 1 round balloon, 3 oblong balloons of different colors, 4 oblong balloons of the same color, 3 thin oblong balloons, 4 additional oblong balloons

PROCEDURE: Ask students to help by blowing up balloons. Hold up a round balloon. What orbital does this represent? (an *s*-orbital) (See Figure T4-1a.) Point out that this also represents an *s*-sublevel.

Twist an oblong balloon in the middle for a *p*-orbital. (See Figure T4-1b.) How do you recognize a *p*-orbital? (It has two lobes.) Do the same for two other oblong balloons of different colors. Twist the three together as in Figure T4-1c to make a *p*-sublevel. Then using the three different colored balloons, you can point out the p_x, the p_y, and the p_z, orbitals.

Twist two oblong balloons of the same color in the middle, and put them together for a *d*-orbital. (See Figure T4-1d.) What orbital does this represent? (a *d*-orbital) How many lobes does it have? (4) For the one odd *d*-orbital—d_z2—curl a thin oblong balloon around the center of a twisted oblong balloon. (See Figure T4-1e.) Sometimes students want to know why the fifth *d*-orbital is different in shape from the rest. The answer is: This is the shape of the space that is left in the *d*-sublevel.

For the *f*-orbitals with eight lobes twist four oblong balloons of the same color together at their centers. (See Figure T4-1f.) For the three *f*-orbitals in Figure T4-1g, curl two thin oblong balloons around the center of a twisted oblong balloon.

Models of the orbitals can be purchased from many of the supply houses. However, balloons seem to be a better choice because they are cheap and they require student involvement. You could make up sets of balloons, and have students work in small groups and let each student group make all the orbitals.

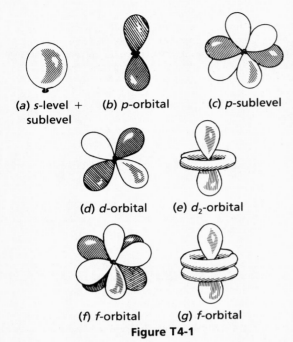

(a) s-level + sublevel (b) p-orbital (c) p-sublevel

(d) d-orbital (e) d_2-orbital

(f) f-orbital (g) f-orbital

Figure T4-1

There is one problem with balloon models, or any orbital model, which is they make it appear that there is a definite point at which the orbital ends. Be sure to emphasize to students that each orbital does not have a definite point at which it ends but the probability of finding a given electron beyond a certain area becomes less and less likely as one moves further out.

Another problem is that after the shapes of the orbitals with accompanying explanations have been presented, students are often confused as to how they all fit together in the atom. Use explanations, chalkboard diagrams, layered transparencies (to show how different sublevels overlap one another) and analogies to help them see how these orbitals fit together to make up sublevels, energy levels and finally the electron cloud. The infant toy that consists of barrels within barrels can be used as an illustration.

An excellent analogy to help students understand (1) the 3-D nature of the electron cloud and (2) that energy levels, sublevels and orbitals are subdivisions of the electron cloud follows. Use an apartment building analogy to represent the electron cloud. Each floor represents an energy level with the first floor being closest to the ground, as the first energy level is closest to the nucleus. Each floor is subdivided into apartments, which represent sublevels; and each apartment is made up of rooms representing orbitals. Apartments have different numbers of rooms: one-room efficiency—s sublevel, three room—p sublevel, five room—d sublevel, and the luxurious seven room—the f sublevel. All rooms have one or two electrons "living" in them.

References and Resources

Books and Periodicals

Bent, Henry. "Should Orbitals Be X-rated in Beginning Chemistry Courses?" *Journal of Chemical Education*, Vol. 61, No. 5,

4.3 Electron Configurations

Students are often puzzled as to how one can be sure that the two electrons in a given orbital are able to spin in opposite directions. They ask the question: If the first electron is spinning in a clockwise direction and the entering one is also spinning in a clockwise direction, how can it go into the orbital. (Of course, we know electrons do not enter atoms one at a time.) The question is good in that it points out how two electrons in the same orbital always manage to spin in opposite directions as is stated in Pauli's exclusion principle. The following demonstration clears this dilemma up nicely.

Demonstration 4: Orbital Electron Pairs Spin in Opposite Directions

PURPOSE: To show how electrons in the same orbital are always able to spin in opposite directions

MATERIALS: 1 boiled peeled egg, 2 nails, 1 narrow strip of paper with arrows going in one direction, 2 small pieces of paper marked N and S, 2 straight pins, tape

PROCEDURE: Put one nail in each end of the boiled egg. (See Figure T4-2) Attach the piece of paper marked N to one nail and S to the other nail. This is to represent the north and south poles of the spinning electrons. (Spinning electrons produce magnetic fields and thus act like miniature magnets with north and south poles. This helps cut down on the repulsion of the negative particles.)

Attach the narrow strip of paper with arrows around the center of the egg using straight pins and spin it in the direction indicated by the arrows. Then show students that if the spinning electron simply flips over it is spinning in the opposite direction.

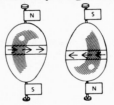

Figure T 4-2

Figuring out the electron configurations of atoms is very difficult for many students; so do several examples and have them do many practice problems. Use the three different methods given in the text for electron configurations. When students are figuring out electron configurations using the diagonal rule, they often want to write the sublevels in the order that they fill rather than keeping each set of sublevels together in their energy levels. Emphasize that, in this text, sublevels must be filled in the order designated by the diagonal rule, but sublevels in the same energy level must be kept together. Other texts do it in order of filling. In the end, the main concern is the number of electrons in the outermost energy level as is shown by the electron-dot notation.

May 1984, pp. 421–3.
Burdman, Clark S. "The Significance of Arrows in Orbital Diagrams," *Journal of Chemical Education*, Vol. 63, No. 4, April 1986, p. 320.

Festa, Roger R. "Wolfgang Pauli (1900–1958): A Brief Anecdotal Biography," *Journal of Chemical Education*, Vol. 58, No. 3, March 1981, pp. 273–4.

Haendler, Bianca L. "Presenting the Bohr Atoms," *Journal of Chemical Education*, Vol. 59, No. 5, May 1982, pp. 372–6.

Lamb, William. "Idea Bank—An Alternative Method for Flame Tests," *The Science Teacher*, Vol. 51, No. 3, March 1984, p. 61.

Morwick, James. "Should Orbitals Be Taught in High School?" *Journal of Chemical Education*, Vol. 56, No. 4, April 1979, pp. 262–3.

Rickard, Douglas. "A Model of Uncertainty," *Journal of Chemical Education*, Vol. 63, No. 10, Oct. 1986, p. 833.

Wiger, George and Melvin Dutton. "A Lecture Demonstration Model of the Quantum Mechanical Atom," *Journal of Chemical Education*, Vol. 58, No. 10, Oct. 1981, pp. 801–2.

Zuhav, Gary. *The Dancing Wu Li Masters*, New York: Bantam Books, 1979.

Audiovisual Resources

Chemistry 103–Dissecting the Atom, sound filmstrip, Learning Arts, 1988.
Electron Configuration, VHS cassette, Learning Arts, 1988.

Computer Software

Atomic Structure Series #3, Apple 48K. Queue, 1988. Several programs relating to this chapter.
Fee, Richard. *Electron Configurations*, Apple 48K. Chemistry According to ROF, 1988.
General Chemistry 1A—Topic 2: Atomic Theory and Structure, Apple 128K (4 disks). Queue, 1988.
Project Seraphim, 1988, Apple II, Disks-Vol. 1. Disk no. 6, Vol. 2, Disk no. 2

Answers and Solutions

Questions

1. Because light and electrons in atoms have some common properties, discoveries about the nature of light led to a revolutionary view of the nature of matter and the atom.

2. During the 19th century, electrons were pictured as particles and light was pictured as waves. The 20th-century concept describes both electrons and light as having a dual wave-particle nature.

3. *(a)* Examples include X rays, ultraviolet light, visible light, infrared light, microwaves, and radio waves. *(b)* Electromagnetic radiation moves through a vacuum at a constant speed of 3.0×10^{10} cm/s. *(c)* In a vacuum, wavelength and frequency are inversely proportional; as wavelength decreases, frequency increases, and vice versa. The equation relating them is $c = \lambda\nu$, where c is the speed of light, λ is wavelength, and ν is frequency.

4. *(a)* A continuous spectrum is one in which all wavelengths within a given range are included. *(b)* When sunlight passes through a prism, it is spread out into an array of colors according to wavelength.

5. *(a)* The electromagnetic spectrum consists of all electromagnetic radiation arranged according to increasing wavelength. *(b)* Shorter wavelengths are measured in nanometers, whereas longer wavelengths are measured in centimeters or meters. *(c)* The SI unit for frequency is the hertz. One hertz is equal to one wave per second.

6. *(a)* The frequency range of microwaves is roughly 10^{11} to 10^{12} Hz; *(b)* the wavelength range of ultraviolet light is roughly 10^{1} to 10^{2} nm; the frequency of visible light is approximately 10^{15} Hz and its wavelength ranges from 400 to 700 nm.

7. The colors of light in the visible spectrum, in order of increasing frequency, are red, orange, yellow, green, blue, and violet.

8. The wave theory could not explain the emission of visible light by matter, the absorption of light by matter, and the photoelectric effect.

9. *(a)* The photoelectric effect is the emission of electrons by certain metals when light shines onto them. *(b)* The German physicist Max Planck is generally credited with this contribution.

10. *(a)* $E = h\nu$ where E is the energy of the photon, ν is the frequency of the light emitted, and h is Planck's constant. *(b)* Energy and frequency are directly proportional.

11. The theory of light that best explains each is: *(a)* the wave theory *(b)* the particle theory *(c)* the particle theory.

12. *(a)* Atoms can be made to emit electromagnetic radiation through heating, which causes their energies to increase. Then, when the atoms return to their original energy states, the added energy is given off in the form of electromagnetic radiation. *(b)* This phenomenon is seen in the production of colored light in neon signs and in fireworks.

13. The ground state is the state of lowest energy of an atom, whereas an excited state is one in which an atom has a higher energy than it does in its ground state.

14. In a line emission spectrum, such as that emitted by excited hydrogen atoms, only photons of certain energies are emitted. These produce the individual lines observed in the hydrogen emission spectrum. In a continuous spectrum, such as that emitted by the sun, photons encompassing a broad range of energies are emitted, so that countless lines corresponding to these energies are produced. The result appears as a solid band of light.

15. *(a)* The most prominent regions in the hydrogen spectrum are the ultraviolet, visible, and infrared regions. *(b)* The groups of spectral lines are referred to as the Lyman, Balmer, and Paschen series, respectively.

16. According to Bohr, a line emission spectrum is produced when an electron drops from a higher-energy orbit to one with lower energy, emitting a photon whose energy is equal to the difference in energy between the orbits.

17. The Lyman series results when electrons drop from various higher-energy levels to the ground-state energy level; the Balmer series results when electrons drop from higher-energy levels to the second energy level; in the Paschen series, the drop is from higher-energy levels to the third energy level of hydrogen.

18. Bohr's model of the hydrogen atom did not explain the spectra of atoms that have more than one electron, nor did it fully explain the chemical behavior of atoms.

19. *(a)* De Broglie proposed that electrons, like light, have a dual wave-particle nature. Since waves within a space have only certain frequencies, he reasoned that if electrons were thought of as waves confined to the space around an atomic nucleus, this could explain why only certain energies of electrons are possible. *(b)* Schrödinger devised equations that treated electrons around nuclei as waves and, thus, laid the foundation for modern quantum theory.

20. The quantum theory describes mathematically the wave properties of electrons and other very small particles.

21. *(a)* An orbital is a three-dimensional region about the nucleus in which there is a high probability that a particular electron is located. *(b)* Orbitals can be pictured as clouds showing the region of probable electron locations. The sizes and shapes of electron clouds depend on the energies of the electrons that occupy them.

22. Whereas Bohr's model worked only for the hydrogen atom, Schrödinger's mathematical model applies to all atoms. The essential difference between the two models involves the issue of certainty. Bohr described definite orbits occupied by electron particles, whereas Schrödinger treated electrons as waves having a certain probability of being found at various distances from the nucleus, in what he called orbitals. The two models are similar from the standpoint of energy, in that Schrödinger's orbitals, like Bohr's orbits, associate the energy of an electron with its location relative to the nucleus, and the production of specific levels of electromagnetic radiation with electrons dropping from higher-energy to lower-energy orbitals. The most probable location for the electron of hydrogen, as described by Schrödinger, is at a distance from the nucleus exactly equal to that of Bohr's lowest-energy orbit.

23. *(a)* Principal quantum numbers indicate the main energy levels surrounding a nucleus. *(b)* The principal quantum number is symbolized by n. *(c)* The term "shells" is used to refer to the main energy levels surrounding a nucleus. *(d)* As the value of n increases, the distance of the main energy levels from the nucleus increases, and their energy increases.

24. *(a)* The orbital quantum number indicates the shape of an orbital. *(b)* Sublevels or subshells are the regions within main energy levels that are occupied by the differently shaped orbitals. *(c)* The number of possible orbital shapes in each main energy level is equal to the value of the principal quantum number. *(d)* The first four orbital shapes are designated by the symbols s, p, d, and f.

25. The responses are: *(a)* 1 orbital, an s orbital *(b)* 2 orbitals, s and p orbitals *(c)* 3 orbitals, s, p, and d orbitals *(d)* 4 orbitals, s, p, d, and f orbitals *(e)* 7 orbitals.

26. *(a)* The magnetic quantum number indicates the orientation of an orbital about the nucleus. *(b)* The number of possible orientations for the s, p, d, and f orbitals are 1, 3, 5, and 7, respectively. *(c)* Subscripts are used to indicate the orientation of the orbital in terms of the x, y, and z axes of a three-dimensional coordinate system centered on the nucleus. The p_x notation refers to a p orbital along the x axis, p_y indicates a p orbital along the y axis, and p_z indicates a p orbital along the z axis.

27. *(a)* The number of orbitals within each main energy level is equal to the square of the principal quantum number, n.

(b) The third main energy level contains nine orbitals; the fifth contains 25.

28. *(a)* The spin quantum number indicates which of the two possible orientations an electron has in an orbital. *(b)* There are two possible values for this number, $+\frac{1}{2}$, and $-\frac{1}{2}$.

(c) The spin quantum number is significant because each orbital can hold no more than two electrons, which must have opposite spins. This limitation thus also determines the maximum number of electrons that can occupy each main energy level.

29. The number of electrons for each main energy level is: *(a)* 2 *(b)* 8 *(c)* 18 *(d)* 32 *(e)* 50 *(f)* 72 *(g)* 98 *(h)* $2n^2$.

30. *(a)* According to the Aufbau principle, an electron occupies the lowest-energy orbital that can receive it. *(b)* In a many-electron atom, the lowest-energy orbital is filled first, and then electrons are added to the orbital with the next-lowest energy, and so on, until all of the electrons in the atom have been placed in orbitals.

31. *(a)* Hund's rule states that orbitals of equal energy are each occupied by one electron before any orbital is occupied by a second electron. *(b)* By placing as many single electrons as possible in separate orbitals in the same energy level, electron-electron repulsion is minimized and more-favorable arrangements result because of lower energy.

32. *(a)* According to the Pauli exclusion principle, no two electrons in the same atom can have the same four quantum numbers. *(b)* The two different values of the spin quantum number permit two electrons of opposite spins to occupy the same orbital. *(c)* Two electrons in the same orbital would have the same principal, orbital, and magnetic quantum number, but would differ in spin quantum number.

33. *(a)* Electron-dot notation shows only those electrons in the highest, or outermost main energy level. *(b)* The highest occupied energy level is the electron-containing main energy level with the highest principal quantum number. *(c)* Inner-shell electrons are electrons not in the highest occupied energy level.

34. The highest occupied energy level in each of the indicated atoms is the: *(a)* 1st *(b)* 2nd *(c)* 3rd *(d)* 4th *(e)* 5th.

35. The electron-dot symbols are: *(a)* $.\dot{Z}\,.$ *(b)* $:\dot{Z}:$ *(c)* $:\dot{Z}:$ *(d)* $.\dot{Z}:$

36. *(a)* Oxygen contains eight electrons. *(b)* Its atomic number is 8. *(c)* The highest occupied energy level is the second main energy level. *(d)* Its orbital notation is

$$\underset{1s}{\uparrow\downarrow}\ \underset{2s}{\uparrow\downarrow}\ \underset{2p}{\uparrow\downarrow}\ \underset{2p}{\uparrow}\ \underset{2p}{\uparrow}$$

(e) Its electron dot symbol is $:\dot{O}\,.$. *(f)* It contains two inner-shell electrons. *(g)* They are located in the $1s$ orbital.

37. The indicated responses are *(a)* 15 *(b)* 15 *(c)* third *(d)*

$$\underset{1s}{\uparrow\downarrow}\ \underset{2s}{\uparrow\downarrow}\ \underset{2p}{\uparrow\downarrow}\ \underset{2p}{\uparrow\downarrow}\ \underset{2p}{\uparrow\downarrow}\ \underset{3s}{\uparrow\downarrow}\ \underset{3p}{\uparrow}\ \underset{3p}{\uparrow}\ \underset{3p}{\uparrow}$$

(e) $.\dot{P}:$ *(f)* 10 *(g)* in $1s$, $2s$, and $2p$ orbitals.

38. *(a)* The [Ne]$3s^2$ configuration denotes an element consisting of the Ne electron configuration with the $3s^2$ electrons added, for a full configuration of $1s^22s^22p^63s^2$. *(b)* The designated element is Mg.

39. The shorthand electron configurations are: *(a)* [Ne]$3s^23p^5$ *(b)* [Ar]$4s^2$.

40. *(a)* The noble gases are the Group-18 elements helium, neon, argon, krypton, xenon, and radon. *(b)* A noble-gas configuration is an outer main energy level fully occupied by eight electrons.

41. The indicated electron configurations and electron-dot symbols are as follows:

41. *(a)* $1s^2 2s^2 2p^6 3s^1$ [Ne]$3s^1$ .Na·

    (b) $1s^2 2s^2 2p^6 3s^2 3p^1$ [Ne]$3s^2 3p^1$.Àl·

    (c) $1s^2 2s^2 2p^6 3s^2 3p^3$ [Ne]$3s^2 3p^3$ .P̈.

    (d) $1s^2 2s^2 2p^6 3s^2 3p^6$ [Ne]$3s^2 3p^6$:Är:

    (e) $1s^2 2s^2 2p^6 3s^2 3p^6 3d^{10} 4s^2 4p^5$ [Ar]$3d^{10} 4s^2 4p^5$:Ḃr:

    (f) $1s^2 2s^2 2p^6 3s^2 3p^6 3d^{10} 4s^2 4p^6 5s^2$ [Kr]$5s^2$ .Sr·

42. The indicated elements are as follows: *(a)* B *(b)* F *(c)* Mg *(d)* Si *(e)* Cl *(f)* K *(g)* Fe.

43. The number of elements contained in each of the first six periods of the Periodic Table are 2, 8, 8, 18, 18, and 32, in that order.

44. The indicated elements are as follows: *(a)* Be *(b)* N *(c)* C *(d)* N *(e)* B *(f)* O *(g)* Ne.

45. The indicated responses are as follows: *(a)* 2 *(b)* 8 *(c)* 3 *(d)* 3 *(e)* Ar *(f)* 6 *(g)* $3s^2$ *(h)* 18 *(i)* 2 *(j)* P *(k)* 2d *(l)* K *(m)* 10 *(n)* 5 *(o)* 3.

Problems

1. *(a)* $\dfrac{3.0 \times 10^{10}\text{ cm}}{\text{s}} \times \dfrac{\text{m}}{10^2\text{ cm}} \times \dfrac{\text{km}}{10^3\text{ m}} = 3.0 \times 10^5\text{ km/s}$

  (b) $\dfrac{3.0 \times 10^5\text{ km}}{\text{s}} \times \dfrac{60\text{ s}}{\text{min}} \times \dfrac{60\text{ min}}{\text{h}} = 1.1 \times 10^9\text{ km/h}$

  (c) $\dfrac{1.1 \times 10^9\text{ km}}{\text{h}} \times \dfrac{24\text{h}}{\text{day}} \times \dfrac{7\text{ days}}{\text{wk}} = 1.8 \times 10^{11}\text{ km/wk}$

2. $\dfrac{3.25 \times 10^{-16}\text{ J} \cdot \text{cal}}{4.184\text{ J/cal}} = 7.77 \times 10^{-17}\text{ cal}$

3. $\dfrac{4.50\text{ mol photons} \times 6.02 \times 10^{23}\text{ photons}}{\text{mol photons}}$

    $= 2.71 \times 10^{24}\text{ photons}$

4. $\dfrac{5.75 \times 10^{15}\text{ photons}}{6.02 \times 10^{23}\text{ photons/mol}} = 9.55 \times 10^{-9}\text{ mol}$

5. *(a)* $\dfrac{1215\text{ Å} \times 1 \times 10^{-8}\text{ cm}}{\text{Å}} = 1.215 \times 10^{-5}\text{ cm}$

  (b) $\dfrac{1.215 \times 10^{-5}\text{ cm}}{10^2\text{ cm/m}} = 1.215 \times 10^{-7}\text{ m}$

6. $V = l \times w \times h$

  $V = \dfrac{15.0\text{ cm} \times 60.0\text{ mm} \cdot \text{cm}}{10\text{ cm}} \times \dfrac{0.400\text{ dm} \times 10\text{ cm}}{\text{dm}}$

  $V = 360.\text{ cm}^3$

  $V = \dfrac{m}{V} = \dfrac{145\text{ g}}{360.\text{ cm}^3} = 0.403\text{ g/cm}^3$

7. $\dfrac{1.25\text{ mol Mg} \times 24.3\text{ g Mg}}{\text{mol Mg}} = 30.4\text{ g Mg}$

8. *(a)* $\dfrac{0.456\text{ g N} \times \text{mol N}}{14.0\text{ g N}} = 0.0326\text{ mol N}$

  (b) $\dfrac{0.0326\text{ mol N} \cdot 6.02 \times 10^{23}\text{ atoms N}}{\text{mol N}}$

    $= 1.96 \times 10^{22}\text{ atoms N}$

9. *(1)* $\dfrac{7.25\text{ g Cl} \times \text{mol Cl}}{35.5\text{ g Cl}} = 0.204\text{ mol Cl}$

  (2) $\dfrac{0.204\text{ mol Cl} \times \text{mol Na}}{\text{mol Cl}} = 0.204\text{ mol Na}$

  (3) $\dfrac{0.204\text{ mol Na} \times 23.0\text{ g Na}}{\text{mol Na}} = 4.69\text{ g Na}$

Application Questions

1. Although the wave theory does explain how light, as a form of energy, can knock electrons loose from a metal surface, it cannot explain why light of only a certain minimum frequency causes the photoelectric effect. Light below that minimum frequency can shine on a metal for a prolonged period without causing this effect, whereas even a few photons of light of the necessary frequency can deliver enough energy to initiate the photoelectric effect.

2. Planck proposed that when a hot object loses energy, it does not do so continuously, as would be expected if the radiation were in the form of waves; instead, the object radiates energy in small, specific amounts he called quanta. Planck proposed the following relationship between a quantum of energy and the frequency of radiation: $E = h\nu$ where E is the energy of a photon, ν is the frequency of the light emitted, and h is Planck's constant.

3. When subjected to high-voltage electric current of the right energy, hydrogen atoms absorb photons of specific energies and move from their ground state to excited states. When the atoms return from the excited state to the ground state, they emit photons corresponding to the same amounts of energy that were absorbed. When passed through a prism, this light is separated into four narrow lines, each of a different color, that make up the hydrogen emission spectrum.

4. Since the line emission spectrum for hydrogen indicates that only photons of specific energies are emitted, and since the energies of these photons is equal to the difference in energy between the initial and final states of the atom, then the allowable energy states available to a hydrogen atom must correspond to specific energy levels.

5. In order of increasing energy, the colors of light in the visible spectrum are red, orange, yellow, green, blue, and violet.

6.

Principal Quantum Number	Number of Orbital Shapes	Total No. of Orbitals per Energy Level	Total No. of Electrons Per Energy Level
(n)	(n)	(n^2)	$(2n^2)$
1	1	1	2
2	2	4	8
3	3	9	18
4	4	16	32
5	5	25	50
6	6	36	72
7	7	49	98
8	8	64	128
n	n	n^2	$2n^2$

7. According to the Pauli exclusion principle, no two electrons in the same atom can have the same set of four quantum

only in terms of spin quantum numbers, and since only two possibilities exist for spin quantum numbers, the maximum number of electrons that can thus occupy an orbital is limited to two.

8. In many-electron atoms, the 4s orbital has a lower energy than does the 3d orbital. Thus, according to the Aufbau principal, the 4s orbital will fill before the 3d orbitals.

9. The order of filling is: 1s, 2s, 2p, 3s, 3p, 4s, 3d, 4p, 5s, 4d, 5p, 6s, 4f, 5d, 6p, 7s, 5f, 6d, 7p.

10. The shorthand electron configurations are as follows:
(a) $[Ar]3d^14s^2$ (b) $[Ar]3d^64s^2$ (c) $[Ar]3d^{10}4s^2$
(d) $[Ar]3d^{10}4s^24p^3$ (e) $[Kr]4d^{10}5s^25p^2$ (f) $[Kr]4d^{10}5s^25p^6$
(g) $[Xe]5d^16s^2$ (h) $[Xe]4f^{14}5d^26s^2$ (i) $[Xe]4f^{14}5d^{10}6s^2$
(j) $[Xe]4f^{14}5d^{10}6s^26p^2$ (k) $[Xe]4f^{14}5d^{10}6s^26p^5$ (l) $[Rn]6d^17s^2$
(m) $[Rn]5f^{14}6d^17s^2$.

11. The electron-dot symbols are: (a) .Zn˙ (b) .Ṡn˙ (c) :Ẍe:
(d) .Ṗḃ˙ (e) :Ȧṫ:

12. The unusual configurations of chromium and copper are due to the lower energy associated with d sublevels that are half-filled in chromium and completely filled in copper.

13. The indicated responses are as follows: (a) 17 (b) Se (c) Sc (d) Fe (e) Sb (f) 32 (g) La (h) Br (i) Pb (j) 8 (k) 3 (l) 18 (m) Bi (n) Sr.

Application Problems

1. $c = \lambda \nu$

$$\lambda = \frac{c}{\nu}$$

$$\lambda = \frac{2.998 \times 10^8 \text{ m/s}}{7.500 \times 10^{12} \text{ /s}} = 3.997 \times 10^{-5} \text{ m}$$

2. $c = \lambda \nu$

$$\nu = \frac{c}{\lambda}$$

$$\nu = \frac{2.998 \times 10^{10} \text{ cm/s}}{4.257 \times 10^{-7} \text{ cm}} = 7.043 \times 10^{16} \text{/s or Hz}$$

3. $E = h\nu$
$\nu = 6.62 \times 10^{-34} \text{ J} \cdot \text{s} \times 3.55 \times 10^{17}\text{/s}$
$\nu = 2.35 \times 10^{-16} \text{ J}$

4. $E = h\nu$

$$\nu = \frac{E}{h}$$

$$\nu = \frac{1.55 \times 10^{-24} \text{ J}}{6.62 \times 10^{-34} \text{ J} \cdot \text{s}} = 2.34 \times 10^9\text{/s or Hz}$$

5. Solving each equation for ν yields:

$$\nu = \frac{c}{\lambda}$$

$$\nu = \frac{E}{h}$$

Equating the two expressions,

$$\frac{c}{\lambda} = \frac{E}{h}$$

$$E = \frac{hc}{\lambda}$$

6. $E = \frac{hc}{\lambda}$

$$E = \frac{6.62 \times 10^{-34} \text{ J} \cdot \text{s} \times 3.00 \times 10^8 \text{ m/s}}{4.25 \times 10^{-8} \text{ m}}$$

$$= 4.67 \times 10^{-18} \text{ J}$$

7. Since the speed of light (c) is 3.00×10^8 m/s,

$$\frac{8.00 \times 10^7 \text{ km}}{3.00 \times 10^8 \text{ m/s}} \times \frac{10^3 \text{m}}{\text{km}} = 2.67 \times 10^2$$

[Note that the frequency of the radiowave did not enter into the calculations.]

8. (a) $E = h\nu$

$$\nu = \frac{E}{h}$$

$$\nu = \frac{3.37 \times 10^{-19} \text{ J}}{6.62 \times 10^{-34} \text{ J} \cdot \text{s}} = 5.09 \times 10^{14}\text{/s or Hz}$$

(b) $\lambda = \frac{c}{\nu} = \frac{3.00 \times 10^8 \text{ m/s}}{5.09 \times 10^{14}\text{/s}}$
$= 5.89 \times 10^{-7} \text{ m}$

9. Recalling that 1 Å $= 10^{-8}$ cm,

$$1.00 \times 10^{-8} \text{ Å} \times \frac{10^{-8} \text{ cm}}{\text{Å}} = 1.00 \times 10^{-16} \text{ cm}$$

Substituting into $E = \frac{hc}{\lambda}$

$$E = \frac{6.62 \times 10^{-34} \text{ J} \cdot \text{s} \times 3.00 \times 10^{10} \text{ cm/s}}{1.00 \times 10^{-16} \text{ cm}}$$

$$= 1.99 \times 10^{-7} \text{ J}$$

Teacher's Notes

Chapter 4

Arrangement of Electrons in Atoms

INTRODUCTION

Chapter 3 showed how the ancient Greek idea of indivisible atoms was replaced by a model of atoms made up of nuclei surrounded by electrons. The present chapter atomic model and on how electrons are arranged around nuclei. This electron arrangement explains the organization of the Periodic Table and accounts for chemical reactivity and for how atoms combine in chemical compounds.

LOOKING AHEAD

As you study this chapter, practice writing the electron configurations for the individual elements. Learn the number of outer electrons in atoms of elements from different regions in the Periodic Table. Much of the rest of your study of chemistry will be based on an understanding of the electron configurations of the elements.

SECTION PREVIEW

4.1 Refinements of the Atomic Model
4.2 Quantum Numbers and Atomic Orbitals
4.3 Electron configurations

All light, including the neon that illuminates the Ginza District of Tokyo, results from charged particles, such as electrons, moving from a high energy level to a low energy level.

4.1 Refinements of the Atomic Model

In the preceding chapter, you followed the evolution of the atomic theory up to the Rutherford model, which consists of a nucleus surrounded by electrons. Although it was an improvement over previous models, Rutherford's model was incomplete. It did not explain how negatively charged electrons could fill the space surrounding the positively charged nucleus. Because oppositely charged particles attract each other, what prevented the electrons from being drawn into the nucleus?

In order to understand how a new atomic model was developed that did not have this and other shortcomings, you need to know something about the nature of light. At this point, you may wonder what light has to do with atoms. As you will see, light and electrons in atoms have some properties in common. Discoveries about the nature of light led directly to a revolutionary view of the nature of matter and of the atom.

Wave–Particle Nature of Light

At the time of Rutherford, electrons were pictured as particles, and light was pictured as waves. Many properties of electrons and light can be explained by these models. A revolution occurred in science in the early 1900s when it was discovered that electrons have certain wavelike properties and that light has certain particle-like properties. To fully explain all observations, it is necessary to describe electrons and light as having a dual wave–particle nature. If you find this difficult to accept and understand, you are no different than many scientists during the period when the idea was introduced. The wave–particle model has withstood many experimental tests and is now well-established. No doubt it will remain so until someone comes up with an observation of either light or electrons that the wave–particle model fails to explain.

Electromagnetic radiation Visible light is one kind of electromagnetic radiation. Some other kinds of electromagnetic radiation are X rays, ultraviolet and infrared light, microwaves, and radio waves. **Electromagnetic radiation** *is a form of energy that exhibits wavelike behavior as it travels through space.* You are familiar with waves moving through water. A duck sitting on the surface moves up and down as waves pass through the water beneath it. The waves pass by, though the duck remains in place.

Electromagnetic radiation moves through a vacuum at a constant speed of 3.0×10^{10} centimeters per second (cm/s) . The same value may be used for the speed of light in air, which is largely a vacuum. According to present theories, this is the fastest speed possible in nature.

Electromagnetic radiation has the measurable wave properties of wavelength and frequency. One wavelength is illustrated in Figure 4-1 as the distance from the base of one wave to the base of the next. **Wavelength** (λ) *is the distance between corresponding points on adjacent waves.* For radiation of a specific energy, this distance is the same throughout the entire succession of waves.

- Discuss the dual wave–particle nature of light.
- Explain the mathematical relationship among the velocity, wavelength, and frequency of electromagnetic radiation.
- Discuss the significance of the line emission spectrum of hydrogen to the model of atomic structure.
- Describe the Bohr model of the hydrogen atom.
- Distinguish between an orbit and an orbital.

Whether electrons and light behave as matter, by showing particle characteristics, or behave as energy, exhibiting wave-like properties, depends upon the type of experiment being carried out. For more detail, refer to *The Dancing Wu Li Masters* by Zuhav.

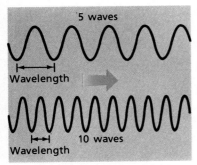

Figure 4-1 Wavelength and frequency are inversely related.

STUDY HINT

λ is lower case Greek lambda; ν is lower case Greek nu.

STUDY HINT

Inverse proportion was discussed in Section 2.4.

Figure 4-2 A prism separates white light into its component colors, creating a rainbow effect.

Frequency (ν) *is defined as the number of waves that pass a given point in a specific amount of time, usually one second.* The higher the frequency of water waves, the more often the duck sitting on the surface will bob up and down. Frequency and wavelength are mathematically related to each other. When waves are closer together, as shown at the bottom of Figure 4-1, more of them pass a given point in the same amount of time (because of light's constant speed).

The relationship between frequency and wavelength is given by the following equation,

$$c = \nu\lambda$$

where c is the speed of light, λ is the wavelength, and ν is the frequency. Because c is the same for all electromagnetic radiation, the product $\lambda\nu$ is a constant, and λ is inversely proportional to ν. In other words, as the wavelength of light decreases, the frequency increases (see Figure 4-1) and vice versa.

When sunlight passes through a prism (Figure 4-2), it is spread out according to its wavelength. The resulting array of colors is an example of a **continuous spectrum,** *a spectrum in which all wavelengths within a given range are included.* At either end of the visible spectrum are other types of

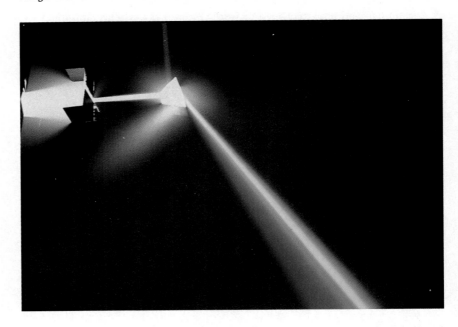

radiation that we cannot see. Visible light is only a small part of the total electromagnetic spectrum, which is shown in Figure 4-3. The **electromagnetic spectrum** *consists of all electromagnetic radiation, arranged according to increasing wavelength.* The right-hand vertical scale in Figure 4-3 gives wavelength (λ) in units of length based on the meter. For visible light and radiation of shorter wavelengths, the unit is the nanometer. For radiation of longer wavelengths, it is the centimeter or the meter. The left-hand vertical scale gives the frequency (ν) in the SI unit for frequency, the *hertz.* One hertz is equal to 1 wave per second.

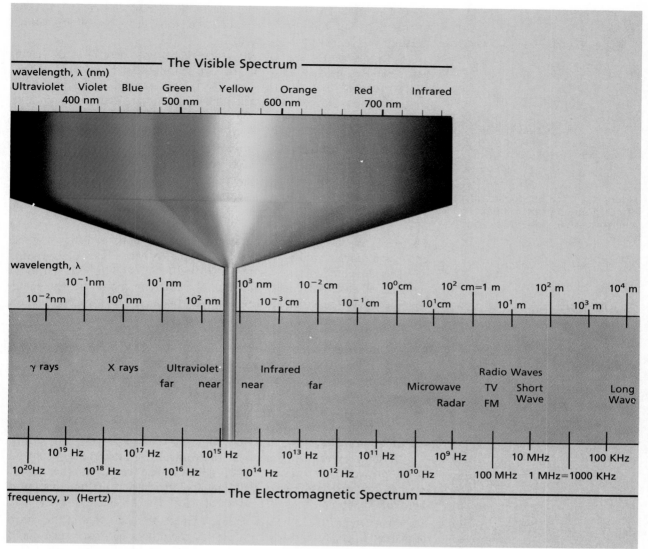

Figure 4-3 Visible light represents only a small portion of the electromagnetic spectrum.

Light as particles As the twentieth century began, scientists encountered two properties of light that could not be explained by wave theory. One involved the emission of visible light by matter, and the other the absorption of light by matter.

When an object made of a material that does not burn, such as iron, is heated, it gives off light. First, a red glow appears. As the temperature (and therefore the energy added) increases, the red glow changes to yellow and then to white. We sometimes speak of this condition as "white hot." The wave theory incorrectly predicted that extremely hot bodies would emit mainly (invisible) ultraviolet light.

A problem was also encountered in explaining the **photoelectric effect**, *which is the emission of electrons by certain metals when light shines on them*. It was agreed that light, being a form of energy, could knock loose an electron from a metal surface. But wave theory could not explain why only

To help students understand a *quantum*, use the following analogy: If a soda costs 45 cents and you put 40 cents in a soda machine, you will not get the soda when you push the button. Only if the required "packet" of money is put in, will you get a soda.

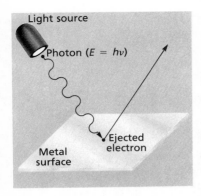

Figure 4-4 The photoelectric effect. Some metals will emit electrons when light shines on them.

Figure 4-5 Hydrogen atoms are excited when high-voltage electric current is passed through a glass tube containing hydrogen gas. The lavender glow is characteristic of hydrogen gas.

light of a certain frequency or higher caused the photoelectric effect. A very intense light with too low a frequency could shine on a metal all day without causing the photoelectric effect.

Both problems were solved when light was described as having particle-like properties. In 1900, the German physicist Max Planck (1858–1947) proposed that when a hot object loses energy, it did not do so continuously, as would be expected if the radiation were in the form of waves. Instead, the object radiated energy in small, specific amounts that he called *quanta*. A **quantum** *is a finite quantity of energy that can be gained or lost by an atom.* A **photon** *is a quantum of light.* Photons are thought of as particles of radiation. Radiation is emitted and absorbed only in whole numbers of photons. Planck proposed the following relationship between a quantum of energy and the frequency of radiation:

$$E = hv$$

where E is the energy of a photon, v is the frequency of the radiation emitted, and h is a fundamental physical constant now known as Planck's constant.

The value of Planck's constant, 6.626×10^{-34} J · s (joule-seconds), is the same for all electromagnetic radiation. The equation shows that the energy and frequency of radiation are directly proportional. In other words, when the frequency increases, the energy of the radiation increases, and vice versa.

Soon, it was proposed that the absorption of particles of light, each with a specific energy, could explain the photoelectric effect. The effect will not occur if the frequency, and therefore the energy of each photon, is too low to dislodge an electron. Even a few photons of light of the necessary frequency, however, deliver enough energy to produce the photoelectric effect (Figure 4-4).

Gradually, evidence accumulated that light can be described as both particles and waves.

In some experiments, such as those involving interference as in the Desktop Investigation on page 102, light acts like waves. In other experiments, such as those involving the photoelectric effect or the emission of light by a hot object, it acts like particles. The acceptance of the dual wave–particle nature of light soon led scientists to a new view of the nature of matter. In addition, the work of Planck and others led to an explantion of an observed property of the hydrogen atom—the hydrogen spectrum.

When atoms in the gaseous state are heated, their potential energy increases. Then, almost immediately, the atoms return to their original energy states, giving off the added energy in the form of electromagnetic radiation as they do so. The production of colored light in neon signs and in fireworks are familiar illustrations of this process. Visible light or other electromagnetic radiation is emitted when an atom passes from a state of higher potential energy to a state of lower potential energy. *The state of lowest energy of an atom is its* **ground state**. *A state in which it has a higher potential energy than it has in its ground state is an* **excited state**. Excited hydrogen atoms can be produced by passing a high-voltage electric current through hydrogen gas. When this happens, the gas glows with a characteristic lavender color, as shown in Figure 4-5. When a narrow beam of this lavender light is passed through a prism, the light is separated into four narrow lines, each of a different color. Figure 4-6 shows this hydrogen spectrum, which is called a line

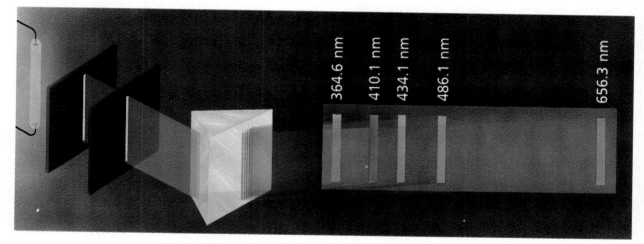

Figure 4-6 A bright line spectrum is produced from hydrogen atoms.

spectrum, or a line-emission spectrum. Each of the colored lines is produced by light of a different wavelength.

As previously discussed, light of a particular wavelength has a definite frequency ($\nu = c/\lambda$), and thus, it also has a definite energy ($E = h\nu$). Each line in the hydrogen spectrum must therefore be produced by emission of photons with certain energies. This would be the result if excited hydrogen atoms could only have specific potential energies. Whenever an excited hydrogen atom drops from such a specific excited state to its ground state or to a low-energy excited state, it emits a photon. The energy of this photon ($E = h\nu$) is equal to the difference in energy between the initial state and the final state, as illustrated in Figure 4-7.

During the late nineteenth and early twentieth centuries, additional series of lines were discovered in the ultraviolet and infrared regions of the hydrogen spectrum. The wavelengths of some of the lines in the hydrogen spectrum—known as the Lyman, Balmer, and Paschen series, after their discoverers—are shown in Figure 4-8. There appeared to be a mathematical relationship between the various energy states of a hydrogen atom. The problem facing scientists was to provide a model of the hydrogen atom that could account for these energy states and, therefore, for the hydrogen-atom spectrum.

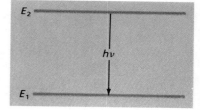

Figure 4-7 When an excited atom with energy $E2$ returns to energy $E1$, it releases a photon having energy $E2 - E1 = E = h\nu$.

Figure 4-8 Representative lines in the hydrogen spectrum. The small letter below each line indicates which of the energy-level transitions produces it.

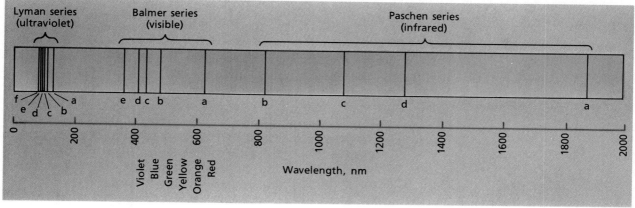

Einstein, using Planck's work, suggested the particle-like behavior of light. He used this premise for his work on the photoelectric effect. For this work, not for his theory of relativity, he received the 1921 Nobel Prize in physics.

Desktop Investigation

The Wave Nature of Light: Interference

Question
Does light show the wave property of interference when a beam of light is projected through a pinhole onto a screen?

Materials
scissors
manila folders
thumb tack
masking tape
aluminum foil
white posterboard
    or cardboard
flashlight

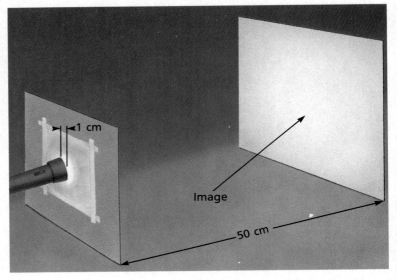

1 cm

Image

50 cm

Procedure
Record all your observations.
1. To make the pinhole screen, cut an 20-cm × 20-cm square from a manila folder. In the center of the square, cut a 2-cm square hole. Cut a 1-cm × 7-cm square of aluminum foil. Using a thumbtack, make a pinhole in the center of this foil square. Tape the aluminum foil over the 2-in square hole, making sure the pin-

hole is centered as shown in the diagram above.
2. Use white posterboard to make a projection screen 35-cm × 35-cm.
3. In a dark room, center the light beam from a flashlight on the pinhole. Hold the flashlight about 1 cm from the pinhole. The pinhole screen should be about 50 cm from the projection

screen, as shown in the diagram. Adjust the distance to form a sharp image on the projection screen.

Discussion
1. Did you observe interference patterns on the screen?
2. As a result of your observations, what do you conclude about the nature of light?

Bohr Model of the Hydrogen Atom

The problem of explaining the hydrogen-atom spectrum was solved in 1913 by the Danish physicist Niels Bohr (1885–1962). Bohr, who had worked with Rutherford, proposed a model of the hydrogen atom as follows: The single electron of the hydrogen atom can circle the nucleus only in allowed paths or *orbits*. When the electron is in each of these allowed orbits, the atom has a definite, fixed energy. Its lowest total energy (kinetic plus potential energy) occurs when it is in the orbit closest to the nucleus. This orbit is separated from the nucleus by a relatively large empty space, where the electron cannot exist. The orbit therefore provides the outer dimension of the hydrogen atom in its ground state. The total energy of the electron increases as it moves into orbits that are successively farther from the nucleus. The highest energy level corresponds to the state in which the electron is sufficiently far from the nucleus that it is no longer attracted by it.

The various electron orbits or atomic energy levels in Bohr's model can be compared to the rungs of a ladder. When you are standing anywhere on a ladder, your feet are on one rung or another. You cannot stand between rungs. The potential energy corresponds to standing on the first rung, the second rung, and so forth. Your potential energy cannot correspond to standing in mid-air between two rungs. The potential energy of your body with respect to the ground is thus subdivided into small, definite amounts—the energy is *quantized*. In the same way, an electron can be in one orbit or another, but not in-between.

How can this model of the hydrogen atom explain the observed spectral lines? While in an orbit, the electron neither gains nor loses energy, But it can move to a higher energy orbit by gaining an amount of energy exactly equal to the difference in energy between the higher energy orbit and the initial lower energy orbit. When a hydrogen atom is in an excited state, its electron is in a higher energy orbit.

Line emission spectra can then be explained as follows: *A **line spectrum** is produced when an electron drops from a higher-energy orbit to a lower energy orbit*. As the electron drops, a photon is emitted that has an energy ($E = h\nu$) equal to the difference in energy between the initial higher energy orbit and the final lower energy orbit. The absorption and emission of radiation according to the Bohr model of the hydrogen atom is illustrated in Figure 4-9.

By using the measured values for the mass and charge of the electron, together with Planck's constant, Bohr was able to calculate the energies that an electron would have in the allowed energy levels for the hydrogen atom. This enabled him to show mathematically how the various spectral series of hydrogen could be produced. Bohr's calculated values agreed exactly with the experimentally observed values for the lines in each series. For example, the Lyman spectral series was shown to be the result of electrons dropping from various higher-energy levels to the ground-state energy level. The origins of several of the series of lines in the hydrogen atom spectrum are shown in Figure 4-10.

The success of Bohr's model of the hydrogen atom in explaining observed spectral lines led many scientists to conclude that a similar model could be applied to all atoms. Scientists soon recognized, however, that Bohr's approach did not explain the spectra of atoms with more than one electron and that it did not fully explain the chemical behavior of atoms.

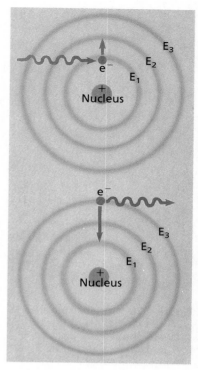

Figure 4-9 According to the Bohr model of the hydrogen atom, the electron can only be in orbits corresponding to the energies, E_1, E_2, E_3, and so forth.

Spectroscopy

Spectroscopy is a useful tool for studying the structure of atoms and the composition of matter in general. By studying the emission spectra of certain substances as described below, the identity of atoms in a substance can be determined.

Use of a spectroscope   The **spectroscope** is an instrument that separates light into a spectrum that can be examined. One type of spectroscope consists of a glass prism and a collimator tube that focuses a narrow beam of light rays

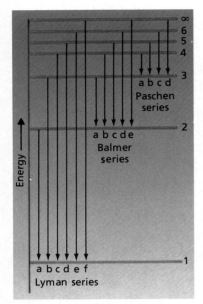

Figure 4-10 An electron energy-level diagram for hydrogen showing the energy transitions for the Lyman, Balmer, and Paschen spectral series.

on the prism. It also has a small telescope for examining the light that passes through the prism. When white light passes through a triangular prism, a band of colors called a continuous spectrum appears. This effect is caused by the unequal bending of light of different wavelengths as it enters and emerges from the prism. A continuous spectrum is shown in the top band of Figure 4-11.

Origin of spectral lines Light rays of different frequencies are bent different amounts when passed through a prism. Light energy of the shortest wavelengths (highest frequencies) is bent most, forming the deep violet color seen at one end of the visible spectrum. Light energy of the longest wavelengths (lowest frequencies) is bent least, forming the deep red color characteristic of the other end of the visible spectrum. Between these two extremes there is a gradual blending from one color to the next. It is possible to recognize six elementary colors: red, orange, yellow, green, blue, and violet.

When substances are heated or energized sufficiently, electrons are raised to higher energy levels by heat energy. When these electrons fall back into the lower energy levels available to them, energy is released. The energy released by any substance has wavelengths characteristic of that substance. Different excited atoms produce different emission spectra. For example, vaporized sodium atoms produce a spectrum consisting of two narrow yellow lines that are very close together (seen in the ordinary spectroscope as a single yellow line). Potassium atoms produce two red lines and a violet line. Lithium atoms yield intense red and yellow lines and weak blue and violet lines. The spectra produced by excited atoms of different elements are as distinct as fingerprints.

Bright lines in the visible portion of the spectrum account for the flame coloration produced by certain metals. The color seen is the combination of emitted light of different wavelengths.

Quantum Model of The Atom

To many scientists, the biggest shortcoming of Bohr's theory was that the very idea of the quantum appeared to contradict common sense. Why could the electron in the hydrogen atom orbit the nucleus only in a relatively small number of allowed paths? Why weren't there limitless orbits of slightly different energies? Scientists searched for some insight into the nature of matter that would explain why atomic energy states are quantized.

The key to a general atomic theory that explains the quantization of electron energies was provided in 1924 by the French physicist Louis de Broglie (1892–1987). Inspired by the dual, wave–particle nature of light, de Broglie proposed that electrons might also have a wave–particle nature. De Broglie pointed out that Bohr's proposed quantized electron orbits were consistent with known wave behavior. For example, waves confined in a space have only certain frequencies. Reflected waves of all other frequencies destroy themselves by destructive interference and disappear. If electrons are thought of as waves confined to the space around an atomic nucleus, this could explain why only certain frequencies, and therefore only certain energies, of electrons are possible. De Broglie's suggestion that electrons have wave properties was soon confirmed by experiments demonstrating the interfer-

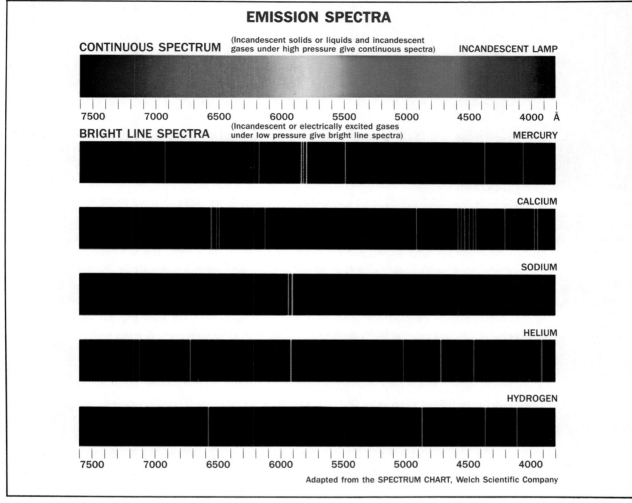

EMISSION SPECTRA

CONTINUOUS SPECTRUM (Incandescent solids or liquids and incandescent gases under high pressure give continuous spectra) INCANDESCENT LAMP

7500 7000 6500 6000 5500 5000 4500 4000 Å

BRIGHT LINE SPECTRA (Incandescent or electrically excited gases under low pressure give bright line spectra) MERCURY

CALCIUM

SODIUM

HELIUM

HYDROGEN

7500 7000 6500 6000 5500 5000 4500 4000

Adapted from the SPECTRUM CHART, Welch Scientific Company

Figure 4-11 Each element's atoms will produce a unique emission spectrum when heated.

ence between electron waves. For instance, when light waves pass through a small opening, like a pinhole in the Desktop Investigation on page 102, the transmitted light waves are bent, or diffracted. Passing electrons through the narrow spaces between atoms in crystalline materials produced diffraction patterns like those previously known only for light waves. (Figure 4-11)

The Austrian physicist Erwin Schrödinger (1887–1961) was impressed by de Broglie's hypothesis that electrons have wave properties. In 1926, Schrödinger devised an equation that treated electrons moving around nuclei as waves. This now-famous equation, known as the Schrödinger wave equation, laid the foundation for modern quantum theory. **Quantum theory** *describes mathematically the wave properties of electrons and other very small particles.*

Schrödinger's wave equation is a mathematical model that applies to all atoms, unlike Bohr's model, which worked only for the hydrogen atom. The quantization of electron energies is a natural outcome of the Schrödinger equation. Only specific energies provide solutions to the equation.

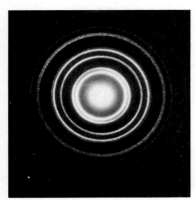

Figure 4-12 Diffraction pattern produced by passing a beam of electrons through many small crystals of aluminum oriented at random.

To develop the idea of probable location, use the analogy of a teacher who constantly moves about the classroom while teaching. At any given hour during a school day, there is a high probability that the teacher will be found somewhere in that assigned room, but exactly where in the room is hard to determine. The same is true of electrons in their orbitals.

Bohr's model described definite orbits occupied by electron particles. In contrast, Schrödinger's model treats electrons as waves that have only a certain probability of being found at various distances from the nucleus. In Schrödinger's model, the nucleus is surrounded not by orbits, but by what are described as *orbitals*.

Strictly speaking, an orbital is a mathematical description of the region within which an electron of specific energy can be found. *An* **orbital** *is a three-dimensional region about the nucleus in which a particular electron can be located.* Orbitals can be thought of as clouds showing the region of probable electron locations. The electron-cloud picture allows us to relate orbitals to something familiar and to make drawings like those in Figure 4-13. The sizes and shapes of electron clouds depend on the energies of the electrons that occupy them.

In the Bohr atomic model, electrons of increasing energy occupy orbits further and further from the nucleus. Atomic orbitals also fall into similar quantized regions of increasing energy, which are referred to as *main energy levels*.

The idea that an electron could not be "pinpointed" was Schrödinger's revolutionary contribution to the model of the atom. The Bohr model had attempted to explain electron behavior as though electrons were physical particles. It made sense to think of the electron as being at a particular point on a precisely specified orbit. But if electrons are thought of as waves, then Schrödinger's electron-cloud or probability approach is easier to accept. After all, waves are not physical bodies that can be precisely located.

This view of matter contradicts our ordinary experience of reality. Many scientists accepted the quantum theory of the atom only after the physicist Werner Heisenberg (1901–1976) proposed his uncertainty principle, discussed in the Chemistry Notebook on the following page.

In many other ways, Schrödinger's atomic model is similar to that of Bohr. From the standpoint of energy, for example, Schrödinger's orbitals can be treated very much like Bohr's orbits. The closer an orbital is to the nucleus, the lower the total energy of an electron in that orbital. To jump from a lower-energy orbital to a higher-energy orbital, an electron must absorb a quantity of energy precisely equal to the energy difference between the two orbitals. When an electron drops from a higher-energy orbital to a lower-energy orbital, it gives off electromagnetic radiation. The energy of the emitted radiation is precisely equal to this energy difference. Finally, the most probable location for the single electron in a hydrogen atom is at a distance from the nucleus exactly equal to that of Bohr's lowest-energy orbit.

Figure 4-13 Two ways of showing electrons around atoms. In *(a)* the probability of finding the electron is proportional to the density of the cloud. In *(b)* the surface within which the electron can be found a certain percentage of the time, i.e., 90% is shown.

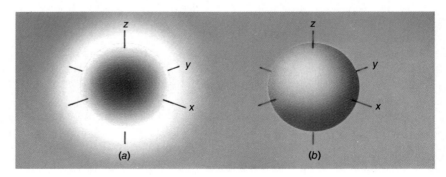

Chemistry Notebook

Heisenberg's Uncertainty Principle

The *uncertainty principle* was first stated in 1927 by the German physicist Werner Heisenberg (1901–1976). According to Heisenberg, it is not possible to measure precisely both the velocity and the position of an electron at the same time. (Velocity is the speed and direction of motion of an object.) The following baseball analogy may help you understand this principle, which is widely accepted by scientists today.

At the crack of the bat, an outfielder can predict almost where a fly ball is going to land, run to that spot, and be ready to catch it. The outfielder instinctively predicts the path that the ball will follow by observing its position after it is hit. The outfielder sees the ball by means of photons reflected off the ball and into his eyes. The photons do not affect the path of the ball because the ball is much more massive than the photons striking it.

Now suppose that a submicroscopic outfielder is trying to catch an electron. This imaginary outfielder also observes objects by means of reflected photons. But unlike a baseball, an electron is so small that its path is significantly changed each time it is struck by a photon. The

outfielder would be able to tell only where the electron had been when it was struck by the last photon. The velocity, and thus, the path of the electron could not be predicted.

If the submicroscopic outfielder tried to avoid this problem by using lower

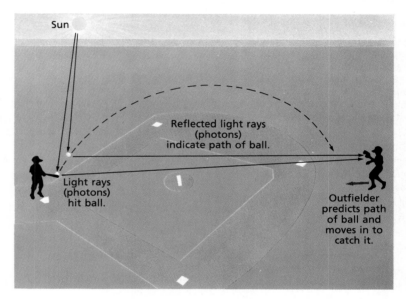

Sun

Reflected light rays (photons) indicate path of ball.

Light rays (photons) hit ball.

Outfielder predicts path of ball and moves in to catch it.

energy light so as not to disturb the path of the electron, a new difficulty would appear. Low energy light would have such a low frequency and long wavelength that it would not be possible for the outfielder to tell the exact direction from which the reflected light came. (You may have noticed that is is much easier to locate the sources of a

high frequency, like that of a flute, than that of a low-frequency sound like that of a ban drum.) The outfielder would therefore not know where the electron was when it reflected the light.

As illustrated in the baseball analogy, it is not possi-

ble to measure both the position and the path of an electron. Since the path of an electron depends partly on its energy, scientists cannot measure both the energy and the position of an electron. ***Heisenberg's uncertainty principle** states that it is not possible to know both the velocity and the position of a particle at the same time.*

Section Review

1. What was the major limitation of Rutherford's model of the atom?
2. Write and label the equation that relates the speed, wavelength, and frequency of electromagnetic radiation.
3. Define *(a)* electromagnetic radiation *(b)* wavelength *(c)* frequency *(d)* quantum *(e)* photon.
4. What is meant by the dual wave–particle nature of light?
5. Describe the Bohr model of the hydrogen atom.

4.2 Quantum Numbers and Atomic Orbitals

This section introduces the terminology, symbols, and rules used to associate individual electrons with specific atomic orbitals. The concepts introduced here will be used again and again in later chapters as you learn more about the characteristic properties of matter.

Quantum Numbers

More than one set of numbers is needed to describe the quantization of electron energies. **Quantum numbers** *are numbers that specify the properties of atomic orbitals and of their electrons.* The first three sets of quantum numbers are derived from the Schrödinger equation. They indicate the region occupied by a given orbital in terms of *(1)* distance from the nucleus, *(2)* orbital shape, and *(3)* orbital position with respect to the three-dimensional x, y, and z axes. These three quantum numbers are whole numbers and are related to each other. The fourth quantum number has only two values and is needed to specify one of two possible orientations of an electron within an orbital. As you read the following descriptions of the quantum numbers, refer to the appropriate columns in Table 4-1.

Principal quantum number *The* **principal quantum number,** *symbolized by n, indicates the main energy levels surrounding a nucleus.* These energy levels, sometimes referred to as *shells*, are the quantum-theory equivalent of the Bohr orbits. Values of n are whole numbers only—1, 2, 3, and so on. The main energy level with $n = 1$ is the one closest to the nucleus. As n increases, the distance of the main energy levels from the nucleus increases and their energy increases. The known elements utilize main energy levels with $n = 1$ to 7 in their ground states.

Orbital quantum number *The* **orbital quantum number** *indicates the shape of an orbital.* Within each main energy level beyond the first, orbitals with different shapes occupy different regions. The regions are referred to as *sublevels*, or *subshells*.

As shown in the second column of Table 4-1, the number of subshells, or possible orbital shapes, in each main energy levels is equal to the value of the principal quantum number. The first four orbital quantum numbers (0,1,2,3) are designated in ascending order by the letters s, p, d, and f (the s sublevel

TABLE 4-1 QUANTUM NUMBER RELATIONSHIPS IN ATOMIC STRUCTURE

Principal Quantum Number: Main Energy Level (n)	Orbitals (Sublevels): n Orbital Shapes; n Sublevels	Number of Orbitals per Sublevel	Number of Orbitals per Main Energy Level (n^2)	Number of Electrons per Sublevel	Number of Electrons per Main Energy Level ($2n^2$)
1	s	1	1	2	2
2	s	1	4	2	8
	p	3		6	
3	s	1	9	2	18
	p	3		6	
	d	5		10	
4	s	1	16	2	32
	p	3		6	
	d	5		10	
	f	7		14	

is the lowest in energy; the p sublevel is higher in energy than the s sublevel; and so on.)

The s orbitals are spherical, the p orbitals have dumbbell shapes, and the d orbitals (with one exception) have four lobes. These are all shown in Figure 4-14. The f orbital shapes are too complex to treat here.

In the first energy level ($n=1$), only one sublevel is allowed. The first energy level therefore has only an s sublevel. In the second energy level ($n=2$), two different sublevels are possible. The second energy level thus has s and p orbitals. The third energy level ($n=3$) has three sublevels, and therefore has s, p, and d orbitals. The fourth energy level ($n=4$) has four possible sublevels and therefore has s, p, d, and f orbitals. In an nth main energy level, orbitals of n shapes are possible.

Each sublevel or type of orbital is designated by the principal quantum number followed by the letter of the sublevel. For example, the $1s$ sublevel is the s sublevel in the first main energy level, the $2s$ sublevel is the s orbital in the second main energy level. A $4d$ orbital is an orbital in the d sublevel of the fourth main energy level. How would you designate the p sublevel in the third main energy level (see margin)? How many other sublevels are in the same main energy level with this one (see margin)?

STUDY HINT

Principal quantum number	Types of orbitals
1	$1s$
2	$2s$, $2p$
3	$3s$, $3p$, $3d$
4	$4s$, $4p$, $4d$, $4f$

Figure 4-14 s, p and d orbitals have different shapes.

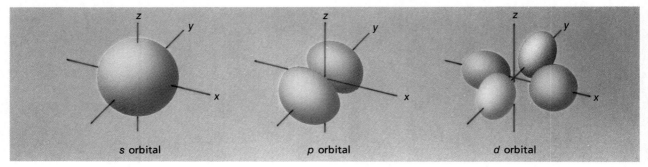

s orbital   p orbital   d orbital

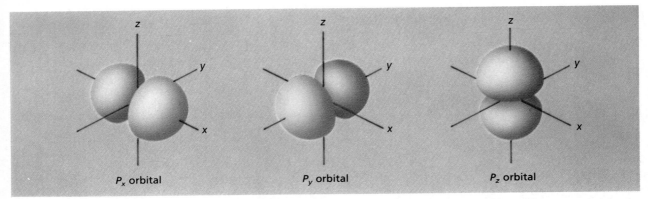

P_x orbital P_y orbital P_z orbital

Figure 4-15 There are three orientations for *p* orbitals.

Magnetic quantum number   *The* **magnetic quantum number** *indicates the orientation of an orbital about the nucleus.* There is only one possible orientation for an *s* orbital, which is a sphere centered on the nucleus, like the orbitals drawn in Figure 4-14. There is therefore only one *s* orbital in each *s* sublevel. The *p* orbitals can have three different orientations. The lobes extend along the *x*, *y*, and *z* axes of a three-dimensional (Cartesian) coordinate system, as shown in Figure 4-15. Each *p* sublevel therefore contains three different *p* orbitals. These usually are designated as p_x, p_y, and p_z orbitals, where the subscripts *x*, *y*, and *z* indicate the locations of the *p* orbitals.

There are five different *d* orbitals (Figure 4-16) in each *d* sublevel, as shown in Table 4-1. There are seven different *f* orbitals in each *f* sublevel.

The combined effect of the orbital and magnetic quantum numbers is that, with increasing main energy levels there are larger numbers of orbitals. Table 4-1 should convince you that the number of orbitals in each main

To help students understand how all parts of the electron cloud fit together and its 3-D nature, use the analogy of an apartment building to represent the electron cloud. Each floor is an energy level, each apartment is a sublevel, and each room is an orbital with one or two people (electrons) living in it.

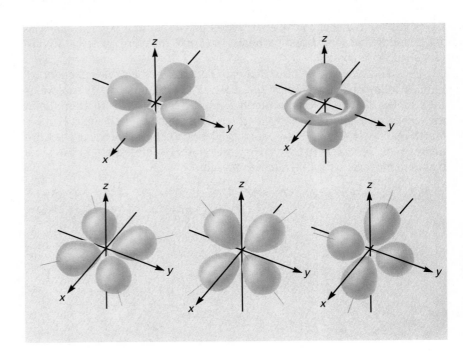

Figure 4-16 There are five orientations for *d* orbitals

energy level equals the square of the principal quantum number. What is the total number of sublevels and orbitals in the second energy level and of what types are they (see margin)?

Spin quantum number   *The* **spin quantum number**, *has only two possible values—(+½, −½)—which indicate two possible states of an electron in an orbital.* Of all the quantum numbers, electron spin has the most familiar physical significance. You can picture an electron spinning on an imaginary axis in much the same way that the earth spins on its axis, thus generating its magnetic field. Each electron thus behaves as though it were a tiny bar magnet with a north and south pole.

Magnetism   All magnetic phenomena stem from the motions of electrons about the nuclei of atoms. *Diamagnetism* is the property of a substance whereby it is weakly repelled by a magnetic field. The magnetic field causes the electrons in any substance to move in such a way that they produce an opposing magnetic field that is repelled by the original magnetic field. *Paramagnetism* is a weak attraction between magnetic fields and substances whose atoms have a more or less uneven electron distribution around their nuclei. This attraction is slightly stronger than the repulsive diamagnetism, so that paramagnetic substances are weakly attracted by magnetic fields. Oxygen is among the common substances that are paramagnetic. *Ferromagnetism*, such as that shown by ordinary bar magnets, is a strong form of paramagnetism, resulting from all of the atoms in the ferromagnetic substance becoming oriented with their like magnetic poles in the same direction.

Electrons in Each Main Energy Level

The rules that govern the location of electrons in the ground state of atoms of every element are discussed in the next section. The first three quantum numbers completely indicate the energy level and shape of each orbital. The spin quantum number is significant, as you will see, because each orbital can hold no more than two electrons, which must have opposite spins. This limitation determines the maximum number of electrons that can occupy each main energy level, as shown in Table 4-1.

The first main energy level has only a single *s* orbital, and therefore, will hold only 2 electrons. The second main energy level can accommodate 2 electrons in its single *s* orbital and 2 electrons in each of its three *p* orbitals, for a total occupancy of 8 electrons. Can you explain how the third main energy level can hold 18 electrons (see margin)? Examine Table 4-1 to convince yourself that the maximum number of electrons in each main energy level is equal to $2n^2$.

Section Review

1. Define   *(a)* main energy levels   *(b)* quantum numbers.
2. *(a)* What general information about atomic orbitals is provided by the four sets of quantum numbers?   *(b)* List the four sets of quantum numbers.
3. Identify and explain the meanings associated with the possible values for each of the four sets of quantum numbers.

Point out that in the first energy level, the *s*-orbital, the *s*-sublevel, and the first energy level are the same thing.

STUDY HINT
Filled third main energy level ($n = 3$) has 2 e⁻ in 3s orbital, 6 e⁻ in 3p orbitals, 10 e⁻ in 3d orbitals.
$2 e^- + 6 e^- + 10 e^- = 18 e^-$. $2n^2 = 2 \times (3)^2 = 2 \times 9 = 18$

1. *(a)* The quantized regions of increasing energy into which atomic orbitals fall   *(b)* Sets of numbers that specify the properties of atomic orbitals and their electrons.
2. *(a)* The distance of an orbital from the nucleus, the shape of the orbital, its position with respect to the three-dimensional axes, and the state of the electrons within that orbital   *(b)* The principal, orbital, magnetic, and spin quantum numbers.
3. (1) Principal quantum number *n* (1, 2, 3, and so on): distance of the main energy level from the nucleus (2) Orbital quantum number *(s, p, d,* and *f)*: various orbital shapes (3) Magnetic quantum number (the subscripts *x, y,* and *z*): orbital orientation with respect to a three-dimensional coordinate system (4) Spin quantum number ($+\frac{1}{2}$ and $-\frac{1}{2}$): the two possible electron states within an orbital

4.3 Electron Configurations

- State the Aufbau principle, Hund's rule, and the Pauli exclusion principle.
- Describe the arrangement of electrons around the atoms of any element using orbital notation, electron ✕ configuration notation or electron-dot notation.
- Build up the electron configuration for atoms of any element, given the atomic number or identity of the element.
- Describe the noble ✕ gas configuration and write it for any noble gas.

Electron configuration *is the arrangement of electrons in atoms.* Atoms of different elements have different numbers of electrons. Therefore, there is a distinctive electron configuration for atoms of each element. Electrons in atoms, like all systems in nature, tend to assume arrangements that have the lowest possible energies. A few simple rules, combined with the quantum number relationships discussed in Section 4.2, permit assignment of these stable, or ground-state, electron configurations.

In this section you will learn how to build up electron configurations for the ground state of any particular atom. The process begins with the hydrogen atom, with its single electron. Electrons are added one by one according to the rules. As you follow the changing electron configurations with increasing atomic number, you will begin to discover the relationship between electron configurations and the Periodic Table. Real atoms are not, however, built up by adding protons and electrons one at a time, and it is wise to remember this.

Rules Governing Electron Configurations

The first rule shows the order in which vacant orbitals will be occupied. According to the **Aufbau** ('building up' in German) **principle**, *an electron occupies the lowest-energy orbital that can receive it.* In hydrogen atoms, the electron is in the orbital with the lowest energy of all orbitals—the $1s$ orbital. The $2s$ and $2p$ orbitals are next highest in energy and will be filled in that order—$2s$, then $2p$.

In multielectron atoms, with the third main energy level ($n = 3$), sublevels in different main energy levels begin to overlap as shown in Figure 4-17. The order of increasing energy for the sublevels is summarized on the lefthand axis. Note, for example, that the $4s$ sublevel is lower in energy than the $3d$ sublevel. In building up the electron configurations for atoms of different elements,

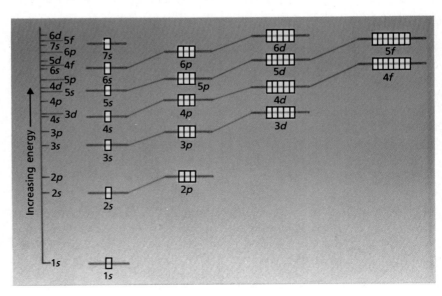

Figure 4-17 Relative energies if an atom's sublevels. At the third main energy level, overlap begins among sublevels of other main energy levels.

therefore, the $4s$ orbital fills before any electrons enter the $3d$ orbitals. Once the $3d$ orbitals are fully occupied by 10 electrons, which sublevel will be occupied next (see margin)?

The second rule requires placing as many single electrons as possible in separate orbitals in the same sublevel. In this way, electron–electron repulsion is minimized and the electron arrangements have a lower energy. According to **Hund's rule**, *orbitals of equal energy are each occupied by one electron before any is occupied by a second electron.* Applying this rule shows, for example, that one electron will enter each of the three p orbitals in a main energy level before a second electron enters any of them. What is the maximum number of unpaired electrons in a d sublevel (see margin)? The third rule is the one we referred to in noting the importance of the spin quantum number. According to the **Pauli exclusion principle**, *no two electrons in the same atom can have the same four quantum numbers.* Different combinations of the first three quantum numbers apply to each different orbital. The two values of the spin quantum number permit two electrons of opposite spins to occupy the same orbital.

Representing Electron Configurations

Three methods, or notations, are used to indicate electron configurations. Two of these notations indicate the ground-state electron configuration. The third indicates only the outermost electron arrangement, which is the most important in the formation of chemical compounds. These three notations are illustrated for atoms of hydrogen (H) and helium (He), the two elements of Period 1.

The single electron in ground-state hydrogen is in the lowest-energy orbital, the $1s$ orbital. The electron can have either one of its two possible spin states. Helium has two electrons, which are paired in the lowest-energy orbital, the $1s$ orbital.

Orbital notation   In orbital notation, an unoccupied orbital is represented by a line ___. An orbital containing one electron is represented as $\uparrow$___ (or $\downarrow$___). An orbital containing two electrons is represented as $\uparrow\downarrow$___, showing the electrons paired and with opposite spins. The lines are labeled with the principal quantum number and subshell letter. The orbital notations for hydrogen and helium are, therefore,

$$\text{H} \quad \underset{1s}{\underline{\uparrow}} \qquad \text{He} \quad \underset{1s}{\underline{\uparrow\downarrow}}$$

Electron-configuration notation   Electron-configuration notation eliminates the lines and arrows of orbital notation. Instead, the number of electrons in a sublevel is shown by adding superscripts to the sublevel designation. The hydrogen configuration is represented by $1s^1$. The superscript ^{1} indicates that one electron is present in the $1s$ orbital. The helium configuration is represented by $1s^2$, the superscript ^{2} indicating that there are two electrons in the $1s$ orbital.

Electron-dot notation   Electron-dot notation shows only electrons in the highest, or outermost, main energy level. The **highest occupied energy level** *is the electron-containing main energy level with the highest principal* *quantum number.* For hydrogen and helium atoms, there is only one occupied energy level.

Dots representing the number of electrons in the highest occupied energy level are placed around the symbol for the element. The electron-dot symbols for hydrogen and helium are

$$\text{H}· \qquad \text{He}:$$

Helium atoms have a fully occupied $1s$ sublevel, therefore, the next electron must enter the next highest vacant sublevel, which is the $2s$ sublevel (see Figure 4-17). For lithium, with atomic number 3 and three electrons, the electron dot symbol is

$$\text{Li}·$$

The two electrons in the $1s$ sublevel are no longer in the outermost main energy level. They have become **inner-shell electrons**, *which are electrons not in the highest occupied energy level.* For atoms of all elements beyond helium, the atomic symbol in electron-dot notation represents the nucleus plus any inner-shell electrons.

Additional dots after the first can be added in several different arrangements. In this book, for elements of the second or higher periods, a second dot will be placed at the left, a third dot at the top, and a fourth dot at the bottom. This arrangement does not always correspond to the ground-state electron configuration. Instead, it provides a clue about how the atom will react with other atoms (see Chapter 6). The dot arrangements for five, six, seven, and eight electrons are also shown in the margin.

Sample Problem 4.1

The electron configuration of boron is $1s^2 2s^2 2p^1$. *(a)* How many electrons are present in a boron atom? What is the atomic number of boron? Write the orbital notation for boron. *(Ans.)* From the electron-configuration notation for boron, each atoms contain $2 + 2 + 1 = 5$ electrons. Since the number of protons equals the number of electrons in a neutral atom, boron has 5 protons and, thus, has an atomic number of 5. To write out the orbital notation, we first draw the lines representing orbitals:

$$\overline{\quad}_{1s} \quad \overline{\quad}_{2s} \quad \overline{\quad}_{2p_x} \quad \overline{\quad}_{2p_y} \quad \overline{\quad}_{2p_z}$$

Next, we add arrows showing the electron locations. The first two electrons occupy $n = 1$ energy level and fill the $1s$ orbital:

$$\overline{\uparrow\downarrow}_{1s} \quad \overline{\quad}_{2s} \quad \overline{\quad}_{2p_x} \quad \overline{\quad}_{2p_y} \quad \overline{\quad}_{2p_z}$$

The next three electrons enter the $n = 2$ main-energy level. As shown by the electron-configuration notation, two electrons occupy the $2s$ orbital and one, a p orbital.

$$\overline{\uparrow\downarrow}_{1s} \quad \overline{\uparrow\downarrow}_{2s} \quad \overline{\uparrow}_{2p_x} \quad \overline{\quad}_{2p_y} \quad \overline{\quad}_{2p_z}$$

(b) Write the electron-dot symbol for boron. How many inner-shell electrons does boron have? In which orbital are they located? *(Ans.)* A boron atom has three electrons in its outermost occupied energy level, the $n = 2$ level. The electron-dot symbol for boron, is therefore,

$$·\dot{\text{B}}·$$

The two remaining electrons in a boron atom are inner-shell electrons in the $1s$ orbital.

Sample Problem 4.1 *continued*

Elements of the Second and Third Periods

Second-period elements As shown above, the first main energy level is filled in the first-period elements, hydrogen and helium. The ground-state configurations in Table 4-2 illustrate how the Aufbau principle, Hund's rule, and the Pauli exclusion principle are applied in the second period. (A diagram that aids in applying the Aufbau principle is given in Figure 4-18.)

According to the Aufbau principle, the first added electron enters the s sublevel in the second main energy level. Thus, lithium (Li) has the $1s^2 2s^1$ configuration. The fourth electron in beryllium (Be) atoms must pair up in the $2s$ sublevel because this sublevel is of lower energy than the $2p$ sublevel.

With the $2s$ sublevel filled, the $2p$ sublevel, which has three vacant orbitals of equal energy, can be occupied. Hund's rule applies here, as clearly shown in the orbital notation. Each of the three p orbitals is occupied in turn by a single electron in boron (B), then carbon (C), then nitrogen (N) atoms.

The next electron must either pair up with another electron in one of the $2p$ orbitals or enter the third main energy level, where it could be alone in an orbital. The Aufbau principle shows that the $2p$ orbitals must be filled before higher-energy orbitals are occupied. The Pauli exclusion principle shows that the next electron can pair up in a $2p$ orbital by having the opposite spin from the first electron. Oxygen (O) atoms, therefore, have the configuration $1s^2 2s^2 2p^4$. This notation does not directly show electron pairing in the $2p$ sublevel. Based on the rules, however, we know that of four electrons in a p sublevel, two are paired and two are not.

The remaining two $2p$ orbitals are filled in fluorine (F) and neon (Ne) atoms, respectively. Atoms like those of neon, which have the s and p sublevels of their highest main energy level filled with eight electrons, are said to have an *octet* of electrons. Note in the Periodic Table inside the back cover than neon is the last element in the second period.

Third-period elements After the outer octet is filled in neon atoms, the filling sequence begins again. The next electron enters the s sublevel in the $n = 3$ main energy level. Thus, sodium (Na) atoms have the configuration $1s^2 2s^2 2p^6 3s^1$. Compare this sodium atom configuration with that of neon atoms in Table 4-2. The first 10 electrons in a sodium atom have the neon configuration, $1s^2 2s^2 2p^6$. To simplify writing the notation, the symbol for neon, enclosed in square brackets, is used to represent the complete neon

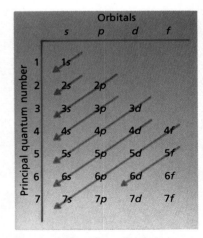

Figure 4-18 The spinning motion of the electron generates a magnetic field. The result is that the electron acts as a tiny bar magnet with its own north and south pole.

TABLE 4-2 **ELECTRON CONFIGURATIONS OF ATOMS OF SECOND-PERIOD ELEMENTS SHOWING THREE NOTATIONS**

Element	Symbol	Orbital Notation 1s	2s	2p	Electron-Configuration Notation	Electron-Dot Notation
Lithium	Li	⇅	↑	__ __ __	$1s^2 2s^1$	Li·
Beryllium	Be	⇅	⇅	__ __ __	$1s^2 2s^2$	·Be·
Boron	B	⇅	⇅	↑ __ __	$1s^2 2s^2 2p^1$	·Ḃ·
Carbon	C	⇅	⇅	↑ ↑ __	$1s^2 2s^2 2p^2$	·Ċ·
Nitrogen	N	⇅	⇅	↑ ↑ ↑	$1s^2 2s^2 2p^3$	·N̈:
Oxygen	O	⇅	⇅	⇅ ↑ ↑	$1s^2 2s^2 2p^4$	:Ö:
Fluorine	F	⇅	⇅	⇅ ⇅ ↑	$1s^2 2s^2 2p^5$	:F̈:
Neon	Ne	⇅	⇅	⇅ ⇅ ⇅	$1s^2 2s^2 2p^6$	:N̈e

configuration, $[Ne] = 1s^2 2s^2 2p^6$. This permits writing $[Ne]3s^1$ for the sodium configuration.

The last element in the third period is argon (Ar). Like neon, the element above it in Group 18, argon has an outermost octet of electrons, $[Ne]3s^2 3p^6$. In fact, each Group 18 element has an electron octet in its highest energy level (see Periodic Table inside back cover). *The Group 18 elements (helium, neon, argon, krypton, xenon, and radon) are called the* **noble gases.** The symbol for each is used, as just described for neon, to simplify the electron configuration notation for elements that follow it. *A* **noble gas configuration** *is an outer main energy level fully occupied by eight electrons.*

Sample Problem 4.2

(a) Write the electron-dot notation for silicon (Si). How many inner-shell electrons does a silicon atom contain? *(Ans.)* The configuration of silicon is $[Ne]3s^2 3p^2$. (See Table 4-2.) A silicon atom has four electrons in the highest energy level, and therefore the electron dot symbol is

$$·\dot{Si}·$$

All of the electrons represented by [Ne] are inner-shell electrons. The total number, as shown in Table 4-3, is 10.

(b) Identify the element with atoms that have five electrons in the *p* sublevel in their third main energy level. What is the total number of electrons in the third main energy level in these atoms? Write the electron-dot symbol for this element. Name the element in the second period with the same number of outermost electrons. *(Ans.)* Table 4-3 shows that a chlorine atom, with configuration $[Ne]3s^2 3p^5$, has five electrons in its 3p sublevel. There are seven electrons in the third main energy level. Therefore, chlorine has the electron-dot symbol

$$:\ddot{Cl}:$$

Fluorine (F), as shown in Table 4-2, also has seven outermost electrons.

Sample Problem 4.2 *continued*

TABLE 4-3   ELECTRON CONFIGURATIONS OF ATOMS OF THIRD-PERIOD ELEMENTS

Name	Symbol	Atomic Number	1s	2s	Number of Electrons in Sublevels 2p	3s	3p	Shorthand Notation
Sodium	Na	11	2	2	6	1		$[Ne]3s^1$
Magnesium	Mg	12	2	2	6	2		$[Ne]3s^2$
Aluminum	Al	13	2	2	6	2	1	$[Ne]3s^23p^1$
Silicon	Si	14	2	2	6	2	2	$[Ne]3s^23p^2$
Phosphorus	P	15	2	2	6	2	3	$[Ne]3s^23p^3$
Sulfur	S	16	2	2	6	2	4	$[Ne]3s^23p^4$
Chlorine	Cl	17	2	2	6	2	5	$[Ne]3s^23p^5$
Argon	Ar	18	2	2	6	2	6	$[Ne]3s^23p^6$

Elements of the Fourth and Fifth Periods

The electron configurations of atoms in the fourth-period elements are shown in Table 4-4. The period begins by filling the $4s$ orbital, the empty orbital of lowest energy. Thus, the first element in the fourth period is potassium (K), which has the electron configuration $[Ar]4s^1$. The next element is calcium (Ca), which has the electron configuration $[Ar]4s^2$. The electron-dot symbols for potassium and calcium are

$$K\cdot \qquad \cdot Ca\cdot$$

With the $4s$ sublevel filled, the $4p$ and $3d$ sublevels have the next available vacant orbitals. Figures 4-17 and 4-18 show that the $3d$ sublevel is lower in energy than the $4p$ sublevel, and therefore the five $3d$ orbitals are next to be filled in the ten elements from scandium (Z = 21) to zinc (Z = 30).

Scandium (Sc) has the electron configuration $[Ar]3d^14s^2$. Titanium (Ti), has the configuration $[Ar]3d^24s^2$, and vanadium (V), has the configuration $[Ar]3d^34s^2$. To this point, three electrons with the same spin have been added to three separate *d* orbitals, as expected from Hund's rule.

TABLE 4-4 ELECTRON CONFIGURATION OF ATOMS OF ELEMENTS IN THE FOURTH PERIOD

Name	Symbol	Atomic Number	Number of Electrons in Sublevels Above 2p					Electron Configuration Notation
			3s	3p	3d	4s	4p	
Potassium	K	19	2	6		1		$[Ar]4s^1$
Calcium	Ca	20	2	6		2		$[Ar]4s^2$
Scandium	Sc	21	2	6	1	2		$[Ar]3d^14s^2$
Titanium	Ti	22	2	6	2	2		$[Ar]3d^24s^2$
Vanadium	V	23	2	6	3	2		$[Ar]3d^34s^2$
Chromium	Cr	24	2	6	5	1		$[Ar]3d^54s^1$
Manganese	Mn	25	2	6	5	2		$[Ar]3d^54s^2$
Iron	Fe	26	2	6	6	2		$[Ar]3d^64s^2$
Cobalt	Co	27	2	6	7	2		$[Ar]3d^74s^2$
Nickel	Ni	28	2	6	8	2		$[Ar]3d^84s^2$
Copper	Cu	29	2	6	10	1		$[Ar]3d^{10}4s^1$
Zinc	Zn	30	2	6	10	2		$[Ar]3d^{10}4s^2$
Gallium	Ga	31	2	6	10	2	1	$[Ar]3d^{10}4s^24p^1$
Germanium	Ge	32	2	6	10	2	2	$[Ar]3d^{10}4s^24p^2$
Arsenic	As	33	2	6	10	2	3	$[Ar]3d^{10}4s^24p^3$
Selenium	Se	34	2	6	10	2	4	$[Ar]3d^{10}4s^24p^4$
Bromine	Br	35	2	6	10	2	5	$[Ar]3d^{10}4s^24p^5$
Krypton	Kr	36	2	6	10	2	6	$[Ar]3d^{10}4s^24p^6$

The 1s, 2s and 2p sublevels are filled in atoms of each element shown.
The configuration for these filled sublevels is $1s^22s^22p^6$.

Somewhat surprisingly, chromium (Cr), has the electron configuration $[Ar]3d^54s^1$. Not only did the added electron go into the fourth $3d$ orbital, but an electron was also moved from the $4s$ orbital into the fifth $3d$ orbital, leaving the $4s$ orbital with a single electron. The electron configuration of chromium atoms is contrary to the Aufbau principle, but it is thought to be of lower energy than that resulting from the Aufbau principle.

A chromium atom with the $[Ar]3d^54s^1$ configuration has six orbitals that each contain only one electron. All six electrons have the same spin. This is very stable according to Hund's rule. Moreover, the $3d$ and $4s$ sublevels are exactly half-filled, and special stability is associated with half-filled (and also completely filled) sublevels. These two factors are apparently strong enough to decrease the total energy of this configuration.

Manganese (Mn) has the electron configuration $[Ar]3d^54s^2$. The added electron goes to the lower-energy $4s$ orbital, completely filling this orbital while leaving the $3d$ orbital still half-filled. Beginning with the next element, electrons pair up in d orbitals, giving iron (Fe) the configuration $[Ar]3d^64s^2$; cobalt (Co), $[Ar]3d^74s^2$; and nickel (Ni), $[Ar]3d^84s^2$. Next is copper, with the electron configuration $[Ar]3d^{10}4s^1$. An electron moves from the $4s$ orbital to pair with the electron in the fifth $3d$ orbital. This gives the $3d$ sublevel a completely filled stable configuration and the $4s$ orbital a half-filled, stable configuration.

In zinc (Zn) atoms, the $4s$ sublevel is filled to give the electron configuration $[Ar]3d^{10}4s^2$. In atoms of the next six elements, electrons add one by one to the three $4p$ orbitals. According to Hund's rule, one electron is added to each of the three $4p$ orbitals before electrons are paired in any $4p$

The Aufbau Principal does not work for atoms of all elements; some elements that deviate are Cr, Cu, Ag, Au, Pd, Pt, and Mo.

orbital. The electron-dot notations for the atoms of the elements from zinc to krypton (Kr) are

$$\cdot Zn\cdot \quad \cdot \overset{\cdot}{G}a\cdot \quad \cdot \overset{\cdot}{G}e\cdot \quad \cdot \overset{\cdot}{A}s\colon \quad \colon \overset{\cdot}{S}e\colon \quad \cdot \overset{\cdot\cdot}{B}r\colon \quad \colon \overset{\cdot\cdot}{K}r\colon$$

In the 18 elements of the fifth period, sublevels fill in the same order as in elements of the fourth period. Successive electrons are added first to the 5s orbital, then to the 4d orbitals, and finally to the 5p orbitals, as shown in Table 4-5. There are occasional deviations from the predicted configurations here also. The deviations differ from those for fourth-period elements; in each case it is believed that the preferred configuration has a lower energy.

TABLE 4-5 ELECTRON CONFIGURATIONS OF ATOMS OF ELEMENTS IN THE FIFTH PERIOD

Name	Symbol	Atomic Number	Number of Electrons in Sublevels above 3d					Electron Configuration Notation
			4s	4p	4d	5s	5p	
Rubidium	Rb	37	2	6		1		$[Kr]5s^1$
Strontium	Sr	38	2	6		2		$[Kr]5s^2$
Yttrium	Y	39	2	6	1	2		$[Kr]4d^15s^2$
Zirconium	Zr	40	2	6	2	2		$[Kr]4d^25s^2$
Niobium	Nb	41	2	6	4	1		$[Kr]4d^45s^1$
Molybdenum	Mo	42	2	6	5	1		$[Kr]4d^55s^1$
Technetium	Tc	43	2	6	5	2		$[Kr]4d^55s^2$
Ruthenium	Ru	44	2	6	7	1		$[Kr]4d^75s^1$
Rhodium	Rh	45	2	6	8	1		$[Kr]4d^85s^1$
Palladium	Pd	46	2	6	10			$[Kr]4d^{10}$
Silver	Ag	47	2	6	10	1		$[Kr]4d^{10}5s^1$
Cadmium	Cd	48	2	6	10	2		$[Kr]4d^{10}5s^2$
Indium	In	49	2	6	10	2	1	$[Kr]4d^{10}5s^25p^1$
Tin	Sn	50	2	6	10	2	2	$[Kr]4d^{10}5s^25p^2$
Antimony	Sb	51	2	6	10	2	3	$[Kr]4d^{10}5s^25p^3$
Tellurium	Te	52	2	6	10	2	4	$[Kr]4d^{10}5s^25p^4$
Iodine	I	53	2	6	10	2	5	$[Kr]4d^{10}5s^25p^5$
Xenon	Xe	54	2	6	10	2	6	$[Kr]4d^{10}5s^25p^6$

All sublevels up to 3d are filled. The configuration for these filled sublevels is $1s^22s^22p^63s^23p^63d^{10}$.

Sample Problem 4.3

(a) Write both the complete and shorthand electron configurations for iron (Fe). *(Ans.)* The complete configuration of iron is $1s^22s^22p^63s^23d^64s^2$, whereas its shorthand notation is $[Ar]3d^64s^2$.

(b) How many electron-containing orbitals are in an atom of iron? How many of these orbitals are filled? How many unpaired electrons are indicated in the electron-configuratrion notation of an iron atom? *(Ans.)* An iron atom contains 1 1s orbital, 1 2s orbital, 3 2p orbitals, 1 3s orbital, 3 3p orbitals, 5 3d orbitals, and 1 4s orbital for a total of 15. All except four of the 3d orbitals are filled for a total of 11.

Practice Problems

1. *(a)* Write the complete and shorthand electron configuration for iodine (I). Write its electron-dot symbol. How many inner-shell electrons does an iodine atom contain? *(Ans.)* $1s^22s^22p^63s^23p^6$ $3d^{10}4s^24p^64d^{10}5s^25p^5$, $[Kr]4d^{10}5s^25p^5$, $\colon\overset{\cdot\cdot}{I}\colon$, 46

(b) How many electron-containing orbitals are in an atom of iodine? How many of these orbitals are filled? How many unpaired electrons are indicated in the electron-configuration notation of an iodine atom?
(Ans.) 27, 26, 1

2. *(a)* Write the shorthand electron configuration for tin (Sn). Write its electron-dot symbol. How many unpaired electrons are indicated in the electron-configuration notation for tin?
(Ans.) $[Kr]4d^{10}\,5s^25p^2$, ·S̈n·, 2

(b) How many electron-containing *d* orbitals are there in an atom of tin? Name the element in the fourth period whose atoms have the same number of highest energy level electrons as tin. Write the electron-dot symbol for that element.
(Ans.) 10, Ge, ·G̈e·

Elements of the Sixth and Seventh Periods

The sixth period, consisting of 32 elements, is much longer than the others. To build up electron configurations for elements of this period, electrons are added first to the 6s orbital, in cesium (Cs) and barium (Ba) atoms. Then, an electron is added to a 5d orbital in lanthanum (La) atoms.

With the next element, cerium (Ce), the 4f orbitals begin to fill, giving cerium atoms the $[Xe]4f^{1}5d^{1}6s^2$ configuration. Electrons are then added to the 4f orbitals in the next 13 elements. The 5d orbitals are next filled and the period is completed by filling the 6p orbitals. Because the 4f and the 5d orbitals are very close in energy, numerous unexpected variations in configurations occur as these orbitals are filling. The electron configurations of the sixth-period elements can be found in the Periodic Table inside the back cover.

The seventh period is incomplete and consists largely of unstable, radioactive elements. These will be discussed in Chapter 29.

Sample Problem 4.4

(a) Write the complete electron configuration of a rubidium atom using the electron-configuration notation and the shorthand electron-configuration notation.
(Ans.) $1s^22s^22p^63s^23p^63d^{10}4s^24p^65s^1$ $[Kr]\,5s^1$

(b) Write the electron-dot symbol for rubidium and identify the elements in the second, third, and fourth periods that have the same number of highest energy level electrons. *(Ans.)* Rubidium has one electron in its highest energy level (the fifth). Its electron-dot symbol is, therefore RB·. The elements with the same outermost configuration are, in the second period, lithium (Li); in the third period, sodium (Na); and in the fourth period, potassium (K).

Practice Problems

1. *(a)* Write both the complete and shorthand-electron configurations for a barium atom.
    (Ans.) $1s^2\,2s^22p^63s^23p^63d^{10}4s^24p^64d^{10}5s^25p^66s^2$, $[Xe]6s^2$

   (b) Write the electron-dot symbol for barium and identify the elements in the second, third, fourth, and fifth periods that have the same number of highest-energy level electrons.
    (Ans.) ·Ba·, Be, Mg, Ca, and Sr

2. *(a)* Identify the element whose electron-dot symbol contains five dots in its fifth and highest-energy level. Write its shorthand electron-configuration notation. *(Ans.)* Sb, $[Kr]4d^{10}5s^25p^3$

   (b) Name the fourth-period element whose atoms have seven dots in their electron-dot notation. Which element would have that same number of outermost electrons in its sixth main-energy level? *(Ans.)* Br, At

Section Review

1. *(a)* What is electron configuration? *(b)* What three principles guide the development of the electron configuration of an atom?
2. What three methods are used to represent the arrangement of electron in atoms?
3. What is an octet of electrons?
4. Write both the complete and shorthand electron configurations, as well as the orbital and electron-dot notation for *(a)* C *(b)* Ne *(c)* S.
5. Identify each of the indicated elements: *(a)* $1s^22s^22p^63s^23p^3$
   (b) $[Ar]4s^1$ *(c)* contains four electrons in its third and outer main energy level *(d)* contains one paired and three single electrons in its electron-dot symbol in the fourth and outer main-energy level
   (e) the first element to contain a *d* electron.

Chapter Summary

- Quantum theory was developed early in this century to explain such observations as the photoelectric effect and the emission spectrum of hydrogen.
- Quantum theory states that electrons can exist in atoms only at specified energy levels called orbitals.
- When an electron moves from one orbital to an orbital with lower energy, a photon is emitted whose energy equals the energy difference between the two orbitals.
- An electron in an atom can move from one orbital to a higher energy orbital within the atom only by absorbing an amount of energy exactly equal to the difference between the two orbitals.
- The four quantum numbers that describe the properties of electrons in atomic orbitals are the principal quantum number, the orbital quantum number, the magnetic quantum number, and the spin quantum number.
- Electrons occupy atomic orbitals in the ground state of an atom according to the Aufbau principle, Hund's rule, and the Pauli exclusion principle.
- Electron configurations can be written using three different types of notations. They are orbital notation, electron-configuration notation, and electron-dot notation.
- Some electron configurations of larger atoms such as chromium do not strictly follow the Aufbau principle, but the entire grounds ✕ state configuration that results is thought to be of a lower energy than that predicted by the Aufbau principle.

1. *(a)* The arrangement of electrons in atoms *(b)* The Aufbau principle, Hund's rule, and the Pauli exclusion principle
2. Orbital notation, electron-configuration notation, and electron-dot notation
3. The filling of the *s* and *p* sublevels of an atom's highest main energy level with eight electrons
4. The indicated responses are as follows:

   (a) $1s^22s^22p^2$, $[He]2s^22p^2$

   $$\frac{\downarrow}{1s} \quad \frac{\downarrow}{2s} \quad \frac{\uparrow\downarrow}{2p} \quad \frac{\uparrow\downarrow}{2p} \quad \frac{}{2p}, \quad \cdot\dot{C}\cdot$$

   (b) $1s^22s^22p^6$, $[H]2s^22p^2$

   $$\frac{\uparrow\downarrow}{1s} \quad \frac{\uparrow\downarrow}{2s} \quad \frac{\uparrow\downarrow}{2p} \quad \frac{\uparrow\downarrow}{2p} \quad \frac{\uparrow\downarrow}{2p}, \quad :\ddot{Ne}:$$

   (c) $1s^22s^22p^63s^23p^4$, $[Ne]3s^23p^4$

   $$\frac{\uparrow\downarrow}{1s} \quad \frac{\uparrow\downarrow}{2s} \quad \frac{\uparrow\downarrow}{2p} \quad \frac{\uparrow\downarrow}{2p} \quad \frac{\uparrow\downarrow}{2p}$$

   $$\frac{\uparrow\downarrow}{3s} \quad \frac{\uparrow\downarrow}{3p} \quad \frac{\uparrow}{3p} \quad \frac{\uparrow}{3p}, \quad :\dot{S}:$$

5. The indicated elements are: *(a)* P *(b)* K *(c)* Si *(d)* As *(e)* Sc

Chapter 4 *Review*

Arrangement of Electrons in Atoms

Vocabulary

Aufbau principle
continuous spectrum
diamagnetic
electromagnetic
radiation
electromagnetic
spectrum
electron configuration
excited state
frequency
ground state
highest occupied energy
level
Hund's rule
inner-shell electrons
line emission spectrum

magnetic quantum
number
orbital
orbital quantum
number
Pauli exclusion
principle
photoelectric effect
photon
principle quantum
number
quantum
quantum numbers
quantum theory
spin quantum number
wavelength

Questions

1. What is the relationship between light and electrons in atoms?
2. Distinguish between the 19th and 20th century theories regarding the composition of both electrons and light.
3. *(a)* List five examples of electromagnetic radiation. *(b)* What is the speed of all forms of electromagnetic radiation in a vacuum? *(c)* Define frequency and wavelength?
4. *(a)* What is a continuous spectrum? *(b)* Cite a familiar example of such a spectrum.
5. *(a)* What is the electromagnetic spectrum? *(b)* What units are used to denote wavelength? *(c)* Define the SI unit for frequency.
6. What is *(a)* the approximate frequency range of microwaves? *(b)* the wavelength range of ultraviolet light; the frequency and wavelength range of visible light?
7. List the colors of light in the visible spectrum in order of increasing frequency.
8. In the 20th century, what three properties of light were encountered that could not be explained by wave theory?

9. *(a)* What is the photoelectric effect? *(b)* Who is credited with explaining this effect?
10. *(a)* Write and label the components of the equation developed by Planck. *(b)* How are energy and frequency related?
11. Which theory of light, the wave or particle, best explains: *(a)* the interference of light *(b)* the photoelectric effect *(c)* the emission of energy by a hot object?
12. *(a)* How can atoms be made to emit electromagnetic radiation? *(b)* Cite two illustrations of this phenomenon.
13. Distinguish between the ground state and the excited state of an atom.
14. Distinguish between a line emission spectrum and a continuous spectrum.
15. *(a)* What are the three most prominent regions in the hydrogen spectrum? *(b)* What names are given to these groups of spectral lines?
16. According to the Bohr model of the hydrogen atom, how is a line emission spectrum produced?
17. How does the Bohr model explain the production of the Lyman, Balmer, and Paschen spectral series for hydrogen?
18. Cite the two major shortcomings of the Bohr model of the atom.
19. Cite the major contributions of each of the following toward the development of the quantum theory: *(a)* de Broglie *(b)* Schrödinger.
20. What is the quantum theory?
21. *(a)* What is an orbital? *(b)* Describe an orbital in terms of an electron cloud.
22. Cite the major similarities and differences between Schrödinger's model of the atom and that proposed by Bohr.
23. *(a)* What is the principal quantum number? *(b)* How is it symbolized? *(c)* What are shells? *(d)* How does n relate to the energy and distance of a main energy level from the nucleus?
24. *(a)* What information is given by the orbital

quantum number? *(b)* What are sublevels or subshells? *(c)* What is the relationship between the value of n and the number of orbitals in a given main energy level? *(d)* List, in order of increasing energy, the first four orbital quantum numbers.

25. For each of the following values of n, indicate the numbers and types of orbitals possible for that main energy level: *(a)* $n=1$ *(b)* $n=2$ *(c)* $n=3$ *(d)* $n=4$ *(e)* $n=7$ (number only).

26. *(a)* What information is given by the magnetic quantum number? *(b)* How many orientations are possible per subshell for the s, p, d, and f orbitals? *(c)* Explain and illustrate the notation for distinguishing among the different p orbitals in a subshell.

27. *(a)* What is the relationship between n and the total number of orbitals in a main energy level? *(b)* How many total orbitals are contained in the 3rd main energy level? in the 5th?

28. *(a)* What information is given by the spin quantum number? *(b)* What are the possible values for this quantum number? *(c)* What is the overriding significance of the spin quantum number?

29. How many electrons could be contained in each of the following main energy levels if n is: *(a)* 1 *(b)* 2 *(c)* 3 *(d)* 4 *(e)* 5 *(f)* 6 *(g)* 7 *(h)* n?

30. *(a)* What is stated in the Aufbau principle? *(b)* Explain the meaning of this principle in terms of an atom with many electrons.

31. *(a)* What is stated in Hund's rule? *(b)* What is the basis for this rule?

32. *(a)* State the Pauli exclusion principle. *(b)* What is the significance of the spin quantum number? *(c)* How would two electrons in the same orbital compare in terms of the values of their respective four quantum numbers?

33. *(a)* What information is provided by electron-dot notation? *(b)* What is meant by the highest×occupied energy level in an atom? *(c)* What are inner-shell electrons?

34. Determine the highest×occupied energy level in: *(a)* He *(b)* Be *(c)* Al *(d)* Ca *(e)* Sn.

35. Write the electron-dot symbol for the unidentified element Z if it contains the following number of highest×energy level electrons: *(a)* 3 *(b)* 6 *(c)* 8 *(d)* 5.

36. Given the electron configuration for oxygen as $1s^2 2s^2 2p^4$, respond to each of the following: *(a)* How many electrons are in each atom? *(b)* What is the atomic number of this element? *(c)* What is the highest occupied energy level? *(d)* Write its orbital notation. *(e)* Write its electron-dot symbol. *(f)* How many inner-shell electrons does the atom contain? *(g)* In which orbital(s) are these inner-shell electrons located?

37. Repeat Question 36 for an atom of phosphorus with an electron configuration of $1s^2 2s^2 2p^6 3s^2 3p^3$.

38. *(a)* What information is given by the shorthand electron configuration $[Ne]3s^2$? *(b)* What element does this represent?

39. Write the shorthand electron configurations for *(a)* Cl *(b)* Ca.

40. *(a)* What are the noble gases? *(b)* What is a noble×gas configuration?

41. Write both the complete and shorthand electron configurations, as well as the electron-dot symbols, for each of the following: *(a)* Na *(b)* Al *(c)* P *(d)* Ar *(e)* Br *(f)* Sr.

42. Identify each atom on the basis of its electron configuration: *(a)* $1s^2 2s^2 2p^1$ *(b)* $1s^2 2s^2 2p^5$ *(c)* $[Ne]3s^2$ *(d)* $[Ne]3s^2 3p^2$ *(e)* $[Ne]3s^2 3p^5$ *(f)* $[Ar]4s^1$ *(g)* $[Ar]3d^6 4s^2$.

43. List, in order, the number of elements in the first six periods of the Periodic Table.

44. Indicate the second×period element that: *(a)* contains a total of four electrons *(b)* contains a total of five electrons in its highest main energy level *(c)* contains a total of two electrons in the p sublevel *(d)* contains three half-filled p orbitals *(e)* has three dots in its electron-dot symbol *(f)* contains six electrons in its electron-dot symbol *(g)* has a filled highest main energy level.

45. Complete each of the following: *(a)* The maximum number of electrons in an orbital is ____. *(b)* A full second main energy level contains ____ electrons. *(c)* The number of occupied energy levels in Cl is ____. *(d)* The electron-dot symbol for Al consists of ____ dots. *(e)* The third×period element that contains an octet is ____. *(f)* The number of electrons in the electron-dot symbol for S is ____. *(g)* The outer energy×level electron configuration for Mg is ____. *(h)* A full third

energy level contains ____ electrons. *(i)* The orbital notation for S contains ____ unpaired *p* electrons. *(j)* Among the elements Be, C, F, P, and Ar, the one with five dots in its electron-dot symbol is ____. *(k)* Among the sublevels listed (1*s*, 2*p*, 3*p*, 2*d*, 3*s*, 5*f*, and 4*d*), the one that cannot exist is ____. *(l)* The element containing a single 4*s* electron is ____. *(m)* The number of inner-shell electrons in an atom of P is ____. *(n)* The number of electron-containing orbitals in an atom of F is ____. *(o)* The number of unpaired electrons in an atom of As is ____.

Problems

1. Electromagnetic radiation travels at a speed of 3.0×10^{10} cm/s in a vacuum. *(a)* Express this speed in km/s. *(b)* What would it be in km/hr? *(c)* If this value is used as the speed of light in air, how far would light travel in one week?
2. A particular photon of light has an energy of 3.25×10^{-16} J. If 1 calorie = 4.184 J, express this energy in calories.
3. How many protons are in 4.50 mol of photons?
4. How many moles are represented by 5.75×10^{15} photons?
5. The line spectrum of hydrogen contains an ultraviolet line with a wavelength of 1215 Å. If $1 \text{ Å} = 1 \times 10^{-8}$ cm, express the wavelength of their line in *(a)* cm *(b)* m.
6. Find the density of a rectangular object of mass 145 g with a length of 15.0 cm, a width of 60.0 mm, and a height of 0.400 dm.
7. What is the mass of 1.25 mol of Mg atoms?
8. Convert 0.456 g of N into *(a)* moles *(b)* number of atoms.
9. What mass of Na would contain the same number of atoms as 7.25 g of Cl?

Application Questions

1. In what way does the photoelectric effect contradict the wave theory?
2. What was Planck's contribution to the particle theory of light?
3. Explain, in general terms, what occurs in a hydrogen atom when a high-voltage electric current passes through a sample of gaseous hydrogen to produce hydrogen's characteristic line✕emission spectrum.

4. Explain the relationship between hydrogen's line emission spectrum and the subsequent conclusion that only specific energy states must be available to a hydrogen atom.
5. Based on the observed relationships among the frequency, wavelength, and energy of any form of electromagnetic radiation, list the colors of light in the visible spectrum in order of increasing energy.
6. Copy and complete the following table for each value of *n*:

Principal quantum number (*n*)	Number of orbital shapes (*n*)	Total no. of orbitals per energy level (n^2)	Total no. of electrons per energy level ($2n^2$)
1	____	____	____
2	____	____	____
3	____	____	____
4	____	____	____
5	____	____	____
6	____	____	____
7	____	____	____
8	____	____	____
n	____	____	____

7. Explain why an orbital can contain no more than two electrons.
8. In multielectron atoms, why do 4*s* orbitals fill with electrons before 3*d* orbitals?
9. According to Figure 4-18, list the order in which orbitals fill, from the 1*s* to the 7*p* orbital.
10. Write the shorthand electron configurations for each of the following: *(a)* Sc *(b)* Fe *(c)* Zn *(d)* As *(e)* Sn *(f)* Xe *(g)* La *(h)* Hf *(i)* Hg *(j)* Pb *(k)* At *(l)* Ac *(m)* Lr.
11. Write the electron-dot symbols for the following elements: *(a)* Zn *(b)* Sn *(c)* Xe *(d)* Pd *(e)* At.
12. Explain the basis for the "deviant" electron configurations of chromium and copper.
13. Complete each of the following: *(a)* The number of electrons in the outer shell of Br is ____. *(b)* The atom with an outer shell configuration of $3d^{10}4s^24p^4$ is ____. *(c)* The element containing the first *d* electron is ____. *(d)* The first element to contain a pair of electrons in a *d* orbital is ____. *(e)* The element for which the electron-dot symbol consists of five dots in the fifth main-energy level is ____. *(f)* A full fourth energy level consists of ____ electrons. *(g)* The first

element to contain an *f* electron is ____.
(h) The element having the same $s\times$ and $p\times$
electron configuration for the principal quantum number 4 as fluorine has for principal
quantum number 2 is ____. *(i)* Among the
elements given (Ba, Pb, Ca, Zn, and Sr), only
____ has an electron-dot symbol not given by
·X·. *(j)* The total number of orbitals filled in
Cl is ____ . *(k)* The number of half-filled
orbitals in an atom of Co is ____ . *(l)* The
number of electron-containing orbitals in an
atom of Br is ____ . *(m)* The sixth$\times$period
element with the same outer shell configuration
as Sb is ____ . *(n)* The element with exactly
19 filled orbitals is ____ .

Application Problems

1. If the speed of light is 2.998×10^8 m/s, calculate
   the wavelength of light whose frequency is 7.500×10^2 Hz.
2. Determine the frequency of light with a
   wavelength of 4.257×10^{-7} cm.
3. Determine the energy, in joules, of a photon
   whose frequency is 3.55×10^{17} Hz.
4. What is the frequency of a radio wave with an
   energy of 1.55×10^{-24} J?
5. Derive an equation expressing E in terms of h,
   c, and λ, using the two equations $E = h\nu$ and
   $c = \lambda\nu$.
6. What is the energy of a photon of ultraviolet
   light with a wavelength of 4.25×10^{-8} m?
7. How long would it take a radiowave with a
   frequency of 7.25×10^5 Hz to travel from Mars
   to earth if this distance between the two planets
   is approximately 8.00×10^7 km?
8. When sodium is heated, a yellow spectral line
   whose energy is 3.37×10^{-19} J is
   produced. *(a)* What is the frequency of this
   light? *(b)* What is its wavelength?
9. Cobalt-60 ($^{60}_{27}$Co) is an artificial radioisotope that
   is produced in a nuclear reactor for use as a
   gamma-ray source in the treatment of certain
   types of cancer. If the wavelength of the gamma
   radiation from a Co-60 source is 1.00×10^{-8} Å,
   determine its energy.

Enrichment

1. Prepare a report summarizing the techniques
   and results of the major historical efforts
   directed toward determining the speed of light.

2. Prepare a two-column table listing those
   properties of light that can best be explained by
   the wave theory in one column, and those best
   explained by the particle theory in the second
   column. You may want to consult a physics
   textbook for reference.
3. Cite some of Einstein's major contributions to
   modern science.
4. Prepare a list of modern devices associated with
   each region in the electromagnetic spectrum.
5. Trace the evolution of today's microwave oven.
   Cite the major advantages and disadvantages
   associated with its use.
6. Research the distinguishing characteristics of
   AM and FM radiowaves. What considerations
   should a potential radio station owner take into
   account in determining whether to seek AM or
   FM broadcast licensing?
7. Prepare a report concerning the photoelectric
   effect, and cite some of its practical uses.
   Explain the basic operation of each device or
   technique mentioned.
8. Trace the evolution of modern spectroscopy,
   and explain some of its specific uses in the
   identification of unknowns. Include the impact
   of this technique on astronomy.
9. Research the evolution of solar cells, and cite
   some of their uses. What are the advantages
   and disadvantages of using solar cells as an
   energy source?
10. Explain, in general terms, how lasers work, and
    cite some of their modern uses.
11. Louis de Broglie thought that if light has a dual
    wave-particle nature, then matter should have a
    dual wave-particle nature. He said that for elec-
    trons and all other particles; $\lambda = h/mv$ where λ
    is the wavelength of the particle, h is Planck's
    constant, m is the mass of the particle and v is
    the velocity of the particle. Using this informa-
    tion, calculate the wavelength of your body if
    you were walking at a velocity of 2 meters per
    second. Calculate the wavelength of a 10 gram
    bullet moving at 100 meters per second and
    calculate the wavelength of an electron moving
    at 100 meters per second. Using your calcula-
    tions, what can you say about the dual
    wave-particle nature of a person walking, a
    bullet speeding through the air, or an electron?
    Hint: Compare the approximate size of the
    particle with its calculated wavelength.

Chapter 5   The Periodic Law

Chapter Planner

5.1 History of the Periodic Table
Mendeleev and Chemical Periodicity
- ■ Demonstration: 1
  Questions: 1, 2
  Application Question: 1
Moseley and the Periodic Law
- ■ Question: 3
  Application Question: 2
The Modern Periodic Table
- ■ Question: 4
- ■ Desktop Investigation
  Questions: 5–7
  Application Question: 3

5.2 Electron Configuration and the Periodic Table
Groups, Periods, and Blocks of the Periodic Table
- ■ Questions: 8, 9, 11–13
- ■ Question: 10
The s-Block Elements: Groups 1 and 2
- ■ Demonstration: 2
  Questions: 14, 15a, 16a
- ■ Questions: 15b-c, 16b-c, 17, 18

The d-Block Elements: Groups 3–12
- ■ Question: 20a
- ■ Questions: 19, 20b-c, 21, 22
The p-Block Elements: Groups 13–18
- ■ Questions: 23, 24, 25a, 26a, 27a, 28
- ■ Questions: 25b, 26b-c, 27b-c, 29
The f-Block Elements: Lanthanides and Actinides
- ■ Question: 30
- ■ Question: 31
  Application Question: 5

5.3 Electron Configuration and Periodic Properties
Atomic Radii
- ■ Questions: 32, 33a-b, 34–36
- ■ Question: 33c

Ionization Energy
- ■ Demonstration: 3
  Questions: 37, 38
- ■ Question: 39
  Application Questions: 6
- ■ Question: 40
  Application Questions: 7
Electron Affinity
- ■ Question: 41a
- ■ Questions: 41b, 42
- ■ Question: 43
Ionic Radii
- ■ Question: 44
  Application Question: 11
Valence Electrons
- ■ Questions: 45–47
- ■ Question: 48
  Application Question: 8
  Application Questions: 9, 10
Electronegativity
- ■ Questions: 49a, 50, 51
- ■ Question: 49b
  Application Question: 12
  Application Question: 13
Periodic Properties of the d- and f-Block Elements
- ■ Question: 52

Teaching Strategies

Introduction

Placing the emphasis on *organized knowledge*, point out that the Periodic Table is the most useful means of organizing or classifying the elements. Throughout the table, the elements tend to follow patterns in their similarities and differences. Thus, the Periodic Table is a valued source of information both for chemists and for scientists in other fields. Mention to the students that perhaps as they observed you teaching chemistry and answering their questions, they often saw you look at the Periodic Table. Perhaps they thought that you were somehow reading the answer from the table because you knew where to look. Then tell them this is partially true; with a limited amount of knowledge, you can obtain a great deal of information from the Periodic Table. Tell them that they too will be able to do this after completing this chapter.

Possibly you have already introduced students to the Periodic Table as a useful tool. If so, ask: In what ways can you already use the Periodic Table as a tool? Then move on to the other things they will be able to do. On a transparency, show the following outline of information that can be obtained from

the Periodic Table.
A. Determine:
   (1) the number of elements (only on up-to-date tables)
   (2) the types of elements (metals, nonmetals, semiconductors, and noble gases)
   (3) whether elements are natural or synthetic
   (4) the state of the element at a given temperature, usually 20°C, generally indicated by the color of its chemical symbol
   (5) the atomic number and atomic weight of an element
   (6) the number of energy levels in atoms of different elements
   (7) the number of electrons in the outer energy level
   (8) the overall electron configuration of an atom
B. Predict:
   (1) the physical properties of elements
   (2) the chemical properties of elements
   (3) the relative sizes of atoms
   (4) the relative ionization energies and electron affinities of atoms
   Tell the students that this list will continue to grow as more is learned throughout the year.

5.1 History of the Periodic Table

Remind the students that because scientists are always looking for patterns in nature (order in the universe), early chemists were looking for some way to organize the elements. Before the establishment of the Periodic Table, approximately 60 elements were recognized, but no one was able to discern the relationships between them.

Demonstration 1: How Mendeleev Discovered the Periodic Table

PURPOSE: To show the thinking Mendeleev used in figuring out the periodic table

MATERIALS: Transparencies, overhead projector

PROCEDURE: An excellent way to help students understand how Mendeleev made his discoveries is as follows: Prior to class make a transparency with elements known to Mendeleev in the block format shown in Table T5-1

TABLE T5-1						
Li	Be	B	C	N	O	F
Na	Mg	Al	Si	P	S	Cl
K	Ca	not discovered		As	Se	Br

(Note that Ga, Ge, and the noble gases are left out because they had not been discovered yet; the transition elements in series 4 are omitted because they would confuse the issue. Point out to the students that this is a simplified presentation.) Fill in each block with the element's atomic weight (no atomic numbers, for they had not yet been discovered) and some information that helps show relationships between that element and other elements in its family. See examples in Table T5-2:

TABLE T5-2		
Li 7 Soft, Silvery, active metal 1 atom of Li reacts with 1 atom of Cl	**Be 9.4** Hard, silvery, active metal 1 atom of Be reacts with 2 atoms of Cl	**F 19** Pale yellow gas Very strong nonmetallic properties 1 atom of F reacts with 1 atom of K
Na 23 Soft, silvery, very active metal 1 atom of Na reacts with 1 atom of Cl	**Mg 24** Hard, silvery, active metal 1 atom of Mg reacts with 2 atoms of Cl	**Cl 35.5** Yellow-green gas Strong nonmetallic 1 atom of Cl reacts with 1 atom of K

After you have made the transparency, cut the blocks of the 20 elements into individual parts. You now have simulations of Mendeleev's cards. Now mix the blocks so they are in no particular order. In class, tell the students: "Mendeleev listed each of the 63 known elements and their properties on individual cards, which he pinned on his laboratory walls, and started searching for some type of order. Let's see if we can duplicate this process."

Scatter the blocks on the overhead. If there is a Periodic Table in the room, ask the students not to look at it during the exercise. "Let's arrange the elements according to their increasing atomic weights, just as Mendeleev did." While doing this, read the properties on the cards. The students will quickly catch the repetition in the similarity of characteristics.

Be sure to leave spaces for the missing elements just as Mendeleev did. Point out that Mendeleev not only left spaces for these elements but predicted their characteristics. To help them understand how he did this, ask: "What would you predict are the characteristics for Rb? For Sr?" (Point out their position in the Periodic Table.) Mendeleev stuck with his idea of the Periodic Table and not only predicted the existence of missing elements, but accurately predicted their properties.

A simulation that is even more fun, if you have a flair for dramatics, is to announce to the students the day before beginning this chapter that you are going to have a guest speaker the next day. Then, with the aid of the theater department, dress up as Mendeleev. It can be handled in the lecture or interview format. Mendeleev was quite a character, so he lends himself to such a presentation. The students will love it! The publications by Jaffe (1976), McBryde (1974), Trimble (1981), and van Spronsen (1981) shown in "References and Resources" will provide the background necessary for the presentation.

Desktop Investigation: Formulating Your Own Periodic Table

The elements were designated by letters, rather than by their names, so that students would not just look at the present Periodic Table to figure out the answers. To keep the exercise simple, only a limited amount of information is given about each element, and the transition elements in Period 3 are left out. The information given was probably of the type available to Mendeleev; this is why the average atomic masses are whole numbers. All elements listed were known to Mendeleev at the time of his work. You may also suggest to the students that they make up fictitious names and chemical symbols for their elements; they might use their last names, such as Smithium or Jacksonium. They could design their own cards, or you could make photocopies or dittoed cards with the format shown in Table T5-3.

TABLE T5-3
Name of element _____ Average atomic mass _____ Type _____
State _____ Density _____ Boiling point _____ Melting point _____
Other physical properties _____
Chemical formulas of compounds _____ _____ _____
Other useful information _____

As students are working on their cards, you might wish to show them actual samples of the elements, if it is safe to do so, and pictures of the dangerous ones.

After students have completed their Periodic Table and determined the actual names of the elements, they might look up additional information about the element in their textbook, encyclopedia, or the *CRC Handbook of Chemistry and Physics*.

Table T5-4 presents a general format for the Periodic Table of the elements given in this excercise.

TABLE T5-4						
A SIMULATION OF MENDELEEV'S PERIODIC TABLE						
Element G Lithium	Element D Beryllium	Element L Boron	Element Q Carbon	Element P Nitrogen	Element N Oxygen	Element K Fluorine
Element M Sodium	Element B Magnesium	Element E Aluminum	Element I Silicon	Element S Phosphorous	Element J Sulfur	Element C Chlorine
Element A Potassium	Element F Calcium	Element X Not found yet	Element Y Not found yet	Element O Arsenic	Element H Selenium	Element R Bromine

ANSWERS TO DISCUSSION QUESTIONS

1. (a) Atomic numbers were not discovered until about 45 years later. (b) See Table T5-4.
2. Seven families and three periods
3. Element X will be a silver metal with an average atomic mass of about 69–70. When combined with chlorine, it will have the formula XCl_3; when combined with oxygen, it will have the formula X_2O_3. Element Y will be a silver metal with an average atomic mass of about 71–72. When combined with chlorine, it will have the formula YCl_3; when combined with oxygen, it will have the formula YO_2.

5.2 Electron Configuration and the Periodic Table

Demonstration 2: Reactivity of the Alkaline–Earth Metals

PURPOSE: To compare the reactions of water with magnesium and calcium
MATERIALS: Calcium, magnesium ribbons, phenolphthalein, distilled water, 2 medium-sized test tubes, sand paper.
CAUTION: Wear goggles and lab coat at all times.
PROCEDURE: Put 10 mL of water into each of the two test tubes. Add two drops of phenolphthalein to each tube. To one test tube, add a 5-cm piece of freshly sand-papered magnesium ribbon rolled into a coil. To a second test tube, add a piece of calcium about equal to $\frac{1}{2}$ of a pea. (The magnesium reacts very slowly, forming hydrogen gas bubbles and turning the water-phenolphthalein light pink. The calcium is more reactive, fizzes, and turns the water-phenolphthalein dark pink.)
The equations for these reactions are:

$$Mg + 2H_2O \rightarrow Mg(OH)_2 + H_2(g)$$
$$Ca + 2H_2O \rightarrow Ca(OH)_2 + H_2(g)$$

Ask students what they think would happen if a small piece of strontium were added to the water solution. (It would be even more reactive than Ca, but similar in that it turns the water solution dark pink and fizzes rapidly as it produces hydrogen.)

The alkali metals also react with water to form hydrogen and are even more reactive than the alkaline–earth metals. This reactivity often prohibits the use of the alkali metals in class demonstrations. In class discussion, you can compare the reactivities of the alkali metals with the alkaline–earth metals.

Students sometimes have difficulty understanding the definition of d-block elements. A transparency with orbital notation drawings similar to Table T5-5 will help. Discuss additional transition elements, such as Y, Zr, and Nb.

Point out that the electron configurations of these elements are all the same except for the d-sublevel in the next-to-outermost energy level.

TABLE T5 5						
THE d-BLOCK ELEMENTS OR TRANSITION ELEMENTS						
	1s 2s 2p		3s 3p		3d	4s
Mn	⇅ ⇅ ⇅ ⇅ ⇅	⇅ ⇅ ⇅ ⇅	↑ ↑ ↑ ↑ ↑		⇅
Fe	⇅ ⇅ ⇅ ⇅ ⇅	⇅ ⇅ ⇅ ⇅	⇅ ↑ ↑ ↑		⇅
Co	⇅ ⇅ ⇅ ⇅ ⇅	⇅ ⇅ ⇅ ⇅	⇅ ⇅ ↑ ↑		⇅

The sources for three good puzzles that will help familiarize students with the Periodic Table are listed in "References and Resources": Nelson (1984), Nelson (1985), and Wolfgram (1987).

5.3 Electron Configuration and Periodic Properties

Demonstration 3: Ionization Energy

PURPOSE: To show that the farther away an electron in the outer energy level is from the nucleus, the smaller amount of energy needed to remove it
MATERIALS: The teacher and two students
PROCEDURE: One student will act as a nucleus, and the teacher plays the part of the electron in the outer energy level. The second student functions as the ionization energy. For an electron close to the nucleus (1st or 2nd energy level), grip the student's arm close to but below the elbow and have him or her do the same. Hold tight! Then have the second student try to dislodge you by pushing on your shoulder. (Fairly gently—of course!) Change the grip to wrist-to-wrist (3rd and 4th energy level), hand-shake hold for 5th energy level, and hold curled tips of fingers for 6th or 7th energy level. At the 6th or 7th energy level, very little pushing is needed to remove the "electron."

To help students understand + and − electron affinities, liken the concept to buying and selling. Those atoms (like Cl) that really need another electron will pay a lot for it. Those (like Na) that are really not interested in getting another e^-, but that have space in their orbital, will pay a small amount. Those (like Mg) that already have a small amount of stability in their filled s sublevel or with total stability (like Ne) would say: "What do you mean—do I want to buy an e^-? You'll have to *pay* me to take one!"

References and Resources

Books and Periodicals

Ciparick, Joseph. "Element X," *ChemMatters*, Vol. 5, 1987, pp. 8–9.

Fernelius, W. Conard. "Some Reflections on the Periodic Table and Its Use," *Journal of Chemical Education*, Vol. 63, 1986, pp. 263–66.

Guenther, William B. "An Upward View of the Periodic Table," *Journal of Chemical Education*, Vol. 64, 1987, pp. 9–10.

Hogg, John T. "Moseley at Oxford," *Journal of Chemical Education*, Vol. 52, 1975, pp. 325–26.

Jaffe, Bernard. "Mendeleev—Siberia Breeds a Prophet," *Crucibles: The Story of Chemistry*, Dover Publications, New York, 1976.

Krishnan, C.V. "The New Format for the Periodic Table of Elements: Concerns of a High School Chemistry Teacher," *Journal of Chemical Education*, Vol. 64, 1987, p. 558.

Mason, Joan. "Periodic Contractions Among the Elements or On Being the Right Size," *Journal of Chemical Education*, Vol. 65, 1988, pp. 17–20.

McBryde, J.W. "The Periodic Law," *CHEM13*, 1974, p. 2.

Nelson, Bronwyn. "The Elements and Their Organization," *ChemMatters*, Vol. 2, 1984, p. 16.

Nelson, Bronwyn. "Periodically Puzzling," *ChemMatters*, Vol. 3, 1985, p. 16.

Ratzlaff, Becky. "Love Always, Francium," *ChemMatters*, Vol. 3, 1985, p. 13.

Sanderson, R.T. "Principles of Electronegativity, Part 1. General Nature," *Journal of Chemical Education*, Vol. 65, 1988, pp. 112–18.

Saturnelli, Annette. "Setting the Periodic Table," *The Science Teacher*, Vol. 52, 1985, pp. 46–49.

Strong, Frederick. "Revised Atomic Form Periodic Table," *Journal of Chemical Education*, Vol. 62, 1985, p. 456.

Trimble, R. F. "Mendeleev's Discovery of the Periodic Law," *Journal of Chemical Education*, Vol. 58, 1981, p. 28.

van Spronsen, Jan W. "Mendeleev as a Speculator," *Journal of Chemical Education*, Vol. 58, 1981, pp. 790–91.

Wolfgram, Dale. "Family Resemblance," *ChemMatters*, Vol. 5, 1987.

Audiovisual Resources

Chemistry in Today's World Series—Metals and the Halogens (Slide sets + cassette) Educational Images, 1988.

Experiments in Chemistry Series—Chemistry of Columns I & II and Chemistry of Halogens (Filmstrips) Educational Images, 1988.

Periodic Table (VHS) Learning Arts, 1988. "The Periodic Table Videodisk: Reactions of the Elements," Project Seraphim/JCE Software.

Computer Software

Chemistry Courseware—Periodic Table, Apple II 64K, Barclay, 1988.

The Chemistry Help Series—Periodic Table, Apple II 48K, Barclay, 1988.

Chemistry Series—Periodic Table, Apple II 48K, TRS-80 III & IV, 32K, Barclay, 1988.

General Chemistry 1A—Topic 4: The Periodic Table, Apple II 128K, Queue, 1988.

Journal of Chemical Education: Software. "The One Computer Classroom," (Apple II).

"Discoverer: Exploring the Properties of the Chemical Elements," (MS-DOS/PC). Available from Project Seraphim

Periodic Table Seraphim Vol. 3, Disk 6, Project Seraphim, 1988.

Fee, Richard. *Periodic Table*, Apple 48K. Chemistry According to Rof, 1988.

Nagel, Edgar. *The Periodic Table: A Computer-Assisted Lecture Aid*, Apple II 48K, Queue, 1988.

Smith, Stanley. *The Elements*, Apple II 48K and IBM PC 128K, Queue, 1988.

Answers and Solutions

Questions

1. *(a)* Cannizzaro presented a method of measuring atomic masses that eventually led to standard values for those masses. *(b)* Mendeleev created a table in which elements with similar properties were grouped together. He is credited with the discovery of periodicity. *(c)* Moseley's work led to the modern definition of atomic number and to the recognition that atomic number is the basis for the organization of the Periodic Table.

2. *(a)* In Mendeleev's table, elements with similar properties were grouped together. *(b)* Prior to Mendeleev, elements had been arranged in order of increasing atomic mass. *(c)* Mendeleev's procedure left several empty spaces in his Periodic Table, but he was able to predict successfully both the existence and properties of the elements that were later discovered to fit those empty spaces. *(d)* Still unanswered was why all but a few elements could be arranged in order of increasing atomic mass and why chemical periodicity existed.

3. According to the periodic law, when the elements are arranged in order of increasing atomic number, elements with similar properties recur at regular intervals.

4. The Periodic Table is an arrangement of elements in order of their atomic numbers such that elements with similar properties fall in the same column.

5. In order to fit the noble gases into Mendeleev's Periodic Table, it was necessary to create a new group that was placed between what are presently known as Groups 17 and 1.

6. *(a)* The lanthanides are the 14 elements with atomic numbers from 58 (Ce) to 71 (Lu) in Period 6. The actinides are the 14 radioactive elements with atomic numbers from 90 (Th) to 103 (Lr) in Period 7. *(b)* The rare earth elements are the 14 lanthanides.

7. Many tables use the IA-VIIA sequence for numbering the seven long groups and the IIB-VIIB sequence for the shorter groups in the center of the table. More recently, a numbering system in which the groups are designated 1–18, from left to right, has come into use.

8. The lack of chemical reactivity among the noble gases is a result of the fact that the highest energy levels are fully occupied in noble-gas atoms.

9. *(a)* For Groups 1, 2, and 12–18, the outer electron configurations of all members of a group are identical. *(b)* Many of the properties of elements are based on their outermost

electron configurations. Inner-shell electrons have little influence on properties.

10. (a) The length of each period in the table is determined by the number of sublevels that can be filled with electrons. (b) Since the second main energy level holds two electrons in the 2s sublevel and six in the 2p, these eight electrons account for eight second-period elements. In the fourth period, the 4s and 4p sublevels together account for eight electrons, but the additional filling of the 3d sublevel adds 10 more, for a total of 18 electrons and thus 18 elements in that period.

11. The period in which an element appears in the table corresponds to its highest occupied main energy level.

12. The indicated configurations and period designations are:
(a) $[He]2s^1$; 2
(b) $[He]2s^22p^4$; 2
(c) $[Ne]3s^2$; 3
(d) $[Ne]3s^23p^3$; 3
(e) $[Ar]3d^14s^2$; 4
(f) $[Ar]3d^{10}4s^24p^5$; 4
(g) $[Kr]4d^{10}5s^25p^2$; 5.

13. (a) The letter of the block indicates the type of sublevel that is being filled in successive elements of that block. (b) The s-block consists of the elements in Groups 1 and 2, the d-block encompasses Groups 3–12, the p-block is made up of Groups 13–18, and the f-block consists of the elements in Periods 6 and 7 that are located between Groups 3 and 4.

14. The group configuration for the Group 1 elements is ns^1 and that for Group 2 is ns^2.

15. (a) The alkali metals are the elements in Group 1 (Li, Na, K, Rb, Cs, and Fr). (b) They are extremely reactive metals, react with increasing vigor with water (and thus must be stored under kerosene), are silvery in appearance, and are soft enough to be cut with a knife. (c) Sodium is used in nuclear reactors and in sodium-vapor lamps, and cesium is used in photoelectric cells such as those that control supermarket doors.

16. (a) The alkaline–earth metals are the Group 2 elements (Be, Mg, Ca, Sr, Ba, and Ra). (b) These elements are harder, denser, stronger, and have higher melting points than the alkali metals, but are less reactive. (c) Magnesium is used to make strong, lightweight alloys, and compounds of strontium are used in highway flares.

17. The information in each case is: (a) Group 2, Period 2, s-block, Be (b) 1, 3, s-block, Na (c) 2, 5, s-block, Sr (d) 2, 4, s-block, Ca (e) 1, 7, s-block, Fr.

18. (a) $1s^22s^1$; $[He]2s^1$ (b) $1s^22s^22p^63s^2$; $[Ne]3s^2$ (c) $1s^22s^22p^6 3s^23p^64s^1$; $[Ar]4s^1$
(d) $1s^22s^22p^63s^23p^63d^{10}4s^24p^64d^{10}5s^25p^66s^2$; $[Xe]6s^2$

19. (a) The indicated group configurations are:
Group 3: $(n-1)d^1ns^2$
Group 4: $(n-1)d^2ns^2$
Group 5: $(n-1)d^3ns^2$
Group 6: $(n-1)d^4ns^2$
Group 7: $(n-1)d^5ns^2$
Group 8: $(n-1)d^6ns^2$
Group 9: $(n-1)d^7ns^2$
Group 10: $(n-1)d^8ns^2$
Group 11: $(n-1)d^9ns^2$
Group 12: $(n-1)d^{10}ns^2$
(b) The sum of the outer s and d electrons is equal to the group number.

20. (a) The d-block elements are sometimes referred to as transition elements. (b) They are good conductors of electricity, have a high luster when cut or polished, are typically less reactive than alkali metals and alkaline–earth metals, and produce many compounds that are brightly colored. (c) Iron is mixed with various other elements to make steel,

many of the d-block element compounds are used to color glass and/or as paint pigments, tungsten is used for the filaments of incandescent light bulbs, copper is used for electrical wiring, and silver is used in photographic film.

21. The indicated period, block, and group are: (a) Period 4, d-block, Group 4 (b) 4, d-block, 6 (c) 5, d-block, 11 (d) 6, d-block, 10.

22. The indicated outer configurations are:
(a) $3d^54s^2$ (b) $4d^15s^2$ (c) $4f^{14}5d^{10}6s^2$.

23. The indicated group configurations are:
(a) Group 13, ns^2np^1; Group 14, ns^2np^2; Group 15, ns^2np^3; Group 16, ns^2np^4; Group 17, ns^2np^5, and Group 18, ns^2np^6. (b) For p-block elements, the total number of electrons in the highest occupied level is equal to the group number minus 10.

24. The indicated numbers of s and p electrons are: two s electrons and one p electron; two and two, two and three, two and four, two and five, and two and six.

25. (a) The p-block elements are diverse; they consist of nonmetals at the right, metalloids in the middle, and metals at the left. (b) The p-block metals are generally harder and more dense than the s-block metals, but softer and less dense than the d-block metals.

26. (a) The halogens are the elements of Group 17 (F, Cl, Br, I, and At). (b) The halogens are the most reactive of the nonmetals. They react vigorously with most metals to form salts. At room temperature, F_2 and Cl_2 are gases, Br_2 is a red liquid, and I_2 is a purple solid. (c) The halogens are used in bleaches, water purification, photography, insecticides, and plastics.

27. (a) The noble gases are the Group 18 elements (He, Ne, Ar, Kr, Xe, and Rn). (b) These elements are the least reactive of all elements. (c) Since they display so little reactivity, noble gases are used instead of air in containers of materials that would react with oxygen or moisture in air.

28. (a) The metalloids include B, Si, Ge, As, Sb, Se, and Te. (b) The metalloids are mostly brittle solids with electrical conductivities intermediate between those of metals (good conductors) and nonmetals (nonconductors). (c) Semiconducting materials made from metalloid elements are used in computers, tiny tape recorders, and "talking" cash registers.

29. The indicated period, block, and group for each is:
(a) Period 3, p-block, Group 16 (b) 5, p-block, 14 (c) 6, p-block, 17.

30. (a) The f-block consists of the lanthanide elements of Period 6 and the actinide elements of Period 7. (b) The lanthanides are shiny reactive metals used in color-television tubes. The actinides are unstable radioactive elements, many of which are laboratory-made and thus rarely encountered in everyday chemistry.

31. The indicated responses are: (a) p-block, Period 3, Group 13, no name, aluminum, metal, reactive (b) p-block, 4, 18, noble gases, krypton, nonmetal, unreactive (c) d-block, 5, 11, transition elements, silver, metal, slightly reactive (d) f-block, 6, no group number assigned, lanthanides, cerium, metal, reactive.

32. (a) The main group elements are the elements of the s and p blocks. (b) Among the observable trends across the main-group elements are those involving atomic size, ease with which atoms gain and lose electrons, sizes of ions, numbers

of electrons involved in the formation of chemical compounds, and electron-attracting abilities.

33. *(a)* The atomic radius is one half the distance between the nuclei of identical combined atoms. *(b)* There is a gradual decrease in atomic radii across periods. *(c)* As electrons are added to *s* and *p* sublevels in the same main energy level, the increasing positive charge of the nucleus gradually pulls electrons closer to the nucleus, thus resulting in decreasing atomic radii.

34. *(a)* The atomic radii of the main group elements generally increase down a group. *(b)* In the atoms of each element, the outer electrons occupy comparable sublevels in successively higher main energy levels and are therefore increasingly far from the nucleus.

35. The indicated responses are: *(a)* Li (largest), O (smallest) *(b)* Ba, Mg *(c)* K, P *(d)* Sn, Si *(e)* Fr, Li *(f)* Li, Ne *(g)* Cs, N.

36. *(a)* An ion is an atom or group of atoms that has a positive or negative charge. *(b)* Ionization refers to any process that results in the formation of an ion. *(c)* Ionization energy is the energy required to remove one electron from an atom of an element.

37. *(a)* In general, first ionization energies increase across a period and decrease down a group. *(b)* Across a period the increasing nuclear charge more strongly attracts electrons in the same energy level. Down a group, the electrons to be removed from each successive element are in increasingly higher energy levels farther from the nucleus, and are thus more easily removed.

38. *(a)* The first ionization energy is the energy required to remove one electron from an atom, whereas the energies required to remove successive electrons from the resulting ions are referred to as the second, third, etc., ionization energies of that atom. *(b)* The values of successive ionization energies increase. *(c)* This occurs because each successive electron must be removed from a particle with an increasingly larger positive charge.

39. *(a)* The elements in order of decreasing first ionization energies are Ne, F, O, C, Li, and K. *(b)* Among the elements listed, Li would be expected to have the highest second ionization energy because Li^+, like K^+ has a noble gas configuration, but is smaller than K^+.

40. *(a)* Electron affinity is the energy absorbed or released when an electron is added to an atom. *(b)* The electron affinity is negative for an exothermic process, and positive for an endothermic process.

41. The indicated elements are: *(a)* O *(b)* Mg *(c)* P *(d)* I *(e)* Li *(f)* F.

42. In order of decreasing electron affinities, the elements are F, O, C, Li, Na, Rb, and Ne.

43. *(a)* A cation is a positive ion, and an anion is a negative ion. *(b)* Cations are always smaller than the atoms from which they were formed, whereas anions are always larger than the neutral atoms.

44. Chemical compounds form because electrons are lost, gained, or shared between atoms of different elements. *(a)* Valence electrons are the electrons available to be lost, gained, or shared in the formation of chemical compounds.

45. *(b)* Valence electrons are generally located in incompletely filled main energy levels.

46. The number of valence electrons in each group is 1 (Group 1), 2, 3, 4, 5, 6, 7, and 8 (Group 18).

47. The loss or gain and the number of electrons most likely to be involved in each case is: *(a)* lost, 1 *(b)* lost, 2 *(c)* lost, 3 *(d)* gained, 2 *(e)* gained, 1 *(f)* neither lost nor gained, 0.

48. *(a)* Electronegativity is the tendency of an atom in a chemical compound to attract shared electrons. *(b)* The most electronegative element, fluorine, is arbitrarily assigned an electronegativity value of 4.0, and the values for the other elements are calculated in relation to this value.

49. *(a)* The most electronegative group is the halogens, whereas the least electronegative is the alkali metals. *(b)* Fluorine is the most electronegative element, whereas cesium and francium are the least.

50. The indicated elements are: *(a)* O *(b)* Mg *(c)* P *(d)* I *(e)* Li *(f)* F *(g)* N.

51. *(a)* Atomic radii generally decrease across the periods, but remain almost identical down the groups. *(b)* Ionization energies generally increase across the periods with decreasing atomic size, but typically increase down each group. *(c)* In ion formation, electrons in the highest occupied sublevel are removed first; cations typically result. *(d)* Electronegativity values generally increase across periods as radii decrease.

Application Questions

1. The fact that Mendeleev was able to predict successfully both the existence and properties of the elements that were later found to fit the empty spaces in his table confirmed that the placement of elements in terms of similar properties was indeed appropriate.

2. *(a)* The periodicity of elements is illustrated by the pattern of repeating properties seen among the atomic numbers of successive elements within a group. These atomic numbers differ by 8, 8, 18, 18, and 32, in that order. *(b)* The periodicity is based on the arrangement of the electrons around the nucleus.

3. Though both have characteristic *s*-block configurations, they are not metals, since both are colorless, odorless, nonconducting gases. Hydrogen has properties that do not greatly resemble those of elements in any group, although it is sometimes placed in Group 1. Helium, on the other hand, has an ns^2-group configuration like the Group 2 elements, but it is placed in Group 18 with the *p*-block elements because its lack of reactivity (due to the fact that its highest energy level is filled) classifies it as a noble gas rather than as a Group 2 metal.

4. The number of the group in which a *p*-block element appears is 10 greater than the sum of the number of *s* and *p* electrons in the highest occupied level. Thus, subtracting 10 from the group number reveals the sum of *s* and *p* electrons.

5. The *f*-block elements are located between Groups 3 and 4 in Periods 6 and 7 because the 4*f* sublevel is filled before electrons are added to the 5*d* sublevel. Since there are seven *f* orbitals to be filled in each period, there are 14 *f*-block elements per period. Limited space usually accounts for the placement of these elements below the main body of the table.

6. *(a)* The high reactivity of the alkali metals is largely due to the low first ionization energy and thus the relative ease

with which these elements lose electrons and react. *(b)* The noble gases have extremely high first ionization energies and thus do not lose electrons or react easily.

7. The indicated responses are *(a)* Cl *(b)* Na *(c)* Mg.

8. The indicated charge and corresponding noble gas in each is: *(a)* 1+, He *(b)* 3+, He *(c)* 2−, Ne *(d)* 1−, Ne *(e)* 2+, Ne *(f)* 3+, Ne *(g)* 3−, Ar *(h)* 2−, Ar *(i)* 1−, Kr *(j)* 2+, Xe.

9. Ca^+ does not have a noble-gas configuration.

10. K^{2+} is least likely to form.

11. Ca^{2+} would have the smaller radius because of the greater attraction its 20 protons would exert on the 18 electrons.

12. The indicated elements are: *(a)* II *(b)* III *(c)* V *(d)* II *(e)* III *(f)* IV *(g)* II *(h)* IV *(i)* II, I, V, IV, and III *(j)* II *(k)* IV.

13. Choice 4 is false.

Teacher's Notes

Chapter 5 The Periodic Law

INTRODUCTION

This chapter returns to the time before the modern model of the atom had been developed and explores the history of the Periodic Table. The periodic law, which is the basis for the Periodic Table, is one of the greatest achievements of science. It was developed during the nineteenth century by carefully classifying and analyzing the observed properties of the known elements. These efforts revealed a cyclic or *periodic* variation in the properties of the different elements. Using the periodic law you can predict the chemical properties of all the elements.

LOOKING AHEAD

As you study this chapter observe how the scientific method was used to develop the periodic laws. Learn to use the Periodic Table as a tool for obtaining information about the electron configurations and the chemical and physical properties of the various elements.

SECTION PREVIEW

5.1 History of the Periodic Table
5.2 Electron Configuration and the Periodic Table
5.3 Electron Configuration and Periodic Properties

Dirigibles like this one are filled with helium, an element that is both light and inert. Each element is assigned a specific place in the Periodic Table on the basis of its characteristics. Chemists are guided by the Periodic Table in choosing elements for specific purposes.

5.1 History of the Periodic Table

From our present viewpoint, it is hard to imagine the confusion among chemists during the middle of the nineteenth century. By 1860, more than 60 elements had been discovered. Chemists had to learn the properties of these elements as well as those of the many compounds that they formed—a difficult task. To make matters worse, there was no method for accurately determining atomic masses or the numbers of atoms of different elements combined in chemical compounds. Different chemists used different atomic masses for the same elements, which resulted in their determining different compositions for the same compounds. This made it difficult for one chemist to understand the reported results of another. The development of the Periodic Table and an understanding of atomic masses helped eliminate some of this confusion.

Mendeleev and Chemical Periodicity

In September 1860, a group of chemists assembled at the First International Congress of Chemists in Karlsruhe, Germany. The goal of the Congress was to review various scientific matters on which there was little agreement. One topic on the agenda was the measurement of atomic masses. Another was how to determine the composition of compounds by using atomic masses. Chemists hoped that communication on these matters, which had become almost impossible, could be improved.

At the Congress, the Italian chemist Stanislao Cannizzaro (1826–1910) presented a method for measuring atomic masses and interpreting the results of the measurements. Cannizzaro's method enabled scientists to agree on standard values for atomic masses. Scientists then began to search for generalizations. Could any relationships be found among atomic masses and other properties of the elements? Could predictions about such properties be made based on these atomic masses?

Dmitri Mendeleev (1834–1907), a Russian chemist, had heard about the new atomic masses discussed at the Karlsruhe Congress. Mendeleev was writing a chemistry textbook and decided to include the new values. In his book, Mendeleev was hoping to organize the elements according to their properties. He went about this much as you might organize information for a research paper. He placed the name of the known element on a card together with the atomic mass of the element and a list of its observed properties. He then arranged the cards according to various properties and looked for trends or patterns.

Mendeleev notices that when the elements were arranged in order of increasing atomic mass, similar properties appeared at regular intervals. Such a repeating pattern is referred to as "periodic." The second hand of a watch, for example, passes over any given mark at periodic, 60-second intervals.

Mendeleev created a table in which elements with similar properties were grouped together—a periodic table of the elements. His first periodic table, shown in Figure 5-2, was published in 1869.

Note that Mendeleev placed iodine (I, atomic mass 127) after tellurium (Te, atomic mass 128), even though this meant not listing the elements in

In the nineteenth century and for a long time afterward, the masses of atoms were referred to as "atomic weights."

Cannizzaro's method, which requires measuring the densities of gases, is discussed in Chapter 12.

Figure 5-1 Mendeleev is credited with formulating the first periodic table.

Figure 5-2 Mendeleev's first published periodic table. The periods are the vertical columns and the groups are the horizontal rows (the reverse of the modern Periodic Table). The numbers after the symbols of the elements are the atomic masses accepted by Mendeleev. The question marks at atomic mass 45, 68, and 70 were later identified as the elements scandium, gallium, and germanium.

Mendeleev, surprisingly, never won a Nobel Prize; the monumental importance of his work was not recognized in his lifetime. In 1906, the year before his death, he was nominated for the Nobel Prize in chemistry but lost by one vote to Henri Moissan (1852–1907), who discovered fluorine.

Moseley shot high-speed electrons at different metal targes in an evacuated tube, producing X rays of different frequencies. When he graphed the square roots of the frequencies against the numbers representing the elements' positions on the periodic chart, he obtained a straight line. He called the number representing each element's position "the atomic number." Later it was discovered that this number was the number of positive charges or protons found in a neutral atom of that element.

но въ ней, мнѣ кажется, уже ясно выражается примѣнимость выставляемаго мною начала ко всей совокупности элементовъ, пай которыхъ извѣстенъ съ достовѣрностію. На этотъ разъ я и желалъ преимущественно найдти общую систему элементовъ. Вотъ этотъ опытъ:

			Ti=50	Zr=90	?=180.
			V=51	Nb=94	Ta=182.
			Cr=52	Mo=96	W=186.
			Mn=55	Rh=104,4	Pt=197,4
			Fe=56	Ru=104,4	Ir=198.
			Ni=Co=59	Pl=106,6	Os=199.
H=1			Cu=63,4	Ag=108	Hg=200.
	Be=9,4	Mg=24	Zn=65,2	Cd=112	
	B=11	Al=27,4	?=68	Ur=116	Au=197?
	C=12	Si=28	?=70	Su=118	
	N=14	P=31	As=75	Sb=122	Bi=210
	O=16	S=32	Se=79,4	Te=128?	
	F=19	Cl=35,5	Br=80	I=127	
Li=7	Na=23	K=39	Rb=85,4	Cs=133	Tl=204
		Ca=40	Sr=87,6	Ba=137	Pb=207.
		?=45	Ce=92		
		?Er=56	La=94		
		?Yt=60	Di=95		
		?In=75,6	Th=118?		

а потому приходится въ разныхъ рядахъ имѣть различное измѣненіе разностей, чего нѣтъ въ главныхъ числахъ предлагаемой таблицы. Или же придется предполагать при составленіи системы очень много недостающихъ членовъ. То и другое мало выгодно. Мнѣ кажется притомъ, наиболѣе естественнымъ составить

order of increasing atomic mass. In this way, tellurium could be placed in the same group as elements with similar properties such as oxygen (O), sulfur (S), and selenium (Se). Iodine could also, then be placed in the same group as fluorine (F), chlorine (Cl), and bromine (Br), which have properties similar to those of iodine.

Mendeleev's procedure left several empty spaces in his periodic table. In 1871, Mendeleev boldly predicted the existence and properties of the elements that would fill three of these spaces. By 1886, all three had been discovered. Their properties were found to be strikingly similar to those predicted by Mendeleev, as shown in Table 5-1. The success of his predictions persuaded most chemists to accept Mendeleev's periodic table and gave him historical credit as the discoverer of periodicity. Two questions remained, however: *(1)* Why was it that most elements could be arranged in the order of increasing atomic mass, but a few could not? *(2)* What is the reason for chemical periodicity? Careful experimentation, observation, and analysis of data would eventually reveal the answers to these questions as you will discover in the remainder of this chapter.

STUDY HINT

The properties of the Group–1 elements are discussed in Section 5.2 and Chapter 29.

Moseley and the Periodic Law

The first question was not answered until over forty years after Mendeleev's first periodic table had been published. In 1911, the English scientist Henry Gwyn-Jeffreys Moseley (1887–1915) discovered why a few elements could not be arranged according to their atomic masses. Moseley, who was working with Rutherford, performed a series of experiments on 38 different metals. By analyzing his data, he discovered a previously unrecognized pattern: The positive charge of the nucleus increases by one unit from one element to the next when the elements are arranged as they are in the Periodic Table.

To emphasize that the Modern Periodic Table arranges elements according to increasing atomic numbers, rather than masses, have students examine the atomic numbers on the Periodic Table to see if any are out of numerical sequence. Then have them do the same with atomic masses. They will find that Ar and K, Co and Ni, and Te and I are reversed in terms of atomic mass, but all elements are in order in terms of atomic numbers.

TABLE 5-1 PROPERTIES OF ELEMENTS PREDICTED BY MENDELEEV

	Mendeleev's Predictions	Modern Values
	Eka*-aluminum	Gallium (discovered 1875)
Atomic mass	68	69.7
Density (g/cm^3)	5.9	5.9
Melting point	Low	29.8°C
Oxide formula	E_2O_3	Ga_2O_3
Oxide density (g/cm^3)	5.5	5.9
	Eka-boron	Scandium (discovered 1879)
Atomic mass	44	44.96
Oxide formula	E_2O_3	Sc_2O_3
Oxide density (g/cm^3)	3.5	3.9
	Eka-silicon	Germanium (discovered 1886)
Atomic mass	72	72.6
Density (g/cm^3)	5.5	5.3
Oxide formula	EO_2	GeO_2
Oxide density (g/cm^3)	4.7	4.2

*Mendeleev used the prefix eka to designate certain then-unknown elements that he assumed would follow in order beyond a specified element. Eka is delivered from a Sanscrit word meaning "one" or "one beyond."

Moseley's work led to both the modern definition of atomic number and the recognition that atomic number, not atomic mass, is the basis for the organization of the periodic table.

Moseley's studies provided experimental justification for Mendeleev's ordering of the periodic table by properties rather than strictly by atomic mass. For example, the order Mendeleev chose for tellurium and iodine is justified by the atomic numbers of 52 for tellurium and 53 for iodine. A similar reversal of argon (Ar, atomic mass 39.95, atomic number 18) and potassium (K, atomic mass 39.10, atomic number 19) can be seen in the modern periodic table.

Today, Mendeleev's principle of chemical periodicity is correctly stated in what is known as the **periodic law:** *the physical and chemical properties of the elements are periodic functions of their atomic numbers.* In other words,

STUDY HINT
Atomic numbers were introduced in Section 3.3.

STUDY HINT
Locate the noble gases in the Periodic Table inside the back cover.

Helium was discovered in 1868 by Pierre Janssen (1824–1907) as he was studying a solar eclipse. However, helium was not accepted as an element until it was discovered that it also exists on earth.

Ramsay received the 1904 Nobel Prize in Chemistry for his discovery of the noble gases.

STUDY HINT
Scandium and yttrium are sometimes considered rare earth elements since their properties are similar.

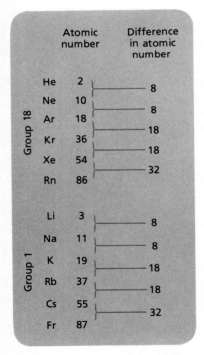

Figure 5-3 Comparison of the differences in atomic numbers for Group 18 and Group 1. The differences between successive elements is 8, 8, 18, 18, and 32. Groups 1, 2, and 13–18 follow the same pattern.

when the elements are arranged in order of increasing atomic number, elements with similar properties recur at regular intervals.

The Modern Periodic Table

The periodic table has undergone extensive change since Mendeleev's time. Chemists have discovered new elements and, in more recent years, synthesized them in the laboratory. Each of the more than forty new elements, however, can be place in a group of other elements with similar properties. *The Periodic Table is an arrangement of the elements in order of their atomic numbers so that elements with similar properties fall in the same column.*

The noble gases   Perhaps the most significant addition to the periodic table came with the discovery of the noble gases. Argon (Ar) was discovered in 1894 by the English physicists John William Strutt (Lord Rayleigh; 1842–1919) and William Ramsay (1852–1916). Although argon makes up about 1 percent of the atmosphere, it had escaped notice because of its total lack of chemical reactivity. Helium (He), another noble gas, had been detected earlier using the spectrum of sunlight. In 1894 Ramsay discovered that helium also exists on earth.

In order to fit argon and helium into the Periodic Table, Ramsay found it necessary to propose a new group. He placed this group between the groups now known as Group 17 (the flourine family) and Group 1 (the Lithium family). In 1898 Ramsay discovered two more noble gases, krypton (Kr) and xenon (Xe). The final noble gas, radon (Rn), was discovered in 1900 by the German scientist Freidrich Ernst Dorn (1848–1916).

The next step in development of the Periodic Table was completed in the early 1900s when the puzzling chemistry of the "rare earth" elements was finally understood. Because the rare earths are so similar in properties, their separation and identification required the efforts of many chemists during the next 100 years. Today, the 14 different rare earth elements are known as the lanthanides. *The **lanthanides, or rare–earth elements,** are the 14 elements with atomic numbersf from 38 (cerium), Ce) to 71 (lutetium, Lu).*

The actinide elements   Another landmark in the development of the Periodic Table was the discovery of the actinide elements, all of which are radioactive. *The **actinide elements** are the 14 elements with atomic numbers from 90 (thorium, Th) to 103 (lawrencium, Lr).* The lanthanides and actinides belong in periods 6 and 7, respectively, or the Periodic Table between the elements of Groups 3 and 4.

Periodicity   Periodicity with respect to atomic number can be observed in any group of elements. Consider, for example, the noble gases of Group 18. The first noble gas is helium, which has an atomic number of 2. The elements following helium in atomic number have completely different properties until the next noble gas, neon (Ne), which has an atomic number of 10, is reached. The remaining noble gases in order of increasing atomic number are argon (atomic number 18), krypton (atomic number 36), xenon (atomic number 54), and radon (atomic number 86). The difference in atomic number between each noble gas is shown in Figure 5-3. Below the noble gases in Figure 5-3 are the elements of Group 1. These are all silvery metals of low density that react strongly with water. As you can see, the differences in atomic number between the Group 1 metals follow exactly the same pattern

Desktop Investigation

Designing Your Own Periodic Table

Question
Using information similar to that available to Mendeleev, can you design your own periodic table?

Materials
index cards

Procedure
1. Write the information available for each element on an index card. The following information is appropriate: a letter of the alphabet (A, B, C, etc.) to identify each element; average atomic mass; state; density; melting point; boiling point; and some readily observable physical properties. Do not write the name of the element on the index card, but keep a separate list of the letters you have assigned to each element.

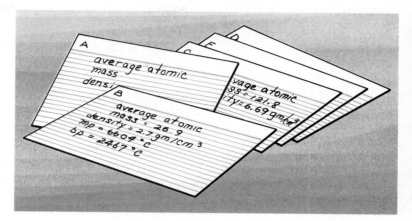

2. Organize the cards for the elements in a logical pattern as you think Mendeleev might have done.

Discussion
1. Keeping in mind that the information you have is similar to that available to Mendeleev in 1869, (a) why are atomic masses given instead of atomic numbers? (b) can you identify each element by name?
2. How many families or groups are in your periodic table? How many periods or series?
3. Predict the characteristics of any missing elements. Once you have finished check your work using your separate list of elements and a periodic table.

---

as for the noble gases. Starting with the first member of Groups 1, 2, and 13–18 the same pattern is repeated: The atomic number of each successive element is 8, 8, 18, 18, and 32 higher than that of the preceding element.

In Section 5.2, you will see that the second mystery presented by Mendeleev's periodic table—the reason for periodicity—is explained by the arrangement of the electrons around the nucleus.

Section Review

1. Who is credited with each of the following: (a) development of a method that led to the determination of standard atomic masses (b) discovery of chemical periodicity (c) development of atomic numbers as the basis for organization of the periodic table.
2. State the periodic law.
3. Name three sets of elements added to the Periodic Table after Mendeleev's time.
4. How do the atomic numbers of the elements within Groups 1, 2, and 13–18 of the Periodic Table differ?

1. (a) Cannizzaro (b) Mendeleev (c) Moseley
2. The physical and chemical properties of the elements are periodic functions of their atomic numbers.
3. The noble gases, the lanthanide elements, and the actinide elements
4. The atomic numbers of successive elements differ by 8, 8, 18, 18, and 32, in that order.

5.2 Electron Configuration and the Periodic Table

SECTION OBJECTIVES

- Describe the relationship between electrons in sublevels and the length of each period.
- Locate and name the four blocks of the Periodic Table. Explain the reasons for these names.
- Discuss the relationship between group configurations and group numbers.
- Describe the location in the Periodic Table and the general properties of the alkali metals, the alkaline earth metals, the halogens, and the noble gases.

The noble gases are known to undergo very few chemical reactions. This lack of chemical reactivity is a result of the fully occupied highest energy levels in noble gas atoms. Electron configurations can be used to explain the chemical reactivity of all of the elements.

Groups, Periods, and Blocks of the Periodic Table

In Chapter 4, you discovered that atoms of fluorine, chlorine, and all other Group 17 elements have identical outer electron configurations. The same is true for Groups 1, 2, and 12–18. In a few cases for Groups 3–12, the arrangements of electrons in the highest occupied energy levels are similar, but not identical for elements in the same group. The outermost electrons play the largest role in determining chemical properties. Inner-shell electrons have little influence on properties of an element because they do not come into contact with their surroundings.

The length of each period in the Periodic Table is determined by the sublevel being filled with electrons as shown in Table 5.2. The first energy level holds only two electrons in its $1s$ sublevel. This accounts for hydrogen and helium in the first period. The second main-energy level holds two electrons in a $2s$ sublevel and six electrons in a $2p$ sublevel. These eight electrons account for the eight elements in the second period. In the eight elements of the third period, $3s$ and $3p$ sublevels are being filled. Filling $3d$ and $4d$ sublevels in addition to s and p sublevels adds 10 elements to the fourth and fifth periods, which therefore include a total of 18 elements each. Filling $4f$ sublevels in addition to s, p, and d sublevels adds 14 elements to the sixth period, for a total of 32 elements in the sixth period. As new artificial elements are created, the 23 known elements in period 7 could be extended to 32. Compare these numbers of elements in each period with the relationships shown in Figure 5-3.

The period of an element can be determined from its electron configuration. For example, arsenic (As) has the configuration $[Ar]3d^{10}$

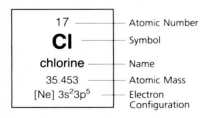

Figure 5-4 Key to the information provided in the Periodic Table in Figure 5-5 and also in the back cover.

TABLE 5-2	RELATIONSHIP BETWEEN PERIOD LENGTH AND SUBLEVELS BEING FILLED IN THE PERIODIC TABLE	
Period Number	Number of Elements in Period	Energy Sublevels in Order of Filling
1	2	$1s$
2	8	$2s$ $2p$
3	8	$3s$ $3p$
4	18	$4s$ $3d$ $4p$
5	18	$5s$ $4d$ $5p$
6	32	$6s$ $4f$ $5d$ $6p$
7	(to date)	$7s$ $5f$ $6d$ etc.

$4s^2 4p^3$. The "4" in "$4p^3$" indicates that the highest occupied energy level is the fourth energy level. Arsenic is therefore in the fourth period in the periodic table.

Refer to the modern Periodic Table in Figure 5-5 and the electron configurations given there, as you read the rest of this chapter.

It is helpful to divide the periodic table into four blocks: the *s* block, the *p* block, the *d* block, and the *f* block, as shown in Figure 5-6. Refer to Table 5-6 as you read about the elements of *s*, *p*, *d*, and *f* blocks in this section.

The name of each block is based on whether an *s*, *p*, *d*, or *f* sublevel is being filled in successive elements of that block.

The *s*-Block Elements: Groups 1 and 2

The elements of the *s* block are all active metals, the Group 1 metals being more active than those of Group 2. Some *s*-block elements are shown in Figure 5-7.

To help students become familiar with the arrangement of the elements on the Periodic Table, give them a blank table so they can fill in information such as chemical symbol, atomic number, and atomic mass of the elements in each group, block, or period being discussed.

STUDY HINT
The properties of the Group 1 metals are discussed in Section 5.2 and Chapter 24.

Figure 5-6 Sublevel blocks of the Periodic Table.

Sublevel Blocks of the Periodic Table

Lanthanide Series

58 Ce	59 Pr	60 Nd	61 Pm	62 Sm	63 Eu	64 Gd	65 Tb	66 Dy	67 Ho	68 Er	69 Tm	70 Yb	71 Lu
90 Th	91 Pa	92 U	93 Np	94 Pu	95 Am	96 Cm	97 Bk	98 Cf	99 Es	100 Fm	101 Md	102 No	103 Lr

Actinide Series

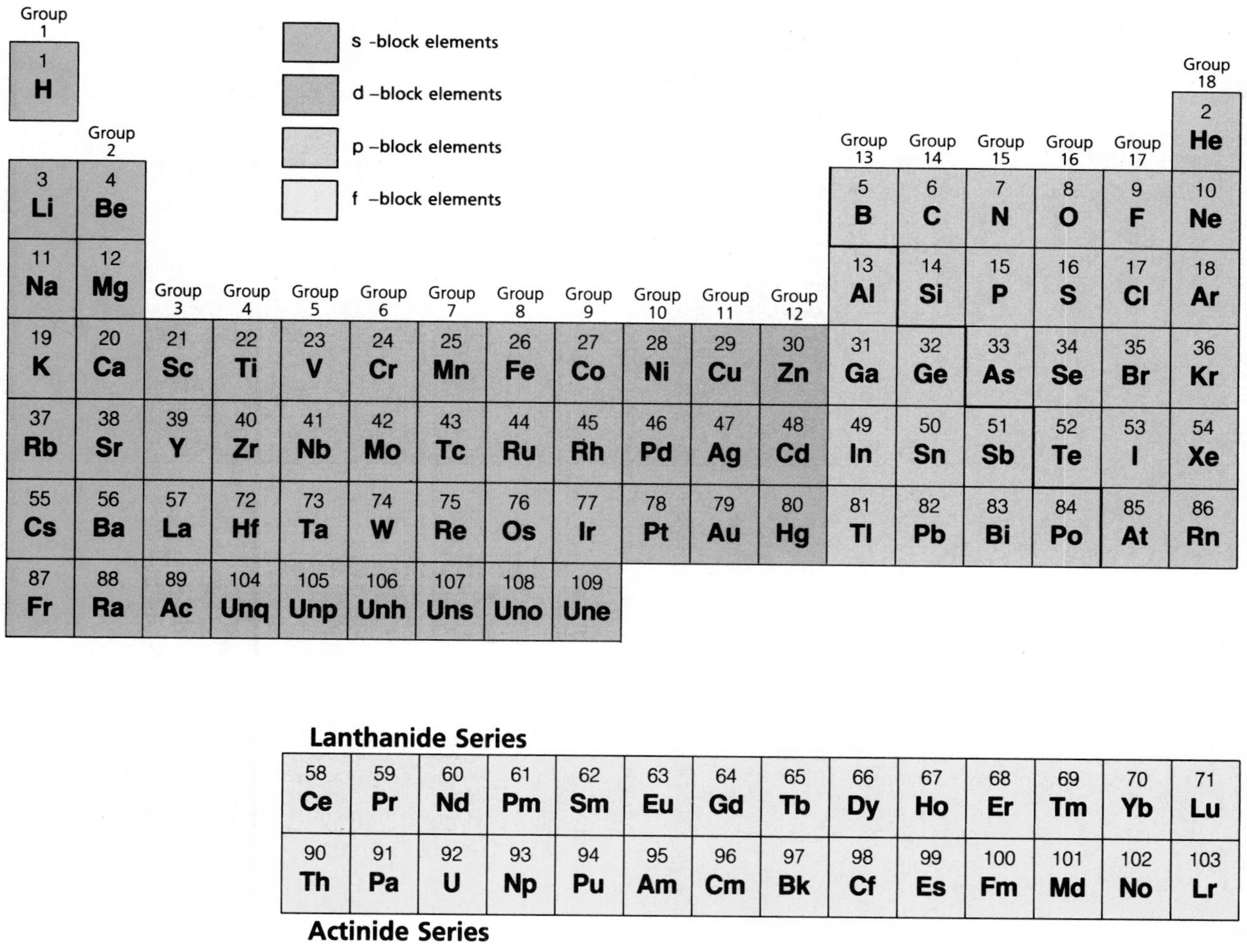

Periodic Table

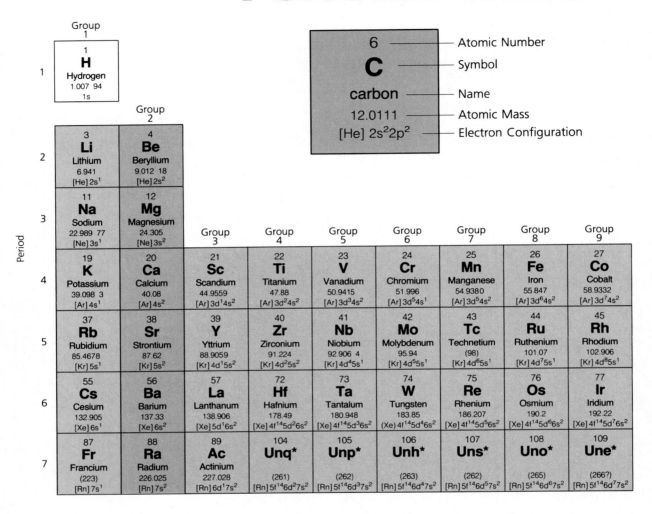

*The systematic names and symbols for elements of atomic number greater than 103 will be used until the approval of trivial names by IUPAC.

Figure 5-5 Periodic Table of the Elements

of the Elements

■ Alkali metals

■ Alkaline earth metals

■ Transition metals

□ Metals

■ Nonmetals

□ Noble gases

			Group 13	Group 14	Group 15	Group 16	Group 17	Group 18
								2 **He** Helium 4.00260 $1s^2$
			5 **B** Boron 10.81 $[He]2s^22p^1$	6 **C** Carbon 12.0111 $[He]2s^22p^2$	7 **N** Nitrogen 14.0067 $[He]2s^22p^3$	8 **O** Oxygen 15.9994 $[He]2s^22p^4$	9 **F** Fluorine 18.998 403 $[He]2s^22p^5$	10 **Ne** Neon 20.179 $[He]2s^22p^6$
Group 10	Group 11	Group 12	13 **Al** Aluminum 26.981 54 $[Ne]3s^23p^1$	14 **Si** Silicon 28.0855 $[Ne]3s^23p^2$	15 **P** Phosphorus 30.973 76 $[Ne]3s^23p^2$	16 **S** Sulfur 32.06 $[Ne]3s^23p^4$	17 **Cl** Chlorine 35.453 $[Ne]3s^23p^5$	18 **Ar** Argon 39.948 $[Ne]3s^23p^6$
28 **Ni** Nickel 58.69 $[Ar]3d^84s^2$	29 **Cu** Copper 63.546 $[Ar]3d^{10}4s^1$	30 **Zn** Zinc 65.39 $[Ar]3d^{10}4s^2$	31 **Ga** Gallium 69.72 $[Ar]3d^{10}4s^24p^1$	32 **Ge** Germanium 72.59 $[Ar]3d^{10}4s^24p^2$	33 **As** Arsenic 74.9216 $[Ar]3d^{10}4s^24p^3$	34 **Se** Selenium 78.96 $[Ar]3d^{10}4s^24p^4$	35 **Br** Bromine 79.904 $[Ar]3d^{10}4s^24p^5$	36 **Kr** Krypton 83.80 $[Ar]3d^{10}4s^24p^6$
46 **Pd** Palladium 106.42 $[Kr]4d^{10}5s^0$	47 **Ag** Silver 107.868 $[Kr]4d^{10}5s^1$	48 **Cd** Cadmium 112.41 $[Kr]4d^{10}5s^2$	49 **In** Indium 114.82 $[Kr]4d^{10}5s^25p^1$	50 **Sn** Tin 118.71 $[Kr]4d^{10}5s^25p^2$	51 **Sb** Antimony 121.75 $[Kr]4d^{10}5s^25p^3$	52 **Te** Tellurium 127.60 $[Kr]4d^{10}5s^25p^4$	53 **I** Iodine 126.905 $[Kr]4d^{10}5s^25p^5$	54 **Xe** Xenon 131.29 $[Kr]4d^{10}5s^25p^6$
78 **Pt** Platinum 195.08 $[Xe]4f^{14}5d^96s^1$	79 **Au** Gold 196.967 $[Xe]4f^{14}5d^{10}6s^1$	80 **Hg** Mercury 200.59 $[Xe]4f^{14}5d^{10}6s^2$	81 **Tl** Thallium 204.383 $[Xe]4f^{14}5d^{10}6s^26p^1$	82 **Pb** Lead 207.2 $[Xe]4f^{14}5d^{10}6s^26p^2$	83 **Bi** Bismuth 208.980 $[Xe]4f^{14}5d^{10}6s^26p^3$	84 **Po** Polonium (209) $[Xe]4f^{14}5d^{10}6s^26p^4$	85 **At** Astatine (210) $[Xe]4f^{14}5d^{10}6s^26p^5$	86 **Rn** Radon (222) $[Xe]4f^{14}5d^{10}6s^26p^6$

63 **Eu** Europium 151.96 $[Kr]4d^{10}4f^75s^25p^66s^2$	64 **Gd** Gadolinium 157.25 $[Kr]4d^{10}4f^75s^25p^65d^16s^2$	65 **Tb** Terbium 158.925 $[Kr]4d^{10}4f^95s^25p^66s^2$	66 **Dy** Dysprosium 162.50 $[Kr]4d^{10}4f^{10}5s^25p^66s^2$	67 **Ho** Holmium 164.930 $[Kr]4d^{10}4f^{11}5s^25p^66s^2$	68 **Er** Erbium 167.26 $[Kr]4d^{10}4f^{12}5s^25p^66s^2$	69 **Tm** Thulium 168.934 $[Kr]4d^{10}4f^{13}5s^25p^66s^2$	70 **Yb** Ytterbium 173.04 $[Kr]4d^{10}4f^{14}5s^25p^66s^2$	71 **Lu** Lutetium 174.967 $[Kr]4d^{10}4f^{14}5s^25p^65d^16s^2$
95 **Am** Americium (243) $[Rn]5f^77s^2$	96 **Cm** Curium (247) $[Rn]5f^76d^17s^2$	97 **Bk** Berkelium (247) $[Rn]5f^97s^2$	98 **Cf** Californium (251) $[Rn]5f^{10}7s^2$	99 **Es** Einsteinium (252) $[Rn]5f^{11}7s^2$	100 **Fm** Fermium (257) $[Rn]5f^{12}7s^2$	101 **Md** Mendelevium (258) $[Rn]5f^{13}7s^2$	102 **No** Nobelium (259) $[Rn]5f^{14}7s^2$	103 **Lr** Lawrencium (260) $[Rn]5f^{14}6d^17s^2$

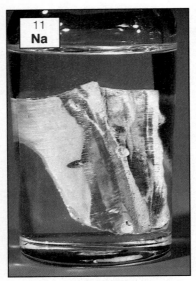

Sodium

Calcium

Figure 5-7 Sodium and calcium are members of the *s*-block of the Periodic Table.

Hydrogen and helium are the simplest and most abundant elements in the universe. Hydrogen makes up 76% of the mass in the universe, and helium makes up 23%.

The highest occupied energy level in atoms of each Group 1 element contains a single electron. For example, the configurations of lithium and sodium are $[He]2s^1$ and $[Ne]3s^1$. As you will see in Section 5-3, the ease with which the single electron is lost helps to make the Group 1 metals very reactive. Using n for the number of the highest occupied energy level, the outer or *group configuration* of the Groups 1 and 2 elements is written ns^1 and ns^2 respectively.

The alkali metals *The elements of Group 1 of the Periodic Table (lithium, sodium, potassium, rubidium, cesium, and francium) are known as the* **alkali metals.** The alkali metals comvine vigorously with many nonmetals. Because they are extremely reactive, alkali metals are not found in nature as free elements. All alkali metals react with water, with increasing vigor for each element down the group, to produce hydrogen gas and aqueous solutions of substances known as alkalis. Alkalis are common and useful chemicals that can, for example, dissolve fats to produce soups. Because of their extreme reactivity with air or moisture, the alkali metals are usually stored under kerosene.

All of the alkali metals have a silvery appearance and are soft enough to cut with a knife. Lithium (Li), sodium (Na), and potassium (K) are less dense than water. Proceeding down the column, alkali metals melt at successively lower temperatures; except for lithium, the alkali metals have melting points lower than the boiling point of water.

Molten sodium is used as a heat-transfer fluid in nuclear reactors, and sodium vapor lamps are efficient light sources. The photoelectric effect occurs readily in cesium. Therefore, cesium (Cs) is used in photoelectric cells, such as those that will open a supermarket door as you approach the door. Compounds of the Group-1 elements sodium and potassium play essential roles in living things, where they are present in body fluids that fill and surround all cells.

The alkaline earth metals *The elements of Group 2 or the Periodic Table (beryllium, magnesium, calcium, strontium, barium, and radium) are called* **the alkaline earth metals.** Atoms of each alkaline earth metal contain a pair of electrons in the outermost s sublevel. Thus, for Group 2, the group configuration is ns^2. The Group 2 metals are all harder, denser, stronger, and have higher melting points than the alkali metals. Although they are less reactive than the alkali metals, the alkaline earth metals are also too reactive to be found in nature as free elements.

Compounds containing calcium are among the most common materials in the earth's crust. The minerals limestone and marble are forms of calcium carbonate, a compound of calcium, carbon (C), and oxygen (O). Compounds containing magnesium (Mg) are also abundant. Sea water contains large quantities of these compounds. Beryllium (Be) is a very rare element. Its source is the gemstone *beryl*. Crystals of colorless or brown beryl are used as the source of metallic beryllium. The light blue-green form of beryl, called aquamarine, and the deep green form, called emerald, are valued for their beauty. Radium (Ra), a naturally occurring radioactive element first discovered in 1898, is extremely rare and is a minor component of uranium ore.

Magnesium is widely used to make strong, lightweight alloys for aircraft, truck bodies, and ladders. The chief use of strontium (Sr) is in fireworks and flares for highway danger signals. Compounds of calcium are extremely important biologically. Teeth and bones consist of calcium compounds.

Hydrogen and helium   The electron configurations of hydrogen and helium are $1s^1$ and $1s^2$, respectively. Neither hydrogen nor helium are metals. Both are colorless, odorless gases. Hydrogen, although it is sometimes placed with Group 1 at the left of the Periodic Table, is a unique element with properties that do not closely resemble those of any group.

Why is helium a member of Group 18 rather than Group 2? Like the Group-2 elements, helium has an ns^2 group configuration. (Helium is placed in Group 18, however, because of its nonreactive nature.) Here, also, electron configuration governs properties. As in the other noble gases, the highest occupied energy level in helium is filled by two rather than eight electrons. As you learned in Chapter 4, there is a special stability associated with a completely filled highest occupied energy level. By contrast, the Group 2 metals have no special stability; their highest occupied energy levels are not filled since they have an empty p sublevel.

Francium, the heaviest and most active alkali metal, is found in minute amounts in uranium minerals. Because it has a half-life of 22 minutes, there is probably less than one ounce of francium to be found at any one time. Francium was discovered in 1939 by Marguerite Perey (1909–1975), who was the first woman to be elected to the French Academy of Sciences.

Sample Problem 5.1

(a) Without looking at the Periodic Table, give the group, period, and block in which the element that has the following electron configuration is located: $[Xe]6s^2$.

(b) Without looking at the Periodic Table, write the electron configuration for the element in the third period in Group 1. Is this element likely to be a more or less active element than that described in part *(a)*?

Solution

(a) The element is in Group 2, as shown by the outer, or group, configuration of ns^2. It is in the sixth period, as shown by the highest principal quantum number of 6, in its configuration. The element is in the *s* block.

(b) A third-period element has filled $1s$, $2s$, and $2p$ sublevels (see Table 5-1). The third main-energy level, with $n = 3$, is its highest occupied energy level. An element in Group 1 has single electron in its highest *s* sublevel (group configuration ns^1). Therefore, this element has the configuration

$$1s^2 2s^2 2p^6 3s^1 \text{ or } [Ne]3s^1$$

Because it is in Group 1 (the alkali metals), this element is likely to be more active than that described in part *(a)*, which is an alkaline earth metal from Group 2.

Practice Problems

1. Without looking at the periodic table give the group, period, and block in which the element with the following shorthand electron configuration is located: $[Kr]5s^1$ *(Ans.)* Group 1, period 5, and *s*-block.

2. *(a)* Without looking at the periodic table, write both the group and complete electron configurations for the element in the fourth period in Group 2.   *(Ans.)* ns^2, $1s^2 2s^2 2p^6 3s^2 3p^6 4s^2$

   (b) Refer to the table to identify the indicated element in part *(c)* and then write its shorthand notation. *(Ans.)* Ca, $[Ar]4s^2$

   (c) How does the reactivity of the element in *(a)* compare with that in Group 1 of the same period?   *(Ans.)* Because this element is in Group 2, it is likely to be less active than that in Group 1 of the same period.

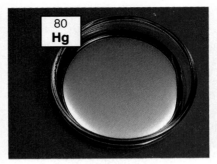

Mercury

Tungsten

Vanadium

Figure 5-8 Mercury, tungsten and vanadium are all members of the *d*-block of the Periodic Table.

Because of their low reactivity, Pd, Pt, and Au are sometimes called the noble metals.

STUDY HINT

The properties of the *d*-block metals are covered further in Chapter 25.

STUDY HINT

Materials that are referred to as *reactive* quickly or easily undergo chemical changes. Materials of low reactivity undergo few chemical changes and do so only with difficulty (often with addition of energy).

STUDY HINT

The formation of negative ions is of little importance in the chemistry of the *d*-block metals, as is true for all metals.

The *d*-Block Elements: Groups 3–12

In the *d*-block elements the filling of the *ns* sublevels in Groups 1 and 2 is followed by the addition of electrons to the *d* sublevels of the main energy level. Some *d*-block elements are shown in Figure 5-8. $(n-1)$. Atoms of the Group 3 elements each have one electron in the *d*-block sublevel corresponding to the $(n-1)$ the energy level just below *n*. This is in addition to the two *ns* electrons from Group 2. For Group 3, the group configuration is the $(n-1)d^1ns^2$. Nine more electrons are added across each period. Atoms of the Group 12 elements hold 10 electrons in the *d* sublevel plus two electrons in the *ns* sublevel. The group configuration for Group 12 is $(n-1)d^{10}ns^2$.

In Groups 4–11, some deviations from orderly *d* sublevel filling occur, as you saw in Chapter 4. As a result, elements in these *d*-block groups, unlike those in *s*-block and *p*-block groups, do not necessarily have identical electron configurations. For example, in Group 10, nickel (Ni) has the electron configuration $[Ar]3d^84s^2$, palladium (Pd) has the configuration $[Kr]4d^{10}5s^0$, and platinum (Pt) has the configuration $[Xe]4f^{14}5d^96s^1$. Notice, however, that the sum of the outer *s* and *d* electrons is equal to the group number. This is true for all *d*-block elements.

The d-block elements are all metals with typical metallic properties and are referred to as **transition elements.** They are good conductors of electricity and have a high luster when freshly cut or polished. The metals of the *d*-block are harder, denser, and with the exception of mercury, have higher melting points than the *s*-block metals. Palladium, platinum, and gold are among the least reactive of all these elements.

The *d*-block metals are typically less reactive than the alkali metals and the alkaline earth metals. Some, such as gold (Au), silver (Ag), and copper in Group 11; palladium (Pd) and platinum (Pt) in Group 10; and iridium (Ir) in Group 9 are commonly found in nature as free elements. These elements are so nonreactive that they do not easily form compounds with other elements.

Perhaps the most important of the *d*-block metals is iron, which is mixed with carbon to make steel. Other elements can be included in the mixture to make different varieties of steel. Some elements used in steel making are vanadium (V), chromium (Cr), manganese (Mn), cobalt (Co), nickel (Ni), molybdenum (Mo), and tungsten (W).

Tungsten has the highest melting point of all the metals and is used for the filaments of incandescent light bulbs. Copper, one of the best electrical conductors, is widely used for electrical wiring. Silver, gold and platinum are well known for their use in jewelry. Silver compounds, which are highly sensitive to light, are essential to photographic film.

Sample Problem 5-2

An element has the electron configuration $[Kr]4d^5 5s^1$. Without looking at the Periodic Table, identify the period, block, and group in which this element is located. Then, consult the Periodic Table and identify this element and the others in its group (excluding the period).

Solution

The number of the highest occupied energy level is 5, showing that the element is in the fifth period. There are 5 electrons in the d sublevel, which is therefore incompletely filled since this sublevel can hold 10 electrons. Therefore, the element is in the d-block. For d-block elements, the sum of the electrons in the ns sublevel and the $(n-1)d$ sublevel is equal to the group number, giving $5 + 1 = 6$. This is the fifth period element in Group 6. The element is molybdenum and the others in Group 6 are chromium and tungsten.

Practice Problems

1. Without looking at the Periodic Table, identify the period, block, and group in which the element with the following electron configuration is located: $[Ar]3d^8 4s^2$. *(Ans.)* 4, d block, and 10
2. *(a)* Without looking at the Periodic Table write the outer electron configuration for the fifth period element in Group 12. *(Ans.)* $4d^{10}5s^2$ *(b)* Refer to the table to identify that element and to write its shorthand electron configuration. *(Ans.)* Cd, $[Kr]4d^{10}5s^2$

Fluorine   **Chlorine**

Bromine

Iodine

Figure 5-9 Fluorine, chlorine, bromine and iodine are all members of p-block of the Periodic Table.

The *p*-Block Elements: Groups 13–18

The p-block elements include Groups 13–18, as shown in Figure 5-9. There are six p-block elements in each period except the first. Because electrons add to a p sublevel only after the s sublevel in the same energy level is filled, atoms of all p-block elements contain two electrons in the ns sublevel. For Group 13 elements, the added electron enters the np sublevel, giving a group configuration of ns^2np^1. Atoms of Group 14 elements each contain two electrons in the p sublevel, giving ns^2np^2 for the group configuration. This pattern continues in Groups 15 through 18. In Group 18 the stable noble gas configuration of ns^2np^6 is reached. Some p-block elements are shown in Figure 5-9.

STUDY HINT

Note that the number of s plus p electrons in atoms of p-block elements is equal to the last digit in the group number.

TABLE 5-3 RELATIONSHIPS AMONG GROUP NUMBERS, BLOCKS, AND ELECTRON CONFIGURATIONS

Group Number	Group Configuration	Block	Comments
1, 2	$ns^{1,\,2}$	s	One or two electrons in ns sublevel
3–12	$(n-1)d^{0-10}ns^{0-2}$	d	Sum of electrons in ns and $(n\text{-}1)d$ levels equals group number
13–18	ns^2np^{1-6}	p	Sum of electrons in ns and np sublevels equals group number minus 10. Number of electrons in np sublevel equals group number minus 12.

STUDY HINT

The properties of the metalloids and aluminum, a *p*-block metal, are discussed in Chapter 26.

For many years, the noble gases were called "the inert gases" because it was thought they would not react to form compounds. However, in 1962 this idea was shown to be incorrect when $XePtF_6$ was formed. Since then, other compounds of Xe, Kr, and Ra have been produced.

STUDY HINT

Carbon and organic chemistry are discussed in Chapter 21 and 22, and the properties of sulfur are discussed in Chapter 27.

STUDY HINT

Chapter 28 is devoted to the chemistry of the halogens.

For atoms of all *p* block elements, the total number of electrons in the highest occupied level is equal to the group number minus 10. For example, bromine, in Group 17, has $17 - 10 = 7$ *s* plus *p* electrons in its highest energy level. Because atoms of all *p*-block elements contain two electrons in the ns sublevel, bromine must have 5 electrons in the *p* sublevel. The electron configuration of bromine (Br) is therefore $[Ar]3d^{10}4s^24p^5$. The relationships among group numbers and electron configurations for all the groups are summarized in Table 5-3.

The properties of elements of the *p* block vary greatly. At its right-hand end, the *p* block includes all of the nonmetals except hydrogen and helium. All seven of the metalloids (boron, silicon, germanium, arsenic, antimony, selenium, and tellurium) are also in the *p* block. To the left and bottom of the block, there are seven *p*-block metals. The location of the nonmetals, metalloids, and metals in the *p* block are shown by distinctive colors in Figure 5-5.

The millions of compounds that contain carbon and hydrogen form the basis of organic chemistry. The nonmetals carbon, nitrogen, oxygen, sulfur, and phosphorus are present in living things and in a great many industrial chemicals. Oxygen also has a central role in the inorganic world. It forms compounds with almost every element and is present in most rocks and minerals.

The elements of Group 17 (fluorine, chlorine, bromine, iodine, and astanine) are known as the **halogens**. The halogens are the most reactive of the nonmetals. They react vigorously with most metals to form the type of compounds known as "salts." The most common salt is ordinary table salt, which contains the halogen chlorine combined with the alkali metal, sodium. As you will see, the reactivity of the halogens is based on the presence of seven electrons, one electron short of the stable noble gas configuration, in their outer energy levels. Fluorine and chlorine are gases at room temperature, bromine is a red liquid, and iodine is a dark purple solid. The halogens are used in bleaches, water purification, photography, insecticides, plastics, and many other practical applications.

The Group-18 elements, the noble gases, are the least reactive of all the elements. They have many uses that take advantage of this lack of reactivity. For example, noble gases are used instead of air in containers with materials that would react with oxygen or moisture in the air.

The metalloids, or semiconducting elements, fall on both sides of a line separating nonmetals and metals in the p block. They are mostly brittle solids with some properties of metals and some of nonmetals. The metalloid elements have electrical conductivity intermediate between that of metals, which are good conductors, and nonmetals, which are nonconductors. The electrical properties of semiconducting materials, made from metalloid elements make them essential components of computers.

The metals of the p block are generally harder and denser than the alkaline earth metals ns^2 block metals, but softer and less dense than the d-block metals. With the exception of bismuth, these metals are sufficiently reactive to be found in nature only in the form of compounds. Once obtained as free metals, however, they are stable in the presence of air. Aluminum is the most important of the p-block metals because it forms strong, lightweight alloys with many uses in building construction and aircraft production.

Sample Problem 5.3

(a) Without looking at the Periodic Table, write the outer electron configuration for the element in the second period of Group 14.

(b) Name this element. Is it a metal, a nonmetal, or metalloid?

Solution

(a) Because the group number is higher than 12, this element must be in the p-block. The total number of electrons in the highest occupied s and p sublevels is therefore $14 - 10 = 4$. With two electrons in the s sublevel, two electrons must also be present in the $2p$ sublevel, giving an outer electron configuration of $2s^2 2p^2$.

(b) This element is carbon (C), which is a nonmetal.

Practice Problems

1. *(a)* Without looking at the Periodic Table, write the outer electron configuration for the third period element in Group 17.   *(Ans.)* $3s^2 3p^5$

   (b) Name the element and identify it as a metal, nonmetal, or metalloid.   *(Ans.)* chlorine, nonmetal

2. *(a)* Without looking at the Periodic Table, identify the period, block, and group of an element with the electron configuration: $[Ar]3d^{10}4s^2 4p^3$ *(Ans. 4, p-block, Group 15.)*

   (b) Name this element and identify it as a metal, nonmetal, or metalloid.   *(Ans.)* arsenic, metalloid

The *f*-Block Elements: Lanthanides and Actinides

In the Periodic Table, the *f*-block elements are wedged between Groups 3 and 4 in Periods 6 and 7. This is because, after lanthanum, the $4f$ sublevel is filled before electrons are added to the $5d$ sublevel. With seven $4f$ orbitals to be filled with two electrons each, there are a total of 14 *f*-block elements between lanthanum (La) and hafnium (Hf) in Period 6. The lanthanides, or "rare earths" (which are not actually rare), are all shiny reactive metals. With modern separation techniques, all are readily available and practical uses have been found for them. For example, many of the phosphors that glow in color television tubes are compounds of lanthanide elements. Figure 5-10 shows two minerals from which lanthanides are recovered.

Bastnasite

Figure 5-10 Thorium, an element in the *f*-block portion of the Periodic Table is contained in the mineral bastnasite shown above.

STUDY HINT
Radioactivity and the creation of new elements by nuclear reactions are discussed in Chapter 29.

There are also 14 *f*-block elements, the actinides, between actinium and element 104 (Unq) in period 7. In these elements the 5*f* sublevel is being filled with 14 electrons. The actinides are all unstable and radioactive. The first four actinides (thorium, TH, through neptunium, Np), have been found naturally on earth. The remaining actinides are known only as manmade elements. Because of their radioactivity, the actinides are not encountered in everyday chemistry.

Sample Problem 5.4

The electron configurations of atoms of four elements are written below. For each, name the block and group in the Periodic Table in which it is located; name the element (consult the Periodic Table); identify it as a metal, nonmetal, or metalloid; and describe it as likely to be of high reactivity or of low reactivity. *(a)* $[Xe]4f^{14}5d^96s^1$ *(b)* $[Ne]3s^23p^5$ *(c)* $[Ne]3s^23p^6$ *(d)* $[Xe]4f^66s^2$

Solution

(a) The 4*f* sublevel is filled with 14 electrons. The 5*d* sublevel is partially with 9 electrons. Therefore, this is a *d*-block element. The element is platinum (Pt) which has a low reactivity.

(b) The incompletely filled *p* sublevel shows this to be a *p*-block element. With a total of 7 electrons in the *ns* and *np* sublevels, this is a Group 17 element, a halogen. The element is chlorine and is likely to be highly reactive.

(c) This element has the noble gas configuration and thus is in Group 18 in the *p* block. The element is argon (Ar), an unreactive nonmetal and a noble gas.

(d) The incomplete 4*f* sublevel shows that it is an *f*-block element and a lanthanide. Group numbers are not assigned to the *f* block. The element is samarium (Sm). The lanthanides are all reactive metals.

Practice Problems

For each of the following, identify the block, period, group, group name (where appropriate), element name, element type (metal, nonmetal, or metalloid), and relative reactivity (high or low):
1. $[He]2s^22p^5$ *(Ans.)* *p* block, 2, 17, halogens, fluorine, nonmetal, highly reactive
2. $[Ar]3d^{10}4s^1$ *(Ans.)* *d* block, 4, 11, transition elements, copper, metal, low reactivity
3. $[Kr]5s^1$ *(Ans.)* *s* block, 5, 1, alkali metals, rubidium, metal, highly reactive

1. They have identical outer electron configurations and thus have similar properties.
2. *s* block, *p* block, *d* block, and *f* block
3. *(a)* alkali metals *(b)* alkaline earth metals *(c)* transition elements *(d)* halogens *(e)* noble gases
4. *s*-block: ns^1 and ns^2 configurations designate elements in Groups 1 and 2, respectively; *p*-block: the ns^2np^1 through ns^2np^6 configurations designate elements in Groups 13 through 18, respectively; *d*-block: $(n-1)d^1ns^2$ through $(n-1)d^{10}ns^2$ configurations denote elements in Groups 3 through 12, respectively

Section Review

1. What is the relationship among the elements within a specific group (for the *s* and *p* block elements)?
2. Into what four blocks can the periodic table be divided to illustrate the relationships between electron configurations and periodic table placement?
3. What name is given to each of the folloing groups of elements on the Periodic Table: *(a)* Group 1 *(b)* Group 2 *(c)* Groups 3–12 *(d)* Group 17   *(e)* Group 18.
4. What is the relationship between group configuration and group number for elements in the *s*, *p*, and *d* blocks?
5. Without looking at the Periodic Table write the outer configuration for the fourth period element in Group 15.
6. Without looking at the Periodic Table identify the period, block, and group of the element: $[Ar]3d^74s^2$.

5. $3d^{10}4s^24p^3$
6. fourth period, *d* block, Group 9

5.3 Electron Configuration and Periodic Properties

Thus far, you have seen that the elements are grouped in the Periodic Table according to their physical and chemical properties and also, as shown in Section 5-2, their electron configurations. In this section, the relationship between the periodic variation in properties and electron configurations will be further explored.

Atomic Radii

Electrons occupy the relatively large region around the nucleus of an atom. The electron cloud determines the size of an atom. The size of an atom is not a definite quantity; however, the boundary is somewhat fuzzy and may vary. For these reasons and also because it is not possible to measure the size of an isolated atom, chemists compare the sizes by measuring the distance between atoms in an element.

The **atomic radius** *is one-half the distance between the nuclei of identical atoms joined in a molecule.* For example, the distance between the nuclei of two iron (Fe) atoms in a sample of solid iron can be measured, as shown in Figure 5-11. The radius of an iron atom is then taken as one-half this distance.

Period trends Figure 5-12 gives the atomic radii of the elements and Figure 5-13 presents this information graphically. There is a gradual decrease in atomic radii across the periods, for example, examine the decrease in atomic radii across the second period from lithium (Li) to neon (Ne).

The trend to smaller atoms across a period is caused by the increasing positive charge of the nucleus. As electrons add to *s* and *p* sublevels in the same main-energy level, they are gradually pulled closer to the more highly charged nucleus. This increased pull results in decreasing atomic radii. The attraction of the nucleus is offset somewhat by repulsion among the increased number of electrons in the same outer energy level, however. As a result, the decrease in radii between neighboring atoms in each period grows smaller, as shown by the graph in Figure 5-13.

Group trends The atomic radii of the main-group elements generally increase as one reads down a group. For example, examine the radii of the Group 1 elements in Figures 5-12 and 5-13. In the atoms of each element, the outer electrons occupy *s* sublevels in successively higher main-energy levels and are therefore farther from the nucleus.

Next, examine the radii of the Group 13 elements. Note that gallium (Ga) atoms have slightly smaller radii than aluminum (Al) atoms, although gallium follows aluminum in the group. Recall the atomic number differences given in Figure 5-3. Unlike aluminum, gallium is preceded in its period by the ten *d*-block elements. The expected increase in radius is overcome in gallium atoms by the higher nuclear charge. Gallium has a slightly smaller atomic radius than zinc (Zn), the element immediately to its left. This decrease is as predicted based on trends across the period.

SECTION OBJECTIVES

- Define atomic and ionic radii, ionization energy, electron affinity, and electronegativity.
- Compare the periodic variations of atomic radii, ionization energy, and electronegativity and state the reasons for these variations.
- Define valence electrons and state how many are present in atoms of each *s* and *p* block elements.
- Describe electron configurations of common ions of elements in the *s* and *p* blocks.
- Compare the atomic radii, ionization energies, and electronegativities of the *d*-block elements with those of the main-group elements.

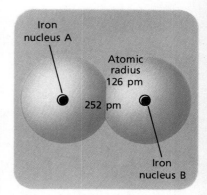

Figure 5-11 Atomic radius is defined as one-half the distance between the nucleus of two identical atoms combined in an element or chemical compound. For iron atoms, the radius is 126 picometers (pm).

Periodic Table of Atomic Radii

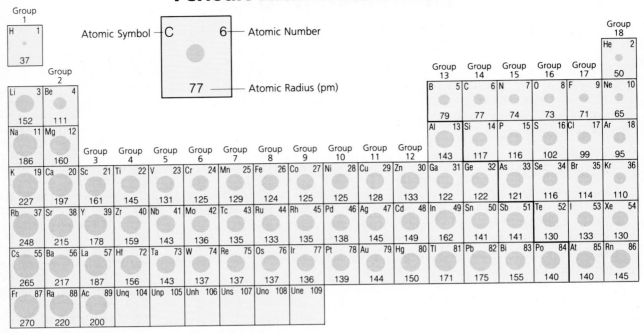

Figure 5-12 Atomic radii decrease from left to right across a period and increase down a group.

Lanthanide Series

Actinide Series

Sample Problem 5.5

(a) Among the elements magnesium (Mg), chlorine (Cl), sodium (Na), and phosphorous (P), which has the largest atomic radius? Explain in terms of periodic table trends.

(b) Among the elements calcium (Ca), beryllium (Be), barium (Ba), and strontium (Sr), which has the largest atomic radius? Explain in terms of periodic table trends.

Solution

(a) These elements are all in the third period. Of the four, sodium has the lowest atomic number and is the first element in the period. Sodium has the largest atomic radius, as expected because atomic radii decrease across the period. *(b)* These elements are all in Group 2. Of the four, barium has the highest atomic number and is furthest down the group. From its position, we would expect it to have the largest atomic radius, and this is correct.

Practice Problems

1. Among the elements Li, O, C, F, and N, identify the one with the: *(a)* largest atomic radius *(Ans.)* Li *(b)* smallest atomic radius. *(Ans.)* F

2. Among the elements Br, At, F, I, and Cl, identify the one with the: *(a)* smallest atomic radius *(Ans.)* F *(b)* largest atomic radius. *(Ans.)* At

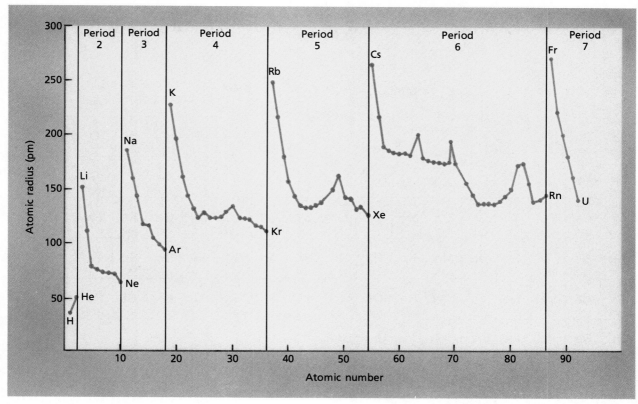

Figure 5-13 Plot of atomic radius versus atomic number

Ionization Energy

An electron can be removed from an atom when enough energy is supplied. Using A as a symbol for an atom of any element, the process can be represented as follows:

$$A + energy \rightarrow A^+ + e^-$$

The A^+ represents an ion of element A with a single positive charge, referred to as a +1 charge. *An* **ion** *is an atom or group of atoms that has a positive or negative charge*. Sodium, for example, readily forms an Na^+ ion. *Any process that results in the formation of an ion is referred to as* **ionization.**

To compare the ease with which atoms of different elements give up electrons, chemists compare ionization energies. *The energy required to remove one electron from an atom of an element is the* **ionization** energy *(or first ionization energy)*. Measurements of ionization energy are made on atoms in the gas phase to eliminate the influence of nearby atoms. Figure 5-14 gives first ionization energies in kilojoules per mole (kJ/mol) for the elements. Figure 5-15 presents this information graphically.

Period trends Examine the ionization energies in Figures 5-14 and 5-15 for the first and last elements in each period. You can see that the Group 1 metals have the lowest first ionization energies: that is, they lose electrons most easily. This ease of electron loss is a major reason for the high reactivity of the Group 1 (alkali) metals. The Group 18 elements, the noble gases, have

First Ionization Energies
kJ/mol

Group 1

1
H
1312

Atomic Number → 6
Symbol → **C**
First Ionization Energy → 1086

Group 1	Group 2											Group 13	Group 14	Group 15	Group 16	Group 17	Group 18
1 **H** 1312																	2 **He** 2372
3 **Li** 520	4 **Be** 900											5 **B** 801	6 **C** 1086	7 **N** 1402	8 **O** 1314	9 **F** 1681	10 **Ne** 2081
11 **Na** 496	12 **Mg** 738	Group 3	Group 4	Group 5	Group 6	Group 7	Group 8	Group 9	Group 10	Group 11	Group 12	13 **Al** 578	14 **Si** 787	15 **P** 1012	16 **S** 1000	17 **Cl** 1251	18 **Ar** 1521
19 **K** 419	20 **Ca** 590	21 **Sc** 631	22 **Ti** 658	23 **V** 650	24 **Cr** 653	25 **Mn** 717	26 **Fe** 759	27 **Co** 758	28 **Ni** 737	29 **Cu** 746	30 **Zn** 906	31 **Ga** 579	32 **Ge** 762	33 **As** 944	34 **Se** 941	35 **Br** 1140	36 **Kr** 1351
37 **Rb** 403	38 **Sr** 550	39 **Y** 616	40 **Zr** 660	41 **Nb** 664	42 **Mo** 685	43 **Tc** 702	44 **Ru** 711	45 **Rh** 720	46 **Pd** 805	47 **Ag** 731	48 **Cd** 868	49 **In** 558	50 **Sn** 709	51 **Sb** 832	52 **Te** 869	53 **I** 1008	54 **Xe** 1170
55 **Cs** 376	56 **Ba** 503	57 **La** 538	72 **Hf** 654	73 **Ta** 761	74 **W** 770	75 **Re** 760	76 **Os** 840	77 **Ir** 880	78 **Pt** 870	79 **Au** 890	80 **Hg** 1007	81 **Tl** 589	82 **Pb** 716	83 **Bi** 703	84 **Po** 812	85 **At** –	86 **Rn** 1038
87 **Fr** –	88 **Ra** 509	89 **Ac** 490	104 **Unq**	105 **Unp**	106 **Unh**	107 **Uns**	108 **Uno**	109 **Une**									

Lanthanide Series

58 **Ce** 528	59 **Pr** 523	60 **Nd** 530	61 **Pm** 536	62 **Sm** 543	63 **Eu** 547	64 **Gd** 592	65 **Tb** 564	66 **Dy** 572	67 **Ho** 581	68 **Er** 589	69 **Tm** 597	70 **Yb** 603	71 **Lu** 523
90 **Th** 590	91 **Pa** 570	92 **U** 590	93 **Np**	94 **Pu**	95 **Am**	96 **Cm**	97 **Bk**	98 **Cf**	99 **Es**	100 **Fm**	101 **Md**	102 **No**	103 **Lr**

Actinide Series

Figure 5-14 Periodic table of the elements showing first ionization energies in kilojoules per mole (kJ/mol)

the highest ionization energies and do not lose electrons easily. As you might guess, the low reactivity of the noble gases is partly based on this difficulty of removing electrons.

The trend across the periods for the main-group elements is toward a general increase in ionization energy. This increase is caused by the increasing nuclear charge, which more strongly attracts electrons in the same energy level. Thus, increasing nuclear charge is responsible for both *increasing* ionization energy and decreasing radii across the periods.

Note that, in general, nonmetals have higher ionization energies than metals. The elements of Group 1, the alkaline-earth metals, have the lowest ionization energies and the elements of Group 13, the noble gases, have the highest ionization energies.

Group trends   Among the main-group elements, the ionization energy generally decreases down the groups as the size of the atoms increases. The electrons removed from atoms of each succeeding element are in higher energy levels. Therefore they are farther from the nucleus and more easily removed. Also, as atomic number increases going down a group, more and more electrons lie between the nucleus and the electrons in the highest

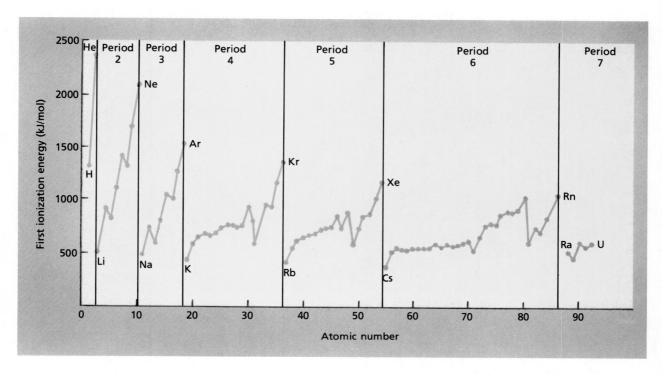

Figure 5-15 Plot of first ionization energy versus atomic number

occupied energy levels. The result is that the outer electrons are shielded from the positive charge of the nucleus. These influences combine to outweigh the attraction for electrons of the increasing nuclear charge.

Formation of positive ions   With sufficient energy, electrons can be removed from ions as well as from neutral atoms. The energies for removal of the second or third electrons from an atom in a gas are referred to as the "second ionization energy," "third ionization energy," and so on.

Table 5-4 shows the first five ionization energies for the elements of Periods 1, 2, and 3. You can see that the second ionization energy is always higher than the first, the third is always higher than the second, and so on. These increases are to be expected. Each successive electron must be removed from an ion with a larger positive charge.

The first ionization energies in Table 5-4 show that removing a single electron from an atom of a Group 18 element is much harder than removing an electron from atoms of other elements in the same period. This special stability of the noble gas electron configuration is also shown by *ions* with the electron configurations of noble gases. In Table 5-4, look at the large increase between the first and second ionization energies of lithium (Li), between the second and third ionization energies of beryllium, between the third and fourth ionization energies of boron (B), and between the fourth and fifth ionization energies of carbon (C). In each case, the jump in ionization energy occurs when the noble gas configuration is reached. For example, the removal of one electron from a lithium atom ([He]$2s^1$) leaves the helium noble gas configuration. The removel of four electron from a carbon atom ([He]$2s^1 2p^2$) also leaves the helium configuration. A bigger table would show that this trend continues across the entire period.

To help students visualize how ionization energies increase with each successive removal of an electron, use the analogy of adults (the protons) looking after energetic children (the electrons). The fewer children per adult, the more control the adults have over the children. With fewer children, more energy is required to take them away.

TABLE 5-4 FIRST FIVE IONIZATION ENERGIES (IN KJ/MOL) FOR ELEMENTS OF PERIODS 1–3

	Period 1		Period 2							
	H	He	Li	Be	B	C	N	O	F	Ne
I	1312	2372	520	900	801	1086	1402	1314	1681	2081
II		5250	7298	1757	2427	2353	2856	3388	3374	3952
III			11 815	14 849	3660	4621	4578	5300	6050	6122
IV				21 007	25 026	6223	7475	7469	8408	9370
V					32 827	37 830	9445	10 990	11 023	12 178

	Period 3							
	Na	Mg	Al	Si	P	S	Cl	Ar
I	496	738	578	787	1012	1000	1251	1521
II	4562	1451	1817	1577	1903	2251	2297	2666
III	6912	7733	2745	3232	2912	3361	3822	3931
IV	9544	10 540	11 578	4356	4957	4564	5158	5771
V	13 353	13 628	14 831	16 091	6274	7031	6540	7238

Sample Problem 5.6

Consider two main-group elements A and B. Element A has an ionization energy of 419 kJ/mol. Element B has an ionization energy of 1000 kJ/mol. For each element, decide if it is more likely to be a metal or a nonmetal, and if it is more likely to be in the *s* block or *p* block. Which element is more likely to form a positive ion in a chemical compound?

Solution
Element A has quite a low ionization energy, meaning that A atoms lose electrons easily. Therefore, element A is most likely to be a metal in the *s* block, because ionization energies increase across the periods. Element B has quite a high ionization energy, meaning that B atoms lose electrons with difficulty. We would expect element B to be a nonmetal, meaning that it must be in the *p* block. Element A is more likely to form a positive ion, because it has a relatively low ionization energy and it is a metal.

Practice Problems
Consider the four hypothetical main-group elements R, S, T, and U with the indicated outer electron configurations: $R—3s^2 3p^5$ $S—3s$ $T—4d^{10}5s^2 5p^5$ $U—4d^{10}5s^2 5p^1$

1. Identify the block location of each hypothetical main-group element and whether each is a metal, nonmetal, or metalloid. *(Ans.)* R is a *p*-block nonmetal, S is an *s*-block metal, T is a *p*-block nonmetal, and U is a *p*-block metal.
2. Which of these elements are in the same *(a)* period? *(Ans.)* S and R, U and T *(b)* group? *(Ans.)* R and T
3. Which would you expect to have *(a)* the highest first ionization energy? *(Ans.)* R *(b)* the lowest first ionization energy? *(Ans.)* S
4. Which of these hypothetical main-group elements would you expect to have the highest second ionization energy? *(Ans.)* S
5. Which is most likely to form a 1+ ion in a chemical compound? *(Ans.)* S

Electron Affinity

Neutral atoms can acquire electrons. The energy involved when an atom gains an electron is referred to as the electron affinity. **Electron affinity** *is the energy change that occurs when an electron is acquired by a neutral atom.* Many atoms readily add electrons and release energy, which classifies this as an exothermic process:

$$A + e^- \rightarrow A^- + \text{energy}$$

By convention, the quantity of energy released in an exothermic process is represented by a negative number.

Some atoms, however, must be "forced" to gain an electron by the addition of energy:

$$A + e^- + \text{energy} \rightarrow A^-$$

The quantity of energy absorbed in such an endothermic process is represented by a positive number. An ion produced in this way is likely to be unstable and to lose the added electron spontaneously.

Figure 5-16 shows the electron affinity in kilojoules per mole for the main-group elements. Figure 5-17 presents these data in graphic form. Note that there are both positive and negative values.

Period trends For each period, halogen atoms (Group 17) gain electrons most easily, as shown by the large negative values for their electron affinities. The ease with which halogen atoms gain electrons is a major reason for the high reactivities of the halogen. Electron affinities become more negative across each period within the *p* block, as electrons add to the same *p* sublevel

Figure 5-16 Periodic table of electron affinities. Values in parentheses are calculated; values not in parentheses are experimental.

Electron Affinities kJ/mol

Key box: 6 — Atomic Number; C — Symbol; −123 — Electron Affinity

Group 1	Group 2	Group 3	Group 4	Group 5	Group 6	Group 7	Group 8	Group 9	Group 10	Group 11	Group 12	Group 13	Group 14	Group 15	Group 16	Group 17	Group 18
1 **H** -73																	2 **He** (21)
3 **Li** -60	4 **Be** (240)											5 **B** -83	6 **C** -123	7 **N** 0	8 **O** -141	9 **F** -322	10 **Ne** (29)
11 **Na** -53	12 **Mg** (230)											13 **Al** (-50)	14 **Si** -120	15 **P** -74	16 **S** -200	17 **Cl** -349	18 **Ar** (35)
19 **K** -48	20 **Ca** (156)	21 **Sc** –	22 **Ti** -38	23 **V** -90	24 **Cr** -64	25 **Mn** –	26 **Fe** -56	27 **Co** -90	28 **Ni** -123	29 **Cu** -123	30 **Zn** –	31 **Ga** (-36)	32 **Ge** -116	33 **As** -77	34 **Se** -195	35 **Br** -325	36 **Kr** (39)
37 **Rb** -47	38 **Sr** (168)	39 **Y** –	40 **Zr** –	41 **Nb** –	42 **Mo** -96	43 **Tc** –	44 **Ru** –	45 **Rh** –	46 **Pd** –	47 **Ag** -126	48 **Cd** –	49 **In** -34	50 **Sn** -121	51 **Sb** -101	52 **Te** -183	53 **I** -295	54 **Xe** (41)
55 **Cs** -46	56 **Ba** (52)	57 **La** –	72 **Hf** –	73 **Ta** -80	74 **W** -50	75 **Re** -14	76 **Os** –	77 **Ir** –	78 **Pt** -205	79 **Au** -223	80 **Hg** –	81 **Tl** -50	82 **Pb** -101	83 **Bi** -101	84 **Po** (-170)	85 **At** (-270)	86 **Rn** (41)
87 **Fr** (-44)	88 **Ra** –	89 **Ac** –	104 **Unq**	105 **Unp**	106 **Unh**	107 **Uns**	108 **Uno**	109 **Une**									

Values in parentheses are calculated; values not in parentheses are experimental.

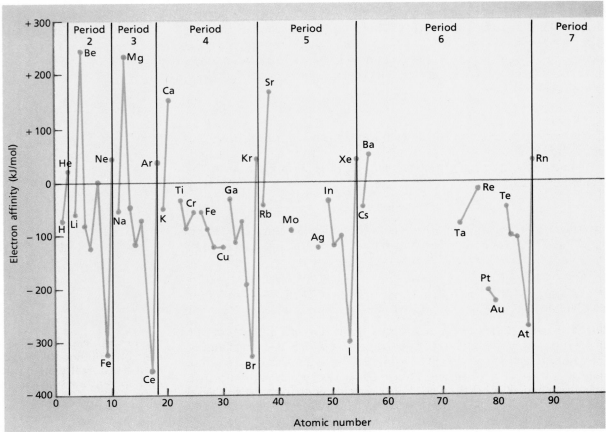

Figure 5-17 Plot of electron affinity versus atomic number. The data are plotted from the experimental and estimated values shown in Figure 5-16.

of atoms with increasing nuclear charge. An exception to this trend occurs between Groups 14 and 15. Compare the electron affinities of carbon ($[He]2s^2 2p^2$) and nitrogen ($[He]2s^2 2p^3$). Adding an electron to a carbon atom gives a half-filled sublevel. This occurs more easily than adding a second electron to an already half-filled p sublevel of a nitrogen atom.

Atoms of Group 2 and Group 18 elements are the least likely to gain electrons least easily, as shown by the positive values for their electron affinities. Here, electrons must be forced into atoms with filled outer sublevels, ns^2 for Group 2 and ns^2np^6 for Group 18 (with the exception of helium, $1s^2$).

Group trends   Trends for electron affinity values within groups are not as regular as ionization energies and they are more difficult to explain. In some main groups, however, especially from the third to the sixth periods, there is a tendency for electrons to add with greater difficulty as one proceeds down the group. You may wonder why the increased nuclear charge down the groups does not increase the attraction for added electrons. Recall that atomic radii also increase as one moves down the group and it is generally more difficult to add electrons to larger atoms than to smaller atoms.

Formation of negative ions   For an isolated ion in the gas phase, it is always more difficult to add a second electron to an already negatively charged ion. Therefore, second electron affinities are all positive. However, ions with negative charges higher than 1 are present in chemical compounds. Electron affinities are measured on isolated atoms in the gas phase. Other factors contribute to the energy balance in the formation of chemical compounds in the liquid or solid states. Certain *p*-block nonmetals tend to form negative ions that have noble gas configurations. The halogens do so by adding one electron. For example, chlorine (Cl, $[\text{Ne}]3s^23p^5$) achieves the configuration of the next highest noble gas, argon (Ar), by adding an electron to give Cl^- ($[\text{Ne}]3s^23p^6$). Atoms of Group 16 elements are present in many compounds as 2- ions. For example, oxygen ($[\text{He}]2s^22p^4$) achieves the configuration of the next highest noble gas, neon, by adding two electrons (O^{2-} $[\text{He}]2s^22p^6$). Nitrogen achieves a neon configuration by adding three electrons to give N^{3-}.

Ionic Radii

Figure 5-18 shows the radii of some of the most common ions of the elements—ions present in many chemical compounds. *An* **ionic radius** *is one-half the diameter of an ion in a chemical compound.*

A positive ion in known as a **cation.** The formation of a cation by the loss of one or more electrons always leads to a decrease in radius. As expected, with the highest energy level electrons removed, the size of the remaining electron cloud is smaller. Also, the remaining electrons are drawn closer to the nucleus by its unbalanced positive charge.

A negative ion is known as an **anion.** The formation of an anion by the addition of one or more electrons always leads to an increase in radius. In the common anions shown in Figure 5-20, enough electrons have been added to

Figure 5-18 The ionic radii of the ions most common in chemical compounds. Cations are smaller, and anions are larger, than the atoms from which they are formed.

Periodic Table of Ionic Radii

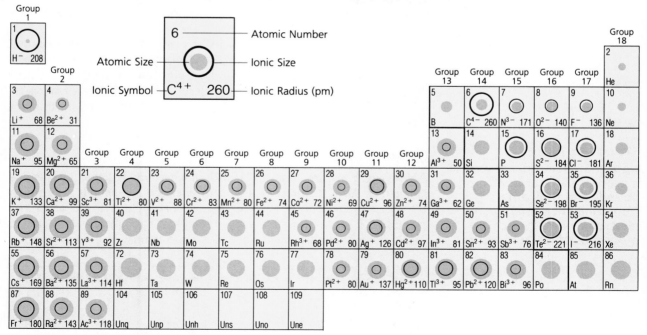

give anions with noble gas configurations. The total positive charge of the nucleus is unchanged. The electron cloud spreads out because there is a smaller attractive force on the electrons and a greater repulsion between them.

Periodic trends   Within each period, the metals at the left form cations and the nonmetals at the right form anions. There is a regular decrease of cation radii with increasing atomic number. This decrease is to be expected as electrons are removed one by one from the highest energy level as the nuclear charge increases.

Beginning with Group 15, anions are more common than cations. There is also a regular decrease of radii because the nuclear charge is increasing and the number of added electrons is decreasing.

Group trends   In proceeding down the groups, the outer electrons for both cations and anions are in higher energy levels, as they are for atoms. Therefore, there is a gradual increase of ionic radii down the groups as there is for atomic radii.

Valence Electrons

As you will learn in detail in Chapter 6, chemical compounds form because electrons are lost, gained, or shared between atoms of different elements. The electrons that interact in this manner are those in the highest energy levels. These are the electrons most subject to the influence of nearby atoms or ions. *The electrons available to be lost, gained, or shared in the formation of chemical compounds are referred to as* **valence electrons.** For example, the electron lost from the $2s$ sublevel of Na to form Na^+ is a valence electron. Valence electrons are often located in incompletely filled main-energy levels.

For main-group elements, electrons in the s and p sublevels of the highest occupied energy level are the valence electrons. The inner electrons are all in filled energy levels and are too closely held by the nucleus to be involved in compound formation.

The Group 1 and Group 2 elements have one and two valence electrons, respectively, as shown in Table 5-5. The elements of Groups 13–18 have a number of valence electrons equal to the group number minus 10. In some cases, both the s and p sublevel valence electrons of the p-block elements are involved in compound formation. In other cases, only the electrons from the p sublevel are involved.

Electronegativity

Valence electrons are responsible for holding atoms together in chemical compounds. Also, the properties of many chemical compounds are strongly influenced by the concentration of negative charge closer to one atom than to another. It is useful to have a measure of how strongly one atom attracts the electrons of another atom when they are combined in a compound.

TABLE 5-5 VALANCE ELECTRONS IN MAIN-GROUP ELEMENTS							
Group number:	1	2	13	14	15	16	17	18
Valence electrons:	1	2	3	4	5	6	7	8

Linus Pauling (1901–), one of the most widely known American chemists, developed the concept of electronegativity, a scale for assigning numerical values to the tendency of an atom to attract electrons. **Electronegativity** is a measure of the power of an atom in a *chemical compound to attract electrons.* The most electronegative element, fluorine, is arbitrarily assigned an electronegativity value of four. Electronegativity values for the other elements are then calculated in relation to this value.

The electronegativities of the elements are shown in Figure 5-19 and plotted in graphical form in Figure 5-20. The graph shows that electronegativity values vary periodically.

Periodic trends Across each period, there is a gradual increase in electronegativity. Nonmetals are generally more electronegative than metals. The active alkali and alkaline earth metals at the left of each period are the least electronegative elements. Their atoms have very little attraction for electrons in chemical compounds. The halogens at the right are the most electronegative elements. Halogen atoms attract electrons in chemical compounds most strongly of all the elements.

STUDY HINT

In general, metals are less electronegative than nonmetals.

Linus Pauling received the 1954 Nobel Prize in Chemistry for his work in molecular structure and chemical bonding (which includes electronegativity). He received the 1962 Nobel Peace Prize for his efforts in support of the Nuclear Test Ban Treaty of 1963. Pauling, along with Marie Curie (1867–1934), is one of the few people to receive two Nobel Prizes.

Figure 5-19 Electronegativities of the elements according to the Pauling scale.

Periodic Table of the Electronegativities

Key: 6 — Atomic Number; C — Symbol; 2.5 — Electronegativity

Group 1	Group 2	Group 3	Group 4	Group 5	Group 6	Group 7	Group 8	Group 9	Group 10	Group 11	Group 12	Group 13	Group 14	Group 15	Group 16	Group 17	Group 18
1 **H** 2.1																	2 **He**
3 **Li** 1.0	4 **Be** 1.5											5 **B** 2.0	6 **C** 2.5	7 **N** 3.0	8 **O** 3.5	9 **F** 4.0	10 **Ne**
11 **Na** 0.9	12 **Mg** 1.2											13 **Al** 1.5	14 **Si** 1.8	15 **P** 2.1	16 **S** 2.5	17 **Cl** 3.0	18 **Ar**
19 **K** 0.8	20 **Ca** 1.0	21 **Sc** 1.3	22 **Ti** 1.5	23 **V** 1.6	24 **Cr** 1.6	25 **Mn** 1.5	26 **Fe** 1.8	27 **Co** 1.8	28 **Ni** 1.8	29 **Cu** 1.9	30 **Zn** 1.6	31 **Ga** 1.6	32 **Ge** 1.8	33 **As** 2.0	34 **Se** 2.4	35 **Br** 2.8	36 **Kr** 3.0
37 **Rb** 0.8	38 **Sr** 1.0	39 **Y** 1.2	40 **Zr** 1.4	41 **Nb** 1.6	42 **Mo** 1.8	43 **Tc** 1.9	44 **Ru** 2.2	45 **Rh** 2.2	46 **Pd** 2.2	47 **Ag** 1.9	48 **Cd** 1.7	49 **In** 1.7	50 **Sn** 1.8	51 **Sb** 1.9	52 **Te** 2.1	53 **I** 2.5	54 **Xe** 2.6
55 **Cs** 0.7	56 **Ba** 0.9	57 **La** 1.1	72 **Hf** 1.3	73 **Ta** 1.5	74 **W** 1.7	75 **Re** 1.9	76 **Os** 2.2	77 **Ir** 2.2	78 **Pt** 2.2	79 **Au** 2.4	80 **Hg** 1.9	81 **Tl** 1.8	82 **Pb** 1.8	83 **Bi** 1.9	84 **Po** 2.0	85 **At** 2.2	86 **Rn** 2.4
87 **Fr** 0.7	88 **Ra** 0.9	89 **Ac** 1.1	104 **Unq**	105 **Unp**	106 **Unh**	107 **Uns**	108 **Uno**	109 **Une**									

Lanthanide Series

58 **Ce** 1.1	59 **Pr** 1.1	60 **Nd** 1.1	61 **Pm** 1.1	62 **Sm** 1.1	63 **Eu** 1.1	64 **Gd** 1.1	65 **Tb** 1.1	66 **Dy** 1.1	67 **Ho** 1.1	68 **Er** 1.1	69 **Tm** 1.1	70 **Yb** 1.1	71 **Lu** 1.2
90 **Th** 1.3	91 **Pa** 1.5	92 **U** 1.7	93 **Np** 1.3	94 **Pu** 1.3	95 **Am** 1.3	96 **Cm** 1.3	97 **Bk** 1.3	98 **Cf** 1.3	99 **Es** 1.3	100 **Fm** 1.3	101 **Md** 1.3	102 **No** 1.3	103 **Lr**

Actinide Series

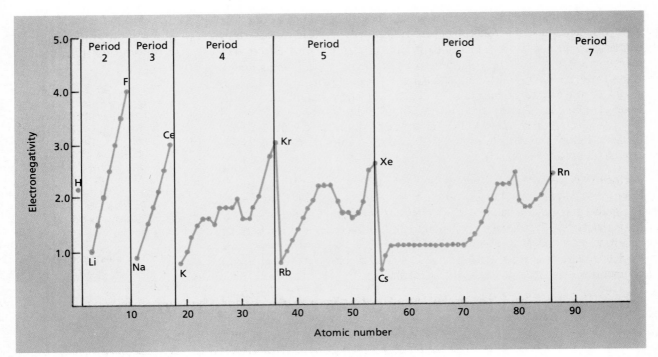

Figure 5-20 Plot of electronegativity versus atomic number for periods 1–6

Group trends Within each main group, electronegativities either decrease down the group or have very similar values. This trend is in keeping with the sizes of the atoms. Larger, more easily ionized atoms attract electrons less strongly than smaller, less easily ionized atoms. Cesium and francium, which have the largest atomic radii and the lowest ionization energies, are the least electronegative elements. Fluorine, the most electronegative element, has a very small radius and a high ionization energy.

Sample Problem 5.7

Among the elements gallium (Ga), bromine (Br), and calcium (Ca), which has the highest electronegativity? Explain in terms of periodic table trends. Will the element of highest electronegativity have the largest or smallest atomic radius among the elements listed?

Solution
The elements are all in period 4. Bromine has the highest atomic number and is furthest to the right in the period. It should have the highest electronegativity, for electronegativity increases across the periods. Bromine should have the smallest atomic radius because smaller atoms have higher electronegativity. And because radii tend to decrease across a period.

Practice Problems
Consider the five hypothetical main group elements V, W, X, Y, and Z with the indicated outer electron configurations:

$$V = 2s^2 2p^5 \qquad W = 4d^{10}5s^2 5p^5 \qquad X = 2s^2 2p^2 \qquad Y = 5d^{10}6s^2 6p^5 \qquad Z = 2s^2 2p^4$$

Sample Problem 5.7 *continued*

1. Identify the block location for each, and then determine which elements are in the same group and which are in the same period. *(Ans.)* All are *p*-block: V, X, and Z are in the same period and V, W, and Y are in the same group.
2. Which would you expect to *(a)* have the highest electron affinity? *(Ans.)* V *(b)* form with a −1 charge ions? *(Ans.)* V, W, Y *(c)* have the highest electronegativity *(Ans.)* V
3. Compare the ionic radius of the typical ion formed by the element W with the radius of its neutral atom. *(Ans.)* The ionic radius would be larger.
4. Which element(s) would be expected to contain seven valence electrons? *(Ans.)* V, W, and Y

Periodic Properties of the *d*- and *f*-Block Elements

Up to now, the fairly regular periodic trends of the main-group elements have been considered. Many properties of the *d*-block elements (which are all metals) vary less and with less regularity than those of the main-group elements. You can see that the curves in Figures 5-13 and 5-15 flatten where the *d*-block elements fall in the middle of periods 4–6.

Atoms of the *d*-block elements contain from zero to two electrons in the highest occupied energy level. Electrons are added to the *d* sublevel of the next highest energy level because it is close in energy to the outermost level for these elements. Electrons in both the *ns* sublevel and the (*n*-1)*d* sublevel are therefore available to interact with their surroundings. As a result, electrons in the incompletely filled *d* sublevels are responsible for many characteristic properties of the *d*-block elements.

Atomic radii The atomic radii of the *d*-block elements generally decrease across the periods. However, this decrease is less than that for the main-group elements because the electrons added to the (*n*-1)*d* sublevel shield the outer electrons from the nucleus. Also, note in Figure 5-13 that the radii dip to a low and then increase slightly across each of the four periods that contain *d*-block elements. As the number of electrons in the *d* sublevel increases, the radii increase because of the repulsion among the electrons.

In the sixth period, the *f*-block elements fall between lanthanum (Group 3) and hafnium (Group 4). Because of the increase in atomic number that occurs from lanthanum to hafnium, the atomic radius of hafnium is actually slightly less than that of zirconium (Zr), which is immediately above it. The radii of elements following hafnium in the sixth period vary with increasing atomic number in the usual manner.

Ionization energy Ionization energies generally increase across the periods with the decreasing size of the atoms, as they do for the main-group elements. In contrast to the decrease down the main groups, however, the first ionization energies of the *d*-block elements increase as one proceeds down each group. The electrons available for ionization in the outer *s* sublevels are apparently less shielded from the increasing nuclear charge by electrons in the incomplete (*n*-1)*d*-sublevels. It is also harder to pull away the outer electrons in atoms of the Period-6 elements, which are about the same size as those of the Period-5 elements but with an increased nuclear charge.

Ion formation and ionic radii   The order in which electrons are removed from all atoms is *exactly the reverse of the order given by the electron-configuration notation*. In other words, electrons in the highest occupied sublevel are always removed first. For the *d*-block elements this means that although newly added electrons are in the *d* sublevels, the first electrons to be removed are those in the highest occupied *s* sublevels. For example, iron (Fe), which has the electron configuration $[Ar]3d^6 4s^2$, first loses a 4s electron to form the Fe^+ ion ($[Ar]3d^6 4s^1$). The Fe^+ ion can then lose the second 4s electron to form the Fe^{2+} ion ($[Ar]3d^6$). The Fe^{2+} ion can then lose a 3d electron to form the Fe^{3+} ion ($[Ar]3d^5$). The iron ions most commonly found in chemical compounds are the Fe^{2+} and the Fe^{3+} ion.

The Group-3 elements form only ions with a +3 charge in compounds Copper forms +1 and +2 ions and silver forms +1 ions. Most other *d*-block elements commonly form +2 ions in compounds although some, such as iron, also form +3 ions. As expected, the cations have smaller radii than the atoms. Comparing +2 ions across the periods shows a decrease in size that parallels the decrease in atomic radii.

Electronegativity   The *d*-block elements all have electronegativity values between 1.1 and 2.4, as shown in Figure 5-20. Only the active metals of Groups 1 and 2 have lower electronegativities. Across the periods there is a general trend for electronegativity values to increase as radii decrease and vice versa. In other words, small atoms tend to attract electrons more strongly than large atoms.

Section Review

1. State the general period and group trends among main-group elements with respect to each of the following: *(a)* atomic radii   *(b)* first ionization energy   *(c)* electron affinity   *(d)* ionic radii   *(e)* electronegativity.
2. Among the main-group elements, what is the relationship between group number and the number of valence electrons among group members?
3. *(a)* In general, how do the periodic properties of the *d*-block elements compare with those of the main-group elements?   *(b)* Explain the comparisons made in *(a)*.

Chapter Summary

- Elements in the Periodic Table with similar chemical properties fall in the same column.
- The rows in the Periodic Table are referred to as *periods*.
- The columns in the Periodic Table are referred to as *groups*.
- As you move one element to the right in the *s, p, d,* and *f* blocks of the Periodic Table, elements in the *s, p, d,* and *f* sublevels respectively gain an electron in their outer sublevel. This same principle holds for substituting *s, p,* and *f* orbitals.
- The Periodic Table can be used to display trends in the following properties of the elements—electron affinity, electronegativity, ionization energy, atomic radii, and ionic radii.

---

1. *(a)* generally decrease across periods and increase down groups   *(b)* generally increase across periods and decrease down groups   *(c)* generally increase across periods among Groups 13–17 and decrease down groups, but numerous exceptions are observed   *(d)* generally decrease across periods and increase down groups   *(e)* gradually increases across periods and either decreases or is very similar in value down groups
2. For Groups 1 and 2, the number of valence electrons is equal to the group number; for Groups 13–18, the number of valence electrons is equal to the group number minus 10.
3. *(a)* vary less, and with less regularity, than main-group elements   *(b)* generally attributed to the presence of electrons in incompletely filled *d* sublevels in these atoms.

Chapter 5 *Review*

The Periodic Law

Vocabulary

actinides
alkali metals
alkaline earth metals
anion
atomic radius
cation
electron affinity
electronegativity
halogens
ion

ionic radius
ionization
ionization energy
lanthanides (rare earth elements)
main-group elements
periodic law
Periodic Table
transition elements
valence electrons

Questions

1. Cite the major contribution of each toward the development of the modern periodic table: *(a)* Cannizzaro *(b)* Mendeleev *(c)* Moseley.
2. *(a)* How did Mendeleev arrange the elements in his first periodic table? *(b)* How was this different from previous tables? *(c)* What dilemma did his newly proposed arrangment present? *(d)* What questions were still left unanswered?
3. Explain the meaning of the periodic law.
4. What is the periodic table?
5. How did the discovery of the noble gases change the appearance of Mendeleev's periodic table?
6. *(a)* Distinguish between the lanthanide and actinide elements by definition and table location. *(b)* What are the rare earth elements?
7. Distinguish between the two systems commonly used for labeling periodic table groups.
8. Why are the noble gases relatively nonreactive?
9. *(a)* How do the electron configurations of the elements within a group compare (for main groups only)? *(b)* What is the relationship between the electron configuration of an element and the properties of that element?
10. *(a)* What determines the length of each period in the table? *(b)* Illustrate this pattern for elements in the second and fourth periods.
11. What is the relationship between the electron

configuration of an element and the period in which that element appears in the table?
12. Write the shorthand electron configuration and indicate the period in which each belongs: *(a)* Li *(b)* 0   *(c)* Mg *(d)* P   *(e)* Sc   *(f)* Br   *(g)* Sn.
13. *(a)* What information is provided by the specific block location of an element? *(b)* Identify, by number, the groups located within each of the four block areas.
14. Using n for the number of the highest occupied energy level, write the group configuration for each group within the s-block.
15. *(a)* Which elements are designated as the alkali metals? *(b)* List four of their characteristic properties. *(c)* Cite two of their common uses.
16. *(a)* Which elements are designated as the alkaline earth metals? (b) How do their characteristic properties compare with those of the alkali metals? *(c)* Cite at least two common uses of alkaline earth metals.
17. Based on the information given below, give the group, period, block, and identity of each element described: *(a)* $[He]2s^2$ *(b)* $[Ne]3s^1$ *(c)* $[Kr]5s^2$ *(d)* $[Ar]4s^2$ *(e)* $[Rn]7s^1$.
18. Without looking at the periodic table write both the group and complete electron configurations for each of the elements described, and then write the shorthand notation for each: *(a)* period 2, Group 1   *(b)* period 3, Group 2 *(c)* period 4, Group 1   *(d)* period 6, Group 2.
19. *(a)* Write the group configuration notation for each d-block group. *(b)* How do these group numbers relate to the number of outer s and d electrons?
20. *(a)* What name is sometimes used to refer to the entire set of d-block elements? *(b)* List four intensive properties of these elements. *(c)* Cite four of their common uses.
21. Without looking at the Periodic Table identify the period, block, and group in which each of the following elements is located: *(a)* $[Ar]\ 3d^2 4s^2$ *(b)* $[Ar]3d^5 4s^1$ *(c)* $[Kr]4d^{10}5s^1$ *(d)* $[Xe]4f^{14}5d^9 6s^1$.

22. Without looking at the periodic table write the expected outer electron configuration for each of the following elements: *(a)* period 4, Group 7 *(b)* period 5, Group 3 *(c)* period 6, Group 12.

23. *(a)* Write the group configuration notation for each *p*-block group. *(b)* How does group number relate to the number of electrons in the highest occupied energy level?

24. Indicate the numbers of *s* and *p*-block electrons in the highest-energy level of the elements in each of Groups 13–18.

25. *(a)* What types of elements make up the *p* block? *(b)* How do the properties of the *p* block metals compare with those in the *s* and *d* blocks?

26. *(a)* Which elements are designated as the halogens? *(b)* List three of their characteristic properties. *(c)* Cite four of their common uses.

27. *(a)* Which elements are designated as the noble gases? *(b)* What is the most significant property of these elements? *(c)* What use is generally made of that property?

28. *(a)* Which elements are metalloids? *(b)* Describe their characteristic properties. *(c)* List some of their typical uses.

29. Without looking at the Periodic Table identify the period, block, and group in which each of the following is located: *(a)* $[Ne]3s^23p^4$ *(b)* $[Kr]4d^{10}5s^25p^2$ *(c)* $[Xe]4f^{14}5d^{10}6s^26p^5$.

30. *(a)* Which elements make up the *f*-block in the periodic table? *(b)* Describe the characteristic properties and primary uses of these elements.

31. For each of the following identify the block, period, group, group name (where appropriate), element name, element type, and relative reactivity: *(a)* $[Ne]3s^23p^1$ *(b)* $[Ar]3d^{10}4s^24p^6$ *(c)* $[Kr]4d^{10}5s^1$ *(d)* $[Xe]4f^26s^2$ *(e)* $[Rn]7s^1$.

32. *(a)* What are the main-group elements? *(b)* What trends can be observed across the various periods within the main-group elements?

33. *(a)* What is meant by atomic radius? *(b)* What trend is observed among the atomic radii of main-group elements across a period? *(c)* How can this trend be explained?

34. *(a)* What trend is observed among the atomic radii of main-group elements down a group? *(b)* How can this trend be explained?

35. Identify the elements with the largest and the smallest atomic radii within each set listed: *(a)* Li, C, and O *(b)* Mg, Ba, and Ca *(c)* Na, P, and K Si, Sn, and I *(e)* the Group-1 elements *(f)* the Period-2 elements *(g)* Li, B, Cs, N, and C.

36. Define each of the following: *(a)* ion *(b)* ionization *(c)* ionization energy.

37. *(a)* How do the first ionization energies of main-group elements vary across a period and down a group? *(b)* Explain the basis for each trend.

38. *(a)* Distinguish among the first, second, and third ionization energies of an atom. *(b)* How do the values of successive ionization energies compare? *(c)* Why does this occur?

39. *(a)* Without looking at the ionization energy table, arrange the elements listed in order of decreasing first ionization energies: *(a)* Li, O, C, K, Ne, and F. *(b)* Which would you expect to have the highest second ionization energy? Why?

40. *(a)* What is electron affinity? *(b)* What signs are associated with electron affinity values and what are their significance?

41. Within each part of question 37 (a–f only), identify the element with the highest electron affinity.

42. Without looking at the electron affinity table, arrange the elements listed in order of *decreasing* electron affinities: C, O, Li, Na, Rb, Ne, and F.

43. *(a)* Distinguish between a cation and an anion. *(b)* How do the sizes of each compare with those of their neutral atoms from which they were formed?

44. What causes chemical compounds to form?

45. *(a)* What are valence electrons? *(b)* Where are such electrons generally located?

46. How many valence electrons are contained within each group of main-group elements?

47. For each of the following groups, indicate whether electrons are more likely to be lost or gained in compound formation and the number of such electrons typically involved: *(a)* Group 1 *(b)* Group 2 *(c)* Group 13 *(d)* Group 16 *(e)* Group 17 *(f)* Group 18.

48. *(a)* What is electronegativity? *(b)* How are electronegativity values assigned to the periodic table elements?

49. *(a)* Identify the most and least electronegative group or groups of elements on the periodic table. *(b)* Identify the most and least

electronegative elements.

50. Within each part of question 37 (a–g), which element would be expected to have the highest electronegativity?

51. What trends are generally observed among d- and f-block elements in terms of *(a)* atomic radii *(b)* ionization energy *(c)* ion formation *(d)* electronegativity across periods?

Application Questions

1. Explain how the presence of empty spaces in Mendeleev's periodic table eventually contributed to chemists' acceptance of that table.

2. *(a)* Explain what is meant by the periodicity of elements in the Periodic Table. *(b)* What is the basis for that periodicity?

3. Explain the group and block placements of hydrogen and helium in the Periodic Table.

4. What information is provided by the number of the group in which a p-block element appears in the Periodic Table?

5. Explain the placement of the f-block elements on the Periodic Table.

6. What is the relationship between first ionization energy and *(a)* the high reactivity of the alkali metals? *(b)* the lack of reactivity of the noble gases?

7. Among the elements Mg, P, Cl, and Na, which would be expected to have the highest: *(a)* first ionization energy *(b)* second ionization energy *(c)* third ionization energy?

8. For each element listed below, determine the charge of the ion most likely to be formed and the identity of the noble gas whose electron configuration is thus achieved: *(a)* Li *(b)* B *(c)* O *(d)* F *(e)* Mg *(f)* Al *(g)* P *(h)* S *(i)* Br *(j)* Ba.

9. Which among the following does not have a noble gas configuration: Na^+, Rb^+, O^{2-}, Br^-, Ca^+, Al^{3+}, S^{2-}?

10. Which of the following ions is least likely to form: I^-, Sr^{2+}, Al^{3+}, Cl^-, O^{2-}, K^{2+}?

11. The two ions K^+ and Ca^{2+} each have 18 electrons surrounding the nucleus. Which would you expect to have the smaller radius? Why?

12. Without referring to specific table values, use the main-group trends to make the requested predictions relative to the neutral atoms I–V described below:
$$\text{I}—1s^2\ 2s^2\ 2p^6\ 3s^2 \qquad \text{IV}—1s^2\ 2s^2\ 2p^5$$
$$\text{II}—1s^2\ 2s^2\ 2p^6\ 3s^1 \qquad \text{V}—1s^2\ 2s^2\ 2p^3$$
$$\text{III}—2s^2\ 2s^2\ 2p^6$$
(a) Which atom (I–V) has the largest atomic radius? *(b)* Which is a noble gas? *(c)* Which has all of its p orbitals half-filled? *(d)* Which would to have the lowest first ionization energy? *(e)* Which would have the highest first ionization energy? *(f)* Which would have the highest electron affinity? *(g)* Which would be expected to have the highest second ionization energy? *(h)* Which would be expected to have the highest electronegativity? *(i)* List the elements in order of increasing ionization energies. *(j)* Which would be expected to most readily form a 1+ ion? *(k)* Which would most readily form a 1− ion?

13. The three elements X, Y, and Z have successive atomic numbers, each increasing by one in the order given. Atoms of element X form stable ions with the formula X^-, and atoms of element Z form stable ions with the formula Z^+. Which of the following statements is false concerning elements X, Y, and Z? *(1)* A neutral atom of element Y has one more electron than a neutral atom of element X, but one less electron than a neutral atom of element Z. *(2)* Element X could be in Group 17. *(3)* Element Z could be in Group 1. *(4)* Elements X, Y, and Z could all be in the same group of the Periodic Table. *(5)* Elements X, Y, and Z could be those elements with atomic numbers 9, 10, and 11, respectively.

Enrichment

1. Prepare a report tracing the evolution of the current periodic table. Cite the chemists involved and the major contribution of each.

2. Write a brief biographic sketch of Dmitri Mendeleev and of his work in developing the first Periodic Table.

3. Describe the experimental procedures used by Henry Gwyn-Jeffreys Moseley in determining the atomic number order of the Periodic Table.

4. Write a report describing the contributions of Glenn Seaborg toward the discovery of many of the actinide elements.

5. Trace the source of the names of each of the actinide elements beyond uranium.

Chapter 6   Chemical Bonding

Chapter Planner

6.1 Introduction to Chemical Bonding
Types of Chemical Bonds
- Questions: 1–6
- Problems: 1, 2
- Application Question: 1
- Application Problems: 1, 2

Why Chemical Bonding Occurs
- Demonstration: 1
- Question: 7

6.2 Covalent Bonding and Molecular Compounds
Formation of Covalent Bonds
- Demonstrations: 2, 3
- Questions: 8, 11
- Questions: 9, 10
- Problem: 3
- Application Questions: 2, 3

The Octet Rule
- Chemistry Notebook

Lewis Structures
- Questions: 12–15
- Problems: 4, 5
- Chemistry Notebook

Multiple Covalent Bonds
- Questions: 16, 19
- Application Problem: 3
- Questions: 17, 18

Polyatomic Ions
- Question: 20
- Application Question: 4
- Application Problem: 4

6.3 Ionic Bonding and Ionic Compounds
- Questions: 21, 22

Formation of Ionic Bonds
- Problem: 6
- Application Problem: 5
- Question: 23

A Comparison of Ionic and Molecular Compounds
- Experiments: 23, 42
- Question: 24

6.4 Metallic Bonding
Formation of Metallic Bonds
- Questions: 25–27

Nature of Metals
- Demonstration: 4
- Application Question: 5

6.5 The Properties of Molecular Compounds
VSEPR Theory
- Demonstration: 5
- Questions: 28–32
- Problem: 7
- Application Question: 6
- Application Problem: 6
- Application Problem: 7

Hybridization
- Questions: 33, 34
- Application Question: 7

Intermolecular Forces
- Questions: 35–37, 39
- Application Question: 9
- Question: 38
- Application Questions: 8, 10

Teaching Strategies

Introduction

Remind students that in Chapter 1 they learned that compounds, and the elements making them up, are vastly different from one another. In Chapters 3 and 4, we dealt with neutral atoms of elements and their configurations. In this chapter, we are going to see what happens to those atoms when they form compounds and *why* this happens, and it lays the basis for formula writing and the naming of compounds.

Your overall goals for this chapter are to discuss *(1)* the three types of bonds found in elements and compounds—ionic, covalent, and metallic bonds, *(2)* why these bonds form, *(3)* the types of particles that result from each type of bond, *(4)* the types of forces these particles exert on one another, and *(5)* the characteristics exhibited by substances with each type of bond.

6.1   Introduction to Chemical Bonding

Students sometimes have difficulty understanding how a compound with a percent composition of 50% can be polar covalent and one with 51% can be ionic. Electronegativity, percent of ionic character, and the 0-4%, 5-50%, 51-99% cut-off points are convenient indicators to give an idea of the type of bonds we are dealing with. To help students understand the concept of chemical bonding, have them visualize the following situation: Take a given space and put black at one end and white at the other. Then, starting with the black, gradually and imperceptively fade it into dark grey, medium grey, lighter grey until it blends into white. Could we draw a line in the middle and say everything to the left is black and everything to the right is white—even though we know it is all different gradations of grey? Of course not! This is similar to what we are dealing with in ionic and covalent bonding. Using these cutoff points helps us to put things in neat and useful compartments that aren't really as clear-cut as they may seem at first.

Most bond formation is an energy-releasing process, or exothermic in nature. The following demonstration involves the formation of an ionic bond, as MgO is made, and a covalent bond, as SO_2 is made.

Demonstration 1: Bond Formation as an Exothermic Reaction
PURPOSE: To show the energy changes that occur in the formation of ionic bonding and covalent bonding
MATERIALS: Bunsen burner, coffee can—3 lb, deflagrating spoon, magnesium ribbon, sulfur, tongs

PROCEDURE: Part 1: (This is a dramatic twist to the burning of magnesium ribbon.) Goggles and a lab coat should be worn. Darken the room as much as possible. CAUTION: Remind students not to look directly at the magnesium ribbon when it ignites. Holding a 6-cm piece of Mg ribbon with tongs, ignite it in a burner flame and immediately put it in a coffee can while it continues to burn. Students will see the reflected light produced as the Mg burns. Describe the energy change that occurred here. (Energy in the form of light was definitely released.) What kind of bond was formed? (Tell students that the magnesium reacts with both the oxygen and the nitrogen in the air, thus two different compounds are formed in this reaction.) Have students figure out that both compounds have ionic bonds by using electronegativities.

$$2Mg + O_2 \rightarrow 2MgO \text{ and } 3Mg + N_2 \rightarrow Mg_3N_2$$

Part 2: Put a small piece of sulfur in a deflagrating spoon, and ignite it under a hood. Students will see that it burns with a blue flame. Describe the energy change that occurred here. (Energy in the form of light was given off.) What kind of bond was formed? (Polar covalent) Remind students that most, but not all, bond formation is exothermic.

6.2 Covalent Bonding and Molecular Compounds

In 1932, Langmuir received the Nobel Prize in chemistry for his work on monomolecular films, which are films that are assumed to have a thickness of one molecule. Using a method similar to his makes it possible to determine the approximate size of a molecule. This is also an excellent demonstration because the size is determined by indirect measurement. Before doing the demonstration, go over the formula for determining the volume of a cylinder, $V = \pi r^2 h$. Point out that if you knew the volume of a cylinder and its diameter, d, you could easily determine its height or thickness, $h = V/\pi r^2$. Do a couple of sample problems determining the thickness of a cylinder when its volume and diameter are known. Hold up a piece of filter paper and ask students if this is a cylinder. This helps develop the idea that cylinders can be very short.

Demonstration 2: Size of a Molecule
PURPOSE: To learn how the size of a molecule can be determined experimentally
MATERIALS: Cafeteria tray (approximately 30 cm × 40 cm), a 1-mL pipet (a 10-mL graduate and an eyedropper can be substituted for the pipet), a plastic ruler, distilled water, 95% ethyl alcohol lycopodium powder or fine talcum powder, oleic acid
PROCEDURE: Prior to class: (1) Clean all equipment several times with detergent, and rinse well with water. *Clean, dry equipment is a must!* (2) Dissolve 5 mL of oleic acid in 95 mL of ethyl alcohol. Make a second solution by dissolving 5 mL of the first solution in 45 mL of ethyl alcohol. The second solution, which will be the one used in the demo, should contain 0.005 mL of oleic acid/mL of solution. To determine the volume of oleic acid/drop, use the pipet and count the number of drops/mL of solution. Volume of oleic acid/drop = 0.005 mL/number of drops per mL of solution. (Assuming 20 drops/mL, the volume of oleic acid/drop would be 0.0025 mL.) (3) Refrigerate sufficient distilled water one day prior to the demonstration. Add enough

distilled water to cover the tray to a depth of 0.5 cm. Put a small amount of lycopodium or talcum powder on a piece of paper; gently blow the powder over the water. Examine the water surface carefully to make sure that there is only a thin dusting of powder spread across the water. Now add one drop of the second oleic acid solution at the center of the tray. Gently tap the tray so that the oleic acid film is spread out over the largest possible area, without being broken. The oleic acid layer should assume a roughly circular shape. If the equipment is dirty, an irregular, star-shaped figure may result. In this case, clean the tray and start over. After waiting several minutes for the alcohol to evaporate, make several measurements of the diameter of the oleic acid film in cm and take an average of these readings. Using 4 = average diameter/2, calculate the thickness of the film; this should turn out to be somewhere in the area of 10^{-7} cm. Assuming it is monomolecular, the thickness will be the diameter of an oleic acid molecule. Even if the demonstration does not turn out perfectly, the students will get the idea.

Demonstration 3: Bond Energies of the Hydrogen Halides
PURPOSE: To help students understand bond energies and, by using models, to show students how to use the Bond Length and Bond Energy Tables to explain why certain reactions take place
MATERIALS: Bond Length and Bond Energy Table, small gumdrops—same color, large gumdrops in 3 colors, round or heavy toothpicks, tape
PROCEDURE: (1) As you propose the following problem to the students, write on the chalkboard the information about each hydrogen halide. (2) As you discuss each molecule, have students assemble a gumdrop-toothpick model of it. Each model of a hydrogen halide should consist of a small gumdrop representing hydrogen and a large gumdrop for the halide. Make each halogen a different color. Break 2 cm off of each end of four toothpicks and tape them together for the H-Cl bond; break 1 cm off each end of two toothpicks and tape for the H-Br bond; and use one full length toothpick for the H-I bond. (See Figure T6-1.) To show bonds breaking, let students break the toothpick bonds and determine which is easiest and hardest to break—indicating which needs the least and most bond energy. Also have students identify how the ease of breaking relates to the length of the toothpick.

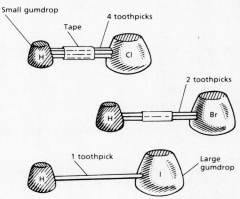

Figure T6-1

(3) A problem that shows differences in bond energies follows: When a heated glass rod is thrust into a test tube of $HCl(g)$, the HCl shows no signs of decomposing. Continuous external heating of the HCl also gives the same results. When a heated glass rod is thrust into a test tube of $HBr(g)$, there is no reaction; but continuous external heating results in the production of red-brown Br_2, indicating the breaking of the H-Br bonds. When a heated glass rod is put into HI, a violet-colored gas is produced, indicating the presence of I_2 vapor due to the breaking of the H-I bonds. Why do you think the bonds of the HI molecules were broken by the hot glass rod and those of the HCl and HBr were not? Hint: Look at Table 6-1 Bond Lengths and Bond Energies. What is the bond energy of the H-Cl bond? Of the H-Br bond? Of the H-I bond? Explain. To introduce the idea of unequal sharing (polar covalent bonds), personalize the atoms a little. You can explain that chlorine promised *faithfully* to share her electron with hydrogen, but she never intended to share *equally*. Chlorine is a pig! (it has a high electronegativity and thus strong pull for e^-). She keeps her electron and hydrogen's a greater percentage of the time. Then she says, "Here, Hydrogen, I'll let you have my electron and yours for just a little while, but I want them right back." Since hydrogen is weaker (weak pull for e^- due to a lower electronegativity), there's nothing he can do. Remind students that what is really true is that the density of the electron cloud is greater near the chlorine and less near the hydrogen.

A spectacular demonstration of the formation of a covalent bond can be found in Shakhashiri, "Reactions of Zinc and Iodine," *Chemical Demonstrations*, Vol. 1, pp. 49–50.

6.3   Ionic Bonding and Ionic Compounds

Use the following drawing on a transparency or chalkboard to help review Lewis structures and as a visual reference for the following explanation of ionic bonding.

A successful means of introducing ionic bonding is with talking atoms. Referring to the above drawing, tell the students that if these atoms could talk, here's what they might say: (*Include a little dramatics.*)

Na: "Drat, if I could get rid of this one stupid electron in my 3rd energy level, the 2nd energy level would be my outermost energy level and I would be a happy atom—I mean chemically stable with a complete outer energy level of eight electrons! I'd sell it cheap!" (*ionization energy*)

Cl: "If I could just get one more electron, I would have a complete outer energy level. I'd be willing to pay lots for it!" (*electron affinity*)

[Na is walking along and meets Cl.]

Na: "Hi, Chlorine, do you know anyone who would like an electron?"

Cl: "Do I ever! I'll take it. It's all I need to complete my outer energy level!"

Na: "Well, then, here it is, Chlorine. Now I'm positively charged, and I feel so stable!" (Using the energy level diagram, show why this is so. Count the number of electrons and compare it to the number of protons in the nucleus.)

Cl: "Thanks, Sodium. I'll take good care of it. Now I'm negatively charged, and do I ever feel stable."

Na: "Say, do you know what, Chlorine? Suddenly I feel very attracted to you."

Cl: "Sodium, I feel exactly the same way."

[Na and Cl move together]

Both: "Ah, togetherness!"

You could make this into a student skit. However you do it, when your atoms start talking, your students start listening! To help the students fully visualize what happened in this process, add the following diagram to the first one.

A dramatic demonstration of ionic bonding is found in Shakhashiri, "Reaction of Sodium and Chlorine," *Chemical Demonstrations*, Vol. 1, pp. 61–63.

To help students distinguish between ionic, polar covalent, and nonpolar covalent bonds, recreate the following chart:

Types of Bonds Compared to Candy Bar Exchange and Sharing	
Type of Bond	
Ionic	Generous George has a candy bar he really doesn't want. Grabby Gertie, who already has one candy bar but wants more, gives him a little kiss, and he readily hands it over to her.
Pure or Nonpolar	Considerate Carl has a chocolate bar, and Caring Carol has a caramel nut bar. Carl shares half of his chocolate bar with Carol, and Carol shares half of her caramel nut bar with Carl. EQUAL SHARING!
Polar	Tricky Tony has a chocolate bar, and Sweet Sue has a coconut bar. Tony shares ONLY $\frac{1}{4}$ of his chocolate bar with Sue and keeps $\frac{3}{4}$ for himself. Sue gives Tony $\frac{3}{4}$ of her coconut bar and keeps only $\frac{1}{4}$ for herself. UNEQUAL SHARING!

6.4   Metallic Bonding

Demonstration 4: Structure of Metals
PURPOSE: To use a model to show how one plane of ions in metallic solids slide over another one without encountering resistance or breaking of bonds

MATERIALS: Gumdrops, toothpicks
PROCEDURE: To make the model, construct each layer by connecting 16 gumdrops with toothpicks—producing a layer of 4 gumdrops × 4 gumdrops. Make several layers and stack them together but do not connect the layers. Then you can show the students how the layers slide over one another when a horizontal force is applied to them. Since no resistance or breaking of bonds occurs, the structure does not shatter.

6.5 The Properties of Molecular Compounds

As you discuss the different theories explaining the shapes of molecules, use models to enhance your discussion. The models suggested below not only show shape, but also show position of lone pair electrons.

Demonstration 5: The Shapes of Molecules
PURPOSE: To help understand molecular polarity by using models to show the shapes of molecules and position of lone pair electrons in them
MATERIALS: Small colored marshmallows, large white marshmallows, Red Hots,® toothpicks, 2 oblong balloons
PROCEDURE: (1) To make a water molecule, use a large marshmallow for the oxygen and attach two small ones for the hydrogens, as shown below. To show the position of the two lone pair electrons, moisten four Red Hots®and stick them to the end

of the large marshmallow opposite the hydrogens. The presence of the lone electrons on models should help students better understand how that end can be negative.

(2) For ammonia use one large marshmallow for nitrogen, three small marshmallows for the hydrogens, and two Red Hots®for the lone electron pair.

(3) A very simple way to show the tetrahedral shape of methane is to blow up two oblong balloons, twist them in the center, twist them around one another in the center, and let them flip to their most stable position.

Desktop Investigation: [Note: Models of Chemical Compounds in Chapter 7 could be used here.]

References and Resources

Books and Periodicals

Halpern, March. "A Simple Inexpensive Model for Student Discovery of USEPR," *Journal of Chemical Education*, Vol. 56, No. 8, Aug. 1979, p. 531.

Jense, William B. "Abegg, Lewis, Langmuir and the Octet Rule," *Journal of Chemical Education*, Vol. 61, No. 3, March 1984, pp. 191–200.

Kapelios, S. and A. Mavrides. "Electronegativity: A Mnemonic Rule," *Journal of Chemical Education*, Vol. 64, No. 11, Nov. 1987, p. 941.

Mickey, Charles. "Molecular Geometry," *Journal of Chemical Education*, Vol. 57, No. 3, March 1980, pp. 210–12.

Pauling, Linus. "C.N. Lewis and the Chemical Bond," *Journal of Chemical Education*, Vol. 61, No. 3, March 1984, pp. 201–3.

Sanderson, R.T. "A Needed Replacement for the Customary Description of Chemical Bonding," *Journal of Chemical Education*, Vol. 59, No. 5, May 1982, p. 376.

Sanderson, R.T. "Principles of Electronegativity, Part 1, General Nature 1," *Journal of Chemical Education*, Vol. 65, No. 2, Feb. 1988, pp. 112–8.

Audiovisual Resources

Chem Study—Chemical Bonding, and Shapes and Polarities of Molecules (16mm film or VHS-video) Central Scientific Co., 1988

Chemistry 104—Bonding of Atoms, sound filmstrip, and *Chemistry 109—Forces in Solids, Liquids, and Gases*, VHS cassette or sound filmstrip, Learning Arts, 1988.

Experiments in Chemistry Series—The Size of Molecules, filmstrip, Educational Images, Ltd., 1988.

Introduction to Chemistry—Ionic and Covalent Bonds, filmstrip, Fisher Scientific-EMD, 1988.

Solo-Learn—Polar Covalence, Introduction to Chemical Bonding, and *Bond Type and Properties of Matter*, sound filmstrips, Central Scientific Co., 1988.

Computer Software

Atomic Structure 3, Apple II 64K and TRS-80 48K, Queue, 1988.

General Chemistry 1A—Chemical Bonding and *Solids and Liquids*, 128 K, Queue, 1988.

Chemistry Courseware—Chemical Bonding, Apple II 48K, Barclay, 1988.

Chemistry Series—Bonding in Molecules and *Bonding between Molecules*, Apple II 48K and TRS-80 32K, Barclay, 1988.

Chemistry Help Series—Atomic Structure and Bonding, Apple II 48K, Barclay.

Fee, Richard. *Bonding*, Apple 48K. Chemistry According to ROF, 1988.

SEI Physical Chemistry, Apple 48K and IBM PC, 238K, Queue, 1988.

VideoChem, Apple 64K and IBM PC, 128 K, Queue, 1988.

Answers and Solutions

Questions

1. *(a)* A chemical bond is a link between atoms resulting from the mutual attraction of their nuclei and electrons. *(b)* Chemical bonds are classified by how valence electrons are distributed around the nuclei of the combined atoms.

2. Metals tend to lose electrons to form positive ions, whereas nonmetals tend to gain electrons to form negative ions.

3. The two major bonding types are ionic and covalent. An ionic bond is the chemical bond resulting from electrostatic attraction between positive and negative ions. A covalent bond is a chemical bond resulting from the sharing of electrons between bonded atoms.

4. In general, the greater the electronegativity difference between two atoms is, the more ionic is the bond between them.

5. *(a)* Polar refers to bonds that have an uneven distribution of charge. *(b)* A polar covalent bond is a covalent bond in which there is an unequal attraction for the shared electrons. A nonpolar covalent bond is a covalent bond in which the bonding electrons are shared equally by the bonded atoms.

6. Typical bonding types in each case would be: *(a)* covalent bonds *(b)* metallic bonds *(c)* ionic bonds *(d)* polar covalent bonds.

7. *(a)* In general, atoms will form a chemical bond if their potential energy is lowered by the change. *(b)* Chemical-bond formation is an energy-releasing process.

8. *(a)* A molecule is a group of two or more atoms held together by covalent bonds and able to exist independently. *(b)* A diatomic molecule is a molecule containing two atoms. *(c)* A molecular compound is a compound composed of molecules. *(d)* A chemical formula is a shorthand representation of the composition of a substance, using atomic symbols and numerical subscripts. *(e)* A molecular formula shows the types and numbers of atoms combined in a single molecule.

9. *(a)* Bond length is determined by the distance at which potential energy is at a minimum; it is the point at which balance is attained between attraction and repulsion in a stable covalent bond. *(b)* Bond energies become larger as bond lengths become shorter.

10. The electrons in a covalent bond occupy overlapping orbitals; each electron is free to move around in either of the orbitals, but both generally spend more time in the space between the two nuclei.

11. The formulas are F_2, Cl_2, Br_2, I_2, and At_2.

12. An unshared pair, or lone pair, is a pair of electrons that is not involved in bonding, but instead belongs exclusively to one atom.

13. *(a)* A structural formula indicates the kind, number, arrangement, and bonds of the atoms in a molecule. *(b)* Such a formula is most useful when unshared pairs are not of interest. *(c)* The formula is F-F.

14. The indicated number of valence electrons in each is: *(a)* 1 *(b)* 7 *(c)* 2 *(d)* 6 *(e)* 3 *(f)* 5 *(g)* 4 *(h)* 8.

15. Three rules that are used in writing Lewis structures are: (1) the least electronegative atom is the central atom; (2) hydrogen and fluorine are always connected to only one other atom; and (3) halogen atoms often surround a central atom.

16. A single-covalent bond is a covalent bond between two atoms produced by the sharing of one pair of electrons, such as in H_2O. A double-covalent bond is a covalent bond between two atoms produced by the sharing of two pairs of electrons, such as in C_2H_4. A triple-covalent bond is a covalent bond between two atoms produced by the sharing of three pairs of electrons, such as in N_2.

17. Atoms of carbon and nitrogen often enter into double or triple bonds.

18. *(a)* Multiple bonds are double or triple bonds. *(b)* Triple bonds have the greatest bond energies, followed by double bonds, and then single bonds. Single bonds are longest, followed by double bonds, and then triple bonds.

19. The need for a multiple bond is evident if there are not enough valence electrons to complete octets by adding unshared pairs.

20. *(a)* A polyatomic ion is a charged group of covalently bonded atoms. *(b)* NH_4^+ and ClO_3^- are examples. *(c)* Polyatomic ions are normally found combined in chemical compounds with ions of opposite charge.

21. *(a)* An ionic compound is composed of positive and negative ions combined so that the positive and negative charges are equal. *(b)* Most ionic compounds occur as solids composed of crystals.

22. *(a)* A formula unit is the simplest unit indicated by the formula of an ionic compound. *(b)* One formula unit of calcium fluoride (CF_2) consists of one calcium ion plus two fluoride ions. *(c)* One formula unit would consist of two potassium ions and one sulfide ion (K_2S).

23. *(a)* Lattice energy is the energy released when one mole of an ionic crystalline compound is formed from gaseous ions. *(b)* The greater the lattice energy is, the stronger is the ionic bond.

24. *(a)* Ionic compounds generally have higher melting and boiling points than do molecular compounds, but they do not tend to vaporize as readily at room temperature as do molecular compounds. *(b)* The differences in their properties are generally due to differences in how strongly the individual particles are held together in the compounds. *(c)* Ionic compounds are hard and brittle crystalline solids. In the molten state or dissolved in water they conduct electricity. Many are soluble in water.

25. *(a)* Metals are better conductors of heat. They are more easily deformed than many solid molecular or ionic compounds, and they are electrical conductors in the solid state. *(b)* The unusual electrical conductivity of metals in the solid state is due to the presence of highly mobile electrons.

26. Metal atoms contain a smaller number of valence electrons than of vacant highest-energy-level orbitals, they have low ionization energies, and they have low electronegativities.

27. *(a)* A metallic bond is a chemical bond resulting from the attraction between positive ions and surrounding mobile electrons. *(b)* The strength of a metallic bond varies with the nuclear charge and the number of valence electrons.

28. The properties of molecules depend upon the types and

arrangement of the atoms bonded together in those molecules.

29. (a) The VSEPR theory for predicting molecular geometry is based on the assumption that electrons in molecules repel each other. (b) The letters VSEPR stand for "Valence-Shell Electron-Pair repulsion," and refer to the repulsion between pairs of valence electrons in the highest occupied energy level, or valence shell. (c) The VSEPR theory states that electrostatic repulsion between the electron pairs surrounding an atom causes these pairs to be oriented as far apart as possible.

30. (a) According to the VSEPR theory, molecules are classified according to how many electron pairs surround a central atom. (b) Both would be expected to be linear.

31. According to the VSEPR theory, the following molecular geometries are indicated: (a) linear (b) triangular planar (c) tetrahedral (d) triangular bipyramidal (e) octahedral.

32. (a) The position of unshared electron pairs does not determine the shape of the molecule, although the presence of such pairs may somewhat alter the bond angles. (b) In predicting molecular shapes, double bonds are treated in the same way as single bonds.

33. (a) Hybridization is the mixing of two or more atomic orbitals of similar energy in the same atom to give new orbitals of equal energies. (b) When orbitals hybridize ("hybridize" means "mix."), they combine and become rearranged to form new, identical orbitals.

34. (a) Hybrid orbitals are orbitals of equal energy produced by the combination of two or more orbitals in the same atom. (b) The number of hybrid orbitals produced is always equal to the number of orbitals that have combined.

35. Since the boiling point is the temperature at which adjacent particles in a liquid have enough energy to pull away from each other and enter the gas phase, the boiling point is a measure of the strength of the intermolecular forces within that substance. The higher the boiling point is, the stronger are the intermolecular forces.

36. (a) Intermolecular forces are the forces of attraction between molecules. (b) These forces are generally weaker than those involved in ionic and metallic bonding. (c) The strongest intermolecular forces act between polar molecules.

37. (a) Dipole-dipole forces are forces of attraction between polar molecules. (b) The polarity of a molecule depends upon both the polarity of the individual bonds and on how the bonds are oriented with respect to each other.

38. (a) An induced dipole is one produced in a nonpolar molecule when its electrons are momentarily attracted by a polar molecule. (b) These forces account for the solubility of nonpolar molecular oxygen in water, a process necessary for aquatic life.

39. (a) Hydrogen bonding is the intermolecular attraction between a hydrogen atom bonded to a strongly electronegative atom (F, O, or N) and an unshared pair of electrons on a strongly electronegative atom (F, O, or N). (b) A hydrogen bond is strong because the great electronegativity difference between an atom of H and one of F, O, or N bonded to it makes the bond highly polar. The hydrogen atom thus has a positive charge approaching that of a proton, and this, coupled with the small size of the hydrogen atom, allows it to attract quite strongly and approach closely an unshared pair of electrons on an adjacent molecule.

40. (a) London dispersion forces are intermolecular attractions resulting from the constant motion of electrons and the creation of instantaneous dipoles and induced dipoles. (b) London forces increase with increasing masses of atoms or molecules.

Problems

1.

Bond	Electronegativity difference	Bond type	More negative atom
(a) H and I	2.5 − 2.1 = 0.4	Polar covalent	I
(b) Ba and S	2.5 − 0.9 = 1.6	Polar covalent	S
(c) K and Br	2.8 − 0.8 = 2.0	Ionic	Br
(d) Zn and O	3.5 − 1.6 = 1.9	Ionic	O
(e) At and I	2.5 − 2.2 = 0.3	Nonpolar covalent	I
(f) S and I	2.5 − 2.5 = 0.0	Nonpolar covalent	Identical

2. The order would be K-Br, Zn-O, Ba-S, H-I, At-I, and S-I.

3. (a) Cl

Cl

(b) O

O

(c) P

P

(d) H

F

4. (a) Li· (b) ·Ca· (c) :Cl: (d) :Ö: (e) ·Ċ· (f) ·Ṗ:

(g) ·Äl· (h) :S:

5. (a)

:F:
:F:C:F: or   :F-C-F:
:F: :F:

(b) H:Se:H   or   H-Se-H

(c)

:Ï: :Ï:
:Ï:N:Ï: or   :Ï-N-Ï:

(d)

:Br: :Br:
:Br:Si:Br: or   :Br-Si-Br:
:Br: :Br:

(e)

H   H
H:C:Cl: or   H-C-Cl:
H   H

6. (a) K· + :F: → K⁺ + :F:⁻

(b) ·Mg· + :Br: + :Br: → Mg²⁺ + :Br:⁻ + :Br:⁻

(c) ·Äl· + :Ï: + :Ï: + :Ï: → Al³⁺ + :Ï:⁻ +
:Ï:⁻ + :Ï:⁻

7. *(a)* :I̤: I̤: (A₂, linear) → $:\ddot{I}:\ddot{I}:$ (A$_2$, linear)

(b) A H:B̤r̤: (AB, linear) → H$:\ddot{B}r:$ (AB, linear)

(c) :F̤: F̤:Ö: F̤: (AD₄, tetrahedral)

$$:\ddot{F}:\;\;\;\ddot{O}\;\;\;:\ddot{F}:$$

with F above and below O (AD$_4$, tetrahedral)

(d) Ö: :C: :Ö (AB₂, linear) → $\ddot{O}::C::\ddot{O}$ (AB$_2$, linear)

(e) :C̤l̤ :B: C̤l̤: (AB₃, triangular planar) with :C̤l̤: below

Application Questions

1. Bonds between unlike atoms are never completely ionic and rarely completely covalent because different atoms attract electrons to different extents based on their electronegativities.

2. As two atoms approach, attraction between opposite charges (the protons of each atom for the electrons of the other) causes a decrease in potential energy. Repulsion between like charges (protons repelling protons and electrons repelling electrons) causes an increase in potential energy. At a point where the attraction is at a maximum and potential energy is at a minimum, a stable molecule is formed.

3. At first, the distance between the hydrogen atoms is large enough that they have no influence on each other, and the potential energy at this point is arbitrarily set at zero. As the atoms approach, the mutual attraction between the electrons of each for the protons of the other dominates over the repulsion between the like charges and causes the potential energy to decrease. As the atoms continue to approach, a point is reached where the potential energy is at a minimum. This point, at the bottom of the potential-energy curve, is where a stable hydrogen molecule forms. If the atoms approach more closely, a sharp rise in potential energy occurs because repulsion then exceeds attraction.

4. Resonance refers to the bonding in molecules that cannot be correctly represented by a simple Lewis structure.

5. *(a)* The motion of the free electrons through the solid accounts for the high electrical conductivity of metals. *(b)* The free electrons can very readily transport heat energy throughout the metal. *(c)* Metals are malleable and ductile because metallic bonding is not directional, but uniform throughout the solid. Unlike layers of ionic crystal, planes of ions in a metal can slide past one another without breaking bonds. *(d)* To reflect light and appear lustrous, a material must be able to absorb and re-emit light of many wavelengths. Since the electrons in the electron sea have a wide range of possible energies, metals can absorb and remit light over a wide range of wavelengths.

6. According to the VSEPR theory, the following geometries are indicated: *(a)* triangular pyramidal *(b)* bent *(c)* bent.

7. The carbon atom in CH_4 contains four valence electrons, two paired in the $2s$ orbital and two unpaired in $2p$ orbitals. Hybridization of the $2s$ and three $2p$ orbitals produces four new, identical orbitals of equal energy. The resulting four sp^3 orbitals are oriented tetrahedrally, 109.5° apart, and contain one electron. Each orbital then overlaps with the electron-containing orbital of a hydrogen atom to form four covalent bonds.

8. The direction of the dipole is toward: *(a)* F *(b)* Cl *(c)* Br *(d)* I.

9. The indicated bonds are: *(a)* nonpolar *(b)* polar *(c)* polar *(d)* nonpolar *(e)* polar *(f)* polar.

10. The molecules are: *(a)* polar *(b)* nonpolar *(c)* nonpolar *(d)* polar *(e)* nonpolar.

Application Problems

1. Al–F = 4.0 − 1.5 = 2.5
   K–Br = 2.8 − 0.8 = 2.0
   C–O = 3.5 − 2.5 = 1.0
   I–I = 2.5 − 2.5 = 0.0
   C–N = 3.0 − 2.5 = 0.5
   (a) K–Br *(b)* Al–F *(c)* C–O and C–N *(d)* I–I

2. The order is I–I, C–N, C–O, K–Br, and Al–F.

3. *(a)* $\ddot{O}::\ddot{O}$ or $\ddot{O}=\ddot{O}$

   (b) $:N::N:$ or $:N\equiv N:$

   (c) $\ddot{O}::Si::\ddot{O}$ or $\ddot{O}=Si=\ddot{O}$

   (d) $:C::O:$ or $:C\equiv O:$

   (e) $\ddot{O}::\ddot{S}:\ddot{O}:$ or $\ddot{O}=\ddot{S}-\ddot{O}:$

4. *(a)* $[:\ddot{O}:H]^-$

   (b) $[:\ddot{O}:\ddot{S}:\ddot{O}:]^{2-}$ with $:\ddot{O}:$ above and below S

   (c) $[:\ddot{O}:\ddot{C}:\ddot{O}:]^{2-}$ with $:\ddot{O}:$ above

   (d) $[:\ddot{O}:\ddot{P}:\ddot{O}:]^{3-}$ with $:\ddot{O}:$ above and below

   (e) $[:\ddot{O}:\ddot{N}:\ddot{O}:]^-$ with $:\ddot{O}:$ above

   (f) $[H:\ddot{C}:\ddot{C}:\ddot{O}:]^-$ with H above and below the first C and $:\ddot{O}$ above the second C

5. *(a)* Na· + Na· + :S̤: → Na⁺ + Na⁺ + $:\ddot{S}:^{2-}$

   (b) ·Ca· + :Ö: → Ca²⁺ + $:\ddot{O}:^{2-}$

   (c) ·Al· + ·Al· + :S̤: + :S̤: + :S̤: →
   Al³⁺ + Al³⁺ + $:\ddot{S}:^{2-}$ + $:\ddot{S}:^{2-}$ + $:\ddot{S}:^{2-}$

6. *(a)* :C̤l̤ :S̤: C̤l̤: (AB₂E₂, bent)

   (b) :I̤: I̤ :P̤: I̤: (AB₃E, triangular pyramidal)

   (c) :C̤l̤ :Ö: C̤l̤: (AB₂E₂, bent)

   (d) H :N̤: C̤l̤: (AB₃E, triangular pyramidal) with H below

:Cl:
(e) :Cl :Si: Cl: (AB$_4$, tetrahedral)
:Br:

(f) O: :N: Cl: (AB$_2$E, bent)

7. (a) [: O :N: O :]$^-$ (AB$_3$, triangular planar)
:O:

(b) [H :N: H]$^+$ (AB$_4$, tetrahedral)
H
H

:O:
(c) [: O :S: O :]$^{2-}$ (AB$_4$, tetrahedral)
:O:

(d) [: O :Cl: O :]$^-$ (AB$_2$E$_2$, bent)

Teacher's Notes

Chapter 6 Chemical Bonding

INTRODUCTION
In this chapter, you will learn how individual atoms share or transfer electrons to form chemical bonds. Three types of chemical bonding are described—covalent, ionic, and metallic. You will also learn what determines the shape and behavior of molecules.

LOOKING AHEAD
As you study this chapter, practice drawing Lewis structures for different molecules and ions. Look for the major differences in the properties of molecular compounds, ionic compounds, and metals, and at how intermolecular forces influence the properties of molecular compounds.

SECTION PREVIEW
6.1 Introduction to Chemical Bonding
6.2 Covalent Bonding and Molecular Compounds
6.3 Ionic Bonding and Ionic Compounds
6.4 Metallic Bonding
6.5 The Properties of Molecular Compounds

Giant rolls of aluminum are readied for shipment. Because of a unique type of bonding that exists between metallic atoms, metals are the only solid substances that can be formed into thin sheets without breaking. How many applications of metals can you think of that use this property?

6.1 Introduction to Chemical Bonding

Isolated and electrically neutral atoms are rare in nature. Ordinarily, only noble-gas atoms exist independently. Atoms of other elements are usually combined with each other or with atoms of different elements.

Atoms are held together by electrostatic attraction between positively charged nuclei and negatively charged electrons. This attraction permits two atoms to be held together by a **chemical bond**, *which is a link between atoms that results from the mutual attraction of their nuclei for electrons.* Chemical bonds are classified by the way in which valence electrons are distributed around the nuclei of the combined atoms.

Types of Chemical Bonds

In Chapter 5 we described how atoms of main group elements can gain or lose electrons to form ions with noble-gas electron configurations. Metals tend to lose electrons to form positive ions, and nonmetals tend to gain electrons to form negative ions. Many chemical compounds are composed of ions; in these compounds the chemical bond is an ionic bond. An **ionic bond** *is the chemical bond resulting from electrostatic attraction between positive and negative ions.* In a *purely* ionic bond, one atom has completely given up one or more electrons, and another atom has gained them—as illustrated for two atoms that each have one unpaired electron at the top in Figure 6-1.

Electrostatics is the study of the properties and behavior of stationary electrical charges.

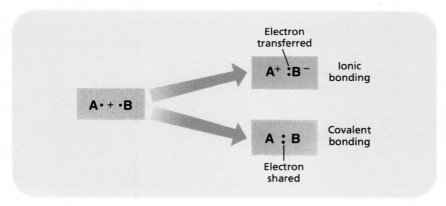

Figure 6-1 In ionic bonding, electrons are transferred from one atom to another, and positive and negative ions are formed. In covalent bonding, an electron pair is shared between two atoms.

In the second major type of chemical bonding, called covalent bonding, neither bonding atom completely loses or gains an electron or electrons. A **covalent bond** *is a chemical bond resulting from the sharing of electrons between two atoms.* A covalent bond in which two electrons are shared is represented by a pair of electron dots, as shown at the bottom right-hand corner in Figure 6-1. In a purely covalent bond, the shared electrons are "owned" equally by the two atoms.

Chemical bonds between unlike atoms are never completely ionic and rarely completely covalent. Bonds can be anywhere in the range between these two extremes, depending upon how strongly the bonded atoms attract electrons.

Electronegativity was introduced in Section 5.3, and electronegativity values are given in Figure 5-20.

Linus Pauling, who received the 1954 Nobel Prize in chemistry for his work on molecular structure, was the first chemist who felt that chemical bonds could show degrees of both a covalent and ionic nature. This led to the development of electronegativity values.

Figure 6-2 The variation of bond type with percent ionic character and the electronegativity difference between the bonded atoms.

The degree to which bonds are ionic or covalent can be estimated by comparing the electronegativities of the bonded atoms. The more two atoms differ in electronegativity, the more ionic the bond between them. In other words, the electrons spend more time close to the bonded atom that attracts them more strongly and hence cause that atom to partially resemble an anion and the other atom, a cation.

Figure 6-2 can be used to classify bonds according to electronegativity differences. The electronegativity of one bonded atom is subtracted from that of the other. For example, the electronegativity difference between a cesium (Cs) atom and a fluorine (F) atom is $4.0 - 0.7 = 3.3$. According to Figure 6-2, a cesium–fluorine bond is an ionic bond. In fact, it is one of the most highly ionic bonds known.

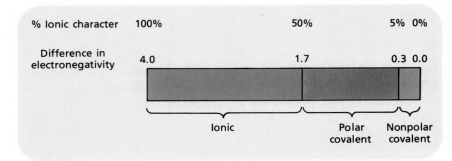

Bonds that have an ionic character of 50% or less are classified as covalent bonds. A bond between identical atoms is completely covalent. Hydrogen, for example, exists in nature not as isolated atoms, but as *pairs* of atoms held together by covalent bonds, H:H. The hydrogen–hydrogen bond has 0% ionic character. It is a **nonpolar-covalent bond**, *a covalent bond in which the bonding electrons are shared equally by the bonded atoms, with a resulting balanced distribution of electrical charge.* Bonds having 0%–5% ionic character, corresponding to electronegativity differences of roughly 0 to 0.3, are generally considered nonpolar-covalent bonds. For example, because the electronegativity difference between hydrogen (H) and boron (B) is 0.1, they form a bond that is essentially nonpolar.

In bonds with significantly different electronegativities, the electrons are more attracted to the more electronegative atom. Such bonds are **polar**, *meaning that they have an uneven distribution of charge.* Covalent bonds having 5%–50% ionic character are classified as polar. A **polar-covalent bond** *is a covalent bond in which the united atoms have an unequal attraction for the shared electrons.*

Nonpolar and polar-covalent bonds are compared in the sketches in Figure 6-3 of the electron density in hydrogen–hydrogen and hydrogen–chlorine bonds. Hydrogen and chlorine atoms combine to produce the compound known as hydrogen chloride (HCl). The electronegativity difference between chlorine and hydrogen atoms is $3.0 - 2.1 = 0.9$, indicating formation of a polar-covalent bond. The electrons in this bond spend more of their time near the more electronegative chlorine atom than near the hydrogen atom, as indicated in Figure 6-3b. Consequently, the chlorine end of the bond has a relative surplus of electrons and a partial negative charge,

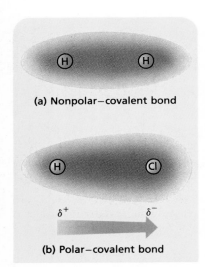

Figure 6-3 Comparison of the electron density in *(a)* a nonpolar, hydrogen–hydrogen bond and *(b)* a polar, hydrogen–chlorine bond. Because a chlorine atom is more electronegative than a hydrogen atom, the electron density in the hydrogen–chlorine bond is greater around the chlorine atom.

indicated by writing δ−. The hydrogen end of the bond then has an equal partial positive charge, δ+.

In addition to ionic and covalent bonding, there is a third major type of bonding—metallic bonding. In solid or liquid metals, metal atoms give up electrons, as in ionic compounds. The liberated electrons however, are free to move throughout the material, rather than being held in place in negative ions.

In general, atoms of nonmetals form covalent bonds with each other, atoms of metals form metallic bonds with each other, and atoms of nonmetals form ionic bonds with atoms of nonmetals. There are many exceptions, however. One common and important exception is the formation of polar-covalent bonds between metals and nonmetals that do not differ greatly in electronegativity.

> An alloy is a solid-solution or combination of metals. Alloys are discussed in Section 14-1.

Sample Problem 6.1

Use electronegativity differences and Figure 6.2 to classify bonds between sulfur and the following elements: hydrogen, cesium, and chlorine. Which atom in each bond will be more negative? The electronegativity of sulfur is 2.5. The more electronegative end of the atom in each bond will be the atom with the larger electronegativity.

Bond from sulfur to	Electronegativity difference	Bond type	More negative atom
hydrogen	2.5 − 2.1 = 0.4	polar covalent	sulfur
cesium	2.5 − 0.7 = 1.8	ionic	sulfur
chlorine	3.0 − 2.5 = 0.5	polar covalent	chlorine

Practice Problems

Use electronegativity differences and Figure 6.2 to classify bonds between chlorine and the following elements: *(a)* calcium *(b)* aluminum *(c)* silver *(d)* bromine
Indicate the more negative atom in each bond

Bond from chlorine to	Electronegativity difference	Bond type	More negative atom
(a) Ca	3.0 − 1.0 = 2.0	ionic	Cl
(b) Al	3.0 − 1.5 = 1.5	polar covalent	Cl
(c) Ag	3.0 − 1.9 = 1.1	polar covalent	Cl
(d) Br	3.0 − 2.8 = 0.2	nonpolar covalent	Cl

Why Chemical Bonding Occurs

The fascinating variety of materials in the world around us is possible because chemical bonds unite atoms of the elements in so many different combinations. Some elements are found in nature only in chemical compounds. Even the atoms of elements that can be found in nature in uncombined form—such as oxygen and nitrogen in the air, or gold and copper in the earth's crust—do not exist as independent atoms. They are bonded together.

The spontaneity of chemical reactions is discussed further in Chapter 18.

If their potential energy is lowered by the change, two atoms will form a chemical bond. Throughout nature, changes that decrease potential energy are favored. Books fall off desks, and the result is lower potential energy for the books. Most atoms have lower potential energy in bonds than as independent atoms.

Chemical-bond formation is often an energy-releasing process. Experiments also show that the reverse—breaking chemical bonds—is often an energy-absorbing process. Atoms separated by breaking a chemical bond have a higher total potential energy than when they were bonded. Whether or not a given chemical reaction occurs spontaneously is partly dependent on whether or not forming new bonds in the products produces enough energy to break bonds in the reactants.

Covalent, ionic, and metallic bonding all decrease the potential energy of the combined atoms. The next section discusses how this occurs in covalent bond formation.

Section Review

1. What are the three major types of bonds?
2. What is the role of electronegativity in the determination of the ionic or covalent character of a bond?
3. What type of bond would be expected between *(a)* H and F *(b)* Cu and S *(c)* I and Br?
4. List the three bonding pairs referred to in the previous question in order of increasing ionic characters.

1. Ionic, covalent, and metallic
2. The degree to which bonds are ionic or covalent can be estimated from their electronegativity difference
3. *(a)* ionic *(b)* polar covalent *(c)* nonpolar covalent
4. Cl—Br, Cu—S, and H—F

Eight elements—H_2, N_2, O_2, F_2, Cl_2, Br_2, I_2, and At_2—have diatomic molecules. An easy way to remember them is that seven of them form a right angle on the Periodic Table.

6.2 Covalent Bonding and Molecular Compounds

Atoms held together by covalent bonds stay together as a unit and can be separated only by a chemical reaction in which the bonds are broken. A **molecule** *is a group of two or more atoms held together by covalent bonds and able to exist independently.*

The hydrogen–hydrogen and hydrogen–chlorine covalent bonds discussed above and represented in Figure 6-3 hold the atoms together in these molecules. On Earth all hydrogen exists as molecules in which two hydrogen atoms are held together by one covalent bond. *A molecule containing two atoms is called a* **diatomic molecule.** The hydrogen chloride molecule is also a diatomic molecule.

Each water molecule consists of one oxygen atom joined by a separate covalent bond to each of two hydrogen atoms. A chemical compound whose simplest formula units are molecules is known as a molecular compound.

A **chemical formula** *is a shorthand representation of the composition of a substance using atomic symbols and numerical subscripts.* The chemical formula for the element hydrogen is H_2, the chemical formula for hydrogen chloride is HCl, and the chemical formula for water is H_2O. Because they represent molecular compounds, these formulas are known as molecular formulas. A **molecular formula** *shows the types and numbers of atoms combined in a single molecule.*

SECTION OBJECTIVES

- Explain the relationships among potential energy, distance between approaching atoms, bond length, and bond energy.
- State the octet rule.
- List the six basic steps used in writing Lewis structures.
- Explain how to determine Lewis structures for polyatomic ions or molecules containing multiple bonds.
- Write the Lewis structure for a molecule or polyatomic ion, given the identity of the atoms combined and other specific information.

Anywhere from two atoms to millions of them can be joined together by covalent bonds. They may be atoms of the same or different elements. A diamond crystal, for example, is a giant molecule containing millions of carbon atoms united by covalent bonds. Molecular compounds are everywhere around us. The chemicals of living things are predominantly molecular, and the processes of life primarily involve making and breaking covalent bonds. The molecular combinations of carbon, hydrogen, and oxygen alone account for more than 90% of all known compounds.

Molecules range in size from 62 to 23,000 picometers.

Formation of Covalent Bonds

What is a covalent bond? One way to answer this question is to consider a simple example, the formation of a hydrogen–hydrogen bond. Picture two hydrogen atoms separated by a large enough distance that they have no influence on each other. The potential energy at this point, shown below in Figure 6-4, is arbitrarily set at zero.

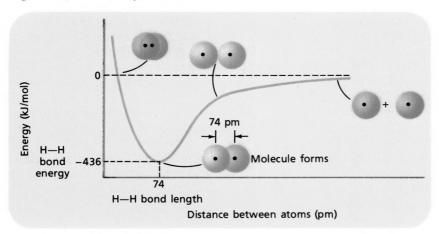

Figure 6-4 The relationship between potential energy and the distance between two hydrogen atoms. The separate atoms, at potential energy = 0, approach each other from right to left. A bond forms at the distance of minimum potential energy.

Each hydrogen atom contains a positively charged proton surrounded by a negatively charged electron in a spherical $1s$ orbital. As the atoms approach, there are several influences on the potential energy, as illustrated in Figure 6-5. By itself repulsion between the like charges of the two protons and the two electrons would cause an increase in potential energy as the atoms approach each other. By itself, attraction between the opposite charges of the proton of each atom and the electron of the other atom would cause a decrease in potential energy as the atoms approach each other.

When the two hydrogen atoms first "see" each other, mutual attraction between the electron of one hydrogen atom and the proton of the other dominates. As shown by the dip in the curve in Figure 6-4, the potential energy decreases. Eventually, a point is reached where the attraction is at a maximum and potential energy is at a minimum. At this point, which is represented by the bottom of the valley in the curve, a stable hydrogen molecule forms. A closer approach of the atoms, shown at the left in Figure 6-4, would cause a sharp rise in potential energy as repulsion becomes increasingly greater than the attraction.

The valley in the curve in Figure 6-4 represents the balance between attraction and repulsion in a stable covalent bond. The two bonded atoms

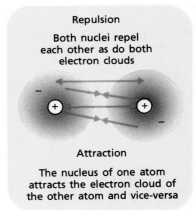

Repulsion

Both nuclei repel each other as do both electron clouds

Attraction

The nucleus of one atom attracts the electron cloud of the other atom and vice-versa

Figure 6-5 Attraction and repulsion between electrons and nuclei of two atoms. If attraction outweighs repulsion, a bond forms at a potential energy minimum. If repulsion outweighs attraction, no bond forms.

vibrate back and forth a bit, but as long as their potential energy remains at this minimum, they are covalently bonded to each other. **Bond length** *is the average distance between two bonded atoms,* that is, the distance of minimum potential energy. The bond length of the hydrogen–hydrogen bond is 74 pm (1 picometer $= 10^{-12}$ m).

The potential-energy change in bond formation is the difference between zero for the isolated atoms and the point of minimum potential energy. This amount of energy, shown in Figure 6-4 as the distance between the zero level and the bottom of the valley, is released in formation of a hydrogen–hydrogen bond. The same amount of energy must be added to separate the bonded atoms. **Bond energy** *is the energy required to break a chemical bond and form neutral atoms.* Bond energies are usually reported for breaking one mole of bonds in isolated molecules in the gaseous state and are given in kilojoules per mole. The addition of 436 kJ of energy is needed to break the hydrogen–hydrogen bonds in one mole of hydrogen molecules.

The general energy relationships described here for formation of a hydrogen–hydrogen bond apply to all covalent bonds. Bond lengths and bond energies vary with the types of atoms that have combined, however, as illustrated by the values in Table 6-1. The energy of a bond between two different types of atoms varies somewhat depending on what other bonds the atoms have formed. As you can see in the table, bond energies become larger as bond lengths become shorter.

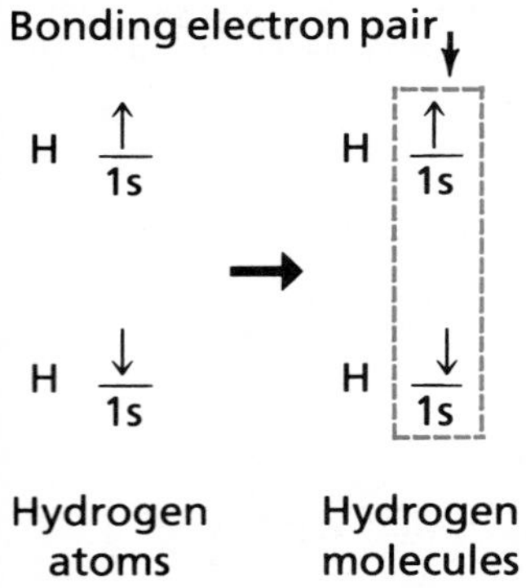

The electrons in a covalent bond are pictured as occupying overlapping orbitals, represented for the hydrogen molecule at the top in Figure 6-6. Each electron is free to move around in either of the orbitals, but both electrons spend more time in the space between the two nuclei than elsewhere. Thus, in the region of orbital overlap, there is an increased electron density. The attraction of the nuclei to the high electron density in this region holds the two atoms together.

By sharing electrons, each atom has achieved the stable, noble-gas configuration of helium, $1s^2$. Orbital notation is used to illustrate this configuration as shown on the left. Noble gases are so-called because they do not react readily with other elements. The notation for each separate atom shows one unpaired electron. A box drawn around the two orbitals containing

Molecules in the gaseous state are not influenced by neighboring molecules because adjacent molecules are far apart.

Bonding electron pair

Hydrogen atoms   Hydrogen molecules

TABLE 6-1	**BOND LENGTHS AND BOND ENERGIES FOR SINGLE COVALENT BONDS**				
Bond	Bond Length (pm)	Bond Energy (kJ/mol)	Bond	Bond Length (pm)	Bond Energy (kJ/mol)
H—H	74	436	C—C	154	347
F—F	128	159	C—N	147	293
Cl—Cl	198	243	C—O	143	351
Br—Br	228	192	C—H	110	414
I—I	266	151	C—Cl	176	330
H—F	92	569	N—N	140	159
H—Cl	127	431	N—H	98	389
H—Br	142	368	O—H	94	464
H—I	161	297			

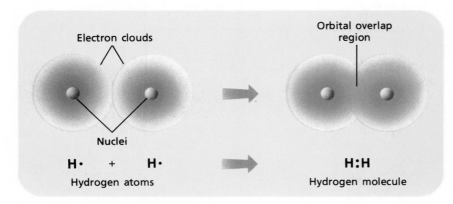

Figure 6-6 Representations of formation of the hydrogen–hydrogen bond in molecular hydrogen (H_2) by *(a)* atomic orbital overlap and *(b)* electron-pair sharing as shown by electron-dot symbols.

these unpaired electrons represents the overlapping orbitals, which are both effectively filled by the shared pair of electrons.

Like hydrogen, the halogens all normally consist of diatomic molecules. The orbital notation for formation of a fluorine molecule from two fluorine atoms is given below. By sharing their unpaired electrons, each fluorine atom in the molecule has its highest energy level completely filled by eight electrons, giving the noble gas configuration of neon $[He]2s^2 2p^6$. The molecular formulas for the diatomic halogen molecules are: fluorine—F_2; chlorine—Cl_2; bromine—Br_2; iodine—I_2; and astatine—At_2.

Fluorine atoms   **Fluorine molecules**

The orbital diagram below shows that in hydrogen chloride, (HCl), the hydrogen achieves a helium configuration by sharing an electron pair, and the chlorine atom achieves the neon configuration. Notice that chlorine, by sharing one electron with hydrogen, has 8 electrons in its outer energy level.

Hydrogen and chlorine atoms   Hydrogen chloride molecule

The Octet Rule

With the exception of helium atoms, each noble gas atom has eight electrons in its highest occupied energy level. In halogen molecules, each atom has achieved a noble-gas configuration by sharing an electron pair (see above). The same is true of hydrogen and of chlorine in hydrogen chloride shown above.

Chemistry Notebook

Exceptions to the Octet Rule

In many molecules, a central atom is surrounded by fewer than an octet of electrons.

Beryllium (Be) and boron (B) from Period 2 of the Periodic Table form such molecules. The Lewis structures for the beryllium fluoride (BeF_2) molecule and for the boron trifluoride (BF_3) molecule, both of which exist only in the gaseous state, are given in Table 6-4 as examples of molecules with AB_2 and AB_3 geometry. In beryllium fluoride, the central beryllium atom is surrounded by only four valence electrons. Each of the beryllium atom's two valence electrons pairs with a valence electron of a fluorine atom to form a total of two covalent bonds. In boron trifluoride, the central boron atom is surrounded by only six valence electrons. Here, there are three valence electrons from boron and one each from the three covalently bonded fluorine atoms.

Beryllium and boron compounds do tend to form additional bonds to complete the octet around the central beryllium or boron atom. For example, when boron trifluoride reacts with ammonia (NH_3), the unshared pair of electrons from the nitrogen atom form a covalent bond between the nitrogen atom and the boron atom as shown below.

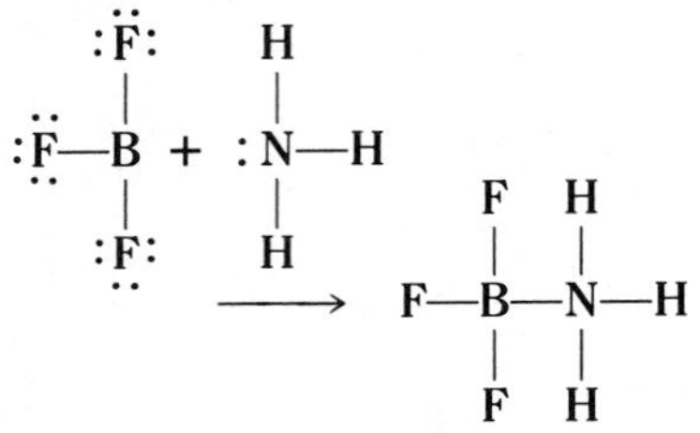

Atoms of elements of the third and higher periods of the periodic table can be surrounded by *more* than four pairs of valence electrons when they combine with atoms of the highly electronegative elements fluorine, oxygen, and chlorine. In these cases of "expanded valence," bonding involves electrons in *d* orbitals as well as in *s* and *p* orbitals. (Such bonding *d* orbitals are not available in Periods 1 and 2.) The Lewis structures for several molecules in which the central atom has expanded valence are shown below.

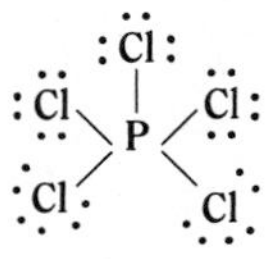

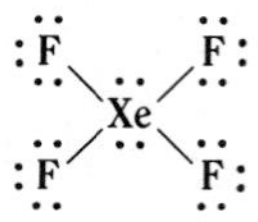

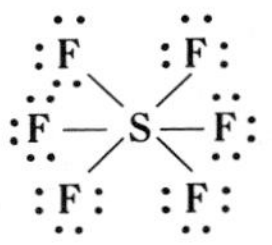

The octet rule is also violated when the total number of valence electrons in a molecule is odd. For example, in nitrogen monoxide (NO), nitrogen has five and oxygen has six valence electrons, a total of eleven.

Gilbert Lewis (1875–1946), an American theoretical chemist, proposed the Octet Rule and electron-dot symbols (also known as Lewis structures) to show arrangement of valence electrons in atoms.

In many compounds, atoms of main group elements form bonds so that each atom has an octet of electrons—four pairs of electrons in the outermost *s* and *p* orbitals. Such bond formation is described by the octet rule: *Chemical compounds tend to form so that each atom, by gaining, losing, or sharing electrons, has an octet of electrons in its highest occupied energy level.* For example, nitrogen, which is in Group 15, has the outer configuration $2s^2 2p^3$. With five electrons in its highest energy level, a nitrogen atom needs three more electrons to complete an octet. Nitrogen forms three covalent bonds in many molecules. The three electrons contributed by other atoms fill the outer energy level.

There are, however, certain exceptions to the octet rule. These are discussed in the Chemistry Notebook above.

Lewis Structures

Electron-dot symbols are used to represent molecules, as illustrated in Figures 6-1 and 6-6. The symbol for a hydrogen *atom* is H. The hydrogen molecule is shown as H:H, where the pair of dots represents the shared electron pair of the covalent bond.

For a halogen molecule, say that of chlorine, Cl_2, the electron-dot symbols of two chlorine atoms are combined to give $:\ddot{C}l:\ddot{C}l:$. Here also the pair of dots in the center represents the covalent bond. Also, each chlorine atom is surrounded by three pairs of electrons that are not shared in bonds. An **unshared pair**, *also called a lone pair, is a pair of electrons that is not involved in bonding, but instead belongs exclusively to one atom.*

The pair of dots representing a covalent bond is often replaced by a long dash, for example H—H instead of H:H or Cl—Cl instead of $:\ddot{C}l:\ddot{C}l:$. These are all **Lewis structures**: *formulas in which atomic symbols represent nuclei and inner-shell electrons, dot-pairs or dashes between two atomic symbols represent electron pairs in covalent bonds, and dots adjacent to only one atomic symbol represent unshared electrons.*

It is common to indicate only the electrons that are shared, using dashes for bonds. A **structural formula** *indicates the kind, number, arrangement, and bonds of the atoms in a molecule.* For example, Cl–Cl and H–Cl are structural formulas. The structural formula of a larger molecule, ethane, shows that the molecule contains two carbon atoms and six hydrogen atoms, arranged so that the carbon atoms are bonded to each other and three hydrogen atoms are bonded to each carbon atom.

$$
\begin{array}{ccc}
H & & H \\
& \diagdown & \diagup \\
H\!-\!\!&C\!-\!C&\!\!-\!H \\
& \diagup & \diagdown \\
H & & H
\end{array}
$$

The Lewis structures for many molecules, especially those composed of main group elements, can be drawn if one knows the composition of the molecule and which atoms are bonded to each other. The following Sample Problem illustrates the basic steps for writing Lewis structures. The molecule described in this problem contains bonds with single shared electron pairs. A single covalent bond, or a **single bond** *is a covalent bond produced by the sharing of one pair of electrons between two atoms.*

Sample Problem 6.4

Draw the Lewis structure of iodomethane (CH_3I), in which the carbon atom is the central atom.

Step 1.   *Determine the type and number of atoms in the molecule*
One carbon atom, one iodine atom, and three hydrogen atoms

Step 2.   *Draw the electron-dot symbols for each type of atom*
Carbon is from Group 14 and has 4 valence electrons; iodine is from Group 17 and has 7 valence electrons; and hydrogen has 1 valence electron.

$$\cdot \dot{C} \cdot \qquad :\ddot{I}: \qquad H\cdot$$

Step 3. *Determine the total number of valence electrons in the atoms to be combined:*

$$C \ 1 \times 4\,e^- = 4\,e^-$$
$$I \ 1 \times 7\,e^- = 7\,e^-$$
$$H \ 3 \times 1\,e^- = \underline{3\,e^-}$$
$$14\,e^-$$

Step 4. *Arrange the atoms to form a skeleton structure for the molecule and connect the atoms by electron-pair bonds*

Usually, the most electropositive atom is central. In this case, carbon and iodine have the same electronegativity, but we know that carbon is the central atom (it usually is).

H
..
H : C : I
..
H

Step 5. *Add unshared pairs of electrons so that each nonmetal atom (except hydrogen) is surrounded by eight electrons*

H   H
.. .. | ..
H : C : I :   or   H—C—I :
.. .. | ..
H   H

Step 6. *Count the electrons in the structure to be sure that the number of valence electrons used equals the number available.*

There are 8 electrons in the four covalent bonds and 6 electrons in the three unshared pairs, giving the correct total of 14 valence electrons.

Practice Problems

1. Draw the Lewis structure of ammonia (NH_3), which is present in household ammonia cleanser. The molecule contains one nitrogen atom and three hydrogen atoms.

   (Ans.) H : N : H   or   H—N—H
   ..   ..
   H   H

2. Draw the Lewis structure of iodine monochloride, which contains one iodine atom and one chlorine atom.

   (Ans.) : I : Cl :   or   : I—Cl :

3. Draw the Lewis structure of silicon tetrafluoride, which contains one silicon atom and four fluorine atoms.

   (Ans.) : F : Si : F :   or   : F—Si—F :
            : F :   : F :

4. Draw the Lewis structure for hydrogen sulfide, which contains two hydrogen atoms and one sulfur atom.

   (Ans.) H : S : H   or   H—S—H

Multiple Covalent Bonds

Atoms of some elements, especially carbon, nitrogen, and oxygen, can share more than one pair of electrons. A double covalent bond, or simply a **double**

bond, *is a covalent bond between two atoms produced by the sharing of two pairs of electrons between two atoms.* In ethane, for example, two electron pairs are simultaneously shared by two carbon atoms:

$$\begin{array}{ccc}
\text{H} & \hspace{2cm} & \text{H} \\
\diagdown & \hspace{2cm} & \diagdown \\
\text{C}::\text{C} & \hspace{0.5cm}\text{or}\hspace{1cm} & \text{C}=\text{C} \\
\diagup\hspace{0.5cm}\diagdown & & \diagup\hspace{0.5cm}\diagdown \\
\text{H}\hspace{1cm}\text{H} & & \text{H}\hspace{1cm}\text{H}
\end{array}$$

A double bond is shown either by two side-by-side pairs of dots or by two parallel dashes. All four electrons in a double bond "belong" to both atoms.

A nitrogen atom, which has five valence electrons, can acquire three electrons to complete an octet by sharing three pairs of electrons with another nitrogen atom.

$$:\!\dot{\text{N}}:\quad+\quad .\dot{\text{N}}:\quad\rightarrow\quad\begin{array}{c}:\text{N}\!:\!:\!\text{N}:\\ \text{or}\\ :\text{N}\!\equiv\!\text{N}:\end{array}$$

A triple covalent bond, or simply a **triple bond**, *is a covalent bond produced by the sharing of three pairs of electrons between two atoms.* Elemental nitrogen, like hydrogen and the halogens, normally exists as diatomic molecules (N_2). The triple bond in nitrogen molecules is nonpolar, as are the single bonds in hydrogen and halogen molecules. The orbital diagrams for the atom for nitrogen and the nitrogen molecule, given in below, illustrate that

Nitrogen molecule

atoms of nonmetals in multiple covalent bonds also follow the octet rule. Each nitrogen atom in the molecule has eight valence electrons, six that it shares with the other nitrogen atom, plus its own unshared pair of electrons.

Carbon forms a number of compounds containing triple bonds. For example, a carbon–carbon triple bond is present in acetylene, familiar for its use in welding and cutting torches.

$$\text{H}:\text{C}:\!:\!:\text{C}:\text{H}\qquad\text{or}\qquad\text{H}\!-\!\text{C}\!\equiv\!\text{C}\!-\!\text{H}$$

Double and triple bonds are referred to as multiple bonds, or multiple covalent bonds. Table 6-2 compares the bond lengths and bond energies for some single, double, and triple bonds. Double bonds in general have higher bond energies and are shorter than single bonds. Triple bonds are yet stronger and shorter.

In writing Lewis structures for molecules that contain carbon, nitrogen, or oxygen, one must remember that multiple bonds between pairs of these atoms are possible. The need for a multiple bond becomes obvious if there are not enough valence electrons to complete octets by adding unshared pairs. Step 6b in the following example illustrates how to deal with this situation.

STUDY HINT

Some common multiple bonds:
$C\!=\!C$, $C\!\equiv\!C$, $C\!\equiv\!N$, $C\!=\!O$, $N\!=\!N$, $N\!=\!O$

TABLE 6-2 COMPARISON OF BOND LENGTHS AND BOND ENERGIES FOR SINGLE AND MULTIPLE COVALENT BOND

Bond	Bond Length (pm)	Bond Energy (kJ/mol)	Bond	Bond Length (pm)	Bond Energy (kJ/mol)
C—C	154	347	C—O	143	351
C=C	134	611	C=O	120	745
C≡C	121	837	C≡O	113	1075
C—N	147	293	N—N	140	159
C≡N	115	891	N=N	125	428
			N≡N	110	946

Formaldehyde is believed to cause cancer and other problems. Because of its bacterial-killing nature, it was used to preserve animals in biology labs. However, now other safer chemicals are sometimes used as preservatives. If formalin-preserved specimens are used, they should be washed in slowly running water for eight hours before being used.

Sample Problem 6.3

Draw the Lewis structure for formaldehyde, which in water solution is used as an antiseptic known as formalin. The formaldehyde molecule has the chemical formula CH_2O.

Step 1.   *Determine the type and number of atoms.*
The formula shows one carbon atom, two hydrogen atoms, and one oxygen atom.

Step 2.   *Draw the electron-dot symbols for each type of atom.*
Carbon from Group 14 has four valence electrons; oxygen, which is in Group 16, has six valence electrons; and hydrogen has only one electron.

$$.\overset{.}{\underset{.}{C}}. \qquad\qquad H\cdot \qquad\qquad :\overset{.}{\underset{.}{O}}:$$

Step 3.   *Determine the total number of valence electrons in the atoms to be combined.*

$$\begin{aligned}
C\ 1 \times 4\,e^- &= 4\,e^-\\
O\ 1 \times 6\,e^- &= 6\,e^-\\
2H\ 2 \times 1\,e^- &= \underline{2\,e^-}\\
&\ 12\,e^-
\end{aligned}$$

Step 4.   *Arrange the atoms to form a skeleton structure for the molecule and connect the atoms by electron-pair bonds.*
The carbon atom must be the central atom.

$$\begin{array}{c} H \\ \!\!\overset{..}{} \\ H:C:O \end{array}$$

Step 5.   *Add unshared pairs of electrons so that each nonmetal atom (except hydrogen) is surrounded by eight electrons.*

$$\begin{array}{c} H \\ \overset{..}{}\ \overset{..}{} \\ H:\overset{..}{C}:\overset{..}{\underset{..}{O}}: \end{array}$$

Step 6a.   *Count the electrons in the Lewis structure to be sure that the number of valence electrons used equals the number available.*
The structure above has 6 electrons in covalent bonds and 8 electrons in four lone pairs, for a total of 14 electrons. The structure has valence electrons.

Step 6b. **If too many electrons have been used, subtract one or more lone pairs from existing bonds (between C, N, or O atoms) until the total number of valence electrons is correct.**
Subtract one lone pair of oxygen electrons from the oxygen–carbon bond. Then move one lone pair of electrons toward the atom that doesn't have a completely filled outershell to obtain the correct structure.

$$\text{H}\underset{..}{:}\text{C}::\underset{..}{\overset{..}{\text{O}}} \quad \text{or} \quad \text{H}\!-\!\overset{\text{H}}{\underset{|}{\text{C}}}\!=\!\underset{..}{\overset{..}{\text{O}}}$$

There are 8 electrons in covalent bonds and 4 electrons in lone pairs, for a total of 12 valence electrons.

Practice Problems

1. Draw the Lewis structure for carbon dioxide (CO_2). *(Ans.)* $\overset{..}{\underset{..}{\text{O}}}\!=\!\text{C}\!=\!\overset{..}{\underset{..}{\text{O}}}$
2. Draw the Lewis structure for hydrogen cyanide, which contains one hydrogen atom, one carbon atom, and one nitrogen atom. *(Ans.)* $\text{H}\!-\!\text{C}\!\equiv\!\text{N}\!:$

Polyatomic Ions

Certain groups of atoms combine to form ions rather than neutral molecules. A **polyatomic ion** *is a charged group of covalently bonded atoms.* The covalent bonds in a polyatomic ion are no different from other covalent bonds. The charge of the ion results from an excess of electrons (negative charge) or a shortage of electrons (positive charge).

The most common positively charged polyatomic ion is the ammonium ion, which contains one nitrogen atom and four hydrogen atoms and has a single positive charge. Its formula is NH_4^+, sometimes written $[NH_4]^+$ to show that the group of atoms as a whole has a charge of +1.

In the ammonium ion, there is a total positive charge of +11 from the seven protons in the nitrogen atom (atomic number 7) plus the four protons in the four hydrogen atoms (atomic number 1). The Lewis structure of the ammonium ion, at the right shows eight valence electrons.

With eight valence electrons and 2 inner-shell electrons ($1s^2$) in the nitrogen atom ($[N]1s^2 2s^2 2p^3$), there is a total negative charge of -10, accounting for the excess +1 charge. Polyatomic ions do not exist in isolation but are found combined in chemical compounds with ions of opposite charge. Electron-dot formulas for the ammonium ion and some common negative polyatomic ions—the nitrate, sulfate, and phosphate ions—are shown below. Most polyatomic ions are negatively charged. To find the Lewis structure for polyatomic ions, follow the usual six steps, but with the following exception. At the end of the third step, add 1 e^- to the total number of valence electrons if the ion is negatively charged. If the ion is positively charged, subtract 1 e^- from the total number of valence electrons.

ammonium ion   nitrate ion   sulfate ion   phosphate ion

Chemistry Notebook

Resonance

Some molecules cannot be represented adequately by a single Lewis structure. An example is the ozone molecule. Ozone is a form of oxygen found in the upper part of the atmosphere. There, it absorbs much of the incoming ultraviolet radiation from the sun, helping to protect people from skin cancer and sunburn. The ozone molecule consists of three atoms of oxygen. Experimental measurements show that ozone is a bent molecule. Following the octet rule, the ozone molecule can be represented by the two Lewis structures shown below. (The double-headed arrow is reserved for use with resonance hybrids.)

Neither structure, however, can be correct by itself because each indicates that the ozone molecule has two different types of O—O bonds—one single and the other double—whereas experiments show that both bonds are identical.

Chemists once speculated that ozone had one of these structures for half the time and the other structure the other half. They thought that it was constantly "resonating" from one structure to the other. Today, scientists say that ozone has a *single* structure that is the *average* of these two. **Resonance** *refers to bonding in molecules that cannot be correctly represented by a single Lewis structure.*

The different Lewis structures that together represent such molecules are called *resonance structures* or *resonance hybrids.*

Another example of a molecule that should be represented by resonance structures is sulfur trioxide. In sulfur trioxide, three resonance structures all obey the octet rule as shown above.

As illustrated by ozone and sulfur trioxide, the atoms in resonance structures all keep the same positions. The only difference is in the distribution of electrons. Lone-pair electrons in one resonance structure become a bonding pair in another. Simultaneously, a bonding pair of electrons in the first resonance structure becomes a lone pair in the other.

Section Review

1. Define *(a)* bond length *(b)* bond energy.
2. State the octet rule.
3. What are Lewis structures?
4. How many pairs of electrons are shared in *(a)* a single bond *(b)* a double bond *(c)* a triple bond?
5. Draw the Lewis structures for molecules consisting of: *(a)* one bromine atom and one iodine atom *(b)* one carbon atom, three hydrogen atoms, and one bromine atom *(c)* one hydrogen atom, one chlorine atom, and two carbon atoms *(d)* the sulfate ion (SO_4^{2-}).

---

1. *(a)* The distance between two bonded atoms, that is, the distance of minimum potential energy *(b)* The energy required to break a chemical bond and form neutral atoms
2. Chemical compounds tend to form so that each atom, by gaining, losing, or sharing electrons, has an octet of electrons in its highest occupied energy level.
3. Formulas in which atomic symbols represent nuclei and inner-shell electrons, dots, or dashes represent electron pairs in covalent bonds, and dots represent unshared
4. *(a)* One pair *(b)* Two pairs *(c)* Three pairs

6.3 Ionic Bonding and Ionic Compounds

Most of the rocks and minerals that make up the earth's crust consist of positive and negative ions held together by ionic bonding. Sodium chloride, which is common table salt, is found in nature as rock salt. It is a typical chemical compound composed of ions. The sodium ion (Na^+) has a charge of +1. The chloride ion (Cl^-) has a charge of -1. These ions combine in a one-to-one ratio—Na^+Cl^-—because this ratio allows each positive charge to be balanced by a negative charge. The chemical formula for sodium chloride is ordinarily written NaCl.

An **ionic compound** *is composed of positive and negative ions combined so that the positive and negative charges are equal.* Most ionic compounds are crystalline solids, such as the familiar crystals of table salt. Some examples of naturally occurring crystalline substances are shown in Figure 6-7. A crystal of any ionic compound is a three-dimensional network of positive and negative ions mutually attracted to one another. As a result, the formula for an ionic compound does not represent an independent unit that can be isolated and examined. It, instead represents the simplest ratio, or smallest numbers, of the combined ions that will give electrical neutrality.

A **formula unit** *is the simplest unit indicated by the formula of any compound.* One formula unit of sodium chloride (NaCl) is one sodium ion plus a chloride ion. The number of ions in one formula unit depends on the charges of the ions combined. For example, the magnesium ion has a charge of +2 (Mg^{2+}). In the ionic compound magnesium chloride, two chloride ions, which each have a -1 charge, must be present for each magnesium ion present in order for neutrality to be maintained. The formula of magnesium chloride is $MgCl_2$. The chemical formula of an ionic compound shows the ratio of the ions present in a sample of the compound of any size. The symbols for a few of the more common ions are given in the margin. How many Al^{3+} and Br^- ions would be in one formula unit of the ionic compound formed by these two types of ions?

SECTION OBJECTIVES

- Compare the meaning of a chemical formula for a molecular compound and one for an ionic compound.
- Discuss the arrangements of ions in a crystal.
- Define lattice energy and explain its significance.
- List and compare the distinctive properties of ionic and molecular compounds.

STUDY HINT

Remember that when describing an ionic compound, you should not refer to "molecules," but rather to formula units

Figure 6-7 Some crystalline minerals, such as aluminum oxide, are ionic compounds.

Formation of Ionic Bonds

Electron-dot structures can be used to demonstrate the changes that take place in the formation of ionic bonds. Ionic compounds do not ordinarily form by combination of isolated atoms, but for the moment consider a sodium atom and a chlorine atom approaching each other. The two atoms are neutral and have one and seven valence electrons, respectively.

$$Na\cdot \qquad\qquad \cdot\ddot{\underset{..}{Cl}}:$$

sodium atom   chlorine atom

We have already seen that atoms of sodium and the other alkali metals readily lose one electron and atoms of chlorine and the other halogens readily gain one electron. Combination of the two atoms to give one formula unit of sodium chloride can be represented as

$$Na\cdot \quad + \quad \cdot\ddot{\underset{..}{Cl}}: \quad \longrightarrow \quad Na^{+} \quad + \quad :\ddot{\underset{..}{Cl}}:^{-}$$

sodium atom   chlorine atom   sodium   chlorine
                                        cation   anion

An electron is transferred from the sodium atom to the chlorine atom. The result is that each atom is transformed into an ion with a noble gas configuration. In the combination of calcium with fluorine, two fluorine atoms are needed to accept the two valence electrons given up by calcium, which is in Group 2.

STUDY HINT

Charges of ions are not usually included in formulas for ionic compounds.

$$\cdot Ca\cdot \quad + \quad \cdot\ddot{\underset{..}{F}}: \quad + \quad \cdot\ddot{\underset{..}{F}}: \quad \longrightarrow \quad Ca^{2+} \quad + \quad :\ddot{\underset{..}{F}}: + :\ddot{\underset{..}{F}}:$$

calcium atom   fluorine   atoms   calcium   fluorine
                                          cation   anions

Electron transfer can also be represented by orbital notation, as illustrated below for sodium and chlorine atoms. The orbital notation shows that the valence electron lost by the sodium atom came from a $3s$ orbital and entered the $3p$ orbital in chlorine that contained a single electron.

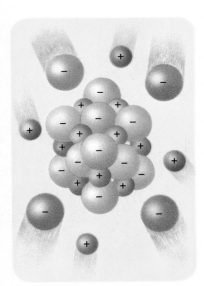

Figure 6-8 Potential energy decreases as ions assume their positions in the orderly array of ions in a crystal.

Recall how hydrogen atoms form diatomic molecules with bond lengths that have a minimum potential energy. In an ionic compound the equivalent potential energy minimum is found in the orderly arrangement of ions in the crystal, as represented in Figure 6-8. According to the laws of electrostatics, like charges repel each other and opposite charges attract each other. The force of attraction or repulsion decreases as the charges move farther apart. In an ionic crystal, there is attraction between oppositely charged ions and between the nuclei and electrons of adjacent atoms. Also, there is repulsion between the valence-shell electrons of adjacent ions. The distance between ions in a crystal and their arrangement represents a balance among all these forces of attraction and repulsion.

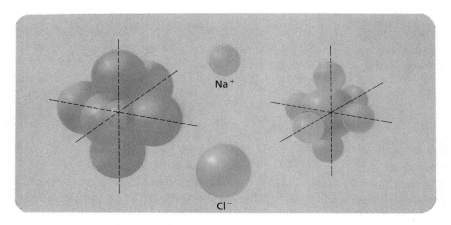

Figure 6-9 The packing of Na$^+$ and Cl$^-$ ions in the NaCl crystal. In the model on the left, the six Cl$^-$ ions are clustered about one Na$^+$ ion that cannot be seen here, but whose location is indicated by the dashed outline.

Within a sodium chloride crystal, as illustrated in Figure 6-9, each sodium cation is surrounded by six chlorine anions and each chlorine anion is also surrounded by six sodium cations. Attraction between the adjacent oppositely charged ions is much greater than repulsion by other ions of the same charge, which are further away. Figure 6-10 shows two views of the arrangement of ions in sodium chloride.

The three-dimensional arrangements of ions and the strength of attraction between them varies with the sizes and charges of the ions and the numbers of ions of different charges. For example, in calcium fluoride (CaF$_2$), which is known commonly as the mineral *fluorite* (Figure 6-11), there are two anions for each cation. Each calcium cation is surrounded by eight

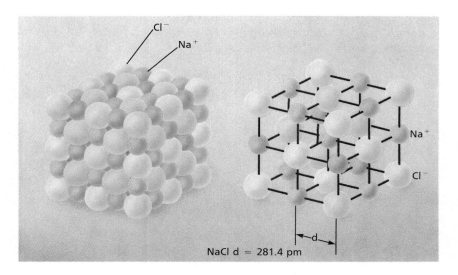

Figure 6-10 Two representations of the sodium and chloride ions in crystalline sodium chloride. The ions *(left)* shown with their electron density regions just touching, as they do in the crystal. An expanded view *(right)* of the positions of the ions in the crystal.

Figure 6-11 Crystals of the mineral fluorite, which is calcium fluoride (CaF_2).

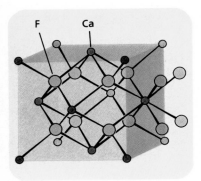

Figure 6-12 The arrangement of the ions in crystalline calcium fluoride (CaF_2).

Figure 6-13 Shattering a crystal. Attraction between positive and negative ions causes planes of ions in a crystalline ionic compound to resist motion. When struck with sufficient force, ions of the same charge approach each other, causing repulsion, and the crystal shatters along the planes.

fluoride anions, and each fluoride ion is surrounded by four calcium cations (see Figure 6-12).

To compare the strengths of ionic bonds, chemists compare the amounts of energy released when isolated ions in a gas come together to form a crystalline solid. **Lattice energy** *is the energy released when one mole of an ionic crystalline compound is formed from gaseous ions.* A few lattice energies are given in the margin. Calculations based on experimental data make it possible to find lattice energy values, which are useful for comparing bond strengths in ionic compounds. Actual ionic compound formation does not always take place by combination of individual ions, however. In a reaction between elements that form an ionic compound, the energy needed to form ions from atoms is supplied by the energy released when the ions form the lower structure of the crystal lattice, which has a relatively low potential energy.

A Comparison of Ionic and Molecular Compounds

The ions in an ionic compound are bound together by ionic bonding and held in place in a crystalline solid. The strong attraction between positive and negative charges is the force that holds ions together in ionic compounds. In a molecular compound, the covalent bonds are quite strong, but the forces of attraction between individual molecules are weaker than the forces of ionic bonding. The differences in the attraction between individual particles in molecular and ionic compounds give rise to different properties in the two types of compounds.

The melting point, boiling point, and hardness of a compound depend upon how strongly the molecules are attracted to each other. Many molecular compounds melt at relatively low temperatures because the forces of attraction between the molecules are not very strong.

Because the forces holding ions together are stronger than those holding molecules together, ionic compounds have generally higher melting and boiling points than do molecular compounds. Also, they do not vaporize as readily at room temperature as many molecular compounds do.

Ionic compounds are hard and brittle crystalline solids. The ions are so positioned that the slightest shifting of one row of ions relative to another causes an enormous buildup of repulsive forces, as shown in Figure 6-13.

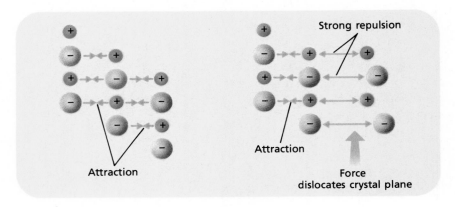

These forces make it difficult for one layer to move relative to another, causing ionic compounds to be hard. If one layer does move, however, the repulsive forces make the layers part completely, causing ionic compounds to be brittle.

In the molten state, or when dissolved in water, ionic compounds are electrical conductors because the ions are free to carry electrical current. In the solid state, the ions cannot move, and the compounds are not electrical conductors. Many ionic compounds are soluble in water. During dissolution, the ions separate from each other and become surrounded by water molecules. Many ionic compounds do not dissolve in water, however, because the attraction of the water molecules cannot overcome the attraction between the ions.

Section Review

1. Give two examples of ionic compounds.
2. Use electron-dot structures to demonstrate the formation of ionic bonds involving *(a)* Li and Cl *(b)* Ca and I.
3. Distinguish between ionic and molecular compounds in terms of the basic units that make up each.
4. When compared to compound B, compound A has lower melting and boiling points, and vaporizes more readily. If one is an ionic compound and the other is molecular, determine the identity of each.

6.4 Metallic Bonding

The properties of metals are quite different from those of either ionic or molecular compounds. These differences suggest that the bonding in metals is also quite different from ionic and covalent bonding. Metals are better conductors of heat and are more easily deformed than many solid molecular or ionic compounds. Most importantly, metals are electrical conductors in the solid state—much better conductors than even molten ionic compounds. This property suggests the presence in solid metals of highly mobile charged particles: the valence-level electrons. Such mobility is not possible in molecular compounds, in which electrons are localized in electron-pair bonds between neutral atoms. Nor is it possible in solid ionic compounds, in which electrons are bound to individual ions and the ions are held in place in crystals.

Formation of Metallic Bonds

Metal atoms contain a smaller number of valence electrons than vacant highest energy level orbitals. In *s*-block and *d*-block metals, there are one or two valence electrons in *s* sublevels, leaving three vacant *p* orbitals in the highest energy levels. In metals of the third and higher periods, there are also vacant *d* orbitals.

Metal atoms also have relatively low ionization energies and electronegativities in common. They easily give up electrons, but do not attract electrons in bonds as strongly as nonmetals. Two adjacent metal atoms consequently, would form at best only a very weak covalent bond.

Some lattice energy values, in kJ/mol:
LiF: -1032
NaF: -922
NaCl: -769
KCl: -718
CsCl: -660

1. Sodium chloride (NaCl) and magnesium chloride (MgCl2)
2. *(a)* Li$^{\cdot}$ + :C̈l: $\longrightarrow$
    Li$^+$ + :C̈l:$^-$

   (b) .Ca$^{\cdot}$ + :Ï: + :Ï: $\longrightarrow$
    Ca^{2+} + :Ï:$^-$ + :Ï:$^-$
3. Ionic compounds consist of positive and negative ions bound together by ionic bonding and held in place in a crystalline solid. Molecular compounds are made up of molecules that move around independently and are capable of independent existence.
4. Compound A is molecular and B is ionic.

Metal atoms do have room in their vacant highest energy level orbitals to share many electrons from surrounding atoms, however. In sodium, for example, the atoms are packed closely together. Each atom is surrounded by eight other sodium atoms. Eight electron pair bonds cannot form. Therefore, each sodium atom ([NE]$3s^1$) is pictured as contributing its one valence electron to a region surrounding the atoms, leaving behind a sodium ion. The electrons are delocalized—they do not belong to any one ion, but are free to move about. All of these valence electrons are shared by all of the atoms. (See Figure 6-14).

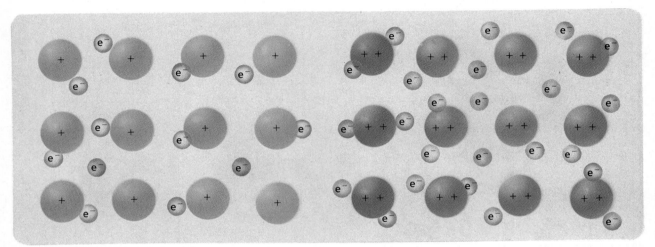

Figure 6-14 Electron-sea model of bonding in metals. The bonding strength increases with the number of electrons contributed by each atom.

A **metallic bond** *is a chemical bond resulting from the attraction between positive ions and surrounding mobile electrons.* The electrons are often referred to as forming an *electron sea.*

The strength of metallic bonding varies with the nuclear charge and the number of valence electrons. To compare bonding strengths in metals, the amounts of heat needed for vaporization are compared. By vaporizing a metal, the bonded atoms in the normal (usually solid) state are converted to individual metal atoms in the gaseous state. The higher the heat of vaporization, therefore, the stronger the metallic bonding.

Some heats of vaporization for metals are given in Table 6-3. Compare, for example, sodium and magnesium. Sodium atoms in the metal give up one electron and magnesium atoms give up two electrons. The attraction between cations and the surrounding electrons should be stronger in magnesium than in sodium because the magnesium cations have twice the charge of the sodium cations and because there are twice as many electrons attracting the cations in magnesium as there are in sodium. The higher heat of vaporization that magnesium has compared to that of sodium shows this to be the case.

Nature of Metals

Whenever you think of metals, you probably think of substances that are lustrous or reflective and that are good conductors of heat and electricity. The electron-sea model of metallic bonding is successful in explaining these intensive properties of metals. The high electrical conductivity is provided by

TABLE 6-3	HEATS OF VAPORIZATION OF SOME METALS (KJ/MOL)		
Row of the Periodic Table		Element	
second	Li 137.7	Be 224	B 540
third	Na 89.6	Mg 132	Al 284
fourth	K 77.08	Ca 153	Sc 306
fifth	Rb 69	Sr 153	Y 390
sixth	Cs 66	Ba 149	La 400

Figure 6-15 This blacksmith relies on the malleability of metals to beat iron into the shape of useful tools.

the motion of the free electrons through the solid. The high thermal conductivity of metals—the rapid transport of heat—also is accounted for by the mobile electrons. When a metal bar is heated at one end, the electrons at that end increase in kinetic energy and rapidly flow through the metal, transporting heat energy with them. In contrast, in molecular or ionic compounds heat can only be passed along stepwise from neighbor to neighbor.

Most metals are also characteristically easy to work and form into desired shapes. Two important characteristics of this type are called *malleability* and *ductility*. **Malleability** *is the state of being able to be shaped or extended by beating with a hammer, rolling, or otherwise exerting physical pressure that results in a change in contour* (see Figure 6-15). **Ductility** *is the state of being able to be drawn, pulled, or extruded through a small opening to produce a wire*. The malleability and ductility of metals are possible because metallic bonding is not directional, but uniform throughout the solid. One plane of ions in a metal can slide past another, without encountering any resistance or breaking any bonds. By contrast, shifting the layers of an ionic crystal causes the bonds to break and the crystal to shatter, as shown in Figure 6-13.

To reflect light and appear lustrous, a material must be able to absorb and remit light of many wavelengths. This occurs in metals because electrons in the electron sea have a wide range of possible energies. As a result, metals can absorb and emit light over a wide range of wavelengths giving them a characteristically shiny appearance.

Section Review

1. Describe the electron-sea model of metallic bonding.
2. What is the relationship between metallic-bonding strength and heat of vaporization?
3. Explain why metals are malleable and ductile, whereas ionic crystals are not.

STUDY HINT
For an introduction to properties of metals, see Section 1.4.

1. Metallic bonding is generally viewed as the result of the mutual sharing of many electrons by many atoms. Each atom contributes its valence electrons to a region surrounding the atoms, leaving behind positive ions. These electrons are then free to move about in what is often called an "electron sea," where thy are shared by all of the atoms.
2. In general, the higher the heat of vaporization, the stronger the metallic bonding.
3. Metallic bonding is not directional but uniform throughout the solid. One plane of ions in a metal can slide past another without any resistance or breaking of bonds as would be the case if layers of an ionic crystal were shifted.

6.5 The Properties of Molecular Compounds

SECTION OBJECTIVES

- Explain the VSEPR theory.
- Predict the shapes of molecules or polyatomic ions using VSEPR.
- Explain how the shapes of molecules are accounted for by hybridization.
- Describe dipole–dipole forces, hydrogen bonding, an induced dipole, and London dispersion forces.
- Explain what determines molecular polarity.

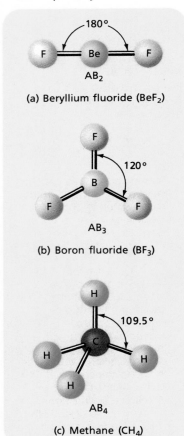

Figure 6-16 Ball and stick models showing the shapes of AB_2, AB_3 and AB_4 molecules according to VSEPR.

The properties of molecules depend not only upon bonding, but also upon the arrangement of the atoms in space—molecular geometry. Understanding molecular geometry is useful, because many properties of molecular compounds are strongly influenced by it. Together with bond polarity, geometry determines molecular polarity. This polarity, in turn, strongly influences the forces that act between molecules in liquids and solids.

Two different, equally successful approaches to predicting molecular geometry are presented in the following sections. It is not unusual for two or more theories or models for the same phenomenon to stand the test of experiment. In such a case, scientists do not hesitate to switch back and forth, using whichever model best fits a given situation.

VSEPR Theory

Diatomic molecules like those of hydrogen (H_2), chlorine (Cl_2), and hydrogen chloride (HCl) must be linear. For more complicated molecules, prediction of geometry requires knowing the locations of electrons in bonds.

Where would you expect the bonding electrons to be in a molecule of beryllium fluoride (F—Be—F)? For each beryllium–flourine bond, the pair of electrons creates a region of high electron density next to the fluorine atom. Because electrons repel each other, the two pairs of bonding electrons stay as far away from each other as possible. The geometry that results is linear. All three atoms lie on a straight line with the two boron–fluorine bonds 180° apart as illustrated in Figure 6-16a.

The VSEPR theory for predicting molecular geometry is based upon the simple assumption that electrons in molecules repel each other. The letters VSEPR stand for "valence-shell, electron-pair repulsion," referring to the repulsion between pairs of bonding valence electrons in the highest occupied energy level (the "valence shell"). **VSEPR theory** *states that electrostatic repulsion between the valence-level electron pairs surrounding an atom causes these pairs to be oriented as far apart as possible.*

To use VSEPR, molecules are classified according to how many electron pairs surround a central atom. Beryllium fluoride is an AB_2 molecule, where A represents the central atom (Be) and B_2 represents the two atoms bonded to the central atom (2F).

What about an AB_3 molecule? Three A-B bonds stay farthest apart by pointing to the corners of an equilateral triangle, giving 120° angles between the bonds. Boron trifluoride (BF_3), shown in Figure 6-16, is found by experiment to have the predicted triangular planar geometry.

The A atoms in AB_4 molecules follow the octet rule and have eight electrons in their highest energy levels. Experiment has shown that methane (CH_4) for example, is a symmetrical molecule with a carbon atom at the center of a regular tetrahedron. Each carbon–hydrogen bond points to one of the four corners of the tetrahedron and all four bond angles are 109.5°, as shown in Figure 6-16. This is exactly the arrangement in which repulsion among the four bonding electron pairs in an AB_4 molecule is at a minimum.

TABLE 6-4 VSEPR AND MOLECULAR GEOMETRY

Molecular Shape		Atoms Bonded to Central Atom	Lone Pairs of Electrons	Type of Molecule	Formula Example	Lewis Structure
Linear		2	0	AB_2	BeF_2	F—Be—F
Bent		2	1	AB_2E	$SnCl_2$	
Triangular planar		3	0	AB_3	BF_3	
Tetrahedral		4	0	AB_4	CH_4	
Triangular pyramidal		3	1	AB_3E	NH_3	
Bent		2	2	AB_2E_2	H_2O	
Triangular bipyramidal		5	0	AB_5	PCl_5	
Octahedral		6	0	AB_6	SF_6	

The shapes of AB_2 to AB_6 molecules are summarized in Table 6-4. The Bs in such molecules need not be single atoms, nor need they be identical. The shape will still be based upon that given in the table. Different sizes of B groups can distort the bond angles, however, making some larger or smaller than those given in the table.

To use Table 6-4 to predict molecular geometry, it is necessary to know the Lewis structure for a molecule and classify it according to the number of bonds surrounding the central atom.

Sample Problem 6.7

Use VSEPR theory to predict the molecular geometry of aluminum trichloride ($AlCl_3$).

First, the Lewis structure must be written for aluminum trichloride ($AlCl_3$) molecules. Aluminum is in Group 13 and has three valence electrons, [·Al·]; chlorine is in Group 17 and has seven valence electrons, [:C̈l:]. The total number of available valence electrons is 24 e^- (3 e^- from aluminum and 21 e^- from chlorine). The following Lewis structure uses all 24 e^-.

$$:\ddot{C}l:\ddot{A}l:\ddot{C}l:$$

As explained in the Chemistry Notebook on page 168, this molecule is an exception to the octet rule. Aluminum trichloride is an AB_3 type of molecule. According to VSEPR theory, therefore, it should have triangular–planar geometry.

Practice Problems
1. Use VSEPR theory to predict the molecular geometry of each of the following: *(a)* HI (*Ans.* linear) *(b)* CBr_4 (Ans. tetrahedral) *(c)* BF_3 *(Ans.)* triangular-planar

Ammonia (NH_3) and water (H_2O) molecules illustrate two more common molecular shapes. The central atoms of these molecules (N and O) are both from *p*-block elements and follow the octet rule. The molecules, however, are not tetrahedral because unshared electron pairs are present.

The Lewis structure of ammonia in Table 6-4 shows that the molecule contains one unshared electron pair. The general formula for molecules like ammonia is AB_3E, where E represents the unshared electron pair. In an ammonia molecule, illustrated in Figure 6-17a, the three hydrogen atoms occupy three corners of a tetrahedron and the unshared pair occupies the fourth corner. The shape of the molecule depends only upon the positions of the four atoms, not on the position of the unshared pair. Consequently, the molecular geometry is that of a pyramid with a triangular base.

The water molecule, as shown by its Lewis structure in Table 6-4, has two unshared electron pairs. It is an AB_2E_2 molecule. Here, the oxygen atom is at the center of a tetrahedron, with two corners occupied by hydrogen atoms

Figure 6-17 The location of bonds and unshared electrons are shown for ammonia *(a)* and water *(b)*. Ball and stick models show the shapes of these molecules.

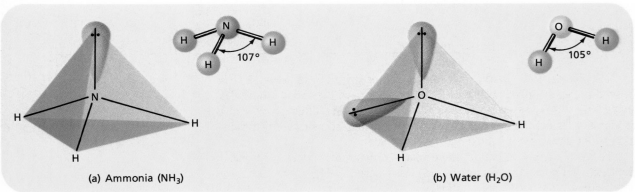

(a) Ammonia (NH_3) (b) Water (H_2O)

and two by the unshared pairs (Figure 6-17b). The shape of the molecule is determined by the three atoms—it is referred to as a "bent" molecule. In Figure 6-17, note that the bond angles in ammonia and water are somewhat less than the 109.5° of a perfect tetrahedron. These angles are decreased because the unshared electron pairs repel electrons more strongly than do bonding electron pairs. Table 6-4 includes an example of an AB_2E type molecule, which results when a central atom forms two bonds and retains one unshared electron pair.

Lewis structures and Table 6-4 can be used together to predict the shapes of many molecules and polyatomic ions. Note that in predicting the shape of a molecule, double bonds are treated in the same way as single bonds.

Sample Problem 6.8

Use VSEPR theory to predict the shape of *(a)* carbon dioxide (CO_2) and *(b)* the chlorate ion($ClO_3{}^-$).

Solution

(a) The Lewis structure of carbon dioxide shows two carbon–oxygen double bonds and no unshared electron pairs on the carbon atom. This is an AB_2-type molecule and is predicted to be linear.

(b) The chlorate ion has three oxygen atoms surrounding a central chlorine atom and, as shown by the Lewis structure, there is an unshared electron pair on the chlorine atom. The chlorate ion is classified as of the AB_3E type and predicted to have triangular–pyramidal geometry, with the three oxygen atoms at the base of the pyramid and the chlorine atom at the top.

Practice Problems
1. Use VSEPR theory to predict the molecular geometry of *(a)* hydrogen cyanide $H{-}C{\equiv}N:$ *(b)* formaldehyde (H_2CO) see Sample Problem 6.5. *(Ans.) (a)* linear, *(b)* triangular pyramidal
2. Use VSEPR theory to predict the molecular geometries of the molecules whose Lewis structures are given below:

(a) $:\ddot{F}-\ddot{S}-\ddot{F}:$ *(Ans.) (a)* bent   *(b)* $:\ddot{Cl}-\overset{\displaystyle}{\underset{\displaystyle :\ddot{Cl}:}{P}}-\ddot{Cl}:$ Cl *(Ans.) (b)* triangular pyramidal

Hybridization

VSEPR theory is an excellent guide to predicting and understanding the shapes of molecules. It does not reveal, however, any relationship between geometry and the orbitals occupied by bonding electrons. For this, a different model of bonding and molecular geometry is needed. One such model is based upon **hybridization**, *which is the mixing of two or more atomic orbitals of similar energies on the same atom to give new orbitals of equal energies.*

Methane (CH_4) provides a good example of how hybridization accounts for bonding and the shape of a molecule. The orbital notation for a carbon atom

$$C \quad \underset{1s^2}{\uparrow\downarrow} \quad \underset{2s^2}{\uparrow\downarrow} \quad \underset{2p^2}{\uparrow \quad \uparrow \quad \underline{}}$$

shows that it has four valence electrons, two in the $2s$ orbital and two in $2p$ orbitals. It is difficult to picture how these four electrons could form four

You might point out that hybrid orbitals are special types of *atomic* orbitals that have energy intermediate between *s*- and *p*-orbitals. An analogy to use might be mixing blue clay (*s* orbital) and yellow clay (*p* orbital) to get green clay (an *sp*, *sp²*, or *sp³* orbital).

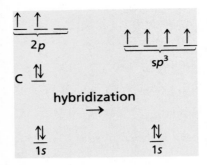

Figure 6-18 sp_3 Hybridization of atomic orbitals on a carbon atom. One s and three p orbitals combine to give four sp^3 hybrid orbitals. As is always the case in hybridization, the hybrid orbitals are at an energy level in between the levels of the orbitals that have combined.

equivalent, tetrahedrally arranged covalent bonds by orbital overlap with orbitals of four other atoms. Two of the electrons are already paired with each other. Also, the $2s$ orbital and the $2p$ orbitals have very different shapes as you learned in Section 4.2.

To account for the bonding in methane, the $2s$ and $2p$ orbitals on the carbon atom are assumed to combine and become rearranged, that is, to *hybridize*. The relative energies of the original and new orbitals on a carbon atom are represented in Figure 6-18. Note that the energy of the hybrid sp^3 orbital is different than that of the $2p$ or $2s$ orbital. Hybridization of the single $2s$ and the three $2p$ orbitals yields four new, identical orbitals.

Hybrid orbitals *are orbitals of equal energy produced by the combination of two or more orbitals on the same atom*. The number of hybrid orbitals produced is always equal to the number of orbitals that have combined. The four orbitals formed by combination of one s orbital and three p orbitals are designated sp^3 hybrid orbitals.

The superscript 3 indicates that three p orbitals were included in the hybridization (the superscript 1 on the s is not written, but understood).

The four sp^3 orbitals are oriented tetrahedrally, 109.5° apart, and account for the tetrahedral molecular geometry around any carbon atom with four single bonds. Each hybrid orbital is occupied by one electron from carbon and can overlap with an electron-containing atomic or hybrid orbital on another atom to form a covalent bond. Such bonding with sp^3 hybrid orbitals is illustrated in Figure 6-19a for methane.

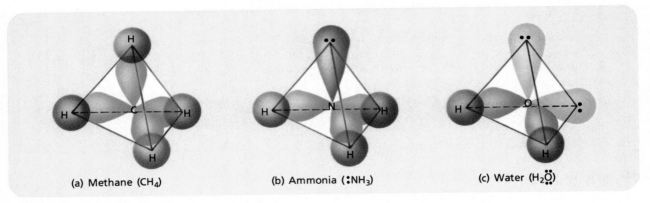

(a) Methane (CH_4) (b) Ammonia (:NH_3) (c) Water ($H_2\overset{\cdot\cdot}{\underset{\cdot\cdot}{O}}$)

Figure 6-19 Bonds formed by overlap of sp_3 orbitals on carbon, nitrogen, and oxygen atoms with $1s$ orbitals on hydrogen atoms. The hybrid orbitals are shown in color.

Hybridization also explains the bonding and geometry of many molecules formed by Group 15 and 16 elements. The sp^3 hybridization of a nitrogen atom (N, $[He]2s^2 2p^3$) yields four hybrid orbitals—one orbital that contains a pair of electrons and three orbitals that each contain an unpaired electron that can form a single bond, as in ammonia (Figure 6-25b). Similarly two of the four sp^3 hybrid orbitals on an oxygen atom ($[He]2s^2 2p^4$) are occupied by two electron pairs, and two are occupied by single electrons that can form single bonds, as in water Figure 6-19c.

The linear and triangular planar geometry of molecules, such as those of beryllium fluoride and boron fluoride (illustrated in Table 6-4), is made possible by hybridization involving only one and two of the available empty p orbitals, respectively. The geometry of sp, sp^2, and sp^3 hybridization is summarized in Table 6-5.

TABLE 6-5 GEOMETRY OF HYBRID ORBITALS

Atomic Orbitals Hybridization	Type of Hybridization	Number of Hybrid Orbitals	Geometry
s, p	sp	2	180° linear
s, p, p	sp²	3	120° planar
s, p, p, p	sp³	4	109.5° tetrahedral

Since Johannes van der Waals studied the attractive forces that exist between molecules of gases and the liquids they formed upon condensation, certain intermolecular forces, dipole–dipole forces, and London dispersion forces are sometimes collectively referred to as van der Waals forces. In 1910, he received the Nobel Prize in physics for this work.

TABLE 6-6 BOILING POINTS AND/BOND TYPES

Bond Type	Substance	bp (1 atm, °C)
noble gas	He	−269
	Ar	−186
nonpolar	H_2	−253
	O_2	−183
	Cl_2	−34
	Br_2	59
molecular	CH_4	−164
	CCl_4	77
	C_6h_6	80
polar	PH_3	−88
	NH_3	−33
	H_2S	−61
molecular	HF	19.5
	HCl	−85
	ICL	97
ionic	NaCl	1413
	MgF_2	2239
metallic	Cu	2567
	Fe	2750
	W	5660

Intermolecular Forces

The boiling point is the highest temperature at which a substance can exist as a liquid. At this temperature, the forces of attraction between particles are not strong enough to overcome the kinetic energy of the particles, which increases with increasing temperature. At the boiling point, adjacent particles in a liquid pull away from each other and enter the gas phase. The strengths of the attraction between particles of various substances, therefore, can be compared by comparing boiling points. The higher the boiling point, the stronger the interparticle forces.

The forces of attraction between molecules are known as **intermolecular forces**. Intermolecular forces vary in strength, but are generally weaker than the bonds that join atoms in molecules, ions in ionic compounds, or metal atoms in solid metals. In Table 6-6, compare the boiling points of the metals and ionic compounds with those of the other substances. You can see that they are much higher, indicating that ionic bonding and metallic bonding are much stronger than intermolecular forces.

Dipole–dipole forces and molecular polarity   The strongest intermolecular forces act between polar molecules. *Equal but opposite charges separated by a short distance create a* **dipole**. The direction of a dipole is represented by an arrow pointing toward the negative pole with a tail crossed to indicate that it lies in the direction of the positive pole. For the hydrogen–chloride molecule, the dipole is indicated as follows, with the negative end at the more electronegative chlorine atom:

H—Cl

Dipole–dipole forces *are forces of attraction between polar molecules.* The negative region in one polar molecule attracts the positive region in adjacent molecules, and so on throughout a liquid or solid. The effect of dipole–dipole forces is seen by comparing the boiling points of iodine chloride (I-Cl) and bromine (Br-Br). The boiling point of polar iodine chloride is 97°C. The boiling point of nonpolar bromine is 59°C. The dipole–dipole forces between molecules of ICl are illustrated schematically in Figure 6-20.

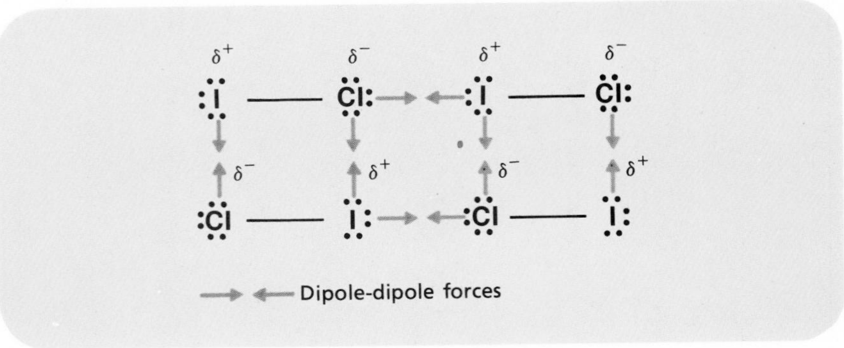

Figure 6-20 Dipole–dipole forces in iodine monochloride. A repulsive force exists where the heads of two arrows meet.

The polarity of diatomic molecules like hydrogen chloride is determined by just one bond. For molecules containing more than two atoms, molecular polarity depends upon both the polarity of the individual bonds and the orientation of bonds.

The water molecule, for example, has two hydrogen–oxygen bonds. The more electronegative oxygen atom is the negative pole of both bonds. The polarities of these two bonds combine to make the water molecule highly polar, as indicated in Figure 6-21. The ammonia molecule is also polar, because the dipoles of the three N–H bonds extend in the same direction. because the dipoles of the three nitrogen-hydrogen bonds extend in the same direction.

A polar molecule can *induce* a dipole in a nonpolar molecule by momentarily attracting its electrons. The result is an intermolecular force that is somewhat weaker than the dipole–dipole force. Such forces account, for example, for the solubility of nonpolar, molecular oxygen in water, a process necessary for aquatic life. The negative pole of a water molecule repels the outer electrons of an adjacent oxygen molecule. The oxygen molecule, then, has an induced positive pole on the side toward the water molecule and an induced negative pole on the opposite side. The result is an attraction to the water molecule.

In some molecules, the individual bond dipoles oppose one another, causing the resulting molecular polarity to be zero. Carbon dioxide and carbon tetrachloride are molecules of this type, as shown in Figure 6-21.

Hydrogen bonding A particularly strong type of dipole–dipole force accounts for the unusually high boiling points of hydrogen-containing compounds, such as hydrogen fluoride (HF), water (H_2O), and ammonia (NH_3). **Hydrogen bonding** *is the intermolecular attraction between a hydrogen atom bonded to a strongly electronegative atom (e.g., fluorine, oxygen, or nitrogen) and an unshared pair of electrons on a strongly electronegative atom (e.g., fluorine, oxygen, or nitrogen).* The large

Linus Pauling was the first chemist to recognize that the hydrogen bond was different in nature and strength from the covalent bond.

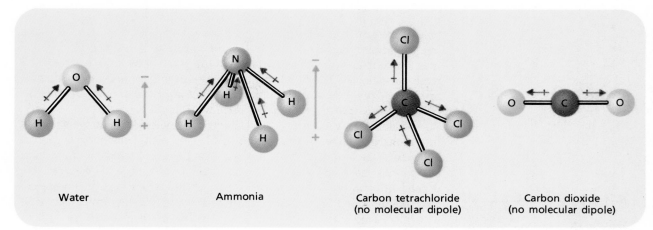

| Water | Ammonia | Carbon tetrachloride (no molecular dipole) | Carbon dioxide (no molecular dipole) |

electronegativity differences between hydrogen atoms and fluorine, oxygen, or nitrogen atoms makes the bonds connecting them highly polar. The hydrogen atom in such a bond has a positive charge approaching that of a proton. Moreover, the small size of the hydrogen atom allows an unshared pair of electrons on adjacent molecule to approach very closely.

Hydrogen bonds are usually represented by dotted lines connecting the hydrogen-bonded hydrogen to the unshared electron pair of the electronegative atom to which it is attracted, as illustrated at right for water:

The effect of hydrogen bonding is shown by comparing the boiling points in Table 6-6 of phosphine (PH_3) and hydrogen-bonded ammonia (NH_3) or hydrogen sulfide (H_2S) and hydrogen-bonded water (H_2O). Hydrogen bonding plays an important role in determining the properties of water, including its high surface tension.

London dispersion forces Even noble-gas atoms and molecules that are nonpolar are attracted to each other when the temperature becomes low enough. The weak intermolecular forces solely responsible for the attraction between such atoms or molecules are known as London dispersion forces. The explanation of these forces was first provided in 1930 by Fritz London (1900–1954).

In any molecule, polar or nonpolar, the electrons are in continuous motion. The result is, that at a given instant, the electron distribution may be

Figure 6-21 Some examples of the effect of molecular shape on molecular polarity. In water and ammonia molecules, the bond polarities all extend in the same direction, causing the molecules to be polar. In the carbon–tetrachloride (CCl_4), and carbon–dioxide (CO_2) molecules, the bond polarities extend symmetrically in different directions and cancel one another.

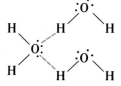

Hydrogen bonding in water

STUDY HINT
The very important role of hydrogen bonding in the properties of water is discussed in Section 13.4.

Figure 6-22 The surface tension of water is the result of hydrogen bonding. The leaf at left is supported by the surface tension of water. At right, water droplets are spherical due to the surface tension of water. If water had zero surface tension, the droplets would not be uniform.

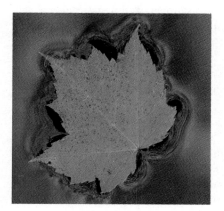

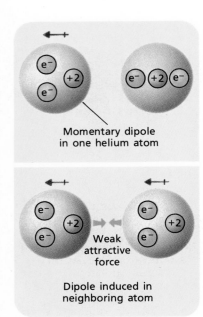

Momentary dipole
in one helium atom

Weak
attractive
force

Dipole induced in
neighboring atom

Figure 6-23 Helium can become temporarily polar under the right conditions. If an instantaneous, temporary dipole develops in a helium atom *(top)*, it can induce a dipole in a neighboring atom *(bottom)*.

1. The spatial arrangement of the atoms bonded together in a molecule

2. The VSEPR theory and/or the Theory of Hybridization

3. *(a)* $\overset{..}{O}::Si::\overset{..}{O}$ (AB$_2$—linear)

   (b) $:\overset{..}{I}:\overset{..}{C}:\overset{..}{I}:$ (AB$_4$—tetrahedral)

4. In sp^3 hybridization, one s and three p orbitals combine and become rearranged to form four new, identical orbitals of equal energies.

5. *(a)* Hydrogen bonding affects the properties of water molecules. *(b)* Such bonding causes water to have an unusually high boiling point and to display high surface tension.

uneven, creating a positive pole in one part of the molecule and a negative pole in another. This instantaneous and temporary dipole can induce a dipole in an adjacent molecule, as illustrated in Figure 6-23. The two molecules are then held together for an instant by the weak attraction between the temporary dipoles. **London dispersion forces** *are intermolecular attractions resulting from the constant motion of electrons and the creation of instantaneous dipoles and induced dipoles.*

London forces act between all atoms and molecules, but are the only intermolecular forces acting among noble-gas atoms and nonpolar molecules. In Table 6-6, it can be seen that noble gases and nonpolar–molecular compounds have the lowest boiling points. Since London forces are dependent upon the motion of electrons, they are stronger as the number of electrons in a molecule is greater. In other words, London forces increase with increasing masses of atoms or molecules. Comparison of the boiling points of the gases helium (He) and argon (Ar), hydrogen (H$_2$) and oxygen (O$_2$), and chlorine (Cl$_2$) and bromine (Br$_2$) shows the effect of increasing mass on increasing London dispersion forces.

Section Review

1. What is meant by molecular geometry?
2. What two theories can be used to predict molecular geometry?
3. Draw the Lewis structure and use the VSEPR theory to predict the molecular geometry of *(a)* SiO$_2$ *(b)* Cl$_4$.
4. Explain what is meant by sp^3 hybridization.
5. *(a)* What type of intermolecular force plays a major role in accounting for various properties of water molecules? *(b)* Explain.

Chapter Summary

- Two or more atoms may form ionic, covalent, or polar-covalent bonds depending on the difference in electronegativity between the atoms.
- The octet rule states that chemical compounds tend to form so that each atom, by gaining, losing or sharing electrons has eight electrons in its highest occupied energy level.
- Double bonds occur when atoms share two pairs of electrons in a covalent bond. Triple bonds occur when atoms share three pairs of electrons in a covalent bond.
- Because of the strong attraction between positively and negatively charged ions, materials containing ionic bonds tend to be harder and more brittle and to have higher boiling points than materials containing covalently bonded atoms.
- The "electron sea" formed in metallic bonding gives metals their properties of high electrical and thermal conductivity, malleability, ductility, and luster.
- VSEPR theory uses the fact that electron pairs strongly repel each other and tend to be oriented as far apart as possible to predict shapes of molecules.
- Hybridization uses the fact that orbitals within an atom can mix to form orbitals of equal energy in order to predict the shapes of molecules.
- Intermolecular forces such as dipole–dipole forces, hydrogen bonding, and London dispersion forces can exist between certain types of molecules.

Chapter 6 *Review*

Chemical Bonding

Vocabulary

bond energy	malleability
bond length	molecular compound
chemical bond	molecular formula
covalent bond	molecule
diatomic molecule	multiple bond
dipole	nonpolar–covalent bond
dipole–dipole forces	octet rule
double bond	polar
ductility	polar-covalent bond
formula unit	polyatomic ion
hybrid orbitals	resonance
hybridization	single bond
hydrogen bonding	structural formula
intermolecular forces	triple bond
ionic bond	unshared pair
ionic compound	van der Waals forces
lattice energy	VSEPR (valence-shell
Lewis structure	electron-pair
London dispersion forces	repulsion) theory

Review Questions

1. *(a)* What is a chemical bond?   *(b)* Upon what basis are the various chemical bonds classified?
2. Among main group elements, what ion formation tendencies are generally observed?
3. Identify and define the two major types of chemical bonding.
4. What is the relationship between electronegativity and the ionic character of a chemical bond?
5. *(a)* What is the meaning of the term polar, as applied to chemical bonding?   *(b)* Distinguish between polar- and nonpolar-covalent bonds.
6. What types of bonds are usually formed between atoms of the following: *(a)* nonmetals *(b)* metals   *(c)* metals with nonmetals of differing electronegativities   *(d)* metals with nonmetals of similar electronegativities.
7. *(a)* In general, what determines whether atoms will form chemical bonds?   *(b)* What energy change is usually associated with bond formation?

8. Define each of the following:
   (a) molecule   *(b)* diatomic molecule
   (c) molecular compound   *(d)* chemical formula   *(e)* molecular formula.
9. *(a)* What determines bond length?   *(b)* How are bond energies and bond lengths related?
10. Describe the general location of the electrons in a covalent bond.
11. Write the molecular formulas for each of the halogens.
12. As applied to covalent bonding, what is meant by an unshared or lone pair of electrons?
13. *(a)* What is a structural formula?   *(b)* When is such a formula generally used?   *(c)* Illustrate the structural formula for F_2.
14. Determine the number of valence electrons contained in an atom of each: *(a)* H   *(b)* F *(c)* Mg   *(d)* O   *(e)* Al   *(f)* N   *(g)* C   *(h)* Ne.
15. List three rules used in drawing Lewis structures.
16. Distinguish among single-, double-, and triple-covalent bonds by defining each and providing an illustration of each type.
17. Name three elements whose atoms form double or triple covalent bonds.
18. *(a)* What are multiple bonds?   *(b)* How do their bond energies and bond lengths compare with those of single bonds?
19. In writing Lewis structures, how is the need for multiple bonds generally determined?
20. *(a)* What is a polyatomic ion?   *(b)* Give two examples of a polyatomic ions.   *(c)* How do such ions normally occur in nature?
21. *(a)* What is an ionic compound?   *(b)* In what form do most ionic compounds occur?
22. *(a)* What is a formula unit?   *(b)* What are the components of one formula unit of CaF_2? *(c)* How many K^+ and S^{2-} ions would be in one formula unit of the ionic compound formed by these ions?
23. *(a)* What is lattice energy?   *(b)* In general, what is the relationship between lattice energy and the strength of an ionic bond?

24. *(a)* In general, how do ionic and molecular compounds compare in terms of melting points, boiling points, and ease of vaporization? *(b)* What accounts for the observed differences in their properties? *(c)* Cite three physical properties of ionic compounds.

25. *(a)* How do the properties of metals differ from those of both ionic and molecular compounds? *(b)* What specific property of metals accounts for their unusual electrical conductivity?

26. What properties of metals contribute to their tendency to form metallic bonds?

27. *(a)* What is a metallic bond? *(b)* What determines the strength of a metallic bond?

28. What two factors generally account for the characteristic properties of molecules?

29. *(a)* What is the basis of the VSEPR theory? *(b)* What do the letters in "VSEPR" represent, and to what do they refer? *(c)* State the VSEPR theory.

30. *(a)* How is the VSEPR theory used to classify molecules? *(b)* What molecular geometry would be expected for F_2 and HF?

31. According to the VSEPR theory, what molecular geometries are associated with the following types of molecules:
(a) AB_2 *(b)* AB_3 *(c)* AB_4 *(d)* AB_5 *(e)* AB_6?

32. Describe the role of each of the following in predicting molecular geometries: *(a)* unshared electron pairs *(b)* double bonds.

33. *(a)* What is hybridization? *(b)* What is the meaning of the term hybridize?

34. *(a)* What are hybrid orbitals? *(b)* What determines the number of hybrid orbitals produced by an atom?

35. What is the relationship between the strength of the intermolecular forces within a substance and the boiling point of that substance?

36. *(a)* What are intermolecular forces? *(b)* How do these forces compare in strength to that in ionic and metallic bonding? *(c)* Where are the strongest intermolecular forces found?

37. *(a)* What are dipole–dipole forces? *(b)* What determines the polarity of a molecule?

38. *(a)* What is meant by an induced dipole? *(b)* What is the everyday importance of this type of intermolecular force?

39. *(a)* What is hydrogen bonding? *(b)* What accounts for its extraordinary strength?

40. *(a)* What are London dispersion forces? *(b)* How are London forces related to atomic and molecular masses?

Review Problems

1. Determine the electronegativity difference, the bond type, and the more negative atom with respect to bonds formed between the following: *(a)* H and I *(b)* Ba and S *(c)* K and Br *(d)* Zn and O *(e)* At and I *(f)* S and I.

2. List the bonding pairs described in Review Problem 1 in order of increasing covalent character.

3. Use orbital notation to illustrate the bonding in each of the following: *(a)* a chlorine molecule *(b)* an oxygen molecule *(c)* a phosphorous molecule *(d)* a molecule of hydrogen fluoride (consisting of one atom of each element).

4. For each of the following, use electron dot notation to illustrate the number of valence electrons present in one atom of each element listed: *(a)* Li *(b)* Ca *(c)* Cl *(d)* O *(e)* C *(f)* P *(g)* Al *(h)* S.

5. Draw Lewis structures for each of the following molecules: *(a)* contains one C and four F atoms *(b)* contains two H and one Se atom *(c)* contains one N and three I atoms *(d)* contains one Si and four Br atoms *(e)* contains one C and three H, and one Cl atom.

6. Use electron-dot structures to demonstrate the formation of ionic bonds involving: *(a)* K and F *(b)* Mg and Br *(c)* Al and I.

7. Draw the Lewis structures and then predict the molecular geometry of each: *(a)* I_2 *(b)* HBr *(c)* CF_4 *(d)* CO_2 *(e)* BCl_3.

Application Questions

1. Can chemical bonds between unlike atoms be purely ionic or purely covalent? Explain.

2. Describe what occurs as two atoms approach each other to form a covalent bond.

3. Use Figure 6-4 to trace the sequence of events involved in the formation of the hydrogen–hydrogen covalent bond.

4. Define the term resonance.

5. Use the electron-sea model of metallic bonding to explain the following properties of metals: *(a)* high electrical conductivity

(b) high thermal conductivity   (c) malleability and ductility   (d) luster or reflectivity.

6. According to the VSEPR theory, what molecular geometries are associated with the following types of molecules?   (a) AB_3E (b) AB_2E_2 (c) AB_2E

7. Use hybridization to explain the bonding in methane (CH_4).

8. For each of the following polar molecules, indicate the direction of the resulting dipole: (a) H—F   (b) H—Cl   (c) H—Br   (d) H—I.

9. Determine whether each of the following bonds would be polar or nonpolar: (a) H—H   (b) H—O   (c) H—F   (d) Br—Br   (e) H—Cl   (f) H—N.

10. On the basis of the individual bond polarity and orientation, determine whether each of the molecules would be polar or nonpolar: (a) H_2O (b) I_2 (c) CH_4 (d) NH_3 (e) CO_2.

Application Problems

1. Given the bonding pairs Al—F, K—Br, C—O, and C—N, determine which bond: (a) has an electronegativity difference of 2.0;   (b) has the greatest ionic character;   (c) is polar-covalent;   (d) is nonpolar-covalent.

2. Among the five bonding arrangements described in the previous question, list these bonds in order of increasing ionic character.

3. Draw Lewis structures for each of the following molecules: (a) contains two oxygen atoms   (b) contains two nitrogen atoms   (c) contains one silicon atom and two oxygen atoms   (d) contains one carbon atom and one oxygen atom   (e) contains one sulfur atom and two oxygen atoms.

4. Draw Lewis structures for each of the following polyatomic ions: (a) the hydroxide ion—one atom of oxygen, one atom of hydrogen, and an overall charge of −1   (b) the sulfate ion—one atom of sulfur, four atoms of oxygen, and an overall charge of −2   (c) the carbonate ion—one atom of carbon, three atoms of oxygen, and an overall charge of −2   (d) the phosphate ion—one atom of phosphorous, four atoms of oxygen, and an overall charge of −3   (e) the nitrate ion—one atom of nitrogen, three atoms of oxygen, and an overall charge of −1   (f) the acetate ion—two atoms of carbon, three atoms of hydrogen, two atoms of oxygen, and an overall charge of −1.

5. Use electron-dot structures to demonstrate the formation of ionic bonds involving: (a) Na and S   (b) Ca and O   (c) Al and S.

6. Draw the Lewis structure for each of the following molecules and then use the VSEPR theory to predict the molecular geometry of each: (a) SCl_2 (b) PI_3 (c) Cl_2O (d) NH_2Cl (e) $SiCl_3Br$ (f) ONCl.

7. Draw the Lewis structure for each of the following polyatomic ions and then use VSEPR theory to identify the ion type and subsequent geometry of each: (a) NO_3^- (b) NH_4^+ (c) SO_4^{2-} (f) ClO_2^-.

Enrichment

1. Prepare and present a brief lesson to an elementary science class illustrating the formation of ionic and covalent bonds. Use models to explain the behavior of the electrons in the bonding process.

2. Identify ten common substances in and around your home, and indicate whether you would expect these substances to contain ionic, covalent, or metallic bonds.

3. Prepare a report on Gilbert Lewis and his major contributions to chemistry.

4. Discuss the historical significance (with respect to science) of sodium chloride (table salt).

5. Prepare a report tracing the development and use of adhesives.

6. Prepare a report concerning the use of sodium nitrite in meat processing. Focus on the current controversy surrounding the use of this substance.

7. Construct molecular models to illustrate the various molecular geometries explained by the VSEPR theory.

8. Prepare a brief paper explaining the cleansing action of shampoo. Focus on the significance of the polarity/nonpolarity of the detergent component.

9. Prepare a report on the work of Linus Pauling.   (a) Discuss his work on the nature of the chemical bond.   (b) Linus Pauling has been an advocate for the use of vitamin C as remedy for colds. Evaluate Pauling's claims. Determine if there is any scientific evidence that indicates whether vitamin C is an effective cold remedy.

CAREERS

Laser Chemist

Laser technology is a rapidly growing field for chemists. Laser applications range from scanning labels at the grocery store to welding detached retinas in eye surgery.

Every compact disk player has a tiny laser that senses information on the surface of the disk and is used to translate it into music. An exciting new application is in the computer industry, where gallium arsenide lasers can serve as extremely precise tools for creating the tiny circuits on computer chips.

Employment in laser chemistry looks promising. If you want to develop new applications for lasers, you need to understand how lasers operate, which requires a background in chemistry and physics. For some technical jobs in laser chemistry, an associate's degree is sufficient. To be a laser chemist, however, you must earn at least a bachelor's degree.

Chemical Technician

Chemical technicians are in high demand these days because of the growth of chemical industries. Before a new chemical product such as an improved pain killer or a solvent is ready for marketing, it must undergo a battery of tests that are usually conducted by technicians.

A chemical technician sets up laboratory equipment such as glassware, balances, burners, and computers according to a chemist's specifications. The technician then mixes chemical solutions, conducts experiments, and

collects data. For example, a technician who works in a laboratory in which materials are tested for corrosion resistance collects and records acids, bases, salts, and other chemicals that are to be used to test the materials. Then, the technician runs tests and collects data on degrees of corrosion.

A chemical technician may specialize in one area, but most are trained to work in a wide range of fields. Strong interests in chemistry and math are helpful to people who wish to become chemical technicians. In most cases, a technician has at least two years of college or technical training. An interest in computers is useful because laboratory data are entered into computers for analysis.

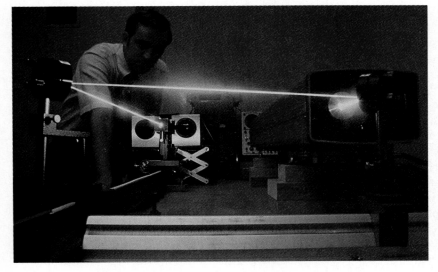

Cumulative *Review*

Unit 2 • Organization of Matter

Questions

1. What did each of the following contribute toward the modern atomic theory:
   (a) Democritus   *(b)* Dalton   *(c)* Thomson
   (d) Millikan   *(e)* Rutherford?   (3.1,2)
2. Define   *(a)* atom   *(b)* nucleus
   (c) isotopes   *(d)* atomic number   *(e)* mass number   *(f)* mole   *(g)* Avogadro's number
   (h) molar mass.   (3.1,2,3)
3. Determine the number of protons, electrons, and neutrons in each of the following isotopes: *(a)* chlorine-37   *(b)* $^{56}_{26}$Fe
   (c) deuterium.   (3.3)
4. *(a)* What is electromagnetic radiation?
   (b) Write the equation that relates the speed, wavelength, and frequency of electromagnetic radiation.   (4.1)
5. Distinguish between   *(a)* the ground state and the excited state of an atom   *(b)* a line emission spectrum and a continuous spectrum
   (c) Bohr's model of the atom and that proposed by Schrödinger.   (4.1)
6. *(a)* What is the quantum theory?   *(b)* List and explain the information provided by the four sets of quantum numbers.   (4.2)
7. *(a)* Identify the three methods used to represent the arrangement of electrons in an atom.
   (b) Illustrate each method for an atom of phosphorus.   (4.3)
8. Cite the chemist credited with   *(a)* the discovery of chemical periodicity   *(b)* the use of atomic number as the basis organizing the Periodic Table.   (5.1)
9. Explain the meaning of the periodic law.   (5.1)
10. What do main-group elements have in common if they belong to the same: *(a)* group
    (b) period   *(c)* block?   (5.1)
11. In what group are each of the following elements found in the Periodic Table: *(a)* alkali metals
    (b) alkaline earth metals   *(c)* transition elements
    (d) halogens   *(e)* noble gases.   (5.2)
12. Without looking at the Periodic Table write the expected outer electron configuration for the element in: *(a)* Period 2, Group 14   *(b)* Period 3, Group 2   *(c)* Period 4, Group 12   *(d)* Period 5, Group 17.   (5.3)
13. Define each of the following: *(a)* ionization energy   *(b)* electron affinity   *(c)* electronegativity   *(d)* cation   *(e)* anion
    (f) valence electrons.   (5.3)
14. *(a)* What is a chemical bond?   *(b)* Identify and define the three major types of chemical bonds.   (6.1,2,3,4)
15. Distinguish between: *(a)* polar– and nonpolar– covalent bonds   *(b)* ionic and molecular compounds (in terms of the basic units comprising each, melting points, boiling points, and ease of vaporization)   *(c)* metals and nonmetals (in terms of their respective ion formation tendencies).   (6.2,3,4)
16. What are Lewis structures?   (6.2)
17. Describe the electron sea model of metallic bonding.   (6.4)
18. Explain the two theories used to predict molecular geometry.   (6.5)

Problems

1. What is the mass in grams of each of the following: *(a)* 2.75 mol Ca   *(b)* 4.50×10^{25} atoms Ag?   (3.3)
2. How many moles of atoms are represented by each of the following quantities of atoms:
   (a) 3.20 g Al
   (b) 1.80×10^{26} atoms of Zn?   (3.3)
3. How many atoms are in each of the following:
   (a) 7.50 mol Pb   *(b)* 0.500 g Au?   (3.3)
4. The average concentration of Na^+ in human blood serum is 3.40 g/L. How many moles of Na^+ would be contained in 2.50 L of serum?   (3.3)
5. Given electronegativity values of 0.9 for Ba and 3.5 for O, determine the bond type expected in BaO.   (6.3)
6. Draw the Lewis structures and then predict the molecular geometry of each: *(a)* CBr_4
   (b) SCl_2 *(c)* BI_3 *(d)* SO_4^{2-} .   (6.2)

Intra-Science
How the Sciences Work Together

The Question in Chemistry: How to Replace Asbestos

When it was established beyond a doubt that asbestos was a health hazard, the call went out for a substitute. The "want ad" might have read something like this:

> Must be noncombustible and resistant to decay, vermin, and most acids. Must withstand continuous heat of up to 500°C and peak temperatures of up to 2000°C. Must have strong, yet flexible fibers that can be spun and woven. Must be cheap.

Asbestos is a general name for six types of rock-forming minerals. The most commonly used form is a fibrous serpentine called chrysolite, which consists of magnesium, silicon, hydrogen, and oxygen.

The Connection to Physics
The most important physical property of asbestos is its fibrous form,

shown in the photograph above, which permits it to be spun into yarn, cloth, or tape. Fibers are removed from crushed rock by equip-

ment that blasts and extracts, separating them from the rock matrix in which they are embedded. The resulting product is fireproof, a good thermal insulator, and has a tensile strength almost as great as some types of steel. Such physical properties led to the use of asbestos in automobile break linings, fire barriers inside walls, insulator blankets around heaters and pipes, and roofing and floor-covering materials. There are more than 200 major industrial applications of asbestos. Replacing it with a safer substance has not been easy. However, dozens of different materials have been developed to meet specific needs.

The asbestos cement used in walls, ceilings, and fire doors is being replaced by cement composites containing glass, steel, and carbon, and by glass-reinforced gypsum and polyester. Pipes that previously might have been made of asbestos cement are now made

of glass-reinforced cement, reinforced concrete, or metal.

Prefabricated asbestos insulating boards are being replaced by boards made of reinforced calcium silicate, reinforced portland cement, rock wool, and ceramic-fiber material. Vermiculite minerals and perlite, a volcanic glass, have proved adequate in insulated boards and in sprayed and loose-fill insulation.

Finding new materials to replace asbestos in brake and clutch linings

resin-impregnated mica and glass fabrics, and thermosetting materials such as polyamide. The choice between the substitutes depends on the heat resistance required.

In floor tiles and roofing materials, asbestos has been replaced by a variety of readily available substances, chiefly cellulose fibers.

The Connection to Health Sciences

The ancient Egyptians and Romans valued asbestos for its fireproofing

from 5 to 100 mm and may remain in the respiratory system—especially the lungs—indefinitely. The disease appears to be more

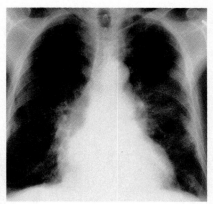

the result of fiber size and shape than of chemical properties. The irritating fibers cause an inflammatory reaction in the lungs, as seen in the X ray above, that results in scarring. If enough scarring occurs, the lungs may become less elastic and breathing more difficult. In the 1960s U.S. scientists found a link between asbestos and cancer. We now know that even a low level of exposure to asbestos is unsafe.

People working directly with asbestos are not the only ones endangered. Dust on clothing worn home from a worksite exposes the entire family to danger. Even the asbestos inside the walls of buildings presents a hazard. As scientists search for acceptable substitutes, the Environmental Protection Agency is working toward a total phaseout of all asbestos products.

One concern was that substitute fibers might be as hazardous as asbestos, but preliminary studies of likely replacements such as glass, vermiculite, and perlite have shown no health hazard. The biggest drawback in the use of these materials is increased cost.

has been a problem. Silicon nitride was used in the brake pads of the *Concorde* airliner prototypes, but was so expensive that in the production planes, reinforced carbon composites were used instead. Steel fibers are cheaper still, but pose an abrasion problem. So do glass fibers.

In the electrical industry, substitutes include oil-impregnated paper and cloth, epoxy resins,

ability, but it did not become commercially important as a building material until the 19th century. Russia and Canada became the world's leading producers. In the 1920s it was discovered that asbestos caused a severe lung disease, which was appropriately named asbestosis.

The disease asbestosis is caused by fibers of asbestos that are inhaled. These fibers range in length

Unit

3

Chapter 7
Chemical Formulas and
Chemical Compounds 200

Chapter 8
Chemical Equations and
Chemical Reactions 232

Chapter 9
Stoichiometry 260

LANGUAGE OF CHEMISTRY

I always liked chemistry and physics classes in school, but wanted a career that would let me use these sciences in a practical way. As a chemical engineer, I can use science to solve real world problems. Now, I help Du Pont's researchers choose the best way to make quality products at the lowest cost.

Chemical engineers use the principles you will learn in this unit to operate chemical plants, making sure the right chemicals are flowing in the right directions, in the right amounts, and at the right temperatures to produce the desired chemical reactions. We use a special language of formulas and equations to communicate about these processes—the language of chemistry.

Robert A. Cantini

R. A. Cantini
Senior Consulting Engineer
E. I. du Pont de Nemours & Company

Petroleum refining plant near Texas City, Texas

Chapter 7 Chemical Formulas and Chemical Compounds

Chapter Planner

7.1 Chemical Names and Formulas
Significance of a Chemical Formula
  ▨ Questions: 1, 3, 4
  ▪ Question: 2
Monatomic Ions
  ▨ Experiment: 8
  Questions: 5–8, 11–13
  ▪ Question: 9
  ▨ Question: 10
  Application Questions: 1, 2
Binary Ionic Compounds
  ▪ Demonstration: 1
  Questions: 14–19
Compounds Containing Polyatomic Ions
  ▨ Questions: 20–24
Binary Molecular Compounds
  ▨ Questions: 25–28
  ▪ Chemistry Notebook
Acids and Salts
  ▨ Demonstration: 2
  Questions: 29–31
  ▪ Desktop Investigation

7.2 Oxidation Numbers
Assigning Oxidation Numbers
  ▨ Demonstration: 3
  Questions: 32–34
Assigning Oxidation Numbers for Formulas and Names
  ▨ Questions: 35, 36

7.3 Using Chemical Formulas
Formula Masses and Molar Masses
  ▨ Questions: 37, 38
  Problems: 1, 2, 4
  Application Problem: 1
  Application Question: 3
Molar Mass as a Conversion Factor
  ▨ Problems: 5–9
  Application Problems: 2, 3
  ▪ Problem: 3
  Application Problems: 4–8

Percent Composition
  ▨ Question: 39
  Problems: 10, 11
  Application Problem: 9
  ▪ Application Problems: 10, 11
  ▨ Problem: 12

7.4 Determining Chemical Formulas
Calculation of Simplest Formulas
  ▨ Experiment: 10
  Questions: 40, 41
  Problems: 13–17
  Application Problems: 12–14
Calculation of Molecular Formulas
  ▨ Question: 42
  Problems: 18, 19
  Application Problems: 15, 16
  ▪ Application Question: 4

Teaching Strategies

Introduction

John Dalton, the English chemist, performed important experiments in the early 19th century involving weight relationships between elements in compounds. To account for the regularities that he observed, Dalton formulated the theory that matter was composed of small units called atoms and that these atoms combined in definite proportions; if the proportion was different, the compound was different. Dalton's work is the basis of Chapter 7.

7.1 Chemical Names and Formulas

An important basic concept in chemistry is presented in this chapter; a chemical compound is represented by a unique chemical formula. The formula consists of symbols for each element in the compound and subscripts which indicate the number of atoms of each element in the compound.

If the symbols are changed, the formula represents a different compound; if the subscripts are changed, the formula repre-

sents a different compound. Students often have difficulty recognizing the importance of maintaining the subscripts once a formula has been determined. You might ask your students to "engrave in stone" correct formulas, or to put boxes around them, so they know that once determined, the formulas cannot be changed. Elements are also represented by symbols and subscripts, but more about that later.

Begin this chapter by bringing to class many compounds and elements for your students to observe. A good collection might be H_2O, H_2O_2, KI, NaI, S_8, Na_2S, Na_2SO_4, Cu_2SO_4, $CuSO_4$, FeO, Fe_2O_3, CH_4, CH_3OH, K_2CrO_7, K_2CrO_4. Have students list their observations of each substance along with its formula. You might distinguish between H_2O and H_2O_2 by adding a few mL of each to a dilute solution of $KMnO_4$. This will not only vividly demonstrate the differences in the chemical properties of H_2O and H_2O_2 but will prepare students for the fact that a single element can exist in various oxidation states.

All of chemistry is encoded in the Periodic Table. It is the job of the chemistry student to learn to decode this information. If you can, give each one of your students a notebook copy of the Periodic Table. The Sargent Welch table is two-sided, and contains excellent information about ionization potential and electronegativity. Require your students to bring this table to class each day. If you cannot supply them with a supplementary table, there is one in the back of their textbook which can be photocopied.

Refer to the Periodic Table constantly in this unit. Point to the class table and indicate that column 1 elements always form +1 ions , column 2 elements always form +2 ions, column 3 elements always form +3 elements and column 17 elements usually form −1 ions. Emphasize that positive ions have more protons than electrons and that negative ions have more electrons than protons. Students manage to construct creative but false theories to explain the charge on particles.

Do a particle count, making use of the Periodic Table to determine the number of protons in the nucleus of each element:

Na	yields	Na +
11 protons		11 protons
11 electrons		10 electrons

Cl$_2$	yields	2Cl$^-$
17 protons	(each Cl atom)	17 protons
17 electrons		18 electrons

Continually remind students that electrons are negatively charged and protons are positively charged. Atoms are neutral: ions are charged particles. The charge on the ion is determined by the balance of positive charges contributed by the protons and the negative charges contributed by the electrons. When an atom loses an electron, the resulting ion is positive because it has more protons than electrons. Students have difficulty with the concept that an atom loses a particle and becomes positive.

All students are familiar with common table salt. Use this compound to illustrate a few points about binary ionic compounds. Ions are charged particles, but the compounds they form are neutral. When writing the formula for the compound, the positive ion is written first and the negative ion, which is written second, has an -ide ending. Refer to the Periodic Table as you are presenting this. Touching the salt and not receiving an electric charge might convince students that the compound is neutral even though it is composed of positive and negative charges. Students will be convinced that the compound does form charged particles if you compare the results of a conductivity test in salt water with the results of a conductivity tester dipped in distilled water. If you also test a sugar solution for conductivity, this demonstration will set the stage for discussing the difference between ionic and molecular compounds.

If space-filled styrofoam models are available, this is an excellent time to use them. NaCl illustrates the ionic nature of the compound as well as showing the difference in size between the positive and negative ions.

If students understand that an ionic compound is composed of charged particles combined to form a neutral substance, they will not have too much trouble writing formulas. Go over the rules carefully. Give the class lots of sample problems. Refer to the Periodic Table as you do that. Diagram the formation of compounds on the chalkboard.

Demonstration 1: The synthesis of a binary iodide compound

PURPOSE: To demonstrate the formation of a compound from two elements
[Note: This reaction will be done quantitatively in Chapter 9. At this time, emphasize the qualitative evidence that a reaction has occurred.]
MATERIALS: 50-mL beaker, 0.5 g of metal—use mossy zinc or coiled copper wire, 0.5 g of iodine crystals, watch glass, ice cube, water bath or heater with temperature control
PROCEDURE: Place 0.5 g of metal in a beaker. Pass the beaker around the classroom. Ask each student to record the properties of the metal. Add iodine crystals to the beaker. Place a watch glass over the beaker and an ice cube on the watch glass. Heat gently over a water bath or on a heater set at a low temperature. Allow to remain for fifteen minutes.

Iodine will sublime resulting in a beautiful purple vapor.
CAUTION: Iodine vapors are toxic—but iodine is the least toxic of all halogens—the ice cube will contain the vapors so that students are not exposed to them. This demonstration should be done in a hood. Some of the vapor will react with the metal. Excess iodine will recrystallize on the bottom of the watch glass. Remove ice cube from watch glass. Discard meltage. Ask one student to carry the watch glass around the room so that students will be able to observe the iodine crystals that have formed. Remove metal with a tweezers and place in a petri dish. Pass around the room so that each student can observe the iodide compound that has formed on the surface of the metal.

A compound has formed from the combination of two elements. The properties of the compound are different than the properties of the elements. A chemical reaction has occurred. In the course of the reaction, electrons have been transferred from the metal to the iodine. Before the iodine received the electrons from the metal, the bond was broken that held the two iodine atoms together into one diatomic elemental molecule. Each iodine atom received one of the metallic electrons.

Write the reaction on the board.

$$Cu + I_2 \rightarrow CuI_2$$

Do a particle count.

Cu	$\rightarrow$ Cu^{+2}	+ 2 electrons
29 protons	29 protons	
29 electrons	27 electrons	

2 electrons	+ I^2	$\rightarrow$ 2I-
	53 protons	53 protons
each atom	53 electrons	54 electrons

$$Cu^{+2} + 2I^- \rightarrow CuI_2$$

Use models to simulate this reaction. The methods and materials for using common materials to make models are described the Desktop Investigation.

Naming compounds can be tricky, but is actually easier than all of the rules seem to indicate. The rules are important and memorization should be encouraged. But, you can try to prevent many of your students' difficulties by showing them how

Chemical Formulas and Chemical Compounds   **200B**

much they already know. They will probably be surprised how much they have learned about naming elements and compounds just through their brief introduction to chemistry in Chapters 1–6. You can encourage them to apply this knowledge by writing formulas on the board and asking for their names. Start with what they may already be familiar with. Also, utilize the tables in the chapter and other resources that you may have. This is an easy way to become familiar with most of the common monatomic and polyatomic ions. Remember that practice makes perfect. Help your students to become familiar with the common names.

A binary compound consists of two elements. To name *binary ionic compounds*, the following rules are used: *(1)* Name the metal first. An *-ide* is added to the stem of the nonmetal (ex. sodium chloride). *(2)* If the metal exhibits more than one oxidation state, a Roman numeral is used to indicate which oxidation state of the metal is present in the compound. The Roman numeral is placed in parentheses following the name of the metal. An *-ide* ending is added to the stem of the nonmetal as in *(1)* above (ex. FeO, iron(II) oxide; Fe_2O_3, iron(III) oxide). To name *binary molecular compounds*, prefixes are used to indicate the number of atoms of the nonmetal. The rules for naming binary ionic compounds are then applied (e. g. PCl_3, phosphorous trichloride; PCl_5, phosphorous pentachloride).

Chemistry is based on observations and problem solving. A lot of memorizing is usually not a good strategy, but learning the names, formulas, and charges of polyatomic ions is the exception. Students have to memorize this, and the sooner the better. One hint that you might give students regarding the charge on ions is that elements in odd numbered columns in the Periodic Table form oxyions that have an odd negative charge (-1 or -3) such as PO_4^{3-}, NO_3^{1-}, NO_2^{1-}, and IO_3^{1-}, and that elements in even numbered columns in the Periodic Table form oxyions that have an even negative charge (-2) such as CO_3^{2-} and SO_4^{2-}.

If students have memorized the table of polyatomic ions, writing formulas for compounds containing polyatomic ions should not be more difficult than writing formulas for binary compounds. Emphasize that parentheses must be placed around the polyatomic ion if two or more exist in the compound. At this point, make sure that students know how to analyze a formula for the correct number of atoms.

CuI_2	1 copper	2 iodine	
NaCl	1 sodium	1 chlorine	
Na_2SO_4	2 sodium	1 sulfur	4 oxygen
$Al_2(SO_4)_3$	2 aluminum	3 sulfur	12 oxygen

Many demonstrations can be done effectively with small quantities of chemicals in 50-mL beakers on an overhead projector projected on a screen. The identity of the material can be included on a transparency prepared for use with 50-mL beakers. The transparency should include a circle for each beaker that you use. Place the beaker on the appropriate circle. You can prepare your master by placing your 50-mL beakers on a piece of paper and tracing around them.

Demonstration 2: Observations of Aqueous Solutions of Acids and Salts
Purpose: To observe some of the properties of aqueous solutions of acids and salts
Materials: 8 50-mL beakers; 0.1-*M* solutions of sulfuric acid,

nitric acid, phosphoric acid, hydrochloric acid; 4 small vials containing sodium sulfate, sodium nitrate, sodium phosphate, sodium chloride; universal indicator
Procedure: Add 20 mL of each acid to a 50-ml beaker placed on an overhead projector with projection on screen. Add 2 drops of universal indicator to each salt solution. Record color. Observe and record physical properties of the salts. Add 20 mL of water to the four empty beakers. Add 0.1 g of each salt to a beaker and stir. Record observations. Add 2 drops of universal indicator to each salt solution. Record color. Compare the color of the indicator in the acid solution to the color of the indicator in the corresponding acid solution. Write the formulas of the acids and the salts.

7.2 Oxidation Numbers

Begin this topic by referring to the Periodic Table. If you are using the Sargent Welch Table, refer to the Pauling electronegative values on page 2. At this point, it is a good idea to expose students to the difference between charge (the actual transfer of electrons) and oxidation state (the distribution of electrons). Do not expect them to thoroughly understand this concept yet. They will review it again in Chapter 20 (and throughout their experiences with chemical reactions). Do make sure they know all of the rules for assigning oxidation numbers.

The various oxidation states of manganese are indicated by a variety of identifying colors. The following demonstration can be done at this time so that students can observe changing oxidation states. Emphasize the formula, the oxidation state of the ion, the oxidation state of the manganese, and the color of the compound. Do not ask students to write equations for these reactions at this time. You can redo this demonstration and discuss the underlying chemical phenomona at that time.

Demonstration 3: Oxidation of alcohols
Purpose: To observe and relate the colors and oxidation states of manganese
Materials: 0.1 *M* $NaHSO_3$ (dissolve 1.04 g $NaHSO_3$ in 100 mL of H_2O), 0.05 *M* $KMnO_4$ (dissolve 0.079 g powdered $KMnO_4$ in 100 mL H_2O, 0.005 *M* $KMnO_4$ (mix 10 mL of solution 2 with 90 mL of H_2O), 0.000 40 *M* $KMnO_4$ (mix 20 mL of solution 3 with 230 mL of H_2O), 6 *M* HCl, 8 *M* NaOH.
Procedure: Add 20 mL of 0.000 40 *M* $KMnO_4$ in each of five 50-mL beakers on an overhead projector. Add the HCl, NaOH, and $NaHSO_3$ by drops as needed to each beaker. Use stirring rods to mix solutions after addition of each reagent.

Drops HCL	Drops NaOH	Drops NaHSO3	Color	Product	Oxidation State of Manganese
0	0	0	purple	MnO_4^{1-}	7
8	0	8	colorless	Mn^{2+}	2
0	0	4	orange	MnO_2	4
0	8	1	green	MnO_4^{2-}	6

It is a bit difficult to obtain the green MnO_4^{2-}. Too little $NaHSO_3$ will produce a blue coloration due to the incomplete reduction of $KMnO_4$. This is really due to absorption of light by two chromophores, MnO_4^{1-} and MnO_4^{2-} leaving blue as the only visible color. If you get this result, add as small a drop as possible and mix. If too much $NaHSO_3$ is added, manganese will be reduced to the orange MnO_2 stage. If this happens, try again. Getting the green Mn^{+6} is well worth the effort.

The attractive force for electrons is referred to as the electronegativity of an element. The greater the electronegativity, the stronger the affinity for electrons. Those elements on the right side of the Periodic Table are more electronegative than those on the left because their electron shells are more nearly complete. Those on the left side have relatively incomplete electron shells and are thus more inclined to donate them. Also, elements lower in the same column of the Periodic Table are larger and, therefore, less electronegative. This is because of the reduced force of attraction between the nucleus (positive charge) and the electrons of the outer energy level. This is due to the increased distance between the nucleus and the outer electron energy level, and because of the shielding effect of core electrons. (See Chapter 5, Section 5.3 regarding electronegativity).

Have students determine the oxidation states of lots of atoms in various ionic states and compounds. Again, show them many different chemicals (e. g. Na_2CrO_4, $Na_2Cr_2O_7$). First determine the oxidation state of the ion. Base this on the fact that Na is always $+1$, and there are two sodium ions in the compounds. Determine the oxidation states of the chromate and dichromate ions. After determining the oxidation state of these ions, and assigning -2 to each oxygen atom, the oxidation state of chromium in each ion can be determined. The same strategy can be used to determine the oxidation states of the central atom in all other polyatomic ions.

Desktop Investigation: Models of Chemical Compounds

Making models of chemical compounds gives students an excellent opportunity to observe the structure of a variety of ionic and molecular substances. Use this investigation to introduce the shapes of molecules, the arrangement of atoms in molecules, the arrangements of ions in crystals, and the differences between ionic and molecular compounds. Students are so accustomed to seeing representations of molecules in two dimensions that it is very important for them to have as many opportunities as possible to see three dimensional representations of these substances. Emphasize that atoms, ionic lattices, and molecules exist in three dimensions, not two. Supplement this Desktop Investigation with a display of other three-dimensional materials. Use styrofoam models if you have them.

This is a good opportunity to use computer displays of molecules.

7.3 Using Chemical Formulas

Once students learn about the combining of ions to form compounds and the oxidation state of atoms, they are ready to start working with the essential quantitative elements used in understanding these compounds. These essential quantitative elements are: *formula mass, molar mass,* and *percent composition.* Formula mass and molar mass are straightforward concepts if correct compound formulas are used. They are essential concepts in order to begin work with chemical reactions. Students who have become familiar with formulas should have little difficulty with this section. Some frustrations may arise as a result of using atomic masses. (These are found in the Appendix, the Periodic Table inside the back cover, and approximate weights are found in the table on the inside front cover.) Emphasis should be given to the use, and the importance, of significant figures. It should also be emphasized that small errors to the hundredth or thousandth of a gram are relatively insignificant. Some students may be greatly relieved to know this. This is especially important when your students begin to work problems, particularly if they have some anxiety already.

Students generally do not have any difficulties, other than formula writing, with these percent composition problems. Once in a while they will make a mathematical error in calculations. A quick and easy way to check this, assuming they have started with the right formula, is to check the resulting percentages to see if they add up to 100%.

Experiment 10, *Water Crystallization and Empirical Formula of a Hydrate,* gives students the opportunity to determine the number of moles of water that are bound to one mole of anhydorus salt.

7.4 Determining Chemical Formulas

Students sometimes have difficulty with this analysis of composition data. You might not want to present this as fundamental material. Three element compounds and molecular formulas that differ from empirical formulas might be considered enrichment. The common mistakes that students make in determining the formula from data are *(1)* to forget to convert percent to grams, and *(2)* to forget to convert grams to moles. Remind students to keep enough significant figures when determining number of moles. This would be an appropriate unit for supplementary computer tutorials.

References and Resources

Books and Periodicals

Fernelius. "Numbers in Chemical Names," *Journal of Chemical Education,* Vol. 59, No. 11, 1982, p. 965.

Kolb, Doris. "Chemical Formulas, Part I: Development," *Journal of Chemical Education,* Vol. 55, No. 1, 1978, p. 44.

Kolb, Doris. "The Chemical Formula, Part II: Determination," *Journal of Chemical Education,* Vol. 55, No. 2, 1978, p. 109.

Meyers. "Electronegativity, Bond Energy, and Reactivity," *Journal of Chemical Education,* Vol. 56, No. 11, 1979, p. 711.

Pearson, Robert and George Arnold. "Manganese Color Reactions," *Journal of Chemical Education,* Vol. 65, 1988, p. 451.

Shakhashiri, Bassam Z. "Reactions of Fe and S," (p. 55) and "Reaction of Al and Br," (p. 68), and "Reaction of Na and Cl," (p. 61), *Chemical Demonstrations,* Vol. 1, Madison: University of Wisconsin Press, 1963. *The Mole Concept,* VHS cassette, TV Ontario.

Computer Software

Fundamental Skills for General Chemistry 3 (ELEMENT, IONS, NOMEN, BALEQ), Apple II, TRS-80, Programs for Learning.

Howbert, J. Jeffrey. *Molecular Animator*, Apple II, IBM, COMPress.

Inorganic Nomenclature II, Apple II, IBM, Intellectual Software.

Michel-Evleth, William H., *Language of Chemistry; The Personal Chemical Nomenclature Learning System*, Apple II, COMPress.

Project Seraphim, Disk #303, NSF Science and Engineering Education, Dept. of Chemistry, Eastern Michigan University, Ypsilanti, MI 48197.

Smith, S.G., Ruth Chabay, and Elizabeth Kean. *Percent Composition*, Apple II, IBM, COMPress.

Topic 3: Formulas, Moles, Stoichiometry, Apple II, Knowledge Factory.

Weyh, John A., Joseph R. Crook, and Les N. Hauge, *Mole Concept*, Version 2.0, IBM, COMPress.

Answers and Solutions

Questions

1. *(a)* Chemical compounds are represented by chemical formulas and chemical names. *(b)* The letter symbols indicate the elements that are present, and the subscripts show how many atoms or ions of each type are combined. *(c)* In addition to the formula, knowledge of the elements, of electron configurations, and of how bonds are formed is needed.

2. The indicated formulas and systematic names are: *(a)* $NaHCO_3$, sodium hydrogen carbonate *(b)* $Mg(OH)_2$, magnesium hydroxide *(c)* $MgSO_4 \cdot 7H_2O$, magnesium sulfate *(d)* $CACO_3$, calcium carbonate *(e)* $NaOH$, sodium hydroxide *(f)* CH_3OH, methanol.

3. Chemical composition is found by analysis or by consulting works that report the results of experiments.

4. A chemical formula can represent one molecule, one mole, or one molar mass of the indicated material.

5. *(a)* The main group elements typically form ions with noble-gas configurations by either gaining or losing electrons. *(b)* The typical charges are: Group 1 = 1+, Group 2 = 2+, Group 13 = 3+, Group 14 = 4+, Group 15 = 3−, Group 16 = 2−, and Group 17 = 1−.

6. *(a)* Monatomic ions are ions formed from a single atom. *(b)* Examples include Na^+, Mg^{2+}, and N^{3-}.

7. The *d*-block elements generally form ions that have 2+, 3+, or, in a few cases, 1+ charges. Several *d*-block elements form common ions that have two different charges.

8. The charges involved are: *(a)* 1 e⁻ lost *(b)* 1 e⁻ gained *(c)* 2 e⁻ lost *(d)* 2 e⁻ gained *(e)* 3 e⁻ lost.

9. The ion most often formed in each case is: *(a)* K^+ *(b)* Ca^{2+} *(c)* S^{2-} *(d)* Cl^- *(e)* Ba^{2+} *(f)* Br^-.

10. The indicated configurations are: *(a)* $Na = [Ne]3s^1$, $Na^+ = [Ne]$ *(b)* $Cl = [Ne]3s^23p^5$, $Cl^- = [Ar]$ *(c)* $Ca = [Ar]4s^2$, $Ca^{2+} = [Ar]$ *(d)* $O = [He]2s^22p^4$, $O^{2-} = [Ne]$.

11. Positive ions are named by the element name followed by *ion*. Negative ions are named by dropping the ending of the element name and adding "-*ide*" to the stem.

12. The indicated formulas are: *(a)* Na^+ *(b)* Al^{3+} *(c)* Cl^- *(d)* N^{3-} *(e)* Fe^{2+} *(f)* Fe^{3+}.

13. The indicated names are: *(a)* potassium ion *(b)* magnesium ion *(c)* aluminum ion *(d)* chloride ion *(e)* oxide ion *(f)* calcium ion *(g)* sodium ion.

14. A binary compound is a compound composed of two different elements.

15. (1) Write the letter symbols for the ions side by side, with the positive ion first. (2) Cross over the charge values to give subscripts. (3) Check the subscipts and write the formula.

16. The indicated formulas are: *(a)* NaI *(b)* CaS *(c)* $ZnCl_2$ *(d)* BaF_2 *(e)* Li_2O *(f)* CrF_2 *(g)* NiO *(h)* Fe_2O_3.

17. The indicated compound names are: *(a)* potassium chloride *(b)* calcium bromide *(c)* magnesium iodide *(d)* sodium fluoride *(e)* silver iodide *(f)* lithium oxide *(g)* aluminum bromide.

18. The Stock system is a method of naming chemical compounds that uses Roman numerals to designate positive ions of different charges formed by the same metal.

19. The indicated names are: *(a)* iron(II) *(b)* iron(III) *(c)* lead(II) *(d)* lead(IV) *(e)* tin(II) *(f)* tin(IV) *(g)* mercury(I).

20. *(a)* Oxanions are polyatomic ions that contain oxygen. *(b)* In such oxyanion pairs, the ion with the larger number of oxygen atoms is named with the -*ate* ending, and the ion with the smaller number of oxygen atoms is named with the -*ite* ending. *(c)* NO_3^- is the nitrate ion; NO_2^- is the nitrite ion.

21. The indicated formulas and charges are: *(a)* NH_4^+ *(b)* $C_2H_3O_2^-$ *(c)* OH^- *(d)* CO_3^{2-} *(e)* SO_4^{2-} *(f)* PO_4 *(g)* Cu^{2+} *(h)* Sn^{2+} *(i)* Fe^{3+} *(j)* Cu^+ *(k)* Hg^+ *(l)* Hg^{2+}.

22. The indicated names are: *(a)* ammonium ion *(b)* chlorate ion *(c)* hydroxide ion *(d)* sulfate ion *(e)* nitrate ion *(f)* carbonate ion *(g)* phosphate ion *(h)* acetate ion *(i)* hydrogen carbonate ion *(j)* chromate ion.

23. The indicated formulas are: *(a)* NaF *(b)* CaO *(c)* K_2S *(d)* $MgCl_2$ *(e)* $AlBr_3$ *(f)* Li_3N *(g)* FeO *(h)* Cu_2S *(i)* $NaOH$ *(j)* NH_4Cl.

24. The indicated names are: *(a)* sodium chloride *(b)* magnesium oxide *(c)* potassium fluoride *(d)* calcium sulfide *(e)* lithium bromide *(f)* potassium hydroxide *(g)* silver nitrate *(h)* iron(II) hydroxide *(i)* tin(II) nitrate *(j)* cobalt(II) nitrate *(k)* iron(III) phosphate *(l)* mercury(I) sulfate *(m)* mercury(II) phosphate.

25. *(a)* The system for naming binary molecular compounds is based on prefixes that indicate the number of atoms of each type that are present. *(b)* The indicated prefixes are *mono-*, *di-*, *tri-*, *tetra-*, *penta-*, and *hexa-*, respectively.

26. The less electronegative element is given first in the names and formulas of binary molecular compounds.

27. The indicated names are: *(a)* carbon dioxide *(b)* carbon tetrachloride *(c)* phosphorous pentachloride *(d)* carbon disulfide *(e)* sulfur dioxide *(f)* phosphorous tribromide *(g)* dinitrogen tetroxide *(h)* carbon tetraiodide *(i)* diarsenic pentoxide *(j)* tetraphosphorous decaoxide.

28. The indicated formulas are: *(a)* CBr_4 *(b)* SiO_2 *(c)* CO *(d)* SO_3 *(e)* P_2O_5 *(f)* As_2S_3.

29. Binary acids, such as HCl and HF, are solutions in water of binary molecular compounds containing hydrogen and, usually, one of the halogens. Oxyacids are acids, such as H_2SO_4 and HNO_3, containing hydrogen, oxygen, and a third element.

30. *(a)* A salt is an ionic compound composed of a cation and of the anion from an acid. *(b)* Two common salts are KCl and $MgSO_4$.

31. The indicated acids are: *(a)* hydrofluoric acid *(b)* hydrobromic acid *(c)* nitric acid *(d)* sulfuric acid *(e)* phosphoric acid.

32. *(a)* Oxidation numbers are assigned numbers that show the general distribution of electrons among bonded atoms. *(b)* Oxidation numbers are useful in naming compounds, in writing formulas, and in studying certain types of chemical reactions.

33. The assigned oxidation numbers are: *(a)* $+1, -1$ *(b)* $+4, -2$ *(c)* $+3, -1$ *(d)* $+4, -2$ *(e)* $+2, -1$ *(f)* $+1, 1$ *(g)* $+1, -1$ *(h)* $+5, -2$ *(i)* $+5, -2$ *(j)* $+1, +5, -1$.

34. The assigned oxidation numbers are: *(a)* $+5, -2$ *(b)* $-3, +1$ *(c)* $+7, -2$ *(d)* $+4, -2$ *(e)* $+5, -2$ *(f)* $+2, -3$ *(g)* $+7, -2$ *(h)* $+2, -2$ *(i)* $+6, -2$.

35. *(a)* carbon(IV) bromide *(b)* Silicon(IV) oxide *(c)* carbon(II) oxide *(d)* Sulfur(VI) oxide *(e)* phosphorous(V) oxide *(f)* arsenic(III) sulfide

36. The indicated formulas are: *(a)* PI_3 *(b)* SCl_2 *(c)* CS_2.

37. *(a)* The formula mass of any compound or polyatomic ion is the sum of the atomic masses of all atoms represented in the formula. *(b)* Formula mass is expressed in atomic mass units (*u*).

38. The molar mass of a compound is the mass, in grams, numerically equal to the formula mass of the compound.

39. *(a)* The percent composition of a compound is the percent by mass of each element in the compound.

   (b) Mass percent of element $= \dfrac{\text{mass of element}}{\text{molar mass of compound}} \times 100\%$

40. A simplest, or empirical, formula consists of the symbols for the elements combined, with subscripts showing the smallest whole-number ratio of the atoms.

41. Finding the simplest formulas from percent-composition requires a knowledge of composition by mass, composition in moles, and the smallest whole-number ratio of atoms.

42. The subscripts in the molecular formula are whole-number multiples of those in the simplest formula.

Problems

1. *(a)* $6 \text{ C atoms} \times \dfrac{12.0111\,u}{\text{C atom}} = 72.0666\,u$

   $12 \text{ H atoms} \times \dfrac{1.007\,94\,u}{\text{H atom}} = 12.095\,u$

   $6 \text{ O atoms} \times \dfrac{15.9994\,u}{\text{O atom}} = 95.9964\,u$

   Formula mass of $C_6H_{12}O_6 = 180.1583\,u$

   (b) $1 \text{ Ca atom} \times \dfrac{40.08\,u}{\text{Ca atom}} = 40.08\,u$

   $4 \text{ C atoms} \times \dfrac{12.0111\,u}{\text{C atom}} = 48.0444\,u$

   $6 \text{ H atoms} \times \dfrac{1.007\,94\,u}{\text{H atom}} = 6.047\,64\,u$

$4 \text{ O atoms} \times \dfrac{15.9994\,u}{\text{O atom}} = 63.9976\,u$

Formula mass of $Ca(C_2H_3O_2)_2 = 158.17\,u$

(c) $1 \text{ N atom} \times \dfrac{14.0067\,u}{\text{N atom}} = 14.0067\,u$

$4 \text{ H atoms} \times \dfrac{1.007\,94\,u}{\text{H atom}} = 4.031\,76\,u$

Formula mass of $NH_4^+ = 18.0385\,u$

(d) $1 \text{ Cl atom} = \dfrac{35.453\,u}{\text{Cl atom}} = 35.453\,u$

$3 \text{ O atoms} = \dfrac{15.9994\,u}{\text{O atom}} = 47.9982\,u$

Formula mass of $ClO_3^- = 83.451\,u$

2. The indicated molar masses are: *(a)* 180.1583 g *(b)* 158.17 g *(c)* 18.0385 g *(d)* 83.451 g.

3. The indicated numbers are *(a)* $1 \text{ mol } K^+$ and $1 \text{ mol } NO_3^-$; 1 mol K, 1 mol N, and 3 mol O *(b)* $2 \text{ mol } Na^+$ and $1 \text{ mol } SO_4^{2-}$; 2 mol Na, 1 mol S, and 4 mol O *(c)* $1 \text{ mol } Ca^{2+}$ and $2 \text{ mol } OH^-$; 1 mol Ca, 1 mol O, and 2 mol H *(d)* $2 \text{ mol } NH_4^+$ and $1 \text{ mol } SO_3^{2-}$; 2 mol N, 8 mol H, 1 mol S, and 3 mol O *(e)* $3 \text{ mol } Ca^{2+}$ and $2 \text{ mol } PO_4^{3-}$; 3 mol Ca, 2 mol P, and 8 mol O *(f)* $2 \text{ mol } Al^{3+}$ and $3 \text{ mol } CrO_4^{2-}$; 2 mol Al, 3 mol Cr, and 12 mol O.

4. *(a)* $1 \text{ mol K} \times \dfrac{39.0983\,\text{g K}}{\text{mol K}} = 39.0983\,\text{g K}$

   $1 \text{ mol N} \times \dfrac{14.0067\,\text{g N}}{\text{mol N}} = 14.0067\,\text{g N}$

   $3 \text{ mol O} \times \dfrac{15.9994\,\text{g O}}{\text{mol O}} = 47.9982\,\text{g O}$

   Molar mass of $KNO_3 = 101.1032\,\text{g}$

   (b) $2 \text{ mol Na} \times \dfrac{22.989\,77\,\text{g}}{\text{mol Na}} = 45.979\,54\,\text{g}$

   $1 \text{ mol S} \times \dfrac{32.06\,\text{g S}}{\text{mol S}} = 32.06\,\text{g}$

   $4 \text{ mol O} \times \dfrac{15.9994\,\text{g O}}{\text{mol O}} = 63.9976\,\text{g}$

   Molar mass of $Na_2SO_4 = 142.04\,\text{g}$

   (c) $1 \text{ mol Ca} \times \dfrac{40.08\,\text{g Ca}}{\text{mol Ca}} = 40.08\,\text{g}$

   $2 \text{ mol O} \times \dfrac{15.9994\,\text{g O}}{\text{mol O}} = 31.8888\,\text{g}$

   $2 \text{ mol H} \times \dfrac{1.007\,94\,\text{g H}}{\text{mol H}} = 2.015\,88\,\text{g}$

   Molar mass of $Ca(OH)_2 = 74.09\,\text{g}$

   (d) $2 \text{ mol N} \times \dfrac{14.0067\,\text{g N}}{\text{mol N}} = 28.0134\,\text{g}$

   $8 \text{ mol H} \times \dfrac{1.007\,94\,\text{g H}}{\text{mol H}} = 8.063\,52\,\text{g}$

   $1 \text{ mol S} \times \dfrac{32.06\,\text{g S}}{\text{mol S}} = 32.06\,\text{g}$

   $3 \text{ mol O} \times \dfrac{15.9994\,\text{g O}}{\text{mol O}} = 46.9982\,\text{g}$

   Molar mass of $(NH_4)_2SO_3 = 116.14\,\text{g}$

(e) $3 \text{ mol Ca} \times \dfrac{40.08 \text{ g Ca}}{\text{mol Ca}} = 120.24 \text{ g}$

$2 \text{ mol P} \times \dfrac{30.973\ 76 \text{ g P}}{\text{mol P}} = 61.947\ 52 \text{ g}$

$8 \text{ mol O} \times \dfrac{15.9994 \text{ g O}}{\text{mol O}} = 127.9952 \text{ g}$

Molar mass of $Ca_3(PO_4)_2 = 310.18 \text{ g}$

(f) $1 \text{ mol Al} \times \dfrac{26.981\ 54 \text{ g Al}}{\text{mol Al}} = 26.981\ 54 \text{ g}$

$3 \text{ mol Cr} \times \dfrac{51.996 \text{ g Cr}}{\text{mol Cr}} = 155.988 \text{ g}$

$12 \text{ mol O} \times \dfrac{15.9994 \text{ g. O}}{\text{mol O}} = 191.9928 \text{ g.}$

Molar mass of $Al_2(CrO_4)_3 = 374.962 \text{ g}$

5. (a) Molar mass of NaCl = 58.443 g

$1.000 \text{ mol NaCl} \times \dfrac{58.44 \text{ g NaCl}}{\text{mol NaCl}} = 58.44 \text{ g NaCl}$

(b) Molar mass of $H_2O = 18.0153 \text{ g}$

$2.000 \text{ mol } H_2O \times \dfrac{18.02 \text{ g } H_2O}{\text{mol } H_2O} = 36.04 \text{ g } H_2O$

(c) Molar mass of $Ca(OH)_2 = 74.09 \text{ g}$

$3.500 \text{ mol } Ca(OH)_2 \times \dfrac{74.09 \text{ g } Ca(OH)_2}{\text{mol } Ca(OH)_2} = 259.3 \text{ g } CaOH_2$

(d) Molar mass of $Ba(NO_3)_2 = 261.34 \text{ g}$

$0.625 \text{ mol } Ba(NO_3)_2 \times \dfrac{261 \text{ g } Ba(NO_3)_2}{\text{mol } Ba(NO_3)_2} = 163 \text{ g } Ba(NO_3)_2$

6. (a) Molar mass of $H_2O = 18.0153 \text{ g}$

$4.50 \text{ g } H_2O \times \dfrac{\text{mol } H_2O}{18.0 \text{ g } H_2O} = 0.250 \text{ mol } H_2O$

(b) Molar mass of $Ba(OH)_2 = 171.34 \text{ g}$

$471.6 \text{ g } Ba(OH)_2 \times \dfrac{\text{mol } Ba(OH)_2}{171.3 \text{ g } Ba(OH)_2}$
$= 2.753 \text{ mol } Ba(OH)_2$

(c) Molar mass of $Fe_3(PO_4)_2 = 357.484 \text{ g}$

$129.68 \text{ g } Fe_3(PO_4)_2 \times \dfrac{\text{mol } Fe_3(PO_4)_2}{357.48 \text{ g } Fe_3(PO_4)_2}$
$= 0.362\ 76 \text{ mol } Fe_3(PO_4)_2$

7. (a) $6.022 \times 10^{23} \text{ units } CO_2 \times \dfrac{\text{mol } KNO_3}{6.022 \times 10^{23} \text{ units } CO_2}$
$= 1.000 \text{ mol } CO_2$

(b) $2.008 \times 10^{23} \text{ units } KNO_3 \times \dfrac{\text{mol } KNO_3}{6.022 \times 10^{23} \text{ units } KNO_3}$
$= 0.3334 \text{ mol } KNO_3$

8. (a) $3.012 \times 10^{23} \text{ units } SO_2 \times \dfrac{64.06 \text{ g } SO_2}{6.022 \times 10^{23} \text{ units } SO_2}$
$= 32.04 \text{ g } SO_2$

(b) $1.506 \times 10^{12} \text{ units } BaCl_2 \times \dfrac{208.24 \text{ g } BaCl_2}{6.022 \times 10^{23} \text{ units } BaCl_2}$
$= 5.207 \times 10^{-10} \text{ g } BaCl_2$

9. (a) $97.45 \text{ g ZnS} \times \dfrac{6.022 \times 10^{23} \text{ units Zn S}}{97.45 \text{ g ZnS}}$

$= 6.022 \times 10^{23} \text{ units ZnS}$

(b) $18.75 \text{ g } Cu(NO_3)_2 \times \dfrac{6.022 \times 10^{23} \text{ units } Cu(NO_3)_2}{187.6 \text{ g } Cu(NO_3)_2}$
$= 6.019 \times 10^{22} \text{ units } Cu(NO_3)_2$

(c) $69.55 \text{ g } PbCl_2 \times \dfrac{6.022 \times 10^{23} \text{ units } PbCl_2}{278.1 \text{ g } PbCl_2}$
$= 1.506 \times 10^{23} \text{ units } PbCl_2$

10. (a) Molar mass of NaCl = 58.443 g

$\% \text{ Na} = \dfrac{22.989\ 77 \text{ g Na}}{58.443 \text{ g NaCl}} \times 100\% = 39.337\% \text{ Na}$

$\% \text{ Cl} = \dfrac{35.453 \text{ g Cl}}{58.443 \text{ g Na}} \times 100\% = 60.663\% \text{ Cl}$

(b) Molar mass of $AgNO_3 = 169.873 \text{ g}$

$\% \text{ Ag} = \dfrac{107.868 \text{ g Ag}}{169.873 \text{ g } AgNO_3} \times 100\% = 63.4992\% \text{ Ag}$

$\% \text{ N} = \dfrac{14.0067 \text{ g N}}{169.873 \text{ g } AgNO_3} \times 100\% = 8.245\ 39\% \text{ N}$

$\% \text{ O} = \dfrac{47.9982 \text{ g O}}{169.873 \text{ g } AgNO_3} \times 100\% = 28.2553\% \text{ O}$

(c) Molar mass of $Mg(OH)_2 = 58.320 \text{ g}$

$\% \text{ Mg} = \dfrac{24.305 \text{ g Mg}}{58.320 \text{ g } Mg(OH)_2} \times 100\% = 41.675\% \text{ Mg}$

$\% \text{ O} = \dfrac{31.9988 \text{ g O}}{58.320 \text{ g } Mg(OH)_2} \times 100\% = 54.868\% \text{ O}$

$\% \text{ H} = \dfrac{2.015\ 88 \text{ g H}}{58.320 \text{ g } Mg(OH)_2} \times 100\% = 3.4566\% \text{ H}$

11. Molar mass of $CuSO_4 \cdot 5H_2O = 249.68 \text{ g}$

$\text{Mass of } H_2O = \dfrac{5 \text{ mol } H_2O}{\text{mol } CuSO_4 \cdot 5H_2O} \times \dfrac{18.0153 \text{ g } H_2O}{\text{mol } H_2O}$

$= 90.0765 \text{ g } H_2O/\text{mol } CuSO_4 \cdot 5H_2O$

$\% \text{ } H_2O = \dfrac{90.0765 \text{ g } H_2O}{249.68 \text{ g } CuSO_4 \cdot 5H_2O} \times 100\% = 36.077\% \text{ } H_2O$

12. (a) $100.0 \text{ g NaCl} \times \dfrac{39.337 \text{ g Na}}{100 \text{ g NaCl}} = 39.34 \text{ g Na}$

(b) $140.0 \text{ g } Mg(OH)_2 \times \dfrac{54.868 \text{ g O}}{100 \text{ g } Mg(OH)_2} = 76.82 \text{ g O}$

13. $63.11 \text{ g Mn} \times \dfrac{\text{mol Mn}}{54.94 \text{ g Mn}} = 1.149 \text{ mol Mn}$

$36.89 \text{ g S} \times \dfrac{\text{mol S}}{32.06 \text{ g S}} = 1.151 \text{ mol S}$

$\dfrac{1.149 \text{ mol Mn}}{1.149} : \dfrac{1.151 \text{ mol S}}{1.149} = 1.000 \text{ mol Mn} : 1.001 \text{ mol S}$

The simplest formula is MnS.

14. $63.50 \text{ g Ag} \times \dfrac{\text{mol Ag}}{107.9 \text{ g Ag}} = 0.5885 \text{ Ag}$

$8.25 \text{ g N} \times \dfrac{\text{mol N}}{14.0 \text{ g N}} = 0.589 \text{ mol N}$

$28.26 \text{ g O} \times \dfrac{\text{mol O}}{16.00 \text{ g O}} = 1.766 \text{ mol O}$

$\dfrac{0.5885 \text{ mol Ag}}{0.5885} : \dfrac{0.589 \text{ mol N}}{0.5885} : \dfrac{1.766 \text{ mol O}}{0.5885}$

1.000 mol Ag : 1.00 mol N : 3.001 mol O

The simplest formula is $AgNO_3$.

15. $52.20 \text{ g C} \times \dfrac{\text{mol C}}{12.01 \text{ g C}} = 4.346 \text{ mol C}$

$13.00 \text{ g H} \times \dfrac{\text{mol H}}{1.008 \text{ g H}} = 12.90 \text{ mol H}$

$34.80 \text{ g O} \times \dfrac{\text{mol O}}{16.00 \text{ g O}} = 2.175 \text{ mol O}$

$\dfrac{4.346 \text{ mol C}}{2.175} : \dfrac{12.90 \text{ mol H}}{2.175} : \dfrac{2.175 \text{ mol O}}{2.175}$

$= 1.998 \text{ mol C} : 5.931 \text{ mol H} : 1.000 \text{ mol O}$

The simplest formula is C_2H_6O.

16. $0.365 \text{ g Na} \times \dfrac{\text{mol Na}}{23.0 \text{ g Na}} = 0.0159 \text{ mol Na}$

$0.221 \text{ g N} \times \dfrac{\text{mol N}}{14.0 \text{ g N}} = 0.0158 \text{ mol N}$

$0.752 \text{ g O} \times \dfrac{\text{mol O}}{16.0 \text{ g O}} = 0.0470 \text{ mol O}$

$\dfrac{0.0159 \text{ mol Na}}{0.0158} : \dfrac{0.0158 \text{ mol N}}{0.0158} : \dfrac{0.0470 \text{ mol O}}{0.0158}$

$= 1.01 \text{ mol Na} : 1.00 \text{ mol N} : 2.97 \text{ mol O}$

The simplest formula is $NaNO_3$.

17. $0.282 \text{ g Mg} \times \dfrac{\text{mol Mg}}{24.3 \text{ g Mg}} = 0.116 \text{ mol Mg}$

$0.188 \text{ g O} \times \dfrac{\text{mol O}}{16.0 \text{ g O}} = 0.0118 \text{ mol O}$

$\dfrac{0.0116 \text{ mol Mg}}{0.0116} : \dfrac{0.118 \text{ mol O}}{0.0116} = 1.00 \text{ mol Mg} : 1.02 \text{ mol O}$

The simplest formula is MgO.

18. (a) Formula mass of $HCO_2 = 45.0178 \, u$

$n = \dfrac{90.0356 \, u}{45.0178 \, u} = 2.000\,00$

Molecular formula $= (HCO_2)_2 = H_2C_2O_4$

(b) Formula mass of $CH_2O = 30.0364 \, u$

$n = \dfrac{180.1584 \, u}{30.0264 \, u} = 6.000\,00$

Molecular formula $= (CH_2O)_6 = C_6H_{12}O_6$

(c) Formula mass of $C_3H_5O_2 = 73.0718 \, u$

$n = \dfrac{146.1436 \, u}{73.0718 \, u} = 2.000\,00$

Molecular formula $= (C_3H_5O_2)_2 = C_6H_{10}O_4$

19. $43.5 \text{ g P} \times \dfrac{\text{mol P}}{31.0 \text{ g P}} = 1.40 \text{ mol P}$

$56.5 \text{ g O} \times \dfrac{\text{mol O}}{16.0 \text{ g O}} = 3.53 \text{ mol O}$

$\dfrac{1.40 \text{ mol P}}{1.40} : \dfrac{3.53 \text{ mol O}}{1.40} = 1.00 \text{ mol P} : 2.52 \text{ mol O}$

The simplest whole-number ratio is 2 mol P : 5 mol O.

Simplest formula $= P_2O_5$

Simplest formula mass $= 141.9445 \, u$

$n = \dfrac{284.0 \, u}{141.9 \, u} = 2.001$

Molecular formula $= (P_2O_5)_2 = P_4O_{10}$

Application Questions

1. The indicated ion is N^{2-}.
2. The indicated ion is Al^{3+}.

3. Formula mass is a general term that applies to all compounds and polyatomic ions, whereas molecular mass correctly applies only to those substances that exist as molecules. Thus, formula mass could apply to both H_2O and NaCl, but molecular mass could apply only to H_2O.

4. Composition data provide only mass percentages and the relative numbers of atoms present in molecules.

Application Problems

1. (a) $1 \text{ Cu atom} \times \dfrac{63.546 \, u}{\text{Cu atom}} = 63.546 \, u$

$1 \text{ S atom} \times \dfrac{32.06 \, u}{\text{S atom}} = 32.06 \, u$

$4 \text{ O atoms} \times \dfrac{15.9994 \, u}{\text{O atom}} = 63.9976 \, u$

$10 \text{ H atoms} \times \dfrac{1.007\,94 \, u}{\text{H atom}} = 10.0794 \, u$

$5 \text{ O atoms} \times \dfrac{15.9994 \, u}{\text{O atom}} = 79.997 \, u$

Formula mass of $CuSO_4 \cdot 5H_2O = 249.68 \, u$

Molar mass of $CuSO_4 \cdot 5H_2O = 249.58 \text{ g}$

(b) $1 \text{ Mg atom} \times \dfrac{24.305 \, u}{\text{Mg atom}} = 24.305 \, u$

$1 \text{ S atom} \times \dfrac{32.06 \, u}{\text{S atom}} = 32.06 \, u$

$4 \text{ O atoms} \times \dfrac{15.9994 \, u}{\text{O atom}} = 63.9976 \, u$

$14 \text{ H atoms} \times \dfrac{1.007\,94 \, u}{\text{H atom}} = 14.111\,16 \, u$

$7 \text{ O atoms} \times \dfrac{15.9994 \, u}{\text{O atom}} = 111.9948 \, u$

Formula mass of $MgSO_4 \cdot 7H_2O = 246.47 \, u$

Molar mass of $MgSO_4 \cdot 7H_2O = 246.47 \text{ g}$

2. (a) $1.00(2 \text{ mol Al} + 3 \text{ mol S} + 12 \text{ mol O}) = 17.0 \text{ mol atoms}$
$17.0(6.022 \times 10^{23} \text{ atoms}) = 1.02 \times 10^{25} \text{ atoms}$

(b) $1.00(1 \text{ mol Mg} + 1 \text{ mol S} + 4 \text{ mol O} + 14.00 \text{ mol H}$
$+ 7 \text{ mol O}) = 27.0 \text{ mol atoms}$
$27.0(6.022 \times 10^{23} \text{ atoms}) = 1.63 \times 10^{25} \text{ atoms}$

(c) $5.00(1 \text{ mol Cu} + 1 \text{ mol S} + 4 \text{ mol O} + 10 \text{ mol H} + 5 \text{ mol O})$
$= 105 \text{ mol atoms}$
$105 \times 6.022 \times 10^{23} \text{ atoms} = 6.32 \times 10^{25}$

3. Molar mass of $Fe_2(SO_4)_3 = 399.87 \text{ g}$

$1.75 \times 10^{-10} \text{ mol Fe}_2(SO_4)_3 \times \dfrac{400 \text{ g Fe}_2(SO_4)_3}{\text{mol Fe}_2(SO_4)_3}$

$= 7.00 \times 10^{-8} \text{ g Fe}_2(SO_4)_3$

4. (a) Molar mass of $NaHCO_3 = 84.0070 \text{ g}$

$4.00 \text{ g NaHCO}_3 \dfrac{\text{mol NaHCO}_3}{84.0 \text{ g NaHCO}_3} = 0.0476 \text{ mol NaHCO}_3$

(b) $0.0476 \text{ mol NaHCO}_3 \times \dfrac{6.02 \times 10^3 \text{ formula units NaHCO}_3}{\text{mol NaHCO}_3}$

$= 2.87 \times 10^{22} \text{ formula units NaHCO}_3$

(c) $2.87 \times 10^{22} \text{ formula units NaHCO}_3 \times \dfrac{6 \text{ atoms}}{\text{formula units}}$

$= 1.72 \times 10^{23} \text{ atoms}$

5. (a) Molar mass of H_2O = 18.0153 g

$$2.500 \text{ mL } H_2O \times \frac{1.000 \text{ g } H_2O}{\text{mL } H_2O} \times \frac{\text{mol } H_2O}{18.02 \text{ g } H_2O}$$

$$= 0.1387 \text{ mol } H_2O$$

(b) $0.1387 \text{ mol } H_2O \times \frac{6.022 \times 10^{23} \text{ molecules } H_2O}{\text{mol } H_2O}$

$$= 8.353 \times 10^{22} \text{ molecules } H_2O$$

6. (a) $9.00 \text{ g } H_2O \times \frac{6.02 \times 10^{23} \text{ units } H_2O}{18.0 \text{ g } H_2O}$

$$= 3.01 \times 10^{23} \text{ units } H_2O$$

$9.00 \text{ g } SiO_2 \times \frac{6.02 \times 10^{23} \text{ units } SiO_2}{60.1 \text{ g } SiO_2}$

$$= 9.01 \times 10^{22} \text{ units } SiO_2 \text{ (Ans.) } 9.00 \text{ g } H_2O$$

(b) $0.25 \text{ mol } AgNO_3 \times \frac{6.0 \times 10^{23} \text{ units } AgNO_3}{\text{mol } AgNO_3}$

$$= 1.5 \times 10^{23} \text{ units } AgNO_3$$

$42.0 \text{ g } Ca(OH)_2 \times \frac{6.02 \times 10^{23} \text{ units } Ca(OH)_2}{74.1 \text{ g } Ca(OH)_2}$

$$= 3.41 \times 10^{23} \text{ units } Ca(OH)_2 \text{ (Ans.) } 42.0 \text{ g } Ca(OH)_2$$

7. (a) $25.0 \text{ g } Pb(NO_3)_2 \times \frac{\text{mol } Pb(NO_3)_2}{331 \text{ g } Pb(NO_3)_2} \times \frac{\text{mol } Na_2CO_3}{\text{mol } PB(NO_3)_2}$

$\times \frac{106 \text{ g } Na_2CO_3}{\text{mol } Na_2CO_3} = 8.01 \text{ g } Na_2CO_3$

(b) $25.0 \text{ g } Pb(NO_3)_2 \times \frac{\text{mol } Pb(NO_3)_2}{331 \text{ g } Pb(NO_3)_2} \times \frac{\text{mol } NH_4Cl}{\text{mol } Pb(NO_3)_2}$

$\times \frac{53.5 \text{ g } NH_4Cl}{\text{mol } NH_4Cl} = 4.04 \text{ g } NH_4Cl$

(c) $25.0 \text{ g } Pb(NO_3)_2 \times \frac{\text{mol } Pb(NO_3)_2}{331 \text{ g } Pb(NO_3)_2} \times \frac{\text{mol } Fe(OH)_3}{\text{mol } Pb(NO_3)_2}$

$\times \frac{107 \text{ g } Fe(OH)_3}{\text{mol } Fe(OH)_3} = 8.08 \text{ g } Fe(OH)_3$

(d) $25.0 \text{ g } Pb(NO_3)_2 \times \frac{\text{mol } Pb(NO_3)_2}{331 \text{ g } Pb(NO_3)_2} \times \frac{\text{mol } Ca_3(PO_4)_2}{\text{mol } Pb(NO_3)_2}$

$\times \frac{310 \text{ g } Ca_3(PO_4)_2}{\text{mol } Ca_3(PO_4)_2} = 23.4 \text{ g } Ca_3(PO_4)_2$

8. (a) $1.00 \text{ mol } KNO_3 \times \frac{6.02 \times 10^{23} \text{ units } KNO_3}{\text{mol } KNO_3}$

$\times \frac{5 \text{ atoms}}{\text{unit } KNO_3} = 3.01 \times 10^{24} \text{ atoms}$

(b) $75.0 \text{ g } O_2 \times \frac{6.02 \times 10^{23} \text{ units } O_2}{32.0 \text{ g } O_2} \times \frac{2 \text{ atoms}}{\text{units } O_2}$

$$= 2.82 \times 10^{24} \text{ atoms}$$

(c) $1000.0 \text{ g } CaCO_3 \times \frac{6.022 \times 10^{23} \text{ units } CaCO_3}{100.09 \text{ g } CaCO_3}$

$\times \frac{5 \text{ atoms}}{\text{unit } CaCO_3} = 3.008 \times 10^{25} \text{ atoms}$

9. (a) Molar mass of $MgSO_4 \cdot 7H_2O$ = 246.47 g

$\% \text{ Mg} = \frac{24.305 \text{ g Mg}}{246.47 \text{ g } MgSO_4 \cdot 7H_2O} \times 100\% = 9.8612\% \text{ Mg}$

$\% \text{ S} = \frac{32.06 \text{ g S}}{246.47 \text{ g } MgSO_4 \cdot 7H_2O} \times 100\% = 13.01\% \text{ S}$

$\% \text{ O} = \frac{175.993 \text{ g O}}{246.47 \text{ g } MgSO_4 \cdot 7H_2O} \times 100\% = 71.406\% \text{ O}$

$\% \text{ H} = \frac{14.1112 \text{ g H}}{246.47 \text{ g } MgSO_4 \cdot 7H_2O} \times 100\% = 5.7253\% \text{ H}$

(b) Molar mass of $C_{17}H_{35}COONa$ (soap) = 306.4663 g

$\% \text{ C} = \frac{216.200 \text{ g C}}{306.4663 \text{ g soap}} \times 100\% = 70.5460\% \text{ C}$

$\% \text{ H} = \frac{35.2779 \text{ g H}}{306.4663 \text{ g soap}} \times 100\% = 11.5112\% \text{ H}$

$\% \text{ O} = \frac{31.9988 \text{ g O}}{306.4663 \text{ g soap}} \times 100\% = 10.4412\% \text{ O}$

$\% \text{ Na} = \frac{22.989 \text{ } 77 \text{ g Na}}{306.4663 \text{ g soap}} \times 100\% = 7.501 \text{ } 570\% \text{ Na}$

10. (a) Molar mass of $Cu(NO_3)_2$ = 187.556 g

$\% \text{ Cu} = \frac{63.546 \text{ g Cu}}{187.556 \text{ g } Cu(NO_3)_2} \times 100\% = 33.881\% \text{ Cu}$

(b) Molar mass of $CuSO_4$ = 159.60 g

$\% \text{ Cu} = \frac{63.546 \text{ g Cu}}{159.60 \text{ g } CuSO_4} \times 100\% = 39.816\% \text{ Cu}$

(c) Molar mass of $CuCl_2$ = 134.452 g

$\% \text{ Cu} = \frac{63.546 \text{ g Cu}}{134.452 \text{ g } CuCl_2} \times 100\% = 47.263\% \text{ Cu}$

(d) Molar mass of $CuC_2H_3O_2$ = 122.591 g

$\% \text{ Cu} = \frac{63.546 \text{ g Cu}}{122.591 \text{ g } CuC_2H_3O_2} \times 100\% = 51.836\% \text{ Cu}$

$CuC_2H_3O_2$ thus has the highest copper content.

11. $\frac{1.15 \text{ g Na}}{2.92 \text{ g sample}} \times 100\% = 39.4\% \text{ Na}$

$\frac{1.77 \text{ g Cl}}{2.92 \text{ g sample}} \times 100\% = 60.6\% \text{ Cl}$

12. $72.35 \text{ g Fe} \times \frac{\text{mol Fe}}{55.85 \text{ g Fe}} = 1.295 \text{ mol Fe}$

$27.65 \text{ g O} \times \frac{\text{mol O}}{16.00 \text{ g O}} = 1.728 \text{ O}$

$\frac{1.295 \text{ mol Fe}}{1.295} : \frac{1.728 \text{ mol O}}{1.295} = 1.000 \text{ mol Fe} : 1.334 \text{ mol O}$

The simplest whole-number ratio is 3 mol Fe : 4 mol O.
The simplest formula is Fe_3O_4.

13. $18.31 \text{ g Ca} \times \frac{\text{mol Ca}}{40.08 \text{ g Ca}} = 0.4568 \text{ mol Ca}$

$32.36 \text{ g Cl} \times \frac{\text{mol Cl}}{35.45 \text{ g Cl}} = 0.9128 \text{ mol Cl}$

$49.33 \text{ g } H_2O \times \frac{\text{mol } H_2O}{18.02 \text{ g } H_2O} = 2.738 \text{ mol } H_2O$

$\frac{0.4568 \text{ mol Ca}}{0.4568} : \frac{0.9128 \text{ mol Cl}}{0.4568} : \frac{2.738 \text{ mol } H_2O}{0.4568}$

$= 1.000 \text{ mol Ca} : 1.998 \text{ mol Cl} : 5.994 \text{ mol } H_2O$

The simplest formula is $CaCl_2 \cdot 6H_2O$.

14. $7.88 \text{ g Al} \times \dfrac{\text{mol Al}}{27.0 \text{ g Al}} = 0.292 \text{ mol Al}$

$14.06 \text{ g S} \times \dfrac{\text{mol S}}{32.06 \text{ g S}} = 0.4386 \text{ mol S}$

$28.06 \text{ g O} \times \dfrac{\text{mol O}}{16.00 \text{ g O}} = 1.754 \text{ mol O}$

$\dfrac{0.292 \text{ mol Al}}{0.292} : \dfrac{0.4386 \text{ mol S}}{0.292} : \dfrac{1.754 \text{ mol O}}{0.292}$

$1.00 \text{ mol Al} : 1.50 \text{ mol S} : 6.01 \text{ mol O}.$

The simplest whole-number ratio is 2 mol Al : 3 mol S : 12 mol O.

The simplest formula is $Al_2S_3O_{12}$.

15. $56.38 \text{ g P} \times \dfrac{\text{mol P}}{30.97 \text{ g P}} = 1.820 \text{ mol P}$

$43.62 \text{ g O} \times \dfrac{\text{mol O}}{16.00 \text{ g O}} = 2.726 \text{ mol O}$

$\dfrac{1.820 \text{ mol P}}{1.820} : \dfrac{2.726 \text{ mol O}}{1.820} = 1.000 \text{ mol P} : 1.498 \text{ mol O}$

The simplest whole-number ratio is 2 mol P : 3 mol O.

Simplest formula = P_2O_3

Simplest formula mass = $109.9458 \ u$

$n = \dfrac{219.8916 \ u}{109.9457 \ u} = 2.000 \ 002$

Molecular formula = $(P_2O_3)_2 = P_4O_6$

16. $18.52 \text{ g Na} \times \dfrac{\text{mol Na}}{22.99 \text{ g Na}} = 0.8056 \text{ mol Na}$

$25.76 \text{ g S} \times \dfrac{\text{mol S}}{32.06 \text{ g S}} = 0.8035 \text{ mol S}$

$4.03 \text{ g H} \times \dfrac{\text{mol H}}{1.01 \text{ g H}} = 3.99 \text{ mol H}$

$51.69 \text{ g O} \times \dfrac{\text{mol O}}{16.00 \text{ g O}} = 3.230 \text{ mol O}$

$\dfrac{0.8056 \text{ mol Na}}{0.8035} : \dfrac{0.8035 \text{ mol S}}{0.8035} : \dfrac{3.99 \text{ mol H}}{0.8035} : \dfrac{3.230 \text{ mol O}}{0.8035}$

$= 1.003 \text{ mol Na} : 1.000 \text{ mol S} : 4.97 \text{ mol H} : 4.02 \text{ mol H}$

The simplest whole-number ratio is
1 mol Na : 1 mol S : 5 mol H : 4 mol O.

Simplest formula = $NaSH_5O_4$

Simplest formula mass = $124.09 \ u$

$n = \dfrac{248.20 \ u}{124.10 \ u} = 2.0000$

Molecular formula = $(NaSH_5O_4)_2 = Na_2S_2H_{10}O_8$

$36.2\% \ H_2O \times 248.20 \text{ g} = 89.8 \text{ g } H_2O$

$89.8 \text{ g } H_2O \times \dfrac{\text{mol } H_2O}{18.0 \text{ g } H_2O} = 4.99 \text{ mol } H_2O = \sim 5 \text{ mol } H_2O$

This leaves 8 − 5 = 3O in the nonhydrate part of the formula.

Final molecular formula = $Na_2S_2O_3 \cdot 5H_2O$

Teacher's Notes

Chapter 7 Chemical Formulas and Chemical Compounds

INTRODUCTION

Unit II, Organization of Matter, took a "micro" view of matter. You saw how chemists explain the properties of matter by a model of atoms composed of electrons, protons, and neutrons. Now you are ready to move on to a "macro" view of matter. The focus will turn to the specific properties of various types of matter. As a foundation, you must learn more about the language of chemistry.

LOOKING AHEAD

The first half of this chapter deals directly with the "words" of the language of chemistry—chemical formulas. As you look at chemical formulas, practice thinking of chemical names. Practice thinking of formulas as you read the names of chemical compounds. The second half of the chapter shows how the relative atomic mass scale and the mole concept are extended to chemical compounds. Chemical formulas relate the masses of chemical compounds to the masses of atoms.

SECTION PREVIEW

7.1 Chemical Names and Formulas
7.2 Oxidation Numbers
7.3 Using Chemical Formulas
7.4 Determining Chemical Formulas

Though they speak different languages, chemists all over the world use the same set of chemical formulas.

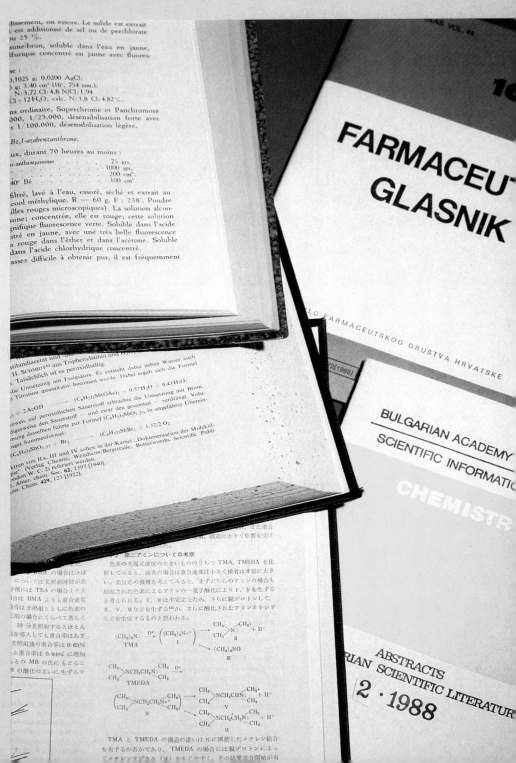

7.1 Chemical Names and Formulas

With all of the compounds of all of the elements to be identified, systematic methods for writing formulas and naming compounds are necessary. In this section you will be introduced to the rules that apply to simple chemical compounds. Table 7-1 lists the common and systematic names for a number of well-known substances. Certain common names such as "milk of magnesia" or "lime" remain in everyday use. As you can see, common names usually give no information about chemical composition.

Significance of a Chemical Formula

Figure 7-1 uses the formulas of octane, a hydrocarbon that is a component of gasoline, and aluminum sulfate, a compound used in dyeing fabrics, to illustrate chemical formulas. Like all hydrocarbons, octane is a molecular compound. One octane molecule is composed of 8 carbon atoms and 18 hydrogen atoms.

Aluminum sulfate contains the sulfate ion and is clearly an ionic compound. Recall that ionic compounds consist of huge numbers of positive and negative ions held together by mutual attraction. The formula therefore represents one *formula unit*—not just a molecule. One formula unit of aluminum sulfate consists of 2 aluminum ions and 3 sulfate ions. Note in the figure how the parentheses are used. They surround the polyatomic ion as a unit. The subscript 3 applies to the entire unit within the parentheses. One formula unit of aluminum sulfate is therefore composed of 2 aluminum atoms, 3 sulfur atoms, and 12 oxygen atoms.

A correctly written chemical formula must represent the known facts about the analytically determined composition of a compound. Care must be

SECTION OBJECTIVES

- Explain the significance of a chemical formula.
- Determine the formula of an ionic compound between any two given ions.
- Explain the two systems for distinguishing different ionic compounds of the same two elements.
- Name an ionic compound, given its formula.
- Using prefixes, name a binary molecular compound from its formula.
- Write the formula of a binary molecular compound, given its name.
- List the names and formulas of the common laboratory acids.

C_8H_{16}, *octene*, is a compound with properties different from octane.

TABLE 7-1 SOME FAMILIAR COMPOUNDS WITH THEIR SYSTEMATIC AND COMMON NAMES

Formula	Systematic Name	Common Name
Hg	mercury	quicksilver
S	sulfur	brimstone
$NaHCO_3$	sodium hydrogen carbonate	baking soda
CaO	calcium oxide	lime
$CaCO_3$	calcium carbonate	limestone, calcite, marble
$MgSO_4 \cdot 7H_2O$	magnesium sulfate hepta hydrate	epsom salts
N_2O	dinitrogen oxide	laughing gas
NaOH	sodium hydroxide	lye, caustic soda
HCl	hydrochloric acid	muriatic acid
$Mg(OH)_2$	magnesium hydroxide	milk of magnesia
NaCl	sodium chloride	table salt
$CaSO_4 \cdot \frac{1}{2}H_2O$	calcium sulfate hemihydrate	plaster of paris
CO_2	carbon dioxide	dry ice (solid CO_2)
CH_3OH	methanol	wood alcohol
CH_3CH_2OH	ethanol	grain alcohol

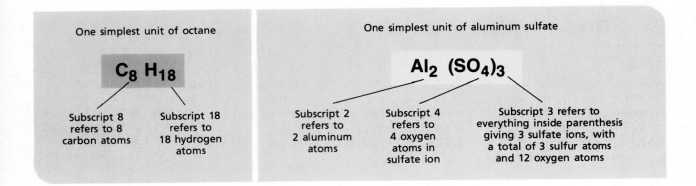

One simplest unit of octane

$$C_8 H_{18}$$

Subscript 8 refers to 8 carbon atoms

Subscript 18 refers to 18 hydrogen atoms

One simplest unit of aluminum sulfate

$$Al_2 (SO_4)_3$$

Subscript 2 refers to 2 aluminum atoms

Subscript 4 refers to 4 oxygen atoms in sulfate ion

Subscript 3 refers to everything inside parenthesis giving 3 sulfate ions, with a total of 3 sulfur atoms and 12 oxygen atoms

Figure 7-1 Interpretation of chemical formulas

taken that subscripts are correct. The two formulas below

$$H_2O \qquad H_2O_2$$
water   hydrogen peroxide

represent different compounds with very different characteristic properties, as illustrated in Figure 7-2.

Chemical formulas are often used to represent amounts of substances. The formula H_2O can represent one molecule of water. As you learned in chapter 3, the mole is equal to 6.02×10^{23}. The formula H_2O can also represent one mole of water molecules. Using the definition of the mole, the mass in grams of one mole of water molecules or one mole of any compound can be calculated. The formula H_2O is also sometimes used to represent one molar mass of water molecules.

$$H_2O$$
one water *molecule*
one *mole* of water molecules
one *molar mass* of water molecules

Section 7.3 is devoted to the quantitative meaning of chemical formulas.

Figure 7-2 The same elements can make different compounds with different chemical properties. Water, (H_2O), *(top)*, is unreactive with green felt cloth, but hydrogen peroxide (H_2O_2) bleaches the green dye in the felt.

Monatomic Ions

Many main group elements form ions with noble-gas configurations by either gaining or losing electrons. The Periodic Table is therefore a guide to the ions formed by these elements. Group 1 metals lose one electron to give +1 ions, such as Na^+. Group 2 metals lose two electrons to give +2 ions such as Mg^{2+}. Of the Group 13 elements, aluminum is the only one commonly found as an ion. Aluminum forms the +3 ion, Al^{3+}. Ions formed from a single atom are known as **monatomic ions**.

With Group 14, the loss of enough electrons to give a noble-gas configuration becomes difficult. The Group 14 metals tin and lead, therefore, form +2 ions in many compounds. These +2 ions result from loss of two of the four valence electrons. Tin and lead also both form some compounds in which all four valence electrons are involved in polar-covalent bonding.

The nonmetals of Groups 15, 16, and 17 form ions by gaining electrons rather than losing electrons. In ionic compounds, nitrogen forms the −3 anion, N^{3-}. The three added electrons plus the five outermost electrons in nitrogen atoms give a completed outermost octet. Similarly, oxygen and sulfur, from Group 16, form −2 ions, and the halogens, from Group 17, form −1 ions. These common ions of main-group elements are shown in Periodic Table form in Table 7-2. Elements from the *d*-block form ions with +2, +3 (or, in a few cases, +1) charges.

The symbols and names of the common monatomic ions are organized according to their charges in Table 7-3. Positive ions are named by the element name followed by "ion" as in

K^+ potassium ion Mg^{2+} magnesium ion

Remind students of the following: Metallic ions contain more protons than electrons and are thus positively charged ions (cations):

Na^+	11 protons	10 electrons
Mg^{2+}	12 protons	10 electrons
Al^{3+}	13 protons	10 electrons

Remind students of the following: The negative ions formed by nonmetals contain more electrons than protons and are thus negatively charged ions (anions):

N^{3-}	15 protons	18 electrons
S^{2-}	16 protons	18 electrons
Cl^{1-}	17 protons	18 electrons

STUDY HINT
The formula of an ionic compound shows the simplest whole-number ratio of ions.

TABLE 7-2   SOME COMMON IONS OF MAIN GROUP ELEMENTS*

Group 1	Group 2	Group 13	Group 14	Group 15	Group 16	Group 17	Group 18
Li^+				N^{3-}	O^{2-}	F^-	
Na^+	Mg^{2+}	Al^{3+}			S^{2-}	Cl^-	
K^+	Ca^{2+}					Br^-	
Rb^+	Sr^{2+}		Sn^{2+}			I^-	
Cs^+	Ba^{2+}	Pb^{2+}					

*All of the ions shown except Sn^{2+} and Pb^{2+} have noble-gas configurations.

TABLE 7-3 SOME COMMON MONATOMIC IONS

+1 Charge		+2 Charge		+3 Charge	
cesium ion	Cs^+	barium ion	Ba^{2+}	aluminum ion	Al^{3+}
copper ion	Cu^+	cadmium ion	Cd^{2+}	chromium(III) ion	Cr^{3+}
lithium ion	Li^+	calcium ion	Ca^{2+}	iron(III) ion	Fe^{3+}
potassium ion	K^+	chromium(II) ion	Cr^{2+}		
rubidium ion	Rb^+	cobalt(II) ion	Co^{2+}		
silver ion	Ag^+	copper(II) ion	Cu^{2+}		
sodium ion	Na^+	iron(II) ion	Fe^{2+}		
		lead(II) ion	Pb^{2+}		
		magnesium ion	Mg^{2+}		
		manganese(II) ion	Mn^{2+}		
		*mercury(I)	Hg_2^{2+}		
		mercury(II) ion	Hg^{2+}		
		nickel(II) ion	Ni^{2+}		
		tin(II) ion	Sn^{2+}		
		zinc ion	Zn^{2+}		

−1 Charge		−2 Charge		−3 Charge	
bromide ion	Br^-	oxide ion	O^{2-}	nitride ion	N^{3-}
chloride ion	Cl^-	sulfide ion	S^{2-}		
fluoride ion	F^-				
iodide ion	I^-				

*The mercury (I) ion exists as two Hg^+ ions joined together by a covalent bond and is written as Hg_2^{2+}.

Negative ions are named by dropping the ending of the element name and adding "ide" to it, as in

F	fluo*rine*	F^-	fluo*ride* ion
N	nit*rogen*	N^{3-}	nit*ride* ion

Binary Ionic Compounds

In ionic compounds, ions must be combined so that the total positive and negative charges are equal. The formulas for many ionic compounds can be written once the identities of the combined ions are known.

Compounds composed of two different elements are known as **binary compounds**. *What is the formula of the binary compound formed by the ions* of magnesium and bromine? Magnesium is in Group 2 and forms the Mg^{2+} ion. Bromine is in Group 17 and forms the Br^- ion. Two bromide ions are required to balance the +2 charge of the magnesium ion. The formula must therefore show one magnesium ion and two bromide ions. It is customary to write the symbol for the positive ion first.

Ions combined: Mg^{2+}, Br^-, Br^- *Chemical formula:* $MgBr_2$

The subscript 1 is understood when no subscript is present and is not written. The subscript 2 shows that two Br^- ions are present in each formula unit. The formula shows that the simplest whole-number ratio of the ions in this compound is 1:2. Note that charges of ions are not usually included in chemical formulas of compounds.

As an aid to writing formulas of ionic compounds, the positive and negative charges can be "crossed over" to give subscripts. The purpose of the "cross-over" is to find the numbers of ions needed to equalize the + and − charges. For example, the formula for the compound formed by the aluminum ion, Al^{3+}, and the oxide ion, O^{2-} is found as follows: *Step 1: Write the symbols for the ions side by side, with the positive ion first.*

$$Al^{3+} \quad O^{2-}$$

Step 2: Cross over the charge values to give subscripts.
Use the value of each charge as the subscript for the other ion.

$$Al_2^{3+} \qquad O_3^{2-}$$

Step 3: Check the subscripts and write the formula. Multiplying the charge by the subscript shows that the charge on two Al^{3+} ions $(2 \times +3 = +6)$ equals the charge on three O^{2-} ions $(3 \times -2 = -6)$. The correct formula is therefore

$$Al_2O_3$$

Ionic compounds are named by combining the names of the positive and negative ions, as illustrated in Figure 7-3. The name of the positive ion is given first. Al_2O_3 is therefore named aluminum oxide.

The names and charges of the ions listed in Table 7-3 should be memorized as an aid to writing and interpreting chemical formulas.

No element commonly forms negative ions of two different charges.

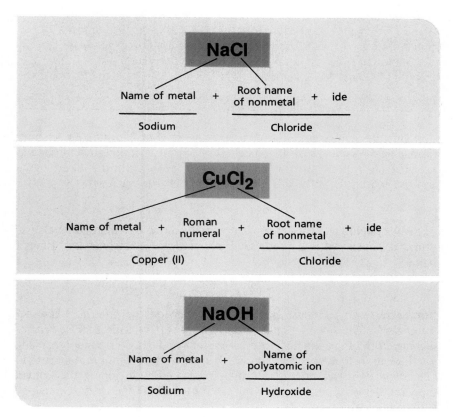

Figure 7-3 Naming ionic compounds

Sample Problem 7.1

Write the formulas for the binary ionic compounds formed between (a) zinc and iodine and (b) zinc and sulfur.

Solution

Step 1.   Write the symbols for the ions side by side, with the positive ion first.
$$(a)\ Zn^{2+}\ I^-\quad (b)\ Zn^{2+}\ S^{2-}$$

Step 2.   Cross over the charge values to give subscripts.
$$(a)\quad Zn_1^{2+}\quad I_2^-\qquad (b)\quad Z_2^{2+}\quad S_2^{2-}$$

Step 3.   Check the subscripts and write the formula.
(a) The subscripts are correct because they give equal total charges of $+2$ ($1 \times +2$) and -2 (2×-1). The subscript 1 is not written, so the formula is ZnI_2. Combining the cation and anion names gives this compound the name *zinc iodide*.
(b) The subscripts are correct because they give equal total charges of $+4$ ($2 \times +2$) and -4 (2×-2). The formula should be written with the simplest whole number ratios of subscripts. Therefore the correct formula is ZnS. This compound is named zinc sulfide.

Practice Problems
1. Write formulas for the binary ionic compounds formed between each of the following:
    (a) potassium and iodine   (Ans.) KI (b) magnesium and chlorine   (Ans.) $MgCl_2$ (c) sodium and sulfur   (Ans.) Na_2S (d) calcium and oxygen   (Ans.) CaO (e) aluminum and sulfur   (Ans.) Al_2S_3.
2. Name the binary ionic compounds indicated by each of the following chemical formulas:
    (a) BaO (Ans.) barium oxide   (b) $AgCl$ (Ans.) silver chloride   (c) KBr (Ans.) potassium bromide
    (d) ZnO (Ans.) zinc oxide

Different names are needed for positive ions of two different charges formed by the same metal. Two different systems for naming such ions are in use. The old system gives the ending *-ous* to the ion with the lower charge and the ending *-ic* to the ion with the higher charge, such as

$$Cu^+\ \text{cpr}ous\ \text{ion}\qquad Cu^{2+}\ \text{cupr}ic\ \text{ion}$$

The new system gives the actual charge on the ion as a Roman numeral. The numeral is enclosed in parentheses and placed immediately after the metal name:

$$Cu^+\ \text{copper(I) ion}\qquad Cu^{2+}\ \text{copper(II) ion}$$

STUDY HINT
To use the Stock system for ionic compounds, you must know which metals form more than one cation.

Nomenclature *is a term that refers to methods of naming chemical compounds.* The use of Roman numerals is part of the *Stock system of nomenclature.* In this system, compounds of metals that form more than one ion (have more than one oxidation state), and therefore combine in more than one proportion with nonmetals, must be named by including the Roman numeral. Names of compounds of metals that form only one ion *do not* include a Roman numeral.

Figure 7-4 Compounds of different ions of the same metal may look different. Compare lead(I) oxide *(left)*, lead(II) oxide *(center)* and lead(III) oxide.

Sample Problem 7.2

Write the formula and give the name in the Stock system for the compound formed by the ions Cr^{3+} and F^-.

Solution

Step 1.   *Write the symbols for the ions side by side, with the positive ion first.*

Step 2.   *Cross over the charge values to give subscripts.*

Step 3.   *Check the subscripts and write the formula.*
The total positive charge is $1 \times +3 = +3$. The total negative charge is $3 \times -1 = -3$. The formula is therefore

$$CrF_3$$

To name this compound correctly, it is necessary to know that chromium forms more than one ion. Since that is the case, this compound is named *chromium (III) fluoride* according to the Stock system.

Practice Problems

1. Write the formula and give the name in the Stock system for the compounds formed between each of the following: *(a)* Cu^{2+} and I^- *(Ans.)* CuI_2, copper(II) iodide   *(b)* Fe^{2+} and O^{2-} *(Ans.)* FeO, iron (II) oxide   *(c)* Pb^{2+} and Cl^- *(Ans.)* $PbCl_2$, lead(II) chloride   *(d)* Hg^{2+} and S^{2-} *(Ans.)* HgS, mercury(II) sulfide   *(e)* Sn^{2+} and F^- *(Ans.)* SnF_2, tin(II) fluoride.
2. Name the following compounds: *(a)* CuO   *(Ans.)* copper(II) oxide   *(b)* CoF_3 *(Ans.)* cobalt(III) fluoride   *(c)* SnI_4 *(Ans.)* tin(IV) iodide   *(d)* FeS   *(Ans.)* iron(II) sulfide   *(e)* Hg_2Cl_2 *(Ans.)* mercury(I) chloride.

TABLE 7-4   SOME POLYATOMIC IONS

+1 Charge		−1 Charge	
Ammonium ion	$NH_4{}^+$	acetate ion	$C_2H_3O_2^-$
		arsenate	$A_5O_4^-$
		chlorate ion	ClO_3^-
		chlorite ion	ClO_2^-
		cyanide ion	CN^-
		hydrogen carbonate ion (bicarbonate ion)	HCO_3^-
		hydrogen sulfate ion	HSO_4^-
		hydroxide ion	OH^-
		hypoiodite	IO^-
		iodate ion	IO_3^-
		nitrate ion	NO_3^-
		nitrite ion	NO_2^-
		oxalate	C_2O_4
		perchlorate ion	ClO_4^-
		permanganate ion	MNO_4^-
		telluride	Te^-

−2 Charge		−3 Charge	
carbonate ion	CO_3^{2-}	phosphate ion	PO_4^{3-}
chromate ion	CrO_4^{2-}	phosphite ion	PO_3^{3-}
dichromate ion	$Cr_2O_7^{2-}$		
peroxide ion	O_2^{2-}		
sulfate ion	SO_4^{2-}		
sulfite ion	SO_3^{2-}		
thiosulfate ion	$S_2O_3^{2-}$		

Compounds Containing Polyatomic Ions

STUDY HINT

The system for naming oxyanions is discussed in Chapter 17.

Hints for relating structure of polyatomic ions to its name: If the ion has an -ide ending, it is monatomic (example: S^{2-} is sulfide). If the ion has an -ate ending or an -ite ending, it is polyatomic and contains oxygen (example: SO_4^{2-} is sulfate, SO_3^{2-} is sulfite).

Some polyatomic ions are listed in Table 7-4. All but the ammonium ion are negatively charged. Note that most of these are *oxyanions*—polyatomic ions that contain oxygen.

In several cases, two different oxyanions are formed by the same two elements. Nitrogen and oxygen, for example, are combined in both NO_3^- and NO_2^-. The ion with the larger number of oxygen atoms is named with the *-ate* ending and the ion with the smaller number of oxygen atoms is named with the *-ite* ending.

$$NO_3^- \qquad NO_2^-$$
$$\text{nitrate ion} \qquad \text{nitrite ion}$$

Compounds containing polyatomic ions are named in the same manner as binary ionic compounds. The cation name is given first and followed by the anion name. For example, the two compounds formed with silver by the nitrate and nitrite anions are named silver nitrate ($AgNO_3$) and silver nitrite ($AgNO_2$). When more than one polyatomic ion is present in a compound, the formula for the ion is surrounded by parentheses as illustrated in Figure 7-1 for aluminum sulfate, $Al_2Cu(SO_4)_3$. This formula says that an aluminum sulfate formula unit has +2 aluminum ions and 3 sulfate ions.

Sample Problem 7.3

Write the formula for tin(IV) sulfate.

Solution

Step 1. *Write the symbols for the ions side by side, with the positive ion first.*

$$Sn^{4+} \ (SO_4)^{2-}$$

Step 2. *Cross over the charge values to give subscripts.*

$$Sn_2^{4+} \ (SO_4)_4^{2-}$$

Step 3. *Check the subscripts and write the formula.*

The total positive charge is $2 \times +4 = +8$. The total negative charge is $4 \times -2 = -8$. The charges are equal. However, the subscripts are not in the smallest possible whole-number ratio. Instead of $2:4$, the ratio $1:2$ should be used. The correct formula is therefore

$$Sn(SO_4)_2$$

Practice Problems

1. Write formulas for each of the following ionic compounds: *(a)* Sodium iodide *(Ans.)* NaI *(b)* calcium chloride *(Ans.)* $CaCl_2$ *(c)* potassium sulfide *(Ans.)* K_2S *(d)* lithium nitrate *(Ans.)* $LiNO_3$ *(e)* copper(II) sulfate *(Ans.)* $CuSO_4$ *(f)* sodium carbonate *(Ans.)* Na_2CO_3 *(i)* barium phosphate *(Ans.)* $Ba_3(PO_4)_2$.
2. Name each of the following ionic compounds using both systems where appropriate: *(a)* Ag_2O *(Ans.)* silver oxide *(b)* $Ca(OH)_2$ *(Ans.)* calcium hydroxide *(c)* $Zn(NO_3)_2$ *(Ans.)* zinc nitrate *(d)* $KClO_3$ *(Ans.)* potassium chlorate *(e)* NH_4OH *(Ans.)* ammonium hydroxide *(f)* $FeCrO_4$ *(Ans.)* iron(II) chromate *(g)* $Pb_3(PO_4)_2$ *(Ans.)* lead(II) phosphate

---

$$\textbf{PBr}_5$$

Prefix needed if less electronegative element contributes more than one atom	+	Name of less electronegative element	+	Prefix needed if more electronegative element contributes more than one atom	+	Root name of more electronegative element	+	ide
		Phosphorus		Pentabromide				

Figure 7-5 Naming binary molecular compounds using prefixes

Binary Molecular Compounds

Compounds between nonmetals are molecular rather than ionic compounds. The old system for naming binary molecular compounds, which is based on prefixes, is illustrated in Figure 7-5. We have already used several names of this type. For CCl_4, the *tetra* in the name carbon *tetra*chloride shows that four chlorine atoms are present. The two oxides of carbon (CO and CO_2) are

The alternate system for naming binary molecular compounds is explained in Section 7.3.

TABLE 7-5
NUMERICAL PREFIXES

Number	Prefix
$\frac{1}{2}$	hemi-
1	mono-
2	di-
3	tri-
4	tetra-
5	penta-
6	hexa-
7	hepta-
8	octa-
9	nona-
10	deca-

The order of nonmetals in formulas is generally P, H, C, S, I, Br, N, Cl, O, F.

The -o or -a at the end of a prefix is often dropped before another vowel, e.g., monoxide, pentoxide.

mon- and *di-* indicate one and two oxygen atoms, respectively.

The least electronegative element is given first in the names and formulas of binary molecular compounds. As a result, oxygen and the halogens are usually given second, as in carbon tetrachloride and the carbon oxides. The order of nonmetals in binary compound names and formulas is generally that listed in the margin.

The prefixes that specify number of atoms (or sometimes other groups) are listed in Table 7-5. These prefixes are used according to the following rules when naming binary molecular compounds:

1. A prefix is used with the name of the first element (the one that is less electronegative) only if more than one atom of that element is present.
2. The second element is named by combining *(a)* a prefix if more than one compound can be formed by the two elements, *(b)* the root of the name of the second element, and *(c)* the ending *-ide*.

With few exceptions the ending *-ide* shows that a compound contains only two elements.

The use of the rules is illustrated by the names of the five oxides of nitrogen, listed in Table 7-6. For example, note the application of rule 1 in the name *nitrogen monoxide* for NO. No prefix is needed with "nitrogen" because only one nitrogen atom is present. A prefix is needed with the single oxygen atom, as specified by rule 2, however, because nitrogen forms more than one oxide. Take a moment to review the prefixes in the other names in Table 7-6.

TABLE 7-6   BINARY COMPOUNDS OF NITROGEN AND OXYGEN

Formula	Prefix-Based Name	Formula	Prefix-Based Name
N_2O	dinitrogen monoxide	NO_2	nitrogen dioxide
NO	nitrogen monoxide	N_2O_4	dinitrogen tetroxide
N_2O_3	dinitrogen trioxide	N_2O_5	dinitrogen pentoxide

Sample Problem 7.4

(a) Give the name for As_2O_5.   *(b)* Write the formula for oxygen difluoride.

Solution

(a) The compound contains two arsenic atoms, so the first word in the name is "*di*arsenic." The five oxygen atoms are indicated by adding the prefix *pent-* to the word "oxide." The complete name is diarsenic pentoxide.

(b) The first symbol in the formula is that for oxygen. Oxygen is first in the name because it is less electronegative than fluorine. Since there is no prefix, there must be only one oxygen atom. The prefix *di-* in *difluoride* shows that there are two fluorine atoms in the molecule. The formula is OF_2.

Practice Problems

1. Name each of the following binary molecular compounds: *(a)* SO_3 *(Ans.)* sulfur trioxide *(b)* CF_4 *(Ans.)* carbon tetrafluoride *(c)* NO_2 *(Ans.)* nitrogen dioxide *(d)* $SiCl_3$ *(Ans.)* silicon trichloride *(e)* P_2O_5 *(Ans.)* diphosphorus pentoxide.
2. Write formulas for each of the following: *(a)* carbon tetraiodide *(Ans.)* CI_4 *(b)* phosphorus trichloride *(Ans.)* PCl_3 *(c)* sulfur dioxide *(Ans.)* SO_2 *(d)* dinitrogen trioxide *(Ans.)* N_2O_3.

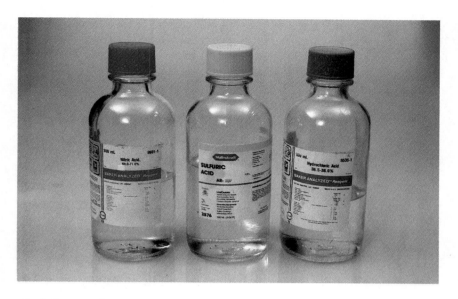

Figure 7-6 Common laboratory acids

1756 Henry Cavendish (1731–1810) discovered hydrogen by experimenting with the reaction of metals with acids.

Acids and Salts

Acids are a distinctive class of chemical compounds about which you will learn much more in later chapters. The acids found on almost every laboratory shelf (Figure 7-6) are either binary acids or oxyacids. Each is a molecular compound that contains one or more hydrogen atoms.

The common binary acids are solutions in water of binary molecular compounds containing hydrogen and one of the halogens—fluorine, chlorine, bromine, or iodine. The binary compound HCl, for example, is named hydrogen chloride. In water solution, HCl is known as *hydrochloric acid*.

Oxyacids *are acids containing hydrogen, oxygen, and a third element.* The common binary and oxyacids are listed in Table 7-7.

Many of the common polyatomic ions are produced by the loss of the hydrogen ions from oxyacids. A few examples of the relationship between oxyacids and oxyanions are the following:

sulfuric acid	H_2SO_4	sulfate ion	SO_4^{2-}
nitric acid	HNO_3	nitrate ion	NO^{3-}
phosphoric acid	H_3PO_4	phosphate ion	PO_4^{3-}

An ionic compound composed of a cation and the anion from an acid is often referred to as a **salt**. Table salt, sodium chloride (NaCl), contains the anion from hydrochloric acid. Calcium sulfate ($CaSO_4$) is a salt containing an anion from sulfuric acid.

CAUTION:
Laboratory acids must be handled with care and according to instructions. They can burn the skin.

TABLE 7-7 COMMON BINARY ACIDS AND OXYACIDS

HF	Hydrofluoric acid	HNO_3	Nitric acid
HCl	Hydrochloric acid	HNO_2	Nitrous acid
HBr	Hydrobromic acid	H_3PO_4	Phosphoric acid
HI	Hydriodic acid	H_2CO_4	Carbonic acid
H_2SO_4	Sulfuric acid	$HC_2H_3O_2$	Acetic acid
H_2SO_3	Sulfurous acid		

Some salts contain anions in which one or more hydrogen atoms from the acid are retained. Such anions are named by adding the word *hydrogen* or the prefix *bi-* to the anion name. The best-known such anion is formed by loss of one hydrogen atom from carbonic acid, H_2CO_3.

$$HCO_3^-$$
hydrogen carbonate ion
bicarbonate ion

Section Review

1. What is the significance of a chemical formula?
2. Write formulas for the compounds formed between (a) aluminum and bromine (b) sodium and oxygen (c) magnesium and iodine (d) Pb^{2+} and O^{2-} (e) Sn^{2+} and I^- (f) Fe^{3+} and S^{2-} (g) Cu^{2+} and NO_3^- (h) NH_4^- and SO_4^{2-}.
3. Name each of the following compounds using the Stock system: (a) NaI (b) MgS (c) CaO (d) K_2S (e) CuBr.
4. Write formulas for each of the following compounds: (a) barium sulfide (b) sodium hydroxide (c) lead(II) nitrate (d) potassium permanganate (e) ferric sulfate.

7.2 Oxidation Numbers

Ionic charges show the distribution of electrons in ionic compounds. Rules have been developed for assigning numbers that similarly show electron distributions for atoms in any type of compound.

Oxidation numbers, also called *oxidation states*, are assigned to the atoms in molecules, including molecular ions, to show the general distribution of electrons among the bonded atoms. In assigning oxidation numbers, "ownership" of the bonding electrons is given to the more electronegative atom in each bond. Unlike ionic charges, oxidation numbers do not have an exact physical meaning. In some cases oxidation numbers are quite arbitrary. However, oxidation numbers are useful in naming compounds, in writing formulas, and, as will be discussed in Chapter 20, in studying certain types of chemical reactions.

Assigning Oxidation Numbers

A comparison among sodium fluoride (NaF), hydrogen fluoride (HF), and water (H_2O) illustrates the basis for oxidation numbers. In sodium fluoride, the sodium and fluorine have ionic charges of +1 and −1. In hydrogen fluoride (HF), the bond is polar, with a partial negative charge on the fluorine atom and a partial positive charge on the hydrogen atom. If HF were an ionic compound in which an electron was fully transferred to the fluorine atom, H would have a +1 charge and F would have a −1 charge. In a water molecule, the oxygen atom is the more electronegative. If H_2O were an ionic compound, the oxygen atom would have a charge of −2 and the hydrogen atoms charges of +1.

Answers to Section Review appear in margin.

1. It is used to represent a chemical compound and indicates the types and numbers of atoms or ions present in the formula unit of that compound.
2. (a) $AlBr_3$ (b) Na_2O (c) MgI_2 (d) PbO (e) SnI_2 (f) Fe_2S_3 (g) $Cu(NO_3)_2$ (h) $(NH_4)_2SO_4$
3. (a) sodium iodide (b) magnesium sulfide (c) calcium oxide (d) potassium sulfide (e) copper(I) bromide (f) iron(II) chloride
4. (a) BaS (b) NaOH (c) $Pb(NO_3)_2$ (d) $KMnO_4$ (e) $Fe_2(SO_4)_3$

Fluorine is the most electronegative element, so it is *assigned* a oxidation state in *every* compound. Oxygen is the second most electronegative element and is given an oxidation state of -2 in most of its compounds. Bonding electrons are assumed to be evenly distributed among atoms of the same element, and therefore, uncombined elements are assigned oxidation numbers of zero.

The complete set of rules for assigning oxidation numbers is as follows:

1. An uncombined element has oxidation number of zero.
2. A monatomic ion has an oxidation number equal to its charge.
3. Fluorine has oxidation number of -1 in all compounds.
4. Oxygen has oxidation number of -2 in almost all compounds (e.g., except in peroxides such as H_2O_2, where it is -1; in superoxides such as KO_2, where it is $-\frac{1}{2}$; and in compounds with fluorine, where it is $+2$).
5. Hydrogen has oxidation number of $+1$ in all compounds except those with metals, in which it has oxidation number -1 (e.g., in sodium hydride, NaH).
6. The more electronegative element in a binary compound is assigned the number equal to the charge it would have if it were an ion.
7. The algebraic sum of the oxidation numbers of all atoms in a neutral compound is zero.
8. The algebraic sum of the oxidation numbers of all atoms in a polyatomic ion is equal to the charge of the ion.

Rules 7 and 8 make it possible to assign oxidation numbers where they are not known. The total of the known and unknown oxidation numbers must satisfy rule 7 or 8, as illustrated in Sample Problem 7.5.

Elements with greater electronegativity tend to be on the right of the Periodic Table and attract electrons, while the elements with the least electronegativity are found on the left and are more likely to donate electrons.

These rules have to be memorized. Tell students that memorization is usually not a good strategy to use in chemistry, that problem solving is. However, occasionally, some information *has* to be memorized.

An *algebraic sum* is a sum of positive and negative numbers.

Sample Problem 7.5

Assign oxidation numbers to each atom in the following compounds or ions. *(a)* UF_6 *(b)* $GeCl_2$ *(c)* ClO_3^- *(d)* H_2SO_4

Solution

(a) In UF_6, fluorine has oxidation number -1 as it does in all compounds in which it is present. Therefore, uranium has oxidation number $+6$.

$$\overset{+6\ -1}{UF_6} \quad \text{Oxidation no. total} = (+6) + (6 \times -1) = 0$$

(b) In $GeCl_2$, chlorine is assigned the oxidation number it would have as an ion, -1, because it is the more electronegative of the two combined elements. Therefore, germanium in this compound has oxidation number $+2$.

$$\overset{+2\ -1}{GeCl_2} \quad \text{Oxidation no. total} = (+2) + (2 \times -1) = 0$$

(c) In ClO_3^-, oxygen has oxidation number -2. The total oxidation number due to the three oxygen atoms is therefore -6. Because the ion has a -1 charge, chlorine must be assigned an oxidation number of $+5$.

$$\overset{+5\ -2}{ClO_3^-} \quad \text{Oxidation no. total} = (+5) + (3 \times -2) = -1$$

(d) In H_2SO_4, according to the rules, hydrogen has oxidation number $+1$ and oxygen has oxidation number -2. With a total positive oxidation number of $+2$ for the two hydrogen atoms, and a total negative oxidation number of -8 for the four oxygen atoms, sulfur must have an oxidation number of $+6$.

$$\overset{+1\ +6\ -2}{H_2SO_4} \quad \text{Oxidation no. total} = (2 \times +1) + (+6) + (4 \times -2) = 0$$

Practice Problems

Assign oxidation numbers to each atom in the following compounds or ions:

1. HCl   *(Ans.)* $+1, -1$;
2. CF_4 *(Ans.)* $+4, -1$;
3. PCl_3 *(Ans.)* $+3, -1$;
4. SO_2 *(Ans.)* $+4, -2$;
5. HNO_3 *(Ans.)* $+1\ +5, -2$;
6. SiO_2 *(Ans.)* $+4, -2$
7. KH   *(Ans.)* $+1, -1$;
8. P_2O_5 *(Ans.)* $+5, -2$
9. $HClO_2$ *(Ans.)* $+1, +3, -2$;)

UF_6 uranium(VI) fluoride

$GeCl_2$ germanium(II) chloride

ClO_3^- chlorate ion

Using Oxidation Numbers for Formulas and Names

Table 7-8 lists the common oxidation numbers for the nonmetals. More extensive list of oxidation numbers are given in Appendix Table A-18. These numbers can sometimes be used in the same manner as ionic charges to determine formulas. Suppose, for example, you want to know the formula of a binary compound between sulfur and oxygen. From the common $+4$ and $+6$ oxidation states of sulfur, you could expect that sulfur might form SO_2 or SO_3. Both are known compounds. As with formulas for ionic compounds, of course, a formula must represent facts. Oxidation numbers alone cannot be trusted to predict whether or not a compound exists.

In Section 7.1 we introduced the use of ionic charges as Roman numerals in the Stock system of nomenclature. This system is actually based upon oxidation numbers. The Roman numeral system of distinguishing between compounds of the same element applies whether or not the compound is ionic. The names for the first three compounds in Sample Problem 7.0 are listed in the margin. These names could be assigned without knowing whether the compounds are molecular or ionic.

The Stock system is an alternative to the prefix system for naming binary molecular compounds. The two oxides of sulfur would be named in the two systems as shown below.

TABLE 7-8   COMMON OXIDATION NUMBERS OF NONMETALS THAT HAVE VARIABLE OXIDATION STATES

Group 14	Carbon	$-4, +4$
Group 15	Nitrogen	$-3, +3, +5$
	Phosphorus	$-3, +3, +5$
Group 16	Sulfur	$-2, +4, +6$
Group 17	Chlorine	$-1, +1, +3, +5, +7$
	Bromine	$-1, +1, +3, +5, +7$
	Iodine	$-1, +1, +3, +5, +7$

Desktop Investigation

Models of Chemical Compounds

Question
Using common materials, can you make models of molecular and ionic compounds?

Materials
red, blue, green, and yellow modeling clay
large and small gumdrops in different colors
toothpicks
centimeter ruler

Procedure
1. Using different colors of clay and toothpicks, make molecular models of the following substances; hydrogen chloride, water, hydrogen peroxide, ammonia, boron trifluoride, carbon dioxide, and sulfur dioxide. As you make the models, check Figure 5-13 in Chapter 5 to make sure the balls of clay represent each atom correctly according to scale. Refer to the diagram at *right* to help you see the way that the atoms are arranged in the different molecules.

2a. Attach gum drops of two different sizes and colors to one another with toothpicks to make an ionic crystal of NaCl similar to the one in the diagram *above*. Use the larger gumdrops for the chlorine ions and the smaller ones for the sodium ions.

2b. Using clay and toothpicks, make models of what you think a formula unit of $FeBr_2$, $FeBr_3$, Cu_2O and CuO might look like. Even though the sizes of Fe and Cu ions are not shown in Figure 5-19 in chapter 5, you can estimate their sizes by comparing with the known sizes of the negative ions.

Discussion
1. How do molecular compounds and ionic compounds differ from one another structurally?

2. Iron(II) bromide and iron(III) bromide are composed of the same elements; the same is true of copper(I) oxide and copper(II) oxide. How could these compounds be differentiated?

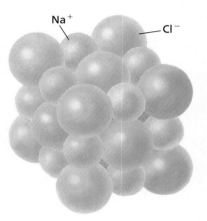

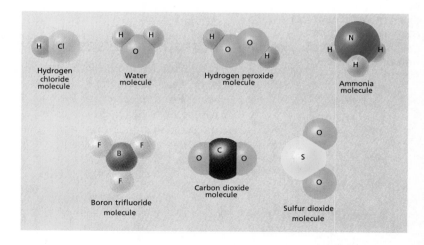

The international body that governs nomenclature has endorsed the Stock system, which is definitely more practical for more complicated compounds. The prefix-based names and the Stock system names are still used interchangeably for many simple compounds. A few more examples of names in both systems are given below.

1. (a) +1, −1 (b) +4, −1
   (c) +1, −2 (d) +3, −1
   (e) +4, −2 (f) +1, −1
   (g) +1, +4, −2 (h) +3,
   −2 (i) +6, −2 (j) +3,
   −2 (k) +5, −2
2. (a) carbon(IV) iodide
   (b) sulfur(VI) oxide
   (c) arsenic(III) sulfide
   (d) nitrogen(VI) oxide

	Old System	Stock System
PCl_3	Phosphorus trichloride	phosphorus(III) chloride
PCl_5	Phosphorus pentachloride	phosphorus(V) chloride
N_2O	Dinitrogen monoxide	nitrogen(I) oxide
NO	Nitrogen monoxide	nitrogen(II) oxide
PbO_2	Lead dioxide	lead(IV) oxide
Mo_2O_3	dimolybdenum trioxide	molybdenum(III) oxide

Section Review

1. Assign oxidation numbers to each atom in the following compounds or ions: *(a)* HF; *(b)* CI_4 *(c)* H_2O *(d)* PI_3 *(e)* CS_2 *(f)* Na_2O_2 *(g)* H_2CO_3 *(h)* NO_2^- *(i)* SO_4^{2-} *(j)* ClO_2^- *(k)* IO_3^-
2. Name each of the following binary molecular compounds according to the Stock system: *(a)* CI_4 *(b)* SO_3 *(c)* As_2S_3 *(d)* NO_3.

7.3 Using Chemical Formulas

The symbols and subscripts in a chemical formula show the numbers and types of combined atoms. The atomic composition, together with the relative atomic masses of these atoms, can be used to find the molar mass of the compound and its percent composition by mass.

In the following paragraphs you will learn how to calculate formula mass, molar mass, and percent composition. These quantities are essential to chemistry. In Section 7.4 you will see how they are applied in the experimental determination of chemical formulas. Chapter 9 expands the use of molar masses to deriving quantitative information about chemical reactions.

Formula Masses and Molar Masses

One hydrogen atom has a mass of 1.007 94 u on the relative atomic mass scale. One oxygen atom has a mass of 15.9994 u on the relative atomic mass scale. From the chemical formula, H_2O, we know that one water molecule is composed of two hydrogen atoms and one oxygen atom. The mass of a water molecule in atomic mass units is found therefore, by summing the masses of the three atoms combined in the molecule.

$$2\,H\,atoms \times \frac{1.007\,94\,u}{H\,atom} = 2.015\,88\,u$$

$$1\,O\,atom \times \frac{15.9994\,u}{O\,atom} = 15.9994\,u$$

$$Formula\ mass\ of\ H_2O\ =\ 18.0153\,u$$

The mass of one water molecule on the atomic mass scale is 18.0153 u.

The procedure illustrated above, applied to any chemical formula, gives the mass of the unit represented by the formula. The unit might be a molecule, a formula unit, or a polyatomic ion. *The **formula mass** of any compound or polyatomic ion is the sum of the atomic masses of all the atoms represented in the formula.*

Figure 7-7 One mole of air (balloon), water (graduated cylinder), $CuSO_4 \cdot 5H_2O$ (blue substance), $COCl_2 \cdot 6H_2O$ (reddish substance) and NaCl (white substance).

The mass of the water molecule is correctly referred to as a molecular mass. Often, the terms *molecular mass* or *molecular weight* are used with the same meaning as *formula mass*. You may see these terms in other books. Strictly speaking, however, the mass of one NaCl formula unit is not a molecular mass because NaCl is an ionic compound. Therefore, we prefer the more general term, and will use "formula mass" throughout this book.

Atomic masses are given to varying numbers of decimal places in the Periodic Table or tables like that in Appendix Table A-6. Therefore, the question arises of how to deal with significant figures in calculating formula masses. In the following problems, we have used atomic masses as given in Appendix Table A-6. The formula masses are rounded off according to the rules for addition of significant figures. If a formula mass is to be used in a further calculation, it is then limited to the number of digits needed in the calculation.

> The formula mass can be rounded off or the atomic masses can be rounded off before the formula mass is found.

Sample Problem 7.6

Find the formula mass of potassium chlorate, $KClO_3$.

Solution

$1 \text{ K atom} \times \dfrac{39.0983 \ u}{\text{K atom}} = 39.0983 \ u; \quad 1 \text{ Cl atom} \times \dfrac{35.453 \ u}{\text{Cl atom}} = 35.453 \ u;$

$3 \text{ O atoms} \times \dfrac{15.9994 \ u}{\text{O atom}} = 47.9982 \ u \quad$ Formula mass of $KClO_3 = 122.550 \ u$

Practice Problems
Find the formula mass of each of the following:
1. H_2SO_4 *(Ans.)* 98.07 *u*   2. $Ca(NO_3)_2$ *(Ans.)* 164.09 *u*   3. PO_4^{-3} *(Ans.)* 94.9714 *u*

In Chapter 3 you learned that the number of atoms in one molar mass of a given element is equal to 6.022×10^{23}. For example, one mole or mass of calcium (Ca) is 40.08 g and contains 6.02×10^{23} atoms. One molar mass of lead (Pb) is 207.2 g and also contains 6.022×10^{23} atoms.

The same relationship applies to compounds. The molar mass of a chemical compound is the mass in grams numerically equal to the formula mass of the compound. This molar mass is equivalent to the sum of the molar masses of all the elements present. The formula mass of water was calculated just above as 18.0153 *u*. The molar mass of water is therefore 18.0153 g. One mole of H_2O molecules contains 2 mol of H atoms and 1 mol of O atoms. The 2 mol of H atoms have a mass of 2.0159 g, and the 1 mol of O atoms has a mass of 15.9994 g. One mole of water weighs 18.0153 g and contains the same number of molecules as there are atoms in one mole of calcium or lead. One mole of any molecular compound contains 6.022×10^{23} molecules, i.e., Avogadro's number of molecules.

One mole of an ionic compound contains one mole of formula units. Potassium chlorate ($KClO_3$) was found in Sample Problem 7.6 to have a formula mass of 122.550 *u*. Taken in grams 3 g is the mass of one mole, one Avogadro's number, of $KClO_3$ formula units. But note that the total number of individual particles in one mole of potassium chlorate is twice

> **1811** Count Amadeo Avogadro (1776–1856) proposed that all gases contain the same number of particles per unit volume. Dalton and Berzelius rejected the proposal, and the idea was not in the mainstream of chemical thought until 1860. Ampère upheld Avogadro's hypothesis.

> **1860** Stanislo Cannizzaro (1826–1910) attended the First International Congress of Chemistry and convinced many attending of the validity of Avogadro's theory. He saw that if Avogadro was correct, molar masses of gases could be determined from their density when compared to the density of hydrogen. The formula for various gasses could then be inferred from the molar mass.

Figure 7-8 Bauxite is 50% AL_2O_3. If all available Al is extracted in the recovery process, how many kg of metallic Al are produced by processing 1.00 metric ton of ore?

Avogadro's number. One mole of $KClO_3$ is composed of one mole of K^+ ions and one mole of ClO_3^- ions. See Figure 7-7.

Several more examples of the meaning of "molar mass" for atoms, molecules, and ionic compounds are given in Figure 7-8.

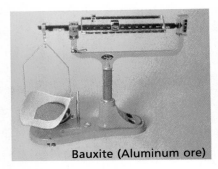

Bauxite (Aluminum ore)

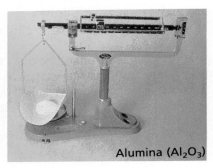

Alumina (Al_2O_3)

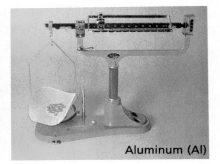

Aluminum (Al)

Sample Problem 7.7

What is the molar mass of barium nitrate, $Ba(NO_3)_2$?

Solution

One mole of barium nitrate contains 1 mol of Ba^{2+} ions and 2 mol of NO_3^- ions, giving a total of 3 mol of ions. One mole of barium nitrate contains 6 mol of oxygen atoms.

$$1 \text{ mol Ba} \times \frac{137.33 \text{ g Ba}}{1 \text{ mol Ba}} = 137.033 \text{ g Ba}$$

$$2 \text{ mol N} \times \frac{14.0067 \text{ g N}}{1 \text{ mol N}} = 28.0134 \text{ g N}$$

$$6 \text{ mol O} \times \frac{15.9994 \text{ g O}}{1 \text{ mol O}} = 95.9964 \text{ g O}$$

Molar mass of $Ba(NO_3)_2 = 261.34 \text{ g}$

Practice Problems

1. Determine the number of moles of ions and the number of moles of atoms present in one mole of each of the following: *(a)* $NaNO_3$ *(Ans.)* 1 mol Na^+ and 1 mol NO_3^-; 1 mol Na, 1 mol N, and 3 mol O *(b)* $Ba(OH)_2$ *(Ans.)* 1 mol Ba^{2+} and 2 mol OH^-; 1 mol Ba, 2 mol O, and 2 mol H.
2. Find the molar mass of each of the compounds listed in the preceding question. *(a)* *(Ans.)* 84.9947 g *(b)* *(Ans.)* 171.34 g *(c)* *(Ans.)* 132.13 g *(d)* *(Ans.)* 262.86 g.

Molar Mass as a Conversion Factor

The molar mass of a compound is a conversion factor that relates mass in grams to number of moles. Conversions of this type for elements and compounds are summarized in Figure 7-9. Molar mass as a conversion factor, you will recall, has the units of grams per mole. To convert a known number of moles to mass in grams the calculation is set up as follows:

$$\text{moles} \times \frac{\text{grams}}{1 \text{ mol}} = \text{mass, in grams}$$

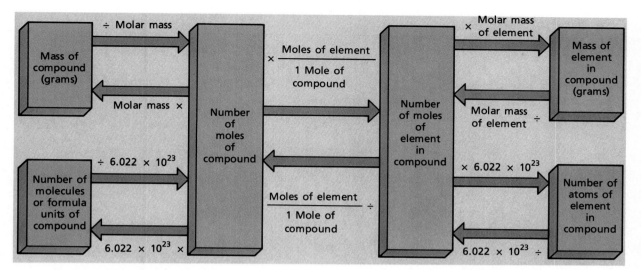

To convert a known mass in grams to number of moles, the mass must be divided by the molar mass. This is done by multiplying by the molar mass in its inverted form so that units are easily cancelled.

$$\text{mass, in grams} \times \frac{1 \text{ mol}}{\text{grams}} = \text{moles}$$

Note that the molar masses used in solving Sample Problems 7.8 to 7.10 have been calculated using all known digits and then rounded to the number of significant figures required by the problem.

Figure 7-9 Relationships between mass, moles, and numbers of molecules or atoms.

18 g/1 mol H_2O means that one mole of water has a mass of 18 grams.

Use of molar mass was introduced in Section 3.3.

Sample Problem 7.8

What is the mass in grams of 2.50 mol of oxygen gas?
Solution

Step 1. Analyze   Given: 2.50 mol O_2
                          Unknown: mass O_2, in grams

Step 2. Plan   moles $O_2 \rightarrow$ mass O_2

The molar mass of O_2 is the conversion factor needed to convert moles to mass.

$$\text{moles } O_2 \times \frac{\text{grams } O_2}{\text{mol } O_2} = \text{grams } O_2$$

Step 3. Compute   First, the molar mass of O_2 must be found:

$$2 \text{ mol O} \times \frac{15.9994 \text{ g}}{1 \text{ mol O}} = 31.9988 \text{ g} = \text{molar mass of } O_2$$

Then, the calculation as in Step 2 can be carried out.

$$2.50 \text{ mol } O_2 \times \frac{32.0 \text{ g } O_2}{1 \text{ mole } O_2} = 80.0 \text{ g } O_2$$

Step 4. Evaluate   The answer is correctly given to three significant figures and is close to an estimated value of 75 (2.5 × 30).

Sample Problem 7.10

Ibuprofen, the active ingredient in many nonprescription pain relievers, has the molecular formula $C_{13}H_{18}O_2$. The molar mass of ibuprofen is 206.2860 g/mol. *(a)* If the tablets in a bottle contain a total of 33 g of ibuprofen, how many moles of ibuprofen are in the bottle? *(b)* How many molecules of ibuprofen are in the bottle? *(c)* What is the total mass in grams of carbon in this much ibuprofen?

Solution

Step 1. Analyze   Given: 33 g of $C_{13}H_{18}O_2$,
                                   molar mass 206.2860 g
                       Unknown: *(a)* moles $C_{13}H_{18}O_2$ *(b)* molecules $C_{13}H_{18}O_2$

Step 2. Plan   *(a)* mass $\rightarrow$ moles   *(b)* moles $\rightarrow$ molecules
(c) moles $C_{13}H_{18}O_2 \rightarrow$ moles C $\rightarrow$ mass C
Part *(a)* requires using the molar mass of $C_{13}H_{18}O_2$ as a conversion factor. Then, knowing the moles of $C_{13}H_{18}O_2$, Avogadro's number is the conversion factor needed in part *(b)* to find the number of molecules. For part *(c)*, the two conversion factors needed are the moles of carbon per mole of $C_{13}H_{18}O_2$ and the molar mass of carbon.

(a) grams $C_{12}H_{18}O_2 \times \dfrac{1 \text{ mol } C_{13}H_{18}O_2}{\text{grams } C_{13}H_{18}O_2} = \text{mol } C_{13}H_{18}O_2$

(b) moles $C_{13}H_{18}O_2 \times \dfrac{6.022 \times 10^{23} \text{molecules}}{1 \text{ mol}} = \text{molecules } C_{13}H_{18}O_2$

(c) moles $C_{13}H_{18}O_2 \times \dfrac{\text{moles C}}{1 \text{ mol } C_{13}H_{18}O_2} \times \dfrac{\text{grams C}}{1 \text{ mol C}} = \text{grams C}$

Step 3. Compute   *(a)* $33 \text{ g } C_{13}H_{18}O_2 \times \dfrac{1 \text{ mol } C_{13}H_{18}O_2}{210 \text{ g } C_{13}H_{18}O_2} = 0.16 \text{ mol } C_{13}H_{18}O_2$

(b) $0.16 \text{ mol } C_{13}H_{18}O_2 \times \dfrac{6.022 \times 10^{23} \text{ molecules}}{1 \text{ mol}} = 9.6 \times 10^{22} \text{ molecules } C_{13}H_{18}O_2$

(c) $0.16 \text{ mol } C_{13}H_{18}O_2 \times \dfrac{13 \text{ mol C}}{\text{mol } C_{13}H_{18}O_2} \times \dfrac{12.01 \text{ g C}}{\text{mol C}} = 25 \text{ g C}$

The bottle contains 0.16 mol of ibuprofen, which is 9.6×10^{22} molecules of ibuprofen. The ibuprofen contains 25 g of carbon.

Step 4. Evaluate   Checking each step shows that the arithmetic is correct, significant figures have been used correctly, and units have cancelled as desired.

Practice Problems

1. What is *(a)* the molar mass of copper(II) nitrate? *(Ans.)* 187.556 g   *(b)* the mass in grams of 6.25 mol of copper(II) nitrate? *(Ans.)* 117 g
2. How many moles are represented by each of the following: *(a)* 11.0 g CO_2 *(Ans.)* 0.250 mol CO_2 *(b)* 6.60 g $(NH_4)_2SO_4$ *(Ans.)* 0.0500 mol $(NH_4)_2SO_4$ *(c)* 4.5 kg $Ca(OH)_2$ *(Ans.)* 61 mol $Ca(OH)_2$.
3. How many molecules are contained in: *(a)* 25.0 g H_2SO_4 *(Ans.)* 1.53×10^{23} molecules H_2SO_4 *(b)* 125 g of sugar $(C_{12}H_{22}O_{11})$ *(Ans.)* 2.20×10^{23} molecules $C_{12}H_{22}O_{11}$.

Chemistry Notebook

Atomic Symbols and Molecular Formulas

It is common to shorten expressions we use repeatedly. The alchemists used symbols to represent the few elements they knew and the common compounds they manipulated in their attempts to turn lead into gold. Unfortunately, their concern for secrecy led each alchemist to use a separate set of symbols. Only a few alchemical symbols became generally accepted.

John Dalton (1766-1844) realized that atomic theory provided a natural way to represent compounds. He invented a set of elemental symbols, and then combined them to represent what he believed were the molecules of various compounds. Thus he represented a molecule of nitrogen(I) oxide (laughing gas) as a simple circle (oxygen) and two circles that each contained a vertical line (nitrogen).

Because it is difficult to create simple, unique figures for all of the elements, Dalton used letters inside circles for some symbols. Jöns Jakob Berzelius (1779-1848) realized that the circles were superfluous, and recommended the one- and two-letter symbols we now use. Since then, we have added subscripts, parentheses, dots for hydrates, and lines representing bonds. To handle complicated molecules, we even resort to three-dimensional perspective. drawings generated by computer.

The standard set of atomic symbols and molecular formulas transcends language barriers and thus can be used internationally.

	Alchemy	Dalton	Present
Oxygen	(Unknown)	○	O
Hydrogen	(Unknown)	⊙	H
Nitrogen	(Unknown)	⦶	N
Sulfur	🜍	Ⓢ	S
Mercury	☿	◌	Hg
Iron	♂	Ⓘ	Fe
Water	≈	⊙○	H_2O
Nitrogen(I) oxide	(Unknown)	⦶○⦶	N_2O

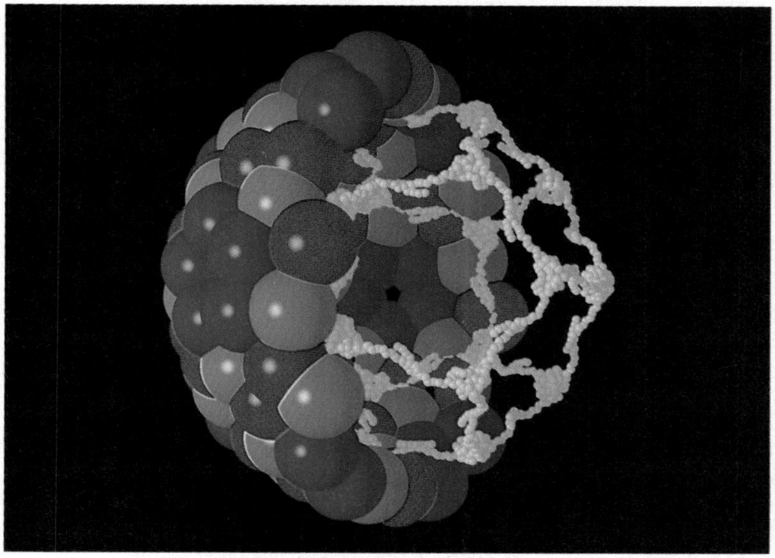

1799 Joseph Proust (1754–1826) experimentally demonstrated that the composition of a compound contained definite proportions by mass of its elements. This specific composition was not dependent on the manner in which the compound was formed.

Percent Composition

It is frequently useful to know the percent composition by mass of a chemical compound compound. The percent of iron in iron(III) oxide (Fe_2O_3) may be used to calculate the mass of iron in an ore. Or the percent of oxygen in potassium chlorate ($KClO_3$) may be useful if it is to be used as a source of it.

*The **percent composition** of a compound is the percent by mass of each element in a compound.* The percent composition is the same, no matter what the size of the sample.

The molar mass of a compound is 100% of the mass represented by the formula. The mass percent of an element is therefore the mass of the element in one molar mass divided by the molar mass or formula mass and multiplied by 100%

$$\frac{\text{mass of element}}{\text{molar mass of compound}} \times 100\% = \%\ \text{of element}$$

Sample Problem 7.11

Find the percent composition of copper(I) sulfide (Cu_2S), which occurs as the copper ore chalcocite.

Solution

Step 1. Analyze   Given: formula, Cu_2S
                     Unknown: percent composition of Cu_2S

Step 2. Plan   formula → molar mass → % of each element

The molar mass of the compound must be found. Then the mass of each element present is used to calculate the percent of each element

Step 3. Compute

$$2\ \text{mol Cu} \times \frac{63.546\ \text{g Cu}}{1\ \text{mol Cu}} = 127.09\ \text{g Cu}$$

$$1\ \text{mol S} \times \frac{32.06\ \text{g S}}{1\ \text{mol S}} = 32.06\ \text{g S}$$

$$\text{Molar mass of } Cu_2S = 159.15\ \text{g}$$

$$\frac{127.09\ \text{g Cu}}{159.15\ \text{g } Cu_2S} \times 100\% = 79.86\%\ \text{Cu}$$

$$\frac{32.06\ \text{g S}}{159.15\ \text{g } Cu_2S} \times 100\% = 20.14\%\ \text{S}$$

Step 4. Evaluate   A good check is to see if the results add up to 100, which they do. (Because of rounding off or experimental errors, the total may not always be exactly 100%.

Sample Problem 7.12

As some compounds crystallize from a water solution, they trap water molecules. Sodium carbonate forms such a hydrate, in which 10 water molecules are present for every formula unit of sodium carbonate. Find the percent of water in sodium carbonate decahydrate, $Na_2CO_3 \cdot 10H_2O$, which has a molar mass of 286 g.

Solution

Step 1. Analyze   Given: chemical formula, $Na_2CO_3 \cdot 10H_2O$
molar mass of $Na_2CO_3 \cdot 10H_2O$
Unknown: percent H_2O

Step 2. Plan   formula → mass H_2O per mole $Na_2CO_3 \cdot 10H_2O$ → % water

The mass of water per mole of sodium carbonate decahydrate must be found using atomic masses and then divided by the mass of 1 mol of $Na_2CO_3 \cdot 10H_2O$.

Step 3. Compute   First, we must find the mass of 10 mol of water:

$$20 \text{ mol H} \times \frac{1.01 \text{ g H}}{1 \text{ mol H}} = 20.2 \text{ g H}$$

$$10 \text{ mol O} \times \frac{16.0 \text{ g O}}{1 \text{ mol O}} = 160 \text{ g O}$$

Mass of H_2O per mole $Na_2CO_3 \cdot 10H_2O = 180$ g

$$\text{Percent mass of } H_2O \text{ per mol } Na_2CO_3 \cdot 10\ H_2O = \frac{180 \text{ g } H_2O}{286 \text{ g } Na_2CO_3 \cdot 10H_2O} \times 100\%$$

$$= 62.9\% \ H_2O$$

Step 4. Evaluate   Checking shows the arithmetic to be correct and units to cancel as desired. The value of the answer is reasonably larger than an estimate of 50% (which would be the answer for 143 g of water).

Practice Problems

1. Find the percent composition of each of the following compounds to five significant figures.
   (a) $PbCl_2$ *(Ans.)* 74.50% Pb and 25.50% Cl
   (b) $Ba(NO_3)_2$ *(Ans.)* 52.548% Ba, 10.719% N, and 36.732% O
2. Find the percent of water in $ZnSO_4 \cdot 7H_2O$ *(Ans.)* 43.860% H_2O
3. Magnesium hydroxide is 54.87% oxygen by mass.
   (a) How many grams of oxygen would be contained in 175 g of the compound? *(Ans.)* 96.0 g O
   (b) How many moles of oxygen does this represent? *(Ans.)* 6.00 mol O

Section Review

1. Determine both the formula mass and molar mass of the compound $(NH_4)_2CO_3$.
2. How many moles of each ion and how many of each atom are present in 1 mol of $(NH_4)_2CO_3$?
3. What is the mass in grams of 3.25 mol $Fe_2(SO_4)_3$?
4. How many moles are present in a 250. g sample of hydrogen nitrate HNO_3?
5. How many molecules of aspirin ($C_9H_8O_4$) are contained in a 100.0 mg tablet of aspirin?
6. Find the percent composition of all of the elements found in ammonium carbonate, $(NH_4)_2CO_3$.

1. Formula mass = 96.086 2 *u*
   Molar mass = 96.086 2 g
2. One mole of $(NH_4)_2CO_3$ contains 2 mol NH_4^+ ion and 1 mol CO_3^{2-} ion. One mole of this compound consists of 2 mol N, 8 mol H, 1 mol C, and 3 mol O atoms.
3. 1.30×10^3 g $Fe_2(SO_4)_3$
4. 3.97 mol HNO_3
5. 3.342×10^{20} molec $C_9H_8O_4$
6. N = 29.154 4%, H = 8.391 96%, C = 12.500 3%, O = 49.953 3%

7.4 Determining Chemical Formulas

- Define simplest formula, and explain how the term applies to ionic and molecular compounds
- Find a simplest formula from either percent or mass composition.
- Explain the relationship between the simplest formula and the molecular formula of a given compound.
- Find a molecular formula from a simplest formula.

When a new substance is prepared in the laboratory or is discovered in nature, its chemical formula is unknown. Qualitative analysis may first be done to determine what elements are present. Next, quantitative analysis may be performed to determine percent composition and the simplest formula. For a molecular compound, molar mass must then be found.

Calculation of Simplest Formulas

The first step in determining a molecular formula is to find the simplest, or empirical, formula of the compound. Composition data make this possible. *The* **simplest formula** *or* **empirical formula** *consists of the symbols for the elements combined, with subscripts showing the smallest whole-number ratio of the atoms.*

Only simplest formulas can be found from composition data. For example, diborane is a compound that is very useful to organic chemists working to synthesize new medicines and other materials. Diborane, which is 78% boron and 22% hydrogen, is known to have the molecular formula B_2H_6. However, the compound BH_3—a short-lived, unstable substance—has the same percent composition as B_2H_6.

For ionic compounds, which do not contain neutral molecules, the simplest formula is the accepted formula. Although the simplest (empirical) formula indicates the ratio of different atoms in a compound, it does not necessarily indicate the actual numbers of atoms present in each molecule. Many molecular compounds have molecular formulas that are whole-number multiples of the empirical formula, as in the case of diborane, which has the empirical formula BH_3 and the molecular formula B_2H_6.

Calculation of a simplest formula requires that composition data, if found as percent, be converted to composition in grams. An easy way to do this is to start with 100 g of the compound. For diborane, for example, the percent composition shows that 100 g contains 78 g of boron and 22 g of hydrogen. Next, the composition in grams is converted to composition in moles by using molar masses. For diborane,

$$78 \text{ g B} \times \frac{1 \text{ mol B}}{10.8 \text{ g B}} = 7.2 \text{ mol B} \qquad 22 \text{ g H} \times \frac{1 \text{ mol H}}{1.01 \text{ g H}} = 22 \text{ mol H}$$

These values give a mole ratio of 7.2 mol B : 22 mol H. The formula could be written $B_{7.2}H_{22}$, but this is not a proper formula with smallest whole-number subscripts. The smallest ratio of atoms (equal to smallest ratio of moles) is found by dividing each molar amount by the smallest number:

$$\frac{7.2 \text{ mol B}}{7.2} : \frac{22 \text{ mol H}}{7.2}$$

$$1.0 \text{ mol B} : 3.1 \text{ mol H}$$

Sometimes the ratio consists of whole numbers, and sometimes, because of rounding off or experimental error, the ratio consists of numbers close to whole numbers. The differences from whole numbers may be ignored and the nearest whole number taken. The calculation shows that the simplest formula of diborane is BH_3.

In summary, finding a simplest formula from percent composition requires finding the following:

> Composition by mass
> Composition in moles
> Smallest whole-number ratio of atoms

Sample Problem 7.13

Analysis shows a compound to contain 26.56% potassium, 35.41% chromium, 38.03% oxygen. Find the simplest formula of this compound.

Solution

Step 1. Analyze   Given: percent composition: 25.56% K, 35.41% Cr, and 38.03% O.
Unknown: simplest formula

Step 2. Plan   % composition → composition in mass → composition in moles → smallest whole-number ratio of atoms

Step 3. Compute   Composition by mass: 26.56 g K, 35.41 g Cr, 38.03 g O

$$\text{Composition in moles: } 26.56 \text{ g K} \times \frac{1 \text{ mol K}}{39.10 \text{ g K}} = 0.6793 \text{ mol K}$$

$$35.41 \text{ g Cr} \frac{1 \text{ mol C}}{52.00 \text{ g Cr}} = 0.6810 \text{ mol Cr}$$

$$38.03 \text{ g O} \times \frac{1 \text{ mol O}}{16.00 \text{ g O}} = 2.377 \text{ mol C}$$

Smallest whole-number ratio of atoms:

$$\frac{0.6793 \text{ mol K}}{0.6793} : \frac{0.6810 \text{ mol Cr}}{0.6793} : \frac{2.377 \text{ mol O}}{0.6793}$$

$$1.00 \text{ mol K} : 1.003 \text{ mol Cr} : 3.499 \text{ mol O}$$

A formula of $KCrO_{3.5}$ would not be correct. When the ratio at this point contains numbers that are not close to whole numbers, the numbers are converted to fractions and the fractions cleared. Here, 3.499 is $3\frac{1}{2}$. Expressing the ratio in halves and clearing fractions gives

$$1 \text{ mol K} : 1 \text{ mol Cr} : 3.5 \text{ mol O}$$

$$1 \text{ mol K} : 1 \text{ mol Cr} : 3\tfrac{1}{2} \text{ mol O}$$

$$2 \text{ mol K} : 2 \text{ mol Cr} : 7 \text{ mol O}$$

The simplest formula of this compound is $K_2Cr_2O_7$.

Step 4. Evaluate   The best evaluation of a problem such as this is to review each step by checking the calculations and the unit cancellations. Each step in this problem appears to be correct.

Sample Problem 7.14

Analysis of a 10.150 g sample of a compound known to contain only phosphorus and oxygen yields 5.717 g of oxygen. What is the simplest formula of this compound?

Solution

Step 1. Analyze

Given: sample mass = 10.150 g
oxygen mass = 5.717 g
Unknown: simplest formula

Step 2. Plan

Composition by mass → composition in moles →
smallest whole-number ratio of atoms

The mass of oxygen is given. The mass of phosphorus is found by subtracting the oxygen mass from the sample mass.

Step 3. Compute

Sample mass − oxygen mass = 10.150 g − 5.717 g = 4.433 g P

Composition by mass: 4.433 g P, 5.717 g O
Composition in moles:

$$4.433 \text{ g P} \times \frac{1 \text{ mol P}}{30.97 \text{ g P}} = 0.1431 \text{ mol P}$$

$$5.717 \text{ g O} \times \frac{1 \text{ mol O}}{16.00 \text{ g O}} = 0.3573 \text{ mol O}$$

Smallest whole-number ratio of atoms:

$$\frac{0.1431 \text{ mol P}}{0.1431} : \frac{0.3573 \text{ mol O}}{0.1431}$$

1.000 mol P : 2.500 mol O
2 mol P : 5 mol O

Step 4. Evaluate

Multiplying the ratio 1.000 mol P : 2.500 mol O by 2 gives the smallest whole-number ratio of 2 : 5. The simplest formula is P_2O_5. Checking each step shows correct use of significant figures and unit cancellation. The formula is reasonable because +5 is a common oxidation state of phosphorus.

Practice Problems

1. A compound is found to contain 63.52% iron and 36.48% sulfur. Find its simplest formula. *(Ans.)* FeS
2. Find the simplest formula of a compound found to contain 32.38% sodium 22.65% sulfur, and 44.99% oxygen. *(Ans.)* Na_2SO_4
3. Analysis of 20.0 grams of a compound containing only calcium and bromine indicates that 4.00 grams of calcium are present. What is the simplest formula of this compound? *(Ans.)* $CaBr_2$.
4. If 3.50 g of iron react with 1.50 g of oxygen, what is the simplest formula of the compound formed? *(Ans.)* Fe_2O_3

Calculation of Molecular Formulas

An experimentally determined simplest formula may or may not be the correct molecular formula. In introducing this section, we found the simplest formula of diborane to be BH_3. Any multiple of BH_3, such as B_2H_6, B_3H_9, B_4H_{12}, and so on represents the same ratio of boron to hydrogen atoms. How can the correct molecular formula be found?

It is not possible to decide which is the correct molecular formula unless the molecular mass of the substance has been determined. The relationship between the simplest formula and the molecular formula is

$$(\text{simplest formula})_x = \text{molecular formula}$$

where x is a whole-number multiple of the simplest formula. Therefore the formula masses have the same relationship

$$(\text{simplest-formula mass})_x = \text{molecular-formula mass}$$

For our diborane example, experiment would show the formula mass of diborane to be 27.67 u. The formula mass for the simplest formula, BH_3, is 13.83 u. Dividing the experimental molar mass by the simplest formula mass gives

$$x = \frac{27.67\ u}{13.83\ u} = 2.000$$

The value of x is 2, showing that the molecular formula is

$$(BH_3)_2 = B_2H_6$$

Sample Problem 7.15

In Sample Problem 7.14, the simplest formula of a compound of phosphorus and oxygen was found to be P_2O_5. Experiment shows that the formula mass of this compound is 283.889 u. What is the molecular formula of this compound? How would it be named in the prefix system and in the Stock system?

Solution

Step 1. Analyze   Given: simplest formula
Unknown: molecular formula

Step 2. Plan
$$x = \frac{\text{molecular-formula mass}}{\text{simplest-formula mass}}$$
$$(\text{simplest formula})_x = \text{molecular formula}$$

Step 3. Compute   The simplest formula mass is found by the usual addition to be 141.945 u.

$$\frac{283.889\ u}{141.945\ u} = 1.99999 \qquad (P_2O_5)_x = P_4O_{10}$$

The compound P_4O_{10} would be named tetraphosphorus decoxide or phosphorus(V) oxide.

Step 4. Evaluate   Checking the arithmetic shows it to be correct.

Practice Problems
1. Determine the molecular formula of a compound having a simplest formula of CH and a formula mass of 78.110 u. *(Ans.)* C_6H_6
2. A sample of a compound with a formula mass of 34.00 u is found to consist of 0.44 g of hydrogen and 6.92 g of oxygen. Find its molecular formula. *(Ans.)* H_2O_2
3. A compound with a formula mass of 42.08 u is found to be 85.64% C and 14.36% H by mass. Find its molecular formula. *(Ans.)* C_3H_6

1. Na_2SO_3
2. Fe_2S_3
3. K_2CrO_4
4. N_2O_5
5. N_2O_4

Section Review

1. A compound is found to contain 36.48 sodium 25.41% sulfur, and 38.02% oxygen. Find its simplest formula.
2. Find the simplest formula of a compound found to contain 53.70% iron and 46.30% sulfur.
3. Analysis of a compound indicates that it contains 1.04 g K, 0.70 g Cr, and 0.82 g O. Find its simplest formula.
4. If 4.04 g of N combine with 11.46 g O to produce a compound with a formula mass of 108.0 u, what is the molecular formula of this compound?
5. The formula mass of a compound is 92 u. Analysis of a sample of the compound indicates that it contains 0.606 g N and 1.390 g O. Find its molecular formula.

Chapter Summary

- Empirical or simplest chemical formulas indicate how many atoms of each type are combined in the simplest unit of a chemical compound.
- In ionic compounds, the charge of each ion in the compound may be used to determine the simplest chemical formula for the compound.
- Positive monatomic ions are named by the element name followed by the word *ion;* negative monatomic ions are named by dropping the ending of the element name and adding *-ide* to it.
- Compounds composed of two different elements are known as binary compounds and are named by combining the names of the positive and negative ions.
- The Stock system of nomenclature is used for compounds of metals that have more than one oxidation state.
- Compounds containing polyatomic ions are named in the same manner as binary ionic compounds.
- In inorganic molecular compounds, oxidation numbers of each element in the compound may be used to determine the compound's simplest chemical formula.
- Oxidation numbers (oxidation states) are used to indicate the general distribution of electrons among bonded atoms and are assigned according to a set of eight rules.
- By knowing the oxidation numbers, names can be assigned to compounds without knowing whether they are ionic or molecular.
- Formula mass, molar mass, and percent composition can be calculated from the chemical formula for a compound.
- The percent composition of a compound is the percent by mass of each element in the compound no matter what the size of the sample.
- Molar mass can be used as a conversion factor between number of moles and mass of a given compound or element.
- A simplest or empirical chemical formula shows the simplest whole number ratio of atoms in a given compound.
- Each molecule of a compound contains a whole-number multiple of the atoms in the simplest chemical formula. In many cases this whole-number multiple is equal to one.

Chapter 7 *Review*

Chemical Formulas and Chemical Compounds

Vocabulary

binary compounds
empirical formula
formula mass
hydroxide
monatomic ions
nomenclature
oxidation numbers

oxidation states
oxide
oxyacids
oxyanions
percent composition
simplest formula

Questions

1. *(a)* How are chemical compounds represented in writing?   *(b)* What information is given by the symbols and subscripts in a chemical formula?   *(c)* What additional information is needed in order to determine the type of compound represented by that formula?
2. Write the chemical formula and systematic name for each of the following: *(a)* baking soda   *(b)* milk of magnesia   *(c)* epsom salts   *(d)* limestone   *(e)* lye   *(f)* wood alcohol.
3. How is the composition of a compound determined?
4. What three different quantitative meanings can be given by a chemical formula?
5. *(a)* How are ions generally formed among the main group elements?   *(b)* What ion charges are typically associated with the majority of the elements in each of the main groups?
6. *(a)* What are monatomic ions?   *(b)* Give three examples of monatomic ions.
7. What ion charges are typically associated with d-block elements?
8. How many electrons are lost or gained in forming each of the following ions: *(a)* Na^+;   *(b)* F^-;   *(c)* Mg^{2+};   *(d)* O^{2-};   *(e)* Al^{3+}?
9. Using only the Periodic Table, write the formula of the ion most typically formed from each of the following elements: *(a)* K   *(b)* Ca   *(c)* S   *(d)* Cl   *(e)* Ba   *(f)* Br.
10. Write the noble-gas electron configurations for each of the following element–ion pairs: *(a)* Na and Na^+ *(b)* Cl and Cl^- *(c)* Ca and Ca^{2+} *(d)* O and O^{2-}.

11. How are monatomic ions named?
12. Write formulas and charges for each of the following: *(a)* sodium ion ion   *(b)* aluminum ion   *(c)* chloride ion   *(d)* nitride ion   *(e)* iron(II) ion   *(f)* iron(III) ion.
13. Name each of the following monatomic ions: *(a)* K^+ *(b)* Mg^{2+} *(c)* Al^{3+} *(d)* Cl^- *(e)* O^{2-} *(f)* Ca^{2+} *(g)* Na^+.
14. What is a binary compound?
15. List the three steps that can be used in writing formulas for ionic compounds.
16. Write formulas for the binary ionic compounds formed between: *(a)* sodium and iodine;   *(b)* calcium and sulfur;   *(c)* zinc and chlorine;   *(d)* barium and fluorine   *(e)* lithium and oxygen;   *(f)* Cr^{2+} and F^-;   *(g)* Ni^{2+} and O^{2-};   *(h)* Fe^{3+} and O^{2-}.
17. Name each of the following binary ionic compounds: *(a)* KCl;   *(b)* $CaBr_2$;   *(c)* MgI_2;   *(d)* NaF;   *(e)* AgI;   *(f)* Li_2O;   *(g)* $AlBr_3$;
18. What is the Stock system of nomenclature?
19. Name each of the following ions according to the Stock system: *(a)* Fe^{2+} *(b)* Fe^{3+} *(c)* Pb^{2+} *(d)* Pb^{4+} *(e)* Sn^{2+} *(f)* Sn^{4+} *(g)* Hg_2^{2+}.   *(g)* Hg_2^{2+}.
20. *(a)* What are oxyanions?   *(b)* How are pairs of oxyanions distinguished by name?   *(c)* Illustrate with oxyanions involving nitrogen.
21. Write the formula and charge for each of the following polyatomic ions: *(a)* ammonium ion   *(b)* acetate ion   *(c)* hydroxide ion   *(d)* carbonate ion   *(e)* sulfate ion   *(f)* phosphate ion   *(g)* $C_u(II)$ ion   *(h)* Sn(II) ion   *(i)* Fe(III) ion   *(j)* copper(I) ion   *(k)* mercury(I) ion   *(l)* mercury(II) ion.
22. Name each of the following ions: *(a)* NH_4^+ *(b)* ClO_3^- *(c)* OH^- *(d)* SO_4^{2-} *(e)* NO_3^- *(f)* CO_3^{2-} *(g)* PO_4^{3-} *(h)* $C_2H_3O_2^-$ *(i)* HCO_3^- *(j)* CrO_4^{2-}.
23. Write formulas for each of the following compounds: *(a)* sodium fluoride   *(b)* calcium oxide   *(c)* potassium sulfide   *(d)* magnesium chloride   *(e)* aluminum bromide   *(f)* lithium nitride   *(g)* iron(II)

(h) copper(I) sulfide *(i)* sodium *(j)* ammonium chloride.

24. Name each of the following ionic compounds using the Stock system: *(a)* NaCl *(b)* MgO *(c)* KF *(d)* CaS *(e)* LiBr *(f)* KOH *(g)* $AgNO_3$ *(h)* $Fe(OH)_2$ *(i)* $Sn(NO_3)_2$ *(j)* $Co(NO_3)_2$ *(k)* $FePO_4$ *(l)* Hg_2SO_4 *(m)* $Hg_3(PO_4)_2$.

25. *(a)* What is the basis for the older naming system used for binary molecular compounds? *(b)* Identify the prefixes including one through six atoms, respectively.

26. In naming and writing formulas for molecular compounds, what determines the order in which the component elements are written?

27. Name the following binary molecular compounds: *(a)* CO_2 *(b)* CCl_4 *(c)* PCl_5 *(d)* CS_2 *(e)* SO_2 *(f)* PBr_3 *(g)* N_2O_4 *(h)* CI_4 *(i)* As_2O_5 *(j)* P_4O_{10}.

28. Write formulas for the following binary molecular compounds: *(a)* carbon tetrabromide; *(b)* silicon dioxide; *(c)* carbon monoxide; *(d)* sulfur trioxide *(e)* diphosphorus pentoxide *(f)* diarsenic trisulfide.

29. *(a)* Distinguish between binary and oxyacids, and give two examples of each.

30. *(a)* What is salt? *(b)* Give two examples.

31. Name each of the following acids: *(a)* HF *(b)* HBr *(c)* HNO_3 *(d)* H_2SO_4 *(e)* H_3PO_4.

32. *(a)* What are oxidation numbers? *(b)* What useful function do oxidation numbers serve?

33. Assign oxidation numbers to each atom in the following compounds: *(a)* HI *(b)* CO_2 *(c)* PBr_3 *(d)* GeS_2 *(e)* OF_2 *(g)* H_2O_2 *(h)* As_2O_5 *(i)* P_4O_{10} *(j)* H_3PO_4.

34. Assign oxidation numbers to each atoms in the following ions: *(a)* NO_3^- *(b)* NH_4^+ *(c)* ClO_4^- *(d)* CO_3^{2-} *(e)* PO_4^{3-} *(f)* MnO_4^- *(g)* $S_2O_3^{2-}$ *(h)* $Cr_2O_7^{2-}$.

35. Name each of the binary molecular compounds in Review Question 28 using the Stock system.

36. Write formulas for each of the following: *(a)* phosphorus(III) iodide *(b)* sulfur(II) chloride *(c)* carbon(IV) sulfide.

37. *(a)* Define formula mass. *(b)* In what use it is formula mass expressed?

38. What is meant by the molar mass of a compound?

39. *(a)* What is meant by the percent composition of a compound? *(b)* Write the formula used to calculate the mass percent of an element in a compound.

40. What is meant by a simplest or empirical formula?

41. What three types of information are needed in order to find simplest formulas from percent composition data?

42. What is the relationship between the simplest formula and the molecular formula of a compound?

Problems

1. Find the formula mass of each of the following: *(a)* glucose, $C_6H_{12}O_6$ *(b)* calcium acetate, $Ca(C_2H_3O_2)_2$ *(c)* the ammonium ion, NH_4^+ *(d)* the chlorate ion, ClO_3^-.

2. Find the molar mass of each substance in Problem 1.

3. Determine the number of moles of ions and then the number of moles of each type of atom present in one mole of each of the following: *(a)* KNO_3 *(b)* Na_2SO_4 *(c)* $Ca(OH)_2$ *(d)* $(NH_4)_2SO_3$ *(e)* $Ca_3(PO_4)_2$ *(f)* $Al_2(CrO_4)_3$.

4. Determine the molar mass of each compound listed in Problem 3.

5. What is the mass in grams of each of the following: *(a)* 1.000 mole NaCl *(b)* 2.000 moles H_2O *(c)* 3.500 moles $Ca(OH)_2$ *(d)* 0.625 moles $Ba(NO_3)_2$?

6. How many moles are represented by each of the following: *(a)* 4.50 g H_2O *(b)* 471.6 g $Ba(OH)_2$ *(c)* 129.68 g $Fe_3(PO_4)_2$?

7. How many moles of each compound are contained in the following: *(a)* 6.022×10^{23} formula units of CO_2 *(b)* 2.008×10^{23} formula units of KNO_3?

8. What is the mass in grams, of each of the following: *(a)* 3.012×10^{23} formula units of SO_2 *(b)* 1.506×10^{12} formula units of $BaCl_2$?

9. How many formula units are contained in each of the following: *(a)* 97.45 g ZnS *(b)* 18.75 g $Cu(NO_3)_2$ *(c)* 69.55 g $PbCl_2$?

10. Find the percent composition of each compound: *(a)* NaCl *(b)* $AgNO_3$ *(c)* $Mg(OH)_2$.

11. Find the percent of water in $CuSO_4 \cdot 5H_2O$.

12. Based on the percent composition information obtained in Problem 10, determine the number of grams of the specified element contained in each of the following: *(a)* Na in 100.00 g NaCl *(b)* O in 140.0 g $Mg(OH)_2$.

13. Find the simplest formula of a compound found to contain 63.11% Mn and 36.89% S?

14. Find the simplest formula of a compound containing 63.50% silver, 8.25% nitrogen, and 28.26% oxygen.

15. Find the simplest formula of a compound found to contain 52.20% carbon, 13.00% hydrogen, and 34.80% oxygen.

16. A 1.338 g sample of a compound contains 0.365 g of Na, 0.221 g of N, and 0.752 g of O. Determine its simplest formula.

17. A 0.470 g sample of a compound consisting of only magnesium and oxygen contains 0.282 of Mg. What is this compound's simplest formula?

18. Determine the molecular formula of the following compounds given the simplest formula and formula mass of each: (a) HCO_2, 90.0356 u; (b) CH_2O, 180.1584 u; (c) $C_3H_5O_2$, 146.1436 u.

19. An oxide of phosphorus is found to contain 43.5% P and 56.5% O. If its formula mass is 284.0 u, find its molecular formula.

Application Questions

1. Among the following ions, which does not have a noble-gas electron configuration: Na^+, S^{2-}, Ca^{2+}, N^{2-}, Cl^-, Ba^{2+}?

2. Among the following ions, which has an electron configuration different from the rest: K^+, S^{2-}, Ca^{2+}, Al^{3+}, Cl^-, and Ar?

3. Distinguish between molecular mass and formula mass as applied to H_2O and NaCl.

4. Explain why composition data can provide only simplest, but not molecular, formulas.

Application Problems

1. Find both the formula mass and molar mass of (a) $CuSO_4 \cdot 5H_2O$; (b) $MgSO_4 \cdot 7H_2O$.

2. How many individual atoms are contained in each of the following: (a) 1.00 mol of $Al_2(SO_4)_3$; (b) 1.00 mol of $MgSO_4 \cdot 7H_2O$; (c) 5.00 mol of $CuSO_4 \cdot 5H_2O$?

3. What is the mass, in grams, of 1.75×10^{-10} moles of iron(III) sulfate?

4. A certain recipe calls for 4.00 g of baking soda (sodium hydrogen carbonate). (a) How many moles of baking soda does this represent? (b) How many molecules are contained in this mass of baking soda? (c) What is the total number of atoms in this amount?

5. (a) How many moles are represented by 2.500 mL of water? (b) How many molecules does this represent?

6. Within each of the following pairs, which contains the greater number of formula units: (a) 9.00 g H_2O or 9.00 g SiO_2 (b) 0.25 mol $AgNO_3$ or 42.0 g $Ca(OH)_2$?

7. Determine the mass of each compound listed below that would contain the same number of moles as 25.0 g $Pb(NO_3)_2$: (a) Na_2CO_3 (b) NH_4Cl (c) $Fe(OH)_3$ (d) $Ca_3(PO_4)_2$.

8. Among the following, which contains the largest total number of atoms: (a) 1.00 mol KNO_3 (b) 75.0 g O_2 (c) 1000.0 g $CaCO_3$

9. Determine the percent composition of each of the following: (a) $MgSO_4 \cdot 7H_2O$ (b) $C_{17}H_{35}COONa$(soap).

10. Among the following, which has the highest copper content? (a) $Cu(NO_3)_2$ (b) $CuSO_4$ (c) $CuCl_2$ (d) $CuC_2H_3O_2$.

11. If 2.92 g of a compound is formed from the reaction of 1.15 g of sodium with chlorine, what is the percent composition of the resulting compound?

12. A compound is found to contain 72.35% Fe and 27.65% O. Find its simplest formula?

13. The percent composition of a particular hydrate is found to be 18.31% Ca, 32.36% Cl, and 49.33% H_2O. What is its empirical formula?

14. A compound contains 7.88 g Al, 14.06 g S, and 28.06 g O. Find its simplest formula.

15. A compound consisting of 56.38% phosphorus and 43.62% oxygen has a formula mass of 2.19.8916 u. Determine its molecular formula.

16. Determine the molecular formula of a hydrate whose formula mass is 248.20 u and whose percent composition is 18.52% Na, 25.76% S, 40.30% H, and 51.49% O. The compound contains 36.2% water of hydration.

Enrichment

1. Survey your house and yard to prepare a list consisting of at least 10 elements, 10 ionic compounds, and 10 molecular compounds that you and your family use on a regular basis. Compare your list with that of your classmates.

2. Read the labels on various brands of multivitamin and multimineral tablets. Using both the common and systematic names, prepare a list of the vitamins contained in the tablets.

Chapter 8 Chemical Equations and Reactions

Chapter Planner

8.1 Chemical Equations
Reading and Writing Chemical Equations
- Demonstration: 1
  Questions: 1–3, 6
- Questions: 4, 7

Significance of a Chemical Equation
- Questions: 5, 10
  Application Questions: 2, 3

Balancing Chemical Equations
- Demonstration: 4
  Questions: 8, 10, 11
  Problems: 1, 2, 5, 6
  Application Problems: 1, 2
- Application Problems: 3, 4
- Problem: 3

8.2 Types of Chemical Reactions
Synthesis Reactions
- Demonstration: 2
  Questions: 12, 13
  Problem: 6
  Application Problems: 5, 6
- Desktop Investigation

Decomposition Reactions
- Demonstration: 5
  Question: 14
  Problem: 7
  Application Problems: 7, 8
- Questions: 15, 16

Single-Replacement Reactions
- Problem: 8
  Application Problems: 9, 10
- Question: 17

Double-Replacement Reactions
- Demonstration: 3
  Question: 9
  Application Problems: 11, 12

Combustion Reactions
- Questions: 18, 19
  Problems: 10, 11
  Application Problems: 13–15
- Technology

8.3 Activity Series of the Elements
- Questions: 20–22
Principles of the Activity Series
- Question: 23
  Application Questions: 4, 16, 17
- Demonstration: 6

Useful Generalizations Based on the Activity Series
- Experiment: 9
  Question: 24
- Question: 27
  Application Question: 5
- Questions: 25, 26
  Application Question: 6
  Application Problem: 18

Teaching Strategies

Introduction

An equation is a shorthand expression for describing a chemical change. Qualitative and quantitative information is contained in it. The equation not only describes what substances are consumed and what new substances are formed in a reaction, but also describes the relative amounts of reactant(s) and product(s).

It is important for students to understand just how powerful equations are. All of the chemical reactions that occur can be described by equations. To help students fully grasp the concept that equations represent actual phenomena that occur in the real world and not just on the pages of textbooks, it is important for them to have many experiences with chemical changes. Many of these can be done in the classroom as demonstrations, Desktop Investigations, or laboratory experiments.

8.1 Chemical Changes

Many chemical changes are part of everyday experience and students should become aware of these reactions: hydrocarbons are oxidized, iron rusts, and nonmetal acids dissolve in water, forming acid rain. Additionally, there is an entire world of biochemical reactivity (see Chapter 23), some of which is neatly locked away in cells, but some of which is evident outside of cells: starch is hydrolyzed to sugar, CaO and CO_2 combine to form the $CaCO_3$ of eggshells, the proteins of our food are broken down to amino acids which are then used to form other proteins which are used in the body. Encourage students to determine from their own observations what substances are consumed and what substances are produced. You can facilitate this with the appropriate use of demonstrations.

Demonstration 1: A Single-Replacement Reaction

PURPOSE: To demonstrate that a metal (zinc) can replace an ion (Cu^{2+}) in solution and that the results of this reaction (elemental Cu) can be determined from easily observable experimental evidence

MATERIALS: 0.5 g of mossy zinc, 20 mL 0.1–M copper(II) sulfate solution, a 100–mm petri dish, an overhead projector

PROCEDURE: Have students examine the zinc ingot and record their observations. (At this point in the year, they will have to be reminded to record all observations carefully.) Pass the zinc around the room so that each student has the opportunity to observe its hardness, luster, approximate size. Ask students to observe and record the color of the copper sulfate solution.

Place a petri dish on an overhead projector. Pour approximately 25 mL of copper sulfate solution into the petri dish until the level of the solution is about 5 mm deep. Place the zinc in the solution. Observe the reaction that will begin immediately. (The zinc will begin to turn black. The color of the solution will begin to fade.)

Cover the petri dish and put aside until the following class. Observe the changes that have occurred. The solution is no longer blue due to reduction of the copper ion; the zinc will be obviously reduced in size; metallic copper will appear.

Encourage students to record all of their observations. Bolster their confidence in their ability to make observations and to interpret them. Although all of the students will see that the solution is no longer blue, they may be hesitant to record that information or to even consider that as a valid observation.

Tell students that the blue color of the solution is an indication of the presence of the copper ion. Alternatively, you can show students several stock solutions of copper(II) compounds and allow them to make that interpretation themselves.

Ask the class to explain the significance of the disappearance of the blue solution. Pass the covered petri dish around the room and ask students to determine what substance has been produced. Emphasize that they do have evidence to make that determination.

When the class has determined that the copper ions have "disappeared," that metallic copper has appeared, and that much of the zinc has "disappeared," write the following half-reactions on the chalkboard.

$$Cu^{2+} + \text{electrons} \rightarrow Cu$$
$$Zn \rightarrow Zn^{2+} + \text{electrons}$$

Although you do not want to formally introduce the concept of oxidation and reduction at this point, it is not too early to discuss:

1. Chemical reactions are based on a rearrangement of electrons.
2. In a chemical reaction, mass must be conserved.
3. In a chemical reaction, charge must be conserved.
4. In a chemical reaction, when electrons move from one substance to another, the number of electrons that leave one substance must be identical to the number of electrons that go to the other substance. This establishes the need to adjust coefficients in chemical equations.

When the following points have been established, write the complete equation on the board.

$$Zn + CuSO_4 \rightarrow ZnSO_4 + Cu$$

Additional demonstrations of single replacement reactions using this procedure can be very helpful. If a copper penny is placed in silver nitrate solution, the copper will replace the silver ion and the penny will become plated with silver. Since silver loses one electron and copper gains two, coefficients will have to be adjusted to balance the equation. A very attractive demonstration based on the Cu/Ag^{2+} reaction is to coil a piece of copper wire and place it upright in silver nitrate solution in a petri dish, covered with an inverted beaker. Glistening particles of the silver will adhere to the copper. This is a logical extension of the $Zn–Cu^{2+}$ demonstration.

A third demonstration could be the generation of hydrogen by the action of zinc on HCl or H_2SO_4. Since hydrogen is diatomic, this will be an opportune time to emphasize the importance of subscripts in representing the composition of elements as well as compounds.

Demonstration 2: The Formation of a Base from a Metal Oxide

PURPOSE: To demonstarate that evidence of a chemical reaction can be detected by indirect means
MATERIALS: Beaker, water, phenopthalein, calcium oxide
PROCEDURE: Add a few drops of phenopthalein to a 50–mL beaker that is half filled with water. Add approximately 0.005 g calcium oxide. [Hints: This simple demonstration can be done in a variety of ways. If you want to use small amounts of material, use a 50–mL beaker placed on an overhead projector. If you want to add a bit of drama to the demonstration, use a large beaker or cylinder placed on a magnetic stirrer. Add the calcium oxide as the stirrer is mixing the reactants at low speed.]

Demonstration 3: A Double-Replacement Reaction

PURPOSE: To demonstrate formation of a precipitate as evidence that chemical reaction has taken place
MATERIALS: $0.1–M$ $PbCl_2$, $0.1–M$ Na_2CrO_4, beaker, test tube, or petri dish
PROCEDURE: Mix together 5 mL of each solution. A precipitate will form immediately.

Double-replacement reactions resulting in the formation of a precipitate can be used extensively in this unit on balancing equations. Actually seeing the reaction take place is a stimulus to students to balance the equation. Continue to ask students how they know that a reaction has taken place.

To entice students to participate in the process, you might ask them to choose the color of the product in the reaction. By mixing a few mL of the appropriate stock solutions in a beaker or test tube, you can convince students that there is beauty and excitement in chemistry.

Color	Compound
White	PbI_2, $PbCl_2$, $BaCo_3$, $BaSO_4$
Yellow	$BaCro_4$, $PbCrO_4$
Blue	$Cu(OH)_2$
Green	$Ni(OH)_2$
Pink	$Co(OH)_2$, $MnCO_3$, $Mn(OH)_2$
Red	$FeSCN$, $SrCrO_4$, $Ag_2(CrO_4)$

Before proceeding to Section 8.2, students should be able to understand the following concepts:

1. A chemical equation represents a chemical reaction.
2. The reactants are written on the left side of the equation and the products are written on the right side of the equation.
3. A chemical equation must be balanced. When an equation is balanced, mass is conserved and charge is conserved.

4. The formulas in a chemical reaction identify the reactants and the products. The subscripts as well as the symbols in the formula are unique to each compound. The subscripts in the formulas cannot be changed: coefficients have to be adjusted to balance the equation.

5. Atoms are rearranged during the course of a chemical reaction. The number of atoms in a single formula unit or molecule of a reactant is not necessarily the same as the number of atoms in a single molecule or formula unit of a product. (Because we emphasize that subscripts cannot be changed in order to balance an equation, some students have the misconception that the subscript must "stay with the symbol" and cannot be changed from reactant to product.)

$$Zn + HCl \rightarrow ZnCl_2 + H_2$$

This equation is a good example of that principle.

6. Even when a molecule of product contains the same number of atoms of a specific element as a molecule of reactant, the atoms are rearranged during the course of the reaction.

$$2H_2 + O_2 \rightarrow 2H_2O$$

Most students will recognize that, when water is formed from hydrogen and oxygen, the bonds between the oxygen are broken. But some will not realize that the bonds between the hydrogen are broken as well. Diagram this reaction on the board using structural formulas and use molecular models to show the structure of the hydrogen molecules, the oxygen molecule, and the water molecule.

8.2 Types of Chemical Reactions

Continue to do at least one simple demonstration each day as you proceed to teach students how to master the skill of equation balancing. Emphasize that the equation represents the chemical reaction. Students will develop a better understanding of this if they continue to have experience with many chemical reactions.

Introduce the concept that the equation contains quantitative information as well as qualitative information. The equation represents the relationship between moles and masses or reactants and products in a chemical reaction.

Translate the following equation into three sentences and have the class join you in chorus:

$$PbCl_2 + Na_2CrO_4 \rightarrow PbCrO_4 + 2NaCl$$

"One formula unit of lead chloride reacts with one formula unit of sodium chromate to produce one formula unit of lead chromate and two formula units of sodium chloride."

"6.02×10^{23} formula units of lead chloride react with 6.02×10^{23} formula units of sodium chromate to produce 6.02×10^{23} formula units of lead chromate and 12.02×10^{23} units of sodium chloride."

"One mole of lead chloride reacts with one mole of sodium chromate to produce one mole of lead chromate and two moles of sodium chloride."

Continue to demonstrate as many of the reactions used in this chapter as possible. Most of your demonstrations will be quick, easy, and qualitative. Also take the time to do one or two quantitative demonstrations.

Demonstration 4: Ratios of Ions

PURPOSE: To determine the formula of a compound and to establish a balanced equation for the reaction

$$Co^{2+}(aq) + OH^-(aq) \rightarrow Co(OH)_2(s)$$

MATERIALS: 0.1 M Co(II) chloride, 0.1 M NaOH, phenopthalein, 24–well microplate or 10–mL test tube, beral micropipets or medicine droppers. (The microplate and beral pipettes are the basic pieces of equipment used for microchemistry techniques. Both types of equipment were developed by biologists and have been utilized by Lewis, Maloney, Russo, and others who have developed student laboratory procedures that require less material and less time than more traditional techniques. Microplates and beral pipettes are available from Flinn.)

PROCEDURE: (1) Add ten drops of cobalt chloride solution to a 10-mL test tube. (Alternatively, you can use a well in a 24-well microplate.) (2) Add 1 drop of phenolpthalein solution to the cobalt chloride solution. (3) Add NaOH solution, one drop at a time, carefully counting drops. The phenolpthalein indicator will turn to red when there is an excess of NaOH. Repeat steps 1–3 five times. If test tubes are used, they will have to be cleaned after each titration. If microplates are used, the five titrations can be done simultaneously in 5 different wells.

Determine the ratio of drops of sodium hydroxide to drops of cobalt chloride in each titration. Find the average ratio of the five titrations and round to the nearest whole number ratio.

The ratio of drops of NaOH to drops of $CoCl_2$ will indicate the relative number of OH ions to cobalt ions in the product. Write the balanced equation for this reaction based on the drop ratio determined above. Compare it to the correct equation for the reaction.

Demonstration 5: The Hydrolysis of H_2O

PURPOSE: To show that the electrolytic decomposition of water produces two gases in the molar proportion of 2:1

MATERIALS: Hoffman apparatus, sodium sulphate, bromthymol blue, a 6 V/12 V d-c power supply

[Note: If platinum electrodes are available, use them in this demonstration in order to get the 2:1 gas volume ratio of hydrogen and oxygen that is predicted. Graphite electrodes can be used for this electrolysis but some adsorption of oxygen on the surface of the anode will interfere with quantitative results.]

PROCEDURE: In a large beaker on a magnetic stirrer prepare a dilute solution of sodium sulfate. Explain to the students that pure water is a very poor conductor of electricity. A small amount of sodium sulfate must be added to the water to conduct the current that will produce the decomposition reaction.

Add a generous amount of bromthymol blue. The water solution is expected to turn green. If your distilled water is not neutral, you may get a yellow or blue solution. Briefly explain the nature of the indicator and the reason for the color that you observe. Explain that you are using the indicator to develop a better understanding of the mechanism of the reaction.

Attach your power supply to your electrodes and begin your electrolysis. A 6 V/12 V battery charger is an excellent source of electrical energy and will give you two distinct reaction rates. Take this opportunity to tell your class that water does not decompose spontaneously but requires an input of energy for the reaction to procede, in this case that energy is electrical energy.

Since heating water results in the formation of a vapor, some students might confuse the physical change associated with adding thermal energy to water with the chemical change associated with adding electrical energy. You can demonstrate the difference experimentally by collecting the gasses produced in electrolysis and the vapor produced by boiling and performing a glowing splint test.

$$\text{Results: } 2H_2O + \text{electrical energy} \rightarrow 2H_2 + O_2$$

As soon as the current is applied, bubbles of hydrogen will form at the cathode and bubbles of oxygen will appear at the anode. The volume of hydrogen produced will be twice the volume of oxygen produced, as predicted.

The solution surrounding the anode will become yellow due to the presence of hydronium ions produced, and the solution surrounding the cathode will turn blue due to the presence of the hydroxide ions produced.

$$\text{Anode: } 6H_2O \rightarrow O_2 + 4H_3O^+(aq) + 4e^-$$
$$\text{Cathode: } 4H_2O + 4e^- \rightarrow 2H_2 + 4OH^-(aq)$$

(Do not show reactions with an unequal number of electrons; that is why you need $4H_2O$ at the cathode.)

$$\text{Overall: } 10H_2O(\ell) \rightarrow 2H_2(g) +$$
$$O_2(g) + 4H_3O^+(aq) + 4OH^-$$

This is not an appropriate time to explain the anode and the cathode reactions in detail. It is sufficient to review the simplified reaction, to state that some acid is produced at the anode and some base is produced at the cathode, and that the reaction will be studied again in more detail. Use this demonstration again when gas laws are introduced, and a third time when oxidation-reduction is formally presented. It will be helpful to students to have some familiarity with this reaction when new concepts are introduced. Chemistry is not a fragmented science: one concept builds on another. Using the same demonstration to present several concepts illustrates that continuity.

8.3 Activity Series of the Elements

Demonstration 6: Reactions in an Activity Series
PURPOSE: To determine the relative ease with which metal atoms lose electrons to form ions in aqueous solutions, to give students an opportunity to determine the products of a reaction from a knowledge of the reactants and from the observations of a reaction, and to introduce students to the concept of rate through observations of slow and fast reactions without any attempt to quantify rate constants
MATERIALS: Metals: Al, Zn, Fe, Co, Ni, Pb, Cu;
Solutions: 0.1–M solutions of the following: $Al(NO_3)_3$, $Zn(NO_3)_2$, $Fe(NO_3)_3$, $Co(NO_3)_2$, $Ni(NO_3)_2$, $Pb(NO_3)_2$, $Cu(NO_3)_2$, and HCl; 49 small containers: 50–mL beakers, petri dishes, or baby food jars
[Many of the chemicals used in this investigation are available in hardware stores and supermarkets so that students can continue this investigation at home. Baby food jars can be used as reaction containers.]
PROCEDURE: Have students prepare a grid as indicated.

Activity Series

TIME	Al/Al^{3+}	Zn/Al^{3+}	Fe/Al^{3+}	Co/Al^{3+}	Ni/Al^{3+}	Pb/Al^{3+}	Cu/Al^{3+}
Three Min	N.R.	N.R.	N.R.	N.R.	N.R.	N.R.	N.R.
Thirty Min							
One Day							

TIME	Al/Zn^{2+}	Zn/Zn^{2+}	Fe/Zn^{2+}	Co/Zn^{2+}	Ni/Zn^{2+}	Pb/Zn^{2+}	Cu/Zn^{2+}
Three Min							
Thirty Min							
One Day							

TIME	Al/Fe^{3+}	Zn/Fe^{3+}	Fe/Fe^{3+}	Co/Fe^{3+}	Ni/Fe^{3+}	Pb/Fe^{3+}	Cu/Fe^{3+}
Three Min							
Thirty Min							
One Day							

TIME	Al/Co^{2+}	Zn/Co^{2+}	Fe/Co^{2+}	Co/Co^{2+}	Ni/Co^{2+}	Pb/Co^{2+}	Cu/Co^{2+}
Three Min							
Thirty Min							
One Day							

TIME	Al/Ni^{2+}	Zn/Ni^{2+}	Fe/Ni^{2+}	Co/Ni^{2+}	Ni/Ni^{2+}	Pb/Ni^{2+}	Cu/Ni^{2+}
Three Min							
Thirty Min							
One Day							

TIME	Al/H^+	Zn/H^+	Fe/H^+	Co/H^+	Ni/H^+	Pb/H^+	Cu/H^+
Three Min							
Thirty Min							
One Day							

TIME	Al/Cu^{2+}	Zn/Cu^{2+}	Fe/Cu^{2+}	Co/Cu^{2+}	Ni/Cu^{2+}	Pb/Cu^{2+}	Cu/Cu^{2+}
Three Min							
Thirty Min							
One Day							

Label containers according to the solutions indicated on the grid sheet. Pour 15 mL of each solution in the appropriately labeled container. If you are doing this as a demonstration, use 50–mL beakers or small petri dishes.

If students are doing this as an activity, test tubes, or baby food jars can be used. This procedure can also be followed using microchem techniques in 2 24-well microplates if 5 mL of solution is used in each well.

Add approximately 0.5 grams of the metal indicated to the appropriately labeled container. Record observations within the first 5 minutes of the reaction, thirty minutes later, and on the following day.

It is not necessary to do all of this investigation at one time. It is important for students to keep an accurate record of the results of the reaction even if you choose to do the demonstration over a period of several weeks.

Students can construct a fairly accurate activity series based on their observations of these reactions by doing the following:

1. By counting the number of reactions in each *column* of the record sheet, tabulate the number of ions that each metal replaces.
2. Based on the number of reactions for each metal, rate each of the metals from the greatest to the least number of ions it will replace. In case of a tie, consider the rate of the reactions.
3. By counting the number of reactions in each *row* of the record sheet, tabulate the number of metals that will replace each ion.
4. Based on the number of reactions for each ion, rate the ions from the highest to the lowest tendency to attract electrons. In case of a tie, consider the rate of the reactions.
5. Compare the two lists.
6. Of the metals tested, which one has the highest tendency to donate electrons to ions tested?
7. Of the metals tested, which one has the lowest tendency to donate electrons to ions tested?
8. Of the ions tested, which one has the highest tendency to attract electrons from the metals tested?
9. Of the ions tested, which one has the lowest tendency to attract electrons from the metals tested?
10. Based on an element's relatively high tendency to donate electrons to ions when that element is in its ionic state, construct an activity series of elements from greatest reducing power to lowest.

References and Resources

Books and Periodicals

Miller, Foll A. "Joseph Priestly, Preeminent Amature Chemist," *Journal of Chemical Education*, Vol. 64, No. 9, 1987, pp. 745–747.

Computer Software

Smith, Stanley G., Ruth Chabay, and Elizabeth Keen. *Chemical Formulas and Equations*, Apple II, IBM, COMPress.

Weyh, John A., Joseph R. Crook, and Les N. Hauge. *Chemical Reactions*, IBM, COMPress.

Answers and Solutions

Questions

1. A chemical equation indicates, with symbols and formulas, the change that occurs in a chemical reaction.
2. A correctly written chemical equation must represent the known facts, contain the correct formulas of the reactants and products, and satisfy the law of conservation of matter.
3. *(a)* A coefficient is a number that appears in front of a formula in a chemical equation. *(b)* The number of atoms of each type in the formula is multiplied by the coefficient.
4. An example of a word equation is: methane + oxygen → carbon dioxide + water. A formula equation is: $CH_4(g) + O_2(g) \rightarrow CO_2(g) + H_2O(g)$. A chemical equation is: $CH_4(g) + 2O_2(g) \rightarrow CO_2(g) + 2H_2O(g)$.
5. Both word and formula equations have only qualitative meaning; they do not express the quantities of reactants used or products formed.
6. *(a)* An aqueous solution is a solution in water. *(b)* A catalyst is a substance that accelerates a chemical reaction but can be recovered unchanged. *(c)* A reversible reaction is a chemical reaction in which the products re-form the original reactants.
7. The indicated formulas are: *(a)* KOH *(b)* $CaNO_3$ *(c)* Na_2CO_3 *(d)* CCl_4 *(e)* $MgBr_2$ *(f)* SO_2 *(g)* $(NH_4)_2SO_4$.
8. Step 1: Identify the names of the reactants and products, and write a word equation. Step 2: Substitute correct formulas for the names of the reactants and products, and write a formula equation. Step 3: Adjust the coefficients in the formula equation so that the law of conservation of matter is satisfied.
9. Four useful guidelines are: (1) balance the different types of atoms one at a time, (2) first balance the types of atoms that appear only once on each side of the equation, (3) balance polyatomic ions that appear on both sides of the equation as single units, and (4) balance H atoms and O atoms last.
10. The number of atoms indicated in each is: *(a)* N = 6 *(b)* H = 4, O = 2 *(c)* H = 4, N = 4, O = 12 *(d)* Ca = 2, O = 4, H = 4 *(e)* Ba = 3, Cl = 6, O = 18 *(f)* Fe = 5, N = 10, O = 30 *(g)* Mg = 12, P = 8, O = 32 *(h)* N = 4, H = 16, S = 2, O = 8 *(i)* Al = 12, C = 18, O = 54.
11. Two common mistakes are neglecting to be sure that each formula is correct and trying to balance an equation by changing subscripts.
12. (1) In synthesis, or composition, reactions, two or more substances combine to form a new substance. Equation: A + X → AX. (2) In decomposition reactions, a single compound undergoes a reaction that produces two or more simpler substances. Equation: AX → A + X. (3) In single-replacement, or displacement, reactions, one element replaces another element in a compound. Equation: A + BX → AX + B or Y + BX → BY + X. (4) In double-

replacement, or ionic, reactions, the ions of two compounds exchange places in an aqueous solution to form two new compounds. Equation: $AX + BY \rightarrow AY + BX$. (5) In combustion reactions, a substance combines with oxygen, releasing a large amount of energy in the form of light and heat. Equation: $C_3H_8(g) + 5O_2(g) \rightarrow 3CO_2(g) + 4H_2O(g)$.

13. The indicated products are: *(a)* an oxide of the element *(b)* an ionic compound *(c)* a metal hydroxide *(d)* a solution of an oxyacid *(e)* a salt.
14. Synthesis and decomposition reactions are opposites.
15. Most decomposition reactions take place only when energy in the form of heat or electricity is added.
16. Electrolysis is the decomposition of a substance by an electric current.
17. *(a)* Most single-replacement reactions take place in aqueous solutions. *(b)* Single replacement reactions usually involve less energy than do synthesis or decomposition reactions.
18. Hydrocarbons are compounds made up of carbon and hydrogen only.
19. The products are carbon dioxide and water.
20. (a) The ability of an element to react (b) The greater the activity of an element, the more readily it reacts.
21. An activity series is a set of elements listed according to the ease with which they undergo certain chemical reactions. The elements are usually listed in an order determined by a tendency to undergo single-replacement reactions. Any element in such a series can replace an element below it.
22. *(a)* The activity series for the halogens lists the elements in the same order as they appear in Group 17 of the Periodic Table. *(b)* In order of decreasing activity, the halogens are: F_2, Cl_2, Br_2, I_2.
23. *(a)* The activity series of metals is based on the ease with which metal atoms lose electrons to form ions in aqueous solution; the higher a metal is in the series, the more easily its atoms lose electrons, and thus the greater the reactivity of that metal. *(b)* The greater the distance between two metals in the activity series is, the more likely it is that one will replace the other in a reaction.
24. (1) An element will replace from a compound in aqueous solution any of those elements below it in the activity series; the larger the interval between the elements in the activity series, the greater is the tendency for the replacement reaction to occur. (2) Any metal above magnesium will replace hydrogen from water. (3) Any metal above cobalt will replace hydrogen from steam. (4) Any metal above hydrogen will react with acids, replacing hydrogen.
25. *(a)* The higher a metal is in the activity series, the more readily it will undergo synthesis with oxygen. *(b)* Any metal above silver reacts with oxygen, forming oxides; those near the top react rapidly. Any metal below mercury forms oxides only indirectly (not by direct reaction with O_2).
26. *(a)* The more active a metal, the more strongly it holds onto oxygen in an oxide, and, therefore, the more strongly the oxide resists decomposition into its elements upon heating. *(b)* Oxides of metals below copper decompose with heat alone, oxides of metals below chromium yield free metals when heated with hydrogen, and oxides of metals above iron resist conversion to free metals when heated with hydrogen.
27. (1) Elements near the top of the activity series are never

found free in nature. (2) Elements near the bottom of the series are often found free in nature.

Problems

1. The indicated word and formula equations are: *(a)* hydrogen + oxygen → water; $H_2(g) + O_2(g) \rightarrow H_2O(g)$ *(b)* copper + silver nitrate → copper(II) nitrate + silver; $Cu(s) + AgNO_3(aq) \rightarrow Cu(NO_3)_2(aq) + Ag(s)$ *(c)* mercury(II) oxide → mercury + oxygen; $HgO(s) \rightarrow Hg(\ell) + O_2(g)$.

2. *(a)* 2, 1, 1 *(b)* 3, 3, 3 *(c)* 5, 10, 5

3. *(a)* $3.0 \text{ mol O} \times \dfrac{4 \text{ mol NH}_3}{3 \text{ mol O}_2} = 4.0 \text{ mol NH}$

   $3.0 \text{ mol O}_2 \times \dfrac{2 \text{ mol N}_2}{3 \text{ mol O}_2} = 2.0 \text{ mol N}_2$

   $3.0 \text{ mol O}_2 \times \dfrac{6 \text{ mol H}_2O}{3 \text{ mol O}} = 6.0 \text{ mol H}_2O$

   (b) $8.0 \text{ mol NH}_3 \times \dfrac{3 \text{ mol O}_2}{4 \text{ mol NH}_3} = 6.0 \text{ mol O}_2$

   $8.0 \text{ mol NH}_3 \times \dfrac{2 \text{ mol N}_2}{4 \text{ mol NH}_3} = 4.0 \text{ mol N}_2$

   $8.0 \text{ mol NH}_3 \times \dfrac{6 \text{ mol H}_2}{4 \text{ mol NH}_3} = 12 \text{ mol H}_2O$

   (c) $1.0 \text{ mol N}_2 \times \dfrac{4 \text{ mol NH}_3}{2 \text{ mol N}_2} = 2.0 \text{ mol NH}_3$

   $1.0 \text{ mol N}_2 \times \dfrac{3 \text{ mol O}_2}{2 \text{ mol N}} = 1.5 \text{ mol O}_2$

   $1.0 \text{ mol N}_2 \times \dfrac{6 \text{ mol H}_2O}{2 \text{ mol N}_2} = 3.0 \text{ mol H}_2O$

   (d) $0.40 \text{ mol H}_2O \times \dfrac{4 \text{ mol NH}_3}{6 \text{ mol H}_2O} = 0.27 \text{ mol NH}_3$

   $0.40 \text{ mol H}_2O \times \dfrac{3 \text{ mol O}_2}{6 \text{ mol H}_2O} = 0.20 \text{ mol O}_2$

   $0.40 \text{ mol H}_2O \times \dfrac{2 \text{ mol N}_2}{6 \text{ mol H}_2O} = 0.13 \text{ mol N}_2$

4. *(a)* $2KBr + BaI_2 \rightarrow 2KI + BaBr_2$
   (b) $2HCl + Mg(OH)_2 \rightarrow MgCl_2 + 2H_2O$
   (c) $2HNO_3 + Ca(OH)_2 \rightarrow Ca(NO_3)_2 + 2H_2O$

5. *(a)* $H_2 + Cl_2 \rightarrow 2HCl$
   (b) $2Al + Fe_2O_3 \rightarrow Al_2O_3 + 2Fe$
   (c) $Pb(C_2H_3O_2)_2 + H_2S \rightarrow PbS + 2HC_2H_3O_2$

6. *(a)* sodium + oxygen → sodium oxide
   $4Na + O_2 \rightarrow 2Na_2O$
   (b) copper + oxygen → copper(I) oxide
   $4Cu + O_2 \rightarrow 2Cu_2O$
   copper + oxygen → copper(II) oxide
   $2Cu + O_2 \rightarrow 2CuO$
   (c) potassium + chlorine → potassium chloride
   $2K + Cl_2 \rightarrow 2KCl$
   (d) magnesium + fluorine → magnesium fluoride
   $Mg + F_2 \rightarrow MgF_2$

7. (a) $2HgO \rightarrow 2Hg + O_2$

(b) $2H_2O(\ell) \xrightarrow[\text{current}]{\text{electric}} 2H_2(g) + O_2(g)$

(c) $2Ag_2O \rightarrow 4Ag + O_2$

(d) $MgCl_2 \xrightarrow[\text{current}]{\text{electric}} Mg + Cl_2$

8. (a) $Zn + Pb(NO_3)_2 \rightarrow Pb + Zn(NO_3)_2$
(b) $2K + Ba(C_2H_3O_2)_2 \rightarrow Ba + 2KC_2H_3O_2$
(c) $2Al + 3NiSO_4 \rightarrow 3Ni + Al_2(SO_4)_3$
(d) $2Na + 2H_2O \rightarrow 2NaOH + H_2$

9. (a) $AgNO_3(aq) + NaCl(aq) \rightarrow NaNO_3(aq) + AgCl(s)$
(b) $Mg(NO_3)_2(aq) + 2KOH(aq) \rightarrow 2KNO_3(aq) + Mg(OH)_2(s)$
(c) $(NH_4)_2S(aq) + ZnCl_2(aq) \rightarrow 2NH_4Cl(aq) + ZnS(s)$
(d) $3LiOH(aq) + Fe(NO_3)_3(aq) \rightarrow 3LiNO_3(aq) + Fe(OH)_3(s)$

10. (a) $CH_4 + 2O_2 \rightarrow CO_2 + 2H_2O$
(b) $2C_3H_6 + 9O_2 \rightarrow 6CO_2 + 6H_2O$
(c) $C_5H_{12} + 8O_2 \rightarrow 5CO_2 + 6H_2O$

11. (a) $H_2 + I_2 \rightarrow 2HI$; synthesis
(b) $2Li + 2HCl \rightarrow 2LiCl + H_2$; single replacement
(c) $Na_2CO_3 \rightarrow Na_2O + CO_2$; decomposition
(d) $2HgO \rightarrow 2Hg + O_2$; decomposition
(e) $Mg(OH)_2 \rightarrow MgO + H_2O$; decomposition
(f) $Cu + Cl_2 \rightarrow CuCl_2$; synthesis
(g) $Ca(ClO_3)_2 \rightarrow CaCl_2 + 3O_2$; decomposition
(h) $2Li + 2H_2O \rightarrow 2LiOH + H_2$; single replacement
(i) $PbCO_3 \rightarrow PbO + CO_2$; decomposition
(j) $CH_4 + 2O_2 \rightarrow CO_2 + 2H_2O$; combustion
(k) $Cu + 2AgNO_3 \rightarrow 2Ag + Cu(NO_3)_2$; single replacement
(l) $2NaF \rightarrow 2Na + F_2$; decomposition
(m) $Mg + 2HNO_3 \rightarrow Mg(NO_3)_2 + H_2$; single replacement
(n) $2AgClO_3 \rightarrow 2AgCl + 3O_2$; decomposition
(o) $F_2 + MgCl_2 \rightarrow Cl_2 + MgF_2$; single replacement
(p) $K_2S + 2HNO_3 \rightarrow 2KNO_3 + H_2S$; double replacement
(q) $Mg + H_2O \rightarrow MgO + H_2$; single replacement

Application Questions

1. According to the law of conservation of matter, the amounts of reactants and products represented in the equation must be adjusted by adding coefficients so that the number and type of atoms is the same on both sides of the equation.
2. (1) A chemical equation gives the smallest numbers of atoms, molecules, or ions that will satisfy the law of conservation of matter, as well as information about moles and masses of reactants and products. (2) The coefficients in chemical equations indicate relative amounts of reactants and products. (3) Since chemical equations represent equalities, they can be interpreted in both the forward and reverse directions.
3. Chemical equations give no indication of whether a reaction will actually occur. They give no information about the speed with which reactions occur. They do not indicate the actual pathway that atoms or ions take in moving from reactants to products.
4. The indicated element in each case is: (a) K (b) Al (c) Cr (d) F_2 (e) Ag (f) Cl (g) Sr (h) F_2.
5. In order of increasing tendency to be found free in nature, the elements are: K, Ba, Na, Al, Fe, Ni, Pb, Cu, Ag, and Au.
6. The indicated element in each case is: (a) Ba (b) Cr (c) Co (d) Sb.

Application Problems

1. (a) $4Al + 3O_2 \rightarrow 2Al_2O_3$
(b) $P_4O_{10} + 6H_2O \rightarrow 4H_3PO_4$
(c) $Fe_2O_3 + 3CO \rightarrow 2Fe + 3CO_2$

2. (a) $Ca(OH)_2 + (NH_4)_2SO_4 \rightarrow CaSO_4 + 2NH_3 + 2H_2O$
(b) $2C_2H_6 + 7O_2 \rightarrow 4CO_2 + 6H_2O$
(c) $2Cu_2S + 3O_2 \rightarrow 2Cu_2O + 2SO_2$
(d) $2NaCl + 2H_2SO_4 + MnO_2 \rightarrow MnSO_4 + Na_2SO_4 + Cl_2 + 2H_2O$

3. (a) LiO_2 is an incorrect formula. $4Li + O_2 \rightarrow 2Li_2O$
(b) H_2Cl_2 is an incorrect formula. $H_2 + Cl_2 \rightarrow 2HCl$
(c) MgO_2 is an incorrect formula, and the equation as written is not balanced. $MgCO_3 \rightarrow MgO + CO_2$
(d) I is an incorrect formula (diatomic I^2), and the equation is not balanced. $2NaI + Cl_2 \rightarrow 2NaCl + I_2$

4. The correct formulas are: (a) Fe_2O_3 (b) NH_3 (c) P_2O_5 (d) NO_2.

5. (a) $2Mg + O_2 \rightarrow 2MgO$ (b) $4Fe + 3O_2 \rightarrow 2Fe_2O_3$ (c) $2Li + Cl_2 \rightarrow 2LiCl$ (d) $Ca + I_2 \rightarrow CaI_2$

6. (a) $2Ba + O_2 \rightarrow 2BaO$ (b) $Mg + Br_2 \rightarrow MgBr_2$ (c) $2Fe + O_2 \rightarrow 2FeO$

7. (a) $2HI \rightarrow H_2 + I_2$ (b) $2H_2O \rightarrow 2H_2 + O_2$ (c) $2NH_3 \rightarrow N_2 + 3H_2$ (d) $CaCO_3 \rightarrow CaO + CO_2$

8. (a) $Mg(OH)_2 \rightarrow MgO + H_2O$ (b) $Pb(OH)_2 \rightarrow PbO + H_2O$ (c) $2LiClO_3 \rightarrow 2LiCl + 3O_2$ (d) $Ba(ClO_3)_2 \rightarrow BaCl_2 + 3O_2$ (e) $Ni(ClO_3)_2 \rightarrow NiCl_2 + 3O_2$

9. (a) $Ca + PbSO_4 \rightarrow Pb + CaSO_4$ (b) $Sr + 2AgNO_3 \rightarrow 2Ag + Sr(NO_3)_2$ (c) $Fe + Ni(NO_3)_2 \rightarrow Ni + Fe(NO_3)_2$ (d) $2Al + 3Pb(NO_3)_2 \rightarrow 3Pb + 2Al(NO_3)_3$ (e) $2Rb + 2H_2O \rightarrow 2RbOH + H_2$

10. (a) $K + NaCl \rightarrow Na + KCl$ (b) $Ca + ZnSO_4 \rightarrow Zn + CaSO_4$ (c) $Fe + Cu(NO_3)_2 \rightarrow Cu + Fe(NO_3)_2$ (d) $2K + 2H_2O \rightarrow 2KOH + H_2$

11. (a) $Cu(NO_3)_2(aq) + Na_2S(aq) \rightarrow 2NaNO_3(aq) + CuS(s)$
(b) $ZnCl_2(aq) + 2LiOH(aq) \rightarrow 2LiCl(aq) + Zn(OH)_2(s)$
(c) $Pb(NO_3)_2(aq) + (NH_4)_2CO_3(aq) \rightarrow 2NH_4NO_3(aq) + PbCO_3(s)$
(d) $Al(NO_3)_3(aq) + 3NaOH(aq) \rightarrow 3NaNO_3(aq) + Al(OH)_3(s)$

12. (a) $ZnBr_2(aq) + K_2S(aq) \rightarrow 2KBr(aq) + ZnS(s)$
(b) $Cu(NO_3)_2(aq) + Ba(OH)_2(aq) \rightarrow Ba(NO_3)_2(aq) + Cu(OH)_2(s)$
(c) $Mn(C_2H_3O_2)_2(aq) + Na_2S(aq) \rightarrow 2NaC_2H_3O_2(aq) + MnS(s)$

13. (a) $C_3H_8 + 5O_2 \rightarrow 3CO_2 + 4H_2O$ (b) $2C_4H_{10} + 13O_2 \rightarrow 8CO_2 + 10H_2O$ (c) $C_5H_{12} + 8O_2 \rightarrow 5CO_2 + 6H_2O$

14. (a) $CH_4 + 2O_2 \rightarrow CO_2 + 2H_2O$ (b) $2C_2H_6 + 7O_2 \rightarrow 4CO_2 + 6H_2O$ (c) $2C_6H_{14} + 19O_2 \rightarrow 12CO_2 + 14H_2O$ (d) $2C_6H_6 + 15O_2 \rightarrow 12CO_2 + 6H_2O$

15. (a) $Zn + S \rightarrow ZnS$; synthesis
(b) $Ca + 2NaNO_3 \rightarrow 2Na + Ca(NO_3)_2$; single replacement
(c) $2AgNO_3 + K_2S \rightarrow 2KNO_3 + Ag_2S$; double replacement
(d) $2NaI \rightarrow 2Na + I_2$; decomposition
(e) $2C_6H_6 + 15O_2 \rightarrow 12CO_2 + 6H_2O$; combustion
(f) $2C_{10}H_{22} + 31O_2 \rightarrow 20CO_2 + 22H_2O$; combustion

16. (a) $2K(s) + CuCl_2(aq) \rightarrow 2KCl(aq) + Cu(s)$
(b) $Zn(s) + Pb(NO_3)_2(aq) \rightarrow Zn(NO_3)_2(aq) + Pb(s)$

(c) $Cl_2(g) + 2KI(aq) \rightarrow 2KCl(aq) + I_2(s)$
(d) no reaction
(e) $Ba(s) + 2H_2O(\ell) \rightarrow Ba(OH)_2(aq) + 2H_2(g)$

17. (a) $2Ca(s) + O_2(g) \rightarrow 2CaO(s)$
(b) $2Ni(s) + O_2(g) \rightarrow 2NiO(s)$
(c) no reaction
(d) $4Al(s) + 3O_2(g) \rightarrow 2Al_2O_3(s)$

18. (a) $2Ag_2O(s) \rightarrow 4Ag(s) + O_2(g)$ (b) No reaction (c) No reaction (d) $CuO(s) + H_2(g) \rightarrow Cu(s) + H_2O(\ell)$

Teacher's Notes

Chapter 8

Chemical Equations and Reactions

INTRODUCTION

You have now been introduced to the vocabulary of chemistry, which includes the symbols for the elements, chemical formulas, and chemical names. In this chapter you will use this vocabulary to write and interpret "chemical sentences"—known in the language of chemistry as chemical equations. You will also discover that learning about a few simple classes of reactions can allow you to predict the products of many chemical reactions.

LOOKING AHEAD

Look for a systematic approach to equation balancing, but also recognize that it is a trial-and-error procedure that will become easier with practice.

SECTION PREVIEW

8.1 Writing Chemical Equations

8.2 Types of Chemical Reactions

8.3 Activity Series of the Elements

In the thermite reaction, iron(III) oxide reacts with aluminum powder to yield aluminim oxide and iron. Temperatures as high as 3500°C are produced by this reaction.

8.1 Chemical Equations

Chemical reactions are described by chemical equations. For example, the formulas of reactants and products can be written in the following form:

$$Fe_2O_3 + 2Al \rightarrow 2Fe + Al_2O_3$$

This equation shows that the *reactants* iron(III) oxide and aluminum yield the *products* iron and aluminum oxide. This strongly exothermic reaction is shown in the photograph on the preceding page. When fully understood, every chemical change can be summarized by a chemical equation.

A **chemical equation** *represents, with symbols and formulas, the reactants and products in a chemical reaction.*

Reading and Writing Chemical Equations

A chemical equation is of value only if it is correct in all details. The requirements outlined below must be met by every chemical equation.

Requirements for chemical equations Whenever you read or write a chemical equation, you must keep the following requirements in mind.

1. The equation must represent the known facts. All the reactants and products must be identified. This information is obtained by chemical analysis in the laboratory or from sources that give the results of experiments.

2. The equation must contain the correct formulas of the reactants and products. To read or write a chemical equation, you must apply what you learned in Chapter 7 about symbols and formulas. Knowledge of the common oxidation states of the elements and methods of writing formulas will enable you to supply formulas for reactants and products if they are not available. Recall that the elements listed in Table 8-1 are found primarily as *diatomic molecules*, such as H_2 and O_2. Each of these elements must be represented in an equation by its molecular formula. Other elements are usually represented by their atomic symbols without any subscripts. In the equation that appears at the beginning of this section, for example, iron (Fe) and aluminum (Al) were represented in this manner.

3. The law of conservation of matter must be satisfied. The law of conservation of matter was introduced in Chapter 1. *Atoms* are neither

SECTION OBJECTIVES

- List three requirements for a correctly written chemical equation.
- Translate chemical equations into sentences.
- Write a word equation and a formula equation, given a description of a chemical reaction.
- List three things you can determine about chemical reactants and products from a chemical equation.
- Balance a formula equation by inspection.

Reactants and products are identified by formulas. All compounds contain more than one element.

Some elements exist as polyatomic molecules. Their formulas must represent this (e.g., S_8, P_4, O_2, N_2, F_2, Cl_2, Br_2, I_2, and H_2).

TABLE 8-1 ELEMENTS THAT NORMALLY EXIST AS DIATOMIC MOLECULES

Element	Symbol	Molecular Formula	Physical State at Room Temperature
hydrogen	H	H_2	gas
nitrogen	N	N_2	gas
oxygen	O	O_2	gas
fluorine	F	F_2	gas
chlorine	Cl	Cl_2	gas
bromine	Br	Br_2	liquid
iodine	I	I_2	solid

created nor destroyed in ordinary chemical reactions. The same number of atoms of each element must appear on each side of a correctly written chemical equation. To equalize numbers of atoms, coefficients are added where necessary. A **coefficient** *is a small whole number that appears in front of a formula in a chemical equation.* Two coefficients of 2 appear in the equation written at the beginning of this section. Placing a coefficient in front of a formula specifies the relative number of moles of the substance.

Word and formula equations The first step in writing a chemical equation is identifying the facts to be represented. It may be helpful to write a **word equation**, *an equation in which the reactants and products in a chemical reaction are represented by words.*

As an example, consider the reaction of methane with oxygen. Methane is the principal component of natural gas. Experiments show that when methane burns in air, it combines with oxygen to produce carbon dioxide and water. Thus, methane and oxygen are the reactants, and carbon dioxide and water are the products.

The word equation for the reaction of methane and oxygen is

$$\text{methane} + \text{oxygen} \rightarrow \text{carbon dioxide} + \text{water}$$

This verbal equation is a brief statement of the experimental facts. A word equation does not give the quantities of reactants used or products formed; thus, it does not tell the whole story. A word equation has only qualitative (descriptive) meaning. The arrow $\rightarrow$ is read as "react to yield," or "yields." The equation above is read, "Methane and oxygen react to yield carbon dioxide and water," or simply, "Methane and oxygen yield carbon dioxide and water."

The next step in writing a chemical equation is to replace the names of the reactants and products with symbols and formulas. Methane (CH_4) is a molecular compound composed of one carbon atom and four hydrogen atoms. Oxygen is one of the elements that exist as diatomic molecules; it is represented as O_2. The correct formulas for carbon dioxide and water are CO_2 and H_2O.

A **formula equation** *represents the reactants and products of a chemical reaction by their symbols or formulas.* The formula equation for the reaction of methane and oxygen is

$$CH_4(g) + O_2(g) \rightarrow CO_2(g) + H_2O(g) \text{ (not balanced)}$$

The *g*'s in parentheses after the four formulas indicate that the corresponding substances are in the gaseous state. Like a word equation, a formula equation gives no information about amounts of reactants or products. A formula equation is also a qualitative statement.

The formula equation meets two of the three requirements for a correct chemical equation by representing the facts and showing the correct symbols and formulas for the reactants and products. To complete the process of writing a correct equation, the law of conservation of matter must be taken into account. The relative molar amounts of reactants and products represented in the equation must be adjusted so that the number and type of atoms is the same on both sides of the equation. This process, called "balancing an equation," is carried out by adding coefficients. Once it is balanced, a formula equation is a correctly written chemical equation.

Look again at the formula equation for the combustion of methane:

$$CH_4(g) + O_2(g) \rightarrow CO_2(g) + H_2O(g) \text{ (not balanced)}$$

Counting hydrogen atoms shows that there are four hydrogen atoms in the reactants but only two in the products. Two hydrogen atoms are therefore needed on the right side of the equation to balance it. They can be added by placing the coefficient *2* in front of the chemical formula H_2O. The number of atoms of each type is multiplied by the coefficient. Thus, $2H_2O$ represents *four* H atoms and *two* O atoms. If no coefficient is written, the coefficient is assumed to be *one*. By placing a *2* in front of the formula for water in the equation for the combustion of methane, we now have the H atoms.

$$CH_4(g) + O_2(g) \rightarrow CO_2(g) + 2H_2O(g) \text{ (partially balanced)}$$

You may have considered changing the subscript in the formula of water from H_2O to H_4O to add two more hydrogen atoms to the right side of the equation, but this would have been a mistake. Subscripts in chemical formulas must not be changed while balancing an equation. When you change the subscripts of a chemical formula, you change the very *identity* of the compound. The compound H_4O is not a product in the combustion of methane (if fact, there is no such compound.) Therefore this cannot be the right way to balance an equation. Inserting coefficients alone is appropriate because, while doing so specifies the relative amounts of reactants and products, it does not change their identity.

We must next count the number of oxygen atoms. There are now four oxygen atoms on the right in the partially balanced equation. Yet there are only two oxygen atoms on the left in the reactants. By placing the coefficient *2* in front of the formula for oxygen on the left side of the equation, we can increase the number of oxygen atoms to four. The correct chemical equation for the combustion of methane—*a balanced formula equation*—can be summarized in Figure 8-1 as

$$CH_4(g) + 2O_2(g) \rightarrow CO_2(g) + H_2O(g)$$

How can you be certain that the coefficents are correct? You could confirm them by carefully carrying out the reaction in the laboratory.

Figure 8-1 The reaction of methane with oxygen to give carbon dioxide and water. The molecular models illustrate that, in a balanced formula equation (as required by the law of conservation of matter), the number of atoms of each element in the reactants equals the number of atoms of each element in the products.

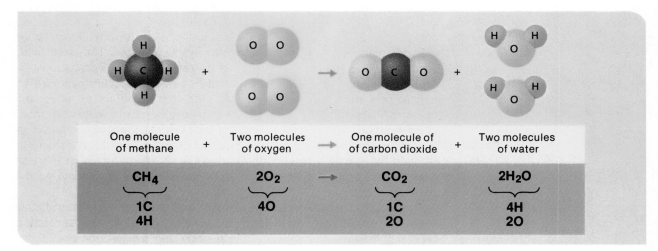

One molecule of methane	+	Two molecules of oxygen	→	One molecule of of carbon dioxide	+	Two molecules of water
CH_4		$2O_2$	→	CO_2		$2H_2O$
1C 4H		4O		1C 2O		4H 2O

TABLE 8-2   SYMBOLS USED IN CHEMICAL EQUATIONS

Symbol	Explanation
$\rightarrow$	"Yields" indicates result of reaction.
$\rightleftarrows$	Used in place of a single arrow for reversible reaction
(s)	A reactant or product in the solid state
$\downarrow$	Alternative to (s); used only for a precipitate (a solid) formed in a solution
(ℓ)	A reactant or product in the liquid state
(aq)	A reactant or product in an aqueous solution (dissolved in water)
(g)	A reactant or product in the gaseous state
$\uparrow$	Alternative to (g); used only for a gaseous product
$\xrightarrow{\Delta}\xrightarrow{\text{heat}}$	Reactants are heated
$\xrightarrow{\text{2 atm}}$	Pressure at which reaction is carried out, in this case 2 atm
$\xrightarrow{\text{pressure}}$	Pressure exceeding normal atmospheric pressure
$\xrightarrow{0°C}$	Temperature at which reaction is carried out, in this case 0°C
$\xrightarrow{MnO_2}$	Formula of catalyst used to alter the rate of the reaction, in this case, manganese dioxide

Sometimes a solid (s) is represented by the letter "c," which comes from the Latin expression for solid food: *cibus firmus.*

Experiments are unnecessary, however. If the values are consistent with the law of conservation of matter, they must be correct. Table 8-2 summarizes symbols that are used in chemical equations. Sometimes a gaseous product is indicated by an arrow pointing upwards, $\uparrow$, instead of (g), as shown above. A downward arrow, $\downarrow$, is sometimes used to show the formation of a precipitate during a reaction in aqueous solution.

The conditions under which a reaction takes place are often indicated by placing information above or below the reaction arrow. The word *heat*, symbolized by a capital Greek delta (Δ) indicates that the reactants must be heated. The specific temperature at which a reaction occurs may also be written over the arrow. For some reactions, it is important to specify the pressure at which the reaction occurs or that the pressure is above normal. Many reactions are speeded up and can take place at lower temperatures in the presence of a catalyst (a substance that changes the rate of a chemical reaction but can be recovered unchanged). To show that a catalyst must be present, the formula for the catalyst (or the word *catalyst*) is written over the reaction arrow.

In many reactions, once products have begun to form, they immediately begin to react to give back the reactants. In other words, the reverse reaction occurs. Depending upon the specific reaction and the reaction conditions, this may occur to a greater or lesser extent. *A reversible reaction is a chemical reaction in which the products re-form the original reactants.* The reversibility of a reaction such as that between iron and steam is indicated by writing two arrows pointing in opposite directions.

$$3Fe(s) + 4H_2O(g) \rightleftharpoons Fe_3O_4(s) + 4H_2(g)$$

In the late 18th century, this reaction was commonly used to prepare hydrogen, by passing the steam through a hot gun barrel.

With a knowledge of all the symbols and formulas used, it is possible·to translate a chemical equation into a sentence. For example, the equation

$$2HgO(s) \xrightarrow{\Delta} 2Hg(\ell) + O_2(g)$$

can be translated in this manner: When heated, solid mercury(II) oxide yields liquid mercury and gaseous oxygen. The equation

$$C_2H_4(g) + H_2(g) \xrightarrow[Pt]{pressure} C_2H_6$$

is read as follows: Under pressure and in the presence of a platinim catalyst, gaseous ethene and hydrogen form gaseous ethane.

Throughout this chapter we will include the symbols for phases (i.e., s, ℓ, and g) in balanced formula equations. You should be able to interpret these symbols when they are used and supply them when the necessary information is available.

Reversible reactions, which are common and important, are discussed in Chapter 19.

The arrow in a chemical reaction is read as "gives", "yields", "produces", or "forms". The plus sign in a chemical reaction does not imply mathematical addition. The + sign on the reactant side of the equation can be read as "reacts with".

Sample Problem 8.1

Write word and formula equations for the chemical reaction(s) occurring when sodium oxide is added to water at room temperature and forms sodium hydroxide (dissolved in the water). Include symbols for phases in the formula equation.

Solution
The word equation must show the reactants, sodium oxide and water, to the left of the arrow and the product, sodium hydroxide, to the right of the arrow.

sodium oxide + water → sodium hydroxide

This word equation is converted to a formula equation by replacing each name with a formula. To do this requires knowing (or looking up) that sodium has the +1 oxidation state, that oxygen usually has the −2 oxidation state, and that a hydroxide ion has a −1 charge.

$$Na_2O + H_2O \rightarrow NaOH \text{ (not balanced)}$$

Adding phase symbols gives

$$Na_2O(s) + H_2O(\ell) \rightarrow NaOH(aq) \text{ (not balanced)}$$

Sample Problem 8.2

Translate the following chemical equation into a sentence:

$$PbCl_2(aq) + Na_2CrO_4(aq) \rightarrow PbCrO_4(s) + 2NaCl(aq)$$

Solution

Each of the reactants on the left side of the chemical equation is an ionic compound and is named according to the rules for such compounds. Both reactants are in aqueous solution. One product is a solid (a precipitate) and the other remains in solution. The meaning of this equation is as follows: Aqueous solution of lead(II) chloride and sodium chromate react to produce a precipitate of lead(II) chromate plus an aqueous solution of sodium chloride.

Practice Problems

1. Write word and formula equations for each of the following chemical reactions. Include symbols for phases when indicated.
   (a) Calcium reacts with sulfur to produce calcium sulfide. *(Ans.)* calcium + sulfur → calcium sulfide; $Ca + S \rightarrow CaS$
   (b) Hydrogen reacts with fluorine to produce hydrogen fluoride. *(Ans.)* hydrogen + fluorine → hydrogen fluoride; $H_2 + F_2 \rightarrow 2HF$
   (c) Aluminum metal (solid) reacts with aqueous zinc chloride to produce zinc metal (solid) and aqueous aluminum chloride. *(Ans.)* aluminum + zinc chloride → zinc + aluminum chloride; $2Al(s) + 3ZnCl_2(aq) \rightarrow 3Zn(s) + 2AlCl_3(aq)$
2. Translate the following chemical equations into sentences:
   (a) $CS_2(s) + 3O_2(g) \rightarrow CO_2(g) + 2SO_2(g)$ *(Ans.)* Solid carbon disulfide reacts with oxygen gas to produce carbon dioxide gas and sulfur dioxide gas.
   (b) $NaCl(aq) + AgNO_3(aq) \rightarrow NaNO_3(aq) + AgCl(s)$ *(Ans.)* Aqueous solutions of sodium chloride and silver nitrate to produce aqueous sodium nitrate and solid silver chloride.

Significance of a Chemical Equation

Chemical equations are of great significance to quantitative work in chemistry. The arrow in a chemical equation (meaning "yields") is like an equals sign, and thus chemical equations are similar to algebraic equations in that both express equalities. Chemical equations can be interpreted quantitatively in three different ways frequently used in chemistry.

1. Amounts of reactants and products. A chemical equation as usually written shows the smallest numbers of atoms, molecules, or ions that will satisfy the law of conservation of matter in a given chemical reaction. Because of the manner in which moles and molar masses are defined, a chemical equation also provides information about moles and masses of reactants and products. Recall that the mole is equal to 6.02×10^{23} atoms, molecules, or formula units. The molar mass of a substance is the mass in grams of one mole of that substance.

The equation for the reaction of hydrogen with chlorine, for example, provides the following information about molecules, moles, and masses of the reactants and products.

$$H_2(g) + Cl_2(g) \rightarrow 2HCl(g)$$

1 molecule of hydrogen reacts with 1 molecule of chlorine to yield 2 molecules of hydrogen chloride

1 mol of hydrogen reacts with 1 mol of chlorine to yield 2 mol of hydrogen chloride

STUDY HINT

You may wish to review the introduction to the mole in Section 3.3.

2 g of hydrogen (1 × molar mass) reacts with 71 g
of chlorine (1 × molar mass) to yield
73 g of hydrogen chloride (2 × molar mass)

The relative number of moles of each substance is given by the coefficients. Going from molecules to moles is like multiplying the equation by Avogadro's number, as shown in Figure 8-2. The relative masses of reactants and products are then known because the molar mass is the mass in grams of one mole.

2. Ratios of reactants and products. The second important interpretation of chemical equations is possible because coefficients indicate relative, not absolute, amounts of reactants and products. From the equation in Figure 8-2, we know that 1 mol of hydrogen reacts with 1 mol of chlorine to produce 2 mol of hydrogen chloride, giving mole ratios of reactants and products of:

$$1 \text{ mol } H_2 : 1 \text{ mol } Cl_2 : 2 \text{ mol } HCl$$

We can therefore see that, for example, 20 mol of hydrogen would react with 20 mol of chlorine to give 40 mol of hydrogen chloride. Similarly, the equation shows the mass ratios of reactants to products. Hydrogen combines with chlorine to form hydrogen chloride in mass ratios of roughly

$$2g \text{ } H_2 : 71g \text{ } Cl_2 : 73g \text{ } HCl$$

Twice as much hydrogen (4g H_2) would therefore react with twice as much chlorine (142g Cl_2) to give twice as much hydrogen chloride (146g HCl), and so on.

3. Reverse reactions. The third important quantitative interpretation of chemical equations is possible because, as for algebraic equations, the equality can be interpreted in the reverse direction as well as in the direction written. The equation in Figure 8-2 shows a reaction in which hydrogen chloride is broken down to give back the elements from which it formed. For example, 2 mol of hydrogen chloride would yield 1 mol of hydrogen plus 1 mol of chlorine. All of the information given is valid for the reverse reaction.

We have discussed several important ways in which chemical equations provide information. There are also several important kinds of information *not* provided by chemical equations, however. They give no indication of whether or not a reaction will actually occur. Chemical equations can be written for reactions that do not necessarily take place. In Sections 8.2 and 8.3,

Mass ratios are derived by multiplying mole ratios by the molar mass of each substance.
1 mol H_2 : 1 mol Cl_2 : 2 mol HCl
(1 × 2) : (1 × 35.5) : (2 × 36.5)
2 g H_2 : 71 g Cl_2 : 73 g HCl

Figure 8-2 The reaction of hydrogen and chlorine to give hydrogen chloride, showing the various ways to interpret a chemical equation.

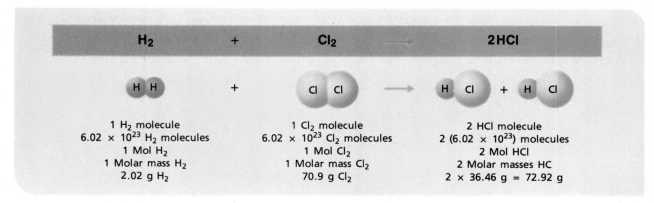

| H_2 | + | Cl_2 | | 2HCl |

1 H_2 molecule	1 Cl_2 molecule	2 HCl molecule
6.02×10^{23} H_2 molecules	6.02×10^{23} Cl_2 molecules	$2 (6.02 \times 10^{23})$ molecules
1 Mol H_2	1 Mol Cl_2	2 Mol HCl
1 Molar mass H_2	1 Molar mass Cl_2	2 Molar masses HC
2.02 g H_2	70.9 g Cl_2	2×36.46 g = 72.92 g

some guidelines are given about types of simple reactions that can be expected to occur. Later chapters will provide further guidelines for other types of reactions. In all cases, experiment forms the basis for the guidelines used in predicting whether or not chemical reactions will occur.

Chemical equations also give no information about the speed with which reactions occur or the actual pathway atoms or ions take in moving from reactants to products. These aspects of chemical reactions are discussed in Chapter 18.

Balancing Chemical Equations

Most of the equations that you will now be expected to balance can be balanced by inspection. The following procedure demonstrates how to master this skill by using a step-by-step approach. The equation for the reaction shown in Figure 8-3 will be used.

 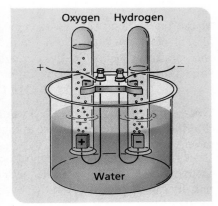

Oxygen   Hydrogen

Water

Figure 8-3 The decomposition of water by an electrical current to give hydrogen and oxygen. Indicators (substances that change color when the reaction occurs) have been added to show that reactions are occurring at the electrodes.

When an electric current is passed through water that has been made slightly conductive, the water molecules breakdown to yield hydrogen and oxygen.

Step 1. Identify the names of reactants and products and write a word equation. The word equation for the reaction shown in Figure 8-3 is

$$\text{water} \rightarrow \text{hydrogen} + \text{oxygen}$$

Step 2. Substitute correct formulas for the names of the reactants and products, and write a formula equation. The formula for water is H_2O (which is correct according to the oxidation number of $+1$ for hydrogen and -2 for oxygen). Both hydrogen and oxygen exist as diatomic molecules and their correct formulas are H_2 and O_2. The formula equation for the breakdown of water is

$$H_2O \rightarrow H_2 + O_2 \text{ (not balanced)}$$

Step 3. Adjust the coefficients in the formula equation so that the law of conservation of mass is satisfied. In other words, balance the formula equation. This is done by trial and error. Coefficients are changed and number of atoms counted until the numbers of each type of atom are the same on both sides of the equation. The trial-and-error method is made easier by use of the following guidelines:

You will learn methods for balancing more complicated equations in Chapter 20.

Water can be made conductive by adding a small amount of salt that dissolves to give ions in solution.

1800 William Nicholson (1735–1815), using a battery designed by Volta, decomposed water into its elements.

- Balance the different types of atoms one at a time.
- First, balance the types of atoms that appear only once on each side of the equation.
- Balance polyatomic ions that appear on both sides of the equation as single units.
- Balance H atoms and O atoms after atoms of all other elements have been balanced.

Checking the formula equation in our example shows that there are two oxygen atoms on the right and only one on the left. To balance oxygen atoms, the number of H_2O molecules must be increased. Placing the coefficient 2 before H_2O gives the necessary 2 oxygen atoms on the left.

$$2H_2O \rightarrow H_2 + O_2 \text{ (partially balanced)}$$

The coefficient 2 in front of H_2O, however, has upset the balance of hydrogen atoms. Placing a coefficient of 2 in front of hydrogen (H_2) on the right gives an equal number of hydrogen atoms (4) on both sides of the equation.

$$2H_2O(\ell) \rightarrow 2H_2(g) + O_2(g)$$

Step 4. Count atoms to be sure that the equation is balanced. By counting, verify that equal numbers of atoms of each type appear on both sides of the arrow.

$$2H_2O(\ell) \rightarrow 2H_2(g) + O_2(g)$$
$$(4H, 2O \rightarrow 4H, 2O)$$

Occasionally at this point, the coefficients are not the smallest possible whole numbers. When this happens, the coefficients should be converted to the smallest possible numbers.

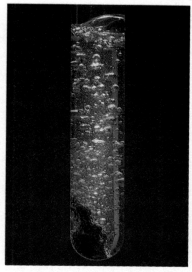

Figure 8-4 The reaction of zinc with aqueous hydrochloric acid to give zinc chloride and hydrogen.

Sample Problem 8.3

The reaction of zinc with aqueous hydrochloric acid produces a solution of zinc chloride and hydrogen gas, as shown in Figure 8-4. Write a balanced chemical equation for this reaction.

Solution

Step 1. Analyze   Write the word equation.

zinc + hydrochloric acid → zinc chloride + hydrogen

Step 2. Plan   Write the formula equation.

$$Zn(s) + HCl(aq) \rightarrow ZnCl_2(aq) + H_2(g) \text{ (not balanced)}$$

Step 3. Compute   Adjust the coefficients. Note that both chlorine and hydrogen appear in only one substance on each side of the equation, but in unequal numbers of atoms. To balance chlorine we must use a coefficient of 2 for HCl. Two molecules of hydrogen chloride, 2HCl, also gives the required two hydrogen atoms on the left. Finally, note that there is one zinc atom on each side in the formula equation, so no further coefficients are needed.

$$Zn(s) + 2HCl(aq) \rightarrow ZnCl_2(aq) + H_2(g)$$

Step 4. Evaluate   Count atoms to check balance.

$$Zn(s) + 2HCl(aq) \rightarrow ZnCl_2(aq) + H_2(g)$$
$$1Zn, 2H, 1Cl \rightarrow 1Zn, 2H, 2Cl$$

The equation is balanced.

Practice Problems

Write word, formula, and balanced chemical equations for each of the following reactions:

1. Sodium reacts with sulfur to produce sodium sulfide.   *(Ans.)* Word: sodium + sulfur → sodium sulfide; Formula: $Na + S \rightarrow Na_2S$; Balanced: $2Na + S \rightarrow Na_2S$
2. The reaction between magnesium and hydrochloric acid produces magnesium chloride and hydrogen.   *(Ans.)* Word: magnesium + hydrochloric acid → magnesium chloride + hydrogen; Formula: $Mg + HCl \rightarrow MgCl_2 + H_2$; Balanced: $Mg + 2HCl \rightarrow MgCl_2 + H_2$
3. Aqueous nitric acid reacts with aqueous magnesium hydroxide to produce aqueous magnesium nitrate and water.   *(Ans.)* Word: nitric acid + magnesium hydroxide → magnesium nitrate + water; Formula: $HNO_3 + Mg(OH)_2 \rightarrow Mg(NO_3)_2 + H_2O$; Balanced: $2HNO_3 + Mg(OH)_2 \rightarrow Mg(NO_3)_2 + 2H_2O$

Balancing chemical equations by inspection will become easier as you continue your study of chemistry and gain experience. Learn to avoid the most common mistakes of *(1)* neglecting to be sure that each formula is correct and *(2)* trying to balance an equation by changing subscripts. Keep in mind that once a formula is correctly written, *subscripts cannot be added, deleted, or changed.*

You should be able to dispense with writing the word equation and the separate steps as you become more experienced. But *do not* dispense with the final step of counting atoms to be sure the equation is balanced.

Sample Problem 8.4

Aluminum carbide (Al_4C_3) reacts with water to produce methane gas and aluminum hydroxide. Write a chemical equation for this reaction.

Solution

The reactants are aluminum carbide and water. The products are methane and aluminum hydroxide.

$$Al_4C_3 + H_2O \rightarrow CH_4 + Al(OH)_3 \text{ (not balanced)}$$

There are 4 Al atoms on the left, which indicates the need for a coefficient of 4 for $Al(OH)_3$.

$$Al_4C_3 + H_2O \rightarrow CH_4 + 4Al(OH)_3 \text{ (partially balanced)}$$

Leaving H and O to balance last, we next balance C atoms. With 3 C atoms on the left, a coefficient of 3 must be added for CH_4 on the right.

$$Al_4C_3 + H_2O \rightarrow 3CH_4 + 4Al(OH)_3 \text{ (partially balanced)}$$

It is best to balance O atoms next, since each appears only once on each side of the equation. On the left, there is 1 O atom. On the right, there are 12 O atoms—3 in each of the 4 $Al(OH)_3$ formula units. A coefficient of 12 for H_2O balances O atoms.

$$Al_4C_3 + 12H_2O \rightarrow 3CH_4 + 4Al(OH)_3$$

This leaves the H atoms to be balanced. There are 24 H atoms on the left (2 in each of 12 water molecules). On the right there are 12 H atoms in methane (4 in each of 3 molecules) plus 12 in aluminum hydroxide (3 in each of 4 formula units) for a total of 24 H atoms on the right. Hydrogen atoms are balanced.

$$Al_4C_3(s) + 12H_2O(\ell) \rightarrow 3CH_4(g) + 4Al(OH)_3(s)$$
$$4Al, 3C, 24H, 12O \rightarrow 4Al, 3C, 24H, 12O$$

The equation is balanced.

Sample Problem 8.5

Aluminum sulfate and calcium hydroxide are used in a water purification process. When added to water, they dissolve and react to produce two insoluble products, aluminum hydroxide and calcium sulfate. These products settle out taking suspended solid impurities with them. Write a balanced chemical equation for the reaction.

Solution

Each of the reactants and products is an ionic compound. Knowing the charges of the ions involved permits writing the formula equation

$$Al_2(SO_4)_3 + Ca(OH)_2 \rightarrow Al(OH)_3 + CaSO_4 \text{ (not balanced)}$$

Calcium appears only once on each side of the equation and is therefore balanced. There are two aluminum atoms on the left and one on the right. Placing a coefficient of 2 in front of $Al(OH)_3$ gives equal number of aluminum atoms on both sides of the equation.

$$Al_2(SO_4)_3 + Ca(OH)_2 \rightarrow 2Al(OH)_3 + CaSO_4 \text{ (partially balanced)}$$

Next, checking SO_4^{2-} ions shows that there are three SO_4^{2-} ions on the left side of the equation and only one on the right side. Placing the coefficient 3 in from of $CaSO_4$ gives an equal number of SO_4^{2-} ions on each side.

$$Al_2(SO_4)_3(aq) + Ca(OH)_2(aq) \rightarrow 2Al(OH)_3(s) + 3CaSO_4(s) \text{ (partially balanced)}$$

There are now three calcium atoms on the right, however, so calcium must be balanced. By placing the coefficient *3* in front of $Ca(OH)_2$, we have an equal number of calcium atoms on each side. This last step also gives an equal number (6) of OH^- ions on both sides of the equation.

$$Al_2(SO_4)_3(aq) + 3Ca(OH)_2(aq) \rightarrow 2Al(OH)_3(s) + 3CaSO_4(s)$$
$$2Al, 3SO_4^{2-}, 3Ca, 6OH^- \rightarrow 2Al, 3SO_4^{2-}, 3Ca, 6OH^-$$

The equation is balanced.

Practice Problems
Write chemical equations for each of the following reactions:
1. Sodium combines with chlorine to produce sodium chloride. *(Ans.)* $2Na + Cl_2 \rightarrow 2NaCl$
2. When metallic copper reacts with aqueous silver nitrate the products are aqueous copper(II) nitrate and metallic silver. *(Ans.)* $Cu(s) + 2AgNO_3(aq) \rightarrow Cu(NO_3)_2(aq) + 2Ag(s)$
3. The reaction between solid iron(III) oxide and carbon monoxide gas produces metallic iron and carbon dioxide gas. *(Ans.)* $Fe_2O_3(s) + 3CO(g) \rightarrow 2Fe(s) + 3CO_2(g)$

1. A word equation uses words to represent the reactants and products in a chemical reaction; a formula equation uses their symbols or formulas; a chemical equation is a balanced formula equation.

2. Sulfuric acid + sodium hydroxide → sodium sulfate + water
   $H_2SO_4(aq) + NaOH(aq) \rightarrow Na_2SO_4(aq) + H_2O(\ell)$

3. Aqueous solutions of nitric acid and magnesium hydroxide react to produce a solution of magnesium nitrate in water.

4. Hydrogen sulfide + oxygen → sulfur dioxide + water
   $H_2S(g) + O_2(g) \rightarrow SO_2(g) + H_2O(g)$
   $2H_2S(g) + 3O_2(g) \rightarrow 2SO_2(g) + 2H_2O(g)$

5. (a) $2NH_4Cl + Ca(OH)_2 \rightarrow CaCl_2 + 2NH_3 + 2H_2O$
   (b) $2C_6H_{14} + 19O_2 \rightarrow 12CO_2 + 14H_2O$

///////

CAUTION
Never look directly at the flame of burning magnesium. This can cause eye damage.

Section Review

1. Distinguish among word, formula, and chemical equations.
2. Write word and formula equations for the reaction in which aqueous solutions of sulfuric acid and sodium hydroxide react to form aqueous sodium sulfate and water.
3. Translate the following chemical equation into a sentence:
   $$2HNO_3(aq) + Mg(OH)_2(aq) \rightarrow Mg(NO_3)_2(aq) + 2H_2O(\ell).$$
4. Write the word, formula, and chemical equations for the reaction between hydrogen sulfide gas and oxygen gas if the products are sulfur dioxide gas and water vapor.
5. Write the chemical equation for each of the following reactions:
   (a) ammonium chloride + calcium hydroxide → calcium chloride + ammonia + water   (b) hexane (C_6H_{14}) + oxygen → carbon dioxide + water.

8.2 Types of Chemical Reactions

Thousands of chemical reactions occur in living systems, in industrial processes, and in chemical laboratories. Based upon observation and analysis, chemical equations can be written to describe these many and varied reactions. Often it is necessary to predict the products formed by the combination of reactants. At this stage in your study of chemistry, you may be given the reactants in certain simple types of reactions and asked to predict the products. You must either remember what the products might be or else be able to reason out the answer from general knowledge about types of reactions.

Memorizing the equations for thousands of chemical reactions would be a very difficult task. It is therefore more useful and realistic to organize reactions according to various similarities and regularities.

There are several different ways of classifying chemical reactions. No single scheme is entirely satisfactory. The classification scheme described in this section provides an introduction to simple chemical reactions and is helpful in predicting the products of many reactions. Five types of chemical reactions are introduced: synthesis, decomposition, single-replacement, double-replacement, and combustion. In later chapters you will be introduced to categories useful in classifying other types of chemical reactions.

Synthesis Reactions

In a **synthisis***, or* **composition***, reaction two or more substances combine to form a new substance.* This type of reaction is represented by the following general equation:

$$A + X \rightarrow AX$$

A and X can be elements or compounds. AX is a compound. The following examples illustrate several kinds of synthesis reactions.

Reactions of elements with oxygen   A simple type of synthesis reaction is the combination of an element with oxygen to produce an oxide of the element. For example, when a thin strip of magnesium metal is placed in a

Bunsen burner flame, it burns with bright white light. When the metal strip is completely burned, only a fine white powder of magnesium oxide is left burning. This chemical reaction, shown in Figure 8-5, is represented by the following equation:

$$2Mg(s) + O_2(g) \rightarrow 2MgO(s)$$

Iron reacts with oxygen to produce two different iron oxides.

$$2Fe(s) + O_2(g) \rightarrow 2FeO(s)$$
$$4Fe(s) + 3O_2(g) \rightarrow 2Fe_2O_3(s)$$

Reactions of two metals   The reaction of two nonmetals forms a covalent compound. For example, hydrogen reacts with oxygen to form water, and sulfur reacts with oxygen to form sulfur dioxide as shown by the equations

$$2H_2(g) + O_2(g) \rightarrow 2H_2O(g)$$
$$S(s) + O_2(g) \rightarrow SO_2(g)$$

Other nonmetals also react with oxygen. For example, when carbon is burned in air, carbon dioxide is produced.

$$C(s) + O_2(g) \rightarrow CO_2(g)$$

Reactions of metals with nonmetals other than oxygen   The reaction of a metal with a nonmetal often produces an ionic compound (a salt). The combination of sodium of this type of synthesis reaction

$$2Na(s) + Cl_2(g) \rightarrow 2NaCl(s)$$

The Group 17 elements, the halogens, undergo synthesis reactions with many different metals. Fluorine, in particular, is so reactive that it combines with almost all metals, for example, with cobalt sodium and uranium

$$Co(s) + F_2(g) \rightarrow CoF_2(s)$$
$$2Na(s) + F_2(g) \rightarrow 2NaF(s)$$
$$U(s) + 3F_2(g) \rightarrow UF_6(g)$$

Sodium fluoride (NaF) is added to municipal water supplies to provide fluoride ions, which help to prevent tooth decay in the people who drink it. Natural uranium is converted to uranium(VI) fluoride (UF_6) as the first step in the production of uranium for use in nuclear power plants. Phosphorus and sulfur are also reactants in many synthesis reactions. For example, the reaction between sodium and molten sulfur is the basis for the sodium–sulfur battery, which has been investigated as a source of power for electric motors for automobiles.

$$2Na(s) + S(\ell) \rightarrow Na_2S(s) + energy$$

Synthesis reactions of oxides   Oxides of active metals react with water to produce metal hydroxides. For example, calcium oxide reacts with water to form calcium hydroxide.

$$CaO(s) + H_2O(\ell) \rightarrow Ca(OH)_2(s)$$

Calcium oxide (CaO), known as lime or quicklime, is manufactured in bulk quantities. The addition of water to lime to produce calcium hydroxide,

Figure 8-5   The synthesis reaction of magnesium and oxygen to give magnesium oxide. $2Mg(s) + O_2(g) \rightarrow 2MgO(s)$.

If the supply of oxygen is limited, carbon monoxide (CO) also forms when carbon is burned in air.

1789 Lavoisier (1743–1794) published *Elementary Treatise on Chemistry* in 1789. His extensive collection of laboratory glassware was carefully illustrated by Mme. Lavoisier. Lavoisier studied the combustion of many substances and was able to conclude that various elements such as mercury and phosphorous increased in mass when burned in air. If the experiment was performed over water in a closed container, one-fifth of the volume of the air was consumed.

Ca(OH)$_2$, known as slaked lime, is a crucial step in the conversion of lime (mixed with other materials) to mortar and cement.

Many nonmetal oxides undergo a synthesis reaction with water to produce a solution of an oxyacid. For example, sulfur trioxide (SO$_3$) combines with water to produce sulfuric acid (H$_2$SO$_4$):

$$SO_3(g) + H_2O(\ell) \rightarrow H_2SO_4(aq)$$

The statue shown in Figure 16-23 was damaged by acid precipitation, which is formed as rain falls through air polluted with SO$_3$.

Certain metal oxides and nonmetal oxides react with each other in synthesis reactions to form salts. For example, calcium sulfate is formed by the reaction of calcium oxide and sulfur trioxide.

$$CaO(s) + SO_3(g) \rightarrow CaSO_4(s)$$

Decomposition Reactions

In a decomposition reaction *a single compound undergoes a reaction that produces two or more simpler substances.* Decomposition reactions are the opposite of synthesis reactions and are represented by the following general equation:

$$AX \rightarrow A + X$$

AX is a compound, and A and X can be elements or compounds.

Most decomposition reactions take place only when energy in the form of heat or electricity is added. Examples of several types of decomposition reactions are given in the following examples.

Decomposition of binary compounds   The simplest kind of decomposition reaction is the decomposition of a binary compound into its elements. When heated, oxides of the less active metals decompose in this manner. For example, oxygen was discovered by Joseph Priestley in 1774 when he heated mercury(II) oxide to produce mercury and oxygen. (Figure 8-7).

$$2HgO(s) \rightarrow 2Hg(\ell) + O_2(g)$$

Electrolysis   *The decomposition of a substance by an electric current is called* **electrolysis**. Passing an electric current through water will decompose it

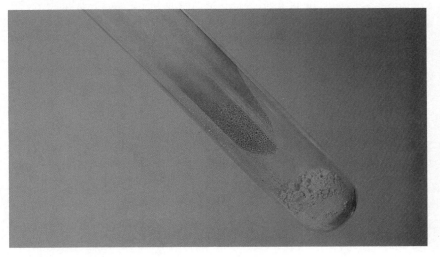

Priestley also discovered NH$_3$, HCl, N$_2$O, NO$_2$, H$_2$S, SO$_2$, and CO. He recognized the O$_2$–CO$_2$ cycles in nature and was a codiscoverer of photosynthesis.

Figure 8-7 The decomposition fo mercury(II) oxide to form mercury metal and oxygen.

$$2HgO(s) \xrightarrow{\Delta} 2Hg(\ell) + O_2(g)$$

into its constituent elements, hydrogen and oxygen, as was shown in Figure 8-3.

$$2H_2O(\ell) \xrightarrow{\text{electric current}} 2H_2(g) + O_2(g)$$

Decomposition of metal carbonates   When a metal carbonate is heated, it breaks down to produce a metal oxide and carbon dioxide gas. For example, calcium carbonate decomposes to calcium oxide and carbon dioxide.

$$CaCO_3(s) \xrightarrow{\Delta} CaO(s) + CO_2(g)$$

Decomposition of metal hydroxides   Most metal hydroxides decompose when heated to give metal oxides and water. For example, calcium hydroxide decomposes to form calcium oxide and water.

$$Ca(OH)_2(s) \xrightarrow{\Delta} CaO(s) + H_2O(g)$$

Sodium hydroxide and potassium hydroxide are exceptions to this generalization.

Decomposition of metallic chlorates   When a metal chlorate is heated it decomposes to a metal chloride and oxygen. For example, potassium chlorate, $(KClO_3)$ decomposes to give potassium chloride and oxygen.

$$2KClO_3(s) \xrightarrow[MnO_2(s)]{\Delta} 2KCl(s) + 3O_2(g)$$

Notice the $MnO_2(s)$ written below the arrow. In this reaction, MnO_2 is a catalyst, which increases the rate of the reaction and also helps the decomposition of $KClO_3$ take place at a lower temperature than would be needed without the presence of MnO_2.

Decomposition of acids   Certain acids decompose to nonmetallic oxides and water. Carbonic acid is unstable and decomposes readily at room temperature to form carbon dioxide and water.

$$H_2CO_3 \rightarrow CO_2(g) + H_2O(\ell)$$

When heated, sulfuric acid decomposes into sulfur trioxide and water.

$$H_2SO_4(aq) \xrightarrow{\Delta} SO_3(g) + H_2O(\ell)$$

Single-Replacement Reactions

In a **single-replacement** *or* **displacement-reaction***, one element replaces a similar element in a compound.* Most single-replacement reactions take place in aqueous solution. The amount of energy involved in this type of reaction is usually smaller than the amount involved in synthesis or decomposition reactions. Single-replacement reactions can be represented by the following general equations:

$$A + BX \rightarrow AX + B$$

or

$$Y + BX \rightarrow BY + X$$

A, B, X, and Y are elements and AX, BX, and BY are compounds.

1807 Sir Humphrey Davy (1778–1829) isolated potassium and sodium by constructing a battery and using it as an energy source for the electrolysis of molten salts. Hydrogen was released when he dropped small pieces of the metals in water.

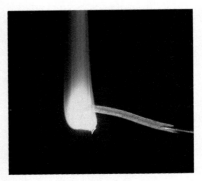

Figure 8-8 The replacement of hydrogen is aqueous hydrochloric acid by magnesium: $Mg(s) + 2HCl(aq) \rightarrow H_2(g) + MgCl_2(aq)$

Replacement of a metal in a compound by a more reactive metal
When metallic aluminum is placed in a solution of iron(II) nitrate, the aluminum atoms replace iron ions, forming metallic iron plus a solution of aluminum nitrate.

$$2Al(s) + 3Fe(NO_3)_2(aq) \rightarrow 3Fe(s) + 2Al(NO_3)_3(aq)$$

Replacement of hydrogen in water by a metal The most active metals (such as those in Group 1) react vigorously with water to produce metal hydroxides and hydrogen. For example, sodium reacts with water to form sodium hydroxide and hydrogen gas.

$$2Na(s) + 2H_2O(\ell) \rightarrow 2\,NaOH(aq) + H_2(g)$$

Less active metals, such as iron, react with steam to form a metal oxide and hydrogen gas.

$$3Fe(s) + 4H_2O(g) \rightarrow Fe_3O_4(s) + 4H_2(g)$$

Replacement of hydrogen in an acid by a metal The more active metals react with certain acids (such as hydrochloric acid and dilute sulfuric acid) to form hydrogen. The metal replaces the hydrogen in the acid, and the reaction products are a metal compound (a salt) and hydrogen gas. For example, when magnesium reacts with hydrochloric acid, as shown in Figure 8-10, the reaction products are hydrogen gas and aqueous magnesium chloride.

$$Mg(s) + 2HCl(aq) \rightarrow H_2(g) + MgCl_2(aq)$$

Replacement of halogens In another type of single-replacement reaction, one halogen replaces another halogen in a compound. Fluorine is the most active halogen, and it can replace any of the other halogens. Each halogen is less active than the one above it in the periodic table. Each halogen can therefore replace those below it, but not those above it in the group (Group 17). Thus, while chlorine can replace bromine in potassium bromide, it cannot replace fluorine in potassium fluoride. The reaction products bromine whereas the combination of flourine and sodium iodide produces sodium flouride and solid iodine.

$$Cl_2(g) + 2KBr(aq) \rightarrow 2KCl(aq) + Br_2(\ell)$$
$$F_2(g) + 2NaI(aq) \rightarrow 2NaF(aq) + I_2(s)$$

Double-Replacement Reactions

In double-replacement reactions *or* ionic reactions, *the ions of two compounds exchange places in an aqueous solution to form two new compounds.* One of the compounds formed is usually a precipitate (a very slightly soluble compound), an insoluble gas that bubbles out of the solution, or a molecular compound, usually water. The other compound is often soluble and remains dissolved in solution. A double-replacement reaction is represented by the following general equation

$$AX + BY \rightarrow AY + BX$$

A, X, B, and Y in the reactants represent ions. AY and BX represent ionic or molecular compounds.

Desktop Investigation

Balancing Equations Using Models

Question
How can molecular models and formula-unit ionic models be used to demonstrate: (1) the rearrangement of atoms in chemical reactions; and (2) the law of conservation of mass as shown by balancing equations?

Materials
large and small gumdrops in at least 4 different colors
toothpicks

Procedure
1. Make models to represent the reactions shown. Use different colors of gumdrops to represent atoms of different elements.

2. Make sure you have the same number of each atom model on each side of the equation. Use toothpicks to connect the "atoms" of your models.

3. Reaction (a) and Reaction (b) in each group are unbalanced equations; balance them using your models. Determine the products for Reaction (c) in each group using models and them complete and balance the equation.

4. Work Reactions (1a) and (5a) using the figures in this book to guide you in model making.

5. Classify each group of chemical reactions listed in (1) through (5).

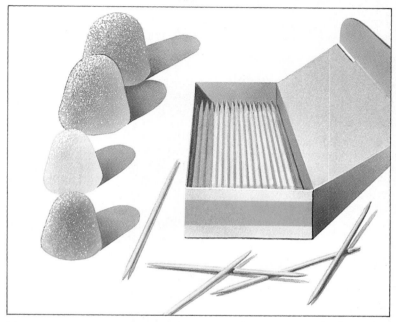

(1) The reactions in this group are classified as ___.
   (a) $H_2 + O_2 \rightarrow H_2O$
   (b) $Mg + O_2 \rightarrow MgO$
      (see Figure 8-5.)
   (c) $BaO + H_2O$

(2) The reactions in this group are classified as ___.
   (a) $H_2CO_3 \rightarrow + CO_2$
   (b) $KClO_3 \rightarrow O_2$
   (c) H_2O electricity $\rightarrow$
      (see Figure 8-3.)

(3) The reactions in this group are classified as ___.
   (a) $Ca + H_2O \rightarrow Ca(OH) + H_2$
   (b) $KI + Br_2 \rightarrow KBr + I_2$
   (c) $Zn + HCl \rightarrow$
      (see Figure 8-4.)

(4) The reactions in this group are classified as ___.
   (a) $AgNO_3 + NcCl \rightarrow AgCl + NaNO_3$
   (b) $FeS + HCl \rightarrow FeCl_2 + H_2S$

   (c) $H_2SO_4 + KOH \rightarrow$
(5) The reactions in this group are classified as ___.
   (a) $CH_4 + O_2 \rightarrow CO_2 + H_2O$
      (see Figure 8-1.)
   (b) $CO + O_2 \rightarrow CO_2$
   (c) $C_3H_8 + O_2 \rightarrow$

Discussion
1. What are the five categories used to classify most reactions? What is the advantage of using this classification method?

2. How does model making in this investigation help illustrate the law of conservation of mass?

3. Some reactions can be placed in more than one category. Can you give an example of such a reaction in this investigation?

Figure 8-9 The double-replacement reaction between lead(II) nitrate and potassium iodide in aqueous solution: $Pb(NO_3)_2(aq) + 2KI(aq) \rightarrow PbI_2(g) + 2KNO_3(aq)$

Formation of a precipitate The formation of a precipitate occurs when the positive ions of one reactant combine with the negative ions of another reactant and form an insoluble or slightly soluble compound. For example, when a solution of potassium iodide is added to a solution of lead(II) nitrate, a yellow precipitate separates from the mixture, as shown in Figure 8-9.

$$Pb(NO_3)_2(aq) + 2KI(aq) \rightarrow PbI_2(s) + 2KNO_3(aq)$$

The precipitate of lead(II) iodide (PbI_2) forms as a result of the very strong attractive forces between the positive Pb^{2+} ions and the negative I^- ions. The other product is the water-soluble salt potassium nitrate (KNO_3). The potassium ions and nitrate ions do not take part in the reaction. Instead, they remain in solution. The principles that help identify which ions form a precipitate and which ions remain in solution are developed in Chapter 15.

Formation of a gas Double-replacement reactions also occur when one of the products is an insoluble gas that bubbles out of the mixture. For example, iron(II) sulfide reacts with aqueous hydrogen chloride to form hydrogen sulfide gas and iron(II) chloride.

$$FeS(s) + 2HCl(aq) \rightarrow H_2S(g) + FeCl_2(aq)$$

Formation of water In some double-replacement reactions, a molecular compound such as water is one of the products. For example,

$$HCl(aq) + NaOH(aq) \rightarrow NaCl(aq) + H_2O(\ell)$$

STUDY HINT

Reactions between acids and hydroxides are discussed in chapters 16 and 17.

Combustion Reactions

In a **combustible reaction**, *a substance combines with oxygen releasing a large amount of energy in the form of light and heat.* The burning of natural gas, propane, and wood are examples of combustion reactions.

When hydrocarbons burn completely, they form carbon dioxide and water. **Hydrocarbons** *are compounds made up of carbon and hydrogen.* For example, the combustion of propane gas, shown in Figure 8-10, produces light and heat with the formation of carbon dioxide and water according to the following equation.

$$C_3H_8(g) + 5O_2(g) \rightarrow 3CO_2(g) + 4H_2O(g)$$

Figure 8-10 The combustion of propane gas.
$C_3H_8(g) + SO_2(g) \rightarrow 3CO_2(g) + 4H_2O(g)$

Section Review

1. List the five types of chemical reactions.
2. Complete and balance each of the following reactions identified by type: *(a)* synthesis: $Li_2O + H_2O \rightarrow$ ___ *(b)* decomposition: $MgClO_3 \rightarrow$ ___ *(c)* single-replacement: $Na + H_2O \rightarrow$ ___ *(d)* double-replacement: $HNO_3 + Ca(OH)_2 \rightarrow$ ___ *(e)* combustion: $C_5H_{12} + O_2 \rightarrow$ ___.
3. Classify each of the following reactions as synthesis, decomposition, single-replacement, double-replacement, or combustion: *(a)* $N_2(g) + 3H_2(g) \rightarrow 2HN_3(g)$ *(b)* $2Li(s) + 2H_2O(\ell) \rightarrow 2LiOH(aq) + H_2(g)$ *(c)* $2NaNO_3(s) \rightarrow 2NaNO_2 + O_2(g)$ *(d)* $2C_6H_{14}(\ell) + 15O_2(g) \rightarrow 12CO_2(g) + 14H_2O(\ell)$ *(e)* $NH_4Cl(s) \rightarrow NH_3(g) + HCl(g)$ *(f)* $BaO(s) + H_2O(\ell) \rightarrow Ba(OH)_2(aq)$ *(g)* $AgNO_3 + NaCl(aq) \rightarrow AgCl(s) + NaNO_3(aq)$.
4. Complete and balance each of the following equations and identify each as to category: *(a)* $Br_2(\ell) + KI(aq) \rightarrow$; *(b)* $Zn(s) + HCl(aq) \rightarrow$; *(c)* $Ca(s) + Cl_2(g) \rightarrow$; *(d)* $NaClO_3(s) \rightarrow$; *(e)* $C_8H_{18}(\ell) + O_2(g) \rightarrow$; *(f)* $CuCl_2(aq) + Na_2S(aq) \rightarrow$.

8.3 Activity Series of the Elements

The ability of an element to react is referred to as the *activity* of the element. The more readily an element reacts with other substances, the greater its activity. *An activity series is a list of elements organized according to the ease with which they undergo certain chemical reactions.* The elements are usually listed in an order determined by single-replacement reactions. The most active element, placed at the top in the series, can replace every element below it. An element further down can replace those elements that follow it, but not those above it.

Flourine is the most active halogen. Activity decreases as one proceeds down this group (Group 7) in the Periodic Table. In discussing single-replacement reactions in Section 8.2, it was noted that each halogen will react to replace those below it. An activity series for the halogens, therefore, lists the elements in the same order as in Group 17. This is shown in Table 8.3 on the following page.

Being able to write a chemical equation does not necessarily mean that a reaction will actually take place. Activity series like those shown in Table 8-3 are useful aids in predicting whether or not certain chemical reactions will occur. For example, based on the activity series for metals in Table 8-3, one would predict that aluminum can replace zinc, and therefore that the following reaction does occur:

$$2Al(s) + 3ZnCl_2(aq) \rightarrow 3Zn(s) + 2AlCl_3(aq)$$

Cobalt, however, cannot replace sodium, and therefore the reaction below does *not* occur:

$$Co(s) + 2NaCl(aq) \longrightarrow 2Na(s) + CoCl_2(aq)$$

---

1. Synthesis, decomposition, single replacement, double replacement, and combustion
2. *(a)* $Li_2O + H_2O \rightarrow 2LiOH$
   (b) $MgClO_3 \rightarrow MgCl_2 + 3O_2$
   (c) $2Na + 2H_2O \rightarrow 2NaOH + H_2$
   (d) $2HNO_3 + Ca(OH)_2 \rightarrow Ca(NO_3)_2 + 2H_2O$
   (e) $C_5H_{12} + 8O_2 \rightarrow 5CO_2 + 6H_2O$
3. *(a)* Synthesis
   (b) Single replacement
   (c) Decomposition
   (d) Combustion
   (e) Decomposition
   (f) Synthesis
   (g) Double replacement
4. *(a)* $Br_2(\ell) + 2KI(aq) \rightarrow 2KBr(aq) + I_2(s)$; single replacement *(b)* $Zn(s) + 2HCl(aq) \rightarrow ZnCl_2(aq) + H_2(g)$; single replacement *(c)* $Ca(s) + Cl_2(g) \rightarrow CaCl_2$; synthesis *(d)* $2NaClO_3(s) \rightarrow 2NaCl(s) + 3O_2(g)$; decomposition *(e)* $2C_8H_{18}(\ell) + 25O_2(g) \rightarrow 16CO_2(g) + 18H_2O(g)$; combustion *(f)* $CuCl_2(aq) + Na_2S(aq) \rightarrow 2NaCl(aq) + CuS(s)$; double replacement

TABLE 8-3 ACTIVITY SERIES OF THE ELEMENTS	
Activity of Metals	
Li	
Rb	
K	Can react with cold
Ba	H_2O and acids,
Sr	replacing hydrogen.
Ca	
Na	
Mg	
Al	Can react with
Mn	steam and acids,
Zn	replacing hydrogen.
Cr	
Fe	
Cd	
Co	Can react with
Ni	acids, replacing
Sn	hydrogen.
Pb	
H_2	
Sb	React with oxygen,
Bi	forming oxides.
Cu	
Hg	
Ag	Fairly unreactive.
Pt	Form oxides only
Au	indirectly.
Activity of Halogen Nonmetals	
F_2	
Cl_2	
Br_2	
I_2	

Principles of the Activity Series

Like many other aids used to predict the products of chemical reactions, activity series are based upon experiment. Some metals (potassium, for example) react vigorously with water and acids, replacing hydrogen to form new compounds. Other metals such as iron or zinc (see Section 8.2) replace hydrogen in acids such as hydrochloric acid but react with water only when it is hot enough to become steam. Nickel will not react with steam but does react with acids, replacing hydrogen. Gold will not react with water (either as a liquid or as steam), or with any single acid to replace hydrogen. (To dissolve gold, one must use a mixture of nitric acid and hydrochloric acid; the gold is dissolved by chlorine, produced by the oxidation of hydrochloric acid by nitric acid.) Such experimental observations led chemists to arrange metals in an activity series.

Metal atoms lose electrons to the positively charged ions of any metal below it in the activity series. The greater the distance between the two elements in the activity series, the more likely the displacement reaction will take place.

Single-replacement reactions of metals occur when, the atoms of one metal give up electrons to the positive ions of another metal in solution. The greater the distance between two metals in the activity series, the more likely it is that one will replace the other.

Consider the reaction between zinc and copper ions in solution, shown in Figure 8-11. A piece of zinc placed in an aqueous solution of copper(II) sulfate becomes coated with copper. These changes show that a reaction does occur, as would be predicted from the activity series.

$$Zn(s) + Cu^{2+}SO_4^{2-} (aq) \rightarrow Zn^{2+}SO_4^{2-} (aq) + Cu(s)$$

The position of zinc in the table indicated the zinc atoms have a greater tendency to lose electrons than do copper atoms. Zinc atoms therefore transfer electrons to the copper ions and form Zn^{2+} ions.

$$Zn \rightarrow Zn^{2+} + 2e^-$$

The copper ions gain the electrons and form Cu metal atoms.

$$Cu^{2+} + 2e^- \rightarrow Cu$$

Notice the SO_4^{2-} ions do not participate in the reaction. During this reaction, the fading of the blue color caused by Cu^{2+} ions in aqueous solution and the coating of zinc by copper can both be observed. The formation of Zn^{2+} ions is also occurring, but it cannot be seen.

When copper is placed in a solution of zinc sulfate ($ZnSO_4$), no reaction is observed. Here, too, the result can be predicted from the activity series. Copper is a less active metal than zinc and does not give up its electrons to Zn^{2+} ions in solution. Copper therefore does not replace zinc ions from a solution of zinc compound, but will replace the ions of a metal below it in the activity series, as illustrated in Figure 8-12 on the following page.

A piece of copper was placed in a solution of silver nitrate. The copper is more active than silver, and replaces silver ions in the compound.

$$Cu(s) + 2Ag^+NO_3^-(aq) \rightarrow Cu^{2+}(aq) + ZNO_3^-(aq) + 2Ag(s)$$

In this reaction, there is also visual evidence of both the gain and loss of electrons as Cu atoms transfer electrons to Ag^+ ions, and the Ag metal crystallizes out of solution. The copper atoms give up electrons and become Cu^{2+} ions, as observed by the blue appearance of the solution.

Useful Generalizations Based on the Activity Series

We have seen that the activity series of metals helps to predict whether or not single replacement reactions of metals with metal ions and of metals with water and acids such as hydrochloric acid will occur. Generalizations 1–4 summarize how the activity series can be used for reactions of these types.

1. An element will replace from a compound in aqueous solution any of those elements below it in the activity series. The larger the interval between the elements in the activity series, the greater the tendency for the replacement reaction to occur.
2. Any metal above magnesium replaces hydrogen from water.
3. Any metal above cobalt replaces hydrogen from steam.
4. Any metal above hydrogen reacts with acids, replacing hydrogen. The synthesis reactions of metals with oxygen also occur more readily, the higher a metal is placed in the activity series. Generalizations 5 and 6 apply to such synthesis reactions.
5. Any metal above silver reacts with oxygen, forming oxides; those near the top react rapidly.
6. Any metal below mercury forms oxides only indirectly (i.e., not by reaction with O_2).

The more active a metal, the more strongly it holds onto oxygen in an oxide and therefore, the more strongly the oxide resists decomposition into its elements upon heating. Generalizations 7–9 summarize how the activity series can be used to predict the outcome of the decomposition of oxides or their conversion to free metals in single replacement reactions with hydrogen such as

$$H_2(g) + NiO(s) \rightarrow Ni(s) + H_2O(\ell)$$

7. Oxides of metals below copper decompose with heat alone.
8. Oxides of metals below chromium yield metals when heated with hydrogen.

Figure 8-11 The replacement of Cu^{2+} ions in aqueous solution by zinc, a metal that is above copper in the activity series:
$Zn(s) + CuSO_4(aq) \rightarrow ZnSO_4(aq) + Cu(s)$

Figure 8-12 The replacement of Ag^+ ions in aqueous solution by copper, a metal that is above silver in the activity series:
$Cu(s) + 2Ag\,NO_3(aq) \rightarrow Cu(NO_3)_2(aq) + 2Ag$

Technology

Hydrazine, a Liquid Rocket Fuel

The Raschig process for forming hydrazine (N_2H_4, a liquid rocket fuel) has three steps:

Step 1 $2NaOH(aq) + Cl_2(aq) \rightarrow NaClO(aq) + NaCl(aq) + H_2O(\ell)$

Step 2 $NaClO(aq) + NH_3(aq) \rightarrow NH_2Cl(aq) + NaOH(aq)$

Step 3 $NH_2Cl(aq) + NH_3(aq) + NaOH(aq) \rightarrow N_2H_4(aq) + NaCl(aq) + H_2O(\ell)$

Notice that one of the products of each step shown above serves as a reactant in the next step and that hydrazine is not formed until the last step. (Additional reactions, known as "side reactions," also occur but are not shown above.)

Sodium hypochlorite (NaClO, the active ingredient in household bleach) is produced in the first reaction. Chloramine (NH_2Cl) is produced in the second. Both are crucial to the Raschig process, but are

consumed immediately after their formation by the next step. This means that they don't have to be transported or stored, which is advantageous because chloramine is dangerous.

The net chemical equation for the Raschig process is the sum of the equations above:

$2NaOH(aq) + Cl_2(g) + 2NH_3(aq) \rightarrow N_2H_4(aq) + 2NaCl(aq) + 2H_2O(\ell)$

The byproducts of the main reaction summarized in the equation above are salt and water. Other byproducts are ammonium chloride and nitrogen, produced by a side reaction between hydrazine and chloramine. This and other side reactions reduce the typical overall yield of hydrazine to 65 percent.

9. Oxides of metals above iron resist conversion to the free metal when heated with hydrogen.

The most active metals are not likely to remain uncombined with other substances for very long. Some are so active that they must be isolated from the air when they are stored. Therefore, generalizations 10 and 11 can be made about the occurrence of free elements in nature.

10. Elements near the top of the series are never found free in nature.
11. Elements near the bottom of the series are often found free in nature.

Sample Problem 8.3

For each of the possible reactions listed below, describe the generalization based on the activity series that applies. Use the generalization to predict whether or not the reaction will take place and what the products will be. *(a)* $Zn(s) + H_2O(\ell) \rightarrow$ ___. *(b)* $Sn(s) + O_2(g) \rightarrow$ ___. *(c)* $Ca(s) + Pb(NO_3)_2 \rightarrow$ ___. *(d)* $Ca(s) + HCl(aq) \rightarrow$ ___. *(e)* $NiO(s) + H_2(g) \rightarrow$ ___.

Solution

(a) This is a reaction between a metal and water—not steam, which would be shown as $H_2O(g)$. A metal more active than magnesium will replace hydrogen from water. Zinc is *above* magnesium according to the activity series. No reaction will occur.

(b) Any metal more active than silver will react with oxygen to form an oxide. Tin is above silver. Therefore, a reaction will occur and the product will be a tin oxide, either SnO or Sn_2O_4.

(c) An element will replace from a compound in aqueous solution any element below it in the activity series. Calcium is above lead and therefore a reaction will occur to produce lead (Pb) and calcium nitrate, $Ca(NO_3)_2$.

(d) A metal more active than hydrogen will replace hydrogen from an acid. Copper is not above hydrogen in the series. Therefore, no reaction will occur.

(e) An oxide of a metal less active than chromium will form the free metal when heated with hydrogen. Therefore, a reaction will occur and the products will be nickel (Ni) and water (H_2O).

Practice Problems

1. Predict whether or not each of the following reactions will occur:

   (a) $Cr(s) + H_2O(l) \rightarrow$ ____   *(Ans.)* No

   (b) $Pt(s) + O_2(g) \rightarrow$ ____.   *(Ans.)* No

   (c) $Pb(s) + HNO_3(aq) \rightarrow$ ____   *(Ans.)* Yes,
       $Pb(s) + 2HNO_3(aq) \rightarrow Pb(NO_3)_2(aq) + H_2(g)$.

   (d) $Mg(s) + H_2O(g) \rightarrow$ ____   *(Ans.)* Yes, Mg(s)
       $+ 2H_2O(g) \rightarrow Mg(OH)_2(aq) + H_2(g)$.

   (e) $BaO(s) \rightarrow$ ____.   *(Ans.)* No

2. Identify the element that replaces hydrogen from water, but cannot replace calcium from its compounds.   *(Ans.)* Na

Section Review

1. In what ways is the activity series useful in predicting chemical behavior?
2. Predict whether or not the following reactions will occur.   (a) $Ni(s) + H_2O(\ell) \rightarrow$ ____   (b) $Br_2(\ell) + KI(aq) \rightarrow$ ____   (c) $Au(s) + HCl(aq) \rightarrow$ ____   (d) $Cd(s) + HCl(aq) \rightarrow$ ____   (e) $Mg(s) + Co(NO_3)_2 \rightarrow$ ____   (f) $FeO(s) \rightarrow$ ____.
3. Write the balanced chemical equation for each of the reactions in Question 2 that can occur.

1. To predict whether or not certain chemical reactions will occur, as well as the products of those reactions that do occur.
2. (a) No   (b) Yes   (c) No   (d) Yes   (e) Yes   (f) No
3. $Br_2(\ell) + 2KI(aq) \rightarrow 2KBr(aq) + I_2(s)$
   $Cd(s) + 2HCl(aq) \rightarrow CdCl_2(aq) + H_2(g)$
   $Mg(s) + Co(NO_3)_2(aq) \rightarrow Mg(NO_3)_2(aq) + Co(s)$

Chapter Summary

- A correct chemical equation will identify all known reactants and products. It contains the correct formulas for the reactants and products and is balanced.
- Synthesis reactions are represented by the general equation $A + X \rightarrow AX$.
- Decomposition reactions are represented by the general equation $AX \rightarrow A + X$.
- Single-replacement reactions are represented by the general equation $A + BX \rightarrow AX + B$.
- Double-replacement reactions are represented by the general equation $AX + BY \rightarrow AY + BX$.
- Activity series list the elements in order of their activity and are useful in predicting whether or not a chemical reaction will occur.
- Chemists have determined the activity series by experiments.

Chapter 8 *Review*

Chemical Equations and Reactions

Vocabulary

activity series
chemical equation
coefficient
combustion reaction
decomposition reaction
double-replacement
   reaction (ionic
   reaction)
electrolysis
formula equation

hydrocarbons
reversible reaction
single-replacement
   reaction
   (displacement
   reaction)
synthesis reaction
   (composition
   reaction)
word equation

Questions

1. What information is provided by a chemical equation?
2. List the three requirements for a correctly written chemical equation.
3. *(a)* What is meant by the term coefficient in relation to a chemical equation? *(b)* How does the presence of a coefficient affect the number of atoms of each type in the formula that it precedes?
4. Give one example each of a word equation, formula equation, and chemical equation.
5. What limitations are associated with the use of both word and formula equations?
6. Define each of the following: *(a)* aqueous solution *(b)* catalyst *(c)* reversible reaction.
7. Write formulas for each of the following compounds: *(a)* potassium hydroxide *(b)* calcium nitrate *(c)* sodium carbonate *(d)* carbon tetrachloride *(e)* magnesium bromide *(f)* sulfur dioxide *(g)* ammonium sulfate.
8. List the there steps used in balancing an equation.
9. What four guidelines are useful in balancing an equation?
10. How many atoms of each type are represented in each of the following? *(a)* $3N_2$ *(b)* $2H_2O$ *(c)* $4HNO_3$ *(d)* $2\ Ca(OH)_2$ *(e)* $3Ba(ClO_3)_2$ *(f)* $5Fe(NO_3)_2$ *(g)* $4Mg_3(PO_4)_2$ *(h)* $2(NH_4)_2$ *(i)* $6Al_2(CO_3)_3$.

11. Cite two common mistakes often encountered when attempting to balance chemical equations.
12. Define and give general equations for the five types of chemical reactions.
13. Complete each of the following general patterns for synthesis reactions by identifying the type of product usually formed: *(a)* element + oxygen *(b)* metal + nonmetal (other than oxygen) *(c)* metal oxide (of active metal) + water *(d)* nonmetal oxide (selected) + water *(e)* metal oxide (selected) + nonmetal oxide (selected).
14. What relationship exists between synthesis and decomposition reactions?
15. How are most decomposition reactions initiated?
16. What is electrolysis?
17. *(a)* In what environment do most single replacement reactions occur? *(b)* How do single-replacement reactions compare with synthesis and decomposition reactions in terms of the amount of energy generally involved?
18. What are hydrocarbons?
19. Complete the general pattern for the complete combustion of hydrocarbons by identifying the types of products usually formed:

$$\text{hydrocarbon + oxygen} \rightarrow$$

20. *(a)* What is meant by the "activity" of an element? *(b)* What is the relationship between the activity of an element and the ease with which it reacts with other substances?
21. *(a)* What is an activity series of elements? *(b)* What is the basis for the ordering of the elements in the activity series?
22. *(a)* What is the relationship between the activity of the halogens and their arrangement on the Periodic Table? *(b)* List the halogens in order of decreasing activity.
23. *(a)* What is the chemical principle upon which the activity series of metals is based? *(b)* What is the significance of the distance between two metals in the activity series?

24. Based on the activity series of metals, summarize the four generalizations that relate to predictions of whether single-replacement reactions of metals with metal ions, and of metals with water and acids, will occur.

25. (a) How does the position of a metal in the activity series affect its ability to react with oxygen through synthesis? (b) What two generalizations (as provided in the text discussion) apply to such synthesis reactions?

26. (a) How does the position of a metal in the activity series affect the tendency of its oxide to undergo decomposition? (b) What three activity series generalizations apply to the ability of metal oxides to decompose or be converted to free metals in single replacement reactions with hydrogen?

27. Based on the activity series, what two generalizations can be made concerning the occurrence of free elements in nature?

Problems

1. Write the word and formula equations for each of the following. Include symbols for phases, such as (s), (g), and (ℓ). (a) Hydrogen gas reacts with oxygen gas to form water vapor. (b) When solid copper reacts with aqueous silver nitrate solution, aqueous copper(II) nitrate and silver metal are produced. (c) When solid mercury(II) oxide is heated, liquid mercury and oxygen gas are produced.

2. Based on the chemical equation given below, complete each of the following:
$Mg(s) + 2HCl(aq) \rightarrow MgCl_2(aq) + H_2(g)$
(a) 1 molecule Mg + ____ molecules HCl → ____ molecules $MgCl_2$ + ____ molecules H_2 (b) ____ molecules Mg + 6 molecules HCl → ____ molecules $MgCl_2$ +____ molecules H_2 (c) ____ molecules Mg + ____ molecules HCl → 5 molecules $MgCl_2$ + ____ molecules H_2.

3. Given the chemical equation below and the number of moles of the indicated reactant or product, determine the number of moles of all remaining reactants and products: $4NH_3 + 3O_2 \rightarrow 2N_2 + 6H_2O$ (a) 3.0 mol O_2 (b) 8.0 mol NH_3 (c) 1.0 mol N_2 (d) 0.40 mol H_2O.

4. Write the formula equation and then the chemical equation for each of the following:

(a) potassium bromide + barium iodide → potassium iodide + barium bromide (b) hydrochloric acid (HCl) + magnesium hydroxide → magnesium chloride + water (c) nitric acid (HNO_3) + calcium hydroxide → calcium nitrate + water.

5. Write chemical equations for each of the following: (a) $H_2 + Cl_2 \rightarrow HCl$ (b) $Al + Fe_2O_3 \rightarrow Al_2O_3 + Fe$ (c) $Pb(C_2H_3O_2)_2 + H_2S \rightarrow PbS + HC_2H_3O_2$.

6. Complete the following synthesis reactions by writing both word and chemical equations for each: (a) sodium + oxygen →____ (b) copper + oxygen → (two possibilities) ____ and ____ (c) potassium + chlorine → ____ (d) magnesium + fluorine → ____.

7. Complete and balance the equation for each of the following decomposition reactions:
(a) $HgO \overset{\Delta}{\rightarrow}$ (b) $H_2O(\ell)$ (electroylsis)
(c) $2Ag_2O \overset{\Delta}{\rightarrow}$ (d) $MgCl_2 \rightarrow$ (electrolysis)

8. Complete and balance the equations for each of the following single replacement reactions:
(a) $Zn + Pb(NO_3)_2 \rightarrow$ ____ (b) $K + Ba(C_2H_3O_2)_2 \rightarrow$ ____ (c) $Al + H_2SO_4 \rightarrow$ ____ (d) $Na + H_2O \rightarrow$ ____.

9. Complete and balance the equation for the following double-replacement reactions, each of which occur, given normal conditions:
(a) $AgNO_3(aq) + NaCl(aq) \rightarrow$ ____
(b) $Mg(NO_3)_2 + KOH (aq) \rightarrow$ ____
(c) $(NH_4)_2S(aq) + ZnCl_2(aq) \rightarrow$ ____
(d) $LiOH(aq) + Fe(NO_3)_3(aq) \rightarrow$ ____.

10. Complete and balance the equation for each of the following combustion reactions: (a) $CH_4 + O_2 \rightarrow$ ____ (b) $C_3H_6 + O_2 \rightarrow$ ____ (c) $C_5H_{12} + O_2 \rightarrow$ ____.

11. Write and balance each of the following equations and then identify each as to type: (a) hydrogen + iodine → hydrogen iodide (b) lithium + hydrochloricacid → lithium chloride + hydrogen (c) sodium carbonate → sodium oxide + carbon dioxide (d) mercury(II) oxide → mercury + oxygen (e) magnesium hydroxide → magnesium oxide + water (f) copper + chlorine → copper(II) chloride (g) calcium chlorate → calcium chloride + oxygen (h) lithium + water → lithium hydroxide + hydrogen (i) lead(II) carbonate → lead(II) oxide + carbon

(j) methane (CH_4) + oxygen → carbon dioxide + water (k) copper + silver nitrate → silver + copper(II) nitrate (l) sodium fluoride → sodium + fluorine (m) magnesium + nitric acid → magnesium nitrate + hydrogen (n) silver chlorate → silver chloride + oxygen (o) fluorine + magnesium chloride → chlorine + magnesium fluoride (p) potassium sulfide + nitric acid → potassium nitrate + hydrogen sulfide (q) magnesium + water → magnesium oxide + hydrogen.

Application Questions

1. What is the relationship between the law of conservation of matter and the process involved in balancing an equation?
2. What are the three quantitative interpretations of chemcial equations?
3. List three kinds of information *not* provided by chemical equations.
4. Based on the activity series of metals and halogens, which element within each pair is more likely to replace the other in a compound? (a) K and Na (b) Al and Ni (c) Bi and Cr (d) Cl_2 and F_2 (e) Au and Ag (f) Cl_2 and I_2 (g) Fe and Sr (h) I_2 and F_2.
5. List the elements given below in order of *increasing* tendency to be found free in nature: Ni, Ba, Cu, Na, Pb, Au, Al, K, Fe, and Ag.
6. Identify the element described in each of the following: (a) The element replaces Sr in its compounds, but not K in its compounds. (b) The element replaces hydrogen from steam, reacts with $Fe(NO_3)_2$, but does *not* react with $Zn(NO_3)_2$. (c) The element replaces hydrogen in acids, cannot replace hydrogen from steam, but reacts with nitrates of Sn, Ni, and Pb. (d) The element reacts with oxygen to form an oxide and replaces Bi in its compounds, but does not react with acids to replace hydrogen.

Application Problems

1. Write chemical equations for each of the following: (a) Aluminum reacts with oxygen to produce aluminum oxide. (b) Phosphoric acid is produced through the reaction between tetraphosphorus decoxide and water. (c) Iron(III) oxide combines with carbon monoxide to produce iron and carbon dioxide.

2. Balance each of the following: (a) $Ca(OH)_2 + (NH_4)_2SO_4 \rightarrow CaSO_4 + NH_3 + H_2O$ (b) $C_2H_6 + O_2 \rightarrow CO_2 + H_2O$ (c) $Cu_2S + O_2 \rightarrow Cu_2O + SO_2$ (d) $NaCl + H_2SO_4 + MnO_2 \rightarrow MnSO_4 + Na_2SO_4 + Cl_2 + H_2O$.
3. The following equations are incorrect in some way. Identify the error and then correctly balance each: (a) $Li + O_2 \rightarrow LiO_2$ (b) $H_2 + Cl_2 \rightarrow H_2Cl_2$ (c) $MgCO_3 \rightarrow MgO_2 + CO_2$ (d) $NaI + Cl_2 \rightarrow NaCl + I$.
4. Each of the following equations is correctly balanced. Determine the subscripts (x and y) needed to complete the specified formulas: (a) $6HCl + Fe_xO_y \rightarrow 2FeCl_3 + 3H_2O$ (b) $3H_2 + N_2 \rightarrow 2N_xH_y$ (c) $4P + 5O_2 \rightarrow 2P_xO_y$ (d) $N_2 + 2O_2 \rightarrow 2N_xO_y$.
5. For each of the following synthesis reactions, identify the missing reactant(s) or product(s) and then balance the resulting equation: (a) $Mg + \underline{\quad} \rightarrow MgO$ (b) $\underline{\quad} + O_2 \rightarrow Fe_2O_3$ (c) $Li + Cl_2 \rightarrow \underline{\quad}$ (d) $Ca + \underline{\quad} \rightarrow CaI_2$.
6. Identify the reactants that could theoretically be combined through synthesis to produce each of the following and then balance the final equation: (a) barium oxide (b) magnesium bromide (c) iron(II) oxide.
7. For each of the following decomposition reactions, identify the missing reactant or product and then balance the resulting equation: (a) $HI \rightarrow H_2 + \underline{\quad}$ (b) $\underline{\quad} \rightarrow H_2 + O_2$ (c) $NH_3 \rightarrow N_2 + \underline{\quad}$ (d) $CaCO_3 \rightarrow CaO + \underline{\quad}$.
8. Identify the compound that could theoretically undergo decomposition to produce the following products and then balance the final equation: (a) magnesium oxide and water (b) lead(II) oxide and water (c) lithium chloride and oxygen (d) barium chloride and oxygen (e) nickel chloride and oxygen.
9. For each of the following single replacement reactions, identify the missing reactant or product, and then balance the resulting equation: (a) $Ca + PbSO_4 \rightarrow Pb + \underline{\quad}$ (b) $Sr + AgNO_3 \rightarrow \underline{\quad} + Sr(NO_3)_2$ (c) $Fe + \underline{\quad} \rightarrow Ni + Fe(NO_3)_2$ (d) $\underline{\quad} + Pb(NO_3)_2 \rightarrow Pb + Al(NO_3)_3$; (e) $Rb + H_2O \rightarrow RbOH + \underline{\quad}$
10. Identify the reactants that could theoretically combine through single replacement to produce

the following products and then balance the final equation: *(a)* sodium and potassium chloride *(b)* zinc and calcium sulfate *(c)* copper and iron(II) nitrate *(d)* potassium hydroxide and hydrogen.

11. For each of the following actual double replacement reactions, identify the missing reactant or product, and then balance the resulting equation: *(a)* $Cu(NO_3)_2(aq) + Na_2S)aq)$ → $NaNO_3(aq) +$ ___ *(b)* $ZnCl_2(aq) + LiOH(aq)$ → ___ $+ Zn(OH)_2(s)$
(c) $Pb(NO_3)_2(aq) +$ ___ → $NH_4NO_3(aq) + PbCO_3(s)$
(d) $Al(NO_3)_3(aq) + NaOH(aq)$ → $NaNO_3(aq)$ + ___.

12. Identify the aqueous reactants that could combine through double replacement to produce the following products actually observed, and then balance each final equation: *(a)* potassium bromide and zinc sulfide; *(b)* barium nitrate and copper(II) hydroxide; *(c)* sodium acetate and manganese(II) sulfide.

13. In each of the following combustion reactions, identify the missing reactant(s), product(s) or both, and then balance the resulting equation: *(a)* $C_3H_8 +$ ___ → ___ $+ H_2O$ *(b)* $C_4H_{10} +$ ___ → ___ $+$ ___ *(c)* ___ $+ 8O_2$ → $5CO_2 + 6H_2O$.

14. Identify the hydrocarbon that could theoretically undergo combustion to produce the specified amounts of the following products, and then balance each equation: *(a)* $CO2 + 2H_2O$ *(b)* $4CO_2 + 6H_2O$ (2 hydrocarbon molecules react) *(c)* $12CO_2 + 14H_2O$ (2 hydrocarbon molecules react) *(d)* $12CO_2 + 6H_2O$ (2 hydrocarbon molecules react).

15. Complete and balance each of the following reactions observed to occur, and then identify each as to type: *(a)* zinc + sulfur → ___ *(b)* calcium + sodium nitrate → ___ *(c)* silver nitrate + potassium sulfide → ___ *(d)* sodium iodide $\overset{\Delta}{\rightarrow}$ ___ *(e)* benzene (C_6H_6) + oxygen → ___ *(f)* decane $(C_{10}H_{22})$ + oxygen → ___.

16. Use the activity series to predict whether each of the following single-replacement reactions will occur. Write the chemical equations for those predicted to occur. *(a)* $K(s) + CuCl_2(aq)$ → ___ *(b)* $Zn(s) + Pb(NO_3)_2(aq)$ → ___ *(c)* $Cl_2(g) + KI(aq)$ → ___

(d) $Cu(s) + FeSO_4(aq)$ → ___ *(e)* $Ba(s) + H_2O(\ell)$ → ___.

17. Use the activity series to predict whether each of the following synthesis reactions will occur and write the chemical equations for those predicted to occur: *(a)* $Ca(s) + O_2(g)$ → ___ *(b)* $Ni(s) + O_2(g)$ → ___ *(c)* $Au(s) + O_2(g)$ → ___ *(d)* $Al(s) + O_2(g)$ → ___.

18. Use the activity series to predict whether each of the following decomposition reactions will occur and write the chemical equations for those predicted to occur: *(a)* $Ag_3O(s) \overset{\Delta}{\rightarrow}$ ___ *(b)* $PbO(s) \overset{\Delta}{\rightarrow}$ ___ *(c)* $NiO(s) \overset{\Delta}{\rightarrow}$ ___ *(d)* $CuO(s) + H_2(g) \overset{\Delta}{\rightarrow}$ ___.

Enrichment

1. Prepare a report on the contributions of Sir Henry Cavendish to chemistry. Include the details surrounding his discovery of hydrogen.

2. Trace the evolution of the current practice of fluoridation of municipal water supplies. What advantages and disadvantages are associated with this practice?

3. Prepare a report describing the academic requirements and general job duties of the analytical chemist. Cite at least five specific types of jobs within the analytical chemistry career field.

4. Cite at least two examples each of everyday substances typically used as reactants to produce desired products through synthesis, decomposition, single replacement, double replacement, and combustion reactions.

5. Trace the development and production of matches, and explain how they work. What role did matches play in shaping life as we know it today?

6. Trace the evolution of electrolysis as a useful technique in today's world. Cite some of its typical products.

7. Explain how a soda-acid fire extinguisher works and determine the chemical equation for the reaction. Check you house and adjacent structures for the location of fire extinguishers and ask your local fire department to verify the effectiveness of the extinguishers.

8. Write a report on candlemaking. Compare the old-fashioned method of making candles with modern techniques.

Chapter 9 Stoichiometry

Chapter Planner

9.1 Introduction to Stoichiometry
Applications of Stoichiometry
- Chemistry Notebook
- Demonstration: 1
- Question: 1

Reaction-Stoichiometry Problems
- Questions: 2–6
- Problem: 1

9.2 Ideal Stoichiometric Calculations
Mole-Mole Calculations
- Problems: 2, 3a-c, 4
- Application Question: 1
- Demonstration: 2
- Application Problem: 1
- Demonstration: 4

Mole-Mass Calculations
- Problems: 3d, 6
- Application Problems: 2, 3, 4a-b
- Application Problem: 4c

Mass-Mole Calculations
- Problems: 5a-b, 7–10
- Application Problems: 5, 6
- Application Problem: 7

Mass-Mass Calculations
- Experiment: 11
- Problems: 5c, 11–13
- Application Problems: 8, 9
- Demonstration: 3

9.3 Limiting Reactants and Percent Yield
Limiting Reactant
- Demonstrations: 5, 6
- Desktop Investigation
- Question: 7
- Problems: 14–17
- Application Problem: 10
- Application Question: 2
- Application Problem: 11

Percent Yield
- Questions: 8, 9
- Problems: 18–20
- Application Questions: 3, 4
- Application Problem: 12

Teaching Strategies

Introduction

This chapter will deal with the mass relationships associated with chemical reactions. You will find that some of your students will master this chapter readily while others will have difficulty with the algebraic manipulations. There are many sample problems with solutions throughout the chapter and many review problems at the end of the chapter. This practice is essential. There are also a variety of interactive computer programs available that can be especially helpful to students who have difficulty analyzing problems.

A two-prong approach can be very effective in this chapter. Here is an opportunity to spark your students' interest as the work gets more quantitative, or to excite them with the phenomena of chemistry. You can develop an enthusiasm for this subject and help them to master stoichiometry by adding several quick demonstrations every day. A majority of the reactions that are the basis of the problems in this chapter can be easily demonstrated to your class. (Some of your students might take the responsibility for this work.) The ideal stoichiometry calculations will be more meaningful to your students if they see these reactions, even if they are not done quantitatively.

The second prong in approaching the topic of stoichiometry is to emphasize a very systematic procedure for solving problems. Emphasize that the first step in solving these problems is to *solve a balanced equation*. The equation describes the chemistry. If you are actually doing the chemistry, students will tend to be more receptive to this idea. Make sure that students include all units in their problems and cancel carefully. If students

have difficulty interpreting problems, have them underline interrogative words. Show them that they can always determine what information is being given and what is being asked for.

9.1 Introduction to Stoichiometry

Priestley is mentioned several times in this chapter. Joseph Priestley was a scientist and a minister. He was introduced to science by Benjamin Franklin. His discoveries included the CO_2 cycle. His friends and colleagues included Dalton, Faraday, Thomas Darwin, whose grandson was Charles Darwin, and Josiah Wedgewood, whose grandaughter was to marry Charles Darwin.

Priestley was forced to leave England because of his radical polital and religious views. He resettled in Northlumberland, Pennsylvania, a small community near Harrisburg. His papers are kept in the Museum of the History of Science at the University of Pennsylvania. There is an excellent article on his life and work in *JCED*, Vol. 65, No. 9, , 1987, pp. 745–7.

Demonstration 1: The production of oxygen by thermal decomposition of an oxide
[This historic reaction, originally performed by Lavoisier, can be repeated in your classroom.]
PURPOSE: To observe chemical change resulting from decomposition of an oxide
MATERIALS: 0.3 g mercury(II) oxide, test tube, Bunsen burner, wooden splint

PROCEDURE: Place 0.3 g mercury(II) oxide in a test tube and heat over a Bunsen burner. CAUTION: This demonstration should only be done under a hood in good working condition. Mercury vapor is toxic. Light a wooden splint and blow out the flame. Quickly hold the split inside the test tube.

The appearance of the powdered material will change as the oxide decomposes into mercury and oxygen. The splint will glow inside the test tube indicating the presence of oxygen.

[Lead(IV) oxide can be used in place of mercury(II) oxide.]

Demonstration 2: Single-Replacement Reaction
[This reaction can be demonstrated in a petri dish placed on an overhead projector and projected on a screen.]
PURPOSE: To show the reaction of a metal in acid
MATERIALS: 100-mm glass petri dish, 0.5 g aluminum, 15 mL 0.1 M H_2SO_4, (Zn and HCl could be used as an alternative.)
PROCEDURE: Pour 15 mL of H_2SO_4 into a petri dish placed on an overhead projector. Add piece a of aluminum.

The aluminum will replace the hydrogen in the acid. Bubbles of hydrogen will evolve as the reaction proceeds.

As an additional demonstration, add a few drops of universal indicator to the acid before you add the aluminum. The indicator will change color from red towards the more neutral color range as the hydrogen ion is consumed.

9.2   Ideal Stoichiometric Calculations

This section focuses on quantitative relationships in chemical reactions. It is important for students to have an opportunity to become familiar with some of the equipment and procedures that are used in stoichiometry. The following demonstration provides that opportunity. Use quantitative methods and record all masses accurately.

Demonstration 3: Formation of Copper Iodide
PURPOSE: To determine the formula of the product formed from the reaction of copper (Cu) and iodine (I_2) and to write the equation for the reaction

$$Cu(s) + I_2(s) \rightarrow CuI_2$$

MATERIALS: Copper wire (1.5–2.0 g), iodine crystals (1.5–2.0 g), 0.1 M sodium thiosulfate solution, acetone, Erlenmeyer or Florence flask, hot plate, balance (0.01 g digital balance works well), watch glass, ice cubes, heat lamp (optional)
PROCEDURE: Reproduce the following data/calculation table on the chalkboard, or give as handouts to your students.

(1) Place a few crystals of iodine in the bottom of a flask. Place the flask on the hot plate and heat gently. A watch glass holding an ice cube should be placed on the mouth of the flask to contain the iodine vapors that will be generated. CAUTION: Iodine vapors are toxic. The ice cube on the watch glass will contain the vapors so that students are not exposed to them. This demonstration should be done in a hood.

(2) Prepare the surface of the copper wire. Make a coil of the copper wire leaving a convenient length of the wire to act as a handle when the wire is lowered into the flask. (3) Lower the copper into the flask containing the iodine vapors and allow it to remain for one minute. (4) Remove the copper wire and immerse it in sodium thiosulfate solution. After one minute, remove the wire from the cleaning solution, and rinse it in distilled water. (5) Immerse the copper wire in acetone. Remove excess acetone, and repeat the acetone immersion two more times. (6) Air dry the prepared copper wire. (7) Determine the mass of the prepared copper wire and record the mass in line 1 of the data table.

(8) Immerse the copper wire in the flask containing the iodine vapors. (9) After two minutes, remove the reacted copper wire from the reaction flask. (10) Hold the copper wire in front of a heat lamp in order to remove excess free iodine. Determine the mass of the wire which is now a mixture of unreacted copper and newly formed copper iodide. Record the mass on line 2 of the data table.

(11) Immerse the copper wire in 0.1 M sodium thiosulfate and rinse in acetone as before. The copper iodide will be removed in this process. (12) Determine the mass of the dried copper wire and record on line 3 of the data table. This is the mass of the unreacted copper.

Repeat steps 8–12 several times if you can. If you do not have enough class time to repeat the procedure, you might give this work to a student to do as a special project.

Copper iodide will form on the surface of the copper after it has been immersed in iodine vapors. The mass of the newly formed copper iodide and the unreacted copper can be determined. All copper iodide is then removed by the sodium thiosulfate. The acetone will remove any excess water leaving pure, unreacted copper. The mass of the residual copper can finally be determined.

Data	
1. Mass of copper	_____
2. Mass of reacted copper and copper iodide	_____
3. Mass of remaining copper	_____

Calculations	
4. Mass of copper iodide (Data line 2 − data line 3)	_____
5. Mass of copper consumed (Data line 1 − data line 3)	_____
6. Mass of iodine in cpd (Data line 4 − data line 5)	_____
7. Moles of copper consumed to form copper iodide	_____
8. Moles of iodine in product	_____
9. Molar ratio copper/iodine	_____

If you have the opportunity to collect data from several trials, determine the mean values of the mass of iodine from all trials and the mean values of the mass of copper from all trials. Base the calculations for the molar ratio of copper to iodine on the mean mass values.

Except for a highly gifted honors class, do not emphasize the calculations at this time.

With this demonstration, as with all of the other demonstrations that accompany this chapter, it is important for you to keep sight of your goals. You are primarily trying to give your students an experimental basis for equation writing. They should be able to develop some intuitive concept of the nature of a chemical reaction. They should recognize the different kinds of evidence for chemical reactions: direct or indirect color change, formation of a precipitate, temperature change, evolution of a gas. They should also start to become familiar with the methods that chemists use to determine the identity of a product; that includes determining molar ratios. The quantitative procedures are not included at this point to teach stoichiometry, although they are good preparation for those manipulations, rather, they are included to give students some actual experiences with fundamental laboratory procedures.

Demonstration 4: An Acid–Base Titration

PURPOSE: To determine the amount of acid required to neutralize a given amount of base

MATERIALS: two burets, burette stand, 25-50 mL 0.1 M NaOH, 25-50 mL 0.1 M HCl, phenolpthalein, 250-mL Erlenmeyer flask

PROCEDURE: HCl will be titrated against NaOH.

Place burettes in burette stand. Fill burette on left with acid and fill burette on right with base. Read both burettes. Add about 20 mL of acid to the Erlenmeyer flask. Read the burette accurately. Add a few drops of phenolpthalein to the acid in the flask. Add NaOH to the flask containing the HCl. Swirl the flask. Continue to add base until the presence of the red color of phenolpthalein indicates that the titration is complete. Read the NaOH burette accurately.

Compare the volume of acid added to the flask to the volume of base added to the flask. Tell students that this procedure allows investigators to determine the number of moles of acid in a solution if the number of moles of base in solution are known. Likewise, if the number of moles of acid in solution are known, the unknown number of moles of base in a solution can be determined.

[Note: The procedure is being done now to aquaint students with the methods of acid/base stoichiometry, not to actually determine the molarity of a base.]

9.3   Limiting Reactants and Percent Yield

Many students have difficulty solving limiting reaction problems. It might be wise to omit this section for your fundamental classes and to spend some additional time working on mass/ mass problems with them.

The Desktop Investigation and the simple demonstration that follows will give students two concrete examples of limiting reactants. Help students grasp the concept that a deficit in the amount of one reactant will limit the amount of product that can be produced whether that product is a batch of cookies, capped bottles, or a chemical compound. After students understand that when one reactant is used up, no more product can be formed. Carefully explain the strategy described in this section for solving limiting reactant problems.

Students often fail to recognize limiting reactant problems. However, if a limiting reactant problem is designated as such, they will be able to solve it successfully. Consider this when you prepare a test that covers this material.

Demonstration 5: How many bottles can a bottle capper cap?

MATERIALS: One dozen bottle caps, ten bottles

PROCEDURE: Invite a student to help you with this demonstration. Ask the students to cap as many bottles as possible. Of course you have a limited supply of bottles. Students never forget this demonstration and you can refer to it while reviewing problems.

Dr. Larry Peck, of Texas A&M University, presented the following demonstration at CAST '86.

Demonstration 6: The Limiting Reactant

PURPOSE: To demonstrate the limiting effect of the amount of magnesium in a reaction

MATERIALS: 3 500-mL flasks, 3 15"-diameter balloons, magnesium turnings, 12 M HCl, water

PROCEDURE: In flask 1, place 25.5 mL of H_2O and 7.5 mL HCl. In flask 2, place 45 mL H_2O and 15 mL HCL. In flask 3, place 22.5 mL H_2O and 37.5 mL HCl. Into each of three balloons, place 2.19 g Mg. Being careful not to spill any Mg into the acid, stretch balloons over the mouth of the flasks, allowing them to dangle over the side of the flasks. Simultaneously hold the balloons up and shake the Mg into the flasks and observe.

Flask 1 contains 0.09 mol Mg and 0.09 mol HCl, so half of the Mg will remain unreacted. Flask 2 contains 0.09 mol Mg and 0.18 mol HCl. All of the Mg reacts because it is a stoichiometric mixture. Balloon 2 will be larger than balloon 1.

Flask 3 contains 0.09 mol Mg and 0.27 mol HCl. In this case, the magnesium is the limiting reactant. Therefore, balloon 3 will be the same size as balloon 2.

References and Resources

Books and Periodicals

Bowman and Schull. "Mysterious Stoichiometry," *Journal of Chemical Education*, Vol. 52, No. 3, 1975, p. 186.

Gizara. "Bridging the Stoichiometric Gap," *ST*, Vol. 48, No. 4, 1981, p. 36.

Maciel. "Chemistry that's Chinese, not Greek, to Students!" *Journal of Chemical Education*, Vol. 57, No. 10, 1980, p. 727.

Morrison. "Emphasis on Stoichiometry and Structure," *Journal of Chemical Education*, Vol. 55, No. 4, 1978, p. 255.

Russo, Tom. "Microchemistry for High School General Chemistry," Educational Concepts, 1986.

Vipond. "Stoichiometry and Chemical Equations", *Science Teacher*, Vol. 43, No. 5, 1978, p. 48.

Computer Software

Chemistry 102 #3: Mole: Its Meaning #4: Mole: Learning by Practice, 1986.

Gold, Marvin. "Principles of Stoichiometry: Part I—Principles of Stoichiometry (Sandwich Analogy), Part II—Stoichiometry in

Chemical Reactions (Ammonia Production)," COM4031A, AII, 48K, 2 disks, COMPress.

Project Seraphim, Disks #305 and #306, NSF Science and Engineering Education, Dept. of Chemistry, Eastern Michigan University, Ypsilanti, MI 48197.

Weyh, J.A., J.R. Crook, and L.N. Hauge. "Chemical Stoichiometry, Version 2.0," COM4010B, I, 256K, CGA, disk, COMPress.

Weyh, J.A., J.R. Crook, and L.N. Hauge. "Mole Concept, Version 2.0," COM 4012B, I, 256K, CGA, disk, COMPress.

Answers and Solutions

Questions

1. *(a)* Composition stoichiometry involves the mass relationships of elements in chemical compounds, whereas reaction stoichiometry involves the mass relationships among reactants and products in a chemical reaction. *(b)* An example of a composition stoichiometry calculation involves the mass/mole relationship. A typical reaction stoichiometry calculation involves the use of a chemical equation to determine the amounts of reactants needed to produce a given amount of product.

2. *(a)* A mole ratio is a conversion factor that relates the number of moles of any two substances involved in a chemical reaction. *(b)* This information is obtained directly from the chemical equation.

3. The coefficients in a chemical equation represent the relative numbers of moles of the substances involved in a reaction.

4. *(a)* $2Ca + O_2 \rightarrow 2CaO$

   1. $\dfrac{2 \text{ mol Ca}}{1 \text{ mol } O_2}$ or $\dfrac{1 \text{ mol } O_2}{2 \text{ mol Ca}}$

   2. $\dfrac{2 \text{ mol Ca}}{2 \text{ mol CaO}}$ or $\dfrac{2 \text{ mol CaO}}{2 \text{ mol Ca}}$

   3. $\dfrac{1 \text{ mol } O_2}{2 \text{ mol CaO}}$ or $\dfrac{2 \text{ mol CaO}}{1 \text{ mol } O_2}$

   (b) $Mg + 2HF \rightarrow MgF_2 + H_2$

   1. $\dfrac{1 \text{ mol } Mg_2}{2 \text{ mol HF}}$ or $\dfrac{2 \text{ mol HF}}{1 \text{ mol Mg}}$

   2. $\dfrac{1 \text{ mol Mg}}{1 \text{ mol } MgF_2}$ or $\dfrac{1 \text{ mol } MgF_2}{1 \text{ mol Mg}}$

   3. $\dfrac{1 \text{ mol Mg}}{1 \text{ mol } H_2}$ or $\dfrac{1 \text{ mol } H_2}{1 \text{ mol Mg}}$

   4. $\dfrac{2 \text{ mol HF}}{1 \text{ mol } MgF_2}$ or $\dfrac{1 \text{ mol MgF}}{2 \text{ mol HF}}$

   5. $\dfrac{2 \text{ mol HF}}{1 \text{ mol } H_2}$ or $\dfrac{1 \text{ mol } H_2}{2 \text{ mol HF}}$

   6. $\dfrac{1 \text{ mol } MgF_2}{1 \text{ mol } H_2}$ or $\dfrac{1 \text{ mol } H_2}{1 \text{ mol } MgF_2}$

5. *(a)* Molar mass is the mass (in grams) of one mole of a substance. *(b)* Molar mass is the conversion factor that relates the mass of a substance to the number of moles of that substance.

6. Ideal stoichiometry calculations do not take into account those factors that can affect the relative amounts of reactants needed or products actually produced in chemical reactions; they deal with the amounts of reactants or products under ideal conditions. Real stoichiometry calculations do take into account actual laboratory or industrial conditions.

7. *(a)* The limiting reactant is the reactant that controls the amount of product formed in a chemical reaction, whereas the excess reactant is the substance that is not used up completely in a reaction. *(b)* The limiting reactant is used up.

8. *(a)* The theoretical yield is the maximum amount of product that can be produced from a given amount of reactant. The actual yield is the measured amount of a product obtained from a reaction. *(b)* The actual yield is less than the theoretical yield.

9. The percent yield is the ratio of the actual yield to the theoretical yield multiplied by 100 percent.

Problems

1. Na_2CO_3: $\dfrac{105.99 \text{ g } Na_2CO_3}{1 \text{ mol } Na_2CO_3}$ or $\dfrac{1 \text{ mol } Na_2CO_3}{105.99 \text{ g } Na_2CO_3}$

   $Ca(OH)_2$: $\dfrac{74.10 \text{ g } Ca(OH)_2}{1 \text{ mol } Ca(OH)_2}$ or $\dfrac{1 \text{ mol } Ca(OH)_2}{74.10 \text{ g } Ca(OH)_2}$

   $NaOH$: $\dfrac{40.00 \text{ g NaOH}}{1 \text{ mol NaOH}}$ or $\dfrac{1 \text{ mol NaOH}}{40.00 \text{ g NaOH}}$

   $CaCO_3$: $\dfrac{100.09 \text{ g } CaCO_3}{1 \text{ mol } CaCO_3}$ or $\dfrac{1 \text{ mol } CaCO_3}{100.09 \text{ g } CaCO_3}$

2. *(a)* $2H_2 + O_2 \rightarrow 2H_2O$

   $5.0 \text{ mol } H_2O \times \dfrac{2 \text{ mol } H_2}{2 \text{ mol } H_2O} = 5.0 \text{ mol } H_2$

   (b) $5.0 \text{ mol } H_2O \times \dfrac{1 \text{ mol } O_2}{2 \text{ mol } H_2O} = 2.5 \text{ mol } O_2$

3. *(a)* $2Fe(OH)_3 + 3H_2SO_4 \rightarrow Fe_2(SO_4)_3 + 6H_2O$

   (b) (1) $5.00 \text{ mol } Fe(OH)_3 \times \dfrac{3 \text{ mol } H_2SO_4}{2 \text{ mol } Fe(OH)_3}$
   $= 7.50 \text{ mol } H_2SO_4$

   (c) (1) $5.00 \text{ mol } Fe(OH)_3 \times \dfrac{1 \text{ mol } Fe_2(SO_4)_3}{2 \text{ mol } Fe(OH)_3}$
   $= 2.50 \text{ mol } Fe_2(SO_4)_3$

   (2) $5.00 \text{ mol } Fe(OH)_3 \times \dfrac{6 \text{ mol } H_2O}{2 \text{ mol } Fe(OH)_3} = 15.0 \text{ mol } H_2O$

   (d) (1) $5.00 \text{ mol } Fe(OH)_3 \times \dfrac{107 \text{ g } Fe(OH)_3}{\text{mol } Fe(OH)_3} = 535 \text{ g } Fe(OH)_3$

   (2) $7.50 \text{ mol } H_2SO_4 \times \dfrac{98.1 \text{ g } H_2SO_4}{\text{mol } H_2SO_4} = 736 \text{ g } H_2SO_4$

   (3) $2.50 \text{ mol } Fe_2(SO_4)_3 \times \dfrac{400 \text{ g } Fe_2(SO_4)_3}{\text{mol } Fe_2(SO_4)_3} = 1.00 \times 10^3 \text{ g } Fe_2(SO_4)_3$

   (4) $15.0 \text{ mol } H_2O \times \dfrac{18.0 \text{ g } H_2O}{\text{mol } H_2O} = 270 \text{ g } H_2O$

535 g + 736 g = 1271 g; 1.00×10^3 g + 270 g = 1270 g

The total masses of the reactants and products are virtually the same (given uncertainties in values), confirming that mass is conserved in a chemical reaction.

4. (a) $2C_2H_6 + 7O_2 \rightarrow 4CO_2 + 6H_2O$

$4.50 \text{ mol } C_2H_6 \times \dfrac{7 \text{ mol } O_2}{2 \text{ mol } C_2H_6} = 15.8 \text{ mol } O_2$

(b) $4.50 \text{ mol } C_2H_6 \times \dfrac{4 \text{ mol } CO_2}{2 \text{ mol } C_2H_6} = 9.00 \text{ mol } CO_2$

$4.50 \text{ mol } C_2H_6 \times \dfrac{6 \text{ mol } H_2O}{2 \text{ mol } C_2H_6} = 13.5 \text{ mol } H_2O$

5. (a) $H_2SO_4 + 2NaOH \rightarrow Na_2SO_4 + 2H_2O$

(b) $0.75 \text{ mol NaOH} \times \dfrac{1 \text{ mol } H_2SO_4}{2 \text{ mol NaOH}} \times \dfrac{98 \text{ g } H_2SO_4}{\text{mol } H_2SO_4} = 37 \text{ g } H_2SO_4$

(c) $0.75 \text{ mol NaOH} \times \dfrac{1 \text{ mol } Na_2SO_4}{2 \text{ mol NaOH}} \times \dfrac{142 \text{ g } Na_2SO_4}{\text{mol } Na_2SO_4} = 53.2 \text{ g } Na_2SO_4$

$0.75 \text{ mol NaOH} \times \dfrac{2 \text{ mol } H_2O}{2 \text{ mol NaOH}} \times \dfrac{18 \text{ g } H_2O}{\text{mol } H_2O} = 14 \text{ g } H_2O$

6. $2Na + Cl_2 \rightarrow 2NaCl$

$25.0 \text{ mol NaCl} \times \dfrac{2 \text{ mol Na}}{2 \text{ mol NaCl}} \times \dfrac{23.0 \text{ g Na}}{\text{mol Na}} = 575 \text{ g Na}$
$\qquad$
$25.0 \text{ mol NaCl} \times \dfrac{1 \text{ mol } Cl_2}{2 \text{ mol NaCl}} \times \dfrac{71.0 \text{ g } Cl_2}{\text{mol } Cl_2} = 888 \text{ g } Cl_2$

7. (a) $Cu + 2AgNO_3 \rightarrow 2Ag + Cu(NO_3)_2$

$2.25 \text{ g Ag} \times \dfrac{\text{mol Ag}}{108 \text{ g Ag}} \times \dfrac{1 \text{ mol } Cu(NO_3)_2}{2 \text{ mol Ag}} = 0.0104 \text{ mol } Cu(NO_3)_2$

(b) $2.25 \text{ g Ag} \times \dfrac{\text{mol Ag}}{108 \text{ g Ag}} \times \dfrac{1 \text{ mol Cu}}{2 \text{ mol Ag}} = 0.0104 \text{ mol Cu}$

$2.25 \text{ g Ag} \times \dfrac{\text{mol Ag}}{108 \text{ g Ag}} \times \dfrac{2 \text{ mol } AgNO_3}{2 \text{ mol Ag}} = 0.0208 \text{ mol } AgNO_3$

8. (a) $Fe_2O_3(s) + 3CO(g) \rightarrow 2Fe(s) + 3CO_2(g)$

$4.00 \text{ kg } Fe_2O_3 \times \dfrac{10^3 \text{ g}}{\text{kg}} \times \dfrac{\text{mol } Fe_2O_3}{160. \text{ g } Fe_2O_3} \times \dfrac{3 \text{ mol CO}}{\text{mol } Fe_2O_3} = 75.0 \text{ mol CO}$

(b) $75.0 \text{ mol CO} \times \dfrac{2 \text{ mol Fe}}{3 \text{ mol CO}} = 50.0 \text{ mol Fe}$

$75.0 \text{ mol CO} \times \dfrac{3 \text{ mol } CO_2}{3 \text{ mol CO}} = 75.0 \text{ mol } CO_2$

9. (a) $2NaOH + CO_2 \rightarrow Na_2CO_3 + H_2O$

$925.0 \text{ g } CO_2 \times \dfrac{\text{mol } CO_2}{44.01 \text{ g } CO_2} \times \dfrac{2 \text{ mol NaOH}}{\text{mol } CO_2} = 42.04 \text{ mol NaOH}$

(b) $42.04 \text{ mol NaOH} \times \dfrac{1 \text{ mol } Na_2CO_3}{2 \text{ mol NaOH}} = 21.02 \text{ mol } Na_2CO_3$

$42.04 \text{ mol NaOH} \times \dfrac{1 \text{ mol } H_2O}{2 \text{ mol NaOH}} = 21.02 \text{ mol } H_2O$

10. (a) $2Al(s) + 3ZnCl_2(aq) \rightarrow 3Zn(s) + 2AlCl_3(aq)$

$13.5 \text{ g Al} \times \dfrac{\text{mol Al}}{27.0 \text{ g Al}} \times \dfrac{3 \text{ mol } ZnCl_2}{2 \text{ mol Al}} = 0.750 \text{ mol } ZnCl_2$

(b) $13.5 \text{ g Al} \times \dfrac{\text{mol Al}}{27.0 \text{ g Al}} \times \dfrac{3 \text{ mol Zn}}{2 \text{ mol Al}} = 0.750 \text{ mol Zn}$

$0.750 \text{ mol Zn} \times \dfrac{2 \text{ mol } AlCl_3}{3 \text{ mol Zn}} = 0.500 \text{ mol } AlCl_3$

11. $CO(g) + 2H_2(g) \rightarrow CH_3OH(g)$

$100.0 \text{ kg } CH_3OH \times \dfrac{10^3 \text{ g}}{\text{kg}} \times \dfrac{\text{mol } CH_3OH}{32.04 \text{ g } CH_3OH} \times \dfrac{1 \text{ mol CO}}{\text{mol } CH_3OH} \times \dfrac{28.01 \text{ g CO}}{\text{mol CO}} = 8.742 \times 10^4 \text{ g CO}$

$100.0 \text{ kg } CH_3OH \times \dfrac{10^3 \text{ g}}{\text{kg}} \times \dfrac{\text{mol } CH_3OH}{32.04 \text{ g } CH_3OH} \times \dfrac{2 \text{ mol } H_2}{\text{mol } CH_3OH} \times \dfrac{2.016 \text{ g } H_2}{\text{mol } H_2} = 1.258 \times 10^4 \text{ g } H_2$

12. $C_6H_{12}O_6 \rightarrow 2C_2H_5OH + 2CO_2$

$25.0 \text{ kg } C_6H_{12}O_6 \times \dfrac{10^3 \text{ g}}{\text{kg}} \times \dfrac{\text{mol } C_6H_{12}O_6}{180. \text{ g } C_6H_{12}O_6} \times \dfrac{2 \text{ mol } C_2H_5OH}{\text{mol } C_6H_{12}O_6} \times \dfrac{46.1 \text{ g } C_2H_5OH}{\text{mol } C_2H_5OH} = 1.28 \times 10^4 \text{ g } C_2H_5OH$

$25.0 \text{ kg } C_6H_{12}O_6 \times \dfrac{10^3 \text{ g}}{\text{kg}} \times \dfrac{\text{mol } C_6H_{12}O_6}{180. \text{ g } C_6H_{12}O_6} \times \dfrac{2 \text{ mol } CO_2}{2 \text{ mol } C_2H_{12}O_6} \times \dfrac{44.0 \text{ g } CO_2}{\text{mol } CO_2} = 1.22 \times 10^4 \text{ g } CO_2$

13. $2NO + O_2 \rightarrow 2NO_2$

(a) $384 \text{ g } O_2 \times \dfrac{\text{mol } O_2}{32.0 \text{ g } O_2} \times \dfrac{2 \text{ mol } NO_2}{1 \text{ mol } O_2} \times \dfrac{46.0 \text{ g } NO_2}{\text{mol } NO_2} = 1.10 \times 10^3 \text{ g } NO_2$

(b) $384 \text{ g } O_2 \times \dfrac{\text{mol } O_2}{32.0 \text{ g } O_2} \times \dfrac{2 \text{ mol } NO}{1 \text{ mol } O_2} \times \dfrac{30.0 \text{ g } NO}{\text{mol } NO} = 720 \text{ g } NO$

14. (a) $2.0 \text{ mol HCl} \times \dfrac{1 \text{ mol NaOH}}{\text{mol HCl}} = 2.0 \text{ mol NaOH}$

Because only 2.0 mol of the 2.5 mol of NaOH are needed to react with the 2.0 mol HCl, the HCl must be the limiting reactant.

(b) $2.5 \text{ mol Zn} \times \dfrac{2 \text{ mol HCl}}{1 \text{ mol Zn}} = 5.0 \text{ mol HCl}$

$6.0 \text{ mol HCl} \times \dfrac{1 \text{ mol Zn}}{2 \text{ mol HCl}} = 3.0 \text{ mol HCl}$

Since only 2.5 mol Zn are available, the Zn is the limiting reactant.

(c) $4.0 \text{ mol Fe(OH)}_3 \times \dfrac{3 \text{ mol } H_2SO_4}{2 \text{ mol Fe(OH)}_3} = 6.0 \text{ mol } H_2SO_4$

$6.5 \text{ mol } H_2SO_4 \times \dfrac{2 \text{ mol Fe(OH)}_3}{3 \text{ mol } H_2SO_4} = 4.3 \text{ mol Fe(OH)}_3$

More than the required 6.0 mol of H_2SO_4 is available, so Fe(OH)$_3$ is the limiting reactant.

15. (a) 2.5 mol NaOH available − 2.0 mol NaOH needed = 0.5 mol excess NaOH (b) 6.0 mol HCl available − 5.0 mol HCl needed = 1.0 mol excess HCl (c) 6.5 mol H_2SO_4 available − 6.0 mol H_2SO_4 needed = 0.5 mol excess H_2SO_4

16. (a) $2.0 \text{ mol HCl} \times \dfrac{1 \text{ mol NaCl}}{1 \text{ mol HCl}} = 2.0 \text{ mol NaCl}$

$2.0 \text{ mol HCl} \times \dfrac{1 \text{ mol } H_2O}{1 \text{ mol HCl}} = 2.0 \text{ mol } H_2O$

(b) $2.5 \text{ mol Zn} \times \dfrac{1 \text{ mol ZnCl}_2}{1 \text{ mol Zn}} = 2.5 \text{ mol ZnCl}_2$

$2.5 \text{ mol Zn} \times \dfrac{1 \text{ mol } H_2}{1 \text{ mol Zn}} = 2.5 \text{ mol } H_2$

(c) $4.0 \text{ mol Fe(OH)}_3 \times \dfrac{1 \text{ mol Fe}_2(SO_4)_3}{2 \text{ mol Fe(OH)}_3} = 2.0 \text{ mol Fe}_2(SO_4)_3$

$4.0 \text{ mol Fe(OH)}_3 \times \dfrac{6 \text{ mol } H_2O}{2 \text{ mol Fe(OH)}_3} = 12.0 \text{ mol } H_2O$

17. (a) $Cu + 2AgNO_3 \rightarrow Cu(NO_3)_2 + 2Ag$

$2.50 \text{ mol Cu} \times \dfrac{2 \text{ mol AgNO}_3}{1 \text{ mol Cu}} = 5.00 \text{ mol AgNO}_3$

$5.50 \text{ mol AgNO}_3 \times \dfrac{1 \text{ mol Cu}}{2 \text{ mol AgNO}_3} = 2.75 \text{ mol Cu}$

The Cu is thus the limiting reactant.

(b) 5.50 mol $AgNO_3$ available − 5.00 mol $AgNO_3$ needed = 0.50 mol excess $AgNO_3$

(c) $2.50 \text{ mol Cu} \times \dfrac{1 \text{ mol Cu(NO}_3)_2}{1 \text{ mol Cu}} = 2.50 \text{ mol Cu(NO}_3)_2$

$2.50 \text{ mol Cu} \times \dfrac{2 \text{ mol Ag}}{1 \text{ mol Cu}} = 5.00 \text{ mol Ag}$

(d) $2.50 \text{ mol Cu(NO}_3)_2 \times \dfrac{188 \text{ g Cu(NO}_3)_2}{\text{mol Cu(NO}_3)_2} = 470 \text{ g Cu(NO}_3)_2$

$5.00 \text{ mol Ag} \times \dfrac{108 \text{ g Ag}}{\text{mol Ag}} = 540. \text{ g Ag}$

18. (a) $\% \text{ yield} = \dfrac{\text{actual yield}}{\text{theoretical yield}} \times 100\% = \dfrac{15.0 \text{ g}}{20.0 \text{ g}} \times 100\% = 75.0\% \text{ yield}$

(b) $\% \text{ yield} = \dfrac{22.5 \text{ g}}{24.0 \text{ g}} \times 100\% = 93.8\% \text{ yield}$

(c) $\% \text{ yield} = \dfrac{4.75 \text{ g}}{5.00 \text{ g}} \times 100\% = 95.0\% \text{ yield}$

19. (a) $\% \text{ yield} = \dfrac{\text{actual yield}}{\text{theoretical yield}} \times 100\%$

$\text{actual yield} = \dfrac{\text{theoretical yield} \times \% \text{ yield}}{100 \%}$

$\text{actual yield} = \dfrac{(12.0 \text{ g})(90.0\%)}{100 \%} = 10.8 \text{ g}$

(b) $\text{actual yield} = \dfrac{(8.50 \text{ g})(70.0\%)}{100 \%} = 5.95 \text{ g}$

(c) $\text{actual yield} = \dfrac{(3.45 \text{ g})(48.0\%)}{100 \%} = 1.66 \text{ g}$

20. $CaCO_3(s) \rightarrow CaO(s) + CO_2(g)$

$25.4 \text{ g CaCO}_3 \times \dfrac{\text{mol CaCO}_3}{100 \text{ g CaCO}_3} \times \dfrac{1 \text{ mol CaO}}{\text{mol CaCO}_3} \times \dfrac{56.0 \text{ g CaO}}{\text{mol CaO}} = 14.2 \text{ g CaO}$ theoretical yield

$\% \text{ yield} = \dfrac{13.2 \text{ g}}{14.2 \text{ g}} \times 100\% = 93.0\% \text{ yield}$

Application Questions

1. (a) Ideal conditions cannot account for certain factors not revealed by chemical equations that can affect the relative amounts of reactants or products actually involved in chemical reactions. Ideal conditions refer to conditions in which complete conversion of all reactants into products takes place. (b) Ideal stoichiometry calculations indicate the maximum amount of product possible before a reaction is carried out in the laboratory.
2. Excess reactants are, typically, those materials that are most readily obtainable and/or least expensive.
3. Actual yields are often less than theoretical yields because: reactions sometimes do not go to completion, impure reactants produce less product than expected, competing side reactions may consume some of the reactants, and some are lost when products collected in impure form are purified.
4. The efficiency of a reaction is expressed in terms of percent yield

$$\frac{\text{actual yield}}{\text{theoretical yield}} \times 100\%$$

Application Problems

1. (a) $4Fe + 3O_2 \rightarrow 2Fe_2O_3$

   (b) $1.50 \text{ mol Fe} \times \dfrac{3 \text{ mol } O_2}{4 \text{ mol Fe}} = 1.12 \text{ mol } O_2$

   (c) $1.50 \text{ mol Fe} \times \dfrac{2 \text{ mol } Fe_2O_3}{4 \text{ mol Fe}} = 0.750 \text{ mol } Fe_2O_3$

2. $AgNO_3 + NaBr \rightarrow AgBr + NaNO_3$

   (a) $4.50 \text{ mol } AgNO_3 \times \dfrac{1 \text{ mol Na Br}}{\text{mol } AgNO_3} \times \dfrac{103 \text{ g NaBr}}{\text{mol NaBr}}$
   $= 464 \text{ g AgBr}$

   (b) $4.50 \text{ mol } AgNO_3 \times \dfrac{1 \text{ mol } NaNO_3}{\text{mol } AgNO_3} \times \dfrac{85.0 \text{ g } NaNO_3}{\text{mol } NaNO_3}$
   $= 382 \text{ g } NaNO_3$

3. $3CCl_4(\ell) + 2SbF_3(s) \rightarrow 3CCl_2F_2(\ell) + 2SbCl_3(s)$

   (a) $15.0 \text{ mol } CCl_2F_2 \times \dfrac{3 \text{ mol } CCl_4}{3 \text{ mol } CCl_2F_2} \times \dfrac{154 \text{ g } CCl_4}{\text{mol } CCl_4}$
   $= 2.31 \times 10^3 \text{ g } CCl_4$

   $15.0 \text{ mol } CCl_2F_2 \times \dfrac{2 \text{ mol } SbF_3}{3 \text{ mol } CCl_2F_2} \times \dfrac{179 \text{ g } SbF_3}{\text{mol } SbF_3}$
   $= 1.79 \times 10^3 \text{ g } SbF_3$

   (b) $15.0 \text{ mol } CCl_2F_2 \times \dfrac{2 \text{ mol } SbCl_3}{3 \text{ mol } CCl_2F_2} \times \dfrac{228 \text{ g } SbCl_3}{\text{mol } SbCl_3}$
   $= 2.28 \times 10^3 \text{ g } SbCl_3$

4. $2C_7H_6O_3(S) + C_4H_6O_3(\ell) \rightarrow 2C_9H_8O_4(s) + H_2O(\ell)$

   (a) $75.0 \text{ mol } C_7H_6O_3 \times \dfrac{2 \text{ mol } C_9H_8O_4}{2 \text{ mol } C_7H_6O_3} \times \dfrac{180.\text{g } C_9H_8O_4}{\text{mol } C_9H_8O_4} \times$
   $\dfrac{Kg \times C_9H_8O_4}{10^3 g \times C_9H_8O_4} = 13.5 \text{ kg } C_9H_8O_4$

   (b) $75.0 \text{ mol } C_7H_6O_3 \times \dfrac{1 \text{ mol } C_4H_6O_3}{2 \text{ mol } C_7H_6O_3} \times \dfrac{10^2 \text{ g } C_4H_6O_3}{\text{mol } C_4H_6O_3} \times$
   $\dfrac{kg \times C_4H_6O_4}{10^3 g \times C_4H_6O_3} = 3.82 \text{ kg } C_4H_6O_3$

   (c) $75.0 \text{ mol } C_7H_6O_3 \times \dfrac{1 \text{ mol } H_2O}{2 \text{ mol } C_7H_6O_3} \times \dfrac{18.0 \text{ g } H_2O}{\text{mol } H_2O} \times$

$\dfrac{\text{mL } H_2O}{\text{g } H_2O} \times \dfrac{\text{L } H_2O}{10^3 \text{ mL } H_2O} = 0.675 \text{ L } H_2O$

5. $H_2SO_4 + 2NaHCO_3 \rightarrow 2CO_2 + Na_2SO_4 + 2H_2O$

   (a) $150.0 \text{ g } H_2SO_4 \times \dfrac{\text{mol } H_2SO_4}{98.08 \text{ g } H_2SO_4} \times \dfrac{2 \text{ mol } NaHCO_3}{\text{mol } H_2SO_4}$
   $= 3.059 \text{ mol } NaHCO_3$

   (b) $3.059 \text{ mol } NaHCO_3 \times \dfrac{2 \text{ mol } CO_2}{2 \text{ mol } NaHCO_3} = 3.059 \text{ mol } CO_2$

   $3.059 \text{ mol } NaHCO_3 \times \dfrac{1 \text{ mol } Na_2SO_4}{2 \text{ mol } NaHCO_3} = 1.530 \text{ mol } Na_2SO_4$

   $3.059 \text{ mol } NaHCO_3 \times \dfrac{2 \text{ mol } H_2O}{2 \text{ mol } NaHCO_3} = 3.059 \text{ mol } H_2O$

6. (a) $4HF + SiO_2 \rightarrow SiF_4 + 2H_2O$

   $45.0 \text{ g HF} \times \dfrac{\text{mol HF}}{20.0 \text{ g HF}} \times \dfrac{1 \text{ mol } SiO_2}{4 \text{ mol HF}} = 0.562 \text{ mol } SiO_2$

   (b) $0.562 \text{ mol } SiO_2 \times \dfrac{1 \text{ mol } SiF_4}{\text{mol } SiO_2} = 0.562 \text{ mol } SiF_4$

   $0.562 \text{ mol } SiO_2 \times \dfrac{2 \text{ mol } H_2O}{\text{mol } SiO_2} = 1.12 \text{ mol } H_2O$

7. $C_{12}H_{25}OH + H_2SO_4 + NaOH \rightarrow C_{12}H_{25}OSO_3Na + 2H_2O$

   $50.0 \text{ T } C_{12}H_{25}OSO_3Na \times \dfrac{10^3 \text{ kg}}{T} \times \dfrac{10^3 \text{ g}}{\text{kg}} \times$
   $\dfrac{\text{mol } C_{12}H_{25}OSO_3Na}{288 \text{ g } C_{12}H_{25}OSO_3Na} \times \dfrac{1 \text{ mol } C_{12}H_{25}OH}{\text{mol } C_{12}H_{25}OSO_3Na}$
   $= 1.74 \times 10^5 \text{ mol } C_{12}H_{25}OH$

   $1.74 \times 10^5 \text{ mol } C_{12}H_{25}OH \times \dfrac{1 \text{ mol } H_2SO_4}{\text{mol } C_{12}H_{25}OH}$
   $= 1.74 \times 10^5 \text{ mol } H_2SO_4$

   $1.74 \times 10^5 \text{ mol } C_{12}H_{25}OH \times \dfrac{1 \text{ mol NaOH}}{\text{mol } C_{12}H_{25}OH}$
   $= 1.74 \times 10^5 \text{ mol NaOH}$

8. (a) $NaHCO_3 + KHC_4H_4O_6 \rightarrow KNaC_4H_4O_6 + CO_2 + H_2O$

   $8.00 \text{ g } KHC_4H_4O_6 \times \dfrac{\text{mol } KHC_4H_4O_6}{188 \text{ g } KHC_4H_4O_6} \times \dfrac{1 \text{ mol } NaHCO_3}{\text{mol } KHC_4H_4O_6} \times$
   $\dfrac{84.0 \text{ g } NaHCO_3}{\text{mol } NaHCO_3} = 3.57 \text{ g } NaHCO_3$

   (v) $3.57 \text{ g } NaHCO_3 \times \dfrac{\text{mol } NaHCO_3}{84.0 \text{ g } NaHCO_3} \times$
   $\dfrac{1 \text{ mol } KNaC_4H_4O_6}{\text{mol } NaHCO_3} \times \dfrac{210. \text{ g } KNaC_4H_4O_6}{\text{mol } KNaC_4H_4O_6} = 8.92 \text{ g } KNaC_4H_4O_6$

   $3.57 \text{ g Na } HCO_3 \times \dfrac{\text{mol } NaHCO_3}{84.0 \text{ g } NaHCO_3} \times \dfrac{1 \text{ mol } CO_2}{\text{mol } NaHCO_3} \times$
   $\dfrac{44.0 \text{ g } CO_2}{\text{mol } CO_2} = 1.87 \text{ g } CO_2$

   $3.57 \text{ g Na } HCO_3 \times \dfrac{\text{mol } NaHCO_3}{84.0 \text{ g } NaHCO_3} \times \dfrac{1 \text{ mol } H_2O}{\text{mol } NaHCO_3} \times$
   $\dfrac{18.0 \text{ g } H_2O}{\text{mol } H_2O} = 0.765 \text{ g } H_2O$

9. $Na_2CO_3 + CaCO_3 + 6SiO_2 \rightarrow Na_2O\cdot CaO\cdot 6SiO_2 + 2CO_2$

$$2500. \text{ bottles} \times \frac{400.0 \text{ g}}{\text{bottle}} \times \frac{\text{mol } Na_2O\cdot CaO\cdot 6SiO_2}{478.6 \text{ g } Na_2O\cdot CaO\cdot 6SiO_2} \times \frac{1 \text{ mol } Na_2CO_3}{\text{mol } Na_2O\cdot CaO\cdot 6SiO_2} \times \frac{106.0 \text{ g } Na_2CO_3}{\text{mol } Na_2CO_3}$$
$$= 2.215 \times 10^5 \text{ g } Na_2CO_3$$

$$2.215 \times 10^5 \text{ g } Na_2CO_3 \times \frac{\text{mol } Na_2CO_3}{106.0 \text{ g } Na_2CO_3} \times \frac{1 \text{ mol } CaCO_3}{\text{mol } Na_2CO_3} \times \frac{100.1 \text{ g } CaCO_3}{\text{mol } CaCO_3} = 2.092 \times 10^5 \text{ g } CaCO_3$$

$$2.215 \times 10^5 \text{ g } Na_2CO_3 \times \frac{\text{mol } Na_2CO_3}{106.0 \text{ g } Na_2CO_3} \times \frac{6 \text{ mol } SiO_2}{\text{mol } Na_2CO_3} \times \frac{60.08 \text{ g } SiO_2}{\text{mol } SiO_2} = 7.533 \times 10^5 \text{ g } SiO_2$$

10. (a) $3H_2SO_4 + 2Al(OH)_3 \rightarrow Al_2(SO_4)_3 + 6H_2O$

$$30.0 \text{ g } H_2SO_4 \times \frac{\text{mol } H_2SO_4}{98.1 \text{ g } H_2SO_4} = 0.306 \text{ mol } H_2SO_4$$

$$25.0 \text{ g } Al(OH)_3 \times \frac{\text{mol } Al(OH)_3}{78.0 \text{ g } Al(OH)_3} = 0.321 \text{ mol } Al(OH)_3$$

$$0.306 \text{ mol } H_2SO_4 \times \frac{2 \text{ mol } Al(OH)_3}{3 \text{ mol } H_2SO_4} = 0.204 \text{ mol } Al(OH)_3$$

The limiting reactant is thus H_2SO_4.

(b) $0.321 \text{ mol } Al(OH)_3 - 0.204 \text{ mol } Al(OH)_3 \times \dfrac{78.0 \text{ g } Al(OH)_3}{\text{mol } Al(OH)_3} = 9.13 \text{ g excess } Al(OH)_3$

(c) $0.306 \text{ mol } H_2SO_4 \times \dfrac{1 \text{ mol } Al_2(SO_4)_3}{3 \text{ mol } H_2SO_4} \times \dfrac{342 \text{ g } Al_2(SO_4)_3}{\text{mol } Al_2(SO_4)_3} = 34.9 \text{ g } Al_2(SO_4)_3$

$$0.306 \text{ mol } H_2SO_4 \times \frac{6 \text{ mol } H_2O}{3 \text{ mol } H_2SO_4} \times \frac{18.0 \text{ g } H_2O}{\text{mol } H_2O} = 11.0 \text{ g } H_2O$$

11. (a) $2N_2H_4 + (CH_3)_2N_2H_2 + 3N_2O_4 \rightarrow 6N_2 + 2CO_2 + 8H_2O$

$$1200. \text{ kg } N_2H_4 \times \frac{10^3 \text{ g } N_2H_4}{\text{kg } N_2H_4} \times \frac{\text{mol } N_2H_4}{32.04 \text{ g } N_2H_4} = 3.745 \times 10^4 \text{ mol } N_2H_4$$

$$1000. \text{ kg } (CH_3)_2N_2H_2 \times \frac{10^3 \text{ g } (CH_3)_2N_2H_2}{\text{kg } (CH_3)_2N_2H_2} \times \frac{\text{mol } (CH_3)_2N_2H_2}{60.10 \text{ g } (CH_3)_2N_2H_2} = 1.664 \times 10^4 \text{ mol } (CH_3)_2N_2H_2$$

$$4500. \text{ kg } N_2O_4 \times \frac{10^3 \text{ g } N_2O_4}{\text{kg } N_2O_4} \times \frac{\text{mol } N_2O_4}{92.01 \text{ g } N_2O_4} = 4.891 \times 10^4 \text{ mol } N_2O_4$$

$$3.745 \times 10^4 \text{ mol } N_2H_4 \times \frac{1 \text{ mol } (CH_3)_2N_2H_2}{2 \text{ mol } N_2H_4} = 1.872 \times 10^4 \text{ mol } (CH_3)_2H_2H_2$$

The compound $(CH_3)_2N_2H_2$ *seems* to limit this reaction thus far.

$$1.664 \times 10^4 \text{ mol } (CH_3)_2N_2H_2 \times \frac{3 \text{ mol } N_2O_4}{1 \text{ mol } (CH_3)_2N_2H_2} = 4.992 \times 10^4 \text{ mol } N_2O_4$$

The N_2O_4 is thus the actual limiting reactant. As such, it is the fuel component first used up.

(b) The amount of water thus produced is:

$$4.891 \times 10^4 \text{ mol } N_2O_4 \times \frac{8 \text{ mol } H_2O}{3 \text{ mol } N_2O_4} \times \frac{18.02 \text{ g } H_2O}{\text{mol } H_2O} \times \frac{\text{kg}}{10^3 \text{ g}} = 2.350 \times 10^3 \text{ kg } H_2O$$

12. $5.00 \text{ kg } NH_3 \times \dfrac{10^3 \text{ g } NH^3}{\text{kg } NH_3} \times \dfrac{\text{mol } NH_3}{17.0 \text{ g } NH_3} \times \dfrac{4 \text{ mol } NO}{4 \text{ mol } NH_3} \times 80\% = 235 \text{ mol } NO$

$$235 \text{ mol } NO \times \frac{2 \text{ mol } NO_2}{2 \text{ mol } NO} \times 80.0\% = 188 \text{ mol } NO_2$$

$$188 \text{ mol } NO_2 \times \frac{2 \text{ mol } HNO_3}{3 \text{ mol } NO_2} \times \frac{63.0 \text{ g } HNO_3}{\text{mol } NHO_3} \times 80.0\% = 6.32 \times 10^3 \text{ g } HNO_3$$

Teacher's Notes

Chapter 9 Stoichiometry

INTRODUCTION

Chapter 8 introduced you to writing chemical equations and many different types of chemical reactions. In this chapter you will explore some ways that chemists use chemical equations. The importance of correctly balancing all chemical equations should become evident.

LOOKING AHEAD

This chapter is concerned mainly with problem solving. Developing a systematic approach to reaction-stoichiometry problems is the key to success. As you work through the problems, try to develop such a system for yourself. Keep in mind that these types of problems are based on real chemical reactions and are used frequently in chemical experiments and manufacturing.

SECTION PREVIEW

9.1 Introduction to Stoichiometry
9.2 Ideal Stoichiometric Calculations
9.3 Limiting Reactions and Percent Yield

This controller-operator in a large chemical plant is adjusting the flow rate of compounds to produce the correct chemical mixture in the desired amount. The chemical manufacturing industry is dependent upon the quantitative principles of stoichiometry.

9.1 Introduction to Stoichiometry

Much of our knowledge of chemistry is based on the careful quantitative analyses of substances involved in chemical reactions. *The branch of chemistry that deals with the mass relationships of elements in compounds and the mass relationships between reactants and products in chemical reactions is known as* **stoichiometry.** The word *stoichiometry* comes from the Greek words *stoichion,* which means element, and *metron,* which means measure.

Applications of Stoichiometry

You are already familiar with one aspect of stoichiometry: **Composition stoichiometry** *involves the mass relationships of elements in chemical compounds.* The calculations involving particle–mass relations in Chapter 3 are examples of composition stoichiometry.

Another aspect of stoichiometry, called reaction stoichiometry, is concerned with chemical reactions. **Reaction stoichiometry** *involves the mass relationships among reactants and products in a chemical reaction.* It is based on chemical equations and the law of conservation of matter. A chemical equation can provide much information. For example, chemical equations indicate the amounts of reactants needed to produce a given amount of product, or the amount of one reactant needed to react with a given amount of another reactant. All reaction-stoichiometry calculations require knowing the chemical equation for the reaction under study.

SECTION OBJECTIVES

- Define *stoichiometry* and distinguish between composition and reaction stoichiometry.
- Define *mole ratio* and describe its role in stoichiometry calculations.
- Give the mole ratio for any two substances in a chemical equation.
- Name the four types of reaction-stoichiometry calculations.

Emphasize that if students know the amount of one substance in a reaction, they can determine the amounts of all of the other substances. They might be given one reactant and be asked to determine the amount of a second reactant needed to complete the reaction, or they might be given the amount of a product and be asked to determine the amount of reactant that is needed to produce it.

Figure 9-1 Calcium carbonate, when heated, decomposes to calcium oxide.

Reaction-Stoichiometry Problems

All of the reaction-stoichiometry problems in this chapter can be classified as one of four types: mole–mole, mole–mass, mass–mole, or mass–mass problems. In each of these types, the given information is the amount of one or more substances involved in a chemical reaction. The amounts are usually expressed in moles or grams. Along with the chemical equation for the reaction, this information permits calculation of the amounts of the other substances involved in the reaction. As you will see in later chapters, you might be given the amount of reactants or products in units other than moles or grams. All stoichiometry problems are solved by converting the given units into moles or grams and then applying the methods described here.

Routes for solving reaction-stoichiometry problems Figure 9-2 shows the pathways or routes by which the four types of problems may be solved. The use of the conversion factors needed to convert between the blocks, namely the molar mass and the mole ratio, are discussed later in this section. Molar masses are needed for mass–mole or mole–mass conversions. Mole ratios are needed for mole–mole conversions between two substances in a chemical reaction.

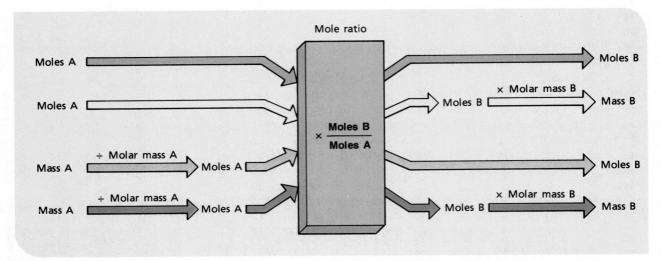

Figure 9-2 Solving the four types of reaction-stoichiometry problems

Remind students to refer to the equation to get the information needed to relate moles of B to moles of A.

1. In *mole–mole problems*, you are given the number of moles of one substance and asked to calculate the number of moles of another substance in the chemical reaction. According to Figure 9-2 the general route is

moles of A → moles of B

2. In *mole–mass problems*, you are given the number of moles of one substance and asked to calculate the mass of another substance in the chemical reaction. According to Figure 9-3, the general route is

moles of A → moles of B → mass of B

3. In *mass–mole problems*, you are given the mass of one substance and asked to calculate the number of moles of another substance in the chemical reaction. According to Figure 9-2 the general route is

mass of A → moles of A → moles of B

4. In *mass–mass problems*, you are given the mass of one substance and asked to calculate the mass of another substance in the chemical reaction. Figure 9-2, the general route is

mass of A → moles of A → moles of B → mass of B

Mole ratio As illustrated in Figure 9-2, each of the four types of reaction-stoichiometry problems requires the use of a mole ratio. A **mole ratio** *is a conversion factor that relates the number of moles of any two substances involved in a chemical reaction.* This information is obtained directly from the chemical equation.

Consider, for example, the chemical equation for the electrolysis of aluminum oxide to produce aluminum and oxygen:

$$2Al_2O_3(\ell) \rightarrow 4Al(s) + 3O_2(g)$$

Recall from Chapter 8 that the coefficients in a chemical equation satisfy the law of conservation of matter and represent the relative numbers of moles of reactants and products. Therefore, 2 mol of aluminum oxide decompose to produce 4 mol of aluminum metal and 3 mol of oxygen gas. These relationships can be expressed in the following mole ratios:

$$\frac{2 \text{ mol } Al_2O_3}{4 \text{ mol } Al} \quad \text{or} \quad \frac{4 \text{ mol } Al}{2 \text{ mol } Al_2O_3}$$

$$\frac{2 \text{ mol } Al_2O_3}{3 \text{ mol } O_2} \quad \text{or} \quad \frac{3 \text{ mol } O_2}{2 \text{ mol } Al_2O_3}$$

$$\frac{4 \text{ mol } Al}{3 \text{ mol } O_2} \quad \text{or} \quad \frac{3 \text{ mol } O_2}{4 \text{ mol } Al}$$

For the decomposition of aluminum oxide, the appropriate mole ratio would be used as a conversion factor to convert a given number of moles of one substance to the corresponding number of moles of another substance. To determine the number of moles of aluminum that can be produced from 13.0 mol of aluminum oxide, the mole ratio needed is that of Al to Al_2O_3.

$$\text{mol } Al_2O_3 \times \frac{\text{mol } Al}{\text{mol } Al_2O_3} = \text{mol } Al$$

$$13.0 \text{ mol } Al_2O_3 \times \frac{4 \text{ mol } Al}{2 \text{ mol } Al_2O_3} = 26.0 \text{ mol } Al$$

Figure 9-3 Industrial production of aluminum via the electrolysis of aluminum oxide

Chemistry Notebook

Scientific Journals: Getting the Word Out

To stay up-to-date on any topic, from racing cars to antique furniture, people read magazines that cover their fields of interest. This is also true for scientists, whose specialty periodicals are called journals. Very recent scientific findings are not contained in books, but in journal articles.

Since there are thousands of journals, a scientist often uses a special index to find articles about a subject. The most widely used index in chemistry is *Chemical Abstracts*, a weekly publication that provides short summaries of articles recently published in all the major disciplines of chemistry. Computers can also be used to search the enormous number of cross-referenced entries in *Chemical Abstracts*.

Scientists write journal articles to gain recognition from their peers in the scientific community and establish that they were the first to find a particular result. Scientists read articles written by other scientists to keep

Chemical Abstracts

abreast of the latest developments in areas of research. For instance, a scientist in the United States working on heat-resistant polymers might read a recent article written by a German scientist who is working in the same field. The United States scientist might gain fresh ideas from the German scientist's work.

Also, the questions and comments that often follow the publication of a journal article are helpful to the author and the readers. Finally, a scientist's results must be verified by others before they are accepted, and this can happen only if others find out about the work and try to duplicate the results.

Because mole ratios are exact, they do not limit the number of significant figures in a calculation. The number of significant figures in the answer is therefore determined only by the number of significant figures of any measured quantities in a particular problem.

Molar mass   Recall from Chapter 7 that the molar mass is the mass, in grams, of one mole of a substance. The molar mass is the conversion factor that relates the mass of a substance to the number of moles of that substance, as illustrated in Figure 9-2. To solve mass–mole problems, mole–mass problems, and mass–mass problems, the appropriate molar masses must therefore be found.

Returning to the previous example, the decomposition of aluminum oxide, what are the molar masses of the substances involved in this reaction? Written as conversion factors, these molar masses are

$$\frac{102.0 \text{ g Al}_2\text{O}_3}{1 \text{ mol Al}_2\text{O}_3} \quad \text{or} \quad \frac{1 \text{ mol Al}_2\text{O}_3}{102.0 \text{ g Al}_2\text{O}_3}$$

$$\frac{27.0 \text{ g Al}}{1 \text{ mol Al}} \quad \text{or} \quad \frac{1 \text{ mol Al}}{27.0 \text{ g Al}} \quad \frac{32.0 \text{ g O}_2}{1 \text{ mol O}_2} \quad \text{or} \quad \frac{1 \text{ mol O}_2}{32.0 \text{ g O}_2}$$

For example, to find the number of grams of aluminum equivalent to 12.0 mol of aluminum, the calculation would be as follows:

$$\text{moles Al} \times \frac{\text{grams Al}}{1 \text{ mol Al}} = \text{grams Al}$$

$$12.0 \text{ mol Al} \times \frac{27.0 \text{ g Al}}{1 \text{ mol Al}} = 324 \text{ g Al}$$

Section Review

1. What is stoichiometry?
2. List the four basic types of reaction-stoichiometry problems.
3. What is the role of mole ratio in reaction stoichiometry problems?
4. For each of the following chemical equations, write all possible mole ratios: (a) $2\text{HgO(s)} \rightarrow 2\text{Hg}(\ell) + \text{O}_2\text{(g)}$ (b) $4\text{NH}_3\text{(g)} + 6\text{NO(g)} \rightarrow 5\text{N}_2\text{(g)} + 6\text{H}_2\text{O}(\ell)$ (c) $2\text{Al(s)} + 3\text{H}_2\text{SO}_4\text{(aq)} \rightarrow \text{Al}_2(\text{SO}_4)_3\text{(aq)} + 3\text{H}_2\text{(g)}$.

1. The branch of chemistry that deals with the mass relationships of elements in compounds, and the mass relationships among reactants and products in chemical reactions
2. Mole-mole, mole-mass, mass-mole, and mass-mass problems
3. Mole ratio is used as a conversion factor to convert a given number of moles of one substance to the corresponding number of moles of another substance.
4. (a) Refer to p. 000 for illustration of mole ratios.

9.2 Ideal Stoichiometric Calculations

In the previous section, four types of reaction-stoichiometry calculations were introduced—mole–mole, mole–mass, mass–mole, and mass–mass calculations. The chemical equation plays a very important part in these calculations, since the mole ratio is obtained directly from it.

Chemical equations help us understand and make predictions about chemical reactions without having to actually work the reactions in the laboratory. Chemical equations do have certain limitations, however. They do not reveal the conditions under which reactions actually occur. Furthermore, as you learned in Chapter 8, chemical equations can be written to describe chemical changes that do not occur. Finally, chemical equations do not tell us how fast reactions occur or the pathways of the reactions.

Factors not revealed by chemical equations can also affect the relative amounts of reactants needed or products actually produced in chemical reactions. These limitations are not taken into account in this section. The reaction-stoichiometry calculations illustrated here are therefore theoretical. They tell us the amounts of reactants needed and the products that are yielded for a given chemical reaction under ideal conditions. By *ideal conditions,* we refer to the complete conversion of all reactants into products. Rarely are

SECTION OBJECTIVES

• Calculate the amount in moles of a reactant or product, given the amount in moles of a different reactant or product.

• Calculate the mass of a reactant or product, given the amount in moles of a different reactant or product.

• Calculate the amount in moles of a reactant or product, given the mass of a different reactant or product.

• Calculate the mass of a reactant or product, given the mass of a different reactant or product.

ideal conditions met in the laboratory or in industry. Theoretical stoichiometry calculations serve a very important function nonetheless. For example, it is useful to know the maximum amount of product possible before a reaction is carried out in the laboratory. Some stoichiometry calculations take into account actual laboratory and industrial conditions to make a more accurate determination of the quantities involved in a reaction.

Solving stoichiometry problems requires practice. Please do not think of the reaction stoichiometry problems presented here as four new types of problems to solve. See them instead as extensions of the composition stoichiometry problems you solved in Chapters 3 and 7. Practice working the sample problems in the rest of this section. You will discover that a logical, systematic approach to solving such problems can be quite successful.

Mole–Mole Calculations

In mole–mole stoichiometry problems, you are asked to calculate the number of moles of one substance that will react with or be produced from another substance. According to Figure 9-2, the route for a mole–mole problem is

$$\text{moles of substance A} \rightarrow \text{moles of substance B}$$

This route requires only one conversion factor—the mole ratio of B to A. To solve this type of problem, simply multiply the *known* quantity (the number of moles of A) by the appropriate conversion factor, which in this case is the mole ratio B/A.

$$\text{given information} \times \text{conversion factor} = \text{unknown information}$$

$$\text{moles A} \times \text{mole ratio of } \frac{\text{A}}{\text{B}} = \text{moles B}$$

Figure 9-4 Lithium hydroxide is used in spacecraft to chemically remove carbon dioxide from the atmosphere. Here, astronaut Jack R. Lousma opens the stowage area for LiOH canisters under a deck of the space shuttle Columbia.

Sample Problem 9.1

In a spacecraft, the carbon dioxide exhaled by astronauts can be removed by its reaction with lithium hydroxide (LiOH) according to the following chemical equation

$$CO_2(g) + 2LiOH(s) \rightarrow Li_2CO_3(s) + H_2O(\ell)$$

How many moles of lithium hydroxide are required to react with 20 mol of CO_2, the average amount exhaled by a person each day?

Solution

Step 1. Analyze
Given: chemical equation amount of CO_2 = 20 mol
Unknown: amount of LiOH in moles

Step 2. Plan
Only one conversion factor is needed—the mole ratio of LiOH to CO_2. This factor can be chosen from the given chemical equation

$$\text{moles of } CO_2 \rightarrow \text{moles of LiOH}$$

Step 3. Compute
Substitute the values in the equation in Step 2, cancel the units, and compute the answer.

$$20 \text{ mol } CO_2 \times \frac{2 \text{ mol LiOH}}{1 \text{ mol } CO_2} = 40 \text{ mol LiOH}$$

Step 4. Evaluate
The answer is correctly given to one significant figure as in the factor 20 mol CO_2. Cancellation of the units in the problem as set up leaves *mol LiOH* as the units for the answer. This is correct since the problem asked for the number of moles of LiOH required.

Practice Problems

1. Ammonia (NH_3) is widely used as a fertilizer and in many household cleaners. How many moles of ammonia are produced when 6 mol of hydrogen gas react with an excess of nitrogen gas? *(Ans.)* 4 mol NH_3
2. The decomposition of potassium chlorate ($KClO_3$) is used as a source of oxygen in the laboratory. How many moles of potassium chlorate are needed to produce 15 mol of oxygen? *(Ans.)* 10 mol $KClO_3$

Mole–Mass Calculations

In mole–mass stoichiometry problems, you are asked to calculate the mass (usually in grams) of a substance that will react with or be produced from a given number of moles of a second substance.

Referring back to Figure 9-3, the route for mole–mass problems is

$$\text{moles A} \rightarrow \text{moles B} \rightarrow \text{mass B}$$

This route requires two conversion factors — the mole ratio of subsance B to substance A and the molar mass of substance B. To solve this kind of problem, you simply multiply the known quantity, the number of moles of A, by the appropriate conversion factors.

$$\text{moles A} \times \text{mole ratio of } \frac{B}{A} \times \text{molar mass of B} = \text{mass of B}$$

STUDY HINT

At this point it would be useful to review balancing chemical equations, in Section 8.1.

When you are told that one reactant is in excess, you do not have to be concerned with its quantity. The reaction is limited by the limiting reactant.

Sample Problem 9.2

In photosynthesis, plants use energy from the sun to combine carbon dioxide and water, forming glucose ($C_6H_{12}O_6$) and oxygen. What mass, in grams, of glucose is produced when 3.00 mol of water react with carbon dioxide?

Solution

Step 1. Analyze   Given: names of reactants and products
                              amount of H_2O = 3.00 mol
                     Unknown: mass of $C_6H_{12}O_6$ produced in grams

Step 2. Plan   moles of $H_2O \rightarrow$ moles $C_6H_{12}O_6 \rightarrow$ mass $C_6H_{12}O_6$

To solve this problem requires writing the chemical equation and finding two conversion factors—the mole ratio of $C_6H_{12}O_6$ to H_2O and the molar mass of $C_6H_{12}O_6$.

$$\text{moles } H_2O \times \frac{\text{moles } C_6H_{12}O_6}{\text{moles } H_2O} \times \frac{\text{grams } C_6H_{12}O_6}{1 \text{ mole } C_6H_{12}O_6} = \text{grams } C_6H_{12}O_6$$

Step 3. Compute   Write the chemical equation for the reaction that occurs during the photosynthesis process.

$$6CO_2(g) + 6H_2O(\ell) \rightarrow C_6H_{12}O_6(s) + 6O_2(g)$$

$$3.00 \text{ mol } H_2O \times \frac{1 \text{ mol } C_6H_{12}O_6}{6 \text{ mol } H_2O} \times \frac{180 \text{ g } C_6H_{12}O_6}{1 \text{ mol } C_6H_{12}O_6} = 90.0 \text{ g } C_6H_{12}O_6$$

Step 4. Evaluate   The answer is correctly rounded off to three significant figures, as in 3.00 mol H_2O. Cancellation of the units in the problem as set up leaves "g $C_6H_{12}O_6$" as the units for the answer. This is correct since the problem asked how many grams of glucose are produced. The answer is reasonable, for it is one-half of 180, and 3/6 = 1/2.

Sample Problem 9.3

What mass, in grams, of carbon dioxide is needed to react with 3.00 mol of H_2O in the photosynthesis reaction described in Sample Problem 9.2?

Solution

Step 1. Analyze   Given: chemical equation
                              amount of H_2O = 3.00 moles
                     Unknown: amount of CO_2 needed in grams = ? g

Step 2. Plan   moles of $H_2O \rightarrow$ moles $CO_2 \rightarrow$ mass CO_2

Two conversion factors are needed—the mole ratio of CO_2 to H_2O and the molar mass of CO_2.

$$3.00 \text{ mol } H_2O \times \frac{6 \text{ mol } CO_2}{6 \text{ mol } H_2O} \times \frac{44.0 \text{ g } CO_2}{1 \text{ mol } CO_2} = 132 \text{ g } CO_2$$

Sample Problem 9.3 *continued*

Step 3. Compute	The chemical equation from Sample Problem 9.2 is $6CO_2(g) + 6H_2O(l) \rightarrow C_6H_{12}O_6(s) + 6O_2(g)$

$$3.00 \text{ mol } H_2O \times \frac{6 \text{ mol } CO_2}{6 \text{ mol } H_2O} \times \frac{44.0 \text{ g } CO_2}{1 \text{ mol } CO_2} = 132 \text{ g } CO_2$$

Step 4. Evaluate	The answer is correctly given to three significant figures as in both factors in the calculation. Cancellation of the units in the problem as set up leaves "g CO_2." This is correct since the problem asked for the amount needed, in grams, of CO_2. The answer is close to an estimate of 120 (3 × 40).

Practice Problems

1. When magnesium burns in air, it combines with oxygen to form magnesium oxide according to the following equation: $2Mg(s) + O_2(g) \rightarrow 2MgOs$. What mass in grams of magnesium oxide is produced from 2.00 mol of magnesium? *(Ans.)* 80.6 g MgO
2. What mass is grams of oxygen combines with 2.00 mol of magnesium in this same reaction? *(Ans.)* 32.0 g O_2

In Sample Problems 9.2 and 9.3, the conversion sequence was from moles of one substance to mass of a second substance. Next, some problems that require conversion from mass into moles are explained.

Mass—Mole Calculations

In mass–mole stoichiometry problems, you are asked to calculate the amount, in moles, of one substance that will react with or be produced from a given mass of another substance. Referring back to Figure 9-3, the route for a mass–mole problem is

$$\text{mass A} \rightarrow \text{moles A} \rightarrow \text{moles B}$$

This route also requires two conversion factors: the molar mass of substance A and the mole ratio of substance B to substance A. To solve this type of problem, simply multiply or divide the known quantity, the mass of A, by the appropriate conversion factors as follows.

$$\frac{\text{mass A}}{\text{molar mass of A}} \times \text{mole ratio of } \frac{B}{A} = \text{moles B}$$

Figure 9-5 The sun's energy drives a reaction in green plants that turns oxygen and water into glucose.

Sample Problem 9.4

The industrial solvent carbon disulfide (CS_2) is produced through the following reaction between coke (C) and sulfur dioxide (SO_2): $C(s) + SO_2 \rightarrow CS_2(l) + CO(g)$ (unbalanced) [If 8.00g of SO_2 reacts, *(a)* how many moles of CS_2 are formed and *(b)* how many moles of CO are formed?]

Solution

Step 1. Analyze	Given:	chemical equation
		mass of SO_2 = 8.00 g
	Unknown:	*(a)* amount of CS_2 produced, in moles
		(b) amount of CO produced, in moles

Step 2. Plan	*(a)* mass SO_2 → moles SO_2 → moles CS_2
	(b) mass SO_2 → moles SO_2 → moles CO

CS_2 to SO_2, and the mole ratio of CO to SO_2. To convert from mass of SO_2 to moles of SO_2, the SO_2 molar mass is used in its inverted form.

moles of SO_2, the SO_2 molar mass is used in its inverted form.

$$(a)\ \text{grams } SO_2 \times \frac{1\ \text{mol } SO_2}{\text{grams } SO_2} \times \frac{\text{moles } CS_2}{\text{moles } SO_2} = \text{moles } CS_2$$

$$(b)\ \text{moles } CO = \text{grams } SO_2 \times \frac{1\ \text{mol } SO_2}{\text{grams } SO_2} \times \frac{\text{moles } CO}{\text{moles } SO_2} = \text{moles } CO$$

Step 3. Compute   The balanced chemical equation is:

$$5C(s) + 2SO_2(g) \rightarrow CS_2(\ell) + 4CO(g)$$

$$(a)\ 8.00\ \text{g } SO_2 \times \frac{\text{mol } SO_2}{64.1\ \text{g } SO_2} \times \frac{1\ \text{mol } CS_2}{2\ \text{mol } SO_2} = 0.0624\ \text{mol } CS_2$$

$$(b)\ 8.00\ \text{g } SO_2 \times \frac{\text{mol } SO_2}{64.1\ \text{g } SO_2} \times \frac{4\ \text{mol } CO}{2\ \text{mol } SO_2} = 0.250\ \text{mol } CO$$

Step 4. Evaluate   The answers are correctly given to three significant figures. Cancellation of units in the two problems set up leaves *mol CS_2* and *mol CO*, respectively. This is correct, since the problem asked for the number of moles of CS_2 and CO.

Practice Problems

Oxygen was discovered by Joseph Priestly in 1774 when he decomposed mercury(II) oxide to its constituent elements by heating it.

1. How many moles of mercury(II) oxide (HgO) are needed to produce 125 g of oxygen (O_2)?
    (Ans.) 7.81 mol HgO
2. How many moles of mercury are produced? *(Ans.)* 7.81 mol Hg

Mass–Mass Calculations

In mass–mass stoichiometry problems, you are asked to calculate the number of grams of one substance that is required to react with or be produced from a given number of grams of a second substance involved in the chemical reaction. For this reason, mass–mass calculations involve a more real-life situation than do mole–mole, mole–mass, and mass–mole calculations. You can never measure moles directly. You are generally required to calculate the number of moles of a substance from its mass, which you can measure in the lab. Mass–mass problems can be viewed as the combination of individual mole–mole, mass–mole, or mole–mass problems. Examine Figure 9-3 to prove this to yourself. The route for solving mass–mass problems is

$$\text{mass A} \rightarrow \text{moles A} \rightarrow \text{moles B} \rightarrow \text{mass B}$$

Three conversion factors are needed to solve mass–mass problems: the molar mass of substance A, the mole ratio B/A, and the molar mass of B.

Sample Problem 9.5

Tin(II) fluoride (SnF_2), or stannous fluoride, is used in some home dental treatment products. It is made by the reaction of tin with hydrogen fluoride according to the equation.

$$Sn(s) + 2HF(g) \rightarrow SnF_2(s) + H_2(g)$$

How many grams of SnF_2 are produced from the reaction of 30.00 g of HF with Sn?

Solution

Step 1. Analyze   Given: chemical equation
amount of HF = 30.00 g
Unknown: amount of SnF_2 produced in grams

mass HF → moles HF → moles SnF_2 → mass SnF_2

Step 2. Plan

$$\text{grams HF} \times \frac{1 \text{ mol HF}}{\text{grams HF}} \times \frac{\text{moles SnF}_2}{\text{moles HF}} \times \frac{\text{grams SnF}_2}{\text{mol SnF}_2} = \text{grams SnF}_2$$

Step 3. Compute

$$30.00 \text{ g HF} \times \frac{1 \text{ mol HF}}{20.01 \text{ g HF}} \times \frac{1 \text{ mol SnF}_2}{2 \text{ mol HF}} \times \frac{156.7 \text{ g SnF}_2}{1 \text{ mol SnF}_2} = 117.5 \text{ g SnF}_2$$

Step 4. Evaluate   The answer is correctly rounded off to four significant figures. Cancellation of the units in the problem as set up correctly leaves *g SnF₂*. The answer is close to an estimated value of 120.

Practice Problems

Laughing gas (nitrous oxide; N_2O) is sometimes used as an anesthetic in dental work. It is produced when ammonium nitrate is decomposed according to the reaction $NH_4NO_3(s) \rightarrow N_2O(g) + H_2O(\ell)$.
1. How many grams of NH_4NO_3 are required to produce 33.0 g of N_2O?   *(Ans.)* 60.0 g NH_4NO_3
2. How many grams of water are produced in this reaction?   *(Ans.)* 27.0 g H_2O

Section Review

1. Balance the equation below, and then, based on the number of moles of each reactant or product given, determine the corresponding number of moles of all remaining reactants and products involved: $NH_3 + O_2 \rightarrow$ $N_2 + H_2O$.   *(a)* 4 mol NH_3 *(b)* 4 mol N_2 *(c)* 4.5 mol O_2
2. Hydrogen gas can be produced through the following unbalanced reaction: $Mg(s) + HCl(aq) \rightarrow MgCl_2(aq) + H_2(g)$.   *(a)* If 2.50 moles of magnesium react completely, balance the chemical equation involved.   *(b)* What mass of HCl is consumed?   *(c)* What mass of each product is produced?
3. Acetylene gas (C_2H_2) is produced as a result of the following reaction: $CaC_2(s) + H_2O(\ell) \rightarrow C_2H_2(g) + Ca(OH)_2(aq)$.   *(a)* If 32.0 grams of CaC_2 are consumed in this reaction, how many moles of H_2 are needed?   *(b)* How many moles of each product would be formed?
4. When sodium chloride reacts with silver nitrate, silver chloride precipitates. What mass of AgCl is produced from 75.0 g of $AgNo_3$?
5. Acetylene gas (C_2H_2), used in welding, produces an extremely hot flame when it burns in pure oxygen according to the reaction: $C_2H_2(g) + O_2(g) \rightarrow$ $CO_2(g) + H_2O(g)$. How many grams of each product are produced when 2.50×10^4 g of C_2H_2 burns completely?

1. $4NH_3 + 3O_2 \rightarrow 2N_2 +$ $6H_2O$ *(a)* 3 mol O_2, 4 mol N_2, 6 mol H_2O *(b)* 8 mol NH^3, 6 mol O_2, 12 mol H_2O *(c)* 6.0 mol NH_3, 3.0 mol N_2, 9.0 mol H_2O *(d)* 0.43 mol NH_3, 0.32 mol O_2, 0.22 mol N_2
2. *(a)* $Mg(s) + 2HCl(aq) \rightarrow$ $MgCl_2(aq) + H_2(g)$ *(b)* 182 g HCl   *(c)* 238 g $MgCl_2$, 5.05 g H_2
3. *(a)* 0.998 mol H_2O *(b)* 0.499 mol C_2H_2, 0.499 mol $Ca(OH)_2$
4. 63.1 g AgCl
5. 8.46×10^4 g CO_2, 1.73×10^4 g H_2O

9.3 Limiting Reactants and Percent Yield

In this section we are going to make the transition from theoretical stoichiometry calculations to real stoichiometry calculations. You will be introduced to two types of stoichiometry calculations that deal with situations common in practical chemistry. In the first, reactants are combined in amounts different from the precise amounts required for complete reaction. In the second, the quantity of product separated after a reaction is finished is less than the maximum quantity possible from the given amounts of reactants.

Limiting Reactant

All the problems that you have seen thus far involved calculations in which you sought to find the amount of a substance that will react with or be produced from a given amount of a second substance. However, in the laboratory a reaction is rarely carried out with exactly required amounts of each of the reactants. In most cases one or more reactants is present in excess; that is, in more than the exact amounts required to react with the given amount of the other reactant according to the chemical equation.

When all of one reactant is used up, no more product can be formed, even if there is more of other reactants available. The substance that is completely used up first in a reaction is called the limiting reactant. *The* **limiting reactant** *is the reactant that controls the amount of product formed in a chemical reaction. The substance that is not used up completely in a reaction is sometimes called the* **excess reactant**.

The concept of the limiting reactant is analogous to the relationship between the number of people that want to take a certain airplane flight and the number of seats available in the airplane. If there are 400 people that want to travel on the flight and only 350 seats are available, then only 350 people can go on the flight. The number of seats on the airplane limits the number of people that can travel. There are 50 people in excess.

You can apply the same reasoning to chemical reactions. Consider the reaction between carbon and oxygen to form carbon dioxide.

$$C(s) + O_2(g) \rightarrow CO_2(g)$$

According to the equation, one mole of carbon reacts with one mole of oxygen to form one mole of carbon dioxide. Suppose you could mix 5 mol of C with 10 mol of O_2 and allow the reaction to take place. There is twice as much oxygen as needed to react with the carbon. In this situation carbon is the limiting reactant and it limits how much CO_2 is formed. Oxygen here is the excess reactant, and 5 mol of O_2 will be unchanged at the end of the reaction. In real laboratory reactions, the reactant in excess is usually one that is easily obtainable and/or least expensive.

Given the amounts of each of two reactants, A and B, how can you decide which is the limiting reactant? You must discover which of the reactants requires more of the other than is available. The following method is recommended. Choose one of the two reactants; let's say you choose reactant

Desktop Investigation

Limiting Reactants in a Cookie Recipe

Question

What are the limiting reactants in a cookie recipe? The limiting reactant sets a limit on the amount of product that can be produced.

Materials

0.5 cup sugar
0.5 cup brown sugar
1 stick (0.5 cup) margarine
1 egg
0.5 tsp salt
1 tsp vanilla
0.5 tsp baking soda
1.5 cup flour
0.75 cup chocolate chips
mixing bowl
wooden spoon
measuring spoons and cups
cookie sheet

Procedure

1. Mix sugars and margarine together until smooth.

2. Add egg, salt, and vanilla and mix well.

3. Stir in baking soda, flour, and chocolate chips. Chill for best results.

4. Make dough into balls 3 cm in diameter; place on ungreased cookie sheet.

5. Bake at 350°F for about 10 minutes, until pale brown. Yields 24 chocolate-chip cookies.

Discussion

Suppose you have only the following materials: a dozen eggs, 4 fluid ounces of vanilla, a pound of salt, a pound of baking soda, 3 cups of chocolate chips, 5 pounds of flour, a pound of margarine, 5 pounds of sugar, and 2 pounds of brown sugar.

1. For each ingredient, calculate how many batches could be prepared if all of that ingredient were consumed. (For example, based on the 12 eggs available, at one egg per batch, 12 batches are possible.)

2. To determine the limiting reactants, identify the ingredient(s) that will produce the fewest batches.

3. How many batches can be produced from these materials?

A. Calculate the amount in moles of the other reactant, B, that is required by A. (This step is similar to solving a mole–mole or mass–mole problem.) Then compare the calculated amount with the amount of B actually available. If more is required than is available, B is the limiting reactant. If less is required, the reactant that you started with (reactant A) is the limiting reactant.

Finding which of two reactants is the limiting reactant is not difficult, but requires a bit of practice. The general method is illustrated in Sample Problems 9.6 to 9.7.

Sample Problem 9.7

Silicon dioxide (quartz) is usually quite stable but reacts with hydrogen fluoride according to the following equation.

$$SiO_2(s) + 4HF(g) \rightarrow SiF_4(g) + 2H_2O(\ell)$$

If 2.0 mol of HF is combined with 4.5 mol of SiO_2, which is the limiting reactant?

Solution

Step 1. Analyze   Given: chemical equation
amount of HF = 2.0 mol
amount of SiO_2 = 4.5 mol
Unknown: limiting reactant

Step 2. Plan   moles of HF → moles of SiO_2 or moles of SiO_2 → moles of HF

The given amount of either reactant is used to calculate the required amount of the other reactant. The calculated amount is then compared with the amount actually available, and the limiting reactant can be identified. We will choose to calculate the moles of SiO_2 required by the given amount of HF.

$$\text{moles HF} \times \frac{\text{moles SiO}_2}{\text{moles HF}} = \text{moles SiO}_2$$

Step 3. Compute $$2.0 \text{ mol HF} \times \frac{1 \text{ mol SiO}_2}{4 \text{ mol HF}} = 0.50 \text{ mol SiO}_2$$

Under ideal conditions, the given amount of HF will require 0.50 mol of SiO_2 for complete reaction. Since this is less than the amount available, the limiting reactant is not SiO_2, but HF.

Step 4. Evaluate   The calculated amount of SiO_2 is correctly given to two significant figures. Since each mole of SiO_2 requires 4 mol of HF, it is reasonable that HF is the limiting reactant, because the molar amount of HF available is less than half that of SiO_2.

Practice Problem

Some rocket engines use a mixture of hydrazine (N_2H_4) and hydrogen peroxide (H_2O_2) as the propellant system. The reaction is given by the equation

$$N_2H_4(\ell) + 2H_2O_2(\ell) \rightarrow N_2(g) + 4H_2O(g)$$

(a) Which is the limiting reactant in this reaction when 0.750 mol of N_2H_4 reacts with 0.500 mol of H_2O_2? *(Ans.)* H_2O_2 *(b)* How much of the excess reactant, in moles, remains unchanged? *(Ans.)* 0.500 mol N_2H_4 *(c)* How much of each product, in moles, is formed? *(Ans.)* 0.250 mol N_2, 1.00 mol H_2O

Sample Problem 9.7

The coating you see on a corroded iron object that has been left in moist soil is black iron oxide (Fe_3O_4). This substance can also be made in the laboratory by the reaction between iron and steam, according to the following equation:

$$3Fe(s) + 4H_2O(g) \rightarrow Fe_3O_4(s) + 4H_2(g)$$

Sample Problem 9.7 *continued*

(a) When 36.0 g of H_2O reacts with 167 g of Fe, which is the limiting reactant? How much of the excess reactant is not used when the reaction is completed?

(b) What mass in grams of black iron oxide is produced?

Solution

Step 1. *Analyze*

Given: chemical equation
mass of H_2O = 36.0 g
mass of Fe = 167.4 g
Unknown: limiting reactant
mass of Fe_3O_4, in grams

Step 2. *Plan*

$$\text{mass } H_2O \rightarrow \text{moles } H_2O; \text{ mass Fe} \rightarrow \text{moles Fe}$$

To determine which is limiting reactant

$$\text{moles } H_2O \rightarrow \text{moles Fe}$$
$$\text{moles limiting reactant} \rightarrow \text{moles } Fe_3O_4 \rightarrow \text{mass } Fe_3O_4$$

(a) It is necessary to convert both given masses to moles. Then, choose one to calculate the needed amount of the other and determine which is the limiting reactant. We have chosen H_2O.

$$\text{grams } H_2O \times \frac{1 \text{ mole } H_2O}{\text{grams } H_2O} = \text{moles } H_2O$$

$$\text{grams Fe} \times \frac{1 \text{ mole Fe}}{\text{grams Fe}} = \text{moles Fe}$$

$$\text{moles } H_2O \times \frac{\text{moles Fe}}{\text{moles } H_2O} = \text{moles Fe}$$

(b) To find the maximum amount of Fe_3O_4 that can be produced, the given number of moles of the limiting reactant must be used in what is simply a mole–mass stoichiometry problem

$$\text{moles limiting reactant} \times \frac{\text{moles } Fe_3O_4}{\text{moles limiting reactant}} \times \frac{\text{grams } Fe_3O_4}{1 \text{ mol } Fe_3O_4} = \text{grams } Fe_3O_4$$

Step 3. *Compute*

$$36.0 \text{ g } H_2O \times \frac{1 \text{ mol } H_2O}{18.0 \text{ g } H_2O} = 2.00 \text{ mol } H_2O$$

$$167 \text{ g Fe} \times \frac{1 \text{ mol Fe}}{55.8 \text{ g Fe}} = 3.00 \text{ mol Fe}$$

$$2.00 \text{ mol } H_2O \times \frac{3 \text{ mol Fe}}{4 \text{ mol } H_2O} = 1.50 \text{ mol Fe}$$

The required 1.50 mol of Fe is less than the 3.00 mol of Fe available, so H_2O is the limiting reactant.

$$2.00 \text{ mol } H_2O \times \frac{1 \text{ mol } Fe_3O_4}{4 \text{ mol } H_2O} \times \frac{231 \text{ g } Fe_3O_4}{1 \text{ mol } Fe_3O_4} = 116 \text{ g } Fe_3O_4$$

Step 4. *Evaluate*

Three significant digits have been correctly carried through each calculation. At each step, units have cancelled correctly. Both questions asked have been answered by identifying the limiting reactant and calculating the mass of product Fe_3O_4. The mass of Fe_3O_4 found in the last step is close to an estimated answer of 115, which is one-half of 230.

Practice Problems
1. Zinc and sulfur react to form zinc sulfide according to the following: $8Zn(s) + S_8(s) \rightarrow 8ZnS(s)$.
   (a) If 2.00 moles of Zn are heated with 1.00 mole of S_8, identify the limiting reactant. *(Ans.)* Zn
   (b) How many moles of excess reactant remain? *(Ans.)* 0.750 mol S_8 remains
   (c) How many moles of the product are formed? *(Ans.)* 2.00 mol ZnS
2. Carbon reacts with steam (H_2O) under certain conditions to produce hydrogen and carbon monoxide.   *(a)* If 2.40 mol of carbon react with 3.10 moles of steam, identify the limiting reactant. *(Ans.)* C   *(b)* How many moles of each product are formed? *(Ans.)* 2.40 mol H_2 and 2.40 mol CO   *(c)* What mass of each product is formed? *(Ans.)* 4.85 g H_2 and 67.2 g CO
3. Aluminim reacts with hydrochloric acid according to the following unbalanced equation: $Al + HCl \rightarrow AlCl_3 + H_2$.   *(a)* If 18 g of Al are combined with 75 g of HCl, which reactant is the limiting reactant? *(Ans.)* Al   *(b)* What mass of each product is formed? *(Ans.)* 89 g $AlCl_3$ and 2.0 g H_2

Percent Yield

The amounts of products calculated in the stoichiometry problems in this chapter so far represent theoretical yields. The **theoretical yield** *is the maximum amount of product that can be produced from a given amount of reactant*. In most chemical reactions, the amount of product obtained is less than the theoretical yield. *The measured amount of a product obtained from a reaction is called the* **actual yield** *of that product.*

Chemists like the one shown in Figure 9-6 strive to vary the conditions of a reaction so that the largest possible actual yield is obtained. There are many reasons why the actual yield may be less than the theoretical yield. Some of the reactant might take part in competing side reactions that reduce the amount of the desired product. Also once a product is formed it is usually

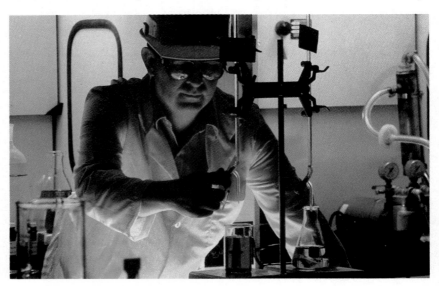

Figure 9-6   One goal of the chemist is to get the maximum possible yield from a reaction

collected in impure form. Often, during the purification process some of the product is lost.

Chemists are usually interested in the efficiency of a reaction. The efficiency is expressed by comparing the actual and theoretical yields. *The percent yield is the ratio of the actual yield to the theoretical yield, multiplied by 100 percent.*

$$\text{percent yield} = \frac{\text{actual yield}}{\text{theoretical yield}} \times 100\%$$

Sample Problem 9.8

Chlorobenzene (C_6H_5Cl) is used in the production of many important chemicals such as aspirin, dyes, and disinfectants. One industrial method of preparing chlorobenzene, used in the chemical industry, is the reaction between benzene (C_6H_6) and chlorine, represented by the equation

$$C_6H_6(\ell) + Cl_2(g) \rightarrow C_6H_5Cl(s) + HCl(g)$$

When 36.8 g of C_6H_6 reacted with an excess of Cl_2, the actual yield of C_6H_5Cl was 38.8 g. What was the percent yield of C_6H_5Cl?

Solution

Step 1. Analyze Given: chemical equation
            mass of C_6H_6 = 36.8 g
            mass of Cl_2 = excess
            actual yield of C_6H_5Cl = 38.8 g
        Unknown: percent yield of C_6H_5Cl

Step 2. Plan First it is necessary to do a mass–mass calculation.

mass $C_6H_6 \rightarrow$ moles $C_6H_6 \rightarrow$ moles $C_6H_5Cl \rightarrow$ mass $C_6H_5Cl \rightarrow$ percent yield

Find the theoretical yield of C_6H_5Cl.

$$\text{grams } C_6H_6 \times \frac{\text{mole } C_6H_6}{\text{grams } C_6H_6} \times \frac{\text{mole } C_6H_5Cl}{\text{mole } C_6H_6} \times \frac{\text{grams } C_6H_5Cl}{1 \text{ mole } C_6H_5Cl}$$
$$= \text{grams } C_6H_5Cl \text{ (theoretical yield)}$$

Then the percent yield can be found.

$$\text{percent yield } C_6H_5Cl = \frac{\text{actual yield}}{\text{theoretical yield}} \times 100\%$$

Step 3. Compute

$$36.8 \text{ g } C_6H_6 \times \frac{1 \text{ mol } C_6H_6}{78.1 \text{ g } C_6H_6} \times \frac{1 \text{ mol } C_6H_5Cl}{1 \text{ mol } C_6H_6} \times \frac{112.6 \text{ g } C_6H_5Cl}{1 \text{ mol } C_6H_5Cl}$$
$$= 53.1 \text{ g } C_6H_5Cl \text{ (theoretical yield)}$$

$$\text{percent yield} = \frac{38.8\text{g}}{53.1\text{g}} \times 100\% = 73.1\%$$

Step 4. Evaluate The answer is correctly rounded off to three significant figures, as in 78.1 g C_6H_6. The units have cancelled correctly. The theoretical yield is close to an estimated value of 50 g (one-half 100 g). The percent yield is close to an estimated value of 70% (70/100 × 100%).

Practice Problems

1. Methanol can be produced through the reaction of CO and H_2 in the presence of a catalyst:

$$CO(g) + 2H_2\ (g) \xrightarrow{\text{catalyst}} CH_3OH(\ell)$$

If 75.0 g of CO reacts to produce 68.4 g CH_3OH, what is the percent yeild of CH_3OH? *(Ans.)* 79.8%

2. Aluminum reacts with excess copper(II) sulfate according to the reaction given below. If 1.85 g of Al react and the percent yield of Cu is 56.6%, what mass of Cu is produced?

Al(s) + $CuSO_4$(aq) → $Al_2(SO_4)_3$(aq) + Cu(s) unbalanced *(Ans.)* 3.70 g

Section Review

1. *(a)* O_2 is the limiting reactant.
   (b) 0.667 mol CS_2 remains
   (c) 0.333 mol CO_2, 0.667 mol SO_2
2. *(a)* H_2O is the limiting reactant.   *(b)* 0.333 mol Mg remains   *(c)* 19.4 g $Mg(OH)_2$, 0.671 g H_2
3. *(a)* The CuO is the limiting reactant   *(b)* 15.9 g Cu
4. 93.8% yield

1. Carbon disulfide burns in oxygen to yield carbon dioxide and sulfur dioxide according to the following chemical equation: $\dot{C}S_2$(s) + $3O_2$(g) → CO_2(g) + $2SO_2$(g).   *(a)* If 1.00 mole of CS_2 is combined with 1.00 mole of O_2, identify the limiting reactant.   *(b)* How many moles of excess reactant remain?   *(c)* How many moles of each product are formed?

2. Metallic magnesium reacts with water to produce magnesium hydroxide and hydrogen gas.   *(a)* If 16.2 g of Mg are combined with 12.0 g of H_2O, what is the limiting reactant?   *(b)* How many moles of the excess reactant are left?   *(c)* How many grams of each product are formed?

3. *(a)* What is the limiting reactant when 19.9 g of CuO react with 2.02 g of H_2 according to the following: CuO + H_2 → Cu + H_2O?   *(b)* How many grams of Cu are produced?

4. Quicklime, CaO, can be prepared by roasting limestone, $CaCO_3$, according to the following reaction: $CaCO_3$(s) → CaO(s) + CO_2(g). When 2.00×10^3 g of $CaCO_3$ are heated, the actual yield of CaO is 1.05×10^3 g. What is the percent yield?

Chapter Summary

- Reaction stoichiometry involves the mass relationships between reactants and products in a chemical reaction.
- There are four types of reaction stoichiometry that may need to be determined in a chemical reaction. They are mole–mole, mole–mass, mass–mole or mass–mass. Each of the four relationships can be found using one of four problem-solving routes.
- A *mole ratio* is the conversion factor that relates the number of moles of any two substances in a chemical reaction. It is one of the quantities that must be known in order to determine any one of the possible relationships being considered.
- In actual reactions the reactants are combined in proportions different from the precise proportions required for complete reaction. Given certain quantities of reactants, the quantity of product is always less than the maximum possible.
- The limiting reactant controls the amount of product formed.
- Percent yield = actual yield ÷ theoretical yield × 100%.

Chapter 9 *Review*

Stoichiometry

Vocabulary

actual yield

composition
  stoichiometry

excess reactant

limiting reactant

mole ratio

percent yield

reaction stoichiometry

stoichiometry

theoretical yield

Questions

1. *(a)* Distinguish between composition stoichiometry and reaction stoichiometry. *(b)* Give one example of each type of calculation.
2. *(a)* Explain the concept of "mole ratio" as used in reaction stoichiometry problems. *(b)* What is the source of this value?
3. What information is given by the coefficients in a chemical equation?
4. For each of the following chemical equations, write all possible mole ratios: *(a)* $2Ca + O_2 \rightarrow 2CaO$ *(b)* $Mg + 2HF \rightarrow MgF_2 + H_2$
5. *(a)* What is molar mass? *(b)* What is its role in reaction stoichiometry?
6. Distinguish between ideal and real stoichiometry calculations.
7. *(a)* Distinguish between the limiting reactant and the excess reactant in a chemical reaction. *(b)* Which is used up?
8. *(a)* Distinguish between the theoretical and actual yields in stoichiometry calculations. *(b)* How do the values of the theoretical and actual yields generally compare?
9. What is meant by the percent yield of a reaction?

Problems

1. Given the chemical equation $Na_2CO_3 + Ca(OH)_2 \rightarrow 2NaOH + CaCO_3$, determine (to two decimal places), the molar masses of all substances involved, and then write these as conversion factors.
2. Hydrogen and oxygen react under a specific set of conditions to produce water according to the following: $2H_2(g) + O_2(g) \rightarrow 2H_2O(g)$.

(a) How many moles of hydrogen would be required in order to produce 5.0 mol of water? *(b)* How many moles of oxygen would be required?

3. Iron (III) hydroxide and sulfuric acid react through the following double replacement reaction: $Fe(OH)_3 + H_2SO_4 \rightarrow Fe_2(SO_4)_3 + H_2O$. *(a)* Balance the equation. *(b)* If 5.00 mol of $Fe(OH)_3$ react, how many moles of H_2SO_4 are required? *(c)* How many of each product are produced? *(d)* Convert each of the molar values in parts *b* and *c* to grams, and then compare the total mass of the reactants with that of the products. What conclusions can you draw?

4. *(a)* If 4.50 mol of ethane (C_2H_6) undergo combustion according to the reaction $C_2H_6 + O_2 \rightarrow CO_2 + H_2O$, how many moles of oxygen are required? *(b)* How many moles of each product are formed?

5. Sulfuric acid reacts with sodium hydroxide according to the following: $H_2SO_4 + NaOH \rightarrow Na_2SO_4 + H_2O$ for this reaction. *(a)* Balance the equation for this reaction. *(b)* What mass of H_2SO_4 would be required to react with 0.75 mol of NaOH? *(c)* What mass of each product is formed?

6. Sodium chloride is produced from its elements through composition. What mass of each reactant would be required to produce 25.0 mol of sodium chloride?

7. Copper reacts with silver nitrate through single replacement. *(a)* If 2.25 g of silver are produced from the reaction, how many moles of copper(II) nitrate are also produced? *(b)* How many moles of each reactant are required in this reaction?

8. Iron ore is generally produced through the following reaction in a blast furnace: $Fe_2O_3(s) + CO(g) \rightarrow Fe(s) + CO_2(g)$. *(a)* If 4.00 kg of Fe_2O_3 are available to react, how many moles of CO are needed? *(b)* How many moles of each product are formed?

9. As early as 1938, the use of NaOH was suggested as a means of removing CO_2 from the cabin atmosphere of a lunar-bound spacecraft according to the following reaction: NaOH + $CO_2 \rightarrow Na_2CO_3 + H_2O$. (a) If the average human body discharges 925.0 g of CO_2 per day, how many moles of NaOH would be needed each day? (b) How many moles of each product would be formed?

10. Aluminum metal reacts with aqueous zinc chloride through single replacement. (a) If 13.5 g of Al react, how many moles of $ZnCl_2$ would be required? (b) How many moles of each product would be formed?

11. Methanol (CH_3OH) is an important industrial compound that is produced from the following reaction: $CO(g) + H_2(g) \rightarrow CH_3OH(g)$. What mass of each reactant would be needed to produce 100.0 kg of methanol?

12. Ethyl alcohol (C_2H_5OH) can be produced by the fermentation of glucose ($C_6H_{12}O_6$) through the following reaction: $C_6H_{12}O_6 \rightarrow C_2H_5OH + CO_2$. If 25.0 kg of glucose react, what mass of each product is formed?

13. Nitrogen combines with oxygen in the atmosphere during lightning flashes to form nitrogen monoxide (NO) which then reacts further with O_2 to produce nitrogen dioxide, NO_2. (a) What mass of NO_2 is formed when NO reacts with 384 g of O_2? (b) How many grams of NO are required to react with this amount of O_2?

14. Given the reactant amounts specified in each chemical equation, determine the limiting reactant in each case:

(a) HCL   + NaOH $\rightarrow$ NaCl + H_2O
   2.0 mol   2.5 mol
(b) Zn   + 2HCl + $ZnCl_2$ + H_2
   2.5 mol   6.0 mol
(c) $2Fe(OH)_3 + 3H_2SO_4 \rightarrow Fe_2(SO_4)_3 + 6H_2O$
   4.0 mol   6.5 mol

15. For each reaction specified in Problem 16 above, determine the number of moles of excess reactant that remains.

16. For each reaction specified in Problem 16 above, calculate the number of moles of each product formed.

17. (a) If 2.50 mol of copper and 5.50 mol of silver nitrate are available to react by sing'.e replacement, identify the limiting reactant.

(b) Determine the number of moles of excess reactant remaining. (c) Determine the number of moles of each product formed.
(d) Determine the mass of each product formed.

18. From theoretical and actual yields of the various chemical reactions given below, calculate the percent yield for each:
(a) theoretical yield = 20.0 g, actual yield = 15.0 g (b) theoretical yield = 24.0 g, actual yield = 22.5 g (c) theoretical yield = 5.00 g, actual yield = 4.75 g.

19. From the theoretical and percentage yields given below, determine the actual yields:
(a) theoretical yield = 12.0 g, percent yield = 90.0% (b) theoretical yield = 8.50 g, percent yield = 70.0% (c) theoretical yield = 3.45 g, percent yield = 48.0%.

20. Calcium carbonate undergoes decomposition through the following reaction:
$$CaCO_3(s) \rightarrow CaO(s) + CO_2(g).$$
If 25.4 g of $CaCO_3$ react to produce 13.2 g of CaO, what is the percent yield of CaO?

Application Questions

1. (a) What is meant by "ideal conditions" relative to stoichiometry calculations? (b) What function do ideal stoichiometry calculations serve?

2. What characteristics are typically observed among those materials that serve as excess reactants in industrial process?

3. Why are actual yields generally less than those calculated theoretically?

4. How do chemists express the efficiency of a reaction?

Application Problems

1. Iron(III) oxide, or rust, is formed from a composition reaction involving its constituent element. (a) Write and balance the corresponding equation. (b) If 1.50 mol of pure iron nails rust completely, how many moles of oxygen are required? (c) How many moles of iron(III) oxide are produced?

2. The double-replacement reaction between silver nitrate and sodium bromide produces silver bromide, a component of photographic film. (a) If 4.50 moles of silver nitrate react, what mass of sodium bromide is required? (b) What mass of each product is formed?

3. The compound freon-12 (CCl_2F_2) is used as a coolant in refrigerators and air conditioners. It is produced through the following reaction: $CCl_4(\ell) + SbF_3(s) \rightarrow CCl_2F_2(\ell) + SbCl_3(s)$. (a) If 15.0 mol of freon-12 are to be produced, what mass of each reactant is required? (b) What mass of $SbCl_3$ is produced?

4. Aspirin ($C_9H_8O_4$) is produced through the following reaction of salicylic acid ($C_7H_6O_3$) and acetic anhydride ($C_4H_6O_3$): $C_7H_6O_3(s) + C_4H_6O_3(\ell) \rightarrow C_9H_8O_4(s) + H_2O(l)$. (a) What mass of aspirin (in kg) could be produced from 75.0 mol of salicylic acid? (b) What mass of acetic anhydride (in kg) would be required? (c) At room conditions, how many liters of water would be formed?

5. In a soda-acid fire extinguisher, concentrated sulfuric acid reacts with sodium hydrogen carbonate to produce carbon dioxide, sodium sulfate and water. (a) How many moles of sodium hydrogen carbonate would be needed to react with 150.0 g of sulfuric acid? (b) How many moles of each product would be formed?

6. Hydrofluoric acid, because it reacts with the silicon dioxide in glass, cannot be stored in glass bottles. (a) If 45.0 g of hydrofluoric acid react with silicon dioxide to produce silicon tetrafluoride and water, how many moles of silicon dioxide are required? (b) How many moles of each product are formed?

7. The detergent sodium dodecanesulfonate ($C_{12}H_{25}OSO_3Na$) is produced through the following reaction: $C_{12}H_{25}OH + H_2SO_4 + NaOH \rightarrow C_{12}H_{25}OSO_3Na + H_2O$. If 50.0 metric tons (1 metric ton = 10^3 kg) of the detergent must be produced, how many moles of each reactant are required?

8. Baking soda ($NaHCO_3$) reacts with cream of tartar ($KHC_4H_4O_6$), according to the reaction: $NaHCO_3 + KHC_4H_4O_6 \rightarrow KNaC_4H_4O_6 + CO_2 + H_2O$. (a) If a recipe calls for 2.00 teaspoons or 8.00 g of cream of tartar, how many grams of baking soda are needed? (b) What mass of each product is formed?

9. Soda-lime glass ($Na_2O \cdot CaO \cdot 6SiO_2$) is produced from the reaction $Na_2CO_3 + CaCO_3 + SiO_2 \rightarrow Na_2O \cdot CaO \cdot 6SiO_2 + CO_2$. If a winery needs 2500 soda-lime glass bottles with a mass of 400.0 g each, what mass of each reactant is needed?

10. Sulfuric acid reacts with aluminum hydroxide by double replacement. (a) If 30.0 g of sulfuric acid react with 25.0 g of aluminum hydroxide, identify the limiting reactant. (b) Determine the mass of excess reactant remaining. (c) Determine the mass of each product formed.

11. The energy used to power one of the Apollo lunar missions was supplied by the following overall reaction: $2N_2H_4 + (CH_3)_2 + 3N_2O_4 \rightarrow 6N_2 + 2CO_2 + 8H_2O$. For that phase of the mission when the lunar module ascended from the surface of the moon, a total of 1200 kg of N_2H_4 were available to react with 1000 kg of $(CH_3)_2N_2H_2$ and 4500 kg of N_2O_4. (a) For this portion of the flight, which of the allocated components was used up first? (b) How much water, in kilograms, was left in the lunar atmosphere through this reaction?

12. The Ostwald Process for producing nitric acid from ammonia consists of the following steps:
$$4NH_3(g) + 5O_2(g) \rightarrow 4NO(g) + 6H_2O(g)$$
$$2NO(g) + O_2(g) \rightarrow 2NO_2(g)$$
$$3NO_2(g) + H_2O(g) \rightarrow 2HNO_3(aq) + NO(g)$$
If the yield in each step is 80.0%, how many grams of nitric acid can be produced from 5.00 kg of ammonia?

Enrichment

1. Prepare a report describing the work of the alchemists. What were their major contributions to modern chemistry? What techniques and procedures were employed by these early chemists?

2. Visit the nearest water-treatment plant and investigate the various chemical reactions employed.

3. Visit any chemical production plants in your area, and prepare a report on the processes being used, the actual chemical reactions being conducted, and the products being made. If any of these materials are toxic, what precautions are being taken?

4. Write a report on the many uses of chemistry in criminal investigations. Describe the work of the forensic chemist.

5. Visit the nearest hospital laboratory and prepare a report on the major use of chemistry in the identification and subsequent control of disease-causing agents.

CAREERS

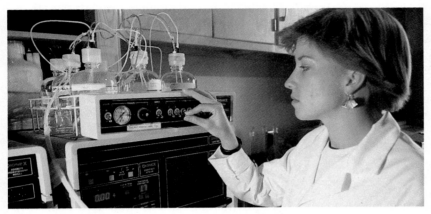

Analytical Pharmaceutical Chemist

Analytical pharmaceutical chemists ensure that the compounds used as medicines (called pharmaceuticals) are safe. This work starts with a study of the composition of the raw materials used. The chemist must look for a variety of impurities, such as halogenated hydrocarbons, arsenic, lead, and mercury.

Sensitive electronic equipment is used to detect trace impurities in medications. In the past, it was difficult to detect 1 g of impurity in a kg of material, but chemists can now detect a few parts per billion.

In recent years, drug products have been labeled with expiration dates that reflect how long a pharmaceutical takes to lose 10% of its potency. Although 90% of a pharmaceutical may remain active, the decomposition products could be toxic. Outdated pharmaceuticals must be discarded. A chemist conducts tests to determine how fast a product decomposes.

Analytical pharmaceutical chemists work for private companies, government agencies, or independent testing laboratories. Technical training through a junior college program is a minimum requirement for more advanced jobs, but most labs require a bachelor's degree in chemistry for such positions. Because working with electronic instrumentation is now part of the job, knowledge of computer programming and electronics is desirable.

Chemical Plant Manager

A plant manager at a chemical manufacturing company carries considerable responsibility. He or she oversees the ordering of materials, makes production schedules, and takes responsibility for plant safety and staff training.

Since the plant manager is ultimately responsible for everything that occurs at the plant, knowledge of chemistry is vital. The manager must be able to answer questions and make decisions quickly. For example, during a chemical process, what side reactions might occur? How might the waste materials be recycled into useful materials? What could go wrong during production? If a chemical spill should occur, how might it affect the environment?

The manager of a chemical plant must have a bachelor's degree in chemistry. Also, chemical manufacturing companies usually require their plant managers to have advanced training in business administration and personnel management.

Cumulative *Review*

Unit 3 • Language of Chemistry

Questions

1. Using only the Periodic Table, write the formula of the ions most typically formed from each of the following: *(a)* Na *(b)* F *(c)* Mg *(d)* Al *(e)* S *(f)* N *(g)* Cl *(h)* K. (7.1)

2. Name each of the following compounds. *(a)* CuCl *(b)* SO_2 *(c)* $AgNO_3$ *(d)* $(NH_4)_2S$ *(e)* SiO_2 *(f)* $Fe_2(SO_4)_3$ *(g)* PCl_3 *(h)* $Sn_3(PO_4)_2$ *(i)* N_2O_3 *(j)* H_2SO_4 *(k)* As_2O_3 (7.1)

3. Assign oxidation numbers to each atom in the following: *(a)* HCl *(b)* PI_3 *(c)* GeO_2 *(d)* As_2S_5 *(e)* H_2SO_4 *(f)* H_3PO_4 *(g)* NO_2^- *(h)* OH^- *(i)* SO_3 *(j)* PO_3^{3-} (7.2)

4. Define or explain each of the following: *(a)* formula mass *(b)* molar mass *(c)* percent composition of a compound *(d)* simplest formula. (7.3)

5. What information is provided by a chemical equation? (8.1)

6. Describe the five basic types of chemical reactions. (8.2)

7. *(a)* Explain what is meant by an activity series of elements. *(b)* Upon what basis are elements ordered in an activity series? *(c)* What function does an activity series serve? (8.3)

8. Define or explain *(a)* molar mass *(b)* percent yield of a reaction. (9.3)

Problems

1. Find the mass in grams of 6.75 mol of the compound $Ca_3(PO_4)_2$. (9.2)

2. *(a)* How many moles are contained in 12.00 g of lead(II) nitrate? *(b)* How many formula units does this mass represent? (9.2)

3. Find the percent composition of $Fe_2(CO_3)_3$. (7.3)

4. Determine the simplest formula of a compound found to contain 51.82% Cu, 19.60% C, 2.46% H, and 26.12% O. (7.4)

5. Determine the molecular formula of a compound consisting of 40.00% C, 6.71% H, and 53.29% O, with a formula mass of 180.1854 *u*. (7.4)

6. Write balanced chemical equations for each of the following and identify each reaction as to type: *(a)* K + HCL → KCl + H_2 *(b)* H_2 + F_2 → HF *(c)* $BaCO_3 \xrightarrow{\Delta} BaO + CO_2$ *(d)* Al + $CuSO_4$ → $Al_2(SO_4)_3$ + Cu *(e)* $Zn(OH)_2$ + H_3PO_4 → $Zn_3(PO_4)_2$ + H_2O *(f)* C_3H_8 + O_2 → CO_2 + H_2O. (8.1,2)

7. Complete and balance each of the following equations for reactions identified by type: *(a)* Synthesis: Ca + Cl_2 → *(b)* Decomposition: $Al(OH)_3 \xrightarrow{\Delta}$ *(c)* Single replacement: Ni + H_3PO_4 → *(d)* Double replacement: $Cu(C_2H_3O_2)_2$ + Li_3PO_4 → *(e)* Combustion: C_2H_2 + O_2 →. (8.1,2)

8. Use the activity series to predict whether each of the following reactions will occur, and write balanced chemical equation for this reaction. *(a)* Zn(s) + $Pb(NO_3)_2$ (aq) → *(b)* Au(s) + O_2(g) → *(c)* Zn(s) + H_2O(g) → *(d)* Zn(s) + H_2O(g) → *(e)* NiO(s) → (8.1,3)

9. Nitric acid reacts with aqueous calcium hydroxide solution by double replacement. *(a)* Write the balanced chemical equation for this reaction. *(b)* If 4.50 moles of nitric acid react, how many of calcium hydroxide are needed? *(c)* How many moles of each product are formed? (8.2, 9.3)

10. Bromine reacts with aluminum iodide by single replacement to produce 7.5000 moles of iodine. *(a)* What mass of each reactant is required? *(b)* What mass of the other product is formed? (8.2, 9.3)

11. If 40.0 grams of Na_2CO_3 undergo decomposition, how many moles of each product are formed? (8.2, 9.3)

12. If 100.0 kg of octane (C_8H_{18}) undergo combustion, what mass of each product is formed? (8.2, 9.3)

13. If 35.0 g of H_3PO_4 react with 65.0 g of $BaCl_2$ by double replacement, what is the limiting reactant? *(b)* How many moles of excess reactant are left? *(c)* What mass of each product is formed? (8.2, 9.3)

The Process in Chemistry: The Amazing Polyhexamethyleneadipamide

The impossible-to-pronounce word in the title above is the proper name of one of the most important discoveries of all time. A skier schussing down a mountain depends on it; so do fishermen casting their lines and drivers driving on the interstate. We now know this discovery as "nylon."

When nylon was in the early stages of development in the 1930s, it was called Fiber 66 or Rayon no. 66, but it soon became apparent that this fiber was so different that its name should be distinct from that of rayon. Suggestions for names included "neosheen," "pontella," and "lustrol." For two years, various committees wrestled with the name problem.

Meanwhile, lab research indicated that Fiber 66 would make women's hosiery that did not run ("no-run" stockings). To make pronunciation easier, "no-run" was changed to nylon, and the new fiber finally had a name.

Nylon is a product of polymerization, a chemical process in which relatively small molecules unite to form very large, chainlike molecules. Natural organic polymers include starch, cellulose, and proteins. Diamonds and graphite are natural inorganic polymers.

Nylon is one type of synthetic polyamide fiber made of four elements—carbon, hydrogen, nitrogen, and oxygen. It is produced by condensation polymerization, in which a diamine such as hexameth-

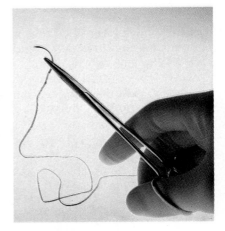

ylene diamine, $H_2N(CH_2)_6NH_2$, reacts with a dibasic acid, $HOOC(CH_2)_4COOH$. As the two molecules join, a molecule of water is formed. The reaction continues, linking additional molecules until a very long chain of polyamides is produced. Different types of nylon are produced by varying the acid and amine building blocks.

The Connection to Health Sciences

During World War II, nylon began replacing twisted silk thread for sutures in surgery, and finely woven nylon fabric served as filters for blood plasma. Nylon's strength and resistance to moisture, mildew, and impact made it ideal for use in parachutes and protective vests.

Today's hospitals have hundreds of nylon products. A plastic wheel, now standard on many wheelchairs, is not only lighter, but can take greater vertical or lateral impact than a wire-spoked wheel. Surgical-staple extractors of nylon resin are made to match the surgeon's hand. The resin provides

the stiffness needed for control and the elongation required for the springlike functioning of the tension arm.

The Connection to Physics

The physical properties of a polymer depend on both the characteristics of its building blocks and its total length. Long chains overlap more and have greater cohesive force than shorter chains. Nylons are strong, durable, elastic, have a low coefficient of friction, and retain these properties over a temperature range of $-100°–110°C$.

Nylon was synthesized first as a fiber, then as a plastic. The first molded nylon products, introduced in 1939, were textile machinery gears. This was a perfect opportunity to demonstrate nylon's potential as an engineering thermoplastic—the gears did not require lubrication. Today, car makers use nylon in many forms. Automotive timing gears, for ex-

ample, are coated with nylon to prolong their life. Because of its high resistance to hydrocarbons, nylon is used for emission-control canisters and is found in electrical connectors, wire jackets, and gears for windshield wipers and speedometers. Glass-reinforced nylons are employed in engine fans, radia-

tor headers, brake-fluid reservoirs, and valve covers.

The Connection to Other Sciences

Nylon's light weight, combined with its strength and durability, make it ideal for people who work outdoors. Packs, clothing, tarpaulins, and tents incorporate nylon fabric. Nylon lines are used by foresters who climb trees and by rock-climbing geologists.

Many field scientists use equipment made of nylon. Biologists use radio collars with nylon bands to track animals in their natural habitats. Ornithologists use mist nets made of nylon to capture birds. The mesh of the nets is very fine and blends in with the background. Flying birds become entangled in the nets and can be removed for weighing and banding, and for parasite studies.

Because it is hydroscopic (it has a propensity to attract water), nylon is often found in humidistats as the mechanism for moisture control. Because nylon resists fire, the electrical industry uses it for plugs, coil forms, writing devices, terminal blocks, wire jacketing, and antenna-mounting devices.

Unit

4

Chapter 10
Representative Gases   288

Chapter 11
Physical Characteristics of Gases   314

Chapter 12
Quantitative Behavior of Gases   342

Chapter 13
Liquids and Solids   370

PHASES OF MATTER

I majored in chemistry in college, but didn't know exactly what sort of chemistry I wanted to spend my life working on. A professor at my college made some laboratory measurements of a chemical reaction that is important to the atmosphere of Jupiter. I was fascinated to hear that chemistry doesn't just take place in test tubes, but in the atmosphere of an entire planet. From then on I was hooked on atmospheric chemistry.

I study the earth's atmosphere and try to understand the influence that human beings may have upon it. I'm especially interested in the depletion of the ozone layer. The polar stratospheric clouds you see at right are important in the chemistry of stratospheric ozone. My work involves experimental projects to gather data on the amount of ozone and other chemicals in the atmosphere, and theoretical studies in which computer models are used to simulate and interpret the data.

Understanding the chemistry of the gases, liquids, and solids of our atmosphere will help us to find ways of protecting it from damaging pollutants.

Susan Solomon
Atmospheric Chemist
National Oceanic and
*   Atmospheric Administration*

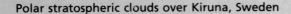

Polar stratospheric clouds over Kiruna, Sweden

Chapter 10 Representative Gases

Chapter Planner

10.1 Oxygen and Ozone
- ▢ Questions: 1–3

Occurrence of Oxygen
- ▪ Question: 4
- ▢ Application Problems: 5, 6

Structure of the Oxygen Molecule
- ▢ Demonstration: 1

Physical Properties of Oxygen
- ▢ Question: 5
- ▪ Application Questions: 1, 2
- ▢ Problem: 6

Chemical Properties of Oxygen
- ▢ Problem: 1
- ▢ Application Problem: 1
- ▪ Application Question: 3
- ▢ Application Question: 4

Uses and Preparation of Oxygen
- ▢ Demonstration: 2
- ▢ Questions: 7, 8
- ▢ Problem: 8

Formation and Properties of Ozone
- ▢ Application Question: 6
- ▢ Application Question: 5

10.2 Hydrogen

Occurrence of Hydrogen
- ▢ Question: 9
- ▢ Application Questions: 7, 8
- ▪ Question: 10
- ▢ Application Questions: 10

Physical Properties of Hydrogen
- ▢ Demonstration: 3

Chemical Properties of Hydrogen
- ▢ Question: 13
- ▢ Application Problem: 3

Uses and Preparation of Hydrogen
- ▢ Question: 14
- ▢ Application Question: 9
- ▢ Application Problem: 2
- ▪ Question: 11
- ▢ Application Problem: 7

10.3 Nitrogen and Ammonia

Occurrence of Nitrogen
- ▢ Questions: 12, 15, 16a

Physical Properties of Nitrogen
- ▢ Question: 16b

Structure and Chemical Properties of Nitrogen
- ▢ Question: 17

Uses and Preparation of Nitrogen
- ▢ Chemistry Notebook Question: 18
- ▢ Application Questions: 11, 12

Occurrence of Ammonia
- ▢ Application Question: 13

Structure and Properties of Ammonia
- ▢ Question: 19
- ▢ Application Question: 14

Uses and Preparation of Ammonia
- ▢ Questions: 20, 21

10.4 Carbon Dioxide and Carbon Monoxide

Occurrence of Carbon Dioxide
- ▢ Question: 23
- ▢ Application Question: 15

Structure of Carbon Dioxide
- ▢ Application Question: 18

Physical Properties of Carbon Dioxide
- ▢ Chemistry Notebook Question: 24a
- ▢ Application Question: 17

Chemical Properties of Carbon Dioxide
- ▢ Questions: 24b, 31

Uses and Preparation of Carbon Dioxide
- ▢ Application Problem: 4
- ▪ Question: 27
- ▢ Application Problem: 8

Structure of Carbon Monoxide
- ▢ Question: 22
- ▪ Question: 28

Properties and Uses of Carbon Monoxide
- ▢ Chemistry Notebook Questions: 25, 26, 29, 32
- ▪ Question: 30

Teaching Strategies

Introduction

This chapter, mainly descriptive chemistry, provides a welcome interlude between the stoichiometry that precedes it and the quantitative application of the gas laws that follows it. A brisk pace is recommended. The text presents more preparation and properties for each gas than can be demonstrated in the time allotted.

10.1   Oxygen and Ozone

Few students are aware of the paramagnetic nature of the oxygen molecule. The traditional double bond between two oxygen atoms needed to complete the octet rule does not take into account the two unpaired electrons. If the materials are available, the following demonstration is dramatic and clearly shows the paramagnetic properties of oxygen.

Demonstration: 1 Paramagnetism in Oxygen

PURPOSE: To show that oxygen molecules are paramagnetic and thus must contain unpaired electrons

MATERIALS: A very strong magnet (the World War II surplus radar magnets are ideal, but hard to find), a dewar flask, liquid nitrogen, large pyrex test tube, a source of oxygen gas

PROCEDURE: (1) Attach a piece of 6-mm glass tubing to a length of rubber tubing. (2) Connect the other end of the rubber tubing to the regulator on the oxygen tank. (3) Fill the dewar flask with liquid nitrogen. (4) Place the large test tube in the dewar flask. (5) Insert the glass tubing in the test tube. Open the valve on the oxygen tank and maintain a very slow flow of oxygen. It will take about 5 minutes to collect about 50 mL of liquid oxygen. (6) Pour some liquid nitrogen between the poles of the magnet. Note that the nitrogen is diamagnetic and flows right through the poles. (7) Remove the test tube containing the liquid oxygen from the dewar flask and note that it has a blue color caused by the unpaired electrons. *(8)* Slowly pour the liquid oxygen between the cooled poles of the magnet. The oxygen will remain suspended between the poles demonstrating the presence of unpaired electrons the resulting paramagnetism.

Demonstration: 2 Preparation of Oxygen
PURPOSE: To produce oxygen

MATERIALS: 3% hydrogen peroxide, 3.0 g OK, 60 mL 3 M H_2SO_4, 600-mL beaker, wooden splints, matches

PROCEDURE: *(1)* Dissolve 3.0 g of potassium permanganate in 60 mL of 3 M sulfuric acid. *(2)* Pour 100 mL 3% hydrogen peroxide into the 600-mL beaker. *(3)* Pour 5 mL of the potassium permanganate solution into the beaker. Note the formation of bubbles and evolution of a gas. *(4)* Light a wooden splint and blow out the flame, leaving a few glowing embers. *(5)* Add another 5-mL portion of permanganate to the beaker and immediately insert the glowing splint. The splint should burst into flames. The procedure may be repeated.

Note that 3% hydrogen peroxide was used. Thirty percent hydrogen peroxide must not be used as it is a very strong oxidizing agent.

10.2   Hydrogen

In discussing the occurrence, preparation, and properties of hydrogen, again take into account the students' previous experiences. Hydrogen makes up 76% of the universe and helium 23%. Point out that the sun utilizes a fusion reaction in which hydrogen is converted to helium to generate the vast amounts of energy on which life on earth depends. We have harnessed the power of the sun in the thermonuclear hydrogen bomb and are working on harnessing the same energy in fusion reactors. Although hydrogen is placed in Group 1 in the Periodic Table, its properties are different from other Group 1 elements.

Demonstration 3: Preparation and Properties of Hydrogen
PURPOSE: To demonstrate the preparation and properties of hydrogen gas

MATERIALS: 30g of mossy zinc, 6 M of hydrochloric acid, 250-mL Erlenmeyer flask and 2-hole rubber stopper, long stem glass funnel or thistle tube, right angle glass bend, rubber tubing, gas-collecting bottles, trough or large pan, matches.

PROCEDURE: Insert funnel and glass bend through the stopper and connect rubber tubing to bend and collecting bottles. Place zinc in the flask and cover with H_2O. Secure stopper to flask with tip of funnel below water line. Add the HCl in 10-mL increments. Collect the hydrogen in the gas-collecting bottles by water displacement.

Collect several 200-mL bottles of hydrogen. The bottles should be inverted and the mouth of the bottle placed on a wet glass plate on the table top. Ask the students to explain the necessity of the inverted bottles. Care must be taken that only hydrogen is present in the bottles. If pure hydrogen is present, the gas will burn quietly at the mouth of the bottle. If a mixture of hydrogen and oxygen is present, the mixture explodes with a loud pop. To demonstrate the difference between burning and supporting combustion, light a small candle. Bring the candle to the mouth of an inverted bottle filled with the hydrogen. The hydrogen burns at the mouth of the bottle with an almost invisible blue flame. Slowly insert the candle into the bottle. The hydrogen extinguishes the flame. Slowly withdraw the candle. The flame at the mouth reignites the candle. Some condensation should have formed on the inside of the bottle. Write the equation for the combustion of hydrogen. Water is a product of the reaction. A very important point should be made at this time. Oxygen gas does not burn, but it does support combustion. Hydrogen gas burns, but it does not support combustion. (Both the fuel and a source of oxygen are needed to burn something.)

Place the mouth of a second bottle of hydrogen over the mouth of an empty (filled with air) bottle. Invert the two. The hydrogen in the bottom bottle will now diffuse into the upper bottle. Separate the two bottles. Invert the lower bottle and place it on the table. Light a candle and bring the flame to the mouth of the inverted bottle that was originally filled with air. A loud pop occurs. Hydrogen diffused into the bottle and mixed with oxygen present in the air. This procedure may be repeated with the remaining bottle. Note the difference between the sound and the rate of the reaction when pure hydrogen burns and when a mixture of hydrogen and oxygen burns. Discuss the von Hindenberg disaster.

After completing the demonstrations, have the students classify them as representing physical or chemical properties of hydrogen. Density and diffusion are physical, while burning and not supporting combustion are chemical.

10.3   Nitrogen and Ammonia

Nitrogen accounts for about 80% of the atmosphere. The triple bond in the nitrogen molecules is strong enough for nitrogen gas to be considered the "poor man's inert or noble gas." This section is best introduced by telling the story about Fritz Haber and Germany's desperate situation in World War I. Explosives at that time used nitrates and nitric acid as major raw materials. Chilean nitrates were an integral part of Germany's war effort. The Allied naval blockade could have deprived Germany of a key raw material and hastened the end of the war. Haber, however, had developed the process, now named after him, for producing ammonia from the nitrogen of the air. The ammonia was easily converted to the nitrates needed for the production of explosives. The very air we breathe became the source of German gunpowder. Many social issues may be discussed at this time. The importance of nitrogen in fertilizers, in feeding underdeveloped nations, in explosives, nitrogen fixation, and in air and water pollution. The students may be given an assignment to find articles that touch on these points.

A number of methods for preparing oxygen-free nitrogen are available. Placing moistened steel wool on a bottle of air inverted in a pan of water may be the simplest. Liquid nitrogen may be obtained from a college, hospital, doctor's office, or welding supply store. Dewar flasks for transporting the nitrogen may be leased from chemical suppliers. Should liquid nitrogen be available, a number of demonstrations would serve as an introduction to the gas laws. Students are fascinated by the liquid nitrogen.

When introducing ammonia, be sure to distinguish between ammonia and ammonium ion. Writing the formulas on the board and displaying molecular models of each will help.

10.4 Carbon Dioxide and Carbon Monoxide

Introduce the study of carbon dioxide by having a piece of dry ice on the desk. You may place it in a large graduated cylinder filled with water to which an acid–base indicator has been added. One of the universal indicators works well.

The "greenhouse effect" and its importance afford you another opportunity to begin discussing carbon dioxide and its properties. Encourage students to bring in articles that discuss the greenhouse effect.

Bubbling carbon dioxide or blowing one's breath into a beaker with limewater is a traditional demonstration for showing the presence of carbon dioxide. You may extend the demonstration by continuing to bubble carbon dioxide into the limewater. After a short period of time, the precipitate dissolves and the limewater is once again clear. Most students will not have seen this latter demonstration. As discussed in the text, the excess carbon dioxide reacts with the calcium carbonate precipitate and forms the soluble calcium–hydrogen carbonate. You may go on and explain how this system is important in cave formation and within the blood.

Write equations and discuss each of the processes for producing carbon dioxide. Most of the processes should have been encountered in earlier science courses.

Carbon monoxide may be introduced while discussing the toxic effects of carbon dioxide. Both are poisonous, but react differently. In discussing the structure of carbon monoxide, you may wish to mention that the structure is isoelectronic with that of nitrogen gas.

You may end the section by pointing out that, despite all of its dangerous features, carbon monoxide serves as a key resource in producing more complex organic compounds. Its role may become more important as petroleum supplies dwindle.

Desktop Investigation: Properties of Carbon Dioxide

If you wish your students to carry this out at home, the following modifications can be made: A 4-ounce juice glass can be substituted for a 150-mL beaker, a pint jar for a 400-mL beaker, and a soda straw for the piece of glass tubing. 1/4 cup of baking soda can be substituted for 50 g of baking soda, and 1/4 cup of vinegar for 50 mL of vinegar. (These are not exact equivalents, but handy amounts to substitute for good results.)

Limewater is made by adding 5.0 g of calcium hydroxide to 1.0 L of distilled water. Shake and allow the undissolved solids to settle. Decant or filter the saturated solution to another bottle.

ANSWERS TO DISCUSSION QUESTIONS

1. Carbon dioxide can be poured from one beaker to another because it is denser than air and thus can displace the air.

2. One experiment that might be proposed, but does not work when attempted, is to bubble exhaled air through limewater for 15 minutes. There is not enough carbon dioxide to redissolve the calcium carbonate precipitate. However, this is a good hypothetical solution to the problem. One that does work is bubbling exhaled air through two different containers of 50 mL of limewater for one minute. Allow the containers to stand until the precipitate settles out. Carefully decant the supernatant liquid and add 100 mL of distilled water to one container and 100 mL of club soda to the second container. (Club soda is a supersaturated solution of carbon dioxide after the bottle is opened.) Stir each container 2 to 3 minutes and observe that the precipitated calcium carbonate dissolves in the club soda, but not in the distilled water.

3. (a) $NaHCO_3(aq) + HC_2H_3O_2(aq) \rightarrow NaC_2H_3O_2(aq) + H_2O(\ell) + CO_2(g)$
   (b) $Ca^{+2}(aq) + 2OH^-(aq) \rightarrow CaCO_3(s) + H_2O(\ell)$

References and Resources

Books and Periodicals

Baugh, Mark. "What's So Bad About Oxygen?," *ChemMatters*, Vol. 4, No. 1, Feb. 1986, pp. 10–11.

Borgford, Christie L. and Lee R. Summerlin. "Gases," *Chemical Activities: A Sourcebook for Science Teachers*, Washington, D.C.: The American Chemical Society, 1988.

Davenport, Derek A. "The Back Burner—Joseph Priestley and the All-American Lunch," *ChemMatters*, Vol. 1, No. 1, Feb. 1983, pp. 14–15.

Dinga, Gustav P. "Hydrogen: The Ultimate Fuel and Energy Carrier," *Journal of Chemical Education*, Vol. 65, No. 8, Aug. 1986, pp. 688–691.

Linner, Marilyn. "Hydrogen and Helium," *ChemMatters*, Vol. 3, No. 3, Oct. 1985, pp. 4–7.

Martins, George F. "Percent Oxygen in Air," *Journal of Chemical Education*, Vol. 64, No. 9, Sept. 1987, pp. 809–810.

Ponnamperuma, Cyril. "The Atmosphere of Planet Earth," *ChemMatters*, Vol. 1, No. 1, Feb. 1983, pp. 10–12.

Shakashiri, Bassam Z. *Chemical Demonstrations*, Vol. 2, Madison: University of Wisconsin Press, 1985.

Summerlin, Lee R., et al. "Gases," *Chemical Demonstrations— A Sourcebook for Teachers*, Vol. 1, Second Edition, Washington, D.C.: American Chemical Society, 1988.

Summerlin, Lee R., et al. "Gases," *Chemical Demonstrations— A Sourcebook for Teachers*, Vol. 2, Washington, D.C.: American Chemical Society, 1987.

Audiovisual Resources

"Out of the Air: 1," part of the series *Chemistry in Action*, VHS cassette, color, 20 min., Films for the Humanities, Inc.

"Out of the Air: 2," part of the series *Chemistry in Action*, VHS cassette, color, 20 min., Films for the Humanities, Inc.

Computer Software

Bauder, Donald. *Haber Tech: A Simulation of the Industrial Synthesis of Ammonia.* Apple, Queue. (Principles of chemical equilibrium, reaction kinetics, formulas, equations, and gas laws)

Edens, R. and K. Shaw. *HABER: Ammonia Synthesis*, Apple, CONDUIT. (Production of ammonia by the Haber process)

Answers and Solutions

Questions

1. *(a)* Oxygen exists as both ordinary oxygen gas (O_2) and ozone (O_3). *(b)* Oxygen gas is necessary for the basic life process of aerobic respiration, in which it combines with molecules from food to produce the energy needed for life. Ozone, found mostly in the upper atmosphere, absorbs most of the ultraviolet portion of the light radiated from the sun to the earth.

2. *(a)* An allotrope is one of the two or more forms of an element that have the same physical state. *(b)* Allotropy is the existence of two or more allotropes of an element. *(c)* Ordinary oxygen (O_2) and ozone (O_3) are allotropes.

3. Chlorofluorocarbons (CFCs), used in aerosols and refrigerants, may be damaging the ozone layer.

4. Oxygen composes 49.5% of the total mass of the earth's crust, waters, and atmosphere; it makes up 20% of the earth's atmosphere by volume; 47% of the earth's crust is oxygen by mass.

5. At normal temperature and pressure, pure oxygen is a colorless, odorless, tasteless gas that is slightly denser than air; at normal atmospheric pressure, liquid oxygen boils at $-183.0°C$ and freezes at $-218.4°C$.

6. *(a)* An oxide is a compound containing the simple oxide anion, which has an oxidation number of -2. A peroxide is a compound containing the O_2^{2-} anion, in which oxygen has the oxidation number -1. A superoxide is a binary compound containing the O_2^- anion, in which oxygen has the uncommon oxidation number $-\frac{1}{2}$. *(b)* The bond character in an oxide depends on the difference in electronegativities between oxygen and the element with which it is combined; the greater the difference in electronegativities is, the more ionic, or less covalent, is the bond.

7. *(1)* Decomposition of hydrogen peroxide:
   $2H_2O_2(aq) \rightarrow 2H_2O(\ell) + O_2(g)$
   (2) Oxidation of water by sodium peroxide:
   $2Na_2O_2(s) + 2H_2O(\ell) \rightarrow 4NaOH(aq) + O_2(g)$
   (3) Thermal decomposition of potassium chlorate:
   $2KClO_3(s) \rightarrow 2KCl(s) + 3O_2(g)$
   (4) Electrolysis of water: $2H_2O(\ell) \rightarrow 2H_2(g) + O_2(g)$
   (5) Liquefaction of air: This method is the most common industrial method for preparing oxygen. Ordinary air is liquefied by compression and cooling, and then nitrogen is separated from the oxygen by fractional distillation.

8. Electrolysis of water is the decomposition of water into hydrogen and oxygen by the passage of a direct current of electricity through the water.

9. *(a)* The most abundant element in the universe is hydrogen. *(b)* On earth, free elemental hydrogen exists as a gas consisting of diatomic H_2 molecules. In combined form, hydrogen exists mainly as water.

10. The indicated forms of hydrogen are: *(a)* elemental hydrogen in the plasma state *(b)* liquid hydrogen *(c)* metallically bonded liquid hydrogen, consisting of protons and mobile electrons.

11. The major commercial source of hydrogen is the decomposition of hydrocarbons.

12. *(a)* Nitrogen is the major component of the earth's atmosphere. *(b)* The most important compound of this element is ammonia. *(c)* Ammonia is used as a fertilizer, as a refrigerant, and as an intermediate in the production of such compounds as nitric acid, nylon, and antimalarial drugs.

13. *(a)* The element most frequently involved in compound-formation is hydrogen. *(b)* With most nonmetals, hydrogen reacts to form molecular compounds with single covalent bonds of varying polarities; with the least electronegative metals, it combines to form ionic hydrides.

14. Hydrogen can be prepared: *(1)* from acids by replacement: $Zn(s) + H_2SO_4(aq) \rightarrow ZnSO_4(aq) + H_2(g)$ *(2)* from water by replacement: $2Na(s) + 2H_2O(\ell) \rightarrow 2NaOH(aq) + H_2(g)$ *(3)* from the electrolysis of water: $2H_2O(\ell) \rightarrow 2H_2(g) + O_2(g)$ *(4)* from the decomposition of hydrocarbons: $CH_4(g) + H_2O(g) \rightarrow CO(g) + 3H_2(g)$

15. Nitrogen is present in the DNA molecules that control heredity and it is a major component of all amino acids.

16. *(a)* The earth's atmosphere is approximately 80% molecular nitrogen, N_2, by volume. *(b)* Elemental nitrogen is a colorless, odorless, and tasteless gas that is slightly less dense than air. At temperatures lower than $-195.8°C$ at 1 atm, it condenses to a colorless liquid; it freezes to a white solid at $-209.9°C$.

17. *(a)* Using energy from ultraviolet light from the sun, nitrogen oxides can react with unburned hydrocarbons from automobile exhaust or with oxygen in the air to produce ozone and other irritating chemicals that together are referred to as photochemical smog. *(b)* The converter helps reduce air pollution by deoxidizing nitrogen oxides to nitrogen and by completely oxidizing hydrocarbons to carbon dioxide and water.

18. *(a)* Nitrogen fixation is the process through which atmospheric nitrogen gas (N_2) is converted into the nitrogen compounds that can be used by plants. *(b)* Since most natural nitrogen fixation is carried out by bacteria on the roots of legumes, such plants can restore nitrogen to soil that has become depleted of it by other crops.

19. *(a)* The Lewis structure is:

$$H \ddot{:}\underset{\ddot{}}{\overset{\cdot\cdot}{N}}\colon H$$
$$H$$

(b) The ammonia molecule is polar covalent, and, as a result of hydrogen bonding, ammonia has higher boiling and freezing points than expected. Ammonia is a colorless gas, has a strong odor, is less dense than air, and is extremely soluble in water.

20. Ammonia is used to revive fainting victims; as household ammonia, it is a useful grease-cutting agent; it is important as a base and as an intermediate in various organic reactions; it is also used in the production of artificial fertilizers and of explosives.

21. *(a)* Ammonia is continually produced naturally by the fixation of atmospheric nitrogen and by the decay of organic matter. *(b)* Industrially, ammonia is produced by decomposition of ammonium compounds: *(1)* $Ca(OH)_2(s) + 2NH_4Cl(s)$

→ CaCl$_2$(s) + 2NH$_3$(g) + 2H$_2$O(g) (2) the Haber process: N$_2$(g) + 3H$_2$(g) ⇌ 2NH$_3$(g)

22. (a) Carbon monoxide is a compound highly toxic to animals, whereas carbon dioxide is nontoxic. Both carbon monoxide and carbon dioxide are key ingredients used by plants in photosynthesis. (b) The CO$_2$ molecule is linear and nonpolar, with an oxygen atom on either side of the carbon atom. The C−O bonds in CO$_2$ are represented by resonance structures, but for most purposes can be considered double bonds. In CO, the bonding between the carbon and oxygen atoms is a complicated mixture of four resonance structures. The CO molecule is linear and polar.

23. (a) Photosynthesis is the process by which plants use energy from the sun to combine carbon dioxide with water to produce glucose and oxygen.
(b) 6CO$_2$(g) + 12H$_2$O(ℓ) + energy (sunlight)
→ C$_6$H$_{12}$O$_6$(s) + 6O$_2$(g) + 6H$_2$O(ℓ)

24. Carbon dioxide is a colorless gas with a faintly irritating odor and a sour taste. It is denser than air, condenses to liquid below 31°C and at pressures above 72.9 atm. At normal pressure, solid CO$_2$ changes directly to gas. CO$_2$ is stable and does not support combustion. It decomposes in the presence of burning magnesium. It combines with water to form carbonic acid, and hydroxides to form carbonates.

25. Solid carbon dioxide, dry ice, is used as a deep-freezing refrigerant; liquid carbon dioxide is used in fire extinguishers; when produced through the action of various leavening agents, such as yeast or baking powder, bubbles of carbon dioxide cause the bread dough to rise and produce a lighter, less dense bread; carbon dioxide is forced into soft drinks to produce effervescence.

26. The indicated reactions are:
(1) Burning carbon-containing substances:
C(combined) + O$_2$(g) → CO$_2$(g)
(2) Reaction of steam and natural gas:
CH$_4$(g) + 2H$_2$O(g) → 4H$_2$(g) + CO$_2$(g)
(3) Fermentation of sugar:
C$_6$H$_{12}$O$_6$(aq) → 2C$_2$H$_5$OH(aq) + 2CO$_2$(g)
(4) Heating carbonates:
CaCO$_3$(s) → CaO(s) + CO$_2$(g)
(5) Reactions of acids and carbonates:
CaCO$_3$(s) + 2HCl(aq) → CaCl$_2$(aq) + H$_2$CO$_3$(aq)
H$_2$CO$_3$(aq) → H$_2$O(ℓ) + CO$_2$(g)
(6) Respiration and decay: Carbon dioxide is a natural product of these two processes.

27. Respiration is the reaction by which foods, usually carbohydrates, are oxidized in plant and animal cells to release energy. Decay is the breakdown of the tissue of dead plants and animals into simpler substances.

28. Carbon monoxide is produced naturally by decaying plants, by live algae, by the oxidation of atmospheric methane, and by volcanic action.

29. (a) The average concentration is 0.5 ppm. (b) This is 0.01 of the level considered toxic. (c) In the United States, as much as 90% of the human-produced carbon monoxide comes from the incomplete combustion of carbon-containing fuels in automobile engines.

30. Carbon monoxide is a colorless, odorless, tasteless gas; it is slightly less dense than air, and only slightly soluble in water.

31. (a) Carbon dioxide does not support combustion of most substances. (b) It decomposes into carbon and oxygen.
2Mg(s) + CO$_2$(g) → 2MgO(s) + C(s)
(c) When dissolved in water, a small amount of the dissolved carbon dioxide unites with water to form carbonic acid.
H$_2$O(ℓ) + CO$_2$ ⇌ H$_2$CO$_3$(aq)
(d) It reacts to form a carbonate, which may or may not precipitate out depending upon the identity of the positive ion.
Ca^{+2}(aq) + 2OH$^-$(aq) + CO$_2$(g) → CaCO$_3$(s) + H$_2$O(ℓ)

32. (a) The carbon monoxide acts as a reducing agent.
Fe$_2$O$_3$(s) + 3CO(g) → 2Fe(s) + 3CO$_2$(g)
(b) When burned, carbon monoxide is oxidized to carbon dioxide.
2CO(g) + O$_2$(g) → 2CO$_2$(g)
(c) The reduction of carbon monoxide produces methanol.
CO(g) + 2H$_2$(g) → CH$_3$OH(ℓ)

Problems

1. The indicated responses are:
(a) Ba = +2, O = −2, oxide (b) Ca = +2, O = −2,
oxide (c) Rb = +1, O = −½, superoxide (d) Al
= +3, O = −2, oxide (e) Na = +1, O = −1, peroxide
(f) Mg = +2, O = −2, oxide (g) = +1, O = −1,
peroxide (h) K = +1, O = −½, superoxide.

2. (a) 2 mol O × $\dfrac{15.9994 \text{ g O}}{\text{mol O}}$ = 31.9988 g O

(b) 3 mol O × $\dfrac{15.9994 \text{ g O}}{\text{mol O}}$ = 47.9982 g O

(c) 2 mol H × $\dfrac{1.00794 \text{ g H}}{\text{mol H}}$ = 2.01588 g H

(d) 2 mol N × $\dfrac{14.0067 \text{ g N}}{\text{mol N}}$ = 28.0134 g N

(e) 1 mol N × $\dfrac{14.0067 \text{ g N}}{\text{mol N}}$ = 14.0067 g N

3 mol H × $\dfrac{1.00794 \text{ g H}}{\text{mol H}}$ = 3.023 82 g H

Molar mass = 17.0305 g

(f) 1 mol C × $\dfrac{12.0111 \text{ g C}}{\text{mol C}}$ = 12.0111 g C

2 mol O × $\dfrac{15.9994 \text{ g O}}{\text{mol O}}$ = 31.9988 g O

Molar mass = 44.0099 g

(g) 1 mol C × $\dfrac{12.0111 \text{ g C}}{\text{mol C}}$ = 12.0111 g C

1 mol O × $\dfrac{15.9994 \text{ g O}}{\text{mol O}}$ = 15.9994 g O

Molar mass = 28.0105 g

3. (a) 5.0 mol H$_2$O × $\dfrac{18.0 \text{ g H}_2\text{O}}{\text{mol H}_2}$ = 90 g H$_2$O

(b) 0.25 mol NH$_3$ × $\dfrac{17.03 \text{ NH}_3}{\text{mol NH}_3}$ = 4.3 g NH$_3$

(c) 3.01×10^{23} molec $CO_2 \times \dfrac{\text{mol } CO_2}{6.022 \times 10^{23} \text{ molec } CO_2} \times$

$\dfrac{44.009 \text{ g } CO_2}{\text{mol } CO_2} = 22.0 \text{ g } CO_2$

4. (a) $25.0 \text{ g HgO} \times \dfrac{\text{mol HgO}}{216.59 \text{ g HgO}} = 0.115 \text{ mol HgO}$

(b) $6.5 \times 10^{-5} \text{ g } H_2 \times \dfrac{\text{mol } H_2}{2.01588 \text{ g } H_2} = 3.2 \times 10^{-5} \text{ mol } H_2$

5. (a) $\dfrac{15.9994 \text{ g O}}{18.0153 \text{ g } H_2O} \times 100\% = 88.8101\% \text{ O}$

(b) $\dfrac{1.007\,94 \text{ g H}}{36.461 \text{ g HCl}} \times 100\% = 2.7644\% \text{ H}$

(c) $\dfrac{14.0067 \text{ g N}}{17.0305 \text{ g } NH_3} \times 100\% = 82.2448\% \text{ N}$

(d) $\dfrac{3 \text{ mol O} \times \dfrac{15.9994 \text{ g O}}{\text{mol O}}}{159.692 \text{ g } Fe_2O_3} \times 100\% = 30.0567\% \text{ O}$

(e) $\dfrac{2 \text{ mol H} \times \dfrac{1.007\,94 \text{ g H}}{\text{mol H}}}{98.07 \text{ g } H_2SO_4} \times 100\% = 2.056\% \text{ H}$

(f) $\dfrac{3 \text{ mol O} \times \dfrac{15.9994 \text{ g O}}{\text{mol O}}}{101.1032 \text{ g } KNO_3} \times 100\% = 47.4745\% \text{ O}$

(g) $\dfrac{2 \text{ mol N} \times \dfrac{14.0067 \text{ g N}}{\text{mol N}}}{80.0434 \text{ g } NH_4NO_3} \times 100\% = 34.9978\% \text{ N}$

(h) $\dfrac{4 \text{ mol H} \times \dfrac{1.007\,94 \text{ g H}}{\text{mol H}}}{16.0429 \text{ g } CH_4} \times 100\% = 25.1311\% \text{ H}$

6. $\dfrac{\text{L } O_2}{1.429 \text{ g O}} \times \dfrac{2 \text{ mol O}}{\text{mol } O_2} \times \dfrac{15.9994 \text{ g O}}{\text{mol O}} = 22.39 \dfrac{\text{L } O_2}{\text{mol } O_2}$

7. $38.67 \text{ g K} \times \dfrac{\text{mol K}}{39.0983 \text{ g K}} = 0.9890 \text{ mol K}$

$13.85 \text{ g N} \times \dfrac{\text{mol N}}{14.0067 \text{ g N}} = 0.9888 \text{ mol N}$

$47.48 \text{ g O} \times \dfrac{\text{mol O}}{15.9994 \text{ g O}} = 2.968 \text{ mol O}$

$\dfrac{0.9890 \text{ mol K}}{0.9888} : \dfrac{0.9888 \text{ mol N}}{0.9888} : \dfrac{2.968 \text{ mol O}}{0.9888}$

$1.00 \text{ mol K} : 1.00 \text{ mol N} : 3.00 \text{ mol O}$

Simplest formula is KNO_3.

$2H_2O(\ell) \rightarrow 2H_2(g) + O_2(g)$

$90.0 \text{ g } H_2O \times \dfrac{\text{mol } H_2O}{18.0153 \text{ g } H_2O} \times \dfrac{2 \text{ mol } H_2}{2 \text{ mol } H_2O} \times \dfrac{2.01588 \text{ g } H_2}{\text{mol } H_2}$

$= 10.1 \text{ g } H_2$

$90.0 \text{ g } H_2O \times \dfrac{\text{mol } H_2O}{18.0153 \text{ g } H_2O} \times \dfrac{1 \text{ mol } O_2}{2 \text{ mol } H_2O} \times \dfrac{31.9988 \text{ g } O_2}{\text{mol } O_2}$

$= 79.9 \text{ g } O_2$

Application Questions

1. The singly bonded oxygen atoms contain two unpaired electrons whose spins do not cancel. This accounts for the attraction of oxygen by a magnet.

2. (a) The two products are nitrogen and oxygen. (b) They can be separated through fractional distillation, in which the lower-boiling-point nitrogen can be separated from the higher-boiling-point oxygen through repeated evaporations and condensations.

3. The indicated responses are: (a) Ba = +2, O = −2, oxide (b) Ca = +2, O = −2, oxide (c) Rb = +1, O = $-\frac{1}{2}$, superoxide (d) Al = +3, O = −2, oxide (e) Na = +1, O = −1, peroxide (f) Mg = +2, O = −2, oxide (g) H = +1, O + −1, peroxide (h) K = +1, O = $-\frac{1}{2}$, superoxide

4. The temperature of the iron and the presence of water are the major factors. At room temperature, moist iron unites slowly with oxygen to form Fe_2O_3; however red-hot steel wool burns brilliantly when plunged into pure oxygen to produce Fe_3O_4.

5.

6. (a) Ozone is produced from oxygen in a two-step process: (1) energy is added to molecular oxygen to produce atomic oxygen, and (2) each free oxygen atom combines with an oxygen molecule to produce O_3.
(b) $3O_2(g) + \text{energy} \rightarrow 2O_3(g)$

7. Unlike the other Group 1 elements, hydrogen exists in the form of diatomic H_2 molecules, and as such is a poor conductor of heat and electricity. Unlike other Group 1 elements, it is a gas and is colorless, odorless, and tasteless.

8. Free hydrogen atoms are extremely reactive, and thus combine readily with each other to form hydrogen molecules or with other elements to form various compounds.

9. (a) Liquid vegetable oils to which hydrogen has not been added are nonhydrogenated. If hydrogen has been added to such oils, the result is a solid or semisolid fat, which is then said to be hydrogenated (b) Nonhydrogenated oils.

10. The nearly 100 other elements were produced from hydrogen nuclei by nuclear fusion in the cores of the stars.

11. (a) Because of the high energy required to break the nitrogen–nitrogen triple bond, it is hard to separate the nitrogen atoms of the N_2 molecule so that they can enter into new combinations. (b) Since it is relatively nonreactive, many food items are packaged in the presence of nitrogen rather than air to prevent the interaction of these products with the oxygen in air, since this interaction often leads to loss of flavor. Due to its nontoxic and nonpolluting characteristics, nitrogen is used as a propellant in aerosol cans.

12. (a) Various oxides of nitrogen formed as a result of combustion in automobile engines can combine with water in the atmosphere to produce acid anhydrides that sometimes fall to earth as acid rain. (b) The catalytic action of ultraviolet light from the sun can cause nitrogen oxides to react with unburned hydrocarbons from automobile exhausts with oxygen in the air to produce ozone and other irritating chemicals collectively referred to as photochemical smog.

13. Ammonia, which may have been abundant in the early atmosphere of the earth, is now converted to other substances in living organisms as fast as it is produced. Also because it is so low in mass, ammonia can escape from the earth's gravitational field to a significant extent.

14. The ammonia molecule is a polar–covalent triangular pyramid.

$$H - H - H$$
$$|$$
$$H$$

15. The "greenhouse effect" is the trapping of reflected energy by the carbon dioxide in the atmosphere. Increased levels of carbon dioxide in the atmosphere (through the increased burning of carbon-containing fuels, or the destruction of forests that normally consume carbon dioxide) could cause an increase in global temperature that could melt the polar ice caps and flood sea-level coastal cities. Other areas could become dry and unproductive.

16. Dry ice, or solid carbon dioxide, has a very low temperature ($-78.5°C$) and at atmospheric pressure, passes directly from the gaseous to the solid state without melting.

17. In a liquid carbon dioxide fire extinguisher, a stream of carbon dioxide snow cools the burning material and also shuts off the supply of oxygen needed for combustion.

18. Carbon dioxide molecules are linear and thus nonpolar. The carbon–oxygen bonds are shorter than ordinary carbon–oxygen double bonds. The molecule has the following resonance hybrid structures:

$$^-:\!\ddot{O}\!: \quad C\!:\!:\!O\!:^+$$

$$:\!\ddot{O}\!: :\!C\!: :\!\ddot{O}\!: \longleftarrow \quad \longrightarrow :\!\ddot{O}\!: :\!C\!: :\!\ddot{O}\!:$$

$$^+:\!O\!: :\!C\!: \quad \ddot{O}\!:^-$$

Application Problems

1. (a) $2Mg(s) + O_2(g) \rightarrow 2MgO(s)$
   (b) $Cs(s) + O_2(g) \rightarrow CsO_2(s)$
   (c) $H_2(g) + O_2(g) \rightarrow H_2O_2(\ell)$

2. (a) $Zn(s) + 2HF(aq) \rightarrow ZnF_2(aq) + H_2(g)$
   (b) $Mg(s) + H_2SO_4(aq) \rightarrow MgSO_4(aq) + H_2(g)$
   (c) $2Fe(s) + 3H_2SO_4(aq) \rightarrow Fe_2(SO_4)_3(aq) + 3H_2(g)$
   (d) $Mg(s) + 2HCl(aq) \rightarrow MgCl_2(aq) + H_2(g)$
   (e) $2Rb(s) + 2HOH(\ell) \rightarrow 2RbOH(aq) + H_2(g)$
   (f) $Mg(s) + 2HOH(\ell) \rightarrow Mg(OH)_2(aq) + H_2(g)$

3. (a) $2H_2(g) + O_2(g) \rightarrow 2H_2O(g)$; $H = +1, O = -2$
   (b) $H_2(g) + Cl_2(g) \rightarrow 2HCl(g)$; $H = +1, Cl = -1$

(c) $2Na(s) + H_2(g) \rightarrow 2NaH(s)$; $Na = +1$; $H = -1$
(d) $Ba(s) + H_2(g) \rightarrow BaH_2(s)$; $Ba = +2$, $H = -1$

4. (a) $MgCO_3(s) \rightarrow MgO(s) + CO_2(g)$
   (b) 1. $MgCO_3(s) + 2HCl(aq) \rightarrow MgCl_2(aq) + H_2CO_3(aq)$
      2. $H_2CO_3(aq) \rightarrow H_2O(\ell) + CO_2(g)$

5. $52.8 \text{ g Al} \times \dfrac{\text{mol Al}}{26.981\,54 \text{ g Al}} = 1.96 \text{ mol A}(\ell)$

$47.2 \text{ g O} \times \dfrac{\text{mol O}}{15.9994 \text{ g O}} = 2.95 \text{ mol O}$

$\dfrac{1.96 \text{ mol Al}}{1.96} : \dfrac{2.95 \text{ mol O}}{1.96}$

$1.00 \text{ mol Al} : 1.51 \text{ mol O}$

Multiplying by 2: $2.00 \text{ Al} : 3.02 \text{ mol O}$

Simplest formula is Al_2O_3.

6. (a) $\dfrac{4 \text{ mol O} \times \dfrac{15.9994 \text{ g O}}{\text{mol O}}}{367.1 \text{ g PbMoO}_4} \times 100\% = 17.43\% \text{ O}$

   (b) $\dfrac{11 \text{ mol O} \times \dfrac{15.9994 \text{ g O}}{\text{mol O}}}{273.29 \text{ g Na}_2B_4O_7 \cdot 4H_2O} \times 100\% = 64.40\% \text{ O}$

   (c) $\dfrac{18 \text{ mol O} \times \dfrac{15.9994 \text{ g O}}{\text{mol O}}}{537.5018 \text{ g } 3BeO \cdot Al_2O_3 \cdot 6SiO_2} \times 100\% = 53.5792\% \text{ O}$

7. (a) $Zn(s) + H_2SO_4(aq) \rightarrow ZnSO_4(aq) + H_2(g)$

   $13.1 \text{ g Zn} \times \dfrac{\text{mol Zn}}{65.39 \text{ g Zn}} \times \dfrac{1 \text{ mol H}_2}{\text{mol Zn}} \times \dfrac{2.015\,88 \text{ g H}}{\text{mol H}_2}$

   $= 0.404 \text{ g H}_2$

   (b) $2Na(s) + 2H_2O(\ell) \rightarrow 2NaOH(aq) + H_2(g)$

   $50.0 \text{ g H}_2O \times \dfrac{\text{mol H}_2O}{18.0153 \text{ g H}_2O} \times \dfrac{1 \text{ mol H}_2}{2 \text{ mol H}_2O} \times \dfrac{2.015\,88 \text{ g H}}{\text{mol H}_2}$

   $= 2.80 \text{ g H}_2$

8. (a) $2NaHCO_3(aq) + H_2SO_4(aq) \rightarrow Na_2SO_4(aq) + 2H_2O(\ell) + 2CO_2(g)$

   $1.500 \text{ kg NaHCO}_3 \times \dfrac{10^3 \text{ g}}{\text{kg}} \times \dfrac{\text{mol NaHCO}_3}{84.0070 \text{ g NaHCO}_3} \times$

   $\dfrac{1 \text{ mol H}_2SO_4}{2 \text{ mol NaHCO}_3} \times \dfrac{98.07 \text{ g H}_2SO_4}{\text{mol H}_2SO_4} = 875.6 \text{ g H}_2SO_4$

   (b) $875.6 \text{ g H}_2SO_4 \times \dfrac{\text{mol H}_2SO_4}{98.07 \text{ g H}_2SO_4} \times \dfrac{2 \text{ mol CO}_2}{\text{mol H}_2SO_4} \times$

   $\dfrac{44.0099 \text{ g CO}_2}{\text{mol CO}_2} = 785.9 \text{ g CO}_2$

Chapter 10 Representative Gases

INTRODUCTION

This chapter serves as a bridge between the theoretical and quantitative chemistry of the previous chapters and the chapters of this unit. In the following four chapters, phases of matter will be introduced. The material presented here, some of which is review, emphasizes the important role that certain gases play in our lives.

LOOKING AHEAD

As you study this chapter, pay attention to the descriptions of the physical and chemical properties of the various gases and to how these properties can be explained in terms of the molecular structure of each gas.

SECTION PREVIEW

10.1 Oxygen and Ozone
10.2 Hydrogen
10.3 Nitrogen and Ammonia
10.4 Carbon Dioxide and Carbon Monoxide

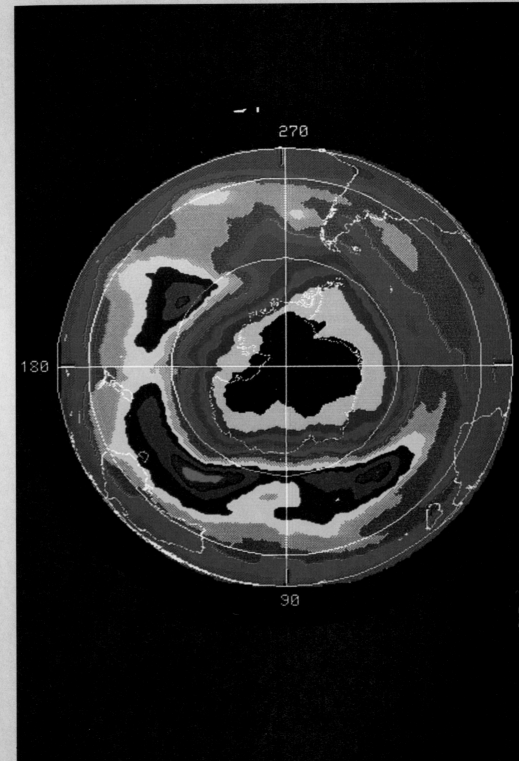

A layer of ozone (O_3) in the upper atmosphere protects us from harmful solar radiation. The increasing size of the hole in the ozone layer is documented by NASA's Nimbus-7 satellite. This satellite plot of ozone distribution in 1987 shows that the hole is nearly half the size of the Antarctic continent.

10.1 Oxygen and Ozone

Two types of gaseous oxygen molecules are present in the earth's atmosphere. One of these molecules is composed of two oxygen atoms and is called, simply, oxygen. The other type of molecule is composed of three oxygen atoms and is called ozone. Oxygen and ozone gas are allotropes. *An* **allotrope** *is one of the two or more forms of an element that exist in the same physical state. The existence of two or more allotropes of an element is called* **allotropy**. Molecular allotropes differ in the arrangement and number of atoms in each molecule and therefore have different physical and chemical properties. To avoid confusion, the term "oxygen" will be used to refer only to ordinary elemental oxygen (O_2), not ozone (O_3).

Oxygen gas is required for the basic life process of aerobic respiration. Respiration occurs in the cells of most living things. In aerobic respiration, oxygen combines with food molecules to produce the energy needed for life. The presence of oxygen in the air is also necessary for the combustion of the various fuels—petroleum, coal, natural gas, and wood—that we use to run our automobiles, generate our electricity, and heat our homes.

In the upper atmosphere, ozone plays an equally crucial role. Ozone absorbs most of the ultraviolet portion of the light radiated from the sun to the earth. Ultraviolet light that reaches the earth's surface causes harmful chemical reactions in plant and animal tissues. Some scientists are concerned about the possible destruction of the ozone layer in the upper atmosphere. This destruction may be caused by chlorofluorocarbons, of CFCs, chemical compounds that escape into the atmosphere during their use in aerosols and refrigerators.

Occurrence of Oxygen

Oxygen is the most abundant element in the earth's crust, waters, and atmosphere. It composes 49.5% of the total mass of these components. Free or uncombined oxygen makes up about 20% of the earth's atmosphere by volume. Free oxygen consists of covalently bonded O_2 molecules. Additional free oxygen is dissolved in the earth's waters, where it is essential for the respiration of aquatic life.

Most of the earth's oxygen exists as combined oxygen, oxygen that has united with other elements to form compounds. For example, water contains about 89% oxygen by mass. The earth's crust is about 47% oxygen by mass. Oxygen is chemically combined with silicon, aluminum, iron, calcium, magnesium, titanium, and other elements to form most of the earth's rocks and minerals.

Structure of the Oxygen Molecule

According to the guidelines introduced in Section 4-3, the Lewis structure for oxygen is

$$\ddot{O}\!:\!\ddot{O}$$

This representation of the oxygen molecule fails to account for certain experimentally determined properties of O_2, however. For example, when

STUDY HINT

CFCs and the ozone layer are discussed in the Intra-Science article, "Pressuring the Ozone Layer" on pages 400-401.

Figure 10-1 Many aerosol sprays, such as deodorants and hair sprays, no longer use chlorofluorocarbons.

liquid oxygen is allowed to flow past a magnet, it is attracted to the magnetic field. As discussed in Section 4.2, a substance that is attracted to a magnetic field is said to be paramagnetic. To be paramagnetic, oxygen would need two unpaired electrons, the spins of which do not cancel. The following resonance hybrid structures of the O_2 molecule better explain this and other observed properties of oxygen:

$$:\ddot{O}:\ddot{O}: \longleftrightarrow :\ddot{O}::\ddot{O}: \longleftrightarrow :O\vdots\vdots O:$$

Physical Properties of Oxygen

At normal temperature and pressure, pure oxygen is a colorless, odorless, tasteless gas that is slightly denser than air. Like all other gases, oxygen can be liquefied. At temperatures lower than −118.6°C and at pressures greater than 49.8 atm (5.02 MPa), it becomes a pale blue liquid. At normal atmospheric pressure, liquid oxygen boils at −183.0°C. When cooled below −218.4°C, liquid oxygen freezes to become a pale-blue crystalline solid. Physical constants of oxygen are listed in Table 10-1.

TABLE 10-1	PHYSICAL PROPERTIES OF OXYGEN (O_2) AND OZONE (O_3)	
	Oxygen (O_2)	Ozone (O_3)
Molecular mass	31.998 8	47.998 2
Boiling point	−183.0°C	−111.9°C
Melting point	−218.4°C	−192.7°C
Density (0°C, 1 atm)	1.429 g/L	2.144 g/L
Bond length (O—O)	0.120 8 nm	0.127 8 nm
Bond angle	−180°	117°

Chemical Properties of Oxygen

Oxygen is one of the most active elements. With the exception of the lighter noble gases, oxygen forms compounds with all other elements. For example, when exposed to air, Group 1 metals combine directly with oxygen to form three types of oxygen compounds called oxides, peroxides, and superoxides. *An* **oxide** *is a compound containing oxygen. The simple oxide anion has an oxidation number of −2.* The bond character in an oxide depends on the difference in electronegativities between oxygen and the element with which it is combined. The greater the difference in electronegativities, the more ionic, or less covalent, the bond.

Oxygen is the second most electronegative element. Because Group 1 and Group 2 metals have low electronegativities, oxides of Group 1 and Group 2 metals are highly ionic. Among the Group 1 metals, only lithium combines directly with oxygen to form an oxide.

$$4Li(s) + O_2(g) \rightarrow 2Li_2O(s)$$

Sodium, which is more reactive than lithium and the Group 2 metals, burns in oxygen to form sodium peroxide. *A* **peroxide** *is a compound containing the O_2^{2-} anion, in which oxygen has the oxidation number −1.*

Important dates in the history of oxygen:
15th century Leonardo da Vinci (Italian artist, 1452–1519) writes that air has several constituents, one of which supports combustion.
1773–1774 Karl Wilhelm Scheele (Swedish chemist, 1742–1786) and Joseph Priestley (English chemist, 1733–1804) independently discover oxygen.
1775–1777 Antoine-Laurent Lavoisier (French chemist, 1743–1794) first recognized oxygen as an element. He developed the modern theory of combustion and demolished the phlogiston theory. Lavoisier coined the name "oxygen" from the Greek *oxys*, "sharp or sour" and *geinomai*, "I produce."

$$2Na(s) + O_2(g) \rightarrow Na_2O_2(s)$$

The still more reactive Group 1 metals—potassium, rubidium, and cesium—burn in oxygen to form superoxides. A **superoxide** *is a binary compound containing the* O_2^- *anion, in which oxygen has the uncommon oxidation number* $-1/2$.

$$K(s) + O_2(g) \rightarrow KO_2(s)$$

The Group 2 metals react spontaneously with pure oxygen to form highly ionic oxides. For example, barium reacts with oxygen to form barium oxide as follows:

$$2Ba(s) + O_2(g) \rightarrow 2BaO(s)$$

Although most of the remaining metals react relatively slowly with oxygen at normal temperatures, many react quickly when heated. Because these metals have electronegativities that are close to the electronegativity of oxygen, the bonding in the resulting oxides is generally polar covalent. In the open air, moist iron reacts with oxygen to form hydrated iron (III) oxide, commonly called rust, in the following reactions.

$$2Fe(s) + O_2(g) + 2H_2O(\ell) \rightarrow 2\ Fe(OH)_2(s)$$

$$4Fe(OH)_2(s) + O_2(g) + 2H_2O(\ell) \rightarrow 4Fe(OH)_3(s)$$

$$2Fe(OH)_3(s) \longrightarrow FeO_3 \cdot H_2O + 2H_2O$$

As shown in Figure 10-2, however, red-hot steel wool burns brilliantly when plunged into pure oxygen. This reaction produces another iron oxide (Fe_3O_4), called *black magnetite*.

$$3Fe(s) + 2O_2(g) \rightarrow Fe_3O_4(s)$$

Black magnetite is a mixed oxide of Fe_2O_3 and FeO and is referred to as an iron(II, III) oxide.

Because they have electronegativities only slightly lower than the electronegativity of oxygen, nonmetals that burn in pure oxygen or in air produce covalently bonded oxides. The combustion of hydrogen and oxygen to produce water, and the combustion of sulfur in air to produce sulfur dioxide, are examples of this type of reaction.

$$2H_2(g) + O_2(g) \rightarrow 2H_2O(\ell)$$

$$S(s) + O_2(g) \rightarrow SO_2(g)$$

Sulfur dioxide is also produced in the combustion of sulfur-containing fuels, such as high-sulfur coal. When SO_2 gas is released into the atmosphere, it is further oxidized and combines with water in the air to form sulfuric acid. The presence of sulfuric acid in the air is a major cause of acid rain. Figure 10-3 shows one harmful effect of acid rain on the environment.

Uses and Preparation of Oxygen

Pure gaseous oxygen has a variety of important applications, some of which are illustrated in Figure 10-4. Oxygen can be used to remove impurities from molten steel. The oxygen gas unites with carbon and other impurities. The resulting compounds escape as gases or solids that float to the surface of the molten steel in the form of removable slag.

Figure 10-2 Iron burns brilliantly in pure oxygen to form the iron oxide Fe_2O_4.

Figure 10-3 Some scientists think that the effects of acid rain caused by the combustion of sulfur-containing fuels may have caused the death of the trees in this forest.

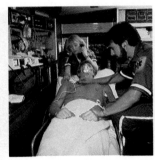

Figure 10-4 *(a)* Impurities are burned off molten steel in a basic-oxygen furnace by a stream of oxygen gas. *(b)* Portable oxygen equipment is used in the emergency treatment of people with breathing difficulties. *(c)* Oxygen in the air kills harmful anaerobic bacteria in a sewage-treatment plant. *(d)* An oxygen–acetylene torch is used to weld steel.

Metals are commonly welded together by the use of an oxygen–acetylene torch. A burning mixture of oxygen gas and acetylene gas can have a temperature as high as 3500°C, which is hot enough to melt most common metals.

The combustion of liquid oxygen and liquid hydrogen provides the energy to lift huge spacecraft into orbit. Oxygen is also used to treat people with breathing difficulties or heart problems and to purify water by killing harmful anaerobic bacteria and by destroying odor-causing substances. In the discussion that follows, some of the most common laboratory and industrial preparations of oxygen are described.

Decomposition of hydrogen peroxide   Oxygen can be prepared safely and conveniently in the laboratory by decomposing hydrogen peroxide in the presence of the catalyst, powdered manganese dioxide. The laboratory apparatus for this reaction is shown in Figure 10-5. The reaction that takes place can be written as follows:

$$2H_2O_2(aq) \xrightarrow[\text{MnO}_2]{\Delta} 2H_2O(\ ) + O_2(g)$$

CAUTION: If you demonstrate this reaction, be sure to use a 3% solution of H_2O_2. A 30% solution is too reactive to be used safely.

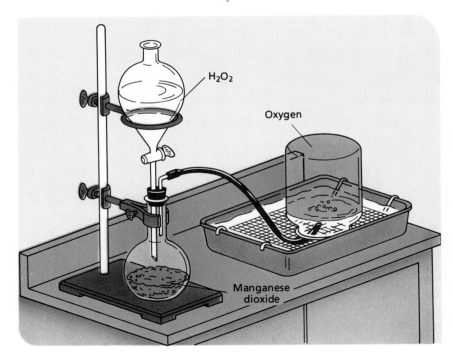

H_2O_2

Oxygen

Manganese dioxide

Figure 10-5 Oxygen can be prepared in the laboratory by the decomposition of hydrogen peroxide in the presence of a catalyst (manganese dioxide). The oxygen is collected by water displacement. This same apparatus can be used to prepare oxygen by allowing water to drop onto sodium peroxide.

Oxidation of water by sodium peroxide When water is allowed to drop onto sodium peroxide in a generator similar to that shown in Figure 10-5, oxygen is liberated as follows:

$$2Na_2O_2(s) + 2H_2O(\ell) \rightarrow 4NaOH(aq) + O_2(g)$$

Thermal decomposition of potassium chlorate When potassium chlorate is mixed with the catalyst manganese dioxide and heated, it decomposes into oxygen and potassium chloride.

$$2KClO_3(s) \xrightarrow[\text{MnO}_2]{\Delta} 2KCl(s) + 3O_2(g)$$

Electrolysis of water Figure 10-6 shows a standard laboratory apparatus used for the electrolysis of water. Electrolysis of water is the decomposition of water into hydrogen gas and oxygen gas by the passage of a direct current of electricity through the water. As the current flows, oxygen gas collects at the positive terminal; at the same time hydrogen gas collects at the negative terminal.

$$2H_2O(\ell) \rightarrow 2H_2(g) + O_2(g)$$

Industrially, this process yields oxygen of very high purity.

Liquefaction of air The most common industrial preparation of oxygen involves the liquefaction of ordinary air by compression and cooling. Nitrogen is then separated from oxygen as follows. When liquid air at $-200°C$ is allowed to warm up, nitrogen, which boils at $-195.8°C$, boils off first. Oxygen, which boils at $-183.0°C$, and a few trace gases remain. This process is called fractional distillation. The oxygen is separated from the trace gases by further fractional distillation and condensation.

Formation and Properties of Ozone

You may already have observed the formation of ozone and not realized it. For example, have you ever smelled an irritating or pungent odor after a lightning storm or near electrical machinery such as a photocopier? If so, what you smelled was probably ozone.

The structure and properties of ozone Like ordinary molecular oxygen (O_2), ozone (O_3) is a gas at room temperature. Oxygen is colorless and odorless, however, and ozone is a poisonous blue gas with a pungent odor. Physical constants of ozone are listed in Table 10-1.

Figure 10-7 compares the shapes of an oxygen molecule and an ozone molecule. The ozone molecule is triatomic and bent. The O—O—O bond angle is about 116.5°. The O—O bond in ozone has a length of 128 pm and an energy of 301 kJ/mol. This bond is shorter and stronger than an O—O single bond, but longer and weaker than an O—O double bond. The ozone molecule is therefore represented by two resonance hybrid structures.

Formation of ozone Ozone is formed from oxygen in a two-step process. First, energy is added to molecular oxygen to produce atomic oxygen. Next,

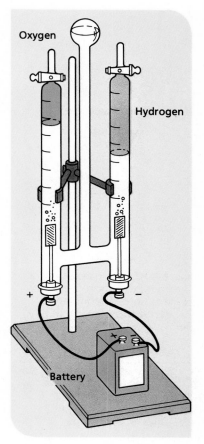

Figure 10-6 Water is decomposed by electrolysis into its constituent elements, oxygen and hydrogen.

STUDY HINT
Resonance is discussed in a Chemistry Notebook in Section 6.2.

CAUTION: Sodium peroxide must be handled carefully. It is a very strong oxidizing agent and a high fire and explosion risk. It should not come in contact with any organic materials, powdered metals, alcohols, acids, or water. It is toxic, irritating, and corrosive to body tissues.

Oxygen molecule

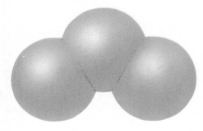

Ozone molecule

Figure 10-7 An oxygen molecule consists of two atoms of oxygen. An ozone molecule consists of three atoms of oxygen.

---

1. *(a)* See definition on page 289. *(b)* Both gases at room temperature
2. Absorbs most of the ultraviolet light radiated from the sun which can harm plant and animal tissue
3. Removes impurities from molten steel, welds metals together, used in spacecraft engines, used to treat people with breathing difficulties or heart problems, and purifies water supplies
4. Ozone is a poisonous blue gas with a pungent odor; oxygen is a colorless and odorless gas and is necessary for life.
5. See "Formation of Ozone," above.

---

each free oxygen atom combines with an oxygen molecule. The overall equation for the formation of ozone is

$$3O_2(g) + \text{energy} \rightarrow 2O_3(g)$$

The energy required in the first step may be provided by ultraviolet radiation, as is the case in the formation of the ozone layer in the upper atmosphere. It can also be provided by electrical discharges, which explains why the smell of ozone can be noticed after lightning storms or near electrical machinery. A typical laboratory ozone generator produces ozone by passing a high-voltage electric current through air or oxygen.

Section Review

1. *(a)* What is an allotrope?   *(b)* Why are oxygen and ozone considered to be allotropes?
2. Why is ozone in the upper atmosphere important to life on earth?
3. Describe five important technological applications of oxygen.
4. Compare the physical properties of oxygen and ozone.
5. Describe the two steps in the formation of ozone and write the overall equation for this process.

10.2 Hydrogen

Hydrogen is the first element in the Periodic Table and the most abundant element in the universe. Is is estimated that 90% of all the atoms in the universe are hydrogen atoms, which constitute 70% of the mass of the universe. The earth's surface environment, however, contains only 0.88% hydrogen by mass. On the earth's surface, free elemental hydrogen exists in negligible quantities as a gas consisting of diatomic hydrogen molecules H_2. The main engines of the space shuttle shown in Figure 10-8 burn liquid hydrogen as fuel. Pure hydrogen, however, is not used as a fuel in ordinary situations, such as heating a home or running a car, because of its tendency to explode. Some scientists speculate that gaseous hydrogen could replace natural gas as a fuel for homes and industries. If better techniques to control its explosiveness are developed, hydrogen gas could be an efficient, nonpolluting source of energy for a "hydrogen economy" of the future.

Occurrence of Hydrogen

Some scientists hypothesize that hydrogen nuclei were the first atomic nuclei formed from subatomic particles as the universe cooled down from the "big bang" that may have marked its origin. Today, the universe still consists mainly of hydrogen, although nearly 100 other elements have been produced by nuclear reactions in the cores of stars. Our sun and the other stars are largely elemental hydrogen in the plasma state.

Very little free elemental hydrogen exists on earth. Some traces of elemental hydrogen are released into the air by volcanoes, but the density of H_2 gas is so low that it rises into the upper atmosphere and escapes into space. The huge amount of water on the earth's surface is 11% hydrogen by mass. In addition, hydrogen is a constituent of many minerals and all living things. Petroleum, natural gas, wood, and soft coal are rich in hydrogen. The combination of hydrogen with oxygen provides much of the heat produced by the combustion of these fuels.

Physical Properties of Hydrogen

Unlike the Group 1 elements, hydrogen ordinarily exists in the form of H_2 molecules. As a typical molecular substance, molecular hydrogen is a poor conductor of heat and electricity.

Under ordinary conditions, hydrogen exists as a colorless, odorless, and tasteless gas. Hydrogen gas—the least dense of all gases—has a density only 7% that of air. (At 0°C and 1 atm, the density of hydrogen is 0.08987 g/L and the density of air is 1.293 g/L.) Because of its low solubility in water, hydrogen can be collected in the laboratory by water displacement, as shown in Figure 10-9. When cooled below −252.9°C, hydrogen gas becomes a liquid. Liquid hydrogen is the least dense of all liquids, having a density only 7% that of liquid water. When cooled below −259.1°C, liquid hydrogen becomes a solid. Solid hydrogen is the least dense solid, being only 8% as dense as liquid water. Some physical properties of hydrogen are listed in Table 10-2.

Figure 10-8 Liquid hydrogen is the fuel used in the space shuttle's main engines. Liquid oxygen is used as the oxidizing agent for the very rapid fuel combustion needed during lift-off.

Figure 10-9 Hydrogen is produced by the action of a metal and dilute acid in the generator. Water vapor carried over with the hydrogen from the generator is removed in the drying tube. The dry hydrogen burns in the bell jar, producing water vapor, which condenses on the cool walls of the jar and drops into a collecting vessel.

TABLE 10-2 PHYSICAL PROPERTIES OF HYDROGEN (H_2)

Molecular mass	2.015 8
Boiling point	−252.8°C
Melting point	−259.1°C
Density (0°C, 1 atm)	0.089 9 g/L
Bond length (H—H)	0.074 611 nm

Chemical Properties of Hydrogen

There are more compounds of hydrogen than of any other element. Hydrogen reacts with both metals and nonmetals.

Reaction with nonmetals   Hydrogen's electronegativity of 2.1 is close to that of most other nonmetals. As a result, hydrogen reacts with nonmetals to form molecular compounds with single covalent bonds. The bonds range in polarity from the nonpolar H—P bond to the highly polar H—F bond.

Although free hydrogen atoms are extremely reactive, ordinary molecular hydrogen is not very reactive at room temperature. A mixture of hydrogen and oxygen must be heated to 800°C or ignited by an electric spark to make the two gases combine. They then combine explosively to form water according to the following equation:

$$2H_2(g) + O_2(g) \rightarrow 2H_2O(g)$$

Hydrogen and chlorine do not combine when they are mixed in the dark. They react explosively in sunlight to form hydrogen chloride, however. Hydrogen chloride can also be prepared by burning a stream of hydrogen gas in chlorine. The equation for the reaction of hydrogen with chlorine is

$$H_2(g) + Cl_2(g) \rightarrow 2HCl(g)$$

Reaction with metals   Hydrogen reacts with the least electronegative metals to form ionic hydrides. *An* **ionic hydride** *is a compound consisting of hydrogen and one other element that is less electronegative.* Each hydrogen atom receives an electron from the less electronegative element to form the hydride ion, H^-. The hydrides of the Group 1 and Group 2 metals are prepared by heating the metals in the presence of hydrogen. The preparations of the white, crystalline, ionic hydrides of sodium and barium are represented by the following equations:

$$2Na(s) + H_2(g) \rightarrow 2NaH(s)$$

$$Ba(s) + H_2(g) \rightarrow BaH_2(s)$$

A **covalent hydrid** *is a compound consisting of hydrogen and one other element that is more electronegative.* For example, hydrogen burns in the presence of bromine to form hydrogen bromide.

$$H_2(g) + Br_2(\ell) \rightarrow 2HBr(g)$$

Uses and Preparation of Hydrogen

Among the important industrial uses of hydrogen is its combination with liquid vegetable oils to produce hydrogenated solid or semisolid fats, such as margarine. Many dietary experts advise against the use of hydrogenated oils, which some research has linked to heart disease.

Hydrogen gas is sometimes used as a reducing (deoxidizing) agent. Its use as a fuel—particularly in fuel cells that produce electricity—is increasing. Hydrogen gas is also used to remove sulfur impurities from petroleum and coal. The hydrogen combines with sulfur to produce hydrogen sulfide gas (H_2S) which can be removed easily. The remainder of this section discusses the common laboratory and commercial preparations of hydrogen.

CAUTION: The mixing of hydrogen and chlorine in a glass tube is a popular but dangerous demonstration because of the possibility of the glass tube shattering. The glass tube should be covered with plastic netting as a safety measure.

Important dates in the history of hydrogen:
1671 Robert Boyle (English physicist and chemist, 1627–1691), along with other 17th century scientists, observed that a flammable gas was produced by the action of dilute sulfuric acid on iron.
1766 Henry Cavendish (English chemist and physicist, 1731–1810) observed that hydrogen gas was lighter than air. He reacted iron, zinc, and tin with several acids, establishing hydrogen's properties.
1781 Cavendish disproved that water is an element by forming it from the combustion of hydrogen in oxygen.
1783 Lavoisier coined the name "hydrogen" from two Greek words, *hydro* and *genes*, meaning "water former."
1878 Sir Joseph Norman Lockyer (English astronomer, 1836–1920) detected hydrogen spectroscopically in the sun's corona.

From acids by replacement The usual laboratory method of preparing hydrogen involves the replacement of hydrogen from acids by metals above it in the activity series. For example, iron, zinc, or magnesium react with hydrochloric acid or sulfuric acid to produce hydrogen. The equations for the chemical reactions of zinc with sulfuric acid (H_2SO_4) and hydrochloric acid (HCl) are

$$Zn(s) + H_2SO_4(aq) \rightarrow ZnSO_4(aq) + H_2(g)$$

$$Zn(s) + 2HCl(aq) \rightarrow ZnCl_2(aq) + H_2(g)$$

In each of these two reactions, an atom of zinc replaces two atoms of hydrogen, which unite to form gaseous molecular hydrogen, H_2.

From water by replacement Group 1 and Group 2 metals react with water, displacing hydrogen and forming a base containing the hydroxide ions. For example, sodium reacts vigorously with water. Each sodium atom replaces one of the hydrogen atoms in a molecule of water, as indicated by the following equation:

$$2Na(s) + 2H_2O(\ell) \rightarrow 2NaOH(aq) + H_2(g)$$

Potassium is more active than sodium and displaces hydrogen from water so vigorously that the heat of the reaction ignites the hydrogen, as shown in Figure 10-10. The reaction is as follows:

$$2K(s) + 2H_2O(\ell) \rightarrow 2KOH(aq) + H_2(g)$$

The Group 2 metal magnesium displaces hydrogen slowly from boiling water, whereas the more reactive Group 2 metal calcium displaces hydrogen from cold water. In each case, each atom of metal displaces two atoms of hydrogen—one from each of two molecules of water—as shown in the following equation:

$$Ca(s) + 2H_2O(\ell) \rightarrow Ca(OH)_2(aq) + H_2(g)$$

Electrolysis of water In the electrolysis of water, water is decomposed to hydrogen and oxygen. Electrolysis is discussed in Section 10.1.

Decomposition of hydrocarbons Hydrocarbons are commonly derived from petroleum or natural gas. They are the major commercial source of hydrogen. If a hydrocarbon such as methane (CH_4) reacts with steam in the presence of a nickel catalyst at a temperature of about 850°C, a mixture of hydrogen and carbon monoxide is produced.

$$CH_4(g) + H_2O(g) \rightarrow CO(g) + 3H_2(g)$$

When the mixture of gaseous products is cooled and compressed, the carbon monoxide liquefies and can be separated from the hydrogen gas. (The reverse of this reaction can be used to synthesize methane.)

Section Review

1. What are some important industrial uses of hydrogen?
2. Discuss the physical properties of hydrogen.
3. How does hydrogen react with *(a)* nonmetals and *(b)* metals? Write one balanced equation for each type of reaction.
4. What is a hydride?

Figure 10-10 Potassium, like other Group 1 metals, reacts vigorously with water. The light for this photograph was produced by dropping a small amount of potassium into a beaker of water.

1. Hydrogen combines with vegetable oils to produce solid or semisolid fats. Hydrogen gas is used as a reducing agent, as a fuel, and to remove impurities from petroleum and coal.
2. Under ordinary conditions, hydrogen exists as a colorless, odorless, and tasteless gas. Hydrogen is the least dense of all gases and has low solubility in water. Molecular hydrogen is a poor conductor of heat and electricity.
3. *(a)* Forms molecular compounds with single covalent bonds
$2H_2(g) + O_2(g) \rightarrow 2H_2O(g)$
$2H_2(g) + Cl_2(g) \rightarrow 2HCl(g)$
(b) Forms ionic hydrides
$2Na(s) + H_2(g) \rightarrow 2NaH(s)$
$Ba(s) + H_2(g) \rightarrow BaH_2(s)$
4. A compound consisting of hydrogen and one other element that is less electronegative

10.3 Nitrogen and Ammonia

SECTION OBJECTIVES

- Discuss the importance and abundance of nitrogen in the atmosphere.
- Compare the structure of molecular nitrogen with that of ammonia.
- Describe the physical and chemical properties of nitrogen and ammonia.
- Discuss the preparation of nitrogen and ammonia.
- Describe the nitrogen cycle.

Nitrogen in the form of gaseous molecular nitrogen (N_2), is the major component of the earth's atmosphere. Although nitrogen in this form is nonreactive, it is important to life on earth. Nitrogen is found in all living things on earth. It is present in the DNA molecules that control heredity. It is also a major component of amino acids, which are the building blocks of proteins. Proteins are present in living tissue and make up the enzymes that control most of the chemical reactions occurring in living things. The nitrogen cycle is described in the Chemistry Notebook on page 300.

The most important nitrogen compound is ammonia (NH_3) which is the chief intermediate between atmospheric nitrogen and the millions of nitrogen-containing compounds. Ammonia is used as a fertilizer, a refrigerant, and as an intermediate in the production of nitric acid, nylon, antimalarial drugs, and other nitrogen-containing compounds. Other uses of nitrogen are illustrated in Figure 10-11.

Occurrence of Nitrogen

The earth's atmosphere consists of approximately 80% molecular nitrogen (N_2) by volume. Combined nitrogen in the form of ammonia, nitrates, nitrites, and nitric acid is widely distributed on earth in living things and in natural deposits of sodium nitrate and potassium nitrate.

Figure 10-11 *(a)* Fruit protected by a "blanket" of nitrogen *(top)*. *(b)* Liquid nitrogen is used in some cryosurgery, such as removing warts *(below)*.

TABLE 10-3 PHYSICAL PROPERTIES OF NITROGEN (N_2)	
Molecular mass	28.013 4
Boiling point	−195.8°C
Melting point	−209.86°C
Density (0°C, 1 atm)	1.251 g/L
Bond length (N—N)	0.109 75 nm

Physical Properties of Nitrogen

Under normal conditions, elemental nitrogen (N_2) is a colorless, odorless, and tasteless gas that is slightly less dense than air. At temperatures lower than −195.8°C at 1 atm, nitrogen gas condenses to a colorless liquid. Liquid nitrogen freezes to a white solid at −209.9°C. Physical properties of nitrogen are listed in Table 10-3.

Structure and Chemical Properties of Nitrogen

Elemental nitrogen normally exists in the form of nonpolar, triple-bonded N_2 molecules. Because of the high energy required to break the nitrogen-nitrogen triple bond, it is hard to separate the nitrogen atoms of the N_2 molecule so that they can enter into new combinations. This explains why elemental nitrogen is relatively nonreactive. The formation of many nitrogen compounds requires the net addition of energy, which is then stored in the compounds. This stored energy has a strong tendency to be released. As a result, these compounds are often unstable and sometimes explosive.

Oxides of nitrogen are also formed during the high-temperature, high-pressure combustion in automobile engines. These oxides can combine with water in the atmosphere to produce acids, that sometimes fall to earth as acid rain. *Using energy from ultraviolet light from the sun, nitrogen oxides can react with unburned hydrocarbons from automobile exhaust or with oxygen in the air to produce ozone and other irritating chemicals that, together, are referred to as* **photochemical smog**. The catalytic converters required on all new cars help reduce this type of air pollution by deoxidizing nitrogen oxides to nitrogen and by completely oxidizing hydrocarbons to carbon dioxide and water.

At high temperatures, nitrogen combines directly with such metals as magnesium, titanium, and aluminum to produce nitrides. Magnesium will burn in nitrogen gas to form Mg_3N_2, magnesium nitride. In nitrides, nitrogen exists in the form of the nitride in ion (N^{3-}).

Figure 10-12 Jupiter's thick atmosphere is composed of swirling clouds of ammonia crystals.

Uses and Preparation of Nitrogen

The major uses of nitrogen are a direct result of its nonreactivity. Many foods are packaged in the presence of pure nitrogen rather than air. This is done to prevent the interaction of these products with the oxygen in air because in some foods, the presence of oxygen results in loss of flavor. Nitrogen is now being used as the propellant in aerosol cans because it is both nontoxic and nonpolluting.

Nitrogen is convenient to use for many of these applications because of its ready availability and ease of production. The industrial production of nitrogen involves the liquefaction of air, as described in Section 10.1.

Nitrogen can be prepared in the laboratory by heating certain ammonia-containing compounds. For example, heating a water solution of ammonium nitrite (NH_4NO_2) produces nitrogen and water according to the following equation:

$$NH_4^+ (aq) + NO_2^- (aq) \rightarrow N_2(g) + 2H_2O(\ell)$$

The extremely high energy of the triple bond in N_2, 941.4 kJ, gives it an inert gas nature. Lithium is the only element to combine with N_2 at room temperature.

Occurrence of Ammonia

Ammonia is abundant in the atmospheres of the outer planets Jupiter, Saturn, and Neptune. In fact, clouds of solid ammonia surround these planets. The top layer of Jupiter's colorful and violent atmosphere as photographed by the *Voyager* spacecraft (Figure 10-12), is made up of ammonia crystals. On Earth, ammonia is continually being produced naturally by the fixation of atmospheric nitrogen and by the decay of organic matter, as described in the Chemistry Notebook "The Nitrogen Cycle" on page 300. As fast as it is produced, however, it is converted into other substances in living organisms.

Structure and Properties of Ammonia

The Lewis structure for ammonia is

$$H:\overset{\cdot\cdot}{N}:H$$
$$\overset{|}{H}$$

As discussed in Chapter 6, the ammonia molecule has the shape of a pyramid. The three hydrogen atoms form the base, the nitrogen atom is in the center,

Chemistry Notebook

The Nitrogen Cycle

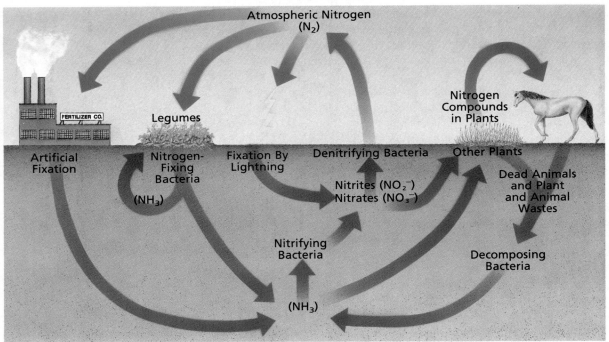

The process by which nitrogen continuously moves back and forth between living things and the atmosphere is called the *nitrogen cycle. Through **nitrogen fixation**, atmospheric nitrogen gas (N_2) is converted into the nitrogen compounds that can be used by plants.*

Most natural nitrogen fixation is carried out by bacteria on the roots of legumes. A legume is a plant belonging to the pea, bean, clover, or alfalfa family. In this case, the nitrogen is fixed in the form of ammonia (NH_3). Ammonia can be incorporated into plants either directly or after having been converted to nitrate ions (NO_3^-) and/or nitrite (NO_2^-) by nitrifying bacteria. Legumes used in crop rotation restore nitrogen to soil that has been depleted of it by other crops. In Asia, the fertility of rice paddies is maintained by free-living, nitrogen-fixing blue-green algae.

A great deal of natural nitrogen fixation also occurs by means of lightning, which causes atmospheric nitrogen and oxygen to react to form nitrate ions. These nitrates are dissolved in rainwater and carried down to the soil.

Animals consume nitrates in plants. Nitrogen compounds in dead plants and animals are then decomposed by bacteria into ammonia compounds.

Nitrogen is fixed artificially on a large scale by the direct combination of nitrogen and hydrogen to form ammonia in the Haber process discussed on page 302. Much of the resulting ammonia is then industrially converted to nitrates. Both ammonia and inorganic nitrates are widely used as artificial fertilizers.

Nitrogen is returned to the atmosphere through the action of denitrifying bacteria, which convert nitrate in the soil to gaseous nitrogen and thus complete the nitrogen cycle.

and the electron pair is at the apex. The ammonia molecule is therefore polar-covalent. In liquid and solid ammonia, hydrogen bonds link the hydrogen atoms of ammonia molecules to the nitrogen atoms of neighboring ammonia molecules. Because of this hydrogen bonding, ammonia has higher boiling and freezing points than would otherwise be expected for a molecule of its molecular mass.

STUDY HINT
Molecular mass is discussed in Chapter 3.

Under normal conditions, ammonia is a colorless gas with such a characteristically strong odor that has traditionally been used to revive fainting victims. However, in high concentration it is very harmful. Ammonia is less dense than air and is extremely soluble in water, with which it forms strong hydrogen bonds. You are probably familiar with household ammonia, which is a useful grease-cutting agent. Some physical properties of ammonia are listed in Table 10-4.

TABLE 10-4 PHYSICAL PROPERTIES OF AMMONIA (NH_3)

Molecular mass	17.030 4
Boiling point	−33.35°C
Melting point	−77.7°C
Density (0°C, 1 atm)	0.771 g/L
Bond length (N—H)	0.100 8 nm
Bond angle (H—N—H)	107.03°

Chemically, ammonia is important as a base and as an intermediate in important natural, laboratory, and industrial organic reactions. At high temperatures, ammonia decomposes into nitrogen (N_2) and hydrogen (H_2). In the presence of a hot platinum-wire catalyst, ammonia also burns in air or oxygen to produce nitrogen and water, as is shown by the following equation:

$$4NH_3(g) + 3O_2(g) \rightarrow 2N_2(g) + 6H_2O(\ell)$$

Uses and Preparation of Ammonia

Ammonia is important in the production of artificial fertilizers. Until about a hundred years ago, farmers commonly added naturally occurring nitrogen compounds to the soil to increase crop production. For example, during the nineteenth century, sodium nitrate from Chile was imported by the United States and European countries for use as fertilizer. Then, in 1913, the German chemist Fritz Haber (1868–1934) invented a process by which nitrogen could be extracted from the air to make ammonia that could be used as an artificial fertlizer.

In 1985, more moles of ammonia were produced than of any other compound. In 1986, nitrogen gas replaced ammonia as the compound most produced. Sulfuric acid continues to be the first in terms of mass.

Decomposition of ammonium compounds In the most common laboratory preparation of ammonia, a mixture of calcium hydroxide and ammonium chloride is heated in a test tube fitted with an L-shaped delivery tube, as shown in Figure 10-13. Ammonia is so soluble in water that, unlike most other gases, it cannot be collected by water displacement. It is collected by the displacement of air in an inverted test tube, instead. The equation for this reaction is

$$Ca(OH)_2(s) + 2NH_4Cl(s) \xrightarrow{\Delta} CaCl_2(s) + 2NH_3(g) + 2H_2O(g)$$

Figure 10-13 When a mixture of ammonium chloride and calcium hydroxide is heated in a test tube, ammonia is given off. The ammonia is collected by the downward displacement of air because it is less dense than air and very soluble in water. Litmus paper turns blue in the presence of ammonia, a base.

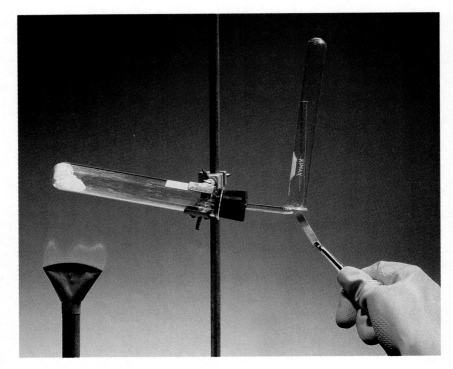

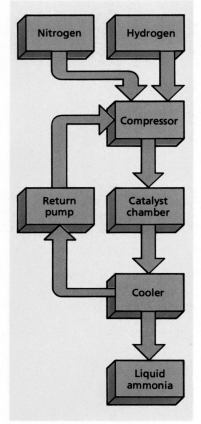

Figure 10-14 (a) A flow chart of the Haber process. Ammonia gas produced in the catalyst chamber is condensed into a liquid in the cooler. The uncombined nitrogen and hydrogen are recirculated through the compressor and catalyst chamber.

Industrial preparation of ammonia   Ammonia is formed as an industrial by-product in the production of coal gas by heating bituminous coal in the absence of air. Most ammonia is produced industrially by the Haber process, however. *The* **Haber process** *is the catalytic synthesis of ammonia from nitrogen gas and hydrogen gas.*

$$N_2(g) + 3H_2(g) \leftrightarrows 2NH_3(g)$$

This reaction is reversible and is carried out at temperatures of 400°–500°C and at pressures as high as 1000 atm (101 MPa) using a catalyst that is a mixture of porous iron and oxides of other metals. Under these conditions, about 40 to 60 percent of the reacting gases are changed to ammonia in each pass through the catalyst chamber, as shown in the chart in Figure 10-14. The ammonia is separated from the unreacted nitrogen and hydrogen by dissolving it in water or by cooling and liquefying it. The unreacted nitrogen and hydrogen are recompressed and again passed over the catalyst. Eventually, all the nitrogen and hydrogen are converted to ammonia in this way. More than 10 billion kilograms of ammonia are produced each year in the United States.

Section Review

1. Explain the importance of nitrogen to life on earth.
2. Compare the structures of nitrogen (N_2) and ammonia (NH_3).
3. Compare the physical properties of nitrogen and ammonia.
4. What is photochemical smog?
5. How is nitrogen prepared industrially?
6. Explain what is meant by the nitrogen cycle.

10.4 Carbon Dioxide and Carbon Monoxide

Depending on the reaction conditions, oxygen and carbon can combine to form either carbon dioxide or carbon monoxide. These two compounds have extremely different properties. Carbon monoxide is a highly toxic compound produced both in nature and as a by-product of many human activities. Carbon dioxide, however, plays a key role in supporting life on the earth. It is one of two ingredients used by plants in photosynthesis. Photosynthesis *is the process by which plants use energy from the sun to combine carbon dioxide with water to produce glucose.* The overall equation for this reaction is

$$6CO_2(g) + 12H_2O(\ell) + \text{energy (sunlight)} \rightarrow$$

$$C_6H_{12}O_6(s) + 6O_2(g)(\ell) + 6H_2O(\ell)$$

In effect, photosynthesis converts the radiant energy from sunlight into stored chemical energy in the glucose molecule. The glucose produced by photosynthesis is converted by living things into a wide variety of other carbon-containing substances and ultimately provides energy for all life processes on earth.

Occurrence of Carbon Dioxide

Carbon dioxide makes up only about 0.03 percent of the earth's atmosphere by volume. This percentage is gradually increasing, however. The increased amounts of carbon dioxide come from the burning of carbon-containing fuels, such as coal. The destruction of forests throughout the world also causes an increase in atmospheric carbon dioxide. These forests remove carbon dioxide from the atmosphere through photosynthesis. Some scientists fear that the increased levels of carbon dioxide may be responsible for a gradual increase in the average temperature of the atmosphere. The earth continuously absorbs and reflects radiant energy from the sun. A certain temperature balance is maintained. Carbon dioxide gas in the atmosphere tends to trap some of the reflected energy. This is known as the "greenhouse effect." Adding carbon dioxide to the atmosphere is like adding a layer of insulation to a house. The house gets warmer.

Even a small increase in global temperature could be harmful. The polar ice caps might melt and raise the level of the oceans, thus flooding coastal cities. Also, areas that are now valuable farmland might become dry and unproductive because of changes in global weather patterns caused by a rise in global temperature.

Structure of Carbon Dioxide

Carbon dioxide molecules are linear, with the two oxygen atoms bonded on opposite sides of the carbon atom, as shown in Figure 10-15. Because of this arrangement, the identical polarities of the individual carbon–oxygen bonds are opposed and the molecule is nonpolar. Scientist have determined experimentally that the carbon–oxygen bonds in carbon dioxide are shorter

C_3O_2, carbon suboxide, is also known. It is a gas (bp 6°C) prepared by the dehydration of malonic acid with P_2O_5 at 300°C.

SECTION OBJECTIVES

- Discuss the importance and relative abundance of carbon dioxide and carbon monoxide in the atmosphere.
- Compare the structure of carbon dioxide with that of carbon monoxide.
- Describe the physical and chemical properties of carbon dioxide and carbon monoxide.

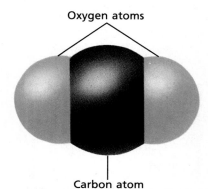

Oxygen atoms

Carbon atom

Figure 10-15 A carbon dioxide molecule is linear and consists of one carbon atom and two oxygen atoms.

1. In DNA molecules; major component of amino acids
2. N_2: nonpolar, triple-bonded molecules. NH_3: covalent molecule in the shape of a pyramid with the three hydrogen atoms forming the base, the nitrogen atom at the center, and an electron pair at the apex
3. N_2: colorless, odorless, tasteless, slightly less dense than air. NH_3: colorless strong odor, less dense than air, extremely soluble in water
4. Ozone and other irritating chemicals produced by the reaction, catalyzed by ultraviolet light, of nitrogen oxides and unburned hydrocarbons from automobile exhaust
5. Liquefaction of air
6. Nitrogen continuously moves back and forth between living things and the atmosphere.

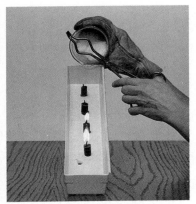

Figure 10-16 Carbon dioxide can be poured down this stair of candles because of its high density and slow rate of diffusion. The candles are extinguished because carbon dioxide gas does not support combustion.

Figure 10-17 Although carbon dioxide does not ordinarily support combustion, a burning magnesium ribbon decomposes CO_2 into carbon and oxygen. The oxygen supports the continued combustion of the magnesium.

than an ordinary carbon–oxygen double bond. Thus, carbon dioxide is often represented by the following four resonance hybrid structures:

$$^- : \overset{..}{O} : C :: O : ^+$$

$$: \overset{..}{O} :: C :: \underset{..}{O} : \longleftrightarrow : \underset{..}{O} :: C :: \overset{..}{O} :$$

$$^+ : O :: C : \underset{..}{\overset{..}{O}} :$$

Each of these structures contributes equally to the actual structure. Such a resonance hybrid has the carbon–oxygen bond distance and energy that are actually observed in carbon dioxide molecules. The Lewis structure shown below is, alone, sufficiently accurate a representation of the carbon dioxide molecule for most purposes.

Physical Properties of Carbon Dioxide

Under ordinary conditions, carbon dioxide is a colorless gas with a faintly irritating odor and a slightly sour taste. The sour taste, characteristic of acids, actually results from the production of carbonic acid when carbon dioxide dissolves in the water in saliva. Carbon dioxide is about 1.5 times as dense as air, which means it can be poured downward into a vessel of air and displace the air. This property of carbon dioxide is illustrated in Figure 10-16, in which the shorter candles have been extinguished by the CO_2 gas poured into the container. Because of its density, carbon dioxide sometimes collects in mines and can cause miners to suffocate.

At temperatures lower than 31°C and at pressures higher than 72.9 atm (7.39 MPa), gaseous carbon dioxide condenses to a liquid. When exposed to normal atmospheric pressure, liquid carbon dioxide evaporates rapidly. Enough heat is lost through evaporation to cause the remaining carbon dioxide to solidify. At normal atmospheric pressure, solid carbon dioxide (dry ice) passes directly to the gaseous state at −78.5°C. Some physical constants of carbon dioxide are listed in Table 10-5.

TABLE 10-5 PHYSICAL PROPERTIES OF CARBON DIOXIDE (CO₂)	
Molecular mass	44.010
Boiling point	−78.5°C (sublimes)
Melting point	−56.6°C (at 5.2 atm)
Density (0°C, 1 atm)	1.977 g/L
Bond length (C—O)	0.120 7 nm
Bond angle (OCO)	180°

Chemical Properties of Carbon Dioxide

Carbon dioxide is a stable gas. It does not burn and does not support combustion. A burning magnesium ribbon, as shown in Figure 10-17, is hot enough to decompose carbon dioxide into carbon and oxygen. The oxygen produced supports the continued combustion of the magnesium. The overall equation for these two reactions is:

$$2Mg(s) + CO_2(g) \rightarrow 2MgO(s) + C(s)$$

When carbon dioxide dissolves in water, a small amount of the dissolved carbon dioxide unites with water to form carbonic acid according to the following equation:

$$H_2O(\ell) + CO_2(aq) \leftrightharpoons H_2CO_3(aq)$$

When carbon dioxide is passed into a water solution of a hydroxide, it reacts to form a carbonate

$$CO_2(g) + 2OH^- \rightarrow CO_3^{2-} + H_2O$$

If the positive ion of the hydroxide forms an insoluble carbonate, it is a precipitate. As shown in the Desktop Investigation, page 000, the test for the presence of carbon dioxide involves the reaction of carbon dioxide with calcium hydroxide, $Ca(OH)_2$, to form a white precipitate of calcium carbonate, $CaCO_3$. The following ionic equation shows this reaction:

$$Ca^{2+}(aq) + 2OH^-(aq) + CO_2(g) \rightarrow CaCO_3(s) + H_2O(\ell)$$

If excess carbon dioxide gas is bubbled through the resulting mixture of water and precipitated calcium carbonate, the precipitate disappears. The excess CO_2 gas reacts with the precipitate and water to form soluble calcium hydrogen carbonate, $Ca(HCO_3)_2$, according to the following ionic equation:

$$CaCO_3(s) + H_2O(\ell) + CO_2(g) \rightarrow Ca^{2+}(aq) + 2HCO_3^-(aq)$$

Figure 10-18 A liquid carbon dioxide fire extinguisher is effective in putting out oil and electrical fires. It cools the burning material and shuts off the supply of oxygen needed for combustion.

Uses and Preparation of Carbon Dioxide

Dry ice, which is used to keep ice cream and other foods frozen during shipping, is solid carbon dioxide. Dry ice is convenient to use for this purpose because it has a very low temperature ($-78.5°C$) and because at atmospheric pressure it does not melt to form a liquid, as does ice. Instead, it changes directly from a solid to a gas without going through the liquid state.

Liquid carbon dioxide fire extinguishers are widely used, especially for putting out oil or electrical fires. Figure 10-18 shows how a stream of liquid carbon dioxide cools the burning material and also shuts off the supply of oxygen needed for burning.

Leavening agents, such as yeast and baking powder, produce bubbles of carbon dioxide gas. These bubbles cause dough to rise and produce a lighter, less dense bread. Soft drinks are carbonated by forcing carbon dioxide into them under pressure. When this pressure is released, as when a soft-drink bottle is opened, the carbon dioxide gas bubbles out of solution.

A common laboratory preparation of carbon dioxide is discussed in the Desktop Investigation on page 307. In numerous reactions, carbon dioxide is a byproduct and thus does not have a specific commercial preparation. The following reactions are examples of commercial sources of carbon dioxide.

Burning carbon-containing substances   Carbon dioxide is one of the products of the complete combustion in oxygen or air of any material that contains carbon in the presence of oxygen or air.

$$C(combined) + O_2(g) \rightarrow CO_2(g)$$

Reaction of steam and natural gas   Carbon dioxide is formed as an industrial by-product when natural gas, which is mostly methane (CH_4), reacts with steam in the presence of metallic oxide catalysts at temperatures between 500°C and 1000°C.

$$CH_4(g) + 2H_2O(g) \xrightarrow[\text{catalyst}]{\Delta} 4H_2O(g) + CO_2(g)$$

The primary purpose of this reaction is to prepare hydrogen for making synthetic ammonia. The carbon dioxide is separated from the hydrogen by dissolving the carbon dioxide in cold water under high pressure.

Fermentation of sugar   Carbon dioxide is a by-product in the production of alcohols from sugars through the process of fermentation. This reaction is catalyzed by enzymes produced by certain types of yeast. The overall equation for the fermentation reaction of glucose ($C_6H_{12}O_6$) to produce ethyl alcohol (C_2H_5OH) is

$$C_6H_{12}O_6(aq) \rightarrow 2C_2H_5OH(aq) + 2CO_2(g)$$

Heating carbonates   When calcium carbonate (in limestone, marble, or seashells, for example) is heated strongly, calcium oxide and carbon dioxide are produced. The calcium oxide produced, known as quicklime or lime, is used for making plaster and mortar. The carbon dioxide produced is a by-product. The equation is

$$CaCO_3(s) \rightarrow CaO(s) + CO_2(g)$$

Reaction of acid and carbonates   One of the simplest ways to produce carbon dioxide is the reaction between vinegar and baking soda, as in the Desktop Investigation on page 307. Another common laboratory preparation of carbon dioxide involves the addition of dilute hydrochloric acid to marble chips. In the first stage of the reaction, the marble and the hydrochloric acid undergo a double displacement reaction:

$$CaCO_3(s) + 2HCl(aq) \rightarrow CaCl_2(aq) + H_2CO_3(aq)$$

In the second stage of the reaction, the unstable carbonic acid decomposes to form water and carbon dioxide.

$$H_2CO_3(aq) \rightarrow H_2O(\ell) + CO_2(g)$$

Geologists make use of the reaction between carbonates and acids to test for the presence of limestone and marble, both of which are mostly $CaCO_3$.

Respiration and decay   Carbon dioxide is produced naturally by the processes of respiration and decay. Respiration is the reaction by which foods, usually carbohydrates, are oxidized in living things to release energy. Decay is the breakdown of the tissue of dead plants and animals into simpler substances.

Occurrence of Carbon Monoxide

Carbon monoxide is a poisonous gas produced naturally by decaying plants, some live algae, the oxidation of atmospheric methane, and volcanoes. As a result, carbon monoxide is a trace constituent of even the least polluted air. The average concentration of carbon monoxide in the atmosphere is about

CO is classified as toxic, while CO_2 is nontoxic but suffocating.

Desktop Investigation

Properties of Carbon Dioxide

Question
How is carbon dioxide made, what are some of its chemical and physical properties, and what is the common test for its presence?

Materials
candle
baking soda (sodium hydrogen carbonate)
vinegar
limewater
matches
balance
100-mL graduated cylinder
three 150-mL beakers
400-mL beaker
stirring rod
plastic straw

Procedure
Record all of your results in a data table. Remove all combustible materials from the work area. Wear safety glasses and an apron.

1. Put the candle into the 150-mL beaker and light it. Put 50 g baking soda into a 400-mL beaker. Slowly add 50 mL vinegar to this second beaker and stir. After foaming slows, carefully pour only the gas formed into the beaker that contains the lighted candle. What gas do you conclude is formed? Why?

2. Repeat the first part of Step 1, but this time do not pour the gas into a beaker containing a candle. Instead, after the reaction has slowed, pour the gas into

50 mL limewater (a saturated solution of calcium hydroxide) in a 150-mL beaker, and stir for one minute. What is the result? Bubble some exhaled air through a plastic straw into another beaker containing 50 mL limewater for one minute. What is the result? What substance does this reaction indicate is present in exhaled breath?

Discussion
1. Why can carbon dioxide gas be poured from one beaker to another?

2. After the reaction between limewater and carbon dioxide is complete, if excess carbon dioxide is bubbled through the mixture of the precipitate and water, the precipitate will redissolve. Can you design an experiment to verify this?

3. Write equations to show (a) the formation of carbon dioxide and (b) the result of bubbling carbon dioxide through limewater.

0.5 parts per million (ppm), which is only about one-hundredth the toxic concentration of carbon monoxide.

Toxic—even lethal—concentrations of carbon monoxide are commonly produced by the incomplete combustion of carbon-containing fuels. In the United States, as much as 90 percent of the human-produced carbon monoxide comes from automobile engines. In a crowded city on a calm day, the street-level concentration of carbon monoxide can reach unhealthy levels. Fortunately, still unidentified natural processes remove carbon monoxide from the atmosphere and prevent its average concentration from approaching a dangerous level.

Figure 10-19 The large number of cars that are concentrated in a city are responsible for most of the high concentration of carbon monoxide found there.

Structure of Carbon Monoxide

The structure of carbon monoxide can be represented by the following resonance structures :

$$^+:C:\ddot{O}:^-$$
$$^-:C:::O:^+ \longleftrightarrow :C::\ddot{O}:$$
$$:C::\ddot{O}:$$

CO is isoelectronic with N_2. Note the similarities in properties.

	CO	N_2
mp(°C)	−205	−210
bp(°C)	−190	−196

Unlike carbon dioxide, these four structures do not contribute equally. The hybrid is estimated to be 10 percent $^+:C:\ddot{O}:^-$, 20 percent each $:C::\ddot{O}:$ and $:C::\ddot{O}:$, and 50 percent $^-:C:::O:^+$. As indicated by this Lewis structure and by the fact that carbon and oxygen differ considerably in electronegativity, the carbon monoxide molecule is polar. The carbon atom is slightly negative and the oxygen atom is slightly positive.

Properties and Uses of Carbon Monoxide

Unfortunately, carbon monoxide is difficult to detect. It is a colorless, odorless, and tasteless gas. It is slightly less dense than air and only slightly soluble in water. Some physical properties of carbon monoxide are shown in Table 10-6.

Chemistry Notebook

Carbon Monoxide Poisoning

Every year, thousands of people are killed by carbon monoxide, which is often produced in toxic concentrations by the incomplete combustion of carbon-containing substances. The major sources of carbon monoxide are automobile engines and faulty home heaters, both of which produce potentially deadly quantities of carbon monoxide. Carbon monoxide is also present in cigarette smoke. In addition, perhaps the majority of deaths resulting from fires are caused by carbon monoxide poisoning. Whatever the source of carbon monoxide, it can have toxic effects, including headaches, nausea, and impairment of vision hearing, and mental functions, whenever its concentration in the surrounding air reaches

50 ppm. Its effects can be lethal when its concentration remains at 1600 ppm for two hours or at 1000 ppm for four hours. To prevent these harmful effects, carbon monoxide should never be released into enclosed or inadequately ventilated spaces.

Carbon monoxide produces its toxic, deadly effect by combining with some of the blood's hemo-

globin. In fact, carbon monoxide binds more readily with hemoglobin than does oxygen. Once combined with carbon monoxide, hemoglobin cannot carry out its usual role of transporting oxygen to the body's cells. This loss of oxygen transport can cause asphyxiation, and ultimately coma or death, for a normally healthy person.

Carbon monoxide acts as a reducing agent in the production of iron, copper, and other metals from their oxides. An example of this is indicated in the following equation describing the reduction of iron(III) oxide:

$$Fe_2O_3(s) + 3CO(g) \rightarrow 2Fe(s) + 3CO_2(g)$$

Carbon monoxide is an important component of artificial industrial fuel gases. When these gases are burned, the carbon monoxide is oxidized to carbon dioxide.

$$2CO(g) + O_2(g) \rightarrow 2CO_2(g)$$

The industrial organic solvent methanol (CH_3OH), which can be used as a paint remover, is prepared by the reduction (hydrogenation) of carbon monoxide.

$$CO(g) + 2H_2(g) \rightarrow CH_3OH(\ell)$$

Carbon monoxide is also used in the production of many other useful organic compounds.

TABLE 10-6 PHYSICAL PROPERTIES OF CARBON MONOXIDE (CO)	
Molecular mass	28.010
Boiling point	−191.5°C
Melting point	−199°C
Density (0°C, 1 atm)	1.250 g/L
Bond length (C—O)	0.113 nm

1. (a) 0.03% (b) About 0.5 parts per million (ppm)
2. The process by which plants use energy from the sun to combine carbon dioxide with water to produce glucose
$6CO_2(g) + 12H_2O(\ell) \rightarrow C_6H_{12}O_6(s) + 6O_2(g) + 6H_2O(\ell)$
3. Carbon dioxide molecules are linear, with two oxygen atoms bonded on opposite sides of the carbon atom. As a result, the molecule is nonpolar. In carbon monoxide, the carbon and oxygen is united by a triple bond. The molecule is polar, with the carbon atom being slightly negative and the oxygen atom slightly positive.
4. Under ordinary conditions, carbon dioxide is a colorless gas with a faintly irritating odor and a slightly sour taste. It is about 1.5 times as dense as air. Carbon monoxide is a colorless, odorless, and tasteless gas that is slightly less dense than air and only slightly soluble in water.
5. Combines in the lungs with hemoglobin. Hemoglobin then cannot transport oxygen to the body's cells.

Section Review

1. *(a)* What is the percentage of carbon dioxide in the earth's atmosphere? *(b)* What is the concentration of carbon monoxide in the earth's atmosphere?
2. What is photosynthesis? Write the equation for this process.
3. Compare the structure of carbon dioxide (CO_2) and carbon monoxide (CO).
4. Compare the physical properties of carbon dioxide and carbon monoxide.
5. How does carbon monoxide interfere with oxygen transport in the body?

Chapter Summary

- Oxygen gas (O_2) and ozone (O_3) are allotropes of elemental oxygen (O).
- An oxide is a compound containing oxygen where oxygen has an oxidation number of -2.
- Oxygen is one of the most active elements and is essential for life.
- Ozone is a poisonous blue gas. A high-altitude layer of ozone in the atmosphere absorbs harmful UV radiation.
- Hydrogen is the most abundant element in the universe.
- Hydrogen reacts with nonmetals forming single covalent bonds. It also reacts with metals to form ionic hydrides.
- Molecular nitrogen (N_2) is the major component of the earth's atmosphere. Nitrogen-containing compounds are found in all living things on earth.
- Ammonia is a colorless gas under normal conditions and is an important ingredient in the production of artificial fertilizers.
- Carbon monoxide is a highly toxic compound. It is used in the manufacture of metals and in the production of some industrial chemicals.
- Carbon dioxide is a colorless gas under normal conditions. It is one of two ingredients used by green plants to produce glucose.
- Tests for the presence of various gases are based on the properties of each gas. Some of these tests are listed in Table 10-7.

TABLE 10-7	TESTS FOR SOME COMMON GASES
H_2	Burns with a pale blue flame to produce water.
CO	Burns with a blue flame to produce CO_2.
CO_2	Forms a white precipitate when bubbled through limewater.
N_2	Does not burn or support combustion. A piece of magnesium ribbon ignited in air will continue to burn in N_2 to produce Mg_3N_2; when water is added, NH_3 is formed, which can be identified by its odor.
O_2	A glowing splint will ignite in O_2. All O_2 will be absorbed when it stands over an aqueous solution of Na_2S.

Chapter 10 *Review*

Representative Gases

Vocabulary

allotrope
allotropy
covalent hydride
Haber process
ionic hydride
nitrogen fixation

oxide
peroxide
photochemical smog
photosynthesis
superoxide

Questions

1. *(a)* What two forms of oxygen are present in the atmosphere? *(b)* How does each affect our everyday lives?
2. *(a)* What is an allotrope? *(b)* What is allotropy? *(c)* Discuss this phenomenon as it applies to oxygen.
3. What types of materials may be damaging the ozone layer in the upper atmosphere?
4. Cite data reflecting the abundance of oxygen in the earth's crust, waters, and atmosphere.
5. Cite at least six physical properties of oxygen, including its boiling and freezing points.
6. *(a)* Distinguish among oxides, peroxides, and superoxides. *(b)* What determines the bond character in an oxide?
7. Identify and write balanced chemical equations (where appropriate) that illustrate five methods used to prepare oxygen.
8. Explain what is meant by the electrolysis of water.
9. *(a)* What is the most abundant element in the universe? *(b)* In what form does it normally exist on earth?
10. Identify the form of hydrogen found in or on each of the following: *(a)* our sun and the stars *(b)* the deep oceans on the surface of Jupiter and Saturn *(c)* middle layers of Jupiter and Saturn.
11. What is the major commercial source of hydrogen?
12. *(a)* Which element is the major component of the earth's atmosphere? *(b)* What is the most important compound in which this element appears? *(c)* List three uses of this compound.

13. *(a)* Which element is involved in the formation of more compounds than any other element? *(b)* How does it combine with most nonmetals; with the least electronegative metals?
14. Identify and write sample balanced equations to illustrate four methods used to prepare hydrogen.
15. Give two examples of the role of nitrogen in the functioning of the human body.
16. *(a)* What is the percent of nitrogen in the earth's atmosphere? *(b)* List five physical properties of elemental nitrogen.
17. *(a)* What is the role of nitrogen in the formation of photochemical smog? *(b)* What is the function of a catalytic converter in a car?
18. *(a)* What is nitrogen fixation? *(b)* Why are legumes often used in crop rotation schemes?
19. *(a)* Draw the Lewis structure for ammonia. *(b)* List at least four properties of ammonia.
20. Cite at least three important uses of ammonia.
21. *(a)* How is ammonia produced in nature? *(b)* industrially? Identify and write balanced equations to illustrate two indentical methods used to prepare ammonia.
22. Compare carbon dioxide and carbon monoxide in terms of (a) their relative effect on living things (b) their molecular structures.
23. *(a)* What is photosynthesis? *(b)* Write the balanced chemical equation that illustrates the reaction.
24. List at least three physical and three chemical properties of carbon dioxide.
25. List four technological applications of carbon dioxide.
26. Identify and write balanced chemical equation (where appropriate) to illustrate six methods used to prepare carbon dioxide.
27. Distinguish between respiration and decay.
28. What four natural processes result in the production of carbon monoxide?
29. *(a)* What is the average concentration of carbon monoxide in the atmosphere? *(b)* How does this compare to the level that is considered

toxic?   (c) In the United States, what is the major source of human-produced carbon monoxide?

30. List five physical properties of carbon monoxide.

31. Describe and write balanced equations (where appropriate) to illustrate the chemical properties of carbon dioxide in each of the following reactions: (a) ability to support combustion of most substances,   (b) response to burning magnesium,   (c) reaction with water,   (d) passage through a water solution of a hydroxide,

32. Describe and write balanced equations (where appropriate) to illustrate the chemical properties of carbon monoxide in each of the following reactions: (a) reaction with metal oxides   (b) combustion (reaction with oxygen)   (c) reduction (reaction with hydrogen).

Problems

1. Determine the oxidation number of each element in the following compounds and identify each compound as an oxide, peroxide, or superoxide: (a) BaO   (b) CaO   (c) RbO_2 (d) Al_2O_3 (e) Na_2O_2 (f) MgO   (g) H_2O_2 (h) KO_2

2. Determine the molar mass of each of the following: (a) O_2 (b) O_3 (c) H_2 (d) N_2 (e) NH_3 (f) CO_2 (g) CO.

3. Determine the mass of each of the following: (a) 5.0 mol H_2O (b) 0.25 mol NH_3 (c) 3.01×10^{23} molecules CO_2.

4. Determine the number of moles of each substance represented by the given mass: (a) 25.0 g HgO   (b) 6.5×10^{-5} g H_2.

5. Determine the percentage by mass (to the nearest 0.1%) of the indicated elements in each of the following compounds: (a) O in H_2O (d) O in Fe_2O_3 (g) N in NH_4NO_3 (b) H in HCl (e) H in H_2SO_4 (h) H in CH_4.   (c) N in NH_3 (f) O in KNO_3

6. The density of oxygen at 0°C and 1.0 atm is 1.429 g/L. Under these conditions, find the molar volume of oxygen for three significant figures.

7. Determine the simplest formula for the compound that consists of 38.67% K, 13.85% N, and 47.48% O.

8. What mass of hydrogen and oxygen are produced through the electrolysis of 90.0 g water?

Application Questions

1. Explain the attraction of the oxygen molecule by a magnet.

2. (a) What two major products result from the liquefaction of air?   (b) How are these two products separated from one another?

3. Determine the oxidation number of each element in the following compounds and then identify each compound as an oxide, a peroxide, or a superoxide: (a) BaO   (b) CaO   (c) RbO_2 (d) Al_2O_3 (e) Na_2O_2 (f) MgO   (g) H_2O_2 (h) KO_2

4. What conditions determine whether the reaction between iron and oxygen will produce Fe_2O_3 or Fe_3O_4?

5. Draw the two resonance forms of ozone.

6. (a) How is ozone produced?   (b) Write the overall equation for the formation of ozone.

7. Though it appears in Group 1 on the Periodic Table, hydrogen differs from the other Group 1 elements in a number of ways. Give at least three differences.

8. Why is most hydrogen found on the earth in combined form, rather than as separate hydrogen atoms?

9. (a) Distinguish between hydrogenated and nonhydrogenated fats.   (b) According to dietary experts, which are better for your health?

10. According to the "big bang" theory, what is the role of hydrogen in the origin of the other elements in the universe?

11. (a) Why is elemental nitrogen relatively nonreactive?   (b) How has this property been put to use?

12. Discuss the role of nitrogen in the production of (a) acid rain   (b) photochemical smog.

13. Explain why ammonia, which is abundant in the atmospheres of Jupiter, Saturn, and Neptune, does not accumulate to any large extent on Earth.

14. Describe and draw the Lewis structure for ammonia.

15. Explain the "greenhouse effect" and its possible effect on the earth's climate.

16. What properties of dry ice make it particularly useful in shipping frozen foods?

17. How does a liquid carbon dioxide fire extinguisher put out a fire?

18. Describe and draw the three resonance hybrid structures of carbon dioxide.

Application Problems

1. Write the balanced chemical equations for the reactions between each of the following: (a) magnesium and oxygen (b) cesium and oxygen (to produce a superoxide) (c) hydrogen and oxygen (to produce a peroxide).
2. Write the balanced chemical equations for the production of hydrogen from each of the following chemical combinations: (a) zinc and hydrofluoric acid (b) magnesium and sulfuric acid (c) iron and sulfuric acid to form ironIII sulfate (d) magnesium and hydrochloric acid (e) rubidium and water (f) magnesium and water.
3. Write the balanced equations for the reactions between hydrogen and each of the following and determine the oxidation numbers of the atoms in each product (a) oxygen (b) chlorine (c) sodium (d) barium.
4. Write the balanced chemical equation for the production of carbon dioxide from each of the following chemical processes: (a) heating of magnesium carbonate (b) reaction of magnesium carbonate and hydrochloric acid.
5. An oxide of aluminum is found to be 52.8% Al. Determine the empirical formula for this compound.
6. Calculate the percentage mass of oxygen in each of the following metallic ores:
   (a) $PbMoO_4$ (b) $Na_2B_4O_7 \cdot 4H_2O$
   (c) $3BeO \cdot Al_2O_3 \cdot 6SiO_2$.
7. Determine the mass of hydrogen that would be produced through each of the following: (a) the replacement reaction between sulfuric acid and 13.1 g zinc (b) from 50.0 g water by single replacement with sodium.
8. (a) How many grams of sulfuric acid are required to react with 1.500 kg sodium hydrogen carbonate in a soda-acid fire extinguisher given the following equation for the reaction? (b) What mass of carbon dioxide is produced in this reaction?

$$2NaHCO_3(aq) + H_2SO_4(aq) \rightarrow$$
$$Na_2SO_4(aq) + 2H_2O(\ell) + 2CO_2(g)$$

Enrichment

1. Prepare a report on the potential impact of the destruction of the ozone layer on plant and animal life.
2. What steps have been taken by the manufacturers of such consumer products as aerosol deodorants and hair sprays to protect the ozone layer? Report current statistics reflecting the impact of these steps.
3. Prepare a report comparing the physical and chemical properties of hydrogenated fats versus nonhydrogenated fats. Discuss the dietary aspects of each.
4. Examine the labels of various food items in your kitchen that contain fats or oils. Indicate which contain hydrogenated, as opposed to nonhydrogenated, fats. Compare your list with those of your classmates and discuss the implications of your findings.
5. Discuss the major uses of oxygen in the space program.
6. Prepare a report concerning the development, operation, and current uses of fuel cells.
7. Explain the "big bang" theory of the origin of the universe. Discuss the role of hydrogen in the formation of most of the other elements in the universe.
8. Prepare a report on the impact of chemical fertilizers in the production of the world's food supply.
9. Write a report on the international efforts to ban the use of chlorofluorocarbons worldwide.
10. Prepare a report (a) on modern alternatives to the Haber process (b) about Haber's role in Germany during World War I.
11. Discuss the factors involved in the production of photochemical smog and the impact of this form of environmental pollution on human life.
12. Prepare a report concerning the disaster in which many people and animals died when carbon monoxide exploded out of Lake Cameroon, in Africa.
13. Research the commercial production of gases such as oxygen, nitrogen and carbon dioxide. What quantities of these gases are produced annually? In what products are they used?
14. Animals are continually converting oxygen and glucose into carbon dioxide and water during respiration. Will the supply of oxygen ever run out? To answer this, research the role of carbon dioxide and oxygen in the animal and plant life on earth. Compare your findings to what you have learned about the nitrogen cycle in this chapter.

Chapter 11 Physical Characteristics of Gases

Chapter Planner

11.1 The Kinetic Theory of Matter
The Kinetic-Molecular Theory of Gases
- ▦ Chemistry Notebook
  Questions: 1–5
- ▦ Question: 6
- ▦ Application Questions: 1, 2
  Application Problem: 1
The Kinetic Theory and the Nature of Gases
- ▦ Demonstration: 1
- ▦ Questions: 7, 8
Deviations of Real Gases from Ideal Behavior
- ▦ Question: 9
- ▦ Application Question: 3

11.2 Qualitative Description of Gases
- ▦ Demonstration: 2
  Question: 10
Pressure and Volume at Constant Temperature
- ▦ Question: 11
Temperature and Volume of Gases—Constant Pressure
- ▦ Question: 12a
- ▦ Question: 12b

Pressure and Temperature
- ▦ Question: 13a
- ▦ Question: 13b
Relationship Between Pressure and Moles
Relationship Between Volume and Moles
- ▦ Question: 14a
- ▦ Question: 14b

11.3 Quantitative Description of Gases
Pressure
- ▦ Demonstrations: 3, 4
  Questions: 15–17, 18a, 21a-b, 22, 23
  Problem: 2
- ▦ Questions: 18b, 19, 20, 21c-d
  Problem: 3
- ▦ Problem: 1
  Application Questions: 3, 5
Boyle's Law: Pressure-Volume Relationship
- ▦ Experiment: 12
  Question: 24
  Problems: 4–6
  Application Problems: 2–4
- ▦ Application Question: 6

Charles' Law: Temperature-Volume Relationship
- ▦ Questions: 25, 26
  Problems: 7–11
- ▦ Application Question: 7
  Application Problem: 5
- ▦ Technology
Gay-Lussac's Law
- ▦ Question: 27
  Problems: 12–14
- ▦ Application Question: 8
  Application Problem: 6
The Combined Gas Law
- ▦ Question: 28
  Problems: 15–19
  Application Problems: 7–9
- ▦ Application Problem: 10
- ▦ Application Problem: 11
Dalton's Law of Partial Pressures
- ▦ Questions: 29, 30
  Problems: 21, 22
  Application Question: 9
  Application Problems: 12, 13
- ▦ Problem: 20

Teaching Strategies

Introduction

Chapter 10 introduced the students to the specific properties of a number of gases. Chapter 11 begins the study of properties common to all gases as well as the underlying assumptions that give rise to the properties. Central to the discussion is the kinetic theory of gases. From the kinetic theory, qualitative relationships among the four properties of all gases are explored prior to formulating the quantitative relationships that exist and which are collectively known as the gas laws.

11.1 The Kinetic Theory of Matter

Before beginning the discussion of the kinetic theory, open a bottle of a strong cologne or perfume, or perhaps release a brief spray of room deodorizer. Have the students raise their hands when they are able to detect the odor. As a pattern develops, have the students try and explain what must be occurring at the molecular level. In short, the invisible particles of gas must be moving through the air.

Be sure that the students understand the concept of an ideal gas, the molecules of which are considered point sources (i.e., occupy no volume), and experience no forces of interaction. The concept of an ideal gas simplifies the model used. Stress that all gases are real, but that they conform to the model of an ideal under certain conditions (high temperature and low pressure).

List the five assumptions of the kinetic theory. Emphasize the meaning of an elastic collision—no friction or energy loss in the collision. Stress that the kinetic energy is directly related to the Kelvin temperature.

Absolute zero is defined as the temperature at which all translational motion stops. Doubling the Kelvin temperature doubles the kinetic energy. Later we shall see that the speed at which gas molecules move is also related to the Kelvin temperature. Qualitatively, at high temperatures gas molecules move faster than at lower temperatures.

Key to the study of gases is the extremely large number of particles that are in any gas sample. Such large samples allow one to make statistically accurate calculations. Many statements discuss the "average kinetic energy;" the statistical average is implied.

Demonstration 1: Kinetic Theory and the Nature of Gases

PURPOSE: To help the students visualize certain characteristic physical properties of gases

MATERIALS: A 50-mL or 100-mL syringe, a candle, one-gallon empty plastic milk container, dropping bottles of conc. HCl and conc. NH_3, a 1-L beaker filled with water

PROCEDURE: *(a) Expansion* Half fill the syringe with air and seal the tip with a serum cap. Pull on the plunger. The air fills the larger volume. *(b) Low Density* Partially fill a balloon with air. Place it in a beaker of water. The balloon floats. *(c) Fluidity* Light a candle and place it on the desk. Stand back about 5 feet and point the open end of a plastic one gallon milk container at the flame. Strike the bottom of the jug. If properly aimed, a pulse of air flows from the container and extinguishes the flame. *(d) Compressibility* Push in on the syringe plunger. The gas compresses and fills a smaller volume. *(e) Diffusion* Place a drop of concentrated ammonia near a drop of concentrated hydrochloric acid on a watch glass. The formation of a white smoke, NH_4Cl, indicates where the diffusing gases combine to form a solid.

SAFETY AND DISPOSAL: Handle the concentrated ammonia and hydrochloric acid carefully. After the reaction, rinse the watch glass with water. The other materials should be saved for further demonstrations.

Ask the students if they know of any substance which, when cooled sufficiently, will not condense and eventually freeze. Van der Waals showed that real gas particles do take up space and do attract each other and the container walls. Conditions which bring molecules close together (very high pressures) and which cause molecules to move slowly enough for the attractive forces to become substantial (low temperatures) lead to nonideal behavior.

Technology: Measuring and Reaching Low Temperatures

The technology feature answers the question sometimes posed by students who realize that a mercury or alcohol thermometer cannot be used to measure very low temperatures because the liquid in the thermometer would freeze. This feature does an excellent job of answering this question.

11.2 Qualitative Description of Gases

Write the four properties of a gas on the chalkboard or overhead projector. Point out that relationships between any two can be studied and described if the remaining two properties are held constant.

Point out that most of the relationships can be predicted from day-to-day experiences. A formula need not be memorized. Use of common sense is the best approach.

Demonstration 2: Gas relationships

PURPOSE: To show the qualitative properties of gases

MATERIALS: a syringe, half-filled with air

PROCEDURE: Demonstrate the particular relationships among the qualitative properties of gases by using the table below. It lists the combinations of variable properties and constant properties and an action, using the half-filled syringe.

Variables	Constants	Action
PV	nT	Squeeze on the plunger of the syringe. *Volume decrease as pressure increases.*
TV	nP	Put the syringe in a beaker of hot water. The plunger moves outward. *Volume increases as the temperature increases.*
PT	nV	If the syringe is heated and the volume expands, the only way to return to the original volume is to exert a larger pressure. *Pressure increases as temperature increases.*
Pn	TV	Add gas to the syringe. Pressure must be added to bring the syringe back to the original volume. *Pressure increases as the number of particles increase.*
Vn	PT	Add gas to the syringe. The volume increases. If the syringe is held vertically, the pressure of the atmosphere and gravity pulling on the syringe plunger remain constant. *Volume increases as the number of particles increases.*

11.3 Quantitative Description of Gases

Based on the qualitative relationships discussed above, students should notice that only two types of relationships were observed: *(1)* as one property was increased, the other property also increased—a direct relationship, and *(2)* as one property was increased, the other property decreased—an inverse relationship. The quantitative formulation of the gas laws is little more than expression the relationships as either direct or inverse.

Demonstration 3: Area and Pressure

PURPOSE: To show the effect of area on the pressure exerted by an object (gravitational force)

MATERIALS: *Modern Chemistry* textbook, thumbtack

PROCEDURE: Place the thumbtack point up on the palm of your hand. Balance the textbook on the point. Now turn the thumbtack point down against the palm of your hand. Ask the students to predict what might happen if you were to place the textbook on the thumbtack. Most will agree that the thumbtack would be driven into your hand. CAUTION: Do NOT actually place the book on the thumbtack. On an overhead projector or chalkboard, show that reducing the size of the volume (the denominator) will increase the size of the fraction (the pressure). The numerator (the gravitational force exerted by the book) remains constant.

Demonstration 4: Atmospheric Pressure and Gases

PURPOSE: To show the difference between atmospheric pressure and the pressure exerted by a gas

MATERIALS: an empty aluminum soda can, beaker, tongs, water, Bunsen burner

PROCEDURE: Hold the can with a pair of beaker tongs, add 10 mL of water to the can, and heat the can over a Bunsen-burner flame. After about one minute of vigorous boiling, turn the can over and place the *mouth* of the can in a beaker of cold water. The can immediately implodes. Perform the demonstration a second time, but this time have your students explain what might be happen during each step of the process. (During the boiling process the air originally in the can is replaced with steam. Placing the mouth of the can in the water causes the steam to condense. A vacuum results. The few vapor molecules left in the can are not able to balance the atmospheric pressure and the walls of the can collapse.)

References and Resources

Books and Periodicals

"Calculating Chemistry—Gases," *ChemMatters*, Vol. 1, No. 1, Feb. 1983, p. 7.

Dolan, Arthur W. "What's the Big Deal About an Empty Bottle? Some Simple Demonstrations with Air," *The Science Teacher*, Vol. 54, No. 7, Oct. 1987, pp. 50–51.

Garde, Ira Batra. "An Easy Approach for Reading Manometers to Determine Gas Pressure: The Analogy of the Child's Seesaw," *Journal of Chemical Education*, Vol. 63, No. 9, Sept. 1986, pp. 796–797.

Rose, Diane. "Charles' Law: Students Develop Their Own Procedure," *Journal of Chemical Education*, Vol. 64, No. 8, Aug. 1987, pp. 712–713.

Shakashiri, Bassam Z. "Physical Behavior of Gases," *Chemical Demonstrations*, Vol. 2, Madison: University of Wisconsin Press, 1985.

Computer Software

Moving Molecules, HRM Software. (Changes in pressure and temperature affect the behavior of molecules in gases, liquids, and solids)

Smith, S.G., R. Chabay, and E. Kean. Introduction to General Chemistry, Disk 7, *Gas Laws*, COMPress. (Ideal Gases, Atmospheric Pressure, Boyle's law, Charles' law, and Pressure—Temperature)

Answers and Solutions

Questions

1. It is based on the idea that particles of matter are always in motion and that this motion has consequences.

2. An ideal gas is an imaginary gas that conforms perfectly to all of the assumptions of the kinetic theory.

3. (1) Gases consist of large numbers of tiny particles that are far apart in comparison to their size. (2) The particles of a gas are in constant motion, moving rapidly in straight lines in all directions. (3) The collisions between particles of the gas and between particles and container walls are elastic collisions. (4) There are no forces of attraction or repulsion between the particles of a gas. (5) The average kinetic energy of the particles of a gas is directly proportional to the Kelvin temperature of the gas.

4. The molecules of gases are much farther.

5. An elastic collision is one in which there is no net energy loss due to friction.

6. *(a)* KE = $\frac{1}{2}mv^2$, where *m* is the mass of the particle, and *v* is its velocity. *(b)* The average velocities and kinetic energies of gas molecules increase with an increase in temperature and decrease with a decrease in temperature.

7. *(a)* Gas molecules are independent particles moving rapidly in all directions, experiencing negligible forces of attraction or repulsion, and thus able to fill the entire volume of any container in which they are enclosed. *(b)* Because the attractive forces between gas particles are negligible, these gas particles are able to glide easily past one another. *(c)* Since the particles in a gas sample are so far apart, there is little mass per unit volume. *(d)* During the compression of a gas, the gas particles, which are initially very far apart, are crowded closer together so as to occupy smaller volumes. *(e)* Since gases consist of particles that are in constant motion, they are able to spread out and mix with one another without stirring and in the absence of circulating currents.

8. *(a)* Diffusion is the spontaneous mixing of the particles of two substances due to their random motion. *(b)* The rate of diffusion of one gas through another is largely dependent upon the speeds of the intermingling gas particles. *(c)* In general, the more massive the gas particle is, the slower is its speed and the more slowly it diffuses through another gas. *(d)* Effusion is the process by which gas particles under pressure pass through a very small opening from one container to another.

9. *(a)* A real gas is a gas that does not completely obey all of the assumptions of the kinetic theory because its particles do occupy space and do attract one another. *(b)* Most real gases behave like ideal gases when their molecules are sufficiently far apart and have sufficiently high kinetic energy. They show the greatest deviations from ideal behavior under high pressures, near their condensation point.

10. The four quantities are its volume, pressure, and temperature, and the quantity or number of molecules.

11. *(a)* A gas in a closed container exerts pressure because its moving molecules hit the walls of the container. *(b)* At constant temperature, the pressure exerted by a fixed quantity of gas increases as the volume decreases.

12. *(a)* As temperature increases, the volume occupied by a fixed quantity of a gas must also increase in order to maintain constant pressure. Conversely, as the temperature of a gas decreases, its volume must also decrease in order to maintain that constant pressure. *(b)* This effect is observed because, at increasingly higher temperatures, the gas molecules move faster and collide with the walls of the container more frequently and with more force. Thus, the only way to maintain constant pressure is to increase volume, thereby reducing the number and frequency of collisions per unit of

wall area and offsetting the greater force of collisions at the higher temperature.

13. *(a)* At constant volume, as temperature increases, so does the pressure exerted by a given quantity of a gas. *(b)* As temperature increases, the average kinetic energy and velocity of the gas molecules increase, so that they experience more frequent and more energetic collisions with the walls of their container. Since the volume occupied by the gas is fixed, the pressure must increase.

14. *(a)* At constant pressure and temperature, the volume of a gas varies directly with the number of molecules. *(b)* As the number of molecules increases, the number of collisions per unit of wall area would increase if volume were held constant. The volume of the container must therefore decrease if the collision rate, and thus the pressure, are held constant. If the number of molecules of gas decreases under the same conditions, the volume must also decrease.

15. The gas laws are simple mathematical relationships between the volume, the temperature, the and pressure of a gas.

16. *(a)* Pressure is defined as the force per unit of surface area. *(b)* Pressure = force/area *(c)* For the same amount of force, the smaller the surface area is, the larger the pressure. *(d)* The newton is the SI unit of force.

17. *(a)* Atmospheric pressure is the pressure exerted by the atmosphere, or blanket of air surrounding the earth. *(b)* The pressure of the atmosphere is due to the weight of the gases composing the atmosphere. The force of gravity holds the atmosphere close to the earth. *(c)* Atmospheric pressure at sea level is equal to the weight of a one-kilogram mass per square centimeter of surface, or 9.8 newtons per square centimeter.

18. *(a)* A barometer is used to measure atmospheric pressure. *(b)* The barometer was introduced by Evangelista Torricelli in 1643.

19. *(a)* According to Torricelli, the height of the mercury in the tube rises or falls until the pressure exerted by its weight is equal to that exerted by the atmosphere. At sea level, the force per unit area exerted by the atmosphere is approximately equal to the force exerted by a column of mercury 760 mm in height. *(b)* The comparable height maintained by a column of water would be about 34 feet. *(c)* The reason for the difference in heights is that mercury is 14 times as heavy as water, so it exerts 14 times more pressure than does a comparable column of water.

20. *(a)* A manometer is a gas-filled flask attached to a U-tube partially filled with mercury and used to measure the pressure of a gas in a closed container. *(b)* The pressure exerted by the gas in the container is given by the difference between the heights of the mercury in the two arms of the U-tube.

21. *(a)* Three commonly used units of pressure are millimeters of mercury (mm Hg), torrs (1 torr = 1 mm Hg), and pascals. *(b)* One atmosphere of pressure is equal to the average sea-level pressure at 0°C, or 760 mm Hg. *(c)* A pascal is defined as the pressure exerted by a force of one newton (N) acting on an area of one square meter; 1 Pa = 1 N/m². *(d)* One standard atmosphere of pressure is defined as 1.01325×10^5 Pa, or 101.325 kPa.

22. *(a)* Such a pressure reading indicates the height of a column of mercury. This is a measure of atmospheric pressure. *(b)* One atmosphere is equal to the average sea-level pres-

sure at 0°C, which is 760 mm Hg. *(c)* The torr is a unit of pressure. One torr is equivalent to the pressure exerted by a column of mercury 1 mm high.

23. *(a)* Standard conditions are 1 atm pressure and 0°C. *(b)* "Standard conditions" or "standard temperature and pressure" is abbreviated as *STP*.

24. *(a)* Boyle's law states that the volume of a fixed mass of gas varies inversely with the pressure at constant temperature. *(b)* The mathematical expressions of Boyle's law are:

$$V = \frac{k}{P} \text{ or } PV = k$$

Thus, since $P_1V_1 = k$ and $P_2V_2 = k$, we have $P_1V_1 = P_2V_2$.

25. *(a)* Charles' law states that the volume of a fixed mass of gas varies directly with the Kelvin temperature at constant pressure.

$$(b) \ V = kT \text{ or } \frac{V}{T} = k \text{ Thus, } \frac{V_1}{T_1} = \frac{V_2}{T_2}$$

26. *(a)* The Celsius equivalent of absolute zero is -273.15°C. *(b)* This temperature is the lowest temperature possible. *(c)* The average kinetic energy of the molecules of a gas is directly proportional to the Kelvin temperature of the gas.

27. *(a)* Gay-Lussac's law states that the pressure of a fixed mass of gas varies directly with the Kelvin temperature at constant volume.

$$(b) \ P = kT \text{ or } \frac{P}{T} = k \text{ Thus, } \frac{P_1}{T_1} = \frac{P_2}{T_2}$$

28. *(a)* $\dfrac{PV}{T} = k$ or $\dfrac{P_1V_1}{T_1} = \dfrac{P_2V_2}{T_2}$

(b) The name "combined gas law" is significant in that each of the individual gas laws is obtained from the combined gas law when the appropriate variable is kept constant.

29. *(a)* The partial pressure of a gas is the pressure that it exerts in a mixture of gases. *(b)* The pressure that each gas exerts in the mixture is independent of the pressures of the other gases present.

30. *(a)* The partial pressure of the dry gas can be determined by subtracting the vapor pressure of the water at the given temperature from the total pressure exerted by the mixture. *(b)* $P_{\text{gas}} = P_{\text{total}} - P_{H_2O}$

Problems

1. *(a)* $14 \times 760 \text{ mm} = 10\ 640 \text{ mm}$

   (b) $\dfrac{1}{1.40}$ or $0.714 \times 760 \text{ mm} = 543 \text{ mm}$

2. *(a)* $1.25 \text{ atm} \times \dfrac{760 \text{ mm Hg}}{\text{atm}} = 950. \text{ mm Hg}$

   (b) $2.48 \times 10^{-3} \text{ atm} \times \dfrac{760 \text{ mm Hg}}{\text{atm}} = 1.88 \text{ mm Hg}$

   (c) $4.75 \times 10^4 \text{ atm} \times \dfrac{760 \text{ mm Hg}}{\text{atm}} = 3.61 \times 10^7 \text{ mm Hg}$

   (d) $7.60 \times 10^6 \text{ atm} \times \dfrac{760 \text{ mm Hg}}{\text{atm}} = 5.78 \times 10^9 \text{ mm Hg}$

3. *(a)* $125 \text{ mm Hg} \times \dfrac{\text{torr}}{1 \text{ mm Hg}} = 125 \text{ torr}$

   (b) $3.20 \text{ atm} \times \dfrac{1.01325 \times 10^5 \text{ Pa}}{\text{atm}} = 3.24 \times 10^5 \text{ Pa}$

(c) $5.38 \text{ kPa} \times \dfrac{\text{atm}}{101.325 \text{ kPa}} \times \dfrac{760 \text{ mm Hg}}{\text{atm}} = 40.4 \text{ mm Hg}$

4. (a) $P_1V_1 = P_2V_2$

$$V_2 = \frac{P_1V_1}{P_2} = \frac{350. \text{ mm} \times 200. \text{ mL}}{700. \text{ mm}} = 100. \text{ mL}$$

(b) $P_1V_1 = P_2V_2$

$$V_1 = \frac{P_2V_2}{P_1} = \frac{0.48 \text{ atm} \times 435. \text{ mL}}{0.75 \text{ atm}} = 2.8 \times 10_2 \text{ mL}$$

(c) $P_1V_1 = P_2V_2$

$$P_1 = \frac{P_2V_2}{V_1} = \frac{(180 \text{ torr})(1.8 \times 10^3 \text{ L})}{2.4 \times 10^5 \text{ L}} = 1.4 \text{ torr}$$

5. $P_1V_1 = P_2V_2$

$$V_2 = \frac{P_1V_1}{P_2} = \frac{325 \text{ mm} \times 240. \text{ mL}}{550 \text{ mm}} = 140 \text{ mL}$$

6. $P_1V_1 = P_2V_2$

$$P_2 = \frac{P_1V_1}{V_2} = \frac{22.5 \text{ kPa} \times 155 \text{ cm}^3}{90.0 \text{ cm}^3} = 38.8 \text{ kPa}$$

7. (a) $0.°C + 273 = 273\text{K}$
   (b) $27°C + 273 = 300\text{K}$
   (c) $-50.°C + 273 = 223\text{K}$
   (d) $-273°C + 273 = 0.\text{K}$

8. (a) $0°C + 273 = 273\text{K}$
   (b) $350.\text{K} - 273 = 77°C$
   (c) $100.\text{K} - 273 = -173°C$
   (d) $20.\text{K} - 273 = -253°C$
   (e) $0.\text{K} - 273 = -273°C$

9. (a) $T_1 = 27°C + 273 = 300.\text{K}; T_2 = 77°C + 273 = 350.\text{K}$

$$\frac{V_1}{T_1} = \frac{V_2}{T_2}$$

$$V_2 = \frac{V_1T_2}{T_1} = \frac{80.0 \text{ mL} \times 350.\text{K}}{300.\text{K}} = 93.3 \text{ mL}$$

(b) $T_2 = 127°C + 273 = 400.\text{K}$

$$\frac{V_1}{T_1} = \frac{V_2}{T_2}$$

$$T_1 = \frac{V_1T_2}{V_2} = \frac{125 \text{ L} \times 400.\text{K}}{85.0 \text{ L}} = 588.\text{K}$$

(c) $T_1 = -33°C + 273 = 240.\text{K}; T_2 = 160.°C + 273$
$= 433\text{K} \quad \dfrac{V_1}{T_1} = \dfrac{V_2}{T_2}$

$$V_1 = \frac{V_2T_1}{T_2} = \frac{54.0 \text{ mL} \times 240.\text{K}}{433\text{K}} = 29.9 \text{ mL}$$

10. $T_1 = 67°C + 273 = 340.\text{K}$

$$\frac{V_1}{T_1} = \frac{V_2}{T_2}$$

$$T_2 = \frac{V_2T_1}{V_1} = \frac{50.0 \text{ mL} \times 340.\text{K}}{140.0 \text{ mL}} = 121\text{K}$$

11. $T_1 = 0.°C \quad 273 = 273\text{K}; T_2 = 130.°C + 273 = 403\text{K}$

$$\frac{V_1}{T_1} = \frac{V_2}{T_2}$$

$$V_2 = \frac{V_1T_2}{T_1} = \frac{275 \text{ mL} \times 403\text{K}}{273\text{K}} = 406 \text{ mL}$$

12. (a) $T_1 = 27°C + 273 = 300.\text{K}$

$$\frac{P_1}{T_1} = \frac{P_2}{T_2}$$

$$T_2 = \frac{P_2T_1}{P_1} = \frac{600. \text{ mm} \times 300.\text{K}}{450. \text{ mm}} = 400. \text{ K}$$

(b) $T_1 = 77°C + 273 = 350.\text{K}; T_2 = 154°C + 237 = 391\text{K}$

$$\frac{P_1}{T_1} = \frac{P_2}{T_2}$$

$$P_2 = \frac{P_1T_2}{T_1} = \frac{1.75 \text{ atm} \times 391\text{K}}{350.\text{K}} = 1.96 \text{ atm}$$

(c) $T_1 = 0.°C + 273 = 273\text{K}; T_2 = -123°C + 273 = 150.\text{K}$

$$\frac{P_1}{T_1} = \frac{P_2}{T_2}$$

$$P_1 = \frac{P_2T_1}{T_2} = \frac{1.00 \text{ atm} \times 273\text{K}}{150.\text{K}} = 1.82 \text{ atm}$$

13. $T_1 = 47°C + 273 = 320\text{K}; T_2 = 77°C + 273 = 350.\text{K}$

$$\frac{P_1}{T_1} = \frac{P_2}{T_2}$$

$$P_2 = \frac{P_1T_2}{T_1} = \frac{250. \text{ mm} \times 350.\text{K}}{320.\text{K}} = 273 \text{ mm}$$

14. $T_1 = 27°C + 273 = 300.\text{K}$

$$\frac{P_1}{T_1} = \frac{P_2}{T_2}$$

$$T_2 = \frac{P_2T_1}{P_1} = \frac{1.125 \text{ atm} \times 300.\text{K}}{0.625 \text{ atm}} = 540.\text{K}$$

15. (a) $T_1 = 27°C + 273 = 300.\text{K}$

$$\frac{P_1V_1}{T_1} = \frac{P_2V_2}{T_2}$$

$$T_2 = \frac{P_2V_2T_1}{P_1V_1} = \frac{2.00 \text{ atm} \times 125 \text{ mL} \times 300.\text{K}}{1.00 \text{ atm} \times 250. \text{ mL}} = 300.\text{K}$$

(b) $T_1 = 77°C + 273 = 350.\text{K}; T_2 = 127°C + 273 = 400.\text{K}$

$$\frac{P_1V_1}{T_1} = \frac{P_2V_2}{T_2}$$

$$P_2 = \frac{P_1V_1T_2}{T_1V_2} = \frac{600. \text{ mm} \times 1.75 \text{ mL} \times 400.\text{K}}{350.\text{K} \times 2.25 \text{ L}} = 533 \text{ mm}$$

(c) $T_1 = 50.°C + 273 = 323\text{K}; T_2 = 70.°C + 273 = 343\text{K}$

$$\frac{P_1V_1}{T_1} = \frac{P_2V_2}{T_2}$$

$$V_1 = \frac{P_2V_2T_1}{P_1T_2} = \frac{820. \text{ mm} \times 3.50 \text{ L} \times 323\text{K}}{740. \text{ mm} \times 343\text{K}} = 3.65 \text{ L}$$

16. $T_1 = 47°C + 273 = 320.K; T_2 = 107°C + 273 = 380.K$

$$\frac{P_1V_1}{T_1} = \frac{P_2V_2}{T_2}$$

$$V_2 = \frac{P_1V_1T_2}{P_2T_1} = \frac{780 \text{ mm} \times 2.20 \text{ L} \times 380.K}{600. \text{ mm} \times 320.K} = 3.4 \text{ L}$$

17. $T_1 = 35°C + 273 = 308K; T_2 = 57°C + 273 = 330.K$

$$\frac{P_1V_1}{T_1} = \frac{P_2V_2}{T_2}$$

$$P_2 = \frac{P_1V_1T_2}{V_2T_1} = \frac{550. \text{ mm} \times 350 \text{ mL} \times 330.K}{425 \text{ mL} \times 308K} = 490 \text{ mm}$$

18. $T_1 = -23°C + 273 = 250.K$

$$\frac{P_1V_1}{T_1} = \frac{P_2V_2}{T_2}$$

$$T_2 = \frac{P_2V_2T_1}{P_1V_1} = \frac{210. \text{ kPa} \times 1.30 \text{ L} \times 250.K}{150. \text{ kPa} \times 1.75 \text{ L}} = 260.K$$

19. $T_1 = 40.°C + 273 = 313K; T_2 = 60.°C + 273 = 333K$

$$\frac{P_1V_1}{T_1} = \frac{P_2V_2}{T_2}$$

$$P_1 = \frac{P_2V_2T_1}{V_1T_2} = \frac{1.40 \text{ atm} \times 1250. \text{ mL} \times 313K}{820. \text{ mL} \times 333K} = 2.01 \text{ atm}$$

20. $P_T = P_{CO_2} + P_{N_2} + P_{O_2}$

$P_{O_2} = P_T - P_{CO_2} - P_{N_2} = 760.000 \text{ mm} - 0.285 \text{ mm}$

$- 593.525 \text{ mm} = 166.190 \text{ mm}$

21. $P_T = P_{O_2} + P_{H_2O}$

$P_{O_2} = P_T - P_{H_2O} = 730.0 \text{ mm Hg} - 17.5 \text{ mm Hg}$

$= 712.5 \text{ mm Hg}$

22. $P_T = P_{gas} + P_{H_2O}$

$P_{gas} = P_T - P_{H_2O} = 742.0 \text{ mm Hg} - 42.2 \text{ mm Hg}$

$= 699.8 \text{ mm Hg}$

Application Questions

1. *(a)* At low temperature, a larger fraction of the molecules have a low velocity, whereas, at the higher temperature, a larger fraction of the molecules have a higher velocity. The peak of the second curve is lower because the velocities are more widely distributed.

   (b) As the temperature is increased further, the resulting distribution should shift farther to the right and peak at an even lower level than was observed in the two previous curves. Thus, a larger fraction of the molecules would have higher velocities.

2. The average kinetic energy values can remain the same because, at the same temperature, the higher the molecular mass of a gas is, the more slowly its molecules move.

3. If all gases behaved ideally, the individual gas particles would never exert the attractive forces on one another that are needed to form liquids or solids.

4. The indicated heights are predicted to be: *(a)* twice that of mercury *(b)* one half that of mercury *(c)* seven times that of mercury.

5. This apparent inconsistency can be explained by the fact that both the pressure due to the column of mercury in the tube and that exerted by the atmosphere on the surface of mercury in the pot can be thought of as being felt at the imaginary surface in the tube that is level with the mercury in the pot outside the tube. Since both forces are acting on the same surface, the size of that surface is not a factor in determining the eventual pressures.

6. *(a)* The volume is halved. *(b)* The pressure is decreased to $\frac{1}{3}$ of the original *(c)* The pressure is increased to four times the original.

7. *(a)* The volume increases by a factor of $\frac{1}{273}$. *(b)* The volume decreases by a factor of $\frac{100}{273}$. *(c)* The volume doubles.

   (d) The volume is reduced to $\frac{1}{4}$ of its original.

8. *(a)* The pressure doubles. *(b)* The pressure is halved. *(c)* The Kelvin temperature triples.

9. *(a)* When a gas is collected by water displacement, it contains a given amount of water vapor, the pressure of which is dependent only on the temperature of the water. Thus, in accordance with Dalton's law of partial pressures, the pressure exerted by the gas alone is the difference between the total pressure exerted by the mixture and that attributable to the water vapor alone. *(b)* When the water levels are the same, the total pressure of the gases inside the bottle is the same as atmospheric pressure, which can then easily be read from a barometer.

Application Problems

1. *(a)* KE = $\frac{1}{2} mv^2$ vs $\frac{1}{2}(2m)(v^2)$, or mv^2
   The KE is thus doubled.
   (b) KE = $\frac{1}{2} mv^2$ vs $\frac{1}{2}(\frac{1}{2}m)(v^2)$, or $\frac{1}{4}mv^2$
   The KE is thus halved.
   (c) KE = $\frac{1}{2} mv^2$ vs $\frac{1}{2}m(2v)^2$, or $2mv^2$
   The KE is thus four times the original value.

2. *(a)* $P_1V_1 = P_2V_2$

   $$V_2 = \frac{P_1V_1}{P_2} = \frac{P_1 \times 450.0 \text{ mL}}{2P_1} = 225.0 \text{ mL}$$

   (b) $P_1V_1 = P_2V_2$

   $$V_2 = \frac{P_1V_1}{P_2} = \frac{P_1 \times 450.0 \text{ mL}}{0.25P_1} = 1800. \text{ mL}$$

3. $P_1V_1 = P_2V_2$

   $$V_2 = \frac{P_1V_1}{P_2} = \frac{(575 \text{ mm})(1.00 \times 10^6 \text{ mL})}{(1.25 \text{ atm})(760 \text{ mm/atm})} = 6.05 \times 10^5 \text{ mL}$$

4. $P_1V_1 = P_2V_2$

   $$P_2 = \frac{P_1V_1}{V_2} = \frac{185 \text{ Pa} \times 365 \text{ mL}}{142 \text{ mL}} = 476 \text{ Pa}$$

5. $T_1 = 75°C + 273 = 348K; T_2 = 30.°C + 273 = 303K$

   $$\frac{V_1}{T_1} = \frac{V_2}{T_2}$$

   $$V_2 = \frac{V_1T_2}{T_1} = \frac{21.5 \text{ m}^3 \times 303K}{348K} = 18.7 \text{ m}^3$$

6. $T_1 = -73°C + 273 = 200.K$

$$\frac{P_1}{T_1} = \frac{P_2}{T_2}$$

$$T_2 = \frac{T_1 P_2}{P_1} = \frac{200.K \times 2P_1}{P_1} = 400.K \text{ or } 127°C$$

7. $T_1 = 18°C + 273 = 291K; T_2 = 25°C + 273 = 298K$

$$\frac{P_1 V_1}{T_1} = \frac{P_2 V_2}{T_2}$$

$$P_2 = \frac{P_1 V_1 T_2}{V_2 T_1} = \frac{1.60 \text{ atm} \times 2.25 \text{ L} \times 298K}{2.45 \text{ L} \times 291K} = 1.50 \text{ atm}$$

8. $T_1 = 17°C + 273 = 290.K$

$$\frac{P_1 V_1}{T_1} = \frac{P_2 V_2}{T_2}$$

$$T_2 = \frac{P_2 V_2 T_1}{P_1 V_1} = \frac{(8.10 \times 10 \text{ Pa})(720. \text{ cm}^3)(290.K)}{(7.75 \times 10 \text{ Pa})(850. \text{ cm}^3)}$$

$$= 257K, \text{ or } -16°C$$

9. $T_1 = 22°C + 273 = 295K; T_2 = -52°C + 273 = 221K$

$$\frac{P_1 V_1}{T_1} = \frac{P_2 V_2}{T_2}$$

$$V_2 = \frac{P_1 V_1 T_2}{P_2 T_1} = \frac{740. \text{ mm} \times 250. \text{ L} \times 221K}{0.750 \text{ atm} \times \frac{760. \text{ mm}}{\text{atm}} \times 295K} = 243 \text{ L}$$

10. $T_1 = 295K$

$$\frac{P_1 V_1}{T_1} = \frac{P_2 V_2}{T_2}$$

$$T_2 = \frac{P_2 V_2 T_1}{P_1 V_1} = \frac{0.475 \text{ atm} \times 400. \text{ L} \times 295K}{740. \text{ mm} \times \frac{\text{atm}}{760. \text{ mm}} \times 250. \text{ L}}$$

$$= 230.K \text{ or } -43°C$$

11. $T_1 = 20.°C + 273 = 293K; T_2 = 0.°C + 273 = 273K$

$$\frac{P_1 V_1}{T_1} = \frac{P_2 V_2}{T_2}$$

$$V_2 = \frac{P_1 V_1 T_2}{P_2 T_1} = \frac{(9.95 \times 10^4 \text{ Pa})(505 \text{ cm}^3)(273K)}{1 \text{ atm} \times \frac{1.01325 \times 10^5 \text{ Pa}}{\text{atm}} \times 293K}$$

$$= 462 \text{ cm}^3$$

$$\frac{462 \text{ cm}^3}{\text{breath}} \times \frac{15.0 \text{ breaths}}{\text{min}} \times \frac{60 \text{ min}}{\text{h}} \times \frac{24 \text{ h}}{\text{day}} \times \frac{\text{m}^3}{10^6 \text{ cm}^3}$$

$$= 9.98 \text{ m}^3$$

12. $P_{\text{gas}} = P_{\text{total}} - P_{H_2O} = 752.0 \text{ mm Hg} - 12.8 \text{ mm Hg}$

$$= 739.2 \text{ mm Hg}$$

Since temperature is constant, $P_1 V_1 = P_2 V_2$.

Thus, $V_2 = \dfrac{P_1 V_1}{P_2} = \dfrac{739.2 \text{ mm Hg} \times 175 \text{ mL}}{770.0 \text{ mm Hg}} = 168 \text{ mL}$

13. $P_{\text{Ar}} = 780.0 \text{ mm Hg} - 23.8 \text{ mm Hg} = 756.2 \text{ mm Hg}$

$T_1 = 25.°C + 273 = 298K; T_2 = 0.°C + 273 = 273K$

$$\frac{P_1 V_1}{T_1} = \frac{P_2 V_2}{T_2}$$

$$V_2 = \frac{P_1 V_1 T_2}{P_2 T_1} = \frac{756.2 \text{ mm} \times 120. \text{ mL} \times 273K}{760. \text{ mm} \times 298K} = 109 \text{ mL}$$

Teacher's Notes

Chapter 11 Physical Characteristics of Gases

INTRODUCTION

Like all matter, gases have kinetic energy. Although gases are made of small, fast-moving particles that are very far apart, scientists have performed experiments that reveal a great deal about the nature of gases and how they behave under a variety of conditions. These experiments have enabled scientists to develop the gas laws, a series of relationships that make it possible to predict the behavior of gases when a change in the temperature, pressure, or volume occurs. In this chapter, you will see how kinetic theory and the gas laws help us to understand the nature of gases.

LOOKING AHEAD

As you study this chapter, note the ways in which a kinetic theory about the behavior of gas molecules can be used to account for the relationships among the various properties of gases.

SECTION PREVIEW

11.1 The Kinetic Theory of Matter
11.2 Qualitative Description of Gases
11.3 Quantitative Description of Gases

Gas particles are usually very far apart. When they are compressed, the volume of a gas can be decreased thousands of times. Gases make excellent propellents, expelling substances like the paint shown here from a pressurized bottle or tank when the pressure is released.

11.1 The Kinetic Theory of Matter

In Chapter 1, matter is described briefly as existing on earth in the form of solids, liquids, and gases. Although it is impossible to observe directly the behavior of individual particles, scientists have been able to study large groups of these particles as they are combined in solids, liquids, or gases.

In the second half of the nineteenth century, scientists developed the kinetic theory of matter to explain the behavior of the atoms and molecules that make up matter. The word *kinetic* comes from the Greek word *kinetikos*, which means "moving." The kinetic theory of matter can be used to explain the properties of gases, liquids, and solids in terms of the energy possessed by the particles of matter and the forces that act between the particles of matter. *The* **kinetic theory** *is based on the idea that particles of matter are always in motion and that this motion has consequences.* In this section, you will study the kinetic theory as it applies to gases.

The Kinetic-Molecular Theory of Gases

The kinetic-molecular theory of gases provides a model of an ideal gas that helps us to understand the behavior of gas molecules and the physical properties of gases. *An* **ideal gas** *is an imaginary gas that conforms perfectly to all of the assumptions of the kinetic theory.*

The kinetic theory of gases is based upon the following five assumptions:

1. Gases consist of large numbers of tiny particles. These particles, usually molecules or atoms, typically occupy a volume about 1000 times larger than that occupied by the same number of particles in the liquid or solid state. Thus, molecules of gases are much farther apart than those of liquids or solids. Most of the volume occupied by a gas is empty space. This accounts for the lower density of gases compared to liquids and solids, and the fact that gases are easily compressible.

2. The particles of a gas are in constant motion, moving rapidly in straight lines in all directions, and thus possess kinetic energy. The kinetic energy of the particles overcomes the attractive forces between them except near the temperature at which the gas condenses and becomes a liquid. Gas particles travel in random directions at high speeds, (See Figure 11-1).

3. The collisions between particles of a gas and between particles and container walls are elastic collisions. An **elastic collision** *is one in which there is no net energy.* Kinetic energy is transferred between two particles during collisions, but the total kinetic energy of the two particles remains the same, at constant temperature.

4. There are no forces of attraction or repulsion between the particles of a gas. You can think of ideal gas molecules as behaving like small billiard balls. They move very fast, and when they collide they do not stick together, but immediately bounce apart.

5. The average kinetic energy of the particles of a gas is directly proportional to the Kelvin temperature of the gas. The kinetic energy of a particle (or any other moving object) is given by the equation

$$\text{K.E.} = \frac{1}{2}mv^2$$

SECTION OBJECTIVES

- State the kinetic theory of matter and describe how it explains certain properties of matter.
- List the five assumptions of the kinetic theory of gases. Define the terms *ideal gas* and *real gas*.
- Describe each of the characteristic properties of gases: expansion, low density, fluidity, compressibility, and diffusion.
- Describe the conditions under which a real gas deviates from ideal behavior.

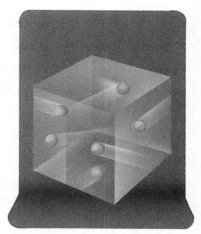

Figure 11-1 Particles of gases are in constant motion, moving rapidly in all directions

Ideal gases do not exist in nature. Molecules of ideal gases are point sources—they occupy no volume, and there are no forces of interaction between molecules. Some gases approach ideality under condition of high temperature and low pressure.

Theories and models are discussed in Chapter 1.

Chemistry Notebook

Molecular Motion

You know that temperature is a measure of the kinetic energy, or energy of motion, of molecules. There are actually three different kinds of molecular motion.

The first kind of motion—and the one you probably tend to think of when the word motion is mentioned—is called "translation." Translation of a molecule is a change in position of the molecule as a whole. The motion can be in any direction, but all parts of the molecule move in the same direction and there is no change in the distances between the atoms within the molecule. (See (a) in the diagram at right.)

In the second type of motion, the molecule as a whole does not change its position, but simply turns "in place." This type of motion is called rotation. (See (b) in the diagram.) Note that different parts of the rotating molecule move in different directions, but that the dis-tances between the atoms within it do not change.

In the third type of motion, the distances between atoms within the molecule may change as bonds stretch or contract. This motion is called vibration. (See (c) in the diagram.) Vibration in more complex molecules can include "bond bending," or changes in the angles between atoms.

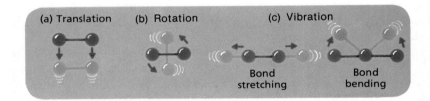

(a) Translation (b) Rotation (c) Vibration Bond stretching Bond bending

All these forms of motion involve kinetic energy. Temperature is actually a direct measure of kinetic energy of translation. Increased temperature can provide energy for rotation or vibration, but is not a measure of the actual kinetic energies of these motions.

As the temperature of a gaseous sample decreases, the translational motion of its molecules decreases. Liquification and solidification occur on further cooling. Even in the solid state, translational motion continues as long as the temperature remains above zero degrees Kelvin. Although the molecules cannot move freely through space in the solid state, they oscillate back and forth, and thus have translational energy. (See (d) in the diagram.) At absolute zero, however, all translational motion ceases. At this point, kinetic energy of translation equals zero and the lowest possible temperature has been reached.

STUDY HINT
Recall that velocity is simply directional speed. It is equal to the displacement (directed distance) divided by time.

where m is the mass of the particle and v is its velocity. Because all the particles of a specific gas have the same mass, their kinetic energies are dependent only on their velocities. The average velocities and kinetic energies of gas molecules will increase with an increase in temperature and decrease with a decrease in temperature.

Figure 11-2 shows the distribution of molecular velocities at two different temperatures. You can see from Figure 11-2 that at the higher temperature, a larger fraction of the molecules has a higher velocity.

Samples of all gases at the *same* temperature have the same average kinetic energy value. Lighter gas molecules, however, such as hydrogen molecules, have higher average velocities than do heavier gas molecules, such as oxygen molecules at the same temperature.

The Kinetic Theory and the Nature of Gases

The kinetic theory applies only to ideal gases. Although ideal gases do not actually exist, the behavior of many gases is close to ideal in the absence of very high pressures or very low temperatures. In the following sections, you will see how characteristic physical properties are explained by the assumptions of the kinetic theory.

Expansiveness   Gases do not have a definite shape or a definite volume. They fill the entire volume of any container in which they are enclosed and assume its shape. A gas transferred from a 1-L vessel to a 2-L vessel will quickly expand to fill the entire 2-L volume. According to the kinetic theory, gas molecules are independent particles moving rapidly in all directions (Assumption 2) without significant attraction or repulsion between them (Assumption 4).

Fluidity   Because the attractive forces between gas particles are negligible (Assumption 4), gas particles glide easily past one another. This ability to flow causes gases to show mechanical behavior similar to that of liquids. Because liquids and gases flow, they are referred to collectively as fluids.

Low density   The density of a substance in the gaseous state is about 1/1000 the density of the same substance in the liquid or solid state because the particles are so much farther apart in the gaseous state (Assumption 1). For example, oxygen gas has a density of .001 429 g/mL at 0°C and 1 atmosphere pressure. As a liquid at −183°C, oxygen has a density of 1.149 g/mL. Figure 11-3 shows the relative volumes of a given amount of oxygen as a gas, a liquid, and a solid.

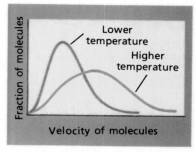

Figure 11-2   Molecular speed distribution in a gas measured at different temperatures.

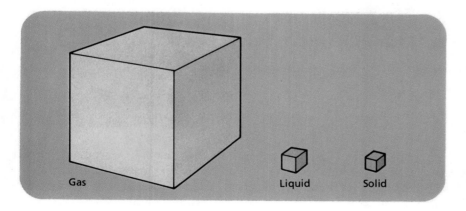

Gas   Liquid   Solid

Figure 11-3   Relative volumes of oxygen as a gas, a liquid, and a solid.

A flat or empty automobile tire is actually full of air at atmospheric pressure.

Compressibility   During the compression of a gas, the gas particles, which are initially very far apart (Assumption 1), are crowded closer together. Under sufficient compression, the volume of a given sample of a gas can be decreased thousands of times. The steel cylinders containing nitrogen, oxygen, or other gases under pressure that are widely used in industry illustrate this point. Such cylinders have an internal volume of about 55 L.

When they are returned "empty" at ordinary pressures, they contain about 55 L of gas (the volume of the cylinder), although when they were delivered "full" they may have had 100 times as many molecules of gas compressed within the same cylinder.

Diffusibility   Gases spread out and mix with one another without stirring and in the absence of circulating currents. If the stopper is removed from a container of ammonia, the presence of this gas, which irritates the eyes, nose, and throat, soon becomes evident. Eventually, the ammonia mixes uniformly with the air in the room, as the random and continuous motion of the ammonia molecules (Assumption 2) carries them throughout the available space. *The spontaneous mixing of the particles of two substances because of their random motion is referred to as* **diffusion.**

Suppose a sample of nitrogen dioxide gas is released into a tube that is already filled with air. As shown in Figure 11-4, brown nitrogen dioxide (NO_2) gas diffuses out of the flask and into the attached tube in a matter of minutes. Because of collisions, diffusion into an enclosed space if slowed, but never prevented, by the presence of other gases in the space.

Figure 11-4   Diffusion of nitrogen dioxide gas.

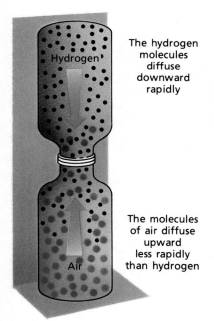

The hydrogen molecules diffuse downward rapidly

The molecules of air diffuse upward less rapidly than hydrogen

Figure 11-5   A bottle filled with hydrogen gas is placed above a bottle filled with air. After several minutes, the molecules of hydrogen have diffused rapidly into the lower jar.

The rate of diffusion of one gas through another depends on three properties of the intermingling gas particles: their speeds, their diameters, and the attractive forces between them. Hydrogen gas diffuses rapidly into other gases at the same temperature because its molecules are smaller and move about with greater speed than the larger, heavier molecules of the other gases, as shown in Figure 11-5.

Whereas **diffusion** is a process by which particles of a gas sample spread out spontaneously and mix with other gases, **effusion** *is a process by which gas particles under pressure pass through a very small opening from one container to another.* The rates of effusion of different gases are directly proportional to their molecular velocities.

Deviations of Real Gases from Ideal Behavior

In 1873, Johannes van der Waals (1837–1923) proposed that real gases deviate from the behavior expected of an ideal gas because *(1)* particles of real gases occupy space, and *(2)* particles of real gases exert attractive forces on each other. *A **real gas** is a gas that does not obey completely all the assumptions of the kinetic theory.*

Most real gases behave like ideal gases when their molecules are sufficiently far apart and have sufficiently high kinetic energy. The kinetic theory does not hold true under all conditions of temperature and pressure, however. Experiments show that gas molecules at very high pressures and/or very low temperatures do not remain far apart. Under such conditions, all gases will deviate considerably from ideal gas behavior. This is illustrated in Figures 11-6 and 11-7.

The kinetic theory is more likely to hold true for gases composed of particles that have little attraction for each other. The noble gases, such as helium (He) and neon (Ne), show essentially ideal gas behavior over a wide range of temperatures and pressures. These gases consist of molecules that are monatomic and thus, nonpolar. Gases such as nitrogen (N_2) and hydrogen (H_2) have nonpolar, diatomic molecules and these are also found to show gas behavior that is close to ideal under many conditions. The more polar its molecules are, the more a gas deviates from ideal gas properties. Substances such as ammonia (NH_3) and water (H_2O) in the gaseous state are composed of highly polar molecules. They deviate from ideal behavior to a large degree.

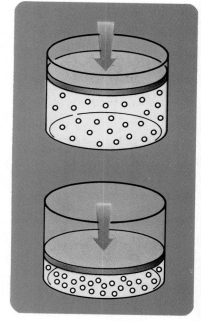

Figure 11-6 Under high pressure, gas molecules move closer together.

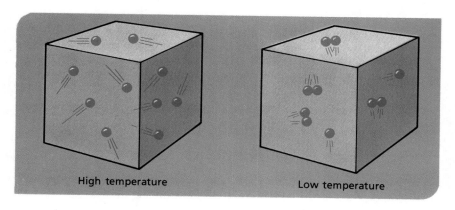

High temperature   Low temperature

Figure 11-7 Particles of gases at high temperatures (left) are far apart. At low temperatures (right), gas particles tend to stick together because the forces of attraction between them exert a greater influence than does their energy of motion.

Section Review

1. Explain each of the following characteristic properties of gases according to the kinetic theory: *(a)* expansiveness *(b)* low density *(c)* fluidity *(d)* compressibility and *(e)* diffusibility.
2. Describe the conditions under which a real gas is most likely to behave ideally.
3. State the two factors that van der Waals proposed to explain why real gases deviate from ideal behavior.
4. Which of the following gases would you expect to deviate the most from ideal behavior? *(a)* He *(b)* O_2 *(c)* H_2 *(d)* H_2O *(e)* N_2 *(f)* HCI *(g)* NH_3.

1. Refer to *The Kinetic Theory and the Nature of Gases,* p.000, for discussion on expansiveness, fluidity, low density, compressibility, and diffusibility.
2. Under conditions of high temperature and low pressure
3. *(1)* Gas particles occupy space. *(2)* Gas particles exert attractive forces on each other.
4. *(d)* H_2O

11.2 Qualitative Description of Gases

SECTION OBJECTIVES

- Explain why the measurable quantities volume, pressure, temperature, and number of molecules of gas are needed to describe properly the state, or condition, of a gas.
- Name the variables that one must hold constant in order to study the (1) pressure–volume, (2) temperature–volume (3) pressure–temperature, (4) pressure–number of molecules, and (5) volume–number of molecules relationships of gases.
- Explain the relationship between the variables in each of the five pairs listed above in terms of the kinetic theory.

New ceramic, metallic oxide superconductors have recently been developed with higher transition temperatures. Liquid nitrogen may be used to cool the material.

P vs V at constant n and T
Where n = moles—the SI unit for quantity of substance

T vs V at constant n and P

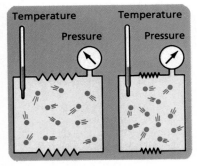

Figure 11-8 At a constant temperature, the pressure that a gas exerts increases as the volume of the gas decreases.

Suppose you have a one-liter bottle of air. How much air do you have? The expression *a liter of air* means little unless the temperature and pressure at which the air is measured are known. A liter of air can be compressed to a few mL of volume; it can also be allowed to expand to fill the volume of an auditorium.

To fully describe the state, or condition, of a gas, you need to use four measurable quantities: *(1)* volume, *(2)* pressure, *(3)* temperature, and *(4)* quantity or number of molecules. Mathematically simple relationships connect these four quantities. If you know the value of any three, you can calculate the value of the fourth. Such a calculated value will be fairly accurate for most gases over a very wide range of conditions.

In the next section, the mathematical relationships between volume, pressure, temperature, and number of molecules of gas will be examined, and examples of how to calculate an unknown value from known information will be given.

Pressure and Volume at Constant Temperature

The pressure exerted by a gas in a container is caused by moving molecules hitting the walls of the container. What happens when a gas in a one-liter container is placed into a one-fourth liter container? There will be four times as many molecules per unit volume and, therefore, four times as many collisions with a given unit of wall area. The pressure will thus increase fourfold.

As represented in Figure 11-8, the pressure exerted by a fixed quantity of gas at constant temperature becomes higher as the volume that the gas occupies becomes smaller. Similarly, the pressure exerted by the same number of gas molecules is lower if the volume available to the gas is larger.

Temperature and Volume of Gases at Constant Pressure

Have you ever watched a hot-air balloon billow out and fill with gas after the flame at its base is lighted? Balloonists such as those in Figure 11-17 are making use of a physical property of gases: if pressure is constant, gases can expand when heated and contract when cooled.

When the temperature increases, the volume occupied by a fixed quantity of gas must increase if the pressure is to remain constant. At the higher temperature, the gas molecules move faster and collide with the walls of the container more frequently and with more force. As illustrated in Figure 11-9, if the volume is increased, the molecules travel farther before reaching the walls. The number of collisions and their frequency per unit of wall area decreases. This smaller frequency of collisions offsets the greater force of collisions at the higher temperature. The pressure thus remains constant.

On the other hand, when the temperature decreases, the volume occupied by a fixed amount of gas must decrease if the pressure is to remain constant. At the lower temperature, the gas molecules move more slowly and they collide with the walls with less force. If the volume is decreased, however, the frequency of collisions per wall area increases, offsetting the force of the collisions. The pressure can therefore remain constant.

Pressure and Temperature

Recall that the temperature of a gas is an indication of the average kinetic energy of its molecules. The higher the temperature of a gas, the more kinetic energy its molecules have, and the more rapidly they move about. We have already seen that the pressure exerted by a gas is due to the collisions of the gas molecules with its container walls. As the temperature increases gas molecules have more kinetic energy and undergo more frequent and more energetic collisions with the walls with more force per collision. As shown in Figure 11-10, if the volume of the same number of molecules remains constant with an increase in temperature, the pressure must increase. It follows that if the temperature is lowered, the pressure, exerted by the same number of molecules at constant volume, is lowered. This is true because the molecules move more slowly and exert less force per collision at lower, temperature.

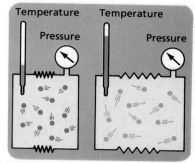

Figure 11-9 At a constant temperature, the pressure that a gas exerts increases as the volume of the gas decreases.

Relationship Between Pressure and Moles

At a given volume and temperature, the number of collisions of gas molecules with the container walls per unit time depends only on the number of molecules of the gas. The more molecules, the more collisions there are per unit time and therefore the higher the pressure. We make use of this property whenever we inflate bicycle or automobile tires. A cylinder that contains twice as many gas molecules as another cylinder has twice the pressure, assuming equal volume and temperature.

Relationship Between Volume and Moles

The volume of a gas depends on its temperature, its pressure, and the number of molecules present. If the temperature and pressure are constant, the volume of a gas varies directly with the number of molecules. As the number of molecules increases, the number of collisions per unit area of the container walls would increase if volume were held constant. The collision rate (and therefore the pressure) is kept constant by an increase in volume. If the number of molecules decreases, and the pressure and temperature are held constant, the volume must decrease.

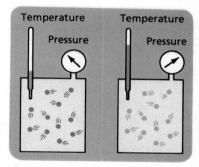

Figure 11-10 At a constant volume, the pressure a gas exerts increases as the temperature of the gas increases.

Section Review

1. Name the four measurable quantities by which the state, or condition, of a gas can be described.
2. Describe qualitatively the relationships between each of the following: *(a)* pressure and volume at constant temperature and number of molecules, *(b)* temperature and volume at constant pressure and number of molecules, *(c)* pressure and temperature at constant volume and number of molecules, and *(d)* pressure and number of molecules at constant volume and temperature.

1. Pressure, volume, temperature, and amount of gas
2. *(a)* Inversely proportional
   (b) Directly proportional
   (c) Directly proportional
   (d) Directly proportional
   (e) Directly proportional

Technology

Measuring and Reaching Low Temperatures

When extremely low temperatures—that is, temperatures below 0.1K—must be measured, ordinary alcohol- or mercury-filled thermometers are useless because the liquid in them freezes before reaching these temperatures. One device for measuring these extremely low temperatures is a nuclear resonance thermometer. This thermometer makes use of the fact that atomic nuclei rotate, or "spin". A strong magnet placed near the sample to be tested causes the nuclei to take on a certain direction of spin. The degree to which the nuclei tend to respond to this influence depends upon how low the kinetic energy of the sample is, and thus is a measure of temperature.

The magnetic spin properties of nuclei are useful not only in measuring very low temperatures, but also in reaching them. In a process called adiabatic demagnetization, the sample is first cooled in a liquid-helium bath in the presence of a powerful magnet, which causes the spinning nuclei in the sample to line up or become magnetized in certain ways. Then, the sample is removed from the bath and insulated so that no heat can flow into it. The sample begins to demagnetize itself, as the nuclei take on their usual spins. The nuclei require energy for this demagnetization and they get it by converting the kinetic energy of their atoms to other forms. Thus, as the kinetic energy falls, so does the temperature. Temperatures as low as 0.000 016K can be reached through this process.

Reaching very low temperatures is important in studying atomic structure and detecting subatomic particles.

11.3 Quantitative Description of Gases

SECTION OBJECTIVES

- Define pressure, explain how it is measured, and state the standard conditions of temperature and pressure.
- State Boyle's law and use it to calculate volume–pressure changes at a fixed temperature.
- Discuss the significance of absolute-zero temperature. Use the Kelvin scale in calculations.

Scientists have been studying the physical properties of gases for hundreds of years. In 1662, Robert Boyle (1627–1691) discovered that the pressure and volume of a gas are related to each other mathematically. The observations of Boyle and others led to development of **gas laws**—simple mathematical relationships between the volume, temperature, and pressure of a gas. These gas laws were established long before the ideal-gas model and kinetic-molecular theory were proposed.

Pressure

Before examining the gas laws, we must first understand *pressure*—how it is defined and how it is measured.

Pressure and Force If we put some air into a rubber balloon, the balloon increases in size because of the pressure created by collisions between the air molecules inside the balloon as they push the walls of the balloon outward. Pressure *is defined as the force per area on a surface:*

$$\text{Pressure} = \frac{\text{force}}{\text{area}}$$

Consider a ballet dancer who has a mass of 51 kg (Figure 11-11). A mass of 51 kg exerts a force of 500 N on the earth's surface. The SI unit for force is the **Newton**, abbreviated N. It is the force that will increase the speed of a 1-kilogram mass by 1 meter per second for every second that it is applied. No matter how the dancer stands, she exerts a force of 500 N against the floor. The pressure she exerts against the floor depends on the area of contact, as shown by the relationship above. When she rests her weight on the soles of both feet, the area of contact with the floor is about 325 cm². The pressure, or force per unit area, when she stands in this manner is 500 N/325 cm², or about 1.5 N per cm². When she stands on her toes, or "on point," as shown in Figure 11-15, the total area of contact of these two toes with the floor is 13 cm². Thus, when she balances in this way, the pressure exerted is equal to 500 N/13 cm²—roughly 38.5 N/cm². And when she stands on one toe, the pressure she exerts is twice that, or 77 N/cm². It is clear that the same force applied to a smaller area results in a greater pressure.

Gas molecules exert pressure on any surface with which they collide. The pressure exerted by a gas depends on the number of molecules present, the temperature, and the volume in which the gas is confined.

The pressure of gases must not be confused with atmospheric pressure —that is, the pressure exerted by the atmosphere. The atmosphere—the blanket of air surrounding the earth—is composed of about 78% nitrogen, 21% oxygen, and 1% other gases, such as carbon dioxide and argon. The pressure of the atmosphere is due to the weight of the gases composing the atmosphere.

Atmospheric pressure at sea level is equal to the weight of a 1-kilogram mass per square centimeter of surface, or 9.8 N per square centimeter. All objects on earth are subjected to the pressure of the atmosphere.

- State Charles' law, and use it to calculate volume—temperature changes at a fixed pressure.
- State Gay-Lussac's law, and use it to calculate pressure—temperature changes at a fixed volume.
- Use the combined gas law to calculate volume—temperature— pressure changes.
- State Dalton's law of partial pressures, and use it to calculate partial pressures and total pressures.

Force is equal to the mass of the object acted upon times the acceleration caused by the force:
force = mass × acceleration.

1 newton (N) = 1 kg m²/s²

STUDY HINT

One newton is equal to the force that must be exerted to accelerate a 1 kg mass at the rate of 1 m/sec².

Figure 11-11 The pressure exerted upon the floor by the ballet dancer is dependent upon the area of contact.

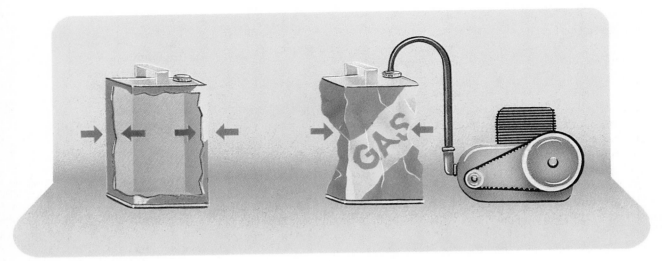

Figure 11-12 A gas can collapses when the air inside it is removed by a vacuum pump.

STUDY HINT

There are two types of manometers. In one type, the mercury column is pushed upward into a tube that contains a vacuum. In another type, the mercury column is pushed upward into a tube that is open to the atmosphere. In the latter type, atmospheric pressure must be taken into account when calculating the pressure exerted by a gas.

To understand the concept of gas pressure and its magnitude, consider the gasoline can shown in Figure 11-12. The "empty" can contains about 5 g of air. The atmosphere exerts a pressure of 9.8 N per square centimeter on the outside of the can. If the can measures 15 cm × 10 cm × 28 cm, it has a total area of 1700 cm^2, and the total force is greater than 1.0 metric tons of weight. To demonstrate the size of the force exerted by the 5 g of air inside the can, we remove the air from the can by using a vacuum pump, as shown in Figure 11-12. In the absence of balancing outward-pressing force exerted by air inside, the force due to atmospheric pressure immediately crushes the can.

Measuring pressure *A device used to measure atmospheric pressure is called a* barometer. The word barometer comes from the Greek *baros*, which means "weight," and *metron*, which means "to measure." The first type of barometer, illustrated in Figure 11-13, was introduced by Evangelista Torricelli (1608–1647) in 1643. Torricelli wondered why water pumps could raise water to a height of only about 34 feet. He thought that the height must be associated with the accepted idea that air had weight. Thus, he reasoned

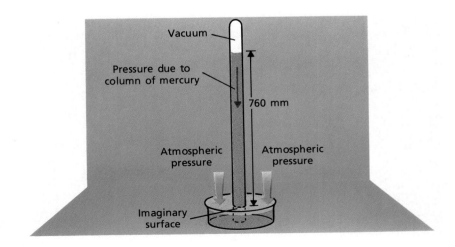

Figure 11-13 Torricelli discovered that the pressure of the atmosphere supported a column of mercury 760 mm above the surface of the mercury in the pot.

that liquid mercury, which is about 14 times as dense as water, could be raised only 1/14 as high as water. To test this idea, Torricelli sealed a long glass tube at one end and filled it with mercury. Holding the open end with his thumb, he inverted the tube into a pot of mercury without allowing any air to enter the tube. When he removed his thumb, the mercury in the tube dropped to a height of about 30 in., or 760 mm, above the surface of the mercury in the pot. He repeated the experiment with tubes of different diameters and lengths longer than 760 mm. In every case, the mercury dropped to a height of about 760 mm.

The space above the mercury in such a tube contains a little mercury vapor, but is nearly a vacuum. The column of mercury in the tube retains its height because the atmosphere exerts a pressure on the surface of the mercury outside the tube. The pressure is transmitted through the mercury to an imaginary surface in the tube that is level with the mercury outside the tube. The mercury in the tube pushes downward because of its weight, which is due to the gravitational force. The mercury in the tube thus falls until the pressure exerted by its weight is equal to the pressure exerted by the atmosphere. The height of the mercury in the tube depends on the pressure—the force per unit area exerted by the atmosphere on the mercury in the pot. The pressure of the atmosphere is proportional to the height of the mercury column supported in the barometer tube. It is this relationship that allows the height of the mercury in a barometer tube to be used to measure atmospheric pressure.

From experiments like those of Torricelli, it is known that at sea level at 0°C, the average pressure of the atmosphere is equal to the pressure that could support a 760-mm column of mercury. At any given place on the earth, the specific atmospheric pressure depends on the elevation and the weather conditions at the time. If the atmospheric pressure is greater than the average at sea level, the height of the mercury column in a barometer will be greater than 760 mm. If the atmospheric pressure is less, the height of the mercury column will be less than 760 mm high.

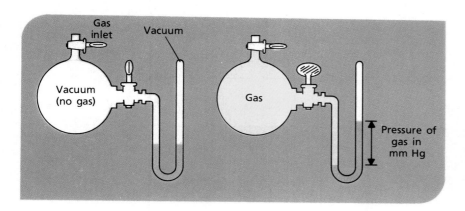

Figure 11-14 In the manometer shown here, the pressure exerted by the gas in the container causes the column of mercury to rise upward. The difference in height of mercury in the two arms of the U-tube can be used to find the gas pressure.

Units of pressure A number of different units are used to measure pressure. Some pressure units are based on force per unit area, some on mass per unit area, and others on the height of a liquid column. Because atmospheric pressure is traditionally measured by a mercury barometer, scientists often express pressure in terms of the height of a column of mercury. *Thus, a common unit of pressure is* **millimeters of mercury, symbolized mm Hg.** The average atmospheric pressure at sea level and 0°C is 760 mm Hg. *To honor Torricelli for his invention of the barometer, a pressure unit named after him, the* **torr,** *has been introduced. By definition, 1 torr = 1 mm Hg.*

High pressures are often measured in units of atmospheres. *One* **atmosphere of pressure (atm)** is defined as being equal to the average sea level pressure at 0° C. Thus, 1 atm = 760 mm Hg or 760 torr.

In the SI system, pressure is expressed by a derived unit named the pascal (Pa) for Blaise Pascal (1623–1662), a French scientist who studied pressure in fluid systems. One pascal is defined as the pressure exerted by a force of one Newton (1 N) acting on an area of one square meter.

$$1 \text{ Pa} = 1 \text{ N/m}^2$$

In many cases it is convenient to express pressure in kilopascals (kPa): 1 kPa = 1000 Pa. The standard atmosphere (1 atm) is equal to 1.01325×10^5 Pa, or 101.325 kPa.

The pressure units used in this book are summarized in Table 11-1. Other common units for expressing pressure include *pounds per square inch* (psi) and *bars* (1 bar = 1×10^5 Pa).

TABLE 11-1 UNITS OF PRESSURE

Unit	Symbol	Definition/Relationship
pascal	Pa	SI pressure unit $1 \text{ Pa} = 1 \text{ kg/(m)} \times s^2 = 1 \text{ N/m}^2$
millimeter of mercury	mm Hg	Pressure that supports a 1-mm mercury column in a barometer
torr	torr	1 torr = 1 mm Hg
atmosphere	atm	Average atmospheric pressure at sea level and 0°C 1 atm = 760 mm Hg = 760 torr $= 1.01325 \times 10^5$ Pa = 101.325 kPa

Sample Problem 11.1

The atmospheric pressure in Denver, Colorado, on the average is 0.830 atm. Express this pressure in *(a)* millimeters of mercury (mm Hg) and *(b)* kilopascals (kPa).

Solution

Step 1. Analyze   Given: $P = 0.830$ atm $\dfrac{760 \text{ mm Hg}}{1 \text{ atm}}$ (definition)

$\dfrac{101.325 \text{ kPa}}{1 \text{ atm}}$ (definition)

Unknown: P in mm Hg; P in kPa

Sample Problem 11.1 *continued*

Step 2. Plan	(a) atm $\rightarrow$ mm Hg (b) atm $\rightarrow$ kPa (a) mm Hg $= atm \times \dfrac{mm\ Hg}{atm}$ (b) kPa $= atm \times \dfrac{kPa}{atm}$
Step 3. Compute	(a) 0.830 atm $\times \dfrac{760\ mm\ Hg}{1\ atm} = 631$ mm Hg (b) 0.830 atm $\times \dfrac{101.325\ kPa}{1\ atm} = 84.1$ kPa
Step 4. Evaluate	Units have canceled to give the desired units, and answers are properly expressed to the correct number of significant digits. The known pressure is roughly 80% of atmospheric pressure. The results are therefore reasonable because each is roughly 80% of 1 atm pressure as expressed in the new units.

Practice Problem

Convert a pressure of 1.75 atm to *(a)* kPa   *(b)* torr.   *(Ans.)* 177 kPa, 1330 torr

Standard temperature and pressure   The volume of a gas depends upon temperature and pressure. Therefore, in order to compare volumes of gases, it is necessary to know the temperature and pressure at which the volumes are measured. *To aid in such comparisons, scientists have agreed upon standard conditions of exactly 1 atm pressure and 0°C. These conditions of standard temperature and pressure are commonly abbreviated* STP.

Boyle's law: Pressure—Volume Relationship

The English chemist and physicist Robert Boyle was the first scientist to make careful measurements that showed the quantitative relationship between the pressures and volumes of gases. He discovered that doubling the pressure on a sample of gas at constant temperature reduces its volume one-half. Tripling the gas pressure reduces its volume to one-third of the original, as illustrated in Figure 11-15. On the other hand, reducing the pressure on a gas by one-half allows the volume of the gas to double.

Data that are representative of pressures and volumes measured for a constant mass of gas at constant temperature are shown in Table 11-2. Notice that the product of the pressure and volume is a constant, k, for each pair of corresponding values. A constant product is found whenever there is an inverse relationship between two variables. Plotting the values of such inversely proportional variables gives a curve like that in Figure 11-16 for volume versus pressure. This volume–pressure relationship is referred to as **Boyle's law:** *the volume of a fixed mass of gas varies inversely with the pressure at constant temperature.*

Mathematically, Boyle's law is expressed as

$$V = k\,\frac{1}{P} \quad \text{or} \quad PV = k$$

STUDY HINT

In a direct relationship, the value of one variable decreases in value as the value of the other variable decreases proportionally. The graph of a direct relationship is linear.

Physical Characteristics of Gases   **327**

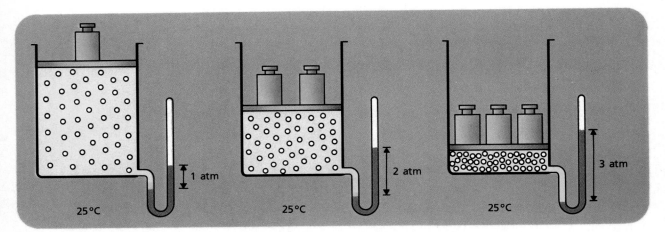

Figure 11-15 At constant temperature, the volume of a gas will be reduced in proportion to increases in pressure.

where k is constant, its value depending only on the quantity of gas and the temperature. If the pressure of a given gas sample at constant temperature changes, the volume will change, but the quantity *pressure times volume* will remain equal to the same value of k. Using P_1 and V_1 to indicate initial conditions and P_2 and V_2 to indicate new conditions,

$$P_1V_1 = k \qquad P_2V_2 = k$$

Combining these two relationships gives the following very useful statement of Boyle's law:

$$P_1V_1 = P_2V_2$$

If three of the four values P_1, V_1, P_2, and V_2 are known, this equation can be used to calculate the fourth. For example, if 1.0 L of gas at 1.0 atm ($V_1 = 1.0$ L, $P_1 = 1.0$ atm) is allowed to expand at constant temperature to 5.0 L($V_2 = 5.0$ L), the new pressure (P_2) is

$$P_2 = \frac{P_1V_1}{V_2}$$

$$= \frac{1.0 \text{ atm} \times 1.0 \text{ L}}{5.0 \text{ L}} = 0.20 \text{ atm}$$

The volume has increased five times and the pressure has decreased to one-fifth the original pressure.

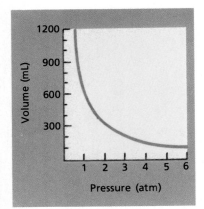

Figure 11-16 The variation of volume with pressure at constant temperature.

TABLE 11-2 PRESSURE–VOLUME DATA (AT CONSTANT MASS AND TEMPERATURE)

Volume (mL)	Pressure (atm)	Pressure × Volume (k) (mL × atm)
1200	0.5	600
600	1.0	600
300	2.0	600
200	3.0	600
150	4.0	600
120	5.0	600
100	6.0	600

Sample Problem 11.2

A sample of oxygen gas collected in the laboratory occupies a volume of 150 mL when its pressure is 720 mm Hg. What volume will the gas occupy at a pressure of 750 mm Hg if the temperature remains constant?

Solution

Step 1. Analyze

Given: V_1 = 150 mL
              P_1 = 720 mm Hg P_2 = 750 mm Hg

Unknown: V_2

Step 2. Plan

$$P_1, V_1, P_2 \rightarrow V_2$$

Rearrange the equation for Boyle's law ($P_1V_1 = P_2V_2$) to obtain V_2.

$$V_2 = \frac{P_1V_1}{P_2}$$

Step 3. Compute

Substitute values for P_1, V_1, and P_2 to obtain the new volume, V_2.

$$V_2 = \frac{P_1V_1}{P_2} = \frac{(720 \text{ mm Hg})(150 \text{ mL})}{750 \text{ mm Hg}} = 144 \text{ mL}$$

Step 4. Evaluate

When the pressure is increased slightly at constant temperature, the volume decreases slightly, as expected. Units cancel to give milliliters, as desired.

Practice Problem

A balloon filled with helium gas has a volume of 500 mL at a pressure of 1 atm. After the balloon is released, it reaches an altitude of 6.5 km, where the pressure is 0.5 atm. Assuming that the temperature has remained the same, what volume does the gas occupy at this height?   *(Ans.)* 1000 mL

Charles' Law: Temperature–Volume Relationship

The quantitative relationship between temperature and volume was first demonstrated by the French scientist Jacques Charles (1746–1823) in 1787. Charles' experiments revealed that all gases expand and contract to extent the same when heated through the same temperature interval. Furthermore, Charles found that the change in volume amounts to $\frac{1}{273}$ of the original volume for each Celsius degree, when the pressure remains the same and the initial temperature is 0°C. Raising the temperature to 1°C causes the gas volume to increase by $\frac{1}{273}$ of the volume it had at 0°C. A 10°C-temperature increase causes the volume to expand by $\frac{10}{273}$ of the original volume at 0°C. If the temperature is increased by 273°C, the volume increases by $\frac{273}{273}$ of the original—that is, the volume doubles.

The same regularity of volume change is observed in cooling a gas at constant pressure. At 0°C, a 1°C decrease in temperature decreases the original volume by $\frac{1}{273}$. At this rate of volume decrease, a gas cooled from 0°C to −273°C would have zero volume. Real gases cannot be cooled to −273°C. Before they reach this temperature, intermolecular forces exceed the kinetic energy of the molecules and the gases condense to liquids or solids.

Figure 11-17 These hot air balloonists are taking advantage of the fact that gases expand upon heating.

If doubling the temperature doubles the volume of a gas, to what temperature must a gas be raised to double its volume if the original sample is at 0°C? This is a nonsense question. Two times zero is zero. Only the kelvin temperature scale may be used.

TABLE 11-3   TEMPERATURE–VOLUME DATA (AT CONSTANT MASS AND PRESSURE)

Temperature (°C)	Volume (mL)
273	1092
100	746
10	566
1	548
0	546
−1	544
−73	400
−173	200
−223	100
−273	0

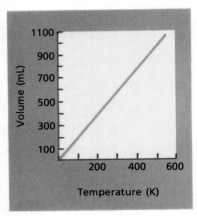

Figure 11-18 The variation in volume with Kelvin temperature at constant pressure.

The data in Table 11-3 illustrate the temperature-volume relationship at constant pressure for a gas sample that has a volume of 546 mL at 0°C. Recall that if the gas is warmed 1°C, it expands $\frac{1}{273}$ its original volume. Each 1°C-temperature change thus causes a volume change of 546 mL/273, or 2 mL in the gas sample mentioned above. Raising the temperature to 100°C increases the volume by $\frac{100}{273}$ of 546 mL, or by 200 mL.

Note that in Table 11-3, the volume does not increase in direct proportion to the Celsius temperature. For example, increasing the temperature tenfold does not cause a proportional increase in the volume.

You learned in Section 2.2 that the Kelvin temperature scale starts at a temperature that corresponds to −273°C, or, more precisely, −273.15°C. Lord Kelvin realized the significance of this temperature, identifying it as the lowest temperature possible. *The temperature −273.15°C is referred to as absolute zero, and is given a value of zero on the Kelvin scale.* This gives the relationship you learned earlier, K = 273.15 + °C. In many calculations, 273.15 is rounded off to 273K.

Because the Kelvin scale has its zero point at the lowest possible temperature, the average kinetic energy of the molecules of a gas is directly proportional to the Kelvin temperature of the gas. As a result, the volume and Kelvin temperature are directly proportional to each other, as shown in Table 11-4. For example, quadrupling the Kelvin temperature from 50K to 200K causes the volume to quadruple. Plotting the volume and Kelvin temperature data of Table 11-4 gives a straight line, as shown in Figure 11-18, which is characteristic of plots of directly proportional variables.

Today, the relationship between the Kelvin temperature of a gas and its volume is known as **Charles' law:** *The volume of a fixed mass of gas varies directly with the Kelvin temperature at constant pressure.* Mathematically, Charles' law may be expressed as

$$V = kT \text{ or } \frac{V}{T} = k$$

where k is a constant, the value of which depends only on the quantity of gas and the pressure, and T is the Kelvin temperature. For the ratio V/T for any set of volume–temperature values (where mass is known), k is always the

TABLE 11-4 VOLUME–TEMPERATURE DATA (AT CONSTANT MASS AND PRESSURE)

Volume (mL)	Celsius Temperature (°C)	Kelvin Temperature (K)	$\dfrac{V}{T(K)}$ (mL/K)
1092	273	546	2
746	100	373	2
566	10	283	2
548	1	274	2
546	0	273	2
544	−1	272	2
400	−73	200	2
100	−223	50	2
0	−273	0	—

same. The relationship thus obtained is a useful expression of Charles' law that can be applied to most volume–temperature problems involving gases.

$$\frac{V_1}{T_1} = \frac{V_2}{T_2}$$

where V_1 and T_1 represent one set of conditions, and V_2 and T_2 represent a different set of conditions. When three of the four values V_1, T_1, V_2 and T_2 are known, this equation can be used to calculate the fourth.

Sample Problem 11.3

A sample of neon gas occupies a volume of 752 mL at 25°C. What volume will the gas occupy at 50°C if the pressure remains constant?

Solution

Step 1. Analyze Given: $V_1 = 752$ mL
  $T_1 = 25°C + 273 = 298K$
  $T_2 = 50°C + 273 = 323K$
  Unknown: V_2

Step 2. Plan $$V_1, T_1, T_2 \rightarrow V_2$$

Because the gas remains at constant pressure, an increase in temperature will cause an increase in volume. Therefore, the gas obeys the Charles' law. It is *very important* to remember that temperature must be expressed in degrees Kelvin in all problems involving Charles' law. To obtain V_2, rearrange the equation for Charles' law.

$$V_2 = \frac{V_1 T_2}{T_1}$$

Step 3. Compute Substitute values for V_1, T_1, and T_2 to obtain the new volume, V_2.

$$V_2 = \frac{V_1 T_2}{T_1} = \frac{752 \text{ mL} \times 323K}{298K} = 815 \text{ mL}$$

Sample Problem 11.3 *continued*

| *Step 4. Evaluate* | As expected, the volume of the gas increases as the temperature increases. Units cancel to yield milliliters, as desired. The answer contains the appropriate number of significant digits, and is reasonably close to an estimated value of 813: |

$$\frac{750 \times 325}{300}.$$

Practice Problem

1. A helium-filled balloon has a volume of 2.75 L at 20°C. The volume of the balloon decreases to 2.46 L after it is placed outside on a cold day. What is the outside temperature?

   (Ans.) 262K, or −11°C

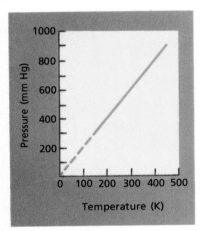

Figure 11-19 The variation in pressure with Kelvin temperature at constant volume.

Gay-Lussac's Law

What would you predict regarding the quantitative relationship between pressure and temperature at constant volume? You have seen that pressure is the result of collisions of molecules with container walls and that the average kinetic energy of molecules is directly proportional to Kelvin temperature. Thus, for a fixed quantity of gas at constant volume, the pressure should be directly proportional to the Kelvin temperature. For every degree of temperature increase or decrease, the pressure of a confined gas increases or decreases respectively by $\frac{1}{273}$ of the volume of the gas at 0°C. Joseph Gay-Lussac (1778–1850) is given credit for recognizing this relationship in 1802.

The data plotted in Figure 11-19 illustrate what is now known as **Gay-Lussac's law:** *The pressure of a fixed mass of gas varies directly with the Kelvin temperature at constant volume.* Mathematically, Gay-Lussac's law is expressed as

$$P = kT \text{ or } \frac{P}{T} = k$$

where k is a constant that depends on the volume of the sample and the quantity of the gas, and T is the temperature in Kelvin. As we have shown for Boyle's and Charles' laws, changes in the pressure or temperature can be found by using the following relationship:

$$\frac{P_1}{T_1} = \frac{P_2}{T_2}$$

when values are known for three of the four quantities involved.

Sample Problem 11.4

The gaseous contents in an aerosol can are under a pressure of 3.00 atm at 25°C. Directions on the can caution the user to keep the can in a place where the temperature does not exceed 52°C. What would the pressure of the gas in the aerosol can be at 52°C?

Sample Problem 11.4 *continued*

Solution

Step 1. Analyze   Given: $P_1 = 3.00$ atm
$T_1 = 25°C + 273 = 298K$
$T_2 = 52°C + 273 = 325K$

Unknown: P_2

Step 2. Plan   Because the gaseous contents remain at the constant volume of the can, an increase in temperature will cause an increase in pressure. The gaseous contents, therefore, obey Gay-Lussac's law.

$$P_1, T_1, T_2 \rightarrow P_2$$

Rearrange Gay-Lussac's law to obtain P_2:

$$P_2 = \frac{P_1 T_2}{T_1}$$

Step 3. Compute   Substitute values for P_1, T_2, and T_1 to obtain the new pressure, P_2.

$$P_2 = \frac{(3.00 \text{ atm})(325K)}{298K} = 3.27 \text{ atm}$$

Step 4. Evaluate   As expected, a temperature increase at constant volume causes the pressure of the contents in the can to increase. Units cancel correctly. The answer contains the proper number of significant digits and is reasonably close to an estimated value of 3.25, calculated as $\frac{3 \times 325}{300}$.

Practice Problem

Before a trip from New York to Boston, the pressure in an automobile tire is 1.8 atm at 20°C. At the end of the trip, the pressure gauge reads 1.9 atm. What is the new Celsius temperature of the air inside the tire? *(Ans.)* 36°C.

The Combined Gas Law

A gas sample often undergoes simultaneous changes in temperature, pressure, and volume. When this happens, three variables must be dealt with at once. Boyle's law, Charles' law and Gay-Lussac's law can be combined into a single expression that describes a relationship that is useful in such situations. *This relationship, known as the* **combined gas law,** *expresses the relationship between pressure, volume, and temperature of a gas when the amount of gas is fixed.* In mathematical form, the combined gas law can be expressed

$$\frac{PV}{T} = k$$

where k is constant and depends on the amount of gas. The combined gas law can also be written as

$$\frac{P_1 V_1}{T_1} = \frac{P_2 V_2}{T_2}$$

As in the individual gas laws, the subscripts indicate two different sets of conditions, and T represents the temperature in degrees Kelvin. From this expression, any one value can be calculated if the other five are known. Note that each of the individual gas laws (Charles' law, Boyle's law, or Gay-Lussac's law) can be obtained from the combined gas law when the appropriate variable is kept constant. When temperature is constant ($T_1 = T_2$), Boyle's law is obtained; when pressure is constant ($P_1 = P_2$), Charles' law is obtained; and when volume is constant ($V_1 = V_2$), Gay-Lussac's law is obtained.

Sample Problem 11.5

A helium-filled balloon has a volume of 50.0 L at 25°C and 820 mm Hg. What volume will it occupy at 650 mm Hg and 10°C?

Solution

Step 1. Analyze   Given: $V_1 = 50.0$ L $T_2 = 10°C + 273 = 283$K
$T_1 = 25°C + 273 = 298$K
$P_2 = 650.$ mm Hg $P_1 = 820.$ mm Hg

Unknown: V_2

Step 2. Plan

$$V_1, T_1, T_2, P_1, P_2 \rightarrow V_2$$

Because the gas is subjected to changes in both temperature and pressure, we must apply the combined gas law. Rearrange the combined gas law,

$$\frac{P_1V_1}{T_1} = \frac{P_2V_2}{T_2}, \text{ to give } V_2 = \frac{P_1V_1T_2}{P_2T_1}$$

Step 3. Compute   Substitute the known values into the equation to obtain a value for V_2:

$$V_2 = \frac{P_1V_1T_2}{P_2T_1} \frac{820 \text{ mm Hg} \times 50.0 \text{ L} \times 283\text{K}}{650 \text{ mm Hg} \times 298\text{K}} = 59.9 \text{ L}$$

Step 4. Evaluate   As expected, the net result of the two changes gives an increase in the volume, from 50.0 L to 59.9 L. The result is expected because the pressure decreases much more that the temperature decreases. Units cancel appropriately. The answer is correctly expressed to three significant digits and is reasonably close to an estimated value of 66.6

$$\frac{800 \times 50 \times 300}{600 \times 300}$$

Practice Problems

1. The volume of a gas is 27.5 mL at 22.0°C and 740 mm Hg pressure. What will be its volume at 15.0°C and 755 mm Hg pressure? *(Ans.)* 26.3 mL
2. A 700-mL gas sample at STP is compressed to a volume of 200 mL and the temperature is increased to 30°C. What is the new pressure of the gas? *(Ans.)* 2.95×10^3 mm Hg

Dalton's Law of Partial Pressures

John Dalton, the English chemist who proposed the atomic theory, also made an important contribution to the study of mixtures of gases. He found that, in the absence of a chemical reaction, the pressure of a mixture of gases is equal to the sum of the individual pressures exerted by each gas alone. For example, suppose we have a 10-L container filled with some oxygen gas that exerts a pressure of 1.0 atm at 0°C. In another 1.0-L container, we have some nitrogen gas that exerts a pressure of 1.0 atm at 0°C. When these gas samples are combined in the same 1.0-L container, the total pressure of the mixture is found to be 2.0 atm at 0°C. The pressure that each gas exerts in the mixture is independent of the other gas or gases present. *The pressure of each gas in a mixture is called the* **partial pressure** *of that gas.* The relationship between the total pressure of a mixture of gases and the individual pressures of the component gases is now known as **Dalton's law of partial pressures:** *The total pressure of a mixture of gases is equal to the sum of the partial pressures of the component gases.* The law holds true regardless of the number of different gases that have been combined. Dalton's law may be expressed as

$$P_T = P_1 + P_2 + P_3 + \ldots$$

where P_T is the total pressure of the mixture and $P_1, P_2, P_3, \ldots$ are the partial pressures of component gases 1, 2, 3, ...

Dalton's law is easily understood in terms of the kinetic-molecular theory. The rapidly moving particles of each gas in a mixture have equal opportunities to collide with the container walls. Therefore, each exerts a pressure independent of that of the other gases present. The total pressure is the result of the total number of collisions per unit of wall area in a given time. (Note that because gas molecules move independently, the combined gas law, as well as Dalton's law, can be applied to gas mixtures.)

Gases collected by water displacement   Gases generated in the laboratory are often collected over water, as shown in Figure 11-20. Dalton's law of partial pressures has a practical application in calculating the volume of gases collected in this manner. A gas collected by water displacement is not pure, but is always mixed with water vapor. This is because water molecules at the liquid surface always evaporate and mingle with the gas molecules. Water vapor, like other gases, exerts a pressure, known as *water vapor pressure.*

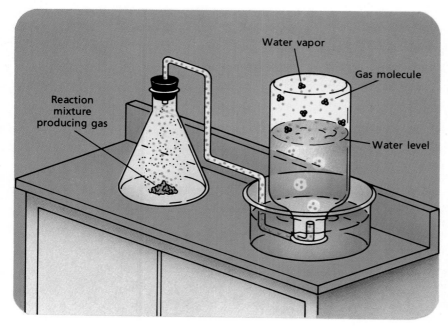

Figure 11-20   Gases are often collected in the laboratory by water displacement.

To determine the total pressure of the gas and water vapor inside the collection bottle, the bottle is raised until the water levels inside and outside the bottle are the same. At this point, the total pressure inside the bottle is the same as the atmospheric pressure. According to Dalton's law of partial pressures,

$$P_T = P_{atm} = P_{gas} + P_{H_2O}$$

To determine the partial pressure of the dry gas collected, one must subtract the vapor pressure of the water at the given temperature from the total pressure. The atmospheric pressure (P_{atm}) at the time of the experiment is read from a barometer in the laboratory. Water vapor pressure varies with temperature. The value of P_{H_2O} at the temperature of the experiment is read from a standard reference table like that given in Table A-11 of this book.

Gases generated in the laboratory can also be collected over mercury. Since mercury does not evaporate markedly at room temperature, it is not necessary to subtract its vapor pressure from the total pressure to determine the pressure of the gas that is being collected.

Sample Problem 11.6

Oxygen from the decomposition of potassium chlorate ($KClO_3$) was collected by water displacement. The barometric pressure and the temperature during the experiment were 731.0 mm Hg and 20.0°C, respectively. What was the partial pressure of the oxygen collected?

Solution

Step 1. Analyze   Given: $P_T = P_{atm} = 731.0$ mm Hg
$P_{H_2O} = 17.5$ mm Hg (vapor pressure of water at 20.0°C, from Table A-11)
$P_T = P_{atm} = P_{O_2} + P_{H_2O}$
Unknown: P_{O_2}

Step 2. Plan $$P_{atm}, P_{H_2O} \rightarrow P_{O_2}$$

The partial pressure of oxygen collected over water must be found by subtracting the partial pressure of water vapor from the atmospheric pressure, according to Dalton's law of partial pressures.

$$P_{O_2} = P_{atm} - P_{H_2O}$$

Step 3. Compute   Substituting values for P_{atm} and P_{H_2O} gives P_{O_2}

$$= 731.0 \text{ mm Hg} - 17.5 \text{ mm Hg} = 713.5 \text{ mm Hg}$$

Step 4. Evaluate   As expected, the oxygen partial pressure is less than atmospheric pressure, but much larger than the partial pressure of water vapor at this temperature. The answer has the appropriate number of places and is reasonably close to an estimated value of 713 (730−17).

Practice Problem

When hydrogen gas is collected over water at 20.0°C, the levels of water inside and outside the gas-collection bottle are the same. The partial pressures of hydrogen was determined to be 742.5 mm Hg. What was the barometric pressure at the time the gas was collected? *(Ans.)* 760.0 mm Hg

Section Review

1. Define pressure and explain how it is measured.
2. Convert the following pressures to pressures in standard atmospheres: *(a)* 151.98 kPa *(b)* 456 mm Hg *(c)* 912 torr.
3. What are the standard conditions for gas measurements?
4. State Boyle's law, Charles' law, and the combined gas law in mathematical terms.
5. A sample of helium gas has a volume of 200.0 mL at 730 mm Hg pressure. What pressure, in atm, is needed to reduce the volume at constant temperature to 50.0 mL?
6. A certain quantity of gas has a volume of 0.750 L at 298K. At what temperature, in degrees Celsius, would this quantity of gas be reduced to 0.500 L at constant pressure?
7. An aerosol can contains gases under a pressure of 4.5 atm at 20.0°C. If the can is left on a hot sandy beach, the pressure of the gases increases to 4.8 atm. What is the Celsius temperature on the beach?
8. Discuss the significance of the absolute-zero temperature.
9. The volume of a sample of methane gas measures 350 mL at 27.0°C and 810 mm Hg. What is the volume in liters at −3.0°C and 650 mm Hg pressure?
10. A certain mass of oxygen was collected over water when potassium chlorate was decomposed by heating. The volume of the oxygen sample collected was 720 mL at 25.0°C and a barometric pressure of 755 mm Hg. What would the volume of the oxygen be at STP?

1. Force per unit area; measured using either a barometer (atmospheric pressure) or a manometer (pressure in a closed container)
2. *(a)* 1.50 atm *(b)* 0.600 atm *(c)* 1.20 atm
3. *(1)* 1 atm pressure *(2)* 273K
4. *(1)* $P_1V_1 = P_2V_2$ *(2)* $V_1/T_1 = V_2/T_2$ *(3)* $P_1V_1/T_1 = P_2V_2/T_2$
5. 3.8 atm
6. −74°C
7. 40.°C
8. Absolute-zero is the temperature at which molecular kinetic motion is at a minimum.
9. 0.393 L
10. 630 mL O_2

Chapter Summary

- The kinetic theory of matter provides an explanation of the properties of gases, liquids, and solids in terms of the forces between the particles of matter and the energy these particles possess.
- The kinetic–molecular theory of gases describes a model of an ideal gas. Although ideal gases do not actually exist, the behavior of most gases is close to ideal under a range of temperatures and pressures.
- Gases consist of large numbers of tiny, fast-moving particles that are far apart in comparison to their size. Their average energy is in direct proportion to the Kelvin temperature of the gas.
- Gases exhibit expansion, fluidity, low density, compressibility, and diffusion.
- Gases are described by four measurable quantities: pressure, volume, temperature, and number of molecules. Relationships among the first three of these four qualities are: Boyle's law—$PV = k$; Charles' law—$V = kT$; Gay-Lussac's law—$P = kT$; and the combined gas law—$PV/T = k$.
- A gas exerts pressure on the walls of its container. In a mixture of gases, the total pressure is equal to the sum of the component gases.
- Conditions of standard temperature and pressure (abbreviated STP) allow us to accurately compare volumes of different gases.
- A barometer measures the pressure of the atmosphere. The pressure of a gas in a closed container can be measured by a manometer.

Chapter 11 *Review*

Physical Behavior of Gases

Vocabulary

absolute zero
atmosphere of
   pressure (atm)
barometer
Boyle's law
Charles' law
combined gas law
Dalton's law of
   partial pressure
diffusion
effusion
elastic collision
fluids

Gay-Lussac's law
ideal gas
kinetic theory
millimeters of mercury
   (mm Hg)
Newton
partial pressure
pressure
real gas
standard temperature
   and pressure (STP)
torr

Questions

1. Upon what two factors is the kinetic theory based?
2. What is an ideal gas?
3. State the five basic assumptions of the kinetic theory.
4. How do gases compare with liquids and solids in terms of the distance between the molecules?
5. What is an elastic collision?
6. *(a)* Write and label the equation that relates the average kinetic energy of the particles in a gas to the velocity of those particles. *(b)* What is the relationship among the temperature, velocity, and kinetic energy of gas molecules?
7. Use the kinetic theory to explain each of the following properties of gases: *(a)* expansion *(b)* fluidity *(c)* low density *(d)* compressibility *(e)* diffusion.
8. *(a)* What is diffusion? *(b)* What factors affect the rate of diffusion of one gas through another? *(c)* What is the relationship between the mass of a gas particle and the rate at which it diffuses through another gas? *(d)* What is effusion?
9. *(a)* What is a real gas? *(b)* When do real gases behave most and least like ideal gases?
10. Name four measurable quantities that must be

specified to fully describe the state or condition of a gas.
11. *(a)* Why does a gas in a closed container exert pressure? *(b)* At constant temperature, what is the relationship between the pressure exerted by a fixed quantity of gas and the volume occupied by that gas?
12. *(a)* At constant pressure, what is the relationship between temperature and the volume occupied by a given quantity of gas? *(b)* Explain the reason for this effect.
13. *(a)* At constant volume, what is the relationship between temperature and the pressure exerted by a given quantity of gas? *(b)* Explain why this effect is observed.
14. *(a)* At constant pressure and temperature, how is the volume occupied by a gas in a closed container related to the number of molecules of gas present? *(b)* Why is this observed?
15. In general terms, what are the gas laws?
16. *(a)* Define pressure. *(b)* Write the equation used to calculate pressure. *(c)* What is the relationship between pressure and the size of the surface area subjected to a given force? *(d)* What is the SI unit for force?
17. *(a)* What is atmospheric pressure? *(b)* Why does the atmosphere exert pressure on the earth? *(c)* In terms of force per unit area, what is the value of atmospheric pressure at sea level?
18. *(a)* What device is used to measure atmospheric pressure? *(b)* Who first introduced this device, and when was it introduced?
19. *(a)* According to Torricelli's experiments, why does a column of mercury in a tube inverted into a pot of mercury retain a height of 760 mm at sea level? *(b)* What comparable height would be maintained by a column of water inverted into a pot of water? *(c)* What accounts for the difference in the heights of the mercury and water columns?
20. *(a)* What is a manometer, and what is it used for? *(b)* How is it read?

21. *(a)* Identify three sets of units typically used to express pressure. *(b)* What is meant by one atmosphere of pressure? *(c)* What is a pascal? *(d)* What is the SI equivalent of one standard atmosphere of pressure?
22. *(a)* What is meant by a pressure reading expressed in millimeters of Hg? *(b)* What is the quantity denoted by one atmosphere of pressure? *(c)* What is the torr?
23. *(a)* What temperature and pressure values are designated as standard conditions? *(b)* How is the term "standard conditions" abbreviated?
24. *(a)* State Boyle's law. *(b)* Write the mathematical equations for Boyle's law.
25. *(a)* State Charles' law. *(b)* Write the mathematical equations that express Charles' law.
26. *(a)* What Celsius temperature represents absolute zero? *(b)* What is the significance of this temperature? *(c)* What is the relationship between Kelvin temperature and the average kinetic energy of the molecules of a gas?
27. *(a)* State Gay-Lussac's law. *(b)* Write the mathematical equations that express that relationship.
28. *(a)* Write the mathematical equations that express the combined gas law. *(b)* What is the significance of the name given to this law?
29. *(a)* Explain what is meant by the partial pressure of each gas within a mixture of gases. *(b)* How do the partial pressures of each of the gases in a mixture affect each other?
30. *(a)* How can the partial pressure of a dry gas collected by water displacement be determined? *(b)* Write the mathematical equation describing this calculation.

Problems

1. If the atmosphere can support a column of mercury 760 mm in height at sea level, what comparable heights of each of the following (in mm) could be supported, given the relative density values cited? *(a)* water, whose density is approximately $\frac{1}{14}$ that of Hg *(b)* a substance with a density 1.40 times that of Hg
2. Convert each of the following into pressure readings expressed in millimeters of mercury: *(a)* 1.25 atm *(b)* 2.48×10^{-3} atm *(c)* 4.75×10^4 atm *(d)* 7.60×10^6 atm.
3. Convert each of the following into the unit specified: *(a)* 125 mm Hg into torr *(b)* 3.20 atm into Pa *(c)* 5.38 kPa into mm Hg.
4. Use Boyle's law to solve for the missing value in each of the following: *(a)* $P_1 = 350$ mm, $V_1 = 200$ mL, $P_2 = 700$ mm, $V_2 = ?$ *(b)* $P_1 = 0.75$ atm, $V_2 = 435$ mL, $P_2 = 0.48$ atm, $V_1 = ?$ *(c)* $V_1 = 2.4 \times 10^5$ L, $P_2 = 180$ torr, $V_2 = 1.8 \times 10^3$ L, $P_1 = ?$
5. If the pressure exerted on a 240. mL sample of hydrogen gas at constant temperature is increased from 325 mm to 550 mm, what will be the final volume of the sample?
6. A flask containing 155 cm^3 of hydrogen was collected under a pressure of 22.5 kPa. What pressure would have been required in order for the volume of the gas to have been 90.0 cm^3 at constant temperature?
7. Convert each of the following Celsius temperatures to Kelvin temperatures: *(a)* 0°C *(b)* 27°C *(c)* −50°C *(d)* −273°C.
8. Convert each of the following Kelvin temperatures to Celsius temperatures: *(a)* 273K *(b)* 350K *(c)* 100K *(d)* 20K *(e)* 0K
9. Use Charles' law to solve for the missing value in each of the following: *(a)* $V_1 = 80.0$ mL $T_1 = 27$°C, $T_2 = 77$°C, $V_2 = ?$ *(b)* $V_1 = 125$ L, $V_2 = 85.0$ L, $T_2 = 127$°C, $T_1 = ?$ *(c)* $T_1 = -33$°C, $V_2 = 54.0$ mL, $T_2 = 160$°C, $V_1 = ?$
10. A sample of air has a volume of 140.0 mL at 67°C. To what temperature must the gas be lowered to reduce its volume to 50.0 mL at constant pressure?
11. At standard temperature a gas has a volume of 275 mL. If the temperature is increased to 130°C, but the pressure is held constant, what is its new volume?
12. Use Gay-Lussac's law to solve for the missing value in each of the following: *(a)* $P_1 = 450.$ mm, $T_1 = 27$°C, $P_2 = 600.$ mm, $T_2 = ?$ *(b)* $P_1 = 1.75$ atm, $T_1 = 77$°C, $T_2 = 154$°C, $P_2 = ?$ *(c)* $T_1 = 0$°C, $P_2 = 1.00$ atm, $T_2 = -123$°C, $P_1 = ?$
13. A sample of hydrogen at 47°C exerts a pressure of 250. mm. If the gas is heated to 77°C at constant volume, what will its new pressure be?
14. To what temperature must a sample of nitrogen at 27°C and 0.625 atm be heated so that its pressure becomes 1.125 atm at constant volume?

15. Use the combined gas law to solve for the missing value in each of the following: (a) P_1 = 1.00 atm, V_1 = 250. mL, T_1 = 27°C, P_2 = atm, V_2 = 125 mL, T_2 = ? (b) P_1 = 600. mm, V_1 = 1.75 L, T_1 = 77°C, V_2 = 2.25 L, T_2 = 127°C, P_2 = ? (c) P_1 = 740. mm, T_1 = 50°C, P_2 = 820. mm, V_2 = 3.50 L, T_2 = 70°C, V_1 = ?

16. A sample of gas at 47°C and 780 mm pressure occupies of volume of 2.20 L. What volume would this gas occupy at 107°C and 600. mm pressure?

17. A 350-mL air sample collected at 35°C has a pressure of 550 mm. What pressure will the air exert if it is allowed to expand to 425 mL at 57°C?

18. A gas measures 1.75 L at −23°C and 150 kPa. At what temperature would the gas occupy 1.30 L at 210 kPa?

19. A sample of oxygen at 40°C occupies 820. mL. If this sample occupies 1250 mL at 60°C and 1.40 atm, what was its original pressure?

20. Three of the primary components of air are carbon dioxide, nitrogen, and oxygen. In a sample containing a mixture of these gases at one atmosphere of pressure, the partial pressures of carbon dioxide and nitrogen are given as P_{CO_2} = 0.285 mm and P_{N_2} = 593.525 mm. What is the partial pressure of oxygen?

21. Determine the partial pressure of oxygen collected by water displacement if the water temperature is 20°C and the total pressure of the gases in the collection bottle is 730.0 mm Hg.

22. A sample of gas is collected over water at a temperature of 35°C when the barometric pressure reading is 742.0 mm Hg. What is the partial pressure of the dry gas?

Application Questions

1. (a) Explain the distribution of molecular velocities in a gas at two different temperatures, as illustrated in Figure 11-2. (b) If the temperature of the gas is increased further, what would you anticipate the resulting kinetic energy curve would look like?

2. Explain how different gases in a mixture can have the same average kinetic energy value, even though the masses of their individual particles differ.

3. If all gases behaved as ideal gases under all conditions of temperature and pressure, there would be no solid or liquid forms of these substances. Explain.

4. In view of Torricelli's findings concerning the heights of mercury and water columns that could be supported by the atmosphere at sea level, what predictions could be made concerning the comparable column heights of each of the following liquids (in relation to that of mercury) if their densities are: (a) half that of mercury (b) twice that of mercury (c) twice that of water.

5. Pressure is defined as force per unit area and yet in his experiments with the barometer, Torricelli found that the diameter of the tub and, correspondingly, the surface area of contact between the mercury in the tube and the mercury in the tub exposed to the atmosphere did not affect the height of mercury that was supported by the atmosphere at sea level. Explain this seemingly inconsistent observation in view of the relationship between pressure and surface areas.

6. According to Boyle's law, at constant temperature, what relative effect would each of the following changes have on the condition specified? (a) effect of doubling pressure on volume (b) effect of tripling volume on pressure (c) effect of reducing volume to ¼ of original on pressure

7. According to Charles' law, what relative effect would each of the following changes have on the condition specified at constant pressure? (a) effect of a 1°C increase on volume (b) effect of a 100°C decrease on volume (c) effect of doubling Kelvin temperature on volume (d) effect of reducing Kelvin temperature to ¼ of original on volume

8. Based on Gay-Lussac's law, at constant volume, what relative effect would each of the following changes have on the condition specified assuming the volume remains constant? (a) effect of doubling Kelvin temperature on pressure (b) effect of halving Kelvin temperature on pressure (c) effect of tripling pressure on Kelvin temperature

9. (a) Explain the relationship between Dalton's law of partial pressures and the procedure used to calculate the volume of a gas collected by

water displacement.   (b) Why is it necessary that the water level inside and outside the gas collection bottle be the same when the total pressure reading is taken?

Application Problems

1. Based on the formula for determining the kinetic energy of gas particles, what quantitative change in kinetic energy of a gas would you expect if: (a) the mass of the gas is doubled, but its velocity remains constant   (b) the mass of the gas is halved, but its velocity remains constant   (c) the velocity of the gas particles is halved, but the mass of the gas remains constant.

2. A gas has a volume of 450.0 mL at standard pressure. If the temperature is held constant, what volume would the gas occupy if the pressure were   (a) doubled   (b) reduced to one-fourth of its original value.

3. If a sample of oxygen that occupies 1.00 $\times$ 10^6 mL at 575 mm is subjected to a pressure of 1.25 atm, what will be the final volume of the sample if the temperature is held constant?

4. A gas in a variable-volume container exerts a pressure of 185 Pa when its volume is 365 mL. If the temperature is held constant, under what pressure would the volume of this gas be 142 mL?

5. A gas storage tank at a large industrial plant is designed to provide fuel under constant pressure. When the plant is shut down on Friday afternoon, the volume of the tank is 21.5 m^3 at 75°C. If the temperature drops to 30°C by Monday morning, what volume will the fuel occupy at that time?

6. If the pressure on a gas at −73°C is doubled but its volume is held constant, what will its final temperature be in degrees Celsius?

7. A balloon filled with air at 1.60 atm pressure and a temperature of 18°C occupies a volume of 2.25 L. If the balloon is left in the sun, so that it expands to 2.45 L at 25°C, what is the pressure of the air inside?

8. A gas at 7.75 $\times$ 10^4 Pa and 17°C occupies a volume of 850. cm^3. At what temperature, in degrees Celsius, would the gas occupy 720. cm^3 at 8.10 $\times$ 10^4 Pa?

9. A meteorological balloon contains 250. L of He at 22°C and 740. mm Hg. If the volume of the balloon can vary according to external conditions, what volume would it occupy at an altitude at which the temperature is −52°C and the pressure is 0.750 atm?

10. The balloon in the previous problem will burst if its volume reaches 400 L. Given the initial conditions specified in that problem, at what temperature, in degrees Celsius, will the balloon burst, if its pressure at that point is 0.475 atm?

11. The normal respiratory rate for a human being is 15 breaths per minute. If the average volume of air for each breath is 505 cm^3 at 20°C and 9.95 $\times$ 10^4 Pa, what volume of air, in cubic meters and at STP, does an individual breathe in one day?

12. A sample of oxygen is collected in a 175-mL container over water at 15°C. If the barometer reads 752.0 mm Hg, what volume would the dry gas occupy at 770.0 mm and 15°C?

13. If 120. mL of argon is collected over water at 25°C and 780.0 mm, compute the volume of the dry agron at STP.

Enrichment

1. Design and conduct a meteorological study to examine the interrelationships among barometric pressure, temperature, humidity, and other pertinent weather variables. Prepare a report explaining your results.

2. Prepare a report tracing the historical development of aerosol products. Identify several products packaged this way and include a discussion of the current status of such packaging.

3. Visit a carbonated-beverage bottling plant to observe the preparation and bottling of these beverages. Report your observations to the class.

4. Prepare a report on the history of scuba diving. Include a discussion of the role of pressure in scuba diving, and of the causes and effects of decompression sickness or "the bends."

5. Prepare a report on the evolution of the modern submarine. Include a discussion of the technology than enables the submarine to withstand the tremendous pressure of the ocean above it. Also report on the equipment used to ensure a sufficient supply of oxygen for submarine crew members.

Chapter 12 Molecular Composition of Gases

Chapter Planner

12.1 Volume-Mass Relationships of Gases
Measuring and Comparing the Volumes of Reacting Gases
- Question: 1

Avogadro's Principle
- Experiment: 13
- Questions: 2–4
- Problem: 1

Molar Volume of Gases
- Experiment: 14
- Question: 5
- Problems: 2–7

Gas Density
- Question: 6
- Problem: 8
- Application Problems: 3, 4a
- Application Problems: 1, 2

Finding Molar Mass From Volume at STP
- Experiment: 15
- Problem: 9
- Application Problem: 4b

- Problems: 10, 11
- Application Problem: 5

12.2 The Ideal Gas Law
Derivation of the Ideal Gas Law
- Questions: 7–9
- Application Question: 1

The Ideal Gas Constant
Using the Ideal Gas Law
- Problems: 12–17
- Application Question: 22
- Application Problems: 6–8, 10–12
- Application Problems: 9, 13
- Application Problem: 14

12.3 Stoichiometry of Gases
- Question: 10
- Problem: 18

Volume-Volume Calculations
- Problems: 19–21
- Application Problem: 15
- Application Problem: 16
- Application Problem: 17

Volume-Mass and Mass-Volume Calculations
- Problems: 22, 26, 27
- Problems: 23–25, 28, 30, 31
- Application Problem: 24
- Problem: 29
- Application Problems: 18–23, 25, 26

12.4 Effusion and Diffusion
Graham's Law of Effusion or Diffusion
- Demonstrations: 1, 2
- Desktop Investigation Questions: 11, 12
- Problem: 32
- Application Problems: 27, 28

Applications of Graham's Law
- Question: 13
- Application Problem: 29
- Technology Question: 14

Teaching Strategies

Introduction

Chapter 11 took the properties of gases and expressed the relationships among pressure, volume, temperature, and number of particles in mathematical form. Chapter 12 continues by developing the relationship between mass and volume, the ideal gas law, stoichiometry of gases, and effusion and diffusion.

12.1 Volume–Mass Relationships of Gases

The introduction presents Dalton's and Gay Lussac's work with combining volumes of gases and with Avogadro's interpretation of the data. It is very important that the students understand the importance of Avogadro's insight. He was able to predict molecular behavior from macroscopic properties and to predict the composition of molecules without being able to see or prove their existence.

Use the chalkboard to lead the students through the reasoning process Avogadro might have used.

Data: Hydrogen gas and oxygen gas react to form water vapor

2 volumes + 1 volume → 2 volumes

If equal volumes at the same temperature and pressure contain the same number of particles, then

2 molecules + 1 molecule → 2 molecules

At this time, the structures (atomic composition) for the molecules of hydrogen, oxygen, and water vapor were not known.

If one assumes that each molecule of (1) oxygen contains a total of X atoms (2) hydrogen contains a total of Y atoms, and (3) water vapor contains a total of Z. atom then from the above molecular equation, one can show that

$$2X \text{ atoms of hydrogen} + Y \text{ atoms of oxygen} \rightarrow 2Z \text{ atoms of water vapor}$$

Now, applying some basic arithmetic, regardless of what X and Z are, 2X and 2Z must be even numbers (two times an odd or an even number gives an even number).

$$\underset{\text{even}}{2X} + Y \rightarrow \underset{\text{even}}{2Z}$$

Y must also be an even number because only an even number can be added to an even number to give an even number. The simplest value of Y would then be two. The oxygen molecule must consist of at least two oxygen atoms. Once one has determined that oxygen is diatomic, one can combine oxygen with other gases molecules and begin to catalog those molecules which must also be diatomic.

Combining volumes may also be related to moles of molecules. Thus the coefficients that balance chemical equations involving gases may be interpreted in terms of molecules, moles of molecules, and volumes of gas.

Since a mole of any gas can be expressed in terms of its molar mass, one can specify the volume of a mole of gas. Thus all gases should occupy the same molar volume at the same temperature and pressure. Students object to this conclusion because large molecules should take up more space than smaller molecules. Remind the students that the distance separating gaseous molecules is very large in comparison to the size of individual molecules. Doubling the size of a molecule still leaves a very large distance between it and the next molecule.

The molar volume of a gas at standard temperature and pressure is 22.4 L. To bring this home to the students, display a 22.4 liter box. If time allows, challenge the students to construct and bring in their own container that measures 22.4 L in volume.

Avogadro's principle is next related to the density of gases. Introduce the topic by having the students reason as follows:

Density is defined as mass/volume, but under conditions of constant temperature and pressure, volume can be related to number of gaseous molecules, thus density is related to mass/number of molecules. If the number of molecules is to be taken to be one mole, then the mass becomes the molar mass, and the original volume related to the molar volume. Thus the density of any gas can be expressed as the ratio of its molar mass to its molar volume (22.4 liters if at STP).

12.2   The Ideal Gas Law

Review Chapter 11, especially the qualitative relationships between the four qualities needed to describe a gas sample—volume, temperature, pressure, and number of particles. Write down the expression of the general gas law developed earlier.

$$\frac{P_1V_1}{n_1T_1} = \frac{P_2V_2}{n_1T_1}$$

Review how each of the gases can be derived from the above general expression if two of the terms are kept constant. Now if the second set of conditions were assumed to be one mole of a gas at STP, the relationship becomes

$$\frac{P_1V_1}{n_1T_1} = \frac{(1 \text{ atm})(22.4 \text{ L})}{(1 \text{ mol})(273K)} = 0.0821 \frac{\text{L} \cdot \text{atm}}{\text{K} \cdot \text{mol}}$$

If the constant calculated is represented by the letter R, then

$$\frac{P_1V_1}{n_1T_1} = R \quad or \quad P_1V_1 = Rn_1T_1$$

This may now be generalized to any set of conditions as

$$PV = nRT$$

Should the students ever forget the value of R, this procedure will help them to calculate it on their own.

The remaining section presents samples of the many problems associated with the ideal gas law. Impress upon the students that the problems studied in Chapter 11 consisted of a gas sample and two sets of conditions. One could predict what might happen to the gas sample if some kind of stress were placed on the gas. The ideal gas law now allows them to calculate one of the variables knowing the other three. In actuality, the above derivation points out that the ideal gas law can be seen as an application of the general gas law. The second set of conditions is always one mole of gas at STP.

The ideal gas law $PV = nRT$ can be rearranged: $\frac{n}{V} = \frac{P}{RT}$

But n/V can be expressed as mass/V if one multiplies n, the number of moles, by the molar mass of the gas. Thus the equation may be rewritten as

$$D = P/RT$$

Another series of problems are once again presented. These problems will give the students the practice in problem solving that will help them in future chapters.

12.3   Stoichiometry of Gases

Calculations studied in Chapter 9 may now be extended to include not only masses and moles of reactants and products, but volumes of each when gases are part of the reaction. A complete flow chart of the process may be presented to the students.

For a reaction A → B

mass A   mass B

moles A → moles B

particles A   particles B
volume A   volume B

If both A and B are gases, Avogadro's principle allows us to directly relate volume and moles. If the coefficients of the equation were 1A and 2B then one mole of A produces two moles of B or 1 L A produces 2 L B.

Students are a bit wary of this simplification at first, and probably will continue to follow the conversion method learned in Chapter 9. Work each type of problem with them and help them to categorize the problem before beginning to work out a solution.

12.4   Effusion and Diffusion

The concluding section on Graham's law combines the derivation, problem solving, a Desktop Investigation, and a Technology feature on the technique of separating isotopes using different diffusion rates.

The text presents the derivation beginning with the definition of kinetic energy. The math is a bit more difficult because students are faced with an inverse square (or square root) relationship. The key qualitative concept is that light gases move more rapidly than heavier slower gases. Average energies are the same.

The analogy of a very slow-moving train and a bullet fired from a rifle might help. The slow-moving train contains a large kinetic energy because of the large mass. The bullet, on the other hand, has a very small mass, but moves very rapidly. Since the velocity term is squared, the kinetic energy of the bullet is also very large. You may wish to bring up other examples from athletics. In football, a smaller lineman will have to be faster than a heavier player if he hopes to compensate for his small mass.

The difference between effusion and diffusion is also very important. Effusion requires a small aperture through which gas will pass. Diffusion involves the mixing of two gases in a container.

Demonstration 1: Rate of Diffusion of Gases

PURPOSE: To compare the rate of diffusion of two gases
MATERIALS: conc. HCl and conc. NH_3(aq), 0.05 to 1-m length of wide bore glass tubing (2 cm), cork or rubber stoppers to fit the glass tubing, cotton, two rings stands and clamps
PROCEDURE: Suspend the piece of glass tubing horizontally on the demonstration table. Take small pieces of cotton and staple them into the ends of the rubber stoppers. (As an alternative, use a cork borer to bore out a hole in the small end of the cork stoppers. The cotton may be placed in the hollow.) Add a few drops of HCl to the cotton on one stopper, and a few drops of NH_3(aq) to the other. Simultaneously insert the two stoppers into the glass tube. Depending on the length of the glass tube, a white ring will form where the ammonia and hydrogen chloride vapors meet. The white ring is solid ammonium chloride.

Conditions in the demonstration are such that Graham's law

will not apply. The ammonia gas will move a greater distance than the hydrogen chloride. You are showing that lighter molecules travel more rapidly. You might wish to calculate the ratio of the square roots of the molar masses of HCl and NH_3 and to predict where the gases should have met should Graham's law have been followed completely.

SAFETY AND DISPOSAL: Observe usual lab precautions when handling the conc. acid and base. Use dropping bottles if possible. Acid and base remaining on the cotton may be rinsed down the sink. The glass may be cleaned by pouring water through it.

If a ready source of hydrogen gas is available, the following is a very dramatic demonstration and may be used to replace the preceding one.

Demonstration 2: Comparative Effusion Rates

PURPOSE: To show that lighter gases effuse more rapidly
MATERIALS: porous unglazed ceramic cup (the kind used in electrochemical cells), one-hole rubber stopper to fit the cup, a 500-mL Erlenmeyer flask, two-hole rubber stopper to fit the flask, glass tubing, 1-L beaker, a source of hydrogen gas
PROCEDURE: Cut one piece of glass tubing approximately 10 cm longer than the length of the flask. Bend the tubing so that the portion that extends beyond the flask is at a 120° angle. Carefully insert the glass tube into the rubber stopper. Insert a straight piece of tubing into the two-hole stopper so that the

end of the tubing does not protrude more than 1 cm through each stopper. Fill the Erlenmeyer flask with water leaving several cm of air in the neck of the flask. Insert the two-hole stopper into the flask. Insert the other stopper into the porous cup.

Collect hydrogen in the inverted 1-L beaker. Carefully place the inverted beaker filled with hydrogen over the porous cup. Immediately water flows from the bent piece of glass tubing. Carefully remove the beaker. Bubbles of air enter the bottom of the flask through the tubing that earlier produced the jet of water. Once again place the beaker over the porous cup. Again water flows from the bent tubing. The second flow is not as vigorous as the first.

Have the students try to explain the demonstration using Graham's law. The competition is between molecules of air (nitrogen and oxygen) and molecules of hydrogen. The hydrogen molecules are so small that they rapidly move through the porous cup into the region in the flask where no hydrogen molecules are present. The added hydrogen molecules increase the pressure in the flask and force the water out the bent tubing. When the beaker is removed, the hydrogen molecules move from the porous cup into the room. A vacuum is left in the flask and the atmospheric pressure pushes air into the flask.

SAFETY AND DISPOSAL: Be careful when handling hydrogen gas. Once constructed, the demonstration apparatus may be stored for later use.

References and Resources

Books and Periodicals

Borgford, Christie L. and Lee R. Summerlin. "Gases," *Chemical Activities: A Sourcebook for Science Teachers*, Washington, D.C.: The American Chemical Society, 1988.

Scott, Arthur F. "The Invention of the Balloon and the Birth of Modern Chemistry," *Scientific American*, Vol. 250, No. 1, Jan. 1984, pp. 126–137.

Audiovisual Resources

Campbell, J. Arthur. *Gas Pressure and Molecular Collisions*, A Chem Study Film, B&W, 21 min., VHS cassette, Central Scientific Co. (Shows the effects of varying the number of molecules per unit of volume and of varying the temperature)

Pimentel, George C. *Gases and How They Combine*, A Chem Study Film, Color, 22 min., VHS cassette, Central Scientific Co. (Simple integer volume ratios lead to Avogadro's hypothesis)

Answers and Solutions

Questions

1. *(a)* At constant temperature and pressure, the volumes of gaseous reactants and products can be expressed as ratios of small whole numbers. *(b)* The law applies only to gas volumes measured at the same temperature and pressure.
2. *(a)* Avogadro's principle states that equal volumes of gases at the same temperature and pressure contain equal numbers of molecules. *(b)* At the same temperature and pressure, the volume of any given gas varies directly with the number of molecules.
3. *(a)* Gas volume is directly proportional to the number of moles of gas at constant temperature and pressure. *(b)* $V = kn$, where n is the number of moles of gas and k is a constant.
4. The coefficients in a chemical equation involving gases give the relative numbers of molecules, numbers of moles, and volumes of reactants and products.
5. *(a)* One mole consists of 6.022×10^{23} molecules. *(b)* The

molar volume of a gas is the volume occupied by one mole of a gas at STP, and has been found to be 22.4 liters. *(c)* At STP, 22.4 L of all gases contain the same number of molecules (6.022×10^{23}), but they have different masses that correspond to their respective molar masses.
6. In order to compare the densities of gases, temperature and pressure must be specified because the volume of a specific number of gas molecules varies with temperature and pressure.
7. *(a)* At constant temperature and molar amount, volume is inversely proportional to pressure. *(b)* At constant pressure and molar amount, volume is directly proportional to the Kelvin temperature. *(c)* At constant pressure and temperature, volume is directly proportional to the number of moles.
8. *(a)* $PV = nRT$
   (b) The ideal gas law relates pressure, volume, temperature, and number of moles for a given sample of gas under any

combination of conditions.

9. *(a)* The ideal gas law applies to any situation in which a single gas sample undergoes no change in conditions and values are known for three of the four variables, P, V, T, and n. *(b)* Care must be taken to match the units of the known quantities and the units of R.

10. *(a)* The molar and volume ratios are the same and are both given by the coefficients in the chemical equation. *(b)* The volume ratios can be used only if all volumes are measured at the same temperature and pressure.

11. *(a)* Diffusion is involved in any movement or mixing of gases. Effusion occurs when molecules of a gas in a container that has a small opening pass through the opening. *(b)* The rates are governed by the velocities of the molecules, which, in turn, are dependent upon the masses of the molecules.

12. *(a)* Graham's law of effusion (or diffusion) states that rates of effusion (or diffusion) of gases at the same temperature and pressure are inversely proportional to their molar masses.

(b) $\dfrac{\text{rate of effusion}_A}{\text{rate of effusion}_B} = \dfrac{\sqrt{\text{molar } m_B}}{\sqrt{\text{molar } m_A}}$

13. $\dfrac{\text{rate of effusion}_A}{\text{rate of effusion}_B} = \dfrac{\sqrt{\text{mol } m_B}}{\sqrt{\text{mol } m_A}} = \dfrac{\sqrt{\text{density}_B}}{\sqrt{\text{density}_A}}$

14. Differences in effusion rates can be used to separate isotopes of differing atomic masses and provide a method for determining molar masses.

Problems

1. *(a)* $\dfrac{1.08 \times 10^{23}\ \text{molec}}{5.00\ \text{L O}_2} \times \dfrac{1\ \text{L O}_2}{\text{L H}_2} \times 5.00\ \text{L H}_2$

$= 1.08 \times 10^{23}\ \text{molec}$

(b) $\dfrac{1.08 \times 10^{23}\ \text{molec}}{5.00\ \text{L O}_2} \times \dfrac{1\ \text{L O}_2}{\text{L CO}_2} \times 5.00\ \text{L CO}_2$

$= 1.08 \times 10^{23}\ \text{molec}$

(c) $\dfrac{1.08 \times 10^{23}\ \text{molec}}{5.00\ \text{L O}_2} \times \dfrac{1\ \text{L O}_2}{\text{L NH}_2} \times 10.00\ \text{L NH}_3$

$= 2.16 \times 10^{23}\ \text{molec}$

2. *(a)* $1.00\ \text{mol O}_2 \times \dfrac{6.02 \times 10^{23}\ \text{molec O}_2}{\text{mol O}_2}$

$= 6.02 \times 10^{23}\ \text{molec O}_2$

(b) $2.50\ \text{mol He} \times \dfrac{6.02 \times 10^{23}\ \text{molec He}}{\text{mol He}}$

$= 1.50 \times 10^{24}\ \text{molec He}$

(c) $0.0650\ \text{mol NH}_3 \times \dfrac{6.02 \times 10^{23}\ \text{molec NH}_3}{\text{mol NH}_3}$

$= 3.91 \times 10^{22}\ \text{molec NH}_3$

(d) $\dfrac{11.5\ \text{g NO}_2}{\text{mol NO}_2} \times \dfrac{\text{mol NO}_2}{46.0\ \text{g NO}_2} \times \dfrac{6.02 \times 10^{23}\ \text{molec NO}_2}{\text{mol NO}_2}$

$= 1.50 \times 10^{23}\ \text{molec NO}_2$

3. *(a)* $2.25\ \text{mol Cl}_2 \times \dfrac{70.9\ \text{g Cl}_2}{\text{mol Cl}_2} = 160\ \text{g Cl}_2$

(b) $3.01 \times 10^{23}\ \text{molec H}_2\text{S} \times \dfrac{\text{mol H}_2\text{S}}{6.02 \times 10^{23}\ \text{molec H}_2\text{S}} \times$

$\dfrac{34.1\ \text{g H}_2\text{S}}{\text{mol H}_2\text{S}} = 17.0\ \text{g H}_2\text{S}$

(c) $25.0\ \text{molec SO}_2 \times \dfrac{\text{mol SO}_2}{6.02 \times 10^{23}\ \text{molec SO}_2} \times \dfrac{64.1\ \text{g SO}_2}{\text{mol SO}_2}$

$= 2.66 \times 10^{-21}\ \text{g SO}_2$

4. *(a)* $1.00\ \text{mol O}_2 \times \dfrac{22.4\ \text{L O}_2}{\text{mol O}_2} = 22.4\ \text{L O}_2$

(b) $3.50\ \text{mol F}_2 \times \dfrac{22.4\ \text{L F}_2}{\text{mol F}_2} = 78.4\ \text{L F}_2$

(c) $0.0400\ \text{mol CO}_2 \times \dfrac{22.4\ \text{L CO}_2}{\text{mol CO}_2} = 0.896\ \text{L CO}_2$

(d) $1.20 \times 10^{-6}\ \text{mol He} \times \dfrac{22.4\ \text{L He}}{\text{mol He}} = 2.69 \times 10^{-5}\ \text{L He}$

5. *(a)* $22.4\ \text{L H}_2 \times \dfrac{\text{mol H}_2}{22.4\ \text{L H}_2} = 1.00\ \text{mol H}_2$

(b) $5.60\ \text{L Cl}_2 \times \dfrac{\text{mol Cl}_2}{22.4\ \text{L Cl}_2} = 0.250\ \text{mol Cl}_2$

(c) $0.125\ \text{L Ne} \times \dfrac{\text{mol Ne}}{22.4\ \text{L Ne}} = 5.58 \times 10^{-3}\ \text{mol Ne}$

(d) $70.0\ \text{mL NH}_3 \times \dfrac{\text{L NH}_3}{10^3\ \text{mL NH}_3} \times \dfrac{\text{mol NH}_3}{22.4\ \text{L NH}_3}$

$= 3.12 \times 10^{-3}\ \text{mol NH}_3$

6. *(a)* $11.2\ \text{L H}_2 \times \dfrac{\text{mol H}_2}{22.4\ \text{L H}_2} \times \dfrac{2.02\ \text{g H}_2}{\text{mol H}_2} = 1.01\ \text{g H}_2$

(b) $2.80\ \text{L CO}_2 \times \dfrac{\text{mol CO}_2}{22.4\ \text{L CO}_2} \times \dfrac{44.0\ \text{g CO}_2}{\text{mol CO}_2} = 5.50\ \text{g CO}_2$

(c) $15.0\ \text{mL SO}_2 \times \dfrac{\text{L SO}_2}{10^3\ \text{mL SO}_2} \times \dfrac{\text{mol SO}_2}{22.4\ \text{L SO}_2} \times \dfrac{64.1\ \text{g SO}_2}{\text{mol SO}_2}$

$= 0.0429\ \text{g SO}_2$

(d) $3.40\ \text{cm}^3\ \text{F}_2 \times \dfrac{\text{mL F}_2}{\text{cm}^3\ \text{F}_2} \times \dfrac{\text{L F}_2}{10^3\ \text{mL F}_2} \times \dfrac{\text{mol F}_2}{22.4\ \text{L F}_2} \times \dfrac{38.0\ \text{g F}_2}{\text{mol F}_2}$

$= 5.77 \times 10^{-3}\ \text{g F}_2$

7. *(a)* $8.00\ \text{g O}_2 \times \dfrac{\text{mol O}_2}{32.0\ \text{g O}_2} \times \dfrac{22.4\ \text{L O}_2}{\text{mol O}_2} = 5.60\ \text{L O}_2$

(b) $3.50\ \text{g CO} \times \dfrac{\text{mol CO}}{28.0\ \text{g CO}} \times \dfrac{22.4\ \text{L CO}}{\text{mol CO}} = 2.80\ \text{L CO}$

(c) $0.0170\ \text{g H}_2\text{S} \times \dfrac{\text{mol H}_2\text{S}}{34.1\ \text{g H}_2\text{S}} \times \dfrac{22.4\ \text{L H}_2\text{S}}{\text{mol H}_2\text{S}} = 0.0112\ \text{L H}_2\text{S}$

(d) $2.25 \times 10^5\ \text{kg NH}_3 \times \dfrac{10^3\ \text{g NH}_3}{\text{kg NH}_3} \times \dfrac{\text{mol NH}_3}{17.0\ \text{g NH}_3} \times$

$\dfrac{22.4\ \text{L NH}_3}{\text{mol NH}_3} = 2.96 \times 10^8\ \text{L NH}_3$

8. *(a)* $\dfrac{28.0\ \text{g}}{\text{mol}} \times \dfrac{\text{mol}}{22.4\ \text{L}} = 1.25\ \text{g/L}$

(b) $\dfrac{70.9 \text{ g}}{\text{mol}} \times \dfrac{\text{mol}}{22.4 \text{ L}} = 3.17 \text{ g/L}$

(c) $\dfrac{28.0 \text{ g}}{\text{mol}} \times \dfrac{\text{mol}}{22.4 \text{ L}} = 1.25 \text{ g/L}$

(d) $\dfrac{64.1 \text{ g}}{\text{mol}} \times \dfrac{\text{mol}}{22.4 \text{ L}} = 2.86 \text{ g/L}$

(e) $\dfrac{17.0 \text{ g}}{\text{mol}} \times \dfrac{\text{mol}}{22.4 \text{ L}} = 0.759 \text{ g/L}$

9. (a) $\dfrac{2.00 \text{ g}}{1.50 \text{ L}} \times \dfrac{22.4 \text{ L}}{\text{mol}} = 29.9 \text{ g/mol}$

(b) $\dfrac{3.25 \text{ g}}{2.30 \text{ L}} \times \dfrac{22.4 \text{ L}}{\text{mol}} = 31.7 \text{ g/mol}$

(c) $\dfrac{0.620 \text{ g}}{250 \text{ mL}} \times \dfrac{10^3 \text{ mL}}{\text{L}} \times \dfrac{22.4 \text{ L}}{\text{mol}} = 55.6 \text{ g/mol}$

10. (a) $\dfrac{P_1 V_1}{T_1} = \dfrac{P_2 V_2}{T_2}$

$V_2 = \dfrac{P_1 V_1 T_2}{P_2 T_1} = \dfrac{740 \text{ mm} \times 250 \text{ mL} \times 273 \text{K}}{760 \text{ mm} \times 300 \text{K}} = 222 \text{ mL}$

(b) $\dfrac{0.205 \text{ g}}{222 \text{ mL}} \times \dfrac{10^3 \text{ mL}}{\text{L}} \times \dfrac{22.4 \text{ L}}{\text{mol}} = 20.7 \text{ g/mol}$

11. First, find the volume of the gas at STP:

$T_1 = 77\,°\text{C} + 273 = 350\text{K}$

$T_2 = 0\,°\text{C} + 273 = 273\text{K}$

$\dfrac{P_1 V_1}{T_1} = \dfrac{P_2 V_2}{T_2}$

$V_2 = \dfrac{P_1 V_1 T_2}{P_2 T_1} = \dfrac{525 \text{ mm} \times 140 \text{ mL} \times 273 \text{K}}{760 \text{ mm} \times 350 \text{K}} = 75.4 \text{ mL}$

Next, find molar mass:

$\dfrac{0.0650 \text{ g}}{75.4 \text{ mL}} \times \dfrac{10^3 \text{ mL}}{\text{L mol}} \times \dfrac{22.4 \text{ L}}{\text{mol}} = 19.3 \text{ g/mol}$

12. (a) $PV = nRT$

$P = \dfrac{nRT}{V} = \dfrac{1.35 \text{ mol} \times 0.0821 \dfrac{\text{L} \cdot \text{atm}}{\text{mol} \cdot \text{K}} \times 320 \text{K}}{2.50 \text{ L}}$

$= 14.2 \text{ atm}$

(b) $PV = nRT$

$P = \dfrac{nRT}{V} = \dfrac{0.86 \text{ mol} \times 0.0821 \dfrac{\text{L} \cdot \text{atm}}{\text{mol} \cdot \text{K}} \times 300 \text{K}}{4.75 \text{ L}}$

$= 4.5 \text{ atm}$

(c) $T = 57\,°\text{C} + 273 = 330.\text{K}$

$P = \dfrac{nRT}{V} = \dfrac{2.15 \text{ mol} \times 0.0821 \times \dfrac{\text{L} \cdot \text{atm}}{\text{mol} \cdot \text{K}} \times 330.\text{K}}{750. \text{mL} \times \dfrac{\text{L}}{10^3 \text{ mL}}} = 77.7 \text{ atm}$

13. (a) $PV = nRT$

$V = \dfrac{nRT}{P} = \dfrac{2.00 \text{ mol} \times 0.0821 \dfrac{\text{L} \cdot \text{atm}}{\text{mol} \cdot \text{K}} \times 300 \text{K}}{1.25 \text{ atm}} = 39.4 \text{ L}$

(b) $T = 37\,°\text{C} + 273 = 310.\text{K}$

$P = 550. \text{ mm Hg} \times \dfrac{\text{atm}}{760 \text{ mm Hg}} = 0.724 \text{ atm}$

$V = \dfrac{nRT}{P} = \dfrac{0.425 \text{ mol} \times 0.0821 \dfrac{\text{L} \cdot \text{atm}}{\text{mol} \cdot \text{K}} \times 310.\text{K}}{0.724 \text{ atm}} = 14.9 \text{ L}$

(c) $T = 57\,°\text{C} + 273 = 330.\text{K}$

$P = 675 \text{ mm Hg} \times \dfrac{\text{atm}}{760 \text{ mm Hg}} = 0.888 \text{ atm}$

$n = 4.00 \text{ g O}_2 \times \dfrac{\text{mol O}_2}{32.0 \text{ g O}_2} = 0.125 \text{ mol O}_2$

$V = \dfrac{nRT}{P} = \dfrac{0.125 \text{ mol O}_2 \times 0.0821 \dfrac{\text{L} \cdot \text{atm}}{\text{mol} \cdot \text{K}} \times 330.\text{K}}{0.888 \text{ atm}} = 3.81 \text{ L}$

14. (a) $PV = nRT$

$n = \dfrac{PV}{RT} = \dfrac{1.06 \text{ atm} \times 1.25 \text{ L}}{0.0821 \dfrac{\text{L} \cdot \text{atm}}{\text{mol} \cdot \text{K}} \times 250.\text{K}} = 0.0646 \text{ mol}$

(b) $T = 27\,°\text{C} + 273 = 300.\text{K}$

$n = \dfrac{PV}{RT} = \dfrac{0.925 \text{ atm} \times 0.80 \text{ L}}{0.0821 \dfrac{\text{L} \cdot \text{atm}}{\text{mol} \cdot \text{K}} \times 250.\text{K}} = 0.036 \text{ mol}$

(c) $T = -50.\,°\text{C} + 273° = 223\text{K}$

$P = 700. \text{ mm Hg} \times \dfrac{\text{atm}}{760 \text{ mm Hg}} = 0.921 \text{ atm}$

$n = \dfrac{PV}{RT} = \dfrac{0.921 \text{ atm} \times 750. \text{ mL} \times \dfrac{\text{L}}{10^3 \text{ mL}}}{0.0821 \dfrac{\text{L} \cdot \text{atm}}{\text{mol} \cdot \text{K}} \times 223\text{K}} = 0.0377 \text{ mol}$

15. (a) $PV = nRT$

$n = \dfrac{PV}{RT} = \dfrac{1.75 \text{ atm} \times 5.60 \text{ L}}{0.0821 \dfrac{\text{L} \cdot \text{atm}}{\text{mol} \cdot \text{K}} \times 250\text{K}} = 0.477 \text{ mol}$

$0.477 \text{ mol O}_2 \times \dfrac{32.0 \text{ g O}_2}{\text{mol O}_2} = 15.3 \text{ g O}_2$

(b) $P = 700. \text{ mm Hg} \times \dfrac{\text{atm}}{760 \text{ mm Hg}} = 0.921 \text{ atm}$

$T = 27\,°\text{C} + 273 = 300.\text{K}$

$n = \dfrac{PV}{RT} = \dfrac{0.921 \text{ atm} \times 3.50 \text{ L}}{0.0821 \dfrac{\text{L} \cdot \text{atm}}{\text{mol} \cdot \text{K}} \times 300.\text{K}} = 0.131 \text{ mol}$

$0.131 \text{ mol NH}_3 \times \dfrac{17.0 \text{ g NH}_3}{\text{mol NH}_3} = 2.23 \text{ g NH}_3$

(c) $P = 625 \text{ mm Hg} \times \dfrac{\text{atm}}{760 \text{ mm Hg}} = 0.822 \text{ atm}$

$T = -53\,°\text{C} + 273 = 220.\text{K}$

$n = \dfrac{PV}{RT} = \dfrac{0.822 \text{ atm} \times 125 \text{ mL} \, \dfrac{\text{L}}{10^3 \text{ mL}}}{0.0821 \dfrac{\text{L} \cdot \text{atm}}{\text{mol} \cdot \text{K}} \times 220.\text{K}} = 5.69 \times 10^{-3} \text{ mol}$

$$5.69 \times 10^{-3} \text{ mol} \times \frac{64.1 \text{ g } SO_2}{\text{mol } SO_2} = 0.365 \text{ g } SO_2$$

16. (a) $n = \dfrac{PV}{RT} = \dfrac{1.08 \text{ atm} \times 1.50 \text{ L}}{0.0821 \dfrac{\text{L}\cdot\text{atm}}{\text{mol}\cdot\text{K}} \times 320.\text{K}} = 0.0617 \text{ mol}$

   (b) Since $n = \dfrac{m}{\text{molar } m}$, then molar $m = \dfrac{m}{n} = \dfrac{3.50 \text{ g}}{0.0617 \text{ mol}}$
   $= 56.7 \text{ g/mol}$

17. (a) $n = \dfrac{PV}{RT} = \dfrac{1.14 \text{ atm} \times 1.12 \text{ L}}{0.0821 \dfrac{\text{L}\cdot\text{atm}}{\text{mol}\cdot\text{K}} \times 280.\text{K}} = 0.0555 \text{ mol}$

   molar $m = \dfrac{0.650 \text{ g}}{0.0555 \text{ mol}} = 11.7 \text{ g/mol}$

   (b) $T = 37°C + 273 = 310.\text{K}$
   $n = \dfrac{PV}{RT} = \dfrac{0.840 \text{ atm} \times 2.35 \text{ L}}{0.0821 \dfrac{\text{L}\cdot\text{atm}}{\text{mol}\cdot\text{K}} \times 310.\text{K}} = 0.0776 \text{ mol}$

   mol $m = \dfrac{1.05 \text{ g}}{0.0776 \text{ mol}} = 13.5 \text{ g/mol}$

   (c) $T = -23°C + 273 = 250.\text{K}$
   $P = 785 \text{ mm Hg} \times \dfrac{\text{atm}}{760 \text{ mm Hg}} = 1.03 \text{ atm}$

   $n = \dfrac{PV}{RT} = \dfrac{1.03 \text{ atm} \times 750. \text{ mL} \times \dfrac{\text{L}}{10^3 \text{ mL}}}{0.0821 \dfrac{\text{L}\cdot\text{atm}}{\text{mol}\cdot\text{K}} \times 250.\text{K}} = 0.0376 \text{ mol}$

   mol $m = \dfrac{0.432 \text{ g}}{0.0376 \text{ mol}} = 11.5 \text{ g/mol}$

18. $H_2(g) + Cl_2(g) \rightarrow 2HCl(g)$
   The missing values are:
   (a) 1, 2 (b) 1, 2 (c) 1, 2.

19. $2CO(g) + O_2(g) \rightarrow 2CO_2(g)$

   (a) $1.0 \text{ L CO} \times \dfrac{1 \text{ L } O_2}{2 \text{ L CO}} = 0.50 \text{ L } O_2$

   (b) $1.0 \text{ L CO} \times \dfrac{2 \text{ L } CO_2}{2 \text{ L CO}} = 1.0 \text{ L } CO_2$

20. $2C_2H_2(g) + 5O_2(g) \rightarrow 4CO_2(g) + 2H_2O(g)$

   (a) $75.0 \text{ L } CO_2 \times \dfrac{2 \text{ L } C_2H_2}{4 \text{ L } CO_2} = 37.5 \text{ L } C_2H_2$

   (b) $75.0 \text{ L } CO_2 \times \dfrac{2 \text{ L } H_2O}{4 \text{ L } CO_2} = 37.5 \text{ L } H_2O$

   (c) $75.0 \text{ L } CO_2 \times \dfrac{5 \text{ L } O_2}{4 \text{ L } CO_2} = 93.8 \text{ L } O_2$

21. $CS_2(\ell) + 3O_2(g) \rightarrow CO_2(g) + 2SO_2(g)$

   $450. \text{ mL } O_2 \times \dfrac{1 \text{ mL } CO_2}{3 \text{ mL } O_2} = 150. \text{ mL } CO_2$

   $450. \text{ mL } O_2 \times \dfrac{2 \text{ mL } SO_2}{3 \text{ mL } O_2} = 300. \text{ mL } SO_2$

22. $2Mg(s) + O_2(g) \rightarrow 2MgO(s)$

   (a) $22.4 \text{ L } O_2 \times \dfrac{\text{mol } O_2}{22.4 \text{ L } O_2} = 1.00 \text{ mol } O_2$

$1.00 \text{ mol } O_2 \times \dfrac{2 \text{ mol MgO}}{1 \text{ mol } O_2} = 2.00 \text{ mol MgO}$

   (b) $11.2 \text{ L } O_2 \times \dfrac{\text{mol } O_2}{22.4 \text{ L } O_2} = 0.500 \text{ mol } O_2$

   $0.500 \text{ mol } O_2 \times \dfrac{2 \text{ mol MgO}}{1 \text{ mol } O_2} = 1.00 \text{ mol MgO}$

   (c) $1.40 \text{ L } O_2 \times \dfrac{\text{mol } O_2}{22.4 \text{ L } O_2} = 0.0625 \text{ mol } O_2$

   $0.0625 \text{ mol } O_2 \times \dfrac{2 \text{ mol MgO}}{1 \text{ mol } O_2} = 0.125 \text{ mol Mg}$

   $0.125 \text{ mol MgO} \times \dfrac{40.3 \text{ g MgO}}{\text{mol MgO}} = 5.04 \text{ g MgO}$

   (d) $4.75 \text{ L } O_2 \times \dfrac{\text{mol } O_2}{22.4 \text{ L } O_2} = 0.212 \text{ mol } O_2$

   $0.212 \text{ mol } O_2 \times \dfrac{2 \text{ mol MgO}}{1 \text{ mol } O_2} = 0.424 \text{ mol MgO}$

   $0.424 \text{ mol MgO} \times \dfrac{40.3 \text{ g MgO}}{\text{mol MgO}} = 17.1 \text{ g MgO}$

   (e) $725 \text{ mL } O_2 \times \dfrac{\text{L } O_2}{10^3 \text{ mL } O_2} \times \dfrac{\text{mol } O_2}{22.4 \text{ L } O_2} = 0.0324 \text{ mol } O_2$

   $0.0324 \text{ mol } O_2 \times \dfrac{2 \text{ mol MgO}}{1 \text{ mol } O_2} = 0.0648 \text{ mol MgO}$

   $0.0648 \text{ mol MgO} \times \dfrac{40.3 \text{ g MgO}}{\text{mol MgO}} = 2.61 \text{ g MgO}$

23. $Fe_2O_3(s) + 3CO(g) \rightarrow 2Fe(s) + 3CO_2(g)$
   $T = 57°C + 273 = 330.\text{K}$
   (a) $PV = nRT$

   $n = \dfrac{PV}{RT} = \dfrac{1.05 \text{ atm} \times 15.0 \text{ L}}{0.0821 \dfrac{\text{L}\cdot\text{atm}}{\text{mol}\cdot\text{K}} \times 330.\text{K}} = 0.581 \text{ mol } (CO_2)$

   $0.581 \text{ mol } CO_2 \times \dfrac{1 \text{ mol } Fe_2O_3}{3 \text{ mol } CO_2} = 0.194 \text{ mol } Fe_2O_3$

   (b) $n = \dfrac{PV}{RT} = \dfrac{1.05 \text{ atm} \times 420. \text{mL} \times \dfrac{\text{L}}{10^3 \text{ mL}}}{0.0821 \dfrac{\text{L}\cdot\text{atm}}{\text{mol}\cdot\text{K}} \times 330.\text{K}} = 0.0163 \text{ mol } (CO_2)$

   $0.0163 \text{ mol } CO_2 \times \dfrac{1 \text{ mol } Fe_2O_3}{3 \text{ mol } CO_2} = 0.005\,43 \text{ mol } Fe_2O_3$

   $0.005\,43 \text{ mol } Fe_2O_3 \times \dfrac{160. \text{ g } Fe_2O_3}{\text{mol } Fe_2O_3} = 0.869 \text{ g } Fe_2O_3$

   (c) $n = \dfrac{PV}{RT} = \dfrac{1.05 \text{ atm} \times 775 \text{ mL} \times \dfrac{\text{L}}{10^3 \text{ mL}}}{0.0821 \dfrac{\text{L}\cdot\text{atm}}{\text{mol}\cdot\text{K}} \times 330.\text{K}} = 0.0300 \text{ mol } (CO_2)$

   $0.0300 \text{ mol } CO_2 \times \dfrac{1 \text{ mol } Fe_2O_3}{3 \text{ mol } CO_2} = 0.0100 \text{ mol } Fe_2O_3$

   $0.0100 \text{ mol } Fe_2O_3 \times \dfrac{160. \text{ g } Fe_2O_3}{\text{mol } Fe_2O_3} = 1.60 \text{ g } Fe_2O_3$

24. $CuO(s) + H_2(g) \rightarrow Cu(s) + H_2O(g)$

(a) $5.60 \text{ L } H_2 \times \dfrac{\text{mol}}{22.4 \text{ L}} = 0.250 \text{ mol } H_2$

(b) $0.250 \text{ mol } H_2 \times \dfrac{1 \text{ mol Cu}}{1 \text{ mol } H_2} = 0.250 \text{ mol Cu}$

(c) $0.250 \text{ mol Cu} \times \dfrac{63.5 \text{ g Cu}}{\text{mol Cu}} = 15.9 \text{ g Cu}$

25. $2KI(aq) + Cl_2(g) \rightarrow 2KCl(aq) + I_2(g)$

(a) $8.50 \text{ L } I_2 \times \dfrac{\text{mol } I_2}{22.4 \text{ L } I_2} = 0.379 \text{ mol } I_2$

(b) $0.379 \text{ mol } I_2 \times \dfrac{2 \text{ mol KI}}{1 \text{ mol } I_2} = 0.758 \text{ mol KI}$

(c) $0.758 \text{ mol KI} \times \dfrac{166 \text{ g KI}}{\text{mol KI}} = 126 \text{ g KI}$

26. $2Fe(OH)_3(s) \rightarrow Fe_2O_3(s) + 3H_2O(g)$

(a) $75.0 \text{ L } H_2O \times \dfrac{\text{mol}}{22.4 \text{ L}} = 3.35 \text{ mol } H_2O$

$3.35 \text{ mol } H_2O \times \dfrac{2 \text{ mol Fe(OH)}_3}{3 \text{ mol } H_2O} = 2.23 \text{ mol Fe(OH)}_3$

$2.23 \text{ mol Fe(OH)}_3 \times \dfrac{107 \text{ g Fe(OH)}_3}{\text{mol Fe(OH)}_3} = 239 \text{ g Fe(OH)}_3$

(b) $3.35 \text{ mol } H_2O \times \dfrac{1 \text{ mol Fe}_2O_3}{3 \text{ mol } H_2O} = 1.12 \text{ mol Fe}_2O_3$

$1.12 \text{ mol Fe}_2O_3 \times \dfrac{160. \text{ g Fe}_2O_3}{\text{mol Fe}_2O_3} = 179 \text{ g Fe}_2O_3$

27. $Fe(s) + H_2SO_4(aq) \rightarrow FeSO_4(aq) + H_2(g)$

$650. \text{ mL } H_2 \times \dfrac{\text{L}}{10^3 \text{ mL}} \times \dfrac{\text{mol}}{22.4 \text{ L}} = 0.0290 \text{ mol } H_2$

$0.0290 \text{ mol } H_2 \times \dfrac{1 \text{ mol FeSO}_4}{1 \text{ mol } H_2} = 0.0290 \text{ mol FeSO}_4$

$0.0290 \text{ mol FeSO}_4 \times \dfrac{152 \text{ g FeSO}_4}{\text{mol FeSO}_4} = 4.41 \text{ g FeSO}_4$

28. $2H_2O(\ell) \rightarrow 2H_2(g) + O_2(g)$

(a) $18.0 \text{ g } H_2O \times \dfrac{\text{mol } H_2O}{18.0 \text{ g } H_2O} = 1.00 \text{ mol } H_2O$

$1.00 \text{ mol } H_2O \times \dfrac{2 \text{ mol } H_2}{2 \text{ mol } H_2O} = 1.00 \text{ mol } H_2$

(b) $4.50 \text{ g } H_2O \times \dfrac{\text{mol } H_2O}{18.0 \text{ g } H_2O} = 0.250 \text{ mol } H_2O$

$0.250 \text{ mol } H_2O \times \dfrac{1 \text{ mol } O_2}{2 \text{ mol } H_2O} = 0.125 \text{ mol } O_2$

(c) $6.00 \text{ g } H_2O \times \dfrac{\text{mol } H_2O}{18.0 \text{ g } H_2O} = 0.333 \text{ mol } H_2O$

$0.333 \text{ mol } H_2O \times \dfrac{2 \text{ mol } H_2}{2 \text{ mol } H_2O} = 0.333 \text{ mol } H_2$

$0.333 \text{ mol } H_2 \times \dfrac{22.4 \text{ L } H_2}{\text{mol } H_2} = 7.46 \text{ L } H_2$

29. $MnO_2(s) + 4HCl(aq) \rightarrow MnCl_2(aq) + Cl_2(g) + 2H_2O(\ell)$

(a) $36.5 \text{ g HCl} \times \dfrac{\text{mol HCl}}{36.5 \text{ g HCl}} = 1.00 \text{ mol HCl}$

$1.00 \text{ mol HCl} \times \dfrac{1 \text{ mol Cl}_2}{4 \text{ mol HCl}} = 0.250 \text{ mol Cl}_2$

(b) $9.12 \text{ g HCl} \times \dfrac{\text{mol HCl}}{36.5 \text{ g HCl}} = 0.250 \text{ mol HCl}$

$0.250 \text{ mol HCl} \times \dfrac{1 \text{ mol Cl}_2}{4 \text{ mol HCl}} = 0.0625 \text{ mol Cl}_2$

$T = 77°C + 273 = 350.\text{K}$

$P = 780. \text{ mm Hg} \times \dfrac{1 \text{ atm}}{760 \text{ mm Hg}} = 1.03 \text{ atm}$

$PV = nRT$

$V = \dfrac{nRT}{P} = \dfrac{0.0625 \text{ mol} \times 0.0821 \dfrac{\text{L} \cdot \text{atm}}{\text{mol} \cdot \text{K}} \times 350.\text{K}}{1.03 \text{ atm}} = 1.74 \text{ L}$

(c) $52.0 \text{ g HCl} \times \dfrac{\text{mol HCl}}{36.5 \text{ g HCl}} = 1.42 \text{ mol HCl}$

$1.42 \text{ mol HCl} \times \dfrac{1 \text{ mol Cl}_2}{4 \text{ mol HCl}} = 0.355 \text{ mol Cl}_2$

$V = \dfrac{nRT}{P} = \dfrac{0.355 \text{ mol} \times 0.0821 \dfrac{\text{L} \cdot \text{atm}}{\text{mol} \cdot \text{K}} \times 350.\text{K} \times \dfrac{10^3 \text{ mL}}{\text{L}}}{1.03 \text{ atm}}$

$= 9.90 \times 10^3 \text{ mL}$

(d) $PV = nRT$

$n = \dfrac{PV}{RT} = \dfrac{1.03 \text{ atm} \times 12.5 \text{ L}}{0.0821 \dfrac{\text{L} \cdot \text{atm}}{\text{mol} \cdot \text{K}} \times 350\text{K}} = 0.448 \text{ mol (Cl}_2)$

$0.448 \text{ mol Cl}_2 \times \dfrac{4 \text{ mol HCl}}{1 \text{ mol Cl}_2} = 1.79 \text{ mol HCl}$

$1.79 \text{ mol HCl} \times \dfrac{36.5 \text{ g HCl}}{\text{mol HCl}} = 65.3 \text{ g HCl}$

(e) $n = \dfrac{PV}{RT} = \dfrac{1.03 \text{ atm} \times 750. \text{ mL} \times \dfrac{\text{L}}{10^3 \text{ mL}}}{0.0821 \dfrac{\text{L} \cdot \text{atm}}{\text{mol} \cdot \text{K}} \times 350.\text{K}}$

$= 0.0269 \text{ mol (Cl}_2)$

$0.0269 \text{ mol Cl}_2 \times \dfrac{4 \text{ mol HCl}}{1 \text{ mol Cl}_2} = 0.108 \text{ mol HCl}$

$0.108 \text{ mol HCl} \times \dfrac{36.5 \text{ g HCl}}{\text{mol HCl}} = 3.94 \text{ g HCl}$

30. $2Al(s) + 6HCl(aq) \rightarrow 2AlCl_3(aq) + 3H_2(g)$

(a) $13.5 \text{ g Al} \times \dfrac{\text{mol Al}}{27.0 \text{ g Al}} = 0.500 \text{ mol Al}$

(b) $0.500 \text{ mol Al} \times \dfrac{3 \text{ mol } H_2}{2 \text{ mol Al}} = 0.750 \text{ mol } H_2$

(c) $0.750 \text{ mol } H_2 \times \dfrac{22.4 \text{ L}}{\text{mol}} = 16.8 \text{ L } H_2$

31. $Na_2SO_3(aq) + 2HCl(aq) \rightarrow 2NaCl(aq) + H_2O(\ell) + SO_2(g)$

(a) $21.0 \text{ g } Na_2SO_3 \times \dfrac{\text{mol } Na_2SO_3}{126 \text{ g } Na_2SO_3} = 0.167 \text{ mol } Na_2SO_3$

$$0.167 \text{ mol Na}_2\text{SO}_3 \times \frac{1 \text{ mol SO}_2}{1 \text{ mol Na}_2\text{SO}_3} = 0.167 \text{ mol SO}_2$$

(b) $0.167 \text{ mol SO}_2 \times \dfrac{22.4 \text{ L SO}_2}{\text{mol SO}_2} = 3.74 \text{ L SO}_2$

32. *(a)* $\dfrac{\text{rate of effusion}_A}{\text{rate of effusion}_B} = \dfrac{\sqrt{\text{mol } m_B}}{\sqrt{\text{mol } m_A}}$

$$\frac{\text{rate}_{H_2}}{\text{rate}_{N_2}} = \frac{\sqrt{28.0 \text{ g/mol}}}{\sqrt{2.02 \text{ g/mol}}} = \sqrt{13.9} = 3.73$$

(b) $\dfrac{\text{rate}_{F_2}}{\text{rate}_{Cl_2}} = \dfrac{\sqrt{71.0 \text{ g/mol}}}{\sqrt{38.0 \text{ g/mol}}} = \sqrt{1.87} = 1.37$

Application Questions

1. From Boyle's law, $V = \dfrac{k}{P}$; from Charles' law $V = kT$; and from Avogadro's principle, $V = kn$. Since a quantity that is proportional to each of two or more other quantities is also proportional to their product:

$$V = \frac{k}{P} \times T \times n;\ V = \frac{kTn}{P}.$$

Replacing k with R, and rearranging, $PV = nRT$ (the ideal gas law).

2. *(a)* The combined gas law is used when a gas changes from one set of *P-V-T* conditions into another. *(b)* The ideal gas law is used to compute one of the four missing quantities (*P*, *V*, *T*, or *n*) when the others are given under the same set of conditions.

Application Problems

1. $4.50 \text{ m}^3 \times \dfrac{10^6 \text{ cm}^3}{\text{m}^3} \times \dfrac{\text{mL}}{\text{cm}^3} \times \dfrac{\text{L}}{10^3 \text{ mL}} \times \dfrac{\text{mol}}{22.4 \text{ L}} \times \dfrac{34.1 \text{ g H}_2\text{S}}{\text{mol}}$

$= 6.85 \times 10^3 \text{ g H}_2\text{S}$

2. $12.0 \text{ g O}_2 \times \dfrac{\text{mol}}{32.0 \text{ g O}_2} \times \dfrac{22.4 \text{ L}}{\text{mol}} \times \dfrac{10^3 \text{ mL}}{\text{L}} \times \dfrac{\text{cm}^3 \text{ O}_2}{\text{mL}}$

$= 8.40 \times 10^3 \text{ cm}^3 \text{ O}_2$

$8.40 \times 10^3 \text{ cm}^3 \text{ O}_2 \times \dfrac{\text{m}^3 \text{ O}_2}{10^6 \text{ cm}^3 \text{ O}_2} = 8.40 \times 10^{-3} \text{ m}^3 \text{ O}_2$

3. $\dfrac{46.0 \text{ g}}{\text{mol}} \times \dfrac{\text{mol}}{22.4 \text{ L}} \times \dfrac{\text{L}}{10^3 \text{ mL}} = 2.05 \times 10^{-3} \text{ g/mL}$

4. *(a)* $D = \dfrac{m}{V} \times \dfrac{4.75 \text{ g}}{2250 \text{ cm}^3} \times \dfrac{\text{cm}^3}{\text{mL}} \times \dfrac{10^3 \text{ mL}}{\text{L}} = 2.11 \text{ g/L}$

(b) $\dfrac{2.11 \text{ g}}{\text{L}} \times \dfrac{22.4 \text{ L}}{\text{mol}} = 47.3 \text{ g/mol}$

5. *(a)* First, find volume at STP:

$T_1 = 50.°\text{C} + 273 = 323\text{K}$

$T_2 = 0.°\text{C} + 273 = 273\text{K}$

$\dfrac{P_1 V_1}{T_1} = \dfrac{P_2 V_2}{T_2}$

$V_2 = \dfrac{P_1 V_1 T_2}{P_2 T_1} = \dfrac{840. \text{ mm Hg} \times 6.51 \text{ m}^3 \times 273\text{K}}{760 \text{ mm} \times 323\text{K}} = 6.08 \text{ m}^3$

Next, find molar mass:

$\dfrac{12.5 \text{ kg}}{6.08 \text{ m}^3} \times \dfrac{10^3 \text{ g}}{\text{kg}} \times \dfrac{\text{m}^3}{10^6 \text{ cm}^3} \times \dfrac{\text{cm}^3}{\text{mL}} \times \dfrac{10^3 \text{ mL}}{\text{L}} \times \dfrac{22.4 \text{ L}}{\text{mol}}$

$= 46.1 \text{ g/mol}$

(b) NO_2, which has a molar mass of 46.0 g/mol, is most likely to be the unknown.

6. First, find P_{gas}:

$P_{gas} = P_{total} - P_{H_2O} = 650.0 \text{ mm} - 17.5 \text{ mm} = 632.5 \text{ mm}$

Then, find volume at STP:

$T_1 = 20.°\text{C} + 273 = 293\text{K}$

$T_2 = 0.°\text{C} + 273 = 273\text{K}$

$\dfrac{P_1 V_1}{T_1} = \dfrac{P_2 V_2}{T_2}$

$V_2 = \dfrac{P_1 V_1 T_2}{P_2 T_1} = \dfrac{632.5 \text{ mm Hg} \times 850. \text{ mL} \times 273\text{K}}{760 \text{ mm Hg} \times 293\text{K}} = 659 \text{ mL}$

Finally, find molar mass:

$\dfrac{0.925 \text{ g}}{659 \text{ mL}} \times \dfrac{10^3 \text{ mL}}{\text{L}} \times \dfrac{22.4 \text{ L}}{\text{mol}} = 31.4 \text{ g/mol}$

7. First, find P_{gas}:

$P_{gas} = 575.0 \text{ mm Hg} - 42.2 \text{ mm Hg} = 532.8 \text{ mm Hg}$

Next, find volume at STP:

$T_1 = 35°\text{C} + 273 = 308\text{K}$

$T_2 = 0.°\text{C} + 273 = 273\text{K}$

$V_2 = \dfrac{P_1 V_1 T_2}{P_2 T_1} = \dfrac{532.8 \text{ mm Hg} \times 500. \text{ cm}^3 \times 273\text{K}}{760 \text{ mm Hg} \times 308\text{K}} = 311 \text{ cm}^3$

Finally, find molar mass:

$\dfrac{0.750 \text{ g}}{311 \text{ cm}^3} \times \dfrac{\text{cm}^3}{\text{mL}} \times \dfrac{10^3 \text{ mL}}{\text{L}} \times \dfrac{22.4 \text{ L}}{\text{mol}} = 54.0 \text{ g/mol}$

8. $T = 40.°\text{C} + 273 = 313\text{K}$

$V = 500. \text{ cm}^3 \times \dfrac{\text{mL}}{\text{cm}^3} \times \dfrac{\text{L}}{10^3 \text{ mL}} = 0.0500 \text{ L}$

$n = 8.25 \text{ g SO}_2 \times \dfrac{\text{mol SO}_2}{64.1 \text{ g SO}_2} = 0.129 \text{ mol SO}_2$

$PV = nRT$

$P = \dfrac{nRT}{V} = \dfrac{0.129 \text{ mol} \times 0.0821 \dfrac{\text{L} \cdot \text{atm}}{\text{mol} \cdot \text{K}} \times 313\text{K}}{0.500 \text{ L}} = 6.63 \text{ atm}$

9. $T = -40.°\text{C} + 273 = 233\text{K}$

$PV = nRT$

$n = \dfrac{PV}{RT} = \dfrac{92.5 \text{ kPa} \times 1500. \text{ mL} \times \dfrac{\text{L}}{10^3 \text{ mL}}}{8.31 \dfrac{\text{L} \cdot \text{kPa}}{\text{mol} \cdot \text{K}} \times 233\text{K}} = 0.0717 \text{ mol}$

10. $P = 625 \text{ mm Hg} \times \dfrac{1 \text{ atm}}{760 \text{ mm Hg}} = 0.822 \text{ atm}$

$n = 4.60 \text{ g CO}_2 \times \dfrac{\text{mol CO}_2}{44.0 \text{ g CO}_2} = 0.105 \text{ mol CO}_2$

$PV = nRT$

$$T = \frac{PV}{nR} = \frac{0.822 \text{ atm} \times 275 \text{ mL} \times \dfrac{L}{10^3 \text{ mL}}}{0.105 \text{ mol} \times 0.0821 \dfrac{L \cdot \text{atm}}{\text{mol} \cdot K}} = 26.2K$$

$$T = 26.2K - 273° = -247°C$$

11. $T = -43°C + 273 = 230.K$

$$P = 240. \text{ mm Hg} \times \frac{1 \text{ atm}}{760 \text{ mm Hg}} = 0.316 \text{ atm}$$

$$\text{mol } m = \frac{mRT}{PV} = \frac{1.06 \times 10^{-3}g \times 0.0821\dfrac{L \cdot \text{atm}}{\text{mol} \cdot K} \times 230.K}{0.316 \text{ atm} \times 14.5 \text{ mL} \times \dfrac{L}{10^3 \text{ mL}}}$$

$$= 4.37 \text{ g/mol}$$

12. $T = 20.°C + 273 = 293K$

$P_{gas} = 420.0 \text{ mm Hg} - 17.5 \text{ mm Hg} = 402.5 \text{ mm Hg}$

$$P = 402.5 \text{ mm Hg} \times \frac{1 \text{ atm}}{760 \text{ mm Hg}} = 0.530 \text{ atm}$$

$$\text{mol } m = \frac{mRT}{PV} = \frac{0.750 \text{ g} \times 0.0821\dfrac{L \cdot \text{atm}}{\text{mol} \cdot K} \times 293}{0.530 \text{ atm} \times 155 \text{ mL} \times \dfrac{L}{10^3 \text{ mL}}} = 220. \text{ g/mol}$$

13. $T = -18°C + 273 = 255K$

$$P = 1650 \text{ mm Hg} \times \frac{1 \text{ atm}}{760 \text{ mm Hg}} = 2.17 \text{ atm}$$

$$\text{mol } m = \frac{DRT}{P} = \frac{\dfrac{3.20 \text{ g}}{L} \times 0.821 \dfrac{L \cdot \text{atm}}{\text{mol} \cdot K} \times 255K}{2.17 \text{ atm}} = 30.9 \text{ g/mol}$$

14. $D = \dfrac{1.40 \text{ g}}{cm^3} \times \dfrac{cm^3}{mL} \times \dfrac{10^3 \text{ mL}}{L} = 1.40 \times 10^3 \text{ g/L}$

$\text{mol } m = \dfrac{DRT}{P}$

$$T = \frac{\text{mol } m}{\dfrac{dR}{D}} = \frac{2.00 \dfrac{g}{\text{mol}} \times 1.30 \times 10^9 \text{ atm}}{1.40 \times 10^3 \dfrac{g}{L} \times 0.0821 \dfrac{L \cdot \text{atm}}{\text{mol} \cdot K}} = 2.26 \times 10^7 K$$

$T = 2.26 \times 10^7 K - 273 = 2.26 \times 10^7 °C$

15. $2NO(g) + O_2(g) \rightarrow 2NO_2(g)$

First, find the volume of O_2 required to react with 350. mL of NO:

$$350. \text{ mL NO} \times \frac{1 \text{ mL } O_2}{2 \text{ mL NO}} = 175 \text{ mL } O_2 \text{ is required}$$

Thus, the O_2 is present in excess.

Therefore, $350. \text{ mL NO} \times \dfrac{2 \text{ mL } NO_2}{2 \text{ mL NO}} = 350. \text{ mL } NO_2$ will be produced.

16. $2C_8H_{18}(g) + 25O_2(g) \rightarrow 16CO_2(g) + 18H_2O(g)$

(a)
$$25.0 \text{ L } C_8H_{18} \times \frac{25 \text{ L } O_2}{2 \text{ L } C_8H_{18}} = 312 \text{ L } O_2$$

$$312 \text{ L } O_2 \times \frac{100 \text{ L air}}{20.9 \text{ L } O_2} = 1.49 \times 10^3 \text{ L air}$$

(b)
$$25.0 \text{ L } C_8H_{18} \times \frac{16 \text{ L } CO_2}{2 \text{ L } C_8H_{18}} = 200. \text{ L } CO_2$$

$$25.0 \text{ L } C_8H_{18} \times \frac{18 \text{ L } H_2O}{2 \text{ L } C_8H_{18}} = 225 \text{ L } H_2O$$

17. $CO(g) + 2H_2(g) \rightarrow CH_3OH(g)$

(a)
$$450. \text{ mL CO} \times \frac{2 \text{ mL } H_2}{1 \text{ mL CO}} = 900. \text{ mL } H_2 \text{ are needed}$$

$$825 \text{ mL } H_2 \times \frac{1 \text{ mL CO}}{2 \text{ mL } H_2} = 412 \text{ mL CO are needed}$$

Thus the CO is present in excess.

(b) The excess amount is 450. mL − 412 mL = 38.0 mL CO

(c)
$$825 \text{ mL } H_2 \times \frac{1 \text{ ml } CH_3OH}{2 \text{ mL } H_2} = 412 \text{ mL } CH_3OH$$

18. $SO_2(g) + H_2O(\ell) \rightarrow H_2SO_3(aq)$

(a) $P = 725 \text{ mm Hg} \times \dfrac{1 \text{ atm}}{760 \text{ mm Hg}} = 0.954 \text{ atm}$

$T = 37°C + 273 = 310K$

$PV = nRT$

$$n = \frac{PV}{RT} = \frac{0.954 \text{ atm} \times 14.0 \text{ L}}{0.0821 \dfrac{L \cdot \text{atm}}{\text{mol} \cdot K} \times 310.K} = 0.525 \text{ mol } H_2O$$

$$0.525 \text{ mol } H_2O \times \frac{18.0 \text{ g } H_2O}{\text{mol } H_2O} = 9.45 \text{ g } H_2O$$

(b)
$$0.525 \text{ mol } H_2O \times \frac{1 \text{ mol } H_2SO_3}{1 \text{ mol } H_2O} = 0.525 \text{ mol } H_2SO_3$$

$$0.525 \text{ mol } H_2SO_3 \times \frac{82.1 \text{ g } H_2SO_3}{\text{mol } H_2SO_3} = 43.1 \text{ g } H_2SO_3$$

19. $CH_4(g) + 2O_2(g) \rightarrow CO_2(g) + 2H_2O(g)$

$P = 730. \text{ mm Hg} \times \dfrac{1 \text{ atm}}{760 \text{ mm Hg}} = 0.961 \text{ atm}$

$T = 20°C + 273 = 293K$

$$n = \frac{PV}{RT} = \frac{0.961 \text{ atm} \times 45.0 \text{ L}}{0.0821 \dfrac{L.\text{atm}}{\text{mol}.K} \times 293K} = 1.80 \text{ mol}$$

$$1.80 \text{ mol } CH_4 \times \frac{1 \text{ mol } CO_2}{1 \text{ mol } CH_4} = 1.80 \text{ mol } CO_2$$

$$1.80 \text{ mol } CO_2 \times \frac{44.0 \text{ g } CO_2}{\text{mol } CO_2} = 79.2 \text{ g } CO_2$$

$$1.80 \text{ mol } CH_4 \times \frac{2 \text{ mol } H_2O}{1 \text{ mol } CH_4} = 3.60 \text{ mol } H_2O$$

$$3.60 \text{ mol } H_2O \times \frac{18.0 \text{ g } H_2O}{\text{mol } H_2O} = 64.8 \text{ g } H_2O$$

20. $2Al(s) + 6NaOH(aq) \rightarrow 3H_2(g) + 2Na_3AlO_3(aq)$

$P_{H_2}H = 740.0 \text{ mm Hg} = 10.9 \text{ mm Hg} = 729.1 \text{ mm Hg}$

$P = 729.1 \text{ mm Hg} \times \dfrac{1 \text{ atm}}{760 \text{ mm Hg}} = 0.959 \text{ atm}$

$T = 12.5°C + 273 = 286K$

$$n = \frac{PV}{RT} = \frac{0.959 \text{ atm} \times 7.50 \text{ L}}{0.0821 \frac{\text{L} \cdot \text{atm}}{\text{mol} \cdot \text{K}} \times 286\text{K}} = 0.306 \text{ mol}$$

$$0.306 \text{ mol H}_2 \times \frac{2 \text{ mol Al}}{3 \text{ mol H}_2} = 0.204 \text{ mol Al}$$

$$0.204 \text{ mol Al} \times \frac{27.0 \text{ g Al}}{\text{mol Al}} = 5.51 \text{ g Al}$$

21. $N_2(g) + 3H_2(g) \rightarrow 2NH_3(g)$

$T = 550.°C + 273 = 823\text{K}$

$$10.0 \text{ kg N}_2 \times \frac{10^3 \text{ g N}_2}{\text{kg N}_2} \times \frac{\text{mol N}_2}{28.0 \text{ g N}_2} = 357 \text{ mol N}_2$$

$$357 \text{ mol N}_2 \times \frac{2 \text{ mol NH}_3}{1 \text{ mol N}_2} = 714 \text{ mol NH}_3$$

$PV = nRT$

$$V = \frac{nRT}{P} = \frac{714 \text{ mol} \times 0.0821 \frac{\text{L} \cdot \text{atm}}{\text{mol} \cdot \text{K}} \times 823\text{K}}{250. \text{ atm}} = 193 \text{ L NH}_3$$

22. $4C_3H_5(NO_3)_3 (\ell) \rightarrow 12CO_2(g) + 6N_2(g) + O_2(g) + 10H_2O(g)$

$$500. \text{ g C}_3\text{H}_5(\text{NO}_3)_3 \times \frac{\text{mol C}_3\text{H}_5(\text{NO}_3)_3}{227 \text{ g C}_3\text{H}_5(\text{NO}_3)_3} = 2.20 \text{ mol C}_3\text{H}_5(\text{NO}_3)_3$$

$$2.20 \text{ mol C}_3\text{H}_5(\text{NO}_3)_3 \times \frac{12 \text{ mol CO}_2}{4 \text{ mol C}_3\text{H}_5(\text{NO}_3)_3} = 6.60 \text{ mol CO}_2$$

$$2.20 \text{ mol C}_3\text{H}_5(\text{NO}_3)_3 \times \frac{6 \text{ mol N}_2}{4 \text{ mol C}_3\text{H}_5(\text{NO}_3)_3} = 3.30 \text{ mol N}_2$$

$$2.20 \text{ mol C}_3\text{H}_5(\text{NO}_3)_3 \times \frac{1 \text{ mol O}_2}{4 \text{ mol C}_3\text{H}_5(\text{NO}_3)_3} = 0.550 \text{ mol O}_2$$

$$2.20 \text{ mol C}_3\text{H}_5(\text{NO}_3)_3 \times \frac{10 \text{ mol H}_2\text{O}}{4 \text{ mol C}_3\text{H}_5(\text{NO}_3)_3} = 5.50 \text{ mol H}_2\text{O}$$

Total number of moles of gases produced = 6.60 mol + 3.30 mol + 0.550 mol + 5.50 mol = 16.0 mol of gases

$$16.0 \text{ mol gas} \times \frac{22.4 \text{ L}}{\text{mol}} = 358 \text{ L of gases}$$

23. $SO_2(g) + 2H_2S(g) \rightarrow 2H_2O(\ell) + 3S(s)$

$$4.50 \times 10^5 \text{ kg S} \times \frac{10^3 \text{ g}}{\text{kg}} \times \frac{\text{mol S}}{32.1 \text{ g S}} = 1.40 \times 10^7 \text{ mol S}$$

$$1.40 \times 10^7 \text{ mol S} \times \frac{1 \text{ mol SO}_2}{3 \text{ mol S}} = 4.67 \times 10^6 \text{ mol SO}_2$$

$$1.40 \times 10^7 \text{ mol S} \times \frac{2 \text{ mol H}_2\text{S}}{3 \text{ mol S}} = 9.33 \times 10^6 \text{ mol H}_2\text{S}$$

$$P = 730. \text{ mm Hg} \times \frac{1 \text{ atm}}{760 \text{ mm Hg}} = 0.961 \text{ atm}$$

$T = 22°C + 273 = 295\text{K}$

$PV = nRT$

$$V_{SO_2} = \frac{nRT}{P} = \frac{4.67 \times 10^6 \text{ mol} \times 0.0821 \frac{\text{L} \cdot \text{atm}}{\text{mol} \cdot \text{K}} \times 295\text{K}}{0.961 \text{ atm}}$$

$$= 1.18 \times 10^8 \text{ L SO}_2$$

$$V_{H_2S} = \frac{nRT}{P} = \frac{9.33 \times 10^6 \text{ mol} \times 0.0821 \frac{\text{L} \cdot \text{atm}}{\text{mol} \cdot \text{K}} \times 295\text{K}}{0.961 \text{ atm}}$$

$$= 2.35 \times 10^8 \text{ L H}_2\text{S}$$

24. $CaC_2(s) + 2H_2O(\ell) \rightarrow C_2H_2(g) + Ca(OH)_2(aq)$

$$3.25 \text{ g CaC}_2 \times \frac{\text{mol CaC}_2}{64.1 \text{ g CaC}_2} = 0.0507 \text{ mol CaC}_2$$

$$0.0507 \text{ mol CaC}_2 \times \frac{1 \text{ mol C}_2\text{H}_2}{1 \text{ mol CaC}_2} = 0.0507 \text{ mol C}_2\text{H}_2$$

$P_{C_2H_2} = 740.0 \text{ mm Hg} - 14.5 \text{ mm Hg} = 725.5 \text{ mm Hg}$

$$725.5 \text{ mm Hg} \times \frac{1 \text{ atm}}{760 \text{ mm Hg}} = 0.955 \text{ atm}$$

$T = 17°C + 273 = 290.\text{K}$

$PV = nRT$

$$V = \frac{nRT}{P} = \frac{0.0507 \text{ mol} \times 0.0821 \frac{\text{L} \cdot \text{atm}}{\text{mol} \cdot \text{K}} \times 290.\text{K} \times \frac{10^3 \text{mL}}{\text{L}}}{0.955 \text{ atm}}$$

$$= 1.26 \times 10^3 \text{ mL}$$

25. $2KOH(aq) + (NH_4)_2SO_4(aq) \rightarrow K_2SO_4(aq) + 2NH_3(g) + 2H_2O(\ell)$

$$15.0 \text{ g KOH} \times \frac{\text{mol KOH}}{56.1 \text{ g KOH}} = 0.267 \text{ mol KOH}$$

$$12.5 \text{ g (NH}_4)_2\text{ SO}_4 \times \frac{1 \text{ mol (NH}_4)_2\text{SO}_4}{132 \text{ g (NH}_4)_2\text{SO}_4} = 0.0947 \text{ mol (NH}_4)_2\text{SO}_4$$

$$0.267 \text{ mol KOH} \times \frac{1 \text{ mol (NH}_4)_2\text{SO}_4}{2 \text{ mol KOH}} = 0.134 \text{ mol (NH}_4)_2\text{SO}_4$$

Since $(NH_4)_2SO_4$ is thus the limiting reagent:

$$0.0947 \text{ mol (NH}_4)_2\text{SO}_4 \times \frac{2 \text{ mol NH}_3}{1 \text{ mol (NH}_4)_2\text{SO}_4} = 0.189 \text{ mol NH}_3$$

$$0.189 \text{ mol NH}_3 \times \frac{22.4 \text{ L}}{\text{mol NH}_3} = 4.23 \text{ L NH}_3$$

26. $Zn(s) + H_2SO_4(aq) \rightarrow ZnSO_4(aq) + H_2(g)$

$$5.00 \text{ g Zn} \times \frac{1 \text{ mol Zn}}{65.4 \text{ g Zn}} = 0.0765 \text{ mol Zn}$$

$$8.00 \text{ g H}_2\text{SO}_4 \times \frac{1 \text{ mol H}_2\text{SO}_4}{98.1 \text{ g H}_2\text{SO}_4} = 0.0815 \text{ mol H}_2\text{SO}_4$$

Since Zn and H_2SO_4 react 1:1, Zn is the limiting reagent.

$$0.0765 \text{ mol Zn} \times \frac{1 \text{ mol H}_2}{1 \text{ mol Zn}} = 0.0765 \text{ mol H}_2$$

$P_{H_2} = 730.0 \text{ mm Hg} - 19.8 \text{ mm Hg} = 710.2 \text{ mm Hg}$

$$710.2 \text{ mm Hg} \times \frac{1 \text{ atm}}{760 \text{ mm Hg}} = 0.934 \text{ atm}$$

$T = 22°C + 273 = 295\text{K}$

$PV = nRT$

$$V = \frac{nRT}{P} = \frac{0.0765 \text{ mol} \times 0.0821 \frac{\text{L} \cdot \text{atm}}{\text{mol} \cdot \text{K}} \times 295\text{K}}{0.934 \text{ atm}} = 1.98 \text{ L}$$

$$\frac{H_2}{Ne} = \frac{\sqrt{\text{mol } m_{Ne}}}{\sqrt{\text{mol } m_{H_2}}} = \frac{\sqrt{20.179 \text{ g/mol}}}{\sqrt{2.0159 \text{ g/mol}}} = 3.16$$

Answers continued on page 854O

Chapter 12 Molecular Composition of Gases

INTRODUCTION

In the last two chapters, you have learned about the properties of several important gases. You have also become acquainted with some of the physical characteristics shared by all gases. In this chapter, you will learn more about the mathematical relationships between the volume, mass, and number of moles of a gas.

LOOKING AHEAD

As you study this chapter, pay attention to the relationships between different variables and to the ways in which ratios and mathematical equations can be used to calculate new information regarding gases.

SECTION PREVIEW

12.1 Volume–Mass Relationships of Gases
12.2 The Ideal Gas Law
12.3 Stoichiometry of Gases
12.4 Effusion and Diffusion

Each SCUBA (self contained underwater breathing apparatus) tank worn by these divers contains a volume of compressed air that would equal as much as 250 times the size of each tank if it were at normal atmospheric pressure. This large supply of air enables divers to stay underwater for extended periods. Technological innovations such as SCUBA let us explore and learn about parts of the Earth that would normally be inhospitable.

12.1 Volume–Mass Relationships of Gases

The volumes taken up by gases under certain conditions can often reveal information about other properties of the gases. In this section, you will examine the relationships that exist among volumes of gases that react with each other. You will also learn about the ways in which volume, mass, density, and molar mass are interrelated.

The first important step toward the recognition of these relationships was taken when early chemists discovered the law of definite composition. In the early 1800s, Dalton developed the atomic theory by applying the law of definite composition to his investigation of the masses of combining substances. At the same time, the French chemist and early balloonist Joseph Louis Gay-Lussac was experimenting with gaseous substances (Figure 12-1). He was particularly interested in measuring and comparing the volumes of gases that react with each other in various chemical reactions.

Gay-Lussac discovered that gases combine in small, whole-number ratios by volume. This simple observation opened the door to solving several problems that had stalled progress in chemistry. It was recognized that some gases, such as oxygen, are composed not of isolated *atoms*, but of diatomic *molecules*. Gay-Lussac's work made possible the comparison of densities and relative molecular masses of gases, a discovery that led to the first correct determination of the relative scales of atomic mass.

Measuring and Comparing the Volumes of Reacting Gases

One of Gay-Lussac's earliest investigations of the volume relationship of gases involved a chemical reaction between hydrogen and oxygen. He observed that, when measured at the same temperature and pressure, 2 L of hydrogen react with 1 L of oxygen to form 2 L of water vapor.

$$\text{hydrogen gas} + \text{oxygen gas} \rightarrow \text{water vapor}$$
$$2\,\text{L} \qquad\qquad 1\,\text{L} \qquad\qquad 2\,\text{L}$$

This equation can be written in a more general form to show the simple and definite 2:1:2 relationship between the volumes of the reactants and product.

$$\text{hydrogen gas} + \text{oxygen gas} \rightarrow \text{water vapor}$$
$$2\,\text{volumes} \qquad 1\,\text{volume} \qquad 2\,\text{volumes}$$

A relationship between volumes like that above applies to any units of volume—2 mL and 1 mL and 2 mL; or 2 m³, 1 m³, and 2 m³. The relative volumes can be of any proportionate size—for example, 4 mL, 2 mL, and 4 mL; or 600 L, 300 L, and 600 L.

In another reaction with gases, Gay-Lussac noticed similar simple and definite proportions by volume.

$$\text{hydrogen gas} + \text{chlorine gas} \rightarrow \text{hydrogen chloride gas}$$
$$1\,\text{L} \qquad\qquad 1\,\text{L} \qquad\qquad 2\,\text{L}$$
$$1\,\text{volume} \qquad 1\,\text{volume} \qquad 2\,\text{volumes}$$

SECTION OBJECTIVES

- State the law of combining volumes.
- State Avogadro's principle, and explain its significance.
- Define standard molar volume of a gas, and use it as a conversion factor to calculate gas masses and volumes.
- Calculate the molar mass and density of a gas by using standard molar volume.

The law of definite composition states that each compound has a definite composition by mass. This law is in Section 1.1.3.

Figure 12-1   Gay-Lussac and Biot making their balloon ascension for scientific observations in 1804.

Joseph-Louis Gay-Lussac (1778–1850), French chemist made his first balloon ascent in 1804.

Gay-Lussac summarized the results of his experiments in 1808 in a statement known today as **Gay-Lussac's law of combining volumes of gases,** which states that *at a constant temperature and pressure, the volumes of gaseous reactants and products can be expressed as ratios of small whole numbers.* This law applies only to gas volumes measured at the same temperature and pressure.

Dalton's atomic theory is discussed in Section 3.1

John Dalton (1766–1844), English chemist; his love of meteorology led to his study of air and properties of gases which naturally led to his atomic theory.

Avogadro's Principle

Recall an important point of Dalton's atomic theory: atoms are indivisible. Dalton also believed that on atom of one element always combines with one atom of another element to form a single particle of the product. He thus thought that gaseous elements exist in the form of isolated single atoms. To explain the volume relationship observed by Gay-Lussac in terms of Dalton's theory would require that atoms be subdivided.

In 1811, Avogadro found a way to explain Gay-Lussac's simple ratios of combining gas volumes without violating Dalton's concepts of indivisible atoms. He rejected Dalton's idea that reactant elements are in monatomic form when they combine to form products and reasoned that molecules could contain more than one atom. Avogadro's explanation, known today as **Avogadro's principle,** states *that equal volumes of gases at the same temperature and pressure contain equal numbers of molecules.* It follows that at the same temperature and pressure, the volume of any given gas varies directly with the number of molecules.

Amedo Avogadro (1776–1856), Count of Quaregna, Italian physicist. Cannizzaro's effort a half century after Avogadro published his findings led to the significance of the work.

Consider the reaction of hydrogen and chlorine to produce hydrogen chloride, illustrated in Figure 12-2.

hydrogen gas + chlorine gas → hydrogen chloride gas
1 volume   1 volume   →   2 volumes

According to Avogadro's principle, equal volumes of hydrogen and chlorine contain the same number of molecules.

Accepting Dalton's idea that matter is composed of indivisible atoms, but rejecting his belief in universally monatomic elements, Avogadro concluded

Figure 12-2 Because hydrogen molecules combine with chlorine molecules in a 1-to-1 ratio to produce two volumes of hydrogen chloride, the molecules of hydrogen and chlorine must be diatomic.

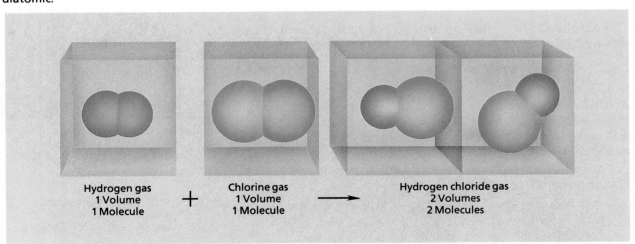

Hydrogen gas
1 Volume
1 Molecule

+

Chlorine gas
1 Volume
1 Molecule

→

Hydrogen chloride gas
2 Volumes
2 Molecules

that the hydrogen and chlorine components must each consist of two or more atoms joined together. The simplest assumption—that hydrogen and chlorine molecules are composed of two atoms each—leads to the following equation for the reaction of hydrogen with chlorine:

$$H_2(g) \quad + \quad Cl_2(g) \quad \rightarrow \quad 2HCl(g)$$

1 molecule	1 molecule	2 molecules
1 volume	1 volume	2 volumes

The simplest hypothetical formula for hydrogen chloride (HCl) indicates that the molecule contains one hydrogen atom and one chlorine atom. The simplest formulas for hydrogen and chlorine must be H_2 and Cl_2, respectively. Determination of the molecular masses of these gases has shown that these formulas are indeed correct.

Avogadro's reasoning applied equally well to the combining volumes and molecular composition of hydrogen and oxygen in the reaction to produce water vapor. The simplest hypothetical formula for oxygen must indicate two oxygen atoms and the simplest possible molecule of water must indicate two hydrogen atoms and one oxygen atom per molecule. Further experiments eventually showed that all of the active gaseous elements except the noble gases normally exist as diatomic molecules.

Using the number of moles to represent numbers of molecules, Avogadro's principle indicates that gas volume is directly proportional to the number of moles of gas (at constant temperature and pressure). The equation for this relationship is

$$V = kn$$

Review direct and inverse proportions.

where n is the number of moles of gas and k is a constant. As illustrated below, the coefficients in a chemical reaction involving gases indicate the numbers of molecules, the numbers of moles, and the relative volumes of reactants and products.

$$2H_2(g) \quad + \quad O_2(g) \quad \rightarrow \quad 2\,H_2O(g)$$

2 molecules	1 molecule	2 molecules
2 moles	1 mole	2 moles
2 volumes	1 volume	2 volumes

Molar Volume of Gases

Recall that one mole contains Avogadro's number of molecules (6.022×10^{23}). One mole of oxygen (O_2) contains 6.022×10^{23} diatomic oxygen molecules and has a mass of 31.9988 g. One mole of hydrogen (H_2) contains the same number of hydrogen molecules but has a mass of 2.01588 g. Helium (He) is a monatomic gas, yet one mole of helium gas contains Avogadro's number of helium atoms and has a mass of 4.00260 g.

Avogadro's number is introduced in Section 3.3.

Because one-mole quantities of all molecular substances contain the Avogadro number of molecules, one mole of any gas contains the same number of molecules as one mole of any other gas. According to Avogadro's principle, each gas will occupy the same volume at the same temperature and pressure (Figure 12-3). Experiments have revealed that one mole of any gas has the same volume at STP as one mole of any other gas. *This volume, known as the **molar volume of a gas,** is the volume occupied by one mole of*

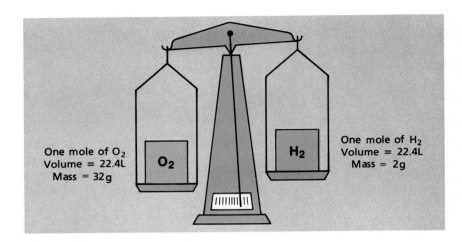

Figure 12-3 One-mole quantities of different gases occupy 22.4 L at STP, yet have different masses.

One mole of O_2
Volume = 22.4L
Mass = 32 g

O_2

H_2

One mole of H_2
Volume = 22.4L
Mass = 2g

Molar volumes at other temperatures

Temp (°C)	Molar volume (L/mole)
25	24.5
100	30.6

a gas at STP, and has been found to be 22.414 L. As illustrated in Figure 12-3, 22.414 L of every gas contains the same number of molecules, but the masses of these volumes are different for different gases (in calculations we will use 22.4 L as the molar volume). The mass of each is equal to the molar mass of the gas—the mass of 1 mole of molecules.

It is often more convenient to measure the volume of a gas than it is to measure its mass. If you remember that molar volume applies only at STP, you can use 1 mol/22.4 L as a conversion factor to find the number of moles, and therefore the mass, of a given volume of a known gas. The molar volume of a gas can be used to find the volume, at STP, of a known number of moles or a known mass of a gas.

Sample Problem 12.1

A chemical reaction is expected to produce 0.0680 mol of oxygen gas. What volume in liters will be occupied by this gas sample at STP?

Solution

Step 1. Analyze

Given: 0.0680 mol of O_2
Unknown: volume of O_2 in liters (at STP)

Step 2. Plan

$$\text{moles of } O_2 \rightarrow \text{volume of } O_2 \text{ (at STP)}$$

The standard molar volume can be used to find the volume of a known molar amount of a gas at STP.

$$\text{moles} \times \frac{22.4 \text{ L}}{\text{mole}} = V \text{ (in liters—L)}$$

Step 3. Compute

$$0.0680 \text{ mol} \times \frac{22.4 \text{ L}}{1 \text{ mol}} = 1.52 \text{ L}$$

Step 4. Evaluate

The answer is close to an estimated value of 1.4, computed as 0.07×20. Units have canceled to yield liters, as desired. The result is correctly carried to three significant figures.

Sample Problem 12.2

A chemical reaction produced 98.0 mL of sulfur dioxide gas (SO_2) at STP. What was the mass (in grams) of the gas produced?

Solution

Step 1. Analyze   Given: $V = 98.0$ mL of SO_2 (at STP)
Unknown: mass of SO_2 in grams

Step 2. Plan SO_2 in mL (at STP) $\rightarrow$ SO_2 volume in L (at STP) $\rightarrow$ SO_2 moles $\rightarrow$ SO_2 mass

$$\text{mL} \times \frac{1 \text{ L}}{1000 \text{ mL}} \times \frac{1 \text{ mol } SO_2}{22.4 \text{ L}} \times \frac{\text{g } SO_2}{1 \text{ mol } SO_2}$$

$$= \text{g } SO_2$$

Step 3. Compute $$98.0 \text{ mL} \times \frac{1 \text{ L}}{1000 \text{ mL}} \times \frac{1 \text{ mol } SO_2}{22.4 \text{ L}} \times \frac{64.1 \text{ g } SO_2}{1 \text{ mol } SO_2}$$

$$= 0.280 \text{ g } SO_2$$

Step 4. Evaluate   The result is correctly limited to three significant figures. Units cancel correctly to give the answer in grams. The known volume is roughly $\frac{1}{200}$ of the molar volume (22400 mL/200 = 112 mL). The answer is reasonable: the mass should be roughly $\frac{1}{200}$ of the molar mass (64 g/200 = 0.32 g).

Practice Problem
What is the mass of 1.33×10^4 mL of oxygen gas (O_2) at STP?   *(Ans.)* 19.0 g

Gas Density

The densities of different gases represent the masses of equal numbers of their respective molecules measured at STP. From the gas laws, you know that the volume of a specific number of gas molecules varies with temperature and pressure. For example, at higher temperatures, a gas tends to occupy more volume and become less dense. If you are to compare the densities of different gases, you must be given the density values for the same temperature and pressure. The density of a gas is usually given in grams per liter at STP. For example, hydrogen has a density at STP of 0.0899 g/L.

As is true for solids and liquids, the density of a gas can be found by dividing the mass of the gas sample by its volume. Unlike solid and liquid densities, however, a gas density at STP can also be found from the molar mass alone. The molar volume is the necessary conversion factor. The relationships in terms of the units are as follows:

$$\text{Density (at STP)} = \frac{\text{molar mass}}{\text{molar volume}} = \frac{\text{g/mol}}{22.4 \text{ L/mol}}$$

$$= \frac{\text{g}}{\text{mol}} \times \frac{1 \text{ mol}}{22.4 \text{ L}}$$

Volume–pressure and volume–temperature relationships are described by Boyle's law and Charles' law, respectively.

Figure 12-4   A balloon filled with helium rises because helium is less dense than air. Why does the balloon filled with carbon dioxide sink?

Sample Problem 12.3

What is the density of CO_2 gas in grams per liter at STP?

Solution

Step 1. Analyze   Given: chemical formula, CO_2
Unknown: density of CO_2 at STP

Step 2. Plan   chemical formula → molar mass → density (at STP)

$$\frac{g}{1\ mol} \times \frac{1\ mol}{22.4\ L} = \frac{g}{L}$$

Step 3: Compute   Result: $\frac{44.0\ g}{1\ mol} \times \frac{1\ mol}{22.4\ L} = 1.96\ g/L$

Step 4. Evaluate   Units cancel correctly and the answer is properly limited to three significant figures. The magnitude of the answer is reasonable for a gas density and is close to an estimated value of 2.0 (44 ÷ 22).

Practice Problem
What is the density of SO_2 gas in grams per liter at STP?   *(Ans.)* 2.86 g/L

Finding Molar Mass from Volume at STP

The combined gas law is discussed in Section 11.3

The molar mass of a gas or any substance that vaporizes easily without decomposing can be determined by use of the molar volume. The density of the gas at STP can be determined by finding either the volume occupied by a given mass of the gas or the mass that occupies a known volume. Then the relationship given above for density, molar mass, and molar volume is used to solve for the molar mass:

$$\text{density (at STP)} = \frac{\text{molar mass}}{\text{molar volume}}$$

$$\text{molar mass} = \text{density (at STP)} \times \text{molar volume}$$

$$= \frac{g}{L} \times \frac{22.4\ L}{mol}$$

Sample Problem 12.4

What is the molar mass of a gas that has a density of 1.28 g/L at STP?

Solution

Step 1. Analyze   Given: density (at STP) = 1.28 g/L
Unknown: molar mass

Step 2. Plan   Route: density (at STP) → molar mass. Because the density is known at STP, the molar volume can be used to find the molar mass, as shown above.

Step 3. Compute   Result: $\frac{1.28\ g}{1\ L} \times \frac{22.4\ L}{1\ mol} = 28.7\ g/mol$

| Step 4. Evaluate | Units cancel correctly. The answer is given to three significant figures and is close to an estimated value of 22, or 1×22. |

Sample Problem 12.5

A 0.519-g gas sample is found to have a volume of 200 mL at STP. What is the molar mass of this gas? Is this gas propane (molar mass = 44.1 g/mol) or butane (molar mass = 58.1 g/mol)?

Solution

| Step 1. Analyze | Given: mass and volume of gas at STP, 0.519 g and 200 mL |
| | Unknown: molar mass |

| Step 2. Plan | Mass/volume (at STP) → density (at STP) → molar mass |

Because the volume is known at STP, the molar volume can be used to find the molar mass. The factor for converting milliliters to liters must be included.

$$\frac{\text{g}}{\text{mL}} \times \frac{1000 \text{ mL}}{1 \text{ L}} \times \frac{22.4 \text{ L}}{1 \text{ mol}} = \frac{\text{g}}{\text{mol}}$$

| Step 3 Compute | $\dfrac{0.519 \text{ g}}{200 \text{ ml}} \times \dfrac{1000 \text{ ml}}{1 \text{ L}} \times \dfrac{22.4}{1 \text{ mol}} = 58.1 \text{ g/mol}$ |

| Step 4. Evaluate | Units cancel correctly and the answer is properly given to three significant figures. The answer has the correct value for one of the two possible gases and is close to an estimated value of 55 ($\frac{0.5}{200} \times 1000 \times 22$). The gas is butane. |

Sample Problem 12.6

A 1.25 g sample of the gaseous product of a chemical reaction was found to have a volume of 350. mL at 20.0°C and 750. mmHg. What is the molar mass of this gas?

Solution

Step 1. Analyze	Given: mass of gas = 1.25 g
	pressure = 750. mm Hg; volume = 350. mL; temperature = 20.0°C
	Unknown: molar mass

| Step 2. Plan | temperature (°C) → temperature (K); pressure, volume, temperature → volume (at STP) → density (at STP) → molar mass |

$$t(°\text{C}) + 273 = T(\text{K})$$

The measured volume is not at STP. Before the molar mass can be found from the density and standard molar volume, therefore, the volume at STP must be found by solving the combined gas law for V_2.

$$V_2(\text{mL}) = \frac{V_1 P_1 T_2}{P_2 T_1}$$

$$\frac{\text{g}}{(\text{mL})} \times \frac{1000 \text{ mL}}{1 \text{ L}} \times \frac{22.4 \text{ L}}{1 \text{ mol}} = \frac{\text{g}}{\text{mol}}$$

Step 3. Compute

$$20°C + 273 = 293 \text{ K}$$

$$V_2 = \frac{350.0 \text{ ml} \times 750 \text{ mm Hg} \times 273\text{K}}{760 \text{ mm Hg} \times 293\text{K}}$$

$$= 322 \text{ mL at STP}$$

$$= \frac{1.25 \text{ g}}{322 \text{ mL}} \times \frac{1000 \text{ mL}}{1 \text{ L}} \times \frac{22.4\text{L}}{1\text{mol}} = 87.0 \text{ g/mol}$$

Step 4. Evaluate Units cancel correctly to give molar mass in grams per mole. The answer is properly limited to three significant digits and is close to an estimated value of 73, or $\frac{1}{300} \times 1000 \times 22$.

Practice Problems

1. What is the molar mass of a gas whose density at STP is 2.08 g/L? *(Ans.)* 46.6 g/mol
2. What is the molar mass of a gas if a 1.39 g sample of the gas has a volume of 375 mL at 22°C and 755 mmHg? *(Ans.)* 89.7 g/mol

1. At constant temperature and pressure, the volumes of gaseous reactants and products can be expressed as ratios of small whole numbers.
2. Equal volumes of gases at the same temperature and pressure contain equal numbers of molecules. This led to the discovery of diatomic molecules and allowed balanced equations with gaseous reactants and products to be interpreted in terms of molecules, moles, and volumes.

SECTION OBJECTIVES

- State the ideal gas law.
- Derive the ideal gas constant and discuss its units.
- Using the ideal gas law, calculate one of the quantities—pressure, volume, temperature, amount of gas—when the other three are known.
- Using the ideal gas law, calculate the molar mass or density of a gas.
- Reduce the ideal gas law to Boyle's law, Charles' law, and Avogadro's principle. Describe the conditions under which each applies.

Section Review

1. Explain Gay-Lussac's law of combining volumes.
2. State Avogadro's principle and explain its significance.
3. Define molar volume.
4. How many moles of oxygen gas are there in 135 L of oxygen at STP?
5. What volume (in mL) at STP will be occupied by 0.0035 mol of methane (CH_4)?
6. What is the density of nitrogen gas in grams per liter at STP?
7. What is the molar mass of a gas if 0.445 g of the gas occupies a volume of 2.80 L at 25°C and 740 mm Hg pressure?

12.2 The Ideal Gas Law

You learned in Section 11.2 that the values of four quantities—pressure, volume, temperature, and the number of moles of a gas—are needed to completely describe a gas sample. A single equation relating pressure, volume, temperature, and number of moles for a given sample of gas under any combination of conditions can be derived experimentally. *The **ideal gas law** is a relationship between pressure, volume, temperature, and the number of moles of a gas.* It can be derived by combining the direct proportionality of volume to the number of moles of a gas (Avogadro's principle) with the P–V–T relationships.

In this section, you will learn about the derivation of the ideal gas law and how it can be used to find unknown information about gas samples.

Derivation of the Ideal Gas Law

The volume of a gas sample varies with pressure, temperature, and the number of moles of gas according to the following laws:

3. The volume occupied by one mole of a gas at STP, or 22.4 L
4. 6.03 moles
5. 78.4 mL
6. 0.625 g/L
7. 3.99 g/mole

Boyle's law: At constant temperature, the volume is inversely proportional to of a given mass of gas the pressure:

$$V = \frac{k}{P}$$

Charles' law: At constant pressure the volume of a given mass of gas is directly proportional to the Kelvin temperature:

$$V = kT$$

Avogadro's principle: At constant pressure and temperature, the volume of a given mass of gas is directly proportional to the number of moles:

$$V = kn$$

Because a quantity that is proportional to each of three separate quantities is also proportional to their product, combining the three gas laws above gives:

$$V = k \times \frac{1}{P} \times T \times n$$

The constant, k, has the same value for all gases whose behavior approaches that of an ideal gas. If we substitute R and k and rearrange the variables, we obtain the equation for the ideal gas law:

$$V = \frac{nRT}{P} \quad or \quad PV = nRT$$

The ideal gas law equation states that the volume of a gas varies directly with the number of moles or molecules of a gas and its Kelvin temperature and varies inversely with its pressure. Under ordinary conditions, most gases exhibit behavior that is close to ideal and the equation can be applied with reasonable accuracy.

A close look will reveal that the ideal gas law reduces to Boyle's law, Charles' law, Gay-Lussac's law, or Avogadro's principle when the appropriate variables are held constant. For example, if n and T are constant, the product nRT is constant because R is also constant. In this case, the ideal gas law reduces to $PV = k$, which is Boyle's law.

The Ideal Gas Constant

In the ideal gas equation, *the constant R is known as the* **ideal gas constant.** Its value depends upon the units chosen for pressure, volume and temperature. Measured values of P, V, T, and n for a gas at conditions that are close to ideal can be used to calculate R. Recall from Section 12.1 that the volume of one mole of an ideal gas at STP (1 atm and 273.15 K) is 22.414 L. Substituting these values and solving the ideal gas law equation for R gives

$$R = \frac{Pv}{nT} = \frac{1 \text{ atm} \times 22.414 \text{ L}}{1 \text{ mol} \times 273.15 \text{K}}$$

$$= 0.082057 \frac{\text{L} \cdot \text{atm}}{\text{mol} \cdot \text{K}}$$

This calculated value of R, usually rounded to 0.0821 L·atm/mol·K, can be used in ideal gas law calculations only when the volume is given in liters, the

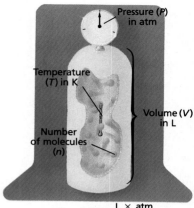

$$R = 0.0821 \frac{\text{L} \times \text{atm}}{\text{mol} \times \text{K}}$$

$$P \times V = n \times R \times T$$

Figure 12-5 The ideal gas law demonstrates the interrelationship between the pressure, volume number of molecules, and temperature of an ideal gas.

Intermediate steps in the derivation

$V = k_1/P \times k_2T \times k_3n = k_1k_2k_3 \times 1/P \times T \times n$

if $R = k_1k_2k_3$, then
$V = R \times 1/P \times T \times n$, or $PV = nRT$

Other values for R are:

R = 8.31 joules/(mole K) = 62.4 liter mm Hg/(mole K) = 1.99 calories/(mole K) = 8.31 volt coulombs/(mole k)

TABLE 12-1　NUMERICAL VALUES OF THE GAS CONSTANT, R

	Units of R	Numerical value of R	Units of P	Units of V	Units of T	Units of n
	$\dfrac{L \cdot mm\ Hg}{mol \cdot K}$	62.4	mm Hg	L	K	mol
$R = \dfrac{PV}{nT}$	$\dfrac{L \cdot atm}{mol \cdot K}$	0.082057	atm	L	K	mol
	$\dfrac{J}{mol \cdot K}$	8.314^a	atm	L	K	mol
	$\dfrac{L \cdot kPa}{mol \cdot K}$	8.314^b	kPa	L	K	mol

(a)　1 L·atm = 101.32 J
(b)　SI units

pressure is given in atmospheres, and the temperature is given in kelvins. Some values of R in other units are given in Table 12-1.

Using the Ideal Gas Law

Finding P, V, T, or n from the ideal gas law　The ideal gas law applies to situations in which a single gas sample undergoes no change in conditions and values are known for three of the four variables—P, V, T, or n. **CAUTION:** Always be sure to match the units of the known quantities and the units of R. We shall be using $R = 0.0821$ L·atm/mol·k. If necessary, volumes must be converted to liters, pressures to atmospheres, temperature to kelvins, and masses to numbers of moles before using the ideal gas law.

Sample Problem 12.7

What is the pressure in atmospheres exerted by a 0.500-mol sample of nitrogen in a 10.0-L container at 298K?

Solution

Step 1. Analyze　　Given:　$V = 10.0$ L; $n = 0.500$ mol; $T = 298$K
　　　　　　　　　　　Unknown: P

Step 2. Plan　　　　　　　　　　　　　　$n, V, T \rightarrow P$

Because the gas sample undergoes no changes in conditions, the ideal gas law can be rearranged and used to find the pressure:

$$P = \frac{nRT}{V}$$

Step 3. Compute　　$P = \dfrac{0.500 \text{ mol} \times (0.0821 \text{ L·atm/mol K}) \times 298\text{K}}{10.0 \text{ L}} = 1.22 \text{ atm}$

Step 4. Evaluate   All units cancel correctly to give the result in atmospheres. The answer is properly limited to three significant figures and is close to an estimated value of 1.5 (computed as $\dfrac{0.5 \times 0.1 \times 300}{10}$).

Sample Problem 12.8

What is the volume in liters occupied by 0.250 mol of oxygen at 20.0°C and 740 mm Hg pressure?

Solution

Step 1. Analyze   Given: $P = 740$ mm Hg; $n = 0.250$ mol; $t(C°) = 20.0°C$
Unknown: V

Step 2. Plan $$P, n, T \rightarrow V$$

The ideal gas law may be rearranged to solve for V for this oxygen sample that undergoes no change in conditions.

$$V = \frac{nRT}{P}$$

To use our chosen value of R (0.0821 L·atm/mol·K) in the equation, the units of P and $t(°C)$ must first be converted into atmospheres and kelvins.

$$P = \text{mm Hg} \times \frac{1 \text{ atm}}{760 \text{ mm Hg}} ; T(K) = °C + 273$$

Step 3. Compute $$P = 740 \text{ mm Hg} \times \frac{1 \text{ atm}}{760 \text{ mm Hg}} = 0.974 \text{ atm}$$

$$T(K) = 20.0°C + 273$$
$$= 293 K$$

$$V = \frac{0.250 \text{ mol} \times 0.0821 \text{ L·atm/mol·K} \times 293 \text{ K}}{0.974 \text{ atm}} = 6.17 \text{ L}$$

Step 4. Evaluate   Units cancel to give liters, as desired. The answer is correctly limited to three significant digits and is close to an estimated value of 6 (calculated as $0.2 \times 0.1 \times 300$).

Sample Problem 12.9

What mass of chlorine (Cl_2), in grams, is contained in a 10.0-L tank at 27°C and 3.50 atm of pressure?

Solution

Step 1. Analyze   Given: identity of gas, Cl_2
$P = 3.50$ atm; $V = 10.0$ L; $t(C°) = 27°C$
Unknown: mass of Cl_2

Step 2. Plan $P, V, T \rightarrow$ moles of Cl_2 $(n) \rightarrow$ mass of Cl_2 in grams

The ideal gas law may be rearranged to solve for n after temperature is converted to kelvins. The number of moles is then converted to grams.

$$n = \frac{PV}{RT}; \ m \text{ (in g)} = n \times \frac{g}{mol}$$

$$T(K) = 27.0°C + 273 = 300 \text{ K}$$

Step 3. Compute

$$n = \frac{3.50 \text{ atm} \times 10.0 \text{ L}}{0.0821 \text{ L·atm/mol·K} \times 300 \text{ K}} = 1.42 \text{ mol}$$

$$\text{mass of } Cl_2 = 1.42 \text{ mol} \times \frac{70.9 \text{ g } Cl_2}{1 \text{ mol}} = 101 \text{ g } Cl_2$$

Step 4. Evaluate   In each calculation, units cancel to leave those that are required. The result of each calculation is correctly given to three significant figures. The answers are close to the estimated values.

Practice Problems

1. What pressure, in atmospheres, is exerted by 0.325 mol of hydrogen gas (H_2) in a 4.08-L container at 35°C?   *(Ans.)* 2.01 atm
2. What is the mass, in grams, of oxygen (O_2) in a 12.5-L container at 45°C and 7.22 atm?   *(Ans.)* 111 g

---

Density is explained in Section 1.2.

Finding molar mass or density from the ideal gas law   You can also use the ideal gas law to calculate the molar mass or density of a gas sample. Suppose that the pressure, volume, and temperature are known for a gas sample of a given mass. The number of moles (n) in the sample can be calculated using the ideal gas law. Then the molar mass (grams per mole) can be calculated by dividing the known mass by the number of moles.

An equation showing the relationship between and among density, pressure, temperature, and molar mass can be derived from the ideal gas law. The number of moles (n) is equal to mass (m) divided by molar mass (mol m).

Substituting $n = m/\text{mol } m$ into $PV = nRT$ gives

$$PV = \frac{mRT}{\text{mol } m} \text{ or mol } m = \frac{mRT}{PV}$$

Density (D) is mass (m) per unit volume (V). Writing this definition in the form of an equation gives $D = \frac{m}{V}$. You can see that $\frac{m}{V}$ appears in the equation on the right just above. Introducing density into that equation gives

$$\text{mol } m = \frac{nRT}{PV} = \frac{DRT}{P}$$

Solving for density gives

$$D = \frac{(\text{mol } m)P}{RT}$$

You can see that the density of a gas varies directly with molar mass and pressure and inversely with kelvin temperature (Figure 12-9).

Sample Problem 12.10

At 28°C and 740 mm Hg pressure, 1.00 L of an unidentified gas has a mass of 5.16 g. What is the molar mass of this gas?

Solution

Step 1. Analyze   Given: $P = 740$ mm Hg; $V = 1.00$ L; $t(C°) = 28$ °C;
$m = 5.16$ g
Unknown: molar mass

Step 2. Plan $P, V, T \rightarrow n \rightarrow$ mol m

Pressure and temperature must be converted to atmospheres and kelvins, respectively. The number of moles equivalent to 5.16 g of the unknown gas is found from the rearranged ideal gas law:

$$n = \frac{PV}{RT}$$

The molar mass (mass of one mole of gas) is then found by dividing the known mass by the number of moles calculated.

$$\text{mol } m = \frac{m \text{ (in g)}}{n(\text{number of moles})}$$

Step 3. Compute $P = 740 \text{ mm Hg} \times \dfrac{1 \text{ atm}}{760 \text{ mm Hg}} = 0.974 \text{ atm}$

$T \text{ (K)} = 28°C + 273 = 301 \text{ K}$

$$n = \frac{0.974 \text{ atm} \times 1.00 \text{ L}}{0.0821 \text{ L·atm/mol·K} \times 301 \text{ K}} = 0.0394 \text{ mol}$$

$$\text{mol } m = \frac{5.16 \text{ g}}{0.0394 \text{ mol}} = 131 \text{ g/mol}$$

Step 4. Evaluate   Units cancel as needed. The answer is correctly given to three significant digits and is close to an estimated value of 125 (calculated as $5 \div 0.04$).

Practice Problems

1. What is the molar mass of a gas if 0.427 g of the gas occupies a volume of 125 mL at 20.0°C and 745 mm Hg pressure?   *(Ans.)* 83.8 g/mol
2. What is the density of a sample of ammonia gas (NH_3) if the pressure is 705 mm Hg and the temperature is 63.0°C?   *(Ans.)* 0.572 g/L

Section Review

1. What is the volume in liters of 2.00 g of CS_2 vapor at 276 mm Hg pressure and 70.0°C?
2. Why must the units P, V, T, and n be appropriate to those of the ideal gas constant in solving problems? What are the units for R if P is in pascals, T is in K, V is in liters, and n is in moles?
3. What is the molar mass of a 1.25 g sample of gas that occupies a volume of 1.00 L at a pressure 730 mm Hg and a temperature of 27.0°C?

if $n_1 = n_2$ and $P_1 = P_2$, then
the equation reduces to

$\dfrac{V_1}{T_1} = \dfrac{V_2}{T_2}$

1. 1.9 L
2. If the units of P, V, T and n are different from those in which R is given, the units will not cancel in the program. The units are L·pascals/K·mole.
3. 32.1 g/mole
4. Under conditions of constant pressure and constant number of moles

$\dfrac{P_1 V_1}{n_1 T_1} = \dfrac{P_2 V_2}{n_2 T_2}$

12.3 Stoichiometry of Gases

SECTION OBJECTIVES

- Explain how Gay-Lussac's law and Avogadro's principle apply to the volumes of gases in chemical reactions
- Use a chemical equation to specify volume ratios for gaseous reactants and/or products
- Use volume ratios, standard molar volume, and the gas laws where appropriate to calculate volumes, masses, or molar amounts of reactants or products in reactions involving gases

The concept of stoichiometry is introduced in Section 9.1.

In Chapter 9, you learned the relationships between moles and masses of reactants and products in chemical reactions. You saw that the numerical coefficients in a balanced chemical equation reveal the relative numbers of moles of any two substances involved in a chemical reaction.

Frequently, one or more reactants or products are gases. When the information known about gases in chemical reactions includes volumes, pressures, and temperatures rather than masses and molar amounts, the combined gas law, the standard molar volume, or the ideal gas law can be used to solve stoichiometric problems involving chemical reactions.

The discoveries of Gay-Lussac and Avogadro described in Section 12.1 can be applied in the stoichiometry of reactions involving gases. For reactants or products that are gases, the coefficients in chemical equations indicate volumes and volume ratios in addition to molar amounts and mole ratios. For example, consider the reaction of carbon monoxide with oxygen to give carbon dioxide:

$$2CO(g) \quad + \quad O_2(g) \quad \rightarrow \quad 2CO_2(g)$$

2 molecules	1 molecule	2 molecules
2 mol	1 mol	2 mol
2 volumes	1 volume	2 volumes

The possible volume ratios can be expressed as:

(1) $\dfrac{2 \text{ vol CO}}{1 \text{ vol O}_2}$ or $\dfrac{1 \text{ vol O}_2}{2 \text{ vol CO}}$ (2) $\dfrac{2 \text{ vol CO}}{2 \text{ vol CO}_2}$ or $\dfrac{2 \text{ vol CO}_2}{1 \text{ vol CO}}$ (3) $\dfrac{1 \text{ vol O}_2}{2 \text{ vol CO}_2}$ or $\dfrac{2 \text{ vol CO}_2}{1 \text{ vol O}_2}$

Volumes can, of course, be compared in this way only if each is measured at the same temperature and pressure.

Volume–Volume Calculations

Volume–volume problems are simple to solve. When the volume of a gas involved in a chemical reaction is known and the volume of another reactant or product must be found, volume ratios like those given above are used exactly like mole ratios. The route for solving the problem is

$$\text{gas volume A} \rightarrow \text{gas volume B}$$

The problem setup is similar to that of the mole-mole calculations given in Section 9.2.

Sample Problem 12.11

The complete combustion of propane (C_3H_8), a gas that is in tanks and used for cooking and heating in some areas (see Figure 12-6), occurs according to the following equation:

$$C_3H_8(g) + 5O_2(g) \rightarrow 3CO_2(g) + 4H_2O(\ell)$$

Assuming all volume measurements are made at the same temperature and pressure, (a) What will be the volume in liters of oxygen required for the complete combustion of 0.350 L of propane; and (b) What will be the volume of carbon dioxide produced in the reaction?

Solution

Step 1. Analyze

Given: balanced chemical equation
                volume of propane $= 0.350$ L
Unknowns: (a) volume of $O_2(g)$ (b) volume of $CO_2(g)$

Step 2. Plan

(a) vol C_3H_8 vol O_2 (b) vol C_3H_8 vol CO_2

Because all volumes are to be compared at the same temperature and pressure, volume ratios can be used like mole ratios to find the unknowns.

Step 3. Compute

(a) $0.350\text{L } C_3H_8 \times \dfrac{5 \text{ L } O_2}{1 \text{ L } C_3H_8} = 1.75 \text{ L } O_2$

(b) $0.350\text{L } \dfrac{3 \text{ L } CO_2}{C_3H_8} \times 1 \text{ L } C_3H_8 = 1.05 \text{ L } CO_2$

Step 4. Evaluate

Each result is correctly given to three significant figures. The answers are reasonably close to estimated values of 2 (calculated as 0.4×5), and 1.2 (calculated as $.04 \times 3$), respectively.

Practice Problem

Assuming all volume measurements are made at the same temperature and pressure, what volume of hydrogen gas is needed to react completely with 4.55 L of oxygen gas to produce water vapor? *(Ans.)* 9.10L

Volume–Mass and Mass–Volume Calculations

In many cases, stoichiometric calculations will involve both masses and gas volumes. Sometimes the volume of a reactant or product is given and the mass of a second gaseous substance is unknown. On the other hand, a mass amount may be known and a volume may be the unknown. The calculations require routes such as

         gas volume A → moles A → moles B → mass B

or

         mass A → moles A → moles B → gas volume B.

To find the unknown information in cases such as these, one must know the conditions under which both the known and unknown gas volumes have been measured. At STP, the conversion factors 1 mol/22.4 L and 22.4 L/mol can be used to convert volume to moles and moles to volume, respectively.

     The ideal gas law is very useful for calculating values at nonstandard conditions. For example, the number of moles of gas together with the known pressure and temperature conditions can be substituted into the ideal gas law and the gas volume calculated directly. If the volume of a gas under nonstandard conditions of temperature and pressure is known, the ideal gas law can be used to calculate the number of moles of the gas. Sample Problem 12.12 is a volume–mass calculation that illustrates use of molar volume as a conversion factor. Sample Problem 12.13 is a mass–volume calculation in

Figure 12-6 Teenage campers in Fairlee, Vermont using a portable propane grill. Propane is often used at trailer parks and at campsites for cooking, water heating, space heating, and refrigeration.

which the ideal gas law is used to find the unknown volume of a gas at nonstandard conditions.

Sample Problem 12.12

Calcium carbonate ($CaCO_3$), which is also known as limestone, is calcined to produce calcium oxide (lime), an industrial chemcial with a wide variety of uses ranging from glass manufacture to sewage treatment (Figure 12-8). The balanced equation for the calcination is

$$CaCO_3(s) \rightarrow CaO(s) + CO_2(g)$$

How many grams of calcium carbonate must be decomposed to produce 5.00 L of carbon dioxide at STP?

Solution

Step 1. Analyze   Given: Balanced chemical equation
                                 desired volume of $CO_2(g)$ produced = 5.00 L (STP)
                        Unknown: mass of $CaCO_3(s)$ in grams

Step 2. Plan   vol $CO_2 \rightarrow$ moles $CO_2 \rightarrow$ moles $CaCO_3 \rightarrow$ mass $CaCO_3$

The known volume is given at STP. Standard molar volume provides the necessary factor to convert the known volume to moles for use with mole ratios from the balanced equation. (Note that volume ratios do not apply here because calcium carbonate is a solid.)

$$L\ CO_2 \times \frac{1\ mol\ CO_2}{22.4\ L\ CO_2} \times \frac{1\ mol\ CaCO_3}{1\ mol\ CO_2} \times \frac{g\ CaCO_3}{1\ mol\ CaCO_3}$$

Step 3. Compute $$(5.00\ L\ CO_2) \times \frac{1\ mol\ CO_2}{22.4\ L\ CO_2} \times \frac{1\ mol\ CaCO_3}{1\ mol\ CO_2} \times \frac{100.0\ g CaCO_3}{1\ mol\ CaCO_3} = 22.3\ g\ CaCO_3$$

Step 4. Evaluate   Units all cancel correctly. The answer is properly given to three significant figures and is close to an estimated value of 25 (computed as $5 \times \frac{1}{20} \times 100$).

Sample Problem 12.13

Tungsten, a metal that is used in light bulb filaments (Fig. 12.9), is produced industrially by the following reaction of tungsten oxide with hydrogen:

$$WO_3(s) + 3H_2(g) \rightarrow W(s) + 3H_2O(l)$$

How many liters of hydrogen at 35°C and 745 mm Hg are needed to react completely with 875 g of tungsten oxide?

Solution

Step 1. Analyze   Given: balanced chemical equation
                                 reactant mass = 875 g $WO_3(s)$
                                 conditions for gaseous reactant: P = 745 mm Hg and t(°C) = 35°C
                        Unknown: volume of H_2 at known non-standard conditions

Sample Problem 12.13 *continued*

Step 2. Plan

$$\text{mass WO}_3 \rightarrow \text{moles WO}_3 \rightarrow \text{moles H}_2 \rightarrow \text{volume H}_2$$

Moles of H_2 are found by converting mass of WO_3 to moles and then using the mole ratio. Because the conditions for the unknown volume are not STP, the ideal gas law is used to find the volume from the calculated number of moles of H_2. Pressure and temperature must therefore be converted first to atmospheres and degrees Kelvin in the usual manner.

Step 3. Compute

$$875 \text{ g WO}_3 \times \frac{1 \text{ mol WO}_3}{231.8 \text{ g WO}_3} \times \frac{3 \text{ mol H}_2}{1 \text{ mol WO}_3} = 11.3 \text{ mol H}_2$$

$$P = 745 \text{ mm Hg} \times \frac{1 \text{ atm}}{760 \text{ mm Hg}} = 0.980 \text{ atm}$$

$$T(k) = 35°C + 273 = 308K$$

$$V = \frac{nRT}{P} = \frac{11.3 \text{ mol} \times 0.0821 \text{ L·atm/mol·K} \times 308K}{0.980 \text{ atm}}$$

$$= 292 \text{ L}$$

Step 4. Evaluate

Unit cancellations are correct in each calculation, as is the use of three significant figures for each result. The answer is reasonably close to an estimated value of 330 (computed as $\frac{11 \times 0.1 \times 300}{1}$).

Practice Problems

1. What mass of sulfur must be used to produce 12.61 L of sulfur dioxide at STP according to the equation

$$S(s) + O_2 \rightarrow SO_2(g) \quad (Ans.) \text{ 19.5 g}$$

2. How many liters of carbon monoxide at 27°C and 188 mm Hg can be produced from the burning of 65.5 g of carbon according to the equation

$$2C(s) + O_2(g) \rightarrow 2CO \text{ (g)} \quad (Ans.) \text{ 544 L}$$

Figure 12-7 The exceptional strength of tungsten at very high temperatures makes it well-suited for use in light-bulb filaments and other industrial applications. The chemical symbol for tungsten, **W**, comes from wolfram, the German name for this element.

Section Review

1. How many liters of ammonia gas can be formed from the reaction of 150L of hydrogen gas, assuming that there is complete reaction of hydrogen with excess nitrogen gas, and that all measurements are made at the same temperature and pressure?
2. How many liters of H_2 at STP can be produced by the reaction of 4.60 g of Na and excess water?
3. How many grams of Na are needed to react with H_2O to liberate 400 mL of H_2 at STP?
4. What volume of oxygen in liters can be collected at 750 mm Hg pressure and 25.0°C when 30.6 g of $KC1O_3$ decomposes by heating?

12.4 Effusion and Diffusion

SECTION OBJECTIVES

- State Graham's law of effusion or diffusion.
- Determine the relative rates of effusion of two gases of known molar masses.

The constant motion of gas molecules causes them to spread out to fill any container in which they are placed. The gradual mixing of two gases due to their spontaneous, random motion is known as **diffusion.** *When a gas is confined in a container that has a very small opening, its molecules will randomly encounter the opening and pass through it in a process called* **effusion** (Figure 12-8). In this section, you will learn how a chemical law describing diffusion and effusion can be used to discover the molar mass of an unknown gas.

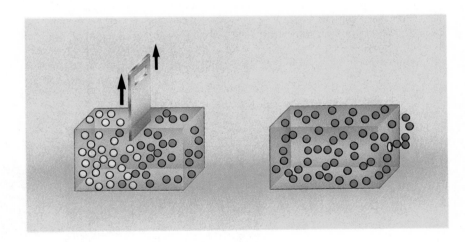

Figure 12-8 Molecules of two gases diffuse into each other after the barrier has been lifted in the box at the left. Gas molecules exit the box on the right by means of effusion.

Graham's Law of Effusion or Diffusion

The rates of effusion and diffusion are governed by the velocities of gas molecules. The velocity of a gas varies inversely with its mass. Lighter molecules move faster than heavier molecules at the same temperature. An experiment demonstrating velocities of the different gases is shown in Figure 12-9.

Recall that the average kinetic energy of the molecules in any gas is determined only by the temperature and is given by $\frac{1}{2}mv^2$. For two different gases, A and B, at the same temperature

Kinetic energy and temperature are discussed in Section 11.1.

See Section 11.1 for a discussion of molecular motion.

Desktop Investigation

Diffusion

Question
How rapidly do gases of different molecular mass diffuse?

Materials
household ammonia
perfume or cologne
two small, wide-mouthed
   plastic containers with lids
   (such as plastic food-
   storage containers)
clock or watch with second
   hand

Procedure
Record all of your results in a data table.

1. Outdoors or in a room separate from the one in which you will carry out the rest of the investigation, pour a small amount (roughly a tablespoon) of ammonia water into a plastic container. Place the lid onto the container. Pour roughly the same amount of perfume or cologne into a second container. Cover it.

2. Take the two samples you have prepared into a large, draft-free large room. Place the two about 12 to 15 feet apart and at the same

height. Remove the lids at as close to the same time as you can, and immediately position yourself midway between the open containers as shown above.

3. Note whether you smell the ammonia or the perfume first. Record how long this takes. Also, record how long it takes the vapor of the other substance to reach you. Air the room after you have finished.

Discussion
1. What do the times that the two vapors took to reach you show about their diffusion rates?

2. The molecular mass of ammonia is about one-tenth the molecular mass of most of the fragrant substances in the perfume or cologne. What does this tell you about the qualitative relationship between molecular mass and diffusion rate?

$$\frac{1}{2} m_A v_A^2 = \frac{1}{2} m_B v_B^2$$

where m_A and m_B represent the molecular masses of gases A and B and v_A and v_B represent their molecular velocities. Multiplying the equation by 2 gives

$$m_A v_A^2 = m_B v_B^2$$

To compare the velocities of the two gases, the equation is first rearranged to give the velocities as a ratio:

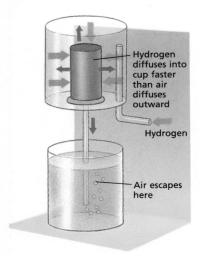

Figure 12-9 The rate of diffusion of hydrogen is greater than that of an average air molecule.

Labels on figure:
- Hydrogen diffuses into cup faster than air diffuses outward
- Hydrogen
- Air escapes here

$$\frac{v_A{}^2}{v_B{}^2} = \frac{m_B}{m_A}$$

The square root of each side of the equation is then taken to give

$$\frac{v_A}{v_B} = \frac{\sqrt{m_B}}{\sqrt{m_A}}$$

This equation shows that the molecular velocities of two different gases are inversely proportional to the square roots of their molar masses.

Because the rates of effusion are directly proportional to molecular velocities, we can express the equation as

$$\frac{\text{rate of effusion of A}}{\text{rate of effusion of B}} = \frac{\sqrt{m_B}}{\sqrt{m_A}}$$

The Scottish chemist Thomas Graham (1805–1869) performed experiments in which he studied the effusion and diffusion of gases. The equation derived above is a mathematical statement of Graham's discoveries and describes the rates of both effusion and diffusion. **Graham's law of effusion** *(or diffusion) states that the rates of effusion (or diffusion) of gases at the same temperature and pressure are inversely proportional to their molar masses.*

Applications of Graham's Law

Graham's experiments dealt with the densities of gases. Because the density of a gas varies directly with its molar mass, it should be clear to you that the square roots of the molar masses in the equation above can be replaced by the square roots of the gas densities. For the rates of effusion of two gases, A and B, the mathematical relationships are

$$\frac{\text{rate of effusion}_A}{\text{rate of effusion}_B} = \frac{\sqrt{m_B}}{\sqrt{m_A}} = \frac{\sqrt{\text{density}_B}}{\sqrt{\text{density}_A}}$$

The mathematical expression for Graham's law can be used in a number of variations, as shown in Table 12-2. In every case, the ratio of some property

---

TABLE 12-2 GRAHAM'S LAW EQUATION

$$\frac{\text{rate of effusion of A}}{\text{rate of effusion of B}} = \frac{\sqrt{\text{molar } m_B}}{\sqrt{\text{molar } m_A}} \qquad \frac{\text{distance traveled by A}}{\text{distance traveled by B}} = \frac{\sqrt{\text{molar } m_B}}{\sqrt{\text{molar } m_A}}$$

$$\frac{\text{molecular velocity of A}}{\text{molecular velocity of B}} = \frac{\sqrt{\text{molar } m_B}}{\sqrt{\text{molar } m_A}} \qquad \frac{\text{rate of effusion of A}}{\text{rate of effusion of B}} = \frac{\sqrt{\text{density}_B}}{\sqrt{\text{density}_A}}$$

$$\frac{\text{effusion time of A}}{\text{effusion time of B}} = \frac{\sqrt{\text{molar } m_A}}{\sqrt{\text{molar } m_B}}$$

Technology

Separating Isotopes

As you know from Graham's law, the rate of effusion of a gas varies inversely with the square root of the molecular mass. The English physicist Francis William Aston (1877–1945) was the first to use this fact to separate isotopes, or atoms of the same atomic number but different atomic mass. Aston placed naturally occurring neon gas, a mixture of mostly neon-20 and neon-22, into porous clay pipes. Atoms of the lower-mass isotope, neon-20, tended to effuse more readily and rapidly through the pores than did atoms of the higher-mass neon-22. The effused gas, which was richer in neon-20, was separated from the gas that remained in the tube. By repeating the process, increasingly pure samples of neon-20 and neon-22 were produced.

Later, this method was improved upon by use of a "cascade" arrangement. In this method, shown in the figure above, the gas is passed into a series of separate porous containers. The portions of the gas that are effused are passed on to containers farther along and uneffused portions are returned to containers farther back. The process is repeated over and over, and isotopes

are progressively separated and concentrated.

This method is used to collect the isotope uranium-235, which is useful in nuclear-energy production. The U-235 makes up only about 0.7% of all uranium, the rest being mostly U-238. The uranium that is to undergo separation occurs in the form of a sample of uranium hexafluoride gas (UF_6), which contains both isotopes. The $^{235}UF_6$, being less massive than the $^{238}UF_6$, effuses more readily. After more than a thousand repetitions of the basic effusion step, the lighter isotope can be separated out with a high degree of purity.

The effusion process separates isotopes of many different elements. In most cases, isotopes of different radioactivity are separated for use in nuclear energy, health sciences, geology, and analytical chemistry.

(for example, rate of effusion, molecular velocity, and so on) is equal to the square root of a ratio of the molar masses or densities.

The experiment illustrated in Figure 12-10 demonstrates the relationship between distance traveled and molecular mass. When ammonia (NH_3) molecules and hydrogen chloride (HCl) molecules diffuse toward each other from opposite ends of a glass tube as shown, a white ring will form at the point where the two gases meet. The white ring is solid ammonium chloride (NH_4Cl) and it forms at the end of the tube closest to the HCl. The lighter NH_3 molecules (molecular mass 17.03 u) diffuse faster than the heavier HCl molecules (molecular mass 36.46 u), as shown by the longer distance traveled by the NH_3 molecules before the two gases meet.

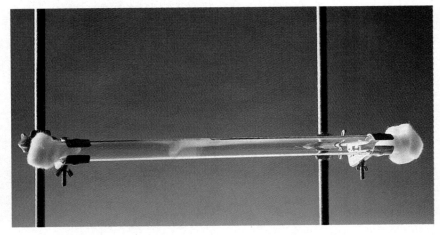

Figure 12-10 Plugs of absorbant cotton moistened with ammonia (NH) and hydrogen chloride (HCl) have been placed at opposite ends of the glass tube several minutes earlier. Why does the white ring of ammonium chloride form at the end closest to the HCl?

The principle expressed by Graham's law and its variations has a number of applications. For example, differences in the rates of effusion of gases are used to separate light isotopes from heavy isotopes (as discussed in Technology, above). Graham's law also provides a method for determining molar masses. The rate of effusion of a gas of known molar mass can be used compared to the effusion rate of an unknown gas at the same temperature and pressure. The unknown molar mass can then be calculated using Graham's law.

Sample Problem 12.14

Compare the rates of effusion of hydrogen and oxygen at the same temperature and pressure.

Solution

Step 1. Analyze   Given: identities of two gases, H_2 and O_2
Unknown: relative rates of effusion

Step 2. Plan   molar mass ratio → rate of effusion ratio

The ratio of the rates of effusion of two gases at the same temperature and pressure can be found from Graham's law.

$$\frac{\text{rate of effusion}_A}{\text{rate of effusion}_B} = \frac{\sqrt{m_B}}{\sqrt{m_A}}$$

Step 3. Compute	$\dfrac{\text{rate of effusion of } H_2}{\text{rate of effusion of } O_2} = \dfrac{\sqrt{m_{O_2}}}{\sqrt{m_{H_2}}} = \dfrac{\sqrt{32}}{\sqrt{2}} = \dfrac{5.7}{1.4} = 4.1$
	Hydrogen diffuses 4.1 times faster than oxygen.
Step 4. Evaluate	The result is correctly limited to two significant figures and is reasonably close to an estimated value of 4.3 (calculate $\sqrt{36}/\sqrt{2}$).

Practice Problem

A sample of hydrogen effuses through a porous container 8.94 times faster than an unknown gas. What is the molar mass of the unknown gas? *(Ans.)* 161 g/mol

Section Review

1. Compare diffusion and effusion.
2. State Graham's law of effusion, and explain its meaning.
3. Calculate the molar mass of a gas that effuses at 1.59 times the effusion rate of carbon dioxide.

Chapter Summary

- Gay-Lussac's law of combining volumes of gases states that the volumes of reacting gases and their products can be expressed as ratios of small whole numbers when they are measured at the same temperature and pressure.
- Avogadro's principle states that equal volumes of gases at the same temperature and pressure contain equal numbers of molecules. At constant temperature and pressure, the volume of a gas is thus directly proportional to the number of moles of the gas.
- Gay-Lussac's law and Avogadro's principle can be used to show that molecules of the active elemental gases are diatomic.
- The volume occupied by one mole of any gas at STP is called the standard molar volume. The standard molar volume of all gases is experimentally determined to be 22.4 L at standard temperature and pressure.
- Charles' law, Boyle's law, and Avogadro's principle can be combined to create the ideal gas law, a useful equation that can be used to describe a gas sample under any combination of conditions.
- The ideal gas law is stated mathematically as $PV = nRT$. It can be used to calculate the pressure, volume, temperature, or number of moles of a gas sample when three of these four variables are known.
- The combined gas law, the standard molar volume, and the ideal gas law can be used to solve stoichiometry calculations involving gases. These calculations may be those in which the known and unknown quantities are both volumes and those in which the known and unknown quantities are volume and mass or molar amount.
- Graham's law of diffusion (or effusion) states that the rates of diffusion (or effusion) of gases at the same temperature and pressure are inversely proportional to their molar masses. Graham's law reflects the fact that less massive molecules diffuse or effuse faster than do more massive ones.

1. Effusion is the process by which gases confined in a container pass through a very small opening in the container. Diffusion is the process by which two gases mix together in a container.
2. Graham's law of effusion (or diffusion) states that the rates of effusion (or diffusion) of gases at the same temperature and pressure are inversely proportional to their molar masses.
3. $M = 17.4$ g/mole

Chapter 12 *Review*

Molecular Composition of Gases

Vocabulary

Avogadro's principle
diffusion
effusion
Gay-Lussac's law of
    combining volumes
    of gases

Graham's law of effusion
    (or diffusion)
ideal gas constant
ideal gas law
standard molar volume
    of a gas

Questions

1. *(a)* State Gay-Lussac's law of combining volumes of gases. *(b)* What restrictions are associated with the use of this law?
2. *(a)* State Avogadro's principle. *(b)* At the same temperature and pressure, what is the relationship between the volume of a gas and the number of molecules present?
3. According to Avogadro, *(a)* what is the relationship between gas volume and the number of moles of gas at constant temperature and pressure? *(b)* What is the mathematical expression denoting this relationship?
4. In a chemical reaction involving gases, what information is given by the coefficients in the equation?
5. *(a)* How many molecules are contained in one mole of any substance? *(b)* What is the molar volume of a gas? *(c)* What is the relationship between the number of molecules and the mass of 22.4 L of different gases at STP?
6. When stating gas density values, why is it necessary that the corresponding temperature and pressure be specified?
7. Briefly summarize the gas behavior described in each of the following: *(a)* Boyle's law *(b)* Charles' law *(c)* Avogadro's principle.
8. *(a)* Write the equation for the ideal gas law. *(b)* What relationship is expressed in the ideal gas law?
9. *(a)* In what situation does the ideal gas law apply? *(b)* What precautions involving units must be taken in using this law?
10. *(a)* In a balanced chemical equation, what is the relationship between the molar ratios and the volume ratios of gaseous reactants and products? *(b)* What restriction applies to the use of the volume ratios in solving stoichiometry problems?
11. *(a)* Distinguish between diffusion and effusion. *(b)* At a given temperature what factor determines the rates at which different molecules undergo these processes?
12. *(a)* State Graham's law of effusion or diffusion. *(b)* Write the mathematical statement of this law.
13. Write the composite expression relating the rates of effusion, molar masses, and densities of two different gases, A and B.
14. What are some of the applications of the principle expressed in Graham's law?

Problems

1. If a 5.00-L sample of O_2 at a given temperature and pressure contains 1.08×10^{23} molecules of O_2, how many molecules would be contained in each of the following at the same temperature and pressure? *(a)* 5.00 L H_2 *(b)* 5.00 L CO_2 *(c)* 10.00 L NH_3
2. How many molecules are contained in each of the following? *(a)* 1.00 mol O_2 *(b)* 2.50 mol He *(c)* 0.0650 mol NH_3 *(d)* 11.5 g NO_2
3. Find the mass of each of the following: *(a)* 2.25 mol Cl_2 *(b)* 3.01×10^{23} molecules H_2S *(c)* 25.0 molecules SO_2
4. What is the volume, in liters, of each of the following at STP? *(a)* 1.00 mol O_2 *(b)* 3.50 mol F_2 *(c)* 0.0400 mol CO_2 *(d)* 1.20×10^{-6} mol He
5. How many moles are contained in each of the following at STP? *(a)* 22.4 L N_2 *(b)* 5.60 L Cl_2 *(c)* 0.125 L Ne *(d)* 70.0 mL NH_3
6. Find the mass, in grams, of each of the following at STP: *(a)* 11.2 L H_2 *(b)* 2.80 L CO_2 *(c)* 15.0 mL SO_2 *(d)* 3.40 cm³ F_2
7. Find the volume, in liters, of each of the following at STP: *(a)* 8.00 g O_2 *(b)* 3.50 g CO *(c)* 0.0170 g H_2S *(d)* 2.25×10^5 kg NH_3

8. Find the density of each of the following gases in grams per liter at STP. *(a)* N_2 *(b)* Cl_2 *(c)* CO *(d)* SO_2 *(e)* NH_3

9. Based on the masses and corresponding volumes of each of the gases listed below that were collected at STP, find their molar masses. *(a)* 2.00 g and 1.50 L *(b)* 3.25 g and 2.30 L *(c)* 0.620 g and 250. mL

10. A sample of a gas collected at 300K and 740. mm Hg pressure has a mass of 0.805 g and occupies a volume of 250 mL. *(a)* What volume would this gas occupy at STP? *(b)* What is its molar mass?

11. Find the molar mass of a gas if 140. mL of the gas collected at 77°C and 525 mm Hg pressure has a mass of 0.0650 g.

12. Calculate the pressure, in atmospheres, exerted by each of the following:
(a) 2.50 L of HF containing 1.35 mol at 320K
(b) 4.75 L of NO_2 containing 0.86 mol at 300K
(c) 750. mL of CO_2 containing 2.15 mol at 57°C

13. Calculate the volume, in liters, occupied by each of the following:
(a) 2.00 mol of H_2 at 300K and 1.25 atm
(b) 0.425 mol of NH_3 at 37°C and 550. mm Hg
(c) 4.00 g of O_2 at 57°C and 675 mm Hg

14. Determine the number of moles of gas contained in each of the following. *(a)* 1.25 L at 250K and 1.06 atm *(b)* 0.80 L at 27°C and 0.925 atm *(c)* 750. mL at $-50°C$ and 700. mm Hg

15. Find the mass of each of the following.
(a) 5.60 L of O_2 at 1.75 atm and 250K
(b) 3.50 L of NH_3 at 700. mm Hg and 27°C
(c) 125 mL of SO_2 at 625 mm Hg and $-53°C$

16. A 1.50 L sample of an unknown gas collected at 320K and 1.08 atm pressure is found to have a mass of 3.50 g. *(a)* How many moles of the gas are contained in this sample? *(b)* What is the molar mass of the gas?

17. Find the molar mass of each gas measured at the specified conditions: *(a)* 0.650 g occupies 1.12 L at 280K and 1.14 atm *(b)* 1.05 g occupies 2.35 L at 37°C and 0.840 atm *(c)* 0.432 g occupies 750. mL at $-23°C$ and 785 mm Hg

18. Balance the chemical equation $H_2(g) + Cl_2(g) \rightarrow HCl(g)$ and then complete each of the following interpretations. *(a)* 1 molecule $H_2(g)$ + ___ molecule(s) $Cl_2(g) \rightarrow$___

molecule(s) HCl(g) *(b)* 1 mole $H_2(g)$ + ___ mole(s) $Cl_2(g) \rightarrow$ ___ mole(s) HCl(g) *(c)* 1 volume $H_2(g)$ + ___ volume(s) $Cl_2(g) \rightarrow$ ___ volume(s) HCl(g)

19. Carbon monoxide reacts with oxygen to produce carbon dioxide. If 1.0 L of carbon monoxide reacts with oxygen, *(a)* how many liters of oxygen are required? *(b)* how many liters of carbon dioxide are produced?

20. Acetylene gas (C_2H_2) undergoes combustion to produce carbon dioxide and water vapor. *(a)* How many liters of C_2H_2 are required to produce 75.0 L of CO_2? *(b)* What volume of H_2O is produced? *(c)* What volume of O_2 is required?

21. If liquid carbon disulfide reacts with 450. mL of oxygen to produce the gases carbon dioxide and sulfur dioxide, what volume of each product is produced?

22. Balance the chemical equation $Mg(s) + O_2(g) \rightarrow MgO(s)$ and then, based on the quantity of reactant or product given, determine the corresponding quantities of the specified reactants or products, assuming that the system is at STP: *(a)* 22.4 L O_2 = ___ mol $O_2 \rightarrow$ mol MgO *(b)* 11.2 L O_2 = ___ mol $O_2 \rightarrow$ mol MgO *(c)* 1.40 L O_2 = ___ mol $O_2 \rightarrow$ mol MgO = ___ g MgO *(d)* 4.75 L O_2 = ___ mol $O_2 \rightarrow$ mol MgO = ___ g MgO *(e)* 725 mL O_2 = ___ mol $O_2 \rightarrow$ mol MgO = ___ g MgO

23. Balance the chemical equation $Fe_2O_3(s) + CO(g) \rightarrow Fe(s) + CO_2(g)$ and then, based on the quantity of reactant or product given, determine the corresponding quantities of the specified reactants or products, assuming that the system is at 1.05 atm and 57°C. *(a)* ___mol $Fe_2O_3 \rightarrow$ ___ mol CO_2 = 15.0 L CO_2 *(b)* ___ g Fe_2O_3 = ___ mol $Fe_2O_3 \rightarrow$ ___ mol CO_2 = 420. mL CO_2 *(c)* ___ g Fe_2O_3 = ___ mol $Fe_2O_3 \rightarrow$ ___ mol CO_2 = 775 mL CO_2

24. Assume that 5.60 L of $H_2(g)$ at STP react with CuO(s) according to the following reaction: $CuO(s) + H_2(g) \rightarrow Cu(s) + H_2O(g)$ *(a)* How many moles of $H_2(g)$ react? *(b)* How many moles of Cu(s) are produced? *(c)* How many grams of Cu(s) are produced?

25. Assume that 8.50 L of $I_2(g)$ are produced according to the following reaction that takes place at STP: $2KI(aq) + Cl_2(g) \rightarrow 2KCl(aq) + I_2(g)$ *(a)* How many moles of $I_2(g)$ are

produced? *(b)* How many moles of KI(aq) were used? *(c)* How many grams of KI(aq) were used?

26. Solid iron(I) hydroxide(s) decompose(s) to produce iron(I) oxide and water vapor. If 75.0 L of water vapor are produced at STP, *(a)* how many grams of iron(I) hydroxide were used? *(b)* how many grams of iron(I) oxide are produced?

27. If 650. mL of hydrogen gas is produced through a replacement reaction involving solid iron and sulfuric acid (H_2SO_4) at STP, how many grams of iron(ll) sulfate are also produced?

28. Balance the chemical equation $H_2O(l) \rightarrow H_2(g) + O_2(g)$ and then, based on the quantity of reactant or product given, determine the corresponding quantities of the requested reactants or products at STP. *(a)* 18.0 g H_2O = __ mol $H_2O \rightarrow$ __ mol H_2 *(b)* 4.50 g H_2O = __ mol $H_2O \rightarrow$ __ mol O_2 *(c)* 6.00 g H_2O = __ mol $H_2O \rightarrow$ __ mol H_2 = __ L H_2 (separated and at STP)

29. Balance the chemical equation $MnO_2(s) + HCl(aq) \rightarrow MnCl_2(g) + H_2O(\ell)$ and then, based on the quantity of reactant or product given, determine the corresponding quantities of the requested reactants or products if the system is at 780. mm Hg and 77°C. *(a)* 36.5 g HCl = __ mol HCl $\rightarrow$ __ mol Cl_2 *(b)* 9.12 g HCl = __ mol HCl __ mol Cl_2 = __ L Cl_2 *(c)* 52.0 g HCl = __ mol HCl __ mol Cl_2 = __ mL Cl_2 *(d)* __ g HCl = __ mol HCl $\rightarrow$ __ mol Cl_2 = 12.5 L Cl_2 *(e)* __ g HCl = __ mol HCl $\rightarrow$ __ mol Cl_2 = 750. mL Cl_1

30. Assume that if 13.5 g of Al(s) react with HCl(aq) according to the following balanced equation at STP:
$2Al(s) + 6HCl(aq) \rightarrow 2AlCl_3(aq) + 3H_2(g)$
(a) How many moles of Al react? *(b)* How many moles of H_2 are produced? *(c)* How many liters of H_2 at STP are produced?

31. Sodium sulfite (aq) reacts with hydrochloric acid to produce aqueous sodium chloride, liquid water, and sulfur dioxide gas. If 21.0 g of sodium sulfite are consumed at STP, *(a)* how many moles of sulfur dioxide are produced? *(b)* how many liters of sulfur dioxide are produced?

32. Quantitatively compare the rates of effusion of the following pairs of gases at the same temperature and pressure: *(a)* hydrogen and nitrogen *(b)* fluorine and chlorine.

Application Questions

1. Explain the basis upon which Boyle's law, Charles' law, and Avogadro's principle can be combined to give the ideal gas law.

2. Explain the specific kind of problem situation in which each of the following would apply: *(a)* the combined gas law *(b)* the ideal gas law.

Application Problems

1. Find the mass of 4.50 m³ of H_2S at STP.

2. What volume, in cubic centimeters and cubic meters, would be occupied by 12.0 g of O_2 at STP?

3. Find the density of NO_2 in grams per milliliter at STP.

4. An unknown gas has a mass of 4.75 g and occupies a volume of 2250 cm³ at STP. *(a)* Find its density, in grams per liter, at STP. *(b)* Find its molar mass.

5. A 6.51 m³ container is filled with an unknown gas at 50°C and 840. mm pressure. The gas has a mass of 12.5 kg. *(a)* What is the molar mass of the gas at STP? *(b)* Among the gases listed below, which is likeliest to be the unknown gas—H_2, N_2, O_2, NH_3, SO_2, NO_2 CO_2, or H_2S?

6. Find the molar mass of a gas at STP if 850. mL of the gas collected over water at 650.0 mm Hg and 20°C has a mass of 0.925 grams.

7. A 500. cm₃ flask contains 0.750 g of a gas collected over water at 35°C and 575.0 mm Hg. What is the molar mass of this gas?

8. What pressure, in atmospheres, would be exerted by 9.25 g of SO_2 at 40°C if it is contained in a flask having a volume of 500. cm³?

9. How many moles of gas will be contained in a 1500−mL flask at −40°C and a pressure of 92.5 kPa?

10. At what temperature, in degrees Celsius, would 4.60 g of CO_2 occupy 275 mL at 625 mm Hg pressure?

11. Use the derived molar mass equation $\left(\text{mol } m = \dfrac{mRT}{PV} \right)$ to determine the molar mass of an unknown gas if 1.06×10^{-3} g occupies 14.5 mL at −43°C and 240. mm Hg pressure.

12. Find the molar mass of a gas sample collected

over water at 20°C if 0.750 g occupies 155 mL at 420.0 mm Hg pressure.

13. If the density of an unknown gas is 3.20 g/L at −18°C and 1650 mm Hg, what is the molar mass of this gas?

14. One method for estimating the temperature of the center of the sun is based on the assumption that the center consists of gases that have an average molar mass of 2.00 g/mol. If the density of the center is 1.40 g/cm^3 at a pressure of 1.30×10^9 atm, calculate its temperature, in degrees Celsius.

15. If 350 mL of nitric oxide reacts with 225 mL of oxygen, what volume of nitrogen dioxide is produced?

16. If air is 20.9% oxygen by volume, (a) How many liters of air are needed to complete the combustion of 25.0 L of octane vapor (C_8H_{18}) (b) what volume of each product is produced?

17. Methanol (CH_3OH) is made by causing carbon monoxide and hydrogen gases to react at high temperature and pressure. If 450. mL of CO and 825. mL of H_2 are mixed, (a) which reactant is present in excess? (b) how much of that reacting remains? (c) what volume of CH_3OH is produced?

18. Balance the equation $SO_2(g) + H_2O(\ell) \rightarrow H_2SO_3(aq)$. If 14.0 L of $SO_2(g)$ at 725 mm Hg and 37°C react, (a) how many grams of $H_2O(\ell)$ are required? (b) how many grams of $H_2SO_3(aq)$ are produced?

19. If 45.0 L of natural gas, which is essentially methane (CH_4), undergoes complete combustion at 730 mm Hg and 20°C, how many grams of each product are formed?

20. How many grams of aluminum metal must react with aqueous sodium hydroxide to produce 7.50 L of hydrogen collected over water at 12.5°C and 740.0 mm Hg if the other product is aqueous Na_3AlO_3?

21. A modified Haber process for making ammonia is conducted at 550°C and 250 atm. If 10.0 kg of nitrogen is used, what volume of ammonia is produced?

22. When nitroglycerin, $C_3H_5(NO_3)_3$, explodes, the products are carbon dioxide, nitrogen, oxygen, and water vapor. If 500 g of nitroglycerin explodes at STP, what is the total volume, at STP, for all gases produced?

23. The principal source of sulfur is in the form of deposits of free sulfur occurring mainly in volcanically active regions. The sulfur was initially formed by the reaction between the two volcanic vapors SO_2 and H_2S to form $H_2O(\ell)$ and S(s). What volume of each gas, at 730. mm Hg and 22°C, was needed to form a sulfur deposit of 4.50×10^5 kg on the slopes of a volcano in Hawaii?

24. A 3.25-g sample of solid calcium carbide (CaC_2) reacts with water to produce acetylene gas (C_2H_2) and aqueous calcium hydroxide. If the acetylene was collected over water at 17°C and 740.0 mm Hg, how many milliliters of acetylene were produced?

25. If 15.0 g of potassium hydroxide reacts with 12.5 g of ammonium sulfate to produce potassium sulfate, ammonia, and liquid water, what volume of ammonia is produced at STP?

26. If 5.00 g of Zn reacts with 8.00 g of H_2SO_4 by single replacement, what volume of hydrogen (collected over water at 22°C and 730.0 mm Hg) is produced?

27. What is the ratio of the velocities of hydrogen molecules and neon atoms at the same temperature and pressure?

28. At a certain temperature and pressure, chlorine molecules have a velocity of 0.0380 m/sec. What is the velocity of sulfur dioxide molecules under the same conditions?

29. A sample of helium effuses through a porous container 6.50 times faster than does unknown gas X. What is the molar mass of the unknown gas?

Enrichment

1. Explain the processes involved in the liquefaction of gases. What substances that are gases under normal room conditions are typically used in the liquid form? Why?

2. Research the relationship between explosives and the establishment of Nobel Prizes. Prepare a report that describes your findings.

3. Prepare a report on the production and varied uses of acetylene gas.

4. What were the major contributions of Thomas Graham to the study of gases?

5. Over a typical day, record every instance in which you encounter the diffusion or effusion of gases (example: smelling perfume as your mother walks by).

Chapter 13 Liquids and Solids

Chapter Planner

13.1 Liquids
Kinetic Theory Description of the Liquid State
- Question: 2
- Question: 1

Properties of Liquids and the Particle Model
- Questions: 3, 4
  Application Questions: 1, 2
- Desktop Investigation
  Questions: 5–9
- Technology

13.2 Solids
Kinetic Theory Description of the Solid State

Properties of Solids and the Particle Model
- Questions: 10–12
- Question: 13

Crystalline Solids
- Questions: 14, 16
  Application Question: 3
- Question: 17
- Technology
  Question: 15

13.3 Changes of State
Equilibrium
- Questions: 18–21
- Application Question: 6
- Application Questions: 4, 5

Equilibrium Vapor Pressure of a Liquid
- Question: 22
  Application Question: 7
- Question: 23

Boiling
- Demonstration: 1
  Questions: 24, 25
  Application Question: 8
- Application Question: 9

Freezing and Melting
- Demonstration: 2
  Experiment: 16
  Questions: 26–29
  Application Question: 10
- Application Question: 11

Phase Diagrams
- Question: 30
  Application Questions: 12, 13
- Question: 31
  Application Questions: 14–16

13.4 Water
Structure of Water
- Question: 32
  Application Questions: 17, 18

Physical Properties of Water
- Experiment: 17
  Question: 33
  Problems: 1, 2
- Application Questions: 19, 20
- Application Problems: 1a, 2a
- Problems: 3–6
  Application Problems: 1b, 2b, 3

Teaching Strategies

Introduction

After studying the kinetic theory as it applies to gases in Chapters 10–12, this chapter moves on to the liquid and the solid states. Liquids and solids are discussed first, followed by the energy changes that accompany changes between states. Then the most common and most important liquid, water, is examined in detail.

13.1 Liquids

Begin by reviewing the kinetic theory of gases. Summarize those aspects that change for liquids and solids. These latter two phases have molecules so close together that they are incompressible. Kinetic energy is restricted, and intermolecular forces become significant. These are the changes that give liquids and solids their properties.

You can illustrate the incompressibility of liquids by drawing 25 mL of water into a syringe and squeezing the plunger. (Make sure there is no air present, and do not squeeze too hard.)

Explain that automobile brakes function because of the incompressibility of the brake fluid. When air is trapped in the line, the brake pedal goes to the floor and no pressure is applied to the brakes. One has to pump the brakes, reducing the volume of air in the line, before the brakes begin to hold.

To show evaporation and boiling (and to provide some data for a later discussion of phase changes) half fill a 250-mL beaker with warm water. Place a cold watch glass on the beaker. Some of the water vapor condenses. Remind students that evaporation is an endothermic process.

Draw some of the warm water into the 25-mL syringe. Make sure no air is present. Pull back on the syringe and the water boils. Do not go into detail at this time. Repeat the demonstration and have the students note that reducing the pressure by pulling back on the syringe causes the water to boil. You may have them touch the syringe to note that the water is not at 100°C, and yet it boiled.

Remind the students that all of these macroscopic properties are the result of interactions of molecules, which in turn can be traced back to the number and arrangement of valence electrons.

13.2   Solids

As you introduce solids, note that intermolecular forces are so strong that only vibrational kinetic energy is evident. Quickly move through each of the properties discussed in the text and contrast them with the same properties in the liquid state. Some students may have a difficult time distinguishing between crystalline and amorphous solids and supercooled liquids and solids. Have examples of each on the desk. Copper sulfate and alum crystals, as well as minerals such as galena, may be used as examples of crystalline solids. A plastic bottle is a good example of an amorphous solid.

Crystal models and styrofoam spheres are excellent illustrations of crystalline solids and the concept of unit cells. Instead of asking students to memorize specific crystal structures, stress the idea that the forces of attraction determine the type of crystal. The unit cell concept may be demonstrated with eight styrofoam balls in a simple cubic packing. With a sharp knife, cut off the parts that belong to a single cell.

Have examples of each crystal type available on the desk. If mineral samples of quartz, mica, and asbestos are available, you can explain the importance of three-dimensional, planar, and linear bonding in each. A candle may be used to illustrate the covalent molecular crystal.

Use models of nonpolar molecules (methane, hydrogen, iodine) and polar molecules (HCl, NH_3, H_2O) to explain the difference in properties. This is a good time to review intermolecular forces of attraction and remind students of the importance of hydrogen bonding.

13.3   Changes of State

To emphasize the equilibrium concept, use the analogy of a sports team. If basketball is used, one notes that during the course of a game there are always 10 players on the court and another fixed number of players on the bench. As the game progresses, players enter the court from the bench while at the same time an equal number of players leave the court and return to the bench. During the course of the game, the actual players on the court may not be the same ones who began the game, but the number on the court has not varied.

Football, soccer, and baseball also work. If students are familiar with ice hockey, players substitute on the fly with no break in the action. The sports team analogy emphasizes the idea that a system in equilibrium must be a closed system, i.e. the total number of particles must be constant.

Equilibrium as it applies to changes of state should be presented with the students giving much of the information while the teacher catalogs and makes inferences. An uncovered beaker half-filled with water serves as a good starting point. Students will observe that with time the level of the water drops. It evaporates. Particles of water leave the liquid phase and enter the gaseous phase. This can be diagramed on an overhead projector or chalkboard. If a cover is placed on the beaker, the particles that enter the gaseous phase remain in the beaker, colliding with the walls, and eventually return to the liquid phase. At some point, the number of particles leaving the surface and entering the gaseous state equals the number striking the surface and returning to the liquid state. Equilibrium has been established, but the process continues on the molecular level with no apparent external macroscopic change. Pose this question: "What would happen if the water were heated?" Most students would agree that more particles would evaporate. Relate this to the kinetic theory. Temperature is directly related to the kinetic energy of the particles. If the volume of the container were reduced, as in a syringe, students would agree that with less space some of the particles in the gaseous state would condense. Diagram this condition on the board.

Liquid and solid deodorizers may be used to reinforce the concept of vapor pressure. Kinetic theory explains what is happening on a molecular level. Even the solid **deodorizer** must have molecules escaping from the surface, since an odor can be detected. Use the graphs in the text to show the relationship between the observed physical phenomena and the graphical interpretation of the phenomena.

The following demonstration may help to bring all of these ideas together.

Demonstration 1: Phase Change

PURPOSE: To allow the students to examine a phase change from liquid to vapor

MATERIALS: Zip-lock sandwich bag, can of dry spray room deodorizer

PROCEDURE: Open the zip-lock bag. Place the nozzle of the spray can inside the bag. Closing the mouth of the bag around the can, hold down the nozzle for approximately three to five seconds and spray the deodorizer into the bag. The bag gets very cold. Seal the bag. You should have captured several milliliters of a very cold liquid in the bag.

Holding the corner of the bag that contains the liquid causes the liquid to boil vigorously. Several bags may be prepared and passed around the room.

As the liquid boils, the volume of the zip-lock bag increases. Pressure may build until the bag begins to leak. Your students should be able to observe the large difference between the volume of a particular mass of liquid and the volume of the same mass of vapor.

Condensation of water vapor from the air may be demonstrated by allowing the liquid in the bag to flow from one corner to the opposite corner. One should be able to observe a trail of ice on the outside of the bag that follows the path taken by the liquid butane.

Any dry spray deodorizer will work. With the removal of halogenated hydrocarbons (freons) as propellants because of their adverse effect on the ozone layer, manufacturers of aerosol sprays had to find another propellant. Most manufacturers are now using butane. Many of the labels do not list butane as an ingredient, but they do indicate that the product is flammable. Butane boils at 0.5°C. At room pressure, butane is a gas. It can be liquefied at room temperature by exerting a greater than atmospheric pressure on the gas. Once liquefied it will remain a liquid until the pressure is lowered: in this case, by pressing the nozzle and allowing some of the gas to escape. Since evaporation is an endothermic process, the container and the target toward which the spray is directed get very cold. When sprayed into the zip-lock bag, the temperature drops below −5°C, and the butane liquefies at room pressure.

The bag and liquid also may be used to qualitatively show the effect of the external pressure on the boiling point. As the bag fills with gas, the liquid does not boil as readily. Opening the bag and allowing the gas to escape reduces the pressure and the boiling point of the liquid. You may wish to make the demonstration quantitative by collecting some of the liquid in a syringe. You may more easily demonstrate the effect of pressure on the boiling point.

The demonstration is a simple one, but it fascinates most students. Be sure to warn them of the danger of spraying the contents near a flame or near electrical equipment. Directing the spray toward exposed skin may cause the equivalent of frostbite.

Dispose of the deodorizer according to the instructions on the can. The zip-locks bags may simply be opened and the fragrance allowed to escape into the room.

After working with the bags or syringes, the concept of molar heat of vaporization may be introduced. Refer once again to the difference between intensive and extensive properties. By agreeing on one mole of material, one obtains an intensive property that can be tabulated and made available to other chemists.

Experiment 5 in *Lab Experiments in Modern Chemistry* discusses freezing and melting in terms of entropy. The students will not cover this concept until later in the course, but you may introduce the topic, using the following very simple demonstration.

Demonstration 2: Phase Changes in Ice and Water

PURPOSE: To introduce the concepts of heat of fusion and heat of vaporization and to observe the temperature behavior of a system undergoing phase changes

MATERIALS: A 250-mL beaker half-filled with an ice and water mixture; a thermometer; and a heat source

PROCEDURE: Stir the ice-water mixture with the thermometer, record the temperature and begin warming the beaker. The temperature remains constant even though heat is flowing into the beaker. Have the students explain what is occurring based on the kinetic theory. (The energy is being used to overcome the hydrogen bonds and other attractive forces present in the ice molecules.)

Once the ice has melted, the temperature begins to rise. Again have the students explain the temperature increase. (Now water molecules are beginning to show in increase in kinetic

energy—vibration, rotation, translation—manifested by the temperature increase.)

At the boiling point, the thermometer again seems to malfunction because it no longer records a temperature increase even though heat flows into the system. Once again have the students describe what might be happening on the molecular level. (The energy this time is used to overcome the attractive forces between liquid molecules and to give them enough energy to enter the gaseous state.)

The students need to be aware that heat is being transferred in both sets of phase changes although the temperature does not change. These energy changes on the molecular level are what the text calls heat of fusion and heat of vaporization.

If a thermistor connected to a computer is available, the same demonstration may be carried out with the change in temperature shown on the computer monitor. Students are able to see in real time that the temperature no longer changes at each of the phase changes.

Observe the usual safety precautions when warming a liquid.

The concept of phase diagram, triple point, and critical temperature and pressure for water may now be introduced. The difference between the boiling point and normal boiling point needs to be clarified. The syringe half-filled with warm water may be used once again to show that water will boil under reduced pressure. Glass "hand boilers" or "love meters," which may be found in novelty stores and catalogs, are also good illustrations of this phenomenon. The normal boiling point is the temperature at which a liquid boils when the pressure is exactly one atmosphere.

13.4   Water

The closing section on water gives you an opportunity to review the concepts presented in this chapter, as well as concepts involving electronic structures, bonding, electronegativity, polarity of bonds, and molecular shapes.

Use molecular models to explain some of the unique properties of water: less dense in solid state than in liquid, a very high heat of vaporization, and a very large temperature range in which it is a liquid.

Encourage students to bring in newspaper and magazine articles that focus on the importance of water in our daily lives.

Answers and Solutions

Questions

1. Liquids are least common because they can exist only within a relatively narrow range of temperatures and pressures.
2. Like gases, liquids can consist of ions, atoms, or molecules; the arrangement and behavior of those particles distinguish liquids from gases. Like gas particles, liquid particles are in constant motion, but they are closer together and lower in energy than gas particles. The attractive forces between liquid particles are more effective than those between gas particles. The movement of the particles in a liquid is more limited than that of gas particles. The entropy of liquids is generally less than that of gases, owing to the lower kinetic

energy, higher interactive forces, and lower mobility of the liquid particles.
3. *(1)* Liquids have a definite volume because of the relatively strong interactive forces between the particles. *(2)* Liquids exhibit fluidity, thus necessitating the use of containers to hold them. Liquids are fluid because the particles are not bound together in fixed positions but can move about constantly. *(3)* Liquids have a relatively high density because of the close packing of their particles. *(4)* Liquids are relatively incompressible because of the close packing of their particles. *(5)* Liquids have the ability to dissolve other substances because of the motion of the particles and the attractive forces that those particles may exert on the particles of the substance to be dissolved. *(6)* Liquids have the

ability to diffuse throughout any other substance that can dissolve them because their particles exhibit constant random motion. *(7)* Liquids exhibit surface tension because of the attractive forces between their particles. *(8)* Liquids have a tendency to evaporate and to boil. Liquids can evaporate when the surface particles acquire enough energy to overcome the interactive forces that bind them to the liquid, thus allowing them to escape into the gas phase. When all of the liquid particles can overcome those attractive forces, the liquid is able to boil. *(9)* Liquids have a tendency to solidify when they are sufficiently cooled so that the kinetic energy of the particles becomes low enough for the attractive forces to hold them together in an orderly arrangement.

4. A fluid is a substance that can flow and therefore conform to the outline of its container.

5. When one substance is added to another so that, after mixing, only one physical state is observed, that substance has been dissolved.

6. Surface tension is a force that tends to pull adjacent parts of a liquid's surface together, making the surface less penetrable by solid bodies.

7. Evaporation is the process by which particles escape from the surface of a nonboiling liquid and enter the gaseous state. Vaporization is any process by which a liquid or solid changes to a gas.

8. Water that falls to the earth in the form of rain and snow originally evaporated from the earth's oceans, lakes, and rivers. Human perspiration cools our bodies when evaporated into the surrounding air.

9. Freezing is the physical change of a liquid to a solid.

10. *(a)* The two types of solids are crystalline and amorphous solids. *(b)* Crystalline solids include salt crystals and snowflakes; amorphous solids include glass and plastics.

11. A crystal is a substance in which the particles are arranged in an orderly, geometric, repeating pattern.

12. *(1)* Solids can maintain a definite shape without a container. Crystalline solids are geometrically regular, whereas amorphous solids are not. *(2)* Solids have definite volumes that change only slightly with changes in temperature or pressure; this property is due to the close packing of the solid particles. *(3)* Solids are nonfluid because they are held in relatively fixed positions. *(4)* Whereas crystalline solids have definite melting points, some amorphous solids do not. Instead, as they are heated, amorphous solids such as glass and plastic gradually soften to become thick, sticky liquids. Solids melt at temperatures at which the kinetic energies of their particles can overcome the interactive forces between them. *(5)* Solids have high densities, due to the close packing of their particles. *(6)* Solids are incompressible, due to the already close packing of their particles. *(7)* Solids have a slow rate of diffusion, due to the low mobility of their particles.

13. A supercooled liquid is a substance that retains certain liquid properties even at temperatures at which it appears to be solid. Glass, since it softens to become a thick, sticky liquid upon being heated and because of its ability to flow, is sometimes classified as a supercooled liquid.

14. A crystal lattice is the total three-dimensional array of points that describes the arrangement of the particles of a crystal. A unit cell is the smallest portion of the crystal

lattice that reveals the three-dimensional pattern of the entire lattice. *(a)* Rubber, glass, plastics, synthetic fibers, amorphous forms of metal alloys *(b)* Answers will vary.

15. The seven basic crystalline systems by shape are isometric or cubic, tetragonal, hexagonal, trigonal, orthorhombic, monoclinic, and triclinic.

16. *(a)* The ionic crystal lattice consists of positive and negative ions arranged in a characteristic regular pattern. In covalent network crystals, the lattice sites contain single atoms, each of which is covalently bonded to its nearest neighboring atoms. The metallic crystal lattice consists of positive metal cations surrounded by valence electrons that are donated by the metal atoms and belong to the crystal as a whole. The attraction between the metal cations and the electrons binds the crystal together. The crystal lattice of covalent molecular crystals consists of covalently bonded molecules held together by van der Waals forces. If these molecules are nonpolar, the bonding is because of weak dispersion forces; however, if these molecules are polar, they are held together by both dispersion forces and stronger dipole-dipole forces. *(b)* Ionic crystals are hard and brittle, have high melting points, and are good insulators. Covalent network crystals are hard and brittle, have high melting points, and are nonconductors. Metallic crystals have high electric conductivity, but their melting points vary greatly. Covalent molecular crystals have low melting points, are relatively soft and easily vaporized, and are good insulators.

17. Metallic glasses are amorphous forms of metal alloys prepared when thin films of melted metal are cooled very rapidly so that the metal atoms do not have time to arrange themselves in a crystalline pattern.

18. *(a)* A phase is any part of a system that has uniform composition and properties. *(b)* A system is a sample of matter being studied.

19. *(a)* Equilibrium is a dynamic condition in which two opposing physical or chemical changes occur at equal rates in the same system. *(b)* Condensation is the process by which a gas changes to a liquid. *(c)* Concentration is the number of particles per unit volume.

20. *(a)* In evaporation, a liquid absorbs heat energy from its surroundings to become a vapor: liquid + heat energy → vapor. *(b)* In condensation, a vapor gives off heat energy to its surroundings to become a liquid: vapor → liquid + heat energy.

21. *(a)* Le Chatelier's principle states that if any of the factors determining an equilibrium are changed, the system will adjust itself in a way that tends to minimize that change. *(b)* Le Chatelier's principle enables one to predict how a change in the conditions of a system at equilibrium will effect the equilibrium.

22. *(a)* The equilibrium vapor pressure of a liquid is the pressure exerted by a vapor in equilibrium with its corresponding liquid at a given temperature. *(b)* The equilibrium vapor pressure of water increases indirectly with increasing temperature.

23. The indicated equilibrium vapor pressures are: *(a)* 50 mm Hg *(b)* 375 mm Hg *(c)* 460 mm Hg *(d)* 375 mm Hg *(e)* 90 mm Hg.

24. The boiling point of a liquid is the temperature at which its equilibrium vapor pressure is equal to atmospheric pressure.

The normal boiling point of a liquid is the temperature at which its equilibrium vapor pressure equals 760 mm Hg.

25. The molar heat of vaporization is the heat energy required to vaporize one mole of liquid at its boiling point.

26. *(a)* Freezing is the physical change of a liquid to a solid. *(b)* Freezing: liquid → solid + heat energy. *(c)* During freezing, entropy decreases.

27. *(a)* Melting is the physical change of a solid to a liquid. *(b)* solid + heat energy → liquid *(c)* During melting, entropy increases.

28. Molar heat of fusion is the heat energy required to melt one mole of a solid at its melting point.

29. *(a)* Sublimation is the change of state directly from a solid to a gas. *(b)* Dry ice and iodine are examples.

30. A phase diagram is a graph of temperature versus pressure that indicates the conditions under which various states of a particular substance exist.

31. The normal freezing point of a substance is the temperature at which the solid and liquid are in equilibrium at 1 atm pressure.

32. A water molecule consists of two atoms of hydrogen and one atom of oxygen united by polar covalent bonds. The angle between the two hydrogen–oxygen bonds is about $105°$.

33. At room temperature, pure water is a transparent, odorless, tasteless, and almost colorless liquid. It freezes and melts at $0°C$ at 1 atm; its molar heat of fusion is 6.008 kJ/mol; water expands 11% in volume as it freezes; it has its maximum density at $4°C$; at a pressure of 1 atm, it boils at $100°C$; its molar heat of vaporization is 40.58 kJ/mol.

Problems

1. *(a)* $5.00 \text{ mol } H_2O \times \dfrac{40.79 \text{ kJ}}{\text{mol } H_2O} = 204 \text{ kJ}$

   (b) $45.0 \text{ g } H_2O \times \dfrac{\text{mol } H_2O}{18.0 \text{ g } H_2O} \times \dfrac{40.79 \text{ kJ}}{\text{mol } H_2O} = 102 \text{ kJ}$

   (c) $8.45 \times 10^{50} \text{ molec } H_2O \times \dfrac{\text{mol } H_2O}{6.02 \times 10^{23} \text{ molec } H_2O} \times$ $\dfrac{40.79 \text{ kJ}}{\text{mol } H_2O} = 5.73 \times 10^{28} \text{ kJ}$

2. *(a)* $12.75 \text{ mol } H_2O \times \dfrac{6.008 \text{ kJ}}{\text{mol } H_2O} = 76.60 \text{ kJ}$

   (b) $6.48 \times 10^{20} \text{ kg } H_2O \times \dfrac{10^3 \text{ g } H_2O}{\text{kg } H_2O} \times \dfrac{\text{mol } H_2O}{18.0 \text{ g } H_2O} \times$ $\dfrac{6.008 \text{ kJ}}{\text{mol } H_2O} = 2.16 \times 10^{23} \text{ kJ}$

3. *(a)* $\dfrac{0.999\,13 \text{ g } H_2O}{\text{cm}^3 H_2O} \times \dfrac{\text{cm}^3 H_2O}{\text{mL } H_2O} \times 525.00 \text{ mL } H_2O$
   $= 524.54 \text{ g } H_2O$

   (b) $524.54 \text{ g } H_2O \times \dfrac{\text{mol } H_2O}{18.015 \text{ g } H_2O} = 29.117 \text{ mol } H_2O$

   (c) $29.117 \text{ mol } H_2O \times \dfrac{6.022 \times 10^{23} \text{ molec } H_2O}{\text{mol } H_2O}$
   $= 1.753 \times 10^{25} \text{ molec } H_2O$

4. $\dfrac{36.5 \text{ kJ}}{0.433 \text{ mol}} = 84.3 \text{ kJ/mol}$

5. *(a)* $71.8 \text{ g} \times \dfrac{1 \text{ mol}}{259.0 \text{ g}} = 0.277 \text{ mol}$

   (b) $\dfrac{4.307 \text{ kJ}}{0.277 \text{ mol}} = 15.5 \text{ kJ/mol}$

6. *(a)* $83.2 \text{ kJ} \times \dfrac{1 \text{ mol}}{3.811 \text{ kJ}} = 21.8 \text{ mol}$

   (b) $\dfrac{5519 \text{ g}}{21.8 \text{ mol}} = 253 \text{ g/mol}$

Application Questions

1. A steel needle can rest on a water surface because the surface tension of the water makes the surface less penetrable by solid bodies. A boat floats in water because it displaces a volume of water whose weight is less than that of the boat.

2. Our bodies are cooled by the evaporation of water-based perspiration that accumulates on our skin. On a humid day, water collects on the skin faster than it can evaporate, because of the high rate of condensation of water from the atmosphere. On a dry day, however, the rate of condensation is lower than the rate of evaporation, and the water does not collect. Thus, one is more comfortable on a hot, dry day than on a hot, humid day.

3. Diamond is a three-dimensional covalent network solid in which the three-dimensional nature of the covalent bonding accounts for the extreme hardness. Graphite, on the other hand, is a planar network solid in which the covalent bonding occurs in only two dimensions. The molecular sheets of graphite are held together only by van der Waals forces. These intersheet bonds are relatively weak and allow the adjacent sheets to slide past each other readily, accounting for the softness of graphite.

4. The indicated responses are:
   (a) higher *(b)* higher *(c)* higher *(d)* lower

5. The indicated responses are:
   (a) increased *(b)* increased *(c)* increased *(d)* decreased

6. With increasing temperature, the rate of evaporation of a liquid increases, thus increasing the number of particles in the vapor phase. This disturbs the liquid-vapor equilibrium and causes an increase in the rate of condensation. Once the equilibrium is reestablished, however, a greater number of particles are in the vapor phase than there were before, thus increasing the vapor pressure.

7. The stronger the attractive forces among the liquid particles are, the smaller the percentage of liquid particles is that can escape to the vapor phase at any given temperature. Since the equilibrium vapor pressure of the liquid is dependent on the relative number of particles in the vapor phase, the pressure above the liquid that has strong attractive forces will be less than that above the liquid in which the comparable forces are weaker.

8. *(a)* As atmospheric pressure increases, so does the boiling point of a liquid, because the liquid must be heated to a higher temperature before its equilibrium vapor pressure is

equal to atmospheric pressure. *(b)* Both the liquid and its vapor have the same constant temperature during boiling. *(c)* The energy from the constant supply of heat is used to overcome the attractive forces between the liquid molecules during the liquid-to-gas change, rather than to increase the kinetic energy of the liquid or vapor molecules, and is stored in the vapor as potential energy.

9. The greater the attraction between the liquid particles is, the greater is the energy that is required to overcome that attraction, and the higher is the molar heat of vaporization.

10. The energy that is continually being removed during freezing is actually a loss of potential energy resulting from the closer approach of particles that attract each other and not of kinetic energy that would be reflected in the temperature of the system.

11. As the degree of attraction between the solid particles increases, more heat must be absorbed by the melting solid in order to overcome the attractive forces and separate the solid particles, resulting in an increase in the molar heat of fusion.

12. According to Le Châtelier's principle, the addition of heat to the ice-water system pushes the equilibrium to the right (toward liquid water), whereas the removal of heat shifts it to the left. As long as both ice and liquid water are present, the addition or removal of heat is simply reflected in the phase change that involves potential energy, and not in the kinetic energy of the overall system.

13. The indicated phase(s) are:
(a) solid and vapor *(b)* solid and liquid *(c)* liquid and vapor *(d)* solid, liquid, and vapor.

14. *(a)* More liquid would vaporize. *(b)* More vapor would condense. *(c)* More vapor would condense. *(d)* More liquid would vaporize.

15. *(a)* More solid would sublime. *(b)* More vapor would solidify. *(c)* More vapor would solidify. *(d)* More solid would vaporize.

16. Decrease the pressure.

17. Methane is a nonpolar molecule and does not undergo hydrogen bonding. Water, on the other hand, is polar and undergoes extensive hydrogen bonding. Higher temperatures are required in order to overcome the resultant attractive forces and convert water from a liquid into a vapor.

18. *(a)* The molecules in ice are hydrogen-bonded in a hexagonal pattern in which the space between the atoms is empty, accounting for the relatively low density of ice. As the ice is heated and melts, the rigid hydrogen bonds among the ice molecules break down. This enables the liquid molecules to crowd closer together, so that they occupy a smaller volume, allowing the liquid water to become more dense than ice. *(b)* Above 4°C, the increased energy of the liquid water molecules causes them to overcome more effectively the molecular attractions between them and then spread out, so that above 4°C the density of water decreases with increasing temperature.

19. The steam radiator makes use of the fact that water has a relatively high molar heat of vaporization. When the steam, which has absorbed a great deal of heat energy during vaporization, reaches the radiator, it condenses and thus releases that large amount of heat into the living area of the home.

20. Yes. A container of water can freeze and boil at the same time if it is attached to a vacuum pump and the pressure

over the water is greatly reduced. At its triple point, the water will boil and its temperature will fall to its freezing point.

21. Water is often called the universal solvent because of its ability to dissolve so many polar substances.

22. Boiling chips are small pieces of inert material that are added to a solution to prevent "bumping" or violent bubbling during boiling.

Application Problems

1. Ice:
$$5.00 \text{ cm}^3 \times \frac{0.917 \text{ g}}{\text{cm}^3} \times \frac{6.02 \times 10^{23} \text{ molecules}}{18.0 \text{ g}}$$
$$= 1.53 \times 10^{23} \text{ molecules}$$

Liquid water:
$$5.00 \text{ cm}^3 \times \frac{1.00 \text{ g}}{\text{cm}^3} \times \frac{6.02 \times 10^{23} \text{ molecules}}{18.0 \text{ g}}$$
$$= 1.67 \times 10^{23} \text{ molecules}$$

The liquid water sample thus contains $1.67 \times 10^{23} - 1.53 \times 10^{23}$, or 1.4×10^{22}, more molecules than does the sample of ice.

2. From the problem above, the liquid water contains 1.67×10^{23} molecules.

Water vapor:
$$5.00 \text{ cm}^3 \times \frac{1 \text{ L}}{1000 \text{ cm}^3} \times \frac{1 \text{ mol}}{22.4 \text{ L}} = 2.23 \times 10^{-4} \text{ mol}$$
$$2.23 \times 10^{-4} \text{ mol} \times 6.02 \times 10^{23} \text{ molecules/mol} = 1.34 \times 10^{20} \text{ molecules}$$

The liquid water thus contains $1.67 \times 10^{23} - 1.34 \times 10^{20}$, or $1.67 \times 10^{23} - 0.00134 \times 10^{23}$, which equals (allowing for significant figures) 1.67×10^{23} more molecules. Thus, the $H_2O(g)$ contains a negligible number of molecules when compared with the $H_2O(\ell)$.

3. Liquid water:
$$100. \text{ cm}^3 \times \frac{1.00 \text{ g}}{\text{cm}^3} \times \frac{1 \text{ mol}}{18.0 \text{ g}} \times 6.008 \text{ kJ/mol} = 33.4 \text{ kJ}$$
(released when the liquid water is frozen)

Steam:
$$33.4 \text{ kJ} \times \frac{1 \text{ mol}}{40.58 \text{ kJ}} = 0.823 \text{ mol (required to release 33.4 kJ}$$
when condensed)
$$T = 100°C + 273 = 373 \text{ K}$$
$$PV = nRT$$
$$V = \frac{nRT}{P} = \frac{0.823 \text{ mol} \times 0.0821 \frac{\text{L} \cdot \text{atm}}{\text{mol} \cdot \text{K}} \times 373 \text{ K}}{1.00 \text{ atm}}$$
$$= 25.2 \text{ L}$$
$$0.823 \text{ mol} \times \frac{18.0 \text{ g}}{1 \text{ mol}} = 14.8 \text{ g}$$

A considerably larger volume of steam is thus required to release the same amount of heat. However, the mass of this larger volume of steam is considerably less than that of the liquid water.

Teacher's Notes

Chapter 13 Liquids and Solids

INTRODUCTION

In Chapters 11 and 12, you took a close look at the nature of gases and examined their physical behavior both on a large scale and at the atomic and molecular levels. You will now take a similar look at two other states of matter—liquids and solids. The chapter concludes with a discussion of the physical behavior of all three states of water.

LOOKING AHEAD

As you study this chapter, look for ways in which the observable physical properties of liquids and solids can be explained in terms of the arrangement and motion of their particles and the attractive forces between them.

SECTION PREVIEW

13.1 Liquids
13.2 Solids
13.3 Changes of State
13.4 Water

Optical fibers like these are made of glass and coated with a protective plastic to preserve their strength. These thin filaments can be used to carry information over long distances or to transmit images of a patient's internal body parts to the eye of a physician. Glasses are amorphous solids. They have some of the properties of solids and some of the properties of liquids.

13.1 Liquids

The water in the breakers crashing on an ocean beach and the molten lava rushing down the sides of a volcano are both examples of matter in the liquid state. When you think of the earth's oceans, lakes, and rivers, the fluids in the bodies of all living things and the many liquids you use every day, it is hard to believe an important fact about liquids—they are probably the *least* common state of matter in the universe. Liquids are less common than solids, gases, or plasmas because the liquid state of any substance can exist only within a relatively narrow range of temperatures and pressures.

In this section, you will examine the properties of the liquid state and compare them with those of the solid state and the gaseous state. These properties will be discussed in terms of a model of liquids at the submicroscopic level.

Kinetic-Theory Description of the Liquid State

How do we describe a liquid? One way to describe a liquid is to note that a liquid is a form of matter that has no definite shape, but instead takes the shape of its container. If you were to observe various types of liquids, you could include other properties in your definition of a liquid.

In order to accurately describe liquids and understand their properties, it is helpful to recall the kinetic molecular theory and the particle model of matter. To do this, think of liquids in terms of the motion, arrangement, and attractive forces of particles. Figure 13-1 is a diagram of liquid particles and the analogous model of gases discussed in Chapter 11.

Like gases and solids, liquids can be seen as consisting of particles—ions, atoms, or molecules. What makes a substance a liquid is not the nature of its particles, but the arrangement and behavior of those particles.

Like gas particles, liquid particles are in constant motion. Liquid particles are closer together and lower in energy than gas particles, however, and therefore the attractive forces between them are more effective than those between gas particles. This attraction between liquid particles is caused by the interactive forces discussed in Chapter 6: dipole–dipole intermolecular forces, London forces, and the hydrogen bonding. Remember that these dispersion forces are not very effective in gases because their particles are widely separated. Because the particles of a liquid are more closely packed than those of a gas, the movement of particles in a liquid is limited.

Liquids are more ordered, or less disordered, than gases. This is because of the effective interactive forces and the resulting decrease in mobility of the liquid particles. The kinetic energy of liquid particles is high enough so that they are not bound together as are solid particles, however. Particles of liquids can change position with other liquid particles, although they tend to do so more slowly and less often than do gas particles.

Properties of Liquids and the Particle Model

In many ways, the liquid state is intermediate between the solid state and the gaseous state. At normal atmospheric pressure (760 mm Hg or 101.325 kPa), most solids pass through the liquid state before turning to gas as their

SECTION OBJECTIVES

- Describe the movement of liquid particles according to the kinetic theory.
- Discuss the properties of liquids in terms of the particle model.

STUDY HINT

Temperature and pressure are discussed in Chapters 1, 11, and 12.

The earth receives just enough energy from the sun to maintain most of the water on earth in the liquid state.

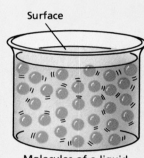

Molecules of a liquid

Molecules of a gas

Figure 13-1 This diagram compares the particle models of liquids (top) and gases (bottom).

temperature is increased. Most gases also pass through the liquid state before solidifying as their temperature is decreased. Some properties of liquids closely resemble those of solids, whereas others closely resemble those of gases.

Now you will take a closer look at some properties of liquids. These properties will be discussed in terms of the particle model of liquids.

Definite volume   Unlike gases, which expand indefinitely unless enclosed by a container, liquids have a definite volume. This volume changes only slightly with changes in temperature or pressure.

Fluidity   Like gases, but unlike solids, liquids can flow. According to the particle model of liquids, the particles are not bound together in fixed positions, but can move about constantly. This particle mobility explains why liquids and gases are referred to as fluids. *A* **fluid** *is a substance that can flow and therefore conform to the outline of its container.* Although most liquids naturally flow downhill because of gravity, flow can be in other directions as well. For example, liquid helium at temperatures near absolute zero has the unusual property of being able to flow uphill. Because they are fluids, liquids must be enclosed in containers. Unlike a gas, which presses against all the walls of its container, however, a liquid can have only one free surface. Its other surfaces must be supported by the container walls.

Relatively high density   At normal atmospheric pressure, most substances are thousands of times denser in their liquid state than in their gaseous state. This density is a result of the closer packing of liquid particles, as described by the particle model. Most substances are slightly (about 10%) less dense in their liquid state than in their solid state. As will be discussed in Section 13.4, water is one of the few substances that decreases in density when it solidifies.

At the same temperature and pressure, different liquids can differ in density by factors of almost 20. A dramatic demonstration of these differences in density is shown in Figure 13-2. The densities of the five liquids shown differ so greatly that the liquids settle out in layers.

Relative incompressibility   When liquid water at 20°C is compressed by a pressure of 1000 atm (roughly 100 MPa), its volume decreases by only 4%. This behavior of liquid water is typical of all liquids and is similar to the behavior of solids. In contrast, the volume of a gas under a pressure of 1000 atm would be only 1/1000 of its volume at normal atmospheric pressure. Liquids are much less compressible than gases because liquid particles are more closely packed together than gas particles. Liquids are used in the braking systems of motor vehicles, as shown in Figure 13-4, because of their relative incompressibility and because they have the ability to transmit pressure equally in all directions.

Dissolving ability   *When one substance is added to another so that, after mixing, only one physical state is observed, the substance has been* **dissolved.** For example, after you add a teaspoon of solid sugar crystals to a glass of liquid water and stir thoroughly, you no longer see solid sugar crystals. The mixture of sugar and water appears to be all liquid. Liquids are the most common dissolving agents and among these, liquid water is the most

Molar volumes of liquids
$10^1 - 10^2$ mL
Molar volumes of gases
$10^4 - 10^5$ mL
Since $D \div m/V$ then
$D_{liq} = 1000 \times D_{gas}$

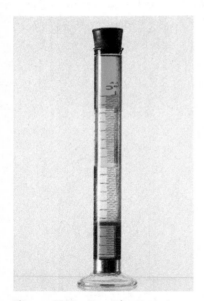

Figure 13-2   From bottom to top, the liquids mercury, glycerin, water, oil and alcohol settle into different layers. Which of these five liquids has the highest density and which has the lowest?

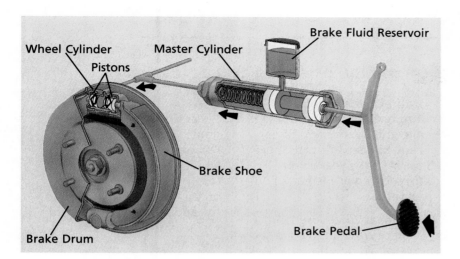

Wheel Cylinder

Master Cylinder

Brake Fluid Reservoir

Pistons

Brake Shoe

Brake Drum

Brake Pedal

Figure 13-3 This diagram shows how liquids are used in the braking systems of automobiles.

generally useful. The dissolving ability of liquids is explained by the motion of their particles and the attractive forces they may exert on the particles of the substance being dissolved.

Ability to diffuse As described in Chapter 11, gases diffuse and mix with the particles of other gases. Liquids also diffuse, as shown in Figure 13.4. Any liquid will gradually diffuse throughout any other liquid in which it can be dissolved. As in gases, diffusion in liquids occurs because of the constant, random motion of particles. Yet diffusion is much slower in liquids than in gases. This is because each forward-moving particle of closely packed liquid soon strikes another particle, changing its path. Also, the attractive forces between the particles of a liquid impede their movement. As the temperature of a liquid is increased, diffusion occurs more rapidly because the average kinetic energy and, therefore, the average speed of the particles is increased.

STUDY HINT
The dissolving process is discussed in Chapter 14.

Figure 13-4 Like gases, the two liquids in this beaker diffuse. The liquid food coloring and water gradually diffuse to form a uniform mixture.

Desktop Investigation

Surface Tension of Liquids and Water Solutions

Question

Do different liquids have different degrees of surface tension; what effect do dissolved substances have on the surface tension of water?

Materials

water
glycerin
cooking oil
rubbing alcohol (70% isopropyl alcohol)
table salt (sodium chloride)
liquid detergent
250-mL beaker
10- and 100-mL graduated cylinders
evaporating dish
tweezers
needle
eye dropper
stirring rod
wax paper

Procedure

Remember to wear safety glasses and an apron. Record all of your results in a data table.

1. Add 70 mL of water to an evaporating dish. Using tweezers, carefully place a needle on the surface of the water, as shown in the illustration. What happens? Empty the dish and dry it and the needle and tweezers. Repeat this procedure using *(a)* glycerin and *(b)* cooking oil instead of water.

2. Repeat Procedure 1, but replace water in the dish with the following solutions. Make sure that you clean and dry the dish, needle, and tweezers between each procedure. *(a)* 60 mL of water + plus 10 mL of alcohol *(b)* 50 mL of water + 20 mL alcohol *(c)* 70 mL water + a few grains of

table salt *(d)* 70 mL water + 1 g table salt *(e)* 70 mL water + 1 drop of liquid detergent.

3. In your data table, include the answers to these questions: *(a)* Does the needle rest on the surface of the liquid or does it sink? *(b)* How long does it remain on the surface? *(c)* Describe the appearance of the liquid surface with the needle resting on it.

4. Place three drops of each liquid or solution used in Procedures 1 and 2 together on a large piece of wax paper. Describe the shape of each drop. Is it rounded or flat? Be sure to use a different area of the wax paper for each different liquid or solution.

Discussion

1. Which liquids have the strongest surface tension? The weakest? How do you know?

2. How do dissolved substances affect the surface tension of water?

3. Did you find any relationship between the ability of a needle to rest on the surface of a liquid or solution and the shapes of the drops of the different liquids or solutions?

Surface tension **Surface tension** *a property common to all liquids, is a force that tends to pull adjacent parts of a liquid's surface together, thus making the surface less penetrable by solid bodies.* For example, the surface tension of water can keep a steel needle on the water's surface, as illustrated in the Desktop Investigation above. Surface tension is a result of the attractive forces between particles of a liquid as described by the particle model of matter discussed in Section 11.1. It gives a taut appearance to a liquid's free surface and acts to minimize its area. As a result of surface tension, liquid droplets tend to take on a spherical shape because a sphere has the smallest possible surface area for a given volume. An example of this phenomenon is shown in Figure 13-5.

Capillary action, the tendency of a liquid to rise up a narrow tube, is closely related to surface tension. Capillary action allows liquids to rise up a piece of slender glass tubing or a straw. It is the result of the attractive forces between particles of a liquid as well as forces of attraction between the liquid particles and the surface of the glass. Capillary action is responsible for the meniscus, the concave liquid surface that you see in a straw, test tube, or drinking glass.

Tendency to evaporate and to boil *The process by which a liquid or solid changes to a gas is* **vaporization.** Evaporation is a form of vaporization. **Evaporation** *is the process by which particles escape from the surface of a nonboiling liquid and enter the gaseous state.*

A small amount of liquid bromine has just been added to the bottle shown in Figure 13-6. Within a few minutes, the air above the liquid bromine is reddish. Some bromine molecules have escaped from the surface of the liquid and gone into the gaseous state, becoming bromine vapor, which mixes with the air. A similar phenomenon occurs when you apply cologne to your wrist. Within seconds, you become aware of the cologne's odor. Cologne scent molecules evaporate from your skin and diffuse through the air to your nose.

Evaporation can be explained by the kinetic energy of the particles in a liquid. At any instant, the separate particles of a liquid have different kinetic energies. Some particles have higher-than-average energies and move faster. Other particles have lower-than-average energies and move slower. Surface particles with higher-than-average energies may overcome the interactive forces that bind them to the liquid and "escape" into the gaseous state.

Evaporation is a crucial process in nature. All of the water that falls on the earth in the form of rain and snow originally evaporated from the earth's oceans, lakes, and rivers. The evaporation of perspiration plays an important role in keeping us cool. Perspiration, which is mostly water, cools us by absorbing body heat when it evaporates.

Boiling is the change of a liquid to bubbles of vapor that appear throughout the liquid when the equilibrium vapor pressure of the liquid equals the atmospheric pressure. The temperature at which a liquid boils is called the *boiling point* of that liquid; boiling is related to evaporation. For most liquids, the boiling point is higher than room temperature. Both evaporation and boiling are discussed in detail in Section 13.3

Tendency to solidify When a liquid is cooled, the average energy of its particles may become low enough that attractive forces hold the particles together in an orderly arrangement. The substance then becomes a solid. The physical change of a liquid to a solid is called *freezing*. Perhaps the best-known example of freezing is the change of liquid water to solid water—or ice—at 0°C. This property of water is discussed more fully in Section 13.4. Another familiar example of the tendency of liquids to solidify is the freezing of liquid candle wax at room temperature. All liquids freeze, although not necessarily at temperatures we encounter in our everyday lives. Ethyl alcohol, for example, freezes at −112°C.

Figure 13-5 This photo shows that, like all liquids, milk forms spherically shaped drops as a result of surface tension.

Figure 13-6 Liquid bromine will gradually diffuse into the air above its surface.

Technology

Liquid Crystals

The unique properties of liquid crystals have intrigued scientists since the discovery of these crystals nearly 100 years ago. Today, liquid-crystal displays (LCDs) are widely used in pocket calculators, digital watches, and clocks.

Liquid crystals derive their name from the fact that they have properties characteristic of both liquids and solids. In the liquid-crystal state, substances can transmit and reflect light in the same way that crystals do, while at the same time taking on the form of a liquid.

There are three types of liquid crystals: nematic, smectic, and cholesteric. Each is defined according to its molecular arrangement. In nematic liquid crystals, the cigar-shaped molecules are parallel to each other.

The attractive forces between molecules in nematic crystals are strong enough for the molecules to move together as a group. Because their molecules change directions in response to an electric field, nematic liquid crystals are most often used in LCDs. The numbers and letters in the LCD are made up of small glass tubes filled with liquid crystals. As an electric charge passes

through the glass, the crystals rotate in the direction of the resulting electric field. When the electric field is not present, the display is lighted. When the field is present, light is blocked in certain areas resulting in a numeric display.

In smectic liquid crystals, the molecules are arranged in planes or layers. The molecules of cholesteric liquid crystals are also arranged in layers. Because they are sensitive to temperature changes, both smectic and cholesteric liquid crystals are used in thermometers.

---

1. See the discussions of kinetic theory and the particle model at the beginning of this section.
2. See the discussion of properties of liquids.
3. Refer to the coverage of the particle model earlier in this section.
4. See "Tendency to evaporate and to boil."

Section Review

1. Describe the liquid state according to the kinetic theory and particle model.
2. List the properties of liquids.
3. How does the particle model explain the following properties of liquids: *(a)* their relatively high density *(b)* their dissolving ability *(c)* their ability to evaporate?
4. Compare vaporization, evaporation, and boiling.

13.2 Solids

"Solid as a rock" is a common expression that intuitively describes a solid—something that is hard or unyielding, and that has a definite shape and volume. Many things besides rocks are solids; in fact, solids are far more common than liquids.

In this section, you will examine the properties of the solid state and compare them with those of the liquid and gaseous states. As with liquids and gases, the properties of solids are explained in terms of the kinetic theory.

Kinetic Theory Description of the Solid State

Like gases and liquids, solids consist of particles. The particles in a solid may be ions, atoms, or molecules. Figure 13-7 compares the structure of solids, liquids, and gases. The kinetic energy of the particles in solids is lower than that of the particles in gases and liquids. Because they have less kinetic energy than gases and liquids, solid particles have less motion than liquid particles. They also have much less motion than gas particles. As shown in Figure 13-7, the particles of a solid do not undergo overall changes in position, but rather move back and forth about fixed positions. Solid particles are more closely packed than liquid or gas particles, and therefore intermolecular forces between particles are even more important in solids. Dipole-dipole attractions, London forces, and hydrogen bonding are all much stronger in solids than in liquids or gases. These forces help to hold solid particles in their positions. Because solid particles are held in relatively fixed positions, solids are more ordered than liquids and much more ordered than gases. The importance of order and disorder in physical and chemical changes is discussed in Chapter 18.

Properties of Solids and the Particle Model

The properties of solids can be explained in terms of the particle model of solids—that is, in terms of the arrangement and motion of solid particles and the attractive forces between them. There are two types of solids: crystalline solids and amorphous solids. *Most solids are* **crystalline solids**—*they consist of crystals. A* **crystal** *is a substance in which the particles are arranged in an orderly, geometric, repeating pattern.* Examples of crystalline solids are shown in Figure 13-8. Other solids, including glass and plastics, are noncrystalline and are called *amorphous solids. An* **amorphous solid** *is one in which the particles are arranged randomly.* These two types of solids are discussed further in Section 13.3.

Definite shape   Unlike liquids and gases, all solids can maintain a definite shape without a container. In addition, crystalline solids are geometrically regular. Even the fragments of a shattered crystalline solid have distinct geometric shapes. On the other hand, while amorphous solids do maintain a definite shape (at least temporarily), they do not have the distinct geometric shapes of crystalline solids. Instead, they can have nearly any shape. For example, glass can be molded into any shape, and, if it is shattered, its fragments can have a wide variety of irregular shapes.

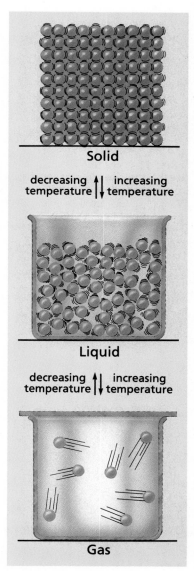

Figure 13-7 This diagram shows particles of a solid, a liquid, and a gas.

Figure 13-8 Amethyst (a), calcite (b), fluorite (c), and azurite (d) are examples of crystals.

Definite volume Like liquids but unlike gases, all solids have a definite volume. The volume of a solid changes only slightly with a change in temperature or pressure. Solids have definite volume because their particles are packed closely together. There is very little empty space into which the particles can be compressed, and the particles do not have a tendency to move apart. Unlike liquids and gases, all the surfaces of a solid can be free.

Nonfluidity Unlike liquids and gases, crystalline solids generally do not flow. This is because their particles, as described by the particle model, are held in relatively fixed positions. Amorphous solids also tend to retain their shapes. Some amorphous solids do flow, although usually at a very slow rate. A distinctive group of amorphous solids is the glasses, which are made by cooling certain molten materials in such a manner that they do not crystallize but remain in an amorphous state.

Definite melting point At a given pressure, a pure crystalline solid melts abruptly at a particular temperature. *The temperature at which a solid becomes a liquid is its* **melting point.** At this temperature, the kinetic energies of the particles within the solid overcome the interactive forces between them, allowing the particles to break out of their positions in the crystal. In contrast, amorphous solids like glass and plastics have no definite melting point. Instead they gradually soften as they are heated, to become thick, sticky liquids. *Because of this, and also because of their ability to flow, amorphous solids such as glass are sometimes classified as* **supercooled liquids;** *these are substances that retain certain liquid properties even at temperatures at which they appear to be solid.* These properties exist because the particles in amorphous solids are not arranged in an orderly, uniform crystal structure, but instead are arranged randomly, like the particles in a liquid. Unlike the particles in a true liquid, however, the particles in amorphous solids are not constantly changing their positions. An amorphous solid can be thought of as being much like a liquid that has been cooled enough so that it no longer flows.

High density   In general, most substances are at their densest in the solid state, which is typically slightly denser than the liquid state. Solids are denser than liquids or gases because the particles of a solid are more closely packed than those of a liquid or a gas. Solid hydrogen is the least dense solid; it has a density about 1/260 that of the densest element, osmium (Os).

Incompressibility   Solids are even less compressible than liquids. For all practical purposes, solids can in fact be considered incompressible. Wood and cork are solids that *seem* compressible, but they are not. They contain pores that are filled with air. When subjected to intense pressure, the pores are compressed, not the wood or cork.

Slow rate of diffusion   If a zinc (Zn) plate and a copper (Cu) plate are clamped together for several months, a small amount of diffusion between the particles of zinc and copper will occur. This observation shows that diffusion does occur in solids. The rate of diffusion is millions of times slower in solids than in liquids, however.

Crystalline Solids

Earlier, you learned the definition of the two types of solids: crystalline and amorphous. Now you will examine types of crystalline solids more closely. Later in this section, you will read more about amorphous solids.

Classification of crystals by arrangement and shape   Crystalline solids exist either as single crystals or as groups of crystals fused together. Figure 13-8 shows four different types of crystals. Scientists can determine the arrangement of the particles of a crystal by mathematical analysis of the crystal's diffraction pattern; a unique pattern that results when X rays or other forms of electromagnetic radiation pass around the edges of particles.

The total three-dimensional array of points that describes the arrangement of the particles of a crystal is called a **crystal lattice.** Such a lattice is shown in Figure 13-9. *The smallest portion of a crystal lattice that reveals the three-dimensional pattern of the entire lattice is called a* **unit cell.** Each crystal lattice is constructed of many unit cells packed solidly together. As a result of this construction, a crystal has the same symmetry as its unit cell. Unit cells can have any one of seven types of symmetry, a characteristic that enables scientists to classify crystals by shape. This classification by shape is a part of the science of crystallography. Any crystal can be placed into one of the seven crystalline systems shown in Figure 13-10.

Binding forces in crystals   Crystal-lattice structures can also be described in terms of the types of particles making up the crystals and the types of chemical bonding between them. According to this method of classifying crystal lattices, there are four types of crystals, which are listed in Table 13-1. You may wish to refer to this table as you read the following discussion.

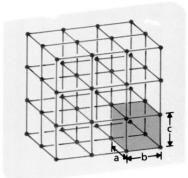

Figure 13-9   The colored cube at the lower right of this crystal lattice is a unit cell.

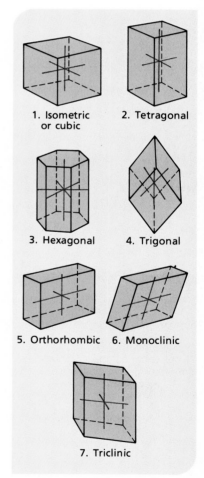

1. Isometric or cubic
2. Tetragonal
3. Hexagonal
4. Trigonal
5. Orthorhombic
6. Monoclinic
7. Triclinic

Figure 13-10   This schematic diagram shows the seven basic crystal systems.

TABLE 13-1   MELTING POINTS AND BOILING POINTS OF REPRESENTATIVE CRYSTALLINE SOLIDS

Type of Substance	Substance	Melting Point (°C)	Boiling Point (1 atm, °C)
ionic	NaCl	801	1413
	MgF_2	1266	2239
covalent network	$(SiO_2)_x$	1610	2230
metallic	Hg	−39	357
	Cu	1083	2567
	Fe	1535	2750
	W	3410	5660
nonpolar	H_2	−259	−253
covalent	O_2	−218	−183
molecular	CH_4	−182	−164
	CCl_4	−23	77
	C_6H_6	6	80
polar	NH_3	−78	−33
covalent	H_2O	0	100
molecular			

Almost all that we know about crystal structure has been learned from X-ray diffraction studies.

1. Ionic crystals. As discussed in Chapter 6, the ionic crystal lattice consists of positive and negative ions arranged in a characteristic regular pattern. The ions can be monatomic or polyatomic. Generally, ionic crystals are formed when Group 1 or Group 2 metals combine with Group 16 or Group 17 nonmetals or the nonmetallic polyatomic ions. The strong binding forces between the positive and negative ions in the crystal lattice cause ionic crystals to be hard and brittle, to have high melting points, and to be good insulators.

2. Covalent network crystals. In covalent network crystals, the lattice sites contain single atoms, each of which is covalently bonded to its nearest neighboring atoms. The covalent bonding extends throughout a network that includes a very large number of atoms. Three-dimensional covalent network solids, including diamond (C_x), shown in Figure 13-11, quartz $(SiO_2)_x$, silicon carbide $(SiC)_x$, and many oxides of transition metals, are essentially giant molecules. The subscript x in these formulas indicates simply that the chain of components extends indefinitely. These network solids are very hard and brittle. They have rather high melting points, and are nonconductors or semiconductors. Planar network solids, such as graphite (C_x), shown in Figure 13-12, consist of molecular sheets. Each sheet can be thought of as a single molecule consisting of atoms held together by many covalent bonds. The adjacent separate sheets of molecules, in turn, are held together by van der Waals forces between atoms in the different sheets. Graphite is commonly used as a lubricant because the weak intersheet bonds of this planar network solid allow adjacent sheets to slide past each other readily. For this same reason, graphite is used as pencil "lead." As you write, a thin layer of graphite is deposited onto the paper.

Graphite conducts electricity. Graphite rods are used in flashlight batteries.

3. Metallic crystals. As discussed in Chapter 6, the metallic crystal lattice consists of positive ions of the metal surrounded by a cloud of valence electrons, that are donated by the metal atoms and belong to the crystal as a whole. The binding force is the attraction between the positive ions and the electrons. At any instant, each electron and the positive metallic ions immediately surrounding it are attracting each other electrostatically. The freedom of the valence electrons to migrate throughout the crystal lattice explains the high electric conductivity of metals. As you can see from Table 13-1, the melting points of different metallic crystals vary greatly.

4. Covalent molecular crystals. The crystal lattice of a covalent molecular substance consists of covalently bonded molecules that are held to one another by van der Waals forces. If the molecules are nonpolar—for example, hydrogen (H_2), iodine (I_2), methane (CH_4), and benzene (C_6H_6)—then the bonding between molecules, resulting from London dispersion forces, is weak. Crystals of the noble gases are held together by the same type of weak forces. In this case, because the noble gases are monatomic, these forces act between individual atoms. If the molecules in a covalent molecular crystal are polar—for example, water (H_2O) and ammonia (NH_3)—they are held together by both dispersion forces and the somewhat stronger dipole-dipole forces. The forces that hold either polar or nonpolar molecules together in the crystal lattice are much weaker than the covalent chemical bonds that bind the atoms within the molecules . Covalent molecular crystals thus have low melting points, are easily vaporized, are relatively soft, and are good insulators. Ice crystals, the most familiar molecular crystals, are discussed in the next section.

Figure 13-11 The three-dimensional structure of diamond, a covalent network crystal.

Amorphous solids   Noncrystalline solids, such as rubber, glass, plastics, and synthetic fibers, are called *amorphous solids*. The word "amorphous" comes from the Greek for "without shape (*a* [without] + *morphe* [form])." Unlike crystals, amorphous solids do not have a regular, natural shape, but instead take on whatever shape is imposed on them. The particles making up amorphous solids are not arranged in a uniform crystal lattice, but are arranged randomly, somewhat like the particles in a liquid.

There are hundreds of types of plastic and glass, which have thousands of important applications. Glass is used for everything from fiberglass automo-

Technology

Superconductors

In early 1987, the scientific community was excited by the discovery of a new class of superconducting materials that offer no resistance to an electric current when cooled to 90K. This temperature can be reached conveniently using inexpensive liquid nitrogen (boiling point, 77K). Previously discovered superconductors functioned at temperatures that could be reached only by using liquid helium (boiling point, 4.2K), which is expensive and inconvenient to handle.

Many of the new superconductors are in the family of materials known as ceramic oxides. They have the empirical formula $YBa_2Cu_3O_x$ and are related to the perovskite class of minerals. In a perfect crystal of this class, x (the number of oxygen atoms) = 9. The supercon-

ductors are imperfect crystals, however, in which the domain of x extends from approximately 6.5 to 7.2.

The new superconductors may someday be used to transmit electricity hundreds of miles without loss of energy in the form of heat. They could also be made into powerful electromagnets that could suspend rail-

road trains in mid-air above their tracks. These magnetically levitated trains could travel without friction, reaching very high speeds. Powerful electromagnets could also produce the strong magnetic fields required to confine the hot plasmas that are needed to generate energy by controlled nuclear fusion.

bile bodies to optical fibers that transmit telephone conversations by means of light waves. Amorphous forms of metal alloys can be prepared by very rapid cooling of thin films of melted metal. The cooling occurs so rapidly that the metal atoms do not have time to arrange themselves in a crystalline pattern. The resulting "metallic glasses" are strong, flexible, and much more resistant to corrosion than crystalline alloys of the same composition.

Section Review

1. Describe the solid state according to kinetic theory and particle model.
2. How does the particle model explain   (a) the definite melting point   (b) the definite volume   (c) the high density of solids?
3. What is the difference between amorphous and crystalline solids?
4. What are the four types of crystals?

13.3 Changes of State

Matter on earth can exist in any of three states—gas, liquid, or solid—and can change from one state to another. Table 13-2 lists the possible changes of state. In this section, you will examine these changes of state and the factors that govern them.

TABLE 13-2 POSSIBLE CHANGES OF STATE

Change of State	Name	Example
solid → liquid	melting	ice → water
solid → gas	sublimation	dry ice → CO_2 gas
liquid → solid	freezing	water → ice
liquid → gas	vaporization	liquid bromine → bromine vapor
gas → liquid	condensation	water vapor → water

Equilibrium

In chemistry, **equilibrium** *is a dynamic condition in which two opposing physical or chemical changes occur at equal rates in the same closed system.* As an analogy, think of a busy department store at holiday time. When the doors are first opened in the morning, customers begin to enter the store. As time passes, some customers finish their shopping and leave the store, while new customers enter. At this point, the store is at a kind of equilibrium. The number of customers inside the store remains constant as long as new customers enter the store at the same rate that other customers leave. As closing time approaches, people leave the store at a faster rate than other people enter; equilibrium no longer exists.

Equilibrium is a very important chemical concept. Here you will learn about it in relation to changes of state. In Chapter 20, you will study equilibrium in terms of chemical reactions.

Equilibrium and changes of state   Consider the evaporation of a liquid in a closed container in which, initially, there is a vacuum over the liquid, as *shown in* Figure 13-13(*a*). Assume that the contents of the container maintain the same temperature as the surroundings, which is normal room temperature (approximately 22°C). The liquid and the container are the **system,** that is, *the sample of matter being studied.* Initially, there is a single phase inside this system—the liquid phase. *A* **phase** *is any part of a system that has uniform composition and properties.*

A molecule at the upper surface of the liquid can overcome the attraction of neighboring molecules and evaporate if its energy is high enough. Molecules that have entered the vapor phase behave as typical gas molecules. Some of these vapor molecules move down toward the liquid surface and condense, however, or reenter the liquid phase. **Condensation** *is the process by which a gas changes to a liquid.*

As long as the temperature and surface area of the liquid remain constant, the rate at which its molecules enter the vapor phase remains constant. The rate at which molecules pass from the vapor phase to the liquid

- Explain the relationship between equilibrium and changes of state.
- Predict changes in equilibrium using Le Châtelier's Principle.
- Explain what is meant by equilibrium vapor pressure.
- Describe the processes of boiling, freezing, melting, and sublimation.
- Interpret phase diagrams.

A chemical equilibrium is a closed system, however, a system in which the total number of particles is a constant. The store represents an open system because new customers are entering; the total number of customers is not a constant, but an ever increasing number.

This is one of several definitions of concentration. Solubilities are expressed in g solute/100 g H_2O, molar concentrations in moles solute/liter of solution, and molar concentrations in moles solute/ kilogram solvent.

interactive forces holding them together and the particles can break out of their positions in the crystal. *(b)* The solid particles are packed closely together. *(c)* The particles of a solid are more closely packed than those of a liquid or a gas.

3. The particles of amorphous solids are not arranged in a uniform crystal lattice, but are arranged randomly, somewhat like the particles in a liquid.
4. ionic crystals, covalent network crystals, metallic crystals, covalent molecular crystals

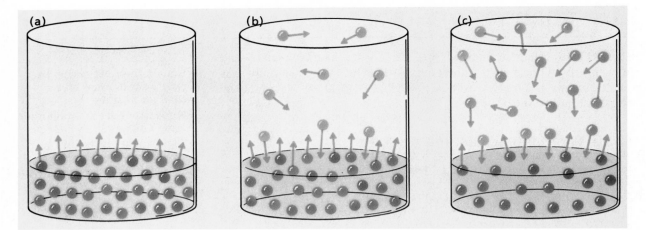

Figure 13-13 A liquid-vapor equilibrium developing in a closed system. *(a)* At first there is only liquid present but molecules begin to evaporate. *(b)* Evaporation continues at the same rate. There is a relatively low concentration of vapor. Vapor molecules are condensing to liquid, however. The rate of condensation is lower than the rate of evaporation. *(c)* The concentration of vapor has increased to the point at which the rates of evaporation and condensation have become equal. Equilibrium has been reached. The amounts of liquid and of vapor now remain constant.

phase also depends upon the concentration of molecules in the vapor phase, however. **Concentration** *is the number of particles per unit volume.* Initially, this concentration and therefore, the rate of condensation are zero, as shown in Figure 13-13(a). As time passes and evaporation continues, the concentration of molecules in the vapor phase increases. This increase in concentration results in an increase in the rate at which molecules reenter the liquid phase, as shown in Figure 13-13(b).

Eventually, the rate at which molecules condense equals the rate at which molecules evaporate, as shown in Figure 13-13(c). In other words, equilibrium has been reached.

Whenever a liquid changes to a vapor, as it does when it evaporates, it absorbs heat energy from its surroundings. Evaporation can therefore be represented as

$$\text{liquid} + \text{heat energy} \rightarrow \text{vapor}$$

Whenever a vapor condenses, it gives off heat energy to its surroundings. Condensation can therefore be represented as

$$\text{vapor} \rightarrow \text{liquid} + \text{heat energy}$$

The liquid-vapor equilibrium can be represented as

$$\text{liquid} + \text{heat energy} \rightleftharpoons \text{vapor}$$

The "double-yields" sign in the equation above represents a reaction at equilibrium. This means that the reaction can be read in either direction. The forward reaction is represented when the equation is read from left to right: liquid + heat energy → vapor. The reverse reaction is represented when the equation is read from right to left vapor → liquid + heat energy.

Le Châtelier's principle   In 1888, the French chemist Henri Louis Le Châtelier (1850–1936) developed an important principle that enables scientists to predict how a change in the conditions of a system at equilibrium will affect the equilibrium. **Le Châtelier's principle** *can be stated as follows: If any of the factors determining an equilibrium are changed, the system will adjust itself in a way that tends to minimize that change.* This principle applies to all kinds of equilibria, both physical and chemical.

Le Châtelier's principle can be used to predict how the liquid-vapor equilibrium in the system just discussed changes when the temperature of the system is increased from 20°C to 50°C. The equilibrium of the system can be represented as

$$\text{liquid} + \text{heat energy} \rightleftarrows \text{vapor}$$

According to Le Châtelier's principle, the system will adjust itself to counteract an increase in temperature. In this case, the forward reaction is endothermic—it absorbs energy. This reaction will be favored in order to counteract the rise in temperature. The forward reaction temporarily proceeds at a faster rate than the reverse reaction, until finally a new equilibrium is reached. At this new, higher-temperature equilibrium, the concentration of vapor is higher than it was at the lower-temperature equilibrium. That is, the equilibrium system has shifted to the right, favoring the products. At the new equilibrium point, the reverse reaction, condensation, occurs at a faster rate than it did at the old equilibrium point so that the rates of the reverse and forward reaction are equal.

Suppose, on the other hand, that the temperature of the system in equilibrium at 20°C is lowered to 5°C. According to Le Châtelier's principle, the system will adjust itself to counteract the decrease in temperature. Here the reverse reaction will be favored because it is reaction is exothermic and releases energy. Equilibrium is shifted to the left and reestablished at 5°C with a vapor concentration less than that at 20°C.

Suppose the mass of the liquid and the temperature of this equilibrium system remain the same, but that the volume of the system suddenly increases. What happens to the equilibrium? Initially, the concentration of molecules in the vapor phase decreases because the same number of vapor-phase molecules now occupy a larger volume. Because the concentration of vapor molecules decreases, fewer vapor molecules strike the liquid surface and return to the liquid phase at any given time. The rate of condensation therefore decreases. The rate of evaporation, which remains the same as it was before, is now higher than the rate of condensation. As a result, an additional net evaporation of liquid counteracts the imposed decrease in vapor concentration by increasing the number of vapor molecules. When equilibrium is reestablished, the concentration of molecules in the vapor phase is the same as in the system of smaller volume. Because there was a net movement of molecules from the liquid phase to the vapor phase, the number of molecules in the liquid phase has decreased. The equilibrium has therefore shifted to the right.

The system will never be able to reestablish the original equilibrium conditions. A system partially counteracts the stress before reestablishing equilibrium.

Equilibrium Vapor Pressure of a Liquid

The vapor molecules in equilibrium with a liquid in a closed system exert a pressure proportional to the vapor concentration. *The pressure exerted by a vapor in equilibrium with its corresponding liquid at a given temperature is called the* **equilibrium vapor pressure** *of the liquid.* Solids can also have an equilibrium vapor pressure, since some of the solid can change directly to vapor. Figure 13-14 shows the apparatus used to measure the equilibrium vapor pressure of a solid and of a liquid at a variety of temperatures.

When the equilibrium vapor pressure of water is graphed, as in Figure 13-15, the result is a vapor pressure-vs.-temperature curve for liquid water.

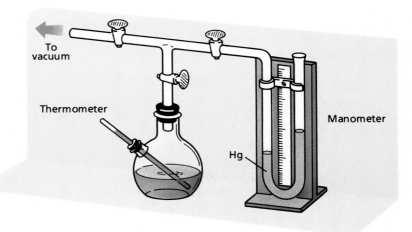

Figure 13-14 This apparatus can be used to measure the equilibrium vapor pressure of a liquid or a solid.

Liquid water can exist in equilibrium with water vapor only up to a temperature of 374.1°C. As discussed later in this chapter, neither liquid water nor solid water can exist at temperatures above 374.1°C. The curve in Figure 13-15 shows that the equilibrium vapor pressure of water increases as temperature increases, although not by direct proportion. This finding is consistent with Le Châtelier's principle since the concentration of water vapor, and therefore the equilibrium vapor pressure of water, increases in response to temperature increases. At any given temperature, water vapor exerts a specific vapor pressure.

The increase in equilibrium vapor pressure with increasing temperature can be explained in terms of the particle models for the liquid and gaseous states. Increasing the temperature of a liquid increases its average kinetic energy, which in turn increases the number of its molecules that have enough energy to escape from the liquid phase into the vapor phase. This increased evaporation rate increases the concentration of molecules in the vapor phase, thus disturbing the liquid-vapor equilibrium. But the increased concentration of molecules in the vapor phase then increases the rate at which molecules reenter the liquid phase. Soon equilibrium is reestablished, but at a higher equilibrium vapor pressure.

Every liquid has a specific equilibrium vapor pressure at a given temperature. This is because all liquids have characteristic forces of attraction between their particles. The stronger these attractive forces are, the smaller is the percentage of liquid particles that can escape to the vapor phase at any given temperature, and therefore, the lower is the equilibrium vapor pressure. Volatile liquids, which evaporate readily, have relatively weak forces between their particles. Ether is a typical volatile liquid. Nonvolatile liquids, which evaporate slowly, have relatively strong attractive forces between particles. Molten ionic compounds are examples of nonvolatile liquids.

Be careful not to confuse the equilibrium vapor pressure of a liquid with the pressure of a vapor that is not in equilibrium with its corresponding liquid. The equilibrium vapor pressure of a liquid depends only on temperature. The pressure of a vapor that is not in equilibrium with the corresponding liquid obeys the gas laws, like any other gas, and is inversely proportional to its volume at constant temperature, according to Boyle's law.

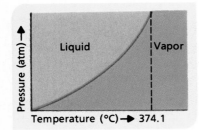

Figure 13-15 This graph shows the vapor pressure curve for water.

Boiling

Equilibrium vapor pressures can be used to explain the common phenomenon of boiling. **Boiling** *is the conversion of a liquid to a vapor, within the liquid as well as at its surface when the equilibrium vapor pressure of the liquid is equal to the atmospheric pressure.*

A liquid does not boil if its equilibrium vapor pressure is lower than the atmospheric pressure. As the temperature of the liquid is increased, the equilibrium vapor pressure also increases. Finally, a temperature is reached at which the equilibrium vapor pressure is equal to the atmospheric pressure. This temperature is known as the boiling point. *The* **boiling point** *of a liquid is the temperature at which the equilibrium vapor pressure of the liquid is equal to the atmospheric pressure.* The lower the atmospheric pressure is, the lower is the boiling point; the higher the atmospheric pressure is, the higher is the boiling point. This is the reason why foods take longer to cook at high elevations where atmospheric pressures are lower than at sea level. This principle is the basis for the operation of the pressure cooker shown in Figure 13-16.

The pressure cooker increases the boiling temperature of the water inside by allowing a certain amount of steam pressure to build up over the surface of the boiling water. In contrast, a device called a vacuum evaporator brings about boiling at lower-than-normal temperatures. Vacuum evaporators are used to remove water from milk and sugar solutions. Under reduced pressure, the water boils away at a temperature low enough to avoid scorching the milk or sugar. This process is used to prepare evaporated milk and sweetened condensed milk.

At normal atmospheric pressure (760 mm Hg, or 101.3 kPa [0.1 M Pa]), the boiling point of water is exactly 100°C. This temperature is known as the normal boiling point of water. It takes longer to hard-boil an egg in Denver, Colorado, than it does in Boston, Massachusetts, however, because water boils at a lower temperature in Denver. Food cooks faster in a pressure cooker because water boils at a higher temperature. The normal boiling points of water and some other liquids are shown in the graph in Figure 13-17. Note that the normal boiling point of each liquid is the temperature at which the liquid's equilibrium vapor pressure equals 760 mm Hg.

If you were to measure carefully the temperature of a boiling liquid and its vapor, you would find that both are at the same constant temperature. As you know from experience, heat must be continuously added in order to keep a liquid boiling. A pot of boiling water stops boiling almost immediately after it is removed from the stove. Since the temperature and, therefore, the average kinetic energy of the particles in a system at the boiling point remain constant despite the continuous addition of heat, where does the added heat energy go? It is used to overcome the attractive forces between liquid molecules during the liquid-to-gas change, and is stored in the vapor as potential energy. *The amount of heat energy required to vaporize one mole of liquid at its boiling point is known as the liquid's* **molar heat of vaporization.** The magnitude of the molar heat of vaporization is a measure of the attraction between the particles of the liquid. The stronger this attraction is, the more energy is required to overcome it, and, therefore, the higher is the molar heat of vaporization. Each liquid has a characteristic molar heat of vaporization. Compared to other liquids, water has an unusually

The strength of the attractive forces is independent of temperature. Higher temperatures with resultant higher kinetic energies make these forces less effective.

Figure 13-16 A pressure cooker makes use of elevated pressure to raise boiling point and shorten cooking time.

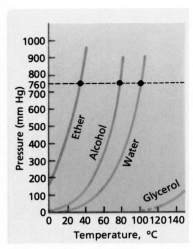

Figure 13-17 The boiling point of each of these liquids is the temperature at which the liquid's equilibrium vapor pressure equals 760 mm Hg (shown here as a dotted line).

high molar heat of vaporization. This property, which you will learn more about in the next section, makes water a very effective cooling agent. When water evaporates from your skin, the escaping molecules carry a great deal of heat away with them.

Freezing and Melting

The physical change of a liquid to a solid is called freezing. Freezing involves a loss of energy, in the form of heat, by the liquid:

$$\text{liquid} \rightarrow \text{solid} + \text{heat energy}$$

In the case of a pure substance, this change occurs at constant temperature. *The normal* freezing point *is the temperature at which the solid and liquid are in equilibrium at 1 atm (101.3kPa). pressure.* This observation indicates that the particles of the liquid and the solid have the same average kinetic energy. Therefore, the energy loss during freezing is a loss of potential energy that was present in the liquid. At the same time as energy decreases, there is a significant increase in order, since the solid state of a substance is much more ordered than the liquid state, even at the same temperature.

The reverse physical change, melting, also occurs at constant temperature. **Melting** *is the physical change of a solid to a liquid.* The **melting point** is the temperature at which the transition from the solid phase to the liquid phase occurs. During melting, the melting solid continuously absorbs heat:

$$\text{solid} + \text{heat energy} \rightarrow \text{liquid}$$

The amount of heat energy required to melt one mole of solid at its melting point is its **molar heat of fusion.** The heat absorbed by the melting solid increases the potential energy of the solid as its particles are pulled apart in opposition to the attractive forces tending to hold them together. At the same time, there is a significant decrease in order.

For pure, crystalline solids, the melting point and the freezing point are the same. At equilibrium at the melting or freezing point, melting and freezing proceed at equal rates according to the following general equilibrium equation:

$$\text{solid} + \text{heat energy} \rightleftarrows \text{liquid}$$

At normal atmospheric pressure, the temperature of an ice–liquid–water system will remain at 0°C as long as both ice and liquid water are present, no matter what the surrounding temperature. As predicted by Le Châtelier's principle, adding heat to an ice–liquid–water system shifts the equilibrium to the right, increasing the proportion of liquid water and decreasing that of ice. Only after all the ice has melted will the addition of heat increase the temperature of the system.

Under sufficiently low temperature and pressure conditions, a liquid cannot exist. Under such conditions, a solid substance exists in equilibrium with its vapor instead of its liquid:

$$\text{solid} + \text{heat energy} \rightleftarrows \text{vapor}$$

The change of state from a solid directly to a gas is known as **sublimation.** Among the common substances that sublime at ordinary temperature are dry ice and iodine. Ordinary ice sublimes slowly at temperatures lower than its melting point (0°C). This is how a thin layer of snow can eventually disappear, despite the fact that the temperature remains below 0°C. Frost-free refrigerators make use of sublimation by periodically slightly warming their freezer compartments, and then using a blower to remove molecules of water as they sublime.

Phase Diagrams

The temperature and pressure conditions under which a substance exists in a given state can be conveniently summarized in a phase diagram. *A **phase diagram** is a graph of temperature versus pressure that indicates the conditions under which gaseous, liquid, and solid phases of a particular substance exist.* A phase diagram also reveals how the states of a system change with changing temperature or pressure.

Figure 13-18 shows the phase diagram for water over a range of temperature and pressure conditions. Note the three curves, AB, AD, and AC. Each point on curve AC indicates the simultaneous temperature and pressure conditions under which liquid water and water vapor can exist together at equilibrium. Similarly, curve AD indicates the temperature and pressure conditions under which ice and liquid water are in equilibrium. Curve AB indicates the temperature and pressure conditions under which ice and water vapor coexist at equilibrium. Point A is called the triple point of water. *The **triple point** of a substance indicates the temperature and pressure conditions at which the solid, liquid, and vapor of the substance can coexist at equilibrium.* Point C is called the critical point of water. *The **critical point** of a substance indicates the **critical temperature** (T_c), the temperature above which the substance cannot exist in the liquid state. The critical point also indicates the **critical pressure** (P_c), the lowest pressure required for the substance to exist as a liquid at the critical temperature.*

The phase diagram in Figure 13-18 also indicates the normal boiling point and the normal freezing point of a substance. As shown by the slope of line AD, ice melts at a higher temperature with decreasing pressure. As the

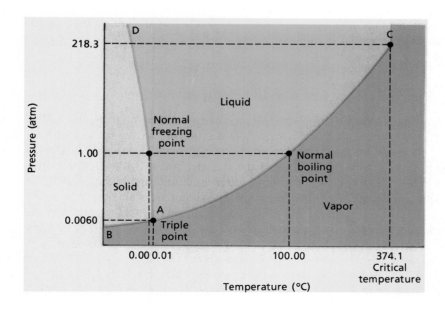

Figure 13-18 The phase diagram for water. Water vapor cannot be converted to a liquid beyond the critical temperature (374.1°C), no matter how high the pressure exerted on the vapor. The critical pressure (218.3 atm) is the lowest pressure required for the substance to exist as a liquid at the critical temperature.

temperature drops below the triple point, the temperature of sublimation decreases with decreasing pressure. The freeze-drying of foods is accomplished by vaporizing the ice and lowering the pressure after freezing.

Section Review

1. What is equilibrium?
2. State Le Châtelier's principle.
3. What happens when a liquid-vapor system at equilibrium experiences an increase in temperature? What happens when it experiences a decrease in temperature?
4. What is the equilibrium vapor pressure of a liquid?
5. What is the boiling point of a liquid?
6. In the phase diagram for water, what is *(a)* the triple point and *(b)* the critical point?

13.4 Water

SECTION OBJECTIVES

- Describe the structure of a water molecule.
- Discuss the physical properties of water and explain how these are determined by the structure of water.

Water is a familiar substance in all three physical states: solid (ice), liquid (liquid water), and gas (water vapor). On earth, water is by far the most abundant liquid. Oceans, rivers, and lakes cover about 75 percent of the earth's surface and significant quantities of water are frozen in glaciers. Water is an essential component of all organisms; 70 percent to 90 percent of the mass of living things is water. The chemical reactions of life processes take place in water, and water is frequently also a reactant or product in such reactions. In order to understand better the importance of water, you will take a closer look first at its structure and then at its properties.

Structure of Water

As discussed in Chapter 6, water molecules consist of two atoms of hydrogen and one atom of oxygen united by polar covalent bonds. Studies of the crystal structure of ice indicate that the water molecule is bent, with a structure that can be represented as follows:

$$O$$
$$H \overset{\diagup\quad\diagdown}{\underset{105°}{}} H$$

The angle between the two hydrogen-oxygen bonds is about 105°. This is close to the angle expected for sp^3 hybridization of the oxygen-atom orbitals.

Because of the difference in electronegativity between the strongly electronegative oxygen atom and the less electronegative hydrogen atoms, the H—O covalent bonds in the water molecule are highly polar. The electrons in each of the covalent bonds spend more of their time close to the oxygen nucleus than the hydrogen nucleus, giving the oxygen portion of the water molecule a partial negative charge and leaving the hydrogen portions with partial positive charges. Since the two polar-covalent bonds do not lie in a straight line, the water molecule as a whole is polar.

The water molecules in solid or liquid water are linked together by intermolecular hydrogen bonds. The number of molecules bonded together

STUDY HINT

Orbitals, hybridization, and chemical bonding are treated in Chapters 4 and 6.

Δe.n. = e.n.(oxygen) − e.n.(hydrogen)
Δe.n. = 3.5 − 2.1 = 1.4

decreases with increasing temperature because increases in kinetic energy makes such bond formation difficult. Nevertheless, there are usually from four to eight molecules per group in liquid water, as shown in Figure 13-19. If it were not for the formation of these molecular groups, water would be a gas at room temperature. In contrast, nonpolar molecules such as methane, CH_4, which are similar in size and mass to water molecules, do not undergo hydrogen bonding and are gases at room temperature.

Ice consists of H_2O molecules arranged in the hexagonal latticework shown in Figure 13-19. The spaces between atoms in this pattern are filled with air and account for the relatively low density of ice. As ice is heated, the increased energy of the atoms and molecules causes them to move and vibrate more vigorously. When the melting point of ice is reached, the energy of the atoms and molecules is so great that the rigid open lattice structure of the ice crystals breaks down. The ice turns into liquid water.

The hydrogen bonds between molecules of liquid water at 0°C are longer and more flexible than those between molecules of ice at the same temperature. Because of the flexibility of hydrogen bonds and because the rigid lattice structure of ice has broken down, groups of water molecules tend to crowd together and liquid water is denser than ice.

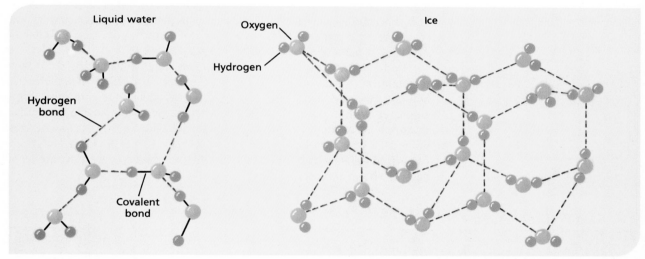

Figure 13-19 The structure of liquid water (left) and ice (right).

As water is warmed from 0°C, the breakage of some of the hydrogen bonds enables water molecules to crowd closer together. At the same time, the increased energy of the water molecules (owing to the rise of temperature) allows them to overcome molecular attractions more effectively and gradually spread apart. Water molecules are as tightly packed as possible at 4°C; hence, we say that water has its maximum density at 4°C. At temperatures above 4°C, the increasing energy of the water molecules causes them to move farther and farther apart as temperatures continue to rise. Thus, the density of water steadily decreases at temperatures above 4°C.

As temperature approaches the boiling point, groups of water molecules must absorb enough energy to break up into single molecules before the water boils. As a result, the normal boiling point of water, 100°C, is relatively high, and a large amount of heat is needed to vaporize the water.

Physical Properties of Water

At room temperature, pure water is a transparent, odorless, tasteless, and almost colorless liquid. Any observable odor or taste is caused by impurities such as dissolved minerals, liquids, or gases.

As shown by its phase diagram in Figure 13-18, at a pressure of 1 atm (0.1 Mpa), water freezes and melts at 0°C). The molar heat of fusion of ice, 6.008 kJ/mol, is relatively large compared to the molar heats of fusion of other solids. Water has the unusual property of expanding 11% in volume as it freezes because its molecules form a rigid latticework, as described earlier. As a result, ice at 0°C has a density of about 0.917 g/cm^3, compared with a density of about 1.00 g/cm^3 for liquid water at 0°C.

This is why ice floats in liquid water. The low density of ice is particularly important in the case of large bodies of water. If ice did not float, the water in lakes and ponds of temperate climates would freeze solid during the winter, killing nearly all the living things in them.

When liquid water at 0°C is warmed, its density increases until its temperature reaches 4°C. At this temperature, water molecules are as tightly packed as they can possibly be. Thus, water is densest at 4°C. This behavior is also unusual compared with that of most other substances. Above 4°C, water behaves more typically by gradually expanding as its temperature increases. This is because of molecular motion due to an increase in kinetic energy that accompanies the rise in temperature. At this temperature, one milliliter of water has a mass of almost exactly one gram.

Under a pressure of 1 atm (101.3 kPa), water boils at 100°C. At this temperature, water's molar heat of vaporization is 40.79 kJ/mol. Both the boiling point and the molar heat of vaporization of water are quite high compared with those of nonpolar substances with comparable molecular masses, such as methane (CH_4). The high molar heat of vaporization makes water useful for household heating systems. Many homes are heated by steam (vaporized water) which condenses in radiators, releasing great quantities of heat, as shown in Figure 13-20.

Sample Problem 13.1

How much heat energy is absorbed when 47.0 g of ice melts at its given melting point (i.e., at STP)? How much heat energy is absorbed when this same mass of liquid water boils?

Solution

Step 1. Analyze Given: mass = 47.0 g H_2O(s)
mass = 47.0 g H_2O(l)
Unknown: heat energy

Step 2. Plan

$$\text{mass} \rightarrow \text{mol} \rightarrow \text{heat energy}$$
$$\frac{g}{g/mol} = \text{mol}$$

mol × molar heat of fusion or vaporization = heat energy
mol × 6.008 kJ/mol = kJ (on melting)
mol × 40.79 kJ/mol = kJ (on vaporizing or boiling)

Step 3. Compute

$$\frac{47.0 \text{ g}}{18.0 \text{ g/mol}} = 2.61 \text{ mol}$$

2.61 mol × 6.008 kJ/mol = 15.7 kJ (on melting)
2.61 mol × 40.79 kJ/mol = 106 kJ (on vaporizing or boiling)

Step 4. Evaluate   Units have canceled appropriately. The answers have the proper number of significant digits and are reasonable close to estimated values of 18 (3 × 6) and 120 (3 × 40), respectively.

Practice Problems

1. What quantity of heat energy is released when 506 g of liquid water freezes?   *(Ans.)* 169 kJ
2. What mass steam is required to release 4.97×10^5 kJ of heat energy on condensation? *(Ans.)* 2.20×10^5 g

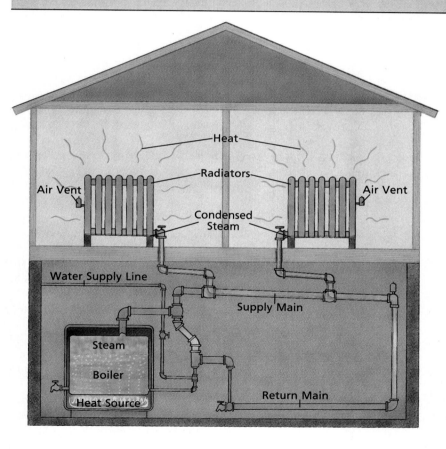

Figure 13-20 A steam heating system is an efficient way to heat a home because of the great amount of heat contained in the steam.

Section Review

1. Why is a water molecule polar?
2. How is the polarity of water responsible for some of water's unique characteristics.?
3. Describe the arrangement of molecules in liquid water and in ice.
4. Why does ice float? Why is this phenomenon important?

Chapter Summary

- The particles of a liquid are closer together and lower in energy than those of a gas. They are more ordered than those of a gas and less ordered than those of a solid.
- Liquids have a definite volume, a fairly high density, and are relatively incompressible. Like gases, liquids can flow and are thus considered to be fluids. Liquids can dissolve other substances.
- Liquids have the ability to diffuse. Unlike gases or solids, they exhibit surface tension and can evaporate or boil. A liquid is said to freeze when it becomes a solid.
- The particles of a solid have less kinetic energy than those of a liquid or gas. They move less than particles of liquids or gases and are more densely packed.
- Solids have a definite shape and may be crystalline or amorphous. They have a definite volume and are nonfluid. High density, incompressibility, extremely slow rates of diffusion, and definite melting points are additional characteristics of solids.
- A crystal lattice is the total three-dimensional array of points that describes the arrangement of the particles of a crystal. A crystal can be classified into one of seven crystalline systems on the basis of the three-dimensional shape of its unit cells. A crystal can also be described as one of four types on the basis of the kinds of particles it contains and the types of chemical bonding between its particles.
- Amorphous solids do not have a regular shape, but instead take on whatever shape is imposed on them.
- A liquid in a closed system will gradually reach a liquid–vapor equilibrium as the rate at which molecules condense equals the rate at which they evaporate.
- When two opposing physical or chemical changes occur at equal rates in the same closed system, such a system is said to be at equilibrium. Le Châtelier's principle states that if any of the factors determining an equilibrium are changed, the system will adjust itself in a way that tends to minimize that change.
- The pressure exerted by a vapor in equilibrium with its corresponding liquid at a given temperature is its equilibrium vapor pressure. A liquid begins to boil when the equilibrium vapor pressure of the liquid is equal to the atmospheric pressure. The amount of heat energy required to vaporize one mole of liquid at its boiling point is known as the liquid's molar heat of vaporization.
- A liquid is said to freeze when it changes to a solid. Freezing involves a loss of energy, in the form of heat, by the liquid. Melting is the physical change of a solid to a liquid. The amount of heat required to melt one mole of solid at its melting point is its molar heat of fusion.
- Water is a polar-covalent compound. The water molecule is bent in shape and has a partial negative charge near its oxygen atom and a partial positive charge near each of its hydrogen atoms. This unique structure is responsible for water's relatively high molar heat of fusion, boiling point and molar heat of vaporization. It is also the reason why water expands upon freezing and is able to float.

Chapter 13 *Review*

Liquids and Solids

Vocabulary

amorphous solid
boiling
boiling point
concentration
condensation
critical point
critical pressure
critical temperature
crystal
crystal lattice
crystalline solid
dissolved
equilibrium
equilibrium vapor
  pressure
evaporation
fluid

freezing
freezing point
Le Châtelier's principle
melting
melting point
molar heat of fusion
molar heat of vaporization
phase
phase diagram
sublimation
supercooled liquid
surface tension
system
triple point
unit cell
vaporization

Questions

1. What property of liquids accounts for the fact that they are the least common state of matter?
2. Compare the particle model of liquids with that of gases.
3. List nine properties of liquids, and briefly explain each in terms of the particle model of liquids.
4. What is a fluid?
5. What does it mean to say a substance has been dissolved?
6. What is surface tension?
7. Distinguish between evaporation and vaporization.
8. Give two reasons why evaporation is such a crucial process in nature.
9. Define "freezing."
10. *(a)* What are the two types of solids? *(b)* Give two examples of each type.
11. What is a crystal?
12. List seven properties of solids and explain each in terms of the particle model of solids.
13. Why is glass sometimes classified as a supercooled liquid?

14. Distinguish between a crystal lattice and a unit cell. *(a)* List four common examples of amorphous solids. *(b)* Cite some of the uses of glass.
15. Name the seven basic crystalline systems.
16. *(a)* List and describe the four types of crystals in terms of the nature of and type of bonding between their component particles. *(b)* What physical properties are associated with each type of crystal?
17. What are metallic glasses?
18. Define *(a)* phase *(b)* system.
19. Define *(a)* equilibrium *(b)* condensation *(c)* concentration.
20. Describe the energy change that takes place and write the corresponding equation illustrating *(a)* evaporation *(b)* condensation.
21. *(a)* What is Le Châtelier's principle? *(b)* How can it be used?
22. *(a)* What is meant by the equilibrium vapor pressure of a liquid? *(b)* What is the relationship between the temperature of water and its equilibrium vapor pressure?
23. Using Figure 13-28, determine the approximate equilibrium vapor pressure of each of the following at the specified temperature: *(a)* water at 40°C *(b)* water at 80°C *(c)* ether at 20°C *(d)* alcohol at 60°C *(e)* glycerol at 140°C.
24. What is meant by the boiling point of a liquid?
25. Define molar heat of vaporization.
26. *(a)* Define freezing. *(b)* Write the equation that illustrates this process. *(c)* What entropy change occurs during freezing?
27. *(a)* Define *melting*. *(b)* Write the equation that illustrates this process. *(c)* What entropy change occurs during melting?
28. Define molar heat of fusion.
29. *(a)* What is sublimation? *(b)* Give two examples of common substances that sublime at ordinary temperatures.
30. What is a phase diagram?
31. What is meant by the normal freezing point of a substance?

32. Describe the basic structure of a water molecule.
33. List at least eight physical properties of water.

Problems

1. The standard molar heat of vaporization for water is 40.79 kJ/mol. How much energy would be required to vaporize each of the following? *(a)* 5.00 mol H_2O *(b)* 45.0 g H_2O *(c)* 8.45 x 10^{50} molecules H_2O
2. The molar heat of fusion for water is 6.008 kJ/mol. How much energy would be required to melt each of the following? *(a)* 12.75 mol ice *(b)* 6.48 x 10^{20} kg ice.
3. *(a)* If water has a density of 0.999 13 g/cm^3 at 15°C, find the mass of 525.00 mL of water at this temperature. *(b)* How many moles of water does this represent? *(c)* How many molecules are contained in this mass of water?
4. Calculate the molar heat of vaporization of a substance given that 0.433 mol of the substance absorbs 36.5 kJ of energy when it is vaporized.
5. Given that a substance has a molar mass of 259.0 g/mol and 71.8 g of the substance absorbs 4.307 kJ when it melts, *(a)* calculate the number of moles in the 71.8-g sample. *(b)* calculate the molar heat of fusion.
6. *(a)* Calculate the number of moles in a liquid sample of a substance that has a molar heat of fusion of 3.811 kJ/mol, given that the sample releases 83.2 kJ when it freezes. *(b)* Calculate the molar mass of this substance if the mass of the sample is 5519 g.

Application Questions

1. Distinguish between the phenomena involved when a steel needle rests on a water surface and when a boat floats in water.
2. Explain the role of the humidity of the air in determining our relative comfort on a hot day.
3. Both diamond and graphite consist only of carbon atoms, yet these materials have very different properties. Explain the basic reasons for these observed differences.
4. The following liquid-vapor system is at equilibrium at a given temperature.

$$\text{liquid + heat energy} \rightleftarrows \text{vapor}$$

If the temperature is increased and equilibrium is established at the higher temperature, how does the final value of each of the following compare with its initial value? (In each case, answer either: higher, lower or the same.) *(a)* the rate of evaporation *(b)* the rate of condensation *(c)* the final concentration of vapor molecules *(d)* the final number of liquid molecules

5. Given a liquid-vapor equilibrium system, decide whether the temperature of that system should be increased or decreased in order to make each of the following changes in the process of establishing equilibrium at the new temperature: *(a)* an increased final rate of evaporation *(b)* an increased final concentration of vapor *(c)* an increased final rate of condensation *(d)* an increased final number of liquid molecules.
6. Explain why the vapor pressure of a liquid increases with increasing temperature.
7. Explain the relationship between the attractive forces between the particles in a liquid and the equilibrium vapor pressure of that liquid.
8. *(a)* Explain the relationship between atmospheric pressure and the actual boiling point of a liquid. *(b)* During continual boiling at this pressure, what is the relationship between the temperature of the liquid and that of its vapor? *(c)* How can this phenomenon be explained?
9. Explain the relationship between the magnitude of the molar heat of vaporization of a liquid and the degree of attraction between the particles of that liquid.
10. During the freezing of a substance, energy is continually being removed from that substance, yet the temperature of the liquid-solid system remains constant. Explain this phenomenon.
11. Explain the relationship between the molar heat of fusion of a solid and the degree of attraction between its particles.
12. At normal atmospheric pressure, the temperature of an ice-water system will remain at 0°C as long as both ice and liquid water are present, no matter what the surrounding temperature. Explain this in terms of Le Châtelier's principle.
13. Based on Figure 13-29, what phase(s) of water is (are) present at each of the following conditions: *(a)* for any corresponding temperature-pressure reading on curve AB

(b) for any corresponding temperature-pressure reading on curve AD *(c)* for any corresponding temperature-pressure reading on curve AC *(d)* at point A.

14. Given a sample of water at any point on curve AC in Figure 13-29, what effect would each of the following changes have on that sample? *(a)* increasing the temperature at constant pressure *(b)* increasing the pressure at constant temperature *(c)* decreasing the temperature at constant pressure *(d)* decreasing the pressure at constant temperature.

15. Given a sample of water at any point on curve AB in Figure 13-29, what effect would each of the following changes have on that sample? *(a)* increasing the temperature at constant pressure *(b)* increasing the pressure at constant temperature *(c)* decreasing the temperature at constant pressure; *(d)* decreasing the pressure at constant temperature.

16. Given a sample of water at any point on curve AD in Figure 13-29, how could more of the liquid water in that sample be converted into a solid without changing the temperature?

17. Explain why methane (CH_4), which is similar in size and mass to water, is a gas at room temperature, whereas water is a liquid at that temperature.

18. *(a)* Why is liquid water more dense than ice? *(b)* Why is liquid water at temperatures between 0°C and 4°C more dense than liquid water above 4°C?

19. What is the principle upon which the use of the steam radiator is based?

20. Is it possible for a container of water to simultaneously freeze and boil? Describe an experimental set-up that could demonstrate this.

21. What liquid is often called "the universal solvent"? Why?

22. What are boiling chips? Explain what they do and how they work.

Application Problems

1. Which contains more molecules of H_2O—5.00 cm^3 of ice at 0°C or 5.00 cm^3 of liquid water at 0°C? How many more?

2. Which contains more molecules of H_2O—5.00 cm^3 of liquid water at 0°C or 5.00 cm^3 of water vapor at STP? How many more?

3. What volume and mass of steam at 100°C and 1.00 atm would, when condensed, release the same amount of heat as 100 cm^3 of liquid water when frozen? What do you note, qualitatively, about the relative volumes and masses of steam and liquid water required to release the same amount of heat?

Enrichment

1. Compile separate lists of crystalline and amorphous solids found in your home. Compare your lists with those of your classmates.

2. Design an experiment to grow crystals of various safe, common household materials. Record the conditions under which each type of crystal is best grown.

3. Write a brief report about the glass-making process. If possible, observe a glassblower at work and record the step-by-step procedures involved.

4. Prepare a report tracing the origin and development of the steam radiator as a home heating device.

5. What are some household examples of sublimation? Make a list of household solids that will sublime under conditions that might be encountered in a home.

6. Describe the use of different kinds of fluids in industrial hydraulic systems. What characteristics make fluids useful for hydraulic applications?

7. Prepare a report about the adjustments that must be made when cooking and baking at high elevations. Collect instructions for high elevation adjustments from the packages of prepared mixes at the grocery store. Explain why changes must be made in recipes that will be prepared at high elevations. Check your library for cookbooks containing information about food preparation at high elevations.

8. How is capillary action important to plants (hint: think of very large plants, such as trees).

9. Look up the molar heat of vaporization of water and compare it to that of several other familiar liquids. How does the molar heat of vaporization compare to that of isopropyl alcohol (rubbing alcohol)?

10. Consult reference materials at the library and prepare a report on the process of freeze-drying, describing its history and its applications.

CAREERS

Occupations in the Compressed-Gas Industries

The major method of producing pure gases is by compressing and cooling air until it liquefies. This liquid is then fractionally distilled to extract the desired gases. The process requires special equipment but generally does not require chemical expertise from the equipment operators.

The liquefied gases are extremely cold. People who work with compressed gases must understand the changes of pressure and volume that occur as the gases warm and the changes that occur in materials that come in contact with cold, liquified gases. Pipes can shatter, since many metals get brittle when they are cold. Joints will leak if

they have not been sealed with materials that can withstand extreme temperature changes.

The manufacture and handling of nonatmospheric gases require chemical knowledge as well as common sense. The chemical reaction used to produce the gas must be carefully monitored to avoid potentially dangerous side reactions. For example, chlorine gas reacts with nearly any metallic surface it touches. Leaks of carbon monoxide can lead to fatalities because, in its early stages, CO poisoning is easily confused with the onset of a bad cold. Plans must be developed for coping with spills and leaking pipes.

Opportunities abound in the compressed-gas industries for people with a background or interest in chemistry. A college degree is usually not necessary, but some technical training is essential.

Chemical Patent Attorney

Students who have an interest in both the law and chemistry may find that chemical patent law is an attractive career. Chemical patent attorneys are becoming increasingly important to industry.

In addition to doing the painstaking work needed to establish a valid patent, chemical patent attorneys help decide whether a product is sufficiently different to justify filing a patent application. Deciding if an idea is new enough to patent or whether it infringes on existing patents is often difficult. Even if a chemical compound is already known, it may be possible to patent a new method of making it.

To become a chemical patent attorney, you must study chemistry as part of your undergraduate college program and then obtain a law degree.

Cumulative *Review*

Unit 4 • Phases of Matter

Questions

1. *(a)* What are allotropes? *(b)* Illustrate the phenomenon of allotropy with oxygen. *(c)* What is the significance of the two allotropes of oxygen in our everyday lives? (10.1)
2. Compare the physical properties of oxygen and ozone. (10.1)
3. Identify each of the following: *(a)* the most abundant element in the earth's crust, waters, and atmosphere *(b)* the industrial process by which most ammonia is produced *(c)* the most abundant element in the universe *(d)* the second most electronegative element *(e)* a triplybonded diatomic gaseous molecule found in the earth's atmosphere *(f)* the molecular substance with a pungent odor often produced during lightning storms *(g)* the gas involved in the creation of the greenhouse effect *(h)* gaseous form of dry ice *(i)* the gas required in photosynthesis. (10.1, 2, 3, 4)
4. Explain the importance of nitrogen to life on earth. (10.3)
5. Compare carbon dioxide and carbon monoxide in terms of *(a)* their effects on living things *(b)* their molecular structures *(c)* their everyday uses. (10.4)
6. *(a)* Why do real gases deviate from ideal gas behavior? *(b)* among the gases listed (Br_2, He, N_2, HCl, O_2, Ne, H_2O, H_2, and NH_3), which are most likely to display ideal gas behavior? *(c)* Which are least likely to behave ideally? (11.1)
7. Describe the liquid state according to the particle model of liquids and cite at least six properties of liquids. (13.1)
8. Describe the solid state according to the particle model of solids and cite at least six properties of solids. (13.2)
9. Distinguish between crystalline and amorphous solids. (13.2)
10. *(a)* What is meant by the equilibrium vapor pressure of a liquid? *(b)* How is it related to the temperature of the liquid? (13.1)

Problems

1. Convert each of the pressure readings into the unit specified: *(a)* 1.75 atm into mm Hg *(b)* 1140 mm Hg into atm *(c)* 250 mm Hg into torr *(d)* 0.500 Pa into mm Hg. (11.2)
2. If a 1.40 L sample of nitrogen exerts a pressure of 0.75 atm at a specified temperature, what pressure would it exert if confined to a volume of 500. mL at that same temperature? (11.3)
3. A sample of hydrogen occupies a volume of 820 mL at 30°C. What volume will the gas occupy at 60°C if the pressure remains constant? (11.3)
4. 1.60 L sample of ammonia at 25°C is compressed to 0.450 L at constant pressure. What is its final temperature in °C? (11.3)
5. A sample of chlorine at 22°C exerts a pressure at 420. mm. If the gas is heated to 77°C at constant volume, what will its new pressure be? (11.3)
6. A sample of helium at 27°C and 700. mm pressure occupies a volume of 650. mL. What volume would the gas occupy at 57°C and 740. mm pressure? (11.3)
7. If 325 mL of hydrogen is collected over water at 25°C and 775.0 mm, compute the volume of the dry hydrogen at STP. (12.2)
8. What volume, in liters, would be occupied by 0.250 mol N_2 at STP? (12.2)
9. Find the mass, in grams, of 8.40 L CO_2 at STP. (12.3)
10. Find the density of NO_2 in grams per liter at STP. (12.3)
11. A 0.420 g sample of a gas has a volume of 186 mL at STP. Find the molar volume. (12.3)
12. What is the pressure, in atmospheres, exerted by 0.86 mol of N_2 in a 4.75 L container at 27°C? (12.2)
13. If 4.50 g of water undergo decomposition at STP, what volume of each product is formed? (12.3)
14. If 45.0 L of natural gas, which is essentially methane (CH_4), undergoes combustion at 730. mm Hg and 20°C, what mass, in grams, of each product is formed? (12.3)

Intra-Science
How the Sciences Work Together

The Problem in Chemistry: Preserving the Ozone Layer

In 1988 a giant U.S. chemical company announced that it would voluntarily phase out production of all chlorofluorocarbons (CFCs). The reason: to protect the earth's ozone shield.

It was a significant announcement. Although it had been suspected for some time that CFCs were damaging the ozone layer, and although an international accord in 1987 had called for a freeze on production levels, E.I. du Pont de Nemours & Company, a chemical concern, was more than merely complying with the agreement's terms. The chemical company was ready to reduce, rather than merely freeze, its level of CFC production.

Why is ozone so important?

No substance is more crucial to life than ordinary molecular oxygen (O_2), which makes up about 20 percent of the volume of the lower atmosphere, where it is crucial to respiration. Ozone (O_3) is a different molecular form of the element oxygen, and is equally important to life. A layer of ozone in the upper atmosphere guards living things on earth against deadly solar radiation. CFCs, such as those used as propellants in aerosol cans, are believed to be threatening this protective layer.

Ironically, ozone is a pollutant at the earth's surface—a major ingredient of photochemical smog. Scientists are trying to minimize the amount of ozone in the lower atmosphere and at the same time trying to preserve the ozone layer in the upper atmosphere.

Ozone is a bluish gas that is produced by the process in which O_2 is broken up into atomic oxygen, followed by the combination of the single atoms of oxygen (O) with O_2 to produce O_3. On the earth's surface, the production of ozone commonly involves an electrical spark that passes through air or pure oxygen. The pungent smell that can be noticed near a spark-producing electrical apparatus or during a lightning storm is that of ozone.

The Connection to Biology and Health Sciences
Because of its strong oxidizing character, ozone is toxic to animals, including humans. Los Angeles smog can contain up to 0.5 parts per million (ppm) of ozone, a concentration sufficiently high to irritate eyes and respiratory tracts. This ozone is formed by the action of sunlight on oxygen in the presence of impurities in the air. At the altitudes flown by intercontinental aircraft, cabin levels of ozone can reach 1.2 ppm unless reduced by carbon filters and ozone-decomposing equipment.

Although ozone itself is toxic to life, its presence in the upper atmosphere protects life because ozone absorbs ultraviolet radiation at a wavelength range of 240–320 nanometers (a nanometer is one billionth of a meter— 10^{-9} m).

Ultraviolet radiation can penetrate 60 feet or more into the ocean and kill one-celled organisms that are at the base of the ocean food chain. The loss of significant amounts of these organisms could result in a collapse of the ocean ecosystem.

Ultraviolet radiation also kills the surface cells of plants. A 15-percent drop in atmospheric ozone could result in $3 billion worth of crop damage in the United States.

Sunburn, premature aging of the skin, and skin cancer are consequences of increased ultraviolet radiation. The Environmental Protection Agency estimates that a 1- percent increase in ultraviolet radiation could cause the incidence of skin cancer to increase by 2 percent.

The Connection to Physics

The production of ozone in the upper atmosphere and at the earth's surface depends upon the absorption of ultraviolet light. In the first step of the production of ozone in the upper atmosphere, ultraviolet light, which has a wavelength lower than 240 nm, is absorbed by O_2. This absorbed light is changed into an equivalent amount of energy, which causes the oxygen molecule to break apart into two oxygen atoms. Ozone is then produced by the reaction of O with O_2.

Once ozone molecules have been formed, they absorb ultraviolet light in the 240−320 nm range. The energy released from this absorbed light causes each ozone molecule to decompose to produce a molecule O_2 and an atom of O. These two species then recombine to form a new molecule of ozone. In this way, the ozone layer is maintained to perform its shielding role.

On the earth's surface, nitrogen dioxide (NO_2), which is produced in automobile engines by the combination of atmospheric oxygen with atmospheric nitrogen, absorbs ultraviolet light with a wavelength lower than 395 nm. The released energy causes the NO_2 molecule to decompose to NO and O. The resulting atomic oxygen can combine with O_2 to form ozone or can combine with unburned hydrocarbons from automobile exhaust to produce a variety of irritating compounds.

The Connection to Earth Science

The British scientist Joseph Farman began monitoring the levels of ozone over Antarctica in 1957 using instruments positioned at Halley Bay. In 1985 he shocked the scientific community with his report that there had been a decline in these ozone levels every year since 1978. His results were confirmed by measurements made from NASA's *Nimbus-7* satellite. Scientists immediately sought an explanation for this "ozone hole."

Today there is evidence that this hole may be caused by human-made chemicals such as CFCs. Because they are nontoxic and chemically inert, these compounds have been widely used as refrigerants and as propellants in aerosol cans. When these substances are released to the atmosphere, however, their inertness ensures that they are not readily destroyed. As a consequence, they eventually find their way into the upper atmosphere.

Scientists hypothesize that, once in the upper atmosphere, CFCs are decomposed by ultraviolet light to produce chlorine, which destroys ozone. Recent experiments in which a NASA aircraft collected air samples high over Antarctica revealed a chlorine level 500 times greater than normal.

Unit 5

Chapter 14
Solutions   404

Chapter 15
Ions in Aqueous Solutions   438

Chapter 16
Acids and Bases   460

Chapter 17
Acid–Base Titrations and pH   490

SOLUTIONS AND THEIR BEHAVIOR

*My high school chemistry teacher, John St.
Sure, hooked me on chemistry with a class
dominated by his twin loves: pranks, and
technical excellence. His perennial stock of
anecdotes about his former students who had
made good in science or engineering delivered
a powerful message about the impotence of mere
enthusiasm unsupported by hard work.*

*My present work in analytical chemistry is
devoted to developing better techniques for
measuring trace amounts of substances in
mixtures known as solutions. The aim is to have
better tools to measure and understand our
environment.*

*Solutions are an important part of our
environment. For example, the oxygen and
carbon dioxide dissolved in oceans, lakes, and
rivers allow aquatic life to flourish. Aquatic
plants liberate tremendous amounts of oxygen
as they break down carbon dioxide during
respiration. This helps maintain a life-sustaining
balance between oxygen and carbon dioxide in
our atmosphere.*

Janet G. Osteryoung

Janet G. Osteryoung
Analytical Chemist
State University of New York at Buffalo

Coral reef in the Red Sea

Chapter 14 Solutions

Chapter Planner

14.1 Types of Mixtures
- Question: 1
- **Application Question: 1**
Solutions
- Demonstration: 1
- Questions: 2, 3
Suspensions
- Question: 4
Colloids
- Demonstration: 2
- Desktop Investigation
- Questions: 5–7
- Application Question: 2

14.2 The Solution Process
Factors Affecting the Rate of Dissolving
- Question: 8
Solubility
- Demonstrations: 3, 5
- Questions: 9–12, 18
- **Application Problems: 1, 2**
Factors Affecting Solubility
- Demonstration: 4
- Experiment: 18
- Questions: 13, 14, 17

- Questions: 15, 16, 19, 20
- Application Questions: 3–5
- Application Question: 6
Heats of Solution
- Demonstration: 6
- Questions: 21, 22
- Application Question: 7
- **Questions: 24, 25**
- Application Question: 8a-d
- Question: 23
- Application Question: 8e-g

14.3 Concentrations of Solutions
Percent by Mass
- Question: 26
- Problem: 1
- Application Problem: 3
- **Application Question: 9a**
Molarity
- Question: 27
- Problems: 2–7
- **Application Question: 9b**
- Application Problem: 6
- Application Problems: 7–9

Molality
- Demonstration: 7
- Question: 28
- Problems: 8–14
- Application Problems: 10, 11
- **Application Questions: 9c, 10**
- Application Problem: 12

14.4 Colligative Properties of Solutions
Vapor-Pressure Lowering
- Questions: 29a, 30, 31
- Technology
Freezing-Point Depression
- Questions: 29b, 32, 33
- Problems: 15–17
- Application Question: 11
- **Application Problems: 13, 15, 16**
- Application Problem: 14
Boiling-Point Elevation
- Questions: 29c, 34, 35
- Problems: 18, 19
Determination of Molar Mass of a Solute
- Demonstration: 8
- Problems: 20, 21
- **Application Problems: 17, 18**

Teaching Strategies

Introduction

Introduce this chapter by telling students they are going to learn why salt is added to ice in an ice-cream freezer, why the temperature goes up on a candy thermometer as you cook the candy, how to remove stains from clothing, why soda fizzes when the cap is removed from the bottle, the types of foods that are colloids, and other exciting things.

The overall objectives for this chapter are *(1)* to study solutions and determine how they can be distinguished from heterogeneous mixtures—colloids and suspensions—which are sometimes mistaken for solutions *(2)* to find out what determines whether or not one substance will dissolve in another and what determines the rate at which they will dissolve *(3)* to learn how to make solutions of specific concentrations *(4)* to determine what happens to the temperature of a solvent when a solute is dissolved in it and why, and *(5)* what happens to the freezing and boiling points of solvents when solutes are dissolved in them.

14.1 Types of Mixtures

The following demonstration involves making two common solutions with seemingly illogical but interesting results!

Demonstration 1: 1 + 1 Does Not Always = 2

PURPOSE: To determine the volume of a solution made up of equal volumes of its components

MATERIALS: 2 250-mL beakers, Celsius thermometer, distilled water, 95% ethanol, 4 100-mL graduates, stirring rod, sucrose

PROCEDURE: Fill four 100-mL graduates with 50 mL of ethyl alcohol, 50 mL of distilled water, 50 mL of sucrose, and 50 mL of hot water (about 70°C). Ask your students what they think the volume of each of the following solutions will be when *(1)* you pour the 50 mL of ethyl alcohol into the graduate containing 50 mL of water, and *(2)* you mix the 50 mL of sucrose and 50 mL of hot water in a beaker? Carry out the two experiments, and determine the volume of the resulting mixtures. Be sure the students are aware of the original volumes of the components of the mixture. (The alcohol and water mixture will be about 92 mL and the sucrose and water solution about 75 mL.) Ask why they think the volumes are smaller? (Possible explanation: The smaller water molecules can slip into the spaces between the bigger alcohol and sucrose molecules. Also, spaces existed between the sucrose crystals.)

You could try similar experiments with other substances that are highly soluble in one another.

Demonstration 2: Common Solutions, Colloids, and Suspensions

PURPOSE: To determine whether certain common substances found around the home are solutions, colloids, or suspensions

MATERIALS: 2 bottles of club soda in a clear glass bottle, lemon-lime soda in a clear glass bottle, 70% ethyl or isopropyl alcohol, fruit gelatin, iodine crystals, milk of magnesia, oil and vinegar salad dressing, starch liquid remaining after macaroni or noodles are cooked, tap water, slide projector, 6 medium test tubes

PROCEDURE: Fill each of the six test tubes three-fourths full of club soda, 70% alcohol, milk of magnesia, starch liquid, fruit gelatin (prepare ahead and refrigerate), tap water. Put the milk of magnesia in several hours prior to the demonstration as it takes awhile for it to settle out.

Remove the labels from the unopened soda bottles and the oil and vinegar dressing. Examine the contents of the test tubes plus the unopened club soda, lemon-line soda, and the oil and vinegar dressing, and classify them as *(1)* suspensions or *(2)* solutions or colloids. (The milk of magnesia, the oil and vinegar salad dressing, and the club soda in the test tube are all suspensions because they settle upon standing. Actually, the gas in the soda rises. The starch liquid is cloudy but does not settle out; thus, it is a colloid.)

Check the remaining clear mixtures by shining the beam of a projector on them in a darkened room to see if they show the Tyndall effect. (The fruit gelatin should show the Tyndall effect; hopefully, the tap water will not. The remaining substances are solutions.)

Add 2 or 3 crystals of iodine to the 70% alcohol solution, and check it for the Tyndall effect. (negative result—solution)

Desktop Investigation Solutions, Colloids, Suspensions

If students wish to do this lab at home, the following substitutions can be made for equipment and measuring devices: pint jars for beakers, small juice glasses for test tubes, 1 cup for 250 mL, 1 tablespoon—15 mL, 1 teaspoon—5 mL, 1 tbsp—12 g sucrose, 1 tsp—3 g starch, $1\frac{1}{2}$ tsp—5 g clay, $\frac{1}{2}$ tsp—2 g borax, $\frac{1}{2}$ package—3 g gelatin. If juice glasses are used for test tubes, they should be filled about $\frac{2}{3}$ full.

Have students make a data table to record the results of their experiment in. The data table (with answers) might look similar to the one below.

DATA TABLE				
Water Mixture of	Appearance	Separation occurs	Path of light visible	Conclusion
sucrose	clear	no	no	solution
starch	slightly cloudy	no	yes	colloid
clay	two layers	yes	—	suspension
food coloring	clear	no	no	solution
sodium borate	clear	no	no	solution
oil	two layers	yes	—	colloid
gelatin	clear-cloudy	no	yes	

ANSWERS TO DISCUSSION QUESTIONS

1. The water mixtures of sucrose, food coloring, and sodium borate are solutions; those of starch and gelatin are colloids; and those of clay and oil are suspensions.

2. Solutions are clear, do not settle out or separate on standing, and do not show a visible light beam path. Colloids are sometimes clear or cloudy, do not settle out or separate on standing, but the path of a light beam is always clearly visible in them. Suspensions settle out or separate on standing.

14.2   The Solution Process

Demonstration 3: Solution Equilibrium

PURPOSE: To explain solution equilibrium using a common saturated solution

MATERIALS: Glass, stirring rod, ice, sucrose, tea bag, water

PROCEDURE: Have a glass of iced tea with too much sugar added to it; thus, the excess is on the bottom of the glass. Think back to our discussion on physical equilibrium in Chapter 13. Is any of the sugar on the bottom dissolving after you have stirred the tea thoroughly? What happens every time another sugar molecule dissolves? Why is this another example of physical equilibrium? What is the relationship between a saturated solution and solution equilibrium? What is wrong with the definition "A saturated solution is one that cannot dissolve any more solute"? (The definition makes a saturated solution sound like a static situation rather the continual movement between particles as they take turns dissolving and recrystallizing.)

The following demonstration can be saved and used from year to year.

Demonstration 4: Solubility of Gases in Liquids

PURPOSE: To show the effects of temperature and pressure on the solubility of gases in liquids

MATERIALS: 4 12-oz bottles of club soda, 3 5×5 cm glass bends, glass plate, 3 1-hole stoppers to fit soda bottles, 3 rubber delivery tubes, 3 1-liter soda bottles, 2 1-liter beakers, pneumatic trough (or a plastic dishpan), permanent felt tip marker, ice, Bunsen burner, ring stand and ring, wire gauze with ceramic center

PROCEDURE: Part 1: Chill one can of soda several hours before the experiment. Ten minutes before class, start to heat 350 mL of water in a 1-L beaker, and fill the trough and the three 1-L soda bottles with water.

When you are ready to start, cover the mouth of each bottle with a glass plate, and carefully invert it in the trough. Have two students help hold the bottles. Assemble all the apparatus as shown in Figure T14-1 so that the stopper assembly can be put into the soda bottles as soon as they are opened and gas collection started immediately. Fill the second beaker with ice.

Open the one chilled bottle and the two room-temperature bottles of soda, and put the stopper assemblies in immediately. Place the chilled bottle in a beaker with ice and one of the room temperature bottles in the beaker containing the hot (not boiling) water. Leave the other one at room temperature. At the end of fifteen minutes, have students note the differences of the water levels in each gas-collecting bottle. How do you explain the results? (The solubility of gases in liquids decreases as the temperature increases. Therefore, you would expect soda in the hot water to produce the most carbon dioxide and the one in the ice, the least.)

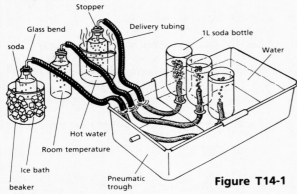
Figure T14-1

Part 2: Have students observe an unopened bottle of soda. Open the bottle, and have students note what occurs. (The soda in the unopened bottle looks like water. After opening, little bubbles of gas immediately appear.) How do you account for the differences? (Opening soda releases the pressure and thus decreased the solubility of the gas in the liquid.)

Demonstration 5: Supersaturated Solutions
PURPOSE: To show the properties of a supersaturated solution
MATERIALS: Sodium thiosulfate pentahydrate, 4 medium test tubes, 4 #3 solid rubber stoppers, 250-mL beaker, ring stand and ring, wire gauge with ceramic center, Bunsen burner
PROCEDURE: Using a water bath, heat 4 or 5 test tubes half filled with sodium thiosulfate pentahydrate ($Na_2S_2O_3 \cdot 5H_2O$) until the crystals are completely melted forming a colorless liquid. Allow the test tubes and their contents to cool to room temperature in an upright position, and then stopper to prevent contamination. To reestablish equilibrium, drop a small crystal of $Na_2S_2O_3 \cdot 5H_2O$ into the supersaturated solution, and crystals start to form immediately. Let students *lightly* touch the sides of the tube to observe the exothermic nature of this recrystallization. Stopper the tube and save for next year. Extra samples should be made because sometimes samples recrystallize on cooling. The extras assure that you will have a successful sample.

For a dramatic variation of this demonstration, refer to Shakhashiri's "Crystallization from Supersaturated Solutions of Sodium Acetate," *Chemical Demonstrations*, Vol. 1, p. 27.

Demonstration 6: Heats of Solution
PURPOSE: To show the effects of solutes having positive and negative heats of solution on the temperature of the solutions they form with water
MATERIALS: Ammonium chloride, calcium chloride, lithium chloride, potassium nitrate, 4 medium test tubes, stirring rod, distilled water
PROCEDURE: Put 10 mL of water into each of four test tubes. Have students note the temperature of the water both by touching the outside of the test tubes and by a thermometer reading. Add 5 g of potassium nitrate to the first tube, 5 g of lithium chloride to the second, 5 g of ammonium chloride to the third, and 5 g of calcium chloride to the last. Have students note the changes in temperature and relate these changes to the heats of formations on Table 14-5 in the text. (KNO_3 and NH_4Cl both caused a decrease in temperature as would be expected since they both have a $+\triangle H_s$, while LiCl, having a $-\triangle H_s$, and $CaCl_2$ both caused an increase in temperature. Ask on the basis of the data gathered from the demonstration, would calcium chloride, which is not listed on the table, have a positive or negative $\triangle H_s$? (Since $CaCl_2$ gave off heat, it would have a $-\triangle H_s$.)

A practical application of heat of solution follows: Hot and cold packs, which are sometimes used for treating injuries when hot water or ice is not available, are perfect examples of heat of solution put to a practical use. They consist of an outer plastic bag containing a dry chemical plus an inner pouch containing water and a dye. The inner pouch must be firmly squeezed to release the water so that it can mix with the dry chemical and produce heat or cold. In the cold packs the dry chemical is ammonium nitrate and in the hot ones, calcium chloride. If you wish to use them in a demonstration, they are available from medical supply outlets or sports stores. Your athletic department might even have them! **CAUTION:** If you use them in a demonstration, the hot packs produce enough heat to cause burns, and the cold ones should not have prolonged contact with the body, so handle with care.

14.3   Concentrations of Solutions

A little consumer economics which applies percent by mass solutions to everyday life follows: If you have a choice of buying two different brands of bleach costing different amounts, you should check the labels, if the % of NaClO is the same, buy the cheaper brand. (Throughout chemistry, emphasize reading labels. As students learn more chemistry, they are capable of becoming better consumers.)

The following demonstration is a detailed version of the one suggested in the TE annotation on page 423.

Demonstration 14.7: Molar and Molal Solutions
PURPOSE: To show how molar solutions and molal solutions differ from one another
MATERIALS: Decigram balance, permanent felt tip marker, 100-mL graduated cylinder, 2 2-L plastic soda bottles, sucrose, water
PROCEDURE: Prior to class, weigh out two 1-mol samples (342 g) of sucrose. Fill one soda bottle with 1000 mL of water. Mark the level of the water with the marking pen, and transfer the water to the second bottle. Again, mark the level of the water, discard the water.

In class, have the students calculate the mass of one mole of sucrose. Tell them the preweighed sample of sucrose has a mass of 342 g. Ask how many moles of sucrose are present? (1 mol) Put the 1 mol of sucrose in one bottle, and add enough warm water (not boiling) to reach the 1-L mark on the bottle. Ask what the volume of the solution is? (1 L) What would you call this solution? (a 1-molar solution)

Weigh the second empty plastic bottle on the decigram balance. Then fill it with water to the 1-L mark and weigh it again. Have students calculate the mass of the water. What is the mass of the water? (1000 g or 1 kg)

Remind students that the reason 1 L of water has a mass of 1 kg is because water has a density of 1 g/mL at 4°C. (The warm water would vary slightly from this if precise measuring devices were used. However, the devices used here are not very precise.) Further, point out that 1 kg of most liquids does NOT occupy a volume of 1 L. For example, 1 kg of ethanol (density 0.789 g/mL) occupies a volume of 1.27 L while 1 kg of glycerol (density 1.26 g/mL) occupies a volume of 0.794 L. Now add the 1 mol of sucrose to the 1 kg of water and shake until dissolved. Have students estimate the volume of the solution. (About 1.3 L) Ask, what would you call this solution? (a 1-molal solution) What is the difference between a 1-*M* and a 1-*m* solution? (A molar solution always has a volume of 1L. The volume of a molal solution depends upon the density of the solvent and the amount of solute used; thus, their final volumes vary greatly.)

14.4 Colligative Properties of Solutions

For an activity that will help your students understand why salt is added to ice in an ice-cream freezer and at the same time make you popular with them (and maybe even other people in the school) do "Team Ice Cream" in *Chemical Activities, Teacher Edition*, by Borgford and Summerlin, American Chemical Society, Washington, D.C., 1988, pp. 299–301.

Demonstration 14.8: Determination of the Molecular Weight of Urea

PURPOSE: To determine the molecular weight of urea experimentally
MATERIALS: 125-mL Erlenmeyer flask, 2-L pan, Celsius thermometer, ice, urea, rock salt, distilled water
PROCEDURE: Dissolve 10 g of urea in 50 mL of water in an Erlenmeyer flask. Submerge the flask in a pan of chipped-ice and salt mixture (1 L chipped ice and 100 mL rock salt—add more salt if a lower temperature is needed). Stir the mixture continuously with the thermometer until freezing is almost complete. At that point, note the temperature. (The temperature should be about −6.2°C.) Have students calculate the molar mass of urea. (Should be about 60 g/mol)

References and Resources

Books and Periodicals

Borgford, Christie and Lee Summerlin. "Mayonnaise: An Edible Emulsion," *Chemical Activities, Teacher Edition*, Washington, D.C.: American Chemical Society, 1988, pp. 292–3.

Davenport, Derek. "When Push Comes to Shove: Disturbing the Equilibrium" *ChemMatters*, Vol. 3, No. 1, Feb. 1985, pp. 15–6.

"Gas Laws and Scuba Diving," *ChemMatters*, Vol. 1, No. 1, Feb. 1983, pp. 4–6.

Goldsmith, Robert H. "A Simple Tyndall Effect Experiment," *Journal of Chemical Education*, Vol. 65, No. 7, July 1988, p. 623.

Grosser, Arthur E. "Emulsified Sauces," and "Ice Cream and Sherbert," *The Cookbook Decoder*, New York: Warner Books, 1981, pp. 166–174.

Nordstrom, Brian H. "The Effect of Polarity on Solubility," *Journal of Chemical Education*, Vol. 65, No. 8, Aug. 1988, pp. 969–7.

Russel, Joan M. "Simple Models for Teaching Equilibrium and LeChâtelier's Principle," *Journal of Chemical Education*, Vol. 65, No. 10, Oct. 1988, pp. 871–2.

Shakhashiri, Bassam Z. "Chemical Cold Pack," and "Chemical Hot Pack," "Heat of Solution of LiCl," *Chemical Demonstrations*, Vol. 1, Madison Wisconsin: University of Wisconsin Press, 1983.

Talesnick, Irwin, ed. "Solubility in Different Solvents," Idea 404, *Idea Bank Collation*, Kingston, Ontario: Science Supplies and Services Co. Ltd., 1984.

Audiovisual Resources

Chemistry 110—Looking at the Solution and *Chemistry 111—Molarity of Solutions*, sound filmstrip or VHS cassette, Learning Arts, 1988.

Chemistry: Part 4, "Solutions," VHS cassette or film, Coronet Films.

Computer Software

Chemistry Help Series-Solutions, Apple II 48K, Barclay School Supplies, 1988.

Chemistry Courseware–Solutions, Apple II 64K and SEI Solutions, Apple 64K and IBM PC 128K, Queue, 1988.

Fee, Richard. *Liquids, Solids, Solutions*, Apple 48K, Chemistry According to ROF, 1988.

Project Seraphim, 1988 Volume 5, Disk 2, Apple II, 1988.

Smith, S.G., R. Chabray, and E. Kean. *Solutions*, Apple 48K and IBM PC 128K, Queue, 1988.

Answers and Solutions

Questions

1. *(a)* Mixtures are classified according to the sizes of the particles that are present. *(b)* The three types of mixtures are solutions, suspensions, and colloids.

2. *(a)* The term *soluble* means "capable of being dissolved." *(b)* A solution is a homogeneous mixture of two or more

substances in a single phase. *(c)* Its two components are the solvent, or dissolving medium, and the solute, or substance dissolved. *(d)* Two examples of solutions are sugar in water and salt in water.

3. *(a)* Electrolytes are substances that dissolve in water to give solutions that conduct electric current. Nonelectrolytes are substances that dissolve in water to give solutions that do not conduct electric current. *(b)* Sodium chloride is an electrolyte; sugar is a nonelectrolyte. *(c)* Generally, electrolytes are soluble ionic compounds, although some highly polar molecular compounds are also electrolytes because their molecules separate into ions when dissolved in water.

4. *(a)* A suspension is a heterogeneous mixture of a solvent-like substance with particles that slowly settle out. *(b)* A solution differs from a suspension in that the particles of the former are much smaller, remain mixed indefinitely (under constant conditions), and cannot be filtered out.

5. *(a)* A colloid is a mixture that contains particles of a size intermediate between those in solutions and those in suspensions. *(b)* The colloidal particles make up the dispersed phase, whereas the solvent-like phase is the dispersing medium.

6. *(a)* Mayonnaise is an emulsion. *(b)* Egg yolk is an emulsifying agent. *(c)* Jelly is a gel.

7. *(a)* The Tyndall effect is the scattering of light by colloidal particles dispersed in a transparent medium. *(b)* The visibility of a headlight beam on a foggy night represents an illustration of this effect. *(c)* This phenomenon can be used to distinguish between a solution and a colloid.

8. *(a)* The rate of dissolving can be increased by increasing the surface area of the sugar, agitating the solution, and heating the water. *(b)* Increasing the surface area and agitating the solution both result in greater contact between the solvent and the solute surface, and thus provide more opportunities for the solute particles to be attracted away from the solute surface and into the solution. Heating the solvent results in collisions that are greater in number and higher in energy, causing the solute molecules to leave the surface at a higher rate.

9. *(a)* Solution equilibrium is the state in which the opposing processes of the dissolving and crystallizing of a solute occur at equal rates. *(b)* The factors that determine the equilibrium point are the nature of the solute, the nature of the solvent, and the temperature. In the case of a gaseous solution, pressure is also a factor.

10. *(a)* A saturated solution is one that contains the maximum possible amount of dissolved solute at solution equilibrium under the existing conditions. *(b)* A residual quantity of undissolved solute remains in contact with the solution. *(c)* An unsaturated solution is one that contains less solute than does a saturated solution under the existing conditions.

11. *(a)* The solubility of a substance is the amount of that substance that is dissolved at solution equilibrium in a specific amount of solvent at a specified temperature. *(b)* Temperature and pressure (for gases) must be specified.

12. There is no relationship between rate of dissolving and solubility.

13. Three factors that affect solubility are the type of solvent and solute, pressure, and temperature.

14. *(a)* The rule of thumb for predicting solubility is "like dissolves like." *(b)* In general, the rule means that polar substances dissolve in polar solvents, and nonpolar substances dissolve in nonpolar solvents.

15. *(a)* The solubility of a gas in a liquid is directly proportional to its pressure above the liquid. *(b)* This is a statement of Henry's law. *(c)* The amount of gas that will dissolve increases.

16. *(a)* Effervescence is the rapid escape of a gas from a liquid in which it is dissolved. *(b)* During bottling, carbon dioxide is forced into solution at 5–10 atm pressure, and the bottle or can is sealed. When the cap is removed, the pressure is reduced to 1 atm, and some of the dissolved carbon dioxide escapes as gas bubbles.

17. *(a)* As the temperature increases, the solubility of a gas usually decreases. *(b)* In general, as temperature increases, the solubility of a solid in a liquid increases.

18. A supersaturated solution is a solution that contains more dissolved solute than does a saturated solution under the same conditions.

19. The indicated solubility levels are: *(a)* 80 g/100 g H_2O *(b)* 110 g/100 g H_2O *(c)* 36 g/100 g H_2O.

20. The indicated temperature readings are: *(a)* 15°C *(b)* 50°C *(c)* 32°C.

21. *(a)* The heat of solution is the overall energy change during solution formation. *(b)* Heat of solution is expressed in kilojoules per mole of solute dissolved in a specified amount of solvent. *(c)* Positive values indicate changes that absorb heat, whereas negative values indicate changes that release heat.

22. Endothermic: Solute + solvent + heat → solution
Exothermic: Solute + solvent → solution + heat

23. In order for solution formation to occur, solute–solute attraction must be broken up (which requires energy), solvent–solvent attraction must be broken up (which requires energy), and then solute–solvent attraction must be formed (which releases energy).

24. *(a)* An increase in temperature increases the solubility of the solid. *(b)* An increase in temperature decreases its solubility.

25. The solubility of gases in liquids generally decreases with increasing temperature because the dissolving process involving gases is usually exothermic and is therefore opposed by imposed increases in temperature.

26. *(a)* A dilute solution contains a relatively small amount of solute in a solvent, whereas a concentrated solution contains a relatively large amount of solute in a solvent. *(b)* These terms are not directly related to solubility; a saturated solution of a substance that is not very soluble could, in fact, be very dilute.

27. *(a)* A 4-*M* solution contains 4 moles of solute dissolved in enough solvent to make 1 liter of solution. *(b)* A 0.75-*M* solution contains 0.75 mole of solute dissolved in enough solvent to make 1 liter of solution.

28. *(a)* A 1-*m* solution contains 1 mole of dissolved solute per 1 kg of solvent. *(b)* A 0.5-*m* solution contains 0.5 mole of dissolved solute per 1 kg of solvent.

29. *(a)* The presence of the nonvolatile solute lowers the vapor pressure of the solvent in which it is dissolved. *(b)* The freezing point of the resulting solution is lower than that of

the pure solvent. *(c)* The boiling point of the solution is higher than that of the pure solvent.

30. *(a)* A colligative property is one that depends on the number of solute particles, but is independent of their nature. *(b)* Examples of such properties include vapor pressure, freezing point, and boiling point.

31. Changes in colligative properties are directly proportional to the molal concentration of molecular solutes.

32. The freezing-point depression of a solvent is the decrease in the freezing point caused by the presence of a solute.

33. The molal freezing-point constant (K_f) is equal to the freezing-point depression of the solvent in a 1-molal solution of a nonvolatile, molecular solute. *(b)* The value of this constant for water is $-1.86°C/molal$.

34. The boiling-point elevation of a solvent is the increase in the boiling point caused by the presence of a solute.

35. *(a)* The molal boiling-point constant is equal to the boiling-point elevation of the solvent in a 1-molal solution of a nonvolatile, molecular solute. *(b)* The value of this constant for water is $0.51°C/molal$.

Problems

1. *(a)* % by mass $= \dfrac{10.0 \text{ g sugar}}{10.0 \text{ g sugar} + 40.0 \text{ g water}} \times 100\% = 20.0\%$ sugar

   (b) % by mass $= \dfrac{0.63 \text{ g AgNO}_3}{0.63 \text{ g AgNO}_3 + 12.00 \text{ g H}_2\text{O}} \times 100\% = 5.0\%$ AgNO$_3$

2. *(a)* $\dfrac{2.5 \text{ mol NaOH}}{1.0 \text{ L}} = 2.5 \, M$ NaOH   *(b)* $\dfrac{4.0 \text{ mol BaCl}_2}{1.5 \text{ L}} = 2.7 \, M$ BaCl$_2$

3. *(a)* $\dfrac{2.5 \text{ mol HCl}}{1 \text{ L}} \times 1.0 \text{ L} = 2.5$ mol HCl   *(b)* $\dfrac{1.25 \text{ mol Al(NO}_3)_3}{1 \text{ L}} \times 150.0 \text{ mL} \times \dfrac{\text{L}}{10^3 \text{ mL}} = 0.188$ mol Al(NO$_3$)$_3$

4. *(a)* $\dfrac{3.50 \text{ mol H}_2\text{SO}_4}{\text{L}} \times 1.00 \text{ L} \times \dfrac{98.1 \text{ g H}_2\text{SO}_4}{\text{mol H}_2\text{SO}_4} = 343$ g H$_2$SO$_4$

   (b) $\dfrac{1.75 \text{ mol Ba(NO}_3)_2}{\text{L}} \times 2.50 \text{ L} \times \dfrac{261 \text{ g Ba(NO}_3)_2}{\text{mol Ba(NO}_3)_2} = 1140$ g Ba(NO$_3$)$_2$

   (c) $\dfrac{0.250 \text{ mol KOH}}{\text{L}} \times 500. \text{ mL} \times \dfrac{\text{L}}{10^3 \text{ mL}} \times \dfrac{56.1 \text{ g KOH}}{\text{mol KOH}} = 7.01$ g KOH

5. *(a)* $\dfrac{20.0 \text{ g NaOH}}{2.00 \text{ L}} \times \dfrac{\text{mol NaOH}}{40.0 \text{ g NaOH}} = \dfrac{0.250 \text{ mol NaOH}}{\text{L}} = 0.250 \, M$ N$_2$OH

   (b) $\dfrac{14.0 \text{ g NH}_4\text{OH}}{150. \text{ mL}} \times \dfrac{\text{mol NH}_4\text{OH}}{35.0 \text{ g NH}_4\text{OH}} \times \dfrac{10^3 \text{ mL}}{\text{L}} = \dfrac{2.67 \text{ mol NH}_4\text{OH}}{\text{L}} = 2.67 \, M$ NH$_4$OH

   (c) $\dfrac{32.7 \text{ g H}_3\text{PO}_4}{500 \text{ mL}} \times \dfrac{\text{mol H}_3\text{PO}_4}{98.0 \text{ g H}_3\text{PO}_4} \times \dfrac{10^3 \text{ mL}}{\text{L}} = \dfrac{0.667 \text{ mol H}_3\text{PO}_4}{\text{L}} = 0.667 \, M$ H$_3$PO$_4$

6. *(a)* $\dfrac{2.20 \text{ mol NaOH}}{\text{L}} \times 65.0 \text{ mL} \times \dfrac{\text{L}}{10^3 \text{ mL}} = 0.143$ mol NaOH

   (b) 0.143 mol NaOH $\times \dfrac{40.0 \text{ g NaOH}}{\text{mol NaOH}} = 5.72$ g NaOH

7. $\dfrac{26.42 \text{ g (NH}_4)_2\text{SO}_4}{50.00 \text{ mL}} \times \dfrac{\text{mol (NH}_4)_2\text{SO}_4}{132.2 \text{ g (NH}_4)_2\text{SO}_4} \times \dfrac{10^3 \text{ mL}}{\text{L}} = \dfrac{3.997 \text{ mol (NH}_4)_2\text{SO}_4}{\text{L}} = 3.997 \, M$

8. *(a)* $\dfrac{4.5 \text{ mol Mg(NO}_3)_2}{2.5 \text{ kg H}_2\text{O}} = 1.8 \, m$ Mg(NO$_3$)$_2$ *(b)* $\dfrac{0.625 \text{ mol K}_2\text{SO}_4}{850. \text{ g H}_2\text{O}} \times \dfrac{10^3 \text{ g}}{\text{kg}} = 0.735 \, m$ K$_2$SO$_4$

9. *(a)* $\dfrac{0.45 \text{ mol CaSO}_4}{1 \text{ kg H}_2\text{O}} \times 2.5 \text{ kg H}_2\text{O} = 1.1$ mol CaSO$_4$

   (b) $\dfrac{3.25 \text{ mol Ba(NO}_3)_2}{1 \text{ kg H}_2\text{O}} \times 750 \text{ mL H}_2\text{O} \times \dfrac{1 \text{ g H}_2\text{O}}{1 \text{ mL H}_2\text{O}} \times \dfrac{1 \text{ kg H}_2\text{O}}{10^3 \text{ g H}_2\text{O}} = 2.44$ mol Ba(NO$_3$)$_2$

10. *(a)* $\dfrac{4.50 \text{ mol H}_2\text{SO}_4}{\text{kg H}_2\text{O}} \times 1.00 \text{ kg H}_2\text{O} \times \dfrac{98.1 \text{ g H}_2\text{SO}_4}{\text{mol H}_2\text{SO}_4} = 441$ g H$_2$SO$_4$

   (b) $\dfrac{1.00 \text{ mol HNO}_3}{\text{kg H}_2\text{O}} \times 2.00 \text{ kg H}_2\text{O} \times \dfrac{63.0 \text{ g HNO}_3}{\text{mol HNO}_3} = 126$ g HNO$_3$

   (c) $\dfrac{3.50 \text{ mol MgCl}_2}{\text{kg H}_2\text{O}} \times 0.450 \text{ kg H}_2\text{O} \times \dfrac{95.2 \text{ g MgCl}_2}{\text{mol MgCl}_2} = 150. \text{ g MgCl}_2$

11. (a) $\dfrac{294.3 \text{ g H}_2\text{SO}_4}{1.000 \text{ kg H}_2\text{O}} \times \dfrac{\text{mol H}_2\text{SO}_4}{98.08 \text{ g H}_2\text{SO}_4} = \dfrac{3.001 \text{ mol H}_2\text{SO}_4}{\text{kg H}_2\text{O}} = 3.001 \, m \text{ H}_2\text{SO}_4$

(b) $\dfrac{63.0 \text{ g HNO}_3}{0.250 \text{ kg H}_2\text{O}} \times \dfrac{\text{mol HNO}_3}{63.0 \text{ g HNO}_3} = \dfrac{4.00 \text{ mol HNO}_3}{\text{kg H}_2\text{O}} = 4.00 \, m \text{ HNO}_3$

(c) $\dfrac{10.0 \text{ g NaOH}}{300. \text{ g H}_2\text{O}} \times \dfrac{\text{mol NaOH}}{40.0 \text{ g NaOH}} \times \dfrac{10^3 \text{ g H}_2\text{O}}{\text{kg H}_2\text{O}} = \dfrac{0.833 \text{ mol NaOH}}{\text{kg H}_2\text{O}} = 0.833 \, m \text{ NaOH}$

12. $\dfrac{17.1 \text{ g C}_{12}\text{H}_{22}\text{O}_{11}}{275 \text{ g H}_2\text{O}} \times \dfrac{\text{mol C}_{12}\text{H}_{22}\text{O}_{11}}{342 \text{ g C}_{12}\text{H}_{22}\text{O}_{11}} \times \dfrac{10^3 \text{ g H}_2\text{O}}{\text{kg H}_2\text{O}} = \dfrac{0.182 \text{ mol C}_{12}\text{H}_{22}\text{O}_{11}}{\text{kg H}_2\text{O}} = 0.182 \, m \text{ C}_{12}\text{H}_{22}\text{O}_{11}$

13. $\dfrac{\text{kg H}_2\text{O}}{0.500 \text{ mol Ca(NO}_3)_2} \times \dfrac{\text{mol Ca(NO}_3)_2}{164 \text{ g Ca(NO}_3)_2} \times 75.5 \text{ g Ca(NO}_3)_2 = 0.921 \text{ kg H}_2\text{O}$

14. $\dfrac{1.25 \text{ mol C}_6\text{H}_{12}\text{O}_6}{\text{kg H}_2\text{O}} \times \dfrac{180. \text{ g C}_6\text{H}_{12}\text{O}_6}{\text{mol C}_6\text{H}_{12}\text{O}_6} \times 750 \text{ g H}_2\text{O} \times \dfrac{\text{kg H}_2\text{O}}{10^3 \text{ g H}_2\text{O}} = 169 \text{ g C}_6\text{H}_{12}\text{O}_6$

15. (a) $\Delta T_f = K_f m = \dfrac{-1.86°\text{C} \cdot \text{kg H}_2\text{O}}{\text{mol C}_6\text{H}_{12}\text{O}_6} \times \dfrac{1.50 \text{ mol C}_6\text{H}_{12}\text{O}_6}{\text{kg H}_2\text{O}} = -2.79°\text{C}$

(b) $\Delta T_f = K_f m = \dfrac{-1.86°\text{C} \cdot \text{kg H}_2\text{O}}{\text{mol C}_{12}\text{H}_{22}\text{O}_{11}} \times \dfrac{\text{mol C}_{12}\text{H}_{22}\text{O}_{11}}{342 \text{ g C}_{12}\text{H}_{22}\text{O}_{11}} \times \dfrac{171 \text{ g C}_{12}\text{H}_{22}\text{O}_{11}}{1 \text{ kg H}_2\text{O}} = -0.930°\text{C}$

(c) $\Delta T_f = K_f m = \dfrac{-1.86°\text{C} \cdot \text{kg H}_2\text{O}}{\text{mol C}_{12}\text{H}_{22}\text{O}_{11}} \times \dfrac{\text{mol C}_{12}\text{H}_{22}\text{O}_{11}}{342 \text{ g C}_{12}\text{H}_{22}\text{O}_{11}} \times \dfrac{77.0 \text{ g C}_{12}\text{H}_{22}\text{O}_{11}}{400. \text{ g H}_2\text{O}} \times \dfrac{10^3 \text{ g H}_2\text{O}}{\text{kg H}_2\text{O}} = -1.05°\text{C}$

16. (a) $\Delta T_f = K_f m$.

Therefore, $m = \dfrac{\Delta T_f}{K_f} = \dfrac{-0.930°\text{C} \cdot \text{mol sugar}}{-1.86°\text{C} \cdot \text{kg H}_2\text{O}} = 0.500 \, m \text{ sugar}$ (b) $m = \dfrac{-3.72°\text{C} \cdot \text{mol sugar}}{-1.86°\text{C} \cdot \text{kg H}_2\text{O}} = 2.00 \, m \text{ sugar}$

(c) $m = \dfrac{-8.37°\text{C} \cdot \text{mol sugar}}{-1.86°\text{C} \cdot \text{kg H}_2\text{O}} = 4.50 \, m \text{ sugar}$

17. (a) $\Delta T_f = K_f m = \dfrac{-1.86 \text{ C} \cdot \text{kg H}_2\text{O}}{\text{mol C}_6\text{H}_{12}\text{O}_6} \times \dfrac{\text{mol C}_6\text{H}_{12}\text{O}_6}{180. \text{ g C}_6\text{H}_{12}\text{O}_6} \times \dfrac{20.0 \text{ g C}_6\text{H}_{12}\text{O}_6}{250. \text{ g H}_2\text{O}} \times \dfrac{10^3 \text{ g H}_2\text{O}}{\text{kg H}_2\text{O}} = -0.827°\text{C}$

(b) $\Delta T_f = T_f \text{solution} - T_f \text{solvent}$; therefore, $T_f \text{solution} = \Delta T_f - T_f \text{solvent} = -0.827°\text{C} - 0.00°\text{C} = -0.827°\text{C}$

18. (a) $\dfrac{0.51°\text{C} \cdot \text{kg H}_2\text{O}}{\text{mol C}_6\text{H}_{12}\text{O}_6} \times \dfrac{2.5 \text{ mol C}_6\text{H}_{12}\text{O}_6}{\text{kg H}_2\text{O}} = 1.3°\text{C}$

(b) $\dfrac{0.51°\text{C} \cdot \text{kg H}_2\text{O}}{\text{mol C}_6\text{H}_{12}\text{O}_6} \times \dfrac{\text{mol C}_6\text{H}_{12}\text{O}_6}{180. \text{ g C}_6\text{H}_{12}\text{O}_6} \times \dfrac{3.20 \text{ g C}_6\text{H}_{12}\text{O}_6}{1.00 \text{ kg H}_2\text{O}} = 0.0091°\text{C}$

(c) $\dfrac{0.51°\text{C} \cdot \text{kg H}_2\text{O}}{\text{mol C}_{12}\text{H}_{22}\text{O}_{11}} \times \dfrac{\text{mol C}_{12}\text{H}_{22}\text{O}_{11}}{342 \text{ g C}_{12}\text{H}_{22}\text{O}_{11}} \times \dfrac{20.0 \text{ g C}_{12}\text{H}_{22}\text{O}_{11}}{500. \text{ g H}_2\text{O}} \times \dfrac{10^3 \text{ g H}_2\text{O}}{\text{kg H}_2\text{O}} = 0.060°\text{C}$

19. (a) $\Delta T_b = T_b \text{solution} - T_b \text{solvent} = 100.25°\text{C} - 100.00°\text{C} = 0.25°\text{C}$

$\Delta T_b = K_b m$

$m = \dfrac{\Delta T_b}{K_b} = \dfrac{0.25°\text{C}}{0.51°\text{C}° \cdot \text{kg H}_2\text{O/mol solute}} = 0.49 \, m \text{ solute}$

(b) $\Delta T_b = 101.53°\text{C} - 100.00°\text{C} = 1.53°\text{C}$

$\Delta T_b = K_b m$

$m = \dfrac{\Delta T_b}{Kb} = \dfrac{1.53°\text{C}}{0.51°\text{C} \cdot \text{kg H}_2\text{0/mol solute}} = 3.0 \, m \text{ solute}$

(c) $\Delta T_b = 102.805°\text{C} - 100.000°\text{C} = 2.805°\text{C}$

$\Delta T_b = K_b m$

$m = \dfrac{\Delta T_b}{K_b} = \dfrac{2.805°\text{C}}{0.51°\text{C} \cdot \text{kg H}_2\text{O/mol solute}} = 5.5 \, m \text{ solute}$

20. $\Delta T_f = K_f m$

$m = \dfrac{\Delta T_f}{K_f} = \dfrac{-1.5°\text{C} \cdot \text{mol solute}}{-7.27°\text{C} \cdot \text{kg phenol}} = 0.21 \, m$ $\dfrac{\text{kg phenol}}{0.21 \text{ mol solute}} \times \dfrac{5.00 \text{ g solute}}{150. \text{ g phenol}} \times \dfrac{10^3 \text{ g phenol}}{\text{kg phenol}} = 160 \text{ g solute/mol solute}$

21. $T_b = T_b \text{solution} - T_b \text{solvent} = 185.5°\text{C} - 184.4°\text{C} = 1.1°\text{C}$

$\Delta T_b = K_b m$

$m = \dfrac{\Delta T_b}{K_b} = \dfrac{3.22°\text{C} \cdot \text{kg aniline}}{\text{mol solute}} \times \dfrac{7.55 \text{ g solute}}{125. \text{ g aniline}} \times \dfrac{10^3 \text{ g aniline}}{\text{kg aniline}} \times \dfrac{1}{1.1°\text{C}} = 180 \text{ g solute/mol solute}$

Application Questions

1. In an electrolyte solution, the positive and negative ions that make up the original compound separate and become surrounded by water molecules. They are free to move, making it possible for an electric current to pass through the solution. When a nonelectrolyte dissolves in water, no mobile charged particles are present, and the resulting solution cannot conduct a current.

2. Shine a beam of light through the mixture. If it is a solution, no scattering of light will be observed. If it is a colloid, the light will be scattered. A suspension will settle out.

3. Water molecules are highly polar. They can attract the positively and negatively charged regions of the polar molecules of the solute and draw those molecules away from each other.

4. The two substances display too little similarity in bonding, polarity, and intermolecular forces to dissolve in each other.

5. According to Le Châtelier's principle, increasing the pressure of the solute gas above the solution places a stress on the established equilibrium. In order to partially offset that stress, gas molecules begin to enter the solution at an increasing rate. Eventually, when the new equilibrium is reached, a greater number of molecules can be found in the solution than were present prior to the time the pressure was increased.

6. Hot water should be used because at a higher temperature, more of the suspended air molecules, which generally cause the cloudiness, will have achieved the speed required to leave the solvent.

7. (a) When the dissolving process is exothermic, the total heat content of the solution is lower than that of its separate components; when the process is endothermic, the heat content of the solution is higher. (b) When the process is exothermic, the solution warms; when it is endothermic, the solution cools. (c) When the solution formation process is exothermic, the heat of solution is negative; when it is endothermic, the heat of the solution is positive. (d) When the process is exothermic, increased temperatures usually decrease solubility; when it is endothermic, increased temperatures generally increase solubility. Most solids have increased solubility at increased temperatures.

8. (a) $AgNO_3(s) + H_2O(\ell) + 22.8$ kJ/mol solution (b) The dissolving process is endothermic; crystallization is exothermic. (c) The temperature drops. (d) The two rates are equal. (e) The rate of dissolving will initially increase at a faster pace than the rate of crystallization because the dissolving process is endothermic and will serve to reduce the stress placed on the system. (f) The amount will increase if additional solute is available. (g) The rate of crystallizing will increase to relieve the stress, and the solubility of the solute will then decrease.

9. (a) Percent by mass would be used for practical purposes when only the mass of the solute, and not its molar mass, is needed. (b) Molarity would be used when it is important to know the molar mass of solute in a given volume of solution. (c) Molality is preferred when it is important to know the relative numbers of solute and solvent particles.

10. First, determine the number of grams of $Mg(NO_3)_2$ needed to prepare the solution:

$$\frac{1.50 \text{ mol } Mg(NO_3)_2}{\text{kg } H_2O} \times \frac{148 \text{ g } Mg(NO_3)_2}{\text{mol } Mg(NO_3)_2} \times 350. \text{ g } H_2O$$
$$\times \frac{1 \text{ kg } H_2O}{10^3 \text{g } H_2O} = 77.7 \text{ g } Mg(NO_3)_2$$

Next, measure out 350. g of H_2O. Finally, dissolve the 77.7 g $Mg(NO_3)_2$ in the 350. g of H_2O.

11. Ethylene glycol lowers the freezing point of the water in the radiator, thereby protecting the cooling system at low temperatures.

Application Problems

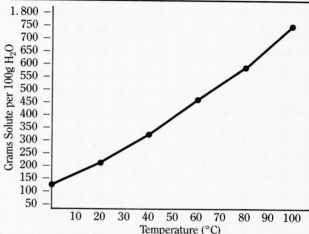

(a) Solubility increases with increasing temperature. (b) At 30°C, solubility is approximately 250 g/100 g H_2O; at 55°C, solubility is approximately 380 g/100 g H_2O; at 75°C, solubility is approximately 540 g/100 g H_2O. (c) This solubility would be observed at approximately 35°C. (d) The solution would be unsaturated; the solution would be saturated.

2. Solubility at 60°C = 110 g; at 10°C = 20 g. The amount of precipitate is therefore 110 g − 20 g = 90 g.

3. (a) % by mass = $\dfrac{\text{solute mass}}{\text{solute mass} + \text{solvent mass}} \times 100\%$

Let x = solute mass (mass of NaCl)

$$10.0\% = \frac{x}{x + 50.0 \text{ g}} \times 100\%$$
$$\frac{10.0\%}{100\%} = \frac{x}{x + 50.0 \text{ g}}$$
$$0.100 = \frac{x}{x + 50.0 \text{ g}}$$
$$0.100(x + 50.0 \text{ g}) = x$$
$$0.100\,x + 5.00 \text{ g} = x$$
$$5.00 \text{ g} = 0.900x$$
$$5.56 \text{ g} = x$$
∴ solute mass = 5.56 g NaCl

(b) Let x = mass of KOH

$$2.4\% = \frac{x}{x + 8.5 \text{ g}} \times 100\%$$
$$\frac{2.4\%}{100\%} = \frac{x}{x + 8.5 \text{ g}}$$

$$0.024 = \frac{x}{x + 8.5\text{ g}}$$

$$0.024(x + 8.5\text{ g}) = x$$

$$0.024\,x + 0.204\text{ g} = x$$

$$0.204\text{ g} = 0.976x$$

$$0.21\text{ g} = x$$

$$\therefore \text{ mass of KOH} = 0.21\text{ g KOH}$$

4. $\dfrac{12.0\text{ mol HCl}}{\text{L}} \times \dfrac{100\text{ mL}}{2.00\text{ L}} \times \dfrac{\text{L}}{10^3\text{ mL}} = \dfrac{0.600\text{ mol HCl}}{\text{L}} = 0.600\text{ } M \text{ HCl}$

5. $\dfrac{\text{L } 16.0\text{ } M \text{ HNO}_3}{16.0\text{ mol HNO}_3} \times 750.\text{ mL dilute solution} \times \dfrac{\text{L dilute solution}}{10^3\text{ mL dilute solution}} \times \dfrac{0.500\text{ mol HNO}_3}{\text{L dilute solution}} \times \dfrac{10^3\text{ mL } 16.0\text{ } M \text{ HNO}_3}{\text{L } 16.0\text{ } M \text{ HNO}_3}$

$= 23.4\text{ mL } 16.0\text{ } M \text{ HNO}_3$

6. $\dfrac{\text{L}}{0.54\text{ mol AgNO}_3} \times \dfrac{10^3\text{ mL}}{\text{L}} \times \dfrac{1\text{ mol AgNO}_3}{170.\text{ g AgNO}_3} \times 0.34\text{ AgNO}_3 = 3.7\text{ mL}$

7. $2\text{H}_3\text{PO}_4 + 3\text{Ca(OH)}_2 \rightarrow \text{Ca}_3(\text{PO}_4)_2 + 6\text{H}_2\text{O}$

$\dfrac{6.00\text{ mol H}_3\text{PO}_4}{\text{L}} \times 750.\text{ mL} \times \dfrac{\text{L}}{10^3\text{ mL}} = 4.50\text{ mol H}_3\text{PO}_4$

$4.50\text{ mol H}_3\text{PO}_4 \times \dfrac{1\text{ mol Ca}_3(\text{PO}_4)_2}{2\text{ mol H}_3\text{PO}_4} \times \dfrac{310.\text{ g Ca}_3(\text{PO}_4)_2}{\text{mol Ca}_3(\text{PO}_4)_2} = 698\text{ g Ca}_3(\text{PO}_4)_2$

$4.50\text{ mol H}_3\text{PO}_4 \times \dfrac{6\text{ mol H}_2\text{O}}{2\text{ mol H}_3\text{PO}_4} \times \dfrac{18.0\text{ g H}_2\text{O}}{\text{mol H}_2\text{O}} = 243\text{ g H}_2\text{O}$

8. $3\text{H}_2\text{SO}_4 + 2\text{Al(OH)}_3 \rightarrow \text{Al}_2(\text{SO}_4)_3 + 6\text{H}_2\text{O}$

$\dfrac{2.50\text{ mol Al(OH)}_3}{\text{L}} \times 250.\text{ mL} \times \dfrac{\text{L}}{10^3\text{ mL}} = 0.625\text{ mol Al(OH)}_3$

$0.625\text{ mol Al(OH)}_3 \times \dfrac{3\text{ mol H}_2\text{SO}_4}{2\text{ mol Al(OH)}_3} \times \dfrac{\text{L}}{18.0\text{ mol H}_2\text{SO}_4} \times \dfrac{10^3\text{ mL}}{\text{L}} = 52.1\text{ mL (of } 18.0\text{ } M \text{ H}_2\text{SO}_4)$

9. Since 1 mol Ag NO$_3$ yields 1 mol Ag, mol AgNO$_3$ = mol Ag. Thus,

$\dfrac{0.250\text{ g Ag}}{75.0\text{ mL}} \times \dfrac{1\text{ mol Ag}}{108\text{ g Ag}} \times \dfrac{10^3\text{ mL}}{1\text{ L}} = 0.0309\text{ } M \text{ (AgNO}_3 \text{ solution)}$

10. $\dfrac{1.75\text{ mol C}_2\text{H}_5\text{OH}}{1\text{ kg H}_2\text{O}} \times 250.\text{ g H}_2\text{O} \times \dfrac{\text{kg H}_2\text{O}}{10^3\text{ g H}_2\text{O}} \times \dfrac{46.0\text{ g C}_2\text{H}_5\text{OH}}{\text{mol C}_2\text{H}_5\text{OH}} = 20.1\text{ g C}_2\text{H}_5\text{OH}$

11. $\dfrac{\text{kg H}_2\text{O}}{0.450\text{ mol NaCl}} \times \dfrac{1\text{ L H}_2\text{O}}{1\text{ kg H}_2\text{O}} \times \dfrac{1\text{ mol NaCl}}{58.5\text{ g NaCl}} \times 65.0\text{ g NaCl} = 2.47\text{ L H}_2\text{O}$

12. $\dfrac{3.00\text{ mol HNO}_3}{\text{kg H}_2\text{O}} \times \dfrac{2.25\text{ kg H}_2\text{O}}{5.75\text{ kg H}_2\text{O}} = 1.17\text{ } m \text{ HNO}_3$

13. $\Delta T_f = K_f m$

$m = \dfrac{\Delta T_f}{K_f}$

$m = \dfrac{-20°\text{C}}{-1.86°\text{C} \cdot \text{kg H}_2\text{O/mol C}_2\text{H}_4(\text{OH})_2}$

$m = 10.8\text{ mol C}_2\text{H}_4(\text{OH})_2/\text{kg H}_2\text{O}$

$m = 10.8\text{ } m \text{ C}_2\text{H}_4(\text{OH})_2$

$\dfrac{10.8\text{ mol CH}_2\text{H}_4(\text{OH})_2}{\text{kg H}_2\text{O}} \times 500.\text{ g H}_2\text{O} \times \dfrac{\text{kg}}{10^3\text{ g}} \times \dfrac{62.0\text{ g C}_2\text{H}_4(\text{OH})_2}{\text{mol C}_2\text{H}_4(\text{OH})_2} = 335\text{ g C}_2\text{H}_4(\text{OH})_2$

14. $T_f = K_f m$

$K_f = \dfrac{T_f}{m} = \dfrac{-1.90°\text{C}}{m}$

$K_f = \dfrac{-1.90°\text{C}}{\dfrac{7.24\text{ g C}_2\text{Cl}_4\text{H}_2}{115\text{ g benzene}} \times \dfrac{\text{mol C}_2\text{H}_2\text{Cl}_4}{168\text{ g C}_2\text{H}_2\text{Cl}_4} \times \dfrac{10^3\text{ g benzene}}{\text{kg benzene}}} = -5.07°\text{C}/m$

15. $\Delta T_f = K_f m = \dfrac{-40.00°C \cdot \text{kg camphor}}{\text{mol solute}} \times \dfrac{1.500 \text{ g solute}}{35.00 \text{ g camphor}} \times \dfrac{\text{mol solute}}{125.0 \text{ g solute}} \times \dfrac{103 \text{ g camphor}}{\text{kg camphor}} = -13.71°C$

New freezing point:
$\Delta T_f = T_f \text{solution} - T_f \text{solvent}$
$T_f \text{solution} = \Delta T_f + T_f \text{solvent} = -13.71°C + 178.40°C = 164.69°C$

16. $\Delta T_f = K_f m$

$m = \dfrac{\Delta T_f}{K_f} = \dfrac{-6.51°C}{-1.86°C \cdot 1 \text{ kg H}_2\text{O/mol solute}} = 3.50 \, m \text{ solute}$

$\Delta T_b = K_b m$

$= \dfrac{0.51°C}{m} \times 3.50 \, m = 1.8°C$

Since $\Delta T_b = T_b \text{solution} - T_b \text{solvent}$, then
$T_b \text{solution} = \Delta T_b + T_b \text{solvent} = 1.8°C + 100.00°C = 101.8°C$

17. $\Delta T_f = K_f m$

$m = \dfrac{\Delta T_f}{K} = \dfrac{-0.568°C}{-1.79°C \cdot \text{kg ether/mol solute}} = 0.317 \text{ mol solute/kg ether}$

$\dfrac{\text{kg ether}}{0.317 \text{ mol solute}} \times \dfrac{21.2 \text{ g solute}}{740. \text{ g ether}} \times \dfrac{10^3 \text{ g ether}}{\text{kg ether}} = 90.4 \text{ g solute/mol solute}$

18. *(a)* $\Delta T_f = K_f m$

$m = \dfrac{\Delta T_f}{K_f} = \dfrac{-0.372°C \cdot \text{mol solute}}{-1.86°C \cdot \text{kg H}_2\text{O}} = 0.200 \, m \text{ solute}$

$\dfrac{\text{kg H}_2\text{O}}{0.200 \text{ mol solute}} \times \dfrac{9.00 \text{ g solute}}{250. \text{ g H}_2\text{O}} \times \dfrac{10^3 \text{ g H}_2\text{O}}{\text{kg H}_2\text{O}} = 180. \text{ g/mol solute}$

(b) The molar masses of CH_3OH, C_6H_6, C_2H_5OH, $C_6H_{12}O_6$, and $C_{12}H_{22}O_{11}$ are 32.0, 78.0, 46.0, 180.0, and 342.0 g/mol, respectively. Since $C_6H_{12}O_6$ is the only substance with a molar mass that is close to the experimentally determined value, $C_6H_{12}O_6$ most likely represents the solute.

Teacher's Notes

Chapter **14** Solutions

INTRODUCTION

In previous chapters you have learned about the elements and the compounds they form. It has been assumed that these elements and compounds have been pure. However, most substances we encounter are mixtures of different elements or compounds. This chapter will discuss some of the important properties of mixtures and how they differ from pure compounds.

LOOKING AHEAD

Think of examples from your own experience of the different types of mixtures you will read about. Look for properties of mixtures that make them different from pure substances.

SECTION PREVIEW

14.1 Types of Mixtures
14.2 The Solution Process
14.3 Concentrations of Solutions
14.4 Colligative Properties of Solutions

These Guajiro men of South America are harvesting salt in the briny waters off the northernmost tip of Columbia. The salt that is dissolved in the ocean waters remains behind when the water evaporates. Solutions of salts and other substances in the human body are the site of reactions necessary for life. Many chemical reactions can occur only in solution.

14.1 Types of Mixtures

It is unlikely that you will encounter many substances of high purity. Almost all the objects or materials that you see, touch, eat, drink, or use are made of mixtures.

It is easy to determine that some materials are mixtures. For example, soil is a mixture of various substances, including small rocks and decomposed animal and plant matter. You can see this by picking up some soil in your hand and taking a close look. Milk, on the other hand, does not appear to be a mixture, but in fact it is. Milk is comprised principally of fats, proteins, milk sugar, and water. Looking at milk under a microscope, you can see round droplets of fat measuring from 1 to 10 micrometers in diameter. Also, irregularly shaped casein (protein) particles about 0.2 micrometers wide can be seen. Both milk and soil are examples of heterogeneous mixtures since their composition is not uniform throughout. Soil, milk, and other heterogeneous mixtures are shown in Figure 14-1.

SECTION OBJECTIVES

- Distinguish between heterogeneous and homogeneous mixtures.
- Distinguish between electrolytes and nonelectrolytes.
- Compare the properties of suspensions, colloids, and solutions.

Other everyday examples of heterogeneous mixtures are tossed salad, buttered popcorn, vegetable soup, and chocolate-almond candy bars.

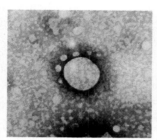

Sugar and water form a mixture in which the sugar molecules are interspersed among the water molecules. To distinguish the components of the mixture you would have to inspect the mixture molecule by molecule.

Above the molecular scale, the mixture of sugar and water is uniform throughout and is considered to be a homogeneous mixture.

Mixtures differ according to the size of the particles in the mixture. Based on this, mixtures are separated into suspensions, colloids and solutions. Solutions contain the smallest particles and suspensions contain the largest. Colloids contain intermediate size particles.

Figure 14-1 Four different mixtures *(left to right)* a handful of soil, milk magnified 500 times, nuts and bolts, whole wheat bread

Solutions

Suppose a lump of sugar is dropped into a glass of water. You know from experience that the sugar will dissolve. Sugar is described as "soluble in water." *By **soluble** we mean capable of being dissolved.*

What happens as the sugar dissolves? The lump gradually disappears as sugar molecules leave the surface of their crystals and intermingle with water molecules. Eventually all of the sugar molecules become uniformly distributed among the water molecules, as indicated by the equally sweet taste of any part of the mixture. Examination of a sugar–water mixture, even under a microscope, would not reveal any particles of sugar. Such a mixture is called a "solution." *A **solution** is a homogeneous mixture of two or more substances in a single phase.* In a solution, atoms, molecules, or ions are thoroughly

Solutions are homogeneous mixtures. Suspensions and colloids are heterogeneous mixtures.

Have you ever looked at a movie projector beam in a dark room and seen the suspended dust particles in the beam of light?

mixed, with the result that the mixture has the same composition and properties throughout.

The nature of solutions In the simplest type of solution, such as a sugar–water solution, the molecules of one substance are distributed around the molecules of another substance. *The dissolving medium in a solution is called the **solvent,** and the substance dissolved in a solution is called the **solute.*** In the sugar–water solution, water is the solvent and sugar is the solute.

Substances that dissolve in water are divided into two types according to whether they yield molecules or ions in solution. When an ionic compound dissolves, the positive and negative ions separate from each other and become surrounded by water molecules. These solute ions are free to move, making it possible for an electric current to pass through the solution. *A substance that dissolves in water to give a solution that conducts electric current is called an **electrolyte.*** Sodium chloride (NaCl), is an electrolyte, as is any soluble ionic compound. Certain highly polar molecular compounds, such as hydrogen chloride (HCl), are also electrolytes because the molecules split into ions (H_3O^+ and Cl^-) when dissolved in water.

By contrast, a solution containing neutral solute molecules does not conduct electric current because no mobile charged particles are available. *A substance that dissolves in water to give a solution that does not conduct an electric current is called a **nonelectrolyte.*** Sugar is a nonelectrolyte. An experiment that demonstrates the difference between electrolytes and nonelectrolytes is shown in Figure 14-2.

Figure 14-2 Apparatus for testing the conductivity of solutions. If the bulb is lit, the solution conducts electricity. A solution of copper(II) sulfate *(right)* conducts electricity, while a sugar solution *(left)* does not.

Types of solutions Solutions are formed in all three states—the gaseous, liquid, and solid states. The nine possible combinations of gases, liquids, and solids in solutions are summarized in Table 14-1. In each example, one component is designated as the solvent and one as the solute. Occasionally, however, these terms have little meaning. For example, in a 50%–50% solution of ethyl alcohol (C_2H_5OH), and water (H_2O), it would be difficult to say which is the solvent and which is the solute.

Figure 14-3 A brass bell. Brass is an alloy of copper and zinc.

TABLE 14-1 TYPES OF SOLUTIONS

Solute	Solvent	Example
gas	gas	air
gas	liquid	soda water
gas	solid	hydrogen in platinum
liquid	gas	water vapor in air
liquid	liquid	alcohol in water
liquid	solid	mercury in copper
solid	gas	sulfur vapor in air
solid	liquid	sugar in water
solid	solid	copper in nickel

In a solution, the dissolved particles are so small that they cannot be seen. They remain mixed with the solvent indefinitely, so long as the existing conditions remain unchanged. If a liquid solution is poured through a filter paper, both the solute and the solvent will pass through the paper. The solute particle dimensions are those of atoms, molecules, and ions—from about 0.1 to 1 nm. It is useful to break solutions into three main types: gaseous, liquid, and solid solutions.

1. Gaseous solutions. Molecules in a gas are far apart and in constant motion. When two or more gases are mixed, therefore, the molecules quickly become uniformly intermingled. Although we do not commonly describe them as such, all mixtures of gases are solutions.

The molecules of any liquid or solid that evaporate enter the gas phase and mix with the molecules there. In a sense, air is a solvent for all such evaporated substances.

2. Liquid solutions. Soft drinks are solutions of a gas, carbon dioxide, in water in which solid flavoring materials are also dissolved. Automobile radiators contain solutions of ethylene glycol, a liquid, in water. Vinegar is acetic acid dissolved in water. Water-soluble solids are present in solutions that we use for washing dishes and clothing.

3. Solid solutions. Many metal objects are actually solid solutions of metal called alloys. **Alloys** *are solid solutions in which the atoms of two or more metals are uniformly mixed.* By properly choosing the proportions of each metal in the alloy, many desirable properties can be obtained. For example, alloys can have higher strength, greater resistance to corrosion and higher melting points than pure elemental metals. (Figure 14-3)
Gas–solid solutions form when small atoms of elements like hydrogen, boron, or nitrogen can intermingle with metal atoms. Hydrogen dissolves especially well in transition metals such as platinum and palladium.

In solutions of substances that are mutually soluble in each other, such as isopropyl alcohol and water or nitrogen and oxygen, the substance that is present in the larger amount is called the solvent.

Water is considered to be the universal solvent. Can you think why?

Some examples of common alloys are sterling silver—92% silver and 8% copper, jewelry gold—gold and copper (percentages vary depending upon the number of karats), white gold—gold containing nickel and possibly tin, zinc, or copper, and the amalgam for filling teeth—silver and mercury.

Desktop Investigation

Solutions, Colloids, Suspensions

Question
How can you distinguish between solutions, colloids, and suspensions?

Materials
water
sucrose (table sugar)
soluble starch
clay
red food coloring
sodium borate (borax)
cooking oil
plain gelatin
7 beakers, 400 mL
flashlight

Procedure
1. Record all your observations in a data table. Remember to wear safety glasses and an apron. Prepare seven mixtures, each containing 250 mL of water and one of the following substances: 12 g of sucrose, 3 g of starch, 5 g of clay, 2 mL of food coloring, 2 g of sodium borate, 50 mL of oil, and 3 g of gelatin. To make the gelatin mixture, first soften the gelatin with 65 mL of cold water, and then add 185 mL of boiling water.

2. Observe the seven mixtures and their characteristics. What is the appearance of each mixture after stirring? Which mixtures separate and which remain mixed after stirring?

3. Transfer 10 mL of each mixture that does not separate after stirring to individual test tubes and shine a flashlight on the mixture in a dark room. Is the path of the light beam clearly visible in the mixture? (See the diagram.)

Discussion
1. Using the information on your data table, decide whether each mixture is a solution, a colloid, or a suspension.
2. How can solutions, colloids, and suspensions be distinguished?

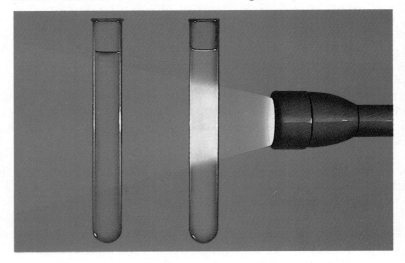

Suspensions

At the opposite extreme from the solute particles in solutions are particles so large that they can be seen and will settle out unless the mixture is constantly stirred or agitated. Think of a jar of muddy water. If left undisturbed, particles of soil collect on the bottom of the jar. Such a mixture is a **suspension**—*a heterogeneous mixture of a solvent-like substance with particles that slowly settle out.* The soil particles are huge and much more dense when compared to the water molecules. Their interaction with the water molecules is not strong enough to prevent them from falling to the bottom of the container. Particles over 1000 nm in diameter—1000 or more times larger than atoms, molecules, or ions—form suspensions. The particles in suspension can be separated from the mutiphase heterogeneous mixture by passing the mixture through a filter.

Colloids

Particles that are intermediate in size between those in solutions and suspensions form mixtures known as colloidal dispersions, or simply colloids. Particles between 1 nm and 1000 nm in diameter form colloids. After large particles settle out of muddy water, the water is often still cloudy because colloidal particles remain dispersed in the water. Pouring the cloudy mixture through a filter will not remove the colloidal particles, and the mixture will remain cloudy. The particles in a colloidal dispersion are small enough to be kept in suspension by the constant movement of the surrounding molecules.

Figure 14-4 The jar at the *(left)* contains a mixture of water and sodium chloride. The jar at right contains a mixture of gelatin in water. Which mixture is a colloidal suspension and which is a solution?

Colloids form in all phases. The colloidal particles make up the *dispersed phase,* and the solventlike phase is called the *dispersing medium.* Examples of the various types of colloids are given in Table 14-2. Note that some familiar terms, such as *emulsion* or *foam,* refer to specific types of colloids. For example, mayonnaise is an emulsion of oil droplets in water; the egg yolk in it acts as an emulsifying agent, which helps to keep the oil droplets dispersed.

TABLE 14-2	COLLOIDAL SUSPENSIONS
Class of Colloid	Phases
sol	solid dispersed in liquid
gel	solid network extending throughout liquid
liquid emulsion	liquid dispersed in liquid
foam	gas dispersed in liquid
aerosol	
smoke	solid dispersed in gas
fog	liquid dispersed in gas
smog	solid and liquid dispersed in gas
solid emulsion	liquid dispersed in solid

Many colloids appear homogeneous because the individual particles cannot be seen with the naked eye. The particles are, however, large enough to scatter light. You have probably noticed how a headlight beam becomes visible on a foggy night. This effect, known as the Tyndall effect occurs only when light is scattered by colloidal particles dispersed in a transparent medium. The Tyndall effect can be used to distinguish between a solution and a colloid, as demonstrated in Figure 14-4.

The distinctive properties of solutions, colloids, and suspensions are summarized in Table 14-3. The individual particles of a colloid can be detected under a microscope if a bright light is cast on the specimen at a right angle. The particles, which appear as tiny specs of light are seen to move rapidly in a random motion. This motion is due to bombardment by rapidly moving molecules and is called Brownian motion after its discoverer, Robert Brown (1773–1858). Brown did not know what caused this motion and suggested that the particles were alive.

Thomas Graham, an English chemist, started the study of colloids in 1861. Since the first colloids were gelatin or glue, he used the Greek word *kolla,* meaning glue, as the root for their name.

TABLE 14-3 PROPERTIES OF SOLUTIONS, COLLOIDS, AND SUSPENSIONS

Solutions	Colloids	Suspensions
Homogeneous	Heterogeneous	Heterogeneous
0.1–1 nm solute particles: atoms, ions, molecules	1–1000 nm dispersed particles: particles or large molecules	Over 1000 nm suspended particles: large particles or aggregates
Does not separate on standing	Does not separate on standing	Particles settle out
Cannot be separated by filtration	Cannot be separated by filtration	Can be separated by filtration
No light scattering	Scatter light (Tyndall effect)	May scatter light, but not transparent

1. (a) Suspension
   (b) Solution   (c) Colloid
2. (a) Heterogeneous
   (b) Homogeneous
3. (a) Electrolyte   (b) Non-electrolyte. Electrolyte solutions conduct electricity. Non-electrolytic solutions do not.
4. A suspension. Unless constantly agitated, it will separate. Let the two mixtures stand until the sand settles out of the water and the sugar remains mixed. Note that light is not scattered by the transparent liquid phase.

- List and explain three factors that influence the rate of dissolving of a solid in a liquid.
- Explain solution equilibrium and distinguish among saturated, unsaturated, and supersaturated solutions.
- Explain the meaning of "like dissolves like" in terms of polar and nonpolar substances.
- List the three interactions that contribute to the heat of solution, and explain what causes dissolving to be exothermic or endothermic.
- Compare the effects of temperature and pressure on solubility.

Section Review

1. Classify the following as either suspensions, colloids, or solutions. Explain. (a) sand mixed with water (b) brass doorknob (c) atmospheric cloud.
2. Classify the following as either a heterogenous or homogenous mixture. Explain. (a) orange juice (b) tap water.
3. What type of solution does (a) an ionic solute form in water? (b) a non-ionic solute form in water? How can you distuinguish between these two types of solutions?
4. In what type of mixture is it easiest to separate the component substances? Why? State one way to prove that a mixture of sugar and water is a solution and a mixture of sand and water is not.

14.2 The Solution Process

Factors Affecting the Rate of Dissolving

Suppose it is a hot day, you are preparing a glass of iced tea, and you are in a hurry. The only available sugar is in large cubes, and the tea solution has already cooled to room temperature. How can you speed up dissolving the sugar in the tea? Which kitchen equipment shown in Figure 14-5 will be of help?

Increasing the surface area of the solute   We noted earlier that sugar dissolves as sugar molecules leave their crystal surface and mingle with water molecules. The same is true for any solid solute in a liquid solvent: molecules or ions of the solute are attracted away from the surface and into solution.

If dissolution occurs at the surface, clearly it can be speeded up if the surface area is increased. Is crushing the sugar cubes one of your suggestions? In general, the more finely divided a substance is the greater the surface area per unit mass and the more quickly it will dissolve.

Agitating the solution   Dissolved solute tends to build up in the solvent close to the solid. Stirring or shaking helps to disperse the solute particles

Figure 14-5 Using the kitchen utensils shown, how would you increase the rate of dissolving of sugar in iced tea?

and brings fresh solvent into contact with the surface. Thus, the effect of stirring is similar to that of crushing a solid—contact between the solvent and the solute surface is increased.

Heating the solvent Most likely you have noticed that sugar and many other materials dissolve more quickly in warm water than in cold water. As the temperature of the solvent increases, solvent molecules move around faster-their average kinetic energy increases. Therefore, collisions at the higher temperatures between the solvent molecules and the solute are more frequent and of higher energy than at the lower temperature. This causes solute molecules to leave the surface at a faster rate.

Figure 14-6 illustrates these three methods of increasing the dissolving rate.

Solubility

If you add spoonful after spoonful of sugar to tea—even if it is powdered sugar added to hot tea with vigorous stirring—a point will come at which no more sugar will dissolve. For every solute–solvent combination at a given temperature, there is a limit to the amount of solute that can be dissolved.

When solid sugar is first dropped into the water, sugar molecules begin to depart from the solid surface and move about at random in the liquid solvent. Eventually, enough sugar molecules are in solution that some collide with the solid surface and crystallize (return to the solid crystal). If the amount of solid sugar present—is less than the limit that can dissolve, sugar molecules will leave the solid surface faster than they reenter it. All of the sugar molecules present will eventually be in solution. If more sugar is present than can dissolve, the rates of sugar molecules leaving and entering

A way to help students understand the constantly changing nature of a saturated solution: If you *could* tie red tags on all the molecules in pile of undissolved sugar in a saturated sugar solution, after a while there would be sugar molecules with red tags in solution and sugar molecules without tags in the undissolved sugar due to the fact that they are continually dissolving and recrystallizing. Can we tie red tags on molecules? In a way—we can tag molecules by using radioactive atoms in them.

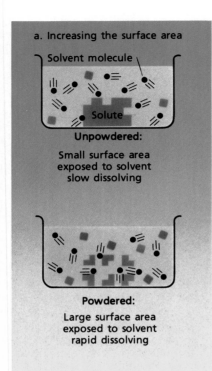

a. Increasing the surface area

Solvent molecule

Solute

Unpowdered:

Small surface area
exposed to solvent
slow dissolving

Powdered:

Large surface area
exposed to solvent
rapid dissolving

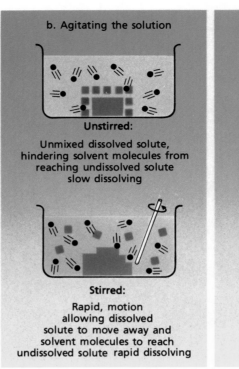

b. Agitating the solution

Unstirred:

Unmixed dissolved solute,
hindering solvent molecules from
reaching undissolved solute
slow dissolving

Stirred:

Rapid, motion
allowing dissolved
solute to move away and
solvent molecules to reach
undissolved solute rapid dissolving

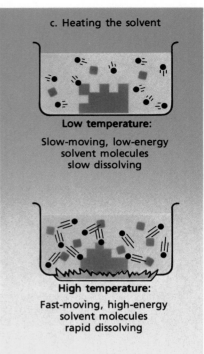

c. Heating the solvent

Low temperature:

Slow-moving, low-energy
solvent molecules
slow dissolving

High temperature:

Fast-moving, high-energy
solvent molecules
rapid dissolving

Figure 14-6 Three methods for increasing the rate at which a solute dissolves

Figure 14-7 A saturated solution contains the equilibrium concentration of solute under existing conditions, as shown in this bottle.

the sugar crystal surface will become equal before all of the sugar has dissolved. In this case, a dynamic physical equilibrium is established.

$$\text{solute} + \text{solvent} \rightleftharpoons \text{solution}$$

Solution equilibrium *is the physical state in which the opposing processes of dissolving and crystallizing of a solute occur at equal rates.* The point at which equilibrium is reached for any solute–solvent combination is difficult to predict precisely and depends upon the nature of the solute, the nature of the solvent, and the temperature.

A solution that contains the maximum amount of dissolved solute is described as a **saturated solution**. (How can you tell that the solution pictured in Figure 14-7 is saturated?) At 20°C, 204 g of sugar is the maximum amount that will dissolve at equilibrium in 100 g of water. If solid sugar is added to the solution just described, it falls to the bottom and does not appear to dissolve because an equilibrium will be established between molecules leaving and entering the solid phase. If more water is added to the saturated solution, then more sugar will dissolve in it. *A solution that contains less solute than a saturated solution under the existing conditions is an* **unsaturated solution.**

When a hot, saturated solution of sugar is allowed to cool, the excess solute usually, but not always, separates as the solution cools. If this does not occur, a **supersaturated solution**—*a solution that contains more dissolved solute than a saturated solution under the same conditions*—is produced. A supersaturated solution is not likely to remain supersaturated. Crystals will eventually form in it until equilibrium is reestablished at the lower temperature. A supersaturated solution can be made by preparing a saturated solution of sodium thiosulfate ($Na_2S_2O_3$) or sodium acetate ($NaC_2H_3O_2$) in hot water, filtering the hot solution, and allowing the filter solution to stand undisturbed as it cools. Dropping a small crystal of the solute into the supersaturated solution ("seeding") or disturbing the solution will cause a rapid formation of crystals by the excess solute.

The **solubility** *of a substance is the amount in grams of that substance required to form a saturated solution with 100 g of solvent at a specified temperature.* We say that the solubility of sugar is 204 g per 100 g of water at 20°C. The temperature must be specified because solubility varies with temperature. For gases, the pressure must also be specified. Solubilities must be determined experimentally; and they vary widely, as illustrated for three compounds in Figure 14-8. Solubility values are recorded in chemical handbooks and are usually given as grams of solute per 100 g of solvent or per 100 mL of solvent at a given temperature.

Figure 14-8 A comparison of the masses of three common solutes that can be dissolved in 100 g of water at 0°C. Convert these quantities to moles of solute per 100 g of water and compare.

In the preceding paragraphs, we listed factors that speed up the rate of dissolving. The speed at which a solid dissolves is *unrelated* to solubility. The maximum amount of solute that enters into solution and reaches equilibrium is always the same under the same conditions.

Factors Affecting Solubility

The extent to which a given solute dissolves in a solvent depends on the identity of the solute and solvent and also on the existing conditions of pressure and temperature.

Gases lose energy when changing to the liquid phase.

Types of solvents and solutes   Sodium chloride is highly soluble in water, but gasoline is not. On the other hand, gasoline mixes readily with benzene (C_6H_6), but sodium chloride does not. What is the basis for such differences in solubility?

"Like dissolves like" is a roughly useful rule for predicting whether or not one substance dissolves in another. The necessary "alikeness" depends on the type of bonding, the polarity or nonpolarity of molecules, and the intermolecular forces that are found in the two mixed substances.

Water molecules are strongly attracted to each other because they interact by hydrogen bonding. According to the "like dissolves like" rule, water should be a good solvent for any substance composed of polar molecules. The molecules are more likely to be drawn away from each other by polar water molecules than by a nonpolar solvent. The positively and negatively charged regions in polar solute molecules are attracted to the oppositively charged regions of the water molecules as diagrammed in Figure 14-9. Molecules that can take part in hydrogen bonding are especially soluble. The large solubility of sugar is aided by the presence in each sugar molecule of six O—H bonds that can form hydrogen bonds with water molecules.

You might wish to review the sections on molecular polarity and intermolecular forces in Chapter 6.

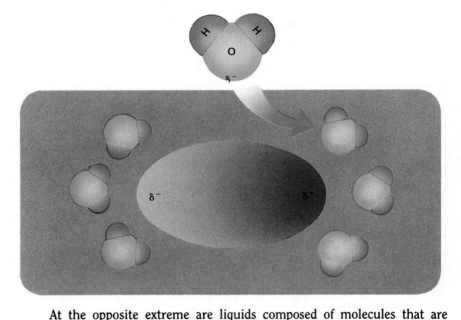

Figure 14-9 Polar molecules are hydrated as they dissolve in water.

See Dalton's law of partial pressure: Section 11.3

At the opposite extreme are liquids composed of molecules that are nonpolar and do not form hydrogen bonds. Carbon tetrachloride (CCl_4) molecules are symmetrical and despite the polar C—Cl bonds, are nonpolar. A carbon tetrachloride molecule can exert only relatively weak London forces on other molecules. Would we expect water or sodium chloride to dissolve in carbon tetrachloride? No, we would not. There is too little similarity in bonding, polarity, and intermolecular forces.

Substances that are not soluble in each other are **immiscible.** Carbon tetrachloride and water are examples of immiscible substances as are oil and water. When you shake a bottle of salad dressing, oil droplets become dispersed in the water. As soon as you stop shaking, the strong attraction between the water molecules squeezes out the oil droplets, and separate layers form.

Substances that are mostly nonpolar, however—such as fats, oils and greases—are quite soluble in carbon tetrachloride. Carbon tetrachloride was once widely used as a household cleaning agent. It is now considered too toxic for such use.

Gasoline contains mainly nonpolar hydrocarbons and is also an excellent solvent for fats, oils, and greases. The major intermolecular forces acting between the nonpolar molecules are the relatively weak London forces.

The "like dissolves like" rule predicts that ethanol (ethyl alcohol, C_2H_5OH) should be soluble in water, and the prediction is correct. Ethanol molecules are polar because of the single electronegative oxygen atom and also can form hydrogen bonds with water molecules through the O—H bond. The intermolecular forces among water and ethanol molecules are so similar that unlimited quantities of these two liquids mix together freely. *Two substances that are mutually soluble in all proportions are said to be* **miscible**. Nonpolar substances with very similar intermolecular forces, such as gasoline and benzene, are also completely miscible. The nonpolar molecules experience no strong forces of attraction or repulsion and therefore mingle freely with little change in energy.

Ethanol is intermediate in polarity between water and carbon tetrachloride (Figure 14-10) and is not as good a solvent for polar or ionic substances as water. Sodium chloride is only slightly soluble in ethanol. On the other hand, ethanol is a better solvent than water for less polar substances because the molecule has a region containing only nonpolar carbon-hydrogen bonds

$$\text{Nonpolar region} \quad \underset{\overset{|}{H}\ \overset{|}{H}}{\overset{\overset{H\quad H}{|\quad |}}{H—C—C—O—H}} \quad \text{polar region}$$

Ether ($C_2H_5OC_2H_5$), for example, has one electronegative atom but no possibility for hydrogen bonding. It is completely miscible with ethanol but only partially miscible with water.

Variations on this are to be expected, however. The small, nonpolar carbon tetrachloride molecules are able to dissolve in the somewhat polar ethanol, for example. The ions in some ionic compounds are so strongly attracted to each other that the compounds are only very slightly soluble in water.

Pressure Changes in pressure have very little effect on the solubilities of liquids or solids in liquid solvents. Gas solubilities in liquids, however, always increase with increasing pressure.

When a gas is in contact with the surface of a liquid, gas molecules can enter the liquid. As the amount of dissolved gas increases, some molecules begin to escape and reenter the gas phase. An equilibrium is eventually established between the rates at which gas molecules enter and leave the liquid phase. As long as this equilibrium is undisturbed, the solubility of the gas in the liquid is unchanged.

$$\text{gas} + \text{solvent} \rightleftharpoons \text{solution}$$

Increasing the pressure of the solute gas above the solution is a stress on the equilibrium. Molecules collide with the liquid surface more often. The increase in pressure is partially offset by an increase in the rate of gas molecules entering the solution. In turn, the increase in the amount of dissolved gas causes an increase in the rate at which molecules escape from the liquid surface and become vapor. Eventually, equilibrium is restored at a higher gas solubility. As expected from Le Chatelier's principle, an increase in gas pressure has caused the equilibrium to shift so that fewer molecules are in the gas phase.

The solubility of a gas in a liquid is directly proportional to the partial pressure of that gas on the surface of the liquid. This is a statement of **Henry's**

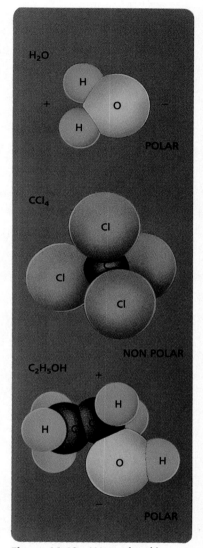

Figure 14-10 Water *(top)* is polar. Carbon tetrachloride *(middle)* is nonpolar. Ethanol *(bottom)* is intermediate in polarity between water and carbon tetrachloride.

law, named after the English chemist William Henry (1775–1836). Henry's law applies to gas-liquid solutions at constant temperature.

Recall that when a mixture of gases is confined in a space of constant volume at a constant temperature, each gas exerts the same pressure it would if it alone occupied the space. Assuming that the gases do not react in any way when in solution, each gas dissolves to the extent it would if no other gases were present.

The increased solubility of gases with increased pressure is put to use in making carbonated beverages. At the bottling plant, carbon dioxide gas (CO_2) is forced into solution in flavored water at a pressure of 5–10 atm. The gas-in-liquid solution is then sealed in bottles or cans. When the cap is removed, the pressure is reduced to 1 atm, and some of the carbon dioxide escapes as gas bubbles. *The rapid escape of a gas from a liquid in which it is dissolved is known as* **effervescence** (Figure 14-11).

Temperature You have already learned that temperature affects the rates at which solutes dissolve. Temperature can also have a large effect on the amounts of solute that dissolve in a given amount of solvent.

First, let's consider gas solubility. Increasing the temperature usually decreases gas solubility. When you first heat water and before boiling begins, you can often see small bubbles. These bubbles are formed by the gases that were in solution in the colder water and are being driven out. As the temperature increases, the average kinetic energy of the molecules in solution increases. A greater number of solute molecules are able to escape from the attraction of solvent molecules and return to the gas phase. Therefore, at higher temperatures equilibrium is reached with fewer gas molecules in solution. (Figure 14-12)

The effect of temperature on the solubility of solids in liquids is more difficult to predict than the effect on gas solubility. Most often, increasing the temperature increases the solubility of solids. However, an equivalent temperature increase may result in a large increase in solubility in one case and a slight increase in another. In Table 14-4 and Figure 14-13, compare the effect of temperature on the solubilities of potassium nitrate (KNO_3) and

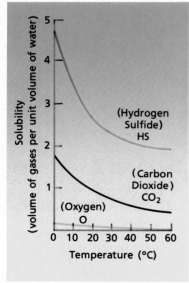

Figure 14-12 The solubility of gases in liquid decreases with increasing temperature.

TABLE 14-4 SOLUBILITY OF SOLUTES AS A FUNCTION OF TEMPERATURE
Grams of solute per 100 grams of H_2O

Substance	Temperature, °C					
	0°	20°	40°	60°	80°	100°
$AgNO_3$	122	216	311	440	585	733
$Ba(OH)_2$	1.67	3.89	8.22	20.9	101	—
$C_{12}H_{22}O_{11}$	179	204	238	287	362	487
$Ca(OH)_2$	0.189	0.173	0.141	0.121	0.094	0.07
$Ce_2(SO_4)_3$	20.8	10.1	—	3.87	—	—
KCl	28.0	34.2	40.1	45.8	51.3	56.3
KI	128	144	162	176	192	206
KNO_3	13.9	31.6	61.3	106	167	245
Li_2CO_3	1.54	1.33	1.17	1.01	0.85	0.72
NaCl	35.7	35.9	36.4	37.1	38.0	39.2
$NaNO_3$	73	87.6	102	122	148	180
$Yb_2(SO_4)_3$	44.2	37.5 at 10°	17.2 at 30°	10.4	6.4	4.7
CO_2(gas at SP)	0.335	0.169	0.0973	0.058	—	—
O_2(gas at SP)	0.00694	0.00537	0.00308	0.00227	0.00138	0.00

sodium chloride (NaCl). About 14 g of potassium nitrate will dissolve in 100 g of water at 0°C. The solubility of potassium nitrate increases by more than 150 g KNO_3/100 g, H_2O when the temperature is raised to 80°C. Under similar circumstances, the solubility of sodium chloride increases by only about 2 g NaCl/100 g H_2O. In some cases, solubility of a solid *decreases* with an increase in temperature. For example, between 0°C and 80°C the solubility of cerium sulfate, $Ce(SO_4)_2$, decreases by about 14 g. Here, too, the size of the change varies.

Heats of Solution

The formation of a solution is accompanied by an energy change. If you dissolve some potassium iodide, KI, in water, you will find that the outside of the container becomes cold to your touch. But if you dissolve some lithium chloride (LiCl), in the same way, the outside of the container will be hot to your touch. The formation of a solid–liquid solution can apparently either absorb heat (KI in water) or release heat (LiCl in water).

The amount of heat energy absorbed or released when a solute dissolves in a specific amount of solute is the **heat of solution**. It is given in kilojoules per mole of (kJ/mole) of solute dissolved in a specified amount of solvent. Table 14-5 lists some heats of solution. You can see that there are both positive values (a change that absorbs heat) and negative values (a change that releases heat). Endothermic—positive heat of solution:

$$solute + solvent + heat \longrightarrow solution$$

Exothermic — negative heat of solution:

$$solute + solvent \longrightarrow solution + heat$$

During solution formation, solvent and solute particles experience changes in the forces attracting them to other particles. There are three different interactions that contribute to the heat of solution. Before

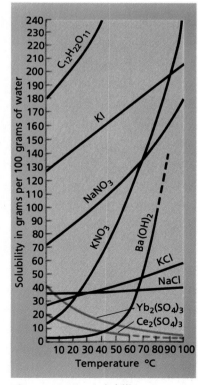

Figure 14-13 Solubility curves

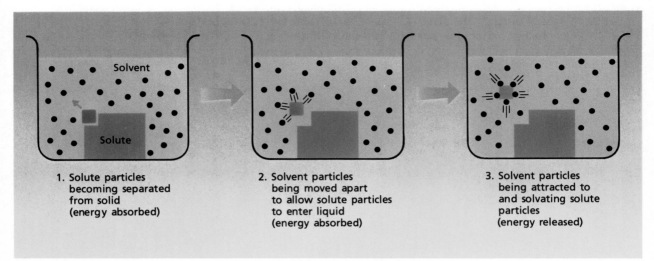

1. Solute particles becoming separated from solid (energy absorbed)

2. Solvent particles being moved apart to allow solute particles to enter liquid (energy absorbed)

3. Solvent particles being attracted to and solvating solute particles (energy released)

Figure 14-14 Three steps in solution formation

The colligative properties of solutions of electrolytes are discussed in Chapter 15.

dissolving begins, solvent molecules are held together by intermolecular forces (solvent–solvent attraction). In the solute, molecules are held together by intermolecular forces (solute–solute attraction). During solution formation, these forces are disrupted. Energy is required to separate solute molecules and solvent molecules from their neighbors.

In the solution, particles acquire new neighbors and solute–solvent attraction takes over. *A solute molecule surrounded by solvent molecules (as in Figure 14-9) is said to be* **solvated.** Intermolecular attraction brings the solvent and solute molecules together, and energy is released in the process.

Solution formation can, thus be pictured as the result of the three interactions summarized in Figure 14-14.

1. Solute–solute attraction is broken up, requiring energy.
2. Solvent–solvent attraction is broken up, requiring energy.
3. Solute–solvent attraction is formed, releasing energy.

The heat of solution is positive (heat is absorbed) when the sum of *(1)* and *(2)* is greater than *(3).* The heat of solution is negative (heat is released) when the sum of *(1)* and *(2)* is less than *(3).* Here is the basis for the effect of increasing temperatures on the solubilities of solids: In a saturated solution,

The amount of heat energy absorbed or released when a solute dissolves in a specific amount of solute is the **heat of solution**. It is given in kilojoules

Figure 14-15 Two beakers of water *(left).* Salt added to the left beaker raises its temperature *(right).* Why?

there is a dynamic equilibrium between undissolved and dissolved solute. Le Chatelier's principle applies to this situation. Adding energy by raising the temperature of the solution causes the rate of dissolving to change in the direction that consumes energy. If dissolving is endothermic, more solute dissolves:

$$\text{solute} + \text{solvent} + \text{heat} \rightleftharpoons \text{solution} + \text{heat}$$

We saw that heating usually decreases the solubility of a gas, so dissolving of gases must usually be exothermic. The heat of solution values in Table 14-5 show this to be correct.

TABLE 14-5   HEATS OF SOLUTION (kJ/mol solute in 200 moles H₂O [(s) = solid, (ℓ) = liquid, (g) = gas at STP]

Substance	Heat of Solution	Substance	Heat of Solution
$AgNO_3(s)$	+22.77	$KOH(s)$	−54.59
$CO_2(g)$	−19.91	$LiCl(s)$	−35.0
$CuSO_4(s)$	−67.81	$Li_2CO_3(s)$	−12.8
$CuSO_4 \cdot 5H_2O(s)$	+11.51	$MgSO_4 \cdot 7H_2O(s)$	+15.9
$HC_2H_3O_2(\ell)$	−1.59	$NaCl(s)$	+4.27
$HCl(g)$	−74.26	$NaNO_3(s)$	+21.1
$HI(g)$	−29.39	$NaOH(s)$	−41.6
$H_2SO_4(\ell)$	−74.30	$Na_2SO_4 \cdot 10H_2O(s)$	+78.53
$KCl(s)$	+17.58	$NH_3(g)$	−34.7
$KClO_3(s)$	+42.03	$NH_4Cl(s)$	+16.2
$KI(s)$	+21.4	$NH_4NO_3(s)$	+25.5
$KNO_3(s)$	+35.7		

In the gaseous state, molecules are so far apart that there are virtually no intermolecular forces of attraction between them. Therefore, the solute–solute interaction has little effect on the heat of solution of a gas. Energy is released when a gas dissolves in a liquid because attraction between solute gas and solvent molecules outweighs the energy needed to separate solvent molecules.

This section has shown that intermolecular forces can explain factors affecting the rate of dissolving, variation in the solubility of a substance in a particular solvent and the energy changes that occur during the solution process.

Section Review

1. Why can you expect a packet of sugar to dissolve faster in hot tea than iced tea?
2. Explain how you would prepare a saturated solution of sugar in water. How would you then make it a super saturated solution?
3. Explain why ethanol will dissolve in water and carbon tetrachloride will not.
4. When a solute molecule is solvated, is heat released or absorbed?
5. A warm and cold bottle of soda pop are opened. Which will effervesce more?

1. The molecules in hot tea move around faster than in ice tea. Therefore, there are more collisions between the tea and sugar molecules in the surface of the sugar grain. Sugar molecules leave the surface of the grains faster in hot tea than in cold tea.
2. Continually add sugar to hot water until undissolved grains remain at the bottom of the container. This is a saturated solution of sugar and water. Set the saturated solution aside until it cools to rooms temperature. Once it cools, it will be a supersaturated solution of sugar and water.
3. Ethanol and water are each polar molecules. The negatively charged region of the ethanol molecules are attracted to the positively charged region of the water molecules and vice-versa. Carbon tetrachloride is nonpolar.
4. Intermolecular attraction brings the solvent and solute molecules together. Energy is released in this process.
5. The warm bottle because the dissolved CO_2 gas is less soluble in warm liquid than in cold liquid.

14.3 Concentration of Solutions

SECTION OBJECTIVES

- Define concentration using molarity, molality, and percent by mass.
- Given the concentration of a solution, find the amount of solute in a given amount of solution.
- Given the concentration of a solution, find the amount of solution that contains a given amount of solute.

Explain that *dilute* and *concentrated* are very indefinite terms by using the example of strong and weak coffee. When people talk about strong coffee, they really mean concentrated coffee; weak would be dilute. What is concentrated coffee to one person may be dilute to another and vice-versa. In science, solutions have to be able to be duplicated exactly to be useful.

Liquid bleach is 5.25% $NaClO$, vinegar is 5–6% $HC_2H_3O_2$, and medicinal hydrogen peroxide is 3% H_2O_2.

The **concentration** *of a solution is a measurement of the amount of solute in a given amount of solvent or solution.* Some medications are solutions of drugs—a one-teaspoonful dose at the correct concentration might cure the condition; the same dose in the wrong concentration might kill the patient.

In this section, we introduce three different ways of expressing the concentrations of solutions: molarity, molality, and percent by mass.

Sometimes solutions are referred to as "dilute" or "concentrated." These are relative terms. "Dilute" means that there is a relatively small amount of solute in a solvent. "Concentrated," on the other hand, means that there is a relatively large amount of solute in a solvent. Note that these terms are unrelated to the degree to which a solution is saturated. A saturated solution of a substance that is not very soluble might be very dilute.

Percent by Mass

Percent by mass of a solute in solution in the number of grams of solute dissolved in 100 g of solution. For example, the percent by mass of a solution made from 10 g of sodium hydroxide (NaOH) dissolved in 90 g of water is found as follows:

$$\text{Percent by mass} = \frac{\text{mass of solute}}{\text{mass of solute} + \text{mass of solvent}} \times 100\%$$

$$= \frac{10 \text{ g NaOH}}{10 \text{ g NaOH} + 90 \text{ g H}_2\text{O}} \times 100\%$$

$$= 10\% \text{ NaOH (by mass)}$$

Note that the mass of the solution is the sum of the masses of the solute and solvent. Percent concentration is common for solutions used for practical purposes—household or industrial cleaning, killing pests in the garden, medical applications. Percent concentration by mass is based only on the mass of a solute and is unrelated to its chemical formula or molar mass.

Sample Problem 14.1

A solution of sodium chloride is prepared by dissolving 5 g of salt in 550 g of water. What is the concentration of this solution given as percent by mass?

Solution

Step 1. Analyze   Given: mass of solute and identity = 5 g NaCl

                             mass of solute and identity = 550 g H_2O

              Unknown: percent by mass concentration of solution

Step 2. Plan   solvent mass, solute mass → solution mass → percent solute by mass

The mass of solute must be divided by the mass of the solution and multiplied by 100%.

$$\frac{\text{mass of solute}}{\text{mass of solute} + \text{mass of solvent}} \times 100\% = \% \text{ solute (by mass)}$$

Sample Problem 14.1 *continued*

Step 3. Compute
$$\frac{5 \text{ g NaCl}}{5 \text{ g NaCl} + 550 \text{ g H}_2\text{O}} \times 100\% = 0.9\% \text{ NaCl}$$

Step 4. Evaluate The answer is rounded to one significant digit because it is limited by the given mass of the solute. If 5 g of solute were dissolved in 500 g of solvent, a 1% solution would be produced. It is reasonable, therefore, that 5 g of solute in somewhat more solvent should give a solution of somewhat smaller concentration.

Practice Problems
1. What is the percent by mass of a solution prepared by dissolving 4.0 g of acetic acid in 35 g of water? *(Ans.)* 10% acetic acid
2. Determine the percent by mass of a solution prepared by dissolving 32.5 g of glucose in 155 g of water. *(Ans.)* 17.3% glucose

Molarity

Molarity *is the number of moles of solute in one liter of solution.* To find the molarity of a solution you must know the molar mass of the solute.

For example, a "one-molar" solution of sodium hydroxide (NaOH) contains one mole of NaOH in every liter of solution. The symbol for molarity in *M* and the concentration of the preceding solution is written 1 *M* NaOH.

One mole of sodium hydroxide (NaOH) has a mass of 40.0 g. This quantity of sodium hydroxide dissolved in enough water to make exactly 1.00 L of solution gives a 1-*M* solution. If 20.0 g of NaOH, which is 0.500 mol, is dissolved in 1.00 L of solution, a 0.500-*M* NaOH solution is produced:

$$\text{molarity} = \frac{\text{number of moles of solute}}{\text{number of liters of solution}}$$

$$= \frac{0.500 \text{ mol NaOH}}{1.00 \text{ L}}$$

$$= 0.500 \, M \text{ NaOH}$$

If twice the molar mass of NaOH (80.0 g) is dissolved in enough water to make 1 L of solution, a 2-*M* solution is produced. The molarity of any solution can be calculated by dividing the number of moles of solute by the number of liters of solution.

Note that a one-molar solution is *not* made by adding 1 mol of solute to 1 L of *solvent*. In such a case, the final total volume would be slightly different from 1 L. Instead, 1 mol of solute is first dissolved in less than 1 L of solvent. Then, the resulting solution is carefully diluted with more solvent to bring the *total volume* to 1 L, as illustrated in Figure 14-16.

Another method for describing the concentration of a solution, called *molality*, will be discussed following sample problems on molarity. Molarity is useful when the quantity of solute participating in a chemical reaction taking place in solution is of interest. Any required molar quantity of a solute can be selected by measuring out the appropriate volume of a solution of known molarity.

Figure 14-16 Preparation of a 1–molar solution. *Clockwise from top right:* 1–mole of solute is added to a 1–liter volumetric flask; enough water is then added to dissolve the solute; finally water is added to bring the volume of the solution to 1 liter.

Sample Problem 14.2

What is the molarity of 3.50 L of solution that contains 90.0 g of sodium chloride (NaCl)?

Solution

Step 1. Analyze   Given: solute mass and solute formula $= 90.0$ g M NaCl
                              solution volume $= 3.50$ L of solution
                    Unknown: molarity of the solution

Step 2. Plan   You only need one conversion (mass to moles of solute) to arrive at your answer.

$$\text{mass of solute} \rightarrow \text{number of moles of solute} \rightarrow \text{molarity}$$

$$\text{mass solute (g)} \times \frac{1}{\text{molar mass of solute (g/mol)}} \times \frac{1}{\text{volume of solution (L)}}$$

$$= \text{molarity of solution}$$

Step 3. Compute $90.0 \text{ g NaCl} \times \dfrac{1 \text{ mole NaCl}}{58.5 \text{ g NaCl}} \times \dfrac{1}{3.5 \text{ L of solution}} = \dfrac{0.440 \text{ mole NaCl}}{1 \text{ L of solution}} = 0.440 \, M \text{ NaCl}$

Step 4. Evaluate   Because each factor involved is limited to three significant digits, the answer should have three significant digits, which it does. The units cancel correctly to give the desired moles of solute per liter of solution, which is the unit of molarity. Because the numerical calculation looks roughly like 90/200 (slightly less than 0.5), the final answer seems reasonable.

Sample Problem 14.3

How many moles of HCl are present in 0.8 L of a 0.5-M HCl solution?

Solution

Step 1. Analyze   Given: volume of solution = 0.8 L
                              molarity of solution = 0.5 M HCl
                    Unknown: moles of HCl in given volume

Step 2. Plan   The molarity indicates the moles of solutes that are in one liter of solution.
                    Given the volume of the solution, the number of moles of solute can then be found.

$$\text{molarity, volume} \rightarrow \text{moles solute}$$

$$\text{molarity (moles of solute/liter of solution)} \times \text{volume (1 liter of solution)}$$
$$= \text{moles of solute}$$

Step 3. Compute $\dfrac{0.5 \text{ moles HCl}}{1 \text{ L of solution}} \times 0.8 \text{ L of solution} = 0.4 \text{ moles HCl}$

Step 4. Evaluate   The answer is correctly given to one significant digit. The units cancel correctly
                    to give the desired label, moles. There should be less than 0.5 moles HCl, since
                    less that 1 L of solution was used.

Practice Problems

1. What is the molarity of a solution composed of 5.85 g of potassium iodide (KI) dissolved in enough water to make 0.125 L of solution?
   (Ans.) 0.282 M KI
2. How many moles of H_2SO_4 are present in 0.500 L of a 0.150-M H_2SO_4 solution?
   (Ans.) 0.0750 mol

Molality

Molality *is the concentration of a solution expressed in moles of solute per kilogram of solvent.* A solution that contains 1 mole of solute, ammonia (NH_3), for example, dissolved in exactly 1 kg of solvent, is a "one-molal" solution. The symbol for molality is m, and the concentration of this solution is written 1 m NH_3.

One mole of ammonia has a molar mass of 17.0 g, and 17.0 g of NH_3 dissolved in 1 kg of water gives one-molal ammonia solution. If 8.50 g of ammonia, which is one-half mole of ammonia, is dissolved in exactly 1 kg of water, a 0.500 m NH_3 solution is prepared.

$$\text{molality} = \frac{\text{no. of moles solute}}{\text{mass of solvent (kg)}} = \frac{0.500 \text{ mol } NH_3}{1 \text{ kg } H_2O} = 0.500 \ m \ NH_3$$

If 34.0 g of ammonia, which is 2 mol, is dissolved in 1 kg of water, a 2 m solution of NH_3 is produced. The molality of any solution can be found by dividing the number of moles of solute by the mass in kilograms of the solvent in which it is dissolved. Note that if grams of solvent are known, the quantity of solvent must be converted to kilograms by multiplying by the factor 1 kg/10^3 g.

To help students understand the difference between molarity and molality, make a 1-M solution of sucrose by adding enough water to 342 g of sucrose to make a liter of solution. Then add 342 grams of sucrose to 1 kg of water, and determine the volume of this 1-m solution. It should form approximately 1.3 L of solution.

Later in this chapter, you will see that concentrations are given in molality when studying properties of solutions related to vapor pressure and temperature changes. Molality and percent composition by mass are useful since they do not change their value with changes in temperature. Molarity can vary with temperature because solution volume varies with temperature.

For a dilute solution, where water is the solvent, molarity and molality have similar values. To see why this is true, imagine a certain number of moles of a solvent, A, dissolved in a given amount of water. The molarity and molality of this solution are, respectively;

$$M = \frac{\text{moles of A}}{\text{volume of solution}} \qquad m = \frac{\text{moles of A}}{\text{mass of solution}}$$

When there are relatively few particles of solute compared to solvent in a solution, the volume and mass of the solution are essentially the volume and mass of the solute. Also, water has a density of 1kg/L. The mass of a sample of water, in kilograms, is numericially equal to its volume in liters. Therefore, for a very dilute aqueous solution, the volume of solution is approximately equal to the mass of the solution. Since the numerators of the above expression for m and M are approximately equal and their denominators are equal, the expressions are approximately equal.

Sample Problem 14.4

A solution contains 17.1 g of sucrose (table sugar, $C_{12}H_{22}O_{11}$) dissolved in 125 g of water. Find the molal concentration of this solution.

Solution

Step 1. Analyze   Given: solute mass and formula = 17.1 g $C_{12}H_{22}O_{11}$
sovent mass and identity = 125 g water

Step 2. Plan   To find molality you must solve for moles of solute per kilogram of solvent. The given mass of solute must be converted to moles and the mass in grams of solvent must be converted to kilograms.

mass solute, mass in grams of solvent → molar mass of solute, mass in kilograms of solvent → molality

$$\frac{\text{mass solute (g solute)}}{\text{mass solvent (g solvent)}} \times \frac{1}{\text{molar mass of solute}} \left(\frac{\text{mol solute}}{\text{g solute}}\right) \times \frac{10^3 \text{ g solvent}}{\text{kg solvent}}$$

$$= \text{molality} \left(\frac{\text{mol solute}}{\text{kg solvent}}\right)$$

Step 3. Compute

$$\frac{17.1 \text{ g } C_{12}H_{22}O_{11}}{125 \text{ g water}} \times \frac{1 \text{ mol } C_{12}H_{22}O_{11}}{342 \text{ g } C_{12}H_{22}O_{11}} \times \frac{10^3 \text{ g water}}{1 \text{ kg } H_2O} =$$

$$\frac{0.400 \text{ mol } C_{12}H_{22}O_{11}}{\text{kg } H_2O} = 0.400 \; m \; C_{12}H_{22}O_{11}$$

Step 4. Evaluate   The answer is correctly given to three significant digits. A numerical estimate of $(2 \times 10^4) \div (4 \times 10^4)$ indicates an answer of approximately 0.5. The label, *mole solute/kg solvent is correct for molality, the desired measurement.* surement.

Sample Problem 14.5

A solution of iodine (I_2) in carbon tetrachloride (CCl_4) is used when iodine is needed for certain chemical tests. How much iodine must be added to prepare a 0.480-*m* solution of iodine in CCl_4 if 100.0 g of CCl_4 are used?

Solution

Step 1. Analyze   Given: molality of solution and solute formula = 0.480 *m* I_2
mass of solvent and identity = 100 g CCl_4
Unknown: mass of solute

Step 2. Plan   The molality gives you the moles of solute, which can be converted to the mass of solute when you determine the molar mass.

$$\text{molality} \rightarrow \text{molar mass of solute} \rightarrow \text{mass of solute}$$

$$\text{molality} \left(\frac{\text{moles of solute}}{\text{kg solvent}}\right) \times \text{molar mass of solute} \left(\frac{\text{g solute}}{\text{moles of solute}}\right)$$

$$\times \text{mass solvent (kg solvent)} = \text{mass solute (g solute)}$$

Step 3. Compute $$\frac{0.480 \text{ mol } I_2}{\text{kg } CCl_4} \times \frac{254 \text{ g } I_2}{1 \text{ mol } I_2} \times 100 \text{ g } CCl_4 \times \frac{1 \text{ kg } CCl_4}{10^3 \text{ g } CCl_4} = 12.2 \text{ g } I_2$$

Step 4. Evaluate   The answer has three significant digits, as do the factors in the calculation. The units cancel, leaving the desired label, g I_2. A numerical estimate of 125/10, or 12.5, would support the computation.

Practice Problems

1. What is the molality of a solution composed of 2.55 g of acetone, $(CH_3)_2CO$, dissolved in 200 g of water? *(Ans.)* 0.219 *m* acetone
2. What quantity, in grams, of methanol (CH_3OH) is required to prepare a 1.244-*m* solution in 400. g of water?   *(Ans.)* 3.12 g CH_3OH
3. How many grams of $AgNO_3$ are needed to prepare a 0.125-*m* solution in 250 mL of water? *(Ans.)* 5.31 g $AgNO_3$
4. What is the molality of a solution containing 18.2g HCl and 250 g of water? *(Ans.)* 2.00 *m*

Section Review

1. What concentration units give the number of molecules of solute in a given volume of solution?
2. A gas–gas solution with a fixed amount of solute and solvent by mass is heated and increases volume. Which units of concentration will not change? Why?
3. Five grams of sugar, $C_6H_{22}O_{11}$, are dissolved in one liter of water. What is the concentration of this solution in percent by mass, molality, and molarity?
4. Two solutions of sugar, $C_6H_{22}O_{11}$, and water are in different bottles. One is a one-mol*ar* solution and the other is a one mol*al* solution. In percent, by mass, which solution is more concentrated?

1. Molarity
2. Molality and percentage by mass. Unlike molarity, molality and percentage by mass do not depend on volume.
3. By mass: 5%, molality: 0.01 *m*, molarity: 0.01 *M* (providing the mixture forms 1 L of solution).
4. A 1-*M* solution would have 243 g of sugar plus just enough water to make 1 L of solution; the 1-*m* solution would have 342 g of sugar plus 1 kg (1 L at 4°C) of water. Since both solutions contain the same amount of sugar, the 1-*M* solution, having less water,

would be more concentrated.

14.4 Colligative Properties of Solutions

SECTION OBJECTIVES

- List three colligative properties and describe how each is caused.
- Write the expression for freezing-point depression and boiling-point elevation. Define and give the units for each term in the expressions.
- Given appropriate information, calculate freezing-point depression, boiling-point elevation, or solution molality.
- Describe an experimental method for determining molar mass using a colligative property.
- Calculate molar mass from freezing-point depression or boiling-point elevation data.

Imagine that it is a very cold day in the middle of winter, like that pictured in Figure 14-17. A steady flow of cars and trucks moves down the highway. The operation of these vehicles is dependent on the steady flow of water in their cooling systems. How can this water remain fluid? What is the purpose of the salt that has been sprinkled on the sidewalk and road? A scientific understanding of the behavior of solutions helps us to continue our daily activities despite the winter cold. Using the properties of solutions, we can counteract nature by lowering the freezing point of water.

Figure 14-17 The cooling systems of most cars depend on liquid water. How can these cars operate at temepratures below the freezing point of water?

Vapor-Pressure Lowering

To understand how the freezing point of water is lowered, we must go back to our study of vapor pressure—the pressure caused by molecules that have escaped from the liquid phase to the gaseous phase. Vapor pressure is a measure of the tendency of molecules to escape from a liquid or a solid. Experiments show that the vapor pressure of a solvent containing a nonvolatile solute is lower than the vapor pressure of the pure solvent at any temperatures. *By* **nonvolatile,** *we mean that a substance that has little tendency to become a gas under the existing conditions.*

Simplest formulas and molecular formulas were discussed in Chapter 4.

Figure 14-18 illustrates the effect a nonvolatile solute has on the pressure of a solvent. Compare the top curve for the pure solvent with the middle curve for a dilute solution of a nonvolatile solute. Vapor pressure has been lowered, and the solution remains liquid over a larger temperature range. At the left, you can see that the freezing point of the solution (the point where the two curves intersect) is lower than that of the pure water. At the right, you can see that the boiling point of the solution is higher than that of pure water.

Have you noticed we have said nothing about the identity of the solute? That is because it does not matter. Each solute particle at the liquid surface has the same effect, whether it is an ion or a molecule and whatever its composition. *A property that depends on the number of solute particles but*

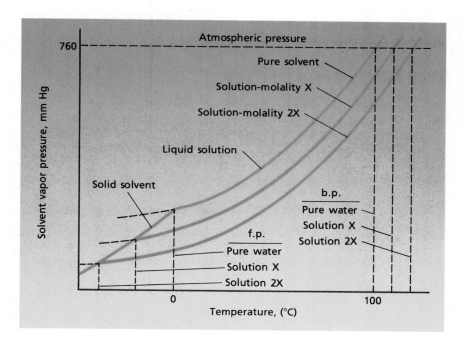

Figure 14-18 Vapor pressure of water as a function of temperature plotted for pure water and water solutions of molalities *X* and *2X.* Freezing point depressions and elevations are proportional to the molal concentrations of the solutions.

is independent of their nature is called a **colligative property.** In Figure 14-18, you can see that adding a nonvolatile solute has lowered the vapor pressure of the solvent, lowered the freezing point of the solution, and raised the boiling point of the solution—these are all colligative properties.

In studying colligative properties, solution concentrations are given in molality, which indicates the relative numbers of solute and solvent particles and does not change with temperature. Our discussion of colligative properties in this chapter is limited to dilute solutions of molecular solutes—that is, nonelectrolytes. Solutions of equal molality of any molecular solute in the same solvent have equal vapor-pressure lowering and other colligative properties. As represented by the bottom curve in Figure 14-18, if the molality is doubled, the vapor-pressure lowering, freezing-point depression, and boiling-point elevation are all doubled. These results show that colligative properties are directly proportional to the molal concentration of a molecular solute.

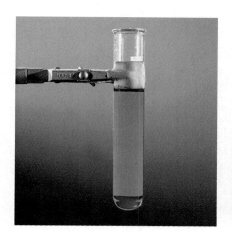 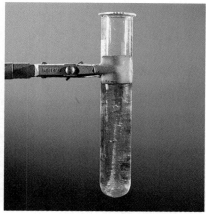

Figure 14-19 A solution of purple dye and water *(left)* is frozen *(right).* The water (solvent) freezes but the dye (solute) does not.

Fresh Water From the Sea

Approximately 97% of the earth's water is contained in its oceans. Human beings have tried for centuries to develop an efficient, inexpensive method to desalinate seawater so that it can be used for drinking and cooking.

One of the oldest techniques for removing salt from water is distillation. Even today, distillation provides about three-fourths of the world's desalinated water. In saltwater distillation, the water is heated and vaporized, and the vapor is then condensed into freshwater. Though effective, distillation is an expensive way to desalinate water.

Newer methods use semipermeable membranes to separate salt from water. One of these is reverse osmosis, in which pressure is used to force salty water against a membrane. The salt is left behind as the wa-

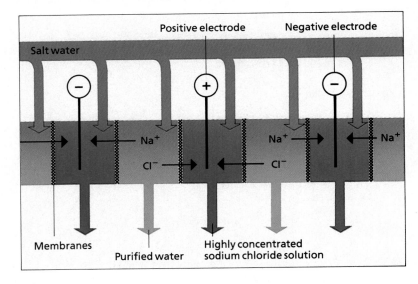

ter passes through the membrane. Another method that uses membranes is electrodialysis. (See the diagram.) In electrodialysis, An electric current is sent through the salt water, drawing the positive and negative ions of salt through separate membranes. Pure water remains between the two membranes.

Membrane processes are less effective than distillation methods of desalination and are usually used to desalinate brackish inland water, rather than seawater. The Yuma Desalting Plant on the United States-Mexico border, for example, uses reverse osmosis to purify water from the Colorado River for irrigation.

Freezing-Point Depression

The ability of motor vehicles to start and run on a cold winter day depends upon the addition of ethylene glycol or other antifreeze substances to the cooling water. All antifreezes act by lowering the vapor pressure of the solution and, as a result, they lower the freezing point.

When a solution containing nonvolatile solute freezes, only pure solvent solidifies, so the solid vapor pressure is the same as that of the pure solid solvent. At the freezing point there is an equilibrium between a solid and a liquid, and their vapor pressures are equal. The curve at the left in **Figure 14-18** shows solid–vapor equilibrium. You can see that the solution must be at a lower temperature before the vapor pressures of the solid and of the solution become equal.

The freezing point of a 1-molal solution of any molecular solute in water is found by experiment to be lowered by 1.86°C. That is, when one mole of a nonelectrolyte is dissolved in 1 kg of water, the freezing point of the solution is −1.86°C. This can be written as −1.86°C/molal. This value is referred to as a **molal freezing-point constant**—*the freezing-point depression of the solvent in a 1-molal solution of a nonvolatile, molecular solute.* Each solvent has its own characteristic molal freezing-point depression constant.

The values of K_f for some common solvents are given in Table 14-6. These values are most accurate for dilute solutions. In addition, these values are most accurate for solutions at 1 atmosphere of pressure. A very slight variation is introduced in the value of K_f and K_b for pressure other than 1 atmosphere.

TABLE 14-6   MOLAL FREEZING-POINT AND BOILING-POINT CONSTANTS

Solvent	Normal f.p. (°C)	Molal f.p. constant, K_f (°C/molal)	Normal b.p. (°C)	Molal b.p. constant, K_b (°C/molal)
acetic acid	16.6	3.90	118.5	3.07
acetone	−94.8	—	56.00	1.71
aniline	−6.1	5.87	184.4	3.22
benzene	5.48	5.12	80.15	2.53
carbon disulfide	−111.5	3.80	46.3	2.34
carbon tetrachloride	−22.96	—	76.50	5.03
ethanol	−114.5	—	78.26	1.22
ether	−116.3	1.79	34.42	2.02
naphthalene	80.2	6.9	218.0	5.65
phenol	40.9	7.27	181.8	3.56
water	0.00	1.86	100.0	0.51

Candy thermometers are based on the boiling-point elevation of water by sugar. As the water boils away in making candy, the solution becomes more concentrated and its temperature increases. The temperature indicates when you should stop cooking your candy.

Because molality is equal to moles of solute per kg of solvent, the units for K_f, for example, for water can be expressed as follows:

$$K_f = \frac{-1.86 \text{ °C}}{\text{mole solute/kg H}_2\text{O}} = \frac{-1.86 \text{ °C} \cdot \text{kg H}_2\text{O}}{\text{mol solute}}$$

The freezing-point depression in a dilute solution of a molecular solute, as illustrated in Figure 14-18, is directly proportional to the molal concentration of the solution: $\Delta T_f \propto m$. In other words, if the molal concentration is doubled, the freezing-point depression is doubled. The proportionality constant for this relationship is K_f, giving

$$\Delta T_f = K_f m \qquad \Delta T_f = K_f m$$

with K_f in the units shown above and m expressed in moles solute/kg solvent ΔT_f is found in degrees Celsius. Sample Problems 14-6 and 14-7 illustrate the use of this expression.

Sample Problem 14.6

What is the freezing-point depression of water in a solution of 17.1 g of sucrose, $(C_{12}H_{22}O_{11})$, and 200 g of water?

Solution

Step 1. Analyze   Given: solute mass and chemical formula = 17.1 g $C_{12}H_{22}O_{11}$
                    solvent mass and identity = 200 g water
       Unknown: freezing point depression

Step 2. Plan   The molal-freezing point constant for water can be found in Table 14-6. Using the formula for freezing-point depression, $\Delta T_f = K_f m$, the following route can be used.

solute mass → molality of solution, molar freezing-point constant → freezing-point depression

$$\Delta T_f = K_f m$$

$$K_f \left(\frac{\text{in °C kg solvent}}{\text{mol solute}} \right) \times \frac{\text{mass of solute (g)}}{\text{mass of solvent (g)}} \times \frac{\text{(mol of solute)}}{\text{molar mass solute (g)}} \times \frac{10^3 \text{ g solvent}}{\text{kg solvent}}$$
$$= \Delta T_f \text{ (°C)}$$

Step 3. Compute $\dfrac{-1.86\text{°C kg water}}{\text{mol } C_{12}H_{22}O_{11}} \times \dfrac{17.1 \text{ g } C_{12}H_{22}O_{11}}{200.\text{ g water}} \times \dfrac{1 \text{ mol } C_{12}H_{22}O_{11}}{342 \text{ g } C_{12}H_{22}O_{11}} \times \dfrac{10^3 \text{ g water}}{\text{kg water}}$

$$= -0.465 \text{ °C}$$

Step 4. Evaluate   As shown, the units cancel correctly to give the answer in °C. The answer is correctly given to three significant digits because there is no factor in the calculation with fewer significant digits.

Sample Problem 14.7

A water solution containing an unknown quantity of a molecular solute is found to have a freezing point of -0.23°C. What is the molal concentration of the solution?

Solution

Step 1. Analyze   Given: solvent identity = water
                    freezing point of solution = -0.23°C
       Unknown: molality of the solution

Step 2. Plan   Look up the value of K_f, the molal-freezing-point constant for water. Recall the equation for freezing-point depression.

$$\Delta T_f = K_f m \qquad \text{Solve for molality, } m.$$

$$m = \frac{\Delta T_f}{K_f}$$

Then the route is: freezing point, temperature change → molality

Step 3. Compute $$-0.23\text{°C} \div \frac{(-1.86\text{°C})}{1m} = 0.12 \, m$$

Step 4. Evaluate   As shown by the unit cencellation, the answer gives the molality, as desired. The answer is properly limited to two significant digits by the value of T_f.

Sample Problem 14.7 *continued*

Practice Problems

1. A solution consists of 10.3 g of glucose ($C_6H_{12}O_6$), dissolved in 250 g of water, What is the freezing-point depression of the solvent? *(Ans.)* −0.425°C

2. In a laboratory experiment, the freezing point of a water solution of glucose is found to be −0.325°C. What is the molal concentration of this solution? *(Ans.)* 0.174 m

Boiling-Point Elevation

The boiling point of a liquid is the temperature at which its vapor pressure of the liquid is equal to the prevailing atmospheric pressure. Therefore, a change in either the atmospheric pressure or the vapor pressure of the liquid will cause a corresponding change in the boiling point.

The **boiling-point elevation** *of a solvent is the increase in the boiling point caused by a solute and is represented by* ΔT_b.

The boiling-point elevation of a 1-molal solution of any molecular solute in water has been found by experiment to be 0.51°C. That is, when one mole of a nonelectrolyte, such as sugar, is dissolved in 1 kg of water, the boiling point of the solution is found to be 100.51°C. For different solvents, the boiling-point elevations of 1-molal solutions have different values. *The* **molal-boiling-point constant** *is the boiling-point elevation of the solvent in a 1-molal solution of a nonvolatile, molecular solute.* For water, the molal-boiling-point constant is

Figure 14-20 Many recipes call for adding salt to water. Why?

$$K_b = \frac{0.51°C}{\text{mole solute/kg } H_2O} = 0.51°C \text{ kg } H_2O/\text{mol solute}$$

Some other values for K_b are included in Table 14-6. Like the freezing-point constants, these values are the most accurate for dilute solutions.

The direct proportionality between boiling-point elevation and the molality of a solution of a nonvolatile molecular solute is given by

$$\Delta T_b = K_b m$$

When K_b is expressed in °C/(mole solute/kg solvent) and m is expressed in moles of solute/kg of solvent, K_b is the boiling-point elevation in °C. This expression can be used in problem solving to find ΔT_b if m is known and to find m if ΔT_b is known in the same ways illustrated for freezing-point depression.

Determination of Molar Mass of a Solute

Because colligative properties depend only on the number of solute particles and not on their identity, boiling point elevation and freezing point depression can be used to determine the molar mass of soluble, nonvolatile electrolytes.

An experiment is performed in which a known mass of a molecular solute is dissolved in a known mass of a solvent. It must be a with which the solute does not react chemically. The freezing-point depression or boiling-point elevation of the solvent is then measured. This experimental value is used to calculate the molality of the solution. Then, the calculated molality and the known solute mass are used to find the molar mass of the solute. The molality, mass of solute, and mass of solvent are related to the molar mass of the solute as follows:

$$\text{molality} = \frac{\text{mass of solute (g)}}{\text{molar mass of solvent} \times \text{mass of solvent (kg)}}$$

or

$$\text{molar mass of solute} = \frac{\text{mass of solute (g)}}{\text{molality} \times \text{mass of solvent (kg)}}$$

The equations that express the relationship between the molality and the change in boiling and freezing points can be expanded to include the factor of molar mass.

$$\Delta T_b = K_b\, m \quad \text{or} \quad m = \frac{\Delta T_b}{K_b} \quad \text{and} \quad \Delta T_f = K_f m \quad \text{or} \quad m = \frac{\Delta T}{K_f}$$

Thus,

$$\frac{\text{mass of solute (g)}}{\text{molar mass of solute} \times \text{mass of solvent (kg)}} = \frac{\Delta T_b}{K_b}$$

and

$$\frac{\text{mass of solute (g)}}{\text{molar mass of solute} \times \text{mass of solvent (kg)}} = \frac{\Delta T_f}{K_f}$$

Rearranging gives expressions for the molar mass of solute

$$\text{molar mass of solute} = \frac{\text{mass of solute (g)} \times K_b}{\Delta T_b \times \text{mass of solvent (kg)}}$$

and

$$\text{molar mass of solute} = \frac{\text{mass of solute (g)} \times K_f}{\Delta T_f \times \text{mass of solvent (kg)}}$$

This method is illustrated in Sample Problem 14.8.

Sample Problem 14.8

In order to determine the molar mass of sulfur, 1.8 g of sulfure (S), is dissolved in 100 g of naphthalene ($C_{10}H_8$): $K_f = -6.9°C/\text{molal}$; Table 14-6). The freezing-point depression of the naphthalene $C_{10}H_8$), was thereby determined. The value found for ΔT_f was $-0.48°C$. What is the molar mass of sulfur? Suggest a molecular formula for the element sulfur.

Solution

Step 1. Analyze   Given: mass and identity of solute = 1.8 g S
mass and identity of solvent = 100 g $C_{10}H_8$
value of $\Delta T_f = -0.48°C$
value of $K_f = -6.9°C/\text{molal}$
Unknown: molar mass (and molecular formula of S) of S

Step 2. Plan   mass of solute, mass of solvent, freezing point depression of solvent, $K_f \rightarrow$
molar mass of sulfur, possible molecular formula.

$$\text{molar mass of sulfur (g/mol)} = \frac{\text{mass of solute (g S)} \times k_f\ (°C \cdot kg\ C_{10}H_8/\text{mol S})}{\Delta T_f\ (°C) \times \text{mass of solvent (kg } C_{10}H_8)}$$

Sample Problem 14.8 *continued*

Sample Problem 14.8 *continued*

Step 3. Compute	$$\frac{1.8\,\text{g S} \times (-6.9°\text{C kg C}_{10}\text{H}_8/\text{mol S})}{-0.48°\text{C} \times 100\,\text{g C}_{10}\text{H}_8} \times \frac{1000\,\text{g C}_{10}\text{H}_8}{1\,\text{kg C}_{10}\text{H}_8} = 260\,\text{g S/mol S}$$

Knowing that the atomic mass of sulfur is 32 g/mol, this value for the molar mass indicated that the molecular formula for sulfur is
S_8 (260 g/mol ÷ 32 g/mol = 8).

Step 4. Evaluate In each expression in Step 3, the units cancel appropriately to yield the desired label, g/mol. The calculated answer is limited to two significant digits. Numerically, a predicted answer would be 280, which is close to the final answer.

Practice Problems

1. When 1.56 g of an unknown, nonvolatile solute is dissolved in 200 g of water, the freezing-point depression of the solvent is $-0.453°$C. Determine the molar mass of the solute.   *(Ans.)* 32.0 g/mol
2. If 1.84 g of a molecular solute is dissolved in 150 g of acetic acid, the boiling point of the solvent is elevated 0.60°C. What is the molar mass of the solute?   *(Ans.)* 63 g/mol

Section Review

1. What colligative properties are responsible for the following phenomena and how are they caused?   *(a)* Antifreeze keeping water flowing in a car's cooling system when the air temperature is below freezing.
   (b) Ice melting on sidewalks after salt has been spread on them.
2. Two moles of a nonelectrolyte solute are added to 1 kg of an unknown solvent. The solution is observed to freeze at 7.9°C below its normal freezing point. What is molal-freezing-point constant of the unknown solvent? Can you guess the identity of the solvent?
3. A solution of methanol and water contains 1.56 g of methanol dissolved in 200 g of water. Describe how you would determine the molar mass of methanol given the above materials.

Chapter Summary

- Most substances we encounter are mixtures.
- A mixture is classified as either a solution, a suspension or a colloid depending on the size of the particles in the mixture.
- A solute will dissolve at a rate that depends on the surface area of the solute, how vigorously the solution is mixed, and the temperature of the solvent.
- The solubility of a substance indicates how much of that substance will dissolve in a specific amount of solvent under certain conditions.
- The solubility of a substance depends on the temperature.
- The solubilities of gases in liquids increases with inceased pressure.
- The overall energy change in solution formation is called the heat of solution.
- The concentration of a solution can be stated in the three different ways: percent by mass, molarity and molality.
- Colligative properties of solutions depend only on the number of solute particles present.

Chapter 14 *Review*

Solutions

Vocabulary

alloy
boiling-point elevation
colligative property
colloid
concentration
effervescence
electrolyte
freezing-point
depression
heat of solution
Henry's law
immiscible
miscible
molal boiling-point
    constant
molal freezing-point
    constant

molarity
percent by mass
nonelectrolyte
nonvolatile substance
saturated solution
solubility
soluble
solute
solution
solution equilibrium
solvated
solvent
supersaturated solution
suspension
unsaturated solution

Questions

1. *(a)* How are mixtures classified?   *(b)* Identify the three types of mixtures.
2. *(a)* What is meant by the term *soluble*? *(b)* What is a *solution*?   *(c)* Identify and define the two components of a solution.   *(d)* Give two examples of common solutions.
3. *(a)* Distinguish between electrolytes and nonelectrolytes.   *(b)* Give one example of each.   *(c)* Generally, what kinds of substances are electrolytes?
4. *(a)* What is a suspension?   *(b)* How does it differ from a solution?
5. *(a)* What is a colloid?*(b)* Identify its two phases.
6. Indentify a common colloid that is *(a)* an emulsion   *(b)* an emulsifying agent   *(c)* a gel.
7. *(a)* What is the Tyndall effect?   *(b)* Cite a common example of this effect.   *(c)* Identify one application of this effect.
8. *(a)* List three ways in which the rate at which sugar dissolves in water can be increased. *(b)* Explain the reason for the effect of each.

9. *(a)* What is solution equilibrium?   *(b)* What factors determine the point at which a given solute–solvent combination reaches equilibrium?
10. *(a)* What is a saturated solution?   *(b)* What visible evidence indicates that a solution is at its saturation point?   *(c)* What is an unsaturated solution?
11. *(a)* What is meant by the solubility of a substance?   *(b)* What condition(s) must be specified in citing solubility levels?
12. What is the relationship between the rate at which a substance dissolves and the solubility of that substance?
13. List three factors that affect the solubility of a substance.
14. *(a)* What rule of thumb is useful for predicting whether or not one substance dissolves in another?   *(b)* In general, what does the rule mean in terms of polar and nonpolar solutes and solvents?
15. *(a)* How does pressure affect the solubility of a gas in a liquid?   *(b)* What law is a statement of this relationship?   *(c)* If the pressure of a gas above a liquid is increased, what happens to the amount of the gas that will dissolve in the liquid, if all other conditions remain constant?
16. *(a)* What is meant by the term *effervescence*? *(b)* Why does a bottle or can of carbonated soda effervesce when opened?
17. What is the effect of temperature on the solubility of: *(a)* a gas in a liquid?   *(b)* a solid in a liquid?
18. What is a supersaturated solution?
19. Based on Figure 14-13, determine the solubility of each of the following in grams of solute per 100 grams H_2O.   *(a)* $NaNO_3$ at 10°C *(b)* KNO_3 at 60°C   *(c)* $NaCl$ at 50°C
20. Based on Figure 14-13, at what temperature would each of the following solubility levels be observed?   *(a)* 140 g KI/100 g H_2O *(b)* 115 g $NaNO_3$/100 g H_2O *(c)* 50 g KNO_3/100 g H_2O.
21. *(a)* What is the heat of solution?   *(b)* In what unit is it expressed?   *(c)* What is the

significance of the positive and negative signs associated with heats of solution?

22. Write the general equations describing endothermic and exothermic dissolving processes.

23. List the three interactions that contribute to the heat of solution and indicate the energy change that must occur in each in order for solution formation to take place.

24. How does an increase in temperature affect the solubility of a solid with *(a)* a positive heat of solution? *(b)* a negative heat of solution?

25. Why does an increase in temperature generally decrease the solubility of a gas in a liquid?

26. *(a)* Distinguish between dilute and concentrated as related to solution strength. *(b)* How do these terms relate to solubility?

27. Express in words what is meant by a: *(a)* 4-M solution *(b)* 0.75-M solution.

28. Express in words what is meant by a: *(a)* 1-m solution. *(b)* 0.5-m solution.

29. How does the presence of a nonvolatile solute affect each of the following properties of the solvent into which the solute is dissolved? *(a)* vapor pressure *(b)* freezing point *(c)* boiling point.

30. *(a)* What is a colligative property? *(b)* List three examples of colligative properties.

31. What is the relationship between colligative properties of a solution and solute concentrations?

32. What is the freezing-point depression of a solvent?

33. *(a)* What is meant by the molal freezing-point constant? *(b)* What is the value of this constant for water?

34. What is the boiling-point elevation of a solvent?

35. *(a)* What is meant by the molal boiling-point constant? *(b)* What is the value of this constant for water?

Problems

1. Find the percent by mass of each of the following solutions:
   (a) 10.0 g of sugar in 40.0 g of water
   (b) 0.63 g of $AgNO_3$ in 12.00 g of water.

2. Determine the molarity of each of the following solutions:
   (a) 2.5 mol NaOH in 1.0 L solution
   (b) 4.0 mol $BaCL_2$ in 1.5 L solution.

3. How many moles of each solute would be required to prepare each of the following solutions?
   (a) 1.0 L of a 2.5-M HCl solution
   (b) 150.0 mL of a 1.25 M $Al(NO_3)_3$ solution

4. Determine the number of grams of solute needed to make solutions of the following volumes and concentrations:
   (a) 1.00 L of a 3.50-M solution of H_2SO_4
   (b) 2.50 L of a 1.75-M solution of $Ba(NO_3)_2$
   (c) 500 mL of a 0.250-M solution of KOH.

5. Determine the molarity of each of the following solutions.
   (a) 20.0 g NaOH in enough H_2O to make 2.00 L of solution
   (b) 14.0 g NH_4OH in enough H_2O to make 150. mL of solution
   (c) 32.7 g H_3PO_4 in enough H_2O to make 500. mL of solution.

6. *(a)* How many moles of NaOH are contained in 65.0 mL of a 2.20-M solution of NaOH in H_2O? *(b)* How many grams of NaOH does this represent?

7. What is the molarity of a solution made by dissolving 26.42 g of $(NH_4)_2SO_4$ in enough H_2O to make 50.00 mL of solution?

8. Determine the molality of each of the following solutions.
   (a) 4.5 mol $Mg(NO_3)_2$ in 2.5 kg H_2O
   (b) 0.625 mol K_2SO_4 in 850. kg H_2O.

9. How many moles of each solute would be required to prepare each of the following solutions?
   (a) 0.45-m $CaSO_4$ in 2.5 kg H_2O
   (b) 3.25-m $Ba(NO_3)_2$ in 750. mL H_2O.

10. Determine the number of grams of solute needed to make each of the following solutions:
    (a) a 4.50-m solution of H_2SO_4 in 1.00 kg H_2O
    (b) a 1.00-m solution of HNO_3 in 2.00 kg H_2O
    (c) a 3.50-m solution of $MgCl_2$ in 0.450 kg H_2O.

11. Determine the molality of each of the following solutions:
    (a) 294.3 g H_2SO_4 in 1.000 kg H_2O
    (b) 63.0 g HNO_3 in 0.250 kg H_2O
    (c) 10.0 g NaOH in 300. g H_2O.

12. A solution is prepared by dissolving 17.1 g of sucrose ($C_{12}H_{22}O_{11}$), in 275 g of H_2O. What is the molality of that solution?

13. How many kilograms of H_2O must be added to 75.5 g of $Ca(NO_3)_2$ to form a 0.500-m solution?

14. How many grams of glucose ($C_6H_{12}O_6$), must be added to 750. g of H_2O to make a 1.25-*m* solution?

15. Determine the freezing-point depression of H_2O in each of the following solutions.
    (a) a 1.50-*m* solution of $C_6H_{12}O_6$
    (b) 171 g of $C_{12}H_{22}O_{11}$ (sucrose) in 1.00 kg H_2O
    (c) 77.0 g of $C_{12}H_{22}O_{11}$ in 400. g H_2O.

16. Determine the molality of each solution of sugar in water given the following freezing-point depressions. (a) $-0.930°C$
    (b) $-3.72°C$ (c) $-8.37°C$.

17. A solution contains 20.0 g of $C_6H_{12}O_6$ in 250. g of water. (a) What is the freezing-point depression of the solvent? (b) What is the freezing point of the solvent?

18. Determine the boiling-point elevation of H_2O in each of the following solutions.
    (a) a 2.5-*m* solution of $C_6H_{12}O_6$
    (b) 3.20 g $C_6H_{12}O_6$ in 1.00 kg H_2O
    (c) 20.0 g $C_{12}H_{22}O_{11}$ in 500 g H_2O.

19. Determine the molality of each water solution given the following boiling points:
    (a) 100.25°C (c) 101.53°C (c) 102.805°C.

20. What is the molar mass of a molecular solute when 5.00 g of the solute dissolved in 150. g of phenol lowers the freezing point 1.5°C?

21. Determine the molar mass of an unknown nonelectrolytic solute when 7.55 g of the solute dissolved in 125 g of aniline raises the boiling point to 185.5°C.

Application Questions

1. Explain why electrolytes can conduct electric current, whereas nonelectrolytes cannot conduct.

2. Given an unknown mixture consisting of two or more substances, explain one technique that could be used to determine whether that mixture is a true solution, a colloid, or a suspension.

3. Explain why water is a good solvent for any substance composed of polar molecules.

4. Why are carbon tetrachloride and water immiscible?

5. Use Le Chatelier's principle to explain the effect of pressure on the solubility of a gas in a liquid.

6. In order to produce clear ice cubes as opposed to cloudy ones, should hot or cold water by used? Explain your answer.

7. Compare endothermic and exothermic dissolving processes in terms of: (a) the heat content of each solution relative to the combined heat content of the initial components of each (b) the change in solution temperature that occurs during the dissolving process (c) the sign of the heat of solution (d) the effect of increased temperature on the solubility of most solids dissolved in a liquid.

8. The heat of solution for $AgNo_3$ is +22.8 kJ/mole. (a) Write the equation that represents the dissolving of $AgNO_3$ in water. (b) What energy change is associated with the dissolving process the crystallizing process? (c) As $AgNO_3$ dissolves, what change occurs in the temperature of the solution? (d) When the system is at equilibrium, how do the rates of dissolving and crystallizing compare? (e) If the solution is then heated, how will rates of dissolving and crystallizing be affected? Why? (f) How will the increased temperature affect the amount of solute that can be dissolved? (g) If the solution is allowed to reach equilibrium and then cooled, how will the system be affected?

9. Under what circumstances would solution concentrations be expressed in terms of:
    (a) percent by mass (b) molarity
    (c) molality.

10. Explain, step-by-step, the procedure that you would use in the laboratory to prepare a 1.50-*m* solution of $Mg(NO_3O_2$ in 350. g of H_2O.

11. Why is ethylene glycol (antifreeze) generally added to motor vehicle radiators?

Application Problems

1. Plot a solubility graph for $AgNO_3$ from the following data, with grams of solute (by increments of 50) per 100 grams of H_2O on the vertical axis, and with temperature in °C on the horizontal axis.

Grams solute per 100 g H_2O	Temperature, °C
122g	0°
216g	20°
311g	40°
440g	60°
585g	80°
733g	100°

(a) How does the solubility of $AgNO_3$ vary with the temperature of the water? (b) Estimate the solubility of $AgNO_3$ at 30°C, 55°C, and 75°C (c) At what temperature would the solubility of $AgNO_3$ be 275 g per 100 g of H_2O? (d) If 100 g of $AgNO_3$ were added to 100 g of H_2O at 10°C, would the resulting solution be saturated or unsaturated? What would occur if 325 g of $AgNO_3$ were added to 100 g of H₂0 at 35°C?

2. If a saturated solution of KNO_3 in 100 g of H_2O at 60°C is cooled to 10°C, approximately how many grams of the solute would precipitate out of the solution? (Use Figure 14-13)

3. Find the mass of each solute present in the given mass of each of the following solutions. (a) 50.0 g of a 10.0% by mass NaCl solution (b) 8.5 g of a 2.4% by mass KOH solution.

4. If 100 mL of a 12.0-M HCl solution is diluted to 2.00 L, what is the molarity of the final solution?

5. How many milliliters of 16.0-M HNO_3 would be required to prepare 750 mL of a 0.500-M solution?

6. How many milliliters of 0.54 M $AgNO_3$ would contain 0.34 g of pure $AgNO_3$?

7. What mass of each product results if 750 mL of 6.00-M H_3PO_4 react according to the following?
$$H_3PO_4 + Ca(OH)_2 \rightarrow Ca_3(PO_4)_2 + H_2O$$

8. How many milliliters of 18.0-M H_2SO_4 would be required to react with 250 mL of 2.50 M $Al(OH)_3$ if the products are aluminum sulfate and water?

9. If 75.0 mL of a $AgNO_3$ solution react with enough Cu to produce 0.250 g of Ag by single replacement, what was the molarity of the initial $AgNO_3$ solution if $Cu(NO_3)_2$ is the other product?

10. A solution of ethyl alcohol (C_2H_5OH) in water is 1.75 m. How many grams of C_2H_5OH are contained per 250. g of water?

11. In how many liters of water should 65.0 g of NaCl be dissolved to make a 0.450-m solution?

12. If a 3.00-m solution of HNO_3 containing 2.25 kg H_2O is diluted so that the resulting solution contains 5.75 kg H_2O, what is the molality of that final solution?

13. How many grams of antifreeze, $C_2H_4(OH)_2$, would be required per 500 g of water to prevent the water from freezing down to −20.0°C?

14. Pure benzene (C_6H_6), freezes at 5.45°C. A solution containing 7.24 g of $C_2Cl_4H_2$ in 115 g of benzene (specific gravity = 0.879) freezes at 3.55°C. Based on these data, what is the molal freezing-point constant for benzene?

15. The freezing-point of pure camphor is 178.40°C; its molal freezing-point constant is −40.00°C/molal. If 1.500 g of a solute having a molecular weight of 125.0 g were dissolved in 35.00 g of camphor, what would be the resulting freezing point of the camphor?

16. Find the boiling point of an aqueous solution that freezes at −6.51°C.

17. The molal freezing-point constant for ether is −1.79°C/molal. When 21.2 g of a solute is dissolved in 740.g of ether, the freezing point of the ether is lowered by 0.568°C. What is the molar mass of that solute?

18. A solution containing 9.00 g of a solute dissolved in 250. g of water freezes at −0.372°C. (a) What is the molar mass of the solute? (b) Among the following, which formula most likely represents the solute: CH_3OH, C_6H_6, C_2H_5OH, $C_6H_{12}O_6$, or $C_{12}H_{22}O_{11}$?

Enrichment

1. The absorptive properties of charcoal make it a versatile substance for use around the home. Explore some of those uses and explain the specific role of charcoal in each.

2. Ask your instructor to provide you with an unknown and then conduct an experiment that would enable you to identify it in the laboratory on the basis of its solubility curve.

3. Try to make a supersaturated solution of sodium acetate, $NaC_2H_3O_2$.

4. Ethylene glycol was originally intended to be placed in automobile radiators for use as an antifreeze. This substance is generally retained in those radiators year-round, however. What function does this substance serve during the summer months?

5. If your laboratory is suitably equipped, obtain a water-soluble unknown from your instructor and attempt to determine its molar mass by using the freezing-point depression approach.

6. Find out how much salt a large northern city, such as New York or Chicago uses on its streets in a typical winter.

Chapter 15 Ions in Aqueous Solutions

Chapter Planner

15.1 Ionic Compounds in Aqueous Solution

Theory of Ionization
- ▨ Demonstration: 1
  Questions: 1, 2, 4
- ▪ Question: 3

Dissolving Ionic Compounds
- ▨ Demonstration: 2
  Questions: 4–6, 9, 10
  Problems: 1, 2
  Application Question: 2
- ▪ Application Question: 1
- ▨ Questions: 7, 8
  Problem: 3

Solubility Equilibria
- ▨ Experiment: 19
  Questions: 5, 6, 11, 12
  Application Problems: 1b, 6a-d
- ▪ Chemistry Notebook
  Question: 4
- ▨ Application Question: 3
  Application Problem: 1a

15.2 Molecular Electrolytes
- ▨ Demonstration: 3
  Questions: 13–18
  Application Question: 4
- ▪ Question: 19

15.3 Properties of Electrolyte Solutions

Conductivity of Solutions
- ▨ Demonstration: 4, 5
  Questions: 21–23
  Application Questions: 6, 7
- ▪ Question: 20
  Application Question: 5

Colligative Properties of Electrolyte Solutions
- ▨ Questions: 24, 25
  Problems: 7–9, 10
  Application Problems: 2, 3
- ▪ Chemistry Notebook
  Question: 26
  Problems: 11, 12
  Application Questions: 8, 9
  Application Problem: 5
- ▨ Application Problems: 6e-f, 7–9

Teaching Strategies

Introduction

Remind students that Chapter 14 dealt mainly with nonelectrolytes and that this chapter deals almost exclusively with electrolytes. It is a difficult and highly theoretical chapter. Go slowly, making sure students understand one concept before proceeding to the next.

The overall objectives for this chapter are *(1)* to study the Theory of Ionization from a historical and modern-day viewpoint *(2)* to determine the differences between ionic and polarcovalent electrolytes and learn how to differentiate between the two in the laboratory *(3)* to learn how to write equations for dissociation, ionization, and reactions involving ions, and *(4)* to discuss the effects of electrolytes on electric conductivity as well as on freezing and boiling points of solutions.

15.1 Ionic Compounds in Aqueous Solution

Students often confuse the terms *conductor*, *nonconductor*, *electrolyte*, and *nonelectrolyte*. Point out that conductors and nonconductors can be pure substances (elements and com-

pounds) or mixtures, but electrolytes and nonelectrolytes are the solutes of water solutions. Thus, either a solution of NaCl and water would be a conductor, but the NaCl would be an electrolyte, or a solution of sugar and water would be a nonconductor, but the sugar would be the nonelectrolyte. These distinctions are important to understanding concepts throughout the chapter. To help students visualize these differences, put Table T15-1 (shown below) on a transparency. This transparency will be very helpful throughout the chapter, and students will be able to understand it better as future sections are discussed.

CONDUCTORS			
Pure Substances		**Mixtures**	
Elements	Compounds	Alloys	Electrolytic Solution
All metals Cu, Ag, Fe	All ionic compounds in liquid state $NaBr(l)$, $KNO_3(l)$	Stainless steel, Sterling silver	Water solutions of NaBr, HCl, NH_3

NONCONDUCTORS		
Pure Substances		**Mixtures**
Elements	Compounds	Nonelectrolytic Solutions
All non-metals S, I_2, P_4	All covalent compounds in liquid state HBR $(\ell)Al_2Cl_6(\ell)$ All solid compounds sucrose(s), NaBr(s), Al_2 Br_2(s), KNO_3(s)	Water solutions of sucrose isopropyl alcohol, ethyl alcohol, glycerin

To help lay a foundation for this chapter, do the following demonstration as an interest-catching introduction.

Demonstration 1: Conductivity of Pure Substances and Solutions

PURPOSE: To help students differentiate between conductors, nonconductors, electrolytes, and nonelectrolytes

MATERIALS: Acetic acid, glacial; ammonia water, concentrated; ethanol (95% denatured will be an acceptable substitute for absolute ethanol); hydrochloric acid, concentrated; glycerin; sodium chloride; sodium hydroxide; sucrose; water, distilled; ammeter, 0-200-milliampere; 2 batteries, 6-volt; 10 beakers, 250-mL; bulb, 6.3-V/150-mA, clips, alligator; copper wire, No. 20 insulated; lamp holder, threaded; popsicle stick, wooden or plastic; switch, push button

PROCEDURE: *(a)* Assemble a set-up like the one shown in Figure T15-1. It is very simple and inexpensive to make. All parts can be obtained from your local electronics store. All connections are made with No. 20 insulated Cu wire.

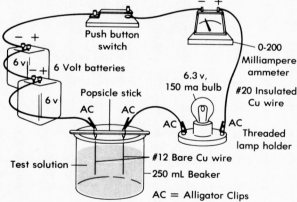

(b) Make up 0.1-*m* solutions of each of the following substances by dissolving the indicated amounts in 200 mL of distilled water: sucrose—7 g; glycerin—2 mL; sodium chloride—1.2 g; ammonia—1.5 mL of concentrated NH_4OH; acetic acid—1.2 mL glacial acetic acid; hydrochloric acid—1.8 mL concentrated HCl; and NaOH—0.8 g.

(c) Measure out 200 mL of the following pure substances — distilled water, glycerol, and ethanol.

(d) First test the pure substances and then the solutions, using the following steps: *(1)* immerse the electrodes of the conductivity apparatus in 200 mL of each material being tested; *(2)* close the switch; *(3)* check to see if the bulb lights up and if

so, whether it is bright or dim; *(4)* take a reading from the ammeter; *(5)* open the switch; *(6)* remove the material being tested; *(7)* rinse the electrodes with distilled water and dry with a paper towel after each immersion. *Note*: The bulb does not light up for some poor conductors, but the degree of conductivity can be noted on the ammeter. After each test ask: Is this a conductor or a nonconductor? Would it be an electrolyte or a nonelectrolyte? (Answers will vary. The pure substances can be only conductors or nonconductors. After students identify the solutions as conductors or nonconductors, have them identify the solutes as electrolytes or nonelectrolytes.)

This section offers an excellent opportunity to emphasize some characteristics of outstanding scientists and at the same time show that scientists, too, are human beings. We often view scientists as older, experienced individuals, but Arrhenius was just a young man working on his doctorate in chemistry. He was obviously a creative thinker—a man far ahead of his time who believed in himself. He received only discouragement from his chemistry professor, Cleve, and his other professors at the university. When he presented his revolutionary theory to Cleve, the response was, "You have a new theory? That is very interesting. Goodbye." When he persisted in using the theory of ionization for his dissertation, he was severely punished with a fourth-class rating—the lowest given. This could have ended his career in chemistry. However, he wrote to several scientists in Europe seeking help.

Finally, Ostwald, a German chemist, came to Arrhenius's rescue by actually making the long trip to Stockholm to give his support. Twenty years later, Arrhenius was properly rewarded with the Nobel Prize for his brilliant work. Students love this story! Ask your students how they would evaluate Cleve's reaction in terms of scientific thinking. (According to one version, Cleve explained his reason for failing to listen to Arrhenius: He had heard so many theories of ionization—all of which were soon proven wrong or inadequate—that he thought this was just another unacceptable student theory.) To add a little dramatics, and at the same time to help students understand part three of the theory, hold up a beaker containing a solution of Na_2SO_4 and stress the fact that it contains billions and billions of electrically charged particles (+ and − ions). Ask if anyone is brave enough to stick a finger into the beaker to find out which, if any, of the charged particles is in excess? Are there more positive ions than negative ions? Since there are twice as many positive ions as negative ones in this solution, why don't you get shocked? Can you give an example of a solution that has twice as many negative ions as positive ones and an example of a solution with an equal number of positive and negative ions? ($Mg(NO_3)_2$ and $SrCO_3$) In each case, what is true about the number of positive and negative charges?

Emphasize that dissociation occurs only when ionic solutes are dissolved in water. Ask how someone can determine whether a compound had ionic bonds? (If it is a binary compound, use the electronegativity difference and percent of ionic character rules. If it is a compound containing a polyatomic ion, it will probably be ionic, unless it is an acid. In the laboratory, see if the compound conducts an electric current when fused.)

Demonstration 2: How Dissociation Takes Place
PURPOSE: To show how water molecules cluster around an ion and pull it into the solution by separating it from the other ions

MATERIALS: Large gumdrops—two colors, small gumdrops—two colors, toothpicks

PROCEDURE: To make a model of a sodium chloride crystal, attach large gumdrops (for chlorine ions) to small ones (for sodium ions) with toothpicks using Figure T15-2a as a guide. Have a few of the gumdrops on one corner loosely attached for easy removal. Using Figure T15-2b as a guide, make water molecules using small gumdrops for oxygen and 1/4 of a small gumdrop for hydrogen. Using the Na^+Cl^- model and the models of water molecules, show how water molecules cluster around a sodium ion; how they separate the Na^+ from the other ions, and pull it into the solution. Repeat with a Cl^-. The water molecules can be attached to the ions by using small pieces of toothpicks. While demonstrating with the models, ask how a water molecule would approach an Na^+ (with its negative end —O). What about Cl^- (with its positive end—H)?

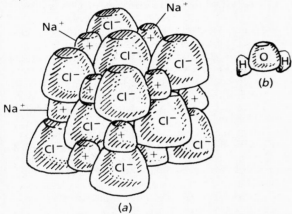

(a)

Students may have difficulty writing ionic equations showing dissociation. (They don't have as many problems until ionization is introduced; then they forget what to do.) Giving them a general equation, having them write several sample equations, and asking questions helps alleviate these problems. General equation: $P^+N^-(s) \rightarrow P^+(aq) + N^-(aq)$. Why are + and − charges written in P^+N^-? (To show that the compound is composed of ions) How do you know the compound is a solid, (s)? (Because of the strong forces of attraction between particles of ionic compounds, they all exist as solids at room temperature.) Why do you write (aq) next to the ions on the product side? (To show that they are hydrated) Now give students several sample problems, for example: Write an ionic equation to show the dissociation of calcium nitrate.

15.2 Molecular Electrolytes

Demonstration 3: How Ionization Takes Place

PURPOSE: To show how water molecules cluster around a polar molecule and pull it apart into ions during the solution process

MATERIALS: Modeling clay, four colors

PROCEDURE: Using modeling clay, make a model of hydrogen chloride and water. (You could use gumdrops also.) Use Figure T15-3 as a guide. After you have clustered the water molecules around the hydrogen chloride molecule, have the water molecules pull it apart into ions. Have students explain how this was different from Demonstration 2.

Demonstration 4: The Difference between Dissociation and Ionization

PURPOSE: To show how dissociation and ionization can be differentiated in the laboratory

MATERIALS: Acetic acid; glacial; potassium nitrate; water, distilled; 3 beakers, 250-mL; porcelain evaporating dish; conducting apparatus used for Demonstration 1

PROCEDURE: Part I Dissociation. To show that ionic compounds are conductors in the liquid state, and electrolytes when dissolved in water, use the simulated fused KNO_3 (described below),

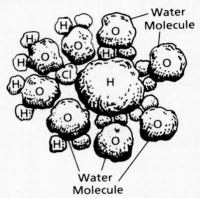

and a 0.1-M solution of KNO_3 (made by dissolving 2.0 g in 200 mL of water) using the conductivity apparatus from Demonstration 1.

Because of the possible danger involved in fusing the low-melting compound, KNO_3, to show it as a conductor in the liquid state, use a simulation. Dissolve a few crystals of KNO_3 in 100 mL of distilled water in an evaporating dish (should give about 75-milliamp reading on ammeter). Tell students that this is what the test for fused KNO_3 would look like.

Why does the fused (melted) KNO_3 conduct an electric current? (The heat energy allows the potassium ions and the nitrate ions to separate from one another and move around; thus, they are able to conduct electricity.) Help students write equations to show what happened when KNO_3 was melted and when it was dissolved in water.

Fused: $KNO_3(s) \rightarrow K^+(\ell) + NO_3^-(\ell)$

Water Solution: $KNO_3(s) \rightarrow K^+(aq) + NO_3^-(aq)$

Part II Ionization. To show that certain covalent compounds that are nonconductors in the liquid state become electrolytes when placed in water, test 200 mL of glacial acetic acid and 200 mL of 0.1-M acetic acid (1.2 mL of glacial acetic acid/200 mL of distilled water) with the conductivity apparatus.

CAUTION: Glacial acid must be handled with care. If you do not have glacial acetic acid, a good alternative to simulate the demonstration is using 200 mL of distilled water in place of the glacial acetic acid and 200 mL of white vinegar in place of the 0.1-M solution.

Why won't glacial acid, which is the most pure acetic acid, conduct an electric current? (In its pure state, its particles are molecules; ions must be present in the liquid state for a substance to conduct an electric current.) Why does the solution of acetic acid conduct an electric current? (During ionization, the water molecules pull the acetic acid molecules apart into hydronium ions and acetate ions.)

To help students write ionic equations for ionization reactions, use the general formula: $PN(s,\ell,g) + H_2O \rightarrow P \cdot H_2O^+ + N^-(aq)$.

Why are no charges written in PN? (PN is a molecule; no ions exist. Students often ask: *(1)* How do I know what state PN is in? (Use these rules: hydrogen halides—HF, HCl, HBr, and HI—are gases at room temperature. Most covalent salts such as the aluminum halides are solids.) *(2)* Why is water included in the reactive side for ionization reactions and not dissociation reactions? (Probably because it is needed to form hydronium ions when acids ionize. To be consistent, therefore, it is included in all ionization reaction equations.)

After dissociation and ionization have been completely covered, it is very important to help students differentiate between the two processes by asking questions such as: What type of particles does a solute that dissociates have before it is dissolved in water? How could you prove in the lab that ions exist? (Fuse the solute, and test it for electrical conductivity.) What do the water molecules do to these + and − ions when the solute is dissolved? (The water molecules separate the ions from one another. The ions then form individual hydrated ions.) What type of particles does a solute that ionizes have before it is dissolved in water? What do the water molecules do to these polar molecules? (Pull them apart into ions.) What kind(s) of solute particles would you expect to find in a solution in which ionization has taken place? (+ and − ions and molecules.) A thought question: A substance is tested in the lab and found not to conduct an electric current in its solid state, but it conducts one in its liquid state. When this substance is dissolved in water, will the solution conduct an electric current? Explain.

15.3 Properties of Electrolyte Solutions

Demonstration 5: Strong and Weak Electrolytes

PURPOSE: To show how to differentiate between strong and weak electrolytes on the basis of electric conductivity

MATERIALS: 0.1-*M* solutions of acetic acid, ammonia water; hydrochloric acid; sodium chloride; sodium hydroxide; 5 beakers, 250-mL; conductivity apparatus from Demonstration 1

PROCEDURE: Using the conductivity apparatus, retest the following 0.1-*M* solutions from Section 14.1: NaCl, HCl, NaOH, $HC_2H_3O_2$, and NH_4OH. This time have students pay close attention to the readings from the ammeter. Which of these solutions are strong electrolytes? weak electrolytes? (Strong—NaCl, HCl, and NaOH; weak—$HC_2H_3O_2$ and NH_4OH.) In the weak electrolyte $HC_2H_3O_2$, what particles would you expect to find in solution? (H_3O^+, $C_2H_3O_2^-$, and $HC_2H_3O_2$ molecules) Which of these would be in excess? (The molecules) In the strong electrolyte HCl, what particles would be in excess? (The H_3O^+ and the Cl^-) To emphasize that compounds with covalent bonds can be strong or weak electrolytes: What do the electrolytes HCl and $HC_2H_3O_2$ have in common? How are they different from one another? (They both ionize and form H_3O^+. HCl ionizes to a large extent and thus is a strong electrolyte; $HC_2H_3O_2$ ionizes to a small extent and thus is a weak electrolyte.)

To help students understand that water solutions of electrolytes never completely ionize or dissociate and that the more dilute the solution, the greater the degree of ionization or dissociation, use a hypothetical problem such as: At what temperature would a 1-*M* solution of H_2SO_4 freeze assuming ionization to be 60%? Write an equation to determine the number of particles formed from each molecule. It works best if you start with an equation showing the ionization of H_2SO_4 and then rewrite it using 100-H_2SO_4 molecules. $H_2SO_4 + 2H_2O \rightarrow 2H_3O^+ + SO_4^{2-}$. Remind students that 60% ionization means that 60 out of every hundred molecules ionize, therefore:

$$100H_2SO_4 + 120H_2O \rightarrow 120H_3O^+ + 60SO_4^{2-} + 40H_2SO_4.$$

100 molecules → 120 ions (+ ions) + 60 (−ions) + 40 molecules *or* 100 molecules → 220 particles *or* 1 molecule → 2.2 particles. Therefore, the freezing point of the solution will be lowered 2.2 times as much as a nonelectrolyte of the same molality *or* the freezing point = $0 - 2.2 \times 1.86 = 4.09°C$.

After discussing the Debye-Hückel theory, reemphasize that even in solutions of ionic solutes, there are two kinds of particles: *(1)* free (+ or −) ions and *(2)* groups of ions or ion clusters. Each ion cluster acts like a single particle in the way that it effects freezing-point depressions and boiling-point elevations.

References and Resources

Books and Periodicals

Jaffe, Bernard. "Svante Arrhenius," *Crucibles: The Story of Chemistry*, New York: Dover Publications, 1976, pp. 140–154.

Kaufman, George B. "Svante August Arrhenius, Swedish Pioneer in Physical Chemistry," *Journal of Chemical Education*, Vol. 65, No. 5, May 1988, pp. 437–8.

Kaufman, George B. and John Baxter. "Hydrated Cations in the General Chemistry Course," *Journal of Chemical Education*, Vol. 58, No. 4, April 1988, pp. 349–53.

Monroe, Manus and Karl Abrams. "A Perspective on Solubility Rules," *Journal of Chemical Education*, Vol. 61, No. 10, Oct. 1984, p. 885.

Parrott, Mary E. "Electrography: A Metal Detective Story," *The Science Teacher*, Vol. 50, No. 5, May 1983, pp. 52–53.

Talesnic, Irwin. "Solubility," *Idea Bank Collation*, Idea 248, Kingston, Ontario: Science Supplies and Services Co. Ltd., 1984.

Tanis, David. "Underground Sculpture," *ChemMatters*, Vol. 2, No. 1, Feb. 1984, pp. 10–11.

Audiovisual Resources

Computer Software General Chemistry 1A—Solutions, 128K, Queue, 1988.

Experiments in Chemistry—Ionization, filmstrip, Educational Images, Ltd.

Introduction to Chemistry—Ionization and Dissociation in Solution, filmstrip, Fisher-Scientific-EMD, 1988.

Answers and Solutions

Questions

1. Electrolytes are substances that conduct electricity when dissolved in water. Nonelectrolytes are substances that do not conduct electricity in aqueous solutions.

2. Faraday suggested that electricity converted dissolved electrolytes into charged particles, which he called ions, that conducted the electric current by moving through the solution. Arrhenius, on the other hand, assumed that electrodes and the passage of current were not necessary for ion formation. He concluded that positive and negative ions were produced by the ionization of molecules in aqueous solutions and that the solution as a whole contained equal numbers of positive and negative charges.

3. The theory of ionization assumes that: (1) electrolytes in solution exist in the form of ions (2) an ion is an atom or a group of atoms that carries an electric charge (3) in the aqueous solution of an electrolyte, the total charge of the positive ions equals the total charge of the negative ions.

4. Since water molecules are polar, their charged ends attract the ions in the ionic compound and bring them into solution. The energy released when ions are attracted to and associate with the charged ends of water molecules also affects the heat of solution of ionic compounds in water.

5. When NaCl crystals come into contact with water molecules, the positive ends of the water molecules are attracted to the Cl^- ions, and the negative ends are attracted to the Na^+ ions. This attraction draws the ions away from the crystal surface and into solution, where each ion is surrounded by water molecules.

6. (a) Hydration is the solution process in which water is the solvent. (b) The number of water molecules surrounding a hydrated ion depends mainly upon the size and charge of the ion.

7. (a) Heats of solution are determined by the magnitude of the solute–solute, solvent–solvent, and solute–solvent interactions. (b) The heat of solution of an ionic compound is the overall result of the energy changes occurring during the separation of ions from the crystal, the separation of water molecules from each other, and the hydration of the separated ions. Energy is required for the first two processes, but is released in the third.

8. (a) Heat of hydration is the energy released when ions become surrounded by water molecules. (b) More energy is usually released in the hydration of smaller ions than of larger ones, and in the hydration of ions with higher charges.

9. (a) Dissociation is the separation of ions that occurs when an ionic compound dissolves. (b) The 1-M solution of KCl contains one mole each of K^+ and Cl^- ions; the BaF_2 solution contains one mole of Ba^{2+} and two moles of F^- ions; the $Mg(NO_3)_2$ solution contains one mole of Mg^{2+} ions and two moles of NO_3^- ions.

10. The indicated responses are: (a) soluble (b) soluble (c) insoluble (d) insoluble (e) insoluble (f) soluble (g) soluble (h) insoluble (i) insoluble (j) insoluble (k) insoluble.

11. (a) Spectator ions are ions that do not take part in a chemi-

cal reaction and are present in solution both before and after the reaction. (b) K^+(aq) and NO_3^-(aq) are the spectator ions.

12. A net ionic equation includes only those compounds and ions that undergo a chemical change in a reaction, and does not include spectator ions.

13. Both molecular and ionic compounds can form electrolytic solutions by forming ions in solution. However, unlike ionic compounds, in which all particles of the compound are ions, many of the particles of a molecular-compound remain as uncharged molecules in solution.

14. When molecules with polar–covalent bonds are dissolved in water, the dipoles of the water molecules and the oppositely charged regions of the solute molecules are attracted to each other such that one or more bonds in the solute molecules are broken. The more electronegative atom retains the bonding electron pair and forms a negative ion; the less electronegative atom forms a positive ion.

15. (a) Ionization refers to the formation of ions from solute molecules by the action of the solvent; it essentially means the creation of ions where there were none. (b) When an ionic compound dissolves through dissociation, the ions that were already present and associated with each other in the crystal become dissociated from each other. On the other hand, when a molecular compound dissolves in a polar solute, ions are formed where none existed in the undissolved compound.

16. The indicated equation is: $HCl(g) + H_2O(\ell) \rightarrow H_3O^+(aq) + Cl^-(aq)$

17. (a) The hydronium ion is produced through the transfer of a proton (H^+) to a water molecule, to which the proton becomes covalently bonded. (b) Its formula is H_3O^+.

18. (a) A strong electrolyte is a solute that is present in aqueous solution entirely as hydrated ions; to whatever extent such substances dissolve in water, they yield only ions. A weak electrolyte is an ionizing solute that yields a relatively low concentration of ions in aqueous solution. (b) Hydrogen chloride and silver chloride are examples of strong electrolytes, whereas hydrogen fluoride and acetic acid are weak electrolytes.

19. In general, the extent to which a solute ionizes in solution depends on the strength of the bonds within the molecules of the solute. If the strength of the bonds within the solute molecule is less than the attractive forces of the water dipoles, then the covalent bond breaks and the molecule is separated into ions.

20. (a) In order for an aqueous solution to be conductive, its solute must be capable of transporting an electric charge. Whereas the ions of electrolytes are charge carriers, the neutral molecules of nonelectrolytes are not, and such substances are thus nonconductive. (b) The degree to which an electrolyte solution conducts is related to the solute ion concentration; the higher the concentration of ions in an aqueous solution is, the more easily a solution conducts electric current.

21. Strong and weak molecular electrolytes differ in the degree

of ionization, whereas dilute and concentrated solutions differ in the amount of solute dissolved in a given quantity of a solvent.

22. *(a), (b), (d),* and *(f)* would conduct.

23. The indicated equation is: $H_2O(\ell) + H_2O(\ell) \rightleftarrows H_3O^+(aq) + OH^-(aq)$

24. *(a)* Nonvolatile electrolytes in aqueous solutions lower the freezing point and raise the boiling point sometimes two, three, or more times as much as nonvolatile nonelectrolytes in solutions of the same molality. *(b)* Both the freezing-point depression and the boiling-point elevation of a solvent by a solute depend on the relative numbers of dissolved particles of solute present, and not on their identity. When electro-lytes dissolve, they produce two, three, or more moles of solute particles for each mole of compound dissolved. Such substances would thus be expected to have two, three, or more times the effect on colligative properties in solutions of the same molality.

25. As the concentration of the aqueous nonvolatile solution decreases, the molal freezing-point depression and boiling-point elevation approach a whole number multiple of K_f and K_b, respectively. However, the result is never quite what it is predicted to be.

26. The more strongly the ions attract each other, the smaller is the apparent degree of ionization based on colligative-property measurements.

Problems

1. *(a)* $KI(s) \rightarrow K^+(aq) + I^-(aq)$;
   (b) $NaNO_3(s) \rightarrow Na^+(aq) + NO_3^-(aq)$;
   (c) $MgCl_2(s) \rightarrow Mg^{2+}(aq) + 2Cl^-(aq)$;
   (d) $Na_2SO_4(s) \rightarrow 2Na^+(aq) + SO_4^{2-}(aq)$;
   (e) $Al_2(CO_3)_3(s) \rightarrow 2Al^{3+}(aq) + 3CO_3^{2-}(aq)$.

2. *(a)* 1 mol K^+, 1 mol I^-; 2 mol ions *(b)* 1 mol Na^+; 1 mol NO_3^- 2 mol ions *(c)* 1 mol Mg^{2+}; 2 mol Cl^-; 3 mol ions *(d)* 2 mol Na^+; 1 mol SO_4^{2-}; 3 mol ions *(e)* 2 mol Al^{3+}; 3 mol CO_3^{2-}; 5 mol ions

3. *(a)* $Sr(NO_3)_2(s)\ \rightarrow Sr^{2+}(aq)\ + 2NO_3^-(aq)$
       0.50 mol   0.50 mol   1.0 mol
       1.5 mol solute ions
   (b) $Na_3PO_4(s)\ \rightarrow 3Na^+(aq)\ + PO_4^{3-}(aq)$
       0.50 mol   1.5 mol   0.50 mol
       2.00 mol solute ions
   (c) $K_2S(s)\ \rightarrow 2K^+(aq)\ + S^{2-}(aq)$
       0.275 mol   0.550 mol   0.275 mol
       0.825 mol solute ions
   (d) $Al_2(SO_4)_3(aq) \rightarrow 2Al^{3+}(aq)\ + 3SO_4^{2-}(aq)$
       0.15 mol   0.30 mol   0.45 mol
       0.75 mol solute ions

4. *(a)* $Co_2S_3(s) \rightleftarrows 2Co^{3+}(aq) + 3S^{2-}(aq)$
   (b) $PbS(s) \rightleftarrows Pb^{2+}(aq) + S^{2-}(aq)$
   (c) $PbCrO_4(s) \rightleftarrows Pb^{2+}(aq) + CrO_4^{2-}(aq)$

5. *(a)* $Fe(NO_3)_2(aq) + K_2S(aq) \rightarrow FeS(s) + 2KNO_3(aq)$
   (b) $HgBr_2(aq) + Na_2S(aq) \rightarrow HgS(s) + 2NaBr(aq)$
   (c) $2NaOH(aq) + MgCl_2(aq) \rightarrow 2NaCl(aq) + Mg(OH)_2(s)$
   (d) $2NaCl(aq) + Mg(NO_3)_2(aq) \rightarrow 2NaNO_3(aq) + MgCl_2(aq)$

6. *(a)* *(1)* $AgNO_3(aq) + NaI(aq) \rightarrow AgI(s) + NaNO_3(aq)$
          (balanced decimal equation)
       (2) $Ag^+(aq) + NO_3^-(aq) + Na^+(aq) + I^-(aq) \rightarrow$
          $AgI(s) + Na^+(aq) + NO_3^-(aq)$
          (overall ionic equation)
       (3) Spectator ions: Na^+ and NO_3^-
          Possible precipitate: AgI
       (4) $Ag^+(aq) + I^-(aq) \rightarrow AgI(s)$
          (net ionic equation)
   (b) *(1)* $HgCl_2(aq) + K_2S(aq) \rightarrow HgS(s) + 2KCl(aq)$
       (2) $Hg^{2+}(aq) + 2Cl^-(aq) + 2K^+(aq) + S^{2-}(aq) \rightarrow HgS(s) + 2K^+(aq) + 2Cl^-(aq)$
       (3) Spectator ions: Cl^- and K^+
          Possible precipitate: HgS
       (4) $Hg^{2+}(aq) + S^{2-}(aq) \rightarrow HgS(s)$
   (c) *(1)* $2Al(NO_3)_3(aq) + 3Ba(OH)_2(aq) \rightarrow 2Al(OH)_3(s) + 3Ba(NO_3)_2(aq)$
       (2) $2Al^{3+}(aq) + 6NO_3^-(aq) + 3Ba^{2+}(aq) + 6OH^-(aq) \rightarrow 2Al(OH)_3(s) + 3Ba^{2+}(aq) + 6NO_3^-(aq)$
       (3) Spectator ions: NO_3^- and Ba^{2+} Possible precipitate: $Al(OH)_3$

(4) $Al^{3+}(aq) + 3OH^-(aq) \rightarrow Al(OH)_3(s)$

(d) *(1)* $3CuCl_2(aq) + 2(NH_4)_3PO_4(aq) \rightarrow Cu_3(PO_4)_2(s) + 6NH_4Cl(aq)$

   (2) $3Cu^{2+}(aq) + 6Cl^-(aq) + 6NH_4^+(aq) + 2PO_4^{3-}(aq) \rightarrow$
   $Cu_3(PO_4)_2(s) + 6NH_4^+(aq) + 6Cl^-(aq)$

   (3) Spectator ions: NH_4^+ and Cl^-
   Possible precipitate: $Cu_3(PO_4)_2$

   (4) $3Cu^{2+}(aq) + 2PO_4^{3-}(aq) \rightarrow Cu_3(PO_4)_2(s)$

(e) *(1)* $2FeI_3(aq) + 3CaCrO_4(aq) \rightarrow Fe_2(CrO_4)_3(s) + 3CaI_2(aq)$

   (2) $2Fe^{3+}(aq) + 6I^-(aq) + 3Ca^{2+}(aq) + 3CrO_4{}^{2-}(aq) \rightarrow$
   $Fe_2(CrO_4)_3(s) + 3Ca^{2+}(aq) + 6I^-(aq)$

   (3) Spectator ions: I^- and Ca^{2+}
   Possible precipitate: $Fe_2(CrO_4)_3$

   (4) $2Fe^{3+}(aq) + 3CrO_4^{2-}(aq) \rightarrow Fe_2(CrO_4)_3(s)$

7. $\Delta T_f = K_f m$

(a) $\dfrac{-1.86°C \cdot kg\ solvent}{mol\ particles} \times \dfrac{1.00\ mol\ KI}{kg\ solvent} \times \dfrac{2\ mol\ particles}{1\ mol\ KI} = -3.72°C$

(b) $\dfrac{-1.86°C \cdot kg\ solvent}{mol\ particles} \times \dfrac{1.00\ mol\ CaCl_2}{kg\ solvent} \times \dfrac{3\ mol\ particles}{1\ mol\ CaCl_2} = -5.58°C$

(c) $\dfrac{-1.86°C \cdot kg\ solvent}{mol\ particles} \times \dfrac{1\ mol\ Ba(NO_3)_2}{kg\ solvent} \times \dfrac{3\ mol\ particles}{1\ mol\ Ba(NO_3)_2} = -5.58°C$

(d) $\dfrac{-1.86°C \cdot kg\ solvent}{mol\ particles} \times \dfrac{1\ mol\ C_{12}H_{22}O_{11}}{kg\ solvent} \times \dfrac{1\ mol\ particles}{1\ mol\ C_{12}H_{22}O_{11}} = -1.86°C$

(e) $\dfrac{-1.86°C \cdot kg\ solvent}{mol\ particles} \times \dfrac{1\ mol\ Al_2(SO_4)_3}{kg\ solvent} \times \dfrac{5\ mol\ particles}{1\ mol\ Al_2(SO_4)_3} = -9.30°C$

8. $\Delta T_f = K_f m$

$\dfrac{-1.86°C \cdot kg\ solvent}{mol\ ions} \times \dfrac{0.015\ mol\ AlCl_3}{kg\ solvent} \times \dfrac{4\ mol\ ions}{1\ mol\ AlCl_3} \rightarrow -0.11°C$

or $\dfrac{-1.86°C}{m\ ions} \times 0.015\ m\ AlCl_3 \times \dfrac{4\ m\ ions}{1\ m\ AlCl_3} \rightarrow -0.11°C$

9. $\Delta T_f = K_f m$

$\dfrac{-1.86°C \cdot kg\ H_2O}{mol\ ions} \times \dfrac{mol\ NaCl}{58.5\ g\ NaCl} \times \dfrac{85.0\ g\ NaCl}{450.\ g\ H_2O} \times \dfrac{2\ mol\ ions}{1\ mol\ NaCl} \times \dfrac{10^3\ g\ H_2O}{kg\ H_2O} = -12.0°C$

$\Delta T_f = T_f solution - T_f solvent$
$T_f solution = \Delta T_f + T_f\ solvent = -12.0°C + 0.0°C = -12.0°C$

10. $\dfrac{0.51°C \cdot kg\ H_2O}{mol\ ions} \times \dfrac{mol\ BaCl_2}{208\ g\ BaCl_2} \times \dfrac{25.0\ g\ BaCl_2}{0.150\ kg\ H_2O} \times \dfrac{3\ mol\ ions}{1\ mol\ BaCl_2} = 1.2°C$

$\Delta T_b = T_b solution - T_b solvent$
$T_b solution = \Delta T_b + T_b\ solvent = 1.2°C + 100.0°C = 101.2°C$

11. $\Delta T_b = K_b m$

$m = \dfrac{\Delta T_b}{K_b} = \dfrac{0.65°C \cdot mol\ ions}{0.51 \cdot kg\ H_2O} \times \dfrac{1\ mol\ KI}{2\ mol\ ions} = 0.64\ m\ KI$

$\dfrac{m\ ions}{0.51°C} \times \dfrac{1\ m\ KI}{2m\ ions} \times 0.65°C = 0.64\ m\ KI$

12. $\Delta T_f = T_f solution - T_f solvent = -2.65°C - 0.00°C = -2.65°C$

$\Delta T_f = K_f m$

$m = \dfrac{\Delta T_f}{K_f} = \dfrac{-2.65°C \cdot mol\ ions}{-1.86°C \cdot kg\ H_2O} \times \dfrac{mol\ Ba(NO_3)_2}{3\ mol\ ions} = 0.475\ m\ Ba(NO_3)_2$

Application Questions

1. Upon crystallization from aqueous solutions, such ionic compounds as $CuSO_4$ and Na_2CO_3 retain a characteristic number of water molecules in forming the crystal lattice. Each resulting crystalline compound, or hydrate, retains a specific number of water molecules. In the case of $CuSO_4$, five water molecules are retained; in Na_2CO_3, 10 are retained.

2. The dissolving process is endothermic because the temperature drop indicates that heat is being removed from the solvent to dissolve the ionic solid. An increase in the

solution temperature would indicate an exothermic process because this would reflect a loss of heat by the solute.

3. Electrolytes play an important role in the functioning of the human nervous system. The alternate stimulation and resting of nerve cells allows for the alternate passage/nonpassage of Na^+ ions through nerve cell membranes, with a resulting change in voltage across those membranes, due to changing ion concentrations. This repeating process accounts for the transmission of impulses along nerve cells.

4. As a pure liquid, hydrogen chloride consists of polar molecules and not of ions. When dissolved in a nonpolar solvent, no ions are formed. However, when HCl is dissolved in water, the attraction between the polar hydrogen chloride and the water dipoles is strong enough to break the H-Cl bond and form hydrogen and chloride ions. Due to the presence of the H^+ and Cl^- ions, the resulting solution can thus conduct electricity.

5. When the battery is wired into the conductivity apparatus, the electrodes serve as conductors that make electric contact with the test solution. In order for current to flow through the complete circuit, the test solution must provide a conducting path between the electrodes that effectively closes the switch to complete the circuit. A good conductor does result in a closed switch, so that current flows through the circuit, whereas a nonconducting solution effectively provides an open switch, such that no current is capable of flowing. Such an apparatus can thus be used to distinguish between substances in terms of their relative conductivities.

6. A concentrated solution of a weak electrolyte contains a relatively large quantity of solute dissolved in a given amount of solvent, but that solute is only slightly ionized. On the other hand, a dilute solution of a strong electrolyte contains a relatively small quantity of solute dissolved in a given amount of solvent, but that solute is ionized to a relatively large degree.

7. Although pure water is considered to be a nonconductor of electric current, tap water generally contains a high enough concentration of dissolved ions to make it a significantly better conductor than pure water. As such, tap water would conduct current if an operating electrical appliance were to be dropped into it, and a considerable shock would be imparted to anyone in contact with the water.

8. The Debye-Hückel theory accounts quantitatively for the attraction between dissociated ions of ionic solids in dilute water solutions. According to this theory, each solute ion is surrounded, on the average, by more ions of opposite charge than of like charge. This clustering effect hinders the movements of solute ions, such that a cluster of hydrated ions may act as a single unit rather than as an individual ion. Thus, the effective concentration of each ion is less than expected on the basis of the number of ions known to be present.

9. One way is to determine the extent to which the solution conducts electricity, since the greater the number of ions in a given volume of a solution is, the greater is its expected conductivity. The second way is to measure the lowering of the freezing point caused by an ionized solute in a measured amount of solvent, since that value is also dependent upon the number of ions in solution.

Application Problems

1. *(1)* $Mg_3(PO_4)_2(aq) + 3Pb(NO_3)_2(aq) \rightarrow Pb_3(PO_4)_2(s) + 3Mg(NO_3)_2(aq)$

   (2) $3Mg^{2+}(aq) + 2PO_4^{3-}(aq) + 3Pb^{2+}(aq) + 6NO_3^-(aq) \rightarrow Pb_3(PO_4)_2(s) + 3Mg^{2+}(aq) + 6NO_3^-(aq)$

   (3) $3Pb^{2+}(aq) + 2PO_4^{3-}(aq) \rightarrow Pb_3(PO_4)_2(s)$

   $$13.0 \text{ g Mg}_3(PO_4)_2 \times \frac{\text{mol Mg}_3(PO_4)_2}{263 \text{ g Mg}_3(PO_4)_2} = 0.0494 \text{ mol Mg}_3(PO_4)_2$$

   $$0.0494 \text{ mol Mg}_3(PO_4)_2 \times \frac{1 \text{ mol Pb}_3(PO_4)_2}{1 \text{ mol Mg}_3(PO_4)_2} = 0.0494 \text{ mol Pb}_3(PO_4)_2$$

   $$0.0494 \text{ mol Pb}_3(PO_4)_2 \times \frac{812 \text{ g Pb}_3(PO_4)_2}{\text{mol Pb}_3(PO_4)_2} = 40.1 \text{ g Pb}_3(PO_4)_2$$

2. $\Delta T_f = K_f m$

   $$\frac{-1.86°C \cdot \text{kg H}_2O}{\text{mol ions}} \times \frac{0.250 \text{ mol NaCl}}{1.00 \text{ kg H}_2O} \times \frac{2 \text{ mol ions}}{1 \text{ mol NaCl}} = -0.930°C$$

   $\Delta T_f = T_f\text{solution} - T_f\text{solvent}$
   $T_f\text{solution} = \Delta T_f + T_f\text{solvent}$
   $= -0.930°C + 0.000°C = -0.930°C$

3. The effect on freezing point is dependent upon the number and concentration of dissolved particles:

   (a)   0.01 mol NaI → 0.02 mol ions
   (b)   0.01 mol $CaCl_2$ → 0.03 mol ions
   (c)   0.01 mol K_3PO_4 → 0.04 mol ions
   (d)   0.01 mol $C_6H_{12}O_6$ → 0.01 mol molecules
   (e)   0.01 mol $Al_2(SO_4)_3$ → 0.05 mol ions

   Since an increasing concentration in particles causes a corresponding increase in the change of the freezing point, then the order is: *(d), (a), (b), (c), (e)*.

4. $$\frac{0.0500 \text{ mol Na}_2CO_3 \cdot 10H_2O}{\text{kg H}_2O} \times \frac{286 \text{ g Na}_2CO_3 \cdot 10H_2O}{1 \text{ mol Na}_2CO_3 \cdot 10H_2O} \times 500. \text{ g H}_2O \times \frac{\text{kg H}_2O}{10^3 \text{ g H}_2O} = 7.15 \text{ g Na}_2CO_3 \cdot 10H_2O$$

5. $\dfrac{\text{L solution}}{0.250 \text{ mol HCl}} \times \dfrac{1 \text{ mol HCl}}{36.5 \text{ g HCl}} \times 3.00 \text{ g HCl} \times \dfrac{10^3 \text{mL solution}}{\text{L solution}} = 329 \text{ mL solution}$

6. *(a)* $3NH_4OH(aq) + Al(NO_3)_3(aq) \rightarrow 3NH_4NO_3(aq) + Al(OH)_3(s)$

   (b) $3NH_4^+(aq) + 3OH^-(aq) + Al^{3+}(aq) + 3NO_3^-(aq) \rightarrow$
   $3NH_4 + (aq) + 3NO_3^-(aq) + Al(OH)_3(s)$

   (c) Spectator ions: $NH_4 +$ and NO_3^-
   Precipitate: $Al(OH)_3$

   (d) $Al^{3+}(aq) + 3OH^-(aq) \rightarrow Al(OH)_3(s)$

   (e) $3NH_4OH(aq) + Al(NO_3)_3(aq) \rightarrow 3NH_4NO_3(aq) + Al(OH)_3(s)$
       3 mol   1 mol   3 mol   1 mol

   This is a limiting-reactant problem. The next step is to determine the limiting reactant.

   $\dfrac{0.250 \text{ mol } NH_4OH}{\text{L solution}} \times 150. \text{ mL solution} \times \dfrac{\text{L solution}}{10^3 \text{ mL solution}} = 0.0375 \text{ mol } NH_4OH$

   $\dfrac{0.400 \text{ mol } Al(NO_3)_3}{\text{L solution}} \times 125 \text{ mL solution} \times \dfrac{\text{L solution}}{10^3 \text{ mL solution}} = 0.0500 \text{ mol } Al(NO_3)_3$

   $0.0375 \text{ mol } NH_4OH \times \dfrac{1 \text{ mol } Al(NO_3)_3}{3 \text{ mol } NH_4OH} = 0.0125 \text{ mol } Al(NO_3)_3$

   (f) Since NH_4OH is the limiting reactant

   $0.0125 \text{ mol } Al(OH)_3 \times \dfrac{78.0 \text{ g } Al(OH)_3}{1 \text{ mol } Al(OH)_3} = 0.975 \text{ g } Al(OH)_3$

   $0.0500 \text{ mol } Al(NO_3)_3 \text{ available} - 0.0125 \text{ mol } Al(NO_3)_3 \text{ required} = 0.0375 \text{ mol } AL(NO_3)_3 \text{ remains in excess.}$

7. $\Delta Tf = K_f m$

   $m = \dfrac{\Delta T_f}{K_f} = \dfrac{-3.72°C \cdot \text{mol ions}}{-1.86°C \cdot \text{kg } H_2O} \times \dfrac{1 \text{ mol } MgSO_3}{2.00 \text{ mol ions}} = 1.00 \, m \text{ } MgSO_3$

   Since $1.00 \text{ mol } MgSO_3 \cdot ?H_2O = 212 \text{ g/mol}$ and $1 \text{ mol } MgSO_3 = 104 \text{ g/mol}$, then there is 108 g of H_2O in the hydrate in the above solution.

   $108 \text{ g } H_2O \times \dfrac{\text{mol } H_2O}{18.0 \text{ g } H_2O} = 6 \text{ mol } H_2O$

   Therefore, the formula is $MgSO_3 \cdot 6H_2O$.

8. $\Delta T_f = K_f m$

   $\dfrac{-1.86°C}{m \text{ ions}} \times 1.00 \, m \text{ } MgI_2 \times \dfrac{3 \, m \text{ ions}}{1 \, m \text{ } MgI_2} = -5.58°C$

   According to the Debye-Hückel theory, there are probably many clusters of hydrated ions acting as a single ion, thereby effectively reducing the concentration of the individual ions. It is also possible that an MgI^+ species may exist due to strong solute–solute attraction. Also, there may not be enough water to hydrate all of the ions.

9. $\Delta T_f = T_f \text{solution} - T_f \text{solvent} = -1.15°C - 0.25°C = -1.40°C$
   $\Delta T_f = K_f m$

   $m = \dfrac{-1.40°C \cdot \text{mol ions}}{-1.86°C \cdot \text{kg } H_2O} \times \dfrac{1 \text{ mol } KNO_3}{2 \text{ mol ions}} = 0.376 \, m \text{ } KNO_3$

Chapter 15 Ions in Aqueous Solutions

INTRODUCTION

The solutions you learned about in detail in Chapter 14 contained neutral solute particles. This chapter describes the unique properties of solutions that contain charged solute particles, or ions.

LOOKING AHEAD

As you read this chapter, look for ways in which the theory of ions in aqueous solutions has been changed in order to explain new experimental observations.

SECTION PREVIEW

15.1 Ionic Compounds in Aqueous Solution
15.2 Molecular Electrolytes
15.3 Properties of Electrolyte Solutions

The salt truck is a familiar winter sight in cold regions of the U.S. Salt lowers the freezing point of water and melts the ice on roads.

15.1 Ionic Compounds in Aqueous Solution

Ionic compounds are soluble only in solvents that have polar molecules, such as water. Some of the properties of solutions containing ionic compounds are different from solutions containing only nonionic compounds. In order to understand the unique properties of ionic solutions, we will look briefly at the theories and observations of Michael Faraday (1791–1867) and Svante Arrhenius (1859–1927). Their work forms the basis of the modern theory of ions in aqueous solution.

Theory of Ionization

Michael Faraday was a self-taught English physicist who did many experiments with electricity and matter in the early nineteenth century. Little was then known of the structure of the atom, but Faraday became convinced that atoms are "associated with electrical powers, to which they owe their most striking qualities." Faraday coined the words *electrolyte* for substances that conduct electricity when dissolved in water and *nonelectrolyte* for substances that do not conduct electricity in aqueous solution. Faraday suggested that electricity converted dissolved electrolytes into charged particles, which he called *ions*. The ions then conducted the electric current by moving through the solution. *Ion* comes from the Greek word meaning "to go."

Examination of the freezing-point lowering and other colligative (binding) properties of electrolytes, however, gave puzzling results. In 1887, the Swedish chemist Svante Arrhenius introduced the theory of ionization. Arrhenius began with a different assumption than Faraday's. Arrhenius assumed that electrodes and the passage of current were not necessary for ion formation and he described many properties of electrolytes to show that this assumption was correct. He concluded that ions were produced by the "ionization" of molecules in aqueous solution.

Arrhenius considered the ions to be electrically charged. The solution as a whole contained equal numbers of positive and negative charges. He also considered ionization to be complete only in very dilute solutions. In more concentrated solutions, Arrhenius believed ions to be in equilibrium with solute molecules that were not ionized.

As chemists gained a better understanding of crystals and water molecules, some of the original concepts developed by Arrhenius were either modified or replaced. Most importantly, the explanation of the behavior of electrolytes in concentrated solutions was brought into agreement with what had been learned about ionic compounds. It is a great tribute to Arrhenius that his original theory served for so long. Present knowledge of the crystalline structure of ionic compounds was not available to him when, at the age of 24, he wrote his thesis on ionization.

From earlier chapters in this book, you have already learned more about the structure of the atom and chemical bonds than Arrhenius knew when he proposed his theory of ionization. You have seen that *before* they dissolve, solid ionic compounds contain equal amounts of positive and negative charge. You know that all ionic compounds have characteristic lattice structures that

SECTION OBJECTIVES

- Describe the solution process for an ionic compound.
- Define *hydration* and write equations representing ion hydration.
- Explain what contributes to the heat of solution of an ionic compound and what determines whether the solution process is exothermic or endothermic.
- Write equations for dissolving soluble ionic compounds in water.
- Write equations for solubility equilibria of very slightly soluble ionic compounds.
- Predict whether a precipitate will form when solutions of soluble ionic compounds are combined.
- Write net ionic equations for precipitation reactions.

STUDY HINT

The relationship between electrical current and chemical reactions is explored in chapter 20.

The colligative properties of electrolyte solutions are discussed in Section 15.6.

Arrhenius developed the Theory of Ionization at the age of 24 as his doctoral dissertation. His professors at the University of Upsala thought it was a radical idea and granted him a doctor of philosophy degree with the lowest possible rating. Twenty years later he was granted the Nobel Prize in chemistry for his creative work.

The electrical neutrality of an ionic solution is an example of the law of conservation of change.

STUDY HINT
A review of Section 6.3 will be helpful at this point.

The general term for solute–solvent interaction is solvation (Section 14.2).

are determined by the relative sizes and charges of their ions. In solution, these ions simply separate from each other. Certain polar molecular compounds are also electrolytes because they are converted into ions as they dissolve. In this chapter we examine the solution process for ionic compounds and molecular electrolytes. We also explore some of the properties of solutions of such compounds.

Dissolving Ionic Compounds

The polarity of water molecules plays an important role in the formation of solutions of ionic compounds in water. The charged ends of water molecules attract the ions in the ionic compounds and bring them into solution. The energy released when an ion is attracted to a water molecule also affects the heat of solution of the ionic compound in water.

The solution process for ionic compounds Suppose we were to drop a few crystals of sodium chloride into a beaker of water. At the crystal surfaces, water molecules come into contact with Na^+ and Cl^- ions. The positive ends of the water molecules are attracted to Cl^- ions; the negative ends are attracted to Na^+ ions. The attraction between water molecules and the ions is strong enough to draw the ions away from the crystal surface and into solution, as illustrated in Figure 15-1. *This solution process with water as the solvent is referred to as* **hydration.** The ions are said to be *hydrated.* The attraction between ions and water molecules is strong enough that each ion in solution remains surrounded by water molecules. As hydrated ions diffuse into the solution, other ions are exposed and attracted away from the crystal surface. In this manner the entire crystal gradually dissolves, and hydrated ions become uniformly distributed in the solution.

The number of water molecules surrounding each hydrated ion depends mainly upon the size and charge of the ion. Some ions in water solution are known to be associated with a specific number of water molecules. For others, the degree of hydration varies. A dynamic equilibrium is established between hydrated ions and water molecules. For Na^+, for example, this equilibrium can be represented as

Figure 15-1 Table salt, sodium chloride, dissolves as water hydrates its sodium and chloride ions.

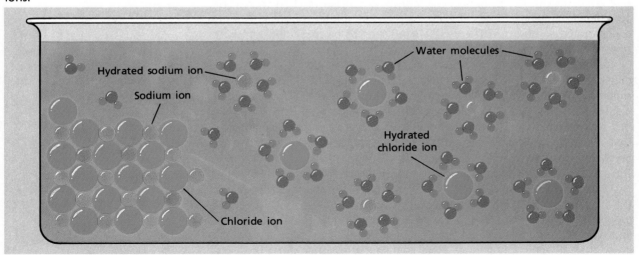

Hydrated sodium ion

Sodium ion

Water molecules

Hydrated chloride ion

Chloride ion

$$Na^+ + nH_2O(\ell) \rightleftharpoons [Na(H_2O)_n]^+$$

where n is the number of water molecules.

If the water is evaporated from a solution of sodium chloride, the solution process is reversed. The Na^+ and Cl^- ions begin to reunite and form a crystal when enough water molecules have evaporated for the solution to become saturated. Unlike sodium chloride, some ionic substances form crystals that incorporate water molecules. Certain anions and/or cations hold onto water molecules as they return to the crystal surface. Such crystalline compounds, known as *hydrates,* retain specific numbers of water molecules and are represented by formulas such as $CuSO_4 \cdot 5H_2O$. When a crystalline hydrate is dissolved in water, the water of hydration returns to the solvent. The solution is no different, therefore, from that of a nonhydrated crystalline compound—it contains hydrated ions and water (Figure 15-2).

Figure 15-2 In the two photos at left, 1 mol of $CuSO_4 \cdot 5H_2O$ (250 g) is added to a quantity of water to produce a 1-m solution. In the two photos at right, 1 mol of $CuSO_4$ is added to a quantity of water to make a 1-m solution. Both solutions that result are similar in all respects. Each molecule of $CuSO_4$ contributes 5 molecules of water to the solution. Therefore, less water is needed to make a 1-m aqueous solution using $CuSO_4 \cdot 5H_2O$.

Heat of solution for ionic compounds The same three interactions (solute–solute, solvent–solvent, and solute–solvent) contribute to the heats of solution of both ionic compounds and molecular compounds. The heat of solution of an ionic compound is the overall result of the energy changes occurring during the separation of ions from the crystal, the separation of water molecules from each other, and the hydration of the separated ions. Energy is required to pull ions away from the attraction of other ions. Energy is also required to separate polar water molecules.

By contrast, hydration is a process in which opposite charges are drawn together. Therefore, energy is always released in hydration. *The energy released when ions become surrounded by water molecules is called the* **heat of hydration.** More energy is usually released in the hydration of smaller ions than large ones. For example, compare the radii and heat of hydration of the lithium, sodium, and chloride ions:

Ion	Radius	Heat of Hydration
Li^+	68 pm	-523 kJ/mol
Na^+	95 pm	-418 kJ/mol
Cl^+	181 pm	-361 kJ/mol

Also, more energy is released in the hydration of ions with higher charges, as illustrated by comparing the radius and heat of hydration of the magnesium ion with the values above for Li^+ and Na^+:

Mg^{2+} radius: 65 pm   heat of hydration: -1949 kJ/mol

The Mg^{2+} ion is about the same size as the Li^+ ion, but has a heat of hydration almost four times larger.

STUDY HINT

The interactions that contribute to the heat of solution are shown in Figure 14-14.

Whether dissolving an ionic compound *releases* heat (exothermic process) or *absorbs* heat (endothermic process) depends upon the relative amounts of energy needed *(1)* to separate the ions in the crystal from each other and *(2)* to hydrate the ions. The energy contribution from separating water molecules is always the same.

As illustrated in Figure 15-3, if more energy is released during ion hydration than is needed to separate ions and water molecules, the solution process is exothermic. If less energy is released during ion hydration than is needed to separate ions and water molecules, the solution process is endothermic.

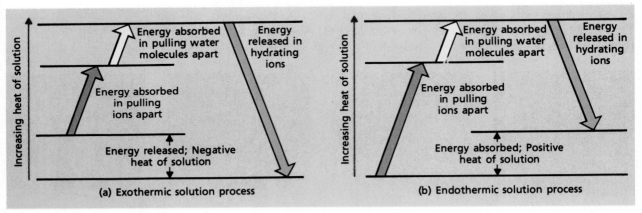

(a) Exothermic solution process   (b) Endothermic solution process

Figure 15-3 During the dissolution of an ionic compound, if more energy is released in hydrating ions than is used to pull them apart, *(a)* the process produces heat or is exothermic. When the reverse is true, *(b)* heat is absorbed, and the solution process is endothermic.

Exothermic solution process: negative heat of solution
Endothermic solution process: positive heat of solution

A simple research project that students could run would be to see how the dissolving of a hydrate and its anhydrous form affects the temperature of the resulting solution. Will both forms lower it or raise it? If so, how much? Or if one raises it, will the other lower it?

Emphasize that in dissociation, the ions already exist; all the water molecules have to do is to separate them.

The heats of solution of several ionic compounds are included in Table 14-5. Heat is released in the hydration of lithium chloride, but is absorbed in the hydration of sodium chloride. The difference results from the larger heat of hydration of the small Li^+ ion. Ammonium nitrate (NH_4NO_3), which is composed of polyatomic ions of larger radii than either sodium or chloride ions, has a much larger positive heat of solution (25.4 kJ/mol) than that of sodium chloride (4.27 kJ/mol). Dissolving a compound like ammonium nitrate absorbs so much heat from the solvent that the solvent temperature drops noticeably. The temperature change that occurs as ammonium nitrate is dissolved has been put to use in cold packs for the treatment of athletic injuries.

Dissociation   *The separation of ions that occurs when an ionic compound dissolves is called* **dissociation.** The equation representing the dissociation of sodium chloride is:

$$NaCl(s) \rightarrow Na^+(aq) + Cl^-(aq)$$

As usual, (s) indicates a solid species and (aq) indicates a species in an aqueous solution. This type of equation, like all others, must be balanced.

$$CaCl_2(s) \rightarrow Ca^{2+}(aq) + 2Cl^-(aq)$$

Whenever ions in solution are studied, one must be aware of the number of ions produced per formula unit. One formula unit of calcium chloride gives three ions in solution, compared with two ions in solution for each formula unit of sodium chloride. A 1-*M* solution of sodium chloride contains 1 mol of Na^+ ions and 1 mol of Cl^- ions.

$$NaCl(s) \rightarrow Na^+(aq) + Cl^-(aq)$$
$$\text{1 mol} \qquad \text{1 mol} \qquad \text{1 mol}$$

A 1-M solution of calcium chloride contains 1 mol of Ca^{2+} ions and 2 mol of Cl^- ions—a total of 3 mol of ions.

$$CaCl_2(s) \rightarrow Ca^{2+}(aq) + 2Cl^-(aq)$$
$$\text{1 mol} \qquad \text{1 mol} \qquad \text{2 mol}$$

Sample Problem 15.1

Write the equation for dissolving aluminum sulfate, $Al_2(SO_4)_3$, in water. How many moles of aluminum ions and sulfate ions are produced by dissolving 1 mol of aluminum sulfate? What is the total number of moles of ions produced by dissolving 1 mol of aluminum sulfate?

Solution

Step 1. Analyze Given: solute amount and formula = 1 mol $Al_2(SO_4)_3$
solvent identity = water

Unknown: *(a)* equation for dissociation of solute in solvent
(b) moles of aluminum ions and sulfate ions
(c) total number of moles of solute ions produced

Step 2. Plan Solute amount and formula →
balanced dissociation equation → mole relationship → total moles of solute ions

Step 3. Compute *(a)* $Al_2(SO_4)_3(s) \rightarrow 2Al^{3+}(aq) + 3SO_4^{2-}(aq)$
(b) 1 mol   2 mol   3 mol
(c) 2 mol Al^{3+} + 3 mol SO_4^{2-} = 5 mol of solute ions

Step 4. Evaluate The ion species (Al^{3+} and SO_4^{2-}) are correctly identified. Since one formula unit of $Al_2(SO_4)_3$ produces 5 ions, then 1 mol of $Al_2(SO_4)_3$ must also produce 5 mol of ions.

Practice Problems

1. Write the equation for dissolving ammonium chloride in water. How many moles of ammonium ion are produced when one mole of ammonium chloride is dissolved?
   (Ans.) $NH_4Cl(s) \rightarrow NH_4^+(aq) + Cl^-(aq)$, 1 mol NH_4^+
2. Write the equation for dissolving sodium sulfide in water. What is the total number of moles of ions formed by dissolving one mole of sodium sulfide?   *(Ans.)* $Na_2S(s) \rightarrow 2Na^+(aq) + S^{2-}$, 3 mol of ions
3. Write the equation for dissolving barium nitrate, $Ba(NO_3)_2$. How many moles of barium and nitrate ions are produced by dissolving 0.5 mol of barium nitrate?
   (Ans.) $Ba(NO_3)_2(s) \rightarrow Ba^{2+}(aq) + 2NO_3^-(aq)$, 0.5 mol Ba^{2+} and 1 mol NO_3^-

Solubility Equilibria

Ionic compounds have different degrees of solubility. No ionic compound is completely insoluble, however. Even the least soluble ionic compounds dissociate to some extent when placed in a polar solvent such as water.

Very slightly soluble ionic compounds   When a very slightly soluble ionic compound is placed in water, an equilibrium is established between the solid compound and its ions in solution, as shown in Figure 15-7. The result is the formation of a saturated solution. Some concentration of positive and

Predicting the concentrations of ions in solution in equilibrium with slightly soluble ionic compounds is explained in Section 19.4.

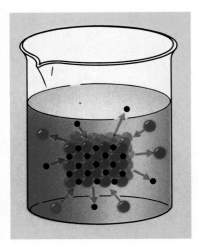

Figure 15-4 When a very slightly soluble ionic compound is placed in water, ions entering and leaving solution will reach equilibrium. At that point, the solution is called *saturated.*

Point out to students that some saturated solutions can be exceedingly dilute!

Rough rules of solubility
soluble: >1 g/100 g H_2O
slightly soluble:
0.1–1 g/100 g H_2O
insoluble: <0.1 g/100 g H_2O

STUDY HINT
Examples of double-replacement reactions are given in Section 8.2.

negative ions will be present in solution whenever water is in contact with such a compound. The equilibrium situation is represented by using the equilibrium arrow in dissociation equations like those previously written for completely soluble compounds. The solubility equilibria for three ionic compounds of low solubility are:

$$AgCl(s) \rightleftarrows Ag^+(aq) + Cl^-(aq)$$
$$Fe(OH)_3(s) \rightleftarrows Fe^{3+}(aq) + 3OH^-(aq)$$
$$Ag_2S(s) \rightleftarrows 2Ag^+(aq) + S^{2-}(aq)$$

Precipitation reactions   Both potassium chloride (KCl) and silver nitrate ($AgNO_3$) are quite soluble in water. Their solubilities at 20°C are 4.5 mol/L for KCl and 13 mol/L for $AgNO_3$. Their dissociation equations are:

$$KCl(s) \rightarrow K^+(aq) + Cl^-(aq)$$
$$AgNO_3(s) \rightarrow Ag^+(aq) + NO_3^-(aq)$$

What is likely to happen when solutions of potassium chloride and silver nitrate are combined? If a precipitate forms, a double-replacement reaction will occur. The two possible precipitates are potassium nitrate (KNO_3) and silver chloride (AgCl). To decide whether a precipitate can form, we must know the solubilities of these two compounds. Just above we saw that silver chloride has a low solubility, 1×10^{-5} mol/L at 20°C. To discover whether potassium nitrate (or any other compound with which we are unfamiliar) is a possible precipitate, we must consult a table of solubilities.

As listed in the margin, compounds are referred to as "soluble" if more than 1 g dissolves in 100 g of water, "slightly soluble" if 0.1–1 g dissolves in 100 g of water, and "insoluble" if the solubility is less than 0.1 g in 100 g of water. Consulting Table 15-1, we notice that KNO_3 is soluble in water.

We can therefore predict that, when solutions of potassium chloride and silver nitrate are combined, potassium nitrate will not precipitate and silver chloride will. When silver and chloride ions in solution are brought together in large enough concentrations, their attraction to each other is greater than their attraction to surrounding water molecules. In such a situation, precipitation occurs, as illustrated in Figure 15-5. Crystals of AgCl form in the solution and settle to the bottom of the container.

The balanced formula equation for the double-replacement reaction between potassium chloride and silver nitrate in solution is used to show that silver chloride precipitates and potassium nitrate remains in solution.

$$KCl(aq) + AgNO_3(aq) \rightarrow KNO_3(aq) + AgCl(s)$$

Net ionic equations   Double-replacement reactions and other reactions of ions in aqueous solution are usually represented by what are known as "net ionic equations." To write a net ionic equation, one first writes an equation in which all soluble compounds are shown as dissociated ions in solution. For the precipitation of silver chloride described above, the ionic equation is

$$K^+(aq) + Cl^-(aq) + Ag^+(aq) + NO_3^-(aq) \rightarrow$$
$$AgCl(s) + K^+(aq) + NO_3^-(aq)$$

Note that the complete formula for silver chloride is written because it is so slightly dissociated that it is considered insoluble. Potassium ion (K^+) and nitrate ion (NO_3^-) appear on both sides of this equation, showing that they

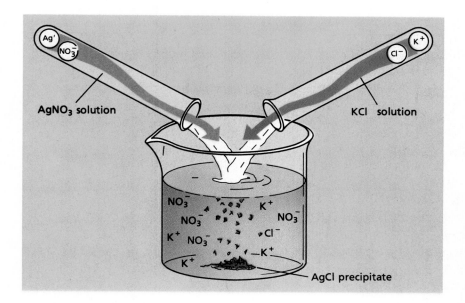

Only dissociated substances are written as ions in net ionic equations. Gases, pure liquids, or solids, if present, are represented by complete formulas.

TABLE 15-1
SOLUBILITY OF IONIC COMPOUNDS

1. Common sodium, potassium, and ammonium compounds are soluble in water.
2. Common nitrates, acetates, and chlorates are soluble.
3. Common chlorides are soluble except those of silver, mercury(I), and lead. [Lead(II) chloride is soluble in hot water.]
4. Common sulfates are soluble except those of calcium, barium, strontium, and lead.
5. Common carbonates, phosphates, and silicates are insoluble except those of sodium, potassium, and ammonium.
6. Common sulfides are insoluble except those of calcium, barium, strontium, magnesium, sodium, potassium, and ammonium.

have not undergone any chemical change and are still present in their original form. *Ions that do not take part in a chemical reaction and are found in solution both before and after the reaction are referred to as* **spectator ions**.

A **net ionic equation** *includes only those compounds and ions that undergo a chemical change in a reaction in an aqueous solution and does not include spectator ions.* To convert an ionic equation into a net ionic equation, the spectator ions are removed from both sides of the equation. Eliminating the K^+ and NO_3^- ions from the equation above gives

$$K^+(aq) + Cl^-(aq) + Ag^+(aq) + NO_3^-(aq) \rightarrow$$
$$AgCl(s) + K^+(aq) + NO_3^-(aq)$$
$$Ag^+(aq) + Cl^-(aq) \rightarrow AgCl(s)$$

This net ionic equation applies to any reaction in which a precipitate of silver chloride forms when the ions are combined in solution. Additional examples of precipitation reactions and net ionic equations are given in Figure 15-6.

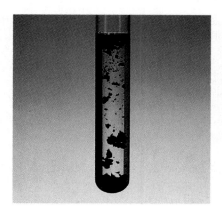

Figure 15-6 Precipitation reactions. Silver nitrate solution added to potassium iodide solution yields a dark precipitate of silver iodide *(left)*.
$Ag^+(ag) + I^-(ag) \rightarrow AgI(s)$
Lead nitrate solution added to sodium chromate solution yields yellow a precipitate of lead chromate *(right)*.
$Pb^{2+}(ag) + CrO_4^{2-}(ag) \rightarrow PbCrO_4(s)$

Sample Problem 15.2

Identify the precipitate that will form when aqueous solutions of zinc nitrate and ammonium sulfide are combined. Write the formula equation, ionic equation, and net ionic equation for the reaction.

Solution

Step 1. Analyze   Given: identity of reactants: zinc nitrate and ammonium sulfide
                              reaction medium: aqueous solution
                     Unknown: *(a)* identity of the precipitate   *(c)* ionic equation
                              (b) formula equation   *(d)* net ionic equation

Step 2. Plan   consult solubility table to determine identity of precipitate → formula
                     equation → ionic equation → net ionic equation
                     $Zn(NO_3)_2(aq) + (NH_4)_2S(aq) → ZnS(?) + NH_4NO_3(?)$

Step 3. Compute   *(a)* Consulting Table 15-1 reveals that zinc sulfide is not one of the soluble
                          sulfides and is, therefore, the precipitate. Ammonium nitrate is soluble
                          according to the table.
                     (b) $Zn(NO_3)_2(aq) + (NH_4)_2S(aq) → ZnS(s) + 2NH_4NO_3(aq)$
                     (c) $Zn^{2+}(aq) + 2NO_3^-(aq) + 2NH_4^+(aq) + S^{2-}(aq) →$
                          $ZnS(s) + 2NH_4^+(aq) + 2NO_3^-(aq)$
                     (d) According to the complete ionic equation above, the ammonium and nitrate ions
                          do not change during the reaction and are, therefore, spectator ions. The net
                          ionic equation is: $Zn^{2+}(aq) + S^{2-}(aq) → ZnS(s)$.

Step 4. Evaluate   The formulas for the reactants have been correctly written. Assuming a double-
                     replacement reaction, the products have been correctly identified. On
                     rechecking the table, zinc sulfide is the insoluble product. This is reasonable
                     because most nitrates are very soluble and most sulfides are not.

Practice Problems

1. Will a precipitate form if solutions of potassium sulfate and barium nitrate are combined? If so, write the net ionic equation for the reaction. *(Ans.)* Yes; $Ba^{2+}(aq) + SO_4^{2-}(aq) → BaSO_4(s)$
2. Will a precipitate form if solutions of barium chloride and sodium sulfate are combined? If so, identify the spectator ions and write the net ionic equation. *(Ans.)* Yes; Na^+ and Cl^-; $Ba^{2+}(aq) + SO_4^{2-}(aq) → BaSO_4(s)$
3. Write the net ionic equation for the precipitation of nickel(II) sulfide. *(Ans.)* $Ni^{2+}(aq) + S^{2-}(aq) →$

Section Review

1. How does the polarity of water molecules affect the solvation process of the ionic compound sodium chloride (NaCl) in water?
2. Is hydration an endothermic or an exothermic process?
3. Write the equation for dissolving $Ba(NO_3)_2$ in water. How many moles of barium ions and nitrate ions are produced by dissolving 0.5 mol of barium nitrate?
4. Will a precipitate form if solutions of magnesium acetate and strontium chloride are combined? If so, write the net ionic equation for the reaction.

Chemistry Notebook

Electrolytes in Your Body

The human body is mostly an electrolytic solution with water as the solvent and potassium ions (K^+), sodium ions (Na^+), calcium ions (Ca^{2+}), and organic ions as the electrolytes. The concentration of these ions serves an important role in the functioning of your nervous system.

Positive- and negative-ion concentrations in solution in the body are controlled by "ion pumps" and cell membranes that at one moment allow ions to pass through and at other moments bar passage to those same ions.

For instance, when a nerve cell is stimulated, Na^+ ions are pumped inside the nerve cell membrane, causing a higher concentration of positively charged ions inside the cell than outside the cell. When a nerve is not

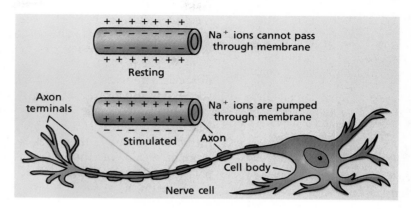

stimulated, or is in the resting state, there is a higher concentration of Na^+ ions outside the cell than inside. Unlike the stimulated cell membranes, the resting cell membrane will not allow Na^+ ions to pass through.

Stimulation of one end of the nerve cell causes a change in voltage across the cell membrane. This is because the concentrations of positive and negative ions has changed. The voltage

change disturbs the adjacent part of the nerve cell, causing Na^+ ions to be pumped inside the adjacent part of the cell. This happens repeatedly in adjacent sections of the cell and the impulse travels along the nerve cell.

Communication between nerve cells uses a different mechanism but once the next cell is stimulated, Na^+ ions are pumped inside the cell and the process is restarted.

15.2 Molecular Electrolytes

Molecular compounds, as well as ionic compounds, can form electrolytic solutions. Unlike ionic compounds in solution, however, many of the particles of a molecular electrolyte in solution may remain uncharged.

The solution process for molecular electrolytes   When a molecule containing one or more polar bonds is dissolved in water, the dipoles of the water molecules and the oppositely charged regions of the solute molecules are attracted to each other. For nonelectrolytes, these forces of attraction draw the molecules into solution. For molecular electrolytes, the attraction of the water molecules is strong enough to overcome the strength of one or more covalent bonds and one or more bonds in the solute molecule break.

STUDY HINT

A review of the discussion of covalent bonds in Section 6.2 and the discussion of intermolecular forces in Section 6.5 will be helpful at this point.

To explain ionization, use a tug-of-war analogy. The water molecules actually *tear* the polar solute molecules apart into ions by pulling on them on opposite sides until they break apart. You might even develop this into a little skit.

STUDY HINT

Acids, an important class of chemical compounds, are discussed in Chapters 16 and 17.

The bonding electron pair stays with the more electronegative atom, forming a negative ion. The less electronegative atom is left with a positive charge, forming a positive ion.

$$A—B \rightarrow A^+ + :B^-$$

The formation of ions from solute molecules by the action of the solvent is called **ionization.** The more general meaning of this term is the creation of ions where there were none. Note that the terms *dissociation* and *ionization* refer to different processes. When an ionic compound dissolves, the ions that were already present and associated with each other in the crystal become dissociated from each other. When a molecular compound dissolves in a polar solute, ions are formed where none existed in the undissolved compound.

Hydrogen chloride (HCl) is a molecular compound that ionizes in aqueous solution. It is one of a series of compounds of hydrogen with the members of Group 17 (known as halogens), which include fluorine, chlorine, bromine, and iodine. The hydrogen halides are all molecular compounds with polar, single covalent bonds. All are gases and very soluble in water.

Experiments show that pure liquefied hydrogen chloride (b.p.$-84.9°C$) does not conduct an electric current. A solution of hydrogen chloride in a nonpolar solvent such as benzene (C_6H_6) is also nonconductive. A solution of hydrogen chloride in water, by contrast, does conduct an electric current. The attraction between the polar H—Cl bond and the water-molecule dipoles is strong enough to break the bond. Hydrogen and chloride ions in solution are formed by the ionization of the hydrogen molecule (Figure 15-7):

$$H:Cl \rightarrow H^+(aq) + :Cl^-(aq)$$

Like all ions in aqueous solution, the ions formed by a molecular solute are hydrated. The heat released during the hydration of the ions provides the energy needed to break the covalent bonds.

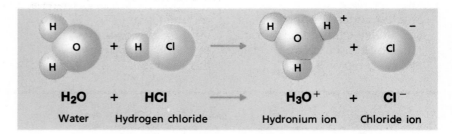

Figure 15-7 When a hydrogen chloride molecule ionizes in water, its hydrogen ion bonds covalently to a water molecule. A hydronium ion and a chloride ion are formed.

Hydrogen chloride in aqueous solution is known as hydrochloric acid. The hydrogen chloride molecules are 100% ionized—the solution contains only ions. The properties of the solution are distinctly different from those of hydrogen chloride gas or liquid. Many of these properties are characteristic of all acids that produce hydrogen ions in aqueous solution.

The hydronium ion The hydrogen ion (H^+) is a bare hydrogen nucleus—usually only a proton. As such it is much smaller than any other ion. The H^+ ion attracts other molecules or ions so strongly that it does not normally have any independent existence. The ionization of hydrogen chloride in water is better described as a chemical reaction in which a proton is transferred directly from HCl to a water molecule, where it becomes covalently bonded to oxygen. The process is represented in Figure 15-7. *The H_3O^+ ion is known as the* **hydronium** *ion*. The hydronium ion has three equivalent covalent bonds between oxygen and hydrogen atoms and an overall charge of $+1$.

Students understand the H_3O^+ formula for a hydronium ion better if you first write it as $H^+ \cdot H_2O$.

The equation for the reaction between hydrogen chloride and water to form hydrochloric acid is often written, as follows, to show the formation of the hydronium ion rather than H^+:

$$HCl(g) + H_2O(\ell) \rightarrow H_3O^+(aq) + Cl^-(aq)$$

The hydration of the H^+ ion to form the hydronium ion is a very favorable reaction. The energy released makes a large contribution to the energy needed to ionize a molecular solute. Many compounds classified as molecular electrolytes contain hydrogen and form H_3O^+ in an aqueous solution.

Strong and weak electrolytes Hydrogen chloride (HCl), hydrogen bromide (HBr), and hydrogen iodide (HI) are 100% ionized in dilute aqueous solution. The product solutions are known as hydrochloric, hydrobromic, and hydroiodic acids, respectively. *Any solute that, like the hydrogen halides, is present in dilute aqueous solution entirely as hydrated ions is referred to as a* **strong electrolyte**. Hydrogen chloride, many other acids, and ionic compounds are strong electrolytes. The distinguishing feature of strong electrolytes is that, to whatever extent they dissolve in water, they yield only ions. A very slightly soluble ionic compound such as silver chloride (AgCl) dissociates to give a very low concentration of ions in water solution. Because Ag^+ and Cl^- are the only solute species in solution, however, silver chloride is considered to be a strong electrolyte.

Certain metal halides that are molecular compounds are also strong electrolytes. Aluminum chloride has the formula Al_2Cl_6 in the liquid and vapor phases and is a very poor conductor as a liquid. The Al^{3+} ion, however, has a strong tendency to become hydrated, and the hydration energy contributes to breaking the aluminum–chlorine bonds. An aqueous solution of aluminum chloride is a good conductor, showing that the compound is a strong electrolyte that ionizes as follows:

$$Al_2Cl_6(\ell) + 12H_2O(\ell) \rightarrow 2[Al(H_2O)_6]^{3+}(aq) + 6Cl^-(aq)$$

Hydrogen fluoride (HF) dissolves in water to give an acid solution known as hydrofluoric acid. For hydrogen fluoride, the process of dissolving and the properties of the solution are somewhat different than for the other hydrogen halides. Fluorine atoms are the smallest halogen atoms. The hydrogen–fluorine bond is not only highly polar, but also much stronger than the bonds between hydrogen and the other halogens. When hydrogen fluoride dissolves, some molecules ionize. But the reverse reaction—the transfer of H^+ ions back to F^- ions to form hydrogen fluoride molecules—also takes place. The result is an equilibrium in which the concentration of dissolved hydrogen fluoride molecules remains high and the concentration of H_3O^+ and F^- ions remains low.

$$HF(aq) + H_2O(\ell) \rightleftharpoons H_3O^+(aq) + F^-(aq)$$

In an aqueous solution, the majority of the HF molecules are present as dissolved HF molecules.

Most organic acids (all of which have carbon in their formulas), including acetic, are weak electrolytes.

A **weak electrolyte** *is a solute that yields a relatively low concentration of ions in aqueous solution.* Hydrogen fluoride is a weak electrolyte. Acetic acid is another. In a 0.1-*M* solution of acetic acid ($HC_2H_3O_2$), only 1% of the acetic acid molecules ionize.

$$HC_2H_3O_2(aq) + H_2O(\ell) \rightleftarrows H_3O^+(aq) + C_2H_3O_2^-(aq)$$

Ammonia (NH_3) is a molecular compound that is a gas at ordinary temperatures and is extremely soluble in water. The aqueous solution of ammonia conducts electric current poorly, showing that the concentrations of ions is low. Ammonia is a weak electrolyte that reacts with water to give a low concentration of ammonium (NH_4^+) and hydroxide (OH^-) ions.

$$NH_3(g) + H_2O(\ell) \rightleftarrows NH_4^+(aq) + OH^-(aq)$$

Ammonia solutions in water were once called ammonium hydroxide solutions because of the presence of the ammonium and hydroxide ions. The name most likely persists because it is etched into many old reagent bottles. Actually, it is inappropriate because the compound NH_4OH has never been found to exist. It is probably better to call its solution ammonia water— $NH_3 \cdot H_2O$.

In general, the extent to which a solute ionizes in solution depends on the strength of the bonds within the molecules of the solute and the strength of attraction to solvent molecules. If the strength of the bonds within the solute molecule is less than the attractive forces of the water dipoles then the covalent bond breaks and the molecule is separated into ions.

Section Review

1. Because the H—F bond is much stronger than the H—Cl bond, the attraction of water molecules is not sufficient to prevent recombination of H^+ and F

2. Electronegativity: atoms with higher electronegativity will attract electrons and form negative ions; those with lower electronegativity will form positive ions.

3. The hydronium ion is formed when a hydrogen ion (H+) or, a bare proton, is attracted to the negative (or oxygen) end of the molecule: its net charge is +1.

1. Explain why HCl is a strong electrolyte and HF is a weak electrolyte.
2. What property of the atoms in a polar–covalent compound determines whether they will gain or lose an electron when they are ionized by a polar solvent.
3. How is a hydronium ion formed from water molecules and what is its net charge?

15.3 Properties of Electrolyte Solutions

- Explain why electrolytes in solution conduct electricity. Write an equation that describes the ionization process in pure water.
- Explain the expected and experimentally observed colligative properties of electrolyte solutions.
- Calculate the expected freezing-point depression of an electrolytic solution.

For an aqueous solution to be conductive, its solute must be capable of transporting an electric charge. The ions of an electrolyte are charge carriers, and the magnitude of the current that an electrolyte solution conducts is related to the solute ion concentration.

Conductivity of Solutions

Figure 15-8 shows apparatus for testing the conductivity of solutions. The test circuit includes a lamp, wire connected to a power source, and a pair of electrodes connected in series. The electrodes are conductors that make electric contact with the test solution. For a current to pass through the lamp

Chemistry Notebook

Boiling Liquids

When a solid is dissolved in a liquid, the vapor pressure of the liquid is lowered because the surface area from which molecules can evaporate is reduced. The vapor pressure is given by Raoult's law:

$$P = xP^{\circ}$$

where x is the mole fraction of the solvent in the solution and P° is the vapor pressure of the pure solvent.

What happens when both solute and solvent are volatile liquids? Each acts as solute to the other and reduces the vapor pressure. If the intermolecular forces between substances A and B are about the same as the A−A and B−B forces, the total vapor pressure can be

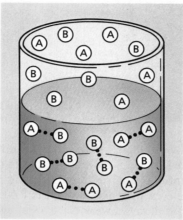

predicted by using Raoult's law twice:

$$P = x_A P_A^{\circ} + x_B P_B^{\circ}$$

The boiling point of a liquid mixture can be estimated by finding the temperature at which this equation yields a value of P equal to the atmospheric pressure.

When the A−B forces are weaker than the A−A or B−B forces, both molecules escape from a solution more easily than from a pure liquid. Mixtures of such liquids have a higher vapor pressure than predicted by the equation, and therefore have a lower boiling point than expected.

Hydrogen bonds between water and ethanol cause their mixtures to vaporize less readily than expected and to have higher boiling points. A mixture that is 95% ethanol and 5% water cannot be separated by simple distillation; such a mixture is called an *azeotrope*.

---

filament, the test solution must provide a conducting path between the two electrodes. A nonconducting solution is, in effect, an open switch between the electrodes. There can be no current in the circuit.

If the liquid tested is a good conductor of electric current, the lamp glows brightly when the switch is closed and the meter registers a substantial current in the circuit. For a liquid that is only a moderate conductor, however, the lamp is not as brightly lit and the meter registers a smaller current. If a liquid is a poor conductor, the lamp does not glow at all and the meter registers only a feeble current. For a nonconductor, no current will show on the meter.

Whether from a strong electrolyte or a weak one, the higher the concentration of ions in an aqueous solution, the more easily a solution will conduct electric current. To compare strong and weak electrolytes, the conductivity of solutions of equal concentration must be compared. The contrast between the conductivity of acetic acid (a weak electrolyte) and hydrochloric acid (a strong electrolyte) is shown in Figure 15-8.

Reference to a "strong electrolyte" or a "weak electrolyte" must not be confused with reference to a solution as "concentrated" or "dilute." Strong

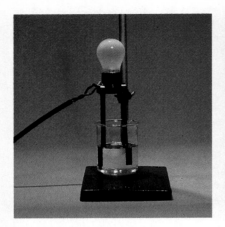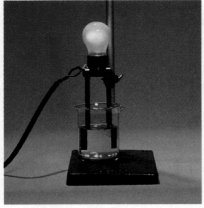

Figure 15-8 The brightness of the lamp filament indicates comparative conductivities of solutions tested. Distilled water *(left)* is a nonconductor. Dilute acetic acid, a solution of a weak electrolyte, is a poor conductor. Dilute hydrochloric acid, a solution of a strong electrolyte, is a good conductor.

In everyday usage, the terms "strong" and "weak" are synonymous with concentrated and dilute, i.e., strong or weak coffee. Be sure to emphasize that the terms "strong" and "weak" are used in a very different way in chemistry; they refer to the ability of solutes to produce ions, not the concentration of the solution.

The hydronium ion is a hydrated proton.

and weak molecular electrolytes differ in the degree of ionization. Concentrated and dilute solutions differ in the amount of solute dissolved in a given quantity of a solvent. Hydrochloric acid is always a strong electrolyte. This is true even in a solution that is 0.000 01 M—very dilute. By contrast, acetic acid is always a weak electrolyte, even in a 5-M solution.

The ionization of water   If pure water is placed in the apparatus shown in Figure 15-8, the light will not shine at all. On the basis of this experiment, water would be judged a nonconductor and a nonelectrolyte.

You might be wondering why we are constantly cautioned about the danger of dropping hair dryers or other electric appliances into the bathtub. The reason is that tap water commonly contains a high enough concentration of dissolved ions to make it a significantly better conductor than pure water. More sensitive tests for conductivity than those using light bulbs show that even very pure water has a slight, but measurable, conductivity. At ordinary temperatures, approximately two water molecules out of one billion are ionized. The ionization occurs, as illustrated in Figure 15-9, when one water molecule gives up a proton (a hydrogen ion) to another water molecule to form a *hydronium ion*. This process is:

$$H_2O(\ell) + H_2O(\ell) \rightleftarrows H_3O^+(aq) + OH^-(aq)$$

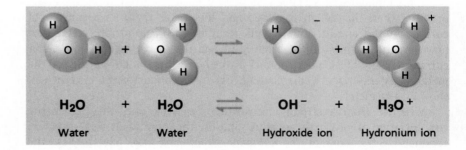

Figure 15-9 The ionization of pure water, a very weak electrolyte. Out of one billion water molecules, only two will ionize.

Water is a *very* weak electrolyte and it is such a poor conductor that, except in special circumstances, it can be considered a nonconductor. In the next chapter, however, you will see that the ionization of water, although slight, plays an important role in the chemistry of acids and bases.

Colligative Properties of Electrolyte Solutions

Colligative properties depend on the relative numbers of dissolved particles and not on their identity. We have already discussed how nonvolatile molecular solutes affect the solvent by lowering the vapor pressure, depressing the freezing point, and elevating the boiling point. Early investigators were puzzled by experiments in which certain substances depressed the freezing point or elevated the boiling point more than expected. Knowing more about dissociation and ionization than Michael Faraday did, you can predict how these properties change when the solute is an electrolyte.

Freezing-point depression in electrolyte solutions The molal freezing-point constant (K_f) for water of $-1.86°C/m$ is derived from experiments with nonelectrolytes in water solution. Electrolytes have a greater influence on the freezing points of solvents than do nonelectrolytes. A 0.1-m solution of sodium chloride lowers the freezing point nearly twice as much as a 0.1-m solution of sugar. A 0.1-m solution of potassium sulfate (K_2SO_4) or calcium chloride ($CaCl_2$) lowers the freezing point nearly three times as much as a 0.1-m solution of sugar. The effect on boiling points is similar. Nonvolatile electrolytes in aqueous solutions raise the boiling point nearly two, three, or more times as much as nonelectrolytes in solutions of the same molality.

Consider that five sugar molecules dissolve to produce five particles in solution. Five hydrogen chloride molecules, however, dissolve to produce ten particles in solution. This is the basis for the results of experiments on the colligative properties of electrolytes: electrolytes produce two or three or more moles of solute particles for each mole of compound dissolved. For the same molal concentrations of sugar and hydrogen chloride, we would expect the effect on colligative properties to be twice as large for hydrochloric acid as for sugar.

What about barium nitrate, $Ba(NO_3)_2$? Each formula unit of barium nitrate yields three ions in solution.

$$Ba(NO_3)_2(s) \rightarrow Ba^{2+}(aq) + 2NO_3^-(aq)$$

We would expect a $Ba(NO_3)_2$ solution of a given molality to lower the freezing point of its solvent three times as much as a nonelectrolyte solution of the same molality.

Sample Problem 15.3

What is the expected change in the freezing point of water in a solution of 125 g of barium nitrate, $Ba(NO_3)_2$, in 1.00 kg of water?

Solution

Step 1. Analyze.   Given: solute mass and formula = 125 g $Ba(NO_3)_2$
                              solvent mass and identity = 1.00 kg water
                              $\Delta T_f = K_f m$
               Unknown: expected freezing-point depression

Step 2. Plan.   mass of solvent,   molality in terms of   expected freezing-
                  mass of solute $\rightarrow$ molality $\rightarrow$ number of ions in solution $\rightarrow$ point depression

$$\frac{\text{mass of solute (g)}}{\text{mass of solvent (kg)}} \times \frac{1 \text{ mol solute}}{1 \text{ molar mass of solute (g)}} = \text{molality of solution} \left(\frac{\text{mol}}{\text{kg}}\right)$$

$$\text{molality of solution} \left(\frac{\text{mol}}{\text{kg}}\right) \times \text{molality conversion} \left(\frac{\text{ions}}{\text{mol}}\right) \times K_f \left(\frac{°C \cdot \text{kgH}_2\text{O}}{\text{mol}}\right)$$

$$= \text{expected freezing-point depression}$$

This problem is similar to Sample Problem 14.6 except that we are using an ionic solute rather than a nonionizing molecular solute. The number of particles in solution will therefore equal the number of ions of the solute.

Step 3. Compute

$$\frac{125 \text{ g Ba(NO}_3)_2}{1 \text{ kg H}_2\text{O}} \times \frac{1 \text{ mol Ba(NO}_3)_2}{261 \text{ g Ba(NO}_3)_2} = 0.479 \frac{\text{mol Ba(NO}_3)_2}{\text{kg H}_2\text{O}}$$

$$\frac{0.479 \text{ mol Ba(NO}_3)_2}{\text{kg H}_2\text{O}} \times \frac{3 \text{ mol ions}}{\text{mol Ba(NO}_3)_2} \times \frac{-1.86°C \cdot \text{kg H}_2\text{O}}{\text{mol ions}} = -2.67°C$$

Step 4. Evaluate

The units cancel properly to give the desired answer in degrees celsius. The answer is correctly given to three significant digits. Noticing that the mass of the solute is approximately one-half of its molar mass, and would give 1.5 mol of ions in the 1 kg of solvent, the estimated answer of $1.5 \times -1.86°C = -2.8°C$ supports our computation.

Practice Problems

1. What is the expected freezing-point depression for a solution that contains 2.0 mol of sodium chloride dissolved in 1.0 kg of water: *(Ans.)* $-7.4°C$
2. What is the expected freezing-point depression of water for a solution that contains 150 g of sodium chloride dissolved in 1.0 kg of water? *(Ans.)* $-9.5°C$

Table 15-2 presents data on the freezing points of solutions of electrolytes that yield two or three ions per formula unit. The actual freezing-point lowering was measured for each solute at the five concentrations shown. The molal freezing-point depressions listed were calculated by dividing the observed freezing-point depression by the molality of the solution.

For an ionic compound that gives two ions per formula unit, we would predict the freezing-point lowering to be $2 \times 1.86°C = 3.72°C$ for a one-molal solution. Examine the data in the table for potassium chloride (KCl). As the concentration of the solution decreases, the molal freezing-point depression comes closer and closer to the expected value of $3.72°C$. But it does not quite reach it even in a 0.001-m solution.

For each solute in Table 15-2, the situation is the same. The more dilute the solution, the closer the molal freezing-point lowering comes to a whole-number multiple of that for nonelectrolytic solute.

Colligative-property measurements can be used to find the percent ionization of weak electrolytes.

Apparent degree of ionization For many years, colligative property measurements like those in Table 15-2 prevented chemists from deciding whether a compound was completely ionized in a water solution or was a weak electrolyte. For example, based on the freezing-point lowerings in Table 15-2, KCl is apparently 93% ionized and $MgSO_4$ is apparently 60% ionized in 0.1-m solutions in water.

TABLE 15-2 MOLAL FREEZING-POINT DEPRESSIONS FOR AQUEOUS SOLUTIONS OF IONIC SOLUTES

Solute	Concentration (in moles solute/kg H₂O)				
	0.1	0.05	0.01	0.005	0.001

Values are observed freezing-point depressions in °C for concentrations indicated divided in each case by the molality.

$2K_f = 2 \times 1.86°C = 3.72°C$

Solute	0.1	0.05	0.01	0.005	0.001
$AgNO_3$	3.32	3.42	3.60	—	—
KCl	3.45	3.50	3.61	3.65	3.66
KNO_3	3.31	3.43	3.59	3.64	—
LiCl	3.52	3.55	3.60	3.61	—
$MgSO_4$	2.25	2.42	2.85	3.02	3.38

$3K_f = 3 \times 1.86°C = 5.58°C$

Solute	0.1	0.05	0.01	0.005	0.001
$BaCl_2$	4.70	4.80	5.03	5.12	5.30
$Ba(NO_3)_2$	4.25	—	5.01	—	5.39
$CaCl_2$	4.83	4.89	5.11	—	—
$Cd(NO_3)_2$	5.08	—	5.20	5.28	—
$CoCl_2$	4.88	4.92	5.11	5.21	—

Today it is recognized that attractive forces exist between ions in aqueous solution. The more strongly the ions attract each other, the smaller the apparent degree of ionization based on colligative-property measurements. Ions in more concentrated solutions attract each other more strongly because they are closer together. The attraction between hydrated ions in solution is small compared to those in the crystalline solid. Forces of attraction do interfere with the movements of the aqueous ions, even in quite dilute solutions, however. Only in very dilute solutions is the average distance between the ions large enough and the attraction between ions small enough

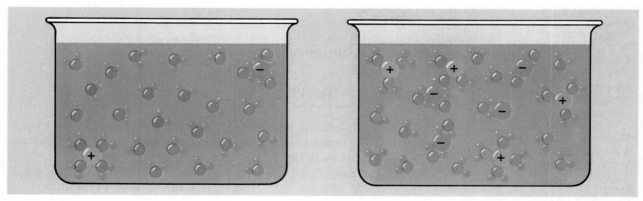

Figure 15-10 In concentrated solution *(right)*, ions are closer together and attract each other more strongly. As a result, the compound in a concentrated solution appears to be less ionized.

for the solute ions to move about freely. This is illustrated in Figure 15-10 and in Table 15-3, which lists the observed freezing-point depressions of increasingly dilute solutions of sodium chloride, the calculated freezing-point depressions per mole of sodium chloride, and the apparent degree of ionization at each concentration.

Peter Debye, a Dutch-born American physical chemist, is most well-known for his work with electrolytes, ionization, and dissociation. He explained the role of water molecules in the process of dissolving an ionic solid. In 1936, he was awarded the Nobel Prize in chemistry for his research on dipole moments and molecular structure.

TABLE 15-3 INFLUENCE OF CONCENTRATION ON FREEZING POINT OF AQUEOUS SOLUTIONS OF NaCl

Concentration of NaCl (m)	Freezing-point depression (°C)	Freezing-point depression/molal NaCl (°C/m)
0.100	0.346	3.46
0.0100	0.0361	3.61
0.001 00	0.003 66	3.66
0.0001 00	0.000 372	3.72

In 1923, Peter Debye and E. Hückel introduced theory to account for the attraction between dissociated ions of ionic solids in dilute aqueous solutions. According to this theory, each solute ion is surrounded, on average, by more ions of opposite charge than of like charge. This clustering effect hinders the movements of solute ions. A cluster of hydrated ions may act as a single unit rather than as individual ions. Thus, the effective concentration of each ion or its "activity" is less than expected on the basis of the number of ions known to be present. A solution dilute enough that the ions have the expected activity is referred to as an "ideal solution."

An ion with an activity coefficient of one is a freely moving particle in an ideal solution.

Ions of higher charge attract other ions very strongly. They therefore have lower activity and lower effective concentrations than ions with smaller charge. Note in Table 15-2 that the molal freezing-point depressions for magnesium sulfate ($MgSO_4$), with its +2 and −2 charged ions, are lower than those for the compounds with +1 and −1 charged ions.

In concentrated solutions, ion activity may be further decreased by a shortage of solvent molecules. In such a case, clusters of unhydrated ions may act as single solute units.

Section Review

1. How would the apparatus in Figure 15-15 perform if acetic acid were used as a test solution? If hydrochloric acid were tested?
2. Compare the conductivity of tap water and distilled water.
3. Why is rock salt used during the making of homemade ice cream?

1. With dilute acetic acid, the light would be dim; with dilute hydrochloric acid, it would be bright.
2. Distilled water is such a poor conductor it is considered to be a nonconductor; tap water, being a very dilute solution of various mineral electrolytes, is a fairly good conductor.
3. Rock salt is added to the ice in the freezer to lower its freezing point; the lower temperature causes the ice cream to freeze.

Chapter Summary

- The heat of solution of an ionic compound is a combination of the energy needed: to *(1)* separate the solutions *(2)* separate the solvent molecules; and *(3)* hydrate the solute ions.
- When two different ionic solutions are mixed, a precipitate may form if any one of the possible compounds formed by the solute ions is insoluble.
- A molecular compound with polar–covalent bonds may ionize in a water solution if the attraction of the polar water molecules is strong enough to overcome the strength of the polar–covalent bonds of the solute molecules.
- Solutes that produce ions in solution are called electrolytes. Electrolytes have a greater effect on the freezing and boiling points of solvents than do nonelectrolytes.

Chapter 15 *Review*

Ions in Aqueous Solutions

Vocabulary

dissociation
heat of hydration
hydration
hydronium ion
ionization

net ionic equations
spectator ions
strong electrolyte
weak electrolyte

Review Questions

1. Distinguish between electrolytes and nonelectrolytes on the basis of the electrical conductivity of their aqueous solutions.
2. Compare and contrast the explanations offered by Michael Faraday and Svante Arrhenius concerning the behavior of electrolytes.
3. List the three basic assumptions of the modern theory of ionization.
4. Explain how the polarity of water molecules affects the formation of solutions of ionic compounds in water.
5. Describe the solution process for ionic compounds like NaCl.
6. *(a)* Define the term hydration. *(b)* What determines the number of water molecules surrounding each hydrated ion?
7. *(a)* What interactions contribute to the heats of solution of ionic and molecular compounds? *(b)* Name the three energy changes and the direction of each, in the determination of the heat of solution of ionic compound.
8. *(a)* Define heat of hydration. *(b)* What is the effect of the size charge of an ion on its heat of hydration?
9. *(a)* Define dissociation. *(b)* How many moles of ions are contained in a 1-*M* solution of KCl? of BaF$_2$? of Mg(NO$_3$)?
10. Use Table 15-1 to predict the relative solubility of each of the following compounds (indicate soluble or insoluble for each). *(a)* KCl *(b)* NaNO$_3$ *(c)* AgCl *(d)* BaSO$_4$ *(e)* Ca$_3$(PO$_4$)$_2$ *(f)* Pb(ClO$_3$)$_2$ *(g)* (NH$_4$)$_2$S *(h)* PbCl$_2$ (in cold water) *(i)* FeS *(j)* Al$_2$(Co$_3$)$_3$ *(k)* Hg$_2$Cl$_2$
11. *(a)* What are spectator ions in a chemical reaction? *(b)* Identify the spectator ions in the reaction between KCl and AgNO$_3$ in an aqueous solution.
12. What is a net ionic equation?
13. How do molecules and ionic compounds compose in terms of the electrolytic solutions formed by each?
14. Describe the solution process for molecular electrolytes.
15. *(a)* What is ionization? *(b)* Distinguish between ionization and dissociation.
16. Write the chemical equation for the reaction between HCl and water to form hydrochloric acid.
17. *(a)* How is the hydronium ion produced? *(b)* Write the formula for the hydronium ion.
18. *(a)* Define and distinguish between strong and weak electrolytes? *(b)* Give two examples of each type.
19. What is the relationship between the molecular bond strength of a solute and the extent to which that solute ionizes in solution.
20. *(a)* What determines whether an aqueous solution will be conductive? *(b)* What determines the magnitude of the current that an electrolyte solution conducts?
21. Distinguish between the two sets of terms *strong* and *weak*, as opposed to *dilute* and *concentrated*, when used to describe electrolyte solutions.
22. If aqueous solutions of each of the following were tested with the apparatus shown in Figure 15-8, which would you expect to conduct electricity? *(a)* NaCl *(b)* KCl *(c)* C$_{12}$H$_{22}$O$_{11}$ *(d)* Mg(NO$_3$)$_2$ *(e)* CCl$_4$ *(f)* Al$_2$(SO$_4$)$_3$
23. Write the equation for the ionization of water.
24. *(a)* Compare the effects of nonvolatile electrolytes with the effects of nonvolatile nonelectrolytes on the freezing and boiling points of solvents into which they are dissolved *(b)* Why are such differences observed?
25. What is the relationship between the concentration of an aqueous nonvolatile electrolyte solution and the observed mold

freezing-point depression and boiling-point elevation?

26. What is the relationship between the attractive forces between ions in aqueous solutions and the apparent degree of ionization of those ions?

Problems

1. Write the equation for the dissolving of each of the following ionic compounds in water: *(a)* KI *(b)* $NaNO_3$ *(c)* $MgCl_2$ *(d)* Na_2SO_4 *(e)* $Al_2(CO_3)_3$.

2. For the compounds listed in the previous problem, determine the number of moles of each ion produced, as well as the total number produced when 1 mole of each compound dissolves in water.

3. Write the equation for dissolving each of the following in water and then indicate the total number of moles of solutions formed: *(a)* 0.50 mol strontium nitrate *(b)* 0.50 mol sodium phosphate *(c)* 0.275 mol potassium sulfide *(d)* 0.15 mol aluminum sulfate.

4. Involving each of the following slightly soluble compounds, write the equations for the solubility equilibria: *(a)* $Co_2S_3(s)$ *(b)* $PbS(s)$ *(c)* $PbCrO_4(s)$

5. Using Table 15-1, unite the balanced chemical (term used in Chapters 8 and 9) equation that represents each of the following possible double replacement reactions occurring in aqueous solutions. Indicate likely precipitates with the (s) designation: *(a)* *i*ron(II) nitrate*(aq)* + potassium sulfide*(aq)* → *(b)* *m*ercury(II) bromide*(aq)* + sodium sulfide*(aq)* → *(c)* *m*agnesium chloride*(aq)* + sodium hydroxide*(aq)* → *(d)* *m*agnesium nitrate*(aq)* + sodium chloride*(aq)* →

6. Use Table 15-1 to write the balanced chemical equation, to write the overall ionic equation, to identify the spectator ions and possible precipitates, and to write the net ionic equation for each of the following reactions: *(a)* silver nitrate*(aq)* + sodium iodide*(aq)* → *(b)* mercury(II) chloride*(aq)* + potassium sulfide*(aq)* → *(c)* aluminum nitrate*(aq)* + barium hydroxide*(aq)* → *(d)* copper(II) chloride(aq) + ammonium phosphate*(aq)* → *(e)* iron(III) iodide*(aq)* + calcium chromate*(aq)* →

7. Given 1.00-*m* aqueous solutions of each of the following substances, what is the expected change in the freezing point of the solvent? *(a)* KI *(b)* $CaCl_2$ *(c)* $Ba(NO_3)_2$ *(d)* $C_{12}H_{22}O_{11}$ *(e)* $Al_2(SO_4)_3$

8. What is the anticipated change in the freezing point of an aqueous solution that is 0.015 *m* $AlCl_3$?

9. What is the freezing point of a solution containing 85.0 g of NaCl dissolved in 450. g of water?

10. Determine the boiling point of a solution made by dissolving 25.0 g of barium chloride dissolved in 0.150 kg of water.

11. The change in the boiling point of an aqueous solution of potassium iodide is 0.65°C. Determine the apparent molal concentration of potassium iodide.

12. The freezing point of an aqueous solution of barium nitrate is −2.65°C. Determine the apparent molal concentration of barium nitrate.

Application Questions

1. Explain the basis for the H_2O molecules in the formulas of such compounds as $CuSO_4·5H_2O$ and $Na_2CO_3·10H_2O$.

2. An ionic solid dissolves in water. The thermometer in the solution indicates a temperature drop of 10°C. Is the dissolving process for this substance endothermic or exothermic? Explain.

3. Discuss one function of electrolytes in the human body.

4. Explain why hydrogen chloride as a pure liquid, or in solution in a nonpolar solvent, does not conduct an electric current but when dissolved in water, the resulting hydrogen chloride solution does conduct.

5. Explain how the test solution in a conductivity apparatus serves as a switch in the circuit and how such an apparatus can be used as a measure of conductivity.

6. Explain the difference between a concentrated solution of a weak electrolyte and a dilute solution of a strong electrolyte.

7. Generally speaking, water is a nonconductor and a nonelectrolyte; yet we are constantly concerned about dropping hair dryers or other electrical appliances into the bathtub. Explain the seeming inconsistency in this appropriate warning.

8. Explain the Debye-Hückel theory.

9. In what two ways can the apparent degree of ionization of a solution be determined?

Application Problems

1. Magnesium phosphate and lead(II) nitrate react in aqueous solutions by double replacement. If 13.0 g of magnesium phosphate react, what is the maximum amount of precipitate that could be found? Write the balanced chemical equation, the overall ionic equation, and the net ionic equation for this reaction.

2. Calculate the predicted freezing point of 1.00 kg of H_2O to which 0.250 mole of NaCl has been added.

3. Given 1.0 kg of 0.01-m aqueous solutions of each of the following, arrange these solutions in order beginning with the one that is expected to cause the least change in the freezing point of the solution: (a) NaI (b) $CaCl_2$ (c) K_3PO_4 (d) $C_6H_{12}O_6$ (glucose) (e) $Al_2(SO_4)_3$.

4. How many grams of sodium carbonate decahydrate crystals, $Na_2CO_3 \cdot 10H_2O$, must be dissolved in 500. g of H_2O to prepare a 0.0500-m solution?

5. How many milliliters of a 0.250-M HCl solution would contain 3.00 g of pure HCl?

6. Given that 150. mL of 0.250M NH_4OH reacts with 125 mL of 0.400M $Al(NO_3)_3$, complete each of the following exercises: (a) Write the balanced chemical equation for the reaction. (b) Write the ionic equation for the reaction. (c) Identify the spectator ions and the insoluble product. (d) Write the net ionic equation. (e) Find the maximum number of grams of precipitate that could be produced through this reaction. (f) Determine the number of moles of excess reactant that remains.

7. A solution made by dissolving hydrated $MgSO_3$ crystals in 1.00 kg of water lowered the freezing point of water by 3.72°C. Determine the chemical formula of this solute if its molar mass is 212 grams.

8. Experimental data for a 1.00-m MgI_2 aqueous solution indicate an actual change in the freezing point of water of −4.78°C. Determine the predicted change in the freezing point of water. Suggest reasons for any discrepancies that may exist between the experimental and the predicted values.

9. (a) You are conducting a freezing point determination in the laboratory using an aqueous solution of KNO_3. The observed freezing point of the solution is −1.15°C. Using a pure water sample, you recorded the freezing point of the pure solvent on the same thermometer at 0.25°C. Determine the molal concentration of KNO_3. Assume that there is no loss in ion activity according to the Debye-Hückel theory. (b) You are not satisfied with the result in part a of this question because you suspect that you cannot ignore the effect of ion interaction. You take a 10.00-mL sample of the solution. After following excellent laboratory procedures, you obtain a mass of 0.415 g KNO_3. Determine the actual molal concentration of KNO_3 and the percentage difference between the predicted and the actual concentrations of KNO_3.

Enrichment

1. Assemble an apparatus similar to the one shown in Figure 15-1. Secure several unknown aqueous solutions of equal molality from your instructor and use the apparatus to distinguish the electrolytes from the nonelectrolytes. Among those identified as electrolytes, rank their relative strengths as conductors, from good conductors down to poor conductors. Have your instructor check your results.

2. Using equal volumes of the unknown solutions from the preceding problem, explain how you could use the freezing-point depression concept to distinguish the electrolytes from the non-electrolytes. Explain further how you could determine the number of ions contained per molecule among the solutes identified as electrolytes. Design and conduct an experiment to test your theories.

3. Research the role of electrolytes and electrolytic solutions in your body. Find out how electro-lytes work in the functioning of nerves and muscles. What are some of the health problems that can arise fom an imbalance of electrolytes in body fluids?

4. Research the role of electrolytes in industry and manufacturing. What processes require the use of electrolytic solutions? Why are electrolytic solution important in these processes?

5. Use the work of Michael Faraday and Svante Arrhenius to explain as an example and the scientific method.

Chapter 16 Acids and Bases

Chapter Planner

16.1 Acids
General Properties of Aqueous Acids
- Demonstration: 1
  Question: 1
Definitions of Acids
- Questions: 2–5, 9, 10
  Application Questions: 1–3
Some Common Acids
- Question: 6
  Problem: 1
- Question: 7
  Application Question: 4
Names and Structures of the Common Acids
- Question: 12

16.2 Bases and Acid-Base Reactions
General Properties of Aqueous Bases
- Demonstration: 2
  Questions: 13, 14
- Question: 11

Definitions of Bases and Acid-Base Reactions
- Demonstration: 3
  Desktop Investigation
  Chemistry Notebook
  Questions: 15–19
  Problems: 2–4
  Application Questions: 5–7
  Application Problem: 2
- Application Question: 11
  Application Problem: 1
Types of Bases
- Question: 20

16.3 Relative Strengths of Acids and Bases
Brønsted Acid-Base Pairs
- Demonstration: 4
  Questions: 21, 22
Relative Strengths of Acids and Bases in Chemical Reactions
- Question: 23
  Application Question: 8
- Application Question: 9
  Application Question: 10

16.4 Oxides, Hydroxides, and Acids
Basic and Acidic Oxides
- Question: 25
  Problem: 5
Amphoteric Oxides and Hydroxides
- Demonstration: 5
  Question: 24
Hydroxides, Acids, and Periodic Trends
- Experiment: 43
  Questions: 26, 27

16.5 Chemical Reactions of Acids, Bases, and Oxides
- Demonstration: 6
  Questions: 29, 30
  Problems: 6–9, 11
- Questions: 28, 31
  Problem: 10
  Application Problem: 3
- Application Problems: 4–7

Teaching Strategies

Introduction

Introduce this chapter by pointing out that many familiar substances will be discussed. Some are edible while others are poisonous and corrosive to the skin. Acids, and salts are found in foods, cleaning supplies, batteries, and even some medicines. The students will learn where acids and bases are found naturally. They will be able to read product labels to check for acids and bases.

The overall objectives for this chapter are to help students (1) recognize acids and bases from their formulas and name them; (2) learn the properties, preparations, reactions, and uses of acids and bases; and (3) differentiate between traditional, Brønsted, and Lewis acids and bases and to know how to determine their strengths.

16.1 Acids

Use the following demonstration to introduce acids and identify some of their properties.

Demonstration 1: Identifying Acids and Their Properties

PURPOSE: To show how acids can be identified using litmus paper and how they can be identified from their formulas as inorganic and organic acids

MATERIALS: Red and blue litmus paper, hydrochloric-acid dilute, nitric-acid dilute, phosphoric-acid dilute, sulfuric-acid dilute, vinegar, mustard, pickle juice, orange juice, grapefruit juice, lemon juice, cranberry juice, grape juice, sour milk or buttermilk, yogurt, lemon-lime soda, fruit-flavored gelatin in liquid form, milk, distilled water, 4 250-mL beakers, 4 stirring rods, 4 reagent bottles

PROCEDURE: CAUTION: Wear goggles, gloves, and a lab coat during the preparation and performance of this demonstration. Remember, *always* add concentrated acid to water slowly with stirring. Work with concentrated HCl under the fume hood.

The concentration of the dilute hydrochloric, nitric, phosphoric, and sulfuric acids can vary. If you do not already have them prepared, you can make each of the dilute acids by dissolving 5 mL of the desired concentrated acid in 100 mL of water in a 250-mL beaker and store in a glass-stoppered reagent bottle.

Inorganic Acids: Tell the students the first group of acids are examples of inorganic acids. Pour a small amount of dilute hydrochloric acid in a 50-mL beaker. Write the name and the formula of the acid. Then test it with red and blue litmus paper. Repeat for nitric, hydrochloric, and phosphoric acids. Ask your students how they could determine if a substance is an inorganic acid by using litmus paper. (If blue litmus turns red it is an acid.) Tell students that the color change from blue to red indicates that the substance is an acid. Red litmus will stay red in neutral substances like water. Blue litmus will remain blue in neutral substances. Again ask, How do you think you can recognize an inorganic acid from its formula? (One or more hydrogens are the first element in the formula.)

Organic Acids: Using red and blue litmus paper, test the following substances to determine if they contain acids: vinegar, mustard, pickle juice, orange juice, grapefruit juice, lemon juice, cranberry juice, sour milk or buttermilk, yogurt, lemon-lime soda, fruit-flavored gelatin in liquid form, and milk. After each substance is tested, write its name, the name of the acid it contains, and its formula. Acetic acid, CH_3COOH, is in vinegar, mustard, and pickle juice.

Citric acid,

$$\begin{array}{c} \text{COOH} \\ | \\ CH_2-C-CH_2, \\ | \quad | \quad | \\ \text{COOH OH COOH} \end{array}$$

is in the juice of

oranges, lemons, grapefruits, gelatins, and most sodas. Lactic acid, $CH_3CHCOOH$, in sour milk, buttermilk, and yogurt.

$$\begin{array}{c} | \\ \text{OH} \end{array}$$

Benzoic acid is in cranberry juice.

All the substances will turn blue litmus paper red except milk. Continue with the following questions. "Do organic acids affect litmus in the same way as inorganic acids?" (Yes, all acids turn blue litmus red.) "Since neither blue nor red litmus change color in milk, what does this mean?" (Milk is a neutral substance.) "If you needed sour milk, how could you make it quickly?" (Add an acid such as vinegar or lemon juice.) "What would happen if milk was used in place of water in making fruit gelatin?" (The citric acid in the gelatin would curdle the milk just like any other acid.) "Have you tasted all the foods we tested today? What taste do all the foods containing acids have in common?" (All foods containing acids have a sour taste.) Tell students that a sour taste is characteristic of all acids, not just organic acids.

CAUTION: Do not allow students to taste the foods, and remind them the taste test is not acceptable in the lab.

Ask the students to compare the list of inorganic acids and their formulas and the organic acids and their formulas and ask: "How can you tell the difference between an inorganic acid and an organic acid by looking at their formulas?" (Organic acids have carbon in their formulas and have one or more–COOH groups.) Organic acids will be discussed in more detail in Chapter 22.

16.2 Bases and Acid–Base Reactions

Introduce bases in a manner similar to acids by using the following demonstration.

Demonstration 2: Identifying Bases in Common Substances

PURPOSE: To show how bases can be identified using litmus paper and how they can be identified by their formulas

MATERIALS: Household ammonia, drain cleaner, red and blue litmus paper, lime, lye, liquid antacid medication, milk of magnesia, oven cleaner, distilled water, 3 150-mL beakers

PROCEDURE: CAUTION: Wear goggles, gloves, and a lab coat. Many of these substances are very caustic. Have students wear goggles and aprons, and make sure all containers are tightly closed.

Test each of the substances with red and blue litmus paper. If it is in solid form, dissolve a small amount in distilled water in a beaker before testing. After testing, have the students read the label to try to determine which base it contains. Write the name of the product, the base it contains, and the formula of the base. Students should observe that all the substances will turn red litmus paper blue and conclude that bases turn red litmus blue. They should also note that bases contain one or more hydroxide ions in their formulas. Ammonia water, however, is an exception; the equation $NH_3(g) + H_2O \rightarrow NH_4^+(aq) + OH^-(aq)$ shows the solution contains hydroxide ions.

Sometimes a substance tests positive for a base when no base is present. This is because it contains a salt with basic properties. A good example of this is automatic dishwashing detergent, which contains Na_2CO_3.

Tell students they will learn why this happens in Chapter 17.

Demonstration 3: Neutralization Reactions

PURPOSE: To show that a reaction between an acid and a base forms a salt and water

MATERIALS: 1-M sodium hydroxide solution, 1 M hydrochloric acid, calcium hydroxide, red and blue litmus paper, distilled water, 4 150-mL beakers, 2 eyedroppers, stirring rod, 2 watch glasses

PROCEDURE: CAUTION: Wear goggles, gloves, and lab coat. Work with concentrated HCl under a hood.

Make the 1-M solution of sodium hydroxide by dissolving 4.0 g of NaOH in enough distilled water to make 100 mL of solution and the 1-M solution of HCl by dissolving 8.6 mL of concentrated HCl in enough distilled water to make 100 mL of solution. Test the HCl and NaOH solutions with red and blue litmus paper. (The HCl has no affect on the red but turns the blue litmus red, and the NaOH has no affect on the blue but turns the red litmus blue.)

To a 150–mL beaker, add 10 mL of HCl solution and 9.5 mL of NaOH solution. Stir thoroughly and test the mixture with red and blue litmus paper. (The red stays red, and the blue turns red.) Continue to add NaOH, one drop at a time, stirring after each addition and checking with red and blue litmus paper until the blue stays blue and the red stays red. Then ask, "Why doesn't the litmus paper change color?" (Because the solution is neutral—there are no hydronium or hydroxide ions present; thus they must have reacted to form water.) Put 2 mL of the neutral solution in a watch glass, and allow the water to evaporate until the next day. Have students observe the white residue. "What is the white substance on the evaporating dish?" (NaCl) "Can you write an equation for the reaction?"

$$NaOH + HCl \rightarrow H_2O + NaCl$$

The following reaction is a common neutralization reaction

used to sweeten acid soil by adding calcium hydroxide (commonly referred to as lime or slaked lime) to the soil. Add 1 g of solid calcium hydroxide to a beaker containing 5 mL of distilled water. Test the mixture with litmus paper. (It will turn red litmus paper blue. Slowly add dilute hydchloric acid to the mixture until almost all the Ca(OH) has reacted. (The white solid Ca(OH) appears to dissolve.) When there is still a tiny amount of un-reacted Ca(OH), test the solution with litmus. (Blue should stay blue, and red should turn blue.) At that point, start adding the HCl drop by drop until all the Ca(OH) is gone. Test again. Red litmus should stay red and blue should stay blue. Add 2 mL of the neutral solution to an evaporating dish and allow it to evaporate until the next day. A white residue of calcium chloride should remain on the evaporating dish. Have students write an equation to show what happened.

$$Ca(OH)_2 + 2HCl \rightarrow CaCl_2 + 2H_2O.$$

Interesting note: The word *alkali* comes from two words that mean *plant ashes*. Wood ashes contain potash (K_2CO_3) which was one of the first alkalies known in ancient times. Two others were soda (Na_2CO_3), made by evaporating alkaline water; and lime (CaO), made by roasting seashells. Caustic soda (NaOH) and caustic potash (KOH), made by roasting seashells. Caustic soda (NaOH) and caustic potash (KOH) were made by reacting soda or potash with lime.

People who have acid soil and a wood-burning stove could use their wood ashes to neutralize their soil. Potassium carbonate is another example of a salt with basic properties.

A good opportunity to review the difference between dissociation and ionization is in the discussion of what determines the strength of bases as compared to what determines the strength of acids. Point out that all acids before they are placed in water have polar molecules; therefore, they ionize in water, and the degree to which they ionize determines their strength. However, most bases (unlike acids) are ionic compounds that dissociate in water, which is why solubility in water is what determines their strength. (NH_3 is the only exception.)

Desktop Investigation: Acids and Bases in Your Kitchen

Preparing cabbage juice to use as an indicator is as simple as dropping wedges of red cabbage into a blender with a cup of water and processing it. Pouring off the water yields about one cup of indicator. In lieu of cabbage juice, the old standby is litmus paper. Suggest to your students that they experiment with other deeply colored juices. Elderberry juice, for example, is deep red when acidic and green when basic.

A discussion of this investigation in class could include the observation that most people enjoy sour, salty, or sweet tastes, but not bitter tastes. Most bases taste bitter, so it's unlikely that any of their foods will be strongly basic.

ANSWERS TO DISCUSSION QUESTIONS:
1. Most cleaning products are basic. There are a few glass cleaning products that contain acetic acid or HCl.
2. Most foods are neutral. Some foods are slightly acidic, such as carbonated soda, lemon juice, vinegar, mayonnaise, and pickle juice.
3. Most strongly basic products such as dishwasher detergent, oven cleaner, and drain cleaner carry consumer warnings.

16.3 Relative Strengths of Acids and Bases

In the discussion of Brønsted–Lowry acids and bases, it is pointed out that unlike traditional acids and bases, Brønsted–Lowry acid–base reactions do not need water to take place. The following demonstration shows this very well.

Demonstration 4: Reaction of a Brønsted–Lowry Acid and Base

PURPOSE: To show a Brønsted–Lowry acid–base reaction that takes place in the absence of water
MATERIALS: Concentrated ammonia water, concentrated hydrochloric acid, plastic storage bag with a snap and seal top (1 gallon capacity), 2 250-mL beakers, 10-mL graduate
PROCEDURE: CAUTION: Wear goggles, gloves, and a lab coat.

Using a fume hood, add 2 mL of concentrated hydrochloric acid to a 250-mL beaker. To a second beaker, add 2 mL of concentrated ammonia water. Place both beakers in a plastic storage bag and seal the top. Within a short time, the two volatile liquids will produce gases, and the area where the gases meet and react will appear white because of the formation of white solid ammonium chloride. Ask students to write the equation for the reaction, and identify the acid, the base, the conjugate acid, the conjugate base, and the conjugate acid–base pairs.

$$HCl(g) + NH_3(g) \rightarrow NH_4^+ + Cl^-$$

acid base conjugate acid conjugate base

conjugate acid–base pair

16.4 Oxides, Hydroxides, and Acids

It is sometimes difficult for students to understand the definitions of amphoteric oxides and hydroxides. The following demonstration makes it very clear.

Demonstration 5: Which Hydroxide Is Amphoteric?

PURPOSE: To learn how to recognize amphoteric hydroxides
MATERIALS: 0.1-M aluminum sulfate solution, dilute ammonia water, dilute hydrochloric acid, 0.1-M iron(III) chloride solution, 2-M sodium hydroxide solution, distilled water, 2 125-mL dropping bottles, 100-mL graduate, 10-mL graduate, 3 reagent bottles, stirring rod, 4 medium test tubes
PROCEDURE: CAUTION: Wear goggles, gloves, and a lab coat.

(1) Make the solutions listed in materials by dissolving: 1 g $Al_2(SO_4)_3 \cdot 18H_2O$/100 mL water in a dropping bottle; 20 mL concentrated HCl/80 mL water in a reagent bottle; 20 mL concentrated $NH_3 \cdot H_2O$/80 mL water in a reagent bottle; 1 g $Fe(OH)_3 \cdot 6H_2O$/100 mL water in a dropping bottle; and 8 g NaOH/100 mL water in a reagent bottle.

(2) Add 6 mL of aluminum sulfate solution to each of two test tubes. To each test tube add 3 mL of dilute ammonia water. What is the result? (A pinkish-white precipitate forms.) Write an equation for the reaction:

$$Al_2(SO_4)_3 + 6 NH_3 \cdot H_2O \rightarrow$$
$$2Al(OH)_3(s) + 3 (NH_4)_2(SO_4)_3$$

Allow the precipitate to settle and decant the clear liquid off the top. To the precipitate in one test, add dilute HCl while stirring until all the Al(OH)$_3$ has reacted. The Al(OH)$_3$ will appear to dissolve. Repeat with the second tube using the NaOH solution instead of the HCl. What happened in each case? (The Al(OH)$_3$ disappeared.) Did it dissolve or react chemically? (It reacted chemically.) Since the Al(OH) reacted with both the HCl and the NaOH, what kind of hydroxide is it? (An amphoteric hydroxide.) Write chemical equations to show what happened.

$$Al(OH)_3 + 3HCl \rightarrow AlCl_3 + 3H_2O$$
$$Al(OH)_3 + NaOH \rightarrow NaAlO_2 + 2H_2O$$

(3) Repeat Step 2, substituting FeCl$_3$ solution for the Al$_2$(SO$_4$)$_3$ solution. In this case a brown precipitate forms. The equation for the reaction is:

$$FeCl_3 + 3NH_3 \cdot H_2O \rightarrow Fe(OH)_3(s) + 3NH_4Cl$$

The addition of the HCl to the Fe(OH)$_3$ caused it to disappear and form a yellow liquid, but the NaOH had no effect. Thus, Fe(OH)$_3$ is not an amphoteric hydroxide. The equations for the reactions are:

$$Fe(OH)_3 + 3\ HCl \rightarrow FeCl_3 + 3H_2O$$
$$Fe(OH)_3 + NaOH \rightarrow \text{no reaction}$$

16.5 Chemical Reactions of Acids, Bases, and Oxides

Many buildings and sculptures are made of marble, which is a form of CaCO$_3$. Some fancy trims, outdoor furniture, and most automobiles contained iron or an iron alloy. Acid precipitation is composed of sulfuric and/or nitric acids. The following demonstration shows what happens when acid precipitation comes in contact with marble and iron.

Demonstration 6: Some Effects of Acid Precipitation
PURPOSE: To show the effects of acid on a carbonate and a metal and to relate these reactions to acid precipitation
MATERIALS: Marble–calcium carbonate, limewater, iron nail, dilute sulfuric acid, wooden splint, 2 beakers, reagent bottle, dropping bottle, filter paper, funnel, 3 large test tubes, rubber stopper to fit one test tube, glass bend (5 × 20 cm)
PROCEDURE: Make dilute sulfuric acid by slowly adding 15 mL of concentrated sulfuric acid to 85 mL of water in a beaker. When cool, store in a reagent bottle. Make limewater by adding 0.5 g of Ca(OH)$_2$ to 100 mL of water in a beaker. Stir thoroughly and filter into a 125-mL dropping bottle.

Put a nail into a test tube, and cover with sulfuric acid. Let the reaction continue until the next day. Bubbles of gas will form in the acid. By the next day the nail is partly "eaten" into by the acid. The equation for the reaction is:

$$Fe + H_2SO_4 \rightarrow FeSO_4 + H_2(g)$$

Put 3 or 4 chips of marble into a test tube. To a second test tube, add 20 mL of limewater. Assemble as shown in Figure T16–1. When the assembly is ready, take the rubber stopper out of the test tube of dilute sulfuric acid, and restopper. Bubbles of carbon dioxide will start to form.

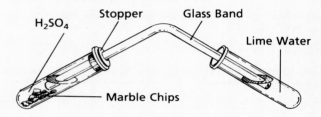

Figure T-16-1

As the CO$_2$ bubbles through the limewater, a white precipitate forms, which is a positive test for CO$_2$. The equations for the reactions are:

Marble and acid: $CaCO_3 + H_2SO_4 \rightarrow CaSO_4(s) + H_2O + CO_2(g)$

Limewater and CO$_2$: $Ca(OH)_2 + CO_2 \rightarrow CaCO_3(s) + H_2O.$

How do the acid reactions compare with similar ones involving acid precipitation? Since acid precipitation is not as concentrated as the acid used in the lab, the reactions involving acid precipitation are slower. But over a period of years, serious damage occurs to objects made of marble or iron.

References and Resources

Books and Periodicals

Borgford, Christie L., and Lee R. Summerlin. "Flower Pigments as Acid–Base Indicators," pp. 92–94; and "Distillation of Vinegar," pp. 266–68. *Chemical Activities, Teacher Edition*, Washington, D.C.: American Chemical Society, 1988.

Bybee, Roger. "Acid Rain: What's the Forecast?," *The Science Teacher*, Vol. 51, 1984, pp. 36–40, 45–47.

Bybee, Roger, *et al.* "The Acid Rain Debate," *The Science Teacher*, Vol. 51, 1984, pp. 50–55.

Charola, A. Elena. "Acid Rain Effects on Stone Monuments," *Journal of Chemical Education*, Vol. 64, 1987, pp. 436–37.

D'Orso, Jennifer, and Diane Burnett. "Mood Lipstick," *Chem-Matters:* Vol. 3, 1985, p. 12.

Jensen, William B. "Acids and Bases: Ancient Concepts in Modern Science," *ChemMatters*, Vol. 1, 1983, pp. 14–15.

Kauffman, George B. "The Brønsted–Lowry Acid–Base Concept," *Journal of Chemical Education*, Vol. 65, 1988, pp. 28–31.

Keever, Diane. "Acids and Bases," *Science Teacher*, Vol. 53, 1986, p. 173.

Kolb, Doris. "Chemical Principles Revisited: Acids and Bases," *Journal of Chemical Education*, Vol. 55, 1976, pp. 459–62.

MCA Staff. "An Acid Can Be Basic," *Journal of Chemical Education*, Vol. 56 , 1979, pp. 529–31.

Naylor, Robert W., and David Blackman. "A Physical Model to Demonstrate Acid–Base Conjugate Pairs," *Journal of Chemical Education*, Vol. 58, 1981, p. 1017.

Shakhashiri, Bassam Z. "Dehydration of Sugar by Sulfuric Acid,"

Chemical Demonstrations, Vol. 1, Madison: University of Wisconsin Press, 1983, pp. 77–78.

Shakhashiri, Bassam Z. "Antacids," *ChemMatter*, Vol. 1, 1983, p. 6.

Summerlin, Lee R., Christie L. Borgford, and Julie B. Ealy. "White Wine or Grape Juice," pp. 169–70; and "Disappearing Ink," *Chemical Demonstrations*, Vol. 2, 2nd ed, Washington, D.C.: American Chemical Society, 1988.

Talesnick, Irwin. "Addition of Acid to Water," Idea No. 89; "Excess Stomach Acid Treatment," Idea No. 52. *Idea Bank Collation: A Handbook for Science Teachers*, Kingston, Ontario, Canada: A17 Science Supplies and Services Co. Ltd., 1984.

Tanis, David. "Underground Sculpture," *ChemMatters*, Vol. 2, 1984, pp. 10–11.

Audiovisual Resources

Acid Rain, VHS, and *Special Topics in Chemistry, Volume I-Solutions, Acids, and Bases*, VHS, Random House Media, Stamford, CT, 1989.

All About Acids and Bases, VHS or sound filmstrip; *Acid Rain*, sound filmstrip; and *Acid Rain: History and Dangers*, VHS, Learning Arts, Wichita, KS, 1988–89.

Newton, David. *Acids and Bases*, Visual Masters, Walch Science, Portland, Maine, 1988.

Computer Software

Chemistry Series (J & S Software) Acid Base Theories, Apple II, 48K; and The Chemistry Help Series—Acids & Bases, Apple II, 48K, Barclay School Supplies, Brooklyn, NY, 1988.

Experiments in Chemistry, Apple II, 48K and IBM PC, 128K; *Chemistry Simulations I—The Manufacture of Sulphuric Acid*, Apple II, 48K, Queue, Bridgeport, Conn. 1988.

Fee, Richard O. *Acids, Bases, Salts 1 and 2*, Apple II, 48K, Chemistry According to ROF, LaGrange, IL, 1988.

Answers and Solutions

Questions

1. Lemons, oranges, and grapefruit contain citric acid, apples contain malic acid, vinegar contains acetic acid, sour milk contains lactic acid, wine contains tartaric acid, and many carbonated beverages contain phosphoric acid.

2. *(a)* According to the traditional definition, an acid is a chemical compound that contains hydrogen and ionizes in aqueous solution to form ions. *(b)* Traditional acids, like concentrated nitric acid (HNO_3), are molecular compounds that have the presence of ionizable hydrogen atoms in common. In concentrated solutions they may be poor conductors of electricity because they are only slightly ionized. However, in dilute aqueous solutions they are highly ionized and thus conductive because each has a hydrogen atom that is either attracted or repressed by the water dipoles.

3. *(a)* According to Brønsted, an acid is a molecule or ion that is a proton donor. *(b)* The Brønsted definition includes all traditional acids in aqueous solutions since they donate protons to water, but other species that donate protons are also included. Additionally, according to the Brønsted definition, the proton donation need not occur in aqueous solution.

4. *(a)* A Lewis acid is an electron-pair acceptor. *(b)* Unlike the traditional and Brønsted definitions, the Lewis definition extends the acid classification to include substances that do not contain protons. It is the broadest definition, applying to any substance that can combine with another by accepting an electron pair to form a covalent bond.

5. *(a)* Strong acids are strong electrolytes that dissociate 100% in aqueous solutions, whereas weak acids are weak electrolytes. *(b)* HCl and HNO_3 are examples of strong acids, and HF and H_3PO_4 are examples of weak acids.

6. Monoprotic acids, such as HCl, donate only one proton per molecule; diprotic acids, such as H_2SO_4, donate two protons per molecule; triprotic acids, such as H_3PO_4, donate three protons per molecule.

7. *(a)* Organic acids are organic compounds that contain an acid group, the carboxyl group (-COOH); mineral acids are acids made from minerals. *(b)* Acetic acid is an example of an organic acid, and sulfuric acid is an example of a mineral acid.

8. *(a)* Binary acids are acids that are compounds of hydrogen and a second element; oxyacids are acids that are compounds of hydrogen, oxygen, and a third element. Binary acids are named using the prefix *hydro-* followed by the root of the name of the second element, and then the suffix *-ic*. In naming oxyacids, a prefix and a suffix are added to the root name of the third element. Generally, one acid in the series is defined as the *-ic* acid and named accordingly, and then the acid with one more oxygen per molecule than the *-ic* acid is named by adding the prefix *per-*. The one containing one less oxygen per molecule than the *-ic* acid is given the suffix *-ous*, and then the acid containing two fewer oxygen atoms per molecule than the *-ic* acid has the prefix *hypo-* in addition to the suffix *-ous*. *(b)* Examples of binary acids are HF, HCl, and HBr. Typical oxyacids include $HClO_3$, HNO_3, and H_2SO_4.

9. The names of the acids indicated are:
   (a) hydrochloric acid *(b)* hydroiodic acid *(c)* nitric acid *(d)* sulfuric acid *(e)* chloric acid *(f)* hydrofluoric acid *(g)* nitrous acid *(h)* perchloric acid *(i)* phosphoric acid *(j)* bromic acid *(k)* hypochlorous acid *(l)* carbonic acid.

10. The indicated formulas are:
    (a) HF *(b)* HNO_3 *(c)* HI *(d)* $HClO_3$ *(e)* HBr *(f)* H_3PO_4 *(g)* HClO *(h)* HNO_2 *(i)* H_2SO_4 *(j)* H_3PO_3

11. Hydroxyl groups are covalently bonded -OH groups such as those found in oxyacids. Hydroxide ions are written as OH^- and are found in ionic compounds known as hydroxides.

12. The acids produced in the greatest quantity in the United States are sulfuric acid (H_2SO_4), phosphoric acid (H_3PO_4), nitric acid (HNO_3), hydrochloric acid (HCl), and acetic acid ($HC_2H_3O_2$). Sulfuric acid is the number-one industrial chemical in terms of volume production per year. It is a dense oily liquid with a high boiling point used in the production of fertilizers, in petroleum refining, in metallurgy, and in automobile batteries.

    Phosphoric acid is the second-largest volume industrial acid, and, in pure form, it is a low-melting solid. It is used in making fertilizers and animal feed, and as a flavoring agent in beverages. Nitric acid is the third largest volume industrial acid. In pure form it is a volatile, unstable liquid. It is important in the manufacture of fertilizers, explosives, rubber, plastics, dyes, and drugs. Hydrochloric acid is 23rd in volume produced. In concentrated form it is 37.2 percent hydrogen chloride (by mass) in water. It is used in the pickling of iron to remove surface impurities, as a cleaning agent, in food processing, and in the activation of oil wells. It is also produced in the stomach as an aid to digestion. Acetic acid is 32nd in volume produced and is a clear colorless liquid. It is produced by the fermentation of malt, barley, and fruit juices. It is found in household vinegar and is used industrially to make plastics and as a solvent.

13. Four common bases are: household ammonia (NH_3), used in general cleaning; lye or sodium hydroxide (NaOH), a component of products used to unclog drains; milk of magnesia, or magnesium hydroxide [$Mg(OH)_2$], used as an antacid; and aluminum hydroxide [$Al(OH)_3$], used as an antacid.

14. *(a)* Litmus, phenolphthalein, and methyl orange are good acid-base indicators. *(b)* Blue litmus changes from blue to pink in an acid and from pink to blue in a base; phenolphthalein is colorless in an acid solution, but red in a basic solution; methyl orange is red in an acid solution and yellow in a basic solution.

15. *(a)* The traditional definition of a base is a substance that contains hydroxide ions (OH^-) and dissociates to give hydroxide ions in aqueous solutions. *(b)* Sodium hydroxide is a traditional base in that it contains a metal cation (Na^+) and the hydroxide ion (OH^-). It is water soluble and dissociates completely to form Na^+(aq) and OH^-(aq).

16. *(a)* An alkaline solution is one that contains OH^- from a soluble base. *(b)* A salt is an ionic compound composed of a metal cation of an aqueous base and the anion from an aqueous acid.

17. *(a)* A neutralization reaction is the reaction of hydronium ions and hydroxide ions to form water molecules.
    (b) H_3O^+(aq) + OH^-(aq) → $2H_2O(\ell)$

18. *(a)* A Brønsted base is molecule or ion that is a proton acceptor. *(b)* In such a reaction, protons are transferred from one reactant (the acid) to another (the base).

19. *(a)* A Lewis base is an electron-pair donor. *(b)* A Lewis acid-base reaction is the formation of one or more covalent bonds between an electron-pair donor and an electron-pair acceptor.

20. *(a)* The strength of a base depends on its solubility and on the concentration of OH^- ions provided in solution. It is unrelated to the number of hydroxide ions per mole of the undissolved compound. *(b)* Sodium hydroxide is strongly basic, calcium hydroxide is moderately basic, and aqueous ammonia is weakly basic.

21. *(a)* A conjugate base is the particle that remains after an acid has given up a proton.

    Example: HF(aq) + $H_2O(\ell)$ → H_3O(aq) + F^-(aq)
              acid   conjugate base

    The F^- is the conjugate base of HF. *(b)* A conjugate acid is the particle formed by the addition of a proton to a base.

    Example: HF(aq) + $H_2O(\ell)$ → H_3O^+(aq) + F^-(aq)
                        base   conjugate acid

    The H_3O is the conjugate acid of H_2O.

22. *(a)* The stronger an acid, the weaker its conjugate base. *(b)* The stronger a base, the weaker its conjugate acid.

23. *(a)* Proton-transfer reactions favor the production of the weaker acid and the weaker base. *(b)* The extent of a proton-transfer reaction depends upon the relative strengths of the acids and the bases involved. For such a reaction to approach completion, the reactants must be much stronger as an acid and a base than the products.

24. *(a)* Amphoteric substances are species that can react as either an acid or a base. *(b)* Sulfuric acid. In the first ionization of H_2SO_4, the HSO_4^- is the conjugate base of H_2SO_4 since it can accept a proton from H_3O^+ in the reverse reaction. However, in a dilute H_2SO_4 solution, the HSO_4 ionizes further to form the SO_4^{2-} ion, so it acts like an acid in this situation.

25. *(a)* Basic anhydrides are oxides that react with water to form alkaline solutions, whereas acid anhydrides are oxides that react with water to form an acid. *(b)* Acid anhydrides are oxides of nonmetals, whereas basic anhydrides are oxides of active metals.

26. *(a)* In general, the active metals form basic oxides, and the nonmetals form acid anhydrides. The elements that lie between these typically form amphoteric oxides. *(b)* The bonding in amphoteric oxides is intermediate between ionic and covalent bonding, and, as a result, these oxides have behavior intermediate between that of acidic and basic oxides. A typical amphoteric oxide, aluminum oxide, is only slightly soluble in water and dissolves in both acidic and alkaline solutions.

27. The nature of the compound is dependent upon the type of bonding to the oxygen atom in the OH group, the size and electronegativity of the atom bonded to the OH group, and the number of other oxygen atoms bonded to the atom connected to the OH group. In general, when the bond to the OH group is ionic and involves metal atoms with large radii and low electronegativities, the resulting compound will be basic. For the compound to be acidic, the bond must be highly polar such as that formed by small, highly electronegative nonmetals. Additionally, the greater the number of other oxygen atoms bonded to the atom connected to the OH group, the more acidic the compound.

28. *(a)* Nonoxidizing acids provide hydronium ions that react with metals above hydrogen in the activity series to produce salts and hydrogen gas. On the other hand, in the reactions of oxidizing acids with metals throughout the entire activity series, the entire acid molecule is likely to be involved.

(b) Hydrochloric acid and dilute sulfuric acid are nonoxidizing acids, and hot concentrated sulfuric acid and nitric acid are typical oxidizing acids.

29. Salts can be produced from the reaction of acids with many metals, the reaction of acids with metal oxides, the reaction of acids with carbonates, the reaction of hydroxides with nonmetal oxides, and the reaction of metal oxides with nonmetal oxides.

30. The probable nature of each is: (a) acid (b) base (c) acid (d) acid (e) base (f) base (g) acid (h) base (i) acid (j) acid (k) base (l) acid (m) base.

Problems

1. (a) (1) $H_2SO_4(aq) + H_2O(\ell) \rightarrow H_3O^+(aq) + HSO_4^-(aq)$
   (2) $HSO_4^-(aq) + H_2O(\ell) \rightleftharpoons H_3O^+(aq) + SO_4^{2-}(aq)$
   (b) The degree of ionization in the first stage is much greater than that in the second.

2. (a) $HCl(aq) + KOH(aq) \rightarrow KCl(aq) + H_2O(\ell)$
   (b) $H_3O^+(aq) + Cl^-(aq) + K^+(aq) + OH^-(aq) \rightarrow K^+(aq) + Cl^-(aq) + 2H_2O(\ell)$
   (c) $H_3O^+(aq) + OH^-(aq) \rightarrow 2H_2O(\ell)$

3. (a) $H_3PO_4(aq) + 3NaOH(aq) \rightarrow Na_3PO_4(aq) + 3H_2O(\ell)$
   (b) $3H_3O^+(aq) + PO_4^{3-}(aq) + 3Na^+(aq) + 3OH^-(aq) \rightarrow 3Na^+(aq) + PO_4^{3-}(aq) + 6H_2O(\ell)$
   (c) $H_3O^+(aq) + OH^-(aq) \rightarrow 2H_2O(\ell)$
   (was 3)   (was 3)   (was 6)

4. (a) $HCl(aq) + NaOH(aq) \rightarrow NaCl(aq) + H_2O(\ell)$
   Ionic = $H_3O^+(aq) + Cl^-(aq) + Na^+(aq) + OH^-(aq) \rightarrow Na^+(aq) + Cl^-(aq) + 2H_2O(\ell)$
   Net = $H_3O^+(aq) + OH^-(aq) \rightarrow 2H_2O(\ell)$
   (b) $HNO_3(aq) + KOH(aq) \rightarrow KNO_3(aq) + H_2O(\ell)$
   Ionic = $H_3O^+(aq) + NO_3^-(aq) + K^+(aq) + OH^-(aq) \rightarrow K^+(aq) + NO_3^-(aq) + 2H_2O(\ell)$
   Net = $H_3O^+(aq) + OH^-(aq) \rightarrow 2H_2O(\ell)$
   (c) $Ca(OH)_2(aq) + H_2SO_4(aq) \rightarrow CaSO_4(aq) + 2H_2O(\ell)$
   Ionic = $Ca^{2+}(aq) + 2OH^-(aq) + 2H_3O^+(aq) + SO_4^{2-} \rightarrow Ca^{2+}(aq) + SO_4^{2-}(aq) + 4H_2O(\ell)$
   Net Ionic = $H_3O^+(aq) + OH^-(aq) \rightarrow 2H_2O(\ell)$
   (d) $3Mg(OH)_2(aq) + 2H_3PO_4(aq) \rightarrow Mg_3(PO_4)_2(aq) + 6H_2O(\ell)$
   Ionic = $3Mg^{2+}(aq) + 6OH^-(aq) + 6H_3O^+(aq) + 2PO_4^{3-} \rightarrow 3Mg^{2+}(aq)$ Net Ionic = $H_3O^+(aq) + OH^-(aq) \rightarrow H_2O(\ell)$

5. Since MgO reacts with water according to the following equation, it is a basic anhydride by definition.
   $MgO(s) + H_2O(\ell) \rightarrow Mg(OH)_2(s)$

6. (a) $Zn(s) + 2HCl(aq) \rightarrow ZnCl_2(aq) + H_2(g)$
   $Zn(s) + 2H_3O(aq) \rightarrow Zn^{2+}(aq) + H_2(g) + 2H_2O(\ell)$
   (b) $2Al(s) + 3H_2SO_4(aq) \rightarrow Al_2(SO_4)_3(aq) + 3H_2(g)$
   $2Al(s) + 6H_3O^+(aq) \rightarrow 2Al^{3+}(aq) + 3H_2(g) + 6H_2O(\ell)$

7. (a) $MgO(s) + H_2SO_4(aq) \rightarrow MgSO_4(aq) + H_2O(\ell)$
   (b) $CaO(s) + 2HCl(aq) \rightarrow CaCl_2(aq) + H_2O(\ell)$
   (c) $Al_2O_3(s) + 6HNO_3(aq) \rightarrow 2Al(NO_3)_3(aq) + 3H_2O(\ell)$
   (d) $3ZnO(s) + 2H_3PO_4(aq) \rightarrow Zn_3(PO_4)_2(aq) + 3H_2O(\ell)$

8. (a) $BaCO_3(s) + 2HCl(aq) \rightarrow BaCl_2(aq) + H_2O(\ell) + CO_2(g)$
   (b) $MgCO_3(s) + 2HNO_3(aq) \rightarrow Mg(NO_3)_2(aq) + H_2O(\ell) + CO_2(g)$
   (c) $Na_2CO_3(s) + H_2SO_4(aq) \rightarrow Na_2SO_4(aq) + H_2O(\ell) + CO_2(g)$
   (d) $3CaCO_3(s) + 2H_3PO_4(aq) \rightarrow Ca_3(PO_4)_2(aq) + 3H_2O(\ell) + 3CO_2(g)$

9. (a) $CO_2(g) + 2KOH(aq) \rightarrow K_2CO_3(aq) + H_2O(\ell)$
   (b) $CO_2(g) + KOH(aq) \rightarrow KHCO_3(aq)$

10. (a) $CaO(s) + CO_2(g) \rightarrow CaCO_3(s)$
    (b) $6\,BaO(s) + P_4O_{10}(s) \rightarrow 2\,Ba_3(PO_4)_2(s)$
    (c) $SrO(s) + SO_3(g) \rightarrow SrSO_4(s)$

11. (a) $Mg(s) + 2HCl(aq) \rightarrow MgCl_2(aq) + H_2(g)$
    (b) $CaO(s) + 2HNO_3(aq) \rightarrow Ca(NO_3)_2(aq) + H_2O(\ell)$
    (c) $2HNO_3(aq) + K_2CO_3(aq) \rightarrow 2KNO_3(aq) + CO_2(g) + H_2O(\ell)$
    (d) $CO_2(g) + Mg(OH)_2(aq) \rightarrow MgCO_3(s) + H_2O(\ell)$
    (e) $BaO(s) + CO_2(g) \rightarrow BaCO_3(s)$

Application Questions

1. (a) Brønsted (b) Brønsted (c) Brønsted (d) Lewis (e) Brønsted (f) Lewis (g) Lewis

2. The strength of an acid depends on the degree of ionization in aqueous solution, not on the amount of hydrogen in

the molecule. Hydrochloric acid, which ionizes completely, is considered to be a strong acid, while phosphoric acid, which ionizes only slightly, is considered weak.

3. Pure hydrogen-chloride gas consists of covalently bonded molecules, with no ions present. When hydrogen chloride is dissolved in a nonpolar solvent, still no hydrogen ions are produced. However, in a dilute aqueous solution, the hydrogen chloride becomes highly ionized because of the hydrating action of the water dipole. This results in the production of hydronium ions, H_3O^+, which account for the acidic properties of the resulting solution.

4. Sulfuric acid is involved in many technical and manufacturing processes. Typically, the more highly industrialized a particular country, the greater the level of sulfuric acid consumed by that country in the course of its everyday industrial activities. Thus, sulfuric acid production can actually serve as a measure of the degree of industrialization and economic activity of that country.

5. Sodium hydroxide is a traditional base because it contains hydroxide ions and dissociates to give hydroxide ions in aqueous solution. However, because it is not a proton acceptor, it does not fit the Brønsted definition. The OH^- ion it produces in solution is, in fact, defined as the Brønsted base.

6. The indicated responses are: (a) traditional, HCl is the acid, KOH is the base (b) Lewis, H^+ is the acid, NH_3 is the base (c) Brønsted, HCl is the acid, NH_3 is the base (d) traditional, H_3PO_4 is the acid, NaOH is the base (e) Brønsted, H_2O is the acid, NH_3 is the base (f) Lewis, NH_3 is the base, Ag^+ is the acid.

7. (1) Test it with litmus; if blue litmus turns pink, it's an acid, whereas if pink litmus turns blue, it's a base. If no change is observed, it may be a salt. (2) Test it with phenolphthalein; if phenolphthalein stays colorless, it may be an acid or a salt but if it turns red, then it is a base. (3) Test it with methyl orange; if the methyl orange turns red, it's an acid; if it turns yellow, it's a base. If it remains unchanged, it may be a salt. (NOTE: Do not attempt to taste or touch this substance.)

8. (a) HCl (acid) and Cl^- (conjugate base); H_2O (base) and H_3O^+ (conjugate acid)
(b) HCl (acid) and Cl^- (conjugate base); NH_3 (base) and NH_4^+ (conjugate acid)
(c) H_2O (acid) and OH^- (conjugate base); NH_3 (base) and NH_4^+ (conjugate acid)
(d) H_2SO_4 (acid) and HSO_4^- (conjugate base); H_2O (base) and H_3O^+ (conjugate acid) (e) $HC_2H_3O_2$ (acid) and $C_2H_3O_2^-$ (conjugate base); H_2O (base) and H_3O^+ (conjugate acid) (f) H_2O (acid) and OH^- (conjugate base); HCO_3^- (base) and H_2CO_3 (conjugate acid) (g) HCN (acid) and CN^- (conjugate base); SO_4^{2-} (base) and HSO_4^- (conjugate acid)

9. The indicated responses are: (a) HNO_3 (b) H_2S (c) HS^- (d) NO_3^-.

10. By definition, strong acids and bases are substances that are in solution. To remain highly ionized, the corresponding conjugate base and acid, respectively, are apparently too weak to compete successfully with the initial acid and base; otherwise, the initial reactants would not be classified as strong. Thus, the strength of an acid or base is rated in terms of the extent to which the corresponding conjugate base or acid remains ionized in solution.

11. Heartburn generally results when the gastric juices in the stomach splash up into the unprotected esophagus. Since these gastric juices contain hydrochloric acid and enzymes, they can effectively be neutralized by bases such as those contained in antacids. Some react with the hydrochloric acid to produce carbon dioxide, various salts, and water; others produce only salts and water. Antacids containing mixtures of hydroxides are generally the most effective.

Application Problems

1. $2H_3PO_4(aq) + 3Mg(OH)_2(aq) \rightarrow Mg_3(PO_4)_2(aq) + 6H_2O(\ell)$
Ionic $= 6H_3O+(aq) + 2PO_4^{3-}(aq) + 3Mg^{2+}(aq) + 6OH^-(aq) \rightarrow 3Mg^{2+}(ag) + 2PO_4^{3-}(ag) + 12H_2O(\ell)$
Net Ionic $= H_3O^+(ag) + OH^-(ag) \rightarrow 2H_2O(1)$

2. (a) $CO_2(g) + H_2O(\ell) \rightarrow H_2CO_3(aq)$
(b) $SO_2(g) + H_2O(\ell) \rightarrow H_2SO_3(aq)$
(c) $CaO(s) + H_2O(\ell) \rightarrow Ca(OH)_2(s)$
(d) $SO_3(g) + H_2O(\ell) \rightarrow H_2SO_4(aq)$

3. $Ca(s) + 2HCl(aq) \rightarrow CaCl_2(aq) + H_2(g)$
$Ca(s) + 2H_3O^+(aq) \rightarrow Ca^{2+}(aq) + H_2(g) + 2H_2O(\ell)$

4. $SO_2 + \frac{1}{2}O_2 \rightarrow SO_3$

$3.5 \times 10^8 \text{ kg SO}_2 \times \dfrac{10^3 \text{ g}}{\text{kg}} \times \dfrac{\text{mol}}{64.1 \text{ g}} = 5.46 \times 10^9 \text{ mol SO}_2$

$5.46 \times 10^9 \text{ mol SO}_2 \times \dfrac{1 \text{ mol SO}_3}{1 \text{ mol SO}_2} = 5.46 \times 10^9 \text{ mol SO}_3$

$SO_3 + H_2O \rightarrow H_2SO_4$

$5.46 \times 10^9 \text{ mol SO}_3 \times \dfrac{1 \text{ mol H}_2SO_4}{1 \text{ mol SO}_3} = 5.46 \times 10^9 \text{ mol H}_2SO_4$

$5.46 \times 10^9 \text{ mol H}_2SO_4 \times \dfrac{98.1 \text{ g}}{\text{mol}} \times \dfrac{\text{kg}}{10^3 \text{ kg}} = 5.36 \times 10^8 \text{ kg H}_2SO_4$

5. (a) $Zn(s) + H_2SO_4(aq) \rightarrow ZnSO_4(aq) + H_2(g)$

$$\frac{6.00 \text{ mol } H_2SO_4}{L} \times \frac{L}{10^3 \text{ mL}} \times 100. \text{ mL} = 0.600 \text{ mol } H_2SO_4$$

$$0.600 \text{ mol } H_2SO_4 \times \frac{1 \text{ mol } ZnSO_4}{1 \text{ mol } H_2SO_4} = 0.600 \text{ mol } ZnSO_4$$

$$0.600 \text{ mol } ZnSO_4 \times \frac{161.5 \text{ g}}{\text{mol}} = 96.9 \text{ g } ZnSO_4$$

(b) $$0.600 \text{ mol } H_2SO_4 \times \frac{1 \text{ mol } H_2}{1 \text{ mol } H_2SO_4} = 0.600 \text{ mol } H_2$$

$$0.600 \text{ mol } H_2 \times \frac{22.4 \text{ L}}{\text{mol}} = 13.4 \text{ L } H_2$$

6. $CaCO_3(s) + 2HCl(aq) \rightarrow CO_2(g) + CaCl_2(aq) + H_2O(\ell)$

(a) $$1500. \text{ mL } CO_2 \times \frac{L}{10^3 \text{ mL}} \times \frac{\text{mol}}{22.4 \text{ L}} = 0.066\,96 \text{ mol } CO_2$$

$$0.066\,96 \text{ mol } CO_2 \times \frac{1 \text{ mol } CaCO_3}{1 \text{ mol } CO_2} = 0.066\,96 \text{ mol } CaCO_3$$

$$0.066\,96 \text{ mol } CaCO_3 \times \frac{100.1 \text{ g } CaCO_3}{\text{mol } CaCO_3} = 6.703 \text{ g } CaCO_3$$

(b) $$0.069\,96 \text{ mol } CaCO_3 \times \frac{2 \text{ mol } HCl}{1 \text{ mol } CaCO_3} = 0.1339 \text{ mol } HCl$$

$$0.1339 \text{ mol } HCl \times \frac{L}{2.00 \text{ mol } HCl} = 0.066\,95 \text{ L or } 66.95 \text{ mL}$$

Teacher's Notes

Chapter 16 Acids and Bases

INTRODUCTION

The properties of acids and the related class of compounds known as bases are examined in this chapter and the next. You will see that acids and bases play important roles in nature, in the chemistry laboratory, and in the chemical industry.

LOOKING AHEAD

The word "acid" was probably familiar to you before you studied chemistry. As you read this chapter, compare what you already know about acids with the scientific description of acids given here. Also, as you read about bases, see if you can name household items that are bases. Relate the uses of these household items to the properties of bases.

SECTION PREVIEW

16.1 Acids
16.2 Bases and Acid–Base Reactions
16.3 Relative Strengths of Acids and Bases
16.4 Oxides, Hydroxides, and Acids
16.5 Chemical Reactions of Acids, Bases, and Oxides

Lime is being added to this lake in the central Adirondack region of New York to reduce its acidity.

16.1 Acds

How many foods can you think of that are sour? Chances are that almost all the foods on your list (such as those in Figure 16-1) owe their sour taste to an acid. Lemons, oranges, and grapefruit contain *citric acid*. Apples contain *malic acid*. Vinegar, which can be produced by the fermentation of hard cider or grape juice, contains *acetic acid*. Almost all fruits contain some kind of acid. Sour milk contains *lactic acid*. Tartaric acid is present in wine, and phosphoric acid imparts a tart flavor to many carbonated beverages.

General Properties of Aqueous Acids

Acids were first recognized as a distinct class of compounds by the sourness and other common properties of their aqueous solutions. These properties are as follows:

1. Acids have a sour taste. You are familiar with the taste of acids in the foods mentioned above. These foods contain weak acids in solution in water. Solid acids, such as citric acid that is often found in candy, and acetylsalicyclic acid, which is aspirin, taste sour as they dissolve in saliva. Many laboratory acids, especially in concentrated solutions, are very corrosive (they destroy skin and clothing) and are powerful poisons.

2. Acids contain hydrogen, and some react with active metals to liberate hydrogen gas (H_2). Metals above hydrogen in the activity series of metals (see Section 8.3) undergo single-replacement reactions with certain acids to produce hydrogen gas.

3. Acids change the color of dyes known as "acid-base indicators." For example, litmus dye, which is extracted from lichens, is an acid-base indicator. If a strip of paper impregnated with blue litmus is dipped into an acid solution, it turns pink, as demonstrated in Figure 16-2.

4. Acids react with bases to produce salts and water. When chemically equivalent amounts of acids and bases react, the three properties described above are "neutralized"—they disappear. The reaction products are water and an ionic compound called a salt.

5. Acids are electrolytes. Aqueous solutions of acids conduct electric current. (See Figure 16-3.)

Definitions of Acids

Traditional acids   Arrhenius included a description of acids in his theory of ionization. He suggested that the characteristic properties of acids

SECTION OBJECTIVES

- List five general properties of aqueous acids.
- Define and give an example of a traditional acid, a Brønsted acid, and a Lewis acid.
- Explain the difference between a strong acid and a weak acid, and give an example of each.
- Explain and use the systems for naming common binary acids and oxyacids.
- Write the electron-dot structures for common oxyacids.
- Name five acids commonly found in industry and/or the laboratory, and describe their nature (strong or weak, the number of ionizable hydrogen atoms, whether inorganic or organic, etc.).

Figure 16-1   Common acidic foods. If a food tastes sour, it probably is acidic.

Figure 16-2   Acids can change the color of many materials. Here, lemon juice, an acid, changes the color *(left to right)* of tea and the acid-base indicator litmus paper.

Figure 16-3 An automobile battery contains acid.

described above are caused by the presence of the hydrogen ion (H^+) in aqueous solution.

The observations of Arrhenius are the basis for the **traditional definition of an acid**: *a chemical compound that contains hydrogen and ionizes in aqueous solution to form hydrogen ions.* Traditional acids are molecular compounds that have the presence of ionizable hydrogen atoms in common. Their water solutions are known as *aqueous acids.*

As explained in our earlier study of the hydration of ions (Section 15.2), the hydrogen ion in aqueous solution is best represented as H_3O^+, the hydronium ion. Using the hydronium ion, the formation of, for example, nitric acid solution is shown by the following equation:

$$HNO_3(\ell) + H_2O(\ell) \rightarrow H_3O^+(aq) + NO_3^-(aq)$$
$$\text{nitric acid}$$

Brønsted acids   In 1923, J. N. Brønsted, a prominent Danish chemist, introduced an acid definition that emphasized the role of the proton.

A **Brønsted acid** is a molecule or ion that is a proton donor. In the equations above and below you can see that HNO_3 and HCl donate protons to H_2O to form H_3O^+. In aqueous solution all traditional acids donate protons to water and are, therefore, Brønsted acids also. In addition, other species that donate protons are included in the category of Brønsted acids. The proton donation need not occur in aqueous solution.

Hydrogen chloride dissolved in ammonia transfers protons to the solvent much as it does in water.

$$HCl(g) + H_2O(\ell) \rightarrow H_3O^+(aq) + Cl^-(aq)$$
$$HCl(g) + NH_3(\ell) \rightarrow NH_4^+(aq) + Cl^-(aq)$$

A proton is transferred from the hydrogen chloride molecule to the ammonia molecule to form the ammonium ion, just as a proton is transferred to the water molecule to form the hydronium ion. The similarity of these reactions is very clear when electron-dot formulas are used.

$$\text{H:}\overset{..}{\underset{..}{\text{Cl}}}\text{:} \;+\; \text{H:}\overset{..}{\underset{\underset{\text{H}}{|}}{\text{O}}}\text{:} \;\longrightarrow\; \text{H:}\overset{..}{\underset{\underset{\text{H}}{|}}{\text{O}}}\text{:H}^+ \;+\; \text{:}\overset{..}{\underset{..}{\text{Cl}}}\text{:}^-$$

$$\text{H:}\overset{..}{\underset{..}{\text{Cl}}}\text{:} \;+\; \text{H:}\overset{..}{\underset{\underset{\text{H}}{|}}{\text{N}}}\text{:H} \;\longrightarrow\; \text{H:}\overset{\overset{\text{H}^+}{}}{\underset{\underset{\text{H}}{|}}{\text{N}}}\text{:H} \;+\; \text{:}\overset{..}{\underset{..}{\text{Cl}}}\text{:}^-$$

In both cases, hydrogen chloride is a Brønsted acid, even though no hydronium ions form in the second reaction.

Water is also a Brønsted acid. In the reaction with ammonia, for example,

$$H_2O(\ell) + NH_3(g) \rightleftarrows NH_4^+(aq) + OH^-(aq)$$

the water molecule donates a proton to the ammonia molecule.

In the Brønsted scheme, the hydronium ion is also an acid and not just an ion produced by an acid. As will be illustrated shortly, the hydronium ion is the actual proton donor in reactions of aqueous acid solutions.

Lewis acids   A third definition extends the acid classification to include substances that do not contain protons. This definition of acids was

STUDY HINT

The solvent–solute attraction is so strong between water and some acids that the acid molecules "break" into ions.

The typical hydrogen ion, (H^+) is an atomic nucleus containing only one proton. Other isotopes of hydrogen do contain one or more neutrons in their nuclei.

Often two or more scientists who are working independently come up with the same idea at the same time. This was the case when Brønsted and T. M. Lowry developed their acid–base concept. It might be interesting for students to try to find out why Brønsted is often given more credit for the idea than Lowry.

The proton, or hydrogen ion, is always a Lewis acid because, having no electrons, it can "accept" an electron pair.

introduced in 1923 by G. N. Lewis, the same American chemist whose name we give to electron-dot structures. Lewis wanted to avoid a definition of acids that depended upon the presence of any particular element. His definition emphasizes the role of electron pairs in chemical reactions. *A **Lewis acid** is an electron-pair acceptor.*

The Lewis definition is the broadest of the three acid definitions we have given. It applies to any substance that can combine with another by accepting an electron pair to form a covalent bond. The bare proton is a Lewis acid in the reaction with ammonia to form ammonium ion. In fact, the proton is a Lewis acid in all reactions in which it forms a covalent bond.

$$\text{H}^+ \ + \ \text{H}:\overset{\cdot\cdot}{\underset{\text{H}}{\text{N}}}:\text{H} \ \longrightarrow \ \left[\text{H}:\overset{\overset{\text{H}}{\cdot\cdot}}{\underset{\text{H}}{\text{N}}}:\text{H}\right]^+$$

$$\text{H}^+ \ + \ :\text{NH}_3 \ \longrightarrow \ \text{NH}_4{}^+$$

A compound in which a central atom has formed three covalent bonds reacts as a Lewis acid by forming an additional covalent bond to complete an electron octet. Boron trifluoride, for example, is an excellent Lewis acid. It forms a fourth covalent bond with many molecules and ions in the manner shown below for the reaction with fluoride ion:

$$\overset{\cdot\cdot}{\underset{\cdot\cdot}{:}\text{F}:}\quad \qquad \qquad \left[\ \overset{\cdot\cdot}{\underset{\cdot\cdot}{:}\text{F}:}\ \right]^-$$

$$:\overset{\cdot\cdot}{\underset{\cdot\cdot}{\text{F}}}:\text{B} \ + \ :\overset{\cdot\cdot}{\underset{\cdot\cdot}{\text{F}}}:{}^- \ \longrightarrow \ :\overset{\cdot\cdot}{\underset{\cdot\cdot}{\text{F}}}:\text{B}:\overset{\cdot\cdot}{\underset{\cdot\cdot}{\text{F}}}:$$

$$\overset{\cdot\cdot}{\underset{\cdot\cdot}{:}\text{F}:}\quad \qquad \qquad \qquad \overset{\cdot\cdot}{\underset{\cdot\cdot}{:}\text{F}:}$$

$$\text{BF}_3(aq) \ + \ \text{F}^-(aq) \ \longrightarrow \ \text{BF}_4^-(aq)$$

The Lewis definition of acids applies to species in all phases. For example, boron trifluoride is a Lewis acid in the gas-phase combination with ammonia:

$$:\overset{\cdot\cdot}{\underset{\cdot\cdot}{\text{F}}}:\quad \quad \text{H} \qquad \qquad :\overset{\cdot\cdot}{\underset{\cdot\cdot}{\text{F}}}:\text{H}$$

$$:\overset{\cdot\cdot}{\underset{\cdot\cdot}{\text{F}}}:\text{B} \ + \ :\overset{\cdot\cdot}{\underset{\text{H}}{\text{N}}}:\text{H} \ \longrightarrow \ :\overset{\cdot\cdot}{\underset{\cdot\cdot}{\text{F}}}:\text{B}:\overset{}{\text{N}}:\text{H}$$

$$:\overset{\cdot\cdot}{\underset{\cdot\cdot}{\text{F}}}:\quad \quad \text{H} \qquad \qquad :\overset{\cdot\cdot}{\underset{\cdot\cdot}{\text{F}}}:\text{H}$$

$$\text{BF}_3(g) \ + \ \text{NH}_3(g) \ \longrightarrow \ \text{F}_3\text{BNH}_3(g)$$

Types of aqueous acids   A general reference to an "acid" usually is a reference to either a substance that produces hydronium ions in water or a solution containing hydronium ions. All of the pure acids are molecular electrolytes. Their molecules are sufficiently polar so that the water molecules attract away one or more of their hydrogen ions to leave behind anions, negatively charged ions.

An acid that is a strong electrolyte and dissociates 100% in aqueous solution is a **strong acid**. Hydrogen chloride is a strong acid. The common strong acids are listed in Table 16-1. Each of the strong acids—perchloric acid ($HClO_4$), hydrochloric acid (HCl), and nitric acid (HNO_3)—ionize completely to give up one proton. *An acid that can donate one proton per molecule is known as a* **monoprotic acid**.

Figure 16-4   J.N. Brønsted *(top)* and G.N. Lewis.

TABLE 16-1 COMMON AQUEOUS ACIDS

Strong Acids		Weak Acids	
$HClO_4$	$\rightarrow H^+ + ClO_4^-$	HSO_4^-	$\rightleftarrows H^+ + SO_4^{2-}$
HCl	$\rightarrow H^+ + Cl^-$	H_3PO_4	$\rightleftarrows H^+ + H_2PO_4^-$
HNO_3	$\rightarrow H^+ + NO_3^-$	HF	$\rightleftarrows H^+ + F^-$
H_2SO_4	$\rightarrow H^+ + HSO_4^-$	$HC_2H_3O_2$	$\rightleftarrows H^+ + C_2H_3O_2^-$
		H_2CO_3	$\rightleftarrows H^+ + HCO_3^-$
		H_2S	$\rightleftarrows H^+ + HS^-$
		HCN	$\rightleftarrows H^+ + CN^-$
		HCO_3^-	$\rightleftarrows H^+ + CO_3^{2-}$

Sulfuric acid (H_2SO_4) is a **polyprotic acid**—*an acid that can donate more than one proton per molecule.* Sulfuric acid ionizes in two stages. In its first ionization, sulfuric acid is a strong acid and is completely converted to hydrogen sulfate ions (HSO_4^-).

$$H_2SO_4(aq) + H_2O(\ell) \rightarrow H_3O^+(aq) + HSO_4^-(aq)$$

The hydrogen sulfate ion is a weak electrolyte, however, and establishes the following equilibrium in solution:

$$HSO_4^-(aq) + H_2O(\ell) \rightleftarrows H_3O^+(aq) + SO_4^{2-}(aq)$$

> As a general rule, the less concentrated the aqueous acid solution is, the greater the percentage of ionization. This percentage of ionization in very dilute solutions may range from an almost negligible amount to 100 percent.

The extent to which this second ionization occurs depends on the amount of dilution: the more dilute the solution, the more hydrogen sulfate ions will convert to sulfate ions. All stages of ionization of a polyprotic acid occur in the same solution. Sulfuric-acid solutions may therefore contain H_3O^+, HSO_4^-, and SO_4^{2-} ions. The ionization of HSO_4^- ions may be complete only in very dilute solutions, giving the following overall reaction:

$$H_2SO_4(aq) + 2H_2O(\ell) \rightarrow 2H_3O^+(aq) + SO_4^{2-}(aq)$$

This equation is the sum of the two stages of ionization of sulfuric acid. In a complete neutralization reaction, sulfuric acid *is an acid that can donate two protons per molecule and is therefore known as a* **diprotic acid**.

Many acids, such as HSO_4^-, are weak electrolytes. In Section 15.2 we showed that hydrogen fluoride is also a weak electrolyte. In aqueous solution, both the ionization of HF and the reverse reaction occur simultaneously:

$$HF(aq) + H_2O(\ell) \rightleftarrows H_3O^+(aq) + F^-(aq)$$

> The more dilute the solution, the more highly ionized are weak electrolytes.

Acids that are weak electrolytes are known as **weak acids.** The aqueous solution of a weak acid contains hydronium ions, anions, and dissolved acid molecules. The common weak acids are listed in Table 16-1. There is a wide variation in the degree that equivalent amounts of different weak acids convert to ions.

Phosphoric acid is a **triprotic acid**—*an acid able to donate three protons per molecule.*

$$H_3PO_4(aq) + H_2O(\ell) \rightleftarrows H_3O^+(aq) + H_2PO_4^-(aq)$$
$$H_2PO_4^-(aq) + H_2O(\ell) \rightleftarrows H_3O^+(aq) + HPO_4^{2-}(aq)$$
$$HPO_4^{2-}(aq) + H_2O(\ell) \rightleftarrows H_3O^+(aq) + PO_4^{3-}(aq)$$

Notice that phosphoric acid is a weak acid in each step of its ionization. The strength of an acid depends on the degree of ionization in aqueous solution, not on the amount of hydrogen in the molecule. A solution of phosphoric acid may contain H_3O^+, H_3PO_4, $H_2PO_4^-$, HPO_4^{2-}, and PO_4^{3-}. As with most polyprotic acids, the concentration of ions formed in the first ionization is the greatest, with lesser concentrations of the respective ions during each succeeding ionization.

Acetic acid is a weak acid that is commonly represented as $HC_2H_3O_2$. Its molecule contains four hydrogen atoms, but only one is ionizable. Acetic acid ionizes slightly in water to give hydronium ions and acetate ions ($C_2H_3O_2^-$).

$$HC_2H_3O_2(aq) + H_2O(\ell) \rightleftharpoons H_3O^+(aq) + C_2H_3O_2^-(aq)$$

Acetic acid is an **organic acid**—*an organic compound that contains an acid group, the carboxyl group (—COOH).* An alternate formula for acetic acid, CH_3COOH, shows that it is the hydrogen atom in the carboxyl group that is "acidic" and forms the hydronium ion, as shown below:

$$CH_3COOH(aq) \rightleftharpoons H_3O^+(aq) + CH_3COO^-(aq)$$

Except for acetic acid, all of the common acids listed in Table 16-1 are inorganic acids. For centuries, four of these substances have been known as *acids. Because they were made from minerals, sulfuric acid, nitric acid, hydrochloric acid, and phosphoric acid are referred to as* **mineral acids**. Each plays an important role in the chemical industry.

Names and Structures of the Common Acids

Some acids contain only two elements and are called *binary acids*; others, such as oxyacids, contain three elements.

Binary acids   A **binary acid** *is an acid that is a compound composed of hydrogen and a second element.* The common inorganic acids are either binary acids or oxyacids. The hydrogen halides—HF, HCl, HBr, HI—are all binary acids. The second element in a binary acid can be a nonmetal or a metalloid. The strength of a binary acid depends upon the polarity of the bond between hydrogen and the second element, and the ease with which the bond can be broken.

The name of a binary acid begins with the prefix *hydro-*. The root of the name of the second element follows this prefix. The name then ends with the suffix *-ic*. This scheme is illustrated by the examples given in Table 16-2. Each of the pure compounds listed in Table 16-2 is a gas. The water solutions of these compounds are known by the names given in the righthand column.

STUDY HINT
Organic compounds are carbon compounds, mostly derivatives of hydrocarbons. Chapters 21–23 are devoted to the chemistry of organic compounds.

The structure of the carboxyl group is

Inorganic compounds include all compounds that do not contain carbon.

TABLE 16-2   NAMES OF BINARY ACIDS

Formula	Name of Pure Substance	Name of Acid
HF	hydrogen fluoride	hydrofluoric acid
HCl	hydrogen chloride	hydrochloric acid
HBr	hydrogen bromide	hydrobromic acid
HI	hydrogen iodide	hydriodic acid
H_2S	hydrogen sulfide	hydrosulfuric acid

Oxyacids Sulfuric acid (H_2SO_4) is an example of an **oxyacid**—*an acid that is a compound of hydrogen, oxygen, and a third element.* An oxyacid is named by adding a prefix and a suffix to the root name of the third element.

The formulas and names of the series of oxyacids of chlorine illustrate the general method for naming such acids.

HClO	*hypo*-chlor-*ous* acid	-ic acid minus two O atoms
$HClO_2$	chlor-*ous* acid	-ic acid minus one O atoms
$HClO_3$	chlor-*ic* acid	
$HClO_4$	*per*-chlor-*ic acid*	-ic acid plus one O atom

The names are determined by defining one acid as the *-ic* acid, which is named by adding *-ic* to the root for the third element. For the chlorine oxyacids, this is chloric acid ($HClO_3$). An acid with one more oxygen atom per molecule than the *-ic* acid is named by adding the prefix *per-*, giving per-chlor-ic acid ($HClO_4$). The prefix *per-* is a contraction of *hyper-*, which means *above*.

The chlorine acid containing one less oxygen atom per molecule than the *-ic* acid is given the suffix *-ous*, which becomes chlor-ous acid, $HClO_2$. The chlorine acid that contains two fewer oxygen atoms per molecule than the *-ic* acid, has the prefix *hypo-* in addition to the suffix *-ous*, which becomes hypo-chlor-ous acid, HClO. The prefix *hypo-* means *below*.

To use the schemes for naming oxyacids, it is necessary only to know the formula and the name of the acid named with *-ic* in any series. You should memorize the following formulas and names for these members of the most common series of oxyacids: $HClO_3$ (chloric acid), H_2SO_4 (sulfuric acid), HNO_3 (nitric acid), H_3PO_4 (phosphoric acid), $HBrO_3$ (bromic acid). The names of some common oxyacids are also given in Table 16-3. The names of the polyatomic ions formed by the loss of all ionizable hydrogen atoms are given in the righthand column. The anion names have the same roots and prefixes

TABLE 16-3 NAMES OF OXYACIDS AND OXYANIONS

Formula	Name of acid	Name of anion
$HC_2H_3O_2$	acetic acid	acetate
H_2CO_3	carbonic acid	carbonate
HIO_3	iodic acid	iodate
HClO	hypochlorous acid	hypochlorite
$HClO_2$	chlorous acid	chlorite
$HClO_3$	chloric acid	chlorate
$HClO_4$	perchloric acid	perchlorate
HNO_2	nitrous acid	nitrite
HNO_3	nitric acid	nitrate
H_3PO_3	phosphorous acid	phosphite
H_3PO_4	phosphoric acid	phosphate
H_2SO_3	sulfurous acid	sulfite
H_2SO_4	sulfuric acid	sulfate

as the acid names. In the anion names, the ending -ic is changed to -ate and the ending -ous is changed to -ite.

All oxyacids are molecular electrolytes that contain one or more oxygen–hydrogen bonds. *The covalently bonded —OH group is referred to as a* **hydroxyl group***.* These —OH groups are not to be confused with hydroxide ions (OH⁻) which are found in the ionic compounds known as hydroxides, such as sodium hydroxide (NaOH).

Figure 16-5 shows the electron-dot formulas of the four oxyacids of chlorine. Notice that each oxygen atom is bonded to the chlorine atom and that each hydrogen atom is bonded to an oxygen atom. Aqueous solutions of these molecules are acids because O—H bonds are broken as the proton is attracted away by water molecules.

The ionizable hydrogen atoms in all polyprotic oxyacids are present in the hydroxyl groups only. For example, examine the structures of sulfuric acid and phosphoric acid in Figure 16-6. Four oxygen atoms surround the central sulfur and phosphorus atoms. The two ionizable hydrogen atoms are bonded to two oxygen atoms. The major exception to this structural pattern is phosphorous acid (H_3PO_3), which is a diprotic acid because one hydrogen atom is bonded directly to the phosphorus atom and is not ionizable.

Some Common Acids

Sulfuric acid, phosphoric acid, and nitric acid are among the ten chemicals manufactured in the largest volume each year in the United States. The major application of each is in the manufacture of agricultural products, principally fertilizers. Sulfuric acid is by far the number-one industrial chemical. It is a versatile chemical with many applications, and over 30 million metric tons are made each year in the United States. The quantity of sulfuric acid produced by a country can be taken as a measure of the size of its economy. Phosphoric acid is usually the second-largest volume industrial acid, and nitric acid the third.

Hydrochloric acid and acetic acid are manufactured in smaller quantities. Their different properties and costs render these two acids useful in ways that differ from those of the "big three" acids. Sulfuric acid, nitric acid, hydrochloric acid, and acetic acid are, however, all important laboratory chemicals.

The properties of each of these common acids as they are supplied in concentrated solutions are summarized in Table 16-4.

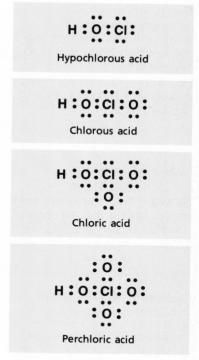

Figure 16-5 Electron-dot formulas of the four oxyacids of chlorine.

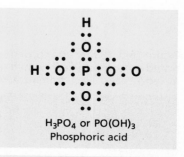

H_3PO_4 or $PO(OH)_3$
Phosphoric acid

H_2SO_4 or $SO_2(OH)_2$
Sulfuric acid

Figure 16-6 Electron-dot formulas of sulfuric acid *(top)* and phosphoric acid.

TABLE 16-4	PROPERTIES OF COMMERCIAL CONCENTRATED ACID SOLUTIONS		
Acid name	Percent by mass (%)	Molarity (mol/L)	Density (g/mL)
Sulfuric acid	98.0%	18	1.84
Phosphoric acid	85.5%	14.8	1.70
Nitric acid	70.4%	15.9	1.42
Hydrochloric acid	37.2%	12.1	1.19
Acetic acid	99.8%	17.4	1.05

The evolution of heat may be so great when sulfuric acid is diluted in water that the solution boils. It is often necessary to cool the solution during the dilution process to avoid spattering and "boiling over."

Figure 16-7 Never add water to an acid. To dilute an acid safely, place a clean, dry glass rod into a container of water and decant the acid along the rod as shown.

Sulfuric acid   In addition to its use in fertilizer manufacture, sulfuric acid is used in large quantities in petroleum refining and metallurgy. Sulfuric acid is also essential to a vast number of industrial processes, including the production of metals, paper, paint, dyes, detergents, and many chemical raw materials. It is the "battery acid" of automobile batteries.

Pure sulfuric acid is a dense, oily liquid that has a high boiling point. Concentrated sulfuric acid contains 98% sulfuric acid by mass combined with water. When concentrated sulfuric acid is diluted with water, a large amount of heat is released during the formation of hydrates, such as $H_2SO_4 \cdot H_2O$ and $H_2SO_4 \cdot 2H_2O$. The safe technique for diluting sulfuric acid is shown in Figure 16-7.

Because of its affinity for water, concentrated sulfuric acid is an effective dehydration (water-removing) agent. It can be used to remove water from gases with which it does not react. Sugar and other organic compounds that contain a 2-to-1 ratio of hydrogen to water are also dehydrated by sulfuric acid. The reaction of sulfuric acid with sugar is quite dramatic, as shown in Figure 16-8. Skin contains organic compounds that are attacked by concentrated sulfuric acid, and it can cause serious burns.

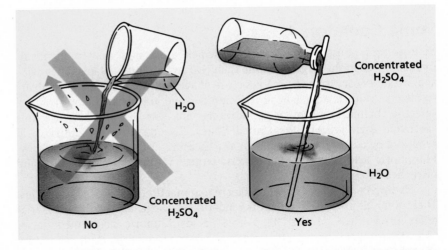

Figure 16-8 Sugar and concentrated sulfuric acid *(left)*, when mixed *(center)* react dramatically.

Phosphoric acid   Pure phosphoric acid is a low-melting solid (mp 42°C). Commercially, phosphoric acid is available as "syrupy phosphoric acid," a viscous liquid that is 85 percent by mass phosphoric acid in water. Dilute phosphoric acid has a pleasant but sour taste and is not toxic. It is used as a flavoring agent in beverages and as a cleaning agent for dairy equipment.

Like nitrogen, phosphorus is an essential element for plants and animals. The bulk of all phosphoric acid produced each year is used directly in impure form for the manufacture of fertilizers and animal feed. Phosphoric acid is also important in the manufacture of detergents, ceramics, and phosphorus-containing chemicals.

Nitric acid   Pure nitric acid is a volatile, unstable liquid rarely used in industry or in laboratories. Concentrated nitric acid contains about 70 percent by mass of the acid dissolved in water and is more stable. While pure nitric acid solutions are colorless, upon standing they gradually become yellow because of slight decomposition to form brown nitrogen–dioxide gas.

$$4HNO_3(aq) \rightarrow 4NO_2(g) + O_2(g) + 2H_2O(\ell)$$

This acid has a suffocating odor and can cause serious burns to the skin. Small amounts of spilled nitric-acid stain the skin yellow.

Nitric acid is used in making explosives (many of which are nitrogen-containing compounds), rubber, chemicals, plastics, dyes, and drugs.

Hydrochloric acid   Dilute and concentrated hydrochloric acids are common laboratory acids and are useful whenever it is necessary to add acid to an aqueous solution. Hydrochloric acid is also produced in the stomach, where secretions contain about 0.4 percent hydrochloric acid that aids in digestion.

Industrially, hydrochloric acid is important for *pickling* of iron and steel, (the immersion of metals in acid solutions to remove surface impurities). It is also used in industry as a general cleaning agent, in food processing, in the activation of oil wells, in recovering magnesium from sea water, and in the production of other chemicals.

Acetic acid   Pure acetic acid is a clear, colorless liquid with a pungent odor at room temperature. It forms crystals in a cold room, since its melting point is 17°C. Acetic acid is available as *glacial* acetic acid, which is 99.8 percent acetic acid by mass. Dilute acetic acid is by mass a 37 percent solution of acetic acid in water. The fermentation of malt, barley, and fruit juices produces acetic acid in vinegars that may retain the flavor of the starting material. Household vinegar contains 4 to 8 percent acetic acid.

Acetic acid is important industrially as a reactant in the synthesis of many chemicals used in making plastics and as a solvent. It is a raw material in the production of food supplements (e.g. lysine, an essential amino acid) and is used as a fungicide.

Section Review

1. What are the five general properties of aqueous acids?
2. Which definition of the three acid definitions is the broadest? Why?
3. *(a)* Why are strong acids also strong electrolytes?   *(b)* Will a strong electrolyte also be a strong acid?
4. Name the following acids: *(a)* HBrO   *(b)* HBrO$_2$ *(c)* HBrO$_3$.

As a reminder, when you are in the laboratory and spill chemicals on your skin, you should rinse the affected area with water immediately and thoroughly. In the case of an acid, if a "tingling" sensation remains, this is your signal to continue rinsing the affected area.

Many proteins (those containing aromatic amino acids) turn yellow in the presence of nitric acid. This explains why small amounts of nitric acid stain skin yellow.

Technical grade, concentrated hydrochloric acid is sold in hardware stores under the name muriatic acid.

1. Sour taste; contain hydrogen, which is given off when certain acids react with metals; change litmus from blue to red, and methyl orange from orange to red; react with bases to produce salts and water; electrolytes.
2. The Lewis definition, because it applies to any substance that can combine with another by accepting an electron pair to form a covalent bond
3. *(a)* Strong acids dissociate 100% in aqueous solutions and thus are strong electrolytes. *(b)* Only if it is an acid to begin with
4. *(a)* Hypobromous acid *(b)* Bromous acid   *(c)* Bromic acid

16.2 Bases and Acid–Base Reactions

SECTION OBJECTIVES

- List five general properties of aqueous bases.
- Define and give an example of a traditional base, a Brønsted base, and a Lewis base.
- Define and recognize traditional, Brønsted, and Lewis acid-base reactions.
- Explain the difference between a strong base and a weak base, and give an example of each.

Figure 16-9 Common household materials that contain bases

Figure 16-10 A base will change the color of an acid–base indicator. Here, the base ammonium hydroxide changes an indicator paper from yellowish to a vivid blue.

General Properties of Aqueous Bases

Several substances that have long been known as bases are commonly found in homes (Figure 16-9). Bases are excellent cleaning agents because they react with fats and oils and convert them to water-soluble substances. Household ammonia is an ammonia–water solution that is very useful for all types of general cleaning. Sodium hydroxide (NaOH), known by the common name *lye*, is present in some commercial products used to unclog sink drains and to clean ovens. Milk of magnesia is a suspension in water of magnesium hydroxide, $Mg(OH)_2$, which is not very water-soluble. It is used as an antacid (a neutralizer of excess hydrochloric acid in the stomach), a laxative, and as an antidote for ingestion of strong acids. Aluminum hydroxide, $Al(OH)_3$, is an antacid in medications such as Maalox and Di-Gel.

All of these common household products are bases of the type identified by Arrhenius in his study of ionization. The distinctive properties associated with such bases are as follows:

1. Bases have a bitter taste. As you may have noticed if you have ever gotten soap into your mouth, solutions of bases have a bitter taste. The stronger basic cleaning agents are caustic—they attack the skin and can cause severe burns. Each year, numerous tragic accidents occur when small children handle drain cleaners and similar substances that have been carelessly stored within easy reach.

2. Dilute aqueous solutions of bases feel slippery to the skin. You encounter this property of aqueous bases whenever you wash with soap.

3. Bases, like acids, change the color of dyes known as acid-base indicators. Aqueous bases cause litmus dye to change color from pink to blue, as shown in Figure 16-10. This is the opposite of the color change caused by aqueous acids. Many dyes are acid-base indicators because they display different colors in acid and base solutions. Phenolphthalein (feen-ol-thal-een) is colorless in the presence of acids and red in the presence of bases. Methyl orange is red in acid solutions and yellow in base solutions.

4. Bases react with acids to produce salts and water. When chemically equivalent quantities of acids and bases react, the previous three properties of bases are "neutralized"—they disappear. The reaction products are salts and water.

5. Bases are electrolytes. Aqueous solutions of bases conduct electricity.

Definitions of Bases and Acid–Base Reactions

For each of the three definitions of acids given in Section 16.1, there is a related definition of a base and an acid–base reaction.

Traditional bases and neutralization reactions   The general properties of bases listed above are based on the Arrhenius definitions of acids and bases. *The* **traditional definition of a base** *is a substance that contains hydroxide ions* (OH^-) *and dissociates to give hydroxide ions in aqueous solution.*

Traditional bases are all ionic compounds containing metal cations (positively-charged ions) and the hydroxide anion (OH^-). Sodium hydroxide (NaOH) and potassium hydroxide (KOH) are common laboratory bases. They are water soluble and dissociate as shown by the following equations:

$$NaOH(s) \rightarrow Na^+(aq) + OH^-(aq)$$
$$KOH(s) \rightarrow K^+(aq) + OH^-(aq)$$

The aqueous solution of a traditional base contains only metal cations and hydroxide ions. All such bases are known as strong bases because they are strong electrolytes and their ions are completely dissociated in solution.

A solution that contains OH^- from a soluble base is referred to as **alkaline**. A highly concentrated solution of a strong base is strongly alkaline. Large amounts of heat are evolved when strong bases are dissolved, and the process must be carried out with care.

The word *alkaline* comes from *alkali*, an old name for substances with the general properties of bases. In Chapter 17 it is explained that aqueous alkaline solutions contain a higher concentration of OH^- than H_3O^+.

Neutralization reactions   A traditional acid-base reaction is the reaction in aqueous solution between an acid that produces H_3O^+ and a hydroxide-producing base. Suppose a solution containing 1 mol of sodium hydroxide is added to a solution containing 1 mol of hydrochloric acid as in Figure 16-11. The formula equation is:

$$HCl(aq) + NaOH(aq) \rightarrow NaCl(aq) + H_2O(\ell)$$

Hydrochloric acid, sodium hydroxide, and sodium chloride are all strong electrolytes. The ionic and net ionic equations are therefore:

$$H_3O^+(aq) + Cl^-(aq) + Na^+(aq) + OH^-(aq) \rightarrow$$
$$Na^+(aq) + Cl^-(aq) + 2H_2O(\ell)$$
$$H_3O^+(aq) + OH^-(aq) \rightarrow 2H_2O(\ell)$$

The Na^+ and Cl^- are spectator ions. The only participants in the reaction are the hydronium ion and the hydroxide ion. Now we can begin to understand why the properties associated with acids and bases disappear when equivalent amounts react with each other. The H_3O^+ and OH^- ions are fully converted to water. *A* **neutralization** *is the reaction of hydronium ions and hydroxide ions to form water molecules.*

The net ionic equation for any reaction between a traditional acid and a traditional base shows only H_3O^+, OH^-, and H_2O. Consider another

STUDY HINT

The formula equations show that neutralization reactions are double-replacement reactions, which are introduced in Chapter 8.

Figure 16-11   When hydrochloric acid reacts with lye (sodium hydroxide), they produce table salt (sodium chloride) and water.

Figure 16-12 The color of a salt depends on the elements that compose it. Pictured here are four different salts, each a different color. The salts are, *clockwise from bottom,* copper nitrate, ammonium chromate, nickel sulfate, and potassium ferricyanide.

Set out several examples of salts, such as NaCl (white), FeCl₃·6H₂O (yellow), and Cu(NO₃)2·3H₂O (green). Remind students that some salts are very poisonous.

example—the reaction of a solution containing one mole of sulfuric acid with a solution containing two moles of potassium hydroxide.

$$H_2SO_4(aq) + 2KOH(aq) \rightarrow K_2SO_4(aq) + 2H_2O(\ell)$$
$$2H_3O^+(aq) + SO_4^{2-}(aq) + 2K^+(aq) + 2OH^-(aq) \rightarrow$$
$$2K^+(aq) + SO_4^{-2}2(aq) + 4H_2O(\ell)$$
$$H_3O^+(aq) + OH^-(aq) \rightarrow 2H_2O(\ell)$$

The potassium and sulfate ions remain in solution. If the solvent water is evaporated, these ions are no longer separated from each other by water molecules. The cations and anions come together in a crystal, and the solid salt, potassium sulfate (K_2SO_4), precipitates as the solvent evaporates. *A* **salt** *is an ionic compound composed of a metal cation of an aqueous base and the anion from an aqueous acid.* Salts vary in solubility, and a slightly soluble salt would precipitate during a neutralization reaction (see Figure 16-12).

Brønsted bases; Brønsted acid-base reactions Earlier, we used the reaction of hydrogen chloride with aqueous ammonia to illustrate the definition of a Brønsted acid. In this reaction, ammonia is a **Brønsted base**—*a molecule or ion that is a proton acceptor:*

$$\underset{\substack{\text{Brønsted} \\ \text{acid}}}{HCl(g)} + \underset{\substack{\text{Brønsted} \\ \text{base}}}{NH_3(aq)} \rightarrow NH_4^+(aq) + Cl^-(aq)$$

The formation of ammonium chloride as a fine solid powder can be seen in the air when HCl and NH₃ mingle as gases (Figure 16-13).

$$HCl(g) + NH_3(g) \rightarrow NH_4Cl(s)$$

Gaseous ammonia is also a Brønsted base in its reaction with water:

$$\underset{\substack{\text{Brønsted} \\ \text{acid}}}{H_2O(\ell)} + \underset{\substack{\text{Brønsted} \\ \text{base}}}{NH_3(g)} \rightarrow NH_4^+(aq) + OH^-(aq)$$

Figure 16-13 When hydrochloric acid is dissolved in an ammonia solution *(top)*, a polar HCl molecule donates a proton to an NH₃ molecule, forming an ammonium ion. NH4+, and a chloride ion, Cl−. HCl and NH₃ react to form the white cloud of ammonium chloride, NH₄Cl.

In this reaction, water is a proton donor and, therefore, is a Brønsted acid. The reactions of HCl and H_2O with NH_3 are representative Brønsted acid-base reactions. In a **Brønsted acid–base reaction**, *protons are transferred from one reactant (the acid) to another (the base)*.

The traditional hydroxide bases such as NaOH are not, strictly speaking, Brønsted bases because, as compounds, they are not proton acceptors. The OH^- ion produced in solution is in each case the Brønsted base.

Lewis bases and Lewis acid–base reactions   *A* **Lewis base** *is an electron-pair donor*. An anion is a Lewis base in any reaction in which it forms a covalent bond by donating an electron pair. The reaction between boron trifluoride and the fluoride anion is a typical Lewis acid–base reaction:

$$BF_3(aq) + :\overset{..}{\underset{..}{F}}:^-(aq) \rightarrow BF_4^-(aq)$$

A **Lewis acid–base reaction** *is the formation of one or more covalent bonds between an electron-pair donor and an electron-pair acceptor*.

Ammonia is a Lewis base in all reactions in which it forms a covalent bond; for example,

$$H^+(aq) + :NH_3(aq) \rightarrow H—NH_3^+(aq), \text{ or } NH_4^+(aq)$$
$$Ag^+(aq) + 2:NH_3(aq) \rightarrow [H_3N—Ag—NH_3]^+(aq) \text{ or } [Ag(NH_3)_2]^+(aq)$$

A comparison of the three acid–base definitions is given in Table 16-5.

Types of Bases

The bases present in aqueous alkaline solutions may be ionic hydroxides or compounds like ammonia that are Brønsted bases because they produce hydroxide ions by accepting protons from water molecules.

The hydroxides of the active metals in Groups 1 and 2 in the Periodic Table are all considered strong bases because they are strong electrolytes. Those that are water soluble can produce strongly alkaline solutions. The alkalinity of aqueous solutions of metal hydroxides depends upon their solubility and the concentration of OH^- ions produced in solution. It is unrelated to the number of hydroxide ions per mole of the undissolved compound.

Like sodium and potassium hydroxides, calcium hydroxide, $Ca(OH)_2$, is an ionic compound and a strong base that dissolves by the dissociation of its ions. Calcium hydroxide is only slightly soluble in water, however. Calcium hydroxide therefore produces a lower concentration of hydroxide ions in aqueous solution than might be expected from its chemical formula.

$$Ca(OH)_2(s) \rightleftharpoons Ca^{2+}(aq) + 2OH^-(aq)$$

The solubilities of a number of metal hydroxides, all traditional bases, are given in Table 16-6. Most are not very soluble, including many hydroxides of the heavy, *d*-block metals, as shown in Figure 16-14 for iron(III) hydroxide.

We have seen that ammonia is a weak electrolyte.

$$NH_3(aq) + H_2O(\ell) \rightleftharpoons NH_4^+(aq) + OH^-(aq)$$

The concentration of OH^- ions in an ammonia solution is relatively low, and ammonia is a *weak base*. The strong, water-soluble bases (sodium hydroxide, potassium hydroxide, and ammonia) are the most common laboratory bases.

TABLE 16-5 ACID-BASE SYSTEMS	
Arrhenius	
Acid	H^+ or H_3O^+ producer
Base	OH^- producer
Brønsted	
Acid	proton (H^+) donor
Base	proton (H^+) acceptor
Lewis	
Acid	electron-pair acceptor
Base	electron-pair donor

TABLE 16-6 SOLUBILITY OF METAL HYDROXIDES

	Hydroxide	Solubility g/100 g H_2O at 20 °C
soluble (>1 g/100 g H_2O)	KOH	112
	NaOH	109
	LiOH	12.4
	$Ba(OH)_2$	3.89
slightly soluble 0.1–1 g/O	$Ca(OH)_2$	0.173
insoluble (<0.1 g/100 g H_2O)	$Pb(OH)_2$	0.016
	$Mg(OH)_2$	0.0009
	$Sn(OH)_2$	0.0002
	$Zn(OH)_2$	negligible
	$Cu(OH)_2$	negligible
	$Al(OH)_3$	negligible
	$Cr(OH)_3$	negligible
	$Fe(OH)_3$	negligible

Many organic compounds that contain nitrogen atoms are also weak bases. For example, aniline, a component of many dyes, is a weak base:

$$C_6H_5NH_2(aq) + H_2O(\ell) \rightleftharpoons C_6H_5NH_3^+(aq) + OH^-(aq)$$

Anions that have a strong attraction for protons are also weak Brønsted bases. Such anions can produce alkaline solutions in the absence of other substances with which they might ordinarily react. For example, the anion of acetic acid, the acetate ion, is a weak base:

$$C_2H_3O_2^-(aq) + H_2O(\ell) \rightarrow HC_2H_3O_3(aq) + OH^-(aq)$$

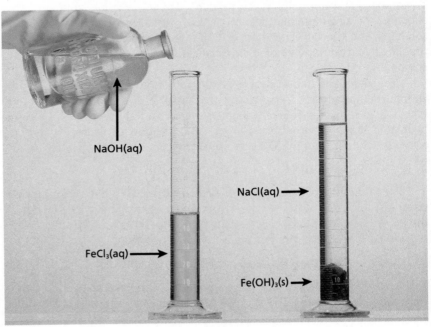

Figure 16-14 The hydroxides of heavy metals are practically insoluble in water as shown in this reaction.

Section Review

1. What are the five general properties of aqueous bases? Name some common substances that fit one or more of these properties.
2. Given the following reactions: $NaOH(s) \rightarrow Na^+(aq) + OH^-(aq)$, $HF + H_2O \rightarrow F^- + H_3O^+$, $H^+ + NH_3 \rightarrow NH_4^+$, which reaction involves a traditional base, a Brønsted base, and a Lewis base? Explain.

16.3 Relative Strengths of Acids and Bases

Any molecule or ion capable of donating a proton is considered to be a Brønsted acid. Any molecule or ion capable of accepting a proton is considered to be a Brønsted base. Thus, Brønsted acids and bases are interconverted by gaining or losing protons.

Brønsted Acid–Base Pairs

The Brønsted definitions of acids and bases provide a basis for the study of proton-transfer reactions. Any molecule or ion capable of *donating* a proton is considered to be an *acid*, while any capable of *accepting* a proton is considered to be a *base*. Suppose that a Brønsted acid gives up a proton. The remainder of the acid is, in theory, capable of adding back that proton. This remaining molecule or ion can act as a base and is known as a conjugate base.

A **conjugate base** *is the species that remains after an acid has given up a proton.* The fluoride ion is the conjugate base of hydrogen fluoride.

$$\underset{\text{acid}}{HF(aq)} + H_2O(\ell) \rightleftharpoons H_3O^+(aq) + \underset{\text{conjugate base}}{F^-(aq)}$$

The acetate ion is the conjugate base of acetic acid.

$$\underset{\text{acid}}{HC_2H_3O_2(aq)} + H_2O(\ell) \rightleftharpoons H_3O^+(aq) + \underset{\text{conjugate base}}{C_2H_3O_2^-(aq)}$$

In these reactions, the water molecule is a Brønsted base that accepts a proton to form H_3O^+, which is an acid. A **conjugate acid** *is the species formed by the addition of a proton to a base.* Thus, the hydronium ion is the conjugate acid of water.

$$HF(aq) + \underset{\text{base}}{H_2O(\ell)} \rightleftharpoons \underset{\text{conjugate acid}}{H_3O^+(aq)} + F^-(aq)$$

In every Brønsted acid-base reaction, therefore, there are two acid-base pairs, known as conjugate acid-base pairs.

$$\underset{\text{acid}}{HF(aq)} + \underset{\text{base}}{H_2O(\ell)} \rightleftharpoons \underset{\text{acid}}{H_3O^+(aq)} + \underset{\text{base}}{F^-(aq)}$$

The conjugate acid–base pairs are *(1)* HF and F^- and *(2)* H_2O and H_3O^+. In every conjugate acid–base pair, the acid has one more proton than its conjugate base.

1. Bitter taste; feel slippery to the skin; change litmus from pink to blue, phenolphtalein from colorless to red, and methyl orange from orange to yellow; react with acids to produce salts and water; electrolytes
2. The NaOH is the traditional base because it yields OH^- ions in a water solution. H_2O is the Brønsted base because it is the proton acceptor, and NH_3 is the Lewis acid because it is the electron-pair donor.

SECTION OBJECTIVES

- Define conjugate acid, conjugate base, and conjugate acid-base pair.
- Write the formula for the conjugate acid of a base and for the conjugate base of an acid.
- Explain why the conjugate base of a strong acid is a weak base and why the conjugate acid of a strong base is a weak acid.
- Explain why proton-transfer reactions favor the production of the weaker acid and the weaker base.
- Define an amphoteric substance, and give an example.

Point out to students that they can recognize conjugate acid-base pairs because one member of the pair has either one more or one less H^+ in its formula than has the other member. Remind them that H^+ is a proton.

Desktop Investigation

Acids and Bases in Your Kitchen

Question
Which substances in the kitchen are acids and which are bases?

Materials
fresh red cabbage
tap water
foods and beverages, such as mayonnaise, baking powder, baking soda, white vinegar, cider vinegar, lemon juice, soft drinks, mineral water, and milk
cleaning products, such as dishwashing liquid, dishwasher detergent, laundry detergent, laundry stain remover, fabric softener, and bleach
blender
strainer
small containers
teaspoon
small beakers

Procedure
Remember to wear safety glasses and an apron. Record all your results in a data table.

1. To make an inexpensive and colorful indicator for acid/base detection, make

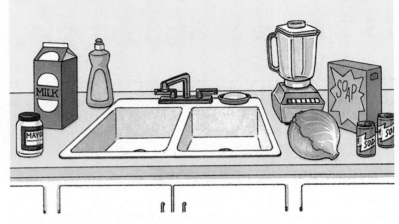

red-cabbage juice. To do this, drop wedges of red cabbage into a blender with a cup of water. Process. Remove the cabbage from the water by straining. Save the water.

2. Assemble foods, beverages, and cleaning products to be tested.

3. If the substance being tested is a liquid, pour about 5 mL into a small container. If it is a solid, place a teaspoonful into a small beaker, moisten it with about 5 mL of water, wait a few minutes for some of the sample to dissolve, and then pour only

the liquid into a small container.

4. Add a drop or two of the red-cabbage juice to the solution being tested and note the color. The solution will turn red if it is a strong acid and green if it is a strong base.

Questions
1. Are cleaning products acids, bases, or neutral?

2. What are acid/base characteristics of foods and beverages?

3. Did you find consumer warning labels on bases or on acidic products?

On the other hand, hydrochloric acid is completely ionized. The HCl molecule gives up protons very readily and is therefore a strong acid. It follows that the Cl^- ion has little tendency to attract and retain a proton. Consequently, the Cl^- ion is a very weak base.

$$HCl(g) + H_2O(\ell) \rightarrow H_3O^+(aq) + Cl^-(aq)$$
$$\text{acid} \qquad \text{base} \qquad \text{acid} \qquad \text{weak base}$$

This observation leads to an important conclusion that follows naturally from the Brønsted definitions of acids and bases: *the stronger an acid, the weaker its conjugate base; the stronger a base, the weaker its conjugate acid.*

Relative Strengths of Acids and Bases in Chemical Reactions

Let's compare perchloric acid ($HClO_4$) and acetic acid ($HC_2H_3O_2$). Perchloric acid is a stronger acid than hydrochloric acid and is highly ionized. Even in concentrated solutions, the reaction to the right is virtually complete:

$$HClO_4(aq) + H_2O(\ell) \rightarrow H_3O^+(aq) + ClO_4^-(aq)$$

stronger acid   stronger base   weaker acid   weaker base

The ClO_4^- ion is the conjugate base of $HClO_4$. It is too weak a base to compete successfully with H_2O as a base in acquiring a proton. The H_3O^+ ion is the conjugate acid of H_2O. It is too weak an acid to compete successfully in donating a proton with the $HClO_4$. There is little tendency, therefore, for the reaction to proceed to the left to recombine the $HClO_4$ and H_2O molecules.

In a comparable solution of acetic acid,

$$HC_2H_3O_2(aq) + H_2O(\ell) \rightleftarrows H_3O^+(aq) + C_2H_3O_2^-(aq)$$

weaker acid   weaker base   stronger acid   stronger base

the H_3O^+ ion concentration is much lower: acetic acid is a weak acid. The $HC_2H_3O_2$ molecule does not compete very successfully with H_3O^+ ions in donating protons to a base. The H_2O molecules do not compete successfully with $C_2H_3O_2^-$ ions in accepting protons. The H_3O^+ ion is the stronger acid, and the $C_2H_3O_2^-$ ion is the stronger base. Thus, the reaction is more favorable from *right to left* as written above.

Note that in each reaction, the favored direction is toward the weaker acid and weaker base. This obervation leads to a second important conclusion based upon the Brønsted definitions: *proton-transfer reactions favor the production of the weaker acid and the weaker base.*

When a Brønsted acid and a Brønsted base are brought together in aqueous solution, a proton-transfer reaction can occur. The extent of the reaction depends upon the relative strengths of the acids and bases involved. For a proton-transfer reaction to approach completion, the reactants must be much stronger as an acid and a base than the products.

By comparing many different acids and bases, a table of relative acid and base strengths like Table 16-7 can be assembled. Note that the strongest acid listed ($HClO_4$) has the weakest conjugate base, ClO_4^-. The strongest base, the hydride ion (H^-), has the weakest conjugate acid (H_2). A violent proton-transfer reaction could result from bringing together the strongest acid and the strongest base in certain proportions. Such a reaction would be highly exothermic and dangerous. In fact, the reaction between hydride ion and water, a much weaker acid than perchloric acid, is quite vigorous, as shown in Figure 16-15.

To help students determine which direction will be favored in a reaction, they may follow three steps:
1) Identify all compounds in the reaction as either acids or bases.
2) Using Table 16-10, determine which of the two acids is the stronger and which of the two bases is the stronger.
3) The side that is favored is the one having the weaker acid and the weaker base. The weaker acid and the weaker base will always be on the same side.

TABLE 16-7   RELATIVE STRENGTHS OF ACIDS AND BASES

		Conjugate Acid	Formula	Conjugate Base	Formula		
Increasing Acid Strength ↑	Decreasing Acid Strength ↓	perchloric	$HClO_4$	perchlorate ion	ClO_4^-	Decreasing Base Strength ↑	Increasing Base Strength ↓
		hydrogen iodide	HI	iodide ion	I^-		
		hydrogen bromide	HB_r	bromide ion	Br^-		
		hydrogen chloride	HCl	chloride ion	Cl^-		
		nitric	HNO_3	nitrate ion	NO_3^-		
		sulfuric	H_2SO_4	hydrogen sulfate ion	HSO_4^-		
		hydronium ion	H_3O^+	water	H_2O		
		hydrogen sulfate ion	HSO_4^-	sulfate ion	SO_4^{2-}		
		phosphoric	H_3PO_4	dihydrogen phosphate ion	$H_2PO_4^-$		
		hydrogen fluoride	HF	fluoride ion	F^-		
		acetic	$HC_2H_3O_2$	acetate ion	$C_2H_3O_2^-$		
		carbonic	H_2CO_3	hydrogen carbonate ion	HCO_3^-		
		hydrogen sulfide	H_2S	hydrosulfide ion	HS^-		
		ammonium ion	NH_4^+	ammonia	NH_3		
		hydrogen carbonate ion	HCO_3^-	carbonate ion	CO_3^{2-}		
		water	H_2O	hydroxide ion	OH^-		
		ammonia	NH_3	amide ion	NH_2^-		
		hydrogen	H_2	hydride ion	H^-		

Note in the table that water and hydrogen sulfate ion (HSO_4^-) can be either acids or bases. For example, in the first ionization of sulfuric acid,

$$H_2SO_4(aq) + H_2O(\ell) \rightarrow HSO_4^-(aq) + H_3O^+(aq)$$

acid	base	base	acid

the HSO_4^- ion is a base—the conjugate base of H_2SO_4. It could accept a proton from H_3O^+ in the reverse reaction; however, HSO_4^- is such a weak acid (Table 16-7) that this reaction occurs only very slightly. In a dilute

Figure 16-15 Calcium hydride reacts violently with water, releasing hydrogen gas.

sulfuric acid solution such as the one described below, the HSO_4^- ion ionizes further, yielding the sulfate ion and the hydronium ion shown in the following reaction.

$$HSO_4^-(aq) + H_2O(\ell) \rightleftarrows SO_4^{2-}(aq) + H_3O^+(aq)$$
$$\text{acid} \qquad \text{base} \qquad \text{base} \qquad \text{acid}$$

Here the HSO_4^- ion is a Brønsted acid. *Any species that can react as either an acid or a base is described as* **amphoteric**.

Section Review

1. The following reaction takes place:

$$H_2CO_3 + H_2O \rightleftarrows HCO_3^- + H_3O^+$$

   (a) Label the conjugate acid–base pairs.   *(b)* Label reactants and products shown in the reaction above as *proton donor or proton acceptor* and either acidic or basic.
2. Label reactants and products in the reaction described in the previous question as either stronger or weaker acids or bases.   *(b)* Determine which direction, forward or reverse, is favored in the reaction shown in the previous question.

16.4 Oxides, Hydroxides, and Acids

An oxide is a binary compound of oxygen and one other element. Oxides are known for almost all of the elements. The bonding in these compounds extends over the entire range from ionic to covalent. The relationships among oxides, hydroxides, and acids are discussed in this section.

Basic and Acidic Oxides

Oxides of the active metals are ionic compounds that contain oxide ions, O^{2-}. The oxide ion from any soluble oxide reacts immediately with water to form hydroxide ions.

$$O^{2-}(aq) + H_2O(\ell) \rightarrow 2OH^-(aq)$$

The active metals of Periodic Table Groups 1 and 2 form ionic oxides that react vigorously with water. A large amount of heat is released, and the reaction product is a metal hydroxide. For example,

$$Na_2O(s) + H_2O(\ ) \rightarrow 2NaOH(aq)$$

To the extent that the hydroxide formed is water soluble, the result is the production of an alkaline solution.

An oxide that reacts with water to form an alkaline solution is referred to as a basic oxide, or a **basic anhydride**. The term anhydride means "without water." Each basic oxide can be pictured as being formed by the removal of water from a hydroxide base, which is the reverse of the reactions of the oxides with water shown above. Oxides of the less active metals, such as magnesium, can be prepared by thermal decomposition to drive water off from their hydroxides.

$$Mg(OH)_2(s) \xrightarrow{\Delta} MgO(g) + H_2O(g)$$

Hydroxides of the very active metals of Group 1 are so stable that they do not decompose in this manner.

Nonmetals to the right in the Periodic Table form molecular oxides. For example, sulfur forms two oxides that are gases:

$$SO_2 \qquad SO_3$$
sulfur dioxide   sulfur trioxide

In reactions typical of nonmetal oxides, each of the sulfur oxides reacts with water to form an oxyacid.

$$SO_2(g) + H_2O(\ell) \rightarrow H_2SO_3(aq)$$
sulfurous acid
$$SO_3(g) + H_2O(\ell) \rightarrow H_2SO_4(aq)$$
sulfuric acid

An oxide that reacts with water to form an acid is known as an acidic oxide, or an **acid anhydride**. As with the basic anhydrides, each acid anhydride can be pictured as being formed by the removal of water from the appropriate oxyacid, which is the reverse of the reactions of the oxides with water shown above. Sulfuric acid, for example, decomposes in this manner when heated.

$$H_2SO_4(aq) \xrightarrow{\Delta} H_2O(\ell) + SO_3(g)$$

Amphoteric Oxides and Hydroxides

Table 16-8 lists in Periodic-Table form some common oxides of main group elements. You can see that the active metal oxides are basic and that the nonmetal oxides are acidic. Between these lies a group of oxides, the amphoteric oxides, with intermediate properties. The bonding in amphoteric oxides is intermediate between ionic and covalent bonding. As a result, oxides of this type have behavior intermediate between that of acidic oxides and basic oxides.

Aluminum oxide (Al_2O_3) is a typical amphoteric oxide. It is only slightly soluble in water and dissolves in both acidic and alkaline solutions. With hydrochloric acid, aluminum oxide forms a salt and water.

$$Al_2O_3(s) + 6HCl(aq) \rightarrow 2AlCl_3(aq) + 3H_2O(\ell)$$

With aqueous sodium hydroxide, aluminum oxide forms a soluble ionic compound that contains aluminate ions, AlO_2^-. We will use the AlO_2^- formula rather than the more precise hydrated aluminate formula, $Al(OH)_4^-$.

$$Al_2O_3(s) + 2NaOH(aq) \rightarrow 2NaAlO_2(aq) + H_2O(\ell)$$

Next, consider aluminum hydroxide. In solution, the Al^{3+} ion is surrounded by six water molecules, $[Al(H_2O)_6]^{3+}$. As aqueous base is added to an Al^{3+} solution, the strongly basic OH^- ions attract protons from three of the water molecules of the hydrated aluminum ion. Aluminum hydroxide

When industries burn coal that contains sulfur, sulfur dioxide is often released into the atmosphere. Some of the sulfur dioxide is further converted into sulfur trioxide in the atmosphere. These two sulfur oxides then combine with rain and other forms of precipitation to form sulfurous and sulfuric acid, which contribute to acid precipitation.

Acidic oxides react with bases. Basic oxides react with acids.

Chemistry Notebook

"Acid Stomach" and Antacids

Digestion begins in the stomach, where gastric juices (hydrochloric acid and enzymes) split the food molecules. The stomach itself is not digested because it is lined with a protective mucus coating. The esophagus, however, is not protected by mucus. When the contents of the stomach splash up into it, you feel the sensation commonly called "heartburn." Sustained anxiety can cause hypersecretion of gastric juices, sometimes leading to peptic ulcers.

To relieve heartburn, people often take antacids, which contain bases. Some antacids produce carbon dioxide as they react with stomach acid:

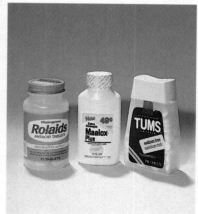

$$NaHCO_3 + HCl \rightarrow$$
$$NaCl + H_2O + CO_2$$
$$MgCO_3 + 2HCl \rightarrow$$
$$MgCl_2 + H_2O + CO_2$$

The more effective antacids contain hydroxides. Although insoluble, they neutralize acid on contact:

$$Mg(OH)_2 + 2HCl \rightarrow$$
$$MgCl_2 + 2H_2O$$
$$Al(OH)_3 + 3HCl \rightarrow AlCl_3 + H_2O$$

They have minor side effects that are often desirable:

magnesium hydroxide is a laxative, and aluminum hydroxide helps alleviate diarrhea. Over-the-counter heartburn remedies are often mixtures of hydroxides, to balance the side effects.

separates as a white jelly-like precipitate that includes the remaining water of hydration.

$$[Al(H_2O)_6]^{3+}(aq) + 3OH^-(aq) \rightarrow [Al(H_2O)_3(OH)_3](s) + 3H_2O(\ell)$$

If the addition of base continues, the precipitate dissolves with the formation of hydrated aluminate ions.

$$[Al(H_2O)_3(OH)_3](s) + OH^-(aq) \rightarrow [Al(H_2O)_3(OH)_4]^-(aq)$$

Writing this equation without the water, as is customary,

$$Al(OH)_3(s) + OH^-(aq) \rightarrow [Al(OH)_4]^-(aq)$$

shows clearly that aluminum hydroxide is soluble in aqueous base. In this reaction, aluminum hydroxide acts as an acid in the presence of a strong base.

Aluminum hydroxide is also soluble in aqueous acid. Written with and without the water of hydration, the equations are:

$$Al(H_2O)_3(OH)_3(s) + 3H_3O^+(aq) \rightarrow Al(H_2O)_6^{3+}(aq) + 3H_2O(\ell)$$
$$Al(OH)_3(s) + 3H_3O^+(aq) \rightarrow Al^{3+}(aq) + 6H_2O(\ell)$$

The three hydroxide ions in the aluminum hydroxide accept three protons

Water of hydration is usually omitted in net ionic equations.

from hydronium ions to form three water molecules. In this reaction, hydrated aluminum hydroxide reacts as a base in the presence of a strong acid.

Compare the equations for the reactions of $Al(OH)_3$ with acid and base with those above for the reactions of aluminum oxide with acid and base. These equations show that aluminum hydroxide, like aluminum oxide, is an amphoteric compound. In the presence of a strong acid, it acts as a base; in the presence of a strong base, it acts as an acid. Each of the elements that form an amphoteric oxide (Table 16-8) also forms an amphoteric hydroxide.

TABLE 16-8   PERIODICITY OF ACIDIC AND BASIC OXIDES OF MAIN GROUP ELEMENTS

Group Number						
1	2	13	14	15	16	17
Li_2O basic	BeO amphoteric	B_2O_3 acidic	CO_2 acidic	N_2O_5 acidic		
Na_2O basic	MgO basic	Al_2O_3 amphoteric	SiO_2 acidic	P_4O_{10} acidic	SO_3 acidic	Cl_2O acidic
K_3O basic	CaO basic	Ga_2O_3 amphoteric	GeO_2 amphoteric	As_4O_6 amphoteric	SeO_3 acidic	
Rb_7O basic	SrO basic	In_2O_3 basic	SnO_2 amphoteric	Sb_4O_6 amphoteric	TeO_3 acidic	I_2O_5 acidic
Ca_2O basic	BaO basic	Tl_2O_3 basic	PbO_2 amphoteric	Bi_2O_3 basic		

Hydroxides, Acids, and Periodic Trends

Compounds that contain OH groups may be acidic, amphoteric, or basic. When the bond to the oxygen atom in the OH group is ionic, as in the oxides and hydroxides of the active metals, the compound is basic. The metal atoms that form basic hydroxides have large radii, low ionization energies, and low electronegativity. Electrons are easily attracted away from these metal atoms to form cations.

What determines whether a molecular compound containing OH groups will be acidic or amphoteric? For the compound to be acidic, it must be possible for a water molecule to attract away a hydrogen atom from an OH group. This occurs more easily when the O—H bond is very polar.

$$\overset{\delta^-}{O}\text{—}\overset{\delta^+}{H} \qquad \overset{\delta^-}{:\overset{..}{O}}\underset{\diagdown H}{\overset{\diagup \,H\,\delta^+}{}}$$

Any feature of a molecule that increases the polarity of the O—H bond increases the acidity of a molecular compound. The small, more highly electronegative atoms of nonmetals at the upper right in the Periodic Table form compounds with acidic hydroxyl groups, like the acids shown in Figures 16-5 and 16-6. These nonmetals do not form amphoteric hydroxides. Atoms of intermediate electronegativity and size tend to form amphoteric oxides and hydroxides. Notice the location of the amphoteric oxides in Table 16-8.

The acidic or basic behavior of a compound is also influenced by the number of other oxygen atoms bonded to the atom connected to the OH group. An increase in the number of such oxygen atoms is likely to cause the compound to be more acidic. The electronegative oxygen atoms draw electron density away from the O—H bond and make it more polar. In other words, the higher the oxidation number of the central atom, the more acidic the compound. It may not always be apparent from the formula or structure of a compound whether it has acidic, amphoteric, or basic properties. For example, chromium forms three different compounds containing OH groups:

basic	*amphoteric*	*acidic*
$Cr(OH)_2$	$Cr(OH)_3$	H_2CrO_4
chromium(II) hydroxide	chromium(III) hydroxide	chromic acid

Consider also the compounds shown in Figure 16-16. Acetic acid is a weak acid. Notice that in acetic acid, a second oxygen atom is bonded to the same carbon atom connected to the OH group. Ethanol is neither an acid nor a base in aqueous solution.

Section Review

1. Using the following equation
    $$CaO + H_2O \rightarrow Ca^{2+} + OH_2^-$$
    determine whether CaO is a basic or acidic anhydride.
2. Using the following equation
    $$CO_2 + 2H_2O \rightarrow HCO_3^- + H_3O^+$$
    determine whether CO_2 is a basic or acidic anhydride.
3. Explain why Group 1 metals tend to form basic anhydrides.
4. Explain how the presence of oxygen atoms in a compound may make the compound more acidic.

16.5 Chemical Reactions of Acids, Bases, and Oxides

Compounds with acidic properties and compounds with basic properties often react with each other. The acid-base reactions of traditional, Brønsted, and Lewis acids and bases were illustrated in Section 16.2.

In this section we list some other common reactions of traditional acids and bases as well as oxides. Even though several of these reactions were included in Chapter 8, it is worthwhile to review them here and recognize the similarities among them. One major similarity is that, in each type of reaction listed, one of the products is a salt.

1. Acids react with many metals. Solutions of acids, such as hydrochloric acid and dilute sulfuric acid, react with metals above hydrogen in the metal activity series. (See Table 8.3.) Single-replacement reactions occur

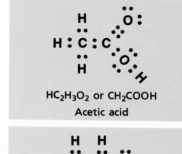

Figure 16-16 Acetic acid *(top)* and ethanol *(bottom)*

1. A basic anhydride, because OH⁻ ions are produced
2. An acidic anhydride, because HO⁺ is produced
3. The active metals of Group 1 form ionic oxides that react immediately with water to form metallic hydroxide. An oxide that reacts with water to form a solution containing hydroxide ions is referred to as a basic anhydride.
4. The acidic nature of a compound is influenced by the number of other oxygen atoms connected to the OH group. An increase in the number of oxygen atoms in a compound causes it to be more acidic because the electronegative oxygen atoms pull the electron density away from the O—H

SECTION OBJECTIVES

- Describe four types of reactions involving aqueous acids.
- Describe two types of reactions involving aqueous bases.
- Describe three common types of reactions involving oxides.
- Predict the products of reactions involving aqueous acids, aqueous bases, and oxides.
- List eight types of reactions in which salts can be produced.

Figure 16-17 When concentrated nitric acid oxidizes metallic copper, it produces the reddish-brown gas nitrogen dioxide and a green solution of copper nitrate *(left)*. Zinc replaces hydrogen in sulfuric acid *(right)*. As can be seen by the formation of hydrogen gas bubbles.

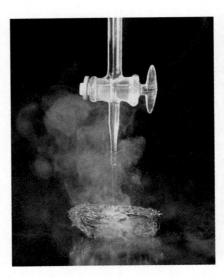 

with salts and hydrogen gas being produced. The single-replacement reaction between zinc and dilute sulfuric acid is typical.

$$Zn(s) + H_2SO_4(aq) \rightarrow ZnSO_4(aq) + H_2(g)$$

Hydrochloric acid and dilute sulfuric acid are nonoxidizing acids. Their reactions are those expected of the hydronium ion acting alone. The distinction between nonoxidizing and oxidizing acids is necessary because oxidizing acids react differently. Rather than just providing hydronium ions that react, the entire acid molecule is likely to be involved in the reactions of oxidizing acids. For example, when hot concentrated sulfuric acid (an oxidizing acid) reacts with zinc, some of the sulfuric acid is converted to hydrogen sulfide (H_2S). The other products are zinc sulfate and water.

$$4Zn(s) + 5H_2SO_4(conc.) \rightarrow 4ZnSO_4(aq) + H_2S(g) + 4H_2O(\ell)$$

Notice that hot, concentrated sulfuric acid is an oxidizing acid and that dilute sulfuric acid is *not* an oxidizing acid.

The most common oxidizing acids are hot concentrated sulfuric acid, nitric acid, and the chlorine-containing oxyacids. Nitric acid produces several different products depending upon the acid concentration and the reactants. In both of the following reactions, nitric acid is an oxidizing acid. With dilute nitric acid, the nitrogen-containing product is nitrogen monoxide (NO).

$$3Cu(s) + 8HNO_3(aq) \rightarrow 3Cu(NO_3)_2(aq) + 2NO(g) + 4H_2O(\ell)$$

With concentrated nitric acid, the nitrogen-containing product is nitrogen dioxide (NO_2). (See Figure 16-17.)

$$Cu(s) + 4HNO_3(conc.) \rightarrow Cu(NO_3)_2(aq) + 2NO_2(g) + 2H_2O(\ell)$$

Solutions of oxidizing acids react with metals both above and below hydrogen in the activity series.

2. Acids react with metal oxides. Metal oxides are either basic or amphoteric. In the reaction between an acid and a metal oxide, the metal oxide acts as a basic substance. The products are a salt and water—the same as those in a traditional acid-base reaction. For example, copper oxide reacts with dilute sulfuric acid to produce copper sulfate and water.

$$CuO(s) + H_2SO_4(dil.) \rightarrow CuSO_4(aq) + H_2O(\ell)$$

Figure 16-18 When a piece of chalk, made primarily of calcium carbonate, is placed into a container of hydrochloric acid, it produces the salt calcium chloride, water, and carbon dioxide. The bubbles are carbon dioxide. The yellow color results from a dye in the chalk.

You may have recognized that this reaction follows the pattern of a double-replacement reaction (Chapter 8).

3. Acids react with carbonates. The products of the reaction of an acid with a carbonate are a salt, water, and carbon dioxide, which bubbles out of the solution as in Figure 16-18. With calcium carbonate the reaction is:

$$CaCO_3(s) + HCl(aq) \rightarrow CaCl_2(aq) + CO_2(g) + H_2O(\ell)$$

This reaction provides a simple test for carbonates. A sea shell or other material placed in an acid solution gives the fizzy production of CO_2 if a carbonate is present. The reaction of acids with carbonates plays a role in the deterioration of marble buildings and statues, as shown in Figure 16-19.

4. Hydroxides react with nonmetal oxides. You have seen that nonmetal oxides are acid anhydrides. The reaction of a hydroxide base with a nonmetal oxide is another acid-base reaction. The product is a salt or a salt and water, depending upon the relative quantities of reactants and their identity. For example, two moles of sodium hydroxide and one mole of carbon dioxide form sodium carbonate and water.

$$CO_2(g) + 2NaOH(aq) \rightarrow Na_2CO_3(aq) + H_2O(\ell)$$

One mole of sodium hydroxide and one mole of carbon dioxide produce sodium hydrogen carbonate.

$$CO_2(g) + NaOH(aq) \rightarrow NaHCO_3(aq)$$

Similar reactions between carbon dioxide and calcium hydroxide (Figure 16-20) are as follows:

$$CO_2(g) + Ca(OH)_2(aq) \rightarrow CaCO_3(s) + H_2O(\ell)$$
$$2CO_2(g) + Ca(OH)_2(aq) \rightarrow Ca(HCO_3)_2(aq)$$

5. Metal oxides react with nonmetal oxides. An oxygen-containing salt can be produced by the reaction between a basic metal oxide and an acidic nonmetal oxide. Water is not involved. Instead, the dry oxides are mixed and heated. Metal carbonates, phosphates, and sulfates are typical of salts that can be produced in this manner.

$$MgO(s) + CO_2(g) \rightarrow MgCO_3(s)$$
$$6CaO(s) + P_4O_{10}(s) \rightarrow 2Ca_3(PO_4)_2(s)$$
$$CaO(s) + SO_3(g) \rightarrow CaSO_4(s)$$

Figure 16-19 Acid precipitation has eroded the features of this marble statue.

Figure 16-20 Limewater, left, is a clear aqueous solution of calcium hydroxide, Ca(OH)₂(aq). When carbon dioxide is added *(center)*, CaCO₃(s) forms, making the water cloudy.

1. (a) Metal + acid → salt + hydrogen
   (b) Metallic oxide + acid → salt + water
   (c) (Metallic) carbonate + acid → salt + carbon dioxide + water
   (d) Nonmetal oxide + hydroxide → salt + water
   (e) Metallic oxide + nonmetallic oxide → salt

2. (a) $Ca(s) + 2HNO_3(aq) \rightarrow Ca(NO_3)_2(aq) + H_2(g)$
   (b) $2CO_2(g) + Ba(OH)_2(aq) \rightarrow Ba(HCO_3)_2(aq)$
   (c) $MgCO_3(s) + 2HCl(aq) \rightarrow MgCl_2(aq) + CO_2(g) + H_2O(\ell)$
   (d) $CaO(s) + H_2SO_4(dil.) \rightarrow CaSO_4(aq) + H_2O(\ell)$
   (e) $BaO(s) + SO_3(g) \rightarrow BaSO_4(s)$
   (f) $CO_2(g) + Sr(OH)_2(aq) \rightarrow SrCO_3(s) + H_2O(\ell)$

Section Review

1. Classify the following reactions as one of the five types of reactions discussed in this section. Identify the components of each reaction. *(a)* $Zn(s) + H_2SO_4(conc.) \rightarrow ZnSO_4(aq) + H_2(g)$ *(b)* $CuO(s) + H_2SO_4(dil) \rightarrow CuSO_4(aq) + H_2O(\ell)$ *(c)* $Na_2CO_3(s) + 2HCl(aq) \rightarrow 2NaCl(aq) + H_2O(\ell) + CO_2(g)$ *(d)* $CO_2(g) + Mg(OH)_2(aq) \rightarrow MgCO_3(s) + H_2O(\ell)$ *(e)* $CaO(s) + SO_3(g) \rightarrow CaSO_4(s)$.

2. Complete and balance each of the following: *(a)* $Ca(s) + HNO_3(aq) \rightarrow$ *(b)* $2CO_2(g) + Ba(OH)_2(aq) \rightarrow$ *(c)* $MgCO_3(s) + HCl(aq) \rightarrow$ *(d)* $CaO(s) + H_2SO_4(dil) \rightarrow$ *(e)* $BaO(s) + SO_3(g) \rightarrow$ *(f)* $CO_2(g) + Sr(OH)_2(aq) \rightarrow$.

Chapter Summary

- In general, acids have a sour taste, contain hydrogen, and react with active metals. They change the color of acid–base indicators, and react with bases to produce salts and water and are electrolytes.
- By definition, a traditional acid contains hydrogen and ionizes in aqueous solution to form hydrogen ions, a Brønsted acid is a proton donor and a Lewis acid is an electron-pair acceptor.
- Acids are described as monoprotic, diprotic, or triprotic in accordance with whether they dontate one, two, or three protons per molecule, respectively, in aqueous solutions.
- In general, bases have a bitter taste, feel slippery to the skin in dilute aqueous solutions, change the color of dyes known as acid–base indicators, react with acids to produce salts and water, and are electrolytes.
- By definition, a traditional base contains hydroxide (OH^-) ions and dissociates to give hydroxide ions in aqueous solution. A Brønsted base is a proton acceptor and a Lewis base is an electron-pair donor.
- The strength of an acid or a base is determined by the extent to which it dissociates in aqueous solutions.
- In every Brønsted acid–base reaction, there are two acid-base pairs known as conjugate acid–base pairs.
- A strong acid has a weak conjugate base; a strong base has a weak conjugate acid.
- Proton-transfer reactions favor the production of weaker acids and bases.
- Basic and acidic anhydrides react with water to give alkaline and acidic solutions, respectively.
- Metals that are highly reactive form metal oxides that are basic. Nonmetal oxides are acidic. Oxides of elements between the active metals and nonmetals in the Periodic Table are amphoteric.
- The acidic or basic behavior of a molecule containing OH groups may depend on the electronegativity of other atoms in the molecule, and the number of oxygen atoms bonded to the atom connected to the OH group.
- In addition to the acid–base reactions of traditional Brønsted and Lewis acids and bases, acids and bases and oxides can react in the following five ways: *(1)* acids react with metals *(2)* acids react with metal oxides *(3)* acids react with carbonates *(4)* hydroxides react with non-metal oxides *(5)* metal oxides react with nonmetal oxides.

Chapter 16 *Review*

Aqueous Acids and Bases

Vocabulary

acid, Brønsted
acid, Lewis
acid, traditional
acid anhydride
acid–base reaction, Brønsted
acid–base reaction, Lewis
alkaline
amphoteric
base, Brønsted
base, Lewis
base, traditional
basic anhydride
binary acid

conjugate acid
conjugate base
diprotic acid
hydroxyl group
mineral acids
monoprotic acid
neutralization
organic acid
oxyacid
polyprotic acid
salt
strong acid
triprotic acid
weak acid

Questions

1. List five common foods, each containing a different acid and name the acid in each.
2. *(a)* What is the traditional definition of an acid?   *(b)* Describe a typical acid according to this definition.
3. *(a)* What is the Brønsted definition of an acid?   *(b)* How does it differ from the traditional definition?
4. *(a)* What is the Lewis definition of an acid? *(b)* How does this definition compare with the traditional and Brønsted definitions?
5. *(a)* What distinguishes strong acids from weak acids?   *(b)* Give two examples of each.
6. Distinguish among and give one example of a monoprotic, a diprotic, and a triprotic acid.
7. *(a)* Distinguish between organic and mineral acids.   *(b)* Give one example of each.
8. *(a)* Distinguish between binary acids and oxyacids in terms of their component elements and their naming systems.   *(b)* Give three examples of each.
9. Name each of the following acids:
   (a) HCl   *(b)* HI   *(c)* HNO_3 *(d)* H_2SO_4
   (e) $HClO_3$ *(f)* HF   *(g)* HNO_2 *(h)* $HClO_4$
   (i) H_3PO_4 *(j)* $HBrO_3$ *(k)* HClO   *(l)* H_2CO_3.
10. Write formulas for each of the following

acids: *(a)* hydrofluoric acid   *(b)* nitric acid   *(c)* hydriodic acid   *(d)* chloric acid   *(e)* hydrobromic acid   *(f)* phosphoric acid   *(g)* hypochlorous acid   *(h)* nitrous acid   *(j)* phosphorous acid.
11. Distinguish between hydroxyl groups and hydroxide ions.
12. Identify and describe the characteristic properties of the five industrial acids produced in the largest quantity in the United States and cite some of the typical uses of each.
13. Write the names and formulas of four common aqueous bases found in the home and cite at least one use for each.
14. *(a)* Name three substances that can be used as acid–base indicators.   *(b)* Explain the unique characteristics of each.
15. *(a)* What is a traditional base?   *(b)* Describe a typical base according to this definition.
16. Define *(a)* alkaline solution   *(b)* salt.
17. *(a)* What is a neutralization reaction?   *(b)* Write the net ionic equation that results .
18. *(a)* What is a Brønsted base?   *(b)* What occurs in a typical Brønsted acid–base reaction?
19. *(a)* What is a Lewis base?   *(b)* What is a Lewis acid–base reaction?
20. *(a)* What determines the strength of a base?   *(b)* Give one example each of solutions that are strongly, moderately, and weakly basic.
21. Define and give an equation to illustrate:
   (a) a conjugate base   *(b)* a conjugate acid.
22. What is the relationship between
   (a) the strength of an acid and that of its conjugate base   *(b)* the strength of a base and that of its conjugate acid?
23. *(a)* What trend can be observed concerning the favored direction of proton-transfer reactions *(b)* What determines the extent to which a proton-transfer reaction occurs?
24. *(a)* What is meant by the term amphoteric? *(b)* Give an example of a substance with amphoteric characteristics.
25. *(a)* Distinguish between basic and acid anhydrides.   *(b)* In terms of their chemical

makeup, how do they differ?

26. (a) In general, which Periodic-Table elements form acidic, basic, and amphoteric oxides? (b) Describe some of the properties of a typical amphoteric oxide.

27. What properties generally indicate whether compounds containing OH groups will be acidic, amphoteric, or basic?

28. (a) Distinguish between oxidizing and nonoxidizing acids in terms of their reactions with metals. (b) Identify two acids of each type.

29. List five types of reactions that can be used to produce salts.

30. Identify each of the following solutions as an acid or a base: (a) turns blue litmus pink (b) has a slippery feel (c) is present in grapefruit juice (d) reacts with NaOH to form NaCl and H_2O (e) supplies OH^- ions when dissolved in H_2O (f) phenolphthalein turns red in its presence (g) has a sour taste and may be corrosive (h) reacts with H_2SO_4 to form $MgSO_4$ and H_2O (i) donates protons (j) reacts with metals to liberate hydrogen gas (k) accepts protons (l) causes a seashell placed in it to fizz from the release of CO_2 gas (n) causes methyl orange to change to yellow.

Problems

1. (a) Write the two-step reaction that describes the ionization of sulfuric acid in a dilute aqueous solution. (b) How do the degrees of ionization in the two steps compare?

2. Dilute HCl(aq) and KOH(aq) are mixed in chemically equivalent quantities. Write the (a) chemical equation for the neutralization reaction, (b) the ionic equation, and (c) the net ionic equation.

3. Repeat Problem 2 with H_3PO_4(aq) and NaOH(aq).

4. Complete the following neutralization reactions. Balance each and then write the complete ionic and net ionic equation for each: (a) HCl(aq) + NaOH(aq) → (b) HNO_3(aq) + KOH(aq) → (c) $Ca(OH)_2$(aq) + H_2SO_4(aq) (very dilute) → (d) $Mg(OH)_2$(aq) + H_3PO_4(aq) (very dilute) →

5. Write an equation that illustrates that MgO is a basic anhydride.

6. Write the chemical equation and net ionic equation for each of the following reactions involving nonoxidizing acids: (a) Zn(s) + HCl(aq) → (b) Al(s) + H_2SO_4(aq) →

7. Write the balanced chemical equation for each of the following reactions between the aqueous acid solution and metal oxide indicated: (a) MgO(s) + H_2SO_4(aq) → (b) CaO(s) + HCl(aq) → (c) Al_2O_3(s) + HNO_3(aq) → (d) ZnO(s) + H_3PO_4(aq) →

8. Write the balanced chemical equation for each of the following reactions between an acid and a carbonate: (a) $BaCO_3$(s) + HCl(aq) → (b) $MgCO_3$(s) + HNO_3(aq) → (c) Na_2CO_3(s) + H_2SO_4 → (d) $CaCO_3$(s) + H_3PO_4(aq) →

9. Write the balanced chemical equation for the reaction involving the following hydroxide and nonmetal oxide: (a) CO_2(g) + 2 KOH(aq) → (b) CO_2(g) + KOH (aq) →

10. Write the balanced chemical equation for the reactions involving the following metal oxides, and nonmetal oxides: (a) CaO(s) + CO_2(g) → (b) BaO(s) + P_4O_{10}(s) → (c) SrO(s) + SO_3(g) →

11. Write the balanced chemical equation that corresponds to the five types of salt-producing reactions described in this chapter: (a) Mg(s) + HCl(aq) → (b) CaO(s) + HNO_3(aq) → (c) HNO_3(aq) + K_2CO_3(aq) → (d) CO_2(g) + $Mg(OH)_2$(aq) → (e) BaO(s) + CO_2(g) →

Application Questions

1. In each equation below, the substance underlined is acting as an acid according to at least one of the three accepted definitions (traditional, Brønsted, and Lewis). Identify the appropriate definition for each. (a) $\underline{HCl}$(g) + $H_2O(\ell)$ → H_3O^+(aq) + Cl^-(aq) (b) $\underline{HCl}$(g) + $NH_3(\ell)$ → NH_4^+(aq) + Cl^-(aq) (c) NH_3(g) + $\underline{H_2O}(\ell)$ → NH_4^+(aq) + OH^-(aq) (d) $\underline{BF_3}$(g) + F^-(aq) → BF_4^-(aq) (e) $\underline{H_2SO_4}(\ell)$ + $H_2O(\ell)$ → H_3O^+(aq) + HSO_4^-(aq) (f) $\underline{AlCl_3}$(aq) + Cl^-(aq) → $AlCl_4^-$(aq) (g) $\underline{H^+}$(aq) + $\underline{H_2O}(\ell)$ → H_3O^+(aq)

2. Explain why H_3PO_4, which contains three hydrogen atoms per molecule, is a weak acid, whereas, HCl, which contains only one hydrogen atom per molecule, is a strong acid.

3. Explain why both pure HCl gas and HCl dissolved in a nonpolar solvent exhibit no acidic

properties in the traditional sense, whereas HCl(aq) does exhibit such properties.

4. Explain how sulfuric acid production serves as a measure of a country's economy.

5. Sodium hydroxide is a traditional base, but it does not qualify as a Brønsted base. Explain.

6. Identify each of the following as traditional, Brønsted, or Lewis acid–base reactions, and label the acid and base in each:
   (a) $HCl(aq) + KOH(aq) \rightarrow KCl(aq) + H_2O(\ell)$
   (b) $H^+(aq) + NH_3(aq) \rightarrow NH_4^+(aq)$
   (c) $HCl(aq) + NH_3(g) \rightarrow NH_4Cl(s)$
   (d) $H_3PO_4(aq) + 3NaOH(aq) \rightarrow Na_3PO_4(aq) + 3H_2O(\ell)$ (e) $H_2O(\ell) + NH_3(g) \rightarrow NH_4^+(aq) + OH^-(aq)$ (f) $2NH_3(aq) + Ag^+(aq) \rightarrow [Ag(NH_3)_2]^+(aq)$

7. Asked to determine whether an unknown solution is an acid, base, or salt, explain three safe tests you could perform.

8. For each reaction listed, identify the proton donor (acid), the proton acceptor (base), and label each conjugate acid–base pair:
   (a) $HCl(aq) + H_2O(\ell) \rightarrow H_3O^+(aq) + Cl^-(aq)$
   (b) $HCl(aq) + NH_3(g) \rightarrow NH_4^+(aq) + Cl^-(aq)$
   (c) $NH_3(g) + H_2O(\ell) \rightleftarrows NH_4^+(aq) + OH^-(aq)$
   (d) $H_2SO_4(\ell) + H_2O(\ell) \rightleftarrows H_3O^+(aq) + HSO_4^-(aq)$ (e) $HC_2H_3O_2(aq) + H_2O(\ell) \rightleftarrows H_3O^+(aq) + C_2H_3O_2^-(aq)$
   (f) $HCO_3^-(aq) + H_2O(\ell) \rightleftarrows H_2CO_3(aq) + OH^-(aq)$ (g) $HCN(aq) + SO_4^{2-}(aq) \rightarrow HSO_4^-(aq) + CN^-(aq)$.

9. Based on the information given in Table 16-7, determine the following relative to HF, H_2S, HNO_3, and $HC_2H_3O_2$: (a) the strongest acid (b) the weakest acid (c) the strongest conjugate base among those produced by the acids listed (d) the weakest conjugate base among the four produced.

10. Explain why the conjugate base of a strong acid is a weak base and the conjugate acid of a strong base is a weak acid.

11. Explain how antacids relieve "heartburn."

Application Problems

1. Write the chemical equation, the ionic equation, and the net ionic equation for the neutralization reaction involving aqueous solutions of H_3PO_4 and $Mg(OH)_2$.

2. Write the equation that illustrates each of the following: (a) that carbon dioxide is the acid anhydride of carbonic acid (b) that sulfur dioxide is the acid anhydride of sulfurous acid (c) that calcium oxide is the basic anhydride of calcium hydroxide (d) that sulfur trioxide is the acid anhydride of sulfuric acid.

3. Write the chemical and net ionic equations for the reaction between Ca(s) and HCl(aq).

4. Acid rain is the term generally used to describe precipitation that is more acidic than normal. Among many of its potential causes may be the formation of sulfuric and nitric acids from the various sulfur and nitrogen oxides produced in volcanic eruptions, forest fires, and thunderstorms. In a typical volcanic eruption, for example, 3.50×10^8 kg of SO_2 may be produced. If this amount of SO_2 were converted to H_2SO_4 according to the two-step process given below, how much H_2SO_4 (in kg) would be produced from such an eruption?
   $$SO_2 + \tfrac{1}{2}O_2 \rightarrow SO_3$$
   $$SO_3 + H_2O \rightarrow H_2SO_4$$

5. Zinc reacts with 100.mL of 6.00 M aqueous sulfuric acid through single replacement. (a) How much zinc sulfate (in grams) is produced? (b) How many liters of hydrogen gas would be released at STP?

6. A seashell (largely calcium carbonate) placed in a solution of HCl resulted in the production of 1500. mL of dry CO_2 gas at STP. The other products are $CaCl_2$ and H_2O. (a) Based on this information, how many grams of $CaCO_3$ were consumed in the reaction? (b) What volume of 2.00-M HCl solution was used in this reaction?

Enrichment

1. Antacids are designed to neutralize excess hydrochloric acid secreted by the stomach during digestion. Carbonates, bicarbonates, and hydroxides serve as the active ingredients in bringing about the neutralization reaction in the most widely used antacids. Examine the labels of several common antacids and identify the active ingredient(s).

2. Develop and test a laboratory procedure for determining the length of time required for a given dose of antacid to neutralize various volume-concentration combinations of HCl. Repeat for several different antacids.

3. Research the three acid–base theories of Arrhenius, Brønsted, and Lewis.

Chapter 17 Acid-Base Titration and pH

Chapter Planner

17.1 Concentration Units for Acids and Bases
Chemical Equivalents
■ Questions: 1–3
  Problems: 1, 2
  Application Question: 1
Normality
■ Questions: 4, 5
  Application Problems: 1–5
■ Problems: 3, 4

17.2 Aqueous Solutions and the Concept of pH
Self-Ionization of Water
■ Demonstration: 1
  Questions: 6, 8a-e
  Problems: 5–7

The pH Scale
■ Chemistry Notebook
  Questions: 7, 8f-g, 9
Calculations Involving pH
■ Experiment: 20
  Problems: 8–15
  Application Questions: 2, 3
  Application Problem: 6
■ Application Question: 4

17.3 Acid-Base Titrations
Indicators
■ Demonstrations: 2–4
  Desktop Investigation
  Questions: 14–16
■ Question: 17
  Application Question: 5

The Principle of Titration
■ Questions: 11, 12
Molarity and Titration
■ Experiments: 21, 22
  Problems: 16–19
  Application Problem: 8
■ Application Problems: 7, 9, 10
Normality and Titrations
■ Question: 13
  Problems: 20, 21

Teaching Strategies

Introduction

This chapter requires a great deal of mathematical computation. However, even without doing the calculations, many of the concepts can be grasped. The concept of pH is an important tool for students to understand. Most students will be able to appreciate the importance of a scale to measure acidity, and will understand that pH values below 7 are acidic, 7 is neutral, and values above 7 are basic. However, many will have difficulty with the exponential values and will require help in relating them to pH. Students will enjoy the opportunity to test some commercial substances. They will quickly recognize that they can rank antacid tablets with respect to the amount of acid they can neutralize even without doing calculations. They will like the challenge of testing a product such as vinegar to see if the ingredients conform to the label. The calculations will require some explanation.

The concept of normality is often used in acid–base titrations, and also in other titration situations, such as for oxidation–reduction reactions. However, all of the calculations for titrations can be carried out using the molarity of the solutions instead of normality. Many teachers will want to skip Section 17.1, *Concentration Units for Acids and Bases*, which teaches the concepts of normality and equivalents, and begin with Section 17.2, *Aqueous Solutions and the Concept of pH*. In Section 17.3, *Acid–Base Titrations*, the last part of the section, which deals with normality and titrations, can also be omitted.

17.1 Concentration Units for Acids and Bases

Before students can understand the concept of equivalents, they must have a thorough understanding of the mole concept. Review the concepts of moles and molarity of solutions. Discuss the preparation of solutions such as 1 M NaCl, and 1 M sucrose, and make up solutions of these substances.

Discuss equivalents with the idea that equivalents are amounts of substances that completely react with each other, with no excess of either reactant. Develop the definitions of one equivalent of an acid and one equivalent of a base in proton transfer reactions. Write net ionic equations for the reactions of various strong acids and strong bases: $H^+(aq) + OH^-(aq) \rightarrow H_2O(\ell)$, and point out that in these reactions it is only the hydrogen and hydroxide ions that are taking part in the reaction, so it is logical to have a concentration unit which deals with just these substances. Have students calculate the equivalent mass of various acids and bases such as HCl, H_2SO_4, HNO_3, H_3PO_4, $Ba(OH)_2$, KOH.

Define the term normality. Discuss normality of solutions in terms of the H_3O^+ or OH^- molarity, then show how this is the same as equivalents per liter. Use volumetric flasks of various sizes, and ask students how to calculate the mass of acids and bases needed to make up solutions of various normalities. Be sure that students can use the equation: $N =$ equivalents/V(liters), and that they can derive the relationship: equivalents $= N \times V$(liters) and can use it properly.

Illustrate dilution operation for preparing weaker concentrations from a 1 N solution of various acids or bases. Develop the dilution relationship: $N_1V_1 = N_2V_2$. Illustrate that milliliters may be used instead of liters in this equation if both V_1 and V_2 are in milliliters. Have students read the data (percent HCl and density) on a label from a bottle of concentrated HCl. Have them calculate the mass of HCl in each milliliter of the concentrated acid, the molarity and the normality. Compare molarity and normality for various acid solutions. Work through the sample problems with the students.

17.2 Aqueous Solutions and the Concept of pH

Write the equation for self-ionization of water. Demonstrate this reaction using molecular models. Use both red and blue litmus paper in water to show that it is neutral, even though it does contain some H_3O^+ ions and OH^- ions. Using the chalkboard or overhead projector, go through the calculations in the text. Illustrate the use of square brackets to represent concentration in moles per liter: $[H_3O^+] = [OH^-] = 1.0 \times 10^{-7}$ mol/L. Give examples of H_3O^+ concentration and OH^- concentrations for acidic and basic solutions. Show how the ion product for water is calculated. Discuss the effect of the increase of temperature on the amount of ionization of water and thus on the ion product.

Illustrate the use of the ion product of water to calculate H_3O^+ and OH^- concentrations in solutions where the concentrations of strong acid or strong base are given. Show how the exponent for the hydronium ion concentration can be used to indicate an acid (smaller negative exponent than -7) or base (larger negative exponent than -7).

Write these values out in decimal form so that the students can relate the negative exponent to the size of the value. Define pH and show how much easier it is to describe hydronium ion concentration by pH rather than by a decimal value. Use the equation $pH = -\log[H_3O^+]$ to calculate the pH of pure water. Discuss the magnitude of the value of pH for acidic and basic solutions. Stress the idea that a pH change of 2 pH units means a change in hydronium and hydroxide ions concentrations of 10^2, or 100 fold. A pH change of 5 means a hydronium ion concentration change of 10^5, or 10 000 fold.

Many students will be intimidated by the use of a logarithmic function and feel that understanding pH is beyond their capabilities. Use the expression $[H_3O^+] = 10^{-pH}$ to help them reach an understanding of the relation between the size of the negative exponent, the value of pH, and the concentration that it expresses. Work with hydronium ion concentrations that are integral powers of 10. More able students should be able to handle the calculation of pH when H_3O^+ concentration is not an integral power of 10, but they will need practice and help in figuring out how to use their calculators. Go through some of the calculations with them as they push each key on their calculators.

Calculate the pH of solutions of strong acids and bases from the molar concentration. Determine both hydronium ion and hydroxide ion concentrations in these. Some students might find the concept of pOH useful: $pOH = -\log[OH^-]$. The relation between pH and pOH is: $pH + pOH = 14$.

Ask students to bring samples of various substances and check their pH using pH indicator paper. Alternatively, give them some pH paper, and ask them to report on the pH of various substances they find at home. Calculate the hydronium and hydroxide ion concentrations of these substances based on the pH values, and discuss the acidity of basicity or the products.

Chemistry Notebook: Acid Rain and Lakes
Ask students to find current articles which deal with the topic of acid rain. Discuss the source of the acids that cause acid rain. Ask students to calculate the hydronium ion concentration of a lake when the pH is 5.0 or 4.5

Demonstration 1: Conductivity of Water
PURPOSE: To show the small amount of conductivity in distilled water
MATERIALS: Conductivity detector using a neon bulb or a 25-W bulb, or a sensitive conductivity meter (from the physics department), distilled water, 6 M HCl
PROCEDURE: Wear goggles. Be careful to avoid shocks if you use a light-bulb conductivity detector. Demonstrate that distilled water has a measurable, but very small conductivity (sufficient to light the neon bulb, but not the 25-W bulb). Add a few drops of HCl solution to illustrate the change when a strong acid ionizes in water. Relate the low conductivity to the self-ionization of water.

17.3 Acid–Base Titrations

Show as many indicators as are available to the students, and demonstrate their color changes. Discuss the transition interval. Illustrate the four types of acid–base combinations and discuss the choice of indicator for each situation. Explain the mechanism by which indicators operate in terms of the weak acid dissociating in water. Demonstrate the use of pH indicator paper, and compare the color changes with universal indicator. Mix several indicators together to create a universal indicator: for example, combine methyl red, methyl orange, bromphenol blue and pehnolphthalein. Extract indicators from natural products such as red cabbage, tea, blueberries, red roses, or from any flower petal and test for their color changes and transition intervals. Use a pH meter to determine the pH of a solution and compare the pH with that obtained using pH indicator paper or other indicators.

Discuss titration as a common technique used by analytical chemists to find the quantity of a substance present in a sample. Demonstrate the technique of titration using burets and common acids and bases. Discuss the necessity for a standard substance in the titration procedure, and define a standard solution. Mention the characteristics of substances which make them acceptable as standards for an acid or base, and show how a solution can be "standardized" using such a substance.

The calculations for titrations can be carried out using concentrations expressed as normality or molarity. If molarity is selected as the concentration unit for the acid and base, show how titration calculations are made. Point out the four steps necessary: find the moles of substance used in the titration, find mole ratio of acid and base from a balanced equation, find the moles of the unknown substance, and calculate the molarity of the unknown solution. Work through the sample problems in the text with the students. Steps 1, 3 and 4 can be combined in one sequence of unit conversion factors, using molarity as the conversion between moles and volume of a substance.

The choice of normality as a concentration unit simplifies the titration calculation, but the concept of normality is more difficult to understand than that of molarity. Demonstrate the use of the equation for calculation using normality: $V_aN_a = V_bN_b$. Work through the computations shown in the text.

There are a number of substances that students might like to analyze themselves by means of acid-base titrations. Some suggestions are to determine the amount of acid in orange juice, to analyze a vitamin C tablet to find the amount of ascorbic acid present, or to determine the amount of ammonia in a household cleaning product.

Testing a commercial product helps students appreciate that chemistry has relevance in their lives. Ask them to bring various brands of antacid tablets to analyze. Try to choose tablets that are white so that the indicator color is easy to see. Students can rank the tablets in relation to the amount of acid they neutralize by adding a few mL of cabbage juice indicator to each crushed tablet, which will give a distinct green color, and then dropping 5% white vinegar (0.837 M acetic acid) onto each tablet. After calibrating a dropper for drops per 10 mL, count and record the number of drops required to change the color of the indicator to red, its acid color. A fairly close measure of the moles of acid neutralized by the tablets can be obtained. Check the labels to see what substances are used in antacid tablets, and write the neutralization reactions. Many tablets will contain an acid–base buffer to keep the pH fairly constant.

A fairly close measure of the moles of acid neutralized by the tablets. Check the labels to see what substances are used in antacid tablets, and write the neutralization reactions. Many tablets will contain an acid-base buffer to keep the pH fairly constant.

The cabbage juice indicator can be obtained by boiling red cabbage in water. It should be stored in a refrigerator when not in use. It will keep for about a week. If cabbage juice is not available, use an indicator that changes pH in the acid range, such as methyl orange or methyl red.

Desktop Investigation: Testing for Acid Rain

The topic of acid rain is one that most students have heard of, but many will be surprised to find that they can determine if the rain in their area is acidic by means of a simple technique. Ask students to collect rain from the area where they live and at their school and compare the pH of rainwater from different sites. Make a chart of the pH of the rainwater in the local area.

Demonstration 2: Behavior of Indicators

PURPOSE: To show the color changes that indicators undergo at various pH values

MATERIALS: solutions of several indicators: methyl orange, methyl red, bromphenol blue, phenolphthalein, thymolphthalein, universal indicator, or other indicators made from flowers, red cabbage, blueberries, etc.; HCl solution, 0.1-M; NaOH solution, 0.1-M; graduated cylinder, 100-mL; beakers and test tubes; distilled water

PROCEDURE: Wear goggles. Pour 100 mL of 0.1-M HCl into a beaker and label it ph 1. Combine 10 mL of 0.1-M HCl and 90 mL of distilled water, and put into a beaker marked pH 2. Take 10 mL of this solution (0.01-M HCl), mix with 90 mL of distilled water to form a solution with pH 3. Continue the sequential dilution, forming solutions with pH values of 4, 5, and 6. Use distilled water for pH 7. Dilute the 0.1 M NaOH solution in a similar fashion to prepare solutions with pH values from 13 to 8. Pour small amounts of these standard pH solutions into a series of test tubes marked with the proper pH, and add the same amount of indicator to each. Note the pH where the color change occurs and the transition interval for the indicators.

Demonstration 3: Indicator Changes with Carbon Dioxide

PURPOSE: To provide a dramatic demonstration of indicator color changes

MATERIALS: 12 large beakers, 6 different indicators such as phenolphthalein, bromphenol blue, universal indicator, methyl orange, etc. dry ice, distilled water, 0.1 M NH_3(aq)

PROCEDURE: Wear goggles. Fill each beaker about three-fourths full of distilled water. Add phenolphthalein to two beakers and add enough NH_3(aq) to cause a definite color change to pink. Add the same amount of NH_3(aq) to all the other beakers. Set the beakers out in pairs of two, and add a different indicator to each. Then place a chunk of dry ice in one beaker of each pair. The dry ice will sublime, forming clouds of water vapor, and the beaker will appear to be boiling. The slow production of acid will gradually cause the indicators to change color. The second beaker in each set is a control to show the color of the indicator in basic solution. The reaction is:

$$CO_2(g) + H_2O(\ell) \rightleftarrows H_2CO_3(aq) \rightleftarrows H^+(aq) + HCO_3^-(aq)$$

Demonstration 4: A Breathtaking Color Change

PURPOSE: To show how carbon dioxide from one's breath can cause an indicator color change

MATERIALS: flasks, 250-mL or 500-mL; long soda straws, 6 M NH_3(aq), phenol red indicator solution or bromthymol blue indicator solution

PROCEDURE: Fill the flasks about two-thirds full of distilled water. Add one drop of 6 M NH_3(aq) and sufficient indicator to give an obvious color. Give student volunteers goggles to wear, and ask them to "blow the color away." Be sure that they blow into the flask, and do not drink the liquid. The carbon dioxide from their breath will form an acidic solution that will cause a color change in the indicator:

$$CO_2(g) + H_2O(\ell) \rightleftarrows H_2CO_3(aq) \rightleftarrows H^+(aq) + HCO_3^-(aq)$$

Demonstration 5: pH Change of an Acid–Base Titration

PURPOSE: To show how pH changes during an acid–base titration

MATERIALS: pH meter, buret, beaker, 0.1 M HCl, 0.1 M NaOH

PROCEDURE: Wear goggles. Carefully measure 25.0 mL of 0.1 M HCl into the beaker. Connect the pH meter, and record the pH of the solution. Continue to record the pH as 0.1 M NaOH is added and graph the resulting titration curve. Point out the very large change in pH that results when one drop of titrant is added at the equivalence point.

References and Resources

Borgford, Christie L. and Lee R. Summerlin, *Chemical Activities*. American Chemical Society, 1988.

Forster, M. "Plant Pigments as Acid-Base Indicators," *Journal of Chemical Education*, Vol. 55, No. 2, 1978, p. 107.

Kolb, D. "The pH Concept," *Journal of Chemical Education*," Vol. 56, No. 1, 1979, p. 49.

Mebane, R. and T. Rybolt. "Edible Acid—Base Indicators," *Journal of Chemical Education*, Vol. 62, No. 4, 1985, p. 285.

Schultz, C.W. and S.I. Spannuth. *"Comparison of Strong and Weak* Titration Curves," *Journal of Chemical Education*, Vol. 56, No. 3, 1979, p. 194.

Audiovisual Resources

"Acid—Base Indicators," CHEM Study 16mm film, Ward's Modern Learning Aids. Shows how a universal indicator is made and determines equilibrium constant for four indicators.

"Indicators and pH," University of Akron, Audio Visual Services.

Ward's Solo-Learn: 78–0700, "Introduction to Chemical Equivalents", Ward's Modern Learning Aids. A programmed film strip to teach chemical equivalents.

WCC 804A "Using Simple Logarithms, Logs and Antilogs," WCC 804B "Using Simple Logarithms, Multiplication and Division," WCC 813A "Titration Techniques, Part I," WCC 897 "Dilution and Titration Problems," WCC 815 "pH Measurements," WCC 866 "Titration Curves and Indicators," WCC 883 "pH, pOH and Buffers," filmstrips/cassettes, Prentice Hall Media.

Answers and Solutions

Questions

1. *(a)* Chemical equivalents are the quantities of substances that have the same combining capacity in chemical reactions. *(b)* One equivalent of an acid is the quantity that supplies one mole of protons; one equivalent of a base is the quantity that accepts one mole of protons or supplies one mole of OH^- ions.

2. *(a)* The mass of one equivalent of an acid is equal to the molar mass of that acid divided by the number of moles of H_3O^+ ions that can be supplied by one mole of the acid. *(b)* One equivalent of a monoprotic acid, such as HCl, is equal to one mole of that acid. One equivalent of a diprotic acid, such as H_2SO_4, is equal to one-half mole of the acid. One equivalent of a triprotic acid, such as H_3PO_4, is equal to one-third mole of the acid.

3. *(a)* The mass of one equivalent of a base is equal to the molecular weight of that base divided by the number of OH^- ions contained in each mole of that base. *(b)* One mole of KOH supplies of one equivalent of OH^- ions; and one mole of $Al(OH)_3$ supplies three equivalents of OH^- ions.

4. Molality is the concentraiton of a solution expressed in moles of solute per kilogram of solvent. The molarity of a solution is an expression of the number of moles of solute per liter of solution. The normality of a solution is an expression of the number of equivalents of solute per liter of solution.

5. *(a)* One equivalent of solute per liter of solution *(b)* Three equivalents of solute per liter of solution *(c)* 0.5 equivalents of solute per liter of solution

6. Hydronium ion concentration in moles per liter, or molar hydronium ion concentration

7. *(a)* The common logarithm of the reciprocal of the hydronium ion concentration *(b)* $pH = -\log[H_3O^+]$ *(c)* The power to which ten must be raised to give the number. (Example: the log of 10^7 is 7.)

8. *(a)* Neutral *(b)* Basic *(c)* Neutral *(d)* Acidic *(e)* Neutral *(f)* Acidic *(g)* Basic

9. *(a)* 2.8 *(b)* 3.1 *(c)* 4.2 *(d)* 6.5 *(e)* 3.1 *(f)* 5.8 *(g)* 2.3 *(h)* 10.5 *(i)* 2.0

10. *(a)* In a neutralization reaction, the OH^- ion acquires a proton from the H_3O^+ ion to form a molecule of water. Complete neutralization has taken place when the H_3O^+ ions and OH^- ions are present in equal numbers. *(b)* $H_3O^+ + OH^- \rightarrow 2H_2O$.

11. *(a)* Titration is the controlled addition of the measured amount of a solution of known concentration required to react completely with a measured amount of solution of unknown concentration. *(b)* Titration is an important laboratory procedure often used in analytical chemistry because it provides a sensitive means of determining the relative volumes of acidic and basic solutions that are chemically equivalent. If the concentration of one solution is known, the concentration of the other solution can be calculated through the use of this technique.

12. *(a)* The end point of a titration is the equivalence point, or the point where equal amounts of H_3O^+ and OH^- ions are present. *(b)* At the end point of the titration, a very rapid change in pH occurs. *(c)* If the indicator selected changes color within the pH range in which the end point of the titration is reached, then it can provide visible evidence that neutralization has occurred.

13. *(a)* The use of normality allows concentrations to be expressed directly in terms of equivalents of solute; and, since solutions of the same normality are chemically equivalent, a simple relationship thus exists between volumes and normalities of solutions used in titration. *(b)* $V_a N_a = V_b N_b$

14. *(a)* The transition interval of an indicator is the pH range over which the indicator color change occurs. *(b)* The choice of the indicator is based on the suitability of its transition interval for a given acid—base reaction.

15. (1) Strong-acid strong-base: pH is about 7, and litmus or bromthymol blue would be suitable indicators. (2) Strong-acid weak-base: pH is below 7, and methyl orange is a suitable indicator. (3) Weak-acid strong-base: pH is above 7, and phenolphthalein is a suitable indicator. (4) Weak-acid strong-base: pH may be either above or below 7 depending upon the reactant that is stronger. No indicator is satisfactory.

16. Through the use of a pH meter

17. The pH is approximately 8.

Problems

1. (a) $1 \text{ equiv HBr} = \dfrac{80.9 \text{ g HBr}}{\text{mol HBr}} \times \dfrac{1 \text{ mol HBr}}{1 \text{ mol } H_3O^+} = 80.9 \text{ g HBr/mol } H_3O^+$

   (b) $1 \text{ equiv } H_2S = \dfrac{34.1 \text{ g } H_2S}{\text{mol } H_2S} \times \dfrac{1 \text{ mol } H_2S}{2 \text{ mol } H_3O^+} = 17.0 \text{ g } H_2S\text{/mol } H_3O^+$

   (c) $1 \text{ equiv } H_3PO_4 = \dfrac{98.0 \text{ g } H_3PO_4}{\text{mol } H_3PO_4} \times \dfrac{1 \text{ mol } H_3PO_4}{3 \text{ mol } H_3O^+} = 32.7 \text{ g } H_3PO_4\text{/mol } H_3O^+$

2. (a) $1 \text{ equiv LiOH} = \dfrac{23.9 \text{ g LiOH}}{\text{mol LiOH}} \times \dfrac{1 \text{ mol LiOH}}{1 \text{ mol } OH^-} = 23.9 \text{ g LiOH/mol } OH^-$

   (b) $1 \text{ equiv } Sr(OH)_2 = \dfrac{122 \text{ g } Sr(OH)_2}{\text{mol } Sr(OH)_2} \times \dfrac{1 \text{ mol } Sr(OH)_2}{2 \text{ mol } OH^-} = 61.0 \text{ g } Sr(OH)_2\text{/mol } OH^-$

   (c) $1 \text{ equiv } Al(OH)_3 = \dfrac{78.0 \text{ g } Al(OH)_3}{1 \text{ mol } Al(OH)_3} \times \dfrac{1 \text{ mol } Al(OH)_3}{3 \text{ mol } OH^-} = 26.0 \text{ g } Al(OH)_3\text{/mol } OH^-$

3. (a) $1.00 \text{ L} \times \dfrac{2.00 \text{ equiv } HNO_3}{L} \times \dfrac{1 \text{ mol } HNO_3}{1 \text{ equiv } HNO_3} \times \dfrac{63.0 \text{ g } HNO_3}{\text{mol } HNO_3} \times \dfrac{\text{mL conc. acid}}{1.42 \text{ g conc. acid}} \times \dfrac{100. \text{ g conc. acid}}{69.5 \text{ g } HNO_3} = 128 \text{ mL conc. acid}$

   (b) $150. \text{ mL} \times \dfrac{L}{10^3 \text{ mL}} \times \dfrac{0.750 \text{ equiv } HNO_3}{L} \times \dfrac{1 \text{ mol } HNO_3}{1 \text{ equiv } HNO_3} \times \dfrac{63.0 \text{ g } HNO_3}{\text{mol } HNO_3} \times \dfrac{\text{mL conc. acid}}{1.42 \text{ g conc. acid}} \times \dfrac{100. \text{ g conc. acid}}{69.5 \text{ g } HNO_3}$

   $= 7.18 \text{ mL conc. acid}$

   (c) $250. \text{ mL} \times \dfrac{L}{10^3 \text{ mL}} \times \dfrac{0.500 \text{ equiv } HNO_3}{L} \times \dfrac{1 \text{ mol } HNO_3}{1 \text{ equiv } HNO_3} \times \dfrac{63.0 \text{ g } HNO_3}{\text{mol } HNO_3} \times \dfrac{\text{mL conc. acid}}{1.42 \text{ g conc. acid}} \times \dfrac{100. \text{ g conc. acid}}{69.5 \text{ g } HNO_3}$

   $= 7.98 \text{ mL conc. acid}$

4. (a) $\dfrac{36.5 \text{ g HCl}}{\text{mol HCl}} \times \dfrac{1 \text{ mol HCl}}{1 \text{ equiv}} \times \dfrac{100 \text{ g conc. acid}}{37.5 \text{ g HCl}} \times \dfrac{\text{mL conc. acid}}{1.19 \text{ g conc. acid}} = 81.8 \text{ mL conc. acid/equiv}$

   (b) $\dfrac{81.8 \text{ mL conc. acid}}{\text{equiv}} \times \dfrac{1.00 \text{ equiv}}{L \text{ dil soln}} \times 2.00 \text{ L dil soln} = 164 \text{ mL conc acid}$

   (c) $\dfrac{81.8 \text{ mL conc acid}}{\text{equiv}} \times \dfrac{0.750 \text{ equiv}}{L \text{ dil soln}} \times 3.00 \text{ L dil soln} = 184 \text{ mL conc acid}$

5. (a) $\dfrac{0.0100 \text{ mol HCl}}{L} \times \dfrac{1 \text{ mol } H_3O^+}{1 \text{ mol HCl}} = 0.0100 \text{ M } H_3O^+$

   (b) $\dfrac{1.65 \times 10^{-4} \text{ mol } HNO_3}{L} \times \dfrac{1 \text{ mol } H_3O^+}{1 \text{ mol } HNO_3}$

   $= 1.65 \times 10^{-4} \text{ M } H_3O^+$

   (c) $\dfrac{0.54 \text{ mol NaOH}}{L} \times \dfrac{1 \text{ mol } OH^-}{1 \text{ mol NaOH}} = 0.54 \text{ M } OH^-$

   (d) $\dfrac{1.54 \times 10^{-3} \text{ mol KOH}}{L} \times \dfrac{1 \text{ mol } OH^-}{1 \text{ mol KOH}}$

   $= 1.54 \times 10^{-3} \text{ M } OH^-$

6. $[H_3O^+][OH^-] = 1 \times 10^{-14} \text{ M}^2$

   $[H_3O^+] = \dfrac{1 \times 10^{-14} \text{ M}^2}{[OH^-]}$

   Now substitute for $[OH^-]$ in (a) – (d).

   (a) $[H_3O^+] = \dfrac{1 \times 10^{-14} \text{ M}^2}{1 \times 10^{-7} \text{ M}} = 1 \times 10^{-7} \text{ M}$

   (b) $[H_3O^+] = \dfrac{1 \times 10^{-14} \text{ M}^2}{1 \times 10^{-4} \text{ M}} = 1 \times 10^{-10} \text{ M}$

   (c) $[H_3O^+] = \dfrac{1 \times 10^{-14} \text{ M}^2}{1 \times 10^{-12} \text{ M}} = 1 \times 10^{-2} \text{ M}$

   (d) $[H_3O^+] = \dfrac{1 \times 10^{-14} \text{ M}^2}{1 \times 10^{-3} \text{ M}} = 1 \times 10^{-11} \text{ M}$

7. $[H_3O^+][OH^-] = 1 \times 10^{-14} \text{ M}^2$

   $[OH^-] = \dfrac{1 \times 10^{-14} \text{ M}^2}{[H_3O^+]}$

   Now substitute for $[H_3O^+]$ in (a) – (d).

   (a) $[OH^-] = \dfrac{1 \times 10^{-14} \text{ M}^2}{1 \times 10^{-7} \text{ M}} = 1 \times 10^{-7} \text{ M}$

   (b) $[OH^-] = \dfrac{1 \times 10^{-14} \text{ M}^2}{1 \times 10^{-9} \text{ M}} = 1 \times 10^{-5} \text{ M}$

   (c) $[OH^-] = \dfrac{1 \times 10^{-14} \text{ M}^2}{1 \times 10^{-1} \text{ M}} = 1 \times 10^{-13} \text{ M}$

   (d) Since $[H_3O^+] = [OH^-]$ was given,
   $[OH^-][OH^-] = 1 \times 10^{-14} \text{ M}^2$
   $[OH^-]^2 = 1 \times 10^{-14} \text{ M}^2$
   $[OH^-] = 1 \times 10^7 \text{ M}$

8. $pH = -\log [H_3O^+]$
   (a) $pH = -\log (1 \times 10^{-7}) = -(-7) = 7$ (c) $pH = -\log (1 \times 10^{-12}) = -(-12) = 12$
   (b) $pH = -\log (1 \times 10^{-3}) = -(-3) = 3$ (d) $pH = -\log (1 \times 10^{-5}) = -(-5) = 5$

9. $pH = -\log [H_3O^+]$
   (a) $pH = -\log (0.01) = -\log (1 \times 10^{-2}) = -(-2) = 2$ (c) $pH = -\log (0.000\,01) = -\log (1 \times 10^{-5}) = -(-5) = 5$
   (b) $pH = -\log (0.001) = -\log (1 \times 10^{-3}) = -(-3) = 3$ (d) $pH = -\log (0.0001) = -\log (1 \times 10^{-4}) = -(-4) = 4$

10. (a) Since $pH = -\log [H_3O^+]$, first find $[H_3O^+]$.
    $[H_3O^+][OH^-] = 1 \times 10^{-14} M^2$

    $[H_3O^+] = \dfrac{1 \times 10^{-14} M^2}{[OH^-]} = \dfrac{1 \times 10^{-14} M^2}{1 \times 10^{-6} M} = 1 \times 10^{-8} M$

    Next, find pH.
    $pH = -\log [H_3O^+] = -\log (1 \times 10^{-8}) = -(-8) = 8$

    (b) $[H_3O^+] = \dfrac{1 \times 10^{-14} M^2}{[OH^-]} = \dfrac{1 \times 10^{-14} M^2}{1 \times 10^{-9} M} = 1 \times 10^{-5} M$

    $pH = -\log [H_3O^+] = -\log (1 \times 10^{-5}) = -(-5) = 5$

    (c) $[H_3O^+] = \dfrac{1 \times 10^{-14} M^2}{[OH^-]} = \dfrac{1 \times 10^{-14} M^2}{1 \times 10^{-2} M} = 1 \times 10^{-12} M$

    $pH = -\log [H_3O^+] = -\log (1 \times 10^{-12}) = -(-12) = 12$

    (d) $[H_3O^+] = \dfrac{1 \times 10^{-14} M^2}{[OH^-]} = \dfrac{1 \times 10^{-14} M^2}{1 \times 10^{-7} M} = 1 \times 10^{-7} M$

    $pH = -\log [H_3O^+] = -\log (1 \times 10^{-7}) = -(-7) = 7$

11. $[H_3O^+][OH^-] = 1 \times 10^{-14} M^2$

    $[H_3O^+] = \dfrac{1 \times 10^{-14} M^2}{[OH^-]}$

    Now continue with parts (a) – (c) and then solve for pH.

    (a) $[H_3O^+] = \dfrac{1 \times 10^{-14} M^2}{0.01 M} = \dfrac{1 \times 10^{-14} M^2}{1 \times 10^{-2} M} = 1 \times 10^{-12} M$

    $pH = -\log [H_3O^+] = -\log (1 \times 10^{-12}) = -(-12) = 12$

    (b) $[H_3O^+] = \dfrac{1 \times 10^{-14} M^2}{0.001 M} \times \dfrac{1 \times 10^{-14} M^2}{1 \times 10^{-3} M} = 1 \times 10^{-11} M$

    $pH = -\log [H_3O^+] = -\log (1 \times 10^{-11}) = -(-11) = 11$

    (c) $[H_3O^+] = \dfrac{1 \times 10^{-14} M^2}{0.0001 M} \times \dfrac{1 \times 10^{-14} M^2}{1 \times 10^{-4} M} = 1 \times 10^{-10} M$

    $pH = -\log [H_3O^+] = -\log (1 \times 10^{-10}) = -(-10) = 10$

12. (a) $pH = -\log [H_3O^+] = -\log (2.0 \times 10^{-5})$
    $= -(\log 2.0 + \log 10^{-5}) = -(0.30 + (-5)) = -(0.30{-}5) =$
    $= -(-4.7) = 4.7$

    (b) $pH = -\log [H_3O^+] = -\log (4.7 \times 10^{-7}) = -(\log 4.7 + \log 10^{-7})$
    $= -(0.67 + (-7)) = -(0.67{-}7) = -(-6.3) = 6.3$

    (c) $pH = -\log [H_3O^+] = -\log (3.8 \times 10^{-3}) =$
    $-(\log 3.8 + \log 10^{-3}) = -(0.58 + (-3)) =$
    $-(0.58{-}3) = -(-2.4) = 2.4$

13. $pH = -\log [H_3O^+]$
    $\log [H_3O^+] = -pH$
    $[H_3O^+] = $ antilog $(-pH)$
    (a) $[H_3O^+] = $ antilog $(-3) = 10^{-3} M$ or $1 \times 10^{-3} M$

    (b) $[H_3O^+] = $ antilog $(-7) = 10^{-7} M$ or $1 \times 10^{-7} M$
    (c) $[H_3O^+] = $ antilog $(-11) = 10^{-11} M$ or $1 \times 10^{-11} M$
    (d) $[H_3O^+] = $ antilog $(-5) = 10^{-5} M$ or $1 \times 10^{-5} M$

14. $pH = -\log [H_3O^+]$
    $\log [H_3O^+] = -pH$
    $[H_3O^+] = $ antilog $(-pH)$
    When $[H_3O^+]$ is known, you can find $[OH^-]$ by substituting into the equation: $[H_3O^+][OH^-] = 1 \times 10^{-14} M^2$, which becomes

    $[OH^-] = \dfrac{1 \times 10^{-14} M^2}{[H_3O^+]}$

    (a) $[H_3O^+] = $ antilog $(-pH) = $ antilog $(-7) = 10^{-7} M$

    $[OH^-] = \dfrac{1 \times 10^{-14} M^2}{10^{-7} M} = 1 \times 10^{-7} M$

    (b) $[H_3O^+] = $ antilog $(-pH) = $ antilog $(-11) = 10^{-11} M$

    $[OH^-] = \dfrac{1 \times 10^{-14} M^2}{10^{-11} M} = 1 \times 10^{-3} M$

    (c) $[H_3O^+] = $ antilog $(-pH) = $ antilog $(-4) = 10^{-4} M$

    $[OH^-] = \dfrac{1 \times 10^{-14} M^2}{10^{-4} M} = 1 \times 10^{-10} M$

    (d) $[H_3O^+] = $ antilog $(-pH) = $ antilog $(-6) = 10^{-6} M$

    $[OH^-] = \dfrac{1 \times 10^{-14} M^2}{10^{-6} M} = 1 \times 10^{-8} M$

15. $pH = -\log [H_3O^+]$ $\log [H_3O^+] = -pH$ $[H_3O^+] = $ antilog $(-pH)$
    (a) $[H_3O^+] = $ antilog $(-pH) = $ antilog $(-4.23) = $ antilog $(0.77{-}5) = 5.9 \times 10^{-5} M$
    (b) $[H_3O^+] = $ antilog $(-pH) = $ antilog $(-7.65) = $ antilog $(0.35{-}8) = 2.2 \times 10^{-8} M$
    (c) $[H_3O^+] = $ antilog $(-pH) = $ antilog $(-9.48) = $ antilog $(0.52{-}10) = 3.3 \times 10^{-10} M$

16. (a) $HCl + NaOH \rightarrow NaCl + H_2O$
    1 mol 1 mol   1 mol   1 mol

    1.0 mol HCl $\times \dfrac{1 \text{ mol NaOH}}{1 \text{ mol HCl}} = 1.0$ mol NaOH

    (b) $HNO_3 + KOH \rightarrow KNO_3 + H_2O$
    1 mol   1 mol   1 mol   1 mol

    0.75 mol KOH $\times \dfrac{1 \text{ mol HNO}_3}{1 \text{ mol KOH}} = 0.75$ mol HNO_3

    (c) $2HF + Ba(OH)_2 \rightarrow BaF_2 + 2H_2O$
    2 mol   1 mol   1 mol   2 mol

    0.20 mol HF $\times \dfrac{1 \text{ mol Ba(OH)}_2}{2 \text{ mol HF}} = 0.10$ mol $Ba(OH)_2$

    (d) $3H_2SO_4 + 2Al(OH)_3 \rightarrow Al_2(SO_4)_3 + 6H_2O$
    3 mol   2 mol   1 mol   6 mol

    0.60 mol $Al(OH)_3 \times \dfrac{3 \text{ mol H}_2\text{SO}_4}{2 \text{ mol Al(OH)}_3} = 0.90$ mol H_2SO_4

17. $H_2SO_4 + 2KOH \rightarrow K_2SO_4 + 2H_2O$
   1 mol   2 mol   1 mol   2 mol

$$\frac{0.0250 \text{ mol } H_2SO_4}{L} \times 15.0 \text{ mL} \times \frac{L}{10^3 \text{ mL}} = 0.000\ 375 \text{ mol } H_2SO_4 \text{ used}$$

$$0.000\ 375 \text{ mol } H_2SO_4 \times \frac{2 \text{ mol KOH}}{1 \text{ mol } H_2SO_4} = 0.000\ 750 \text{ mol KOH reacted}$$

$$\frac{0.000\ 750 \text{ mol KOH}}{10.0 \text{ mL}} \times \frac{10^3 \text{ mL}}{L} = 0.0750 \text{ mol KOH/L} = 0.0750\,M \text{ KOH}$$

18. $2HNO_3 + Ba(OH)_2 \rightarrow Ba(NO_3)_2 + 2H_2O$
   2 mol   1 mol   1 mol   2 mol

$$\frac{0.0175 \text{ mol } Ba(OH)_2}{L} \times 12.5 \text{ mL} \times \frac{L}{10^3 \text{ mL}} = 0.000\ 219 \text{ mol } Ba(OH)_2 \text{ used}$$

$$0.000\ 219 \text{ mol } Ba(OH)_2 \times \frac{2 \text{ mol } HNO_3}{1 \text{ mol } Ba(OH)_2} = 0.000\ 438 \text{ mol } HNO_3 \text{ reacted}$$

$$\frac{0.000\ 438 \text{ mol } HNO_3}{14.5 \text{ mL}} \times \frac{10^3 \text{ mL}}{L} = 0.0302 \text{ mol } HNO_3/L = 0.0302-M \text{ HNO}_3$$

19. $2H_3PO_4 + 3Ba(OH)_2 \rightarrow Ba_3(PO_4)_2 + 6H_2O$
   2 mol   3 mol   1 mol   6 mol

$$\frac{0.0230 \text{ mol. } H_3PO_4}{L} \times 13.4 \text{ mL} \times \frac{L}{10^3 \text{ mL}} = 0.000\ 308 \text{ mol } H_3PO_4 \text{ used}$$

$$0.000\ 308 \text{ mol } H_3PO_4 \times \frac{3 \text{ mol } Ba(OH)_2}{2 \text{ mol } H_2PO_4} = 0.000\ 462 \text{ mol } Ba(OH)_2 \text{ reacted}$$

$$\frac{0.000\ 462 \text{ mol } Ba(OH)_2}{22.5 \text{ mL}} \times \frac{10^3 \text{ mL}}{L} = 0.0205 \text{ mol } Ba(OH)_2/L = 0.0205\,M \text{ Ba(OH)}_2$$

20. $V_a N_a = V_b N_b$

$$V_b = \frac{V_a N_a}{N_b} = \frac{(20.0 \text{ mL})(0.75\,N)}{0.50\,N} = 30. \text{ mL}$$

21. $V_a N_a = V_b N_b$

$$N_b = \frac{V_a N_a}{V_b} = \frac{(35 \text{ mL})(0.10\,N)}{21.8 \text{ mL}} = 0.16\,N$$

Application Questions

1. The mass of one equivalent of a diprotic acid such as H_2SO_4 depends upon the degree to which the acid is neutralized in a chemical reaction. If two moles of H_3O^+ ion are donated, then one equivalent of H_2SO_4 is the same as one-half mole of the acid. However, if only one mole of H_3O^+ ions is donated, then one equivalent is the same as one mole of the acid.

2. (a) 0 (b) 0.18 (c) 0.45 (d) 0.57 (e) 0.66 (f) 0.81 (g) 0.93 (h) 1 (i) −4 (j) −12

3. (a) 4.57 (b) 1.53 (c) 2.94 (d) 4.57 (e) 5.39 (f) 6.42 (g) 8.77 (h) 10 (i) 10^3 (j) 10^{-6} (k) 10^{-11}

4. Aqueous solutions of HCl below 1 M are considered to be completely ionized. Thus, the molarities of such solutions are directly indicative of the $[H_3O^+]$. However, for acids such as $HC_2H_3O_2$ and H_2CO_3, the ionization is not complete. As a result, no direct relationship exists between molarity and the $[H_3O^+]$.

5. In titrations other than those involving strong acids and strong bases, the end point may occur at pH values other than 7 because the aqueous solutions of the resulting salts may themselves be slightly acidic or slightly basic depending upon the composition of the salt.

Application Problems

1. (a) $\dfrac{1.0 \text{ equiv } HNO_3}{L} \times 1.0 \text{ L} = 1.0 \text{ equiv } HNO_3$

(b) $\dfrac{2.5 \text{ equiv NaOH}}{L} \times 1.0 \text{ L} = 2.5 \text{ equiv NaOH}$

(c) $\dfrac{1.0 \text{ equiv } Mg(OH)_2}{L} \times 2.0 \text{ L} = 2.0 \text{ equiv } Mg(OH)_2$

(d) $\dfrac{3.0 \text{ equiv HCl}}{L} \times 1.5 \text{ L} = 4.5 \text{ equiv HCl}$

(e) $\dfrac{1.20 \text{ equiv } K_2CO_3}{L} \times 250. \text{ mL} \times \dfrac{L}{10^3 \text{ mL}}$
$= 0.300 \text{ equiv } K_2CO_3$

2. (a) $\dfrac{1.0 \text{ equiv NaOH}}{1.0 \text{ L}} = \dfrac{1.0 \text{ equiv NaOH}}{L} = 1.0\,N \text{ NaOH}$

(b) $\dfrac{2.5 \text{ equiv } HNO_3}{5.0 \text{ L}} = \dfrac{0.50 \text{ equiv } HNO_3}{L} = 0.50\,N \text{ HNO}_3$

(c) $\dfrac{40.1 \text{ g HF}}{1.00 \text{ L}} \times \dfrac{1 \text{ equiv HF}}{1 \text{ mol HF}} \times \dfrac{1 \text{ mol HF}}{20.0 \text{ g HF}} = \dfrac{2.00 \text{ equiv HF}}{L}$
$= 2.00\,N \text{ HF}$

(d) $\dfrac{74.1 \text{ g Ca(OH)}_2}{3.00 \text{ L}} \times \dfrac{2 \text{ equiv Ca(OH)}_2}{1 \text{ mol Ca(OH)}_2} \times \dfrac{1 \text{ mol Ca(OH)}_2}{74.1 \text{ g Ca(OH)}_2} = \dfrac{0.667 \text{ equiv Ca(OH)}_2}{\text{L}} = 0.667\text{-}N \text{ Ca(OH)}_2$

(e) $\dfrac{10.9 \text{ g H}_3\text{PO}_4}{60.0 \text{ mL}} \times \dfrac{10^3 \text{ mL}}{\text{L}} \times \dfrac{3 \text{ equiv H}_3\text{PO}_4}{\text{mol H}_3\text{PO}_4} \times \dfrac{1 \text{ mol H}_3\text{PO}_4}{98.0 \text{ g H}_3\text{PO}_4} = \dfrac{5.56 \text{ equiv H}_3\text{PO}_4}{\text{L}} = 5.56 \, N \text{ H}_3\text{PO}_4$

3. $\dfrac{1.20 \text{ equiv H}_2\text{SO}_4}{\text{L}} \times 500. \text{ mL} \times \dfrac{\text{L}}{10^3 \text{ mL}} = 0.600 \text{ equiv H}_2\text{SO}_4 \qquad \dfrac{\text{L}}{1.50 \text{ equiv H}_2\text{SO}_4} \times 0.600 \text{ equiv H}_2\text{SO}_4 \dfrac{10^3 \text{ mL}}{\text{L}} = 400. \text{ mL}$

4. $\dfrac{0.200 \text{ mol H}_3\text{PO}_4}{\text{L}} \times 2.50 \text{ L} \times \dfrac{3 \text{ equiv H}_3\text{PO}_4}{1 \text{ mol H}_3\text{PO}_4} = 1.50 \text{ equiv H}_3\text{PO}_4 \qquad \dfrac{\text{L}}{0.750 \text{ equiv H}_3\text{PO}_4} \times 1.50 \text{ equiv H}_3\text{PO}_4 = 2.00 \text{ L}$

5. (a) $\dfrac{1.0 \text{ equiv HF}}{F} \times \dfrac{1 \text{ mol HF}}{1 \text{ equiv HF}} = 1.0 \, M \text{ HF}$

(b) $\dfrac{3.0 \text{ equiv NaOH}}{\text{L}} \times \dfrac{1 \text{ mol NaOH}}{1 \text{ equiv NaOH}} = 3.0 \, M \text{ NaOH}$

(e) $\dfrac{0.75 \text{ equiv NH}_4\text{OH}}{\text{L}} \times \dfrac{1 \text{ mol NH}_4\text{OH}}{1 \text{ equiv NH}_4\text{OH}} = 0.75 \, M \text{ NH}_4\text{OH}$

(c) $\dfrac{2.0 \text{ equiv Mg(OH)}_2}{\text{L}} \times \dfrac{1 \text{ mol Mg(OH)}_2}{1 \text{ equiv Mg(OH)}_2} = 1.0 \, M \text{ Mg(OH)}_2$

(f) $\dfrac{0.50 \text{ equiv Ca(OH)}_2}{\text{L}} \times \dfrac{1 \text{ mol Ca(OH)}_2}{2 \text{ equiv Ca(OH)}_2} = 0.25 \, M \text{ Ca(OH)}_2$

(d) $\dfrac{1.5 \text{ equiv H}_2\text{SO}_4}{\text{L}} \times \dfrac{1 \text{ mol H}_2\text{SO}_4}{2 \text{ equiv H}_2\text{SO}_4} = 0.75 \, M \text{ H}_2\text{SO}_4$

(g) $\dfrac{2.4 \text{ equiv H}_3\text{PO}_4}{\text{L}} \times \dfrac{1 \text{ mol H}_3\text{PO}_4}{3 \text{ equiv H}_3\text{PO}_4} = 0.80 \, M \text{ H}_3\text{PO}_4$

6. (a) $[\text{H}_3\text{O}^+] = \text{antilog} - \text{pH} = \text{antilog}(-2.70) = (0.30 - 3) = 2.0 \times 10^{-3} \, M$

(b) $[\text{H}_3\text{O}^+][\text{OH}^-] = 1 \times 10^{-14} \, M^2$

$[\text{OH}^-] = \dfrac{1 \times 10^{-14} \, M^2}{[\text{H}_3\text{O}^+]} \times \dfrac{1 \times 10^{-14} \, M^2}{2 \times 10^{-3} \, M} = 5 \times 10^{-12} \, M$

(c) Since HNO$_3$ is a monoprotic acid giving 1 mol of H$_3$O$^+$ ions per mol of HNO$_3$ when completely ionized in solution, $2.0 \times 10^{-3} M$ H$_3$O$^+$ was, therefore, provided by a 2.0×10^{-3} M solution.

Thus, the HNO$_3$ solution is $2.0 \times 10^{-3} M$.

$\dfrac{2.0 \times 10^{-3} \text{ mol HNO}_3}{\text{L}} \times 5.50 \text{ L} \times \dfrac{63.0 \text{ g HNO}_3}{\text{mol HNO}_3} = 0.693 \text{ g HNO}_3$

(d) $0.693 \text{ g HNO}_3 \times \dfrac{100. \text{ g conc. acid}}{69.5 \text{ g HNO}_3} \times \dfrac{\text{mL conc. acid}}{1.42 \text{ g conc. acid}} = 0.702 \text{ mL conc. acid}$

7. $\dfrac{0.200 \text{ mol H}_2\text{SO}_4}{\text{L}} \times \dfrac{2 \text{ equiv H}_2\text{SO}_4}{1 \text{ mol H}_2\text{SO}_4} = 0.400 \, N \text{ H}_2\text{SO}_4 \qquad N_b = \dfrac{V_a N_a}{V_b} = \dfrac{12.5 \text{ mL} \times 0.400 \, N}{27.5 \text{ mL}} = \, = 0.182 \, N$

8. $\dfrac{0.250 \text{ mol H}_3\text{PO}_4}{\text{L}} \times \dfrac{3 \text{ equiv H}_3\text{PO}_4}{1 \text{ mol H}_3\text{PO}_4} = 0.750 N \text{ H}_3\text{PO}_4 \qquad V_a \dfrac{V_b N_b}{N_a} = \dfrac{40.0 \text{ mL} \times 0.560 \, N}{0.750 \, N} = 29.9 \text{ mL}$

$\dfrac{0.280 \text{ mol Ba(OH)}_2}{\text{L}} \times \dfrac{2 \text{ equiv Ba(OH)}_2}{1 \text{ mol Ba(OH)}_2} = 0.560 \, N \text{ Ba(OH)}_2$

9. $\dfrac{0.350 \text{ mol H}_2\text{SO}_4}{\text{L}} \times \dfrac{2 \text{ equiv H}_2\text{SO}_4}{1 \text{ mol H}_2\text{SO}_4} = 0.700 \, N \text{ H}_2\text{SO}_4$

$N_b = \dfrac{V_a N_a}{V_b} = \dfrac{18.5 \text{ mL} \times 0.700 \, N}{12.5 \text{ mL}} = 1.04 \, N$

$\dfrac{1.04 \text{ equiv LiOH}}{\text{L}} \times 12.5 \text{ mL} \times \dfrac{\text{L}}{10^3 \text{ mL}} \times \dfrac{1 \text{ mol LiOH}}{1 \text{ equiv LiOH}} \times \dfrac{23.9 \text{ g LiOH}}{\text{mol LiOH}} = 0.311 \text{ g LiOH}$

10. (a) Write the balanced equation, and determine the quantities of both reactants that are present.

$$2\text{HNO}_3 \quad + \quad \text{Ba(OH)}_2 \quad \rightarrow \quad \text{Ba(NO}_3)_2 \quad + \quad 2\text{H}_2\text{O}$$
2 mol   1 mol   1 mol   2 mol

$\dfrac{0.200 \text{ mol HNO}_3}{\text{L}} \times 15.0 \text{ mL} \times \dfrac{\text{L}}{10^3 \text{ mL}} = 0.003 \, 00 \text{ mol HNO}_3$

$\dfrac{0.0750 \text{ mol Ba(OH)}_2}{\text{L}} \times 25.0 \text{ mL} \times \dfrac{\text{L}}{10^3 \text{ mL}} = 0.001 \, 88 \text{ mol Ba(OH)}_2$

Based on the molar ratios established by the balanced equation, 0.003 00 mol HNO$_3$ requires 0.001 50 mol Ba(OH)$_2$. Since there is 0.001 88 mol Ba(OH)$_2$ available, then the Ba(OH)$_2$ is in excess by 0.001 88 mol minus 0.001 50 mol, or 0.000 38 mol.

(b) Since $Ba(OH)_2$ is present in excess, the final solution is basic.

(c) Prior to determining pH, first find $[OH^-]$ and then $[H_3O^+]$.

$$0.000\ 38\ \text{mol Ba(OH)}_2 \times \frac{2\ \text{mol OH}^-}{\text{mol Ba(OH)}_2} = 0.000\ 76\ \text{mol OH}^-$$

This fraction of a mole of OH^- is found in a total volume of 40.0 mL (15.0 mL + 25.0 mL).

$$\frac{0.000\ 76\ \text{mol OH}^-}{40.0\ \text{mL}} \times \frac{10^3\ \text{mL}}{\text{L}} = 0.019\ M\ \text{OH}^-$$

$$[H_3O^+][OH^-] = 1 \times 10^{-14}\ M^2$$

$$[H_3O^+] = \frac{1 \times 10^{-14}\ M^2}{[OH^-]} = \frac{1 \times 10^{-14}\ M^2}{1.9 \times 10^{-2}\ M} = 5.3 \times 10^{-13}\ M$$

$$pH = -\log\ [H_3O^+] = -\log\ (5.3 \times 10^{-13}) = -(0.72 + (-13))$$
$$-(0.72 - 13) = 12.3$$

Teacher's Notes

Chapter 17 Acid–Base Titration and pH

INTRODUCTION

In previous chapters of this unit, the strength and weakness of acids and bases were discussed qualitatively. This chapter introduces the pH scale, a quantitative measure of the strength of acids and bases. Titration, a procedure for determining the concentration and pH of a solution is also introduced.

LOOKING AHEAD

Keep in mind that the major purpose of this chapter is to introduce the analytical methods of titration. In reading about titration in Section 17.3, note how the pH of an acidic or basic solution is not measured directly. It is measured by comparison with a "standard" acidic or basic solution.

SECTION PREVIEW

17.1 Concentration Units for Acids and Bases
17.2 Aqueous Solutions and the Concept of pH
17.3 Acid–Base Titrations

A technician measures the acidity of blood plasma samples.

17.1 Concentration Units for Acids and Bases

In Chapter 14, you learned three ways to express concentration: mass percent, molarity, and molality. Recall that molarity *(M)* is the number of moles of solute per liter of solution and that molality *(m)* is the number of moles of solute per kilogram of solvent. In this section you will learn about a method of expressing concentration that is useful in dealing with acids and bases.

Chemical Equivalents

Solution concentrations can be expressed in a way that allows for simple measurement of solute quantities that are chemically equivalent; that is, quantities of substances that have the same combining capacity in chemical reactions. *These quantities of solutes that have equivalent combining capacity are called* **equivalents** *(equiv).*

Consider the following equations. They show that one mole of HCl and one-half mole of H_2SO_4 are chemically equivalent in neutralization reactions with basic KOH. That is, they both react completely with the same quantity (in this case, one mole) of KOH.

$$HCl\ +\ KOH\ \rightarrow\ KCl\ +\ H_2O$$
$$\text{1 mol}\qquad\text{1 mol}\qquad\text{1 mol}\qquad\text{1 mol}$$

$$\tfrac{1}{2}H_2SO_4\ +\ KOH\ \rightarrow\ \tfrac{1}{2}K_2SO_4\ +\ H_2O$$
$$\tfrac{1}{2}\text{ mol}\qquad\text{1 mol}\qquad\tfrac{1}{2}\text{ mol}\qquad\text{1 mol}$$

Since the mass of one mole of HCl is 36.5 g and the mass of one-half mole of H_2SO_4 is 49.0 g, then 36.5 g of HCl and 49.0 g of H_2SO_4 are chemically

SECTION OBJECTIVES

- Define equivalent and equivalent mass.
- Find the number of equivalents and the equivalent mass for a given acid or base in an acid–base reaction.
- State the relationship between normality and molarity.
- Calculate solution normality from solution concentration for a given acid or base.

STUDY HINT
It is helpful to review molarity, Section 14.3.

Ionization of 1 hydrogen chloride molecule (HCl) produces 1 hydrogen ion (H^+) and 1 chloride ion (Cl^-). Ionization of 1 mole of hydrogen chloride molecules (HCl) produces 1 mole of hydrogen ions (H^+) and 1 mole of chloride ions (Cl^-).

Ionization of 1 hydrogen sulfate molecule produces 2 hydrogen ions (H^+) and 1 sulfate ion (SO_4^{2-}). Ionization of 1 mole of hydrogen sulfate molecules (H_2SO_4) produces 2 moles of hydrogen ions (H^+) and 1 mole of sulfate ions (SO_4^{2-}).

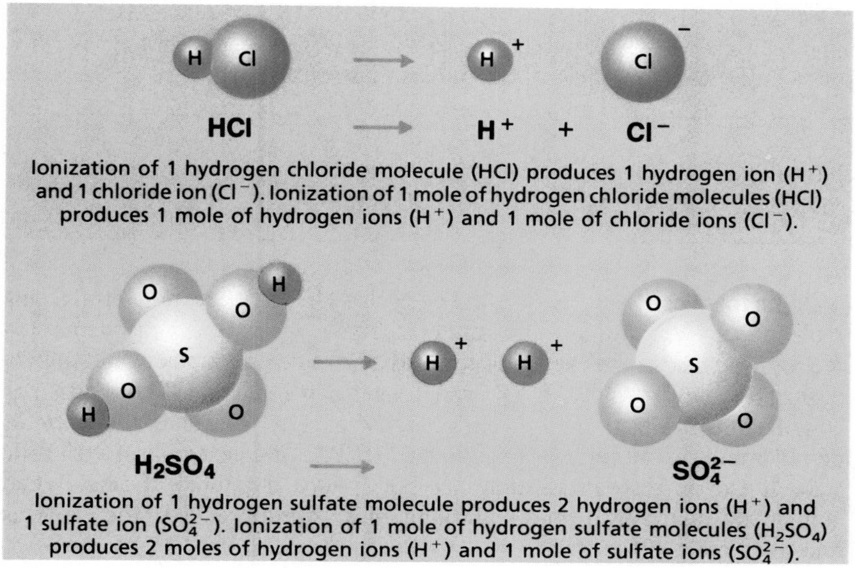

Figure 17-1 A mole of hydrogen sulfide produces twice as many hydrogen ions as a mole of hydrogen chloride.

equivalent to each other in these neutralization reactions. Both of them are also chemically equivalent to 56.1 g (the mass of 1 mole) of KOH in these reactions.

In *proton-transfer reactions, the* **equivalent mass** *of an acid is the quantity (in grams) that supplies one mole of protons.* A method for determining the equivalent mass of an acid in neutralization reactions is shown in Sample Problem 17.1. Note that in each case the mass of *one equivalent* is numerically equal to the mass of one mole of the acid divided by the number of moles of protons (expressed in terms of H^+ or H_3O^+) that one mole of the acid can provide.

Remember, proton-transfer refers to hydrogen ion (H^+) transfer.

Sample Problem 17.1

Determine the mass of one equivalent of HCl and H_2SO_4 in neutralization reactions.

Solution

Step 1. Analyze   Given: identity of substances: HCl and H_2SO_4
Unknown: *(a)* one equivalent mass of HCl; *(b)* one equivalent mass of H_2SO_4

Step 2. Plan   To find one equivalent mass of H_2SO_4 and HCl, the molar mass of each acid must be divided by the number of protons provided per mole of each acid.
molar mass, protons per mole of acid → equivalents

$$\text{molar mass}\left(\frac{g}{mol}\right) \times \frac{\text{mole substance (mol)}}{\text{mole protons (mol } H^+)} = \text{equivalents}\left(\frac{g}{mol\ H^+}\right)$$

Step 3. Compute   (a) $\dfrac{36.5\ g\ HCl}{mol\ HCl} \times \dfrac{1\ mol\ HCl}{1\ mol\ H^+} = 36.5\ g\ HCl/mol\ H^+$

(b) $\dfrac{98.1\ g\ H_2SO_4}{mol\ H_2SO_4} \times \dfrac{1\ mol\ H_2SO_4}{2\ mol\ H^+} = 49.0\ g\ H_2SO_4/mol\ H^+$

Step 4. Evaluate   Since one mole of HCl yields one mole of protons, it is reasonable that the molar mass of HCl is equal to one equivalent mass of the acid. Since one mole of H_2SO_4 yields two moles of protons, it is reasonable that the molar mass of H_2SO_4 is equal to twice the equivalent mass of the acid. One equivalent mass should therefore be equal to only half of one molar mass of H_2SO_4.

Practice Problems
1. Determine the mass of 1.00 equivalent of HNO_3. *(Ans.)* 63.0 g HNO_3/mol H^+
2. Determine the mass of 1.00 equivalent of H_2SO_3. *(Ans.)* 41.0 g H_2SO_3/mol H^+

The three common acids, HCl, HNO_3, and $HC_2H_3O_2$ are monoprotic. One mole of each can supply one mole of H_3O^+ ions. One equivalent of a monoprotic acid therefore is the same as one mole of the acid. One mole of a diprotic acid, such as H_2SO_4, can supply two moles of H_3O^+ ions. For complete neutralization, one equivalent of a diprotic acid is the same as one-half mole of the acid. Phosphoric acid (H_3PO_4) is an example of a triprotic acid. It can furnish three moles of H_3O^+ ions per mole of acid. When completely neutralized, one equivalent of a triprotic acid is the same as one-third mole of the acid.

A similar relationship exists between chemical equivalents and moles of bases. One equivalent of a base is the quantity, in grams, that accepts one mole of protons or supplies one mole of hydroxide (OH^-) ions. Consider again the type of neutralization reaction presented above. Suppose now that KOH is replaced with $Ca(OH)_2$ in the reaction with HCl. You can see, from the following equations, that one-half mole of $Ca(OH)_2$ is equivalent to one mole of KOH.

$$HCl \; + \; KOH \; \rightarrow \; KCl \; + \; H_2O$$
$$\text{1 mol} \quad \text{1 mol} \quad \text{1 mol} \quad \text{1 mol}$$

$$HCl \; + \tfrac{1}{2} Ca(OH)_2 \rightarrow \tfrac{1}{2} CaCl_2 + H_2O$$
$$\text{1 mol} \quad \tfrac{1}{2}\text{ mol} \quad \tfrac{1}{2}\text{ mol} \quad \text{1 mol}$$

Said another way, one mole of $Ca(OH)_2$ is equal to two equivalents of KOH. The equivalent masses of KOH and $Ca(OH)_2$ are found from the molar masses as illustrated for acids in Sample Problem 17.1.

$$\frac{56.1 \text{ g KOH}}{\text{mol KOH}} \times \frac{1 \text{ mol KOH}}{1 \text{ mol OH}^-} = 56.1 \text{ g KOH/mol OH}^-$$

$$\frac{74.1 \text{ g Ca(OH)}_2}{\text{mol Ca(OH)}_2} \times \frac{1 \text{ mol Ca(OH)}_2}{2 \text{ mol OH}^-} = 37.0 \text{ g Ca(OH)}_2/\text{mol OH}^-$$

In many chemical reactions, a diprotic or triprotic acid is not completely neutralized. In such a case, the number of moles of protons supplied per mole of acid is determined by the reaction that the acid undergoes. In fact, in all cases the number of equivalents of a given mass of reactant depends on the specific reaction involved. For example, suppose a solution containing one mole of H_2SO_4 is added to a solution containing one mole of NaOH. The product is the "acid salt" sodium hydrogen sulfate, which can be recovered by evaporating the water solvent. The equation is as follows:

$$H_2SO_4(aq) \; + \; NaOH(aq) \; \rightarrow \; NaHSO_4(aq) \; + \; H_2O(\ell)$$
$$\text{1 mol} \qquad \text{1 mol} \qquad \text{1 mol} \qquad \text{1 mol}$$
$$\text{98.1 g} \qquad \text{40.0 g}$$

Observe that the neutralization of H_2SO_4 is not complete. One mole of H_2SO_4 supplies only one mole of its two moles of protons (H^+) to the base. An acid salt containing HSO_4^- ions is the product. In this reaction, one equivalent of H_2SO_4 is the same as one molar mass, i.e. 98.1 g, of the acid.

Now, suppose that a solution containing one mole of phosphoric acid, H_3PO_4, is added to another containing one mole of NaOH. The reaction is shown below:

$$H_3PO_4(aq) \; + \; NaOH(aq) \; \rightarrow \; NaH_2PO_4(aq) \; + \; H_2O(\ell)$$
$$\text{1 mol} \qquad \text{1 mol} \qquad \text{1 mol} \qquad \text{1 mol}$$
$$\text{98.0 g} \qquad \text{40.0 g}$$

The product is the acid salt sodium dihydrogen phosphate, which can be recovered by evaporation. Since each H_3PO_4 unit in this reaction gives up only one of its three protons, one mole of H_3PO_4 is needed to supply one mole of protons to the base to form the product acid salt. In this reaction also, one equivalent of H_3PO_4 (98.0 g) is the same as one molar mass of this acid.

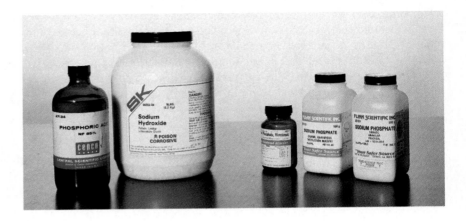

Figure 17-2 When phosphoric acid and sodium hydroxide react, any of three salts may be produced, depending on the relative number of moles of the reactants.

The product $[H_3O^+][OH^-]$ is often referred to as the ionization constant of water (K_w).

Normality

The concentration of solutions can be expressed by stating the quantity of the solute in equivalents rather than in moles. This measure of concentration is called normality (N). *The* **normality** *of a solution is the number of equivalents of solute per liter of solution.* A one-normal (1-N) solution contains one equivalent of solute per liter of solution. A 0.25-N solution contains 0.25 equivalent of solute per liter of solution. Three ways of expressing solution concentration are summarized in Table 17-1.

The normality of an acid or base solution is commonly expressed using the number of H_3O^+ or OH^- ions available for a complete neutralization. One mole of monoprotic hydrogen chloride has a mass of 36.5 g. As a reactant, this quantity of HCl can furnish one mole of protons. Thus, one mole of HCl in one liter of aqueous solution provides one equivalent of protons as H^+, or H_3O^+, ions. The concentration of this solution is 1 N.

TABLE 17-1 METHODS OF EXPRESSING CONCENTRATION OF SOLUTIONS

Name	Symbol	Solute Unit	Solvent Unit	Dimensions
molality	m	mole	kilogram solvent	$\dfrac{\text{mol solute}}{\text{kg solvent}}$
molarity	M	mole	liter solution	$\dfrac{\text{mol solute}}{\text{L solution}}$
normality	N	equivalent	liter solution	$\dfrac{\text{equiv solute}}{\text{L solution}}$

Sample Problem 17.2

Determine the normality of solutions that contain   *(a)* 1 mol H_2SO_4 dissolved in 1 L of solution,   *(b)* 0.01 mol of $Ca(OH)_2$ dissolved in 1 L of solution.

Solution
Step 1. Analyze   Given: identity and amount of solutes, volume of solution
                      Unknown: *(a)* Normality of H_2SO_4 solution;
                               (b) Normality of $Ca(OH)_2$ solution

Step 2. Plan

Equivalents per mol, volume → normality

H_2SO_4 gives two equivalents of H^+ per mole and $Ca(OH)_2$ gives two equivalents of OH^- per mole

$$\frac{\text{equivalents}}{\text{mol}} \times \frac{\text{moles}}{\text{volume of solution}}(L) = \frac{\text{equivalents}}{L} = \text{normality}$$

Step 3. Compute (a)

$$\frac{2 \text{ equivalents}}{1 \text{ mol } H_2SO_4} \times \frac{1 \text{ mol } H_2SO_4}{1L} \times \frac{2 \text{ equivalents}}{1L} = 2 N H_2SO_2$$

(b)

$$\frac{2 \text{ equivalents}}{0.01 \text{ mol } Ca(OH)_2} \times \frac{1 \text{ mol } Ca(OH)_2}{1L} = \frac{0.02 \text{ equivalents}}{1L} = 0.02 N Ca(OH)_2$$

Step 4. Evaluate

Since one mole of H_2SO_4 and one mole of $Ca(OH)_2$ yield 2 mol of H^+ and OH^- ions, respectively, solutions containing 1 mol/L and 0.01 mol/L must be 2 N and 0.02 N.

Practice Problems

1. Determine the normality of a 0.5-L solution containing 2.5 equivalents of H_2SO_3. *(Ans.)* 5 N
2. Determine the normality of a 4-L solution that contains 0.06 mol of $AL(OH)_3$ *(Ans.)* 0.015 N

The Relationship of Normality to Molarity

An advantage in expressing solution concentration in terms of normality rather than molarity is that equal volumes of different solutions of the same normality are chemically equivalent; that is, they contain equal numbers of equivalents. In general, if one mole of a solute is also one equivalent, the molarity and normality of the solution are the same. A 1-*M* HCl solution is thus a 1-*N* solution. If one mole of solute is two equivalents, however, a 1-*M* solution of that solute is 2 *N*. Therefore, a 0.01-*M* H_2SO_4 solution is 0.02 *N*. Similarly, a 0.01-*M* H_3PO_4 solution is 0.03 *N* in a complete neutralization. Solutions of equal normality are chemically equivalent, volume for volume.

Normality is numerically related to molarity by the simple relationship

$$N = nM$$

where n is the number of equivalents per mole.

Sample Problem 17.3

What is the molarity of a 0.06 *N* $Sr(OH)_2$ solution?

Solution

Step 1. Analyze

Given: normality of $Sr(OH)_2$ solution
Unknown: molarity of $Sr(OH)_2$ solution

Step 2. Plan

Normality → equivalents per mole → molarity

$Sr(OH)_2$ has 2 equivalents of OH^- per mole

$$M = \frac{N}{n} = \frac{\text{equivalents}}{L} \times \frac{\text{mole}}{\text{equivalents}}$$

Step 3. Compute

$$\frac{0.06 \text{ equivalents Sr(OH)}_2}{1 \text{ L}} \times \frac{1 \text{ mol Sr(OH)}_2}{2 \text{ equivalents}} = \frac{0.03 \text{ mol Sr(OH)}_2}{1 \text{ L}} = 0.03 \, M \text{ Sr(OH)}_2$$

Step 4. Evaluate   The molarity of a substance that yields 2 equivalents per mole should be one-half its normality.

Practice Problems
1. Determine the normality of a 0.3 M H_3PO_4 solution. *(Ans.)* 0.9 N
2. Compute the molarity of a 0.45 N NaOH solution. *(Ans.)* 0.45 M

1. 40.0 g NaOH/mol OH^-
2. One equivalent is the mass in grams that supplies or accepts one mole of protons. Normality is the number of equivalents per liter.
3. $N = nM$ where n is the no. of equivalents per mole.
4. Equal volumes of the same normality contain the same number of equivalents.

Section Review

1. Find the equivalent mass of NaOH and $BA(OH)_2$ in acid–base reactions.
2. What is the relationship between equivalents and normality?
3. Explain the difference between molarity and normality.
4. Why would normality be a more useful measure of solution concentration when dealing with reactions between two different solutions.

17.2 Aqueous Solutions and the Concept of pH

SECTION OBJECTIVES

- Describe the self-ionization of water.
- Define pH and give the pH of a neutral solution.
- Explain and use the pH scale.
- Given $[H_3O^+]$ or $[OH^-]$, find pH.
- Given pH, find $[H_3O^+]$ or $[OH^-]$.

You are probably used to thinking of water as an inactive medium that can be used to dissolve solutes and permit reaction between them. Actually, however, water has an important chemical property that makes it far from inactive. In the future, you will need to take this property into consideration when you deal with aqueous solutions—especially those containing acids or bases.

Self-Ionization of Water

Water, however pure, is ionized in a process called **self-ionization**. This process, which occurs only to a slight degree, is sometimes referred to as *autoprotolysis*. The conductivity of pure water results from this slight ionization of water molecules.

This fact can be demonstrated by testing water that has been highly purified by several different techniques. Electric conductivity measurements

Figure 17-3   Water is weakly ionized by self-ionization.

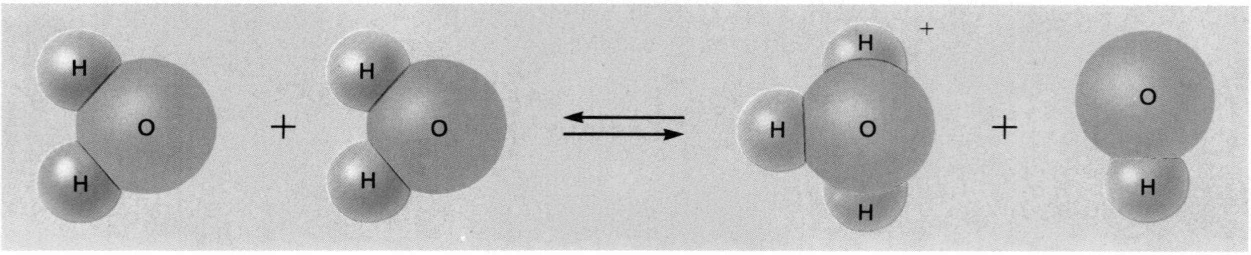

of such pure water show that it is indeed very slightly ionized to form H_3O^- and OH^- ions. Concentrations of these ions in pure water are only one mole of each per 10^7 liters of water at 25°C.

You will find it is useful to express ion concentration as moles per liter rather than as moles per 10^7 liters.

$$\frac{1 \text{ mol } H_3O^+}{10^7 \text{ L } H_2O} = 10^{-7} \text{ mol } H_3O^+/L \text{ } H_2O$$

The concentration of H_3O^+ ions (and OH^- ions) in water at 25°C is thus 10^{-7} M. In percentage, water is 2×10^{-9}% ionized.

Chemists use a standard notation to represent concentration in terms of moles per liter. The symbol or formula of the particular ion or molecule is enclosed in brackets, [ ]. For example, $[H_3O^+]$ means "hydronium ion concentration in moles per liter, or "molar hydronium ion concentration." For the ionic concentrations in water at 25°C, $[H_3O^+]$ is 10^{-7} M and $[OH^-]$ is 10^{-7} M. Because the hydronium and hydroxide ion concentrations are the same, water is neutral: it is neither acidic nor basic. This neutrality prevails in any solution in which $[H_3O^+] = [OH^-]$. Thus, the word *neutral*, when applied to an aqueous solution, does not generally mean that there are no H_3O^+ or OH^- ions present. It means only that the concentrations of these ions are equal.

If the H_3O^+ ion concentration exceeds 10^{-7} mol/L, an aqueous solution is acidic. For example, a solution containing 10^{-5} mol H_3O^+ ion per liter is acidic ($10^{-5} > 10^{-7}$). If the OH^- ion concentration exceeds 10^{-7} mol per liter, the solution is basic. Thus, a solution containing 10^{-4} mol OH^- ions per liter is basic ($10^{-4} > 10^{-7}$).

In an aqueous solution, the following self-ionization equilibrium is taking place.

$$H_2O(\ell) + H_2O(\ell) \rightleftharpoons H_3O^+ + OH^-$$

The product of $[H_3O^+]$ and $[OH^-]$ remains constant in water and dilute aqueous solutions as long as the temperature does not change. According to LeChâtelier's principle, an increase in the concentration of either H_3O^+ or OH^- in an aqueous solution causes a decrease in the concentration of the other ion. Since an increase in either H_3O^+ or OH^- results in an exact proportional decrease in the other ion, it is also true that the *product* of $[H_3O^+]$ and $[OH^-]$ remains constant in water and dilute aqueous solutions. This constant will remain the same for a given temperature; however, a change in temperature results in a different constant. In water and dilute aqueous solutions at 25°C, the following relationship is valid:

$$[H_3O^+]\,[OH^-] = 10^{-7} M \times 10^{-7} M = (10^{-7} M)^2 = 10^{-14} M^2$$

The ionization of water increases slightly with temperature increases. The ion product $[H_3O^+]\,[OH^-]$ of 10^{-14} M^2 is commonly used as a constant within the ordinary range of room temperatures.

When the margin for error is very slim, as in quantitative analysis, preparing standardized solutions is critical.

The pH Scale

The range of solution concentrations encountered by chemists is very wide. It varies from above 10 M to below 10^{-15} M. Concentrations of 1 M or only somewhat less than 1 M are most commonly used, however. As stated above, the product of $[H_3O^+]$ and $[OH^-]$ is a constant at a given temperature. Therefore, if the concentration of either ionic species is known, the concentration of the other species can be determined using this information. For example, the OH^- ion concentration of a 0.01-M NaOH solution is 0.01, or 10^{-2} mole per liter. The H_3O^+ ion concentration of this solution is calculated as follows.

$$[H_3O^+] \, [OH^-] = 10^{-14} \, M^2$$

$$[H_3O^+] = \frac{1 \times 10^{-14} \, M^2}{[OH^-]} = \frac{1 \times 10^{-14} \, M^2}{1 \times 10^{-2} \, M} = 1 \times 10^{-12} \, M$$

Similarly, the H_3O^+ and OH^- concentrations in a 0.0001-M HNO$_3$ solution, which is also 100% ionized, are found as follows.

Sample Problem 17.4

A 0.0001-M solution of HNO$_3$ has been prepared for a laboratory experiment.   *(a)* Calculate the $[H_3O^+]$ of this solution.   *(b)* Calculate the $[OH^-]$.

Solution

Step 1. Analyze   Given: Concentration and identity of solution = 0.0001-M HNO$_3$
Unknown: *(a)* $[H_3O^+]$ *(b)* $[OH^-]$

Step 2. Plan   concentration HNO$_3 \rightarrow$ concentration $H_3O^+ \rightarrow$ concentration OH^-

Since HNO$_3$ is a strong acid, which is essentially 100% ionized in dilute solutions, then the concentration of the hydronium ions is equal to the concentration of the acid: one molecule of acid produces one hydronium ion. Because the ion product of $[H_3O^+] \, [OH^-]$ is a constant, the $[OH^-]$ can easily be determined by using the value for $[H_3O^+]$.

(a) $\underset{\text{1 mol}}{HNO_3} + \underset{\text{1 mol}}{H_2O} \rightarrow \underset{\text{1 mol}}{H_3O^+} + \underset{\text{1 mol}}{NO_3^-}$ (assuming 100% ionization)

$$\text{conc. of HNO}_3 \frac{(\text{mol HNO}_3)}{(\text{L sol'n})} \times \frac{(\text{mol H}_3\text{O}^+)}{(\text{mol HNO}_3)} = \text{conc. of H}_3\text{O}^+ \frac{(\text{mol H}_3\text{O}^+)}{(\text{L sol'n})}$$

(b) $[H_3O^+] \, [OH^-] = 10^{-14} \, M^2$

$$[OH^-] = \frac{10^{-14} \, M^2}{[H_3O^+]}$$

Step 3. Compute   *(a)* $[H_3O^+] = \dfrac{0.0001 \text{ mol HNO}_3}{\text{L solution}} \times \dfrac{1 \text{ mol H}_3\text{O}^+}{1 \text{ mol HNO}_3} = 0.0001 \, M \, H_3O^+ = 10^{-4} \, M \, H_3O^+$

(b) $[OH^-] = \dfrac{10^{-14} \, M^2}{[H_3O^+]} \times \dfrac{10^{-14} \, M^2}{10^{-4} \, M} = 10^{-10} \, M \, OH^-$

Step 4. Evaluate   Since one mole of HNO$_3$ produces one mole of hydronium ions, then a 0.0001-M HNO$_3$ solution will have a 0.0001-M concentration of H_3O^+. The answers are correctly expressed to one significant digit.

Practice Problems

1. Determine the hydronium and hydroxide ion concentrations in a solution that is 10^{-4} M HCl.
(Ans.) $[H_3O^+] = 10^{-4}$ M; $[OH^-] = 10^{-10}$ M

2. Determine the hydronium and hydroxide ion concentrations in a solution which is 0.0010 M HNO_3.
(Ans.) $[H_3O^+] = 1.0 \times 10^{-3}$ M; $[OH^-] = 1.0 \times 10^{-11}$ M

Expressing acidity or basicity in terms of the concentration of H_3O^+ or OH^- can become cumbersome. This is especially true in dilute solutions, whether decimal or scientific notation is used. Because it is more convenient, chemists use a quantity called pH to indicate the hydronium ion concentration of a solution. The letters *pH* stand for the French words *pouvoir hydrogène,* meaning "hydrogen power."

Numerically, the pH of a solution is defined as the negative of the common (base = 10) logarithm of the hydronium ion concentration, $[H_3O^+]$. *The* **pH** *is expressed by the following equation:*

$$\text{pH} = -\log\ [H_3O^+]$$

As you may recall from your mathematics studies, the common logarithm of a number is the power to which 10 must be raised to give the number. For example, the logarithm of 10^7 is $+7$.

Pure water at 25°C contains a 10^{-7} M concentration of H_3O^+. The pH of water is therefore determined as follows:

$$\text{pH} = -\log\ [H_3O^+] = -\log\ (10^{-7}) = -(-7) = 7$$

Suppose the H_3O^+ ion concentration in a solution is greater than that in pure water, as is the case in acidic solutions. Then the concentration of H_3O^+ is higher than 10^{-7} M. Suppose, for example, that its value is 10^{-6} M. The pH of this acidic solution would then be

$$\text{pH} = -\log\ [H_3O^+] = -\log\ (10^{-6}) = -(-6) = 6$$

As you can see, the pH is a number smaller than 7. This is the case for all acidic solutions.

If you suppose, on the other hand, that the H_3O^+ ion concentration is smaller than that in pure water, the pH would then be a number larger than 7. This is the case for all basic solutions. To summarize, *if pH is less than 7, the solution is acidic. If pH is greater than 7, the solution is basic.*

The possible range of pH values of aqueous solutions generally falls between 0 and 14. The pH system is particularly useful in describing the acidity and basicity of solutions that are not too far from neutral. Near-neutral solutions include many food substances and many fluids, such as blood. The pH values of some common substances are given in Table 17-2.

Chemical solutions with pH values below 0 and above 14 can be prepared. For example, the pH of 6-M H_2SO_4 is between 0 and -1. The pH of 3-M KOH solution is approximately 14.5. Only common pH values in the 0–14 range will be considered in this course, however.

There are two basic types of pH problems that will concern you: (1) the calculation of pH when the $[H_3O^+]$ of a solution is known, and (2) the calculation of $[H_3O^+]$ when the pH of a solution is known.

Figure 17-4 The pH of shampoo is adjusted during manufacturing to be slightly acidic or neutral; therefore, it is gentler to the hair.

Figure 17-5 Guess the pH of these common household products. Check your answers using Table 17-2.

There exists an inverse relationship between volume and concentration. If the concentration increases, the volume needed to get the same amount of substance decreases.

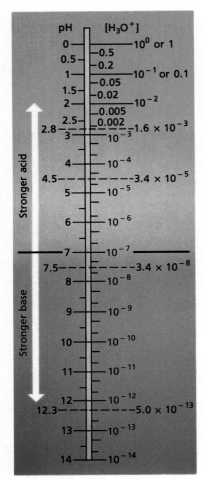

$[H_3O^+]$

10^0 or 1
0.5
0.2
10^{-1} or 0.1
0.05
0.02
10^{-2}
0.005
0.002
1.6×10^{-3}
10^{-3}
10^{-4}
3.4×10^{-5}
10^{-5}
10^{-6}
10^{-7}
3.4×10^{-8}
10^{-8}
10^{-9}
10^{-10}
10^{-11}
10^{-12}
5.0×10^{-13}
10^{-13}
10^{-14}

Stronger acid

Stronger base

Figure 17-6 A comparison of scales showing pH and hydronium ion concentration

TABLE 17-2 APPROXIMATE pH OF SOME COMMON MATERIALS (AT 25 °C)

Material	pH	Material	pH
1.0-N HCl	0.1	bread	5.5
1.0-N H$_2$SO$_4$	0.3	potatoes	5.8
0.1-N HCl	1.1	rainwater	6.2
0.1-N H$_2$SO$_4$	1.2	milk	6.5
gastric juice	2.0	saliva	6.5–6.9
0.01-N H$_2$SO$_4$	2.1	pure water	7.0
lemons	2.3	eggs	7.8
vinegar	2.8	0.1-N NaHCO$_3$	8.4
0.1-N HC$_2$H$_3$O$_2$	2.9	seawater	8.5
soft drinks	3.0	milk of magnesia	10.5
apples	3.1	0.1-N NH$_3$	11.1
grapefruit	3.1	0.1-N Na$_2$CO$_3$	11.6
oranges	3.5	0.1-N NaOH	13.0
cherries	3.6	1.0-N NaOH	14.0
tomatoes	4.2	1.0-N KOH	14.0
bananas	4.6		

Calculations Involving pH

In the simplest of pH problems, the $[H_3O^+]$ of the solution is an integral power of 10, such as 1 M or 0.01 M. You can solve these problems simply by inspection. For example, the pH of a solution in which $[H_3O^+]$ is 10^{-5} M is 5. This is because the negative logarithm of 10^{-5} is equal to the exponent of 10, or -5 in this case. The pH, or negative of the logarithm, is thus $-(-5)$ = 5.

$$pH = -\log [H_3O^+] = -\log (10^{-5}) = -(-5) = 5$$

In general, when $[H_3O^+]$ is an integral power of 10, the pH of a solution is the exponent of the hydronium ion concentration with the sign changed. Sample Problem 17.5 gives an example involving a base.

Sample Problem 17.5

What is the pH of a 0.001-M NaOH solution?

Solution

Step 1. Analyze Given: Identity and concentration of solution = 0.001 M NaOH
Unknown: pH of solution

Step 2. Plan concentration of base → concentration of OH$^-$ → concentration of H$_3$O$^+$ → pH

Since NaOH is completely dissociated when it is dissolved in water, a 0.001-M NaOH solution produces an $[OH^-]$ equal to 0.001-M. The ion product of $[H_3O^+]$ and $[OH^-]$ is a constant, 10^{-14} M^2. By substitution, the $[H_3O^+]$ can be determined. The pH is then determined as in the above problem.

Sample Problem 17.5 *continued*

Step 3. Compute

$$[H_3O^+][OH^-] = 10^{-14}$$

$$[H_3O^+] = \frac{10^{-14}\,M^2}{[OH^-]} = \frac{10^{-14}\,M^2}{(10^{-3}\,M)} = 10^{-11}\,M$$

$$pH = -\log[H_3O^+] = -\log(10^{-11}) = -(-11) = 11$$

Step 4. Evaluate   The answer correctly indicates that NaOH forms a solution with pH > 7, which is basic.

Practice Problems
1. Determine the pH of a 10^{-3}-M HCl solution. *(Ans.)* pH = 3
2. Determine the pH of a 10^{-5}-M HNO$_3$ solution. *(Ans.)* pH = 5
3. Determine the pH of a 10^{-4}-M NaOH solution. *(Ans.)* pH = 10
4. Determine the pH of a 10^{-2}-M KOH solution. *(Ans.)* pH = 12

The preceding problems have hydronium ion concentrations that are integral powers of 10. They are easily solved by inspection. Many problems, however, involve hydrogen ion concentrations that are not integral powers of 10. Solving such problems requires some basic knowledge of logarithms and exponents. Most scientific calculators have a "log" key. You may need to enter the number whose log is to be found and then press the log key, or the procedure may be in the reverse order. (Consult the instructions for your particular calculator. In the absence of a calculator, log tables must be used.) Note that the logarithm given by your calculator may have far more than the necessary number of digits.

The relationship between the pH and $[H_3O^+]$ is shown on the scale of Figure 17-6. This scale can be used to estimate the pH from a known $[H_3O^+]$ value. For example, suppose that you are given that the $[H_3O^+]$ of a solution is $3.4 \times 10^{-5}\,M$ and you are asked to find the pH. Observe that 3.4×10^{-5} lies between 10^{-4} and 10^{-5} on the $[H_3O^+]$ scale of Figure 17-6. The pH of the solution thus must be between 4 and 5. The value can be estimated to a second significant digit as 4.5 on the pH scale of Figure 17-6. You must carry out calculations as to obtain a more precise pH value. However, this reliable estimate of pH helps prevent you from making the sorts of errors that commonly occur in such calculations. Sample Problem 17.6 continues the actual calculation of the pH value for this example.

Point out to the students that pH is an exponential value. This means that if the pH changes by 3 units, the concentration of H_3O^+ changes by 10^3 or 1000 units.

Stress the fact that as the concentration of H_3O^+ increases, the pH decreases.

Sample Problem 17.6

What is the pH of a solution if $[H_3O^+]$ is $3.4 \times 10^{-5}\,M$?

Solution

Step 1. Analyze   Given: Concentration of $H_3O^+ = 3.4 \times 10^{-5}\,M$
Unknown: pH of solution

Step 2. Plan   concentration of $H_3O^+ \rightarrow$ pH

The only difference between this problem and previous pH problems is that you must determine the logarithm of 3.4 by using your calculator.

Step 3. Compute $pH = -\log [H_3O^+] = -\log (3.4 \times 10^{-5}) = -(\log 3.4 + \log 10^{-5}) = -(0.53 + (-5)) = -(0.53 - 5) = -(-4.47) = 4.47$

Step 4. Evaluate   Since the pH of a $10^{-5}-M$ H_3O^+ solution is 5, it is reasonable that a solution with a greater concentration of hydronium ions would be more acidic and have a pH less than 5.

Practice Problems
1. What is the pH of a solution if $[H_3O^+]$ is 6.7×10^{-4} mol/L? *(Ans.)* pH = 3.17
2. What is the pH of a solution with a hydronium ion concentration of $2.5 \times 10^{-2} M$? *(Ans.)* pH = 1.60

You have now learned to calculate the pH of a solution when given its hydronium ion concentration. Now suppose that you are given the pH of a solution instead. How can you determine its hydronium ion concentration? The equation for the pH in terms of the $[H_3O^+]$ is:

$$pH = -\log [H_3O^+]$$

Remember that the base of common logarithms is 10. The antilog of the common logarithm of a number is the power to which 10 must be raised to give that number. For the pH, this would be

$$\log [H_3O^+] = -pH$$

$$[H_3O^+] = \text{antilog } (-pH)$$

$$[H_3O^+] = 10^{-pH}$$

For an aqueous solution that has a pH of 2, for example, the $[H_3O^+]$ is equal to $10^{-2} M$. As you can see, the exponent of 10 is simply the negative of the pH. When the pH is 0, the $[H_3O^+]$ is $1 M$, since 10^0 is 1. Sample Problem 17.7 involves another example of a pH value that is a positive integer. Sample Problem 17.8, which follows, involves a pH that is not an integral number. The procedure for finding an antilog varies with the calculator. There may be an "antilog" key, or a "10^x" key, or several keystrokes may be required. You should familiarize yourself with the method for your own calculator.

Sample Problem 17.7

Determine the hydronium ion concentration of an aqueous solution that has a pH of 4.

Solution
Step 1. Analyze   Given: pH = 4
Unknown: $[H_3O^+]$

Step 2. Plan   pH → hydronium ion concentration

This problem requires that you rearrange the pH equation and solve for the $[H_3O^+]$. You must use antilogs.

Sample Problem 17.7 *continued*

Step 3. Compute

$$pH = -\log [H_3O^+]$$

$$\log [H_3O^+] = -pH$$

$$[H_3O^+] = \text{antilog } (-pH) = \text{antilog } (-4) = 10^{-4} \, M$$

Step 4. Evaluate It follows that if the pH of $10^{-4} \, M \, H_3O^+$ is 4, then a solution with a pH of 4 has a $[H_3O^+]$ equal to $10^{-4} \, M$. This problem is reasonably simple to double check by taking your answer and determining the pH.

Sample Problem 17.8

The pH of a solution is measured and determined to be 7.52.   *(a)* What is the hydronium ion concentration?   *(b)* What is the hydroxide ion concentration?   *(c)* Is the solution acidic or basic?

Solution

Step 1. Analyze Given: pH of the solution = 7.52
Unknown: *(a)* $[H_3O^+]$ *(b)* $[OH^-]$ *(c)* Is solution acidic or basic?

Step 2. Plan

$$pH \rightarrow [H_3O^+] \rightarrow [OH^-]$$

This problem is very similar to previous pH problems. You will need to substitute values into the pH $= -\log [H_3O^+]$ equation and work the antilogs properly. Once the $[H_3O^+]$ is determined, the ion-product constant of $[H_3O^+]$ $[OH^-]$ equals $10^{-14} \, M^2$ may be used. The pH of the solution tells you about the acidity or basicity of the solution.

Step 3. Compute *(a)* $pH = -\log [H_3O^+]$
$\log [H_3O^+] = -pH$
$[H_3O^+] = \text{antilog } (-pH) = \text{antilog } (-7.52) = 10^{-7.52} = 3.0 \times 10^{-8} M$

(b) $[H_3O^+] [OH^-] = 10^{-14} \, M^2$

$$[OH^-] = \frac{10^{-14} \, M^2}{[H_3O^+]} = \frac{10^{-14} \, M^2}{3.0 \times 10^{-8} \, M} = 3.3 \times 10^{-7} \, M$$

(c) A pH of 7.5.2 is slightly greater than a pH of 7. This means that the solution is slightly basic.

Step 4. Evaluate Since the solution is slightly basic, a hydroxide ion concentration slightly larger than $10^{-7} \, M$ is predicted. A hydronium ion concentration slightly less than $10^{-7} \, M$ is also predicted.

Practice Problems
1. The pH of a solution is determined to be 5. What is the hydronium ion concentration of this solution? *(Ans.)* $[H_3O^+] = 10^{-5} \, M$
2. The pH of a solution is determined to be 12. What is the hydronium ion concentration of this solution? *(Ans.)* $[H_3O^+] = 10^{-12} \, M$
3. The pH of an aqueous solution is measured as 1.50.   *(a)* Calculate the $[H_3O^+]$. *(Ans.)* $[H_3O^+] = 3.2 \times 10^{-2} \, M$ *(b)* Determine the $[OH^-]$. *(Ans.)* $[OH^-] = 3.2 \times 10^{-13} \, M$
4. The pH of an aqueous solution is 3.67. Determine $[H_3O^+]$. *(Ans.)* $[H_3O^+] = 2.1 \times 10^{-4} \, M$

Chemistry Notebook

Acid Rain and Lakes

Acid rain is caused by the combustion of fossil fuels, which produces oxides of nitrogen and sulfur that then dissolve in water to form nitric and sulfuric acid. Much of the northeastern United States and Canada is affected by acid rain.

A lake can withstand acid rain for a few years if the local soil is rich in alkaline materials. However, when the substructure is granite,

as is typical in New England, there is little protective buffering action, and the lake is susceptible to permanent

damage. Some adult fish begin to die if the pH drops below 5.0, and no fish can reproduce when it is below 4.5. A lake can have over 50 varieties of plankton at a pH of 7.0, but only 12 species can survive at a pH of 5.0, and none can survive below 4.5. Once the pH of a lake falls below 5.0, it is defined as acidic. Hundreds of lakes in the northeast are classified as acidic.

TABLE 17-3	RELATIONSHIP OF $[H_3O^+]$ TO $[OH^-]$ AND pH (AT 25 °C)				
Solution	$[H_3O^+]$	$[OH^-]$	$[H_3O^+][OH^-]$	pH
0.02-M KOH	5.0×10^{-13}	2.0×10^{-2}	1.0×10^{-14}	12.3
0.01-M KOH	1.0×10^{-12}	1.0×10^{-2}	1.0×10^{-14}	12.0
pure H_2O	1.0×10^{-7}	1.0×10^{-7}	1.0×10^{-14}	7.0
0.001-M HCl	1.0×10^{-3}	1.0×10^{-11}	1.0×10^{-14}	3.0
0.1-M $HC_2H_3O_2$	1.3×10^{-3}	7.7×10^{-12}	1.0×10^{-14}	2.9

Table 17-3 shows the relationship between the hydronium ion and hydroxide ion concentrations, the product of these concentrations, and the pH for several solutions of typical molarities. Since KOH is a soluble ionic compound, it is completely dissociated when forming an aqueous solution. The molarity of each KOH solution indicates directly the $[OH^-]$. Since the product of $[H_3O^+]$ and $[OH^-]$ is the constant $10^{-14}\ M^2$ at 25°C, the $[H_3O^+]$ can be calculated. If the $[H_3O^+]$ is known, the pH can be determined as $-\log [H_3O^+]$.

Hydrochloric acid may be considered to be completely ionized in any aqueous solution that has a concentration below 1 M. The $[H_3O^+]$ concentration is therefore equal to the molar concentrations of the solution. For example, a 0.001-M HCl solution has a $[H_3O^+]$ equal to 0.001 M.

The weakly ionized acetic acid ($HC_2H_3O_2$) solution presents a different problem. Information about the concentrations of acetic acid molecules, hydronium ions, and acetate ions in the equilibrium mixture may be lacking. You can determine the hydronium ion concentration, however, by measuring the pH of the solution experimentally by using a pH meter. You can then calculate the $[H_3O^+]$ as the antilog ($-$pH). (See Sample Problems 17.7 and 17.8)

Section Review

1. What is the concentration of hydronium and hydroxide ions in pure water at 25°C?
2. Why does the pH scale range from 0 to 14 in aqueous solutions?
3. Why does a pH of 7 represent a neutral solution?
4. Identify each of the following solutions as acidic or basic.

   (a) $[H_3O^+] = 1 \times 10^{-3} \dfrac{mol}{L}$ (b) $[OH^-] = 1 \times 10^{-4} \dfrac{mol}{L}$

   (c) pH = 5   (d) pH = 8.

17.3 Acid–Base Titrations

Everything you have learned so far can be applied to the analysis of acidic and basic solutions and the reactions between them. In order for you to apply your knowledge in this way, however, you will need to find out about certain "tools." These "tools" will enable you to get general information on acid–base concentrations virtually at a glance.

Indicators

The chemical "tools" just referred to are the substances called indicators. You have already read about the common acid–base indicator, litmus. As you may recall, litmus is red in acidic solutions and blue in basic solutions.

Indicators *are weak acid or base dyes whose colors are sensitive to pH, or hydronium ion concentration.* Chemists have a wide choice of indicators for use in the study of acids and bases and their reactions. In a given situation they are able to choose one that has different colors at certain pH values and that changes colors over the correct pH range for a particular reaction. *The pH range over which an indicator changes color is called its* **transition interval**. Table 17-4 gives the color changes and transition intervals for a number of common acid–base indicators. As you will see, the choice of

SECTION OBJECTIVES

- Describe how an acid–base indicator functions.
- Explain the procedures involved in carrying out an acid–base titration.
- Calculate the molarity of a solution from titration data.
- Calculate the normality of a solution from titration data.

TABLE 17-4   INDICATOR COLORS

	Color			Transition Interval
Indicator	Acid	Transition	Base	(pH)
methyl violet	yellow	aqua	blue	0.0– 1.6
methyl yellow	red	orange	yellow	2.9– 4.0
bromophenol blue	yellow	green	blue	3.0– 4.6
methyl orange	red	orange	yellow	3.2– 4.4
methyl red	red	buff	yellow	4.8– 6.0
litmus	red	pink	blue	5.5– 8.0
bromothymol blue	yellow	green	blue	6.0– 7.6
phenol red	yellow	orange	red	6.6– 8.0
phenolphthalein	colorless	pink	red	8.2–10.6
thymolphthalein	colorless	pale blue	blue	9.4–10.6
alizarin yellow	yellow	orange	red	10.0–12.0

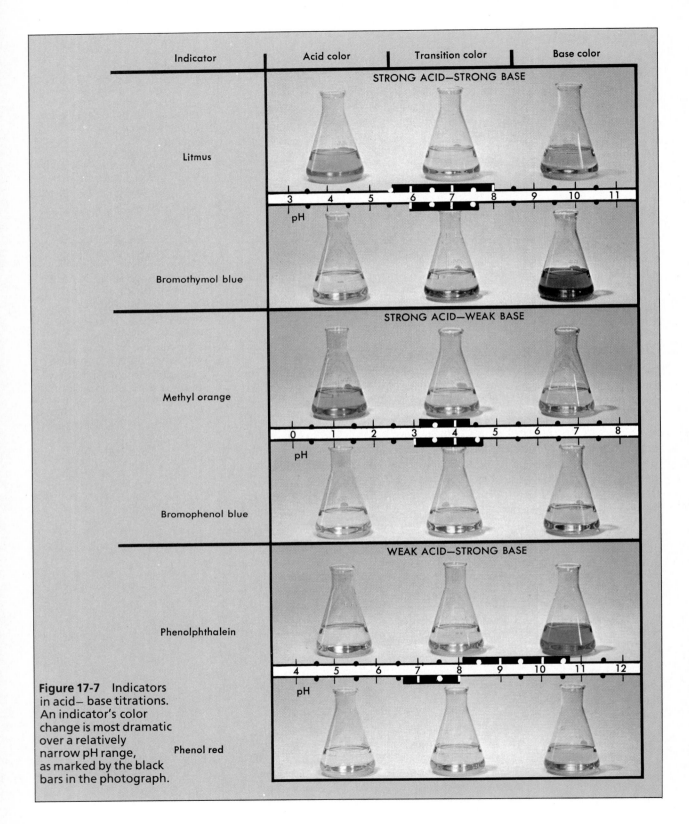

Figure 17-7 Indicators in acid–base titrations. An indicator's color change is most dramatic over a relatively narrow pH range, as marked by the black bars in the photograph.

indicator is based on the suitability of its transition interval for a given acid–base reaction.

Indicators are generally divided into three types, according to the pH at which they change color:

1. Those of the first type change color at about pH 7. Such indicators include litmus. The color-change interval for litmus, however, is inconveniently broad—pH 5.5–8.0. This makes for considerable imprecision when acid–base reactions are being studied. The indicator bromthymol blue performs more satisfactorily in this respect since its transition interval is relatively narrow—pH 6.0–7.6.

Bromthymol blue is yellow in acids, blue in bases, and green in solutions that are almost neutral. Indicators of this general type that undergo transition at about pH 7 are especially useful in the study of neutralization reactions between strong acids and strong bases. The neutralization of strong acids with strong bases produces a salt solution with a pH of approximately 7. The use of this type of indicator makes it easy to observe the point at which such neutralization occurs; that is, the point at which equal numbers of equivalents of acids and bases have combined to form products.

2. Indicators of the second type change color at pH lower than 7. Methyl orange is an example of this type. Such indicators are especially useful in studying neutralization reactions between strong acids and weak bases. These reactions produce salt solutions whose pH is lower than 7.

3. The third type of indicator changes color at pH higher than 7. Phenolphthalein is an example. Such indicators are especially useful when studying neutralization reactions between weak acids and strong bases. These reactions produce salt solutions whose pH is greater than 7.

You may be wondering what sort of indicator is useful when studying reactions between weak acids and weak bases. The surprising answer is "none at all!" The pH of the solution that results from complete neutralization reaction between weak acids and weak bases may be almost any value, depending on the relative strengths of the reactants. Without reasonably complete information, the transition interval of an indicator helps very little in determining whether reactions between such acids and bases are complete.

So-called universal indicators can be made by mixing together several different indicator solutions. The resulting solution can turn a number of colors, depending on the pH of the solution it is being used to test. Strips of paper can be soaked in universal indicator solution to produce what is called pH paper. This paper can take on almost any of the colors of the rainbow and thus, can be a fairly accurate way of distinguishing solutions having different pH values. Figure 17-8 shows the pH scale and the colors for this indicator paper.

The mechanism by which indicators operate is interesting. The indicators themselves are weak acids. In solution, there is an equilibrium between the un-ionized indicator molecule—let's give it the general formula HIn—and the H^+ and In^- ions that are produced when this molecule ionizes.

$$HIn \rightleftharpoons H^+ + In^-$$

The colors that an indicator displays result from the fact that HIn and In^- are different colors. For example, in the case of litmus, HIn is red and In^- is blue.

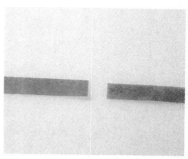

Figure 17-8 When pH paper touches an acidic or basic solution, it changes color. The pH of the solution may be estimated by comparison against the scale on the dispenser.

If you study the equilibrium equation above in terms of Le Châtelier's principle and the Brønsted theory of acids, you will understand how indicators are able to operate the way they do. In sufficiently acidic solutions, the equilibrium is forced to the left by the presence of the many protons (H^+ ions) that the acid produces. Any In^- ions that are present are forced to act as Brønsted bases and thus accept these protons from the acid. The indicator is then present in largely un-ionized form, HIn. It then has its characteristic acid-indicating color, which for litmus is red.

In sufficiently basic solutions, the indicator's equilibrium is forced to the right by the presence of the many hydroxide ions (OH^-) that the base produces. These OH^- ions combine with the H^+ ions that are produced by the indicator and force the indicator molecules to ionize further to try to offset the loss of H^+ ions. The indicator is thus present largely in the form of its anion, In^-. The solution now displays the characteristic base-indicating color of the indicator anion, which for litmus is blue.

Indicators such as those in methyl orange that change color at pH lower than 7 are simply stronger acids than the other types of indicators. They tend to ionize more completely than the others. The anions (In^-) that they produce are weaker Brønsted bases and have less tendency to accept protons from any acid being tested. These indicators therefore do not shift to their un-ionized (HIn) form unless the concentration of H^+ is fairly high. Their color transition occurs at rather low pH. What can you say about the acid strength of indicators, such as phenolphthalein, that undergo transition in the higher pH range? If you said that they are weaker acids, you are correct.

Indicators can be used to compare the pH of different solutions. An indicator added to different solutions may show the same transition color, for example. If so, the solutions are considered to have the same pH. This is the basis for the common colorimetric determination of pH. A measured volume of a suitable indicator is added to each solution whose pH is to be determined. The color is then compared with that of the same indicator in solutions of known pH. By careful color comparison, the pH of a solution can be estimated with a precision of about 0.1 pH unit.

The pH of solutions can be found in ways other than by the use of indicators. Instrumental methods are generally used by chemists to make

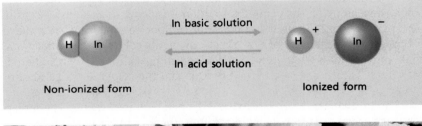

Figure 17-9 An acid pushes the equilibrium of litmus to the left; the nonionized form predominates, and the litmus turns pink. A base pushes the equilibrium to the right; the ionized form predominates, and the litmus turns blue.

Desktop Investigation

Testing for Acid Precipitation

Question
Do you have acid precipitation in your area?

Materials
rainwater
distilled water
500-mL glass containers
thin, transparent plastic
  metric ruler (± 0.1 cm)
hydrion-pH test
  paper: narrow-range,
  ± 0.2–0.3, or pH meter.

Procedure
Record all your results in a data table.

1. Each time it rains, set out five clean glass containers to collect the rainwater. If the rain continues longer than 24 hours, put out new containers at the end of each 24-hour period until the rain stops. (The same procedure can be used with snow if the snow is allowed to melt before measurements are taken. You may need to use larger containers if a heavy snowfall is expected.)

2. After the rain stops or at the end of each 24-hour period, measure with a thin plastic ruler the depth of the water collected to the nearest 0.1 cm, and test the water with the hydrion paper to determine its pH to the nearest 0.2–0.3.

3. Record the following information: (a) the date and time the collection was started; (b) the date and

time the collection was ended; (c) the location where the collection was made (town and state); (d) the amount of rainfall in cm; and (e) the pH of the rainwater.

4. Find the average pH of each collection that you have made for each rainfall, and record it in the data table.

5. Make collections on at least five different days. The more collections you make, the more reliable your data becomes.

6. For comparison, determine the pH of pure water by testing five samples of distilled water with hydrion paper. Record your results in a separate data table and then calculate an average pH

for distilled water.

Discussion
1. What is the pH of distilled water?

2. What is the pH of normal rainwater? How do you explain this?

3. How is acid precipitation defined?

4. What are the drawbacks of using a ruler to measure the height of collected water?

5. Does the amount of rainfall or the time of day the sample is taken have an effect on its pH? Explain any variability among samples.

6. What conclusion can you draw from this investigation? Explain how your data supports your conclusion.

Figure 17-10 pH colorimetry. A sample's pH is determined by adding indicator and comparing its color to those of standard solutions containing indicator.

rapid pH determinations. A pH meter (Figure 17-11) provides a convenient way of measuring the pH of a solution. The pH meter measures the voltage difference between a special electrode and a reference electrode, which are placed in the solution. The voltage changes as the hydronium ion concentration in the solution changes. In an acid–base neutralization reaction, a large change in voltage occurs at the point when equivalent quantities of the two reactants are present in the solution.

The Principle of Titration

The reaction that occurs between an acid and a base is neutralization. Recall that, in a neutralization reaction, the basic OH^- ion acquires a proton from the H_3O^+ ion to form a molecule of water.

$$H_3O^+ + OH^- \rightarrow 2H_2O$$

One mole of hydronium ions (19.0 g) and one mole of hydroxide ions (17.0 g) are chemically equivalent. They combine in a one-to-one ratio. Neutralization occurs when hydronium ions and hydroxide ions are supplied by reactants in equal numbers. The product for $[H_3O^+]$ $[OH^-]$ of 10^{-14} M^2 is a constant for the aqueous acid and base solutions and for the solution that results when acids and bases mix and react. If 0.1 mol of gaseous HCl is dissolved in a liter of water for which $[H_3O^+]$ is 10^{-7} M, the H_3O^+ ion concentration would rise to 0.1, or $10^{-1} M$. Since the product $[H_3O^+]$ $[OH^-]$ remains at 10^{-14} M^2 and since $[H_3O^+]$ rises to 10^{-1} M, the $[OH^-]$ must decrease from $10^{-7} M$ to $10^{-13} M$. The hydroxide ions are removed from the solution by combining with H_3O^+ ions to form water molecules. Almost 10^{-7} mol/L of H_3O^+ ions is also removed in this way. This quantity is only a small

Figure 17-11 A pH meter

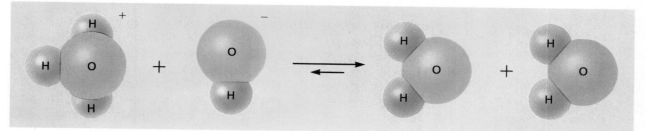

portion (0.0001%) of the 0.1 mol of hydronium ions that are present in each liter of solution.

Now suppose that 0.1 mol (4.0 g) of solid NaOH is added to the liter of 0.1-M HCl solution. The NaOH dissolves and supplies 0.1 mol of hydroxide ions to the solution. Before the reaction proceeds significantly, both [H_3O^+] and [OH^-] are high. Their ion product is momentarily much greater than the constant value of $10^{-14} \, M^2$ for this dilute aqueous solution.

The ion-removal reaction (that is, neutralization) then proceeds quickly. Hydronium and hydroxide ions, which are present in equal numbers, combine until the product [H_3O^+] [OH^-] returns to the constant value of $10^{-14} \, M^2$. This neutralization process is one in which chemically equivalent quantities of hydronium ions and hydroxide ions have combined. Because the salt (NaCl) produced is the product of the neutralization of a strong acid and a strong base, the resulting solution is neither acidic nor basic, but it is neutral.

These preceding examples should help you understand the nature of the chemical reactions that occur between acids and bases as a solution of one of them is progressively added to a solution of the other. This progressive addition of an acid to a base (or a base to an acid) can be done in the laboratory to compare the concentrations of acid and base solutions. In such a case, it is called titration. **Titration** *is the controlled addition and measurement of the amount of a solution of known concentration that is required to react completely with a measured amount of a solution of unknown concentration.*

Titration provides a sensitive means of determining the relative volumes of acidic and basic solutions that are chemically equivalent. If the concentration of one solution is known, the concentration of the other solution can be calculated after determining the volume of one that is required to neutralize a given volume of the other. *A solution that contains a precisely known concentration of a solute is known as a* **standard solution.** Titration is an important laboratory procedure and is often used in analytical chemistry.

Let's consider an example of a titration. Suppose that successive additions of an aqueous base are made to a measured volume of an aqueous acid. Eventually, the acid is neutralized. With the continued addition of base, the solution becomes distinctly basic. The pH has now changed from a low to a high numerical value. If you were to monitor the change in pH, you would find that it occurs slowly at first, rapidly through the neutral point, and slowly again as the solution becomes basic. Typical pH curves for strong-acid/strong-base and weak-acid/strong-base titrations are shown in Figure 17-13.

Figure 17-12 The ion-removal reaction (neutralization). The water molecules produced are only slightly ionized.

The product [H_3O^+][OH^-] is often referred to as the ion-ization constant of water (K_w).

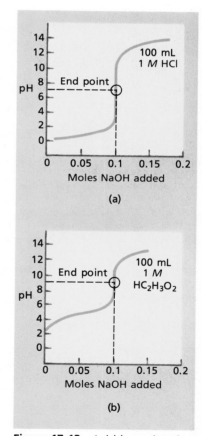

Figure 17-13 Acid-base titration curves: *(a)* strong acid, strong base; *(b)* weak acid, strong base.

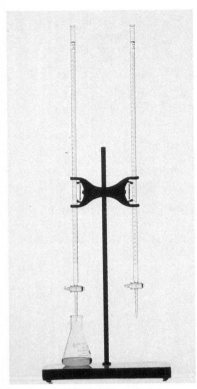

Figure 17-14 A titration stand

Strictly speaking, the "equivalence point" is the place where equivalent quantities of acid and base are present. The "end point" occurs when an indicator or pH meter signals the equivalence point. Ideally, they occur at the same time.

The point at which this very rapid change in pH occurs, at which equivalent quantities of hydronium and hydroxide ions are present, is called the **equivalence point**, *or the end point of the titration.* Any method that shows this abrupt change in pH can be used to detect the end point. Indicators or pH meters may be used for this purpose. At the end point, a pH meter shows a large voltage change.

To determine the concentration of an "unknown" solution, one must titrate it using a "known" solution. To be certain of actually knowing the concentration of a "known" or "standard" solution, however, that solution must be compared to another standard solution whose concentration is well established. The standard solution is first prepared, and its volume is adjusted to the desired concentration. This concentration information is then refined by titrating the solution with a carefully measured quantity of a solution of a highly purified compound. *A highly purified compound, when used in solution to check the concentration of the known solution in a titration is known as a* **primary standard**. The actual concentration of the standard solution is established by this standardizing procedure.

Molarity and Titration

Burets, or graduated glass columns having good precision—like those shown in Figure 17-14—are commonly used to measure solution volumes during titration. Suppose that an aqueous solution of an unknown concentration of the base NaOH is added dropwise to 10.0 mL of a 0.01-M aqueous solution of hydrochloric acid that contains a few drops of a suitable indicator such as bromthymol blue. The addition of drops continues until the end point is reached. (The color of the solution changes from yellow to green.) The readout of the buret that contains the base indicates that 20.0 mL of the NaOH solution is used. How can these titration data be used to determine the molarity of the basic solution?

The equation for the neutralization reaction is

$$HCl(aq) \ + \ NaOH(aq) \ \rightarrow \ NaCl(aq) \ + \ H_2O(\ell)$$
$$1 \text{ mol} \qquad 1 \text{ mol} \qquad 1 \text{ mol} \qquad 1 \text{ mol}$$

The volume and molarity of the HCl solution are known. From these data, the quantity (in moles) of the HCl used for titrating can be determined.

$$\frac{0.01 \text{ mol HCl}}{L} \times 20.0 \text{ mL} \times \frac{1 \text{ L}}{1000 \text{ mL}} = 0.0001 \text{ mol HCl (used)}$$

The balanced chemical equation shows that one mole of NaOH is used for one mole of HCl. In other words, NaOH and HCl show chemical equivalence, mole for mole, in the reaction. The quantity of NaOH used in the titration, therefore, is also 0.0001 mole. From the titration, this quantity of base is furnished by 10.0 mL of the NaOH solution. The molarity of the NaOH solution can now be obtained as follows:

$$\frac{0.0001 \text{ mol NaOH}}{20.0 \text{ mL}} \times \frac{1000 \text{ mL}}{1 \text{ L}} = 0.005 \text{ mol NaOH/L or } 0.005 \ M \text{ NaOH}$$

Suppose the titration is repeated with the same "unknown" NaOH solution but this time with 10.0 mL of a 0.01-M solution of sulfuric acid (the diprotic acid H_2SO_4) as the standard acid solution. When titrated to the end point, the base buret readout indicates that 40.0 mL of the NaOH solution is used. The equation for this reaction is

$$H_2SO_4(aq) + 2NaOH(aq) \rightarrow Na_2SO_4(aq) + 2H_2O(\ell)$$
$$\text{1 mol} \qquad \text{2 mol} \qquad \text{1 mol} \qquad \text{2 mol}$$

The amount of H_2SO_4 in the standard solution used in this titration was.

$$\frac{0.01 \text{ mol } H_2SO_4}{L} \times 10.0 \text{ mL} \times \frac{L}{1000 \text{ mL}} = 0.0001 \text{ mol } H_2SO_4 \text{ (used)}$$

The reaction equation shows that 2 mol of NaOH is required for each 1 mol of H_2SO_4. Therefore, the 40.0 mL of NaOH used to neutralize 0.0001 mol of H_2SO_4 contained 0.0002 mol of NaOH, as shown by the following calculation:

$$0.0001 \text{ mol } H_2SO_4 \times \frac{2 \text{ mol NaOH}}{1 \text{ mol } H_2SO_4} = 0.0002 \text{ mol NaOH}$$

The molarity of the NaOH solution is obtained as follows:

$$\frac{0.0002 \text{ mol NaOH}}{40.0 \text{ mL}} \times \frac{1000 \text{ mL}}{L} = 0.005 \text{ mol NaOH/L or } 0.005 \text{ } M \text{ NaOH}$$

To summarize, the molarity of an aqueous base (or acid) of unknown concentration can be determined by titrating against an aqueous acid (or base) of known concentration. The following four steps are involved:
1. Determine the moles of acid (or base) from the standard solution used during the titration.
2. From a balanced chemical equation, determine the ratio of moles of acid (or base) to base (or acid).
3. Determine the moles of solute of the unknown solution used during the titration.
4. Determine the molarity of the unknown solution. Sample Problem 17.9 illustrates the above steps.

When the calculations for a titration are done by this method, molarity is used and not normality. Steps 2 and 3, a chemical equation and mole ratio conversion, are necessary to make the proper computation.

Sample Problem 17.9

In a titration, 27.4 mL of a standard solution of $Ba(OH)_2$ is added to a 20.0-mL sample of an HCl solution. The concentration of the standard solution is 0.0154 M. What is the molarity of the acid solution?

Solution

Step 1: Analyze Given: volumes, concentration, and identity of standard solution = 27.4 mL of 0.0154 M $Ba(OH)_2$
Unknown: molarity of acid solution.

Step 2: Plan Use the above four steps as follows.
1. Volume of solution used → moles of base used

$$\frac{\text{mol } Ba(OH)_2}{L} \times mL \times \frac{1 \text{ L}}{1000 \text{ mL}} = \text{mol } Ba(OH)_2$$

2. Balanced chemical equation → ratio of moles of acid to moles of base
3. moleration of base used → moles of acid used from unknown solution.

Sample Problem 17.9 *continued*

4. Volume of unknown solution, moles of solute in unknown solution →
molarity of unknown solution

$$\frac{\text{moles of solute in unknown solution}}{\text{volume of unknown solution in mL}} \times \frac{1000 \text{ mL}}{1 \text{ L}}$$

$$= \text{molarity of unknown solution}$$

Step 3: Compute   1. $27.4 \text{ mL} \times \dfrac{0.0154 \text{ mol Ba(OH)}_2}{\text{L}} \times \dfrac{1 \text{ L}}{1000 \text{ mL}} = 4.22 \times 10^{-4} \text{ mol Ba(OH)}_2$

2. $\text{Ba(OH)}_2 + 2\text{HCl} \rightarrow \text{BaCl}_2 + 2\text{H}_2\text{O}$ 1 mol Ba(OH)_2: 2 mol HCl

3. $\dfrac{2 \text{ mol HCl}}{1 \text{ mol Ba(OH)}_2} \times 4.22 \times 10^{-4} \text{ mol Ba(OH)}_2 = 8.44 \times 10^{-4} \text{ mol HCl}$

4. $\dfrac{8.44 \times 10^{-4} \text{ mol HCl}}{20 \text{ mL}} \times \dfrac{1000 \text{ mL}}{1 \text{ L}} = 0.0422 \ M \text{ HCl}$

Step 4: Evaluate   The answer is correctly stated to three significant digits. A 0.0154-M Ba(OH)$_2$ solution is 0.0308 N. We would expect the HCl solution to be greater than 0.0308 N (or in the case of HCl, greater than 0.0308 M) by a factor of roughly 1.4 since about 1.4 times more base solution was used than acid solution. An answer of approximately 0.043 M HCl (0.031 × 1.4) is therefore predicted.

Practice Problems

1. A 15.5-mL sample of 0.215-M KOH solution required 21.2 mL of aqueous acetic acid solution in a titration experiment. Calculate the molarity of the acetic acid solution. *(Ans.)* 0.157 M HC$_2$H$_3$O$_2$
2. By titration, 17.6 mL of aqueous H$_2$SO$_4$ just neutralized 27.4 mL of 0.0165-M LiOH solution. What was the molarity of the aqueous acid solution? *(Ans.)* 0.0128 M H$_2$SO$_4$

Normality and Titrations

Chemists often prefer to express acid and base concentrations in terms of normality rather than molarity. The advantage in doing so is that concentrations are expressed directly in terms of equivalents of solute per liter of solution. Solutions of the same normality are chemically equivalent, milliliter for milliliter. The normality of a given solution is its molarity times a whole-number factor. As we have seen earlier in this chapter, this factor depends on the substance and on the reaction in which the substance is involved.

A very simple relationship exists between volumes and normalities of the solutions used in titration. Take as an example a titration that requires 50.0 mL of a 0.100-N solution of NaOH to reach an end point with 10.0 mL of vinegar, a water solution of acetic acid (HC$_2$H$_3$O$_2$). The number of chemical equivalents of the acetic acid used is the product of the volume of the acid solution used, V_a, and the normality of the acid, N_a.

$$V_a \times N_a = \text{equiv}_a \quad (\text{L} \times \text{equiv/L} = \text{equiv})$$

Similarly, the number of chemical equivalents of NaOH used is the product of the volume of the base solution used, V_b, and the normality of the base, N_b.

$$V_b \times N_b = \text{equiv}_b \quad (\text{L} \times \text{equiv/L} = \text{equiv})$$

At the end point in the titration, equiv_a equals equiv_b . Therefore,

$$V_a N_a = V_b N_b$$

Sample Problem 17.10

A titration requires 50.0 mL of a 0.100-N NaOH solution to reach an end point with 10.0 mL of acetic acid solution. Determine the normal concentration of the acetic acid (vinegar) solution.

Solution

Step 1. Analyze   Given: volume, concentration, and identity of standard solution = 50.0 mL of 0.100 N NaOH
volume, and identity of acid solutions = 10.0 mL acetic acid
Unknown: normality of acetic acid solution

Step 2. Plan   concentration of base, volume of base, volume of acid → concentration of acid

The equation $V_a N_a = V_b N_b$ will be used, substituting appropriately. This means that you should rearrange the equation algebraically to solve for the unknown quantity.

Step 3. Compute
$$V_a N_a = V_b N_b$$
$$N_a = \frac{V_b N_b}{V_a} = \frac{50.0 \text{ mL} \times 0.100 \, N}{10.0 \text{ mL}} = 0.500 \, N$$

Step 4. Evaluate   The answer is correctly given to three significant digits. Since the volume of base required is five times the volume of acid, then the acid is five times more concentrated than the base, or 0.500 N as predicted.

Practice Problems

1. Determine the normal concentration of an unknown acid solution when 17.3 mL of this solution was used in titrating 31.2 mL of a 0.125-N NaOH solution. *(Ans.)* 0.225 N
2. Determine the normality of a NaOH solution when 45.1 mL of a 0.100-N HCl solution was used in titrating 31.0 mL of the base. *(Ans.)* 0.145 N

The acidity of vinegar results from presence of acetic acid. A 1.0-N acetic acid solution contains 1.0 equivalent of acid (H_3O^+) per liter of solution. In this case, 1.0 equivalent equals 1.0 mol (60 g) of acetic acid per liter of solution. The 0.500-N acetic acid solution described in the sample problem above must contain 30 g of $HC_2H_3O_2$ per liter. A liter of vinegar has a mass of approximately 1000 g. The sample of vinegar thus used contains 3.0% acetic acid by mass.

NaOH is a strong base, and $HC_2H_3O_2$ is a weak acid. The pH curve for this titration is shown in Figure 17-14. The end point occurs at a higher pH because the sodium acetate solution that is formed in the titration is slightly basic. Phenolphthalein would be a suitable indicator to use in this titration because it changes at a pH higher than 7. If a weak base and a strong acid are titrated, the end point would occur below a pH of 7 because of the acidic nature of the salt formed. Methyl orange would be a suitable indicator to use in this titration.

1. (a) Methyl red, methyl orange, or bromphenyl blue. (b) Phenol red, phenolphthalein, or thymolphthalein.

2. Standard solutions of HCl, H_2SO_4, or $HC_2H_3O_2$

3. $20.0 \text{ mL} \times \dfrac{0.0100 \text{ mol HCl}}{L} \times$

$\dfrac{1 \text{ L}}{1000 \text{ mL}} =$

2.00×10^{-4} mol HCl

$HCl + NaOH \rightarrow H_2O + NaCl$

2.00×10^{-4} mol HCl $\times$

$\dfrac{1 \text{ mol NaOH}}{1 \text{ mol HCl}} =$

2.00×10^{-4} mol NaOH

$\dfrac{2.00 \times 10^{-4} \text{ mol NaOH}}{30 \text{ mL NaOH}} \times$

$\dfrac{1000 \text{ mL}}{1 \text{ L}} = 6.7 \times 10^{-3} M$

NaOH

4. $V_a N_a = V_b N_b$

$N_a = \dfrac{V_b N_b}{V_a} =$

$\dfrac{20.0 \text{ mL} \times 0.10 \text{ N}}{12 \text{ mL}} =$

0.17 N HCl

Section Review

1. Name an appropriate indicator for testing two types of solutions. *(a)* strong acid and a weak base. *(b)* strong base and a weak acid.

2. You have a solution of an unknown concentration containing NaOH. Name three possible standard solutions that could be used to titrate the NaOH solution.

3. If 20.0 mL of 0.0100-*M* aqueous HCl is required to neutralize 30 mL of an aqueous solution of NaOH, determine the molarity of the NaOH solution.

4. If 20.0 mL of 0.10-*N* NaOH is required to neutralize 12 mL of aqueous HCl solution, what is the normality of the HCl solution?

Chapter Summary

- The quantities of solutes that have equivalent combining capacity in a chemical reaction are called equivalents.
- In a neutralization reaction, one equivalent of an acid provides one mole of H_3O^+ ions and reacts completely with one equivalent of a base, which provides one mole of OH^- ions.
- The equivalent mass of a substance is the mass (in grams) that provides one equivalent in a given chemical reaction.
- A solution with a normality of 1 contains one equivalent mass of solute per liter of solution.
- Pure water undergoes self-ionization to give 1×10^{-7} H_3O^+ ions and 10^{-7} OH^- ions per liter.
- Normality is numerically related to molarity by the simple relationship $N = nM$, where n is the number of equivalents per mole.
- Pure water undergoes self-ionization to give 1×10^{-7} M H_3O^+ ions and $10^{-7} M$ OH^- ions per liter.
- $pH = -\log[H_3O^+]$
- Acids have a pH of less than 7, bases have a pH of greater than 7, and neutral solutions have a pH of 7.
- Indicators are divided into three types: *(1)* those that change color at a pH of 7 and are used to study the neutralization reaction between strong acid and strong bases *(2)* those that change color at a pH lower than 7 and are used to study the neutralization reactions between strong acids and weak bases *(3)* those that change color at pH greater than 7 and are used to study the neutralization reaction between strong bases and weak acids.
- Titration uses a solution of known concentration (referred to as a *standard solution*), to determine the concentration of an "unknown" solution. The solutes in both the "standard" and the "unknown" solution must be known in order to calculate the molar concentration of the unknown solution from the titration data.
- When the molarity and volume of standard solution used in a titration is known, then the molarity of an unknown solution can be found, provided its volume has been measured.
- When a titration is complete, the volume of standard acid solution used (V_a) multiplied by its normality (N_a) is equal to the volume of unknown base solution (V_b) multiplied by its normality (N_b): $V_a N_a = V_b N_b$.

Chapter 17 *Review*

Acid–Base Titration and pH

Vocabulary

end point	pH
equivalence point	primary standard
equivalents	self-ionization (water)
equivalent mass	standard solution
indicator	titration
normality	transition interval

Questions

1. *(a)* What are chemical equivalents? *(b)* In proton-transfer reactions, distinguish between one equivalent of an acid and one equivalent of a base.
2. *(a)* What is the relationship between the protic nature of an acid and the mass of one equivalent of that acid? *(b)* Illustrate with one example each of a monoprotic, diprotic, and triprotic acid.
3. *(a)* What is the relationship between the number of OH^- ions in one mole of a base and the mass of one equivalent of that base? *(b)* Illustrate with one example each of bases that contain one, two, and three moles of OH^- ions per mole.
4. Distinguish among molality, molarity, and normality.
5. Express in words what is meant by a: *(a)* 1-*N* solution *(b)* 3-*N* solution *(c)* 0.5-*N* solution.
6. What is meant by the expression $[H_3O^+]$?
7. *(a)* Define what is meant by the pH of a solution. *(b)* Write the equation for determining pH. *(c)* Explain and illustrate what is meant by the common logarithm of a number.
8. Identify each of the following solutions as acidic, basic, or neutral: *(a)* $[H_3O^+] = 1 \times 10^{-7}$ mole/L; *(b)* $[H_3O^+] = 1 \times 10^{-10}$ mole/L *(c)* $[OH^-] = 1 \times 10^{-7}$ mole/L *(d)* $[OH^-] = 1 \times 10^{-11}$ mole/L *(e)* $[H_3O^+] = [OH^-]$ *(f)* pH = 3 *(g)* pH = 13
9. What is the approximate pH of each of the following common substances? *(a)* vinegar *(b)* apples *(c)* tomatoes *(d)* milk *(e)* grapefruit *(f)* potatoes *(g)* lemons *(h)* milk of magnesia *(i)* gastric juice.

10. *(a)* Explain briefly what occurs in a neutralization reaction. *(b)* Write the net ionic equation for this reaction.
11. *(a)* What is titration? *(b)* How is titration used in the laboratory?
12. *(a)* What is meant by the end point of a titration? *(b)* What can be observed about the pH of a solution at the end point of a titration? *(c)* What is the role of an indicator in the titration process?
13. *(a)* What is the advantage of expressing the concentrations of solutions involved in titration in terms of normality? *(b)* What equation relates the normalities and volumes of solutions involved in titration?
14. *(a)* What is meant by the transition interval of an indicator? *(b)* What is the basis upon which an indicator is selected for a particular titration experiment?
15. For each of the four possible types of acid–base titration combinations, indicate the pH at the end point and one or more suitable indicators for detecting that end point.
16. Other than through the use of indicators, how can the end point of a titration experiment or the pH of a solution be determined?
17. An unknown solution is colorless when tested with phenolphthalein, but turns pink in the presence of the indicator phenol red. Based on this information, what is the approximate pH of this solution?

Problems

1. Determine the mass of one equivalent (in g/mol H_3O^+) to three significant digits for each of the following acids: *(a)* HBr *(b)* H_2S *(c)* H_3PO_4.
2. Determine the mass one equivalent (in g/mol OH^-) to three significant digits for each one of the following bases: *(a)* LiOH *(b)* $Sr(OH)_2$ *(c)* $Al(OH)_3$.
3. A given supply of concentrated nitric acid is 69.5% HNO_3 by mass and has a density of 1.42 g/mL. Based on this information, determine the number of milliliters of concentrated nitric acid

that would be needed to prepare: (a) 1.00 L of 2.00-N HNO_3 solution (b) 150. mL of 0.750-N HNO_3 solution (c) 250. mL of 0.500-N HNO_3 solution.

4. Concentrated hydrochloric acid generally contains 37.5% HCl by mass and has a density of 1.19 g/mL. Based on this information, determine the number of milliliters of concentrated hydrochloric acid that would:
(a) contain 1.00 equivalent of HCl
(b) be required to prepare 2.00 L of 1.00-N HCl solution (c) be needed to prepare 3.00 L of 0.750-N HCl solution.

5. Determine the molar concentrations of the hydronium and hydroxide ions for each of the following solutions: (a) 0.0100-M HCl
(b) 1.65×10^{-4}-M HNO_3 (c) 0.54-M NaOH (d) 1.54×10^{-3}-M KOH.

6. Based on the $[OH^-]$ given for each of the following solutions, determine the corresponding $[H_3O^+]$: (a) $1 \times 10^{-7} M$ (b) $1 \times 10^{-4} M$ (c) $1 \times 10^{-12} M$ (d) $1 \times 10^{-3} M$.

7. Based on the $[H_3O^+]$ given for each of the following solutions, determine the corresponding $[OH^-]$: (a) $1 \times 10^{-7} M$ (b) $1 \times 10^{-9} M$ (c) $1 \times 10^{-1} M$ (d) $1 \times 10^{-5} M$.

8. Given the following $[H_3O^+]$ concentrations, determine the pH of each solution: (a) $1 \times 10^{-7} M$ (b) $1 \times 10^{-3} M$ (c) $1 \times 10^{-12} M$ (d) $1 \times 10^{-5} M$.

9. Determine the pH of each of the following solutions: (a) 0.01-M HCl (b) 0.001-M HNO_3 (c) 0.00001-M HF (d) 0.0001-M HBr.

10. Given the following $[OH^-]$ concentrations, determine the pH of each solution: (a) 1×10^{-6}-M (b) 1×10^{-9}-M (c) 1×10^{-2}-M (d) 1×10^{-7}-M.

11. Determine the pH of each solution:
(a) 0.01-M NaOH (b) 0.001-M KOH
(c) 0.0001-M LiOH.

12. Determine the pH of each solution when given the following $[H_3O^+]$: (a) $2.0 \times 10^{-5} M$ (b) $4.7 \times 10^{-7} M$ (c) $3.8 \times 10^{-3} M$.

13. Given the following pH values, determine the $[H_3O^+]$ for each solution: (a) 3 (b) 7 (c) 11 (d) 5.

14. Given the following pH values, determine the $[OH^-]$ for each solution: (a) 7; (b) 11; (c) 4; (d) 6.

15. Determine $[H_3O^+]$ for each of the solutions with the following pH values: (a) 4.23; (b) 7.65; (c) 9.48.

16. For each of the following acid–base titration combinations, determine the number of moles of the first substance listed that would be required to react with the given quantity of the second: (a) NaOH with 1.0 mole HCl; (b) HNO_3 with 0.75 mole KOH; (c) $Ba(OH)_2$ with 0.20 mole HF; (d) H_2SO_4 with 0.60 mole $Al(OH)_3$.

17. If 15.0 mL of 0.0250-M aqueous H_2SO_4 is required to neutralize 10.0 mL of an aqueous solution of KOH, what is the molarity of the KOH solution?

18. In a titration experiment, a 12.5-mL sample of 0.0175-M $Ba(OH)_2$ just neutralized 14.5 mL of HNO_3 solution. Calculate the molarity of the HNO_3 solution.

19. If 13.4 mL of 0.0230-M H_3PO_4 is required to neutralize 22.5 mL of $Ba(OH)_2$ in a titration experiment, determine the molarity of the $Ba(OH)_2$ solution.

20. What volume of 0.50 N KOH would be needed to neutralize 20.0 mL of 0.75-N HNO_3 solution?

21. If 35 mL of 0.10-N acid solution is neutralized by 21.8 mL of a basic solution, what is the normality of the base?

Application Questions

1. Explain how one equivalent of H_2SO_4 is the same as one mole of the acid (98.1 g) in some reactions, but can be equal to one-half mole of the acid (49.0 g) in other reactions.

2. Using a scientific calculator, determine the logarithm of the following, to two significant figures: (a) 1.00 (b) 1.50 (c) 2.80 (d) 3.75 (e) 4.53 (f) 6.47 (g) 8.52 (h) 10 (i) 10^{-4} (j) 10^{-12}.

3. Use the calculator, where necessary, to determine the antilog of each of the following:
(a) 0.6599 (b) 0.1847 (c) 0.4683 (d) 0.6599 (e) 0.7316 (f) 0.8075 (g) 0.9430 (h) 1 (i) 3 (j) -6 (k) -11.

4. For any aqueous solution of HCl that has a concentration below 1-M, the $[H_3O]$ is indicated directly by the molarity of the solution. For weak acids like as $HC_2H_3O_2$ and H_2CO_3, however, such a direct relationship does not exist. Explain.

5. Explain why the end points in certain titration experiments may occur at pH values other than seven.

Application Problems

1. Determine the number of equivalents needed to make each of the following solutions: *(a)* 1.0 L of a 1.0-*N* solution of HNO_3 *(b)* 1.0 L of a 2.5-*N* solution of NaOH *(c)* 2.0 L of a 1.0-*N* solution of $Mg(OH)_2$ *(d)* 1.5 L of a 3.0-*N* solution of HCl *(e)* 250 mL of a 1.20-*N* solution of K_2CO_3.

2. Determine the normality of each of the following solutions:
   (a) 1.0 equiv NaOH in 1.0 L of total solution
   (b) 2.5 equiv HNO_3 in 5.0 of total solution
   (c) 40.1 g HF in 1.00 L of total solution
   (d) 74.1 g $Ca(OH)_2$ in 3.00 L of total solution
   (e) 10.9 g H_3PO_4 in 60.0 mL of total solution.

3. How many milliliters of 1.50-*N* H_2SO_4 solution would be required to prepare 500 mL of 1.20-*N* H_2SO_4 solution?

4. What volume of 0.750-*N* H_3PO_4 would be required to prepare 2.50 L of 0.200-*M* H_3PO_4 solution?

5. Determine the molarity of each of the following solutions: *(a)* 1.0-*N* HF *(b)* 3.0-*N* NaOH *(c)* 2.0-*N* $Mg(OH)_2$ *(d)* 1.5-*N* H_2SO_4 *(e)* 0.75-*N* NH_4OH *(f)* 0.50-*N* $Ca(OH)_2$ *(g)* 2.4-*N* H_3PO_4.

6. A nitric acid solution is found to have a pH of 2.70. Determine each of the following: *(a)* the $[H_3O^+]$ *(b)* the $[OH^-]$ *(c)* the mass of anhydrous HNO_3 required to prepare 5.50 L of this solution *(d)* the amount, in milliliters, of the concentrated acid needed to prepare the solution in *(c)* (Concentrated nitric acid is 69.5% HNO_3 by mass and has a density of 1.42 g/mL.).

7. If 12.5 mL of 0.200-*M* H_2SO_4 is required to neutralize 27.5 mL of KOH solution, what is the normality of the KOH solution?

8. How many milliliters of 0.250-*M* H_3PO_4 solution would be required to neutralize 40.0 mL of 0.280-*M* $Ba(OH)_2$?

9. If 18.5 mL of p.350-*M* H_2SO_4 neutralizes 12.5 mL of aqueous LiOH solution, how many grams of LiOH were used to make the LiOH solution?

10. In a laboratory experiment, 15.0 mL of 0.200-*M* HNO_3 is mixed with 25.0 mL of 0.0750-*M* $Ba(OH)_2$. Determine each of the following: *(a)* the identity and quantity of the reactant present in excess *(b)* whether the resulting solution is acidic, basic, or neutral *(c)* the pH of the final solution.

Enrichment

1. Many common household items are acids or bases. Based on your experiences with such food items as fruits, fruit juices, vinegar, milk, and soft drinks, as well as other items—cleaning agents, soaps, detergents, shampoos, antacids, and baking soda—try to determine whether each material is acid, base, or neutral; then confirm your conclusion by testing each with litmus paper, pH paper, or another appropriate indicator.

2. Conduct a titration experiment with a variety of brands of orange juice, and determine the percentage of citric acid in each.

3. Design and conduct an experiment to isolate and extract possible acid–base indicators from such potential sources as red cabbage, berries, and flower petals. Test the action of each indicator that you are able to isolate with known acidic, basic, and neutral solutions.

4. Design and conduct an experiment to study the pH of rain in various locations for an extended period of time. Coordinate your data-collection activities with your classmates so that the pH readings will be made in a variety of locations. Try to determine whether any patterns emerge due to location, season, time of day, degree of industrialization in the area, etc.

5. Design and conduct an experiment to determine the pH of a variety of brands of shampoo. Do any patterns emerge in terms of types of shampoos or contents of each, relative to the results?

6. Design and conduct and experiment to test the effectiveness of various brands of antacid. Use the following substances in your experiment: antacids, white vinegar, and red cabbage water. (See Desktop Investigation in Chapter 16 for instructions in making the cabbage water.) Write a report, detailing your methods, results, and conclusions about the relative effectiveness of the different brands of antacid.

CAREERS

Food Chemist

Chemists play an essential role in the preparation and packaging of foods so that they come to you without loss of color, flavor, or texture. Chemists from the U.S. Department of Agriculture and the U.S. Food and Drug Administration monitor ingredients and packaging processes to ensure that the food you eat is safe. Other chemists synthesize new flavor and aroma molecules, adding them to improve the appeal of the food.

Food chemists work in a variety of areas within the food industries. Nutrition chemists design diets for hospitals, and flavor chemists synthesize such essences as "watermelon flavor" for bubble gum. A recent project is the development of calorie-free substitutes for sugar and for fats such as shortening.

What sort of background do you need to be a food chemist? Chemistry and biochemistry are useful basic courses, but, depending on what you wish to focus on, you will need extra courses in cooking, nutrition, and physiology. With a high school diploma and some training as a lab technician, you may qualify as a technician in a food chemistry lab. Food chemists must have a college degree in chemistry.

Aroma Chemist

Aroma chemistry permeates many industries, just as its products permeate the air. The job of an aroma chemist is unusual because it combines science and aesthetics. These are two interests that are sometimes erroneously considered to be incompatible with each other. The aroma chemist, also known as a perfumer, mixes chemicals to develop pleasant fragrances.

Aroma chemists are hired by cosmetic or consumer- products companies to develop new fragrances that have certain characteristics. It usually takes about two years to arrive at a final combination of essential oils—derived from flowers, herbs, and occasionally animals—and base ingredients such as soaps, oils, waxes, or powders. As a consultant, a perfumer may specialize in

one product line or develop fragrances as diverse as those found in cologne, window cleaner, and kitty litter.

An analytical aroma chemist uses chromatography to separate the different aroma-producing compounds in a natural essence. The chemical structure of each compound is then determined by using various types of spectroscopy, including nuclear magnetic resonance. When the structure of each compound is known, it can be synthesized in the aroma laboratory and combined with other compounds in an aromatic mixture.

Most aroma chemists have a bachelor's degree in chemistry. A master's or doctoral degree in biochemistry and an interest in psychology are needed for aroma chemists who wish to understand how people perceive smells.

Cumulative *Review*

Unit 5 • Solutions and Their Behavior

Questions

1. Identify and define *(a)* the three types of mixtures *(b)* the two components of a solution. (14.1)
2. Identify three factors that affect the rate of dissolving of a solid in a liquid. (14.2)
3. Explain the meaning of the "like dissolves like" rule. (14.2)
4. *(a)* What is meant by the concentration of a solution? *(b)* List and define three ways of expressing solution concentration. (14.3)
5. Identify three colligative properties of a solution by the addition of a nonvolatile solute and indicate how each is affected. (14.4)
6. *(a)* Define ionization. *(b)* How does ionization differ from dissociation? (15.1)
7. *(a)* How do nonvolatile electrolytes in aqueous solutions affect the boiliing and freezing points of solvents into which they are placed? *(b)* Explain your answer to part *a*. (15.3)
8. Identify five general properties of aqueous acids. (16.1)
9. *(a)* What determines the strength of an acid? *(b)* Give one example each of a strong and a weak acid. (16.3)
10. In the reaction that follows, identify the conjugate acid-base pairs: $HCl(g) + H_2O(\ell) \rightarrow H_3O^+(aq) + Cl^-(aq)$. (16.2)
11. Distinguish between one equivalent of an acid and one equivalent of a base. (17.1)
12. What type of solution (acidic, basic, or neutral) is indicated by each of the following: *(a)* pH = 2 *(b)* $[H_3O^+] = 10^{-5}M$ *(c)* $[OH^-] = 10^{-2}M$ *(d)* pH = 7. (17.2)
13. Explain the function of the titration process in a chemical laboratory. (17.3)

Problems

1. How many grams of solute are required to prepare 4.50 L of 1.75-*M* solution of nitric acid (HNO_3)? (14.3)
2. If 150.0 mL of 18.0-*M* H_2SO_4 is diluted to 2.50 L, what is the molarity of the solution that results? (14.3)

3. If 75.0 mL of 2.00 *M* HCl react with $Ca(OH)_2$ by double replacement, what mass of each product is formed? (14.3)
4. What is the molality of a solution that contains 32.0 g of KNO_3 dissolved in 0.625 kg of water? (14.3)
5. If 50.0 g of antifreeze, $C_2H_4(OH)_2$, is dissolved in 750. g of water, what would be the new freezing and boiling points of the resulting solution? (14.4)
6. Determine the molar mass of a molecular solute if 3.75 g of the solute dissolved in 250. g of water lowers the freezing point of the solution to $-0.820°C$. (14.4)
7. Write the equation for the solubility equilibrium involving the very slightly soluble compound iron(III) hydroxide. (15.1)
8. Use Table 15-1 to write the balanced chemical equation, write the overall ionic equation, identify the spectator ions and possible precipitates, and write the net ionic equation for the following reactions: *(a)* sodium sulfate(aq) + calcium nitrate(aq) $\rightarrow$ *(b)* potassium carbonate(aq) + iron(III) nitrate(aq) $\rightarrow$. (15.1)
9. Determine both the freezing and boiling points of water that contains 8.20 g of calcium nitrate per 250. g of water. (15.3)
10. Complete each of the following salt-producing reactions and write the balanced chemical equation for each: *(a)* $Ni(s) + HCl(aq) \rightarrow$ *(b)* $FeO(s) + HNO_3(aq) \rightarrow$ *(c)* $MgCO_3(s) + HCl(aq) + \rightarrow$ *(d)* $CO_2(g) + Ba(OH)_2(aq) \rightarrow$ *(e)* $BaO(s) + SO_3(g) \rightarrow$ (16.2)
11. What mass of H_2SO_4 would be required to prepare 250. mL of a 6.00-*N* solution? (17.1)
12. Determine the pH, $[H_3O^+]$, and $[OH^-]$ values for a 10^{-5}-*M* HCL solution. (17.2)
13. If 12.5 mL of 0.250 *N* KOH is required to neutralize 18.5 mL of aqueous H_3PO_4 solution in a titration experiment, determine both the normality and the molarity of the acid solution. (17.3)

The Problem in Chemistry: Detergents and the Environment

Lake Tahoe is growing old before its time. Surrounded by the Sierra Nevada mountains of California, Lake Tahoe *was* like most other geologically young lakes—sparkling clear and cold, with few plants or plankton. In the past few decades, however, runoff containing organic wastes and phosphate

detergents has nourished unwanted plants to the point that they have choked off other plant and animal life. Now, each spring the shores are covered by slimy green algae.

In the mid 1950s, soaps made of plant and animal fatty acids were replaced by synthetic detergents in American washing machines. Detergents give a cleaner, brighter wash than soaps, especially in "hard water," which is high in dissolved minerals.

Soaps reacted with dissolved calcium (Ca) and magnesium (Mg) ions in hard water to form insoluble salts, which leave a white scum on clothing. Detergents, on the other hand, remove dirt and grease and enter the wash water without leaving a residue.

Unfortunately the detergent-containing wastewater that entered the water systems through municipal sewage greatly altered the ecology and accelerated the aging of lakes and estuaries. Detergents have now been reformulated to avoid ecological damage, but much work remains to restore lakes to their original state and to eliminate other sources of water pollution.

The Connection to Physics

The cleansing action of detergents results from their ability not only to break up the oily film that binds dirt particles but also to weaken the cohesion between water molecules so that oil and dirt can enter the wash water. Detergent molecules contain a long, narrow hydrocarbon component that is attached to a group that has a negative charge. Because the hydrocarbon portion of detergent molecules adheres well to oil and because the charged portion of detergent molecules adheres well

to water, oil is pulled off the clothing and into the wash water. Once in the water, each oil droplet is surrounded by a ball of detergent molecules. The hydrocarbon portions of the detergent molecules point inward, where they adhere to the oil molecules, which are also hydrocarbons. The negatively charged portion of the detergent molecules point outward into the water, which is a good solvent for charged molecules.

Once the oil has been taken up by the wash water, the dirt particles can also enter the water. This is facilitated by the action of the detergent in reducing the cohesive force between adjacent water molecules. By reducing the surface tension of water, detergents make it easier for the particles of dirt to push apart water molecules at the water-dirt interface and enter into the wash water, where they remain suspended.

The Connection to Biology and Health Sciences

Unlike soap, many of the first synthetic detergents were not biodegradable. The molecules of these synthetic detergents contained branched chains that could not be broken down into simple inorganic substances through the action of soil bacteria. Consequently, these detergents tended to collect in groundwater and contaminate it.

In contrast, detergents manufactured today are more similar to soaps made from natural fats. In the molecules of both, the hydrocarbon portions have skeletons in which carbon atoms are joined in a straight chain. Since soil bacteria are adapted to ''eating'' straight-chain hydrocarbons, they can prevent detergent molecules from accumulating.

Sodium phosphate pollution is the most severe environmental

problem resulting from synthetic detergents, and it remains unresolved. Sodium phosphates are alkaline salts that form complexes with dissolved metal ions in the wash water so that these ions do not interfere with the process by which oil is emulsified.

When the phosphate concentration of a lake increases, so does the growth of algae and other plants. Plants and animals that

were flourishing are forced out by other species. Lakes that were once clear become covered with a scum of blue-green algae. Decaying plants consume much of the lake's dissolved oxygen, and some species of fish die because there is not enough oxygen to sustain them.

Connection to Earth Science

Lakes are created by glaciers or rivers. Lakes age as sediment is deposited on the bottom and vegetation encroaches from the shore. The maturing of a lake (called eutrophication) normally takes thousands of years. The addition of phosphates greatly accelerates eutrophication by increasing the biomass and the rate of sedimentation. Because of pollution, the eutrophication of Lake Tahoe is occurring in only a few decades rather than thousands of years.

Many lakes, such as Lake Onondaga in New York, were redeemed when efforts were made to keep phosphate-laden wastes from flowing into them. Today, most detergents are free of phosphates. Phosphates have been replaced by safer but less effective substances, such as sodium carbonate.

Unit

6

Chapter 18
Reaction Energy
and Reaction Kinetics   526

Chapter 19
Chemical Equilibrium   562

Chapter 20
Oxidation–Reduction Reactions   596

CHEMICAL REACTIONS

*Growing up in Taiwan after World War II, I was
exposed to the excitement of science transform-
ing human society. But my decision to be a
research chemist can perhaps be traced to the
time when I had a chance to read the biography
of Madame Curie in junior-high school. Her
dedication to scientific research, her excitement
about new discoveries, and her idealistic attitude
convinced me that the life of a research chemist
could be very beautiful.*

*My main research interest is understanding
the exact details of how chemical reactions take
place at the molecular level. Molecular collisions
are responsible for chemical change. Since it is
not possible to see the collisional events of
microscopic atoms and molecules with the
naked eye, we devise various complicated
experimental apparatuses to try to "visualize"
the reactive scattering processes of atoms and
molecules. With tools such as the molecular
beam apparatus and laser, scientists can study
chemical reactions at the molecular and atomic
levels in great detail.*

Yuan T. Lee

Yuan T. Lee
Physical Chemist
University of California, Berkeley

Laser study of flame chemistry

Chapter 18 Reaction Energy and Reaction Kinetics

Chapter Planner

18.1 Thermochemistry
Heat of Reaction
- Demonstration: 1
  Questions: 1–4
  Problems: 1, 2
- Application Questions: 4, 5

Heat of Formation
- Question: 5
- Application Question: 3

Stability and Heat of Formation
- Application Question: 1
- Application Problem: 6

Heat of Combustion
- Technology
  Experiment: 25
  Problems: 5–9
- Question: 6
  Application Question: 2
  Application Problem: 1

Calculating Heats of Reaction
- Experiment: 26
  Questions: 7, 8a
  Problems: 3, 4
- Question: 8b

Bond Energy and Reaction Heat
- Question: 9

18.2 Driving Force of Reactions
Enthalpy and Reaction Tendency
- Application Question: 7

Entropy and Reaction Tendency
- Demonstration: 2
  Application Question: 6

Free Energy
- Questions: 10–15
  Problem: 10
- Application Question: 8
  Application Problems: 3, 4

18.3 The Reaction Process
Reaction Mechanisms
- Application Question: 9

Collision Theory
- Question: 16
  Application Questions: 10, 11

Activation Energy
- Question: 17

The Activated Complex
- Questions: 18–20
  Problems: 10, 11
- Application Problem: 5

18.4 Reaction Rate
Rate-Influencing Factors
- Demonstrations: 3, 4
  Desktop Investigation
  Experiment: 27
  Questions: 21, 22
  Application Question: 13
- Application Question: 12

Rate Laws for Reactions
- Questions: 23–25
  Problem: 12
  Application Problem: 2
- Application Questions: 14–16

Teaching Strategies

Introduction

Three questions confront a chemist studying a chemical reaction. *(1)* Will the reaction proceed as written? *(2)* How long will the reaction take to reach completion? *(3)* How much product will be obtained? Each question is answered by different but related fields of chemistry. Thermodynamics and the calculation of the free energy of the reaction answers the first question. Chemical kinetics or the study of rates of reactions answers the second, and, finally, the study of chemical equilibrium and the law of mass action answers the third. Many texts treat these as separate topics. Here you have the opportunity to show the interrelatedness of these three in one chapter.

Pose the question "What information does a chemist need to know about a chemical system under study?" In answering the question, describe the three fields of chemistry mentioned above and urge them to keep the global picture in mind as they study the individual components. All students, basic to advanced, can grasp the importance of knowing whether or not something will happen, how long will it take, and how much they will get.

Chapter 18 begins with the study of thermochemistry. Stress from the very beginning that chemists are working from the difference in energy between some initial state (e.g., the reactants), and a final state (e.g., the products). Quantities or variables that depend only on the initial and final conditions are known as *state functions*.

18.1 Thermochemistry

The term *heat content* is a confusing one. The term describes the energy stored within molecules. It represents the energy stored within the chemical bonds and, in a sense, measures the stability of the molecules. Explosives are carefully synthesized to contain large amounts of stored energy. This energy is released in the explosion or detonation of the material.

The physical state of the substance described is very important and must be written as part of the chemical reaction. Ask the question "What would cause a more severe burn, 10 g of liquid water at $100°C$ or 10 g of steam (water vapor) at $100°C$?"

(The gaseous water-vapor molecules contain more kinetic energy than the molecules of the liquid. The steam would cause a more severe burn.)

As equations with the heat of reaction are presented, keep stressing that the law of conservation of energy is in effect. The total energy of the reactant side of the equation must equal the total energy of the product side. This total energy includes the heat that is either absorbed (when energy is a reactant) or released (when energy is a product). If the energy term is removed from the balanced equation, the total stored energies are no longer equal. If the heat term were removed from the reactant side, the products would have more energy. If the heat term were removed from the product side, the reactants would have more stored energy. Explain to the students that this stored energy is what is being described as heat content, ΔH. ΔH for a reaction is defined as the difference between the total heat content of the products minus the total heat content of the reactants. If the energy term is removed from the reactant side, ΔH must be positive, because the products have stored within them the energy written as a separate term on the reactant side. ΔH is a negative term for an exothermic reaction. The students must realize that there is no such thing as negative energy.

This is generally a very confusing issue and if it can be clarified early in the chapter, the students will have a less difficult time with changes in entropy and free energy. The sign of ΔH simply indicates heat flow. When positive, heat is flowing into the system and when negative, it is flowing away from the system.

Demonstration 1: Reaction between Hydrogen and Oxygen

PURPOSE: To illustrate thermochemistry by the combination of hydrogen with oxygen

MATERIALS: 2 small balloons filled with approximately 2:1 ratios of hydrogen and oxygen, candle, meter stick, matches, string, tape

PROCEDURE: Fill the balloons appropriately with the two gases and attach them to either end of the teachers desk with string. Discuss the concepts presented in this first part of the chapter in terms of the molecules in each of the balloons. Students are generally disappointed that nothing seems to happen even though the text indicates that hydrogen and oxygen form an explosive mixture and give off a very large amount of energy when they combine to form water. CAUTION: The following safety precautions must be followed: ears must be protected either by placing a piece of tissue in each or by cupping one's hands over the ears; teachers (and administrators) in neighboring classrooms need to be warned about the upcoming explosion; and, if any students are seriously affected by loud noises, they may be excused from the room.

Turn off the room lights and, using a candle taped to the end of a meter stick, touch one of the balloons with the flame. A very loud explosion and a brilliant yellow flash result. Have the students try to explain what they have seen. Many of the concepts that will be used later in the chapter may be drawn from them. Hydrogen and oxygen molecules were colliding in the balloon but not reacting (insufficient activation energy). Energy was needed to initiate the reaction. Although a small amount of heat was used to initiate the reaction, a large amount was given off—the reaction was definitely exothermic. Under room condi-

tions, the reaction would be very slow, not even noticeable. After initiation, the reaction proceeded at an explosive pace.

At this point, introduce the concept of molar heat of formation, an intensive property that can be tabulated. Compare this term with the heat of reaction, an extensive property, presented earlier in the chapter.

Since all of these terms are defined by initial and final states, explain the rationale of assigning the heats of formation of elements in their standard state a value of zero. This does not mean that elements have no energy. It is a relative term that describes the difference in energy between the elements and the molecule produced. If ΔH is positive, the molecule contains more energy than the elements used to produce it. On the other hand, if ΔH is negative, the elements have more energy stored within them than does the molecule produced. Where does the extra energy go? It must be released as heat, sound, light, etc. The reaction must be exothermic.

With this background most students will begin to predict stability based on heats of reactions. If a synthesis process is very endothermic, the product is probably unstable. Manufacturers of explosives and fireworks go to great lengths to store a lot of energy within the chemical bonds of the molecules that make up the fireworks. A newspaper article describing the explosion of a fireworks manufacturing firm brings the concept home.

When calculating the heats of combustion, encourage the students not to take short cuts as they begin to learn the process. Show the students how they must take the tabulated data, change the sign depending on where the molecule occurs in the equation describing the reaction, and how they must multiply the tabulated value by the coefficient from the equation. Much drill at this time will make the presentation of Hess' law less difficult.

When presenting Hess' law, work out the initial examples in detail. First be sure that the students realize how the formulas in the equations must be shifted and then how the energy terms associated with those formulas are added algebraically. Stress that Hess's law is an extremely powerful tool. Energies for reactions that would be next to impossible to carry out in a lab, can be calculated from known heats of formation.

Working with the remaining hydrogen balloon and molecular models of hydrogen gas and oxygen gas, ask the students to explain what must occur before hydrogen and oxygen can react to form water. Explain how splitting apart an oxygen molecule requires energy and therefore is an endothermic process and how uniting the individual hydrogen and oxygen atoms as water molecules is an exothermic process.

18.2 Driving Force of Reactions

At this time, you might "accidentally" drop a book onto the floor. Ask the students why the book fell to floor. All students will acknowledge the presence of gravity. Stress that in a gravitational field such as that found on earth, all objects tend to move towards the center of the earth unless something, such as the floor, stands in the way. Use this analogy to explain that it is a fundamental tendency in nature to move to the most energetically stable position. In the case of a chemical reaction, considering only energy, the most stable state must be the state that contains the least amount of stored energy. Explain how the products of exothermic reactions meet this criteria and how many exothermic reactions are indeed spontaneous.

Pose, but don't answer, the question "Why don't the hydrogen and oxygen molecules in the balloon spontaneously react to form water?" The reaction is certainly exothermic and thus the water molecules must contain less energy than the original hydrogen and oxygen molecules. After posing the question, perform the following demonstration.

Demonstration 2: A Spontaneous Endothermic Reaction

PURPOSE: To introduce the concept of entropy as a driving force for chemical reactions

MATERIALS: Approximately 20 g of barium hydroxide ($Ba(OH)_2 \cdot 8H_2O$) and 10 g of ammonium thiocyanate (NH_4SCN) (Carefully grind the crystals of $Ba(OH)_2 \cdot 8H_2O$ into a fine powder with a mortar and pestle), 250-mL Erlenmeyer flask, solid rubber stopper for the flask, block of wood, distilled water bottle

PROCEDURE: Explain to the students that you will be carrying out a reaction involving two solids. Ask the students to describe what they observe. Show the vials containing the two solids to the students. Appear to clean the block of wood by pouring some distilled water onto the block of wood. The puddle should be about half the diameter of the base of flask.

Pour each of the solids into flask. Stopper the flask and begin to swirl the contents. The solids appear to get wet and begin to form a liquid. As this happens, place the flask on the puddle of water on the block and hold up the block and flask for all to see. Continue to swirl the block and flask. After a minute or so, the flask adheres to the block of wood and a film of ice forms on the outside part of the flask in contact with the liquid. Feign surprise and try to remove the flask from the block of wood. (Don't try too hard. The ice layer binds the flask to wood to such an extent that the flask may break if too much pressure is applied.)

You may pass the block of wood around the room, allowing the students to feel the intense cold. CAUTION: Do not touch the base of the flask. Frostbite might occur if the mixture is placed directly on the skin. For this reason this reaction is not suitable for chemical ice packs.

Open the flask and waft the fumes towards the nose. The odor of ammonia will be detected. Thus two solids reacted to form a solution and a gas. The solution is formed from the water of hydration. The process absorbed so much energy that water in contact with the mixture froze. Explain how the solution and gaseous state contain particles in more random states than the original crystalline solids. Finally, explain that a second tendency in nature is the tendency toward the more random of two states. Such a movement always involves absorption of energy. This is a case of an entropy-driven reaction.

$$Ba(OH)_2 \cdot 8H_2O + 2NH_4SCN \rightarrow$$
$$Ba^{2+} + 2SCN^{-1} + 2NH_3 + 10H_2O$$

Summarize the information gleaned from this demonstration and from the examples of entropy driven reactions in the text by writing the phases of matter on the chalkboard or overhead projector.

lower		gas		more
energy		liquid		random
state		solid		state

Nature tends to move spontaneously to the lower energy state (solid) and to the more random state (gas). The two tendencies oppose each other. Whichever one wins out determines the course of the reaction. With this background, present the chemist's need for another term or function that takes both tendencies into consideration. Gibb's Free Energy, ΔG, is the function.

Point out that exothermic reactions move to lower energy states and that a negative ΔH is associated with exothermic reactions. A positive sign is required for the change in entropy because one moves from a less random to a more random state. Since the temperature of a material determines its physical state, an argument can be made for including temperature with the entropy term.

Define ΔG as the difference between the change in enthalpy (heat content) and the product of the Kelvin temperature and the change in entropy (randomness) of a system. For a reaction to proceed spontaneously as written (from reactant to product *or* left to right), the sign of ΔG must be negative.

When working the problems in the text, point out that tabulated values of ΔH and ΔS have different units. For our beginning study, ΔH is considered independent of temperature and is expressed in kJ/mol. ΔS is dependent on temperature and expressed in J/K·mol. Students frequently forget to convert kJ to J or vice versa.

18.3 The Reaction Process

Return to the hydrogen–oxygen balloon to discuss the reaction process. Have the students once again try to describe what must be taking place on a molecular level. (Molecules are colliding with each other.) Bonds are not being broken because no reaction is evident at room conditions. Once the reaction is initiated, it goes to completion. Here is a reaction driven by the large decrease in enthalpy. All the heat evident in the reaction must have come from the original hydrogen and oxygen molecules. The water molecules must be at a lower enthalpy state. Entropy does not appear to be too important because one has gases as reactants and as a product.

At this point the concepts of reaction mechanism, collision theory, activation energy, and activated complex may be introduced using the hydrogen–oxygen reaction as well as the examples listed in the text. Draw from the students the following requirements: oxygen and hydrogen molecules must split apart so that the atoms can rearrange to form products; the molecules or atoms must come in contact with each other if they are to react; energy is required to split the molecules; as the molecules split and rearrange to form the product molecules, they must pass through some point in which they can return to the original molecules or move on to the product molecules.

18.4 Reaction Rate

Students will agree that the collision theory makes sense. In discussing the rate of reactions and factors that influence the rates, use a traffic analogy to bring the concepts home.

Most students will agree that more accidents (collisions) occur during rush hour when more cars are on the road (greater concentration). At the same time, the collisions during rush hour may not be as severe as collisions at other times because most traffic moves more slowly during rush hour (lower temperature and velocity of molecules). In a collision of a large car with a compact car, the smaller car is damaged more severely (nature of reactants). The analogy can continue with ice on the road acting as a catalyst and how the cars hit (orientation) will, in part, determine how much damage occurs.

A number of demonstrations are available to bring these concepts home to students. Grain-elevator explosions and some coal-mine explosions are due to the very large surface area of the grain residues and coal dust. If lycopodium powder is available, a little may be sprinkled into a Bunsen-burner flame. The resulting flash simulates what happens in a grain-elevator explosion.

Demonstration 3: Concentration vs Reaction Rate

PURPOSE: To show the effect of concentration on the rate of a reaction

MATERIALS: A clean, one-pint paint can and lid, masking tape, short length of rubber hose, matches

PROCEDURE: Punch a hole with a can opener near the bottom of the can. Punch a smaller hole, approximately 1 cm diameter, in the lid. (The size of the upper hole determines the rate of the reaction.) Place the lid on the can and fill the can from the top with methane gas from a gas jet. Attach the short hose to the gas valve and allow the methane to flow until the odor is noticeable. Ask students why the can is filled from the top and not from the bottom (methane is less dense than air). After the can is filled, place a piece of tape over the two holes. The can may then be transported to a safe place, preferably one away from the students and not below any light fixtures.

Explain that you are going to light the methane in the can and the students are to try and explain what occurs. The lit methane burns with a bright yellow flame. It does not burn inside the can because no oxygen is present in the can. The flame remains on the outside of the can as long as the mixture in the can does not have enough oxygen to support combustion. Ask the students why two holes are needed. (Most will reply to replace the burning methane.) Depending on the size of the hole in the lid, reaction continues for several minutes. You may have to turn off the room lights as the flame gets smaller and burns with an almost invisible blue flame. As the flame disappears into the can, an explosion occurs, blowing off the lid. If the lid is pressed on tightly enough, the force of the explosion can send it to the ceiling. CAUTION: It is imperative that you and the students are not near the can at the end of the demonstration.

You should make the can and practice ahead of time. Enlarge the size of the hole in the lid so that the reaction time is between three and four minutes.

Demonstration 4: Temperature vs Rate of Reaction

PURPOSE: To show that temperature determines the rate of a reaction

MATERIALS: Three cyalume light sticks, three 1000-mL beakers: one filled with boiling water, one with water at room temperature, and one with ice water

PROCEDURE: Initiate the cyalume reaction by bending the light sticks, breaking the inner glass ampule, and mixing the reagents. All three light sticks should glow with the same intensity.

Place all three light sticks in the water at room temperature. Note that all give off the same amount of light. Remove one from the beaker and place it in the beaker with the hot water. Remove a second and place it in the beaker with ice water. The brightness of the glow is indicative of the rate of the reaction. Take the brightly glowing light stick from the hot water and place it in the ice water, and do the reverse for the one that was in the ice water. Whichever stick is in the hot water will glow the brightest.

Explain that increasing the temperature also shortens the time the light stick will glow. Most yellow-green light sticks are rated for 12 hours at room temperature. The demonstration is a very dramatic one.

The demonstration books listed in the reference section include a number of other demonstrations that would be appropriate. The Desktop Investigation, *Factors Influencing Reaction Rate*, will allow students to study the factors influencing reaction rates on their own.

Now that the students have a qualitative grasp of the factors that influence reaction rates, explain that chemists need to quantitatively describe the rates of reactions. Rates are always positive numbers. Since reactions ordinarily take place in a series of steps, the slowest step must be the one that determines the overall rate. You may explain this in terms of a dishwashing analogy. If six students decide to wash glassware and two collect the dirty glassware, one washes and rinses and the others dry the glassware, the rate-determining step is the dishwashing step. The only way to increase the rate is to take one of the dryers and have him/her also wash dishes.

The rate-determining step must be determined experimentally and cannot be deduced from the balanced chemical equation. For a reaction to occur, individual steps will involve two-body collisions. Students will agree that even on a busy highway, seldom will three or four cars collide simultaneously. Most multiple car collisions occur in a chain reaction form with no more than two cars colliding at any one time.

Using the examples in the text, write out rate laws for the students.

References and Resources

Books and Periodicals

Borgford, Christie L. and Lee R. Summerlin. "Chemical Energy and Rates of Reaction" *Chemical Activities: A Sourcebook for Science Teachers*, The American Chemical Society, 1988.

Olney, David J. "Some Analogies for Teaching Rates/Equilibrium," *Journal of Chemical Education*, Vol. 65, No. 8, Aug. 1988, pp. 696–697.

Shakashiri, Bassam Z. "Thermochemistry," *Chemical Demon-*

strations, Vol. 2, University of Wisconsin Press, 1985.

Spencer, J.N. and E.S. Holmboe. "Entropy and Unavailable Energy," *Journal of Chemical Education*, Vol. 60, No. 12, Dec. 1983, pp. 1018–1021.

Summerlin, Lee R., et al. "Kinetics," *Chemical Demonstrations—A Sourcebook for Teachers*, Vol. 1, Second Edition, Washington, D.C.: American Chemical Society, 1988.

Summerlin, Lee R., et al. "Kinetics and Equilibrium," *Chemical Demonstrations—A Sourcebook for Teachers*, Vol. 2, Washington, D.C.: American Chemical Society, 1987.

Treptow, Richard S. "LeChâtelier's Principle Applied to the Temperature Dependence of Solubility," *Journal of Chemical Education*, Vol. 61, No. 6, June 1984, pp. 499–502.

Audiovisual Resources

Eyring, Henry. *Introduction to Reaction Kinetics*, A Chem Study Film, Color, 13 min., 16mm film or VHS cassette, Central Scientific Co.

Powell, Richard E. *Catalysis*, A Chem Study Film, Color, 17 min., 16-mm film or VHS cassette, Central Scientific Co.

Answers and Solutions

Questions

1. *(a)* During a chemical reaction, energy is used to break chemical bonds and energy is released when new bonds are formed. *(b)* The overall energy change depends mainly on the relative strengths and numbers of bonds broken and formed. *(c)* Thermochemistry is the study of the changes in heat energy that accompany chemical reactions and physical changes.

2. A calorimeter is a device that measures the heat absorbed or released in a chemical or physical change.

3. *(a)* Higher *(b)* Lower

4. *(a)* A thermochemical equation is an equation that includes the quantity of heat released or absorbed during the reaction as written. *(b)* An enthalpy change is the amount of heat exchanged by a system and its surroundings in a process carried out at constant temperature. *(c)* The heat of reaction is taken as a negative when the heat content of a system decreases. It is taken as positive when the heat content of a system increases.

5. *(a)* Molar heat of formation is the heat released or absorbed when one mole of a compound is formed by combination of its elements. *(b)* Standard states are the states found at atmospheric pressure and, usually, room temperature.

6. *(a)* The heat of reaction is the quantity of heat released or absorbed during a chemical reaction and after the reactants return to their original temperature. The heat of formation is the heat released or absorbed when one mole of a compound is formed by combination of its elements. The heat of combustion is the heat released by the complete combustion of 1 mol of a substance. *(b)* The heat of combustion is defined in terms of 1 mol of reactant. Heat of formation, on the other hand, is defined in terms of 1 mol of product.

7. *(a)* The heat of a reaction is the same no matter how many intermediate steps make up the pathway from reactants to products. *(b)* The general principles for combining thermochemical equations are (1) to reverse the direction and/or (2) multiply the coefficients of the known equations so that when added together they give the desired thermochemical equation.

8. *(a)* ΔH = sum of $\Delta H_f°$ of products − sum of $\Delta H_f°$ of reactants *(b)* $\Delta H_f°$ of compound X = sum of $\Delta H_f°$ of combustion products of X − ΔH_c of X.

9. The change in heat content of a reaction system is related to (1) the change in the number of bonds breaking and forming, and (2) the strengths of these bonds as the reactants form products.

10. *(a)* There is a tendency for processes to occur that lead to a lower energy state, and thus toward greater stability in a system. There is also a tendency for processes to occur that lead to a less orderly, or a more disordered, state. *(b)* The free energy of a system is a simultaneous assessment of both the enthalpy-and-entropy change tendencies. A reaction system proceeds spontaneously in the direction that lowers its free energy.

11. *(a)* A spontaneous chemical reaction is one that takes place without any continuing outside influence, such as heating, cooling, or stirring. *(b)* Most reactions that occur spontaneously in nature are exothermic, following the tendency toward lower energy so as to form more stable structural configurations.

12. *(a)* The free-energy change of a system is defined as the difference between the change in enthalpy and the product of the Kelvin temperature and the entropy change. *(b)* $\Delta G = \Delta H - T\Delta S$

13. Changes of phase in the direction of the solid phase are accompanied by decreases in entropy.

14. A rise in temperature generally favors an increase in entropy.

15. A negative ΔH and a positive ΔS always result in a negative ΔG that is indicative of a spontaneous reaction.

16. *(a)* The collision theory is the set of assumptions regarding collisions and reactions. *(b)* The collision must be energetic enough to supply the necessary activation energy and the colliding molecules must be oriented in a way that favors their efficient interaction.

17. Activation energy is the energy required to transform reactants into the activated complex.

18. *(a)* The reactant particles must undergo an effective collision that raises the internal energies of the reactants to their minimum level for reaction. *(b)* The activated complex occurs at the maximum-energy position along the reaction pathway.

19. The activated complex could re-form the original bonds and separate into the reactant particles or it could form new bonds and separate into product particles.

20. The activation energy required for the endothermic change is greater than that required for the exothermic change by the amount of the heat of reaction of the system.

21. *(a)* The reaction rate is the change in concentration of reactants per unit time as a reaction proceeds. *(b)* The rate

of reaction is measured by the amounts of reactants converted to products in a unit of time. *(c)* Reaction rate depends on the collision frequency of the reactants and on the collision efficiency.

22. Nature of reactants—the specific reactants and the individual bonds involved influence reaction rate Surface area—reaction rate is proportional to the area of contact of the reactant particles. Concentration—in general, an increase in the concentration of one or more of the reactants causes an increase in the rate of the reaction. Temperature—in general, a rise in temperature increases reaction rate because it causes an increase in both collision energy and collision frequency. Presence of a catalyst—the presence of a catalyst increases reaction rate because it provides an alternate pathway or reaction mechanism in which the energy barrier between reactants and products is lowered.

23. The rate-determining step is the step that proceeds at the lowest rate, and thus determines the overall reaction rate.

24. $R = k[A]^n[B]^m \ldots$

25. The law of mass action states that the rate of a chemical reaction is directly proportional to the product of the concentrations of reacting substances, each raised to the appropriate power.

Problems

1. *(a)* $\Delta H = -393.51$ kJ; exothermic
   (b) $\Delta H = -890.31$ kJ; exothermic
   (c) $\Delta H = +176$ kJ; endothermic
   (d) $\Delta H = -44.02$ kJ; exothermic
   (e) $\Delta H = -197.8$ kJ; exothermic

2. *(a)* $H_2(g) + \frac{1}{2}O_2(g) \rightarrow H_2O(\ell) + 285.83$ kJ; exothermic
   (b) $2Mg(s) + O_2(g) \rightarrow 2MgO(s) + 72.3$ kJ; exothermic
   (c) $I_2(s) + 62.4$ kJ $\rightarrow I_2(g)$; endothermic
   (d) $3CO(g) + Fe_2O_3(s) \rightarrow 2Fe(s) + 3CO_2(g) + 24.7$ kJ; exothermic
   (e) $2NO_2(g) + 114.2$ kJ $\rightarrow 2NO(g) + O_2(g)$; endothermic
   (f) $C_2H_4(g) + 3O_2(g) \rightarrow 2CO_2(g) + 2H_2O(\ell) + 1411.0$ kJ; exothermic

3. *(a)* $Ca(s) + Cl_2(g) \rightarrow CaCl_2(s) + 795.0$ kJ ($\Delta H_{reverse} = +795.0$ kJ)
   (b) $2C(s) + H_2(g) + 226.73$ kJ $\rightarrow C_2H_2(g)$ ($\Delta H_{reverse} = -226.83$ kJ)
   (c) $S(s) + O_2(g) \rightarrow SO_2(g) + 296.83$ kJ ($\Delta H_{reverse} = +296.83$ kJ)

4. *(a)* Reaction: $CaCO_3(s) \rightarrow CaO(s) + CO_2(g)$

   $CaCO_3(s) \rightarrow Ca(s) + C(s) + \frac{3}{2}O_2(g)$; $\Delta H = +1206.9$ kJ

   $Ca(s) + \frac{1}{2}O_2(g) \rightarrow CaO(s)$; $\Delta H = -635.5$ kJ

   $C(s) + O_2(g) \rightarrow CO_2(g)$; $\Delta H = -393.51$ kJ
   $CaCO_3(s) \rightarrow CaO(s) + CO_2(g)$; $\Delta H = +177.9$ kJ
   $\Delta H = [\Delta H_f°(CaO) + \Delta H_f°(CO_2)] - [\Delta H_f°(CaCO_3)]$
   $\Delta H = [(-635.5 \text{ kJ}) + (-393.51 \text{ kJ})] - [-1206.9 \text{ kJ}]$
   $\Delta H = +177.9$ kJ

   (b) Reaction: $Ca(OH)_2(s) \rightarrow CaO(s) + H_2O(g)$
   $Ca(OH)_2(s) \rightarrow Ca(s) + O_2(g) + H_2(g)$; $\Delta H = +986.6$ kJ

   $Ca(s) + \frac{1}{2}O_2(g) \rightarrow CaO(s)$; $\Delta H = -635.5$ kJ

   $H_2(g) + \frac{1}{2}O_2(g) \rightarrow H_2O(g)$; $\Delta H = -241.82$ kJ
   $Ca(OH)_2(s) \rightarrow CaO(s) + H_2O(g)$; $\Delta H = +109.3$ kJ
   $\Delta H = [\Delta H_f°(CaO) + \Delta H_f°(H_2O)] - \Delta H_f°(Ca(OH)_2]$
   $\Delta H = [(-635.5 \text{ kJ}) + (-241.82 \text{ kJ})] - [-986.6 \text{ kJ}]$
   $\Delta H = +109.3$ kJ

   (c) Reaction: $Fe_2O_3(s) + 3CO(g) \rightarrow 2Fe(s) + 3CO_2(g)$

   $Fe_2O_3(s) \rightarrow 2Fe(s) + \frac{3}{2}O_2(g)$; $\Delta H = +824.2$ kJ
   $3CO(g) \rightarrow 3C(s) + \frac{3}{2}O_2(g)$; $3[\Delta H = +110.53$ kJ]
   $3C(s) + 3O_2(g) \rightarrow 3CO_2(g)$; $3[\Delta H = -393.51$ kJ]
   $Fe_2O_3(s) + 3CO(g) \rightarrow 2Fe(s) + 3CO_2(g)$; $\Delta H = -24.7$ kJ
   $\Delta H = [\Delta H_f°(Fe) + 3\Delta H_f°(CO_2)] - [\Delta H_f°(Fe_2O_3) + 3\Delta H_f°(CO)]$
   $\Delta H = [0 + 3(-393.51 \text{ kJ})] - [(-824.2 \text{ kJ}) + 3(-110.53 \text{ kJ})]$
   $\Delta H = -24.7$ kJ

5. *(a)* Reaction: $C_2H_6(g) + \frac{7}{2}O_2(g) \rightarrow 2CO_2(g) + 3H_2O(\ell)$

   $C_2H_6(g) \rightarrow 2C(s) + 3H_2(g)$; $\Delta H = +84.68$ kJ
   $2C(s) + 2O_2(g) \rightarrow 2CO_2(g)$; $2[\Delta H = -393.51$ kJ]

   $3H_2(g) + \frac{3}{2}O_2(g) \rightarrow 3H_2O(\ell)$; $3[\Delta H = -285.83$ kJ]

   $C_2H_6(g) + \frac{7}{2}O_2(g) \rightarrow 2CO_2(g) + 3H_2O(\ell)$; $\Delta H = -1559.83$ kJ
   $\Delta H = [2\Delta H_f°(CO_2) + 3\Delta H_f°(H_2O)] - [\Delta H_f°(C_2H_6) + \Delta H_f°(O_2)]$
   $\Delta H = [2(-393.51 \text{ kJ}) + 3(-285.83 \text{ kJ})] - [-84.68 \text{ kJ} + 0]$
   $\Delta H = -1559.83$ kJ

   (b) Reaction: $C_6H_6(\ell) + \frac{15}{2}O_2(g) \rightarrow 6CO_2(g) + 3H_2O(\ell)$

   $C_6H_6(\ell) \rightarrow 6C(s) + 3H_2(g)$; $\Delta H = -48.50$ kJ
   $6C(s) + 6O_2(g) \rightarrow 6CO_2(g)$; $6[\Delta H = -393.51$ kJ]

   $3H_2(g) + \frac{3}{2}O_2(g) \rightarrow 3H_2O(\ell)$; $3[\Delta H = -285.83$ kJ]

   $C_6H_6(\ell) + \frac{15}{2}O_2(g) \rightarrow 6CO_2(g) + 3H_2O$; $\Delta H = -3267.05$ kJ
   $\Delta H = [6\Delta H_f°(CO_2) + 3\Delta H_f°(H_2)] - [\Delta H_f°(C_6H_6) + \Delta H_f°(O_2)]$
   $\Delta H = [6(-393.51 \text{ kJ}) + 3(-285.83 \text{ kJ})] - [+48.50 \text{ kJ} + 0]$
   $\Delta H = -3267.05$ kJ

6. $C(s) = -393.51$ kJ/mol; $CH_4(g) = -890.31$ kJ/mol;

$$SO_2(g) = \frac{-197.8 \text{ kJ}}{2 \text{ mol}} = -98.9 \text{ kJ/mol}$$

7. *(a)* $C_3H_8(g) + 5O_2(g) \rightarrow 3CO_2(g) + 4H_2O(\ell) + 2219.9$ kJ
   Reaction: $3C(s) + 4H_2(g) \rightarrow C_3H_8(g)$
   $3C(s) + 3O_2(g) \rightarrow 3CO_2(g); 3[\Delta H = -393.51 \text{ kJ}]$
   $4H_2(g) + 2O_2(g) \rightarrow 4H_2O(\ell); 4[\Delta H = -285.83 \text{ kJ}]$
   $3CO_2(g) + 4H_2O(\ell) \rightarrow C_3H_8(g) + 5O_2(g); \Delta H = +2219.9 \text{ kJ}$
   $3C(s) + 4H_2(g) \rightarrow C_3H_8(g); \Delta H = -104.0 \text{ kJ}$
   $\Delta H_f°(C_3H_8) = [3\Delta H_f°(CO_2) + 4\Delta H_f°(H_2O)] - [\Delta H_c(C_3H_8)]$
   $\Delta H_f°(C_3H_8) = 3(-393.51 \text{ kJ/mol}) + 4(-285.83 \text{ kJ/mol}) - (-2219.9 \text{ kJ/mol})$
   $\Delta H_f°(C_3H_8) = -104.0 \text{ kJ/mol}$

   (b) $C_2H_6(g) + \frac{7}{2}O_2(g) \rightarrow 2CO_2(g) + 3H_2O(\ell) + 1559.8$ kJ

   Reaction: $2C(s) + 3H_2(g) \rightarrow C_2H_6(g)$
   $2C(s) + 2O_2(g) \rightarrow 2CO_2(g); 2[\Delta H = -393.51 \text{ kJ}]$

   $3H_2(g) + \frac{3}{2}O_2(g) \rightarrow 3H_2O(\ell); 3[\Delta H = -285.83 \text{ kJ}]$

   $2CO_2(g) + 3H_2O(\ell) \rightarrow C_2H_6(g) + \frac{7}{2}O_2(g); \Delta H = +1559.8 \text{ kJ}$

   $2C(s) + 3H_2(g) \rightarrow C_2H_6(g); \Delta H = -84.7 \text{ kJ}$
   $\Delta H_f°(C_2H_6) = [2\Delta H_f°(CO_2) + 3\Delta H_f°(H_2O)] - [\Delta H_c°(C_2H_6)]$
   $\Delta H_f°(C_2H_6) = 2(-393.51 \text{ kJ/mol}) + 3(-285.83 \text{ kJ/mol}) - (-1559.8 \text{ kJ/mol})$
   $\Delta H_f°(C_2H_6) = -84.7 \text{ kJ/mol}$

   (c) $C_2H_2(g) + \frac{5}{2}O_2(g) \rightarrow 2CO_2(g) + H_2O(\ell) + 1299.6$ kJ

   Reaction: $2C(s) + H_2(g) \rightarrow C_2H_2(g)$
   $2C(s) + 2O_2(g) \rightarrow 2CO_2(g); 2[\Delta H = -393.51 \text{ kJ}]$

   $H_2(g) + \frac{1}{2}O_2(g) \rightarrow H_2O(\ell); \Delta H = -285.83 \text{ kJ}$

   $2CO_2(g) + H_2O(\ell) \rightarrow C_2H_2(g) + \frac{5}{2}O_2(g); \Delta H = +1299.6 \text{ kJ}$

   $2C(s) + H_2(g) \rightarrow C_2H_2(g); \Delta H = +266.8 \text{ kJ}$
   $\Delta H_f°(C_2H_2) = [2\Delta H_f°(CO_2) + \Delta H_f°(H_2O)] - [\Delta H_c(C_2H_2)]$
   $\Delta H_f°(C_2H_2) = 2(-393.51 \text{ kJ/mol}) + (-285.83 \text{ kJ/mol}) - (-1299.6 \text{ kJ/mol})$
   $\Delta H_f°(C_2H_2) = +226.8 \text{ kJ/mol}$

   (d) $C_6H_6(\ell) + \frac{15}{2}O_2(g) \rightarrow 6CO_2(g) + 3H_2O(\ell) + 3267.5$ kJ

   Reaction: $6C(s) + 3H_2(g) \rightarrow C_6H_6(\ell)$
   $6C(s) + 6O_2(g) \rightarrow 6CO_2(g); 6[\Delta H = -393.51 \text{ kJ}]$

   $3H_2(g) + \frac{3}{2}O_2(g) \rightarrow 3H_2O(g); 3[\Delta H = -285.83 \text{ kJ}]$

   $6CO_2(g) + 3H_2O(\ell) \rightarrow C_6H_6(\ell) + \frac{15}{2}O_2(g); \Delta H = +3267.5 \text{ kJ}$

   $6C(s) + 3H_2(g) \rightarrow C_6H_6(\ell); \Delta H = +49.0 \text{ kJ}$
   $\Delta H_f°(C_6H_6) = [6\Delta H_f°(CO_2) + 3\Delta H_f°(H_2O)] - [\Delta H_c(C_6H_6)]$
   $\Delta H_f°(C_6H_6) = 6(-393.51 \text{ kJ/mol}) + 3(-285.83 \text{ kJ/mol}) - (-3267.5 \text{ kJ/mol})$
   $\Delta H_f°(C_6H_6) = +49.0 \text{ kJ/mol}$

8. *(a)* $H_2(g) + \frac{1}{2}O_2(g) \rightarrow H_2O(\ell) + (-\Delta H_c)$

   $\Delta H_f°(H_2) = \Delta H_f°(H_2O) - \Delta H_c(H_2)$
   $\Delta H_c(H_2) = \Delta H_f°(H_2O) - \Delta H_f°(H_2)$
   $\Delta H_c(H_2) = (-285.83 \text{ kJ/mol}) - (0)$
   $\Delta H_c(H_2) = -285.83 \text{ kJ/mol}$

   (b) $S(s) + O_2(g) \rightarrow SO_2(g) + (-\Delta H_c)$
   $\Delta H_f°(S) = \Delta H_f°(SO_2) - \Delta H_c(S)$
   $\Delta H_c(S) = \Delta H_f°(SO_2) - \Delta H_f°(S)$
   $\Delta H_c(S) = (-296.83 \text{ kJ/mol}) - (0.00 \text{ kJ/mol})$
   $\Delta H_c(S) = -296.83 \text{ kJ/mol}$

9. $\Delta H_c(C_2H_5OH) = [2\Delta H_f°(CO_2) + 3\Delta H_f°(H_2O)] - [\Delta H_f°(C_2H_5OH)]$
   $\Delta H_c(C_2H_5OH) = [2(-393.51 \text{ kJ/mol}) + 3(-285.83 \text{ kJ/mol})] - [-277.0 \text{ kJ/mol}]$
   $\Delta H_c(C_2H_5OH) = -1367.5 \text{ kJ/mol}$

10. (a) $\Delta G = \Delta H - T\Delta S$
$\Delta G = (+125 \text{ kJ}) - (293\text{K})(0.0350 \text{ kJ/K})$
$\Delta G = +125 \text{ kJ} - 10.3 \text{ kJ}$
$\Delta G = 115 \text{ kJ}$; not spontaneous

   (b) $T = 127°\text{C} + 273 = 400\text{K}$
$\Delta G = \Delta H - T\Delta S$
$\Delta G = (-85.2 \text{ kJ}) - (400\text{K})(0.125 \text{ kJ/K})$
$\Delta G = -85.2 \text{ kJ} - 50.0 \text{ kJ}$
$\Delta G = 135.2 \text{ kJ}$; spontaneous

   (c) $T = 500°\text{C} + 273 = 773\text{K}$
$\Delta G = \Delta H - T\Delta S$
$\Delta G = (-275 \text{ kJ}) - (773\text{K})(0.450 \text{ kJ/K})$
$\Delta G = -275 \text{ kJ} - 348 \text{ kJ}$
$\Delta G = -623 \text{ kJ}$; spontaneous

11. (a)
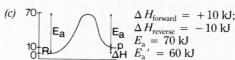
$\Delta H_{\text{reverse}} = -30 \text{ kJ}$
$E_a = 50 \text{ kJ}$

   (b)
$\Delta H_{\text{reverse}} = +30 \text{ kJ}$
$E_a' = 50 \text{ kJ}$

12. $R = k[\text{A}]^2[\text{B}]$. Doubling A increases the rate to 4 times its original rate. Doubling B increases the rate to twice its original rate.

13. (a)
$\Delta H_{\text{forward}} = +60 \text{ kJ}$;
$\Delta H_{\text{reverse}} = -60 \text{ kJ}$
$E_a = 80 \text{ kJ}$
$E_a' = 20 \text{ kJ}$

   (b)
$\Delta H_{\text{forward}} = -40 \text{ kJ}$;
$\Delta H_{\text{reverse}} = +40 \text{ kJ}$
$E_a = 20 \text{ kJ}$
$E_a' = 60 \text{ kJ}$

   (c)
$\Delta H_{\text{forward}} = +10 \text{ kJ}$;
$\Delta H_{\text{reverse}} = -10 \text{ kJ}$
$E_a = 70 \text{ kJ}$
$E_a' = 60 \text{ kJ}$

Application Question

1. A large amount of energy is released when a compound with a high negative heat of formation is formed. The same amount of energy is required to cause such a compound to decompose into its separate elements. This energy must be supplied by an external source. Such compounds are very stable. Once they start, the reactions forming them usually proceed spontaneously and vigorously.

2. Fuels are usually chosen for their energy content, the energy released per unit consumed, not the temperature of the flame. Cost is the prime criterion in most comparisons; transportation, ease of use, and cleanliness are also considerations.

3. A positive heat of formation suggests that compound A is very unstable and may react or decompose explosively, while the $-\Delta H_f$ for compound B indicates that this compound is very stable.

4. If the temperature of the gases remained constant throughout the mixing, the heat content did not change.

5. The self-mixing of the two gases is accompanied by an increase in entropy, a driving force in natural processes.

6. An endothermic reaction in which the entropy change is favorable, and the increase in entropy (expressed as $T\Delta S$) exceeds the unfavorable energy change, will proceed spontaneously as it is accompanied by a decrease in free energy. In such cases, the free energy change is negative.

7. If an exothermic reaction is accompanied by a decrease in entropy such that the unfavorable change in entropy, expressed as $T\Delta S$, exceeds the favorable change in energy content, the reaction will not be spontaneous. This is to say that the free energy change is positive.

8. The temperature of a system is the dominant factor in determining the relative importance of the tendencies toward lower energy and higher entropy. At low temperatures, $T\Delta S$ is generally small in comparison to ΔH, so the reaction proceeds as the ΔH dictates. However, at high temperatures, when ΔS is positive, the $T\Delta S$ factor may be large enough to exceed the ΔH value and thus the entropy factor dictates the spontaneity of the reaction.

9. Since most reactions actually take place in a sequence of one-step processes, it is unlikely that the burning of ethane takes place in this single step.

10. As the formation of a single activated complex by the simultaneous collision of five molecules is highly improbable, the reaction mechanism must involve some sequence of simple steps.

11. $NO_2 + O \rightarrow NO + O_2$

12. (a) This is the pathway by which the potential energy of the reactants is raised to the minimum level needed for effective collisions. (b) The maximum-energy region of the minimum-energy pathway for reaction is reached when the activated complex is formed.

13. In such areas, the powdered nature of the dry combustible sawdust or grain allows for large surface areas of contact with oxygen so that the presence of even a spark could initiate a very rapid reaction that could result in an explosion.

14. Measure the pressure of the reaction system. Since two moles of gas react to form one mole of gas product, the pressure will decrease as the reaction proceeds.

15. (a) The rate is increased. (b) The rate is doubled. (c) The rate is cut in half. (d) The rate is increased. (e) The rate is increased.

16. The coefficients correspond to the powers *only* for reactions that follow a simple one-step pathway and occur at the molecular level *exactly* as written.

Application Problems

1. (a) $6C(s) + 6O_2(g) \rightarrow 6CO_2(g)$; $6[\Delta H_f^\circ = -393.51 \text{ kJ/mol}]$

   $7H_2(g) + \frac{7}{2}O_2(g) \rightarrow 7H_2O(\ell)$; $7[\Delta H_f^\circ = -285.83 \text{ kJ/mol}]$

   $6CO_2(g) + 7H_2O(\ell) \rightarrow C_6H_{14}(\ell) + \frac{19}{2}O_2(g)$; $\Delta H = +4163.1 \text{ kJ}$

   $6C(s) + 7H_2(g) \rightarrow C_6H_{14}(\ell)$; $\Delta H = -198.77 \text{ kJ}$
   $\Delta H_f^\circ = [6(-393.51 \text{ kJ}) + 7(-285.83 \text{ kJ})] - [-4163.1 \text{ kJ}]$
   $\Delta H_f^\circ = [-2361.06 \text{ kJ} - 2000.81 \text{ kJ}] + [4163.1 \text{ kJ}]$
   $\Delta H_f^\circ = -198.77 \text{ kJ}$

   (b) $12C(s) + 12O_2(g) \rightarrow 12CO_2(g)$; $12[\Delta H_f^\circ(CO_2) = -393.51 \text{ kJ/mol}]$

   $11H_2(g) + \frac{11}{2}O_2(g) \rightarrow 11H_2O(\ell)$; $11[\Delta H_f^\circ(H_2O) = -285.83 \text{ kJ/mol}]$

   $12CO_2(g) + 11H_2O(\ell) \rightarrow C_{12}H_{22}O_{11}(s) + 12O_2(g)$; $\Delta H = +5640.9 \text{ kJ}$

   $12C(s) + 11H_2(g) + \frac{11}{2}O_2(g) \rightarrow C_{12}H_{22}O_{11}(s)$; $\Delta H_f^\circ = -2225.35 \text{ kJ/mol}$

   $\Delta H_f^\circ = [12(-393.51 \text{ kJ}) + 11(-285.83 \text{ kJ})] - [-5640.9 \text{ kJ}]$
   $\Delta H_f^\circ = [-4722.12 \text{ kJ} - 3144.13 \text{ kJ}] - [5640.9 \text{ kJ}]$
   $\Delta H_f^\circ = -2225.35 \text{ kJ}$

2. (a) $R = k[A][B]^2$

   (b) $k = \dfrac{R}{[A][B]^2} = \dfrac{2.0 \times 10^{-4} \text{ mol/L} \cdot \text{min}}{(0.20 \text{ mol/L})(0.20 \text{ mol/L})^2} = 2.5 \times 10^{-2} \dfrac{L^2}{\text{mol}^2 \cdot \text{min}}$

   (c) $R = k[A][B]^2 = 2.5 \times 10^{-2} \dfrac{L^2}{\text{mol}^2 \cdot \text{min}} \times 0.30 \text{ mol/L} \times (0.30 \text{ mol/L})^2 = 6.8 \times 10^{-4} \dfrac{\text{mol}}{L \cdot \text{min}}$

3. $T = 25°C + 273 = 298K$
   $\Delta G = \Delta H - T\Delta S$
   $\Delta G = (-393.51 \text{ kJ}) - (298K)(0.003\ 00 \text{ kJ/K})$
   $\Delta G = -394. \text{ kJ}$; spontaneous

4. $T = 27°C + 273 = 300K$
   $\Delta G = \Delta H - T\Delta S$
   $\Delta G = (-74.8 \text{ kJ}) - (300K)(-0.0809 \text{ kJ/K})$
   $\Delta G = -50.5 \text{ kJ}$; spontaneous

5. (a) 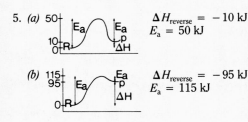 $\Delta H_{\text{reverse}} = -10 \text{ kJ}$
   $E_a = 50 \text{ kJ}$

   (b) $\Delta H_{\text{reverse}} = -95 \text{ kJ}$
   $E_a = 115 \text{ kJ}$

   (c) $\Delta H_{\text{forward}} = -40 \text{ kJ}$
   $E_a' = 70 \text{ kJ}$

6. $S(s) + O_2(g) \rightarrow SO_2(g)$; $\Delta H = -296.83 \text{ kJ/mol}$

   $SO_2(g) + \frac{1}{2}O_2(\ell) \rightarrow SO_3(g)$; $\Delta H = -98.91 \text{ kJ/mol}$

   $SO_3(g) + H_2O(\ell) \rightarrow H_2SO_4(\ell)$; $\Delta H = -132.4 \text{ kJ/mol}$

   $H_2(g) + \frac{1}{2}O_2(g) \rightarrow H_2O(\ell)$; $\Delta H = -283.83 \text{ kJ/mol}$

   $S(s) + 2O_2(g) + H_2(g) \rightarrow H_2SO_4(\ell)$; $\Delta H = -812.0 \text{ kJ/mol}$

Chapter 18 Reaction Energy and Reaction Kinetics

INTRODUCTION

In Chapters 8 and 9, stoichiometry and methods for balancing chemical equations were presented to provide an understanding of chemical changes. This chapter also deals with chemical change. The focus here is on energy changes involved when substances react, the actual processes that take place during reactions, and the rates at which reactions occur.

LOOKING AHEAD

As you study this chapter, pay special attention to the quantitative aspects of chemical reactions. Note the ways that energy, concentration, and time values are related mathematically and how simple mathematical methods are used to solve for unknown quantities.

SECTION PREVIEW

18.1 Thermochemistry
18.2 Driving Force of Reactions
18.3 The Reaction Process
18.4 Reaction Rate

An explosion is a dramatic example of a chemical reaction that results in the release of energy.

18.1 Thermochemistry

The role of energy in chemical reactions is of primary interest. Virtually every chemical reaction is accompanied by the absorption or the release of energy. To compare the quantities of energy associated with chemical reactions, the reactions are compared under clearly defined conditions. When this is done, a characteristic absorption or release of energy can be assigned to each.

During a chemical reaction, energy is used to break chemical bonds and energy is released when new bonds form. The overall energy change depends mainly on the relative strengths and numbers of bonds broken and formed.

Most often chemical reactions absorb or release energy in the form of heat. You learned in Chapter 13 that heat is also absorbed or released in physical changes, such as melting ice or condensing a vapor. **Thermochemistry** *is the study of the changes in heat energy that accompany chemical reactions and physical changes.*

Heat of Reaction

The heat absorbed or released in a chemical or physical change is measured in a device known as a **calorimeter.** For example, if a mixture of hydrogen and oxygen is ignited, water is formed and heat energy is released explosively. Known quantities of reactants are sealed in the reaction chamber, which is immersed in a known quantity of water in an insulated vessel. Therefore, the heat given off (or absorbed) during the reaction is determined from the temperature change of the known mass of surrounding water. For example, if the reaction is exothermic, the surrounding water will become warmer.

The quantity of heat given up during the formation of water from H_2 and O_2 is proportional to the quantity of water formed. No heat is supplied externally, except to ignite the mixture. The heat that is released comes from the reactants as they change into products. The heat content of the product water must be less than the heat content of the reactants before ignition. The equation for this reaction is written as

$$2H_2(g) + O_2(g) \rightarrow 2H_2O(g)$$

which expresses the fact that when 2 mol of hydrogen gas at room temperature are burned, 1 mol of oxygen gas is used, and 2 mol of water vapor are formed.

Writing the heat released as a reaction product gives the equation

$$2H_2(g) + O_2(g) \rightarrow 2H_2(g) + 483.6 \text{ kJ}$$

This is a **thermochemical equation,** *an equation that includes the quantity of heat released or absorbed during the reaction as written.* The quantity of heat for this or any reaction depends on the amounts of reactants and products. Forming twice as much water vapor would release 2×483.6 kJ. Forming one-half as much water would release $1/2 \times 483.6$ kJ. The thermochemical equations for these reactions are

$$4H_2(g) + 2O_2(g) \rightarrow 4H_2O(g) + 967.2 \text{ kJ}$$
$$H_2(g) + \tfrac{1}{2}O_2(g) \rightarrow H_2O(g) + 241.8 \text{ kJ}$$

Fractional coefficients are sometimes needed in thermochemical equations.

- Explain heat of reaction, heat of formation, and enthalpy.
- Solve problems involving heats of reaction, heats of formation, and heats of combustion.
- Explain the concept of bond energy and its relationship to heat of reaction.

STUDY HINT

One calorie of heat energy is absorbed by each gram of liquid water as its temperature increases by 1°C. The joule (J), rather than the calorie, is the SI unit of heat energy. One calorie equals exactly 4.184 J. Thus, a 1°C increase in the temperature of 1 g of water is brought about by the absorption of 4.184 J. One kilojoule (kJ) is equal to 1000 J.

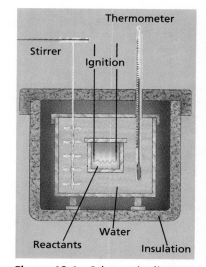

Figure 18-1 Schematic diagram of an ignition calorimeter used for measuring nutritional calories.

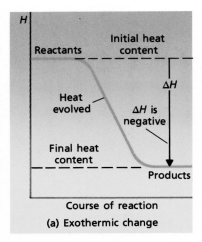

(a) Exothermic change

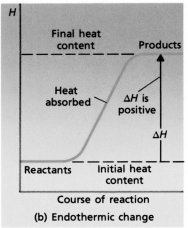

(b) Endothermic change

Figure 18-2 Change in heat content during a chemical reaction.

Δ always signifies the difference between a final state and the initial state. If energy is involved, a positive value indicates absorption of energy, and a negative value indicates a release of energy.

The **heat of reaction** *is the quantity of heat released or absorbed during a chemical reaction.* One way to think of heat of reaction is as the difference between the stored heat energy, or heat content, of the reactants and the products. As pictured in Figure 18-2a, in an exothermic reaction, the heat evolved is the difference between the higher heat content of the reactants and the lower heat content of the products. The situation is reversed when products have a higher heat content than reactants, as is the case in an endothermic reaction. The reverse reaction absorbs the same amount of heat that is released by the forward reaction. This is to be expected, for the difference between the heat content of reactants and products is unchanged. For the endothermic decomposition of water vapor, the thermochemical equation is

$$2H_2O(g) + 483.6 \text{ kJ} \rightarrow 2H_2(g) + O_2(g)$$

The states of reactants and products must always be included in thermochemical equations because they influence the overall amount of heat exchanged. The heat for the decomposition of water, for example, would be larger if we started with ice.

Chemical reactions are usually carried out in open vessels. Volume may change, but the pressure is equal to atmospheric pressure and does not change. The heat of a reaction under these conditions is represented by ΔH. The H is the symbol for a quantity called **enthalpy,** *the heat content of a system at constant pressure.* However, it is not very useful to talk just about heat content or enthalpy because we have no way to measure the enthalpy of a system. Only *changes* in enthalpy can be measured. The Greek letter Δ (a capital "delta") stands for "change in." Therefore, ΔH is read as the "change in enthalpy" or the "enthalpy change." An **enthalpy change** *is the amount of heat absorbed or lost by a system in a process carried out at constant pressure.*

Thermochemical equations are usually written by including ΔH. In an exothermic reaction, ΔH is always given a minus sign.

$$2H_2(g) + O_2(g) \rightarrow 2H_2O(g) \qquad \Delta H = -483.6 \text{ kJ}$$

For an endothermic reaction, ΔH is always given a positive value (the + sign is not usually written).

$$2H_2O \rightarrow 2H_2(g) + O_2(g) \qquad \Delta H = 483.6 \text{ kJ}$$

The sign convention is arbitrary, but it is logical. The heat of reaction is taken as negative when the heat content of a system decreases (see Figure 18-2a). It is taken as positive when the heat content of a system increases (see Figure 18-2b).

Heat of Formation

Doing an experiment with a calorimeter every time a heat of reaction is needed is not very practical. Instead, the results of many hundreds of such measurements are recorded in chemistry reference books. By consulting these data and, if necessary, combining them as illustrated later in this section, chemists are able to calculate heats of reaction for many common reactions.

The formation of water from hydrogen and oxygen is a composition reaction—the formation of a compound from its elements. Thermochemical data are often recorded as the heats of such composition reactions. The **molar heat of formation** *is the heat released or absorbed when one mole of a compound is formed by combination of its elements.*

To make comparisons meaningful, heats of formation are given for the standard states of reactants and products—these are the states found at atmospheric pressure and, usually room temperature. Thus, the standard state of water is liquid, not gas or solid. The standard state of iron is solid, not molten. To signify that a value represents measurements on substances in their standard states, a ° sign is added to the symbol, giving $\Delta H°$ for the standard heat of a reaction. Adding an $_f$ as in $\Delta H°_f$ further indicates a standard heat of formation.

Some standard heats of formation are given in Appendix Table A-16. Each is the heat of reaction for the composite reaction for *one mole* of the compound listed. The thermochemical equation to accompany each heat of reaction shows the formation of one mole of the compound from the elements in their standard states. For liquid water, sodium chloride, and sulfur dioxide, for example, the equations are

$$H_2(g) + \tfrac{1}{2}O_2(g) \rightarrow H_2O(\ell) \qquad \Delta H°_f = -285.9 \text{ kJ}$$
$$Na(s) + \tfrac{1}{2}Cl_2(g) \rightarrow NaCl(s) \qquad \Delta H°_f = -411.10 \text{ kJ}$$
$$S(s) + O_2(g) \rightarrow SO_2(g) \qquad \Delta H°_f = -296.83 \text{ kJ}$$

Elements in their standard states are *defined* as having $\Delta H°_f = 0$. A negative $\Delta H°_f$ indicates a substance that is more stable than the free elements.

Stability and Heat of Formation

You can see in Appendix Table A-16 that the majority of the heats of formation are negative.

A large amount of energy is released when a compound with a high negative heat of formation is formed. The same amount of energy is required to cause such a compound to decompose into its separate elements. This energy must be supplied by an external source. Such compounds are very stable. Once they start, the reactions forming them usually proceed spontaneously and vigorously. For example, the ΔH_f of carbon dioxide is -393.51 kJ per mole of gas produced.

Compounds with relatively high values of heats of formation (that is, values that are positive or only slightly negative) are relatively unstable. For example, hydrogen sulfide (H_2S) has a heat of formation of -20.63 kJ/mol. It is not very stable and decomposes when heated. Hydrogen iodide (HI) is a colorless gas that decomposes somewhat when stored at room temperature. It has a relatively high heat of formation of $+26.48$ kJ/mol. As it decomposes, violet iodine vapor becomes visible throughout the container of the gas.

A compound with a high positive heat of formation is very unstable and may react or decompose explosively. For example, ethyne (C_2H_2) reacts explosively with oxygen. Nitrogen triiodide and mercury fulminate decompose explosively. The formation reactions of such compounds store a great deal of energy within them. Mercury fulminate ($HgC_2N_2O_2$) has a heat formation of $+270$ kJ/mol. Its instability makes it useful as a detonator for explosives.

Figure 18-3 Hydrogen iodide gas *(top)* decomposes into hydrogen and iodine. Iodine vapor is violet.

Figure 18-4 The combustion of wood releases all its stored energy. Ashes have no stored energy.

Heat of Combustion

Whether used in a furnace, automobile, or rocket, fuels must be energy-rich. The products of their combustion reaction with oxygen are energy-poor substances. In these combustion reactions, the energy yield may be very high. The products of the chemical action may be of little importance compared to the quantity of heat energy given off.

The combustion of 1 mol of graphite, which is pure carbon (C), yields 393.51 kJ of heat energy.

$$C(s) + O_2(g) \rightarrow CO_2(g) \qquad \Delta H = -393.51 \text{ kJ}$$

TABLE 18-1	HEAT OF COMBUSTION	
Substance	Formula	ΔH_c
hydrogen (g)	H_2	−285.9
carbon (graphite) (s)	C	−393.51
carbon monoxide (g)	CO	−283.0
methane (g)	CH_4	−890.31
ethane (g)	C_2H_6	−1559.8
propane (g)	C_3H_8	−2219.9
butane (g)	C_4H_{10}	−2878.5
pentane (g)	C_5H_{12}	−3536.1
hexane (ℓ)	C_6H_{14}	−4163.1
heptane (ℓ)	C_7H_{16}	−4811.2
octane (ℓ)	C_8H_{18}	−5450.5
ethene (ethylene) (g)	C_2H_4	−1411.0
propene (propylene) (g)	C_3H_6	−2051
ethyne (acetylene) (g)	C_2H_2	−1299.6
benzene (ℓ)	C_6H_6	−3267.5
toluene (ℓ)	C_7H_8	−3909

ΔH_c = heat of combustion of the given substance. All values of ΔH_c are expressed as kJ/mol of substance oxidized to H_2O and/or CO_2 (g) at constant pressure and 25°C. (s) = solid, (ℓ) = liquid, (g) = gas.

Technology

Fuels and Energy Content

Acetylene burns with an unusually hot flame. Why isn't it always used for fuel? Fuels are usually chosen for their energy content, the energy released per unit mass consumed, not the temperature of the flame. The "unit consumed" can be kilograms, moles, or liters of fuel. However, cost is the prime criterion in most comparisons; consumers want the most joules per dollar.

Coal is almost pure carbon. When burned, it releases 32.76 kJ/g. Coal is inexpensive for industrial use. It is inconvenient and messy for use in individual houses, however, and expen-

sive since the cost for home delivery has to be included.

Natural gas, which is principally methane, is easy to

transport or pipe into homes. Methane releases 55.50 kJ/g when it burns, which is greater than the heat released by carbon. Natural gas is also cleaner and easier to use than coal, but it is more expensive in many industrial situations.

So why isn't acetylene always used as fuel? It releases 49.92 kJ/g when burned, less than natural gas. It is also expensive to use, since pressurized oxygen is needed to burn it efficiently. Moreover, actylene is more dangerous to handle than natural gas and it must be manufactured since it is not a component of natural gas or oil.

The heat released by the complete combustion of 1 mol of a substance is called the **heat of combustion** *of the substance.* Heat of combustion is defined in terms of *1 mol of reactant*. Heat of formation, on the other hand, is defined in terms of *1 mol of product*. The general heat-of-reaction notation—ΔH—applies to heats of combustion as well. The notation ΔH_c is used to refer specifically to heat of combustion, however, as shown in Table 18-1. A more extensive list is in Appendix Table A-17.

Carbon dioxide and water are the products of the complete combustion of organic compounds containing carbon, hydrogen, and oxygen. Knowing this, the thermochemical equation for the combustion of any compound listed in the Tables can be written by balancing the equation for the reaction of one mole of the compound. For propane, for example,

$$C_3H_8(g) + 5O_2(g) \rightarrow 3CO_2(g) + 4H_2O(\ell) \qquad \Delta H_c = -2219.9 \text{ kJ}$$

Calculating Heats of Reaction

Thermochemical equations can be rearranged, terms can be cancelled, and the equations can be added to give heats for reactions not included in the data tables. The basis for this type of calculation is known as **Hess's law:** *The heat of a reaction is the same no matter how many intermediate steps make up the pathway from reactants to products.* The energy difference between

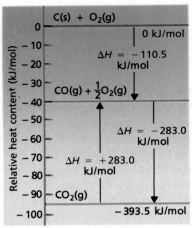

Figure 18-5 Heat of formation diagram for carbon dioxide and carbon monoxide

reactants and products, which is illustrated in Figure 18-2, is independent of the route taken to get from one to the other. In fact, measured heats of reaction can be combined to calculate heats of reaction that are difficult or impossible to actually measure.

To demonstrate how to apply Hess's law, we will work our way through the calculation of several heats of reaction. To start with a simple one, suppose you need to know the heat of reaction for the decomposition of ice to give hydrogen and oxygen. The available data are the ΔH_f° of liquid water as recorded in the table, and the heat of fusion, or melting, of ice, which is 6.0 kJ/mol.

$$H_2(g) + \tfrac{1}{2}O_2(g) \rightarrow H_2O(\ell) \qquad \Delta H^\circ_f = -285.9 \text{ kJ}$$
$$H_2O(s) \rightarrow H_2O(\ell) \qquad \Delta H = 6.0 \text{ kJ}$$

The reaction for which ΔH is desired is

$$H_2O(s) \rightarrow H_2(g) + \tfrac{1}{2}O_2(g) \qquad \Delta H = ?$$

The general principles for combining thermochemical equations are *(1)* to reverse the direction and/or *(2)* multiply the coefficients of the known equations so that when added together they give the desired thermochemical equation. In this case, all that is needed is to reverse the formation equation for water, remembering to change the sign of ΔH_f from negative to positive, and add the equations:

$$
\begin{array}{lll}
H_2O(\ell) & \rightarrow H_2(g) + \tfrac{1}{2}O_2(g) & \Delta H = 285.9 \text{ kJ} \\
H_2O(s) & \rightarrow H_2O(\ell) & \Delta H = 6.0 \text{ kJ} \\
\hline
H_2O(s) & \rightarrow H_2(g) + \tfrac{1}{2}O_2(g) & \Delta H = 291.9 \text{ kJ}
\end{array}
$$

The decomposition of ice to hydrogen and oxygen requires 291.9 kJ per mole of ice.

As a somewhat more complex example, consider the heat of a reaction used to prepare ultra-pure silicon for the electronics industry:

$$2Mg(s) + SiCl_4(\ell) \rightarrow Si(s) + 2MgCl_2(s) \qquad \Delta H = ?$$

The available information is the heats of formation of the reactant and product:

$$
\begin{array}{lll}
Si(s) + 2Cl_2(g) & \rightarrow & SiCl_4(\ell) \quad \Delta H^\circ_f = -687 \text{ kJ} \\
Mg(s) + Cl_2(g) & \rightarrow & MgCl_2(s) \quad \Delta H^\circ_f = -641 \text{ kJ}
\end{array}
$$

These equations can be combined to give the desired ΔH. The first equation must be reversed because $SiCl_4$ is a reactant, not a product, in the new equation. The second equation need not be reversed, because $MgCl_2$ is a product. However, this equation must be multiplied by 2 because 2MgCl are needed in the new equation. With this done, as shown below, the equal number of unwanted Cl_2 molecules on opposite sides of the equation cancel to give the desired thermochemical equation

$$
\begin{array}{lll}
SiCl_4(\ell) & \rightarrow Si(s) + 2Cl_2(g) & \Delta H = 687 \text{ kJ} \\
2Mg(s) + 2Cl_2(g) & \rightarrow 2MgCl_2(s) \; \Delta H = 2 \times -641 \text{ kJ} & = -1282 \text{ kJ} \\
\hline
2Mg(s) + SiCl_4(\ell) & \rightarrow Si(s) + 2MgCl_2(s) & \Delta H = -595 \text{ kJ}
\end{array}
$$

After doing many calculations of the type just illustrated, you would recognize that there is a general relationship between the heat of a reaction and the heats of formation of the reactants and products.

$$\Delta H = \text{Sum of } \Delta H_f^\circ \text{ of products} - \text{Sum of } \Delta H_f^\circ \text{ of reactants}$$

To find the necessary sums, the ΔH°_f values for each reactant and product must be multiplied by the coefficients of each in the desired equation and the signs adjusted as necessary for any changes of direction. For the reaction of Mg with $SiCl_4$, applying this equation gives

$$\Delta H = [0 + (2 \times -641 \text{ kJ})] - [0 + (+687 \text{ kJ})] = \text{ kJ}$$

Note the use of the assigned value of zero for the heats of formation of elements in their standard states.

Sample Problem 18.1

Calculate the heat of reaction for the combustion of nitrogen monoxide gas (NO) to form nitrogen dioxide gas (NO_2). Use the heat-of-formation data in Appendix Table A-16. Solve by combining the known thermochemical equations. Verify the result by using the general equation for finding heats of reaction from heats of formation.

Solution

Step 1. Analyze
Given: $\frac{1}{2}N_2(g) + \frac{1}{2}O_2(g) \rightarrow NO(g)$ $\Delta H_f^\circ = 90.25 \text{ kJ/mol}$

$\frac{1}{2}N_2(g) + O_2(g) \rightarrow NO_2(g)$ $\Delta H_f^\circ = 33.18 \text{ kJ/mol}$

Unknown: ΔH for $NO(g) + \frac{1}{2}O_2(g) \rightarrow NO_2(g)$

Step 2. Plan
ΔH_f° (NO), ΔH_f° (NO_2) $\rightarrow \Delta H_c$ (NO). The ΔH requested can be found by adding the ΔHs of the component reactions as specified in Hess's law such that $NO(g) + \frac{1}{2}O_2(g) \rightarrow NO_2(g)$ is the net result. Thus, reversing the first reaction and the sign of its ΔH yields $NO(g) \rightarrow \frac{1}{2}N_2(g) + \frac{1}{2}O_2(g)$ $\Delta H = -90.25$ kJ. Retaining the second, $\frac{1}{2}N_2(g) + O_2(g) \rightarrow NO_2(g)$ $\Delta H = +33.18$ kJ

Step 3. Compute
$$NO(g) \rightarrow \tfrac{1}{2}N_2(g) + \tfrac{1}{2}O_2(g)\ \Delta H = -90.25.\ \text{kJ}$$
$$\frac{\tfrac{1}{2}N_2(g) + O_2(g) \rightarrow NO_2(g)\ \Delta H = +33.18\ \text{kJ}}{NO(g) + \tfrac{1}{2}O_2(g) \rightarrow NO_2(g)\ \Delta H = -57.07\ \text{kJ}}$$

(Note the cancellation of the $\frac{1}{2}N_2(g)$ and the partial cancellation of the $O_2(g)$.)

Step 4. Evaluate
The unnecessary reactants and products cancel to give the desired equation. Verification: $\Delta H = \Delta H_f(NO_2) - [\Delta H_f(NO) + 0] = 33.18 \text{ kJ/mol} - 90.25 \text{ kJ/mol} = -57.07 \text{ kJ/mol}$

Practice Problem
1. Calculate the heat of reaction for the combustion of propane gas (C_3H) to form $CO_2(g)$ + $H_2O(\ell)$. Write all equations involved. *(Ans.)* $\Delta H = -2219.9$ kJ.

By the creative combination of available data, many types of thermochemical calculations are possible. Let's consider one in which a heat of formation and a heat of combustion are combined. When carbon is burned in a limited supply of oxygen, carbon monoxide is produced. In this reaction, carbon a limited supply of air, carbon monoxide is produced: $C(s) + O_2(g) \rightarrow CO_2(g)$. In this reaction, carbon is probably first oxidized to carbon dioxide. Then the carbon dioxide is reduced with carbon to give the carbon monoxide: $C(s) + CO_2(g) \rightarrow 2CO(g)$. Because these two reactions occur simultaneously, it is not possible to measure directly the heat of formation of $CO(g)$ from $C(s)$ and $O_2(g)$.

$$C(s) + \tfrac{1}{2}O_2(g) \rightarrow CO(g) \qquad \Delta H^{\circ}_f = ?$$

However, we do know the heat of formation of carbon dioxide and the heat of combustion of carbon monoxide.

$$
\begin{aligned}
C(s) + O_2(g) &\rightarrow CO_2(g) & \Delta H^{\circ}_f &= -393.5 \text{ kJ} \\
CO(g) + \tfrac{1}{2}O_2(g) &\rightarrow CO_2(g) & \Delta H_c &= -283.0 \text{ kJ}
\end{aligned}
$$

Reversing the second equation and adding gives the desired heat of formation of carbon monoxide.

$$
\begin{aligned}
C(s) + O_2(g) &\rightarrow CO_2(g) & \Delta H^{\circ} &= -393.5 \text{ kJ} \\
CO_2(g) &\rightarrow CO(g) + \tfrac{1}{2}O_2(g) & \Delta H &= 283.0 \text{ kJ} \\
\hline
C(s) + \tfrac{1}{2}O_2(g) &\rightarrow CO(g) & \Delta H &= -110.5 \text{ kJ}
\end{aligned}
$$

Figure 18-6 A possible mechanism for the water-gas reaction

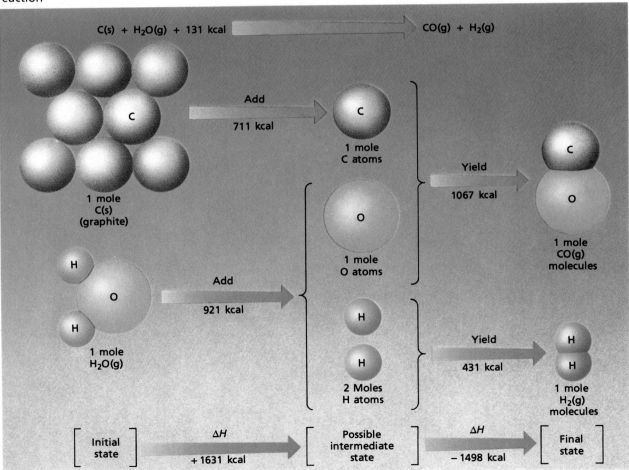

The combined use of heats of formation and heats of combustion to find unknown heats of formation can also be generalized. The following equation applies: heat of formation of compound X = sum of formation of products of combustion of compound X − heat of combustion of compound X.

Sample Problem 18.2

Calculate the heat of formation of pentane (C_5H_{12}) using the information on heats of formation and on heats of combustion in Appendix Table A-16 and Table 18-1. Solve by combining the known thermochemical equations. Verify the result by using the general equation for finding heats of formation from heats of combustion.

Solution

Step 1. Analyze　Given:

$$C(s) + O_2(g) \rightarrow CO_2(g) \qquad\qquad \Delta H_f^\circ = -393.51 \text{ kJ/mol}$$
$$H_2(g) + \tfrac{1}{2}O_2(g) \rightarrow H_2O(\ell) \qquad\qquad \Delta H_f^\circ = -285.83 \text{ kJ/mol}$$
$$C_5H_{12}(g) + 8O_2(g) \rightarrow 5CO_2(g) + 6H_2O(\ell) \quad \Delta H_c = -3536.1 \text{ kJ/mol}$$

Unknown: $\Delta H_f(C_5H_{12})$ for $5C(s) + 6H_2(g) \rightarrow C_5H_{12}(g)$

Step 2. Plan

$$\Delta H_f^\circ(CO_2),\ \Delta H_f^\circ(H_2O),\ \Delta H_c(C_5H_{12}) \rightarrow \Delta H_f(C_5H_{12})$$

To combine the equations given above according to Hess's law, the equation for combustion of C_5H_{12} must be reversed, the equation for formation of CO_2 must be multiplied by 5 to give 5C as reactant, and the equation for formation of H_2O must be multiplied by 6 to give $6H_2$ as reactant.

Step 3. Compute

$$5C(s) + 5O_2(g) \rightarrow 5CO_2(g) \qquad\qquad 5\Delta H_f = -1967.6 \text{ kJ}$$
$$6H_2(g) + 3O_2(g) \rightarrow 6H_2O(g) \qquad\qquad 6\Delta H_f = -1715.0 \text{ kJ}$$
$$5CO_2(g) + 6H_2O(\ell) \rightarrow C_5H_{12}(g) + 8O_2(g) \quad \Delta H = 3536.1 \text{ kJ}$$

$$\overline{5C(s) + 6H_2(g) \rightarrow C_5H_{12}(g) \qquad\qquad \Delta H = -146.5 \text{ kJ}}$$

Step 4. Evaluate　The unnecessary reactants and products cancel to give the correct equation. The answer is reasonably close to an estimated value of −145 (−1970 kJ/mol − 1715 kJ/mol = −3540 kJ/mol).

Verification:

$$\Delta H_f(C_5H_{12}) = 5\Delta H_f(CO_2) + 6\Delta H_f(H_2O) - \Delta H_c(C_5H_{12})$$
$$= 5(-393.51 \text{ kJ/mol}) + 6(-285.83 \text{ kJ/mol}) - (-3536.1 \text{ kJ/mol})$$
$$= -146.5 \text{ kJ/mol}$$

Practice Problems

1. Calculate the heat of formation of butane, C_4H_{10}, using the information in Appendix Table A-16 and Table 18-1. Write out the solution according to Hess's law.　*(Ans.)* −124.7 kJ/mol
2. Calculate the heat of combustion of 1 mole of nitrogen, N_2, to form NO_2 using Appendix Table 14-16. (Hint: Rearrange the general equation for ΔH_f of compound X to solve for ΔH_c.)　*(Ans.)* +66.36 kJ/mol

Bond Energy and Reaction Heat

The change in heat content of a reaction system is related to *(1)* the change in the number of bonds breaking and forming, and *(2)* the strengths of these bonds as the reactants form products. The reaction for the formation of water gas shown below can be used to illustrate these relationships.

STUDY HINT
Thermodynamics is the study of heat and other forms of energy, and of heat-energy transfer.

1. Heat content of a system at constant pressure
2. (a) Heat released or absorbed by a chemical reaction
   (b) Difference in enthalpy or heat content of products and reactants
3. The more negative the heat of formation, the more stable the compound
4. Equations and heats of reaction are rearranged and added algebraically to give the desired reaction.

SECTION OBJECTIVES

- Explain the relationship between enthalpy change and the tendency of a reaction to occur.
- Explain the relationship between entropy change and the tendency of a reaction to occur.
- Discuss the concept of free energy and explain how the value of this quantity is calculated and interpreted.

The two oxygen-to-hydrogen bonds of each steam molecule must be broken, as must the carbon-to-carbon bonds of the graphite. Carbon-to-oxygen and hydrogen-to-hydrogen bonds must be formed. Recall that energy is absorbed when bonds break and is released when bonds form.

A hypothetical series of steps for the water-gas reaction is illustrated in Figure 18-6. Observe that a total heat input of 1632 kJ is required to break the bonds of 1 mol of graphite and 1 mole of steam. The formation of bonds in 1 mol of CO and 1 mol of H_2 releases 1498 kJ of heat energy. The net effect is that 134 kJ of heat must be supplied to the system from an external source. This quantity agrees quite closely with the experimental value of 131.3 kJ of heat input per mole of carbon used.

Section Review

1. What is meant by "enthalpy?"
2. (a) What is meant by "heat of reaction?" (b) How is heat of reaction related to the enthalpy, or energy content, of reactants?
3. Describe the relationship between a compound's stability and its heat of formation.
4. How can thermochemical equations be combined to calculate ΔH values according to Hess's law?

18.2 Driving Force of Reactions

Whether a reaction occurs and how it occurs are questions that have always concerned chemists. The answers to these questions involve a study of reactions conducted within the framework of the laws of thermodynamics. This is the realm of physical chemistry and physics. In this section, the concepts that chemists label collectively as the "driving force" of chemical reactions will be examined.

Enthalpy and Reaction Tendency

It is not surprising to see a ball roll down an incline unaided; it is just what would be expected. Potential energy is given up in this process, and the ball attains a more stable state at a lower energy level. It would be surprising, however, if the ball rolled up the incline by and of itself. From such experiences with nature, a basic rule for natural processes becomes apparent: There is a tendency for processes to occur that lead to a lower energy state. This tendency in nature is toward greater stability (resistance to change) in a system.

The great majority of chemical reactions in nature are exothermic. As the reactions proceed, energy is liberated and the products have less energy than the original reactants. With the above rule in mind, one would expect exothermic reactions to occur spontaneously; that is, that they would have the ability to proceed without the assistance of an external agent.

It should follow that endothermic reactions, in which engery is absorbed, would not occur spontaneously but would proceed only with the assistance of an external agent. Certainly energy must be expended on the ball to roll it up the incline. At the top of the incline the ball's potential energy is high and its stability is low.

Some endothermic reactions *do* take place spontaneously, however. The production of water gas involves such a reaction. Steam is passed over white-hot coke (impure carbon), and the reaction proceeds spontaneously (Figure 18-7). It is not driven by any activity outside the reacting system. However, the process is endothermic, showing that the product gases, CO and H_2, collectively have a higher heat content than do the reactants, steam and carbon. The enthalpy change is positive, and therefore unfavorable, yet the reaction proceeds spontaneously.

$$H_2O(g) + C(s) \rightarrow CO(g) + H_2(g) \qquad \Delta H = +131.3 \text{ kJ}$$

A word about what chemists mean by *spontaneous* is in order. A spontaneous chemical reaction takes place without any continuing outside influence, such as heating, cooling, or stirring. A spontaneous change will keep going, once it gets started. As used in chemistry, *spontaneous* has nothing to do with speed. Some "spontaneous" chemical reactions are actually quite slow.

Entropy and Reaction Tendency

If a process is not driven by decreasing energy, what does drive it? To answer this question, consider a physical process that proceeds by and of itself with no energy change. Figure 18-8 shows two identical flasks connected by a valve. One flask is filled with ideal gas A and the other with ideal gas B. The entire system is at room temperature.

When the valve is opened, the two gases mix spontaneously until they are distributed evenly throughout the two containers. The gases will remain in this state indefinitely, with no tendency to separate. The temperature does not rise but remains constant throughout the mixing process. Thus, no exothermic process can have taken place. The total heat content, cannot have changed to a lower level. Clearly, the self-mixing process, since it is spontaneous, must be caused by a driving force other than the enthalpy-change tendency.

Another endothermic, but spontaneous, process is melting. An ice cube melts spontaneously at room temperature as heat is transferred from the

STUDY HINT
Natural processes tend toward maximum disorder.

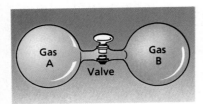

Figure 18-8 The mixing of gases may occur without an energy change.

STUDY HINT
The bonding of water is discussed in Section 13.4.

Gibbs Free Energy was named for Josiah Willar Gibbs, (1939–1903). He was a professor of mathematical physics at Yale University from 1871–1903. He laid the foundations of many areas of thermodynamics.

warm air to the cool ice and to the ice water formed. The well-ordered structure of the ice crystal (the solid phase) is lost, and the less orderly liquid phase of higher energy content is formed.

What is observed here is a tendency for the ice to move into the less orderly liquid phase. This is a tendency for processes to occur that lead to a less orderly or a more disordered state. This tendency in nature is toward greater disorder in a system. A disordered system is one that lacks a regular arrangement of its parts. *The property that describes the disorder of a system is called* **entropy** *(S).* The more disordered or more random a system, the higher is its entropy. Liquid water has higher entropy than ice. A mixture of gases has higher entropy than a pure gas.

Thus, processes in nature are driven in two ways: toward lowest energy and toward highest entropy. When these two oppose each other, the dominant factor determines the direction of the spontaneous change. In the steam-plus-hot-carbon reaction, the temperature is high enough that the entropy factor overcomes the unfavorble energy-change factor. The spontaneous endothermic reaction occurs.

A system that can go from one state to another without an enthalpy change does so by becoming more disordered. Entropy increases, for example, when a gas expands, a solid or liquid dissolves, or the number of particles in a system increases.

Free Energy

The overall drive to spontaneous change, as you have seen, depends on two factors—the tendency toward lowest enthalpy and the tendency toward highest entropy. A function of the state of a reaction system has been defined to relate the energy and entropy factors at a given temperature. *This combined enthalpy–entropy function is called the* **free energy** *of the system.* It is the function that simultaneously assesses both the enthalpy-change and entropy-change tendencies. In this sense, a process proceeds spontaneously in the direction that lowers the free energy of a system.

The net driving force of a chemical or physical change is called the free-energy change (ΔG) of the system. When the enthalpy change and the entropy change oppose each other, the direction of the net driving force is determined by the factor having the greater influence.

At a constant pressure and temperature, the **free-energy change,** ΔG, *of a system is defined as the difference between the change in enthalpy,* ΔH, *and the product of the Kelvin temperature and the entropy change, which is defined as* $T\Delta S$.

$$\Delta G = \Delta H - T\Delta S$$

In the equation above, ΔG is the change in free energy of the system, ΔH is the change in the enthalpy (heat content), T is the temperature (expressed as kelvins), and ΔS is the change in entropy. The product $T\Delta S$ and the quantities ΔG and ΔH all have the same dimensions, usually measured as kilojoules per mole. The dimensions of ΔS for use in this equation are usually kJ/mol · K

A chemical reaction proceeds spontaneously if it is accompanied by a decrease in free energy, that is, if the free energy of the products is lower than that of the reactants. In such a case, the free energy change in the system

(ΔG) is negative. In practical terms, free energy is the quantity of energy available to do useful work.

In exothermic reactions, ΔH has a negative value; in endothermic reactions, its value is positive. From the expression for ΔG, observe that the more negative ΔH is, the more negative ΔG is likely to be. Reaction systems that change from a high enthalpy state to a low one tend to proceed spontaneously. The more positive ΔS is, the more negative ΔG is likely to be. Thus, systems that change from well-ordered states to highly disordered states also tend to proceed spontaneously. Both of these tendencies in a given reaction system are assessed simultaneously in terms of the free-energy change—ΔG.

In some processes, ΔH is negative and ΔS is positive. Here, regardless of the relative magnitudes of ΔH and ΔS, the process will proceed spontaneously because ΔG is negative. Processes in which ΔH and ΔS have the same signs are more common.

Note that the sign of ΔG can be either positive or negative depending on the temperature T. The temperature of the system is the dominant factor that determines the relative importance of the tendencies toward lower energy and the higher entropy.

Consider the water-gas reaction

$$H_2O(g) + C(s) \rightarrow CO(g) + H_2(g) \qquad \Delta H = +131.3 \text{ kJ}$$

Here, a negative free-energy change requires that $T\Delta S$ be positive and greater than 131.3 kJ. This can be so if T is large, or if ΔS is large and positive, or both. Consider first the magnitude and sign of ΔS. Entropy increases (that is, ΔS is positive) when water gas is produced. In order to understand why, recall that the solid phase is well-ordered and the gaseous phase is random. One of the reactants, carbon, is a solid, and the other, steam, is a gas. In this case, however, both products are gases. The change from the orderly solid phase to the random gaseous phase involves an increase in disorder and thus an increase in entropy.

In general, a change from a solid to a gas will proceed with an increase in entropy. If all reactants and products are gases, an increase in the number of product particles increases entropy.

At low temperatures, whether ΔS is positive or negative, the product $T\Delta S$ may be small compared to ΔH. In such cases, the $T\Delta S$ term is nearly negligible, and the reaction proceeds as the enthalpy change predicts.

For the water gas reaction, measurements show that ΔS is $+0.134$ kJ/mol K at 25°C (298 K). Thus

$$T\Delta S = 298 \text{ K} \times 0.134 \, \frac{\text{kJ}}{\text{mol·K}} = 39.9 \text{ kJ/mol}$$

and

$$\Delta G = \Delta H - T\Delta S = +131.3 \text{ kJ/mol} - 39.9 \text{ kJ/mol}$$
$$+91.4 \text{ kJ/mol}$$

Since ΔG has a positive value, the reaction that produces water gas is not spontaneous at the relatively low temperature of 25°C.

Increases in temperature tend to favor increases in entropy. When ΔS is positive, a high temperature gives $T\Delta S$ a large positive value. Certainly, at a high enough temperature $T\Delta S$ will be larger than ΔH, and ΔG will have a negative value.

The water-gas reaction occurs spontaneously at the temperature of white-hot carbon, approximately 900°C. Strictly speaking, the values of ΔH and ΔS vary slightly with different temperature. However, they are assumed to remain about the same at the reaction temperature of 900°C. Therefore the values can be used to calculate an approximate value of ΔG for this temperature. Retaining the value of ΔH as $+131.3$ kJ/mol, $T\Delta S$ is determined at 1173K (900°C) to be

$$\left(1173 \text{ K} \times 0.134 \frac{\text{kJ}}{\text{mol·K}}\right) = 157 \text{ kJ/mol}$$

$$\Delta G = \Delta H - T\Delta S = +131.3 \frac{\text{kJ}}{\text{mol}} - \left(\frac{157 \text{ kJ}}{\text{mol}}\right)$$

$$-26 \text{ kJ/mol}$$

The free-energy change for the water-gas reaction is negative at 900°C, and that is why the reaction is spontaneous at this temperature.

1. Large negative value of ΔH
2. A measure of randomness. A positive change in entropy is spontaneous.
3. Crystalline solid to a solution of the solid in water. Change of phase from solid to liquid to gas. Decomposition of large molecules into smaller ones.

Sample Problem 18.3

For the reaction $NH_4Cl(s) \rightarrow NH_3(g) + HCl(g)$, at 25°C, $\Delta H = 176$ kJ and $\Delta S = 0.285$ kJ/K. Calculate ΔG and decide if this reaction will be spontaneous in the forward direction at 25°C.

Solution

Step 1. Analyze Given: $\Delta H = 176$ kJ at 25°C
$\Delta S = 0.285$ kJ/K at 25°C

Unknown: ΔG at 25°C

Step 2. Plan $\Delta S, \Delta H, T \rightarrow \Delta G$

The value of ΔG can be calculated according to the equation

$$\Delta G = \Delta H - T\Delta S$$

Step 3. Compute $\Delta G = 176 \text{ kJ} - (298 \text{ K} \times 0.285 \text{ kJ/K})$
$= 176 \text{ kJ} - 84.9 \text{ kJ}$
$= 91 \text{ kJ}$

Step 4. Evaluate The answer is reasonably close to an estimated value of 110 [200 − (300 × .03)]. The positive value of ΔG shows that this reaction is not spontaneous at 25°C.

Sample Problem
1. In the reaction above, why does the entropy increase? *(Ans.)* Gas formed from solid
2. What should be the effect of increasing temperature on the value of ΔG for this reaction? *(Ans.)* At a large enough value of T, ΔG changes from positive to negative.

Section Review

4. $\Delta G = \Delta H - T\Delta S$. Calculated from the difference of the change in enthalpy and the products of kelvin temperature change in entropy of a system
5. A reaction for which $\Delta G < 0$ is spontaneous

1. What kind of enthalpy change tends to produce a spontaneous reaction?
2. What is entropy and how does it relate to spontaneity of reactions?
3. List several changes that result in entropy increase.
4. Define *free energy* and explain how it is calculated.
5. Explain the relationship between free energy and spontaneity of reactions.

18.3 The Reaction Process

Up to now, this chapter has dealt with aspects of chemical reactions that are independent of the actual routes by which reactions occur. For example, enthalpy change (ΔH) and entropy change (ΔS) depend only on the initial and final states of a reaction, irrespective of any intermediate processes that may have taken place. This section explores the energy pathway that a reaction follows and the changes that take place on the molecular level when substances interact.

SECTION OBJECTIVES

• Explain the concept of reaction mechanisms.

• Use the collision theory to interpret chemical reactions.

• Define *activated complex.*

• Relate activation energy and heat of reaction.

Reaction Mechanisms

Chemical reactions involve breaking existing chemical bonds and forming new ones. The relationships and arrangements of atoms in the products are different from those in the reactants. Colorless hydrogen gas consists of pairs of hydrogen atoms bonded together as diatomic molecules, H_2. Violet-colored iodine vapor is also diatomic, consisting of pairs of iodine atoms bonded together as I_2 molecules. A chemical reaction between these two gases at elevated temperatures produces hydrogen iodide (HI) a colorless gas. Hydrogen iodide molecules, in turn, tend to decompose and re-form hydrogen and iodine molecules. The reaction equations are

$$H_2(g) + I_2(g) \rightarrow 2HI(g)$$

and

$$2HI(g) \rightarrow H_2(g) + I_2(g)$$

Such equations indicate only what molecular species disappear as a result of the reactions and what species are produced. They do not show the pathway by which either reaction proceeds. That is, they do not show the step-by-step sequence of reactions by which the overall chemical change occurs. *Such a sequence of steps is called the* **reaction mechanism.**

A chemical system can usually be examined before a reaction occurs or after the reaction is over. Such an examination reveals nothing about the mechanism by which the action proceeded, however. For most chemical reactions, only the reactants that disappear and final products that appear are known. In other words, only the net chemical change is directly observable.

Sometimes, however, chemists are able to devise experiments that reveal the sequence of steps in a reaction mechanism. Each reaction step is usually a simple process. The equation for each step represents the *actual* atoms, ions, or molecules that participate in that step. Complicated chemical reactions take place in a sequence of simple steps. Even a reaction that appears from its balanced equation to be a simple process may actually occur in a sequence of steps.

The reaction between hydrogen gas and bromine vapor to produce hydrogen bromide gas is an example of a **homogeneous reaction,** *a reaction whose reactants and products exist in a single phase* —in this case, the gas phase. This reaction system is also an example of a homogeneous chemical system. In such a system, all reactants and products in all intermediate steps are in the same phase. The overall chemical equation for this reaction is

$$H_2(g) + Br_2(g) \rightleftharpoons 2HBr(g)$$

It might appear that one hydrogen molecule reacts with one bromine molecule to form the product in a simple, one-step-process. However, chemists have found that the reaction follows a more complex pathway. The initial forward reaction is the splitting of bromine molecules into bromine atoms.

$$(1) \quad Br_2 \rightleftarrows 2Br$$

The bromine atoms, in turn, initiate a sequence of two steps that can repeat themselves.

$$(2) \quad Br + H_2 \rightleftarrows HBr + H$$
$$(3) \quad H + Br_2 \rightleftarrows HBr + Br$$

Observe that the sum of steps 2 and 3 gives the net reaction as shown in the overall equation above.

The equation

$$H_2(g) + I_2(g) \rightleftarrows 2HI(g)$$

represents another homogeneous chemical system. Figure 18-9 illustrates the initial reactants and final products identified by this equation. For many years chemists thought this reaction was a simple one-step process. They assumed it involved interaction of two molecules, H_2 and I_2, in the forward direction and two HI molecules in the reverse reaction.

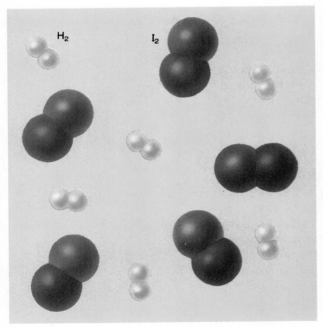

Initial State

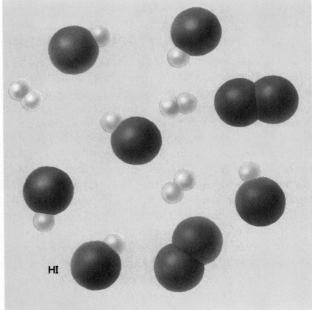

Final State

Figure 18-9 The initial and final states of a mixture of hydrogen gas and iodine vapor, which react to form hydrogen iodide gas.

Experiments eventually showed, however, that a direct reaction between H_2 and I_2 does not take place. Chemists proposed alternative mechanisms for the reaction. The steps in each reaction mechanism had to add up to give the overall equation. Note that two of the species—I and H_2I—in the mechanism steps do not appear in the net equation. *Such species that appear in some steps, but not in the net equation are known as* **intermediates.** The first possible mechanism has a two-step pathway:

$$(1) \qquad I_2 \rightleftharpoons 2I$$
$$\underline{(2) \quad 2I + H_2 \rightleftharpoons 2HI}$$
$$I_2 + H_2 \rightleftharpoons 2HI$$

The second possible mechanism has a three-step pathway:

$$(1) \qquad I_2 \rightleftharpoons 2I$$
$$(2) \quad I + H_2 \rightleftharpoons H_2I$$
$$\underline{(3) \quad H_2I + I \rightleftharpoons 2HI}$$
$$I_2 + H_2 \rightarrow 2HI$$

In studying reaction mechanisms, chemists know that it is easier to prove what does *not* happen than to pin down what happens. No matter how simple a balanced equation, the reaction pathway may be complicated and difficult to determine. Some chemists believe that simple, one-step reaction mechanisms are unlikely, and that nearly all reactions are complicated.

Collision Theory

In order for reactions to occur between substances, their particles (molecules, atoms, or ions) must collide. Further, these collisions must result in interactions. *The set of assumptions regarding collisions and reactions is known as* **collision theory.** Chemists use this theory to interpret many facts they observe about chemical reactions.

From the kinetic theory you know that the molecules of gases are continuously in random motion. The molecules have kinetic energies ranging from very low to very high values. The energies of the greatest portion are near the average value for all the molecules in the system. When the temperature of a gas is raised, the average kinetic energy of the molecules is increased as is shown in Figure 18-10.

Consider what might happen on a molecular scale in one step of a homogeneous reaction system. We will analyze a proposed first step in the decomposition of hydrogen iodide.

$$HI + HI \rightleftharpoons H_2 + 2I$$

According to the collision theory, the two HI molecules must collide in order to react. Further, they must collide while favorably oriented and with enough energy to disrupt the bonds of the molecules. If they do so, a reshuffling of bonds leads to the formation of the products: one H_2 molecule and two I atoms.

If collision is too gentle, the old bonds are not disrupted and new ones are not formed. The two molecules simply rebound from each other unchanged. This effect is illustrated in Figure 18-11(a).

Similarly, a collision in which the reactant molecules are poorly oriented has little effect. The angles and distances between certain atoms in the molecules are not those necessary for new bonds to form. The colliding molecules rebound without changing. A poorly oriented collision is shown in Figure 18-11(b).

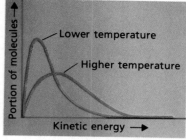

Figure 18-10 Energy distribution among gas molecules at two different temperatures

Rebounding occurs because of the electron–electron repulsion.

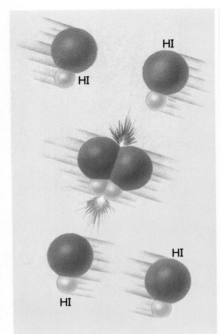

(a) Collision too gentle

(b) Collision in poor orientation

(c) Effective collision

Figure 18-11 Possible collision patterns for HI molecules.

Thus, collision theory provides two reasons why a collision between reactant molecules may fail to produce a new chemical species: *(1)* the collision is not energetic enough to supply the required activation energy, and *(2)* the colliding molecules are not oriented in a way that enables them to react with each other. We will now look at what is necessary for an effective collision that produces a new chemical species.

Activation Energy

When a collision is suitably oriented and violent enough, an interpenetration of the electron clouds of the appropriate atoms in the colliding molecules occurs. Then, old bonds can be broken and new bonds can form. See Figure 18-11(c). Only collisions that occur with sufficient energy can be effective.

Consider the reaction for the formation of water from the diatomic gases oxygen and hydrogen. The heat of formation is quite high: $\Delta H = -285.8$ kJ/mol at 25°C. The free energy change is also large: $\Delta G = -241.0$ kJ/mol. Why, then, do hydrogen and oxygen not combine spontaneously and immediately to form water when they are mixed at room temperature?

Hydrogen and oxygen gases exist as diatomic molecules. By some reaction mechanism, the bonds of these molecular species must be broken in order for new bonds between oxygen and hydrogen atoms to be formed. Bond breaking is an endothermic process, and bond forming is exothermic. Even though the net process is exothermic, it appears that an initial energy kick is needed to start the reaction.

It might help to think of the reactants as lying in an energy trough. Before they can react to form water, they must be lifted from this trough even though the energy content of the product is lower than that of the reactants. Once the reaction is started, the energy released is enough to sustain the reaction by activating other molecules. Thus, the reaction rate keeps increasing. It is finally limited only by the time required for reactant particles to acquire the energy and make contact. Energy from a flame or a spark discharge, or the energy associated with high temperatures or radiations may start reactants along the pathway of reaction. This activation-energy concept is illustrated in Figure 18-12.

The reverse reaction is the decomposition of water molecules. The product water lies at an energy level that is lower than the one from which the reactants were lifted. The water molecules must be lifted from this deeper energy trough before they can decompose to form oxygen and hydrogen. The activation energy needed to start this endothermic reaction is greater than that required for the original exothermic change. The difference equals the amount of heat of reaction, ΔH, released in the original reaction. See Figure 18-13.

In the energy diagram shown in Figure 18-14, the difference in height of the two level portions of the diagram is compared to the energy of reaction, ΔH. Energies E_a and E_a' represent the activation energies of the forward and reverse reactions, respectively. These quantities are comparable to the heights of the pass above the high and low valleys.

The Activated Complex

The possession of high kinetic energy does not alone make molecules unstable. When molecules collide, some of this energy is converted into internal potential energy within the colliding molecules. If enough energy is converted, the molecules may be activated.

When particles collide with energy at least equal to the activation energy for the species involved, their interpenetration can disrupt existing bonds. New bonds can then form. In the brief interval of bond disruption and bond formation, the collision complex is said to be in a transition state. Some sort of partial bonding exists in this transitional structure. *A transitional structure that results from an effective collision and that persists while old bonds are breaking and new bonds are forming is called an* **activated complex**.

An activated complex is formed when an effective collision raises the internal energies of the reactants to their minimum level for reaction. (See Figure 18-14.) Both forward and reverse reactions go through the same activated complex. A bond that is broken in the activated complex for the forward reaction must be re-formed in the activated complex for the reverse reaction. Observe that the activated complex occurs at the maximum-energy position along the reaction pathway. In this sense, the activated complex defines the activation energy for the system. **Activation energy** *is thus the energy required to transform the reactants into the activated complex.*

In its brief existence, the activated complex has partial bonding that is characteristic of both reactant and product. In this state, it may respond to either of two possibilities: 1. It may re-form the original bonds and separate into the reactant particles or, 2. It may form new bonds and separate into product particles. Usually, the formation of products is just as likely as the formation of reactants. Do not confuse the activated complex with the relatively stable intermediate products of different steps of a reaction

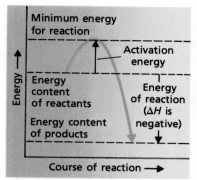

Figure 18-12 Pathway of an exothermic reaction.

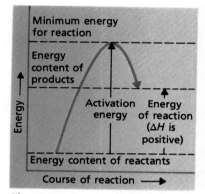

Figure 18-13 Pathway of an endothermic reaction.

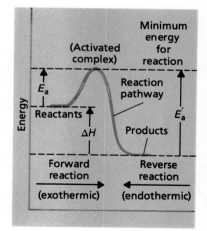

Figure 18-14 Activation energies for the forward reaction E_a and the reverse reaction E_a' and the change in internal energy ΔH in a reversible reaction.

Figure 18-15 Possible activated complex configuration, which could form either 2HI or H2 + 2I

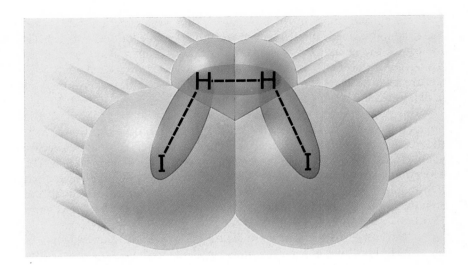

mechanism. The activated complex unlike intermediate products, is a very-short-lived molecular complex in which bonds are in the process of being broken and formed.

A possible configuration of the activated complex in the hydrogen iodide reaction is shown in Figure 18-15. The broken lines represent some sort of partial bonding. This transitional structure may produce HI molecules or an H_2 molecule and two I atoms.

Figure 18-16 shows the energy profile for the hydrogen iodide (HI) decomposition, which occurs at elevated temperatures of 400–500°C. Of the 183-kJ activation energy, 171 kJ is available for excitation of the hydrogen (H_2) and iodide (I_2) molecules. Of this amount, 149 kJ is used in producing iodine atoms.

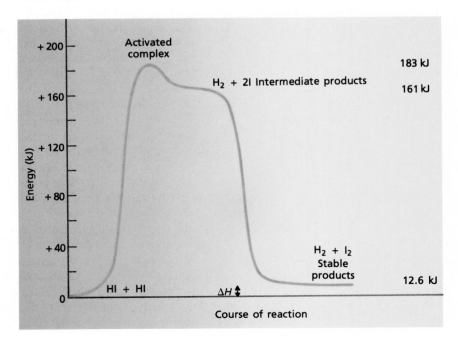

Figure 18-16 An energy profile. Results of kinetic studies of the decomposition of hydrogen iodide at high temperatures

Sample Problem 18.4

Copy the energy diagram below and label the reactants, products, ΔH, E_a, and E_a'. Determine the value of $\Delta H_{forward}$, $\Delta H_{reverse}$, E_a, and E_a'.

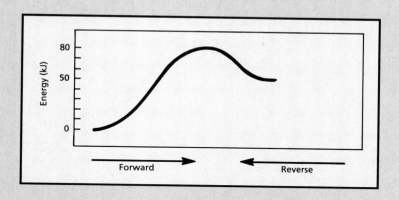

Solution

The reactant energy level is always at the left-hand end of such a curve and the product energy level is always at the right-hand end. The heat of reaction, ΔH, is the difference between these two energy levels. The activation energy, which differs in the forward and reverse directions, is the difference between the reactant or product energy level, and the peak in the curve—the minimum energy needed to achieve effective reaction.

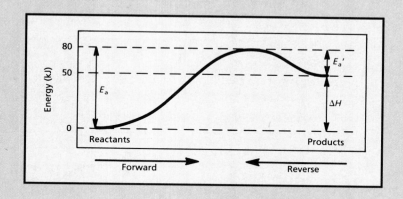

$$\Delta H_{forward} = 50 \text{ kJ} - 0 \text{ kJ} = +50 \text{ kJ}$$
$$\Delta H_{reverse} = 0 \text{k J} - 50 \text{ kJ} = -50 \text{ kJ}$$
$$E_a \qquad\quad = 80 \text{ kJ} - 0 \text{ kJ} = 80 \text{ kJ}$$
$$E_a' \qquad\quad = 80 \text{ kJ} - 50 \text{ kJ} = 30 \text{ kJ}$$

Sample Problem 18.3 *continued*

Practice Problems

1. Repeat the sample problem for the following energy diagram:

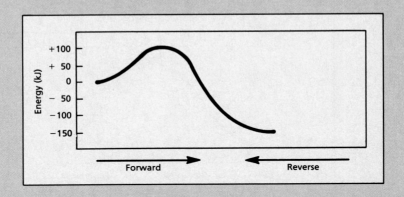

(Ans.)

$\Delta H_{forward}$ = −150 kJ
$\Delta H_{reverse}$ = +150 kJ
E_a' = 100 kJ
E_a' = 250 kJ

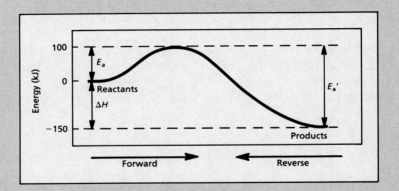

2. Draw and label the energy diagram for a reaction for which $\Delta H = 30$ kJ/mol and $E_a = 40$ kJ/mol. Place the reactants at energy level zero. Give values of $\Delta H_{forward}$, $\Delta H_{reverse}$, and E_a'.

(Ans.)

$\Delta H_{forward}$ = +30 kJ
$\Delta H_{reverse}$ = −30 kJ
E_a' = 10 kJ

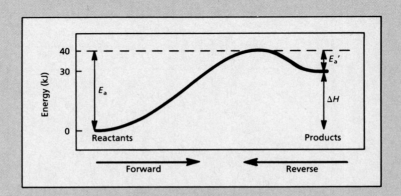

Section Review

1. What is meant by "reaction mechanism"?
2. What factors determine whether a molecular collision produces a reaction?
3. What is an activated complex?
4. What is activation energy?
5. How is activation energy related to the energy of reaction?

18.4 Reaction Rate

Chemists are interested in learning how rapidly reactants change into products in chemical reactions. *The change in concentration of reactants per unit time as a reaction proceeds is called the* **reaction rate.** *The branch of chemistry that is concerned with reaction rates and reaction mechanisms is called* **chemical kinetics.** As you will see as you read this concluding section of the chapter, the study of rates deals with the factors that affect rate and with the mathematical expressions that reveal the specific dependencies of rate upon concentration.

Rate-Influencing Factors

The rates of chemical reactions vary widely. Some reactions are over in an instant, and others take months or years to complete. The rate of reaction is measured by the amounts of reactants converted to products in a unit of time. Two conditions are necessary for reactions (other than simple decompositions) to occur at all. First, particles must come into contact. Second, this contact must result in appropriate interaction. Thus, the rate of a reaction depends on the collision frequency of the reactants and on the collision efficiency. An efficient collision is one with enough energy for activation and in which the reactant molecules are favorably oriented.

Changing conditions may affect either the collision frequency or the collision efficiency. Any such change affects the reaction rate. Five important factors influence the rate of chemical reaction.

Nature of reactants   Substances vary greatly in their tendencies to react. For example, hydrogen combines vigorously with chlorine under certain conditions. Under the same conditions, it may react only feebly with nitrogen. Sodium and oxygen combine much more rapidly than iron and oxygen under similar conditions. Platinum and oxygen do not combine directly. Atoms, ions, and molecules are the particles of substances that react. Bonds are broken and other bonds are formed in reactions. The rate of reaction depends on the particular reactants and bonds involved.

Surface area   Solid zinc reacts with aqueous hydrochloric acid and hydrogen gas according to the following equation.

$$Zn(s) + 2HCl(aq) \rightarrow ZnCl_2(aq) + H_2(g)$$

This reaction occurs at the surface of the zinc cube. A cube of zinc measuring 1 cm on each edge presents only 6 cm^2 of contact area to the solvent. The same amount of zinc in the form of a fine powder might provide a contact area on the order of 10^4 times the original area. Consequently, the reaction rate of the powdered solid is much higher.

A lump of coal burns slowly when kindled in air. The rate of burning can be increased by breaking the lump into smaller pieces, exposing new surfaces. If the piece of coal is powdered and then ignited while suspended in air, it burns explosively. Nickel in large pieces shows no noticeable oxidation in air, but finely powdered nickel reacts vigorously and spectacularly.

The reactions just mentioned are examples of heterogeneous reactions. **Heterogeneous reactions** *involve reactants in two different phases*. Such reactions can occur only when the two phases are in contact. Thus, the surface area of a solid (or liquid) reactant is an important rate consideration. Gases and dissolved particles do not have surfaces in the sense just described. In heterogeneous reactions, the reaction rate is proportional to the area of contact of the reaction substances.

Some chemical reactions between gases actually take place at the solid walls of the container. Others may occur at the surface of a solid or liquid catalyst. If the products are also gases, such a reaction system presents a problem of language. It is a heterogeneous reaction because it takes place at a surface between two phases, gas and solid. On the other hand, it is a homogeneous reaction because all the reactants and all the products are in one phase.

Concentration Suppose a small lump of charcoal is heated in air until combustion begins. If it is then lowered into a bottle of pure oxygen, the reaction proceeds at a much higher rate. A substance that oxidizes in air reacts more vigorously in pure oxygen, as shown in Figure 18-17. The partial pressure of oxygen in air is approximately one-fifth of the total pressure. Pure oxygen at the same pressure as the air has five times the concentration of oxygen molecules.

This charcoal oxidation is a heterogeneous reaction system in which one reactant is a gas. Not only does the reaction rate depends not only on the amount of exposed charcoal surface, but also on the concentration of the gas.

Figure 18-17 Carbon burns faster in oxygen *(right)* than in air because air is only about one-fifth oxygen.

Homogeneous reactions may involve reactants in gaseous or liquid solutions. The concentration of gases changes with pressure according to Boyle's law. In liquid solutions, the concentration of reactants changes if either the quantity of solute or the quantity of solvent is changed. Solids and liquids are practically noncompressible. Thus, it is not possible to change the concentration of pure solids and pure liquids to any measurable extent.

In homogeneous reaction systems, reaction rates depend on the concentration of the reactants. From collision theory, a rate increase might be expected if the concentration of one or more of the reactants is increased. Lowering the concentration should have the opposite effect. The specific effect of concentration changes on reaction rate, however, must be determined experimentally.

Increasing the concentration of substance A in reaction with substance B could increase the reaction rate or have no effect on it. The effect depends on the particular reaction. One cannot tell from the balanced equation for the net reaction how the reaction rate is affected by a change in concentration of reactants. Chemists account for these differences in behavior in terms of the reaction mechanisms.

> This is a key point: a balanced equation does not determine the reaction pathway.

Complex chemical reactions may take place in a series of simple steps. Instead of a single activated complex, there may be several activated complexes in sequence along the reaction pathway. Of these steps, the one that proceeds at the lowest rate determines the overall reaction rate. *This lowest-rate step is called the* **rate-determining step** *for the chemical reaction.*

> Emphasize that *lowest-rate* means *slowest* step.

Temperature   The average kinetic energy of the particles of a substance is proportional to the temperature of the substance. Collision theory explains why a rise in temperature increases the rate of chemical reaction. According to this theory, a decrease in temperature lowers the reaction rate for both exothermic and endothermic reactions.

Near room temperature, the rates of many reactions roughly double or triple with a 10°C rise in temperature. This rule of thumb should be used with caution, however. The actual rate increase with a given rise in temperature must be determined experimentally.

Large increases in reaction rate are caused partly by the increase in collision frequency of reaction particles. However, for a chemical reaction to occur, the particles must also collide with enough energy to cause them to react. At higher temperatures, more particles possess enough energy to form the activated complex when collisions occur. In other words, more particles have the necessary activation energy. Thus, a rise in temperature produces an increase in collision energy and collision frequency.

Presence of catalysts   Some chemical reactions proceed quite slowly. Frequently, their reaction rates can be increased dramatically by the presence of a catalyst. *A* **catalyst** *is a substance that changes the rate of a chemical reaction without itself being permanently consumed. The action of a catalyst is called* **catalysis.** Catalysts do not appear among the final products of reactions they accelerate. They may participate in one step along a reaction pathway and be regenerated in a later step. In large-scale and cost-sensitive reaction systems, catalysts are recovered and reused.

> The collision energy is a more important factor than the collision frequency.

A catalyst that is in the same phase as all the reactants and products in a reaction system is called a **homogeneous catalyst.** *When its phase is* different from that of the reactants, it is called a **heterogeneous catalyst.** Metals are often used as heterogeneous catalysts. The catalysis of many reactions is promoted by adsorption of reactants on the metal surfaces, which has the effect of increasing the concentration (and therefore the amount of contact) of the reactants.

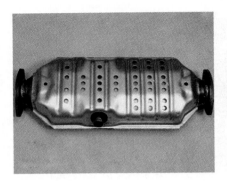 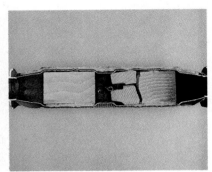

Figure 18-18 An automobile's catalytic converter makes some chemicals in exhaust less toxic.

STUDY HINT

Enzymes are examples of biological catalysts. Each enzyme accelerates a specific chemical process, enabling the process to be completed typically within only a few hours. The same process uncatalyzed might require weeks for completion. See Section 23.2.

A tunnel through a mountain is analogous to a catalyst in a chemical reaction. The tunnel provides a new, easier path across the mountain. The starting and ending points are still the same.

Such catalytic actions are hindered by the presence of inhibitors. *An* **inhibitor** *is a substance that slows down a process.* Inhibitors, once referred to as negative catalysts, can interfere with the surface chemistry of heterogeneous catalysts and thus inhibit and eventually destroy the catalytic action. An example is the "poisoning" of metallic-oxide catalysts in the catalytic converters of automobile exhaust systems by leaded gasoline.

A catalyst provides an alternative in pathway or reaction mechanism in which the potential-energy barrier between reactants and products is changed. The catalyst may be effective in forming an alternative activated complex that requires a lower activation energy, as suggested in the energy profiles of Figure 18-19.

Rate Laws for Reactions

The influence of reactant concentrations on reaction rate in a given chemical system has been mentioned. Measuring this influence is a challenging part of kinetics. Chemists determine the relationship between the rate of a reaction and the concentration of one reactant by first keeping the concentrations of other reactants and the temperature of the system constant. Then the reaction rate is measured for various concentrations of the reactant in question. A series of such experiments reveals how the concentration of each reactant affects the reaction rate.

The following homogeneous reaction is used as an illustration. The reaction is carried out in a vessel of constant volume and at an elevated constant temperature.

$$2H_2(g) + 2NO(g) \rightarrow N_2(g) + 2H_2O(g)$$

Here, four moles of reactant gases produce three moles of product gases. Thus, the pressure of the system diminishes as the reaction proceeds. The rate of the reaction can therefore be determined by measuring the change of pressure in the vessel with time.

Suppose a series of experiments is conducted using the same initial concentration of nitrogen monoxide but different initial concentrations of

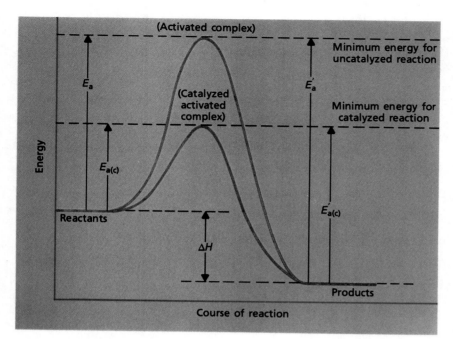

Figure 18-19 If a reaction is catalyzed, the change in potential energy may differ from that of the uncatalyzed reaction.

hydrogen. The initial reaction rate is found to vary directly with the hydrogen concentration. That is, doubling the concentration of H_2 doubles the rate, and tripling the concentration of H_2 triples the rate. Therefore,

$$R \propto [H_2]$$

Here R is the reaction rate and $[H_2]$ is the concentration of hydrogen in moles per liter. The $\propto$ is a symbol that is read "is proportionate to."

Now suppose the same initial concentration of hydrogen is used but the initial concentration of nitrogen monoxide is varied. The initial reaction rate is found to increase fourfold when the NO concentration is doubled and ninefold when the concentration of NO is tripled. Thus, the reaction rate varies directly with the square of the nitrogen monoxide concentration.

$$R \propto [NO]^2$$

Since R is proportional to $[H_2]$ and to $[NO]^2$, it is proportional to their product.

$$R \propto [H_2]\,[NO]^2$$

By introduction of an appropriate proportionality constant k, the expression becomes an equality.

$$R = k[H_2][NO]^2$$

An equation that relates reaction rate and concentrations of reactants is called the **rate law** *for the reaction.* It is applicable for a specific reaction at a given temperature. A rise in temperature causes an increase in the rate of nearly all reactions. The value of k usually increases as the temperature increases.

Collision theory indicates that the frequency of collisions between particles increases as the concentration of these particles is raised (Figure

Rate laws MUST be determined experimentally.

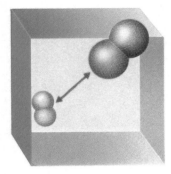

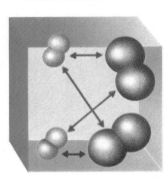

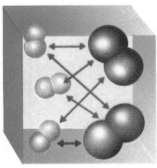

Figure 18-20 Under constant conditions, the collision frequency increases with the concentration of each reactant.

18-20). The reaction rate for any step in a reaction pathway is directly proportional to the frequency of collisions between the particles involved. It is also directly proportional to the collision efficiency.

Consider a single step in a reaction pathway. Suppose one molecule of gas A collides with one molecule of gas B to form two molecules of substance C. The equation for the step is

$$A + B \rightarrow 2C$$

One particle of each reactant is involved in each collision. Thus, doubling the concentration of either reactant will double the collision frequency. It will also double the reaction rate *for this step*. Therefore, the rate for the step is directly proportional to the concentration of A and of B. The rate law for the step becomes

$$R = k[A][B]$$

Now suppose the reaction is reversible. In the reverse step, two molecules of C must collide to form one molecule of A and one of B.

$$2C \rightarrow A + B$$

Thus the reaction rate for this reverse step is directly proportional to $[C] \times [C]$. The rate law for the step is

$$R = k[C]^2$$

A simple relationship may or may not exist between the chemical equation for a reaction and the rate law for that reaction. Observe that the power to which the molar concentration of each reactant or product is raised in the rate laws above corresponds to the coefficient for the reactant or product in the balanced equation. Such a relationship does not always hold, however. It holds if the reaction follows a simple one-step that is, if the reaction occurs at the molecular level exactly as written. A reaction may proceed in a series of steps, however, and in such reactions the rate law may be simply that for the slowest (rate-determining) step. The rate law may also be more complex, depending on the mechanism. Thus, the rate law for a reaction must be determined experimentally. It cannot be written based only on the balanced equation for the net reaction.

Again consider the reaction between hydrogen and nitrogen monoxide. The balanced equation for the net reaction is

$$2H_2(g) + 2NO(g) \rightarrow N_2(g) + 2H_2O(g)$$

Suppose this reaction occurs in a single step involving a collision between two H_2 molecules and two NO molecules. Doubling the concentration of either reactant should then quadruple the collision frequency and the rate. The rate law should then be proportional to $[H_2]^2$ as well as $[NO]^2$. It would take the form

$$R = k[H_2]^2[NO]^2$$

This equation is not in agreement with experimental results, and the assumed single-step reaction mechanism cannot be correct.

The actual experimentally determined rate law for the net reaction is

$$R = k[H_2][NO]^2$$

Desktop Investigation

Factors Influencing Reaction Rate

Question

How do the type of reactants, surface area of reactants, concentration of reactants, temperature of reactants, and catalysts affect the rates of chemical reactions?

Materials

Bunsen burner
paper ash
copper foil
graduate-10mL
magnesium ribbon
matches
paper clip
2 potatoes
sandpaper
steel wool
2 sugar cubes
white vinegar
sheet of zinc
6 test tubes, 16 x 150 mm
tongs

Procedure

Remove all combustible material from the work area. Wear safety glasses and an apron. Record all your results in a data table.

1. Add 10 mL of vinegar to each of three test tubes. To one test tube, add a 3-cm piece of magnesium ribbon to a second, a 3-cm zinc strip, and to a third 3 cm copper strip. (All metals should be the same width.) If necessary, sandpaper the metals until shiny. (See drawing below.)

2. Using tongs, hold a paper clip in the hottest part of the burner flame for 30 seconds. Repeat with a ball of

steel wool 2 cm in diameter. (See drawing at upper right.)

3. To one test tube, add 10 mL of vinegar; to a second, 5 mL of vinegar plus 5 mL of water; to a third, 2.5 mL of vinegar plus 7.5 mL of water. To each of the three test tubes add a 3 cm piece of magnesium ribbon.

4. Select two potatoes that are approximately equal in shape and mass. Bake one at 300°F until it is thoroughly cooked. Bake the second at 350°F until it is thoroughly cooked.

5. Using tongs, hold a sugar cube and try to ignite it with a match. Then try to ignite it in a burner flame. Rub paper ash on a second cube and try to ignite it with a match.

Discussion

1. What are the rate-influencing factors in Procedures 1, 2, 3, 4, and 5?

2. What were the results from each of the five parts of the experiment? How do you interpret these results?

The reaction pathway must therefore, consist of more than one step. The slowest step, thought to be the collision of one H_2 molecule with two NO molecules, is rate-determining. The rate law found experimentally for the net reaction is the rate expression of this rate-determining step.

The general form for the rate law of reactions is

$$R = k[A]^n[B]^m \ldots$$

Here R is the reaction rate, k is the rate constant, and [A] and [B]... represent the molar concentrations of reactants. The n and m are the respective powers to which the concentrations are raised, *based on experimental data*. Again it should be emphasized that one cannot assume that the coefficients in the balanced equation for a net reaction are the exponents in the rate law for the reaction.

The dependence of reaction rate on concentration of reactants was first recognized as a general principle by two Norwegian chemists, Cato Maximilian Guldberg (1836–1902) and Peter Waage (1833–1900), in 1867. They stated their discovery of this dependence as the **law of mass action:** *the rate of a chemical reaction is directly proportional to the product of the concentrations of reacting substances, each raised to the appropriate power.* This principle is interpreted in terms of the rate law for a chemical system in the modern science of chemical reaction kinetics.

1. The rate at which chemical reactions occur
2. (1) Nature of reactants
   (2) Surface area
   (3) Concentration
   (4) Temperature
   (5) Presence of a catalyst

Sample Problem 18.5

A reaction involving reactants X and Y was found to occur by a one-step mechanism: $X + 2Y \rightarrow XY_2$. Write the rate law for this reaction and then determine the effect of each of the following on the reaction rate: *(a)* doubling the concentration of X *(b)* doubling the concentration of Y *(c)* using one-third the concentration of Y.

Solution
Because the equation represents a single-step mechanism, the rate law can be written from the equation (otherwise it could not be). The rate will vary directly with the concentration of X, which has no coefficient in the equation, and will vary directly with the square of the concentration of Y, which has the coefficient of 2: $R = k[X][Y]^2$.
(a) Doubling the concentraction of X will double the rate ($R = k[2X][Y]^2$).
(b) Doubling the concentration of Y will increase the rate fourfold (Rate $= k[X][2Y]^2$.
(c) Using one-third the concentration of Y will reduce the rate to one-ninth of its original value ($R = k[X][\frac{1}{3}Y]^2$).

Practice Problem
The rate of reaction involving L, M, and N is found to double if the concentration of L is doubled, to increase eightfold if the concentration of M is doubled, and to double if the concentration of N is doubled. Write the rate law for this reaction. *(Ans.)* $R = k[L][M]^3[N]$

3. A substance that provides an alternative lower energy pathway for a reaction. It lowers the activation energy.
4. An equation that relates the reaction rate to the concentrations of reactants. The rate law can be based on the balanced equation only if the reaction takes place in a single step. If the reaction requires more than one step, the rate law is based on the concentrations of the reactants involved in the slowest step.

Section Review

1. What is studied in the branch of chemistry that is known as chemical kinetics?
2. List the five important factors that influence the rate of chemical reactions.
3. What is a catalyst? Explain the effect of a catalyst on the rate of chemical reactions. How does a catalyst influence the activation energy required by a particular reaction?
4. What is meant by a rate law for a chemical reaction? Explain the conditions under which a rate law can be written from a chemical equation. When can a rate law not be written from a single step?

Chapter Summary

- Thermochemistry is the study of the changes in heat energy that accompany chemical reactions and physical changes.
- A thermochemical equation is an equation that includes the quantity of heat released or absorbed during the reaction as written.
- The heat of reaction is the quantity of heat released or absorbed during a chemical reaction.
- Enthalpy is the heat content of a system at constant pressure.
- An enthalpy change is the amount of heat absorbed or lost by a system in a process carried out at constant pressure.
- The heat of formation is negative for exothermic reactions and positive for endothermic reactions.
- Compounds with highly negative heats of formation tend to be stable; compounds with highly positive or only slightly negative heats of formation tend to be unstable.
- The standard molar heat of formation is the heat absorbed or released in formation of one mole of a compound from the elements in their standard states at 25°C.
- The heat released in a combustion reaction is called the heat of combustion.
- Heats of reaction can be found by using heats of formation of reactants and products.
- The heat of formation of a compound may be found by adding together the heats of formation of the products of combustion of the compound, and then subtracting the heat of combustion of the compound.
- The change in the heat content, or enthalpy of a system is related to the number of bonds breaking, forming the number of bonds, and the strength of the bonds.
- Entropy is a measure of the disorder of a system.
- Free-energy change combines the effects of entropy and enthalpy changes and temperatures of a system, and is a measure of the overall drive to spontaneous change.
- A reaction will occur spontaneously if it causes a favorable change (decrease) in free energy. It will not occur if it causes an increase in free energy.
- The step-by-step process by which an overall chemical reaction occurs is called the reaction mechanism.
- In order for chemical reactions to occur, the particles of the reactants must collide.
- Activation energy is needed to loosen bonds so molecules can become reactive.
- An activated complex is formed when an effective collision between molecules of reactants raises their internal energies to the minimum level for a reaction to occur.
- The rate of reaction is influenced by the following factors: nature of reactants, surface area, temperature, concentration of reactants; and the presence of catalysts and inhibitors.
- The rates at which chemical reactions occur can sometimes be experimentally measured and expressed in terms of mathematical equations called rate laws.

Chapter 18 *Review*

Reaction Energy and Reaction Kinetics

Vocabulary

activated complex	heterogeneous catalyst
activation energy	heterogeneous reaction
calorimeter	homogeneous catalyst
catalysis	homogeneous reaction
catalyst	inhibitor
chemical kinetics	intermediate
collision theory	law of mass action
enthalpy	molar heat
enthalpy change	of formation
entropy	rate-determining step
free-energy	rate law
free-energy change	reaction mechanism
heat of combustion	reaction rate
heat of reaction	thermochemical equation
Hess's law	thermochemistry

Questions

1. *(a)* What role does energy play in a chemical reaction?   *(b)* What determines the energy change associated with a given reaction?   *(c)* What is thermochemistry?
2. What is a calorimeter?
3. How does the heat content of the products of a reaction system compare with the heat content of the reactants when the reaction is *(a)* endothermic   *(b)* exothermic?
4. *(a)* What is a thermochemical equation?   *(b)* What is an enthalpy change?   *(c)* What is the significance of the sign of ΔH?
5. *(a)* Define molar heat of formation.   *(b)* What is meant by the standard states of reactants and products?
6. *(a)* Distinguish between heats of reaction, formation, and combustion.   *(b)* Upon what basis are heats of formation and combustion defined?
7. *(a)* State Hess's law.   *(b)* Cite the two principles used in writing combinations of thermochemical equations.
8. Write the equations that can be used to calculate   *(a)* heat of reaction from heats of formation   *(b)* heat of formation from heats of combustion.
9. What factors affect the value of ΔH in a reaction system?
10. *(a)* Cite the two basic rules that govern the tendency for natural processes to occur.   *(b)* How is the free energy of a system related to these two reaction tendencies?
11. *(a)* What is meant by a spontaneous reaction?   *(b)* What types of reactions typically occur spontaneously in nature? Why?
12. *(a)* Explain what is meant by the free-energy change of a sytem.   *(b)* Write and label the equation used to calculate this value.
13. What kind of entropy change is observed in changes of phase in the direction of the solid?
14. How does an increase in temperature affect the entropy of a system?
15. According to the ΔG equation, what combination of ΔH and ΔS values always results in a spontaneous reaction?
16. *(a)* What is the collision theory?   *(b)* According to this theory, what two conditions must be met in order for a collision between reactant molecules to be effective in producing new chemical species?
17. Define activation energy in terms of the formation of an activated complex.
18. *(a)* What condition must be met in order for an activated complex to result from the collision of reactant particles?   *(b)* Where, in terms of energy, does the complex occur along a typical reaction pathway?
19. Within the brief period of time during which an activated complex exists, what two possibilities may result?
20. In a reversible reaction, how does the activation energy required for the exothermic change compare with the activation energy required for the endothermic change?
21. *(a)* What is meant by the concept of reaction rate?   *(b)* How is reaction rate measured?   *(c)* Upon what two factors is reaction rate

dependent?

22. Identify the five factors that influence the rate of chemical reactions, and describe the effect of each.

23. What is meant by the rate-determining step for a chemical reaction?

24. Write the general equation that expresses the rates of many chemical reactions, and label the various factors.

25. State the law of mass action.

Problems

1. For each reaction listed below, determine the ΔH and type of reaction (endothermic or exothermic):
   (a) $C(s) + O_2(g) \rightarrow CO_2(g) + 393.51$ kJ
   (b) $CH_4(g) + 2O_2(g) \rightarrow CO_2(g) + 2H_2O(\ell)$ $+ 890.31$ kJ
   (c) $CaCO_3(s) + 176$ kJ $\rightarrow CaO(s) + CO_2(g)$
   (d) $H_2O(g) \rightarrow H_2O(\ell) + 44.02$ kJ
   (e) $2SO_2(g) + O_2(g) \rightarrow 2SO_3(g) + 197.8$ kJ

2. Rewrite each reaction below with the ΔH value in its appropriate position and identify that reaction as endothermic or exothermic:
   (a) $H_2(g) + \frac{1}{2}O_2(g) \rightarrow H_2O(\ell)$;
   $\Delta H = -285.83$ kJ
   (b) $2Mg(s) + O_2(g) \rightarrow 2MgO(s)$;
   $\Delta H = -72.3$ kJ
   (c) $I_2(s) \rightarrow I_2(g)$; $\Delta H = +62.4$ kJ
   (d) $3CO(g) + Fe_2O_3(s) \rightarrow 2Fe(s) + 3CO_2(g)$;
   $\Delta H = -24.7$ kJ
   (e) $2NO_2(g) \rightarrow 2NO(g) + O_2(g)$;
   $\Delta H = +114.2$ kJ
   (f) $C_2H_4(g) + 3O_2(g) \rightarrow 2CO_2(g) + 2H_2O(\ell)$;
   $\Delta H = -1411.0$ kJ

3. Use Table 18-1 and Appendix Table A-16 to write the reaction illustrating the formation of each of the following compounds from its elements. Write the ΔH as part of each equation and indicate the ΔH for the reverse reaction: (a) $CaCl_2(s)$ (b) $C_2H_2(g)$ (ethyne or acetylene)   (c) $SO_2(g)$

4. Use heat of formation data given in Table 18-1 and Appendix Table A-16 to calculate the heat of reaction for each of the following. Solve each by combining the known thermochemical equations. Verify each result by using the general equation for finding heats of reaction from heats of formation.   (a) $CaCO_3(s) \rightarrow$ $CaO(s) + CO_2(g)$ (b) $Ca(OH)_2(s) \rightarrow CaO(s) +$

$H_2O(g)$ (c) $Fe_2O_3(s) + 3CO(g) \rightarrow 2Fe(s)$ $+ 3CO_2(g)$

5. Calculate the heats of reaction for combustion reactions in which each of the following are the reactants and $CO_2(g)$ and and $H_2O(\ell)$ are the products. Solve each by combining the known thermochemical equations using $\Delta H_f°$ values in Table 18-1 and Appendix Table A-16. Verify each result by using the general equation for finding heats of reaction from heats of formation.
   (a) $C_2H_6(g) + O_2(g) \rightarrow$
   (b) $C_6H_6(\ell) + O_2(g) \rightarrow$

6. Refer to Problems 1(a), 1(b), and 1(e) to determine th heats of combustion of $C(s)$, $CH_4(g)$, and $SO_2(g)$.

7. Write the thermochemical equations for the complete combustion of 1 mole of each of the following to form carbon dioxide and water. (See Table 18-1 for heats of combustion.) Calculate the heat of formation of each by combining the known thermochemical equations and then verify each by using the general equation for finding heats of formation from heats of combustion: (a) propane   (b) ethane   (c) ethyne   (d) benzene.

8. Using heats-of-formation data, calculate the heat of reaction for the combustion of one mole of each of the following: (a) hydrogen gas (assuming that the product is $H_2O(\ell)$)   (b) sulfur.

9. The heat of formation of ethanol, C_2H_5OH, is -277.0 kJ/mol at 25°C. Calculate the heat of combustion of one mole of ethanol, assuming that the products are $CO_2(g)$ and $H_2O(\ell)$.

10. Based on the following values, compute ΔG values for each reaction, and predict whether the reaction will occur spontaneously.
    (a) $\Delta H = +125$ kJ, $T = 293$ K, $\Delta S = 0.0350 \dfrac{\text{kJ}}{\text{K}}$
    (b) $\Delta H = -85.2$ kJ, $T = 127$°C, $\Delta S = 0.125 \dfrac{\text{kJ}}{\text{K}}$
    (c) $\Delta H = -275$ kJ, $T = 500$°C, $\Delta S = 0.450 \dfrac{\text{kJ}}{\text{K}}$

11. Draw and label energy diagrams depicting the following reactions, and determine all remaining values. Place the reactants at energy level zero.
    (a) $\Delta H_{\text{forward}} = +30$ kJ $E_a' = 20$ kJ
    (b) $\Delta H_{\text{forward}} = -30$ kJ $E_a = 20$ kJ

12. A reaction involving reactants A and B is found

to occur in the one-step mechanism: $2A + B \rightarrow A_2B$. Write the rate law for this reaction and predict the effect of doubling the concentration of each reactant on individually the overall reaction rate.

13. For each of the energy diagrams provided below, label the reactants, products, ΔH, E_a, and E_a'. Also determine the values of ΔH for the forward and reverse reactions, and of E_a, and E_a'.

(a)

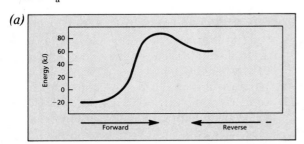

(b)

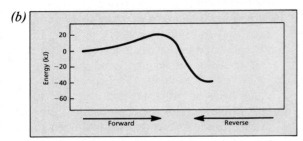

(c)

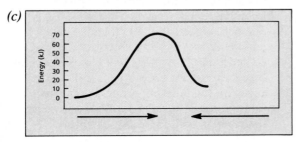

Application Questions

1. Explain why the heat of formation of a compound is an indication of its stability.
2. Explain the relative importance, in fuel selection of the level of energy contained in each unit mass of the reactants and products, as opposed to their specific ΔH_c values per mole of their flame temperature. What other factors should also be considered?
3. Compound A has a ΔH of $+87.3$ kJ/mol, and B has a $\Delta H_f°$ of -340.0 kJ/mol. What do these

$\Delta H_f°$ values imply about the relative stabilities of these compounds? Explain.

4. Two flasks, each containing a different gas at room temperature, are connected so that the two gases mix. What kind of evidence would indicate that the gases experienced no change in energy content during the mixing?
5. Explain why the gases in Application Question 4 exhibit a tendency to mix.
6. Explain the circumstances under which an endothermic reaction is spontaneous.
7. Explain the circumstances under which an exothermic reaction is nonspontaneous.
8. Explain the relationship between temperature and the spontaneity of reactions.
9. Would you expect the following reaction to represent the mechanism by which ethane (C_2H_6) burns? Why or why not?
$$C_2H_6(g) + \tfrac{7}{2}O_2(g) \rightarrow 2CO_2(g) + 3H_2O(g)$$
10. The balanced equation for a rapid homogeneous reaction between two gases is as follows: $4A + B \rightarrow 2C + 2D$. Since the simultaneous collision of four molecules of one reactant with one molecule of the other reactant is extremely improbable, what would you assume about the nature of the reaction mechanism for this reaction system?
11. The decomposition of nitrogen dioxide $2NO_2 \rightarrow 2NO + O_2$ occurs in a two-step sequence at elevated temperatures. The first step is $NO_2 \rightarrow NO + O$. Predict a possible second step that, when combined with the first step, gives the complete reaction.
12. How can you justify calling the reaction pathway that is shown in Figure 18-14 the minimum-energy pathway for reaction? (b) What significance is associated with the maximum-energy region of this minimum-energy pathway?
13. Explain why there is a danger of explosion in such places as saw mills and grain elevators where large amounts of dry, powdered combustible materials are present.
14. What property would you measure in order to determine the reaction rate for the following reaction? Justify your choice.
$$2NO_2(g) \rightarrow N_2O_4(g)$$
15. In a thought experiment, suppose two moles of hydrogen gas and one mole of iodine vapor are passed simultaneously into a one liter flask. The

rate law for the forward reaction is $R=k[I_2][H_2]$. What is the effect on the rate of the forward reaction if *(a)* the temperature is increased *(b)* one mole of iodine vapor is added *(c)* one mole of hydrogen is removed *(d)* the volume of the flask is reduced (assume this is possible)

16. Under what circumstances do the coefficients in a balanced chemical equation correspond to the powers to which the molar concentrations of the respective reactants or products are raised in a rate law?

Application Problems

1. Use heat of combustion and product heats of formation data to write all component reactions and calculate the heat of formation of each of the following. Verify the results by using an appropriate mathematical equation. *(a)* hexane $(C_6H_{14}(\ell))$ *(b)* sucrose $(C_{12}H_{22}O_{11}(s))$.

2. A chemical reaction is expressed by the balanced chemical equation $A + B \rightarrow C$. Three reaction rate experiments yield the following data:

Experiment number	Initial [A]	Initial [B]	Initial rate of formation of C
1	0.20 M	0.20 M	2.0×10^{-4} M/min
2	0.20 M	0.40 M	8.0×10^{-4} M/min
3	0.40 M	0.40 M	1.6×10^{-3} M/min

(a) Determine the rate law for the reaction. *(b)* Calculate the value of the specific rate constant. *(c)* If the initial concentrations of both A and B are 0.30 M, at what initial rate is C formed?

3. The ΔS for the reaction below is $0.00300 \frac{kJ}{K}$ at 25°C. Calculate the ΔG for this reaction and determine whether it will occur spontaneously at 25°C.

$$C(s) + O_2(g) \rightarrow CO_2(g) + 393.51 \text{ kJ}$$

4. When graphite reacts with hydrogen at 27°C, ΔH is -74.8 kJ and ΔS is $-0.0809 \frac{kJ}{K}$.

Will this reaction occur spontaneously?

5. Draw and label energy diagrams depicting the following reactions, and determine all remaining values. Place the reactants at energy level zero.
(a) $\Delta H_{forward} = 10$ kJ $E_a' = 40$ kJ
(b) $\Delta H_{forward} = 95$ kJ $E_a' = 20$ kJ
(c) $\Delta H_{reverse} = 40$ kJ $E_a = 30$ kJ

6. Calculate the heat of formation of $H_2SO_4(\ell)$ from the ΔH values for the combustion of sulfur to $SO_2(g)$ ($\Delta H = -296.83$ kJ/mol), the oxidation of SO_2 to $SO_3(g)$ ($\Delta H = -98.91$ kJ/mol), and the solution of SO_3 in $H_2O(\ell)$ (heat of solution $= -132.4$ kJ/mol) to give $H_2SO_4(\ell)$ at 25°C. Solve by combining the known thermochemical equations.

Enrichment

1. Prepare a report on the invention and changing design of calorimeters. Include information on how each type is used.

2. Carry out library research on the history of the study of heat in physical and chemical processes, and on the way in which such study led to the discovery of energy and the law of conservation of energy.

3. Keep track of your food caloric intake for one day. Convert the total number of food Calories to joules (1 food Cal = 1000 cal = 4184 J). Compare your totals with those of your classmates and members of your family. How does your caloric intake compare with that recommended by nutritionists?

4. Take note of various everyday processes that occur on a given day. Record whether the processes are chemical or physical and whether they seem to be exothermic or endothermic.

5. Carry out library research to obtain information on alternative measurement units used to express values of heat and energy. Also, find out how the quantities relate to SI metric units. Include information specifically on English units, such as the British Thermal Unit (BTU), and on typical BTU ratings of household appliances. Calculate how these ratings would be expressed in joules instead.

6. Design a simple calorimeter investigation to determine the molar heat of fusion of water. Use of the following materials: a large Styrofoam cup with cover, a thermometer, a balance, water at room temperature, and an ice cube. Allow your teacher to review your design. Then carry out the investigation and write up a laboratory report, including your calculations and a comparison of your quantitative results with known values. Attempt to account for any disagreements between the experimental and actual values.

Chapter 19 Chemical Equilibrium

Chapter Planner

19.1 The Nature of Chemical Equilibrium
Reversible Reactions
- ▢ Questions: 1, 2
  Application Question: 2
- ■ Application Question: 1

Equilibrium, a Dynamic State
- ▢ Questions: 3, 4, 6
- ■ Question: 5

The Equilibrium Constant
- ▢ Experiment: 29
  Questions: 7–9
  Problems: 1–3, 5
  Application Questions: 3, 4
- ■ Application Problem: 1

19.2 Shifting Equilibrium
- ▢ Technology
  Experiment: 28
  Questions: 10–21
  Application Questions: 5, 7, 8, 10, 11

- ■ Application Question: 6

Reactions that Run to Completion
- ▢ Questions: 22, 23
  Problem: 6
- ▢ Application Question: 12

Common-Ion Effect
- ▢ Application Question: 13

19.3 Equilibrium of Acids, Bases, and Salts
Ionization Constant of a Weak Acid
- ▢ Application Question: 14
- ■ Question: 24
- ▢ Application Problems: 2, 3

Ionization Constant of Water
- ▢ Question: 25
  Problem: 7
- ■ Application Question: 15

Hydrolysis of Salts
- ▢ Questions: 26, 27

19.4 Solubility Equilibrium
Solubility Product
- ▢ Experiment: 30
  Questions: 28–30
  Problems: 9–11
  Application Question: 16
- ■ Application Problems: 4, 5

Calculating Solubilities
- ▢ Problems: 8, 12, 13
  Application Problems: 6, 7
- ▢ Problems: 14, 15

Precipitation Calculations
- ▢ Application Problems: 8–11

Teaching Strategies

Introduction

In the preceding chapter, students were able to determine whether or not a particular reaction would take place as written by calculating the free energy, $\triangle G$, of the reaction. In the same chapter, the students were able to predict the rate of the reaction. Now they will be able to determine the amount of product that can be expected by applying the law of mass action and developing the concepts underlying the equilibrium constant.

The chapter on chemical equilibrium is one that students may find to be very difficult. Part of the difficulty is due to the mathematical basis of the problems, but the initial presentation of the concept may also add to the problem. The students have already been introduced to the key element of the chapter—the statement of the law of mass action. If the students are brought to realize that all of the different types of equilibrium problems they will encounter are different applications of the same law of mass action, they will probably fare better.

Begin with a restatement of the law of mass action for a general equation:
$$nA + mB . . . \rightleftarrows xC + yD + . . .$$

$$K = \frac{[C]^x[D]^y . . .}{[A]^n[B]^m . . .}$$

Since a system at equilibrium is being discussed, the temperature must be a constant and the system must be closed.

Under these conditions, pure solids and liquids that occur as reactants and products do not appear in the law of mass action expression because their "concentrations" are really expressions of density which is at constant temperature.

$$[\ ] = mol/liter = mol/volume = mass/volume = density$$

Liquid water is a special case because one liter of water has a mass of approximately 1000 g or 55.5 mol/L. In most cases where water is involved in the reaction, so little water is actually used that the value 55.5 mol/L remains constant.

Only variables are retained in the law of mass action expression. Constants are incorporated into the constant, K.

If substances **A**, **B**, **D**, and **E** are gases, K_c would describe an equilibrium in which concentration is expressed in mol/L, while K_p would use partial pressures.

$$3H_2(g) + N_2(g) \rightleftarrows 2NH_3(g)$$

$$K_c = \frac{[NH_3(g)]^2}{[H_2(g)]^3[N_2(g)]}$$

$$K_p = \frac{P[NH_3(g)]^2}{P[H_2(g)]^3 P[N_2(g)]}$$

If **A** is an acid, then K becomes K_a. The same applies if A is base, K becomes K_b.

$$HC_2H_3O_2(aq) \rightleftarrows H^+(aq) + C_2H_3O_2^-(aq)$$

$$K_a = \frac{[H^+(aq)][C_2H_3O_2^-(aq)]}{[HC_2H_3O_2(aq)]}$$

or $HC_2H_3O_2(aq) + H_2O(\ell) \rightleftarrows H_3O^+(aq) + C_2H_3O_2^-(aq)$

$$K_a = \frac{[H_3O^+(aq)][C_2H_3O_2^-(aq)]}{[HC_2H_3O_2(aq)]}$$

In the case of a weak base,

$$NH_3(aq) + H_2O(\ell) \rightleftarrows NH_4^+(aq) + OH^-(aq)$$

$$K_b = \frac{[NH_4^+(aq)][OH^-(aq)]}{[NH_3(aq)][H_2O(\ell)]}$$

$$K[H_2O(\ell)] = \frac{[NH_4^+(aq)][OH^-(aq)]}{[NH_3(aq)]}$$

$$K_b = \frac{[NH_4^+(aq)][OH^-(aq)]}{[NH_3(aq)]}$$

If **A** represents water, $H_2O(\ell)$, then K becomes K_w.

$$H_2O(\ell) \rightleftarrows H^+(aq) + OH^-(aq)$$

$$K = \frac{[H^+(aq)][OH^-(aq)]}{[H_2O(\ell)]}$$

$$K[H_2O(\ell)] = [H^+(aq)][OH^-(aq)]$$
$$K_w = [H^+(aq)][OH^+(aq)]$$

or $H_2O(\ell) + H_2O(\ell) \rightleftarrows H_3O^+(aq) + OH^-(aq)$

$$K = \frac{[H^+(aq)][OH^-(aq)]}{[H_2O(\ell)]^2}$$

$$K[H_2O(\ell)]^2 = [H_3O^+(aq)][OH^-(aq)]$$
$$K_w = [H_3O^+(aq)][OH^-(aq)]$$

If **A** represents a slightly soluble ionic solid, K becomes K_{sp}.

$$AgCl(s) \rightleftarrows Ag^+(aq) + Cl^-(aq)$$

$$K = \frac{[Ag^+(aq)][Cl^-(aq)]}{[AgCl(s)]}$$

$$K[AgCl(s)] = [Ag^+(aq)][Cl^-(aq)]$$
$$K_{sp} = [Ag^+(aq)][Cl^-(aq)]$$

By initially outlining the chapter and how the same definition will be applied, you will make the students' task of trying to understand the concept less difficult.

Students who go on to take more advanced courses in chemistry will approach the same topic but base it on activities and activity coefficients. These should not be discussed at the introductory level. In the more advanced courses, students will learn of unit activities rather than constant densities.

As many demonstrations as are possible should be incorporated into the lectures and classroom discussions. Problem solving techniques are also very important. Students must work as many practice, review, and end-of-chapter problems as are needed to make them feel comfortable with them.

The demonstration manuals mentioned throughout this text unit have many demonstrations that are suitable and which may be used to supplement the ones described below.

19.1 The Nature of Chemical Equilibrium

Review physical equilibrium, dynamic equilibrium, equilibrium vapor pressure, Le Châtelier's principle, and solution equilibrium. Also review the role of free-energy change and the spontaneity of a reaction for the water–gas reaction. You may wish to use the data in that section for ΔS and ΔH and have the students calculate the temperature at which a closed system would reach equilibrium (about 980K).

Discuss Lavoisier's mercury(II) oxide bell-jar experiment and have students write the equation for this experiment and for equilibrium. Define chemical equilibrium and discuss the balance of the driving force of energy change, ΔH, and the driving force of entropy change, ΔS.

Refer to Chapter 11 and discuss equilibrium vapor pressure (and saturated solutions) to show that for a given closed system, the particular equilibrium state depends upon temperature and, for systems involving gases, also depends on the pressure.

Discuss the ionization of dilute acetic acid as an example of a reaction reaching equilibrium far to the left since acetic acid is a weak acid. Discuss the dissociation of NaCl or the ionization of HCl as reactions that go far to the right. Discuss the chemist's problem of producing desirable products in terms of ammonia synthesis. Have students restate the five factors that determine the rate of chemical reactions.

Refer to the introduction in Chapter 11 and further develop the concept of the equilibrium concept. Much practice on the part of the students is needed. Follow the development in the text. Begin with the hypothetical equations and move to the real equations given. Have the students write out equilibrium constant expressions until they are comfortable with the practice. Point out that if concentrations are known, the equilibrium constant may be calculated, and that if the equilibrium constant is known, equilibrium concentrations can be calculated from the initial concentrations and the stoichiometry of the reaction.

The Technology feature on *Swimming Pool Chemistry* gives you the opportunity to discuss another equilibrium system.

19.2 Shifting Equilibrium

Raise the problem of how chemists may displace an equilibrium in order to increase the desired production. Review and restate Le Châtelier's principle. The text discusses the effect of changes in concentration, of changes in pressure where gases are involved, and the effect of temperature. Stress with the students that the only change that will affect the numerical value of the equilibrium constant is a change in temperature. An increase in temperature will favor the endothermic reaction. In all other cases, a new set of equilibrium concentrations or pressure will be established, but the numerical value of the equilibrium constant will be the same as before any changes occurred. Students will have a difficult time visualizing this. They should be led carefully through a description of the changes that occur on the molecular level and then have these changes related to the equilibrium expression that describes them. They should note the numerator (concentration of products) and the denominator (concentration of reactants). Catalysts deserve special attention. They affect the rate of a reaction, not the net amount of product. Catalysts have no effect on the numerical value of the equilibrium constant.

Explain to the students that there are signs that will help them predict whether a given reaction will run to completion. Two signs are readily apparent: *(1)* the formation of a gas and *(2)* the formation of a precipitate. A third is more difficult to see: *(3)* the formation of a slightly ionized product. Each of these should be demonstrated.

The section concludes with a discussion of the common ion effect. Treat it as just another case of the application of Le Châtelier's principle. Most students will be able to predict what should happen.

19.3   Equilibrium of Acids, Bases, and Salts

A more detailed look at one specific type of equilibrium is now discussed. Refer back to your initial lecture. The one common aspect of the discussion is the dissociation of a slightly ionized hydrogen containing a water soluble compound—a weak acid. Because all equations have this in common, K_{eq} is now referred to as K_a.

As you proceed with your explanation, have your students note the difference in relative concentrations of reactants and products when K_a is very small. Very little product is present in the system. This will help you later when you broach the subject of buffers—systems in which both reactant and product (acid anion) concentrations are of the same magnitude. Buffers may be introduced as another case of the common ion effect.

Water is a very important case. Impress upon the students that it is so important that the equilibrium constant describing the dissociation of water has its own symbol, K_w. Refer back to your introductory lecture and point out, once again, that the water system is just another case of the application of the law of mass action.

The remaining sections discuss the hydrolysis of salts, anion hydrolysis, and cation hydrolysis. One method of helping students recall and organize the information is to have them reconstruct the acid and base from which the salt might have been formed. Using NaCl as an example, the cation, Na^+, is always derived from the base, and the anion, Cl^-, from the acid. To predict the base and acid, add the proper number of OH^-'s to the cation and the proper number of H^+'s to the anion. In the case of NaCl, the base must have been NaOH and the acid HCl. Since the salt formed from the neutralization of a strong acid and a strong base is neutral, NaCl is a neutral salt. Each of your predictions should be tested with an acid–base indicator. Use the following table as a mnemonic.

SA = strong acid   SB = strong base
wA = weak acid   wB = weak base

SB + SA → neutral salt (NaCl, or Na_2SO_4)
wB + SA → acid salt (NH_4Cl)
SB + wA → basic salt (NaF)
wB + wA → determined by the stronger of the two.

(One needs to know the K_a and K_b.)

For the mnemonic to be of any use, students must be aware of the common strong acids and strong bases.

The anion hydrolysis constant should again be discussed in terms of just another application of the law of mass action. When introducing cation hydrolysis, note the special importance of ions with a charge of $+2$ and $+3$. Ions of such solutions will always test acidic.

19.4   Solubility Equilibrium

Impress upon the students that no substance is completely insoluble. Certain ranges of solubility have been established. It is very important that after you derive the solubility product expression (again as a special case of the more general law of mass action), you point out to the students the general class of problems most frequently associated with solubility products: calculating the K_{sp} from solubility data, calculating solubilities given the K_{sp} for a reaction, and determining whether or not a precipitate will form, given the K_{sp} and initial concentration of the reactants.

References and Resources

Books and Periodicals

Kauffman, George B. "The Brønsted-Lowry Acid-Base Concept," *Journal of Chemical Education*, Vol. 65, No. 1, Jan. 1988, pp. 28–29.
Russell, Joan M. "Simple Models for Teaching Equilibrium and Le Châtelier's Principle," *Journal of Chemical Education*, Vol. 65, No. 10, Oct. 1988, pp. 871–872.

Audiovisual Resources

Campbell, J. Arthur. *Acid-Base Indicators*, A Chem Study Film, Color, 19 min., VHS cassette, Central Scientific Co. (Behavior of acid-base indicators according to the proton donor-acceptor theory)
Pimentel, George C. *Equilibrium*, A Chem Study Film, Color, 24 min., 16mm film or VHS cassette, Central Scientific Co. (The dynamic nature of equilibrium is examined)

Computer Software

"Chemical Equilibrium Series," Apple, Cambridge Development Laboratory.

Answers and Solutions

Questions

1. *(a)* A reversible reaction is one in which the products can react to re-form the reactants. *(b)* $2Hg(s) \rightleftarrows 2Hg(\ell) + O_2(g)$

2. At equilibrium, the rate of the forward reaction is equal to the rate of the reverse reaction.

3. The term dynamic refers to a state in which two opposing processes are occurring simultaneously at the same rate.

4. Two examples are water in contact with its saturated vapor in a closed vessel, and sugar crystals in contact with a saturated sugar solution.

5. *(a)* When an ionic compound such as NaCl is dissolved in water, the resulting solution consists of hydrated Na^+ and Cl^- ions dispersed throughout. When enough NaCl is added to produce a saturated solution, a solution equilibrium is established in which the rate of association of ions re-forming the NaCl crystals equals the rate of dissociation of ions from the crystals such that: $Na^+Cl^-(s) \rightleftarrows Na^+ (aq) + Cl^- (aq)$.
   (b) When a polar-covalent compound, such as acetic acid $(HC_2H_3O_2)$, is combined with water, H_3O^+ and $C_2H_3O_2^-$ ions are formed. As pairs of these ions rejoin, acetic acid molecules are re-formed so that ionic equilibrium is established between the non-ionized molecules and their hydrated ions such that: $HC_2H_3O_2(aq) + H_2O(\ell) \rightleftarrows H_3O^+(aq) + C_2H_3O_2^- (aq)$.

6. *(a)* In the process of establishing equilibrium, the forward reaction occurs to a greater extent than the reverse such that there is a higher concentration of products than of reactants. *(b)* The reverse reaction occurrs to a greater extent in the process of establishing equilibrium so that there is a higher concentration of reactants than products. *(c)* The concentration of the products of the forward reaction is greater than that of the reactants of that reaction. *(d)* The concentration of the reactants of the forward reaction is greater than that of the products of that reaction. *(e)* This refers to the forward reaction. *(f)* This refers to the reverse reaction.

7. At the start, A and B are at their maximum concentrations, and no C and D exist. As A and B collide and react to form C and D, the concentrations of A and B decrease, while those of C and D increase. As the amounts of C and D increase, the rate at which they combine to re-form A and B increases, while the rate at which A and B combine decreases. Eventually, the two rates become equal and equilibrium is established. At that point, the concentrations of A, B, C, and D remain constant as long as conditions remain the same.

8. *(a)* $K = \dfrac{[C]^x[D]^y \cdots}{[A]^n[B]^m \cdots}$
   (b) The value of K provides information about the extent to which the reactants are converted into the products of a given reaction at the specified temperature.

9. *(a)* Neither is favored. *(b)* The reverse reaction is favored. *(c)* The forward reaction is favored.

10. *(a)* Le Châtelier's principle states that if a system at equilibrium is subjected to a stress, the equilibrium is shifted in the direction that relieves the stress. *(b)* Le Châtelier's principle applies to all dynamic equilibria: physical, chemical, and ionic.

11. *(a)* (1) A change in concentration causes the equilibrium to shift in the direction that relieves the imposed stress. An increase in the concentration of any component favors the reaction that consumes that component. (2) A change in pressure affects an equilibrium system that involves gases. An increase in pressure favors the reaction that produces the fewer number of gas molecules since the result would be a reduction in volume. (3) Temperature changes affect equilibrium systems in accordance with the endothermic or exothermic nature of the forward and reverse reactions. An increase in temperature favors the endothermic reaction since it absorbs the additional heat. Likewise, a decrease on temperature favors the exothermic reaction. *(b)* Only temperature changes affect the value of K.

12. *(a)* Forward *(b)* Reverse *(c)* Forward *(d)* Reverse

13. *(a)* Higher *(b)* Higher *(c)* Lower *(d)* Lower

14. *(a)* Forward *(b)* Reverse

15. *(a)* Higher *(b)* Lower

16. The concentrations of pure substances in the solid and liquid phrases can be removed from the equilibrium constant expression because the equilibrium concentrations of such substances are not changed by adding or removing quantities of these materials. Since concentration is density-dependent, the densities of substances in these phrases is constant regardless of the total amounts present.

17. *(a)* Reverse *(b)* Forward

18. *(a)* Increased temperature would reduce the amount of NO_2, while decreased temperature would increase the level of NO_2. *(b)* Increased temperature would reduce the value of K since the reverse reaction would be favored, while decreased temperature would increase the value of K.

19. A catalyst increases the rates of both the forward and reverse reactions at equilibrium, but does so to an equal extent such that the relative equilibrium amounts, and thus the value of K, are not affected.

20. *(a)* Forward *(b)* Forward *(c)* Neither *(d)* Forward *(e)* Reverse *(f)* Neither *(g)* Neither *(h)* Reverse *(i)* Neither

21. *(a)* HCl would increase; K would remain the same *(b)* HCl would decrease; K would remain the same *(c)* HCl would not change; K would remain the same *(d)* HCL would increase; K would increase *(e)* HCl would decrease; K would remain the same *(f)* HCl would not change; K would remain the same *(g)* HCl would not change; K would remain the same *(h)* HCl would decrease; K would decrease *(i)* HCl would not change; K would remain the same.

22. The extent to which reacting ions are removed from solution depends on the solubility of the compound formed and the degree of ionization, if the compound is soluble.

23. (1) A precipitate is formed: $Ag^+(aq) + Cl^- (aq) \rightarrow AgCl(s)$.
    (2) A gaseous product is formed: $H_3O^+(aq) + HCO_3^- (aq) \rightarrow 2H_2O(\ell) + CO_2(g)$.
    (3) A slightly ionized product is formed: $H_3O^+(aq) + OH^-(aq) \rightarrow 2H_2O$.

24. Since buffers can guard against changes in pH, they are used in medicines to protect the body against large pH changes that could otherwise lead to serious disturbances of normal body functions or even death.

25. *(a)* $K_w = [H_3O^+][OH^-]$ *(b)* $K_w = 10^{-14}$

26. (1) Salts of strong acids and strong bases—Neither the cations of a strong base nor the anions of a strong acid hydrolyze appreciably in aqueous solutions. Such salts yield neutral aqueous solutions with a pH of 7. NaCl is an example. (2) Salts of weak acids and strong bases—The aqueous solutions of such salts are basic. The anions of the salt (the negative ions from the weak acid) undergo hydrolysis, while the cations of the salt (the positive ions from the strong base) do not. $NaC_2H_3O_2$ is an example. (3) Salts of strong acids and weak bases—The aqueous solutions of such salts are acidic. The cations of the salt (the positive ions from the weak base) undergo hydrolysis, while the anions do not. NH_4Cl is an example. (4) Salts of weak acids and weak bases—In aqueous solutions of such salts, both ions of the dissolved salt are hydrolyzed extensively and the solution can be acidic, basic, or neutral, depending on the salt involved. $NH_4C_2H_3O_2$ is an example.

27. Anion hydrolysis results in aqueous solutions that are basic, while cation hydrolysis produces acidic solutions.

28. In general, a substance is said to be soluble if the solubility is greater than 1 g per 100 g of water, and insoluble if its solubility is less than 0.1 g per 100 g of water. Substances whose solubilities fall between these limits are termed slightly soluble.

29. Salts that are sparingly soluble in water, and generally described as insoluble, are typically involved.

30. If the ion product is less than the value of the given K_{sp}, the solution is unsaturated. However, if the ion product exceeds the K_{sp}, the solution is saturated and precipitation occurs until the ion concentrations decrease to equilibrium values.

Problems

1. *(a)* $K = \dfrac{[C]^3[D]}{[A][B]^2}$ *(b)* $K = \dfrac{[C]^2[D]^3}{[A]^4[B]}$

   (c) $K = \dfrac{[NH_3]^2}{[N_2][H_2]^3}$ *(d)* $K = \dfrac{[CO_2][NO]}{[CO][NO_2]}$

2. *(a)* $K = \dfrac{[C]}{[A][B]} = \dfrac{4.0}{(2.0)(3.0)} = 0.67$

   (b) $K = \dfrac{[F][G]^3}{[D][E]^2}$ *(c)* $K = \dfrac{[NH_3]^2}{[N_2][H_2]^3}$

   $K = \dfrac{[1.8][1.2]^3}{[1.5][2.0]^2}$ $K = \dfrac{[0.62]^2}{[0.45][0.14]^3}$

   $K = 0.52$ $K = 310$

3. $K = \dfrac{[H_2O]^2[Cl_2]^2}{[HCl]^4[O_2]} = \dfrac{(5.8 \times 10^{-2})^2(5.8 \times 10^{-2})^2}{(1.2 \times 10^{-3})^4(3.8 \times 10^{-4})}$

   $K = 1.43 \times 10^{10}$

4. $K = \dfrac{[NH_3]^2}{[N_2][H_2]^3}$

   $[N_2] = [NH_3]^2/K[H_2]^3$

$$[N_2] = \dfrac{(1.23 \times 10^{-4})^2}{(6.59 \times 10^{-3})(2.75 \times 10^{-6})^3}$$

$[N_2] = 1.10 \times 10^{13}$ mol/L

5. *(a)* $K = \dfrac{[Zn^{2+}]}{[Ag^+]^2}$ *(b)* $K = [Pb^{2+}][I^-]^2$

   (c) $K = \dfrac{[HCN][OH^-]}{[CN^-]}$ *(d)* $K = [Cd^{2+}][S^{2-}]$ *(e)* $K = \dfrac{1}{[O_2]^3}$

6. *(a)* $Pb^{2+}(aq) + 2NO_3^-(aq) + 2Na^+(aq) + 2Cl^-(aq) \rightarrow$
   $2Na^+(aq) + 2NO_3^-(aq) + PbCl_2(s)$
   $Pb^{2+}(aq) + 2Cl^-(aq) \rightarrow PbCl_2(s)$

   (b) $Ba^{2+}(aq) + 2NO_3^-(aq) + 2K^+(aq) + SO_4^{2-}(aq) \rightarrow$
   $2K^+(aq) + 2NO_3^-(aq) + BaSO_4(s)$
   $Ba^{2+}(aq) + SO_4^{2-}(aq) \rightarrow BaSO_4(s)$

   (c) $Ag^+(aq) + C_2H_3O_2^-(aq) + Li^+(aq) + Cl(aq) \rightarrow Li^+(aq)$
   $+ C_2H_3O_2^-(aq) + AgCl(s)$
   $Ag^+(aq) + Cl^-(aq) \rightarrow AgCl(s)$

   (d) $6Na_2^+(aq) + 2PO_4^{3-}(aq) + 3Cu^{2+}(aq) + 3SO_4^{2-}(aq) \rightarrow$
   $6Na^+(aq) + 3SO_4^{2-}(aq) + Cu_3(PO_4)_2(s)$
   $3Cu^{2+}(aq) + 2PO_4^{3-}(aq) \rightarrow Cu_3(PO_4)_2(s)$

   (e) $Ba^{2+}(aq) + 2Cl^-(aq) + 2Na^+(aq) + SO_4^{2-}(aq) \rightarrow$
   $2Na^+(aq) + 2Cl^-(aq) + BaSO_4(s)$
   $Ba^{2+}(aq) + SO_4^{2-}(aq) \rightarrow BaSO_4(s)$

7. $K_w = [H_3O^+][OH^-]$

   $[H_3O^+] = \dfrac{K_w}{[OH^-]} = \dfrac{10^{-14}}{10^{-5}} = 10^{-9}$

   It is a base.

8. *(a)* $\dfrac{36.0 \text{ g NaCl}}{100.\text{g}} \times \dfrac{10^3 \text{ g}}{L} \times \dfrac{\text{mol NaCl}}{58.4 \text{ g NaCl}} = 6.16$ mol NaCl/L

   (b) $\dfrac{0.0241 \text{ g PbSO}_4}{100. \text{ g}} \times \dfrac{10^3 \text{ g}}{L} \times \dfrac{\text{mol PbSO}_4}{303 \text{ g PbSO}_4} = 7.95 \times$
   10^{-4} mol PbSO$_4$/L

   (c) $\dfrac{1.84 \times 10^{-11} \text{ g Cu}_2\text{S}}{100.\text{g}} \times \dfrac{10^3 \text{ g}}{L} \times \dfrac{\text{mol Cu}_2\text{S}}{159 \text{ g Cu}_2\text{S}} = 1.16 \times$
   10^{-12} mol Cu$_2$S/L

9. $GH = [G^+] = [H^-] = 0.0427$ mol/L
   $K_{sp} = [G^+][H^-] = (0.0427)^2 = 1.82 \times 10^{-3}$

10. $[E^{2+}] = [J^{2-}] = 8.45 \times 10^{-6}$ mol/L
    $K_{sp} = [E^{2+}][J^{2-}] = (8.45 \times 10^{-6})^2 = 7.14 \times 10^{-11}$

11. *(a)* $\dfrac{1.2 \times 10^{-3} \text{ g CaCO}_3}{100. \text{ g}} \times \dfrac{10^3 \text{ g}}{L} \times \dfrac{\text{mol CaCO}_3}{1.0 \times 10^2 \text{ g CaCO}_3}$
    $= 1.2 \times 10^{-4}$ mol CaCO$_3$/L
    $CaCO_3(s) \rightleftharpoons Ca^{2+}(aq) + CO_3^{2-}(aq)$
    $K_{sp} = [Ca^{2+}][CO_3^{2-}] = (1.2 \times 10^{-3})(1.2 \times 10^{-3})$
    $= 1.4 \times 10^{-6}$

    (b) $\dfrac{2.4 \times 10^{-4} \text{ g BaSO}_4}{100. \text{ g}} \times \dfrac{10^3 \text{ g}}{L} \times \dfrac{\text{mol B}_2\text{SO}_4}{233 \text{ g BaSO}_4}$
    $= 1.0 \times 10^{-5}$ mol B$_2$SO$_4$/L
    $BaSO_4(s) \rightleftharpoons Ba^{2+}(aq) + SO_4^{2-}(aq)$
    $K_{sp} = [Ba^{2+}][SO_4^{2-}] = (1.0 \times 10^{-5})(1.0 \times 10^{-5})$
    $= 1.0 \times 10^{-10}$

(c) $\dfrac{0.173 \text{ g Ca(OH)}_2}{100. \text{ g}} \times \dfrac{10^3 \text{ g}}{\text{L}} \times \dfrac{\text{mol Ca(OH)}_2}{74.1 \text{ g Ca(OH)}_2}$

$= 2.33 \times 10^{-2} \text{ mol Ca(OH)}_2/\text{L}$

$\text{Ca(OH)}_2(s) \rightleftarrows \text{Ca}^{2+}(aq) + 2\text{OH}^-(aq)$

$K_{sp} = [\text{Ca}^{2+}][\text{OH}^-]^2$

$K_{sp} = (2.33 \times 10^{-2})(2(2.33 \times 10^{-2}))^2 = 5.06 \times 10^{-5}$

12. $K_{sp} = [\text{M}^{2+}][\text{N}^{2-}] = 8.1 \times 10^{-6}$

Solubility $= [\text{M}^{2+}] = [\text{N}^{2-}] = K_{sp} = \sqrt{8.1 \times 10^{-6}}$

$= 2.8 \times 10^{-3} \text{ mol/L}$

13. (a) $\text{AgBr}(s) \rightleftarrows \text{Ag}^+(aq) + \text{Br}^-(aq)$, $K_{sp} = [\text{Ag}^+][\text{Br}^-]$

Since $[\text{Ag}^+] = [\text{Br}^-]$, $K_{sp} = [\text{Ag}^+]^2$

$[\text{Ag}^+] = \sqrt{K_{sp}} = \sqrt{5.0 \times 10^{-13}} = 7.1 \times 10^{-7}$

Solubility AgBr $= 7.1 \times 10^{-7} \text{ mol/L}$

(b) $\text{CoS}(s) \rightleftarrows \text{Co}^{2+}(aq) + \text{S}^{2-}(aq)$

$K_{sp} = [\text{Co}^{2+}][\text{S}^{2-}]$

Since $[\text{Co}^{2+}] = [\text{S}^{2-}]$, $K_{sp} = [\text{Co}^{2+}]^2$

$[\text{Co}^{2+}] = \sqrt{K_{sp}} = \sqrt{3.0 \times 10^{-26}} = 1.7 \times 10^{-13}$

Solubility CoS $= 1.7 \times 10^{-13} \text{ mol/L}$

14. (a) $\text{PbSO}_4(s) \rightleftarrows \text{Pb}^{2+}(aq) + \text{SO}_4^{2-}(aq)$

(b) $K_{sp} = [\text{Pb}^{2+}][\text{SO}_4^{2-}]$

(c) Since $0.0500M$ $\text{Pb(NO}_3)_2$ is $0.0500M$ Pb^{2+},

$\dfrac{0.0500 \text{ mol Pb}^{2+}}{\text{L}} \times 25.0 \text{ mL} \times \dfrac{\text{L}}{10^3 \text{ mL}}$

$= 1.25 \times 10^{-3} \text{ mol Pb}^{2+}$

Since $0.0400M$ Na_2SO_4 is $0.0400M$ SO_4^{2-},

$\dfrac{0.0400 \text{ mol SO}_4^{2-}}{\text{L}} \times 25.0 \text{ mL} \times \dfrac{\text{L}}{10^3 \text{ mL}}$

$= 1.00 \times 10^{-3} \text{ mol SO}_4^{2-}$

For Pb^{2+}, $\dfrac{1.25 \times 10^{-3} \text{ mol Pb}^{2+}}{50.0 \text{ mL}} \times \dfrac{10^3 \text{ mL}}{\text{L}}$

$= 2.50 \times 10^{-2} \text{ mol Pb}^{2+}/\text{L}$

For SO_4^{2-}, $\dfrac{1.00 \times 10^{-3} \text{ mol SO}_4^{2-}}{50.0 \text{ mL}} \times \dfrac{10^3 \text{ mL}}{\text{L}}$

$= 2.00 \times 10^{-2} \text{ mol SO}_4^{2-}/\text{L}$

(d) $[\text{Pb}^{2+}][\text{SO}_4^{2-}] = (2.50 \times 10^{-2})(2.00 \times 10^{-2})$

$= 5.00 \times 10^{-4}$

(e) K_{sp} for PbSO_4 at 25°C $= 1.8 \times 10^{-8}$

Since the ion product of 5.00×10^{-4} exceeds the given K_{sp}, a precipitate of PbSO_4 will form.

15. Since the ion product, 1.9×10^{-5}, exceeds the K_{sp} for CaCO_3, a precipitate will form.

Application Questions

1. Since reaction rate depends on both collision frequency and efficiency, the only occasion for reactant collisions in a heterogeneous system is at the surface area of contact of the reactants. In a homogeneous system, many more opportunities exist for contact, thus the concentration of reactant particles available for this contact is a major factor.

2. Place the steam and iron in a closed reaction chamber and heat. The products of the forward reaction hydrogen and magnetic iron oxide, cannot escape from the reaction chamber, and so can react to form steam and iron. When the chamber is opened, all four substances will be present.

3. (a) K_3 (b) K_1

4. By understanding the principles of chemical equilibrium, an industrial chemist can apply various stresses to equilibrium systems that result in the increased production of desired products.

5. To the equilibrium system continually add more H_2 and F_2 while removing the HF as it is formed.

6. Once equilibrium is established, such changes have an equal effect on the numerator for and denominator of the equilibrium constant expression so that the numerical rate is not affected.

7. High pressure would shift the equilibrium to the right since the forward reaction converts three molecules into two, thereby effectively reducing the volume and relieving the stress imposed by the increased pressure.

8. High pressure would shift the equilibrium toward the increased production of methanol.

9. The yield of methanol would be increased by lowering the temperature of the system since the reaction to the right is exothermic. A rise in temperature would act as a stress on the system, and the equilibrium would shift in the direction of the endothermic reaction to relieve the stress. This would reduce the concentration of methanol.

10. (a) High reactant concentrations, high pressure, low temperature (b) High reactant concentrations, pressure not relevant, low temperature (c) High reactant concentrations, pressure not relevant, high temperature (d) High reactant concentrations, high pressure, low temperature (e) High reactant concentrations, low pressure, high temperature

11. (a) Low (b) Low (c) High (d) Low (e) High

12. (a) The rate of the reaction increases decidedly. (b) Air is one-fifth oxygen; thus, pure oxygen at the same pressure as the air will provide 5 times the number of collisions between the reactants.

13. Acetic acid is a weak electrolyte and is slightly ionized in water solution. Equilibrium is quickly established between acetic acid molecules and hydronium and acetate ions in the solution: $\text{HC}_2\text{H}_3\text{O}_2 + \text{H}_2\text{O} \rightleftarrows \text{H}_3\text{O}^+ + \text{C}_2\text{H}_3\text{O}_2^-$. Sodium acetate is an ionic salt and is therefore completely ionized in water solution. When added to the acetic acid solution, the concentration of the acetate ion is increased, driving the equilibrium to the left. This removes hydronium ions from solution and increases the concentration of un-ionized acetic acid molecules. Thus the solution is less acidic and has a higher pH value. At equilibrium, the concentration of molecular acetic acid is high and the concentrations of H_3O^+ and $\text{C}_2\text{H}_3\text{O}_2^-$ ions are low. Equilibrium is established early, before the forward reaction has progressed very far. The low concentrations of ions in the equilibrium system show that acetic acid is a weak electrolyte.

15. (a) $\text{NH}_3(aq) + \text{H}_2\text{O}(\ell) \rightleftarrows \text{NH}_4^+(aq) + \text{OH}^-(aq)$

$K_b = \dfrac{[\text{NH}_4^+][\text{OH}^-]}{[\text{NH}_3]}$

(b) The value of K_b is small, showing that there are comparatively few NH_4^+ and OH^- ions at equilibrium.

16. Although a saturated solution contains the maximum concentration of solute possible at a given temperature in

equilibrium with an undissolved substance, the actual concentration of dissolved solute may be high or low, depending on the solubility of that solute. Thus, saturated does not necessarily mean concentrated.

Application Problems

1. Since K forward $= 40.0$, K reverse $= \dfrac{1}{40.0} = 0.0250$

2. $[H_3O^+] = $ antilog $(-3.14) = 7.24 \times 10^{-4}$
$[H_2O] = [X^-] = 7.24 \times 10^{-4}$
$[HX] = 0.100 - 0.000\ 072\ 4 = 0.100$
$K_a = \dfrac{[H_3O^+][X^-]}{[HX]} = \dfrac{(7.24 \times 10^{-4})^2}{0.100} = 5.24 \times 10^{-6}$

3. $K_h = \dfrac{K_w}{K_a} = \dfrac{1.00 \times 10^{-14}}{K_a\ 2.69 \times 10^{-9}} = 3.72 \times 10^{-6}$

4. $X_2Y(s) \rightleftarrows 2X^+(aq) + Y^{2-}(aq)$
$K_{sp} = [X^+]^2[Y^{2-}]$
$[X^+] = 2. \times 5.54 \times 10^{-7} mol/L = 1.11 \times 10^{-6} mol/L$
$[Y^{2-}] = 5.54 \times 10^{-7} mol/L$
$K_{sp} = (1.11 \times 10^{-6})^2 (5.54 \times 10^{-7})$
$= 6.83 \times 10^{-19}$

5. $T_3U_2(s) \rightleftarrows 3T^{2+}(aq) + 2U^{3-}(aq)$
$K_{sp} = [T^{2+}]^3[U^{3-}]^2$
$[T^{2+}] = 3\ (3.77 \times 10^{-20}) mol/L = 1.13 \times 10^{-19} mol/L$
$[U^{3-}] = 2\ (3.77 \times 10^{-20}) mol/L = 7.54 \times 10^{-20} mol/L$
$K_{sp} = (1.13 \times 10^{-19})^3(7.54 \times 10^{-20})^2 = (1.44 \times 10^{-57})(5.69 \times 10^{-39}) = 8.19 \times 10^{-96}$

6. $VW_2(s) \rightleftarrows V^{2+}(aq) + 2W^-(aq)$
$K_{sp} = [V^{2+}][W^-]^2 = 4.0 \times 10^{-9}$
If X represents solubility of VW_2 then
$[V^{2+}] = [X]$ and
$[W^-]^2 = [2X]^2$ such that
$K_{sp} = [X][2X]^2$
$4.0 \times 10^{-9} = (X)(4X^2)$
$4.0 \times 10^{-9} = 4X^3$
$1.0 \times 10^{-9} = X^3$
$1.0 \times 10^{-3} = X$

7. $AgI(s) \rightleftarrows Ag^+(aq) + I^-(aq)$
$K_{sp} = [Ag^+][I^-]$
$8.3 \times 10^{-17} = (2.7 \times 10^{-10})(I^-)$
$[I^-] = \dfrac{8.3 \times 10^{-17}}{2.7 \times 10^{-10}} = 3.1 \times 10^{-7} mol/L$

8. $Ca(OH)_2(s) \rightleftarrows Ca^{2+}(aq) + 2OH^-(aq)$
$K_{sp} = [Ca^{2+}][OH^-]^2$
For Ca^{2+}, $0.35\ L \times \dfrac{4.4 \times 10^{-3}\ mol\ Ca^{2+}}{L} = 1.5 \times 10^{-3}\ mol\ Ca^{2+}$
$\dfrac{1.5 \times 10^{-3}\ mol\ Ca^{2+}}{(0.35\ L + 0.17\ L)} = 2.9 \times 10^{-3}\ mol\ Ca^{2+}/L$
$0.17\ L \times \dfrac{3.9 \times 10^{-4}\ mol\ OH^-}{L} = 6.6 \times 10^{-5}\ mol\ OH^-/L$
$\dfrac{6.6 \times 10^{-5}\ mol\ OH^-}{(0.35\ L + 0.17\ L)} = 0.000\ 13\ mol\ OH^-/L$
Trial product: $[Ca^{2+}][OH^-]^2 = (6.6 \times 10^{-5})(1.3 \times 10^{-4})^2 = 1.1 \times 10^{-12}$
Since the trial product is less than the K_{sp} of 5.5×10^{-6}, no precipitate is formed.

9. $CaSO_4(s) \rightleftarrows Ca^{2+}(aq) + SO_4^{2-}(aq)$

$K_{sp} = [Ca^{2+}][SO_4^{2-}]$

For Ca^{2+}: $\dfrac{0.0040 \text{ mol } Ca^{2+}}{L} \times 70.0 \text{ mL} \times \dfrac{L}{10^3 \text{ mL}} = 2.8 \times 10^{-4} \text{ mol } Ca^{2+}$

$\dfrac{2.8 \times 10^{-4} \text{ mol } Ca^{2+}}{100.0 \text{ mL}} \times \dfrac{10^3 \text{ mL}}{L} = 2.8 \times 10^{-3} \text{ mol } Ca^{2+}/L$

For SO_4^{2-}: $\dfrac{0.010 \text{ mol } SO_4^{2-}}{L} \times 30.0 \text{ mL} \times \dfrac{L}{10^3 \text{ mL}} = 3.00 \times 10^{-4} \text{ mol } SO_4^{2-}$

$\dfrac{3.00 \times 10^{-4} \text{ mol } SO_4^{2-}}{100.0 \text{ mL}} \times \dfrac{10^3 \text{ mL}}{L} = 3.00 \times 10^{-3} \text{ mol } SO_4^{2-}/L$

Trial product: $[Ca^{2+}][SO_4^{2-}] = (2.8 \times 10^{-3})(3.00 \times 10^{-3}) = 8.4 \times 10^{-6}$

Since the trial product is less than the K_{sp} of 9.1×10^{-6}, no precipitate is formed.

10. $AgCl(s) \rightleftarrows Ag^+(aq) + Cl^-(aq)$

$K_{sp} = [Ag^+][Cl^-]$

For Ag^+: $1.70 \text{ g } AgNO_3 \times \dfrac{108 \text{ g } Ag^+}{170. \text{ g } AgNO_3} = 1.08 \text{ g } Ag^+$

$1.08 \text{ g } Ag^+ \times \dfrac{\text{mol } Ag^+}{108 \text{ g } Ag^+} = 0.0100 \text{ mol } Ag^+$

$\dfrac{0.0100 \text{ mol } Ag^+}{200. \text{ mL}} \times \dfrac{10^3 \text{ mL}}{L} = 0.0500 \text{ mol } Ag^+/L$

For: Cl^-: $14.5 \text{ g } NaCl \times \dfrac{23.0 \text{ g } Na^+}{58.4 \text{ g } NaCl} = 5.72 \text{ g } Na^+$

$5.72 \text{ g } Na^+ \dfrac{\text{mol } Na^+}{23.0 \text{ g } Na^+} = 0.249 \text{ mol } Na^+$

$\dfrac{0.249 \text{ mol } Na^+}{200. \text{ mL}} \times \dfrac{10^3 \text{ mL}}{L} = 1.24 \text{ mol } Cl^-/L$

Trial product: $[Ag^+][Cl^-] = (0.0500)(1.24) = 6.20 \times 10^{-2}$

Since the trial product exceeds the K_{sp} of 1.8×10^{-10}, a precipitate does form.

11. $Fe(OH)_3(s) \rightleftarrows Fe^{3+}(aq) + 3OH^-(aq)$

$K_{sp} = [Fe^{3+}][OH^-]^3$

For Fe^{3+}: $2.50 \text{ g } Fe(NO_3)_3 \times \dfrac{55.8 \text{ g } Fe^{3+}}{242 \text{ g } Fe(NO_3)_3} = 0.576 \text{ g } Fe^{3+}$

$0.576 \text{ g } Fe^{3+} \times \dfrac{\text{mol } Fe^{3+}}{55.8 \text{ g } Fe^{3+}} = 1.03 \times 10^{-2} \text{ mol } Fe^{3+}$

$\dfrac{1.03 \times 10^{-2} \text{ mol } Fe^{3+}}{100. \text{ mL}} \times \dfrac{10^3 \text{ mL}}{L} = 1.03 \times 10^{-1} \text{ mol } Fe^{3+}/L$

For OH^-: 1.00×10^{-20} M NaOH is 1.00×10^{-20} M OH^-

Trial product: $[Fe^{3+}][OH^-]^3 = (1.03 \times 10^{-1})(1.00 \times 10^{-20})^3 = 1.03 \times 10^{-61}$

Since the trial product is less than the K_{sp} of 1.5×10^{-36}, no precipitate is formed.

Chapter 19 Chemical Equilibrium

INTRODUCTION

In the preceding chapter, you learned some general information about the energy of chemical reactions and about the rates at which they occur. In this chapter, you will focus on the relationships among reactants and products in reactions that have come to equilibrium. You will also learn the effect of changing conditions on equilibrium systems.

LOOKING AHEAD

As you study this chapter, keep in mind that reactions are occurring even in systems in which no visible process is taking place. Pay attention to the changes that occur in these underlying reactions when equilibrium is disturbed.

SECTION PREVIEW

19.1 The Nature of Chemical Equilibrium
19.2 Shifting Equilibrium
19.3 Equilibria of Acids, Bases, and Salts
19.4 Solubility Equilibrium

These stalactites and stalagmites were formed by a disruption of chemical equilibrium—the open air of the cave changed soluble calcium hydrogen carbonate into insoluble calcium carbonate.

19.1 The Nature of Chemical Equilibrium

In earlier chapters, you learned about the equilibrium that can exist between certain physical processes. In the first section of this chapter, you will investigate the nature of the equilibrium that can exist between chemical processes. Then you will examine the mathematical relationships that govern equilibrium.

Reversible Reactions

The products formed in a chemical reaction may react to re-form the original reactants. *A chemical reaction in which the products can react to re-form the reactants is called a* **reversible reaction**. An example of such a reaction is shown in Figure 19-1. Essentially, all chemical reactions are considered to be reversible under suitable conditions. Theoretically, every reaction pathway can be traversed in two directions, forward and reverse. Some reverse reactions occur or are driven less easily than others, however. For example, a temperature near 3000°C is required for water vapor to decompose into measurable quantities of elemental hydrogen and oxygen.

Mercury(II) oxide decomposes when heated strongly.

$$2HgO(s) \xrightarrow{\Delta} 2Hg(\ell) + O_2(g)$$

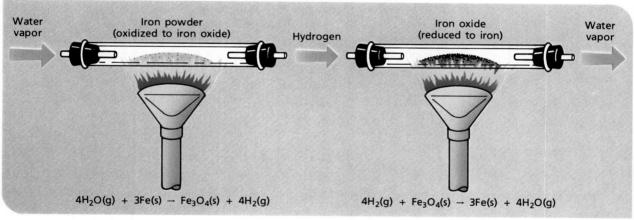

Water vapor — Iron powder (oxidized to iron oxide) — Hydrogen — Iron oxide (reduced to iron) — Water vapor

$$4H_2O(g) + 3Fe(s) \rightarrow Fe_3O_4(s) + 4H_2(g)$$

$$4H_2(g) + Fe_3O_4(s) \rightarrow 3Fe(s) + 4H_2O(g)$$

Figure 19-1 A reversible reaction. Water vapor (steam) passed over the hot iron is reduced to hydrogen, and the iron is oxidized to iron(II) and iron(III) oxide. Hydrogen passed over the hot iron oxide is oxidized to water vapor, and the iron oxide is reduced to iron.

Mercury and oxygen combine to form mercury(II) oxide when heated gently.

$$2Hg(\ell) + O_2(g) \xrightarrow{\Delta} 2HgO(s)$$

Suppose mercury(II) oxide is heated in a closed container from which neither the mercury nor the oxygen can escape. Once decomposition has begun, the mercury and oxygen released can recombine to form mercury(II) oxide again. Thus, both reactions can proceed at the same time. Under these conditions, the rate of the composition reaction will eventually equal that of the decomposition reaction. Mercury and oxygen will combine to form mercury(II) oxide at the same rate as mercury(II) oxide decomposes into mercury and oxygen. The amounts of mercury(II) oxide, mercury, and oxygen can be expected to remain constant as long as these conditions persist. At this point, a state of equilibrium has been reached between the two chemical reactions. Both reactions continue, but there is no net change in the composition of the system. The equation for the reaction at equilibrium is written as follows to show the overall reversible reaction:

$$2HgO(s) \rightleftarrows 2Hg(\ell) + O_2(g)$$

Chemical equilibrium *is a state of balance in which the rates of opposing chemical reactions are exactly equal.*

Equilibrium, A Dynamic State

STUDY HINT

Physical processes, such as evaporation and condensation, that can give rise to a state of equilibrium are discussed in Chapter 13.

In systems that are in equilibrium, opposing processes occur at the same time and at the same rate. For example, as you read in Chapter 14, the evaporation of a liquid in a closed vessel and the condensation of its saturated vapor can proceed at equal rates. The resulting equilibrium vapor pressure is a characteristic of the liquid at the prevailing temperature.

If an excess of sugar is placed in water, sugar molecules go into solution. Some of these molecules in turn separate from solution and rejoin the crystals. At saturation, molecules of sugar are crystallizing at the same rate that crystal molecules are dissolving.

The preceding are examples of physical equilibria. The opposing physical processes occur at exactly the same rate. Such an equilibrium, whether chemical or physical, is a dynamic state in which two opposing processes proceed simultaneously at the same rate.

Compounds such as sodium chloride are composed of ions. Their aqueous solutions consist of hydrated solute ions dispersed throughout the water solvent. When an excess of sodium chloride is placed in water, a saturated solution eventually results. A solution equilibrium is established in which the rate of association of ions reforming the crystals equals the rate of dissociation of ions from the crystals. The equilibrium for this reversible dissociation process is shown in the ionic equation

$$Na^+Cl^-(s) \rightleftarrows Na^+(aq) + Cl^-(aq)$$

In a chemical equation, formulas written to the left of the arrow are assumed to be the reactants, and the reaction proceeds from left to right.

Many chemical reactions are reversible under ordinary conditions of temperature and concentration. They may reach a state of equilibrium unless at least one of the substances involved escapes or is removed. In some cases, however, the forward reaction is nearly completed before the reverse reaction rate becomes high enough to equal it and establish equilibrium. Here, the products of the forward reaction ($\rightarrow$) are favored, meaning that at equilibrium, there is a higher concentration or amount of products than of reactants. The favored reaction that produces this situation is referred to as a "reaction to the right" because the convention for writing chemical reactions is that *left-to-right* is forward and *right-to-left* is reverse. An example of such a system is

$$HBr + H_2O \leftrightarrows H_3O^+ + Br^-$$

Note the inequality of the two arrow lengths.

In still other cases, the forward reaction is barely under way when the rate of the reverse reaction becomes equal to that of the forward reaction, and equilibrium is established. In these cases, the products of the reverse reaction (←) are favored and the original reactants are formed. That is, at equilibrium, there is a higher concentration of reactants than of products. The favored reaction that produces this situation is referred to as the "reaction to the left." An example of such a system is

$$H_2CO_3(aq) + H_2O \leftrightharpoons H_3O^+ + HCO_3^-$$

In still other cases, both forward and reverse reactions occur to nearly the same extent before chemical equilibrium is established. Neither reaction is favored and considerable concentrations of both reactants and products are present at equilibrium. An example is

$$H_3PO_4(aq) + H_2O \leftrightharpoons H_3O^+ + H_2PO_4^-$$

Chemical reactions are ordinarily used to convert available reactants into more-desirable products. Chemists try to produce as much of these products as possible from the reactants used. Chemical equilibrium may seriously limit the possibilities of a seemingly useful reaction. In dealing with equilibrium systems, it is important to remember the conditions that influence reaction rates: *(1)* the nature of reactants; *(2)* the surface area; *(3)* the concentration of reactants; *(4)* the temperature; and *(5)* the presence of a catalyst. Any factor that disturbs the rate of the forward or the reverse reaction upsets the equilibrium. In Section 19-2, you will see how a reaction at equilibrium responds to such stresses. But first, it is important to understand the quantitative nature of chemical equilibrium.

The only one we cannot change in the lab is *(1)* the nature of the reactants.

STUDY HINT
The factors that influence reaction rate are discussed in Section 18.4.

The Equilibrium Constant

Many chemical reactions are reversible and reach a state of equilibrium before the reactants are completely changed into products. Once at equilibrium, the concentrations of products and reactants remain constant.

Suppose two substances, A and B, react to form products C and D. In turn C and D react to produce A and B. Under appropriate conditions, equilibrium occurs for this reversible reaction. The equation for this hypothetical equilibrium reaction is

$$A + B \rightleftarrows C + D$$

Initially, the concentrations of C and D are zero, and those of A and B are maximum. With time, as shown in Figure 19-2, the rate of the forward reaction decreases as A and B are used up. Meanwhile, the rate of the reverse reaction increases as C and D are formed. As these two reaction rates become equal, equilibrium is established. The individual concentrations of A, B, C, and D undergo no further change if conditions remain the same.

At equilibrium, the ratio of the mathematical product [C] × [D] to the mathematical product [A] × [B] for this reaction has a definite value at a given temperature. It is known as the equilibrium constant of the reaction and is designated by the letter K. Thus,

$$K = \frac{[C] \times [D]}{[A] \times [B]}$$

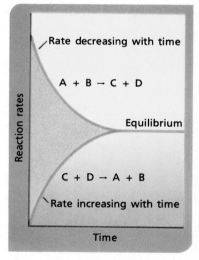

Figure 19-2 Reaction rates for the hypothetical equilibrium reaction system $A + B \rightleftarrows C + D$. The rate of the forward reaction is represented by curve A + B. Curve C + D represents the rate of the reverse reaction. At equilibrium, the two rates are equal.

Observe that the concentrations of substances on the right side of the chemical equation are given in the numerator. These substances are the products of the forward reaction. The concentration of substances on the left side of the chemical equation found are in the denominator. These substances are the reactants of the forward reaction. Concentrations of reactants and products are given in moles per liter. The constant K is independent of the initial concentrations. It is, however, dependent on the temperature of the system.

The value of K for a given equilibrium reaction at a given temperature is important to chemistry. It shows the extent to which the reactants are converted into the products of the reaction. If K is equal to 1, the products of the concentrations in the numerator and denominator have the same value, and therefore, at equilibrium, these are roughly equal amounts of reactants and products. If the value of K is very small, the forward reaction occurs only very slightly before equilibrium is established, and the reactants are favored. A large value of K indicates an equilibrium in which the original reactants are largely converted to products. The numerical value of K for a particular equilibrium system is obtained experimentally. The chemist must analyze the equilibrium mixture and determine the concentrations of all substances.

A value of $K = 10$ is considered very large.

Suppose the balanced equation for an equilibrium reaction has the general form

$$nA + mB + ... \rightleftarrows xC + yD + ...$$

The equilibrium constant relationship at equilibrium (for this generalized reaction system) becomes

$$K = \frac{[C]^x[D]^y...}{[A]^n[B]^m...}$$

In general, then, *the **equilibrium constant** K is the ratio of the mathematical product of the concentrations of substances formed at equilibrium to the mathematical product of the concentrations of reacting substances, each concentration being raised to the power that is the coefficient of that substance in the chemical equation. The equation for K is sometimes referred to as the **chemical equilibrium expression or law.***

To illustrate, suppose a reaction system at equilibrium is shown by the equation

$$3A + B \rightleftarrows 2C + 3D$$

The equilibrium constant K is given by the chemical equilibrium expression

$$K = \frac{[C]^2[D]^3}{[A]^3[B]}$$

Note that the exponents in the expression correspond to the coefficients in the chemical equation.

Now consider an actual example. The reaction between H_2 and I_2 vapor in a closed flask at an elevated temperature is easy to follow by simply observing the rate at which the violet color of the iodine vapor diminishes (Figure 19-3). Suppose this reaction runs to completion with respect to iodine. If so, the color must disappear entirely, since the product, hydrogen iodide, is a colorless gas.

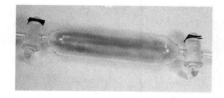

The color does not disappear entirely, however, because the reaction is reversible. Hydrogen iodide decomposes to re-form hydrogen and iodine. The rate of this reverse reaction increases as the concentration of hydrogen iodide builds up. The concentrations of hydrogen and iodine decrease, meanwhile, as they are used up. The rate of the forward reaction decreases accordingly.

As the rates of the opposing reactions become equal, an equilibrium is established. A constant intensity of the color violet indicates that equilibrium exists between hydrogen, iodine, and hydrogen iodide. The net chemical equation for the reaction system at equilibrium is

$$H_2(g) + I_2(g) \rightleftarrows 2HI(g)$$

Thus, the chemical equilibrium expression is

$$K = \frac{[HI]^2}{[H_2][I_2]}$$

Chemists have carefully measured the concentrations of H_2, I_2, and HI in equilibrium mixtures at various temperatures. In some experiments, the flasks were filled with hydrogen iodide at known pressure. The flasks were held at fixed temperatures until equilibrium was established. In other experiments, hydrogen and iodine were the original substances.

Experimental data, together with the calculated values for K, are listed in Table 19-1. Experiments 1 and 2 began with hydrogen iodide. Experiments 3 and 4 began with hydrogen and iodine. Note the close agreement obtained for the numerical values of the equilibrium constant in each case.

Figure 19-3 The production of hydrogen iodide gas from gaseous hydrogen and iodine. The violet color of iodine gas *(left)* becomes fainter *(center)* as the reaction consumes the iodine. The violet will not disappear but will reach a constant intensity when the reaction reaches equilibrium *(right)*.

TABLE 19-1 TYPICAL EQUILIBRIUM CONCENTRATION OF H_2, I_2, AND HI IN MOLE/LITER AT 425°C

Experiment	$[H_2]$	$[I_2]$	$[HI]$	$K = \dfrac{[HI]^2}{[H_2][I_2]}$
(1)	0.4953×10^{-3}	0.4953×10^{-3}	3.655×10^{-3}	54.46
(2)	1.141×10^{-3}	1.141×10^{-3}	8.410×10^{-3}	54.33
(3)	3.560×10^{-3}	1.250×10^{-3}	15.59×10^{-3}	54.62
(4)	2.252×10^{-3}	2.336×10^{-3}	16.85×10^{-3}	53.97

The equilibrium constant K for this equilibrium reaction system at 425°C has the average value of 54.34. This value for K should hold for any system of H_2, I_2, and HI at equilibrium at this temperature. If the calculation for K yields a different result, there must be a reason. Either the H_2, I_2, and HI system has not reached equilibrium or the system is not at 425°C.

The balanced chemical equation for an equilibrium system yields the expression for the equilibrium constant. The data in Table 19-1 show that the validity of this expression is confirmed when the actual values of the equilibrium concentrations of reactants and product are determined experimentally. The values of K are calculated from these concentrations. No information concerning the kinetics of the reacting systems is required.

Once the value of the equilibrium constant is known, the equilibrium constant expression can be used to calculate concentrations of reactants or products at equilibrium. Suppose a system at equilibrium at 425°C is found to contain 1.5 mol/L each of H_2 and I_2. To find the concentration of HI in this system the chemical equilibrium expression can be rearranged

Note that the numerical value of K depends on the coefficients used to balance the chemical equation. Doubling the coefficients squares the numerical value of K.

$$K = \frac{[H_2][I_2]}{[HI]}$$

$$[HI] = \frac{[H_2][I_2]}{K}$$

and then solved, using the known K value.

$$[HI] = \frac{1.5 \times 1.5}{54.35}$$
$$= 0.41 \text{ mol/L}$$

Sample Problem 19.1

An equilibrium mixture of H_2, I_2, and HI gases at 425°C is determined to consist of 4.5647×10^{-3} mol/L of H_2, 0.7378×10^{-3} mol/L of I_2 and 13.544×10^{-3} mol/L of HI. What is the equilibrium constant for the system at this temperature?

Solution

Step 1. Analyze   Given: $[H_2] = 4.5647 \times 10^{-3}$ mol/L
$[I_2] = 7.378 \times 10^{-4}$ mol/L
$[HI] = 1.354\ 4 \times 10^{-2}$ mol/L

Unknown: K

Step 2. Plan $[H_2], [I_2], [HI] \rightarrow K$

The balanced chemical equation is $H_2(g) + I_2(g) \rightleftarrows 2HI(g)$

The chemical equilibrium expression is $K = \dfrac{[HI]^2}{[H_2][I_2]}$

Step 3. Compute $$K = \frac{(1.354\ 4 \times 10^{-2})^2}{(4.5647 \times 10^{-3})(7.378 \times 10^{-4})} = 54.47$$

Step 4. Evaluate   The answer has the correct number of significant digits and is close to an estimated value of 56, $\dfrac{(14)^2}{(5)(0.7)}$

Practice Problems

1. An equilibrium mixture at 425°C is determined to consist of 1.83×10^{-3} mol/L of H_2, 3.13×10^{-3} mol/L of I_2, and 17.7×10^{-2} mol/L of HI. Calculate the equilibrium constant K.   *(Ans.)* 54.7

2. An equilibrium mixture at 425°C consists of 4.79×10^{-4} mol/L of H_2, 4.79×10^{-4} mol/L or I_2, and 3.53×10^{-3} mol/L of HI. Determine the equilibrium constant.   *(Ans.)* 54.3

Section Review

1. What is meant by "chemical equilibrium"?
2. What is an equilibrium constant?
3. How does the value of an equilibrium constant relate to the relative quantities of reactants and products at equilibrium?
4. What is meant by a chemical equilibrium expression?
5. Write the chemical equilibrium expression for the reaction $4HCl(g) + O_2(s) \rightleftarrows 2Cl_2(g) + 2H_2O(g)$
6. At equilibrium at 2500K, $[HCl] = 6.25$ mol/L and $[H_2] = [Cl_2] = 0.45$ mol/L for the reaction $H_2(g) + Cl_2(g) \rightleftarrows 2HCl(g)$. Find the value of K.

19.2 Shifting Equilibrium

In systems that have attained chemical equilibrium, opposing reactions are proceeding at equal rates. Any change that alters the rates of these reactions disturbs the original equilibrium. The system then seeks a new equilibrium state. By shifting an equilibrium in the desired direction, chemists can often increase production of important industrial chemicals. In this section, you will read about such shifts in equilibrium.

Le Châtelier's principle (Section 13.3) provides a means of predicting the influence of disturbing factors on equilibrium systems. As you may recall, Le Châtelier's principle states: *If a system at equilibrium is subjected to a stress, the equilibrium is shifted in the direction that relieves the stress.* This principle holds for all kinds of dynamic equilibria, physical as well as chemical. In applying Le Châtelier's principle to chemical equilibrium, three important stresses will be considered.

1. Change in concentration. From collision theory, an increase in the concentration of a reactant causes an increase in collision frequency. A reaction resulting from these collisions should then proceed at a faster rate. Consider the hypothetical reaction

$$A + B \rightleftarrows C + D$$

An increase in the concentration of A will shift the equilibrium to the right. Both A and B will be used up faster, and more of C and D will be formed. The equilibrium will be reestablished with a lower concentration of B. The equilibrium has shifted in such direction as to reduce the stress caused by the increase in concentration of A.

Similarly, an increase in the concentration of B drives the reaction to the right. An increase in either C or D shifts the equilibrium to the left. A decrease in the concentration of C or D has the same effect as an increase in the concentration of A or B, that is, it will shift the equilibrium to the right.

Changes in concentration have no effect on the value of the equilibrium constant. This is because such changes have an equal effect on the numerator and the denominator of the chemical equilibrium expression. Thus, all concentrations give the same numerical ratio for the equilibrium constant when equilibrium is reestablished.

2. Change in pressure. A change in pressure can affect only equilibrium systems in which gases are involved. According to Le Châtelier's principle, if

5. $K = \dfrac{[Cl_2]^2[H_2O]^2}{[HCl]^4[O_2]}$

6. $K = \dfrac{[HCl]^2}{[H_2][Cl_2]} = \dfrac{(6.25)^2}{(0.45)(0.45)} = 190$

By shifting an equilibrium one means that the balance is shifted in favor of either the forward or the reverse reaction.

Technology

Swimming Pool Chemistry

Killing bacteria in swimming pools is the job of the highly effective disinfectant, hypochlorous acid (HOCl). Tiny HOCl molecules are deadly to bacteria because they can easily penetrate cell walls and oxidize proteins. Too much HOCl, however, encourages algae growth and irritates eyes.

HOCl can be produced by bubbling corrosive chlorine gas through water, but this requires special equipment and is impractical for home pools. Solid sodium hypochlorite (NaClO, the active ingredient in household bleach) is a safer way to generate HOCl. The reaction is

$$NaClO(s) + H_2O(\ell) \rightleftharpoons$$
$$HOCl(aq) + NaOH(aq)$$

Commercial pool products often contain calcium hypochlorite, $Ca(ClO)_2$, which reacts similarly:

$$Ca(ClO)_2 + 2H_2O(\ell) \rightleftharpoons$$
$$2HOCl(aq) + Ca(OH)_2$$

These equations show that adding sodium or calcium hypochlorite to a pool initially produces a basic solution, due to the sodium or calcium hydroxide. The HOCl also affects the pH, since it can dissociate to make the solution acidic:

$$HOCl(aq) \rightleftharpoons H^+(aq)$$
$$+ ClO^-(aq)$$

As the pool is used, sunlight causes the ClO^- ion to be lost in other reactions, and the HOCl is consumed in the destruction of bacteria. The

pH changes are therefore unpredictable. The pH in a pool should be checked daily and adjusted by adding HCl to lower it or $NaHSO_4$ to raise it. The ideal pH range is 7.2–7.8.

If moles of gaseous reactants and products are equal, pressure has no effect on equilibrium concentrations.

the pressure on an equilibrium system is increased, the reaction is driven in the direction that relieves the pressure.

The Haber process for catalytic synthesis of ammonia from its elements illustrates the influence of pressure on an equilibrium system.

$$N_2(g) + 3H_2(g) \rightleftharpoons 2NH_3(g)$$

The equation indicates that 4 molecules of the reactant gases form 2 molecules of ammonia gas. Suppose the equilibrium mixture is subjected to an increase in pressure. This pressure can be relieved by the reaction that produces fewer gas molecules and, therefore, a smaller volume. Thus the stress is *lessened* by the formation of ammonia. Equilibrium is shifted toward the right, favoring the production of ammonia. High pressure is desirable in this industrial process. Figure 19-4 illustrates the effect of pressure on this equilibrium system.

An increase in the partial pressure of H_2 or N_2 in the reaction vessel would shift the equilibrium to the right as this increase in pressure effects an increase in concentration of the resultant gas. Similarly, a decrease in the partial pressure of NH_3 gas in the reaction vessel would also shift the

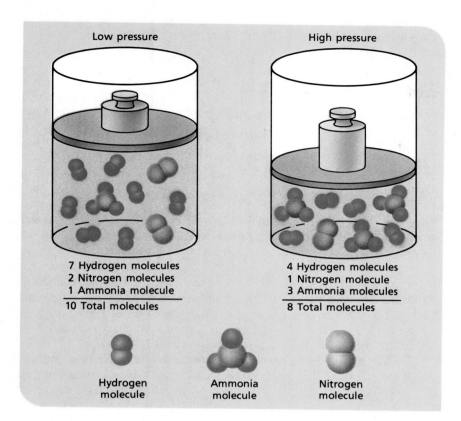

Figure 19-4 Increased pressure results in a higher yield of ammonia because the equilibrium shifts in the direction that produces fewer molecules.

Low pressure

High pressure

7 Hydrogen molecules
2 Nitrogen molecules
1 Ammonia molecule
_____
10 Total molecules

4 Hydrogen molecules
1 Nitrogen molecule
3 Ammonia molecules
_____
8 Total molecules

Hydrogen
molecule

Ammonia
molecule

Nitrogen
molecule

equilibrium to the right. A change in the partial pressure (and thus the concentration) of at least one of the three reactant gases is necessary for the equilibrium position to be affected.

The introduction of an inert gas, such as helium into the reaction vessel for the synthesis of ammonia, would increase the total pressure in the vessel. However, both the partial pressures and the concentrations of the reactant gases present would not be altered by this kind of pressure change. Therefore, increasing pressure by adding a gas that is not a reactant cannot affect the equilibrium position of the reaction system.

In the Haber process, the ammonia produced is continuously removed by condensation to liquid. This condensation removes most of the product from the gas phase in which the reaction occurs. The change in concentration also tends to shift the equilibrium to the right. The Haber process is illustrated in Figure 19-5.

Many chemical processes involve heterogeneous reactions, reactions in which the reactants and products are in different phases. The *concentrations* in equilibrium systems of pure substances in solid and liquid phases are not changed by adding or removing quantities of the substance. This is because concentration is density-dependent, and the density of these phases is constant, regardless of the total amounts present. *A pure substance in a condensed phase, solid or liquid, can thus be removed from the expression for the equilibrium constant.* This is done by substituting the number *1* for its concentration, which is assumed to remain unchanged in the equilibrium system.

An inert gas does NOT alter the equilibrium.

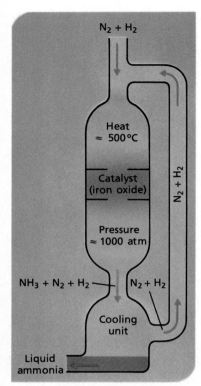

Figure 19-5 The Haber process *(top)* for the production of ammonia and a bottle of ammonia *(bottom)*, a household cleaning agent. Application of Le Châtelier's principle to the equilibrium system, $N_2(g) + 3H_2 \rightarrow 2NH_3(g) + 92$ kJ, suggest that high pressure and low temperature favors the yield of ammonia. Why, then is a moderately high temperature used in this process?

Consider the equilibrium system given by the equation

$$CaCO_3(s) \rightleftarrows CaO(s) + CO_2(g)$$

Carbon dioxide is the only substance in the system subject to changes in concentration. Since it is a gas, the forward (decomposition) reaction is favored by a low pressure. The expression for the equilibrium constant is

$$K = \frac{[CaO][CO_2]}{[CaCO_3]} = \frac{[1][CO_2]}{[1]} = [CO_2]$$

In the reaction

$$CO(g) + H_2O(g) \rightleftarrows CO_2(g) + H_2(g)$$

there are equal numbers of molecules of gaseous reactants and gaseous products. Pressure change cannot produce a shift in the equilibrium of this system. Pressure change thus has no effect on the reaction.

Obviously, an increase in pressure on confined gases has the same effect as an increase in the concentrations of these gases. Recall that changes in concentration have no effect on the value of the equilibrium constant. Changes in pressure likewise do not affect the value of the equilibrium constant. They can shift the equilibrium position, but affect the numerator and denominator of the mass-action expression equally.

3. Change in temperature. Chemical reactions are either exothermic or endothermic. Reversible reactions are exothermic in one direction and endothermic in the other. The effect of changing the temperature of an equilibrium mixture depends on which of the opposing reactions is endothermic.

According to Le Châtelier's principle, addition of heat shifts the equilibrium so that heat is absorbed. This favors the endothermic reaction. The removal of heat favors the exothermic reaction. A rise in temperature increases the rate of any reaction. In an equilibrium system, however, the rates of the opposing reactions are raised unequally. Thus, *the value of the equilibrium constant for a given system is affected by the temperature.*

The synthesis of ammonia is exothermic.

$$N_2(g) + 3H_2(g) \leftrightarrows 2NH_3(g) + 92 \text{ kJ}$$

A high temperature is not desirable because it favors the decomposition of ammonia, the endothermic reaction. At low temperatures, however, the forward reaction is too slow to be commercially useful. The temperature used represents a compromise between kinetic and equilibrium requirements. It is high enough that equilibrium is established rapidly but low enough that the equilibrium concentration of ammonia is significant. Moderate temperature (about 500°C) and very high pressure (700–1000 atmospheres) produce a satisfactory yield of ammonia.

The reactions of the system are also accelerated by a suitable catalyst. Note, however, that catalysts have no effect on relative equilibrium amounts, but only on the rates at which equilibrium is reached. This is because catalysts increase the rates of forward and reverse reactions in a system to an equal extent. They therefore do not affect K.

Data on several equilibrium systems are listed in Table 19-2. The table also gives numerical values of the equilibrium constants at various temperatures. A very small K value means that the equilibrium mixture consists mainly of the substances on the left of the equation. If the value of K is large, the equilibrium mixture consists mainly of the substances on the right of the equation.

An increase in temperature always favors the endothermic reaction.

TABLE 19-2   EQUILIBRIUM CONSTANTS

Equilibrium System	Temperature (°C)	Value of K
$N_2(g) + 3H_2(g) \rightleftarrows 2NH_3(g)$	350	2.66×10^{-2}
$N_2(g) + 3H_2(g) \rightleftarrows 2NH_3(g)$	450	6.59×10^{-3}
$N_2(g) + 3H_2(g) \rightleftarrows 2NH_3(g)$	727	2.37×10^{-3}
$H_2(g) + I_2(g) \rightleftarrows 2HI(g)$	350	66.9
$H_2(g) + I_2(g) \rightleftarrows 2HI(g)$	425	54.4
$H_2(g) + I_2(g) \rightleftarrows 2HI(g)$	1123	45.9
$2H_2(g) + S_2(g) \rightleftarrows 2H_2S(g)$	477	9.39×10^{-5}
$H_2(g) + CO_2(g) \leftrightharpoons H_2O(g) + CO(g)$	1727	4.40
$2H_2O(g) \leftrightharpoons 2H_2(g) = O_2(g)$	1727	5.31×10^{-10}
$2CO(g) + O_2 \rightleftarrows 2CO_2(g)$	727	2.24×10^{22}
$C(s) + CO_2(g) \times 2CO(g)$	1123	14.1
$Cu(s) + 2Ag^+(aq) \leftrightharpoons Cu^{2+}(aq) + 2Ag(s)$	25	2×10^{15}
$I_2(s) \leftrightharpoons 2I(g)$	727	3.76×10^{-5}
$2O_3(g) \rightleftarrows 3O_2(g)$	1727	2.54×10^{12}
$N_2(g) \rightleftarrows 2N(g)$	1000	1.31×10^{-31}

Reactions That Run to Completion

Many reactions are easily reversible under suitable conditions. A state of equilibrium may be established unless one or more of the products escapes or is removed. An equilibrium reaction may be driven in the preferred direction by application of Le Châtelier's principle.

Some reactants appear to be converted irreversibly into products if reactions proceed, on their own, in the forward direction. For example, potassium chlorate decomposition is essentially irreversible: no one has found a method of recombining potassium chloride and oxygen directly once the potassium chlorate decomposes. Sugar is decomposed into carbon and water by the application of heat. No single-step method of recombining these products is yet known.

Many compounds are formed by the reaction, to completion, of ions in solutions. Consider mixing of solutions of soluble ionic compounds. Two pairings of ions are possible. In some cases, these pairings do not occur. If dilute solutions of sodium chloride and potassium bromide are mixed, for example, no reaction occurs. The resulting solution merely contains a mixture of Na^+, K^+, Cl^-, and Br^- ions. Association of ions occurs, in this case, only if enough water is evaporated to cause crystals to separate from solution. The result is then a mixture of NaCl, KCl, NaBr, and KBr.

With some combinations of ions, however, reactions do occur. Such reactions appear to run to completion in the sense that the ions are almost completely removed from solution. The extent to which reacting ions are removed from solution depends on: *(1)* the solubility of the compound formed and *(2)* the degree of ionization, if the compound is soluble. Thus, a product that escapes as a gas or is precipitated as a solid or is only slightly ionized effectively removes from solution the bulk of the reacting ions that make it up. Consider some specific examples of situations in which such ionic reactions run to completion.

1. Formation of a gas. Unstable substances formed as products of ionic reactions decompose spontaneously. An example is carbonic acid, the acid in carbonated water such as club soda, which yields a gas as a decomposition product.

$$H_2CO_3(aq) \rightarrow H_2O(\ell) + CO_2(g)$$

Carbonic acid is produced in the reaction between sodium–hydrogen carbonate and hydrochloric acid, as shown by the equation

$$NaHCO_3(aq) + HCl(aq) \rightarrow NaCl(aq) + H_2CO_3(aq)$$

The aqueous solution of HCl contains H_3O^+ and Cl^- ions. A careful examination of the reaction suggests that the HCO_3^- ion, a Brønsted base, acquires a proton from the H_3O^+ ion, a Brønsted acid; Na^+ and Cl^- ions are merely spectator ions. The following equations may be more appropriate for this chemical reaction.

$$H_3O^+ + HCO_3^- \rightarrow H_2O(\ell) + H_2CO_3(aq)$$
$$H_2CO_3(aq) \rightarrow H_2O(\ell) + CO_2(g)$$

The net ionic equation is, thus, simply

$$H_3O^+ + HCO_3^- \rightarrow 2H_2O(\ell) + CO_2(g)$$

This reaction runs practically to completion because one of the products (CO_2) escapes as a gas.

The reaction between iron(II) sulfide and hydrochloric acid also goes virtually to completion because a gaseous product escapes. The reaction equation is

$$FeS(s) + 2HCl(aq) \rightarrow FeCl_2(aq) + H_2S(g)$$

The net ionic equation is

$$FeS(s) + 2H_3O^+(aq) \rightarrow Fe^{2+}(aq) + H_2S(g) + 2H_2O(\ell)$$

The hydrogen sulfide formed is only moderately soluble and is largely given off as a gas. The iron(II) chloride shown in the first equation could be obtained on evaporation of the water.

2. Formation of a precipitate. When solutions of sodium chloride and silver nitrate are mixed, a white precipitate of silver chloride immediately forms.

$$Na^+ + Cl^- + Ag^+ + NO_3^- \rightarrow Na^+ + NO_3^- + Ag^+Cl^-(s)$$

If chemically equivalent amounts of the two solutes are used, only Na^+ ions and NO_3^- ions remain in solution in appreciable quantities. All but a meager portion of the Ag^+ ions and Cl^- ions combine and separate from the solution as a precipitate of AgCl. This is because silver chloride is only very sparingly soluble in water. It separates by precipitation from what turns out to be a

Figure 19-6 When a sodium chloride solution is poured into a solution of silver nitrate, silver chloride, a white precipitate, is formed.

saturated solution of its particles. The reaction thus effectively runs to completion because an "insoluble" product is formed.

The only reaction is between the silver ions and chloride ions. Omitting the spectator ions, Na^+ and NO_3^-, the net ionic equation is simply

$$Ag^+(aq) + Cl^-(aq) \rightarrow Ag^+Cl^-(s)$$

Crystalline sodium nitrate could be recovered by evaporation of the water solvent.

3. *Formation of a slightly ionized product.* Neutralization reactions between H_3O^+ ions from aqueous acids and OH^- ions from aqueous bases result in the formation of water molecules, which are only slightly ionized. A reaction between HCl and NaOH illustrates this process. Aqueous HCl supplies H_3O^+ ions and Cl^- ions to the solution, and aqueous NaOH supplies Na^+ ions and OH^- ions.

$$H_3O^+ + Cl^- + Na^+ + OH^- \rightarrow Na^+ + Cl^- + 2H_2O$$

Neglecting the spectator ions, the net ionic equation is simply

$$H_3O^+(aq) + OH^-(aq) \rightarrow 2H_2O(\ell)$$

Because it is only slightly ionized, the water exists almost entirely as covalently bonded molecules. Thus, insofar as they are initially present in equal amounts, hydronium ions and hydroxide ions are almost entirely removed from the solution. Thus, the reaction effectively runs to completion because the product is only slightly ionized.

Common-Ion Effect

Suppose hydrogen chloride gas is bubbled into a saturated solution of sodium chloride. As the hydrogen chloride dissolves in sufficient quantity, sodium chloride separates as a precipitate. Since chloride ions but not sodium ions are common to both solutes, the concentration of chloride ions is initially increased, and that of sodium ions is not. In accordance with Le Châtelier's principle, sodium chloride crystals form, relieving the stress of added chloride. The concentration of sodium ions in the solution is lowered along with the total concentration of chloride ions; the overall amount of sodium chloride in solution is likewise lowered. Thus, increasing the concentration of chloride ions has the effect of decreasing the concentration of sodium ions. *This phenomenon, in which the addition of an ion common to two solutes brings about precipitation or reduced ionization is called the* **common-ion effect.** The initial equilibrium in this example is

$$Na^+Cl^-(s) \rightleftarrows Na^+(aq) + Cl^-(aq)$$

Additions of hydrogen chloride disturb this equilibrium and drive the reaction to the left. When the reaction is forced to the left, sodium chloride precipitates. This reduces the concentration of both the sodium and chloride ions in solution.

The common-ion effect is also observed when one ion species of a weak electrolyte is added in excess to a solution. Acetic acid ($HC_2H_3O_2$) is such an electrolyte. A 0.1-M $HC_2H_3O_2$ solution is only about 1.4% ionized. The ionic equilibrium is shown by the equation

$$HC_2H_3O_2(aq) + H_2O \leftrightarrows H_3O^+ + C_2H_3O_2^-$$

The common-ion effect is an application of LeChâtelier's principle.

Figure 19-7 One product of the reaction between hydrocholoric acid and sodium hydroxide is water.

Small additions of sodium acetate (an ionic salt that is completely dissociated in water solution) to a solution containing acetic acid greatly increase the acetate-ion concentration. The equilibrium shifts in the direction that uses up acetate ions in accordance with Le Châtelier's principle. More molecules of acetic acid are formed and the concentration of hydronium ions is reduced. In general, the addition of a salt with an ion common to the solution of a weak electrolyte reduces the ionization of the electrolyte. A 0.1-M $HC_2H_3O_2$ solution that contains no sodium acetate has a pH of 2.9. A solution 0.1-M concentrations of both acetic acid and sodium acetate has a pH of 4.6, indicating lower $[H^+]$ and thus lowered acetic acid ionization.

Section Review

1. Name three ways in which chemical equilibrium can be disturbed.
2. Describe three situations in which ionic reactions go to completion.
3. Describe the common-ion effect.

19.3 Equilibria of Acids, Bases, and Salts

In Chapters 16 and 17, you read about the important classes of compounds called acids, bases, and salts. At the end of the preceding section of this chapter, you studied the effect of common ions on acid ionization. In this section you will look at the equilibrium-related characteristics of acidic and basic water solutions and at the behavior of salts dissolved in water.

Ionization Constant of a Weak Acid

About 1.4% of the solute molecules in a 0.1-M acetic acid solution are ionized at room temperature. The remaining 98.6% of the $HC_2H_3O_2$ molecules remain un-ionized. Thus, the solution contains three species of particles in equilibrium—$HC_2H_3O_2$ molecules, H_3O^+ ions, and $C_2H_3O_2^-$ ions.

At equilibrium, the rate of the forward reaction

$$HC_2H_3O_2 + H_2O \rightarrow H_3O^+ + C_2H_3O_2^-$$

is equal to the rate of the reverse reaction

$$H_3O^+ + C_2H_3O_2^- \rightarrow H_2O + HC_2H_3O_2$$

The equilibrium equation is

$$HC_2H_3O_2 + H_2O \rightleftarrows H_3O^+ + C_2H_3O_2^-$$

The equilibrium constant for this system expresses the equilibrium ratio of ions to molecules. From the equilibrium equation for the ionization of acetic acid,

$$K = \frac{[H_3O^+][C_2H_3O_2^-]}{[HC_2H_3O_2][H_2O]}$$

At the 0.1-M concentration, water molecules greatly exceed the number of acetic acid molecules. Without introducing a measurable error, one can assume that the mole concentration of H_2O molecules remains constant in such a solution. Thus the product $K[H_2O]$ is constant (since both K and $[H_2O]$ are constant).

$$K[H_2O] = \frac{[H_3O^+][C_2H_3O_2^-]}{[HC_2H_3O_2]}$$

By setting $K[H_2O] = K_a$

$$K_a = \frac{[H_3O^+][C_2H_3O_2^-]}{[HC_2H_3O_2]}$$

The term K_a is called the **acid-ionization constant.**

The equilibrium equation for the typical weak acid HB is

$$HB(aq) + H_2O(\ell) \rightleftarrows H_3O^+(aq) + B^-(aq)$$

From this equation, the expression for K_a can be written in the general form

$$K_a = \frac{[H_3O^+][B^-]}{[HB]}$$

How can the numerical value of the ionization constant K_a for acetic acid at a specific temperature be determined? First, the equilibrium concentrations of H_3O^+ ions, $C_2H_3O_2^-$ ions, and $HC_2H_3O_2$ molecules must be known. The ionization of a molecule of $HC_2H_3O_2$ in water yields one H_3O^+ ion and one $C_2H_3O_2^-$ ion. These concentrations can therefore be found experimentally by measuring the pH of the solution.

Suppose that precise measurements in an experiment show the pH of a 0.1000-M solution of acetic acid to be 2.876 at 25°C. The numerical value of K_a for $HC_2H_3O_2$ at 25°C can be determined as follows:

$$[H_3O^+] = [C_2H_3O_2^-] = 10^{-2.876}\,\frac{\text{mol}}{\text{L}}$$

$$[H_3O^+] = [C_2H_3O_2] = \text{antilog}\,(-2.876) = \text{antilog}\,(0.124 - 3)$$
$$= 1.33 \times 10^{-3}$$

$$[HC_2H_3O_2]_{\text{final}} = 0.1000\,\text{mol/L} - 0.00133\,\text{mol/L}$$
$$= [HC_2H_3O_2]_{\text{initial}} - [H_3O^+] = 0.0987\,\text{mol/L}$$

$$K_a = \frac{[H_3O^+][C_2H_3O_2^-]}{[HC_2H_3O_2]}$$

$$K_a = \frac{(1.33 \times 10^{-3})^2}{9.87 \times 10^{-2}} = 1.79 \times 10^{-5}$$

An increase in temperature causes the equilibrium to shift according to Le Châtelier's principle. Thus K_a has a new value for each temperature. At constant temperature, an increase in the concentration of $C_2H_3O_2^-$ ions, through the addition of $NaC_2H_3O_2$, also disturbs the equilibrium. This disturbance causes a decrease in $[H_3O^+]$ and an increase in $[HC_2H_3O_2]$. Eventually, the equilibrium is reestablished with the same value of K_a. There is a higher concentration of nonionized acetic acid molecules and a lower concentration of H_3O^+ ions, however. Changes in the hydronium-ion concentration and in pH go together. In this example, the reduction in $[H_3O^+]$ means an increase in the pH of the solution. Ionization data and constants for some dilute acetic acid solutions at 25°C are given in Table 19-3 (shown below).

TABLE 19-3   IONIZATION CONSTANT OF ACETIC ACID

Molarity	% Ionized	$[H_3O^+]$	$[HC_2H_3O_2]$	K_a
0.1000	1.33	0.00133	0.09867	1.79×10^{-5}
0.0500	1.89	0.000945	0.04906	1.82×10^{-5}
0.0100	4.17	0.000417	0.009583	1.81×10^{-5}
0.0050	5.86	0.000293	0.004707	1.81×10^{-5}
0.0010	12.60	0.000126	0.000874	1.72×10^{-5}

A solution containing both a weak acid and a salt of the acid can react with either an acid or a base. The pH of the solution remains nearly constant even when small amounts of acids or bases are added. Suppose an acid is added to a solution of the weak acid, acetic acid, and its salt, sodium acetate ($NaC_2H_3O_2$). Acetate ions react with the hydronium ions to form nonionized acetic acid molecules.

$$C_2H_3O_2^-(aq) + H_3O^+(aq) \rightarrow HC_2H_3O_2(aq) + H_2O(\ell)$$

The hydronium-ion concentration and the pH of the solution remain practically unchanged.

Suppose a small amount of a base is added to the original solution. The OH^- ions of the base react with and remove hydronium ions, to form nonionized water molecules. Acetic acid molecules ionize and restore the equilibrium concentration of hydronium ions.

$$HC_2H_3O_2(aq) + H_2O(\ell) \rightarrow H^+(aq) + C_2H_3O_2^-(aq)$$

The pH of the solution again remains practically unchanged.

A solution of a weak base containing a salt of the base behaves in a similar manner. The hydroxide-ion concentration (and the pH) of the solution remain essentially constant with small additions of acids or bases. Suppose a base is added to an aqueous solution of ammonia that also contains ammonium chloride. Ammonium ions remove the added hydroxide ions to form nonionized water molecules.

$$NH_4^+(aq) + OH^-(aq) \rightarrow NH_3(aq) + H_2O$$

If a small amount of an acid is added to the solution instead, hydroxide ions from the solution remove the added hydronium ions to form nonionized water molecules. Ammonia molecules in the solution then ionize and restore the equilibrium concentration of hydronium ions and the pH of the solution.

$$NH_3(aq) + H_2O(\ell) \rightarrow NH_4^+(aq) + OH^-(aq)$$

The common ion salt in each of these solutions acts as a "buffer" against significant changes in pH in the solution. *Such solutions that can resist changes in pH are referred to as* **buffered solutions.**

Buffer action has many important applications in chemistry and physiology (see Figure 19-8). Many medicines are buffered to prevent large and potentially damaging changes in pH. Human blood is naturally buffered to maintain a pH of about 7.3. This is essential because large changes in pH would lead to serious disturbances of normal body functions or even death.

Figure 19-8 When acetic acid is buffered with sodium acetate *(left)*, the addition of a small amount of hydrochloric acid *(right)* does not affect the solution's pH as shown by the unchanging color of litmus paper.

Ionization Constant of Water

Pure water is a very poor conductor of electricity because it is only very slightly ionized according to the equation

$$H_2O(\ell) + H_2O(\ell) \rightleftharpoons H_3O^+(aq) + OH^-(aq)$$

The degree of ionization is slight. Equilibrium is quickly established with a very low concentration of H_3O^+ and OH^- ions.

Conductivity experiments with pure water at 25°C show that the concentrations of H_3O^+ and OH^- ions are both 10^{-7} moles per liter (see Figure 19-10). The expression for the equilibrium constant is

$$K = \frac{[H_3O^+][OH^-]}{[H_2O]^2}$$

A liter of water at 25°C contains

$$\frac{997 \text{ g } H_2O}{18.0 \text{ g/mol}} = 55.4 \text{ mol } H_2O$$

This concentration of water molecules, 55.4 mol/L, remains practically the same in all dilute solutions.

Figure 19-9 Many medicines and personal products are buffered to protect the body from possibly harmful pH changes.

Figure 19-10 Water is a poor conductor as shown by the high reading on the ohmeter *(left)*, because it only ionizes to a very slight degree. Addition of HCl creates a solution with higher conductivity as indicated by the low reading on the ohmeter *(right)*. An ohmeter measures resistivity which is the inverse of conductivity.

Thus, both $[H_2O]^2$ and K in the above equilibrium expression are constants. Their product is the ion-product constant for water. As you learned in Chapter 17, this ion product is the product of the molar concentrations of the H_3O^+ and OH^- ions. The ion-product constant is symbolized by K_w. Thus,

$$K_w = [H_3O^+][OH^-]$$

At 25°C,

$$K_w = 10^{-7} \times 10^{-7} = 10^{-14}$$

The product, K_w, of the molar concentrations of H_3O^+ and OH^- ions has this constant value not only in pure water but in all water solutions at 25°C. An acid solution with a pH of 4 has a $[H_3O^+]$ of 10^{-4} mol/L and thus, has a $[OH^-]$ of 10^{-10} mol/L (since $10^{-4} \times 10^{-10} = 10^{-14}$). A basic solution with a pH of 8 has an $[H_3O^+]$ concentration of 10^{-8} mol/L and thus has a $[OH^-]$ concentration of 10^{-6} mol/L.

Hydrolysis of Salts

When a salt is dissolved in water, the solution might be expected to be neutral. The aqueous solutions of some salts, such as NaCl and KNO_3, are indeed neutral; such solutions have a pH of 7.

Some other salts form aqueous solutions that are either acidic or basic, however. As salts, they are completely dissociated in aqueous solution; but, unlike salts whose aqueous solutions are neutral, their ions react chemically with the water solvent. The ions of a salt that do react with the solvent are said to hydrolyze in water solution. **Hydrolysis** *is a reaction between water and ions of a dissolved salt.*

Salts can be placed in four general categories, depending on their hydrolysis properties.

1. Salts of strong acids and strong bases. Recall that a salt is one product of an aqueous acid-base neutralization reaction. The salt consists of the positive ions (cations) of the base and negative ions (anions) of the acid. For example,

$$HCl(aq) + NaOH(aq) \rightarrow Na^+Cl^-(aq) + H_2O(\ell)$$

HCl(aq) is a strong acid, and NaOH(aq) is a strong base. Equivalent quantities of these two reactants yield a neutral aqueous solution of Na^+ and Cl^- ions; the pH of the solution is 7. Neither the Na^+ cation (the positive ion of the strong base) nor the Cl^- anion (the negative ion of the strong acid) undergoes hydrolysis in water solutions.

Similarly, KNO_3 is the salt of the strong acid HNO_3 and the strong base KOH. Measurements show that the pH of an aqueous KNO_3 solution is always very close to 7. Neither the cation of a strong base nor the anions of a strong acid hydrolyze appreciably in aqueous solutions. (See Figure 19-11a.)

2. Salts of weak acids and strong bases. The aqueous solutions of such salts are basic. Anions of the dissolved salt are hydrolyzed in the water solvent and the pH of the solution is raised, indicating that the hydroxide-ion concentration has increased. Observe that the anions of the salt that undergo hydrolysis are the negative ions forming the weak acid. The cations of the salt (the positive ions from the strong base) do not hydrolyze appreciably. An example of such a salt is sodium acetate. (See Figure 19-11b.)

STUDY HINT

The concept of strong and weak acids and bases is dealt with in Section 16.3. Recall that strong acids unlike weak ones, tend to ionize nearly completely aqueous solutions.

Figure 19-11 The acidity or basicity of solutions of various salts are tested using pH paper: (a) NaCl, (b) NaC₂H₃O₂, (c) NH₄Cl, and (d) NH₄C₂H₃O₂.

3. *Salts of strong acids and weak bases.* The aqueous solutions of such salts are acidic. Cations of the dissolved salt are hydrolyzed in the water solvent, and the pH of the solution is lowered, indicating that the hydronium-ion concentration has increased. Observe that, in this case, the cations of the salt that undergo hydrolysis are the positive ions from the weak base. The anions of the salt (the negative ions form the strong acid) do not hydrolyze appreciably. An example of such a salt is ammonium chloride (NH_4Cl). (See Figure 19-11c.)

4. *Salts of weak acids and weak bases.* In the aqueous solutions of such salts, both ions of the dissolved salt are hydrolyzed extensively and the solution can be either neutral, acidic, or basic, depending on the salt dissolved. If both ions of such a salt in aqueous solution are hydrolyzed equally, the solution remains neutral. Ammonium acetate ($NH_4C_2H_3O_2$) is such a salt. (See Figure 19-11d.) In cases in which both the acid and the base are very weak indeed, the salt may undergo essentially complete decomposition to the products of hydrolysis. For example, when aluminum sulfide is placed in water, both a precipitate and a gas are formed. The reaction is

$$Al_2S_3(s) + 6H_2O(\ell) \rightarrow 2Al(OH)_3(s) + 3H_2S(g)$$

Both products are sparingly soluble in water and are removed from solution.

Hydrolysis often has important effects on the properties of solutions. Sodium carbonate (Na_2CO_3), known as washing soda, is widely used as a cleaning agent because of the basic properties of its water solution. (See Figure 19-12.) Sodium–hydrogen carbonate, known as baking soda, forms a mildly alkaline solution in water and has many practical uses. Through the study of hydrolysis, you can now better understand a fact you were given in Chapter 17: the end point of a neutralization reaction can occur at a pH other than 7.

Now it is time to examine hydrolysis in more detail. First, you will learn more about anion hydrolysis, and then you will study cation hydrolysis.

Anion hydrolysis The cations of the salt of a weak acid and a strong base do not hydrolyze in water. Only anion hydrolysis occurs in an aqueous solution of such a salt. In the Brønsted sense, the anion of the salt is the conjugate base (proton acceptor) of the weak acid. These ions are basic enough to remove protons from some water molecules (proton donors) to form OH^- ions. An equilibrium is established in which the net effect of the anion hydrolysis is an increase in the hydroxide-ion concentrations, $[OH^-]$, of the solution. Using the arbitrary symbol B^- for the anion, this hydrolysis reaction is represented by the general anion hydrolysis equation that follows:

$$B^-(aq) + H_2O(\ell) \rightleftharpoons HB(aq) + OH^-(aq)$$

Figure 19-12 A polarized light photomicrograph of sodium carbonate (washing soda) crystals (9.5x).

Figure 19-13 A polarized light photomicrograph of sodium bicarbonate (baking soda) crystals (130x).

In the reaction to the right, the anion B^- acquires a proton from the water molecule to form the weak acid HB and hydroxide ion OH^-.

The equilibrium constant for an ion hydrolysis reaction is sometimes called **hydrolysis constant,** K_h. From this general anion hydrolysis equation, the expression for K_h is

$$K_h = \frac{[HB][OH^-]}{[B^-]}$$

You can see that a K_h has the same form as the general chemical-equilibrium expression, but without $[H_2O]$, which is essentially constant in dilute solutions. The hydrolysis constant K_h is also expressed by the ratio of the ion-product constant K_w for water to the ionization constant K_a for the weak acid.

$$K_h = \frac{K_w}{K_a}$$

The validity of this expression can be demonstrated in the following way. Since

$$K_w = [H_3O^+][OH^-] \quad \text{and} \quad K_a = \frac{[H_3O^+][B^-]}{[HB]}$$

Then,

$$\frac{K_w}{K_a} = \frac{[H_3O^+][OH^-]}{\dfrac{[H_3O^+][B^-]}{[HB]}} \neq \frac{[HB][OH^-]}{[B^-]} = K_h$$

Suppose sodium carbonate is dissolved in water and the solution is tested with litmus paper. The solution turns red litmus blue. The solution contains a higher concentration of OH^- ions than does pure water, and is basic.

The sodium ions (Na^+) in sodium carbonate do not undergo hydrolysis in aqueous solution, but the carbonate ions (CO_3^{2-}) react as a Brønsted base. A CO_3^{2-} anion acquires a proton from a water molecule to form the slightly ionized hydrogen carbonate ion, HCO_3^-, and the OH^- ion.

$$CO_3^{2-}(aq) + H_2O(\ell) \rightleftarrows HCO_3^-(aq) + OH^-(aq)$$

The OH^- ion concentration increases until equilibrium is established. Consequently, the H_3O^+ ion concentration decreases since the product $[H_3O^+][OH^-]$ remains equal to the ionization constant K_w of water at the temperature of the solution. Thus the pH is *higher* than 7, and the solution is basic. In general, then, salts formed from weak acids and strong bases hydrolyze in water to form basic solutions.

Cation hydrolysis   The anions of the salt of a strong acid and a weak base do not hydrolyze in water. Only cation hydrolysis occurs in an aqueous solution of such a salt.

An aqueous solution of ammonium chloride, NH_4Cl turns blue litmus paper red. This color change indicates that hydrolysis has occurred and the solution contains an excess of H_3O^+ ions. Chloride ions, the anions of the salt, show no noticeable tendency to hydrolyze in aqueous solution. Ammonium ions, the cations of the salt, donate protons to water molecules according to the equation

$$NH_4^+(aq) + H_2O(\ell) \rightleftarrows H_3O^+(aq) + NH_3(aq)$$

Equilibrium is established with an increased H_3O^+ ion concentration. The pH is lower than 7, and the solution is acidic.

The metallic cations of some salts are hydrated in aqueous solution. These hydrated cations may donate protons to water molecules. Because of this cation hydrolysis, the solution becomes acidic. For example, aluminum chloride forms the hydrated cations

$$Al(H_2O)_6^{3+}$$

Copper(II) sulfate in aqueous solution forms the light-blue hydrated cations

$$Cu(H_2O)_4^{2+}$$

These ions react with water to form hydronium ions as follows:

$$Al(H_2O)_6^{3+} + H_2O(\ell) \leftrightharpoons Al(H_2O)_5OH^{2+} (aq) + H_3O^+(aq)$$
$$Cu(H_2O)_4^{2+} + H_2O(\ell) \leftrightharpoons Cu(H_2O)_3OH^+ (aq) + H_3O^+(aq)$$

Cations such as $Cu(H_2O)_3OH^+$ ions may experience a secondary hydrolysis to a slight extent

$$Cu(H_2O)_3OH^+(aq) + H_2O(\ell) \leftrightharpoons Cu(H_2O)_2(OH)_2(s) + H_3O^+(aq)$$

The resulting hydrated copper(II) hydroxide is not very soluble. The scummy appearance of reagent bottles in which copper(II) salt solutions are stored for a long time is caused by this slight secondary hydrolysis of $Cu(H_2O)_3OH^+$ cations. (See Figure 19-14.) Expressions for the hydrolysis constant K_h for cation acids are similar to those of anion bases. In general, salts formed from strong acids and weak bases hydrolyze in water to form acidic solutions.

Section Review

1. What is meant by an acid-ionization constant?
2. How is an acid-ionization equilibrium expression written?
3. What is meant by the term buffered solution?
4. What is meant by the ion-product constant for water? What is the value of this constant?
5. What is hydrolysis? Compare cation and anion hydrolysis.

19.4 Solubility Equilibrium

Chemical-equilibrium considerations can be applied to solubilities of various substances. As you may recall from Chapter 14, solubility refers to the maximum amounts of a substance that will dissolve per unit volume of solution under various conditions. Solubility is generally expressed in units of moles of solute per liter of solution. In this section solubility is expressed in terms of equilibrium expressions, which are then used to predict whether precipitation occurs when solutions of various substances are combined.

Solubility Product

A saturated solution contains the maximum amount of solute possible at a given temperature in equilibrium with an undissolved excess of the

Figure 19-14 When stored for a long time, a solution of copper(II) sulfate develops a scummy appearance. The $Cu(H_2O)_3OH^+$ cations hydrolyze, forming insoluble hydrated copper(II) hydroxide.

1. The equilibrium constant for the dissociation of a weak acid, K_a
2. $K_a = \dfrac{[H_3O^+][B^-]}{[HB]}$
3. Solutions that can resist changes in pH
4. The product of the hydrogen ion and hydroxide ion concentration. $K_w = 1.0 \; !x! \; 10^{-14}$
5. The reaction between water and ions of a dissolved salt. Cation hydrolysis produces an acid solution; anion hydrolysis produces a basic solution.

Nothing is truly "insoluble." Amounts dissolving may only be too small to measure.

substance. A saturated solution is not necessarily a concentrated solution. The concentration may be high or low, depending on the solubility of the solute.

A general rule is often used to express solubilities qualitatively. By this rule, a substance is said to be *soluble* if the solubility is *greater than* 1 g per 100 g of water. It is said to be *insoluble* if the solubility is *less than* 0.1 g per 100 g of water. Substances whose solubilities fall between these limits are described as *slightly soluble.*

Substances usually referred to as *insoluble* are, in fact, very sparingly soluble. An extremely small quantity of such a solute saturates the solution. Equilibrium is established with the undissolved excess remaining in contact with the solution. Equilibria between sparingly soluble solids and their saturated solutions are especially important in analytical chemistry.

You have observed that silver chloride precipitates when Ag^+ and Cl^- ions are placed in the same solution. Silver chloride is so sparingly soluble in water that it is described as insoluble. Its solution reaches saturation at a very low concentration of its ions. All Ag^+ and Cl^- ions in excess of this concentration eventually separate as an AgCl precipitate.

The equilibrium principles developed in this chapter apply to all saturated solutions of sparingly soluble salts. Consider the equilibrium system in a saturated solution of silver chloride containing an excess of the solid salt. The equilibrium equation is

$$\mathrm{AgCl(s)} \rightleftarrows \mathrm{Ag^+(aq)} + \mathrm{Cl^-(aq)}$$

The equilibrium constant is expressed as

$$K = \frac{[\mathrm{Ag^+}][\mathrm{Cl^-}]}{[\mathrm{AgCl}]}$$

Because the concentration of a pure substance in the solid or liquid phase remains constant, adding more solid AgCl to this equilibrium system does not change the concentration of the undissolved AgCl present (see Figure 19-15). Thus [AgCl] in the equation is set equal to 1, giving what is called the solubility-product constant K_{sp}. By combination of the two constants, this equation becomes

$$K[\mathrm{AgCl}] = [\mathrm{Ag^+}][\mathrm{Cl^-}]$$

$$K_{sp} = [\mathrm{Ag^+}][\mathrm{Cl^-}]$$

This equation is the solubility-equilibrium expression for the reaction. It expresses the fact that the solubility-product constant K_{sp} of AgCl is the product of the molar concentrations of its ions in a saturated solution.

Calcium fluoride is another sparingly soluble salt. The equilibrium in a saturated CaF_2 solution is given by the following equation:

$$\mathrm{CaF_2(s)} \rightleftarrows \mathrm{Ca^{2+}(aq)} + \mathrm{2F^-(aq)}$$

The solubility-product constant is given by

$$K_{sp} = [\mathrm{Ca^{2+}}][\mathrm{F^-}]^2$$

Notice that this constant is the product of the molar concentration of Ca^{2+} ions and the molar concentration of F^- ions squared as required by the general chemical equilibrium expression.

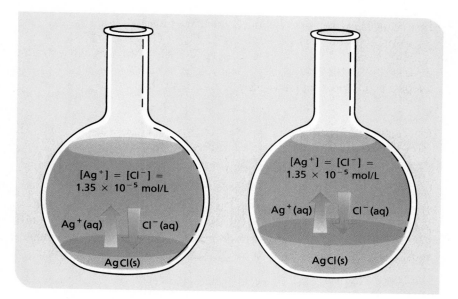

Similar equations apply to any sparingly soluble salt having the general formula M_aX_b. The equilibrium system in a saturated solution is shown by

$$M^aX^b(s) \leftrightarrows aM^{b+} + bX^{a-}$$

The solubility-product constant is expressed by

$$K_{sp} = [M^{b+}]^a[X^{a-}]^b$$

The **solubility-product constant** *of a substance is the product of the molar concentrations of its ions in a saturated solution, each raised to the power that is the coefficient of that ion in the chemical equations.*

From solubility data listed in Appendix Table A-16, 1.94×10^{-4} g of AgCl saturates 100 g of water at 20°C. One mole of AgCl has a mass of 143.4 g. The saturation concentration (solubility) of AgCl can therefore be expressed in moles per liter of water, which is roughly equal to moles per liter of solution:

$$\frac{1.94 \times 10^{-4} \text{ g AgCl}}{100. \text{ g H}_2\text{O}} \times \frac{1 \text{ g H}_2\text{O}}{1 \text{ mL H}_2\text{O}} \times \frac{10^3 \text{ mL}}{1 \text{ L}} \times \frac{1 \text{ mol AgCl}}{143.4 \text{ g AgCl}}$$

$$= 1.35 \times 10^{-5} \text{ mol/L}$$

The equilibrium equation is

$$\text{AgCl(aq)} \rightleftarrows \text{Ag}^+(\text{aq}) + \text{Cl}^-(\text{aq})$$

Silver chloride dissociates in solution, contributing equal numbers of aqueous Ag^+ and Cl^- ions. The ion concentrations in the saturated solution, are therefore 1.35×10^{-5} mol/L

$$[Ag^+] = 1.35 \times 10^{-5}$$
$$[Cl^-] = 1.35 \times 10^{-5}$$

and

$$K_{sp} = [Ag^+][Cl^-]$$
$$K_{sp} = (1.35 \times 10^{-5})(1.35 \times 10^{-5})$$
$$K_{sp} = (1.35 \times 10^{-5})^2$$
$$K_{sp} = 1.82 \times 10^{-10}$$

This result is the solubility-product constant of AgCl at 20°C.

From Appendix Table A-16, the solubility of CaF_2 at 25°C is 1.7×10^{-3} g/100 g of water. Expressed in moles per liter as before, this concentration becomes 2.2×10^{-4} mol/L.

The equilibrium equation for a saturated solution of CaF_2 is

$$CaF_2(s) \rightleftarrows Ca^{2+}(aq) + 2F^-(aq)$$

CaF_2 dissociates in solution to yield twice as many F^- ions as Ca^{2+} ions. The ion concentrations in the saturated solution are

$$[Ca^{2+}] = 2.2 \times 10^{-4}$$
$$[F^-] = 2(2.2 \times 10^{-4}) = 4.4 \times 10^{-4}$$

Note carefully that at equilibrium at 25°C $[Ca^2]$ equals the solubility of 2.2×10^{-4} mol/L, but $[F^-]$ equals twice the solubility, or 4.4×10^{-4} mol/L. The number of moles of positive and negative ions per mole of compound must always be accounted for in using K_{sp}s and solubilities. Therefore

$$K_{sp} = [Ca^{2+}][F^-]^2$$
$$K_{sp} = (2.2 \times 10^{-4})(4.4 \times 10^{-4})^2$$
$$K_{sp} = (2.2 \times 10^{-4})(1.9 \times 10^{-7})$$
$$K_{sp} = 4.2 \times 10^{-11}$$

Thus, the solubility-product constant of CaF_2 is 4.2×10^{-11} at 25°C.

If the ion product $[Ca^{2+}][F^-]^2$ is less that the value for K_{sp} at a particular temperature, the solution is unsaturated. If the ion product is greater than the value for K_{sp}, CaF_2 precipitates. This precipitation reduces the concentrations of Ca^{2+} and F^- ions until equilibrium is established.

It is difficult to measure very small concentrations of a solute with precision. For this reason, solubility data from different sources may result in slightly different values of K_{sp} for a substance. Thus, calculations of K_{sp} ordinarily should be limited to two significant figures. Representative values of K_{sp} at 25°C for some sparingly soluble compounds are listed in Table 19-4.

Sample Problem 19.2

Calculate the solubility-product constant K_{sp} for copper(I) chloride (CuCl) given that the solubility of this compound at 25°C is 1.08×10^{-2} g/100 g H_2O.

Solution

Step 1. Analyze   Given: solubility of CuCl = 1.08×10^{-2} g CuCl/100. g H_2O
Unknown: K_{sp}

Step 2. Plan   solubility in g/100. g H_2O → solubility in mol/L
solubility in mol/L, solubility equilibrium
expression → K_{sp}

$$\frac{g\ CuCl}{100.\ g\ H_2O} \times \frac{1\ g\ H_2O}{1\ mL\ H_2O} \times \frac{10^3\ mL}{1\ L} \times \frac{1\ mol\ CuCl}{g\ CuCl} = \text{solubility in mol/L}$$

$$CuCl(s) \rightleftarrows Cu^+(aq) + Cl^-(aq)$$
$$[Cu^+] = [Cl^-] = \text{solubility in mol/L}$$
$$K_{sp} = [Cu^+][Cl^-]$$

Step 3. Compute   Since the molar mass of CuCl is 99.0 g/mol,

$$\text{solubility in mol/L} = \frac{1.08 \times 10^{-2}\text{ g}}{100\text{ g}} \times \frac{10^3\text{ g}}{\text{L}} \times \frac{\text{mol CuCl}}{99.0\text{ g CuCl}} = 1.09 \times 10^{-3}\text{ mol/L}$$

$$[Cu^+]\quad[Cl^-] = 1.09 \times 10^{-3}\text{ mol/L}$$

$$K_{sp} = (1.09 \times 10^{-3})\,(1.09 \times 10^{-3}) = 1.19 \times 10^{-6}$$

Step 4. Evaluate   The answer contains the proper number of significant digits, is approximately equal to an estimated value of 1.2×10^{-6}, $(1.1 \times 10^{-3}) \times (1.1 \times 10^{-3})$, and is also close the actual K_{sp} value given in Table 19-4.

Practice Problem

Calculate the solubility-product constant, K_{sp}, of lead(II) chloride ($PbCl_2$), which has a solubility of 1.0 g/100 g H_2O at 25°C.   *(Ans.)* 1.85×10^{-4}

Calculating Solubilities

Solubility-product constants are computed from very careful measurements of solubilities and other solution properties. Once known, the solubility product is very helpful in determining the solubility of a sparingly soluble salt.

Suppose you wish to know how much barium carbonate ($BaCO_3$) can be dissolved in one liter of water solution at 25°C. From Table 19-4, K_{sp} for $BaCO_3$ has the numerical value 1.2×10^{-8}. The solubility equation is written as follows:

$$BaCO_3(s) \rightleftharpoons Ba^{2+}(aq) + CO_3{}^{2-}(aq)$$

Knowing the value for K_{sp} gives

$$K_{sp} = [Ba^{2+}][CO_3{}^{2-}] = 1.2 \times 10^{-8}$$

Therefore, $BaCO_3$ dissolves until the product of the molar concentrations of Ba^{2+} and $CO_3{}^{2-}$ equals 1.2×10^{-8}.

The solubility-equilibrium equation shows that Ba^{2+} ions and CO_3^{2-} ions enter the solution in equal numbers as the salt dissolves. Thus,

$$[Ba^{2+}] = [CO_3{}^{2-}] = \text{solubility of } BaCO_3$$
$$K_{sp} = 1.2 \times 10^{-8} = (\text{solubility of } BaCO_3)^2$$
$$\text{solubility of } BaCO_3 = \sqrt{1.2 \times 10^{-8}}$$
$$\text{solubility of } BaCO_3 = 1.1 \times 10^{-4}\text{ mol/L}$$

The solubility of $BaCO_3$ is 1.1×10^{-4} mol/L. Thus, the solution concentration is 1.1×10^{-4} M for Ba^{2+} ions and for $CO_3{}^{2-}$ ions.

Sample Problem 19.3

Calculate the solubility of silver acetate ($AgC_2H_3O_2$) in mol/L, given the K_{sp} value for this compound listed in Table 19-4.

TABLE 19-4 SOLUBILITY-PRODUCT CONSTANTS K_{sp} AT 25°C

Salt	Ion Product	K_{sp}	Salt	Ion Product	K_{sp}
$AgC_2H_3O_2$	$[Ag^+][C_2H_3O_2^-]$	2.5×10^{-3}	Cu_2S	$[Cu^+]^2[S^{2-}]$	2.5×10^{-48}
$AgBr$	$[Ag^+][Br^-]$	5.0×10^{-13}	CuS	$[Cu^{2+}][S^{2-}]$	6.3×10^{-36}
Ag_2CO_3	$[Ag^+]^2[CO_3^{2-}]$	8.1×10^{-12}	FeS	$[Fe^{2+}][S^{2-}]$	6.3×10^{-18}
$AgCl$	$[Ag^+][Cl^-]$	1.8×10^{-10}	Fe_2S_3	$[Fe^{3+}]^2[S^{2-}]^3$	1.4×10^{-85}
AgI	$[Ag^+][I^-]$	8.3×10^{-17}	$Fe(OH)_3$	$[Fe^{3+}][OH^-]^3$	1.5×10^{-36}
Ag_2S	$[Ag^+]^2[S^{2-}]$	1.6×10^{-49}	HgS	$[Hg^{2+}][S^{2-}]$	1.6×10^{-52}
$Al(OH)_3$	$[Al^{3+}][OH^-]^3$	1.3×10^{-33}	$MgCO_3$	$[Mg^{2+}][CO_3^{2-}]$	3.5×10^{-8}
$BaCO_3$	$[Ba^{2+}][CO_3^{2-}]$	1.2×10^{-8}	$Mg(OH)_2$	$[Mg^{2+}][OH^-]^2$	1.5×10^{-11}
$BaSO_4$	$[Ba^{2+}][SO_4^{2-}]$	1.1×10^{-10}	MnS	$[Mn^{2+}][S^{2-}]$	2.5×10^{-13}
CdS	$[Cd^{2+}][S^{2-}]$	8.0×10^{-27}	$PbCl_2$	$[Pb^{2+}][Cl^-]^2$	1.9×10^{-4}
$CaCO_3$	$[Ca^{2+}][CO_3^{2-}]$	1.4×10^{-8}	$PbCrO_4$	$[Pb^{2+}][CrO_4^{2-}]$	2.8×10^{-13}
CaF_2	$[Ca^{2+}][F^-]^2$	4.2×10^{-11}	$PbSO_4$	$[Pb^{2+}][SO_4^{2-}]$	1.8×10^{-8}
$Ca(OH)_2$	$[Ca^{2+}][SO_4^{2-}]$	5.5×10^{-6}	PbS	$[Pb^{2+}][S^{2-}]$	8.0×10^{-28}
$CaSo_4$	$[Ca^{2+}][SO_4^{2-}]$	9.1×10^{-6}	SnS	$[Sn^{2+}][S^{2-}]$	1.2×10^{-25}
CoS	$[Co^{2+}][S^{2-}]$	3.0×10^{-26}	$SrSO_4$	$[Sr^{2+}][SO_4^{2-}]$	3.2×10^{-7}
Co_2S_3	$[Co^{3+}]^2[S^{2-}]^3$	2.6×10^{-124}	ZnS	$[Zn^{2+}][S^{2-}]$	1.6×10^{-24}

Sample Problem 9.3 *continued*

Solution

Step 1. Analyze   Given: $K_{sp} = 2.5 \times 10^{-3}$

Unknown: solubility of $AgC_2H_3O_2$

Step 2. Plan

$$K_{sp} \rightarrow [Ag^+] = \text{solubility of } AgC_2H_3O_2$$
$$K_{sp} = [Ag^+][C_2H_3O_2^-]$$
$$\text{Since } [Ag^+] = [C_2H_3O_2^-]$$
$$K_{sp} = [Ag^+]^2$$
$$[Ag^+] = \sqrt{K_{sp}}$$
$$\text{solubility of } AgC_2H_3O_2 = [Ag^+]$$

Step 3. Compute   Solubility of $AgC_2H_3O_2 = [Ag^+] = \sqrt{2.5 \times 10^{-3}} = 5.0 \times 10^{-2}$ mol/L

Step 4. Evaluate   The answer has the proper number of significant digits and is equal to an estimated value of 5.0×10^{-2} ($\sqrt{25 \times 10^{-4}}$)

Practice Problem

Calculate the solubility of cadmium sulfide (CdS) in moles/liter, given the K_{sp} value listed in Table 19-4. (*Ans.*) 8.9×10^{-14} mol/L

Precipitation Calculations

In the example used earlier in this section of calculating solubility, $BaCO_3$ served as the source of both Ba^{2+} and CO_3^{2-} ions. Since each mole of $BaCO_3$ yields one mole of Ba^{2+} and one mole of CO_3^{2-}, the concentrations of the two ions were equal. However, the equilibrium condition does not require that the two ion concentrations are equal. Equilibrium will still be established so that the ion product $[Ba^{2+}][CO_3^{2-}]$ not exceed the value of K_{sp} for the system.

Suppose unequal quantities of $BaCl_2$ and $CaCO_3$ are added to water. If the ion product $[Ba^{2+}][CO_3{}^{2-}]$ exceeds the K_{sp} of $BaCO_3$, a precipitate of $BaCO_3$ forms. Precipitation continues until the ion concentrations decrease to the point where $[Ba^{2+}]$ $[CO_3^{2-}]$ equals the $Ksp\ell$.

Substances differ greatly in their tendencies to form precipitates when mixed in moderate concentrations. The photos in Figure 19-16 show the behavior of some negative ions in the presence of certain metallic ions. Note that some of the combinations have produced precipitates and some have not.

The solubility product can be used to predict whether a precipitate forms when two solution are mixed.

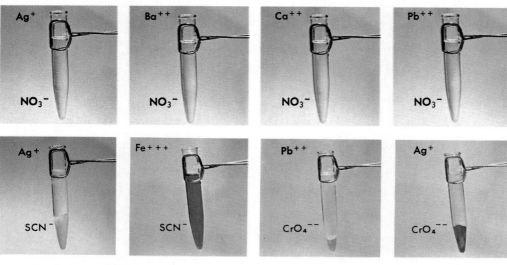

Figure 19-16

Sample Problem 19.4

Will a precipitate form if 20.0 mL of 0.010 M $BaCl_2$ is mixed with 20.0 mL of 0.0050 M Na_2SO_4?

Solution

Step 1. Analyze

Given: concentration of $BaCl_2$ = 0.010 M,
volume of $BaCl_2$ = 20.0 mL
concentration of Na_2SO_4 = 0.0050 M,
volume of Na_2SO_4 = 20.0M

Unknown: whether a precipitate forms

Step 2. Plan

Concentration, volume → ion product → precipitate? The two possible new pairings of ions are NaCl and $BaSO_4$. Of these, $BaSO_4$ is a sparingly soluble salt. It will precipitate if the ion product $[Ba^{2+}][SO_4^{2-}]$ in the combined solutions exceeds K_{sp} for $BaSO_4$. From the table of solubility products (Table 19-4), the K_{sp} is found to be 1.1×10^{-10}. The solubility equilibrium equation is

$$BaSO_4(s) \leftrightharpoons Ba^{2+}\ (aq) + SO_4^{2-}(aq)$$

The solubility equilibrium expression is

$$K_{sp} = [Ba^{2+}][SO_4^{2-}] = 1.1 \times 10^{-10}$$

First $[Ba^{2+}]$ and $[SO_4^{2-}]$ in the new solution must be found. Then the ion product is calculated and compared with the K_{sp}.

Step 3. Compute   Mole quantities of Ba^{2+} and SO_4^{2-} ions:

$$0.020 \text{ L} \times \frac{0.010 \text{ mol } Ba^{2+}}{L} = 0.000\ 20 \text{ mol } Ba^{2+}$$

$$0.020 \text{ L} \times \frac{0.0050 \text{ mol } SO_4^{2-}}{L} = 0.000\ 10 \text{ mol } SO_4^{2-}$$

Total volume of solution containing Ba^{2+} and SO_4^{2-} ions:

$$0.020 \text{ L} + 0.020 \text{ L} = 0.040 \text{ L}$$

Ba^{2+} and SO_4^{2+} ion concentrations:

$$\frac{0.000\ 10 \text{ mol } SO_4^{2-}}{0.040 \text{ L}} = 2.5 \times 10^{-3} \text{ mol } SO_4^{2-} \text{ L}$$

$$\frac{0.000\ 10 \text{ mol } SO_4{}^{2-}}{0.040 \text{ L}} = 2.5 \times 10^{-3} \text{ mol } SO_4{}^{2-}\text{L}$$

Trial value of ion product:

$$[Ba^{2+}][SO_4^{-2}] = (5.0 \times 10^{-3})(2.5 \times 10^{-3})$$
$$= 1.2 \times 10^{-5}$$

The ion product is much greater than K_{sp} ($K_{sp} = 1.1 \times 10^{-10}$), so precipitation occurs.

Step 4. Evaluate   The answer contains the appropriate number of significant digits and is close to an estimated value of 1×10^{-5}, $(5 \times 10^{-3})(2 \times 10^{-3})$; since $10^{-5} > 10^{-10}$, precipitation should occur.

Practice Problems
1. Does a precipitate form when 100 mL of 0.0025-M AgNO$_3$ and 150 mL of 0.0020-M NaBr solutions are mixed?   (*Ans.*) AgBr-ion product = 1.2×10^{-6}, K_{sp}(AgBr) = 5.0×10^{-13}; AgBr precipitates.

2. Does a precipitate form when 20 mL of 0.038 M Pb(NO$_3$)$_2$ and 30 mL of 0.018-M KCl solutions are mixed?   (*Ans.*) PbCl$_2$ ion product = 1.8×10^{-6}) K_{sp}(PbCl$_2$) = 1.9×10^{-4}; PbCl$_2$ does *not* precipitate.

The solubility-product principle can be very useful when applied to solutions of sparingly soluble substances. It *cannot* be applied very successfully to solutions of moderately soluble or very soluble substances. Also many solubility-product constants are known only roughly because of difficulties involved in solubility measurements. Sometimes it is necessary to consider two equilibria simultaneously. For example, if either ion hydrolyzes, the salt will be more soluble than predicted by using only the solubility product constant. The solubility product is also sensitive to changes in solution temperature to the extent that the solubility of the dissolved substance is affected by such changes. All of these factors limit the conditions under which the principle can be applied.

Section Review

1. What is a solubility-product constant? How are such constants determined?
2. How are solubility-product constants used to calculate solubilities?
3. What is an ion product?
4. How are calculations to predict possible precipitation carried out?
5. What is the value of K_{sp} for Ag_2SO_4 if 5.40 g is soluble in 1 L of water?
6. Determine whether a precipitate will form if 20.0 mL of 1.00×10^{-7} M $AgNO_3$ is mixed with 20.0 mL of 2.00×10^{-9} M NaCl at 25°C.

Chapter Summary

- A reaction system in which the forward and reverse reactions occur simultaneously and at the same rate is said to be in *equilibrium*. Both reactions continue, but there is no net change in the composition of the system.
- At equilibrium, the ratio of the product of the mole concentrations of substances formed to the product of the mole concentrations of reactants, each raised to the appropriate power, has a definite numerical value, K, which is the equilibrium constant at a given temperature. For values of K greater than 1, the products of the forward reaction are favored. For values of K less than 1, the reactants of the reverse reaction are favored.
- Any change that alters the rate of either the forward or reverse reaction disturbs the equilibrium of the system. According to Le Châtelier's principle, the equilibrium is shifted in the direction that relieves the stress.
- Catalysts increase the rates of forward and reverse reactions equally, and they do not shift an equilibrium or change the value of K.
- The common-ion effect is recognized when a solute containing ions like those of a reactant in an equilibrium system is added to the system. Le Châtelier's principle explains the response of the system to the stress.
- The equilibrium a weak for the ionization constant of a weak acid HB is
  $$Ka \neq \frac{[H_3O^+][Bi]}{[HB]} \quad K_a = K[H_2O].$$
- Salts formed from strong bases and weak acids produce aqueous solutions that are basic because of *anion hydrolysis*.
- Salts formed from strong acids and weak bases produce aqueous solutions that are acidic because of *cation hydrolysis*.
- Salts formed from strong acids and strong bases do not hydrolyze in water, and their solutions are neutral.
- Salts formed from weak acids and weak bases may produce neutral, acidic, or basic solutions, depending on the relative amounts of cation and anion hydrolysis. They may also hydrolyze completely in water solution.
- Ions of salts that are very sparingly soluble form saturated aqueous solutions at extremely low concentrations. The solubility-equilibrium expression for such salts yields a useful constant (K_{sp})—the solubility-product constant—which equals the product of the mole concentrations of solute ions in the saturated solution raised to appropriate exponential powers.

1. The product of the molar concentrations of the ions in a saturated solution, each raised to the appropriate power. They are determined from very careful measurements of solubilities and other solution properties.
2. If the K_{sp} and equation are known, solubilities may be calculated from the K_{sp} expression. (See Sample problem 19.5)
3. The product of the concentration of ions present in a solution each raised to the appropriate power
4. If the ion product exceeds the numerical value of K_{sp}, a precipitate will form.
5. 2.07×10^{-5}
6. Since the trial product (5.00×10^{-17}) is less than the K_{sp} of 1.8×10^{-11}, no precipitate is formed.

Chapter 19 *Review*

Chemical Equilibrium

Vocabulary

acid-ionization constant
buffered solution
chemical equilibrium
chemical-equilibrium
  expression
common-ion effect

equilibrium constant
hydrolysis
hydrolysis constant
reversible reaction
solubility-product
  constant

Questions

1. *(a)* What is meant by a reversible reaction? *(b)* Give an example of such a reaction.
2. Describe chemical equilibrium in terms of reaction rates.
3. What is the meaning of the term *dynamic* as applied to an equilibrium system?
4. Give two examples of physical equilibria.
5. Explain what occurs in the establishment of an equilibrium system involving *(a)* an ionic compound *(b)* a polar–covalent compound.
6. Explain what is meant by each of the following expressions: *(a)* the forward reaction is favored *(b)* the reverse reaction is favored *(c)* the products are framed *(d)* the reactants are favored *(e)* the reaction to the right *(f)* the reaction to the left.
7. Explain how the concentrations of A, B, C, and D change from the time when A and B are first combined to the point at which equilibrium is established given $A + B \rightleftarrows C + D$.
8. *(a)* Write the general expressions for an equilibrium constant based on the equation: $nA + mB+ . . . \rightleftarrows xC + yD$. *(b)* What information is provided by the value of K for a given equilibrium system at a specified temperature?
9. In general, which reaction is favored (forward, reverse, or neither) if the value of K at a specified temperature is *(a)* equal to 1 *(b)* very small *(c)* very large?
10. *(a)* State Le Chatelier's principle. *(b)* To what kinds of equilibria does it apply?

11. *(a)* Name three types of stress that can be placed on an equilibrium system and describe the impact of varying levels of each. *(b)* Which of these stresses affect the value of K?
12. Predict whether each of the following concentration changes would favor the forward or reverse reaction:
$$2HI(g) \rightleftarrows H_2(g) + I_2(g)$$
(a) addition of HI *(b)* addition of H_2 *(c)* removal of I_2 *(d)* removal of HI.
13. What will be the effect of each of the following changes on the new equilibrium concentration of H_2O?
$$4HCl(g) + O_2(g) \rightleftarrows 2H_2O(g) + 2Cl_2(g)$$
(a) addition of HCl *(b)* removal of Cl_2 *(c)* removal of O_2 *(d)* addition of Cl_2
14. Predict whether each of the following pressure changes would favor the forward or reverse reaction:
$$2NO(g) + O_2(g) \rightleftarrows 2NO_2(g)$$
(a) increased pressure *(b)* decreased pressure.
15. Based on the equilibrium system in Question 14, what would be the effect of a and b on the concentration of NO_2 at the new equilibrium? equilibrium?
16. In heterogeneous reaction systems, what type of substances can be removed from the equilibrium constant expression? Why?
17. Predict whether each of the following temperature changes would favor the forward or reverse reaction:
$$2NO(g) + O_2(g) \rightleftarrows 2NO_2(g) + 113 \text{ kJ}$$
(a) increased temperature *(b)* decreased temperature.
18. *(a)* Based on the equilibrium system in Question 17, what would be the effect of a and b on the concentration of NO_2 at the new equilibrium? *(b)* How would a and b affect the value of K?
19. Explain the effect of a catalyst on an equilibrium system.
20. Predict the effect of each of the following on the indicated equilibrium system in terms of

which reaction will be formed (forward, reverse, or neither):

$$H_2(g) + Cl_2(g) \rightleftarrows 2HCl(g) + 184 \text{ kJ}$$

(a) addition of Cl_2 (b) removal of HCl (c) increased pressure (d) decreased temperature (e) removal of H_2 (f) decreased pressure (g) addition of a catalyst (h) increased temperature (i) decreasing system volume.

21. How would parts a through i of Question 20 affect the new equilibrium concentration of HCl and the value of K at the new equilibrium?

22. What two fractions determine the extent to which reacting ions are removed from solution?

23. Identify the three conditions under which ionic reactions can run to completion and write an equation for each.

24. How are buffers important to the human body?

25. (a) Write the ion-product constant expression for water. (b) What is the value of this constant?

26. List and distinguish among the four general categories of salts, based on their hydrolysis properties.

27. Distinguish between anion and cation hydrolysis in terms of the type of solution resulting from each.

28. What rule of thumb is used to distinguish among soluble, insoluble, and slightly soluble substances?

29. What is the major solubility characteristic of those types of substances typically involved in solubility equilibrium systems?

30. What is the realtionship between K_{sp} and the product of the ion concentrations in terms of determining whether a solution of those ions is saturated?

Problems

1. Write the equilibrium-constant expression for each of the following systems: (a) A(g) + 2B(g) $\rightleftarrows$ 3C(g) + D(g) (b) 4A(g) + B(g) $\rightleftarrows$ 2C(g) + 3D(g) (c) $N_2(g)$ + $3H_2(g)$ $\rightleftarrows$ $2NH_3(g)$ (d) CO(g) + $NO_2(g)$ $\rightleftarrows$ $CO_2(g)$ + NO(g).

2. Determine the value of the equilibrium constant for each reaction below. If the equilibrium concentrations are found to be those specified: (a) A + B $\rightleftarrows$ C; A = 2.0 mol/L; B = 3.0 mol/L; C = 4.0 mol/L. (b) D + 2E

= F + 3G; D = 1.5 mol/L; E = 2.0 mol/L; F = 1.8 mol/L; G = 1.2 mol/L. (c) $N_2(g)$ + $3H_2(g)$ $\rightleftarrows$ $2NH_3(g)$; N_2 = 0.45 mol/L; H_2 = 0.14 mol/L; NH_3 = 0.62 mol/L.

3. An equilibrium mixture at a specific temperature is found to consist of 1.2 × 10^{-3} mol/L HCl, 3.8 × 10^{-4} mol/L O_2, 5.8 × 10^{-2} mol/L H_2O, and 5.8 × 10^{-2} mol/L Cl_2 according to the following: 4HCl(g) + $O_2(g)$ $\rightleftarrows$ $2H_2O(g)$ + $2Cl_2(g)$ Determine the value of the equilibrium constant for this system.

4. At 450°C the value of the equilibrium constant for the following system is 6.59 × 10^{-3}. If $[NH_3]$ = 1.23 × 10^{-4} mol/L + $[H_2]$ = 2.75 × 10^{-6} mol/L at equilibrium, determine the concentration of N_2 at that point.
$$N_2(g) + 2H_2(g) \rightleftarrows 2NH_3(g)$$

5. Write equilibrium-constant expressions for each of the following heterogeneous systems:
(a) Zn(s) + $2Ag^+(aq)$ $\rightleftarrows$ $Zn^{2+}(aq)$ + 2 Ag(s)
(b) $PbI_2(s)$ $\rightleftarrows$ $Pb^{2+}(aq)$ + $2I^-(aq)$
(c) $CN^-(aq)$ + $H_2O(\ell)$ $\rightleftarrows$ HCN(aq) + $OH^-(aq)$
(d) CdS(s) $\rightleftarrows$ $Cd^{2+}(aq)$ + $S^{2-}(aq)$ (e) 4Fe(s) + $3O_2(g)$ $\rightleftarrows$ $2Fe_2O_3(s)$.

6. Under suitable conditions, each of the following double-replacement reactions results in the formation of a precipitate. Write both the overall and the net ionic reactions for each and identify the precipiate by the label(s).
(a) $Pb(NO_3)_2(aq)$ + NaCl(aq) $\rightarrow$
(b) $Ba(NO_3)_2(aq)$ + $K_2SO_4(aq)$ $\rightarrow$
(c) $AgC_2H_3O_2(aq)$ + LiCl(aq) $\rightarrow$
(d) $Na_3PO_4(aq)$ + $CuSO_4(aq)$ $\rightarrow$
(e) $BaCl_2(aq)$ + $Na_2SO_4(aq)$ $\rightarrow$

7. Based on the K_w for water, what is the $[H_3O^+]$ of a solution with a pH of 9? What type of solution is it?

8. Express the solubility information given for each substance below in moles per liter:
(a) NaCl = 36.0 g/100. g H_2O at 20°C
(b) $PbSO_4$ = g/100. g H_2O at 25°C
(c) Cu_2S = 1.84 × 10^{-11} g/100. g H_2O at 25°C.

9. The ionic substance, GH ionizes to form G^+ and H^- ions. The solubility of GH is 0.0427 mol/L. What is the value of the solubility V product constant, K_{sp}?

10. The ionic substances EJ ionizes to form E^{2+} and J^{2-} ions. The solubility of EJ is 8.45 × 10^{-6} mol/L. What is the value of the solubility-product constant?

11. Calculate the solubility-product constant K_{sp} for each of the following, based on the solubility information provided: (a) $C_2CO_3 = 1.2 \times 10^{-3}$ g/100. g H_2O at 20°C   (b) $BaSO_4 = 2.4 \times 10^{-4}$ g/100. g H_2O at 20°C   (c) $Ca(OH)_2 = 0.173$ g/100. g H_2O at 20°C

12. Calculate the solubility of a substance MN that ionizes to form M^{2+} and N^{2-} ions, given that $K_{sp} = 8.1 \times 10^{-6}$.

13. Use the K_{sp} values given in Table 19-4 to evaluate the solubility of each of the following in moles per liter: (a) AgBr   (b) CoS.

14. Complete each of the following relative to the reaction that occurs when 25.0 mL of 0.0500 M $Pb(NO_3)_2$ is combined with 25.0 mL of 0.0400 M Na_2SO_4 if equilibrium is reached at 25°C: (a) Write the solubility-equilibrium equation at 25°C.   (b) Write the solubility-equilibrium expression for the net reaction.   (c) Determine the number of moles and then the molar concentration of each in the solubility-equilibrium expression.   (d) Determine whether a precipitate will form.

15. Follow the same sequence of steps outlined in Problem 13 to determine whether a precipitate of $CaCO_3$ will form if 0.10 L of 0.025 M $Ca(NO_3)_2$ is mixed with 0.10 L of 0.0031 M Na_2CO_3 at 25°C. (See Table 19-4 for K_{sp} values).

Application Questions

1. Explain why surface area of reactants is of particular importance in rate considerations in heterogeneous systems, while reactant concentrations are of more importance in such considerations in homogeneous solutions.

2. The reaction between steam and iron is reversible. Steam passed over hot iron produces magnetic iron oxide (Fe_3O_4) and hydrogen. Hydrogen passed over hot magnetic iron oxide reduces it to iron and forms steam. Under what general conditions could this reversible reaction could be brought to a state of equilibrium?

3. The equilibrium constants for four different reactions are given below:
$K_1 = 4.2 \times 10^{-6}$
$K_2 = 3.1 \times 10^{-2}$
$K_3 = 1.8 \times 10^4$
$K_4 = 27$
In which case are   (a) the products favored to

the greatest extent   (b) the reactants favored to the greatest extent?

4. Why is a knowledge of the principles of chemical equilibrium important to an industrial chemist?

5. What concentration stresses can be placed on the following equilibrium system so that the maximum level of HF can be produced?
$$H_2(g) + F_2(g) \rightleftharpoons 2HF(g)$$

6. Changes in the concentrations of the reactants and products at equilibrium have no impact on the value of the equilibrium constant. Explain.

7. What relative pressure (high or low) would result in the production of the maximum level of CO_2 according to the following? Explain.
$$2CO(g) + O_2(g) \rightleftharpoons 2CO_2(g)$$

8. What relative pressure (high or low) would increase the yield of methanol in the following equilibrium system?

9. $CO(g) + 2H_2(g) \rightleftharpoons CH_3OH(g)$ methanol
What relative temperature (high or low) would increase the yield of methanol in Application Question 8 if the formed reaction has a $\Delta H = -101$ kJ. Explain.

10. What relative conditions (reactant concentrations, pressure, and temperature) would favor a high equilibrium concentration of the underlined substance in each of the following equilibrium systems?   (a) $2CO(g) + O_2(g) \rightleftharpoons 2CO_2(g) + 167$ kJ   (b) $Cu^{2+}(aq) + 4NH_3(g) \rightleftharpoons Cu(NH_3)_4^{2+}(aq) + 42$ kJ   (c) $2HI(g) + 12.6$ kJ $\rightleftharpoons H_2(g) + I_2(g)$ (d) $4HCl(g) + O_2(g) \rightleftharpoons 2H_2O(g) + 2Cl_2(g) + 113$ kJ   (e) $H_2O(\ell) + 42$ kJ $\rightleftharpoons H_2O(g)$

11. At what relative temperature (high or low) would each equilibrium system in Application Question 10 exhibit its highest equilibrium-constant value?

12. A combustion reaction proceeding in air under standard pressure is transferred to an atmosphere of pure oxygen under the same pressure.   (a) What effect would you observe?   (b) How can you account for this effect?

13. The pH of a solution containing both acetic acid and sodium acetate is higher than that of a solution containing the same concentration of acetic acid alone. Explain.

14. The ionization constant K_a for acetic acid is 1.8×10^{-5} at 25°C. Explain the significance of this value.

15. *(a)* From the development of K_a described in Section 19-3, show how you would express an ionization constant K_b for the weak base NH_3. *(b)* In this case, $K_b = 1.8 \times 10^{-5}$. What is the significance of this numerical value to equilibrium?

16. A saturated solution is not necessarily a concentrated solution. Explain.

Application Problems

1. The value of the equilibrium constant for the reaction below is 40.0 at a specified temperature. What would be the value of that constant for the reverse reaction under the same conditions? $H_2(g) + I_2(g) \rightleftarrows 2HI(g)$

2. The pH of a 0.100-M solution of an acid HX is 3.14. Calculate the value of K_a.

3. A weak acid has a K_a value of 2.69×10^{-9}. Calculate the value of hydrolysis constant K_h. (Assume that $K_w = 1.00 \times 10^{-14}$.)

4. The ionic substance X_2Y ionizes to form X^+ and Y^{2-} ions. The solubility of X_2Y is 5.54×10^{-7} mol/L. What is the value of the solubility-product constant?

5. The ionic substance T_3U_2 ionizes to form T^{2+} and U^{3-} ions. The solubility of TU_2 is 3.77×10^{-20} mol/L. What is the value of the solubility product constant?

6. Calculate the solubility in moles per liter of a substance VW_2 that ionizes to form V^{2+} and W^{-1} ions, given that $K_{sp} = 4.0 \times 10^{-9}$.

7. A saturated solution of AgI at 20°C contains 2.7×10^{-10} mol/L Ag^+. Find its I^- concentration.

8. Calculate whether a precipitate will form if 0.35 L of 0.0044 M $Ca(NO_3)_2$ and 0.17 L of 0.00039 M NaOH are mixed at 25°C. (See Table 19-4 for K_{sp} values.)

9. Determine whether a precipitate will form if 70.0 mL of a 0.0040-M $Ca(NO_3)_2$ solution is mixed with 30.0 mL of a 0.010-M Na_2SO_4 solution at 25°C.

10. Determine whether a precipitate will form if 1.70 g of solid $AgNO_3$ and 14.5 g of solid NaCl are dissolved in 200. mL of water to form a solution at 25°C.

11. If 2.50 g of solid $Fe(NO_3)_3$ are added to 100. mL of a 1.00×10^{-20}-M NaOH solution, will a precipitate form?

Enrichment

1. Carry out library research to learn more about applications of chemical equilibrium principles in industry. If possible, arrange a visit to an industrial plant in which such principles are applied.

2. Use reference works to investigate one or more processes that take place within the human body involving the maintenance of chemical balances. Find out how these chemical balances may differ from and actually oppose equilibrium states.

3. Make up a gridlike chart illustrating combinations of various ions. Show which combinations tend to form precipitates and which do not.

4. Many chemical processes that take place in the kitchen, such as hard-boiling an egg, involve the use of varied temperatures to increase reaction rates and produce virtually irreversible changes in substances. Observe such processes and make a list of them. Then, make use of a reference work on food chemistry to find out what is occurring in each case.

5. Investigate the chemical equilibria of oceans. Recent research indicates that the composition of oceans is probably controlled by equilibria involving seawater and marine minerals.

6. Find out, and report on how a disruption of chemical equilibrium is responsible for the growth of crystals.

7. Research the role of chemical equilibria in chromatography. Distinguish between the various chromatographic methods and find out the particular each method. Write a report of your findings.

8. Carry out library research on the formation of stalagmites and stalagtites in caves. (See photograph on first page of this chapter.) Investigate the equilibria processes involved in the formation of stalagmites and stalagtites.

9. *(a)* Investigate the life and work of the French scientist Henri Le Châtelier. How did he apply the scientific method to arrive at the principle which is named after him? What other important contributions did Le Châtelier make to science and technology? *(b)* Carry out library research on Le Châtelier's principle. Explain how the principle can be determined using the laws of thermodynamics which you studied in Chapter 18.

Chapter 20 Oxidation-Reduction Reactions

Chapter Planner

20.1 The Nature of Oxidation and Reduction

Oxidation
- Demonstration: 1
  Question: 2a
  Reduction
- Demonstration: 2
  Questions: 1, 2b
  Problem: 1

Oxidation and Reduction as a Process
- Desktop Investigation
  Chemistry Notebook
  Experiment: 31
  Questions: 3–5
  Problem: 2
  Application Questions: 1, 2

20.2 Balancing Redox Equations

Oxidation Number Method
- Problems: 3–6
  Application Problems:
  1–5

Ion-Electron Method
- Question: 6
  Problems: 7, 8
  Application Problems: 6, 7
- Application Problem: 8

20.3 Oxidizing and Reducing Agents

Chemical Equivalent of Oxidizing and Reducing Agents
- Experiments: 32, 33
  Questions: 8–12
  Application Questions: 3, 4
- Application Questions: 5, 8

20.4 Electrochemistry

Electrochemical Cells
- Demonstrations: 3–5
  Question: 13

Electrolytic Cells
- Demonstration: 6
  Questions: 15–17
- Application Question: 6

Rechargeable Cells
- Application Question: 7

Electrode Potentials
- Questions: 14, 18–21
  Problems: 9, 10
  Application Problems: 10–12
- Application Problem: 9

Teaching Strategies

Introduction

This chapter provides a good opportunity to review the year's work. The spontaneity or nonspontaneity of redox reactions provides a good time to review thermodynamics. Concentration effects on voltage provide a good review of equilibrium and Le Châtelier's theory. A comparison of redox potentials with reference to the Periodic Table can stimulate discussion of electron structure and periodicity. The oxidation of alcohols and alkenes to acids, aldehydes, and ketones is a good introduction to organic chemistry. Balancing redox equations will add a new dimension to stoichiometry and chemical reactivity. Since Avogadro's number was established by Ampère, using electrochemical techniques, and since 1 Faraday equals 1 mol of electrons, students can develop an additional understanding of mole theory in this unit. There are many interesting demonstrations and lab investigations that can be done with this chapter.

20.1 The Nature of Oxidation and Reduction

Here is an opportunity to emphasize that many chemical reactions involve a redistribution of electrons. In ionic species, we assume that an electron has been removed or added to an atom resulting in a protonic charge $(+)$ or an electronic charge $(-)$. In a molecular species, oxidation numbers are assigned according to the polarity of covalent bonds. For example, in water (H-O-H) the electron density of the O-H bond is greater towards the oxygen. Therefore, the oxygen is assigned an oxidation state of negative one, and the hydrogen is assigned an oxidation state of $+1$. The periodate ion, IO_4^-, has a charge of -1 because it

has gained an electron from the environment and the total number of electrons exceeds the total number of protons by one (85 protons, 86 electrons). In this polyatomic ion, the electron density of the I-O bond is greater towards the oxygen. Therefore, each oxygen is assigned an oxidation state of -2, and the central iodine is assigned an oxidation state of $+7$. If IO_4^- is reduced to IO_3^-, the iodine will be assigned an oxidation state of $+5$.

The emphasis in this section is on the chemical process of oxidation and reduction. Although oxidation-reduction plays a part in the simplest syntheses, decompositions, and single-replacement reactions, and these reactions have been studied in class all year, some students will still need to have the underlying concepts of oxidation-reduction clarified. *Oxidation* results from a *loss of electrons*. *Reduction* results from a *gain of electrons*. Even though it seems that they should understand this concept, some students will have to be reminded of the fact that electrons are negative and that is why a *gain* leads to a *reduction* of oxidation state.

$KMnO_4$ is an excellent oxidizing agent that goes through a series of color changes as it is reduced. The variety of colors produced in this reduction illustrates that a single element can have many oxidation states. Repeat Demonstration 3 on manganese described in the Chapter 7 Teaching Strategies. Incorporate this into your lesson explaining the rules that govern the assignment of oxidation numbers.

Make sure that students know how oxidation numbers are assigned. Bring out samples of various compounds that contain polyatomic ions when you do this.

Demonstration 1: The Oxidation of I⁻(aq) by MnO_4^-(aq)

PURPOSE: To observe evidence of a spontaneous redox reaction between a polyatomic ion and a monoatomic ion in aqueous solution

MATERIALS: 100-mm petri dish, 7 mL 0.1 M KI, 7-mL 0.01 M $KMnO_4$, 1 M HCl, starch (optional)

PROCEDURE: Add approximately 7 mL 0.1 M KI to a petri dish on an overhead projector. Add 0.01 M $KMnO_4$ to the same petri dish and mix the two solutions with a stirring rod. Add a few drops of 1.0 M HCl to the solutions. Stir.

The reaction will not proceed until the acid is added. Within a few minutes of adding HCl, the purple of the $KMnO_4$ will begin to fade. The appearance of an orange-brown precipitate indicating the presence of MnO_2 will develop along with the red-brown solution indicative of dissolved I_2. Spraying with starch or adding a few grains of soluble powdered starch will give positive evidence for the presence of iodine.

This simple demonstration illustrates many of the topics covered in this chapter. Students will observe that redox reactions are not limited to monoatomic ions and metallic elements. Additionally, they will see elements in a variety of oxidation states and they will observe that a reduction is accompanied by an oxidation. In this chapter, they will learn that redox reactions accompanied by the loss or gain of oxygen by a polyatomic ion will require the presence of acid or base. This point is illustrated by the above demonstration since the addition of acid (or base) is a requirement for the spontaneity of this reaction.

Equation:

$$2\ MnO_4^{-1}(aq) + 6I^-(aq) + 8H^+(aq)$$
$$\rightarrow 2MnO_2 + 3I_2 + 4H_2O$$

Many of the reactions demonstrated before in Chapters 7, 8, and 9 can be repeated now with emphasis on the transfer of electrons from reducing agent to oxidizing agent. Repeat some of the Activity Series reactions of Chapter 8. This time, refer to the orbital origin and destination of the electrons being transferred.

$Ni(s) \rightarrow Ni^{2+}(aq) + 2e^-$ (from s orbital at 4th energy level)

$2e^-$ (to s orbital at 4th energy level) $+ Cu^{+2}(aq) \rightarrow Cu(s)$

The electrolysis of water and the oxidation of alcohols can be used effectively to illustrate redox phenomena.

Desktop Investigation: Redox Reactions

This investigation gives students an opportunity to observe oxidation and reduction of some familiar household items. Make sure to schedule this investigation so that students can follow the reactions over a twenty-four hour period. It is a good time to talk about rust and the dangers of allowing steel bridges to go unpainted. This is a timely topic since we have just become aware of the fact that many older bridges are decaying because their steel girders have not been cared for properly. Experiment 32, *Investigating the Factors that Influence the Corrosion of Iron*, also illustrates this material.

20.2 Balancing Redox Equations

Balancing equations is one of the primary tools used in chemistry. Accounting for electron shifts in half-reactions can simplify

balancing procedures. Because students have only been exposed to simple reactions, they might resist learning redox methods for balancing. Give your students a reaction that appears to be really difficult initially, then help them to balance it with the methods carefully outlined in this section. Make sure to point out that mass and charge must be conserved; that the number of electrons lost by the reducing agent must equal the number of electrons gained by the oxidizing agent. Most of the reactions in this chapter include hydrogen or hydroxyl ions from water when needed. If they are not specified, hydrogen or hydroxyl ions from water can always be added if they are required to conserve the number of oxygen atoms or the charge in a reaction.

20.3 Oxidizing and Reducing Agents:

Actually doing an electrochemical titration or diagramming it on the chalkboard is a good way to introduce the concept of equivalent mass. Make sure to point out that the equivalent mass of an element depends on the reaction in which it is participating. If Fe^0 is oxidized to Fe^{+3}, its equivalent mass is 55.8/3 g. If Fe is reduced from Fe^{+3} to Fe^{+2}, its equivalent mass is 55.8/1 g.

Equivalent Mass Addition

The reduction of a permanganate ion with oxalic acid can be demonstrated qualitatively or quantitatively. In this reaction, the permanganate ion is reduced to manganese dioxide. The manganese is reduced from +7 to +4; therefore, the equivalent weight of potassium permanganate is calculated by dividing the molar mass (158) of potassium permanganate by 3.

$$3H_2C_2O_4 + 2MnO_4^{-1} \rightarrow$$
$$6CO_2 + 2OH^- + 2H_2O + 2MnO_2$$

Emphasize that the equivalent weight of a substance is dependent on the number of electrons lost or gained in a specific reaction. If potassium permanganate had been reduced to MnO_4^{-2}, the manganese is reduced from +7 to +6; therefore, the equivalent mass of potassium permanganate in this reaction would be 158.

Demonstration 2: The Reaction of Permanganate Ion with Oxalic Acid

PURPOSE: To show the reduction of the permanganate ion

MATERIALS: Overhead projector and screen, 50-mL beaker, 0.300 M oxalic acid, 0.01 M potassium permanganate, recently prepared (within the previous two weeks)

PROCEDURE: Add 10 mL of each solution to a 50-mL beaker on an overhead projector.

The characteristic purple color of the permanganate ion will fade within a few minutes as the ion is reduced. You can use this reaction qualitatively to introduce equivalent mass. This reaction is also the basis of an interesting rate study described in HRM's "Experiments in Colorimetry," which could be assigned to a student who would like to do some independent work.

Potentiometric titrations are usually not used as class laboratory investigations in a first year course, but could be used effectively as a demonstration or as an independent study. The $Fe^{2+}/Cr_2O_7^{2-}$ and the Fe^{2+}/Ce^{4+} systems can be studied with the help of the HRM software package "Experiments in Chemistry," Experiment 15.

The reactions involved can serve as illustrations for equivalent mass calculations.

$$6Fe^{+2} + Cr_2O_7{}^{2-} + 14H^+ \rightarrow$$
$$6Fe^{3+} + 2Cr^{3+} + 7H_2O$$

$$Fe^{+2}(aq) + Ce^{+4}(aq) \rightarrow$$
$$Fe^{3+} + Ce^{3+}$$

In both reactions, iron is the reducing agent. One mole of iron donates one mole of electrons to the oxidizing agent; therefore, the equivalent mass of iron is the molar mass of iron (55.8) divided by 1.

In the chromate reaction, each mole of chromium atoms receives 3 electrons, therefore, the equivalent mass of chromium in this reaction is the molar mass of chromium (52.0) divided by 3.

In the cerium reaction, since each mole of cerium receives one mole of electrons, the equivalent mass of cerium is the molar mass of cerium (140) divided by 1.

20.4   Electrochemistry

While chemistry students can acquire the ability to handle the algebraic manipulations concerning electrochemical cells, they often cannot relate that algebraic ability to understanding the fundamental chemical principles or the applications of electrochemistry. The best way to make sure that students can relate algebraic manipulations to chemical phenomena is to continue to present demonstrations and hands-on experiences in tandem with presentation of theoretical material.

The microchemical techniques that have been developed by Maloney, Lewis, Russo, and others enable you to introduce experiments that require very small amounts of material and reduced laboratory time. Microcomputer-based laboratory approaches developed by Tinker and by others allow students to convert the computer into a sophisticated scientific instrument capable of tracking redox reactions.

Electrical energy is released from the spontaneous flow of electrons from a substance that has less affinity for electrons to a substance that has a greater affinity. The greater the difference in attraction, the greater the energy released. This chemical process is the basis of the electrochemical cell. If electrons are going to flow from a substance that has greater affinity for electrons to a substance that has less affinity for electrons, the reaction will not be spontaneous and energy will have to be added to the chemical system. This process is the basis of the electrolytic cell.

An understanding of the fundamental process of electron flow is necessary for an understanding of electrochemistry. The flow of electrons from one substance to another can be demonstrated with single-replacement reactions in which one metal replaces a metallic ion in solution with the concommitant reduction of the ion to its neutral state. Select a few of the reactions used in the Chapter 8 Activity Series demonstration to illustrate the chemical phenomona underlying potential differences and electric current.

Students will be in a better position to understand electrochemical and electrolytic cells if they have an unambiguous understanding of a cell reaction. Once that firm foundation is laid, they can apply their understanding of electrochemistry to batteries, photovoltaic cells, electroplating and semiconductors.

After you have demonstrated the spontaneous reduction and oxidation that occurs in a simple replacement reaction, explain that the two half reactions, oxidation and reduction, have to be physically separated for the energy released to be harnessed to do work.

Ronald Ulrich has adapted an idea (from an article by Earl Zwicker) in developing two unique demonstrations based on the principles of potential difference, current, and redox reactions that will engage your students' interest and enthusiasm while utilizing electrochemical energy to do work.

Demonstration 3: Bio-Clock
PURPOSE: To show that electrochemical reactions can take place in nature
MATERIALS: small battery clock, the type that sticks on the dash of your car, 3 short connecting wires, soldering gun, solder, flux, 2 copper wires or flattened $1/8''$ copper strips, 2 zinc strips ($1/4'' \times 5$ cm) (or a zinc-coated nail), 2 apples
PROCEDURE: The clock can be separated by inserting a screwdriver into the groove and twisting. Notice where the + part of the battery is connected. Use a small Phillips screwdriver to remove the retaining metal piece. Solder the copper wires to the appropriate poles + and −. Solder one of the wires to one of the copper electrodes, the other to one of the zinc electrodes. Solder one end of the remaining wire to a copper strip, the other end to a zinc strip. Insert the two electrodes coming from the clock into two different apples. Insert the remaining wire (zinc and copper) next to the electrodes in the apples approximately $1/4''$ apart (the copper next to the zinc and the zinc next to the copper). Set the clock according to the directions that came with it. The clock is surprisingly accurate.

Demonstration 4: Screaming Lemon
PURPOSE: To show the electrochemical properties of a lemon
MATERIALS: Piezo sounder, 1.5-24 V d-c, 20 mA consumption; 1 zinc strip, $1/4''$ x 5 cm, 1 copper strip, $1/4'' \times 5$ cm, two connecting wires, soldering iron, solder, flux, lemon
PROCEDURE: Solder the + and − ends of the Piezo Sounder to separate connecting wires. Solder the + wire to the copper strip. Solder the − wire to the zinc strip. Place the two strips approximately $1/4''$ apart into the lemon. The "Screamin' Lemon" is beyond belief.

In addition to presenting these demonstrations, build a Daniell Cell, which is based on the spontaneous Zn/Cu^{2+} reaction that the class has seen before.

Demonstration 5: The Daniell Electrochemical Cell
PURPOSE: To build an electrochemical cell that is based on the spontaneous Zn/Cu^{2+} reaction and to measure the potential difference between the zinc and copper half-cells
MATERIALS: copper and zinc strips, approximately 4 x 150 mm; steel wool; 250-mL beaker; porous cup, 35 mm x 80 mm; 0.1 M-solutions of copper nitrate (or copper sulfate) and zinc nitrate; voltmeter, 0-5 V − / + 0.01 V; leads with alligator clips
PROCEDURE: Clean the metal strips with steel wool. Fill the porous cup 3/4 full with 0.1-M copper ion solution. Place the porous cup in the beaker and fill the beaker 3/4 full with 0.1 M zinc nitrate. Place the copper strip in the copper solution

and the zinc strip in the zinc solution. Connect the leads from the voltmeter to the two electrodes. If no reading is recorded on the voltmeter after adjusting the sensitivity, reverse the leads. Record the voltage.

A transparent voltmeter placed on an overhead projector will allow students to observe the measurable potential difference between the zinc and copper half-cells. Since students know from previous experience that the zinc is oxidized and the copper is reduced, they are in a good position to understand the terms *anode* (the electrode where oxidation takes place) and *cathode* (the electrode where reduction takes place). Once students are introduced to the Daniell cell, they can understand the chemistry fundamental to all batteries.

Encourage students to translate their understanding of electrochemical cells to electrolytic cells. In an electrochemical cell, electrons flow spontaneously, free energy is released, and that energy can be measured and utilized. In an electroytic cell, energy is required to force electrons "uphill"; that is, to force electrons from a source that has a higher affinity for electrons to an electrode that has a lower affinity.

Carefully define the locations at which important events occur. The electrolysis of potassium iodide can attract students' attention and will illustrate the electrode reactions.

Demonstration 6: The Electrolysis of Potassium Iodide

PURPOSE: To illustrate electrode reactions and to define the locations at which electrochemical events occur

MATERIALS: 100-mm petri dish, overhead projector, screen, 0.1 *M* KI, 6 V/12 V battery charger, two electrodes, phenopthalein, starch solution

EQUATIONS:

$$Anode: 2I^- \rightarrow I_2 + 2e^-$$
$$I^- + I_2 \rightarrow I_3^- \text{ (brown color)}$$
$$Cathode: 2e^- + 2H_2O \rightarrow H_2(g) + 2OH^-$$

PROCEDURE: Place the petri dish on the overhead projector. Add 20 mL 0.1-*M* KI solution. Add a few drops of phenolpthalein,

and a few drops of starch solution to the KI solution. Attach electrodes to insulated leads of battery charger. Turn battery charger to 12 V (6 V if you want a slower reaction). Hold electrodes in KI solution.

I_2 can be detected at the anode by adding starch and detecting the characteristic blue-black color that develops. H_2 bubbles can be detected at the cathode. OH^- can be detected in solution near the cathode by adding a few drops of phenolpthalein.

Electrons are being removed from the iodide ions at the anode where they are oxidized to I_2. The electrons are being pumped from the anode, causing it to be positively charged. These electrons are being forced through an external circuit to the cathode where the hydrogen ions of the water successfully compete for the electrons with the sodium ions. Reduction of the hydrogen ions takes place at the cathode, which is the negative electrode because electrons have been pumped to it by the external energy source.

In both the electrolytic cell and the electrochemical cell, a closed circuit and electroneutrality must be maintained during the electrochemical reaction.

In both cases, the requirements for continuous conductivity in the internal circuit and electroneutrality in the cells are fulfilled by migrations of ions. In an electrolytic cell, when a salt solution, such as potassium iodide, is electrolyzed, there is a migration of the cations Na^+ and H^+ and of the anions I^- and OH^-. In an electrochemical cell, a pathway must be provided for electrons to flow from the anode to the cathode. This pathway is provided by the external circuit, usually a copper wire. A pathway must also be provided for ions to flow between the electrolyte solutions which surround the electrodes. This is the internal circuit. The pathway can be provided by building one half-cell in a porous cup and placing that within the other half-cell. Alternatively, the pathway can be provided by an electrolyte bridge connecting two half-cells that are physically separated. The traditional bridge consists of a glass tube packed with glass wool and an ammonium nitrate solution. Hot dogs or pickles also make unique conducting devices.

References and Resources

Books and Periodicals

Baca and Lewis. "Electrochemistry in a Nutshell," *Journal of Chemical Education*, Vol. 55, No. 12, 1978, p. 804.

Campbell, J.A. "Kinetics—Early and Often," *Journal of Chemical Education*, Vol. 40, No. 11, 1963, pp. 578–583.

Ensman, E., T.R. Hacker, and R.A.D. Wentworth. "Vegetable Voltage and Fruit 'Juice': An Electrochemical Demonstration," *Journal of Chemical Education*, Vol. 65, No. 8, 1988, p. 727.

Gilbert, "Electrical Energy From Cells — A Corridor Demonstration," *Journal of Chemical Education*, Vol. 57, No. 3, 1980, p. 216.

Heldeman, Stephen, "The Electrolysis of Water: An Improved Demonstration Procedure," *Journal of Chemical Education*, Vol. 63, No. 9, 1986, p. 809.

Kam, T.T. "On the Nernst Equations of Mercury-Mercuric Oxide Half-Cell Electrodes," *Journal of Chemical Education*, Vol. 60, No, 2, 1983, p. 133.

Koch, Klaus R. "Oxidation by Mn_2O_7: An Impressive Demonstration of the Powerful Oxidizing Property of Dimanganeseheptoxide," *Journal of Chemical Education*, Vol. 59, No. 11, 1982, pp. 973–974.

Lipeles, Enid S. "The Chemical Contributions of Amadeo Avogadro," *Journal of Chemical Education*, Vol. 60, No. 2, 1983, pp. 127–128.

MacBeath, Marie E. and Andrew L. Richardson, "Tomato Juice Rainbow: A Colorful and Instructive Demonstration," *Journal of Chemical Education*, Vol. 63, No. 12, 1986, pp. 1092–1094.

Maloney, Barb and Robert Lewis. "Electrolysis of Potassium Iodide," unpublished, taken from Manjkow, Joseph and Dana Levine, "Electrodeposition of Nickel on Copper," *Journal of Chemical Education*, Vol. 63, No. 9, 1986, p. 809.

Microscaled Chemistry Workshop, ISTA, 1987.

Ophardt, Charles E. "Redox Demonstrations and Descriptive Chemistry Part 2: Halogens," *Journal of Chemical Education*, Vol. 64, No. 9, 1987, p. 807.

Parker, "50–Minute Experiment, An Appropriate Determination of the Avogadro Constant," *Journal of Chemical Education,* Vol. 57, No. 10, 1980, p. 735.

Pearson, Robert S. "Manganese Color Reactions," *Journal of Chemical Education,* Vol. 65, No. 5, 1988, pp. 451-452.

Russo, Tom, Microchemistry for High School General Chemistry, Educational Corp., 1985, (esp. Module 9: Electrochemistry).

"State of the Art Symposium: Electrochemistry," *Journal of Chemical Education,* Vol. 60, No. 4, 1983.

Stevens, George H. "Teaching Suggestions for Oxidation-Reduction," excerpted from *Journal of Chemical Education,* Feb. 1979.

Stroud, Linda M. and Denise Creech, "Effects of NO_2 and SO_2 on Plant Pigments," presented at the Southeast Regional Meeting of NSTA, Dec. 1988.

Ulrich, Roanald, "Bio Clock" and "Screamin, Lemon," adapted from Earl Zwicke (ed.), *The Physics Teacher,* Jan. 1988.

Weissman, "Batteries: The Workhorses of Chemical Energy," *CHEM* , Vol. 45, Oct. 1972.

Williams, L. Pearce, "Andre-Marie Ampère," *Scientific American,* Vol. 260, No. 1, 1989, pp. 90–97.

Computer Software

"Experiments in Colorimetry," Apple II, HRM.

Tinker, Robert F. and Diana Malone, "Experiment 12: Basic EMF Measurements – Fundamentals of Electrochemistry," Apple II, IBM, COMPress.

"Topic 7: REDOX-Acids and Bases," Apple II, Knowledge Factory.

"Topic 7: Electrochemistry Part I: Oxidation/Reduction, Part II: The Nernst Equation, The Relationship between delta E, delta G, and K, and the Activity Series," Apple II, Knowledge Factory.

Weyh, John A., Joseph R. Crook, and Les N. Hauge, "Oxidation-Reduction Reactions," IBM, COMPress.

Answers and Solutions

Questions

1. *(a)* Oxidation processes are reactions in which the atoms or ions of an element attain a more positive (or less negative) oxidation state. Reactions in which the atoms or ions of an element attain a more negative (or less positive) oxidation state are described as reduction processes.
   (b) Oxidation: $Na \rightarrow Na^+ + e^-$
   Reduction: $Cl_2 + 2e^- \rightarrow 2Cl^-$

2. *(a)* An oxidized species is one that undergoes a change in oxidation state in a positive mathematical direction. *(b)* One that undergoes a change in oxidation state in a negative sense is said to be reduced.

3. An oxidation-reduction reaction is a chemical process in which elements undergo a change in oxidation number.

4. *(a)*, *(b)*, *(c)*, *(f)*, *(g)*, *(i)*

5.

Equation	Oxidized	Reduced
(a)	Na	Cl_2
(b)	C	O_2
(c)	O^{2-}	H^+
(f)	O^{2-}	Cl^{5+}
(g)	H_2	Cl_2
(i)	Zn	Cu^{2+}

6. In the oxidation-number method, the electron shift is balanced for the oxidized and reduced species by balancing the two electronic equations that represent the shift, and then the coefficients of all reactants and products are adjusted to balance the overall equation. In the ion-electron method, the participating species are identified and then separate oxidation and reduction equations are balanced for *both* atoms and charge, and then added together to give a balanced net ionic equation.

7. An oxidizing agent is a substance that attains a more negative oxidation state during an oxidation-reduction reaction whereas a reducing agent is one that attains a more positive oxidation state.

8. *(a)* Lithium is the most active reducing agent of all the common elements. *(b)* Due to their large atomic radii, the Group-1 metals have weak attraction for their valence electrons and form positive ions readily, thus making those lost electrons available to be gained by other substances that then undergo reduction. *(c)* Fluorine is the most active oxidizing agent among the common elements.

9. *(a)* Strongest = Ca; weakest = Cl^-
   (b) Strongest = Al; weakest = Br^-
   (c) Strongest = Na; weakest = F^-

10. *(a)* Strongest = NO_3^-; weakest = K^+
    (b) Strongest = Cl_2; weakest = Zn^{2+}
    (c) Strongest = F_2; weakest = Li^+

11. *(a)* Oxidizing agent = Cl_2; reducing agent = Na *(b)* Oxidizing agent = O_2; reducing agent = C *(c)* Oxidizing agent = H_2; reducing agent = O^{2-} *(f)* Oxidizing agent = Cl^{5+}; reducing agent = O^{2-} *(g)* Oxidizing agent = Cl_2; reducing agent = H_2 *(i)* Oxidizing agent = Cu^{2+}; reducing agent = Zn

12. *(a)* Yes *(b)* Yes *(c)* No *(d)* Yes *(e)* No

13. Electrochemistry is the branch of chemistry that deals with the electricity-related applications of redox reactions.

14. *(a)* The electrode potential is the potential difference between an electrode and its solution. *(b)* A half-reaction that occurs at one electrode in an electrochemical cell. *(c)* A half-cell is one portion of a voltaic cell, consisting of a metal electrode in contact with a solution of its ions.

15. *(a)* In an electrochemical cell, the redox reaction is spontaneous and the system of electrodes and electrolytes is used to transmit the electrical energy produced by the reaction. In an electrolytic cell the redox reaction is not spontaneous and thus current is used to drive the reaction through the electrodes and electrolytes used. *(b)* In an electrochemical cell, the anode is the negative electrode, while the cathode is the positive electrode. In an electrolytic cell, the anode is the positive electrode, while the cathode is the negative electrode.

16. The specific products are determined by the relative ease with which the different particles present can be oxidized or reduced.
17. (a) Electroplating is an electrolytic process in which a metal is deposited on a surface. (b) The object to be plated serves as the cathode, while the plating metal is the anode.
18. (a) The potential difference between the two electrodes is a measure of the energy required to move a certain electric charge between the electrodes. (b) This difference can be measured by a voltmeter connected across the two electrodes, and is given in volts (V).
19. (a) A voltaic cell is an arrangement that is capable of generating an electric current in an external circuit connected between two electrodes. (b) A half-cell is one of the two portions of a voltaic cell, each consisting of a metal electrode in contact with a solution of its ions. (c) The electrode potential is the potential difference between an electrode and its solution. (d) A standard electrode potential is a half-cell potential measured relative to a potential of zero for the standard hydrogen electrode.
20. Whereas the potential difference across a voltaic cell is easily measured, there is no direct way to measure an individual electrode potential. By assigning an arbitrary potential to the standard reference electrode, a specific potential can then be determined for the other electrode of the complete cell made by combining the two half-cells involved.
21. (a) A given half-cell potential is an indication of the tendency of the substance to undergo reduction. (b) The larger the value of the potential for a given half-reaction, the greater is the tendency for it to undergo reduction and the less likely it is to undergo oxidation. The smaller the potential, the less the tendency for the substance to undergo reduction and the greater is its tendency to undergo oxidation instead.

Problems

1. (a) Oxidation: $K \rightarrow K^+ + e^-$
   (b) Reduction: $S + 2e^- \rightarrow S^{2-}$
   (c) Oxidation: $Mg \rightarrow Mg^{2+} + 2e^-$
   (d) Oxidation: $2F^- \rightarrow F_2 + 2e^-$
   (e) Oxidation: $H_2 \rightarrow 2H^+ + 2e^-$
   (f) Reduction: $O_2 + 4e^- \rightarrow O^{2-}$
   (g) Reduction: $Fe^{3+} + e^- \rightarrow Fe^{2+}$
   (h) Oxidation: $Mn^{2+} \rightarrow Mn^{7+} + 5e^-$

2. (a) $\overset{0}{H_2}$ (b) $\overset{+1-2}{H_2O}$ (c) $\overset{0}{Al}$ (d) $\overset{+2-2}{MgO}$ (e) $\overset{+3\ -2}{Al_2 S_3}$
   (f) $\overset{+1+5-2}{HNO_3}$ (g) $\overset{+2+6-2}{H_2SO_4}$
   (h) $\overset{+2-2+1}{Ca(OH)_2}$ (i) $\overset{+2+5-2}{Fe(NO_3)_2}$ (j) $\overset{0}{O_2}$ (k) $\overset{+1+7-2}{KMnO_4}$
   (l) $\overset{+1+4-2}{Na_2SO_3}$ (m) $\overset{+5-2}{ClO_3^-}$
   (n) $\overset{+6-2}{SO4_2^{2-}}$ (o) $\overset{-3+1}{NH_4}$ (p) $\overset{+5-2}{PO_4^{3-}}$ (q) $\overset{+1\ +6\ -2}{Na_2Cr_2O_7}$
   (r) $\overset{-3+1}{(NH_4)_3}\ \overset{+5-2}{PO_4}$

3. (a) $Fe + HCl \rightarrow FeCl_2 + H_2$ (b) $\overset{0}{Fe} + \overset{+1-1}{HCl} \rightarrow \overset{+2-1}{FeCl_2} + \overset{0}{H_2}$
   Fe is oxidized; H^+ is reduced

(c) Oxidation: $\overset{0}{Fe} \rightarrow \overset{+2}{Fe} + 2e^-$
   Reduction: $\overset{+1}{2H} + 2e^- \rightarrow \overset{0}{H_2}$
   (d) $Fe + 2HCl \rightarrow FeCl_2 + H_2$

4. (a) $\overset{0}{Li} + \overset{+1\ -2}{H_2O} \rightarrow \overset{+1-2+1}{LiOH} + \overset{0}{H_2}$
   $2[\overset{0}{Li} \rightarrow \overset{+1}{Li^+} + e^-]$
   $\overset{+1}{2H} + 2e^- \rightarrow \overset{0}{H_2}$
   $2Li + 2H_2O \rightarrow 2LiOH + H_2$

   (b) $\overset{0}{Al} + \overset{+1-1}{HCl} \rightarrow \overset{+3-1}{AlCl_3} + \overset{0}{H_2}$
   $2[\overset{0}{Al} \rightarrow \overset{+3}{Al^{3+}} + 3e^-]$
   $3[\overset{+1}{2H} + 2e^- \rightarrow \overset{0}{H_2}]$
   $2Al + 6HCl \rightarrow 2AlCl_3 + 3H_2$

   (c) $\overset{+1-2}{H_2S} + \overset{+1+5-2}{HNO_3} \rightarrow \overset{+1+6-2}{H_2SO_4} + \overset{+4\ -2}{NO_2} + \overset{+1-2}{H_2O}$
   $8[\overset{+5}{N} + e^- \rightarrow \overset{+4}{N}]$
   $\overset{-2}{S^{2-}} \rightarrow \overset{+6}{S} + 8e^-$
   $H_2S + 8HNO_3 \rightarrow H_2SO_4 + 8NO_2 + 4H_2O$

   (d) $\overset{0}{Cu} + \overset{+1+5-2}{HNO_3} \rightarrow \overset{+2+5-2}{Cu(NO_3)_2} + \overset{+2-2}{NO} + \overset{+1-2}{H_2O}$
   $3[\overset{0}{Cu} \rightarrow \overset{+2}{Cu^{2+}} + 2e^-]$
   $2[\overset{+5}{N} + 3e^- \rightarrow \overset{+2}{N}]$
   $3Cu + 8HNO_3 \rightarrow 3Cu(NO_3)_2 + 2NO + 4H_2O$

   (e) $\overset{+1+5-2}{KClO_3} \rightarrow \overset{+1-1}{KCl} + \overset{0}{O_2}$
   $2[\overset{+5}{Cl} + 6e^- \rightarrow \overset{-1}{Cl^-}]$
   $3[\overset{-2}{2O} \rightarrow \overset{0}{O_2} + 4e^-]$
   $2KClO_3 \rightarrow 2KCl + 3O_2$

   (f) $\overset{+1-2+10}{KOH} + \overset{0}{Cl_2} \rightarrow \overset{+1+5-2}{KClO_3} + \overset{+1-1}{KCl} + \overset{+1\ -2}{H_2O}$
   $\overset{0}{Cl_2} \rightarrow \overset{+5}{2Cl} + 10e^-$
   $5[\overset{0}{Cl_2} + 2e^- \rightarrow \overset{-1}{2Cl^-}]$
   $6KOH + 3Cl_2 \rightarrow KClO_3 + 5KCl + 3H_2O$

   (g) $\overset{0}{S} + \overset{+1+5-2}{HNO_3} \rightarrow \overset{+4-2}{SO_2} + \overset{+2-2}{NO} + \overset{+1-2}{H_2O}$
   $3[\overset{0}{S} \rightarrow \overset{+4}{S} + 4e^-]$
   $4[\overset{+5}{N} + 3e^- \rightarrow \overset{+2}{N}]$
   $3S + 4HNO_3 \rightarrow 3SO_2 + 4NO + 2H_2O$

   (h) $\overset{0}{As_4} + \overset{+1+5-2}{HNO_3} \rightarrow \overset{+1+5-2}{HAsO_3} + \overset{+2-2}{NO} + \overset{+1-2}{H_2O}$
   $3[\overset{0}{As_4} \rightarrow \overset{+5}{4As} + 20e^-]$
   $20[\overset{+5}{N} + 3e^- \rightarrow \overset{+2}{N}]$
   $3As_4 + 20HNO_3 \rightarrow 12HAsO_3 + 20NO + 4H_2O$

5. $\overset{0}{Zn} + \overset{+1+5-2}{HNO_3} \rightarrow \overset{+2+5-2}{Zn(NO_3)_2} + \overset{-3+1+5-2}{NH_4NO_3} + \overset{+1-2}{H_2O}$

$\overset{+5}{N} + 8e^- \rightarrow \overset{-3}{N}$

$4[\overset{0}{Zn} \rightarrow 4\overset{+2}{Zn^{2+}} + 2e^-]$

$4Zn + 10HNO_3 \rightarrow 4Zn(NO_3)_2 + NH_4NO_3 + 3H_2O$

6. $\overset{0}{Zn} + \overset{+1+6-2}{Na_2CrO_4} + \overset{+1-2+1}{NaOH} \rightarrow \overset{+1+2-2}{Na_2ZnO_2} + \overset{+1+3-2}{NaCrO_2} + \overset{+1-2}{H_2O}$

$2[\overset{+6}{Cr} + 3e^- \rightarrow 2\overset{+3}{Cr}]$

$3[\overset{0}{Zn} \rightarrow \overset{2+}{Zn^{2+}} + 2e^-]$

$3Zn + 2Na_2CrO_4 + 4NaOH \rightarrow$
$3Na_2ZnO_2 + 2NaCrO_2 + 2H_2O$

7. (a) $K + H_2O \rightarrow KOH + H_2$

(b) $\overset{0}{K} + \overset{+1-2}{H_2O} \rightarrow \overset{+1}{K^+} + \overset{-2+1}{OH^-} + \overset{0}{H_2}$

(c) $\overset{+1-2}{2H_2O} + 2e^- \rightarrow \overset{0}{H_2} + 2OH^-$

(d) $\overset{0}{K} \rightarrow \overset{+1}{K^+} + e^-$

(e) $2H_2O + 2e^- \rightarrow H_2 + 2OH^-$

$2K \rightarrow 2\overset{+1}{K^+} + 2e^-$

(f) $2K + H_2O \rightarrow 2KOH + H_2$

8. (a) $HI + HNO_2 \rightarrow NO + I_2 + H_2O$

$\overset{+1}{H^+} + \overset{-1}{I^-} + \overset{+1}{H^+} + \overset{+3-2}{NO_2^-} \rightarrow \overset{+2-20}{NO} + \overset{}{I_2} + \overset{+1-2}{H_2O}$

$2\overset{-1}{I^-} \rightarrow \overset{0}{I_2} + 2e^-$

$2[\overset{+3-2}{NO_2^-} + e^- + 2H^+ \rightarrow \overset{+2-2}{NO} + H_2O]$

(b) $FeCl_3 + H_2S \rightarrow FeCl_2 + HCl + S$

$\overset{+3}{Fe^{3+}} + 3\overset{-1}{Cl^-} + \overset{+1-2}{H_2S} \rightarrow \overset{+2}{Fe^{2+}} + 2\overset{-1}{Cl^-} + \overset{+1}{H^+} + \overset{-1}{Cl^-} + \overset{0}{S}$

$2[\overset{+3}{Fe^{3+}} + e^- \rightarrow \overset{+2}{Fe^{2+}}]$

$\overset{+1-2}{H_2S} \rightarrow \overset{0}{S} + 2e^- + 2H^+$

$2FeCl_3 + H_2S \rightarrow 2FeCl_2 + 2HCl + S$

(c) $SbCl_5 + KI \rightarrow KCl + I_2 + SbCl_3$

$\overset{+5-1}{SbCl_5} + \overset{+1-1}{K^+I^-} \rightarrow \overset{+1}{K^+} + \overset{-1}{Cl^-} + \overset{0}{I_2} + \overset{+3-1}{SbCl_3}$

$\overset{+5-1}{SbCl_5} + 2e^- \rightarrow \overset{+3-1}{SbCl_3} + 2\overset{-1}{Cl^-}$

$2\overset{-1}{I^-} \rightarrow \overset{0}{I_2} + 2e^-$

$SbCl_5 + 2KI \rightarrow 2KCl + I_2 + SbCl_3$

(d) $Ca(OH)_2 + NaOH + ClO_2 + C \rightarrow$
$NaClO_2 + CaCO_3 + H_2O$

$\overset{+2}{Ca^{2+}} + \overset{-2+1}{OH^-} + \overset{+1}{Na^+} + \overset{-2+1}{OH^-} + \overset{+4-2}{ClO_2} + \overset{0}{C} \rightarrow$

$\overset{+1}{Na^+} + \overset{+3-2}{ClO_2^-} + \overset{+2}{Ca^{2+}} + \overset{+4-2}{CO_3^{2-}} + \overset{+1-2}{H_2O}$

$\overset{0}{C} + 6\overset{-2+1}{OH^-} \rightarrow \overset{+4-2}{CO_3^{2-}} + 4e^- + 3H_2O$

$4[\overset{+4-2}{ClO_2} + e^- \rightarrow \overset{+3-2}{ClO_2^-}]$

$Ca(OH)_2 + 4NaOH + 4ClO_2 + C \rightarrow 4NaClO_2 +$
$CaCO_3 + 3H_2O$

9. (a)
	$E° = -0.34\ V$
$\overset{0}{Cu} \rightarrow \overset{+2}{Cu^{2+}} + 2e^-$	$E° = -0.34\ V$
$2[\overset{+1}{Ag^+} + e^- \rightarrow \overset{0}{Ag}]$	$E° = +0.80\ V$
$Cu + 2Ag^+ \rightarrow Cu^{2+} + 2Ag$	$E° = +0.46\ V$

(b)
$\overset{0}{Cd} \rightarrow \overset{+2}{Cd^{2+}} + 2e^-$	$E° = +0.40\ V$
$\overset{+2}{Co^{2+}} + 2e^- \rightarrow \overset{0}{Co}$	$E° = -0.28\ V$
$Cd + Co^{2+} \rightarrow Cd^{2+} + Co$	$E° = +0.12\ V$

(c)
$Ni^{2+} + 2e^- \rightarrow Ni$	$E° = -0.23\ V$
$2[Na \rightarrow Na^+ + e^-]$	$E° = +2.71\ V$
$2Na + Ni^{2+} \rightarrow 2Na^+ + Ni^0$	$E° = +2.48\ V$

(d)
$Br_2 + 2e^- \rightarrow 2Br^-$	$E° = +1.06\ V$
$2I^- \rightarrow I_2 + 2e^-$	$E° = -0.54\ V$
$Br_2 + 2I^- \rightarrow 2Br^- + I_2$	$E° = +0.52\ V$

(e)
$2[Au^{3+} + 3e^- \rightarrow Au]$	$E° = +1.42\ V$
$3[Mg \rightarrow Mg^{2+} + 2e^-]$	$E° = +2.38\ V$
$3Mg + 2Au^{3+} \rightarrow 2Au + 3Mg^{2+}$	$E° = +3.80\ V$

10. (a)
$Mg \rightarrow Mg^{2+} + 2e^-$	$E° = +2.38\ V$
$Sn^{2+} + 2e^- \rightarrow Sn$	$E° = -0.14\ V$
$Mg + Sn^{2+} \rightarrow Mg^{2+} + Sn$	$E° = +2.24\ V$

The reaction does occur spontaneously.

(b)
$3[K \rightarrow K^+ + e^-]$	$E° = +2.92\ V$
$Al^{3+} + 3e^- \rightarrow Al$	$E° = -1.71\ V$
$3K + Al^{3+} \rightarrow 3K^+ + Al$	$E° = +1.21\ V$

The reaction does occur spontaneously.

(c)
$2[Li^+ + e^- \rightarrow Li]$	$E° = -3.04\ V$
$Zn \rightarrow Zn^{2+} + 2e^-$	$E° = +0.76\ V$
$2Li^+ + Zn \rightarrow 2Li + Zn^{2+}$	$E° = -2.28\ V$

The reaction does not occur spontaneously.

(d)
$Cu \rightarrow Cu^{2+} + 2e^-$	$E° = -0.34\ V$
$Cl_2 + 2e^- \rightarrow 2Cl^-$	$E° = +1.36\ V$
$Cu + Cl_2 \rightarrow Cu^{2+} + 2Cl^-$	$E° = +1.02\ V$

The reaction does occur spontaneously.

(e)
$Zn \rightarrow Zn^{2+} + 2e^-$	$E° = +0.76\ V$
$Fe^{2+} + 2e^- \rightarrow Fe$	$E° = -0.41\ V$
$Zn + Fe^{2+} \rightarrow Zn^{2+} + Fe$	$E° = +0.35\ V$

The reaction does occur spontaneously.

Application Questions

1. In a reaction system, in order for particles to *gain* electrons and undergo *reduction*, other particles must lose electrons and thus undergo *oxidation*. In order for both changes to occur, the two processes must take place simultaneously so that the total increase in oxidation number equals the total decrease.

2. The assignment of oxidation numbers to such species allows for a determination of the partial loss or gain of electrons in compounds that are not ionic.

3. The oxidized substance, since it has lost electrons, has made those electrons available to another substance so that the other substance can be reduced. The reduced substance, in gaining electrons, is thus able to serve as an oxidizing agent in that it receives the electrons lost by the other substance in the process of that other substance being oxidized.

4. Since fluorine is the most highly electronegative atom, it forms negative ions readily and thus is the most active oxidiz-

ing agent. However, because of its strong attraction for its own electrons, the fluoride ion does not easily give them up and is thus a weak reducing agent.

5. (a) Yes (b) Yes (c) No (d) Yes (e) No (f) Yes

6. In an electrochemical cell the redox reaction is spontaneous and, as such, is able to provide electrical energy. In an electrolytic cell the redox reaction is nonspontaneous and thus requires external energy to drive it.

7. In rechargeable cells, oxidation-reduction reactions characteristic of electrochemical cells are used as a source of electrical energy through the conversion of chemical energy into electrical energy during the discharge cycle. When such cells are being charged, electric energy is converted to chemical energy by the oxidation-reduction reactions characteristic of electrolytic cells.

8. The metals at the top of the table are strong reducing agents. They readily give up electrons to oxidized substances appearing below them. This property diminishes, moving down the table, to the extent that metals near the bottom give up electrons only in the presence of strong oxidizing agents.

Application Problems

1. (a)
$$\overset{+3-1}{AuCl_3} + \overset{+3-2}{Sb_2O_3} + \overset{+1-2}{H_2O} \rightarrow \overset{0}{Au} + \overset{+5-2}{Sb_2O_5} + \overset{+1-1}{HCl}$$

$$3[\overset{+3}{Sb} \rightarrow \overset{+5}{Sb} + 2e^-]$$

$$2[\overset{+3}{Au} + 3e^- \rightarrow \overset{0}{Au}]$$

(Since $\frac{3}{2}$ would be required as the coefficient of both Sb_2O_3, the coefficients in the two electronic equations are doubled before being placed in the overall equation.)

$$4AuCl_3 + 3Sb_2O_3 + 6H_2O \rightarrow 4Au + 3Sb_2O_5 + 12HCl$$

(b)
$$\overset{+1+7-2}{KMnO_4} + \overset{+1+6-2}{H_2SO_4} + \overset{+1-1}{KBr} \rightarrow \overset{+1+6-2}{K_2SO_4} + \overset{+2+6-2}{MnSO_4} + \overset{0}{Br_2} + \overset{+1-2}{H_2O}$$

$$2[\overset{+7}{Mn} + 5e^- \rightarrow \overset{+2}{Mn^{2+}}]$$

$$5[\overset{-1}{2Br^-} \rightarrow \overset{0}{Br_2} + 2e^-]$$

$$2KMnO_4 + 8H_2SO_4 + 10KBr \rightarrow 6K_2SO_4 + 2MnSO_4 + 5Br_2 + 8H_2O$$

2.
$$\overset{+1+4-2}{K_2CO_3} + \overset{0}{Br_2} \rightarrow \overset{+1-1}{KBr} + \overset{+1+5-2}{KBrO_3} + \overset{+4-2}{CO_2}$$

$$\overset{0}{Br_2} \rightarrow \overset{+5}{2Br} + 10e^-$$

$$5[\overset{0}{Br} + 2e^- \rightarrow \overset{-1}{2Br^-}]$$

$$3K_2CO_3 + 3Br_2 \rightarrow 5KBr + KBrO_3 + 3CO_2$$

3.
$$\overset{+1+7-2}{KMnO_4} + \overset{+1+4-2}{Na_2SO_3} + \overset{+1+6-2}{H_2SO_4} \rightarrow \overset{+1+6-2}{K_2SO_4} + \overset{+2+6-2}{MnSO_4} + \overset{+1+6-2}{Na_2SO_4} + \overset{+1-2}{H_2O}$$

$$2[\overset{+7}{Mn} + 5e^- \rightarrow \overset{+2}{Mn}]$$

$$5[\overset{+4}{S} \rightarrow \overset{+6}{S} + 2e^-]$$

$$2KMnO_4 + 5Na_2SO_3 + 3H_2SO_4 \rightarrow K_2SO_4 + 2MnSO_4 + 5Na_2SO_4 + 3H_2O$$

4. (a)
$$\overset{+1+5-2}{HNO_3} + \overset{0}{Cu} \rightarrow \overset{+2+5-2}{Cu(NO_3)_2} + \overset{+4-2}{NO_2} + \overset{+1-2}{H_2O}$$

$$2[\overset{+5}{N} + e^- \rightarrow \overset{+4}{N}]$$

$$\overset{0}{Cu} \rightarrow \overset{+2}{Cu} + 2e^-$$

$$Cu + 4HNO_3 \rightarrow Cu(NO_3)_2 + 2NO_2 + 2H_2O$$

(b)
$$4H^+(aq) + 2\overset{+5}{NO_3^-} + 2e^- \rightarrow 2\overset{+4}{NO_2} + 2H_2O$$

$$\overset{0}{Cu} \rightarrow \overset{+2}{Cu^{2+}} + 2e^-$$

$$Cu + 4H^+(aq) + 2NO_3^- \rightarrow Cu^{2+} + 2NO_2 + 2H_2O$$

or $Cu + 4HNO_3 \rightarrow Cu(NO_3)_2 + 2NO_2 + 2H_2O$

5. (a)
$$\overset{+1+6-2}{H_2SO_4} + \overset{0}{Zn} \rightarrow \overset{+2+6-2}{ZnSO_4} + \overset{+1-2}{H_2S} + \overset{+1-2}{H_2O}$$

$$\overset{+6}{S} + 8e^- \rightarrow \overset{-2}{S}$$

$$4[\overset{0}{Zn} \rightarrow \overset{+2}{Zn} + 2e^-]$$

$$4Zn + 5H_2SO_4 \rightarrow 4ZnSO_4 + H_2S + 4H_2O$$

(b)
$$10H^+(aq) + \overset{+1}{SO_4^{2-}} + 8e^- \rightarrow \overset{+1-2}{H_2S} + \overset{+1-2}{4H_2O}$$

$$4[\overset{0}{Zn} \rightarrow \overset{+2}{Zn^{2+}} + 2e^-]$$

$$4Zn + 10H^+(aq) + SO_4^{2-} \rightarrow 4Zn^{2+} + H_2S + 4H_2O$$

or $4Zn + 5H_2SO_4 \rightarrow 4ZnSO_4 + H_2S + 4H_2O$

6. $\overset{+1}{8H^+}(aq) + \overset{+5}{2NO_3^-} + 6e^- \rightarrow \overset{+2}{2NO} + \overset{+1-2}{4H_2O}$

$$\dfrac{\overset{+2}{3[Zn \rightarrow Zn^{2+} + 2e^-]}}{3Zn + 8H^+(aq) + 2NO_3^- \rightarrow 3Zn^{2+} + 2NO + 4H_2O}$$

or $3Zn + 8HNO_3 \rightarrow 3Zn(NO_3)_2 + 2NO + 4H_2O$

7. $\overset{+7}{16H^+}(aq) + 2MnO_4^- + 10e^- \rightarrow \overset{+2}{2Mn}^{2+} + 8H_2O$

$$\dfrac{\overset{+2}{10[Fe^{2+}} \rightarrow \overset{+3}{Fe^{3+}} + e^-]}{16H^+(aq) + 2MnO_4^- + 10Fe^{2+} \rightarrow 2Mn^{2+} + 10Fe^{3+} + 8H_2O}$$

or

$2KMnO_4 + 10FeSO_4 + 8H_2SO_4 \rightarrow K_2SO_4 + 2MnSO_4 + 5Fe_2(SO_4)_3 + 8H_2O$

8. *(a)* $3[Na \rightarrow Na^+ + e^-]$
$Cr^{3+} + 3e^- \rightarrow Cr$
Overall: $3Na + Cr^{3+} \rightarrow 3Na^+ + Cr$
(b) $Pb \rightarrow Pb^{2+} + 2e^-$
$2[Ag^+ + e^- \rightarrow Ag]$
Overall: $Pb + 2Ag^+ \rightarrow Pb^{2+} + 2Ag$
(d) $F_2 + 2e^- \rightarrow 2F^-$
$2I^- \rightarrow I_2 + 2e^-$
Overall: $F_2 + 2I^- \rightarrow 2F^- + I_2$
(f) $2[Al \rightarrow Al^{3+} + 3e^-]$
$3[Pb^{2+} + 2e^- \rightarrow Pb]$
Overall: $2Al + 3Pb^{2+} \rightarrow 2Al^{3+} + 3Pb$

9.

	Actual Values	Hypothetical Values
$Br_2 + 2e^- \rightleftarrows 2Br\text{-}$	$+1.06$ V	$(1.06 \text{ V} - 0.54 \text{ V}) = 0.52$ V
$I_2 + 2e^- \rightleftarrows 2I^-$	$+0.54$ V	0 V
$Al^{3+} + 3e^- \rightleftarrows Al$	-1.71 V	$(-1.71 \text{ V} - 0.54 \text{ V}) - 2.25$ V

(a) $+0.52$ V *(b)* -2.25 V
(c) original $= +1.06$ V $- 0.54$ V $= +0.52$ V
new $= +0.52$ V $- 0 = +0.52$ V
No change would be expected and computations verify that none would occur.

10.

$Ni \rightarrow Ni^{2+} + 2e^-$	$E° = +0.23$ V
$2[Ag^+ + e^- \rightarrow Ag]$	$E° = +0.80$ V
$Ni + 2Ag^+ \rightarrow Ni^{2+} + 2Ag$	$E° = +1.03$ V

The Ni would gradually enter the solution as Ni^{2+} ions.

11. *(a)*

$2[Al \rightarrow Al^{3+} + 3e^-]$	$E° = +1.71$ V
$3[Zn^{2+} + 2e^- \rightarrow Zn]$	$E° = -0.76$ V
$2Al + 3Zn^{2+} \rightarrow 2Al^{3+} + 3Zn$	$E° = +0.95$ V

The Al spoon would begin to disintegrate as its atoms entered the solution as Al^{3+} ions.

(b)

$3[Zn \rightarrow Zn^2 + 2e^-]$	$E° = +0.76$ V
$2[Al^{3+} + 3e^- \rightarrow Al]$	$E° = -1.71$ V
$3Zn + 2Al^{3+} \rightarrow 3Zn^{2+} + 2Al$	$E° = -0.95$ V

The Zn strip could be used since the negative $E°$ value indicates that no reaction would occur.

12.

$2[Al \rightarrow Al^{3+} + 3e^-]$	$E° = +1.71$ V
$3[Sn^{2+} + 2e^- \rightarrow Sn]$	$E° = -0.14$ V
$2Al + 3Sn^{2+} \rightarrow 2Al^{3+} + 3Sn$	$E° = +1.57$ V

The Al container could not be used to store the $Sn(NO_3)_2$ solution because the two substances would react. Al would enter the solution as Al^{3+} ions, causing the container to disintegrate.

Teacher's Notes

Chapter 20 Oxidation–Reduction Reactions

INTRODUCTION

This chapter deals with a large and varied class of chemical reactions. Many of these reactions, which were first introduced in Chapter 6, involve electron transfer. As you will see, they have many important applications in industry and everyday life.

LOOKING AHEAD

As you study this chapter, pay special attention to the role of electrons in chemical reactions. Be aware of changes in electric charge and the numbers of electrons involved in each chemical reaction.

SECTION PREVIEW

20.1 The Nature of Oxidation and Reduction
20.2 Balancing Redox Equations
20.3 Oxidizing and Reducing Agents
20.4 Electrochemistry

During steelmaking, impurities in iron are removed by oxidation. Many other industrial processes also involve oxidation—reduction reactions

20.1 The Nature of Oxidation and Reduction

In Chapter 7, you first learned about oxidation numbers, or numbers that are assigned to atoms and ions as a way of keeping track of electrons. Table 20-1 summarizes the set of rules by which oxidation numbers are assigned. In this section, you will learn to apply your knowledge of these numbers to an analysis of chemical reactions.

Oxidation

Reactions in which the atoms or ions of an element attain a more positive (or less negative) oxidation state were described in Section 7.2 as oxidation processes. A species that undergoes a change in oxidation state in a positive mathematical direction is said to be **oxidized**.

The combustion of metallic sodium in an atmosphere of chlorine gas illustrates the oxidation process (Figure 20-1). The product is sodium chloride, and the sodium–chlorine bond is ionic. The equation for this reaction is

$$2Na(s) + Cl_2(g) \rightarrow 2NaCl(s)$$

In this reaction, each sodium atom loses an electron and becomes a sodium ion.

$$Na \rightarrow Na^+ + e^-$$

The oxidation state of sodium has changed from the 0 state of the atom (Rule 1, Table 20-1) to the more positive +1 state of the ion (Rule 2). The sodium atom is said to be *oxidized* to the sodium ion. This change in oxidation state is indicated by the oxidation numbers placed above the symbol of the atom and the ion. For each sodium atom,

$$\overset{0}{Na} \rightarrow \overset{+1}{Na^+} + e-$$

A second example of an oxidation process occurs when hydrogen burns in chlorine to form molecular hydrogen chloride gas. The hydrogen–chlorine bond is covalent. The equation for this reaction is

$$H_2(g) + Cl_2(g) \rightarrow 2HCl(g)$$

By Rule 1, the oxidation state of each hydrogen atom in the hydrogen molecule is 0. By Rule 5, the oxidation state of the combined hydrogen in the HCl molecule is +1. Because the hydrogen atom has changed from the 0 to the more positive +1 oxidation state, this change is an oxidation process.

$$\overset{0}{H_2} + Cl_2 \rightarrow 2\overset{+1}{HCl}$$

Reduction

Reactions in which the atoms or ions of an element attain a more negative (or less positive) oxidation state are recognized as **reduction** *processes. A species that undergoes a change in oxidation state in a negative sense is said to be* **reduced**. Consider the behavior of chlorine in the above reaction

- Define oxidation and reduction.
- Assign oxidation number to reactant and product species.
- Explain what is meant by oxidation–reduction reaction, (redox reaction).

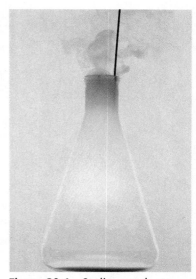

Figure 20-1 Sodium and chlorine react violently.

The reduced chlorine atom has 17 protons and 18 electrons.

with sodium and hydrogen. With sodium, each chlorine atom acquires an electron and becomes a chloride ion. The oxidation state of chlorine changes from the 0 state of the chlorine atom (Rule 1) to the more negative -1 state of the chloride ion (Rule 2). The chlorine atom is reduced to the chloride ion. For each chlorine atom,

$$\overset{0}{Cl} + e^- \rightarrow \overset{-1}{Cl^-}$$

In the reaction of chlorine with hydrogen, the pair of electrons shared by the hydrogen and chlorine atoms in the hydrogen-chloride molecule is not shared equally. Rather, the pair of electrons is more strongly attracted to the chlorine atom because of its higher electronegativity. By Rules 5 and 6, chlorine is assigned an oxidation number of -1. Thus, the chlorine atom is changed from the 0 to the -1 oxidation state, a reduction process.

$$\overset{0}{H_2} + \overset{-1}{Cl_2} \rightarrow 2HCl$$

Assignment of oxidation numbers to the atoms of covalent molecular species in this somewhat arbitrary way is useful in studying reactions. The numbers allow determination of the partial loss or gain of electrons in compounds that are not ionic. For the hydrogen-chloride molecule, the oxidation number of the hydrogen atom is $+1$, and that of the chlorine atom is -1.

$$\overset{0}{H_2} + \overset{0}{Cl_2} \rightarrow 2\overset{+1\,-1}{HCl}$$

Decreases in charge or in oxidation number can be seen in terms of complete or partial gains of electrons. Reduction can thus be understood as a process in which electrons are gained.

Oxidation and Reduction as a Process

If oxidation occurs during a chemical reaction, then reduction must occur simultaneously. Furthermore, the amount of oxidation must match the amount of reduction. In other words, the total increase in oxidation number of the species being oxidized must equal the total decrease in oxidation number of

The number of electrons lost from the particles that have been oxidized must be gained by the particles that are reduced.

the species being reduced. *Any chemical process in which elements undergo a change in oxidation number is an* **oxidation–reduction reaction**. The name is often shortened to "redox" reaction.

Observe that as metallic sodium burns in chlorine gas, two sodium atoms are oxidized to Na^+ ions as one diatomic chlorine molecule is reduced to two Cl^- ions. Elemental sodium and elemental chlorine are chemically equivalent atom for atom and ion for ion. That is, for every electron lost by a sodium atom, one electron is gained by a chlorine atom.

$$\overset{0}{2Na} \rightarrow \overset{+1}{2Na^+} + 2e^- \quad \text{(oxidation)}$$
$$\overset{0}{Cl_2} + 2e^- \rightarrow \overset{-1}{2Cl^-} \quad \text{(reduction)}$$
$$\overline{\overset{0}{2Na} + \overset{0}{Cl_2} \rightarrow \overset{+1}{Na} + \overset{-1}{Cl^-}} \quad \text{(redox reaction)}$$

Suppose a scheme is devised to recover metallic aluminum from an aluminum salt. In this scheme, sodium atoms are oxidized to Na^+ ions, and Al^{3+} ions are reduced to aluminum atoms. For these oxidation and reduction processes to be chemically equivalent, three Na^+ ions must be formed for each Al^{3+} ion reduced.

$$\overset{0}{3Na} \rightarrow \overset{+1}{3Na^+} + 3e^- \quad \text{(oxidation)}$$
$$\overset{+3}{Al^{3+}} + 3e^- \rightarrow \overset{0}{Al} \quad \text{(reduction)}$$
$$\overline{\overset{0}{3Na} + \overset{+3}{Al^{3+}} \rightarrow \overset{+1}{3Na^+} + \overset{0}{Al}} \quad \text{(redox reaction)}$$

Observe that in the combined equation for each example, both electric charge and atoms are conserved, and that the oxidation and reduction processes are indeed equivalent.

Many of the reactions studied in elementary chemistry involve oxidation–reduction processes. Of the three examples considered so far in this chapter, two are composition reactions. (Figure 20-2 shows another composition redox reaction.) The third example is a replacement reaction. Not all composition reactions are oxidation–reduction reactions, however. For example, sulfur-dioxide gas (SO_2) dissolves in water to form an acidic solution containing a low concentration of sulf*ous* acid (H_2SO_3).

$$\overset{+4-2}{SO_2} + \overset{+1-2}{H_2O} \rightarrow \overset{+1+4-2}{H_2SO_3}$$

Observe that the oxidation states of all elemental species remain unchanged in this composition reaction. It is therefore not a redox reaction.

When a solution of sodium chloride is added to a solution of silver nitrate, an ion-exchange reaction occurs, and a white precipitate of silver chloride separates.

$$\overset{+1}{Na^+} + \overset{-1}{Cl^-} + \overset{+1}{Ag^+} + \overset{+5-2}{NO_3^-} \rightarrow \overset{+1}{Na^+} + \overset{+5-3}{NO_3^-} + \overset{+1\ -1}{Ag^+Cl^-}$$

or, more simply, by omitting spectator ions,

$$\overset{+1}{Ag^+}(aq) + \overset{-1}{Cl^-}(aq) \rightarrow \overset{+1\ -1}{Ag^+Cl^-}(s)$$

Figure 20-2 The combustion of antimony in chlorine is an oxidation–reduction reaction.

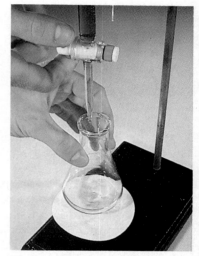

Figure 20-3 As $KMnO_4$ solution is added to an acidic solution of $FeSO_4$, Fe^{2+} ions are oxidized to Fe^{3+} ions, and the red MnO_4^- ions are changed to colorless Mn^{2+} ions. When all Fe^{2+} ions are oxidized, MnO_4^- ions are no longer changed to colorless Mn^{2+} ions. Thus the first faint appearance of the MnO_4^- color indicates the end point of the titration.

Desktop Investigation

Redox Reactions

Question
How do you recognize oxidation–reduction reactions in the laboratory?

Materials
aspirin, 325 mg tablet
liquid bleach, 5.25% NaClO solution
bunsen burner
forceps
hydrogen peroxide
manganese dioxide
stirring rod
steel wool
2 test tubes, 16 x 150 mm
5 test tubes, 25 x 150 mm
test tube clamp
white vinegar, 6% $HC_2H_3O_2$
water, distilled
wooden splints

Procedure
CAUTION: Goggles and apron should be worn during this investigation. Record all your results in a data table.

1. (a) Put 10 mL of hydrogen peroxide in a 16 x 150 mm test tube and add a small amount of manganese dioxide (equal to about half a pea). What is the result? (b) After the reaction contin-

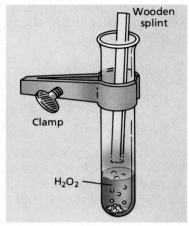

Wooden splint

Clamp

H_2O_2

ues for 1 minute, test for oxygen by inserting a glowing wooden splint into the test tube. What is the result? (See diagram.)

2. Using forceps, heat a small roll of steel wool (about 3 cm in length) in the burner flame. What is the result?

3. (a) Repeat procedure 1(a). (b) Heat a small roll of steel wool in the burner flame and immediately insert it into the test tube. What is the result?

4. Using 5 large test tubes, add (a) 10 mL of distilled water to the first test tube; (b) 10 mL of bleach to the

second test tube (c) 10 mL of white vinegar to the third test tube (d) 9.5 mL of distilled water, 5 drops of bleach, and 5 drops of vinegar to the fourth test tube (stir) (e) 1 aspirin tablet and 10 mL of distilled water to the fifth test tube (stir). Then add a 2.5 cm ball of steel wool to each of the five test tubes. Note the speed of reaction and color changes. Observe the reactions over a 24 hour-period. What are the results?

Discussion
1. Write general equations to show what happened in the reactions.

2. Assign oxidation numbers to the elements in the general equations for 1(a), 2, 4(b), and 4(c), and write electronic equations to show which substance was oxidized and which was reduced.

3. How do you identify the oxidizing reaction and the reducing reaction? Give two examples.

4. Write a good conclusion for this experiment.

Peroxides are the only compounds in which oxygen has an oxidation state of -1 instead of -2. (peroxide ion: O_2^{2-}, hydrogen peroxide is H_2O_2: H—O—O—H.)

The oxidation state of each monatomic ion remains unchanged. This reaction is not an oxidation–reduction reaction.

Redox reactions sometimes involve polyatomic ions. An example is the redox reactions of permanganate ion (MnO_4^-), in which Mn has an oxidation number $+7$ (Rules 4 and 8). Under certain conditions, this MnO_4^- ion is changed to the *manganese(II) ion*, Mn^{2+}, as shown in Figure 20-3. Under other conditions, the MnO_4^- ion is changed to the nearly colorless manganate ion, MnO_4^{2-}, in which MnO_4^{2-} has an oxidation number $+6$ (Rules 4 and 8).

To correctly balance the equation for a redox reaction requires knowing all changes in oxidation state that occur in the reaction. The substance that is oxidized and the substance that is reduced must be identified. In the following electronic equations, the substances on the left are oxidized.

$$\overset{0}{Na} \rightarrow \overset{+1}{Na^+} + e^-$$
$$\overset{0}{Fe} \rightarrow \overset{+2}{Fe^{2+}} + 2e^-$$
$$\overset{+2}{Fe^{2+}} \rightarrow \overset{+3}{Fe^{3+}} + e^-$$
$$\overset{-1}{2Cl^-} \rightarrow \overset{0}{Cl_2} + 2e^-$$

The difference between oxidation numbers indicates the change in oxidation state. The chloride ion is assigned the oxidation number -1, a negative oxidation state. Oxidation results in the change of the oxidation numbers in a positive mathematical direction. The change from -1 to 0 is a change in a positive mathematical direction, as is the change from 0 to $+1$, from 0 to $+2$, or from $+2$ to $+3$.

In the electronic equations that follow, the substances on the left are reduced.

$$\overset{+1}{Na^+} + e^- \rightarrow \overset{0}{Na}$$
$$\overset{0}{Cl_2} + 2e^- \rightarrow \overset{-1}{2Cl^-}$$
$$\overset{+3}{Fe^{3+}} + e^- \rightarrow \overset{+2}{Fe^{2+}}$$
$$\overset{+2}{Cu^{2+}} + 2e^- \rightarrow \overset{0}{Cu}$$
$$\overset{0}{Br_2} + 2e^- \rightarrow \overset{-1}{2Br^-}$$

In the third equation, the iron(III) ion, oxidation number $+3$, is reduced to the iron(II) state, with oxidation number $+2$. The reduction results in the change of the oxidation number in a negative direction. The change from $+3$ to $+2$ is a change in a negative direction, as is the change from $+1$ to 0, 0 to -1, $+3$ to $+2$, and from $+2$ to 0.

Reactants and products in redox reactions are not limited to monatomic ions and uncombined elements. Molecular compounds or polyatomic ions containing elements that have more than one nonzero oxidation state can also be oxidized and reduced. For example, sulfur dioxide when (SO_2) is converted to sulfur dioxide (SO_3),

$$\cdots + \overset{+4}{SO_2} \rightarrow \overset{+6}{SO_3} + \cdots$$

sulfur is oxidized. (Usually we refer to the oxidation or reduction of the entire molecule or ion, e.g., "sulfur dioxide is oxidized to sulfur trioxide.") In a reaction in which permanganate ion (MnO_4^-) ion is converted to manganese dioxide (MnO_2),

$$\cdots + \overset{+7}{MnO_4^-} \rightarrow \overset{+4}{MnO_2} + \cdots$$

manganese is reduced.

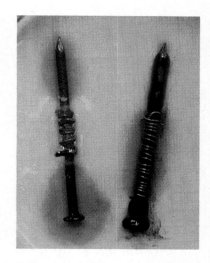

Figure 20-4 Oxidation–reduction reactions. Top photo, zinc wrapped around iron nail, iron reduced; copper wrapped around iron nail, iron oxidized. Bottom photo, when a strip of zinc is immersed in an aqueous solution of copper (II) sulfate, what substance is oxidized? What is reduced?

Species with oxidation states of $+7$ always are good oxidizing agents.

SECTION OBJECTIVES

- Explain what must be conserved in redox equations.
- Balance redox equations by using the oxidation-number method.
- Balance redox equations by using the ion-electron method.

Section Review

1. What is oxidation? Give an example of this process.
2. What is reduction? Give an example.
3. How are oxidation numbers assigned?
4. What is meant by "redox reaction"?
5. Which of the following are oxidation–reduction reactions? Among those that are, identify the substance oxidized and that reduced.
   (a) $SO_3(g) + H_2O(\ell) \rightarrow H_2SO_4(aq)$ (b) $2KNO_3(s) \rightarrow 2KNO_2(s) + O_2(g)$
   (c) $H_2(g) + CuO(s) \rightarrow Cu(s) + H_2O(\ell)$
6. For each redox reaction identified in the previous question, write the electronic equations for the substances oxidized and reduced in each.

20.2 Balancing Redox Equations

A principal use of oxidation numbers is in balancing equations for oxidation–reduction reactions. These reactions are recognized by the fact that changes in oxidation states occur in reactants that undergo oxidation or reduction. In the equation-balancing process, accommodation is made for the conservation of charge as well as for the conservation of atoms. Equations for simple redox reactions can be balanced by inspection, which you learned to do in Chapter 8. Most redox equations, however, require more systematic methods.

The use of a pair of electronic equations to achieve electron balance in a redox equation has been demonstrated in two examples in Section 20.1. Two of these methods are illustrated here. The first method is simply an extension of the general balancing procedure you already know how to use. The second scheme is commonly used for ionic equations.

Oxidation-Number Method

In the oxidation-number method the electron shift is balanced for the particles oxidized and the particles reduced by balancing the pair of electronic equations that represents this shift. Then the coefficients of all reactants and products are adjusted to achieve a balanced equation.

The *oxidation-number method* (or, as it is sometimes called, the *electron-transfer method*) of balancing a redox equation includes the steps listed below. Observe that *Steps 2, 3,* and *4* represent additions to the steps you are accustomed to using in equation-balancing. *Step 1: Write the skeleton equation for the reaction.* To do this, you must know the reactants and products and must represent each by the correct formula. *Step 2: Assign oxidation numbers to all atoms in all species and determine what is oxidized and what is reduced. Step 3: Write the electronic equation for the oxidation process and the electronic equation for the reduction process. Step 4: Adjust the coefficients in both electronic equations so that the number of electrons lost equals the number gained. Step 5: Place these coefficients into the skeleton equation. Step 6: Supply the proper coefficients for the rest of the equation to satisfy the conservation of atoms.*

These steps are illustrated by application first to a very simple oxidation–reduction equation that is easily balanced by inspection. Gaseous hydrogen sulfide burns in air to form sulfur dioxide and water. These facts are used to write the skeleton equation.

Step 1: $$H_2S + O_2 \rightarrow SO_2 + H_2O$$

Step 2: Next, oxidation numbers are assigned. Changes in oxidation numbers indicate that sulfur is oxidized from the -2 state to the $+4$ state; oxygen is reduced from the 0 state to the -2 state. The oxidation number of hydrogen remains the same; it therefore plays no part in the primary action of oxidation-reduction.

$$\overset{+1\ -2}{H_2S} + \overset{0}{O_2} \rightarrow \overset{+4-2}{SO_2} + \overset{+1\ -2}{H_2O}$$

Step 3: The change in oxidation state of sulfur requires the loss of 6 electrons: $(-2) - (+4) = -6$. The change in oxidation state of oxygen requires the gain of 2 electrons: $(0) - (-2) = +2$. The electronic equations for these two reactions are as follows.

$$\overset{-2}{S} \rightarrow \overset{+4}{S} + 6e^- \quad \text{(oxidation)}$$
$$\overset{0}{O} + 2e^- \rightarrow \overset{-2}{O} \quad \text{(reduction)}$$

Free oxygen is diatomic. Thus 4 electrons must be gained during the reduction of a molecule of free oxygen.

$$\overset{0}{O_2} + 4e^- \rightarrow \overset{-2}{2O}$$

Step 4: The coefficients of the two electronic equations are now adjusted so that the number of electrons lost in the oxidation of sulfur is equal to the number gained in the reduction of oxygen. The smallest number of electrons common to both equations is 12. To show the gain and loss of 12 electrons in the two equations, the oxidation equation is multiplied by 2, and the reduction equation is multiplied by 3.

$$\overset{-2}{2S} \rightarrow \overset{+4}{2S} + 12e^-$$
$$\overset{0}{3O_2} + 12e^- \rightarrow \overset{-2}{6O}$$

Step 5: Hence, the coefficients of H_2S and SO_2 are both 2, and the coefficient of O_2 is 3. Observe that the $\overset{-2}{6O}$ is divided between the two products SO_2 and H_2O. The 6 O atoms are accounted for by placing the coefficient 2 in front of each product formula including coefficients in the skeleton equation gives

$$2H_2S + 3O_2 \rightarrow 2SO_2 + 2H_2O$$

Step 6: The coefficients of the equation are adjusted in the usual way to satisfy the law of conservation of atoms. In this case, no further adjustments are needed; the equation is already balanced for both atoms and charges.

$$2H_2S + 3O_2 \rightarrow 2SO_2 + 2H_2O$$

As a second example, a redox equation that is slightly more difficult to balance will be used. This reaction occurs between manganese dioxide and hydrochloric acid (Figure 20-5). Water, manganese(II) chloride, and chlorine gas are formed.

Figure 20-5 The oxidation–reduction reaction between hydrochloric acid and manganese dioxide produces water, manganese(II) chloride and chlorine gas.

Steps 1 and 2: The skeleton equation is written and oxidation numbers are assigned to the elements in the reaction.

$$\overset{+4-2}{MnO_2} + \overset{+1-1}{HCl} \rightarrow \overset{+1-2}{H_2O} + \overset{+2-1}{MnCl_2} + \overset{0}{Cl_2}$$

They show that $\overset{+4}{Mn}$ is reduced to $\overset{+2}{Mn^{2+}}$, and some of the $\overset{-1}{Cl^-}$ is oxidized to $\overset{0}{Cl}$. Hydrogen and oxygen do not take part in the oxidation–reduction reaction.

Step 3: The electronic equations are

$$\overset{-1}{2Cl^-} \rightarrow \overset{0}{Cl_2} + 2e^- \qquad \text{(oxidation)}$$

$$\overset{+7}{Mn} + 5e^- \rightarrow \overset{+2}{Mn^{2+}} \qquad \text{(oxidation)}$$

Steps 4 and 5: The numbers of electrons lost and gained are the same. The coefficients are transferred to the skeleton equation, which becomes

$$MnO_2 + 2HCl \rightarrow H_2O + MnCl_2 + Cl_2$$

Step 6: The complete equation can now be balanced by inspection. Two additional molecules of HCl are needed to supply the two Cl^- ions of the $MnCl_2$. This balancing requires 2 molecules of water. These water molecules also account for the 2 oxygen atoms of the MnO_2. The final equation reads

$$MnO_2 + 4HCl \rightarrow 2H_2O + MnCl_2 + Cl_2$$

The equations for both of these examples could have been balanced without the use of electronic equations to balance electron shifts. The sample problem below involves a more complicated redox reaction, one in which use of electronic equations is necessary.

Sample Problem 20.1

The oxidation–reduction reaction between hydrochloric acid and potassium permanganate yields water, potassium chloride, manganese(II) chloride, and chlorine gas. Write the balanced equation.

Solution

We will balance this equation by using the stepwise, oxidation number method.

Step 1. $\qquad$ $HCl(aq) + KMnO_4(aq) \rightarrow H_2O(\ell) + KCl(aq) + MnCl_2(aq) + Cl_2(g)$

Step 2. $\qquad$ $\overset{+1\ -1}{HCl} + \overset{+1\ +7-2}{KMnO_4} \rightarrow \overset{+1\ -2}{H_2O} + \overset{+1\ -1}{KCl} + \overset{+2-1}{MnCl_2} + \overset{0}{Cl_2}$

Step 3. $\qquad$ Chlorine is oxidized from -1 to 0 and manganese is reduced from $+7$ to $+2$. Note that in the electronic equations we include only the symbol for the element oxidized or reduced, a coefficent showing how many atoms of an element are oxidized or reduced, and the number of electrons needed to bring about the oxidation number change.

$$\overset{-1}{2Cl^-} \rightarrow \overset{0}{Cl_2} + 2e^- \qquad \text{(oxidation)}$$

$$\overset{+7}{Mn} + 5e^- \rightarrow \overset{+2}{Mn^{2+}} \qquad \text{(oxidation)}$$

Sample Problem 20.1 *continued*

Step 4.   To equalize the number of electrons, the first equation must be multiplied by 5 and the second by 2, to give 10e$^-$ in both equations:

$$\overset{-1}{10Cl^-} \rightarrow \overset{0}{5Cl_2} + 10e^-$$

$$\overset{+7}{2Mn} + 10e^- \rightarrow \overset{+2}{2Mn^{2+}}$$

Step 5. $10HCl + 2KMnO_4 \rightarrow H_2O + KCl + 2MnCl_2 + 5Cl_5$

Step 6.   By inspection, $2KMnO_4$ produces $2KCl$ and $8H_2O$. Now $2KCl$ and $2MnCl_2$ call for 6 additional molecules of HCl. The balanced equation becomes

$$16HCl(aq) + 2KMnO_4(aq) \rightarrow 8H_2O(\ell) + 2KCl(aq) + 2MnCl_2(aq) + 5Cl_2(g)$$

Inspection reveals that the equation is indeed balanced for atoms and for overall oxidation number. Balancing a redox equation should always be completed by an evaluation step. Check to be sure that the number of atoms of each element is the same on both sides of the equation and that the total oxidation number on each side equals zero (for a formula equation).

Practice Problem

1. A pure solution of nitric acid (HNO_3) is colorless but on standing becomes yellow because of slight decomposition that forms brown nitrogen dioxide, oxygen, and water. Using the oxidation-number method, balance the redox equation for this decomposition reaction.

(*Ans.*) $\overset{+5\ -2}{HNO_3}(aq) \rightarrow \overset{+4}{NO_2}(g) + \overset{0}{O_2}(g) + \overset{+1\ -2}{H_2O}(\ell)$

          $\overset{+5}{4N} + 4e^- \rightarrow \overset{+4}{4N} \qquad \overset{-2}{2O} \rightarrow \overset{0}{O_2} + 4e^-$

          $4HNO_3(aq) \rightarrow 4NO_2(g) + O_2(g) + 2H_2O(\ell)$

Ion-Electron Method

In this alternative scheme (also called the half-reaction method), the participating species are determined, as before, by assignment of oxidation numbers. Separate oxidation and reduction equations are balanced for both atoms and charge (adjusted so that equal numbers of electrons are gained and lost) and added together to give a balanced net ionic equation.

To compare the two balancing schemes, in Sample Problem 20.2, the ion-electron method is now used to balance the same equation balanced in Sample Problem 20.1 by the oxidation-number method. Sample Problem 20.2 shows the steps for the ion-electron method.

Sample Problem 20.2

Write a balanced equation for the reaction in Sample Problem 20.1 using the ion-electron method.

Solution
Step 1: Write the skeleton equation.

$$HCl(aq) + KMnO_4(aq) \rightarrow H_2O(\ell) + KCl(aq) + MnCl_2(aq) + Cl_2(g)$$

Step 2: Write the ionic equation.

$$H^+ + Cl^- + K^+ + MnO_4^- \rightarrow H_2O + K^+ + Cl^- + Mn^{2+} + 2Cl^- + Cl_2$$

Step 3: Assign oxidation numbers and retain only those species that include an element that changes oxidation state.

$$\overset{-1}{Cl^-} + \overset{+7}{MnO_4} \rightarrow \overset{+2}{Mn^{2+}} + \overset{0}{Cl_2}$$

Step 4: Write the equation for the reduction; balance it for both atoms and charge. Note the difference in this step from the oxidation number method. Entire formulas for the species oxidized or reduced are included (not just the atoms).

$$\overset{+7}{MnO_4} \rightarrow \overset{+2}{Mn^{2+}}$$

Observe from the initial equation that oxygen forms water. Account for the 4 oxygen atoms on the left by adding $4H_2O$ on the right.

$$MnO_4^- \rightarrow Mn^{2+} + 4H_2O$$

This addition now requires that $8H^+(aq)$ be added on the left, so that H atoms are balanced.

$$\underset{(-1)}{MnO_4^-} + \underset{(+8)}{8H^+(aq)} \rightarrow \underset{(+2)}{Mn^{2+}} + \underset{(0)}{4H_2O}$$

The equation is now balanced for atoms but not for charge; $5e^-$ must be added on the left side to balance the charge.

$$\underset{(-1)}{MnO_4^-} + \underset{(+8)}{8H^+(aq)} + \underset{(-5)}{5e^-} \rightarrow \underset{(+2)}{Mn^{2+}} + \underset{(0)}{4H_2O}$$

The reduction equation is now balanced for both atoms and charge.

Step 5: Write the equation for the oxidation and balance for atoms and charge.

$$\underset{(-2)}{\overset{-1}{2Cl^-}} \rightarrow \underset{(0)}{\overset{0}{Cl_2}}$$

To balance charge, $2e^-$ must be added on the right.

$$\underset{(-2)}{2Cl^-} \rightarrow \underset{(0)}{Cl_2} + \underset{(-2)}{2e^-}$$

Step 6: Multiply the oxidation and reduction equations by coefficients so that electrons gained equals electrons lost and add the equations. The reduction equation must be multiplied by 2 and the oxidation equation by 5 so that the number of electrons gained in reduction is equal to that lost in oxidation ($2e^- \times 5e^- = 5e^- \times 2e^- = 10e^-$)

$$2[MnO_4^- + 8H^+(aq) + 5e^- \rightarrow Mn^{2+} + 4H_2O]$$
$$5[2Cl^- \rightarrow Cl_2 + 2e^-]$$
$$\overline{2MnO_4^- + 16H^+(aq) + 10Cl^- + 10e^- \rightarrow 2Mn^{2+} + 8H_2O + 5Cl_2 + 10e^-}$$

The two $10e^-$ terms cancel, and the ion equation is balanced.

$$2MnO_4^- + 16H^+(aq) + 10Cl^- \rightarrow 2Mn^{2+} + 8H_2O + 5Cl_2$$

Step 7: Add whatever species are needed to balance the formula equation. You must include one K^+ ion for each MnO_4^- ion, and six more Cl^- ions must be added so there is one for each $H^+(aq)$ ion. The formula equation may then be expressed as follows:

$$2KMnO_4(aq) + 16HCl(aq) \rightarrow 2MnCl_2(aq) + 2KCl(l) + 8H_2O(l) + 5Cl_2(g)$$

Checking shows that atoms and oxidation numbers are balanced.

Practice Problems

1. Copper reacts with hot, concentrated sulfuric acid to form copper(II) sulfate, sulfur dioxide, and water. Balance the equation using *(a)* the oxidation-number method and *(b)* the ion-electron method.

(Ans.) (a)
$$\overset{+6}{S} + 2e^- \rightarrow \overset{+4}{S}$$
$$\overset{0}{Cu} \rightarrow \overset{+2}{Cu^{+2}} + 2e^-$$
$$Cu + 2H_2SO_4 \rightarrow CuSO_4 + SO_2 + 2H_2O$$

(b)
$$4H^+(aq) + \overset{+6}{SO_4^{2-}} + 2e^- \rightarrow \overset{+4}{SO_2} + 2H_2O$$
$$\overset{0}{Cu} \rightarrow \overset{+2}{Cu^{2+}} + 2e^-$$
$$\overline{Cu + 2H_2SO_4 \rightarrow CuSO_4 + SO_2 + 2H_2O}$$

2. The products of a reaction between nitric acid and potassium iodide are potassium nitrate, iodine, nitrogen monoxide, and water. Balance the equation *(a)* using the oxidation number method and *(b)* the ion-electron method.

(Ans.) (a)
$$\overset{+5}{2N} + 6e^- \rightarrow \overset{+2}{2N}$$
$$\overset{-1}{6I^-} \rightarrow \overset{0}{3I_2} + 6e^-$$
$$8HNO_3 + 6KI \rightarrow 6KNO_3 + 3I_2 + 2NO + 4H_2O$$

(b)
$$8H^+(aq) + \overset{+5}{2NO_3^-} + 6e^- \rightarrow \overset{+2}{2NO} + 4H_2O$$
$$\overset{-1}{6I^-} \rightarrow \overset{0}{3I_2} + 6e^-$$
$$\overline{8H^+(aq) + 2NO_3^- + 6I^- \rightarrow 3I_2 + 2NO + 4H_2O}$$

or

$$8HNO_3 + 6KI \rightarrow 6KNO_3 + 3I_2 + 2NO + 4H_2O$$

Section Review

1. What two quantities must be conserved in redox equations?
2. What steps are involved in the oxidation-number method of equation-balancing?
3. What steps are involved in the ion-electron method of equation-balancing?
4. Use the oxidation number method to balance the following oxidation reduction reaction: $I_2O_5 + CO \rightarrow I_2 + CO_2$
5. Use the ion–electron method to balance the following oxidation–reduction reaction: $Na_2SnO_2 + Bi(OH)_3 \rightarrow Bi + Na_2SnO_3 + H_2O$

1. Charge and number of atoms
2. Refer to Section 20.2 for steps 1 through 6.
3. Refer to Section 20.2 for steps 1 through 7.
4. $I_2O_5 + 5CO \rightarrow I_2 + 5CO_2$
5. $3Na_2SnO_2 + 2Bi(OH)_3 \rightarrow 2Bi + 3Na_2SnO_3 + 3H_2O$

20.3 Oxidizing and Reducing Agents

An **oxidizing agent** *attains a more negative oxidation state during an oxidation–reduction reaction.* A **reducing agent** *attains a more positive oxidation state.*

It follows that the substance oxidized is also the reducing agent, and the substance reduced is the oxidizing agent. In other words, in every redox reaction, there is one reducing agent and one oxidizing agent. Study the oxidation–reduction terms presented in Table 20-2. Refer to them in this section.

The relatively large atoms of the sodium family of metals make up Group 1 of the Periodic Table. These atoms have weak attraction for their valence electrons and form positive ions readily. They are very active reducing agents. According to electrochemical measurements, the lithium atom is the most active reducing agent of all the common elements. The lithium ion, on the other hand, is the weakest oxidizing agent of the common ions. The electronegativity scale suggests that Group-1 metals starting with lithium should become progressively more active reducing agents. Except for lithium, this is true. A possible basis for the unusual activity of lithium is discussed in Chapter 24.

Atoms of the halogen family (Group 7 of the Periodic Table) have a strong attraction for electrons. They form negative ions readily and are very active oxidizing agents. The fluorine atom is the most highly electronegative atom. It is also the most active oxidizing agent among the elements. Because of its strong attraction for its own electrons, the fluoride ion is the weakest reducing agent.

The protons in the fluorine nucleus exert a strong attractive (coulombic) force on the electrons.

The following species make excellent oxidizing agents because the central atoms in these polyatomic ions have high oxidation states: IO_4^-, ClO_4^-, MnO_4^-, NO_3^-, CrO_4^{2-}, and $Cr_2O_7^-$. The electron density of the covalent bonds that exists between the oxygen and the central atom is closer to the oxygen.

TABLE 20-2   OXIDATION–REDUCTION TERMINOLOGY

Term	Change in Oxidation Number	Change in Electron Population
oxidation	in a positive direction	loss of electrons
reduction	in a negative direction	gain of electrons
oxidizing agent	in a negative direction	gains electrons
reducing agent	in a positive direction	loses electrons
substance oxidized	in a positive direction	loses electrons
substance reduced	in a negative direction	gains electrons

In Table 20-3 many familiar substances are arranged according to their activity as oxidizing and reducing agents. The left column of each pair shows the relative abilities of some metals to displace other metals from their compounds. Such displacements which you first read about in Chapter 8, are oxidation–reduction processes. Zinc, for example, appears above copper. Thus, zinc is the more active reducing agent and displaces copper ions from solutions of copper compounds (Figure 20-6).

TABLE 20-3 RELATIVE STRENGTH OF OXIDIZING AND REDUCING AGENTS

Reducing Agents	Oxidizing Agents	Reducing Agents	Oxidizing Agents
Li Strong	Li^+ Weak	Cu	Cu^{2+}
K	K^+	I^-	I_2
Ca	Ca^{2+}	MnO_4^{2-}	MnO_4^-
Na	Na^+	Fe^{2+}	Fe^{3+}
Mg	Mg^{2+}	Hg	Hg_2^{2+}
Al	Al^{3+}	Ag	Ag^+
Zn	Zn^{2+}	NO_2^-	NO_3^-
Cr	Cr^{3+}	Br^-	Br_2
Fe	Fe^{2+}	Mn^{2+}	MnO_2
Ni	Ni^{2+}	SO_2	H_2SO_4 (conc)
Sn	Sn^{2+}	Cr^{3+}	$Cr_2O_7^{2-}$
Pb	Pb^{2+}	Cl^-	Cl_2
H_2	H_3O^+	Mn^{2+}	MnO_4^-
H_2S	S	F^- Weak	F_2 Strong

$$Zn(s) + Cu^{2+}(aq) \rightarrow Zn^{2+}(aq) + Cu(s)$$
$$\overset{0}{Zn} \rightarrow \overset{+2}{Zn^{2+}} + 2e^- \quad \text{(oxidation)}$$
$$\overset{+2}{Cu^{2+}} + 2e^- \rightarrow \overset{0}{Cu} \quad \text{(reduction)}$$

The copper(II) ion, on the other hand, is a more active oxidizing agent than the zinc ion.

Nonmetals and some important ions are included in the series in Table 20-3. Any reducing agent is oxidized by the oxidizing agents below it. Observe that F_2 displaces Cl^-, Br^-, and I^- ions from their solutions. Cl_2 displaces Br^- and I ions, Br_2 displaces I^- ions. The equation for the displacement of Br^- by Cl_2 is

$$Cl_2 + 2Br^-(aq) \rightarrow 2Cl^-(aq) + Br_2$$
$$\overset{-1}{2Br^-} \rightarrow \overset{0}{Br_2} + 2e^- \quad \text{(oxidation)}$$
$$\overset{0}{Cl_2} + 2e^- \rightarrow \overset{-1}{2Cl^-} \quad \text{(reduction)}$$

Permanganate ions (MnO_4^-) and dichromate ions ($Cr_2O_7^{2-}$) are important oxidizing agents. They are used mainly in the form of their potassium salts. In neutral or mildly basic solutions, $\overset{+7}{Mn}$ in permanganate ion is reduced to $\overset{+4}{Mn}$ in MnO_2. If a solution is strongly basic, manganate ion containing $\overset{+6}{Mn}$ is formed. In acid solutions, the $\overset{+7}{Mn}$ in permanganate ion is reduced to manganese(II) ion, Mn^{2+}, and $\overset{+6}{Cr}$ in dichromate ion is reduced to chromium(III) ion, Cr^{3+}.

Peroxide ions (O_2^{2-}) have a relatively unstable covalent bond between the two oxygen atoms. The electron-dot formula is written as

$$\left[:\overset{..}{\underset{..}{O}}:\overset{..}{\underset{..}{O}}:\right]^{2-}$$

Figure 20-6 Zinc displaces copper ions from a copper sulfate solution. Metallic copper precipitates. This reaction is the basis for the electrochemical cell.

Chemistry Notebook

Not-So-Blue Jeans

If you want to fade your new blue jeans, rough them up a little with fine sandpaper before you wash them. This works because the blue jeans are colored by a special procedure that ensures their fading.

Most fabrics are colored by a thorough dye soaking, and the color becomes permanent. The cotton yarn that will be made into denim, however, is passed quickly through a solution of indigo (in a pale yellow soluble form, $C_6H_{12}N_2O_2$) so that only its surface gets wet. The core of the yarn never gets dyed, even with several passes through the indigo bath. Gradually the indigo

reacts with oxygen in the air and changes to a navy blue compound that is insoluble in water as shown in the following reaction.

$$2C_6H_{12}N_2O_2 + O_2 \rightarrow 2C_{16}H_{10}O_2 + 2H_2O$$

Jeans fade partly because some indigo washes out, but mostly because the thread surface gets worn off, exposing the white core. Rubbing your jeans with sandpaper just speeds up the process.

Figure 20-7 The antiseptic hydrogen peroxide

This structure represents an intermediate state of oxidation between free oxygen and oxides. The oxidation number of oxygen in the peroxide structure is −1.

Hydrogen peroxide (H_2O_2) is often used as an antiseptic (Figure 20-7). It decomposes by the oxidation *and* reduction of its oxygen. The products are water and molecular oxygen.

$$\overset{-1}{H_2O_2} + \overset{-1}{H_2O_2} \rightarrow 2\overset{-2}{H_2O} + \overset{0}{O_2}(g)$$

In this decomposition, half of the oxygen in the peroxide sample is *reduced* (from −1 to −2 oxidation state) to the oxide, forming water, and half is *oxidized* (from −1 to 0) to gaseous oxygen. *Such a process, in which a substance acts as both an oxidizing and a reducing agent and is self-oxidizing and self-reducing is called* **auto-oxidation**. Impurities in a water solution of hydrogen peroxide may catalyze this process.

Because of its oxidizing properties, hydrogen peroxide is used to bleach natural and artificial fibers and to purify gas and liquid waste materials produced by industrial plants. It is also used in uranium mining and metal extraction processes and as a starting material for preparing other compounds in which oxygen has an oxidation number of −1.

Section Review

1. Describe the chemical activity of the alkali in metals and of the halogens on the basis of oxidizing and reducing strength.
2. What is meant by "auto-oxidation?"
3. Would Ca be oxidized by Cu^{2+}? Explain.
4. Would Cl_2 be reduced by I^-? Explain.

20.4 Electrochemistry

Oxidation–reduction reactions, like all reactions, involve energy changes. Since these reactions involve electron transfer, the net *release* or net *absorption* of energy can occur in the form of electrical energy rather than heat energy. As you will learn in this section, this property allows for a great many practical applications of redox reactions and also makes possible quantitative prediction and comparison of the oxidizing and reducing abilities of different substances. *The branch of chemistry that deals with these electricity-related applications of oxidation–reduction reactions is called* **electrochemistry**.

Electrochemical Cells

Oxidation–reduction reactions involve a transfer of electrons from the substance oxidized to the substance reduced. Sometimes, such reactions occur spontaneously and produce undesirable effects, such as the rusting of iron (Figure 20-8). Rust causes a tremendous amount of damage to metal structures every year. In light of the high cost of repairs and dwindling supplies of metal ores, research into rust prevention is a growing enterprise.

However, some spontaneous redox reactions can be used as sources of electric energy. If the reactants are in contact, the energy released during the electron transfer is in the form of heat. If the reactants are separated solutions and are connected externally by a wire conductor, the transfer of electrons takes place through the wire and the energy is released in the form of electrical energy. Solutions that conduct electricity, as you learned in Chapter 15, are classified as electrolytes. *A* **conductor** *used to establish electric contact with a nonmetallic part of a circuit, such as electrolytes, is known as an* **electrode**. *A system of electrode and electrolyte in which a spontaneous redox reaction is known as an* **electrochemical cell**. The transfer of electrons through a metallic conductor constitutes an electric current. Although, under these conditions, in an electrochemical cell, a part of the energy released during the electron transfer does appear as heat, the remainder is in fact available as electrical energy.

The dry cell (Figure 20-11) is a common source of electrical energy in the laboratory, as well as in countless devices from flashlights to portable radios. Small dry cells are familiar as so-called "flashlight batteries". A zinc cup serves as the negative electrode. A carbon rod serves as the positive electrode. The carbon electrode is surrounded by a mixture of manganese dioxide and powdered carbon. The electrolyte is a moist paste of ammonium chloride containing some zinc chloride. These components of the dry cell are illustrated in the schematic diagram of Figure 20-10.

SECTION OBJECTIVES

- Describe the nature of electro-chemical cells.
- Describe the nature of electro-lytic cells.
- Discuss the electrolysis of water.
- Discuss the electrolysis of aqueous salt solutions.
- Explain the process of electro-plating.
- Discuss the oxidation of re-chargeable cells.
- Explain, use, and carry out cal-culations involving standard electrode potentials.

Figure 20-8 Rust, the reddish corrosion product formed by electrochemical interaction between iron and atmospheric oxygen.

Figure 20-9 Many common batteries are zinc dry cells

When the external circuit is closed, zinc atoms are oxidized at the negative (zinc) electrode.

$$\overset{0}{Zn} \rightarrow \overset{+2}{Zn^{2+}} + 2e^-$$

Electrons flow through the external circuit to the positive (carbon) electrode. If manganese dioxide were not present, hydrogen gas would be formed at the carbon electrode by the reduction reaction shown in the following equation.

$$\overset{+1}{2NH_4} + 2e^- \rightarrow 2NH_3 + \overset{0}{H_2(g)}$$

Such a reaction that occurs at one electrode in an electrochemical cell is known as a **half-reaction**. However, manganese oxidizes the hydrogen gas to form water. Thus, manganese, rather than hydrogen, is reduced at the positive (carbon) electrode.

$$\overset{+4}{2MnO_2} + 2NH_4 + 2e^- \rightarrow \overset{+3}{Mn_2O_3} + 2NH_3 + H_2O$$

The ammonia that is formed reacts with Zn^{2+} ions to form complex $Zn(NH_3)_4^{2+}$ ions.

The electrode at which oxidation occurs is called the **anode**, *and the electrode at which reduction occurs is called the* **cathode**. In this context, because oxidation occurs at the negative electrode of an electrochemical cell (spontaneous reactions), the negative electrode of such a cell is an anode. Because reduction occurs at the positive electrode, the positive electrode of such a cell is a cathode. The zinc electrode (oxidation) of Figure 20-10 is the anode, and the carbon electrode (reduction) is the cathode of the zinc-carbon dry cell.

Electrolytic Cells

Some oxidation–reduction reactions do not occur spontaneously. Such reactions can be driven by means of electrical energy, however. *The process in which an electric current is used to drive an oxidation–reduction reaction is called* **electrolysis**. *An* **electrolytic cell** *is a system of electrodes and electrolytes in which a current is used to drive a nonspontaneous redox reaction.* A battery or other direct-current source is connected across the cell. One electrode, which is connected to the negative terminal of the battery, acquires a negative charge. The other electrode, connected to the positive terminal of the battery, acquires a positive charge.

Ion migration in the cell is responsible for the transfer of electric charge between electrodes. Positively charged ions migrate, or move, toward the negative electrode. Negatively charged ions migrate toward the positive electrode.

In an electrolytic cell, as in an electrochemical cell, reduction occurs at the cathode. In an electrolyte cell, however, this reduction occurs at the negative electrode. So, in this type of cell, the cathode is negatively charged. Electrons are removed from the cathode by the electrolyte in this process. Oxidation occurs at the anode, which in this case is the positive electrode, the electrode that acquires electrons from the electrolyte in the process. The chemical reactions at the electrodes complete the electric circuit between an external source of electric current and the cell. The closed-loop path

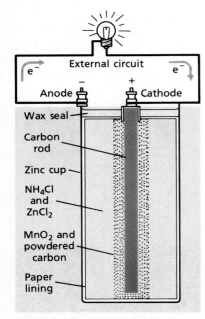

Figure 20-10 In a zinc dry cell, zinc is oxidized to Zn^{2+} at the negative electrode and manganese(IV) is reduced to manganese(III) at the positive electrode.

1800 Count Alessandro Volta (1745–1827) produced the first electric battery by stacking silver and zinc disks in a "pile," separating them by cardboard pads soaked in salt water, and attaching a metal wire at each end.

for electric current allows energy to be transferred from the external source to the electrolytic cell. This energy drives the electrode reactions in the electrolytic cell.

transferred from the external source to the electrolytic cell. This energy drives the electrode reactions in the cell.

Electrolysis of water   In the decomposition of water by electrolysis, energy is transferred from the energy source to the decomposition products. The reaction is endothermic, and the amount of energy required is 285.9 kJ/mol of water decomposed. The overall reaction is

$$2H_2O(\ell) + 571.8 \text{ kJ} \rightarrow 2H_2(g) + O_2(g)$$

Hydrogen gas is given up at the cathode (the reduction site), and oxygen gas is given up at the anode (the oxidation site).

A suitable electrolysis cell is shown in Figure 20-11. It consists of two inert platinum electrodes immersed in water. A very small amount of an electrolyte (such as H_2SO_4), is added to provide adequate conductivity. The electrodes are connected to a battery that supplies the electrical energy to drive the decomposition reaction forward.

The electric current in the external circuit consists of a flow of electrons. The electrode of the cell connected to the negative electrode of the battery acquires an excess of electrons and becomes the cathode of the electrolytic cell. The other electrode of the cell, which is connected to the positive electrode of the battery, loses electrons to the battery and becomes the anode of the electrolytic cell. The battery can be thought of as an electron pump supplying electrons to the cathode of the electrolytic cell and recovering electrons from the anode of the cell at equal rates.

Reduction occurs at the cathode, where, it is believed, water molecules acquire electrons directly. Hydrogen gas is the product of this electrochemical reaction.

$$\text{cathode reaction (reduction): } 2\overset{+1}{H_2}O(\ell) + 2e^- \rightarrow 2OH^-(aq) + \overset{0}{H_2}(g)$$

The OH^- ion concentration in the solution around the cathode rises.

Oxidation occurs at the anode; oxygen gas is the product. There are SO_4^{2-} ions, OH^- ions, and water molecules in the region about the anode. The OH^- ion concentration is quite low, and this ion is not likely to appear in the anode reaction. The SO_4^{2-} ions, also at low concentration, are more difficult to oxidize than are water molecules. For these reasons, chemists believe that water molecules are oxidized by giving up electrons directly to the anode.

$$\text{anode reaction (oxidation): } 6\overset{-2}{H_2O}(\ell) \rightarrow 4H_3O^+(aq) + \overset{0}{O_2}(g) + 4e^-$$

The H_3O^+ ion concentration in the solution around the anode rises.

The overall reaction is the sum of the net cathode and anode reactions. The equation for the cathode reaction is doubled because equal numbers of electrons must be involved in the oxidation and reduction processes.

$$\text{cathode: } 4H_2O(\ell) + 4e^- \rightarrow 4OH^-(aq) + 2H_2(g)$$

$$\text{anode: } 6H_2O(\ell) \rightarrow 4H_3O^+(aq) + O_2(g) + 4e^-$$

$$\text{cell: } 10H_2O(\ell) \rightarrow 2H_2(g) + O_2(g) + 4H_3O^+(aq) + 4OH^-(aq)$$

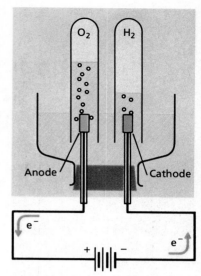

Figure 20-11   The electrolysis of water. In electrolytic cells, reduction occurs at the negative electrode, and oxidation occurs at the positive electrode.

1832 Michael Faraday (1791–1867) described the following laws:
1. The mass of substance liberated at an electrode during electrolysis is proportional to the quantity of electricity driven through the solution.
2. The mass liberated by a given quantity of electricity is proportional to the atomic weight of the element liberated and inversely proportional to the valence of the element liberated.

Figure 20-12 Industrial chlorine production units

The solution around the cathode becomes basic because of the production of OH^- ions. The solution around the anode becomes acidic because of the production of H_3O^+ ions. Because of the ordinary mixing tendency in the solution, these ions can be expected to diffuse together eventually and form water. If this neutralization is complete, the net electrolysis reaction is

cell neutralization

cell: $\quad 10H_2O(\ell) \rightarrow 2H_2(g) + O_2(g) + 4H_3O^+(aq) + 4OH^-(aq)$

neutralization: $4H_3O^+(aq) + 4OH^-(aq) \rightarrow 8H_2O(\ell)$

net: $\quad\quad\quad\quad\quad 2H_2O(\ell) \rightarrow 2H_2(g) + O_2(g)$

Reduction at the cathode lowers the oxidation state of hydrogen from $+1$ to 0; the oxidation state of oxygen at the cathode remains -2. At the anode, oxidation raises the oxidations state of oxygen from -2 to 0; the oxidation state of hydrogen at the anode remains $+1$.

Electrolysis of aqueous salt solutions The electrode products from the electrolysis of aqueous salt solutions are determined by the relative ease with which the different particles present can be oxidized or reduced. In the case of aqueous NaCl, for example (see Figure 20-12), Na^+ ions are more difficult to reduce at the cathode than H_2O molecules or H_3O^+ ions. Since a solution of NaCl is neutral, the H_3O^+ ion concentration remains very low (10^{-7} mol/L). Molecules of H_2O are therefore the particles that are reduced at the cathode.

Cathode reaction (reduction):

$$2\overset{+1}{H_2}O(\ell) + 2e^- \rightarrow 2OH^-(aq) + \overset{0}{H_2}(g)$$

Thus, Na^+ ions remain in solution, and hydrogen gas is released.

In general, metals that are easily oxidized form ions that are difficult to reduce. Thus, hydrogen gas, not sodium metal, is the cathode product in the preceding reaction. On the other hand, metals such as copper, silver, and gold are difficult to oxidize, and they form ions that are easily reduced. Aqueous solutions of their salts give up the metal at the cathode.

The choice for anode reaction in the aqueous NaCl electrolysis lies between Cl^- ions and H_2O molecules. The Cl^- ions are more easily oxidized, so Cl_2 gas is produced at the anode.

Anode reaction (oxidation):

$$2\overset{-1}{Cl}^-(aq) \rightarrow \overset{0}{Cl_2}(g) + 2e^-$$

Adding the cathode and anode equations gives the net reaction for the cell.

Net: $2H_2O(\ell) + 2Cl^-(aq) \rightarrow 2OH^-(aq) + H_2(g) + Cl_2(g)$

The electrolytic solution gradually changes from aqueous NaCl to aqueous NaOH as the electrolysis continues, as long as the Cl_2 gas is continuously removed from the cell.

Aqueous Br^- ions and I^- ions are oxidized electrolytically in the same way as Cl^- ions. Their solutions give the free halogen at the anode. On the other hand, aqueous solutions of negative ions that do not participate in the oxidation reaction (NO_3^- ions, for example) give O_2 gas at the anode.

Electroplating   Inactive metals form ions that are more easily reduced than hydrogen. *This fact makes possible an electrolytic process called* **electroplating**, *in which a metal is deposited on a surface.*

An electroplating cell contains a solution of a salt of the plating metal. It has an object to be plated (the cathode) and a piece of the plating metal (the anode). A silverplating cell, illustrated in Figure 20-13, contains a solution of a soluble silver salt and a silver anode. The cathode is the object to be plated. The silver anode is connected to the positive electrode of a battery or to some other source of direct current. The object to be plated is connected to the negative electrode. Silver ions are reduced at the cathode of the cell when electrons flow through the circuit.

$$\overset{+1}{Ag^+} + e^- \rightarrow \overset{0}{Ag}$$

Silver atoms are oxidized at the anode.

$$\overset{0}{Ag} \rightarrow \overset{+1}{Ag^+} + e^-$$

Silver ions are removed from the solution at the cathode and deposited as metallic silver. Meanwhile, metallic silver is removed from the anode as ions. This action maintains the Ag^+ ion concentration of the solution. Thus, in effect, silver is transferred from the anode to the cathode of the cell.

Rechargeable Cells

A rechargeable cell is a cell that, in effect, combines the oxidation–reduction chemistry of both the electrochemical cell and the electrolytic cell, which are compared in Figure 20-14. When the rechargeable cell is used as a source of

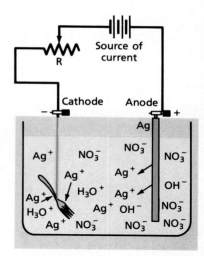

Figure 20-13   An electrolytic cell (top) used for silver plating. This door knocker is silver-plated.

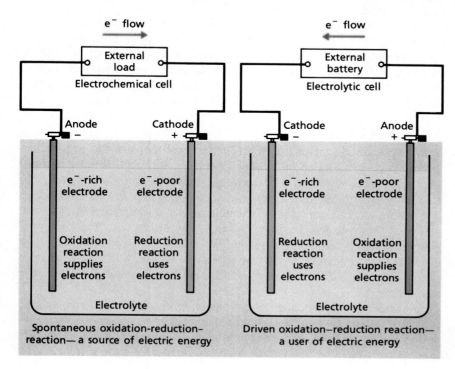

Figure 20-14   A comparison of electrochemical and electrolytic cells

Figure 20-15 The rechargeable cells of a car battery produce electricity from reactions between lead(IV) oxide, lead, and sulfuric acid.

electrical energy (during the so-called discharge cycle), chemical energy is converted to electrical energy by the oxidation–reduction reactions characteristic of an electrochemical cell. When the cell is being recharged (charge cycle), electric energy is converted to chemical energy by the oxidation–reduction reactions of an electrolytic cell.

The standard 12-volt automobile battery consists of six lead(IV) oxide–lead–sulfuric acid rechargeable cells that are connected in series. The battery is charged by the conversion of electrical energy from an external source to stored chemical energy which occurs by an oxidation–reduction reaction in which each cell acts as an electrolytic cell. While the battery is used as a source of electrical energy, chemical energy stored during the charging cycle is converted to electrical energy by the reverse oxidation–reduction reaction, in which each cell acts as an electrochemical cell. In an automobile engine, these charge and discharge modes are regulated automatically to maintain the state of charge in the battery while the engine is in operation.

Electrode Potentials

You have learned that oxidation–reduction reactions are the sum of two distinct processes: oxidation, in which electrons are supplied to the system, and reduction, in which electrons are acquired from the system. In electrochemical cells, these reactions take place at the separate electrodes.

As the electrochemical-cell reaction begins, a difference in electrical potential, or voltage, develops between the electrodes. This potential difference, which can be measured by a voltmeter connected across the two electrodes, is a measure of the energy required to move a certain electric charge between the electrodes. Potential difference is measured in units called volts (V).

Consider the electrochemical cell shown in Figure 20-16. A strip of zinc is placed in a solution of $ZnSO_4$, and a strip of copper is placed in a solution of $CuSO_4$. *The two solutions are separated by a porous partition that permits ions to pass but that otherwise prevents mixing of the solutions. This*

The battery voltage is produced by a spontaneous electrochemical reaction. If the system is closed, that is, if reactions are reversible by providing electrons, the battery can be recharged.

Ions pass from one electrolytic half-cell to another so that overall electroneutrality is maintained in solution throughout the process.

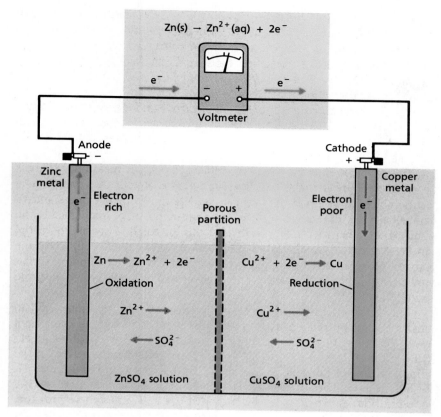

$$Zn(s) \rightarrow Zn^{2+}(aq) + 2e^-$$

Voltmeter

Anode −

Cathode +

Zinc metal

Copper metal

Electron rich

Porous partition

Electron poor

$$Zn \longrightarrow Zn^{2+} + 2e^-$$

$$Cu^{2+} + 2e^- \longrightarrow Cu$$

Oxidation

Reduction

$$Zn^{2+} \longrightarrow$$

$$Cu^{2+} \longrightarrow$$

$$\longleftarrow SO_4^{2-}$$

$$\longleftarrow SO_4^{2-}$$

$ZnSO_4$ solution

$CuSO_4$ solution

Figure 20-16 In a zinc-copper cell, zinc is oxidized and copper is reduced.

arrangement, which is capable of generating an electron current in an external circuit connected between the electrodes, is called a **voltaic cell**.

In the two electrode reactions, the zinc electrode acquires a negative charge relative to the copper. The copper electrode becomes positively charged relative to the zinc. This fact reveals that zinc atoms have a stronger tendency to release electrons and enter the solution as ions than do copper atoms. Zinc is said to be more active, or more easily oxidized, than copper.

In the cell shown in Figure 20-16, then, the reaction at the surface of the zinc electrode is an oxidation.

$$\overset{0}{Zn}(s) \rightarrow \overset{+2}{Zn^{+2}}(aq) + 2e^-$$

The reaction at the surface of the copper electrode is a reduction.

$$\overset{+2}{Cu^{2+}}(aq) + 2e^- \rightarrow \overset{0}{Cu}(s)$$

As Zn^{2+} ions form, electrons accumulate on the zinc electrode, giving it a negative charge. As Cu atoms form, electrons are removed from the copper electrode, leaving it with a positive charge. Electrons also flow through the external circuit from the zinc electrode to the copper electrode. Here they replace the electrons removed, enabling Cu^{2+} ions to undergo reduction to Cu atoms. Thus, in effect, electrons are transferred from Zn atoms through the external circuit to Cu^{2+} ions. The overall reaction can be written as

$$Zn(s) + Cu^{2+}(aq) \rightarrow Zn^{2+}(aq) + Cu(s)$$

Oxidation-Reduction Reactions **617**

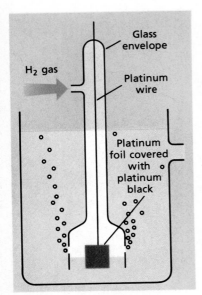

Figure 20-17 A hydrogen electrode, the standard reference electrode for measuring electrode potentials. The electrode surface in contact with the solution is actually a layer of hydrogen adsorbed onto the surface of the platinum black.

A voltmeter connected across the Cu–Zn voltaic cell measures the potential difference. This difference is about 1.1 V when the solution concentrations of Zn^{2+} and Cu^{2+} ions are each at a one molar concentration (1 M).

A voltaic cell consists of two portions, called half-cells, each consisting of a metal electrode in contact with a solution of its ions. The reaction taking place is the half-reaction for that cell.

The potential difference between an electrode and its solution is known as its electrode potential. The sum of the electrode potentials for two half-reactions roughly equals the potential difference measured across the complete voltaic cell.

The potential difference across a voltaic cell is easily measured. There is no way to measure an individual electrode potential directly, however. A value for the potential of a half-reaction can be determined by use of a standard half-cell along with it as a reference. An arbitrary potential is assigned to the standard reference electrode. Relative to this potential, a specific potential can be determined for the other electrode of the complete cell. Electrode potentials are expressed as reduction potentials. These reduction potentials provide a reliable indication of the tendency of a substance to undergo reduction.

Chemists use a hydrogen electrode immersed in a 1-M solution of H^+(aq) ions as a standard reference electrode. This practice provides a convenient way to examine the relative tendencies of metals to react with aqueous hydrogen ions [H^+(aq) or H_3O^+(aq)]. It is responsible for the activity series of metals listed in Table 8-3. A hydrogen electrode is shown in Figure 20-17. It consists of a platinum electrode dipping into an acid solution of 1-M concentration and surrounded by hydrogen gas at 1 atmosphere pressure. This standard hydrogen electrode is assigned a potential of zero volts. The half-cell reaction is

$$\overset{0}{H_2}(g) \rightleftarrows 2\overset{+1}{H^+}(aq) + 2e^-$$

Since the potential of the hydrogen electrode is arbitrarily set at 0 V, the potential difference across the complete cell is attributed entirely to the electrode of the other half-cell. *A half-cell potential measured relative to a potential of zero for the standard hydrogen electrode is called a **standard electrode potential**,* symbolized by $E°$.

Suppose a complete cell consists of a zinc half-cell and a standard hydrogen half-cell, as in Figure 20-18. The potential difference across the cell measures the standard electrode potential (symbol: $E°$) of the zinc electrode relative to the hydrogen electrode (the zero reference electrode). It is found to be -0.76 V. The value is relative because the standard electrode potential of any electrode is given a negative value if this electrode has a negative charge relative to the standard hydrogen electrode. Electrons flow through the external circuit from the zinc to the hydrogen electrode. There, H^+(aq) ions are reduced to H_2 gas.

This reaction indicates that the tendency for Zn^{2+} ions to be reduced to Zn atoms is 0.76 volt less than the tendency for H^+(aq) ions to be reduced to H_2. The half-reaction (as a reduction) is

$$\overset{+2}{Zn^{2+}} + 2e^- \rightarrow \overset{0}{Zn} \qquad E° = -0.76 \text{ V}$$

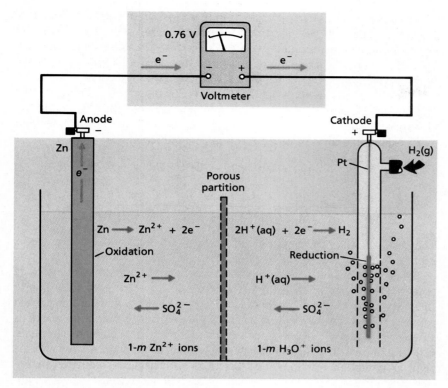

Figure 20-18 The electrode potential of a zinc half-cell is measured by coupling it with a standard hydrogen electrode.

This half-reaction has less tendency to occur than the half-reaction

$$\overset{+1}{2H^+}(aq) + 2e^- \rightarrow \overset{0}{H_2}(g)$$

by 0.76 V. This statement also means that the half-reaction (as an oxidation)

$$\overset{0}{Zn}(s) \rightarrow \overset{+2}{Zn^{2+}}(aq) + e^- \quad E° = +0.76\ V$$

has a greater tendency to occur by 0.76 volt than

$$\overset{0}{H_2}(g) \rightarrow \overset{+1}{2H^+}(aq) + 2e^-$$

Observe that the sign of the electrode potential is reversed when the Zn half-cell reaction is reversed and written as an oxidation.

A copper half-cell coupled with the standard hydrogen electrode gives a potential difference measurement of +0.34 V. This measurement indicates that Cu^{2+}(aq) ions are more readily reduced than H^+(aq) ions. *The value is positive because the standard electrode potential (E°) of an electrode is given a positive value if this electrode has a positive charge relative to the standard hydrogen electrode.* The half-reaction for copper (reduction) is

$$\overset{+2}{Cu^{2+}}(aq) + 2e^- \rightarrow \overset{0}{Cu}(s) \quad E° = +0.34\ V$$

This reaction has a greater tendency to occur than

$$\overset{+1}{2H^+}(aq) + 2e^- \rightarrow \overset{0}{H_2}(g)$$

by 0.34 V.

If Cu^{2+}, Zn^{2+}, and H^+ were competing for electrons, Cu^{2+} would get them because it exerts a stronger attractive force on the electrons.

Two observations can be made from these measurements: *(1)* Zinc has a greater tendency to yield electrons than does hydrogen by 0.76 V; and *(2)* Hydrogen has a greater tendency to yield electrons than does copper by 0.34 V. Taken together, these potentials indicate that zinc has a greater tendency toward oxidation than does copper by 1.10 V (0.76 V + 0.34 V).

How do these potentials apply to the Zn–Cu voltaic cell of Figure 20-16? The potential difference across the complete cell is obtained by adding the electrode potentials of the two half-reactions; the Zn half-reaction is written as an oxidation and the Cu half-reaction as a reduction.

$$\overset{0}{Zn}(s) \rightarrow \overset{+2}{Zn^{2+}}(aq) + 2e^- \qquad E° = +0.76 \text{ V}$$

$$\overset{+2}{Cu^{2+}}(aq) + 2e^- \rightarrow \overset{0}{Cu}(s) \qquad E° = +0.34 \text{ V}$$

$$Zn(s) + Cu^{2+}(aq) \rightarrow Zn^{2+}(aq) + Cu(s) \quad E° = +1.10 \text{ V}$$

The positive sign of the potential difference shows that the reaction proceeds spontaneously to the right.

Half-reactions for some common electrodes and their standard electrode potentials are listed in Table 20-4. These reactions are arranged according to their standard electrode potentials ($E°$) relative to a standard hydrogen reference electrode. All electrode reactions are written as reduction reactions to the right, and all electrode potentials are given as reduction potentials. Half-reactions with positive reduction potentials occur spontaneously to the right as reduction reactions. Half-reactions with negative reduction potentials occur spontaneously to the left as oxidation reactions. When a half-reaction is written as an oxidation reaction, the sign of the electrode potential is changed.

TABLE 20-4 STANDARD ELECTRODE POTENTIALS (AS REDUCTION POTENTIALS)

	Standard Electrode Potential $(E)°$ (in volts)		Standard Electrode Potential $(E)°$ (in volts)
$F_2 + 2e^- \rightleftarrows 2F^-$	+2.87	$Fe^{3+} + 3e^- \rightleftarrows Fe$	−0.04
$MnO_4^- + 8H^+(aq) + 5e^- \rightleftarrows Mn^{2+} + 4H_2O$	+1.49	$Pb^{2+} + 2e^- \rightleftarrows Pb$	−0.13
$Au^{+3} + 3e^- \rightleftarrows Au$	+1.42	$Sn^{2+} + 2e^- \rightleftarrows Sn$	−0.14
$Cl_2 + 2e^- \rightleftarrows 2Cl^-$	+1.36	$Ni^{2+} + 2e^- \rightleftarrows Ni$	−0.23
$Cr_2O_7^{2-} + 14H^+(aq) + 6e^- \rightleftarrows 2Cr^{3+} + 7H_2O$	+1.33	$Co^{2+} + 2e^- \rightleftarrows Co$	−0.28
$MnO_2 + 4H^+(aq) + 2e^- \rightleftarrows Mn^{2+} + 2H_2O$	+1.21	$Cd^{2+} + 2e^- \rightleftarrows Cd$	−0.40
$Br_2 + 2e^- \rightleftarrows 2Br^-$	+1.06	$Fe^{2+} + 2e^- \rightleftarrows Fe$	−0.41
$Hg^{2+} + 2e^- \rightleftarrows Hg$	+0.85	$S + 2e^- \rightleftarrows S^{2-}$	−0.51
$Ag^+ + e^- \rightleftarrows Ag$	+0.80	$Cr^{3+} + 3e^- \rightleftarrows Cr$	−0.74
$Hg_2^{2+} + 2e^- \rightleftarrows 2Hg$	+0.80	$Zn^{2+} + 2e^- \rightleftarrows Zn$	−0.76
$Fe^{3+} + e^- \rightleftarrows Fe^{2+}$	+0.77	$Al^{3+} + 3e^- \rightleftarrows Al$	−1.71
$MnO_4^- + e^- \rightleftarrows MnO_4^{2-}$	+0.56	$Mg^{2+} + 2e^- \rightleftarrows Mg$	−2.38
$I_2 + 2e^- \rightleftarrows 2I^-$	+0.54	$Na^+ + e^- \rightleftarrows Na$	−2.71
$Cu^{2+} + 2e^- \rightleftarrows Cu$	+0.34	$Ca^{2+} + 2e^- \rightleftarrows Ca$	−2.76
$Cu^{2+} + e^- \rightleftarrows Cu^+$	+0.16	$Ba^{2+} + 2e^- \rightleftarrows Ba$	−2.90
$S + 2H^+(aq) + 2e^- \rightleftarrows H_2S(aq)$	+0.14	$K^+ + e^- \rightleftarrows K$	−2.92
$2H^+(aq) + 2e^- \rightleftarrows H_2$	0.00	$Li^+ + e^- \rightleftarrows$	−3.04

The magnitude of the electrode potential measures the tendency of the reduction half-reaction to occur as the equation is written in Table 20-4. The half-reaction at the bottom of the column has the least tendency toward reduction (adding electrons). Stated in another way, it has the greatest tendency to occur as an oxidation (yielding electrons). The half-reaction at the top of the column has the greatest tendency to occur as a reduction. Thus it has the least tendency to occur as an oxidation.

The higher a half-reaction's position is in the column, the greater is the tendency for its reduction reaction to occur. The lower a half-reaction is in the column, the greater is the tendency for the oxidation reaction to occur. For example, potassium (K) has a large negative electrode potential and a strong tendency to form K^+ ions. Thus, potassium is a strong reducing agent. Fluorine has a large positive electrode potential and a strong tendency to form F^- ions. Fluorine, then, is a strong oxidizing agent. If you compare the listings in Table 20-3 with those in Table 20-4, you will see that the former is a shortened and inverted form of the latter (inverted since activity series of elements are traditionally listed in descending order of reducing strength of metals).

Standard electrode potentials can be used to predict the direction in which an electrochemical reaction proceeds spontaneously. Sample Problem 20.3 illustrates the method that is followed. *Note that when a half-reaction is multiplied by a constant, the $E°$ value is not multiplied, but remains the same.* When a half-reaction is reversed, however, the sign of $E°$ is changed.

Sample Problem 20.3

Two half-cells, one containing Fe^{3+} and Fe and the other containing Ag^+ and Ag are connected to form a voltaic cell. Use Table 20-4 to determine the direction of spontaneous reaction and the value of the electrode potential ($E°$).

Solution

Step 1. Analyze   Given: Fe^{3+} — Fe half-cell: $E° = -0.04$ V
                                        Ag^+ — Ag half cell: $E° = +0.80$ V
               Unknown: reaction direction, $E°$

Step 2. Plan   half-cells → half-reaction → overall reaction → $E°$

Step 3. Compute   From the $E°$ values of the half-reaction, the combination most likely to result in a spontaneous reaction is for the Ag^+–Ag half-cell to undergo reduction ($E° = +0.85$ V) and for the Fe–Fe^{3+} half-cell to undergo oxidation ($E° = +0.04$) since this is the combination that will result in a positive $E°$ value.

$$Fe \rightarrow Fe^{3+}(aq) + 3e^- \qquad\qquad E° = +0.04 \text{ V}$$
$$3[Ag^+(aq) + e^- \rightarrow Ag(s)] \qquad\qquad E° = +0.80 \text{ V}$$
$$\overline{Fe(s) + 3Ag^+(aq) \rightarrow Fe^{3+}(aq) + 3Ag(s) \quad E° = +.84 \text{ V}}$$

The relative $E°$ value confirms that the overall reaction does proceed spontaneously as written since a positive value results.

Step 4. Evaluate   The answer contains the proper number of digits and its sign corresponds to that of the $E°$ of a spontaneous electrochemical reaction.

Practice Problems
Given the following half-cells, determine the overall electrochemical reaction that proceeds spontaneously and the value of $E°$.
1. $Ni^{2+}-Ni$, $Fe^{3+}-Fe^{2+}$ *(Ans.)* $Ni + 2Fe^{3+} \rightarrow Ni^{2+} + 2Fe^{2+}$, $E° = +1.00$ V
2. $Hg^{2+}-Hg$, $Cr^{3+}-Cr$ *(Ans.)* $3Hg^{2+} + 2Cr \rightarrow 3Hg + 2Cr^{3+}$, $E° = +1.59$

1. A system of electrodes and electrolytes in which a spontaneous redox reaction is used as a source of electrical energy
2. A system of electrodes and electrolytes in which a current is used to drive a nonspontaneous redox reaction
3. Refer to "Electrolysis of water" for discussion.
4. Refer to "Electrolysis of aqueous salt solutions" for discussion.
5. Refer to "Rechargeable Cells" for explanation.
6. A cell that combines the redox chemistry of both electrochemical and electrolytic cells
7. The potential difference between an electrode and its solution; its magnitude measures the tendency of reduction half-reactions to occur
8. $K + Na^+ \rightarrow K^+ + Na$
   $E° = +0.21$ V

Section Review

1. What is an electrochemical cell?
2. What is an electrolytic cell?
3. Discuss briefly the electrolysis of water.
4. Discuss briefly the electrolysis of an aqueous salt solution.
5. Explain the process of electroplating.
6. What is a rechargeable cell?
7. What is electrode potential, and how is it used to calculate information about an electrochemical reaction?
8. Given the Na^+-Na and K^+-K half cells, determine the overall electrochemical reaction that proceeds spontaneously and the $E°$ value.

Chapter Summary

- In the oxidation-number method for balancing equations, oxidation numbers are assigned to all atoms taking part in the reaction. Electronic equations representing the shift in electrons are then written and balanced. Coefficients from the balanced electronic equations are then used to balance the redox equation.
- In the ion-electron method for balancing equations, separate oxidation and reduction equations are balanced for atoms and charge. A balanced net ionic equation is then obtained.
- In redox reactions, the substance that is *reduced* acts as an *oxidizing* agent because it *acquires* electrons from the substance oxidized.
- The substance that is *oxidized* in redox reactions is a *reducing* agent because it supplies the electrons to the substance reduced.
- Some oxidation–reduction reactions occur spontaneously and may be used as sources of electrical energy when arranged in electrochemical cells. Other oxidation–reduction reactions that are not spontaneous can be driven by an external source of electric energy in a process called electrolysis.
- The oxidation or reduction reaction between an electrode and its electrolyte in an electrochemical cell is called a half-reaction.
- The potential difference between the electrode and its solution is called the electrode potential. The sum of the electrode potentials of the two half-reactions of a voltaic cell is roughly equal to the potential difference across the cell.
- Standard electrode potentials, which are measured relative to a standard hydrogen half-cell, indicate the relative strengths of substances as oxidizing and reducing agents.

Chapter 20 *Review*

Oxidation–Reduction Reactions

Vocabulary

anode
auto-oxidation
cathode
electrochemical cell
electrochemistry
electrode
electrode potential
electrolysis
electrolytic cell
electroplating
half-cell
half-reaction

oxidation
oxidized
oxidizing agent
oxidation-reduction
  reaction
redox reaction
reduced
reducing agent
reduction
standard electrode
potential
voltaic cell

Questions

1. *(a)* Distinguish between the processes of oxidation and reduction. *(b)* Write an equation to illustrate each.
2. Explain what is meant when a substance is *(a)* oxidized *(b)* reduced.
3. What is an oxidation-reduction or redox reaction?
4. Which of the following are redox reactions?
   (a) $2Na + Cl_2 \rightarrow 2NaCl$ *(b)* $C + O_2 \rightarrow CO_2$
   (c) $2H_2O \rightarrow 2H_2 + O_2$ *(d)* $NaCl + AgNO_3 \rightarrow$
   $AgCl + NaNO_3$ *(e)* $NH_3 + HCl \rightarrow NH_4^+ + Cl^-$
   (f) $2KClO_3 \rightarrow 2KCl + 3O_2$ *(g)* $H_2 + Cl_2 \rightarrow 2HCl$
   (g) $H_2 + Cl_2 \rightarrow 2HCl$ *(h)* $H_2SO_4 + 2KOH \rightarrow$
   $K_2SO_4 + 2H_2O$ *(i)* $Zn + CuSO_4 \rightarrow ZnSo_4 + Cu$.
5. For each oxidation-reduction reaction in Question 4, identify: the substance oxidized; and the substance reduced.
6. Distinguish between the two methods described for balancing redox equations.
7. Distinguish between oxidizing and reducing agents.
8. *(a)* Identify the most active reducing agent among all common elements. *(b)* Why are all of the elements in its Periodic Table group very active reducing agents? *(c)* Identify the most active oxidizing agent among the common elements.
9. Based on Table 20-3, identify the strongest and weakest reducing agents among the substances listed within each of the following groupings:
   (a) Ca, Ag, Sn, Cl^- *(b)* Fe, Hg, Al, Br^-
   (c) F^-, Pb, Mn^{2+}, Na
10. Based on Table 20-3, identify the strongest and weakest oxidizing agents among the substances listed within each of the following groupings:
    (a) Cr^{3+}, Cu^{2+}, NO_3^-, K+ *(b)* Cl_2, S, Zn^{2+}, Ag^+ *(c)* Li^+, F_2, Ni^{2+}, Fe^{3+}
11. For each oxidation-reduction reaction determined as such in Question 4, identify *(a)* the oxidizing agent *(b)* the reducing agent.
12. Use Table 20-3 to respond to each of the following: *(a)* Would Al be oxidized by Ni^{2+}? *(b)* Would CU be oxidized by Ag^+? *(c)* Would PB be oxidized by Na^+? *(d)* Would F_2 be reduced by Cl^-? *(e)* Would Br_2 be reduced by Cl^-?
13. What is studied in the area of electrochemistry?
14. Define: *(a)* electrode potential *(b)* half-reaction *(c)* half-cell.
15. Distinguish between an electrochemical cell and an electrolytic cell in terms of *(a)* the nature of the reaction involved *(b)* the signs of the two electrodes?
16. What determines the specific products that result from the electrolysis of an aqueous salt solution?
17. *(a)* What is electroplating? *(b)* Distinguish between the nature of the anode and cathode in such a process.
18. *(a)* Explain what is meant by the potential difference between the two electrodes in an electrochemical cell. *(b)* How, and in what units, is this potential difference measured?
19. Define: *(a)* voltric cell *(b)* half-cell *(c)* electrode potential *(d)* standard electrode potential
20. The standard hydrogen electrode is assigned an electrode potential of 0.00 V. Explain why this voltage is assigned.

21. What information is provided by the electrode potential of a given half-cell? (b) What does the relative value of the potential of a given half-reaction indicate about its oxidation-reduction tendency?

Problems

1. Each of the following atom/ion pairs undergoes the oxidation number change indicated below. For each pair, determine whether oxidation or reduction has occurred, and then write the electronic equation indicating the corresponding number of electrons lost or gained:
   (a) $K \rightarrow K^+$ *(b)* $S \rightarrow S^{2-}$ *(c)* $Mg \rightarrow Mg^{2+}$
   (d) $F^- \rightarrow F_2$ *(e)* $H_2 \rightarrow H^+$ *(f)* $O_2 \rightarrow O^{2-}$
   (g) $Fe^{3+} \rightarrow Fe^{2+}$ *(h)* $Mn^{2+} \rightarrow Mn^{7+}$.

2. Determine the oxidation number of each atom indicated in the following: *(a)* H_2 *(b)* H_2O
   (c) Al *(d)* MgO *(e)* Al_2S_3 *(f)* HNO_3
   (g) H_2SO_4 *(h)* $Ca(OH)_2$ *(i)* $Fe(NO_3)_2$ *(j)* O_2
   (k) $KMnO_4$ *(l)* Na_2SO_3 *(m)* ClO_3^- *(n)* SO_4^{2-}
   (o) NH_4^+ *(p)* PO_4^{3-} *(q)* $Na_2Cr_2O_7$
   (r) $(NH_4)_3PO_4$.

3. Balance the oxidation-reduction equation below by using the oxidation-number method in response to each requested step:
   $$Fe + HCl \rightarrow FeCl_2 + H_2$$
   (a) Write the skeleton equation for the reaction. *(b)* Assign oxidation numbers to all atoms and determine what is oxidized and what is reduced. *(c)* Write separate electronic equations for the oxidation and reduction processes, and balance each of them so that the number of electrons lost is equal to the number gained. *(d)* Place the coefficients from the two electronic equations into the skeleton equation and then balance the overall equation so that atoms are conserved.

4. Use the oxidation-number method as outlined in Problem 3 to balance each of the following redox equations:
   (a) $Li + H_2O \rightarrow LiOH + H_2$
   (b) $Al + HCl \rightarrow AlCl_3 + H_2$
   (c) $H_2S + HNO_3 \rightarrow H_2SO_4 + HO_2 + H_2O$
   (d) $Cu + HNO_3 \rightarrow Cu(NO_3)_2 + NO_2 + H_2O$
   (e) $KClO_3 \rightarrow KCl + O_2$
   (f) $KOH + Cl_2 \rightarrow KCl_3 + KCl + H_2O$
   (g) $S + HNO_3 \rightarrow SO_2 + NO + H_2O$
   (h) $As_4 + HNO_3 \rightarrow HA_sO_3 + NO + H_2O$

5. Zinc reacts with nitric acid to form zinc nitrate,

ammonium nitrate, and H_2O. Balance the redox equation by the oxidation-number method.

6. Zinc reacts with sodium chromate in a sodium hydroxide solution to form Na_2ZnO_2, $NaCrO_2$, and H_2O. Balance the redox equation using the oxidation-number method.

7. Balance the oxication-reduction equation below by using the ion-electron method in response to each requested step: $K + H_2O \rightarrow KOH + H_2$.
   (a) Write the skeleton equation for the reaction.
   (b) Write the net ionic equation and assign oxidation numbers to all atoms to determine what is oxidized and what is reduced.
   (c) Write the equation for the reduction and balance it for both atoms and charge.
   (d) Write the equation for the oxidation and balance it for both atoms and charge.
   (e) Adjust the oxidation and reduction equations by multiplying the coefficients as needed so that electrons lost equals electrons gained and add the two resulting equations.
   (f) Add species as necessary to balance the overall formula equation.

8. Use the ion-electron method in Problem 7 to balance each of the reactions below.
   (a) $HI + HNO_2 \rightarrow NO + I_2 + H_2O$
   (b) $FeCl_3 + H_2S \rightarrow FeCl_2 + HCl + S$
   (c) $SbCl_5 + KI \rightarrow KCl + I_2 + SbCl_3$
   (d) $Ca(OH)_2 + NaOH + ClO_2 + C \rightarrow NaClO_2 + CaCO_3 + H_2O$

9. Given the following half-cells, determine the overall electrochemical reaction that proceeds spontaneously and the value of $E°$: *(a)* $Cu^{2+} - Cu, Ag^+ - Ag$ *(b)* $Cd^{2+} - Cd, Co^{2+} - Co$ *(c)* $Na^+ - Na, Ni^{2+} - Ni$ *(d)* $I_2 - I^-, Br_2 - Br^-$ *(e)* $Mg^{2+} - Mg, Au^{3+} - Au$

10. Predict whether each of the following reactions will occur spontaneously as written by determining the $E°$ value for potential reaction. Write and balance the overall equation for each reaction that does occur. *(a)* $Mg + Sn^{2+} \rightarrow$ *(b)* $K + Al^{3+} \rightarrow$ *(c)* $Li^+ + Zn \rightarrow$ *(d)* $Cu + Cl_2 \rightarrow$ *(e)* $Zn + Fe^{2+} \rightarrow$

Application Questions

1. If oxidation occurs in a chemical reaction, reduction must occur simultaneously. Explain.

2. Explain the significance of assigning oxidation numbers to the atoms of covalent molecules.

3. In an oxidation-reduction reaction, the sub-

stance oxidized serves as the reducing agent, while the substance reduced serves as the oxidizing agent. Explain.

4. Fluorine is the most active oxidizing agent among the elements, while the fluorine ion is the weakest reducing agent. Explain this observation.

5. Based on Table 20-3, predict whether each of the following reactions should occur: *(a)* Na + Cr^{3+} → *(b)* Pb + Ag^+ → *(c)* Ni + Mg^{2+} → *(d)* F_2 + I^- → *(e)* Cl_2 + F^- → *(f)* Al + Pb^{2+} →

6. While an electrochemical cell can be viewed as a type of battery, the electrolytic cell requires an external energy source such as a battery to drive it. Explain the distinction between these two types of cells.

7. Explain how the oxidation-reduction chemistry of both the electrochemical cell and the electrolytic cell are combined in the chemistry of rechargeable cells

8. The active metals in Table 20-3, down to magnesium, replace hydrogen from liquid water. Magnesium and succeeding metals replace hydrogen from steam. Metals near the bottom of the list do not replace hydrogen from liquid water or steam. How can the data given in the table help you explain this behavior?

Application Problems

1. Balance each of the following oxidation-reduction reactions by using the oxidation number method: *(a)* $AuCl_3$ + Sb_2O_3 + H_2O → Au + Sb_2O_5 + HCl *(b)* $AuCl_3$ + H_2SO_4 + KBr → K_2SO_4 + $MnSO_4$ + Br_2 + H_2O

2. Potassium carbonate and bromine react to form potassium bromide, potassium bromate, and carbon dioxide. Balance the equation using the oxidation-number method.

3. Potassium permanganate, sodium sulfite, and sulfuric acid react to form potassium sulfate, manganese(II) sulfate, sodium sulfate, and water. Balance the equation for this oxidation-reduction reaction by the oxidation-number method.

4. Concentrated nitric acid reacts with copper to form copper(II) nitrate, nitrogen dioxide, and water. Balance the equation using *(a)* the oxidation-number method; *(b)* the ion-electron method.

5. Hot, concentrated sulfuric acid reacts with zinc to form zinc sulfate, hydrogen sulfide, and water. Balance the equation, using *(a)* the oxidation-number method; *(b)* the ion-electron method.

6. Balance the equation for the following redox reaction by the ion-electron method.
zinc + nitric acid → zince nitrate + nitrogen monoxide + water

7. Balance the equation for the following redox reaction by the ion-electron method.
$KMnO_4$ + $FeSO_4$ + H_2SO_4 → K_2SO_4 + $MnSO_4$ + $Fe_2(SO_4)_3$ + H_2O

8. For each reaction predicted to occur in Application Question 5, complete the half-reactions and balance the overall equation for the reaction based on the information provided in Table 20-3.

9. Suppose chemists had chosen to make the I_2 + $2e^-$ ⇌ $2I^-$ half-cell the standard electrode and assigned it a potential of zero(0) volts. *(a)* What would be the $E°$ value for the Br_2 + $2e^-$ ⇌ $2Br^-$ half-cell? *(b)* What would be the $E°$ value for the Al^{3+} + $3e^-$ ⇌ Al half-cell? *(c)* How much charge would be observed in the $E°$ value for the reaction involving Br_2 + I^- using the I_2 half-cell as the standard?

10. If a strip of Ni is dipped into a solution of $AgNO_3$, what would be expected to occur? Explain, using $E°$ values and equations.

11. *(a)* What would happen if an aluminum spoon were used to stir a solution of $Zn(NO_3)_2$? Explain using $E°$ values. *(b)* Could a strip of Zn be used to stir a solution of $Al(NO_3)_3$? Explain, using $E°$ values.

12. Can a solution of $Sn(NO_3)_2$ be stored in an aluminum container? Explain, using $E°$ values.

Enrichment

1. Take an inventory of the type of batteries used in your house. Find out the voltage supplied by each battery and what type of electrochemical reaction each uses. See if you can provide a reason why each type of electrochemical reaction is used in each case.

2. Go to the library and find out what you can on the electroplating industry in the United States. What are the top three metals used for plating and how many metric tons of each are used in the United States each year for electroplating?

CAREERS

Industrial Technician

Thousands of chemical products—dyes, cleaning agents, drugs, and fuels—are produced every day. In most industries, the bulk of the work that keeps production running smoothly is done by industrial technicians.

Industrial technicians run the equipment that is used to make chemical products. A technician receives shipments of raw materials from other manufacturers, measures and mixes the materials, and runs production equipment. In electronic and pharmaceutical industries, in which the product must be kept sterile or dust-free, the technician controls remote or robotic equipment. Other technicians monitor temperature and pressure gauges during chemical processing. Mechanical skills help a technician spot and solve problems such as jamming or slippage, which may cause equipment to malfunction.

Most industrial technicians have a high school diploma or an associate's degree certifying their competence in a specific technical area. The ability to work well with others is important for industrial technicians who wish to become supervisors.

Environmental Chemist

An environmental chemist determines the identity and level of pollutants in various parts of our ecosystem.

To determine the composition of soil and water samples, environmental chemists use spectroscopy, titration, and other tools of analytical chemistry. Chemists who are interested in air pollution study the ozone layer, health hazards presented by leaded gasoline, and acid precipitation.

Analysis of waste is another major area of environmental work. Hazardous waste is often improperly disposed. Sometimes it is dumped into open pits, where rain leaches soluble materials into nearby streams. Environmental chemists are needed to analyze problems and to propose treatments to prevent further pollution.

Many job opportunities exist for chemists interested in environmental issues: in the field gathering samples, in small businesses devoted to testing materials, in industries concerned about their toxic wastes, and at government research agencies. Most environmental chemists have at least a bachelor's degree in chemistry, though a high school diploma is sufficient for some technical jobs.

Cumulative *Review*

Unit 6 • Chemical Reactions

Questions

1. Write the thermochemical equations, including ΔH_f°, for the formation of *(a)* zinc sulfate *(b)* nitrogen dioxide, *(c)* barium nitrate. (18.1)
2. Consult the Appendix Table of ΔH_f values and decide which of the calcium compounds listed is *(a)* most stable *(b)* least stable. (18.1)
3. Write the equation for the net driving force of a chemical reaction and identify each of the factors and how they influence spontaneity. (18.2)
4. For $2Ag(s) + Cl_2(g) \rightarrow 2AgCl(s)$, at 25°C $\Delta S = -0.116$ kJ. Explain this entropy change. (18.2)
5. For the following reaction mechanism, *(a)* explain what happens in each step *(b)* identify the intermediate step *(c)* identify the rate-determining step *(d)* give the overall reaction:

   $$A_2 \rightarrow 2A \quad \text{fast}$$
   $$2A + B_2 \rightarrow 2AB \quad \text{slow}$$ (18.3)
6. Write the chemical equilibrium expression for *(a)* $N_2H_4(g) = N_2(g) + 2H_2(g)$ *(b)* $S(s) + 3F_2(g) = SF_6(g)$ *(c)* $2H_2(g) + O_2(g) = H_2O(\ell)$. (19.1)
7. For each of the following in aqueous solution, write a chemical equation and the equilibrium constant expression: *(a)* Ag_2S, a slightly soluble salt *(b)* HF, a weak acid *(c)* SO_3^{2-}, which undergoes hydrolysis. (19.4)
8. *(a)* What happens to [HA] if [A^-] in a solution of weak acid HA is increased? *(b)* Name this effect. (19.2)
9. Will the following solutions be acid, basic, or neutral? *(a)* $AlCl_3$ *(b)* $NaNO_3$ *(c)* $NaHCO_3$. (19.3)
10. Assign oxidation numbers to each element and classify the following as redox or nonredox reactions: *(a)* $Ca(OH)_2(ag) + CO_2(g) \rightarrow CaCO_3(s) + H_2O(\ell)$ *(b)* $3Na(s) + P(s) \rightarrow Na_3P(s)$ *(c)* $NH_3(ag) + ClO^- \rightarrow N_2H_4(g) + Cl^- + H_2O(\ell)$. (20.1)
11. For each redox reaction in Question 10 identify the *(a)* substance oxidized *(b)* substance reduced *(c)* oxidizing agent *(d)* reducing agent. (20.3)

12. What is the difference between the chemical reactions in a flashlight battery and in an electrolytic cell? (20.4)
13. Define standard electrode potential and explain why it is needed. (20.4)
14. Would you expect chromium (Cr) to be oxidized in a reaction with *(a)* sodium ion (NA^+) *(b)* an aqueous acid (H_3O^+) *(c)* iron(III) ion (Fe^{3+})? (20.4)

Problems

1. Calculate the heat of formation of octane ($C_8H_{18}(\ell)$), using ΔH_F and ΔH_c values in Tables 18-1 and 18-2. *(a)* Solve by combining the known thermochemical equations. *(b)* Verify the result by using the general equation for finding heats of formation from heats of combustion. (18.1)
2. Determine whether a reaction occuring at 25°C with $\Delta H = -165.4$ kJ and $\Delta S = 0.0275$ kJ/K will occur spontaneously. (18.2)
3. Draw and label the energy diagram for a reaction in which the reactants start at energy level = 0 kJ, $\Delta H_{forward} = +10$ kJ, and $E_a' = 30$ kJ. Determine $\Delta H_{reverse}$ and E_a'. (18.3)
4. Determine whether or not a precipitate will form if 10.0 mL of $1.00 \times 10^{-6} M$ $Al(NO_3)_3$ is mixed with 10.0 mL of $2.00 \times 10^{-5} M$ KOH at 25°C. (19.4)
5. Use the oxidation number method tp balance the oxidation–reduction equation given below:
   $$KMnO_4(aq) + KBr(aq) + H_2SO_4(aq) \rightarrow Br_2(\ell) + K_2SO_4(aq) + MnS_4(aq) + H_2O(\ell)$$ (20.2)
6. Use the ion-electron method to balance the redox equation given below: (20.2)
   $$HNO_3(aq) + Bi_2S_3(aq) \rightarrow Bi(NO_3)_3(aq) + NO(g) + S(s) + H_2O(\ell)$$
7. Predict whether the reaction between $Zn(s)$ and $Ag^+(aq)$ will occur spontaneously by determining the possible $E°$ value. If the reaction does occur, write and balance the overall equation. (20.4)

Intra-Science
How the Sciences Work Together

The Process in Chemistry: Chemiluminescence

As small children, we learn (sometimes painfully) that most sources of light—flames and incandescent lights, for example—are dangerously hot. When we get older, we are fascinated to discover that some substances are luminescent—that is, they emit light when they are at normal temperatures. The most surprising type of luminescence is chemiluminescence, which is the reaction of two or more substances in complete darkness to produce light without becoming hot.

If hydrogen peroxide is added to a solution of the organic compound luminol, a chemical reaction produces a beautiful greenish blue luminescence. This continues until one or both of the reagents have been chemically consumed. Most chemiluminescent reactions are the result of oxidations in which an organic compound—such as luminal—is oxidized by oxygen, hydrogen peroxide, or ozone.

One familiar example of this is the operation of a Bunsen burner. The blue color of the innermost cone of a Bunsen burner flame is the result of a chemiluminescent reaction between oxygen from the air and natural gas. The innermost part of the flame is relatively cool. The heat of a Bunsen burner flame is produced when the products of this chemiluminescent reaction are further oxidized above the upper tip of the blue cone.

The Connection to Physics
Photons of light—particles of radiant energy—are produced when an atom or molecule changes from an excited state to a lower-energy state. This change involves an elec-

tron moving from an atomic or molecular orbital that is relatively far from the corresponding nucleus (where it has a relatively high energy) to an orbital closer to the nucleus (where it has a lower energy). The energy of the emitted photon is equal to the energy difference between the higher-energy and lower-energy orbital. The energy of the emitted photon determines whether the corresponding light is colored and, if so, what its color is.

Where the types of light production differ is in the method used to produce excited atoms or molecules. In ordinary incandescent lights and in flames, this excitation is caused by heat. In luminescence, however, atoms or molecules are excited without becoming hot. Sonoluminescence is a type of luminescence in which electrons of atoms and molecules are excited

by high-frequency sound waves. In fluorescence and phosphorescence, atoms or molecules are excited by absorbing light. In fluorescence, the excited species re-emit light almost immediately (within 1/100 000 000 of a second). In phosphorescence, this re-emission occurs relatively slowly. Consequently, a phosphorescent object may glow for many minutes in the dark. In chemiluminescence, chemical reactions produce molecules that are in excited states to begin with. Light is emitted as these molecules drop to their lowest energy state.

The Connection to Biology
Long before the general phenomenon of chemiluminescence was recognized and understood, people had been fascinated by the light-producing ability of fireflies and glowworms (the larvae of fireflies),

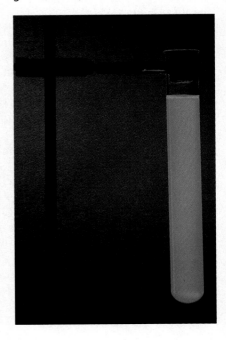

and by foxfire (the glow produced by certain fungi). This emission of light by living organisms is called bioluminescence.

During the last thirty years, it has been recognized that bioluminescence is a type of chemiluminescence. In each case of bioluminescence, a particular biochemical substance called luciferin is oxidized, usually by oxygen, in a reaction catalyzed by an enzyme called luciferase. Each bioluminescing species has its own type of luciferin and luciferase.

Many of the bioluminescing species are marine bacteria and fungi. The flashlight fish, found in the Pacific and Indian oceans, uses luminous bacteria, which it harbors in its light organ for a variety of purposes. It communicates with other flashlight fish by blinking—opening and closing a lid over the light organ. Feeding only on dark nights, the flashlight fish uses this light to find and capture its prey. It

can also confuse and evade its predators by flashing the light on and off while zigzagging away.

The small crustacean *Cypridina*, found in Japanese waters, makes luciferin and luciferase in separate glands. When it is attacked, it squirts them into water, where they react, producing a ball of light. *Cypridina* can use this technique to escape predators, which will pursue the ball of light while the crustacean escapes.

During mating season, the firefly uses its light to attract the opposite sex. Each species of firefly has its own particular light signal so that it will not attract a firefly of a different species.

The Connection to Health Sciences

The bioluminescent reaction that occurs in the firefly requires adenosine triphosphate (ATP) in addition to luciferin and luciferase. Because even extremely dim light can be detected and measured, a mixture of luciferin and luciferase provides a sensitive method for detecting ATP, a crucial energy-carrying molecule present in most living cells. This bioluminescent reaction is used in cancer research to detect the quantity of ATP in cells under different conditions.

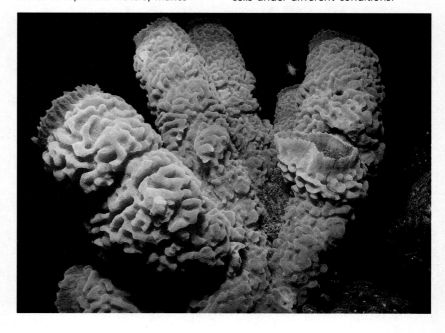

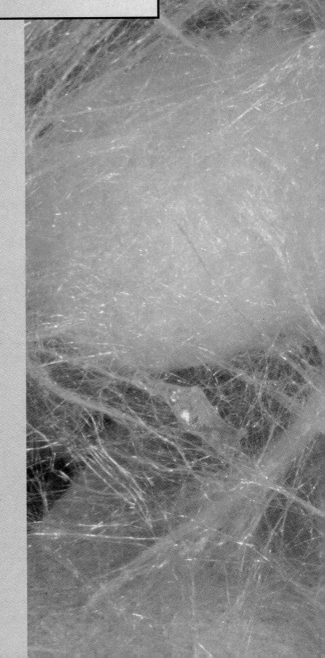

Unit
7

Chapter 21
Carbon and Hydrocarbons   632

Chapter 22
Substituted Hydrocarbons   664

Chapter 23
Biochemistry   684

CARBON AND ITS COMPOUNDS

I vividly remember washing and breaking many dishes during my childhood. If allowed to continue, we would have been eating off paper napkins. In time, however, my parents and I gained a great appreciation for the unbreakable nature of plastic cups and dinnerware.

While in high school, I was introduced to organic polymer chemistry and the notion that most plastics are the result of adding together a number of small molecules. This concept is comparable to building a long train from individual cars. Presently, I am synthesizing new, highly flexible, impact-resistant polymers that will be used in applications ranging from transparent airplane windows to high-gloss automotive paint.

Much of the effort in developing new polymers is toward mimicking the properties of natural substances. The subtle beauty of silk, for instance, has been valued in clothing for almost 5000 years. Silk fibers are made from the cocoons of silkworm larvae. Today, fabrics containing polymers such as polyesters can be produced with many of the prized qualities of silk.

Van I. W. Stuart
Organic/Polymer Chemist
General Electric Corporation

Silkworm spinning cocoon

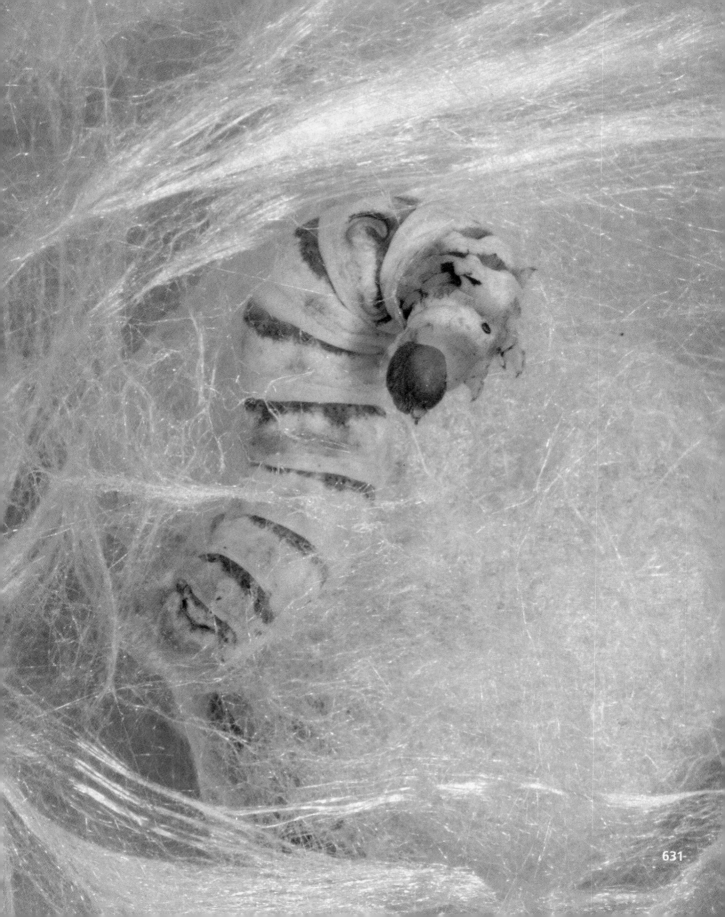

631

Chapter 21 Carbon and Hydrocarbons

Chapter Planner

21.1 Abundance and Importance of Carbon
Structure and Bonding of Carbon
■ Demonstration: 1
Experiment: 24
Questions: 1–3
Allotropic Forms of Carbon
■ Questions: 4, 5, 7
■ Questions: 6, 8, 9
Amorphous Forms of Carbon
■ Questions: 10, 11
■ Questions: 12, 13

21.2 Organic Compounds
■ Questions: 14, 15
Structural Formulas and Bonding
■ Demonstration: 4
Questions: 16–18

Differences Between Organic and Inorganic Compounds
■ Demonstration: 3
Question: 19

21.3 Hydrocarbons
■ Demonstration: 2
Questions: 23, 24
■ Chemistry Notebook
Alkanes
■ Questions: 25–27, 35
■ Chemistry Notebook
Questions: 27, 29
Alkenes
■ Experiment: 40
Questions: 31–33
■ Question: 30

Alkynes
■ Questions: 34, 36
Benzene and Aromatic Hydrocarbons

21.4 Representative Hydrocarbons and Polymers
Natural Gas and Petroleum
■ Questions: 20–22
Petroleum Substitutes
■ Question: 38
Rubber
■ Questions: 39–44

Teaching Strategies

Introduction

Start this chapter by pointing out that almost everything that will be discussed will be something that touches the students' lives in some way. Some of the substances you will discuss are: methane (the main component of natural gas); propane (bottled gas, which many of them use to heat their homes); gasoline (used to drive their cars); and kerosene (used to drive big diesel trucks, which deliver many of the commodities they buy in grocery stores, department stores, and at service stations). Many of the plastics and rubber they use are made from some component of crude oil or petroleum.

At the same time, much of the chapter will seem very foreign and different from the chemistry that has been discussed previously. Even though structural formulas have already been covered, the idea that every compound has its own individual structural formula and that many compounds have the same chemical formulas is difficult for students to grasp. Emphasize that similarity in structure enables many compounds to be grouped together and that similarity causes them to react in a similar way. In this chapter and the following chapter, students will see how over 4 000 000 compounds can be placed in a relatively small number of groups on the basis of similarity in structure. After they understand the basis for naming, they will begin to see the pattern for naming all classes of organic compounds. It is important to stress that, although organic chemis-

try may at first seem to be very different from inorganic chemistry, the same principles apply to both. Because everything is so new and different, it will take a little extra time at the beginning to lay the ground work for this unit.

The overall objectives for this chapter are to help the students: (1) differentiate between the different allotropic and amorphous forms of carbon, know where they are found or how they are made, and know their uses; (2) to know the differences between inorganic and organic compounds; (3) differentiate between alkanes, alkenes, alkynes, and aromatics on the basis of their structural formulas (both two-dimensional and three-dimensional and their names); (4) know the natural sources or preparations of the different hydrocarbon classes, and their uses and reactions; and (5) to understand that natural gas and petroleum are natural sources of hydrocarbons and that they provide raw materials for making many polymers such as plastics and synthetic rubber.

21.1 Carbon

This demonstration is more helpful to the students if they make models along with you.

Demonstration 1: Bonding of Carbon Atoms
Purpose: To demonstrate the bond angles of carbon atoms and

how they are able to link with one another to form chains, rings, plates, and networks

MATERIALS: Gumdrops, scissors, tape, toothpicks, transparencies, enlarged photocopied drawings of Figure T21-1

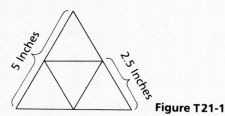

Figure T21-1

PROCEDURE: Prior to class, reproduce the drawing in Figure T21-1 on a sheet of paper $8\frac{1}{2} \times 11''$. You will be able to get four per page. Make as many photocopies as needed, and make transparencies using the photocopies of the drawings.

Have the students work in groups of two. First make a transparent model of a tetrahedron by cutting out the large equilateral triangle and bending up the sides of the tetrahedron. Tape two of the sides together, and leave one side open to insert the toothpick-gumdrop model of the carbon atom with its four covalent bonds. (See Figure T21-2.) Students may have to reinsert toothpicks a few times to get them at the correct angles so they go to the corners of the tetrahedron. Now the students have a model from which to work when making models through out this chapter.

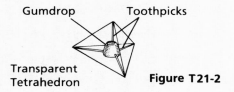

Gumdrop   Toothpicks

Transparent
Tetrahedron   **Figure T21-2**

Using the first model as a guide, make additional carbon atoms with their bonds and join them in chains and rings, or even plates and networks, to understand how they join with one another. (See Figure 21.3.) Save these models for further use in this chapter.

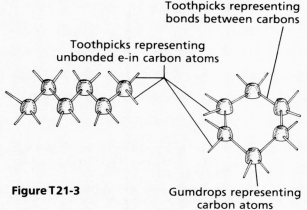

Toothpicks representing
bonds between carbons

Toothpicks representing
unbonded e-in carbon atoms

Figure T21-3

Gumdrops representing
carbon atoms

This demonstration lays the groundwork for understanding structural formulas of organic compounds.

As you discuss diamonds, you might use a diamond phonograph needle to scratch a piece of glass to illustrate its hardness. With certain cuts of fake diamonds it is sometimes difficult even for jewelers to determine if they are fake or not. The ultimate test is to discover if it can be scratched by another diamond. If it can be scratched, it is not a real diamond; if it cannot, it is real!

As you discuss graphite, exhibit a sample of powdered graphite and different substances or objects containing graphite, such as graphite lubricants and greases, graphite electrodes, pencil lead, and rackets or fishing rods made of graphite fibers. If you made gumdrop models of graphite plates in Demonstration 1 you can use them to show how these plates slip over one another with ease. This explains why graphite feels greasy and is a good lubricant.

When you discuss coal, exhibit different types of coal. Along with the discussion of destructive distillation, show the different products of destructive distillation with their original form. For example, a piece of wood with a charcoal briquet, coal with coke, and a bone with boneblack. Remind students that they will be making charcoal in the lab. After they have completed the lab, you might ask them to design a similar lab to make coke.

21.2 Organic Compounds

It is very important that all discussions in this section and future sections dealing with the structure of organic molecules be accompanied by three-dimensional models that can be translated into two-dimensional structural formulas. Many types of commercial model kits are available. In addition to the ball-and-stick model kits, which represent all atoms as being the same size, newer, more sophisticated ones are now being produced with balls of varying sizes. The ball-and-stick models make it easier to see the arrangement of the atoms in the molecules; however, in reality the atoms are much closer together. The compact models are better to use, as they more accurately represent the atoms in the molecule.

Most molecular model kits are fairly expensive. If your school can afford them, it is a good idea to have a ball-and-stick set and a compact set for demonstration purposes.

Let the students work along with you, building their own models using different colored gumdrops for different atoms, toothpicks for single-covalent bonds, and pipecleaners for double and triple bonds.

There are several advantages to using gumdrop models: *(1)* They are cheap enough that the students can build their own models. *(2)* You can quickly see if the students understand the three-dimensional nature of the discussion. *(3)* Gumdrops come in at least two sizes, so larger ones can be used for large atoms and smaller ones for small atoms. *(4)* Students can make ball-and-stick gumdrop models first and then push the gumdrops together into the compact form, which is the best representative of the molecule. Not even the commercial models can do this!

Demonstration: 2 Building Models and Writing Structural Formulas of C_5H_{12} Isomers
PURPOSE: To help students develop a better understanding of isomers and structural formulas by building three-dimensional models of the three isomers having the formula C_5H_{12} and writing their structural formulas using the models

MATERIALS: Kits composed of 15 large gumdrops of one color, 36 small gumdrops of a different color, 51 toothpicks, the gumdrop-toothpick model of a carbon atom with its four hybrid covalent bonds from Demonstration 1, a commercial molecular model kit (optional)

PROCEDURE: If possible, have students work along with you using their own kits. Using the model from Demonstration 1 as a guide for bond angles, first build a continuous chain of five carbon atoms, letting the large gumdrops represent the carbon atoms. (See Figure T21-4.) Next attach small gumdrops, representing hydrogens, to all toothpicks with unattached ends. (See Figure T21-5.) Then write the structural formula for the compound.

Figure T21-4

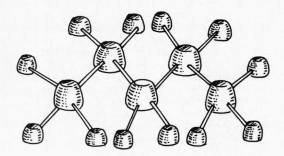

Figure T21-5

Point out that the zig-zag nature of the three-dimensional model is represented as a straight-chain in the structural formula. Thus, in some ways the structural formula is not a good representation of the model or the actual molecule. To make the ball- and-stick model into a compact model, push the gumdrops together. (See Figure T21-6.)

Figure T 21-6

Next make a model of the C_5H_{12} isomer having the structural formula:

By using the model and turning it in different directions, you can help students see that the following three formulas are the same as the first formula:

and

Last, make a model of the isomer with the structural formula,

Demonstration 3: Differences Between Organic and Inorganic Compounds

PURPOSE: To show how organic compounds differ from inorganic compounds on the basis of *(1)* solubility in polar and nonpolar solvents, and *(2)* rate of reaction

MATERIALS: Ammonium chloride, dilute ammonia water, corn-starch, 0.5-*M* iron(III) chloride solution, kerosene, mineral oil, napthalene, potassium nitrate, saliva, 0.1-*M* silver nitrate solution, 0.1-*M* sodium chloride solution, sucrose, distilled water, dry yeast, 2 125-mL beakers, burner, ringstand and iron ring, stirring rod, 8 rubber stoppers to fit large test tubes, 4 medium test tubes(15 × 150 mm), 8 large test tubes(25 × 150 mm), wire gauze

PROCEDURE: *(1)* Add 20 mL of water to each of the four large tubes. Check for solubility of each of the following substances in water by adding the indicated amount to the water: 0.5 g of NH_4Cl, 0.5 g KNO_3, 0.5 g of napthalene, and 5 mL of kerosene. Stopper and shake each test tube and its contents. Which substances dissolved in the water? NH_4Cl and KNO_3. Why did this happen? NH_4Cl and KNO_3 are inorganic compounds; inorganic compounds generally dissolve in water. Napthalene and kerosene are organic compounds that do not dissolve in water. *(2)* Repeat part *(1)* but substitute mineral oil for the water. Results? The NH_4Cl and KNO_3 did not dissolve, but the naphthalene and the kerosene did dissolve. Why? Napthalene and kerosene are organic compounds and mineral oil is an organic liquid; organic substances dissolve in other organic substances. (Like dissolves like: nonpolar solvents dissolve nonpolar substances; polar solvents dissolve polar and ionic substances.) Thus, inorganic compounds such as NH_4Cl and KNO_3 do not dissolve in organic solvents such as mineral oil. *(3)* Mix 5 mL of $FeCl_3$ solution with 5 mL of ammonia water in a medium test tube. Result? A brown precipitate forms immediately. *(4)* Mix 5 mL of NaCl solution with 1 mL of silver nitrate solution in a medium test tube. Result? A white precipitate forms immediately. *(5)* Add dry yeast and a small pea and 1 g of sucrose to 40 mL of water in a beaker and stir. Result? Nothing immediately, but by the next day gas bubbles are forming, indicating a reaction is taking place. *(6)* Before class, dissolve 2 g of cornstarch in 25 mL of cold water in a beaker and gently heat while stirring until it thickens. Cool. During class, add about 1 mL of saliva to the cool, thick starch mixture. Result? Nothing happens immediately, but in 5-10 minutes it changes to a thin, watery mixture. How do you explain these results? The organic reactions in parts *(5)* and *(6)* proceed at much slower rates than the inorganic reactions in parts *(3)* and *(4)*.

21.3 Hydrocarbons

To introduce this section, display several models of alkanes, alkenes, and alkynes. Remember, if you use gumdrop models, sections of pipecleaners work very nicely for the flexibility needed for double and triple bonds. By using models, students will be able to quickly see the differences and similarities between the hydrocarbon classes.

Demonstration 4: Tests for Alkanes and Alkenes

PURPOSE: To show how to differentiate between alkanes and alkenes on the basis of the reactivity of the single bond in alkanes and double bonds in alkenes

MATERIALS: 1,1,2-trichloro-1,2,2-trifluoroethane, cyclohexane, cyclohexene, iodine, potassium permanganate, sodium hydroxide, 4-50 mL beakers, 2 stirring rods, overhead projector

PROCEDURE: CAUTION: Wear goggles, gloves, and a lab coat or lap apron during the preparation and performance of this lab. NaOH and any solution containing NaOH is extremely corrosive to the skin. A great deal of heat is given off when NaOH is added to water. Cyclohexane and cyclohexene are extremely flammable. Store under a fume hood until ready to use, and return to the hood as soon as the demonstration is completed.

Make the basic potassium permanganate solution by: *(1)* adding 4 g of NaOH to enough water to make 100 mL of solution; and *(2)* right before doing the lab, add 0.16 g of $KMnO_4$ to the NaOH solution. (Make sure the solution is cool before adding the $KMnO_4$.)

Put 10 mL of cyclohexane and 10 mL of cyclohexene in two separate 50-mL beakers. Place each beaker on an overhead projector, and project on a screen. Add 3 mL of the basic $KMnO_4$ solution to the cyclohexane and 3 mL to the cyclohexene. Stir both solutions thoroughly. If no double bonds are present, the solution will remain purple; if double or triple bonds are present, it will turn brown. (Note: Depending on the purity of the alkene, the double-bonded compounds may turn green instead of brown. Either color change indicates the presence of an unsaturated double- or triple-bond.) Results: The cyclohexane and basic $KMnO_4$ solution will remain purple, but in the presence of the cyclohexene the color will change to brown (or possibly green). See equations below:

$$RH + MnO_4^- \rightarrow \text{No reaction}$$

$$+ MnO_4^- \rightarrow \text{No reaction}$$

$$R\text{-}CH=CH\text{-}R + MnO_4^- \rightarrow R\text{-}\underset{OH}{C}H\text{-}\underset{OH}{C}H\text{-}R + \underset{brown}{MnO_2} \text{ (or } \underset{green}{MnO_4^{2-}})$$

$$+ MnO_4^- \rightarrow$$

What do the final colors indicate? (Cyclohexane has no double or triple bonds and cyclohexene has at least one double or triple bond. Of course, the name tells us it is a double bond.)

Using the fume hood, make the iodine solution by adding 3 g of iodine to 100 mL of 1,1,2-trichloro-1,2,2-trifluoroethane. Put 10 mL of cyclohexane and 10 mL of cyclohexene in two separate 50-mL beakers. Place each beaker on an overhead projector, and project on a screen. Add 3 mL of the iodine solution to the cyclohexane and 3 mL to the cyclohexene. Stir both solutions thoroughly. If no double bonds are present, the solution will not change color; if double bonds are present, it will turn colorless.

What do the final colors indicate? (Cyclohexane has no double or triple bonds, and cyclohexene does.)

The equations for the reactions follow:

$$\text{cyclohexane} + I_2 \rightarrow \text{No reaction}$$

$$\text{cyclohexene} + I_2 \rightarrow \text{diiodocyclohexane}$$

21.4 Representative Hydrocarbons and Polymers

The societal issue of plastic pollution is receiving a great deal of attention. The problems with plastic pollution are increasing yearly. Environmentalists point out some of the serious problems already plaguing our society:

(1) Plastics make up 30% of the volume of solid wastes in city garbage. Plastics, along with paper, are the fastest growing types of garbage. *(2)* All types of garbage are on the increase and places to put them on the decrease. *(3)* Most plastics are not biodegradable and will be in the environment for years. *(4)* Plastic pollution in the ocean is killing fish, turtles, birds, and mammals.

Students could research the problem and do short reports and position papers on it. Some solutions might be: *(1)* developing more biodegradable plastics, *(2)* recycling plastic by encouraging people to return plastic containers with bottle deposits and developing more ways to reuse plastic, *(3)* plastic consumption.

You might wish to set a day aside to present some information on the problem to the class. This is definitely a problem that students will have to deal with in their lifetimes. Three good sources are: Seymour (1989), Thayer (1989), and Weisskopf (1988).

References and Resources

Books and Periodicals

Alper, Joseph. "Polymers," *ChemMatters*, Vol. 4, 1986, pp. 4–7.

Carraher, Charles E. Jr., and Raymond B. Seymour. "Polymer Structure—Organic Aspects (Definitions)," *Journal of Chemical Education*, Vol. 65, 1988, pp. 314–18.

Jaffe, Bernard. "Woehler: Urea without a Kidney," *Crucibles: The Story of Chemistry*, Dover Publications, New York, 1976, pp. 108–24.

Kaufman, George B. "Wallace Hume Carothers and Nylon, The First Completely Synthetic Fiber," *Journal of Chemical Education*, Vol. 65, 1988, pp 803–7.

Marsella, Gail. "Silly Putty," *ChemMatters*, Vol. 4, 1986, pp. 15–17.

Polymer Chemistry: A Teaching Package for Pre-College Teachers, National Science Teachers Association, Washington, D.C., 1986.

Scheinberg, Stephen P. "Tyvek," *ChemMatters*, Vol. 4, 1986, pp. 8–10.

Seymour, Raymond B. "Alkanes: Abundant, Pervasive, Important, and Essential," *Journal of Chemical Education*, Vol. 66, 1989, pp. 59–63.

Shakhashiri, Bassam Z. "Combustion of Methane, Polystyrene," *Chemical Demonstrations: A Handbook for Teachers of Chemistry*, Vol. 1, Madison: The University of Wisconsin Press, 1983, pp. 241–42.

Shakhashiri, Bassam Z. "Determination of the Molecular Mass of the Gas from a Butane Lighter," *Chemical Demonstrations: A Handbook for Teachers of Chemistry*, Vol. 2, Madison: The University of Wisconsin Press, 1985, pp. 48–50.

Summerlin, Lee R., Christie L. Borgford, and Julie B. Ealy. "The Nonpolar Disk Game," p. 36; "Producing Methane Gas," p. 93; and "Alkanes versus Alkenes: Reaction of the Double Bond," pp. 94–95. *Chemical Demonstrations: A Sourcebook for Teachers*, Vol. 2, Washington, D.C.: American Chemical Society, 1988.

Summerlin, Lee R., and James L. Ealy, Jr. "Production of a Gas: Acetylene," p. 16; and " Synthetic Rubber," pp. 176–77. *Chemical Demonstrations: A Sourcebook for Teachers*, Vol. 1, Washington, D.C.: American Chemical Society, 1988.

Thayer, Ann M. "Solid Waste Concerns Spur Plastic Recycling Efforts," *Chemical & Engineering News*, Vol. 67, 1989, pp. 7–15.

Vernon, William. "Chocolate Chip Petroleum," *The Science Teacher*, Vol. 55, 1988, pp. 105–6.

Weisskopf, Michael. "Plastic Reaps a Grim Harvest in the Oceans of the World," *Smithsonian*, Vol. 18, 1988, pp. 58–67.

Wittcoff, Harold. "Nonleaded Gasoline: Its Impact on the Chemical Industry," *Journal of Chemical Education*, Vol. 1987, pp. 773–76.

Wood, Clair G. "Dissolving Plastic," *ChemMatters*, Vol. 5, 1987, pp. 12–14.

Young, Jay. "The Interrupted Party," *ChemMatters*, Vol. 2, 1984, pp. 4–5.

Audiovisual Resources

Chemistry 120—Organic Chemistry (sound filmstrips) Learning Arts, Wichita, KS, 1988.

Chemistry in Today's World Series—Carbon (Slides and cassette) Educational Images, Elmira, NY, 1988.

Chemistry Transparencies—Crystal structure of a diamond and graphite, Models of methane, ethane and propane, plus nine others, Frey Scientific, Mansfield, OH, 1988–89.

Chemistry Study Cards—Organic Chemistry: Organic Nomenclature and Organic Reactions, Frey Scientific, Mansfield, OH, 1988–9.

Organic Chemistry I and II. TV Ontario, Chapel Hill, NC, 27516.

Computer Software

Flash, Patrick, and Victor Bendall. *Organic Reaction Chemistry*, Apple II, 64K, Queue, Bridgeport, CN, 1988.

Groves, Paul. *An Introduction to Polymerization*, Apple II, National Science Teachers Association, Washington, D.C.,1985.

Smith, Stanley. *Alkanes and Alkenes and Arenes*, Apple II 48K and IBM PC 256K, Queue, Bridgeport, CN, 1988.

Chemistry Series (J & S Software) Organic Chemistry, Apple II, 48K, Barclay, Brooklyn, NY, 1988.

Organic Nomenclature 1, Apple II, 64K and TRA-80 48K, Queue, Bridgeport, CN, 1988.

SEI Organic Chemistry, Apple II, 64K and IBM PC, 128K, Queue, Bridgeport, CN, 1988.

Isomers, Disks AP301 & IB 301; *Octane*, Disks AP803 & IB803; *Refinery*, Disk AP806; *Polymerization*, Disk AP705; and *Organic Nomenclature*, Disk AP705, Project Seraphim, Ypsilanti, MI, 1988. AP = Apple II; IB = IBM PC.

Answers and Solutions

Questions

1. *(a)* Organic chemistry is the study of carbon compounds. *(b)* Because of the large number of carbon compounds
2. Toward the four vertices of a regular tetrahedron, with the center of the atom at the center of the tetrahedron
3. Carbon atoms join readily with the atoms of other elements. Carbon atoms also link together in chains, rings, plates, and networks.
4. *(a)* Allotropy is the existence of an element in two or more forms in the same physical phase. *(b)* Diamond and graphite
5. Its hardness
6. Heat is conducted through diamond by transfer of energy of vibration from atom to atom. This process is efficient because carbon atoms have a small mass and are strongly bonded to one another. The diamond structure has no mobile electrons; thus diamond is a nonconductor of electricity.
7. The carbon atom layers in graphite slide over one another easily, accounting for the softness. The atoms within the layers are strongly bonded and are difficult to pull apart in the direction of the layer. Hence, carbon fibers are strong.
8. *(a)*

(b) Evidence indicates that all the bonds in graphite are the same. None of the structures has an independent existence.

The actual structure is a resonance hybrid of the written structures. The carbon-carbon bonds are intermediate in character between single and double bonds.

9. The flat scalelike plates slide over one another easily. Graphite is not soluble, subject to corrosion, or affected by heat.
10. Coal is a solid, rocklike material that contains at least 50% carbon and burns readily.
11. *(a)* By long-time decomposition of buried vegetation at high temperature and pressure in the absence of air *(b)* Peat, lignite, subbituminous coal, bituminous coal, and anthracite *(c)* The length of decomposition time and increasingly higher temperatures and pressures
12. *(a)* Destructive distillation is the process of decomposing materials by heating them in a closed container without access to air or oxygen. *(b)* The process is destructive only in the sense that it breaks up or decomposes substances; it is not actually destructive, however, since many new and useful substances are formed by this process.
13. Because of its very large internal surface area
14. Carbon atoms link together by means of covalent bonds, and the same atoms may be arranged in several different isomeric structures.
15. *(a)* A compound composed only of carbon and hydrogen *(b)* CH_4 *(c)* C_2H_2
16. *(a)* A pair of shared electrons *(b)* A pair of shared electrons
17. *(a)* Compounds having the same molecular formula but different structures

(b)

18. The exact number and types of atoms in a molecule and how the atoms are bonded to each other
19. Most organic compounds do not dissolve in water; many inorganic compounds do so. Organic compounds are decomposed by heat more easily than most inorganic compounds. Organic reactions generally proceed at much slower rates than inorganic reactions. Organic reactions are greatly affected by reaction conditions. Organic compounds exist as molecules consisting of atoms joined by covalent bonds.

20. (a) Natural gas is a mixture of hydrocarbon gases and vapors. Up to about 97% of natural gas is methane. The remainder consists of hydrocarbons whose molecules contain from two to seven carbon atoms. (b) Petroleum is a complex mixture of hydrocarbons that varies greatly in composition from place to place. The hydrocarbon molecules in petroleum contain from one to more than 50 carbon atoms and are of several different types.

21. (a) The process of evaporation followed by condensation of the vapors in a separate vessel (b) Fractional distillation is a method of separating the components of a mixture on the basis of differences in their boiling points. (c) It is fairly easy to separate the lower members of the alkane series. However, it is usually only possible to separate the substances with higher boiling points into mixtures of compounds with similar boiling points.

22. (a) Motor fuel (b) Candles, waterproofing, home canning

23. (a) C_nH_{2n+2} (b) C_nH_{2n} (c) C_nH_{2n-2}

24. (a) C_7H_{16} (b) C_7H_{14} (c) C_7H_{12}

25. (a) A series in which adjacent members differ by a constant unit (b) Methane, ethane, propane, butane (c) By using the Greek or Latin prefixes that show the number of carbon atoms in the molecule and the suffix -ane

26. (a) A generalized formula for an alkane (b) The generalized formula for an alkyl group

27. $NaC_2H_3O_2(s) + NaOH(s) \rightarrow CH_4(g) + Na_2CO_3(s)$

28. (a) $CH_4 + 2O_2 \rightarrow CO_2 + 2H_2O$
    (b) $C_2H_4 + 3O_2 \rightarrow 2CO_2 + 2H_2O$
    (c) $2C_2H_2 + 5O_2 \rightarrow 4CO_2 + 2H_2O$
    (d) $2C_4H_6 + 11O_2 \rightarrow 8CO_2 + 6H_2O$

29. $CH_4 + Cl_2 \rightarrow CH_3Cl + HCl$; $CH_3Cl + Cl_2 \rightarrow CH_2Cl_2 + HCl$; $CH_2Cl_2 + Cl_2 \rightarrow CHCl_3 + HCl$; $CHCl_3 + Cl_2 \rightarrow CCl_4 + HCl$

30. (a) Ethene (b) Thermal cracking of petroleum

31. *Cracking* is a process by which complex organic molecules are broken up into simpler molecules. *Thermal cracking* is done by heat alone. *Catalytic cracking* involves the use of both heat and a catalyst.

32.

33. (a) The small molecule from which a polymer is formed (b) A molecule formed by combination of two identical molecules (c) A very large molecule consisting of many repeating structural units (d) The process of joining together of a very large number of monomer molecules; the formation of a polymer

34. It is used in the synthesis of complex organic compounds and in the oxyacetylene torch.

35. (a) They refer to the bonding in hydrocarbon molecules. Saturated molecules have only single bonds between carbon atoms, and unsaturated molecules have at least one double or triple bond. (b) They refer to solutions. A saturated solution is one in which the concentration of solute is the maximum possible under existing conditions. An unsaturated solution is one in which the concentration of solute is less than the maximum possible under existing conditions.

36. Because any moisture in the air would react with the calcium carbide and liberate ethyne

37. They have structures that consist, respectively, of two and three benzene molecules joined together.

38. (1) Mining and transportation to the surface, where it is heated in large ovens to drive out the oil (2) Underground heating and collection of the oil

39. By adding formic acid

40. The repeating units are joined in a zigzag chain

41. By mixing the rubber with sulfur and various other chemicals

42. (a) Organic catalysts and other organic chemicals (antioxidants) (b) Organic catalysts speed vulcanization. The organic chemicals slow the aging process caused by oxygen in the air.

43. The heating of rubber with other materials, always including sulfur, to a definite temperature for a definite time to improve its properties

44. It is little affected by oils and greases.

Problems

1. $1.00 \text{ c} \times \dfrac{200 \text{ mg}}{1c} \times \dfrac{1 \text{ g}}{1000 \text{ mg}} \times \dfrac{1 \text{cm}^3}{3.51 \text{ g}} = .0570 \text{ cm}^3$

2. (a) $100.0 \text{ g} \times \dfrac{1 \text{ mol}}{44.0 \text{ g}} = 2.272 \text{ mol } CO_2$

   (b) $2.272 \text{ mol} \times \dfrac{6.02 \times 10^{23} \text{ molecules}}{\text{mol}} = 1.368 \times 10^{24} \text{molecules } CO_2$

   (c) $2.272 \text{ mol} \times \dfrac{22.4 \text{ L}}{1 \text{ mol}} = 50.9 \text{ L } CO_2$

3. Molecular mass $C_{12}O_9 = 288.1$

   $C = \dfrac{144.1 \text{ g C}}{288.1 \text{ g } C_{12}O_9} \times 100\% = 50.02\% \text{ C}$

   $O = \dfrac{144.0 \text{ g O}}{288.1 \text{ g } C_{12}O_9} \times 100\% = 49.98\% \text{ O}$

4. (a) $Fe_2O_3 + 3CO \rightarrow 2Fe + 3CO_2$
   　　 1 mol　　3 mol　　2 mol　　3 mol

   $5.00 \text{ mol CO} \times \dfrac{1 \text{ mol} Fe_2O_3}{3 \text{ mol CO}} = 1.67 \text{ mol } Fe_2O_3$

   (b) $5.00 \text{ mol CO} \times \dfrac{2 \text{ mol Fe}}{3 \text{ mol CO}} = 3.33 \text{ mol Fe}$

   (c) $5.00 \text{ mol CO} \times \dfrac{3 \text{ mol } CO_2}{3 \text{ mol CO}} = 5.00 \text{ mol } CO_2$

5. $ZnO + CO \rightarrow Zn + CO_2(g)$

   $15.0 \text{ g ZnO} \times \dfrac{1 \text{ mol ZnO}}{81.4 \text{ g ZnO}} \times \dfrac{1 \text{ mol CO}}{1 \text{ mol ZnO}} \times \dfrac{28.0 \text{ g CO}}{1 \text{ mol CO}}$

   $= 5.16 \text{ g CO}$

6. (a) $15.0 \text{ g ZnO} \times \dfrac{1 \text{ mol ZnO}}{81.4 \text{ g ZnO}} \times \dfrac{1 \text{ mol Zn}}{1 \text{ mol ZnO}} \times \dfrac{65.4 \text{ g Zn}}{1 \text{ mol Zn}}$

   $= 12.1 \text{ g Zn}$

   (b) $15.0 \text{ g ZnO} \times \dfrac{1 \text{ mol ZnO}}{81.4 \text{ g ZnO}} \times \dfrac{1 \text{ mol } CO_2}{1 \text{ mol ZnO}} \times \dfrac{22.4 \text{ L } CO_2}{1 \text{ mol } CO_2}$

   $= 4.13 \text{ L } CO_2$

7. $2NaHCO_3 + H_2SO_4 \rightarrow Na_2SO_4 + 2H_2O + 2CO_2$

   $900. \text{ g NaHCO}_3 \times \dfrac{1 \text{ mol NaHCO}_3}{84.0 \text{ g NaHCO}_3} \times \dfrac{1 \text{ mol } H_2SO_4}{2 \text{ mol NaHCO}_3} \times$

$$\frac{98.1\text{g H}_2\text{SO}_4}{1\text{ mol H}_2\text{SO}_4} = 526\text{ g H}_2\text{SO}_4$$

8. $900.\text{ g NaHCO}_3 \times \dfrac{1\text{mol NaHCO}_3}{84.0\text{ g NaHCO}_3} \times \dfrac{2\text{ mol CO}_2}{2\text{ mol NaHCO}_3} \times$

   $\dfrac{22.4\text{ L CO}_2}{1\text{ mol CO}_2} = 240.\text{ L CO}_2$

9. $12.0\text{ gal} \times \dfrac{4\text{ qt}}{1\text{ gal}} \times \dfrac{946\text{ mL}}{1\text{ qt}} \times \dfrac{0.692\text{ g}}{\text{mL}} \times \dfrac{1\text{ kg}}{1000\text{ g}} = 31.4\text{ kg}$

10. Molecular mass $c_6H_5SO_3H = 158.17\text{ g}$

    $\dfrac{(32.1)\text{S}}{(158.1)\text{C}_6\text{H}_5\text{SO}_3\text{H}} \times 100\%\ C_6H_5SO_3H = 20.3\%\ S$

11. $CaC_2 + 2H_2O \rightarrow C_2H_2 + Ca(OH)_2$

    $2.5\text{ L C}_2\text{H}_2 \times \dfrac{1\text{ mol C}_2\text{H}_2}{22.4\text{ L C}_2\text{H}_2} \times \dfrac{1\text{ mol CaC}_2}{1\text{ mol C}_2\text{H}_2} \times \dfrac{64.1\text{ g CaC}_2}{1\text{ mol CaC}_2}$

    $= 7.2\text{ g CaC}_2$

Application Questions

1. Ionic bonds are usually stronger than covalent bonds, and more energy (higher temperature) is needed to break them.

2.

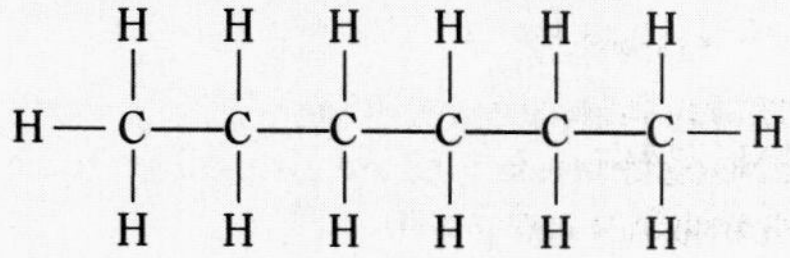

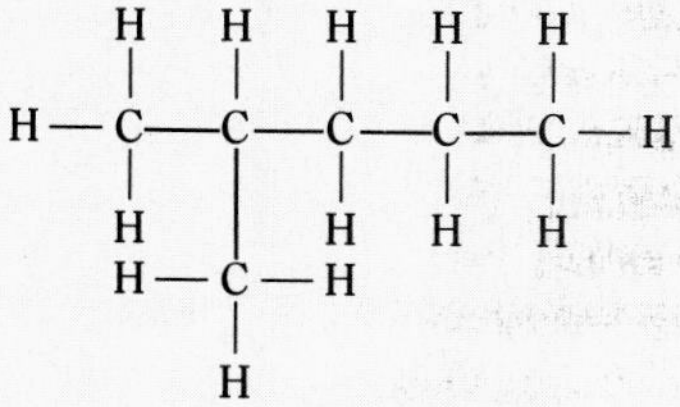

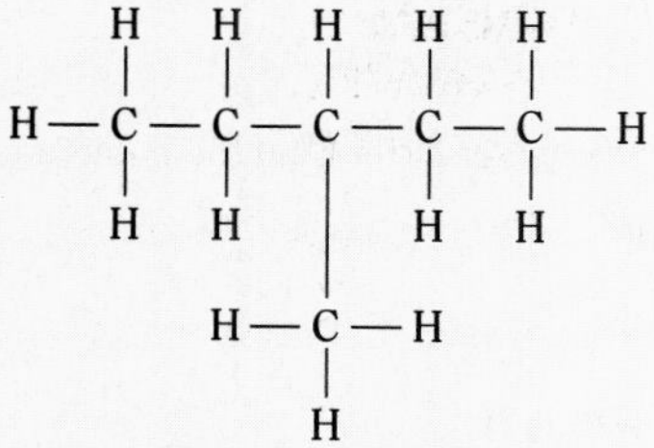

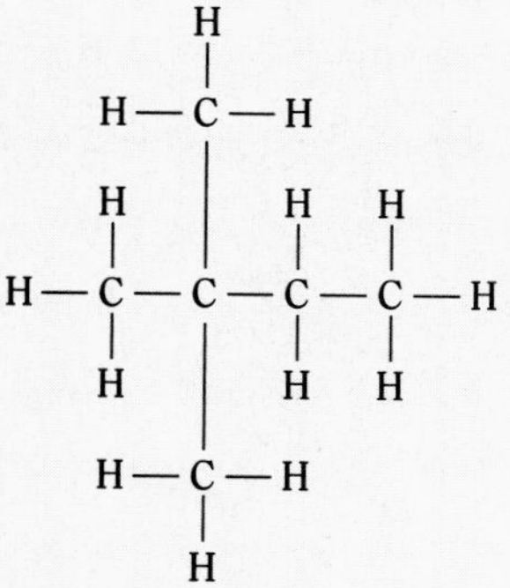

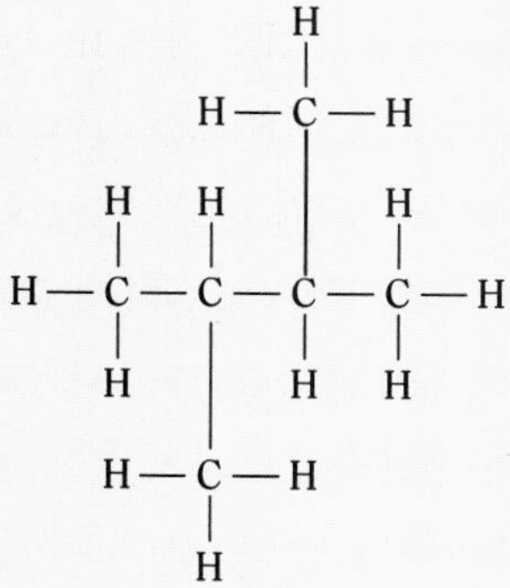

3. $2C_{10}H_{22} + 31O_2 \rightarrow 20CO_2 + 22H_2O$

4.

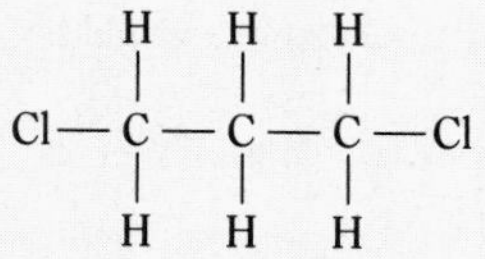

5. 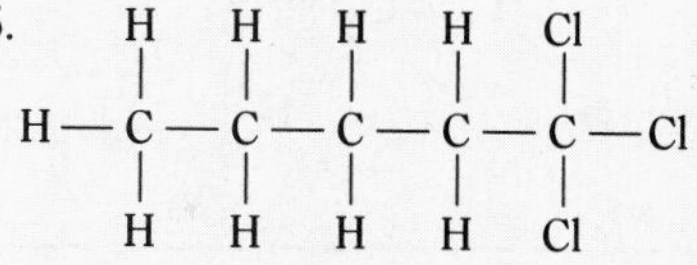 1,1,1-trichloropentane

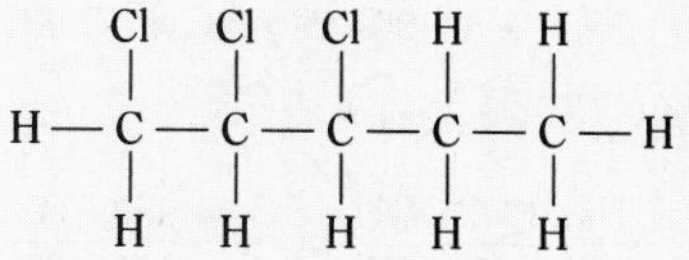

   1,2,3-trichloropentane

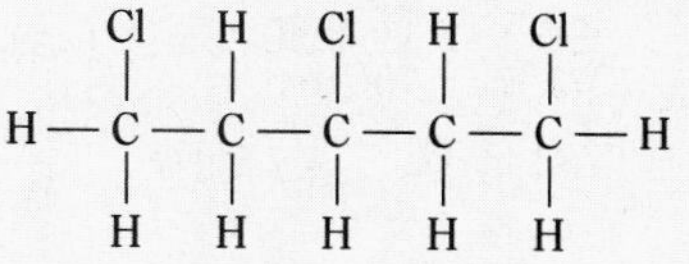

   1,3,5-trichloropentane

6. Almost all carbon compounds contain hydrogen. In addition, hydrogen is contained in acids, hydroxides, and ammonium compounds and other inorganic compounds.

7.

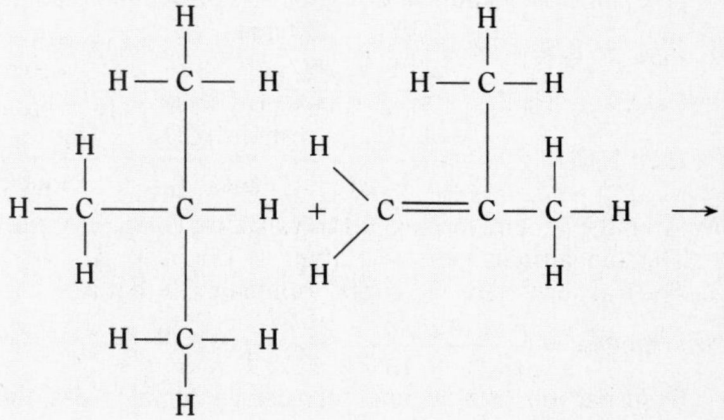

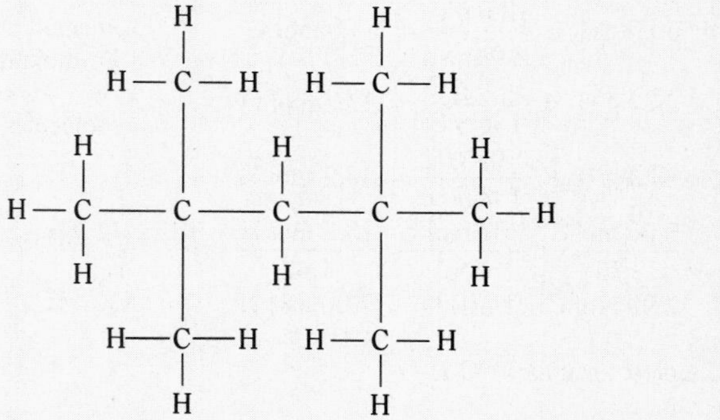

8.

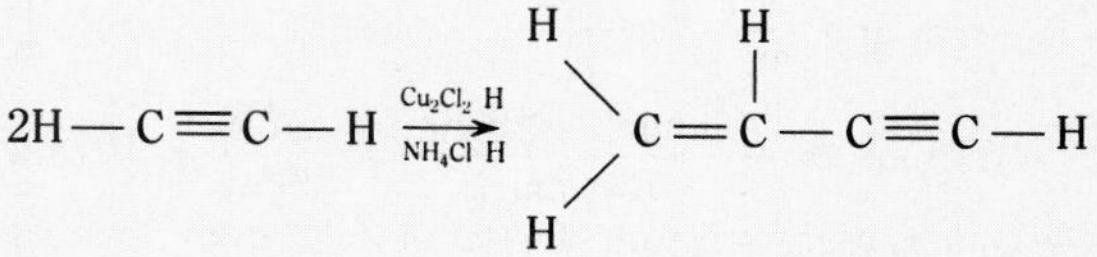

9. *(a)* Yes *(b)* Yes The structural formula would be:

(c) The six hydrogen atoms and the four carbon atoms would all lie in the same plane.

Application Problems

1. $(12.000\ 00\ u + 13.003\ 35\ u + 14.003\ 24\ u)/3 = 13.002\ 20\ u.$

2. CO, $+2$; CO_2, $+4$; C_3O_2, $+4/3$; $C_{12}O_9$, $+3/2$.

3. $HCOOH \rightarrow H_2O + CO$

$$= 250.\ g\ HCOOH \times \frac{1\ mol\ HCOOH}{46.0\ g\ HCOOH} \times \frac{1\ mol\ CO}{1\ mol\ HCOOH} \times \frac{28.0\ g\ CO}{1\ mol\ CO} = 152\ g\ CO$$

4. $152\ g\ CO \times \dfrac{22.4\ L}{28.0\ g} \times \dfrac{300.K}{273\ K} \times \dfrac{760.mm}{750\ mm} \times = 135\ L\ CO$

5. $52.9\ g\ C = \dfrac{1\ mol\ C}{12.0\ g\ C} = 4.41\ mol\ C$ $47.1\ g\ O \times \dfrac{1\ mol\ O}{16.0\ g\ mol} = 2\ 94\ mol\ O$ $\dfrac{4.41\ mol\ C}{2.94} : \dfrac{2.94 mol\ O}{2.94}$

   1.5 mol C : 1 mol O
   3 mol C : 2 mol O
   Simplest formula = C_3O_2
   $C_3O_2 + O_2 \rightarrow CO_2$ (not balanced) $C_3O_2 + 2O_2 \rightarrow 3CO_2$ (balanced according to conservation of atoms)
   Since the balanced equation shows 2 moles of O_2 for each mole of C_3O_2 and the volume relations are also known to be 2 volumes of O_2 for each volume of C_3O_2, C_3O_2 is proved to be the molecular formula by Avogadro's principle.

6. $V_{STP} = 2.50\ L \times \dfrac{273K}{295K} \times \dfrac{735\ mm}{760\ mm} = 2.24\ L$ at STP

   $CaCO_3 + 2HCl \rightarrow CaCl_2 + H_2O + CO_2(g)$

   $2.24\ L\ CO_2 \times \dfrac{1\ mol\ CO_2}{22.4\ L\ CO_2} \times \dfrac{1\ mol\ CaCO_3}{1\ mol\ CO_2} \times \dfrac{100.1\ g\ CaCO_3}{1\ mol\ CaCO_3} = 10.0\ g\ CaCO_3$

7. From the general formula for the alkanes, C_nH_{2n+2}, the part of the formula mass that is hydrogen $= 1.0(2n + 2)$.
   The total formula mass $= 12.0(n) + 1.0(2n + 2)$.
   The fractional part of the total formula mass that is:

   hydrogen $= \dfrac{1.0(2n + 2)}{12(n) + 1.0(2n + 2)\ 1} = \dfrac{2n + 2}{14n + 2} = \dfrac{n + 1}{7n + 1}$.

   By inspection, as n assumes increasing integral values, the denominator will become larger faster than the numerator, and the value of the fraction will decrease. The percentage of hydrogen therefore decreases.

8. $60.0\ g\ C \times \dfrac{12.0\ g\ C}{1\ mol\ C} = 5.00\ mol\ C$ Formula mass of $C_3H_8O = 60.\ u$

   $26.7\ g\ O = \dfrac{16.0\ g\ O}{1\ mol\ C} = 1.67\ mol\ O$ Formula mass of $(C_3H_8O)_x = 60.\ u$

   $13.3\ g\ H \times \dfrac{1.01\ g\ H}{1\ mol\ H} = 13.2\ mol\ H$ $x = 60./60.0 = 1.00$

   Molecular formula $= (C_3H_8O)_1$ or C_3H_8O

   $\dfrac{5.00\ mol\ C}{1.67} : \dfrac{1.67\ mol\ O}{1.67} : \dfrac{13.2\ mol\ H}{1.67}$

   2.99 mol C: 1 mol O 7.90 mol H

Simplest formula = C_3H_8O

9. $V_{STP} = V_1 \times \dfrac{P_1}{P_{STP}} \times \dfrac{T_{STP}}{T_1}$

   $= 115\ mL \times \dfrac{738\ mm - 20\ mm}{760\ mm} \times \dfrac{273K}{295K} = 101 mL$

   (20 mm is vapor pressure of water at 22°C from Appendix A.)

Chapter 21 Carbon and Hydrocarbons

INTRODUCTION

Carbon exhibits some unusual characteristics shown by no other element. More than 95 percent of all known compounds contain carbon. The chemistry of carbon compounds, organic chemistry, is an important branch of chemistry.

In this chapter, we consider the chemistry of carbon, the backbone of all organic compounds.

LOOKING AHEAD

As you study this chapter, examine the structure and bonding characteristics of carbon. Pay special attention to the classification and systematic naming of organic compounds. Next, identify the specific hydrocarbons that occur in natural gas and petroleum and note possible substitutes for petroleum as a source of energy.

SECTION PREVIEW

21.1 Abundance and Importance of Carbon
21.2 Organic Compounds
21.3 Hydrocarbons
21.4 Representative Hydrocarbons and Polymers

Hydrocarbons provide us with energy to travel. Carbon-hydrogen compounds are used in countless areas of our lives.

21.1 Abundance and Importance of Carbon

Carbon is found in nature both as an element and in the combined form. Although ranks about seventeenth in abundance by mass among the elements in the earth's crust, it is exceedingly important because of its universal presence in living matter. Carbon is present in body tissues and in the foods you eat. It is found in common fuels such as coal, petroleum, natural gas, synthetic gas, and wood.

Structure and Bonding of Carbon

Carbon, the first member of Group 14, exhibits mostly nonmetallic properties. In its ground state, a carbon atom has an electronic configuration of $1s^2\,2s^2\,2p^2$. The two $1s$ electrons are tightly bound to the nucleus. The two $2s$ electrons and the two $2p$ electrons are the valence electrons. Carbon atoms show a very strong tendency to share electrons and form covalent bonds. This electron sharing usually has the effect of producing a stable highest-energy-level octet around a carbon atom.

The possession of four valence electrons makes it possible for a carbon atom to form four covalent bonds. These bonds are directed in space toward the four corners of a regular tetrahedron. As shown in Figure 21-1, the nucleus of the carbon atom is at the center of the tetrahedron. During a chemical reaction, all four of the carbon valence electrons are usually involved in bonding through sp, sp^2, or sp^3 hybridization. As shown in Figure 21-2, the ground-state electron configuration can be changed by the promotion of one of the $2s$ electrons to the unoccupied orbital of the $2p$ energy level. This change allows hybridization to produce four sp^3 orbitals that are equivalent. Each of these orbitals can form a single covalent bond. Carbon also forms double and triple covalent bonds.

Because the property of forming covalent bonds is so strong in carbon atoms, they join readily with other elements with similar electronegativities, such as hydrogen, oxygen, nitrogen, sulfur, phosphorus, and the halogens. They also link together with other carbon atoms in chains, rings, plates, and networks.

SECTION OBJECTIVES

- List the sources of carbon.
- Recognize the importance of carbon in chemistry.
- Discuss the structure and bonding of carbon atoms.
- Identify the allotropic forms of carbon.
- Name several amorphous forms of carbon.
- Describe the production, properties, and uses of various forms of carbon.

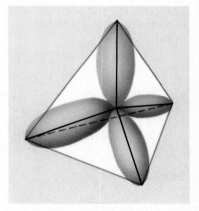

Figure 21-1 A carbon atom can form four covalent bonds and may be shown as a tetrahedron.

The role of carbon in living organisms is discussed in Chapter 23.

Hybridization is explained in Section 6.5.

The ability of carbon atoms to form chains by bonding to one another is known as *catenation*.

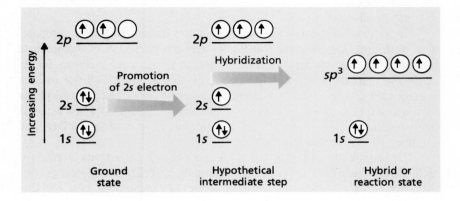

Figure 21-2 Carbon has four sp^3 hybrid orbitals.

Allotropes are two or more forms of the same element that exist in the same physical state and that have different chemical and physical properties.

Borazon is the second-hardest substance known. It is formed when boron nitride, BN, is exposed to very high pressures, creating a tetrahedral structure similar to diamond.

Diamond and graphite are also discussed in Section 13.2.

Figure 21-3 An uncut diamond *(left)* lacks the sparkle and brilliance of a cut diamond (right).

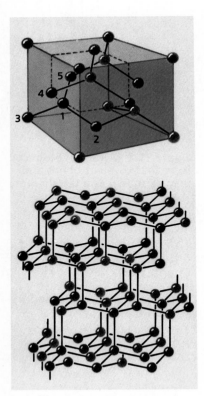

Figure 21-4 In diamond, each carbon atom is bonded to four tetrahedrally oriented carbon atoms.

Allotropic Forms of Carbon

Carbon occurs in two well-known solid allotropic forms, diamond and graphite. These two forms have dramatically different properties. In the form of **diamond**, *carbon is a colorless, crystalline solid with an extremely high density of 3.514 g/cm³*. **Graphite** *is a soft, black, crystalline form of carbon that is a fair conductor of electricity.* Carbon is able to form such unusual allotropes because its atoms can join together in many different ways, each arrangement resulting in a different compound.

Diamond   Over 95% of the world's natural diamonds are mined in Africa. Diamonds form when certain types of molten rock cool under conditions of extremely high temperature and pressure. When mined, diamonds do not have the shape or sparkle of gemstones. It is the art of cutting and polishing that gives diamonds their brilliant appearance, as shown in Figure 21-3. Because of the high refractive index of diamond, white light passing through the diamond crystal is broken into its various colors, which are dispersed.

Synthetic diamonds are chemically identical to natural diamonds, but are produced in the laboratory. They are prepared by subjecting graphite and a metal that acts as solvent and catalyst to extremely high pressure (55,000 atm) and high temperature (2000°C) for several hours.

Diamond is the hardest material known. It is the densest form of carbon—about 3.5 times as dense as water. Both its hardness and density are due to its structure. Figure 21-4 shows that carbon atoms in diamond are covalently bonded in a strong, compact fashion. Note that each carbon atom is tetrahedrally oriented to its four nearest neighbors. The distances between the carbon nuclei are 154 pm. This type of structure gives the crystal its great strength in all three dimensions.

The rigidity of its structure gives diamond its hardness. The compactness, resulting from the small distances between nuclei, gives diamond its high density. The covalent network structure of diamond accounts for its extremely high melting point (greater than 3550°C). Since all the valence electrons are used in forming covalent bonds, none can migrate. This explains why diamond is a nonconductor of electricity. Because of its extreme hardness, the major industrial uses of diamond are in cutting, drilling, and grinding.

The unique structure of diamond also explains its extremely efficient ability to conduct heat. A perfect single diamond crystal conducts heat greater than five times as readily as silver or copper, which are the best metallic conductors. In diamond, heat is conducted by the transfer of energy of vibration from one carbon atom to the next. In a perfect single diamond crystal, this process is very efficient because the carbon atoms have a small mass. The forces binding the atoms together are strong and can easily transfer vibratory motion among the atoms.

Graphite Graphite is nearly as remarkable for its softness as diamond is for its hardness. It is easily crumbled and feels greasy characteristics readily explained by its structure. The carbon atoms in graphite are arranged in layers of thin hexagonal plates, as shown in Figure 12-5. The distance between the centers of adjacent carbon atoms within a layer is 142 pm. This distance is less than the distance between adjacent carbon atoms in diamond. The distance between the centers of atoms in adjacent layers is 335 pm, however.

Figure 21-5 shows the bonding within a layer of graphite. Each carbon atom in a layer is bonded to only three other carbon atoms in that layer. This bonding consists of single and double covalent bonds between carbon atoms. When presented in this fashion, three different equivalent patterns appear. In each of these, some carbon-carbon bonds are single and others are double. There is, however, no experimental evidence that the bonds in a layer of graphite are of these two distinct types. On the contrary, the evidence indicates that the bonds are all the same. The layers of graphite have a resonance structure in which the carbon-carbon bonds are intermediate in character between single and double bonds. Each layer in graphite is a strongly bonded covalent network structure. As with diamond, this structure gives graphite a high melting point (3570°C). The strong bonds between the carbon atoms within a layer make graphite difficult to pull apart in the direction of the layer. As a result, carbon fibers, in which the carbon is in the form of graphite, are very strong.

The layers of carbon atoms in graphite are too far apart for the formation of covalent bonds between them. They are held together by weak London dispersion forces. The weak attraction between layers accounts for the softness of graphite and its greasy feel as one layer slides over another.

The average distance between carbon atoms in graphite is greater than the average distance in diamond; thus, graphite has a lower density. The mobile electrons in each carbon-atom layer make graphite a fairly good conductor of electricity, even though it is a nonmetal. Like diamond, graphite does not dissolve in any ordinary solvent. Similarly, it forms carbon dioxide when burned in oxygen.

Uses of graphite Graphite is a good lubricant, because it is not soluble, resists corrosion, and is not affected by heat. It is sometimes mixed with petroleum jelly or motor oil to form graphite lubricants, and can be used for lubricating machine parts that operate at temperatures that are too high for the usual petroleum lubricants. Graphite leaves a gray streak or mark when it is drawn across a sheet of paper. In making "lead" pencils (the name *lead* is merely poetic), graphite is powdered, mixed with clay, and then formed into sticks. The hardness of a pencil depends on the relative amount of clay that is used.

The most important use of synthetic graphite is in electrodes for electric-arc furnaces for making steel. Synthetic graphite electrodes are also

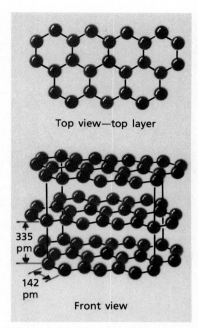

Top view—top layer

335 pm

142 pm

Front view

Figure 21-5 The arrangement of carbon atoms in graphite allows adjacent layers to slide past each other. Each carbon atom is bonded to three other carbon atoms.

A solvent is a dissolving medium. A catalyst is a substance that increases the rate of a chemical reaction without itself being permanently changed.

Figure 21-6 Carbon fibers give this graphite fishing rod its strength.

used in the electrolysis of salt water for making chlorine and sodium hydroxide. Graphite does not react with acids, bases, or organic and inorganic solvents. These properties make it useful for a variety of processes and equipment in the food, chemical, and petroleum industries. Graphite is also used in nuclear reactors, as described in Chapter 30.

If certain synthetic fibers are combined with plastic resins and heated under pressure, they become carbon fibers. Carbon fibers are less dense than steel, but are stronger and stiffer. They are used in aircraft floor decking, wings, and wing flaps and in weather and communication satellites. In sporting goods, carbon fibers are used to make golf club shafts, tennis rackets, fishing rods (see Figure 21-6), and bicycle frames. Carbon fibers have been implanted in severely torn ligaments and tendons to help the natural process of reconstruction.

Amorphous Forms of Carbon

When substances that contain combined carbon are decomposed by heat, they leave **amorphous carbon**, *black, carbon-containing residues that have no crystalline shape*. Examples of amorphous carbon are coal, coke, charcoal, petroleum coke, and activated carbon. Studies of these substances' structures, made by x-ray diffraction, reveal that the various forms of so-called amorphous carbon contain regions in which the carbon atoms have an orderly arrangement. In charcoal, for example, the carbon atoms are arranged somewhat as they are in a layer of graphite.

Coal Coal *is a natural dark-brown or black solid that consists mostly of amorphous carbon with various organic and some inorganic compounds*. It is a solid, rocklike material that burns readily. Coal contains at least 50 percent carbon and varying amounts of moisture, plus volatile (easily evaporated) materials and noncombustible mineral matter (ash). It is our most plentiful fossil fuel.

Geologists believe that coal was formed about 300 million years ago during the Carboniferous Period, a time often referred to as "The Coal Age." Much of the North American land mass was then covered with shallow swamps filled with dense forests of ferns, horsetails, and primitive seed plants (see Figure 21-7). The moist, warm climate of the time supported luxurious plant growth. Tree ferns, giant club mosses, and other vegetation growing in these swamps supplied the material for the coal deposits we use today.

Peat *is a form of partially carbonized plant material that can be used as fuel*. Peat bogs are found extensively in Pennsylvania, Michigan, Wisconsin, and other northern states. The bogs in which it is found contain mosses, sedges, and other forms of vegetation that undergo partial decomposition in swampy land in the almost complete absence of oxygen. Peat burns with a smoky flame, and its heat content is rather low because the vegetable material within it becomes only partially converted to carbon and hydrocarbons. Peat contains a high percentage of moisture. In Ireland and Sweden, peat is used for domestic fuel. In the United States, peat is used mainly as a soil conditioner, not as fuel.

In some areas, upheavals of parts of the earth's crust buried extensive deposits of slowly decaying plant material during the Carboniferous Period. Once buried, this material was subjected to increased temperatures and

Figure 21-7 Lush vegetation during the Coal Age was the source of the coal, natural gas, and petroleum we use today.

pressures over millions of years. An early stage of carbon-containing material that formed under these conditions is **lignite**, *a low grade form of coal sometimes called "brown coal" because of its brownish-black color.* Lignite also burns with a smoky flame, and although its heat content is higher than that of peat, it is still significantly below that of the other coals. The moisture content of lignite is high.

The next stage of coal formation is subbituminous coal, a form of coal that is found extensively in Wyoming and other Western states. **Subbituminous coal** *is a form of coal that has a higher percentage of carbon and a lower percentage of moisture than lignite.* Subbituminous coal also has a higher heat content than lignite. It can be used for industrial heating, as in electric generating stations.

Bituminous coal *is another form of coal that appears to have been subjected to greater heat and pressure than subbituminous coal.* The percentage of carbon is high; the moisture content is low. Bituminous coal has a high heat content. It is widely distributed throughout the Appalachian region, the Midwest, and the Rocky Mountain states. It is used for home and industrial heating and for making coke and coal by-products.

Anthracite *is believed to have been subjected to greater temperatures and pressures than other coals.* It contains the least amount of volatile matter of all coals and has the highest percentage of carbon. Most anthracite is found in Pennsylvania. Anthracite coal has a higher heat content than any other coal.

Coke When bituminous coal is heated in a test tube, a flammable gas escapes that burns readily if mixed with air and ignited. A tarlike liquid also condenses on the upper walls of the test tube. If the heating is continued until all the volatile material is driven off, coke is left as a residue. **Coke** *is the solid residue left after volatile materials are removed from bituminous coal.*

Commercially, coke is prepared by destructive distillation of bituminous coal in by-product coke ovens, as shown in Figure 21-8. **Destructive distillation** *is a process by which a complex carbon-containing material, such as coal, is heated in a closed container in the absence of air or oxygen.* The volatile products are separated into coal gas, ammonia (NH_3), and coal tar. Coal gas can be used as a fuel. Ammonia is used in making fertilizers.

Deep coal mining is a hazardous occupation, with one out of every 1000 underground miners killed each year and 30% of long-time coal miners developing "black lung" disease.

Coal gas is composed mainly of the three combustible gases—methane, hydrogen, and carbon monoxide. Coal tar can be separated by distillation into benzene, toluene, xylenes, naphthalene, phenol, and creosols.

Figure 21-8 Coke is produced by destructive distillation of bituminous coal in by-product coke ovens.

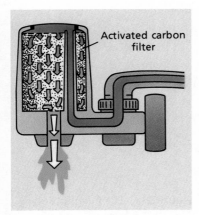

Figure 21-9 Activated carbon is used in some home water-purification systems, such as this water faucet filter unit.

Adsorption is the adhesion of particles of gases, liquids, or solids to the surfaces of a solid.

1. Graphite, diamond, charcoals, soot, living matter, natural gas, coal, wood, petroleum, metal carbonates
2. It is present in all living matter.
3. The ground-state electron configuration can be changed by the promotion of one of the $2s$ electrons to the unoccupied orbital of the $2p$ energy level. This change produces 4 sp^3 orbitals that are equivalent.

Coal tar can be separated by distillation into materials used to make drugs, dyes, and explosives. The black pitch that remains after distillation of coal tar is used to surface roads.

About 60 million tons of coke are produced each year in the United States. Coke is a gray, porous solid that is harder and denser than charcoal. Coke burns with little flame, has a high heat content, and is a valuable fuel.

It is an excellent reducing agent and is widely used in obtaining iron, tin, copper, and zinc from oxide ores containing these metals. Coke readily reduces the metals in these oxides to yield relatively pure metals.
relatively pure metals.

Charcoal   Destructive distillation of wood yields several gases that can be burned. Wood also produces methanol (wood alcohol—CH_3OH)), acetic acid (CH_3COOH), and other volatile products. *The residue that remains after the destructive distillation of wood is* **charcoal**. Charcoal is prepared commercially by heating wood in retorts, closed containers in which substances are distilled or decomposed by heat.

Charcoal is a porous, black, brittle solid. It is odorless and tasteless. It is denser than water, but it often adsorbs enough gas to make it float on water. The ability of charcoal to adsorb a large quantity of gas is perhaps its most remarkable physical property. One cubic centimeter of freshly prepared willow charcoal adsorbs about 90 cc of ammonia gas. Charcoal has been used to adsorb toxic waste products from the blood of persons suffering from liver failure, from drug overdose, and from accidental swallowing of poisonous carbon–hydrogen compounds. It is also used in aquarium filters and to remove ozone from the cabin atmosphere of jet airplanes.

Petroleum coke   **Petroleum coke** *is a form of amorphous carbon that is produced by the destructive distillation of an oily residue that is a remnant of petroleum refining.* Rods of petroleum coke are converted to synthetic graphite for use as positive electrodes in dry cells. They are also used in the production of aluminum by electrolysis.

Activated carbon   **Activated carbon** *is a form of amorphous carbon that is prepared in a way that gives it a very large internal surface area.* This large surface area makes activated carbon exceptionally useful for the adsorption of liquids and gases. Activated carbon can be made from a variety of carbon-containing materials. Such a material is first destructively distilled and the resultant carbon is then treated with steam or carbon dioxide at about 100°C. These two processes produce a very porous form of carbon, which contains a very large internal surface area. This form of activated carbon can remove odors from the air circulated by large air conditioning systems. It is also used in municipal and industrial water treatment and in home water faucet units (see Figure 21-9) to adsorb harmful materials as well as objectionable odors and tastes.

Section Review

1. List several sources of carbon both in free and combined form.
2. What makes carbon an important element in the study of chemistry?
3. Explain how a carbon atom forms four covalent bonds when its valence configuration in the ground-state is $2s^2 2p^2$.

21.2 Organic Compounds

Organic compounds have one thing in common—they all contain carbon atoms. With few exceptions, *carbon-containing compounds are the subject of the study of* **organic chemistry**. Ionic compounds such as sodium carbonate (Na_2CO_3), and the oxides of carbon (CO, CO_2) are considered inorganic compounds.

The number of possible carbon compounds is virtually unlimited. Over 4 million natural and synthetic organic compounds are known, and about one hundred thousand new ones are isolated or synthesized each year. The fact that carbon is the backbone of more compounds than all of the other elements combined suggests that it has some unique characteristics. There are two major reasons for the occurrence of so many organic compounds.

1. Carbon atoms have the ability to link up with each other to form a virtually limitless number of chain, branched-chain, and ring structures of countless shapes and dimensions. This makes possible molecules in which thousands of carbon atoms are bonded to another. Not only are carbon atoms linked by single covalent bonds, but they are sometimes linked by double or triple covalent bonds. Also, carbon atoms form bonds with a large number of other elements.

2. Carbon atoms represented by a specific molecular formula may be arranged in several different ways, which creates compounds with different properties and structures. For example, the molecular formula C_2H_6O represents the molecular formula for both ethyl alcohol and dimethyl ether. Such compounds have the same molecular formula but different structures. *Compounds that have the same molecular formula but different structures are called* **isomers**. As the number of carbon atoms increases in a molecular formula, the number of possible isomers increases rapidly. For example, there are 18 isomers with the molecular formula C_8H_{18}, 35 with the molecular formula C_9H_{20}, and 75 with the molecular formula $C_{10}H_{22}$. Calculations show that with the molecular formula of just 40 carbon atoms and 82 hydrogen atoms ($C_{40}H_{82}$), there are theoretically 69 491 178 805 831 isomers. To distinguish one isomer from another, we must use structural formulas, the topic of our next discussion.

Although the number of possible organic compounds is virtually unlimited, the study of organic compounds can be organized into two major divisions: hydrocarbons and substituted hydrocarbons, the subject of Chapters 21 and 22.

Hydrocarbons *are the simplest organic compounds, composed only of carbon and hydrogen.* They are grouped into several families with altogether different characteristics—namely alkanes, alkenes, alkynes, and aromatic hydrocarbons. These families are based mainly on the type of bonding between carbon atoms. **Substituted hydrocarbons** *are those hydrocarbons that are classified according to the functional groups they contain. A* **functional group** *is an atom or group of atoms bonded to a hydrocarbon and is responsible for the specific properties of the resulting organic compound.* In some cases, the functional groups are substituted for H atoms in hydrocarbons, and in other cases for C atoms, thus the name substituted hydrocarbons. Because organic compounds with the same functional group

The structure for ethyl alcohol is

The structure for dimethyl ether is

Give a few examples of functional groups, such as the hydroxyl group in ethyl alcohol and isopropyl alcohol, or the formyl group in acetaldehyde and propionaldehyde, or the carboxyl group in acetic acid and butyric acid. This will help lay the groundwork for the next chapter.

Chemistry Notebook

Chlorohydrocarbons

Chlorohydrocarbons are hydrocarbons in which chlorine atoms are substituted for one or more of the hydrogen atoms. A wide variety of chlorohydrocarbon compounds were used extensively in industry and agriculture until it was found that use of many of these compounds may cause cancer, damage the kidney and liver, and result in ailments of the circulatory and respiratory systems.

The members of one group of chlorohydrocarbons, the chlorinated forms of methane, were widely used until their health hazards were recognized. Carbon tetrachloride (tetrachloromethane) was used in dry cleaning because it dissolves fats, oils, and greases. Chloroform (trichloromethane) is still used as a solvent in many paint strippers. Methyl chloride (chloromethane)

was stored under pressure as a liquid and then sprayed onto skin as a local anaesthetic.

Many other chlorohydrocarbons are known to be dangerous to living organisms. DDT (dichlorodipenyltrichloroethane) is a highly effective pesticide, but is now banned in many parts of the world. DDT becomes increasingly concentrated as it moves up the food chain and accumulates in animal tissues. It poisons fish and many other living organisms and interferes with bird reproduction. PCBs (polychlorinated bipenyls) were used in the United States as coolants and insulating fluids for electrical equipment. A herbicide called Agent Orange was used as a defoliant in Vietnam by the United States during its participation in the Vietnam War. A byproduct of the manufac-

ture of PCBs and Agent Orange, the dioxins are a group of chlorinated aromatic hydrocarbons that are known to cause a wide variety of health problems. The town of Times Beach, Missouri, shown in the photograph below was abandoned in 1983 because of dioxin contamination of the soil.

or groups behave similarly, we are able to classify them in terms of their functional groups.

Structural Formulas And Bonding

The molecular formula of a compound shows the exact number of atoms of each type of element present in a molecule. While the formula for sulfuric acid (H_2SO_4) gives enough information for most purposes in inorganic chemistry, a molecular formula such as that for sucrose ($C_{12}H_{22}O_{11}$) is not satisfactory in organic chemistry because there are over 100 different compounds that have the same molecular formula as sucrose.

For this reason, organic chemists use structural formulas to represent organic compounds. *A **structural formula** indicates the exact number and types of atoms present in a molecule and also shows how the atoms are bonded* to each other. *The various isomers with a specific molecular formula are referred to as* **structural isomers**.

Structural isomers may exist for hydrocarbons with four or more carbon atoms. For example, careful study of the molecular formula C_4H_{10} reveals the existence of two compounds that exhibit different properties. When we attempt to write structural formulas for C_4H_{10} we notice that atoms can be arranged to give two different structures.

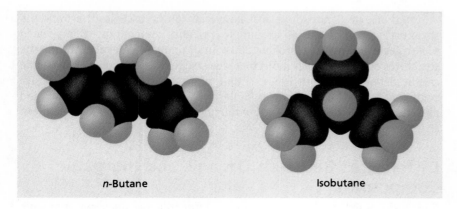

These structural formulas represent molecules of two different compounds because the atoms are linked together in different ways, as shown in Figure 21-10. These two compounds are isomers of one another. The dash (—) in the structural formulas represents a pair of shared electrons forming a covalent bond. The *n* in the name *n*-butane stands for normal, which indicates a straight-chain structural formula—the four carbon atoms are in a continuous chain. The chain may be bent or twisted, but it is continuous. The formula of isobutane shows a continuous chain of three carbon atoms, with the fourth carbon atom attached to the second carbon of the chain.

![n-Butane (left) and Isobutane (right) space-filling molecular models]

n-Butane   Isobutane

Figure 21-10 *N*-butane *(left)* and isobutane *(right)* are isomers of butane.

Determination of an organic structural formula   There are two different isomers of organic compounds having a molecular weight of 46. In each, carbon represents 52.2% of the molecular weight, hydrogen, 13%, and oxygen, 34.8%. One compound is a colorless liquid that boils at 78°C. The other is a colorless gas that condenses to liquid at −25°C under one atmosphere of pressure (760 mm Hg). Each has its own distinctive odor. How can their structural formulas be determined?

From the percentage composition, you can calculate the empirical formula, which is C_2H_6O. Since this empirical formula has a formula weight of 46, it must also be the molecular formula of each compound. From what you have already learned about bonding, there are only two ways in which two

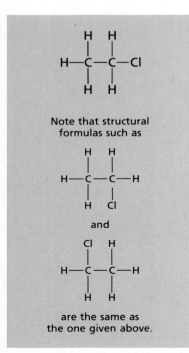

Note that structural formulas such as

are the same as the one given above.

Figure 21-11 The structural formula for ethyl chloride (C_2H_5Cl) is written as shown here.

CAUTION
Many organic compounds are flammable and poisonous. Quite a few cause or are suspected of causing cancer. Some organic reactions are rapid and highly exothermic. A student should not perform any experiments with organic compounds without detailed laboratory directions, and then only with adequate ventilation and other safety precautions and under the supervision of an experienced instructor.

carbon atoms, six hydrogen atoms, and a single oxygen atom can combine, as shown below.[1]

Structure A   Structure B

Now the problem is to match these structures to the two compounds. If each compound is tested for reaction with metallic sodium, only the liquid compound reacts. In this reaction, hydrogen gas is given off. The amount of hydrogen given off is equal to one-sixth of the hydrogen that the compound contains. This evidence indicates that in the molecules of the liquid, one of the six hydrogen atoms is bonded in a way that is different from the others. Structure A is indicated.

Next, you discover that the liquid reacts with phosphorus trichloride and gives a product with the molecular formula C_2H_5Cl. In this reaction, chlorine has replaced both a hydrogen atom and an oxygen atom. Only one structural formula for C_2H_5Cl can be written, as shown in Figure 21-11. You may assume that the chlorine atom occupies the same position as the oxygen and hydrogen atoms that it replaced. Structure A is again indicated. Additional evidence can be found to indicate that the liquid does indeed have Structure A. This liquid substance is ethanol, or ethyl alcohol—C_2H_5OH. The gaseous substance has the other structural formula and is dimethyl ether—CH_3OCH_3.

Methods similar to those just described can be used to determine the structural formulas of other simple organic compounds. Complicated molecules are generally broken down into simpler molecules. From the structures of these simpler molecules, the structure of the complex molecule can be determined. Sometimes, simple molecules of known structure are combined to produce a complex molecule. A comparison of chemical and physical properties of compounds of unknown structure with those of known structure is sometimes helpful.

Differences Between Organic and Inorganic Compounds

The basic laws of chemistry are the same for organic and inorganic compounds. The behavior of organic compounds is somewhat different from that of inorganic compounds, however. Some of the most important differences are:

1. Most organic compounds do not dissolve in water. Many inorganic compounds will dissolve more or less readily in water. Organic compounds generally dissolve in other organic liquids, such as alcohol, chloroform, ether, or carbon tetrachloride.

2. Organic compounds are decomposed by heat more easily than most inorganic compounds. The decomposition (charring) of sugar when it undergoes moderate heating is familiar. Such charring upon heating is often a test for organic substances. By contrast, many inorganic compounds, such

TABLE 21-1 COMPARISON OF AN ORGANIC COMPOUND AND AN INORGANIC COMPOUND		
Compound	Benzene (C_6H_6)	Table salt (NaCl)
formula	C_6H_6	NaCl
melting point	5.5°C	801°C
boiling point	80.1°C	1413°C
bonding	covalent	ionic
solubility in H_2O	insoluble	soluble
solubility in gasoline	soluble	insoluble

as common salt (sodium chloride), are simply vaporized without decomposition when exposed to high temperatures.

3. Organic reactions generally proceed at much slower rates than inorganic reactions. Such organic reactions often require hours or even days for completion. (Some organic reactions, such as those in living cells, may take place with great speed, however.) Most inorganic reactions occur almost as quickly as solutions of the reactants are brought together.

4. Organic reactions are greatly affected by reaction conditions. Many inorganic reactions follow well-known patterns. This makes it possible for you to learn about many inorganic reactions by studying the general types of reactions explained in Chapter 8. There are some general types of organic reactions, too. But changing the temperature or pressure or the nature of a catalyst can alter the identity of the products formed. The same organic reactants can form different products, depending on reaction conditions.

5. Organic compounds exist as molecules consisting of atoms joined by covalent bonds. Many inorganic compounds have ionic bonds.

6. Organic compounds may exist as isomers. Inorganic compounds do not exhibit isomerism.

Section Review

1. Give two major characteristics of the carbon atoms that make possible the occurrence of millions of organic compounds.
2. What is meant by the term *isomer*?
3. Define each of the following terms: *(a)* hydrocarbons *(b)* structural formula.
4. How many structural isomers have the molecular formula C_5H_{12}?
5. List several ways in which organic compounds are different from inorganic compounds.

21.3 Hydrocarbons

Because they are the basic structures from which other organic compounds are derived, the hydrocarbons are studied first. Hydrocarbons are grouped into several different series of compounds. These groupings are based mainly on the type of bonding between carbon atoms.

1. *The **alkanes** (al-kaynes) are straight-chain or branched-chain hydrocarbons whose carbon atoms are connected by single covalent bonds:*

1. Ability to link up with each other to form a virtually limitless (theoretically infinite) number of chain, branch-chain, and ring structures of countless shapes and dimensions. May be arranged in several different ways, which gives compounds with different properties and structures
2. Compounds that have the same molecular formula, but different structures
3. *(a)* Compounds composed only of carbon and hydrogen *(b)* Indicates the exact number and types of atoms present in a molecule and how the atoms are bonded to each other
4. Three
5. Most organic compounds do not dissolve in water, while many inorganic compounds do. Organic compounds are decomposed by heat more easily than most inorganic compounds. Organic reactions are slow, while inorganic reactions are fast. Organic reactions are greatly affected by reaction conditions, while inorganic reactions follow well-known patterns. Organic compounds consist of atoms joined by covalent bonds, while many inorganic compounds have ionic bonds. Organic compounds may exist as isomers, while many inorganic compounds do not exhibit isomerism.

SECTION OBJECTIVES

- Classify hydrocarbons into several different groups and give examples of each.
- Draw structural formulas for hydrocarbon molecules if given their molecular formulas.

- Give systematic names of hydrocarbons if given their structural formulas.
- Discuss sources of alkanes and describe preparation methods of alkenes and ethylene.
- Describe reactions of alkanes, alkenes, alkynes, and benzene.
- Explain why alkanes and aromatic hydrocarbons react by substitution and alkenes and alkynes react by addition.
- Name and draw the structural formulas of aromatic hydrocarbons.

2. *The* **alkenes** *(al-keens) are straight- or branches-chain hydrocarbons in which two carbon atoms in each molecule are connected by a double covalent bond:*

3. *The* **alkynes** *(al-kynes) are straight- or branched-chain hydrocarbons in which two carbon atoms in each molecule are connected by a triple covalent bond:*

$$H—C≡C—H$$

5. *The* **aromatic hydrocarbons** *are hydrocarbons with six-membered carbon rings that exhibit resonance.* These structures are sometimes represented by alternate single and double covalent bonds in six-membered carbon rings.

Alkanes

This series of organic compounds is sometimes called *the paraffin series because* **paraffin wax** *is a mixture of hydrocarbons of this series.* The word *paraffin* means "little attraction." Compared to the hydrocarbons of other series, the alkanes have low chemical reactivity. This stability results from their single covalent bonds. Because they have only single covalent bonds in each molecule, the alkanes are known as saturated hydrocarbons. **Saturated hydrocarbons** *are those in which each carbon atom in the molecule forms four single covalent bonds with other atoms.*

Table 21-2 lists a few members of the alkane series. The names of the first four members of this series follow no system. Beginning with pentane, the first part of each name is a Greek or Latin numerical prefix. This prefix indicates the number of carbon atoms. The name of each member ends in *-ane,* the same as the name of the series.

If you examine the formulas for successive alkanes, you will see a clear pattern. Each member of the series differs from the preceding one by the —CH$_2$ group.

TABLE 21-2　N-ALKANES

Alkane	Formula	Melting Point (°C)	Phase at Room Temperature	Number of Possible Isomers
Methane	CH_4	−183	gas	1
Ethane	C_2H_6	−172	gas	1
Propane	C_3H_8	−187	gas	1
Butane	C_4H_{10}	−135	gas	2
Pentane	C_5H_{12}	−130	liquid	3
Hexane	C_6H_{14}	−94	liquid	5
Heptane	C_7H_{15}	−91	liquid	9
Octane	C_8H_{18}	−57	liquid	18
Nonane	C_9H_{20}	−54	liquid	35
Decane	$C_{10}H_{22}$	−30	liquid	75

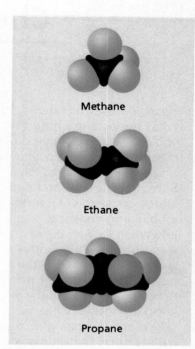

Figure 21-12　Methane *(top)*, ethane *(center)*, and propane *(bottom)* are the first three alkanes.

Compounds that differ in this fashion belong to a homologous series. A **homologous series** *is one in which adjacent members differ by a constant unit.* It is not necessary to remember the formulas for all members of a homologous series. Instead, a general formula, such as C_nH_{2n+2} for the alkanes, can be used to determine the formulas. Suppose a member of this series has 30 carbon atoms in its molecules. To find the number of hydrogen atoms, multiply 30 by 2, then add 2. The formula is $C_{30}H_{62}$.

Structures and systematic names of alkanes　Names of the millions of organic compounds are derived from a systematic method developed by the International Union of Pure and Applied Chemistry (IUPAC). Systematic names can be derived from structural formulas. At the same time, structural formulas can be drawn from systematic names of organic compounds.

Each of the first three alkanes can have only one molecular structure, as shown by the formulas below. See the models in Figure 21-12.

methane　　　　ethane　　　　propane

Butane (C_4H_{10}), has two structural isomers

n-butane　　　　2-methylpropane

Chemistry Notebook

Octane Ratings

Have you ever wondered what the octane rating that you see on a gasoline pump means? The octane rating of a fuel is a measure of its burning efficiency and anti-knock properties. An automobile engine works properly when its fuel burns smoothly. If a fuel ignites spontaneously before the piston is in the proper position, power is lost and a knocking sound is heard.

Heptane is a straight-chain hydrocarbon that burns with a lot of knocking, so it is given an arbitrary octane rating of 0. Isooctane (also called 2,2,4-trimethylpentane) is a branched hydrocarbon that burns smoothly without knocking and is given a rating of 100. A gasoline mixture with an octane rating of 85 causes the same amount of knocking as a mixture of 85% isooctane and 15% heptane.

Until recently, small amounts of lead tetraethyl were added to fuels. This chemical inhibits knocking and thus increases the octane rating of a fuel. Unfortunately, lead tetraethyl directly contributes to air pollution and also ruins the catalytic converters that are installed on new cars to meet the reduced emissions requirement of the Clean Air Amendments of 1970. Other branched or ring hydrocarbons are now added to unleaded fuels to increase their octane ratings. This causes gasoline to be more expensive but the high performance coupled with a reduction in air pollution makes the extra cost worthwhile.

The basic part of the systematic name of an organic compound is the name of the longest carbon chain in the molecule. The straight-chain molecule is n-butane. Recall that *n* stands for normal—a structural formula in a continuous chain. The branched-chain molecule is named 2-methylpropane.

The longest continuous carbon chain in the second molecule shown above is three carbon atoms long, as in propane. One hydrogen atom attached to the second carbon atom in propane is replaced by the —CH_3 group. This is a substitution group called the methyl group and is indicated by the following structure:

$$H—\overset{\displaystyle H}{\underset{\displaystyle H}{C}}—H$$

The **methyl group** *is methane with one of the hydrogen atoms removed. It is an example of an alkyl group. An* **alkyl group** *is simply the neutral species formed when one hydrogen atom is removed from the parent alkane molecule.*

Table 21-3 lists a few alkyl groups. The methyl and other alkyl groups are neutral—they have the same number of protons and electrons. Carbon atoms in the main chain of the molecule are numbered. This numbering begins at the end of the molecule giving the carbon atoms with substitution groups the smallest numbers. Thus in the name 2-methylpropane, the 2 refers to the number of the carbon atom on which there is a substitution. Methyl is the substitution group. Propane is the parent hydrocarbon. The branched-chain butane is also commonly known as isobutane. The prefix *iso-* is needed to represent the alkyl group

$$CH_3-\underset{\underset{CH_3}{|}}{\overset{\overset{H}{|}}{C}}-$$

This alkyl group, which sometimes is written as $-CH(CH_3)_2$, is derived from propane when one hydrogen atom is removed from the second carbon atom. The name that follows the prefix *iso-* indicates the number of carbon atoms.

There are three possible pentanes—C_5H_{12}. They are named: *n*-pentane, 2-methylbutane, and 2,2-dimethylpropane. Why is the name 2-methylbutane used rather than 3-methylbutane?

substitutions. The prefix *dimethyl* shows that the two substitutions are both methyl groups. Propane is the parent hydrocarbon. Which of the pentanes also has also the name *isopentane*?

Just as the $-CH_3$ group derived from methane is the *methyl* group, the $-C_2H_5$ group that is derived from ethane is the *ethyl* group, and $-C_3H_7$ derived from propane is the *n*-propyl group. The *n*-butyl group ($-C_4H_{19}$), is derived from butane. The formula—C_5H_{11} is usually called the *amyl* group rather than the *pentyl* group. Other groups are given names following the general rule of dropping the *-ane* suffix and adding *-yl*. The alkyl groups of the first ten alkanes are listed in Table 21.4. The symbol R— is frequently used to represent an alkyl group in a formula.

Sources and uses of alkanes   Alkanes are generally found in petroleum and natural gas. It is fairly easy to separate the smaller members of the alkane series individually from petroleum and natural gas by fractional distillation, a technique that will be described in Section 21-4. The alkanes with higher boiling points are usually separated into mixtures with similar boiling points.

Methane is a colorless, nearly odorless gas that forms about 90% of natural gas. Pure methane can be separated from the other components of natural gas. Chemists sometimes prepare small amounts of methane in the laboratory by heating soda lime, which contains sodium hydroxide, with sodium acetate.

$$NaC_2H_3O_2(s) + NaOH(s) \rightarrow CH_4(g) + Na_2CO_3(s)$$

Ethane (C_2H_6) is a colorless gas that occurs in natural gas and is a product of petroleum refining. It has a higher melting point and boiling point

TABLE 21-3 ALKYL GROUPS	
Alkyl Group	Formula
methyl	CH_3^-
ethyl	$C_2H_5^-$
propyl	$C_3H_7^-$
butyl	$C_4H_9^-$
pentyl (amyl)	$C_5H_{11}^-$
hexyl	$C_6H_{13}^-$
heptyl	$C_7H_{15}^-$
octyl	$C_8H_{17}^-$
nonyl	$C_9H_{19}^-$
decyl	$C_{10}H_{21}^-$

Any straight-chained alkane that has a hydrogen replaced by a methyl group on the second carbon has the "iso-" prefix. For example: isopentane, isohexane, and isoheptane.

Coal can be converted into methane and other combustible gases using modern coal gasification processes that involve passing steam and oxygen through heated coal. The CO and H that are produced are then converted to methane in the presence of a catalyst. The result is a synthetic "natural" gas (SNG) with a composition and heat value similar to natural gas.

than methane. These properties are related to ethane's higher molecular weight.

Reactions of alkanes   **Combustion:** Because the alkanes make up a large proportion of our gaseous and liquid fuels, their most important reaction is combustion. Methane burns with a bluish flame.

$$CH_4 + 2O_2 \rightarrow CO_2 + 2H_2O$$

Ethane and other alkanes also burn in air to form carbon dioxide and water vapor.

$$2C_2H_6 + 7O_2 \rightarrow 4CO_2 + 6H_2O$$

Substitution: The alkanes react with halogens such as chlorine or bromine. In reactions such as these, one or more atoms of a halogen are substituted for one or more atoms of hydrogen. Therefore, the products are called substitution products. In the reaction shown below and in Figure 21-14 on page 646, an atom of bromine is substituted for one atom of hydrogen.

$$\begin{array}{c} H \\ | \\ H-C-H \\ | \\ H \end{array} + Br_2 \longrightarrow \begin{array}{c} H \\ | \\ H-C-Br \\ | \\ H \end{array} + HBr$$

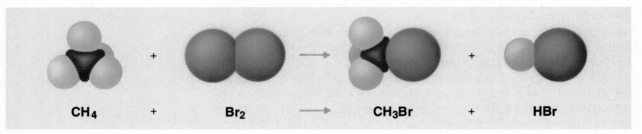

CH₄   +   Br₂   ⟶   CH₃Br   +   HBr

Figure 21-13   An atom of bromine is substituted for an atom of hydrogen during this substitution reaction.

By supplying additional molecules of the halogen, a halogen atom can be substituted for each of the hydrogen atoms. Carbon tetrachloride (CCl_4), dichlorodifluoromethane (CCl_2F_2), and Teflon [$F(CF_2)_xF$], are examples of halogen-substituted alkanes.

Preparation of hydrogen: Propane reacts with steam in the presence of a nickel catalyst at a temperature of about 850°C.

$$C_3H_8 + 6H_2O \rightarrow 3CO_2 + 10H_2$$

To separate the carbon dioxide from the hydrogen, the carbon dioxide under pressure is easily dissolved in water, thus allowing the hydrogen to be collected.

Cycloalkanes   The alkanes you have seen so far have been composed of open-ended chains of carbon atoms. Carbon and hydrogen atoms can also be arranged in a ring or cyclic structure. The rings may contain three, four, five, or a larger number of carbon atoms. Aromatic hydrocarbons, those that contain six-membered carbon rings, form an important class of compounds as you will see at the end of this section.

The physical and chemical properties of cyclic hydrocarbons are quite similar to those of their related open-chain hydrocarbons. The names of cycloalkanes are derived by adding the prefix *cyclo-* to the name of the open-chain molecule that has the same number of carbon atoms.

Alkenes

The alkenes, sometimes called the olefin series, are distinguished by a double covalent bond between two carbon atoms. Thus, the simplest alkene must have two carbon atoms. Its structural formula is

$$\begin{array}{c} H \\ \diagdown \\ H \diagup \end{array} C = C \begin{array}{c} H \\ \diagup \\ \diagdown H \end{array}$$

and its name is ethene.

The name of an alkene comes from the name of the alkane with the same number of carbon atoms. Simply substitute the suffix *-ene* for the suffix *-ane*. Since ethane is the alkane with two carbon atoms, the alkene with two carbon atoms is named ethene. (This substance is also commonly called *ethylene*.) The general formula for the alkenes is C_nH_{2n}.

Ethene is the hydrocarbon commercially produced in greatest quantity in the United States. It is also the organic chemical industry's most important starting material.

Preparation of alkenes   Cracking alkanes: The commercial method of producing alkenes is by cracking petroleum. **Cracking** *is a process in which complex organic molecules are broken up into simpler molecules.* This process involves the action of heat and sometimes involves a catalyst.

Cracking that uses heat alone is known as thermal cracking. During cracking, alkanes decompose in several ways and produce a variety of alkenes. A simple example is the thermal cracking of propane, which proceeds in either of two ways in nearly equal proportions.

$$C_3H_8 \xrightarrow{\;460\;°C\;} C_3H_6 + H_2$$

$$C_3H_8 \xrightarrow{\;460\;°C\;} C_2H_4 + CH_4$$

Alkenes (especially ethene), smaller alkanes, and hydrogen are typical products of the thermal cracking of alkanes. They can be separated, purified, and used as starting materials for making other organic compounds.

The cracking process that involves heat and requires a catalyst is catalytic cracking. The high-boiling fractions from petroleum distillation are catalytically cracked to produce smaller, lower-boiling hydrocarbons useful in gasoline. The catalysts used most frequently are oxides of silicon and aluminum. The cracking reactions produce smaller alkanes and alkenes with highly branched structures. Aromatic hydrocarbons are also formed.

Dehydration of alcohols: Ethene can be prepared in the laboratory by dehydrating (that is, removing water from) ethyl alcohol. Hot concentrated sulfuric acid is used as the dehydrating agent.

$$C_2H_5OH \xrightarrow[170\;°C]{H_2SO_4} C_2H_4 + H_2O$$

Reactions of alkenes   Addition: *An organic compound that has one or more double or triple covalent bonds in each molecule is said to be unsaturated.* It is chemically possible to add other atoms directly to such molecules to form molecules of a new compound in what are known as addition reactions. For example, hydrogen atoms can be added to an alkene in the presence of a finely divided nickel catalyst. This reaction produces the corresponding alkane. Converting the carbon–carbon double bond to a single bond provides two new bond positions for the hydrogen atoms.

$$\underset{H}{\overset{H}{>}}C=C\overset{H}{\underset{H}{<}} \quad + \quad H_2 \quad \xrightarrow{\text{Ni}} \quad H-\overset{\overset{\displaystyle H}{|}}{\underset{\underset{\displaystyle H}{|}}{C}}-\overset{\overset{\displaystyle H}{|}}{\underset{\underset{\displaystyle H}{|}}{C}}-H$$

Halogen atoms can be added to alkene molecules. For example, two bromine atoms added to ethene form 1,2-dibromoethane as shown in Figure 21-14.

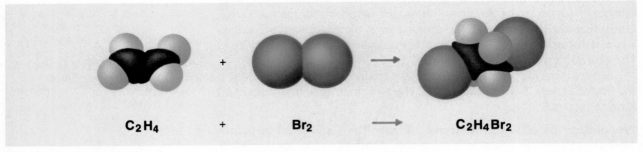

C_2H_4 + Br_2 $\longrightarrow$ $C_2H_4Br_2$

Figure 21-14 An addition reaction occurs when bromine atoms are added to ethene.

$$\underset{H}{\overset{H}{>}}C=C\overset{H}{\underset{H}{<}} \quad + \quad Br_2 \quad \longrightarrow \quad H-\overset{\overset{\displaystyle Br}{|}}{\underset{\underset{\displaystyle H}{|}}{C}}-\overset{\overset{\displaystyle Br}{|}}{\underset{\underset{\displaystyle H}{|}}{C}}-H$$

When the double bond between the carbon atoms is converted to a single bond, there is one bond position available for each bromine atom.

The name of the product is 1,2-*dibromoethane*. The basic part of the name, ethane, is that of the related alkane with two carbon atoms. *Dibromo-* refers to the two bromine atoms that have been substituted for hydrogen atoms in ethane. The 1,2- means that one bromine atom is bonded to the first carbon atom and the other is bonded to the second carbon atom. An isomer, 1,1-dibromoethane, has the formula

$$Br-\overset{\overset{\displaystyle Br}{|}}{\underset{\underset{\displaystyle H}{|}}{C}}-\overset{\overset{\displaystyle H}{|}}{\underset{\underset{\displaystyle H}{|}}{C}}-H$$

A molecule of a hydrogen halide, such as hydrogen bromide, can be added to an alkene molecule. What is the name of the resultant product?

$$\underset{H}{\overset{H}{>}}C=C\overset{H}{\underset{H}{<}} \quad + \quad HBr \quad \longrightarrow \quad H-\overset{\overset{\displaystyle H}{|}}{\underset{\underset{\displaystyle H}{|}}{C}}-\overset{\overset{\displaystyle Br}{|}}{\underset{\underset{\displaystyle H}{|}}{C}}-H$$

Polymerization: A **polymer** *is a very large molecule composed of repeating structural units.* For example, molecules of ethene join together, or polymerize, at about 15°C and 20 atmospheres pressure in the presence of a catalyst. The resulting large molecules have molecular weights of about 30,000. This polymerized material is called *polyethylene. The small molecules that join together to make a polymer are known as* **monomers.** Ethene is the monomer for polyethylene. During **polymerization,** *which is the reaction of monomers to form polymers,* the double bond in ethene becomes a single bond. Thus, the repeating unit in polyethylene is

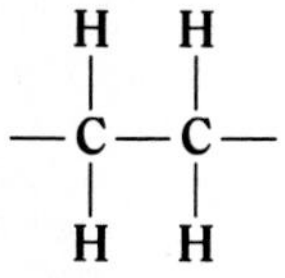

$$
\begin{array}{ccc}
 & \text{H} & \text{H} \\
 & | & | \\
-\text{C} & - & \text{C}- \\
 & | & | \\
 & \text{H} & \text{H}
\end{array}
$$

and polyethylene is represented by $H(CH_2)_x H$, where x is a large number. Many monomers join together to make the polymer, which means "many units." Polyethylene is used in film for packaging, in electric insulation, in a variety of containers, and in tubing.

Alkylation: In **alkylation,** *gaseous alkanes and alkenes are combined.* For example, 2-methylpropane and 2-methylpropene combine in the presence of a sulfuric acid or anhydrous hydrogen fluoride catalyst. The result is a highly branched octane, 2,2,4-trimethylpentane, an important component of gasoline.

Combustion: The alkenes burn in oxygen. For example,

$$C_2H_4 + 3O_2 \rightarrow 2CO_2 + 2H_2O$$

Butadiene: an important alkadiene *The* **alkadienes** *(al-kah-dy-eens) are straight- or branched-chain hydrocarbons that have two double covalent bonds between carbon atoms in each molecule.* An example of an alkadiene is the substance butadiene shown here

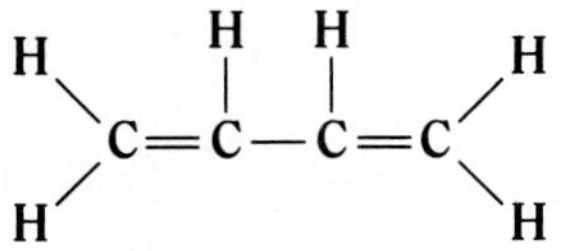

$$
\begin{array}{ccccc}
\text{H} & & \text{H} & \text{H} & & \text{H} \\
\diagdown & & | & | & & \diagup \\
 & \text{C}=\text{C} & - & \text{C}=\text{C} & \\
\diagup & & & & \diagdown \\
\text{H} & & & & \text{H}
\end{array}
$$

The *-ene* suffix indicates a double bond. The *-diene* suffix indicates two double bonds. The names of the alkadienes are derived in much the same way as those of the other hydrocarbon series. Butadiene must, therefore, have four carbon atoms and contain two double bonds in each molecule.

Actually, this structure is 1,3-butadiene, since the double bonds follow the first and third carbon atoms; the isomer 1,2-butadiene is so uncommon that 1,3-butadiene is usually called simply butadiene.

Butadiene is prepared by cracking petroleum fractions containing butane. It is used in the manufacture of SBR rubber, the most common type of synthetic rubber.

Alkynes

The alkynes are distinguished by a triple covalent bond between two carbon atoms. This series is sometimes called the acetylene series because the

Figure 21-15 Calcium carbide and water, shown in the beaker, react together to produce ethyne. This reaction was used in miner's lamps like the one above to provide fuel for light.

simplest alkyne has the common name acetylene. It has two carbon atoms, with the structural formula

$$H—C\equiv C—H$$

The names of the alkynes are derived from the names of the alkanes that have the same number of carbon atoms. The suffix -yne is substituted for -ane. Hence the systematic name for acetylene, the simplest alkyne, is ethyne. The general formula for the alkynes is

$$C_nH_{2n-2}$$

Ethyne, C_2H_2, is a minor component of the atmospheres of Jupiter and Saturn. Ethyne molecules have also been detected in interstellar space.

Preparation of ethyne

From calcium carbide: Ethyne, a colorless gas, can be prepared by the action of water on calcium carbide (CaC_2). Calcium carbide is made from limestone ($CaCO_3$) in a series of operations. First, the limestone is heated in a kiln (oven) and CaO is produced.

$$CaCO_3(s) \rightarrow CaO(s) + CO_2(g)$$

The calcium oxide is then heated with coke at 2000°C in an electric resistance furnace.

$$CaO(s) + 3C(s) \rightarrow CaC_2(s) + CO(g)$$

Calcium carbide is an ionic compound with the electron-dot structure

$$Ca^{2+}$$
$$^-:C:::C:^-$$

When it reacts with water, hydrogen replaces the calcium in the calcium carbide structure. The bonding of hydrogen and carbon in ethyne is covalent.

$$CaC_2(s) + 2H_2O(\ell) \rightarrow C_2H_2(g) + Ca(OH)_2(aq)$$

The preparation of ethyne from calcium carbide is both a commercial and laboratory method. An increasingly large percentage of the ethyne used commercially is obtained by the cracking of hydrocarbons at very high temperatures.

By partial oxidation of methane: This is an alternate commercial

$$6CH_4 + O_2 \xrightarrow{1500°C} 2C_2H_2 + 2CO + 10H_2$$

The source of methane is petroleum refining. The carbon monoxide and some of the hydrogen can be used to produce methanol. The balance of the hydrogen can be used as a fuel to maintain the temperature required for the reaction.

Reactions of ethyne Combustion: Ethyne burns in air with a very smoky flame. Carbon, carbon dioxide, and water vapor are the products of combustion. With special burners, the combustion to carbon dioxide and water vapor is complete.

$$2C_2H_2 + 5O_2 \rightarrow 4CO_2 + 2H_2O$$

The oxyacetylene welding torch burns ethyne in the presence of oxygen and reaches a temperature of over 3200°C.

Halogen addition: Ethyne is more unsaturated than ethene because of the triple bond. It is chemically possible to add two molecules of bromine to an ethyne molecule to form 1,1,2,2-tetrabromomethane.

$$H-C\equiv C-H \;+\; 2Br_2 \longrightarrow \underset{\underset{Br\;\;Br}{|\;\;\;|}}{\overset{\overset{Br\;\;Br}{|\;\;\;|}}{H-C-C-H}}$$

Dimerization: A principal use of ethyne is in the synthesis of more complex organic compounds. For example, two molecules of ethyne may combine and form the dimer, vinylacetylene. *A dimer is a molecule formed by the combination of two simple molecules.* This dimerization is brought about by passing ethyne through a water solution of copper(I) chloride and ammonium chloride (NH_4Cl). The solution acts as a catalyst.

$$2H-C\equiv C-H \xrightarrow[NH_4Cl]{Cu_2Cl_2} \underset{H}{\overset{H}{>}}C=\overset{\overset{H}{|}}{C}-C\equiv C-H$$

vinylacetylene

The —CH = CH₂ *group is the* **vinyl group**. Vinylacetylene is the basic raw material for producing Neoprene, a synthetic rubber.

Benzene and Aromatic Hydrocarbons

The aromatic hydrocarbons are generally obtained from coal tar and petroleum. The aromatic hydrocarbons are a group of hydrocarbons characterized by the presence of a benzene ring or related structure.

Benzene (C_6H_6) may be represented by the following resonance formula in which the two structures shown contribute equally.

The bonds in benzene are neither single bonds nor double bonds. Instead, each bond is a resonance hybrid bond. All of the carbon–carbon bonds in the molecule are the same. As a result, benzene and other aromatic hydrocarbons are not as unsaturated as the alkenes.

Because of the resonance structure of benzene, the benzene ring is sometimes written as

Benzene, C₆H₆

Toluene, C₇H₈

Figure 21-16 Aromatic hydrocarbons such as these are characterized by the presence of six-membered carbon rings.

///////

CAUTION
Benzene is poisonous. Its vapors are harmful to breathe and are very flammable. It is now recommended that benzene *not* be used in school laboratories.

Prolonged exposure to benzene may destroy the bone marrow's ability to make blood cells and may also cause cancer.

The two main sources of benzene are petroleum and coal. Benzene is synthesized from petroleum alkanes and from the distillation of coal tar.

The resonance nature of its structure is understood. The —C_6H_5 group derived from benzene is the *phenyl group*.

Benzene is produced commercially from petroleum. It is a flammable liquid that is used as a solvent. Benzene is used in manufacturing many chemical products, including dyes, drugs, and explosives. Benzene has a strong, yet fairly pleasant, aromatic odor. It is less dense than water and only very slightly soluble in water.

Reactions of benzenes   Halogenation: In a process known as **halogenation,** *benzene reacts with a halogen to give a halogen-substituted benzene.* For example, benzene reacts with bromine in the presence of iron to form the substitution product bromobenzene, or phenyl bromide.

$$\text{benzene} + Br_2 \xrightarrow{\text{Fe}} \text{bromobenzene} + HBr$$

Further treatment causes the successive substitution of other bromine atoms for hydrogen atoms. With complete substitution, hexabromobenzene is produced.

Nitration: In a process known as *nitration,* an —NO_2 group is added to a benzene when it is treated with nitric acid plus another acid. For example, nitrobenzene is produced by treating benzene with concentrated nitric and sulfuric acids.

$$\text{benzene} + HNO_3 \xrightarrow{H_2SO_4} \text{nitrobenzene} + H_2O$$

Friedel-Crafts reaction: In the **Friedel-Crafts reaction,** *an alkyl group can be introduced into the benzene ring by using an alkyl halide in the presence of anhydrous aluminum chloride.*

$$\text{benzene} + RCl \xrightarrow{AlCl_3} \text{alkylbenzene} + HCl$$

Other aromatic hydrocarbons   Several other aromatic hydrocarbons are shown in Figure 21-16. Most or all aromatic compounds contain one or more benzene rings. Toluene, or methylbenzene, is obtained from petroleum.

The xylenes or dimethylbenzenes, $C_6H_4(CH_3)_2$, are a mixture of three liquid isomers. The xylenes are used as starting materials for the production of polyester fibers, films, and bottles.

Ethylbenzene is produced by the Friedel-Crafts reaction of benzene and ethene in the presence of hydrogen chloride.

Ethylbenzene is treated with a catalyst of mixed metallic oxides to eliminate hydrogen and produce styrene. Styrene is used along with butadiene in making SBR synthetic rubber.

Styrene may be polymerized to polystyrene, a tough, transparent plastic. Polystyrene molecules have molecular weights of about 500,000. Styrofoam, a porous form of polystyrene, is used as a packaging and insulating material and for making throwaway beverage cups.

Section Review

1. List the basic features that characterize each of the following:
   (a) alkanes (b) alkenes (c) alkynes.
2. Draw all of the structural formulas that can represent C_4H_{10}.
3. Give the systematic name for each of the compounds whose formulas appear in Question 2.
4. Describe preparation methods for alkanes, alkenes, and ethyne.
5. Consider alkanes, alkenes, alkynes, and aromatic hydrocarbons. Which ones react by addition and which ones react by substitution? Explain why.
6. Give the structural formula of (a) butadiene (b) nitrobenzene (c) ethyl benzene

1. (a) Saturated hydrocarbons in straight or branched chains
   (b) Straight- or branched-chain hydrocarbons that contain a double C—C bond (C=C)
   (c) Straight- or branched-chain hydrocarbons that contain a triple C—C (C≡C)
2. See p. 641
3. n-butane; 2-methylpropane
4. Alkanes are produced by fractional distillation from petroleum and natural gas. Alkenes are produced by cracking petroleum or dehydration of alcohols. Ethyne is produced by partial oxidation of methane or by action of water on calcium carbide.
5. Alkenes and alkynes react by addition because they contain multiple bonds, double and triple bonds respectively. Alkanes and aromatic hydrocarbons react by substitution.
6. (a) See p. 647. (b) See p. 650. (c) See p. 651.

21.4 Representative Hydrocarbons and Polymers

The remarkable development of organic chemistry during the last two centuries has given rise to many industrial applications that have raised the quality of life. The identification and modification of many natural products has played a key role in the development of new chemical products.

Natural Gas and Petroleum

Natural gas is a mixture of hydrocarbon gases and vapors found in porous formations in the earth's crust. It is mostly methane (CH_4) and frequently is

SECTION OBJECTIVES

- Recognize that natural gas and petroleum are complex mixtures of hydrocarbons
- Describe the formation of natural gas and petroleum.
- Identify products of the fractional distillation of petroleum.
- Identify several energy sources as substitutes for petroleum.
- Recognize rubber as an important hydrocarbon and describe the purpose of vulcanization.
- Describe the structure and uses of neoprene and SBR.

found with crude oil. Crude oil is **petroleum**, *a complex mixture of different hydrocarbons that varies greatly in composition.* The hydrocarbon molecules in petroleum contain from one to more than 50 carbon atoms.

Like coal, natural gas and petroleum were formed from the decay of plants and other organisms living in swampy areas millions of years ago. These plant remains became trapped beneath subsequent layers of new plant growth and slowly decomposed in the absence of oxygen. Later changes in the earth's surface subjected these carbon-rich plant remains to intense heat and pressure, resulting in natural gas and petroleum.

The processing of natural gas   Up to about 97 percent of natural gas is methane (CH_4). Mixed with it are other hydrocarbons whose molecules contain between two and seven carbon atoms. These different hydrocarbons have different boiling points, and thus can be separated using a method called fractional distillation, described below for petroleum.

Hydrocarbons with three or four carbon atoms per molecule are sometimes separated and used for fuel as "bottled gas." Those hydrocarbons with five to seven carbon atoms per molecule are liquids at ordinary temperature. Their vapors can be condensed and used as solvents or in gasoline.

The refining of petroleum   Petroleum is refined by using fractional distillation to separate crude oil into different portions, each with unique properties suitable for certain uses. Distillation is evaporation followed by condensation of the vapors in a separate vessel. In **fractional distillation** *components of a mixture with different boiling points are collected separately.* No attempt is made to separate the petroleum into individual hydrocarbons as is done for the lower alkanes in natural gas. Instead, portions

The hydrocarbons in petroleum are alkanes, cycloalkanes, and aromatics.

The methane in natural gas is still produced today by the same process that produced it millions of years ago. Methane production occurs in swamps, mud, sewage, and reclaimed garbage dumps.

TABLE 21-4   SUMMARY OF FRACTIONAL DISTILLATION OF PETROLEUM

Portion	No. of C atoms per molecule	Boiling point range (°C)	Uses
gas	C_1 to C_5	−164–30	fuel; hydrogen, gasoline by alkylation
petroleum ether	C_5 to C_7	20–100	solvent; dry cleaning
gasoline	C_5 to C_{12}	30–200	motor fuel
kerosene	C_{12} to C_{16}	175–275	fuel
fuel oil diesel oil	C_{15} to C_{18}	250–400	furnace fuel; diesel engine fuel; cracking
lubricating oils greases petroleum jelly	C_{16}	350	lubrication
paraffin wax	C_{20}	melts 52–57	candles; waterproofing; home canning
pitch tar		residue	road construction
petroleum coke		residue	electrodes

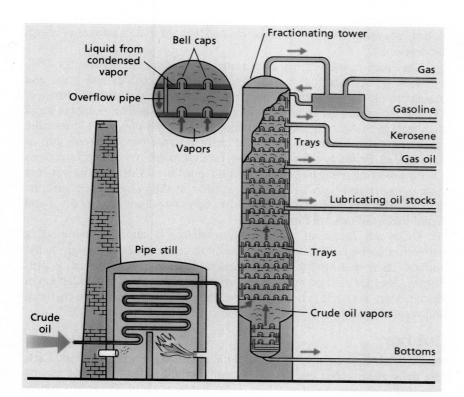

Bell caps

Liquid from condensed vapor

Overflow pipe

Vapors

Fractionating tower

Gas

Gasoline

Kerosene

Trays

Gas oil

Lubricating oil stocks

Trays

Crude oil vapors

Bottoms

Pipe still

Crude oil

Figure 21-17 A pipe still and fractionating tower are shown here in cross-section.

that distill between certain temperature ranges are collected in separate receivers. Table 21-4 summarizes the characteristics of the different portions obtained during the fractional distillation of petroleum.

Petroleum refining is carried out in a pipe still and a fractionating tower, as shown in Figure 21-17. The crude oil is heated to about 370°C in the pipe still. At this temperature, nearly all the components of the crude oil are vaporized. The hot vapors are then discharged into the fractionating tower at a point near its base. Here, the portions with the highest condensation temperatures condense and are drawn off to collecting vessels. Portions with lower condensation temperatures continue to rise in the tower. As they rise, they are gradually cooled. In this way, the various portions reach their condensation temperatures at different levels. As they condense, the liquids collect in shallow troughs that line the inside of the tower.

Pipes carry away the overflow of condensed liquids from the troughs. The fraction that will become gasoline, together with the more volatile portions of the petroleum, passes as a gas from the top of the tower and is then liquefied in separate condensers. The liquids and remaining gaseous fractions will be subjected to other processes depending on what products are required. Figure 21-18 shows a petroleum refinery.

Petroleum Substitutes

Approximately 73 percent of the energy used in the United States comes from natural gas and petroleum. Within a few decades the known supplies of these important raw materials may be exhausted so it is essential that supplies of

Figure 21-18 The fractionating towers of this refinery are used to separate petroleum into its different components.

natural gas and petroleum be conserved and that alternative sources of energy be developed.

Coal may possibly serve as a source for petroleum substitutes. Two well-known methods that can be used to convert coal to fuel gases and petroleum-like liquids. The Fischer-Tropsch process starts with a mixture of CO and H_2 formed by combining steam and hot coke. The resultant gases then react at suitable pressures and temperatures in the presence of a metallic catalyst to form a variety of straight-chain hydrocarbons. The amount of unsaturated products varies with the catalyst used. The Bergius process involves reacting finely powdered coal suspended in oil with hydrogen at very high pressure and high temperature, usually in the presence of a metallic catalyst. The complex, many-ring compounds in coal are broken down and liquid hydrocarbons are formed.

Oil shale, found in some western states, is another source of a petroleumlike liquid. In one process, the oil shale is mined and brought to the surface. There it is heated in large ovens to drive out the oil, which is then refined. Another method involves the underground heating and collection of the oil. More research is needed to develop economical production methods and resolve possible environmental problems.

Rubber

Rubber is an elastic hydrocarbon material obtained from rubber trees. Each tree daily yields about one ounce of a milky fluid called *latex* that can be collected as shown in Figure 21-19. Latex contains about 35 percent rubber in colloidal suspension. When formic acid is added to latex, the rubber separates out as a curdlike mass. After being washed and dried, it is shipped to market as large sheets of crude rubber.

The simplest formula for rubber is $(C_5H_8)_x$. The "x" indicates a very large number. The C_5H_8 unit (shown) is a monomer of rubber, which is a polymer made up of many C_5H_8 units. The units are joined in a zigzag chain and this acocunts for rubber's elasticity.

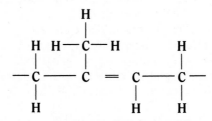

rubber structural unit

Compounding of rubber For commercial processing, raw rubber is mixed with a number of other materials in large batches. The ingredients in these batches vary according to the products to be made. Sulfur, however, is always one of the ingredients.

After mixing, the product is shaped either in a mold or from thin sheets. The whole mass is then vulcanized. **Vulcanization** *is a treatment process in which rubber is heated with sulfur.* During vulcanization, the sulfur atoms form bonds between adjacent rubber molecules. This makes rubber more elastic, less sticky, and longer wearing.

Figure 21-19 A worker taps a rubber tree to collect latex.

Neoprene—a synthetic rubber and its uses In 1910, scientists first produced hydrocarbon synthetic rubber by polymerizing isoprene (C_5H_8). This synthetic product was too expensive and was inferior to natural rubber from plantations in the East Indies. In 1931, a successful synthetic rubber called neoprene appeared on the market. **Neoprene** *is a synthetic rubber that is resistant to heat and flames, as well as to oil and chemicals.*

Hydrogen chloride can be added to vinylacetylene to yield chloroprene:

$$\underset{\text{vinylacetylene}}{\overset{\displaystyle H \quad H \qquad\qquad}{\underset{\displaystyle H}{C}=\underset{}{C}-C\equiv C-H}} \;+\; HCl \;\rightarrow\; \underset{\text{chloroprene}}{\overset{\displaystyle H \quad H \qquad\quad H}{\underset{\displaystyle H \qquad\; Cl \;\; H}{C}=C-C=C}}$$

Figure 21-20 Neoprene suits protect surfers from cold water.

The catalytic polymerization of chloroprene yields the unit shown below.

Oils and greases cause natural rubber to swell and rot; they have little effect on neoprene. Neoprene can hence be used in gasoline delivery hoses and in other objects (as shown in Figure 21-20) that must be flexible while resisting the action of hydrocarbons or temperature extremes.

$$-\underset{\displaystyle H}{\overset{\displaystyle H}{C}}-\underset{}{\overset{\displaystyle Cl}{C}}=\underset{}{\overset{\displaystyle H}{C}}-\underset{\displaystyle H}{\overset{\displaystyle H}{C}}-$$

neoprene structural unit

SBR—a synthetic rubber and its uses SBR, Styrene Butadiene Rubber, is a good all-purpose synthetic rubber that can replace natural rubber for most purposes. Radial automobile tires are made from a mixture of about half natural rubber and half synthetic rubber. SBR is used often as the synthetic-rubber component in automobile tire treads because it resists wear better than other synthetic rubbers (see Figure 21-21). It is made by churning butadiene and styrene together in soapy water. The churning is carried out at 5°C, using a catalyst. This causes the chemicals to polymerize and form SBR. The addition of an acid causes the rubber to separate in curdlike masses, which are washed and dried. A possible structural unit is shown in the structural formula below:

SBR structural unit

Figure 21-21 A worker inspects a block of synthetic rubber before it is packaged for shipment.

The remaining SBR, about 25 percent of total consumption, is used in a large number of products including hoses, seals, footwear, coated fabrics, and wire and cable insulation. A large quantity of SBR is sold as latex to be applied as carpet backing.

1. Methane
2. Probably formed by the decay of plants and animals living millions of years ago. Because of the changes in the earth's surface, these plants and animal residues were trapped in rock formations where they slowly decomposed under intense heat and pressure in the absence of atmospheric oxygen.
3. Petroleum ether, gasoline, kerosene, fuel oil, diesel oil, lubricating oils, petroleum jelly, paraffin wax, pitch tar, petroleum coke
4. Nuclear fission, nuclear fusion, solar energy, biomass energy
5. During vulcanization, which is heating of rubber with sulfur, the rubber becomes more elastic and loses its sticky qualities. Neoprene is resistant to heat, cold, and flames, as well as to oil and chemicals.

Section Review

1. What is the predominant constituent of natural gas?
2. How were natural gas and petroleum formed?
3. Name some of the products of the fractional distillation of petroleum..
4. Identify some of the possible energy sources in the future.
5. Describe vulcanization and compare the characteristics of vulcanized rubber to those of neoprene.

Chapter Summary

- Carbon is an important constituent of many substances found in the natural world. A carbon atom contains four valence electrons and forms four covalent bonds with other atoms. Carbon atoms bond to other carbon atoms and to atoms of other elements with similar electronegativities.
- Diamond and graphite are two allotropic forms of carbon with dramatically different crystalline structures. Amorphorus forms of carbon, such as coal, coke, and charcoal, lack a definite crystalline shape.
- Chemists refer to carbon-containing compounds as organic compounds. An almost infinite variety of organic compounds exists because of the carbon atom's ability to join with other carbon atoms and with other molecules. Carbon atoms can link up with hundreds or even thousands of other carbon atoms to form chains, branched chains, or ring structures.
- Carbon atoms represented by a specific molecular formula may be arranged in several different ways, creating compounds with different properties and structures. Compounds that have the same molecular formula but different structures are called isomers.
- Organic compounds that contain only carbon and hydrogen are called hydrocarbons. Hydrocarbons that contain functional groups in addition to atoms of carbon and hydrogen are called substituted hydrocarbons and are discussed in Chapter 22.
- Hydrocarbons are classified into four different groups based on differences in their carbon–carbon bonds. These four groups are the alkanes, alkenes, alkynes, and the aromatic hydrocarbons. Each group exhibits characteristic chemical properties.
- Alkanes are straight-chain or branched-chain hydrocarbons whose carbon atoms are connected by single covalent bonds. Alkenes are straight- or branched-chain hydrocarbons in which two carbon atoms in each molecule are connected by double covalent bond. Alkynes are straight- or branched-chain hydrocarbons in which two carbon atoms in each molecule are connect by a triple covalent bond. The aromatic hydrocarbons contain one or more benzene rings or related structures.
- Many hydrocarbons play important roles in energy and industry. Natural gas, petroleum, and coal are examples of hydrocarbons that are important sources of energy. Petroleum is refined by using fractional distillation to separate crude oil into different portions. Rubber, a natural hydrocarbon obtained from trees, can be chemically compounded to increase its strength. Synthetic forms of rubber such as neoprene and SBR are frequently used in place of natural rubber because they are more resistant to chemicals and extreme temperatures.

Chapter 21 *Review*

Carbon and Hydrocarbons

Vocabulary

activated carbon	hydrocarbon
alkadienes	isomer
alkanes	lignite
alkenes	methyl group
alkyl group	monomer
alkylation	neoprene
alkynes	nitration
amorphous carbon	organic chemistry
anthracite	paraffin wax
aromatic hydrocarbons	peat
bituminous coal	petroleum
charcoal	petroleum coke
coal	polyethylene
coke	polymer
cracking	polymerization
destructive distillation	saturated
diamond	structural formula
dimer	structural isomers
fractional distillation	subbituminous coal
Friedel-Crafts reaction	substituted hydrocarbons
functional groups	sulfonation
graphite	unsaturated
halogenation	vinyl group
homologous series	vulcanization

Questions

1. *(a)* What is organic chemistry? *(b)* Why is organic chemistry a separate branch of chemistry?
2. How are the four covalent bonds of carbon atom oriented in sp^3 hybridization?
3. There are several times as many carbon compounds as noncarbon compounds. What property of carbon atoms makes this possible?
4. *(a)* Define allotropy. *(b)* Name two allotropic forms of carbon.
5. What property of diamond determines most of its industrial uses?
6. Diamond is an excellent conductor of heat but a nonconductor of electricity. Why?
7. Graphite is soft, yet carbon fibers in which the carbon is in the form of graphite are very strong. Why?
8. *(a)* Draw structural formulas for the three types of graphite layers. *(b)* How is the actual structure of graphite an example of resonance?
9. List several properties of graphite that make it useful as a lubricant.
10. What is coal?
11. *(a)* Explain how geologists believe coal was formed. *(b)* In what sequence were the various types of coal probably formed? *(c)* What conditions probably determined this sequence?
12. *(a)* Define destructive distillation. *(b)* Comment on the descriptive accuracy of this term.
13. Why is activated carbon a useful adsorbent?
14. Why are there so many carbon compounds?
15. *(a)* What is a hydrocarbon? Give the formula of *(b)* the simplest saturated hydrocarbon *(c)* the unsaturated hydrocarbon having the fewest atoms per molecule.
16. *(a)* In an electron-dot formula, what does : represent? *(b)* In a structural formula, what does the symbol " —" represent?
17. *(a)* What are isomers? *(b)* Illustrate using structural formulas for the simplest alkene having isomers.
18. What information is provided by a properly written structural formula?
19. List five important differences between organic and inorganic compounds.
20. Describe the composition of *(a)* natural gas *(b)* petroleum.
21. Define *(a)* distillation *(b)* fractional distillation. *(c)* How effectively are the components of petroleum separated by fractional distillation?
22. What use is made of the petroleum fraction *(a)* having 5 to 12 carbon atoms per molecule *(b)* having a melting point range of 52°C to 57°C?
23. What is the general formula for the members of the *(a)* alkane series *(b)* alkene series *(c)* alkyne series?

24. A hydrocarbon molecule contains seven carbon atoms. Give its formula if it is
    (a) an alkane   *(b)* an alkene   *(c)* an alkyne.
25. *(a)* What is a homologous series?   *(b)* Name the first four members of the alkane series.
    (c) How are the other members of the alkane series named?
26. *(a)* What does the formula RH represent?
    (b) What does the symbol "R—" represent?
27. Write the balanced equation for the laboratory preparation of methane.
28. Write a balanced equation for the complete combustion of   *(a)* methane   *(b)* ethene
    (c) ethyne   *(d)* butadiene.
29. Write equations for the step-by-step substitution of each of the hydrogen atoms in methane by chlorine
30. *(a)* Which hydrocarbon is commercially produced in greatest quantity in the United States?   *(b)* What method is used for commercial production of this hydrocarbon?
31. Define: cracking, thermal cracking, catalytic cracking.
32. Butene reacts with hydrogen in the presence of a nickel catalyst. Write a structural-formula equation for the reaction.
33. What is   *(a)* a monomer   *(b)* a dimer
    (c) a polymer   *(d)* polymerization?
34. Give two uses for ethyne.
35. *(a)* What do the terms saturated and unsaturated mean when applied to hydro-carbons?   *(b)* What other meaning do these terms have in chemistry?
36. Why is calcium carbide sold in airtight cans?
37. How are naphthalene and anthracene molecules related structurally to benzene molecules?
38. Describe two methods of obtaining oil from oil shale.
39. How is rubber separated out from latex?
40. Why is rubber elastic?
41. How is rubber compounded?
42. *(a)* What kinds of additives are used in making rubber goods?   *(b)* Why?
43. Define vulcanization.
44. What advantage does neoprene have over rubber?

Problems

1. The jewelers' mass unit for diamond is the carat. By definition 1 carat equals exactly 200 mg.

What is the volume of a 1.00-carat diamond? The density of diamond is 3.51 g/cm^3.
2. For 100.0 g of carbon dioxide, calculate
   (a) the number of moles   *(b)* the number of molecules   *(c)* the volume of STP.
3. A complex oxide of carbon has the molecular formula $C_{12}O_9$. Determine its percentage composition.
4. *(a)* How many moles of iron (III) oxide can be reduced by the carbon in 5.00 moles of carbon monoxide, according to the equation:
   $Fe_2O_3 + 3CO \rightarrow 2Fe + 3CO_2$?
   (b) How many moles of iron are produced?
   (c) How many moles of carbon dioxide are produced?
5. How many grams of carbon monoxide are needed to react with 15.0 g of zinc oxide to produce elemental zinc ($ZnO + CO \rightarrow Zn + CO_2$)?
6. In Problem 5, *(a)* How many grams of zinc are produced?   *(b)* What is the volume in liters at STP of the carbon dioxide gas produced?
7. How many grams of $H_2SO_4 \rightarrow$ are required for the reaction with 900 g of sodium hydrogen carbonate in a soda–acid fire extinguisher?

   $2NaHCO_3 + H_2SO_4 \rightarrow Na_2SO_4 + 2H_2O + 2CO_2$

8. Calculate the number of liters of carbon dioxide at STP given off during the discharge of the fire extinguisher of Problem 7.
9. What is the mass in kilograms of 12.0 gallons of gasoline? Assume the gasoline is isooctane, which has a density of 0.692 g/mL.
10. Calculate the percentage of sulfur in benzene-sulfonic acid.
11. How many grams of calcium carbide are required for the production of 2.5 L of ethyne at STP?

Application Questions

1. Why are organic compounds with covalent bonds usually less stable to heat than inorganic compounds with ionic bonds?
2. Draw structural formulas for the five isomers of hexane.
3. When decane ($C_{10}H_{22}$) burns completely, carbon dioxide and water vapor are formed. Write the balanced equation for this reaction.
4. The compound 1,3-dichloropropane is a colorless liquid that boils at 120.4°C under standard

pressure. Draw the structural formula for this compound.

5. Draw structural formulas for any three isomers of trichloropentane, and name the isomers whose formulas you draw.

6. The element that appears in the greatest number of compounds is hydrogen. The element forming the second greatest number of compounds is carbon. Why are there more hydrogen compounds than carbon compounds?

7. The organic compounds 2-methylpropane and 2-methylpropene react to form 2,2,4-trimethylpentane. Write the structural formula equation for the reaction.

8. Ethyne molecules dimerize to form vinyl acetylene. Write the structural formula equation for this reaction.

9. *(a)* Is it geometrically possible for the four hydrogen atoms attached to the end carbon atoms in the 1,3-butadiene molecule to lie in the same plane? *(b)* If carbon–hydrogen bonds on adjacent singly bonded carbon atoms tend to repel each other, would it be likely for all six hydrogen atoms to lie in the same plane? *(c)* If they do, what is their relation to the plane of the carbon atoms?

Application Problems

1. On a mythical planet in a far-off galaxy, the element carbon consists of equal numbers of atoms of C-12, atomic mass exactly 12 u; C-13, atomic mass, 13.00335 u; and C-14, atomic mass, 14.00324 u. What is the atomic mass of carbon on this make-believe planet?

2. The stable oxides of carbon have the molecular formulas CO, CO_2, C_3O_2, and $C_{12}O_9$. What is the oxidation number of carbon in each oxide?

3. How many grams of carbon monoxide can be obtained by the dehydration of 250 g of formic acid by sulfuric acid?

4. How many liters of dry carbon monoxide are produced in the reaction described in the previous problem if the temperature is 27°C and the pressure is 750 mm Hg?

5. A gaseous compound contains 52.9% carbon and 47.1% oxygen. One volume of this gas reacts with two volumes of oxygen and yields three volumes of carbon dioxide. Knowing that oxygen molecules are diatomic, determine the molecular formula of this compound.

6. How many grams of calcium carbonate are needed to prepare 2.50 L of dry carbon dioxide at 22°C and 735 mm Hg pressure by the reaction between calcium carbonate and hydrochloric acid?

7. As the number of carbon atoms in an alkane molecule increases, does the percentage of hydrogen increase, decrease, or remain the same?

8. A compound consists of 60.0% carbon, 26.7% oxygen, and 13.3% hydrogen. Its formula mass is 60 u. What are the possible structures for molecules of this chemical compound?

9. A volume of ethene (115 mL) is collected by water displacement at 22°C and 738 mm pressure. What is the volume of the dry ethene at STP?

Enrichment

1. Perform library research to learn about fibers used in medicine. Summarize your findings in a written report.

2. Investigate the organic material from which bulletproof vests are made and describe to your class the structure for a segment of a molecule of that material.

3. Read the article entitled "What Compound was Discovered as a Result of an Insurance Claim?" by Maureen M. Julian, *Journal of Chemical Education*, October, 1981, p. 793. Summarize this article for your classmates.

4. *Chemical and Engineering News* publishes a list once a year of the top fifty chemicals. Find out how many of the chemicals on this year's list are hydrocarbons and report their identity to the class.

5. Consult reference materials in the library to learn about the ways in which petroleum refining can damage the environment. Describe some of the steps that the petroleum industry has taken to reduce or eliminate some of the environmentally undesirable side effects of petroleum refining.

6. Consult reference materials at the library and read about products made from hydrocarbons. Keep a list of the number of petroleum-related products you use in a single day.

7. Do library research and write a paper on the health hazards of some of the aromatic hydrocarbons.

Chapter 22 Substituted Hydrocarbons

Chapter Planner

22.1 Alcohols
■ Question: 1b
■ Application Problems: 4

Preparation of Alcohols
■ Desktop Investigation
Application Question: 6
■ Application Problems: 1, 2

Reactions of Alcohols
■ Demonstration: 1
Application Questions: 9–11
■ Question: 11
■ Questions: 2a, 10

Properties, Preparation, and Uses of Alcohols
■ Demonstration: 2
Questions: 6–9
Application Questions: 7, 8
■ Problems: 3, 4
Application Problem: 5

22.2 Halocarbons
■ Question: 1a
Application Question: 1a
■ Problem: 1

Preparation of Halocarbons
■ Demonstration: 3
Application Questions: 2, 4

Reactions of Halocarbons
■ Application Question: 3

Properties, Preparation, and Uses of Several Halocarbons
■ Questions: 3–5
Application Questions: 1c, 5

22.3 Ethers
■ Question: 1c

Preparation of Ethers
Reactions, Properties, and Uses of Ethers

22.4 Aldehydes and Ketones
■ Question: 1d-e
Application Question: 1d-e

Preparation and Uses of Aldehydes
■ Question: 12

Preparation and Uses of Ketones
■ Question: 13
■ Demonstration: 4

22.5 Carboxylic Acids
■ Question: 1f
Application Question: 1f
■ Application Question: 13a

Preparation of Carboxylic Acids
■ Question: 14
Application Question: 12

Reactions of Carboxylic Acids
■ Experiment: 39
Question: 15
■ Question: 2b
■ Problems: 2, 5

Preparation, Properties, and Uses of Carboxylic Acids
■ Application Problem: 3

22.6 Esters
■ Question: 1g
■ Application Question: 13b

Preparation of Esters
■ Demonstration: 5
Application Questions: 1g, 14

Reactions of Esters
■ Demonstration: 6
Question: 1h
■ Questions: 2c, 17
Application Question: 15
■ Question: 16

Teaching Strategies

Introduction

Introduce this chapter by pointing out that many familiar substances will be discussed, such as wood alcohol, rubbing alcohol, cleaning fluids, formaldehyde, vanilla, acetone, vinegar, citric acid, and flavorings. They will learn how many of these substances are made or where they are found in nature. They will be able to read and understand product labels.

The objectives for this chapter are to help students demonstrate the following abilities as they work with substituted hydrocarbons: *(1)* identify and name organic compounds from their structural formulas and write their structural formulas from their names; *(2)* know some preparations and reactions for each class of organic compounds and be able to write equations using structural formulas for these reactions; and *(3)* know the properties of each class of organic compounds and their uses.

It is very helpful to continue to use models and to have students make models as you cover each class of organic compounds.

22.1 Alcohols

Introduce this section by exhibiting several substances containing different alcohols, such as windshield cleaning fluid (methanol or methyl alcohol), liquid medicines (ethanol or ethyl alcohol), rubbing alcohol (2–propanol or isopropyl alcohol), antifreeze (1-, 2– ethanediol or ethylene glycol), and hand lotion (1-,2-,3-propanetriol or glycerol). Have students read the labels to determine what alcohols are present. Write the structural formula for each compound. See if the students can write the general formula for alcohols by determining what the structural formulas for alcohols have in common.

Demonstration 1: Oxidation of Alcohols
PURPOSE: To show the oxidation of alcohols and how this reaction helps to differentiate between primary or secondary alcohols and tertiary alcohols
MATERIALS: 1-butanol (n-butyl alcohol), 2-butanol (sec-butyl

alcohol), 2-methyl-2-propanol (tert-butyl alcohol), potassium permanganate, distilled water, 3 50-mL beakers, overhead projector

PROCEDURE: CAUTION: Wear goggles, gloves, and a lab coat or lab apron during the preparation and performance of this demonstration. Make a solution of potassium permanganate by adding 0.16 g of $KMnO_4$ to 100-mL of water. Alcohols are flammable; make sure all flames are extinguished in the room.

Put 10 mL of each of the following alcohols into three separate beakers: 1-butanol, 2-butanol, and 2-methyl-2-propanol. Place each beaker on an overhead projector and project on a screen. Add 3 mL of $KMnO_4$ to each alcohol and gently swirl the two liquids. (The 1-butanol and 2-butanol will turn brown because oxidation takes place. The 2-methyl-2-propanol will remain purple. Tertiary alcohols cannot be oxidized except under extreme conditions.)

The reactions may take several minutes before they turn brown. They may slowly turn red and finally brown. This is a good time to remind students that organic reactions take place more slowly than inorganic reactions.

Ask the students why the 1- and 2-butanol changed color. (They were oxidized by the $KMnO_4$.) What do they think the products of these reactions were? (1-butanol produced an aldehyde and 2-butanol, a ketone.) Ask the students to write equations for these reactions and the names of the organic compounds that were formed.

$$-\overset{|}{\underset{|}{C}}-\overset{|}{\underset{|}{C}}-\overset{|}{\underset{|}{C}}-\overset{|}{\underset{|}{C}}-OH + O \rightarrow -\overset{|}{\underset{|}{C}}-\overset{|}{\underset{|}{C}}-\overset{|}{\underset{|}{C}}-\overset{|}{\underset{|}{C}} = O + H_2O$$

$(KMnO_4)$ butanol or butynaldehyde

Atomic oxygen is supplied by the $KMnO_4$.

$$-\overset{|}{\underset{|}{C}}-\overset{|}{\underset{OH}{C}}-\overset{|}{\underset{|}{C}}-\overset{|}{\underset{|}{C}} + O \rightarrow -\overset{|}{\underset{|}{C}}-\overset{|}{\underset{|}{C}}-\overset{|}{\underset{O}{C}} \ \ C + H_2O$$

$(KMnO_4)$ methyl ethyl ketone or 2-butanone

Ask students how tertiary alcohols may be distinguished from primary or secondary alcohols. (Tertiary alcohols, unlike primary and secondary alcohols, are not oxidized by $KMnO_4$ in the presence of a base.) The color change of the $KMnO_4$ from purple to brown indicates that oxidation has occurred.

Demonstration 2: How Is a Super Ball Made?

PURPOSE: To make a polymer and to show that silicon also forms polymers.

MATERIALS: Sodium silicate (water glass) solution, 95% ethyl alcohol, 25-mL graduated cylinder, 5-ounce paper cup, plastic sandwich bag, wooden splint

PROCEDURE: CAUTION: Wear safety goggles, a lab coat or lab apron, and disposable plastic gloves.

Put 24 mL of sodium silicate solution in the paper cup. While stirring the sodium silicate solution with a wooden splint, add 6 mL of ethanol. Continue to stir until a solid substance is formed. Put the polymer in the palm of one hand. Gently press it with the palms of both your hands until a ball that does not crumble is formed. This takes a little time and patience. Occasionally moisten the ball by letting a small amount of water from a faucet run over it. When the ball no longer crumbles, bounce it! Store the ball in a plastic bag. If it starts to crumble, it can be reformed.

The formula for the final product is:

$$\underset{C_2H_5}{\overset{C_2H_5}{\underset{|}{Si}}}\overset{O}{-}\underset{C_2H_2}{\overset{C_2H_5}{\underset{|}{Si}}}\overset{O}{-}\underset{C_2H_5}{\overset{C_2H_5}{\underset{|}{Si}}}\overset{O}{-}\underset{C_2H_4}{\overset{C_2H_5}{\underset{|}{Si}}}$$

Water is also formed. The liquid given off as the polymer is pressed is excess alcohol and a little water.

If students make their own super balls, you might have them devise additional experiments to see: (1) if the size of the ball has any effect on the height of its bounce; (2) the effect of different concentrations of ethyl alcohol; or (3) what the difference is between using isopropyl alcohol and ethyl alcohol.

22.2 Halocarbons

To introduce halocarbons, make a table on a transparency similar to the following:

NAME(S)	USES AND COMMENTS	STRUCTURAL FORMULA
Trichloromethane (Chloroform)	Previously used as an industrial solvent and anesthetic; use now limited because considered toxic and carcinogenic	$Cl-\overset{H}{\underset{Cl}{C}}-Cl$
Difluorodichloromethane (Freon-12) 2,-dichloro-1,1,2,2-tetra- chloroethane (Freon-114)	Coolant in refrigerators and air conditioners. Not toxic but is damaging earth's ozone layer	$Cl-\overset{F}{\underset{F}{C}}-Cl$ $Cl-\overset{F}{\underset{F}{C}}-\overset{F}{\underset{F}{C}}-Cl$
1,4-dichlorobenzene	Insecticide, moth balls	Cl—◯—Cl
Polychlorinated biphenyls (PCBs)	Multitude of uses: hydraulic fluids, making paint and plastics, insulating fluid in transformers; may be most widespread chemical pollutant	Cl—◯—◯—Cl with Cl substituents
Polybromonated biphenyls (PBBs)	Formerly, flame retardant for clothing	Br—◯—◯—Br with Br substituents
Triiodomethane Iodoform	Antiseptic powder	$H-\overset{H}{\underset{H}{C}}-I$
Vinyl chloride	Monomer for polyvinyl chloride plastic	$\overset{H}{\underset{H}{C}}=\overset{H}{\underset{Cl}{C}}$
tetrafluoroethane	Monomer for teflon polymer	$\overset{F}{\underset{F}{C}}=\overset{F}{\underset{F}{C}}$
Dichlorodiphenyl-trichloroethane (DDT)	Insecticide; harmful to many animals; very long-lasting in the environment; illegal in U.S.	Cl—◯—CH—◯—Cl, $Cl-\overset{|}{\underset{Cl}{C}}-Cl$
Hexachlorocyclohexane Benzenehexachloride (BHC)	Insecticide; long-lasting in the environment; causes environmental pollution	Cl-substituted cyclohexane

Ask students what these compounds have in common, and what their general formula is. All have a carbon chain, frequently with hydrogens, and one or more halogens located on the chain. Their general formula is RX.

As is apparent from the chart and from the discussion of freon in this chapter, many of the halocarbons are environmental pollutants that cause health problems that will affect life in the future. This would be a good place to assign reports (for extra credit or as a class project) on some of the problems caused by halocarbons.

Demonstration 3: Making a Halocarbon

PURPOSE: To make a halocarbon by adding a halogen to an alkene

MATERIALS: 1,1,2-trichloro−1,2,2, trifluoroethane, cyclohexene, iodine, 50mL beaker, overhead projector

PROCEDURE: CAUTION: Wear goggles, gloves, and a lab coat or lab apron during the preparation and performance of this lab. Cyclohexene is extremely flammable. Store under a fume hood until ready to use, and return to hood as soon as the demonstration is completed.

Using a fume hood, add 3 g of iodine to 100 mL of 1,1,2-trichloro − 1,2,2-trifluoroethane. Put 10 mL of cyclohexene into a 50-mL beaker. Put the beaker on the overhead projector to a screen. Add 3 mL of the iodine solution to the cyclohexene and gently swirl. (The color disappears, indicating the addition of the iodine to the double bonds of the cyclohexene and the formation of a halocycloalkane. Ask students to write an equation for the reaction and name the halocarbon.

1,2 diiodocyclohexane

22.3 Ethers

Remind students that there are many different types of ethers. Put the structural formulas and names for several different

ethers on the board and ask the students to derive their general formula, ROR.

Emphasize that ethers are used mostly as solvents, though many people believe their main use is as anesthetics.

Remind students that ethers (like alcohols) burn easily because part of the oxygen needed for combustion is in the compound itself. Those that are volatile (the ethers with the smaller number of carbons in their chains) explode readily as they evaporate and mix with the air. An example equation for the combustion of methyl ethyl ether is:

$$2CH_3OC_2H_5 + 9O_2 \rightarrow 6CO_{2(g)} + 8H_2O(g)$$

22.4 Aldehydes and Ketones

Introduce this section by showing the students the general formulas for aldehydes and ketones and some simple examples of each. Display some common flavorings, spices, or odors whose flavor or odor is result of an aldehyde or ketone.

Demonstration 4: Fehling's Test for Aldehydes

PURPOSE: To show that aldehydes give a positive test with Fehling's solution

MATERIALS: Acetone, copper(II) sulfate pentahydrate, ethanol, glucose, potassium sodium tartrate (Rochelle salts), sodium hydroxide, distilled water, 400-mL beaker, 3 250-mL Erlenmeyer flasks, 3 #5 rubber stoppers, 3 medium test tubes

PROCEDURE: CAUTION: Wear goggles, gloves, and a lab coat during the preparation and performance of this lab. Acetone and ethanol are extremely flammable; be sure there are no flames in the room.

(1) Dissolve 7 g of copper(II) sulfate pentahydrate in 100 mL of distilled water in an erlenmeyer flask and insert stopper. In a second flask, dissolve 35 g of potassium sodium tartrate and 10 g of sodium hydroxide in 100 mL of distilled water and stopper. To a third flask, dissolve 5 g of glucose in 100 mL of distilled water. (2) Fill the beaker half full of water, and heat to boiling on a hot plate. To each of three test tubes, add 3 mL of copper sulfate solution and 3 mL of basic potassium sodium tartrate solution. To one test tube, add 1 mL of the glucose solution; to the second, 1 mL of acetone; and to the third, 1 mL of ethanol. Put all three test tubes in the boiling water bath for 5 minutes. (3) The formation of a brick-red precipitate is a positive test for an aldehyde. Glucose will give a positive test; the other two will not. Glucose contains an aldehyde group. The equation for the reaction is:

22.5   Carboxylic Acids

Introduce carboxylic acid by writing the names and formulas of several common substances that contain carboxylic acids, such as vinegar (acetic acid), CH_3COOH; yogurt or buttermilk (lactic acid),

$$CH_3CH\text{-}COOH;$$
$$|$$
$$OH$$

oranges, lemons, grapefruits (citric acid),

$$OH$$
$$|$$
$$HOOCCH_2\text{—}C\text{—}CH_2COOH;$$
$$|$$
$$COOH$$

rhubarb and sauerkraut (oxalic acid), $HOOC\text{-}COOH$;

green apples (malic acid), $HOOC\ CH_2\ CH\ COOH$;
$$|$$
$$OH$$

and cranberries (benzoic acid), —$COOH$.

Ask students to locate the functional group that they have in common and write the general formula for carboxylic acids, RCOOH.

The following demonstration shows a reaction of an organic acid and makes a good lead-in for the next unit.

Demonstration 5: More Good Flavor-Making Esters

PURPOSE: To show how common flavorings are made
MATERIALS: Glacial acetic acid, butyric acid, decanoic acid, salicylic acid, concentrated sulfuric acid, n-amyl alcohol, iso-amyl alcohol, ethyl alcohol, methyl alcohol, octyl alcohol, 7 medium size test tubes, test tube rack, test tube holder, hot plate, 400-mL beaker, 7 stirring rods, filter paper, 7 watch glasses
PROCEDURE: CAUTION: Wear a face shield, gloves, and a lab coat during the preparation and performance of this lab. Glacial acetic acid and concentrated sulfuric acid are corrosive. Butyric acid is an irritant and has a horrible smell that remains with you for hours if spilled on clothing or skin. Extinguish all flames in the room.

Fill the beaker half full of water, and start to heat it to boiling on the hot plate. Number the test tubes. Use the chart to determine which acid and alcohol to put into each test tube. Add 2 mL of each alcohol and 2 mL of each liquid acid to each test tube as indicated on the chart. If the acid is a solid, add 1 g. Add 1 mL of concentrated sulfuric acid to each test tube. To mix the contents of the test tube, tap the bottom of the tube with your finger. Place the test tubes in the boiling water bath for a few minutes.

Cut a piece of filter paper into seven strips and place on seven watch glasses. Dip a clean stirring rod into each test tube and wet the filter paper by touching the stirring rod to the paper. Have the students try to identify the odor of the ester.

Tube	Acid	Alcohol	Odor of Ester	Name of Ester
1	butyric	n-amyl	apricot	n-amyl butyrate
2	butyric	ethyl	pineapple	ethyl butyrate
3	acetic	iso-amyl	banana	iso-amyl acetate
4	acetic	ethyl	apple	ethyl acetate
5	acetic	octyl	orange	octyl acetate
6	salicylic	methyl	wintergreen	methyl salicylate
7	decanoic	ethyl	grape	ethyl decanoate

22.6   Esters

Demonstration 6: Testing for Double Bonds in Oils

PURPOSE: To test for double bonds in cooking oils
MATERIALS: 1,1,2-trichloro-1,2,2-trifluoroethane, cooking oil, iodine, 50-mL beaker, overhead projector
PROCEDURE: CAUTION: Wear goggles, gloves, and a lab coat during the preparation and performance of this lab.

Put 10 mL of cooking oil into a 50-mL beaker. Put the beaker on an overhead projector and project on a screen. Add 3 mL of the iodine solution to the oil and stir thoroughly. (The color disappears, indicating the addition of the iodine to the double bonds of the oil. This shows that cooking oil has double bonds and thus is an unsaturated fat. The equation that indicates what happens with one oil follows:

References and Resources

Books and Periodicals

Asimov, Isaac. *The World of Carbon*, Collier Books, New York, 1962.

Borgford, Christie L., and Lee R. Summerlin. "'Slime': A Shimmery, Clear Fluid Polymer," pp. 87–88; "Making a Super Ball," pp. 89–91; "Preparing a Detergent," pp. 152–54; and "Artificial Flavorings and Fragrances," pp. 237–39. *Chemical Activities*, Teacher's Edition, American Chemical Society, Washington, D.C., 1988.

Breedlove, C. H. "Vanilla," *ChemMatters*, Vol. 6, 1988, pp. 8–9.

Casassa, E. Z. "The Gelation of Polyvinyl Alcohol with Borax," *Journal of Chemical Education*, Vol. 57, 1986, pp. 57–60.

Doheny, Anthony, and G. Marc Loudon. "The Effect of Free Radical Stability on the Rate of Bromination of Hydrocarbons," *Journal of Chemical Education*, Vol. 57, 1980, pp. 507–8.

Friedstein, Harriet G. "Basic Concepts of Culinary Chemistry," *Journal of Chemical Education*, Vol. 60, 1983, pp. 1037–38.

Gough, Michael. "Dioxin: From Plant Hormones to Agent Orange," *ChemMatters*, Vol. 6, 1988, pp. 9–12.

Gough, Michael. "Dioxin: Past War. Future Risk?" *ChemMatters*, Vol. 6, 1988, pp. 15–19.

Nagel, Mirian C. "Peroxides Can Be Treacherous," *Journal of Chemical Education*, Vol. 61, 1984, pp. 250–51.

Schultz, Emeric. "Pop-and-Sniff Experimentation," *Journal of Chemical Education*, Vol. 64, 1987, pp. 797–98.

Wood, Clair G. "Dissolving Plastic," *ChemMatters*, Vol. 5, 1987, pp. 12–14.

Zurer, Pamela S. "Studies on Ozone Destruction Expand Beyond Antarctic," *Chemical and Engineering News*, Vol. 66, 1988, pp. 16–25.

Audiovisual Resources

Chemistry 120—Organic Chemistry (Sound Filmstrips) Learning Arts, Wichita, KS, 1988.

Chemistry Study Cards—Organic Chemistry:Organic Nomenclature and Organic Reactions, Frey Scientific, Mansfield, OH, 1988–89.

The Ozone Layer: How Important Is It? (VHS) Random House Media, Stamford CT, 1988.

Special Topics in Chemistry, Vol. II Organic Chemistry and Biochemistry, (VHS) Random House Media, Stamford, CT, 1988.

Computer Software

Flash, Patrick and Victor Bendall. *Organic Reaction Chemistry*, Apple II, 64K, Queue, Bridgeport, CT, 1988.

Smith, Stanley. *Substitution reactions, Alcohols, Aldehydes and Ketones*, and *Carboxylic Acids*, Apple II 48K and IBM PC 256K, Queue, Bridgeport, CT, 1988.

Chemistry Series (J & S Software) Organic Chemistry, Apple II, 48K, Barclay, Brooklyn, NY, 1988.

General Chemistry 1B, Topic 9: Organic Chemistry, Apple II, 128K, Queue, Bridgeport, CT, 1988.

SEI Organic Chemistry, Apple II, 64K and IBM PC 128K, Queue, Bridgeport, CT, 1988.

Answers and Solutions

Questions

1. *(a)* RX *(b)* ROH *(c)* ROR′ *(d)* RCHO *(e)* RCOR′ *(f)* RCOOH *(g)* RCOOR *(h)* RCOOCH$_2$ RCOOCH R′COOCH$_2$

2. *(a)* Sulfation is the addition of sulfur to alcohols. *(b)* Esterification is the reaction of an acid with an alcohol to produce an ester. *(c)* Saponification is the hydrolysis of fats or oils using a solution of a strong hydroxide.

3. *(a)* It is an excellent nonpolar solvent. *(b)* It is used for dry-cleaning fabrics, removing grease from metals, extracting oils from seeds, and as an industrial solvent.

4. *(a)* Good ventilation must be provided. *(b)* The vapors are poisonous.

5.

$$\begin{array}{ccccccccccc}
& H & & H & & H & & H & & H & & H \\
& | & & | & & | & & | & & | & & | \\
- & C & - & C & - & C & - & C & - & C & - & C & - \\
& | & & | & & | & & | & & | & & | \\
& H & & Cl & & H & & Cl & & H & & Cl \\
\end{array}$$

6. Alcohols do not yield hydroxide ions in water; alcohols are molecular and hydroxides are ionic compounds.

7. It destroys the cells of the optic nerve, causing blindness, eventually death.

8. Denatured alcohol is a mixture composed principally of ethanol to which poisonous and nauseating materials have been added to make it unfit for consumption.

9. It is hygroscopic.

10. *(a)* $2Na + 2H_2O \rightarrow 2NaOH + H_2(g)$ *(b)* $2Na + 2CH_3OH \rightarrow 2CH_3ONa + H_2(g)$

11. $4\frac{1}{2}$

12. For making adhesives for plywood and particle board, and as resins for plastics

13. As a solvent in the manufacture of acetate rayon, as a solvent for ethyne in storage tanks, for cleaning metals, for removing stains, and for preparing synthetic organic chemicals

14. The apple cider is permitted to ferment to hard cider. The ethanol in the hard cider is oxidized to acetic acid. Oxygen of the air acts as the oxidizing agent. The reaction is catalyzed by enzymes from certain bacteria.

15. *(a)* $H_2C_2O_4 + H_2O \rightleftarrows H_3O^+ + HC_2O_4^-$
    $HC_2O_4^- + H_2O \rightleftarrows H_3O^+ + C_2O_4^-$;
    (b) Six moles

16. Dehydration of alcohols to ethers or alkenes; esterification

17. *(a)* They are esters of glycerol and long-carbon-chain acids.

(b) Oils are liquids at room temperature; fats are solids.
(c) Oils usually contain more unsaturated hydrocarbon chains than fats. The long-carbon-chain acids with double bonds produce esters having lower melting points.

Problems

1. $1 \text{ C atom} \times \dfrac{12.0111 \text{ u}}{\text{C atom}} = 12.0111 \text{ u}$

   $3 \text{ Cl atoms} \times \dfrac{35.453 \text{ u}}{\text{Cl atom}} = 106.359 \text{ u}$

   $1 \text{ F atom} \times \dfrac{18.9984 \text{ u}}{\text{F atom}} = 18.9984 \text{ u}$

   Formula mass $CCl_3F = 137.369 \text{ u}$

2. $V_a N_a = V_b N_b.$ $N_b = V_a N_a / V_b.$

   $N_b = \dfrac{35.0 \text{ mL} \times 0.150 - N}{45.0 \text{ mL}} = 0.117 - N$

3. Molar mass $C_3H_5(OH)_3 = 92.0 \text{ g}$

   $\dfrac{0.400 \text{ mol } C_3H_5(OH)_3}{\text{kg } H_2O} \times \dfrac{92.0 \text{ g } C_3H_5(OH)_3}{\text{mol } C_3H_5(OH)_3} \times 0.250 \text{ kg } H_2O$
   $= 9.20 \text{ g}$

4. $0.000°C - (0.400 \, m \times 1.86 \, C°/\text{molal}) = -0.744°C$
   $100.00°C + (0.400 \, m \times 0.51 \, C°/\text{molal}) = 100.20°C$

5. Molar mass $(COOH)_2 = 90.0357 \text{ g}$

   Equiv. mass $(COOH)_2 = \dfrac{\text{molar mass }(COOH)_2}{2} = \dfrac{90.0357 \text{ g}}{2}$
   $= 45.0179 \text{ g}$

Application Questions

1.

(a) Cl — C — Cl with Cl below (dichloromethane structure)

(b) H — C — C — C — H with O, H, O below

(c) H — C — C — O — C — C — H

(d) H — C — C with =O and H

(e) H — C — C — C — C — H with =O

(f) H — C — C with O—H

(g) H — C with =O, O — C — C — H

2. *(a)* $C_2H_6 + Br_2 \rightarrow C_2H_4Br + HBr$ *(b)* $C_2H_4 + HBr$
   $\rightarrow C_2H_5Br$ *(c)* $C_2H_5OH + HBr \rightarrow C_2H_5Br + H_2O.$

3. Alkyl halides react with water solutions of strong hydroxides to yield alcohols and the halide ion. Alkyl halides react with benzenes to form alkyl benzenes and hydrogen chloride. Preparation of dichlorodifluoromethane from carbon tetrachloride and hydrofluoric acid.

4. $CH_4 + 4Cl_2 \rightarrow CCl_4 + 4HCl$

   $CCl_4 + 2HF \xrightarrow{\text{catalyst}} CCl_2F_2 + 2HCl.$

5. Freons in the stratosphere are believed to react with and deplete the ozone that protects the earth's surface from excess ultraviolet radiation from the sun.

6.

(a) $C=C$ (ethene structure) $+ H—O—H \xrightarrow{H_3O^+}$
   H — C — C — O — H

(b) H — C — C — Cl $+ O—H^- \rightarrow$
   H — C — C — O — H $+ Cl^-$

(c) $C_{12}H_{22}O_{11} + H—O—H \rightarrow$
   $4 \, O=C=O + 4H —$ C — C — O — H

(d) H — C — O — H $+ C=O + 2H—H \rightarrow$
   H — C — C — O — H $+ H—O—H$

7. As the formula mass increases, increasing masses are required per unit mass of water to produce the same freezing-point depression. The order of increasing formula mass is: methanol, ethanol, and ethylene glycol. However, in this same order, the boiling points increase. Although a smaller mass of methanol is required for a freezing-point depression, it evaporates most readily. A larger mass of ethylene glycol is required, but it does not evaporate appreciably.

8. It has a much higher boiling temperature than pure water.

9. $C_3H_7OH + HI \rightarrow C_3H_7I + H_2O$

10. $2C_3H_7OH \xrightarrow{H_2SO_4} C_3H_7OC_3H_7 + H_2O$

11. (a) $CH_3CHOHCH_2CH_3 + O$ (from oxidizing agent) $\rightarrow$
$CH_3COCH_2CH_3 + H_2O$
(b) Methyl ethyl ketone

12. (a) $CH_3OH + CO \xrightarrow{catalyst} CH_3COOH$
(b) $C_2H_5OH + O_2 \rightarrow CH_3COOH + H_2O$ (catalyzed by bacterial enzymes). Equations for the stepwise oxidation of ethanol to acetaldehyde and then to acetic acid could also be given.

13. (a) The carbon–oxygen bonds in formic acid are a single and a double bond because they have different lengths. (b) The carbon–oxygen bonds in sodium formate are alike because they have the same length. (c) The carbon–oxygen bonds in sodium formate are resonance hybrid bonds intermediate between single and double covalent bonds.

14. $C_4H_9OH + CH_3COOH \xrightarrow{H_2SO_4} CH_3COOC_4H_9 + H_2O$

15. $(C_{17}H_{35}COO)_3C_3H_5 + 3NaOH \rightarrow 3C_{17}H_{35}COONa + C_3H_5(OH)_3$

Application Problems

1. Molar mass of $C_{12}H_{25}OSO_2ONa = 288.38$ g

$$\% \text{ C} = \frac{144.13 \text{ g C}}{288.38 \text{ g C}} \times 100\% = 49.98\% \text{ C}$$

$$\% \text{ H} = \frac{25.199 \text{ g C}}{288.38 \text{ g C}} \times 100\% = 8.74\% \text{ H}$$

$$\% \text{ O} = \frac{63.998 \text{ g O}}{288.38 \text{ g C}} \times 100\% = 22.19\% \text{ O}$$

$$\% \text{ S} = \frac{32.06 \text{ g S}}{288.38 \text{ g C}} \times 100\% = 11.12\% \text{ S}$$

$$\% \text{ Na} = \frac{22.990 \text{ g Na}}{288.38 \text{ g C}} \times 100\% = 7.97\% \text{ Na}$$

2. (a) $54.5 \text{ g C} \times \dfrac{1 \text{ mol C}}{12.0 \text{ g}} = 4.54 \text{ mol C}$

$9.1 \text{g H} \times \dfrac{1 \text{ mol H}}{1.0 \text{ g H}} = 9.1 \text{ mol H}$

$36.4 \text{ g O} \times \dfrac{1 \text{ mol O}}{16.0 \text{ g O}} = 2.28 \text{ mol O}$

$\dfrac{4.54}{2.28} \text{ mol C} : \dfrac{9.1}{2.28} \text{ mol H} : \dfrac{2.28}{2.28} \text{ mol O}$

$1.99 \text{ mol C} : 3.99 \text{ mol H} : 1 \text{ mol O}$

Simplest formula $= C_2H_4O$

(c) molar mass of $C_2H_4O = 44$ g
molar mass of $(C_2H_4O)_x = 88$ g
$x = 88g/44g = 2$
molecular formula $= C_4H_8O$
The compound could be ethyl acetate, $CH_3COOC_2H_5$.

(b) $\Delta T_b = K_b m$

$$m = \frac{78.87°C - 78.26°C}{1.22°C \text{ Kg ethanol/mol solute}} = 0.50 \text{ mol solute/kg ethanol}$$

$$\text{molar mass solute} = \frac{11.0 \text{ g solute}}{(0.50 \text{ mol/solute/kg ethanol})(1 \text{ kg}/10^3\text{g})(250 \text{ g ethanol})} = 88 \text{ g}$$

3. $\text{pH} = -\log [H_3O^+] = -\log(9.4 \times 10^{-4}) = -(\log 9.4 + \log 10^{-4}) = -(0.97 - 4) = 3.03$

4. Molar mass $C_2H_5OH = 46.0$ g
$$\frac{0.750 \text{ mol } C_2H_5OH}{\text{L } H_2O} \times \frac{46.0 \text{ g } C_2H_5OH}{\text{mol } C_2H_5OH} \times 500 \text{ mL} \times \frac{1 \text{ L}}{1000 \text{ mL}} \times \frac{1 \text{ mL}}{0.789 \text{ g}} = 21.9 \text{ mL}$$

5. (a) Molar mass $C_2H_4(OH)_2 = 62.0$ g. Using one liter of ethylene glycol (1432 g) and one liter of water (1000 g), the mass ratio is
$$\frac{1432 \text{ g } C_2H_4(OH)_2}{1000 \text{ g } H_2O} = \frac{23.1 \text{ mol } C_2H_4(OH)_2}{1 \text{ kg } H_2O} = 23.1 \text{ m}$$

$0.0°C - (23.1 \text{ } m \times 1.86C°/m) = -43.0°C$, the theoretical freezing point.
(b) Freezing-point-depression calculations are based on a relationship that holds experimentally only for dilute solutions.

Chapter 22 Substituted Hydrocarbons

INTRODUCTION

In Chapter 21 you learned about the structure and bonding characteristics of carbon and the chemistry of hydrocarbons. Here you will examine the chemistry of *substituted hydrocarbons*— carbon compounds formed by replacing one or more hydrogen atoms of a hydrocarbon with some other atom or group of atoms.

LOOKING AHEAD

As you study this chapter, carefully examine the structural formulas of the various functional groups, and associate their structures to the properties that their parent molecules exhibit. Pay particular attention to the methods of preparation of the various substituted hydrocarbons and the reactions they undergo. Relate the properties that they exhibit to their useful applications.

SECTION PREVIEW

22.1 Alcohols
22.2 Halocarbons
22.3 Ethers
22.4 Aldehydes and Ketones
22.5 Carboxylic Acids
22.6 Esters

The plastic coating on these compact disks is just one of many products made from substituted hydrocarbons.

22.1 Alcohols

Alcohols *are substituted hydrocarbons that contain one or more hydroxyl units (—OH).* They are derived from saturated or unsaturated hydrocarbons by replacing a hydrogen atom with a hydroxyl group. The boiling points of alcohols are generally higher than those of their corresponding hydrocarbons because their molecules join together through the formation of hydrogen bonds between the hydrogen atom of the hydroxyl group of one molecule and the oxygen atom of the hydroxyl group of another molecule. Alcohols are good solvents and are useful in the preparation of other organic compounds.

Common names are used frequently for lower members of the alcohol series. A common name is derived by adding the word *alcohol* to the name of the alkyl group attached to the hydroxyl group. The IUPAC system is used to name the more complex alcohols. The systematic name of alcohols is derived by dropping the final *-e* of the alkane corresponding to the longest continuous chain of carbon atoms and adding *-ol*. The parent chain is numbered from the end nearer the hydroxyl group. The position of the hydroxyl group is often indicated by a number in front of the longest parent chain. Two common alcohols and their common and systematic names are

$$\overset{3}{C}H_3\overset{2}{C}H_2\overset{1}{C}H_2OH \qquad \overset{3}{C}H_3\overset{2}{C}H\overset{1}{C}H_3$$
$$\qquad\qquad\qquad\qquad\qquad\;\; |$$
$$\qquad\qquad\qquad\qquad\qquad OH$$

| n-propyl alcohol | isobutyl alcohol |
| 1-propanol | 2-proponal |

The chemical behavior of alcohols depends on the nature of the carbon atom to which the hydroxyl group is attached. Alcohols are classified as primary, secondary, or tertiary depending on the number of carbons bonded to the carbon atom bearing the —OH group. If one carbon is bonded to this carbon atom, the alcohol is primary; if two carbons are bonded, it is secondary; if three carbons are bonded, it is tertiary. An example of each type of alcohol is shown in Figure 22-1.

Preparation of Alcohols

1. Hydration of alkenes. Alcohols are formed when alkenes react with water in the presence of hydronium ions that are furnished by the catalytic action of H_2SO_4 or H_3PO_4. The reaction of ethene that is used commercially to produce an alcohol is

$$\underset{H}{\overset{H}{\diagdown}}C=C\underset{H}{\overset{H}{\diagup}} \;+\; H—O—H \xrightarrow[400°C,\;70\;atm]{H_3PO_4} H—\underset{\underset{H}{|}}{\overset{\overset{H}{|}}{C}}—\underset{\underset{H}{|}}{\overset{\overset{H}{|}}{C}}—OH$$

This reaction produces C_2H_5OH, an alcohol known as *ethanol* or ethyl alcohol. Ethanol is commercially produced in operations such as the one shown in Figure 22-2.

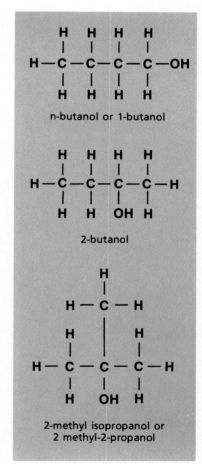

Figure 22-1 Alcohols may be classified as primary *(top)*, secondary *(middle)*, or tertiary *(bottom)*.

Figure 22-2 This milling complex in Iowa produces ethonol from corn for use in fuel.

The reaction between sodium and ethanol is slower than the reaction between sodium and water. The reaction rate of alcohols decreases as the number of carbons in the alcohol molecule increases.

The covalently bonded hydroxyl group (—OH) found in alcohols must not be confused with the ionically bonded hydroxide ion (OH⁻). The hydroxyl group is a neutral species. Unlike hydroxide ions, the hydroxyl group does not exist as a stable ion in water solution.

2. Hydrolysis of alkyl halides. Alkyl halides, which are covered in Section 22.2, react with water solutions of strong hydroxides to form the corresponding alcohols.

$$CH_3CH_2Cl + OH^- \rightarrow CH_3CH_2OH + Cl^-$$
ethyl chloride   ethanol

The halide ion (in this case, Cl^-) is displaced by the hydroxide ion.

Reactions of Alcohols

1. With sodium. Sodium reacts vigorously with ethanol, releasing hydrogen. A second product of the reaction is sodium ethoxide, C_2H_5ONa. This compound is recovered as a white solid after the excess ethanol is evaporated. This reaction is similar to the reaction of sodium with water.

$$2\ \text{H—C—C—OH} + 2\ \text{Na} \longrightarrow 2\ \text{H—C—C—ONa} + \text{H}_2$$

2. With HX. Alcohols react with concentrated water solutions of hydrogen halides, particularly hydrobromic acid, HBr, and hydriodic acid, HI, acid. These reactions produce alkyl halides. Sulfuric acid is used as a dehydrating agent to remove the water produced by the reaction. In the example shown below, ethanol reacts with hydrobromic acid to produce alkyl bromide.

$$\text{H—C—C—OH} + \text{HBr} \longrightarrow \text{H—C—C—Br} + \text{H}_2\text{O}$$

3. Dehydration. Depending on reaction conditions, ethanol dehydrated by hot, concentrated sulfuric acid yields either diethyl ether ($C_2H_5OC_2H_5$), which is covered in Section 22.3, or ethene (C_2H_4).

$$2\ \text{H—C—C—OH} \longrightarrow \text{H—C—C—O—C—C—H} + \text{H}_2\text{O}$$

$$\text{H—C—C—OH} \longrightarrow \text{C=C} + \text{H}_2\text{O}$$

4. Oxidation. Alcohols that have the hydroxyl group attached to the end carbon can be oxidized by hot copper(II) oxide. The product of such an oxidation is an aldehyde, RCHO.

$$\text{R—C—OH} + \text{CuO} \longrightarrow \text{R—C=O} + \text{H}_2\text{O} + \text{Cu}$$

Low-molecular-mass alcohols such as methanol are flammable and burn readily in air:

$$2 CH_3OH + 3 O_2 \rightarrow 2 CO_2 + 4 H_2O$$

5. Sulfation of long-chain alcohols. Under certain conditions, sulfur may be added to alcohols. This addition of sulfur is called **sulfation**. 1-dodecanol ($C_{12}H_{25}OH$) is commonly called lauryl alcohol. It is obtained from coconut oil by hydrogenation and partial decomposition. Lauryl alcohol may be *sulfated* by treatment with sulfuric acid and then neutralized with sodium hydroxide. This process yields sodium lauryl sulfate ($C_{12}H_{25}OSO_2ONa$), a very effective cleaning agent and an ingredient in many shampoos.

$$C_{12}H_{25}OH + H_2SO_4 \rightarrow C_{12}H_{25}OSO_2OH + H_2O$$

$$C_{12}H_{25}OSO_2OH + NaOH \rightarrow C_{12}H_{25}OSO_2ONa + H_2O$$

Properties, Preparation, and Uses of Alcohols

1. Methanol. Methanol (CH_3OH) is a colorless liquid with a rather pleasant odor. It has a low density and boils at 64°C. *It is very poisonous, even when used externally.* If taken internally in small quantities, it causes blindness by destroying the cells of the optic nerve. Larger quantities cause death. Methanol is a good fuel and burns with a hot, smokeless flame.

All alcohols burn, even the solid ones. However, the shorter chain alcohols C to C are much more volatile, and therefore their vapors ignite more readily.

All alcohols with one hydroxyl group are poisonous. Even breathing methanol for long periods of time can cause blindness and death. Ethanol is the least toxic of all the alcohols, but in large doses, even it can cause death.

Figure 22-3 Ethanol is produced commercially in large fermentation tanks *(left)*. Both methanol and ethanol can be added to unleaded gasoline to form gasohol, a fuel that is used chiefly in automobile and truck engines *(right)*.

Methanol is produced by the catalytic hydrogenation of carbon monoxide under pressure:

$$CO + 2 H_2 \rightarrow CH_3OH$$

Methanol is used as a solvent and is added to ethanol to make it unfit for drinking. It serves as a starting material for preparing other organic compounds such as formaldehyde and acetic acid.

2. Ethanol. Ethanol (C_2H_5OH) is a colorless liquid that has a characteristic odor and a sharp, biting taste. It boils at 78°C, freezes at −117°C, and burns with a nearly colorless flame. Ethanol is a good solvent for many organic compounds that are insoluble in water.

Methanol is often called wood alcohol; this is because it is made as a byproduct of the destructive distillation of wood.

Chemistry Notebook

Polyunsaturated Fats and Pour-on Margarines

We hear a lot these days about switching from saturated to polyunsaturated fats as part of a proper diet. Fats and oils are glycerides: glycerol molecules (see facing page) with one or more of the —OH groups replaced by a long-chain carboxylic acid (see Section 22.5). If these acids have only single bonds between the carbon atoms in the chain, they are considered saturated because they contain the maximum possible number of hydrogen atoms. If they have one carbon-carbon double bond, they are considered unsaturated. Polyunsaturated fats have carbon chains with more than one double bond. Polyunsaturated fats have low melting points because their chains bend at their double bonds and thus do not pack tightly together. Those that are

liquid at room temperature may be called oils.

Polyunsaturated fats react readily at the double-bond sites and so are digested by our bodies more easily than saturated fats. Saturated fats may tend to be deposited in our arteries, leading to increased risk of high blood pressure, heart attacks, and strokes. Animal products usually contain saturated fats, such as lard and butter, that

are solid at room temperature. Plant products generally contain unsaturated oils, such as corn oil and olive oil, that are liquid at room temperature. Changing from animal fats to vegetable oils may reduce health risks.

Margarine was first developed as a cheap replacement for butter, but new awareness of the dangers of saturated fats has caused increasing numbers of consumers to use margarines high in polyunsaturated oils, such as corn oil instead of butter. These polyunsaturated margarines and spreads are made by blending two or more fats or oils to produce a product of the desired consistency; some are even available in liquid forms. The bread and butter of yesterday may become tomorrow's bread and oil.

Ethanol is the alcohol in alcoholic beverages. It is a type of drug known as a depressant and slows down the activity of the central nervous system. Excessive use of alcohol over prolonged periods can damage the liver and cause a wide variety of other physical and psychological problems. In the United States, the production and distribution of ethanol is controlled by the federal government, and ethanol intended for use in alcoholic beverages if taxed. Ethanol that is manufactured for other uses is denatured, or made unfit to drink. Denatured alcohol is a mixture composed primarily of ethanol and other poisonous and nauseating materials that have been added to make it unfit for consumption.

The first alcoholic beverages produced by humans were fermented. The ethanol in modern alcoholic beverages is produced by fermentation or distillation. Fruits, grains, or vegetables can be used as a source of sugar.

Large quantities of ethanol are produced by hydrating ethene or by fermentation. The latter method occurs when yeast is added to a dilute solution of sugar at room temperature, or slightly warmer. In nature, ripe fruit often serves as the sugar source for fermentation. The yeast cells begin to grow and divide and, in the absence of oxygen, carry out fermentation. Yeast cells produce sucrase and zymase—two enzymes that act as catalysts in changing the sugar into ethyl alcohol and carbon dioxide.

$$C \text{ sucrose } C_{12}H_{22}O_{11} + H_2O \rightarrow 4C_2H_5OH + 4CO_2$$

Although both processes are used in producing industrial alcohol, the hydration of ethene is less expensive than fermentation.

3. Ethylene glycol. The compound ethylene glycol, $C_2H_4(OH)_2$, has the structural formula

Figure 22-4 Ethylene glycol makes an excellent automotive antifreeze.

Ethylene glycol is used extensively as a "permanent" antifreeze in automobile radiators. Its boiling point (198°C) is so much higher than that of water that it does not readily evaporate or boil away. Ethylene glycol is poisonous; care should be taken to ensure that any spilled ethylene glycol is not left standing, for it is lethal to pets.

4. Glycerol. Glycerol, or glycerin, $C_3H_5(OH)_3$, has the structural formula

Figure 22-5 Glycerol is an ingredient of these clear soaps.

It is a colorless, odorless, slow-flowing liquid with a sweet taste. It has a low vapor pressure and is hygroscopic (it uptakes water readily). It is used in making synthetic resins for paints and in cigarettes to keep the tobacco moist. It is also used in the manufacture of cellophane, nitroglycerin, and some soaps. Glycerol is an ingredient of cosmetics and drugs and is used in many foods and beverages.

Section Review

1. Write the structural formulas for all alcohols having the molecular formula C_4H_9OH. Classify each as primary, secondary, or tertiary, and give the systematic name for each.
2. Why do alcohols have higher boiling points than the corresponding hydrocarbons?
3. How are alcohols prepared?
4. Name a specific alcohol and describe its properties, preparation, and uses.

1. See examples of structural formulas and rules for systematic naming of alcohols given in this section.
2. The alcohol molecules are associated through the formation of hydrogen bonds.
3. Hydration of alkenes or hydrolysis of alkyl halides.
4. Accept answers as given in the subsection "Properties, Preparation, and Uses of Alcohol."

22.2 Halocarbons

SECTION OBJECTIVES

- Draw structural formulas of halocarbons and give their names.
- Describe methods of preparation of halocarbons.
- Write equations for the reactions of halocarbons.
- Give examples of some special halocarbons, their properties, preparation, and uses.

Halogens are discussed in Chapter 28.

Halocarbons *or* **alkyl halides** *are alkanes in which one or more halogen atoms—fluorine, chlorine, bromine, or iodine—are substituted for one or more hydrogen atoms in a hydrocarbon.* Because R— is often used to represent an alkyl group and —X any halogen, a halocarbon or alkyl halide may be represented by RX. The common name of a halocarbon is derived by adding the name of the halide—fluoride, chloride, bromide, or iodide—to the name of the alkyl group attached to the halide. The systematic name indicates the halogen by the prefixes *fluoro-, chloro-, bromo-,* and *iodo-*. The name consists of the name of the parent alkane preceded by the prefix of the halogen with a number to indicate its position. Two halides and their names are as follows:

$$\overset{1}{C}H_3\overset{2}{C}H_2\overset{3}{C}H_2Br$$

n-propyl bromide, or
1-bromo propane

$$\overset{1}{C}H_3\overset{2}{C}H\overset{3}{C}H_2\overset{4}{C}H\overset{5}{C}H_2\overset{6}{C}H_3$$
$$\quad\;\; | \qquad\quad |$$
$$\quad\;\; CH_3 \qquad Cl$$

4-chloro-2-methyl hexane

Halocarbons are rare in nature, but many have been synthesized in the laboratory. They are easily converted to many other classes of organic compounds. Many of the halocarbons are excellent solvents. Other halocarbon compounds have been used as anesthetics, insecticides, aerosol propellants, fire extinguishers, and refrigerants.

Preparation of Halocarbons

1. Direct halogenation. Recall from your reading in Section 21.3 that halogens react with alkanes to form substitution products. For example, under suitable conditions halogen atoms can be substituted for each of the four hydrogen atoms in methane. This reaction occurs in four stages:

$$CH_4 + X_2 \rightarrow CH_3X + HX$$
$$CH_3X + X_2 \rightarrow CH_2X_2 + HX$$
$$CH_2X_2 + X_2 \rightarrow CHX_3 + HX$$
$$CHX_3 + X_2 \rightarrow CX_4 + HX$$

2. From alkanes and alkynes. Recall from Section 21.3 that alkenes and alkynes react with halogens or hydrogen halides to form halocarbons.

3. From alcohols. You learned earlier in this chapter that reactions of alcohols with hydrogen halides yield the corresponding halocarbons.

$$ROH + HX \rightarrow RX + H_2O$$

Reactions of Halocarbons

Halocarbons react with many molecules and ions. In some of these reactions, the halogen atom is replaced by an atom of another element. For example, halocarbons react with water solutions of strong hydroxides to yield alcohols and the halide ion. The hydroxyl group from the hydroxide is substituted for the halogen atom in the halocarbon.

$$RX + OH^- \rightarrow ROH + X^-$$

Figure 22-6 Insecticides may contain chlorinated hydrocarbons. Here, insecticides are being used to control mosquitoes in Aswan, Egypt.

The Friedel-Crafts reaction [Section 21.3] is an alkyl halide substitution reaction. Here, the phenyl group (C_6H_5—) is substituted for the halogen atom in the alkyl halide

Properties, Preparation, and Uses of Several Halocarbons

The chemical commonly known as carbon tetrachloride (CCl_4) is **tetrachloromethane**—*a colorless, volatile, nonflammable liquid, and an excellent nonpolar solvent.*

Carbon tetrachloride is prepared commercially by the direct chlorination of methane:

$$CH_4 + 4Cl_2 \rightarrow CCl_4 + 4HCl$$

Carbon tetrachloride is sometimes used for dry-cleaning fabrics, removing grease from metals, as a spot remover for clothes, and extracting oils from seeds. Its vapors are poisonous. Its most important use is in the preparation of a family of polyhalogenated derivatives of methane and ethane known as the Freon halocarbons.

Dichlorodifluoromethane (CCl_2F_2) is the most important member of the Freon halocarbons. It is an odorless, nontoxic, nonflammable, easily liquefied gas that is prepared from carbon tetrachloride and hydrofluoric acid with antimony compounds as catalysts.

$$CCl_4 + 2HF \xrightarrow{\text{catalyst}} CCl_2F_2 + 2HCl$$

Because it has a low boiling point and is nontoxic and nonflammable, dichlorodifluoromethane is used as a refrigerant in mechanical refrigerators and air conditioners (Figure 22-7) and as a propellant in aerosol products. It has also been used as a bubble-making agent in the manufacture of plastic foams.

Once released into the air, Freon halocarbons rise into the stratosphere, where ultraviolet radiation from the sun decomposes them and releases free halogen atoms. The halogen atoms catalyze the destruction of the ozone in the ozone layer, which protects us from cancer-causing ultraviolet radiation. Consequently, the production of the Freon halocarbons has been limited, and their use as propellants in aerosol cans has been discontinued in the United States.

Tetrafluoroethene (C_2F_4) can be polymerized to form a product that has a structure in which the following structural unit occurs again and again, forming long chains:

Carbon tetrachloride is thought to be a cancer-causing agent; it is toxic if taken internally, absorbed by the skin, or inhaled.

CAUTION:
There must be good ventilation when carbon tetrachloride is used. Its use in high school laboratories should be avoided.

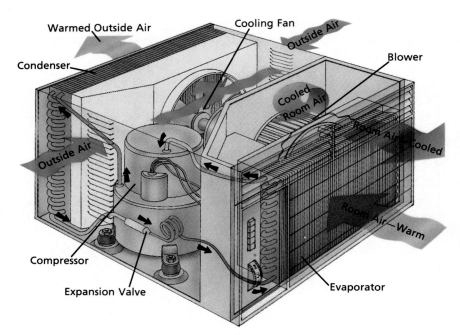

Figure 22-7 An air conditioner uses recirculating Freon gas. The gas exits the compressor under high pressure and condenses to a liquid, giving up heat to the outdoors. It then passes through an expansion valve and enters the evaporator, where it becomes a gas and gets extremely cold. A fan pulls warm room air across the cold evaporator coils and blows cool air back into the room.

useful where heat-resistant, nonlubricated moving parts are needed and in many items as a "nonsticking" surface.

Ethylene dichloride ($ClCH_2CH_2Cl$) also called 1,2-dichloroethane dichloride, is made by chlorinating ethene, as described in Section 21.3. This liquid is used as a solvent and for making adhesives. Most of it, however, is converted to vinyl chloride:

$$Cl-\overset{\overset{\displaystyle H}{|}}{\underset{\underset{\displaystyle H}{|}}{C}}-\overset{\overset{\displaystyle H}{|}}{\underset{\underset{\displaystyle H}{|}}{C}}-Cl \xrightarrow{\text{heat}} \quad \overset{\displaystyle H}{\underset{\displaystyle H}{}}C=C\overset{\displaystyle H}{\underset{\displaystyle Cl}{}} \quad + \quad HCl$$

Vinyl chloride is then polymerized to form polyvinyl chloride, which has the structural unit

$$-\overset{\overset{\displaystyle H}{|}}{\underset{\underset{\displaystyle H}{|}}{C}}-\overset{\overset{\displaystyle H}{|}}{\underset{\underset{\displaystyle Cl}{|}}{C}}-$$

Polyvinyl chloride has a molecular weight of about 1,500,000. It is used for plastic pipe, plastic siding for buildings, film, wire, and insulation.

Section Review

1. Write structural formulas for all halocarbons having the molecular formula C_4H_9Cl, and give the systematic name for each.
2. Write an equation for the preparation of each of the following halocarbons: *(a)* 2-chloropropane *(b)* dichloromethane.
3. Write an equation showing the preparation of ethyl alcohol from the reaction of a halocarbon with the water solution of a strong base.

1. See examples of structural formulas and rules for systematic naming of halocarbons given in this section.

2. *(a)*

$$CH_3\overset{\overset{\displaystyle H}{|}}{C}=\overset{\overset{\displaystyle H}{|}}{C}-H + HCl \rightarrow$$

$$CH_3\overset{\overset{\displaystyle H}{|}}{\underset{\underset{\displaystyle Cl}{|}}{C}}-C\overset{\displaystyle H}{\underset{\displaystyle H}{}}$$

(b) $CH_3Cl + Cl_2 \rightarrow CH_2Cl_2 + HCl$

3. $CH_3CH_2Cl + OH^- \rightarrow CH_3CH_2OH + Cl^-$

22.3 Ethers

Ethers *are a group of compounds in which two hydrocarbon groups are bonded to the same atom of oxygen.* They may be represented by the general formula ROR′, in which R and R′ indicate two alkyl groups. These alkyl groups may or may not be the same.

Common names are usually used when ethers are named. They are usually named by giving the name of each alkyl group first, then the word "ether." Some common ethers and their names are shown below.

$$CH_3OCH_2CH_3 \qquad CH_3CH_2OCH_2CH_3$$
$$\text{methyl ethyl ether} \qquad \text{diethyl ether or ethyl ether}$$

If both alkyl groups are the same, the prefix *di-* is usually used but the use of the prefix is not necessary.

Preparation of Ethers

1. Dehydration of alcohols. Ethers can be prepared by dehydration of alcohols as described in Section 22.1. For example, diethyl ether can be prepared by heating ethanol and sulfuric acid to 140°C.

$$2\ CH_3CH_2OH \xrightarrow{H_2SO_4} CH_3CH_2OCH_2CH_3 + H_2O$$

2. Williamson synthesis. In this method, the reactants are the sodium salt of an alcohol and a halocarbon.

$$RO^-Na^+ + R'X \rightarrow ROR' + Na^+X^-$$

This method is used to prepare mixed ethers. In a mixed ether, the R and R′ groups are different.

Reactions, Properties, and Uses of Ethers

Ethers are relatively unreactive. However, when ethers are treated with hot concentrated mineral acids (HI, HBr), they are cleaved:

$$CH_3OCH_2CH_3 + HI \rightarrow CH_3I + CH_3CH_2OH$$
$$CH_3CH_2OH + HI \rightarrow CH_3CH_2I + H_2O$$

The properties of ethers generally resemble those of the corresponding hydrocarbons. Like alcohols, ethers are extremely flammable. Ethers are only slightly soluble in water and are mainly used as solvents for many organic compounds. Diethyl ether slows down the operation of the central nervous system, hence its former widespread use as an anesthetic. One of the methyl butyl ethers, $(CH_3)_3COCH_3$, is used in gasoline to increase the octane rating.

Section Review

1. Draw structural formulas for each of the following ethers:
   (a) dimethyl ether   *(b)* ethyl n-propyl ether   *(c)* ethyl isopropyl ether.
2. Name each of the following ethers:
   (a) $CH_3CH_2CH_2OCH_2CH_2CH_3$ *(b)* $CH_3CH_2OCH_2CH_2CH_2CH_3$
3. Write an equation showing how dimethyl ether can be prepared from the dehydration of methanol.

1. *(a)* CH_3OCH_3
   (b) $CH_3CH_2OCH_2CH_2CH_3$

   (c) $CH_3CH_2OCHCH_3$
               |
              CH_3

2. *(a)* n-propyl ether or di-n-propyl ether
   (b) methyl n-butyl ether

3. $2CH_3OH \xrightarrow{H_2SO_4} CH_3OCH_3$
   $+ \quad H_2O$

22.4 Aldehydes and Ketones

Aldehydes and *ketones are substituted hydrocarbons characterized by the presence of the* **carbonyl group**.

$$\underset{/}{\overset{\backslash}{C}}{=}O$$

If one of the bonds of the carbonyl group is attached to a hydrogen atom, the compound is an **aldehyde**. *In* **aldehydes**, *the carbon atom in the carbonyl group is attached to at least one hydrogen atom. The functional group of an aldehyde is*

$$\underset{H}{\overset{\backslash}{\underset{/}{C}}}{=}O$$

known as the **formyl group**. The other bond of the carbonyl carbon may bond to either a hydrogen or an alkyl group. The general formula for an aldehyde is

$$\underset{H}{\overset{R}{\underset{/}{C}}}{=}O$$

When both bonds of the carbonyl group are attached to carbon atoms, the compounds are **ketones**. The general formula for ketones is

$$\underset{R'}{\overset{R}{\underset{/}{C}}}{=}O$$

The alkyl groups R and R′ may be the same or different. Because both aldehydes and ketones contain the carbonyl group, their properties are usually similar.

Simple aldehydes are referred to by common names. The systematic name is derived by substituting *-al* for the final *-e* in the parent hydrocarbon chain name. Some aldehydes and their common and systematic names are:

Figure 22-8 Acetone, an aldehyde, is the active ingredient in nail polish remover.

$H-C{\overset{\displaystyle O}{\underset{\displaystyle H}{\big\langle}}}$	$CH_3C{\overset{\displaystyle O}{\underset{\displaystyle H}{\big\langle}}}$	$CH_3CH_2CH_2C{\overset{\displaystyle O}{\underset{\displaystyle H}{\big\langle}}}$
formaldehyde	acetaldehyde	butyraldehyde
or	or	or
methanal	ethanal	butanal

The simpler ketones are usually referred to by common names. The names are derived by naming the alkyl groups bonded to the carboxyl group and adding the word *ketone*. Systematic names are given usually for very complex ketones. The suffix-*one* is substituted for the final *-e* in the name of the parent hydrocarbon chain. Three ketones and their names are:

$$\begin{array}{ccc}
\overset{\overset{\displaystyle O}{\|}}{CH_3- C -CH_3} & \overset{\overset{\displaystyle O}{\|}}{CH_3- C -CH_2CH_3} & \overset{\overset{\displaystyle O}{\|}}{CH_3CH_2- C -CH_2CH}
\end{array}$$

acetone   methyl ethyl ketone   diethyl ketone
or
dimethyl ketone

Many natural substances are aldehydes and ketones. Some of these include flavors such as cinnamon, vanilla, almond, jasmine, clove oil, and the scents of many perfumes. Some of the sex hormones and some vitamins are aldehydes and ketones.

Figure 22-9 Vanilla beans *(left)* and lemons *(right)* each owe their characteristic flavors to a specific aldehyde.

Preparation and Uses of Aldehydes

The general method of preparing aldehydes, the oxidation of alcohols, was described in Section 22.1. A common process of this type involves passing methanol vapor and a regulated amount of air over heated copper. Formaldehyde (HCHO) is produced.

$$2Cu + O_2 \rightarrow 2CuO$$
$$CH_3OH + CuO \rightarrow HCHO + H_2O + Cu$$

The commercial preparation of formaldehyde requires the catalytic action of silver or an iron-molybdenum oxide. At room temperature, formaldehyde is a gas with a pungent, distinctive odor. Dissolved in water, it makes an excellent disinfectant. It is also used to preserve biological or medical specimens. The largest uses of formaldehyde are in making certain types of adhesives for plywood and particleboard and producing resins for plastics.

Preparation And Uses Of Ketones

Ketones can be prepared from alcohols that do not have the hydroxyl group attached to an end-carbon atom. For example, acetone (CH_3COCH_3) is prepared by the mild oxidation of 2-propanol ($CH_3CHOHCH_3$) with potassium dichromate ($K_2Cr_2O_7$) in water solution as the oxidizing agent.

The molecular structures of some of the sex hormones are discussed in Section 23.4.

Figure 22-10 The pungent odor of formaldehyde is familiar to those who have dissected biological specimens.

$$\underset{\substack{|\ \ \ \ |\ \ \ |\\ H\ \ \ H\ \ H}}{H-\overset{\displaystyle H}{\underset{\displaystyle H}{C}}-\overset{\displaystyle OH}{\underset{}{C}}-\overset{\displaystyle H}{\underset{}{C}}-H} + \text{O (from oxidizing agent)} \longrightarrow \underset{\substack{|\ \ \ \ \ \ \ \ |\\ H\ \ \ \ \ \ \ H}}{H-\overset{\displaystyle H}{\underset{}{C}}-\overset{\displaystyle O}{\underset{}{C}}-\overset{\displaystyle H}{\underset{}{C}}-H} + H_2O$$

Acetone is a colorless, volatile liquid that is widely used as a solvent.

Section Review

1. Identify each of the following compounds as an aldehyde or a ketone, and give the name for each:

(a) $CH_3CH_2CH_2C\overset{\displaystyle O}{\underset{\displaystyle H}{\diagup}}$

(b) $CH_3\underset{\displaystyle CH_3}{\overset{}{CH}}\overset{\displaystyle O}{\overset{\|}{C}}-CH_2CH_3$

2. Write an equation that represents a method of preparing formaldehyde.
3. Describe the uses of formaldehyde and acetone.

22.5 Carboxylic Acids

Carboxylic acids contain the *carboxyl group.*

$$\overset{\displaystyle O}{\overset{\|}{-C}}-OH$$

Carboxylic *or* **organic acids** are *hydrocarbons in which two of the hydrogen atoms on an end carbon atom have been replaced by a doubly bonded oxygen, and the third hydrogen on the end carbon atom has been replaced by a hydroxyl group,* —OH. The general formula for a carboxylic acid can be written as

$$R-\overset{\displaystyle O}{\overset{\|}{C}}-OH$$

The R could be an H atom, designating the simplest carboxylic acid

$$H-\overset{\displaystyle O}{\overset{\|}{C}}-OH$$

The carboxyl group can also be written —COOH.

known as formic acid, or it could be an alkyl group. Carboxylic acids are highly polar, and their molecules form stable hydrogen bonds. As a result, organic acids have higher boiling points than alcohols of corresponding molecular masses.

Because organic acids were known before a systematic nomenclature was developed, many of them are known by common names derived from a Greek or Latin name indicating the sources from which they were originally obtained. For example, formic acid was originally obtained from the distillation of ants and hence derives its name from the Latin word for ant, *formica*. The systematic system for naming organic acids uses the suffix *-oic* in place of the final *-e* of the alkane name and adds the word *acid*. Hence, acetic acid (CH_3COOH) is given the systematic name ethanoic acid. Butyric acid ($CH_3CH_2CH_2COOH$) is assigned the systematic name, butanoic acid.

Preparation of Carboxylic Acids

Carboxylic acids are most commonly prepared by the oxidation of primary alcohols, or by the oxidation of aldehydes.

$$RCH_2OH \xrightarrow{\text{oxidation}} R-\overset{\overset{\displaystyle O}{\|}}{C}-OH \quad R-\overset{\overset{\displaystyle O}{\|}}{C}-H \xrightarrow{\text{oxidation}} R-\overset{\overset{\displaystyle O}{\|}}{C}-OH$$

When oxidized, primary alcohols and aldehydes each produce acids with the same number of carbon atoms per molecule.

Commercially, carboxylic acids can be prepared from the reaction of CO and the alcohol possessing one fewer carbon atoms than the desired acid. The reaction is carried out under pressure and in the presence of dicobalt octacarbonyl, acting as a catalyst.

$$R-OH + CO \xrightarrow{CO_2(CO)_8} R-\overset{\overset{\displaystyle O}{\|}}{C}-OH$$

$$CH_3CH_2OH + CO \xrightarrow{CO_2(CO)_8} CH_3CH_2COOH$$

Reactions of Carboxylic Acids

1. Ionization. This reaction involves the hydrogen atom bonded to an oxygen atom in the carboxyl group. It is this hydrogen atom that ionizes in water solution, giving carboxylic acids their acid properties. The hydrogen atoms bonded to carbon atoms in these acids *never* ionize in water solution.

$$HCOOH + H_2O \rightleftarrows H_3O^+ + HCOO^-$$
$$CH_3COOH + H_2O \rightleftarrows H_3O^+ + CH_3COO^-$$

These equilibria yield low H_3O^+ ion concentrations. Therefore, carboxylic acids are generally weak acids.

2. Neutralization. An organic acid may be neutralized by a hydroxide. A salt is formed in a reaction similar to that undergone by inorganic acids.

$$CH_3COOH + NaOH \rightarrow CH_3COONa + H_2O$$

3. Esterification. In this reaction, an acid reacts with an alcohol to produce a substance known as an ester. We will study this reaction in Section 22.6.

Preparation, Properties, and Uses of Carboxylic Acids

1. Acetic acid. Acetic acid, CH_3COOH, can be produced by the catalytic oxidation of acetaldehyde. Very pure acetic acid can be prepared by reacting methanol with carbon monoxide at 180°C and 30–40 atm pressure in the presence of a catalyst.

Figure 22-11 Nylon, shown here being wound onto a stirring rod, is produced from the polymerization of adipic acid (a carboxylic acid) and hexamethylene-diamine.

$$CH_3OH + CO \xrightarrow{\text{catalyst}} CH_3COOH$$

Concentrated acetic acid is a colorless liquid that is a good solvent for some organic chemicals. It is used for making cellulose acetate, a basic material of many fibers and films.

Cider vinegar is made from apple cider that has fermented to "hard cider." The ethanol in hard cider is slowly oxidized in the presence of the oxygen in the air. This oxidation produces acetic acid. The reaction is catalyzed by enzymes produced by certain species of bacteria.

$$C_2H_5OH + O_2 \rightarrow CH_3COOH + H_2O$$

Pure acetic acid has a penetrating odor and produces burns if in contact with the skin. Acetic acid is the most commercially important organic acid. It is used extensively in the dyeing of textiles, and in the production of cellulose acetate, a major component of synthetic fibers. It is also an excellent solvent for organic compounds.

2. Formic acid. Formic acid, HCOOH, is prepared from sodium hydroxide solution and carbon monoxide under pressure. This reaction yields sodium formate (HCOONa).

$$NaOH + CO \rightarrow HCOONa$$

If sodium formate is carefully heated with sulfuric acid, formic acid distills off:

$$HCOONa + H_2SO_4 \rightarrow HCOOH + NaHSO_4$$

Formic acid is found in nature in stinging nettle plants and in certain ants. It is used as an acidifying agent in the textile industry.

3. Oxalic acid. Oxalic acid, $(COOH)_2$, is the first member of the dicarboxylic acid series. **Dicarboxylic acids** *are acids that have two carboxyl groups; they can be said to be dibasic.* Oxalic acid is prepared by heating sodium formate, and then acidifying sodium oxylate, the product of the reaction shown below.

$$2\ H-\overset{\overset{\displaystyle O}{\|}}{C}-O^-Na^+ \xrightarrow{\text{NaOH}} Na^+O^--\overset{\overset{\displaystyle O}{\|}}{C}-\overset{\overset{\displaystyle O}{\|}}{C}-O^-Na^+ + H_2$$

Oxalic acid occurs in plants such as rhubarb and sorrel (wildflowers in the genus *Oxalis.* It is also found in urine and guano. Oxalic acid is used commercially for dyeing, calico printing, and in laundries for the removal of stains and ink spots.

Section Review

1. Write the structural formula for each of the following compounds:
   (a) pentanoic acid   *(b)* butyric acid   *(c)* isobutyric acid
   (d) 2-methyl butanoic acid.
2. Name each of the following compounds:

(a) $CH_3CH_2CH_2CH_2CH_2CH_2C\begin{smallmatrix}\diagup O \\ \diagdown OH\end{smallmatrix}$

(b) $CH_3\underset{\underset{\textstyle CH_3}{|}}{CH}CH_2C\begin{smallmatrix}\diagup O \\ \diagdown OH\end{smallmatrix}$

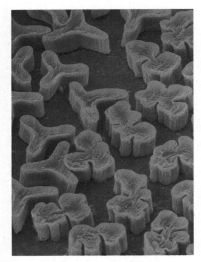

Figure 22-12 Acetic acid is used to convert some of the hydroxyl (–OH) groups of cellulose into ester groups, resulting in cellulose acetate fibers, shown here in cross section.

1. *(a)*

$$CH_3CH_2CH_2CH_2\overset{\overset{\displaystyle O}{\|}}{C}-OH$$

(b)

$$CH_3CH_2CH_2\overset{\overset{\displaystyle O}{\|}}{C}-OH$$

(c)

$$CH_3\underset{\underset{\textstyle CH_3}{|}}{CH}\overset{\overset{\displaystyle O}{\|}}{C}-OH$$

Guano is manure produced by sea birds or bats. It is often used as fertilizer.

(d)

$$CH_3CH_2\underset{\underset{\textstyle OH_2}{|}}{CH}\overset{\overset{\displaystyle O}{\|}}{C}-OH$$

2. *(a)* heptanoic acid
   (b) 3-methylbutanoic acid
   (c) 2-chlorobutanoic acid
   (d) propanoic acid

(c) $CH_3CH_2CHCOOH$
         |
         Cl

(d) CH_3CH_2COOH

3. Why do carboxylic acids have higher boiling points than alcohol with corresponding molecular masses?
4. Show the ionization reaction of acetic acid.

3. Carboxylic acids are highly polar and their molecules form stable hydrogen bonds. As a result, they have higher boiling points than alcohols of corresponding molecular masses.
4. $CH_3COOH + H_2O \leftrightarrow$
$CH_3COO^- + H_3O^+$

22.6 Esters

Esters are similar to carboxylic acids, but have an alkyl group in place of one of the hydrogen atoms of the carboxyl group.

An **ester** *may be represented by the general formula* $RC\overset{O}{\overset{\|}{—}}OR'$ *in which R' is always an alkyl group and R can be either a hydrogen atom or an alkyl group.* Although the molecules of an ester are polar, unlike those of acids, they do not form hydrogen bonds with one another. As a result, esters have much lower boiling points than acids with corresponding molecular masses. Ester molecules are able to form hydrogen bonds with water molecules and are somewhat soluble in water. Esters are excellent solvents for organic compounds. Many esters occur naturally and are responsible for the distinctive odors and flavors of fruits and flowers. Esters are used in perfumes and artificial flavoring.

Esters are easy to name. The name of the alkyl group (designated R' in the general formula above) is given first and then is followed by the name of the acid that is represented by the carboxylic group before the replacement of the hydrogen atom by the alkyl group. The second part of the ester name is obtained by replacing the *-ic* of the acid name by *-ate*. The structural formulas and names of some common esters are as follows:

$$H—\overset{O}{\overset{\|}{C}}—OCH_2CH_3$$

ethyl methonate
or
ethyl formate

$$CH_3\overset{O}{\overset{\|}{C}}—OCH_2CH_3$$

ethyl ethanoate
or
ethyl acetate

$$CH_3CH_2CH_2\overset{O}{\overset{\|}{C}}—OCH_2CH_3$$

ethyl-n-butanoate

SECTION OBJECTIVES

- Draw structural formulas of esters and give their names.
- Describe properties of esters.
- Describe esterification
- Describe reactions of esters and define saponification.

Figure 22-13 The different flavors and aromas of pineapple, banana, orange, and apple are a result of different esters.

Preparation of Esters

An ester is produced when an organic acid reacts with an alcohol in a reaction called an **esterification reaction.** The reaction usually takes place in the presence of an inorganic acid, which acts as a catalyst, and can be represented by the general equation.

$$R—\overset{O}{\overset{\|}{C}}—OH + R'OH \underset{}{\overset{H_2SO_4}{\rightleftarrows}} R—\overset{O}{\overset{\|}{C}}—OR' + H_2O$$

As an example, ethyl acetate is the ester formed when acetic acid and ethanol react as shown below.

$$CH_3COOH + C_2H_5OH \xrightarrow{H_2SO_4} CH_3COOC_2H_5 + H_2O$$

Reactions of Esters

Esters are neutral compounds and, as such, they do not exhibit acidic or basic properties. They undergo reactions in which the —OR' group is replaced by another group. Esters can be decomposed by hydrolysis into the alcohol and acid from which they were derived. This hydrolysis may occur in the presence of dilute acid or metallic hydroxide solutions.

1. Acidic hydrolysis. When the hydrolysis of an ester is catalyzed by an acid such as HC1, esters yield a carboxylic acid and an alcohol.

$$CH_3H_2\overset{\displaystyle O}{\overset{\displaystyle \|}{C}}\!-\!OCH_2CH_2 + H_2O \xrightarrow[\text{catalyst}]{H^+} CH_3\overset{\displaystyle O}{\overset{\displaystyle \|}{C}}\!-\!OH + CH_3CH_2OH$$
$$\text{acetic acid} \qquad \text{ethanol}$$

This reaction is the reverse of esterification. In reality, the reverse reaction takes place to an appreciable extent, and the reaction shown above does not go to completion.

2. Alkaline hydrolysis. When the hydrolysis of an ester is carried out in basic solution, the ester is converted to its alcohol and a salt of its acid.

$$R\!-\!\overset{\displaystyle O}{\overset{\displaystyle \|}{C}}\!-\!OR' + NaOH \longrightarrow R\!-\!\overset{\displaystyle O}{\overset{\displaystyle \|}{C}}\!-\!ONa + R'OH$$

Fats *and* **oils** *from plants or animals are esters of glycerol and long carbon chain acids.* The carbon chains of the acids usually contain from 12 to 20 carbon atoms. The structure of a fat or oil can be represented as

$$\begin{array}{l} RCOOCH_2 \\ | \\ R'COOCH \\ | \\ R''COOCH_2 \end{array}$$

R, R', and R″ are saturated or unsaturated long-chain-hydrocarbon groups.

The only difference between a fat and an oil is the physical state of each at room temperature. Oils are liquids at room temperature; fats are solids. Oils usually contain more unsaturated hydrocarbon chains than fats. This is because long-carbon-chain acids with double bonds produce esters that have lower melting points and thus are liquid at room temperature. When natural fats are treated with sodium hydroxide, they are converted to glycerol and the sodium salt of the fatty acids. This process results in soap, the sodium salt of the fatty acids. *The hydrolysis of fats or oils using a solution of a strong hydroxide is called* **saponification.** The term *saponification* was originally used to describe the alkaline hydrolysis of fats or oils in the manufacture of soaps.

Figure 22-14 Polyesters are polymers containing many ester linkages. When a carboxylic acid with two carboxyl groups reacts with an alcohol with two hydroxyl groups, esterification may occur at both ends of each molecule, resulting in polyester.

The alkaline hydrolysis of fatty acid esters is represented by the general equation

RCOOCH$_2$
|
R'COOCH + 3NaOH → RCOONa + R'COONa + R"COONa + C$_3$H$_5$(OH)$_3$
|
R"COOCH$_2$

Soaps are generally made by hydrolyzing fats and oils with water heated to about 250°C in a closed container at high pressure. The long-chain carboxylic acids thus produced are neutralized with sodium hydroxide. This neutralization yields a mixture of the sodium salts that we recognize as soap.

Section Review

1. Give the structural formulas of the following compounds. (a) methyl acetate (b) ethyl methanoate (c) isopropyl butanoate (d) n-propyl acetate.
2. Give an example of an esterification reaction and write the equation for it.
3. Explain why esters have lower boiling points than acids with corresponding molecular masses.
4. Define *saponification* and write an equation to represent the reaction.

Chapter Summary

- Substituted hydrocarbons are hydrocarbons in which an atom or a group of atoms substitute for one or more of the hydrogen atoms.
- Alcohols are alkanes in which one or more hydroxyl groups have been substituted for an equal number of hydrogen atoms. The general formula is ROH.
- Alcohols are prepared by the hydration of alkenes and from alkyl halides by substitution. They react with sodium and with hydrogen halides and also undergo dehydration, oxidation, and sulfation, reactions.
- A halocarbon is an alkane in which a halogen atom is substituted for a hydrogen atom. The general formula for a halocarbon is RX.
- Halocarbons are prepared by the reaction of halogens with alkanes, alkenes, or alkynes and by the reaction of hydrogen halides with alkenes, alkynes, and alcohols. Many reactions of halocarbons are substitution reactions.
- Ethers are organic oxides with the general formula ROR'. They can be prepared by dehydrating alcohols.
- Aldehydes have the general formula RCHO. They are prepared by the controlled oxidation of alcohols having a hydroxyl group on a carbon at the end of the molecule.
- Ketones have the general formula RCOR'. They are prepared by oxidizing alcohols that do not have the hydroxyl group attached to the carbon at the end of the molecule.
- The general formula for carboxylic acids is RCOOH. They can be prepared by the oxidation of alcohols and aldehydes.
- Esters give fruits their characteristic flavors and odors. Their general formula is RCOOR'. Saponification is the hydrolysis of a fat using a solution of strong hydroxide and yields a mixture of sodium salts that makes up soap.

1. (a)

(b)

(c)

(d)

2. Accept all reactions between an alcohol and an organic acid that give the appropriate ester according to the general equation

3. Ester molecules do not form hydrogen bonding with one another, unlike organic acids. As a result, esters have lower boiling points than the acids with corresponding molecular masses.

4. Hydrolysis of fats and oils using a solution of a strong hydroxide. Accept any reaction that is represented by the general equation

Chapter 22 *Review*

Substituted Hydrocarbons

Vocabulary

alcohol	esterification
aldehyde	ether
alkyl halide	fat
carbonyl group	formyl group
carboxyl group	halocarbon
carboxylic acid	ketone
denatured alcohol	oil
dicarboxylic acid	saponification
ester	sulfation

Questions

1. Write the general formula for a(n) *(a)* alkyl halide *(b)* alcohol *(c)* ether *(d)* aldehyde *(e)* ketone *(f)* carboxylic acid *(g)* ester *(h)* fat.
2. Define *(a)* sulfation *(b)* esterification *(c)* saponification.
3. *(a)* What is an important physical property of carbon tetrachloride? *(b)* What uses of carbon tetrachloride depend on this property?
4. *(a)* What precautions must be taken when using carbon tetrachloride? *(b)* Why?
5. Draw a structural formula showing the portion of a polymer formed from three molecules of vinyl chloride.
6. Alcohols and metallic hydroxides both contain —OH groups. Why are their chemical properties so different?
7. How does methanol affect the body?
8. What is denatured alcohol?
9. Why is glycerol an ingredient in many moisturizing skin lotions?
10. Write the balanced formula equation for the reaction of sodium with *(a)* water *(b)* methanol.
11. How many molecules of oxygen are required for the complete combustion of one molecule of 2-propanol, $CH_3CHOHCH_3$? This compound is commonly called isopropyl alcohol.
12. Give two uses for formaldehyde.
13. Name some uses for acetone.
14. How is vinegar made from apple cider?
    (a) For carboxylic acids, define ionization, neutralization, and esterification. *(b)* Write a general equation for each type of reaction for a carboxylic acid, RCOOH.
15. *(a)* Write equations for the step-by-step complete ionization of oxalic acid. *(b)* How many moles of sodium hydroxide are required for the complete neutralization of 3 mol of oxalic acid?
16. List the types of reactions given in this chapter in which sulfuric acid is used as a dehydrating agent.
17. *(a)* How are fats and oils alike? *(b)* How do they differ? *(c)* Why do they differ?

Problems

1. Calculate the formula mass of trichlorofluoromethane.
2. A sodium hydroxide solution (45.0 mL) exactly neutralizes 35.0 mL of 0.150-*N* formic acid solution. What is the normality of the sodium hydroxide solution?
3. Calculate the number of grams of glycerol that must be dissolved in 0.250 kg of water in order to prepare a 0.400-*m* solution.
4. Calculate the freezing point and boiling point of the water of Problem 3.
5. How many grams of diprotic oxalic acid, $(COOH)_2$, are required to prepare 1.75 L of 0.200-*N* acid?

Application Questions

1. Draw structural formulas for *(a)* trichloromethane (chloroform) *(b)* 1,3-dihydroxypropane *(c)* diethyl ether *(d)* acetaldehyde *(e)* methylethylketone *(f)* acetic acid *(g)* ethyl formate.
2. Write an equation for the preparation of ethyl bromide starting with *(a)* ethane *(b)* ethene *(c)* ethanol.

3. Describe the types of alkyl halide substitution reactions given in this chapter.
4. Starting with methane, chlorine, and hydrogen fluoride, write equations for the preparation of dichlorodifluoromethane.
5. How do Freon halocarbons affect the stratosphere?
6. Using structural formula equations, show how ethanol may be prepared from (a) ethene (b) ethyl chloride (c) sugar ($C_{12}H_{22}O_{11}$) (d) methanol.
7. On the basis of formula mass and boiling point, compare the advantages of methane, ethanol, and ethylene glycol as antifreezes.
8. Why is a 50%–50% mixture of ethylene glycol and water better than water alone as a radiator fluid for extremely hot summer driving conditions?
9. Write an equation for the preparation of propyl iodide from propyl alcohol.
10. Write an equation for the preparation of propyl ether from propyl alcohol.
11. (a) Write an equation for the mild oxidation of 2-butanol, $CH_3CHOHCH_2CH_3$. (b) Name the product formed.
12. Write a formula equation showing the preparation of acetic acid from (a) methanol; (b) ethanol.
13. (a) How do the two carbon–oxygen bonds in formic acid compare? (b) How do the two carbon–oxygen bonds in sodium formate compare? (c) Account for the difference.
14. The ester n-butyl acetate is to be prepared from n-butyl alcohol, C_4H_9OH, and acetic acid. Write the equation for the preparation.
15. Write a balanced formula equation for the saponification with NaOH of a fat having the formula $(C_{17}H_{35}COO)_3C_3H_5$.

Application Problems

1. What is the percentage composition (to two decimal places) of $C_{12}H_{25}OSO_2ONa$, sodium lauryl sulfate?
2. A compound is found to contain 54.5% carbon, 9.1% hydrogen, and 36.4% oxygen. (a) Determine the empirical formula. (b) The compound is soluble in ethanol. If 11.0 g of the compound dissolved in 250 g of ethanol boils at 78.87°C, what is the molecular weight of the compound?

(c) What is the molecular formula?
3. The hydronium ion concentration in 0.05M acetic acid is 9.4×10^{-4} mol/L. What is the pH of this solution.
4. What volume of ethanol must be diluted with water to prepare 500 mL of 0.750M solution? The density of ethanol is given as 0.789 g/mL.
5. The density of ethylene glycol is 1.432 g/mL. (a) Calculate the theoretical freezing point of the water in a 50% (by volume) solution of ethylene glycol. (b) The actual freezing point of such a solution is about −37°C. Account for any difference from your calculation.

Enrichment

1. Consult current science magazines to learn the extent to which the concentration of ozone in the upper atmosphere has changed recently, including the "ozone hole." Learn how scientists explain these changes.
2. Examine some newspaper or popular magazine stories about the ozone hole. In your opinion, are the descriptions fair and accurate?
3. Find out the meaning of "PCBs," the structural formula of a PCB, and the properties that made PCBs useful in electrical equipment such as transformers. Then find out why the use of PCBs in transformers is now banned.
4. Explore the reasons for and against the addition of methyl or ethyl alcohol to gasoline ("gasohol"). Explain why you would support or oppose the use of alcohol in gasoline.
5. Consult reference materials at the library and gather information about the different methods that various human cultures have used to synthesize ethanol.
6. Find out how commercial breweries and distilleries manufacture alcohol.
7. Consult reference materials at the library and learn about the dangers that formaldehyde poses to human health.
8. Describe the discovery and uses of DDT. Learn why DDT has been banned in the United States.
9. Learn about the effects of dioxin on human health. Describe the link between dioxin and "Agent Orange," the herbicide used by the United States during our participation in the Vietnam War.

Chapter 23   Biochemistry

Chapter Planner

23.1 The Chemistry of Life
Inorganic Molecules and Water
- ▨ Questions: 1–4

Macromolecules
- ▨ Questions: 9, 10
- ▨ Application Problem: 8

23.2 Proteins
Amino Acids
- ▨ Experiment: 41
- ▨ Application Question: 10

Protein Structure
- ▨ Questions: 11, 12a
- Application Question: 5
- ▨ Question: 5
- Application Problem: 3

Enzymes
- ▨ Demonstration: 1
- Question: 13
- Problem: 2
- Application Question: 2
- ▨ Application Question: 8
- Application Problem: 4

23.3 Carbohydrates
Simple Sugars
- ▨ Question: 8
- ▨ Application Questions: 6, 9

Disaccharides and Polysaccharides
- ▨ Question: 12b
- ▨ Demonstration: 2
- Desktop Investigation
- Application Questions: 7, 11–13

23.4 Lipids
- ▨ Problem: 4
- Application Questions: 1, 4

Fatty Acids
Waxes
- ▨ Question: 6

Fats and Oils
Phospholipids
Steroids

23.5 Nucleic Acids
- ▨ Question: 12c

Nucleotides
- ▨ Questions: 7, 15

The Genetic Code
- ▨ Problems: 1, 3

DNA Replication
- ▨ Question: 14

RNA Synthesis
Protein Synthesis
Genetic Engineering
- ▨ Technology

23.6 Biochemical Pathways
- ▨ Demonstration: 3

Glycolysis
- ▨ Application Problem: 2

Teaching Strategies

Introduction

This chapter is less analytical than most of the other chapters in this book. Many of your students may be excited by the topics covered. However, some of your students, especially those who are quantitatively oriented, may have difficulty with the volume of conceptual detail and memorization. Remind them that some of the most interesting research problems of the coming decade will be in the area of biochemistry, particularly in molecular genetics, membrane structure, and protein structure.

The first step in teaching this chapter is classifying the four types of macromolecules that are essential to life: proteins, carbohydrates, lipids, and nucleic acids. Since the structure of these molecules is so important to their function, it would be helpful to use molecular models and computer modeling displays, so that students can develop visual images of their complex geometries.

23.1 The Chemistry of Life

The evolution of life from small components is a fascinating story. You may wish to assign library research projects on this topic. Students might not be aware of the fact that the earth's atmosphere has changed from a reducing atmosphere to an oxidizing atmosphere. The precursors of proteins were

smaller and less complex molecules that contained carbon and nitrogen. Based on the Urey-Miller experiments, in which an attempt was made to simulate earth's early atmosphere, it was thought that the small protein precursors were amino acids. There is now a growing body of evidence, however, that the immediate precursors of proteins were cyano-based compounds, molecules consisting of highly unsaturated chains of carbon and nitrogen that form polypeptides when they come in contact with water. Interestingly, space explorations searching for signs of extraterrestrial life have detected molecules that are identical to the proposed polycyanides.

Whatever those prehistoric molecules were, it is an indisputable fact that water played a large role in the evolution of life and it continues to play an essential role in its maintenance. The chemistry of life occurs in essentially aqueous solutions.

The chemical reactions in our bodies take place in intra- and extra-cellular fluids that are very much like sea water. Water is also a participant in the condensation and hydrolysis reactions that are important in every metabolic pathway.

23.2 Proteins

Proteins are found almost everywhere in the body and have a broad range of functions—hormonal, enzymatic, and structural (see Table 23-1 in the student text). A good way to understand proteins is to know their composition and structure. Proteins have primary, secondary, tertiary, and quarternary levels of structure. Their functionality is determined by their structure, which can be altered by changes in the pH, temperature, and salinity of the protein's environment.

The basic building blocks of proteins are amino acids. There are at least 150 naturally occurring amino acids. Only 20 of these, however, are used to build genetically coded proteins. Although the relative number of specific amino acids as well as their sequence in the protein chain differs from protein to protein, proteins in all living systems are composed of the same 20 amino acids.

Demonstration 1: Bromolain: A Proteolytic Enzyme

PURPOSE: To demonstrate the proteolytic activity of an enzyme and to investigate the factors that will destroy that activity
MATERIALS: 3 packages of gelatin, fresh pineapple, four 1000-mL beakers
PROCEDURE: (1) Cut the pineapple in small pieces—approximately 1 cm x 1 cm in size. (2) Divide the class into three groups. One group (the control group) will not add anything to the gelatin. One group will add fresh pineapple to the gelatin. One group will add boiled pineapple to the gelatin. (3) Prepare the gelatin according to the instructions on the package.

The gelatin prepared without pineapple will gel normally. The gelatin with boiled pineapple will also gel. The gelatin prepared with fresh pineapple, however, will not gel.

Bromolain is a mucoproteolytic agent found in fresh pineapple. It takes its name from the Bromeliad family, of which the pineapple is a member. Gelatin is a protein that absorbs hot water and forms a gel as the water cools. Bromolain can break down gelatin, as well as many other proteins. Boiling the pineapple before addition to the gelatin denatures the proteolytic enzyme, preventing it from interfering with the properties of the gelatin.

Canned pineapple may be substituted for boiled pineapple in this demonstration because the canning process subjects the pineapple to heat and pressure.

23.3 Carbohydrates

Students may have heard the terms sucrose, glucose, and fructose, but may be confused about their role in the diet. Overhead transparencies are useful in distinguishing the disacharride sucrose from the monosaccharides glucose and fructose.

Demonstration 2: Lugol's Test for Starch

PURPOSE: To detect the presence of starch using a simple chemical test
MATERIALS: five test tubes, Lugol's solution, 1% glucose solution, 1% fructose solution, 1% sucrose solution, 1% starch solution, distilled water
PROCEDURE: (1) Label five test tubes as follows: glucose, fructose, sucrose, starch, and water. (2) Place 3 mL of each solution in the corresponding test tube. (3) Add 3 mL of Lugol's solution to each test tube.

Lugol's solution contains I_2, which stains starch dark blue. The contents of the test tube containing the starch will turn dark blue immediately after addition of the Lugol's solution.

When the red-brown Lugol's solution is added to the other test tubes, no color change will be detected.

You may want to ask students to bring foods to class to test for the presence of starch. A positive test will be found in foods such as potatoes, crackers, bread, and cake. Some students may be surprised to find that sodas and candies do not contain starch.

Desktop Investigation: Enzyme Action on Starch

Nine grams of cornstarch is equal to one tablespoon. It is very important that students use a fresh, unused splint with each tube to avoid contamination.

DATA TABLE		
STARCH + SALIVA AND WATER	STARCH + BOILED SALIVA AND WATER	STARCH + VINEGAR AND WATER
Starch changes from semisolid to liquid	No change	No change

ANSWERS TO DISCUSSION QUESTIONS

1. The amylase changed the starch into a simpler form: the sugar maltose, which is soluble in water.

$$\text{starch} \xrightarrow{\text{amylase}} \text{maltose}$$

2. The heat and vinegar inactivated the amylase so that it had no effect on the starch. High temperatures (55°C or higher) and extremes in pH cause proteins to undergo total denaturation. Since enzymes (including amylase) are proteins, the heat and pH changes denatured the amylase. The reaction cannot be catalyzed and thus does not take place.

3. Since 1mL of vinegar was added to one test tube containing saliva, 1 mL of water must be added to the other two so they will have equal volumes of liquid. The investigation is thus a controlled one, with only one variable.

4. The enzyme amylase, acting as a digestive catalyst, causes the starch to be broken down into smaller maltose units. High temperatures (100°C) and extremes in acid pH (which is caused by vinegar) denatures the enzyme and prevents the reaction from taking place.

23.4. Lipids

The term "polyunsaturated" is now a common household word, though most students do not know that the term refers to the long hydrocarbon chain in fatty acids. If the chain contains one or more double bonds, the fatty acid is not saturated with hydrogen and is referred to as unsaturated. Lipids of animal origin tend to be saturated and are solids at room temperature. Lipids derived from plants and fish are usually unsaturated and are liquids at room temperature. In general, people whose diets contain unsaturated fats suffer less from deposits on their arteries than people whose diets contain large amounts of saturated fats.

Students may find the subject of neurotransmitters particularly interesting. If there is sufficient time, you could discuss in detail a single neurotransmitter, such as serotonin.

Serotonin regulates several body mechanisms, including mood, alertness, and carbohydrate intake. In turn, the amount of serotonin produced is regulated by the proportion of carbohydrate in the diet. A high proportion of carbohydrate stimulates the rate at which tryptophan, an amino acid precursor, is converted to serotonin.

This type of biofeedback mechanism is common in regulating biochemical pathways in the body. These mechanisms work for most of us most of the time; sometimes they don't. Carbohydrate Craving Obesity is a syndrome that results from a severe aberration in the carbohydrate/serotonin feedback system. People with this syndrome suffer from both mood and appetite disorders. They tend to have bouts of depression in late afternoon or evening and eat large amounts of carbohydrate-rich snacks to compensate.

23.5   Nucleic Acids

Reading about the history of the discovery of DNA may help your students to understand how scientists synthesize information of many types and from many sources.

The discovery of the structure of DNA in 1953 is an example of a situation in which the function of a molecule could not be fully understood until the structure of the molecule was known. The discovery of the structure of DNA was not an isolated event. Following Friedrich Miescher's initial discovery of DNA, it took over one hundred years to elucidate its structure. The well-known double-helix model developed by Watson and Crick required the supporting evidence of the x-ray crystallography of Rosalind Franklin.

Students can read about the latest developments in genetic engineering in the popular press, even on the pages of the *Wall Street Journal*. Never before has biotechnology so quickly found its way into the marketplace. Many ethical questions have been raised: What role should animals have in genetic research? At what point is it appropriate to use a drug on humans? Should we use fetal tissue for experiments in cloning? Could we create a new microorganism that could endanger the human race?

These questions should be discussed in the classroom. Encourage students to bring relevant press releases to class. Think of asking a colleague in the humanities to be a cofacilitator in a discussion of these issues.

23.6   Biochemical Pathways

Emphasize to your students that energy is required for the growth and maintenance of cells. In a healthy individual, energy is extracted from food molecules when the molecules are broken down in the cell to smaller fragments. The energy released from the breakdown of molecules is conserved in the nucleotide; ATP is consumed.

The energy in macromolecules has to come from somewhere. The ultimate source of energy is the sun. Green plants, algae, and some bacteria can capture the energy of the sun and convert light into chemical energy. The energy is stored in macromolecules produced by these photosynthesizing organisms and is utilized by consumers, organisms not capable of utilizing light directly for an energy source.

Demonstration 3: Oxidation of Methylene Blue
PURPOSE: To demonstrate the conversion of light energy into chemical energy through the reduction of an organic dye
MATERIALS: petri dish, 100 mm; overhead projector; black poster board, 150 mm x 150 mm; 10 g $FeSO_4$; 3 mL 3 M H_2SO_4; 500 mL H_2O; methylene blue
PROCEDURE: *(1)* Prepare an aqueous solution of methylene blue by adding 25 mg of methylene blue to 5 mL of water. This solution may be stored for several weeks. *(2)* Prepare an acidified solution of ferrous sulfate by adding 10 g of ferrous sulfate to 500 mL of water. Add 3 mL of 3 M sulfuric acid. This solution must be prepared on the day of the demonstration. Ferrous sulfate oxidizes easily to ferric sulfate, which would destroy the reducing properties of the iron solution. *(3)* Place a petri dish on an overhead projector. The projector light should be off. Position a piece of black poster board so that one-half of the petri dish is shielded from the light. *(4)* Pour 50 mL of the ferrous sulfate solution into the petri dish. *(5)* Add 4 to 6 drops of methylene blue to the ferrous sulfate. The solution should have a deep blue color. If it does not, add more methylene blue, one drop at a time, until the solution is deep blue. *(6)* Turn on the projector light. The solution in the half of the projector that is exposed to light will become colorless as the methylene blue is photochemically reduced. *(7)* Remove the black poster board. The solution in the protected half of the petri dish will still be deep blue. Color will begin to disappear as the remaining half of the solution absorbs light.

Remind students that light is the ultimate source of energy in living systems. It is absorbed by producers that store the energy from the sun in a variety of macromolecules that are passed along the food chain to consumers. Although students know about this process from their work in biology, they sometimes continue to think of heat as the source of all energy-absorbing reactions. The photochemical reduction of methylene blue clearly demonstrates that light can supply the necessary energy for a chemical reaction.

An analysis of the chemical structure of the oxidized and reduced form of methylene blue illustrates the relationship between form and function in organic molecules. This relationship has many analogies in living systems.

References and Resources

Books and Periodicals

Butler, P.J.G., and A. Klug. "The Assembly of a Virus," *Scientific American*, Vol. 239, 1978, pp. 62–69.
Dickerson, Richard E., and Irving Geis. *The Structure and Action of Proteins*, Benjamin Cummings Publishing Co., 1969.
Ferguson, Lloyd N. "Bio-Organic Mechanisms II: Chemorecep-

tion," *Journal of Chemical Eduction*, Vol. 58, 1981, pp. 456.
Freifelder, David. *Molecular Biology and Biochemistry: Problems and Applications*, W.H. Freeman and Co., 1978.
Goldsmith, Robert H. "A Tale of Two Sweeteners," *Journal of Chemical Education*, Vol. 64, 1987, p. 954.
The Hastings Center, 360 Broadway, Hastings-on-Hudson, New York 10706. Publications available to teachers at a nominal cost.

MacBeath, Marie E., and Andrew L. Richardson. "Tomato Juice Rainbow, A Colorful and Instructive Demonstration," *Journal of Chemical Education*, Vol. 63, 1986, pp. 1092–1095.

Ophardt, Charles E., and Paul F. Krause. "Blood Buffer Demonstration," *Journal of Chemical Education*, Vol. 60, 1983, p. 493.

Perutz, M.F. "The Hemoglobin Molecule," *Scientific American*, Vol. 211, 1964. pp. 64–76.

Piel, Jonathan, ed. "What Science Knows About AIDS: A Single Topic Issue," *Scientific American*, Vol. 259, 1988.

Portugal, Franklin H. and Jack S. Cohen. *A Century of DNA*, The MIT Press, 1977.

Sayre, A. "Rosalind Franklin and DNA," W.W. Norton, 1975.

Stent, G.S., and R. Calendar. *Molecular Genetics: An Introductory Narrative*, W.H. Freeman and Co., 1971.

Stroud, Robert M. "A Family of Protein-Cutting Proteins," *Scientific American*, Vol. 231, 1974, pp. 74–88.

Watson, James D. "The Double Helix," *Atheneum*, 1968.

Wurtmann, Richard and Judith Wurtmann. "Carbohydrates and Depression," *Scientific American*, Vol. 260, 1989, p. 68.

Computer Software

"Blood Sugar," ALB4605A, AII, 48K, Disk, available from Queue.

"Chemicals of Life II: Water, Carbohydrates, and Lipids," IBM02PC, I, Disk, available through Queue.

"Chemicals of Life III: Protein and Nucleic Acids," IBM03PC, I, Disk, available from Queue.

Currie, James and G. Scott Owen. "Molec," COM4084A, AII, 48K, Disk, available from Queue.

"Enzyme Investigations," Biology Explorations, HRM Software, HRM543A, available from Queue.

"Gene Machine," Genes and Heredity, HRM Software, HRM537A, AII, Disk, HRM537C, HRM537T, available from Queue.

Kleinsmith, Lewis J. "Protein Synthesis/Condon," COM4206A, Apple 128K, Disk; COM4206B, IBM, 256K, Disk; COM4206J, PB-Jr., 256K, Disk; available from Queue.

"Metabolic Pathways," ALB4614A, AII, 48K, Disk, available from Queue.

Zimmerman, S. Scott, "Stimulation of Hemoglobin Function," COM4200A (AII,48K,Disk), available from Queue.

Answers and Solutions

Questions

1. N_2, CO_2, H_2O
2. C, H, N, O
3. Nearly all biochemical reactions take place in solution in water. Water makes up most of the mass of every living organism.
4. Organic compounds are all but the smallest compounds of carbon. All other chemical compounds are inorganic
5. $R_1\text{—CHCOOH} + R_2\text{—CHCOOH} \rightarrow R_1\text{—CHC—N—CHCOOH}$ with NH_2, NH_2, and the product having NH_2, H, and the group $\overset{O}{\overset{\|}{C}}\ \overset{R_2}{|}$

6. Main use of waxes in living organisms is waterproofing the exterior.
7. Sugar (ribose or deoxyribose), phosphate group, nitrogen base
8. A monosaccharide (sugar, carbohydrate)
9. A fatty acid (lipid)
10. Carbohydrates, proteins, lipids, nucleic acids
11. It denatures it, altering its three-dimensional structure. Sufficient heat will destroy the protein completely and prevent it from functioning.
12. *(a)* Amino acids
   (b) monosaccharides
   (c) nucleotides
13. An enzyme is a protein that catalyzes a biochemical reaction.
14. Nucleotides form hydrogen bonds between pairs of bases so the nucleotide sequence in one strand corresponds to the sequence in its partner. A nucleotide containing adenine base pairs to one containing thymine, and a nucleotide containing cytosine base pairs to one containing guanine.
15. Adenosine triphosphate (ATP)

Problems

1. DNA is the genetic material. The information it contains specifies the sequence of amino acids in all the proteins in the body. This determines the genetic characteristics of the body and is the information that the organism will pass on to its offspring when it reproduces.
2. The enzyme in beaker B was probably denatured by heat.
3. TTTAGCCGT
4. 108 g of fat

Application Questions:

1. Fatty acids (oleic acids); waxes (beeswax, covering on a leaf; fats and oils, margarine, butter, olive oil, peanut oil, corn oil); steroids (estrogen, cortisone, testosterone)
2. Enzyme action is specific because an enzyme binds its substrate in a "lock and key" fashion and each enzyme can bind only one or a few different substrate molecules.
3. Fat is produced mainly by animals, carbohydrate by plants. Fat contains more energy than the same weight of carbohydrate so it has taken more calories to produce. We produce most of our fatty foods by growing plants and then feeding them to animals which convert them to fat. This is more expensive than just growing plants which produce most of our carbohydrate food.
4. Cholesterol: part of the structure of all biological membranes
   Estrogen: female sex hormone
   Testosterone: male sex hormone
5. Enzymes: catalyze chemical reactions
   Structural proteins: form part of the structure of a body or cell

Hormones: carry messages that regulate bodily functions
Blood transport proteins: carry things in the blood

6.

or

7. Cellulose, chitin
8. pH, temperature

9. Energy (food) supply, food storage, structural support of a body or cell
10. Hormones, transport, structure support
11. Carbohydrate
12. Glycogen is the carbohydrate that stores energy in the liver of animals.
13. Cellulose is the structural carbohydrate that strengthens plant cell walls.
14. Starch. Potatoes.

Application Problems:

1. Life depends on huge molecules held together by carbon–carbon bonds. Life based on silicon would be impossible because silicon–silicon macromolecules would fall apart in the presence of oxygen. Oxygen is vital to life. It is produced by plants during photosynthesis, which is indirectly responsible for the production of all the food on earth. Also, many metabolic reactions would be impossible if oxygen became stably bound to macromolecules, as it would if they were based on silicon.
2. The sugar in ATP is ribose; in a DNA nucleotide it is deoxyribose.
3. When an egg is cooked, the protein in it is denatured, meaning that it loses its three-dimensional structure, although the chain of amino acids may remain the same.
4. Glucose. It is readily digested by the body.

Teacher's Notes

Chapter 23 Biochemistry

INTRODUCTION

Plants and animals synthesize some of the carbon compounds discussed in the last chapter. They also build these and other carbon compounds into very large molecules that are found only in living things. These large carbon molecules make the difference between being alive and not being alive. In this chapter we look at some of the molecules of life and at some of the chemical reactions they undergo.

LOOKING AHEAD

In studying this chapter, note that the chemistry of life is based on two things: large molecules with skeletons of carbon atoms, and water. Living things use energy to build up and break down some of the biggest molecules in the world.

SECTION PREVIEW

23.1 The Chemistry of Life
23.2 Proteins
23.3 Carbohydrates
23.4 Lipids
23.5 Nucleic Acids
23.6 Biochemical Pathways

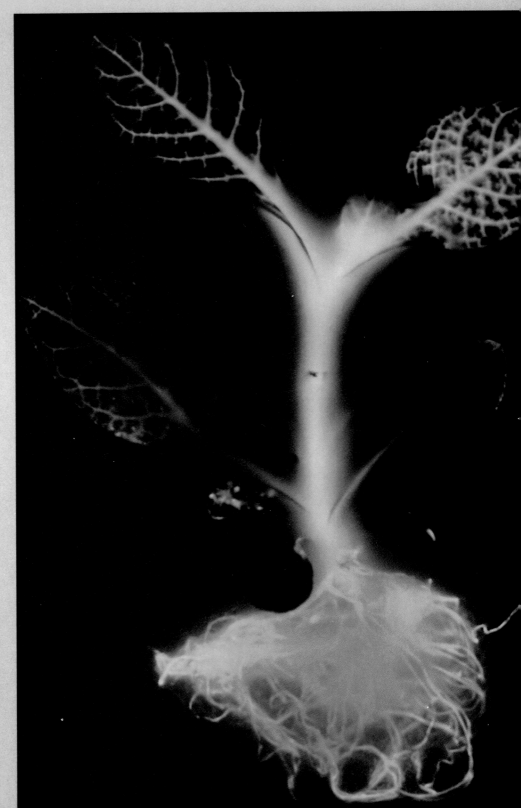

The cells of this tobacco plant contain a gene transplanted from fireflies. The same gene that makes fireflies light up on a summer evening causes the plant to glow in the dark. Biochemists may soon be able to transfer human genes.

23.1 The Chemistry of Life

The chemistries of plants, animals, bacteria, and fungi are impressive in two ways. First, these living things are more than half water. Second, what is not water is mainly composed of organic molecules that, as you have seen, all contain carbon. The biggest molecules on earth are organic. One kind of organic molecule is DNA. When it is straightened out, a molecule of DNA from a fruit fly is nearly 2 cm long—ten times as long as the fly itself!

We live surrounded by organic molecules; we touch them when we touch our noses or a clump of grass or a spider. All these things are alive, and they are kept alive by reactions between organic molecules. **Biochemistry** *is the study of the chemical reactions that go on in living things.* As people learn more about biochemistry, they are learning how to produce some of the molecules found in living things. We can now make substances such as hormones, vaccines, and even genes, in the laboratory—with the assistance of living organisms.

Inorganic Molecules and Water

The first living things evolved more than three billion years ago from the chemical compounds that were common on earth at that time, such as water, ammonia and carbon dioxide. These substances contain the elements hydrogen, oxygen, carbon, and nitrogen. Living things today are made up almost entirely of these elements. In addition, they contain small amounts of 18 more of the 92 natural elements.

Water's main role in life is to act as a solvent. The chemical reactions in our bodies take place in a dilute solution that is much like sea water (Figure 23-1). The solution contains the ions of salts, such as potassium phosphate (K^+ and PO_4^{3-}), magnesium bicarbonate (Mg^{2+} and HCO_3^{2-}), and sodium chloride (Na^+ and Cl^-) dissolved in water. All substances that travel through the body, such as oxygen, hormones, and food, travel dissolved in a salt solution, and most biochemical reactions take place in this solution. When we digest food, store fat, or move our muscles, dozens of reactions take place between dissolved compounds. Water makes up most of the mass of every living organism. Chemists have spent many years discovering how to imitate the body's solution of salts in water. As a result of this research, many body parts can now be kept alive outside the body, bathed in salt solution. These body parts may be kidneys or hearts waiting to be transplanted, cells to be used for research, or eggs to be fertilized outside the body.

Macromolecules

Large organic molecules are called **macromolecules** *(the prefix* macro *means large),* and they do many jobs in the body. Some of them are structural molecules that provide support in the bodies of plants and animals. The muscles that hold our bodies up and the wood that holds up a tree, are made of macromolecules. Other macromolecules store and release energy. Still others carry information that the body needs.

Most of the macromolecules essential to life belong to four main classes, three of which we know as types of food:

SECTION OBJECTIVES

- Understand how the chemistry of living things is unique.
- List the four elements that are most common in living organisms. List four main classes of biological macromolecules.
- Describe the formation of a macromolecule.

The chemistry of carbon is introduced in Section 21.1.

Although oxygen was present in CO_2, the atmosphere did not contain free oxygen molecules. A reducing atmosphere existed until organisms evolved that were capable of carrying on photosynthesis.

The chemical nature of the fluid inside our bodies is discussed in Section 15.1.

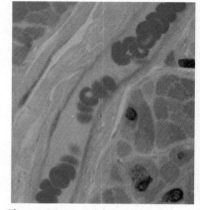

Figure 23-1 Without water, there can be no life. These cells are in a solution of salts in water.

Proteins are chains of many amino acids that are linked together. The three-dimensional structure of proteins is very important to their function.

Lipids are the primary components of cell walls. They form bilayers with the hydrocarbon tails on the inside and the phosphate heads on the outside.

The carboxyl group is discussed in Section 22.5.

Polarity is discussed in Section 6.5.

The behavior of water as a solvent is discussed in Section 13.4 and Section 14.2.

1. The study of the chemical reactions that go on in living things
2. Structural, energy storage and release, information storage
3. carboxyl group (— COOH), hydroxyl group (— OH), amino group (NH$_2$), carbonyl group (C = O)

1. *Proteins*. Hair, tendons, ligaments, and silk are made of protein. Other proteins act as hormones, carry things around the body, and fight infections. Enzymes are proteins that control the body's chemical reactions.
2. *Carbohydrates*. Sugars, starches, and cellulose are carbohydrates.
3. *Lipids*. Fats, oils, waxes, and steroids are lipids, nonpolar substances, which do not dissolve in water.
4. *Nucleic acid*. The nucleic acids are DNA and RNA. In most organisms, DNA is used to store hereditary information, and RNA helps to assemble proteins.

Functional groups   Organic molecules come in a vast variety of types and sizes. To make sense of this diversity, biochemists examine the functional groups that organic molecules contain. As you learned in Chapter 22, functional groups are clusters of atoms that behave in particular ways. The functional groups of a molecule determine the molecule's chemical properties. For instance, any compound containing a carboxyl group (—COOH) is acidic because the carboxyl group releases hydrogen ions in solution:

$$—COOH \rightarrow —COO^- + H^+$$

As another example, C—H bonds are nonpolar. Therefore, a part of a molecule containing only carbon and hydrogen is also nonpolar. Such parts are not soluble in water.

Other functional groups contain polar bonds, particularly O—H, N—H, and C=O. Functional groups containing polar bonds are soluble in water. Macromolecules are so large that part of the molecule may be insoluble in water while the rest is soluble. Functional groups determine a molecule's solubility and determine what kinds of chemical reactions it will take part in.

Section Review

1. What is biochemistry?
2. List three functions that different macromolecules perform in the body.
3. List four different functional groups found in organic molecules.

23.2 Proteins

Proteins contain the elements carbon, oxygen, hydrogen, nitrogen, and usually some sulfur. They perform many different functions in the body (Table 23-1). **Proteins** *are macromolecules formed by condensation reactions between monomers called amino acids.*

Amino Acids

Every **amino acid** *has a carboxyl group (—COOH) and an amino group (—NH$_2$), both attached to the same carbon atom, which is also attached to a hydrogen (—H).* The reason one amino acid differs from another is that the fourth bond of this carbon is attached to a functional group (called an R group) that differs from one amino acid to another (Figure 23-2). The proteins of all organisms contain the same twenty common amino acids.

TABLE 23-1 SOME FUNCTIONS OF PROTEINS

Type	Example	Function
Enzymes	Amylase	Converts starch to glucose
	DNA polymerase I	Repairs DNA molecules
Structural proteins	Keratin	In hair, nails, hoofs
	Collagen	In tendons, cartilage
Hormones	Insulin	Regulate glucose metabolism
	Vasopressin	Slows water loss by kidneys
Transport proteins	Hemoglobin	Carries O_2 in blood

Insulin was one of the first proteins for which the protein structure was determined. Its amino acid sequence was discovered in the early 1950s.

Protein Structure

When a protein forms, the first step is a condensation reaction joining two amino acids by a **peptide bond**, *a type of bond that joins amino acids in proteins*. A peptide bond is formed by removing an OH from the carboxyl (—COOH) group of one amino acid and the H from the amino (—NH₂) group of another (Figure 23-2). Further peptide bonds join more amino acids to the chain. The finished peptide chain usually contains 100–300 amino acids. Each protein has its own unique sequence of amino acids. A complex organism has at least several thousand different proteins, each with a special structure and function.

Proteins are made up of long, unbranched chains of amino acids, but the final protein is quite a complex structure. This is because peptide chains fold up to form various structures. The most common of these structures is a coil formed by part of the chain and known as an alpha helix.

Each peptide chain may also link up with other chains to form the final protein. For instance, *insulin*, a hormone that helps the body to regulate the level of sugar in the blood, is made up of two linked chains. The chains are held together by S—S bonds between sulfur atoms in two cysteine amino acids (Figure 23-3). Insulin is one of the smaller proteins, containing only 51 amino acids. In contrast, hemoglobin, which carries oxygen in the blood, is a large protein consisting of four long chains with complicated three-dimensional structures. Proteins can be made to lose their shape by gentle heating or various chemical treatments. When they are returned to more normal surroundings, they fold or coil up again and join together, re-forming their original structure.

Changing even one amino acid can change a protein's structure and what it does. For example, the difference between normal hemoglobin and the hemoglobin that causes sickle cell anemia is just two amino acids.

Figure 23-2 The general structure of amino acids consists of an amino and a carboxyl group, both attached to the same carbon atom, which is also bonded to a hydrogen atom (above). This carbon's fourth bond is attached to an R group, which may be nonpolar, polar, or ionic. A peptide bond (below) forms by a condensation reaction between two amino acids. A peptide bond is one between the carboxyl carbon of one amino acid and the amino nitrogen of another.

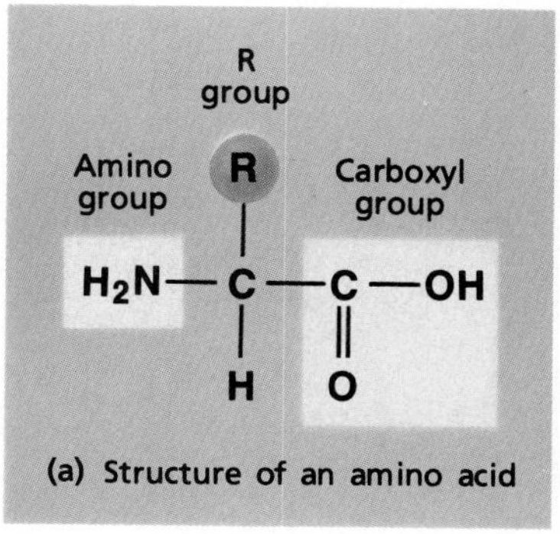

(a) Structure of an amino acid

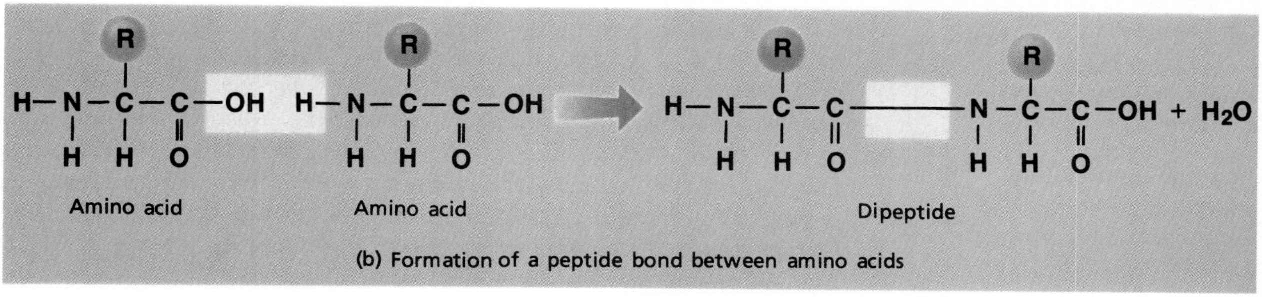

(b) Formation of a peptide bond between amino acids

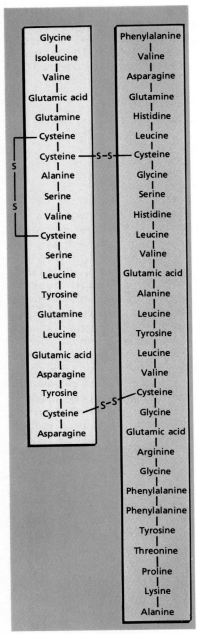

Figure 23-3 The amino acid sequence of the hormone insulin. This small protein is made up of two short polypeptide chains joined by (yellow) "sulfur bridges," attachments between suifur-containing cytosine amino acids.

Enzymes

Nearly all the chemical reactions in living things are controlled by **enzymes**, *protein molecules that act as catalysts.* A catalyst is a substance that speeds up a chemical reaction. *An enzyme is rather like a robot on an assembly line, performing the same task over and over on a particular set of parts called* **substrates.** The molecules produced by an enzyme reaction are called the reaction's *products.* Most enzymes have names ending with *-ase* and are named according to their specific substrates and the kinds of reactions they catalyze.

How does an enzyme speed up a reaction? Each enzyme combines with its substrates. An enzyme can bind only one or a few types of molecules and hold them in the correct position for the reaction to occur. Parts of the enzyme pull on the substrates' bonds and loosen them. This lowers the energy of activation needed for the reaction to occur. As a result, the reaction occurs readily.

Vitamins, minerals, and coenzymes Some enzymes cannot bind to their substrates without the help of additional molecules. These may be *minerals* such as ions of calcium or iron, or *helper molecules called* **coenzymes** that play accessory roles in enzyme catalyzed reactions. Many vitamins are coenzymes or parts of coenzymes.

Vitamins are organic molecules that we cannot manufacture and hence need to eat in small amounts. You can see why we need vitamins and minerals in our diet—to permit our enzymes to work. You can also see why we need only small amounts of them. Minerals and coenzymes are not destroyed in biochemical reactions. Like enzymes, coenzymes and minerals can be used over and over again.

Factors that affect enzyme activity Various factors, notably temperature and pH, affect the rates of reactions catalyzed by enzymes. Many enzymes are affected by the pH of the surrounding solution. Most proteins work best in a solution of approximately neutral pH. (Most of our body cells have a pH of 7.4.) Some enzymes work in acidic or basic environments, however. Pepsin, the collective term for the digestive enzymes found in the human stomach, works best at the very acidic pH of about 1.5. Cells lining the stomach secrete hydrochloric acid (HCl) to produce this acid environment. When food travels down the digestive tract, it carries these enzymes out of the stomach into the intestine. Here the stomach enzymes stop working because the intestine contains sodium bicarbonate that raises the pH to about 8. The digestive enzymes that work in the intestine are formed by the pancreas and work best at pH 8.

Most chemical reactions, including enzyme reactions, speed up as it gets hotter. On the other hand, *high temperatures (above about 60°C) destroy, or* **denature** *proteins by breaking up their three-dimensional structure.* Denaturation is the loss of configuration in an enzyme. When we cook an egg or cook an egg or a piece of meat, we denature the proteins in the food. Heating preserves food by denaturing the enzymes of organisms that cause decay. Milk is pasteurized by heating it to denature enzymes that would turn it sour. Refrigeration and freezing preserve food by cooling, which slows down the reactions of enzymes that cause decay.

Section Review

1. Name the monomers of proteins and draw a peptide bond.
2. What are the main elements found in proteins?
3. Why do temperature changes affect the activity of proteins?

1. Amino acids. See Figure 23-2 for an example of a peptide bond.
2. Carbon, oxygen, hydrogen, nitrogen, usually some sulfur.
3. Can alter the three-dimensional structure of a protein.

23.3 Carbohydrates

SECTION OBJECTIVES

- Name four carbohydrates.
- Describe the monomers from which carbohydrates are made.
- List three functions of carbohydrates in living organisms.
- Describe how carbohydrates are synthesized.

Carbohydrates *are the sugars, starches, and related compounds.* Some carbohydryates supply energy. Others are structural molecules, building materials that shape the body. Carbohydrates tend to be polar, and thus will associate with water.

Simple Sugars

The monomers of carbohydrates are **monosaccharides**, *or simple sugars, such as fructose and glucose. A monosaccharide contains carbon, hydrogen, and oxygen in about a 1:2:1 ratio* (CH_2O).

Monosaccharides provide the body with ready energy. Sugary foods are quick-energy foods because sugars are easily converted by enzymes into molecules that the body can use for energy. Carbohydrates are transported in our blood mainly in the form of the six-carbon monosaccharide glucose, which the cells of the body take up and use for energy as they need it.

Carbohydrates are a good source of energy. If we eat more of them than our bodies can immediately use for energy, carbohydrates are then converted into fats to be stored in the body for later use.

Disaccharides and Polysaccharides

Two monosaccharides may be joined together to form one **disaccharide** (Figure 23-4). Sucrose (table sugar) and lactose (milk sugar) are disaccharides. A disaccharide can be hydrolyzed into the monosaccharides that formed it. *By a series of condensation reactions, many monosaccharides can be joined to form a polymer called a* **polysaccharide** *(commonly known as a* **complex carbohydrate**).

It is the repetition and chemical stability of these polysaccharides that gives them their physical stability and, thus, their ability to form hard shells, bones, and fibers.

Three important polysaccharides made of glucose monomers are glycogen, starch, and cellulose. *Animals store energy in* **glycogen**. The liver and muscles remove glucose from the blood and condense it into glycogen, which can later be hydrolyzed back into glucose and used to supply energy as needed.

Starch *is the polysaccharide that is used to store energy in plants.* Starch consists of two kinds of glucose polymers. A plant hydrolyzes its starch into

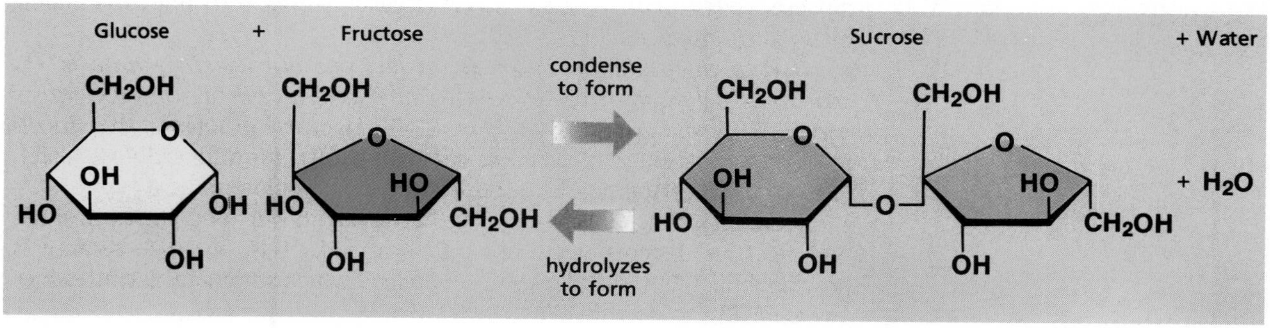

Figure 23-4 Formation of a disaccharide. A condensation reaction between the monosaccharides glucose and fructose produces the disaccharide sucrose (table sugar) and water. This reaction is reversible by hydrolysis.

Desktop Investigation

Enzyme Action on Starch

Questions
What effect does the enzyme *amylase* have on the digestion of starch? What effect do heat and acids have on enzyme action?

Materials
cornstarch
distilled water
saliva
vinegar
beaker, 250 mL
stirring rod
6 test tubes (14 × 100 mm)
wax pencil
Bunsen burner
ringstand with ring
wire gauze-ceramic center
3 wooden splints

Procedure
Record all your results in a data table. Remember to wear safety glasses and an apron.

1. Add 9 g of cornstarch to a beaker containing 125 mL of cold distilled water. While stirring continuously, heat the mixture until it has boiled about two minutes.

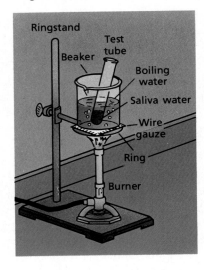

Ringstand
Beaker
Test tube
Boiling water
Saliva water
Wire gauze
Ring
Burner

2. Fill three test tubes with 5 mL each of the cornstarch mixture and allow them to cool. Label each tube near the rim with a wax pencil.

3. Spit equal amounts of saliva into each of 3 other test tubes. To two of these test tubes, add 1 mL of distilled water. Add 1 mL of vinegar to the third test tube. Label each tube appropriately.

4. Boil one of the test tubes containing saliva and water in a water bath for five minutes, as shown in the diagram.

5. Add the heated water and saliva to one of the test tubes that contains cornstarch. Add the unheated water and saliva to the second test tube that contains cornstarch. Add the vinegar and saliva to the third test tube that contains cornstarch. Label each tube appropriately. Stir each of these three test tubes with a new, unused wooden splint.

Discussion
1. What effect did amylase in the saliva have on the starch?

2. What effect did the heat and vinegar have on the amylase action? Explain.

3. Why was it necessary to add 1 mL of water to the test tubes containing saliva?

4. Write a good conclusion for this investigation.

---

glucose. The plant uses some of this glucose for energy, and some as a building material to produce more cells.

The main building material made from glucose in plants is the structural polysaccharide **cellulose**, *which is probably the most common organic compound on earth* (Figure 23-5). Hydroxyl groups on the glucose monomers link different cellulose chains together, forming cellulose fibers. (The fibers of cotton consist almost entirely of cellulose.)

Humans cannot digest cellulose. Cellulose in our food passes through our digestive systems unchanged. It is important in our diet because it provides bulk (often called *fiber* or *roughage*), which stimulates the intestines and keeps things moving along.

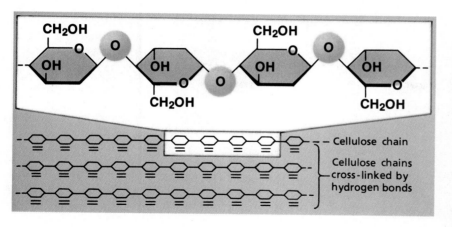

Another structural polysaccharide is *chitin*, which is composed of glucose with a nitrogen-containing group attached. Chitin is an important component of the "armor plating" of crabs, spiders, insects, and their relatives.

Section Review

1. What is the difference between a disaccharide and a polysaccharide?
2. How are two monosaccharides joined to form a disaccharide?
3. Name four carbohydrates.
4. What is cellulose?

23.4 Lipids

Lipids *are a varied group of organic compounds that share one property: they are not very soluble in water.* This is because lipids contain a high proportion of C—H bonds. They dissolve in nonpolar organic solvents, such as ether, chloroform, and benzene, but not in water, which is polar. Most lipids are too small to be called macromolecules and they are not made of monomers, as other biological molecules are.

Lipids contain carbon, hydrogen, and oxygen, but the proportion of oxygen in lipids is much lower than in carbohydrates. Some lipids contain other elements, such as phosphorus and nitrogen.

Because they are insoluble in water, lipids are the main compounds in biological membranes. Membranes divide living things into many compartments, such as cells, each containing a particular salt solution and set of reactions.

Lipids are used as energy stores because a given weight of lipid can store much more energy than the same weight of carbohydrate or protein (Table 23-2). This is undoubtedly why lipids are used as food reserves in the body. Without lipid reserves, for instance, the migrations of many birds would be impossible. A small bird may almost double its body weight in the fall as it lays down fat reserves for the long flight. To carry the same energy reserve in the carbohydrate glycogen, it would have to get so fat that it could not fly.

Figure 23-5 In cellulose, glucose monomers are linked into long, unbranched chains (left). Cellulose fibers are deposited in layers to form rigid plant cell walls that help to provide structural support for the entire plant (right).

1. A disaccharide is formed from two monosaccharides joined together. A polysaccharide consists of many

monosaccharides joined together.
2. By means of a condensation reaction
3. Glycogen, starch, cellulose, chitin
4. A large polysaccharide that consists of many glucose monomers joined together. It provides structural support for plants and is probably the most common organic compound on earth.

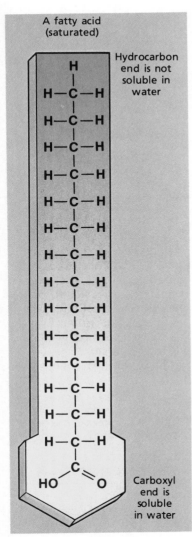

A fatty acid
(saturated)

Hydrocarbon end is not soluble in water

Carboxyl end is soluble in water

Figure 23-6 The carboxyl group in the fatty acid shown above is at the bottom. The rest of the molecule is a nonpolar, hydrocarbon chain, insoluble in water. This is saturated because the carbon atoms of the hydrocarbon chain contain as many hydrogens as they can hold.

TABLE 23-2 ENERGY YIELD OF TYPES OF FOOD		
Class	Composition	Energy Yield
Carbohydrates	CH_2O	3.8 kilocalories/gram
Lipids	CHO	9.3 kilocalories/gram
Proteins	CHON(S)	3.1 kilocalories/gram

Fatty Acids

Fatty acids are the simplest lipids. A **fatty acid** *consists of a chain of carbon and hydrogen atoms, with a carboxyl group at one end (Figure 23-6). If all carbon atoms of the hydrocarbon chain are "filled" with as many hydrogens as they can possibly hold, the fatty acid is said to be* **saturated.** *A fatty acid with one or more double bonds in its hydrocarbon chain is said to be* **unsaturated** *because it could hold more hydrogens if a double bond were broken and two hydrogen atoms were attached to each carbon instead of one.*

The bonds in a carboxyl group are polar, and so the carboxyl end of a fatty acid attracts water molecules. The carbon-hydrogen bonds of a lipid's hydrocarbon chain are nonpolar, however. The polar end will dissolve in water, and the other end will dissolve in nonpolar organic compounds. This behavior means that fatty acids will form membranes when they are dropped into water. It also gives soaps and detergents their cleaning power.

Fatty acids are usually combined with other molecules to form substances such as **glycolipids** *(carbohydrate + lipid) or* **lipoproteins** *(lipid + protein).* They are also parts of more complex lipids.

Waxes

A **wax** *is a lipid with one very long fatty acid and one very long carbon-hydrogen chain with an alcohol group (C—OH) in it.* These long chains make waxes solid at room temperature, with high melting points. Waxes form a waterproof coating on the outer surfaces of many organisms, for example on skin, hair, feathers, and leaves (Figure 23-7). People make polishes and candles from waxes, and use another wax, lanolin, to make cosmetics.

Fats and Oils

Fats and oils are molecules formed by the esterification of three fatty acids with a molecule of the alcohol glycerol, as described in Section 22.6. Fats are solid at room temperature, whereas oils are liquid. **Saturated fats** *contain only saturated fatty acid groups.* **Unsaturated fats** *contain at least one fatty acid group with double bonds. Polyunsaturated fats are fats and oils with double bonds in their fatty acid chains.* There is some evidence that eating large amounts of saturated fat may increase the risk of heart attacks, so some cooking oils and margarine are advertised as "polyunsaturated." These are made of oils extracted from plants because oils generally contain more unsaturated fatty acids than do fats, and oils are more common in plants than in animals. (Examples are olive, corn, and peanut oils).

Americans would probably be healthier if we ate less fat and oil. Part of the reason for this is that many people are overweight and a given weight of lipid provides more calories (energy) than the same weight of carbohydrate or protein. In addition, if the body contains too much of certain types of lipids, it may be more susceptible to some diseases. Fats perform many vital functions in the body, and some vitamins are soluble in fat but not in water. We need some fat in our diet.

Phospholipids

Phospholipids are the most important structural lipids because they are the chief lipids in biological membranes. **Phospholipids** *are lipids in which one (or two) of the fatty acids is replaced by a phosphate group.*

Steroids

Steriods *differ from other lipids in structure. A steroid is a lipid with a "skeleton" structure consisting of four carbon rings.* The most common steroid is cholesterol, shown in Figure 23-8, which has received some bad publicity because deposits of cholesterol in the body may cause gallstones or heart disease. **Cholesterol** *is a steroid that makes up part of every animal cell membrane and of the brain and nerves,* however. It is also the raw material from which steroid hormones are produced. We could not live without it. A buildup of cholesterol and other materials within the arteries makes it difficult for the arteries to expand and contract properly and blood moves through them with difficulty. Diets high in cholesterol increase the risk of heart disease.

Hormones *are chemical messengers that "communicate" between different parts of the body.* Steroid hormones include cortisone, as well as sex hormones from the ovaries and testes—estrogens and testosterone. These hormones control reproduction. They cause the differences between male and female bodies and affect sexual behavior. Synthetic derivatives of testosterone are sometimes taken by athletes and weightlifters to promote muscle growth. Recent research, however, has shown that these steroid drugs can have serious side effects, including shrinking of the testes, breast development, and decreased sperm production in men; facial and body hair and menstrual irregularity in women; and high blood pressure in both sexes.

Figure 23-7 Wax. Flakes of wax form a waterproof covering on the surface of a leaf and prevent the plant losing water into the surrounding dry air.

Hormones are basically regulatory in nature because they control certain biochemical activities in the body. An example is insulin, which is secreted by the pancreas in response to an increase in blood sugar levels. Insulin facilitates the transport of glucose into cells. Diabetes results from a deficiency of insulin.

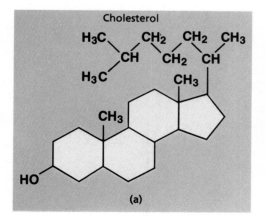

(a)

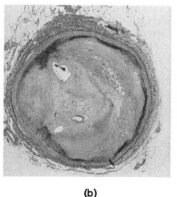

(b)

Figure 23-8 Cholesterol (left) is an important component of animal cell membranes. The basic skeleton of carbon rings (color) is the same in all steroids, but other steroids have other groups attached to this skeleton. The accumulation of cholesterol on artery walls (right) may eventually block the passage of blood.

Biochemistry **693**

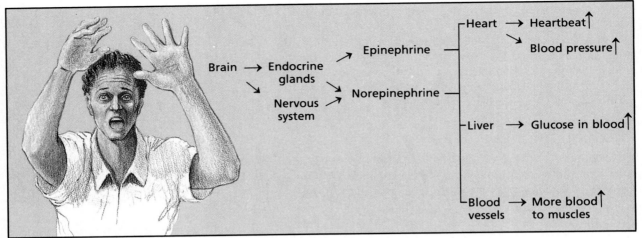

Figure 23-9 Fight or flight. When the brain registers excitement, glands and neurons produce the hormones epinephrine and norepinephrine, which travel around the body and get it ready for action.

Not all hormones are steroids. Many are proteins or amines. *(An **amine** is a base derived from ammonia in which one or more of the three hydrogens is replaced by an alkyl or aryl group.)* For instance, epinephrine (also called adrenalin) is an amine. **Epinephrine** *is the hormone enacted when you experience excitement. It speeds up the heart, increases the blood supply to the muscles, and generally gets the body ready for "flight or fight"* (Figure 23-9).

All the hormones mentioned so far are secreted by endocrine glands, special glands such as the thyroid, pancreas, and testes, which secrete the hormones they produce into the bloodstream. Until recently, biologists gave the name "hormone" only to substances produced by endocrine glands. But it is now becoming obvious that the body is full of all kinds of chemical

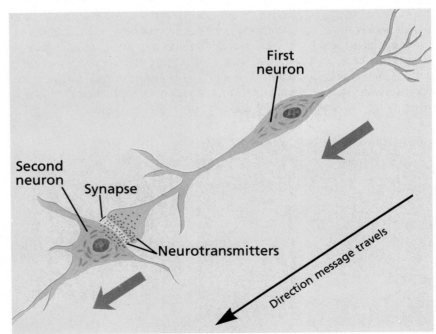

Figure 23-10 A synapse. The first neuron and part of a second neuron are shown. When the message in the first neuron reaches the synapse, neurotransmitter is released from the first neuron and travels across the synapse to stimulate the second neuron.

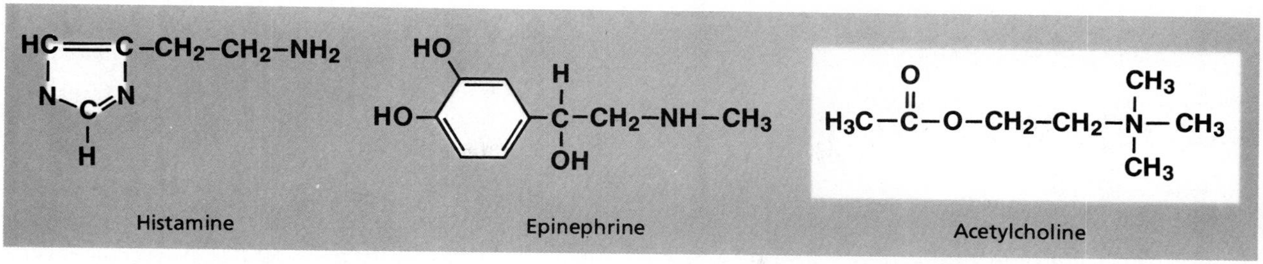

| Histamine | Epinephrine | Acetylcholine |

Figure 23-11 The chemical structure of three neuro-transmitters that are amines.

messengers carrying information from one place to another. These include local chemical messengers (short range chemical messengers not produced by endocrine glands) and nerotransmitters. *Neuro* means "nervous," and *the cells of the nervous system are called* **neurons**. When a message travels along a neuron, it travels as an electrical signal (which is why messages travel round the nervous system so rapidly). When a message reaches the end of a neuron, it is faced with a **synapse**—*a space between one neuron and the next* (Figure 23-10). *A chemical messenger called a* **neurotransmitter** *transmits the message from one neuron to the next across the synapse.* Several examples of neurotransmitters are shown in Figure 23-11.

Section Review

1. What is a lipid?
2. What is the difference between a fat and an oil?
3. Describe what is meant by the terms *saturated* and *unsaturated?*
4. Why are all biological membranes made up largely of lipids?
5. Are all hormones lipids?
6. What is a neurotransmitter?

1. A type of organic compound insoluble in water, contains a high proportion of C—H bonds
2. Fats are solid at room temperature, oils are liquids.
3. Saturated means that the carbons hold as many hydrogens as possible. Unsaturated means that there are one or more double bonds in the hydrocarbon chain.
4. Because they are generally insoluble in water.
5. No. Many are proteins or amines.
6. A chemical that carries a message from one neuron to the next.

23.5 Nucleic Acids

Nucleic acids *are macromolecules that transmit genetic information.* Nucleic acids include the largest molecules made by organisms. There are two kinds. **Deoxyribonucleic acid (DNA)** *is the material that contains the genetic information that all organisms pass on to their offspring during reproduction.* This information includes instructions for making proteins and also for making the other nucleic acid, *ribonucleic acid* (RNA). **Ribonucleic acid (RNA)** *assists in protein synthesis by helping to coordinate the process of protein assembly.*

Nucleotides

Nucleotides *are the monomers of nucleic acids.* A nucleotide has three parts: one or more phosphate groups, a sugar containing five carbon atoms, and a ring-shaped nitrogen base (Figure 23-12). RNA nucleotides contain the simple sugar ribose. DNA nucleotides contain deoxyribose (ribose stripped of one oxygen atom). Cells contain nucleotides with one, two, or three phosphate groups attached.

Besides being the monomers of nucleic acids, several nucleotides play other roles. For example, **adenosine triphosphate (ATP)** *is the nucleotide that supplies the energy for many chemical reactions.*

- Describe the monomers of nucleic acids.
- Describe the structure of DNA.
- Describe the genetic code.
- Understand how DNA controls which proteins are synthesized.

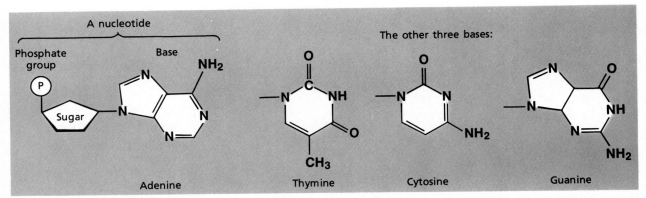

A nucleotide

Phosphate group

Base

NH₂

Sugar

Adenine

The other three bases:

Thymine

Cytosine

Guanine

Figure 23-12 DNA is made up of nucleotides. One nucleotide unit, containing the base adenine, is shown at the left. In other DNA nucleotides, adenine is replaced by one of three other bases: thymine, cytosine, or guanine.

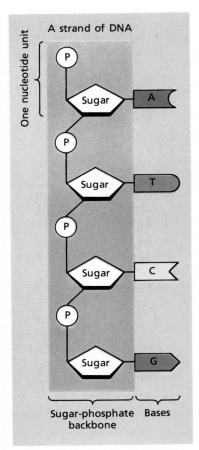

A strand of DNA

One nucleotide unit

P

Sugar — A

P

Sugar — T

P

Sugar — C

P

Sugar — G

Sugar-phosphate backbone   Bases

Figure 23-13 A strand of DNA is is made up of nucleotides.

When enzymes join nucleotides to form nucleic acids, the phosphate group of one nucleotide is linked to the sugar of the next. This forms a long backbone of sugars and phosphates, with the bases sticking out at one side (Figure 23-13). This is what many RNA and some DNA molecules look like. But some nucleic acids coil up or join together to form more complicated structures held together by hydrogen bonds.

The bases in nucleotides attract each other in pairs, a phenomenon known as **base-pairing**. DNA, for instance, is made up of four different nucleotides—those containing the bases adenine (A), thymine (T), guanine (G), and cytosine (C). Adenine forms more hydrogen bonds with thymine. Similarly, cytosine and guanine bond together. This *base-pairing* holds DNA nucleotides, and the DNA strands they lie in, together in pairs (Figure 23-14). *Paired strands of DNA then coil up to form a structure called a* **double helix**.

The Genetic Code

Base-pairing between nucleotides is one of the characteristics of DNA that permits it to store information. We can think of DNA's information as rather like a computer program: it tells the body what to do when certain things happen. Information in DNA, like computer information, is stored in code. But DNA is much more impressive than any computer. It stores more information in a smaller space than the most advanced gallium arsenide chip, it tells its owners how to reproduce themselves, and it even shows some of them how to learn to build computers!

DNA does all this by coding the information needed to build proteins. The genetic code is simple: a string of three DNA nucleotides codes for an amino acid. For instance, the nucleotide string TTC stands for the amino acid lysine. To read the DNA code, you find the string of nucleotides *TAC*, which means "start here," and read the nucleotides after the C in threes. For instance:

<u>TAC</u>   <u>TTC</u>   <u>CAA</u>   <u>CCT</u>   <u>TCG</u>

Means: START lysine valine glycine serine

This gives you the code for a chain of amino acids—in other words, the code for a protein. DNA contains coded information that tells the body what proteins to make. Some of the proteins are blood proteins or structural proteins. But the most important proteins are enzymes because these determine which chemical reactions go on in the body. The reason you are

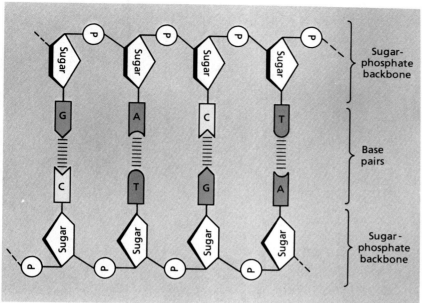

Figure 23-14 Two strands of nucleotides make up each DNA molecule. Here, a pair of strands, each four nucleotides long, is shown. The strands of both sugar-phosphate "backbones" lie on the outside of the molecule, with the nitrogen-containing bases meeting in the middle. Each base is held to a base on the opposite strand by hydrogen bonds (dotted lines). Adenine on one strand always pairs with thymine on the other, and cytosine always pairs with guanine.

chemically different from any other human being (unless you have an identical twin) is that you contain collections of enzymes different from anyone else.

A **gene** *is a length of DNA that codes for a particular protein.* Different living organisms contain different genes.

DNA Replication

DNA codes for protein structure, but it cannot actually make proteins. This is where base-pairing comes in. Remember that an A nucleotide on one DNA strand always bonds to a T nucleotide on the opposite strand, and a C bonds to a G. One strand of a DNA double helix has all the information needed to build a second strand because it consists of a string of nucleotides, each bearing one of the four bases A, T, C, or G. For instance, if the single strand is:

$$T—A—C—T—T—C—C—A—A—C—C—T—T—C—G$$

the missing second strand must be:

$$A—T—G—A—A—G—G—T—T—G—G—A—A—G—C$$

It is easy to synthesize the second strand because a free A nucleotide, floating around in the cell, will form hydrogen bonds with a T nucleotide in the single DNA strand. So the correct nucleotides line up opposite their base pairs, and the enzymes that build DNA merely have to join them together (Figure 23-15).

This is how DNA replicates (duplicates) itself. The two strands of the molecule separate. Each single strand then becomes the base upon which a new paired strand is built. DNA replication is vital to growth. Each of us starts with one set of genes (DNA) in the single fertilized egg cell from which we develop. But your body contains billions of cells. Each cell needs a copy of

The human body contains about 25 billion kilometers of the DNA double helix.

A sequence of DNA can be spliced into bacterial chromosomes that are replicated every time the bacteria divide. Millions of copies of that gene can be made and harvested.

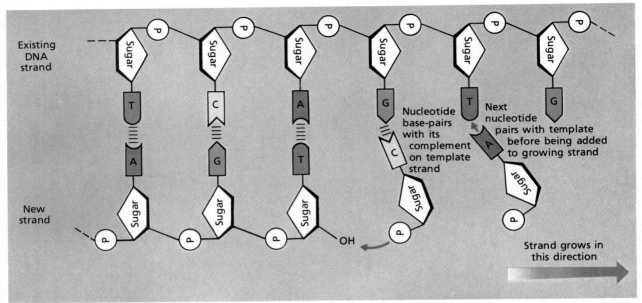

Figure 23-15 Replication of DNA. The double-stranded DNA "unzips" along the hydrogen bonds between the strands, producing single strands. At 1, a nucleotide containing cytosine forms hydrogen bonds with a complementary nucleotide, containing guanine, on the single strand. At 2, the enzyme DNA polymerase attaches the incoming cytosine nucleotide by its phosphate group to the sugar at the free end of the new, growing strand.

In the figure, the following labels appear:

- Existing DNA strand
- New strand
- Nucleotide base-pairs with its complement on template strand
- Next nucleotide pairs with template before being added to growing strand
- Strand grows in this direction

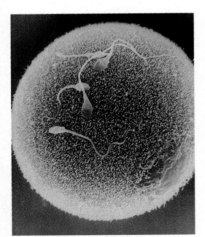

Figure 23-16 A human egg is shown here surrounded by sperm cells.

that DNA so that it has the information to make you you. Every time a new cell forms, it gets a complete copy of all your genes, made by duplicating the DNA.

There is another time when DNA must be duplicated—reproduction. When we pass on genes to our children, we do it by sending a copy of our genes to each sperm or egg we produce (Figure 23-16). All organisms, even the simplest ones, must transfer genetic information to their offspring. Plants pass on their genetic information in their eggs and pollen grains. Bacteria do it by dividing into two and leaving a copy of their DNA in each new bacterium.

The extraordinary process of DNA synthesis continues throughout an organism's life span. Remarkably, mistakes almost never occur. *An error in the sequence of DNA in a gene is known as* **mutation**. *Agents that can cause mutations, such as X rays or certain chemicals, are known as* **mutagens**. Some forms of cancer are caused by mutations.

RNA Synthesis

In addition to making copies of DNA, base-pairing permits DNA to make short RNA molecules containing the same information as parts of the DNA molecule. In this case, the RNA molecules do not stay bound to the DNA strands where they were produced. They leave the DNA and go off into the rest of the cell, carrying genetic messages to the locations where their information is required for protein synthesis.

Protein Synthesis

Proteins are made by structures inside the cell called **ribosomes**, *which are collections of RNA and protein molecules that act as the site of protein synthesis.* Ribosomes find out which proteins they are to make from RNA molecules called messenger RNAs. *A* **messenger RNA** *carries the coded information for a particular protein from DNA to a ribosome.* The ribosome decodes the RNA and makes the correct protein.

Technology

Toward Human Gene Transplants

Recent research indicates that we will one day be able to cure many inherited human diseases by gene transplants. For instance, beta thalassemia is a hereditary disease in which the blood cells produce defective hemoglobin. Victims suffer from severe anemia (lack of hemoglobin) and usually die in infancy. In one experiment, a human gene for hemoglobin was injected into fertilized eggs of mice with thalassemia. The eggs developed into adult mice with normal hemoglobin and passed the transplanted gene on to their offspring. In another experiment, scientists transplanted a rat growth hormone into mice, producing mice that are genetically identical to their normal counterparts except for their size (see photo).

The likelihood that we will one day be able to transplant genes from one human to another opens up many possibilities—both good and bad. It would mean that we could produce people with any genetic makeup we chose. We are not, and may never will be, morally ready to cope with genetic control over the human population. Scientists believe that organizations such as churches and medical societies should start to consider the moral and ethical implications of genetic engineering now, before the techniques are ready to be used.

Genetic Engineering

The biochemistry of genes, known as **molecular genetics***, is one of the most exciting and fast-moving branches of science.* Although we still have much to learn about genes, molecular genetics has already given rise to a new technology. We can isolate a desired gene and grow millions of copies of it. We can analyze these copies to find out the nucleotide sequence of the gene. We can find out the sequence of amino acids in the corresponding protein, and we can compare the structures of different genes. In several cases, we have even transferred functioning genes into the cells of plants and animals. Someday, we may be able to correct genetic defects with gene transplants. Current research in genetic diseases is seeking to find the precise locations of the one or more genes responsible for a disease.

We can also make DNA to order. Many laboratories now have "gene machines" that can be programmed to produce short strands of DNA in any desired sequence. This tailor-made DNA is a useful tool for the study of DNA. It can also be used in protein synthesis experiments. By changing the genetic code so as to eliminate particular amino acids from a protein, biochemists can determine how the amino acids affect the function of the protein as a whole.

These feats are all part of **genetic engineering**, *the deliberate manipulation of genetic material.* Its applications, today or in the future, include making safer vaccines by engineering a weaker version of the disease-causing agent; producing chemicals synthesized by bacteria; producing enzymes for industry; cleaning up toxic wastes from industry, oil spills, and pesticide accidents; replacing defective genes; and creating pest-resistant plants and animals.

Mapping the human genome   The availability of techniques for analyzing DNA makes it possible to think about mapping the entire human genome—analyzing human DNA so that we would know the relative positions of all human genes. In 1986, the United States committed itself to this formidable task. To print the nucleotide sequence for the entire human genome would require the equivalent of 200 Manhattan telephone books. It would take a large number of researchers years of time to assemble all this information. The effort would also require a lot of expensive equipment, both high-technology machinery to analyze the DNA, and powerful computers to handle the resulting data.

Many scientists question the value of committing so much money and talent to such a large project. Would we be doing it simply because it is now technically feasible? These people feel we would be better off working as in the past: focusing on specific genetic defects, pinpointing the responsible genes, and studying these genes and their neighbors to discover ways to alleviate the suffering of people born with those genetic defects.

Section Review

1. What is a gene?
2. Chemically, how does one organism differ from another?
3. How is genetic information coded in DNA?
4. What is the function of RNA?

23.6 Biochemical Pathways

Biochemistry is much more than the study of macromolecules and their reactions. It is the study of the chemical reactions that allow organisms to do what they do. For example, what chemical reactions permit living organisms to respond to stimuli—by exploring, running away, or rolling into a ball?

The general answer to these questions is that living things use energy to power a vast number of chemical reactions. These reactions permit all living organisms to act in ways that nonliving things cannot. Every living cell is a busy biochemical factory. Here monomers are converted from one form to another, new macromolecules are built and old ones dismantled, and the energy required to run all this is extracted from food molecules.

Thermodynamics is the study of the energy reactions required to perform chemical reactions. Biochemical thermodynamics is the study of the energy reactions in biochemical processes.

1. A length of DNA that codes for a particular protein
2. Each organism contains DNA that is chemically unique
3. The information in DNA is used to specify the sequence of amino acids in a protein.
4. It helps to coordinate the process of protein synthesis by carrying the coded information for a particular protein to a ribosome.

SECTION OBJECTIVES

- Define metabolism.
- State how one reaction is linked to the next in a biochemical pathway.
- Describe the importance of ATP in metabolic pathways.

Insulin is an example of a hormone that has been synthesized using genetic engineering.

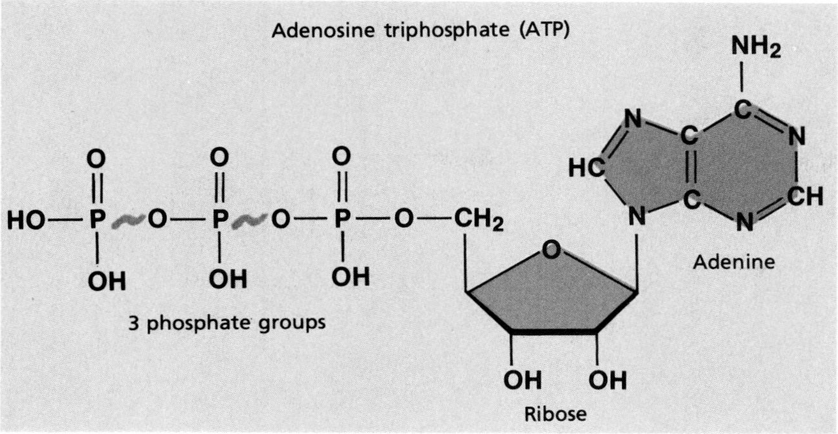

Adenosine triphosphate (ATP)

$$HO-\underset{\underset{OH}{|}}{\overset{\overset{O}{\|}}{P}}\sim O-\underset{\underset{OH}{|}}{\overset{\overset{O}{\|}}{P}}\sim O-\underset{\underset{OH}{|}}{\overset{\overset{O}{\|}}{P}}-O-CH_2$$

3 phosphate groups

NH₂

Adenine

Ribose

OH   OH

Figure 23-17 Adenosine triphosphate (ATP) is made up of the base adenine, the sugar ribose, and three phosphate groups (left). Considerable energy is released when the bond attaching the last phosphate group (red squiggle) is hydrolyzed, converting the ATP into adenosine disphosphate, ADP.

All of an organism's biochemical reactions are collectively called its **metabolism** and its reactions are organized into **metabolic pathways**, *chemically-linked steps in which a series of enzymes converts substrate molecules, step by step, into a final product.* For example, in animals one pathway converts the six-carbon sugar glucose into the five-carbon sugar ribose, which is needed to make nucleotides. This conversion uses four different enzymes. Other metabolic pathways require more or fewer enzymes.

Glycolysis

As an example of a metabolic pathway, let us consider **glycolysis**, *one of a group of metabolic pathways by which animals turn their food into an energy supply.* The energy supply produced by most energy-producing pathways is the nucleotide, adenosine triphosphate (ATP) (Figure 23-17). When the bond holding the last of ATP's three phosphate groups is hydrolyzed, considerable energy is released:

$$\underset{\substack{\text{adenosine}\\\text{triphosphate}}}{ATP} + \underset{\text{water}}{H_2O} \rightarrow \underset{\substack{\text{adenosine}\\\text{diphosphate}}}{ADP} + \underset{\substack{\text{phosphate}\\\text{ion}}}{PO_4^{3-}} + \text{energy}$$

The energy released by this reaction can be used to power reactions that cannot occur without an input of energy, such as the synthesis of macromolecules.

Cells can produce ATP by breaking down many different food molecules. One of the food molecules most commonly used to produce ATP is glucose, which is the substrate for glycolysis. Glycolysis is summarized in Figure 23-18. It consists of nine different reactions, each catalyzed by a different enzyme. Glycolysis adds to the amount of energy in the body by producing four molecules of ATP for each molecule of glucose that is broken down. However, note that glycolysis also uses up two existing molecules of ATP. This is because some of the reactions in the glycolysis series do not proceed spontaneously but are energetically "uphill." They need an input of energy from ATP to make them happen. One of the energy-using reactions is the first one, catalyzed by the enzyme hexokinase:

$$\text{Glucose} + \text{ATP} \xrightarrow{\text{hexokinase}} \text{glucose-6-phosphate} + \text{ADP}$$

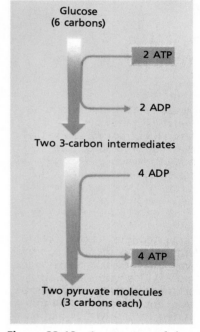

Glucose
(6 carbons)

2 ATP

2 ADP

Two 3-carbon intermediates

4 ADP

4 ATP

Two pyruvate molecules
(3 carbons each)

Figure 23-18 A summary of the nine reactions that make up glycolysis. Energy is needed to convert glucose into two 3-carbon molecules. The energy is supplied by a reaction that converts ATP to ADP. In the second half of glycolysis, two molecules of pyruvate are formed. These reactions release energy that is used to attach phosphate groups to 4 molecules of ADP, converting them into ATP.

Ask students to predict what would happen if a mistake occurred in the genetic code for hexokinase.

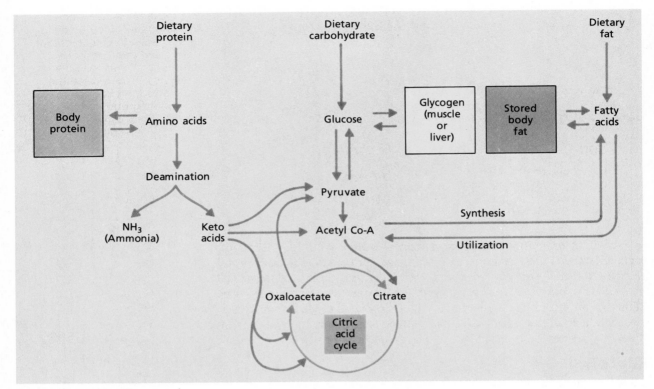

Figure 23-19 Interconnected metabolic pathways. Proteins, fats, and carbohydrates in the diet may become part of the body (circles) or may be processed to release energy. The most usual way for a cell to obtain energy is via glycolysis (from glucose to pyruvate, center). Pyruvate is usually converted into acetyl CoA to enter another metabolic pathway, the citric acid cycle, where more ATP is synthesized. Fatty acids and some keto acids made from amino acids may be converted into acetyl CoA and enter this pathway. Other keto acids enter as pyruvate or as molecules in the citric acid cycle. Both proteins and carbohydrates can also follow pathways that result in accumulation of body fat.

The pyruvate molecules that are the products of glycolysis are themselves the substrates for other metabolic pathways. Some of these generate still more ATP. Others use the pyruvate as substrates for building macromolecules.

Glucose is not the only food molecule that we use to produce energy. Energy comes from fat, proteins, and other carbohydrates. Figure 23-19 gives you an idea of the complex ways in which some metabolic pathways connect. For example, it shows that if we eat fat, the fat is hydrolyzed to fatty acids. These form the substrate for a metabolic pathway that produces a substance called acetyl Co-A. Acetyl Co-A, in turn, is a substrate for the pathway called the citric acid cycle. One product of the citric acid cycle is oxaloacetate that can be broken down into pyruvate. Pyruvate can be used as a monomer to build the storage polysaccharide glycogen. So this series of pathways, involving many different enzymes, has converted fat in our diet into a carbohydrate stored in the liver.

Organizing Metabolism

The various metabolic pathways interconnect in a complex pattern. Furthermore, each substrate occurs in very low concentrations. These tiny amounts of dozens of substrates must all be pushed into the right pathways when the body needs them, with no shortages or surpluses. How can these biochemical traffic patterns be controlled?

To bring order into the chaos, different metabolic pathways often occur in different areas. For instance, all the enzymes for one metabolic pathway may be arranged together in a group so that one enzyme's product is it neighbor's substrate. As the first enzyme releases its product, the second one binds the molecule before it can get away.

Membranes provide a second way to organize metabolic pathways. Many membranes separate compartments containing particular kinds of molecules and pathways.

Metabolic pathways can also be controlled by negative feedback control of a process by its product. When the pathway works too fast, too much product builds up. Some of the product molecules attach to molecules of the pathway's first enzyme and switch them off (just as hot air switches off the thermostat on a heater). This stops the entire pathway. Later, as the product is used up by other pathways, product molecules unbind from the start-up enzymes, and the enzymes go back to work.

Section Review

1. Define metabolism.
2. How does glycolysis contribute to a cell's energy store?

Chapter Summary

- Living organisms are made up of water, inorganic ions, and organic molecules. Organic macromolecules are polymers, each made up of many monomers.
- Large organic molecules are called macromolecules. Most of them fall into four main groups: carbohydrates, lipids, nucleic acids, and proteins.
- Proteins are made of monomers called amino acids. Different proteins have different sequences of amino acids. When linked together, these form polypeptides that have unique shapes.
- Enzymes are protein catalysts. Each enzyme catalyzes reactions between specific substrates. Enzymes enable organisms to carry out chemical reactions quickly. The activity of enzymes if affected by pH and temperature.
- Carbohydrates and lipids are composed mainly of carbon, hydrogen, and oxygen. Carbohydrates include monosaccharids, disaccharides, and polysaccharides. Some carbohydrates and lipids are important energy-storage compounds that may be broken down to release energy.
- Carbohydrates tend to be polar and associate with water, but lipids are nonpolar and so do not dissolve in water. Lipids include fatty acids, waxes, fats, oils, phospholipids, and steroids.
- Nucleic acids, such as DNA and RNA, and proteins play vital roles in directing an organism's growth, activity, and reproduction. DNA stores genetic information. RNA helps translate this information into proteins. Enzymes and structural proteins are examples of important proteins.
- Organisms extract useful energy from the energy stored in the chemical bonds of food molecules. The energy so released is used to make the cell's supply of ATP. ATP, in turn, donates the energy to various energy-requiring processes, such as metabolic reactions or production of new polymers.
- Glycolysis is an example of a metabolic pathway. During glycolysis, glucose is broken down to two molecules of pyruvate. Many other metabolic pathways feed into glycolysis, enabling cells to use many organic compounds other than glucose as food sources to generate usable energy in the form of ATP.

1. A term used to collectively describe all of an organism's biochemical reactions
2. It begins the process of glucose breakdown by converting glucose into pyruvate, which can then serve as a substrate for other metabolic pathways.

Chapter 23 *Review*

Biochemistry

Vocabulary

adenosine
 triphosphate (ATP)
amine
amino acid
base-pairing
biochemistry
carbohydrate
cellulose
chitin
cholesterol
coenzyme
complex carbohydrate
denaturation
deoxyribonucleic acid (DNA)
disaccharide
double helix
enzyme
epinephrine
fatty acid
gene
genetic engineering
glycogen
glycolipid
glycolysis
hormone
lipid
lipoprotein
macromolecule

messenger RNA
metabolic pathways
metabolism
molecular genetics
monosaccharide
mutagen
mutation
neuron
neurotransmitter
nucleic acid
nucleotide
peptide bond
phospholipid
polysaccharide
polyunsaturated fat
product
protein
ribonucleic acid (RNA)
ribosome
saturated fat
starch
steroid
substrate
synapse
unsaturated
unsaturated fat
wax

Questions

1. What are some of the inorganic compounds that supply the elements needed by living things to build their organic compounds?
2. What four elements make up the bulk of living organisms?
3. How is water important to life?
4. How do organic and inorganic compounds differ from one another?
5. Write a reaction showing the formation of a peptide bond between two amino acids.
6. What is the function of waxes in living things?

7. List the three components that make up a nucleotide.
8. To what class of organic molecules does the following belong?

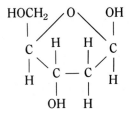

9. To what class of organic molecules does the following belong?

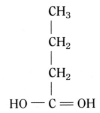

10. Name the four main classes of biological molecules.
11. How does heat affect a protein?
12. Name the monomers that make up
    (a) proteins   *(b)* carbohydrates   *(c)* nucleic acids.
13. What is an enzyme?
14. How does the order of nucleotides in one strand of DNA reflect the order of nucleotides in the other strand?
15. Name a nucleotide that is not part of a nucleic acid.

Problems

1. Explain the function of DNA as fully as you can.
2. You have a solution of an enzyme. You put half of it into each of two beakers containing identical substrate at equal concentrations. After waiting a while, you test both solutions and find

that the substrate in beaker A has been changed but that the substrate in beaker B has not been acted on by the enzyme. Suddenly you notice that beaker B has been sitting on a hot plate with the switch turned to "high." Why did the enzyme in beaker B probably not work?

3. A fragment of DNA has the nucleotide sequence AAATCGGCA. If another strand of DNA paired with this fragment, what would be its nucleotide sequence?

4. If you consume 1000 kilocalories of carbohydrate, how many grams of fat would you have to eat to supply the same amount of energy?

Application Questions

1. List the four classes of lipids and give an example of each.
2. Why is the action of an enzyme specific, meaning that each enzyme catalyzes reactions involving only one or a few substrates?
3. Why do foods that are rich in fat tend to be costly compared to the cost of carbohydrates?
4. Name two lipids with different functions and state their uses in the body.
5. List three functions of different proteins.
6. Draw the molecular structure of glucose.
7. Name two structural carbohydrates.
8. Name two factors that can alter the activity of an enzyme.
9. List three functions of carbohydrates.
10. Many proteins are enzymes. Name some other functions that proteins perform in the body.
11. To what group of macromolecules does glycogen belong?
12. What is the function of glycogen?
13. What is the function of cellulose?
14. What type of macromolecule is used to store energy in plants? Name a plant that is useful to humans because of its ability to produce large amounts of this macromolecule.

Application Problems

1. Science fiction tales sometimes feature life forms that are based on silicon rather than carbon. Silicon is much more abundant on earth than carbon. Like carbon, its atoms can bond to four other atoms. Bonds between two silicon atoms are unstable in the presence of O_2, but bonds between silicon and oxygen atoms are extremely stable and difficult to break. Is this the reason life is based on carbon not on silicon? Why or why not?

2. How does ATP differ from a nucleotide in a DNA molecule?

3. Explain what happens to the protein in an egg when the egg is cooked.

4. A person who is too sick to eat may be given liquid nourishment intravenously. What carbohydrate is used for this purpose? Why?

Enrichment

1. Find an article in a newspaper or magazine about recent biochemical research. Use this chapter and a biochemistry text from the library to make notes for a talk in which you explain this research to your classmates.

2. How do you imagine genetic engineering might affect your life? How might it affect the next generation?

3. What inorganic compound is the source the carbon used to build macromolecules? (You may need a biology book to answer this one.)

4. When you cut an apple or banana, phenol oxidase enzymes in the injured areas quickly begin a "wound reaction," which results in the cut surfaces turning brown. Many cooks sprinkle lemon juice on sliced fruit to prevent it from discoloring in this way. Why does this work?

5. Keep a written record of all the foods you consume in a single day. Classify the foods as proteins, carbohydrates, or lipids.

6. Consult recent magazines and scientific journals in your library to learn the most current reports on the project to map the human genome. Share your findings with your classmates.

7. After your class has read more about genetic engineering, have a debate on this new area of scientific research.

8. Compile a list of chemicals that are considered to be mutagens. What mutagenic substances might you encounter in your daily life in your town or city?

9. Prepare a report on the dangers posed by the use of anabolic steroids to improve athletic performance.

10. Make a list of some of the products in the grocery store that are fats or oils. Look up information in the library about each item. Which are saturated fats? Which are unsaturated? Group the products according to whether they are considered healthy or unhealthy.

CAREERS

Agricultural Chemist

From seed to harvest, plants need the care and protection that chemists help provide. It's the agricultural chemist who develops new techniques to improve the yield and quality of crops. Seeds are often treated with pesticides to protect them until germination occurs. Herbicides keep down the growth of weeds that tend to crowd seedlings. And fertilizers provide missing nutrients.

If working outdoors appeals to you, you may consider

employment with the U.S. Department of Agriculture, doing field tests and consulting with farmers about their crops, animals, and water supplies . A high school diploma is required to work as a caretaker of plants and animals. Job advancement requires an associate's degree from a junior college, including as many chemistry and biology courses as possible. For a future in biotechnology, you need a bachelor's degree in chemistry, botany, ecology, molecular biology, genetics, or microbiology.

Biochemist

One of the most fascinating fields of science is biochemistry. A biochemist, such as the one pictured at upper right, works with substances that are present in living organisms. He or she might explore the pathways of an organism's metabolism or discover how closely related organisms share chemical similarities.

Perhaps the most active area of biochemistry is genetic engineering. Biochemists working in this area assist in "redesigning" DNA, a molecule in living cells that stores hereditary information and allows an organism to pass character traits from one generation to the next. Biochemists use chemical techniques to identify specific

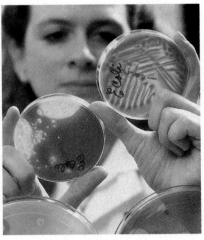

regions of DNA. These regions can then be isolated and added to the DNA within cells of a different organism, changing the organism's genetic characteristics. The resultant cells contain hybrid DNA and can be grown in the laboratory. In this way, cells can be given the instructions to manufacture a product that they would not produce under natural conditions. Biochemists can then use these techniques to produce valuable products, such as human insulin.

A few examples of the many research areas for biochemists are endocrinology, the study of hormones; oncology, the study of cancer; and cell biology, which includes study of the chemical reactions within cells.

A biochemist must have a master's or doctoral degree in chemistry or biology.

Cumulative *Review*

Unit 7 • Carbon and Its Compounds

Questions

1. Name the three major natural sources of organic compounds. (21.1)
2. Describe the forces acting in graphite between carbon atoms in each layer and between the layers. (21.1)
3. Write the formulas for any two organic compounds and any two carbon-containing inorganic compounds. (21.1)
4. Two solid compounds, A and B, are found to have the following properties: compound A, flammable and burns in air with a smoky flame, soluble in ether, not soluble in water; compound B, not flammable, not soluble in ether, soluble in water. Which is most likely organic and most likely inorganic? (21.2)
5. Identify each of the following compounds as as alkane, an alkene, or an alkyne:
   (a) $CH_3CH_2C\!\equiv\!CCH_3$ *(b)* $CH_3CH_2CH_2CH_3$
   (c) $CH_2\!=\!CH_2$ *(d)* CH_4.   (21.3)
6. Identify each of the following compounds as an alkane, an alkene, or an alkyne: *(a)* C_3H_6
   (b) $C_{10}H_{22}$ *(c)* C_3H_8 *(d)* C_5H_{10} (21.3)
7. By drawing the structural formulas, determine the possible isomers of pentadiene. (21.2)
8. *(a)* What is the name given to a reaction of the following type?

   $$CH_3CH_3 + Br_2 \rightarrow CH_3CH_2Br + HBr$$

   (b) What is the name given to a reaction of the following type?

   $$\overset{\displaystyle Br \quad Br}{\underset{\textstyle |\quad\;\; |}{CH_3CH = CH_2 + Br_2 \rightarrow CH_3CH\ CH_2}} \quad (21.3)$$

9. Name the monomer that is polymerized to produce neoprene. Write the structure of this monomer and of the repeating unit in neoprene. (21.4)
10. Draw the structural formula of a carbonyl group and a carboxyl group. (22.4, 22.5)
11. Write the formula for each of the following compounds: *(a)* trifluoroacetic acid

(b) dimethyl ether   *(c)* hexanoic acid
(d) 2-methylbutanol   *(e)* formic acid
(f) acetaldehyde.   (22.2, 22.3, 22.4, 22.5)

12. Name each of the following compounds and choose the one most likely to be used as a gasoline additive, a flavoring, a grease solvent, the monomer for a common polymer, a skin moisturizer: *(a)* CCl_4 *(b)* C_2H_5OH
    (c) $C_3H_5(OH)_3$ *(d)* $CH_2\!=\!CHCl$
    (e) $C_3H_7COOC_2H_5$.   (22.1, 22.2, 22.4, 22.5)
13. Define condensation and hydrolysis reactions as they occur in living bodies, and explain their roles.   (23.2, 22.5)
14. Name the four classes of macromolecules essential to life and indicate which class includes   *(a)* fats   *(b)* sucrose   *(c)* DNA
    (d) enzymes   *(e)* cartilage   *(f)* polysaccharides   *(g)* cholesterol.   (23.1)
15. What is an enzyme and what is an enzyme substrate?   (23.2)
16. What is the function of a coeyzyme? What is the name given to coenzymes that we must include in our diet?   (23.2)
17. *(a)* To what major class of biochemical compounds does glucose belong?   *(b)* Is glucose a monosaccharide or a disaccharide?   *(c)* What are two functions of glucose in the body?   (23.3)
18. *(a)* What is the function of a biochemical membrane?   *(b)* Of what class of biochemical compounds are membranes composed?   (23.4)
19. Describe neurotransmitter function.   (23.4)
20. *(a)* What do the abbreviations DNA and RNA stand for?   *(b)* What is the very important function of these molecules in living things?   (23.5)
21. *(a)* Define metabolism.   *(b)* What are metabolic pathways?   *(c)* What role is played by the metabolic pathway known as "glycolysis"? (23.6)
22. *(a)* Name the nucleotide that is the energy storing product of glycolysis.   *(b)* Describe the chemical reaction by which the stored energy is released from this nucleotide.   (23.6)

Intra-Science
How the Sciences Work Together

The Process in Chemistry: Custom-Designed Plants

Custom-designed plants that are able to flourish in salty soil, resist drought, or survive deadly diseases are gradually becoming a reality. Tobacco plants that are resistant to herbicides are being developed. Strawberry plants sprayed with genetically altered bacteria are able to resist frost damage.

Ever since agriculture began, about 10 thousand years ago, farmers have tried to create improved breeds of plants. Parent plants with desirable characteristics were bred to each other to produce superior offspring. It was not until the mid 1800s, however, that a systematic study of inheritance

patterns was made. Gregor Mendel, an Austrian monk, clarified the basis of heredity with his pea-plant experiments. Mendel's greatest contribution was the discovery that hereditary characteristics are determined by units (now called genes) rather than by a blending of parental characteristics.

A second major step in genetics was Watson and Crick's descrip-

tion of the DNA (deoxyribonucleic acid) molecule. Their work in the 1950s began a revolution in genetics, leading to the creation of artificial genes and new forms of life.

The molecular basis of heredity is the DNA molecule. DNA is made of long chains of nucleotides. A nucleotide consists of a molecule of the sugar deoxyribose, a phosphate group, and one of four bases—adenine, thymine, cytosine, or quinine. The sequence of bases acts as a code that regulates the sequence of amino acids in a protein. DNA controls the characteristics of an organism by controlling the synthesis of specific proteins.

Every plant has thousands of genes (a gene is a section of DNA). By changing the genes or inserting new ones, plants with new characteristics—custom-designed plants—can be created.

The Connection to Physics
The chemical nature of genes already present in an organism can be altered with radiation. Ultraviolet light, alpha-particles, beta-particles, high-energy electrons,

X rays, and other types of high-energy radiation can dislodge one or more electrons from the DNA of a gene. The ionized molecule is more reactive than normal and may combine with other substances. A nucleotide may be added or deleted, and this changes the sequence of bases. As a result, the sequence of amino acids in the

inserted. This process, electroporation, has been used to transfer new genes into corn cells. The biggest stumbling block has been the difficulty of regenerating whole plants from genetically altered cells. If this problem can be solved, electroporation will enable scientists to transfer genes between unrelated species. More nutritious corn, or wheat that can survive floods, may be produced.

The Connection to Biology

Viruses and bacteria are used to transport new genes into the cells of plants. Bacteria of the genus *Agrobacterium* are most commonly used. These bacteria normally cause diseases in plants by inserting a plasmid, a ring of bacterial DNA, into plant cells. The plasmid

the synthesis of an enzyme, EPSP synthetase, which is needed for making three essential amino acids. A mutant gene that produces a slightly different EPSP synthetase enzyme was inserted into cells of tobacco plants, using *Salmonella* bacteria. When fields of genetically altered tobacco plants are sprayed by glyphosate, the weeds are killed. The tobacco plants survive because they are still able to produce the essential amino acids.

Plant pathologists isolated the

gene of bacteria that causes ice crystals to form and destroy plant cells. They altered the gene that codes for the synthesis of the protein and inserted it into bacteria. The new bacteria crowd out the old frost-forming type of bacteria, leaving the plants undamaged.

Scientists also learned how to form artificial genes by joining nucleotides together in long chains. The ability to make artificial genes, coupled with techniques to introduce genetic material into plant cells, opened the possibility of creating new life forms.

protein produced by the gene will be different from the protein normally produced. Such a change in the amino acid sequence could improve the organism in some way. Conversely, the change could render the protein nonfunctional and kill the organism.

To insert foreign genes into organisms, genetic engineers use electricity to create tiny openings in the membranes of plant cells through which the genes can be

codes for the synthesis of a protein that causes plant cells to grow abnormally. Molecular biologists inserted a desirable gene into the bacterial plasmid. When the plant cell is infected by the bacterium, it also receives a copy of the desirable gene.

This process has been used to give tobacco plants greater resistance to herbicides that are used to kill weeds. The herbicide glyphosate kills plants by preventing

Unit

8

Chapter 24
The Metals of Groups 1 and 2   712

Chapter 25
The Transition Metals   728

Chapter 26
Aluminum and the Metalloids   750

Chapter 27
Sulfur and Its Compounds   764

Chapter 28
The Halogens   778

DESCRIPTIVE CHEMISTRY

Growing up in Butte, Montana, one of the major historical copper-producing areas in North America, one cannot help being interested in the mining and processing of copper ores. It wasn't until college, however, that I fully appreciated the chemical complexity of many of the modern processes. I am now a process metallurgist in a plant that uses a new method to treat previously untreatable gold ores.

In this unit you will learn about some of the more useful chemical elements, how they are recovered from nature, and how the science of chemistry is used to turn them into valuable products.

William J. Janhunen, Jr.
Process Metallurgist
Homestake Mining Company / McLaughlin Mine

Copper mine near Ruth, Nevada

Chapter 24 The Metals of Groups 1 and 2

Chapter Planner

24.1 The Alkali Metals
Structure and Properties
  ▪ Demonstration: 1
  Question: 1
Chemical Behavior
  ▪ Question: 8, 9
  ▪ Demonstration 2
  Chemistry Notebook
  ▪ Problems: 1–3

24.2 The Alkaline-Earth Metals
Structure and Properties
  ▪ Demonstration: 3
  ▪ Demonstration: 4

**Properties of Alkaline-Earth
Metal Compounds**
  ▪ Demonstrations: 5–10
  Experiment: 34
  ▪ Chemistry Notebook

**24.3 Sodium: One of the Active
Metals**
Preparation of Sodium
Properties and Uses of Sodium
  ▪ Demonstration: 11
  Questions: 2 , 3

**Important Compounds of
Sodium**
  ▪ Question: 6
  Application Questions: 2,
  5
  ▪ Questions: 5, 10
  Application Question: 4
  ▪ Demonstration: 12
  Question: 4
  Application Question: 7
  Application Problem: 1, 3

Teaching Strategies

Introduction

This is a descriptive chapter dealing with the alkali metals and the alkaline-earth metals. The emphasis is on the group properties of these elements and how they relate to the electron configurations and periodic properties of atoms discussed earlier. The chemistry of sodium is described in some detail, including a discussion of the preparation methods, major uses, and important industrial compounds.

24.1 The Alkali Metals

In discussing the properties of the alkali metals, keep in mind the general properties of the family of elements and relate them to the electron configurations of this group.

Students often confuse the properties of the elemental substances with the properties of the compounds formed by the elements. Help them keep in mind this distinction when discussing their properties. The elements are extremely reactive, and the compounds are much less reactive. Display samples of sodium and potassium stored under kerosene and show samples of some of the common compounds of the elements. Show samples of sodium chloride, sodium hydrogen carbonate, sodium hydroxide, potassium chloride, potassium iodide, and potassium carbonate. This will help students appreciate the difference in reactivity between the elements and the compounds.

To help students understand the concept of the unit cell, exhibit a model made from styrofoam balls or gumdrops and toothpicks of a body-centered cubic lattice. Discuss the mobility of free electrons and related properties .

Contrast the low melting points, low densities, and softness of alkali metals with the properties of the more familiar metals iron, nickel, copper, silver, and gold. Contrast the soft metallic lattice with a rigid salt crystal, showing that bending a metallic crystal forces rows of atoms to move past other identical rows of atoms, while bending a salt crystal would force positive and negative ions to move past one another.

Have students observe the long liquid temperature range (melting point to boiling point) of the alkali metals, which is indicative of the strength of the metallic bond. Ask students to supply details of electronic configuration and general trends in properties. Help them understand that the Group 1 ions have the stable electron configurations of the preceding noble gases and that this is why the ions have different chemical properties from the element. Review the trends in atomic and ionic sizes and ionization energies. Discuss metallic and ionic radii. Discuss the electron configuration of negative ions such as Na^- and the oxidation states of the anions.

Explain why free alkali metals are such vigorous reducing agents. Remind students that a good reducing agent is itself readily oxidized, losing electrons to the substance that is reduced. Show the relationship between the ease of oxidation and the ionization energy. Explain the unusual behavior of lithium in terms of small ion size, the consequent high charge density, and the hydration energy.

Demonstration 1: Flame Tests
PURPOSE: To show the colors produced when alkali metal compounds are excited in a flame
MATERIALS: Device made from 2-liter plastic soft drink bottle, cork, burner, and spray bottle; Assemble as shown in Figure T 24-1.
PROCEDURE: Spray solutions of alkali metal compounds into hole in bottle. Observe color in flame. Use solutions of NaCl, KCl, and Li_2CO_3. Compare the colors when NaCl, Na_2SO_4, KCl, and K_2SO_4 are used to verify that the color is produced by the alkali metal ion and not the negative ion. With this device the color is sufficiently long-lasting that students can observe the line spectra using an inexpensive hand spectroscope.

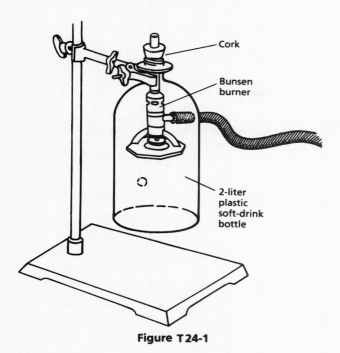

Figure T 24-1

Demonstration 2: Reaction of Sodium with Water.

PURPOSE: To show the reactivity of sodium with water
MATERIALS: Beaker, sodium-lead alloy (90% lead, 10% sodium, obtainable from J. T. Baker Chemical Co., Phillipsburg, NJ), water, phenolphthalein, overhead projector
PROCEDURE: Wear safety goggles. Drop small pieces of sodium-lead alloy into a beaker of water containing some phenolphthalein on the overhead projector. This alloy is a convenient and safe source for reacting small amounts of sodium. Hydrogen gas is evolved, and the pink color of the solution shows that sodium hydroxide solution has also formed. Explain to students that the lead is a "moderator."

Chemistry Notebook: Group 1 Ions in Medication and Diet

Discuss with the students the effects of salt concentration in cells as a means of regulating water retention of cells by the process of osmosis. Show containers of "No-Salt Salt" for seasoning food, and have the students read the labels to determine the contents. Most students will be aware of the fact that perspiration removes much sodium chloride from the body. Discuss the need to replace sodium chloride after strenuous exertion.

24.2   The Alkaline-Earth Metals

Compare and contrast the properties of the elements of Group 2 with those of Group 1. Students should recognize that this group of metals is also a very reactive family, though not as reactive as the Group 1 metals. Point out that the alkaline-earth metals are, like the alkali metals, never found free in nature but are always in a combined form. Display samples of the metals magnesium and calcium. Discuss the fact that these do not have

to be stored under kerosene. Explain the difference in reactivity in terms of the energy required for these atoms to lose two electrons while the alkali metals need only lose one electron to form compounds.

Contrast the greater densities, hardness, and higher melting and boiling points of the alkaline-earth metals with the properties of the alkali metals. Explain the differences in terms of metallic crystal makeup. The alkaline-earth metal 2 + ions are more strongly attracted to the electron "gas" that permeates the metal than are the alkali metal 1 + ions. Compare the sizes of the atoms of Groups 1 and 2.

Display some of the common compounds of the alkaline-earth metals. You could show samples of $CaCO_3$, $Ca(OH)_2$, CaO, MgO, $Mg(OH)_2$, and $Ba(OH)_2$. Write the electron configurations of the elements, and discuss the reason for the large increase in ionization energy when the elements lose their third electrons. Use the concept of ionization energy to explain the increase in reactivity as one moves down the column in the Periodic Table.

Demonstration 3: Reactivity of Magnesium and Calcium with Water

PURPOSE: To compare the reactivity of magnesium and calcium with water
MATERIALS: Calcium metal, magnesium metal, two beakers, water, phenolphthalein
PROCEDURE: Wear goggles. Place a small piece of magnesium metal in a beaker that contains water and a few drops of phenolphthalein solution. Put a piece of calcium metal in another beaker with water and phenolphthalein solution. Observe that the calcium reacts fairly rapidly with evolution of hydrogen gas and formation of slightly soluble $Ca(OH)_2$, as shown by the phenolphthalein turning pink. The magnesium does not react. Relate the difference in reactivity to their positions in the Periodic Table.

Demonstration 4: Combination of Magnesium with Nitrogen

PURPOSE: To show that magnesium metal can react with nitrogen.
MATERIALS: Magnesium ribbon, crucible and crucible cover, red litmus paper, beaker, ring stand, triangle, burner
PROCEDURE: Wear goggles. Make a coil of a 5-cm piece of magnesium ribbon and place it in a crucible. Cover it loosely with the crucible cover. Support it on a ring stand and triangle, and heat it until the magnesium reacts. Allow it to cool. Examine the product and note the presence of white MgO and traces of dark Mg_3N_2. Add water, and the Mg_3N_2 will react to form NH_3 gas, which can be detected if a piece of moist red litmus is held over the crucible. Add some MgO to water in a beaker. $Mg(OH)_2$ will form, which can also be detected by the litmus paper.

Demonstration 5: Formation of Hydroxides from Alkaline-Earth Oxides

PURPOSE: To show that oxides of Ca, Mg, and Ba form hydroxides with water
MATERIALS: CaO, MgO, BaO, water, litmus paper, test tubes
PROCEDURE: Add the oxides to water, test with litmus paper to show that the solutions are basic. Write the equations for the reactions. Observe that the hydroxides are not very soluble.

Demonstration 6: Flame Tests with Alkaline-Earth Compounds

PURPOSE: To show the colors produced when alkaline-earth metal compounds are excited in a flame

MATERIALS: Device made from 2-liter plastic soft drink bottle, cork, burner, and spray bottle (See Demonstration 1); solutions of alkaline-earth compounds such as magnesium chloride, calcium chloride, barium chloride, and strontium chloride

PROCEDURE: Spray solutions of alkaline-earth compounds into the hole in the bottle and observe the characteristic colors in the flame

Demonstration 7: Reaction of Milk of Magnesia with Acid

PURPOSE: To show the gradual neutralization of acid by $Mg(OH)_2$

MATERIALS: Milk of Magnesia tablet, 600-mL beaker, distilled water, 1.0 M HCl, universal indicator, magnetic stirrer

PROCEDURE: Place 300 mL of distilled water in the beaker on a magnetic stirrer. Add 6 drops of 1.0 M HCl and 24 drops of the universal indicator. Drop in one tablet of Milk of Magnesia. Observe the color changes. The pH of the solution slowly changes from pH 4 to pH 10 as the tablet gradually neutralizes the acid. The indicator changes color from red to orange to yellow to green to blue and then violet.

Demonstration 8: Limewater Test for Carbon Dioxide

PURPOSE: To show how carbon dioxide precipitates calcium carbonate in limewater solution

MATERIALS: Saturated solution of calcium hydroxide (limewater), drinking straw

PROCEDURE: Have a student blow his or her breath into a solution of limewater. (Caution them not to drink the solution.) Observe the formation of the precipitate of $CaCO_3$. Write the equation for the reaction.

Demonstration 9: Hard Water

PURPOSE: To show the effects of hard water on soap and detergent solutions

MATERIALS: Distilled water, tap water, hard water containing a little calcium chloride and magnesium chloride, soap solution, detergent solution, test tubes

PROCEDURE: Place equal amounts of distilled water, tap water, and hard water in separate test tubes. Add soap solution, a few drops at a time, to each test tube and shake until a stable lather occurs. Observe that: distilled water lathers readily and appears clear; tap water forms a cloudy solution and needs more soap to form a lather; and hard water forms the most scum and requires the greatest amount of soap for a lather. Repeat with detergent instead of soap solution. Carry out Demonstration 10 to show how the hard water can be treated.

Demonstration 10: Softening Hard Water

PURPOSE: To show ways of softening hard water

MATERIALS: Hard water solutions from Demonstration 9, Na_2CO_3, soap solution

PROCEDURE: Add Na_2CO_3 to the hard water solution used in Demonstration 9 to precipitate $CaCO_3$ and $MgCO_3$. Repeat

the test with soap solution to see how much is needed to form a stable lather. Compare these results with those of the unsoftened water in Demonstration 9.

Chemistry Notebook: Strontium: Fireworks and Fallout

Students are always fascinated by fireworks. If you have done flame-test demonstrations, you can ask them to discuss different substances that will give various colors to fireworks.

Point out that strontium is very similar to calcium in its chemistry. Growing children absorb more calcium than adults and are thus more susceptible to having strontium-90 deposited in bones. Remind them that blood is produced in bone marrow, and discuss diseases of the blood and blood-forming tissues, such as leukemia.

24.3 Sodium: One of the Active Metals

Students may need some help in understanding the equations given for the various reactions. Explain the oxidation-reduction reaction that occurs in the Downs cell. Write the formulas for the ions oxide, O^{2-}, peroxide, O_2^{2-}, and superoxide, O_2^-. Discuss the uses of sodium. Explain production of tetraethyl lead from ethyl chloride and a sodium-lead alloy. Display and operate a sodium lamp. (Your physics department may have one.) Discuss the use of sodium as a heat-transfer agent in atomic reactors.

Display rock salt, table salt, and halite crystals. Examine NaCl crystals with a hand lens or microscope and compare them with a model of NaCl face-centered structure. Show that when rock salt crystals break, they fracture so that the faces are flat and the angles between the faces are 90°.

Exhibit forms of NaOH—flake, stick, and pellet—along with containers of commercial lye. Have students read the labels on cans of lye, which provide valuable information about the preparation of solutions, the uses of NaOH, and antidotes. Exhibit a can of any product used to clear clogged drains that contains NaOH in excess of 50 percent.

Demonstration 11: Dissolution of NaOH and Reaction of NaOH and Aluminum

PURPOSE: To show the exothermic nature of the dissolution of sodium hydroxide and the reaction that occurs between sodium hydroxide solution and aluminum

MATERIALS: Beaker, thermometer, sodium hydroxide pellets, aluminum foil

PROCEDURE: Wear safety goggles. Prepare approximately 3 M NaOH by adding 12 g of NaOH pellets to 92 mL of water. Measure the temperature change as the dissolution progresses. (Remember that NaOH is caustic. Wash off any spills with copious amounts of water.) Add a few pieces of aluminum foil to the warm solution to show how the aluminum dissolves with the formation of hydrogen gas. Write the equation for the reaction. Discuss the use of aluminum in drain cleaners containing NaOH. Explain why oven cleaners can be used on porcelain or stainless steel ovens and utensils, but not on aluminum.

Display a chalkboard replica or transparency of the flow diagram of the Solvay process, and write out the appropriate equations. Exhibit commercial varieties of baking soda and washing soda.

Demonstratration 12: Preparation of Potassium Nitrate

PURPOSE: To show the preparation of potassium nitrate and sodium chloride by crystallization from solution.

MATERIALS: Sodium nitrate, potassium chloride, filter paper, funnel

PROCEDURE: To 25 mL of water add 12 g $NaNO_3$ and 10 g KC1.

Heat to complete solution and evaporate to one-half volume. Let the solution cool. Filter out the crystals (mostly sodium chloride). Examine with a hand lens or microscope, and compare with NaCl crystals. Evaporate filtrate further until a second crop of crystals forms (mostly KNO_3). Filter and examine these with a lens or microscope to see their crystal form, comparing with KNO_3.

References and Resources

Books and Periodicals

Alyea, Hubert N. "Burning Magnesium in Steam." *Tested Demonstrations in Chemistry*, 6th ed., 1965, p. 142. Reprint from the *Journal of Chemical Education*. Published by the Division of Chemicial Education of the American Chemical Society.

Alyea, Hubert N. "Removal of Hardness and Deionization of Water by the Ion Exchange Method." *Tested Demonstrations in Chemistry*, 6th ed., 1965, p. 219. Reprint from the *Journal of Chemical Education*. Published by the Division of Chemical Education of the American Chemical Society.

Dandy, Alan J. "Chemistry of Cement." *CHEM*, Vol. 51, 1978, p. 13.

Gouge, Edward M. "A Flame Test Demonstration Device," *Journal of Chemical Education*, Vol. 65, 1988, p. 544.

Hall, Christopher. "On the History of Portland Cement," *Journal of Chemical Education*, Vol. 53, 1976, p. 222.

Hall, Christopher. "Chemistry of Fossilization," *Journal of Chemical Education*, Vol. 53, 1976, p. 270.

Summerlin, Lee R., Christie L. Borgford, and Julie B. Ealy. "Making Sodium Chloride from Sodium and Chlorine," *Chemical Demonstrations, A Sourcebook for Teachers*, Vol. 2, American Chemical Society, 1987, p. 56.

Summerlin, Lee R., Christie L. Borgford, and Julie B. Ealy. "Burning Magnesium in Carbon Dioxide." *Chemical Demonstrations: A Sourcebook for Teachers*, Vol. 2, American Chemical Society, 1987, p. 58.

Summerlin, Lee R., Christie L. Borgford, and Julie B. Ealy. "Reduction of Sand with Magnesium." *Chemical Demonstrations: A Sourcebook for Teachers*, Vol. 2, Americal Chemical Society, 1987, p. 193.

Audiovisual Resources

"Chemical Families," CHEM Study 16mm film, Ward's Modern Learning Aids. Shows the elements potassium, lithium, and sodium reacting with a number of nonmetals

"Chemicals from NaCl: 1" video, Films for the Humanities and Sciences, FFH1110F. Shows production of sodium hydroxide and chlorine by electrolysis of salt

"Chemicals from NaCl: 2" video, Films for the Humanities and Sciences, FFH1111F. Shows production of sodium metal from salt

"The Electrolysis of Brine": 16 mm film, PPG Industries, Inc., Free-loan film available from Modern Talking Picture Service. Shows production of chlorine and sodium hydroxide and their importance

"Salt—The Essence of Life": 16mm film, Salt Institute, Free-loan film available from Modern Talking Picture Service. Shows history, production, and uses of salt

Computer Software

"The Periodic Table Videodisc: Reactions of the Elements": CAV-type videodisc, Journal of Chemical Education Software, 1989. Shows each element in its most common forms; reactions with air, water, acids, bases; and some common applications

Answers and Solutions

Questions

1. The atoms of Group 1 elements have electron configurations of the preceding noble gases plus one additional electron in the next higher energy level. The ions have the electron configurations of the preceding noble gases.

2. Sodium is prepared by electrolysis of its fused chloride. Potassium is now prepared by the reaction of metallic sodium and potassium chloride.

3. Uses: in making tetraethyl lead for antiknock gasoline; as a heat-transfer agent; in making dyes; in making sodium-vapor lamps

4. Caustic substances, such as sodium hydroxide, are capable of changing some types of animal and vegetable matter into soluble materials by chemical means. Caustic substances have destructive effects on skin, hair, and wool. "Corrosive" is a more general term for substances that eat away by chemical action, such as the action of an acid on a metal.

5. *(a)* Common salt, limestone, and coal *(b)* Coal tar, sodium carbonate, sodium hydrogen carbonate, and calcium chloride

6. Molasses contains acids that react with baking soda to form carbon dioxide gas.

7. Large deposits of potassium chloride, crystallized with magnesium and calcium compounds, are found in Texas

and New Mexico. Some potassium compounds are extracted from Searles Lake in California.

8. *(a)* Under kerosene *(b)* Because they react vigorously with water

9. A clean platinum wire is dipped into a solution of either a sodium or a potassium compound and then placed in an oxidizing flame. The sodium atoms yield a yellow flame. The potassium atoms yield a violet flame.

10. $CaCO_3(s) \rightarrow CaO(s) + CO_2(s)$
$CaO(s) + H_2O(\ell) \rightarrow Ca(OH)_2(s)$
$NH_4(aq) + OH^-(aq) \rightarrow NH_3(g) + H_2O(\ell)$

11. *(a)* Consists of monopositive ions of the metal, is permeated by an electron "gas" consisting of relatively free valence electrons *(b)* A body-centered cubic structure in which each metallic ion is surrounded by eight nearest neighbors at the corners of the cube.

Problems

1. *(a)* 15.0 g   X
$$2NaOH + H_2SO_4 \rightarrow Na_2SO_4 + 2H_2O$$
2 mol   1 mol
1 mol NaOH = 40.0 g   1 mol H_2SO_4 = 98.1 g
$$X = \frac{15.0 \text{ g NaOH}}{40.0 \text{ g/mol}} \times \frac{1 \text{ mol } H_2SO_4}{2 \text{ mol NaOH}} \times \frac{98.1 \text{ g}}{\text{mol}} = 18.4 \text{ g } H_2SO_4$$

(b) 15.0 g   X
$$2KOH + H_2SO_4 \rightarrow K_2SO_4 + 2H_2O$$
2 mol   1 mol
1 mol KOH = 56.1 g   1 mol H_2SO_4 = 98.1 g
$$X = \frac{15.0 \text{ g KOH}}{56.1 \text{ g/mol}} \times \frac{1 \text{ mol } H_2SO_4}{2 \text{ mol KOH}} \times \frac{98.1 \text{ g}}{\text{mol}} = 13.1 \text{ g } H_2SO_4$$

2. 1.50 kg   1.50 kg   X
$$NaNO_3 + KCl \rightarrow KNO_3 + NaCl$$
1 mol   1 mol   1 mol
1 mol $NaNO_3$ = 85.0 g   1 mol KCl = 74.6 g   1 mol KNO_3 = 101.1 g

Because the reactants combine in a 1:1 mole ratio, KCl is in excess. The mass of KNO_3 produced will depend on the mass of $NaNO_3$ that is used.

$$X = \frac{1.50 \text{ kg NaNO}_3}{85.0 \text{ g/mol}} \times \frac{1 \text{ mol KNO}_3}{\text{mol NaNO}_3} \times \frac{101.1 \text{ g}}{\text{mol}} = 1.78 \text{ kg KNO}_3$$

3. *(a)* $X = 50.0 \text{ g Na}_2CO_3 \times \frac{\text{mol}}{106 \text{ g}} \times \frac{1 \text{ mol CO}_2}{\text{mol Na}_2CO_3} \times \frac{22.4 \text{ L}}{\text{mol}} = 10.6 \text{ L CO}_2$

(b) $X = 50.0 \text{ g NaHCO}_3 \times \frac{\text{mol}}{84.0 \text{ g}} \times \frac{1 \text{ mol CO}_2}{\text{mol NaHCO}_3} \times \frac{22.4 \text{ L}}{\text{mol}} = 13.3 \text{ L CO}_2$

(c) $X = 50.0 \text{ g K}_2CO_3 \times \frac{\text{mol}}{138 \text{ g}} \times \frac{1 \text{ mol CO}_2}{\text{mol K}_2CO_3} \times \frac{22.4 \text{ L}}{\text{mol}} = 8.12 \text{ L CO}_2$

(d) $X = 50.0 \text{ g KHCO}_3 \times \frac{\text{mol}}{100 \text{ g}} \times \frac{1 \text{ mol CO}_2}{\text{mol KHCO}_3} \times \frac{22.4 \text{ L}}{\text{mol}} = 11.2 \text{ L CO}_2$

Application Questions

1. Sodium compounds are cheaper per gram of reactant.
2. The salt contains magnesium chloride, which is very deliquescent.
3. The increase in volume between sodium and potassium atoms is proportionally greater than the increase in mass between them.
4. Because the sodium hydrogen carbonate produced is practically insoluble under the conditions of the reaction

5. Carbonate ions undergo hydrolysis with water to form hydrogen carbonate ions and hydroxide ions, which turn the litmus paper blue.

6. Lithium hydroxide. More hydroxide ions are obtained per kilogram.

7. Common-ion effect: the presence of $NaC_2H_3O_2$ greatly increases the $[C_2H_3O_2^-]$ and shifts the $HC_2H_3O_2 + H_2O \rightleftarrows H_3O^+ + C_2H_3O_2^-$ equilibrium to the left. The $[H_3O^+]$ decreases and the pH rises.

Application Problems

1. $X = \dfrac{1.00 \times 103 \text{ kg Na}_2\text{CO}_3 \times 117 \text{ g NaCl}}{106 \text{ g Na}_2\text{CO}_3} = 1.10 \times 10^3 \text{ kg NaCl}$

2. $\dfrac{1.1 \text{ g}}{500 \text{ mL}} \times \dfrac{\text{mol}}{40.0 \text{ g}} \times \dfrac{10^3 \text{ mL}}{\text{L}} = 0.055 \text{ mol/L} = 0.055\ m$

   $[\text{H}_3\text{O}^+] = \dfrac{1.0 \times 10^{-14} \text{ mol}^2/\text{L}^2}{5.5 \times 10^{-2} \text{ mol/L}} = 1.8 \times 10^{-13} \text{ mol/L}$

   $\text{pH} = {}^-\log[\text{H}_3\text{O}^+] = {}^-\log(1.8 \times 10^{-13}) = {}^-12.74$

Teacher's Notes

Chapter 24 The Metals of Groups 1 and 2

INTRODUCTION

In this chapter you will take a close look at the occurrence, methods of preparation and extraction, and the trends in physical, chemical, and atomic properties of the alkali metals (Group 1) and the alkaline-earth metals (Group 2). The emphasis will be on important chemical properties and the roles of the metals and their compounds in industrial and biological processes.

LOOKING AHEAD

As you study this chapter, examine the properties of the alkali metals and the alkaline-earth metals, in terms of the outer energy level electron configuration of their atoms. Pay particular attention to the ionization energies and compare the chemical behavior of the alkali metals to that of the alkaline-earth metals. Relate the properties of the metals and their compounds to their uses in industrial biological processes.

SECTION PREVIEW

24.1 The Alkali Metals
24.2 The Alkaline-Earth Metals
24.3 Sodium

Sodium vapor lamps line Lake Shore Drive in Chicago.

24.1 The Alkali Metals

The group of elements at the left of the Periodic Table includes lithium, sodium, potassium, rubidium, cesium, and francium. These elements are the alkali metals, sometimes called the Sodium family of elements. Because of their great chemical reactivity, none exist in nature in the elemental state. They are found in natural compounds as monatomic ions with a +1 charge. Sodium and potassium are plentiful, lithium is fairly rare, and rubidium and cesium are rarer still. Francium does not exist as a stable element. Only trace quantities have been produced in certain nuclear reactions.

Structure and Properties

The Group-1 elements possess metallic characteristics to a high degree. Each has a silvery luster, is a good conductor of electricity and heat, and is ductile and malleable. These metals are relatively soft and can be cut with a knife. The properties of the Group-1 metals are related to their characteristic metallic crystal lattice structures, which are described in Chapter 13.

The crystal lattice consists of metallic ions with a +1 charge. The lattice is built around a body-centered cubic unit cell in which each metallic ion is surrounded by eight nearest neighbors at the corners of the cube. A model of the body-centered cubic unit cell is shown in Figure 24-1. Valence electrons are not localized to one atom but are free to permeate the lattice structure. These "free" electrons belong to the metallic crystal as a whole. The mobility of the free electrons in Group-1 metals accounts for their high thermal and electric conductivity and silvery luster as well.

The softness, ductilty, and malleability of the alkali metals are explained by the binding force in the metallic crystal lattice. Their low melting points and densities, together with their softness, distinguish these elements from the more familiar common metals. Some of the properties of the alkali metals are shown in Table 24-1.

The atoms of each element in Group-1 have a single electron in their highest energy level. Lithium atoms have the electron configuration $1s^2 2s^1$. All other Group-1 elements have next-to-highest energy levels containing eight electrons. An ion formed by removing the single-valence electron has the stable electron configuration of the preceding noble gas. For example, the sodium ion has the electron configuration $1s^2 2s^2 2p^6$. This matches the configuration of the neon atom. The electron configuration of the potassium ion, $1s^2 2s^2 2p^6 3s^2 3p^6$, matches the argon atom.

It is also possible to add an electron to a Group 1 atom to produce an anion having an electron pair in the *s* orbital of the highest energy level. An example is the Na^- anion which has the electron configuration $1s^2 2s^2 2p^6 3s^2 3p^6$. Here the oxidation number of sodium is -1. Similar anions of potassium, rubidium, and cesium have also been produced.

Chemical Behavior

The very active free metals of the Sodium family are obtained by reducing the +1 ions in their natural compounds. The metals are vigorous reducing agents. They have a weak attraction for their valence electrons.

Sodium makes up about 2.8% by mass of the earth's crust, potassium 2.6%, lithium 0.003%, rubidium 0.000 3%, and cesium 0.000 07%.

- Identify the alkali-metal elements
- List some of the properties of the alkali metal elements.
- Relate the properties of alkali metals to their characteristic metallic crystal lattice structures.
- Explain certain observed physical properties of the Group-1 elements in terms of their atomic properties.
- Explain why none of the alkali metals exist free in nature.

An alkali is a substance that forms a base and can neutralize an acid.

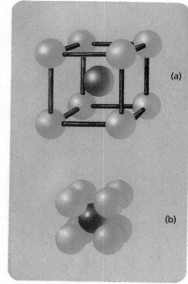

(a)

(b)

Figure 24-1 Body-centered cubic unit cell.

The Group-1 elements form hydroxides whose water solutions are extremely alkaline. Thus the Group-1 elements are "alkali metals."

STUDY HINT
Crystalline structure is covered in section 13.2

The Metals of Group 1 and 2   **713**

TABLE 24-1 PROPERTIES OF THE GROUP 1 METALS

	Li	Na	K	Rb	Cs	Fr
Atomic number	3	11	19	37	55	87
Atomic mass	6.941	22.99	39.10	85.47	132.9	223*
Electron Configuration	[He]2s^1	[Ne]3s^1	[Ar]4s^1	[Kr]5s^1	[Xe]6s^1	[Rn]$^-$
Melting point (°C)	186	97.5	63.65	38.89	28.5	27
Boiling point (°C)	1326	889	774	688	690	677
Density (g/cm^3)	0.534	0.971	0.862	1.53	1.87	——
Ionization energy (kJ/mol)	519	498	418	403	376	——
Electronegativity	1.0	0.9	0.8	0.8	0.7	0.7
Atomic radius (pm)	152	186	231	244	262	——
Ionic radius (pm)	60	95	133	148	169	——
Oxidation number	+1	+1	+1	+1	+1	——
Crystal structure**	bcc	bcc	bcc	bcc	bcc	——
Hardness (Mohs scale)	0.6	0.4	0.5	0.3	——	——
Flame color	bright red	yellow	violet	purple	reddish violet	——

*atomic mass of most stable isotope   **body centered cubic

STUDY HINT

You may wish to review ionization energies in Section 5.3

All alkali metals have relatively low first ionization energies, separated by a relatively large energy gap from the second ionization energy, as shown in Table 24-2. This suggests the presence of one easy-to-remove electron and accounts for the +1 charge of ions formed by the Group-1 elements. Ionization energy decreases as the atom size increases going down the group. As shown in Figure 24-2, this decreasing energy with increasing size shows that it is easier to remove electrons from the highest energy levels of the heavier atoms. If you examine the net energy of reactions in which electrons are given up to other elements, however, you find that the lithium atom is the strongest reducing agent. Although it may be harder to remove an electron from the lithium atom, more energy is given back by the subsequent interaction of the lithium ion and its surroundings than by the larger alkali-metal ions with their surroundings.

Handling and storing the alkali metals is difficult because of their chemical behavior. They are usually stored in kerosene or some other liquid hydrocarbon because they react vigorously with water.

TABLE 24-2 FIRST AND SECOND IONIZATION ENERGIES AND THE SIZE OF ATOMS OF SELECTED GROUP-1 METALS

Element	Size of Atoms (pm)	First Ionization Energy (kJ/mol)	(eV)	Second Ionization Energy (kJ/mol)	(eV)
Li	152	519	5.34	7297	75.7
Na	186	498	5.14	4656	47.4
K	227	418	4.34	3071	31.9
Rb	248	403	4.18	2653	27.6
Cs	265	376	3.89	2423	25.1

Chemistry Notebook

Group 1 Ions in Medication and Diet

The elements of Group 1 are important to the diet and in medicine because they form ionic compounds that usually dissolve well in water. If such ions are incorporated into large molecules, those molecules can then be dissolved and carried easily through the bloodstream. One example is sodium pentothal, an anesthetic popularly known as "truth syrum."

Sodium and potassium ions are critical nutrients because they facilitate the transmission of nerve impulses and control the amount of water retained by cells.

Physicians often advise people with heart problems or high blood pressure to go on low-sodium diets to reduce the volume of water in their bodies. Diuretic drugs help to release excess water, but cause the loss of potas-

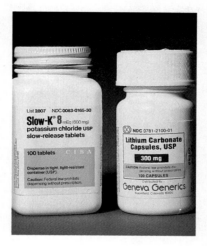

sium ions. Potassium-rich foods or supplements are sometimes needed to replace the loss.

The distribution of positively charged potassium ions and sodium ions and negatively charged organic ions across nerve cell membranes is an essential factor in the operation of the nervous

system. This uneven distribution of ions creates a voltage across these membranes. When a nerve cell is stimulated, the ion concentration changes. This leads to a change in voltage across the nerve cell membrane. The resulting voltage change initiates the transmission of a nerve impulse.

In medical use, lithium carbonate is an ionic compound that is often prescribed to reduce nervous tremors, a symptom of Parkinson's disease. Frequently, it is referred to only as "lithium." Lithium carbonate is also the most effective drug for treating manic depressive disorder, a mental illness. The way it works is not well understood, but it is believed that lithium ions affect the way nerve cells communicate.

A small piece of potassium dropped into water, will react explosively, releasing H_2 to form a strongly basic hydroxide solution. The heat of reaction ignites the hydrogen that is produced.

$$2K(s) + 2H_2O(\ell) \rightarrow 2K^+ + 2OH^-(aq) + H_2(g)$$

All of the ordinary compounds of the alkali metals are ionic, including their hydrides. Only lithium forms the oxide directly with oxygen. Sodium forms the peroxide instead, and the higher metals tend to form superoxides of the form $M^+O_2^-$. The ordinary oxides can be prepared indirectly, however. These oxides are basic anhydrides. They form hydroxides in water.

Nearly all the compounds of the alkali metals are quite soluble in water. The alkali-metal ions are colorless, have little tendency to hydrolyze in water solution, or to form polyatomic ions.

Alkali-metal peroxides and superoxides react with water to form oxygen gas and alkali-metal hydroxide solutions. This property is put to use in self-con-

A basic anhydride is a substance that will react with water to form a base.

Caution
The reaction of potassium with water is highly exothermic and can lead to an explosion. Do not attempt to use this reaction as an experiment or a demonstration.

Figure 24-2 First ionization energies (shown in kJ/mol) decreases with increasing atomic radii for the atoms of Group-1 elements.

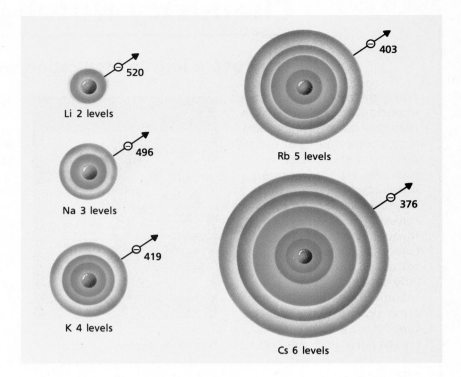

Li 2 levels — 520

Na 3 levels — 496

K 4 levels — 419

Rb 5 levels — 403

Cs 6 levels — 376

1. Lithium, sodium, potassium, rubidium, cesium, and francium
2. Silvery luster, good conductor of electricity and heat, ductile and malleable
3. Valance electrons form an electron "gas" that permeates the lattice structure. The mobility of the electrons accounts for the high thermal and electric conductivity, as well as the silvery luster. The softness, ductility, and malleability are explained by the binding force in the metallic crystal.
4. Ionization energy decreases as the atom size increases, making it easier to remove electrons from the highest energy levels of the heavier atoms. The outer energy-level electron is attracted less by the nuclear positive charge as the size of the atom gets larger.
5. They have a weak attraction for their valence electrons and therefore are easily oxidized to +1 ions.

tained breathing appartus where breath is passed through a filter compartment containing KO_2. The moisture from the breath reacts with KO_2 to form oxygen gas. Carbon dioxide dissolves in the resulting hydroxide solution.

Compounds of the more important alkali metals are easily identified by flame tests. **Flame tests** are carried out by dipping a clean platinum wire into a solution containing a salt of the metal to be tested, then placing the wire in an oxidizing flame. Their compounds impart characteristic colors to a Bunsen flame. Sodium compounds color the flame yellow. Lithium compounds give a flame a red (carmine) color.

Potassium imparts a fleeting violet color to a Bunsen flame. The color comes from potassium atoms. The presence of sodium, however, masks the violet color of potassium in the flame. Potassium is detected in a mixture of sodium and potassium compounds by observing the colored flame through cobalt-blue glass. This glass filters out the yellow of the sodium flame, allowing the violet potassium flame to show clearly.

Section Review

1. Name the alkali metals.
2. Describe several properties of the alkali metals.
3. How are the properties of the Group 1 elements related to their characteristic metallic crystalline lattice structures?
4. Explain why it is easier to remove electrons from the heavier alkali atoms than it is from lithium.
5. Explain why none of the alkali metals exist in an uncombined state in nature.

24.2 The Alkaline-Earth Metals

The elements of Group 2 of the Periodic Table are beryllium, magnesium, calcium, strontium, barium, and radium are known as the alkaline-earth metals.

Early chemists used the term "earth" for nonmetallic substances, such as the Group-2 metal oxides, that were practically insoluble in water and unchanged by strong heating. The Group-2 metal oxides give an alkaline reaction with water. Thus, these metal oxides have historically been called *alkaline earths,* and the metals *alkaline-earth metals.* Like the alkali metals, they are never found as free elements in nature. The metals must be recovered from their natural compounds. Many of their compounds are insoluble or slightly soluble and are found in the earth's crust. Their carbonates, phosphates, silicates, and sulfates are the most important deposits.

Beryllium and magnesium are commercially important light metals. In their chemical behavior, they resemble the corresponding alkali metals, lithium and sodium. Radium is important because it is radioactive. Radioactivity and other properties of radium are discussed in Chapter 30. The remaining three elements of Group 2—calcium, strontium, and barium—have similar properties. They are considered to be the typical alkaline-earth metals.

Structure and Properties

Each alkaline-earth element has two valence electrons beyond the stable configuration of the preceding noble gas. All form doubly charged ions of the M^{2+} type. Ions of the alkaline-earth metals thus have stable noble-gas structures. Their chemistry, like that of the alkali metals of Group 1, is generally uncomplicated.

The attraction between the metal ions and the electron "gas" of the alkaline-earth metal crystals is stronger than in the alkali metals. The *alkaline -earth metals* are therefore denser and harder than the corresponding alkali family metals, and they have higher melting and boiling points.

The atoms and ions of the alkaline-earth metals are smaller than those of the corresponding alkali metals because of their higher nuclear charge. For example, the magnesium ion (Mg^{2+}) has the same electron configuration as the sodium ion, Na^+. This configuration is $1s^2 2s^2 p^6$. The nuclear charge of Mg^{2+} is +12, and Na^+ has a nuclear charge of +11. The higher nuclear charge of the Mg^{2+} ion attracts electrons more strongly and results in a smaller ionic radius. Some properties of Group-2 metals are listed in Table 24-1.

In Section 24.1 you learned that ionization energies of Group 1 metals decrease down the group as atomic size increases. This same relationship is seen in Group 2. The smaller atoms hold their outer electrons more securely than do the larger atoms.

Recall that ionization energy measures the tendency of an isolated atom to hold a valence electron. The first ionization energy relates to the removal of one electron from the neutral atom. The second ionization energy is that required to remove an electron from the ion with a +1 charge (the atom that

SECTION OBJECTIVES

- Identify the alkaline-earth metals and some of their sources.
- List some of the properties of the alkaline-earth metals.
- Name some alkaline-earth metal compounds and describe their properties.

Remind students that Group-1 metal oxides are all water soluble, forming bases in solution. They are also unchanged by strong heating.

One interesting use of beryllium is that it can be alloyed with copper to form a hard metal that is used for electrical contacts because it doesn't form sparks.

Figure 24-1 Beryllium processed in Utah. This metal produces alloys that are extremely elastic.

The alkaline-earth metals form 2+ ions that will have a greater attraction for the electron "gas" than the alkali metal 1+ ions.

TABLE 24-1 PROPERTIES OF GROUP-2 METALS

	Be	Mg	Ca	Sr	Ba	Ra
Atomic number	4	12	20	38	56	88
Atomic mass	9.012	24.31	40.08	87.62	137.33	226.0
Electron configuration	$[He]2s^2$	$[Ne]3s^2$	$[Ar]4s^2$	$[Kr]5s^2$	$[Xe]6s^2$	$[Rn]7s^2$
Melting point (°C)	1280	650	839	770	725	700
Boiling point (°C)	2970	1120	1484	1380	1640	1140
Density (g/cm³)	1.85	1.74	1.55	2.54	3.51	5
Atomic radius (pm)	111	160	197	215	217	220
Ionic radius (pm)	31	65	99	113	135	—
First ionization energy (kJ/mol)	900	736	590	548	502	510
Second ionization energy (kJ/mol)	1757	1459	1145	1058	958	979
Electronegativity	1.5	1.2	1.0	1.0	1.0	1.0
Oxidation number	+2	+2	+2	+2	+2	+2
Crystal structure	hexagonal	hexagonal	cfc	cfc	cfc	—
Hardness (Mohs scale)	4	2.0	3	1.8	1.5	—
Flame Color	—	—	brick red	crimson	green	—

has already had one valence electron removed). This second ionization energy is always higher than the first because the particle from which the electron is removed now has a positive charge.

If a third electron were to be removed from a Group 2 atom, it would come from the stable inner electron configuration (the noble-gas structure) of the +2 ion. The predicted energy requirement would be much higher. In fact, the energies required for a third level of ionization of alkaline-earth elements are very high. They exceed the energies usually available in chemical reactions. The only forms observed are thus the +2 ions of these elements. The first, second, and third ionization energies of the Group-2 elements are listed in Table 24-2.

Properties of Alkaline-Earth Metal Compounds

The alkaline-earth metal form hydrides, oxides or peroxides, and halides similar to those of the alkali metals. Almost all hydrides are ionic and contain H^- ions. An exception is BeH_2. Binary compounds of beryllium have fairly

TABLE 24-2 IONIZATION ENERGIES OF THE GROUP-2 ATOMS

	First Ionization Energy (kJ/mol)	Second Ionization Energy (kJ/mol)	Third Ionization Energy (kJ/mol)
Be	900	1757	14849
Mg	736	1450	7732
Ca	590	1145	4912
Sr	548	1058	4205 (?)
Ba	502	958	3556 (?)
Ra	510	979	—

Chemistry Notebook

Strontium: Fireworks and Fallout

Few people today have even heard of the element strontium (Sr), even though they have probably seen it in use many times. The bright red color in fireworks displays is due to the presence of strontium sulfate in the fireworks' fuel. Luminous paints sometimes contain strontium sulfide, which is prepared by heating strontium hydroxide with sulfur.

In the 1960s concern about the effects of a certain strontium isotope was a major political issue.

The isotope was released into the environment during atmospheric testing of nuclear weapons.

Natural strontium is a mixture of four stable isotopes. An unstable isotope, strontium-90, is produced by nuclear explosions.

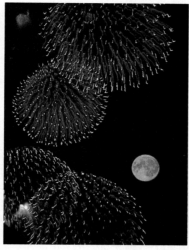

$$^{235}U + n \rightarrow {}^{90}Sr + {}^{143}Xe + 3n$$

Of the various components of nuclear fallout, the strontium-90 isotope presents the greatest hazard to people. When it enters the human body, strontium-90 is deposited in bones and can cause serious illnesses.

Concern about the possibility of strontium-90 finding its way into children's milk contributed to the Nuclear Test Ban Treaty of 1963, which prohibited atmospheric testing of nuclear weapons. Now, nuclear weapons tests are conducted deep underground.

---

strong covalent bonds and react with water to release hydrogen gas and form basic hydroxide solutions. Calcium hydride is often used as a laboratory source of hydrogen.

The oxides of the alkaline-earth metals have very high melting points. CaO and MgO are used as heat-resistant (refractory) materials. Beryllium oxide is amphoteric; otherwise, the oxides are basic. Recall, from Section 16.3, that any species that can react as either an acid or a base is *amphoteric*.

With the exception of beryllium oxide, the oxides are essentially ionic, but with more covalent character than the alkali-metal oxides. Strontium and barium form peroxides with oxygen, probably because of the large size of their ions.

The hydroxides are formed by adding water to the oxides. Except for Be(OH)$_2$, which is amphoteric, the hydroxides dissociate in water solution to yield OH$^-$ ions. Hydroxides above barium are only slightly soluble in water, the solubility increasing with the metallic ion's size. Solutions of these hydroxides have low concentrations of OH$^-$ ions.

The reactions are:
$$BeH_2(s) + 2H_2O \rightarrow 2H_2(g) + Be(OH)_2$$
$$CaH_2(s) + 2H_2O \rightarrow 2H_2(g) + Ca(OH)_2$$

Beryllium and its compounds are extremely toxic.

Be(OH)$_2$ can dissolve in both acid and base:
$$Be(OH)_2(s) + 2H^+(aq) \rightarrow Be^{2+}(aq) + 2H_2O$$
$$Be(OH)_2(s) + 2OH^-(aq) \rightarrow [Be(OH)_4]^{2-}(aq)$$

1. Beryllium, magnesium, calcium, strontium, barium, radium
2. In the earth's crust as carbonates, phosphates, silicates, and sulfates
3. Form ions with an +2 oxidation state. Denser and harder than the corresponding alkali metals and have higher melting and boiling points. Each atom bears a nuclear charge one greater than that of an atom of the corresponding alkali metal; thus, the two outer-level electrons in the

SECTION OBJECTIVES

- List some of the sources of sodium.
- Describe the preparation, properties, and uses of sodium.
- Name some sodium compounds and describe their preparation and uses.

alkaline-earth metals are held more firmly than the single outer-level electron of the alkali metals. The ionization energies decrease down the group as the atomic sizes increase.

4. Form hydrides, oxides or peroxides, and halides similar to those of the alkali metals. The oxides are essentially ionic but with more covalent character than the alkali-metal oxides.

5. Forms oxides that are amphoteric while the oxides of the other Group-2 metals are basic. Beryllium oxides are covalent while the others are mostly ionic in character. Unlike other Group-2 metal hyroxides, beryllium hydroxide is not soluble and is amphoteric.

Figure 24-4 Elemental sodium is produced by the electrolysis of melted sodium chloride in a Downs cell. Chlorine is a by-product.

Section Review

1. Name the alkaline-earth metals.
2. Where are the alkaline-earth metals found in nature?
3. What are some of the properties of the alkaline-earth metals?
4. How does the reactivity of the alkaline-earth metals compare to that of the alkali metals?
5. How do the properties of beryllium compare to those of the other Group-2 metals?

24.3 Sodium: One of The Active Metals

Metallic sodium is never found free in nature. Compounds containing the Na^+ ion and sodium complexes are found in soil, natural waters, and in plants and animals, however. Because of the solubility of sodium compounds, it is almost impossible to find a sodium-free material. Vast quantities of sodium chloride are present in sea water and rock salt deposits. There are important deposits of sodium nitrate in Chile and Peru. The carbonates, sulfates, and borates of sodium are found in dry lake beds.

Preparation of Sodium

Sir Humphry Davy (1778–1829) prepared metallic sodium in 1807 by the electrolysis of fused sodium hydroxide. Today *sodium is prepared by the electrolysis of fused sodium chloride with an apparatus called the* **Downs cell** (Figure 24-4).

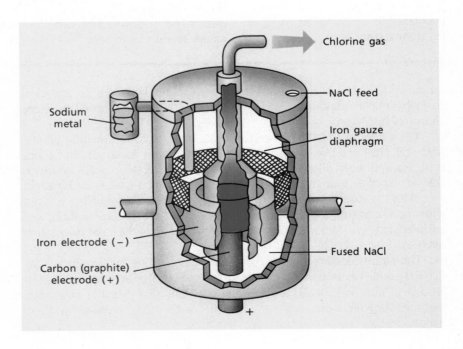

The process used in the Downs-cell apparatus requires the mixing of sodium chloride with calcium chloride. This results in the lowering of the melting point of the sodium chloride from 80l°C to 580°C and in the simultaneous production of liquid sodium and chlorine gas. The liquid metallic sodium is collected under oil and is kept separate from the chlorine gas by an iron-gauze diaphragm.

The sodium is recovered by reducing Na^+ ions. This occurs at the negative electrode of the Downs cell. Each sodium ion acquires an electron from this electrode to form a neutral sodium atom.

$$Na^+ + e^- \rightarrow Na(\ell)$$

Chloride ions are oxidized at the positive electrode . Each chloride ion loses an electron to this electrode to form a neutral chlorine atom .

$$Cl^- \rightarrow Cl + e^-$$

But two chlorine atoms form the diatomic molecule of elemental chlorine gas. The net reaction at the positive electrode is

$$2Cl^- \rightarrow Cl_2(g) + 2e^-$$

In the overall cell reaction, electron-transfer balance is maintained between the electrodes. Two Na^+ ions are reduced for each Cl_2 molecule formed. The cell reaction can be written

negative electrode: $2Na^+ + 2e^- \rightarrow 2Na(\ell)$
positive electrode: $\underline{\hspace{2em} 2Cl^- \rightarrow Cl_2(g) + 2e^-}$
cell: $2Na^+ + 2Cl^- \rightarrow 2Na(\ell) + Cl_2(g)$

Properties and Uses of Sodium

Sodium is a silvery-white, lustrous metal that tarnishes rapidly when exposed to air. It is very soft and has a lower density than water and a low melting point. A pellet of sodium dropped into water melts from the heat of the vigorous exothermic reaction that occurs. This reaction yields hydrogen gas and a strongly basic solution.

$$2Na(s) + 2H_2O(\ell) \rightarrow 2NA^+(aq) + 2OH^-(aq) + H_2(g)$$

When exposed to air, sodium unites with oxygen to form sodium peroxide (Na_2O_2). If sodium is supplied in excess, some sodium oxide (Na_2O) can be produced along with the bulk product, Na_2O_2. This production of Na_2O is a result of the strong reducing character of sodium atoms. Sodium oxide can also be formed by heating NaOH with sodium.

$$2NaOH + 2Na \rightarrow 2Na_2O + H_2(g)$$

The superoxide of sodium, NaO_2, can be prepared indirectly. Superoxides have the O_2^- ions with one unpaired electron.

Sodium reacts with all aqueous acids. It burns in an atmosphere of chlorine gas, and directly forms sodium chloride.

As shown in Figure 24-5, a flame test of sodium compounds reveals a strong yellow color characteristic of vaporized sodium atoms. This yellow flame commonly identifies sodium. For comparison, Figure 24-5 also shows flame test for salts of other Group-1 elements.

Caution
Never add an alkaline metal to water, an explosion can result.

Sodium is a good reducing agent because it loses its outer electron readily. It is easily oxidized to the +1 ion, causing other substances to be reduced.

Figure 24-5 Flame tests of salts of (from l. to r.) the elements lithium, sodium, potassium, rubidium.

Much of the sodium produced in the United States has been used in making tetraethyl lead, an antiknock additive for gasoline. This use is decreasing, however, as the production of leaded gasoline is being phased out to reduce the amount of lead in the atmosphere. Sodium is also used as a heat-transfer agent, in making dyes and organic compounds, and in sodium-vapor lamps.

Important Compounds of Sodium

Among the most important compounds of sodium are sodium chloride, sodium hydroxide, and sodium carbonate. Representative sodium compounds and their uses are shown in Table 24-3.

Sodium chloride *Sodium chloride is found in sea water, in salt wells, and in deposits of* **rock salt,** *known as* **halite,** *which is mined in many places in the world.*

Pure sodium chloride is not deliquescent (tending to undergo gradual dissolution by the absorbtion of moisture from the air). Magnesium chloride is very deliquescent, however, and is usually present in sodium salt as an

Some nuclear power plants use liquid sodium as the agent to carry the heat generated in the reactor core to a secondary coil where the heat is used to turn water to steam to drive a turbine. Sodium's good thermal conductivity, large liquid temperature range, and high heat capacity make it ideal for this use.

Sodium chloride makes up 2.8% by mass of the ocean.

Figure 24-6 Native crystal formations of sodium chloride, known as halite, in Death Valley, California.

TABLE 24-3 REPRESENTATIVE SODIUM COMPOUNDS

Chemical Name	Common Name	Formula	Color	Uses
sodium tetraborate, decahydrate	borax	$Na_2B_4O_7 \cdot 10H_2O$	white	as a water softener; in making glass; as a flux
sodium carbonate, decahydrate	washing soda	$Na_2CO_3 \cdot 10H_2O$	white	as a water softener; in making glass
sodium hydrogen carbonate	baking soda	$NaHCO_3$	white	as a leavening agent in baking
sodium cyanide	prussiate of soda	$NaCN$	white	to destroy vermin; to extract gold from ores; in silver and gold plating; in case-hardening steel
sodium hydride	(none)	NaH	white	in cleaning scale and rust from steel forgings and castings
sodium nitrate	Chile saltpeter	$NaNO_3$	white or colorless	as a fertilizer
sodium peroxide	(none)	Na_2O_2	yellowish white	as an oxidizing and bleaching agent; as a source of oxygen
sodium phosphate, decahydrate	TSP	$Na_3PO_4 \cdot 10H_2O$	white	as a cleaning agent; as a water softener
sodium sulfate, decahydrate	Glauber's salt	$Na_2S O_4 \cdot 10H_2O$	white or colorless	in making glass; as a cathartic in medicine
sodium thiosulfate, pentahydrate	hypo	$Na_2S_2O_3 \cdot 5H_2O$	white or colorless	as a fixer in photography; as an antichlor
sodium sulfide	(none)	Na_2S	colorless	in the preparation of sulfur dyes; for dyeing cotton; to remove hair from hides

impurity. The presence of magnesium chloride explains why table salt becomes wet and sticky in damp weather.

Sodium chloride is essential in the diets of humans and animals and is present in certain body fluids. Perspiration contains considerable amounts of sodium chloride.

Sodium chloride is the cheapest compound of sodium and is used as a starting material in the production of most other sodium compounds. Some sodium compounds are most easily prepared directly from metallic sodium, however.

Sodium hydroxide   Most commercial sodium hydroxide is produced by electrolysis of an aqueous sodium chloride solution. This electrolysis of aqueous NaCl is different from that of fused NaCl in the Downs cell. Chlorine gas is produced at the positive electrode, but hydrogen gas (instead of metallic

People sometimes put rice in salt shakers to absorb moisture in humid weather.

sodium) is produced at the negative electrode. The solution, meanwhile, becomes aqueous NaOH. Evidence indicates that water molecules acquire electrons from the negative electrode and are reduced to H_2 gas and OH^- ions.

The cell reaction for the electrolysis of an aqueous $NaC\ell$ solution is

electrode $(-)$ $2H_2O + 2e^- \rightarrow H_2(g) + 2OH^-(aq)$

electrode $(+)$ $2Cl^- \rightarrow Cl_2(g) + 2e^-$

cell: $2H_2O + 2Cl^- \rightarrow H_2(g) + Cl_2(g) + 2OH^-(aq)$

As the Cl^- ion concentration diminishes, the OH^- ion concentration increases. The Na^+ ion concentration remains unchanged during electrolysis. Thus the solution is converted from aqueous NaCl to aqueous NaOH.

Sodium hydroxide, commonly known as lye, converts some types of animal and vegetable matter into soluble materials by chemical action. It is very **caustic,** *meaning that it has destructive effects on skin, hair, and wool.*

Sodium hydroxide is a white crystalline solid. It is marketed in the form of flakes, pellets, and sticks. It is very deliquescent and dissolves in the water that it removes from the air. It reacts with carbon dioxide from the air, producing sodium carbonate. Its water solution is strongly basic.

Sodium hydroxide reacts with fats, forming soap and glycerol. One of its important uses, therefore, is in making soap. It is also used in the production of rayon, cellulose film, and paper pulp, and in petroleum refining.

Sodium carbonate and the Solvay process   Almost all of the sodium carbonate produced in the United States comes from the mining of natural sodium carbonate deposits located in Wyoming or from the evaporation of brine from Searles Lake, California.

Sodium carbonate produced in the United States and much of that produced in the rest of the world is manufactured by the Solvay process. *The* Solvay process *uses brine (salty water) and limestone to produce sodium carbonate.* This process, a classic example of efficiency in chemical production, was developed in the 1860's by Ernest Solvay (1838–1922), a Belgian industrial chemist. We will describe the Solvay process here because it illustrates many chemical principles.

A flow diagram of the Solvay process is shown in Figure 24-7. The raw materials for the Solvay process are common salt, limestone, and coal. The salt is pumped as brine from salt wells. The thermal decomposition of limestone yields the carbon dioxide and calcium oxide needed in the process.

$$CaCO_3(s) \rightarrow CaO(s) + CO_2(g) \qquad \text{(Equation l)}$$

Coal is converted into coke, gas, coal tar, and ammonia by destructive distillation. The coke and gas are used as fuel in the plant. The ammonia is also used in the process, and the coal tar is sold as a useful by-product.

To begin the process, a cold saturated solution of sodium chloride is further saturated with ammonia and carbon dioxide. The following reactions occur:

$$CO_2(g) + H_2O \rightarrow H_2CO_3$$

$$H_2CO_3 + NH_3(g) \rightarrow NH_4^+(aq) + HCO_3^-(aq)$$

$$\text{Net: } CO_2(g) + NH_3(g) + H_2O \rightarrow NH_4^+(aq) + HCO_3^-(aq)$$

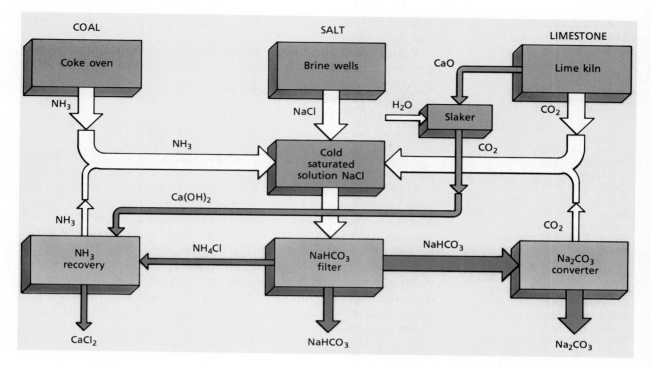

Figure 24-7 A flow diagram of the Solvay process.

The HCO_3^- ions form in a solution that has a high concentration of Na^+ ions. Because sodium hydrogen carbonate is only slightly soluble in this cold solution, it precipitates.

$$Na^+(aq) + HCO_3^-(aq) \rightarrow NaHCO_3(s)$$

The solution that remains contains NH_4^+ ions and Cl^- ions. The precipitated sodium hydrogen carbonate is filtered and dried. It is either sold as baking soda or converted into sodium carbonate by thermal decomposition.

$$2NaHCO_3(s) \rightarrow Na_2CO_3(s) + H_2O(g) + CO_2(g)$$

The dried sodium carbonate is an important industrial chemical called **soda ash,** *which is used to manufacture a wide variety of materials including glass, chemicals, paper, and detergents.*

The ammonia used in the process is more valuable than the sodium carbonate of sodium hydrogen carbonate. Hence, it must be recovered and used over again if the process is to be profitable.

The calcium oxide from Equation 1 is slaked, or crumbled, by adding water. Calcium hydroxide is formed in this reaction.

$$CaO(s) + H_2O(\ell) \rightarrow Ca(OH)_2(s)$$

The calcium hydroxide is added to the solution containing NH_4^+ ions and Cl^- ions to release the ammonia from the ammonium ion.

$$NH_4^+ + OH^-(aq) \rightarrow NH_3(g) + H_2O$$

The solution now contains mainly Ca^{2+} ions and Cl^- ions. Neither of these ions is recycled into the process. As a by-product, calcium chloride has some

Sodium carbonate is also called washing soda.

Some of the excess calcium chloride is spread on streets on icy days to help melt the ice.

1. Sea water, rock salt deposits, dry lakes, soil, natural waters, plants, and animals
2. Sodium chloride is mixed with calcium chloride to lower its melting point from 801°C to 580°C. It is then placed in the Downs cell, and metallic sodium is prepared by the electrolysis of the fused salt. Liquid sodium is formed by reduction of Na+ ions at the negative electrode and is collected under oil.
3. Silvery-white, lustrous metal that tarnishes when exposed to air; very soft, lower density than water, low melting point; unites with oxygen to form sodium peroxide when exposed to air; reacts vigorously with water to produce hydrogen gas and a strongly basic solution. Most sodium is used in making tetraethyl lead; also used in making dyes and organic compounds
4. (a) Found in sea water, salt wells, and in deposits of rock salt; can be prepared directly from metallic sodium and chlorine. Used as a starting material in the production of most other sodium compounds
(b) Produced by electrolysis of an aqueous sodium chloride solution. Used in the manufacture of soaps, rayon, cellulose film, and paper pulp, and in petroleum refining
(c) Mined from natural sodium carbonate deposits; often prepared by the Solvay process, most important use is the production of glass.

use as an inexpensive dehydrating agent. The supply generally exceeds the demand, however.

Baking soda *is composed of sodium hydrogen carbonate.* It is the main ingredient of baking powders. Mixed with acid, it releases CO_2 bubbles that expand dough as it bakes.

The water solution of sodium carbonate is mildly basic because of hydrolysis of CO_3^{2-} ions.

$$CO_3^{2-} + H_2O \rightleftarrows HCO_3^- + OH^-$$

Nuclear reactor coolant Alkali metals, in particular sodium, have properties that could possibly make nuclear power plants less costly. Nuclear power plants generate electricity using heat transferred from the reactor by a substance known as a coolant. Most nuclear reactors now use steam or boiling water as a coolant. However steam and boiling water must be kept under high pressure (70 atm to 150 atm) at efficient operating temperatures (over 500°C). Building reactors that can contain such high pressures adds to the cost of construction and maintenance. Unlike water, liquid sodium can reach efficient operating temperatures (900°C) at pressures of less than 2 atm. However since water reacts violently with sodium, extreme care must be taken to avoid any contact between the liquid sodium and water used in other parts of the liquid-sodium cooled reactor.

Section Review

1. Name several sources of sodium.
2. Describe the preparation of sodium by the electrolysis of sodium chloride in a Downs cell.
3. What are some of the properties and uses of the alkali metal sodium?
4. Give the preparation and uses of each of the following compounds: *(a)* sodium chloride   *(b)* sodium hydroxide   *(c)* sodium carbonate.

Chapter Summary

- Group 1 elements do not exist as free elements in nature; they are found only in natural compounds as monpositive ions.
- The elements of Group 1 in the Periodic Table are among the most chemically active metals.
- The elements of Group 1 are referred to collectively as the Sodium family and as the alkali metals.
- Group-1 elements are characterized by a single highest-energy-level electron added to the preceding noble-gas structure. These elements have relatively low ionization energies.
- Group-1 elements are soft, silvery-white metals that are good conductors of heat and electricity.
- The elements of Group 2 are the metals beryllium, magnesium, calcium, strontium, barium, and radium. They are known as the alkaline-earth metals.
- The alkaline-earth metals are characterized by two valence electrons beyond the preceding noble-gas structure. They are metals denser and harder than the alkali metals of Group 1 and have higher melting points.

Chapter 24 *Review*

The Metals of Groups 1 and 2

Vocabulary

alkali metals	Downs cell
alkaline-earth	soda ash
metals	Solvay process
baking soda	halite
caustic	lye

Questions

1. How do the electron configurations of Group 1 atoms and ions differ?
2. How are sodium and potassium prepared?
3. List three uses for metallic sodium.
4. Distinguish *caustic* from *corrosive*.
5. Identify: *(a)* the raw materials used in the Solvay process *(b)* the products and by-products of the process.
6. What leavening action occurs when molasses and baking soda are used in baking cookies?
7. What are the principal sources of potassium compounds in the United States?
8. *(a)* How are metallic sodium and potassium stored in the laboratory stockroom? *(b)* Why must they be stored in this manner?
9. Describe the flame tests for sodium and potassium.
10. Write three equations for the recovery of ammonia in the Solvay process.
11. Describe: *(a)* the metallic crystal lattice of the alkali metals *(b)* the unit cell upon which this lattice is built.

Problems

1. *(a)* How many grams of sulfuric acid in water solution can be neutralized by 15.0 g of sodium hydroxide *(b)* 15.0 g of potassium hydroxide?
2. Given 1.50 kg of sodium nitrate and 1.50 kg of potassium chloride, how many grams of potassium nitrate can be produced by reacting these two substances?
3. If sodium carbonate decahydrate sells for 50 cents per kilogram, what is the anhydrous sodium carbonate worth per kilogram?

3. How many liters of carbon dioxide can be liberated from 50.0 g of each of the following: *(a)* Na_2CO_3 *(b)* $NaHCO_3$ *(c)* K_2CO_3 *(d)* $KHCO_3$?

Application Questions

1. Why are sodium compounds more frequently used than potassium compounds?
2. Why does table salt become sticky in damp weather, although pure sodium chloride is not deliquescent?
3. Explain why potassium has a lower density than sodium, although it consists of heavier atoms.
4. In the Solvay process, the reaction between sodium chloride and ammonium hydrogen carbonate runs to completion. Why?
5. Why does a solution of sodium carbonate in water turn red litmus paper blue?
6. Suppose you had a tremendous quantity of acid that had to be neutralized and that NaOH, KOH, and LiOH were all available at the same price per kilogram. Which of these three would you use? Why?
7. A 0.1-*M* solution of $HC_2H_3O_2$ is found to have a pH of 2.9. A solution of 0.1-*M* $NaC_2H_3O_2$ is added to the acetic acid solution and the pH rises. Explain.

Application Problems

1. How many kilograms of sodium chloride are required to produce 1.00 metric ton of anhydrous sodium carbonate?
2. A solution is prepared by dissolving 1.1 g NaOH in water and diluting it to 500 mL. What is the pH of the solution?

Enrichment

1. Research and write a report explaining the development of fireworks.
2. Investigate the advantages of hydrometallurgy over pyrometallurgy and make a short presentation to your class.

Chapter 25 The Transition Metals

Chapter Planner

25.1 The Transition Elements
  ■ Question: 1
General Properties of the Transition Elements
  ■ Demonstration: 1
  Experiment: 35
Transition Metal Groups
Variable Oxidation States
  ■ Questions: 2, 6
The Formation of Colored Compounds
  ■ Demonstrations: 2, 3
The Formation of Complex Ions
  ■ Application Question: 2

25.2 Iron, Cobalt, and Nickel
General Characteristics
  ■ Desktop Investigation
Occurrence of Iron
  ■ Demonstration: 4
  Question: 3
  ■ Application Problems: 1, 2
Pure Iron
  ■ Demonstration: 5
  Application Questions: 3, 4
  ■ Problems: 1–3, 5, 6

25.3 Copper, Silver, and Gold
General Characteristics
  ■ Questions: 4, 5
  Application Problems: 3, 4
  ■ Problem: 3
  Application Question: 1
Copper
  ■ Demonstrations: 6–8
  Application Question: 5

Teaching Strategies

Introduction

Students generally find the study of the transition elements to be of great interest because they are already familiar with many of the elements. This chapter discusses the electron arrangements of the transition metals. The general properties of the elements are studied, based on their electron configurations. A basis is given for the similarities among the transition elements. A comparison is made with the alkali metals and the alkaline-earth metals. The iron and copper families are studied in detail, emphasizing the occurrence and preparation of iron and copper and some of their compounds. The chemistry of the coordination complexes of the transition metals is also examined.

25.1 The Transition Elements

It is helpful to students to develop a chart of the electron configurations of all the elements in Period 4. Point out that the transition metals are adding electrons to the next-to-highest energy level, close to the outermost level in energy. Mention the names of the transition elements and ask students if they know of substances formed from the various elements.

Compare the properties of the transition elements with those of the alkali metals and alkaline-earth metals. Emphasize the metallic properties of the transition elements: their shiny appearance, malleability, ductility, and good conduction of electricity. Show that all the metals have one or two electrons in their highest energy levels. Note the reduced reactivity of the transi-tion elements. Explain their greater hardness and brittleness and higher melting and boiling points in terms of the covalent bonding resulting from the sharing of d electrons.

Map the melting points of the transition elements and show that they follow a trend—all have high melting points, increasing and then decreasing as electrons begin to half fill and then totally fill d orbitals. Show that these properties increase as one moves down the Periodic Table and that tungsten has the highest melting point of all. Point out the very high densities of osmium, iridium, and platinum. Have students calculate the mass of one cubic foot of osmium (634 kg or 1400 lb). Note the unusual behavior of mercury, the only metal that is a liquid at room temperature.

Define paramagnetism. Draw electron filling diagrams for iron, cobalt, and nickel to show the unpaired electrons in d orbitals.

Stress the horizontal similarities found in the transition metal groups.

Make a long-form version of the Periodic Table by cutting out the lanthanide and actinide series and pasting it in between Groups 2 and 3. Use this to discuss the electron configurations of the lanthanide and actinide series.

Discuss various oxidation states of the transition metals. Manganese can have oxidation states of $+2$, $+3$, $+4$, $+6$, and $+7$. Write out the electron configurations for these. Note that elements beyond manganese seldom have higher oxidation states.

Display samples of transition metal compounds that show a range of colors. Include compounds with various oxidation states of the same element, such as $MnSO_4$, MnO_2, and $KMnO_4$, and point out that the color is related to the oxidation state.

Carry out some demonstrations involving complex ion formation. Write the equations for the reactions and analyze the formulas to show the metal ion and the complexing ions. Determine the charge on the complex ion from the charges on the metal ion and the complexing groups.

Demonstration 1: Paramagnetism of Transition Metal Compounds

PURPOSE: To demonstrate that many transition metal compounds are paramagnetic
MATERIALS: Pharmaceutical gelatin capsules, thread, strong magnet, soda-straw balance or any other sensitive balance, transition metal compounds such as $MnSO_4$, $FeSO_4$, $CoSO_4$, MnO_2, $KMnO_4$, Mn, and nontransition compounds such as NaCl, $CaCO_3$
PROCEDURE: Fill gelatin capsules with the compounds to be tested and suspend them from the balance. Bring the magnet under them and observe that the paramagnetic substances are pulled downward into the magnet, while the diamagnetic substances are not. Compare the effect of the magnet on substances with various numbers of unpaired electrons. For example, compare $MnSO_4$, which has five unpaired electrons, with MnO_2, which has three.

Demonstration 2: Colored Objects Absorb Light

PURPOSE: To demonstrate that colored objects reflect the color that is observed and absorb their complementary colors
MATERIALS: Flashlight, colored filters (colored cellophane, probably available from physics lab), construction paper in white and in various colors
PROCEDURE: In a darkened room, shine white light on white paper. Observe that white is reflected. Put a blue filter over a flashlight and shine on white paper. Now blue is reflected. Shine blue filter light on blue paper and again blue is reflected. Shine blue filtered light on yellow (the complement of blue) paper, and the paper appears black, showing that yellow absorbs the blue light. Reverse the colors and try with other combinations of complementary colors.

Demonstration 3: Dehydration of Copper(II) Sulfate Pentahydrate Crystals

PURPOSE: To show the color change of copper(II) sulfate pentahydrate crystals when dehydrated and then hydrated
MATERIALS: Fine crystals of copper(II) sulfate pentahydrate, 250-mL beaker, watch glass, hot plate
PROCEDURE: Place one to two grams of copper(II) sulfate pentahydrate crystals in the beaker. Cover with watch glass. Place on hot plate on medium setting. Observe water droplets on watch glass as crystals lose their water of hydration. Remove watch glass and continue to heat until the colorless anhydrous crystals are formed. Allow to cool. Add a few drops of water and observe the reappearance of the blue color as the crystals absorb water. Also note that heat is evolved as the hydration occurs.

Desktop Investigation: Paramagnetism of Nickel

Even the most sophisticated of students will play with iron filings on a plastic sheet with a magnet underneath.

The colorful, single displacement (replacement) reaction described in this Desktop Investigation is quite fascinating, especially when students observe that the "dust particles" formed respond to a magnet.

Many students believe that iron is the only magnetic substance. Three common metals, however, are magnetic: iron, cobalt, and nickel. Nickel-copper alloys were used in coins for many years because of the low cost of nickel and the durability of copper. However, the United States has decreased the amount of nickel in coins enough so that the "nickel" no longer has enough of that metal in it to respond to a magnetic field.

ANSWERS TO DISCUSSION QUESTIONS
1. Nothing happened with nickel(II) nitrate or aluminum near the magnet.
2. The nickel dust responded to the magnetic field, just as iron filings would.
3. A single displacement (or replacement) reaction took place. The aluminum displaced the nickel from the compound.
4. No, the United States has decreased the amount of nickel in five-cent coins. Some older U.S. coins are attracted by a magnet.

25.2 Iron, Cobalt, and Nickel

Point out the location of the metals iron, cobalt, and nickel in the Periodic Table and call attention to the horizontal similarities. Write out their electron configurations and compare the properties of the elements. Discuss the fact that iron can form a +3 ion readily, leaving it with a half-filled $3d$ electron sublevel. Show the electron configurations of the +2 ions of iron, cobalt, and nickel.

Show samples of the three metals and some of their alloys. Discuss ferromagnetism and the domain explanation. Test the metals and their alloys for attraction to a magnet.

Discuss the abundance of iron. List the various iron ores and their formulas. Give the various equations for the reactions in the blast furnace, pointing out the oxidation and reduction reactions that occur.

Emphasize steel production as a process in which impurities in blast furnace iron are removed by oxidation. The sources of oxygen are the elemental gas or the oxygen in the iron ore. Give the equations for the oxidation process and for slag formation.

Exhibit the oxides of iron and discuss their uses. Show a sample of $FeSO_4 \cdot 7H_2O$. Write the equation for its oxidation and explain how the addition of sulfuric acid and pieces of iron keep the Fe^{2+} in the reduced state.

Show a sample of $FeCl_3 \cdot 6H_2O$ and discuss its uses. Explain the equations for the hydrolysis of Fe^{3+} ions.

Review the complex ions $Fe(CN)_6^{4-}$ and $Fe(CN)_6^{3-}$, pointing out the difference in oxidation states of the iron. Show solutions of $K_3Fe(CN)_6$ and of $K_3Fe(CN)_6$. Explain the tests for the presence of the Fe^{2+} and the Fe^{3+} ions.

Demonstration 4: Rusting of Iron

PURPOSE: To discover the source of oxygen in rusting of iron
MATERIALS: Test tube, beaker, steel wool, water
PROCEDURE: Put a loose wad of steel wool into the bottom of a test tube. Moisten the steel wool with water. Place the test tube

upside down in a beaker containing some water. Note that the test tube will have the steel wool at the top and will be filled almost entirely with air that is trapped by the water in the beaker. After a day or two, check to see that rust has formed on the steel wool. Did the oxygen needed to form the rust come from the water or from the air? The raised level of water inside the test tube indicates that the air provided the oxygen to form the rust.

Demonstration 5: Finding Iron in Cereal

PURPOSE: To show the presence of iron in iron-enriched cereals
MATERIALS: Iron-enriched cereal, beaker, magnetic stirrer and stirring magnet, water
PROCEDURE: Add water to some iron-fortified cereal and stir until the cereal becomes soggy. Place a magnetic stirring bar in the cereal, and stir on a magnetic stirrer for 15-30 minutes. Remove the stirring bar and note the slivers of metallic iron covering the magnet. Discuss the reaction of metallic iron with the hydrochloric acid in the stomach to produce Fe^{2+} ions.

25.3   Copper, Silver, and Gold

Discuss the fact that the metals of the Copper family are not easily oxidized and are among the few metals that occur in the free state. Exhibit samples of the metals and of their ores.

Discuss the steps in copper metallurgy and write the equations for the reactions that occur.

Point out that Cu^{2+} compounds are the most important copper compounds. Show samples of Cu^{2+} compounds such as $CuSO_4 \cdot 5H_2O$, CuO, and $CuCO_3$. Discuss the hydrolysis of Cu^{2+} in water, and test the pH of water solutions of Cu^{2+}. Ask students to list the uses of copper. Explain the tests to show the presence of the Cu^{2+} ion.

Demonstration 6: Copper Plating

PURPOSE: To demonstrate plating of copper from solution
MATERIALS: Solution of $CuSO_4$, carbon electrodes, small nine-volt battery, beaker, wires and clips
PROCEDURE: Place the electrodes in the beaker with the $CuSO_4$ solution. Connect them to the terminals of the battery. Observe the formation of metallic copper on the cathode. Write the equation for the reduction of Cu^{2+} ions. Try using other substances for electrodes.

Demonstration 7: Floating Pennies

PURPOSE: To show the relative reactivity of copper and zinc
MATERIALS: Pennies minted after 1982, hydrochloric acid ($6M$), sharp file, beaker
PROCEDURE: Wear goggles. Use the file to make a sharp scratch in the edge of a penny on two opposite sides. The scratch must cut through the copper coating and expose the zinc layer underneath. Place several of these pennies in the hydrochloric acid and allow to react overnight. The copper will not react with the acid, but the zinc will, forming hydrogen gas and zinc chloride. The bubbles of hydrogen gas will collect inside the pennies and cause them to float. This is a demonstration that is a real attention getter!

Demonstration 8: Changing Pennies to Silver and Gold.

PURPOSE: To show the formation of a coating of zinc on copper, and the formation of a bronze alloy of copper and zinc, making pennies appear to be "silver" and "gold"
MATERIALS: Pennies minted before 1982, $6M$ NaOH solution, granular zinc, evaporating dish, hot plate, steel wool, crucible tongs, burner.
PROCEDURE: Wear goggles. Place about 5 g of granular zinc in an evaporating dish. Add enough $6M$ NaOH solution to cover the zinc and fill the dish about one-third full. Heat on a hot plate until near boiling. Clean a copper penny with steel wool. Hold the penny with crucible tongs by opposite edges in the dish for 3-4 minutes, until it appears "silver" in color as a zinc coating forms. Remove it, rinse with water, and blot dry. Prepare several of these "silver" pennies. Hold one of the "silver" pennies by opposite edges with crucible tongs in the flame of a burner. A gold color forms almost immediately as the zinc coating alloys with the copper. Remove the coin after 3-5 seconds, and wash and dry it. These pennies make great awards for students!

Dispose of Zn-NaOH solution by neutralizing it with acid and pouring it down the drain with large amounts of water. Spontaneous fires have occurred when a Zn-OH mixture was discarded in the trash with paper.

References and Resources

Books and Periodicals

Bonatti, Enrico. "Sea-floor Ore," *Scientific American*, Vol. 240, 1979, p. 95.

Bonatti, Enrico. "Chemistry of Iron, Raw Materials, and Steel," American Iron and Steel Institute, 150 West 42nd St., New York, NY 10017. Copies are provided free to schools.

Maddin, Robert, James D. Muhly, and Tamara S. Wheeler. "How the Iron Age Began," *Scientific American*, Vol. 237, 1977, p. 122.

Shakhashiri, Bassam Z., Glen E. Dureen, and Lloyd G. Williams. "Paramagnetism and Color of Liquid Oxygen; A Lecture Demonstration," *Journal of Chemical Education*, Vol. 57, 1980, p. 363.

Summerlin, Lee R., and James L. Ealy Jr. "The Silver Mirror Reaction," *Chemical Demonstrations, A Sourcebook for Teachers*, Vol. 1, American Chemical Society, 1985 p. 91.

Summerlin, Lee R., and James L. Ealy Jr. "Oxidation States of Manganese: Mn^{7+}, Mn^{6+}, Mn^{4+}, and Mn^{2+}," *Chemical Demonstrations, A Sourcebook for Teachers*, Vol. 1, American Chemical Society, 1985, p. 99.

Summerlin, Lee R., and James L. Ealy Jr. "Oxidation States of Vanadium: Reduction of V^{5+} to V^{2+}," *Chemical Demonstrations, A Sourcebook for Teachers*, Vol. 1, American Chemical Society, 1985, p. 108.

Summerlin, Lee R., Christie L. Borgford, and James L. Ealy, Jr. "Transition Metals and Complex Ions," *Chemical Demonstra-*

tions, *A Sourcebook for Teachers*, Vol. 2, American Chemical Society, 1987, pp. 69–86.

Viswanadham, Puligandla. "An Inexpensive Guoy Balance for Magnetic Susceptibility Determination," *Journal of Chemical Education*, Vol. 55, 1978, p. 54.

Audiovisual Resources

"Steel from Inland": 16-mm film or video, #13878, free-loan film from Modern Talking Picture Service. Shows how iron ore, coal, and limestone become steel

"Mining and the Environment": 16-mm film or videocassette, #16269, Free-Loan Film from Modern Talking Picture Service, 5000 Park Street North, St. Petersburg, FL 33709.

Shows the operation of a copper and silver mine in Montana

"Vanadium—A Transition Element": CHEM Study 16-mm film, Ward's Modern Learning Aids. Shows different oxidation states of vanadium, electronic structure, and complex ion formation

Computer Software

"The Periodic Table Videodisc: Reactions of the Elements": CAV type videodisc, Journal of Chemical Education Software, 1989. Shows action sequences and still shots of each element in its most common forms; reactions with air, water, acids, and bases; some common applications

Answers and Solutions

Questions

1. The metallic character of transition elements is attributed to the small number of electrons in the outermost energy sublevel of their atoms.
2. Yellow, the complement of blue light
3. (a) It is reduced. (b) It is purified.
4. All are found as the native metal. They are soft, dense, and inactive metals. They all have an oxidation state of $+1$ in some compounds.
5. In moist air, copper forms a basic copper carbonate coating. Sulfur dioxide in the air may also combine with copper, forming a green basic copper sulfate.
6. (a) 0 (b) $+4$ (c) $+2$ (d) $+7$ (e) $+6$.

Problems

1. $Fe + 2HCl \rightarrow FeCl_2 + H_2$

$165 \text{ g Fe} \times \dfrac{1 \text{ mol Fe}}{55.8 \text{ g Fe}} \times \dfrac{1 \text{ mol FeCl}_2}{1 \text{ mol Fe}} \times \dfrac{127 \text{ g FeCl}_2}{1 \text{ mol FeCl}_2}$
$= 376 \text{ g FeCl}_2$

2. $2FeCl_2 + Cl_2 \rightarrow 2FeCl_3$

$376 \text{ g FeCl}_2 \times \dfrac{1 \text{ mol FeCl}_2}{127 \text{ g FeCl}_2} \times \dfrac{2 \text{ mol FeCl}_3}{2 \text{ mol FeCl}_2} \times \dfrac{162 \text{ g FeCl}_3}{1 \text{ mol FeCl}_3}$
$= 480. \text{ g FeCl}_3$

3. Formula mass of limonite $= 373.4 \text{ g/mol}$

$\dfrac{223.4 \text{ g Fe}}{373.43 \text{ g limonite}} \times 100\% = 59.83\% \text{ Fe}$

4. Each Ag atom forms 1 formula unit of $AgNO_3$.

$100. \text{ g Ag} \times \dfrac{1 \text{ mol Ag}}{108 \text{ g Ag}} \times \dfrac{1 \text{ mol AgNO}_3}{1 \text{ mol Ag}} \times \dfrac{170. \text{ g AgNO}_3}{1 \text{ mol AgNo}_3}$
$= 157 \text{ g AgNO}_3$

5. $\dfrac{112 \text{ g Fe}}{160 \text{ g Fe}_2O_3} \times 87.0\% \text{ Fe}_2O_3 = 60.9\% \text{ Fe}$

6. $1.0 \times 10^6 \text{ metric tn} \times 0.04 = 4.0 \times 10^4 \text{ metric tn}$

Application Questions

1. The metals used in early times were largely the ones that occur in the native state. Those difficult to extract from their ores have been obtained within the last century because an increased knowledge of chemistry made their extraction possible.
2. A polyatomic ion is a charged structure made up of two or more atoms covalently bonded. A complex ion is a charged structure composed of a central metal ion combined with a specific number of polar molecules or ions.
3. Iron(II) ions gradually oxidize to the $+3$ state, and the blue color characteristic of Fe ions present in two different oxidation states in the $KFeFe(CN)_6 \cdot H_2O$ product develops.
4. $K_4Fe(CN)_6$ forms the blue precipitate $KFeFe(CN)_6 \cdot H_2O$ with iron (III) ions. $K_3Fe(CN)_6$ forms the same blue precipitate with iron(II) ions.
5. Enough gold and silver are recovered to pay for the process

Application Problems

1. Each mole of $CaCO_3$ yields 1 mole of CaO and each mole of CaO reacts with 1 mole of SiO_2.

$(1.0 \times 10^6 \text{ metric tn})(0.08) = 8 \times 10^4 \text{ metric tn SiO}_2$

$8.0 \times 10^4 \text{ metric tn SiO}_2 \times \dfrac{10^6 \text{g SiO}_2}{\text{metric tn SiO}_2} \times \dfrac{1 \text{ mol SiO}_2}{60. \text{ g SiO}_2} \times$

$\dfrac{1 \text{ mol CaCO}_3}{1 \text{ mol SiO}_2} \times \dfrac{100. \text{ g CaCO}_3}{1 \text{ mol CaCO}_3} \times \dfrac{1 \text{ metric tn CaCO}_3}{10^6 \text{g CaCO}_3}$

$= 1.3 \times 10^5 \text{ metric tn CaCO}_3$

2. *(a)* $Fe_2O_3 + 3CO \rightarrow 2Fe + 3CO_2$

$1.0 \times 10^6 \text{ metric tn ore} \times 0.87 \dfrac{\text{metric tn Fe}_2O_3}{\text{metric tn ore}}$
$= 8.7 \times 10^5 \text{ metric tn Fe}_2O_3$

$8.7 \times 10^5 \text{ metric tn Fe}_2O_3 \times \dfrac{10^6 \text{g Fe}_2O_3}{\text{metric tn Fe}_2O_3} \times \dfrac{1 \text{ mol Fe}_2O_3}{160. \text{ Fe}_2O_3} \times$

$\dfrac{3 \text{ mol CO}}{1 \text{ mol Fe}_2O_3} \times \dfrac{28 \text{ g CO}}{1 \text{ mol CO}} \times \dfrac{1 \text{ metric tn CO}}{10^6 \text{g CO}}$

$= 4.6 \times 10^5 \text{ metric tn CO}$

(b) One mol of C is required to produce each mole of CO_2. A second mole of C is required to convert each mole of CO_2 to 2 mol of CO. Therefore, overall, 1 mol of C is required for each mole of CO.

$$4.6 \times 10^5 \text{ metric tn CO} \times \frac{10^6 \text{g CO}}{\text{metric tn CO}} \times$$

$$\frac{1 \text{ mol CO}}{28 \text{ g CO}} \times \frac{1 \text{ mol C}}{1 \text{ mol C}} \times \frac{12 \text{ g C}}{1 \text{ mol C}} \times \frac{1 \text{ metric tn C}}{10^6 \text{g C}}$$

$$= 2.0 \times 10^5 \text{ metric tn C}$$

3. $\overset{0}{Ag} + \overset{+5}{H}NO_3 \rightarrow \overset{+1}{Ag}NO_3 + \overset{+2}{N}O + H_2O$

$\overset{0}{Ag} \rightarrow \overset{+1}{Ag} + e^-$

$\overset{+5}{N} + 3e^- \rightarrow \overset{+2}{N}$

$3Ag + 4HNO_3 \rightarrow 3AgNO_3 + NO + 2H_2O$

4. $\overset{0}{Cu} + H_2\overset{+6}{S}O_4 \rightarrow \overset{+2}{Cu}SO_4 + \overset{+4}{S}O_2 + H_2O$

$\overset{0}{Cu} \rightarrow \overset{+2}{Cu} + 2e^-$

$\overset{+6}{S} + 2e^- \rightarrow \overset{+4}{S}$

$Cu + 2H_2SO_4 \rightarrow CuSO_4 + SO_2 + 2H_2O$

Teacher's Notes

Chapter 25 The Transition Metals

INTRODUCTION

The transition elements have widely varying and interesting properties and are used to construct bridges and buildings and to make appliances, utensils, and other common products. In this chapter, we cover only a few of the most commonly seen properties of the transition elements. Next, we take a brief look at the chemistry of their complex ions. We will examine iron, cobalt, and nickel, and the process by which iron ore is converted into the metal. We will also discuss the chemistry of copper, silver, and gold.

LOOKING AHEAD

As you study this chapter, review the electron configuration of the transition elements and compare and contrast their properties to those of the Group 1 and Group 2 metals. Pay particular attention to the presence and behavior of *d*-orbital electrons and the chemistry of complex ions.

SECTION PREVIEW

25.1 The Transition Elements
25.2 Iron, Cobalt, and Nickel
25.3 Copper, Silver, and Gold

Humans have long valued precious metals for their scarcity and beauty. This Asante umbrella top exhibits the brilliance of gold, one of the transition metals.

25.1 The Transition Elements

Beginning with Period 4, ten subgroups of elements intervene between Group 2 and Group 13 they are called the **transition elements**. As the atomic number of the transition elements increases, there is an increase in the number of electrons in the next-to-highest energy level. This inner building of electronic structure among the transition elements can be interpreted as an interruption in the regular increase of highest-energy-level electrons from Group 1 to Group 18 across a period.

The transition elements (Figure 25-1), include gold, silver, and platinum, some of the most important elements in valuable jewelry. This category of elements also includes the iron used as the structural backbone of our buildings and bridges and the copper in millions of miles of wire bringing electricity to every part of the country.

Transition Elements

Group 2		Group 3	Group 4	Group 5	Group 6	Group 7	Group 8	Group 9	Group 10	Group 11	Group 12	Group 13
12 **Mg** 24.305												13 **Al** 26.98154
20 **Ca** 40.08		21 **Sc** 44.956	22 **Ti** 47.88	23 **V** 50.9415	24 **Cr** 51.996	25 **Mn** 54.938	26 **Fe** 55.847	27 **Co** 58.9332	28 **Ni** 58.69	29 **Cu** 63.546	30 **Zn** 65.38	31 **Ga** 69.72
38 **Sr** 87.62		39 **Y** 88.906	40 **Zr** 91.22	41 **Nb** 92.906	42 **Mo** 95.94	43 **Tc** (98)	44 **Ru** 101.07	45 **Rh** 102.906	46 **Pd** 106.42	47 **Ag** 107.8682	48 **Cd** 112.41	49 **In** 114.82
56 **Ba** 137.33		57 **La** 138.906	72 **Hf** 178.49	73 **Ta** 180.948	74 **W** 183.85	75 **Re** 186.207	76 **Os** 190.2	77 **Ir** 192.22	78 **Pt** 195.08	79 **Au** 196.967	80 **Hg** 200.59	81 **Tl** 204.383
88 **Ra** 226.025		89 **Ac** 227.028	104 **Unq** (261)	105 **Unp** (262)	106 **Unh** (263)	107 **Uns** (262)	108 **Uno** (265)	109 **Une** (266?)				

Remember that the periods of the Periodic Table are horizontal rows; groups are vertical columns.

Figure 25-1 Transition metals are characterized by the buildup of d–sublevel electrons in the next-to-highest energy level.

General Properties of the Transition Elements

As you might expect, the properties of the elements remain similar going down each main group. Going horizontally across Periods 1–3 from Group 1 to Group 17, however, the properties change steadily from metallic to nonmetallic. In the transition region, horizontal similarities may be as great as (and in some cases greater than) similarities going down a subgroup.

All transition elements are metals. The metallic character is related to the presence of no more than one or two electrons in the highest energy levels of their atoms. These metals are generally harder and more brittle than the Group 1 and Group 2 metals. Their hardness is partially due to the covalent bonding resulting from the sharing of *d* electrons. Their melting points are higher than those of the Group 1 and Group 2 metals. Mercury (atomic number 80) is a familiar exception. It is the only metal that is liquid at room temperature. The increase in the melting points of the elements up through vanadium and chromium may be attributed to additional covalent bond formation resulting from the larger number of unpaired electrons. The transition metals are good conductors of heat and electricity. Such conduction is characteristic of a metallic crystal lattice permeated by an electron gas.

Tungsten, with a melting point of 3180°C, has the highest melting point of all the metals. The densest metal is osmium, which has a density of 22.4 g/mL.

The only other element that is a liquid at room temperature is bromine.

Desktop Investigation

Paramagnetism of Nickel

Question
Can you demonstrate the magnetic character of a fourth-row transition element?

Materials
nickel(II) nitrate
aluminum foil
distilled water
five-cent coins of various
  ages
magnet
100-mL beaker
scissors
steel wool

Procedure
Record your results in a data table.
1. Test the solid nickel(II) nitrate and the aluminum foil for any response to a magnetic field.
2. Dissolve 2–3 g of nickel(II) nitrate in 10 mL of distilled water.
3. Cut a piece of aluminum foil about 4 cm × 4 cm; gently buff it with steel wool to expose a fresh surface.

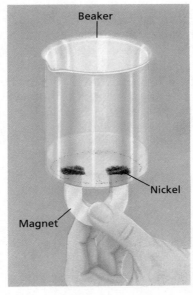

4. Place the foil into the nickel(II) solution and allow it to react overnight.
5. The solid that forms is elemental nickel. Test it for its ability to respond to a magnetic field by moving the beaker over the magnet.
6. Test five-cent coins for their response to a magnetic field.

Discussion
1. What happened when the solid nickel(II) nitrate and aluminum were placed near the magnet?
2. What happened when the newly formed nickel was brought near the magnet?
3. What general type of chemical reaction took place?
4. Is there very much nickel in a five-cent coin? Are there differences between coins of different ages?

The chemical properties of transition elements are varied. Electrons of the next-to-highest d sublevel as well as those of the highest energy level may become involved in compound formation. Most transition metals exhibit variable oxidation states in forming compounds, and most of their compounds have color. They have a pronounced tendency to form complex ions. Many of their compounds are attracted into a magnetic field (a property called paramagnetism).

Transition metal similarities   The number of electrons in the highest energy level of the transition elements remains fairly constant. It never exceeds two electrons as the number of electrons in the d sublevel increases across a period. Similarities often appear in a horizontal sequence of transition elements as well as vertically through a group. In some cases, horizontal similarity may exceed vertical similarity. The first five groups (Groups 3 and 7) are those headed by Sc, Ti, V, Cr, and Mn. Generally, similarities among these elements are recognized within each group. The next three groups (Groups 8–10) are headed by Fe, Co, and Ni. Here the similarities across each period are greater than those down each group. Iron, cobalt, and nickel resemble each other more than do iron, ruthenium, and osmium, the members of Group 8. The final two groups are headed by Cu and Zn. Here again, the greatest similarity is within each group.

Transition Metal Groups

The electron configuration of each transition element is shown in the Periodic Table inside the back cover. Compare the electron configuration of the Group 2 metals with those of the corresponding metals of the first transition group, the scandium (Sc) group. What differences do you observe? Now compare the electron configurations of elements across the first transition series.

There are minor irregularities in the 3d-electron buildup at Cr (atomic number 24) and at Cu (atomic number 29). These apparent irregularities and those in the other periods of the transition elements are attributed to the extra stability associated with half-filled and completely filled sublevels.

The transition elements occupy the long periods in the Periodic Table. The choice of the first and last elements in the series depends somewhat on the definition used. Based on d- and s-electron configurations, the transition elements consist of the ten groups between Group 2 and Group 13. The elements of the first transition series are the most important, and they are the most abundant. Refer also to Appendix Table A-9 to see the electron arrangements of the transition elements.

Variable Oxidation States

Variable oxidation states are common among the transition metals. Some form different compounds in which they exhibit several oxidation states. The energies of the outermost d and s electrons do not differ greatly. The energies to remove these electrons are relatively low.

The d sublevel has five available orbitals that can hold ten electrons. The s sublevel has one orbital, and it can hold two electrons. Electrons occupy the d orbitals singly as long as unoccupied orbitals of similar energy exist. The s electrons and one or more d electrons can be used in bonding.

In metals of the first transition series, several 4s and 3d electrons may be transferred to or shared with other substances. Thus, several oxidation states become possible. Remember that the maximum oxidation state is limited by the total number of 4s and 3d electrons present.

Table 25-1 gives the metals' common oxidation states and 3d- and 4s-electron configurations. It uses orbital notations introduced in Section 4.3. In these notations ___ represents an unoccupied orbital. The symbol $\underline{\uparrow}$ represents an orbital occupied by one electron, and $\underline{\uparrow\downarrow}$ represents an orbital with an electron pair.

Observe that the maximum oxidation state increases to +7 for manganese and then decreases abruptly beyond manganese. The difficulty in forming the higher oxidation states increases toward the end of the row

Figure 25-2 These nodules, found on the floor of the Pacific Ocean, contain nickel, copper, cobalt and manganese. They might prove to be an important source of strategic transition metals.

Figure 25-3 Chromium, magnesium, and nickel are ingredients of stainless steel, used to make durable rugged items such as wheelchairs.

Figure 25-4 "Setting out" the Cathedral of St. John the Divine. Zinc templates are used to trace the shape of each stone onto rough limestone blocks.

Figure 25-5 Rhodochrosite is a rose-red mineral that consists primarily of manganese carbonate ($MnCo_3$).

because of the general increase in ionization energy with atomic number. The higher oxidation states generally involve covalent bonding.

Manganese atoms lose two $4s$ electrons to become Mn^{2+} ions. Higher oxidation states involve $3d$ electrons. In the permanganate ion, MnO_{4-}, manganese is in the $+7$ oxidation state, covalently bonded with the oxygen atoms.

When the electron configuration includes both $3d$- and $4s$-valence electrons, the $3d$ electrons are lower in energy than the $4s$ electrons. Thus the first electron removed in an ionizing reaction is the one most loosely held, a $4s$ electron. This fact is illustrated by the step-by-step ionization of titanium. The ground-state configurations of the valence electrons are:

$$
\begin{array}{ll}
\text{Ti} & 3d^2 4s^2 \\
\text{Ti}^+ & 3d^2 4s^1 \\
\text{Ti}^{2+} & 3d^2 4s^0 \\
\text{Ti}^{3+} & 3d^1 \\
\text{Ti}^{4+} & 3d^0
\end{array}
$$

TABLE 25-1   OXIDATION STATES OF FIRST-ROW TRANSITION ELEMENTS

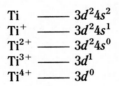

Element	Electron Configuration $1s^2 2s^2 2p^6 3s^2 3p^6$	3d	4s	Common Oxidation States
scandium	Each transition	↑ — — — —	↓↑	+3
titanium	element has all	↑ ↑ — — —	↓↑	+2 +3 +4
vanadium	these sublevels	↑ ↑ ↑ — —	↓↑	+2 +3 +4 +5
chromium	filled in an	↑ ↑ ↑ ↑ ↑	↑	+2 +3 +6
manganese	argon	↑ ↑ ↑ ↑ ↑	↓↑	+2 +3 +4 +6 +7
iron	structure	↓↑ ↑ ↑ ↑ ↑	↓↑	+2 +3
cobalt		↓↑ ↓↑ ↑ ↑ ↑	↓↑	+2 +3
nickel		↓↑ ↓↑ ↓↑ ↑ ↑	↓↑	+2 +3
copper		↓↑ ↓↑ ↓↑ ↓↑ ↓↑	↑	+1 +2
zinc		↓↑ ↓↑ ↓↑ ↓↑ ↓↑	↓↑	+2

Remind students of Hund's rule from Chapter 4.

The Formation of Colored Compounds

A striking property of many compounds of transition metals is their color. Most colored inorganic compounds involve elements of the transition region of the Periodic Table, as shown in Figure 25-6. The color of the compounds and their solutions may vary depending on what other ions or groups are associated with the transition element. In general, these colored compounds are thought to have some electrons that are easily excited by selectively absorbing part of the energy of white light.

The color of transition metal compounds is associated with partially filled d orbitals. Note that Zn^{2+} compounds, which have completely filled d orbitals, are colorless. Compounds of Sc^{3+}, in which there are no d electrons, are also colorless.

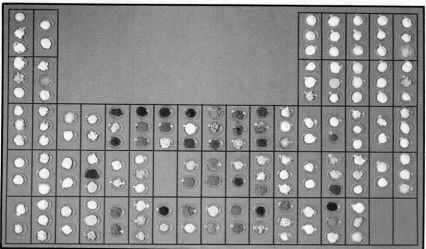

Figure 25-6 Most compounds of transition metals are colored. The color depends on the metal, its oxidation state and the anion with which it is combined.

As shown in Figure 4-3, the continuous spectrum of white light is composed of red, orange, yellow, green, blue, indigo, and violet colors of light. If light energy of a discrete band of wavelengths corresponding to a color region is absorbed by a substance, the remaining light reflected to the eye is the complement of the color absorbed. The energies of the wavelengths of white light that are absorbed are the energies required to raise d electrons to higher energy states.

Copper(II) sulfate pentahydrate crystals are blue when viewed in white light. Their aqueous solutions are also blue. They are blue because the energy of wavelengths corresponding to yellow light is absorbed. The light reflected (or transmitted) is blue light, the complement of yellow light. When these crystals are heated, the water of hydration is given up, and the anhydrous salt is neither blue nor crystalline. It is a white powder. If water is added to the anhydrous powder, it turns blue. Because many sulfate compounds are colorless, we can conclude that it must be the hydrated copper(II) ion, $Cu(H_2O)_4^{2+}$, that is blue.

The Formation of Complex Ions

The NO_3^- ion, SO_4^{2-} ion, and the NH_4^+ ion are examples of polyatomic ions. They are charged particles made up of more than a single atom. These familiar ions are covalent structures. Their net ionic charge is the algebraic sum of the oxidation numbers of all the atoms present. Because of their small size and their stability, they behave just like single-atom ions, and go through many reactions unchanged.

Polyatomic ions are described in Section 6.23.

Figure 25-7 Crystals of hydrated sulfates of nickel (green), iron (tan), cobalt (pink), and copper (blue).

Complex ions can be considered to be examples of Lewis acid-base reactions. They occur when a coordinate covalent bond forms as complexing agents donate electron pairs to a central metal ion.

Figure 25-8 *(below)* Solution of $CuSO_4$ in concentrated HCl; $[CuCl_4]^{2-}$ (aq) ions give this solution a light green color *(a)*. If H_2O is added to this solution, H_2O molecules replace Cl^- ions in the $[CuCl_4]^{2-}$ complex and the solution turns dark green *(b)*. Anhydrous $CuSO_4$ is dissolved in water; $[Cu(H_2O)_4]^{2+}$ (aq) ions give this solution a light blue color *(c)*. Addition of NH_3(aq) causes the displacement of H_2O molecules by NH_3 molecules to form $[Cu(NH_3)_4]^{2+}$ (aq) ions, giving the solution a deep blue color *(d)*.

The name **complex ion** *is ordinarily restricted to an ionic species composed of a central metal ion combined with a specific number of polar molecules or ions. Complex ions vary greatly in stability, but none is as stable as the common polyatomic ions mentioned above.*

The most common complex ions are formed by ions of the transition metals combined with chemical species such as chloride ions (Cl^-), ammonia molecules (NH_3), water molecules (H_2O), and cyanide ions (CN^-). The transition metal cation is always the central ion in the complex. *The number of units covalently bonded to, or coordinated with, the central ion is the* **coordination number** *for that complex ion.* For example, the Fe^{2+} ion in the $Fe(CN)_6^{4-}$ complex has a coordination number of 6.

The hydrated Cu^{2+} ion, $Cu(H_2O)_4^{2+}$, discussed earlier in this section, is a complex ion. The Cu^{2+} ion is coordinated with 4 molecules of water of crystallization. It is this ionic species that crystallizes from an aqueous sulfate solution as the compound $CuSO_4 \cdot 5H_2O$.

When an excess of concentrated ammonia-water solution is added to an aqueous solution of Cu^{2+} ions (a $CuSO_4$ solution, for example), the light-blue color of the solution changes to a deep-blue color. The soluble $Cu(NH_3)_4^{2+}$ complex has been formed.

$$Cu(H_2O)_4^{2+} + 4NH_3(aq) \rightarrow Cu(NH_3)_4^{2+} + 4H_2O$$
$$\text{light blue} \qquad\qquad\qquad \text{deep blue}$$

TABLE 25-2 SOME COMMON COMPLEX IONS

Coordinating Atom	Coordinating Group	Complex Ions	Color
N	H :N̈:H Ḧ	$Ag(NH_3)_2^+$ $Ni(NH_3)_4^{2+}$ $Co(NH_3)_6^{3+}$	— blue blue
C	:C:::N:⁻	$Fe(CN)_6^{4-}$ $Fe(CN)_6^{3-}$	yellow red
S	:Ö: :S̈:S̈:Ö: :Ö:	$Ag(S_2O_3)_2^{3-}$	—
N	:S̈::C::N̈:⁻	$FeSCN^{2+}$	red

Zinc and cadmium form similar complex ions with ammonia. Table 25-2 gives the formulas and colors of some typical complex ions formed by transition metals.

If the ammonia-water solution is added to a solution of silver nitrate, the complex $Ag(NH_3)_2^+$ ion is formed.

$$Ag^+ + 2NH_3 \longrightarrow Ag(NH_3)_2^+$$

$$Ag^+ + 2 \begin{matrix} H \\ :N:H \\ H \end{matrix} \longrightarrow \left[\begin{matrix} H & & H \\ H:N:Ag:N:H \\ H & & H \end{matrix} \right]^+$$

Brackets are used in this expression simply to enclose the electron-dot formula and indicate that the +1 charge applies to the ion as a whole. They do not indicate a concentration expression.

Because complex ions are unstable to some extent, the reaction is reversible. The more stable the complex ion, the more quickly equilibrium is established between the reactions involving the formation and dissociation of the complex ions. At equilibrium the equation describing the decomposition of $Ag(NH_3)^+$ is

$$Ag(NH_3)_2^+ \rightleftarrows Ag^+ + 2NH_3$$

The equilibrium constant has the usual form.

$$K = \frac{[Ag^+][NH_3]^2}{[Ag(NH_3)_2^+]} \quad 7 \times 10^{-8}$$

In cases of complex ion equilibria, the equilibrium constant is called an *instability constant*. The larger the value of K, the more unstable is the complex ion.

Point out to students that these equations are written to show the complex ion breaking apart. Thus, the equilibrium constant is called the "instability" constant. For the reaction written in the reverse direction, the equilibrium constant is called the "formation" constant.

When metallic ions in solution form complexes, they are, in effect, removed from the solution as simple ions. Thus the formation of complex ions increases the solubility of metallic ions. The solubility of a precipitate may also be increased if cations of the substance form complex ions. For example, the formation of chloride complexes may increase solubility where, by common ion effect, a decrease would be expected.

Consider the equilibrium of sparingly soluble AgCl.

$$AgCl(s) \rightleftarrows Ag^+(aq) + Cl^-(aq)$$

Equilibrium constants are discussed in Section 19.1.

The common ion effect is discussed in Section 19.2.

A small addition of Cl^- ions does shift the equilibrium to the left and lower the solubility as expected. If a large concentration of Cl^- ions is added, however, the solid AgCl dissolves. This apparent contradiction is explained by the formation of $AgCl_2^-$ complex ions.

The stability of complex ions is indicated by the small value of the equilibrium constants, K, as shown in Table 25-3. The smaller the value of K, the more stable is the complex ion and the more effectively it holds the central metallic ion as part of the soluble complex.

TABLE 25-3 STABILITY OF COMPLEX IONS

Complex Ion	Equilibrium Reaction	K
$Ag(NH_3)_2^+$	$Ag(NH_3)_2^+ \rightleftharpoons Ag^+ + 2NH_3$	7×10^{-8}
$Co(NH_3)_6^{2+}$	$CO(NH_3)_6^{2+} \rightleftharpoons Co^{2+} + 6NH_3$	1×10^{-5}
$Co(NH_3)_6^{3+}$	$Co(NH_3)_6^{3+} \rightleftharpoons Co^{3+} + 6NH_3$	2×10^{-34}
$Co(NH_3)_6^{4+}$	$Cu(NH_3)_4^{2+} \rightleftharpoons Cu^{2+} + 4NH_3$	3×10^{-13}
$Ag(CN)_2^-$	$Ag(CN)_2^- \rightleftharpoons Ag^+ + 2CN^-$	2×10^{-19}
$Au(CN)_2^-$	$Au(CN)_2^- \rightleftharpoons Au^+ + 2CN^-$	5×10^{-39}
$Cu(CN)_3^{2-}$	$Cu(CN)_3^{2-} \rightleftharpoons Cu_+ + 3CN^-$	1×10^{-35}
$Fe(CN)_6^{4-}$	$Fe(CN)_6^{4-} \rightleftharpoons Fe^{2+} + 6CN^-$	1×10^{-35}
$Fe(CN)_6^{3-}$	$Fe(CN)_6^{3-} \rightleftharpoons Fe^{3+} + 6CN^-$	1×10^{-42}
$FeSCN^{2+}$	$FeSCN^{2+} \rightleftharpoons Fe^{3+} + SCN^-$	8×10^{-3}
$Ag(S_2O_3)_2^{3-}$	$Ag(S_2O_3)_2^{3-} \rightleftharpoons Ag^+ + 2S_2O_3^{2-}$	6×10^{-14}

Section Review

1. Where are the transition elements found in the periodic table?
2. What are some of the properties of the transition elements?
3. How do the transition elements differ from alkali metals and alkaline-earth metals?
4. What are the common oxidation states of the elements in the first transition series?

25.2 Iron, Cobalt, and Nickel

Iron, cobalt, and nickel, are elements of the first transition series. Each is the first member of a group of transition metals (Groups 8–10) that bears its name, as indicated in Figure 25-1. The similarities among iron, cobalt, and nickel are greater than those within each of the groups they head. Thus iron, cobalt, and nickel are referred to as the Iron family. The remaining six members of these three groups have properties similar to platinum and are considered to be members of the Platinum family. They are called *noble metals* because they show little chemical activity. They are rare and expensive.

General Characteristics

Iron is the most important of these elements. Alloys of iron, cobalt, and nickel are important structural metals. Some properties of the Iron family are listed in Table 25-4.

SECTION OBJECTIVES

- Identify the members of the Iron Family, and describe their general characteristics.
- Describe the natural occurrence, refining methods, and compounds of iron.
- Outline the metallurgy of iron and steel.
- Describe the reactions and uses of Fe^{2+} and Fe^{3+} ions.

TABLE 25-4   THE IRON FAMILY

Element	Atomic Number	Atomic Mass	Electron Configuration	Oxidation Numbers	Melting Point (°C)	Boiling Point (°C)	Density (g/cm³)
iron	26	55.847	2,8,14,2	+2, +3	1535	2750	7.87
cobalt	27	58.9332	2,8,15,2	+2, +3	1495	2870	8.9
nickel	28	58.69	2,8,16,2	+2, +3	1453	2732	8.90

The Iron family is located in the middle of the transition elements. The next-to-highest energy level in these elements is incomplete: the $3d$ sublevels of iron, cobalt, and nickel contain only 6, 7, and 8 electrons, respectively, instead of the full number, 10. Each can exhibit the +2 and +3 oxidation states. The Fe^{2+} ion is easily oxidized to the Fe^{3+} ion by air and other oxidizing agents, but the +3 oxidation state occurs only rarely in cobalt and nickel. The electron configurations of iron, cobalt, and nickel sublevels are shown in Table 25-5. Notice that, in this table, paired and unpaired electrons of the $3d$ and $4s$ sublevels are represented by paired and unpaired dots. Compare this table with the electron configuration shown in Figure 25-1.

TABLE 25-5   ELECTRON CONFIGURATIONS OF THE IRON FAMILY

Sublevel	1s	2s	2p	3s	3p	3d	4s
Maximum number of electrons	2	2	6	2	6	10	2
iron	2	2	6	2	6	:....	:
cobalt	2	2	6	2	6	::...	:
nickel	2	2	6	2	6	:::..	:

The two $4s$ electrons are removed easily, as is usually the case with metals. This removal forms the Fe^{2+}, Co^{2+}, or Ni^{2+} ions. In the case of iron, one $3d$ electron is also easily removed because the five remaining $3d$ electrons make a half-filled sublevel. When two $4s$ and one $3d$ electrons are removed, the Fe^{3+} ion is formed. It becomes increasingly difficult to remove a $3d$ electron from cobalt and nickel. This increasing difficulty is partly explained by the higher nuclear charges. Another factor is that neither $3d$ sublevel would be left at the filled or half-filled stage. Neither Co^{3+} nor Ni^{3+} ions are common. Cobalt atoms in the +3 oxidation state occur in complexes, however.

All three metals of the Iron family have a strong magnetic property. This property is commonly known as *ferromagnetism* because of the unusual extent to which it is possessed by iron. Cobalt is strongly magnetic, whereas nickel is the least magnetic of the three. The ferromagnetic nature of these metals is thought to be related to the similar spin orientations of their unpaired $3d$ electrons. Table 25-5 shows that atoms of iron have 4 unpaired $3d$ electrons, atoms of cobalt have 3 unpaired $3d$ electrons, and atoms of nickel have 2 unpaired $3d$ electrons.

Each spinning electron acts like a tiny magnet. Electron pairs are formed by two electrons spinning in opposite directions. The electron magnetisms of such a pair of electrons neutralize each other. In Iron family metals, groups of atoms may be aligned and form small magnetized regions called *domains*. Ordinarily, magnetic domains within the metallic crystals point in random directions. In this way, they cancel one another so that the net magnetism is zero. A piece of iron becomes magnetized when it is acted upon by an outside force that for both paired and unpaired electrons aligns the domains in the same direction.

Occurrence of Iron

Iron is the fourth element in abundance by mass in the earth's crust. Nearly 5 percent of the earth's crust is iron. It is the second most abundant metal, following aluminum. The known magnetic nature of the earth itself and the fact that meteors are known to contain iron suggests that the earth's core may consist mainly of iron.

Unfortunately, much of the iron in the crust of the earth cannot be removed profitably. Although iron is found in many minerals, only those from which iron can be profitably recovered by practical methods are considered to be iron ores.

Great deposits of *hematite* (Fe_2O_3) once existed in the Lake Superior region of the United States. They were the major sources of iron in this country for many years. These rich ore deposits have been depleted, and the iron industry now depends on medium- and low-grade ores, the most abundant of which is *taconite*.

Taconite has an iron content of roughly 25–50% in a chemically complex mixture. The main iron minerals that are present are hematite (Fe_2O_3) and magnetite (Fe_3O_4). Modern blast furnaces that are designed to reduce iron ore to iron require ores containing well above 60% iron. Thus, the raw ores must be concentrated by removing much of the waste materials. This is done at the mines before the ores are transported to the blast furnaces or smelters.

Taconite ores are first crushed and pulverized and then concentrated by a variety of complex methods. The concentrated ores are hardened into pellets for shipment to the smelters. Ore concentrates containing 90% iron are produced by the chemical removal of oxygen during the pelletizing treatment, as shown in Figure 25-9.

Iron is the most widely used of all metals. More iron is consumed than 14 times all other metals together.

Figure 25-9 Enrichment of low-grade iron ores. Almost all iron ore mined in the United States is concentrated before it is shipped to a blast furnace.

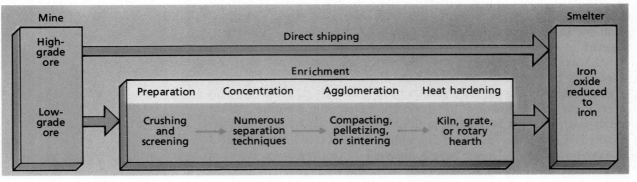

The blast furnace *Iron ore is converted to iron in a giant structure called a* **blast furnace**. The charge placed in the blast furnace consists of iron ore concentrate, coke, and a flux, which causes mineral impurities in the ore to melt more readily. The proper proportions for the charge are calculated by analyzing the raw materials. Usually the flux is limestone, because silica or sand is the most common impurity in the iron ore. Some iron ores contain limestone as an impurity; in such cases, the flux added is sand. A blast of hot air, sometimes enriched with oxygen, is forced into the base of the furnace, as shown in Figure 25-10.

Iron oxide is reduced in the blast furnace to iron. *The earthly impurities that remain are called* **slag**. *The iron recovered from a blast furnace is called* **pig iron**.

What happens inside the blast furnace when iron is produced? The actual chemical changes that occur are complex and are not entirely understood. First, the coke is ignited by the blast of hot air, and some of it burns to form carbon dioxide.

$$C(s) + O_2(g) \rightarrow CO_2(g)$$

When the oxygen is introduced, the carbon dioxide comes in contact with other pieces of hot coke and is reduced to carbon monoxide gas.

$$CO_2(g) + C(s) \rightarrow 2CO(g)$$

The carbon monoxide thus formed is actually the reducing agent that reduces the iron oxide to metallic iron.

$$Fe_2O_3(s) + 3CO(g) \rightarrow 2Fe(\ell) + 3CO_2(g)$$

This reduction probably occurs in steps as the temperature increases toward the bottom of the furnace. Some possible steps are

$$Fe_2O_3 \rightarrow Fe_3O_4 \rightarrow FeO \rightarrow Fe$$

The white-hot liquid iron collects in the bottom of the furnace and is removed every four or five hours. It may be cast in molds as pig iron or converted directly to steel.

In the central region of the furnace, the limestone decomposes to calcium oxide and carbon dioxide.

$$CaCO_3(s) \rightarrow CaO(s) + CO_2(g)$$

The calcium oxide combines with silica to form a calcium silicate slag.

$$CaO(s) + SiO_2(s) \rightarrow CaSiO_3(\ell)$$

This glassy slag melts readily and collects at the bottom of the furnace. Since it has a much lower density than liquid iron, it floats on top of the iron, thus preventing the reoxidation of the iron. The melted slag is tapped off every few hours and usually is thrown away.

Steel production The relatively high carbon content of iron recovered from the blast furnace makes it very hard and brittle. Phosphorus and sulfur are two other impurities. The phosphorus makes pig iron brittle at low temperatures. The sulfur makes it brittle at high temperatures.

The conversion of iron to steel is essentially a purification process in which impurities are removed by oxidation. This purification process is

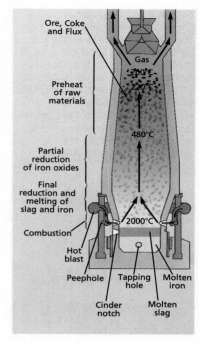

Figure 25-10 A sectional view of a blast furnace.

Note that the oxidation number of iron in this sequence goes from Fe^{3+} in Fe_2O_3, to a combination of Fe^{3+} and Fe^{2+} in Fe_3O_4, to Fe^{2+} in FeO, and finally to Fe^0 in elemental Fe.

Figure 25-11 Steelmaking by the basic oxygen process. The furnace receives a charge of scrap steel *(upper left)* and a charge of molten iron from a blast furnace *(upper right)*. Oxygen is blown into the furnace and flux is added *(lower right)*. Molten steel is released from a tap hole *(lower left)* and transferred to a ladle. Alloying substances are then added according to precise specifications.

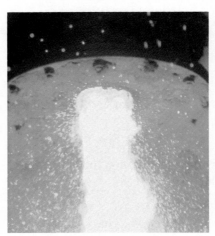

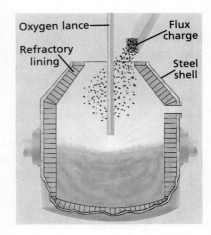

carried out in a furnace at a high temperature. Several steps of this process are shown in Figure 25-11. Scrap steel that is being recycled is sometimes added, and iron ore may be added to supply oxygen for oxidizing the impurities. Oxygen gas is blown in for very rapid oxidation. Limestone or lime is included to form a slag with nongaseous oxides. Near the end of the process, selected alloying substances are added. By using different kinds and quantities of these added substances, the steel is given different desired properties.

The impurities in the pig iron are oxidized in the following way:

$$3C + Fe_2O_3 \rightarrow 3CO(g) + 2Fe$$
$$3Mn + Fe_2O_3 \rightarrow 3MnO + 2Fe$$
$$12P + 10Fe_2O_3 \rightarrow 3P_4O_{10} + 20Fe$$
$$3Si + 2Fe_2O_3 \rightarrow 3SiO_2 + 4Fe$$
$$3S + 2Fe_2O_3 \rightarrow 3SO_2(g) + 4Fe$$

Carbon and sulfur compounds escape as gases. The limestone flux decomposes as in the blast furnace.

$$CaCO_3 \rightarrow CaO + CO_2(g)$$

Figure 25-12 These marble-sized pellets are produced by direct reduction of iron ore.

Calcium oxide and the oxides of the other impurities react to form slag.

$$P_4O_{10} + 6CaO \rightarrow 2Ca_3(PO_4)_2$$
$$SiO_2 + CaO \rightarrow CaSiO_3$$
$$MnO + SiO_2 \rightarrow MnSiO_3$$

Pure Iron

Pure iron is a metal that is seldom encountered. It is silver-white, soft, ductile, tough, and does not tarnish readily. It melts at 1535°C. Commercial iron contains carbon and other impurities that alter its properties. Cast iron melts at approximately 1150°C. All forms of iron corrode, or rust, in moist air, so iron is not a self-protective metal. The rust that forms is brittle and scales off, exposing the metal underneath to further corrosion. Iron does not rust in dry air or in water that is free of dissolved oxygen.

Iron is only moderately active chemically. Even so, it corrodes more extensively than more active metals such as zinc and aluminum. These metals form oxide coatings that adhere to the metal surface and protect it from further corrosion. The rusting of iron is a complicated (and not fully understood) electrochemical process that involves water, air, and carbon dioxide.

Rusting begins when water containing CO_2 in solution comes into contact with iron. Such a solution is acidic and reacts with iron to form Fe^{2+} ions. The iron(II) ions are oxidized to the iron(III) state as hydrated Fe_2O_3 (rust) by oxygen. Iron is made rust-resistant in several ways: (1) by alloying it with chromium, copper, or nickel; (2) by treating it to form a coating of Fe_3O_4, which adheres to and protects the surface; (3) by painting it, which protects the surface as long as the paint adheres and is not chipped or scratched; (4) by dip-coating it with a self-protective metal (for example galvanizing with zinc); (5) by plating it with nickel or chromium; or (6) by coating it with a material like porcelain.

Although dilute acids generally react readily with iron, concentrated nitric acid does not react with it, and concentrated sulfuric acid has little effect on iron. If fact, dipping iron into concentrated nitric acid makes the iron passive, or inactive, with other chemicals. Strong hydroxides do not react with iron.

Three oxides of iron Of the three oxides of iron, iron(II) oxide, FeO, is of little importance. It oxidizes rapidly when exposed to air to form iron(III) oxide, Fe_2O_3. This oxide is the important ore of iron. It is used as a cheap pigment for red paint and for grinding and polishing glass lenses and mirrors. Magnetic iron oxide known as magnetite, (Fe_3O_4), is an important ore. It is composed of Fe_2O_3 and FeO. Thus, it may be considered to be iron(II, III) oxide, and its formula can be written $FeO \cdot Fe_2O_3$.

Reactions of the Fe^{2+} ion Hydrated iron(II) sulfate, $FeSO_4 \cdot 7H_2O$, is the most useful compound of iron in the +2 oxidation state. It is used as a reducing agent and in medicine for iron tonics. Iron(II) sulfate can be prepared by the action of dilute sulfuric acid on iron. The crystalline hydrate loses water of hydration when exposed to air and turns brown because of oxidation. Iron(II) sulfate in solution is gradually oxidized to the iron(III) state by dissolved oxygen. The brown precipitate of basic iron(III) sulfate that

Figure 25-13 A worker inspects pipes at an iron oxide plant in West Virginia.

Galvanized iron is familiar as the metal used to make garbage cans and buckets.

The passivity of iron is easily destroyed by a scratch or by shock.

forms is evidence of this change:

$$4FeSO_4 + O_2 + 2H_2O \rightarrow 4Fe(OH)SO_4(s)$$

The Fe^{2+} ions can be kept in the reduced state by making the solution acidic with sulfuric acid and adding pieces of iron. Hydrated iron(II) ammonium sulfate, $Fe(NH_4)_2(SO_4)_2 \cdot 6H_2O$, is a better source of Fe^{2+} ions in the laboratory because it is stable in contact with air. Iron(II) salts are readily oxidized to iron(III) salts by the corresponding acid and an oxidizing agent. In the case of the nitrate, nitric acid meets both requirements.

$$3Fe(NO_3)_2 + 4HNO_3 \rightarrow 3Fe(NO_3)_3 + NO(g) + 2H_2O$$

Reactions of the Fe^{3+} ion Hydrated iron(III) chloride, $FeCl_3 \cdot 6H_2O$, is the most useful compound of iron in the +3 oxidation state. It is used as a mordant for dyeing cloth and as an oxidizing agent. The yellow crystalline hydrate is deliquescent. The hydrated Fe^{3+} ion, $Fe(H_2O)_6^{3+}$, is pale violet in color. This color usually is not seen, however, because of hydrolysis, which yields hydroxide complexes that are yellow-brown in color. The hydrolysis of Fe^{3+} ions in water solutions of its salts gives solutions that are acidic.

$$Fe^{3+} + 2H_2O \rightleftarrows FeOH^{2+} + H_3O^+$$
$$FeOH^{2+} + 2H_2O \rightleftarrows Fe(OH)_2^+ + H_3O^+$$
$$Fe(OH)_2^+ + 2H_2O \rightleftarrows Fe(OH)_3 + H_3O^+$$

The hydrolysis is extensive when it occurs in boiling water. A blood-red colloidal suspension of iron(III) hydroxide is formed. This colloidal suspension can be produced by adding a few drops of $FeCl_3$ solution to a flask of boiling water.

Iron(III) ions are removed from solution by adding a solution containing hydroxide ions. A red-brown jelly-like precipitate of iron(III) hydroxide is formed.

$$Fe^{3+} + 3OH^- \rightarrow Fe(OH)_3(s)$$

If water is evaporated, red Fe_2O_3 remains.

Tests for iron ions Potassium hexacyanoferrate(II), $K_4Fe(CN)_6$ (also called potassium ferrocyanide), is a light yellow, crystalline salt. It contains the complex hexacyanoferrate(II) ion (ferrocyanide ion), $Fe(CN)_6^{4-}$. The iron is in the +2 oxidation state. The hexacyanoferrate(II) ion forms when an excess of cyanide ions is added to a solution of an iron(II) salt.

$$6CN^- + Fe^{2+} \rightarrow Fe(CN)_6^{4-}$$

Suppose KCN is used as the source of CN^- ions and $FeCl_2$ as the source of Fe^{2+} ions. The equation is

$$6KCN + FeCl_2 \rightarrow K_4Fe(CN)_6 + 2KCl$$

The iron of the $Fe(CN)_6^{4-}$ ion can be oxidized by chlorine to the +3 state. This reaction forms the hexacyanoferrate(III) ion (ferricyanide ion) $Fe(CN)_6^{3-}$.

$$2Fe(CN)_6^{4-} + Cl_2 \rightarrow 2Fe(CN)_6^{3-} + 2Cl^-$$

As the iron is oxidized from the +2 oxidation state to the +3 state, the chlorine is reduced from the 0 state to the −1 state.

Solutions containing cyanide ions are deadly poisons. Safety precautions should be followed when handling these solutions.

Figure 25-14 Potassium hexacyanoferrate(II) (left) and potassium hexacyanoferrate(III) (right) are salts containing iron.

The (II) following hexacyanoferrate tells you it is the Fe^{2+} ion in the complex. The (III) tells you it is the Fe^{3+} ion.

Using the potassium salt as the source of the $Fe(CN)_6^{4-}$ ions, the empirical equation is

$$2K_4Fe(CN)_6 + Cl_2 \rightarrow 2K_3Fe(CN)_6 + 2KCl$$

$K_3Fe(CN)_6$, potassium hexacyanoferrate(III) (known also as potassium ferricyanide), is a dark red crystalline salt.

Intense colors are observed in most compounds that have an element present in two different oxidation states. When iron(II) ions and hexacyanoferrate(III) ions are mixed and when iron(III) ions and hexacyanoferrate(II) ions are mixed, the same intense blue substance is formed.

$$Fe^{2+} + K^+ + Fe(CN)_6^{3+} + H_2O \rightarrow K\overset{+2}{Fe}\overset{+3}{Fe}(CN)_6 \cdot H_2O(s)$$

and

$$Fe^{3+} + K^+ + Fe(CN)_6^{4+} + H_2O \rightarrow K\overset{+3}{Fe}\overset{+2}{Fe}(CN)_6 \cdot H_2O(s)$$

In both reactions, the precipitate contains iron in the +2 and +3 oxidation states, and the same intense blue color is seen.

Iron(II) ions, Fe^{2+}, and hexacyanoferrate(II) ions, $Fe(CN)_6^{4-}$, form a white precipitate, $K_2FeFe(CN)_6$. Note that both Fe ions are in the same +2 oxidation state. This precipitate remains white if the oxidation of Fe^{2+} ions is prevented. Of course, on exposure to air, it begins to turn blue. Iron(III) ions (Fe^{3+}) and hexacyanoferrate(III) ions, $Fe(CN)_6^{3-}$, give a brown solution. Here, both Fe ions are in the same +3 oxidation state. The reactions given in the two preceding equations provide ways to detect the presence of iron in each of its two oxidation states.

1. Test for the Fe^{2+} ion. If a few drops of $K_3Fe(CN)_6$ solution is added to a solution of iron(II) sulfate, the characteristic intense blue precipitate $KFeFe(CN)_6 \cdot H_2O$ forms. Two-thirds of the potassium ions and all of the sulfate ions are merely spectator ions. The net reaction is

$$Fe^{2+} + K^+ + Fe(CN)_6^{3-} + H_2O \rightarrow KFeFe(CN)_6 \cdot H_2O(s)$$

2. Test for the Fe^{3+} ion. When a few drops of $K_4Fe(CN)_6$ solution are added to a solution of iron(III) chloride, the characteristic blue precipitate $KFeFe(CN)_6 \cdot H_2O$, is formed. The net reaction is

$$Fe^{3+} + K^+ + Fe(CN)_6^{4-} + H_2O \rightarrow KFeFe(CN)_6 \cdot H_2O(s)$$

Potassium thiocyanate, KSCN, provides another good test for the Fe^{3+} ion. It is often used to confirm the $K_4Fe(CN)_6$ test. A blood-red solution results from the formation of the complex $FeSCN^{2+}$ ion.

Section Review

1. What elements make up the Iron family?
2. Describe some of the properties of the Iron family.
3. Where does iron occur and what are some of its compounds?
4. How are limestone impurities removed from an iron ore?
5. What is the primary purpose of the reactions that occur during the conversion of iron to steel?
6. Write chemical equations to illustrate the reactions of Fe^{2+} and Fe^{3+}.
7. What precipitate serves to identify the presence of Fe^{2+} ion?

Blueprints: The basis for the blueprinting process is a photochemical reaction. Blueprint paper is first treated with a solution of iron(III) ammonium citrate and potassium hexacyanoferrate(III) and allowed to dry in the dark. A drawing on tracing paper is laid over the blueprint paper and exposed to light. Where light strikes the paper, Fe^{3+} ions are reduced to Fe^{2+} ions. The paper is dipped into water and the Fe^{2+} ions react with the hexacyanoferrate(III) ions to form the blue color in the paper. Where drawing lines covered the paper, no reduction of Fe^{3+} ions occurs and the paper rinses out white. Thus, a blueprint has white lines on blue paper.

1. Iron, cobalt, and nickel
2. Metals with low reactivity; can all show the +2 and +3 oxidation states; have high melting and boiling points; all form complex ions; all have strong magnetic properties. Ferromagnetism is the result of the similar spin orientations of their 3d electrons.
3. In the earth's crust, in meteors. Accept any compounds mentioned in Section 25-2.
4. By adding sand as the flux
5. Purification process.
6. See equations in Section 25-2.
7. Blue precipitate of $KFeFe(CN)_6$ when $K_3Fe(CN)_6$ is added to a solution.

25.3 Copper, Silver, and Gold

Group 11, sometimes known as the Copper family, consists of copper, silver, and gold. All three metals appear below hydrogen in the electrochemical series. They are not easily oxidized and often occur in nature in the free, or native, state. Because of their pleasing appearance, durability, and scarcity, these metals have been highly valued since the time of their discovery. All have been used in ornamental objects and coins throughout history.

SECTION OBJECTIVES

- Identify the elements of the Copper family, and describe their general characteristics.
- Describe the natural occurrence of and refining methods for copper and name compounds of copper and their uses.
- Describe the properties and uses of metallic copper.

Gold can be made into a very thin, flexible film that is used to coat statues or picture frames.

The major use of silver is in photographic film. A thin, light-sensitive coating on photographic film contains silver bromide. When it is exposed to light, a few of the silver ions are reduced to

General Characteristics

The atoms of copper, silver, and gold have a single electron in their highest energy levels. Thus they often form compounds in which they exhibit the $+1$ oxidation state, and thereby resemble the Group 1 metals of the Sodium family.

Each metal of the Copper family has 18 electrons in the next-to-highest energy level. The d electrons in this next-to-highest energy level have energies that differ only slightly from the energy of the outer s electron. Thus, one or two of these d electrons can be removed with relative ease. For this reason, copper and gold often form compounds in which they exhibit the $+2$ and $+3$ oxidation states, respectively. In silver, the $+2$ oxidation state is reached only under extreme oxidizing conditions.

Copper, silver, and gold are very dense, ductile, and malleable. They are classed as heavy metals along with other transition metals in the central region of the Periodic Table. Some important properties of each metal are shown in Table 25-6.

TABLE 25-6   THE COPPER FAMILY

Element	Atomic number	Atomic mass	Electron configuration	Oxidation numbers	Melting point (°C)	Boiling point (°C)	Density (g/cm³)
copper	29	63.546	2,8,18,1	+1, +2	1083.4	2567	8.96
silver	47	107.868	2,8,18,18,1	+1	961.9	2212	10.5
gold	79	196.967	2,8,18,32,18,1	+1, +3	1064.4	2807	19.3

Copper

metallic silver. These ions sensitize the tiny silver bromide crystals in which they occur. Developing the film causes additional silver ions in the sensitized crystals to be reduced. An image is formed of dark areas of silver and light areas of unchanged AgBr in the unexposed areas. The unreacted AgBr is then removed so the developed film will no longer be light sensitive.

Copper, alloyed with tin in the form of bronze, has been in use for over 5000 years. Native copper deposits lie deep underground and are difficult to mine. Most copper is thus obtained from its ores by either treating ores with extremely high temperatures or subjecting ores to aqueous solution at relatively low temperatures.

Sulfide ores of copper yield most of the supply of this metal. Chalcocite (Cu_2S), chalcopyrite ($CuFeS_2$), and bornite (Cu_3FeS_3), are the major sulfide ores. Malachite, $Cu_2(OH)_2CO_3$, and azurite, $Cu_3(OH)_2(CO_3)_2$, are basic carbonates of copper. Malachite is a rich green, and azurite is a deep blue. Besides serving as ores of copper, fine specimens of these minerals are sometimes polished for use as ornaments or in making jewelry.

Copper and its recovery The carbonate ores of copper are washed with a dilute solution of sulfuric acid, forming a solution of copper(II) sulfate. The metallic copper is then recovered by electrolysis. High-grade carbonate ores are heated in air to convert them to copper(II) oxide. The oxide is then reduced with coke, yielding metallic copper as the final product in the process.

 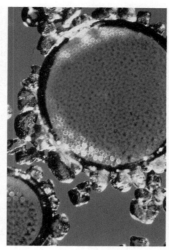

Figure 25-15 The oil-flotation process *(left)* for concentrating copper ore. Air bubbles carry ore particles to the surface. A photomicrograph *(right)* of a flotation bubble from the oil-flotation process

The sulfide ores are usually low-grade and require concentration before they can be refined profitably. The concentration is accomplished by oil-flotation which is illustrated in Figure 25-15. Earthly impurities associated with the ore are wetted by water, and the ore is wetted by oil. Air is blown into the mixture to form a froth. The oil-wetted ore floats to the surface in the froth. This treatment changes the concentration of the ore from about 2% copper to as high as 30% copper. Metallic copper is recovered from the concentrated ore by a process shown diagrammatically in Figure 25-16.

Figure 25-16 A flow diagram of the copper refining process

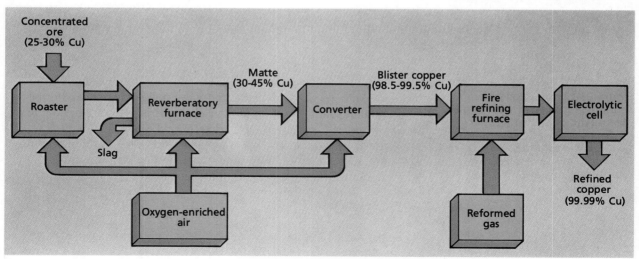

The concentrated ore is partially roasted to form a mixture of Cu_2S, FeS, FeO, and SiO_2. This mixture is known as calcine. The roasting process uses oxygen-enriched air and yields high-quality sulfur dioxide. This gas is converted to sulfuric acid, as described in Section 27.2. Calcine is fused with limestone in a furnace. Part of the iron is removed as a silicate slag. The rest of the iron, together with the copper, forms a mixture of sulfides known as copper matte. Copper matte is processed in a reverberatory furnace, and the end product contains 40% copper.

The melted matte is further refined in a converter that is supplied with oxygen-enriched air as the oxidizing agent. Sulfur from the sulfides as well as impurities of arsenic and antimony are removed as volatile oxides. Most of the iron is removed as slag. Some of the copper(I) sulfide is converted into copper(I) oxide. The copper(I) oxide then reacts with more copper(I) sulfide, forming metallic copper and sulfur dioxide. The following equations show the chemical reactions involved in this process.

$$2Cu_2S + 3O_2 \rightarrow 2Cu_2O + 2SO_2(g)$$
$$2Cu_2O + Cu_2S \rightarrow 6Cu + SO_2(g)$$

The molten copper is cast as blister copper of 98.5—99.5% purity. As the copper cools, dissolved gases escape and form blisters, hence the name. Impurities remaining are iron, silver, gold, and sometimes zinc. In the fourth step of the copper-refining process, blister copper is further purified in a fire-refining furnace or by electrolytic refining.

Electrolytic refining of copper Unrefined copper contains fairly large amounts of silver and gold. Thus the cost of its refining is offset by the recovery of these precious metals. Copper is largely used for making electric conductors, and very small amounts of impurities greatly increase the electric resistance.

In electrolytic refining, shown in Figure 25-17, sheets of pure copper are used as the negative electrodes in electrolytic cells. Large plates of impure copper are used as the positive electrodes. The electrolyte is a solution of copper(II) sulfate in sulfuric acid. A direct current at low voltage is used to operate the cell. During the electrolysis, copper and the other metals in the positive electrode that are above copper in the electrochemical series are oxidized. They enter the solution as ions.

$$Cu \rightarrow Cu^{2+} + 2e^-$$
$$Fe \rightarrow Fe^{2+} + 2e^-$$
$$Zn \rightarrow Zn^{2+} + 2e^-$$

At the low voltages used, the less active silver and gold are not oxidized and do not go into solution. As the positive electrode is used up, they fall to the bottom of the cell as a sludge and are recovered easily.

You may expect the various positive ions of the electrolyte to be reduced at the negative electrode. But H_3O^+ ions, Fe^{2+} ions, and Zn^{2+} ions all require higher voltages than Cu^{2+} ions for reduction. At the low potential maintained across the cell, only Cu^{2+} ions are reduced.

$$Cu^{2+} + 2e^- \rightarrow Cu$$

Of all the metals present, only copper plates out on the negative electrode. Electrolytic copper is 99.99% pure.

Figure 25-17 Electrolytic refining of copper. In these large electrolytic cells, positive electrodes of impure copper yield Cu2+ ions in solution. The Cu2+ ions are plated out on pure copper negative electrodes. Electolytically refined copper is 99.99% pure.

Properties and uses of copper and its compounds Copper is a soft, ductile, malleable, red metal with a density of 8.96 g/cm^3. It is second to silver in electric conductivity.

Copper compounds, in which copper exhibits the +1 oxidation state, are relatively unimportant as compared with those in which copper exhibits the +2 oxidation state. The best known copper(I) compounds are complexes or slightly soluble solids. These include CuI, $CuCN$, $CuCl$, and $CuBr$. Heated in air, copper forms a black coating of copper(II) oxide, CuO. Metallic copper and most copper compounds color a bunsen flame green.

Copper forms copper(II) salts that dissociate in water to give blue solutions. The color is characteristic of the hydrated copper(II) ion, $Cu(H_2O)_4^{2+}$. Adding an excess of ammonia to solutions containing this ion produces the deeper-blue complex ion, $Cu(NH_3)_4^{2+}$. Aqueous solutions of copper(II) salts are weakly acidic because of the mild hydrolysis of the Cu^{2+} ion.

$$Cu(H_2O)_4^{2+} + H_2O \rightarrow Cu(H_2O)_3OH^+ + H_3O^+$$

An excess of sulfur vapor forms a blue-black coating of copper(I) sulfide on hot copper. In moist air, copper tarnishes and forms a protective coating. This coating is a green basic carbonate, $Cu_2(OH)_2CO_3$. Sulfur dioxide in the air may also combine with copper. If so, a green basic sulfate, $Cu_4(OH)_6SO_4$, is produced. The green color seen on copper roofs is caused by the formation of these compounds.

Copper(II) compounds are much more common than copper(I) compounds. Copper(II) oxide is used to change alternating current to direct current. It is also used as an oxidizing agent in chemical laboratories.

Hydrated copper(II) sulfate, $CuSO_4 \cdot 5H_2O$, called blue vitriol, is an important copper compound. It is used to kill algae in reservoirs and to make certain pesticides. It is also used in electroplating and in preparing other copper compounds.

Because copper stands below hydrogen in the activity series, it does not replace hydrogen from acids. Thus it is not acted on by nonoxidizing acids such as hydrochloric acid and dilute sulfuric acid except very slowly when oxygen is present. The oxidizing acids, such as nitric and hot concentrated sulfuric, react vigorously with copper. Such reactions produce the corresponding copper(II) salts.

Because of its high thermal and electrical conductivity, and its ductility, malleability, and resistance to corrosion, copper has many uses. Copper is used in the manufacture of electrical wire, various instruments, transmission lines, and electrical machinery. It is also used in plumbing, roofing, screens, and gutters. A large amount of copper is used to make various alloys, including brass and bronze. The alloys are stronger, less expensive than copper, and easy to work with.

Tests for the Cu^{2+} ion A dilute solution of a copper(II) salt changes to a very deep blue color when an excess of ammonia is added. This color change is caused by the formation of complex $Cu(NH_3)_4^{2+}$ ions.

$$Cu^{2+} + 4NH_3 \rightarrow Cu(NH_3)_4^{2+}$$

The addition of $K_4Fe(CN)_6$ to a solution containing Cu^{2+} ions produces a red precipitate of $Cu_2Fe(CN)_6$. This precipitate is copper(II) hexacyanoferrate(II),

Gold is the third-best electrical conductor.

This black coating of copper(II) oxide is familiar to anyone who has cooked in a pot that has a copper-coated bottom.

The green coating is called a patina.

Brass is an alloy of copper and zinc. Bronze is an alloy of copper and tin.

1. Copper, silver, gold
2. Accept any properties shown in Table 25-8.
3. Found in sulfide ores, namely, chalcocite, Cu_2S; chalcopyrite, $CuFeS_2$; bornite, Cu_3FeS_3; malachite, $Cu_2(OH)_2CO_3$; and azurite, $Cu_3(OH)_2(CO_3)_2$. The recovery process depends on the ore. Accept any answers dealing with the recovery and refining processes discussed under the headings "Copper and its recovery" and "Electrolytic refining of copper."
4. CuO is used to change alternating current to direct current and as an oxidizing agent in chemical laboratories. $CuSO_4 \cdot H_2O$ is used to kill algae in reservoirs, to make certain pesticides, in electroplating, and in preparing other copper compounds.
5. High thermal conductivity, electrical conductivity, ductility, malleability, and resistance to corrosion. Used in the manufacture of electrical wire, instruments, machinery, plumbing, roofing, screens and gutters, and various alloys.

also known as copper(III) ferrocyanide. If copper is present in a borax bead formed in an oxidizing flame, a clear blue color appears on cooling. The hot bead is green. A bead formed in a reducing flame is colorless when hot and an opaque red when cool.

Section Review

1. What elements constitute the Copper family?
2. Describe some properties that the elements of the Copper family exhibit.
3. Where does copper occur and how is it refined?
4. Name some of the important compounds of copper and describe some of their uses.
5. What properties make copper a very useful element? What are some of its uses?

Chapter Summary

- The transition metals consist of ten groups interposed between Group 2 and Group 13 in the Periodic Table. They are in this position because they represent an interruption of the regular increase in the number of highest-energy-level electrons from Group 1 to Group 18 in the long periods of the table.
- Chemical properties of transition metals are varied. Electrons of the next-to-highest d sublevel and the highest energy level may become involved in the formation of compounds.
- Most of the transition metals show different oxidation states in their compounds. They have strong tendencies to form complex ions. Many of their compounds are paramagnetic and are colored.
- With minor, apparent irregularities, the transition metals have two highest energy level electrons. In the first row of transition elements, chromium and copper have one $4s$ electron. The availability of $3d$ electrons tends to increase the number of oxidation states up to manganese in the middle of the row, after which the number of common oxidation states for each element decreases.
- The color of the compounds of transition elements depends to some extent on the ions or polyatomic groups associated with the element. The colors result from the ease with which d electrons can be excited by absorption of discrete quantities of light energy.
- Transition metals form many complex ions, with the transition element cation serving as the central ion in the complex. Certain ionic or molecular species are able to coordinate with the central ion to form the complex ion.
- The Iron family consists of iron, cobalt, and nickel. These metals are ferromagnetic.
- Iron forms compounds in two oxidation state, $+2$ and $+3$. Iron(II) ions are easily oxidized to the iron(III) state. Iron(II) compounds are sometimes used as reducing agents. Iron(III) compounds are used as oxidizing agents.
- The Copper family or Group 11, consists of copper, silver, and gold. All are found as native metals in nature. Most copper used in commerce and industry is recovered from its sulfide ores by reduction and electrolytic refining.

Chapter 25 *Review*

The Transition Metals

Vocabulary

blast furnace

complex ion

coordination number

pig iron

slag

transition elements

Questions

1. Describe the structural similarity that determines the metallic character of the transition elements?
2. Copper(II) sulfate crystals are blue. To what color does the light energy absorbed by the crystals correspond?
3. Chemically, what happens to iron *(a)* in a blast furnace; *(b)* during steel making?
4. List five points of similarity among copper, silver, and gold.
5. Why does the copper trim on roofs frequently acquire a green surface?
6. What is the oxidation number of manganese in *(a)* MN *(b)* MNO_2 *(c)* $MnSO_4 \cdot 4H_2O$ *(d)* $KMnO_4$ *(e)* $BaMnO_4$?

Problems

1. How much iron(II) chloride can be produced by adding 165 g of iron to an excess of hydrochloric acid?
2. How much iron(III) chloride can be prepared from the iron(II) chloride in Problem 1 if more hydrochloric acid is added and air is blown through the solution? (Assume 1 mol of $FeCl_3$ is formed for each mole of $FeCl_2$.)
3. What is the percentage of iron in a sample of limonite, $2Fe_2O_3 \cdot 3H_2O$ to a figure of two decimal places?
4. How many grams of silver nitrate can be obtained by adding 100 g of pure silver to an excess of nitric acid?
5. A sample of hematite ore contains Fe_2O_3 87.0%, silica 8.0%, moisture 4.0%, other impurities 1.0%. What is the percentage of iron in the ore sample?
6. What will be the loss in mass when 1.0×10^6 metric tons of the ore in Problem 5 are heated to 200°C.

Application Questions

1. Why were some metals used in very early times, while many of the other metals were obtained and used only within the last century?
2. Distinguish between a polyatomic ion and a complex ion.
3. When potassium hexacyanoferrate(II) is added to a solution of iron (II) sulfate, a white precipitate forms that gradually turns blue. Explain the color change.
4. How can you detect an iron(II) and an iron(III) compound if both are present in the same solution?
5. Why does it usually pay to refine copper by electrolysis?

Application Problems

1. How much limestone will be needed to combine with the silica in 1.0×10^6 metric tons of the ore described in Problem 5?
2. *(a)* How much carbon monoxide is required to reduce 1.0×10^6 metric tons of the ore in Problem 5? *(b)* How much coke must be supplied to meet this requirement? (Assume the coke to be 100% carbon.)
3. Silver reacts with dilute nitric acid to form silver nitrate, water, and nitrogen monoxide. Write the balanced formula equation.
4. Copper reacts with hot concentrated sulfuric acid to form copper(II) sulfate, sulfur dioxide, and water. Write the balanced equation.

Enrichment

1. Investigate and report on the uses of ferromagnetic oxides in magnetic recording tape.
2. Do library research and investigate how silver halides are used in the photographic process. Summarize your findings in a report.
3. Because of its central role as an oxygen carrier in living things, hemoglobin has been studied extensively. Investigate how the iron in the heme group binds oxygen and report your findings to the class.

Chapter 26 Aluminum and the Metalloids

Chapter Planner

26.1 Introduction to Aluminum and Metalloids
The Nature of Aluminum
■ Question: 1
The Nature of Metalloids
■ Demonstration: 2
    Application Question: 3
■ Problem: 4

26.2 Aluminum
Recovery of Aluminum
■ Question: 7
    Application Question: 2
■ Technology
Properties of Aluminum
■ Demonstration: 1
    Experiment: 36
    Questions: 2, 3
    Application Question: 1
■ Problem: 3
The Thermite Reaction
■ Problem: 1
Some Compounds of Aluminum
■ Problem: 2

26.3 Representative Metalloids
Silicon
■ Demonstration: 3
    Questions: 4, 5
    Application Questions: 4, 5
Arsenic
■ Question: 6
■ Chemistry Notebook
■ Application Problem: 1
Antimony

Teaching Strategies

Introduction

The importance of aluminum as an industrial metal and the expanding use of semiconductors should make this chapter relevant to the students. Emphasis is placed on the properties of the metalloids, which are intermediate between those of the metals and nonmetals. Aluminum is a metal, but it is included here because of its position in the Periodic Table.

The metalloids occupy a diagonal location in the Periodic Table, and their properties follow this diagonal relationship. Some of the properties and reactions of boron, silicon, arsenic, and antimony are examined.

26.1 Introduction to Aluminum and Metalloids

When discussing the properties and reactions of these elements, ask students to keep in mind which properties are typically metallic and which are nonmetallic. Physical properties of metals include shininess, malleability and ductility, and good conduction of heat and electricity. Chemical properties include low ionization energies, and formation of positive ions and ionic compounds.

Have the students identify the metalloids in the Periodic Table. Make comparisons of the physical properties of these elements: their melting points, boiling points, appearances, densities, oxidation states, and ionization energies. Determine which element has the smallest temperature range for the liquid

phase (boron) and how this relates to the nonmetallic behavior of boron. Discuss other generalizations in this section, such as variations from metallic to nonmetallic in a period from left to right, and the increase in metallic properties going down a group. Use these generalizations to account for the diagonal similarities. Discuss the characteristics of the hydroxide, oxide, and hydride of aluminum.

Section 26.2 Aluminum

Identify aluminum as a metal by its physical properties. Exhibit aluminum foil, wire, and other aluminum objects. Ask students to name items that contain aluminum and discuss the abundance of aluminum.

You may wish students to report on the history and discovery of aluminum and the coincidence of the Hall-Heroult discovery. Exhibit a set of charts and materials such as bauxite, refined alumina, clay, cryolite, and carbon electrodes. Review the chemical equations for oxidation and reduction that take place in the electrolytic cell used in the production of aluminum.

Discuss the invisible coating of aluminum oxide on aluminum metal and the advantage of this type of "built-in" corrosion protection. Explain the reaction for the hydrolysis of aluminum ions in solution. Test the pH of a water solution of an aluminum salt. Discuss the fact that aluminum will readily react with acids such as hydrochloric acid, forming aluminum ions and hydrogen gas, but will not react with oxidizing acids such as nitric acid. Show that the amphoteric aluminum will also react with bases to form the aluminate ion and hydrogen gas.

Discuss the thermite reaction and the enormous amount of heat released by this process. Explain how this reaction can be used to form molten iron that can be used for welding.

Point out the usefulness of substances containing aluminum oxide: the abrasives corundum and emery, the gems ruby and sapphire, and the refractory substances such as crucibles.

Ask students to write formulas for various compounds that can be classed as alums. The octahedral alum crystals are easy to grow from saturated solutions, and some students may like to see how large a crystal they can produce from a seed crystal of potassium aluminum sulfate dodecahydrate suspended by a thread in a saturated alum solution.

Technology: Can a Can Become a Plane?

Most students know that aluminum can be recycled, but they will be surprised to find that the recycled metal cannot be used for all purposes. However, recycled aluminum can be purified and made into new products at a fraction of the cost of obtaining aluminum from its ore. Discuss the ecological advantages of recycling and different types of substances that can be economically recycled.

Demonstration 1: Aluminum Displacement of Copper

PURPOSE: To demonstrate the relative reactivities of aluminum and copper

MATERIALS: Copper(II) chloride crystals, tall graduated cylinder, water, strip of heavy aluminum foil

PROCEDURE: Dissolve 4-5 teaspoons of the copper(II) chloride in about 500 mL of water and put it in a graduated cylinder. Twist the copper into an interesting shape and drop it into the solution. Observe the changes that occur. (The foil becomes coated with a dark substance—metallic copper. The green-blue color of the solution fades. The aluminum metal dissolves.) The reaction is exothermic. This demonstration provides a number of observations that students can make.

$$2Al(s) + 3Cu^{2+}(aq) \rightarrow 2Al^{3+}(aq) + 3Cu(s)$$

26.3 Representative Metalloids

Boron is an element that is not very familiar to most students. Some will have heard of boric acid and borax, but they may not know that these contain boron. Examine the properties of boron and show that they are typical of a metalloid.

Discuss boron carbide and boron nitride. Emphasize the covalent network bonding that must occur to form these extremely hard substances. Point out the formulas of the tetraborate ion, $B_4O_7^{2-}$, and borate, BO_3^{3-}. In both of these, the oxidation state of boron is $+3$. Have students read labels on laundry products that use boron compounds to determine which substances are used.

Discuss the properties of silicon, again noting that they are typical of a metalloid. Compare the properties of carbon bonded in diamond to those of silicon, which has a diamond-type structure. Point out that the larger silicon atoms cannot form the double bonds needed for a graphite-like structure. In silicon semiconductor chips, each silicon atom is tetrahedrally bonded to four other silicon atoms in a covalent network.

Emphasize the fact that silicon forms silicon-oxygen bonds more readily than silicon-silicon bonds, and that the silicates that make up so much of the earth's crust contain these silicon-oxygen bonds. Have students look up some of the types of silicates that might be found locally. Ask them to report on the manufacture of glass.

Discuss the meaning of the term *polymer*. Compare polymeric compounds of silicones to those of hydrocarbons. Note that the silicones are very heat resistant and are used as lubricants and oils.

Compare the properties of arsenic and antimony and their compounds. Discuss the uses of their compounds.

Chemistry Notebook: Gallium Arsenide, Lasers, and Computer Chips

The quantity of information etched into a computer chip will astound most students. They will easily appreciate the complexity of the processes that must be employed to produce the chips. Suggest that students look up other methods that have been used to produce these circuits and that they find out what information is etched into the chips.

Demonstration 2: Borax Bead Tests

PURPOSE: To show the colors of various metallic oxides and salts when fused with borax

MATERIALS: Platinum or nichrome wire, burner, powdered borax (sodium tetraborate), compounds such as iron(II) sulfate, cobalt nitrate, manganese(IV) oxide, nickel(II) nitrate

PROCEDURE: Wear goggles. Make a loop about 3 mm in diameter in the end of the wire. Heat the wire to redness in a flame. While it is still hot, dip it into powdered borax and then hold the wire and the adhering borax in the upper part of a flame. Repeat until a clear, rounded, glass-like bead is formed. Heat the bead again and, while it is hot, touch it to a tiny speck of one of the compounds.

Heat again. A colored transparent bead should form. Note the colors formed with the various substances tested. If the bead is opaque, too much of the salt was used and the test should be repeated with a new bead. To remove the bead from the wire, heat it, and flick it off while it is hot. If necessary, heat, dip in cold water, and rub it off.

Demonstration 3: A Silicate Garden

PURPOSE: To demonstrate the growth of insoluble silicates

MATERIALS: Wide-mouth container (a small fish bowl is a good choice), sodium silicate solution (also called water glass), various colored small crystals such as iron(III) chloride (brown), nickel(II) nitrate (dark green), copper(II) chloride (bright green), cobalt(II) chloride (dark blue), and zinc nitrate (white)

PROCEDURE: Dilute 1 part sodium silicate to 4 parts water. Fill the container with the solution. Drop in 3-5 small crystals (matchhead size).

Observe that the crystals seem to grow like stalagmites within a few seconds. As the salts are placed in the solution, a semipermeable membrane composed of insoluble silicate is formed around the crystal. Because the concentration is greater inside the membrane, water enters the membrane by osmosis. This causes the membrane sack to break. It ruptures upward where the pressure is least. A new membrane forms and the crystal grows. This is a real attention-getting demonstration.

References and Resources

Books and Periodicals

Garrett, Alfred B. "The Flash of Genius-Discovery of the Transistor," *Journal of Chemical Education*, Vol. 40, 1963, p. 302.

Kolb, Doris, and Kenneth E. Kolb. "Chemistry of Glass," *Journal of Chemical Education*, Vol. 56, 1979, p. 604.

Sarquis, Mickey. "Arsenic, Old Myths," *Journal of Chemical Education*, Vol. 56, 1979.

Shakhashiri, Bassam Z. "Thermite Reaction," *Chemical Demonstrations*, Vol. 1, University of Wisconsin Press, 1983, p. 85.

Summerlin, Lee R., Christie L. Borgford, and Julie B. Ealy, "Reduction of Sand with Magnesium," *Chemical Demonstrations, A Sourcebook for Teachers*, Vol. 2, American Chemical Society, 1987, p. 193.

Summerlin, Lee R., Christie L. Borgford, and Julie B. Ealy, "Recycling Aluminum," *Chemical Demonstrations, A Sourcebook for Teachers*, Vol. 2, American Chemical Society, 1987, p. 54.

Summerlin, Lee R., and James L. Ealy, Jr., "Silicate Garden," *Chemical Demonstrations, A Sourcebook for Teachers*, Vol. 1, American Chemical Society, 1985, p. 30.

Audiovisual Resources

"Aluminum": video FFH1108F, Films for the Humanities and Sciences. Describes the extraction of aluminum from bauxite and the production of various alloys.

The Aluminum Association, Educational Services. Free or inexpensive materials available, including wall charts, "The Aluminum Film Catalog," listing motion pictures and filmstrips, and resource guides for teachers.

Organosilicon Chemistry Course (slides, cassette, chart, etc.) from Dow Corning Corporation.

Computer Software

"The Periodic Table Videodisc: Reactions of the Elements": CAV type videodisc, Journal of Chemical Education Software, 1989. Shows action sequences and still shots of each element in its most common forms, and reacting with air, water, acids, bases, and some common applications.

Answers and Solutions

Questions

1. Elemental aluminum is metallic, having familiar properties of metals. However, such chemical properties as the amphoterism of aluminum hydroxide, $Al(OH)_3$, the tendency to form the negative aluminate ion, $Al(OH)_4^-$, and formation of covalent bonds in some compounds are metalloidal.
2. Many clays, rocks, and minerals; principal ore is bauxite.
3. Silver – white in color; density 2.70 g per cm^3; ductile; malleable; not as strong as copper, brass, or steel; good conductor of electricity; can be welded, spun, or cast; difficult to solder
4. *(a)* Oxides of aluminum *(b)* Al_2O_3 *(c)* As abrasives
5. Aluminum oxide is very high melting. It can be dissociated by being dissolved in fused cryolite.
6. To produce a carbon-free metal. Also because aluminum is a more powerful reducing agent
7. $KAl(SO_4)_2 \cdot 12H_2O$; $NaAl(SO_4)_2 \cdot 12H_2O$; $KFe(SO_4)_2 \cdot 12H_2O$; $KCr(SO_4)_2 \cdot 12H_2O$
8. Second among the elements of the earth's crust
9. Arsenic dust and vapor and most compounds of arsenic are very toxic.

Problems

1. $2Al + Fe_2O_3 \rightarrow Al_2O_3 + 2Fe$

$$10.0 \text{ kg Fe} \times \frac{1 \text{ kg mol} \cdot \text{Fe}}{55.8 \text{ kg Fe}} \times \frac{2 \text{ kg mol Al}}{2 \text{ kg mol Fe}} \times \frac{27.0 \text{ kg Al}}{1 \text{ kg mol Al}}$$

$$= 4.84 \text{ kg Al}$$

$$10.0 \text{ kg Fe} \times \frac{1 \text{ kg mol Fe}}{55.8 \text{ kg Fe}} \times \frac{1 \text{ kg mol Fe}_2O_3}{2 \text{ kg mol Fe}} \times$$

$$\frac{160. \text{ kg Fe}_2O_3}{1 \text{ kg mol Fe}_2O_3} = 14.3 \text{ kg Fe}_2O_3$$

2. $NaAl(SO_4)_2 \cdot 12H_2O$, molar mass $= 458$ g
$27.0 \text{ g} \div 458 \text{ g} \times 100\% = 5.90\%$ Al

3. $2Al + 2NaOH + 6H_2O \rightarrow 2NaAl(OH)_4 + 3H_2$

$$50.0 \text{ g Al} \times \frac{1 \text{ mol Al}}{27.0 \text{ g Al}} = 1.85 \text{ mol Al}$$

$$100. \text{ g NaOH} \times \frac{1 \text{ mol NaOH}}{40.0 \text{ g NaOH}} = 2.5 \text{ mol NaOH}$$

By inspection: NaOH is in excess

$$1.85 \text{ mol Al} \times \frac{3 \text{ mol H}_2}{2 \text{ mol Al}} \times \frac{22.4 \text{ L}}{1 \text{ mol}} = 62.2 \text{ L H}_2$$

4. $Ca_2B_6O_{11} \cdot 5H_2O$, molar mass 411.0 g 6B, 64.8 g
$64.8 \text{ g} \div 411.0 \text{ g} \times 100\% = 15.8\%$ boron

Application Questions

1. *(a)* $2Al(s) + 6H_3O^+ + 6H_2O \rightarrow 2Al(H_2O)_6^{3+} + 3H_2(g)$
(b) $2Al(s) + 2OH^- + 6H_2O \rightarrow 2Al(OH)_4^- + 3H_2(g)$
(c) The protective oxide coating isolates the aluminum from the water.
2. Negative electrode: $Al^{3+} + 3e^- \rightarrow Al$
Positive electrode: $2O^{2-} \rightarrow O_2 + 4e^-$
3. The aluminum cannot be produced economically from the clay.
4. Availability of cheap electricity, usually generated by waterpower.
5. *(a)* Small atomic radius *(b)* Ionization energy intermediate between metals and nonmetals *(c)* electronegativity of about 2 *(d)* Form covalent bonds *(e)* Low electric conductivity at

ordinary temperatures; electric conductivity increases with an increase in temperature

(a) A compound containing a chain of alternate silicon and
6. oxygen atoms, with organic groups attached to the silicon

atoms *(b)* As a lubricant, as a varnish for the windings of electric motors, as a water repellent, and in automobile and furniture polishes

7. Because of the hydrocarbon groups attached to the silicon atoms

Application Problems

1. $96.2 \text{ g As} \times \dfrac{1 \text{ mol As}}{74.9 \text{ g As}} = 1.28 \text{ mol As}$

$3.85 \text{ g H} \times \dfrac{1 \text{ mol H}}{1.01 \text{ g H}} = 3.81 \text{ mol H}$

$\dfrac{1.28 \text{ mol As}}{1.28} : \dfrac{3.81 \text{ mol H}}{1.28}$

$1.00 \text{ mol As} : 2.97 \text{ mol H}$
Simplest formula, AsH_3

$\dfrac{3.48 \text{ g}}{1L} \times \dfrac{22.4L}{1 \text{ mol}} = 77.9 \text{ g/mol}$

molar mass $AsH_3 = 77.9454$
The molecular formula is AsH_3.

Teacher's Notes

Chapter 26 Aluminum and the Metalloids

INTRODUCTION

In this chapter you will study the chemistry of the metalloids, elements that are found in a diagonal pattern spanning four groups in the Periodic Table. Although it is considered a metal, aluminum is included in this chapter because of its position in the Periodic Table.

LOOKING AHEAD

As you study this chapter, investigate why the metalloids exhibit intermediate characteristics compared to those of the metals and nonmetals. Pay particular attention to the properties and bonding of these elements, and relate these properties to the nature and uses of their compounds.

SECTION PREVIEW

26.1 Introduction to Aluminum and Metalloids
26.2 Aluminum
26.3 Representative Metalloids

Remind students of "metallic" properties: shiny surface, malleability, ductility, good conduction of heat and electricity, low ionization energy, forms positive ions easily, forms ionic compounds.

Its characteristic of low density makes aluminum an ideal material for the construction of aircraft. It is only one-third as dense as steel.

26.1 Introduction to Aluminum and Metalloids

SECTION OBJECTIVES

- Name the elements known as metalloids and locate their position in the Periodic Table.
- Explain the ways in which aluminum can be related to the metalloids.
- Describe the general nature of the metalloids, and list specific properties of the metalloid elements.

Some elements are neither distinctly metallic nor distinctly nonmetallic. Their properties are intermediate between those of metals and nonmetals. As a group, these elements are called *metalloids* or *semimetals*. In the Periodic Table, they occupy a diagonal region from the upper center toward the lower right as shown in Figure 26-1.

The metalloids are the elements boron, silicon, germanium, arsenic, antimony, selenium, and tellurium. Although aluminum is not included in the metalloids, it is included in this chapter because of its unique position in the Periodic Table relative to the metalloids.

The Nature of Aluminum

Elemental aluminum is distinctly metallic and possesses many of the familiar properties of metals. But aluminum forms the negative aluminate ion, $Al(OH)_4^-$. Its hydroxide, $Al(OH)_3$ is amphoteric. The oxide of aluminum is ionic, yet AlH_3 and $AlCl_3$ contain covalent bonds. Aluminum is so resistant to oxidation that it can be used as cookingware on hot stove burners. Yet it is easily oxidized in sodium hydroxide solution to yield hydrogen and sodium aluminate. In these respects, aluminum is similar to the metalloids.

Amphoteric hydroxides dissolve in both acids and bases:

$$Al(OH)_3(s) + 3H^+(aq) \rightarrow Al^{3+}(aq) + 3H_2O$$

$$Al(OH)_3(s) + OH^-(aq) \rightarrow Al(OH)_4-(aq)$$

The Nature of Metalloids

Silicon, arsenic, and antimony are typical metalloids and are discussed in Section 26.3. Table 26-1 lists some properties of these elements together with those of aluminum and the other metalloids.

Group 9	Group 10	Group 11	Group 12	Group 13	Group 14	Group 15	Group 16	Group 17	Group 18
									2 **He** 4.002 60
				5 **B** 10.81	6 **C** 12.011	7 **N** 14.006 7	8 **O** 15.999 4	9 **F** 18.9984	10 **Ne** 20.179
				13 **Al** 26.981 54	14 **Si** 28.085 5	15 **P** 30.9738	16 **S** 32.06	17 **Cl** 35.453	18 **Ar** 39.948
27 **Co** 58.933 2	28 **Ni** 58.69	29 **Cu** 63.546	30 **Zn** 65.38	31 **Ga** 69.72	32 **Ge** 72.59	33 **As** 74.921 6	34 **Se** 78.96	35 **Br** 79.904	36 **Kr** 83.80
45 **Rh** 102.906	46 **Pd** 106.42	47 **Ag** 107.868	48 **Cd** 112.41	49 **In** 114.82	50 **Sn** 118.69	51 **Sb** 121.75	52 **Te** 127.60	53 **I** 126.905	54 **Xe** 131.29
77 **Ir** 192.22	78 **Pt** 195.08	79 **Au** 196.967	80 **Hg** 200.59	81 **Tl** 204.383	82 **Pb** 207.2	83 **Bi** 208.98	84 **Po** (209)	85 **At** (210)	86 **Rn** (222)

Figure 26-1 The metalloids are the elements *boron, silicon, germanium, arsenic, antimony, selenium,* and *tellurium.* Aluminum is a metal but it has some unique properties that relate it to the metalloids.

TABLE 26-1 PROPERTIES OF ALUMINUM AND METALLOIDS

Element	Atomic Number	Electron Configuration	Oxidation States	Melting Point (°C)	Boiling Point (°C)	Density (g/cm³)	Atomic Radius (pm)	First Ionization Energy (kJ/mol)
boron	5	2,3	+3	2300	2550	2.34	79	801
aluminum	13	2,8,3	+3	660	2467	2.70	143	578
silicon	14	2,8,4	+2, +4, −4	1410	2355	2.33	117	787
germanium	32	2,8,18,4	+2, +4, −4	937	2830	5.32	122	762
arsenic	33	2,8,18,5	+3, +5, −3	sublimes		5.73	121	944
selenium	34	2,8,18,6	+2, +4, +6, −2	217	685	4.5	116	941
antimony	51	2,8,18,18,5	+3, +5, −3	631	1750	6.69	141	832
tellurium	52	2,8,18,18,6	+2, +4, +6, −2	450	990	6.24	130	869

The amphoterism of aluminum hydroxide is discussed in detail in Section 16.4

A semiconductor is a material that conducts electricity better than an insulator but not as well as a conductor. Semiconductors are used extensively in modern computer and other important electronic devices.

The metalloids in general have atomic radii and ionization energies intermediate between those of the metals and the nonmetals. The electronegativities of these elements are also intermediate, most being approximately 2. The metalloids form covalent bonds in many compounds. Their most characteristic property is electrical conductivity that is low at ordinary temperatures, but increases with rising temperature. (The conductivity of metals decreases with rising temperature.)

Germanium is a moderately rare element. With the development of the transistor, germanium became important as a semiconductor material. It is chemically similar to silicon, the element above it in Group 14. Germanium is more metallic than arsenic just to its right in period four. The major oxidation state of germanium is +4.

Selenium and tellurium, members of Group 16, are somewhat rarer elements than germanium. Both have also become valuable because of their use in electronic devices. Selenium has several allotropes, including red, gray, and black forms. Black selenium, the commercially available form, is brittle but has a metallic luster. Selenium is a photoconductor—its ability to conduct electricity increases when light shines on it. This property is utilized in photocells and in creating the image in photocopying machines such as Xerox copiers.

Selenium has positive oxidation states of +2, +4, and +6. Like sulfur, just above in the group, it forms a dioxide, SeO_2, and an oxyacid, H_2SeO_4. Selenium also forms dihalides, such as $SeCl_2$, and selenides, such as Na_2Se, which contain the selenide ion, Se^{2-}. Selenium and its compounds are used in metallurgy, glass and ceramics, pigments, and medicines. As expected, the metal-like characteristics of Group-16 elements increase down the group. Tellurium is more metallic than selenium above it and less metallic than polonium below it. This gradation in properties down the group is illustrated by the change in odor of the hydrogen compounds. Hydrogen oxide (water) is odorless. Hydrogen sulfide has the offensive odor of rotten eggs. The odor of hydrogen selenide is even more offensive, and that of hydrogen telluride is the foulest.

Although it can be amorphous, tellurium is more commonly a brittle, silvery, metal-like crystalline substance. It is classed as a semiconductor, and its chemistry is typically metalloidal. It appears with oxygen in both tellurite

$(TeO_3{}^{2-})$ and tellurate $(TeO_4{}^{2-})$ ions. In these ions, tellurium shows the $+4$ and $+6$ oxidation states, respectively. Tellurium combines covalently with oxygen and the halogens, to give compounds in which the $+2$, $+4$, and $+6$ oxidation states are observed. It forms tellurides (-2 oxidation state) with such elements as gold, hydrogen, and lead. In fact, tellurium is the only element combined with gold in nature.

Section Review

1. Which elements are known as metalloids and where they are located in the Periodic Table?
3. What general characteristics make aluminum similar to the metalloids?
4. Descibe some of the important properties of the metalloids.
5. Describe some of the properties of selenium and tellurium.

26.2 Aluminum

Aluminum, atomic number 13, is the second member of Group 13. This group is headed by boron and includes gallium, indium, and thallium. All are typically metallic except boron, which is classed as a metalloid. The chemistry of boron differs from that of aluminum and the other Group 13 elements mainly because of the small size of the boron atom. Its chemistry resembles that of silicon and germanium more than it does that of aluminum. Much of the chemistry of aluminum is similar to that of the corresponding Group 2 metals, magnesium and beryllium.

Aluminum is a low-density metal; it is about one-third as dense as steel. The pure metal is used in chemical processes, in electronics, for forming jewelry, and as foil wrapping. Aluminum alloys are used in structural and industrial applications.

Aluminum is the most abundant metal in the earth's crust. Although it is found in many clays, rocks, and minerals, the only source for economical commercial production is **bauxite,** *a hydrated aluminum oxide ore, is the source of aluminum.* Bauxite containing 40–60% aluminum oxide is required for present aluminum-recovery technology. Major foreign sources for the recovery of bauxite are Australia, Jamaica, Surinam, and Guyana. Bauxite is also mined in Georgia, Alabama, Tennessee, and Arkansas. *Anhydrous aluminum oxide is called* **alumina.**

Recovery of Aluminum

In what is known as the **Hall process** *aluminum is extracted by electrolyzing anhydrous aluminum oxide (refined bauxite) dissolved in molten* **cryolite,** $Na_3 AlF_6.$ The process requires a temperature slightly below 1000°C.

In the electrolytic cell, an iron box lined with graphite serves as the negative electrode, graphite rods serve as the positive electrode, and molten cryolite–aluminum oxide is the electrolyte. The cryolite is melted in the cell, and the aluminum oxide dissolves as it is added to the molten cryolite. The operating temperature of the cell is above the melting point of aluminum (660°C). The aluminum metal becomes liquid and sinks to the bottom of the cell where it is easily drawn off.

1. Boron, silicon, germanium, arsenic, antimony, selenium, tellurium. They occupy a diagonal region from the upper center to the lower right in the Periodic Table.
2. Although elemental aluminum is distinctly metallic, it forms some compounds that exhibit ionic character and others in which it is covalently bonded. Furthermore, its compound, $Al(OH)_3$, is amphoteric.
3. Accept properties listed in Table 26-1 and properties given in the subsection "The Nature of Metalloids."
4. Accept properties of selenium and tellurium from Table 26-1 and those given in the subsection "The Nature of Metalloids."

SECTION OBJECTIVES

- Cite the occurrence, recovery, and uses of pure aluminum.
- Describe the properties of aluminum.
- Describe the *thermite reaction* and discuss its uses.
- Give the properties and uses of some important aluminum compounds.

This process for extracting aluminum from its ores was independently discovered in 1886 by two 23-year-old men, Charles Hall in the United States and Paul Heroult in France. Before this time, pure aluminum was very rare and expensive. In the 1860s, Napoleon III of France used a fork made of aluminum which was considered more precious than the gold and silver tableware his guests used.

Technology

Can A Can Become A Plane?

Aluminum recycling has become increasingly common. What happens to the recycled aluminum cans? Can they end up as part of an airplane? Probably not, because the metal recovered from aluminum cans is not pure aluminum.

Until recently, titanium was found in recycled aluminum because the paint on aluminum cans contained titanium oxide. The metal ions accumulated in the recycled aluminum, building up to a high enough level to affect the quality of the aluminum being produced. Now cans are painted with organic dyes. Even so, recycled aluminum cans still contain small amounts of silicon (from glass), iron, and cop-

per. These elements can stick together in particles as large as 10–50 micrometers. Such colloid particles can cause the recycled aluminum to

crack when being rolled into sheets. Recycled aluminum cans cannot be used, therefore, for critical purposes such as aircraft parts.

The electrode reactions are complex and are not understood completely. Aluminum is reduced at the negative electrode. The graphite rods that serve as a positive electrode are gradually converted to carbon dioxide. This fact suggests that oxygen is formed at the positive electrode by oxidation of the O^{2-} ion. The following equations for the reaction mechanism summarize the oxidation–reduction processes in the cell.

$$\text{negative electrode:} \quad 4Al^{3+} + 12e^- \rightarrow 4Al$$
$$\text{positive electrode:} \quad 6O^{2-} \rightarrow 3O_2 + 12e^-$$
$$3C + 3O_2 \rightarrow 3CO_2(g)$$

Another important source of aluminum is scrap metal, which provides a valuable source of aluminum at a much lower cost than that of the electrolysis of aluminum oxides.

Properties of Aluminum

Aluminum has a density of 2.70 g/cm^3 and a melting point of 660°C. It is ductile and malleable but is not as strong as brass, copper, or steel. Only silver, copper, and gold are better conductors of electricity. Aluminum can be welded, cast, or spun and can be soldered only by using a special solder.

Aluminum is a very active metal. The surface is always covered with a thin layer of aluminum oxide, which is not affected by air or moisture; aluminum is a self-protective metal that resists corrosion. At high temperatures, the metal combines with oxygen, releasing intense light and heat. Photoflash lamps contain aluminum foil or fine wire in an atmosphere of oxygen.

Aluminum is a very good reducing agent but is not as active as the Group 1 and Group 2 metals.

$$Al \rightarrow Al^{3+} + 3e^-$$

The Al^{3+} ion is quite small and carries a large positive charge. The ion hydrates vigorously in water solution and is usually written as the hydrated ion, $Al(H_2O)_6{}^{3+}$. Water solutions of aluminum salts are generally acidic because of the hydrolysis of $Al(H_2O)_6{}^{3+}$ ions.

$$Al(H_2O)_6{}^{3+} + H_2O \rightarrow Al(H_2O)_5OH{}^{2+} + H_3O{}^+$$

Water molecules are amphoteric but are very weak proton donors or acceptors. Water molecules that hydrate the Al^{3+} ion give up protons more readily, however. This increased activity results from the repulsion effect of the highly positive Al^{3+} ion. In the hydrolysis shown above, a water molecule succeeds in removing one proton from the $Al(H_2O)_6{}^{3+}$ ion.

Hydrochloric acid reacts with aluminum to form aluminum chloride and release hydrogen. The net ionic equation is

$$2Al(s) + 6H_3O^+ + 6H_2O \rightarrow 2Al(H_2O)_6{}^{3+} + 3H_2(g)$$

Nitric acid does not react readily with aluminum because of aluminum's protective oxide layer.

In basic solutions, aluminum forms aluminate ions, $Al(OH)_4{}^-$, and releases hydrogen.

$$2Al(s) + 2OH^- + 6H_2O \rightarrow 2Al(OH)_4{}^- + 3H_2(g)$$

If the base is sodium hydroxide, sodium aluminate, $NaAl(OH)_4$, is the soluble product, and hydrogen is given up as a gas.

$$2Al(s) + 2NaOH + 6H_2O \rightarrow 2NaAl(OH)_4 + 3H_2(g)$$

The oxide surface coating protects aluminum under normal atmospheric conditions and must be removed before the metal underneath can react with hydronium ions or with hydroxide ions. Hydroxide ions dissolve this oxide layer more readily than do hydronium ions. This explains why aluminum reacts more readily with hydroxides than with acids.

The Thermite Reaction

When a mixture of powdered aluminum and an oxidizing agent such as iron(III) oxide is ignited, the aluminum reduces the oxide to the free metal. This reduction is rapid and violent and yields a tremendous amount of heat. The sudden release of this heat energy produces temperatures from 3000 to 3500°C, enough to melt the iron. *Such a reaction between aluminum and the oxide of a less active metal is called the* **thermite reaction**. *The mixture of powdered aluminum and iron (III) oxide is called thermite.*

The Al_2O_3 coating on the aluminum is transparent and invisible.

Ductile: Capable of being drawn into wire. Malleable: Capable of being rolled into sheets

Solder is any metal alloy used when melted for joining or patching metal parts or surfaces.

The Al^{3+} ion is so small and highly charged that it has a very great charge density. This means that it will strongly attract electrons. The water molecules bonded to the aluminum ion have their electrons pulled so strongly toward the aluminum that the bonds between oxygen and hydrogen are weakened. Hydrogen ions readily break away.

Because HNO_3 is an oxidizing acid, it enhances the protective Al_2O_3 layer on the aluminum surface. Nitric acid can even be stored in aluminum containers. Nonoxidizing acids such as HCl will react with aluminum.

The formation of aluminum oxide is strongly exothermic. The reaction releases 1669 kJ of heat per mole of aluminum oxide formed. The heat of formation of iron(III) oxide is 820 kJ per mole. In the thermite reaction, the amount of heat released per mole of aluminum oxide formed equals the difference between these values. The net thermite reaction is considered to be the sum of these two separate reactions.

$$2Al + \tfrac{3}{2}O_2 \rightarrow Al_2O_3 \qquad \Delta H = -1669 \text{ kJ}$$
$$Fe_2O_3 \rightarrow 2Fe + \tfrac{3}{2}O_2 \qquad \Delta H = +820 \text{ kJ}$$
$$\overline{2Al + Fe_2O_3 \rightarrow 2Fe + AlO_3 \qquad \Delta H = -849 \text{ kJ}}$$

It is not practical to use aluminum to produce cheaper metals by the reduction of their oxides. The thermite reaction is often used to produce small quantities of carbon-free metal, however. A more important use of this reaction is to reduce metallic oxides that are not readily reduced with carbon. Chromium, manganese, titanium, tungsten, and molybdenum can be recovered from their oxides by the thermite reaction. All these metals are used in making alloy steels.

The very high temperature produced by the thermite reaction makes it useful in welding. Large steel parts, such as propeller shafts and rudder posts on ships or the crankshafts of heavy machinery, are repaired by thermite welding, as shown in Figure 26-2.

Some Compounds of Aluminum

Because of its amphoteric properties, aluminum and its oxide react both with acids and bases to form compounds with many useful properties. In this section, we limit our discussion to only a few aluminum compounds and their uses.

Aluminum oxide **Bauxite,** the chief ore of aluminum, contains the hydrated oxide Al_2O_3. **Corundum** *or* **emery** *is a natural oxide of this metal that is useful as abrasives.* Emery is the abrasive of emery boards, emery paper, emery cloth, and emery grinding wheels.

The thermite reaction provides a convenient way to weld railroad tracks because no heavy tanks of gas or electrical equipment are needed.

Figure 26-2 The very high temperature produced by the thermite reaction makes it useful in welding large steel parts, as in shipbuilding.

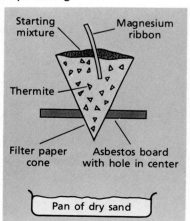

Starting mixture — Magnesium ribbon

Thermite —

Filter paper cone — Asbestos board with hole in center

Pan of dry sand

Anhydrous aluminum oxide, usually a white powder called **alumina**, *is a very inert substance having an extremely high melting point of 2027°C*. It is used in crucibles for high-temperature reactions.

Pure crystalline alumina, Al_2O_3, is clear and colorless. *The **ruby** and **sapphire** gemstones shown in Figure 26-3 are composed of aluminum oxide contaminated with traces of certain impurities.* The presence of a trace quantity of chromium imparts the brilliant ruby-red color to a gem-quality crystal of Al_2O_3, yielding one of the most valuable gemstones found in nature whereas traces of titanium yield the blue sapphire.

Synthetic rubies and sapphires are produced by fusing aluminum oxide with the required trace contaminants in the flame of an oxyhydrogen blowtorch. As synthetic gemstones, they are used in jewelry. Because of the hardness of aluminum oxide, these synthetic crystals are also used as bearings (jewels) in watches and other precision instruments and as dies for drawing wires.

Scientists are finding new ways to strengthen and stiffen structural materials. One important development involves **fiber composites**, *monocrystalline strands or fibers of one substance embedded in and held in place by some other material*. An aluminum rod containing very hard, strong sapphire "whiskers" is quite unlike ordinary soft, ductile aluminum. Such a structure is six times stronger than aluminum and twice as stiff. These fiber composites enable engineers to greatly improve the strength-to-weight ratio of structural materials.

Alundum is a material made by melting and resolidifying Al_2O_3. It is used for making grinding wheels and other abrasives. It is also found in crucibles, funnels, tubing, and other pieces of laboratory equipment.

The alums   **Alums** *are common compounds that usually contain aluminum and have the general formula $M^+M^{3+}(SO_4)_2 \cdot 12H_2O$.* The M^+ can be any one of several monopositive ions but is usually Na^+, K^+, or NH_4^+. The M^{3+} can be any one of a number of tripositive ions such as Al^{3+}, Cr^{3+}, Co^{3+}, Fe^{3+}, or Mn^{3+}. Potassium aluminum sulfate, $KAl(SO_4)_2 \cdot 12H_2O$, is the most common alum. It is easy to see from this formula why alums are sometimes called double sulfates.

All alums, also called aluminum salts, have the same crystal structure. This crystal-structure requirement explains why some monopositive and tripositive ions do not form alums. The alum crystals are hydrated structures

Figure 26-3   Both synthetic and natural sapphires (left) and rubies (right) are composed of aluminum oxide, Al_2O_3. The blue color of sapphire is due to a trace of titanium. The red color of a ruby is due to a trace of chromium.

---

in which six water molecules are coordinated to each metallic ion. This coordination number of six for both the monopositive and tripositive ions may restrict the kinds of metallic ions that can form alums.

Alums are used for water purification, as mordants, and in paper manufacture. $NaAl(SO_4)_2 \cdot 12H_2O$ is used as the acid compound in one type of baking powder.

Section Review

1. Where is aluminum found in nature, and how is it recovered?
2. What are some of the uses of aluminum and aluminum alloys?
3. List the physical and chemical properties of aluminum.
4. Describe the thermite reaction and list some important uses.
5. Name some aluminum compounds and their uses.

26.3 Representative Metalloids

Recall that the metalloids are elements that exhibit both metallic and nonmetallic properties to some extent. *Those elements and compounds that are characterized by a slight electric conductivity that increases with temperature are known as* **semiconductor**. This property is the reason for the use of metalloids in the electronics industry. Semiconductors permit the manufacture of tiny electronic devices such as transistors. A single computer memory chip, like those shown in Figure 26-4, may contain over a million transistors.

In this section, you will study the properties and uses of some of the metalloids.

Silicon

Silicon is as important in the mineral world as is carbon in living systems. Silicon dioxide and silicates make up about 87 percent of the material of the earth's crust. Silicon does not occur free in nature. It is found in many minerals in the form of SiO_2, as shown in Figure 26-5.

Properties of silicon Silicon atoms have four valence electrons. Silicon crystallizes with a tetrahedral-bond arrangement similar to that of carbon atoms in diamond. Atoms of silicon also have small atomic radii and tightly held electrons. Thus their ionization energy and electronegativity are fairly high. Silicon is a metalloid. It forms bonds with other elements that are essentially covalent. The electric conductivity of silicon is similar to that of boron; it, too, is a semiconductor. Unlike carbon, silicon forms only single bonds. It forms silicon–oxygen bonds more readily than silicon–silicon or silicon–hydrogen bonds. However, much of its chemistry is similar to that of carbon. Silicon has much the same role in mineral chemistry as carbon has in organic chemistry.

Boron and silicon atoms have roughly similar small radii. This similarity allows them to be substituted for one another in glass, even though boron is a member of Group 13 and silicon of Group 14.

Because the relative size of atoms frequently defines properties, elements in diagonal positions in the Periodic Table (such as boron-silicon-arsenic)

Figure 26-4 Chunks of raw poly-silicon are processed into silicon ingots, which are sliced and polished to form wafers. The silicon wafers can then be used to make semiconductor integrated circuits.

Extremely pure silicon that contains a trace of arsenic is used to make **n**-type semiconductors. Arsenic has one more valence electron than silicon, and this extra electron is a negative charge carrier that can diffuse through the crystal. A **p**-type semiconductor contains a trace of boron. With one less valence electron, a positive electron "hole" is formed, which can also diffuse through the crystal.

often have similar properties. In general, atom size decreases to the right from Groups 1 to 18. It increases down each group. This balancing effect results in atoms of similar size along diagonals in the Periodic Table.

Elemental silicon, of about 98 percent purity, is used in the production of alloys of iron, spring steel, copper, and bronze. It is also used in the manufacture of **silicones**, *polymeric organosilicon compounds composed of silicon, carbon, oxygen, and hydrogen.* Ultrapure silicon is used in microcomputers, transistors, and solar cells.

Silicones Silicon resembles carbon in the ability of its atoms to form chains. Unlike carbon chains, which are formed by carbon–carbon bonds, silicon chains are usually formed by silicon–oxygen bonds. In *silicones,* the silicon chain is bound together with oxygen atoms. Hydrocarbon groups are attached to the silicon atoms. Thus the silicones are part organic and part inorganic. By using different hydrocarbon groups, a variety of silicones can be produced. One silicone chain has the structure

$$\left[\begin{array}{c} H \\ | \\ H-C-H \quad H-C-H \\ | \\ -Si-O-Si- \\ | \\ H-C-H \quad H-C-H \\ | \\ H \end{array}\right]_x$$

The silicones are not greatly affected by heat. They have very good electric insulating properties and are water repellent. Water does not penetrate cloth treated with a silicone. Some silicones have the character of oils or greases and can be used as lubricants. Silicone varnishes are used to coat wires for the windings of electric motors. This insulation permits the electric motor to operate at high temperatures without developing short circuits. Silicones are used in automobile and furniture polishes.

Figure 26-5 Silicon, the second most abundant element. Its abundance is due to the wide distribution in nature of silicon dioxide (quartz rocks) and many different silicates.

Figure 26-6 Amethyst is a form of silicon dioxide (SiO2) crystals.

Arsenic

Arsenic occurs in nature principally in the form of sulfide minerals such as orpiment, As_2S_3, realgar, As_4S_4, and arsenopyrite, FeAsS. It is also found in the mineral andetite, As_2O_3, and in metal arsenides such as $FeAs_2$ and $CoAs_2$.

Arsenic is allotropic and exists in two distinct crystalline forms, known as yellow arsenic and metallic arsenic. In the yellow form, arsenic is present as As_4 tetrahedral molecules. In the metallic form, it is present as sheets of atoms. Yellow arsenic is soft, has a density of 2.03 g/cm^3, and is metastable—it passes rapidly to the metallic form by heating or by the action of light. Metallic arsenic is a brittle, gray solid with density of 5.73 g/cm^3. Although it is a good conductor of heat, it is a very poor conductor of electricity.

Chemically, arsenic may act as a metal and form oxides and chlorides. It may also act as a nonmetal and form acids. When heated in moist air, arsenic sublimes and forms As_2O_3. Metallic arsenic has few commercial uses except as a hardening agent in alloys. A trace of arsenic is used to improve the hardness and sphericity of lead shots. Compounds of arsenic are used mostly for their toxicity as weed killers, sheep and cattle dips, and insecticides.

Antimony

Antimony is found in nature principally as the black sulfide ore, stibnite, Sb_2S_3. Both the metal and the ore have been used since antiquity for making vases and ornaments and in cosmetics. Antimony, like arsenic, exhibits the property of allotropy. It exists in the yellow and metallic forms. Yellow antimony is a soft solid with a density of 5.3 g/cm^3. This form of antimony is very unstable, and it exists only at very low temperatures. Metallic antimony is a dense, brittle, silver-white metalloid with a bright metallic luster and a density of 6.7 g/cm^3. It is less active than arsenic. When strongly heated in air, antimony forms a white oxide, Sb_4O_6.

An alloy of lead and antimony is used for the plates in storage batteries. This alloy is stronger and more resistant to acids than lead alone. Tartar emetic, or potassium antimony tartrate, $KSbOC_4H_4O_6$, is used as a mordant in the dyeing of cotton goods.

Metalloids in computers Silicon is a primary ingredient in the integrated circuits of computers. When precise quantities of the metalloids arsenic, antimony, or boron are added to high purity silicon, the electrical conductivity

The symbol for antimony is based on its Latin name, *stibium*.

CAUTION
The soluble compounds of antimony are almost as toxic as those of arsenic.

Figure 26-7 *Solar cells* (below) which are high purity silicon crystals treated with certain metalloids convert light to electricity. A diagram of a typical solar cell is shown at right.

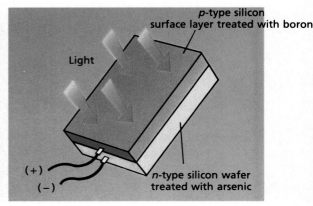

Chemistry Notebook

Gallium Arsenide, Lasers, and Computer Chips

Gallium arsenide (GaAs) is a semiconductor similar to silicon, except that its crystals can be made to "lase"—to emit light of a single wavelength. By including some aluminum arsenide, a GaAs laser can be tuned to other wavelengths, as applications demand. The GaAs laser acts like a heat gun, focusing infrared light into an extremely small region for nanosecond intervals. This technique is extraordinarily useful for the manufacture of computer chips.

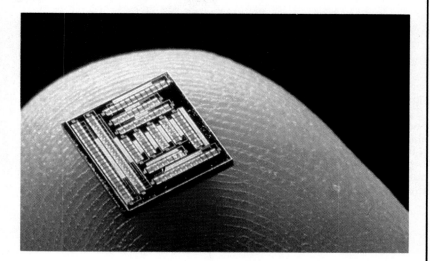

The procedure used now to make computer chips is to cover a wafer of silicon with a stencil that shows the wiring layout needed. The wafer is etched to expose a clean surface, and then a thin layer of metal is deposited into the etched lines. The stencil is lifted and the microscopic metal circuit is sealed with a polymer coating.

Gallium arsenide lasers will soon be used to "draw" a circuit on a silicon wafer that is surrounded by a gaseous mixture of tungsten hexafluoride and hydrogen. The energy of the laser will cause these gases to react to form tungsten metal, but only in the microscopic region where the laser is focused. Essentially, this makes the laser act like a pencil, drawing metal circuit lines wherever it traces. In the microchip industry, such a laser pencil will allow much greater flexibility in the design of microcircuits.

---

of the silicon sample can be changed dramatically depending on the voltage that is applied. This voltage-sensitive conductivity was the basis for the operation of vacuum tubes which you may have seen in an old radio or television. Circuits made of silicon have many of the same electronic properties of the old vacuum tubes but can operate using a tiny fraction of the power and space required by the vacuum tubes. This is how electronic engineers pack so much computing power into today's computers.

Section Review

1. Identify some properties of silicon and discuss its uses.
2. What are silicones and how are they used?
3. In what ways are antimony and arsenic the same? In what ways are they different?
4. Name some of the uses of arsenic and antimony.

Most arsenic compounds are used for their toxicity as weed killers, animal dips, and insecticides. Antimony alloys with lead are used to reduce friction in machinery and for the plates in storage batteries. Another antimony alloy is used as a mordant.

1. Accept any properties given under the heading "Properties of silicon."
2. Synthetic compounds composed of silicon, carbon, oxygen, and hydrogen. Used as water repellents, lubricants, varnishes to coat wires, and in automotive and furniture polishes
3. Both exhibit the property of allotropy, and their nonmetallic and metallic forms exhibit similar properties. Arsenic is more active than antimony.
4. Arsenic, when heated in air, forms the oxide As_2O_3, and antimony, the oxide Sb_4O_6.

Figure 26-8 Gallium arsenide (GaAs) as well as silicon is an important ingredient in advanced electronic devices. The computer chip shown here are composed largely of GaAs. A single computer memory chip may contain over a million transistors.

Chapter Summary

- Metalloids have properties intermediate between those of metals and nonmetals. They occupy a diagonal region of the periodic table from upper center toward lower right. The metalloids include boron, silicon, germanium, arsenic, antimony, selenium, and tellurium.
- Although certain of its properties tend to place aluminum among the metalloids, it is considered a metal. It forms negative aluminate ions and some covalent halides.
- Aluminum is recovered by electrolysis of the aluminum oxide dissolved in molten cryolite. It is used as a low-density structural metal when alloyed with selected elements. Powdered aluminum mixed with iron(3) oxide and subsequently ignited reduces the oxide to the free metal in a rapid, violent combustion reaction known as the thermite reaction.
- Some natural or synthetic oxides of aluminum are very hard and are used as abrasives.
- Alums are double sulfates with the general formula $M^+M^{3+}(SO_4)_2 \cdot 12H_2O$.
- Silicon forms strong bonds with oxygen. Silicon–oxygen lattice chains are similar to carbon–carbon chains. Much of the chemistry of the element silicon is similar to that of carbon.
- Silicon chains bound together with oxygen atoms and having hydrocarbon groups attached to the silicon atoms are called silicones. Some silicones have properties similar to organic oils and greases.
- Arsenic and antimony are used in alloys. Arsenic is a hardening agent for lead. Antimony and lead are used in "antifriction" alloys and battery-plate alloys.
- Aluminum is distinctly metallic in many of its properties. It is always covered with an oxide surface coating that protects it from corrosion by air and moisture.
- Silicon mixed with precise amounts of the metalloids such as antimony, arsenic, or boron has electronic properties that can be used to construct integrated circuits.

Chapter 26 *Review*

Introduction to Aluminum and Metalloids

Vocabulary

alumina	Hall process
bauxite	ruby
corundum	sapphire
cryolite	semiconductor
emery	thermite
fiber composite	thermite reaction

Questions

1. Why is aluminum, a familiar structural metal, included in the study of metalloids?
2. In what materials does aluminum occur in nature?
3. What are the important physical properties of aluminum?
4. *(a)* What is the chemical composition of corundum and emery? *(b)* Write the formulas for these substances. *(c)* What is their important use?
5. Why must aluminum oxide be dissolved in molten cryolite before it can be decomposed by electricity?
6. Why are certain metallic oxides reduced with aluminum rather than with carbon?
7. Write the chemical formulas for four different alums.
8. How does silicon rank in abundance among the elements?
9. Why must extreme care be used in handling arsenic and its compounds?

Problems

1. How much aluminum and how much iron(III) oxide must be used in a thermite mixture to produce 10.0 kg of iron for a welding job?
2. What is the percentage of aluminum in sodium alum that crystallizes with 12 molecules of water of hydration?
3. How many liters of hydrogen can be prepared by the reaction of 50.0 g of aluminum metal and 100. g of sodium hydroxide in solution?
4. Calculate the percentage of boron in colemanite, $Ca_2B_6O_{11} \cdot 5H_2O$.

Application Questions

1. What reaction occurs when aluminum is placed in *(a)* hydrochloric acid solution *(b)* sodium hydroxide solution? *(c)* Why is there no reaction in neutral water?
2. Write equations to show the net positive and negative electrode reactions during the electrolysis of aluminum oxide.
3. Why is bauxite imported from the West Indies and South America when almost any clay bank in the United States contains aluminum?
4. What geographic conditions affect the location of plants for the production of aluminum from purified bauxite?
5. Describe the properties characteristic of metalloids such as arsenic and silicon in terms of *(a)* atomic radius *(b)* ionization energy
6. *(c)* electronegativity *(d)* types of bonds formed *(e)* electric conductivity.
7. *(a)* What is a silicone? *(b)* What are some of the important uses for silicones?
8. Why would you expect silicones to be water repellent?

Application Problems

1. A compound contains 96.2% arsenic and 3.85% hydrogen. Its vapor is found to have a density of 3.48 g/L at STP. What is the molecular formula of the compound?

Enrichment

1. Investigate the chemistry of insulators and semiconductors and write a report.
2. Perform library research and write a report on the current state of optical fibers. How are optical figures used? How might you or your classmates encounter optical fibers in everyday life?
3. Describe the role of silicon in the formation of minerals. Conduct library research to discover why minerals such as quartz are found in a variety of colors.

Chapter 27 Sulfur and Its Compounds

Chapter Planner

27.1 Elemental Sulfur
The Production of Sulfur
- ■ Questions: 1–4
- ▨ Problem: 1
  Application Questions: 1, 2

Physical Properties of Elemental Sulfur
- ■ Demonstrations: 1, 2
  Question: 6
  Application Questions: 3–5
- ■ Question: 7
- ▨ Application Problem: 1

Chemical Properties of Elemental Sulfur
- ▨ Problems: 2, 4

Uses of Elemental Sulfur
- ▨ Question: 8

27.2 Important Compounds of Sulfur
Sulfur Dioxide
- ■ Questions: 5, 9, 10, 12, 14c–e
  Application Questions: 6–8
- ■ Chemistry Notebook
  Questions: 11, 13
  Application Questions: 10–12
- ▨ Problems: 5, 7
  Application Question: 9
  Application Problems: 4, 5, 7

Sulfuric Acid
- ■ Demonstration: 3
  Questions: 16–20
  Application Question: 13
- ■ Question: 15
- ▨ Problems: 3, 6
  Application Question: 14
  Application Problems: 2, 3, 6

Teaching Strategies

Introduction

This chapter deals with the physical and chemical properties of sulfur and the chemistry of two important sulfur compounds: sulfur dioxide and sulfuric acid. The pollution of the atmosphere by sulfur dioxide and the formation of acid precipitation is a major environmental concern and will be of interest to students. Since more sulfuric acid is used each year than any other single chemical substance, its preparation and some of its uses should be discussed.

27.1 Elemental Sulfur

Exhibit samples of elemental sulfur in various forms such as flowers of sulfur and roll sulfur. Show minerals of sulfur such as gypsum and iron pyrite. If iron pyrite is available, point out the golden crystals that led to the name "fool's gold." Note the oxidation state of the sulfur in pyrite, gypsum, and other available minerals.

Discuss the Frasch process of obtaining sulfur. Explain how water can be heated to a temperature of 170°C when it is under great pressure. Refer to Figure 27-1 to explain the process for pumping the molten sulfur from underground.

Have students note the color and lack of odor of sulfur. Drop some sulfur in water to show the relative density of sulfur and its lack of solubility in water. Relate the solubility to the nonpolar nature of the sulfur molecule.

Make a ball-and-stick model of the S_8 molecule to show the ring structure. Discuss rhombic sulfur and monoclinic sulfur as allotropes containing the S_8 structure though stacked in different ways.

Discuss the behavior of sulfur when heated. Show how *lambda*-sulfur consists of S_8 molecules in the liquid state. Break open the ball-and-stick model to show what happens upon further heating, and explain the viscosity of *mu*-sulfur in terms of the tangling of the open chains of sulfur. Explain how further heating shortens the chains so that the viscosity decreases. Amorphous sulfur forms when the short chains are cooled so rapidly that the chains and rings cannot reform. In time, the bonds slowly form so that S_8 molecules are again present.

Discuss the gases SO_2 and SO_3, pointing out the oxidation states of +4 and +6 for sulfur in these molecules.

Show that elemental sulfur is used in some substances. Ask students to name chemical compounds that contain sulfur.

Demonstration 1: Plastic Sulfur
PURPOSE: To show some of the allotropic forms of sulfur
MATERIALS: Powdered sulfur, large test tube, beaker, water, burner, tongs
PROCEDURE: Wear goggles. Work in a hood because sulfur dioxide gas may be produced. Place about 1 inch of sulfur in the test tube. Observe its properties. Heat the sulfur until it melts into an orange liquid (*lambda*-sulfur). Continue to strongly heat

the test tube until a thick, dark red liquid forms (*mu*-sulfur). Quickly pour the molten sulfur into a beaker of cold water. Observe the formation of the brown, rubbery, plastic sulfur (amorphous sulfur). Allow the sulfur to cool for several minutes, remove with tongs, and note its elasticity. On standing, the plastic sulfur will revert back to its original form.

Demonstration 2: A Chemical Sunset
PURPOSE: To show the formation of colloidal sulfur
MATERIALS: Glass container of approximately 2 L (an empty acid bottle works well), $Na_2S_2O_3 \cdot 5H_2O$, 6 M hydrochloric acid, slide projector, water
PROCEDURE: Wear goggles. Dissolve approximately 15 g of $Na_2S_2O_3 \cdot 5H_2O$ in 2 L of water in the container. In a darkened room, shine the beam from a slide projector through the container onto a wall or screen. Pour about 10 mL of 6 M HCl into the container, and quickly swirl it to mix. Observe the color changes in the container and on the wall or screen. The reaction slowly produces colloidal sulfur that scatters light as it is being formed and produces a sequence of colors from yellow to red on the screen, just as the colors of a natural sunset are produced when light is scattered by dust particles in the atmosphere.

$$2H^+(aq) + S_2O_3^{2-}(aq) \rightarrow H_2S_2O_3(aq)$$

$$H_2S_2O_3 \rightarrow H_2SO_3 + S(\text{colloidal})$$

27.2   Important Compounds of Sulfur

Discuss sulfur dioxide as both a chemical substance with great utility and as a byproduct of combustion that is an environmental hazard. Explain the equations that lead to the formation of SO_2. Discuss the oxidation states of sulfur in SO_2 and in the sulfite ion (both +4). Show that sulfur dioxide is the "anhydride" of sulfurous acid, H_2SO_3, because in both compounds sulfur has the same oxidation state.

Ask students to look at labels of substances such as dried fruits, cereals with fruit, and wines for evidence of treatment with sulfur dioxide. Discuss treatment of fresh fruits and vegetables with SO_2 to preserve them. Discuss papermaking, in which sulfites are used to dissolve the lignin, the sticky substance that holds the wood fibers together.

Acid precipitation is an important ecological issue. Assign students reports based on current articles. Discuss the strengths of the acids formed when carbon dioxide, sulfur dioxide, sulfur trioxide, and nitrogen dioxide dissolve in rainwater, and how each of these affect the pH of the precipitation.

Sulfuric acid is sometimes called the king of chemicals because of the enormous quantities of it that are consumed. Show students a reagent bottle of H_2SO_4 and have them read the label and note properties of the acid. Swirl the acid so they can observe the oily viscosity. Write the equations showing the stepwise ionization of the acid. Discuss the fact that the first ionization goes essentially to completion, and the equilibrium constant for the second ionization shows that it is about 10% ionized. Write formulas for salts of sulfuric acid such as Na_2SO_4 and $NaHSO_4$, explaining that it is possible to form both sulfate and hydrogen sulfate compounds.

Discuss the reactivity of sulfuric acid in three ways: as a hydrogen ion donor, as an oxidizing agent, and as a dehydrating agent. This variety of reactions makes it a very useful chemical substance. Write equations showing each of these types of behavior.

The majority of sulfuric acid in the United States is used to make fertilizers. Discuss the need for the enormous quantity of fertilizer that is produced each year. Show a lead storage battery that uses sulfuric acid as an electrolyte and discuss the fact that the density of the solution is used as a measure of the state of the battery.

Chemistry Notebook: The Smell of Sulfur Compounds
Students might be surprised to think that the sense of smell is a chemical sense, and that the combination of atoms in the molecule can cause odors to form. Sulfur compounds are notorious for having bad smells. Discuss the benefit of adding compounds with strong odors to natural gas to allow gas leaks to be detected.

Demonstration 3: Dehydration of Sucrose
PURPOSE: To demonstrate the dehydrating ability of sulfuric acid
MATERIALS: Sucrose (table sugar), 100-mL beaker, concentrated H_2SO_4
PROCEDURE: Wear goggles. Work in a hood. Fill a small beaker about one-third full of sucrose. Set the beaker in a tray. Carefully pour 5-10 mL of concentrated sulfuric acid over the sugar. Do NOT stir. Stand back and watch. The sugar will start to char as the acid dehydrates the sugar; the carbon that remains forms a large "snake," along with steam and smoke. When cool, remove the carbon snake with tongs because it will contain some sulfuric acid. Rinse it thoroughly with water before discarding.

References and Resources

Books and Periodicals

Likens, Gene E. "Acid Rain," *Scientific American*, Vol. 241, 1979, p. 43.

Martin, J.A., P. Baudot, J.L. Manal, and M.F. Lejaille. "Synthesis of Sulfuric Acid by the Contact Process," *Journal of Chemical Education*, Vol. 52, 1975, p. 188.

Mohnen, Volker A. "The Challenge of Acid Rain," *Scientific American*, Vol. 259, 1988, p. 259.

Shakhashiri, Bassam Z. "Heat of Dilution of Sulfuric Acid." *Chemical Demonstrations*, Vol. 1, University of Wisconsin Press, 1983, p. 17.

Sheldon, Richard P. "Phosphate Rock," *Scientific American*, Vol. 246, 1982, p. 45.

Summerlin, Lee R. and James L. Ealy, Jr. "Exothermic Reaction: Sodium Sulfite and Bleach," *Chemical Demonstrations, A Sourcebook for Teachers*, Vol. 1, American Chemical Society, 1985.

Summerlin, Lee R., Christie L. Borgford, and Julie B. Ealy. "Acid Rain," *Chemical Demonstrations, A Sourcebook for Teachers*, Vol. 2, American Chemical Society, 1987, p. 165.

"Sulfuric Acid": film, University of Akron, Audio Visual Services.
"Sulfur and Its Compounds (2nd ed)": film and video, #1436, Coronet Films & Video.

Audiovisual Resources

"Sulfur": slides and cassette. Educational Images Ltd. Allotropes of sulfur, sulfur dioxide, and sulfuric acid discussed.
"Sulfur Dioxide and Sulfurous Acid": film, University of Akron, Audio Visual Services.

Computer Software

"The Periodic Table Videodisc: Reactions of the Elements." CAV type videodisc, Journal of Chemical Education Software, 1989. Shows action sequences and still shots of each element in its most common forms; reacting with air, water, acids, bases; and some common applications

Answers and Solutions

Questions

1. As the free element or combined with other elements in sulfides and sulfates
2. Between 150 and 600 meters underground in Texas and Louisiana near the Gulf of Mexico
3. *(a)* To melt the sulfur *(b)* To force the melted sulfur and water mixture to the surface
4. From pyrite, FeS_2, and from the purification of coke oven gas, smelter gases, petroleum, or natural gas
5. *(a)* Odorless *(b)* A suffocating, choking odor
6. *(a)* FeS_2 *(b)* Slow flowing *(c)* One of the two or more different forms of an element in the same physical phase *(d)* Ease of flowing *(e)* Having neither definite form nor structure
7. This was the result of a change in the sulfur from amorphous sulfur back to rhombic sulfur.
8. From the manufacture of sulfuric acid, black gunpowder, medicines, fungicides; in the vulcanization of rubber
9. From volcanic gases and mineral waters, from the burning of fuels with sulfur as an impurity, and from the roasting of sulfide ores
10. $S(s) + O_2(g) \rightarrow SO_2(g)$
    $2\,ZnS(s) + 3\,O_2(g) \rightarrow 2\,ZnO(s) + 2SO_2$
11. *(a)* Upward displacement of air *(b)* Its solubility in water and its density greater than air
12. Making sulfuric acid
13. *(a)* $SO_2(aq) + NaOH(aq) \rightarrow NaHSO_3(aq)$
    (b) $SO_2(aq) + 2NaOH(aq) \rightarrow Na_2SO_3(aq) + H^2O(\ell)$
14. *(a)* About 5.0 *(b)* Because normal rainwater contains dissolved carbon dioxide and organic acids of natural origin *(c)* Rainwater having a pH below about 5.0 *(d)* Oxides of nitrogen and of sulfur *(e)* Acid rain may harm plant and animal life
15. *(a)* The contact process *(b)* Because the mixture of sulfur dioxide and oxygen comes in contact with the catalyst
16. By slowly adding sulfuric acid to water while constantly stirring
17. Dilute sulfuric acid is a strong acid. Hot, concentrated sulfuric acid is an oxidizing agent. Concentrated sulfuric acid is a dehydrating agent.
18. The manufacture of fertilizers
19. To remove oxides from the surface of iron or steel before the metal is plated or coated with enamel
20. Sulfuric acid acts as a dehydrating agent to absorb the water formed in the reaction between the nitric acid and the glycerol, thus causing the reaction to proceed more readily.

Problems

1. $49.4 \text{ m}^3 \times \dfrac{10^6 \text{cm}^3}{1 \text{ m}^3} \times \dfrac{2.07 \text{ g}}{1 \text{ cm}^3} \times \dfrac{1 \text{ kg}}{10^3 \text{ g}} \times \dfrac{1 \text{ metric tn}}{10^3 \text{ kg}}$
   $= 102$ metric tn

2. Na_2S: $\dfrac{2.5 - 0.9}{2.5} \times 100\% = 64\%$

   MgS: $\dfrac{2.5 - 1.2}{2.5} \times 100\% = 52\%$

   Al_2S_3: $\dfrac{2.5 - 1.5}{2.5} \times 100\% = 40\%$

   SiS_2: $\dfrac{2.5 - 1.8}{2.5} \times 100\% = 28\%$

   P_4S_{10}: $\dfrac{2.5 - 2.1}{2.5} \times 100\% = 16\%$

   SCl_2: $\dfrac{3.0 - 2.5}{3.0} \times 100\% = 17\%$

3. Molar mass of $H_2SO_4 = 98.1$g
   $\dfrac{2.02 \text{ g H}}{98.1 \text{ g } H_2SO_4} \times 100\% \ H_2SO_4 = 2.06\%$ H
   $\dfrac{32.1 \text{ g S}}{98.1 \text{ g } H_2SO_4} \times 100\% \ H_2SO_4 = 32.7\%$ S
   $\dfrac{64.0 \text{ g O}}{98.1 \text{ g } H_2SO_4} \times 100\% \ H_2SO_4 = 65.2\%$ O

4. 100.0% cpd $- 86.6\%$ Pb $= 13.4\%$ S
   $86.6 \text{ g Pb} \times \dfrac{1 \text{ mol Pb}}{207 \text{ g Pb}} = 0.418$ mol Pb
   $13.4 \text{ g S} \times \dfrac{1 \text{ mol S}}{32.1 \text{ g S}} = 0.417$ mol S   Pb:S $= 0.418{:}0.417 = 1{:}1$
   Empirical formula, PbS

5. $S + O_2 \rightarrow SO_2$
   $2.5 \text{ kg S} \times \dfrac{10^3 \text{ g S}}{1 \text{ Kg S}} \times \dfrac{1 \text{ mol S}}{32.1 \text{ g S}} \times \dfrac{1 \text{ mol } SO_2}{1 \text{ mol S}} \times \dfrac{64.1 \ SO_2}{1 \text{ mol } SO_2}$
   $\times \dfrac{1 \text{ Kg } SO_2}{10^3 \text{ g } SO_2} = 5.0 \text{ kg } SO_2$

6. For each mole of ZnS roasted, one mole of H_2SO_4 may be theoretically produced.
   $450. \text{ metric tn ZnS} \times \dfrac{10^6 \text{ kg}}{\text{metric ton}} \times \dfrac{1 \text{ mol ZnS}}{97.5 \ ZnS} \times \dfrac{1 \text{ mol } H_2SO_4}{1 \text{ mol ZnS}}$
   $\times \dfrac{98.1 \text{ g } H_2SO_4}{1 \text{ mol } H_2SO_4} \times \dfrac{1 \text{ Kg } H_2SO_4}{10^3 \text{ g } H_2SO_4} = 4.53 \times 10^5 \text{ kg } H_2SO_4$

7. $Na_2SO_3 + H_2SO_4 \rightarrow H_2O + SO_2(g) + Na_2SO_4$

$5.00 \text{ L } SO_2 \times \dfrac{1 \text{ mol}}{22.4 \text{ L}} \times \dfrac{1 \text{ mol } Na_2SO_3}{1 \text{ mol } SO_2} \times \dfrac{126.1 \text{ g } Na_2SO_3}{1 \text{ mol } Na_2SO_3}$

$= 28.1 \text{ g } Na_2SO_3$

Application Questions

1. To recover sulfur deposits located below a quicksand layer, which cannot be recovered by conventional mining methods

2. *(a)* Water heated under pressure to above its normal boiling point *(b)* To be at a high enough temperature to melt the sulfur

3. *(a)* S_8 *(b)* It would complicate an equation by making it necessary to multiply all the coefficients by 8.

4. At the melting point, sulfur is a pale-yellow liquid that flows easily. As it is heated further, it becomes more viscous; above about 160°C, it hardly pours. The color at this time has changed from light yellow to reddish-brown to almost black. Near the boiling point the fluidity increases again. The fluidity changes are the result of changing from S_8 rings of sulfur to long, entangled chains of sulfur atoms and then back to smaller groups of atoms. The color changes are due to the increasing number of unshared electrons.

5. Endothermic. Monoclinic sulfur is stable only at a higher temperature than rhombic.

6. Reduction of sulfuric acid:
$Cu + 4H_3O^+ + SO_4^{2-} \rightarrow Cu^{2+} + 6H_2O + SO_2$
Decomposition of a sulfite:
$SO_3^{2-} + 2H_3O^+ \rightarrow 3H_2O + SO^2$

7. *(a)* The bonding situation in substances whose bond properties cannot be satisfactorily represented by any single formula using the electron-dot notation system and keeping the octet rule
(b)

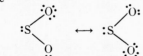

8. Easy to liquefy. Only atm pressure at room temperature is required to liquefy sulfur dioxide.

9. *(a)* NO_2 is slightly soluble; CO_2 is moderately soluble; SO_2 is very soluble. *(b)* The extent of solubility and degree of reactivity are similar.

10. It is used as a preservative, for bleaching, and for preparing paper pulp.

11. *(a)* NO and NO_2 get into the air as combustion products of fuels. NO reacts with oxygen in the air to form NO_2. NO_2 reacts with water vapor in the air to form nitric acid. *(b)* SO_2 is formed by the combustion of sulfur impurities in fuels. SO_2 dissolves in drops of water in the air, and the solution is oxidized by oxygen in the air to sulfuric acid.

12. Student answer

13. It has a high temperature and is also a powerful oxidizing and dehydrating agent.

14. *(a)* $\overset{-2}{H_2S} + \overset{+4}{SO_2} \rightarrow \overset{0}{S} + H_2O$

$2(\overset{-2}{S} \rightarrow \overset{0}{S} + 2e^-)$

$\overset{+4}{S} + 4e^- \rightarrow \overset{0}{S}$

$2H_2S(g) + SO_2(g) \rightarrow 3S(s) + 2H_2O(\ell)$

(b) $\overset{0}{I_2} + \overset{+4}{Na_2SO_3} + H_2O \rightarrow \overset{-1}{NaI} + \overset{+6}{H_2SO_4}$

$2(\overset{0}{I} + e^- \rightarrow \overset{-1}{I^-})$

$\overset{+4}{S} \rightarrow \overset{+6}{S} + 2e^-$

$I_2(aq) + Na_2SO_3(aq) + H_2O(\ell) \rightarrow 2NaI(aq) + H_2SO_4(aq)$

(c) $\overset{0}{Hg} + \overset{+6}{H_2SO_4} \rightarrow \overset{+2}{HSO_4} + \overset{+4}{SO_2} + H_2O$

$\overset{0}{Hg} \rightarrow \overset{+2}{Hg^{2+}} + 2e^-$

$\overset{+6}{S} + 2e^- \rightarrow \overset{+4}{S}$

$Hg(\ell) + 2H_2SO_4(aq) \rightarrow HgSO_4(aq) + SO_2(g) + 2H_2O(\ell)$

(d) $\overset{-2}{Cu_2S} + \overset{0}{O_2} \rightarrow \overset{-2}{Cu_2O} + \overset{+4}{SO_2}$

$\overset{-2}{S^{2-}} \rightarrow \overset{+4}{S} + 6e^-$

$3(\overset{0}{O} + 2e^- \rightarrow \overset{-2}{O^{2-}})$

$Cu_2S + \tfrac{3}{2}O_2 \rightarrow Cu_2O + SO_2$

$2Cu_2S(s) + 3O_2(g) \rightarrow 2Cu_2O(s) + 2SO_2(g)$

(e) $\overset{+4}{SO_2} + H_2O + \overset{+7}{KMnO_4} \rightarrow \overset{+2}{MnSO_4} + \overset{+6}{} \overset{+2}{K_2SO_4} + \overset{+6}{H_2SO_4}$

$5(\overset{+4}{S} \rightarrow \overset{+6}{S} + 2e^-)$

$2(\overset{+7}{Mn} + 5e^- \rightarrow \overset{+2}{Mn^{2+}})$

$5SO_2(g) + 2H_2O(\ell) + 2KMnO_4(aq) \rightarrow 2MnSO_4(aq) + K_2SO_4(aq) + 2H_2SO_4(aq)$

Application Problems

1. $31.972\,07\,u \times 0.950\,18 = 30.379\,u$
$32.971\,46\,u \times 0.007\,50 = 0.247\,u$
$33.967\,86\,u \times 0.042\,15 = 1.432\,u$
$35.967\,09\,u \times 0.000\,17 = \underline{0.006\,u}$
$\hspace{4cm} 32.064\,u,$ atomic mass

Atomic mass $= 32.064$

2. From one mole of sulphur, one mole of sulfuric acid may be theoretically produced.

$0.995 \times 50.0 \text{ kg S} \times \dfrac{10^3 \text{g S}}{1 \text{ Kg S}} \times \dfrac{1 \text{ mol S}}{32.1 \text{ S}} \times \dfrac{1 \text{ mol } H_2SO_4}{1 \text{ mol S}} \times$

$\dfrac{98.1 \text{ g } H_2SO_4}{1 \text{ mol } H_2SO_4} \times \dfrac{1 \text{ kg } H_2SO_4}{10^3 \text{ g } H_2SO_4} = 152 \text{ kg } H_2SO_4$

3. *(a)* $Fe + H_2SO_4 \rightarrow FeSO_4 + H_2$

$70.0 \text{ kg Fe} \times \dfrac{10^3 \text{g } FeSO_4}{1 \text{ g } FeSO_4} \times \dfrac{1 \text{ mol Fe}}{55.8 \text{ kg Fe}} \times \dfrac{1 \text{ mol } FeSO_4}{1 \text{ kg mol Fe}} \times$

$\dfrac{151.9 \text{ g } FeSO_4}{1 \text{ mol } FeSO_4} \times \dfrac{1 \text{ Kg Fe } SO_4}{10^3 \text{g Fe } SO_4} = 191 \text{ kg } FeSO_4$

(b) $70.0 \text{ kg Fe} \times \dfrac{10^3 \text{g Fe}}{1 \text{ kg Fe}} \times \dfrac{1 \text{ mol Fe}}{55.8 \text{ Fe}} \times \dfrac{1 \text{ mol } H_2SO_4}{1 \text{ mol Fe}} \times$

$\dfrac{98.1 \text{ g } H_2SO_4}{1 \text{ mol } H_2SO_4} \times \dfrac{1 \text{ Kg } H_2SO_4}{10^3 \text{g } H_2SO_4} \times \dfrac{100\% \text{ } H_2SO_4}{95\% \text{ } H_2SO_4} = 130. \text{ kg acid}$

4. $Cu + 2H_2SO_4 \rightarrow CuSO_4 + 2H_2O + SO_2$

$100. \text{ g } H_2SO_4 \times \dfrac{1 \text{ mol } H_2SO_4}{98.1 \text{ g } H_2SO_4} = 1.02 \text{ mol } H_2SO_4$

$100. \text{ g Cu} \times \dfrac{1 \text{ mol Cu}}{63.5 \text{ g Cu}} = 1.57 \text{ mol Cu}$

The copper is in excess.

$100. \text{ g } H_2SO_4 \times \dfrac{1 \text{ mol } H_2SO_4}{98.1 \text{ g } H_2SO_4} \times \dfrac{1 \text{ mol } SO_2}{2 \text{ mol } H_2SO_4} \times$

$\dfrac{22.4 \text{ L}}{1 \text{ mol}} = 11.4 \text{ L } SO_2$

5. $4FeS_2 + 11O_2 \rightarrow 2Fe_2O_3 + 8SO_2(g)$

$1500. \text{ kg } FeS_2 \times \dfrac{10^3 \text{ g}}{1 \text{ kg}} \times \dfrac{1 \text{ mol } FeS_2}{120.0 \text{ g } FeS_2} \times \dfrac{8 \text{ mol } SO_2}{4 \text{ mol } FeS_2} \times$

$\dfrac{22.4 \text{ L}}{1 \text{ mol}} \times \dfrac{298 \text{ K}}{273 \text{ K}} \times \dfrac{760. \text{ mm}}{740. \text{ mm}} = 6.28 \times 10^5 \text{ L } SO_2$

6. $[H_3O^+] = \text{antilog} (-pH) = \text{antilog} (-2.1) = 7.9 \times 10^{-3} \text{ mol/L}$

$[OH^-] = \dfrac{1.0 \times 10^{-14} \text{ mol}^2/\text{L}^2}{7.9 \times 10^{-3} \text{ mol/L}} = 1.3 \times 10^{-12} \text{ mol/L}$

7. $[Ba^{2+}][SO_4^{2-}] = 1.1 \times 10^{-10}$
$0.01[SO_4^{2-}] = 1.1 \times 10^{-10}$
$[SO_4^{2-}] = 1.1 \times 10^{-8}$

Teacher's Notes

Chapter 27 Sulfur and Its Compounds

INTRODUCTION
Sulfur is one of the most important nonmetallic elements. In this chapter, we will study its occurrence, its physical and chemical properties, and some of its uses. We will also discuss the structure, formation, and uses of two of the most important sulfur compounds, sulfur dioxide and sulfuric acid.

LOOKING AHEAD
As you study this chapter, examine carefully the different allotropes of sulfur. Pay particular attention to the various physical forms of sulfur and interpret its properties in terms of its structure.

SECTION PREVIEW
28.1 Elemental Sulfur
28.2 Important Compounds of Sulfur

Of all sulfur produced, 90 percent is used by factories producing sulfuric acid, the leading industrial chemical in the United States.

27.1 Elemental Sulfur

Sulfur is one of the elements known since ancient times. It occurs in nature as the free element or combined with other elements in sulfides and sulfates. It is commonly found in the form of pale, yellow crystals. In the United States, huge deposits of nearly pure sulfur occur between 150 and 600 meters underground in Texas and Louisiana, near the Gulf of Mexico.

The Production of Sulfur

The sulfur beds in Texas and Louisiana are as much as 60 meters thick. A layer of quicksand between the surface of the ground and the sulfur makes it difficult to sink a shaft and mine the sulfur by ordinary methods.

The American chemist Herman Frasch (1851–1914) developed a method of obtaining the sulfur without sinking a shaft. *The **Frasch process** is a sulfur-mining process that uses a complex system of pipes to pump superheated water (170°C) under pressure into the sulfur deposit.* **Superheated water** *is water that has been heated under pressure to above the boiling point without conversion to steam.* The hot water then melts the sulfur, which is forced to the surface by compressed air.

Some of the elemental sulfur produced in the United States is now obtained from *a compound of iron and sulfur called* **pyrite** *(FeS₂)* and from the purification of coke oven gas, smelter gases, petroleum, and natural gas.

Physical Properties of Elemental Sulfur

Common sulfur is a yellow, odorless solid. It is practically insoluble in water and is twice as dense as water. It dissolves readily in carbon disulfide and less readily in carbon tetrachloride.

Allotropes of sulfur Sulfur exists in several different solid and liquid allotropic forms. These forms are produced by different arrangements of groups of sulfur atoms.

An ancient name for sulfur is "brimstone," which means burning stone. Sulfur burns with a bright blue flame, forming the choking compound sulfur dioxide. It is often found in burning volcanos. This led people to think that the fires of hell must be like burning sulfur, brimstone.

Superheated water is steam that has been heated above its saturation temperature.

Figure 27-1 Sulfur is found in many forms in nature.

Most people don't realize that sulfur itself is odorless. It is the compounds of sulfur that have the bad odors.

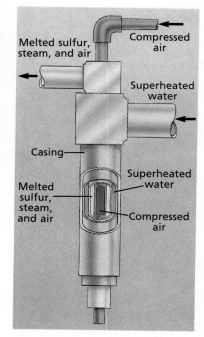

Figure 27-2 The Frasch sulfur-mining process uses a system of nested pipes.

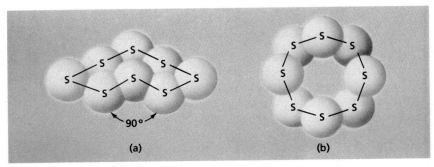

(a)   (b)

1. *Rhombic sulfur (S_α).* **Rhombic sulfur** *is the form of solid sulfur that is stable at ordinary temperatures.* Electron diffraction measurements and x-ray analysis give evidence that a crystal cell consists of sixteen S_8 molecules. These molecules consist of eight sulfur atoms bonded to each other by single covalent bonds in a puckered-ring structure, as shown in Figure 27-3. Rhombic crystals are shown in Figure 27-4.

2. *Monoclinic sulfur (S_β).* **Monoclinic sulfur** *is a second form of solid sulfur that is less stable than rhombic sulfur.* In this form, the eight-membered rings of sulfur are arranged in a monoclinic crystal pattern as shown in Figure 27-5. This allotropic form is prepared by first melting some sulfur in a crucible at as low a temperature as possible. Next the sulfur is allowed to cool slowly, and crystals form. When such crystals cool below 95°C, they gradually change back into the rhombic form. The density of monoclinic sulfur is 1.96 g/cm³.

3. *λ-sulfur (Lambda-sulfur).* **Lambda-sulfur** *is the liquid allotropic form of sulfur produced at temperatures just above the melting point of 112.8 °C.* It flows easily and has a straw-yellow color. At this temperature, the eight-membered rings of sulfur atoms slip and roll over one another easily, giving this form of sulfur its fluidity.

4. *μ-sulfur (Mu-sulfur).* **Mu-sulfur** *is an allotropic form of sulfur that forms at temperatures above 160°C.* When sulfur is heated to approximately this temperature, it darkens to a reddish liquid and then turns almost black. The melted sulfur becomes so viscous that it does not flow. The heat gives enough energy to the sulfur atoms to break some of the eight-membered rings, forming chains of sulfur atoms as shown in Figure 27-6. When a ring of sulfur atoms breaks open, the sulfur atoms on either side of the break are each left with an unshared electron. These sulfur atoms form bonds with similar sulfur atoms from other open rings. In this way, long chains of μ-sulfur are formed. The dark color of μ-sulfur is due to the greater absorption of light by electrons from the broken ring structure.

The high viscosity of μ-sulfur is explained by the tangling of the sulfur-atom chains. As the temperature is raised above about 190°C, however, these chains break up into smaller groups of atoms. The mass then flows more easily. The color becomes still darker because more free electrons, which absorb more light, are produced by the breaking of chains.

Sulfur vapor, produced when sulfur boils at 444.6°C, also consists of S_8 molecules. If sulfur vapor is heated to a higher temperature, these molecules gradually dissociate into S_6 and then S_2 molecules. Monatomic molecules of sulfur are produced at very high temperatures.

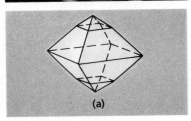

(a)

Figure 27-4 Rhombic crystals of sulfur are stable at room temperature.

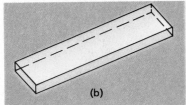

(b)

Figure 27-5 Notice the long, needlelike shape of monoclinic sulfur crystals.

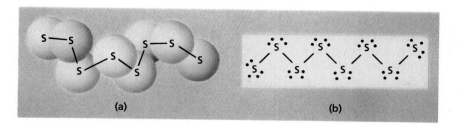
(a)   (b)

   5. Amorphous sulfur. **Amorphous sulfur** *is a rubbery, dark brown or black mass made by pouring boiling sulfur into cold water.* When sulfur is boiling, the long tangled chains of μ-sulfur have largely broken down, and the molten sulfur is fluid again. Eight-membered rings of sulfur atoms and chains of sulfur atoms are in equilibrium. When this boiling mixture is suddenly cooled, the chains of μ-sulfur have no time to reform into rings. Instead, amorphous sulfur is produced.

   A mass of amorphous sulfur soon loses its elasticity and becomes hard and brittle. For amorphous sulfur at room temperature, the successive changes into allotropic forms occur in reverse order. Eventually the amorphous sulfur again acquires the S_8 arrangement of stable rhombic sulfur.

Remind students that amorpohous means without crystalline form.

Chemical Properties of Elemental Sulfur

At room temperature, sulfur is not very active chemically. When heated, sulfur combines with oxygen to produce sulfur dioxide.

$$S(s) + O_2(g) \rightarrow SO_2(g)$$

Traces of sulfur trioxide (SO_3) also form when sulfur burns in the air. Sulfur can be made to combine with nonmetals such as hydrogen, carbon, and

TABLE 27-1	SOME PROPERTIES OF ELEMENTAL SULFUR
Property	**Sulfur**
Symbol	S
Atomic number	16
Atomic mass	32.064
Electron configuration	[Ne]$3s^23p^4$
Atomic radius (pm)	104
Ionic radius of S^{2-} (pm)	182
Oxidation numbers	−2 to +6
Electronegativity	2.5
Ionization energy (kJ/mol)	1004
Melting point (°C)	112.8
Boiling point (°C)	444.6
Density (g/cm³)	2.07[a]
	1.96[b]
Color of element	yellow

[a] density for rhombic sulfur
[b] density for monoclinic sulfur

Figure 27-7 Molten sulfur (bottom) becomes viscous and dark, but reverts to rhombic form (top) upon cooling.

chlorine. Such compounds are formed with difficulty and are not very stable however. Hydrogen sulfide (H_2S) has a structure resembling that of water, but unlike water, is a gas at room temperature. It is a deadly poison. Fortunately, its terrible odor can be detected at extremely low concentrations (see the Chemistry Notebook, "The Smell of Sulfur Compounds," in Section 27.2).

The formulas SO_3, SO_2, and H_2S indicate that sulfur can have oxidation numbers of +6 or +4 when combined with oxygen and −2 when combined with hydrogen. Electron-dot formulas for these compounds are shown below. The actual molecules of sulfur trioxide and sulfur dioxide are resonance hybrids of these possible structures shown.

resonance structures of
sulfur trioxide

resonance structures of   hydrogen
sulfur dioxide   sulfide

At least ten other oxides of sulfur are known, but only sulfur trioxide (SO_3) and sulfur dioxide (SO_2) are commercially important. The preparation and uses of sulfur dioxide are discussed in Section 27.2.

Sulfur combines directly with all metals except gold and platinum. Powdered zinc and sulfur combine vigorously, as shown in Figure 27-8. The heat produced when iron filings and sulfur unite causes the whole mass to glow red hot. Copper unites with the vapor of boiling sulfur to form CuS.

Uses of Elemental Sulfur

About 90% of the sulfur produced is used to make sulfuric acid. Sulfur is also used in making sulfur dioxide, carbon disulfide, and other sulfur compounds. Matches, fireworks, and black gunpowder all contain either sulfur or sulfur

Figure 27-8 Zinc dust and sulfur react dramatically in the laboratory.

Chemistry Notebook

Medicines Made from Sulfur

The medicinal properties of sulfur have been known for centuries. Sulfur can be used as an antiseptic and to kill parasites such as the mites that burrow in the skin of cattle, sheep, and humans to cause the skin disease called scabies.

Large deposits of sulfur in Sicily were mined by Roman slaves and prisoners. The workers seldom lived long because the mines were quite dangerous. The sulfur could be ignited easily, causing fires and the suffocating fumes of sulfur dioxide. The sulfur mines also contained hydrogen sulfide, a flammable, poisonous gas.

Three forms of pure sulfur are used in pharmacy and veterinary medicine. Sulfur that has been purified by sublimation is known as flowers of sulfur because the sulfur forms flowerlike pat-

terns. Washed sulfur is made by treating sublimed sulfur with ammonia to dissolve impurities and to remove traces of acid. Precipitated sulfur, also known as milk of sulfur, is made by boiling sulfur with lime and precipitating the sulfur.

Sulfur is also an important ingredient in a group of sulfonamide compounds known as sulfa drugs. Sulfa drugs, discovered in the late 1930s, were the first drugs to be proven safe and effective against many common bacterial infections.

compounds. Sulfur is used in the preparation of certain dyes, medicines, and **fungicides**, *compounds used to kill fungi.*

Sulfur is also used in making practical forms of rubber. Untreated rubber lacks strength, becomes brittle at cool temperatures, and tacky at warm temperatures. In a process called *vulcanization*, rubber is heated with sulfur causing the polymer chains of rubber to become cross linked. This makes the rubber stronger and flexible through a wider temperature range.

Section Review

1. In what form does sulfur occur in nature?
2. Where is sulfur found in the United States?
3. What are some of the physical and chemical properties of sulfur?
4. Name some of the uses of elemental sulfur.

1. In huge deposits underground. Occurs in combined form with other elements in sulfides and sulfates
2. In Texas and Louisiana, near the Gulf of Mexico
3. Accept all answers that are given in the subsections "Physical Properties of Elemental Sulfur" and "Chemical Properties of Elemental Sulfur."
4. In the manufacture of many sulfur compounds, including sulfuric acid, sulfur dioxide, and carbon disulfide. Also used in matches, fireworks, black gunpowder, medicines, and fungicides

27.2 Important Compounds of Sulfur

SECTION OBJECTIVES

- Cite the common occurrences and preparation of sulfur dioxide.
- Describe the physical properties, chemical properties, and uses of sulfur dioxide.
- Identify the contact process as the major method for the preparation of sulfuric acid.
- Describe the physical properties, chemical properties, and uses of sulfuric acid.

Because it is used in so many industrial processes, sulfur is essential to the chemical industry and our national economy. The variety and complexity of sulfur compounds is quite large. Several sulfur compounds are shown in Table 27-2. Two of the most important sulfur compounds are sulfur dioxide and sulfuric acid.

Sulfur Dioxide

Sulfur dioxide is most harmful to individuals who have respiratory problems and to older people and children. High concentrations of sulfur dioxide in areas such as London, England, have been known to cause many deaths.

Sulfur dioxide is a colorless gas sometimes found in large amounts in the atmosphere of densely populated urban areas. Although sulfur dioxide occurs in some volcanic gases and in some mineral waters, its primary source is the burning of coal and fuel oil that contains sulfur as an impurity. As these fuels are burned, the sulfur combines with oxygen to form sulfur dioxide. Coal and fuel oil of low sulfur content produce less sulfur dioxide air pollution than similar fuels of high sulfur content. In the United States and many other countries, environmental protection laws limit the amount of sulfur that can be emitted by coal-burning power plants.

The roasting of sulfide ores converts the sulfur in the ore to sulfur dioxide. In modern smelting plants, this sulfur dioxide is converted to sulfuric acid.

Preparation of sulfur dioxide   *1. By burning sulfur.* The simplest way to prepare sulfur dioxide is to burn sulfur in air or in pure oxygen.

$$S(s) + O_2(g) \rightarrow SO_2(g)$$

TABLE 27-2   COMMON COMPOUNDS OF SULFUR AND SOME OF THEIR PROPERTIES

Oxidation Number of Sulfur	Compound	Formula	Structure	Density	Melting Point (°C)	Boiling Point (°C)
−2	Hydrogen sulfide	H_2S		1.539 g/L	−85.5	−60.7
−4	Sulfur dioxide	SO_2		2.927 g/L	−72.7	−10
+6	Sulfur trioxide	SO_3		1.97 g/mL	16.83	44.8
+6	Sulfuric acid	H_2SO_4		1.841 g/mL	10.36	33

2. By roasting sulfides. Huge quantities of sulfur dioxide are produced by *heating sulfide ores to convert the sulfides to oxides, a process known as* **roasting**. The roasting of zinc sulfide ore is typical.

$$2ZnS(s) + 3O_2(g) \rightarrow 2ZnO(s) + 2SO_2(g)$$

Sulfur dioxide is a by-product in this operation.

3. By the reduction of sulfuric acid. In one laboratory method of preparing this gas, copper is heated with concentrated sulfuric acid, as shown in Figure 27-9.

$$Cu(s) + 2H_2SO_4(aq) \rightarrow CuSO_4(aq) + 2H_2O(\ell) + SO_2(g)$$

4. By the decomposition of sulfites. In this second laboratory method, sulfur dioxide is formed by the action of a strong acid on a sulfite, as shown in Figure 28-11.

$$Na_2SO_3(aq) + H_2SO_4(aq) \rightarrow Na_2SO_4(aq) + H_2O(\ell) + SO_2(g)$$

Physical properties of sulfur dioxide Pure sulfur dioxide is a colorless gas with a suffocating, choking odor. It is more than twice as dense as air and is very soluble in water. It is one of the easiest gases to liquefy, becoming liquid at room temperature under a pressure of about 3 atmospheres.

Chemical properties of sulfur dioxide *1. It is an acid anhydride.* Sulfur dioxide dissolves in water and also reacts with the water.

$$SO_2(g) + 2H_2O(\ell) \rightleftharpoons H_3O^+(aq) + HSO_3^-(aq)$$
$$HSO_3^-(aq) + H_2O(\ell) \rightleftharpoons H_3O^+(aq) + SO_3^{2-}(aq)$$

These reactions partly account for the high solubility of sulfur dioxide in water. The solution is commonly called "sulfurous acid" although no significant amount of H_2SO_3 exists in the solution.

2. It is a stable gas. Sulfur dioxide does not burn. With a suitable catalyst and at a high temperature, it can be oxidized to sulfur trioxide.

$$2SO_2(g) + O_2(g) \rightleftharpoons 2SO_3(g)$$

Uses for sulfur dioxide *1. For making sulfuric acid.* In the chemical industry, great quantities of sulfur dioxide are oxidized to sulfur trioxide. The sulfur trioxide is then combined with water, forming sulfuric acid.

2. As a preservative. Dried fruits such as apricots and prunes are treated with sulfur dioxide, which acts as a preservative. Sulfiting agents are also used to prolong the fresh appearance of salad bar fruits and vegetables. Some individuals are extremely allergic to these sulfur compounds and must avoid them.

3. For bleaching. The weakly acid water solution of sulfur dioxide does not harm the fibers of wool, silk, straw, or paper. It can thus be used to bleach these materials.

4. In preparing paper pulp. Sulfur dioxide in water reacts with limestone to form calcium hydrogen sulfite, $Ca(HSO_3)_2$. Wood chips are heated in calcium–hydrogen sulfite solution as a first step in making paper (Figure 27-10). The hot solution dissolves the lignin, which binds the cellulose fibers of wood together. The cellulose fibers are left unchanged and are processed to form paper.

Figure 27-9 The reduction of concentrated sulfuric acid by copper produces sulfur dioxide.

Laboratory methods of production of chemicals are not always the most economical, but are easiest to carry out.

The SO_3 is dissolved in H_2SO_4, rather than in water because it dissolves much more quickly in H_2SO_4.

Figure 27-10 Sulfur dioxide is used to make paper pulp.

Chemistry Notebook

The Smell of Sulfur Compounds

Nearly every compound of sulfur has a distinctive, unpleasant smell. Everyone is familiar with the pungent, nauseating odor of rotten or overcooked eggs. The culprit responsible for this memorable smell is hydrogen sulfide (H_2S).

Idling in traffic with the car windows open is another instance in which we are aware of hydrogen sulfide, the production of which is catalyzed by the platinum in the catalytic converter. Not only is H_2S repugnant, it is highly toxic. Luckily, our noses can detect it at low concentrations, so we can escape before the danger level is exceeded. Unfortunately, the phenomenon of smell fatigue is known to have caused the deaths of several researchers working on sulfur chemistry. They work "just another minute," then no longer smell the gas

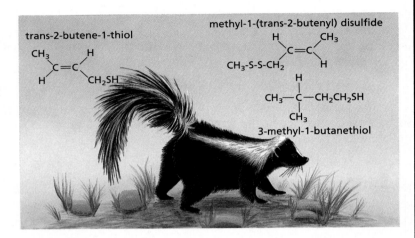

and mistakenly think that it is no longer present.

Two odorous groups of organic sulfur compounds are mercaptans, which contain an —SH group, and sulfides. Our sense of smell can detect these compounds in extremely low concentrations. Dimethyl sulfide, $(CH_3)_2S$, is often used as an additive to give a distinct odor to natural gas because of its chemical stability. Even

smellier and more often used is *t*-butyl mercaptan (systematically named 2-methyl-2-propanethiol). It can be detected at levels of one molecule in a billion molecules of air or natural gas. Both of these organic sulfur compounds have an extremely unpleasant skunk-like odor. In fact, the skunk's defensive spray is a 4:3:3 mixture of the three sulfurous compounds shown above.

Sulfur-containing compounds are also the cause of the odor and taste in onions, garlic, and horseradish.

Acid precipitation   **Acid precipitation** *forms when oxides of sulfur or nitrogen enter the air and dissolve in airborne water vapor, producing acids* that form as a result lower the pH of precipitation below the average normal pH of 5.6. Sulfur dioxide is produced when sulfur-containing coal is burned or when sulfide ores are smelted. Although the waste gases of newer industrial plants are treated to remove sulfur dioxide, millions of metric tons of it still enter the atmosphere annually. Sulfur dioxide reacts with oxygen and water vapor in the air to form sulfuric acid.

Oxides of nitrogen, principally nitrogen monoxide (NO) and nitrogen dioxide (NO_2) get into the air as combustion products of automobile engines and some generating processes for electricity. Nitrogen monoxide reacts with oxygen in the air to form nitrogen dioxide, which then reacts with water vapor in the air to form nitric acid (HNO_3).

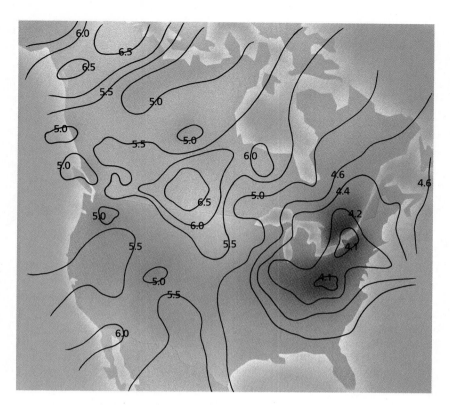

Figure 27-11 This map shows the variation in pH of the precipitation in North America. The areas of lowest precipitation pH correspond to areas within and downwind of heavy industrialization and urbanization.

Acid precipitation and its effects are also discussed in "Acid Precipitation and Lakes," Chemistry Notebook in Section 17.2. Students can test for acid precipitation in the Desktop Investigation, "Testing for Acid Precipitation," in Section 17.3.

Sulfur oxide and nitrogen oxides can be carried many miles through the atmosphere by the prevailing winds. A local acid-precipitation problem becomes widespread as pollutants travel across state and national boundaries. Figure 27-11 shows the pHs of rainfall in North America.

Sulfuric Acid

Fertilizer, steel, and petroleum industries consume huge amounts of sulfuric acid each year. Over 3×10^{10} kg of sulfuric acid, the so-called "king of chemicals," is produced each year in the United States.

Preparation of sulfuric acid Sulfuric acid is produced in the United States today by the **contact process**, *a method in which sulfur dioxide is prepared by burning sulfur or by roasting iron pyrite (FeS₂)*. Impurities that might combine with and ruin the catalyst used in the process are then removed from the sulfur dioxide gas. The purified sulfur dioxide is mixed with air and passed through heated iron pipes containing a catalyst, usually divanadium pentoxide (V_2O_5). This close "contact" of the sulfur dioxide and the catalyst gives the contact process its name. Sulfur dioxide and oxygen from the air are both absorbed on the surface of the catalyst where they react to form sulfur trioxide. The sulfur trioxide is then dissolved in approximately 98% sulfuric acid with which it combines readily to form pyrosulfuric acid ($H_2S_2O_7$).

$$SO_3(g) + H_2SO_4(\ell) \rightarrow H_2S_2O_7(\ell)$$

When diluted with water, pyrosulfuric acid yields sulfuric acid.

$$H_2S_2O_7(\ell) + H_2O(\ell) \rightarrow 2H_2SO_4(\ell)$$

Figure 27-12 These trees in the high elevations of the Great Smoky Mountains have been damaged by acid precipitation.

Sulfur and Its Compounds **773**

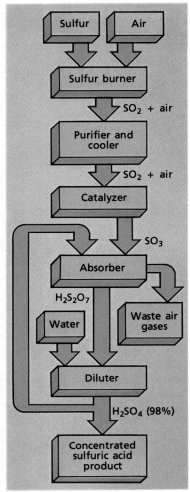

Figure 27-13 This diagram summarizes the contact process for manufacturing sulfuric acid.

The acid converts the $Ca_3(PO_4)_2$ to the more soluble $Ca(H_2PO_4)_2$.

CAUTION:
Sulfuric acid burns the flesh severely. The burns are the result of its dehydrating action on the skin. Handling sulfuric acid requires constant care to prevent its contact with the skin.

Very pure, highly concentrated sulfuric acid is produced by the contact process, diagrammed in Figure 27-13.

In terms of tonnage, sulfuric acid ranks first among the substances produced by the United States chemical industry. Each year, over 37 million metric tons are prepared.

Chemical properties of sulfuric acid   The properties of sulfuric acid make it useful in a variety of ways:

1. As an acid. Sulfuric acid is a diprotic acid. It ionizes in dilute water solution in two stages:

$$H_2SO_4 + H_2O \rightleftarrows H_3O^+ + HSO_4^-$$
$$HSO_4^- + H_2O \rightleftarrows H_3O^+ + SO_4^{2-}$$
$$K_a = 1.26 \times 10^{-2}$$

At 25°C, 0.1-M H_2SO_4 is completely ionized in the first stage and about 10% ionized in the second stage. Sulfuric acid reacts with hydroxides to form hydrogen sulfates and sulfates. It also reacts with metals and with the oxides of metals. Dilute sulfuric acid is more highly ionized than cold, concentrated sulfuric acid. Thus the dilute acid reacts more vigorously than the cold, concentrated acid with metals above hydrogen in the oxidizing and reducing agents series.

2. As an oxidizing agent. Hot, concentrated sulfuric acid is a vigorous oxidizing agent. The sulfur is reduced from the +6 oxidation state to the +4 or −2 oxidation state. The extent of the reduction depends on the strength of the acid and on the reducing agent used.

3. As a dehydrating agent. The strong attraction of sulfuric acid for water makes it a very good **dehydrating agent.** Gases that do not react with sulfuric acid can be dried by bubbling them through the concentrated acid. Sulfuric acid is such an active dehydrating agent that it takes hydrogen and oxygen directly from certain substances in the proportion in which they are found in water. For example, it dehydrates cellulose, $(C_6H_{10}O_5)_n$, and sucrose, $(C_{12}H_{22}O_{11})$ in this manner, leaving the carbon uncombined. The equation for this process with sucrose is

$$C_{12}H_{22}O_{11} + 11H_2SO_4 \rightarrow 12C + 11H_2SO_4 \cdot H_2O$$

When sulfuric acid is poured on sucrose, the sucrose rapidly decomposes and turns into a foamy, black carbon mass.

In some commercial chemical processes, water is formed as a byproduct. The production of nitroglycerin, $C_3H_5(NO_3)_3$, is such a process.

$$C_3H_5(OH)_3 + 3HNO_3 \rightarrow C_3H_5(NO_3)_3 + 3H_2O$$

Sulfuric acid can be used as a dehydrating agent for the reaction above, absorbing the water as fast as it is formed and maintaining the reaction rate.

Uses of sulfuric acid   About 60% of the sulfuric acid produced in the United States is used to make **superphosphate** *fertilizer, a mixture of phosphate compounds that have been treated with sulfuric acid to increase their solubility.*

Sulfuric acid is also used in making phosphoric and other acids, various sulfates, and many chemicals. The iron and steel industries use sulfuric acid to remove oxides from the surface of iron or steel before the metal is plated or coated with an enamel. In petroleum refining, sulfuric acid is used to remove certain organic impurities.

Sulfuric acid serves as a dehydrating agent in the production of smokeless powder and nitroglycerin. It is used in making photographic film, nitrocellulose plastics, and rayon, paints and pigments, cellophane, and thousands of other commercial articles. The electrolyte in lead storage batteries is dilute sulfuric acid.

Section Review

1. Cite two common occurrences of sulfur dioxide.
2. List several physical and chemical properties of sulfur dioxide.
3. What are some of the uses of sulfur dioxide?
4. Write a balanced chemical equation for the preparation of sulfuric acid.
5. List some physical and chemical properties of sulfuric acid.
6. Identify several uses of sulfuric acid.

Chapter Summary

- Sulfur occurs in nature as a free element or combined with other elements in sulfides and sulfates. The Frasch process is used to mine sulfur in Texas and Louisiana. It uses a system of pipes to pump superheated water under pressure into the sulfur deposit. Molten sulfur is then forced to the surface by compressed air. Elemental sulfur is also obtained from pyrite and from the purification of coke-oven gas, smelter gases, petroleum, and natural gas.

- Common sulfur is a yellow, odorless solid that is practically insoluble in water. It is soluble in carbon disulfide and in carbon tetrachloride. Many sulfur compounds have unpleasant odors.

- Sulfur exists in several allotropic forms. The solid allotropes are rhombic, monoclinic, and amorphous sulfur. The liquid allotropes are λ-sulfur (Lambda-sulfur) and μ-sulfur (Mu-sulfur). Mu-sulfur is a viscous, dark brown form of molten sulfur that does not flow.

- Sulfur dioxide is a dense, suffocating gas that is easily liquefied and extremely soluble in water. Water solutions of sulfur dioxide are weakly acidic.

- Sulfur dioxide is used for making sulfuric acid and sulfites. It is also used for bleaching, as a preservative, and in preparing paper pulp.

- Oxides of sulfur and nitrogen react with water vapor in the air to form sulfuric acid and nitric acid. These airborne acids can cause acid precipitation.

- Sulfuric acid is made from sulfur dioxide, oxygen, and water by the contact process. It is a dense, oily, colorless liquid that is a strong acid and must be handled carefully. In dilute form, it acts as an acid. When hot and concentrated, it is a vigorous oxidizing agent. It is also a good dehydrating agent.

- Sulfuric acid is one of the most important chemicals used in industry. More tons of sulfuric acid are used each year than any other chemical. It is used in fertilizer manufacture, petroleum refining, explosives production, the making of lead storage batteries and a great many other products.

1. Volcanic gases, mineral waters
2. Colorless gas, choking odor, twice as dense as air, very soluble in water, easily liquefiable; reacts with water, oxidizes to sulfur trioxide at high temperature and in the presence of a suitable catalyst
3. Accept all answers given under the heading "Uses for sulfur dioxide."
4. $H_2S_2O_7(l) + H_2O(l) \rightarrow 2H_2SO_4(l)$
5. Accept answers given under the headings "Physical properties of sulfuric acid" and "Chemical properties of sulfuric acid."
6. Accept all uses given under the heading "Uses of sulfuric acid."

Chapter 27 *Review*

Sulfur and Its Compounds

Vocabulary

acid precipitation
amorphous sulfur
contact process
dehydrating agent
Frasch process
fungicide
lambda-sulfur

monoclinic sulfur
mu-sulfur
pyrite
rhombic sulfur
roasting
superheated water
superphosphate

Questions

1. In what forms does sulfur occur in nature?
2. What is the location of the deposits of elemental sulfur in the United States?
3. In the Frasch process, what is the purpose of *(a)* the superheated water? *(b)* the compressed air?
4. In addition to Frasch process sulfur, what other sources of sulfur are important in the United States?
5. Describe the odor of *(a)* sulfur *(b)* sulfur dioxide.
6. Define *(a)* pyrite *(b)* viscous *(c)* allotrope *(d)* fluidity *(e)*amorphous.
7. A pupil prepared some nearly black amorphous sulfur in the laboratory. When it was examined the following week, it had become brittle and much lighter in color. Explain.
8. List several uses for elemental sulfur.
9. Sulfur dioxide is an important air pollutant. Where does it come from?
10. Write balanced equations for two commercial methods of preparing sulfur dioxide.
11. *(a)* In the laboratory preparation of sulfur dioxide, what method of collecting the gas is used? *(b)* What properties of sulfur dioxide are considered in making this decision?
12. What is the most important use of sulfur dioxide?
13. Write balanced equations to show the formation from a water solution of sulfur dioxide and sodium hydroxide of *(a)* sodium hydrogen sulfite *(b)* sodium sulfite.

14. *(a)* What is the pH of normal rainwater? *(b)* Why is the pH not 7.0? *(c)* What is acid rain? *(d)* What are the principal causes of acid rain? *(e)* Why is acid rain harmful?
15. *(a)* What process is used for the manufacture of sulfuric acid in the United States? *(b)* How did the process get its name?
16. How is sulfuric acid diluted safely?
17. Give three properties of sulfuric acid.
18. What is the most important use for sulfuric acid in the United States?
19. For what purpose is sulfuric acid used in the iron and steel industries?
20. Why are both nitric acid and sulfuric acid used in making nitroglycerin?

Problems

1. A railroad gondola car has a capacity of 49.4 m^3. How many metric tons of sulfur does it contain when loaded to capacity? The density of sulfur is 2.07 g/cm^2.
2. Common sulfides of Period 3 elements are Na_2S, MgS, Al_2S_3, SiS_2, P_4S_{10}, and SCl_2. What is the percentage of ionic character of the bonds in each of these compounds?
3. Calculate the percentage composition (to three significant figures) of H_2SO_4.
4. A compound of lead and sulfur contains 86.6% lead. Determine the empirical formula of the compound.
5. How many kilograms of sulfur dioxide can be produced by burning 2.5 kg of pure sulfur?
6. A lead smelter processes 450. metric tons of zinc sulfide (ZnS) each day. If no sulfur dioxide is lost, how many kilograms of sulfuric acid could be made in the plant daily?
7. How many grams of sodium sulfite are needed to produce 5.00 L of sulfur dioxide at STP by reaction with sulfuric acid?

Application Questions

1. Why was the Frasch process for obtaining sulfur developed?

2. (a) What is superheated water? (b) Why must the water used in the Frasch process be superheated?

3. (a) Give the molecular formula for rhombic sulfur. (b) Why do you suppose this molecular formula is not usually used in equations?

4. As sulfur is heated from the melting point to the boiling point, it undergoes changes in color and fluidity. Describe these changes and explain why they occur.

5. Is the change from rhombic sulfur to monoclinic sulfur exothermic or endothermic? Explain.

6. Write balanced net ionic equations for two laboratory preparations of sulfur dioxide.

7. (a) Define resonance. (b) Draw electron-dot formulas to demonstrate the resonance structure of sulfur dioxide.

8. Is sulfur dioxide easy or difficult to liquefy? Explain.

9. (a) Using data from Appendix Table 11, compare the solubility in water of CO_2, NO_2, and SO_2. (b) What do these solubilities indicate about the extent to which these oxides react with water?

10. What uses are there for sulfur dioxide other than the manufacture of sulfuric acid?

11. (a) How is the nitric acid in acid precipitation formed? (b) How is the sulfuric acid formed?

12. What is being done now about the environmental problem of acid precipitation?

13. Give two reasons why boiling concentrated sulfuric acid burns the flesh.

14. Balance the following oxidation-reduction equations: (a) $H_2S(g) + SO_2(g) \rightarrow S(s) + H_2O(\ell)$ (b) $I_2(aq) + Na_2SO_3(aq) + H_2O(\ell) \rightarrow NaI(aq) + H_2SO_4(aq)$ (c) $Hg(\ell) + H_2SO_4(aq) \rightarrow HgSO_4(aq) + SO_2(g) + H_2O(\ell)$ (d) $Cu_2S(s) + O_2(g) \rightarrow Cu_2O(s) + SO_2(g)$ (e) $SO_2(g) + H_2O(\ell) + KMnO_4(aq) \rightarrow MnSO_4(aq) + K_2SO_4(aq) + H_2SO_4(aq)$.

Application Problems

1. Natural sulfur consists of 95.018% S-32 (31.97207 u), 0.750% S-33 (32.97146 u), 4.215% S-34 (33.96786 u), and 0.017% S-36 (35.96709 u). Calculate the atomic mass of sulfur.

2. How many kilograms of sulfuric acid can be prepared from 50.0 kg of sulfur that is 99.5% pure?

3. (a) If 70.0 kg of scrap iron is added to a large vat of dilute sulfuric acid, how many kilograms of iron(II) sulfate can be produced? (b) How many kilograms of 95% sulfuric acid are required?

4. A laboratory technician has on hand 100 g of copper but only 100 g of H_2SO_4. How many liters of sulfur dioxide at STP can the technician prepare?

5. How many liters of sulfur dioxide at 25°C and 740 mm pressure can be produced by roasting 1500 kg of iron pyrite, FeS_2?

6. A 0.01-N solution of sulfuric acid has a pH of 2.1. Calculate $[H_3O^+]$ and $[OH^-]$.

7. For barium sulfate, $K_{sp} = 1.1 \times 10^{-10}$. If a barium–chloride solution is 0.01 M, what is the smallest sulfate ion concentration that can be detected by precipitation?

Enrichment

1. Research London smog and Los Angeles smog in the library. Compare the histories of the air pollution problems of these two cities and report your findings to the class.

2. Read the article entitled "Air: An Atmosphere of Uncertainty" by Noel Grove in the April 1987 *National Geographic* magazine. Summarize the article for the class.

3. Describe recent research efforts by chemists and chemical engineers to reduce the amount of sulfur dioxide released during the combustion of coal, fuel oil, and gasoline.

4. Could sulfuric acid be produced from the sulfur dioxide obtained from the gases released by coal combustion at electric power plants? Consult reference materials to learn about the feasibility of such a possibility. Find out if any electric power plants are currently equipped to produce sulfuric acid in this manner.

5. Contact state agencies, universities, or library references in your area to learn if your area has acid precipitation. What is the approximate pH of precipitation in your community? Can you identify the nearest sources of sulfur oxides or nitrogen oxides in your area?

6. Consult references in the library and write a report about the history of human uses of sulfur.

7. Ask several veterinarians for the names and purposes of sulfur-containing medicines or drugs that they use to treat animals in their veterinary practice.

Chapter 28 The Halogens

Chapter Planner

28.1 The Halogen Family
General Characteristics of the Halogens
- ▪ Experiment: 37
- Question: 3
- Application Question: 1
- ▪ Question: 1

The Physical Phases of the Halogens
- ▪ Question: 2
- Application Questions: 3, 10
- ▪ Application Question: 11

28.2 Fluorine
Preparation and Properties of Fluorine
- ▪ Questions: 4, 17a
- Application Question: 2

Compounds of Fluorine and Their Uses
- ▪ Question: 5
- ▪ Application Question: 4

28.3 Chlorine
Preparation of Chlorine
- ▪ Question: 6
- Application Question: 6
- ▪ Demonstration: 1
- ▪ Problem: 4
- Application Problem: 3

Physical and Chemical Properties of Chlorine
- ▪ Questions: 7, 8
- Application Question: 7
- ▪ Application Question: 5
- ▪ Problems: 1, 3
- Application Problems: 1, 2, 4

Uses of Chlorine
- ▪ Questions: 9, 17b-c
- ▪ Technology
- ▪ Problem: 5
- Application Questions: 12, 13
- Application Problem: 5

An Important Chlorine Compound: Hydrogen Chloride
- ▪ Questions: 10, 11

28.4 Element Bromine
Preparation of Bromine
- ▪ Question: 12
- Application Question: 8

Physical and Chemical Properties of Bromine
- ▪ Question: 13

Compounds of Bromine and Their Uses
- ▪ Question: 14
- ▪ Problem: 2

28.5 Iodine
Preparation of Iodine
Physical and Chemical Properties of Iodine
- ▪ Question: 16
- Application Question: 9

Uses of Iodine and Its Compounds
- ▪ Desktop Investigation Question: 17d-f

Teaching Strategies

Introduction

The chemistry of the halogens follows very regular trends. As one moves down the column of the halogens, one can predict many of the physical and chemical properties with a fair degree of certainty. This chapter gives students a good opportunity to make predictions based on their knowledge of electron configurations, atomic and ionic sizes, electronegativities, electron affinities, and ionization energies.

28.1 The Halogen Family

An introduction to the halogen family might include discussing a number of important and familiar compounds that contain halogens. Ask students to name as many as they can. Show samples of, for example, salt, iodized salt, tincture of iodine, Teflon plastic, fluoride toothpaste, liquid bleach, and photographic film.

Have students develop a table of properties of the halogens based on their knowledge of periodic trends among the elements. Discuss the difference between the halogen atoms,

halogen molecules, and halide ions. Use electron dot structures to show these differences and to show why the halogens occur as diatomic molecules.

Note the trend in melting and boiling points with increasing atomic number. Explain the effect of increased van der Waals forces in the larger molecules. Note that the intensity of the colors of the halogens also follows a periodic trend.

28.2 Fluorine

Fluorine can react with more elements than any other element. Explain the reactivity of fluorine in terms of its high electronegativity. Emphasize the reasons for the high electronegativity and reactivity. Discuss the problem of finding containers to store fluorine gas.

Most students know that fluorides are added to toothpastes and drinking water to help prevent tooth decay.

Mention other fluorine compounds, such as Teflon and Freons. Students may want to read about Teflon in the Intra-Science feature that appears after this unit. Discuss the problems associated with the interaction of Freons with ozone and the implications of the depletion of the earth's ozone layer. An Intra-Science article about the ozone layer may be found after Unit 4.

28.3 Chlorine

Exhibit chlorides, table salt, iodized salt, salt substitutes, rock salt, halite crystals, and a model of NaCl. Show the equations for the laboratory production of chlorine, identifying the oxidation and reduction steps of each reaction.

Discuss the physical properties of chlorine. Most students will be familiar with the odor of chlorine from the familiar odor of chlorine bleach or a swimming pool. Point out the reactivity of chlorine, which is predicted by its high electron affinity and electronegativity.

Obtain bottles of swimming-pool disinfectants and commercial bleaches and ask students to examine their labels. Students may be interested in finding out the manner in which local water supplies are disinfected.

Explain the equations for the production of hydrogen chloride. Point out that hydrochloric acid is a water solution of the gas hydrogen chloride. Show students a bottle of reagent hydrochloric acid, and ask them to read the contents and the warnings on the label. Discuss the fact that the stomach contains 0.16 M hydrochloric acid to help digest foods. Remind students that chloride compounds are generally very soluble in water.

Technology: PVC Pipes

The ingenious technology involved in this application of polyvinyl chloride pipe should appeal to the students. Discuss the term "polymer" and show that the plastic pipe contains the vinyl chloride grouping, $(-CH_2-CHCl-)$, repeated many times. Other common forms of polyvinyl chloride are in credit cards, phonograph records, and floor tile.

One problem associated with polyvinyl chloride is its disposal. When this plastic is incinerated, hydrogen chloride gas can form, causing an environmental problem.

Demonstration 1: Electrolysis of Salt Solutions

PURPOSE: To show the formation of chlorine and hydrogen gases from the electrolysis of salt water

MATERIALS: U-tube, small 9-volt battery, carbon electrodes (can be pencil lead), wire and wire clips, sodium chloride, phenolphthalein

PROCEDURE: Perform this demonstration in a hood or with good ventilation for only a short time. Dissolve the salt in water, add a few drops of the phenolphthalein solution, and fill the U-tube. Use the wire and clips to attach the carbon electrodes to the battery, and put one in each side of the U-tube. Note the formation of chlorine gas at the anode and hydrogen gas at the cathode (negative terminal). Also note that hydroxide ions form at the cathode and cause the phenolphthalein solution to turn pink. Write the equations for the half-reactions that occur:

$$2Cl^-(aq) \rightarrow Cl_2(g) + 2e^-$$
$$2H_2O + 2e^- \rightarrow H_2(g) + 2OH^-(aq)$$

28.4 Bromine

Bromine is not as familiar to students as fluorine and chlorine. This is understandable since it is far less abundant. Discuss the physical properties of bromine, and show how they continue the trend shown by the halogens.

Desktop Investigation: Silver Halides

Photosensitive substances are interesting to students. You might suggest that students lay a wire gauze over their precipitates when they expose them to light. The grids of the wire gauze will show up as a "photo" on their precipitates.

To make 0.1-M AgNO$_3$ solution, dissolve 17 g of AgNO$_3$ in enough water to make a liter of solution. The AgNO$_3$ solution must be stored in a brown bottle to prevent its decomposition. This is a very sensitive reaction; if necessary you can use less silver nitrate to make your solution and still get good results. To make 0.2-M NaCl, make a liter of solution using 12 g of NaCl. To make the NaBr solution use 21 g of NaBr. To make the NaI solution, use 30 g of NaI.

Be sure to caution students to be careful with silver nitrate. If spilled on hands or clothing it causes stains that do not appear for several hours. The stains may be permanent on clothing and last for days on hands.

ANSWERS TO DISCUSSION QUESTIONS
1. The reactions in Procedure 1 all formed precipitates. However, they were different colors. AgCl is white, AgBr is yellow, and AgI is yellowish green. In Procedure 2, all the silver halides become dark when exposed to light. However, the rate at which the change takes place varies. AgI was the fastest, and AgCl was the slowest.
2. $NaCl + AgNO_3 \rightarrow AgCl(s) + NaNO_3$
   $NaBr + AgNO_3 \rightarrow AgBr(s) + NaNO_3$
   $NaI + AgNO_3 \rightarrow AgI(s) + NaNO_3$

   $$2AgCl \xrightarrow{\text{light}} 2Ag + Cl_2(g)$$
   $$2AgBr \xrightarrow{\text{light}} 2Ag + Br_2$$
   $$2AgI \xrightarrow{\text{light}} 2Ag + I_2$$
3. AgI is used to make fast film, AgCl to make slow film.
4. Silver halides are made by reacting a sodium salt of a halide with silver nitrate. The ionic reaction goes to completion because of the formation of the insoluble silver halide. All silver halides are light-sensitive and turn dark when exposed to light. As silver iodide is the most light-sensitive, it decomposes first. As silver chloride is the least light-sensitive, it breaks down last.

28.5 Iodine

Show crystals of iodine, and comment on their metallic appearance. Discuss the fact that the dividing line between metals and nonmetals comes close to iodine. Predict the properties of iodine based on the periodic trends shown by the other metals.

Show that iodine does not dissolve well in water, but dissolves readily in water that contains iodide ions. Iodine dissolves readily in alcohol, forming a brown solution. Tincture of iodine is an alcohol solution of iodine that is used as a disinfectant. In nonpolar solvents, iodine forms a pink-purple color.

References and Resources

Books and Periodicals

Shakhashiri, Bassam A. "Photochemical Reaction of Hydrogen and Chlorine," p. 121; "Reaction of Sodium and Chlorine," p. 61; "Reaction of Antimony and Chlorine," p. 64; "Reaction of Iron and Chlorine," p. 66; "Reaction of Aluminum and Bromine," p. 68; "Reaction of Zinc and Iodine," p. 49; "Reaction of Metals and Hydrochloric Acid, p. 25. *Chemical Demonstrations*, Vol. 1, University of Wisconsin Press, 1983.

Summerlin, Lee R., and James Ealy. "Preparation of Chlorine Gas from Laundry Bleach," *Chemical Demonstrations, A Sourcebook for Teachers*, Vol. 1, American Chemical Society, 1985, p. 13.

Summerlin, Lee R., Christie L. Borgford, and Julie B. Ealy. "Electrolysis of Potassium Iodide," p. 196; "Colorful Effects of Hydrochloric Acid Dilution," p. 175; "Halogens Compete for Electrons," p. 60. *Chemical Demonstrations, A Sourcebook for Teachers*, Vol. 2, American Chemical Society, 1987.

Audiovisual Resources

"Bromine—Element from the Sea": *CHEM* Study 16 mm film, Ward's Modern Learning Aids. Shows production of elemental bromine

"Chemical Families": *CHEM* Study 16 mm film, Ward's Modern Learning Aids. Shows halogens reacting with alkali metals and with phosphorus, and the reaction of xenon with fluorine

"A Research Problem: Inert(?) Gas Compounds": *CHEM* Study 16 mm film, Ward's Modern Learning Aids. Shows the research to find the first compound of krypton and fluorine

"The Halogens (2nd ed.)": film or video, #1643, Coronet Films and Video.

Computer Software

"The Periodic Table Videodisc: Reactions of the Elements": CAV type videodisc, Journal of Chemical Education Software, 1989. Shows action sequences and still shots of each element in its most common forms; reactions with air, water, acids, bases; and some common applications

Answers and Solutions

Questions

1. *(a)* The name given to the family of elements having seven valence electrons; "salt producer" *(b)* An atom of fluorine, chlorine, bromine, iodine, or astatine *(c)* A diatomic molecule of fluorine, chlorine, bromine, iodine, or astatine *(d)* A singly charged negative ion of fluorine, chlorine, bromine, iodine, or astatine; a fluoride, chloride, bromide, iodide, or astatide ion

2. *(a)* Almost exclusively as ions *(b)* They are active elements with high electronegativities.

3. *(a)* Increase *(b)* Increase *(c)* Decrease *(d)* Decrease *(e)* Decrease *(f)* Increase *(g)* Increase *(h)* Increase *(i)* Remains constant

4. *(a)* Carbon steel *(b)* The iron becomes coated with iron fluoride, which resists the action of fluorine.

5. It is nonflammable, nontoxic, and has a moderately high critical temperature and critical pressure.

6. *(a)* Sodium chloride *(b)* Sodium

7. When inhaled in small quantities, chlorine affects the nose and throat membranes. If inhaled in larger quantities, it may cause death.

8. *(a)* Hydrogen *(b)* When a paraffin candle burns in chlorine, the chlorine unites with the hydrogen of the paraffin and leaves the carbon uncombined.

9. For bleaching, as a disinfectant, and for making compounds

10. *(a)* More dense than air and very soluble in water *(b)* Upward displacement of air

11. *(a)* Cleaning of metals to remove oxide coatings and other forms of tarnish *(b)* It is essential in the process of digestion.

12. *(a)* $2Br^-(aq) + Cl_2(g) \rightarrow 2Cl^-(aq) + Br_2(\ell)$ *(b)* Displacement reaction *(c)* Chlorine is more active than bromine and displaces it from its compounds.

13. Dark, reddish-brown volatile liquid, three times as dense as water, and moderately soluble in water. Readily soluble in carbon tetrachloride, carbon disulfide, and in water solutions of bromides.

14. CH_3Br, soil fumigant; $C_2H_4Br_2$, in antiknock gasoline. $AgBr$, in making photographic film; $CaBr_2$, in solution as a high-density fluid in oil well drilling

15. At atmospheric pressure, iodine sublimes; that is, it passes directly from the solid to the vapor phase.

16. $:\ddot{\underset{..}{I}}. + e^- \rightarrow :\ddot{\underset{..}{I}}:^-$

17. *(a)* A corrosion-resistant alloy composed principally of nickel and copper *(b)* The operation in which color is partially or wholly removed from a colored material *(c)* The compound $Ca(ClO)Cl$ *(d)* Chemical agent that destroys microorganisms but not bacterial spores *(e)* A substance that checks the growth or action of microorganisms on or in the body *(f)* A substance that removes dirt

Problems

1. *(a)* $\dfrac{3.214 \text{ g}}{L} \times 5.00 \text{ L} \times \dfrac{\text{mol}}{71.0 \text{ g}} = 0.226 \text{ mol}$

   (b) $0.226 \text{ mol Cl}_2 \times \dfrac{6.02 \times 10^{23} \text{ molecules}}{1 \text{ mol}} = 1.36 \times 10^{23} \text{ molecules Cl}_2$

2. **Molar mass of $C_2H_4Br_2$** = 187.9 g/mol

$$\frac{159.8 \text{ g Br}}{187.9 \text{ g } C_2H_4Br_2} \times 100\% \ C_2H_4Br_2 = 85.0\%$$

3. *(a)* $Cl_2 + Ca(OH)_2 \rightarrow Ca(ClO)Cl + H_2O$

$$= 200. \text{ g Ca (ClO)Cl} \times \frac{1 \text{ mol Ca(ClO)Cl}}{127.1 \text{ g Ca(ClO)Cl}} \times \frac{1 \text{ mol Ca(OH)}_2}{1 \text{ mol Ca(ClO)Cl}} \times \frac{74.1 \text{ g Ca(OH)}_2}{1 \text{ mol Ca(OH)}_2} = 117 \text{ g Ca(OH)}_2$$

(b) $200. \text{ g Ca (ClO)Cl} \times \dfrac{1 \text{ mol Ca(ClO)Cl}}{127.1 \text{ g Ca(ClO)Cl}} \times \dfrac{1 \text{ mol Cl}_2}{1 \text{ mol Ca(ClO)Cl}} \times \dfrac{71.0 \text{ Cl}_2}{1 \text{ mol Cl}_2} = 112 \text{ g Cl}_2$

4. $2 \text{ NaCl} \xrightarrow{\text{elect}} 2\text{Na} + Cl_2 \text{(g)}$

$$= 375 \text{ g NaCl} \times \frac{1 \text{ mol NaCl}}{58.5 \text{ g NaCl}} \times \frac{1 \text{ mol Cl}_2}{2 \text{ mol NaCl}} \times \frac{22.4 \text{ L}}{1 \text{ mol}} = 71.8 \text{ L Cl}_2$$

5. $\text{Zn} + Cl_2\text{(g)} \rightarrow ZnCl_2$

$$0.500 \text{ L Cl}_2 \times \frac{1 \text{ mol}}{22.4 \text{ L}} \times \frac{1 \text{ mol ZnCl}_2}{1 \text{ mol Cl}_2} \times \frac{136.4 \text{ g ZnCl}_2}{1 \text{ mol ZnCl}_2} = 3.04 \text{ g ZnCl}_2$$

Application Questions

1. *(a)* $2Cl^- + F_2 \rightarrow Cl_2 + 2F^-$ *(b)* No reaction *(c)* $2I^- + F_2 \rightarrow I_2 + 2F^-$ *(d)* $2Br^- + Cl_2 \rightarrow Br_2 + 2Cl^-$ *(e)* No reaction *(f)* No reaction

2. There are no elements more electronegative than fluorine; therefore, fluorine is assigned a negative oxidation state in all compounds.

3. *(a)* The ionic radius increases because of the increase in number of energy levels occupied. *(b)* The first ionization energy decreases because the first electron removed is increasingly farther from the nucleus. *(c)* The melting point becomes higher because of greater dispersion interaction forces between the molecules with the greater number of electrons. *(d)* Electronegativity decreases because as the atoms increase in size, the attraction for an electron becomes less. *(e)* The principal oxidation number remains the same because the outer electron configuration of each halogen is the same.

4. *(a)* and *(b)*
$$\overset{+1-2}{2H_2O} + \overset{+2-1}{2XeF_2} \rightarrow \overset{0}{2Xe} + \overset{0}{O_2} + \overset{+1-1}{4HF}$$

5. Because otherwise they would react with one another

6. $\overset{-1}{4HCl}\text{(aq)} + Ca(\overset{+1}{Cl}O)_2\text{(s)} \rightarrow CaCl_2\text{(aq)} + \overset{0}{2Cl_2}\text{(g)} + 2H_2O(\ell)$
Chloride ions in hydrochloric acid are oxidized. Chlorine in calcium hypochlorite is reduced.

7. *(a)* Greenish-yellow *(b)* The color of chlorine is imparted to it. *(c)* Yes, the color disappears. *(d)* The chlorine reacts with water to form hypochlorous acid and hydrochloric acid, which are colorless.

8. *(a)* $\overset{+4}{MnO_2} + 4H\overset{-1}{Br} \rightarrow \overset{+2}{MnBr_2} + 2H_2O + \overset{0}{Br_2}$
(b) $2Na\overset{-1}{Br} + 2H_2SO_4 + \overset{+4}{MnO_2} \rightarrow Na_2SO_4 + \overset{+2}{MnSO_4} + 2H_2O + \overset{0}{Br_2}$
(c) See *(a)* and *(b)* above. In both reactions, bromine is oxidized and manganese is reduced.
(d) They are essentially the same.

9. *(a)* Bluish black *(b)* Dark brown *(c)* Violet *(d)* Violet

10. *(a)* HF, HCl, HBr, HI *(b)* The polarity of the bonds decreases as one goes from HF to HI because the electronegativity differences become smaller. *(c)* $HF\text{(g)} + H_2O(\ell) \rightleftarrows H_3O^+\text{(aq)} + F^-\text{(aq)}; HCl\text{(g)} + H_2O(\ell) \rightarrow H_3O^+\text{(aq)} + Cl^-\text{(aq)}; HBr\text{(g)} + H_2O(\ell) \rightarrow H_3O^+\text{(aq)} + Br^-\text{(aq)}; HI\text{(g)} + H_2O(\ell) \rightarrow H_3O^+\text{(aq)} + I^-\text{(aq)}$

11. *(a)* HF: The concentration of hydrogen-bonded hydronium ion—fluoride ion pairs is much higher than the concentrations of either separate hydronium ions and fluoride ions or hydrogen fluoride molecules. HCl, HBr, and HI: The concentrations of hydronium and halide ions are much higher than the concentration of hydrogen halide molecules.
(b) Hydrogen-bonded hydronium ion-fluoride ion pairs are more stable than either separated hydronium ions and fluoride ions or hydrogen fluoride molecules. Hydrogen chloride, hydrogen bromide, and hydrogen iodide molecules are less stable than hydronium ions and the corresponding halide ions.

12. Aluminum is not as highly electropositive a metal as sodium and magnesium.

13. Student answer

Application Problems

1. Let x = fractional part due to Cl-35; $1-x$ = fractional part due to Cl-37.
$34.96885 x + 36.96590(1-x) = 35.453$
$x = 0.7576$ or 75.76% Cl-35
$1-0.7576 + 02424$ or 24.24% Cl-37

2. $Cl_2 + 2HBr \rightarrow 2HCl + Br_2$
$\triangle H = \{[2 \text{ mol HCl} \times \triangle H_f\text{(HCl)}] + \triangle H_f\text{(Br}_2)\} - \{\triangle H_f\text{(Cl}_2) + [2 \text{ mol HBr} \times \triangle H_f\text{(HBr)}]\}$
$= \{(2 \text{ mol HCl} \times \dfrac{-92.31 \text{ kJ}}{1 \text{ mol HCl}}) + O\}$
$- \{O + (2 \text{ mol HBr} \times \dfrac{-36.40 \text{ kJ}}{\text{mol HBr}})\} = -111.8 \text{ kJ}$

3. $5 \times 250 \text{ mL} \times \dfrac{1 \text{ L}}{10^3 \text{ mL}} \times \dfrac{273 \text{ K}}{295 \text{ K}} \times \dfrac{730 \text{ mm}}{760 \text{ mm}} = 1.11 \text{ L Cl}_2$ at STP

$4\text{HCl} + \text{MnO}_2 \rightarrow \text{MnCl}_2 + 2\text{H}_2\text{O} + \text{Cl}_2\text{(g)} = 36.5 \text{ g}$

$1.11 \text{ L Cl}_2 \times \dfrac{1 \text{ mol}}{22.4 \text{ L}} \times \dfrac{4 \text{ mol HCl}}{1 \text{ mol Cl}_2} \times \dfrac{36.5 \text{ g HCl}}{1 \text{ mol HCl}} \times \dfrac{100\% \text{ acid}}{38\% \text{ HCl}} \times \dfrac{1 \text{ mol}}{1.20 \text{ g}} = 15.9 \text{ mL conc. HCl}$

$1.11 \text{ L Cl}_2 \times \dfrac{1 \text{ mol}}{22.4 \text{ L}} \times \dfrac{1 \text{ mol MnO}_2}{1 \text{ mol Cl}_2} \times \dfrac{86.9 \text{ g MnO}_2}{1 \text{ mol MnO}_2} = 4.31 \text{ g MnO}_2$

4. $p\text{V} = n\text{RT}$
$n = p\text{V/RT}$
$n = (1 \text{ atm} \times 5.00 \text{ L})/(0.082057 \text{ L atm/mol } K \times 273 \text{ } K)$
$n = 0.223 \text{ mol}$
At STP chlorine does not behave exactly as an ideal gas. This is probably because at STP, chlorine is so near its condensation temperature of $-34.6°\text{C}$.

5. 1 mol CaCl_2 = 111.1 g; 1 mol CaCl_2 yields 3 mol of ions.

$\dfrac{62.5 \text{ g CaCl}_2}{500. \text{ g H}_2\text{O}} \times \dfrac{10_3 \text{ g}}{1 \text{ kg}} \times \dfrac{\text{mol CaCl}_2}{111.1 \text{ g CaCl}_2} \times 3 \times \dfrac{^{-}1.86°\text{C}}{\text{molal}} = ^{-}6.28° \text{ C}$

Teacher's Notes

Chapter 28 The Halogens

INTRODUCTION

In this chapter you will study the chemistry of the halogens, a family of five nonmetals with closely related properties. The properties of the halogens are interpreted in terms of the periodic atomic properties such as electronegativity, ionization energy, and atomic radius. Emphasis will be placed on the occurrence, preparation, physical properties, chemical properties, and the uses of the halogens and some of their compounds.

LOOKING AHEAD

As you study this chapter, try to explain the properties and behavior of the halogens in terms of the fundamental concepts developed in earlier chapters. Pay particular attention to the periodic atomic properties such as electronegativity, ionization energy, and atomic radius.

SECTION PREVIEW

28.1 The Halogen Family
28.2 Fluorine
28.3 Chlorine
28.4 Bromine
28.5 Iodine

Photographic films are coated with silver halides, which undergo a light-induced reaction to form an image.

28.1 The Halogen Family

The members of the Halogen family are the colorful, active, nonmetallic elements fluorine, chlorine, bromine, iodine, and astatine. *These elements constitute Group 17 of the Periodic Table and are referred to as* **halogens.** They are named *halogens* from the Greek words meaning *salt producer* because they form ionic salts with the alkali metals, salts that are among the most abundant soluble salts found in nature. Samples of the first four of these elements are shown in Figure 28-1.

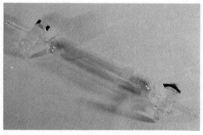

Figure 28-1 The halogens are colorful elements. Fluorine *(top)* is a pale yellow gas, chlorine a greenish-yellow gas, bromine a reddish-brown liquid, and iodine a bluish-black crystal.

General Characteristics of the Halogens

Some of the properties of the halogens are shown in Table 28-1. The atoms of each halogen have seven electrons in the highest energy level. The addition of one electron to a halogen atom converts it to a halide ion having an outer octet of electrons.

$$:\ddot{Cl}: \;+\; e^- \;\longrightarrow\; :\ddot{Cl}:^-$$

Using a chlorine atom as an example, the halogens are all active elements that have high electronegativities. They are very rarely found free in nature. In the elemental state they exist as covalent, diatomic molecules.

Fluorine has the smallest atoms in the Halogen family. It has a strong attraction for electrons. Fluorine is the most highly electronegative element and the most active nonmetal. Because of these properties, fluorine cannot be prepared from its compounds by chemical oxidation. It must be prepared by electrolysis.

The other halogens, with increasingly larger atoms, are less electronegative than fluorine. As a result, the smaller, lighter halogens can replace and oxidize the larger, heavier halogens from compounds. Astatine is a synthetic radioactive halogen.

STUDY HINT

Reactions involving the replacement of halogens are described in Section 8.2

Students might expect that hydrogen fluoride would be ionic because of the great difference in electronegativity between hydrogen and fluorine. However, in the gas phase the polar molecules become hydrogen-bonded to one another, forming a chain-like structure of HF units rather than an ionic crystal. In aqueous solution the negative fluorines bond readily to hydrogen atoms in water molecules, and the molecule does not dissociate.

Table 28-1 clearly shows regular change in properties that occurs in this family. This change in properties proceeds from the smallest and lightest atoms to the largest and heaviest.

Each of the halogens combines with hydrogen. The great electronegativity difference between hydrogen and fluorine explains why hydrogen fluoride molecules are so polar that they associate by hydrogen bonding. The remaining hydrogen halides have smaller electronegativity differences and do not show this property. All hydrogen halides are colorless gases that are ionized in water solution. Hydrofluoric acid is a weak acid because hydronium ions are strongly hydrogen bonded to fluoride ions. The other binary halogen acids are highly dissociated, strong acids.

TABLE 28-1 PROPERTIES OF THE HALOGEN ATOMS

	Fluorine	Chlorine	Bromine	Iodine	Astatine
Atomic number	9	17	35	53	85
Atomic mass	18.998	35.453	79.904	126.905	(210)*
Electron configuration	$[He]2s^22p^5$	$[Ne]\,3s^23p^5$	$[Ar]3d^{10}4s^24p^5$	$[Kr]4d^{10}5s^25p^5$	$[Xe]ff^{14}5d^{10}6s^26p^5$
Ionization energy (kJ/mol)	1680	1254	1139	1113	897
Electron affinity (kJ/mol)	−335	−356	−335	−305	
Bond energy X_2 (kJ/mol)	159	243	192	151	80**
Electronegativity	4.0	3.0	2.8	2.5	2.1
Atomic radius (pm)	72	99	114	133	
Ionic radius of X^- (pm)	133	181	196	220	
Melting point (°C)	−219.62	−100.98	7.2	113.5	302
Boiling point (°C)	−188.14	−34.6	58.78	184.35	337**
Density	1.70 g/L	3.21g/L	3.12 g/mL	4.94 g/cm³	
Oxidation numbers	−1	−1 to +7	−1 to +7	−1 to +7	−1 to +7**
Reduction potential (V)	2.87	1.36	1.07	0.54	
Color of element	pale yellow	yellow-green	reddish-brown	dark violet vapor;	
Standard phase	gas	gas	liquid	dark metallic solid solid	

* Atomic mass in parentheses is that of the most stable isotopes.
**Estimated

The Physical Phases of the Halogens

STUDY HINT

Half-lives are covered in Section 29.2.

If you examine closely the data in Table 28-1, you will find that, under normal conditions, fluorine and chlorine are gases, bromine is a liquid, and iodine is a solid. Astatine is a radioactive element; its longest-lived isotope, astatine-210, has a half-life of 8.3 hours. Because a cyclotron must be used to synthesize astatine, it is very expensive to produce, and its chemistry has not been investigated extensively.

We can use our knowledge of periodicity to make certain predictions about astatine, however. Using trends of the periodic properties of the other halogens, we would expect astatine to have the smalles ionization energy among the halogens, and the highest melting point and boiling point. We would also predict astatine would be a dark-colored solid, that it would have a greater density than the other halogens, and that it would be less soluble in water than the other halogens.

Under ordinary conditions, all of the halogens exist as diatomic molecules. In these molecules, sharing a single pair of electrons gives each halogen atom a stable noble-gas electronic configuration. In the liquid and solid states, the molecules are held together by intermolecular forces (van der Waals forces). As the atomic number increases, the intermolecular forces tend to increase. We should thus expect an increase in the intermolecular forces between halogen molecules as we go from top to bottom within the Halogen family in the Periodic Table. The ionization energies of the halogen elements decrease, as would be expected, with increasing atomic numbers. Because iodine is the largest and most easily ionized of the naturally occuring halogens, the intermolecular forces between iodine molecules are the strongest. The strength of the intermolecular forces between halogen molecules explains the increase in the melting and boiling points from fluorine to iodine. Iodine molecules have 106 electrons, while bromine molecules have 70. The intermolecular force should thus be greater between iodine molecules than between bromine molecules, which explains the higher melting and boiling points for iodine than bromine.

Section Review

1. Which elements are known as the halogens, and what are their symbols?
2. List the general characteristics of the halogens.
3. Why does fluorine exhibit properties that are different than those of the other halogens?

28.2 Fluorine

Because of its high electronegativity and this tendency to form compounds, fluorine is never found free in nature. In the earth's crust, fluorine exists mainly in the form of the minerals feldspar, fluorapatite, $C_5(PO_4)_3F$, and cryolite, Na_3AlF_6.

Preparation and Properties of Fluorine

Fluorine is prepared by electrolyzing a mixture of potassium fluoride and hydrogen fluoride. A steel and Monel metal electrolytic cell with a positive carbon electrode is used. **Monel metal** is a corrosion-resistant alloy composed principally of nickel and copper. The fluoride coating that forms on these metals protects them from further reaction.

Fluorine is the most active nonmetallic element. It unites with hydrogen explosively. It forms compounds with all elements except helium, neon, and argon. As the most electronegative element, fluorine is assigned an oxidation number of -1 in all compounds. There are no known positive oxidation states of fluorine. It forms salts known as fluorides. Fluorine reacts with gold and platinum slowly. Special carbon-steel containers are used to transport fluorine. These containers become coated with iron fluoride, which resists further action.

1. Fluorine, F; chlorine, Cl; bromine, Br; iodine, I; astatine, At
2. Reactive, high electronegativities, high ionization energies, high electron affinities, low melting and boiling points, colorful, exist as covalent diatomic molecules in the elemental state
3. It has the smallest atoms in the halogen family and has a strong attraction for electrons. As a result, it cannot be prepared from its compounds by chemical reduction. Also, because of the great electronegative difference between hydrogen and fluorine, hydrogen-fluorine molecules are so polar that they associate by hydrogen bonding. The other hydrogen halides do not exhibit this property.

SECTION OBJECTIVES

- Cite the common occurrences of fluorine.
- Describe the properties and preparation of fluorine.
- Identify commercial uses of fluorine compounds.

Nothing forms compounds with helium, neon, or argon.

Since no other element pulls harder on electrons than fluorine, it cannot lose electrons or form positive oxidation states in ordinary chemical reactions.

Compounds of Fluorine and Their Uses

The mineral fluorspar (CaF_2) is often used in preparing fluorine compounds. Sodium fluoride is used as a poison in pesticides for roaches, rats, and other vermin. A trace of sodium fluoride, or the less expensive sodium hexafluorosilicate (Na_2SiF_6), is added to drinking water in many areas to help prevent tooth decay. The fluoride ion concentration is carefully regulated to be 1 ppm. Fluorides have also been added to some toothpastes.

One of the freons, dichlorodifluoromethane (CCl_2F_2) is used as a refrigerant. It is odorless, nonflammable, and nontoxic. Its critical temperature is is 111.7°C and critical pressure is 39.4 atm. In producing aluminum, melted cryolite Na_3AlF_6 is used as a solvent for aluminumm oxide. Uranium is changed to uranium hexafluoride gas (UF_6) in the process for separating the uranium isotopes.

Fluorine combines with the noble gases krypton, xenon, and radon. Krypton difluoride (KrF_2) is prepared by passing electricity through a krypton-fluorine mixture. This process is carried out at the temperature of liquid nitrogen. Krypton difluoride is a volatile white solid that decomposes slowly at room temperature.

Three fluorides of xenon—XeF_2, XeF_4, and XeF_6—have been made. All are colorless crystalline solids at ordinary temperatures. XeF_6 is the most highly reactive. Each of the three compounds reacts with hydrogen to produce elemental xenon and hydrogen fluoride. With water, XeF_2 produces xenon, oxygen, and hydrogen fluoride. The other two fluorides react with water to yield xenon trioxide (XeO_3) a white, highly explosive solid.

Hydrofluoric acid (HF) is used as a catalyst in producing high-octane gasoline. It is also used in making synthetic cryolite for aluminum production. For many years hydrofluoric acid has been used for etching glass. Glassware is given a frosty appearance by exposing it to hydrogen fluoride fumes, as shown in Figure 28-2.

Section Review

1. What are the principal minerals in which fluorine is found?
2. What is the reason that fluorine does not exhibit positive oxidation states?
3. Describe the method used for the preparation of fluorine.
4. Name two fluorine compounds and identity their commercial uses.

28.3 Chlorine

Chlorine is rarely found uncombined in nature. Elemental chlorine is found in small amounts in some volcanic gases. Chlorides of sodium, potassium, and magnesium are fairly abundant. Common table salt, sodium chloride, is a widely distributed compound. It is found in sea water, in underground brines, and in rock salt deposits. Sodium chloride is the commercial source of chlorine.

You may be familiar with the use of chlorine for maintaining clean water in swimming pools. In fact chlorine is used worldwide for the purification of drinking water.

Figure 28-2 Hydrofluoric acid is used to etch glass.

1. Feldspar; fluorapatite, $Ca_5(PO_4)_3F$; cryolite, Na_3AlF_6
2. The position of fluorine in the top of Table 20-00 indicates that it is the strongest common oxidizing agent known. This means that fluoride ions cannot be oxidized by other chemical oxidizing agents.
3. Fluorine is prepared by electrolyzing a mixture of potassium fluoride and hydrogen fluoride.
4. Accept all answers given in the subsection "Compounds of Fluorine and Their Uses."

SECTION OBJECTIVES

- Cite the principal occurrences of chlorine.
- Describe methods for the preparation of chlorine.
- Relate the properties of chlorine to its important commercial uses.
- Describe the preparation, properties, and uses of hydrogen chloride.

Preparation of Chlorine

The element chlorine was first isolated in 1774 by Carl Wilhelm Scheele, the codiscoverer of oxygen. The preparation of elemental chlorine involves the oxidation of chloride ions. Strong oxidizing agents are required. The following three methods can be used to isolate chlorine.

1. By the electrolysis of sodium chloride. Chlorine is most often prepared by the electrolysis of a saturated water solution of sodium chloride. At this concentration, hydrogen from the water is released at the negative electrode, and chlorine is set free at the positive electrode. The hydrogen and chlorine gases are kept separate from each other and from the solution by asbestos partitions. The sodium and hydroxide ions remaining in the solution are recovered as sodium hydroxide.

$$2NaCl(aq) + 2H_2O(\ell) \xrightarrow{\text{(electricity)}} 2NaOH(aq) + H_2(g) + Cl_2(g)$$

2. By the oxidation of hydrogen chloride. This method involves heating a mixture of manganese dioxide and concentrated hydrochloric acid. The manganese oxidizes half of the chloride ions in the reacting HCl to chlorine atoms. Manganese is reduced during the reaction from the +4 oxidation state to the +2 state.

$$MnO_2(s) + 4HCl(aq) \rightarrow MnCl_2(aq) + 2H_2O(\ell) + Cl_2(g)$$

This was the method used by Scheele in first preparing chlorine. It is a useful laboratory preparation. An alternative process involves heating a mixture of manganese dioxide, sodium chloride, and sulfuric acid, as shown in Figure 28-3.

$$2NaCl(s) + 2H_2SO_4(aq) + MnO_2(s) \rightarrow$$
$$Na_2SO_4(aq) + MnSO_4(aq) + 2H_2O(\ell) + Cl_2(g)$$

3. By the action of hydrochloric acid on calcium hypochlorite. Hydrochloric acid added drop by drop to calcium hypochlorite powder releases chlorine and forms calcium chloride and water.

$$4HCl(aq) + Ca(ClO)_2(s) \rightarrow CaCl_2(aq) + 2Cl_2(g) + 2H_2O(\ell)$$

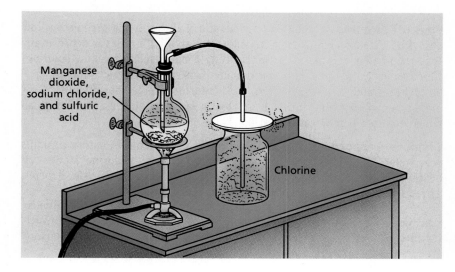

Manganese dioxide, sodium chloride, and sulfuric acid

Chlorine

Most chlorine compounds are soluble, so chlorine is found in sea water and deposits from sea water. Fluorine compounds are generally insoluble, so fluorine is found in mineral ores.

Figure 28-3 Chlorine may be prepared by heating a mixture of manganese dioxide, sodium chloride, and sulfuric acid. Chlorine is collected by upward displacement of air. The preparation should be performed under a hood.

Technology

PVC Pipes

Lead was once used extensively for water pipes and tanks because it was inexpensive and slow to corrode. When the hazards of lead poisoning were recognized, clay was used instead. As our cities age, replacement of those aging clay pipes has become a major problem. How can the pipes be replaced without interrupting the water and sewage service? Engineers developed an interesting process using PVC.

PVC stands for "polyvinyl chloride," a polymeric form of $CH_2—CH—CH_2Cl$. This plastic is durable, cheap, and easy to manufacturer into pipes that do not corrode. The PVC pipes are created

from a continuous strip of PVC with special clamping edges. The PVC is fed as a spiral directly into the old pipes, forming a sealed tube that is flexible enough to

expand up against the old pipe. This lightweight PVC tube is supported by the old pipes, so it doesn't have to be as thick as other kinds of pipe.

Physical and Chemical Properties of Chlorine

Chlorine is a toxic, highly reactive element. Its chemical properties, however, make it an extremely useful element.

Physical Properties   At room temperature, chlorine is a greenish-yellow gas with a disagreeable, suffocating odor. It is about 2.5 times as dense as air. It is moderately soluble in water, forming a pale greenish-yellow solution. Chlorine is easily liquefied and is usually marketed in steel cylinders.

When inhaled in small quantities, chlorine affects the mucous membranes of the nose and throat. If inhaled in larger quantities, chlorine is so poisonous that it may cause death.

Chemical Properties   The third energy level of a chlorine atom contains seven electrons. There are many reactions by which chlorine atoms can acquire an additional electron and complete the third-energy-level octet. Three reactions involving chlorine are given as examples.

1. Reaction with metals and metalloids. Most metals and metalloids react with chlorine directly. For example, when powdered antimony is sprinkled into a jar of moist chlorine, the two elements combine spontaneously, emitting sparks. Antimony trichloride is formed.

The odor of chlorine is familiar to everyone as the odor of chlorine bleaches and of swimming-pool disinfectant.

$$2Sb(s) + 3Cl_2(g) \rightarrow 2SbCl_3(s)$$

In a similar manner, hot metallic sodium burns in chlorine to form sodium chloride. Chlorine combines directly with such metals as copper, iron, zinc, and arsenic, if they are heated slightly.

2. Reaction with hydrogen. If hydrogen and chlorine are mixed in the dark, no reaction occurs. But such a mixture explodes violently if it is heated or exposed to sunlight. The heat or sunlight provides the activation energy. A jet of hydrogen that is burning in air will continue to burn if it is inserted into a bottle of chlorine.

$$H_2(g) + Cl_2(g) \rightarrow 2HCl(g)$$

Chlorine does not support the combustion of wood or paper. A paraffin candle, however, continues to burn in chlorine with a smoky flame, as shown in Figure 28-4. In this reaction the hydrogen of the paraffin combines with the chlorine, forming hydrogen chloride. The carbon is left uncombined.

3. Reaction with water. A freshly prepared solution of chlorine in water has a greenish-yellow color. If it stands in sunlight for a few days, both the color and the strong chlorine odor will disappear. The chlorine combines with the water to form hypochlorous acid and hydrochloric acid. Hypochlorous acid is unstable and decomposes into hydrochloric acid and oxygen.

$$2H_2O\,(\ell) + 2Cl_2(g) \rightarrow 2HClO(aq) + 2HCl(aq)$$
$$2HClO(aq) \rightarrow 2HCl(aq) + O_2(g)$$

The instability of hypochlorous acid makes chlorine water a good oxidizing agent.

Uses of Chlorine

1. For bleaching. **Bleaching** is the operation by which color is partially or wholly removed from colored materials. Household bleaching solution is usually a solution of sodium hypochlorite. It is made by reacting liquid chlorine with sodium hydroxide solution. Chlorine is used as a bleaching agent by the pulp and paper industry.

2. As a disinfectant. Since moist chlorine is a good oxidizing agent, it destroys bacteria. **Chlorinated lime** *is Ca(ClO)Cl.* Large quantities of Ca(ClO)Cl, are used as a disinfectant *which is a chemical agent that destroys microorganisms,* but not bacterial spores. In city water systems, billions of gallons of water are treated with chlorine to kill disease-producing bacteria. The water in swimming pools is usually treated with chlorine. As part of the sewage purification process, chlorine is sometimes used to kill bacteria in sewage.

3. For making compounds. Because chlorine combines directly with many substances, it is used to produce a variety of compounds. Among these are aluminum chloride (Al_2Cl_6), used as a catalyst; carbon tetrachloride (CCl_4), a solvent; 1,1,2-trichloroethene ($ClHC=CCl_2$), a cleaning agent; and **polyvinyl chloride** which is *the polymer of vinyl chloride (CH_2CHCl).* Polyvinyl chloride is replacing steel and cast iron as the material of choice in plumbing and electrical pipe. Polyvinyl chloride has the advantage of being lightweight, strong, and free from corrosion.

Seventy percent of chlorine produced is used for the production of organic-chloride compounds such as polyvinyl chloride.

Figure 28-4 A paraffin candle will burn in chlorine. The hydrogen of the paraffin combines with the chlorine to form hydrogen chloride. The smoke is the carbon from the wax, which is left uncombined.

CAUTION
Chlorine destroys silk or wool fibers. Never use commercial bleaches containing hypochlorites on silk or wool.

Figure 28-5 Chlorine is used to kill microorganisms in swimming pools.

Figure 28-6 Large quantities of hydrochloric acid are used to clean metals.

STUDY HINT

Some of these processes are the chlorination of hydrocarbons described in Chapter 21.

1. Usually in the combined state as chlorides of sodium, potassium, and magnesium. Sodium chloride is found in salt brines underground, in rock-salt deposits, and in sea water as sodium chloride. Elemental chlorine is found in some volcanic gases.
2. Accept all answers given in the subsection "Preparation of Chlorine."
3. Accept all answers given in the subsection "Uses of Chlorine."
4. By treating sodium chloride with concentrated sulfuric acid
5. Colorless gas, sharp penetrating odor, denser than air, soluble in water. Hydrochloric acid reacts with metals above hydrogen in the activity series and with oxides of metals; it is used in preparing chlorides and in cleaning metals.

An Important Chlorine Compound: Hydrogen Chloride

Hydrogen chloride ranks in the top 30 among industrial chemicals in production. Hydrogen chloride dissolved in pure water is sold under the name hydrochloric acid. Its nontechnical name is muriatic acid. Both hydrogen chloride and hydrochloric acid are industrial chemicals of major importance.

Preparation of hydrogen chloride   In the laboratory, hydrogen chloride can be prepared by treating sodium chloride with concentrated sulfuric acid.

$$NaCl(s) + H_2SO_4(\ell) \rightarrow NaHSO_4(s) + HCl(g)$$

This same reaction, carried out at a higher temperature, is used commercially. If more NaCl is used at this higher temperature a second molecule of HCl can be produced per molecule of H_2SO_4.

$$2NaCl(s) + H_2SO_4(\ell) \rightarrow Na_2SO_4(s) + 2HCl(g)$$

Another commercial preparation involves the direct union of hydrogen and chlorine. Both elements are obtained by the electrolysis of concentrated sodium chloride solution.

By far the most important commercial source of hydrogen chloride results from the formation of HCl as a by-product of chlorination processes.

Physical and chemical properties of hydrogen chloride   Hydrogen chloride is a colorless gas with a sharp, penetrating odor. It is denser than air and extremely soluble in water. One volume of water at 0°C dissolves more than 500 volumes of hydrogen chloride at standard pressure. Hydrogen chloride makes a fog with moist air. It is so soluble that it condenses water vapor from the air into tiny drops of hydrochloric acid.

Hydrogen chloride is a stable compound that does not burn. Some vigorous oxidizing agents react with it to form water and chlorine. The water solution of hydrogen chloride is hydrochloric acid. Concentrated hydrochloric acid contains about 38 percent hydrogen chloride by weight and is about 1.2 times as dense as water. Hydrochloric acid reacts with metals above hydrogen in the activity series and with the oxides of metals. It neutralizes hydroxides, forming salts and water.

Uses of hydrochloric acid   Hydrochloric acid is used in preparing certain chlorides and in cleaning metals. This cleaning involves removing oxides and other forms of tarnish.

Some hydrochloric acid is essential in the process of digestion. The concentration of hydrochloric acid in gastric juice is $0.16 M$.

Section Review

1. In what state is chlorine found in nature?
2. Describe a method for the preparation of chlorine, and write a balanced chemical equation that shows the reaction.
3. List one commercial use of chlorine and explain its use in terms of its properties.
4. How is hydrogen chloride prepared in the laboratory?
5. List some of the properties of hydrogen chloride and its uses as a water solution.

28.4 Bromine

Bromine is far less abundant in the earth's crust than the first two members of the Halogen family. It occurs mainly as potassium, magnesium, and sodium bromides in sea water, in underground brines, and in salt deposits. Water from brine wells in Michigan and Arkansas have relatively high concentrations of bromine and are a major source of the halogen. Seawater, which contains relatively low concentrations of bromine (about 85 parts per million), is not an economical source for this element. Bromine is toxic and reacts readily with many elements. It is the only nonmetal that is liquid under standard conditions.

Preparation of Bromine

In the laboratory, bromine can be prepared by using manganese dioxide, sulfuric acid, and sodium bromide as shown in Figure 28.7.

$$2NaBr(s) + MnO_2(s) + 2H_2SO_4(aq) \rightarrow$$
$$Na_2SO_4(aq) + MnSO_4(aq) + 2H_2O(\ell) + Br_2(g)$$

This is similar to the preparation of chlorine as discussed earlier in this section.

The commercial production of bromine from salt-well brines depends on the ability of chlorine to displace bromide ions from solution. Chlorine displaces bromide ions because it is more highly electronegative than bromine.

$$2Br^-(aq) + Cl_2(g) \rightarrow 2Cl^-(aq) + Br_2(\ell)$$

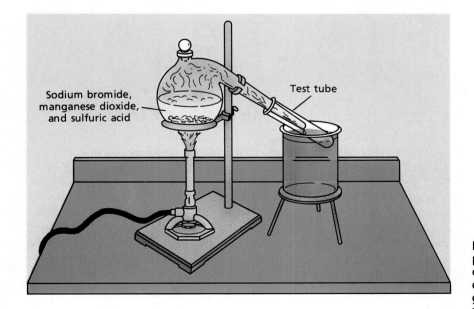

Sodium bromide, manganese dioxide, and sulfuric acid

Test tube

Figure 28-7 Bromine can be prepared by heating a mixture of sodium bromide, manganese dioxide, and sulfuric acid in a glass-stoppered retort. This preparation should be performed under a hood.

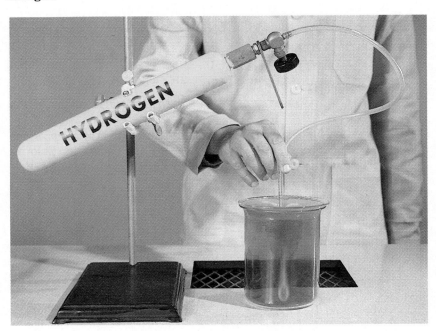
Physical and Chemical Properties of Bromine

Bromine is a dark, reddish-brown liquid that is about three times as dense as water. It evaporates readily, forming a vapor that burns the eyes and throat and has a very disagreeable odor. The name "bromine" is derived from the Greek word *bromos*, meaning stench. Bromine is moderately soluble in water. The reddish-brown water solution used in the laboratory is known as bromine water. Bromine dissolves readily in carbon tetrachloride and carbon disulfide and in water solutions of bromines.

Bromine unites directly with hydrogen to form hydrogen bromide, as shown in Figure 28-8. It combines with most metals to form bromides. When it is moist, bromine is a good bleaching agent. A water solution of bromine is a strong oxidizing agent that forms hydrobromic acid and oxygen in sunlight.

Figure 28-8 Hydrogen burns in an atmosphere of bromine vapor. Hydrogen bromide is formed.

Compounds of Bromine and Their Uses

Complex bromine-containing organic compounds are added to polyurethane and polystyrene plastics to make them flame retardant. Methyl bromide (CH_3Br) is a useful soil fumigant. Improved crop yields result from such treatment. Ethylene bromide ($C_2H_4Br_2$) increases the efficiency of lead tetraethyl in antiknock gasoline. The use of bromine in this application is becoming more limited since the United States Environmental Protection Agency has initiated a phaseout of the use of lead in gasoline. This phaseout started in the early 1980's. Ethylene bromide was formerly used to control insects in grain. This use has been stopped because ethylene bromide is believed to cause cancer.

Silver bromide (AgBr) is a yellowish solid. It is highly photosensitive, *meaning it is sensitive to light rays* and is widely used in making photographic film. A solution of calcium bromide ($CaBr_2$) is used as a

Desktop Investigation

Silver Halides

Question
How do you make silver ha-lides, and what effect does light have on these halogen compounds?

Materials
sodium chloride, 0.2-*M* solution
sodium bromide, 0.2-*M* solution
sodium iodide, 0.2-*M* solution
silver nitrate, 0.1-*M* solution
three test tubes, 16 × 150 mm
medicine dropper
filter paper
ring stand with iron ring
beaker, 150 mL

Procedure
Remove all combustible materials from the work area. Remember to wear safety glasses and apron. **Caution:** Do not allow AgNO₃ to come in contact with your skin or clothing.
1. Add 5 mL of sodium chlo-ride solution to the first test

tube. Add 5 mL of sodium bromide solution to the sec-ond test tube. Add 5 mL of sodium iodine solution to the third test tube. To each of the three test tubes, add 10 drops of silver nitrate so-lution. What is the result?
2. Immediately filter the contents of each test tube. (See diagram.) If necessary, flush the test tube with

small amounts of water from a wash bottle to remove all of the precipitate. Unfold the filter papers and expose their contents to direct sun-light for several minutes. What is the result? Note which of the three test tubes shows a change first and which last.

Discussion
1. In what ways were the reaction in Procedures 1 and 2 *(a)* similar and *(b)* differ-ent?
2. Write equations for all the reactions in Procedures 1 and 2.
3. Silver halides are used in black-and-white film. Which silver halide is probably used to make fast film (film that requires relatively little light)? Which is used to make slow film (film that requires a relatively large amount of light)?
4. Write a good conclusion for this experiment.

---

high-density fluid in oil-well drilling. The bromides of sodium and potassium are used in medicine as sedatives. Such medicines should not be used unless prescribed by a physician. Certain bromine compounds may be used as disinfectants. Many organic bromine compounds are useful intermediates in reactions forming amine, phenols, or ethers.

Section Review

1. Write a balanced chemical equation for the preparation of bromine in the laboratory.
2. What are some of the properties of bromine?
3. Name two compounds of bromine and identify their commercial uses.
4. Describe some ways in which bromine compounds are known to be dangerous to human health.

1. $2NaBr(s) + MnO_2(s) + 2H_2SO_4(aq) \rightarrow Na_2SO_4(aq) + MnSO_4(aq) + 2H_2O(\ell) + Br_2(g)$
2. Accept all answers given in the subsection "Physical and Chemical Properties of Bromine," and in Table 29-1.
3. Accept all answers given in the subsection "Compounds of Bromine and Their Uses."
4. See "Compounds of Bromine and Their Uses."

28.5 Iodine

Iodine, like bromine, is far less abundant in the earth's crust than are fluorine and chlorine. Iodine occurs principally as sodium iodate and periodate impurities in sodium nitrate deposits found in Chile. The largest source of iodine in the United States is underground brine from oil wells in California.

Preparation of Iodine

The laboratory preparation of iodine is similar to that of chlorine and bromine. An iodide is heated with manganese dioxide and sulfuric acid, as shown in Figure 28-9.

$$2NaI(s) + MnO_2(s) + 2H_2SO_4(aq) \rightarrow Na_2SO_4(aq) + MnSO_4(aq)$$
$$+ 2H_2O(\ell) + I_2(g)$$

The iodine is driven off as a vapor and can be condensed as a solid on the walls of a cold dish or beaker.

Figure 28-9 Iodine may be prepared by heating sodium iodide and manganese dioxide with sulfuric acid. It should be done under a hood.

Physical and Chemical Properties of Iodine

Iodine is a bluish-black crystalline solid. When heated, it sublimes (vaporizes without melting) and produces a violet-colored vapor. The odor of this vapor is irritating, resembling that of chlorine. Iodine is very slightly soluble in water. It is much more soluble in water solutions of sodium or potassium iodide. With these solutions it forms complex I_3^- ions. It dissolves readily in alcohol, forming a dark-brown solution. It is very soluble in carbon disulfide and carbon tetrachloride, giving a rich violet color.

Iodine is active chemically, though less so than either bromine or chlorine. It combines with metals to form iodides. It can also form compounds with all nonmetals except sulfur, selenium, and the noble gases. Iodine reacts with organic compounds just as chlorine and bromine do. But since the carbon-iodine bond is not strong, many of the compounds formed decompose readily.

Uses of Iodine and Its Compounds

Iodine is most useful as a disinfectant. Iodine that is "complexed" (loosely bonded) with certain organic compounds is used as an antiseptic on the skin. *An antiseptic is a substance that checks the growth of microorganisms on or in the body.* Detergents with which iodine is complexed make good combination cleaning and sanitizing agents. Iodine and some iodine compounds are used as catalysts in certain organic reactions.

Silver iodide is a light-sensitive compound used in photographic film. Potassium iodide is added to table salt to provide the iodine necessary for proper nutrition.

Section Review

1. Write a balanced chemical equation for the preparation of iodine.
2. What are some of the properties of elemental iodine?
3. Describe some of the uses of iodine and its compounds.

Chemistry Notebook

Discovering How to Discover Elements

The search for new elements is as old as chemistry, but in the early days was not systematic. In 1807, Sir Humphry Davy built a powerful battery (see photograph), ran current through molten "caustic pot-ash" (potassium hydroxide), and produced a new element, potassium. Because "caustic soda" (sodium hydroxide) was chemically similar, he treated it similarly and discovered sodium. Within a year, he had discov-ered four other elements!

Later in the same century, Dmitri Mendeleev discovered the Periodic Law: properties of elements occur in cycles. He believed that some elements had not been discovered because of gaps in some cycles. Mendeleev predicted several properties of these missing elements, and eventually he was shown to be correct.

Modern atomic theory correlates the properties of elements and their position on the periodic table with their electronic structure. This allows chemists to predict the properties of unknown elements with high precision. Such predictions help us to confirm that a new element has been discovered. For example, the elements technetium and promethium have been identified in the light of certain stars.

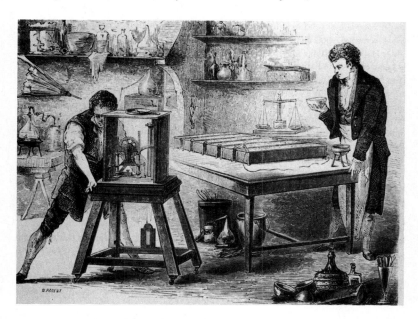

Chapter Summary

- The Halogen family consists of the highly electronegative elements fluorine, chlorine, bromine, iodine, and astatine.
- Fluorine compounds are used *(1)* as refrigerants *(2)* to help prevent tooth decay *(3)* in aluminum production *(4)* for separating uranium isotopes *(5)* in gasoline production and *(6)* for etching glass. Chlorine is used for bleaching, disinfecting, and making chlorine-containing compounds.
- Bromine compounds are used *(1)* to make plastics flame retardant *(2)* in agriculture *(3)* in antiknock gasoline *(4)* in oil-well drilling *(5)* as disinfec-tants *(6)* in medicine and *(7)* in photography.
- Iodine is used as a disinfectant and in organic synthesis. Iodides are used in photography and as nutrients.

Chapter 28 *Review*

The Halogen Family

Vocabulary

antiseptic
chlorinated lime
disinfectant
halogen

Monel metal
photosensitive
polyvinyl chloride

Questions

1. Define: *(a)* halogen *(b)* halogen atom *(c)* halogen molecule *(d)* halide ion.
2. *(a)* In what form are the halogens found in nature, as atoms, molecules, or ions? *(b)* Why?
3. As one goes down the Halogen family on the Periodic Table, how do the members vary in *(a)* atomic radius *(b)* ionic radius *(c)* first ionization energy *(d)* electron affinity *(e)* electronegativity *(f)* melting point *(g)* boiling point *(h)* density *(i)* principal oxidation number?
4. *(a)* What kind of container is used for fluorine? *(b)* What chemical property makes this use possible?
5. What properties make dichlorodifluoromethane useful as a refrigerant?
6. *(a)* What compound is the commercial source of chlorine? *(b)* For what other element is this compound the commercial source?
7. What effects does chlorine have on the body?
8. *(a)* For which does chlorine have greater attraction, carbon or hydrogen? *(b)* What experimental evidence can you give to support your answer?
9. What are some commercial uses of chlorine?
10. *(a)* List the physical and chemical properties of hydrogen chloride that must be considered in choosing a method of collecting this gas in the laboratory. *(b)* Which method of collection would you choose for a gas with this combination of properties?
11. *(a)* Describe an industrial use for hydrochloric acid. *(b)* What is the function of hydrochloric acid in the human body?

12. *(a)* Write the ionic equation for the reaction involved in extracting bromine from salt-well brines. *(b)* What type of reaction is this? *(c)* Why does this reaction occur?
13. Describe the important physical properties of bromine.
14. Write formulas for two organic and two inorganic bromine compounds and give a use for each.
15. What phase change does iodine undergo that the other halogens do not?
16. Write an equation using electron-dot symbols for the formation of an iodide ion from an iodine atom.
17. Define the following terms: *(a)* Monel metal *(b)* bleaching *(c)* chlorinated lime *(d)* disinfectant *(e)* antiseptic *(f)* detergent.

Problems

1. *(a)* Given that chlorine has a density of 3.214 g/L how many moles of Cl_2 are there in 5.00 L of the gas at STP? *(b)* How many Cl_2 molecules is this?
2. What is the mass percentage of bromine in ethylene bromide ($C_2H_4Br_2$)?
3. Chlorine reacts with calcium hydroxide to produce bleaching powder, Ca(ClO)Cl, and water. *(a)* What mass of calcium hydroxide is required for making 200 g of bleaching powder? *(b)* What mass of chlorine is also required?
4. How many liters of chlorine at STP can be obtained from 375 g of sodium chloride by electrolysis?
5. How many grams of zinc chloride can be produced from 0.500 L of chlorine at STP?

Application Questions

1. If a reaction occurs, complete and balance the following ionic equations:
   (a) $Cl^- + F_2 \rightarrow$ _____.
   (b) $F^- + Br_2 \rightarrow$ _____.
   (c) $I^- + F_2 \rightarrow$ _____.

(d) $Br + Cl_2 \rightarrow$ _____.
(e) $Cl^- + I_2 \rightarrow$ _____.
(f) $Br^- + I_2 \rightarrow$ _____.

2. All the halogens except fluorine exhibit positive oxidation states. Why is fluorine different?

3. On the basis of atomic or molecular structure, explain the variations with increasing atomic number, if any, the halogens show in (a) ionic radius (b) first ionization energy (c) melting point (d) electronegativity (e) principal oxidation number.

4. Water reacts with xenon difluoride and yields xenon, oxygen, and hydrogen fluoride. (a) Write an equation for this reaction. (b) Assign oxidation numbers to each element and balance the equation by a method used to balance oxidation-reduction equations.

5. In the commercial electrolysis of sodium chloride solution, the products chlorine, hydrogen, and sodium hydroxide are kept separated from each other. Why?

6. Write the balanced formula equation for the lab preparation of chlorine by the action of hydrochloric acid on calcium hypochlorite. Assign oxidation numbers. Tell what is oxidized and what is reduced.

7. (a) What is the color of freshly prepared chlorine water? (b) Why? (c) Does the color change as chlorine water is exposed to sunlight? (d) Explain.

8. Write the equation for the laboratory preparation of bromine from (a) manganese dioxide and hydrobromic acid (b) manganese dioxide, sodium bromide, and sulfuric acid. (c) Assign oxidation numbers in both equations and tell which element is oxidized and which element is reduced. (d) How do these reactions compare?

9. Compare the colors of (a) solid iodine (b) iodine in alcohol (c) iodine in carbon tetrachloride (d) iodine vapor.

10. Hydrogen forms binary compounds with each of the four common halogens. (a) Write the formulas for these compounds. (b) How do the hydrogen-halogen bonds in these compounds vary in polarity? (Use electronegativity data to support your answer.) (c) Each hydrogen halide reacts with water. Write equations for the reactions.

11. The chemical reactions between water molecules and molecules of the hydrogen halides are reversible.   (a) Qualitatively, at equilibrium, what are the relative concentrations of the particles involved? (b) What does this indicate about the relative stability of the hydrogen halide molecules compared with the stability of the ions that can be formed from them?

12. Sodium chloride and magnesium chloride are ionic compounds, but aluminum chloride is molecular. Why?

13. The improved health that results from using chlorine to kill disease-producing bacteria in municipal water supply systems has been recognized for many years. But chlorine can also react with natural substances in water to produce chloroform ($CHCl_3$), which causes cancer in laboratory animals. In what ways can scientists deal constructively with this dilemma?

Application Problems

1. Natural chlorine consists of only the two isotopes Cl-35 (34.96885 u) and Cl-37 (36.96590 u). If the atomic mass is 35.453, what are the percentages of these two isotopes in naturally occurring chlorine?

2. Calculate the heat of reaction for:

$$\tfrac{1}{2}Cl_2 + HBr \rightarrow HCl + \tfrac{1}{2}Br_2$$

from (a) bond energy data (Table 6-4) (b) heat of formation data (Appendix Table 14).

3. A laboratory experiment requires five 250-mL bottles of chlorine measured at 22°C and 730 mm pressure. What volume of 38% hydrochloric acid (density 1.20 g/mL) and what mass of manganese dioxide will be required?

4. Using the ideal gas equation (Section 12.2), calculate the number of moles of Cl_2 in 5.00 L of the gas at STP. Why does your answer differ from that obtained in Review Problem 1?

5. A solution contains 62.5 g of $CaCl_2$ in 500. g of water. What is the theoretical freezing point of the water of the solution?

Enrichment

1. Investigate such factors as the hardness of water, temperature, alkalinity, and amount of sunlight a swimming pool receives and explain the chemistry involved. Report your findings to your class.

Metallurgist and Materials Scientist

Metallurgy is the science of metals, alloys, and their properties. Materials science is a combination of metallurgy and the study of modern ceramic, plastic, and composite materials. Both of these fields offer exciting opportunities.

If cars are your passion, then designing custom alloys for automobile parts is a career possibility. If bicycles are more your speed, why not improve the linkages and frame while keeping the materials as light as possible? Space technology is another rewarding area for future work as a metallurgist or materials scientist. The challenge is to keep spacecraft light to reduce fuel costs—yet strong enough to survive meteor impact—and sufficiently heat-resistant to withstand the heat of a fiery reentry through the atmosphere.

To work with metals and alloys, a metallurgist must understand the chemical and physical nature of metals. This requires a strong background in chemistry and engineering. A similar background is needed for understanding the complex nature of plastics and composites. In the current job market, a bachelor's degree in chemistry or metallurgy (with a minor in chemistry) is the minimum requirement. A doctorate enables you to conduct research on the new materials that will be needed for tomorrow's technology.

Chemical Supplier

Chemical suppliers must be salespersons, regulation and transportation experts, and applied chemists all at the same time. Marketing a chemical product requires good sales techniques, communications skills, and knowledge of the products. A technical background is helpful in answering questions about which products to use for a specific purpose and about possible side reactions and impurities.

Packaging and transporting chemicals also requires chemical knowledge. The shelf life of each chemical

being handled, stored, or shipped must be considered when arranging production and shipping dates. Chemical reactions with air, light, or other chemicals must be considered. Transportation of chemicals is subject to strict government regulation. Packaging must be clearly labeled with the name and hazardous nature of the chemical. Procedures in case of accidents, such as spills, should be clearly listed on the label.

An associate's degree, together with the ability to deal with people and to sell products, can get you an entry-level job with a chemical supplier. A bachelor's degree in business with a minor in chemistry can lead to a job as a salesperson or as a technical representative. The people most in demand have a master's degree in business administration with an undergraduate degree in chemistry.

Cumulative *Review*

Unit 8 • Descriptive Chemistry

Questions

1. Give the Periodic Table group number (or numbers) for the following: *(a)* metalloids *(b)* alkali metals *(c)* alkaline-earth metals *(d)* noble metals *(e)* halogens. (24, 25, 26, 27)
2. Among potassium, beryllium, and magnesium, which has the *(a)* highest first ionization energy *(b)* largest atomic radius *(c)* smallest atomic radius *(d)* most soluble hydroxide *(e)* highest concentration in sea water? (24)
3. Give the common names for the following compounds: *(a)* $NaHCO_3$ *(b)* $NaOH$ *(c)* Na_2CO_3 *(d)* KOH. (24)
4. Name the two alkali metal ions and two alkaline earth ions essential in human body fluids. (24)
5. Identify the properties by which alkali metals differ from most other metals. (24.1)
6. Identify the oxidation and reduction products and write the cell reactions for the *(a)* electrolysis of aqueous sodium chloride *(b)* electrolysis of molten sodium chloride. (24.3)
7. For the Solvay process, identify the raw materials, the principle product, and the by-products that have commercial value. (24.3)
8. List some of the properties in which the alkaline earth metals differ significantly from the alkali metals. (24)
9. Beryllium is more likely to form covalent bonds than the other alkaline-earth metals. Why? (24.2)
10. *(a)* Define a self-protective metal. *(b)* Which of the following metals are self-protective: aluminum, magnesium, iron, sodium, copper? (24, 25,)
11. *(a)* Name four characteristic properties of transition metals and their compounds. *(b)* What is most responsible for these properties? (25.1)
12. Write *(a)* the equation for reduction of iron(III) oxide in a blast furnace and *(b)* the equation for formation of slag in a blast furnace. (25.2)
13. Give names and formulas for the three oxides of iron. (25.2)
14. Describe what happens during *(a)* oil flotation *(b)* oxidation of copper matte *(c)* electrolytic refining in processing copper ores. (25.3)
15. *(a)* Name the seven metalloids. *(b)* Identify their most characteristic property. *(c)* In what industry is this characteristic property of the metalloids essential? (26.1)
16. Name the following compounds and identify which is an abrasive, a bleach, colored, an antiseptic: *(a)* $CrCl_3$ *(b)* $NaOCl$ *(c)* H_3BO_3 *(d)* BN. (28.3, 4)
17. Among the elements Al, Se, Cl, and S, which would you expect to be *(a)* in a photocell *(b)* in a swimming pool *(c)* the best electrical conductor *(d)* the most electronegative *(e)* the most allotropic *(f)* found naturally as a free element *(g)* produced by electrolysis? (26, 27, 28)
18. Describe the properties of silicates, silicon, and silicones. (26.3)
19. *(a)* Show by three equations conversion of SO_2 to H_2SO_4. *(b)* Name the H_2SO_4 production process. (27.2)
20. *(a)* Name three industries that are major consumers of H_2SO_4. *(b)* Name three useful chemical properties of H_2SO_4. (27.2)
21. *(a)* What gases are responsible for acid rain? *(b)* What are sources of these gases? *(c)* What do these gases form in the atmosphere? *(d)* Give the pH of acid rain. (27.2)
22. List four of the major uses of the compound sulfur dioxide. (27.2)
23. Choose the halogen(s) with the *(a)* largest atomic radius *(b)* weakest HX acid *(c)* highest electronegativity *(d)* light-sensitive silver salt *(e)* industrially most important HX acid *(f)* lowest ionization energy. (28)
24. Name three uses for chlorine. (28.3)
25. Complete each equation and identify the name (where appropriate) and use of each of the following reactions: *(a)* $Fe_2O_3 + 3CO \rightarrow$ *(b)* $2Br^- + Cl_2 \rightarrow$ *(c)* $2Cu_2S + 3O_2 \rightarrow$ *(d)* $2Al + Fe_2O_3 \rightarrow$ (25.2, 3; 26.2; 28.4)

Intra-Science
How the Sciences Work Together

The Process in Chemistry: The Miracle Fluorocarbon Resin

The *Guinness Book of World Records* says that the most slippery material in the world is a fluorocarbon resin that was invented more than 50 years ago. Two smooth, flat surfaces of this resin can slide past each other more easily than can two equivalent surfaces of any other material.

The name of the substance is Teflon.* It is best known for keeping eggs from sticking to frying pans, but it does far more. It has been to the moon. It helps prevent corrosion on the Statue of Liberty. It insulates wires. It coats lightbulbs so that the glass doesn't shatter when they break. It is used for gaskets, seals, and bearings in cars.

Teflon was discovered by accident during research into refrigeration gases in 1938. Its molecular structure is based on a chain of carbon atoms, the same as all polymers. Unlike some other fluoropolymers, Teflon has a chain completely surrounded by fluorine atoms. The bond between carbon and fluorine is very strong, and the fluorine atoms shield the vulnerable carbon chain. This unusual structure gives Teflon its unique properties. In addition to being extremely slippery, it is inert to almost every known chemical. It is useful over a remarkably wide temperature range, from −250 to 300°C.

The Connection to Physics
It took scientists years to understand Teflon's chemical properties and work out production methods before exploring the potential physical applications. It was not commercially available until 1948 and did not become widely used until the late 1950s. As technology advances, engineers are finding more and more uses for Teflon.

The biggest single use of Teflon today is not in kitchen pans, although this is a widespread commercial application, but in electrical

wire and cable. In modern office buildings, miles of telephone and computer cables are insulated with Teflon. It also coats the logging cables that carry instruments to measure geological characteristics at various depths in oil wells.

Teflon keeps Detroit Lions football fans comfortable as they sit under the 10-acre fiberglass-fabric roof of the Pontiac Silverdome, which is coated with translucent and flame-resistant Teflon. It also helps preserve the Statue of Liberty. When the statue was restored in the early 1980s, Teflon replaced shellac-soaked asbestos as a galvanic corrosion insulator and lubricant between the copper skin and the inner stainless-steel framework.

*Teflon is a registered trademark of DuPont Company.

perspiration escapes through the membrane, but wind, snow, and rain cannot penetrate from the outside.

Other applications include soil- and stain-repellent products, laboratory equipment, pumps, films, molded parts, food-packaging machinery, pipes to measure groundwater contamination, solar collectors, straps, and fasteners. Sometimes, Teflon replaces other materials, simply because it can do the job better. As the material in pipes for systems monitoring groundwater, for instance, it

When the U.S. put a person on the moon, Teflon was part of the space suits, fuel tanks, linings for liquid oxygen systems, heat shields, and reentry nose cone. It is useful to the space program because it doesn't get brittle and remains flexible at low temperatures.

The Connection to Health Sciences

Teflon is almost completely inert, and therefore serves as an ideal material for surgical implants. More than a million people are functioning with artificial arteries or veins made of a material containing Teflon. It replaces knee ligaments, and heart patches, and improves suture materials because its inertness makes it compatible with living tissue.

Teflon powder has been injected directly into the vocal cords of people who have lost their voices as a result of extreme vocal cord relaxation. The fluoropolymer acts as an inert substance that supports, or fills out, the cords.

The Connection to Other Sciences

Teflon's versatility has led to its use in hundreds of everyday objects. Sandwiched as a membrane between two layers of nylon, Teflon helps make outdoor gear both waterproof and breathable. It is used in rainwear, ski clothing, running shoes and suits, and military uniforms for wet, cold weather. The membrane has nine billion pores per square inch. Each pore is 20 000 times larger than a molecule of water vapor. Because people sweat vapor, not drops,

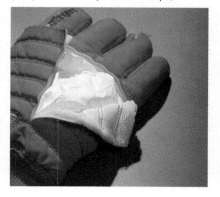

proved superior to glass, PVC (polyvinyl chloride), and stainless steel. It is the best available substance for protecting water samples against contamination, and it can withstand many years of exposure to chemicals, bacteria, and moisture.

The chemistry of fluorocarbon resins makes them difficult and expensive to put together. Because it is expensive, Teflon is used only where nothing else works, or where it is clearly superior to less expensive materials.

NUCLEAR REACTIONS

As a young man growing up in New York City, I was always fascinated by science fiction stories about space and traveling to other worlds. This made me wonder about the forces that would have to be understood and conquered to reach other planets and stars. My early curiosity convinced me to choose a career in physics so that I could learn about the fundamental nature of matter.

This unit introduces the concepts of nuclear reactions. Nuclear chemistry is the study of the interactions between atomic nuclei. The forces holding the nucleus together are much greater than those that hold electrons around atoms. Consequently, nuclear reactions involve a great deal more energy than chemical reactions.

For centuries, the way in which energy is produced by the sun and stars remained a mystery to scientists. It has only been during this century that we have discovered that this energy is released through the fusion of light-weight atoms. To study and understand the nuclear interactions involved in a fusion reaction, elaborate measuring instruments are needed. As a physicist working on the development of a fusion energy reactor, I am responsible for producing these "eyes and ears."

Donald H. Priester
Plasma Physicist
United States Department of Energy

Nuclear fusion is the source of the sun's energy.

Chapter 29 Nuclear Chemistry

Chapter Planner

29.1 The Composition and Structure of the Nucleus
Mass Defect and Nuclear Binding Energy
- Application Question: 11
Relationship Between Nuclear Stability and the Neutron/Proton Ratio
- Application Questions: 11, 12
Types of Nuclear Reactions
- Question: 11
  Application Question: 13
- Problem: 4, 5
- Application Problem: 3, 4

29.2 The Phenomenon of Radioactivity
Naturally Occurring Radioactive Nuclides
- Experiment: 38
  Questions: 1–9
  Problems: 1–3
  Application Questions: 1–10
- Chemistry Notebook
- Application Problems: 1, 2
Artificially Induced Radioactive Nuclides
- Questions: 12–14
  Application Questions: 14–16

29.3 Applications of Radioactivity
Radioactive Dating
- Question: 10
Radiosotope Uses in Medicine as Tracers
- Chemistry Notebook
  Question: 15

29.4 Energy from the Nucleus
Nuclear Fission
- Questions: 16–19
  Application Question: 17
Nuclear Fusion
- Question: 20

Teaching Strategies

Introduction

This important topic is often omitted since it is covered in the last chapter of the text. Spend at least one day teaching your students to balance nuclear reactions. Some of the societal issues suggested by this topic can be covered earlier in the year. If you maintain a bulletin board that covers recent developments in science, you will definitely have a few items that include medical uses of radioactivity, nuclear power plants, and particle accelerators.

29.1 The Composition and Structure of the Nucleus

Students have studied chemical reactions, all of which involve some rearrangement of particles outside of the nucleus. Now they will study a rearrangement of nuclear particles. You have emphasized that an element gets its identity from the number of protons in its nucleus; now students will see that the identity of elements can change. Compare the relative energy changes in a physical change, chemical reaction, and nuclear reaction. Calculate the binding energy released from a mass defect using Einstein's equation so they will have some understanding of the source of this enormous quantity of energy.

29.2 The Phenomenon of Radioactivity

The emphasis in this section is on balancing nuclear reactions. Students will be pleased to learn that this is not as difficult as it sounds. Those students with greater scholastic abilities will master the skill with little effort. If some of your students are going to be taking standardized examinations in chemistry,

make sure that they have some experience with nuclear equations and half-life determinations.

Spend some time reviewing the sample half-life problems and the sample nuclear equations. Students who have an advanced knowledge of mathematics will recognize that radioactive half-lives are an example of first-order rate reactions.

Other students should not be required to do complicated calculations, but can get an intuitive understanding of half-lives from some concrete examples. You can fill a one-liter graduate with water and sequentially remove half of the water. Another concrete example of half-life that students do not quickly forget is to dismiss the class by "half-class." A student will realize that there is always part of the class left after each "half-class" dismissal.

Make the following points to students who are learning to balance nuclear reactions:
(1) Mass numbers of reactants and products must balance.
(2) Atomic numbers of reactants and products must balance.
(3) The identity of products can be determined by comparing the atomic number of products with those on the Periodic Table.

The notation used to describe a nuclide might not be obvious to some students. Take a few minutes to explain that the subscripts and superscripts should be written on the *left* of the symbol: the subscript refers to the number of positively charged particles, and the superscript refers to the number of protons and neutrons.

Remind students that there are many isotopes that undergo radioactive decay. The complete list is in the *CRC Handbook of Chemistry and Physics*. Using the *Handbook* as a source for equations that you balance in class will give you the opportunity to emphasize its usefulness.

If your school has a Geiger counter, use it to demonstrate the background count and the increased count when the counter is exposed to a radioactive source. Devices designed for radon determinations are available from most hardware stores at a modest price.

29.3 Applications of Radioactivity

Never before has there been a discovery in science that offered such potential for both destruction and progress as the discovery of radioactivity. Allow your classroom to become a forum for a discussion of these societal issues. In anticipation of this topic, encourage your students to bring newspaper and magazine articles to class. Initiate a "nuclear" bulletin board several weeks before you begin this chapter. Clippings could refer to such diverse topics as CAT scanners, PET scanners, radiation therapy, tracer studies, carbon-dating to determine the age of archeological finds (the Shroud of Turin, for example), particle accelerators, nuclear power plants, disposal of nuclear waste, nuclear armaments, and ongoing effects of survivors of nuclear disasters. Help your students recognize that they will deal with issues related to the use of radioactivity for the rest of their lives.

Students will know that the medical hazards of radioactivity in the environment are a constant concern. They may not, however, be aware of the use of radionuclides in medicine. If time permits, you may wish to discuss the developments in nuclear medicine that improve diagnostic methods and treatment of a variety of illnesses.

Section 29.4 Energy from the Nucleus

Your students might be interested in knowing some of the research that led to the discovery of fission and eventually, to the nuclear reactor.

Several important discoveries of induced radioactivity preceded the discovery of fission. The first artificial nucleus was made by Lord Rutherford in 1914. He bombarded nitrogen with high-speed particles and produced oxygen $^-17$.

$$^{14}_{7}\text{N} + ^{4}_{2}\text{He} \rightarrow ^{17}_{8}\text{O} + ^{1}_{1}\text{H} + \text{energy}$$

In 1934, Frédéric and Irène Joliot-Curie discovered artificial radioactivity: the radioactive disintegration of a nuclide produced by an induced nuclear reaction. They bombarded aluminum with alpha particles, which produced an unstable form of phosphorous, which then decayed, emitting positrons and previously unknown radioisotopes.

Concurrent research with artificial radioactivity was being conducted by Enrico Fermi in Rome and by Lise Meitner, Otto Hahn, and Fritz Strassman in Berlin. Both groups bombarded uranium with neutrons, which produced new elements with atomic numbers greater than uranium. The heavy nuclides decayed rapidly, producing a mixture of unidentified elements. Lise Meitner was forced to leave Germany for Sweden in 1938. In Sweden she received reports of ongoing research from Otto Hahn. Based on these data, she concluded that the bombarded uranium split, forming two elements that were each approximately half of the mass of the uranium. Careful calculations indicated that some mass had disappeared. Meitner postulated that the mass was converted into energy according to Einstein's theory $E = Mc^2$.

$$^{235}_{92}\text{U} + ^{1}_{0}\text{N} \rightarrow \begin{array}{l} \text{fission} \\ \text{fragments} + \text{neutrons} + \text{energy} \end{array}$$

While Meitner was interpreting the results of nuclear fission in Sweden, Enerico Fermi fled from Rome and continued his research in the United States. In 1942, Fermi and his team produced the first artificially created, self-sustained chain reaction at the University of Chicago. Fermi's work was a critical step in harnessing nuclear energy for the first atomic bomb.

Controlled fission of radioactive material supplies the energy for nuclear power plants. Although a tremendous amount of energy can be produced in a nuclear reactor, there are potential dangers in the large amounts of radioactive wastes that accumulate in spent fuel rods.

Because a fusion reactor would not accumulate heavy radioactive isotopes, it would be a much safer way to harness the energy of nuclear reactions; however, the technology to accomplish this is still being developed. Present fusion reactors require more energy than they produce.

References and Resources

Books and Periodicals

Alazrak, Naomi P., and Fred S. Mishkin, eds. *Fundamentals of Nuclear Medicine*, The Society of Nuclear Medicine, Inc., 1984.

Asimov, Isaac. *The Neutrino*, Avon Books, 1983.

Atwood, Charles H. "Nuclear Diagnosis," *ChemMatters*, Vol. 3, 1985, p. 4.

Atwood, Charles H. "Chernobyl — What Happened?," *Journal of Chemical Education*, Vol. 65, 1988, p. 1037

Bodner, George. "Binding Energy and Atomic Weight Calculations," *Journal of Chemical Education*, Vol. 55, 1978, p. 598.

Champlin, Richard. "With the Chernobyl Victims," *Miami Herald*, July 13, 1986.

Colburn, Don. "Radiation: What It Is, What It Does," *Washington Post Magazine Section*, May 7, 1986. Reprinted in SIRS-Medical Sciences, 1986.

Downey, Daniel M., and Glen Simolunas. "Measurement of Radon in Indoor Air — A Laboratory Exercise" *Journal of Chemical Education*, Vol. 65, 1988, p. 1042.

Felia, Arthur. "The Role of Chemistry in Positron Emission Tomography," *Journal of Chemical Education*, Vol. 65, 1988. p. 655.

Fundamental Particles and Interactions Chart Committee. "Fundamental Particles," *Physics Today*, Vol. 26, 1988, p. 556.

Hartt, Kenneth. "Beta Decay," *Physics Today*, Vol. 26, 1988, p. 471.

Mann, Charles C., and Robert P. Crease, "Waiting for Decay," *Science*, 1986.

The Nucleus, Inc. "Radon Issues and Answers" booklet. 761 Emory Valley Road, Oak Ridge, TN, 37830-2561. Available free.

Raloff, Janet. "The Complete Breeder?" *Science News*, Vol. 127, Jan. 26, 1985.

Simmons, Janien. "NMR's Magnetic Appeal," *Yale Scientific*, Winter, 1985.

Suillen, Michael. "The Paradox of Antimatter," *Science Digest*, Feb. 1985.

Sutton, Christine. "Neutrons Attack Cancer," *New Scientist*, Sept. 19, 1985.

Taubes, Gay. "The Ultimate Theory of Everything," *Discover*, April, 1985.

Weast, Ronert C., Melvin J. Astle, and William H. Beyer, eds. *CRC Handbook of Chemistry and Physics*, 69th ed., CRC Press, Inc., 1988.

Wolke, Robert L. "Marie Curie's Doctoral Thesis: Prelude to a Nobel Prize," *Journal of Chemical Education*, 1988, p. 561.

Audiovisual Resources

HUMAN RELATIONS MEDIA, Nuclear Weapons: Concepts and Controversies (Two-part filmstrip, filmstrip on videocassette) (Award-Winner: Society for Technical Communication). 175 Tompkins Ave. Pleasantville, NY 10570-9973).

Computer Software

Donald L. Pavia, Atomic Structure 2, Apple II's TRS-80, Compress.

Queue, Chemistry Package II, Radioactive Decay, Apple II's.

Queue, General Chemistry 1B, Topic 8, Nuclear Chemistry, Apple II's.

Queue, The Physical Science Programs, Radioactivity, Apple II's.

Queue, Physics Simulations Package 1, Radioactivity Decay, Apple II's.

Queue, Laboratory Simulations in Atomic Physics (includes Thomson and Milliken simulations).

Answers and Solutions

Questions

1. *(a)* The spontaneous breakdown of an unstable atomic nucleus with the release of particles and rays *(b)* In 1896 by Henri Becquerel *(c)* By accident. Becquerel found that uranium compounds give off invisible rays that affect a photographic plate in the same way light does.

2. *(a)* Pitchblende is four times as radioactive as the amount of uranium it contains warrants. *(b)* Polonium, and radium, two elements much more radioactive than uranium, were present in pitchblende.

3. Radium is more than 1 000 000 times as radioactive as the same mass of uranium. Polonium is 5000 times as radioactive as the same mass of radium.

4. The length of time during which half of a given number of atoms of a radioactive nuclide decay.

5. The nucleus

6. *(a)* In luminous paint *(b)* The beta particles emitted by Pm-147 are not as hazardous, and the nuclide has a much shorter half-life.

7. The atomic number decreases by two, which now identifies the nuclide, and its mass number decreases by four.

8. The atomic number increases by one, which now identifies the nuclide, and its mass number remains the same.

9. *(a)* Radon is continually being produced by the decay of uranium-238, which is widely distributed in the earth's crust. *(b)* It is used in the treatment of disease and to help in predicting earthquakes. *(c)* It may be a health hazard if it collects in airtight energy-efficient homes.

10. *(a)* By measuring the amount of parent nuclide and daughter nuclides in the sample *(b)* That radioactive substances decay at known rates, that these rates have not changed during the existence of the mineral, and that there has been no gain or loss of parent or daughter nuclides except by radioactive decay

11. Radioactive decay: The nucleus releases an alpha or beta particle and gamma rays, forming a slightly lighter, more stable nucleus. Nuclear disintegration: A nucleus is bombarded with alpha particles, protons, deuterons, neutrons, or other particles. The now unstable nucleus emits a proton or a neutron and becomes more stable. Fission: A very heavy nucleus splits to form medium-weight nuclei. Fusion: Lightweight nuclei combine to form heavier, more stable nuclei

12. $^{9}_{4}\text{Be} + ^{4}_{2}\text{He} \rightarrow ^{12}_{6}\text{C} + ^{1}_{0}\text{n}$

13. Because they do not carry any charge. They are not attracted or repelled by a nucleus and can easily penetrate it.

14. It may go through the atom, or it may be caught by the nucleus. It may cause the nucleus to disintegrate.

15. To treat certain forms of cancer, for diagnostic purposes, for test purposes, in food preservation, as tracers

16. Uranium-238, 99.3%. Uranium-235, 0.7%. Uranium-234, insignificant traces

17. The $^{235}_{92}\text{U}$ nucleus captures a slow neutron. This produces a highly unstable nucleus that immediately undergoes fission.

18. *(a)* A reaction in which the material or energy that starts the reaction is also one of the products *(b)* A device in which the controlled fission of radioactive material produces new radioactive substances and energy *(c)* The amount of radioactive material required to sustain a chain reaction

19. Student answer

20. A fusion reaction in which four hydrogen nuclei combine to form a helium nucleus

Problems

1. For three-fourths to decay requires 2 half-lives. 2 × 18.72 da = 37.44 da

2. 27.0 hr ÷ 6.75 hr/half-life = 4.00 half-lives. After 4 half-lives, one-sixteenth remains.

3. 4806 yr ÷ 1602 yr/half-life = 3.00 half-lives. After 3 half-lives, one-eighth remains. 0.250 g/8 = 0.0312 g

4. [3(1.007276 u) + 4(1.008665 u)] − 7.01436 u = 0.04213 u

5. [10(1.007276 u) + 10(1.008665 u)] − 19.98695 u = 0.17246 u

Application Questions

1. Beyond bismuth
2. The radiation easily penetrates the black paper and affects the film.
3. The radioactive material increases the ionization of the surrounding air molecules and permits the electroscope to be discharged more rapidly.
4. The Geiger counter
5. *(a)* Chemical form, low temperature, high pressure, a varying electric field, and interaction of the decaying nucleus and its valence electrons *(b)* Slight
6.

Particle	Identity	Mass number	Charge	Speed	Penetrating ability
alpha particle	helium nucleus	4	+2	1/10 the speed of light	low penetrating ability
beta particle	electron	0	−1	near the speed of light	100 times the penetrating ability of alpha particles

7. *(a)* High-energy electromagnetic waves *(b)* They are produced when nuclear particles undergo transitions in nuclear energy levels. *(c)* They are the most penetrating of the radiations.
8. $^{210}_{84}Po \rightarrow ^{206}_{82}Pb + ^{4}_{2}He$
9. $^{210}_{82}Pb \rightarrow ^{210}_{83}Bi + ^{0}_{-1}e$
10. $^{232}_{90}Th \rightarrow ^{228}_{88}Ra + ^{4}_{2}He$
    $^{228}_{88}Ra \rightarrow ^{228}_{89}Ac + ^{0}_{-1}e$
    $^{228}_{89}Ac \rightarrow ^{228}_{90}Th + ^{0}_{-1}e$
    $^{228}_{90}Th \rightarrow ^{224}_{88}Ra + ^{4}_{2}He$
11. *(a)* The elements with lowest and highest mass numbers have smaller binding energies per nuclear particle. The elements with intermediate mass numbers have the largest binding energies per nuclear particle. *(b)* The greater the binding energy per nuclear particle, the more stable the nucleus.

12. The ratio of neutrons to protons and the even-odd nature of the number of protons and the number of neutrons in a nucleus affect its stability.
13. Each type of nuclear reaction results in more stable nuclei because the products are nuclei with greater binding energy per nuclear particle.
14. *(a)* Rutherford *(b)* in 1919 *(c)* $^{14}_{7}N + ^{4}_{2}He \rightarrow ^{17}_{8}O + ^{1}_{1}H$
15. *(a)* By bombarding lithium with high-speed protons and comparing the actual energy release with Einstein's prediction *(b)* J.D. Cockcroft and E.T.S. Walton, English
16. $^{238}_{92}U$ is bombarded with neutrons to produce the unstable isotope of uranium, $^{239}_{92}U$. This emits a beta particle to form neptunium, $^{239}_{93}Np$. This in turn emits a beta particle to form plutonium, $^{239}_{94}Pu$.
17. The fission of $^{235}_{92}U$ is started by a neutron. During the fission process, several neutrons are emitted. Some of these initiate fission in other $^{235}_{92}U$ nuclei, which in turn produce other neutrons.
18. It is constructed by blocks of natural uranium alternated with blocks of graphite that serve as a moderator. Control rods and a cooling apparatus are also incorporated.
19. Scientists must find ways to confine the nuclear fuel and make it hot enough and dense enough long enough for fusion to occur. The energy produced must be greater than the energy required.
20. Student answer
21. Student answer

Application Problems

1. 7.646 da ÷ 3.823 da/half-life = 2.000 half-lives. After 2 half-lives, one-fourth remains. 0.0500 g × 4 = 0.200 g
2. 10^5 yr ÷ 24 100 yr/half-life = approximately 4 half-lives. After 4 half-lives, one-sixteenth remains. 100 g/16 = 6.25 g. Approximately 6 g remain.
3. 0.042 13 *u*/7 nuclear particles = 0.006 019 *u*/nuclear particle. 0.006019 *u* × 931 MeV/*u* = 5.60 MeV.
4. [92(1.007276 *u*) + 146(1.008665 *u*)] − 238.0003 *u* = 1.9342 *u*. 1.9342 *u*/238 nuclear particles = 0.008 1269 *u*/nuclear particle. 0.008 1269 *u* × 931 MeV/*u* = 7.57 MeV

Teacher's Notes

Chapter 29 Nuclear Chemistry

INTRODUCTION

Chemical reactions involve only the electronic structures of atoms. We have paid little attention so far to atomic nuclei other than to establish the masses of atoms and to note that their positive charge determines the number of electrons in a neutral atom. In this chapter you will study the nature and reaction of nuclei and the characteristics and uses of the different forms of radiation.

LOOKING AHEAD

As you study this chapter, pay particular attention to the structure and properties of nuclei and the different kinds of nuclear reactions. Contrast the source and magnitude of the energies liberated in nuclear transformations to those in chemical reactions. Investigate the effect of radiation on biological systems and the great usefulness of radioisotopes.

SECTION PREVIEW

29.1 The Composition and Structure of the Nucleus
29.2 The Phenomenon of Radioactivity
29.3 Applications of Radioactivity
29.4 Energy from the Nucleus

Nuclear power plants such as this one in Pennsylvania use the fission of uranium nuclei to produce energy.

29.1 The Composition and Structure of the Nucleus

SECTION OBJECTIVES

- Distinguish between chemical and nuclear reactions.
- Define mass defect and nuclear binding energy.
- Discuss the factors affecting the stability of atomic nuclei.
- Identify four types of nuclear reactions.

In your study of atomic structure you investigated experiments which revealed the general composition of the atomic nucleus. Recall that all nuclei except for ordinary hydrogen (protium) contain two fundamental particles—protons and neutrons. Nuclei of protium contain only a proton. *The protons and neutrons are referred to as* **nucleons.** The nucleus is quite small (about 10^{-13} cm) compared to the size of the atom (about 10^{-8} cm). The nucleons are therefore packed tightly together. Because of this close packing, positive charges repel each other. An attractive force is thus required to hold the positively charged protons in the tiny nucleus. *The force that keeps the nucleons together is called the* **strong nuclear force.** This force is the powerful short-range force of attraction between, respectively, proton and proton, proton and neutron, and neutron and neutron. The strong nuclear force affects only the nucleons, and is distinct from the electrostatic force or gravitational force.

Recall that all atoms of the same element have the same atomic number, that is, the same number of protons. The atoms of a given element can have different mass numbers, however. In other words, because the mass number represents the total number of nucleons in a given nucleus, the atoms of a given element can have different numbers of neutrons. Because the nuclear properties of an atom depend on the number of protons and neutrons in its nucleus, we must distinguish between different isotopes of an element.

Recall that in chemical reactions the atoms and ions are rearranged to form new substances simply by the breaking and making of chemical bonds. The atoms are neither created nor destroyed. Only the outer electrons of atoms are affected in chemical reactions. Recall also that chemical reactions involve the release or absorption of certain amounts of energy, and that the rates of reactions are affected by conditions of pressure, temperature, concentration, and catalysts. In contrast, in nuclear changes, isotopes of one element are converted to isotopes of another element as the contents of the nucleus, the nucleons, change. In contrast to chemical reactions, however, the amounts of energies involved in nuclear reactions are huge, and the reactions are not affected by conditions of pressure, temperature, or catalysts.

Up to this point you have not considered reactions in which the nuclei of atoms undergo a change. You are now ready to investigate in more detail, the structure, properties, composition, and reactions of atomic nuclei.

Nucleus comes from the Latin word *nux*, meaning 'nut' or 'seed.'

The concept of neutrons and protons presents a simplified, but workable, model of the nucleus. The Standard Model of Fundamental Particles and Interactions presents a more detailed model of the nucleus that is based on recent experiments in the field of particle physics.

Compare the free energy changes associated with a physical, chemical, and nuclear reaction.

Mass Defect and Nuclear Binding Energy

On the atomic mass scale, the isotope of carbon with six protons and six neutrons in its nucleus is defined as having an *atomic mass* of exactly 12 u. On this scale, a $^4_2$He nucleus has a mass of 4.0015 u. The mass of a proton is 1.0073 u and the mass of a neutron is 1.0087 u. A $^4_2$He nucleus contains two protons and two neutrons. You might expect its mass to be the combined mass of these four particles — 4.0320 u [2(1.0073 u) + 2(1.0087 u = 4.0320 u]. Note, however, that there is a *difference* of 0.0305 u between the measured mass (4.0015 u) and the calculated mass, 4.0320 u, of a $^4_2$He nucleus.

Calculate the mass of 1 u in grams.
1 hydrogen atom has a mass of 1.00 u
1 mol of hydrogen atoms has a mass of 1.00 g
1 mol of hydrogen contains 6.02 × 10²³ atoms
1.00 g/6.02 × 10²³ atoms = mass of 1 atom in grams = mass of 1 u
1.00 u = 1.66 × 10⁻²⁴ g

This difference in mass is called the nuclear mass defect. *The **nuclear mass defect** is the difference between the mass of a nucleus and the sum of the masses of its constituent particles.* The mass defect is converted into energy units by using Einstein's equation, $E = mc^2$. The energy calculated in this manner is the energy released when a nucleus is formed from the particles that compose it. This energy is generally referred to as the **nuclear binding energy,** *which is the energy released when a nucleus is formed from its constituent particles.* This energy must be supplied to a nucleus to separate it into its constituent particles.

Calculations of binding energies of the atoms of the elements show that the lightest and the heaviest elements have the smallest binding energies per nuclear particle. Elements having intermediate atomic weights have the greatest binding energies per nuclear particle, as shown in Figure 29-1. The elements with the greatest binding energies per nuclear particle are those with the most stable nuclei. The nuclei of the lightest and heaviest elements are therefore less stable than the nuclei of elements having intermediate atomic weights.

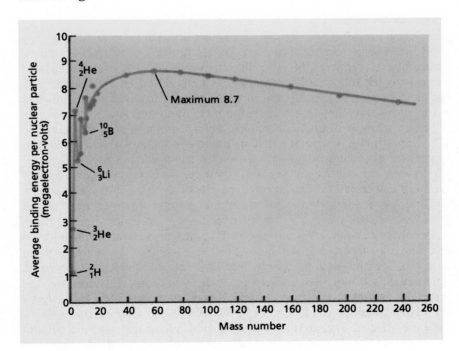

Figure 29-1 This graph shows the relationship between binding energy per nuclear particle and mass number.

A thermodynamic comparison can be made between the stability of a nucleus and the stability of a compound. Elements containing nuclei with the greatest binding energy per nuclear particle are the least likely to decay. Compounds that are formed with the greatest release of energy are the least likely to decompose.

Relationship Between Nuclear Stability and the Neutron/Proton Ratio

The stability of atomic nuclei is affected by the ratio of the neutrons to protons that compose them. By plotting the number of protons against the number of neutrons in the nuclei, a belt-like graph is obtained as shown in Figure 29-2. Among atoms having low atomic numbers, the most stable nuclei are those with a neutron-to-proton ratio of 1:1. Nuclei with a greater number of neutrons than protons have lower binding energies and are less stable. Many properties of nuclear particles indicate that energy levels exist

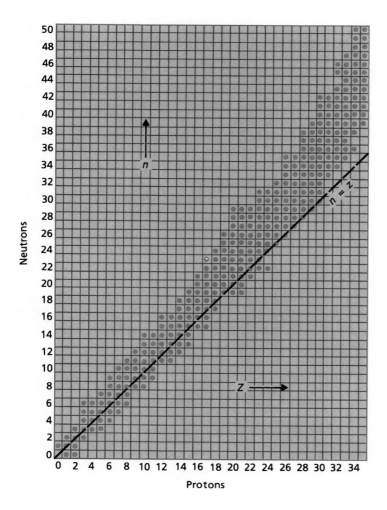

Figure 29-2 The relationship between the number of neutrons and protons in atomic nuclei is shown here.

within the atomic nucleus. In nuclei with an equal number of neutrons and protons, the particles apparently occupy the lowest energy levels in the nucleus. In this way, they give it stability. In low-atomic-number nuclei that contain an excess of neutrons over protons, however, some of the neutrons seem to occupy higher energy levels. This reduces the binding energy and consequently lowers the stability of the nucleus.

As the atomic number increases, stable nuclei have a neutron-to-proton ratio increasingly greater than 1:1. For example, $^{127}_{53}\text{I}$, the stable isotope of iodine, has a $\frac{\text{neutron}}{\text{proton}}$ ratio of $\frac{74}{53}$ (about 1.40:1). The stable end product of the uranium-238 decay series is $^{206}_{82}\text{Pb}$. It has a $\frac{\text{neutron}}{\text{proton}}$ ratio of $\frac{124}{82}$ (about 1.51:1).

The stability of a nucleus depends *also* on the even-odd relationship of the number of protons and neutrons. Most stable nuclei (157 of them) have even numbers of both protons and neutrons. There are 55 stable nuclei having an even number of protons and an odd number of neutrons. Fifty stable nuclei have an odd number of protons and an even number of neutrons. Only four stable nuclei having odd numbers of both protons and neutrons are known.

1. Chemical: No new elements are formed; only electrons outside the nucleus are involved; reactions absorb or release small amounts of energy; reaction rates are affected by pressure, temperature, concentration, and catalysts. Nuclear: New elements are formed; particles inside the nucleus are involved; reactions involve huge amounts of energy; reaction cannot be affected by pressure, temperature, or catalysts.

2. *(a)* The difference between the mass of a nucleus and the sum of the masses of its constituent particles. *(b)* The energy released when a nucleus is formed from its constituent particles.

3. The most stable nuclei are those whose neutron-to-proton ratio is 1:1. Nuclei with a greater number of neutrons than protons have lower binding energies and are less stable.

4. Radioactive decay, nuclear disintegration, fission, fusion

SECTION OBJECTIVES

- Identify the unusual properties of radioactive nuclides.
- Define *half-life* and give an example.
- Name the three types of radioactive emissions and write their nuclear symbols.
- Complete and balance nuclear reactions.
- Define the term *transmutation* and distinguish between naturally occurring and induced radioactive nuclides.
- Identify transuranium elements as the result of the bombardment of nuclei by neutrons.

Fission comes fro the Latin word *fissionem,* which means 'splitting.'

Fusion comes from the Latin word *fusionem,* which means 'a pouring out.'

Types of Nuclear Reactions

Because of the difference in stability of different nuclei, there are four types of nuclear reactions. In each type a small amount of the mass of the reactants is converted into energy, forming products of greater stability.

1. **Radioactive decay** *refers to the emission of an alpha particle, a beta particle, or gamma radiation and the formation of a slightly lighter and more stable nucleus.*

2. A nucleus is bombarded with alpha particles, protons, deuterons (deuterium nuclei, ^2_1H), neutrons, or other particles. *The unstable nucleus emits a proton or a neutron and becomes more stable in a process called* **nuclear disintegration.**

3. **Fission** *refers to the process in which a very heavy nucleus splits to form medium-weight nuclei.*

4. **Fusion** *refers to the process in which lightweight nuclei combine to form heavier, more stable nuclei.*

Section Review

1. Summarize the difference between chemical and nuclear changes.
2. What is the meaning of *(a)* mass defect, and *(b)* binding energy?
3. How is nuclear stability related to the neutron/proton ratio?
4. Name four types of nuclear reactions.

29.2 The Phenomenon of Radioactivity

In 1896 Henri Becquerel (bek-*rel*) (1852–1908) was studying the properties of uranium compounds. He was particularly interested in the ability of these compounds to **fluoresce** or *give off visible light after being exposed to sunlight.* By accident Becquerel found that, whether they flouresce or not, all uranium compounds give off invisible rays. He discovered that these rays penetrate the lightproof covering of a photographic plate and affect the film as if it had been exposed to light rays directly. Figure 29-4 shows an example of this phenomenon. Substances that give off such invisible rays are radioactive, and the property is called radioactivity. **Radioactivity** *is the process whereby an unstable nucleus forms a more stable nucleus by the release of high energy particles and radiation.*

Becquerel was very interested in the source of radioactivity. At his suggestion, Marie Curie (1867–1934) and Pierre Curie (1859–1906) began to investigate the properties of uranium and its ores. They soon learned that uranium and uranium compounds are only mildly radioactive. They also discovered that one uranium ore, pitchblende, has four times the amount of radioactivity expected on the basis of its uranium content.

The Curies discovered two new radioactive metallic elements in pitchblende in 1898. These elements, *polonium* and *radium*, account for the high radioactivity of pitchblende. Radium is more than a million times as radioactive as the same mass of uranium. Polonium is five thousand times as radioactive as the same mass of radium.

Figure 29-4 A fragment of metallic uranium, one of the radioactive elements *(top)*. A photograph produced when radiation from the same fragment of uranium penetrated the light-tight wrappings of a photographic plate *(bottom)*.

The heaviest member of Group 16, the oxygen family, is polonium. It is a silvery metal, confirming the trend from nonmetallic to metallic properties with increasing atomic mass shown by families of active elements at the right of the Periodic Table. Radium, also a silvery metal, is the heaviest member of Group 2, the calcium family. Its chemical properties are similar to those of barium. Some properties of the most abundant nuclides of these radioactive elements are given in Table 29-1.

Radium is always found in uranium ores although it is present in only minute amounts. For instance, there is less than 1 gram of radium in 6×10^3 kilograms of pitchblende. It never occurs in these ores in a proportion greater than 1 part of radium to 3×10^6 parts of uranium. The production of radium bromide (its usual commercial form) from uranium ore is a long, difficult, and costly procedure. Radium was produced commercially in Europe, North America, and Australia from 1904 until 1960. Today, there still remains an ample supply for its limited medicinal and industrial uses. Also, some of the more recently produced artificial isotopes such as ^{60}Co are being used instead of radium.

TABLE 29-1   PROPERTIES OF SOME RADIOACTIVE NUCLIDES

Nuclide	Electron Configuration	Principal Oxidation Numbers	Melting Point (°C)	Boiling Point (°C)	Color	Density	Atomic Radius (Å)	Ionic Radius (Å)	Decay Type	Half-life
U-238	2,8,18,32,21,9,2	+4,+6	1132	3818	silver metal	19.0 g/cm³	1.42	0.93(+4)	α	4.468×10^9y
Th-232	2,8,18,32,18,10,2	+4	1750	4790	silver metal	11.7 g/cm³	1.65	0.99(+4)	α	1.41×10^{10}y
Ra-226	2,8,18,32,18,8,2	+2	700	1140	silver metal	5(?) g/cm³	2.20	1.40(+2)	α	1.60×10^3 Y
Rn-222	2,8,18,32,18,8	0	−71	−61.8	colorless gas	9.73 g/L	1.43(?)		α	3.8235 d
Po-210	2,8,18,32,18,6	−2,+2, +4, +6	254	962	silver metal	9.32g/mL	1.46	2.30(−2)	α	138.38 d

Chemistry Notebook

Radium—Cause of and Cure for Cancer

Radium is dangerous because it emits alpha particles, beta particles, and gamma radiation. Since it is chemically similar to calcium and magnesium, it can be incorporated into body tissue.

Alpha particles from radium compounds are responsible for the luminous glow of older watch dials. Years ago, the numerals were painted by hand on watch faces so they would glow in the dark. To paint the tiny numerals, factory workers licked their paintbrushes into fine points. It was nearly 20 years before anyone realized that alpha radiation from the radium paint penetrated tongues and mouth linings and eventually

caused mouth cancers. Modern watches have LEDs (light-emitting diodes), eliminating the need for radium dials.

Ironically, gamma rays emitted by radium can make the element helpful. Radium

has been used in the treatment of some forms of cancer, although it must be handled with extreme care. Marie Curie, co-discoverer with her husband, of radium was unaware of its dangers, and died from leukemia.

Naturally Occurring Radioactive Nuclides

Of the elements known to Becquerel in 1896, testing by the Curies showed that only uranium and thorium were radioactive. The Curies discovered two more radioactive elements, polonium and radium. Since that time, many other natural radioactive nuclides have been identified. Some of these are listed in Table 29-2. All nuclides of the elements beyond bismuth (Bi, at. no. 83) in the Periodic Table are radioactive. Only polonium (Po, at. no. 84), radon (Rn, at. no. 86), radium (Ra, at. no. 88), actinium (Ac, at. no. 89), thorium (Th, at. no. 90), protactinium (Pa, at. no. 91), and uranium (U, at. no. 92) have any natural radioactive nuclides. The remainder of the elements beyond bismuth have only radioactive nuclides that have been artificially produced.

Because of their radioactivity, these nuclides have several unusual properties. Some of these properties are:

1. They affect the light-sensitive emulsion on a photographic film. Photographic film can be wrapped in heavy black paper and stored in the dark. This will protect it from ordinary light, but radiations from radioactive nuclides penetrate the wrapping and affect the film in the same way that light does when the film is exposed. When the film is developed, a black spot shows

TABLE 29-2   REPRESENTATIVE NATURAL RADIOACTIVE NUCLIDES WITH ATOMIC NUMBERS UP TO 83

Nuclide	Abundance Natural Element (%)	Half-life (years)	Nuclide	Abundance in Natural Element (%)	Half-life (years)
$^{40}_{19}K$	0.0118	1.3×10^9	$^{130}_{52}Te$	34.49	8×10^{20}
$^{48}_{20}Ca$	0.185	$> 10^{18}$	$^{180}_{74}W$	0.135	$> 1.1 \times 10^{15}$
$^{64}_{30}Zn$	48.89	$> 8 \times 10^{15}$	$^{182}_{74}W$	26.4	$> 2 \times 10^{17}$
$^{70}_{30}Zn$	0.62	$> 10^{15}$	$^{183}_{74}W$	14.4	$> 1.1 \times 10^{17}$
$^{87}_{37}Rb$	27.85	4.8×10^{10}	$^{186}_{74}W$	28.4	$> 6 \times 10^{15}$
$^{113}_{48}Cd$	12.26	$> 1.3 \times 10^{15}$	$^{190}_{78}Pt$	0.0127	6.9×10^{11}
$^{116}_{48}Cd$	7.58	$> 10^{17}$	$^{192}_{78}Pt$	0.78	10^{15}
$^{124}_{50}Sn$	5.98	$> 2 \times 10^{17}$	$^{198}_{78}Pt$	7.19	$> 10^{15}$
$^{123}_{51}Sb$	42.75	$> 1.3 \times 10^{16}$	$^{196}_{80}Hg$	0.146	$> 1 \times 10^{14}$
$^{123}_{52}Te$	0.87	1.2×10^{13}	$^{209}_{83}Bi$	100	$> 2 \times 10^{18}$

up on the negative where the invisible radiation struck it. The rays from radioactive nuclides penetrate paper, wood, flesh, and thin sheets of metal.

2. They produce an electric charge in the surrounding air. The radiation from radioactive nuclides ionizes the molecules of the gases in the air. These ionized molecules conduct electric charges away from the knob of a charged electroscope, thus discharging it. In early studies, radioactivity was measured by the rate at which it discharged an electroscope. In Geiger counters, which are used today, radiation ionizes low pressure gas in a tube. Electricity thus passes through the tube for an instant. The passage of electricity registers as a "click" in a set of earphones.

3. They produce fluorescence with certain other compounds. A small quantity of radium bromide added to zinc sulfide causes the sulfide to glow. Since the glow is visible in the dark, the mixture has been used in making luminous paint. A radioactive promethium compound is now used instead of radium bromide.

4. Their radiations have special physiological effects. The radiation from radium can destroy the germinating power of seed. It can kill bacteria or animals. People who work with radium may be severely burned by the rays it emits. Such burns heal slowly and can be fatal. Controlled radiations from a variety of radioactive nuclides are used in the treatment of cancer and certain skin diseases, however.

5. They undergo radioactive decay. The atoms of all radioactive nuclides steadily decay into simpler atoms as they release radiation. For example, one-half of any number of radium-226 atoms decays into simpler atoms in 1602 years. One-half of what remains, or one-fourth of the original atoms, decays in the next 1602 years. One-half of what is left, or one-eighth of the original atoms, decays in the next 1602 years, and so on. This period of 1602 years is called the half-life of radium-226. **Half-life** *is the length of time during which half of a given number of atoms of a radioactive nuclide decays.* Each radioactive nuclide has its own half-life.

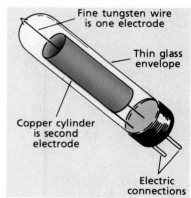

Fine tungsten wire is one electrode

Thin glass envelope

Copper cylinder is second electrode

Electric connections

Figure 29-5 A diagram showing the construction of a Geiger-Müller counter tube. Radiation passing through the tube ionizes the gas it contains and enables current to flow.

Radiation can be absorbed by DNA of cells, resulting in the alteration of genetic material. This can result in long-term deleterious effects. The ultraviolet radiation of the sun can cause the same damage as the radiation coming from decaying nuclides.

Depending on its chemical form, the half-life of a nuclide varies slightly. For example, beryllium-7 decays at very slightly different rates when it is elemental, in BeO or as Be^{2+} in an HCl(aq) solution. It is found in solid form as $Be_2P_2O_7$ The decay rate is also affected by large changes in the environment. Some of these changes occur at very low temperatures, high pressure, or large variations in a surrounding electric field. Somewhat larger alteration of the half-life of certain nuclides occurs when the decaying nucleus and its valence electrons interact.

Sample Problem 29.1

Phosphorus-32, $^{32}_{15}P$, a radioactive isotope of phosphorus, has a half-life of 14.3 years. How many grams remain after 57.2 years if you have 4.0 g of the isotope at the beginning?

Solution

Step 1. Analyze

Given: original amount of phosphorus-32 = 4.0 g
           half-life of phosphorus-32 = 14.3 years
           time elapsed = 57.2 years
Unknown: amount of phosphorus-32 remaining after 57.2 years

Step 2. Plan

In order to determine the number of grams of phosphorus-32 remaining, we must first find the number of half-lives that have passed. Then, the amount of phosphorous-32 is determined by reducing the original amount by ½ for every half-life that has passed.

$$\text{number of half-lives} = \text{time elapsed (years)} \times \left(\frac{1 \text{ half-life}}{14.3 \text{ years}}\right)$$

$$\text{original amount of phosphorus-32 remaining} \times \frac{1}{2} \text{ for each half-life}$$

$$= \text{amount of phosphorus-32 remaining}$$

Step 3. Compute

$$4.00 \text{ half-lives} = 57.2 \text{ years} \times \frac{1 \text{ half-life}}{14.3 \text{ years}} = 4 \text{ half-lives}$$

$$4.0 \text{ g} \times \frac{1}{2} \times \frac{1}{2} \times \frac{1}{2} \times \frac{1}{2} = 0.25 \text{ g}$$

Step 4. Evaluate

A period of 57.2 years is four half-lives for phosphorus-32. That means that only $\frac{1}{16}\left(\frac{1}{2} \times \frac{1}{2} \times \frac{1}{2} \times \frac{1}{2}\right)$ of the original amount will remain. In other words, at the end of one half-life, 2.0 g of phosphorus-32 remains; 1.0 g remains at the end of two half-lives; 0.50 g remains at the end of three half-lives; and 0.25 g remains at the end of four half-lives.

Practice Problems
1. How many years will be needed for the decay of $\frac{15}{16}$ of a given amount of radium-226? *(Ans.)* 6408 years
2. The half-life of radon-222 is 3.823 days. After what time will only one-fourth of a given amount of radon remain? *(Ans.)* 7.646 days
3. The half-life of polonium-210 is 138.4 days. What fraction remains after 415.2 days? *(Ans.)* $\frac{1}{8}$

Nature of radiation   The radiation given off by radioactive nuclides can be separated into three different kinds of particles and rays.

1. The α (alpha particles are helium nuclei. Their mass is nearly four times that of a protium atom. They have a charge of +2 and move at speeds that are approximately one-tenth the speed of light. Because of their relatively low speed, they have low penetrating ability. A thin sheet of aluminum foil or a sheet of paper stops them. They burn flesh and ionize air easily, however.

2. The β (beta) particles are electrons. They travel at speeds close to the speed of light, with penetrating ability about 100 times greater than that of alpha particles.

3. The γ (gamma) rays are high-energy electromagnetic waves. They are the same kind of radiation as visible light, but are of much shorter wavelength and higher frequency. Gamma rays are produced when nuclear particles undergo transitions in nuclear-energy levels. They are the most penetrating of the radiations given off by radioactive nuclides. Alpha and beta particles are seldom, if ever, given off simultaneously from the same nucleus. Gamma rays, however, are often produced along with either alpha or beta particles.

STUDY HINT
Electrons are described in Section 3.2.

Alpha and beta rays were named by Ernest Rutherford.

STUDY HINT
Electromagnetic radiation is described in Section 4.1.

TABLE 29-2   CHARACTERISTICS OF ALPHA, BETA, AND GAMMA EMISSIONS

Particle	Symbol	Composition	Charge	Penetrating Power
Alpha, α	^4_2He	2 protons 2 neutrons	2+	low
Beta, β	$^0_{-1}e$	electron	1−	100 times greater than that of alpha
Gamma, γ	$^0_0\gamma$	electromagnetic waves	0	very great

Figure 29-6 shows the effect of a powerful magnetic field on the complex radiation given off by a small particle of radioactive material. The field is perpendicular to the plane of the paper. The heavy alpha particles are slightly deflected in one direction. The lighter beta particles are deflected more sharply in the opposite direction. The gamma rays, being uncharged, are not affected by the magnet.

Radioactive nuclides such as uranium-238, radium-226, and polonium-210 decay spontaneously, yielding energy. A long series of experiments has shown that this energy results from the decay of their nuclei. Alpha and beta particles are the products of such nuclear decay. Certain heavy nuclei break down spontaneously into simpler and lighter nuclei, releasing enormous quantities of energy.

At first it was believed that radioactive nuclides did not lose mass and would give off energy forever. More careful investigation proved that radioactive materials do lose mass slowly, however. The presence of electrons and helium nuclei among the radiations is evidence for the loss of mass.

A series of related radioactive nuclides   All naturally occurring radioactive nuclides with atomic numbers greater than 83 belong to one of

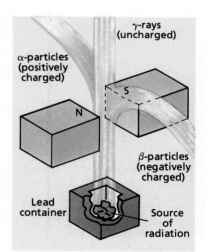

Figure 29-6   The effect of a magnet on the different types of radiations. The north pole of the magnet is toward the reader, and the south pole is away from the reader.

three series of related nuclides. *The heaviest nuclide of each series is called the* **parent nuclide.** *The nuclides produced by the decay of the parent nuclides are called* **daughter nuclides.** The parent nuclides are uranium-238, uranium-235, and thorium-232. The decay series of uranium-238 contains radium-226 and is traced in the following example. The various nuclear changes are charted in Figure 29-7.

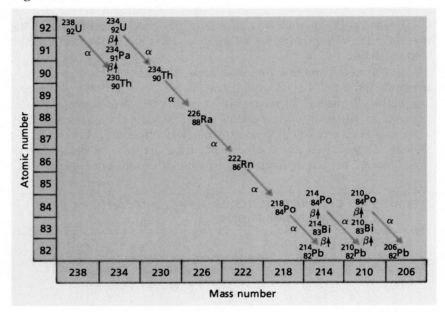

Figure 29-7 The parent nuclide of the uranium decay series is $^{238}_{92}U$. The final nuclide of the series is $^{206}_{82}Pb$.

The nucleus of a uranium-238 atom contains 92 protons (the atomic number of uranium is 92). It has a mass number (number of protons + number of neutrons) of 238. As this nucleus decays, it emits an alpha particle, which becomes an atom of helium when its positive charge is neutralized. An alpha particle has a mass number of 4. Since it contains two protons, it has an atomic number of 2. The remainder of the uranium nucleus thus has an atomic number of 90 and a mass number of 234. This daughter nuclide is an isotope of thorium. See the left portion of Figure 29-7. A **transmutation reaction** has taken place. A transmutation *is a change in the identity of a nucleus because of a change in the number of its protons.* The *nuclear equation* for this transmutation reaction can be written as

$$^{238}_{92}U \rightarrow \, ^{234}_{90}Th + \, ^{4}_{2}He$$

The above equation is a **nuclear equation,** *an equation in which only nuclei are represented.* In each symbol the superscript is the mass number and the subscript is the atomic number. Alpha particles are represented as helium nuclei ($^{4}_{2}He$). The total of the mass numbers on the left side of the equation must equal the total of the mass numbers on the right side of the equation. The total of the atomic numbers on the left side must equal the total of the atomic numbers on the right.

The half-life of $^{234}_{90}Th$ is about 24 days. It decays by giving off beta particles. The loss of a beta particle from a nucleus increases the number of positive charges in the nucleus (the atomic number) by one. The beta particle is believed to be formed by the change of a neutron into a proton and beta particle (electron). Since the mass of the lost beta particle is so small that it may be neglected, the mass number of the resulting nuclide stays the same. See the left portion of Figure 29-7.

$$^{234}_{90}\text{Th} \rightarrow {}^{234}_{91}\text{Pa} + {}^{0}_{-1}\text{e}$$

The symbol ${}^{0}_{-1}\text{e}$ represents an electron with an atomic number of -1 and a mass number of 0. One isotope of protactinium is ${}^{234}_{91}\text{Pa}$. This nuclide decays by releasing beta particles to produce ${}^{234}_{92}\text{U}$.

$$^{234}_{91}\text{Pa} \rightarrow {}^{234}_{92}\text{U} + {}^{0}_{-1}\text{e}$$

The ${}^{234}_{92}\text{U}$ nuclide decays by giving off alpha particles.

$$^{234}_{92}\text{U} \rightarrow {}^{230}_{90}\text{Th} + {}^{4}_{2}\text{He}$$

The resulting isotope of thorium also emits alpha particles, forming radium-226.

$$^{230}_{90}\text{Th} \rightarrow {}^{226}_{88}\text{Ra} + {}^{4}_{2}\text{He}$$

You can see now why ores of uranium contain radium. Radium is one of the products of the decay of uranium atoms. The half-lives of ${}^{238}_{92}\text{U}$ and ${}^{226}_{88}\text{Ra}$ determine the proportion of uranium atoms to radium atoms in uranium ores.

The decay of ${}^{226}_{88}\text{Ra}$ proceeds according to the chart shown in Figure 29-7. The ${}^{226}_{88}\text{Ra}$ nuclide decays by giving off alpha particles, forming radon-222. The nuclear equation is

$$^{226}_{88}\text{Ra} \rightarrow {}^{222}_{86}\text{Rn} + {}^{4}_{2}\text{He}$$

Radon-222 is a radioactive noble gas. It is collected in tubes and used for the treatment of disease. Variations in the amount of radon in soil gases near geologic faults may help in predicting earthquakes. The accumulation of radon in airtight, energy-efficient homes may be a health hazard. The source of the radon is the decay of uranium, which is widely distributed in the earth's crust.

The ${}^{222}_{86}\text{Rn}$ nuclei are unstable and have a half-life of about four days. They decay by giving off alpha particles.

$$^{222}_{86}\text{Rn} \rightarrow {}^{218}_{84}\text{Po} + {}^{4}_{2}\text{He}$$

The remaining atomic number and mass number changes shown on the decay chart are also explained in terms of the particles given off. When it loses alpha particles, ${}^{210}_{84}\text{Po}$ forms ${}^{206}_{82}\text{Pb}$. This is a stable, nonradioactive isotope of lead. Thus a series of spontaneous transmutations begins with ${}^{238}_{92}\text{U}$, passes through ${}^{226}_{88}\text{Ra}$, and ends with the formation of stable ${}^{206}_{82}\text{Pb}$.

Figure 29-8 The variation in the amount of radon in soil gases near geologic faults helps in predicting earthquakes. The monitoring station shown in the photos measures and records the amount of radon.

A chemical change results in a change in the identity of a substance by changing the arrangement of atoms. A nuclear change results in change in the identity of a substance by changing the identity of atoms.

Sample Problem 29.2

Identify the product X that balances the following nuclear reaction: ${}^{212}_{84}\text{Po} \rightarrow {}^{4}_{2}\text{He} + X$.

Solution
The sum of the mass numbers on the left-hand side of the equation must equal the sum of the mass number on the right-hand side of the equation. Furthermore, the sum of the atomic numbers of the

Sample Problem 29.2 *continued*

reactants and products must be equal to maintain electrical balance. Here the mass number and atomic number on the left-hand side of the equation are 212 and 84, respectively. Thus, the mass number and atomic number of the product X must be 208 and 82, respectively, to have the same sums on both sides of the equation. Thus, the equation is balanced as follows:

$$^{212}_{84}\text{Po} \rightarrow \, ^{4}_{2}\text{He} + \, ^{208}_{82}\text{Pb}$$

Practice Problems

Complete the following nuclear equations (if necessary, consult Figure 29-8):

1. $^{218}_{84}\text{Po} \rightarrow \, ... + \, ^{4}_{2}\text{He}.$ (*Ans.*) $^{218}_{84}\text{Po} \rightarrow \, ^{214}_{82}\text{Pb} + \, ^{4}_{2}\text{He}$

2. $^{214}_{82}\text{Pb} \rightarrow \, ^{214}_{83}\text{Bi} + \, ...$ (*Ans.*) $^{214}_{82}\text{Pb} \rightarrow \, ^{214}_{83}\text{Bi} + \, ^{0}_{-1}\text{e}$

3. $^{214}_{83}\text{Bi} \rightarrow \, ... + \, ...$ (*Ans.*) $^{214}_{83}\text{Bi} \rightarrow \, ^{214}_{84}\text{Po} + \, ^{0}_{-1}\text{e}$

4. $^{214}_{84}\text{Po} \rightarrow \, ... + \, ...$ (*Ans.*) $^{214}_{84}\text{Po} \rightarrow \, ^{210}_{82}\text{Pb} + \, ^{4}_{2}\text{He}$

In a nuclear reaction, mass number and atomic number are conserved.

Stable nuclei from radioactive decay   The release of an alpha particle from a radioactive nucleus such as $^{238}_{92}\text{U}$ decreases the mass of the nucleus. The resulting lighter nucleus has higher binding energy per nuclear particle. The release of an alpha particle decreases the number of protons and neutrons in a nucleus *equally and also by an even number*.

Beta particles are released when neutrons change into protons, as during the decay of $^{234}_{90}\text{Th}$. This change lowers the neutron-to-proton ratio toward the value found for stable nuclei of the same mass number. Both alpha-particle release and beta-particle release yield a product nucleus that is more stable than the original nucleus. (See Figure 29-7.)

Artificially Induced Radioactive Nuclides

1935: Iréne and Frédéric Joliot-Curie were awarded the Nobel Prize in chemistry for their transmutation (nuclear disintegration) reactions.

In 1934, Madame Curie's daughter Irène Joliot-Curie (1897–1956) and her husband Frédéric Joliot-Curie (1900–1958) discovered that stable atoms can be made radioactive by artificial means. This occurs when they are bombarded with deuterons or neutrons. Radioactive isotopes of all the elements have been prepared. For example, radioactive $^{60}_{27}\text{Co}$ can be produced from natural nonradioactive $^{59}_{27}\text{Co}$ by slow-neutron bombardment. The nuclear equation is

$$^{59}_{27}\text{Co} + \, ^{1}_{0}\text{n} \rightarrow \, ^{60}_{27}\text{Co}$$

Radiation from $^{60}_{27}\text{Co}$ consists of beta particles and gamma rays.

Radioactive $^{32}_{15}\text{P}$ is prepared by bombardment of $^{32}_{16}\text{S}$ with slow neutrons.

$$^{32}_{16}\text{S} + \, ^{1}_{0}\text{n} \rightarrow \, ^{32}_{15}\text{P} + \, ^{1}_{1}\text{H}$$

The radiation from $^{32}_{15}\text{P}$ consists of only beta particles.

The first artificial nuclear disintegration   After scientists discovered how uranium and radium undergo natural decay and transmutation, they worked to produce artificial transmutations. They had to find a way to add protons to a nucleus of an atom of an element, converting it to the nucleus of an atom of a different element. In 1919, Rutherford produced the first artificial nuclear disintegration. His method involved bombarding nitrogen with alpha particles from radium. He obtained protons (hydrogen nuclei) and

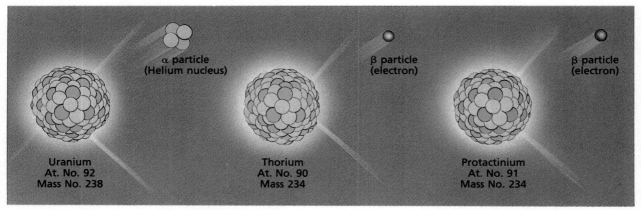

α particle
(Helium nucleus)

β particle
(electron)

β particle
(electron)

Uranium
At. No. 92
Mass No. 238

Thorium
At. No. 90
Mass 234

Protactinium
At. No. 91
Mass No. 234

a stable isotope of oxygen. See Figure 29-10. This nuclear disintegration is represented by the following equation:

Figure 29-9. This diagram shows successive alpha and beta particle emissions in the decay of $^{238}_{92}U$.

$$^{14}_{7}N + ^{4}_{2}He \rightarrow ^{17}_{8}O + ^{1}_{1}H$$

Neutron emission in some nuclear disintegrations You have already learned that neutrons were discovered by Chadwick in 1932. He first detected them in an experiment that involved bombarding beryllium with alpha particles.

$$^{9}_{4}Be + ^{4}_{2}He \rightarrow ^{12}_{6}C + ^{1}_{0}n$$

The symbol for a neutron is $^{1}_{0}n$. This symbol indicates a particle with zero atomic number (no protons) and a mass number of 1. The reaction described above proved that neutrons were a second type of particle in the nuclei of atoms.

Proofs of Einstein's equation In 1932, the two English scientists J. D. Cockcroft (1897–1967) and E. T. S. Walton (b. 1903) experimentally proved Einstein's equation, $E = mc^2$. They bombarded lithium with high-speed protons. Alpha particles and an enormous amount of energy were produced.

$$^{7}_{3}Li + ^{1}_{1}H \rightarrow ^{4}_{2}He + ^{4}_{2}He + energy$$

There is a loss of matter in this reaction. One lithium nucleus (mass 7.0144 u) was hit by a proton (mass 1.0073 u). These particles formed two alpha

Figure 29-10 This diagram shows the historic nuclear disintegration experiment performed by Rutherford.

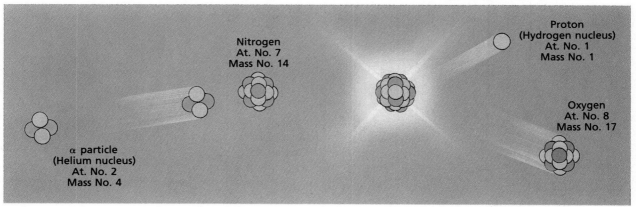

Nitrogen
At. No. 7
Mass No. 14

Proton
(Hydrogen nucleus)
At. No. 1
Mass No. 1

α particle
(Helium nucleus)
At. No. 2
Mass No. 4

Oxygen
At. No. 8
Mass No. 17

particles (helium nuclei), each having a mass of 4.0015 u. Calculation shows that there is a loss of 0.0187 u. [(7.0144 u + 1.0073 u) − 2(4.0015 u)]. Cockcroft and Walton found that the energy released very nearly equaled that predicted by Einstein for such a loss in mass. Additional experiments have further supported Einstein's equation.

Neutrons as "bullets" Before the discovery of neutrons in 1932, alpha particles and protons were used to study atomic nuclei. But alpha particles and protons are charged particles. It requires great quantities of energy to "fire" these charged "bullets" into a nucleus. Their positive charge causes them to be repelled by the positive nuclear charge. The various kinds of particle accelerators were developed to give charged "bullets" enough energy to overcome this repelling force.

When accelerated positive particles strike a target material, usually lithium or beryllium, neutrons are produced. Neutrons have no charge. Thus there is no repelling force, and they can easily penetrate the nucleus of an atom. Some fast neutrons may go through an atom without causing any change in it. Other fast neutrons may cause nuclear disintegration. Slow neutrons, on the other hand, are sometimes trapped by a nucleus. This nucleus then becomes unstable and may break apart. Fast neutrons are slowed down by passage through materials composed of elements of low atomic weight. Examples are deuterium oxide or graphite.

Artificial elements from neutron bombardment The $^{238}_{92}\text{U}$ nuclide is the most plentiful isotope of uranium. When hit by slow neutrons, a $^{238}_{92}\text{U}$ nucleus may capture a neutron. This capture produces the nucleus of an atom of an unstable isotope of uranium, $^{239}_{92}\text{U}$. This nucleus emits a beta particle and, in so doing, becomes the nucleus of an atom of an artificial radioactive element, neptunium. Neptunium has atomic number 93. The nuclide formed has the symbol $^{239}_{93}\text{Np}$

$$^{238}_{92}\text{U} + {}^{1}_{0}\text{n} \rightarrow {}^{239}_{92}\text{U}$$

$$^{239}_{92}\text{U} \rightarrow {}^{239}_{93}\text{Np} + {}^{0}_{-1}\text{e}$$

Neptunium is itself an unstable element. The nucleus of a neptunium atom gives off a beta particle. This change produces the nucleus of an atom of still another artificial element, plutonium, atomic number 94. This nuclide has the symbol $^{239}_{94}\text{Pu}$.

$$^{239}_{93}\text{Np} \rightarrow {}^{239}_{94}\text{Pu} + {}^{0}_{-1}\text{e}$$

Neptunium and plutonium were the first artificial *transuranium* elements. Transuranium elements are those with more than 92 protons in their nuclei. As this is written, 18 artificially prepared transuranium elements have been reported. In addition to neptunium (Np) and plutonium (Pu), there are americium (Am), curium (Cm), berkelium (Bk), californium (Cf), einsteinium (Es), fermium (Fm), mendelevium (MD), nobelim (No), and lawrencium (Lr). Elements 104 through 109 have only systematic names. All these elements are prepared by bombarding the nuclei of uranium or other complex atoms with neutrons, alpha particles, or other "nuclear bullets." The nuclear reactions for the synthesis of several transuranium elements are shown in Table 29-4.

TABLE 29-4   REACTIONS FOR THE FIRST PREPARATION OF TRANSURANIUM ELEMENTS

Atomic Number	Name	Symbol	Nuclear Reaction
93	neptunium	Np	$^{238}_{92}U + ^{1}_{0}n \rightarrow ^{239}_{93}Np + ^{0}_{-1}e$
94	plutonium	Pu	$^{238}_{92}U + ^{2}_{1}H \rightarrow ^{238}_{93}Np + 2^{1}_{0}n$
			$^{238}_{93}Np \rightarrow ^{238}_{94}Pu + ^{0}_{-1}e$
95	americium	Am	$^{239}_{94}Pu + 2^{1}_{0}n \rightarrow ^{241}_{95}Am + ^{0}_{-1}e$
96	curium	Cm	$^{239}_{94}Pu + ^{4}_{2}He \rightarrow ^{242}_{96}Cm + ^{1}_{0}n$
97	berkelium	Bk	$^{241}_{95}Am + ^{4}_{2}He \rightarrow ^{243}_{97}Bk + 2^{1}_{0}n$
98	californium	Cf	$^{242}_{96}Cm + ^{4}_{2}He \rightarrow ^{245}_{98}Cf + ^{1}_{0}n$
99	einsteinium	Es	$^{238}_{92}U + 15^{1}_{0}n \rightarrow ^{253}_{99}Es + 7^{0}_{-1}e$
100	fermium	Fm	$^{238}_{92}U + 17^{1}_{0}n \rightarrow ^{255}_{100}Fm + 8^{0}_{-1}e$
101	mendelevium	Md	$^{253}_{99}Es + ^{4}_{2}He \rightarrow ^{256}_{101}Md + ^{1}_{0}n$
102	nobelium	No	$^{246}_{96}Cm + ^{12}_{6}C \rightarrow ^{254}_{102}No + 4^{1}_{0}n$
103	lawrencium	Lr	$^{252}_{98}Cf + ^{10}_{5}B \rightarrow ^{258}_{103}Lr + 4^{1}_{0}n$
104	unnilquadium (USSR)	Unq	$^{242}_{94}Pu + ^{22}_{10}Ne \rightarrow ^{260}_{104}Unq + 4^{1}_{0}n$
104	unnilquadium (US)	Unq	$^{249}_{98}Cf + ^{12}_{6}C \rightarrow ^{257}_{104}Unq + 4^{1}_{0}n$
105	unnilpentium (US)	Unp	$^{249}_{98}Cf + ^{15}_{7}N \rightarrow ^{260}_{105}Unp + 4^{1}_{0}n$
106	unnilhexium (US)	Unh	$^{249}_{98}Cf + ^{18}_{8}O \rightarrow ^{263}_{106}Unh + 4^{1}_{0}n$
107	unnilseptium (USSR)	Uns	$^{209}_{83}Bi + ^{54}_{24}Cr \rightarrow ^{261}_{107}Uns + 2^{1}_{0}n$
108	unniloctium (W. Germany)	Uno	$^{208}_{82}Pb + ^{58}_{26}Fe \rightarrow ^{265}_{108}Uno + ^{1}_{0}n$
109	unnilennium (W. Germany)	Une	$^{209}_{83}Bi + ^{58}_{26}Fe \rightarrow ^{266}_{109}Une + ^{1}_{0}n$

Section Review

1. What are some of the unusual properties that nuclides exhibit because of their radioactivity?
2. Define the term *half-life* and give the half-life of any four radioactive isotopes.
3. What fraction of a given sample of a radioactive nuclide will remain after four half-lives?
4. Write balanced nuclear equations for the following nuclear transmutations:
   (a) Uranium-233 undergoes alpha decay
   (b) Uranium-239 undergoes beta decay.
5. Complete and balance the following nuclear equations:

   (a) $^{187}_{75}Re + \underline{\quad} \rightarrow ^{188}_{75}Re + ^{1}_{1}H$

   (b) $^{176}_{71}Lu + \underline{\quad} \rightarrow ^{177}_{71}Lu + ^{0}_{0}\gamma$

   (c) $^{212}_{84}Po \rightarrow ^{4}_{2}He + \underline{\quad}$

   (d) $^{32}_{14}Si \rightarrow ^{32}_{15}P + \underline{\quad}$

6. What is meant by the term *transmutation?*
7. How do you distinguish between a naturally occurring radioactive nuclide and an artificially induced radioactive nuclide?

particles and rays. Induced radioactive nuclide: produced as a result of bombardment of

a stable nucleus with particles such as neutrons, alpha particles, or beta particles

1. Affect the light-sensitive emulsion on a photographic film; produce an electric charge in the surrounding air; produce fluorescence with certain compounds; radiations have special physiological effects; undergo radioactive decay

2. The length of time during which half of a given number of atoms of a radioactive nuclide decays. Accept the half-lives of any isotopes given in Table 30-1 and Table 30-2.

3. One-sixteenth. ($\frac{1}{2} \times \frac{1}{2} \times \frac{1}{2} \times \frac{1}{2}$)

4. *(a)* $^{233}_{92}U \rightarrow ^{4}_{2}He + ^{229}_{90}Th$
   (b) $^{239}_{92}U \rightarrow ^{0}_{-1}e + ^{239}_{93}Np$

5. *(a)* $^{187}_{75}Re + ^{2}_{1}H \rightarrow ^{188}_{75}Re + ^{1}_{1}H$
   (b) $^{176}_{71}Lu + ^{1}_{0}n \rightarrow ^{177}_{71}Lu + ^{0}_{0}\gamma$
   (c) $^{212}_{84}Po \rightarrow ^{4}_{2}He + ^{208}_{82}Pb$
   (d) $^{32}_{14}Si \rightarrow ^{32}_{15}P + ^{0}_{-1}e$

6. A change in the identity of a nucleus because of a change in the number of its protons

7. Naturally occurring radioactive nuclide: spontaneous breakdown with release of

29.3 Applications of Radioactivity

SECTION OBJECTIVES

- Describe the use of half-life to determine the age of an object.
- Give examples of the use of radioisotopes in medicine.
- Explain how radioisotopes can be used as tracers.

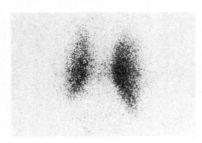

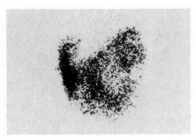

Figure 29-11 Iodine–131, a radioactive tracer, is absorbed by the human thyroid, an organ that secretes hormones and is found in the neck. The I–131 can be detected by a linear photo scanner. A normal pattern of I–131 absorption *(top)* indicates a healthy thyroid. A clumpy absorption pattern *(bottom)* indicates thyroid cancer.

STUDY HINT

The atomic mass scale based on carbon-12 is described in Section 3.3

Stable and radioactive isotopes of a given element behave the same in both chemical and physical processes. Thus, scientists make use of radioactive nuclides in a variety of fields.

Radioactive Dating

The age of any mineral containing radioactive substances can be estimated with a fair degree of accuracy. Such an estimate is based on the fact that radioactive substances decay at known rates. It is also assumed that these rates have not changed during the existence of the mineral and that there has been no gain or loss of parent or daughter nuclides except by radioactive decay. The mineral is analyzed to determine the amount of long-lived parent nuclide and the amounts of shorter-lived daughter nuclides in the sample. Then, by calculation, scientists can determine how long it must have taken for these amounts of daughter nuclides to have been produced. This time is assumed to be the age of the mineral. By this method, the oldest known minerals on earth, found in western Australia in 1983, have been estimated to be between 4.1 and 4.2 billion years old. Dust from sites of moon landings has been found to be about 4.6 billion years old. The ages of moon rocks range from 3.2 to 4.6 billion years.

The age of more recent potassium-containing minerals, 50 thousand to 50 million years old, is determined quite accurately by the proportion of potassium to argon they contain. Some nuclei of $^{40}_{19}K$ decay by capturing an orbital electron to form $^{40}_{18}Ar$. So, over time, the proportion of argon to potassium in the mineral increases and is used to establish the mineral's age.

Some carbon atoms involved in the oxygen–carbon dioxide cycle of living plants and animals are radioactive. Radioactive $^{14}_{6}C$ is continuously being produced from $^{14}_{7}N$ atoms in the atmosphere. This change is brought about by the action of *cosmic rays*. **Cosmic rays** *are protons and other nuclei of very high energy that come to the earth from outer space.* Some of the carbon atoms combine with oxygen in the upper atmosphere, forming carbon dioxide which diffuses down to earth and is absorbed by living things. When living things die, the oxygen—carbon dioxide cycle ceases to continue in them. They no longer replace carbon atoms in their cells with other carbon atoms. Thus, the level of radioactivity produced by the radioactive carbon in a given amount of nonliving material slowly diminishes.

Carbon from a wooden beam taken from the tomb of an Egyptian pharaoh yields about half the radiation of carbon in living trees. The half-life of a $^{14}_{6}C$ atom is about 5730 years. Thus the age of dead wood with half the radioactivity of living wood is about 5730 years. The use of $^{14}_{6}C$ dating has also been applied to the study of bone proteins. Dating by this method can presently be used with objects up to about 50 000 years old. More sensitive measuring techniques are under development, however.

Radioisotope Uses in Medicine and as Tracers

Radioactive phosphorus, radioactive cobalt, and some other radioactive nuclides are used to treat certain forms of cancer. Radioactive drugs are used for diagnostic purposes and for test purposes in blood and tissue samples. The gamma radiation from $^{60}_{27}\text{Co}$ can be used to preserve food. It also kills bacteria that spoil food and insects that infest food. Many radioactive nuclides are used as **radioactive tracers,** *radioactive atoms incorporated in substances so that movement of substances can be followed by radiation detectors.* By using them, scientists can determine the course of chemical reactions, the cleaning ability of detergents, the wearing ability of various products, the efficiency of fertilizers, the flow of fluids through pipelines, and the movement of sand along sea coasts. Many new radioactive nuclides are made by slow-neutron bombardment in the nuclear reactor at Oak Ridge, Tennessee.

Section Review

1. Describe how carbon-14 is used to determine the age of an ancient object.
2. How are radioisotopes used in medicine?
3. Explain why radioactivity might be used in chemical reactions.

29.4 Energy from the Nucleus

In nuclear reactions in which a heavy nucleus splits into lighter nuclei and several neutrons, large amounts of energy are produced. Similarly, large amounts of energy are generated when light nuclei combine to form heavier nuclei.

Nuclear Fission

The element uranium exists as three naturally occurring isotopes: $^{238}_{92}\text{U}$, $^{235}_{92}\text{U}$, and $^{234}_{92}\text{U}$. Most uranium is the nuclide $^{238}_{92}\text{U}$. Only 0.7% of natural uranium is $^{235}_{92}\text{U}$. Nuclide $^{234}_{92}\text{U}$ occurs in only the slightest traces. When $^{235}_{92}\text{U}$ is bombarded with slow neutrons, each atom may capture one of the neutrons. This extra neutron in the nucleus makes it very unstable. Instead of giving off an alpha or beta particle, as in other radioactive changes, the nucleus splits into medium-weight parts. Neutrons are usually produced during this *fission*. There is a small loss of mass, which appears as a great amount of energy. One equation for the fission of $^{235}_{92}\text{U}$ is

$$^{235}_{92}\text{U} + ^{1}_{0}\text{n} \rightarrow ^{138}_{56}\text{Ba} + ^{95}_{36}\text{Kr} + 3^{1}_{0}\text{n} + \text{energy}$$

The atomic mass of $^{235}_{92}\text{U}$ is slightly greater than 235 u. The atomic masses of the unstable barium and krypton isotopes are slightly less than 138 u and 95 u, respectively. The masses of the reactants and the masses of the products thus are not equal. Instead, about 0.2 u of mass is converted to energy for each uranium atom undergoing fission. Plutonium, made from $^{238}_{92}\text{U}$, also undergoes fission to produce more neutrons when bombarded with slow neutrons.

Uranium-238 atoms exist in very small amounts in many minerals where they undergo spontaneous fission. These fissions have left tracks that can be made visible under a microscope by etching. The number of tracks in a given area and the amount of $^{238}_{92}\text{U}$ in the specimen are used to determine the age of

1. When living things die, the accumulation of carbon-14 in their bodies stops. By comparing the amount of radioactive carbon remaining in ancient objects such as wood with that in present-day objects, a determination of age can be made.
2. To treat certain forms of cancer and for tests of blood and tissue samples
3. Used as tracers to determine the course of chemical reactions

Lise Meitner (1878–1968) discovered nuclear fission in collaboration with her colleague Otto Hahn. In 1944, the Nobel Prize in physics was awarded for the discovery of the fission of uranium to Otto Hahn alone.

Chemistry Notebook

Measuring Radiation Exposure

In 1987 the world was reminded of the potentially deadly nature of nuclear radiation when several people in a Brazilian city were killed, and many were severely injured by their exposure to radioactive cesium that had been contained in an abandoned piece of medical equipment. The biological hazard posed by nuclear radiation and by X rays stems from their ionizing quality, which can cause chemical changes in organic matter.

The unit that measures the dosage of ionizing radiation is the *roentgen*, which is the amount of radiation that produces 2×10^9 ion pairs when it passes through 1 cm^3 of dry air. Ionizing-radiation damage to human tissue is measured in terms of the rem (roentgen-equivalent-man). One rem is the quantity of ionizing radiation that does as much damage to human tissue as is done by 1 roentgen of approximately 200 kilovolts of X radiation.

One-time doses of 100–400 rem of ionizing radiation produce radiation sickness, the 400 rem doses are fatal in half the cases. Lower radiation doses can cause cellular damage that ultimately can lead to cancer—often about 25 years after exposure to the radiation—or can damage reproductive cells, possibly leading to harmful mutations. Current guidelines suggest that most people should be exposed to doses of no more than 0.005 rem/year (500 millirems/year) and that radiation workers limit themselves to an exposure of 5 rems/year. The table below contains the average annual dose rates from a variety of common sources.

ANNUAL DOSE RATES FROM COMMON SIGNIFICANT SOURCES OF HIGH-ENERGY (IONIZING) RADIATION EXPOSURE			
Source	Exposed Group	Body Portion Exposed	Average Dose Rate millirems/yr
MEDICAL X RAYS			
Medical diagnosis	Adult patients	Bone marrow	103
Dental diagnosis	Adult patients	Bone marrow	3
ATMOSPHERIC WEAPONS TESTS	Total population	Whole body	4-5
COMMERCIAL NUCLEAR POWER PLANTS (effluent releases)	Population within 10 miles	Whole body	<<10
CONSUMER PRODUCTS			
Building materials	Population in brick and masonry buildings	Whole body	7
Television receivers	Viewing populations	Gonads	0.2-1.5
AIRLINE TRAVEL (COSMIC RADIATION)	Passengers	Whole body	3
	Crew members and flight attendants	Whole body	160

the crystal. This method, called fission-track dating, can establish the age of materials from a few decades old to as old as the solar system.

Nuclear chain reaction   The fissions of $^{235}_{92}$U and $^{239}_{92}$Pu can produce chain reactions. One neutron causes the fission of one $^{235}_{92}$U nucleus. Two or three neutrons given off when this fission occurs can cause the fission of other $^{235}_{92}$U nuclei. Again neutrons are emitted. These can cause the fission of still other $^{235}_{92}$U nuclei. This is a chain reaction. *A **chain reaction** is one in which the material that starts the reaction is also one of the products.* It continues until all the $^{235}_{92}$U atoms have split or until the neutrons fail to strike $^{235}_{92}$U nuclei. See Figure 29-13.

Nuclear reactor   *A **nuclear reactor** is a device in which the controlled fission of radioactive material produces new radioactive substances and energy.* One of the earliest nuclear reactors was built at Oak Ridge, Tennessee, in 1943. This reactor uses natural uranium. It has a lattice-type structure with blocks of graphite forming the framework. Spaced between the blocks of graphite are rods of uranium, encased in aluminum cans for protection. *Control rods* are inserted into the lattice. **Control rods** *are composed of neutron-absorbing steel and are used to limit the number of free neutrons.* The reactor is air-cooled.

The rods of uranium or uranium oxide are the nuclear fuel for the reactor. The energy released in the reactor comes from changes in the uranium nuclei. **Graphite** *is the moderator which slows down the fast neutrons produced by fission.* By doing so, it makes them more readily captured by a nucleus and thus more effective for producing additional nuclear changes. The

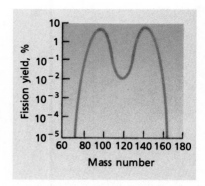

Figure 29-12   The yield of various nuclides of uranium-235. The products vary in mass number from 72 to 161. Observe the low probability of fission products of nearly equal mass numbers.

Figure 29-13   An illustration of a nuclear chain reaction.

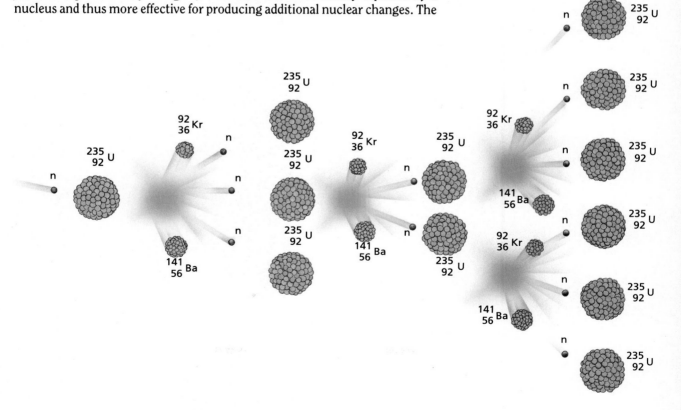

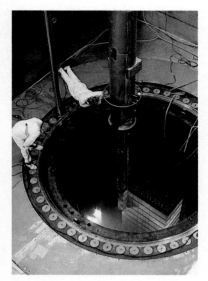

Figure 29-14 A nuclear power plant

mass of uranium in such a reactor is important. *The amount of uranium needed to provide the* number of neutrons needed to sustain a chain reaction is called the **critical mass.**

Two types of reactions occur in the fuel in such a reactor. Neutrons cause $^{235}_{92}U$ nuclei to undergo fission. The fast neutrons from this fission are slowed down as they pass through the graphite. Some strike other $^{235}_{92}U$ nuclei and continue the chain reaction. Other neutrons strike $^{238}_{92}U$ nuclei, starting the changes that finally produce plutonium. Great quantities of heat energy are released. For this reason the reactor must be cooled continuously by blowing air through tubes in the lattice. The rate of the reaction is controlled by the insertion or removal it the neutron-absorbing control rods. This type of reactor is now used to produce radioactive isotopes.

In nuclear power plants, the reactor is the source of heat energy. Pressurized water is both the moderator and the coolant. The heat from the reactor absorbed by the pressurized water is used to produce steam. This steam turns the turbines, which drive the electric generators. Present problems with nuclear power plant development include location and environmental requirements, safety of operation, and plant construction costs. The procurement and enrichment of uranium, as well as methods for storing or reprocessing used nuclear reactor fuel, are also being studied.

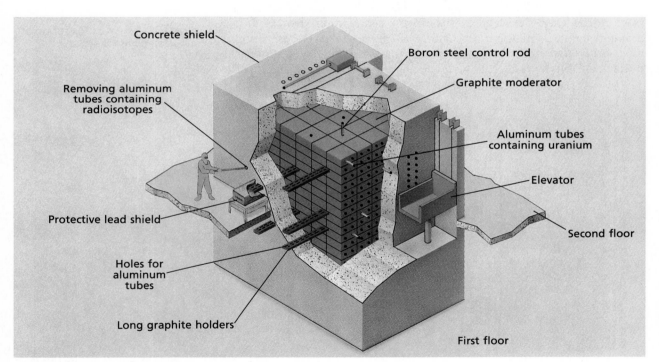

Figure 29-15 A cutaway of the Oak Ridge reactor

Nuclear Fusion

Nuclear stability can be increased by combining light-weight nuclei into heavier nuclei in a process is defined as nuclear fusion.

Fusion reactions are the source of the sun's energy. There are two series of such reactions going on in the sun it is believed. One series occurs at the very hot center of the sun, and the other takes place in the cooler outer portion of the sun. These two reactions proceed by different pathways. Their net effect, however, is the combination of four hydrogen nuclei into a helium nucleus. A loss of mass occurs and a tremendous amount of energy is released. More energy is released per gram of fuel in a fusion reaction than in a fission reaction.

Current research indicates that fusion reactions may be controlled. Scientists are attempting to find ways to confine the nuclear fuel, usually ionized deuterium and tritium, and at the same time, make it hot enough and dense enough long enough for fusion to occur. Magnetic fields are usually used to confine the ionized fuel because, at the initial temperature required (about 10^8K), no known material container will do. Experiments using high-powered laser light or beams of particles, such as protons or lithium ions, as ways to start the fusion reaction are being conducted. If fusion reactions are to be a practical source of energy, the energy given off must be greater than the energy required to heat and compress the ionized fuel.

Section Review

1. Distinguish between *nuclear fission* and *nuclear fusion*.
2. Define *chain reaction*.
3. What is the function of the following in a nuclear reactor: *(a)* nuclear fuel *(b)* control rods *(c)* moderator *(d)* cooling fluid.

Chapter Summary

- Protons and neutrons, called nucleons, are held together strongly by a short-range force of attraction known as the nuclear strong forces.
- Radioactivity is the spontaneous breakdown of an unstable atomic nucleus with the release of particles and rays.
- Radioactive nuclides and their compounds have several unusual properties: *(1)* they affect the light-sensitive emulsion in a photographic film *(2)* they produce an electric charge in the surrounding air *(3)* they produce fluorescence with certain other compounds *(4)* their radiations have special physiological effects and *(5)* they undergo radioactive decay.
- The half-life of a radioactive nuclide is the length of time that it takes for half of a given number of atoms of the nuclide to decay.
- The radiation given off by radioactive nuclides consists of three kinds of particles and rays: *(1)* alpha particles, which are helium nuclei *(2)*beta particles, which are electrons *(3)* gamma rays, which are high-energy X rays.
- The age of certain minerals and of carbon-containing materials can be estimated by the amounts of radioactive nuclides they contain.
- The difference between the sum of the masses of the nucleons in a nucleus and the actual mass of a nucleus is the nuclear mass defect.
- The four types of reactions that produce more stable nuclei are: *(1)* radioactive decay, *(2)* nuclear disintegration, *(3)* fission, and *(4)* fusion.
- A nuclear reactor is a device in which the controlled fission of radioactive material produces new radioactive substances and heat energy.
- A fusion reaction is one in which lightweight nuclei are combined into heavier nuclei. Fusion reactions produce the sun's heat and light.

Figure 29-16 Glass pellet containing deuterium and tritium: The target contains over 100 atmospheres of deuterium-tritium fuel ready to be fused by the heat from a powerful laser.

1. Nuclear fission is a neutron reaction in which a heavy nucleus splits into medium-weight fragments. Nuclear fusion is a process in which light-weight nuclei combine into heavier nuclei.
2. The material or energy that starts the reaction is also one of the products.
3. *(a)* Provides the energy in a nuclear reactor *(b)* used to absorb neutrons *(c)* slows down fast-moving neutrons produced by fission *(d)* fluid used to cool a reactor in which great amounts of heat energy are released

Chapter 29 *Review*

Nuclear Chemistry

Vocabulary

alpha particle	nuclear binding energy
beta particle	nuclear disintegration
chain reaction	nuclear equation
control rod	nuclear mass defect
cosmic ray	nuclear reactor
critical mass	nucleons
daughter nuclide	parent nuclide
deuteron	radioactive decay
fission	radioactive tracer
fluoresce	radioactivity
fusion	strong nuclear force
gamma ray	transmutation
half-life	transuranium element
moderator	

Questions

1. *(a)* What is radioactivity? *(b)* When and by whom was radioactivity discovered? *(c)* How was the discovery made?
2. *(a)* Why did Marie and Pierre Curie suspect that there was a radioactive element other than uranium in pitchblende? *(b)* What did their subsequent discoveries show?
3. Compare the radioactivity of uranium, radium, and polonium.
4. Define *half-life.*
5. What is the source of the alpha or beta particles emitted by a radioactive nuclide?
6. *(a)* What practical use was formerly made of the fluorescence produced in zinc sulfide by a radium compound? *(b)* Why is a promethium-147 compound now used instead of a radium compound?
7. What changes in atomic number, identity, and mass number occur when a radioactive nuclide gives off a beta particle?
8. What changes in atomic number, identify, and mass number occur when a radioactive nuclide gives off a beta particle?
9. *(a)* Radon-222 has a half-life of less than four days, yet it is continually found in very low concentrations in the earth's crust, surface waters, and atmosphere. Why? *(b)* In what ways is radon useful? *(c)* In what way may radon be harmful?
10. *(a)* Describe how the age of a radioactive mineral is estimated. *(b)* What assumptions are necessary?
11. Name and briefly describe the four types of nuclear reactions that produce more stable nuclei.
12. Write the nuclear equation for the reaction by which Chadwick discovered neutrons.
13. Neutrons are more effective than protons or alpha particles for bombarding atomic nuclei. Why?
14. What may happen to a neutron that is fired at the nucleus of an atom?
15. List five different uses of artificially radioactive nuclides.
16. Name the naturally occurring isotopes of uranium and indicate their relative abundance.
17. How is the fission of a $^{235}_{92}U$ nucleus produced?
18. Define: *(a)* chain reaction *(b)* nuclear reactor *(c)* critical mass.
19. What factors are currently affecting the construction of nuclear power plants?
20. Describe the reaction that produces the sun's energy.

Problems

1. The half-life of thorium-27 is 18.72 days. How many days are required for three-fourths of a given amount to decay?
2. The half-life of $^{234}_{91}Pa$ is 6.75 hours. How much of a given amount remains after 27.0 hours?
3. After 4806 years, how much of an original 0.250 g of $^{226}_{88}Ra$ remains?
4. The mass of a $^{7}_{3}Li$ nucleus is 7.01436 u. Calculate the nuclear mass defect. The atomic masses of nuclear particles are given in Section 3.2.
5. Calculate the nuclear mass defect of $^{20}_{20}Ne$ if its nucleus has a mass of 19.98695 u.

Application Questions

1. Where on the Periodic Table are most of the radioactive nuclides located?
2. Why can the radiation from a radioactive material affect photographic film even though the film is well wrapped in black paper?
3. Explain how a radioactive material can change the discharge rate of a charged electroscope.
4. What instrument for studying radioactivity has replaced the electroscope? Explain how it works.
5. *(a)* What conditions cause variations in the decay rate of a radioactive material? *(b)* Are these variations slight, moderate, or great?
6. Make a chart that compares the following properties of alpha particles and beta particles: identity, mass number, charge, speed, and penetrating ability.
7. *(a)* What are gamma rays? *(b)* How do scientists believe they are produced? *(c)* How does their penetrating ability compare with that of alpha particles and beta particles?
8. Write the nuclear equation for the release of an alpha particle by $^{210}_{84}$Po.
9. Write the nuclear equation for the release of a beta particle by $^{210}_{82}$Pb.
10. The parent nuclide of the thorium decay series is $^{232}_{90}$Th. The first four decays are: alpha particle emission, beta particle emission, beta particle emission, alpha particle emission. Write the nuclear equations for this series of emissions.
11. *(a)* How does binding energy per nuclear particle vary with mass number? *(b)* How does binding energy per nuclear particle affect the stability of a nucleus?
12. Describe two ways in which the number of protons and the number of neutrons in a nucleus affect its stability.
13. Explain how each type of nuclear reaction produces more stable nuclei.
14. *(a)* Who produced the first artificial nuclear disintegration? *(b)* How many years ago? *(c)* Write the equation for this reaction.
15. *(a)* How was Einstein's equation for the relationship between matter and energy, $E = mc^2$, proved to be correct? *(b)* Give the names and nationality of the scientists responsible for the proof.
16. Explain the nuclear changes that occur when $^{239}_{94}$Pu is produced from $^{238}_{92}$U.
17. How does the fission of $^{235}_{92}$U produce a chain reaction?
18. How is a uraniun-graphite reactor constructed?
19. What problems must be overcome before energy-producing controlled fusion reactions are a reality?
20. To what extent has the use of radiation replaced the use of chemical fumigants and insecticides in food preservation?
21. What progress is being made in devising environmentally safe methods for storing radioactive wastes?

Application Problems

1. The half-life of $^{222}_{86}$Rn is 3.823 days. What was the original mass of $^{222}_{86}$Rn if 0.0500 g remains after 7.646 days?
2. The half-life of $^{239}_{94}$Pu is 24 100 years. Of an original mass of 100 g, approximately how much remains after 10^5 years?
3. Calculate the nuclear mass defect per nuclear particle for $^{7}_{3}$Li using the answer to Problem 4. Convert the mass in u to binding energy in megaelectron volts by using the relationship $1\ u = 931$ Mev.
4. Calculate the binding energy per nuclear particle of $^{238}_{92}$U in Mev if the atomic mass of its nucleus is 238.000 3 u.

Enrichment

1. Research *positron-emission tomography,* a medical tool diagnosis of many diseases. Investigate the chemistry involved in positron-emission transaxial tomography.
2. Investigate the history of the Manhattan Project.
3. Write a report on the findings in May 1972, at the nuclear fuel processing plant at Pierrelatte, France, regarding "Nature's Fission Reactor" at the Oklo mine in the Gabon Republic of West Africa.
4. Find out about the various fusion-energy research projects that are being conducted in the United States and other parts of the world. How close are the researchers to finding a method that will economically produce energy? What obstacles must still be overcome?
5. Research the history of the discovery of x-rays. How was the scientific method used in this discovery?

APPENDIX A

Quantity	Symbol	Value
Atomic mass unit	u	$1.660\,565\,5 \times 10^{-27}$ kg
		$5.485\,803 \times 10^{-4}$ u
Avogadro number	N_A	$6.022\,045 \times 10^{23}$/mol
Electron rest mass	m_e	$9.109\,534 \times 10^{-31}$ kg
		$5.485\,9 \times 10^{-4}$ u
Molar gas constant	R	$8.205\,68 \times 10^{-2}$ L atm/mol K
		$8.314\,41$ J/mol K
Molar volume of ideal gas at STP	V_m	$22.413\,83$ L/mol
Neutron rest mass	m_n	$1.674\,954\,3 \times 10^{-27}$ kg
		$1.008\,665$ u
Normal boiling point of water	T_b	373.15 K = $100.0\,°C$
Normal freezing point of water	T_b	273.15 K = $0.00\,°C$
Planck's constant	h	$6.626\,176 \times 10^{-34}$ J s
Proton rest mass	m_p	$1.672\,648\,5 \times 10^{-27}$ kg
		$1.007\,277$ u
Speed of light in a vacuum	c	$2.997\,924\,58 \times 10^{8}$ m/s
Temperature of triple point of water		273.16 K = $0.01\,°C$

TABLE A–2 HEAT OF COMBUSTION

Substance	Formula	Phase	ΔH_c
hydrogen	H_2	g	-285.9
graphite	C	s	-393.5
carbon monoxide	CO	g	-283.0
methane	CH_4	g	-890.31
ethane	C_2H_6	g	$-1\,559.8$
propane	C_3H_8	g	$-2\,219.9$
butane	C_4H_{10}	g	$-2\,878.5$
pentane	C_5H_{12}	g	$-3\,536.1$
hexane	C_6H_{14}	l	$-4\,163.1$
heptane	C_2H_{16}	l	$-4\,811.2$
octane	C_8H_{18}	l	$-5\,450.5$
ethene (ethylene)	C_2H_4	g	$-1\,411.0$
propene (propylene)	C_3H_6	g	$-2\,051.0$
ethyne (acetylene)	C_2H_2	g	$-1\,300.0$
benzene	C_6H_6	l	$-3\,267.5$
toluene	C_7H_8	l	$-3\,908.7$
naphthalene	$C_{10}H_8$	s	$-5\,153.9$
anthracene	$C_{14}H_{10}$	s	$-7\,163.0$
methanol	CH_3OH	l	-726.34
ethanol	C_2H_5OH	l	$-1\,366.8$
ether	$(C_2H_5)_2O$	l	$-2\,751.1$
formaldehyde	CH_2O	g	-570.78
glucose	$C_6H_{12}O_6$	s	$-2\,803.0$
sucrose	$C_{12}H_{22}O_{11}$	s	$-5\,640.9$

ΔH_c = heat of combustion of the given substance. All values of ΔH_c are expressed as kJ/mol of substance oxidized to $H_2O(l)$ and/or $CO_2(g)$ at constant pressure and $25\,°C$. s = solid, l = liquid, g = gas.

TABLE A-3 THE ELEMENTS—THEIR SYMBOLS, ATOMIC NUMBERS, AND ATOMIC MASSES

The more common elements are printed in color.

Name of element	Symbol	Atomic number	Atomic mass	Name of element	Symbol	Atomic number	Atomic mass
actinium	Ac	89	227.028	neon	Ne	10	20.179
aluminum	Al	13	26.981 54	neptunium	Np	93	237.048
americium	Am	95	[243]	nickel	Ni	28	58.69
antimony	Sb	51	121.75	niobium	Nb	41	92.9064
argon	Ar	18	39.948	nitrogen	N	7	14.0067
arsenic	As	33	74.9216	nobelium	No	102	[259]
astatine	At	85	[210]	osmium	Os	76	190.2
barium	Ba	56	137.33	oxygen	O	8	15.9994
berkelium	Bk	97	[247]	palladium	Pd	46	106.42
beryllium	Be	4	9.012 18	phosphorus	P	15	30.973 76
bismuth	Bi	83	208.980	platinum	Pt	78	195.08
boron	B	5	10.81	plutonium	Pu	94	[244]
bromine	Br	35	79.904	polonium	Po	84	[209]
cadmium	Cd	48	112.41	potassium	K	19	39.0983
calcium	Ca	20	40.08	praseodymium	Pr	59	140.908
californium	Cf	98	[251]	promethium	Pm	61	[145]
carbon	C	6	12.0111	protactinium	Pa	91	231.036
cerium	Ce	58	140.12	radium	Ra	88	226.025
cesium	Cs	55	132.905	radon	Rn	86	[222]
chlorine	Cl	17	35.453	rhenium	Re	75	186.207
chromium	Cr	24	51.996	rhodium	Rh	45	102.906
cobalt	Co	27	58.9332	rubidium	Rb	37	85.4678
copper	Cu	29	63.546	ruthenium	Ru	44	101.07
curium	Cm	96	[247]	samarium	Sm	62	150.36
dysprosium	Dy	66	162.50	scandium	Sc	21	44.9559
einsteinium	Es	99	[252]	selenium	Se	34	78.96
erbium	Er	68	167.26	silicon	Si	14	28.0855
europium	Eu	63	151.96	silver	Ag	47	107.868
fermium	Fm	100	[257]	sodium	Na	11	22.989 77
fluorine	F	9	18.998 403	strontium	Sr	38	87.62
francium	Fr	87	[223]	sulfur	S	16	32.06
gadolinium	Gd	64	157.25	tantalum	Ta	73	180.948
gallium	Ga	31	69.72	technetium	Tc	43	[98]
germanium	Ge	32	72.59	tellurium	Te	52	127.60
gold	Au	79	196.967	terbium	Tb	65	158.925
hafnium	Hf	72	178.49	thallium	Tl	81	204.383
helium	He	2	4.002 60	thorium	Th	90	232.038
holmium	Ho	67	164.930	thulium	Tm	69	168.934
hydrogen	H	1	1.007 94	tin	Sn	50	118.71
indium	In	49	114.82	titanium	Ti	22	47.88
iodine	I	53	126.905	tungsten	W	74	183.85
iridium	Ir	77	192.22	unnilennium	Une	109	[266?]
iron	Fe	26	55.847	unnilhexium	Unh	106	[263]
krypton	Kr	36	83.80	unniloctium	Uno	108	[265]
lanthanum	La	57	138.906	unnilpentium	Unp	105	[262]
lawrencium	Lr	103	[260]	unnilquadium	Unq	104	[262]
lead	Pb	82	207.2	unnilseptium	Uns	107	[261?]
lithium	Li	3	6.941	uranium	U	92	238.029
lutetium	Lu	71	174.967	vanadium	V	23	50.9415
magnesium	Mg	12	24.305	xenon	Xe	54	131.29
manganese	Mn	25	54.9380	ytterbium	Yb	70	173.04
mendelevium	Md	101	[258]	yttrium	Y	39	88.9059
mercury	Hg	80	200.59	zinc	Zn	30	65.39
molybdenum	Mo	42	95.94	zirconium	Zr	40	91.224
neodymium	Nd	60	144.24				

A value given in brackets denotes the mass number of the most stable or most common isotope. The atomic masses of most of these elements are believed to have an error no greater than ±1 in the last digit given.

TABLE A–4 COMMON ELEMENTS

Name	Symbol	Approx. mass	Common ox. nos.	Name	Symbol	Approx. mass	Common ox. nos.
aluminum	Al	27.0	+3	magnesium	Mg	24.3	+2
antimony	Sb	121.8	+3, +5	manganese	Mn	54.9	+2, +4, +7
arsenic	As	74.9	+3, +5	mercury	Hg	200.6	+1, +2
barium	Ba	137.3	+2	nickel	Ni	58.7	+2
bismuth	Bi	209.0	+3	nitrogen	N	14.0	−3, +3, +5
bromine	Br	79.9	−1, +5	oxygen	O	16.0	−2
calcium	Ca	40.1	+2	phosphorus	P	31.0	+3, +5
carbon	C	12.0	+2, +4	platinum	Pt	195.1	+2, +4
chlorine	Cl	35.5	−1, +5, +7	potassium	K	39.1	+1
chromium	Cr	52.0	+2, +3, +6	silicon	Si	28.1	+4
cobalt	Co	58.9	+2, +3	silver	Ag	107.9	+1
copper	Cu	63.5	+1, +2	sodium	Na	23.0	+1
fluorine	F	19.0	−1	strontium	Sr	87.6	+2
gold	Au	197.0	+1, +3	sulfur	S	32.1	−2, +4, +6
hydrogen	H	1.0	−1, +1	tin	Sn	118.7	+2, +4
iodine	I	126.9	−1, +5	titanium	Ti	47.9	+3, +4
iron	Fe	55.8	+2, +3	tungsten	W	183.8	+6
lead	Pb	207.2	+2, +4	zinc	Zn	65.4	+2

TABLE A–5 COMMON IONS

Name	Symbol	Name	Symbol
aluminum	Al^{+3}	lead(II)	Pb^{+2}
ammonium	NH_4^+	magnesium	Mg^{+2}
barium	Ba^{+2}	mercury(I)	Hg_2^{+2}
calcium	Ca^{+2}	mercury(II)	Hg^{+2}
chromium(III)	Cr^{+3}	nickel(II)	Ni^{+2}
cobalt(II)	Co^{+2}	potassium	K^+
copper(I)	Cu^+	silver	Ag^+
copper(II)	Cu^{+2}	sodium	Na^+
hydronium	H_3O^+	tin(II)	Sn^{+2}
iron(II)	Fe^{+2}	tin(IV)	Sn^{+4}
iron(III)	Fe^{+3}	zinc	Zn^{+2}
acetate	$C_2H_3O_2^-$	hydrogen sulfate	HSO_4^-
bromide	Br^-	hydroxide	OH^-
carbonate	CO_3^{-2}	hypochlorite	ClO^-
chlorate	ClO_3^-	iodide	I^-
chloride	Cl^-	nitrate	NO_3^-
chlorite	ClO_2^-	nitrite	NO_2^-
chromate	CrO_4^{-2}	oxide	O^{-2}
cyanide	CN^-	perchlorate	ClO_4^-
dichromate	$Cr_2O_7^{-2}$	permanganate	MnO_4^-
fluoride	F^-	peroxide	O_2^{-2}
hexacyanoferrate(I)	$Fe(CN)_6^{-4}$	phosphate	PO_4^{-2}
hecacyanoferrate(II)	$Fe(CN)_6^{-3}$	sulfate	SO_4^{-2}
hydride	H^-	sulfide	S^{-2}
hydrogen carbonate	HCO_3^-	sulfite	SO_3^{-2}

	Sublevels	1s	2s	2p	3s	3p	3d	4s	4p	4d	4f	5s	5p	5d	5f	6s	6p	6d	6f	7s
1	hydrogen	1																		
2	helium	2																		
3	lithium	2	1																	
4	beryllium	2	2																	
5	boron	2	2	1																
6	carbon	2	2	2																
7	nitrogen	2	2	3																
8	oxygen	2	2	4																
9	fluorine	2	2	5																
10	neon	2	2	6																
11	sodium	2	2	6	1															
12	magnesium	2	2	6	2															
13	aluminum	2	2	6	2	1														
14	silicon	2	2	6	2	2														
15	phosphorus	2	2	6	2	3														
16	sulfur	2	2	6	2	4														
17	chlorine	2	2	6	2	5														
18	argon	2	2	6	2	6														
19	potassium	2	2	6	2	6		1												
20	calcium	2	2	6	2	6		2												
21	scandium	2	2	6	2	6	1	2												
22	titanium	2	2	6	2	6	2	2												
23	vanadium	2	2	6	2	6	3	2												
24	chromium	2	2	6	2	6	5	1												
25	manganese	2	2	6	2	6	5	2												
26	iron	2	2	6	2	6	6	2												
27	cobalt	2	2	6	2	6	7	2												
28	nickel	2	2	6	2	6	8	2												
29	copper	2	2	6	2	6	10	1												
30	zinc	2	2	6	2	6	10	2												
31	gallium	2	2	6	2	6	10	2	1											
32	germanium	2	2	6	2	6	10	2	2											
33	arsenic	2	2	6	2	6	10	2	3											
34	selenium	2	2	6	2	6	10	2	4											
35	bromine	2	2	6	2	6	10	2	5											
36	krypton	2	2	6	2	6	10	2	6											
37	rubidium	2	2	6	2	6	10	2	6			1								
38	strontium	2	2	6	2	6	10	2	6			2								
39	yttrium	2	2	6	2	6	10	2	6	1		2								
40	zirconium	2	2	6	2	6	10	2	6	2		2								
41	niobium	2	2	6	2	6	10	2	6	4		1								
42	molybdenum	2	2	6	2	6	10	2	6	5		1								
43	technetium	2	2	6	2	6	10	2	6	5		2								
44	ruthenium	2	2	6	2	6	10	2	6	7		1								
45	rhodium	2	2	6	2	6	10	2	6	8		1								
46	palladium	2	2	6	2	6	10	2	6	10										
47	silver	2	2	6	2	6	10	2	6	10		1								
48	cadmium	2	2	6	2	6	10	2	6	10		2								
49	indium	2	2	6	2	6	10	2	6	10		2	1							
50	tin	2	2	6	2	6	10	2	6	10		2	2							
51	antimony	2	2	6	2	6	10	2	6	10		2	3							
52	tellurium	2	2	6	2	6	10	2	6	10		2	4							
53	iodine	2	2	6	2	6	10	2	6	10		2	5							
54	xenon	2	2	6	2	6	10	2	6	10		2	6							

	Sublevels	1s	2s	2p	3s	3p	3d	4s	4p	4d	4f	5s	5p	5d	5f	6s	6p	6d	6f	7s
55	cesium	2	2	6	2	6	10	2	6	10		2	6			1				
56	barium	2	2	6	2	6	10	2	6	10		2	6			2				
57	lanthanum	2	2	6	2	6	10	2	6	10		2	6	1		2				
58	cerium	2	2	6	2	6	10	2	6	10	2	2	6			2				
59	praseodymium	2	2	6	2	6	10	2	6	10	3	2	6			2				
60	neodymium	2	2	6	2	6	10	2	6	10	4	2	6			2				
61	promethium	2	2	6	2	6	10	2	6	10	5	2	6			2				
62	samarium	2	2	6	2	6	10	2	6	10	6	2	6			2				
63	europium	2	2	6	2	6	10	2	6	10	7	2	6			2				
64	gadolinium	2	2	6	2	6	10	2	6	10	7	2	6	1		2				
65	terbium	2	2	6	2	6	10	2	6	10	9	2	6			2				
66	dysprosium	2	2	6	2	6	10	2	6	10	10	2	6			2				
67	holmium	2	2	6	2	6	10	2	6	10	11	2	6			2				
68	erbium	2	2	6	2	6	10	2	6	10	12	2	6			2				
69	thulium	2	2	6	2	6	10	2	6	10	13	2	6			2				
70	ytterbium	2	2	6	2	6	10	2	6	10	14	2	6			2				
71	lutetium	2	2	6	2	6	10	2	6	10	14	2	6	1		2				
72	hafnium	2	2	6	2	6	10	2	6	10	14	2	6	3		2				
73	tantalum	2	2	6	2	6	10	2	6	10	14	2	6	3		2				
74	tungsten	2	2	6	2	6	10	2	6	10	14	2	6	4		2				
75	rhenium	2	2	6	2	6	10	2	6	10	14	2	6	5		2				
76	osmium	2	2	6	2	6	10	2	6	10	14	2	6	6		2				
77	iridium	2	2	6	2	6	10	2	6	10	14	2	6	7		2				
78	platinum	2	2	6	2	6	10	2	6	10	14	2	6	9		1				
79	gold	2	2	6	2	6	10	2	6	10	14	2	6	10		1				
80	mercury	2	2	6	2	6	10	2	6	10	14	2	6	10		2				
81	thallium	2	2	6	2	6	10	2	6	10	14	2	6	10		2	1			
82	lead	2	2	6	2	6	10	2	6	10	14	2	6	10		2	2			
83	bismuth	2	2	6	2	6	10	2	6	10	14	2	6	10		2	3			
84	polonium	2	2	6	2	6	10	2	6	10	14	2	6	10		2	4			
85	astatine	2	2	6	2	6	10	2	6	10	14	2	6	10		2	5			
86	radon	2	2	6	2	6	10	2	6	10	14	2	6	10		2	6			
87	francium	2	2	6	2	6	10	2	6	10	14	2	6	10		2	6			1
88	radium	2	2	6	2	6	10	2	6	10	14	2	6	10		2	6			2
89	actinium	2	2	6	2	6	10	2	6	10	14	2	6	10		2	6	1		2
90	thorium	2	2	6	2	6	10	2	6	10	14	2	6	10		2	6	2		2
91	protactinium	2	2	6	2	6	10	2	6	10	14	2	6	10	2	2	6	1		2
92	uranium	2	2	6	2	6	10	2	6	10	14	2	6	10	3	2	6	1		2
93	neptunium	2	2	6	2	6	10	2	6	10	14	2	6	10	4	2	6	1		2
94	plutonium	2	2	6	2	6	10	2	6	10	14	2	6	10	6	2	6			2
95	americium	2	2	6	2	6	10	2	6	10	14	2	6	10	7	2	6			2
96	curium	2	2	6	2	6	10	2	6	10	14	2	6	10	7	2	6	1		2
97	berkelium	2	2	6	2	6	10	2	6	10	14	2	6	10	9	2	6			2
98	californium	2	2	6	2	6	10	2	6	10	14	2	6	10	10	2	6			2
99	einsteinium	2	2	6	2	6	10	2	6	10	14	2	6	10	11	2	6			2
100	fermium	2	2	6	2	6	10	2	6	10	14	2	6	10	12	2	6			2
101	mendelevium	2	2	6	2	6	10	2	6	10	14	2	6	10	13	2	6			2
102	nobelium	2	2	6	2	6	10	2	6	10	14	2	6	10	14	2	6			2
103	lawrencium	2	2	6	2	6	10	2	6	10	14	2	6	10	14	2	6	1		2
104	unnilquadium	2	2	6	2	6	10	2	6	10	14	2	6	10	14	2	6	2		2?
105	unnilpentium	2	2	6	2	6	10	2	6	10	14	2	6	10	14	2	6	3		2?
106	unnilhexium	2	2	6	2	6	10	2	6	10	14	2	6	10	14	2	6	4		2?
107	unnilseptium	2	2	6	2	6	10	2	6	10	14	2	6	10	14	2	6	5		2?
108	unniloctium	2	2	6	2	6	10	2	6	10	14	2	6	10	14	2	6	6		2?
109	unnilennium	2	2	6	2	6	10	2	6	10	14	2	6	10	14	2	6	7		2?

TABLE A–7   WATER-VAPOR PRESSURE

Temperature (°C)	Pressure (mm Hg)	Temperature (°C)	Pressure (mm Hg)	Temperature (°C)	Pressure (mm Hg)
0.0	4.6	19.5	17.0	27.0	26.7
5.0	6.5	20.0	17.5	28.0	28.3
10.0	9.2	20.5	18.1	29.0	30.0
12.5	10.9	21.0	18.6	30.0	31.8
15.0	12.8	21.5	19.2	35.0	42.2
15.5	13.2	22.0	19.8	40.0	55.3
16.0	13.6	22.5	20.4	50.0	92.5
16.5	14.1	23.0	21.1	60.0	149.4
17.0	14.5	23.5	21.7	70.0	233.7
17.5	15.0	24.0	22.4	80.0	355.1
18.0	15.5	24.5	23.1	90.0	525.8
18.5	16.0	25.0	23.8	95.0	633.9
19.0	16.5	26.0	25.2	100.0	760.0

TABLE A–8   DENSITY OF GASES AT STP

Gas	Density (g/L)	Gas	Density (g/L)
air, dry	1.2929	hydrogen	0.0899
ammonia	0.771	hydrogen chloride	1.639
carbon dioxide	1.977	hydrogen sulfide	1.539
carbon monoxide	1.250	methane	0.716
chlorine	3.214	nitrogen	1.251
dinitrogen monoxide	1.977	nitrogen monoxide	1.340
ethyne (acetylene)	1.171	oxygen	1.429
helium	0.1785	sulfur dioxide	2.927

TABLE A–9   DENSITY OF WATER

Temperature (°C)	Density (g/mL)	Temperature (°C)	Density (g/mL)
0	0.999 84	24	0.997 30
2	0.999 94	26	0.996 78
3.98 (maximum)	0.999 973	30	0.995 65
4	0.999 97	40	0.992 22
6	0.999 94	50	0.988 04
8	0.999 85	60	0.983 20
10	0.999 70	70	0.977 77
14	0.999 24	80	0.971 79
16	0.998 94	90	0.965 31
20	0.998 20	100	0.958 36

TABLE A–10 SOLUBILITY OF GASES IN WATER

Volume of gas (reduced to STP) that can be dissolved in 1 volume of water at the temperature (°C) indicated.

Gas	0°	10°	20°	60°
air	0.029 18	0.022 84	0.018 68	0.012 16
ammonia	1130	870	680	200
carbon dioxide	1.713	1.194	0.878	0.359
carbon monoxide	0.035 37	0.028 16	0.023 19	0.014 88
chlorine	4.54	3.148	2.299	1.023
hydrogen	0.021 48	0.019 55	0.018 19	0.016 00
hydrogen chloride	512	475	442	339
hydrogen sulfide	4.670	3.399	2.582	1.190
methane	0.055 63	0.041 77	0.033 08	0.019 54
nitrogen	0.023 54	0.018 61	0.015 45	0.010 23
nitrogen dioxide	0.073 81	0.057 09	0.047 06	0.029 54
oxygen	0.048 89	0.038 02	0.031 02	0.019 46
sulfur dioxide	79.789	56.647	39.374	—

TABLE A–11 SOLUBILITY CHART

S = soluble in water. A = soluble in acids, insoluble in water. P = partially soluble in water, soluble in dilute acids.
I = insoluble in dilute acids and in water. a = slightly soluble in acids, insoluble in water. d = decomposes in water.

	acetate	bromide	carbonate	chlorate	chloride	chromate	hydroxide	iodide	nitrate	oxide	phosphate	silicate	sulfate	sulfide
aluminum	S	S	—	S	S	—	A	S	S	a	A	I	S	d
ammonium	S	S	S	S	S	S	—	S	S	—	A	—	S	S
barium	S	S	P	S	S	A	S	S	S	S	A	S	a	d
calcium	S	S	P	S	S	S	P	S	S	P	P	P	P	P
copper(II)	S	S	—	S	S	—	A	—	S	A	A	A	S	A
hydrogen	S	S	—	S	S	—	—	S	S	—	S	I	S	S
iron(II)	S	S	P	S	S	—	A	S	S	A	A	—	S	A
iron(III)	S	S	—	S	S	A	A	A	S	A	P	—	P	d
lead(II)	S	S	A	S	S	A	P	P	S	P	A	A	P	A
magnesium	S	S	P	S	S	S	A	S	S	A	P	A	S	d
manganese(II)	S	S	P	S	S	—	A	S	S	A	P	I	S	A
mercury(I)	P	A	A	S	a	P	—	A	S	A	A	—	P	I
mercury(II)	S	S	—	S	S	P	A	P	S	P	A	—	d	I
potassium	S	S	S	S	S	S	S	S	S	S	S	S	S	S
silver	P	a	A	S	a	P	—	I	S	P	A	—	P	A
sodium	S	S	S	S	S	S	S	S	S	S	S	S	S	S
strontium	S	S	P	S	S	P	S	S	S	S	A	A	P	S
tin(II)	d	S	—	S	S	A	A	S	d	A	A	—	S	A
tin(IV)	S	S	—	—	S	S	P	d	—	A	—	—	S	A
zinc	S	S	P	S	S	P	A	S	S	P	A	A	S	A

TABLE A–12 SOLUBILITY OF COMPOUNDS

Solubilities are given in grams of anhydrous solute that can be dissolved in 100 g of water at the temperature (°C) indicated.

Compound	Formula	0°	20°	60°	100°
aluminum sulfate	$Al_2(SO_4)_3$	31.2	36.4	59.2	89.0
ammonium chloride	NH_4Cl	29.4	37.2	55.3	77.3
ammonium nitrate	NH_4NO_3	118	192	421	871
ammonium sulfate	$(NH_4)_2SO_4$	70.6	75.4	88	103
barium carbonate	$BaCO_3$	$0.0016^{8°}$	$0.0022^{18°}$	—	0.0065
barium chloride	$BaCl_2$	31.2	35.8	46.2	59.4
barium hydroxide	$Ba(OH)_2$	1.67	3.89	20.94	$101.40^{80°}$
barium nitrate	$Ba(NO_3)_2$	4.95	9.02	20.4	34.4
barium sulfate	$BaSO_4$	0.000 115	0.000 24	—	0.000 413
cadmium sulfate	$CdSO_4$	75.4	76.6	81.8	60.8
calcium acetate	$Ca(C_2H_3O_2)_2$	37.4	34.7	32.7	29.7
calcium carbonate	$CaCO_3$	—	0.0012	—	0.002
calcium fluoride	CaF_2	$0.0016^{18°}$	$0.0017^{25°}$	—	—
calcium hydrogen carbonate	$Ca(HCO_3)_2$	16.15	16.60	17.50	18.40
calcium hydroxide	$Ca(OH)_2$	0.189	0.173	0.121	0.076
calcium sulfate	$CaSO_4$	0.176	$0.209^{30°}$	0.205	0.162
cerium sulfate	$Ce_2(SO_4)_3$	21.4	9.84	3.87	—
cesium nitrate	$CsNO_3$	9.33	23.0	83.8	197
copper(II) chloride	$CuCl_2$	68.6	73.0	96.5	120
copper(II) sulfate	$CuSO_4$	23.1	32.0	61.8	114
lead(II) chloride	$PbCl_2$	0.67	1.00	1.94	3.20
lead(II) nitrate	$Pb(NO_3)_2$	37.5	54.3	91.6	133
lithium chloride	$LiCl$	69.2	83	98.4	128
lithium sulfate	Li_2SO_4	36.1	34.8	32.6	29.9
magnesium hydroxide	$Mg(OH)_2$	—	$0.0009^{18°}$	—	—
magnesium sulfate	$MgSO_4$	22.0	33.7	54.6	68.3
mercury(I) chloride	Hg_2Cl_2	0.000 14	0.0002	$0.0007^{40°}$	—
mercury(II) chloride	$HgCl_2$	3.63	6.57	16.3	61.3
potassium aluminum sulfate	$KAl(SO_4)_2$	3.00	5.90	24.8	109
potassium bromide	KBr	53.6	65.3	85.5	104
potassium chlorate	$KClO_3$	3.3	7.3	23.8	56.3
potassium chloride	KCl	28.0	34.2	45.8	56.3
potassium chromate	K_2CrO_4	56.3	63.7	70.1	75.6
potassium iodide	KI	128	144	176	206
potassium nitrate	KNO_3	13.9	31.6	106	245
potassium permanganate	$KMnO_4$	2.83	6.34	22.1	—
potassium sulfate	K_2SO_4	7.4	11.1	18.2	24.1
silver acetate	$AgC_2H_3O_2$	0.73	1.05	1.93	$2.59^{80°}$
silver chloride	$AgCl$	0.000 07	0.000 194	$0.0005^{50°}$	0.002
silver nitrate	$AgNO_3$	122	216	440	733
sodium acetate	$NaC_2H_3O_2$	36.2	46.4	139	170.15
sodium chlorate	$NaClO_3$	79.6	95.9	137	204
sodium chloride	$NaCl$	35.7	35.9	37.1	39.2
sodium nitrate	$NaNO_3$	73.0	87.6	122	180
sucrose	$C_{12}H_{22}O_{11}$	179.2	203.9	287.3	487.2
ytterbium sulfate	$Yb_2(SO_4)_3$	44.2	$22.2^{30°}$	10.4	4.7

ΔH_f is heat of formation of the given substance from its elements. All values of ΔH_f are expressed as kJ/mol at 25°C.
Negative values of ΔH_f indicate exothermic reactions. s = solid, ℓ = liquid, g = gas.

Substance	Phase	ΔH_f	Substance	Phase	ΔH_f
aluminum oxide (α)	s	−1675.7	lead (II) nitrate	s	−451.97
ammonia	g	−46.11	lead (II) sulfate	s	−919.94
ammonium chloride	s	−314.43	lithium chloride	s	−408.78
ammonium sulfate	s	−1180.85	lithium nitrate	s	−481.58
barium chloride	s	−860.06	lithium sulfate	s	−1434.4
barium nitrate	s	−991.86	magnesium chloride	s	−641.83
barium sulfate	s	−1465.2	magnesium oxide	s	−601.83
benzene	g	+82.93	magnesium sulfate	s	−1278.2
benzene	ℓ	+48.50	manganese (IV) oxide	s	−520.03
calcium carbonate	s	−1206.9	manganese (II) sulfate	s	−1065.2
calcium chloride	s	−795.0	mercury (I) chloride	s	−265.22
calcium hydroxide	s	−986.6	mercury (II) chloride	s	−224.3
calcium nitrate	s	−937.22	mercury (II) fulminate	s	+270
calcium oxide	s	−635.5	mercury (II) nitrate	s	−392
calcium sulfate	s	−1432.69	mercury (II) oxide (red)	s	−90.84
carbon (diamond)	s	+1.90	methane	g	−74.81
carbon (graphite)	s	0.00	nitrogen dioxide	g	+33.18
carbon dioxide	g	−393.51	nitrogen monoxide	g	+90.25
carbon disulfide	g	+117.07	dinitrogen monoxide	g	+82.05
carbon disulfide	ℓ	+89.70	dinitrogen pentoxide	g	+11.3
carbon monoxide	g	+110.53	dinitrogen pentoxide	s	−43.1
carbon tetrachloride	ℓ	−95.81	dinitrogen tetroxide	g	+9.16
copper (II) nitrate	s	−302.9	oxygen (O_2)	g	0.00
copper (II) oxide	s	−157.3	ozone (O_3)	g	+142.7
copper (II) sulfate	s	−771.36	diphosphorus pentoxide	s	−2984.0
copper (I) sulfide	s	−79.5	potassium bromide	s	−392.17
copper (II) sulfide	s	−53.1	potassium chloride	s	−435.87
ethane	g	−84.68	potassium hydroxide	s	−425.85
ethyne (acetylene)	g	+226.73	potassium iodide	s	−327.65
hydrogen (H_2)	g	0.00	potassium nitrate	s	−492.71
hydrogen bromide	g	−36.40	potassium sulfate	s	−1433.69
hydrogen chloride	g	−92.31	silicon dioxide (quartz)	s	−910.94
hydrogen fluoride	g	−271.1	silver acetate	s	−399
hydrogen iodide	g	+26.48	silver chloride	s	−127.07
hydrogen oxide (water)	g	−241.82	silver nitrate	s	−124.39
hydrogen oxide (water)	ℓ	−285.83	silver sulfide	s	−32.6
hydrogen peroxide	g	−136.11	sodium bromide	s	−361.1
hydrogen peroxide	ℓ	−187.78	sodium chloride	s	−411.01
hydrogen sulfide	g	−20.63	sodium hydroxide	s	−426.73
iodine (I_2)	s	0.00	sodium nitrate	s	−467.9
iodine (I_2)	g	+62.44	sodium sulfate	s	−1484.49
iron (III) chloride	s	−399.49	sulfur dioxide	g	−296.83
iron (III) oxide	s	−824.2	sulfur trioxide	g	−395.72
iron (II, III) oxide	s	−1118.4	tin (IV) chloride	ℓ	−511.3
iron (II) sulfate	s	−928.4	zinc nitrate	s	−483.7
iron (II) sulfide	s	−100.0	zinc oxide	s	−358.28
lead (II) oxide	s	−218.07	zinc sulfate	s	−982.82
lead (IV) oxide	s	−277.4	zinc sulfide (sphalerite)	s	−205.98

TABLE A–14 PROPERTIES OF COMMON ELEMENTS

Name	Form/color at room temperature	Density (g/cm³)	Melting point (°C)	Boiling point (°C)	Common oxidation numbers
aluminum	silv metal	2.70	660.4	2467	+3
antimony	silv metal	6.69	630.7	1750	+3, +5
argon	colorless gas	1.784*	−189.2	−185.7	0
arsenic	gray metal	5.73	817 (28 atm)	613 (sublimes)	+3, +5
barium	silv metal	3.5	725	1640	+2
beryllium	gray metal	1.85	1280	2970	+2
bismuth	silv metal	9.75	271.3	1560	+3
boron	blk solid	2.34	2079	2550 (sublimes)	+3
bromine	red-br liquid	3.12	−7.2	58.8	−1, +5
calcium	silv metal	1.55	839	1484	+2
carbon	diamond	3.51	3700	4200	+2, +4
	graphite	2.26	3620 (sublimes)	4200	
chlorine	grn-yel gas	3.214*	−101.0	−34.6	−1, +5, +7
chromium	silv metal	7.19	1860	2672	+2, +3, +6
cobalt	silv metal	8.9	1495	2870	+2, +3
copper	red metal	8.96	1083.4	2567	+1, +2
fluorine	yel gas	1.696*	−219.6	−188.1	−1
germanium	gray metalloid	5.32	937.4	2830	+4
gold	yel metal	19.3	1064.4	2807	+1, +3
helium	colorless gas	0.1785*	−272.2 (26 atm)	−268.9	0
hydrogen	colorless gas	0.08988*	−259.1	−252.9	−1, +1
iodine	bl-blk solid	4.93	113.5	184.4	−1, +5
iron	silv metal	787	1535	2750	+2, +3
lead	silv metal	11.4	327.5	1740	+2, +4
lithium	silv metal	0.534	180.5	1347	+1
magnesium	silv metal	1.74	649	1090	+2
manganese	silv metal	7.3	1244	1962	+2, +4, +7
mercury	silv liquid	13.5	−38.8	356.6	+1, +2
neon	colorless gas	0.8999*	−248.7	−246.0	0
nickel	silv metal	8.90	1453	2732	+2
nitrogen	colorless gas	1.251*	−209.9	−195.8	−3, +3, +5
oxygen	colorless gas	1.429*	−218.4	−183.0	−2
phosphorus	yel solid	1.82	44.1	280	+3, +5
platinum	silv metal	21.4	1772	3800	+2, +4
plutonium	silv metal	19.8	641	3232	+3, +4, +5, +6
potassium	silv metal	0.862	63.6	774	+1
radium	silv metal	5(?)	700	1140	+2
radon	colorless gas	9.73*	−71	−61.8	0
silicon	gray solid	2.33	1410	2355	+4
silver	silv metal	10.5	961.9	2212	+1
sodium	silv metal	0.971	97.8	882.9	+1
strontium	silv metal	2.54	769	1384	+2
sulfur	yel solid	2.07	112.8	444.7	−2, +4, +6
tin	silv metal	7.31	232.0	2270	+2, +4
titanium	silv metal	4.54	1660	3287	+3, +4
tungsten	gray metal	19.3	3410	5660	+6
uranium	silv metal	19.0	1132	3818	+4, +6
xenon	colorless gas	5.89*	−111.9	−107	0
zinc	silv metal	7.13	419.6	907	+2

*Densities of gases are given in grams/liter at STP.

APPENDIX B

Logarithms

Logarithms are exponents. They are useful when working with calculations in which numbers vary over a large range. The logarithm of a number is the exponent or the power to which the number must be raised in order to obtain that number. For example, if $x = zy$, then by raising z to the y power, x is obtained. This can be said mathematically as: y is the log of x using base z or; $y = \log^z x$. For example, if 10 is used as the base, then $100 = 10^2$ or 2 is the log of 100 using base 10.

Logarithms that use 10 as the base are called common logarithms. Only common logarithms are used in this book so it is not necessary to write 10 as the base when using common logarithms. For example, in the expression, $2 = \log 100$, it is understood that base 10 is being used.

Logarithms are exponents and follow the laws of exponents in multiplication and division. The logarithm of a product is the sum of the logarithms of the factors.

$$\log(a \times b) = \log(a) + \log(b)$$

The logarithm of a quotient is the logarithm of the dividend minus the logarithm of the divisor.

$$\log(a/b) = \log(a) - \log(b)$$

A logarithm is composed of two parts: the characteristic, or whole-number part; and the mantissa, or decimal part. For example, the logarithm of 219. is 2.340, where 2 is the characteristic and .340 is the mantissa.

To find the logarithm of a number, it is helpful to use the fact that

$$\log(a \times 10^b) = \log(a) + \log(10^b) = \log(a) + b$$

To illustrate this we will find the logarithm of 219 using four steps.

Step 1. Write the number in scientific notation.
$$219. = 2.19 \times 10^2$$

Step 2. Then take the log of both sides of this equation.

$$\log(219.) = \log(2.19) \times \log(10^2) = \log(2.19) + 2$$

Step 3. Find the log of 2.19 by consulting a logarithm table (Table B-1). Find 2.1 in the left column. Then, move to the right until you are under the column labeled 9. The log of 2.19 is .3404.

Step 4. Since we know the log of 2.19, we can fill in the missing information from Step 2 as follows.

$$\log(219) = 2 + .3404 = 2.34$$

In many cases you are given the log of a number and are asked to find the number. The number whose log is given is called the antilog. How can you find an antilog? The antilog of a number is the number obtained by raising ten to the power indicated by the number. For example, the antilog of 2 is 100; that, is 10 raised to the power 2 is 100, or;

$$\text{antilog}(2) = 100$$

To find the antilog of a mixed number or fraction such as 10.32 or 0.189, the 4 steps shown above must essentially be reversed. This is illustrated in the following sample problem.

Sample Problem
Find the antilog of 1.32.

Solution
Step 1. Write the number as the sum of a fraction and a whole number.

$$1.32 = 0.32 + 1$$

Step 2. Take the antilog of the fraction and the whole number.

$$\text{antilog}(1.32) = \text{antilog}(0.32 + 1)$$

Step 3. The antilog of 1 is 10; remember that $10 = 10^1$. Find the antilog of 0.32 by looking up 0.32 in the body of Table B-1. It is found in the column labeled 9 and the row labeled 2.0. The antilog of 0.32 is 2.09.

Step 4. Since the antilog of the sum of two numbers is equal to the product of the two antilogs, obtain the final antilog by multiplying the antilogs of the two numbers.

$$10 \times 2.09 = 20.9$$

TABLE B–1 COMMON LOGARITHMS

N	0	1	2	3	4	5	6	7	8	9	N	0	1	2	3	4	5	6	7	8	9
1.0	.0000	.0043	.0086	.0128	.0170	.0212	.0253	.0294	.0334	.0374	5.5	.7404	.7412	.7419	.7427	.7435	.7443	.7451	.7459	.7466	.7474
1.1	.0414	.0453	.0492	.0531	.0569	.0607	.0645	.0682	.0719	.0755	5.6	.7482	.7490	.7497	.7505	.7513	.7520	.7528	.7536	.7543	.7551
1.2	.0792	.0828	.0864	.0899	.0934	.0969	.1004	.1038	.1072	.1106	5.7	.7559	.7566	.7574	.7582	.7589	.7597	.7604	.7612	.7619	.7627
1.3	.1139	.1173	.1206	.1239	.1271	.1303	.1335	.1367	.1399	.1430	5.8	.7634	.7642	.7649	.7657	.7664	.7672	.7679	.7686	.7694	.7701
1.4	.1461	.1492	.1523	.1553	.1584	.1614	.1644	.1673	.1703	.1732	5.9	.7709	.7716	.7723	.7731	.7738	.7745	.7752	.7760	.7767	.7774
1.5	.1761	.1790	.1818	.1847	.1875	.1903	.1931	.1959	.1987	.2014	6.0	.7782	.7789	.7796	.7803	.7810	.7818	.7825	.7832	.7839	.7846
1.6	.2041	.2068	.2095	.2122	.2148	.2175	.2201	.2227	.2253	.2279	6.1	.7853	.7860	.7868	.7875	.7882	.7889	.7896	.7903	.7910	.7917
1.7	.2304	.2330	.2355	.2380	.2405	.2430	.2455	.2480	.2504	.2529	6.2	.7924	.7931	.7938	.7945	.7952	.7959	.7966	.7973	.7980	.7987
1.8	.2553	.2577	.2601	.2625	.2648	.2672	.2695	.2718	.2742	.2765	6.3	.7993	.8000	.8007	.8014	.8021	.8028	.8035	.8041	.8048	.8055
1.9	.2788	.2810	.2833	.2856	.2878	.2900	.2923	.2945	.2967	.2989	6.4	.8062	.8069	.8075	.8082	.8089	.8096	.8102	.8109	.8116	.8122
2.0	.3010	.3032	.3054	.3075	.3096	.3118	.3139	.3160	.3181	.3201	6.5	.8129	.8136	.8142	.8149	.8156	.8162	.8169	.8176	.8182	.8189
2.1	.3222	.3243	.3263	.3284	.3304	.3324	.3345	.3365	.3385	.3404	6.6	.8195	.8202	.8209	.8215	.8222	.8228	.8235	.8241	.8248	.8254
2.2	.3424	.3444	.3464	.3483	.3502	.3522	.3541	.3560	.3579	.3598	6.7	.8261	.8267	.8274	.8280	.8287	.8293	.8299	.8306	.8312	.8319
2.3	.3617	.3636	.3655	.3674	.3692	.3711	.3729	.3747	.3766	.3784	6.8	.8325	.8331	.8338	.8344	.8351	.8357	.8363	.8370	.8376	.8382
2.4	.3802	.3820	.3838	.3856	.3874	.3892	.3909	.3927	.3945	.3962	6.9	.8388	.8395	.8401	.8407	.8414	.8420	.8426	.8432	.8439	.8445
2.5	.3979	.3997	.4014	.4031	.4048	.4065	.4082	.4099	.4116	.4133	7.0	.8451	.8457	.8463	.8470	.8476	.8482	.8488	.8494	.8500	.8506
2.6	.4150	.4166	.4183	.4200	.4216	.4232	.4249	.4265	.4281	.4298	7.1	.8513	.8519	.8525	.8531	.8537	.8543	.8549	.8555	.8561	.8567
2.7	.4314	.4330	.4346	.4362	.4378	.4393	.4409	.4425	.4440	.4456	7.2	.8573	.7579	.8585	.8591	.8597	.8603	.8609	.8615	.8621	.8627
2.8	.4472	.4487	.4502	.4518	.4533	.4548	.4564	.4579	.4594	.4609	7.3	.8633	.8639	.8645	.8651	.8657	.8663	.8669	.8675	.8681	.8686
2.9	.4624	.4639	.4654	.4669	.4683	.4698	.4713	.4728	.4742	.4757	7.4	.8692	.8698	.8704	.8710	.8716	.8722	.8727	.8733	.8739	.8745
3.0	.4771	.4786	.4800	.4814	.4829	.4843	.4857	.4871	.4886	.4900	7.5	.8751	.8756	.8762	.8768	.8774	.8779	.8785	.8791	.8797	.8802
3.1	.4914	.4928	.4942	.4955	.4969	.4983	.4997	.5011	.5024	.5038	7.6	.8808	.8814	.8820	.8825	.8831	.8837	.8842	.8848	.8854	.8859
3.2	.5051	.5065	.5079	.5092	.5105	.5119	.5132	.5145	.5159	.5172	7.7	.8865	.8871	.8876	.8882	.8887	.8893	.8899	.8904	.8910	.8915
3.3	.5185	.5198	.5211	.5224	.5237	.5250	.5263	.5276	.5289	.5307	7.8	.8921	.8927	.8932	.8938	.8943	.8949	.8954	.8960	.8965	.8971
3.4	.5315	.5328	.5340	.5353	.5366	.5378	.5391	.5403	.5416	.5428	7.9	.8976	.8982	.8987	.8993	.8998	.9004	.9009	.9015	.9020	.9025
3.5	.5441	.5453	.5465	.5478	.5490	.5502	.5514	.5527	.5539	.5551	8.0	.9031	.9036	.9042	.9047	.9053	.9058	.9063	.9069	.9074	.9079
3.6	.5563	.5575	.5587	.5599	.5611	.5623	.5635	.5647	.5658	.5670	8.1	.9085	.9090	.9096	.9101	.9106	.9112	.9117	.9122	.9128	.9133
3.7	.5682	.5694	.5705	.5717	.5729	.5740	.5752	.5763	.5775	.5786	8.2	.9138	.9143	.9149	.9154	.9159	.9165	.9170	.9175	.9180	.9186
3.8	.5798	.5809	.5821	.5832	.5843	.5855	.5866	.5877	.5888	.5899	8.3	.9191	.9196	.9201	.9206	.9212	.9217	.9222	.9227	.9232	.9238
3.9	.5911	.5922	.5933	.5944	.5955	.5966	.5977	.5988	.5999	.6010	8.4	.9243	.9248	.9253	.9258	.9263	.9269	.9274	.9279	.9284	.9289
4.0	.6021	.6031	.6042	.6053	.6064	.6075	.6085	.6096	.6107	.6117	8.5	.9294	.9299	.9304	.9309	.9315	.9320	.9325	.9330	.9335	.9340
4.1	.6128	.6138	.6149	.6160	.6170	.6180	.6191	.6201	.6212	.6222	8.6	.9345	.9350	.9555	.9360	.9365	.9370	.9375	.9380	.9385	.9390
4.2	.6232	.6243	.6253	.6263	.6274	.6284	.6294	.6304	.6314	.6325	8.7	.9395	.9400	.9405	.9410	.9415	.9420	.9425	.9430	.9435	.9440
4.3	.6335	.6345	.6355	.6365	.6375	.6385	.6395	.6405	.6415	.6425	8.8	.9445	.9450	.9455	.9460	.9465	.9469	.9474	.9479	.9484	.9489
4.4	.6435	.6444	.6454	.6464	.6474	.6484	.6493	.6503	.6513	.6522	8.9	.9494	.9499	.9504	.9509	.9513	.9518	.9523	.9528	.9533	.9538
4.5	.6532	.6542	.6551	.6561	.6571	.6580	.6590	.6599	.6609	.6618	9.0	.9542	.9547	.9552	.9557	.9562	.9566	.9571	.9576	.9581	.9586
4.6	.6628	.6637	.6646	.6656	.6665	.6675	.6684	.6693	.6702	.6712	9.1	.9590	.9595	.9600	.9605	.9609	.9614	.9619	.9624	.9628	.9633
4.7	.6721	.6730	.6739	.6749	.6758	.6767	.6776	.6785	.6794	.6803	9.2	.9638	.9643	.9647	.9652	.9657	.9661	.9666	.9671	.9675	.9680
4.8	.6812	.6821	.6830	.6839	.6848	.6857	.6866	.6875	.6884	.6893	9.3	.9685	.9689	.9694	.9699	.9703	.9708	.9713	.9717	.9722	.9727
4.9	.6902	.6911	.6920	.6928	.6937	.6946	.6955	.6964	.6972	.6981	9.4	.9731	.9736	.9741	.9745	.9750	.9754	.9759	.9763	.9768	.9773
5.0	.6990	.6998	.7007	.7016	.7024	.7033	.7042	.7050	.7059	.7067	9.5	.9777	.9782	.9786	.9791	.9795	.9800	.9805	.9809	.9914	.9818
5.1	.7076	.7084	.7093	.7101	.7110	.7118	.7126	.7135	.7143	.7152	9.6	.9823	.9827	.9832	.9836	.9841	.9845	.9850	.9854	.9859	.9863
5.2	.7160	.7168	.7177	.7185	.7193	.7202	.7210	.7218	.7226	.7235	9.7	.9868	.9872	.9877	.9881	.9886	.9890	.9894	.9899	.9903	.9908
5.3	.7243	.7251	.7259	.7267	.7275	.7284	.7292	.7300	.7308	.7316	9.8	.9912	.9917	.9921	.9926	.9930	.9934	.9939	.9943	.9948	.9952
5.4	.7324	.7332	.7340	.7348	.7356	.7364	.7372	.7380	.7388	.7396	9.9	.9956	.9961	.9965	.9969	.9974	.9978	.9983	.9987	.9991	.9996

GLOSSARY

A

absolute zero the lowest possible temperature; $-273.15°C$ or $0°K$ (330)

accuracy the closeness of a measurement to the true or accepted value of the quantity measured (46)

acid anhydride an oxide that reacts with water to form an acid (480)

acid, Brønsted a molecule or ion that is a proton donor (462)

acid ionization constant the equilibrium constant for the ionization of an acid; K_A (577)

acid, Lewis an electron-pair acceptor (463)

acid precipitation precipitation having a pH below about 5.0 (765)

acid, traditional a chemical compound that contains hydrogen and ionizes in aqueous solution to form hydrogen ions (462)

acid-base reaction, Brønsted the transfer of protons from one reactant (the acid) to another (the base) (473)

actinides the 14 elements with atomic numbers from 90 (thorium, Th) through 103 (lawrencium, Lr) (130)

activated carbon a form of amorphous carbon that is prepared in a way that gives it a very large internal surface area (638)

activated complex a transitional structure that results from an effective collision and that persists while bonds are breaking and new bonds are forming (545)

activation energy the energy needed to transform reactants into an activated complex (545)

activity series a list of elements organized according to the ease with which they undergo certain chemical reactions (251)

actual yield the measured amount of a product obtained from a reaction (276)

adenosine triphosphate the nucleotide that supplies the energy for many metabolic reactions; ATP (695)

alcohol a substituted hydrocarbon that contains one or more hydroxyl (—OH) units (665)

aldehyde a substituted hydrocarbon with a carbonyl group that has one bond attached to a hydrogen atom (674)

alkadienes straight- or branched-chain hydrocarbons in which two carbon atoms in each molecule are connected by two double covalent bonds (651)

alkali metals the elements of Group 1 of the Periodic Table (lithium, sodium, potassium, rubidium, cesium, and francium) (136)

alkaline a solution that contains the hydroxide anion (OH^-) from a soluble base (471)

alkaline earth metals the elements of Group 2 of the Periodic Table (beryllium, magnesium, calcium, strontium, barium, and radium) (136)

alkanes straight- or branched-chain hydrocarbons whose atoms are connected by single covalent bonds (643)

alkenes straight- or branched-chain hydrocarbons in which two carbon atoms in each molecule are connected by a double covalent bond (644)

alkyl group the neutral species formed when one hydrogen atom is removed from an alkane molecule (647)

alkyl halide an alkane in which one or more halogen atoms are substituted for a hydrogen atom; a halocarbon (670)

alkylation the process by which gaseous alkanes are combined with unsaturated hydrocarbons by heat in the presence of a catalyst (651)

alkynes straight- or branched-chain hydrocarbons in which two carbon atoms in each molecule are connected by a triple covalent bond (644)

allotrope one of the two or more forms of an element that have the same physical state (289)

allotropy the existence of two or more allotropes of an element (289)

alloy a solid solution in which the atoms of two or more metals are uniformly mixed (407)

alpha particle a helium nucleus emitted from the nucleus of a radioactive nuclide (809)

alumina anhydrous aluminum oxide, Al_2O_3 (757)

amine an ammonia base in which one or more of the three hydrogens is replaced by an alkyl or aryl group (694)

amino acid organic molecules containing nitrogen in the form of NH_2 and a carboxyl group (COOH) bonded to the same carbon atom

amorphous carbon a carbon-containing residue, such as charcoal or activated carbon, that is black and has no crystalline shape (636)

amorphous solid a solid, such as plastic, in which the particles are arranged randomly (77)

amorphous sulfur a rubbery, dark brown or black mass made by pouring boiling sulfur into cold water (767)

amphoteric any species that can react as either an acid or a base (478)

anion a negative ion (151)

anode the electrode at which oxidation occurs in an electrochemical cell (612)

anthracite a coal believed to have been subjected to greater temperatures and pressures than other coals (637)

antiseptic a substance that checks the growth or action of microorganisms on or in the body (790)

aqueous solution a solution in which water is the solvent (18, 236)

aromatic hydrocarbons hydrocarbons with six-membered carbon rings that exhibit resonance (644)

atmosphere of pressure (atm) average atmospheric pressure at sea level and $0°C$; 1 atm = 760 mm Hg or 760 torr (321)

atom the smallest particle of an element that can exist either alone or in combination with other atoms (75)

atomic mass unit a unit of mass that is exactly $\frac{1}{12}$ the mass of the carbon-12 atom or $1.60\ 565 \times 10^{-24}$ g (85)

atomic number the number of protons in the nucleus of an atom of an element (83)

atomic radius one-half of the distance between the nuclei of identical atoms joined in a molecule (143)

atomic structure the identity and arrangement of smaller particles within atoms (75)

Aufbau principle the rule that an electron occupies the lowest-energy orbital that can receive it (112)

auto-oxidation a process in which a substance acts as both an oxidizing and a reducing agent and is self-oxidizing and self-reducing (610)

average atomic mass the weighted average of the atomic masses of the naturally occurring isotopes of an element (86)

Avogadro's number $6.022\ 045 \times 10^{23}$; the number of particles in 1.0 mole of a pure substance (88)

Avogadro's principle equal volumes of gases at the same temperature and pressure contain equal numbers of molecules (344)

B

baking soda sodium hydrogen carbonate, $NaHCO_3$ (726)

barometer a device used to measure atmospheric pressure (324)

base, Brønsted a molecule or ion that is a proton acceptor (472)

base, traditional a substance that contains hydroxide ions (OH^-) and dissociates in water to give hydroxide ions (470)

base-pairing phenomenon in which the bases in nucleotides attract each other in pairs (696)

basic anhydride an oxide that reacts with water to form an alkaline solution (480)

bauxite a hydrated aluminum oxide ore (753)

beta particle an electron emitted from the nucleus of a radioactive nuclide (809)

binary acid an acid composed of hydrogen and a second element (467)

binary compounds compounds composed of two different elements (204)

biochemistry the study of chemical reactions that take place within living things (685)

bituminous coal a form of coal that appears to have been subjected to greater heat and pressure than subbituminous coal (637)

blast furnace a tall, cylindrical chamber in which iron oxide is reduced using coke, limestone, and a blast of hot air (739)

boiling   the change of a liquid to a vapor, within the liquid and at its surface, when the equilibrium vapor pressure of the liquid equals the atmospheric pressure (375)

boiling point   the temperature at which the equilibrium vapor pressure of a liquid is equal to the atmospheric pressure, (387)

boiling-point elevation   the increase in the boiling point of a solvent caused by a solute (431)

bond energy   the energy required to break a chemical bond and form neutral atoms (166)

bond length   the distance between two bonded atoms (166)

Boyle's law   the volume of a fixed mass of gas varies inversely with pressure at constant temperature (327)

buffered solutions   solutions that can resist changes in pH (579)

C

calorie   the quantity of heat required to raise the temperature of 1 g of water through one Celsius degree, equivalent to 4.184 J (43)

calorimeter   a device used to measure heats of reactions (527)

carbohydrate   organic compounds containing hydrogen and oxygen in a ratio of approximately $2:1$; includes sugars, starches, and related compounds (689)

carbonyl group   a group with the formula $C{=}O$ (674)

carboxyl group   a group with the formula COOH (676)

carboxylic acid   a hydrocarbon in which two of the hydrogen atoms on a terminal carbon atom have been replaced by a doubly bonded oxygen, and the third hydrogen atom has been replaced by a hydroxyl group, —OH (676)

catalysis   activity by a catalyst (551)

catalyst   a substance that speeds up a chemical reaction but is not permanently consumed in the reaction (551)

cathode   the electrode at which reduction occurs in an electrochemical cell (612)

cation   a positive ion (151)

caustic   having destructive effects on skin, hair, and wool (724)

cellulose   a polysaccharide of glucose monomers with hydroxyl links; the main building material for plants (690)

chain reaction   a reaction in which the material or energy that starts the reaction is also one of the products (819)

changes of state   changes between the gaseous, liquid, or solid states (14)

charcoal   the residue that remains after the destructive distillation of wood (638)

Charles' law   the volume of a fixed mass of gas varies directly with the Kelvin temperature at constant pressure (330)

chemical bond   a link between atoms resulting from the mutual attraction of their nuclei for electrons (161)

chemical change or chemical reaction   any change or reaction in which one or more substances are converted into different substances with different characteristic properties (15)

chemical compound   a pure substance that can be decomposed into two or more simpler substances by an ordinary chemical change (22)

chemical equation   a representation, with symbols and formulas, of the reactants and products in a chemical reaction (233)

chemical equilibrium   a state of balance in which the rates of opposing chemical reactions are equal (564)

chemical equilibrium expression   the equation for the equilibrium constant K (566)

chemical formula   a shorthand representation of the composition of a substance using atomic symbols and numeric subscripts (165)

chemical kinetics   the branch of chemistry that is concerned with reaction rates and reaction mechanisms (549)

chemical property   the ability of a substance to undergo a change that alters its identity (14)

chemistry   the study of the composition and structure of materials and the changes they undergo (5)

chitin   a polysaccharide of repeating glucose units with a nitrogen-containing group that composes the "armor" of insects and other invertebrates and is also produced by fungi (691)

chlorinated lime   Ca(ClO)Cl; a disinfectant (785)

chlorohydrocarbons   hydrocarbons in which chlorine atoms are substituted for one or more of the hydrogen atoms (640)

cholesterol   the most common steroid, making up part of every animal cell membrane, brain, and nerve (693)

coal   a natural dark-brown or black solid that consists mostly of amorphous carbon with various organic and some inorganic compounds (636)

coefficient   a small whole number that appears in front of a formula in a chemical equation (234)

coenzyme   a molecule that helps an enzyme bind to its substrate (688)

coke   the solid residue left behind after volatile materials are removed from bituminous coal (637)

colligative property   a property of a system that depends on the number of particles present but is independent of their nature (427)

collision theory   the set of assumptions regarding collisions of particles and the reactions that result from those collisions (543)

colloid   a mixture consisting of particles intermediate in size between those in solutions and those in suspensions (409)

combined gas law   relationship between pressure, volume, and temperature of a gas when the amount of gas is fixed (333)

combustion reaction   a reaction in which a large amount of light and heat energy is produced (250)

common-ion effect   the phenomenon in which the addition of an ion common to two solutes brings about precipitation or reduced ionization (576)

complex carbohydrate   a polymer formed by joining many monosaccharides; a polysaccharide (689)

complex ion   an ionic species composed of a central metal ion combined with a specific number of polar molecules or ions (734)

composition stoichiometry   calculations involving the mass relationships of elements in chemical compounds (261)

concentration   a measure of the amount of solute in a given amount of solvent or solution (420)

concentration   the number of particles per unit volume (384)

condensation   the process by which a gas changes to a liquid (383)

conjugate acid   the species formed by the addition of a proton to a base (475)

conjugate base   the species that remains after an acid has given up a proton (475)

contact process   a method in which sulfur dioxide is prepared by burning sulfur or by roasting iron pyrite, FeS_2, in contact with the catalyst divanadium pentoxide, V_2O_5 (773)

continuous spectrum   a spectrum that includes all wavelengths within a given range (98)

control rod   a rod of neutron-absorbing material used to regulate the reaction in a nuclear reactor (819)

conversion factor   a ratio of two different units that can be used to convert from one unit to the other (34)

coordination number   the number of units covalently bonded to, or coordinated with, the central ion (734)

copper matte   an intermediate product in copper refining, containing copper sulfides, iron sulfides and other sulfides (746)

corundum   a natural oxide of aluminum (756)

cosmic rays   protons and other nuclei of very high energy that come to earth from outer space (816)

covalent bond   a chemical bond resulting from the sharing of electrons between two atoms (161)

covalent hydride   a compound consisting of hydrogen and one more electronegative element (296)

cracking   a process in which complex organic molecules are broken up into simpler molecules (649)

critical mass   the mass of uranium that provides the number of neutrons needed to sustain a chain reaction (820)

critical point   indicates the critical temperature and critical pressure of a substance (390)

critical pressure   the lowest pressure required for a substance to exist as a liquid at the critical temperature; P_c (390)

critical temperature   the temperature above which a substance cannot exist in the liquid state no matter how high the pressure exerted T_c (390)

cryolite Na_3AlF_6, used in refining aluminum (753)

crystal   a substance in which the particles are arranged in an orderly, geometric, repeating pattern (377)

crystal lattice   the total three-dimensional array of points that describes the arrangement of the particles of a crystal (379)

crystalline solid   a solid consisting of crystals (377)

D

Dalton's law of partial pressures the total pressure of a mixture of gases is equal to the sum of the partial pressures of the component gases (335)

daughter nuclide a nuclide that is the product of the sequential radioactive decay of a parent nuclide (810)

decomposition reaction a reaction in which a compound produces two or more simpler substances (246)

degree Celsius the unit of temperature on the Celsius scale (41)

degree Kelvin the unit of temperature on the Kelvin scale; the fundamental SI unit for temperature (41)

dehydrating agent a substance that removes water from a material (774)

denaturation destruction of proteins by breaking up their three-dimensional structure with high temperatures (688)

denatured alcohol a mixture composed principally of ethanol and other poisonous and nauseating materials that have been added to make it unfit for consumption (668)

density a material's mass divided by its volume, or its mass per unit volume (37)

deoxyribonucleic acid nucleic acid of all organisms that contains the genetic information passed on to their offspring; DNA (695)

destructive distillation a process by which a complex carbon-containing material, such as coal, is heated in a closed container in the absence of air or oxygen (637)

deuterons deuterium nuclei (812)

diamagnetic a substance that is weakly repelled by a magnetic field (111)

diamond the form of carbon that is a colorless crystalline solid with an extremely high density of 3.514 g/cm^3 (634)

diatomic molecule a molecule containing two atoms (164)

diffusion the process in which constant movement causes molecules of two gases to mix together (360)

diffusion the gradual mixing of two substances because of their spontaneous, random motion (318)

dimer a molecule formed by the combination of two simpler molecules (635)

dimerization the process that forms a dimer (653)

dipole equal but opposite charges separated by a short distance (188)

dipole–dipole forces forces of attraction between polar molecules (189)

diprotic acid an acid that can donate two protons per molecule (466)

directly proportional two variables that have a constant ratio (58)

disaccharide two monosaccharides joined together (689)

disinfectant a chemical agent that destroys microorganisms but not bacterial spores (785)

dissociation the separation of ions that occurs when an ionic compound dissolves (442)

dissolved describes a substance that, when mixed with another, is no longer of a different physical state (372)

domain small magnetized regions formed by groups of properly aligned atoms of ferromagnetic substances (738)

double bond a covalent bond produced by sharing two pairs of electrons between two atoms (170)

double helix structural shape of a DNA molecule, consisting of two nucleotide strands wound around each other (697)

double-replacement reaction (ionic reactions) a type of reaction in which the ions of two compounds exchange places in an aqueous solution to form two new compounds (248)

Downs cell apparatus that prepares elemental sodium by the electrolysis of fused sodium chloride (720)

ductility the state of being able to be drawn, pulled, or extruded through a small opening to produce a wire (182)

E

effervescence the rapid escape of a gas from a liquid in which it is dissolved (416)

effusion a process by which gas particles under pressure pass through a very small opening from one container to another (318)

effusion in a container with a very small opening, the process by which gas molecules pass through the opening as they randomly encounter it (360)

elastic collision a collision between gas particles or between gas particles and container walls in which there is no net energy loss due to friction (315)

electrochemical cell a system of electrode and electrolyte in which a spontaneous redox reaction is used as a source of electrical energy (611)

electrochemistry the branch of chemistry that deals with the electricity-related applications of redox reactions (611)

electrode a conductor used to establish electric contact with a non-metallic part of a circuit, such as a solution of an electrolyte (611)

electrode potential the difference in potential between an electrode and the solution it rests in (618)

electrolysis the process in which an electric current is used to drive oxidation-reduction reactions (613); also the decomposition of a substance by an electric current (246)

electrolyte a substance that dissolves in water to give a solution that conducts electric current (406)

electrolytic cell a system of electrodes and electrolyte in which a current is used to drive a nonspontaneous redox reaction (612)

electromagnetic radiation a form of energy that exhibits wavelike behavior as it travels through space (97)

electromagnetic spectrum all electromagnetic radiation, arranged according to increasing wavelength (98)

electron affinity the energy change that occurs when an electron is acquired by a neutral atom (149)

electron configuration the arrangement of electrons in atoms (112)

electronegativity a measure of the ability of an atom in a chemical compound to attract electrons (153)

electrons negatively charged, subatomic particles (76)

electroplating an electrolytic process in which a metal is deposited on a surface (697)

element a substance that cannot be separated into other substances by any ordinary chemical change (22)

emery a natural oxide of aluminum, used as an abrasive (756)

empirical formula the symbols for the elements of a compound with subscripts showing the smallest whole-number ratio of the atoms (224)

end point the point in a titration at which a very rapid change in pH occurs, at which equivalent quantities of hydronium and hydroxide ions are present (512)

endothermic describes processes that absorb heat (19)

energy the ability to cause change or to do work (12)

enthalpy the heat content of a substance at constant pressure (528)

enthalpy change the amount of heat absorbed or lost by a system in a process carried out at constant pressure (528)

entropy the property that describes the disorder of a system; S (538)

enzyme a protein molecule that acts as a catalyst (688)

epinephrine an amine hormone that speeds up the heart, increases the blood supply to the muscles, and generally gets the body ready for "flight or fight" (694)

equilibrium a dynamic condition in which two opposing physical or chemical changes occur at equal rates in the same closed system (383)

equilibrium constant the ratio of the mathematical product of the concentrations of the products in a reversible reaction, at equilibrium, to the mathematical product of the concentrations of the reactants, each concentration being raised to the power that is the coefficient of the number of moles of that substance in the chemical equation; K (566)

equilibrium vapor pressure the pressure exerted by a vapor in equilibrium with its corresponding liquid at a given temperature (386)

equivalent mass of acid or base for an acid, the quantity in grams that supplies one mole of protons; for a base, the quantity in grams that supplies one mole of hydroxide ions (492)

equivalents quantities of solutes that have equivalent combining capacity (491)

ester a compound with the formula RCOOR′, in which R′ is always an alkyl group and R is either a hydrogen atom or an alkyl group (679)

esterification the production of an ester by the reaction of an organic acid and an alcohol (679)

ether a compound in which two hydrocarbon groups are bonded to an atom of oxygen (673)

evaporation the process by which particles escape from the surface of a nonboiling liquid and enter the gaseous state (375)

excess reactant the substance that is not used up completely in a reaction (272)

excited state a state in which an atom has a higher potential energy than it has in its ground state (100)

exothermic describes processes that release heat (17)

extensive physical properties properties, including mass, length and volume, that depend on the amount of matter present (14)

F

factor-label method a problem-solving method in which units of measurement are treated in calculations as if they were algebraic factors (34)

fat an ester, of plant or animal origin, of glycerol and long-carbon-chain acids (680)

fatty acids the simplest lipids; a fatty acid is a chain of carbon and hydrogen atoms with a carboxyl group at one end (692)

ferromagnetism the property of certain metals whereby they are strongly attracted by a magnet (737)

fiber composite monocrystalline strands or fibers of one substance embedded in and held in place by another material (757)

fission the process in which a very heavy nucleus splits to form medium-weight nuclei (804)

fluid a substance that can flow and therefore conform to the outline of its container (372); also a liquid or a gas composed of particles that glide easily past one another (317)

fluoresce give off visible light after being exposed to sunlight (804)

formula equation a representation of the reactants and products of a chemical reaction by their symbols or formulas (234)

formula mass the sum of the atomic masses of all atoms represented in the formula of any compound or polyatomic ion (216)

formula unit the simplest unit indicated by the formula of any compound (177)

fractional distillation distillation in which components of a mixture with different boiling points are collected separately (656)

Frasch process a method of obtaining sulfur from the ground by melting the deposit of sulfur with superheated water and then forcing it to the surface with compressed air (765)

free energy the combined enthalpy-entropy function of a system; G (538)

free-energy change the difference between the change in enthalpy, ΔH, and the product of the Kelvin temperature and the entropy change, $T\Delta S$, at a constant pressure and temperature; ΔG (538)

freezing the physical change of a liquid to a solid (375)

freezing point the temperature at which the solid and liquid are in equilibrium at 1 atm (101.3k Pa) pressure (388)

freezing-point depression the decrease in the freezing point of a solvent caused by a solute (428)

frequency the number of waves that pass a given point in a specific amount of time, usually one second (98)

Friedel-Crafts reaction a reaction in which an alkyl group can be introduced into a benzene ring using an alkyl halide in the presence of anhydrous aluminum chloride (654)

functional group an atom or a group of atoms bonded to a hydrocarbon and responsible for the specific properties of the organic compound (639)

fundamental unit a unit that is defined by a physical standard of measurement (32)

fungicide compounds used to kill fungi (769)

fusion the process in which lightweight nuclei combine to form heavier, more stable nuclei (804)

G

gamma rays high-energy electromagnetic waves (809)

gaseous state the state of any matter that has neither a definite shape nor a definite volume (13)

Gay-Lussac's law the pressure of a fixed mass of gas varies directly with the Kelvin temperature at constant volume (332)

Gay-Lussac's law of combining volumes of gases at constant temperature and pressure, the volumes of gaseous reactants and products can be expressed as ratios of small whole numbers (344)

gene a segment of a DNA chain that codes for the structure of a particular protein (697)

genetic engineering the deliberate manipulation of genetic material (700)

glycogen a glucose monomer that stores energy in animals (689)

glycolipids fatty acids combined with carbohydrates (692)

glycolysis part of a group of metabolic pathways by which animals turn their food into an energy supply (701)

graphite a soft, black crystalline form of carbon that is a fair conductor of electricity (634)

Graham's law of effusion (diffusion) the rates of effusion (or diffusion) of gases at the same temperature and pressure are inversely proportional to their molar masses (362)

ground state the state of lowest energy of an atom (100)

groups or families the vertical columns of elements in the Periodic Table, numbered consecutively from 1 to 18 (23)

H

Haber process the catalytic synthesis of ammonia from nitrogen gas and hydrogen gas in a ratio of 1:3 by volume (302)

half-cell one of two portions of a voltaic cell, consisting of a metal electrode in contact with a solution of its ions (618)

half-life the length of time during which half of a given number of atoms of a radioactive nuclide decay (807)

half-reaction the reaction taking place at each electrode in an electrochemical cell (612)

halite rock salt; it is mined throughout the world (722)

Hall process a process in which aluminum is extracted by electrolyzing anhydrous aluminum oxide (refined bauxite) dissolved in molten cryolite (753)

halogen any of the the colorful, active, nonmetallic elements fluorine, chlorine, bromine, iodine, and astatine that constitute Group 17 of the Periodic Table (779)

halogenation a chemical reaction by which a halogen is introduced into a compound by addition or substitution (654)

heat the sum of the kinetic energies of the particles in a sample of matter (40)

heat capacity the amount of heat energy needed to raise the temperature of a given sample of matter by 1°C (44)

heat of combustion heat released by the complete combustion of 1 mole of a substance; defined in terms of 1 mol of reactant (531)

heat of formation molar heat of formation (529)

heat of hydration the energy released when ions become surrounded by water molecules (441)

heat of reaction the quantity of heat released or absorbed during a chemical reaction (528)

heat of solution the overall energy change in solution formation (418)

Henry's law the solubility of a gas in a liquid is directly proportional to the pressure of that gas above the liquid (415)

Hess's law the heat of a reaction is the same no matter how many intermediate steps make up the pathway from reactants to products (531)

heterogeneous mixture a mixture whose composition and properties are not uniform but that differ from point to point in the mixture (19)

heterogeneous reaction a reaction involving reactants in two different phases (550)

heterogenous catalyst a catalyst whose phase is different from that of the reactants (552)

highest occupied energy level the main energy level with the highest principal quantum number that contains at least one electron (113)

homogeneous catalyst a catalyst in the same phase as all the reactants and products in a reaction system (552)

homogeneous mixture a mixture whose composition and properties are uniform throughout the mixture; also called a solution (20)

homogeneous reaction a reaction system that occurs in a single phase (541)

homologous series one in which adjacent members differ by a constant unit (645)

hormones chemicals, such as steroids, proteins, or amines, that carry messages between different parts of the body (693)

Hund's rule orbitals of equal energy are each occupied by one electron before any orbital is occupied by a second electron (113)

hybrid orbitals orbitals of equal energy produced by the combination of two or more orbitals on the same atom (187)

hybridization the production of hybrid orbitals (186)

hydration the solvation process with water as the solvent (440)

hydrocarbons the simplest organic compounds, composed only of carbon and hydrogen (639)

hydrogen bonding the intermolecular attraction between a hydrogen atom bonded to a strongly electronegative atom and an unshared pair of electrons on a strongly electronegative atom (190)

hydrolysis a reaction between water and the ions of a dissolved salt (580)

hydrolysis constant the equilibrium constant for an ion hydrolysis reaction; K_h (582)

hydronium ion the H_3O^+ ion (449)

hydroxyl group the covalently bonded —OH group (465)

hypothesis a testable statement (8)

I

ideal gas an imaginary gas that conforms perfectly to all of the assumptions of the kinetic theory (315)

ideal gas constant the constant R (351)

ideal gas law a relationship between pressure, volume, temperature, and the number of moles of a gas (350)

immiscible not mutually soluble (414)

indicators weak acid or base dyes whose colors are sensitive to hydronium ion concentration (505)

inertia resistance to change in motion (10)

inhibitor a substance that slows down a process (552)

inner-shell electrons electrons that are not in the highest-occupied energy level (114)

instability constant the equilibrium constant for complex-ion equilibria (735)

intensive physical properties properties that do not depend on the amount of matter present; they include melting point, boiling point, density, ductility, malleability, color, crystalline shape, and refractive index (14)

intermolecular forces the forces of attraction between molecules (188)

inversely proportional two variables that have a constant mathematical product (59)

ion an atom or group of atoms that has a positive or negative charge (145)

ionic bond the chemical bond resulting from electrostatic attraction between positive and negative ions (161)

ionic compound a compound of positive and negative ions combined so that the positive and negative charges are equal (176)

ionic hydride a compound consisting of hydrogen and one element that is less electronegative (296)

ionic radius one-half of the diameter of an ion in a chemical compound (151)

ionization the formation of ions from solute molecules by the action of the solvent (448)

ionization energy the energy required to remove one electron from an atom of an element (145)

isomers compounds that have the same molecular formula but different structures (639)

isotopes atoms of the same element that have different masses (81)

J

joule the SI unit of heat energy and of all forms of energy (43)

K

ketone a substituted hydrocarbon with a carbonyl group in which the carbon atom of the carbonyl group is attached to two other carbon atoms (674)

kilogram the SI standard unit for mass (33)

kinetic energy the energy of an object in motion (12)

kinetic theory the idea that particles of matter are always in motion and that this motion has consequences (315)

L

lambda-sulfur the liquid allotropic form of sulfur produced at temperatures just above the melting point of sulfur, 112.8°C (766)

lanthanides or rare-earth elements the 14 elements with atomic numbers from 58 (cerium, Ce) through 71 (lutetium, Lu) (130)

lattice energy the energy released when one mole of an ionic crystalline compound is formed from gaseous ions (179)

law a generalization that describes a wide variety of behaviors in nature (8)

law of conservation of energy energy can be converted from one form to another, but it cannot be created or destroyed in ordinary chemical or physical changes (13)

law of conservation of matter matter cannot be either created or destroyed in ordinary chemical or physical changes (12)

law of definite composition a chemical compound contains the same elements in exactly the same proportions by mass regardless of the size of the sample or source of the compound (22)

law of mass action the rate of a chemical reaction is directly proportional to the product of the concentrations of reacting substances, each raised to the appropriate power (554)

law of multiple proportions if two or more different compounds are composed of the same two elements, the masses of the second element combined with a certain mass of the first element can be expressed as ratios of small whole numbers (73)

Le Châtelier's principle if any of the factors determining an equilibrium are changed, the system will adjust itself in a way that tends to minimize that change (384)

Lewis structure a formula in which atomic symbols represent nuclei and inner-shell electrons, dots or dashes represent electron pairs in covalent bonds, and dots represent unshared electrons (169)

lignite a low-grade form of coal sometimes called "brown coal" because of its brownish-black color (637)

limiting reactant the reactant that controls the amount of product formed in a chemical reaction (272)

line emission spectrum the colored lines produced when an electron drops from a higher-energy orbit to a lower energy orbit (103)

lipids various organic compounds not very soluble in water that have a high proportion of C—H bonds (691)

lipoprotein a fatty acid combined with a protein (692)

liquid state the state of any matter that has a definite volume but an indefinite shape (13)

London dispersion forces intermolecular attractions resulting from the constant motion of electrons and the creation of instantaneous dipoles and induced dipoles (191)

lye sodium hydroxide; converts some types of animal and vegetable matter into soluble materials (724)

M

macromolecules large organic molecules (685)

magnetic quantum number the quantum number that indicates the orientation of an orbital about the nucleus (110)

main-group elements the elements of the s-and p-blocks.

malleability the state of being able to be shaped or extended by beating with a hammer, rolling, or otherwise exerting physical pressure that results in a change in contour (182)

mass a measure of the quantity of matter (11)

mass number the total number of protons and neutrons in the nuclei of an isotope (83)

matter anything that has mass and occupies space (12)

melting point the temperature at which a solid becomes a liquid (378)

messenger RNA the type of RNA that carries the coded information for a particular protein from DNA to a ribosome (698)

metabolic pathway chemically linked steps in which a series of enzymes converts substrate molecules, step by step, into a final product (701)

metabolism all of the biochemical reactions that occur within an organism (701)

metal an element that is a good conductor of heat and electricity (24)

metallic bond a chemical bond resulting from the attraction between positive ions and surrounding mobile electrons (181)

metalloid an element that has some properties characteristic of metals and others characteristic of nonmetals, and is a semiconductor (25)

meter the SI standard unit for length (33)

methyl group methane with one of its hydrogen atoms removed (647)

millimeters of mercury (mm Hg) a common unit of pressure (326)

mineral acids acids made from minerals; include sulfuric acid, nitric acid, hydrochloric acid, and phosphoric acid (467)

miscible mutually soluble (415)

mixture a combination of two or more kinds of matter, each of which retains its own composition and properties (19)

moderator a material that slows down neutrons (819)

molal boiling-point constant the boiling-point elevation of a solvent in a 1 molal solution of a nonvolatile, molecular solute (431)

molality the concentration of a solution expressed in moles of solute per kilogram of solvent (423)

molar heat of formation heat of reaction released or absorbed when 1 mole of a compound is formed by combination of its elements; defined in terms of 1 mol of product (529)

molar heat of fusion the heat energy required to melt one mole of solid at its melting point (388)

molar heat of vaporization the amount of heat energy required to vaporize one mole of liquid at its boiling point (388)

molar mass the mass in grams of one mole of an element or compound (88)

molarity the number of moles of solute in one liter of solution (421)

mole the amount of a substance that contains the same number of particles as there are atoms in exactly 12 g of carbon-12 (88)

mole ratio a conversion factor that relates the number of moles of any two substances involved in a chemical reaction (263)

molecular compound a chemical compound whose simplest formula units are molecules (165)

molecular formula shows the types and numbers of atoms combined in a single molecule (165)

molecular genetics the branch of science that studies the biochemistry of genes (699)

molecule a group of two or more atoms held together by covalent bonds and able to exist independently (164)

monatomic ions ions formed from a single atom (203)

Monel metal an alloy of nickel and copper that is highly resistant to corrosion (781)

monoclinic sulfur a form of solid sulfur in which the eight-membered rings of sulfur are arranged in a monoclinic crystal pattern (766)

monomers the small molecules that join together to make a polymer (651)

monoprotic acid an acid that can donate one proton per molecule (463)

monosaccharides the monomers of carbohydrates, such as fructose and glucose; simple sugars (689)

mu-sulfur an allotropic form of sulfur that forms at temperatures above 160°C (766)

multiple bonds double and triple bonds (171)

N

neoprene a synthetic rubber that is resistant to heat and flames, as well as to oil and chemicals (659)

net ionic equation an equation that includes, for a reaction in aqueous solution, only the compounds and ions that change chemically (445)

neuron a nerve cell (695)

neurotransmitter a chemical messenger that carries a message across a synapse from one neuron to the next (695)

neutralization the reaction of hydronium ions and hydroxide ions to form water molecules (471)

neutrons electrically neutral, subatomic particles found in atomic nuclei (80)

Newton (N) the SI unit for force (323)

nitration a chemical reaction by which an —NO_2 group is introduced into an organic compound (654)

nitrogen fixation the conversion of atmospheric nitrogen gas (N_2) into nitrogen compounds that can be used by plants (300)

noble gas configuration an atomic configuration in which the outer main energy level is fully occupied by eight electrons (116)

noble gases the Group 18 elements, helium, neon, argon, krypton, xenon, and radon (116)

noble metals metals that show little chemical activity, especially toward oxygen (736)

nomenclature methods of naming chemical compounds (206)

nonelectrolyte a substance that dissolves in water to give a solution that does not conduct an electric current (406)

nonmetal an element that is a poor conductor of heat and electricity (25)

nonpolar–covalent bond a covalent bond in which the bonding electrons are shared equally by the bonded atoms, with a resulting balanced distribution of electrical charge (162)

nonvolatile a substance that has little tendency to become a gas under the existing conditions (426)

normality the number of equivalents of solute per liter of solution; N (491)

nuclear binding energy the energy released when a nucleus is formed from its constituent particles (802)

nuclear disintegration the emission of a proton or neutron from a nucleus as a result of bombarding the nucleus with alpha particles, protons, deuterons, or neutrons (804)

nuclear equation an equation representing changes in the nuclei of atoms (810)

nuclear mass defect the difference between the mass of a nucleus and the sum of the masses of its constituent particles (802)

nuclear reactor a device in which controlled fission of radioactive material produces new radioactive substances and energy (819)

nucleic acids macromolecules that transmit genetic information (695)

nucleons the protons and neutrons in an atom (801)

nucleotide the monomer of a nucleic acid (695)

nucleus the positively charged, dense central portion of an atom that contains nearly all of its mass but takes up only an insignificant fraction of its volume (78)

nuclide the general term for any isotope of any element (84)

O

octet rule chemical compounds tend to form so that each atom, by gaining, losing, or sharing electrons, has 8 electrons (an octet) in its highest occupied energy level (168)

oil a liquid ester of glycerol and long carbon chain acids having a plant or animal origin (680)

orbital a three-dimensional region about the nucleus in which a particular electron can be located (106)

orbital quantum number the quantum number that indicates the shape of an orbital (108)

organic acid an organic compound that contains an acid group, the carboxyl group (—COOH) (467)

organic chemistry the chemistry of carbon compounds (632)

oxidation a reaction in which the atoms or ions of an element lose one or more electrons and hence attain a more positive (or less negative) oxidation state (597)

oxidation numbers or oxidation states numbers assigned to the atoms in molecules that show the general distribution of electrons among bonded atoms (212)

oxide a compound containing oxygen in which oxygen has an oxidation number of −2 (290)

oxidized a species that undergoes a change in oxidation state in a positive mathematical direction (597)

oxidizing agent a species that attains a more negative oxidation state during an oxidation-reduction reaction (608)

oxyacids acids containing hydrogen, oxygen, and a third element (211)

oxyanions polyatomic ions that contain oxygen (208)

P

pH the concentration of the hydronium ions of a solution, expressed as the negative of the common logarithm (499)

paraffin series the alkanes (644)

paraffin wax a mixture of hydrocarbons of the paraffin series (644)

parent nuclide the heaviest, most complex, naturally occurring nuclide in a decay series of radioactive nuclides (810)

partial pressure the pressure of each gas in a mixture (335)

Pauli exclusion principle no two electrons in the same atom can have the same four quantum numbers (113)

peat a form of partially carbonized plant material that can be used as fuel (637)

peptide bond a bond joining amino acids in proteins (687)

percent by mass the number of parts by mass or solute per 100 parts by mass of solution (420)

percent composition the percent by mass of each element in a compound (222)

percent yield the ratio of the actual yield to the theoretical yield, multiplied by 100 percent (277)

periodic law the physical and chemical properties of the elements are periodic functions of their atomic numbers (129)

Periodic Table an arrangement of the elements in order of their atomic numbers so that elements with similar properties fall in the same column (130)

periods the horizontal rows of elements in the Periodic Table, numbered 1 to 7 from the top down (23)

peroxide a compound containing the O_2^{2-} anion, in which oxygen has the oxidation number -1 (290)

petroleum A complex mixture of hydrocarbons that varies greatly in composition from place to place (crude oil) (656)

petroleum coke a form of amorphous carbon produced by the destructive distillation of an oily residue that is a remnant of petroleum refining (638)

phase any part of a system that has uniform composition and properties (20)

phase diagram a graph of temperature versus pressure that indicates the conditions under which gaseous, liquid, and solid states of a substance exist (389)

phases portions of matter that have both the same chemical properties and the same physical properties (22)

phenyl group a $C_6H_5^-$ group derived from benzene (654)

phospholipid a lipid in which one or two of the fatty acids are replaced by a phosphate group (693)

photochemical smog ozone and other irritating chemicals in the air that are the result of nitrogen oxides reacting either with unburned hydrocarbons from automobile exhausts or with atmospheric oxygen (299)

photoelectric effect the emission of electrons by certain metals when light shines on them (99)

photon a quantum of light, thought of as particles of radiation (100)

photosensitive sensitive to light rays (788)

photosynthesis the process by which plants use energy from the sun to combine carbon dioxide with water to produce glucose (303)

physical change any change in a property of matter that does not result in a change in identity (14)

physical property a property that can be observed or measured without altering the identity of a material (14)

pig iron iron recovered from a blast furnace (739)

polar a bond that has an uneven distribution of charge (162)

polar–covalent bond a covalent bond in which there is an unequal attraction for the shared electrons (162)

polyatomic ion a charged group of covalently bonded atoms (173)

polyethylene a polymer of ethene (651)

polymer a very large molecule composed of repeating structural units (651)

polymerization the process that produces polymers (651)

polyprotic acid an acid that can donate more than one proton per molecule (466)

polysaccharide a polymer formed by joining many monosaccharides; a complex carbohydrate (689)

polyunsaturated fat fats and oils with double bonds in their fatty acid chains (692)

polyvinyl chloride a polymeric form of vinyl chloride (CH_2CHCl) also called PVC (785)

potential energy energy that an object has because of its position or composition (12)

precipitate a solid that separates from a solution (16)

precision the agreement among the numerical values of a set of measurements of the same quantity made in the same way (46)

pressure force per unit area (323)

primary standard a highly purified compound, when used in solution to check the concentration of the known solution in a titration (512)

principal quantum number the quantum number that indicates the main energy levels surrounding a nucleus; n (108)

products the new substance or substances produced by a chemical reaction or by an enzyme reaction (15, 688)

properties characteristics that enable us to distinguish one kind of matter from another (14)

proteins macromolecules formed by condensation reactions between monomers called amino acids (686)

protons subatomic particles that have a positive charge equal in magnitude to the negative charge of an electron and that are present in atomic nuclei (79)

pure substance a homogeneous sample of matter that has the same composition and characteristic properties throughout (20)

pyrite a compound of iron and sulfur (765)

Q

qualitative information non-numerical information (8)

quantitative information numerical information (8)

quantum a finite quantity of energy that can be gained or lost by an atom (100)

quantum numbers numbers that determine the properties of atomic orbitals and their electrons (108)

quantum theory a mathematical description of the wave properties of electrons and other very small particles (105)

R

radioactive decay the release, by the nucleus, of an alpha or beta particle and gamma rays, forming a slightly lighter and more stable nucleus (804)

radioactive tracer radioactive atoms incorporated in substances so that movement of substances can be followed by radiation detectors (817)

radioactivity the spontaneous breakdown of an unstable nucleus with the release of particles and rays (804)

rate law an equation that relates reaction rate and concentrations of reactants (553)

rate-determining step the lowest-rate step for a chemical reaction (551)

reactant the substance that undergoes a chemical reaction (15)

reaction mechanism the step-by-step sequence of reactions by which the overall chemical change occurs (541)

reaction rate the change in concentration of reactants per unit time as a reaction proceeds (549)

reaction stoichiometry calculations involving the mass relationships among reactants and products in a chemical reaction (201)

real gas a gas that does not obey completely all the assumptions of the kinetic theory (319)

redox reaction oxidation–reduction reaction; any chemical process in which elements undergo a change in oxidation number (599)

reducing agent a species that attains a more positive oxidation state during an oxidation–reduction reaction (608)

reduction a reaction in which the atoms or ions of an element gain one or more electrons and hence attain a more negative (or less positive) oxidation state (597)

resonance the bonding in molecules that cannot be correctly represented by a single Lewis structure (174)

reversible reaction a chemical reaction in which the products can react to re-form the reactants (237)

rhombic sulfur the form of solid sulfur that is stable at ordinary temperatures (766)

ribonucleic acids nucleic acids that assist in protein synthesis by helping to coordinate the process of protein assembly; RNA (695)

ribosomes a structure within the cell, containing RNA and protein that serves as the site for protein assembly (698)

roasting heating in the presence of air (771)

ruby a gemstone of aluminum oxide contaminated with traces of impurities, including chromium (757)

S

salt an ionic compound composed of a metal cation of an aqueous base and the anion from an aqueous acid (472)

saponification the hydrolysis of fats or oils using a solution of a strong-hydroxide (681)

sapphire a gemstone of aluminum oxide contaminated with traces of impurities, including titanium (757)

saturated fat a fatty acid in which all carbon atoms of the hydrocarbon chain are filled with hydrogens (692)

saturated hydrocarbon a hydrocarbon in which each carbon atom in the molecule forms four single covalent bonds with other atoms (644)

saturated solution a solution that contains the maximum amount of dissolved solute possible at solution equilibrium under the existing conditions (412)

scientific method a logical approach to the solution of problems that lend themselves to investigation by observing (7)

scientific notation numbers written in the form $M \times 10^n$ where M is a number greater than or equal to 1 but less than 10 and n is an integer (52)

second the SI standard unit for time (34)

self-ionization the process that ionizes water; autoprotolysis(496)

semiconductor a substance with an electric conductivity between that of a metal and that of an insulator (758)

significant figures all digits in a measurement that are known with certainty, plus one final digit, which is uncertain or estimated (47)

simplest formula an expression of a compound that combines the symbols of the elements with subscripts showing the smallest whole-number ratio of the atoms (224)

single bond a covalent bond produced by sharing one pair of electrons between two atoms (169)

single-replacement reaction (displacement reaction) one element replaces a similar element in a compound (247)

slag an easily melted product of the reaction between the flux and the impurities of an ore; one of the impurities that remain when iron oxide is reduced in a blast furnace (739)

soda ash dried sodium carbonate, Na_2CO_3—an important industrial chemical (725)

solid state the state of any matter that has a definite shape and volume (13)

solubility the amount of a substance that is dissolved at solution equilibrium in a specific amount of solvent at a specified temperature (413)

solubility product constant the product of the molar concentrations of ions of a substance in a saturated solution, each raised to the power that is the coefficient of that ion in the chemical equations; K_{sp} (585)

soluble capable of being dissolved (405)

solute in a solution, the substance that is dissolved in solvent (406)

solution a homogeneous mixture of two or more substances in a single phase (205)

solution equilibrium the physical state in which the opposing processes of dissolving and crystallizing of a solute occur at equal rates (412)

solvated a solute molecule surrounded by solvent molecules (418)

Solvay process the method for industrial production of sodium carbonate from brine and limestone (724)

solvent the dissolving medium in a solution (406)

specific heat the amount of heat energy required to raise the temperature of 1 g of a substance, 1°C (44)

spectator ions ions that do not take part in a chemical reaction and are found in solution both before and after the reaction (445)

spin quantum number the quantum number that indicates two possible orientations of an electron in an orbital (111)

standard electrode potentials a half-cell potential measured relative to a potential of zero for the standard hydrogen electrode (618)

standard molar volume of a gas the volume occupied by one mole of a gas at STP, 22.4L (345)

standard solution a solution that contains a precisely known concentration of a solute (511)

standard temperature and pressure conditions (STP) agreed-upon standard conditions of exactly 1 atm pressure and 0°C (327)

standards of measurement objects or natural phenomena of constant value, easy to preserve and reproduce, and practical in size, that are used to define SI units (31)

starch a plant polysaccharide that is used as energy storage and as building material (689)

steroid a lipid with a "skeleton" structure consisting of four carbon rings (693)

stoichiometry the branch of chemistry dealing with mass relationships of elements in compounds and among reactants and products in chemical reactions (261)

strong acid an acid that is a strong electrolyte and dissociates 100% in aqueous solution (463)

strong electrolyte any solute that is present in dilute aqueous solution entirely as hydrated ions (449)

strong nuclear force the force that keeps the nucleons together in the nucleus (801)

structural formula a formula that indicates the exact number and types of atoms present in a molecule and how the atoms are bonded to each other (640)

structural isomers structural formulas for the various isomers with a specific molecular formula (640)

subbituminous coal a form of coal that has a higher percentage of carbon and a lower percentage of moisture than lignite (637)

sublimation the change of state from a solid directly to a gas (389)

substituted hydrocarbons hydrocarbons that are classified according to the functional groups they contain (639)

substrates the substance or substances on which an enzyme has its effect (688)

supercooled liquid a substance that retains certain liquid properties even at temperatures at which it appears to be solid (378)

superheated water water heated under pressure to a temperature above its normal boiling point (765)

superoxide a binary compound containing the O_2^- anion, in which oxygen has the uncommon oxidation number $-\frac{1}{2}$ (291)

superphosphate the fertilizer calcium phosphate, $Ca_3(PO_4)_2$, which has been increased in solubility (774)

supersaturated solution a solution that contains more dissolved solute than a saturated solution under the same conditions (412)

surface tension a force, common to all liquids, that tends to pull adjacent parts of a liquid's surface together, thus making the surface less penetrable by solid bodies (374)

suspension a heterogeneous mixture of a solvent-like substance with particles that slowly settle out (408)

synapse the space between one neuron and the next (695)

synthesis reaction (composition reaction) a reaction in which two or more substances combine to form a new substance (244)

system the sample of matter being studied (383)

T

temperature a measure of the average kinetic energy of the particles in a sample of matter (40)

theoretical yield the maximum amount of a product that can be produced from a given amount of reactants (276)

theory a broad generalization that explains a body of facts or phenomena (9)

thermite a mixture of powdered aluminum and iron(III) oxide (755)

thermite reaction the reaction by which a metal is prepared from its oxide by reduction with aluminum (755)

thermochemical equation an equation that includes the quantity of heat released or absorbed during the reaction as written (527)

thermochemistry the study of the changes in heat energy that accompany chemical reactions (527)

titration the controlled addition and measurement of the amount of a solution of known concentration that is required to react completely with a measured amount of a solution of unknown concentration (511)

torr a pressure unit named after Torricelli. 1 torr = 1 mm Hg (326)

transition element the ten subgroups of elements that intervene between Group 2 and Group 13, beginning with Period 4 (729)

transition interval the pH range over which an indicator changes color (505)

transmutation a change in the identity of a nucleus because of a change in the number of its protons (810)

transuranium element an element with more than 92 protons in its nucleus (814)

triple bond a covalent bond produced by the sharing of three pairs of electrons between two atoms (171)

triple point indicates the temperature and pressure conditions at which the solid, liquid, and vapor of a substance can coexist at equilibrium (390)

triprotic acid an acid able to donate three protons per molecule (466)

U

unit cell the smallest portion of a crystal lattice that reveals the three-dimensional pattern of the entire lattice (379)

unit of measurement a physical quantity of a defined size (31)

unsaturated describes an organic compound that has one or more double or triple covalent bonds in each molecule (650)

unsaturated solution solution that contains less solute than a saturated solution under the existing conditions (412)

unshared pair a lone pair of electrons that is not involved in bonding but instead belongs exclusively to one atom (169)

V

valence electrons the electrons available to be lost, gained, or shared in the formation of chemical compounds (152)

van der Waals forces forces of attraction between molecules (188)

vaporization the process by which a liquid or solid changes to a gas (375)

variable a quantity that can change in value (58)

VSEPR (valence-shell electron-pair repulsion) theory electrostatic repulsion between the electron pairs surrounding an atom causes these pairs to be oriented as far apart as possible (183)

vinyl group the —CH = CH_2 group (653)

voltaic cell a device in which an electric current can be generated from two solutions separated by a porous partition that permits ions to pass (but that otherwise prevents mixing of the solutions) and which has two electrodes connected in a circuit (617)

volume the amount of space occupied by an object (36)

vulcanization the process by which rubber is heated with sulfur, which bonds with rubber molecules, making the rubber more elastic, less sticky and longer wearing (658)

W

wavelength the distance between corresponding points on adjacent waves (97)

wax a lipid having one long fatty acid and one long carbon–hydrogen chain with an alcohol group (C—OH) in it (692)

weak acid an acid that is a weak electrolyte (466)

weak electrolyte a solute that yields a relatively low concentration of ions in aqueous solution (450)

weight a measure of the earth's gravitational attraction for matter (11)

word equation an equation in which the reactants and products in a chemical reaction are represented by words (234)

INDEX

Page numbers in boldface type refer to figures. Page number preceded by the letter *t* refer to tables.

A

absolute zero, 330
accuracy, 46–47, **46,** 59
acetic acid: anion of, 474; common-ion effect and, 576; conjugate base of, 475; ionization constant of, 577–578, t 578; preparation of, 677–678; properties of, 465, t 465, 469; structure of, 467; uses of, 469, 678, **678**; as weak acid, 477, **483**
acetone, **674,** 675–676
acetylene, 531
acid(s): binary acids, 467, t 467; Brønsted acids, 462, 486, 475–477, 508; in carbon dioxide preparation, 306; chemical reactions of, 483–485, **484, 485,** 486; conjugate, 475–478, t 478; decomposition of, 247; diprotic acids, 466; examples of common acids, 465–469; household, 476; in hydrogen preparation, 298; hydroxyl group, 465; indicators for, 505–510, t 505, **506–508, 510,** 516; ionization constant of, 576–579, t 578, 590; Lewis acids, 462–463, 486; mineral acids, 467; monoprotic acids, 463; organic acids, 467; oxyacids, 464–465, t 464, **465**; pH and, 499, 516; polyprotic acids, 466; properties of, 461, **461,** 486; relative strengths of, 477–478, t 478; strong acids, 463, t 466; traditional, 461–462, 486; triprotic acids, 466; types of, 211; weak acids, 466, t 466
acid anhydride(s), 480
acid-base reaction(s), 473, 486
acid precipitation, 772–773, 775
acid rain, **291,** 299, 504, 509
actinide(s), 130, 141
activated carbon, 638
activated complex, 545–548, **546,** 557
activation energy, 544–545, **545,** 557
activity series, 251–255, t 252, **253**
actual yield, 276
adenosine triphosphate (ATP), 695, 701–703, **701**
adiabatic demagnetization, 322
Agent Orange, 640
alcohol(s), 665–669, **667–667, 669,** 681
aldehyde(s), 674–675, **674, 675,** 681
alkali metal(s), 136, 713–716, **713,** t 714, **716,** 726
alkaline(s), 471
alkaline-earth metal(s), 136–137, 717–719, **717,** t 718, 726
alkane(s), 643–649, **645,** t 645 **648,** 660
alkene(s), 644, 649–651, **650,** 660
alkyl group, 647, t 647
alkyl halide(s), 670
alkylation, 651
alkyne(s), 644, 651–653, **651,** 660
allotrope(s), 289
allotropy, 289
alloy(s), 407, **407**

alpha particle(s), 806, 809, **809,** t 809, 814
alum(s), 757–758, 762
alumina, 753, 757
aluminum: atomic radii of, 143; compounds of, 756–758, **757**; electron configuration of, t 117; properties of, 751, t 752, 753, 754–755, 762; thermite reaction, 755–756, **756,** 762
aluminum chloride, 449, 583
aluminum hydroxide, 481–482
aluminum nitrate, 248
aluminum oxide, 205, 263, 265, 480, 756–757
aluminum recycling, 754
aluminum sulfate, 201
alundum, 757
amine(s), 694, **695**
amino acid(s), 686, **687**
ammonia: aqueous solution of, 450; as base, 472–474; molality and, 423; occurrence of, 299, **299,** 300; preparation of, 301–302, **203**; reaction with hydrogen chloride, 364, **364**; structure and properties of, 184, **184, 186,** 299, 301, t 301; uses of, 301, 310
ammonium chloride, 301, 364, **364,** 472, 582–583, 611
ammonium ion(s)& 173
ammonium nitrite, 299
amorphous carbon, 636–638
amorphous solid(s), 377, 381–382, 394
amorphous sulfur, 767, **767**
ampere, t 32
amphoteric, 478
amphoteric hydroxide(s), 480–483
amphoteric oxide(s), 480–483
analytical chemistry, 7
anhydride(s), 480
aniline, 474
anion(s), 151, 212, 464–465, t 464, 474
anion hydrolysis, 581–582, 590
anode, 612
antacid(s), 481
anthracite, 637
antimony, **25,** t 119, 760–761, 762
antiseptic, 790
aqueous salt solution(s), 614
argon, 116, t 117, 130
Aristotle, 71
aromatic hydrocarbon(s), 644, 653–655, **654,** 660
Arrhenius, Svante, 439, 461–462, 470
arsenic, **25,** 760, 762
arsenopyrite, **25**
astatine, 779, t 780
Aston, Francis William, 363
atm. *See* atmosphere of pressure (atm)
atmosphere of pressure (atm), 326, t 326
atom(s): excited state of, 100–101; Greek ideas on, 71; ground state of, 100–101; nuclear forces in, 80; quantum model of, 104–106; radii of, 143–144, **143–145**; sizes of, 81–82, 92; structure of, 75–82; subdivision of, 75. *See also* electron(s); neutron(s); proton(s)

atomic mass(es), 85–87, t 86, 92, 217, 801
atomic mass unit, 85
atomic nucleus: composition and structure of, 75, 78–80, **78,** 801–804, **802, 803**; discovery of, 77–78, 92; size of, 81–82, 92
atomic number(s), 83, 92, 803
atomic radii, 143–144, **143–145,** 155
atomic structure, 75
atomic symbol(s), 221
atomic theory: Dalton's model of, 71–72, 92, 221, 344; and definite composition of chemical compounds, **72,** 73; and law of conservation of mass, 72–73, **72,** 92; and law of multiple proportions, 73–74; quantum model of, 104–106
ATP. *See* adenosine triphosphate (ATP)
Aufbau principle, 112–113, **115,** 121
auto-oxidation, 610
autoprotolysis, 496
average atomic mass, 85–87, t 86, 92
Avogadro, Amadeo, 88, 217, 344–345
Avogadro's number, 88, 90, **90,** 92
Avogardo's principle, 344–345, **344,** 365
azeotrope(s), 451

B

baking soda, 726
barium hydride, 296
barium oxide, 291
barometer, 324, 325–326, 337
base(s): Brønsted bases, 472–473, **472,** t 473, 475–477, 486; chemical reactions of, 483–485, **484, 485,** 486; conjugate, 475–478, t 478; household, 476; indicators for, 505–510, t 505, **506–508, 510,** 516; Lewis bases, 473, t 473, 486; neutralization reactions, 471–472, **471, 472**; pH and, 499, 516; properties of, 470, **470,** 486; traditional, 486; traditional bases, 470–471, t 473; types of, 473–474, **474,** 486; weak bases, 473–474
base-pairing, 696, **697**
basic anhydride(s), 480
bauxite, **218,** 753, 756
Becquerel, Henri, 804, **805,** 806
benzene(s), 415, 653–654
Bergius process, 658
beryl, 136
beryllium, 115, t 116, 136, 168
beryllium fluoride, 182
Berzelius, Jöns Jakob, 23, 221
beta particle(s), 809, **809,** t 809
binary acid(s), 211, t 211, 467, t 467
binary compound(s), 204–207, **205,** 209–210, **209,** t 210, 246, **246**
biochemistry, 7, 685
bituminous coal, 637
Black, Joseph, 260
black magnetite, 291
blast furnace, 739, **739**
bleaching, 785
block(s), in Periodic Table of Elements, 133–142, **133–136, 138, 139,** t 140, **141**

blue jeans, 610
blueprint(s), 743
Bohr, Niels, 102, 104, 105–106
boiling, 375, 387–388, **387, 388,** 394, 451
boiling point(s): and bond types, t 187, 189,
190; of crystals, t 380; of liquid, 375,
387–388, 394, 451; of solutions, 429, t 429
boiling-point elevation, 431, **431**
bond(s). *See* chemical bond(s)
bond energy: covalent bonds and, 166, t
166; and reaction heat, 535–536, **535,**
557; single and multiple covalent bonds
and, 171, t 172
bond length(s), 166, t 166, 171, t 172
boron, t 116, 168
boron trifluoride, 463
Boyle, Robert, 296, 322, 327
Boyle's law, 327–329 **327, 328,** t 328, 337
bromine compounds of, 788–789; **788;**
preparation of, 787, **787;** properties of,
25, **139,** 140, 375, **375, 779,** t 780, 788;
single-replacement reactions with, 248;
uses of, 788–789, 791
Brønsted, J. N., 462
Brønsted acid(s), 462, 486, 508
Brønsted acid-base reaction(s), 473,
475–477, 486
Brønsted base(s), 472–473, **472,** t 473,
475–477, 486
Brown, Robert, 409
Brownian motion, 409
buffered solution(s), 579, **579**
butadiene, 651
butane, 645

C

calcium, 137
calcium bromide, 788–789
calcium carbonate, 136, 247, 306, 485
calcium chloride, 442–443
calcium fluoride, 177–178, **178,** 585–586
calcium hydroxide, 245–246, 247, 301, 305,
473, 485
calcium hypochlorite, 570
calcium oxide, 245, 247
calcium sulfate, 211, 246
calculator(s), 501–502
calorie(s), 43, t 43, **527**
calorimeter, 527, **527**
cancer, 172, 806, 816, 818
candela, t 32
Cannizzaro, Stanislao, 127, 344
capillary action, 375
carbohydrate(s), 689–691, **689, 691,** 703
carbon: allotropic forms of, 634–636, **634,**
660; amorphous forms of, 636–638, 660;
electron configuration of, t 116; structure
and bonding of, 171, t 172, 633, **633,**
660; uses of, 140
carbon dioxide: chemical properties of,
304–305, 307; desktop investigation, 307;
occurrence of, 303; physical properties of,
304, **304,** t 304, 307; preparation of,
234–235, **235,** 245, 247, 272, 305–306;
structure of, 303–304, **203;** uses of, 303,
305, **305,** 310, 407, 416, **416**
carbon-14, 816
carbon monoxide: occurrence of, 306, 308;
physical properties of, t 309; poisoning,

306, 308, 309; and preparation of hydrogen, 297; structure of, 308; uses of, 303,
309, 310
carbon tetrachloride, 414–415, **415,** 671
carbonate(s): in carbon dioxide preparation,
306; reactions of acids with, 485
carbonic acid, 247, 306, 574
carboxyl group, 676
carboxylic acid(s), 676–678, **677,** 681
catalysis, 551
catalyst(s), 236, 551–552, **553**
cathode, 612
cathode ray(s), 75–77, **76, 77**
cation(s), 151
cation hydrolysis, 582–583, 590
caustic, 724
Cavendish, Henry, 211, 296
cellulose, 690, **691**
Celsius, Anders, 41
Celsius temperature scale, 41–43, **42**
cesium, 25, 136, 154
Chadwick, Sir James, 813
chain reaction, 819, **819**
charcoal, 550–551, 638
Charles, Jacques, 329
Charles' law, 329–332, **329, 330,** t 330,
t 331, 337
chemical(s), 6–7
chemical bond(s), 161–164, **161, 162,** 190.
See also covalent bond(s); ionic bond(s);
metallic bond(s)
chemical change(s). *See* chemical reaction(s)
chemical compound(s): chemical formulas
for, 201–202, t 201, **202;** containing
polyatomic ions, 208–209; definite
composition of, 73; desktop investigation, 215; law of definite composition,
22–23, **22;** law of multiple proportions,
73–74; low and high heats of formation,
529; models of, 215; molar masses and,
216, 217–220, **218;** nomenclature of,
206, **207;** numerical prefixes for, **209,**
t 210; percent composition of, 222–223;
systematic and common names of,
t 201
chemical element(s). *See* element(s)
chemical equation(s): balancing, 240–243,
240, 241, 249; coefficients in, 234; desktop investigation, 249; reading and writing, 233–238, 255; requirements for,
233–234; significance of, 238–240, **239;**
symbols used in, t 236; word and formula
equations, 234–238, **235**
chemical equilibrium, 564. *See also*
equilibrium
chemical equilibrium expression, 566
chemical equivalent(s). *See* equivalent(s)
chemical formula(s): of binary ionic compounds, 204–207, **205,** 228; calculation
of, 224–227; description of, 165; empirical, 224, 228; of familiar compounds,
t 201; formula masses and, 216–217;
molar masses and, **216,** 217–220, **218,**
228; molecular formulas, 226–227;
molecular masses and, 217; oxidation
numbers and, 214–216; percent composition and, 222–223; significance of,
201–202, **202,** 228; simplest, 224, 228
chemical kinetics, 549

chemical properties, 14–15, 27
chemical reaction(s): description of, 15;
desktop investigation, 18; endothermic,
18–19; exothermic, 16–17, **17;** indications of, 16, **16, 17,** 27; spontaneous,
536–537; types of, 244–250
chemical solution(s). *See* solution(s)
chemical symbol(s), 23, t 23, 236–237, t 236
chemistry: branches of, 7, 27; description
of, 5; descriptive, 25–26; in modern
society, **5,** 6, **6**
chitin, 691
chloramine, 254
chlorinated lime, 785
chlorine: bonding and, 162, **162,** 182; compounds of, 786, **786;** electron configurations of, t 117; preparation of, 782, **782;**
properties of, **139,** 779, **779,** t 780,
784–785; single-replacement reactions,
248; uses of, 782, 785, **785,** 791
chlorofluorocarbon(s), 289, **289**
chlorohydrocarbon(s), 640
cholesteric crystal(s), 376
cholesterol, 693, **693**
chromium, 118, t 118
coal, 531, 550, 636–637, t 641, 658
cobalt, 118, t 118, 736–737, t 737
Cockcroft, J. D., 813–814
coefficient(s), 234
coenzyme(s), 688
coke, 637–638
colligative properties: of electrolyte solutions, 453–456, **455,** t 455, t 456; of
solutions, 426–433, **426, 427, 431**
collision theory, 543–544, **544,** 557
colloid(s), 405, 408, 409, **409,** t 409, t 410
combined gas law, 333–334, 337
combustion reaction(s), 250, **250,** 530–531,
530, t 530, 557
common-ion effect, 575–576, 590
complex carbohydrate(s), 689–690
complex ion(s), 733–736, *734,* t 735, t 736,
748
composition reaction(s). *See* synthesis reac-
tion(s) composition stoichiometry, 261
compound(s). *See* binary compound(s);
chemical compound(s); ionic com-
pound(s); molecular compound(s)
computer(s), 760–762
computer chip(s), 761
concentration, 384, 420
concentration: description of, 384, 420;
equilibrium and changes in, 569; and reaction rate, 550–551, **550**
condensation, 383–384
conductor(s), 611
conjugate acid(s), 475–478, t 478
conjugate base(s), 475–478, t 478
conservation of energy, law of, 13, 15, 27
conservation of mass, law of, 72–73,
72, 92
conservation of matter, law of, 12, 15, 27,
233–234
contact process, 773
continuous spectrum, 98, **98,** 104, **105**
control rod(s), 819
conversion factor(s), 34–35, 56
Cooke, Josiah Parsons, Jr., 127
coordination number(s), 734

copper: activity series and, 252–253, **253;** electrode potentials and, 616–620; electrolytic refining of, 746, **746;** electron configuration of, t 118; oxidation-reduction reactions and, 608–609; properties of, 24, 26, **26,** 138, 744, 747; recovery of, 745–746, **745;** tests for copper ion, 747–748; uses of, 744, 747
copper chloride, **205**
copper hydroxide, 583
copper oxide, 484
copper sulfate, **406,** 484, 583, 747
corundum, 756
cosmic ray(s), 816
covalent bond(s): bond lengths and bond energies for, 166, t 166; description of, 161–162, 163; double, 170–171, 190; formation of, 165–167, **165, 167;** Lewis structures and, 169–170; multiple, 170–173, t 172, 190; octet rule, 167–168; single, 169, t 172; triple, 171, 190
covalent hydride(s), 296
covalent molecular crystal(s), 381
covalent network crystal(s), 380
cracking, 649
critical mass, 820
critical point(s), 389
critical pressure, 389
critical temperature, 389
crude oil, 656
cryolite, 753
crystal(s): binding forces in, 379–381, **381;** classification of, 379, **379,** 394; ionic crystals, **176–178;** liquid crystals, 376; melting points and boiling points of, t 380; as solids, 377
crystal lattice, 379, **379**
crystalline solid(s), 377, **378,** 379–381, **379,** t 380, **381,** 394
Curie, Marie, 804, **805,** 806
cycloalkane(s), 648–649

D

Dalton, John, 71–72, 73, 92, 221, 334, 344
Dalton's law of partial pressures, 334–336, **335**
Da Vinci, Leonardo, 290
daughter nuclide(s), 810
Davy, Sir Humphry, 247, 720, 791
DDT, 640
De Broglie, Louis, 104–105
decay, in carbon dioxide preparation, 306
decomposition reaction(s), 246–247, **246,** 255
definite composition, law of, 22–23
dehydrating agent(s), 774
Democritus, 71
denaturation, 688
denatured alcohol, 668
density: of common materials, **37,** 39, t 39; derived SI units and, t 36, 37, 39; desktop investigation, 38
deoxyribonucleic acid (DNA), 695–698, **696-698,** 700, 703
destructive distillation, 637
deuterium, 80, 83, t 84, 821, **821**
diamagnetism, 111
diamond, 380, **381,** 634–635, **634,** 660

diatomic molecule(s), 164, 233, t 233
diborane, 224–227
dicarboxylic acid(s), 678
dichlorodifluoromethane, 671, 782
diet, 715
diffusion, 318, **318,** 360–362, **360, 362,** 364–365, 373, **373**
dimer(s), 653
dimethyl sulfide, 772
dipole(s), 187–188
dipole-dipole forces, 187–188, **188**
diprotic acid(s), 466
direct proportion, 58
disaccharide(s), 689, **689**
disinfectant(s), 785, 790
displacement reaction(s). *See* single-replacement reaction(s)
dissociation, 442–443
dissolving agent(s), 372–373
distillation, 428
DNA. *See* deoxyribonucleic acid (DNA)
Dorn, Freidrich Ernst, 130
double bond(s), 170–171, 190
double helix, 696
double-replacement reaction(s), 248, 250, **250,** 255
Downs cell, 720–721, **720**
dry cells(s), 611, **612**
dry ice, 305
ductility, 181

E

effervescence, 416
effusion, 318, 360–365, **360**
Einstein, Albert, 15, 101
Einstein's equation, 15, 813–814
elastic collision(s), 315
electrochemical cell(s), 611–612, **610, 611,** 622
electrochemistry: electrochemical cells, 611–612, 622; electrode potentials, 616–622, **617–169,** t 620; electrolytic cells, 612–615, **613–615;** rechargeable cells, 615–616
electrode(s), 611
electrode potential(s), 616–622, **617–619,** t 620
electrolysis: of aqueous salt solutions, 614–615; as decomposition reaction, 246–247; description of, 612, 622; in hydrogen preparation, 298; in oxygen preparation, 293; of water, 613–614
electrolyte(s): acids as, 461; apparent degree of ionization of solutions, 454–456, **455;** bases as, 470; coining of word, 439; colligative properties of solutions, 453–454, **455,** t 455, t 456; conductivity of solutions, 450–452, **452;** description of 406; freezing point depression of solutions, 453–454, t 455, t 456; in human bodies, 447; molecular electrolyte(s), 447–450, **448;** strong, 449; weak, 450
electrolyte solution(s), 447–448, **448,** 450–456, t 455, t 456
electrolytic cell(s), 612–615, **613–615**
electromagnetic radiation, 97–98, 100
electromagnetic spectrum, 98, **98, 99**
electron(s): Bohr's theory of, 102–106, **103;** charge and mass of, 76–77, 79; discovery

of, 75–77, 92; highest occupied energy level and, 113–114; inner-shell electrons, 114; in main energy levels, 111; orbitals of, 106, 110, **110,** 121; orbits of, 103, 104; properties of, t 80; uncertainty principle and, 106–107; unshared pair of, 169; valence electrons, 152, t 152; wave properties of, 105–106, **106;** wave-particle model of, 97, 104–105
electron affinity, 149–151, **149, 150**
electron configuration(s): fifth-period elements, 119, t 119; first-period elements, 114, 115; fourth-period elements, 117–119, t 118; notations for, 113–114, 121; and Periodic Table, 132–142, **134–136,** 138, **139,** t 132, t 140, 141; rules governing, 112–113, **112,** 121; second-period elements, 115, t 116; seventh-period elements, 120; sixth-period elements, 120; third-period elements, 115–116, t 117
electron-configuration notation, 113
electron-dot notation, 113–114
electron-dot symbol(s). *See* Lewis structure(s)
electronegativity, 152–154, **153,** 156
electroplating, 615
element(s): activity series of, 251–255, t 252; atomic masses and abundances of, t 86, 87; classes of, 24–25; discovery of, 791; electron configurations of, 115–120, t 116, t 117, t 118, t 119; fifth-period, 119, t 119; first-period, 114, 115; fourth-period, 117–119, t 118; number of, 21–22; reactions wtih oxygen, 244–245; second-period, 115, t 116; seventh-period, 120; sixth-period, 120; symbols for, 23, t 23; third-period, 115–116, t 117. *See also* Periodic Table of Elements
emery, 756
empirical formula(s), 224, 228
end point of titration, 512
endothermic reaction, 18–19, **545**
energy: conservation of, 13, 27; interconversion of matter and energy, 15; kinetic, 12; potential, 12; in starting a car, 12, **13**
enthalpy, 528, 536–537, **537**
enthalpy change, 528
entropy, 537–538, **538,** 557
enzyme(s), 688, 703
epinephrine, 694, **694**
equation(s). *See* chemical equation(s)
equilibrium: change in concentration and, 569; change in pressure and, 569–572, **571, 572;** change in temperature and, 572–573; and changes of state, 383–384, **384;** common-ion effect and, 575–576, 590; description of, 383, 394, 564–565, **565,** 590; equilibrium constant, 565–568, **567,** t 567, 573, t 573, 590; equilibrium vapor pressure, 385–386, **386,** 394; Le Châtelier's principle and, 384–385, 394; reactions that run to completion, 573–575, **575**
equilibrium constant, 565–568, **567,** t 567, 573, t 573, 590
equilibrium vapor pressure, 385–386, **386,** 394
equivalence point, 512

equivalent(s), 491–493, **491,** 516
equivalent mass, 492, 516
ester(s), 679, **679, 680,** 681
esterification, 677, 679–680
ethane, 171, 645
ethanol, 414–415, **415,** 665, **666,** 667–669, **667**
ethene, 649
ether(s), 415, 673, 681
ethyl alcohol, 407, 665
ethylene bromide, 788
ethylene dichloride, 672
ethylene glycol, 407, 669, **669**
ethyne, 652–653
evaporation, 375
excess reactant(s), 272
excited state, 100–101
exothermic reaction, 16–17, **17, 545**
experiment(s), 8
extensive physical properties, 14

F

factor-label method, 34–36, 59, 56
Fahrenheit, Gabriel Daniel, 42
Fahrenheit temperature scale, 42–43, **42**
families of elements, 23
Faraday, Michael, 439
fat(s), 296, 414, 680–681, 692–693
fatty acid(s), 692, **692**
fermentation, 306
ferromagnetism, 111, 737
fertilizer(s), 301
fiber composite(s), 757
fire extinguisher(s), 305, **305**
Fischer-Tropsch process, 658
fission, 75, 804, 817, 819, **819**
flame test(s), 716
fluid(s), 372. *See also* liquid(s); water
fluoresce, 804, **805,** 807
fluorine: compounds of, 782; oxidation number, 213; preparation of, 781; properties of, 115, t 115, **139,** 154, 251, 779, **779,** t 780; reactions involving, 245, 245; uses of, 782, 791
fluorite, 177, **178**
fluorspar, 782
food(s): calories in, 43, t 43; carbon dioxide in, 305, 407, 416, **416;** recipes for, 273, **431**
formaldehyde, 172, 675, **675**
formic acid, 678
formula(s). *See* chemical formula(s); empirical formula(s); molecular formula(s); structural formula(s)
formula equation(s), 234–235, **235**
formula mass(es), 216–217
formula unit(s), 175, 201
formyl group, 674
fractional distillation, 656–657, t 656
francium, 154
Frasch, Herman, 765
Frasch process, 765, **765,** 775
free energy, 538–539, 557
free-energy change, 538–540, 557
freezing, 375–376, 388, 394
freezing point(s), 388, 429, t 429
freezing-point depression, 428–430, 453–454, t 455, t 456
Freon(s), 671, **672**

frequency, 98
Friedel-Crafts reaction, 654, 671
fuel(s), 531
functional group(s), 639–640
fundamental unit(s) of measurement, 32–34, t 32, t 33, 59
fungicide(s), 769
fusion, 804, 820–821, **821**

G

gallium, 25, 143
gamma ray(s), 809, **809,** t 809
gas(es): Avogadro's principle, 344–345, 365; Boyle's law, 327–329, **327, 328,** t 328, 337; changes of state, t 383; Charles' law, 329–332, **329, 330,** t 330, t 331, 337; collection by water displacement, 335–336; combined gas law, 333–334, 337; compressibility of, 317–318; Dalton's law of partial pressures, 334–336, **335;** density of, 347–348, **347;** deviations of real from ideal gases, 319, **319;** diffusibility of, 318, **318;** discoveries of, 130; expansiveness of, 317; fluidity of, 317; formation of, 250, 574–575; Gay-Lussac's law, 332–333, 337, 343–344, 366; ideal gas law, 350–355, **351,** t 352, 365; ideal gases, 315, 319, 337; kinetic theory of, 315–319, **315, 317–319,** 337; low density of, 317; measuring and comparing volumes of, 343–344, 366; molar mass of, 348–350; molar volume of, 345–347, **346,** 365; nature of, 317–318, 337; noble gases, 130, 140; pressure and, 322–327, **323–325,** t 326; pressure and temperature, 321, **321;** pressure and volume at constant temperature, 320; qualitative description of, 320–321; quantitative description of, 322–336, t 326, t 328, t 330, t 331, 337; real gases, 319, **319;** relationship between pressure and moles, 321; relationship between volume and moles, 321, **321;** solubility of, 415–416; solutions of, 407, t 407; stoichiometry of, 356–359; temperature and volume at constant pressure, 320–321, **320;** tests for, t 310; volume-mass and mass-volume calculations, 357–359; volume-volume calculations, 356–357; water-gas reaction, 535–536, **535.** *See also* names of specific gases
gas law(s): Avogadro's principle, 344–345, 365; Boyle's law, 327–329, t 328, 337; Charles' law, 329–332, t 330, t 331, 337; Dalton's law of partial pressures, 334–336; description of, 322; Gay-Lussac's law, 332–333, 337, 343–344, 366; ideal gas law, 350–355, t 352, 365
gaseous state, 13–14
gasoline, 414, 415, 646, 788
Gay-Lussac, Joseph, 332, 343–344, **343**
Gay-Lussac's law, 332–333, **332,** 337, 343–344, 366
Geiger, Hans, 77
Geiger counter(s), 807, **807**
gene(s), 697
genetic engineering, 699–700
germanium, 25, 752, t 752
Gibbs, Josiah Willar, 538

Gibbs Free Energy, 538
glycerol, 669, **669**
glycogen, 689
glycolipid(s), 692
glycolysis, 701, **701, 702,** 703
gold, 24, 138–139, 252, 744, t 744
Graham, Thomas, 409
Graham's law of effusion or diffusion, 360–362, **362,** 364–365, **364**
granite, 20, **20**
graphite, 380, **381,** 634, 635–636, **635, 636,** 660, 819
ground state, 100–101
group(s) of elements, 23
Guldberg, Cato Maximilian, 556

H

Haber, Fritz, 301
Haber process, 302, **302,** 570–571, **572**
Hahn, Otto, 75, 817
half-cell(s), 618
half-life, 807–808, 821
half-reaction(s), 612, 622, 620–621, t 620
halite, 722
Hall, Charles, 753
Hall process, 753
halocarbon(s), 670–672, **670, 672,** 681
halogen(s), 140, 248, 779–781, **779,** t 780, 791
halogenation, 654
heat: description of, 40, **41,** 59; heat capacity, 44; specific heat, 44–45, t 44; symbol for, 236; units of, 43, t 43; water-gas reaction, 535–536, **535**
heat capacity, 44
heat of combustion, 530–531, **530,** t 530, 557
heat of formation, 528–529, **529,** 556–557
heat of hydration, 441
heat of reaction, 527–528, **527, 528,** 531–536, **532,** 556, 557
heat of solution, 417–419, t, 419, 441–442, **442**
Heisenberg, Werner, 106–107
helium, t 84, 113–114, 130, 137, 345
hematite, 738
Henry, William, 416
Henry's law, 415–416
Heroult, Paul, 753
hertz, 98
Hess's law, 531–532
heterogeneous catalyst(s), 552
heterogeneous mixture(s), 19–20
heterogeneous reaction(s), 550
homogeneous catalyst(s), 551–552
homogeneous mixture(s), 20, **20, 21,** 27
homogeneous reaction(s), 541–542
homologous series, 645
hormone(s), 693–695, **694–695**
Hückel, E., 456
Hund's rule, 113
hybrid orbital(s), 186, t 187
hybridization, 185–186, **186,** t 187, 190
hydration, 440
hydrazine, 254
hydrobromic acid, 449

hydrocarbon(s): alkanes, 643–649, t 645, 660; alkenes, 644, 649–651, 660; alkynes, 644, 651–653, 660; aromatic hydrocarbons, 644, 653–655, 660; description of, 250, 643, 660; groups of, 643–644, 660; in hydrogen preparation, 298; industrial and energy uses of, 655–660

hydrochloric acid: formation of, 449; as nonoxidizing acid, 483–484; and preparation of carbon dioxide, 306; and preparation of hydrogen, 297; properties of, 465, t 465, 467, 469; reactions including, 248, **248**, 549, 574, 603–604; as strong acid, 463, 476; uses of, 786, **786**

hydrofluoric acid, 782, **782**

hydrogen: chemical properties of, 296; covalent bonding, 162, **162**, 164–167, **165, 167**; diffusion of, **362**; discovery of, 211, 211; electron configurations of, 113–114, 137; history of, 296; isotopes of, 80–81, 83, t 84; mass of, 216, 345; molecules of, 182; occurrence of, 294–295; physical properties of, 295, t 295; preparation of, 248, **295**, 297, **297**, 648; reaction with chlorine, 785, **785**; reaction with metals, 296, 310; reaction with nonmetals, 296, 310; replacement by metal, 248, **248**; uses of, 294, **295**, 296, 310

hydrogen atom(s): Bohr model of, 102–103, **103**; bonding of, 162, **162**, 164–167, **165, 167**, 189, **189**; line emission spectra of, 103; spectrum of, **99**, 100–103, **103**

hydrogen bonding, 162, **162**, 164–167, **165, 167**, 189, **189**

hydrogen bromide, 449, 788, **788**

hydrogen chloride: as acid, 463, 472; ammonia and, 364, **364**; bonding of, 162, 164, 182; chemical equation for, 238–239, **239**; common-ion effect and, 576; description of, 786; double-replacement reaction and, 250; formation of, 296, 343–345; molecular electrolytes of, 448–449, **448**; oxidation and reduction and, 597–598

hydrogen fluoride, 212, 449–450, 475

hydrogen iodide, 449

hydrogen peroxide, **202**, 292 **292**, 610

hydrogen sulfate, 478

hydrogen sulfide, 250, 575, 772

hydroiodic acid, 449

hydrolysis, 580–583, **581–583**, 590

hydrolysis constant, 582

hydronium ion(s), 448–449, t 504

hydroxide(s), 480–483, 485, t, 504

hydroxyl group, 465

hyberbola, 59

hypochlorous acid, 570

hypothesis, 8

I

ice crystal(s), 391, **391**

ideal gas(es), 315, 319, 337

ideal gas constant, 351–352, t 352

ideal gas law, 350–355, **351**, t 352, 365

immiscible, 414

indicator(s), 505–510, t 505, **506–508, 510**, 516

inertia, 10–11

inhibitor(s), 552

inner-shell electron(s), 114

inorganic chemistry, 7

inorganic compound(s), 642–643, t 643

insecticide(s), 670, **670**

instability constant, 735

insulin, 687

integrated circuit(s), 760–761

intensive physical properties, 14

intermediate(s), 542–543

intermolecular forces, 187–190, **188–190**

inverse proportion, **58**, 59

iodine, **139**, 248, **779**, t 780, 790, **790**, 791

ion(s): complex, 733–736, **734**, t 735, t 736, 748; description of, 145; formation of negative, 150–151; formation of positive, 147–148, 151; in metals, 156; monatomic, 203–204, t 203, t 204; origin of word, 439; spectator ions, 445

ionic bond(s), 161–163, **161, 162**, 176–178, **176, 177**

ionic compound(s): binary, 204–207, **205**; compared with molecular compounds, 178–179, **178**, 190; description of, 175, **175**; dissociation and, 442–443; dissolving of, 440–443, **440–442**, 456; heat of solution for, 441–442, **442**; net ionic equations, 444–445; precipitation reactions, 444; solubility equilibria, 443–446, **444, 445**; solution process for, 440–441, **441**; theory of ionization, 439–440; very slightly soluble, 443–444

ionic crystal(s), 380

ionic hydride(s), 296

ionic radii, 151–152, **151**, 156

ionization: apparent degree of, 454–456, **455**; description of, 145, 448; theory of, 439–440; of water, 452

ionization constant: of water, 579–580; of weak acids, 576–579, t 578, 590

ionization energy: first ionization energies, 147–148, t 148; formation of positive ions, 147–148; group trends, 146–147; of metals, 155; period trends, 145–146, **146, 147**

iron: blast furnace and, 739, **739**; characteristics of, 252, 736–738, t 737; electron configuration of, 118, t 118; occurrence of, 738–741, **738–740**; oxides of, 741–742; pure iron, 741–743, **741, 742**; steel production and, 739–741, **740**; tests for iron ions, 742–743

iron chloride, 250, 575

iron nitrate, 248

iron oxide, 222, 245, 291, 309, 741

iron sulfate, 741–742

iron sulfide, 250, 574

isomer(s), 639

isotope(s): artificial, 85; atomic masses and abundances of, t 86, 87; description of, 72, 81, 92; designation of, 83; of helium, t 84; of hydrogen, 80–81, 83, t 84; relative atomic masses of, 85; separation of, 363

J

Janssen, Pierre, 130

Joliot-Curie, Frederic, 85, 812

Joliot-Curie, Irene, 85, 812

joule, t 36, 43, t 43

journals. *See* scientific journals

K

Kelvin, William Thomson Lord, 41

Kelvin temperature scale, t 32, 41–42, **42**

ketone(s), 674–676, 681

kilogram(s), t 32, 33

kinetic energy, 12

kinetic theory: of gases, 315–319, **315, 317–319**, 337; of liquids, 371; of solids, 377

krypton, t 118, 130

krypton difluoride, 782

L

lambda-sulfur, 764

lanthanide(s), 130, 141

latex, 658

lattice energy, 178

Lavoisier, Antoine-Laurent, 245, 290, 296

law(s), 8–9. *See also* gas law(s); rate law(s); and names of specific laws

law of conservation of energy, 13, 15, 27

law of conservation of mass, 72–73, **72**, 92

law of conservation of matter, 12, 15, 27, 233–234

law of definite composition, 22–23

law of mass action, 556

law of multiple proportions, 73–74

LCDs. *See* liquid-crystal display(s) (LCDs)

Le Châtelier, Henri Louis, 384

Le Châtelier's principle, 384–386, 394, 508, 569–570, 572, 575, 578, 590

lead nitrate, 250, **250**

lead oxide, **207**

Lewis, Gilbert Newton, 114, 168, 462, 463

Lewis acid(s), 462–463, 486

Lewis acid-base reaction(s), 473

Lewis base(s), 473, t 473, 486

Lewis structure(s), 169–170

light: electromagnetic spectrum of, 98, **98, 99**; interference and, 100, 102; as particles, 99–101, **100, 101**; speed of, 97; as waves, 97–98, 102

lignite, 637

lime, 245, 306

limestone, 136

limiting reactant(s), 272–276

line emission spectrum, 101, 103, **105**

lipid(s), 691–695, **692–695**, t 692, 703

lipoprotein(s), 692

liquid(s): boiling of, 387–388, 394, **387, 388**; capillary action of, 375; changes of state, t 383; density of, 372, **372**; diffusion ability, 373, **373**; dissolving ability of, 372–373; equilibrium vapor pressure of, 385–386, **386**, 394; evaporation and boiling, 375, **375**; fluidity of, 372; incompressibility of, 372 **373**; kinetic-theory description of, 371; particle model of, 371–376, **371**; properties of, 371–376, **372 373, 375**, 394; solutions of, 407, t 407; surface tension of, 374–375, **375**; tendency to solidify, 375; volume of, 372. *See also* fluid(s); water

liquid air, 293

liquid crystal(s), 376

liquid state, 13–14

liquid-crystal display(s) (LCDs), 376

liter(s), 37, **37**

lithium, 115, t 116

lithium carbonate, 715
litmus, 461, 470, 505, 506
Lockyer, Joseph Norman, 296
London, Fritz, 189
London dispersion forces, 189–190, **190**
lone pair. *See* unshared pair
lye, 724

M

macromolecule(s), 685–686
magnesium, t 117, 136, 244–245, 299, 304–305, 441
magnesium bromide, 204
magnesium chloride, 175, 248, **248**
magnesium nitride, 299
magnesium oxide, 245, **245**
magnetic quantum number(s), 110–111, **110**
magnetism. *See* ferromagnetism
main-group element(s), valence electrons in, 152, t 152
malachite, **26,** 744
malleability, 181
manganese, 118, t 118, 600–601, 612
manganese dioxide, 292, **292,** 601, 603–604
manometer, **325,** 337
marble, 136, 306
Marsden, Ernest, 77
mass: conservation of, 72–73, **72,** 92; description of, 11, 27; percent by mass of solutions, 420–421; volume-mass and mass-volume calculations for gases, 357–359
mass action, law of, 556
mass number, 83, 92
mass-mass calculation(s), 262, 270–271
mass-mole calculation(s), 262, 269–270, **269**
matter: changes of state, 14, **14;** chemical and physical changes in, 16–19; classification of, **22;** conservation of, 12, 27; gaseous state of, 13–14; interconversion of matter and energy, 15; law of conservation of matter, 233–234; liquid state of, 13–14; mass and weight of, 11; mixtures, 19–20; properties of, 10–11, 14–15, 27; pure substances, 20–23, 27; solid state of, 13–14
measurement: accuracy and precision, 46–47, **46,** 59; calculations involving, 34–36, **35;** derived SI units, 36–39, t 36, **37;** of heat, 43–45, t 43, t 44; SI measurement system, 31–34, **31,** t 32, t 33, 59; significant figures and, 47–52, **47–49,** t 48, t 50; standards of, 31, **31–34;** of temperature, 41–43, **42;** units of measurement, 31–32
medication, 715
Meitner, Lise, 75, 817
melting, 378, 388, 394
melting point(s): of crystals, t 380; description of, 388; of solids, 378
Mendeleev, Dmitri, 127–129, **127, 128,** t 129, 791
mercaptan(s), 772
mercury, 24, 246, **246**
mercury oxide, 563–564
messenger RNA, 698
metabolic pathway(s), 701–703, **701, 702**

metabolism, 701–703
metal(s): atomic radii of, 155; description of, 141; ductility of, 181; electronegativity of, 156; heats of vaporization of, t 181; ionic formation in, 156; ionic radii of, 156; ionization energy of, 155; malleability of, 181; properties of, 24, **24,** 179–181, 190; reaction with acids, 483–484; reaction with chlorine, 784–785; reaction with hydrogen, 296, 310; reaction with nonmetals, 245; reaction of two metals, 245; replacement in compound by more reactive metal, 248; replacement of hydrogen by, 248. *See also* specific types of metals
metal carbonate(s), 247
metal hydroxide(s), 473, t 474
metal oxide(s), 484–485
metallic bond(s): description of, 163, 180; electron sea model of, 180, **180,** 190; formation of, 179–180, **180**
metallic chlorate(s), 247
metallic crystal(s), 380–381
metalloid(s): antimony 760–761, 762; arsenic, 760, 762; description of, 25, **25,** 141; properties of, 751–753, **751,** t 752, 758, 762; reaction with chlorine, 784–785; silicon, 758–762, **759, 760**
meter(s), t 32, 33
methane, 185–186, **186,** 234–235, **235,** 297, 306, 645, 647, 656
methanol, 309, 667
methyl bromide, 788
methyl group, 647
methylpropane, 645
Millikan, Robert A., 76–77, 79
millimeter(s) of mercury (mm Hg), 326, t 326
mineral(s), 688, t 692
mineral acid(s), 467
miscible substance(s), 415
mixture(s): compared with pure substances, 21; description of, 405, **405;** heterogeneous, 19–20, 27; homogeneous, 20, **20, 21,** 27. *See also* colloid(s); solution(s); suspension(s)
mm Hg. *See* millimeter(s) of mercury (mm Hg)
model(s), construction of, 74
moderator, 820
molal boiling-point constant, t 429, 431
molal freezing-point constant, 429, t 429
molality, 423–425
molar heat of formation, 529
molar heat of fusion, 388
molar heat of vaporization, 387, 394
molar mass(es): of chemical compounds, **216,** 217–218, **218;** as conversion factor, **87,** 88–89, **88, 90,** 218–220, **219,** 228; of gases, 348–350, 354–355; of solutes, 431–433
molar volume of a gas, 345–347, **346,** 365
molarity: description of, 421, **422;** formula for, 421–423; and normality, 495–496; and titration, 512–514, 516
mole(s), t 32, 88
mole ratio, 263–264, 278
mole-mass calculation(s), 262, 267–269
mole-mole calculation(s), 262, 266–267, **266**

molecular compound(s): binary, 209–210, **209,** t 210; compared with ionic compounds, 178–179, **178,** 190; hybridization and, 185–186, **186,** t 187, 190; intermolecular forces and, 187–190, **188–190;** naming of, 209–210, **209,** t 210; numerical prefixes for, **209,** t 210; VSEPR theory and, 182–185, **182,** t 183, **184,** 190
molecular electrolyte(s): hydronium ion, 448–449; solution process for, 447–448, **448;** strong and weak electrolytes, 449–450
molecular formula(s): atomic symbols and, 221; calculation of, 226–227; description of, 165
molecular genetics, 699
molecular mass(es), 217
molecular motion, 316
molecular weight, 217
molecule(s), diatomic, 164, 233, t 233
monatomic ion(s), 203–204, t 203, t 204
Monel metal, 781
monoclinic sulfur, 764, **764**
monomer(s), 651
monoprotic acid(s), 463
monosaccharide(s), 689
Moseley, Henry Gwyn-Jeffreys, 129–130
multiple bond(s), 170–173, t 172, 190
multiple proportions, law of, 73–74
mu-sulfur, 764, **765**
mutagen(s), 698
mutation, 698

N

natural gas, 306, 531, 655–656
nematic crystal(s), 376
neon, 115–116, t 116
neoprene, 659
neptunium, 814
net ionic equation(s), 444–445
neuron(s), **694,** 695
neurotransmitter(s), 695, **695**
neutral solution(s), 497
neutralization, 471–472, **471, 472,** 492–493, 516
neutron(s), 75, 80, t 80, 801–804, **803,** 821, 814
Newton, Sir Isaac, 326
Nicholson, William, 240
nickel, 118, t 118, 138, 252, 730, 736–737, t 737
nitration, 654
nitric acid, 463, 465, t 465, 467, 469, 484
nitrogen: binary compounds of, t 210; electron configuration of, t 116; occurrence of, 298; physical properties of, 298, t 298; preparation of, 299; structure and chemical properties of, 298–299; triple bond and, 171; uses of, 140, 298, **298,** 299, 310
nitrogen fixation, 300
nitrogen monoxide& 168
noble gas(es), 116, 130, **130,** 140
noble gas configuration, 116
nomenclature, 206, **207**
nonelectrolyte(s), 406, 439
nonmetal(s): description of, 25–27, **27;** reaction with hydrogen, 296, 310; reaction with metals, 245

nonmetal oxide(s), 485
nonpolar-covalent bond(s), 162
nonvolatile substance(s), 426
normality, 494–496, **494,** t 494, 514–516
nuclear binding energy, 801–802, **802**
nuclear disintegration, 804, 812–813, **813**
nuclear equation, 810–811
nuclear fission, 658, 817, 819, **819**
nuclear forces, 80
nuclear fusion, 820–821, **821**
nuclear mass defect, 802
nuclear reaction(s), 804
nuclear reactor(s), 726, 819–820, **820,** 821
nucleic acid(s), 695–700, **696–698,** 703
nucleon(s), 801, 821
nucleotide(s), 695–696, **696, 697**
nucleus. *See* atomic nucleus
nuclide(s), 84
nylon, **677**

O

octane, 201
octane rating(s), 646
octet rule, 168, 190
oil(s), 296, 414, 680–681, 692–693
oil shale, 658
orbital(s), 106, 110, **110,** 121
orbital notation, 113
orbital quantum number(s), 108–109, **109,**
  t 109
organic acid(s), 467
organic chemistry, 7
organic compound(s): compared with
  inorganic compounds, 642–643, t 643;
  description of, 639–640, 660; structural
  formulas and bonding, 640–642, 660
oxalic acid, 678
oxidation, 597, **597,** t 598, t 608, 622
oxidation number(s): assigning of, 212–214,
  t 598; formulas and names, 214–216;
  method for balancing redox equations,
  602–605; for nonmetals, t 214; uses of,
  213, 228
oxidation states, 213
oxidation-reduction reaction, 598–601,
  599, 601, t 608, **609, 610,** 622
oxide(s): amphoteric, 480–482; basic and
  acidic, 479–480, 486, t 482; chemical
  reactions of, 483–485, **484, 485,** 486;
  description of, 290, 310; synthesis
  reactions of, 245–246
oxidizing agent(s), 608–610, t 608, t 609,
  622
oxyacid(s), 211, t 211, 464–465, t 464, **465**
oxyanion(s), 208, 211
oxygen: binary compounds of, t 210; chemical properties of, 290–291, **291;** electron
  configuration of, 115, t 116; history of,
  290; mass of, 216, 345; molecular structure of, 289–290; occurrence of, 289;
  oxidation number of, 213; physical properties of, 290, t 290; preparation of, 246,
  246, 292–293, **292, 293;** reactions of
  elements with, 244–245, **245;** uses of,
  140, 289, 291–292, **292,** 310
ozone: formation of, 293–294; importance
  of, 289, 310; nitrogen oxides and,
  299; structure and properties of,
  293, **294**

P

palladium, 138
paraffin wax, 644
paramagnetism, 111
parent nuclide(s), 810, **810**
partial pressure, 335
particle model: of liquids, 371–376, **371;** of
  solids, 377–379, **377**
pascal, 326, t 326
Pascal, Blaise, 326
Pauli exclusion principle, 113
Pauling, Linus, 153
PCBs, 640
peat, 636–637, t 641
peptide bond(s), 687
percent by mass, 420–421
percent composition, 222–223
percent yield, 276–278, **276**
perchloric acid, 463, 477
Percy, Marguerite, 137
period(s), 23, 132–133
periodic law, 129–130, 791
Periodic Table of Atomic Radii, 143, **144**
Periodic Table of Electron Affinities, **149**
Periodic Table of Elements: atomic numbers
  and, 83; chart of, **24, 134–135;** description of, 23–24, 27; design of, 131; desktop
  investigation, 131; electron configurations and, 132–142, **132–136,** t 132,
  138, 139, t 140, **141;** history of,
  127–130, **128,** t 129
Periodic Table of Ionic Radii, **151**
Periodic Table of Ionization Energy, 146
Periodic Table of the Electronegativities,
  153
periodicity, 130–131, **130**
permanganate, 609
peroxide(s), 290, 609
pesticide(s), 640
petroleum, 656–658, t 656, **657**
petroleum coke, 638
petroleum substitute(s), 658
pH: calculations involving, 500–504; of
  common materials, t 500; equation for,
  499, 516; indicators and, 505–506, t 505,
  506–508, 508, 510, 516; relationship between hydronium ion and hydroxide ion concentrations and, 504, t 504;
  scale, 497–499 **499, 500;** transition
  interval and, 505–506, t 505
phase(s), 20, 383
phase diagram(s), 389–390, **389**
phenyl group, 654
phospholipid(s), 693
phosphoric acid, 465, t 465, 466–467, 469
phosphorous acid, 465
phosphorus, 26–27, **27,** t 117, 140, 245
photochemical smog, 299
photoelectric cell(s), 136
photoelectric effect, 99–100, 136
photon(s), 100, 121
photosensitive, 788
photosynthesis, 268, 303
physical change(s), 14, 18
physical chemistry, 7
physical properties, 14, 27
pig iron, 739
Planck, Max, 100
Planck's constant, 100

platinum, 138–139
plutonium, 814
polar bond(s), 162
polar-covalent bond(s), 162–163
pollution. *See* acid rain
polonium, 804–805
polyatomic ion(s), 173, 208–209, t 208
polyethylene(s), 651
polymer(s), 651
polymerization, 651
polyprotic acid(s), 466
polysaccharide(s), 689–690
polyunsaturated fat(s), 692
polyvinyl chloride, 784, 785
potassium, 136, 252, 621, 791
potassium bromide, 248
potassium chlorate, 217–218, 222, 247, 293
potassium chloride, 247, 444
potassium fluoride, 248
potassium hydroxide, 247, 471, 474
potassium iodide, 250, **250,** 790
potassium nitrate, 298, 444
potential energy, 12
precipitate(s), 16, **17,** 250, **250,** 575
precipitation, 444, 589
precision, 46–47, **46,** 59
pressure: Boyle's law, 327–329, **327, 328,**
  t 328, 337; combined gas law, 333–334;
  critical pressure, 389; Dalton's law of partial pressures, 334–336, **335;** equilibrium
  and changes in, 569–572, **571, 572;** and
  force, 323–324; Gay-Lussac's law, 332–
  333, **332,** 337; ideal gas law and, 352–
  354; measurement of, 324–325, **235;**
  partial pressure, 335; phase diagrams
  and, 389–390; solubility and, 415–416;
  standard temperature and pressure, 327;
  units of pressure, 325–326, t 326
Priestley, Joseph, 246, 290
primary standard(s), 512
principal quantum number(s), 108, t 109
product(s), in chemical equations, 15,
  233, 688
propane, **357,** 645
proportional relationship(s), 58–59, **58**
protein(s), 686–688, t 687, **687, 688,**
  698, 703
protium, 80, 83, t 84
proton(s), 75, 79, t 80, 801–804, **803,**
  814, 821
Proust, Joseph Louis, 23, 222
pure substance(s), 20, 21–23, 27
PVC pipe(s), 784
pyrite, 765

Q

qualitative information, 8
quantitative information, 8
quantitative problem(s), 55–56, 59
quantum, 100
quantum number(s), 108–111, **109,**
  t 109, 121
quantum theory, 104–106, **106,** 121
quartz, 380
quicklime, 245, 306

R

radiation. *See* electromagnetic radiation
radiation exposure, 818

radioactive dating, 816
radioactive decay, 804, 807, 812
radioactive nuclide(s): artificially induced, 812–814, **813,** t 815; naturally occurring, 806–812, **807,** t 807, **809–811,** t 809; properties of, t 805, 806–808, 821; radiation given off by, 809, **809,** t 809, 821; series of, 809–812, **810;** stable nuclei from radioactive decay, 812
radioactive tracer(s), 817
radioactivity, 804, 821
radioisotope(s), 816–817
radium, 136, 804–806
radon, 130, 811, **811**
Ramsay, William, 130
Rankine Scale, 42
rare-earth element(s), 130
rate law(s), 552–557, **553**
rate-determining step, 551
reactant(s), in chemical equations, 15, 233
reaction(s). *See* chemical reaction(s)
reaction mechanism(s), 541–543, **542, 543,** 557
reaction rate: desktop investigation, 555; factors influencing, 549–552, **550, 552, 553,** 555, 557; rate laws, 552–557, **553**
reaction stoichiometry: mass-mass calculations, 262, 270–271; mass-mole calculations, 262, 269–270, **269;** mole-mass calculations, 262, 267–269; mole-mole calculations, 262, 266–267, **266;** problem-solving techniques, 261–265, **261–263**
real gas(es), 319, **319**
rechargeable cell(s), 615–616, **615, 616**
recipe(s), 273, **431**
redox equation(s): ion-electron method of balancing, 605–607; oxidation-number method of balancing, 602–605, **603**
redox reaction(s), 600–601
reducing agent(s), 608–610, t 608, t 609, 622
reduction, 597–598, t 608, 622
relative atomic mass, 85, 92
replacement reaction(s), 247–248, **248,** 250, **250**
resonance, 174
resonance hybrid(s), 174
respiration, in carbon dioxide preparation, 306
reversible reaction(s), 237, 563–564, **563**
rhombic sulfur, 764, **764**
ribonucleic acid (RNA), 695, 698, 703
ribosome(s), 698
RNA. *See* ribonucleic acid (RNA)
roasting, 770–771
rock salt, 722
rocket fuel, 254
roentgen, 818
rotation, 316
rounding off, in calculations, 49–50, t 50
rubber, 658–659, **658, 659**
ruby, 757, **757**
Rutherford, Ernest, 77–78, 92, 97, 812–813

S

salt. *See* sodium chloride
salt(s): description of, 211, 472, **472;** hydrolysis of, 580–583, **581–583,** 590

saltwater distillation, 428
saponification, 681
sapphire, 757, **757**
saturated fat(s), 692
saturated hydrocarbon(s), 644
saturated solution(s), 412
SBR, 659 scandium, 117, t 118
Scheele, Carl Wilhelm, 290, 783
Schrödinger, Erwin, 105
Schrödinger wave equation, 105
scientific journals, 264
scientific measurement. *See* measurement
scientific method, 7–10, **7–9,** 27
scientific notation: form of, 52–53; mathematics involving, 54; purpose of, 52, 59
sea water, 136, 428
second(s), t 32, 34
selenium, 752, t 752
self-ionization, of water, 496–497, **496,** 516
semiconductor(s), 25
SI measurement system: derived SI units, 36–39, t 36; description of, 31–32, 59; fundamental SI units, 32–34, t 32, **33,** t 33, **34**
significant figure(s), 47–52, **47–49,** t 48, t 50, 59
silicon, 25, t 117, 758–762, **759, 760**
silicon carbide, 380
silicone(s), 759
silver, t 119, 138–139, 744, t 744
silver bromide, 788
silver chloride, **17,** 444, **445,** 449, 575, 584–586, 599
silver halide(s), 789
silver iodide, 790
silver nitrate, **17,** 208, 252–253, 444, **445,** 575, 599
silver nitrite, 208
single bond(s), 169, t 172
single-replacement reaction(s), 247–248, **248,** 255
slag, 739
smectic crystal(s), 376
smog. *See* photochemical smog
soap(s), 680–681
soda ash, 725
sodium: compounds of, 722–726, **722,** t 723, **725;** electron configuration of, 115–116, t 117; formation of metallic bonds, 180; preparation of, 720–721, **720;** properties and uses of, 136, 721–722, **722**
sodium acetate, 413, 576
sodium carbonate, 485, 582, 724–726, **725**
sodium chloride: common-ion effect and, 575–576; description of, 722–723, **722;** dissolving of, 440–441, **440;** as electrolyte, 406; equilibrium and, 564; hydration of, 442; hydrochloric acid and, 211; as ionic compound, 175–177, **177, 205;** oxidation and, 597; oxidation-reduction reaction and, 599; solubility of, **409,** 413, 415; solutions of, **445**
sodium fluoride, 212, 245, 248, 782
sodium hydride, 296
sodium hydroxide, **205,** 247, 248, 420, 421, 471, 473, 485, 723–724
sodium hypochlorite, 254
sodium iodide, 248

sodium nitrate, 298
sodium pentothal, 715
sodium peroxide, 290, 293
sodium thiosulfate, 413
solid(s): amorphous, 377, 381–382, 394; changes of state, t 383; crystalline, 377; density of, 379; diffusion rate of, 379; incompressibility of, 379; kinetic theory description of, 377; melting point of, 378–379; nonfluidity of, 378; particle model of, 377–379, **377;** properties of, 377–379, 394; shape of, 377; solutions of, 407, t 407; volume of, 378
solid state, 13–14
solubility: calculations of, 587–589; description of, 411–413, 433; factors affecting, 413–417, **414–417,** t 417, 433; equilibrium of, 443–446, **444, 445,** 584–590, **585;** solubility-product constant, 585–587, t 588, 590
solubility equilibria, 443–446, **444, 445**
solubility-product constant, 585–587, t 588, 590
soluble, 405
solute(s), 406, 413–414, 431–433
solution(s): aqueous, 496–497; boiling-point elevation, 431, **431;** buffered solutions, 579; colligative properties of, 426–433, **426, 427, 431;** concentration of, 420–425; description of, 20, **20, 21,** 405–406; desktop investigation, 408; electrolyte, 447–448, 450–456, t 455, t 456; equivalents, 491–493, **491,** 516; factors affecting rate of dissolving, 410–411; freezing-point depression, 428–430; formation of, **418;** heats of, 417–419, t 419, 441–442, **442;** molality of, 423–425; molar mass of solute, 431–433; molarity of, 421–423, **422,** 495–496; molecular electrolyte(s), 447–448, **448;** nature of, 406, **406;** neutral, 497; normality, 494–496, **494,** t 494, 516; percent by mass, 420–421; and pH, 496–504; properties of, t 410; saturated, 412; solubility and, 411–417, **411–417,** t 417, 433; solution equilibrium, 412; standard solutions, 511, 516; supersaturated, 412–413; types of, 407, t 407; unsaturated, 412; vapor-pressure lowering, 426–427, **427**
solution equilibrium, 412
solvated, 418
Solvay, Ernest, 724
Solvay process, 724–726, **725**
solvent(s), 406, 413–414
specific heat, 44–45, t 44
spectator ion(s), 445
spectroscope, 103–104
spectroscopy, 103–104, **105**
spectrum, 104. *See also* electromagnetic spectrum; line emisison spectrum
speed: of electromagnetic radiation, 97; formula for, 59, **59**
spin quantum number(s), 111
standard(s), of measurement, 31
standard electrode potential, 618–620, t 620, 622
standard molar volume of a gas, 345–347, **346,** 365

standard solution, 511, 516
standard temperature and pressure (STP), 327, 337
starch, 689–690
steel production, 291, **292,** 739–741, **740**
steroid(s), 693
Stock system of nomenclature, 206, 214–216
stoichiometry: applications of, 261; of gases, 356–359; ideal conditions and, 265–266; limiting reactants and, 272–276; mass-mass calculations, 262, 270–271; mass-mole calculations, 262, 269–270, **269**; mole-mass calculations, 262, 267–269; mole-mole calculations, 262, 266–267, **266**; percent yield, 276–278, **276**; reaction-stoichiometry problems, 261–265, **261–263**
STP. *See* standard temperature and pressure (STP)
strong acid(s), 463, t 466
strong electrolyte(s), 449
strontium, 136–137, 719
structural formula(s), 169, 640–642, **641, 642**
structural isomer(s), 641
Strutt, John William, 130
subatomic particle(s). *See* electron(s); neutron(s); proton(s)
subbituminous coal, 637
sublimation, 388–389
substituted hydrocarbon(s): alcohols, 665–669, 681; aldehydes, 674–675, 681; carboxylic acids, 676–678, 681; description of, 639, 681; esters, 679, 681; ethers, 673, 681; halocarbons, 670–672, 681; ketones, 674–676, 681
substrate(s), 688
sugar(s): disaccharides and polysaccharides, 689–690, **689**; heating of, 22, **22**; simple sugars, 689; solution with water, 20, **21,** 405, 406, 410–412, **411,** 414, 564
sulfur: allotropes of, 765–767, **764, 765,** 775; chemical properties of, 767–768, t 767, **768**; compounds of, 770–775, t 770, **771, 773, 774**; electron configuration of, t 117; mixed with zinc, **17**; physical properties of, **764,** 765–767, **765,** 775; production of, 765, **765,** 775; sythesis reactions with, 245; uses of, 140, 769
sulfur dioxide, 245, 291, 770–771, **771,** 775
sulfur trioxide, 174, 247
sulfuric acid, 246, 247, 297, 464–468, t 465, **468,** 483–484, 773–775
superconductor(s), 382
supercooled liquid(s), 378–379
superoxide(s), 291
supersaturated solution(s), 412–413
surface tension, 374–375, **375**
suspension(s), 405, 408–409, t 410
swimming pool chemistry, 570
synapse, **694,** 695
synthesis reaction(s), 244–246 **245,** 255
system(s), 383

T

taconite, 738
Teflon, 672
tellurium, 752–753, t 752

temperature: absolute zero, 330; Charles' law, 329–332, **329, 330,** t 330, t 331, 337; combined gas law, 333–334; critical pressure, 389; description of, 40, **41,** 59; equilibrium and changes in, 572–573; Gay-Lussac's law, 332–333, **332,** 337; ideal gas law and, 352–354; low temperatures, 322; measurement of, 41–43, **47**; and molecular motion, 316; phase diagrams and, 389–390, **389**; solubility and, 416–417, t 417. *See also* boiling; freezing; melting
tetrachloromethane, 671
tetrafluoroethene, 671–672
theoretical yield, 276
theory, 9
thermite, 755
thermite reaction, 755–756, **756,** 762
thermochemical equation(s), 527
thermochemistry: description of, 527; heat of combustion, 530–531, **530,** t 530, 557; heat of formation, 528–529, **529,** 556–557; heat of reaction, 527–528, **527, 528,** 531–536, **532,** 556, 557
thermometer(s), 41, **47,** 322. *See also* temperature
Thomson, Sir John, 76, 77
tin, t 119
titanium, 117, t 118
titration, 510–516, **511**
torr, 326, t 326
Torricelli, Evangelista, 324–325, 326
transition element(s), 138, 729–731
transition interval, 505–506, t 505
transition metal(s): formation of colored compounds, 733, **733,** 748; formation of complex ions, 733–736, **734,** t 735, t 736, 748; oxidation states of, 731–732, t 732, 748; properties of, 729–731, **729,** 748
translation, 316
transmutation, 810
transmutation reaction, 810
transuranium element(s), 814, t 815
triple bond(s), 171, 190
triple point(s), 389
triprotic acid(s), 466 tritium, 80, 83, t 84, 821, **821**
tungsten, **359**
Tyndall effect, 409

U

uncertainty principle, 106–107
unit(s) of measurement: calculations involving, 34–36; derived SI units, 36–39, t 36; SI measurement system, 31–34, t 32, t 33, 59; symbols for, 31–32
unit cell(s), 379
unsaturated, 650
unsaturated fat(s), 692
unsaturated solution(s), 412
unshared pair, 169
uranium, 245, 817, 819, **819**
uranium hexafluoride, 363
uranium-235, 363

V

valence electron(s), 152
van der Waals, Johannes, 187, 319

van der Waals force(s), 187
vanadium, 117, t 118
vaporization, 375, 388, 394
variable(s), 58
vibration, 316
vitamin(s), 688, t 692
voltaic cell(s), 616–617
volume: derived SI units and, 36–37, t 36, **37**; ideal gas law and, 352–354; volume-mass and mass-volume calculations for gases, 357–359; volume-volume calculations of gases, 356–357
VSEPR theory, 182–185, **182,** t 183, **184,** 190
vulcanization, 658

W

Waage, Peter, 556
Walton, E. T. S., 813–814
water: boiling point of, 387, **388**; distillation of, 428; electrolysis of, 21, **21,** 613–614; equilibrium vapor pressure of, 385–386, **386**; formation of, 250; formula for, 202, **202**; fresh water from the sea, 428; in hydrogen preparation, 298; ionization constant of, 579–580; ionization of, 452; mass of, 216–217; oxidation number of, 212; in oxygen preparation, 293, **293**; phase diagrams of, 389–390, **389**; physical properties of, **52,** 392–393, **393**; reaction with chlorine, 785; self-ionization of, 496–497, **496,** 516; structure of, 184–185, **184, 186,** 390–391, **391,** 394; uses of, 685, **685**; water-gas reaction, 535–536, **535**
wave properties: of electromagnetic radiation, 97; of electrons, 105–106, **106**; of light, 97–98, 102
wavelength, 97, 98
wax, 692, **693**
weak acid(s), 466–467, t 466
weak base(s), 473–474
weak electrolyte(s), 449
weight, 11, 27
Williamson synthesis, 673
word equation(s), 234 xenon, t 119, 130, 782

Z

zero(s): absolute zero, 330; significant figures involving, t 48
zinc: activity series and, 252; electrode potentials of, 616–620; electron configuration of, 118–119, t 118; as oxidizing agent, 608–609; preparation of hydrogen 297; reaction with copper, 252, **253**; reaction with hydrochloric acid, 549–550
zinc chloride, 611
zinc sulfate, 252

Answers to Cumulative Review Questions

1. (a) Science is a very systematic and organized collection of facts. (b) Chemistry is the study of the composition of materials and changes in the composition of materials. (c) The scientific method is a logical approach to the solution of problems that lend themselves to investigation. (d) Inertia is resistance to change in motion. (e) Matter is anything that has mass and occupies space. (f) Energy is the ability to cause change. (g) A characteristic property is a property that does not depend on the size, shape, or mass of a material. (h) A precipitate is a solid that separates from a solution.

2. The five major areas of chemistry are organic chemistry, inorganic chemistry, physical chemistry, biochemistry, and analytical chemistry.

3. The four parts of the scientific method are observing, generalizing, theorizing, and testing. Observing involves such activities as collecting data, measuring, experimenting, and communicating findings to others. Generalizing involves the organization and analysis of accumulated data in an effort to find the relationships among those data. Theorizing involves the construction of models that show relationships among data or events, often leading to the formulation of theories. Testing involves predicting the outcome of experiments based upon hypotheses, laws, or theories, often leading to modifications or abandonment based on newly acquired evidence.

4. (a) Numerical information is quantitative, whereas nonnumerical information in qualitative. (b) Weight is a measure of the earth's attraction for matter, whereas mass is a measure of the quantity of the matter. (c) Kinetic energy is the energy of an object in motion, whereas potential energy is the energy an object has by virtue of its position. (d) Physical properties can be observed or measured without altering the identity of a material, whereas chemical properties can only be observed in the processes that alter the identity of a material. (e) A physical change is any change in a property of matter that does not result in a change in identity, whereas a chemical change is any change in which one or more substances are converted into different substances with different characteristic properties. (f) The substances that undergo a chemical reaction are the reactants, whereas the products are the new substances produced by a chemical reaction. (g) An exothermic process is one in which heat is released, whereas an endothermic process is one in which heat is absorbed. (h) Heterogeneous matter is not uniform in composition and properties, where homogeneous matter is uniform in composition and properties throughout. (i) A pure substance is a homogeneous material that has the same composition and characteristic properties, whatever its source, whereas a mixture is a combination of two or more kinds of matter, each of which retains its own characteristic properties. (j) An element is a substance that cannot be separated into other substances by any ordinary chemical change. A chemical compound is a pure substance that can be decomposed chemically into two or more simpler substances. (k) Groups are the vertical columns of the Periodic Table, whereas periods are the horizontal rows.

5. (a) The law of conservation of matter states that matter cannot be either created or destroyed in ordinary chemical of physical changes. (b) The law of conservation of energy states that energy can be converted from one form to another, but it cannot be either created or destroyed in ordinary chemical or physical changes. (c) The law of definite composition states that every chemical compound has a definite composition by mass.

6. The three states of matter are the solid state (the state of matter characterized by a definite shape and volume), the liquid state (the state of matter characterized by a definite volume, but not a definite shape), and the gaseous state (the state of matter characterized by having neither a definite shape nor a definite volume).

7. Options a, b, c, and e illustrate physical changes, whereas d and f illustrate chemical changes.

8. Observable changes that generally indicate that a chemical reaction has occurred include the evolution of heat and light, the production of a gas, or the formation of a precipitate.

9. The three general classes of elements are the metals, the nonmetals, and the metalloids. The metals are elements that are good conductors of heat and electricity. The nonmetals are the elements that are poor conductors of heat and electricity. The metalloids are those elements that have some properties characteristic of metals and others characteristic of nonmetals, and are semiconductors.

10. The fundamental SI units for length, mass, and time are meter, kilogram, and second, in that order.

11. (a) Temperature is a measure of the average kinetic energy of the particles in a sample of matter. Heat is energy transferred between two systems at different temperatures. (b) Accuracy is the closeness of a measurement to a true or accepted value; precision is agreement among the numerical values of a set of measurements made in the same way. (c) Two variables are directly proportional if dividing one by the other gives a constant value. Two variables are inversely proportional if their product has a constant value.

12. The four steps in the stepwise method of problem are (1) analyze the problem (2) select a route to solve the problem (3) compute the answer (4) evaluate the answer.

13. The most appropriate SI unit(s) would be (a) Millimeters (b) Grams or milligrams (c) Milliliters (d) Kilograms (e) Kilometers (f) Kiloliters or cubic meters.

14. The indicated answers are: (a) 125.096 cm (b) 125.10 cm (c) 125 cm (d) 130 cm (e) 125.1 cm (f) 100 cm.

15. The correct scientific notation is: (a) 4.5×10^3 (b) 3.7×10^5 (c) 5.3×10^{-2} (d) 8.1×10^{-5} (e) 6.5×10^6 (f) 2.06×10^{-5}.

16. The usual, long form for each is: (a) 72 000 (b) 0.000 013 (c) 30 000 000 000.

Solutions to Cumulative Review Problems

1. (a) $3.5 \text{ m} \times \dfrac{100 \text{ cm}}{\text{m}} = 350 \text{ cm}$

(b) $425 \text{ g} \times \dfrac{\text{kg}}{1000 \text{ g}} = 0.425 \text{ kg}$

(c) $7.8 \text{ L} \times \dfrac{1000 \text{ ml}}{\text{L}} = 7800 \text{ ml}$

(d) $1.5 \text{ km} \times \dfrac{1000 \text{ m}}{\text{km}} \times \dfrac{1000 \text{ mm}}{\text{m}}$
$= 1.5 \times 10^6 \text{ mm}$

(e) $12.6 \text{ L} \times \dfrac{1000 \text{ ml}}{\text{L}} \times \dfrac{\text{cm}^3}{\text{ml}}$
$= 1.26 \times 10^4 \text{ cm}^3$

2. Density $= \dfrac{\text{mass}}{\text{volume}}$

$D = \dfrac{m}{v}$

$D = \dfrac{15.0 \text{ g}}{4.5 \text{ cm}^3}$

$D = 3.3 \text{g/cm}^3$

3. $D = \dfrac{m}{v}$

$V = \dfrac{m}{D}$

$V = \dfrac{19.40 \text{ g}}{8.62 \text{ g/cm}^3}$

$V = 2.25 \text{ cm}^3$

4. (a) $T(\text{K}) = t(^\circ\text{C}) = 273$
$T(\text{K}) = 20^\circ\text{C} + 273$
$T(\text{K}) = 293^\circ\text{C}$

(b) $T(\text{K}) = -65^\circ\text{C} + 273$
$T(\text{K}) = 208^\circ\text{C}$

(c) $t(^\circ\text{C}) = T(\text{K}) - 273$
$t(^\circ\text{C}) = 350\text{K} - 273$
$t(^\circ\text{C}) = 77\text{K}$

(d) $t(^\circ\text{C}) = 100\text{K} - 273$
$t(^\circ\text{C}) = -173\text{K}$

5. $350.0 \text{ cal} \times \dfrac{4.184 \text{ J}}{\text{cal}} = 1464 \text{ J}$

6. Ht. $=$ sp. ht. $\times$ m $\times \Delta t$

Ht. $= \dfrac{0.24 \text{ J}}{9^\circ\text{C}} \times 20.0 \text{ g} \times 15^\circ\text{C}$

Ht. $= 72 \text{ J}$

7. (a) $4.2 \times 10^3 + 2.9 \times 10^4 = 0.42$
$\times 10^4 + 2.9 \times 10^4 = 3.32$
$\times 10^4 = 3.3 \times 10^4$

(b) $(3.12 \times 10^5)(8.1 \times 10^{-9})$
$= 25.272 \times 10^{-4} = 2.5 \times 10^{-3}$

(c) $\dfrac{(5.2 \times 10^{-7})(7.0 \times 10^4)}{1.2 \times 10^{-5}}$
$= 30.333 \times 10^2 = 3.0 \times 10^3$

8. (a) $2.40 \text{ m} \times \dfrac{100 \text{ cm}}{\text{m}} = 240. \text{ cm}$

$14.0 \text{ dm} \times \dfrac{10 \text{ cm}}{\text{dm}} = 140. \text{ cm}$

$V = l \times w \times h$
$V = 240. \text{ cm} \times 140. \text{ cm} \times 80.0 \text{ cm}$
$V = 2.69 \times 10^6 \text{ cm}^3$

(b) $2.69 \times 106 \text{ cm}^3 \times \dfrac{\text{ml}}{\text{cm}^3} \times \dfrac{\text{L}}{10^3 \text{ ml}}$

$= 2.69 \times 10^3 \text{ L}$

Answers to Cumulative Review Questions

1. *(a)* Democritus first used the term atom to describe the smallest particles of matter that can exist. *(b)* Dalton proposed an atomic theory that related atoms to the measurable property of mass. *(c)* Thomson's work provided strong support for the hypothesis that cathode rays are negatively charged particles. *(d)* Millikan's work confirmed that the electron has the smallest possible negative electric charge. *(e)* Rutherford discovered the nucleus and described its contents.

2. *(a)* An atom is the smallest particle of an element that can exist either alone or in combination with other atoms. *(b)* The nucleus of an atom is its positively charged dense, central portion, which contains nearly all of its mass but takes up only an insignificant fraction of its volume. *(c)* Isotopes are atoms of the same element that have different masses. *(d)* The atomic number of an element is the number of protons in the nuclei of atoms of that element. *(e)* The mass number of an isotope is the total number of protons and neutrons in its nuclei. *(f)* A mole is the amount of a substance that contains a number of particles equal to the number of atoms in exactly 12 grams of carbon-12. *(g)* Avogadro's number is the number of particles in exactly one mole of a pure substance. *(h)* The molar mass of an element is the mass in grams of that element that is numerically equal to its mass on the relative atomic mass scale.

3. *(a)* Chlorine-37 contains 17 protons, 17 electrons, and 30 neutrons *(b)* $^{56}_{26}$Fe contains 26 protons, 26 electrons, and 30 neutrons *(c)* Deuterium ($^2_1$H) contains 1 proton, 1 electron, and 1 neutron.

4. *(a)* Electromagnetic radiation is a form of energy that exhibits wavelike behavior as it travels through space. *(b)* The equation is $c = \lambda\nu$.

5. *(a)* The ground state is the state of lowest energy of an atom, whereas an excited state is one in which an atom has higher energy than in its ground state. *(b)* In a line emission spectrum only photons of certain energies are emitted so that only individual lines are observed. Whereas in a continuous spectrum, photons encompassing a broad range of energies are emitted so that countless lines corresponding to those energies are produced. *(c)* Bohr's model described definite orbits occupied by electron particles, whereas Schrödinger treated electrons as waves having a certain probability of being found at definite distances from the nucleus in what he called orbitals.

6. *(a)* The quantum theory describes mathematically the wave properties of electrons and other very small particles. *(b)* The four sets of quantum numbers are the principal quantum number, the orbital quantum number, the magnetic quantum number, and the spin quantum number. The principal quantum number indicates the main energy levels surrounding a nucleus. The orbital quantum number indicates the shape of the orbital. The magnetic quantum number indicates the orientation of an orbital about the nucleus. The spin quantum number indicates the two possible orientations of an electron in an orbital.

7. *(a)* The three methods are orbital notation, electron-configuration notation, and electron dot notation. *(b)* Orbital notation:

$$\underset{1s}{\uparrow\downarrow} \quad \underset{2s}{\uparrow\downarrow} \quad \underset{2p}{\uparrow\downarrow \; \uparrow\downarrow \; \uparrow\downarrow} \quad \underset{3s}{\uparrow\downarrow} \quad \underset{3p}{\uparrow \; \uparrow \; \uparrow}$$

Electron-configuration notation:
$1s^2 2s^2 2p^6 3s^2 3p^3$
Electron-dot notation: P

8. The chemists are *(a)* Mendeleev *(b)* Moseley.

9. According to the periodic law, when the elements are arranged in order of increasing atomic number, elements with similar properties recur at regular intervals.

10. *(a)* For main-group elements, the outer electron configurations of all members of a particular group are identical, the result being that such elements have similar properties. *(b)* All elements within a specific period have the same number of main energy levels. *(c)* For elements in the same block, the last electrons added are in the same sublevel of their respective main energy levels.

11. The group numbers are: *(a)* 1 *(b)* 2 *(c)* 3–12 *(d)* 17 *(e)* 18.

12. The expected outer configurations are: *(a)* $2s^2 2p^2$ *(b)* $3s^2$ *(c)* cd$^{10}4s^2$ *(d)* $4d^{10}5s^2 5p^5$.

13. *(a)* Ionization energy is the energy required to remove one electron from an atom of an element. *(b)* Electron affinity is the energy absorbed or released when an electron is added to an atom. *(c)* Electronegativity is the power of an atom in a chemical compound to attract electrons. *(d)* A cation is a positive ion. *(e)* An anion is a negative ion. *(f)* Valence electrons are the electrons available to be lost, gained, or shared in the formation of chemical compounds.

14. *(a)* A chemical bond is a link between atoms, resulting from the mutual attraction of their nuclei for electrons. *(b)* The three major types of bonds are ionic, covalent, and metallic. ionic bonding, but these electrons are free to move throughout the material rather than being held in place in negative ions.

15. *(a)* In polar–covalent bonds there is an unequal attraction for the shared electrons, whereas in nonpolar–covalent bonds, the bonding electrons are shared equally by the bonded atoms, with a resulting balanced distribution of electrical charge. *(b)* Ionic compounds consist of positive and negative ions bound together by ionic bonding and held in place in a crystalline solid. Molecular compounds are made up of molecules that move around independently and are capable of independent existence. Ionic compounds generally have higher melting and boiling points than molecular compounds, but they do not vaporize as readily at room temperature. *(c)* Metals tend to lose electrons to form positive ions, whereas nonmetals tend to gain electrons to form negative ions.

16. Lewis structures are formulas in which atomic symbols represent nuclei and inner-shell electrons, dots or dashes represents electron pairs in covalent bonds, and dots represent unshared electrons.

17. Metallic bonding is generally viewed as the result of the mutual sharing of many electrons by many atoms in which each contributes its valence electrons to a region surrounding the atoms, leaving behind positive ions. These electrons are then free to move about in what is often termed an electron sea, where they are shared by all of the atoms.

18. According to the VSEPR theory (valence-shell electron-pair repulsion), electrons in molecules repel each other. The theory states that electrostatic repulsion between the electron pairs surrounding an atom causes these pairs to be oriented as far apart as possible. The other theory, that of hybridization, refers to the mixing of two or more atomic orbitals of similar energies on the same atom to give new orbitals of equal energies. When orbitals hybridize, they combine and become rearranged to form new, identical orbitals.

Solutions to Cumulative Review Problems

1. (a) $2.75 \text{ mol Ca} \times \dfrac{40.1 \text{ g Ca}}{\text{mol Ca}} = 110. \text{ g Ca}$

   (b) $4.50 \times 10^{25} \text{ atoms Ag} \dfrac{\text{mol Ag}}{6.02 \times 10^{23} \text{ atoms Ag}}$
   $= 74.8 \text{ mol Ag}$

2. (a) $3.20 \text{ g Al} \times \dfrac{\text{mol Al}}{27.0 \text{ g Al}} = 0.119 \text{ mol Al}$

   (b) $1.80 \times 10^{26} \text{ atoms Zn} \dfrac{\text{mol Zn}}{6.02 \times 10^{23} \text{ atoms Zn}}$
   $= 299 \text{ mol Zn}$

3. (a) $7.50 \text{ mol Pb} \dfrac{6.02 \times 10^{23} \text{ atoms Pb}}{\text{mol Pb}}$
   $= 4.52 \times 10^{24} \text{ atoms Pb}$

   (b) $0.500 \text{ g Au} \times \dfrac{6.02 \times 10^{23} \text{ atoms Au}}{197 \text{ g Au}}$
   $= 1.53 \times 10^{21} \text{ atoms Au}$

4. $\dfrac{3.40 \text{ g Na}^+}{\text{L serum}} \times 2.50 \text{ L serum} \times \dfrac{\text{mol Na}^+}{23.0 \text{ g Na}^+}$
   $= 0.148 \text{ mol Na}^+$

5. Electronegativity difference $= 3.5 - 0.9 = 2.6$
   This value is indicative of an ionic bond.

6. (a) Br $:\overset{\displaystyle Br}{\underset{\displaystyle Br}{\ddot{C}}}:$ Br (AB_4—tetrahedral)

   (b) $:\ddot{C}l: \ \ddot{S} \ :\ddot{C}l:$ (AB_2E_2—bent)

   (c) $:\ddot{I}: \ \underset{\displaystyle :\ddot{I}:}{B} \ :\ddot{I}:$ (AB_3—triangular planar)

   (d) $:\ddot{O}: \ \overset{\displaystyle :\ddot{O}:}{\underset{\displaystyle :\ddot{O}:}{S}} \ :\ddot{O}:$ (AB_4—tetrahedral)

Answers to Cumulative Review Questions

1. The indicated formulas are: (a) Na^+ (b) F^- (c) Mg^{2+} (d) Al^{3+} (e) S^{2-} (f) N^{3-} (g) Cl^- (h) K^+.

2. The indicated compounds are: (a) Copper(I) chloride (b) Sulfur dioxide sulfur(IV) oxide (c) Silver nitrate (d) Ammonium sulfide (e) Silicon(IV) oxide (f) Iron(III) sulfate (g) Phosphorus(III) chloride (h) Tin(II) phosphate (i) Nitrogen(III) oxide (j) Sulfuric acid (k) Arsenic(V) oxide.

3. The indicated oxidation numbers are: (a) $+1$, -1 (b) $+3$, -1 (c) $+4$, -2 (d) $+5$, -2 (e) $+1$, $+6$, -2 (f) $+1$, $+5$, -2 (g) $+3$, -2 (h) -2, $+1$ (i) $+4$, -2 (j) $+3$, -2 (k) $+1$, $+6$, -2 (l) $+6$, -2.

4. (a) The formula mass of a compound or polyatomic ion is the sum of the atomic masses of all atoms represented in the formula. (b) The molar mass of a compound is the mass in grams numerically equal to the formula mass of the compound. (c) The percent composition of a compound is the percent by mass of each element in a compound. (d) The simplest formula of a compound consists of the symbols for the elements combined, with subscripts showing the smallest whole-number ratio of the atoms.

5. A chemical equation uses chemical symbols and formulas to indicate the change that occurs in a chemical reaction.

6. The five major types of reactions are synthesis, decomposition, single-replacement, double-replacement, and combustion. In synthesis reactions, two or more substances combine to form a new substance. In decomposition, a single compound undergoes a reaction that produces two or more simpler substances. In single-replacement reactions, one element replaces a similar element in a compound. In double-replacement reactions, the ions of two compounds exchange places in an aqueous solution to form two new compounds. In combustion, a substance combines with oxygen releasing a large amount of energy in the form of light and heat.

7. (a) An activity series is a list of elements according to the ease with which they undergo certain chemical reactions. (b) The elements are usually listed in an order determined by single-replacement reactions such that any element in the series can replace an element below it, but not those above it. (c) An activity series can be used to predict whether or not certain chemical reactions will occur, as well as the products of those reactions that do occur.

8. (a) Molar mass is the mass, in grams, of one mole of a substance. (b) Percent yield is the ratio of the actual yield to the theoretical yield, multiplied by 100 percent. An ionic bond is the chemical bond resulting from the electrostatic attraction between positive and negative ions. A covalent bond is a chemical bond resulting from the sharing of electrons between the two bonded atoms. In metallic bonding the metal atoms give up electrons as in

Solutions to Cumulative Review Problems

1. $3 \text{ mol Ca} \times \dfrac{40.08 \text{ g Ca}}{\text{mol Ca}} = 120.2 \text{ g Ca}$

   $2 \text{ mol P} \times \dfrac{30.973\,76 \text{ g P}}{\text{mol P}} = 61.947\,52 \text{ g P}$

   $8 \text{ mol O} \times \dfrac{15.9994 \text{ g O}}{\text{mol O}} = 127.9952 \text{ g O}$

   Molar mass of $Ca_3(PO_4)_2 = 310.1 \text{ g}$

   $6.75 \text{ mol } Ca_3(PO_4)_2 \times \dfrac{310.1 \text{ g } Ca_3(PO_4)_2}{\text{mol } Ca_3(PO_4)_2}$
   $= 20909 \text{ } Ca_3(PO_4)2$

2. (a) $1 \text{ mol Pb} \times \dfrac{207.2 \text{ g Pb}}{\text{mol Pb}} = 207.2 \text{ g Pb}$

   $2 \text{ mol N} \times \dfrac{14.0067 \text{ g N}}{\text{mol N}} = 28.0134 \text{ g N}$

   $6 \text{ mol O} \times \dfrac{15.9994 \text{ g O}}{\text{mol O}} = 95.9964 \text{ g O}$

   Molar mass of $Pb(NO_3)_2 = 331.2 \text{ g}$

   $12.00 \text{ g } Pb(NO_3)_2 \times \dfrac{\text{mol } Pb(NO_3)_2}{331.2 \text{ g } Pb(NO_3)_2}$
   $= 0.036\,23 \text{ mol } Pb(NO_3)_2$

   (b) $0.036\,23 \text{ mol } Pb(NO_3)_2 \times \dfrac{6.022 \times 10^{23} \text{ units } Pb(NO_3)_2}{\text{mol } Pb(NO_3)_2}$
   $= 2.182 \times 10^{22} \text{ units } Pb(NO_3)_2$

3. $2 \text{ mol Fe} \times \dfrac{55.847 \text{ g Fe}}{\text{mol Fe}} = 111.69 \text{ g Fe}$

   $3 \text{ mol C} \times \dfrac{12.0111 \text{ g C}}{\text{mol C}} = 36.0333 \text{ g C}$

   $9 \text{ mol O} \times \dfrac{15.9994 \text{ g O}}{\text{mol O}} = 143.995 \text{ g O}$

   Molar mass $Fe_2(CO_3)_3 = 291.72 \text{ g}$

   $\% \text{ Fe} = \dfrac{111.69 \text{ g Fe}}{291.72 \text{ g } Fe_2(CO_3)_3} \times 100\% = 38.287\% \text{ Fe}$

   $\% \text{ C} = \dfrac{36.0333 \text{ g C}}{291.72 \text{ g } Fe_2(CO_3)_3} \times 100\% = 12.352\% \text{ C}$

   $\% \text{ O} = \dfrac{143.995 \text{ g O}}{291.72 \text{ g } Fe_2(CO_3)_3} \times 100\% = 49.361\% \text{ O}$

4. $51.82 \text{ g Cu} \times \dfrac{\text{mol Cu}}{63.546 \text{ g Cu}} = 0.8155 \text{ mol Cu}$

   $19.60 \text{ g C} \times \dfrac{\text{mol C}}{12.0111 \text{ g C}} = 1.632 \text{ mol C}$

   $2.4 \text{ g H} \times \dfrac{\text{mol H}}{1.007\,94 \text{ g H}} = 2.44 \text{ mol H}$

   $26.12 \text{ g O} \times \dfrac{\text{mol O}}{15.9994 \text{ g O}} = 1.633 \text{ mol O}$

   $\dfrac{0.8155 \text{ mol Cu}}{0.8155} : \dfrac{1.632 \text{ mol C}}{0.8155} : \dfrac{2.44 \text{ mol H}}{0.8155} : \dfrac{1.633 \text{ mol O}}{0.8155}$

   $1.000 \text{ mol Cu} : 2.001 \text{ mol C} : 2.99 \text{ mol H} : 2.002 \text{ mol O}$

   The simplest formula is $CuC_2H_3O_2$.

5. $40.00 \text{ } u \text{ C} \times \dfrac{\text{mol C}}{12.0111 \text{ } u \text{ C}} = 3.330 \text{ mol C}$

   $6.71 \text{ } u \text{ H} \times \dfrac{\text{mol H}}{1.007\,94 \text{ } u \text{ H}} = 6.66 \text{ mol H}$

$$53.29 \, u \, O \times \frac{\text{mol O}}{15.9994 \, u \, O} = 3.331 \text{ mol O}$$

$$\frac{3.330 \text{ mol C}}{3.330} : \frac{6.66 \text{ mol H}}{3.330} : \frac{3.331 \text{ mol O}}{3.330}$$

$1.000 \text{ mol C} : 2.00 \text{ mol H} : 1.000 \text{ mol O}$

Simplest formula = CH_2O

Simplest formula mass = $30.0264 \, u$

$$n = \frac{180.1854 \, u}{30.0264 \, u} = 6.000 \, 00$$

Molecular formula = $(CH_2O)_6 = C_6H_{12}O_6$

6. (a) $2K + 2HCl \rightarrow 2KCl + H_2$; single replacement
(b) $H_2 + F_2 \rightarrow 2HF$; synthesis (c) $BaCO_3 \, \Delta \rightarrow BaO + CO_2$; decomposition (d) $2Al + 3CuSO_4 \rightarrow Al_2(SO_4)_3 + 3Cu$; single replacement (e) $3Zn(OH)_2 + 2H_3PO_4 \rightarrow Zn_3(PO_4)_2 + 6H_2O$; double replacement (f) $C_3H_8 + 5O_2 \rightarrow 3CO_2 + 4H_2O$.

7. (a) $Ca + Cl_2 \rightarrow CaCl_2$ (b) $2Al(OH)_3 \, \Delta \rightarrow Al_2O_3 + 3H_2O$
(c) $3Ni + 2H_3PO_4 \rightarrow Ni_3(PO_4)_2 + 3H_2$ (d) $3Cu(C_2H_3O_2)_2 + 2Li_3PO_4 \rightarrow 6LiC_2H_3O_2{}^+ Cu_3(PO_4)_2$ (e) $2C_2H_2 + 5O_2 \rightarrow$

8. (a) Yes; $Zn(s) + Pb(NO_3)_2(aq) \rightarrow Zn(NO_3)_2(aq) + Pb(s)$
(b) No
(c) Yes; $Zn(s) + 2H_2O(g) \rightarrow Zn(OH)_2(aq) + 2H_2(g)$
(d) No

9. (a) $2HNO_3(aq) + Ca(OH)_2(aq) \rightarrow Ca(NO_3)_2(aq) + 2H_2O(\ell)$

(b) $4.50 \text{ mol HNO}_3 \times \dfrac{1 \text{ mol Ca(OH)}_2}{2 \text{ mol HNO}_3} = 2.225 \text{ mol Ca(OH)}_2$

(c) $4.50 \text{ mol HNO}_3 \times \dfrac{1 \text{ mol Ca(NO}_3)_2}{2 \text{ mol HNO}_3} = 2.25 \text{ mol Ca(NO}_3)_2$

$4.50 \text{ mol HNO}_3 \times \dfrac{2 \text{ mol H}_2O}{2 \text{ mol HNO}_3} = 4.50 \text{ mol H}_2O$

10. $3Br_2 + 2AlI_3 - 2AlBr_3 + 3I_2$

(a) $7.500 \text{ mol I}_2 \times \dfrac{3 \text{ mol Br}_2}{3 \text{ mol I}_2} \times \dfrac{159.808 \text{ g Br}_2}{\text{mol Br}_2}$

$= 1199 \text{ g Br}_2$

$7.550 \text{ mol I}_2 \times \dfrac{2 \text{ mol AlI}_3}{3 \text{ mol I}_2} \times \dfrac{407.7 \text{ g AlI}_3}{\text{mol AlI}_3} = 2038 \text{ g AlI}_3$

(b) $7.500 \text{ mol I}_2 \times \dfrac{2 \text{ mol AlBr}_3}{3 \text{ mol I}_2} \times \dfrac{266.69 \text{ g AlBr}_3}{\text{mol AlBr}_3}$

$= 1333 \text{ g AlBr}_3$

11. $Na_2CO_3 \, \Delta \rightarrow Na_2O + CO_2$

$40.0 \text{ g Na}_2CO_3 \times \dfrac{\text{mol Na}_2CO_3}{105.988 \text{ g Na}_2CO_3} \times \dfrac{1 \text{ mol Na}_2O}{\text{mol Na}_2CO_3}$

$= 0.377 \text{ mol Na}_2O$

$40.0 \text{ g Na}_2CO_3 \times \dfrac{\text{mol Na}_2 CO_3}{105.9888 \text{ g Na}_2CO_3} \times \dfrac{\text{mol CO}_2}{\text{mol Na}_2CO_3}$

$= 0.377 \text{ mol CO}_2$

12. $2C_8H_{18} + 17O_2 \rightarrow 8CO_2 + 18H_2O$

$100.0 \text{ kg C}_8H_{18} \times \dfrac{10^3 \text{ g}}{\text{kg}} \times \dfrac{\text{mol C}_8H_{18}}{114.2 \text{ g C}_8H_{18}} \times \dfrac{8 \text{ mol CO}}{2 \text{ mol C}_8H_{18}} \times$

$\dfrac{44.0099 \text{ g CO}_2}{\text{mol CO}_2} = 15.2 \text{ kg CO}_2$

$100.0 \text{ kg C}_8H_{18} \times \dfrac{10^3 \text{ g}}{\text{kg}} \times \dfrac{\text{mol C}_8H_{18}}{114.2 \text{ g C}_8H_{18}} \times \dfrac{18 \text{ mol H}_2}{2 \text{ mol C}_8H_{18}} \times$

$\dfrac{18.0153 \text{ g H}_2}{\text{mol H}_2} = 142.0 \text{ kg H}_2O$

13. (a) $2H_3PO_4 + 3BaCl_2 \rightarrow Ba_3(PO_4)_2 + 6 HCl$

$35.0 \text{ g H}_3PO_4 \times \dfrac{\text{mol H}_3PO_4}{97.9952 \text{ g H}_3PO_4} \times \dfrac{3 \text{ mol BaCl}_2}{2 \text{ mol H}_3PO_4}$

$= 0.536 \text{ mol BaCl}_2$

$65.0 \text{ g BaCl}_2 \times \dfrac{\text{mol BaCl}_2}{208.24 \text{ g BaCl}_2} = 0.312 \text{ mol BaCl}_2$

Thus, $BaCl_2$ is the limiting reactant.

(b) $0.312 \text{ mol BaCl}_2 \times \dfrac{2 \text{ mol H}_3PO_4}{3 \text{ mol BaCl}_2} = 0.208 \text{ mol H}_3PO_4$

$35.0 \text{ g H}_3PO_4 \times \dfrac{\text{mol H}_3PO_4}{97.9951 \text{ g H}_3PO_4} = 0.357 \text{ mol H}_3PO_4$

$0.357 \text{ mol H}_3PO_4 - 0.208 \text{ mol H}_3PO_4$

$= 0.149 \text{ mol excess H}_3PO_4$

(c) $0.312 \text{ mol BaCl}_2 \times \dfrac{1 \text{ mol Ba}_3(PO_4)_2}{3 \text{ mol BaCl}_2} \times$

$\dfrac{601.93 \text{ Ba}_3(PO_4)_2}{\text{mol Ba}_3(PO_4)_2} = 62.6 \text{ g Ba}_3(PO_4)_2$

$0.312 \text{ mol BaCl}_2 \times \dfrac{6 \text{ mol HCl}}{3 \text{ mol BaCl}_2} \times \dfrac{36.461 \text{ g HCl}}{\text{mol HCl}}$

$= 22.8 \text{ g HCl}$

Answers to Cumulative Review Questions

1. *(a)* Allotropes are the two or more forms of an element that have the same physical state. *(b)* Oxygen exists in two gaseous forms in the earth's atmosphere—as O_2 (called oxygen) and as O_3 (called ozone). *(c)* Oxygen gas is required for the basic life process of aerobic respiration that occurs in the cells of most living things. Ozone absorbs most of the ultraviolet portion of the light radiated from the sun to the earth in the upper atmosphere.

2. Both oxygen (O_2) and ozone (O_3) are gases at room temperature. However, while oxygen is colorless and odorless, ozone is a poisonous blue gas with a pungent odor. Oxygen has a smaller molar mass, and lower boiling and melting points than ozone. It also has a lower density and a shorter bond length than ozone. The oxygen molecule is linear, whereas the ozone molecule is triatomic and bent with a bond angle of $116.5°$.

3. The indicated responses are *(a)* oxygen *(b)* the Haber process *(c)* hydrogen *(d)* oxygen *(e)* nitrogen *(f)* ozone *(g)* carbon dioxide *(h)* carbon dioxide *(i)* carbon dioxide.

4. Nitrogen is present in the DNA molecules that control heredity and it is a major component of amino acids which are the building blocks of proteins.

5. *(a)* Carbon dioxide is a very critical ingredient used by plants in photosynthesis, whereas carbon monoxide is highly toxic to living things. *(b)* Carbon dioxide consists of one carbon and two doubly-bonded oxygen atoms to produce a linear, nonpolar molecule. Carbon monoxide contains one carbon and one oxygen atom triply-bonded to produce a linear, polar molecule. *(c)* Carbon dioxide, as solid dry ice, is used as a refrigerant; as a liquid, it is used in fire extinguishers; through the action of leavening agents it causes bread to rise; and as a gas, it is forced into soft drinks to produce carbonation. Carbon monoxide has no practical uses.

6. *(a)* Real gases deviate from ideal gas behavior because the particles of real gases do occupy space and do exert attractive forces on each other. *(b)* Those most likely to behave ideally are Br_2, He, N_2, O_2, Ne, and H_2. *(c)* Those least likely to behave ideally are HCl, H_2O, and NH_3.

7. Liquids have no definite shape but take the shape of their containers; they consist of ions, atoms, or molecules; their particles are in constant motion; their particles are closer together and lower in energy than gas particles but the attractive forces among liquid particles are more effective than those among gas particles; their entropy is generally less than that of gases. Liquids have a definite volume, exhibit fluidity, have relative high densities, are relatively incompressible, can dissolve, can diffuse, exhibit surface tension, and can evaporate, boil, and solidify.

8. Like gases and liquids, solids consist of ions, atoms, or molecules; their kinetic energy and thus their motion is lower than that of gases and liquids; the solid particles are more closely packed than those in liquids or gases; the interactive forces among their particles are stronger than those in liquids or gases: and they have the lowest entropy of the three physical states of matter. Solids have a definite shape, definite volume, nonfluidity, definite melting points, high densities, incompressibility, and a slow rate of diffusion.

9. Crystalline solids consist of crystals, substances in which the particles are arranged in an orderly, geometric, repeating pattern. Amorphous solids are solids in which the particles are arranged randomly.

10. *(a)* The equilibrium vapor pressure of a liquid is the pressure exerted by a vapor in equilibrium with its corresponding liquid at a given temperature. *(b)* The equilibrium vapor pressure of a liquid increases with increasing temperature.

Solutions to Cumulative Review Problems

1. *(a)* $1.75 \text{ atm} \times \dfrac{760 \text{ mm Hg}}{\text{atm}} = 1330 \text{ mm Hg}$

   (b) $1140 \text{ mm Hg} \times \dfrac{1 \text{ atm}}{760 \text{ mm Hg}} = 1.50 \text{ atm}$

   (c) $250 \text{ mm Hg} \times \dfrac{\text{torr}}{1 \text{ mm Hg}} = 250 \text{ torr}$

   (d) $0.500 \text{ Pa} \times \dfrac{1 \text{ atm}}{1.013\,25 \times 10^5 \text{ Pa}} \times \dfrac{760 \text{ mm Hg}}{1 \text{ atm}}$
   $= 3.75 \times 10^{-3} \text{ mm Hg}$

2. $P_1V_1 = P_2V_2$

   $P_2 = \dfrac{P_1V_1}{V_2} = \dfrac{0.75 \text{ atm} \times 1.40 \text{ L}}{500. \text{ mL} \times \dfrac{\text{L}}{1000 \text{ mL}}} = 2.1 \text{ atm}$

3. $T_1 = 30°C + 273 = 303\text{K}; T_2 = 60°C + 273 = 333\text{K}$

   $\dfrac{V_1}{T_1} = \dfrac{V_2}{T_2}$

   $V_2 = \dfrac{V_1T_2}{T_1} = \dfrac{830 \text{ mL} \times 333\text{K}}{303\text{K}} = 901 \text{ mL}$

4. $T_1 = 25°C + 273 = 298\text{K}$

   $\dfrac{V_1}{T_1} = \dfrac{V_2}{T_2}$

   $T_2 = \dfrac{V_2T_1}{V_1} = \dfrac{0.450 \text{ L} \times 298\text{K}}{1.60 \text{ L}} = 83.8\text{K}$

5. $T_1 = 22°C + 273 = 295\text{K}; T_2 = 77°C + 273 = 350\text{K}$

   $\dfrac{P_1}{T_1} = \dfrac{P_2}{T_2}$

   $\dfrac{P_2}{P_2} = \dfrac{P_1T_2}{T_1} = \dfrac{420. \text{ mm} \times 350\text{K}}{295\text{K}} = 498 \text{ mm}$

6. $T_1 = 27°C + 273 = 300\text{K}; T_2 = 57°C + 273 = 330\text{K}$

   $\dfrac{P_1V_1}{T_1} = \dfrac{P_2V_2}{T_2}$

   $V_2 = \dfrac{P_1V_1T_2}{T_1T_1} = \dfrac{700. \text{ mm} \times 650. \text{ mL} \times 330\text{K}}{300\text{K} \times 740. \text{ mm}} = 676 \text{ mL}$

7. $P_{H_2} = 775.0 \text{ mm Hg} - 23.8 \text{ mm Hg} = 751.2 \text{ mm Hg}$
$T_1 = 25°C + 273 = 298K$
$T_2 = 0°C + 273 = 273K$

$$\frac{P_1 V_1}{T_1} = \frac{P_2 V_2}{T_2}$$

$$V_2 = \frac{P_1 V_1 T_2}{T_1 P_2} = \frac{775.0 \text{ mm} \times 325 \text{ mL} \times 273K}{298K \times 760 \text{ mm}} = 304 \text{ mL}$$

8. $0.250 \text{ mol N}_2 \times \dfrac{22.4 \text{ L}}{\text{mol}} = 5.60 \text{ L N}_2$

9. $8.40 \text{ L CO}_2 \times \dfrac{\text{mol}}{22.4 \text{ L}} \times \dfrac{44.0 \text{ g CO}_2}{\text{mol CO}_2} = 16.5 \text{ g CO}_2$

10. $\dfrac{46.0 \text{ g NO}_2}{\text{mol NO}_2} \times \dfrac{\text{mol}}{22.4 \text{ L}} = 2.05 \text{ g/L}$

11. $\dfrac{0.420 \text{ g}}{186 \text{ mL}} \times \dfrac{10^3 \text{ mL}}{\text{L}} \times \dfrac{22.4 \text{ L}}{\text{mol}} = 50.6 \text{ g/mol}$

12. $PV = nRT$

$$P = \frac{nRT}{V} = \frac{0.86 \text{ mol} \times 0.0821 \dfrac{\text{L} \cdot \text{atm}}{\text{mol}} \times 300K}{4.75 \text{ L}}$$

$= 4.5 \text{ atm}$

13. $2H_2O(\ell) \rightarrow 2H_2(g) + O_2(g)$

$4.50 \text{ g H}_2O \times \dfrac{\text{mol H}_2O}{18.0 \text{ g H}_2O} = 2.50 \text{ mol H}_2O$

$0.250 \text{ mol H}_2O \times \dfrac{2 \text{ mol H}_2}{2 \text{ mol H}_2O} = 0.250 \text{ mol H}_2$

$0.250 \text{ mol H}_2O \times \dfrac{1 \text{ mol O}_2}{2 \text{ mol H}_2O} = 1.25 \text{ mol O}_2$

$0.250 \text{ mol H}_2 \times \dfrac{22.4 \text{ L}}{\text{mol}} = 5.60 \text{ L H}_2$

$0.125 \text{ mol O}_2 \times \dfrac{22.4 \text{ L}}{\text{mol}} = 2.80 \text{ L O}_2$

14. $CH_4(g) + 2O_2(g) \rightarrow CO_2(g) + 2H_2O(g)$

$P = 730. \text{ mm Hg} \times \dfrac{1 \text{ atm}}{760 \text{ mm Hg}} = 0.961 \text{ atm}$

$T = 20°C + 273 = 293K$

$$n\text{CH}_4 = \frac{PV}{RT} = \frac{0.961 \text{ atom} \times 45.0 \text{ L}}{\dfrac{0.0821 \text{ L} \cdot \text{atm} \times 293K}{\text{mol} \cdot K}} = 1.80 \text{ mol}$$

$1.80 \text{ mol CH}_4 \times \dfrac{1 \text{ mol CO}_2}{1 \text{ mol CH}_4} = 1.80 \text{ mol CO}_2$

$1.80 \text{ mol CO}_2 \times \dfrac{44.0 \text{ g CO}_2}{\text{mol CO}_2} = 79.2 \text{ g CO}_2$

$1.80 \text{ mol CH}_4 \times \dfrac{2 \text{ mol H}_2O}{\text{mol CH}_4} = 3.60 \text{ mol H}_2O$

$3.60 \text{ mol H}_2O \times \dfrac{18.0 \text{ g H}_2O}{\text{mol H}_2O} = 64.8 \text{ g H}_2O$

Answers to Cumulative Review Questions

1. *(a)* The three types of mixtures are solution, suspensions, and colloids. A solution is a homogenous mixture of two or more substances in a single phase. A suspension is a heterogenous mixture of a solvent-like substance with particles that slowly settle out. A colloid is a mixture involving particles that are intermediate in size between those in solutions and those in suspensions. *(b)* The two components of a solution are the solvent, or dissolving medium, and the solute, or substance dissolved.

2. The rate of dissolving can be increased by increasing the surface area of the sugar, agitating the solution, and heating the water.

3. "Like dissolves like" is a useful rule for predicting whether or not one substance dissolves in another. The more alike the solute and solvent in terms of bonding type, polarity or nonpolarity of molecules, and intermolecular forces, the greater the probability that one will dissolve in the other.

4. *(a)* The concentration of a solution is a measurement of the amount of solute in a given amount of solvent or solution. *(b)* Solution concentrations are generally expressed as molarity, molality, and percent by mass. Molarity is the number of moles of solute in one liter of solution. Molality is the concentration of a solution expressed in moles of solute per kilogram of solvent. Percent by mass is the number of parts by mass of solute per 100 parts by mass of solution.

5. Vapor pressure, freezing point, and boiling point are all colligative properties. The addition of a nonvolatile solute lowers the vapor pressure, lowers the freezing point, and raises the boiling point of the solvent into which it is dissolved.

6. *(a)* Ionization refers to the formation of ions from solute molecules by the action of the solvent; it essentially refers to the creation of ions where there were none. *(b)* When ionic compounds dissolve through dissociation, the ions that were *already* present and associated with each other in the crystal become dissociated from each other. However, when molecular compounds dissolve in polar solutes, ions are formed where none existed in the undissolved compounds.

7. *(a)* Nonvolatile electrolytes in aqueous solutions lower the freezing point and raise the boiling point sometimes two, three, or more times as much as nonvolatile nonelectrolytes in solutions of the same molality. *(b)* This effect is observed because, upon dissolving, such substances produce two, three, or more moles of solute particles for each mole of compound dissolved. Since colligative properties depend on relative numbers of dissolved particles of solute present, the effect should be two or more times as great as that observed among nonelectrolytes that do not form such ions.

8. Aqueous acids have a sour taste; acids contain hydrogen, and some react with active metals to liberate hydrogen gas (H_2); acids change the color of dyes known as acid-base indicators; acids react with bases to produce salts and water; and acids are electrolytes.

9. *(a)* The strength of an acid depends on the degree of ionization in aqueous solution. *(b)* HCl is an example of a strong acid because it ionizes completely, whereas HF is a weak acid because it ionizes only slightly in aqueous solution.

10. HCl is the acid, with Cl^- as its conjugate base; H_2O is the base, with H_3O^+ as its conjugate acid.

11. One equivalent of an acid is the quantity, in grams, that supplies one mole of protons, whereas one equivalent of a base is the quantity, in grams, that accepts one mole of protons or supplies one mole of hydroxide (OH^-) ions.

12. The corresponding solution types are *(a)* acidic *(b)* acidic *(c)* basic *(d)* neutral.

13. Titration is used to determine the concentration of an unknown acidic or basic solution through the controlled addition and measurement of the amount of a solution of known concentration required to react completely with a measured amount of the unknown. It is based on the chemical equivalence of the acidic and basic solutions involved.

Solutions to Cumulative Review Problems

1. $\dfrac{1.75 \text{ mol } HNO_3}{L} \times 4.50 \text{ L} \times \dfrac{63.0 \text{ g } HNO_3}{\text{mol } HNO_3} = 496 \text{ g } HNO_3$

2. $\dfrac{18.0 \text{ mol } H_2SO_4}{L} \times \dfrac{150.0 \text{ mL}}{2.50 \text{ L}} \times \dfrac{L}{10^3 \text{mL}} = \dfrac{1.08 \text{ mol } H_2SO_4}{L}$
$= 1.08 \, M \, H_2SO_4$

3. $2HCl + Ca(OH)_2 \rightarrow CaCl_2 + 2H_2O$

$\dfrac{2.00 \text{ mol } HCl}{L} \times 75.0 \text{ mL} \times \dfrac{L}{10^3 \text{ mL}} = 0.150 \text{ mol } HCl$

$0.150 \text{ mol } HCl \times \dfrac{1 \text{ mol } CaCl_2}{\text{mol } HCl} \times \dfrac{74.1 \text{ g } CaCl_2}{\text{mol } CaCl_2}$
$= 11.1 \text{ g } CaCl_2$

$0.150 \text{ mol } HCl \times \dfrac{2 \text{ mol } H_2O}{\text{mol } HCl} \times \dfrac{18.0 \text{ g } H_2O}{\text{mol } H_2O} = 5.40 \text{ g } H_2O$

4. $\dfrac{32.0 \text{ g } KNO_3}{0.625 \text{ kg } H_2O} \times \dfrac{\text{mol } KNO_3}{101 \text{ g } KNO_3} = 0.507 \, m \, KNO_3$

5. $\Delta T_f = K_f m = \dfrac{-1.86°C \cdot \text{kg } H_2O}{\text{mol } C_2H_4(OH)_2} \times \dfrac{50.0 \text{ g } C_2H_4(OH)_2}{750. \text{ g } H_2O}$

$\times \dfrac{10_3}{\text{kg}} \times \dfrac{\text{mol } C_2H_4(OH)_2}{62.0 \text{ g } C_2H_4(OH)_2} = -2.00°C$

New freezing point:
$\Delta T_f = T_f \text{ solution} - T_f \text{ solvent}$
$T_f \text{ solution} = \Delta T_f + T_f \text{ solvent} = -2.00°C + 0°C$
$= -2.00°C$

$\Delta T_b = K_b m = \dfrac{0.51°C \cdot \text{kg } H_2O}{\text{mol } C_2H_4(OH)_2} \times \dfrac{50.0 \text{ g } C_2H_4(OH)_2}{750. \text{ g } H_2O}$

$\times \dfrac{10^3 \text{g}}{\text{kg}} \times \dfrac{\text{mol } C_2H_4(OH)_2}{62.0 \text{ g } C2H_4(OH)_2} = 0.59°C$

New boiling point $= 100°C + 0.59°C = 100.59°C$

6. $\Delta T_f = K_f m$

$m = \dfrac{\Delta T_f}{K_f} = \dfrac{-0.820°C \cdot \text{mol solute}}{-1.86°C \cdot \text{Kg H}_2\text{O}} = 0.441\ m$

$\dfrac{\text{kg H}_2\text{O}}{0.441\ \text{mol solute}} \times \dfrac{3.75\ \text{g solute}}{250.\ \text{g H}_2\text{O}} \times \dfrac{10^3 g}{\text{kg}} = 34.0\ \text{g/mol}$

7. $Fe(OH)_3(s) \rightleftarrows Fe^{3+}(aq) + 3OH^-(aq)$

8. (a) (1) $Na_2SO_4(aq) + Ca(NO_3)_2(aq) \rightarrow CaSO_4(s) + 2NaNO_3(aq)$
   (2) $2Na^+(aq) + SO_4^{2-}(aq) + Ca^{2+}(aq) + 2NO_3^-(aq) \rightarrow$
       $CaSO_4(s) + 2NaNO_3(aq) + 2NO_3^-(aq)$
   (3) Spectator ions: Na^+ and NO_3 –
       Possible precipitate: $CaSO_4$
   (4) $Ca^{2+}(aq) + SO_4^{2-}(aq) \rightarrow CaSO_4(s)$
   (b) (1) $3K_2CO_3(aq) + 2Fe(NO_3)_3(aq) \rightarrow$
       $Fe_2(CO_3)_3(s) + 6KNO_3(aq)$
   (2) $6K^+(aq) + 3CO_3^{2-}(aq) + 2Fe^{3+}(aq) + 6NO_3^-(aq) \rightarrow$
       $Fe_2(CO_3)_3(s) + 6K^+(aq) + 6NO_3^-(aq)$
   (3) Spectator ions: K^+ and NO_3^-
       Possible precipitate: $Fe_2(CO_3)_3$
   (4) $2Fe^{3+}(aq) + 3CO_3^{2-}(aq) \rightarrow Fe_2(CO_3)_3(s)$

9. $\Delta T_f = K_f m$

$\Delta T_f = \dfrac{-1.86°C \cdot \text{kg H}_2\text{O}}{\text{mol ions}} \times \dfrac{\text{mol Ca(NO}_3)_2}{164\ \text{g Ca(NO}_3)_2} \times$

$\dfrac{3\ \text{mol ions}}{\text{mol Ca(NO}_3)_2} \times \dfrac{8.20\ \text{g Ca(NO}_3)_2}{250.\ \text{g H}_2\text{O}} \times \dfrac{10^3\ g}{\text{kg}} = -1.12°C$

$\Delta T_f = T_f\ \text{solution} - T_f\ \text{solvent}$
$T_f\ \text{solution} = \Delta T_f + T_f\ \text{solvent}$

$T_f\ \text{solution} = -1.12°C + 0°C = -1.12°C$
$\Delta T_b = K_b m$

$\Delta T_b = \dfrac{0.51°C \cdot \text{kg H}_2\text{O}}{\text{mol ions}} \times \dfrac{\text{mol Ca(NO}_3)_2}{164\ \text{g Ca(NO}_3)_2} \times \dfrac{3\ \text{mol ions}}{\text{mol Ca(NO}_3)_2} \times$

$\dfrac{8.20\ \text{g Ca(NO}_3)_2}{250.\ \text{g H}_2\text{O}} \times \dfrac{10^3 g}{\text{kg}} = 0.306°C$

$\Delta T_b = T_b\ \text{solution} - T_b\ \text{solvent}$
$T_b\ \text{solution} = \Delta T_b + T_b\ \text{solvent} = 0.306°C + 100°C$
$= 100.306°C$

10. (a) $Ni(s) + 2HCl(aq) \rightarrow NiCl_2(aq) + H_2(g)$
    (b) $FeO(s) + 2HNO_3(aq) \rightarrow Fe(NO_3)_2(aq) + H_2O(\ell)$
    (c) $MgCO_3(s) + 2HCl(aq) \rightarrow MgCl_2(aq) + CO_2(g) + H_2O(\ell)$
    (d) $CO_2(g) + Ba(OH)_2(aq) \rightarrow BaCO_3(s) + H_2O(\ell)$
    (e) $BaO(s) + SO_3(g) \rightarrow BaSO_4(s)$

11. $\dfrac{6.00\ \text{equiv H}_2\text{SO}_4}{L} \times 250.\ \text{mL} \times \dfrac{L}{10^3\ \text{mL}} \times \dfrac{1\ \text{mol H}_2\text{SO}_4}{2\ \text{equiv H}_2\text{SO}_4} \times$

$\dfrac{98.1\ \text{g H}_2\text{SO}_4}{\text{mol H}_2\text{SO}_4} = 73.6\ \text{g H}_2\text{SO}_4$

12. $pH = -\log[H_3O^+] = -\log(10^{-5}) = -(-5) = 5$
$[H_3O^+] = 10^{-5}\ M$ since HCl is a monoprotic acid

Since $[H_3O^+][OH^-] = 10^{-14}\ M^2$, $[OH^-] = \dfrac{10^{-14}\ M^2}{[H_3O^+]}$
$= \dfrac{10^{-14}\ M^2}{10^{-5}\ M} = 10^{-9}\ M$

13. $V_a N_a = V_b N_b$

$N_a = \dfrac{V_b N_b}{V_a} = \dfrac{12.5\ \text{mL} \times 0.250\ N}{18.5\ \text{mL}} = 0.169\ N$

$\dfrac{0.169\ \text{equiv}}{L} \times \dfrac{\text{mol H}_3\text{PO}_4}{3\ \text{equiv H}_3\text{PO}_4} = 0.056\ 3\ M$

Answers to Cumulative Review Questions

1. (a) $Zn(s) + S(s) + 2O_2(g) \rightarrow ZnSO_4(s)$ $\Delta H_f = -982.82$ kJ
   (b) $\frac{1}{2}N_2(g) + O_2(g) \rightarrow NO_2(g)$ $\Delta H_f = 33.18$ kJ
   (c) $Ba(s) + N_2(g) + 3O_2(g) \rightarrow Ba(NO_3)_2$ $\Delta H_f = -991.86$ kJ
2. (a) calcium sulfate, (b) calcium oxide
3. (a) $\Delta G = \Delta H - T\Delta S$; the free energy change, ΔH is the enthalpy change—the more negative the enthalpy change, the more likely that a reaction will be spontaneous; T is the Kelvin temperature—at a high enough temperature a non-spontaneous reaction becomes spontaneous; ΔS is the entropy change—the more positive and larger the entropy change, the more likely a reaction is to be spontaneous, although entropy change is often small relative to enthalpy change.
4. The entropy change is negative because a system becomes less random when a gas is converted to a solid.
5. (a) In the first step one molecule of A_2 breaks down to give two atoms of A. In the second step the two atoms of A collide with one molecule of B_2 to produce two molecules of AB. (b) A is the intermediate. (c) The second step (the slow step) is the rate-determining step. (d) $A_2 + B_2 \rightarrow 2AB$.
6. (a) $K = \dfrac{[N_2][H_2]^2}{[N_2H_4]}$
   (a) $K = \dfrac{[SF_6]}{[F_2]^3}$
   (c) $\dfrac{1}{[H_2]^2[O_2]}$
7. (a) $Ag_2S(s) \rightleftarrows 2Ag^+(aq) + S^{2-}(aq)$
   $K_{sp} = [Ag^{2+}]^2[S^{2-}]$
   (b) $HF(aq) + H_2O(\ell) \rightleftarrows H_3O^+(aq) + F^-(aq)$
   $K_a = \dfrac{[H^3O^+][A^-]}{[HF]}$
   (c) $SO_3^{2-}(aq) + H_2O(\ell) \rightleftarrows HSO_3^{2-}(aq) + OH^-$
   $K_n = \dfrac{[HSO^{3-}][OH^-]}{[SO_3{}^{2-}]}$
8. (a) [HA] increases (b) Common-ion effect
9. (a) Acidic (b) Neutral (c) Basic
10. (a) $\overset{+2-2+1}{Ca(OH)_2(aq)} + \overset{+4-2}{CO_2(g)} \rightarrow \overset{+2+4-2}{CaCO_3(s)} + \overset{+1-2}{H_2O(\ell)}$
    a nonredox reaction
    (b) $\overset{0}{3Na(s)} + \overset{0}{P(s)} \rightarrow \overset{+1-3}{Na_3P(s)}$
    a redox reaction
    (c) $\overset{-3+1}{NH_3(aq)} + \overset{+1-2}{ClO^-} \rightarrow \overset{+2+1}{N_2H_4(g)} + \overset{-1}{Cl^-} + \overset{+1-2}{H_2O(\ell)}$
    a redox reaction

11. (b) (i) Na (ii) P (iii) P (iv) Na (c) (i) NH_3 (ii) ClO^- (iii) ClO^- (iv) NH_3
12. In a flashlight battery a spontaneous electrochemical reaction produces electrical current. In an electrolytic cell a nonspontaneous chemical reaction is caused to occur by an electric current supplied from outside the cell.
13. Standard electrode potential is the potential, in volts, of an electrode measured relative to the standard hydrogen electrode, which is assigned a potential of zero. It is necessary because half-cell potentials cannot be measured, and a standard must be chosen in order to compare electrode potentials.
14. (a) No (b) Yes (c) Yes

Solutions to Cumulative Review Problems

1. Reaction: $8C(s) + 9H_2(g) \rightarrow C_8H_{18}(\ell)$
   (a) $8CO_2(g) + 9H_2O(\ell) \rightarrow$

$C_8H_{18}(\ell) + \frac{25}{2}O_2(g)$	$\Delta H = +5450.5$ kJ
$8C(s) + 8O^2(g) \rightarrow 8CO^2(g)$	$8[\Delta H = -393.51$ kJ$]$
$9H^2(g) + \frac{9}{2}O_2(g) \rightarrow 9H_2O(\ell)$	$9[\Delta H = -285.83$ kJ$]$
$8C(s) + 9H_2(g) \rightarrow C_8H_{18}(\ell)$	$\Delta H = -270.05$ kJ

   (b) $\Delta H_f{}^\circ = [\Delta H_f{}^\circ(CO_2) + \Delta H_f{}^\circ(H_2O)] - [\Delta H_c(C_8H_{18})]$
   $\Delta H_f{}^\circ = [8(-393.51$ kJ$) + 9(-285.83$ kJ$)] - [-5450.5$ kJ$]$
   $\Delta H_f{}^\circ = -270.05$ kJ
2. $\Delta G = \Delta H - T\Delta S$
   $T = 25°C + 273 = 298K$
   $\Delta G = (-165.4$ kJ$) - (298K)(0.0275$ kJ$)$
   $\Delta G = -165.4$ kJ $- 8.20$ kJ
   $\Delta G = -173.6$ kJ
   This reaction will occur spontaneously.
3. $\Delta H_{reverse} = -10$ kJ
   $E_a = 40$ kJ
4. $Al(OH)_3(s) \rightleftarrows Al^{3+}(aq) + 3OH^-(aq)$
   $K_{sp} = [Al^{3+}][OH^-]^3$
   For Al^{3+}; $\dfrac{1.00 \times 10^{-6} \text{ mol } Al^{3+}}{L} \times (10.0$ mL$)$
   $\times \dfrac{L}{10^3 \text{ mL}} = 1.00 \times 10^{-8}$ mol Al^{3+}
   $\dfrac{1.00 \times 10^{-8} \text{ mol } Al^{3+}}{20.0 \text{ mL}} \times \dfrac{10^3 \text{ mL}}{L} = 5.00 \times 10^{-7}$ mol Al^{3+}/L
   For OH^-; $\dfrac{2.00 \times 10^{-5} \text{ mol } OH^-}{L} \times 10.0$ mL $\times \dfrac{L}{10^3 \text{ mL}} =$
   2.00×10^{-7} mol OH^-
   $\dfrac{2.00 \times 10^{-7} \text{ mol } OH^-}{20.0 \text{ mL}} \times \dfrac{10^3 \text{ mL}}{L} = 1.00 \times$
   10^{-5} mol OH^-/L
   Trial product: $[Al^{3+}][OH^-]^3 = [5.00 \times 10^{-7}][1.00 \times 10^{-5}]^3$
   $= 5.00 \times 10^{-22}$
   Since the trial product exceeds the K_{sp} of 1.3×10^{-33}, a precipitate is formed.

5. $\overset{+1+7-2}{KMnO_4} + \overset{+1-1}{KBr} + \overset{+1+6-2}{H_2SO^4} \rightarrow \overset{0}{Br_2} + \overset{+1+6-2}{K_2SO_4} + \overset{+2+6-2}{MnSO_4} +$ $\overset{+1-2}{H_2O}$

$2[\overset{+7}{Mn} + 5e^- \rightarrow \overset{+2}{Mn^{2+}}]$

$5[\overset{-1}{2Br^-} \rightarrow \overset{0}{Br_2} + 2e^-]$

$2KMnO_4 + 10KBr + 8\ H_2SO_4 \rightarrow\ 5Br_2\ + 6K_2SO_4 +$

$2MnSO_4 + 8H_2O$

6. $HNO_3 + Bi_2S_3 \rightarrow\ Bi(NO_3)_3 +\ NO + S\ + H_2O$

$\overset{+1}{H^+} + \overset{+5-2}{NO_3} - + Bi_2S_3 \rightarrow \overset{+3}{Bi^{3+}} + \overset{+5-2}{3NO_3} - + \overset{+2-2}{NO} +$ $\overset{0}{S} + \overset{+1-2}{2H_2O}$

$2[4NO_3^- + 3e^- + 4H^+ \rightarrow NO + 3NO_3^- + 2H_2O]$

$3[S_2^- \rightarrow S + 2e^-]$

$8HNO_3 + Bi_2S_3 \rightarrow 2Bi(NO_3)_3 + 2NO + 3S + 4H_2O$

7.

$Zn \rightarrow \quad Zn^{2+} + 2e^-$		$E° = +0.76\ V$
$2[Ag^+ + \quad e^- \rightarrow \quad Ag]$		$E° = +0.80\ V$
$Zn + 2Ag^+ \quad \rightarrow \quad Zn^{2+} + 2Ag$		$E° = +1.56\ V$

The reaction does occur spontaneously, as evidenced by the positive $E°$ value.

Answers to Cumulative Review Questions

1. Natural gas, petroleum, coal
2. Carbon atoms are joined by equivalent covalent bonds intermediate between single and double bonds; layers are joined by intermolecular forces known as London dispersion forces.
3. CH_4, C_2H_6 . . . any hydrocarbon; $NaCO_3$ (or any carbonate), CaC (or any carbide), $NaCN$ (or any cyanide), CO, CO_2
4. Compound A, organic; compound B, inorganic
5. (a) Alkyne (b) Alkane (c) Alkene (d) Alkane
6. (a) Alkene (b) Alkane (c) Alkane (d) Alkene
7.

8. (a) Substitution (b) Addition
9.

   monomer, chloroprene neoprene repeating unit
10.

11. (a) F_3CCOOH
    (b) CH_3OCH_3
    (c) $CH_3CH_2CH_2CH_2CH_2COOH$
    (d) $CH_3CH(CH_3)CH_2CH_2OH$
    (e) $HCOOH$
    (f) CH_3CHO
12. (a) Carbon tetrachloride, a grease solvent (b) Ethanol, a gasoline additive (c) Glycerol, a skin moisturizer (d) Vinyl chloride, monomer for a common polymer (polytheylene) (e) Ethyl butyrate (ethyl butanoate), a flavoring
13. In condensation, two monomers combine when the loss of a molecule of water produces a bond between them. In hydrolysis, the reverse of condensation, water is added to the bond between two monomers, breaking the bond. In living bodies, macromolecules are formed by condensation reactions, and foods are broken down during digestion by hydrolysis.
14. Proteins, carbohydrates, lipids, nucleic acids; (a) lipids (b) carbohydrates (c) nucleic acids (d) proteins (e) proteins (f) carbohydrates (g) lipids
15. An enzyme is a catalyst for reactions in living things; the reactants in enzyme-catalyzed reactions are called substrates.
16. A coenzyme aids in the binding of an enzyme to its substrates. Vitamins
17. (a) Carbohydrate (b) Monosaccharide (c) Source of energy in cells and carbohydrate monomer
18. (a) Divide living things into separate compartments such as cells. (b) Lipids
19. A neurotransmitter is a chemical messenger that transmits a message across a space (synapse) between cells of the nervous system (neurons).
20. (a) Deoxyribonucleic acid and ribonucleic acid (b) DNA stores the genetic code and RNA delivers the genetic messages to places where they are needed.
21. (a) All of an organism's chemical reactions (b) Reaction pathways in which a series of enzymes allow stepwise conversion of substrates into specific products. (c) Conversion of food into energy supply
22. (a) Adenosine triphosphate (ATP) (b) Hydrolysis of a phosphate group

Answers to Cumulative Review Questions

1. *(a)* Groups 13–16 *(b)* Group 1 *(c)* Group 2 *(d)* Groups 8–10 *(e)* Group 17
2. *(a)* Beryllium *(b)* Potassium *(c)* Beryllium *(d)* Potassium *(e)* Magnesium
3. *(a)* Baking soda *(b)* Lye or caustic soda *(c)* Soda ash *(d)* Potash or caustic potash
4. Alkali metal ions—sodium and potassium; alkaline earth metal ions—calcium and magnesium
5. The alkali metals are more highly reactive, and have lower melting points and densities, and greater softness than other metals.
6. *(a)* Oxidation product is Cl_2 and reduction product is H_2.
   $2H_2O + 2 Cl^- \rightarrow H_2(g) + Cl_2(g) + 2OH^-$
   (b) Oxidation product is Cl_2 and reduction product is sodium metal.
7. Raw metals, coal, salt, and limestone; major product, sodium carbonate; by-products, sodium hydrogen carbonate and calcium chloride
8. Alkaline earth metals have smaller radii, higher ionization energies, and higher melting and boiling points, are denser and harder, and form $+2$ rather than $+1$ ions.
9. Beryllium atoms are smaller than those of the other alkaline-earth metals.
10. *(a)* A self-protective metal forms a non-porous, non-scaling coat of tarnish, thus protecting itself from deterioration. *(b)* Aluminum, magnesium, and copper
11. *(a)* Paramagnetism, variable oxidation sates, formation of complex ions, colored compounds *(b)* Electrons in incompletely fill *d* sublevels can participate in bonding and interact with surroundings.

12. *(a)* $Fe_2O_3 + 3CO \rightarrow 2Fe + 3CO_2$
    (b) $CaO + SiO_2 + SiO_2 \rightarrow CaSiO_3$
13. Iron(II) oxide, FeO; iron(III) oxide, Fe_2O_3; iron(II,III) oxide, Fe_3O_4 or $FeO \cdot Fe_2O_3$
14. *(a)* Low-grade ore is concentrated as the oil-wetted ore floats on the surface of the water. *(b)* During oxidation sulfur is driven from the matte as CO_2 and iron is removed from slag. *(c)* During electrolytic processing, very pure copper is produced and separated from silver and gold, which fall to the bottom of the cell and are collected.
15. *(a)* Boron, silicon, germanium, arsenic, antimony, selenium, tellurium *(b)* Intermediate electrical conductivity that increases with temperature *(c)* Electronics industry
16. *(a)* Chromium(III) chloride, colored *(b)* Sodium hypochlorite, a bleach *(c)* Boric acid, an antiseptic *(d)* Boron nitride, abrasive
17. *(a)* Se *(b)* Cl_2 *(c)* Al *(d)* Cl *(e)* S *(f)* S *(g)* Al and Cl_2
18. Silicates are present in the earth's crust; silicon is an element; silicones are synthetic compounds containing silicon and oxygen chains with hydrocarbon groups bonded to silicon.
19. *(a)* $2SO_2 + O_2 \rightarrow 2SO_3$
    $SO_3 + H_2SO_4 \rightarrow H_2S_2O_7$
    $H_2S_2O_7 + H_2O \rightarrow 2H_2SO_4$
    (b) Contact process
20. *(a)* Fertilizer, steel, petroleum industries *(b)* acid, oxidizing agent, dehydrating agent
21. *(a)* Nitrogen and sulfur oxides *(b)* Coal, gasoline, oil burning, and roasting sulfide ores *(c)* Nitric acid and sulfuric acid *(d)* Normal rain, pH 5.6, acid rain pH lower than 5.0
22. For making sulfuric acid, as a preservative, for bleaching, in making paper pulp
23. *(a)* Iodine *(b)* Fluorine *(c)* Fluorine *(d)* Bromine and iodine *(e)* Chlorine *(f)* Iodine
24. In bleaches, disinfectants, synthesis of chlorine-containing compounds
25. *(a)* $2Fe + 3CO_2$, reduction of iron oxide to give iron in blast furnace *(b)* $Cl^- + Br_2$, commercial production of bromine from brine *(c)* $2Cu_2O + 2SO_2$, roasting copper ore during production of copper *(d)* $2Fe + Al_2O_3$, thermite reaction, welding

Continued from page 342J

27. $\dfrac{Cl_2}{SO_2} = \dfrac{\sqrt{\text{mol } m_{SO_2}}}{\sqrt{\text{mol } m_{Cl_2}}}$

$SO_2 = \dfrac{Cl_2 \sqrt{\text{molar } m_{Cl_2}}}{\sqrt{\text{molar } m_{SO_2}}} = \dfrac{0.0380 \text{ m/s} \sqrt{70.9 \text{ g/mol}}}{\sqrt{64.1 \text{ g/mol}}} = 0.0400 \text{ m/s}$

28. $\dfrac{He}{X} = \dfrac{\sqrt{\text{mol } m_X}}{\sqrt{\text{mol } m_{He}}}$

$\sqrt{\text{mol } m_X} = \dfrac{V_{He} \sqrt{\text{mol } m_{He}}}{V_X}$

Substituting 6.50 for $\dfrac{V_{He}}{V_X}$:

$\sqrt{\text{mol } m_X} = 6.50 \sqrt{\text{mol } m_{He}}$

$\text{molar } m = (6.50 \sqrt{\text{mol } m_{He}})^2 = 42.2 \times \text{mol } m_{He}$

$= 42.2 \times 4.00 \text{ g/mol} = 169 \text{ g/mol}$

Teacher's Notes

Periodic Table

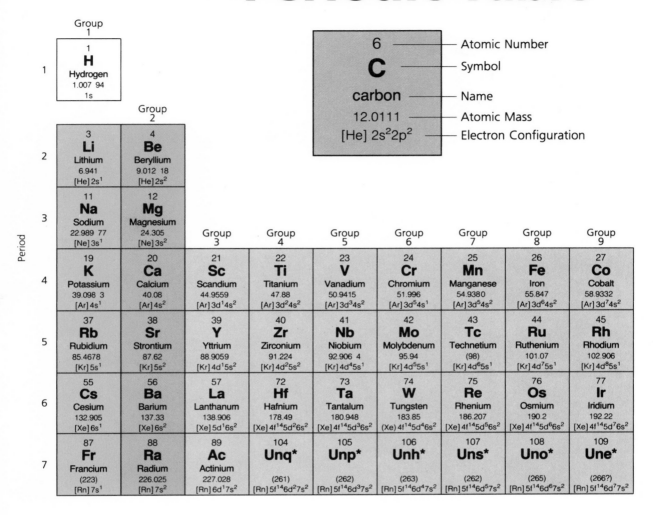

Figure 5-5 Periodic Table of the Elements

*The systematic names and symbols for elements of atomic number greater than 103 will be used until the approval of trivial names by IUPAC.